STANDARD HANDBOOK OF
ENGINEERING CALCULATIONS

OTHER McGRAW-HILL HANDBOOKS OF INTEREST

Avallone & Baumeister • MARKS' STANDARD HANDBOOK FOR MECHANICAL ENGINEERS

Brady & Clauser • MATERIALS HANDBOOK

Considine • PROCESS/INDUSTRIAL INSTRUMENTS AND CONTROLS HANDBOOK

Fink & Beaty • STANDARD HANDBOOK FOR ELECTRICAL ENGINEERS

Fink & Christiansen • ELECTRONICS ENGINEER'S HANDBOOK

Freeman • STANDARD HANDBOOK OF HAZARDOUS WASTE TREATMENT AND DISPOSAL

Ganic & Hicks • THE MCGRAW-HILL HANDBOOK OF ESSENTIAL ENGINEERING
INFORMATION AND DATA

Haines & Wilson • HVAC SYSTEMS DESIGN HANDBOOK

Harris • HANDBOOK OF ACOUSTIC MEASUREMENTS AND NOISE CONTROL

Harris & Crede • SHOCK AND VIBRATION HANDBOOK

Higgins • MAINTENANCE ENGINEERING HANDBOOK

Hodson • MAYNARD'S INDUSTRIAL ENGINEERING HANDBOOK

Juran & Gryna • JURAN'S QUALITY CONTROL HANDBOOK

Karassik et al. • PUMP HANDBOOK

Lingaiah • MACHINE DESIGN DATA HANDBOOK

Maidment • HANDBOOK OF HYDROLOGY

Merritt & Ricketts • BUILDING DESIGN AND CONSTRUCTION HANDBOOK

Perry & Green • PERRY'S CHEMICAL ENGINEER'S HANDBOOK

Rosaler • STANDARD HANDBOOK OF PLANT ENGINEERING

Shigley & Mischke • STANDARD HANDBOOK OF MACHINE DESIGN

Tuma • ENGINEERING MATHEMATICS HANDBOOK

Wadsworth • HANDBOOK OF STATISTICAL METHODS FOR ENGINEERS AND SCIENTISTS

Walsh • MCGRAW-HILL MACHINING AND METALWORKING HANDBOOK

Wang • HANDBOOK OF AIR CONDITIONING AND REFRIGERATION

THIRD EDITION

STANDARD HANDBOOK OF ENGINEERING CALCULATIONS

Tyler G. Hicks, P.E. Editor

International Engineering Associates
Member: American Society of Mechanical Engineers
Institute of Electrical and Electronics Engineers
United States Naval Institute

S. David Hicks Coordinating Editor

International Engineering Associates

Joseph Leto, P.E. Assistant Editor

Consulting Engineer

McGraw-Hill, Inc.

New York San Francisco Washington, D.C. Auckland Bogotá
Caracas Lisbon London Madrid Mexico City Milan
Montreal New Delhi San Juan Singapore
Sydney Tokyo Toronto

Library of Congress Cataloging-in-Publication Data

Standard handbook of engineering calculations / Tyler G. Hicks, editor
; S. David Hicks, coordinating editor.—3rd ed.
    p. cm.
   Includes index.
   ISBN 0-07-028812-7
   1. Engineering mathematics—Handbooks, manuals, etc. I. Hicks,
Tyler Gregory, date.   II. Hicks, S. David.
TA332.S73   1994
620′.00212—dc20
                                      95-11138
                                            CIP

3 4 5 6 7 8 9 0 DOC/DOC 9 0 9 8 7 6 5

ISBN 0-07-028812-7

The sponsoring editor for this book was Harold B. Crawford. This book was set in Caledonia by Techna Type.

Printed and bound by R. R. Donnelley & Sons Company.

This book is printed on acid-free paper.

A Note about This Handbook

The *Standard Handbook of Engineering Calculations*, 3rd edition, is designed to present accurate and authoritative information on engineering calculation procedures in a variety of fields. The editors, advisers, and contributors used every effort possible to assure the accuracy of the procedures, formulas, and constants presented in the handbook. Further, the handbook is written with the understanding that the publisher, the editors, and the contributors are supplying guidelines and not attempting to render professional engineering, legal, accounting, or other professional services. If such services are required, the assistance of a qualified professional should be sought.

CONTENTS

CONTRIBUTORS AND ADVISERS   vii
PREFACE   ix
ACKNOWLEDGMENTS   xiii
HOW TO USE THIS HANDBOOK   xv

SECTION *1*   CIVIL ENGINEERING   1.1

SECTION *2*   ARCHITECTURAL ENGINEERING   2.1

SECTION *3*   MECHANICAL ENGINEERING   3.1

SECTION *4*   ELECTRICAL ENGINEERING   4.1

SECTION *5*   ELECTRONICS ENGINEERING   5.1

SECTION *6*   CHEMICAL AND PROCESS PLANT
                 ENGINEERING   6.1

SECTION *7*   CONTROL ENGINEERING   7.1

SECTION *8*   AERONAUTICAL AND ASTRONAUTICAL
                 ENGINEERING   8.1

SECTION *9*   MARINE ENGINEERING   9.1

SECTION *10*   NUCLEAR ENGINEERING   10.1

SECTION *11*   SANITARY ENGINEERING   11.1

SECTION *12*   ENGINEERING ECONOMICS   12.1

SECTION *13*   ENVIRONMENTAL ENGINEERING   13.1

INDEX FOLLOWS SECTION 13

This handbook is dedicated to the engineering profession throughout the world. Your accomplishments—from the simplest irrigation project that helps crops grow better and feed hungry people, to the latest advances in medical microelectronics, to the frontiers of space exploration—are seldom acknowledged or recognized. The editor is proud and grateful to be a member of this elite group and he hopes this handbook will help every user save time and effort in applying the principles of our profession.

TYLER G. HICKS, P.E.

CONTRIBUTORS AND ADVISERS

In preparing the various sections of this handbook, the following individuals either contributed sections, or portions of sections, or advised the editor or contributors, or both, on the optimum content of specific sections. The affiliations shown are those prevailing at the time of the preparation of the contributed material or the recommendations as to section content.

In choosing the procedures and worked-out problems, these specialists used a number of guidelines, including: (1) What are the most common applied problems that must be solved in this discipline? (2) What are the most accurate methods for solving these problems? (3) What other problems might be met in this discipline? When the answers to these and other related questions were obtained, the procedures and worked-out problems were chosen. Thus, the handbook represents a cross section of the thinking of a large number of experienced practicing engineers, project directors, and educators.

To those who might claim that the use of step-by-step solution procedures and worked-out examples makes engineering "too easy," the editor points out that for many years engineering educators have recognized the importance and value of problem solving in the development of engineering judgment and experience. Problems courses have been popular in numerous engineering schools for many years and are still given in many schools. However, with the greater emphasis on engineering science in most engineering schools, there is less time for the problems courses. The result is that many of today's graduates can benefit from a more extensive study of specific problem-solving procedures.

FREDERICK S. BARTON, *Director*, Hewlett-Packard Limited

HAROLD BECHER, *President*, Strato Missles, Inc.

EDMUND B. BESSELIEVRE, P.E., *Consultant*, Forrest & Cotton, Inc.

ROBERT T. CHIEN, *Professor of Electrical Engineering*, University of Illinois—Urbana-Champaign

ROBERT L. DAVIDSON, *Consulting Engineer*

STEPHEN M. EBER, P.E., Ebasco Services, Inc.

ANDREW W. EDWARDS, *Power Engineer*, Westinghouse Electric Corp.

GERALD M. EISENBERG, *Project Engineering Administrator*, American Society of Mechanical Engineers

CHARLES F. HAFER, P.E.

WILLIAM J. HANNAN, *Vice President*, Research and Technology Development, Itek Optical Systems Corp.

GREGORY T. HICKS, R.A., Gregory T. Hicks and Associates, Architects

S. DAVID HICKS, International Engineering Associates

TYLER G. HICKS, P.E., International Engineering Associates

EDGAR J. KATES, P.E., *Consulting Engineer*

MAX KURTZ, P.E., *Consulting Engineer*

JOSEPH LETO, P.E., *Consulting Engineer*

SAMUEL C. LIND, *Consultant*, United States Atomic Energy Commission

JOSEPH MITTLEMAN, *Consulting Engineer*

JEROME F. MUELLER, P.E., Mueller Engineering Corp.

GEORGE M. MUSCHAMP, *Consulting Engineer*, Honeywell, Inc.

RUFUS OLDENBERGER, *Professor*, Purdue University

ZVI PRIHAR, *Scientific Adviser*, Operations Research, Inc.

JOHN S. REARICK, P.E., *Consulting Engineer*

RAYMOND J. ROARK, *Professor*, University of Wisconsin

JOHN P. ROEDEL, *Research Engineer*, The Boeing Company

HAROLD L. RORDEN, *Consulting Engineer*, American Electric Power Service Corp.

LYMAN F. SCHEEL, *Consulting Engineer*

B. G. A. SKROTZKI, P.E., *Power* Magazine

S. W. SPIELVOGEL, *Piping Engineering Consultant*

MICHAEL K. STAFFORD, *Manager*, Magnetics Department, Disk-Development Division, Dysan Corporation

JOHN C. STERLING, *Consultant*, Newport News Shipbuilding & Drydock Co.

FREDERICK W. SUHR, *Consulting Engineer*, General Electric Co.

BERNARD TICHAZ, *Director*, George G. Sharp Inc.

PREFACE

Three major developments occurred in engineering since publication of the second edition of this handbook. These are: (1) the emergence of environmental engineering as a key discipline in almost every area of civilized life; (2) the broad adoption of personal computers (desktop and laptop) for the solution of repetitive engineering problems; and (3) the continued emphasis on energy conservation in the design and operation of a variety of structures, transportation platforms, and energy generating facilities.

To help engineers everywhere cope with the enormous emphasis on environmental protection, a new section, 13, has been added to the handbook. When used with other applicable sections of the handbook, Section 13 permits any engineer to "get up to speed" in this fastest growing discipline in the engineering field today. Since environmental engineering uses the methods and solutions from a variety of other disciplines—mechanical, civil, electrical, chemical, industrial, architectural, sanitary, nuclear, and control engineering—having key solutions for these specialities in the book, as this handbook does, makes the volume highly valuable to environmental engineers, and all other engineers.

Thousands of environmental regulations from federal, state, county, and city regulators impact engineering design today. Just keeping current with such regulations is an enormous task for the engineer. To help all engineers with this task, the new edition of this handbook includes pertinent environmental data for numerous calculation procedures. Using this information, the engineer can tailor a design to include environmental considerations that make it more acceptable to regulators having control over the project.

The second major development since publication of the second edition of this handbook is the broad adoption of desktop and laptop computers to solve repetitive engineering problems. Using a computer can save many hours of design time. So engineers have widely adopted computer solutions for repetitive problems.

Hundreds of computer programs are available for solving common repetitive engineering problems. Prices of such programs range from a few hundred dollars to several thousand dollars, depending on the program and its developer.

Engineers using such programs often find that the data-entry time requirements are excessive for quick one-off-type calculations. When typical design calculations are needed most, engineers today turn to their electronic calculator and perform the necessary steps to provide the solution desired. But where repetitive calculations are required, the computer program will save time and energy in the usual medium-size or large engineering office. Small engineering offices generally resort to manual calculation for even repetitive procedures because the investment for one or more programs is difficult to justify.

Even where computer programs are used extensively, careful engineers still insist on manually checking results on a random basis to be certain the program is accurate. This checking can be speeded by using any of the procedures given in this handbook. Many engineers have remarked to the editor that they feel safer, knowing that they have manually verified the computer results on a spot-check basis. With liability for engineering designs extending beyond the lifetime of the designer, every engineer seeks the "security blanket" provided by manual verification of the results furnished by a computer program run on a desktop, laptop, or notebook computer. This handbook gives the tools needed for manual verification of some 5,000 calculation procedures.

The third major development since publication of the second edition of this handbook is the greater emphasis on energy conservation in the design and operation of a variety of structures, transportation platforms, and energy-generating facilities. One excellent example of this is the introduction of aero-derivative gas turbines for power-plant expansion, topping service, repowering, and modernization. Highly efficient gas turbines save construction time, reduce space requirements, and allow greater fuel flexibility.

Today's power plant looks much different from those of just 10 years ago. Steam is being generated in heat-recovery steam generators fired by gas-turbine exhaust gases. A variety of unique fuels—biomass, municipal waste, agricultural byproducts, and so on—now generate steam, reducing land pollution, saving fossil fuel, and reducing operating costs.

The third edition of this handbook gives full coverage to the major developments in energy conservation during the last 10 years. Using the data presented, any engineer can prepare a sensible preliminary design ready for further study by specialists in the various fields covered by the design.

With the increased attention on improving the environment and saving energy, engineering and engineers are in the spotlight more than ever before. Why? Because saving the environment and energy all comes down to engineering. Engineers design *not* to pollute the environment and *not* to waste energy. If a pollution problem is discovered, engineers design ways to stop the pollution and remove whatever undesirables may have been produced in the past. So you can say that the modern environmentally conscious, energy-saving world is more dependent than ever before on engineers and engineering. This edition of the handbook recognizes this situation and addresses it in a way that is practical and useful for every engineer producing designs in thirteen different disciplines.

As with the two previous editions, this is a handbook of specific engineering calculation procedures that presents to its users more than five thousand direct and related calculation procedures for solving almost all routine, and many nonroutine, problems met in everyday engineering practice in thirteen important technical fields. These fields of engineering are: aeronautical and astronautical, architectural, chemical and process plant, civil, control, electrical, electronic, economic, marine, mechanical, nuclear, environmental, and sanitary. Having this handbook on the desk or bench, the engineer or scientist will be able to solve most of the applied problems met in daily activities of design, operation, analysis, or economic evaluation anywhere in the world.

The step-by-step *practical and applied* calculation procedures in this handbook are arranged so they can be followed by anyone with an engineering or scientific background. Each worked-out procedure presents *fully explained and illustrated steps* for solving similar problems in design, industrial, research, government, academic, or license-examination situations. For any applied problem, all the handbook user need do is place his or her calculation sheets alongside this handbook and follow the step-by-step procedure line for line to obtain the desired solution for the actual, real-life problem. By following the calculation procedures in this handbook, the engineer, scientist, or technician will obtain accurate results in minimum time with least effort. And the approach and solutions are modern throughout.

The purpose of this handbook is to provide engineers everywhere with specific step-by-step calculation procedures for the most common design and operating problems met in daily practice. While specialists in a given discipline may know of and use more advanced methods, the procedures given in this handbook will produce safe and usable results for the majority of situations met by practicing engineers.

Spanning thirteen separate disciplines from aeronautical and astronautical engineering to environmental engineering, the handbook is useful to these engineers and to architectural, chemical, civil, control, electrical, electronics, marine, mechanical, sanitary, and nuclear engineers as well. The editors and contributors endeavored to present calculation procedures that are the most important and useful in each branch of engineering, and which will give practitioners the greatest help in their daily-work.

Beside being highly useful in the daily practice of engineering, the handbook is valuable to candidates for professional engineering licenses, marine engineering licenses, and a variety of civil service examinations. Many open-book license examinations contain problems that closely resemble those in this handbook.

Further, the handbook is particularly valuable to engineers asked to work outside their area of specialty. Thus, the electrical engineer who has to size a beam to support a motor or a transformer will find a specific calculation procedure for such a choice in Section 1, Civil Engineering. And the mechanical engineer asked to design a lightning protection system for an industrial plant will find the solution in Section 4, Electrical Engineering. The same is true for all the other engineering disciplines covered by this handbook. Each procedure is accompanied by a worked-out design or operating situation, showing each step to take and the reasons for the step. Thus, to the design or operating engineer, this is the one handbook he or she should not be without. This is particularly so for this third edition of the handbook because it covers the major changes and developments that took place since publication of the first edition.

All calculation procedures in this third edition use both systems of units—the United States Customary System (USCS) and SI. Thus, the engineer unfamiliar with SI can learn this world-wide system quickly and easily simply by reviewing pertinent calculation procedures in this handbook using both systems of units. Such a review will give the engineer a better sense of the "size" of particular SI units much more rapidly than isolated study of a listing of conversion factors. In each calculation procedure the SI unit is indicated in parentheses or brackets after the USCS unit. Thus, a 100-ft span is presented as a 100-ft (30.5-m) span, while a 1000-lb/in^2 pressure is presented as a 1000-lb/in^2 (6894-kPa) pressure. The engineer can use the handbook calculation procedure either solely in USCS units, solely in SI units, or a combination of them. Thus, overseas users will find the handbook compatible with their local practice and a valuable tool in helping them save time, money, and effort.

Besides the addition of a completely new section on environmental engineering, this third edition has been thoroughly updated throughout. A few of the steps taken to accomplish this updating include: Substitution of the new beam designations for the previous standard designations used in earlier editions in both Sections 1 and 2; addition of new design procedures for machinery in Section 3; numerous energy-saving procedures were added to Section 3; data and procedures were added on Boolean algebra; much new information and specific procedures were added on heat recovery steam generators and gas turbines used in power generation; quick design procedures are included for a variety of heat exchangers; vessel-emptying time (a tricky calculation) is looked at from several vantage points; plus a variety of other specific, hands-on procedures for rapid and accurate calculation of various designs.

So with the addition of many new calculation procedures, the full metrication of the entire handbook, i.e., *every* calculation procedure, and the comprehensive coverage of environmental design and energy conservation, this is an entirely new handbook. Users will find the handbook coverage in keeping with today's engineering practice throughout the world. This is a handbook every engineer should own so that he or she is ready to solve the many calculation challenges that come to everyone in the profession sooner or later.

In a work of this size—some 1500 pages—with nearly half of every page comprised of mathematical material, errors can occur. For this reason, the editor asks each user of the handbook to call to his attention any errors that are found. The errors will be corrected in the next printing of the handbook.

Further, if a reader feels that one or more important calculation procedures have been excluded from the handbook, the editor would like to have these called to his attention. And if a reader would like to submit a favorite calculation procedure for possible inclusion in the next edition, the editor will be glad to receive the procedure and evaluate it. All accepted procedures will be fully acknowledged as to source and contributor in the next edition. To have a procedure considered, just send the name of the procedure to the editor in care of the publisher, McGraw-Hill Book Company, 1221 Avenue of the Americas, New York, NY 10020, Attn. Professional Books Group. The editor will respond, indicating if he is interested in seeing the full procedure. Please do *not* send the full procedure unless requested to do so by the editor.

Lastly, thank you for using this handbook in your work. I hope it helps you in all fields of modern engineering practice.

TYLER G. HICKS, P.E.

ACKNOWLEDGMENTS

The contributors and advisers consulted hundreds of sources when preparing the material for inclusion in this handbook. Besides using the books and other publications listed as references at the beginning of each major section of the handbook, the contributors and advisers consulted and drew material from technical magazines and journals, trade-association standards, engineering and scientific papers, industrial and engineering catalogs, and a variety of similar publications. Most of these are noted in appropriate places throughout the handbook. Additional acknowledgments, listed in the order received, are given below.

Data and charts credited to the Hydraulic Institute are reprinted from the *Hydraulic Institute Standards*, copyright by the Hydraulic Institute, and from the *Pipe Friction Manual*, copyright by the Hydraulic Institute. Data on diesel engine cooling systems are reprinted from *Marine Diesel Standard Practices*, copyright by the Diesel Engine Manufacturers Association. Data on minimum requirements for plumbing are drawn from the *American Standard National Plumbing Code* with permission of the publisher, The American Society of Mechanical Engineers.

Specific firms, trade associations, and publications that were extremely helpful in supplying data for various sections of the handbook include Martin Marietta Corporation; *Electronic Design* magazine; Dresser Industries Inc.—Dresser Industrial Valve and Instrument Division; Ingersoll-Rand Company; Anaconda American Brass Company; Waterloo Register Division—Dynamics Corporation of America; ITT Hammel-Dahl; *Mechanical Engineering*, a monthly publication of The American Society of Mechanical Engineers; McQuay, Inc.; The G. C. Breidert Co.; Modine Manufacturing Company; Rubber Manufacturers Association; Condenser Service & Engineering Co., Inc.; Armstrong Machine Works; American Air Filter Company; Crane Company; *Machine Design* magazine; The RAND Corporation; Texas Instruments Incorporated; McGraw-Hill Publications Company, McGraw-Hill, Inc.; Morse Chain Company; Grinnell Corporation; General Electric Company; The B. F. Goodrich Company; American Standard Inc.; the American Society of Heating, Refrigerating and Air-Conditioning Engineers; International Engineering Associates; Taylor Instrument Process Control Division of Sybron Corporation; Clark-Reliance Corporation; American Society for Testing and Materials; Acoustical and Insulating Materials Association; W. S. Dickey Clay Manufacturing Co.; Flexonics Division, Universal Oil Products Co.; Dunham-Bush, Inc.; Carrier Air Conditioning Company; National Industrial Leather Association; Worthington Corporation; Goulds Pumps, Inc. Illustrations and problems credited to Carrier Air Conditioning Company are copyrighted by Carrier Air Conditioning Company.

Individuals who were helpful to the editor of this handbook at one or more times during its preparation include Lyman F. Scheel, Consulting Engineer; Jack Jaklitsch, Editor, *Mechanical Engineering*; Spencer A. Tucker, Martin & Tucker; Paul V. DeLuca, Porta Systems

Corp.; Professor Steven Edelglass, Cooper Union; Professor William Vopat, Cooper Union; Professor Theodore Baumeister, Columbia University; Frederick S. Merritt, Consulting Engineer; James J. O'Connor, Editor, *Power* magazine; Nathan R. Grossner, Consulting Engineer; Nicholas P. Chironis, *Product Engineering;* Franklin D. Yeaple, *Product Engineering* and *Design Engineering;* John D. Constance, Consulting Engineer; John R. Miller, Texas Instruments Incorporated; Rupert Le Grand, *American Machinist;* Ronald G. Kogan, United Computing Systems, Inc.; Al Brons, Flexonics Div., Universal Oil Products Company; Carl W. MacPhee, ASHRAE *Guide and Data Book;* Frank P. Anderson, Secretary, Hydraulic Institute; Joseph Mittleman, *Electronics;* Cheryl A. Shaver, E.E., who was a major help in metricating several sections of the handbook; Thomas F. Epley, Editorial Director, U.S. Naval Institute Press; Janet Eyler, *Electronics*; Charles R. Hafer, P.E.; Calvin S. Cronan, *Chemical Engineering* Magazine; Nicholas Chopey, Executive Editor, *Chemical Engineering* Magazine; Joseph C. McCabe, Editor-Publisher, *Combustion* Magazine; Francis J. Lavoie, Managing Editor, *Machine Design* Magazine; Donald E. Fink, Managing Editor—Technical, *Aviation Week & Space Technology* Magazine; Richard J. Zanetti, Editor-in-Chief, *Chemical Engineering* Magazine; Robert G. Schwieger, Editorial Director, *Power* Magazine; Barbara LoSchiavo, Editorial Support, *Machine Design* Magazine; Joseph Leto, P.E., Consulting Engineer; Gerald M. Eisenberg, Project Engineering Administrator, American Society of Mechanical Engineers, who contributed a number of new procedures and ideas; Stephen M. Eber, P.E., Ebasco Services, Inc., who also contributed a number of new procedures and ideas; Jerome Mueller, P.E., Mueller Engineering Corporation, who was most helpful with thoughts on applied calculation procedures; Joseph B. Shanley, Mechanical Engineer, who metricated many illustrations and procedures; and numerous working engineers and scientists in firms and universities in the United States and abroad.

HOW TO USE THIS HANDBOOK

There are two ways to enter this handbook to obtain the maximum benefit from the time invested. The first entry is through the index; the second is through the table of contents of the section covering the discipline, or related discipline, concerned. Each method is discussed in detail below.

Index.   Great care and considerable time were expended on preparation of the index of this handbook so that it would be of maximum use to every reader. As a general guide, enter the index using the generic term for the type of calculation procedure being considered. Thus, for the design of a beam, enter at *beam*(s). From here, progress to the specific type of beam being considered—such as *continuous, of steel*. Once the page number or numbers of the appropriate calculation procedure are determined, turn to them to find the step-by-step instructions and worked-out example that can be followed to solve the problem quickly and accurately.

Contents.   The contents of each section lists the titles of the calculation procedures contained in that section. Where extensive use of any section is contemplated, the editor suggests that the reader might benefit from an occasional glance at the table of contents of that section. Such a glance will give the user of this handbook an understanding of the breadth and coverage of a given section, or a series of sections. Then, when he or she turns to this handbook for assistance, the reader will be able more rapidly to find the calculation procedure he or she seeks.

Calculation Procedures.   Each calculation procedure is a unit in itself. However, any given calculation procedure will contain subprocedures that might be useful to the reader. Thus, a calculation procedure on pump selection will contain subprocedures on pipe friction loss, pump static and dynamic heads, etc. Should the reader of this handbook wish to make a computation using any of such subprocedures, he or she will find the worked-out steps that are presented both useful and precise. Hence, the handbook contains numerous valuable procedures that are useful in solving a variety of applied engineering problems.

One other important point that should be noted about the calculation procedures presented in this handbook is that many of the calculation procedures are equally applicable in a variety of disciplines. Thus, a beam-selection procedure can be used for civil-, chemical-, mechanical-, electrical-, and nuclear-engineering activities, as well as some others. Hence, the reader might consider a temporary neutrality for his or her particular specialty when using the handbook because the calculation procedures are designed for universal use.

Any of the calculation procedures presented can be programmed on a computer. Such programming permits rapid solution of a variety of design problems. With the growing use of low-cost time sharing, more engineering design problems are being solved using a remote terminal in the engineering office. The editor hopes that engineers throughout the world will make greater use of work stations and portable computers in solving applied engineering problems. This modern equipment promises greater speed and accuracy for nearly all the complex design problems that must be solved in today's world of engineering.

To make the calculation procedures more amenable to computer solution (while maintaining ease of solution with a handheld calculator), a number of the algorithms in the handbook have been revised to permit faster programming in a computer environment. Likewise, all the new calculation procedures—of which there are many in this third edition—have their algorithms in programmable form. This enhances ease of solution for any method used—work station, portable computer, or calculator.

SI Usage. The technical and scientific community throughout the world accepts the SI (System International) for use in both applied and theoretical calculations. With such widespread acceptance of SI, every engineer must become proficient in the use of this system of units if he or she is to remain up-to-date. For this reason, every calculation procedure in this handbook is given in both the United States Customary System (USCS) and SI. This will help all engineers become proficient in using both systems of units. In this handbook the USCS unit is generally given first, followed by the SI value in parentheses or brackets. Thus, if the USCS unit is 10 ft, it will be expressed as 10 ft (3 m).

Engineers accustomed to working in USCS are often timid about using SI. There really aren't any sound reasons for these fears. SI is a logical, easily understood, and readily manipulated group of units. Most engineers grow to prefer SI, once they become familiar with it and overcome their fears. This handbook should do much to "convert" USCS-user engineers to SI because it presents all calculation procedures in both the known and unknown units.

Overseas engineers who must work in USCS because they have a job requiring its usage will find the dual-unit presentation of calculation procedures most helpful. Knowing SI, they can easily convert to USCS because all procedures, tables, and illustrations are presented in dual units.

Learning SI. An efficient way for the USCS-conversant engineer to learn SI follows these steps:

1. List the units of measurement commonly used in your daily work.

2. Insert, opposite each USCS unit, the usual SI unit used; Table 1 shows a variety of commonly used quantities and the corresponding SI units.

3. Find, from a table of conversion factors, such as Table 2, the value to use to convert the USCS unit to SI, and insert it in your list. (Most engineers prefer a conversion factor that can be used as a multiplier of the USCS unit to give the SI unit.)

4. Apply the conversion factors whenever you have an opportunity. Think in terms of SI when you encounter a USCS unit.

5. Recognize—here and now—that the most difficult aspect of SI is becoming comfortable with the names and magnitude of the units. Numerical conversion is simple, once you've set up *your own* conversion table. So think pascal whenever you encounter pounds per square inch pressure, newton whenever you deal with a force in pounds, etc.

SI Table for a Mechanical Engineer. Let's say you're a mechanical engineer and you wish to construct a conversion table and SI literacy document for yourself. List the units you commonly meet in your daily work; Table 1 is the list compiled by one mechanical engineer. Next, list the SI unit equivalent for the USCS unit. Obtain the equivalent from Table 2. Then, using Table 2 again, insert the conversion multiplier in Table 1.

Keep Table 1 handy at your desk and add new units to it as you encounter them in your work. Over a period of time you will build a personal conversion table that will be valuable to you whenever you must use SI units. Further, since *you* compiled the table, it will have a familiar and nonfrightening look, which will give you greater confidence in using SI.

Units Used. In preparing the calculation procedures in this handbook, the editors and contributors used standard SI units throughout. In a few cases, however, certain units are still in a state of development. For example, the unit *tonne* is used in certain industries, such as

TABLE 1 Commonly Used USCS and SI Units*

USCS unit	SI unit	SI symbol	Conversion factor—multiply USCS unit by this factor to obtain the SI unit
square feet	square meters	m²	0.0929
cubic feet	cubic meters	m³	0.2831
pounds per square inch	kilopascal	kPa	6.894
pound force	newton	N	4.448
foot pound torque	newton-meter	N·m	1.356
Btu per pound	kilojoule per kilogram	kJ/kg	2.326
gallons per minute	liters per second	L/s	0.06309
Btu per cubic foot	kilojoule per cubic meter	kJ/m³	37.26

*Because of space limitations this table is abbreviated. For a typical engineering practice an actual table would be many times this length.

metalworking. This unit is therefore used in the metalworking section of this handbook because it represents current practice. However, only a few SI units are still under development. Hence, users of this handbook face little difficulty from this situation.

Computer-aided Calculations. Widespread availability of programmable pocket calculators and low-cost microcomputers allows engineers and designers to save thousands of hours of calculation time. Yet each calculation procedure must be programmed, unless the engineer is willing to use off-the-shelf software. The editor—observing thousands of engineers over the years—detects reluctance among technical personnel to use untested and unproven software programs in their daily calculations. Hence, the tested and proven procedures in this handbook form excellent programming input for programmable pocket calculators, microcomputers, minicomputers, and mainframes.

TABLE 2 Typical Conversion Table*

To convert from	To	Multiply by	
square feet	square meters	9.290304	E − 02
foot per second squared	meter per second squared	3.048	E − 01
cubic feet	cubic meters	2.831685	E − 02
pound per cubic inch	kilogram per cubic meter	2.767990	E + 04
gallon per minute	liters per second	6.309	E − 02
pound per square inch	kilopascal	6.894757	
pound force	newton	4.448222	
British thermal unit per square foot	joule per square meter	1.135653	E + 04
British thermal unit per hour	Watt	2.930711	E − 01
British thermal unit per cubic foot	kilojoule per cubic meter	3.725697	E + 01
foot-pound torque	newton-meter	1.355818	
British thermal unit per pound	joule per kilogram	2.326	E + 03

Note: The E indicates an exponent, as in scientific notation, followed by a positive or negative number, representing the power of 10 by which the given conversion factor is to be multiplied before use. Thus, for the square feet conversion factor, 9.290304 × 1/100 = 0.09290304, the factor to be used to convert square feet to square meters. For a positive exponent, as in converting British thermal units per cubic foot to kilojoule per cubic meter, 3.725697 × 10 = 37.25697.

Where a conversion factor cannot be found, simply use the dimensional substitution. Thus, to convert pounds per cubic inch to kilograms per cubic meter, find 1 lb = 0.4535924 kg, and 1 in³ = 0.00001638706 m³. Then, 1 lb/in³ = 0.4535924 kg/0.00001638706 m³ = 27,67990, or 2.767990 E + 04.

*This table contains only selected values. See the U.S. Department of the Interior *Metric Manual*, or National Bureau of Standards, *The International System of Units* (SI), both available from the U.S. Government Printing Office (GPO), for far more comprehensive listings of conversion factors.

A variety of software application programs can be used to put the procedures in this handbook on computer. Typical of these are MathSoft, Algor, and similar programs.

There are a number of advantages for the engineer who programs his or her own calculation procedures, namely: (1) The engineer knows, understands, and approves *every* step in the procedure; (2) there are *no* questionable, unknown, or legally worrisome steps in the procedure; (3) the engineer has complete faith in the result because he or she knows every component of it; and (4) if a variation of the procedure is desired, it is relatively easy for the engineer to make the needed changes in the program, using this handbook as the source of the steps and equations to apply.

Modern computer equipment provides greater speed and accuracy for almost all complex design calculations. The editor hopes that engineers throughout the world will make greater use of available computing equipment in solving applied engineering problems. Becoming computer literate is a necessity for every engineer, no matter which field he or she chooses as a specialty. The procedures in this handbook simplify every engineer's task of becoming computer literate because the steps given comprise—to a great extent—the steps in the computer program that can be written.

SECTION 1

CIVIL ENGINEERING

MAX KURTZ, P.E.
CONSULTING ENGINEER

METRICATED BY

GERALD M. EISENBERG
PROJECT ENGINEERING ADMINISTRATOR
AMERICAN SOCIETY OF MECHANICAL ENGINEERS

PRINCIPLES OF STATICS; GEOMETRIC PROPERTIES OF AREAS 1.7
    Graphical Analysis of a Force System 1.7
    Analysis of Static Friction 1.8
    Analysis of a Structural Frame 1.9
    Graphical Analysis of a Plane Truss 1.10
    Truss Analysis by the Method of Joints 1.12
    Truss Analysis by the Method of Sections 1.14
    Reactions of a Three-Hinged Arch 1.14
    Length of Cable Carrying Known Loads 1.15
    Parabolic Cable Tension and Length 1.17
    Catenary Cable Sag and Distance between Supports 1.17
    Stability of a Retaining Wall 1.18
    Analysis of a Simple Space Truss 1.19
    Analysis of a Compound Space Truss 1.20
    Geometric Properties of an Area 1.22
    Product of Inertia of an Area 1.24
    Properties of an Area with Respect to Rotated Axes 1.25
ANALYSIS OF STRESS AND STRAIN 1.25
    Stress Caused by an Axial Load 1.26
    Deformation Caused by an Axial Load 1.26
    Deformation of a Built-Up Member 1.26
    Reactions at Elastic Supports 1.27
    Analysis of Cable Supporting a Concentrated Load 1.28
    Displacement of Truss Joint 1.29
    Axial Stress Caused by Impact Load 1.29
    Stresses on an Oblique Plane 1.30
    Evaluation of Principal Stresses 1.31

Hoop Stress in Thin-Walled Cylinder under Pressure 1.32
Stresses in Prestressed Cylinder 1.32
Hoop Stress in Thick-Walled Cylinder 1.33
Thermal Stress Resulting from Heating a Member 1.34
Thermal Effects in Composite Member Having Elements in Parallel 1.34
Thermal Effects in Composite Member Having Elements in Series 1.35
Shrink-Fit Stress and Radial Pressure 1.35
Torsion of a Cylindrical Shaft 1.36
Analysis of a Compound Shaft 1.36
STRESSES IN FLEXURAL MEMBERS 1.37
Shear and Bending Moment in a Beam 1.37
Beam Bending Stresses 1.39
Analysis of a Beam on Movable Supports 1.40
Flexural Capacity of a Compound Beam 1.40
Analysis of a Composite Beam 1.41
Beam Shear Flow and Shearing Stress 1.42
Locating the Shear Center of a Section 1.43
Bending of a Circular Flat Plate 1.45
Bending of a Rectangular Flat Plate 1.45
Combined Bending and Axial Load Analysis 1.45
Flexural Stress in a Curved Member 1.46
Soil Pressure under Dam 1.47
Load Distribution in Pile Group 1.48
DEFLECTION OF BEAMS 1.48
Double-Integration Method of Determining Beam Deflection 1.49
Moment-Area Method of Determining Beam Deflection 1.49
Conjugate-Beam Method of Determining Beam Deflection 1.50
Unit-Load Method of Computing Beam Deflection 1.51
Deflection of a Cantilever Frame 1.52
STATICALLY INDETERMINATE STRUCTURES 1.53
Shear and Bending Moment of a Beam on a Yielding Support 1.53
Maximum Bending Stress in Beams Jointly Supporting a Load 1.54
Theorem of Three Moments 1.55
Theorem of Three Moments: Beam with Overhang and Fixed End 1.56
Bending-Moment Determination by Moment Distribution 1.57
Analysis of a Statically Indeterminate Truss 1.58
MOVING LOADS AND INFLUENCE LINES 1.59
Analysis of Beam Carrying Moving Concentrated Loads 1.59
Influence Line for Shear in a Bridge Truss 1.60
Force in Truss Diagonal Caused by a Moving Uniform Load 1.62
Force in Truss Diagonal Caused by Moving Concentrated Loads 1.63
Influence Line for Bending Moment in Bridge Truss 1.64
Force in Truss Chord Caused by Moving Concentrated Loads 1.65
Influence Line for Bending Moment in Three-Hinged Arch 1.66
Deflection of a Beam under Moving Loads 1.67
RIVETED AND WELDED CONNECTIONS 1.68
Capacity of a Rivet 1.68
Investigation of a Lap Splice 1.69
Design of a Butt Splice 1.70
Design of a Pipe Joint 1.71
Moment on Riveted Connection 1.72
Eccentric Load on Riveted Connection 1.72
Design of a Welded Lap Joint 1.74
Eccentric Load on a Welded Connection 1.75
STEEL BEAMS AND PLATE GIRDERS 1.76
Most Economic Section for a Beam with a Continuous Lateral Support under a
    Uniform Load 1.76
Most Economic Section for a Beam with Intermittent Lateral Support under
    Uniform Load 1.76

Design of a Beam with Reduced Allowable Stress 1.78
Design of a Cover-Plated Beam 1.79
Design of a Continuous Beam 1.82
Shearing Stress in a Beam—Exact Method 1.83
Shearing Stress in a Beam—Approximate Method 1.83
Moment Capacity of a Welded Plate Girder 1.84
Analysis of a Riveted Plate Girder 1.84
Design of a Welded Plate Girder 1.85
STEEL COLUMNS AND TENSION MEMBERS 1.88
Capacity of a Built-Up Column 1.89
Capacity of a Double-Angle Star Strut 1.90
Section Selection for a Column with Two Effective Lengths 1.91
Stress in Column with Partial Restraint against Rotation 1.91
Lacing of Built-Up Column 1.92
Selection of a Column with a Load at an Intermediate Level 1.93
Design of an Axial Member for Fatigue 1.93
Investigation of a Beam Column 1.94
Application of Beam-Column Factors 1.95
Net Section of a Tension Member 1.96
Design of a Double-Angle Tension Member 1.96
PLASTIC DESIGN OF STEEL STRUCTURES 1.97
Allowable Load on Bar Supported by Rods 1.98
Determination of Section Shape Factors 1.99
Determination of Ultimate Load by the Static Method 1.100
Determining the Ultimate Load by the Mechanism Method 1.101
Analysis of a Fixed-End Beam under Concentrated Load 1.102
Analysis of a Two-Span Beam with Concentrated Loads 1.102
Selection of Sizes for a Continuous Beam 1.104
Mechanism-Method Analysis of a Rectangular Portal Frame 1.105
Analysis of a Rectangular Portal Frame by the Static Method 1.107
Theorem of Composite Mechanisms 1.108
Analysis of an Unsymmetric Rectangular Portal Frame 1.108
Analysis of Gable Frame by Static Method 1.110
Theorem of Virtual Displacements 1.112
Gable-Frame Analysis by Using the Mechanism Method 1.113
Reduction in Plastic-Moment Capacity Caused by Axial Force 1.114
TIMBER ENGINEERING 1.116
Bending Stress and Deflection of Wood Joists 1.116
Shearing Stress Caused by Stationary Concentrated Load 1.117
Shearing Stress Caused by Moving Concentrated Load 1.117
Strength of Deep Wooden Beams 1.117
Design of a Wood-Plywood Beam 1.118
Determining the Capacity of a Solid Column 1.119
Design of a Solid Wooden Column 1.120
Investigation of a Spaced Column 1.120
Compression on an Oblique Plane 1.121
Design of a Notched Joint 1.121
Allowable Lateral Load on Nails 1.122
Capacity of Lag Screws 1.123
Design of a Bolted Splice 1.123
Investigation of a Timber-Connector Joint 1.124
REINFORCED CONCRETE 1.125
DESIGN OF FLEXURAL MEMBERS BY ULTIMATE-STRENGTH METHOD . . 1.126
Capacity of a Rectangular Beam 1.127
Design of a Rectangular Beam 1.128
Design of the Reinforcement in a Rectangular Beam of Given Size 1.129
Capacity of a T Beam 1.129
Capacity of a T Beam of Given Size 1.130
Design of Reinforcement in a T Beam of Given Size 1.130

Reinforcement Area for a Doubly Reinforced Rectangular Beam 1.131
Design of Web Reinforcement 1.132
Determination of Bond Stress 1.134
Design of Interior Span of a One-Way Slab 1.134
Analysis of a Two-Way Slab by the Yield-Line Theory 1.136
DESIGN OF FLEXURAL MEMBERS BY THE WORKING-STRESS METHOD . . 1.138
Stresses in a Rectangular Beam 1.140
Capacity of a Rectangular Beam 1.141
Design of Reinforcement in a Rectangular Beam of Given Size 1.142
Design of a Rectangular Beam 1.142
Design of Web Reinforcement 1.143
Capacity of a T Beam 1.144
Design of a T Beam Having Concrete Stressed to Capacity 1.145
Design of a T Beam Having Steel Stressed to Capacity 1.146
Reinforcement for Doubly Reinforced Rectangular Beam 1.147
Deflection of a Continuous Beam 1.148
DESIGN OF COMPRESSION MEMBERS BY ULTIMATE-STRENGTH
    METHOD 1.149
Analysis of a Rectangular Member by Interaction Diagram 1.150
Axial-Load Capacity of Rectangular Member 1.151
Allowable Eccentricity of a Member 1.152
DESIGN OF COMPRESSION MEMBERS BY WORKING-STRESS METHOD . . 1.153
Design of a Spirally Reinforced Column 1.153
Analysis of a Rectangular Member by Interaction Diagram 1.154
Axial-Load Capacity of a Rectangular Member 1.156
DESIGN OF COLUMN FOOTINGS 1.157
Design of an Isolated Square Footing 1.157
Combined Footing Design 1.159
CANTILEVER RETAINING WALLS 1.162
Design of a Cantilever Retaining Wall 1.162
PRESTRESSED CONCRETE 1.165
Determination of Prestress Shear and Moment 1.167
Stresses in a Beam with Straight Tendons 1.168
Determination of Capacity and Prestressing Force for a Beam with Straight
    Tendons 1.170
Beam with Deflected Tendons 1.172
Beam with Curved Tendons 1.173
Determination of Section Moduli 1.174
Effect of Increase in Beam Span 1.174
Effect of Beam Overload 1.175
Prestressed-Concrete Beam Design Guides 1.175
Kern Distances 1.176
Magnel Diagram Construction 1.176
Camber of a Beam at Transfer 1.178
Design of a Double-T Roof Beam 1.179
Design of a Posttensioned Girder 1.182
Properties of a Parabolic Arc 1.186
Alternative Methods of Analyzing a Beam with Parabolic Trajectory 1.187
Prestress Moments in a Continuous Beam 1.187
Principle of Linear Transformation 1.188
Concordant Trajectory of a Beam 1.190
Design of Trajectory to Obtain Assigned Prestress Moments 1.191
Effect of Varying Eccentricity at End Support 1.191
Design of Trajectory for a Two-Span Continuous Beam 1.197
Reactions for a Continuous Beam 1.197
DESIGN OF HIGHWAY BRIDGES 1.197
Design of a T-Beam Bridge 1.198
Composite Steel-and-Concrete Bridge 1.200

FLUID MECHANICS 1.204
  HYDROSTATICS 1.204
  Buoyancy and Flotation 1.204
  Hydrostatic Force on a Plane Surface 1.204
  Hydrostatic Force on a Curved Surface 1.206
  Stability of a Vessel 1.206
  MECHANICS OF INCOMPRESSIBLE FLUIDS 1.208
  Viscosity of Fluid 1.208
  Application of Bernoulli's Theorem 1.209
  Flow through a Venturi Meter 1.210
  Flow through an Orifice 1.211
  Flow through the Suction Pipe of a Drainage Pump 1.211
  Power of a Flowing Liquid 1.211
  Discharge over a Sharp-Edged Weir 1.212
  Laminar Flow in a Pipe 1.212
  Turbulent Flow in Pipe—Application of Darcy-Weisbach Formula 1.213
  Determination of Flow in a Pipe 1.215
  Pipe-Size Selection by the Manning Formula 1.215
  Loss of Head Caused by Sudden Enlargement of Pipe 1.215
  Discharge of Looping Pipes 1.216
  Fluid Flow in Branching Pipes 1.217
  Uniform Flow in Open Channel—Determination of Slope 1.217
  Required Depth of Canal for Specified Fluid Flow Rate 1.218
  Alternate Stages of Flow; Critical Depth 1.218
  Determination of Hydraulic Jump 1.220
  Rate of Change of Depth in Nonuniform Flow 1.221
  Discharge between Communicating Vessels 1.221
  Variation in Head on a Weir without Inflow to the Reservoir 1.222
  Variation in Head on a Weir with Inflow to the Reservoir 1.222
  Dimensional Analysis Methods 1.224
  Hydraulic Similarity and Construction of Models 1.225
SURVEYING AND ROUTE DESIGN 1.225
  Plotting a Closed Traverse 1.225
  Area of Tract with Rectilinear Boundaries 1.227
  Partition of a Tract 1.228
  Area of Tract with Meandering Boundary: Offsets at Irregular Intervals . . . 1.230
  Differential Leveling Procedure 1.230
  Stadia Surveying 1.231
  Volume of Earthwork 1.232
  Determination of Azimuth of a Star by Field Astronomy 1.233
  Time of Culmination of a Star 1.235
  Plotting a Circular Curve 1.236
  Intersection of Circular Curve and Straight Line 1.238
  Realignment of Circular Curve by Displacement of Forward Tangent . . . 1.238
  Characteristics of a Compound Curve 1.239
  Analysis of a Highway Transition Spiral 1.241
  Transition Spiral: Transit at Intermediate Station 1.243
  Plotting a Parabolic Arc 1.245
  Location of a Single Station on a Parabolic Arc 1.247
  Location of a Summit 1.247
  Parabolic Curve to Contain a Given Point 1.248
  Sight Distance on a Vertical Curve 1.249
  Mine Surveying: Grade of Drift 1.250
  Determining Strike and Dip from Two Apparent Dips 1.250
  Determination of Strike, Dip, and Thickness from Two Skew Boreholes . . . 1.252
AERIAL PHOTOGRAMMETRY 1.255
  Flying Height Required to Yield a Given Scale 1.255
  Determining Ground Distance by Vertical Photograph 1.257

Determining the Height of a Structure by Vertical Photograph 1.257
Determining Ground Distance by Tilted Photograph 1.259
Determining Elevation of a Point by Overlapping Vertical Photographs 1.261
Determining Air Base of Overlapping Vertical Photographs by Use of Two
   Control Points 1.261
Determining Scale of Oblique Photograph 1.263
SOIL MECHANICS 1.265
Composition of Soil 1.265
Specific Weight of Soil Mass 1.266
Analysis of Quicksand Conditions 1.266
Measurement of Permeability by Falling-Head Permeameter 1.267
Construction of Flow Net 1.267
Soil Pressure Caused by Point Load 1.269
Vertical Force on Rectangular Area Caused by Point Load 1.269
Vertical Pressure Caused by Rectangular Loading 1.270
Appraisal of Shearing Capacity of Soil by Unconfined Compression Test . . . 1.270
Appraisal of Shearing Capacity of Soil by Triaxial Compression Test 1.272
Earth Thrust on Retaining Wall Calculated by Rankine's Theory 1.273
Earth Thrust on Retaining Wall Calculated by Coulomb's Theory 1.274
Earth Thrust on Timbered Trench Calculated by General Wedge Theory . . . 1.275
Thrust on a Bulkhead 1.277
Cantilever Bulkhead Analysis 1.278
Anchored Bulkhead Analysis 1.278
Stability of Slope by Method of Slices 1.280
Stability of Slope by ϕ-Circle Method 1.282
Analysis of Footing Stability by Terzaghi's Formula 1.284
Soil Consolidation and Change in Void Ratio 1.285
Compression Index and Void Ratio of a Soil 1.285
Settlement of Footing 1.286
Determination of Footing Size by Housel's Method 1.287
Application of Pile-Driving Formula 1.287
Capacity of a Group of Friction Piles 1.288
Load Distribution among Hinged Batter Piles 1.288
Load Distribution among Piles with Fixed Bases 1.290
Load Distribution among Piles Fixed at Top and Bottom 1.291

REFERENCES: Crawley and Dillion—*Steel Buildings: Analysis and Design*, Wiley; Bowles—*Structural Steel Design*, McGraw-Hill; ASCE Council on Computer Practices—*Computing in Civil Engineering*, ASCE; American Concrete Institute—*Building Code Requirements for Reinforced Concrete*; American Institute of Steel Construction—*Manual of Steel Construction*; National Forest Products Association—*National Design Specification for Stress-Grade Lumber and Its Fastenings*; Abbett—*American Civil Engineering Practice*, Wiley; Gaylord and Gaylord—*Structural Engineering Handbook*, McGraw-Hill; LaLonde and Janes—*Concrete Engineering Handbook*, McGraw-Hill; Lincoln Electric Co.—*Procedure Handbook of Arc Welding Design and Practice*; Merritt—*Standard Handbook for Civil Engineers*, McGraw-Hill; Timber Engineering Company—*Timber Design and Construction Handbook*, McGraw-Hill; U.S. Department of Agriculture, Forest Products Laboratory—*Wood Handbook (Agriculture Handbook 72)*, GPO; Urquhart—*Civil Engineering Handbook*, McGraw-Hill; Borg and Gennaro—*Advanced Structural Analysis*, Van Nostrand; Gerstle—*Basic Structural Design*, McGraw-Hill; Jensen—*Applied Strength of Materials*, McGraw-Hill; Kurtz—*Comprehensive Structural Design Design Guide*, McGraw-Hill; Roark—*Formulas for Stress and Strain*, McGraw-Hill; Seely—*Resistance of Materials*, Wiley; Shanley—*Mechanics of Materials*, McGraw-Hill; Timoshenko and Young—*Theory of Structures*, McGraw-Hill; Beedle, et al.—*Structural Steel Design*, Ronald; Grinter—*Design of Modern Steel Structures*, Macmillan; Lothers—*Advanced Design in Structural Steel*, Prentice-Hall; Beedle—*Plastic Design of Steel Frames*, Wiley; Canadian Institute of Timber Construction—*Timber Construction*; Scofield and O'Brien—*Modern Timber Engineering*, Southern Pine Association; Dunham—*Theory and Practice of Reinforced Concrete*, McGraw-Hill; Winter, et al.—*Design of Concrete Structures*, McGraw-Hill; Viest, Fountain, and Singleton—*Composite Construction in Steel and Concrete*, McGraw-Hill; Chi and Biberstein—*Theory of Prestressed Concrete*, Prentice-Hall; Connolly—*Design of Prestressed Concrete Beams*, McGraw-Hill; Evans and Bennett—*Pre-stressed Concrete*, Wiley; Libby—*Prestressed Concrete*, Ronald; Magnel—*Prestressed Concrete*, McGraw-Hill; Gennaro—*Computer Methods in Solid Mechanics*, Macmillan;

Laursen—*Matrix Analysis of Structures*, McGraw-Hill; Weaver—*Computer Programs for Structural Analysis*, Van Nostrand; Brenkert—*Elementary Theoretical Fluid Mechanics*, Wiley; Daugherty and Franzini—*Fluid Mechanics with Engineering Applications*, McGraw-Hill; King and Brater—*Handbook of Hydraulics*, McGraw-Hill; Li and Lam—*Principles of Fluid Mechanics*, Addison-Wesley; Sabersky and Acosta—*Fluid Flow*, Macmillan; Streeter—*Fluid Mechanics*, McGraw-Hill; Allen—*Railroad Curves and Earthwork*, McGraw-Hill; Davis, Foote, and Kelly—*Surveying: Theory and Practice*, McGraw-Hill; Hickerson—*Route Surveys and Design*, McGraw-Hill; Hosmer and Robbins—*Practical Astronomy*, Wiley; Jones—*Geometric Design of Modern Highways*, Wiley; Meyer—*Route Surveying*, International Textbook; American Association of State Highways Officials—*A Policy on Geometric Design of Rural Highways;* Chellis—*Pile Foundations*, McGraw-Hill; Goodman and Karol—*Theory and Practice of Foundation Engineering*, Macmillan; Huntington—*Earth Pressures and Retaining Walls*, Wiley; Ritter and Paquette—*Highway Engineering*, Ronald; Scott and Schoustra—*Soil: Mechanics and Engineering*, McGraw-Hill; Spangler—*Soil Engineering*, International Textbook; Teng—*Foundation Design*, Prentice-Hall; Terzaghi and Peck—*Soil Mechanics in Engineering Practice*, Wiley; U.S. Department of the Interior, Bureau of Reclamation—*Earth Manual*, GPO.

Principles of Statics; Geometric Properties of Areas

If a body remains in equilibrium under a system of forces, the following conditions obtain:

1. The algebraic sum of the components of the forces in any given direction is zero.

2. The algebraic sum of the moments of the forces with respect to any given axis is zero.

The above statements are verbal expressions of the *equations of equilibrium*. In the absence of any notes to the contrary, a clockwise moment is considered positive; a counterclockwise moment, negative.

GRAPHICAL ANALYSIS OF A FORCE SYSTEM

The body in Fig. 1*a* is acted on by forces *A*, *B*, and *C*, as shown. Draw the vector representing the equilibrant of this system.

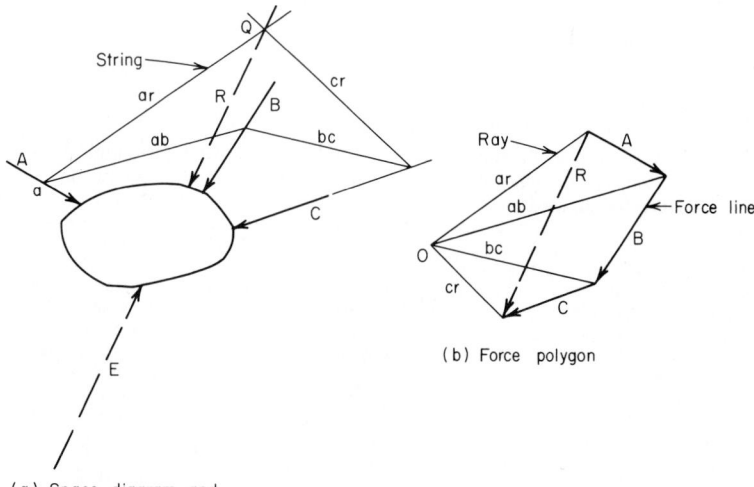

(a) Space diagram and
string polygon

(b) Force polygon

FIG. 1 Equilibrant of force system.

Calculation Procedure:

1. Construct the system force line

In Fig. 1b, draw the vector chain A-B-C, which is termed the *force line*. The vector extending from the initial point to the terminal point of the force line represents the resultant R. In any force system, the resultant R is equal to and collinear with the equilibrant E, but acts in the opposite direction. The equilibrant of a force system is a single force that will balance the system.

2. Construct the system rays

Selecting an arbitrary point O as the pole, draw the rays from O to the ends of the vectors and label them as shown in Fig. 1b.

3. Construct the string polygon

In Fig. 1a, construct the string polygon as follows: At an arbitrary point a on the action line of force A, draw strings parallel to rays ar and ab. At the point where the string ab intersects the action line of force B, draw a string parallel to ray bc. At the point where string bc intersects the action line of force C, draw a string parallel to cr. The intersection point Q of ar and cr lies on the action line of R.

4. Draw the vector for the resultant and equilibrant

In Fig. 1a, draw the vector representing R. Establish the magnitude and direction of this vector from the force polygon. The action line of R passes through Q.

Last, draw a vector equal to and collinear with that representing R but opposite in direction. This vector represents the equilibrant E.

Related Calculations: Use this general method for any force system acting in a single plane. With a large number of forces, the resultant of a smaller number of forces can be combined with the remaining forces to simplify the construction.

ANALYSIS OF STATIC FRICTION

The bar in Fig. 2a weighs 100 lb (444.8 N) and is acted on by a force P that makes an angle of 55° with the horizontal. The coefficient of friction between the bar and the inclined plane is 0.20. Compute the minimum value of P required (a) to prevent the bar from sliding down the plane; (b) to cause the bar to move upward along the plane.

Calculation Procedure:

1. Select coordinate axes

Establish coordinate axes x and y through the center of the bar, parallel and perpendicular to the plane, respectively.

2. Draw a free-body diagram of the system

In Fig. 2b, draw a free-body diagram of the bar. The bar is acted on by its weight W, the force P, and the reaction R of the plane on the bar. Show R resolved into its x and y components, the former being directed upward.

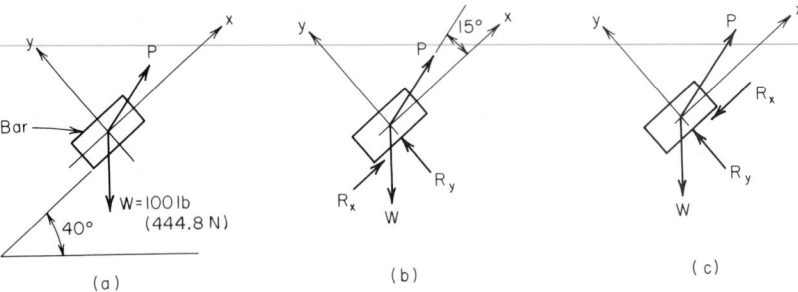

FIG. 2 Body on inclined plane.

3. Resolve the forces into their components

The forces W and P are the important ones in this step, and they must be resolved into their x and y components. Thus

$$W_x = -100 \sin 40° = -64.3 \text{ lb } (-286.0 \text{ N}) \qquad W_y = -100 \cos 40° = -76.6 \text{ lb } (-340.7 \text{ N})$$
$$P_x = P \cos 15° = 0.966P \qquad\qquad\qquad P_y = P \sin 15° = 0.259P$$

4. Apply the equations of equilibrium

Consider that the bar remains at rest and apply the equations of equilibrium. Thus

$$\Sigma F_x = R_x + 0.966P - 64.3 = 0 \qquad R_x = 64.3 - 0.966P$$
$$\Sigma F_y = R_y + 0.259P - 76.6 = 0 \qquad R_y = 76.6 - 0.259P$$

5. Assume maximum friction exists and solve for the applied force

Assume that R_x, which represents the frictional resistance to motion, has its maximum potential value. Apply $R_x = \mu R_y$, where μ = coefficient of friction. Then $R_x = 0.20R_y = 0.20(76.6 - 0.259P) = 15.32 - 0.052P$. Substituting for R_x from step 4 yields $64.3 - 0.966P = 15.32 - 0.052P$; so $P = 53.6$ lb (238.4 N).

6. Draw a second free-body diagram

In Fig. 2c, draw a free-body diagram of the bar, with R_x being directed downward.

7. Solve as in steps 1 through 5

As before, $R_y = 76.6 - 0.259P$. Also the absolute value of $R_x = 0.966P - 64.3$. But $R_x = 0.20R_y = 15.32 \times 0.052P$. Then $0.966P - 64.3 = 15.32 - 0.052P$; so $P = 78.2$ lb (347.8 N).

ANALYSIS OF A STRUCTURAL FRAME

The frame in Fig. 3a consists of two inclined members and a tie rod. What is the tension in the rod when a load of 1000 lb (4448.0 N) is applied at the hinged apex? Neglect the weight of the frame and consider the supports to be smooth.

Calculation Procedure:

1. Draw a free-body diagram of the frame

Since friction is absent in this frame, the reactions at the supports are vertical. Draw a free-body diagram as in Fig. 3b.

With the free-body diagram shown, compute the distances x_1 and x_2. Since the frame forms a 3-4-5 right triangle, $x_1 = 16(4/5) = 12.8$ ft (3.9 m) and $x_2 = 12(3/5) = 7.2$ ft (2.2 m).

2. Determine the reactions on the frame

Take moments with respect to A and B to obtain the reactions:

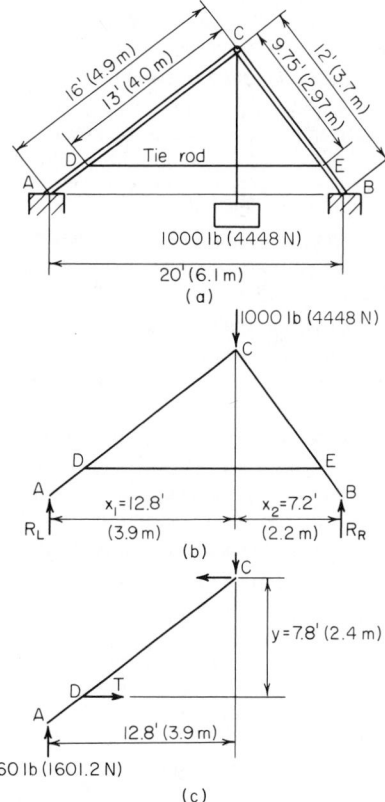

FIG. 3

$$\Sigma M_B = 20R_L - 1000(7.2) = 0 \qquad R_L = 360 \text{ lb } (1601.2 \text{ N})$$
$$\Sigma M_A = 1000(12.8) - 20R_R = 0 \qquad R_R = 640 \text{ lb } (2846.7 \text{ N})$$

3.   Determine the distance y in Fig. 3c

Draw a free-body diagram of member AC in Fig. 3c. Compute $y = 13(3/5) = 7.8$ ft (2.4 m).

4.   Compute the tension in the tie rod

Take moments with respect to C to find the tension T in the tie rod:

$$\Sigma M_C = 360(12.8) - 7.8T = 0 \qquad T = 591 \text{ lb } (2628.8 \text{ N})$$

5.   Verify the computed result

Draw a free-body diagram of member BC, and take moments with respect to C. The result verifies that computed above.

GRAPHICAL ANALYSIS OF A PLANE TRUSS

Apply a graphical analysis to the cantilever truss in Fig. 4a to evaluate the forces induced in the truss members.

Calculation Procedure:

1.   Label the truss for analysis

Divide the space around the truss into regions bounded by the action lines of the external and internal forces. Assign an uppercase letter to each region (Fig. 4).

2.   Determine the reaction force

Take moments with respect to joint 8 (Fig. 4) to determine the horizontal component of the reaction force R_U. Then compute R_U. Thus $\Sigma M_8 = 12R_{UH} - 3(8 + 16 + 24) - 5(6 + 12 + 18) = 0$; so $R_{UH} = 27$ kips (120.1 kN) to the right.

Since R_U is collinear with the force DE, $R_{UV}/R_{UH} = {}^{13}\!/_{24}$, so $R_{UV} = 13.5$ kips (60.0 kN) upward, and $R_U = 30.2$ kips (134.3 kN).

3.   Apply the equations of equilibrium

Use the equations of equilibrium to find R_L. Thus $R_{LH} = 27$ kips (120.1 kN) to the left, $R_{LV} = 10.5$ kips (46.7 kN) upward, and $R_L = 29.0$ kips (129.0 kN).

4.   Construct the force polygon

Draw the force polygon in Fig. 4b by using a suitable scale and drawing vector **fg** to represent force FG. Next, draw vector **gh** to represent force GH, and so forth. Omit the arrowheads on the vectors.

5.   Determine the forces in the truss members

Starting at joint 1, Fig. 4b, draw a line through a in the force polygon parallel to member AJ in the truss, and one through h parallel to member HJ. Designate the point of intersection of these lines as j. Now, vector **aj** represents the force in AJ, and vector **hj** represents the force in HJ.

6.   Analyze the next joint in the truss

Proceed to joint 2, where there are now only two unknown forces—BK and JK. Draw a line through b in the force polygon parallel to BK and one through j parallel to JK. Designate the point of intersection as k. The forces BK and JK are thus determined.

7.   Analyze the remaining joints

Proceed to joints 3, 4, 5, and 6, in that order, and complete the force polygon by continuing the process. If the construction is accurately performed, the vector **pe** will parallel the member PE in the truss.

8.   Determine the magnitude of the internal forces

Scale the vector lengths to obtain the magnitude of the internal forces. Tabulate the results as in Table 1.

9.   Establish the character of the internal forces

To determine whether an internal force is one of tension or compression, proceed in this way: Select a particular joint and proceed around the joint in a clockwise direction, listing the letters

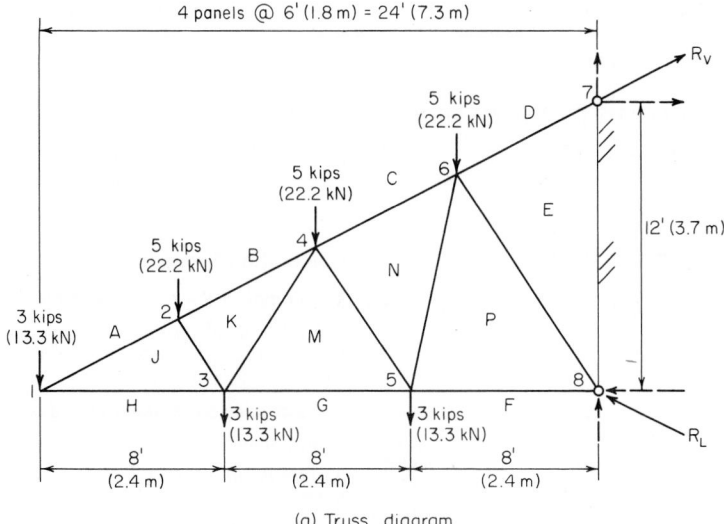

(a) Truss diagram

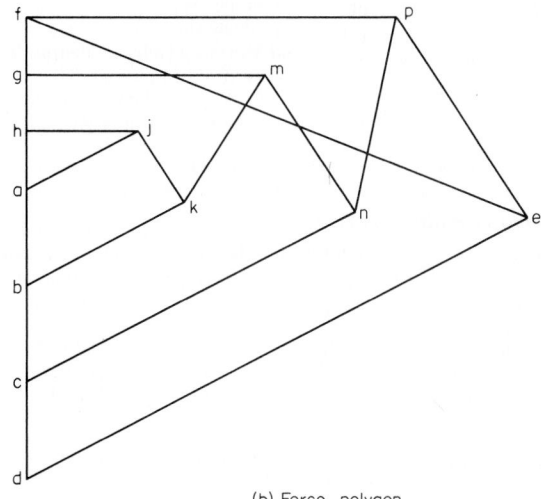

(b) Force polygon

FIG. 4

in the order in which they appear. Then refer to the force polygon pertaining to that joint, and proceed along the polygon in the same order. This procedure shows the direction in which the force is acting at that joint.

For instance, by proceeding around joint 4, *CNMKB* is obtained. By tracing a path along the force polygon in the order in which the letters appear, force *CN* is found to act upward to the right; *NM* acts upward to the left; *MK* and *KB* act downward to the left. Therefore, *CN*, *MK*, and *KB* are directed away from the joint (Fig. 4); this condition discloses that they are tensile forces. Force *NM* is directed toward the joint; therefore, it is compressive.

The validity of this procedure lies in the drawing of the vectors representing external forces while proceeding around the truss in a clockwise direction. Tensile forces are shown with a positive sign in Table 1; compressive forces are shown with a negative sign.

Related Calculations: Use this general method for any type of truss.

TRUSS ANALYSIS BY THE METHOD OF JOINTS

Applying the method of joints, determine the forces in the truss in Fig. 5a. The load at joint 4 has a horizontal component of 4 kips (17.8 kN) and a vertical component of 3 kips (13.3 kN).

TABLE 1 Forces in Truss Members (Fig. 4)

Member	Force	
	kips	kN
AJ	+6.7	+29.8
BK	+9.5	+42.2
CN	+19.8	+88.0
DE	+30.2	+134.2
HJ	−6.0	−26.7
GM	−13.0	−57.8
FP	−20.0	−88.9
JK	−4.5	−20.0
KM	+8.1	+36.0
MN	−8.6	−38.2
NP	+10.4	+46.2
PE	−12.6	−56.0

Calculation Procedure:

1. Compute the reactions at the supports

Using the usual analysis techniques, we find R_{LV} = 19 kips (84.5 kN); R_{LH} = 4 kips (17.8 kN); R_R = 21 kips (93.4 kN).

2. List each truss member and its slope

Table 2 shows each truss member and its slope.

3. Determine the forces at a principal joint

Draw a free-body diagram, Fig. 5b, of the pin at joint 1. For the free-body diagram, assume that the unknown internal forces AJ and HJ are tensile. Apply the equations of equilibrium to evaluate these forces, using the subscripts H and V, respectively, to identify the horizontal and vertical components. Thus ΣF_H = 4.0 +

$AJ_H + HJ = 0$ and $\Sigma F_V = 19.0 + AJ_V = 0$; $\therefore$ AJ_V = −19.0 kips (−84.5 kN); AJ_H = − 19.0/0.75 = −25.3 kips (−112.5 kN). Substituting in the first equation gives HJ = 21.3 kips (94.7 kN).

The algebraic signs disclose that AJ is compressive and HJ is tensile. Record these results in Table 2, showing the tensile forces as positive and compressive forces as negative.

4. Determine the forces at another joint

Draw a free-body diagram of the pin at joint 2 (Fig. 5c). Show the known force AJ as compressive, and assume that the unknown forces BK and JK are tensile. Apply the equations of equilibrium, expressing the vertical components of BK and JK in terms of their horizontal components. Thus $\Sigma F_H = 25.3 + BK_H + JK_H = 0$; $\Sigma F_V = -6.0 + 19.0 + 0.75BK_H - 0.75JK_H = 0$.

Solve these simultaneous equations, to obtain BK_H = −21.3 kips (−94.7 kN); JK_H = − 4.0 kips (−17.8 kN); BK_V = −16.0 kips (−71.2 kN); JK_V = −3.0 kips (−13.3 kN). Record these results in Table 2.

5. Continue the analysis at the next joint

Proceed to joint 3. Since there are no external horizontal forces at this joint, CL_H = BK_H = 21.3 kips (94.7 kN) of compression. Also, KL = 6 kips (26.7 kN) of compression.

6. Proceed to the remaining joints in their numbered order

Thus, for *joint 4*: $\Sigma F_H = -4.0 - 21.3 + 4.0 + LM_H + GM = 0$; $\Sigma F_V = -3.0 - 3.0 - 6.0 + LM_V = 0$; LM_V = 12.0 kips (53.4 kN); LM_H = 12.0/2.25 = 5.3 kips (23.6 kN). Substituting in the first equation gives GM = 16.0 kips (71.2 kN).

Joint 5: $\Sigma F_H = 21.3 - 5.3 + DN_H + MN_H = 0$; $\Sigma F_V = -6.0 + 16.0 - 12.0 - 0.75DN_H - 2.25MN_H = 0$; DN_H = −22.7 kips (−101.0 kN); MN_H = 6.7 kips (29.8 kN); DN_V = −17.0 kips (−75.6 kN); MN_V = 15.0 kips (66.7 kN).

Joint 6: $EP_H = DN_H$ = 22.7 kips (101.0 kN) of compression; NP = 11.0 kips (48.9 kN) of compression.

Joint 7: $\Sigma F_H = 22.7 - PQ_H + FQ_H = 0$; $\Sigma F_V = -8.0 - 17.0 - 0.75PQ_H - 0.75FQ_H = 0$; PQ_H = −5.3 kips (−23.6 kN); FQ_H = −28.0 kips (−124.5 kN); PQ_V = −4.0 kips (−17.8 kN); FQ_V = −21.0 kips (−93.4 kN).

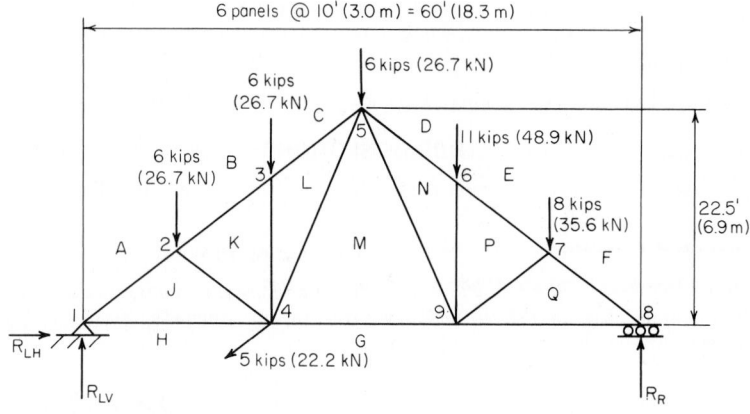

(a) Truss diagram

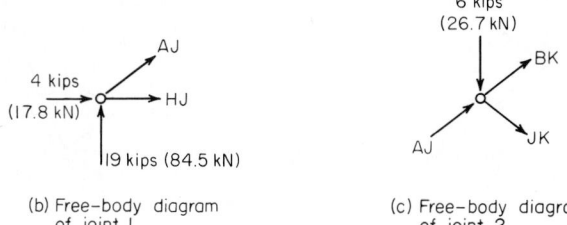

(b) Free-body diagram
of joint 1

(c) Free-body diagram
of joint 2

FIG. 5

TABLE 2   Forces in Truss Members (Fig. 5)

Member	Slope	Horizontal component	Vertical component	Force kips	Force kN
AJ	0.75	25.3	19.0	−31.7	−141.0
BK	0.75	21.3	16.0	−26.7	−118.8
CL	0.75	21.3	16.0	−26.7	−118.8
DN	0.75	22.7	17.0	−28.3	−125.9
EP	0.75	22.7	17.0	−28.3	−125.9
FQ	0.75	28.0	21.0	−35.0	−155.7
HJ	0.0	21.3	0.0	+21.3	+94.7
GM	0.0	16.0	0.0	+16.0	+71.2
GQ	0.0	28.0	0.0	+28.0	+124.5
JK	0.75	4.0	3.0	−5.0	−22.2
KL	∞	0.0	6.0	−6.0	−26.7
LM	2.25	5.3	12.0	+13.1	+58.3
MN	2.25	6.7	15.0	+16.4	+72.9
NP	∞	0.0	11.0	−11.0	−48.9
PQ	0.75	5.3	4.0	−6.7	−29.8

Joint 8: $\Sigma F_H = 28.0 - GQ = 0$; $GQ = 28.0$ kips (124.5 kN); $\Sigma F_V = 21.0 - 21.0 = 0$.
Joint 9: $\Sigma F_H = -16.0 - 6.7 - 5.3 + 28.0 = 0$; $\Sigma F_V = 15.0 - 11.0 - 4.0 = 0$.

7. Complete the computation

Compute the values in the last column of Table 2 and enter them as shown.

TRUSS ANALYSIS BY THE METHOD OF SECTIONS

Using the method of sections, determine the forces in members BK and LM in Fig. 5a.

Calculation Procedure:

1. Draw a free-body diagram of one portion of the truss

Cut the truss at the plane aa (Fig. 6a), and draw a free-body diagram of the left part of the truss. Assume that BK is tensile.

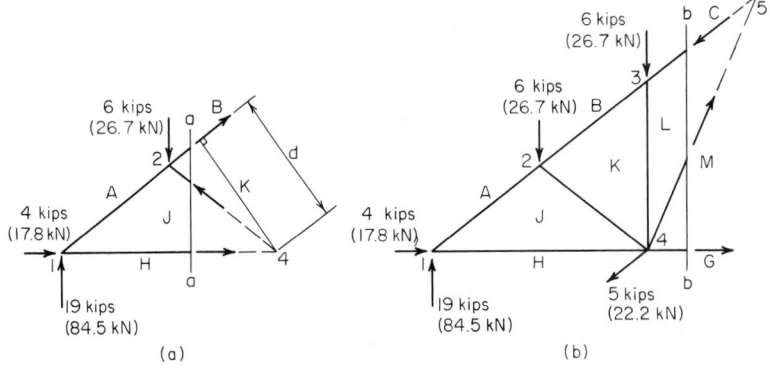

FIG. 6

2. Determine the magnitude and character of the first force

Take moments with respect to joint 4. Since each half of the truss forms a 3-4-5 right triangle, $d = 20(3/5) = 12$ ft (3.7 m), $\Sigma M_4 = 19(20) - 6(10) + 12BK = 0$, and $BK = -26.7$ kips $(-118.8$ kN).

The negative result signifies that the assumed direction of BK is incorrect; the force is, therefore, compressive.

3. Use an alternative solution

Alternatively, resolve BK (again assumed tensile) into its horizontal and vertical components at joint 1. Take moments with respect to joint 4. (A force may be resolved into its components at any point on its action line.) Then $\Sigma M_4 = 19(20) + 20BK_V - 6(10) = 0$; $BK_V = -16.0$ kips $(-71.2$ kN); $BK = -16.0(5/3) = -26.7$ kips $(-118.8$ kN).

4. Draw a second free-body diagram of the truss

Cut the truss at plane bb (Fig. 6b), and draw a free-body diagram of the left part. Assume LM is tensile.

5. Determine the magnitude and character of the second force

Resolve LM into its horizontal and vertical components at joint 4. Take moments with respect to joint 1: $\Sigma M_1 = 6(10 + 20) + 3(20) - 20LM_V = 0$; $LM_V = 12.0$ kips (53.4 kN); $LM_H = 12.0/2.25 = 5.3$ kips (23.6 kN); $LM = 13.1$ kips (58.3 kN).

REACTIONS OF A THREE-HINGED ARCH

The parabolic arch in Fig. 7 is hinged at A, B, and C. Determine the magnitude and direction of the reactions at the supports.

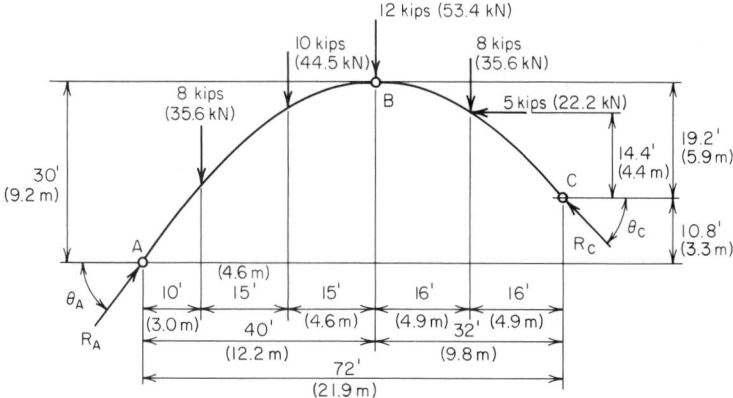

FIG. 7 Parabolic arch.

Calculation Procedure:

1. *Consider the entire arch as a free body and take moments*

Since a moment cannot be transmitted across a hinge, the bending moments at A, B, and C are zero. Resolve the reactions R_A and R_C (Fig. 7) into their horizontal and vertical components.

Considering the entire arch ABC as a free body, take moments with respect to A and C. Thus $\Sigma M_A = 8(10) + 10(25) + 12(40) + 8(56) - 5(25.2) - 72R_{CV} - 10.8R_{CH} = 0$, or $72R_{CV} + 10.8R_{CH} = 1132$, Eq. *a*. Also, $\Sigma M_C = 72R_{AV} - 10.8R_{AH} - 8(62) - 10(47) - 12(32) - 8(16) - 5(14.4) = 0$, or $72R_{AV} - 10.8R_{AH} = 1550$, Eq. *b*.

2. *Consider a segment of the arch and take moments*

Considering the segment BC as a free body, take moments with respect to B. Then $\Sigma M_B = 8(16) + 5(4.8) - 32R_{CV} + 19.2R_{CH} = 0$, or $32R_{CV} - 19.2R_{CH} = 152$, Eq. *c*.

3. *Consider another segment and take moments*

Considering segment AB as a free body, take moments with respect to B: $\Sigma M_B = 40R_{AV} - 30R_{AH} - 8(30) - 10(15) = 0$, or $40R_{AV} - 30R_{AH} = 390$, Eq. *d*.

4. *Solve the simultaneous moment equations*

Solve Eqs. *b* and *d* to determine R_A; solve Eqs. *a* and *c* to determine R_C. Thus $R_{AV} = 24.4$ kips (108.5 kN); $R_{AH} = 19.6$ kips (87.2 kN); $R_{CV} = 13.6$ kips (60.5 kN); $R_{CH} = 14.6$ kips (64.9 kN). Then $R_A = [(24.4)^2 + (19.6)^2]^{0.5} = 31.3$ kips (139.2 kN). Also $R_C = [(13.6)^2 + (14.6)^2]^{0.5} = 20.0$ kips (89.0 kN). And $\theta_A = \arctan (24.4/19.6) = 51°14'$; $\theta_C = \arctan (13.6/14.6) = 42°58'$.

LENGTH OF CABLE CARRYING KNOWN LOADS

A cable is supported at points P and Q (Fig. 8a) and carries two vertical loads, as shown. If the tension in the cable is restricted to 1800 lb (8006 N), determine the minimum length of cable required to carry the loads.

Calculation Procedure:

1. *Sketch the loaded cable*

Assume a position of the cable, such as $PRSQ$ (Fig. 8a). In Fig. 8b, locate points P' and Q', corresponding to P and Q, respectively, in Fig. 8a.

2. *Take moments with respect to an assumed point*

Assume that the maximum tension of 1800 lb (8006 N) occurs in segment PR (Fig. 8). The reaction at P, which is collinear with PR, is therefore 1800 lb (8006 N). Compute the true perpendicular

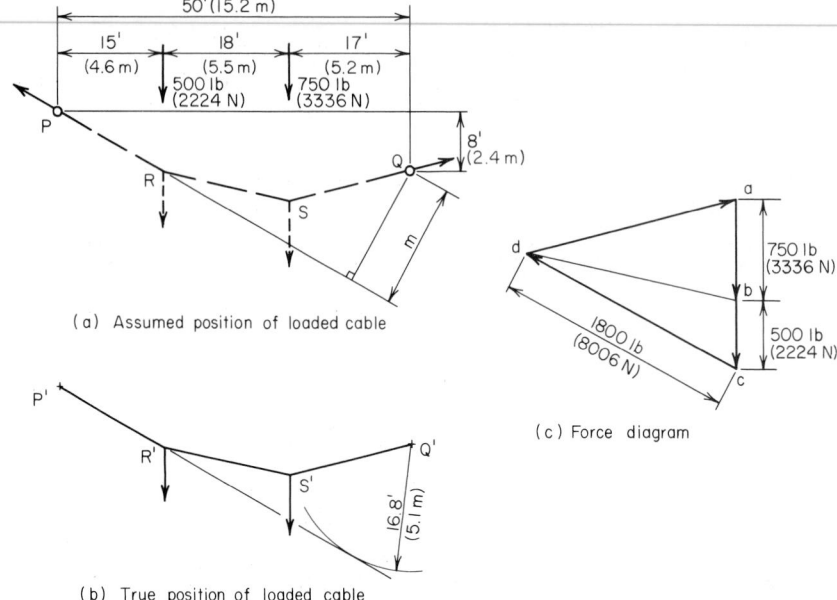

(a) Assumed position of loaded cable

(b) True position of loaded cable

(c) Force diagram

FIG. 8

distance m from Q to PR by taking moments with respect to Q. Or $\Sigma M_Q = 1800m - 500(35) - 750(17) = 0$; $m = 16.8$ ft (5.1 m). This dimension establishes the true position of PR.

3. Start the graphical solution of the problem

In Fig. 8b, draw a circular arc having Q' as center and a radius of 16.8 ft (5.1 m). Draw a line through P' tangent to this arc. Locate R' on this tangent at a horizontal distance of 15 ft (4.6 m) from P'.

4. Draw the force vectors

In Fig. 8c, draw vectors **ab**, **bc**, and **cd** to represent the 750-lb (3336-N) load, the 500-lb (2224-N) load, and the 1800-lb (8006-N) reaction at P, respectively. Complete the triangle by drawing vector **da**, which represents the reaction at Q.

5. Check the tension assumption

Scale **da** to ascertain whether it is less than 1800 lb (8006 N). This is found to be so, and the assumption that the maximum tension exists in PR is validated.

6. Continue the construction

Draw a line through Q' in Fig. 8b parallel to **da** in Fig. 8c. Locate S' on this line at a horizontal distance of 17 ft (5.2 m) from Q.

7. Complete the construction

Draw $R'S'$ and **db**. Test the accuracy of the construction by determining whether these lines are parallel.

8. Determine the required length of the cable

Obtain the required length of the cable by scaling the lengths of the segments in Fig. 8b. Thus $P'R' = 17.1$ ft (5.2 m); $R'S' = 18.4$ ft (5.6 m); $S'Q' = 17.6$ ft (5.4 m); and length of cable = 53.1 ft (16.2 m).

PARABOLIC CABLE TENSION AND LENGTH

A suspension bridge has a span of 960 ft (292.61 m) and a sag of 50 ft (15.2 m). Each cable carries a load of 1.2 kips per linear foot (kips/lin ft) (17,512.68 N/m) uniformly distributed along the horizontal. Compute the tension in the cable at midspan and at the supports, and determine the length of the cable.

Calculation Procedure:

1. *Compute the tension at midspan*

A cable carrying a load uniformly distributed along the horizontal assumes the form of a parabolic arc. In Fig. 9, which shows such a cable having supports at the same level, the tension at midspan

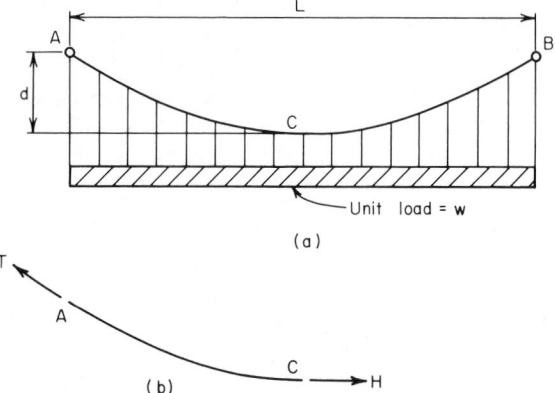

FIG. 9 Cable supporting load uniformly distributed along horizontal.

is $H = wL^2/(8d)$, where H = midspan tension, kips (kN); w = load on a unit horizontal distance, kips/lin ft (kN/m); L = span, ft (m); d = sag, ft (m). Substituting yields $H = 1.2(960)^2/[8(50)]$ = 2765 kips (12,229 kN).

2. *Compute the tension at the supports*

Use the relation $T = [H^2 + (wL/2)^2]^{0.5}$, where T = tension at supports, kips (kN), and the other symbols are as before. Thus, $T = [(2765)^2 + (1.2 \times 480)^2]^{0.5}$ = 2824 kips (12,561 kN).

3. *Compute the length of the cable*

When d/L is 1/20 or less, the cable length can be approximated from $S = L + 8d^2/(3L)$, where S = cable length, ft (m). Thus, $S = 960 + 8(50)^2/[3(960)]$ = 966.94 ft (294.72 m).

CATENARY CABLE SAG AND DISTANCE BETWEEN SUPPORTS

A cable 500 ft (152.4 m) long and weighing 3 pounds per linear foot (lb/lin ft) (43.8 N/m) is supported at two points lying in the same horizontal plane. If the tension at the supports is 1800 lb (8006 N), find the sag of the cable and the distance between the supports.

Calculation Procedure:

1. *Compute the catenary parameter*

A cable of uniform cross section carrying only its own weight assumes the form of a catenary. Using the notation of the previous procedure, we find the catenary parameter c from $d + c = T/w = 1800/3 = 600$ ft (182.9 m). Then $c = [(d + c)^2 - (S/2)^2]^{0.5} = [(600)^2 - (250)^2]^{0.5}$ = 545.4 ft (166.2 m).

2.   Compute the cable sag

Since $d + c = 600$ ft (182.9 m) and $c = 545.4$ ft (166.2 m), we know $d = 600 - 545.4 = 54.6$ ft (16.6 m).

3.   Compute the span length

Use the relation $L = 2c \ln (d + c + 0.5S)/c$, or $L = 2(545.5) \ln (600 + 250)/545.4 = 484.3$ ft (147.6 m).

STABILITY OF A RETAINING WALL

Determine the factor of safety (FS) against sliding and overturning of the concrete retaining wall in Fig. 10. The concrete weighs 150 lb/ft³ (23.56 kN/m³), the earth weighs 100 lb/ft³ (15.71 kN/m³), the coefficient of friction is 0.6, and the coefficient of active earth pressure is 0.333.

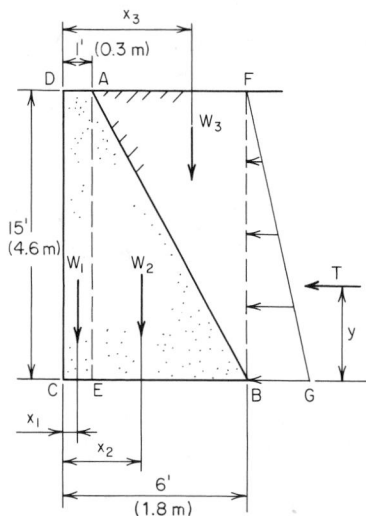

FIG. 10

Calculation Procedure:

1.   Compute the vertical loads on the wall

Select a 1-ft (304.8-mm) length of wall as typical of the entire structure. The horizontal pressure of the confined soil varies linearly with the depth and is represented by the triangle *BGF* in Fig. 10.

Resolve the wall into the elements *AECD* and *AEB*; pass the vertical plane *BF* through the soil. Calculate the vertical loads, and locate their resultants with respect to the toe *C*. Thus $W_1 = 15(1)(150) = 2250$ lb (10,008 N); $W_2 = 0.5(15)(5)(150) = 5625$; $W_3 = 0.5(15)(5)(100) = 3750$. Then $\Sigma W = 11,625$ lb (51,708 N). Also, $x_1 = 0.5$ ft; $x_2 = 1 + 0.333(5) = 2.67$ ft (0.81 m); $x_3 = 1 + 0.667(5) = 4.33$ ft (1.32 m).

2.   Compute the resultant horizontal soil thrust

Compute the resultant horizontal thrust T lb of the soil by applying the coefficient of active earth pressure. Determine the location of T. Thus $BG = 0.333(15)(100) = 500$ lb/lin ft (7295 N/m); $T = 0.5(15)(500) = 3750$ lb (16,680 N); $y = 0.333(15) = 5$ ft (1.5 m).

3.   Compute the maximum frictional force preventing sliding

The maximum frictional force $F_m = \mu(\Sigma W)$, where μ = coefficient of friction. Or $F_m = 0.6(11,625) = 6975$ lb (31,024.8 N).

4.   Determine the factor of safety against sliding

The factor of safety against sliding is FSS $= F_m/T = 6975/3750 = 1.86$.

5.   Compute the moment of the overturning and stabilizing forces

Taking moments with respect to *C*, we find the overturning moment $= 3750(5) = 18,750$ lb·ft (25,406.3 N·m). Likewise, the stabilizing moment $= 2250(0.5) + 5625(2.67) + 3750(4.33) = 32,375$ lb·ft (43,868.1 N·m).

6.   Compute the factor of safety against overturning

The factor of safety against overturning is FSO = stabilizing moment, lb·ft (N·m)/overturning moment, lb·ft (N·m) $= 32,375/18,750 = 1.73$.

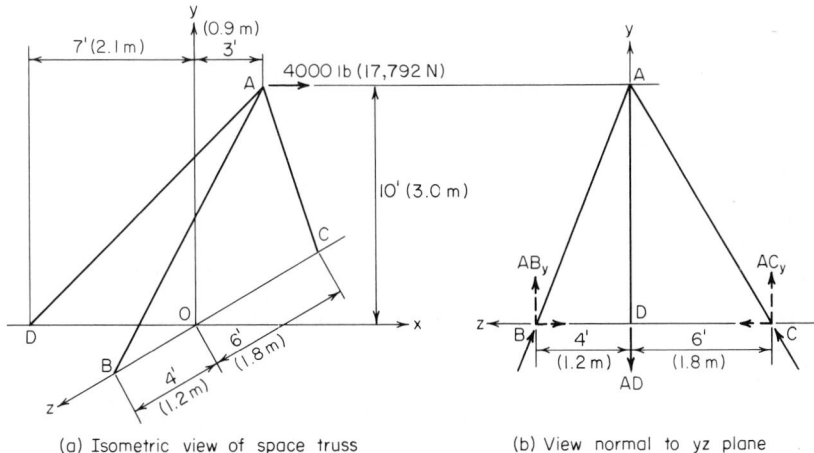

(a) Isometric view of space truss   (b) View normal to yz plane

FIG. 11

ANALYSIS OF A SIMPLE SPACE TRUSS

In the space truss shown in Fig. 11a, A lies in the xy plane, B and C lie on the z axis, and D lies on the x axis. A horizontal load of 4000 lb (17,792 N) lying in the xy plane is applied at A. Determine the force induced in each member by applying the method of joints, and verify the results by taking moments with respect to convenient axes.

Calculation Procedure:

1. Determine the projected length of members

Let d_x, d_y, and d_z denote the length of a member as projected on the x, y, and z axes, respectively. Record in Table 3 the projected lengths of each member. Record the remaining values as they are obtained.

2. Compute the true length of each member

Use the equation $d = (d_x^2 + d_y^2 + d_z^2)^{0.5}$, where d = the true length of a member.

3. Compute the ratio of the projected length to the true length

For each member, compute the ratios of the three projected lengths to the true length. For example, for member AC, $d_z/d = 6/12.04 = 0.498$.

These ratios are termed *direction cosines* because each represents the cosine of the angle between the member and the designated axis, or an axis parallel thereto.

TABLE 3   Data for Space Truss (Fig. 11)

Member	AB		AC		AD	
d_x, ft (m)	3	(0.91)	3	(0.91)	10	(3.05)
d_y, ft (m)	10	(3.0)	10	(3.0)	10	(3.0)
d_z, ft (m)	4	(1.2)	6	(1.8)	0	(0)
d, ft (m)	11.18	(3.4)	12.04	(3.7)	14.14	(4.3)
d_x/d	0.268		0.249		0.707	
d_y/d	0.894		0.831		0.707	
d_z/d	0.358		0.498		0	
Force, lb (N)	−3830	(−17,036)	−2750	(−12,232)	+8080	(+35,940)

Since the axial force in each member has the same direction as the member itself, a direction cosine also represents the ratio of the component of a force along the designated axis to the total force in the member. For instance, let AC denote the force in member AC, and let AC_x denote its component along the x axis. Then $AC_x/AC = d_x/d = 0.249$.

4. Determine the component forces

Consider joint A as a free body, and assume that the forces in the three truss members are tensile. Equate the sum of the forces along each axis to zero. For instance, if the truss members are in tension, the x components of these forces are directed to the left, and $\Sigma F_x = 4000 - AB_x - AC_x - AD_x = 0$.

Express each component in terms of the total force to obtain $\Sigma F_x = 4000 - 0.268AB - 0.249AC - 0.707AD = 0$; $\Sigma F_y = -0.894AB - 0.831AC - 0.707AD = 0$; $\Sigma F_z = 0.358AB - 0.498AC = 0$.

5. Solve the simultaneous equations in step 4 to evaluate the forces in the truss members

A positive result in the solution signifies tension; a negative result, compression. Thus, $AB = 3830$-lb (17,036-N) compression; $AC = 2750$-lb (12,232-N) compression; and $AD = 8080$-lb (35,940-N) tension. To verify these results, it is necessary to select moment axes yielding equations independent of those previously developed.

6. Resolve the reactions into their components

In Fig. 11b, show the reactions at the supports B, C, and D, each reaction being numerically equal to and collinear with the force in the member at that support. Resolve these reactions into their components.

7. Take moments about a selected axis

Take moments with respect to the axis through C parallel to the x axis. (Since the x components of the forces are parallel to this axis, their moments are zero.) Then $\Sigma M_{Cx} = 10AB_y - 6AD_y = 10(0.894)(3830) - 6(0.707)(8080) = 0$.

8. Take moments about another axis

Take moments with respect to the axis through D parallel to the x axis. So $\Sigma M_{Dx} = 4AB_y - 6AC_y = 4(0.894)(3830) - 6(0.831)(2750) = 0$.
The computed results are thus substantiated.

ANALYSIS OF A COMPOUND SPACE TRUSS

The compound space truss in Fig. 12a has the dimensions shown in the orthographic projections, Fig. 12b and c. A load of 5000 lb (22,240 N), which lies in the xy plane and makes an angle of 30° with the vertical, is applied at A. Determine the force induced in each member, and verify the results.

Calculation Procedure:

1. Compute the true length of each truss member

Since the truss and load system are symmetric with respect to the xy plane, the internal forces are also symmetric. As one component of an internal force becomes known, it will be convenient to calculate the other components at once, as well as the total force.

Record in Table 4 the length of each member as projected on the coordinate axes. Calculate the true length of each member, using geometric relations.

2. Resolve the applied load into its x and y components

Use only the absolute values of the forces. Thus $P_x = 5000 \sin 30° = 2500$ lb (11,120 N); $P_y = 5000 \cos 30° = 4330$ lb (19,260 N).

3. Compute the horizontal reactions

Compute the horizontal reactions at D and at line CC' (Fig. 12b). Thus $\Sigma M_{CC'} = 4330(12) - 2500(7) - 10H_1 = 0$; $H_1 = 3446$ lb (15,328 N); $H_2 = 3446 - 2500 = 946$ lb (4208 N).

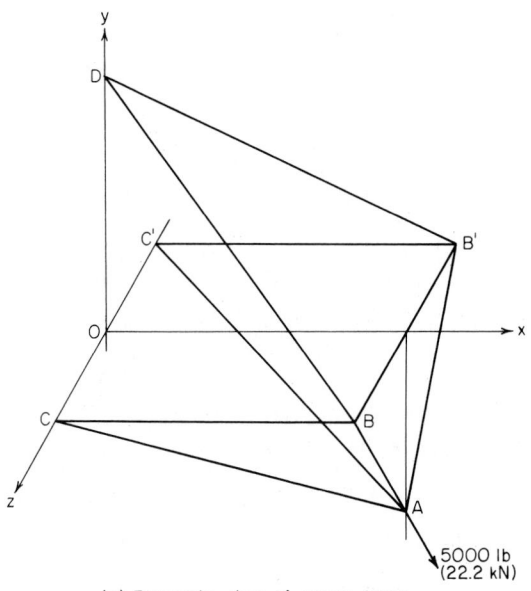

(a) Isometric view of space truss

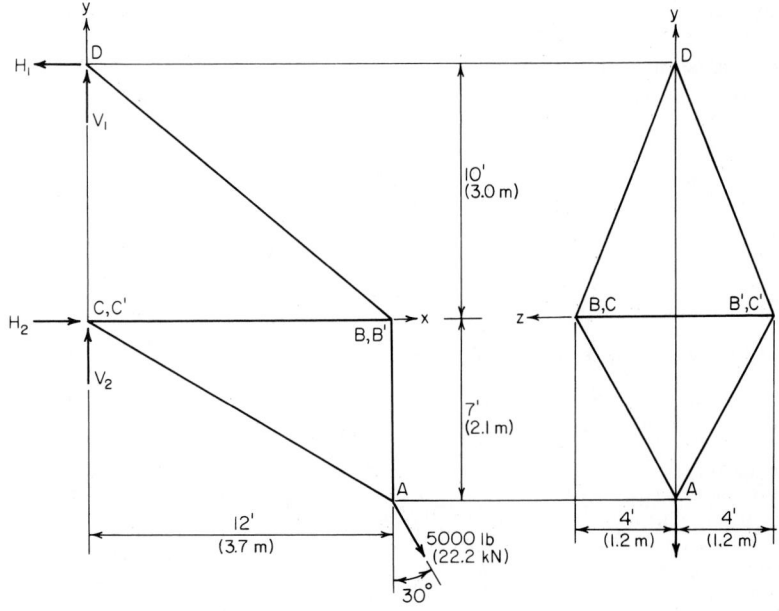

(b) View normal to xy plane

(c) View normal to yz plane

FIG. 12

1.21

4. *Compute the vertical reactions*

Consider the equilibrium of joint D and the entire truss when you are computing the vertical reactions. In all instances, assume that an unknown internal force is tensile. Thus, at joint D: $\Sigma F_x = -H_1 + 2BD_x = 0$; $BD_x = 1723$-lb (7664-N) tension; $BD_y = 1723(10/12) = 1436$ lb (6387 N); likewise, $\Sigma F_y = V_1 - 2BD_y = V_1 - 2(1436) = 0$; $V_1 = 2872$ lb (12,275 N).

For the entire truss, $\Sigma F_y = V_1 + V_2 - 4330 = 0$; $V_2 = 1458$ lb (6485 N).

The z components of the reactions are not required in this solution. Thus, the remaining calculations for BD are $BD_z = 1723(4/12) = 574$ lb (2553 N); $BD = 1723(16.12/12) = 2315$ lb (10,297 N).

5. *Compute the unknown forces by using the equilibrium of a joint*

Calculate the forces AC and BC by considering the equilibrium of joint C. Thus $\Sigma F_x = 0.5H_2 + AC_x + BC = 0$, Eq. a; $\Sigma F_y = 0.5V_2 - AC_y = 0$, Eq. b. From Eq. b, $AC_y = 729$-lb (3243-N) tension. Then $AC_x = 729(12/7) = 1250$ lb (5660 N). From Eq. a, $BC = 1723$-lb (7664-N) compression. Then $AC_z = 729(4/7) = 417$ lb (1855 N); $AC = 729(14.46/7) = 1506$ lb (6699 N).

6. *Compute another set of forces by considering joint equilibrium*

Calculate the forces AB and BB' by considering the equilibrium of joint B. Thus $\Sigma F_y = BD_y - AB_y = 0$; $AB_y = 1436$-lb (6387-N) tension; $AB_z = 1436(4/7) = 821$ lb (3652 N); $AB = 1436 (8.06/7) = 1653$ lb (7353 N); $\Sigma F_z = -AB_z - BD_z - BB' = 0$; $BB' = 1395$-lb (6205-N) compression.

All the internal forces are now determined. Show in Table 4 the tensile forces as positive, and the compressive forces as negative.

7. *Check the equilibrium of the first joint considered*

The first joint considered was A. Thus $\Sigma F_x = -2AC_x + 2500 = -2(1250) + 2500 = 0$, and $\Sigma F_y = 2AB_y + 2AC_y - 4330 = 2(1436) + 2(729) - 4330 = 0$. Since the summation of forces for both axes is zero, the joint is in equilibrium.

8. *Check the equilibrium of the second joint*

Check the equilibrium of joint B by taking moments of the forces acting on this joint with respect to the axis through A parallel to the x axis (Fig. 12c). Thus $\Sigma M_{Ax} = -7BB' + 7BD_z + 4BD_y = -7(1395) + 7(574) + 4(1436) = 0$.

9. *Check the equilibrium of the right-hand part of the structure*

Cut the truss along a plane parallel to the yz plane. Check the equilibrium of the right-hand part of the structure. Now $\Sigma F_x = -2BD_x + 2BC - 2AC_x + 2500 = -2(1723) + 2(1723) - 2(1250) + 2500 = 0$, and $\Sigma F_y = 2BD_y + 2AC_y - 4330 = 2(1436) + 2(729) - 4330 = 0$. The calculated results are thus substantiated in these equations.

GEOMETRIC PROPERTIES OF AN AREA

Calculate the polar moment of inertia of the area in Fig. 13: (*a*) with respect to its centroid, and (*b*) with respect to point A.

TABLE 4 Data for Space Truss (Fig. 12)

Member	AB	AC	BC	BD	BB'
d_x, ft (m)	0 (0)	12 (3.7)	12 (3.7)	12 (3.7)	0 (0)
d_y, ft (m)	7 (2.1)	7 (2.1)	0 (0)	10 (3.0)	0 (0)
d_z, ft (m)	4 (1.2)	4 (1.2)	0 (0)	4 (1.2)	8 (2.4)
d, ft (m)	8.06 (2.5)	14.46 (4.4)	12.00 (3.7)	16.12 (4.9)	8 (2.4)
F_x, lb (N)	0 (0)	1,250(5,560)	1,723 (7,664)	1,723 (7,664)	0 (0)
F_y, lb (N)	1,436 (6,387)	729 (3,243)	0 (0)	1,436 (6,387)	0 (0)
F_z, lb (N)	821 (3,652)	417 (1,855)	0 (0)	574 (2,553)	1,395 (6,205)
F, lb (N)	+1,653 (+7,353)	+1,506 (+6,699)	−1,723 (−7,664)	+2,315 (+10,297)	−1,395 (−6,205)

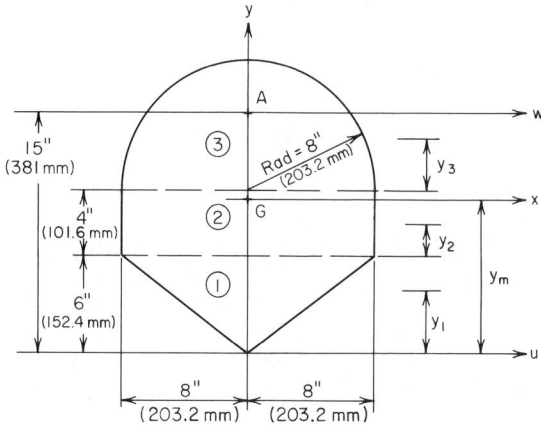

FIG. 13

Calculation Procedure:

1. Establish the area axes

Set up the horizontal and vertical coordinate axes u and y, respectively.

2. Divide the area into suitable elements

Using the American Institute of Steel Construction (AISC) *Manual*, obtain the properties of elements 1, 2, and 3 (Fig. 13) after locating the horizontal centroidal axis of each element. Thus y_1 = ⅔(6) = 4 in (101.6 mm); y_2 = 2 in (50.8 mm); y_3 = 0.424(8) = 3.4 in (86.4 mm).

3. Locate the horizontal centroidal axis of the entire area

Let x denote the horizontal centroidal axis of the entire area. Locate this axis by computing the statical moment of the area with respect to the u axis. Thus

Element	Area, in² (cm²)	×	Arm, in (cm)	=	Moment, in³ (cm³)
1	0.5(6)(16) = 48 (309.7)		4 (10.2)	=	192 (3,158.9)
2	4(16) = 64 (412.9)		8 (20.3)	=	512 (8,381.9)
3	1.57(8)² = 100.5 (648.4)		13.4 (34.0)	=	1,347 (22,045.6)
Total	212.5 (1,371.0)				2,051 (33,586.4)

Then y_m = 2051/212.5 = 9.7 in (246.4 mm). Since the area is symmetric with respect to the y axis, this is also a centroidal axis. The intersection point G of the x and y axes is, therefore, the centroid of the area.

4. Compute the distance between the centroidal axis and the reference axis

Compute k, the distance between the horizontal centroidal axis of each element and the x axis. Only absolute values are required. Thus k_1 = 9.7 − 4.0 = 5.7 in (144.8 mm); k_2 = 9.7 − 8.0 = 1.7 in (43.2 mm); k_3 = 13.4 − 9.7 = 3.7 in (94.0 mm).

5. Compute the moment of inertia of the entire area—x axis

Let I_0 denote the moment of inertia of an element with respect to its horizontal centroidal axis and A its area. Compute the moment of inertia I_x of the entire area with respect to the x axis by applying the transfer equation $I_x = \Sigma I_0 + \Sigma Ak^2$. Thus

Element	I_0, in^4 (dm^4)	Ak^2, in^4 (dm^4)
1	$\frac{1}{36}(16)(6)^3 = \ 96 \ (0.40)$	$48(5.7)^2 = 1560 \ (6.49)$
2	$\frac{1}{12}(16)(4)^3 = \ 85 \ (0.35)$	$64(1.7)^2 = \ 185 \ (0.77)$
3	$0.110(8)^4 = \underline{451} \ \underline{(1.88)}$	$100.5(3.7)^2 = \underline{1376} \ \underline{(5.73)}$
Total	632 (2.63)	3121 (12.99)

Then, $I_x = 632 + 3121 = 3753$ in^4 (15.62 dm^4).

6. Determine the moment of inertia of the entire area—y axis

For this computation, subdivide element 1 into two triangles having the y axis as a base. Thus

Element	I about y axis, in^4 (dm^4)
1	$2(\frac{1}{12})(6)(8)^3 = \ 512 \ (2.13)$
2	$\frac{1}{12}(4)(16)^3 = 1365 \ (5.68)$
3	$\frac{1}{2}(0.785)(8)^4 = \underline{1607} \ \underline{(6.69)}$
	$I_y = 3484 \ (14.5)$

7. Compute the polar moment of inertia of the area

Apply the equation for the polar moment of inertia J_G with respect to G: $J_G = I_x + I_y = 3753 + 3484 = 7237$ in^4 (30.12 dm^4).

8. Determine the moment of inertia of the entire area—w axis

Apply the equation in step 5 to determine the moment of inertia I_w of the entire area with respect to the horizontal axis w through A. Thus $k = 15.0 - 9.7 = 5.3$ in (134.6 mm); $I_w = I_x + Ak^2$
$$= 3753 + 212.5(5.3)^2 = 9722 \text{ in}^4 \ (40.46 \text{ dm}^4).$$

9. Compute the polar moment of inertia

Compute the polar moment of inertia of the entire area with respect to A. Then $J_A = I_w + I_y = 9722 + 3484 = 13,206$ in^4 (54.97 dm^4).

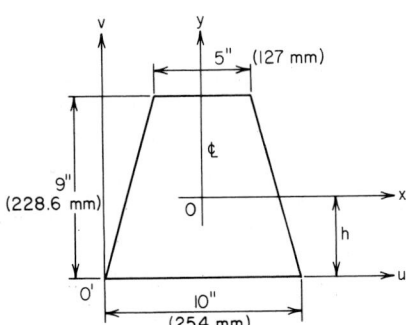

FIG. 14

PRODUCT OF INERTIA OF AN AREA

Calculate the product of inertia of the isosceles trapezoid in Fig. 14 with respect to the rectangular axes u and v.

Calculation Procedure:

1. Locate the centroid of the trapezoid

Using the AISC *Manual* or another suitable reference, we find h = centroid distance from the axis (Fig. 14) = $(9/3)[(2 \times 5 + 10)/(5 + 10)] = 4$ in (101.6 mm).

2. Compute the area and product of inertia P_{xy}

The area of the trapezoid is $A = \frac{1}{2}(9)(5 + 10) = 67.5$ in^2 (435.5 cm^2). Since the area is symmetrically disposed with respect to the y axis, the product of inertia with respect to the x and y axes is $P_{xy} = 0$.

3. Compute the product of inertia by applying the transfer equation

The transfer equation for the product of inertia is $P_{uv} = P_{xy} + Ax_m y_m$, where x_m and y_m are the coordinates of O' with respect to the centroidal x and y axes, respectively. Thus $P_{uv} = 0 + 67.5(-5)(-4) = 1350$ in^4 (5.6 dm^4).

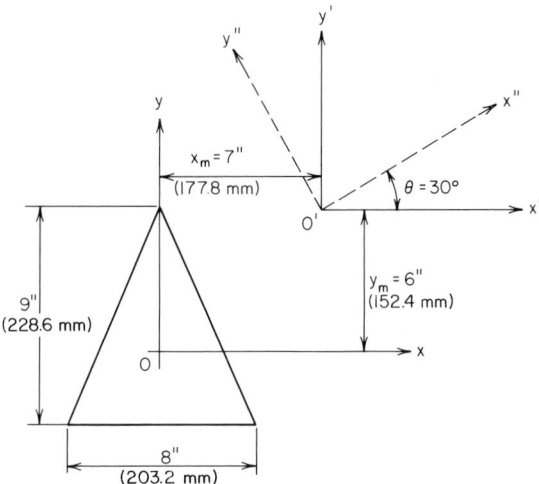

FIG. 15

PROPERTIES OF AN AREA WITH RESPECT TO ROTATED AXES

In Fig. 15, x and y are rectangular axes through the centroid of the isosceles triangle; x' and y' are axes parallel to x and y, respectively; x'' and y'' are axes making an angle of 30° with x' and y', respectively. Compute the moments of inertia and the product of inertia of the triangle with respect to the x'' and y'' axes.

Calculation Procedure:

1. *Compute the area of the figure*

The area of this triangle = 0.5(base)(altitude) = 0.5(8)(9) = 36 in² (232.3 cm²).

2. *Compute the properties of the area with respect to the x and y axes*

Using conventional moment-of-inertia relations, we find $I_x = bd^3/36 = 8(9)^3/36 = 162$ in⁴ (0.67 dm⁴); $I_y = b^3d/48 = (8)^3(9)/48 = 96$ in⁴ (0.39 dm⁴). By symmetry, the product of inertia with respect to the x and y axes is $P_{xy} = 0$.

3. *Compute the properties of the area with respect to the x' and y' axes*

Using the usual moment-of-inertia relations, we find $I_{x'} = I_x + Ay_m^2 = 162 + 36(6)^2 = 1458$ in⁴ (6.06 dm⁴); $I_{y'} = I_y + Ax_m^2 = 96 + 36(7)^2 = 1860$ in⁴ (7.74 dm⁴); $P_{x'y'} = P_{xy} + Ax_my_m = 0 + 36(7)(6) = 1512$ in⁴ (6.29 dm⁴).

4. *Compute the properties of the area with respect to the x'' and y'' axes*

For the x'' axis, $I_{x''} = I_{x'} \cos^2\theta + I_{y'} \sin^2\theta - P_{x'y'} \sin 2\theta = 1458(0.75) + 1860(0.25) - 1512(0.866) = 249$ in⁴ (1.03 dm⁴).

For the y'' axis, $I_{y''} = I_{x'} \sin^2\theta + I_{y'} \cos^2\theta + P_{x'y'} \sin 2\theta = 1458(0.25) + 1860(0.75) + 1512(0.866) = 3069$ in⁴ (12.77 dm⁴).

The product of inertia is $P_{x''y''} = P_{x'y'} \cos 2\theta + [(I_{x'} - I_{y'})/2] \sin 2\theta = 1512(0.5) + [(1458 - 1860)/2]0.866 = 582$ in⁴ (2.42 dm⁴).

Analysis of Stress and Strain

The notational system for axial stress and strain used in this section is as follows: A = cross-sectional area of a member; L = original length of the member; Δl = increase in length; P = axial force; s = axial stress; ϵ = axial strain = $\Delta l/L$; E = modulus of elasticity of material =

s/ϵ. The units used for each of these factors are given in the calculation procedure. In all instances, it is assumed that the induced stress is below the proportional limit. The basic stress and elongation equations used are $s = P/A$; $\Delta l = sL/E = PL/(AE)$. For steel, $E = 30 \times 10^6$ lb/in² (206 GPa).

STRESS CAUSED BY AN AXIAL LOAD

A concentric load of 20,000 lb (88,960 N) is applied to a hanger having a cross-sectional area of 1.6 in² (1032.3 mm²). What is the axial stress in the hanger?

Calculation Procedure:

1. Compute the axial stress

Use the general stress relation $s = P/A = 20,000/1.6 = 12,500$ lb/in² (86,187.5 kPa).

  Related Calculations: Use this general stress relation for a member of any cross-sectional shape, provided the area of the member can be computed and the member is made of only one material.

DEFORMATION CAUSED BY AN AXIAL LOAD

A member having a length of 16 ft (4.9 m) and a cross-sectional area of 2.4 in² (1548.4 mm²) is subjected to a tensile force of 30,000 lb (133.4 kN). If $E = 15 \times 10^6$ lb/in² (103 GPa), how much does this member elongate?

Calculation Procedure:

1. Apply the general deformation equation

The general deformation equation is $\Delta l = PL/(AE) = 30,000(16)(12)/[2.4(15 \times 10^6)] = 0.16$ in (4.06 mm).

  Related Calculations: Use this general deformation equation for any material whose modulus of elasticity is known. For composite materials, this equation must be altered before it can be used.

DEFORMATION OF A BUILT-UP MEMBER

A member is built up of three bars placed end to end, the bars having the lengths and cross-sectional areas shown in Fig. 16. The member is placed between two rigid surfaces and axial loads

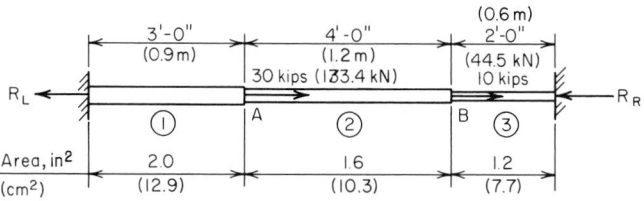

FIG. 16

of 30 kips (133 kN) and 10 kips (44 kN) are applied at A and B, respectively. If $E = 2000$ kips/in² (13,788 MPa), determine the horizontal displacement of A and B.

Calculation Procedure:

1. Express the axial force in terms of one reaction

Let R_L and R_R denote the reactions at the left and right ends, respectively. Assume that both reactions are directed to the left. Consider a tensile force as positive and a compressive force as negative. Consider a deformation positive if the body elongates and negtive if the body contracts.

Express the axial force P in each bar in terms of R_L because both reactions are assumed to be directed toward the left. Use subscripts corresponding to the bar numbers (Fig. 16). Thus, $P_1 = R_L$; $P_2 = R_L - 30$; $P_3 = R_L - 40$.

2. *Express the deformation of each bar in terms of the reaction and modulus of elasticity*

Thus, $\Delta l_1 = R_L(36)/(2.0E) = 18R_L/E$; $\Delta l_2 = (R_L - 30)(48)/(1.6E) = (30R_L - 900)/E$; $\Delta l_3 = (R_L - 40)24/(1.2E) = (20R_L - 800)/E$.

3. *Solve for the reaction*

Since the ends of the member are stationary, equate the total deformation to zero, and solve for R_L. Thus $\Delta l_t = (68R_L - 1700)/E = 0$; $R_L = 25$ kips (111 kN). The positive result confirms the assumption that R_L is directed to the left.

4. *Compute the displacement of the points*

Substitute the computed value of R_L in the first two equations of step 2 and solve for the displacement of the points A and B. Thus $\Delta l_1 = 18(25)/2000 = 0.225$ in (5.715 mm); $\Delta l_2 = [30(25) - 900]/2000 = -0.075$ in (-1.905 mm).

Combining these results, we find the displacement of $A = 0.225$ in (5.715 mm) to the right; the displacement of $B = 0.225 - 0.075 = 0.150$ in (3.81 mm) to the right.

5. *Verify the computed results*

To verify this result, compute R_R and determine the deformation of bar 3. Thus $\Sigma F_H = -R_L + 30 + 10 - R_R = 0$; $R_R = 15$ kips (67 kN). Since bar 3 is in compression, $\Delta l_3 = -15(24)/[1.2(2000)] = -0.150$ in (-3.81 mm). Therefore, B is displaced 0.150 in (3.81 mm) to the right. This verifies the result obtained in step 4.

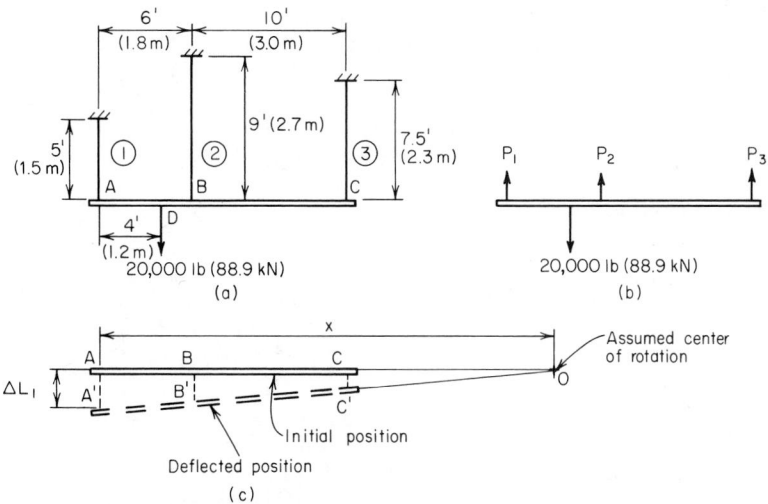

FIG. 17   Bar supported by three hangers.

REACTIONS AT ELASTIC SUPPORTS

The rigid bar in Fig. 17a is subjected to a load of 20,000 lb (88,960 N) applied at D. It is supported by three steel rods, 1, 2, and 3 (Fig. 17a). These rods have the following relative cross-sectional areas: $A_1 = 1.25$, $A_2 = 1.20$, $A_3 = 1.00$. Determine the tension in each rod caused by this load, and locate the center of rotation of the bar.

Calculation Procedure:

1. Draw a free-body diagram; apply the equations of equilibrium

Draw the free-body diagram (Fig. 17b) of the bar. Apply the equations of equilibrium: $\Sigma F_V = P_1 + P_2 + P_3 - 20,000 = 0$, or $P_1 + P_2 + P_3 = 20,000$, Eq. a; also, $\Sigma M_C = 16P_1 + 10P_2 - 20,000(12) = 0$, or $16P_1 + 10P_2 = 240,000$, Eq. b.

2. Establish the relations between the deformations

Selecting an arbitrary center of rotation O, show the bar in its deflected position (Fig. 17c). Establish the relationships among the three deformations. Thus, by similar triangles, $(\Delta l_1 - \Delta l_2)/(\Delta l_2 - \Delta l_3) = 6/10$, or $10\Delta l_1 - 16\Delta l_2 + 6\Delta l_3 = 0$, Eq. c.

3. Transform the deformation equation to an axial-force equation

By substituting axial-force relations in Eq. c, the following equation is obtained: $10P_1(5)/(1.25E) - 16P_2(9)/(1.20E) + 6P_3(7.5)/E = 0$, or $40P_1 - 120P_2 + 45P_3 = 0$, Eq. c'.

4. Solve the simultaneous equations developed

Solve the simultaneous equations a, b, and c' to obtain $P_1 = 11,810$ lb (52,530 N); $P_2 = 5100$ lb (22,684 N); $P_3 = 3090$ lb (13,744 N).

5. Locate the center of rotation

To locate the center of rotation, compute the relative deformation of rods 1 and 2. Thus $\Delta l_1 = 11,810(5)/(1.25E) = 47,240/E$; $\Delta l_2 = 5100(9)/(1.20E) = 38,250/E$.

In Fig. 17c, by similar triangles, $x/(x - 6) = \Delta l_1/\Delta l_2 = 1.235$; $x = 31.5$ ft (9.6 m).

6. Verify the computed values of the tensile forces

Calculate the moment with respect to A of the applied and resisting forces. Thus $M_{Aa} = 20,000(4) = 80,000$ lb·ft (108,400 N·m); $M_{Ar} = 5100(6) + 3090(16) = 80,000$ lb·ft (108,400 N·m). Since the moments are equal, the results are verified.

ANALYSIS OF CABLE SUPPORTING A CONCENTRATED LOAD

A cold-drawn steel wire ¼ in (6.35 mm) in diameter is stretched tightly betwen two points lying on the same horizontal plane 80 ft (24.4 m) apart. The stress in the wire is 50,000 lb/in² (344,700 kPa). A load of 200 lb (889.6 N) is suspended at the center of the cable. Determine the sag of the cable and the final stress in the cable. Verify that the results obtained are compatible.

Calculation Procedure:

1. Derive the stress and strain relations for the cable

With reference to Fig. 18, L = distance between supports, ft (m); P = load applied at center of cable span, lb (N); d = deflection of cable center, ft (m); ϵ = strain of cable caused by P; s_1 and s_2 = initial and final tensile stress in cable, respectively, lb/in² (kPa).

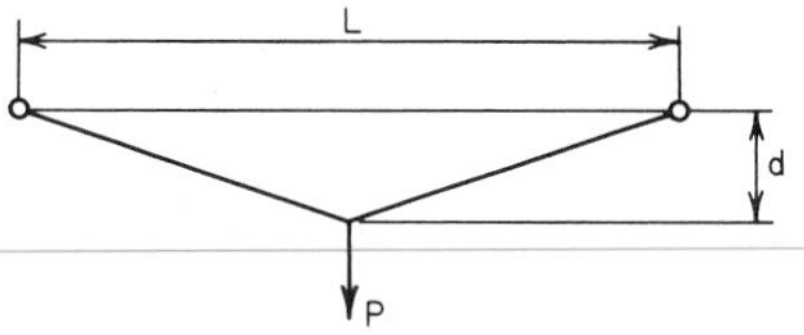

Refer to the geometry of the deflection diagram. Taking into account that d/L is extremely small, derive the following approximations: $s_2 = PL/(4Ad)$, Eq. a; $\epsilon = 2(d/L)^2$, Eq. b.

2. Relate stress and strain

Express the increase in stress caused by P in terms of ϵ, and apply the above two equations to derive $2E(d/L)^3 + s_1(d/L) = P/(4A)$, Eq. c.

FIG. 18 Deflection of cable under concentrated load.

3. Compute the deflection at the center of the cable

Using Eq. c, we get $2(30)(10)^6(d/L)^3 + 50,000d/L = 200/[4(0.049)]$, so $d/L = 0.0157$ and $\therefore d = 0.0157(80) = 1.256$ ft (0.382 m).

4. Compute the final tensile stress

Write Eq. a as $s_2 = [P/(4A)]/(d/L) = 1020/0.0157 = 65,000$ lb/in² (448,110 kPa).

5. *Verify the results computed*

To demonstrate that the results are compatible, accept the computed value of d/L as correct. Then apply Eq. b to find the strain, and compute the corresponding stress. Thus $\epsilon = 2(0.0157)^2 = 4.93 \times 10^{-4}$; $s_2 = s_1 + E\epsilon = 50,000 + 30 \times 10^6 \times 4.93 \times 10^{-4} = 64,800$ lb/in^2 (446,731 kPa). This agrees closely with the previously calculated stress of 65,000 lb/in^2 (448,110 kPa).

DISPLACEMENT OF TRUSS JOINT

In Fig. 19a, the steel members AC and BC both have a cross-sectional area of 1.2 in^2 (7.7 cm^2). If a load of 20 kips (89.0 kN) is suspended at C, how much is joint C displaced?

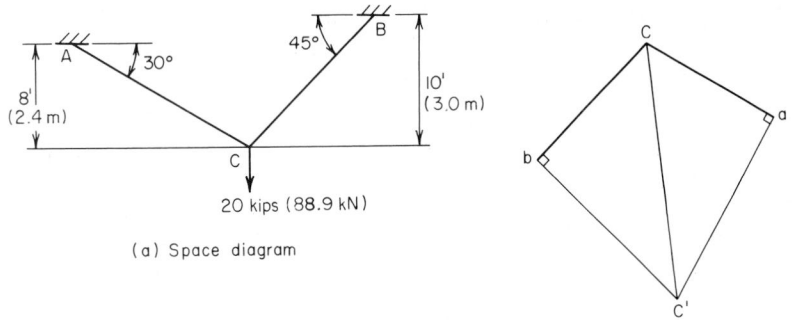

(a) Space diagram

(b) Displacement diagram

FIG. 19

Calculation Procedure:

1. *Compute the length of each member and the tensile forces*

Consider joint C as a free body to find the tensile force in each member. Thus $L_{AC} = 192$ in (487.7 cm); $L_{BC} = 169.7$ in (431.0 cm); $P_{AC} = 14,640$ lb (65,118.7 N); $P_{BC} = 17,930$ lb (79,752.6 N).

2. *Determine the elongation of each member*

Use the relation $\Delta l = PL/(AE)$. Thus $\Delta l_{AC} = 14,640(192)/[1.2(30 \times 10^6)] = 0.0781$ in (1.983 mm); $\Delta l_{BC} = 17,930(169.7)/[1.2(30 \times 10^6)] = 0.0845$ in (2.146 mm).

3. *Construct the Williott displacement diagram*

Selecting a suitable scale, construct the Williott displacement diagram as follows: Draw (Fig. 19b) line Ca parallel to member AC, with $Ca = 0.0781$ in (1.98 mm). Similarly, draw Cb parallel to member BC, with $Cb = 0.0845$ in (2.146 mm).

4. *Determine the displacement*

Erect perpendiculars to Ca and Cb at a and b, respectively. Designate the intersection point of these perpendiculars as C'.

Line CC' represents, in both magnitude and direction, the approximate displacement of joint C under the applied load. Scaling distance CC' to obtain the displacement shows that the displacement of $C = 0.134$ in (3.4036 mm).

AXIAL STRESS CAUSED BY IMPACT LOAD

A body weighing 18 lb (80.1 N) falls 3 ft (0.9 m) before contacting the end of a vertical steel rod. The rod is 5 ft (1.5 m) long and has a cross-sectional area of 1.2 in^2 (7.74 cm^2). If the entire kinetic energy of the falling body is absorbed by the rod, determine the stress induced in the rod.

Calculation Procedure:

1. State the equation for the induced stress

Equate the energy imparted to the rod to the potential energy lost by the falling body: $s = (P/A)\{1 + [1 + 2Eh/(LP/A)]^{0.5}\}$, where h = vertical displacement of body, ft (m).

2. Substitute the numerical values

Thus, $P/A = 18/1.2 = 15$ lb/in^2 (103 kPa); $h = 3$ ft (0.9 m); $L = 5$ ft (1.5 m); $2Eh/(LP/A) = 2(30 \times 10^6)(3)/[5(15)] = 2,400,000$. Then $s = 23,250$ lb/in^2 (160,285.5 kPa).

   Related Calculations: Where the deformation of the supporting member is negligible in relation to the distance h, as it is in the present instance, the following approximation is used: $s = [2PEh/(AL)]^{0.5}$.

STRESSES ON AN OBLIQUE PLANE

The prism $ABCD$ in Fig. 20a has the principal stresses of 6300- and 2400-lb/in^2 (43,438.5- and 16,548.0-kPa) tension. Applying both the analytical and graphical methods, determine the normal and shearing stress on plane AE.

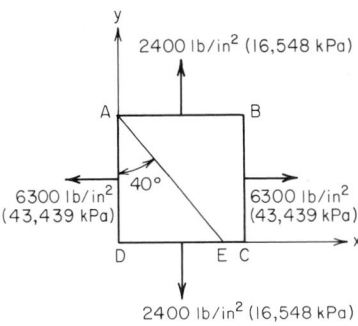

(a) Stresses on prism

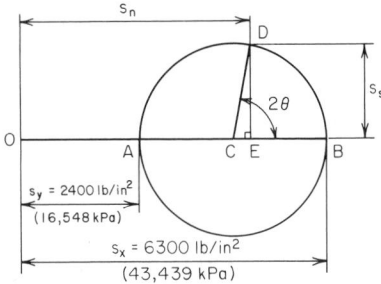

(b) Mohr's circle of stress

FIG. 20

Calculation Procedure:

1. Compute the stresses, using the analytical method

A *principal stress* is a normal stress not accompanied by a shearing stress. The plane on which the principal stress exists is termed a *principal plane*. For a condition of plane stress, there are two principal planes through every point in a stressed body and these planes are mutually perpendicular. Moreover, one principal stress is the maximum normal stress existing at that point; the other is the minimum normal stress.

   Let s_x and s_y = the principal stress in the x and y direction, respectively; s_n = normal stress on the plane making an angle θ with the y axis; s_s = shearing stress on this plane. All stresses are expressed in pounds per square inch (kilopascals) and all angles in degrees. Tensile stresses are positive; compressive stresses are negative.

   Applying the usual stress equations yields $s_n = s_y + (s_x - s_y)\cos^2 \theta$; $s_s = \frac{1}{2}(s_x - s_y)\sin 2\theta$. Substituting gives $s_n = 2400 + (6300 - 2400)0.766^2 = 4690$-lb/in^2 (32,337.6-kPa) tension, and $s_s = \frac{1}{2}(6300 - 2400)0.985 = 1920$ lb/in^2 (13,238.4 kPa).

2. Apply the graphical method of solution

Construct, in Fig. 20b, Mohr's circle of stress thus: Using a suitable scale, draw $OA = s_y$, and $OB = s_x$. Draw a circle having AB as its diameter. Draw the radius CD making an angle of $2\theta = 80°$ with AB. Through D, drop a perpendicular DE to AB. Then $OE = s_n$ and $ED = s_s$. Scale OE and ED to obtain the normal and shearing stresses on plane AE.

Related Calculations: The normal stress may also be computed from $s_n = (s_x + s_y)0.5 + (s_x - s_y)0.5 \cos 2\theta$.

EVALUATION OF PRINCIPAL STRESSES

The prism $ABCD$ in Fig. 21a is subjected to the normal and shearing stresses shown. Construct Mohr's circle to determine the principal stresses at A, and locate the principal planes.

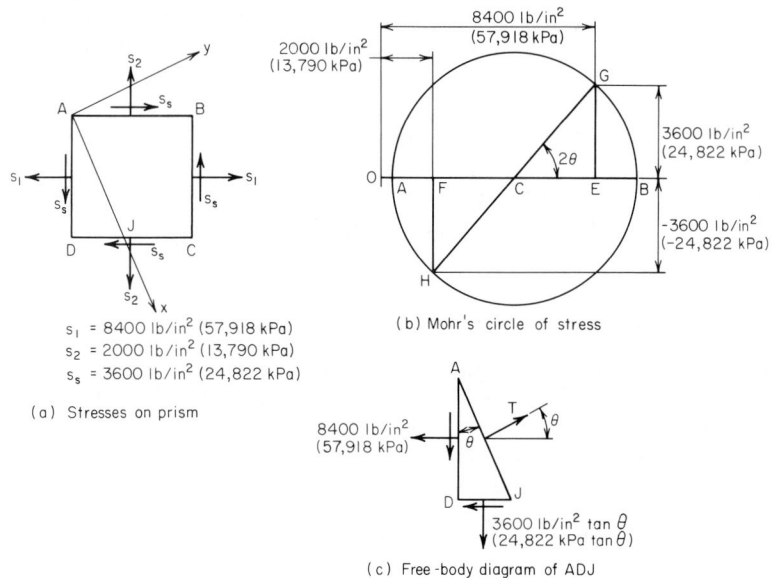

$s_1 = 8400$ lb/in² (57,918 kPa)
$s_2 = 2000$ lb/in² (13,790 kPa)
$s_s = 3600$ lb/in² (24,822 kPa)

(a) Stresses on prism

(b) Mohr's circle of stress

(c) Free-body diagram of ADJ

FIG. 21

Calculation Procedure:

1. Draw the lines representing the normal stresses (Fig. 21b)

Through the origin O, draw a horizontal base line. Locate points E and F such that $OE = 8400$ lb/in² (57,918.0 kPa) and $OF = 2000$ lb/in² (13,790.0 kPa). Since both normal stresses are tensile, E and F lie to the right of O. Note that the construction required here is the converse of that required in the previous calculation procedure.

2. Draw the lines representing the shearing stresses

Construct the vertical lines EG and FH such that $EG = 3600$ lb/in² (24,822.0 kPa), and $FH = -3600$ lb/in² ($-24,822.0$ kPa).

3. Continue the construction

Draw line GH to intersect the base line at C.

4. Construct Mohr's circle

Draw a circle having GH as diameter, intersecting the base line at A and B. Then lines OA and OB represent the principal stresses.

5. Scale the diagram

Scale OA and OB to obtain $f_{max} = 10,020$ lb/in² (69,087.9 kPa); $f_{min} = 380$ lb/in² (2620.1 kPa). Both stresses are tension.

6. Determine the stress angle

Scale angle BCG and measure it as $48°22'$. The angle between the x axis, on which the maximum stress exists, and the side AD of the prism is one-half of BCG.

7. Construct the x and y axes

In Fig. 21a, draw the x axis, making a counterclockwise angle of $24°11'$ with AD. Draw the y axis perpendicular thereto.

8. Verify the locations of the principal planes

Consider ADJ as a free body. Set the length AD equal to unity. In Fig. 21c, since there is no shearing stress on AJ, $\Sigma F_H = T \cos \theta - 8400 - 3600 \tan \theta = 0$; $T \cos \theta = 8400 + 3600(0.45)$ $= 10{,}020$ lb/in^2 (69,087.9 kPa). The stress on $AJ = T/AJ = T \cos \theta = 10{,}020$ lb/in^2 (69,087.9 kPa).

HOOP STRESS IN THIN-WALLED CYLINDER UNDER PRESSURE

A steel pipe 5 ft (1.5 m) in diameter and ⅜ in (9.53 mm) thick sustains a fluid pressure of 180 lb/in^2 (1241.1 kPa). Determine the hoop stress, the longitudinal stress, and the increase in diameter of this pipe. Use 0.25 for Poisson's ratio.

Calculation Procedure:

1. Compute the hoop stress

Use the relation $s = pD/(2t)$, where s = hoop or tangential stress, lb/in^2 (kPa); p = radial pressure, lb/in^2 (kPa); D = internal diameter of cylinder, in (mm); t = cylinder wall thickness, in (mm). Thus, for this cylinder, $s = 180(60)/[2(⅜)] = 14{,}400$ lb/in^2 (99,288.0 kPa).

2. Compute the longitudinal stress

Use the relation $s' = pD/(4t)$, where s' = longitudinal stress, i.e., the stress parallel to the longitudinal axis of the cylinder, lb/in^2 (kPa), with other symbols as before. Substituting yields $s' = 7200$ lb/in^2 (49,644.0 kPa).

3. Compute the increase in the cylinder diameter

Use the relation $\Delta D = (D/E)(s - \nu s')$, where ν = Poisson's ratio. Thus $\Delta D = 60(14{,}400 - 0.25 \times 7200)/(30 \times 10^6) = 0.0252$ in (0.6401 mm).

STRESSES IN PRESTRESSED CYLINDER

A steel ring having an internal diameter of 8.99 in (228.346 mm) and a thickness of ¼ in (6.35 mm) is heated and allowed to shrink over an aluminum cylinder having an external diameter of 9.00 in (228.6 mm) and a thickness of ½ in (12.7 mm). After the steel cools, the cylinder is subjected to an internal pressure of 800 lb/in^2 (5516 kPa). Find the stresses in the two materials. For aluminum, $E = 10 \times 10^6$ lb/in^2 (6.895 $\times$ 10^7 kPa).

Calculation Procedure:

1. Compute the radial pressure caused by prestressing

Use the relation $p = 2\Delta D/\{D^2[1/(t_a E_a) + 1/(t_s E_s)]\}$, where p = radial pressure resulting from prestressing, lb/in^2 (kPa), with other symbols the same as in the previous calculation procedure and the subscripts a and s referring to aluminum and steel, respectively. Thus, $p = 2(0.01)/\{9^2[1/(0.5 \times 10 \times 10^6) + 1/(0.25 \times 30 \times 10^6)]\} = 741$ lb/in^2 (5109.2 kPa).

2. Compute the corresponding prestresses

Using the subscripts 1 and 2 to denote the stresses caused by prestressing and internal pressure, respectively, we find $s_{a1} = pD/(2t_a)$, where the symbols are the same as in the previous calculation procedure. Thus, $s_{a1} = 741(9)/[2(0.5)] = 6670$-lb/in^2 (45,989.7-kPa) compression. Likewise, $s_{s1} = 741(9)/[2(0.25)] = 13{,}340$-lb/in^2 (91,979-kPa) tension.

3. Compute the stresses caused by internal pressure

Use the relation $s_{s2}/s_{a2} = E_s/E_a$ or, for this cylinder, $s_{s2}/s_{a2} = (30 \times 10^6)/(10 \times 10^6) = 3$. Next, compute s_{a2} from $t_a s_{a2} + t_s s_{s2} = pD/2$, or $s_{a2} = 800(9)/[2(0.5 + 0.25 \times 3)] = 2880$-lb/in^2 (19,857.6-kPa) tension. Also, $s_{s2} = 3(2880) = 8640$-lb/in^2 (59,572.8-kPa) tension.

4. Compute the final stresses

Sum the results in steps 2 and 3 to obtain the final stresses: $s_{a3} = 6670 - 2880 = 3790$-lb/in^2 (26,132.1-kPa) compression; $s_{s3} = 13,340 + 8640 = 21,980$-lb/in^2 (151,552.1-kPa) tension.

5. Check the accuracy of the results

Ascertain whether the final diameters of the steel ring and aluminum cylinder are equal. Thus, setting $s' = 0$ in $\Delta D = (D/E)(s - vs')$, we find $\Delta D_a = -3790(9)/(10 \times 10^6) = -0.0034$ in $(-0.0864$ mm$)$, $D_a = 9.0000 - 0.0034 = 8.9966$ in (228.51 mm). Likewise, $\Delta D_s = 21,980(9)/(30 \times 10^6) = 0.0066$ in (0.1676 mm), $D_s = 8.99 + 0.0066 = 8.9966$ in (228.51 mm). Since the computed diameters are equal, the results are valid.

HOOP STRESS IN THICK-WALLED CYLINDER

A cylinder having an internal diameter of 20 in (508 mm) and an external diameter of 36 in (914 mm) is subjected to an internal pressure of 10,000 lb/in^2 (68,950 kPa) and an external pressure of

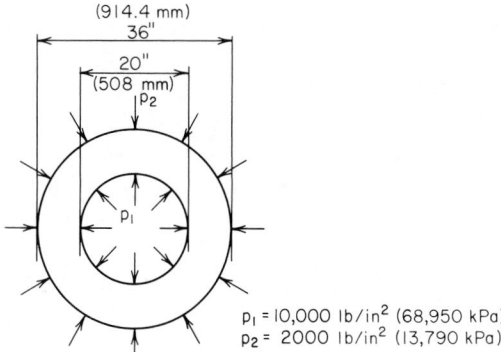

$p_1 = 10{,}000$ lb/in^2 (68,950 kPa)
$p_2 = 2000$ lb/in^2 (13,790 kPa)

FIG. 22 Thick-walled cylinder under internal and external pressure.

2000 lb/in^2 (13,790 kPa) as shown in Fig. 22. Determine the hoop stress at the inner and outer surfaces of the cylinder.

Calculation Procedure:

1. Compute the hoop stress at the inner surface of the cylinder

Use the relation $s_i = [p_1(r_1^2 + r_2^2) - 2p_2 r_2^2]/(r_2^2 - r_1^2)$, where s_i = hoop stress at inner surface, lb/in^2 (kPa); p_1 = internal pressure, lb/in^2 (kPa); r_1 = internal radius, in (mm); r_2 = external radius, in (mm); p_2 = external pressure, lb/in^2 (kPa). Substituting gives $s_i = [10,000(100 + 324) - 2(2000)(324)]/(324 - 100) = 13,100$-lb/in^2 (90,324.5-kPa) tension.

2. Compute the hoop stress at the outer cylinder surface

Use the relation $s_0 = [2p_1 r_1^2 - p_2(r_1^2 + r_2^2)]/(r_2^2 - r_1^2)$, where the symbols are as before. Substituting gives $s_0 = [2(10,000)(100) - 2000(100 + 324)]/(324 - 100) = 5100$-lb/in^2 (35,164.5-kPa) tension.

3.   Check the accuracy of the results

Use the relation $s_i r_1 - s_0 r_2 = [(r_2 - r_1)/(r_2 + r_1)](p_1 r_1 + p_2 r_2)$. Substituting the known values verifies the earlier calculations.

THERMAL STRESS RESULTING FROM HEATING A MEMBER

A steel member 18 ft (5.5 m) long is set snugly between two walls and heated 80°F (44.4°C). If each wall yields 0.015 in (0.381 mm), what is the compressive stress in the member? Use a coefficient of thermal expansion of $6.5 \times 10^{-6}/°F$ ($1.17 \times 10^{-5}/°C$) for steel.

Calculation Procedure:

1.   Compute the thermal expansion of the member without restraint

Replace the true condition of partial restraint with the following equivalent conditions: The member is first allowed to expand freely under the temperature rise and is then compressed to its true final length.

To compute the thermal expansion without restraint, use the relation $\Delta L = cL\Delta T$, where c = coefficient of thermal expansion, /°F (/°C); ΔT = increase in temperature, °F (°C); L = original length of member, in (mm); ΔL = increase in length of the member, in (mm). Substituting gives $\Delta L = 6.5(10^{-6})(18)(12)(80) = 0.1123$ in (2.852 mm).

2.   Compute the linear restraint exerted by the walls

The walls yield $2(0.015) = 0.030$ in (0.762 mm). Thus, the restraint exerted by the walls is $\Delta L_w = 0.1123 - 0.030 = 0.0823$ in (2.090 mm).

3.   Determine the compressive stress

Use the relation $s = E\Delta L/L$, where the symbols are as given earlier. Thus, $s = 30(10^6)(0.0823)/[18(12)] = 11,430$ lb/in^2 (78,809.9 kPa).

THERMAL EFFECTS IN COMPOSITE MEMBER HAVING ELEMENTS IN PARALLEL

A ½-in (12.7-mm) diameter Copperweld bar consists of a steel core ⅜ in (9.53 mm) in diameter and a copper skin $\frac{1}{16}$ in (1.6 mm) thick. What is the elongation of a 1-ft (0.3-m) length of this bar, and what is the internal force between the steel and copper arising from a temperature rise of 80°F (44.4°C)? Use the following values for thermal expansion coefficients: $c_s = 6.5 \times 10^{-6}$ and $c_c = 9.0 \times 10^{-6}$, where the subscripts s and c refer to steel and copper, respectively. Also, $E_c = 15 \times 10^6$ lb/in^2 (1.03×10^8 kPa).

Calculation Procedure:

1.   Determine the cross-sectional areas of the metals

The total area $A = 0.1963$ in^2 (1.266 cm^2). The area of the steel $A_s = 0.1105$ in^2 (0.712 cm^2). By difference, the area of the copper $A_c = 0.0858$ in^2 (0.553 cm^2).

2.   Determine the coefficient of expansion of the composite member

Weight the coefficients of expansion of the two members according to their respective AE values. Thus

$A_s E_s$ (relative) = $0.1105 \times 30 \times 10^6$ =	3315
$A_c E_c$ (relative) = $0.0858 \times 15 \times 10^6$ =	1287
Total	4602

Then the coefficient of thermal expansion of the composite member is $c = (3315c_s + 1287c_c)/4602 = 7.2 \times 10^{-6}/°F$ ($1.30 \times 10^{-5}/°C$).

3.   *Determine the thermal expansion of the 1-ft (0.3-m) section*

Using the relation $\Delta L = cL\Delta T$, we get $\Delta L = 7.2(10^{-6})(12)(80) = 0.00691$ in (0.17551 mm).

4.   *Determine the expansion of the first material without restraint*

Using the same relation as in step 3 for copper *without* restraint yields $\Delta L_c = 9.0(10^{-6}) \times (12)(80) = 0.00864$ in (0.219456 mm).

5.   *Compute the restraint of the first material*

The copper is restrained to the amount computed in step 3. Thus, the restraint exerted by the steel is $\Delta L_{cs} = 0.00864 - 0.00691 = 0.00173$ in (0.043942 mm).

6.   *Compute the restraining force exerted by the second material*

Use the relation $P = (A_c E_c \Delta L_{cs})/L$, where the symbols are as given before: $P = [1,287,000(0.00173)]/12 = 185$ lb (822.9 N).

7.   *Verify the results obtained*

Repeat steps 4, 5, and 6 with the two materials interchanged. So $\Delta L_s = 6.5(10^{-6})(12)(80) = 0.00624$ in (0.15849 mm); $\Delta L_{sc} = 0.00691 - 0.00624 = 0.00067$ in (0.01701 mm). Then $P = 3,315,000(0.00067)/12 = 185$ lb (822.9 N), as before.

THERMAL EFFECTS IN COMPOSITE MEMBER HAVING ELEMENTS IN SERIES

The aluminum and steel bars in Fig. 23 have cross-sectional areas of 1.2 and 1.0 in^2 (7.7 and 6.5 cm^2), respectively. The member is restrained against lateral deflection. A temperature rise of 100°F (55°C) causes the length of the member to increase to 42.016 in (106.720 cm). Determine the stress and deformation of each bar. For aluminum, $E = 10 \times 10^6$, $c = 13.0 \times 10^{-6}$; for steel, $c = 6.5 \times 10^{-6}$.

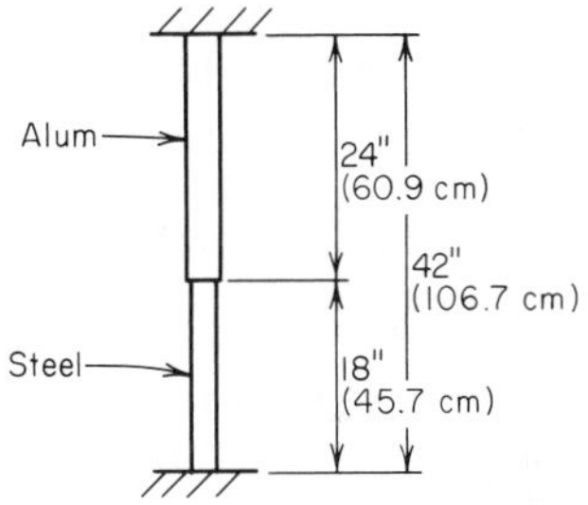

Calculation Procedure:

1.   *Express the deformation of each bar resulting from the temperature change and the compressive force*

The temperature rise causes the bar to expand, whereas the compressive force resists this expansion. Thus, the net expansion is the difference between these two changes, or $\Delta L_a = cL\Delta T - PL/(AE)$, where the subscript a refers to the aluminum bar; the other symbols are the same as

FIG. 23

given earlier. Substituting gives $\Delta L_a = 13.0 \times 10^{-6}(24)(100) - P(24)/[1.2(10 \times 10^6)] = (31,200 - 2P)10^{-6}$, Eq. *a*. Likewise, for steel: $\Delta L_s = 6.5 \times 10^{-6}(18)(100) - P(18)/[1.0(30 \times 10^6)] = (11,700 - 0.6P)10^{-6}$, Eq. *b*.

2.   *Sum the results in step 1 to obtain the total deformation of the member*

Set the result equal to 0.016 in (0.4064 mm); solve for P. Or, $\Delta L = (42,900 - 2.6P)10^{-6} = 0.016$ in (0.4064 mm); $P = (42,900 - 16,000)/2.6 = 10,350$ lb (46,037 N).

3.   *Determine the stresses and deformation*

Substitute the computed value of P in the stress equation $s = P/A$. For aluminum $s_a = 10,350/1.2 = 8630$ lb/in^2 (59,503.9 kPa). Then $\Delta L_a = (31,200 - 2 \times 10,350)10^{-6} = 0.0105$ in (0.2667 mm). Likewise, for steel $s_s = 10,350/1.0 = 10,350$ lb/in^2 (71,363.3 kPa); and $\Delta L_s = (11,700 - 0.6 \times 10,350)10^{-6} = 0.0055$ in (0.1397 mm).

SHRINK-FIT STRESS AND RADIAL PRESSURE

An open steel cylinder having an internal diameter of 4 ft (1.2 m) and a wall thickness of $\frac{5}{16}$ in (7.9 mm) is to be heated to fit over an iron casting. The internal diameter of the cylinder before heating is $\frac{1}{32}$ in (0.8 mm) less than that of the casting. How much must the temperature of the cylinder be increased to provide a clearance of $\frac{1}{32}$ in (0.8 mm) all around between the cylinder and casting? If the casting is considered rigid, what stress will exist in the cylinder after it cools, and what radial pressure will it then exert on the casting?

Calculation Procedure:

1. Compute the temperature rise required

Use the relation $\Delta T = \Delta D/(cD)$, where ΔT = temperature rise required, °F (°C); ΔD = change in cylinder diameter, in (mm); c = coefficient of expansion of the cylinder = 6.5×10^{-6}/°F (1.17×10^{-5}/°C); D = cylinder internal diameter before heating, in (mm). Thus $\Delta T = (3/32)/[6.5 \times 10^{-6}(48)] = 300$°F (167°C).

2. Compute the hoop stress in the cylinder

Upon cooling, the cylinder has a diameter 1/32 in (0.8 mm) larger than originally. Compute the hoop stress from $s = E\Delta D/D = 30 \times 10^{6}(1/32)/48 = 19,500$ lb/in² (134,452.5 kPa).

3. Compute the associated radial pressure

Use the relation $p = 2ts/D$, where p = radial pressure, lb/in² (kPa), with the other symbols as given earlier. Thus $p = 2(5/16)(19,500)/48 = 254$ lb/in² (1751.3 kPa).

TORSION OF A CYLINDRICAL SHAFT

A torque of 8000 lb·ft (10,840 N·m) is applied at the ends of a 14-ft (4.3-m) long cylindrical shaft having an external diameter of 5 in (127 mm) and an internal diameter of 3 in (76.2 mm). What are the maximum shearing stress and the angle of twist of the shaft if the modulus of ridigity of the shaft is 6×10^{6} lb/in² (4.1×10^{4} MPa)?

Calculation Procedure:

1. Compute the polar moment of inertia of the shaft

For a hollow circular shaft, $J = (\pi/32)(D^4 - d^4)$, where J = polar moment of inertia of a transverse section of the shaft with respect to the longitudinal axis, in⁴ (cm⁴); D = external diameter of shaft, in (mm); d = internal diameter of shaft, in (mm). Substituting gives $J = (\pi/32)(5^4 - 3^4) = 53.4$ in⁴ (2222.6 cm⁴).

2. Compute the shearing stress in the shaft

Use the relation $s_s = TR/J$, where s_s = shearing stress, lb/in² (MPa); T = applied torque, lb·in (N·m); R = radius of shaft, in (mm). Thus $s_s = [(8000)(12)(2.5)]/53.4 = 4500$ lb/in² (31,027.5 kPa).

3. Compute the angle of twist of the shaft

Use the relation $\theta = TL/JG$, where θ = angle of twist, rad; L = shaft length, in (mm); G = modulus of ridigity, lb/in² (GPa). Thus $\theta = (8000)(12)(14)(12)/[53.4(6,000,000)] = 0.050$ rad, or 2.9°.

ANALYSIS OF A COMPOUND SHAFT

The compound shaft in Fig. 24 was formed by rigidly joining two solid segments. What torque may be applied at B if the shearing stress is not to exceed 15,000 lb/in² (103.4 MPa) in the steel and 10,000 lb/in² (69.0 MPa) in the bronze? Here $G_s = 12 \times 10^{6}$ lb/in² (82.7 GPa); $G_b = 6 \times 10^{6}$ lb/in² (41.4 GPa).

Calculation Procedure:

1. Determine the relationship between the torque in the shaft segments

Since segments AB and BC (Fig. 24) are twisted through the same angle, the torque applied at the junction of these segments is distributed in proportion to their relative rigidi-

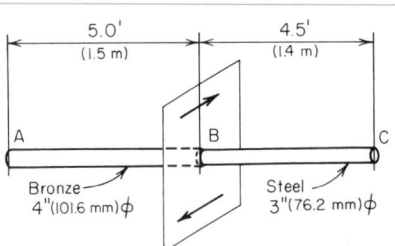

FIG. 24 Compound shaft.

ties. Using the subscripts s and b to denote steel and bronze, respectively, we see that $\theta = T_sL_s/(J_sG_s) = T_bL_b/(J_bG_b)$, where the symbols are as given in the previous calculation procedure. Solving yields $T_s = (5/4.5)(3^4/4^4)(12/6)T_b = 0.703T_b$.

2. Establish the relationship between the shearing stresses

For steel, $s_{ss} = 16T_s/(\pi D^3)$, where the symbols are as given earlier. Thus $s_{ss} = 16(0.703T_b)/(\pi 3^3)$. Likewise, for bronze, $s_{sb} = 16T_b/(\pi 4^3)$, $\therefore s_{ss} = 0.703(4^3/3^3)s_{sb} = 1.67s_{sb}$.

3. Compute the allowable torque

Ascertain which material limits the capacity of the member, and compute the allowable torque by solving the shearing-stress equation for T.

If the bronze were stressed to 10,000 lb/in² (69.0 MPa), inspection of the above relations shows that the steel would be stressed to 16,700 lb/in² (115.1 MPa), which exceeds the allowed 15,000 lb/in² (103.4 MPa). Hence, the steel limits the capacity. Substituting the allowed shearing stress of 15,000 lb/in² (103.4 MPa) gives $T_s = 15,000\pi(3^3)/[16(12)] = 6630$ lb·ft (8984.0 N·m); also, $T_b = 6630/0.703 = 9430$ lb·ft (12,777.6 N·m). Then $T = 6630 + 9430 = 16,060$ lb·ft (21,761.3 N·m).

Stresses in Flexural Members

In the analysis of beam action, the general assumption is that the beam is in a horizontal position and carries vertical loads lying in an axis of symmetry of the transverse section of the beam.

The vertical shear V at a given section of the beam is the algebraic sum of all vertical forces to the left of the section, with an upward force being considered positive.

The bending moment M at a given section of the beam is the algebraic sum of the moments of all forces to the left of the section with respect to that section, a clockwise moment being considered positive.

If the proportional limit of the beam material is not exceeded, the bending stress (also called the flexural, or fiber, stress) at a section varies linearly across the depth of the section, being zero at the neutral axis. A positive bending moment induces compressive stresses in the fibers above the neutral axis and tensile stresses in the fibers below. Consequently, the elastic curve of the beam is concave upward where the bending moment is positive.

SHEAR AND BENDING MOMENT IN A BEAM

Construct the shear and bending-moment diagrams for the beam in Fig. 25. Indicate the value of the shear and bending moment at all significant sections.

Calculation Procedure:

1. Replace the distributed load on each interval with its equivalent concentrated load

Where the load is uniformly distributed, this equivalent load acts at the center of the interval of the beam. Thus $W_{AB} = 2(4) = 8$ kips (35.6 kN); $W_{BC} = 2(6) = 12$ kips (53.3 kN); $W_{AC} = 8 + 12 = 20$ kips (89.0 kN); $W_{CD} = 3(15) = 45$ kips (200.1 kN); $W_{DE} = 1.4(5) = 7$ kips (31.1 kN).

2. Determine the reaction at each support

Take moments with respect to the other support. Thus $\Sigma M_D = 25R_A - 6(21) - 20(20) - 45(7.5) + 7(2.5) + 4.2(5) = 0$; $\Sigma M_A = 6(4) + 20(5) + 45(17.5) + 7(27.5) + 4.2(30) - 25R_D = 0$. Solving gives $R_A = 33$ kips (146.8 kN); $R_D = 49.2$ kips (218.84 kN).

3. Verify the computed results and determine the shears

Ascertain that the algebraic sum of the vertical forces is zero. If this is so, the computed results are correct.

Starting at A, determine the shear at every significant section, or directly to the left or right of that section if a concentrated load is present. Thus V_A at right = 33 kips (146.8 kN); V_B at left = 33 − 8 = 25 kips (111.2 kN); V_B at right = 25 − 6 = 19 kips (84.5 kN); $V_C = 19 − 12 = 7$ kips (31.1 kN); V_D at left = 7 − 45 = −38 kips (−169.0 kN); V_D at right = −38 + 49.2 = 11.2 kips (49.8 kN); V_E at left = 11.2 − 7 = 4.2 kips (18.7 kN); V_E at right = 4.2 − 4.2 = 0.

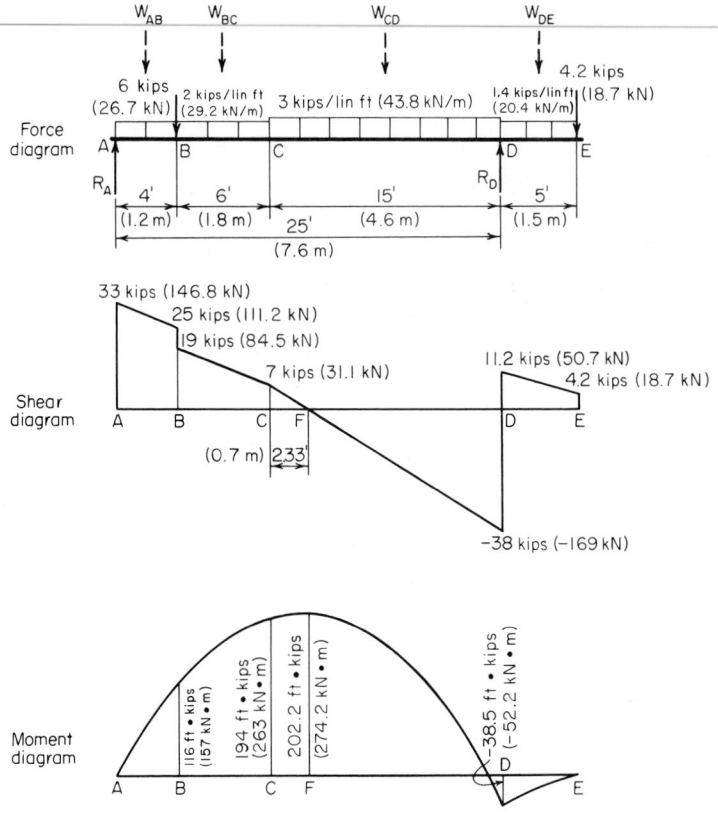

FIG. 25

4. Plot the shear diagram

Plot the points representing the forces in the previous step in the shear diagram. Since the loading betwen the significant sections is uniform, connect these points with straight lines. In general, the slope of the shear diagram is given by $dV/dx = -w$, where w = unit load at the given section and x = distance from left end to the given section.

5. Determine the bending moment at every significant section

Starting at A, determine the bending moment at every significant section. Thus $M_A = 0$; $M_B = 33(4) - 8(2) = 116$ ft·kips (157 kN·m); $M_C = 33(10) - 8(8) - 6(6) - 12(3) = 194$ ft·kips (263 kN·m). Similarly, $M_D = -38.5$ ft·kips (-52.2 kN·m); $M_E = 0$.

6. Plot the bending-moment diagram

Plot the points representing the values in step 5 in the bending-moment diagram (Fig. 25). Complete the diagram by applying the slope equation $dM/dx = V$, where V denotes the shear at the given section. Since this shear varies linearly between significant sections, the bending-moment diagram comprises a series of parabolic arcs.

7. Alternatively, apply a moment theorem

Use this theorem: If there are no externally applied moments in an interval 1-2 of the span, the difference between the bending moments is $M_2 - M_1 = \int_1^2 V\, dx$ = the area under the shear diagram across the interval.

Calculate the areas under the shear diagram to obtain the following results: $M_A = 0$; $M_B =$

$M_A + \frac{1}{2}(4)(33 + 25) = 116$ ft·kips (157.3 kN·m); $M_C = 116 + \frac{1}{2}(6)(19 + 7) = 194$ ft·kips (263 kN·m); $M_D = 194 + \frac{1}{2}(15)(7 - 38) = -38.5$ ft·kips (-52.2 kN·m); $M_E = -38.5 + \frac{1}{2}(5)(11.2 + 4.2) = 0$.

8. Locate the section at which the bending moment is maximum

As a corollary of the equation in step 6, the maximum moment occurs where the shear is zero or passes through zero under a concentrated load. Therefore, $CF = 7/3 = 2.33$ ft (0.710 m).

9. Compute the maximum moment

Using the computed value for CF, we find $M_F = 194 + \frac{1}{2}(2.33)(7) = 202.2$ ft·kips (274.18 kN·m).

BEAM BENDING STRESSES

A beam having the trapezoidal cross section shown in Fig. 26a carries the loads indicated in Fig. 26b. What is the maximum bending stress at the top and at the bottom of this beam?

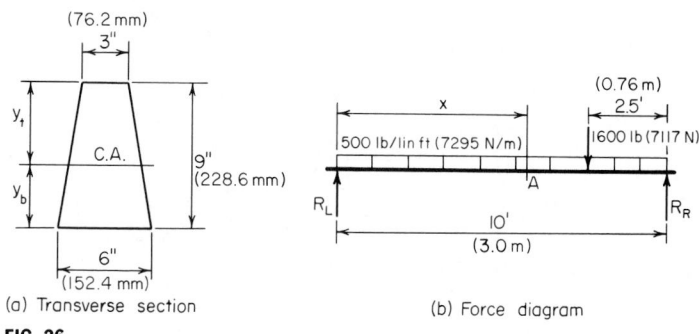

(a) Transverse section   (b) Force diagram

FIG. 26

Calculation Procedure:

1. Compute the left reaction and the section at which the shear is zero

The left reaction $R_L = \frac{1}{2}(10)(500) + 1600(2.5/10) = 2900$ lb (12,899.2 N). The section A at which the shear is zero is $x = 2900/500 = 5.8$ ft (1.77 m).

2. Compute the maximum moment

Use the relation $M_A = \frac{1}{2}(2900)(5.8) = 8410$ lb·ft (11,395.6 N·m) $= 100,900$ lb·in (11,399.682 N·m).

3. Locate the centroidal axis of the section

Use the AISC *Manual* for properties of the trapezoid. Or $y_t = (9/3)[(2 \times 6 + 3)/(6 + 3)] = 5$ in (127 mm); $y_b = 4$ in (101.6 mm).

4. Compute the moment of inertia of the section

Using the AISC *Manual*, $I = (9^3/36)[(6^2 + 4 \times 6 \times 3 + 3^2)/(6 + 3)] = 263.3$ in⁴ (10,959.36 cm⁴).

5. Compute the stresses in the beam

Use the relation $f = My/I$, where f = bending stress in a given fiber, lb/in² (kPa); y = distance from neutral axis to given fiber, in. Thus $f_{top} = 100,900(5)/263.3 = 1916$-lb/in² (13,210.8-kPa) compression, $f_{bottom} = 100,900(4)/263.3 = 1533$-lb/in² (10,570.0-kPa) tension.

   In general, the maximum bending stress at a section where the moment is M is given by $f = Mc/I$, where c = distance from the neutral axis to the outermost fiber, in (mm). For a section that is symmetric about its centroidal axis, it is convenient to use the section modulus S of the section, this being defined as $S = I/c$. Then $f = M/S$.

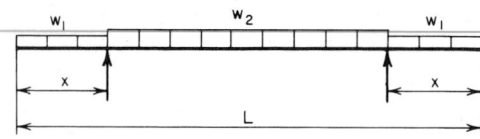

(a) Loads carried by overhanging beam

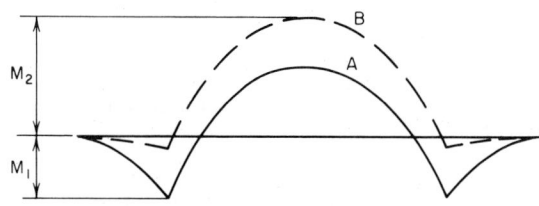

Diagram A: Full load on entire span
Diagram B: Dead load on overhangs ; full load
between supports

(b) Bending-moment diagrams

FIG. 27

ANALYSIS OF A BEAM ON MOVABLE SUPPORTS

The beam in Fig. 27a rests on two movable supports. It carries a uniform live load of w lb/lin ft and a uniform dead load of $0.2w$ lb/lin ft. If the allowable bending stresses in tension and compression are identical, determine the optimal location of the supports.

Calculation Procedure:

1. Place full load on the overhangs, and compute the negative moment

Refer to the moment diagrams. For every position of the supports, there is a corresponding maximum bending stress. The position for which this stress has the smallest value must be identified.

As the supports are moved toward the interior of the beam, the bending moments between the supports diminish in algebraic value. The optimal position of the supports is that for which the maximum potential negative moment M_1 is numerically equal to the maximum potential positive moment M_2. Thus, $M_1 = -1.2w(x^2/2) = -0.6wx^2$.

2. Place only the dead load on the overhangs and the full load between the supports. Compute the positive moment.

Sum the areas under the shear diagram to compute M_2. Thus, $M_2 = \frac{1}{2}[1.2w(L/2 - x)^2 - 0.2wx^2] = w(0.15L^2 - 0.6Lx + 0.5x^2)$.

3. Equate the absolute values of M_1 and M_2 and solve for x

Substituting gives $0.6x^2 = 0.15L^2 - 0.6Lx + 0.5x^2$; $x = L(\overline{10.5}^{0.5} - 3) = 0.240L$.

FLEXURAL CAPACITY OF A COMPOUND BEAM

A W16 × 45 steel beam in an existing structure was reinforced by welding an WT6 × 20 to the bottom flange, as in Fig. 28. If the allowable bending stress is 20,000 lb/in² (137,900 kPa), determine the flexural capacity of the built-up member.

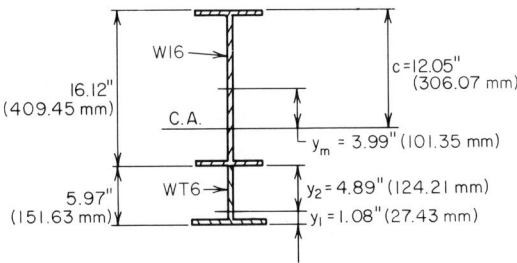

FIG. 28 Compound beam.

Calculation Procedure:

1. *Obtain the properties of the elements*

Using the AISC *Manual,* determine the following properties. For the W16 × 45, $d = 16.12$ in (409.45 mm); $A = 13.24$ in^2 (85.424 cm^2); $I = 583$ in^4 (24,266 cm^4). For the WT6 × 20, $d = 5.97$ in (151.63 mm); $A = 5.89$ in^2 (38.002 cm^2); $I = 14$ in^4 (582.7 cm^4); $y_1 = 1.08$ in (27.43 mm); $y_2 = 5.97 - 1.08 = 4.89$ in (124.21 mm).

2. *Locate the centroidal axis of the section*

Locate the centroidal axis of the section with respect to the centerline of the W16 × 45, and compute the distance c from the centroidal axis to the outermost fiber. Thus, $y_m = 5.89[(8.06 + 4.89)]/(5.89 + 13.24) = 3.99$ in (101.346 mm). Then $c = 8.06 + 3.99 = 12.05$ in (306.07 mm).

3. *Find the moment of inertia of the section with respect to its centroidal axis*

Use the relation $I_0 + Ak^2$ for each member, and take the sum for the two members to find I for the built-up beam. Thus, for the W16 × 45: $k = 3.99$ in (101.346 mm); $I_0 + Ak^2 = 583 + 13.24(3.99)^2 = 793$ in^4 (33,007.1 cm^4). For the WT6 × 20: $k = 8.06 - 3.99 + 4.89 = 8.96$ in (227.584 mm); $I_0 + Ak^2 = 14 + 5.89(8.96)^2 = 487$ in^4 (20,270.4 cm^4). Then $I = 793 + 487 = 1280$ in^4 (53,277.5 cm^4).

4. *Apply the moment equation to find the flexural capacity*

Use the relation $M = fI/c = 20,000(1280)/[12.05(12)] = 177,000$ lb·ft (240,012 N·m).

ANALYSIS OF A COMPOSITE BEAM

An 8 × 12 in (203.2 × 304.8 mm) timber bean (exact size) is reinforced by the addition of a 7 × ½ in (177.8 × 12.7 mm) steel plate at the top and a 7-in (177.8-mm) 9.8-lb (43.59-N) steel channel at the bottom, as shown in Fig. 29a. The allowable bending stresses are 22,000 lb/in^2 (151,690 kPa) for steel and 1200 lb/in^2 (8274 kPa) for timber. The modulus of elasticity of the timber is 1.2×10^6 lb/in^2 (8.274×10^6 kPa). How does the flexural strength of the reinforced beam compare with that of the original timber beam?

Calculation Procedure:

1. *Compute the rigidity of the steel compared with that of the timber*

Let n = the relative rigidity of the steel and timber. Then $n = E_s/E_t = (30 \times 10^6)/(1.2 \times 10^6) = 25$.

2. *Transform the composite beam to an equivalent homogeneous beam*

To accomplish this transformation, replace the steel with timber. Sketch the cross section of the transformed beam as in Fig. 29b. Determine the sizes of the hypothetical elements by retaining the dimensions normal to the axis of bending but multiplying the dimensions parallel to this axis by n.

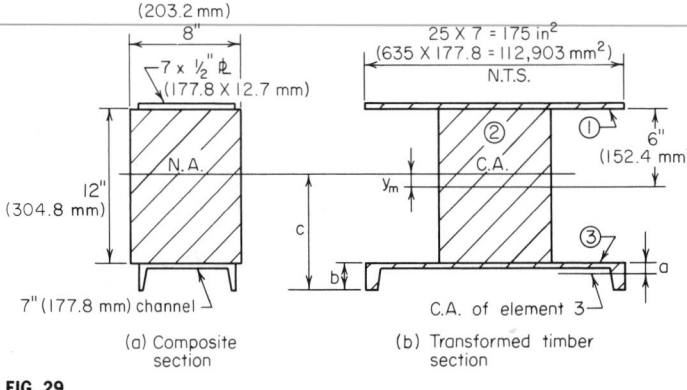

FIG. 29

(a) Composite section

(b) Transformed timber section

3. Record the properties of each element of the transformed section

Element 1: $A = 25(7)(\frac{1}{2}) = 87.5$ in^2 (564.55 cm^2); I_0 is negligible.
Element 2: $A = 8(12) = 96$ in^2 (619.4 cm^2); $I_0 = \frac{1}{12}(8)12^3 = 1152$ in^4 (4.795 dm^4).
Element 3: Refer to the AISC *Manual* for the data; $A = 25(2.85) = 71.25$ in^2 (459.71 cm^2); I_0 $= 25(0.98) = 25$ in^4 (1040.6 cm^4); $a = 0.55$ in (13.97 mm); $b = 2.09$ in (53.09 mm).

4. Locate the centroidal axis of the transformed section

Take static moments of the areas with respect to the centerline of the 8×12 in (203.2 $\times$ 304.8 mm) rectangle. Then $y_m = [87.5(6.25) - 71.25(6.55)]/(87.5 + 96 + 71.25) = 0.31$ in (7.87 mm). The neutral axis of the composite section is at the same location as the centroidal axis of the transformed section.

5. Compute the moment of inertia of the transformed section

Apply the relation in step 3 of the previous calculation procedure. Then compute the distance c to the outermost fiber. Thus, $I = 1152 + 25 + 87.5(6.25 - 0.31)^2 + 96(0.31)^2 + 71.25(6.55 + 0.31)^2 = 7626$ in^4 (31.74 dm^4). Also, $c = 0.31 + 6 + 2.09 = 8.40$ in (213.36 mm).

6. Determine which material limits the beam capacity

Assume that the steel is stressed to capacity, and compute the corresponding stress in the transformed beam. Thus, $f = 22,000/25 = 880$ lb/in^2 (6067.6 kPa) < 1200 lb/in^2 (8274 kPa).

In the actual beam, the maximum timber stress, which occurs at the back of the channel, is even less than 880 lb/in^2 (6067.6 kPa). Therefore, the strength of the member is controlled by the allowable stress in the steel.

7. Compare the capacity of the original and reinforced beams

Let subscripts 1 and 2 denote the original and reinforced beams, respectively. Compute the capacity of these members, and compare the results. Thus $M_1 = fI/c = 1200(1152)/6 = 230,000$ lb·in (25,985.4 N·m); $M_2 = 880(7626)/8.40 = 799,000$ lb·in (90,271.02 N·m); $M_2/M_1 = 799,000/230,000 = 3.47$. Thus, the reinforced beam is nearly 3½ times as strong as the original beam, before reinforcing.

BEAM SHEAR FLOW AND SHEARING STRESS

A timber beam is formed by securely bolting a 3×6 in (76.2 $\times$ 152.4 mm) member to a 6×8 in (152.4 $\times$ 203.2 mm) member (exact size), as shown in Fig. 30. If the beam carries a uniform load of 600 lb/lin ft (8.756 kN/m) on a simple span of 13 ft (3.9 m), determine the longitudinal shear flow and the shearing stress at the juncture of the two elements at a section 3 ft (0.91 m) from the support.

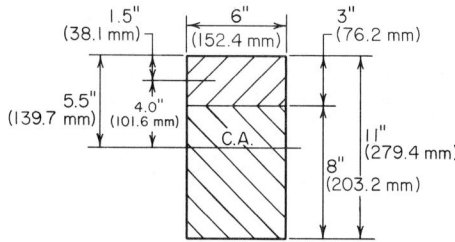

FIG. 30

Calculation Procedure:

1. Compute the vertical shear at the given section

Shear flow is the shearing force acting on a unit distance. In this instance, the shearing force on an area having the same width as the beam and a length of 1 in (25.4 mm) measured along the beam span is required.

Using dimensions and data from Fig. 30, we find $R = \frac{1}{2}(600)(13) = 3900$ lb (17,347.2 N); $V = 3900 - 3(600) = 2100$ lb (9340.8 N).

2. Compute the moment of inertia of the cross section

$$ I = (\tfrac{1}{12})(bd^3) = (\tfrac{1}{12})(6)(11)^3 = 666 \text{ in}^4 \ (2.772 \text{ dm}^4) $$

3. Determine the static moment of the cross-sectional area

Calculate the static moment Q of the cross-sectional area above the plane under consideration with respect to the centroidal axis of the section. Thus, $Q = Ay = 3(6)(4) = 72$ in^3 (1180.1 cm^3).

4. Compute the shear flow

Compute the shear flow q, using $q = VQ/I = 2100(72)/666 = 227$ lb/lin in (39.75 kN/m).

5. Compute the shearing stress

Use the relation $v = q/t = VQ/(It)$, where t = width of the cross section at the given plane. Then $v = 227/6 = 38$ lb/in^2 (262.0 kPa).

Note that v represents both the longitudinal and the transverse shearing stress at a particular point. This is based on the principle that the shearing stresses at a given point in two mutually perpendicular directions are equal.

LOCATING THE SHEAR CENTER OF A SECTION

A cantilever beam carries the load shown in Fig. 31a and has the transverse section shown in Fig. 31b. Locate the shear center of the section.

Calculation Procedure:

1. Construct a free-body diagram of a portion of the beam

Consider that the transverse section of a beam is symmetric solely about its horizontal centroidal axis. If bending of the beam is not to be accompanied by torsion, the vertical shearing force at any section must pass through a particular point on the centroidal axis designated as the *shear,* or *flexural, center.*

Cut the beam at section 2, and consider the left portion of the beam as a free body. In Fig. 31b, indicate the resisting shearing forces V_1, V_2, and V_3 that the right-hand portion of the beam exerts on the left-hand portion at section 2. Obtain the directions of V_1 and V_2 this way: Isolate the segment of the beam contained between sections 1 and 2; then isolate a segment $ABDC$ of the top flange, as shown in Fig. 31c. Since the bending stresses at section 2 exceed those at section 1, the resultant tensile force T_2 exceeds T_1. The resisting force on CD is therefore directed to the

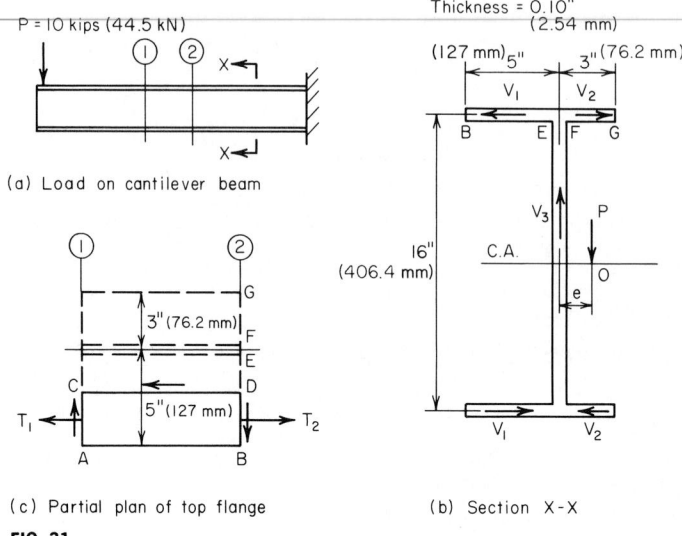

(a) Load on cantilever beam

(c) Partial plan of top flange   (b) Section X-X

FIG. 31

left. From the equation of equilibrium $\Sigma M = 0$ it follows that the resisting shears on AC and BD have the indicated directions to constitute a clockwise couple.

This analysis also reveals that the shearing stress varies linearly from zero at the edge of the flange to a maximum value at the juncture with the web.

2.   Compute the shear flow

Determine the shear flow at E and F (Fig. 31) by setting Q in $q = VQ/I$ equal to the static moment of the overhanging portion of the flange. (For convenience, use the dimensions to the centerline of the web and flange.) Thus $I = \frac{1}{12}(0.10)(16)^3 + 2(8)(0.10)(8)^2 = 137$ in^4 (5702.3 cm^4); $Q_{BE} = 5(0.10)(8) = 4.0$ in^3 (65.56 cm^3); $Q_{FG} = 3(0.10)(8) = 2.4$ in^3 (39.34 cm^3); $q_E = VQ_{BE}/I = 10,000(4.0)/137 = 292$ lb/lin in (51,137.0 N/m); $q_F = 10,000(2.4)/137 = 175$ lb/lin in (30,647.2 N/m).

3.   Compute the shearing forces on the transverse section

Since the shearing stress varies linearly across the flange, $V_1 = \frac{1}{2}(292)(5) = 730$ lb (3247.0 N); $V_2 = \frac{1}{2}(175)(3) = 263$ lb (1169.8 N); $V_3 = P = 10,000$ lb (44,480 N).

4.   Locate the shear center

Take moments of all forces acting on the left-hand portion of the beam with respect to a longitudinal axis through the shear center O. Thus $V_3e + 16(V_2 - V_1) = 0$, or $10,000e + 16(263 - 730) = 0$; $e = 0.747$ in (18.9738 mm).

5.   Verify the computed values

Check the computed values of q_E and q_F by considering the bending stresses directly. Apply the equation $\Delta f = Vy/I$, where Δf = increase in bending stress per unit distance along the span at distance y from the neutral axis. Then $\Delta f = 10,000(8)/137 = 584$ lb/(in$^2 \cdot$in) (158.52 MPa/m).

In Fig. 31c, set $AB = 1$ in (25.4 mm). Then $q_E = 584(5)(0.10) = 292$ lb/lin in (51,137.0 N/m); $q_F = 584(3)(0.10) = 175$ lb/lin in (30,647.1 N/m).

Although a particular type of beam (cantilever) was selected here for illustrative purposes and a numeric value was assigned to the vertical shear, note that the value of e is independent of the type of beam, form of loading, or magnitude of the vertical shear. The location of the shear center is a geometric characteristic of the transverse section.

BENDING OF A CIRCULAR FLAT PLATE

A circular steel plate 2 ft (0.61 m) in diameter and ½ in (12.7 mm) thick, simply supported along its periphery, carries a uniform load of 20 lb/in^2 (137.9 kPa) distributed over the entire area. Determine the maximum bending stress and deflection of this plate, using 0.25 for Poisson's ratio.

Calculation Procedure:

1. Compute the maximum stress in the plate

If the maximum deflection of the plate is less than about one-half the thickness, the effects of diaphragm behavior may be disregarded.

Compute the maximum stress, using the relation $f = (\frac{3}{8})(3 + \nu)w(R/t)^2$, where R = plate radius, in (mm); t = plate thickness, in (mm); ν = Poisson's ratio. Thus, $f = (\frac{3}{8})(3.25)(20)(12/0.5)^2 = 14{,}000$ lb/in^2 (96,530.0 kPa).

2. Compute the maximum deflection of the plate

Use the relation $y = (1 - \nu)(5 + \nu)fR^2/[2(3 + \nu)Et] = 0.75(5.25)(14{,}000)(12)^2/[2(3.25)(30 \times 10^6)(0.5)] = 0.081$ in (2.0574 mm). Since the deflection is less than one-half the thickness, the foregoing equations are valid in this case.

BENDING OF A RECTANGULAR FLAT PLATE

A 2×3 ft (61.0×91.4 cm) rectangular plate, simply supported along its periphery, is to carry a uniform load of 8 lb/in^2 (55.2 kPa) distributed over the entire area. If the allowable bending stress is 15,000 lb/in^2 (103.4 MPa), what thickness of plate is required?

Calculation Procedure:

1. Select an equation for the stress in the plate

Use the approximation $f = a^2b^2w/[2(a^2 + b^2)t^2]$, where a and b denote the length of the plate sides, in (mm).

2. Compute the required plate thickness

Solve the equation in step 1 for t. Thus $t^2 = a^2b^2w/[2(a^2 + b^2)f] = 2^2(3)^2(144)(8)/[2(2^2 + 3^2)(15{,}000)] = 0.106$; $t = 0.33$ in (8.382 mm).

COMBINED BENDING AND AXIAL LOAD ANALYSIS

A post having the cross section shown in Fig. 32 carries a concentrated load of 100 kips (444.8 kN) applied at R. Determine the stress induced at each corner.

Calculation Procedure:

1. Replace the eccentric load with an equivalent system

Use a concentric load of 100 kips (444.8 kN) and two couples producing the following moments with respect to the coordinate axes:

$$M_x = 100{,}000(2) = 200{,}000 \text{ lb} \cdot \text{in (25,960 N} \cdot \text{m)}$$

$$M_y = 100{,}000(1) = 100{,}000 \text{ lb} \cdot \text{in (12,980 N} \cdot \text{m)}$$

2. Compute the section modulus

Determine the section modulus of the rectangular cross section with respect to each axis. Thus $S_x = (\frac{1}{6})bd^2 = (\frac{1}{6})(18)(24)^2 = 1728$ in^3 (28,321.9 cm^3); $S_y = (\frac{1}{6})(24)(18)^2 = 1296$ in^3 (21,241 cm^3).

3. Compute the stresses produced

Compute the uniform stress caused by the concentric load and the stresses at the edges caused by the bending moments. Thus $f_1 = P/A = 100{,}000/[18(24)] = 231$ lb/in^2 (1592.7 kPa); $f_x = M_x/$

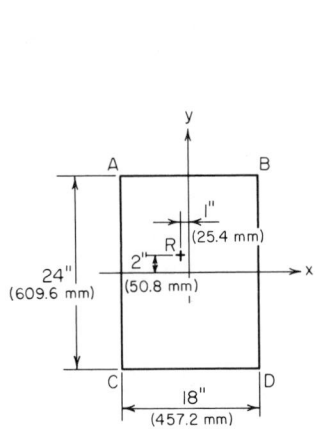

FIG. 32   Transverse section of a post.

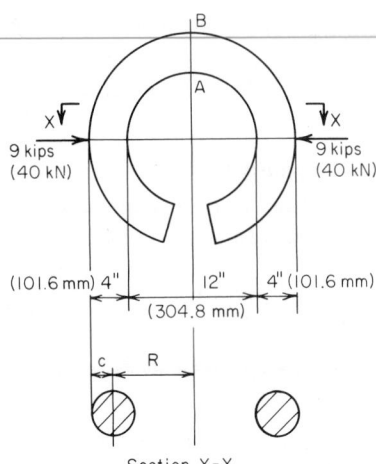

FIG. 33   Curved member in bending.

$S_x = 200,000/1728 = 116 \text{ lb/in}^2$ (799.8 kPa); $f_y = M_y/S_y = 100,000/1296 = 77 \text{ lb/in}^2$ (530.9 kPa).

4.   Determine the stress at each corner

Combine the results obtained in step 3 to obtain the stress at each corner. Thus $f_A = 231 + 116 + 77 = 424 \text{ lb/in}^2$ (2923.4 kPa); $f_B = 231 + 116 - 77 = 270 \text{ lb/in}^2$ (1861.5 kPa); $f_c = 231 - 116 + 77 = 192 \text{ lb/in}^2$ (1323.8 kPa); $f_D = 231 - 116 - 77 = 38 \text{ lb/in}^2$ (262.0 kPa). These stresses are all compressive because a positive stress is considered compressive, whereas a tensile stress is negative.

5.   Check the computed corner stresses

Use the following equation that applies to the special case of a rectangular cross section: $f = (P/A)(1 \pm 6e_x/d_x \pm 6e_y/d_y)$, where e_x and e_y = eccentricity of load with respect to the x and y axes, respectively; d_x and d_y = side of rectangle, in (mm), normal to x and y axes, respectively. Solving for the quantities within the brackets gives $6e_x/d_x = 6(2)/24 = 0.5$; $6e_y/d_y = 6(1)/18 = 0.33$. Then $f_A = 231(1 + 0.5 + 0.33) = 424 \text{ lb/in}^2$ (2923.4 kPa); $f_B = 231(1 + 0.5 - 0.33) = 270 \text{ lb/in}^2$ (1861.5 kPa); $f_C = 231(1 - 0.5 + 0.33) = 192 \text{ lb/in}^2$ (1323.8 kPa); $f_D = 231(1 - 0.5 - 0.33) = 38 \text{ lb/in}^2$ (262.0 kPa). These results verify those computed in step 4.

FLEXURAL STRESS IN A CURVED MEMBER

The ring in Fig. 33 has an internal diameter of 12 in (304.8 mm) and a circular cross section of 4-in (101.6-mm) diameter. Determine the normal stress at A and at B (Fig. 33).

Calculation Procedure:

1.   Determine the geometrical properties of the cross section

The area of the cross section is $A = 0.7854(4)^2 = 12.56 \text{ in}^2$ (81.037 cm²); the section modulus is $S = 0.7854(2)^3 = 6.28 \text{ in}^3$ (102.92 cm³). With $c = 2$ in (50.8 mm), the radius of curvature to the centroidal axis of this section is $R = 6 + 2 = 8$ in (203.2 mm).

2.   Compute the R/c ratio and determine the correction factors

Refer to a table of correction factors for curved flexural members, such as Roark—*Formulas for Stress and Strain*, and extract the correction factors at the inner and outer surface associated with the R/c ratio. Thus $R/c = 8/2 = 4$; $k_i = 1.23$; $k_o = 0.84$.

3. Determine the normal stress

Find the normal stress at A and B caused by an equivalent axial load and moment. Thus $f_A = P/A + k_i(M/S) = 9000/12.56 + 1.23(9000 \times 8)/6.28 = 14{,}820\text{-lb/in}^2$ (102,183.9-kPa) compression; $f_B = 9000/12.56 - 0.84(9000 \times 8)/6.28 = 8930\text{-lb/in}^2$ (61,572.3-kPa) tension.

SOIL PRESSURE UNDER DAM

A concrete gravity dam has the profile shown in Fig. 34. Determine the soil pressure at the toe and heel of the dam when the water surface is level with the top.

Calculation Procedure:

1. Resolve the dam into suitable elements

The soil prism underlying the dam may be regarded as a structural member subjected to simultaneous axial load and bending, the cross section of the member being identical with the bearing surface of the dam. Select a 1-ft (0.3-m) length of dam as representing the entire structure. The weight of the concrete is 150 lb/ft³ (23.56 kN/m³).

Resolve the dam into the elements AED and $EBCD$. Compute the weight of each element, and locate the resultant of the weight with respect to the toe. Thus $W_1 = \frac{1}{2}(12)(20)(150) = 18{,}000$ lb (80.06 kN); $W_2 = 3(20)(150) = 9000$ lb (40.03 kN); $\Sigma W = 18{,}000 + 9000 = 27{,}000$ lb (120.10 kN). Then $x_1 = (\frac{2}{3})(12) = 8.0$ ft (2.44 m); $x_2 = 12 + 1.5 = 13.5$ ft (4.11 m).

2. Find the magnitude and location of the resultant of the hydrostatic pressure

Calling the resultant $H = \frac{1}{2}wh^2 = \frac{1}{2}(62.4)(20)^2 = 12{,}480$ lb (55.51 kN), where w = weight of water, lb/ft³ (N/m³), and h = water height, ft (m), then $y = (\frac{1}{3})(20) = 6.67$ ft (2.03 m).

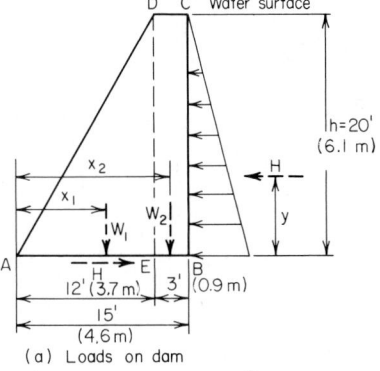

(a) Loads on dam

3. Compute the moment of the loads with respect to the base centerline

Thus, $M = 18{,}000(8 - 7.5) + 9000(13.5 - 7.5) - 12{,}480(6.67) = 20{,}200$ lb·ft (27,391 N·m) counterclockwise.

4. Compute the section modulus of the base

Use the relation $S = (\frac{1}{6})bd^2 = (\frac{1}{6})(1)(15)^2 = 37.5$ ft³ (1.06 m³).

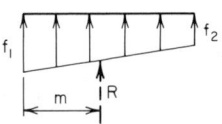

(b) Soil pressure under dam

FIG. 34

5. Determine the soil pressure at the dam toe and heel

Compute the soil pressure caused by the combined axial load and bending. Thus $f_1 = \Sigma W/A + M/S = 27{,}000/15 + 20{,}200/37.5 = 2339$ lb/ft² (111.99 kPa); $f_2 = 1800 - 539 = 1261$ lb/ft² (60.37 kPa).

6. Verify the computed results

Locate the resultant R of the trapezoidal pressure prism, and take its moment with respect to the centerline of the base. Thus $R = 27{,}000$ lb (120.10 kN); $m = (15/3)[(2 \times 1261 + 2339)/(1261 + 2339)] = 6.75$ ft (2.05 m); $M_R = 27{,}000(7.50 - 6.75) = 20{,}200$ lb·ft (27,391 N·m). Since the applied and resisting moments are numerically equal, the computed results are correct.

LOAD DISTRIBUTION IN PILE GROUP

A continuous wall is founded on three rows of piles spaced 3 ft (0.91 m) apart. The longitudinal pile spacing is 4 ft (1.21 m) in the front and center rows and 6 ft (1.82 m) in the rear row. The resultant of vertical loads on the wall is 20,000 lb/lin ft (291.87 kN/m) and lies 3 ft 3 in (99.06 cm) from the front row. Determine the pile load in each row.

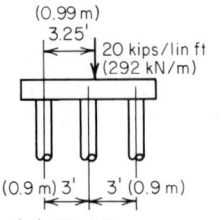

(0.99 m)
3.25'
20 kips/lin ft
(292 kN/m)

(0.9 m) 3'   3' (0.9 m)

(a) Elevation

Calculation Procedure:

1. Identify the "repeating group" of piles

The concrete footing (Fig. 35a) binds the piles, causing the surface along the top of the piles to remain a plane as bending occurs. Therefore, the pile group may be regarded as a structural member subjected to axial load and bending, the cross section of the member being the aggregate of the cross sections of the piles.

Indicate the "repeating group" as shown in Fig. 35b.

2. Determine the area of the pile group and the moment of inertia

Calculate the area of the pile group, locate its centroidal axis, and find the moment of inertia. Since all the piles have the same area, set the area of a single pile equal to unity. Then $A = 3 + 3 + 2 = 8$.

Take moments with respect to row A. Thus $8x = 3(0) + 3(3) + 2(6)$; $x = 2.625$ ft (66.675 mm). Then $I = 3(2.625)^2 + 3(0.375)^2 + 2(3.375)^2 = 43.9$.

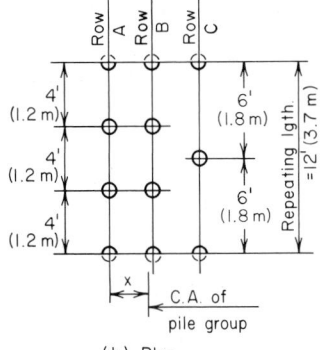

(b) Plan

3. Compute the axial load and bending moment on the pile group

The axial load $P = 20,000(12) = 240,000$ lb (1067.5 kN); then $M = 240,000(3.25 - 2.625) = 150,000$ lb·ft (203.4 kN·m).

4. Determine the pile load in each row

Find the pile load in each row resulting from the combined axial load and moment. Thus, $P/A = 240,000/8 = 30,000$ lb (133.4 kN) per pile; then $M/I = 150,000/43.9 = 3420$. Also, $p_a = 30,000 - 3420(2.625) = 21,020$ lb (93.50 kN) per pile; $p_b = 30,000 + 3420(0.375) = 31,280$ lb (139.13 kN) per pile; $p_c = 30,000 + 3420(3.375) = 41,540$ lb (184.76 kN) per pile.

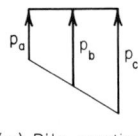

(c) Pile reactions

FIG. 35

5. Verify the above results

Compute the total pile reaction, the moment of the applied load, and the pile reaction with respect to row A. Thus, $R = 3(21,020) + 3(31,280) + 2(41,540) = 239,980$ lb (1067.43 kN); then $M_a = 240,000(3.25) = 780,000$ lb·ft (1057.68 kN·m), and $M_r = 3(31,280)(3) + 2(41,540)(6) = 780,000$ lb·ft (1057.68 kN·m). Since $M_a = M_r$, the computed results are verified.

Deflection of Beams

In this handbook the slope of the elastic curve at a given section of a beam is denoted by θ, and the deflection, in inches, by y. The slope is considered positive if the section rotates in a clockwise direction under the bending loads. A downward deflection is considered positive. In all instances, the beam is understood to be prismatic, if nothing is stated to the contrary.

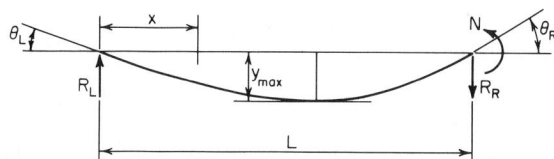

FIG. 36 Deflection of simple beam under end moment.

DOUBLE-INTEGRATION METHOD OF DETERMINING BEAM DEFLECTION

The simply supported beam in Fig. 36 is subjected to a counterclockwise moment N applied at the right-hand support. Determine the slope of the elastic curve at each support and the maximum deflection of the beam.

Calculation Procedure:

1. *Evaluate the bending moment at a given section*

Make this evaluation in terms of the distance x from the left-hand support to this section. Thus $R_L = N/L$; $M = Nx/L$.

2. *Write the differential equation of the elastic curve; integrate twice*

Thus $EI\, d^2y/dx^2 = -M = -Nx/L$; $EI\, dy/dx = EI\theta = -Nx^2/(2L) + c_1$; $EIy = -Nx^3/(6L) + c_1x + c_2$.

3. *Evaluate the constants of integration*

Apply the following boundary conditions: When $x = 0$, $y = 0$; $\therefore c_2 = 0$; when $x = L$, $y = 0$; $\therefore c_1 = NL/6$.

4. *Write the slope and deflection equations*

Substitute the constant values found in step 3 in the equations developed in step 2. Thus $\theta = [N/(6EIL)](L^2 - 3x^2)$; $y = [Nx/(6EIL)](L^2 - x^2)$.

5. *Find the slope at the supports*

Substitute the values $x = 0$, $x = L$ in the slope equation to determine the slope at the supports. Thus $\theta_L = NL/(6EI)$; $\theta_R = -NL/(3EI)$.

6. *Solve for the section of maximum deflection*

Set $\theta = 0$ and solve for x to locate the section of maximum deflection. Thus $L^2 - 3x^2 = 0$; $x = L/3^{0.5}$. Substituting in the deflection equation gives $y_{max} = NL^2/(9EI3^{0.5})$.

MOMENT-AREA METHOD OF DETERMINING BEAM DEFLECTION

Use the moment-area method to determine the slope of the elastic curve at each support and the maximum deflection of the beam shown in Fig. 36.

Calculation Procedure:

1. *Sketch the elastic curve of the member and draw the M/(EI) diagram*

Let A and B denote two points on the elastic curve of a beam. The moment-area method is based on the following theorems:

The difference between the slope at A and that at B is numerically equal to the area of the $M/(EI)$ diagram within the interval AB.

The deviation of A from a tangent to the elastic curve through B is numerically equal to the static moment of the area of the $M/(EI)$ diagram within the interval AB with respect to A. This tangential deviation is measured normal to the unstrained position of the beam.

Draw the elastic curve and the $M/(EI)$ diagram as shown in Fig. 37.

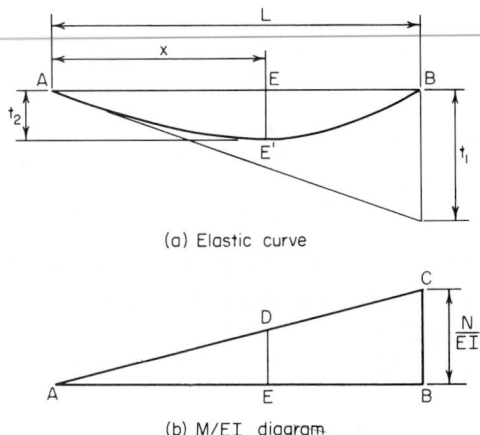

(a) Elastic curve

(b) M/EI diagram

FIG. 37

2. Calculate the deviation t_1 of B from the tangent through A

Thus, $t_1 = $ moment of $\triangle ABC$ about $BC = [NL/(2EI)](L/3) = NL^2/(6EI)$. Also, $\theta_L = t_1/L = NL/(6EI)$.

3. Determine the right-hand slope in an analogous manner

4. Compute the distance to the section where the slope is zero

Area $\triangle AED = $ area $\triangle ABC(x/L)^2 = Nx^2/(2EIL)$; $\theta_E = \theta_L - $ area $\triangle AED = NL/(6EI) - Nx^2/(2EIL) = 0$; $x = L/3^{0.5}$.

5. Evaluate the maximum deflection

Evaluate y_{max} by calculating the deviation t_2 of A from the tangent through E' (Fig. 37). Thus area $\triangle AED = \theta_L = NL/(6EI)$; $y_{max} = t_2 = [NL/(6EI)](2x/3) = [NL/(6EI)][(2L/(3 \times 3^{0.5})] = NL^2/(9EI3^{0.5})$, as before.

CONJUGATE-BEAM METHOD OF DETERMINING BEAM DEFLECTION

The overhanging beam in Fig. 38 is loaded in the manner shown. Compute the deflection at C.

Calculation Procedure:

1. Assign supports to the conjugate beam

If a conjugate beam of identical span as the given beam is loaded with the $M/(EI)$ diagram of the latter, the shear V' and bending moment M' of the conjugate beam are equal, respectively, to the slope θ and deflection y at the corresponding section of the given beam.

At A, the given beam has a specific slope but zero deflection. Correspondingly, the conjugate beam has a specific shear but zero moment; i.e., it is simply supported at A.

At C, the given beam has a specific slope and a specific deflection. Correspondingly, the conjugate beam has both a shear and a bending moment; i.e., it has a fixed support at C.

2. Construct the M/(EI) diagram of the given beam

Load the conjugate beam with this area. The moment at B is $-wd^2/2$; the moment varies linearly from A to B and parabolically from C to B.

3. Compute the resultant of the load in selected intervals

Compute the resultant W_1' of the load in interval AB and the resultant W_2' of the load in the interval BC. Locate these resultants. (Refer to the AISC *Manual* for properties of the complement of a half parabola.) Then $W_1' = (L/2)[wd^2/(2EI)] = wd^2L/(4EI)$; $x_1 = \frac{2}{3}L$; $W_2' = (d/3)[wd^2/(2EI)] = wd^3/(6EI)$; $x_2 = \frac{3}{4}d$.

4. *Evaluate the conjugate-beam reaction*

Since the given beam has zero deflection at B, the conjugate beam has zero moment at this section. Evaluate the reaction R_L' accordingly. Thus $M_B' = -R_L'L + W_1'L/3 = 0$; $R_L' = W_1'/3 = wd^2L/(12EI)$.

5. *Determine the deflection*

Determine the deflection at C by computing M_c'. Thus $y_c = M_c' = -R_L'(L + d) + W_1'(d + L/3) + W_2'(3d/4) = wd^3(4L + 3d)/(24EI)$.

UNIT-LOAD METHOD OF COMPUTING BEAM DEFLECTION

The cantilever beam in Fig. 39a carries a load that varies uniformly from w lb/lin ft at the free end to zero at the fixed end. Determine the slope and deflection of the elastic curve at the free end.

Calculation Procedure:

1. *Apply a unit moment to the beam*

Apply a counterclockwise unit moment at A (Fig. 39b). (This direction is selected because it is known that the end section rotates in this manner.) Let x = distance from A to given section; w_x = load intensity at the given section; M and m = bending moment at the given section induced by the actual load and by the unit moment, respectively.

2. *Evaluate the moments in step 1*

Evaluate M and m. By proportion, $w_x = w(L - x)/L$; $M = -(x^2/6)(2w + w_x) = -(wx^2/6)[2 + (L - x)/L] = -wx^2(3L - x)/(6L)$; $m = -1$.

3. *Apply a suitable slope equation*

Use the equation $\theta_A = \int_0^L [Mm/(EI)]\, dx$. Then $EI\theta_A = \int_0^L [wx^2(3L - x)/(6L)]\, dx = [w/(6L)] \int_0^L (3Lx^2 - x^3)\, dx = [w/(6L)](3Lx^3/3 - x^4/4)|_0^L = [w/(6L)](L^4 - L^4/4)$; thus, $\theta_A = \frac{1}{8}wL^3/(EI)$ counterclockwise. This is the slope at A.

4. *Apply a unit load to the beam*

Apply a unit downward load at A as shown in Fig. 39c. Let m' denote the bending moment at a given section induced by the unit load.

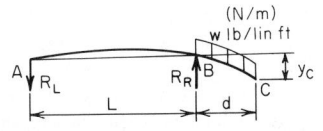

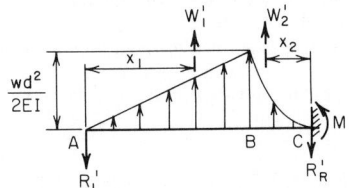

(a) Force diagram of given beam

(b) Force diagram of conjugate beam

FIG. 38   Deflection of overhanging beam.

(a)   Actual load on beam

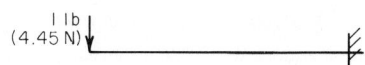

(b) Superimposed moment to find θ_A

(c) Superimposed load to find y_A

FIG. 39

5. Evaluate the bending moment induced by the unit load; find the deflection

Apply $y_A = \int_0^L [Mm'/(EI)]\, dx$. Then $m' = -x$; $EIy_A = \int_0^L [wx^3(3L - x)/(6L)]\, dx = [w/(6L)] \int_0^L x^3(3L - x)\, dx$; $y_A = (11/120)wL^4/(EI)$.

The first equation in step 3 is a statement of the work performed by the unit moment at A as the beam deflects under the applied load. The left-hand side of this equation expresses the external work, and the right-hand side expresses the internal work. These work equations constitute a simple proof of Maxwell's theorem of reciprocal deflections, which is presented in a later calculation procedure.

DEFLECTION OF A CANTILEVER FRAME

The prismatic rigid frame $ABCD$ (Fig. 40a) carries a vertical load P at the free end. Determine the horizontal displacement of A by means of both the unit-load method and the moment-area method.

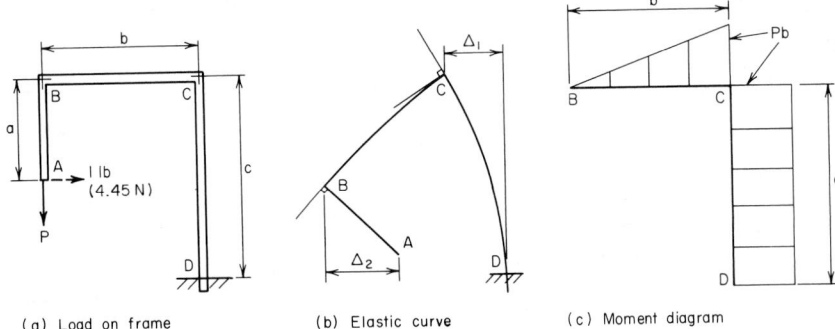

(a) Load on frame   (b) Elastic curve   (c) Moment diagram

FIG. 40

Calculation Procedure:

1. Apply a unit horizontal load

Apply the unit horizontal load at A, directed to the right.

2. Evaluate the bending moments in each member

Let M and m denote the bending moment at a given section caused by the load P and by the unit load, respectively. Evaluate these moments in each member, considering a moment positive if it induces tension in the outer fibers of the frame. Thus:

Member AB: Let x denote the vertical distance from A to a given section. Then $M = 0$; $m = x$.

Member BC: Let x denote the horizontal distance from B to a given section. Then $M = Px$; $m = a$.

Member CD: Let x denote the vertical distance from C to a given section. Then $M = Pb$; $m = a - x$.

3. Evaluate the required deflection

Calling the required deflection Δ, we apply $\Delta = \int [Mm/(EI)]\, dx$; $EI\Delta = \int_0^b Paxdx + \int_0^c Pb(a - x)\, dx = Pax^2/2]_0^b + Pb(ax - x^2/2)]_0^c = Pab^2/2 + Pabc - Pbc^2/2$; $\Delta = [Pb/(2EI)](ab + 2ac - c^2)$.

If this value is positive, A is displaced in the direction of the unit load, i.e., to the right. Draw the elastic curve in hyperbolic fashion (Fig. 40b). The above three steps constitute the unit-load method of solving this problem.

4. Construct the bending-moment diagram

Draw the diagram as shown in Fig. 40c.

5. Compute the rotation and horizontal displacement by the moment-area method

Determine the rotation and horizontal displacement of C. (Consider only absolute values.) Since there is no rotation at D, $EI\theta_C = Pbc$; $EI\Delta_1 = Pbc^2/2$.

6. Compute the rotation of one point relative to another and the total rotation

Thus $EI\theta_{BC} = Pb^2/2$; $EI\theta_B = Pbc + Pb^2/2 = Pb(c + b/2)$. The horizontal displacement of B relative to C is infinitesimal.

7. Compute the horizontal displacement of one point relative to another

Thus, $EI\Delta_2 = EI\theta_B a = Pb(ac + ab/2)$.

8. Combine the computed displacements to obtain the absolute displacement

Thus $EI\Delta = EI(\Delta_2 - \Delta_1) = Pb(ac + ab/2 - c^2/2)$; $\Delta = [Pb/(2EI)](2ac + ab - c^2)$.

Statically Indeterminate Structures

A structure is said to be *statically determinate* if its reactions and internal forces may be evaluated by applying solely the equations of equilibrium and *statically indeterminante* if such is not the case. The analysis of an indeterminate structure is performed by combining the equations of equilibrium with the known characteristics of the deformation of the structure.

SHEAR AND BENDING MOMENT OF A BEAM ON A YIELDING SUPPORT

The beam in Fig. 41a has an EI value of 35×10^9 lb·in^2 (100,429 kN·m^2) and bears on a spring at B that has a constant of 100 kips/in (175,126.8 kN/m); i.e., a force of 100 kips (444.8 kN) will compress the spring 1 in (25.4 mm). Neglecting the weight of the member, construct the shear and bending-moment diagrams.

Calculation Procedure:

1. Draw the free-body diagram of the beam

Draw the diagram in Fig. 41b. Consider this as a simply supported member carrying a 50-kip (222.4-kN) load at D and an upward load R_B at its center.

2. Evaluate the deflection

Evaluate the deflection at B by applying the equations presented for cases 7 and 8 in the AISC *Manual*. With respect to the 50-kip (222.4-kN) load, $b = 7$ ft (2.1 m) and $x = 14$ ft (4.3 m). If y is in inches and R_B is in pounds, $y = 50,000(7)(14)(28^2 - 7^2 - 14^2)1728/[6(35)(10)^9 28] - R_B(28)^3 1728/[48(35)(10)^9] = 0.776 - (2.26/10^5)R_B$.

3. Express the deflection in terms of the spring constant

The deflection at B is, by proportion, $y/1 = R_B/100,000$; $y = R_B/100,000$.

4. Equate the two deflection expressions, and solve for the upward load

Thus $R_B/10^5 = 0.776 - (2.26/10^5)R_B$; $R_B = 0.776(10)^5/3.26 = 23,800$ lb (105,862.4 N).

5. Calculate the reactions R_A and R_C by taking moments

We have $\Sigma M_C = 28R_A - 50,000(21) + 23,800(14) = 0$; $R_A = 25,600$ lb (113,868.8 N); $\Sigma M_A = 50,000(7) - 23,800(14) - 28R_C = 0$; $R_C = 600$ lb (2668.8 N).

6. Construct the shear and moment diagrams

Construct these diagrams as shown in Fig. 41. Then $M_D = 7(25,600) = 179,200$ lb·ft (242,960 N·m); $M_B = 179,200 - 7(24,400) = 8400$ lb·ft (11,390.4 N·m).

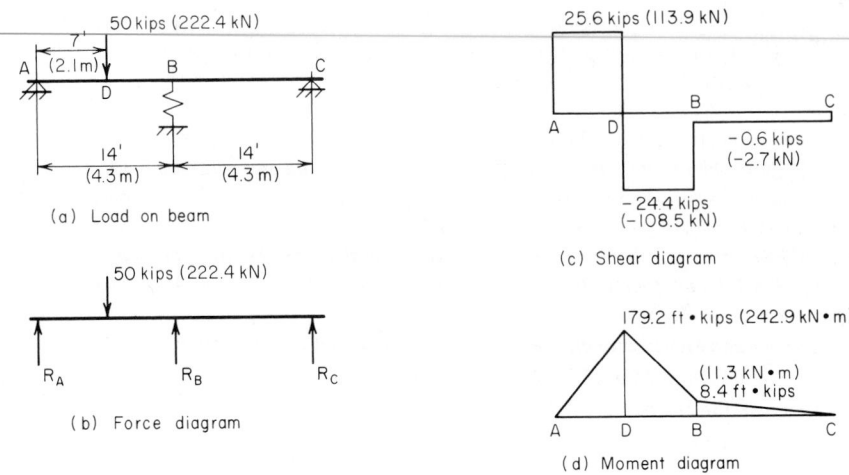

(a) Load on beam

(b) Force diagram

(c) Shear diagram

(d) Moment diagram

FIG. 41

MAXIMUM BENDING STRESS IN BEAMS JOINTLY SUPPORTING A LOAD

In Fig. 42a, a W16 × 40 beam and a W12 × 31 beam cross each other at the vertical line V, the bottom of the 16-in (406.4-mm) beam being ⅜ in (9.53 mm) above the top of the 12-in (304.8-mm) beam before the load is applied. Both members are simply supported. A column bearing on the 16-in (406.4-mm) beam transmits a load of 15 kips (66.72 kN) at the indicated location. Compute the maximum bending stress in the 12-in (304.8-mm) beam.

Calculation Procedure:

1. Determine whether the upper beam engages the lower beam

To ascertain whether the upper beam engages the lower one as it deflects under the 15-kip (66.72-kN) load, compute the deflection of the 16-in (406.4-mm) beam at V if the 12-in (304.8-mm)

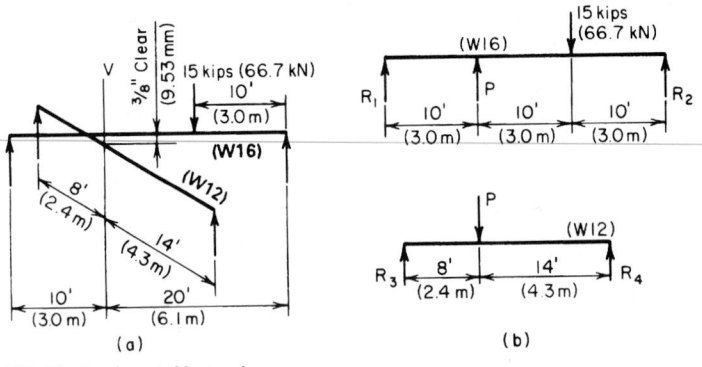

FIG. 42 Load carried by two beams.

beam were absent. This distance is 0.74 in (18.80 mm). Consequently, the gap between the members is closed, and the two beams share the load.

2.   *Draw a free-body diagram of each member*

Let P denote the load transmitted to the 12-in (304.8-mm) beam by the 16-in (406.4-mm) beam [or the reaction of the 12-in (304.8-mm) beam on the 16-in (406.4-mm) beam]. Draw, in Fig. 42b, a free-body diagram of each member.

3.   *Evaluate the deflection of the beams*

Evaluate, in terms of P, the deflections y_{12} and y_{16} of the 12-in (304.8-mm) and 16-in (406.4-mm) beams, respectively, at line V.

4.   *Express the relationship between the two deflections*

Thus, $y_{12} = y_{16} - 0.375$.

5.   *Replace the deflections in step 4 with their values as obtained in step 3*

After substituting these deflections, solve for P.

6.   *Compute the reactions of the lower beam*

Once the reactions of the lower beam are computed, obtain the maximum bending moment. Then compute the corresponding flexural stress.

THEOREM OF THREE MOMENTS

For the two-span beam in Fig. 43a, compute the reactions at the supports. Apply the theorem of three moments to arrive at the results.

Calculation Procedure:

1.   *Using the bending-moment equation, determine M_B*

Figure 43b represents a general case. For a prismatic beam, the bending moments at the three successive supports are related by $M_1L_1 + 2M_2(L_1 + L_2) + M_3L_2 = -\frac{1}{4}w_1L_1^3 - \frac{1}{4}w_2L_2^3 - $

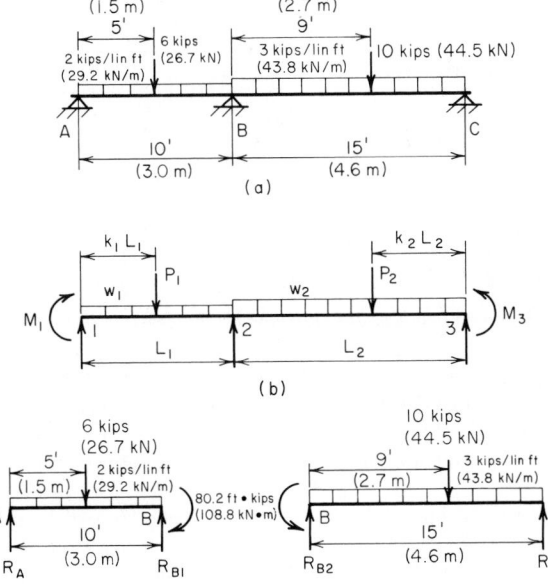

FIG. 43

$P_1L_1^2(k_1 - k_1^3) - P_2L_2^2(k_2 - k_2^3)$. Substituting in this equation gives $M_1 = M_3 = 0$; $L_1 = 10$ ft (3.0 m); $L_2 = 15$ ft (4.6 m); $w_1 = 2$ kips/lin ft (29.2 kN/m); $w_2 = 3$ kips/lin ft (43.8 kN/m); $P_1 = 6$ kips (26.7 N); $P_2 = 10$ kips (44.5 N); $k_1 = 0.5$; $k_2 = 0.4$; $2M_B(10 + 15) = -\frac{1}{4}(2)(10)^3 - \frac{1}{4}(3)(15)^3 - 6(10)^2(0.5 - 0.125) - 10(15)^2(0.4 - 0.064)$; $M_B = -80.2$ ft·kips (-108.8 kN·m).

2. Draw a free-body diagram of each span

Figure 43c shows the free-body diagrams.

3. Take moments with respect to each support to find the reactions

Span AB: $\Sigma M_A = 6(5) + 2(10)(5) + 80.2 - 10R_{B1} = 0$; $R_{B1} = 21.02$ kips (93.496 kN); $\Sigma M_B = 10R_A - 6(5) - 2(10)(5) + 80.2 = 0$; $R_A = 4.98$ kips (22.151 kN).

   Span BC: $\Sigma M_B = -80.2 + 10(9) + 3(15)(7.5) - 15R_C = 0$; $R_C = 23.15$ kips (102.971 kN); $\Sigma M_C = 15R_{B2} - 80.2 - 10(6) - 3(15)(7.5) = 0$; $R_{B2} = 31.85$ kips (144.668 kN); $R_B = 21.02 + 31.85 = 52.87$ kips (235.165 kN).

THEOREM OF THREE MOMENTS: BEAM WITH OVERHANG AND FIXED END

Determine the reactions at the supports of the continuous beam in Fig. 44a. Use the theorem of three moments.

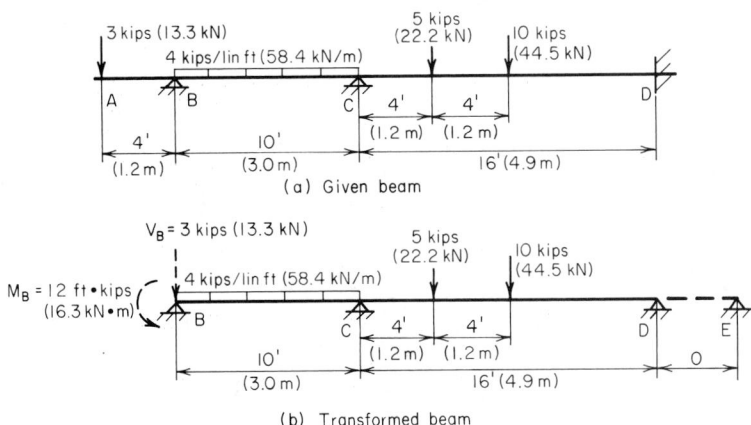

(a) Given beam

(b) Transformed beam

FIG. 44

Calculation Procedure:

1. Transform the given beam to one amenable to analysis by the theorem of three moments

Perform the following operations to transform the beam:

a. Remove the span AB, and introduce the shear V_B and moment M_B that the load on AB induces at B, as shown in Fig. 44b.

b. Remove the fixed support at D and add the span DE of zero length, with a hinged support at E.

   For the interval BD, the transformed beam is then identical in every respect with the actual beam.

2. Apply the equation for the theorem of three moments

Consider span BC as span 1 and CD as span 2. For the 5-kip (22.2-kN) load, $k_2 = 12/16 = 0.75$; for the 10-kip (44.5-kN) load, $k_2 = 8/16 = 0.5$. Then $-12(10) + 2M_C(10 + 16) + 16M_D =$

$-\frac{1}{4}(4)(10)^3 - 5(16)^2(0.75 - 0.422) - 10(16)^2(0.5 - 0.125)$. Simplifying gives $13M_C + 4M_D = -565.0$, Eq. a.

3. Apply the moment equation again

Considering CD as span 1 and DE as span 2, apply the moment equation again. Or, for the 5-kip (22.2-kN) load, $k_1 = 0.25$; for the 10-kip (44.5-kN) load, $k_1 = 0.5$. Then $16M_C + 2M_D(16 + 0) = -5(16)^2(0.25 - 0.016) - 10(16)^2(0.50 - 0.125)$. Simplifying yields $M_C + 2M_D = -78.7$, Eq. b.

4. Solve the moment equations

Solving Eqs. a and b gives $M_C = -37.1$ ft·kips $(-50.30$ kN·m); $M_D = -20.8$ ft·kips $(-28.20$ kN·m).

5. Determine the reactions by using a free-body diagram

Find the reactions by drawing a free-body diagram of each span and taking moments with respect to each support. Thus $R_B = 20.5$ kips (91.18 kN); $R_C = 32.3$ kips (143.67 kN); $R_D = 5.2$ kips (23.12 kN).

BENDING-MOMENT DETERMINATION BY MOMENT DISTRIBUTION

Using moment distribution, determine the bending moments at the supports of the member in Fig. 45. The beams are rigidly joined at the supports and are composed of the same material.

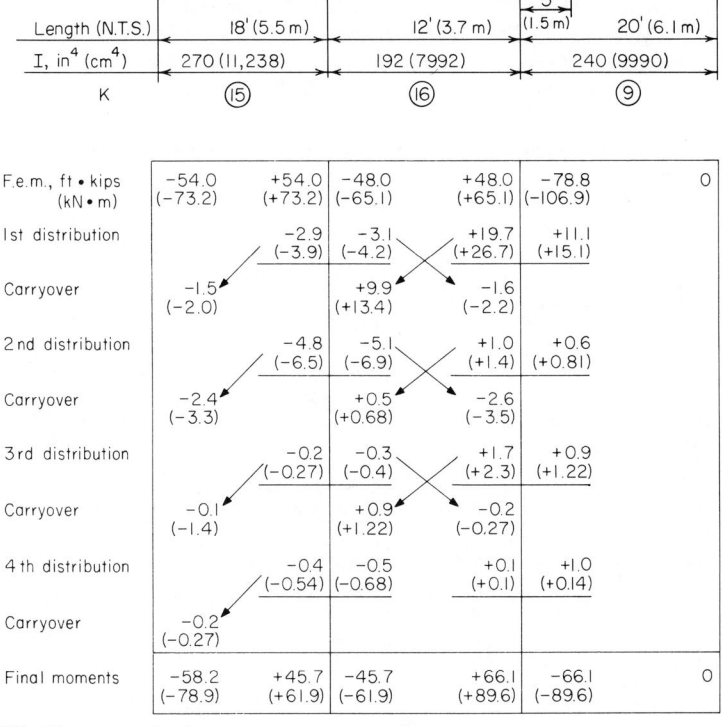

FIG. 45 Moment distribution.

Calculation Procedure:

1.   Calculate the flexural stiffness of each span

Using K to denote the flexural stiffness, we see that $K = I/L$ if the far end remains fixed during moment distribution; $K = 0.75I/L$ if the far end remains hinged during moment distribution. Then $K_{AB} = 270/18 = 15$; $K_{BC} = 192/12 = 16$; $K_{CD} = 0.75(240/20) = 9$. Record all the values on the drawing as they are obtained.

2.   For each span, calculate the required fixed-end moments at those supports that will be considered fixed

These are the *external* moments with respect to the span; a clockwise moment is considered positive. (For additional data, refer to cases 14 and 15 in the AISC *Manual*.) Then $M_{AB} = -wL^2/12 = -2(18)^2/12 = -54.0$ ft·kips (-73.2 kN·m); $M_{BA} = +54.0$ ft·kips (73.22 kN·m). Similarly, $M_{BC} = -48.0$ ft·kips (-65.1 kN·m); $M_{CB} = +48.0$ ft·kips (65.1 kN·m); $M_{CD} = -24(15)(5)(15 + 20)/[2(20)^2] = -78.8$ ft·kips (-106.85 kN·m).

3.   Calculate the unbalanced moments

Computing the unbalanced moments at B and C yields the following: At B, $+54.0 - 48.0 = +6.0$ ft·kips (8.14 kN·m); at C, $+48.0 - 78.8 = -30.8$ ft·kips (-41.76 kN·m).

4.   Apply balancing moments; distribute them in proportion to the stiffness of the adjoining spans

Apply the balancing moments at B and C, and distribute them to the two adjoining spans in proportion to their stiffness. Thus $M_{BA} = -6.0(15/31) = -2.9$ ft·kips (-3.93 kN·m); $M_{BC} = -6.0(16/31) = -3.1$ ft·kips (-4.20 kN·m); $M_{CB} = +30.8(16/25) = +19.7$ ft·kips (26.71 kN·m); $M_{CD} = +30.8(9/25) = +11.1$ ft·kips (15.05 kN·m).

5.   Perform the "carry-over" operation for each span

To do this, take one-half the distributed moment applied at one end of the span, and add this to the moment at the far end if that end is considered to be fixed during moment distribution.

6.   Perform the second cycle of moment balancing and distribution

Thus $M_{BA} = -9.9(15/31) = -4.8$; $M_{BC} = -9.9(16/31) = -5.1$; $M_{CB} = +1.6(16/25) = +1.0$; $M_{CD} = +1.6(9/25) = +0.6$.

7.   Continue the foregoing procedure until the carry-over moments become negligible

Total the results to obtain the following bending moments: $M_A = -58.2$ ft·kips (-78.91 kN·m); $M_B = -45.7$ ft·kips (-61.96 kN·m); $M_C = -66.1$ ft·kips (-89.63 kN·m).

ANALYSIS OF A STATICALLY INDETERMINATE TRUSS

Determine the internal forces of the truss in Fig. 46a. The cross-sectional areas of the members are given in Table 5.

Calculation Procedure:

1.   Test the structure for static determinateness

Apply the following criterion. Let j = number of joints; m = number of members; r = number of reactions. Then if $2j = m + r$, the truss is statically determinate; if $2j < m + r$, the truss is statically indeterminate and the deficiency represents the degree of indeterminateness.

In this truss, $j = 6$, $m = 10$, $r = 3$, consisting of a vertical reaction at A and D and a horizontal reaction at D. Thus $2j = 12$; $m + r = 13$. The truss is therefore statically indeterminate to the first degree; i.e., there is *one* redundant member.

The method of analysis comprises the following steps: Assume a value for the internal force in a particular member, and calculate the relative displacement Δ_i of the two ends of that member caused solely by this force. Now remove this member to secure a determinate truss, and calculate the relative displacement Δ_a caused solely by the applied loads. The true internal force is of such magnitude that $\Delta_i = -\Delta_a$.

2.   *Assume a unit force for one member*

Assume for convenience that the force in *BF* is 1-kip (4.45-kN) tension. Remove this member, and replace it with the assumed 1-kip (4.45-kN) force that it exerts at joints *B* and *F*, as shown in Fig. 46*b*.

3.   *Calculate the force induced in each member solely by the unit force*

Calling the induced force *U*, produced solely by the unit tension in *BF*, record the results in Table 5, considering tensile forces as positive and compressive forces as negative.

4.   *Calculate the force induced in each member solely by the applied loads*

With *BF* eliminated, calculate the force *S* induced in each member solely by the applied loads.

5.   *Evaluate the true force in the selected member*

Use the relation $BF = -[\Sigma SUL/(AE)]/[\Sigma U^2 L/(AE)]$. The numerator represents Δ_a; the denominator represents Δ_i for a 1-kip (4.45-kN) tensile force in *BF*. Since *E* is constant, it cancels. Substituting the values in Table 5 gives $BF = -(-266.5/135.5) = 1.97$ kips (8.76 kN). The positive result confirms the assumption that *BF* is tensile.

6.   *Evaluate the true force in each member*

Use the relation $S' = S + 1.97U$, where $S' = $ true force. The results are shown in Table 5.

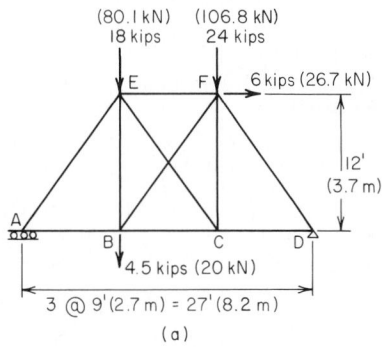

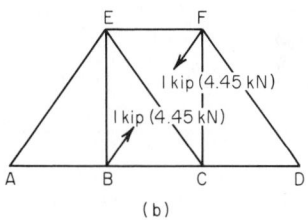

FIG. 46   Statically indeterminate truss.

Moving Loads and Influence Lines

ANALYSIS OF BEAM CARRYING MOVING CONCENTRATED LOADS

The loads shown in Fig. 47*a* traverse a beam of 40-ft (12.2-m) simple span while their spacing remains constant. Determine the maximum bending moment and maximum shear induced in the beam during transit of these loads. Disregard the weight of the beam.

Calculation Procedure:

1.   *Determine the magnitude of the resultant and its location*

Since the member carries only concentrated loads, the maximum moment at any instant occurs under one of these loads. Thus, the problem is to determine the position of the load system that causes the *absolute* maximum moment.

   The magnitude of the resultant *R* is $R = 10 + 4 + 15 = 29$ kips (129.0 kN). To determine the location of *R*, take moments with respect to *A* (Fig. 47). Thus $\Sigma M_A = 29AD = 4(5) + 15(17)$, or $AD = 9.48$ ft (2.890 m).

2.   *Assume several trial load positions*

Assume that the maximum moment occurs under the 10-kip (44.5-kN) load. Place the system in the position shown in Fig. 47*b*, with the 10-kip (44.5-kN) load as far from the adjacent support as the resultant is from the other support. Repeat this procedure for the two remaining loads.

3.   *Determine the support reactions for the trial load positions*

For these three trial positions, calculate the reaction at the support adjacent to the load under consideration. Determine whether the vertical shear is zero or changes sign at this load. Thus, for

TABLE 5   Forces in Truss Members (Fig. 46)

Member	A, in² (cm²)	L, in (mm)	U, kips (kN)	S, kips (kN)	U²L/A	SUL/A	S', kips (kN)
AB	5	108	0	+15.25	0	0	+15.25
	(32.2)	(2,743.2)	(0)	(+67.832)	(0)	(0)	(+67.832)
BC	5	108	−0.60	+15.25	+7.8	−197.6	+14.07
	(32.2)	(2,743.2)	(−2.668)	(+67.832)	(+615.54)	(−15,417.78)	(+62.583)
CD	5	108	0	+13.63	0	0	+13.63
	(32.2)	(2,743.2)	(0)	(+60.626)	(0)	(0)	(+60.626)
EF	4	108	−0.60	−13.63	+9.7	+220.8	−14.81
	(25.8)	(2,743.2)	(−2.668)	(−60.626)	(+756.84)	(+17,198.18)	(−65.874)
BE	4	144	−0.80	+4.50	+23.0	−129.6	+2.92
	(25.8)	(3,657.6)	(−3.558)	(+20.016)	(+1,794.68)	(−10,096.24)	(+12.988)
CF	4	144	−0.80	+2.17	+23.0	−62.5	+0.59
	(25.8)	(3,657.6)	(−3.558)	(+9.952)	(+1,794.68)	(−4,868.55)	(+2.624)
AE	6	180	0	−25.42	0	0	−25.42
	(38.7)	(4,572.0)	(0)	(−113.068)	(0)	(0)	(−113.068)
BF	5	180	+1.00	0	+36.0	0	+1.97
	(32.2)	(4,572.0)	(+4.448)	(0)	(+2,809.18)	(0)	(+8.762)
CE	5	180	+1.00	−2.71	+36.0	−97.6	−0.74
	(32.2)	(4,572.0)	(+4.448)	(−9.652)	(+2,809.18)	(−6,095.82)	(−3.291)
DF	6	180	0	−32.71	0	0	−32.71
	(38.7)	(4,572.0)	(0)	(−145.494)	(0)	(0)	(−145.494)
Total					+135.5	−266.5	
					(+10,580.1)	(−19,280.2)	

position 1: $R_L = 29(15.26)/40 = 11.06$ kips (49.194 kN). Since the shear does not change sign at the 10-kip (44.5-kN) load, this position lacks significance.

Position 2: $R_L = 29(17.76)/40 = 12.88$ kips (57.290 kN). The shear changes sign at the 4-kip (17.8 kN) load.

Position 3: $R_R = 29(16.24)/40 = 11.77$ kips (52.352 kN). The shear changes sign at the 15-kip (66.7-kN) load.

4. Compute the maximum bending moment associated with positions having a change in the shear sign

This applies to positions 2 and 3. The absolute maximum moment is the larger of these values. Thus, for position 2: $M = 12.88(17.76) − 10(5) = 178.7$ ft·kips (242.32 kN·m). Position 3: $M = 11.77(16.24) = 191.1$ ft·kips (259.13 kN·m). Thus, $M_{max} = 191.1$ ft·kips (259.13 kN·m).

5. Determine the absolute maximum shear

For absolute maximum shear, place the 15-kip (66.7-kN) load an infinitesimal distance to the left of the right-hand support. Then $V_{max} = 29(40 − 7.52)/40 = 23.5$ kips (104.53 kN).

When the load spacing is large in relation to the beam span, the absolute maximum moment may occur when only part of the load system is on the span. This possibility requires careful investigation.

INFLUENCE LINE FOR SHEAR IN A BRIDGE TRUSS

The Pratt truss in Fig. 48a supports a bridge at its bottom chord. Draw the influence line for shear in panel cd caused by a moving load traversing the bridge floor.

Calculation Procedure:

1. Compute the shear in the panel being considered with a unit load to the right of the panel

Cut the truss at section YY. The algebraic sum of vertical forces acting on the truss at panel points to the left of YY is termed the *shear* in panel cd.

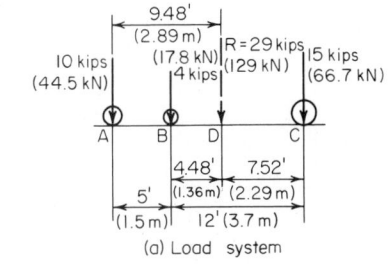

(a) Load system

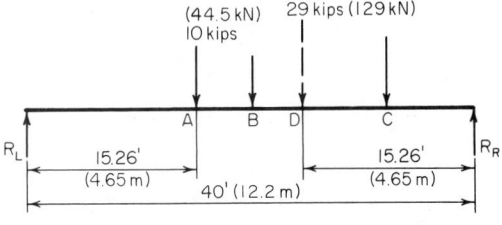

(b) Position 1, for 10-kip (44.5-kN) load

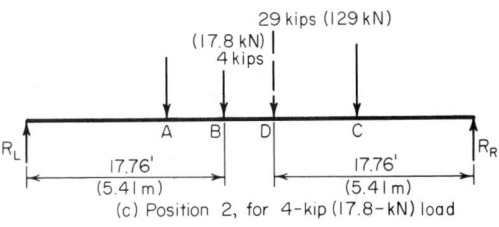

(c) Position 2, for 4-kip (17.8-kN) load

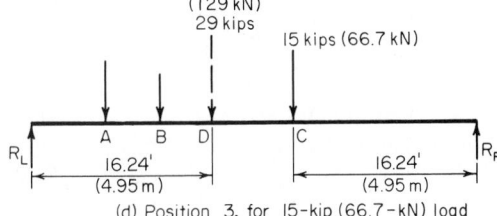

(d) Position 3, for 15-kip (66.7-kN) load

FIG. 47

Consider that a moving load traverses the bridge floor from right to left and that the portion of the load carried by the given truss is 1 kip (4.45 kN). This unit load is transmitted to the truss as concentrated loads at two adjacent bottom-chord panel points, the latter being components of the unit load. Let x denote the instantaneous distance from the right-hand support to the moving load.

Place the unit load to the right of d, as shown in Fig. 48b, and compute the shear V_{cd} in panel cd. The truss reactions may be obtained by considering the unit load itself rather than its panel-point components. Thus: $R_L = x/120$; $V_{cd} = R_L = x/120$, Eq. a.

2.   *Compute the panel shear with the unit load to the left of the panel considered*

Placing the unit load to the left of c yields $V_{cd} = R_L - 1 = x/120 - 1$, Eq. b.

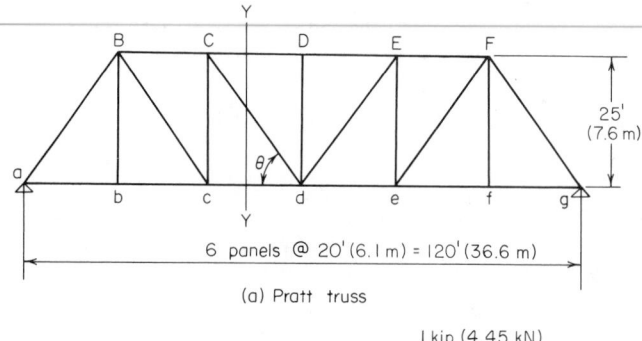

(a) Pratt truss

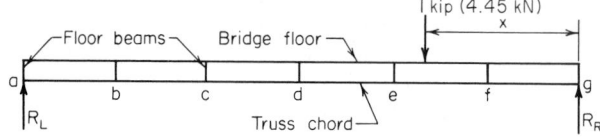

(b) Transmission of load through floor beams

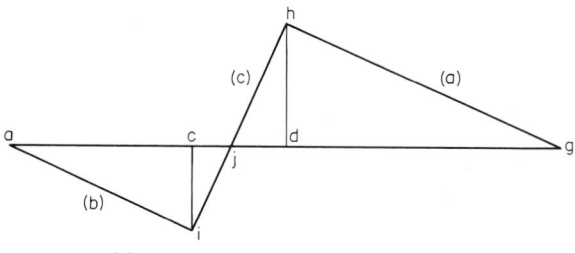

(c) Influence line for shear in panel cd

FIG. 48

3. Determine the panel shear with the unit load within the panel

Place the unit load within panel cd. Determine the panel-point load P_c at c, and compute V_{cd}. Thus $P_c = (x - 60)/20 = x/20 - 3$; $V_{cd} = R_L - P_c = x/120 - (x/20 - 3) = -x/24 + 3$, Eq. c.

4. Construct a diagram representing the shear associated with every position of the unit load

Apply the foregoing equations to represent the value of V_{cd} associated with every position of the unit load. This diagram, Fig. 48c, is termed an *influence line*. The point j at which this line intersects the base is referred to as the *neutral point*.

5. Compute the slope of each segment of the influence line

Line a, $dV_{cd}/dx = 1/120$; line b, $dV_{cd}/dx = 1/120$; line c, $dV_{cd}/dx = -1/24$. Lines a and b are therefore parallel because they have the same slope.

FORCE IN TRUSS DIAGONAL CAUSED BY A MOVING UNIFORM LOAD

The bridge floor in Fig. 48a carries a moving uniformly distributed load. The portion of the load transmitted to the given truss is 2.3 kips/lin ft (33.57 kN/m). Determine the limiting values of the force induced in member Cd by this load.

Calculation Procedure:

1. Locate the neutral point, and compute dh

The force in Cd is a function of V_{cd}. Locate the neutral point j in Fig. 48c and compute dh. From Eq. c of the previous calculation procedure, $V_{cd} = -jg/24 + 3 = 0$; $jg = 72$ ft (21.9 m). From Eq. a of the previous procedure, $dh = 60/120 = 0.5$.

2. Determine the maximum shear

To secure the maximum value of V_{cd}, apply uniform load continuously in the interval jg. Compute V_{cd} by multiplying the area under the influence line by the intensity of the applied load. Thus, $V_{cd} = \frac{1}{2}(72)(0.5)(2.3) = 41.4$ kips (184.15 kN).

3. Determine the maximum force in the member

Use the relation $Cd_{max} = V_{cd}(\csc \theta)$, where $\csc \theta = [(20^2 + 25^2)/25^2]^{0.5} = 1.28$. Then $Cd_{max} = 41.4(1.28) = 53.0$-kip (235.74-kN) tension.

4. Determine the minimum force in the member

To secure the minimum value of V_{cd}, apply uniform load continuously in the interval aj. Perform the final calculation by proportion. Thus, $Cd_{min}/Cd_{max} = \text{area } aij/\text{area} jhg = -(2/3)^2 = -4/9$. Then $Cd_{min} = -(4/9)(53.0) = 23.6$-kip (104.97-kN) compression.

FORCE IN TRUSS DIAGONAL CAUSED BY MOVING CONCENTRATED LOADS

The truss in Fig. 49a supports a bridge that transmits the moving-load system shown in Fig. 49b to its bottom chord. Determine the maximum tensile force in De.

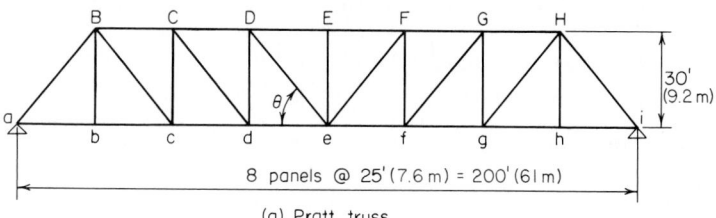

(a) Pratt truss

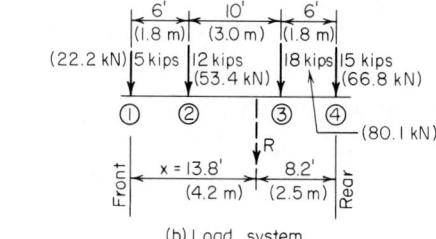

(b) Load system

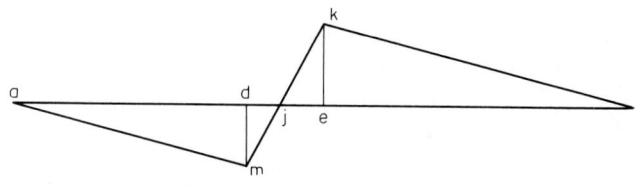

(c) Influence line for shear in panel de

FIG. 49

Calculation Procedure:

1. Locate the resultant of the load system

The force in De (Fig. 49) is a function of the shear in panel de. This shear is calculated without recourse to a set rule in order to show the principles involved in designing for moving loads.

To locate the resultant of the load system, take moments with respect to load 1. Thus, $R = 50$ kips (222.4 kN). Then $\Sigma M_1 = 12(6) + 18(16) + 15(22) = 50x$; $x = 13.8$ ft (4.21 m).

2. Construct the influence line for V_{de}

In Fig. 49c, draw the influence line for V_{de}. Assume right-to-left locomotion, and express the slope of each segment of the influence line. Thus slope of $ik =$ slope of $ma = 1/200$; slope of $km = -7/200$.

3. Assume a load position, and determine whether V_{de} increases or decreases

Consider that load 1 lies within panel de and the remaining loads lie to the right of this panel. From the slope of the influence line, ascertain whether V_{de} increases or decreases as the system is displaced to the left. Thus $dV_{de}/dx = 5(-7/200) + 45(1/200) > 0$; $\therefore$ V_{de} increases.

4. Repeat the foregoing calculation with other assumed load positions

Consider that loads 1 and 2 lie within the panel de and the remaining loads lie to the right of this panel. Repeat the foregoing calculation. Thus $dV_{de}/dx = 17(-7/200) + 33(1/200) < 0$; $\therefore$ V_{de} decreases.

From these results it is concluded that as the system moves from right to left, V_{de} is maximum at the instant that load 2 is at e.

5. Place the system in the position thus established, and compute V_{de}

Thus, $R_L = 50(100 + 6 - 13.8)/200 = 23.1$ kips (102.75 kN). The load at panel point d is $P_d = 5(6)/25 = 1.2$ kips (5.34 kN); $V_{de} = 23.1 - 1.2 = 21.9$ kips (97.41 kN).

6. Assume left-to-right locomotion; proceed as in step 3

Consider that load 4 is within panel de and the remaining loads are to the right of this panel. Proceeding as in step 3, we find $dV_{de}/dx = 15(7/200) + 35(-1/200) > 0$.

So, as the system moves from left to right, V_{de} is maximum at the instant that load 4 is at e.

7. Place the system in the position thus established, and compute V_{de}

Thus $V_{de} = R_L = [50(100 - 8.2)]/200 = 23.0$ kips (102.30 kN); $\therefore$ $V_{de,max} = 23.0$ kips (102.30 kN).

8. Compute the maximum tensile force in De

Using the same relation as in step 3 of the previous calculation procedure, we find $\csc \theta = [(25^2 + 30^2)/30^2]^{0.5} = 1.30$; then $De = 23.0(1.30) = 29.9$-kip (133.00-kN) tension.

INFLUENCE LINE FOR BENDING MOMENT IN BRIDGE TRUSS

The Warren truss in Fig. 50a supports a bridge at its top chord. Draw the influence line for the bending moment at b caused by a moving load traversing the bridge floor.

Calculation Procedure:

1. Place the unit load in position, and compute the bending moment

The moment of all forces acting on the truss at panel points to the left of b with respect to b is termed the *bending moment* at that point. Assume that the load transmitted to the given truss is 1 kip (4.45 kN), and let x denote the instantaneous distance from the right-hand support to the moving load.

Place the unit load to the right of C, and compute the bending moment M_b. Thus $R_L = x/120$; $M_b = 45R_L = 3x/8$, Eq. a.

2. Place the unit load on the other side and compute the bending moment

Placing the unit load to the left of B and computing M_b, $M_b = 45R_L - (x - 75) = -5x/8 + 75$, Eq. b.

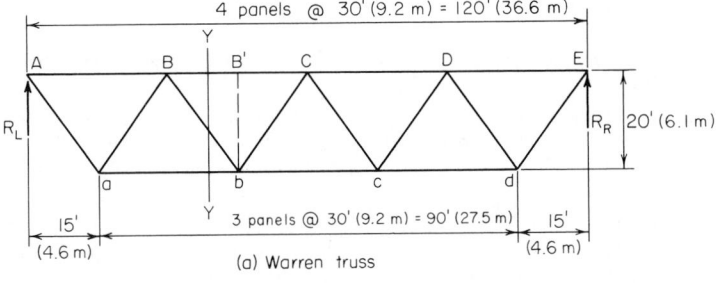

(a) Warren truss

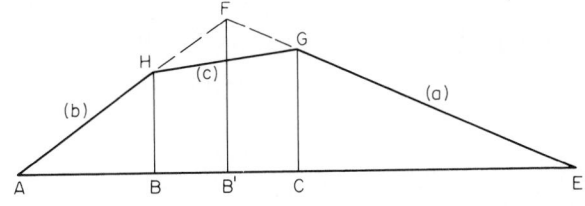

(b) Influence line for bending moment at b

FIG. 50

3.   *Place the unit load within the panel; compute the panel-point load and bending moment*

Place the unit load within panel *BC*. Determine the panel-point load P_B and compute M_b. Thus $P_B = (x - 60)/30 = x/30 - 2$; $M_b = 45R_L - 15P_B = 3x/8 - 15(x/30 - 2) = -x/8 + 30$, Eq. *c*.

4.   *Applying the foregoing equations, draw the influence line*

Figure 50*b* shows the influence line for M_b. Computing the significant values yields $CG = (3/8)(60) = 22.50$ ft·kips (30.51 kN·m); $BH = -(5/8)(90) + 75 = 18.75$ ft·kips (25.425 kN·m).

5.   *Compute the slope of each segment of the influence line*

This computation is made for subsequent reference. Thus, line *a*, $dM_b/dx = 3/8$; line *b*, $dM_b/dx = -5/8$; line *c*, $dM_b/dx = -1/8$.

FORCE IN TRUSS CHORD CAUSED BY MOVING CONCENTRATED LOADS

The truss in Fig. 50*a* carries the moving-load system shown in Fig. 51. Determine the maximum force induced in member *BC* during transit of the loads.

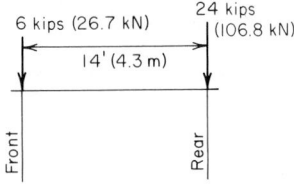

FIG. 51

Calculation Procedure:

1.   *Assume that locomotion proceeds from right to left, and compute the bending moment*

The force in *BC* is a function of the bending moment M_b at *b*. Refer to the previous calculation procedure for the slope of each segment of the influence line. Study of these slopes shows that M_b increases as the load system moves until the rear load is at *C*, the front load being 14 ft (4.3 m) to the left of *C*. Calculate the value of M_b corresponding to this load disposition by applying the computed properties of the influence line. Thus, $M_b = 22.50(24) + (22.50 - 1/8 \times 14)(6) = 664.5$ ft·kips (901.06 kN·m).

2.   Assume that locomotion proceeds from left to right, and compute the bending moment

Study shows that M_b increases as the system moves until the rear load is at C, the front load being 14 ft (4.3 m) to the right of C. Calculate the corresponding value of M_b. Thus, $M_b = 22.50(24) + (22.50 - 3/8 \times 14)(6) = 643.5$ ft·kips (872.59 kN·m). $\therefore M_{b,max} = 664.5$ ft·kips (901.06 kN·m).

3.   Determine the maximum force in the member

Cut the truss at plane YY. Determine the maximum force in BC by considering the equilibrium of the left part of the structure. Thus, $\Sigma M_b = M_b - 20BC = 0$; $BC = 664.5/20 = 33.2$-kips (147.67-kN) compression.

INFLUENCE LINE FOR BENDING MOMENT IN THREE-HINGED ARCH

The arch in Fig. 52a is hinged at A, B, and C. Draw the influence line for bending moment at D, and locate the neutral point.

Calculation Procedure:

1.   Start the graphical construction

Draw a line through A and C, intersecting the vertical line through B at E. Draw a line through B and C, intersecting the vertical line through A and F. Draw the vertical line GH through D.

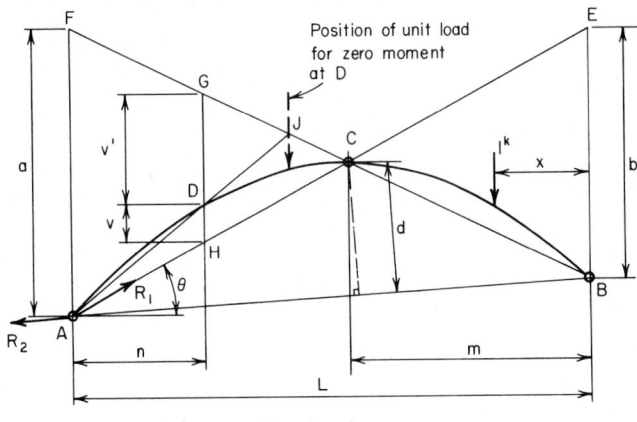

(a) Three-hinged arch

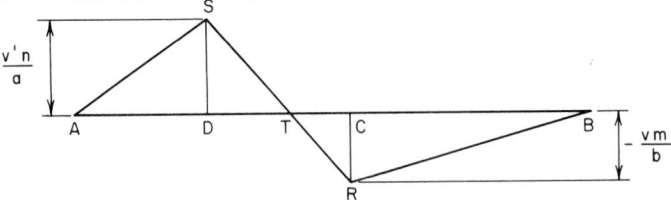

(b) Influence line for bending moment at D

FIG. 52

Let θ denote the angle between AE and the horizontal. Lines through B and D perpendicular to AE (omitted for clarity) make an angle θ with the vertical.

2.   Resolve the reaction into components

Resolve the reaction at A into the components R_1 and R_2 acting along AE and AB, respectively (Fig. 52).

3.   Determine the value of the first reaction

Let x denote the horizontal distance from the right-hand support to the unit load, where x has any value between 0 and L. Evaluate R_1 by equating the bending moment at B to zero. Thus $M_B = R_1 b \cos \theta - x = 0$; or $R_1 = x/(b \cos \theta)$.

4.   Evaluate the second reaction

Place the unit load within the interval CB. Evaluate R_2 by equating the bending moment at C to zero. Thus $M_C = R_2 d = 0$; $\therefore R_2 = 0$.

5.   Calculate the bending moment at D when the unit load lies within the interval CB

Thus, $M_D = -R_1 v \cos \theta = -[(v \cos \theta)/(b \cos \theta)]x$, or $M_D = -vx/b$, Eq. a. When $x = m$, $M_D = -vm/b$.

6.   Place the unit load in a new position, and determine the bending moment

Place the unit load within the interval AD. Working from the right-hand support, proceed in an analogous manner to arrive at the following result: $M_D = v'(L - x)/a$, Eq. b. When $x = L - n$, $M_D = v'n/a$.

7.   Place the unit load within another interval, and evaluate the second reaction

Place the unit load within the interval DC, and evaluate R_2. Thus $M_C = R_2 d - (x - m) = 0$, or $R_2 = (x - m)/d$.

Since both R_1 and R_2 vary linearly with respect to x, it follows that M_D is also a linear function of x.

8.   Complete the influence line

In Fig. 52b, draw lines BR and AS to represent Eqs. a and b, respectively. Draw the straight line SR, thus completing the influence line. The point T at which this line intersects the base is termed the *neutral point*.

9.   Locate the neutral point

To locate T, draw a line through A and D in Fig. 52a intersecting BF at J. The neutral point in the influence line lies vertically below J; that is, M_D is zero when the action line of the unit load passes through J.

The proof is as follows: Since $M_D = 0$ and there are no applied loads in the interval AD, it follows that the total reaction at A is directed along AD. Similarly, since $M_C = 0$ and there are no applied loads in the interval CB, it follows that the total reaction at B is directed along BC. Because the unit load and the two reactions constitute a balanced system of forces, they are collinear. Therefore, J lies on the action line of the unit load.

Alternatively, the location of the neutral point may be established by applying the geometric properties of the influence line.

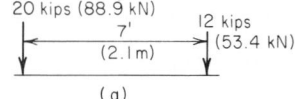

(a)

DEFLECTION OF A BEAM UNDER MOVING LOADS

The moving-load system in Fig. 53a traverses a beam on a simple span of 40 ft (12.2 m). What disposition of the system will cause the maximum deflection at midspan?

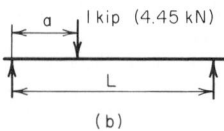

(b)

FIG. 53

Calculation Procedure:

1.   Develop the equations for the midspan deflection under a unit load

The maximum deflection will manifestly occur when the two loads lie on opposite sides of the centerline of the span. In calculating the deflection at midspan caused by a load applied at any point on the span, it is advantageous to apply Maxwell's theorem of reciprocal deflections, which states the following: *The deflection at A caused by a load at B equals the deflection at B caused by this load at A.*

In Fig. 53*b*, consider the beam on a simple span *L* to carry a unit load applied at a distance *a* from the left-hand support. By referring to case 7 of the AISC *Manual* and applying the principle of reciprocal deflections, derive the following equations for the midspan deflection under the unit load: When $a < L/2$, $y = (3L^2a - 4a^3)/(48EI)$. When $a > L/2$, $y = [3L^2(L - a) - 4(L - a)^3]/(48EI)$.

2.   *Position the system for purposes of analysis*

Position the system in such a manner that the 20-kip (89.0-kN) load lies to the left of center and the 12-kip (53.4-kN) load to the right of center. For the 20-kip (89.0-kN) load, set $a = x$. For the 12-kip (53.4-kN) load, $a = x + 7$; $L - a = 40 - (x + 7) = 33 - x$.

3.   *Express the total midspan deflection in terms of x*

Substitute in the preceding equations. Combining all constants into a single term *k*, we find $ky = 20(3 \times 40^2x - 4x^3) + 12[3 \times 40^2(33 - x) - 4(33 - x)^3]$.

4.   *Solve for the unknown distance*

Set $dy/dx = 0$ and solve for *x*. Thus, $x = 17.46$ ft (5.321 m).

For maximum deflection, position the load system with the 20-kip (89.0-kN) load 17.46 ft (5.321 m) from the left-hand support.

Riveted and Welded Connections

In the design of riveted and welded connections in this handbook, the American Institute of Steel Construction *Specification for the Design, Fabrication and Erection of Structural Steel for Buildings* is applied. This is presented in Part 5 of the *Manual of Steel Construction*.

The structural members considered here are made of ASTM A36 steel having a yield-point stress of 36,000 lb/in² (248,220 kPa). (The yield-point stress is denoted by F_y in the *Specification*.) All connections considered here are made with A141 hot-driven rivets or fillet welds of A233 class E60 series electrodes.

From the *Specification*, the allowable stresses are as follows: Tensile stress in connected member, 22,000 lb/in² (151,690.0 kPa); shearing stress in rivet, 15,000 lb/in² (103,425.0 kPa); bearing stress on projected area of rivet, 48,500 lb/in² (334,408.0 kPa); stress on throat of fillet weld, 13,600 lb/in² (93,772.0 kPa).

Let *n* denote the number of sixteenths included in the size of a fillet weld. For example, for a ⅜-in (9.53-mm) weld, $n = 6$. Then weld size = $n/16$. And throat area per linear inch of weld = $0.707n/16 = 0.0442n$ in². Also, capacity of weld = $13,600(0.0442n) = 600n$ lb/lin in ($108.0n$ N/mm).

As shown in Fig. 54, a rivet is said to be in *single shear* if the opposing forces tend to shear the shank along one plane and in *double shear* if they tend to shear it along two planes. The symbols R_{ss}, R_{ds}, and R_b used here to designate the shearing capacity of a rivet in single shear, the shearing capacity of a rivet in double shear, and the bearing capacity of a rivet, respectively, expressed in pounds (newtons).

CAPACITY OF A RIVET

Determine the values of R_{ss}, R_{ds}, and R_b for a ¾-in (19.05-mm) and ⅞-in (22.23-mm) rivet.

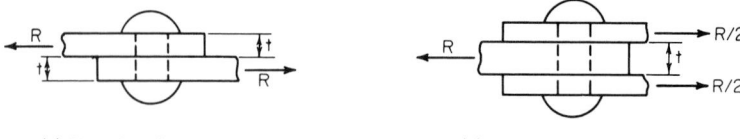

(a) Rivet in single shear   (b) Rivet in double shear

FIG. 54

Calculation Procedure:

1. *Compute the cross-sectional area of the rivet*

For the ¾-in (19.05-mm) rivet, area $= A = 0.785(0.75)^2 = 0.4418$ in² (2.8505 cm²). Likewise, for the ⅞-in (22.23-mm) rivet, $A = 0.785(0.875)^2 = 0.6013$ in² (3.8796 cm²).

2. *Compute the single and double shearing capacity of the rivet*

Let t denote the thickness, in inches (millimeters) of the connected member, as shown in Fig. 54. Multiply the stressed area by the allowable stress to determine the shearing capacity of the rivet. Thus, for the ¾-in (19.05-mm) rivet, $R_{ss} = 0.4418(15,000) = 6630$ lb (29,490.2 N); $R_{ds} = 2(0.4418)(15,000) = 13,250$ lb (58,936.0 N). Note that the factor of 2 is used for a rivet in double shear.

Likewise, for the ⅞-in (22.23-mm) rivet, $R_{ss} = 0.6013(15,000) = 9020$ lb (40,121.0 N); $R_{ds} = 2(0.6013)(15,000) = 18,040$ lb (80,242.0 N).

3. *Compute the rivet bearing capacity*

The effective bearing area of a rivet of diameter d in (mm) $= dt$. Thus, for the ¾-in (19.05-mm) rivet, $R_b = 0.75t(48,500) = 36,380t$ lb (161,709t N). For the ⅞-in (22.23-mm) rivet, $R_b = 0.875t(48,500) = 42,440t$ lb (188,733t N). By substituting the value of t in either relation, the numerical value of the bearing capacity could be obtained.

INVESTIGATION OF A LAP SPLICE

The hanger in Fig. 55a is spliced with nine ¾-in (19.05-mm) rivets in the manner shown. Compute the load P that may be transmitted across the joint.

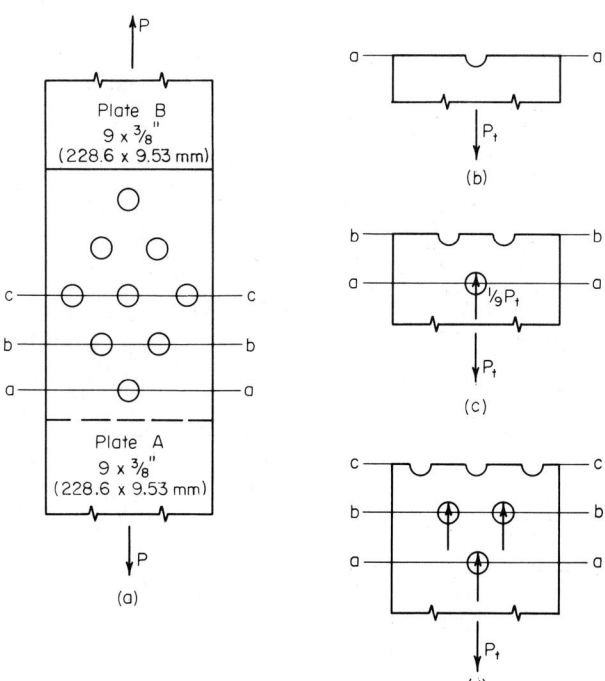

FIG. 55

Calculation Procedure:

1.   Compute the capacity of the joint in shear and bearing

There are three criteria to be considered: the shearing strength of the connection, the bearing strength of the connection, and the tensile strength of the net section of the plate at each row of rivets.

Since the load is concentric, assume that the load transmitted through each rivet is $\frac{1}{9}P$. As plate A (Fig. 55) deflects, it bears against the upper half of each rivet. Consequently, the reaction of the rivet on plate A is exerted *above* the horizontal diametral plane of the rivet.

Computing the capacity of the joint in shear and in bearing yields $P_{SS} = 9(6630) = 59,700$ lb (265,545.6 N); $P_b = 9(36,380)(0.375) = 122,800$ lb (546,214.4 N).

2.   Compute the tensile capacity of the plate

The tensile capacity P_t lb (N) of plate A (Fig. 55) is required. In structural fabrication, rivet holes are usually punched $\frac{1}{16}$ in (1.59 mm) larger than the rivet diameter. However, to allow for damage to the adjacent metal caused by punching, the *effective* diameter of the hole is considered to be $\frac{1}{8}$ in (3.18 mm) larger than the rivet diameter.

Refer to Fig. 55b, c, and d. Equate the tensile stress at each row of rivets to 22,000 lb/in² (151,690.0 kPa) to obtain P_t. Thus, at aa, residual tension $= P_t$; net area $= (9 - 0.875)(0.375) = 3.05$ in² (19.679 cm²). The stress $s = P_t/3.05 = 22,000$ lb/in² (151,690.0 kPa); $P_t = 67,100$ lb (298,460.0 N).

At bb, residual tension $= \frac{8}{9}P_t$; net area $= (9 - 1.75)(0.375) = 2.72$ in² (17.549 cm²); $s = \frac{8}{9}P_t/2.72 = 22,000$; $P_t = 67,300$ lb (299,350.0 N).

At cc, residual tension $= \frac{6}{9}P_t$; net area $= (9 - 2.625)(0.375) = 2.39$ in² (15.420 cm²); $\frac{6}{9}P_t/2.39 = 22,000$; $P_t = 78,900$ lb (350,947.0 N).

3.   Select the lowest of the five computed values as the allowable load

Thus, $P = 59,700$ lb (265,545.6 N).

DESIGN OF A BUTT SPLICE

A tension member in the form of a $10 \times \frac{1}{2}$ in (254.0 × 12.7 mm) steel plate is to be spliced with $\frac{7}{8}$-in (22.23-mm) rivets. Design a butt splice for the maximum load the member may carry.

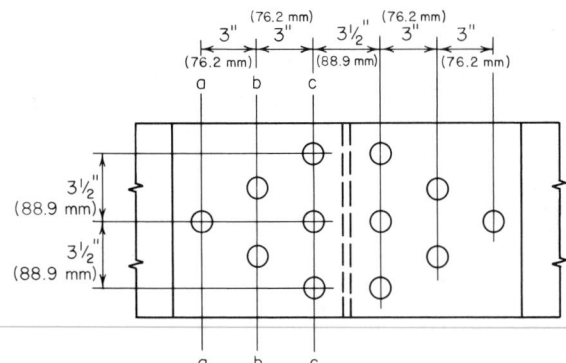

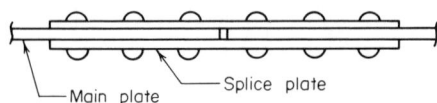

FIG. 56

Calculation Procedure:

1.   *Establish the design load*

In a butt splice, the load is transmitted from one member to another through two auxiliary plates called *cover*, *strap*, or *splice* plates. The rivets are therefore in double shear.

Establish the design load, P lb (N), by computing the allowable load at a cross section having one rivet hole. Thus net area = $(10 - 1)(0.5) = 4.5$ in^2 (29.03 cm^2). Then $P = 4.5(22,000) = 99,000$ lb (440,352.0 N).

2.   *Determine the number of rivets required*

Applying the values of rivet capacity found in an earlier calculation procedure in this section of the handbook, determine the number of rivets required. Thus, since the rivets are in double shear, $R_{ds} = 18,040$ lb (80,241.9 N); $R_b = 42,440(0.5) = 21,220$ lb (94,386.6 N). Then $99,000/18,040 = 5.5$ rivets; use the next largest whole number, or 6 rivets.

3.   *Select a trial pattern for the rivets; investigate the tensile stress*

Conduct this investigation of the tensile stress in the main plate at each row of rivets.

The trial pattern is shown in Fig. 56. The rivet spacing satisfies the requirements of the AISC *Specification*. Record the calculations as shown:

Section	Residual tension in main plate, lb (N)	÷	Net area, in^2 (cm^2)	=	Stress, lb/in^2 (kPa)
aa	99,000 (440,352.0)		4.5 (29.03)		22,000 (151,690.0)
bb	82,500 (366,960.0)		4.0 (25.81)		20,600 (142,037.0)
cc	49,500 (220,176.0)		3.5 (22.58)		14,100 (97,219.5)

Study of the above computations shows that the rivet pattern is satisfactory.

4.   *Design the splice plates*

To the left of the centerline, each splice plate bears against the *left* half of the rivet. Therefore, the entire load has been transmitted to the splice plates at *cc*, which is the critical section. Thus the tension in splice plate = ½(99,000) = 49,500 lb (220,176.0 N); plate thickness required = $49,500/[22,000(7)] = 0.321$ in (8.153 mm). Make the splice plates 10 × ⅜ in (254.0 × 9.53 mm).

DESIGN OF A PIPE JOINT

A steel pipe 5 ft 6 in (1676.4 mm) in diameter must withstand a fluid pressure of 225 lb/in^2 (1551.4 kPa). Design the pipe and the longitudinal lap splice, using ¾-in (19.05-mm) rivets.

Calculation Procedure:

1.   *Evaluate the hoop tension in the pipe*

Let L denote the length (Fig. 57) of the *repeating group* of rivets. In this case, this equals the rivet pitch. In Fig. 57, let T denote the hoop tension, in pounds (newtons), in the distance L. Evaluate the tension, using $T = pDL/2$, where p = internal pressure, lb/in^2 (kPa); D = inside diameter of pipe, in (mm); L = length considered, in (mm). Thus, $T = 225(66)L/2 = 7425L$.

2.   *Determine the required number of rows of rivets*

Adopt, tentatively, the minimum allowable pitch, which is 2 in (50.8 mm) for ¾-in (19.05-mm) rivets. Then establish a feasible rivet pitch. From an earlier calculation procedure in this section, $R_{ss} = 6630$ lb (29,490.0 N). Then $T = 7425(2) = 6630n$; $n = 2.24$. Use the next largest whole number of rows, or three rows of rivets. Also, $L_{max} = 3(6630)/7425 = 2.68$ in (68.072 mm). Use a 2½-in (63.5-mm) pitch, as shown in Fig. 57a.

3. Determine the plate thickness

Establish the thickness t in (mm) of the steel plates by equating the stress on the net section to its allowable value. Since the holes will be drilled, take $^{13}/_{16}$ in (20.64 mm) as their diameter. Then $T = 22,000t(2.5 - 0.81) = 7425(2.5)$; $t = 0.50$ in (12.7 mm); use ½-in (12.7-mm) plates. Also, $R_b = 36,380(0.5) > 6630$ lb (29,490.2 N). The rivet capacity is therefore limited by shear, as assumed.

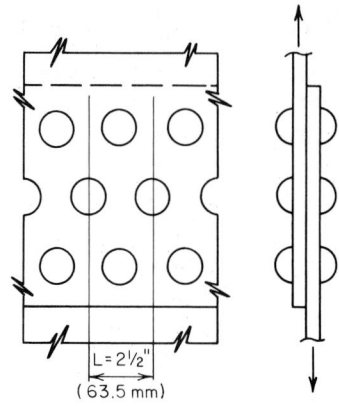

$L = 2½"$
(63.5 mm)

(a) Longitudinal pipe joint

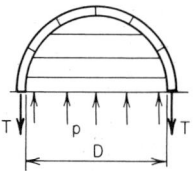

(b) Free-body diagram of upper half
of pipe and contents

FIG. 57

MOMENT ON RIVETED CONNECTION

The channel in Fig. 58a is connected to its supporting column with ¾-in (19.05-mm) rivets and resists the couple indicated. Compute the shearing stress in each rivet.

Calculation Procedure:

1. Compute the polar moment of inertia of the rivet group

The moment causes the channel (Fig. 58) to rotate about the centroid of the rivet group and thereby exert a tangential thrust on each rivet. This thrust is directly proportional to the radial distance to the center of the rivet.

Establish coordinate axes through the centroid of the rivet group. Compute the polar moment of inertia of the group with respect to an axis through its centroid, taking the cross-sectional area of a rivet as unity. Thus, $J = \Sigma(x^2 + y^2) = 8(2.5)^2 + 4(1.5)^2 + 4(4.5)^2 = 140$ in^2 (903.3 cm^2).

2. Compute the radial distance to each rivet

Using the right-angle relationship, we see that $r_1 = r_4 = (2.5^2 + 4.5^2)^{0.5} = 5.15$ in (130.810 mm); $r_2 = r_3 = (2.5^2 + 1.5^2)^{0.5} = 2.92$ in (74.168 mm).

3. Compute the tangential thrust on each rivet

Use the relation $f = Mr/J$. Since $M = 12,000(8) = 96,000$ lb·in (10,846.1 N·m), $f_1 = f_4 = 96,000(5.15)/140 = 3530$ lb (15,701.4 N); and $f_2 = f_3 = 96,000(2.92)/140 = 2000$ lb (8896.0 N). The directions are shown in Fig. 58b.

4. Compute the shearing stress

Using $s = P/A$, we find $s_1 = s_4 = 3530/0.442 = 7990$ lb/in^2 (55,090 kPa); also, $s_2 = s_3 = 2000/0.442 = 4520$ lb/in^2 (29,300 kPa).

5. Check the rivet forces

Check the rivet forces by summing their moments with respect to an axis through the centroid. Thus $M_1 = M_4 = 3530(5.15) = 18,180$ in·lb (2054.0 N·m); $M_2 = M_3 = 2000(2.92) = 5840$ in·lb (659.8 N·m). Then $\Sigma M = 4(18,180) + 4(5840) = 96,080$ in·lb (10,855.1 N·m).

ECCENTRIC LOAD ON RIVETED CONNECTION

Calculate the maximum force exerted on a rivet in the connection shown in Fig. 59a.

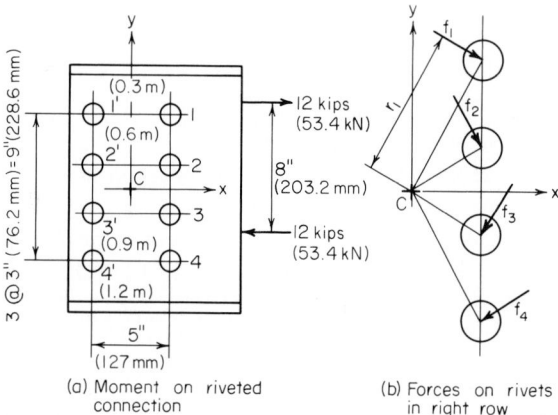

(a) Moment on riveted connection

(b) Forces on rivets in right row

FIG. 58

Calculation Procedure:

1. Compute the effective eccentricity

To account implicitly for secondary effects associated with an eccentrically loaded connection, the AISC *Manual* recommends replacing the true eccentricity with an *effective* eccentricity.

To compute the effective eccentricity, use $e_e = e_a - (1 + n)/2$, where e_e = effective eccentricity, in (mm); e_a = actual eccentricity of the load, in (mm); n = number of rivets in a vertical row. Substituting gives $e_e = 8 - (1 + 3)/2 = 6$ in (152.4 mm).

2. Replace the eccentric load with an equivalent system

The equivalent system is comprised of a concentric load P lb (N) and a clockwise moment M in·lb (N·m). Thus, $P = 15,000$ lb (66,720.0 N), $M = 15,000(6) = 90,000$ in·lb (10,168.2 N·m).

3. Compute the polar moment of inertia of the rivet group

Compute the polar moment of inertia of the rivet group with respect to an axis through its centroid. Thus, $J = \Sigma(x^2 + y^2) = 6(3)^2 + 4(4)^2 = 118$ in^2 (761.3 cm^2).

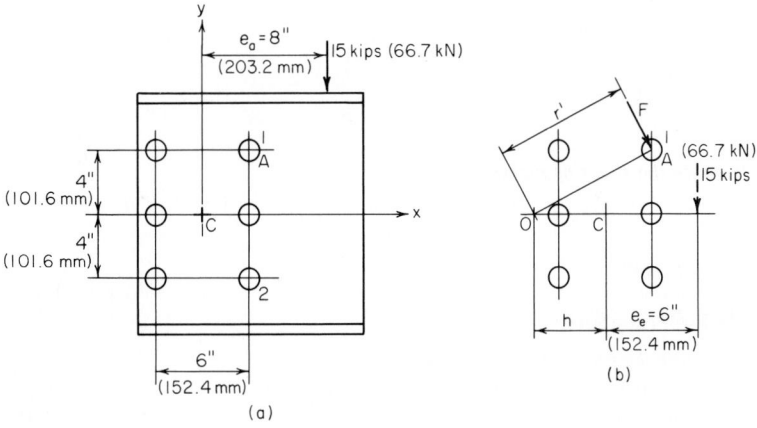

(a)

(b)

FIG. 59

4. Resolve the tangential thrust on each rivet into its horizontal and vertical components

Resolve the tangential thrust f lb (N) on each rivet caused by the moment into its horizontal and vertical components f_x and f_y, respectively. These forces are as follows: $f_x = My/J$ and $f_y = Mx/J$. Computing these forces for rivets 1 and 2 (Fig. 59) yields $f_x = 90,000(4)/118 = 3050$ lb (13,566.4 N); $f_y = 90,000(3)/118 = 2290$ lb (10,185.9 N).

5. Compute the thrust on each rivet caused by the concentric load

This thrust is $f'_y = 15,000/6 = 2500$ lb (11,120.0 N).

6. Combine the foregoing results to obtain the total force on the rivets being considered

The total force F lb (N) on rivets 1 and 2 is desired. Thus, $F_x = f_x = 3050$ lb (13,566.4 N); $F_y = f_y + f'_y = 2290 + 2500 = 4790$ lb (21,305.9 N). Then $F = [(3050)^2 + (4790)^2]^{0.5} = 5680$ lb (25,264.6 N).

The above six steps comprise method 1. A second way of solving this problem, method 2, is presented below.

The total force on each rivet may also be found by locating the *instantaneous center of rotation* associated with this eccentric load and treating the connection as if it were subjected solely to a moment (Fig. 59*b*).

7. Locate the instantaneous center of rotation

To locate this center, apply the relation $h = J/(e_cN)$, where N = total number of rivets and the other relations are as given earlier. Then $h = 118/[6(6)] = 3.28$ in (83.31 m).

8. Compute the force on the rivets

Considering rivets 1 and 2, use the equation $F = Mr'/J$, where r' = distance from the instantaneous center of rotation O to the center of the given rivet, in. For rivets 1 and 2, $r' = 7.45$ in (189.230 mm). Then $F = 90,000(7.45)/118 = 5680$ lb (25,264.6 N). The force on rivet 1 has an action line normal to the radius OA.

DESIGN OF A WELDED LAP JOINT

The 5-in (127.0-mm) leg of a 5 × 3 × ⅜ in (127.0 × 76.2 × 9.53 mm) angle is to be welded to a gusset plate, as shown in Fig. 60. The member will be subjected to repeated variation in stress. Design a suitable joint.

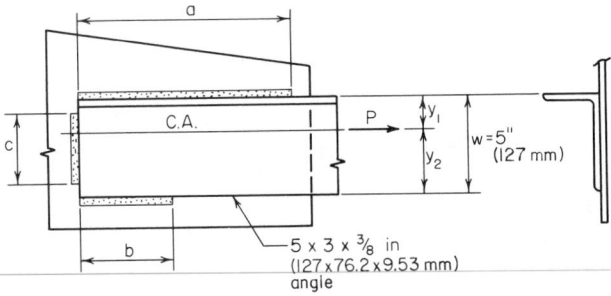

FIG. 60

Calculation Procedure:

1. Determine the properties of the angle

In accordance with the AISC *Specification*, arrange the weld to have its centroidal axis coincide with that of the member. Refer to the AISC *Manual* to obtain the properties of the angle. Thus $A = 2.86$ in^2 (18.453 cm^2); $y_1 = 1.70$ in (43.2 mm); $y_2 = 5.00 - 1.70 = 3.30$ in (83.820 mm).

2.   *Compute the design load and required weld length*

The design load P lb (N) $= As = 2.86(22,000) = 62,920$ lb (279,868.2 N). The AISC *Specification* restricts the weld size to $\frac{5}{16}$ in (7.94 mm). Hence, the weld capacity $= 5(600) = 3000$ lb/lin in (525,380.4 N/m); $L =$ weld length, in (mm) $= P/\text{capacity}$, lb/lin in $= 62,920/3000 = 20.97$ in (532.638 mm).

3.   *Compute the joint dimensions*

In Fig. 60, set $c = 5$ in (127.0 mm), and compute a and b by applying the following equations: $a = Ly_2/w - c/2$; $b = Ly_1/w - c/2$. Thus, $a = (20.97 \times 3.30)/5 - \frac{5}{2} = 11.34$ in (288.036 mm); $b = (20.97 \times 1.70)/5 - \frac{5}{2} = 4.63$ in (117.602 mm). Make $a = 11.5$ in (292.10 mm) and $b = 5$ in (127.0 mm).

ECCENTRIC LOAD ON A WELDED CONNECTION

The bracket in Fig. 61 is connected to its support with a $\frac{1}{4}$-in (6.35-mm) fillet weld. Determine the maximum stress in the weld.

Calculation Procedure:

1.   *Locate the centroid of the weld group*

Refer to the previous eccentric-load calculation procedure. This situation is analogous to that. Determine the stress by locating the instantaneous center of rotation. The maximum stress occurs at A and B (Fig. 61).

Considering the weld as concentrated along the edge of the supported member, locate the centroid of the weld group by taking moments with respect to line aa. Thus $m = 2(4)(2)/(12 + 2 \times 4) = 0.8$ in (20.32 mm).

2.   *Replace the eccentric load with an equivalent concentric load and moment*

Thus $P = 13,500$ lb (60,048.0 N); $M = 124,200$ in·lb (14,032.1 N·m).

3.   *Compute the polar moment of inertia of the weld group*

This moment should be computed with respect to an axis through the centroid of the weld group. Thus $I_x = (1/12)(12)^3 + 2(4)(6)^2 = 432$ in³ (7080.5 cm³); $I_y = 12(0.8)^2 + 2(1/12)(4)^3 + 2(4)(2 - 0.8)^2 = 29.9$ in³ (490.06 cm³). Then $J = I_x + I_y = 461.9$ in³ (7570.54 cm³).

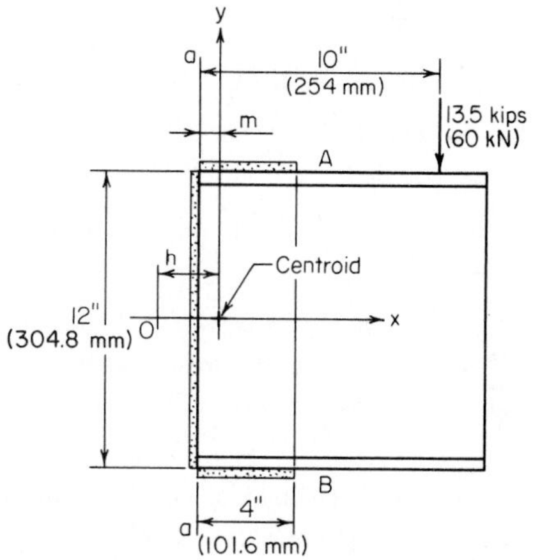

4.   *Locate the instantaneous center of rotation O*

FIG. 61

This center is associated with this eccentric load by applying the equation $h = J/(eL)$, where $e =$ eccentricity of load, in (mm), and $L =$ total length of weld, in (mm). Thus, $e = 10 - 0.8 = 9.2$ in (233.68 mm); $L = 12 + 2(4) = 20$ in (508.0 mm); then $h = 461.9/[9.2(20)] = 2.51$ in (63.754 mm).

5.   *Compute the force on the weld*

Use the equation $F = Mr'/J$, lb/lin in (N/m), where $r' =$ distance from the instantaneous center of rotation to the given point, in (mm). At A and B, $r' = 8.28$ in (210.312 mm); then $F = [124,200(8.28)]/461.9 = 2230$ lb/lin in (390,532.8 N/m).

6.   *Calculate the corresponding stress on the throat*

Thus, $s = P/A = 2230/[0.707(0.25)] = 12,600$ lb/in² (86,877.0 kPa), where the value 0.707 is the sine of 45°, the throat angle.

Steel Beams and Plate Girders

In the following calculation procedures, the design of steel members is executed in accordance with the *Specification for the Design, Fabrication and Erection of Structural Steel for Buildings* of the American Institute of Steel Construction. This specification is presented in the AISC *Manual of Steel Construction*.

Most allowable stresses are functions of the yield-point stress, denoted as F_y in the *Manual*. The appendix of the *Specification* presents the allowable stresses associated with each grade of structural steel together with tables intended to expedite the design. The *Commentary* in the *Specification* explains the structural theory underlying the *Specification*.

Unless otherwise noted, the structural members considered here are understood to be made of ASTM A36 steel, having a yield-point stress of 36,000 lb/in² (248,220.0 kPa).

The notational system used conforms with that adopted earlier, but it is augmented to include the following: A_f = area of flange, in² (cm²); A_w = area of web, in² (cm²); b_f = width of flange, in (mm); d = depth of section, in (mm); d_w = depth of web, in (mm); t_f = thickness of flange, in (mm). t_w = thickness of web, in (mm); L' = unbraced length of compression flange, in (mm); f_y = yield-point stress, lb/in² (kPa).

MOST ECONOMIC SECTION FOR A BEAM WITH A CONTINUOUS LATERAL SUPPORT UNDER A UNIFORM LOAD

A beam on a simple span of 30 ft (9.2 m) carries a uniform superimposed load of 1650 lb/lin ft (24,079.9 N/m). The compression flange is laterally supported along its entire length. Select the most economic section.

Calculation Procedure:

1. *Compute the maximum bending moment and the required section modulus*

Assume that the beam weighs 50 lb/lin ft (729.7 N/m) and satisfies the requirements of a compact section as set forth in the *Specification*.

The maximum bending moment is $M = (1/8)wL^2 = (1/8)(1700)(30)^2(12) = 2,295,000$ in·lb (259,289.1 N·m).

Referring to the *Specification* shows that the allowable bending stress is 24,000 lb/in² (165,480.0 kPa). Then $S = M/f = 2,295,000/24,000 = 95.6$ in³ (1566.88 cm³).

2. *Select the most economic section*

Refer to the AISC *Manual*, and select the most economic section. Use W18 × 55 = 98.2 in³ (1609.50 cm³); section compact. The disparity between the assumed and actual beam weight is negligible.

A second method for making this selection is shown below.

3. *Calculate the total load on the member*

Thus, the total load = $W = 30(1700) = 51,000$ lb (226,848.0 N).

4. *Select the most economic section*

Refer to the tables of allowable uniform loads in the *Manual*, and select the most economic section. Thus use W18 × 55; $W_{allow} = 52,000$ lb (231,296.0 N). The capacity of the beam is therefore slightly greater than required.

MOST ECONOMIC SECTION FOR A BEAM WITH INTERMITTENT LATERAL SUPPORT UNDER UNIFORM LOAD

A beam on a simple span of 25 ft (7.6 m) carries a uniformly distributed load, including the estimated weight of the beam, of 45 kips (200.2 kN). The member is laterally supported at 5-ft (1.5-m) intervals. Select the most economic member (*a*) using A36 steel; (*b*) using A242 steel, having a yield-point stress of 50,000 lb/in² (344,750.0 kPa) when the thickness of the metal is ¾ in (19.05 mm) or less.

Calculation Procedure:

1. Using the AISC allowable-load tables, select the most economic member made of A36 steel

After a trial section has been selected, it is necessary to compare the unbraced length L' of the compression flange with the properties L_c and L_u of that section in order to establish the allowable bending stress. The variables are defined thus: L_c = maximum unbraced length of the compression flange if the allowable bending stress = $0.66f_y$, measured in ft (m); L_u = maximum unbraced length of the compression flange, ft (m), if the allowable bending stress is to equal $0.60f_y$.

The values of L_c and L_u associated with each rolled section made of the indicated grade of steel are recorded in the allowable-uniform-load tables of the AISC *Manual*. The L_c value is established by applying the definition of a *laterally supported* member as presented in the *Specification*. The value of L_u is established by applying a formula given in the *Specification*.

There are four conditions relating to the allowable stress:

Condition	Allowable stress
Compact section; $L' \leq L_c$	$0.66f_y$
Compact section; $L_c < L' \leq L_u$	$0.60f_y$
Noncompact section; $L' \leq L_u$	$0.60f_y$
$L' > L_u$	Apply the *Specification* formula—use the larger value obtained when the two formulas given are applied.

The values of allowable uniform load given in the AISC *Manual* apply to beams of A36 steel satisfying the first or third condition above, depending on whether the section is compact or noncompact.

Referring to the table in the *Manual*, we see that the most economic section made of A36 steel is W16 × 45; W_{allow} = 46 kips (204.6 kN), where W_{allow} = allowable load on the beam, kips (kN). Also, L_c = 7.6 > 5. Hence, the beam is acceptable.

2. Compute the equivalent load for a member of A242 steel

To apply the AISC *Manual* tables to choose a member of A242 steel, assume that the shape selected will be compact. Transform the actual load to an equivalent load by applying the conversion factor 1.38, that is, the ratio of the allowable stresses. The conversion factors are recorded in the *Manual* tables. Thus, equivalent load = 45/1.38 = 32.6 kips (145.0 N).

3. Determine the lightest satisfactory section

Enter the *Manual* allowable-load table with the load value computed in step 2, and select the lightest section that appears to be satisfactory. Try W16 × 36; W_{allow} = 36 kips (160.1 N). However, this section is noncompact in A242 steel, and the equivalent load of 32.6 kips (145.0 N) is not valid for this section.

4. Revise the equivalent load

To determine whether the W16 × 36 will suffice, revise the equivalent load. Check the L_u value of this section in A242 steel. Then equivalent load = 45/1.25 = 36 kips (160.1 N), L_u = 6.3 ft (1.92 m) > 5 ft (1.5 m); use W16 × 36.

5. Verify the second part of the design

To verify the second part of the design, calculate the bending stress in the W16 × 36, using S = 56.3 in³ (922.76 cm³) from the *Manual*. Thus M = $(1/8)WL$ = $(1/8)(45,000)(25)(12)$ = 1,688,000 in · lb (190,710.2 N · m); $f = M/S$ = 1,688,000/56.3 = 30,000 lb/in² (206,850.0 kPa). This stress is acceptable.

DESIGN OF A BEAM WITH REDUCED ALLOWABLE STRESS

The compression flange of the beam in Fig. 62a will be braced only at points A, B, C, D, and E. Using AISC data, a designer has selected W21 × 55 section for the beam. Verify the design.

Calculation Procedure:

1. *Calculate the reactions; construct the shear and bending-moment diagrams*

The results of this step are shown in Fig. 62.

2. *Record the properties of the selected section*

Using the AISC *Manual*, record the following properties of the 21WF55 section: $S = 109.7$ in^3 (1797.98 cm^3); $I_y = 44.0$ in^4 (1831.41 cm^4); $b_f = 8.215$ in (208.661 mm); $t_f = 0.522$ in (13.258

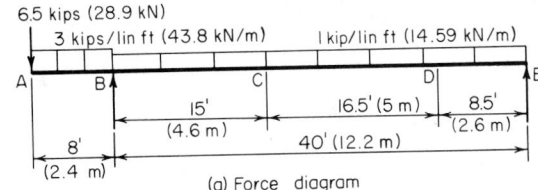

(a) Force diagram

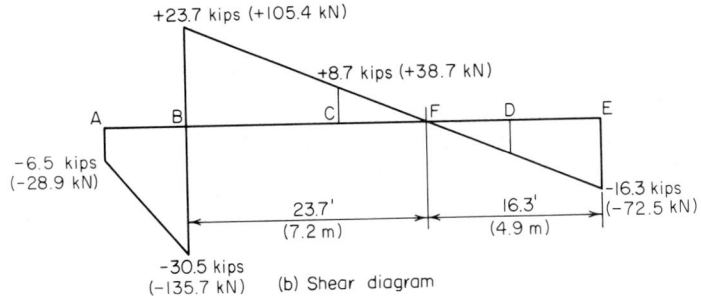

(b) Shear diagram

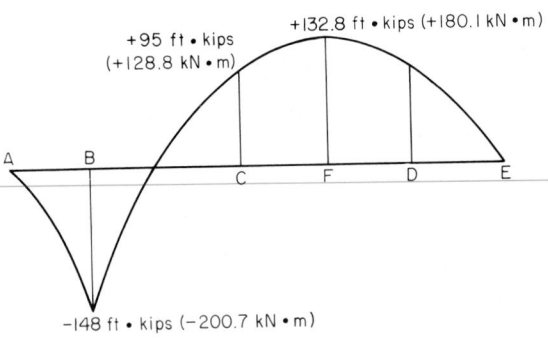

(c) Bending – moment diagram

FIG. 62

mm); d = 20.80 in (528.32 mm); t_w = 0.375 in (9.525 mm); d/A_f = 4.85/in (0.1909/mm); L_c = 8.9 ft (2.71 m); L_u = 9.4 ft (2.87 m).

Since $L' > L_u$, the allowable stress must be reduced in the manner prescribed in the *Manual*.

3. Calculate the radius of gyration

Calculate the radius of gyration with respect to the y axis of a T section comprising the compression flange and one-sixth the web, neglecting the area of the fillets. Referring to Fig. 63, we see A_f = 8.215(0.522) = 4.29 in² (27.679 cm²); (1/6)A_w = (1/6)(19.76)(0.375) = 1.24; A_T = 5.53 in² (35.680 cm²); I_T = 0.5I_y of the section = 22.0 in⁴ (915.70 cm⁴); r = $(22.0/5.53)^{0.5}$ = 1.99 in (50.546 mm).

4. Calculate the allowable stress in each interval between lateral supports

By applying the provisions of the *Manual*, calculate the allowable stress in each interval between lateral supports, and compare this with the actual stress. For A36 steel, the *Manual* formula (4) reduces to f_1 = 22,000 − $0.679(L'/r)^2/C_b$ lb/in² (kPa). By *Manual* formula (5), f_2 = 12,000,000/$(L'd/A_f)$ lb/in² (kPa). Set the allowable stress equal to the greater of these values.

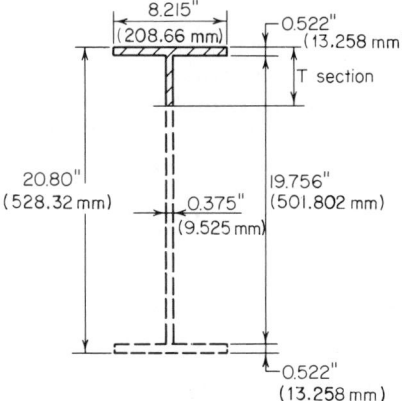

FIG. 63 Dimensions of W21 × 55.

For interval *AB:* L' = 8 ft (2.4 m) < L_c; $\therefore f_{\text{allow}}$ = 24,000 lb/in² (165,480.0 kPa); $f_{\max}$ = 148,000(12)/109.7 = 16,200 lb/in² (111,699.0 kPa)—this is acceptable.

For interval *BC:* L'/r = 15(12)/1.99 = 90.5; M_1/M_2 = 95/(−148) = −0.642; C_b = 1.75 − 1.05(−0.642) + 0.3(−0.642)² = 2.55; $\therefore$ set C_b = 2.3; f_1 = 22,000 − 0.679(90.5)²/2.3 = 19,600 lb/in² (135,142.0 kPa); f_2 = 12,000,000/[15(12)(4.85)] = 13,700 lb/in² (94,461.5 kPa); $f_{\max}$ = 16,200 < 19,600 lb/in² (135,142.0 kPa). This is acceptable.

Interval *CD:* Since the maximum moment occurs within the interval rather than at a boundary section, C_b = 1; L'/r = 16.5(12)/1.99 = 99.5; f_1 = 22,000 − 0.679(99.5)² = 15,300 lb/in² (105,493.5 kPa); f_2 = 12,000,000/[16.5(12)(4.85)] = 12,500 lb/in² (86,187.5 kPa); $f_{\max}$ = 132,800(12)/109.7 = 14,500 < 15,300 lb/in² (105,493.5 kPa). This stress is acceptable.

Interval *DE:* The allowable stress is 24,000 lb/in² (165,480.0 kPa), and the actual stress is considerably below this value. The W21 × 55 is therefore satisfactory. Where deflection is the criterion, the member should be checked by using the *Specification*.

DESIGN OF A COVER-PLATED BEAM

Following the fabrication of a W18 × 60 beam, a revision was made in the architectural plans, and the member must now be designed to support the loads shown in Fig. 64a. Cover plates are to be welded to both flanges to develop the required strength. Design these plates and their connection to the W shape, using fillet welds of A233 class E60 series electrodes. The member has continuous lateral support.

Calculation Procedure:

1. Construct the shear and bending-moment diagrams

These are shown in Fig. 64. Also, M_E = 340.3 ft·kips (461.44 kN·m).

2. Calculate the required section modulus, assuming the built-up section will be compact

The section modulus S = M/f = 340.3(12)/24 = 170.2 in³ (2789.58 cm³).

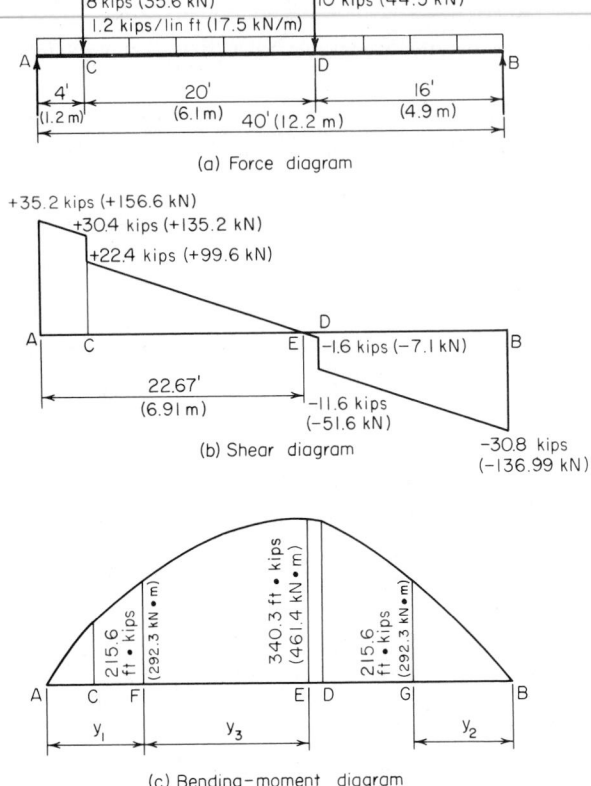

(a) Force diagram

(b) Shear diagram

(c) Bending–moment diagram

FIG. 64

3. Record the properties of the beam section

Refer to the AISC *Manual*, and record the following properties for the W18 × 60; $d = 18.25$ in (463.550 mm); $b_f = 7.56$ in (192.024 mm); $t_f = 0.695$ in (17.653 mm); $I = 984$ in⁴ (40.957 cm⁴); $S = 107.8$ in³ (1766.84 cm³).

4. Select a trial section

Apply the approximation $A = 1.05(S - S_{WF})/d_{WF}$, where A = area of one cover plate, in² (cm²); S = section modulus required, in³ (cm³); S_{WF} = section modulus of wide-flange shape, in³ (cm³); d_{WF} = depth of wide-flange shape, in (mm). Then $A = [1.05(170.2 - 107.8)]/18.25 = 3.59$ in² (23.163 cm²).

    Try 10 × ⅜ in (254.0 × 9.5 mm) plates with $A = 3.75$ in² (24.195 cm²). Since the beam flange is 7.5 in (190.50 mm) wide, ample space is available to accommodate the welds.

5. Ascertain whether the assumed size of the cover plates satisfies the AISC Specification

Using the appropriate AISC *Manual* section, we find $7.56/0.375 = 20.2 < 32$, which is acceptable; $½(10 - 7.56)/0.375 = 3.25 < 16$, which is acceptable.

6. Test the adequacy of the trial section

Calculate the section modulus of the trial section. Referring to Fig. 65a, we see $I = 984 + 2(3.75)(9.31)^2 = 1634$ in⁴ (68,012.1 cm⁴); $S = I/c = 1634/9.5 = 172.0$ in³ (2819.08 cm³). The reinforced section is therefore satisfactory.

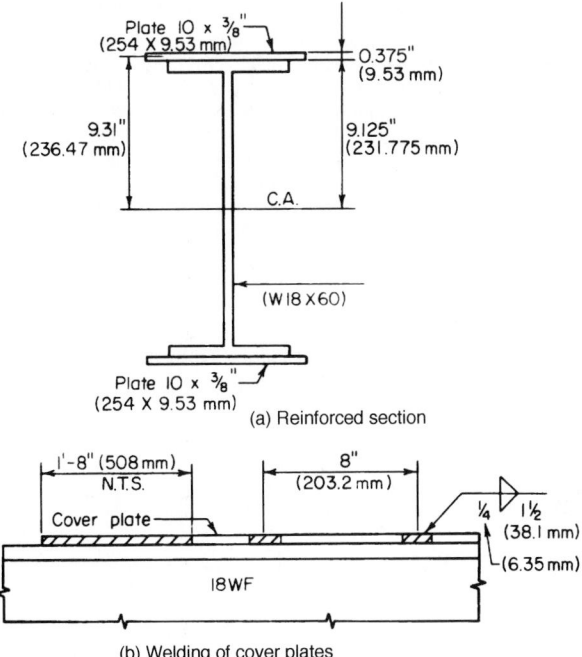

Plate 10 x ⅜"
(254 X 9.53 mm)

0.375"
(9.53 mm)

9.31"
(236.47 mm)

9.125"
(231.775 mm)

C.A.

(W18 X60)

Plate 10 x ⅜"
(254 X 9.53 mm)

(a) Reinforced section

1'-8" (508 mm)
N.T.S.

8"
(203.2 mm)

Cover plate

¼ 1½
(38.1 mm)
(6.35 mm)

18WF

(b) Welding of cover plates

FIG. 65

7. *Locate the points at which the cover plates are not needed*

To locate the points at which the cover plates may theoretically be dispensed with, calculate the moment capacity of the wide-flange shape alone. Thus, $M = fS = 24(107.8)/12 = 215.6$ ft·kips (292.3 kN·m).

8. *Locate the points at which the computed moment occurs*

These points are F and G (Fig. 64). Thus, $M_F = 35.2y_1 - 8(y_1 - 4) - \frac{1}{2}(1.2y_1^2) = 215.6$; $y_1 = 8.25$ ft (2.515 m); $M_G = 30.8y_2 - \frac{1}{2}(1.2y_2^2) = 215.6$; $y_2 = 8.36$ ft (2.548 m).

Alternatively, locate F by considering the area under the shear diagram between E and F. Thus $M_F = 340.3 - \frac{1}{2}(1.2y_3^2) = 215.6$; $y_3 = 14.42$ ft (4.395 m); $y_1 = 22.67 - 14.42 = 8.25$ ft (2.515 m).

For symmetry, center the cover plates about midspan, placing the theoretical cutoff points at 8 ft 3 in (2.51 m) from each support.

9. *Calculate the axial force in the cover plate*

Calculate the axial force P lb (N) in the cover plate at its end by computing the mean bending stress. Determine the length of fillet weld required to transmit this force to the W shape. Thus $f_{mean} = My/I = 215,600(12)(9.31)/1634 = 14,740$ lb/in² (101,632.3 kPa). Then $P = Af_{mean} = 3.75(14,740) = 55,280$ lb (245,885.4 N). Use a ¼-in (6.35-mm) fillet weld, which satisfies the requirements of the *Specification*. The capacity of the weld $= 4(600) = 2400$ lb/lin in (420,304.3 N/m). Then the length L required for this weld is $L = 55,280/2400 = 23.0$ in (584.20 mm).

10. *Extend the cover plates*

In accordance with the *Specification*, extend the cover plates 20 in (508.0 mm) beyond the theoretical cutoff point at each end, and supply a continuous ¼-in fillet weld along both edges in this extension. This requirement yields 40 in (1016.0 mm) of weld as compared with the 23 in (584.2 mm) needed to develop the plate.

11.   *Calculate the horizontal shear flow at the inner surface of the cover plate*

Choose F or G, whichever is larger. Design the intermittent fillet weld to resist this shear flow. Thus $V_F = 35.2 - 8 - 1.2(8.25) = 17.3$ kips (76.95 kN); $V_G = -30.8 + 1.2(8.36) = -20.8$ kips (-92.51 kN). Then $q = VQ/I = 20,800(3.75)(9.31)/1634 = 444$ lb/lin in (77,756.3 N/m).

The *Specification* calls for a minimum weld length of 1.5 in (38.10 mm). Let s denote the center-to-center spacing as governed by shear. Then $s = 2(1.5)(2400)/444 = 16.2$ in (411.48 mm). However, the *Specification* imposes additional restrictions on the weld spacing. To preclude the possibility of error in fabrication, provide an identical spacing at the top and bottom. Thus, $s_{max} = 21(0.375) = 7.9$ in (200.66 mm). Therefore, use a ¼-in (6.35-mm) fillet weld, 1.5 in (38.10 mm) long, 8 in (203.2 mm) on centers, as shown in Fig. 65b.

DESIGN OF A CONTINUOUS BEAM

The beam in Fig. 66a is continuous from A to D and is laterally supported at 5-ft (1.5-m) intervals. Design the member.

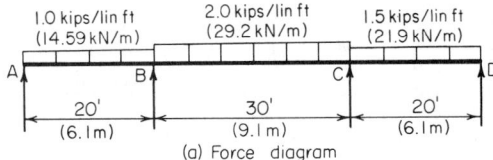

(a) Force diagram

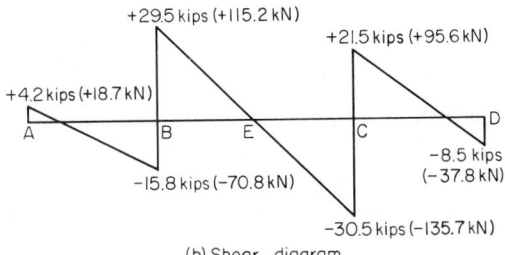

(b) Shear diagram

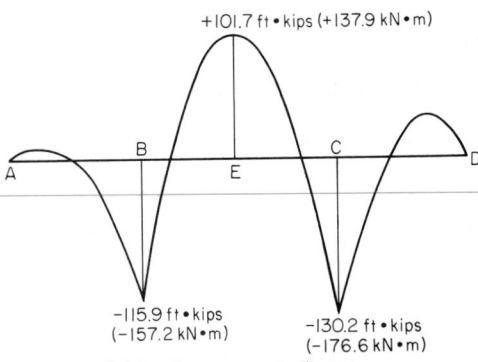

(c) Bending–moment diagram

FIG. 66

Calculation Procedure:

1. Find the bending moments at the interior supports; calculate the reactions and construct shear and bending-moment diagrams

The maximum moments are $+101.7$ ft·kips (137.9 kN·m) and -130.2 ft·kips (176.55 kN·m).

2. Calculate the modified maximum moments

Calculate these moments in the manner prescribed in the AISC *Specification*. The clause covering this calculation is based on the postelastic behavior of a continuous beam. (Refer to a later calculation procedure for an analysis of this behavior.)

Modified maximum moments: $+101.7 + 0.1(0.5)(115.9 + 130.2) = +114.0$ ft·kips (154.58 kN·m); $0.9(-130.2) = -117.2$ ft·kips (-158.92 kN·m); design moment $= 117.2$ ft·kips (158.92 kN·m).

3. Select the beam size

Thus, $S = M/f = 117.2(12)/24 = 58.6$ in³ (960.45 cm³). Use W16 × 40 with $S = 64.4$ in³ (1055.52 cm³); $L_c = 7.6$ ft (2.32 m).

SHEARING STRESS IN A BEAM—EXACT METHOD

Calculate the maximum shearing stress in a W18 × 55 beam at a section where the vertical shear is 70 kips (311.4 kN).

Calculation Procedure:

1. Record the relevant properties of the member

The shearing stress is a maximum at the centroidal axis and is given by $v = VQ/(It)$. The static moment of the area above this axis is found by applying the properties of the WT9 × 27.5, which are presented in the AISC *Manual*. Note that the T section considered is one-half the wide-flange section being used. See Fig. 67.

The properties of these sections are $I_W = 890$ in⁴ (37,044.6 cm⁴); $A_T = 8.10$ in² (52.261 cm²); $t_w = 0.39$ in (9.906 mm); $y_m = 9.06 - 2.16 = 6.90$ in (175.26 mm).

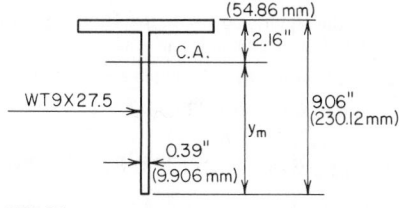

2. Calculate the shearing stress at the centroidal axis

Substituting gives $Q = 8.10(6.90) = 55.9$ in³ (916.20 cm³); then $v = 70,000(55.9)/[890(0.39)] = 11,270$ lb/in² (77,706.7 kPa).

FIG. 67

SHEARING STRESS IN A BEAM—APPROXIMATE METHOD

Solve the previous calculation procedure, using the approximate method of determining the shearing stress in a beam.

Calculation Procedure:

1. Assume that the vertical shear is resisted solely by the web

Consider the web as extending the full depth of the section and the shearing stress as uniform across the web. Compare the results obtained by the exact and the approximate methods.

2. Compute the shear stress

Take the depth of the web as 18.12 in (460.248 mm), $v = 70,000/[18.12(0.39)] = 9910$ lb/in² (68,329.45 kPa). Thus, the ratio of the computed stresses is $11,270/9910 = 1.14$.

Since the error inherent in the approximate method is not unduly large, this method is applied in assessing the shear capacity of a beam. The allowable shear V for each rolled section is recorded in the allowable-uniform-load tables of the AISC *Manual*.

The design of a rolled section is governed by the shearing stress only in those instances where the ratio of maximum shear to maximum moment is extraordinarily large. This condition exists in a heavily loaded short-span beam and a beam that carries a large concentrated load near its support.

MOMENT CAPACITY OF A WELDED PLATE GIRDER

A welded plate girder is composed of a 66 × ⅜ in (1676.4 × 9.53 mm) web plate and two 20 × ¾ in (508.0 × 19.05 mm) flange plates. The unbraced length of the compression flange is 18 ft (5.5 m). If $C_b = 1$, what bending moment can this member resist?

Calculation Procedure:

1.   Compute the properties of the section

The tables in the AISC *Manual* are helpful in calculating the moment of inertia. Thus $A_f = 15$ in^2 (96.8 cm^2); $A_w = 24.75$ in^2 (159.687 cm^2); $I = 42,400$ in^4 (176.481 dm^4); $S = 1256$ in^3 (20,585.8 cm^3).

For the T section comprising the flange and one-sixth the web, $A = 15 + 4.13 = 19.13$ in^2 (123.427 cm^2); then $I = (1/12)(0.75)(20)^3 = 500$ in^4 (2081.1 dm^4); $r = (500/19.13)^{0.5} = 5.11$ in (129.794 mm); $L'/r = 18(12)/5.11 = 42.3$.

2.   Ascertain if the member satisfies the AISC Specification

Let h denote the clear distance between flanges, in (cm). Then: flange, ½(20)/0.75 = 13.3 < 16— this is acceptable; web, $h/t_w = 66/0.375 = 176 < 320$—this is acceptable.

3.   Compute the allowable bending stress

Use $f_1 = 22,000 - 0.679(L'/r)^2/C_b$, or $f_1 = 22,000 - 0.679(42.3)^2 = 20,800$ lb/in^2 (143,416.0 kPa); $f_2 = 12,000,000/(L'd/A_f) = 12,000,000(15)/[18(12)(67.5)] = 12,300$ lb/in^2 (84,808.5 kPa). Therefore, use 20,800 lb/in^2 (143,416.0 kPa) because it is the larger of the two stresses.

4.   Reduce the allowable bending stress in accordance with the AISC Specification

Using the equation given in the *Manual* yields $f_3 = 20,800\{1 - 0.005(24.75/15)[176 - 24,000/(20,800)^{0.5}]\} = 20,600$ lb/in^2 (142,037.0 kPa).

5.   Determine the allowable bending moment

Use $M = f_3 S = 20.6(1256)/12 = 2156$ ft·kips (2923.5 kN·m).

ANALYSIS OF A RIVETED PLATE GIRDER

A plate girder is composed of one web plate 48 × ⅜ in (1219.2 × 9.53 mm); four flange angles 6 × 4 × ¾ in (152.4 × 101.6 × 19.05 mm); two cover plates 14 × ½ in (355.6 × 12.7 mm). The flange angles are set 48.5 in (1231.90 mm) back to back with their 6-in (152.4-mm) legs outstanding; they are connected to the web plate by ⅞-in (22.2-mm) rivets. If the member has continuous lateral support, what bending moment may be applied? What spacing of flange-to-web rivets is required in a panel where the vertical shear is 180 kips (800.6 kN)?

Calculation Procedure:

1.   Obtain the properties of the angles from the AISC Manual

Record the angle dimensions as shown in Fig. 68.

2.   Check the cover plates for compliance with the AISC Specification

The cover plates are found to comply with the pertinent sections of the *Specification*.

3.   Compute the gross flange area and rivet-hole area

Ascertain whether the *Specification* requires a reduction in the flange area. Thus gross flange area = 2(6.94) + 7.0 = 20.88 in^2 (134.718 cm^2); area of rivet holes = 2(½)(1) + 4(¾)(1) = 4.00 in^2 (25.808 cm^2); allowable area of holes = 0.15(20.88) = 3.13. The excess area = hole area −

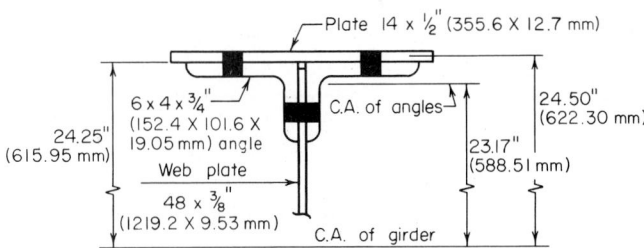

FIG. 68

allowable area $= 4.00 - 3.13 = 0.87$ in² (5.613 cm²). Consider that this excess area is removed from the outstanding legs of the angles, at both the top and the bottom.

4. Compute the moment of inertia of the net section

	in⁴	dm⁴
One web plate, I_0	3,456	14.384
Four flange angles, I_0	35	0.1456
$Ay^2 = 4(6.94)(23.17)^2$	14,900	62.0184
Two cover plates:		
$Ay^2 = 2(7.0)(24.50)^2$	8,400	34.9634
I of gross section	26,791	111.5123
Deduct $2(0.87)(23.88)^2$ for excess area	991	4.12485
I of net section	25,800	107.387

5. Establish the allowable bending stress

Use the *Specification*. Thus $h/t_w = (48.5 - 8)/0.375 < 24,000/(22,000)^{0.5}$; ∴ use 22,000 lb/in² (151,690.0 kPa). Also, $M = fI/c = 22(25,800)/[24.75(12)] = 1911$ ft·kips (2591.3 kN·m).

6. Calculate the horizontal shear flow to be resisted

Here Q of flange $= 13.88(23.17) + 7.0(24.50) - 0.87(23.88) = 472$ in³ (7736.1 cm³); $q = VQ/I = 180,000(472)/25,800 = 3290$ lb/lin in (576,167.2 N/m).

From a previous calculation procedure, $R_{ds} = 18,040$ lb (80,241.9 N); $R_b = 42,440(0.375) = 15,900$ lb (70,723.2 N); $s = 15,900/3290 = 4.8$ in (121.92 mm), where $s =$ allowable rivet spacing, in (mm). Therefore, use a 4¾-in (120.65-mm) rivet pitch. This satisfies the requirements of the *Specification*.

Note: To determine the allowable rivet spacing, divide the horizontal shear flow into the rivet capacity.

DESIGN OF A WELDED PLATE GIRDER

A plate girder of welded construction is to support the loads shown in Fig. 69a. The distributed load will be applied to the top flange, thereby offering continuous lateral support. At its ends, the girder will bear on masonry buttresses. The total depth of the girder is restricted to approximately 70 in (1778.0 mm). Select the cross section, establish the spacing of the transverse stiffeners, and design both the intermediate stiffeners and the bearing stiffeners at the supports.

Calculation Procedure:

1. Construct the shear and bending-moment diagrams

These diagrams are shown in Fig. 69.

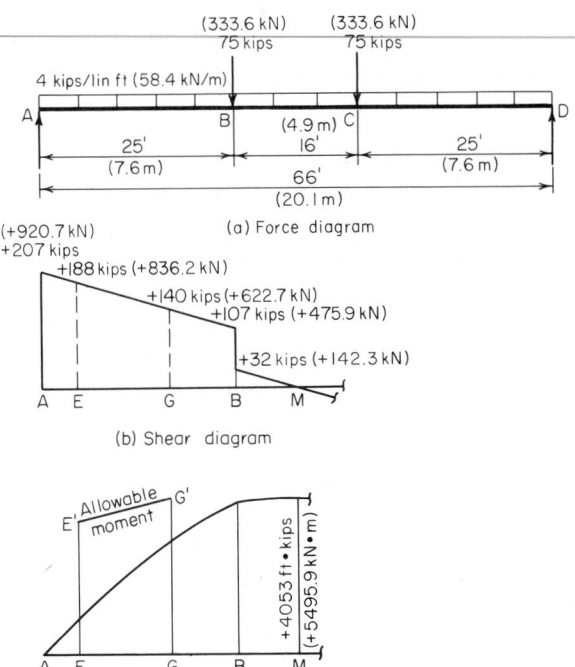

(a) Force diagram

(b) Shear diagram

(c) Bending-moment diagram

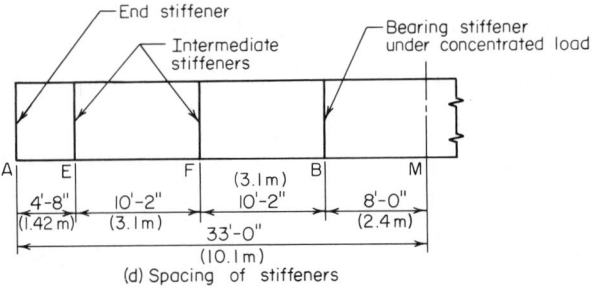

(d) Spacing of stiffeners

FIG. 69

2. Choose the web-plate dimensions

Since the total depth is limited to about 70 in (1778.0 mm), use a 68-in (1727.2-mm) deep web plate. Determine the plate thickness, using the *Specification* limits, which are a slenderness ratio h/t_w of 320. However, if an allowable bending stress of 22,000 lb/in² (151,690.0 kPa) is to be maintained, the *Specification* imposes an upper limit of $24,000/(22,000)^{0.5} = 162$. Then $t_w = h/162 = 68/162 = 0.42$ in (10.668 mm); use a ⁷⁄₁₆-in (11.112-mm) plate. Hence, the area of the web $A_w = 29.75$ in² (191.947 cm²).

3. Select the flange plates

Apply the approximation $A_f = Mc/(2fy^2) - A_w/6$, where y = distance from the neutral axis to the centroidal axis of the flange, in (mm).

Assume 1-in (25.4-mm) flange plates. Then $A_f = 4053(12)(35)/[2(22)(34.5)^2] - 29.75/6 = 27.54$ in^2 (177.688 cm^2). Try $22 \times 1\frac{1}{4}$ in (558.8 × 31.75 mm) plates with $A_f = 27.5$ in^2 (177.43 cm^2). The width-thickness ratio of projection $= 11/1.25 = 8.8 < 16$. This is acceptable.

Thus, the trial section will be one web plate $68 \times \frac{7}{16}$ in (1727 × 11.11 mm); two flange plates $22 \times 1\frac{1}{4}$ in (558.8 × 31.75 mm).

4. Test the adequacy of the trial section

For this test, compute the maximum flexural and shearing stresses. Thus, $I = (1/12)(0.438)(68)^3 + 2(27.5)(34.63)^2 = 77,440$ in^3 (1,269,241.6 cm^3); $f = Mc/I = 4053(12)(35.25)/77,440 = 22.1$ kips/in^2 (152.38 MPa). This is acceptable. Also, $v = 207/29.75 = 6.96 < 14.5$ kips/in^2 (99.98 MPa). This is acceptable. Hence, the trial section is satisfactory.

5. Determine the distance of the stiffeners from the girder ends

Refer to Fig. 69d for the spacing of the intermediate stiffeners. Establish the length of the end panel AE. The *Specification* stipulates that the smaller dimension of the end panel shall not exceed $11,000(0.438)/(6960)^{0.5} = 57.8 < 68$ in (1727.2 mm). Therefore, provide stiffeners at 56 in (1422.4 mm) from the ends.

6. Ascertain whether additional intermediate stiffeners are required

See whether stiffeners are required in the interval EB by applying the *Specification* criteria.

Stiffeners are not required when $h/t_w < 260$ and the shearing stress within the panel is below the value given by either of two equations in the *Specification*, whichever equation applies. Thus $EB = 396 - (56 + 96) = 244$ in (6197.6 mm); $h/t_w = 68/0.438 = 155 < 260$; this is acceptable. Also, $a/h = 244/68 = 3.59$.

In lieu of solving either of the equations given in the *Specification*, enter the table of a/h, h/t_w values given in the AISC *Manual* to obtain the allowable shear stress. Thus, with $a/h > 3$ and $h/t_w = 155$, $v_{\text{allow}} = 3.45$ kips/in^2 (23.787 MPa) from the table.

At E, $V = 207 - 4.67(4) = 188$ kips (836.2 kN); $v = 188/29.75 = 6.32$ kips/in^2 (43.576 MPa) > 3.45 kips/in^2 (23.787 MPa); therefore, intermediate stiffeners are required in EB.

7. Provide stiffeners, and investigate the suitability of their tentative spacing

Provide stiffeners at F, the center of EB. See whether this spacing satisfies the *Specification*. Thus $[260/(h/t_w)]^2 = (260/155)^2 = 2.81$; $a/h = 122/68 = 1.79 < 2.81$. This is acceptable.

Entering the table referred to in step 6 with $a/h = 1.79$ and $h/t_w = 155$ shows $v_{\text{allow}} = 7.85 > 6.32$. This is acceptable.

Before we conclude that the stiffener spacing is satisfactory, it is necessary to investigate the combined shearing and bending stress and the bearing stress in interval EB.

8. Analyze the combination of shearing and bending stress

This analysis should be made throughout EB in the light of the *Specification* requirements. The net effect is to reduce the allowable bending moment whenever $V > 0.6V_{\text{allow}}$. Thus, $V_{\text{allow}} = 7.85(29.75) = 234$ kips (1040.8 kN); and $0.6(234) = 140$ kips (622.7 kN).

In Fig. 69b, locate the boundary section G where $V = 140$ kips (622.7 kN). The allowable moment must be reduced to the left of G. Thus, $AG = (207 - 140)/4 = 16.75$ ft (5.105 m); $M_G = 2906$ ft·kips (3940.5 kN·m); $M_E = 922$ ft·kips (1250.2 kN·m). At G, $M_{\text{allow}} = 4053$ ft·kips (5495.8 kN·m). At E, $f_{\text{allow}} = [0.825 - 0.375(188/234)](36) = 18.9$ kips/in^2 (130.31 MPa); $M_{\text{allow}} = 18.9(77,440)/[35.25(12)] = 3460$ ft·kips (4691.8 kN·m).

In Fig. 69c, plot points E' and G' to represent the allowable moments and connect these points with a straight line. In all instances, $M < M_{\text{allow}}$.

9. Use an alternative procedure, if desired

As an alternative procedure in step 8, establish the interval within which $M > 0.75M_{\text{allow}}$ and reduce the allowable shear in accordance with the equation given in the *Specification*.

10. Compare the bearing stress under the uniform load with the allowable stress

The allowable stress given in the *Specification* is $f_{b,\text{allow}} = [5.5 + 4/(a/h)^2]10,000/(h/t_w)^2$ kips/in^2 (MPa), or, for this girder, $f_{b,\text{allow}} = (5.5 + 4/1.79^2)10,000/155^2 = 2.81$ kips/in^2 (19.374 MPa). Then $f_b = 4/[12(0.438)] = 0.76$ kips/in^2 (5.240 MPa). This is acceptable. The stiffener spacing in interval EB is therefore satisfactory in all respects.

11.   *Investigate the need for transverse stiffeners in the center interval*

Considering the interval BC, $V = 32$ kips (142.3 kN); $v = 1.08$ kips/in^2 (7.447 MPa); $a/h = 192/68 = 2.82 \simeq [260/(h/t_w)]^2$.

The *Manual* table used in step 6 shows that $v_{\text{allow}} > 1.08$ kips/in^2 (7.447 MPa); $f_{b,\text{allow}} = (5.5 + 4/2.82^2)10,000/155^2 = 2.49$ kips/in^2 (17.169 MPa) > 0.76 kips/in^2 (5.240 MPa). This is acceptable. Since all requirements are satisfied, stiffeners are not needed in interval BC.

12.   *Design the intermediate stiffeners in accordance with the* Specification

For the interval EB, the preceding calculations yield these values: $v = 6.32$ kips/in^2 (43.576 MPa); $v_{\text{allow}} = 7.85$ kips/in^2 (54.125 MPa). Enter the table mentioned in step 6 with $a/h = 1.79$ and $h/t_w = 155$ to obtain the percentage of web area, shown in italics in the table. Thus, A_{st} required $= 0.0745(29.75)(6.32/7.85) = 1.78$ in^2 (11.485 cm^2). Try two 4 $\times$ ¼ in (101.6 $\times$ 6.35 mm) plates; $A_{st} = 2.0$ in^2 (12.90 cm^2); width-thickness ratio $= 4/0.25 = 16$. This is acceptable. Also, $(h/50)^4 = (68/50)^4 = 3.42$ in^4 (142.351 cm^4); $I = (1/12)(0.25)(8.44)^3 = 12.52$ in^4 (521.121 cm^4) > 3.42 in^4 (142.351 cm^4). This is acceptable.

The stiffeners must be in intimate contact with the compression flange, but they may terminate 1¾ in (44.45 mm) from the tension flange. The connection of the stiffeners to the web must transmit the vertical shear specified in the *Specification*.

13.   *Design the bearing stiffeners at the supports*

Use the directions given in the *Specification*. The stiffeners are considered to act in conjunction with the tributary portion of the web to form a column section, as shown in Fig. 70. Thus, area of web $= 5.25(0.438) = 2.30$ in^2 (14.839 cm^2). Assume an allowable stress of 20 kips/in^2 (137.9 MPa). Then, plate area required $= 207/20 - 2.30 = 8.05$ in^2 (51.938 cm^2).

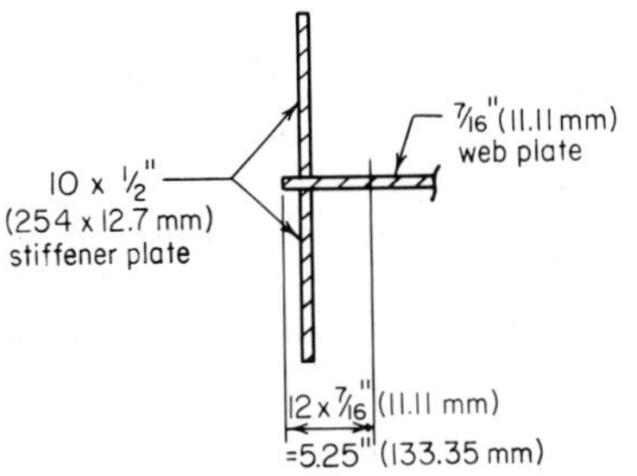

Try two plates 10 $\times$ ½ in (254.0 $\times$ 12.7 mm), and compute the column capacity of the section. Thus, $A = 2(10)(0.5) + 2.30 = 12.30$ in^2 (79.359 cm^2); $I = (1/12)(0.5)(20.44)^3 = 356$ in^4 (1.4818 dm^4); $r = (356/12.30)^{0.5} = 5.38$ in (136.652 mm); $L/r = 0.75(68)/5.38 = 9.5$.

Enter the table of slenderness ratio and allowable stress in the *Manual* with the slenderness ratio of 9.5, and obtain an allowable stress of 21.2 kips/in^2 (146.17 MPa). Then $f = 207/12.30 = 16.8$ kips/in^2 (115.84 MPa) < 21.2 kips/in^2 (146.17 MPa). This is acceptable.

10 x ½"
(254 x 12.7 mm)
stiffener plate

$\frac{7}{16}$"(11.11 mm)
web plate

12 x $\frac{7}{16}$"(11.11 mm)
$=5.25$"(133.35 mm)

FIG. 70   Effective column section.

Compute the bearing stress in the stiffeners. In computing the bearing area, assume that each stiffener will be clipped 1 in (25.4 mm) to clear the flange-to-web welding. Then $f = 207/[2(9)(0.5)] = 23$ kips/in^2 (158.6 MPa). The *Specification* provides an allowable stress of 33 kips/in^2 (227.5 MPa).

The 10 $\times$ ½ in (254.0 $\times$ 12.7 mm) stiffeners at the supports are therefore satisfactory with respect to both column action and bearing.

Steel Columns and Tension Members

The general remarks appearing at the opening of the previous part apply to this part as well.

A column is a compression member having a length that is very large in relation to its lateral dimensions. The *effective* length of a column is the distance between adjacent points of contraflexure in the buckled column or in the imaginary extension of the buckled column, as shown in Fig. 71. The column length is denoted by L, and the effective length by KL. Recommended design values of K are given in the AISC *Manual*.

The capacity of a column is a function of its effective length and the properties of its cross section. It therefore becomes necessary to formulate certain principles pertaining to the properties of an area.

Consider that the moment of inertia I of an area is evaluated with respect to a group of concurrent axes. There is a distinct value of I associated with each axis, as given by earlier equations in this section. The *major* axis is the one for which I is maximum; the *minor* axis is the one for which I is minimum. The major and minor axes are referred to collectively as the *principal* axes.

With reference to the equation given earlier, namely, $I_{x''} = I_{x'} \cos^2 \theta + I_{y'} \sin^2 \theta - P_{x'y'} \sin 2\theta$, the orientation of the principal axes relative to the given x' and y' axes is found by differentiating $I_{x''}$ with respect to θ, equating this derivative to zero, and solving for θ to obtain tan $2\theta = 2P_{x'y'}/(I_{y'} - I_{x'})$, Fig. 15.

The following statements are corollaries of this equation:

1. The principal axes through a given point are mutually perpendicular, since the two values of θ that satisfy this equation differ by 90°.

2. The product of inertia of an area with respect to its principal axes is zero.

3. Conversely, if the product of inertia of an area with respect to two mutually perpendicular axes is zero, these are principal axes.

4. An axis of symmetry is a principal axis, for the product of inertia of the area with respect to this axis and one perpendicular thereto is zero.

Let A_1 and A_2 denote two areas, both of which have a radius of gyration r with respect to a given axis. The radius of gyration of their composite area is found in this manner: $I_c = I_1 + I_2 = A_1 r^2 + A_2 r^2 = (A_1 + A_2)r^2$. But $A_1 + A_2 = A_c$. Substituting gives $I_c = A_c r^2$; therefore, $r_c = r$.

This result illustrates the following principle: If the radii of gyration of several areas with respect to a given axis are all equal, the radius of gyration of their composite area equals that of the individual areas.

The equation $I_x = \Sigma I_0 + \Sigma A k^2$, when applied to a single area, becomes $I_x = I_0 + A k^2$. Then $A r_x^2 = A r_0^2 + A k^2$, or $r_x = (r_0^2 + k^2)^{0.5}$. If the radius of gyration with respect to a centroidal axis is known, the radius of gyration with respect to an axis parallel thereto may be readily evaluated by applying this relationship.

The Euler equation for the strength of a slender column reveals that the member tends to buckle about the minor centroidal axis of its cross section. Consequently, all column design equations, both those for slender members and those for intermediate-length members, relate the capacity of the column to its minimum radius of gyration. The first step in the investigation of a column, therefore, consists in identifying the minor centroidal axis and evaluating the corresponding radius of gyration.

CAPACITY OF A BUILT-UP COLUMN

A compression member consists of two C15 × 40 channels laced together and spaced 10 in (254.0 mm) back to back with flanges outstanding, as shown in Fig. 72. What axial load may this member carry if its effective length is 22 ft (6.7 m)?

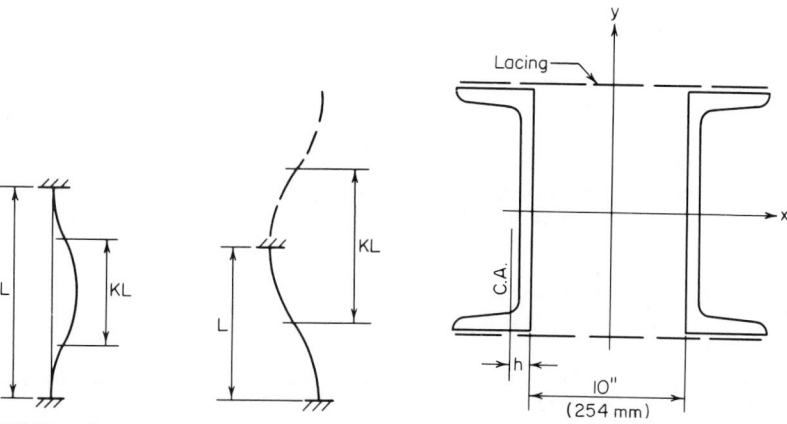

FIG. 71   Effective column lengths.   **FIG. 72**   Built-up column.

Calculation Procedure:

1. Record the properties of the individual channel

Since x and y are axes of symmetry, they are the principal centroidal axes. However, it is not readily apparent which of these is the minor axis, and so it is necessary to calculate both r_x and r_y. The symbol r, without a subscript, is used to denote the *minimum* radius of gyration, in inches (centimeters).

Using the AISC *Manual*, we see that the channel properties are $A = 11.70$ in^2 (75.488 cm^2); $h = 0.78$ in (19.812 mm); $r_1 = 5.44$ in (138.176 mm); $r_2 = 0.89$ in (22.606 mm).

2. Evaluate the minimum radius of gyration of the built-up section; determine the slenderness ratio

Thus, $r_x = 5.44$ in (138.176 mm); $r_y = (r_2^2 + 5.78^2)^{0.5} > 5.78$ in (146.812 mm); therefore, $r = 5.44$ in (138.176 mm); $KL/r = 22(12)/5.44 = 48.5$.

3. Determine the allowable stress in the column

Enter the *Manual* slenderness-ratio allowable-stress table with a slenderness ratio of 48.5 to obtain the allowable stress $f = 18.48$ kips/in^2 (127.420 MPa). Then, the column capacity $= P = Af = 2(11.70)(18.48) = 432$ kips (1921.5 kN).

CAPACITY OF A DOUBLE-ANGLE STAR STRUT

A star strut is composed of two $5 \times 5 \times \frac{3}{8}$ in (127.0 $\times$ 127.0 $\times$ 9.53 mm) angles intermittently connected by $\frac{3}{8}$-in (9.53-mm) batten plates in both directions. Determine the capacity of the member for an effective length of 12 ft (3.7 m).

Calculation Procedure:

1. Identify the minor axis

Refer to Fig. 73*a*. Since p and q are axes of symmetry, they are the principal axes; p is manifestly the minor axis because the area lies closer to p than q.

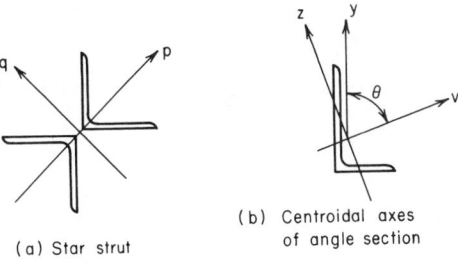

(a) Star strut

(b) Centroidal axes of angle section

FIG. 73

2. Determine r_v^2

Refer to Fig. 73*b*, where v is the major and z the minor axis of the angle section. Apply $I_{x'} = I_{y'} \cos^2 \theta + I_{y'} \sin^2 \theta - P_{x'y'} \sin 2\theta$, and set $P_{vz} = 0$ to obtain $r_y^2 = r_v^2 \cos^2 \theta + r_z^2 \sin^2 \theta$; therefore, $r_v^2 = r_y^2 \sec^2 \theta - r_z^2 \tan^2 \theta$. For an equal-leg angle, $\theta = 45°$, and this equation reduces to $r_v^2 = 2r_y^2 - r_z^2$.

3. Record the member area and computer r_v

From the *Manual*, $A = 3.61$ in^2 (23.291 cm^2); $r_y = 1.56$ in (39.624 mm); $r_z = 0.99$ in (25.146 mm); $r_v = (2 \times 1.56^2 - 0.99^2)^{0.5} = 1.97$ in (50.038 mm).

4. Determine the minimum radius of gyration of the built-up section; compute the strut capacity

Thus, $r = r_p = 1.97$ in (50.038 mm); $KL/r = 12(12)/1.97 = 73$. From the *Manual*, $f = 16.12$ kips/in^2 (766.361 MPa). Then $P = Af = 2(3.61)(16.12) = 116$ kips (515.97 kN).

SECTION SELECTION FOR A COLUMN WITH TWO EFFECTIVE LENGTHS

A 30-ft (9.2-m) long column is to carry a 200-kip (889.6-kN) load. The column will be braced about both principal axes at top and bottom and braced about its minor axis at midheight. Architectural details restrict the member to a nominal depth of 8 in (203.2 mm). Select a section of A242 steel by consulting the allowable-load tables in the AISC *Manual* and then verify the design.

Calculation Procedure:

1.   *Select a column section*

Refer to Fig. 74. The effective length with respect to the minor axis may be taken as 15 ft (4.6 m). Then $K_xL = 30$ ft (9.2 m) and $K_yL = 15$ ft (4.6 m).

The allowable column loads recorded in the *Manual* tables are calculated on the premise that the column tends to buckle about the minor axis. In the present instance, however, this premise is not necessarily valid. It is expedient for design purposes to conceive of a uniform-strength column, i.e., one for which K_x and K_y bear the same ratio as r_x and r_y, thereby endowing the column with an identical slenderness ratio with respect to the two principal axes.

Select a column section on the basis of the K_yL value; record the value of r_x/r_y of this section. Using linear interpolation in the *Manual* Table shows that a W8 × 40 column has a capacity of 200 kips (889.6 kN) when $K_yL = 15.3$ ft (4.66 m); at the bottom of the table it is found that $r_x/r_y = 1.73$.

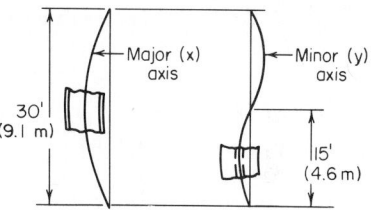

FIG. 74   Effective column lengths.

2.   *Compute the value of K_xL associated with a uniform-strength column, and compare this with the actual value*

Thus, $K_xL = 1.73(15.3) = 26.5$ ft (8.1 m) < 30 ft (9.2 m). The section is therefore inadequate.

3.   *Try a specific column section of larger size*

Trying W8 × 48, the capacity = 200 kips (889.6 kN) when $K_yL = 17.7$ ft (5.39 m). For uniform strength, $K_xL = 1.74(17.7) = 30.8 > 30$ ft (9.39 m > 9.2 m). The W8 × 48 therefore appears to be satisfactory.

4.   *Verify the design*

To verify the design, record the properties of this section and compute the slenderness ratios. For this grade of steel and thickness of member, the yield-point stress is 50 kips/in² (344.8 MPa), as given in the *Manual*. Thus, A = 14.11 in² (91.038 cm²); $r_x = 3.61$ in (91.694 mm); $r_y = 2.08$ in (52.832 mm). Then $K_xL/r_x = 30(12)/3.61 = 100$; $K_yL/r_y = 15(12)/2.08 = 87$.

5.   *Determine the allowable stress and member capacity*

From the *Manual*, $f = 14.71$ kips/in² (101.425 MPa) with a slenderness ratio of 100. Then $P = 14.11(14.71) = 208$ kips (925.2 kN). Therefore, use W8 × 48 because the capacity of the column exceeds the intended load.

STRESS IN COLUMN WITH PARTIAL RESTRAINT AGAINST ROTATION

The beams shown in Fig. 75a are rigidly connected to a W14 × 95 column of 28-ft (8.5-m) height that is pinned at its foundation. The column is held at its upper end by cross bracing lying in a plane normal to the web. Compute the allowable axial stress in the column in the absence of bending stress.

Calculation Procedure:

1.   *Draw schematic diagrams to indicate the restraint conditions*

Show these conditions in Fig. 75b. The cross bracing prevents sidesway at the top solely with respect to the minor axis, and the rigid beam-to-column connections afford partial fixity with respect to the major axis.

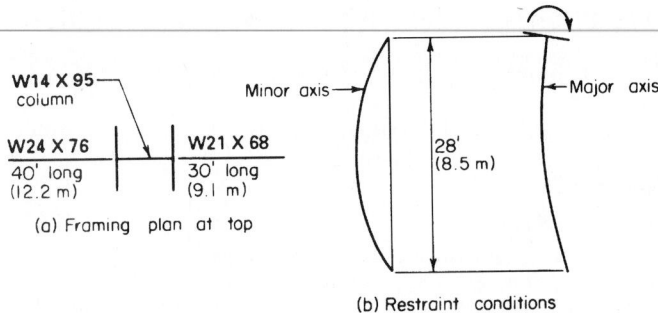

(a) Framing plan at top

(b) Restraint conditions

FIG. 75

2. *Record the I_x values of the column and beams*

| | I_x | |
Section	in⁴	cm⁴
W14 × 95	1064	44,287
W24 × 76	2096	87,242
W21 × 68	1478	61,519

3. *Calculate the rigidity of the column relative to that of the restraining members at top and bottom*

Thus, $I_c/L_c = 1064/28 = 38$. At the top, $\Sigma(I_g/L_g) = 2096/40 + 1478/30 = 101.7$. At the top, the rigidity $G_t = 38/101.7 = 0.37$.

In accordance with the instructions in the *Manual*, set the rigidity at the bottom $G_b = 10$.

4. *Determine the value of K_x*

Using the *Manual* alignment chart, determine that $K_x = 1.77$.

5. *Compute the slenderness ratio with respect to both principal axes, and find the allowable stress*

Thus, $K_xL/r_x = 1.77(28)(12)/6.17 = 96.4$; $K_yL/r_y = 28(12)/3.71 = 90.6$.

Using the larger value of the slenderness ratio, find from the *Manual* the allowable axial stress in the absence of bending $= f = 13.43$ kips/in² (92.600 MPa).

LACING OF BUILT-UP COLUMN

Design the lacing bars and end tie plates of the member in Fig. 76. The lacing bars will be connected to the channel flanges with ½-in (12.7-mm) rivets.

Calculation Procedure:

1. *Establish the dimensions of the lacing system to conform to the AISC Specification*

The function of the lacing bars and tie plates is to preserve the integrity of the column and to prevent local failure.

Refer to Fig. 76. The standard gage in 15-in (381.0-mm) channel = 2 in (50.8 mm), from the AISC *Manual*. Then $h = 14 < 15$ in (381.0 mm); therefore, use single lacing.

Try $\theta = 60°$; then, $v = 2(14) \cot 60° = 16.16$ in (410.5 mm). Set $v = 16$ in (406.4 mm); therefore, $d = 16.1$ in (408.94 mm). For the built-up section, $KL/r = 48.5$; for the single channel, $KL/r = 16/0.89 < 48.5$. This is acceptable. The spacing of the bars is therefore satisfactory.

2. *Design the lacing bars*

The lacing system must be capable of transmitting an assumed transverse shear equal to 2 percent of the axial load; this shear is carried by two bars, one on each side. A lacing bar is classified as a secondary member. To compute the transverse shear, assume that the column will be loaded to its capacity of 432 kips (1921.5 N).

Then force per bar = $\frac{1}{2}(0.02)(432)(16.1/14)$ = 5.0 kips (22.24 N). Also, $L/r \leq 140$; therefore, $r = 16.1/140 = 0.115$ in (2.9210 mm).

For a rectangular section of thickness t, $r = 0.289t$. Then $t = 0.115/0.289 = 0.40$ in (10.160 mm). Set $t = \frac{7}{16}$ in (11.11 mm); $r = 0.127$ in (3.226 mm); $L/r = 16.1/0.127 = 127$; $f = 9.59$ kips/in² (66.123 MPa); $A = 5.0/9.59 = 0.52$ in² (3.355 cm²). From the *Manual*, the minimum width required for ½-in (12.7 mm) rivets = 1½ in (38.1 mm). Therefore, use a flat bar 1½ × $\frac{7}{16}$ in (38.1 × 11.11 mm); $A = 0.66$ in² (4.258 cm²).

3. *Design the end tie plates in accordance with the* Specification

The minimum length = 14 in (355.6 mm); $t = 14/50 = 0.28$. Therefore, use plates 14 × $\frac{5}{16}$ in (355.6 × 7.94 mm). The rivet pitch is limited to six diameters, or 3 in (76.2 mm).

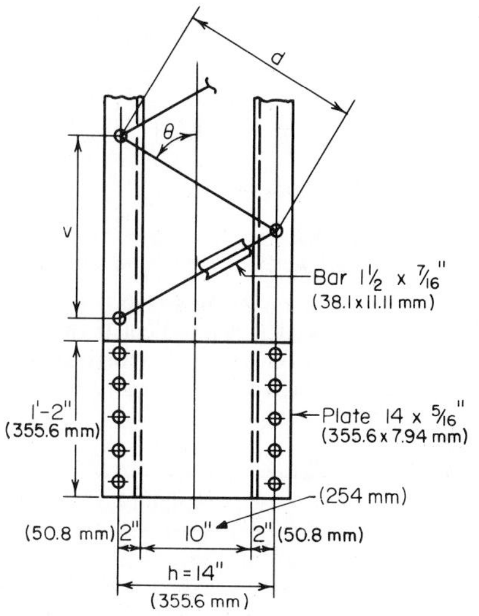

FIG. 76 Lacing and tie plates.

SELECTION OF A COLUMN WITH A LOAD AT AN INTERMEDIATE LEVEL

A column of 30-ft (9.2-m) length carries a load of 130 kips (578.2 kN) applied at the top and a load of 56 kips (249.1 kN) applied to the web at midheight. Select an 8-in (203.2-mm) column of A242 steel, using $K_xL = 30$ ft (9.2 m) and $K_yL = 15$ ft (4.6 m).

Calculation Procedure:

1. *Compute the effective length of the column with respect to the major axis*

The following procedure affords a rational method of designing a column subjected to a load applied at the top and another load applied approximately at the center. Let m = load at intermediate level, kips per total load, kips (kilonewtons). Replace the factor K with a factor K' defined by $K' = K(1 - m/2)^{0.5}$. Thus, for this column, $m = 56/186 = 0.30$. And $K_x'L = 30(1 - 0.15)^{0.5} = 27.6$ ft (8.41 m).

2. *Select a trial section on the basis of the K_yL value*

From the AISC *Manual* for a W8 × 40, capacity = 186 kips (827.3 kN) when $K_yL = 16.2$ ft (4.94 m) and $r_x/r_y = 1.73$.

3. *Determine whether the selected section is acceptable*

Compute the value of K_xL associated with a uniform-strength column, and compare this with the actual effective length. Thus, $K_xL = 1.73(16.2) = 28.0 > 27.6$ ft (8.41 m). Therefore, the W8 × 40 is acceptable.

DESIGN OF AN AXIAL MEMBER FOR FATIGUE

A web member in a welded truss will sustain precipitous fluctuations of stress caused by moving loads. The structure will carry three load systems having the following characteristics:

System	Force induced in member, kips (kN)		No. of times applied
	Maximum compression	Maximum tension	
A	46 (204.6)	18 (80.1)	60,000
B	40 (177.9)	9 (40.0)	1,000,000
C	32 (142.3)	8 (35.6)	2,500,000

The effective length of the member is 11 ft (3.4 m). Design a double-angle member.

Calculation Procedure:

1. Calculate for each system the design load, and indicate the yield-point stress on which the allowable stress is based

The design of members subjected to a repeated variation of stress is regulated by the AISC *Specification*. For each system, calculate the design load and indicate the yield-point stress on which the allowable stress is based. Where the allowable stress is less than that normally permitted, increase the design load proportionately to compensate for this reduction. Let $+$ denote tension and $-$ denote compression. Then

System	Design load, kips (kN)	Yield-point stress, kips/in^2 (MPa)
A	$-46 - \frac{2}{3}(18) = -58 \ (-257.9)$	36 (248.2)
B	$-40 - \frac{2}{3}(9) = -46 \ (-204.6)$	33 (227.5)
C	$1.5(-32 - \frac{3}{4} \times 8) = -57 \ (-253.5)$	33 (227.5)

2. Select a member for system A and determine if it is adequate for system C

From the AISC *Manual*, try two angles $4 \times 3\frac{1}{2} \times \frac{3}{8}$ in (101.6 × 88.90 × 9.53 mm), with long legs back to back; the capacity is 65 kips (289.1 kN). Then $A = 5.34$ in^2 (34.453 cm^2); $r = r_x = 1.25$ in (31.750 mm); $KL/r = 11(12)/1.25 = 105.6$.

From the *Manual*, for a yield-point stress of 33 kips/in^2 (227.5 MPa), $f = 11.76$ kips/in^2 (81.085 MPa). Then the capacity $P = 5.34(11.76) = 62.8$ kips (279.3 kN) > 57 kips (253.5 kN). This is acceptable. Therefore, use two angles $4 \times 3\frac{1}{2} \times \frac{3}{8}$ in (101.6 × 88.90 × 9.53 mm), long legs back to back.

INVESTIGATION OF A BEAM COLUMN

A W12 × 53 column with an effective length of 20 ft (6.1 m) is to carry an axial load of 160 kips (711.7 kN) and the end moments indicated in Fig. 77. The member will be secured against sidesway in both directions. Is the section adequate?

Calculation Procedure:

1. Record the properties of the section

The simultaneous set of values of axial stress and bending stress must satisfy the inequalities set forth in the AISC *Specification*.

The properties of the section are $A = 15.59$ in^2 (100.586 cm^2); $S_x = 70.7$ in^3 (1158.77 cm^3); $r_x = 5.23$ in (132.842 mm); $r_y = 2.48$ in (62.992 mm). Also, from the *Manual*, $L_c = 10.8$ ft (3.29 m); $L_u = 21.7$ ft (6.61 m).

2. Determine the stresses listed below

The stresses that must be determined are the axial stress f_a; the bending stress f_b; the axial stress F_a, which would be permitted in the absence of bending; and the bending stress F_b, which would be permitted in the absence of axial load. Thus, $f_a = 160/15.59 = 10.26$ kips/in^2 (70.742 MPa); $f_b = 31.5(12)/70.7 = 5.35$ kips/in^2 (36.888 MPa); $KL/r = 240/2.48 = 96.8$; therefore, $F_a = 13.38$ kips/in^2 (92.255 MPa); $L_u < KL < L_c$; therefore, $F_b = 22$ kips/in^2 (151.7 MPa). (Although this consideration is irrelevant in the present instance, note that the *Specification* establishes two maximum d/t ratios for a compact section. One applies to a beam, the other to a beam column.)

3. Calculate the moment coefficient C_m

Since the algebraic sign of the bending moment remains unchanged, M_1/M_2 is positive. Thus, $C_m = 0.6 + 0.4(15.2/31.5) = 0.793$.

4. Apply the appropriate criteria to test the adequacy of the section

Thus, $f_a/F_a = 10.26/13.38 = 0.767 > 0.15$. The following requirements therefore apply: $f_a/F_a + [C_m/(1 - f_a/F'_e)](f_b/F_b) \leq 1$; $f_a/(0.6f_y) + f_b/F_b \leq 1$, where $F'_e = 149,000/(KL/r)^2$ kips/in^2 and KL and r are evaluated with respect to the plane of bending.

Evaluating gives $F'_e = 149,000(5.23)^2/240^2 = 70.76$ kips/in^2 (487.890 MPa); $f_a/F'_e = 10.26/70.76 = 0.145$. Substituting in the first requirements equation yields $0.767 + (0.793/0.855)(5.35/22) = 0.993$. This is acceptable. Substituting in the second requirements equation, we find $10.26/22 + 5.35/22 = 0.709$. This section is therefore satisfactory.

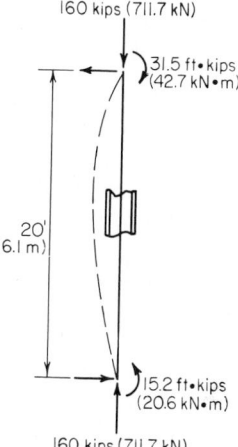

160 kips (711.7 kN)

31.5 ft·kips (42.7 kN·m)

20' (6.1 m)

15.2 ft·kips (20.6 kN·m)

160 kips (711.7 kN)

FIG. 77 Beam column.

APPLICATION OF BEAM-COLUMN FACTORS

For the previous calculation procedure, investigate the adequacy of the W12 × 53 section by applying the values of the beam-column factors B and a given in the AISC *Manual*.

Calculation Procedure:

1. Record the basic values of the previous calculation procedure

The beam-column factors were devised in an effort to reduce the labor entailed in analyzing a given member as a beam column when $f_a/F_a > 0.15$. They are defined by $B = A/S$ per inch (decimeter); $a = 0.149 \times 10^6 I$ in^4 (6201.9I dm^4).

Let P denote the applied axial load and P_{allow} the axial load that would be permitted in the absence of bending. The equations given in the previous procedure may be transformed to $P + BMC_m(F_a/F_b)a/[a - P(KL)^2] \leq P_{\text{allow}}$, and $PF_a/(0.6f_y) + BMF_a/F_b \leq P_{\text{allow}}$, where KL, B, and a are evaluated with respect to the plane of bending.

The basic values of the previous procedure are $P = 160$ kips (711.7 kN); $M = 31.5$ ft·kips (42.71 kN·m); $F_b = 22$ kips/in^2 (151.7 MPa); $C_m = 0.793$.

2. Obtain the properties of the section

From the *Manual* for a W12 × 53, $A = 15.59$ in^2 (100.587 cm^2); $B_x = 0.221$ per inch (8.70 per meter); $a_x = 63.5 \times 10^6$ in^4 (264.31 × 10^3 dm^4). Then when $KL = 20$ ft (6.1 m), $P_{\text{allow}} = 209$ kips (929.6 kN).

3. Substitue in the first transformed equation

Thus, $F_a = P_{\text{allow}}/A = 209/15.59 = 13.41$ kips/in^2 (92.461 MPa), $P(KL)^2 = 160(240)^2 = 9.22 \times 10^6$ kip·in^2 (2.648 × 10^4 kN·m^2), and $a_x/[a_x - P(KL)^2] = 63.5/(63.5 - 9.22) = 1.17$; then $160 + 0.221(31.5)(12)(0.793)(13.41/22)(1.17) = 207 < 209$ kips (929.6 kN). This is acceptable.

4. Substitute in the second transformed equation

Thus, $160(13.41/22) + 0.221(31.5)(12)(13.41/22) = 148 < 209$ kips (929.6 kN). This is acceptable. The W12 × 53 section is therefore satisfactory.

NET SECTION OF A TENSION MEMBER

The 7 × ¼ in (177.8 × 6.35 mm) plate in Fig. 78 carries a tensile force of 18,000 lb (80,064.0 N) and is connected to its support with three ¾-in (19.05-mm) rivets in the manner shown. Compute the maximum tensile stress in the member.

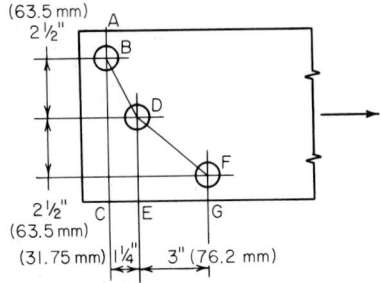

(63.5 mm)
2½"
A
B
D
F
2½"   C   E   G
(63.5 mm)
(31.75 mm) 1¼"   3" (76.2 mm)

FIG. 78

Calculation Procedure:

1. Compute the net width of the member at each section of potential rupture

The AISC *Specification* prescribes the manner of calculating the net section of a tension member. The effective diameter of the holes is considered to be ⅛ in (3.18 mm) greater than that of the rivets.

After computing the net width of each section, select the minimum value as the effective width. The *Specification* imposes an upper limit of 85 percent of the gross width.

Refer to Fig. 78: From B to D, $s = 1.25$ in (31.750 mm), $g = 2.5$ in (63.50 mm); from D to F, $s = 3$ in (76.2 mm), $g = 2.5$ in (63.50 mm); $w_{AC} = 7 - 0.875 = 6.12$ in (155.45 mm); $w_{ABDE} = 7 - 2(0.875) + 1.25^2/[4(2.5)] = 5.41$ in (137.414 mm); $w_{ABDFG} = 7 - 3(0.875) + 1.25^2/(4 \times 2.5) + 3^2/(4 \times 2.5) = 5.43$ in (137.922 mm); $w_{max} = 0.85(7) = 5.95$ in (151.13 mm). Selecting the lowest value gives $w_{eff} = 5.41$ in (137.414 mm).

2. Compute the tensile stress on the effective net section

Thus, $f = 18,000/[5.41(0.25)] = 13,300$ lb/in^2 (91,703.5 kPa).

DESIGN OF A DOUBLE-ANGLE TENSION MEMBER

The bottom chord of a roof truss sustains a tensile force of 141 kps (627.2 kN). The member will be spliced with ¾-in (19.05-mm) rivets as shown in Fig. 79a. Design a double-angle member and specify the minimum rivet pitch.

Calculation Procedure:

1. Show one angle in its developed form

Cut the outstanding leg, and position it to be coplanar with the other one, as in Fig. 79b. The gross width of the angle w_g is the width of the equivalent plate thus formed; it equals the sum of the legs of the angle less the thickness.

2. Determine the gross width in terms of the thickness

Assume tentatively that 2.5 rivet holes will be deducted to arrive at the net width. Express w_g in terms of the thickness t of each angle. Then net area required = $141/22 = 6.40$ in^2 (41.292 cm^2); also, $2t(w_g - 2.5 \times 0.875) = 6.40$; $w_g = 3.20/t + 2.19$.

3. Assign trial thickness values, and determine the gross width

Construct a tabulation of the computed values. Then select the most economical size of member. Thus

t, in (mm)	w_g, in (mm)	$w_g + t$, in (mm)	Available size, in (mm)	Area, in^2 (cm^2)
½ (12.7)	8.59 (218.186)	9.09 (230.886)	6 × 3½ × ½ (152.4 × 88.9 × 12.7)	4.50 (29.034)
⁷⁄₁₆ (11.11)	9.50 (241.300)	9.94 (252.476)	6 × 4 × ⁷⁄₁₆ (152.4 × 101.6 × 11.11)	4.18 (26.969)
⅜ (9.53)	10.72 (272.228)	11.10 (281.940)	None	

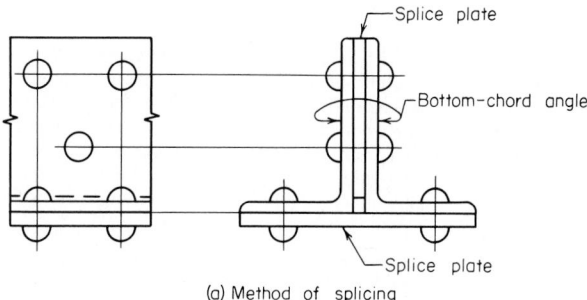

(a) Method of splicing

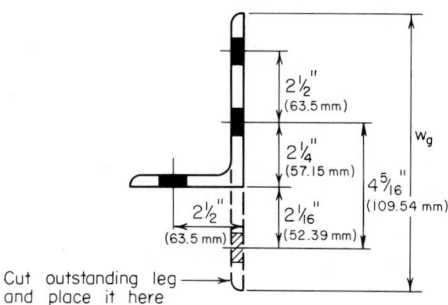

(b) Development of angle for net section

FIG. 79

The most economical member is the one with the least area. Therefore, use two angles 6 × 4 × $\frac{7}{16}$ in (152.4 × 101.6 × 11.11 mm).

4. Record the standard gages

Refer to the *Manual* for the standard gages, and record the values shown in Fig. 79*b*.

5. Establish the rivet pitch

Find the minimum value of s to establish the rivet pitch. Thus, net width required = $\frac{1}{2}[6.40/(7/16)] = 7.31$ in (185.674 mm); gross width = $6 + 4 - 0.44 = 9.56$ in (242.824 mm). Then 9.56 $- 3(0.875) + s^2/(4 \times 2.5) + s^2/(4 \times 4.31) = 7.31$; $s = 1.55$ in (39.370 mm).

For convenience, use the standard pitch of 3 in (76.2 mm). This results in a net width of 7.29 in (185.166 mm); the deficiency is negligible.

Plastic Design of Steel Structures

Consider that a structure is subjected to a gradually increasing load until it collapses. When the yield-point stress first appears, the structure is said to be in a state of *initial yielding*. The load that exists when failure impends is termed the *ultimate load*.

In elastic design, a structure has been loaded to capacity when it attains initial yielding, on the theory that plastic deformation would annul the utility of the structure. In plastic design, on the other hand, it is recognized that a structure may be loaded beyond initial yielding if:

1. The tendency of the fiber at the yield-point stress toward plastic deformation is resisted by the adjacent fibers.

2. Those parts of the structure that remain in the elastic-stress range are capable of supporting this incremental load.

The ultimate load is reached when these conditions cease to exist and thus the structure collapses.

Thus, elastic design is concerned with an allowable *stress*, which equals the yield-point stress divided by an appropriate factor of safety. In contrast, plastic design is concerned with an allowable *load*, which equals the ultimate load divided by an appropriate factor called the *load factor*. In reality, however, the distinction between elastic and plastic design has become rather blurred because specifications that ostensibly pertain to elastic design make covert concessions to plastic behavior. Several of these are underscored in the calculation procedures that follow.

In the plastic analysis of flexural members, the following simplifying assumptions are made:

1. As the applied load is gradually increased, a state is eventually reached at which all fibers at the section of maximum moment are stressed to the yield-point stress, in either tension or compression. The section is then said to be in a state of *plastification*.

2. While plastification is proceeding at one section, the adjacent sections retain their linear-stress distribution.

   Although the foregoing assumptions are fallacious, they introduce no appreciable error.

   When plastification is achieved at a given section, no additional bending stress may be induced in any of its fibers, and the section is thus rendered impotent to resist any incremental bending moment. As loading continues, the beam behaves as if it had been constructed with a hinge at the given section. Consequently, the beam is said to have developed a *plastic hinge* (in contradistinction to a true hinge) at the plastified section.

   The *yield moment* M_y of a beam section is the bending moment associated with initial yielding. The plastic moment M_p is the bending moment associated with plastification.

   The *plastic modulus* Z of a beam section, which is analogous to the section modulus used in elastic design, is defined by $Z = M_p/f_y$, where f_y denotes the yield-point stress. The *shape factor* SF is the ratio of M_p to M_y, being so named because its value depends on the shape of the section. Then SF $= M_p/M_y = f_yZ/(f_yS) = Z/S$.

   In the following calculation procedures, it is understood that the members are made of A36 steel.

ALLOWABLE LOAD ON BAR SUPPORTED BY RODS

A load is applied to a rigid bar that is symmetrically supported by three steel rods as shown in Fig. 80. The cross-sectional areas of the rods are: rods A and C, 1.2 in^2 (7.74 cm^2); rod B, 1.0 in^2 (6.45 cm^2). Determine the maximum load that may be applied, (*a*) using elastic design with an allowable stress of 22,000 lb/in^2 (151,690.0 kPa); (*b*) using plastic design with a load factor of 1.85.

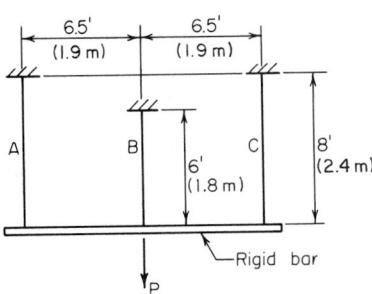

FIG. 80

Calculation Procedure:

1. *Express the relationships among the tensile stresses in the rods*

The symmetric disposition causes the bar to deflect vertically without rotating, thereby elongating the three rods by the same amount. As the first method of solving this problem, assume that the load is gradually increased from zero to its allowable value.

Expressing the relationships among the tensile stresses, we have $\Delta L = s_A L_a/E = s_B L_B/E = s_C L_C/E$; therefore, $s_A = s_C$, and $s_A = s_B L_B/L_A = 0.75 s_B$ for this arrangement of rods. Since s_B is the maximum stress, the allowable stress first appears in rod B.

2. *Evaluate the stresses at the instant the load attains its allowable value*

Calculate the load carried by each rod, and sum these loads to find P_{allow}. Thus $s_B = 22,000$ lb/in^2 (151,690.0 kPa); $s_A = 0.75(22,000) = 16,500$ lb/in^2 (113,767.5 kPa); $P_A = P_C = 16,500(1.2) = 19,800$ lb (88,070.4 N); $P_B = 22,000(1.0) = 22,000$ lb (97,856.0 N); $P_{\text{allow}} = 2(19,800) + 22,000 = 61,600$ lb (273,996.8 N).

Next, consider that the load is gradually increased from zero to its ultimate value. When rod B attains its yield-point stress, its tendency to deform plastically is inhibited by rods A and C because the rigidity of the bar constrains the three rods to elongate uniformly. The structure therefore remains stable as the load is increased beyond the elastic range until rods A and C also attain their yield-point stress.

3. Find the ultimate load

To find the ultimate load P_u, equate the stress in each rod to f_y, calculate the load carried by each rod, and sum these loads to find the ultimate load P_u. Thus, $P_A = P_C = 36,000(1.2) = 43,200$ lb (192,153.6 N); $P_B = 36,000(1.0) = 36,000$ lb (160,128.0 N); $P_u = 2(43,200) + 36,000 = 122,400$ lb (544,435.2 N).

4. Apply the load factor to establish the allowable load

Thus, $P_{\text{allow}} = P_u/\text{LF} = 122,400/1.85 = 66,200$ lb (294,457.6 N).

DETERMINATION OF SECTION SHAPE FACTORS

Without applying the equations and numerical values of the plastic modulus given in the AISC *Manual*, determine the shape factor associated with a rectangle, a circle, and a W16 × 40. Explain why the circle has the highest and the W section the lowest factor of the three.

Calculation Procedure:

1. Calculate M_y for each section

Use the equation $M_y = Sf_y$ for each section. Thus, for a rectangle, $M_y = bd^2f_y/6$. For a circle, using the properties of a circle as given in the *Manual*, we find $M_y = \pi d^3 f_y/32$. For a W16 × 40, $A = 11.77$ in^2 (75.940 cm^2), $S = 64.4$ in^3 (1055.52 cm^3), and $M_y = 64.4f_y$.

2. Compute the resultant forces associated with plastification

In Fig. 81, the resultant forces are C and T. Once these forces are known, their action lines and M_p should be computed.

Thus, for a rectangle, $C = bdf_y/2$, $a = d/2$, and $M_p = aC = bd^2f_y/4$. For a circle, $C = \pi d^2 f_y/8$, $a = 4d/(3\pi)$, and $M_p = aC = d^3f_y/6$. For a W16 × 40, $C = \frac{1}{2}(11.77f_y) = 5.885f_y$.

To locate the action lines, refer to the *Manual* and note the position of the centroidal axis of the WT8 × 20 section, i.e., a section half the size of that being considered. Thus, $a = 2(8.00 - 1.82) = 12.36$ in (313.944 mm); $M_p = aC = 12.36(5.885f_y) = 72.7f_y$.

3 Divide M_p by M_y to obtain the shape factor

For a rectangle, SF $= (bd^2/4)/(bd^2/6) = 1.50$. For a circle, SF $= (d^3/6)/(\pi d^3/32) = 1.70$. For a WT16 × 40, SF $= 72.7/64.4 = 1.13$.

4. Explain the relative values of the shape factor

To explain the relative values of the shape factor, express the resisting moment contributed by a given fiber at plastification and at initial yielding,

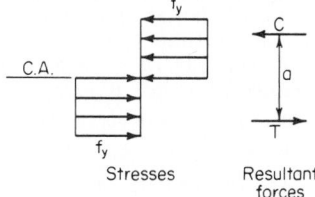

FIG. 81 Conditions at section of plastification.

and compare the results. Let dA denote the area of the given fiber and y its distance from the neutral axis. At plastification, $dM_p = f_y y dA$. At initial yielding, $f = f_y y/c$; $dM_y = f_y y^2 dA/c$; $dM_p/dM_y = c/y$.

By comparing a circle and a hypothetical W section having the same area and depth, the circle is found to have a larger shape factor because of its relatively low values of y.

As this analysis demonstrates, the process of plastification mitigates the detriment that accrues from placing any area near the neutral axis, since the stress at plastification is independent of the position of the fiber. Consequently, a section that is relatively inefficient with respect to flexure has a relatively high shape factor. The AISC *Specification* for elastic design implicitly recognizes the value of the shape factor by assigning an allowable bending stress of $0.75f_y$ to rectangular bearing plates and $0.90f_y$ to pins.

DETERMINATION OF ULTIMATE LOAD BY THE STATIC METHOD

The W18 × 45 beam in Fig. 82a is simply supported at A and fixed at C. Disregarding the beam weight, calculate the ultimate load that may be applied at B (a) by analyzing the behavior of the beam during its two phases; (b) by analyzing the bending moments that exist at impending collapse. (The first part of the solution illustrates the postelastic behavior of the member.)

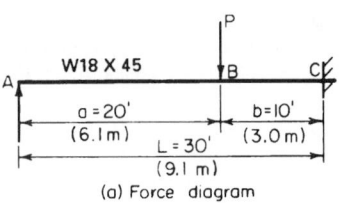

(a) Force diagram

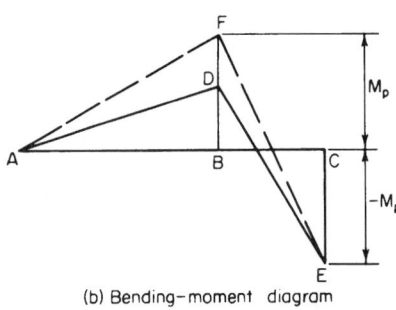

(b) Bending-moment diagram

FIG. 82

Calculation Procedure:

1. Calculate the ultimate-moment capacity of the member

Part a: As the load is gradually increased from zero to its ultimate value, the beam passes through two phases. During phase 1, the *elastic phase*, the member is restrained against rotation at C. This phase terminates when a plastic hinge forms at that end. During phase 2—the *postelastic*, or *plastic, phase*—the member functions as a simply supported beam. This phase terminates when a plastic hinge forms at B, since the member then becomes unstable.

Using data from the AISC *Manual*, we have $Z = 89.6$ in^3 (1468.54 cm^3). Then $M_p = f_y Z = 36(89.6)/12 = 268.8$ ft·kips (364.49 kN·m).

2. Calculate the moment BD

Let P_1 denote the applied load at completion of phase 1. In Fig. 82b, construct the bending-moment diagram ADEC corresponding to this load. Evaluate P_1 by applying the equations for case 14 in the AISC *Manual*. Calculate the moment BD. Thus, $CE = -ab(a + L)P_1/(2L^2) = -20(10)(50)P_1/[2(900)] = -268.8$; $P_1 = 48.38$ kips (215.194 kN); $BD = ab^2(a + 2L)P_1/(2L^3) = 20(100)(80)(48.38)/[2(27,000)] = 143.3$ ft·kips (194.31 kN·m).

3. Determine the incremental load at completion of phase 2

Let P_2 denote the incremental applied load at completion of phase 2, i.e., the actual load on the beam minus P_1. In Fig. 82b, construct the bending-moment diagram AFEC that exists when phase 2 terminates. Evaluate P_2 by considering the beam as simply supported. Thus, $BF = 268.8$ ft·kips (364.49 kN·m); $DF = 268.8 - 143.3 = 125.5$ ft·kips (170.18 kN·m); but $DF = abP_2/L = 20(10)P_2/30 = 125.5$; $P_2 = 18.82$ kips (83.711 kN).

4. Sum the results to obtain the ultimate load

Thus, $P_u = 48.38 + 18.82 = 67.20$ kips (298.906 kN).

5. Construct the force and bending-moment diagrams for the ultimate load

Part b: The following considerations are crucial: The bending-moment diagram always has vertices at B and C, and formation of two plastic hinges will cause failure of the beam. Therefore, the plastic moment occurs at B and C at impending failure. *The sequence in which the plastic hinges are formed at these sections is immaterial.*

These diagrams are shown in Fig. 83. Express M_p in terms of P_u, and evaluate P_u. Thus, $BF = 20R_A = 268.8$; therefore, $R_A = 13.44$ kips (59.781 kN). Also, $CE = 30R_A - 10P_u = 30 \times 13.44 - 10P_u = -268.8$; $P_u = 67.20$ kips (298.906 kN).

Here is an alternative method: $BF = (abP_u/L) - aM_p/L = M_p$, or $20(10)P_u/30 = 50M_p/30$; $P_u = 67.20$ kips (298.906 kN).

This solution method used in part b is termed the *static*, or *equilibrium*, method. As this solution demonstrates, it is unnecessary to trace the stress history of the member as it passes through its successive phases, as was done in part a; the analysis can be confined to the conditions that exist

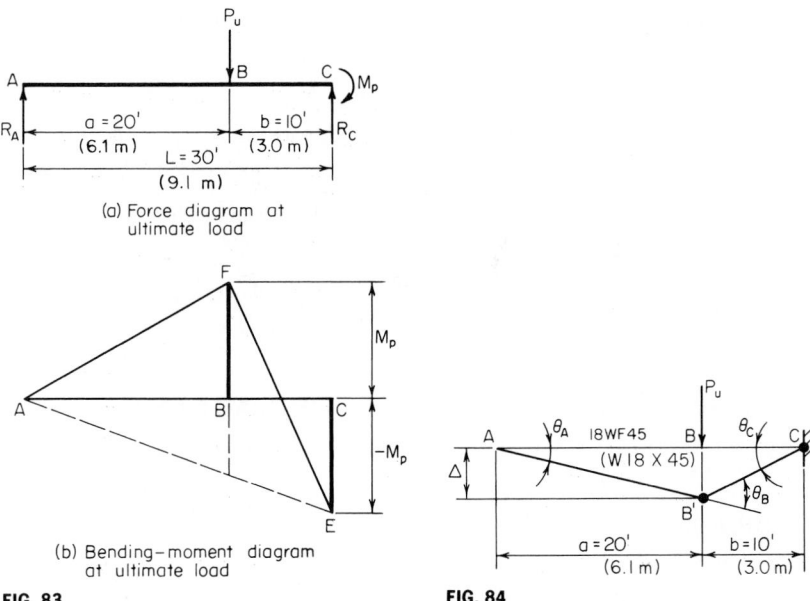

(a) Force diagram at ultimate load

(b) Bending−moment diagram at ultimate load

FIG. 83

FIG. 84

at impending failure. This procedure also illustrates the following important characteristics of plastic design:

1. Plastic design is far simpler than elastic design.

2. Plastic design yields results that are much more reliable than those secured through elastic design. For example, assume that the support at C does not completely inhibit rotation at that end. This departure from design conditions will invalidate the elastic analysis but will in no way affect the plastic analysis.

DETERMINING THE ULTIMATE LOAD BY THE MECHANISM METHOD

Use the mechanism method to solve the problem given in the previous calculation procedure.

Calculation Procedure:

1. Indicate, in hyperbolic manner, the virtual displacement of the member from its initial to a subsequent position

To the two phases of beam behavior previously considered, it is possible to add a third. Consider that when the ultimate load is reached, the member is subjected to an incremental deflection. This will result in collapse, but the behavior of the member can be analyzed during an infinitesimally small deflection from its stable position. This is termed a *virtual* deflection, or displacement.

Since the member is incapable of supporting any load beyond that existing at completion of phase 2, this virtual deflection is not characterized by any change in bending stress. Rotation therefore occurs solely at the real and plastic hinges. Thus, during phase 3, the member behaves as a mechanism (i.e., a constrained chain of pin-connected rigid bodies, or links).

In Fig. 84, indicate, in hyperbolic manner, the virtual displacement of the member from its initial position ABC to a subsequent position $AB'C$. Use dots to represent plastic hinges. (The initial position may be represented by a straight line for simplicity because the analysis is concerned solely with the deformation that occurs *during* phase 3.)

2.   *Express the linear displacement under the load and the angular displacement at every plastic hinge*

Use a convenient unit to express these displacements. Thus, $\Delta = a\theta_A = b\theta_C$; therefore, $\theta_C = a\theta_A/b = 2\theta_A$; $\theta_B = \theta_A + \theta_C = 3\theta_A$.

3.   *Evaluate the external and internal work associated with the virtual displacement*

The work performed by a constant force equals the product of the force and its displacement parallel to its action line. Also, the work performed by a constant moment equals the product of the moment and its angular displacement. Work is a positive quantity when the displacement occurs in the direction of the force or moment. Thus, the external work $W_E = P_u\Delta = P_u a\theta_A = 20P_u\theta_A$. And the internal work $W_I = M_p(\theta_B + \theta_C) = 5M_p\theta_A$.

4.   *Equate the external and internal work to evaluate the ultimate load*

Thus, $20P_u\theta_A = 5M_p\theta_A$; $P_u = (5/20)(268.8) = 67.20$ kips (298.906 kN).

The solution method used here is also termed the *virtual-work*, or *kinematic*, method.

ANALYSIS OF A FIXED-END BEAM UNDER CONCENTRATED LOAD

If the beam in the two previous calculation procedures is fixed at A as well as at C, what is the ultimate load that may be applied at B?

Calculation Procedure:

1.   *Determine when failure impends*

When hinges form at A, B, and C, failure impends. Repeat steps 3 and 4 of the previous calculation procedure, modifying the calculations to reflect the revised conditions. Thus $W_E = 20P_u\theta_A$; $W_I = M_p(\theta_A + \theta_B + \theta_C) = 6M_p\theta_A$; $20P_u\theta_A = 6M_p\theta_A$; $P_u = (6/20)(268.8) = 80.64$ kips (358.687 kN).

2.   *Analyze the phases through which the member passes*

This member passes through three phases until the ultimate load is reached. Initially, it behaves as a beam fixed at both ends, then as a beam fixed at the left end only, and finally as a simply supported beam. However, as already discussed, these considerations are extraneous in plastic design.

ANALYSIS OF A TWO-SPAN BEAM WITH CONCENTRATED LOADS

The continuous W18 × 45 beam in Fig. 85 carries two equal concentrated loads having the locations indicated. Disregarding the weight of the beam, compute the ultimate value of these loads, using both the static and the mechanism method.

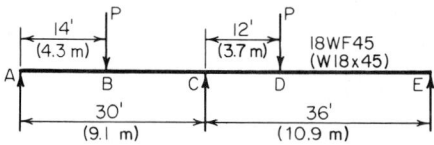

FIG. 85

Calculation Procedure:

1.   *Construct the force and bending-moment diagrams*

The continuous beam becomes unstable when a plastic hinge forms at C and at another section. The bending-moment diagram has vertices at B and D, but it is not readily apparent at which of these sections the second hinge will form. The answer is found by assuming a plastic hinge at B and at D, in turn, computing the corresponding value of P_u, and selecting the lesser value as the correct result. Part a will use the static method; part b, the mechanism method.

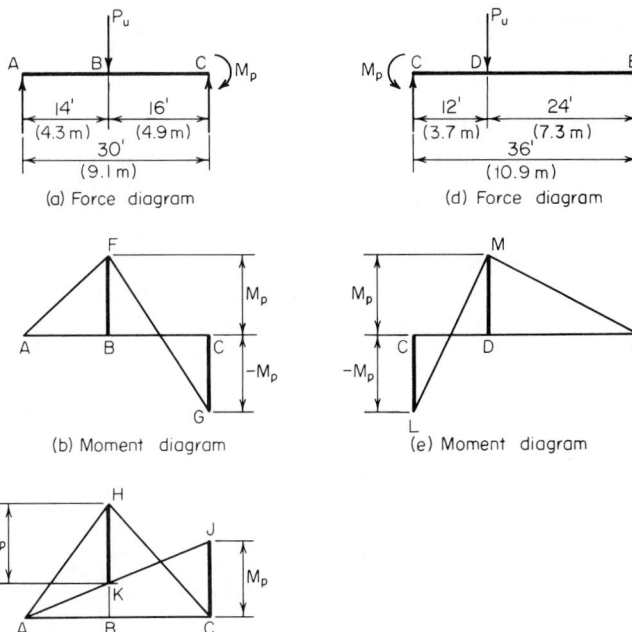

(a) Force diagram

(d) Force diagram

(b) Moment diagram

(e) Moment diagram

(c) Moment diagram by parts

FIG. 86

Assume, for part a, a plastic hinge at B and C. In Fig. 86, construct the force diagram and bending-moment diagram for span AC. The moment diagram may be drawn in the manner shown in Fig. 86b or c, whichever is preferred. In Fig. 86c, ACH represents the moments that would exist in the absence of restraint at C, and ACJ represents, in absolute value, the moments induced by this restraint. Compute the load P_u associated with the assumed hinge location. From previous calculation procedures, $M_p = 268.8$ ft·kips (364.49 kN·m); then $M_B = 14 \times 16P_u/30 - 14M_p/30 = M_p$; $P_u = 44(268.8)/224 = 52.8$ kips (234.85 kN).

2. Assume another hinge location and compute the ultimate load associated with this location

Now assume a plastic hinge at C and D. In Fig. 86, construct the force diagram and bending-moment diagram for CE. Computing the load P_u associated with this assumed location, we find $M_D = 12 \times 24P_u/36 - 24M_p/36 = M_p$; $P_u = 60(268.8)/288 = 56.0$ kips (249.09 kN).

3. Select the lesser value of the ultimate load

The correct result is the lesser of these alternative values, or $P_u = 52.8$ kips (234.85 kN). At this load, plastic hinges exist at B and C but not at D.

4. For the mechanism method, assume a plastic-hinge location

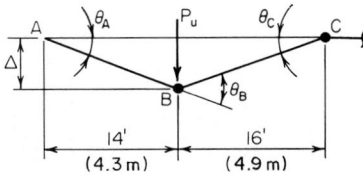

FIG. 87

It will be assumed that plastic hinges are located at B and C (Fig. 87). Evaluate P_u. Thus, $\theta_C = 14\theta_A/16$; $\theta_B = 30\theta_A/16$; $\Delta = 14\theta_A$; $W_E = P_u\Delta = 14P_u\theta_A$; $W_I = M_p(\theta_B + \theta_C) = 2.75M_p\theta_A$; $14P_u\theta_A = 2.75M_p\theta_A$; $P_u = 52.8$ kips (234.85 kN).

5. Assume a plastic hinge at another location

Select C and D for the new location. Repeat the above procedure. The result will be identical with that in step 2.

SELECTION OF SIZES FOR A CONTINUOUS BEAM

Using a load factor of 1.70, design the member to carry the working loads (with beam weight included) shown in Fig. 88a. The maximum length that can be transported is 60 ft (18.3 m).

Calculation Procedure:

1. Determine the ultimate loads to be supported

Since the member must be spliced, it will be economical to adopt the following design:

a. Use the particular beam size required for each portion, considering that the two portions will fail simultaneously at ultimate load. Therefore, three plastic hinges will exist at failure—one at the interior support and one in the interior of each span.

b. Extend one beam beyond the interior support, splicing the member at the point of contraflexure in the adjacent span. Since the maximum simple-span moment is greater for AB than for BC, it is logical to assume that for economy the left beam rather than the right one should overhang the support.

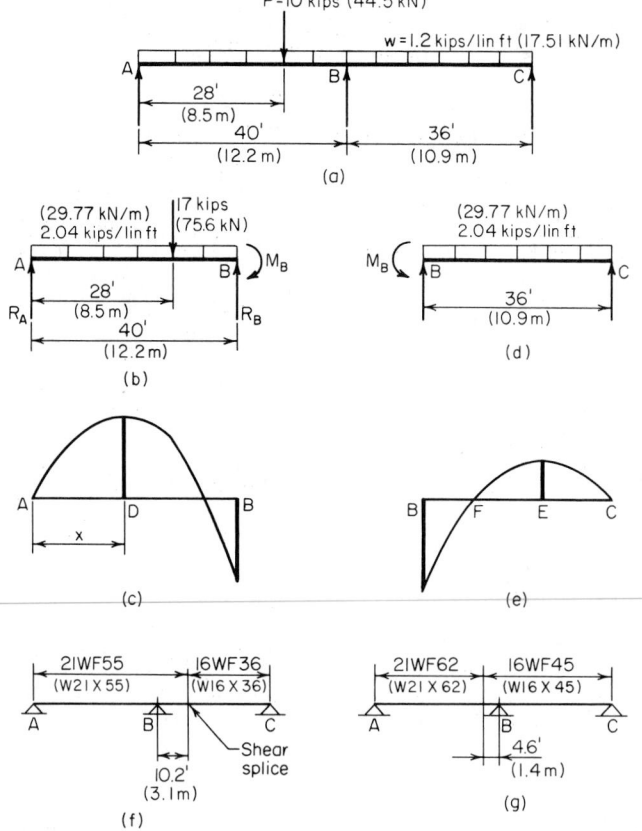

FIG. 88

Multiply the working loads by the load factor to obtain the ulitmate loads to be supported. Thus, $w = 1.2$ kips/lin ft (17.51 kN/m); $w_u = 1.70(1.2) = 2.04$ kips/lin ft (29.77 kN/m); $P = 10$ kips (44.5 kN); $P_u = 1.70(10) = 17$ kips (75.6 kN).

2. Construct the ultimate-load and corresponding bending-moment diagram for each span

Set the maximum positive moment M_D in span AB and the negative moment at B equal to each other in absolute value.

3. Evaluate the maximum positive moment in the left span

Thus, $R_A = 45.9 - M_B/40$; $x = R_A/2.04$; $M_D = \frac{1}{2}R_A x = R_A^2/4.08 = M_B$. Substitue the value of R_A and solve. Thus, $M_D = 342$ ft·kips (463.8 kN·m).

An indirect but less cumbersome method consists of assigning a series of trial values to M_B and calculating the corresponding value of M_D, continuing the process until the required equality is obtained.

4. Select a section to resist the plastic moment

Thus, $Z = M_p/f_y = 342(12)/36 = 114$ in^3 (1868.5 cm^3). Referring to the AISC *Manual*, use a W21 × 55 with $Z = 125.4$ in^3 (2055.31 cm^3).

5. Evaluate the maximum positive moment in the right span

Equate M_B to the true plastic-moment capacity of the W21 × 55. Evaluate the maximum positive moment M_E in span BC, and locate the point of contraflexure. Thus, $M_B = -36(125.4)/12 = -376.2$ ft·kips (-510.13 kN·m); $M_E = 169.1$ ft·kips (229.30 kN·m); $BF = 10.2$ ft (3.11 m).

6. Select a section to resist the plastic moment

The moment to be resisted is M_E. Thus, $Z = 169.1(12)/36 = 56.4$ in^3(924.40 cm^3). Use W16 × 36 with $Z = 63.9$ in^3 (1047.32 cm^3).

The design is summarized in Fig. 88f. By inserting a hinge at F, the continuity of the member is destroyed and its behavior is thereby modified under gradually increasing load. However, the ultimate-load conditions, which constitute the only valid design criteria, are not affected.

7. Alternatively, design the member with the right-hand beam overhanging the support

Compare the two designs for economy. The latter design is summarized in Fig. 88g. The total beam weight associated with each scheme is as shown in the following table.

Design 1	Design 2
55(50.2) = 2,761 lb (12,280.9 N)	62(35.4) = 2,195 lb (9,763.4 N)
36(25.8) = 929 lb (4,132.2 N)	45(40.6) = 1,827 lb (8,126.5 N)
Total 3,690 lb (16,413.1 N)	4,022 lb (17,889.9 N)

For completeness, the column sizes associated with the two schemes should also be compared.

MECHANISM-METHOD ANALYSIS OF A RECTANGULAR PORTAL FRAME

Calculate the plastic moment and the reactions at the supports at ultimate load of the prismatic frame in Fig. 89a. Use a load factor of 1.85, and apply the mechanism method.

Calculation Procedure:

1. Compute the ultimate loads to be resisted

There are three potential modes of failure to consider:

a. Failure of the beam BD through the formation of plastic hinges at B, C, and D (Fig. 89b)

b. Failure by sidesway through the formation of plastic hinges at B and D (Fig. 89c)

c. A composite of the foregoing modes of failure, characterized by the formation of plastic hinges at C and D

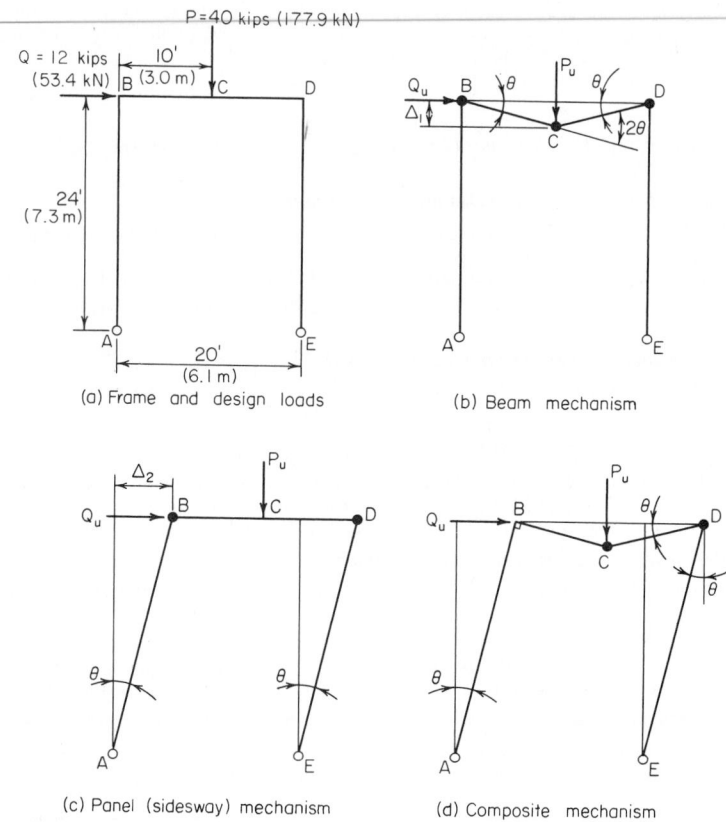

(a) Frame and design loads   (b) Beam mechanism

(c) Panel (sidesway) mechanism   (d) Composite mechanism

FIG. 89

Since the true mode of failure is not readily discernible, it is necessary to analyze each of the foregoing. The true mode of failure is the one that yields the highest value of M_p.

Although the work quantities are positive, it is advantageous to supply each angular displacement with an algebraic sign. A rotation is considered positive if the angle on the interior side of the frame increases. The algebraic sum of the angular displacements must equal zero.

Computing the ultimate loads to be resisted yields P_u = 1.85(40) = 74 kips (329.2 kN); Q_u = 1.85(12) = 22.2 kips (98.75 kN).

2.   *Assume the mode of failure in Fig. 89b and compute M_p*

Thus, Δ_1 = 10θ; W_E = 74(10θ) = 740θ. Then indicate in a tabulation, such as that shown here, where the plastic moment occurs. Include all significant sections for completeness.

Section	Angular displacement	Moment	W_I
A			
B	$-\theta$	M_p	$M_p\theta$
C	$+2\theta$	M_p	$2M_p\theta$
D	$-\theta$	M_p	$M_p\theta$
E	$\cdot\ \cdot$	$\cdot\ \cdot$	$\cdot\ \cdot$
Total			$4M_p\theta$

Then $4M_p\theta = 740\theta$; $M_p = 185$ ft·kips (250.9 kN·m).

3. Repeat the foregoing procedure for failure by sidesway

Thus, $\Delta_2 = 24\theta$; $W_E = 22.2(24\theta) = 532.8\theta$.

Section	Angular displacement	Moment	W_I
A	$-\theta$		
B	$+\theta$	M_p	$M_p\theta$
C			
D	$-\theta$	M_p	$M_p\theta$
E	$+\theta$		
Total			$2M_p\theta$

Then $2M_p\theta = 532.8\theta$; $M_p = 266.4$ ft·kips (361.24 kN·m).

4. Assume the composite mode of failure and compute M_p

Since this results from superposition of the two preceding modes, the angular displacements and the external work may be obtained by adding the algebraic values previously found. Thus, $W_E = 740\theta + 532.8\theta = 1272.8\theta$. Then the tabulation is as shown:

Section	Angular displacement	Moment	W_I
A	$-\theta$		
B			
C	$+2\theta$	M_p	$2M_p\theta$
D	-2θ	M_p	$2M_p\theta$
E	$+\theta$		
Total			$4M_p\theta$

Then $4M_p\theta = 1272.8\theta$; $M_p = 318.2$ ft·kips (431.48 kN·m).

5. Select the highest value of M_p as the correct result

Thus, $M_p = 318.2$ ft·kips (431.48 kN·m). The structure fails through the formation of plastic hinges at C and D. That a hinge should appear at D rather than at B is plausible when it is considered that the bending moments induced by the two loads are of like sign at D but of opposite sign at B.

6. Compute the reactions at the supports

Draw a free-body diagram of the frame at ultimate load (Fig. 90). Compute the reactions at the supports by applying the computed values of M_C and M_D. Thus, $\Sigma M_E = 20V_A + 22.2(24) - 74(10) = 0$; $V_A = 10.36$ kips (46.081 kN); $V_E = 74 - 10.36 = 63.64$ kips (283.071 kN); $M_C = 10V_A + 24H_A = 103.6 + 24H_A = 318.2$; $H_A = 8.94$ kips (39.765 kN); $H_E = 22.2 - 8.94 = 13.26$ kips (58.980 kN); $M_D = -24H_E = -24(13.26) = -318.2$ ft·kips (-431.48 kN·m). Thus, the results are verified.

ANALYSIS OF A RECTANGULAR PORTAL FRAME BY THE STATIC METHOD

Compute the plastic moment of the frame in Fig. 89a by using the static method.

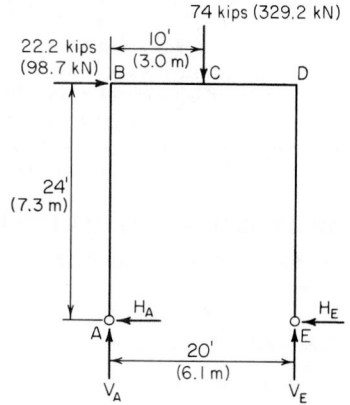

FIG. 90

Calculation Procedure:

1. Determine the relative values of the bending moments

Consider a bending moment as positive if the fibers on the interior side of the neutral plane are in tension. Consequently, as the mechanisms in Fig. 89 reveal, the algebraic sign of the plastic moment at a given section agrees with that of its angular displacement during collapse.

Determine the relative values of the bending moments at B, C, and D. Refer to Fig. 90. As previously found by statics, $V_A = 10.36$ kips (46.081 kN), $M_B = 24H_A$, $M_C = 24H_A + 10V_A$; therefore, $M_C = M_B + 103.6$, Eq. a. Also, $M_D = 24H_A + 20V_A - 74(10)$; $M_D = M_B - 532.8$, Eq. b; or $M_D = M_C - 636.4$, Eq. c.

2. Assume the mode the failure in Fig. 89b

This requires that $M_B = M_D = -M_p$. This relationship is incompatible with Eq. b, and the assumed mode of failure is therefore incorrect.

3. Assume the mode of failure in Fig. 89c

This requires that $M_B = M_p$, and $M_C < M_p$; therefore, $M_C < M_B$. This relationship is incompatible with Eq. a, and the assumed mode of failure is therefore incorrect.

By a process of elimination, it has been ascertained that the frame will fail in the manner shown in Fig. 89d.

4. Compute the value of M_p for the composite mode of failure

Thus, $M_C = M_p$, and $M_D = -M_p$. Substitute these values in Eq. c. Or, $-M_p = M_p - 636.4$; $M_p = 318.2$ ft·kips (431.48 kN·m).

THEOREM OF COMPOSITE MECHANISMS

By analyzing the calculations in the calculation procedure before the last one, establish a criterion to determine when a composite mechanism is significant (i.e., under what conditions it may yield an M_p value greater than that associated with the basic mechanisms).

Calculation Procedure:

1. Express the external and internal work associated with a given mechanism

Thus, $W_E = e\theta$, and $W_I = iM_p\theta$, where the coefficients e and i are obtained by applying the mechanism method. Then $M_p = e/i$.

2. Determine the significance of mechanism sign

Let the subscripts 1 and 2 refer to the basic mechanisms and the subscript 3 to their composite mechanism. Then $M_{p1} = e_1/i_1$; $M_{p2} = e_2/i_2$.

When the basic mechanisms are superposed, the values of W_E are additive. If the two mechanisms do not produce rotations of opposite sign at any section, the values of W_I are also additive, and $M_{p3} = e_3/i_3 = (e_1 + e_2)/(i_1 + i_2)$. This value is intermediate between M_{p1} and M_{p2}, and the composite mechanism therefore lacks significance. But if the basic mechanisms produce rotations of opposite sign at any section whatsoever, M_{p3} may exceed both M_{p1} and M_{p2}.

In summary, a composite mechanism is significant only if the two basic mechanisms of which it is composed produce rotations of opposite sign at any section. This theorem, which establishes a necessary but not sufficient condition, simplifies the analysis of a complex frame by enabling the engineer to discard the nonsignificant composite mechanisms at the outset.

ANALYSIS OF AN UNSYMMETRIC RECTANGULAR PORTAL FRAME

The frame in Fig. 91a sustains the ultimate loads shown. Compute the plastic moment and ultimate-load reactions.

Calculation Procedure:

1. Determine the solution method to use

Apply the mechanism method. In Fig. 91b, indicate the basic mechanisms.

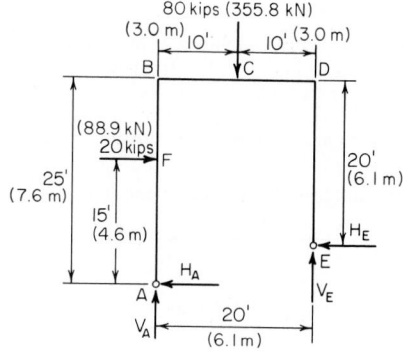

(a) Frame and ultimate loads

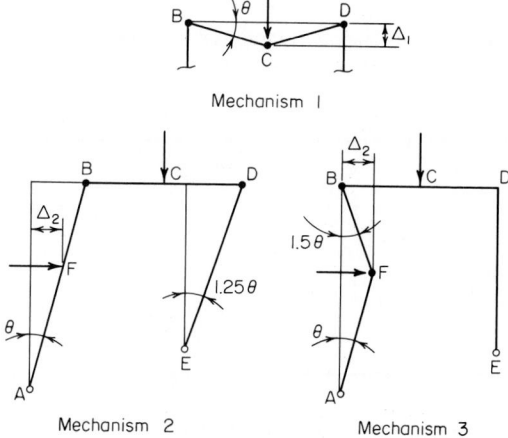

Mechanism 1

Mechanism 2

Mechanism 3

(b) Basic mechanisms

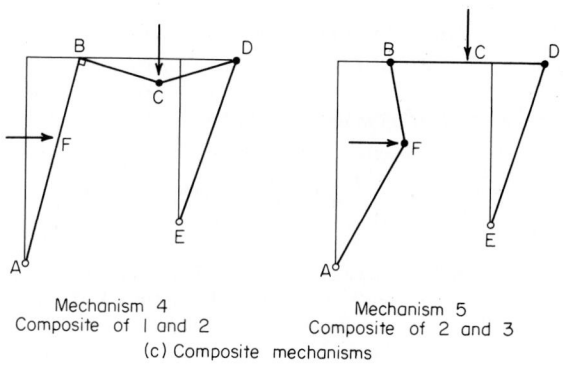

Mechanism 4
Composite of 1 and 2

Mechanism 5
Composite of 2 and 3

(c) Composite mechanisms

FIG. 91

1.109

2. Identify the significant composite mechanisms

Apply the theorem of the previous calculation procedure. Using this theorem, identify the significant composite mechanisms. For mechanisms 1 and 2, the rotations at B are of opposite sign; their composite therefore warrants investigation.

For mechanisms 1 and 3, there are no rotations of opposite sign; their composite therefore fails the test. For mechanisms 2 and 3, the rotations at B are of opposite sign; their composite therefore warrants investigation.

3. Evaluate the external work associated with each mechanism

Mechanism	W_E
1	$80\Delta_1 = 80(10\theta) = 800\theta$
2	$20\Delta_2 = 20(15\theta) = 300\theta$
3	300θ
4	1100θ
5	600θ

4. List the sections at which plastic hinges form; record the angular displacement associated with each mechanism

Use a list such as the following:

Mechanism	Section			
	B	C	D	F
1	$-\theta$	$+2\theta$	$-\theta$	
2	$+\theta$	. .	-1.25θ	
3	-1.5θ		. . .	$+2.5\theta$
4	. .	$+2\theta$	-2.25θ	
5	-0.5θ	. .	-1.25θ	$+2.5\theta$

5. Evaluate the internal work associated with each mechanism

Equate the external and internal work to find M_p. Thus, $M_{p1} = 800/4 = 200$; $M_{p2} = 300/2.25 = 133.3$; $M_{p3} = 300/4 = 75$; $M_{p4} = 1100/4.25 = 258.8$; $M_{p5} = 600/4.25 = 141.2$. Equate the external and internal work to find M_p.

6. Select the highest value as the correct result

Thus, $M_p = 258.8$ ft · kips (350.93 kN · m). The frame fails through the formation of plastic hinges at C and D.

7. Determine the reactions at ultimate load

To verify the foregoing solution, ascertain that the bending moment does not exceed M_p in absolute value anywhere in the frame. Refer to Fig. 91a.

Thus, $M_D = -20H_E = -258.8$; therefore, $H_E = 12.94$ kips (57.557 kN); $M_C = M_D + 10V_E = 258.8$; therefore, $V_E = 51.76$ kips (230.23 kN); then $H_A = 7.06$ kips (31.403 kN); $V_A = 28.24$ kips (125.612 kN).

Check the moments. Thus $\Sigma M_E = 20V_A + 5H_A + 20(10) - 80(10) = 0$; this is correct. Also, $M_F = 15H_A = 105.9$ ft · kips (143.60 kN · m) $< M_p$. This is correct. Last, $M_B = 25H_A - 20(10) = -23.5$ ft · kips (-31.87 kN · m) $> -M_p$. This is correct.

ANALYSIS OF GABLE FRAME BY STATIC METHOD

The prismatic frame in Fig. 92a carries the ultimate loads shown. Determine the plastic moment by applying the static method.

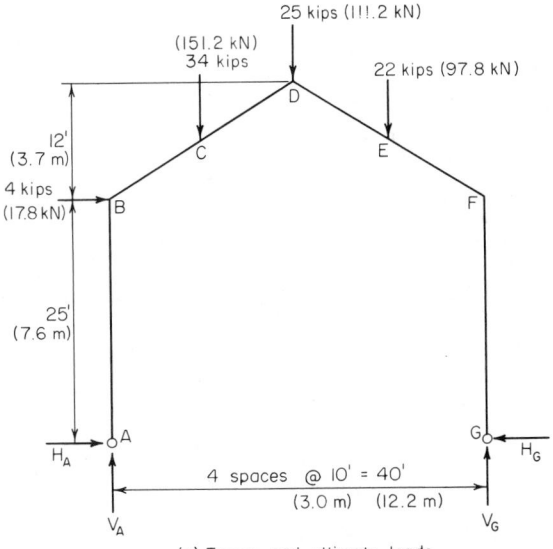

(a) Frame and ultimate loads

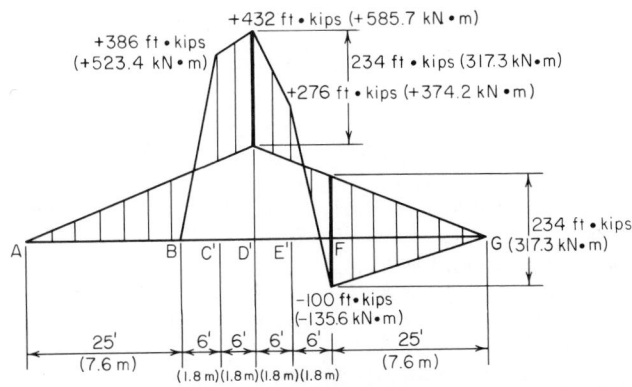

(b) Projected bending–moment diagram

FIG. 92

Calculation Procedure:

1. Compute the vertical shear V_A and the bending moment at every significant section, assuming $H_A = 0$

Thus, $V_A = 41$ kips (182.4 kN). Then $M_B = 0$; $M_C = 386$; $M_D = 432$; $M_E = 276$; $M_F = -100$.

Note that failure of the frame will result from the formation of two plastic hinges. It is helpful, therefore, to construct a "projected" bending-moment diagram as an aid in locating these hinges. The computed bending moments are used in plotting the projected bending-moment diagram.

2. Construct a projected bending-moment diagram

To construct this diagram, consider the rafter BD to be projected onto the plane of column AB and the rafter FD to be projected onto the plane of column GF. Juxtapose the two halves, as shown in Fig. 92b. Plot the values calculated in step 1 to obtain the bending-moment diagram corresponding to the assumed condition of $H_A = 0$.

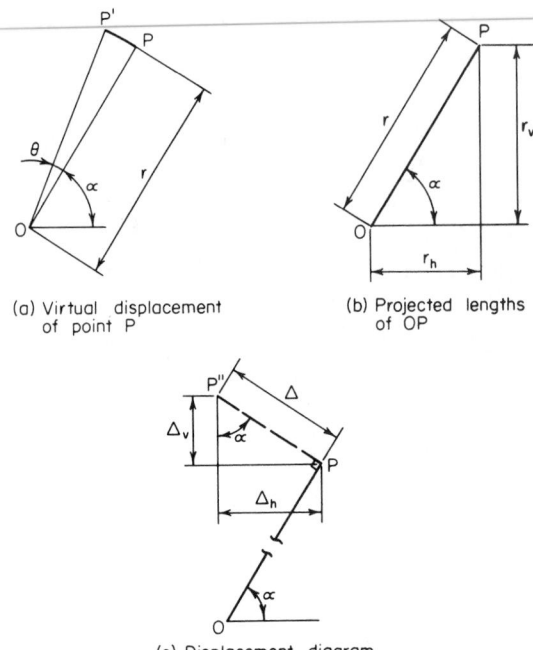

(a) Virtual displacement
of point P

(b) Projected lengths
of OP

(c) Displacement diagram

FIG. 93

The bending moments caused solely by a specific value of H_A are represented by an isosceles triangle with its vertex at D'. The true bending moments are obtained by superposition. It is evident by inspection of the diagram that plastic hinges form at D and F and that H_A is directed to the right.

3.   *Evaluate the plastic moment*

Apply the true moments at D and F. Thus, $M_D = M_p$ and $M_F = -M_p$; therefore, $432 - 37H_A = -(-100-25H_A)$; $H_A = 5.35$ kips (23.797 kN) and $M_p = 234$ ft·kips (317 kN·m).

THEOREM OF VIRTUAL DISPLACEMENTS

In Fig. 93a, point P is displaced along a virtual (infinitesimally small) circular arc PP' centered at O and having a central angle θ. Derive expressions for the horizontal and vertical displacement of P in terms of the given data. (These expressions are applied later in analyzing a gable frame by the mechanism method.)

Calculation Procedure:

1.   *Construct the displacement diagram*

In Fig. 93b, let r_h = length of horizontal projection of OP; r_v = length of vertical projection of OP; Δ_h = horizontal displacement of P; Δ_v = vertical displacement of P.

In Fig. 93c, construct the displacement diagram. Since PP' is infinitesimally small, replace this circular arc with the straight line PP'' that is tangent to the arc at P and therefore normal to radius OP.

2.   *Evaluate Δ_h and Δ_v , considering only absolute values*

Since θ is infinitesimally small, set $PP'' = r\theta$; $\Delta_h = PP'' \sin \alpha = r\theta \sin \alpha$; $\Delta_v = PP'' \cos \alpha = r\theta \cos \alpha$. But $r \sin \alpha = r_v$ and $r \cos \alpha = r_h$; therefore, $\Delta_h = r_v\theta$ and $\Delta_v = r_h\theta$.

These results may be combined and expressed verbally thus: If a point is displaced along a virtual circular arc, its displacement as projected on the u axis equals the displacement angle times the length of the radius as projected on an axis normal to u.

GABLE-FRAME ANALYSIS BY USING THE MECHANISM METHOD

For the frame in Fig. 92a, assume that plastic hinges form at D and F. Calculate the plastic moment associated with this assumed mode of failure by applying the mechanism method.

Calculation Procedure:

1. *Indicate the frame configuration following a virtual displacement*

During collapse, the frame consists of three rigid bodies: ABD, DF, and GF. To evaluate the external and internal work performed during a virtual displacement, it is necessary to locate the instantaneous center of rotation of each body.

In Fig. 94, indicate by dash lines the configuration of the frame following a virtual displacement. In Fig. 94, D is displaced to D' and F to F'. Draw a straight line through A and D intersecting the prolongation of GF at H.

Since A is the center of rotation of ABD, DD' is normal to AD and HD; since G is the center of rotation of GF, FF' is normal to GF and HF. Therefore, H is the instantaneous center of rotation of DF.

2. *Record the pertinent dimensions and rotations*

Record the dimensions a, b, and c in Fig. 94, and express θ_2 and θ_3 in terms of θ_1. Thus, $\theta_2/\theta_1 = HD/AD$; $\therefore \theta_2 = \theta_1$. Also, $\theta_3/\theta_1 = HF/GF = 49/25$; $\therefore \theta_3 = 1.96\theta_1$.

3. *Determine the angular displacement, and evaluate the internal work*

Determine the angular displacement (in absolute value) at D and F, and evaluate the internal work in terms of θ_1. Thus, $\theta_D = \theta_1 + \theta_2 = 2\theta_1$; $\theta_F = \theta_1 + \theta_3 = 2.96\theta_1$. Then $W_I = M_p (\theta_D + \theta_F) = 4.96M_p\theta_1$.

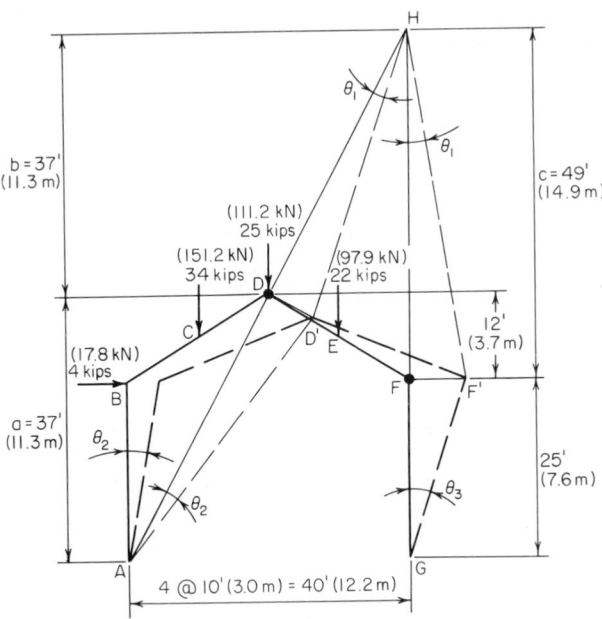

FIG. 94   Virtual displacement of frame.

4.   Apply the theorem of virtual displacements to determine the displacement of each applied load

Determine the displacement of each applied load in the direction of the load. Multiply the displacement by the load to obtain the external work. Record the results as shown:

Section	Load kips	Load kN	Displacement in direction of load ft	Displacement in direction of load m	External work ft·kips	External work kN·m
B	4	17.8	$\Delta_h = 25\theta_2 = 25\,\theta_1$	$7.6\,\theta_1$	$100\,\theta_1$	$135.6\,\theta_1$
C	34	151.2	$\Delta_v = 10\theta_2 = 10\,\theta_1$	$3.0\,\theta_1$	$340\,\theta_1$	$461.0\,\theta_1$
D	25	111.2	$\Delta_v = 20\,\theta_1$	$6.1\,\theta_1$	$500\,\theta_1$	$678.0\,\theta_1$
E	22	97.9	$\Delta_v = 10\,\theta_1$	$3.0\,\theta_1$	$220\,\theta_1$	$298.3\,\theta_1$
Total					$1160\,\theta_1$	$1572.9\,\theta_1$

5.   Equate the external and internal work to find M_p

Thus, $4.96 M_p \theta_1 = 1160\theta_1$; $M_p = 234$ ft·kips (317.3 kN·m).

Other modes of failure may be assumed and the corresponding value of M_p computed in the same manner. The failure mechanism analyzed in this procedure (plastic hinges at D and F) yields the highest value of M_p and is therefore the true mechanism.

REDUCTION IN PLASTIC-MOMENT CAPACITY CAUSED BY AXIAL FORCE

A W10 × 45 beam-column is subjected to an axial force of 84 kips (373.6 kN) at ultimate load. (a) Applying the exact method, calculate the plastic moment this section can develop with respect to the major axis. (b) Construct the interaction diagram for this section, and then calculate the plastic moment by assuming a linear interaction relationship that approximates the true relationship.

Calculation Procedure:

1.   Record the relevant properties of the member

Let P = applied axial force, kips (kN); P_y = axial force that would induce plastification if acting alone, kips (kN) = $A f_y$; M_p' = plastic-moment capacity of the section in combination with P, ft·kips (kN·m).

A typical stress diagram for a beam-column at plastification is shown in Fig. 95a. To simplify the calculations, resolve this diagram into the two parts shown at the right. This procedure is tantamount to assuming that the axial load is resisted by a central core and the moment by the outer segments of the section, although in reality they are jointly resisted by the integral action of the entire section.

From the AISC *Manual*, for a W10 × 45: $A = 13.24$ in² (85.424 cm²); $d = 10.12$ in (257.048 mm); $t_f = 0.618$ in (15.6972 mm); $t_w = 0.350$ in (8.890 mm); $d_w = 10.12 - 2(0.618) = 8.884$ in (225.6536 mm); $Z = 55.0$ in³ (901.45 cm³).

2.   Assume that the central core that resists the 84-kip (373.6-kN) load is encompassed within the web; determine the core depth

Calling the depth of the core g, refer to Fig. 95d. Then $g = 84/[0.35(36)] = 6.67 < 8.884$ in (225.6536 mm).

3.   Compute the plastic modulus of the core, the plastic modulus of the remaining section, and the value of M_p'

Using data from the *Manual* for the plastic modulus of a rectangle, we find $Z_c = \frac{1}{4} t_w g^2 = \frac{1}{4}(0.35)(6.67)^2 = 3.9$ in³ (63.92 cm³); $Z_r = 55.0 - 3.9 = 51.1$ in³ (837.53 cm³); $M_p' = 51.1(36)/$

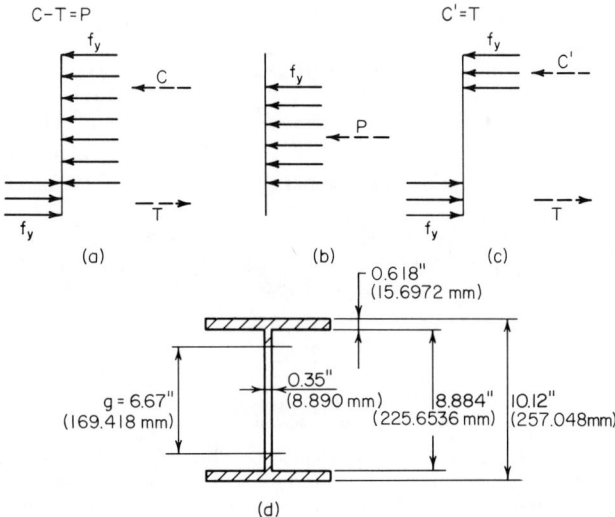

FIG. 95

$12 = 153.3$ ft·kips (207.87 kN·m). This constitutes the solution of part *a*. The solution of part *b* is given in steps 4 through 6.

4.   *Assign a series of values to the parameter g, and compute the corresponding sets of values of P and M_p'*

Apply the results to plot the interaction diagram in Fig. 96. This comprises the parabolic curves *CB* and *BA*, where the points *A*, *B*, and *C* correspond to the conditions $g = 0$, $g = d_w$, and $g = d$, respectively.

The interaction diagram is readily analyzed by applying the following relationships: $dP/dg = f_y t$; $dM_p'/dg = -\frac{1}{2}f_y tg$; $\therefore dP/dM_p' = -2/g$. This result discloses that the change in slope along *CB* is very small, and the curvature of this arc is negligible.

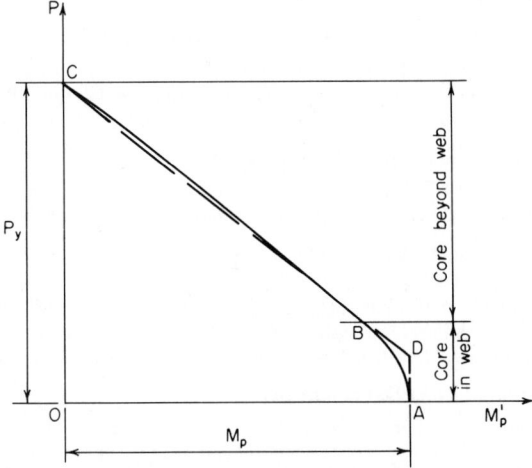

FIG. 96   Interaction diagram for axial force and moment.

5. *Replace the true interaction diagram with a linear one*

Draw a vertical line $AD = 0.15P_y$, and then draw the straight line CD (Fig. 96). Establish the equation of CD. Thus, slope of $CD = -0.85P_y/M_p$; $\therefore P = P_y - 0.85P_y M_p'/M_p$, or $M_p' = 1.18(1 - P/P_y)M_p$.

The provisions of one section of the AISC *Specification* are based on the linear interaction diagram.

6. *Ascertain whether the data are represented by a point on AD or CD; calculate M_p' accordingly*

Thus, $P_y = Af_y = 13.24(36) = 476.6$ kips (2119.92 kN); $P/P_y = 84/476.6 = 0.176$; therefore, apply the last equation given in step 5. Thus, $M_p = 55.0(36)/12 = 165$ ft·kips (223.7 kN·m); $M_p' = 1.18(1 - 0.176)(165) = 160.4$ ft·kips (217.50 kN·m). This result differs from that in part *a* by 4.6 percent.

Timber Engineering

In designing timber members, the following references are often used: *Wood Handbook*, Forest Products Laboratory, U.S. Department of Agriculture, and *National Design Specification for Stress-Grade Lumber and Its Fastenings*, National Forest Products Association. The members are assumed to be continuously dry and subject to normal loading conditions.

For most species of lumber, the true or *dressed* dimensions are less than the nominal dimensions by the following amounts: ⅜ in (9.53 mm) for dimensions less than 6 in (152.4 mm); ½ in (12.7 mm) for dimensions of 6 in (152.4 mm) or more. The average weight of timber is 40 lb/ft³ (6.28 kN/m³). The width and depth of the transverse section are denoted by b and d, respectively.

BENDING STRESS AND DEFLECTION OF WOOD JOISTS

A floor is supported by 3×8 in (76.2 × 203.2 mm) wood joists spaced 16 in (406.4 mm) on centers with an effective span of 10 ft (3.0 m). The total floor load transmitted to the joists is 107 lb/in² (5.123 kN/m²). Compute the maximum bending stress and initial deflection, using $E = 1,760,000$ lb/in² (12,135 kPa).

Calculation Procedure:

1. *Calculate the beam properties or extract them from a table*

Thus, $A = 2⅝(7½) = 19.7$ in² (127.10 cm²); beam weight $= (A/144)$ (lumber density, lb/ft³) $= (19.7/144)(40) = 5$ lb/lin ft (73.0 N/m); $I = (1/12)(2⅝)(7½)^3 = 92.3$ in⁴ (3841.81 cm⁴); $S = 92.3/3.75 = 24.6$ in³ (403.19 cm³).

2. *Compute the unit load carried by the joists*

Thus, the unit load $w = 107(1.33) + 5 = 148$ lb/lin ft (2159.9 N/m), where the factor 1.33 is the width, ft, of the floor load carried by each joist and $5 =$ the beam weight, lb/lin ft.

3. *Compute the maximum bending stress in the joist*

Thus, the bending moment in the joist is $M = (1/8)wL^2 12$, where $M =$ bending moment, in·lb (N·m); $L =$ joist length, ft (m). Substituting gives $M = (1/8)(148)(10)^2(12) = 22,200$ in·lb (2508.2 N·m). Then for the stress in the beam, $f = M/S$, where $f =$ stress, lb/in² (kPa), and $S =$ beam section modulus, in³ (cm³); or $f = 22,200/24.6 = 902$ lb/in² (6219.3 kPa).

4. *Compute the initial deflection at midspan*

Using the AISC *Manual* deflection equation, we see that the deflection Δ in (mm) $= (5/384)wL^4/(EI)$, where $I =$ section moment of inertia, in⁴ (cm⁴) and other symbols are as before. Substituting yields $\Delta = 5(148)(10)^4(1728)/[384(1,760,000)(92.3)] = 0.205$ in (5.2070 mm). In this relation, the factor 1728 converts cubic feet to cubic inches.

SHEARING STRESS CAUSED BY STATIONARY CONCENTRATED LOAD

A 3 × 10 in (76.2 × 254.0 mm) beam on a span of 12 ft (3.7 m) carries a concentrated load of 2730 lb (12,143.0 N) located 2 ft (0.6 m) from the support. If the allowable shearing stress is 120 lb/in² (827.4 kPa), determine whether this load is excessive. Neglect the beam weight.

Calculation Procedure:

1. Calculate the reaction at the adjacent support

In a rectangular section, the shearing stress varies parabolically with the depth and has the maximum value of $v = 1.5V/A$, where V = shear, lb (N).

The *Wood Handbook* notes that checks are sometimes present near the neutral axis of timber beams. The vitiating effect of these checks is recognized in establishing the allowable shearing stresses. However, these checks also have a beneficial effect, for they modify the shear distribution and thereby reduce the maximum stress. The amount of this reduction depends on the position of the load. The maximum shearing stress to be applied in design is given by $v = 10(a/d)^2 v'/\{9[2 + (a/d)^2]\}$, where v = true maximum shearing stress, lb/in² (kPa); v' = nominal maximum stress computed from $1.5V/A$; a = distance from load to adjacent support, in (mm).

Computing the reaction R at the adjacent support gives $R = V_{max} = 2730(12 - 2)/12 = 2275$ lb (10,119.2 N). Then $v' = 1.5V/A = 1.5(2275)/24.9 = 137$ lb/in² (944.6 kPa).

2. Find the design stress

Using the equation given in step 1, we get $(a/d)^2 = (24/9.5)^2 = 6.38$; $v = 10(6.38)(137)/[9(8.38)] = 116$ lb/in² (799.8 kPa) < 120 lb/in² (827.4 kPa). The load is therefore not excessive.

SHEARING STRESS CAUSED BY MOVING CONCENTRATED LOAD

A 4 × 12 in (101.6 × 304.8 mm) beam on a span of 10 ft (3.0 m) carries a total uniform load of 150 lb/lin ft (2189.1 N/m) and a moving concentrated load. If the allowable shearing stress is 130 lb/in² (896.4 kPa), what is the allowable value of the moving load as governed by shear?

Calculation Procedure:

1. Calculate the reaction at the support

The transient load induces the absolute maximum shearing stress when it lies at a certain critical distance from the support rather than directly above it. This condition results from the fact that as the load recedes from the support, the reaction decreases but the shear-redistribution effect becomes less pronounced. The approximate method of analysis recommended in the *Wood Handbook* affords an expedient means of finding the moving-load capacity.

Place the moving load P at a distance of $3d$ or $\frac{1}{4}L$ from the support, whichever is less. Calculate the reaction at the support, disregarding the load within a distance of d therefrom.

Thus, $3d = 2.9$ ft (0.884 m) and $\frac{1}{4}L = 2.5$ ft (0.762 m); then $R = V_{max} = 150(5 - 0.96) + \frac{3}{4}P = 610 + \frac{3}{4}P$.

2. Calculate the allowable shear

Thus, $V_{allow} = \frac{2}{3}vA = \frac{2}{3}(130)(41.7) = 3610$ lb (16,057.3 N). Then $610 + \frac{3}{4}P = 3610$; $P = 4000$ lb (17,792.0 N).

STRENGTH OF DEEP WOODEN BEAMS

If the allowable bending stress in a shallow beam is 1500 lb/in² (10,342.5 kPa), what is the allowable bending moment in a 12 × 20 in (304.8 × 508.0 mm) beam?

Calculation Procedure:

1. Calculate the depth factor F

An increase in depth of a rectangular beam is accompanied by a decrease in the modulus of rupture. For beams more than 16 in (406.4 mm) deep, it is necessary to allow for this reduction in strength by introducing a *depth factor F*.

Thus, $F = 0.81(d^2 + 143)/(d^2 + 88)$, where d = dressed depth of beam, in. Substituting gives $F = 0.81(19.5^2 + 143)/(19.5^2 + 88) = 0.905$.

2.   Apply the result of step 1 to obtain the moment capacity

Use the relation $M = FfS$, where the symbols are as given earlier. Thus, $M = 0.905 \times (1.5)(728.8)/12 = 82.4$ ft·kips (111.73 kN·m).

DESIGN OF A WOOD-PLYWOOD BEAM

A girder having a 36-ft (11.0-m) span is to carry a uniform load of 550 lb/lin ft (8026.6 N/m), which includes its estimated weight. Design a box-type member of glued construction, using the allowable stresses given in the table. The modulus of elasticity of both materials is 1,760,000 lb/in² (12,135.2 MPa), and the ratio of deflection to span cannot exceed 1/360. Architectural details limit the member depth to 40 in (101.6 cm).

	Lumber	Plywood
Tension, lb/in² (kPa)	1,500 (10,342.5)	2,000 (13,790.0)
Compression parallel to grain, lb/in² (kPa)	1,350 (9,308.3)	1,460 (10,066.7)
Compression normal to grain, lb/in² (kPa)	390 (2,689.1)	405 (2,792.5)
Shear parallel to plane of plies, lb/in² (kPa)	. . . .	72° (496.4)
Shear normal to plane of plies, lb/in² (kPa)	. . . .	192 (1,323.8)

° Use 36 lb/in² (248.2 kPa) at contact surface of flange and web to allow for stress concentration.

Calculation Procedure:

1.   Compute the maximum shear and bending moment

Thus, $V = \frac{1}{2}(550)(36) = 9900$ lb (44,035.2 N); $M = \frac{1}{8}(wL^2)12 = \frac{1}{8}(550)(36)^2 12 = 1,070,000$ in·lb (120,888.6 N·m). To preclude the possibility of field error, make the tension and compression flanges alike.

2.   Calculate the beam depth for a balanced condition

Assume that the member precisely satisfies the requirements for flexure and deflection, and calculate the depth associated with this balanced condition. To allow for the deflection caused by shear, which is substantial when a thin web is used, increase the deflection as computed in the conventional manner by one-half. Thus, $M = fI/c = 2fI/d = 2700I/d$, Eq. a. $\Delta = (7.5/48)L^2M/(EI) = L/360$, Eq. b.

Substitute in Eq. b the value of M given by Eq. a; solve for d to obtain $d = 37.3$ in (947.42 mm). Use the permissible depth of 40 in (1016.0 mm). As a result of this increase in depth, a section that satisfies the requirement for flexure will satisfy the requirement for deflection as well.

3.   Design the flanges

Approximate the required area of the compression flange; design the flanges. For this purpose, assume that the flanges will be 5½ in (139.7 mm) deep. The lever arm of the resultant forces in the flanges will be 34.8 in (883.92 mm), and the average fiber stress will be 1165 lb/in² (8032.7 kPa). Then $A = 1,070,000/[1165(34.8)] = 26.4$ in² (170.33 cm²). Use three 2 × 6 in (50.8 × 152.4 mm) sections with glued vertical laminations for both the tension and compression flange. Then $A = 3(8.93) = 26.79$ in² (170.268 cm²); $I_o = 3(22.5) = 67.5$ in⁴ (2809.56 cm⁴).

4.   Design the webs

Use the approximation $t_w = 1.25V/dv_n = 1.25(9900)/[40(192)] = 1.61$ in (40.894 mm). Try two ⅞-in (22.2-mm) thick plywood webs. A catalog of plywood properties reveals that the ⅞-in (22.2-mm) member consists of seven plies and that the parallel plies have an aggregate thickness of 0.5 in (12.7 mm). Draw the trial section as shown in Fig. 97.

5.   Check the bending stress in the member

For simplicity, disregard the webs in evaluating the moment of inertia. Thus, the moment of inertia of the flanges $I_f = 2(67.5 + 26.79 \times 17.25^2) = 16,080$ in⁴ (669,299.448 cm⁴); then the stress $f = Mc/I = 1,070,000(20)/16,080 = 1330 < 1350$ lb/in² (9308.25 kPa). This is acceptable.

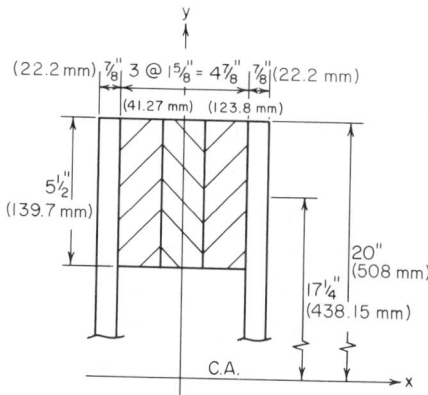

FIG. 97

6.   Check the shearing stress at the contact surface of the flange and web

Use the relation $Q_f = Ad = 26.79(17.25) = 462$ in^3 (7572.2 cm^3). The q per surface $= VQ_f/(2I_f) = 9900(462)/[2(16,080)] = 142$ lb/lin in (24.8 kN/m). Assume that the shearing stress is uniform across the surface, and apply 36 lb/in^2 (248.2 kPa), as noted earlier, as the allowable stress. Then, $v = 142/5.5 = 26$ lb/in^2 (179.3 kPa) < 36 lb/in^2 (248.2 kPa). This is acceptable.

7.   Check the shearing stress in the webs

For this purpose, include the webs in evaluating the moment of inertia but apply solely the area of the parallel plies. At the neutral axis $Q = Q_f + Q_w = 462 + 2(0.5)(20)(10) = 662$ in^3 (10,850.2 cm^3); $I = I_f + I_w = 16,080 + 2(1/12)(0.5)(40)^3 = 21,410$ in^4 (89.115 dm^4). Then $v = VQ/(It) = 9900(662)/[21,410(2)(0.875)] = 175$ lb/in^2 (1206.6 kPa) < 192 lb/in^2 (1323.8 kPa). This is acceptable.

8.   Check the deflection, applying the moment of inertia of only the flanges

Thus, $\Delta = (7.5/384)wL^4/(EI_f) = 7.5(550)(36)^4(1728)/[384(1,760,000)(16,080)] = 1.10$ in (27.94 mm); $\Delta/L = 1.10/[36(12)] < 1/360$. This is acceptable, and the trial section is therefore satisfactory in all respects.

9.   Establish the allowable spacing of the bridging

To do this, compare the moments of inertia with respect to the principal axes. Thus, $I_y = 2(1/12)(5.5)(4.875)^3 + 2(0.5)(40)(2.875)^2 = 433$ in^4 (18,022.8 cm^4); then $I_x/I_y = 16,080/433 = 37.1$.

For this ratio, the *Wood Handbook* specifies that "the beam should be restrained by bridging or other bracing at intervals of not more than 8 ft (2.4 m)."

DETERMINING THE CAPACITY OF A SOLID COLUMN

An 8 × 10 in (203.2 × 254 mm) column has an unbraced length of 10 ft 6 in (3.20 m). The allowable compressive stress is 1500 lb/in^2 (10,342.5 kPa), and $E = 1,760,000$ lb/in^2 (12,135.2 MPa). Calculate the allowable load on this column (*a*) by applying the recommendations of the *Wood Handbook*; (*b*) by applying the provisions of the *National Design Specification*.

Calculation Procedure:

1.   Record the properties of the member; evaluate K; classify the column

Let L = unbraced length of column, in (mm); d = smaller side of rectangular section, in (mm); f_c = allowable compressive stress parallel to the grain in short column of the same species, lb/in^2 (kPa); f = allowable compressive stress parallel to grain in column under investigation, lb/in^2 (kPa).

The *Wood Handbook* divides columns into three categories: short, intermediate, and long. Let K denote a parameter defined by the equation $K = 0.64(E/f_c)^{0.5}$.

The range of the slenderness ratio and the allowable stress for each category of column are as follows: *short column*, $L/d \leq 11$ and $f = f_c$; *intermediate column*, $11 < L/d \leq K$ and $f = f_c[1 - \frac{1}{3}(L/d/K)^4]$; *long column*, $L/d > K$ and $f = 0.274E/(L/d)^2$.

For this column, the area $A = 71.3$ in^2 (460.03 cm^2), using the dressed dimensions. Then $L/d = 126/7.5 = 16.8$. Also, $K = 0.64(1,760,000/1500)^{0.5} = 21.9$. Therefore, this is an intermediate column because L/d lies between K and 11.

2. Compute the capacity of the member

Use the relation capacity, lb (N) $= P = Af = 71.3(1500)[1 - \frac{1}{3}(16.8/21.9)^4] = 94,600$ lb (420,780.8 N). This constitutes the solution to part *a*, using data from the *Wood Handbook*. For part *b*, data from the *National Design Specification* are used.

3. Compute the capacity of the column

Determine the stress from $f = 0.30E/(L/d)^2 = 0.30(1,760,000)/16.8^2 = 1870$ lb/in^2 (12,893.6 kPa). Setting $f = 1500$ lb/in^2 (10,342.5 kPa) gives $P = Af = 71.3(1500) = 107,000$ lb (475,936 N). Note that the smaller stress value is used when the column capacity is computed.

DESIGN OF A SOLID WOODEN COLUMN

A 12-ft (3.7-m) long wooden column supports a load of 98 kips (435.9 kN). Design a solid section in the manner recommended in the *Wood Handbook*, using $f_c = 1400$ lb/in^2 (9653 kPa) and $E = 1,760,000$ lb/in^2 (12,135.2 MPa).

Calculation Procedure:

1. Assume that d = 7.5 in (190.5 mm), and classify the column

Thus, $L/d = 144/7.5 = 19.2$ and $K = 0.64(1,760,000/1400)^{0.5} = 22.7$. This is an intermediate column if the assumed dimension is correct.

2. Compute the required area and select a section

For an intermediate column, the stress $f = 1400[1 - \frac{1}{3}(19.2/22.7)^4] = 1160$ lb/in^2 (7998.2 kPa). Then $A = P/f = 98,000/1160 = 84.5$ in^2 (545.19 cm^2).

Study of the required area shows that an 8×12 in (203.2 $\times$ 304.8 mm) column having an area of 86.3 in^2 (556.81 cm^2) should be used.

INVESTIGATION OF A SPACED COLUMN

The wooden column in Fig. 98 is composed of three 3×8 in (76.2 $\times$ 203.2 mm) sections. Determine the capacity of the member if $f_c = 1400$ lb/in^2 (9653 kPa) and $E = 1,760,000$ lb/in^2 (12,135.2 MPa).

Calculation Procedure:

1. Record the properties of the elemental section

In analyzing a spaced column, it is necessary to assess both the aggregate strength of the elements and the strength of the built-up section. The end spacer blocks exert a restraining effect on the elements and thereby enhance their capacity. This effect is taken into account by multiplying the modulus of elasticity by a *fixity factor F*.

The area of the column $A = 19.7$ in^2 (127.10 cm^2) when the dressed sizes are used. Also, $L/d = 114/2.625 = 43.4$; $F = 2.5$; $K = 0.64(2.5 \times 1,760,000/1400)^{0.5} = 35.9$. Therefore, this is a long column.

2. Calculate the aggregate strength of the elements

Thus, $f = 0.274E/(L/d)^2$ for a long column, or $f = 0.274(2.5)(1,760,000)/(43.4)^2 = 640$ lb/in^2 (4412.8 kPa). $P = 3(19.7)(640) = 37,800$ lb (168,134.4 N).

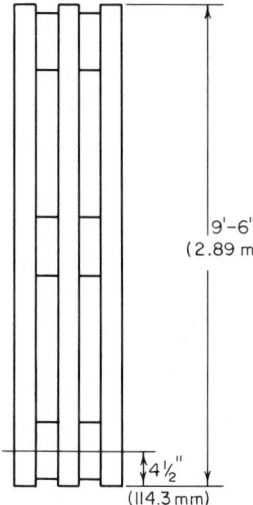

FIG. 98 Spaced column.

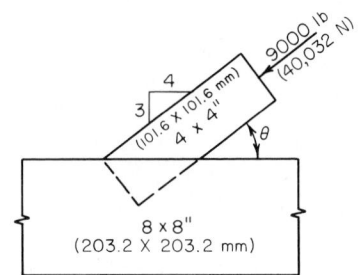

FIG. 99

3. Repeat the foregoing steps for the built-up member

Thus, $L/d = 114/7.5 \times 15.2$; $K = 22.7$; therefore, this is an intermediate column. Then $f = 1400[1 - \frac{1}{3}(15.2/22.7)^4] = 1306$ lb/in² (9004.9 kPa) > 640 lb/in² (4412.8 kPa).
The column capacity is therefore limited by the elements and $P = 37,800$ lb (168,134.4 N).

COMPRESSION ON AN OBLIQUE PLANE

Determine whether the joint in Fig. 99 is satisfactory with respect to bearing if the allowable compressive stresses are 1400 and 400 lb/in² (9653 and 2758 kPa) parallel and normal to the grain, respectively.

Calculation Procedure:

1. Compute the compressive stress

Thus, $f = P/A = 9000/3.625^2 = 685$ lb/in² (4723.1 kPa).

2. Compute the allowable compression stress in the main member

Apply Hankinson's equation: $N = PQ/(P \sin^2 \theta + Q \cos^2 \theta)$, where P = allowable compressive stress parallel to grain, lb/in² (kPa); Q = allowable compressive stress normal to grain; lb/in² (kPa); N = allowable compressive stress inclined to the grain, lb/in² (kPa); θ = angle between action line of N and direction of grain. Thus, $\sin^2 \theta = 0.36$, $\cos^2 \theta = (4/5)^2 = 0.64$; then $N = 1400(400)/(1400 \times 0.36 + 400 \times 0.64) = 737$ lb/in² (5081.6 kPa) > 685 lb/in² (4723.1 kPa). Therefore, the joint is satisfactory.

3. Alternatively, solve Hankinson's equation by using the nomogram in the Wood Handbook

DESIGN OF A NOTCHED JOINT

In Fig. 100, $M1$ is a 4×4, $F = 5500$ lb (24,464 N), and $\phi = 30°$. The allowable compressive stresses are $P = 1200$ lb/in² (8274 kPa) and $Q = 390$ lb/in² (2689.1 kPa). The projection of $M1$ into $M2$ is restricted to a vertical distance of 2.5 in (63.5 mm). Design a suitable notch.

Calculation Procedure:

1. Record the values of the trigonometric functions of ϕ and $\phi/2$

The most feasible type of notch is the one shown in Fig. 100, in which AC and BC bisect the angles between the intersecting edges. The allowable bearing pressures on these faces are therefore identical for the two members.

With $\phi = 30°$, $\sin 30° = 0.500$; $\sin 15° = 0.259$; $\cos 15° = 0.966$; $\tan 15° = 0.268$.

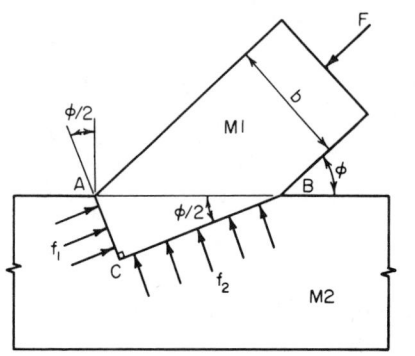

FIG. 100   FIG. 101

2. Find the lengths AC and BC

Express these two lengths as functions of AB. Or, $AB = b/\sin \phi$; $AC = [b \sin (\phi/2)]/\sin \phi$; $BC = [b \cos (\phi/2)]/\sin \phi$; $AC = 3.625(0.259/0.500) = 1.9$ in (48.26 mm); $BC = 3.625(0.966/0.500) = 7.0$ in (177.8 mm). The projection into $M2$ is therefore not excessive.

3. Evaluate the stresses f_1 and f_2

Resolve F into components parallel to AC and BC. Thus, $f_1 = (F \sin \phi)/(A \tan \phi/2)$; $f_2 = (F \sin \phi)[\tan (\phi/2)]/A$, where A = crossectional area of $M1$. Substituting gives $f_1 = 783$ lb/in² (5399 kPa); $f_2 = 56$ lb/in² (386.1 kPa).

4. Calculate the allowable stresses

Compute the allowable stresses N_1 and N_2 on AC and BC, respectively, and compare these with the actual stresses. Thus, by using Hankinson's equation from the previous calculation procedure, $N_1 = 1200(390)/(1200 \times 0.259^2 + 390 \times 0.966^2) = 1053$ lb/in² (7260.4 kPa). This is acceptable because it is greater than the actual stress. Also, $N_2 = 1200(390)/(1200 \times 0.966^2 + 390 \times 0.259^2) = 408$ lb/in² (2813.2 kPa). This is also acceptable, and the joint is therefore satisfactory.

ALLOWABLE LATERAL LOAD ON NAILS

In Fig. 101, the Western hemlock members are connected with six 50d common nails. Calculate the lateral load P that may be applied to this connection.

Calculation Procedure:

1. Determine the member group

The capacity of this connection is calculated in conformity with Part VIII of the *National Design Specification*. Refer to the *Specification* to ascertain the classification of the species. Western hemlock is in group III.

2. Determine the properties of the nail

Refer to the *Specification* to determine the properties of the nail. Calculate the penetration-diameter ratio, and compare this value with that stipulated in the *Specification*. Thus, length = 5.5 in

(139.7 mm); diameter $=$ 0.244 in (6.1976 mm); penetration/diameter ratio $=$ $(5.5 - 1.63)/0.244$ $=$ $15.9 > 13$. This is acceptable.

3. Find the capacity of the connection

Using the *Specification*, find the capacity of the nail. Then the capacity of the connection $=$ P $=$ $6(165)$ $=$ 990 lb (4403.5 N).

CAPACITY OF LAG SCREWS

In Fig. 102, the cottonwood members are connected wtih three ⅝-in (15.88-mm) lag screws 8 in (203.2 mm) long. Determine the load P that may be applied to this connection.

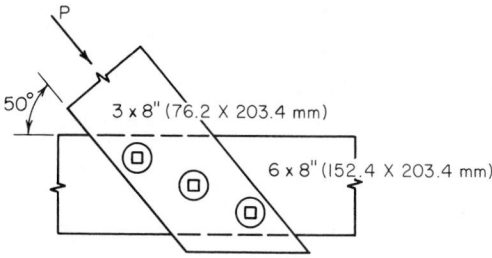

P

50°

3 x 8" (76.2 X 203.4 mm)

6 x 8" (152.4 X 203.4 mm)

FIG. 102

Calculation Procedure:

1. Determine the member group

The *National Design Specification* shows that cottonwood is classified in group IV.

2. Find the allowable screw loads

The *National Design Specification* gives the following values for each screw: allowable load parallel to grain $=$ 550 lb (2446.4 N); allowable load normal to grain $=$ 330 lb (1467.8 N).

3. Compute the allowable load on the connection

Use the Scholten nomogram, or $N = PQ/(P \sin^2 \theta + Q \cos^2 \theta)$, with $\theta = 50°$, and solve as given earlier. Either solution gives $P = 3(395) = 1185$ lb (5270.9 N).

DESIGN OF A BOLTED SPLICE

A 6 $\times$ 12 in (152.4 $\times$ 304.8 mm) southern pine member carrying a tensile force of 56 kips (249.1 kN) parallel to the grain is to be spliced with steel side plates. Design the splice.

Calculation Procedure:

1. Determine the number of bolts, and bolt size, required

Find the bolt capacity from the *National Design Specification*. The *Specification* allows a 25 percent increase in capacity of the parallel-to-grain loading when steel plates are used as side members.

Determine the number of bolts from $n = P/$capacity per bolt, lb (N), where $P =$ load, lb (N). By assuming ⅞-in (22.2-mm) diameter bolts, $n = 56,000/[3940(1.25)] = 11.4$; use 12 bolts. The value 1.25 in the denominator is the increase in bolt load mentioned above.

As a trial, use three rows of four bolts each, as shown in Fig. 103.

2. Determine whether the joint complies with the Specification

Assume ¹⁵⁄₁₆-in (23.8-mm) diameter bolt holes. The gross area of the dressed lumber is 63.25 in² (408.089 cm²). The net area $=$ gross area $-$ area of the bolt holes $= 63.25 - 3(0.94)(5.5) =$

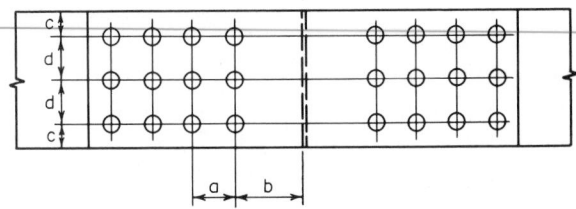

FIG. 103

47.74 in^2 (308.018 cm^2). The bearing area under the bolts = number of bolts [bolt diameter, in (mm)] [width, in (mm)] = 12(0.875)(5.5) = 57.75 in^2 (372.603 cm^2). The ratio of the net to bearing area is 47.74/57.75 = 0.83 > 0.80. This is acceptable, according to the *Specification*. The joint is therefore satisfactory, and the assumptions are usable in the design.

3. Establish the longitudinal bolt spacing

Using the *Specification*, we find a = 4(⅞) = 3.5 in (88.90 mm); b_{min} = 7(⅞) = 6⅛ in (155.58 mm).

4. Establish the transverse bolt spacing

Using the *Specification* gives L/D = 5.5(⅞) = 6.3 > 6. Make c = 2 in (50.8 mm) and d = 3¾ in (95.25 mm).

INVESTIGATION OF A TIMBER-CONNECTOR JOINT

The members in Fig. 104a have the following sizes: A, 4 × 8 in (101.6 × 203.2 mm); B, 3 × 8 in (76.2 × 203.2 mm). They are connected by six 4-in (101.6-mm) split-ring connectors, in the manner shown. The lumber is dense structural redwood. Investigate the adequacy of this joint, and establish the spacing of the connectors.

Calculation Procedure:

1. Determine the allowable stress

The *National Design Specification* shows that the allowable stress is 1700 lb/in^2 (11,721.5 kPa).

2. Find the lumber group

The *Specification* shows this species is classified in group C.

3. Compute the capacity of the connectors

The *Specification* shows that the capacity of a connector in parallel-to-grain loading for group C lumber is 4380 lb (19,482.2 N). With six connectors, the total capacity is 6(4380) = 26,280 lb (116,890 N). This is acceptable.

The *Specification* requires a minimum edge distance of 2¾ in (69.85 mm). The edge distance in the present instance is 3¾ in (95.25 mm).

4. Calculate the net area of member A

Apply the dimensions of the groove, which are recorded in the *Specification*. Referring to Fig. 104b, gross area = 27.19 in^2 (175.430 cm^2). The projected area of the groove and bolt hole = 4.5(1.00) + 0.813(2.625) = 6.63 in^2 (42.777 cm^2). The net area = 27.19 − 6.63 = 20.56 in^2 (132.7 cm^2).

5. Calculate the stress at the net section; compare with the allowable stress

The stress f = load/net area = 26,000/20.56 = 1260 lb/in^2 (8688 kPa). From the *Specification*, the allowable stress is f_{allow} = (⅞)(1700) = 1488 lb/in^2 (10,260 kPa). Also from the *Specification*, f_{allow} = 1650 lb/in^2 (11,377 kPa). The joint is therefore satisfactory in all respects.

6. Establish the connector spacing

Using the *Specification*, apply the recorded values without reduction because the connectors are stressed almost to capacity. Thus, a = 7 in (177.8 mm) and b = 9 in (228.6 mm).

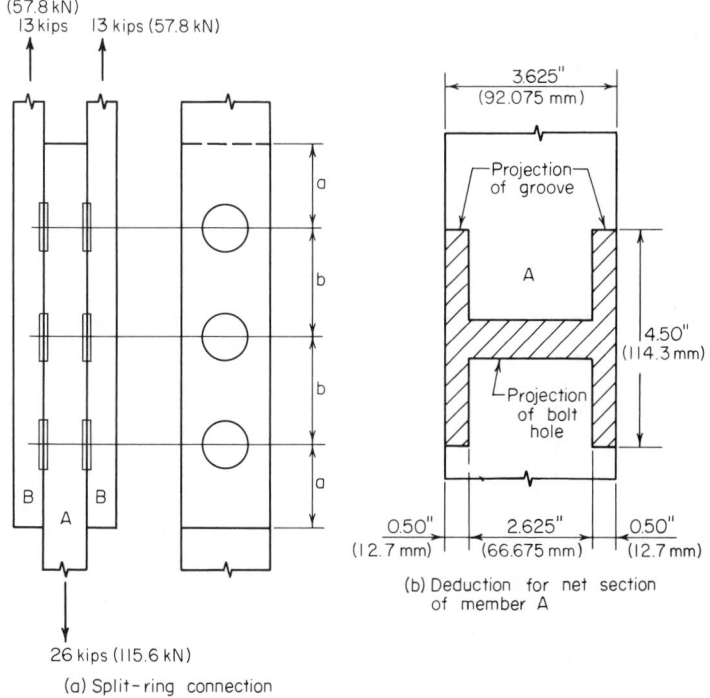

(57.8 kN)
13 kips   13 kips (57.8 kN)

a

b

b

a

B   B
    A

26 kips (115.6 kN)

(a) Split-ring connection

3.625"
(92.075 mm)

Projection
of groove

A

4.50"
(114.3 mm)

Projection
of bolt
hole

0.50"   2.625"   0.50"
(12.7 mm)   (66.675 mm)   (12.7 mm)

(b) Deduction for net section
of member A

FIG. 104

Reinforced Concrete

The design of reinforced-concrete members in this handook is executed in accordance with the specification titled *Building Code Requirements for Reinforced Concrete* of the American Concrete Institute (ACI). The ACI *Reinforced Concrete Design Handbook* contains many useful tables that expedite design work. The designer should become thoroughly familiar with this handbook and use the tables it contains whenever possible.

The spacing of steel reinforcing bars in a concrete member is subject to the restrictions imposed by the ACI *Code*. With reference to the beam and slab shown in Fig. 105, the reinforcing steel is assumed, for simplicity, to be concentrated at its centroidal axis, and the effective depth of the flexural member is taken as the distance from the extreme compression fiber to this axis. (The term *depth* hereafter refers to the *effective* rather than the overall depth of the beam.) For design purposes, it is usually assumed that the distance from the exterior surface to the center of the first row of steel bars is 2½ in (63.5 mm) in a beam with web stirrups, 2 in (50.8 mm) in a beam without stirrups, and 1 in (25.4 mm) in a slab. Where two rows of steel bars are provided, it is usually assumed that the distance from the exterior surface to the centroidal axis of the reinforcement is 3½ in (88.9 mm). The ACI *Handbook* gives the minimum beam widths needed to accommodate various combinations of bars in one row.

In a well-proportioned beam, the width-depth ratio lies between 0.5 and 0.75. The width and overall depth are usually an even number of inches.

The basic notational system pertaining to reinforced concrete beams is as follows: f'_c = ultimate compressive strength of concrete, lb/in^2 (kPa); f_c = maximum compressive stress in concrete, lb/in^2 (kPa); f_s = tensile stress in steel, lb/in^2 (kPa); f_y = yield-point stress in steel, lb/in^2 (kPa); ϵ_c = strain of extreme compression fiber; ϵ_s = strain of steel; b = beam width, in (mm); d = beam depth, in (mm); A_s = area of tension reinforcement, in^2 (cm^2); p = tension-reinforcement ratio, $A_s/(bd)$; q = tension-reinforcement index, pf_y/f'_c; n = ratio of modulus of elasticity of steel to that of concrete, E_s/E_c; C = resultant compressive force on transverse section, lb (N); T = resultant tensile force on transverse section, lb (N).

Where the subscript b is appended to a symbol, it signifies that the given quantity is evaluated at balanced-design conditions.

Design of Flexural Members by Ultimate-Strength Method

In the ultimate-strength design of a reinforced-concrete structure, as in the plastic design of a steel structure, the capacity of the structure is found by determining the load that will cause failure and dividing this result by the prescribed load factor. The load at impending failure is termed the *ultimate load*, and the maximum bending moment associated with this load is called the *ultimate moment*.

Since the tensile strength of concrete is relatively small, it is generally disregarded entirely in analyzing a beam. Consequently, the effective beam section is considered to comprise the reinforcing steel and the concrete on the compression side of the neutral axis, the concrete between these component areas serving merely as the ligature of the member.

The following notational system is applied in ultimate-strength design: a = depth of compression block, in (mm); c = distance from extreme compression fiber to neutral axis, in (mm); ϕ = capacity-reduction factor.

Where the subscript u is appended to a symbol, it signifies that the given quantity is evaluated at ultimate load.

For simplicity (Fig. 106), designers assume that when the ultimate moment is attained at a given section, there is a uniform stress in the concrete extending across a depth a, and that $f_c = 0.85f'_c$, and $a = k_1c$, where k_1 has the value stipulated in the ACI *Code*.

A reinforced-concrete beam has three potential modes of failure: crushing of the concrete, which is assumed to occur when ϵ_c reaches the value of 0.003; yielding of the steel, which begins when f_s reaches the value f_y; and the simultaneous crushing of the concrete and yielding of the steel. A beam that tends to fail by the third mode is said to be in *balanced design*. If the value of p exceeds that corresponding to balanced design (i.e., if there is an excess of reinforcement), the beam tends to fail by crushing of the concrete. But if the value of p is less than that corresponding to balanced design, the beam tends to fail by yielding of the steel.

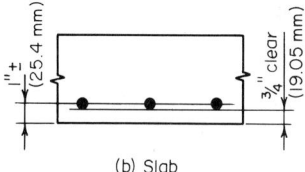

(a) Beam with stirrups

(b) Slab

FIG. 105 Spacing of reinforcing bars.

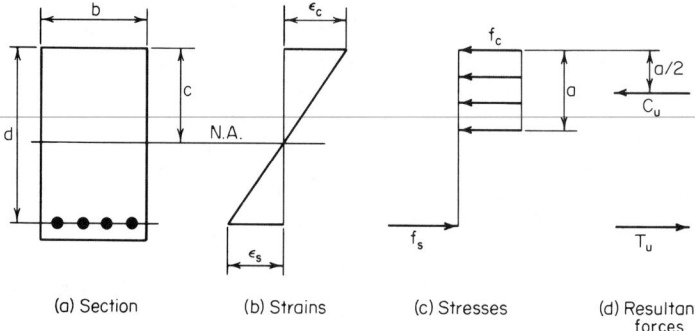

(a) Section   (b) Strains   (c) Stresses   (d) Resultant forces

FIG. 106 Conditions at ultimate moment.

Failure of the beam by the first mode would occur precipitously and without warning, whereas failure by the second mode would occur gradually, offering visible evidence of progressive failure. Therefore, to ensure that yielding of the steel would occur prior to failure of the concrete, the ACI *Code* imposes an upper limit of $0.75p_b$ on p.

To allow for material imperfections, defects in workmanship, etc., the *Code* introduces the capacity-reduction factor ϕ. A section of the *Code* sets $\phi = 0.90$ with respect to flexure and $\phi = 0.85$ with respect to diagonal tension, bond, and anchorage.

The basic equations for the ultimate-strength design of a rectangular beam reinforced solely in tension are

$$C_u = 0.85abf'_c \qquad T_u = A_s f_y \tag{1}$$

$$q = \frac{[A_s/(bd)]f_y}{f'_c} \tag{2}$$

$$a = 1.18qd \qquad c = \frac{1.18qd}{k_1} \tag{3}$$

$$M_u = \phi A_s f_y \left(d - \frac{a}{2} \right) \tag{4}$$

$$M_u = \phi A_s f_y d(1 - 0.59q) \tag{5}$$

$$M_u = \phi bd^2 f'_c q(1 - 0.59q) \tag{6}$$

$$A_s = \frac{bdf_c - [(bdf_c)^2 - 2bf_c M_u/\phi]^{0.5}}{f_y} \tag{7}$$

$$p_b = \frac{0.85k_1 f'_c}{f_y} \frac{87,000}{87,000 + f_y} \tag{8}$$

$$q_b = 0.85k_1 \left(\frac{87,000}{87,000 + f_y} \right) \tag{9}$$

In accordance with the *Code*,

$$q_{\max} = 0.75q_b = 0.6375k_1 \left(\frac{87,000}{87,000 + f_y} \right) \tag{10}$$

Figure 107 shows the relationship between M_u and A_s for a beam of given size. As A_s increases, the internal forces C_u and T_u increase proportionately, but M_u increases by a smaller proportion because the action line of C_u is depressed. The M_u-A_s diagram is parabolic, but its curvature is small. By comparing the coordinates of two points P_a and P_b, the following result is obtained, in which the subscripts correspond to that of the given point:

$$\frac{M_{ua}}{A_{sa}} > \frac{M_{ub}}{A_{sb}} \qquad \text{where } A_{sa} < A_{sb} \tag{11}$$

CAPACITY OF A RECTANGULAR BEAM

A rectangular beam having a width of 12 in (304.8 mm) and an effective depth of 19.5 in (495.3 mm) is reinforced with steel bars having an area of 5.37 in² (34.647 cm²). The beam is made of 2500-lb/in² (17,237.5-kPa) concrete, and the steel has a yield-point stress of 40,000 lb/in² (275,800 kPa). Compute the ultimate moment this beam may resist (*a*) without referring to any design tables and without apply-

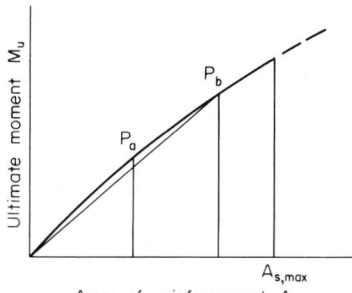

FIG. 107

ing the basic equations of ultimate-strength design except those that are readily apparent; (b) by applying the basic equations.

Calculation Procedure:

1. Compute the area of reinforcement for balanced design

Use the relation $\epsilon_s = f_y/E_s = 40,000/29,000,000 = 0.00138$. For balanced design, $c/d = \epsilon_c/(\epsilon_c + \epsilon_s) = 0.003/(0.003 + 0.00138) = 0.685$. Solving for c by using the relation for c/d, we find $c = 13.36$ in (339.344 mm). Also, $a = k_1c = 0.85(13.36) = 11.36$ in (288.544 mm). Then $T_u = C_u = ab(0.85)f'_c = 11.36(12)(0.85)(2500) = 290,000$ lb (1,289,920 N); $A_s = T_u/f_y = 290,000/40,000 = 7.25$ in^2 (46,777 cm^2); and $0.75A_s = 5.44$ in^2 (35.097 cm^2). In the present instance, $A_s = 5.37$ in^2 (34.647 cm^2). This is acceptable.

2. Compute the ultimate-moment capacity of this member

Thus $T_u = A_sf_y = 5.37(40,000) = 215,000$ lb (956,320 N); $C_u = ab(0.85)f'_c = 25,500a = 215,000$ lb (956,320 N); $a = 8.43$ in (214.122 mm); $M_u = \phi T_u(d - a/2) = 0.90(215,000)(19.5 - 8.43/2) = 2,960,000$ in·lb (334,421 N·m). These two steps comprise the solution to part a. The next two steps comprise the solution of part b.

3. Apply Eq. 10; ascertain whether the member satisfies the Code

Thus, $q_{max} = 0.6375k_1(87,000)/(87,000 + f_y) = 0.6375(0.85)(87/127) = 0.371$; $q = [A_s/(bd)]f_y/f'_c = [5.37/(12 \times 19.5)]40/2.5 = 0.367$. This is acceptable.

4. Compute the ultimate-moment capacity

Applying Eq. 5 yields $M_u = \phi A_sf_yd(1 - 0.59q) = 0.90(5.37)(40,000)(19.5)(1 - 0.59 \times 0.367) = 2,960,000$ in·lb (334,421 N·m). This agrees exactly with the result computed in step 2.

DESIGN OF A RECTANGULAR BEAM

A beam on a simple span of 20 ft (6.1 m) is to carry a uniformly distributed live load of 1670 lb/lin ft (24,372 N/m) and a dead load of 470 lb/lin ft (6859 N/m), which includes the estimated weight of the beam. Architectural details restrict the beam width to 12 in (304.8 mm) and require that the depth be made as small as possible. Design the section, using $f'_c = 3000$ lb/in^2 (20,685 kPa) and $f_y = 40,000$ lb/in^2 (275,800 kPa).

Calculation Procedure:

1. Compute the ultimate load for which the member is to be designed

The beam depth is minimized by providing the maximum amount of reinforcement permitted by the Code. From the previous calculation procedure, $q_{max} = 0.371$.

Use the load factors given in the Code: $w_{DL} = 470$ lb/lin ft (6859 N/m); $w_{LL} = 1670$ lb/lin ft (24,372 N/m); $L = 20$ ft (6.1 m). Then $w_u = 1.5(470) + 1.8(1670) = 3710$ lb/lin ft (54,143 N/m); $M_u = \frac{1}{8}(3710)(20)^212 = 2,230,000$ in·lb (251,945.4 N·m).

2. Establish the beam size

Solve Eq. 6 for d. Thus, $d^2 = M_u/[\phi bf'_cq(1 - 0.59q)] = 2,230,000/[0.90(12)(3000) \times (0.371)(0.781)]$; $d = 15.4$ in (391.16 mm).

Set $d = 15.5$ in (393.70 mm). Then the corresponding reduction in the value of q is negligible.

3. Select the reinforcing bars

Using Eq. 2, we find $A_s = qbdf'_c/f_y = 0.371(12)(15.5)(3/40) = 5.18$ in^2 (33.421 cm^2). Use four no. 9 and two no. 7 bars, for which $A_s = 5.20$ in^2 (33.550 cm^2). This group of bars cannot be accommodated in the 12-in (304.8-mm) width and must therefore be placed in two rows. The overall beam depth will therefore be 19 in (482.6 mm).

4. Summarize the design

Thus, the beam size is 12 $\times$ 19 in (304.8 $\times$ 482.6 mm); reinforcement, four no. 9 and two no. 7 bars.

DESIGN OF THE REINFORCEMENT IN A RECTANGULAR BEAM OF GIVEN SIZE

A rectangular beam 9 in (228.6 mm) wide with a 13.5-in (342.9-mm) effective depth is to sustain an ultimate moment of 95 ft·kips (128.8 kN·m). Compute the area of reinforcement, using f'_c = 3000 lb/in^2 (20,685 kPa) and f_y = 40,000 lb/in^2 (275,800 kPa).

Calculation Procedure:

1. Investigate the adequacy of the beam size

From previous calculation procedures, q_{max} = 0.371. By Eq. 6, $M_{u,max}$ = 0.90 × $(9)(13.5)^2(3)(0.371)(0.781)$ = 1280 in·kips (144.6 kN·m); M_u = 95(12) = 1140 in·kips (128.8 kN·m). This is acceptable.

2. Apply Eq. 7 to evalute A_s

Thus, f_c = 0.85(3) = 2.55 kips/in^2 (17.582 MPa); bdf_c = 9(13.5)(2.55) = 309.8 kips (1377.99 kN); A_s = [309.8 − (309.8^2 − 58,140)$^{0.5}$]/40 = 2.88 in^2 (18.582 cm^2).

CAPACITY OF A T BEAM

Determine the ultimate moment that may be resisted by the T beam in Fig. 108a if f'_c = 3000 lb/in^2 (20,685 kPa) and f_y = 40,000 lb/in^2 (275,800 kPa).

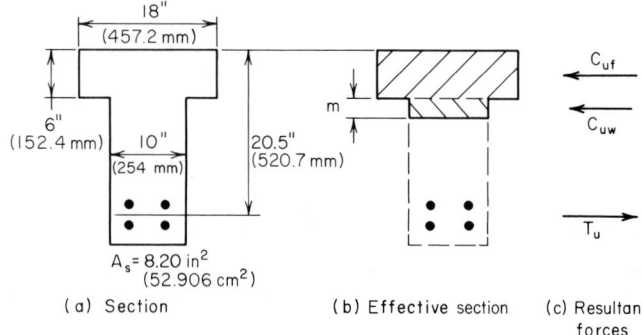

(a) Section   (b) Effective section   (c) Resultant forces

FIG. 108

Calculation Procedure:

1. Compute T_u and the resultant force that may be developed in the flange

Thus, T_u = 8.20(40,000) = 328,000 lb (1,458,944 N); f_c = 0.85(3000) = 2550 lb/in^2 (17,582.3 kPa); C_{uf} = 18(6)(2550) = 275,400 lb (1,224,979 N). Since C_{uf} < T_u, the deficiency must be supplied by the web.

2. Compute the resultant force developed in the web and the depth of the stress block in the web

Thus, C_{uw} = 328,000 − 275,400 = 52,600 lb (233,964.8 N); m = depth of the stress block = 52,600/[2550(10)] = 2.06 in (52.324 mm).

3. Evaluate the ultimate-moment capacity

Thus, M_u = 0.90[275,400(20.5 − 3) + 52,600(20.5 − 6 − 1.03)] = 4,975,000 in·lb (562,075.5 N·m).

4. Determine if the reinforcement complies with the Code

Let b' = width of web, in (mm); A_{s1} = area of reinforcement needed to resist the compressive force in the overhanging portion of the flange, in^2 (cm^2); A_{s2} = area of reinforcement needed to resist the compressive force in the remainder of the section, in^2 (cm^2). Then $p_2 = A_{s2}/(b'd)$; A_{s1} = 2550(6)(18 − 10)/40,000 = 3.06 in^2 (19.743 cm^2); A_{s2} = 8.20 − 3.06 = 5.14 in^2 (33.163 cm^2). Then p_2 = 5.14/[10(20.5)] = 0.025.

A section of the ACI *Code* subjects the reinforcement ratio p_2 to the same restriction as that in a rectangular beam. By Eq. 8, $p_{2,\text{max}}$ = 0.75p_b = 0.75(0.85)(0.85)(3/40)(87/127) = 0.0278 > 0.025. This is acceptable.

CAPACITY OF A T BEAM OF GIVEN SIZE

The T beam in Fig. 109 is made of 3000-lb/in^2 (20,685-kPa) concrete, and f_y = 40,000 lb/in^2 (275,800 kPa). Determine the ultimate-moment capacity of this member if it is reinforced in tension only.

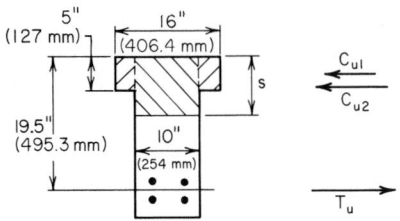

FIG. 109

Calculation Procedure:

1. Compute C_{u1}, $C_{u2,\text{max}}$, and s_{max}

Let the subscript 1 refer to the overhanging portion of the flange and the subscript 2 refer to the remainder of the compression zone. Then f_c = 0.85(3000) = 2550 lb/in^2 (17,582.3 kPa); C_{u1} = 2550(5)(16 − 10) = 76,500 lb (340,272 N). From the previous calculation procedure, $p_{2,\text{max}}$ = 0.0278. Then $A_{s2,\text{max}}$ = 0.0278(10)(19.5) = 5.42 in^2 (34.970 cm^2); $C_{u2,\text{max}}$ = 5.42(40,000) = 216,800 lb (964,326.4 N); s_{max} = 216,800/[10(2550)] = 8.50 in (215.9 mm).

2. Compute the ultimate-moment capacity

Thus, $M_{u,\text{max}}$ = 0.90[76,500(19.5 − 5/2) + 216,800(19.5 − 8.50/2)] = 4,145,000 in·lb (468,300 N·m).

DESIGN OF REINFORCEMENT IN A T BEAM OF GIVEN SIZE

The T beam in Fig. 109 is to resist an ultimate moment of 3,960,000 in·lb (447,400.8 N·m). Determine the required area of reinforcement, using f'_c = 3000 lb/in^2 (20,685 kPa) and f_y = 40,000 lb/in^2 (275,800 kPa).

Calculation Procedure:

1. Obtain a moment not subject to reduction

From the previous calculation procedure, the ultimate-moment capacity of this member is 4,145,000 in·lb (468,300 N·m). To facilitate the design, divide the given ultimate moment M_u by the capacity-reduction factor to obtain a moment M'_u that is not subject to reduction. Thus M'_u = 3,960,000/0.9 = 4,400,000 in·lb (497,112 N·m).

2. Compute the value of s associated with the given moment

From step 2 in the previous calculation procedure, M'_{u1} = 1,300,000 in·lb (146,874 N·m). Then M'_{u2} = 4,400,000 − 1,300,000 = 3,100,000 in·lb (350,238 N·m). But M'_{u2} = 2550(10s)(19.5 − s/2), so s = 7.79 in (197.866 mm).

3. Compute the area of reinforcement

Thus, C_{u2} = $M'_{u2}/(d - \frac{1}{2}s)$ = 3,100,000/(19.5 − 3.90) = 198,700 lb (883,817.6 N). From step 1 of the previous calculation procedure, C_{u1} = 76,500 lb (340,272 N); T_u = 76,500 + 198,700 = 275,200 lb (1,224,089.6 N); A_s = 275,200/40,000 = 6.88 in^2 (174.752 mm).

4. Verify the solution

To verify the solution, compute the ultimate-moment capacity of the member. Use the notational system given in earlier calculation procedures. Thus, C_{uf} = 16(5)(2550) = 204,000 lb (907,392

N); $C_{uw} = 275,200 - 204,000 = 71,200$ lb (316,697.6 N); $m = 71,200/[2550(10)] = 2.79$ in (70.866 mm); $M_u = 0.90\,[204,000\,(19.5 - 2.5) + 71,200(19.5 - 5 - 1.40)] = 3,960,000$ in·lb (447,400.8 N·m). Thus, the result is verified because the computed moment equals the given moment.

REINFORCEMENT AREA FOR A DOUBLY REINFORCED RECTANGULAR BEAM

A beam that is to resist an ultimate moment of 690 ft·kips (935.6 kN·m) is restricted to a 14-in (355.6-mm) width and 24-in (609.6-mm) total depth. Using $f'_c = 5000$ lb/in² and $f_y = 50,000$ lb/in² (344,750 kPa), determine the area of reinforcement.

Calculation Procedure:

1. *Compute the values of q_b, q_{max}, and p_{max} for a singly reinforced beam*

As the following calculations will show, it is necessary to reinforce the beam both in tension and in compression. In Fig. 110, let A_s = area of tension reinforcement, in² (cm²); A'_s = area of

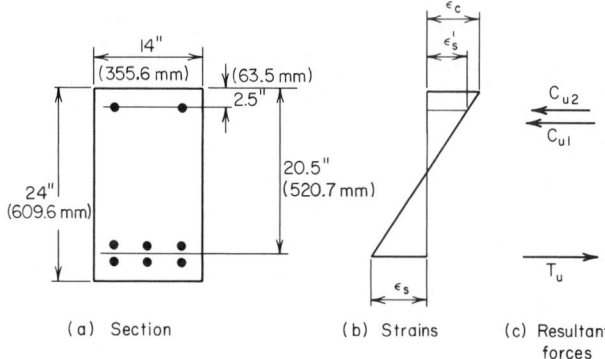

(a) Section   (b) Strains   (c) Resultant
                                              forces

FIG. 110   Doubly reinforced rectangular beam.

compression reinforcement, in² (cm²); d' = distance from compression face of concrete to centroid of compression reinforcement, in (mm); f_s = stress in tension steel, lb/in² (kPa); f'_s = stress in compression steel, lb/in² (kPa); ϵ'_s = strain in compression steel; $p = A_s/(bd)$; $p' = A'_s/(bd)$; $q = pf_y/f'_c$; M_u = ultimate moment to be resisted by member, in·lb (N·m); M_{u1} = ultimate-moment capacity of member if reinforced solely in tension; M_{u2} = increase in ultimate-moment capacity resulting from use of compression reinforcement; C_{u1} = resultant force in concrete, lb (N); C_{u2} = resultant force in compression steel, lb (N).

If $f'_s = f_y$, the tension reinforcement may be resolved into two parts having areas of $A_s - A'_s$ and A'_s. The first part, acting in combination with the concrete, develops the moment M_{u1}. The second part, acting in combination with the compression reinforcement, develops the moment M_{u2}.

To ensure that failure will result from yielding of the tension steel rather than crushing of the concrete, the ACI *Code* limits $p - p'$ to a maximum value of $0.75p_b$, where p_b has the same significance as for a singly reinforced beam. Thus the *Code*, in effect, permits setting $f'_s = f_y$ if inception of yielding in the compression steel will precede or coincide with failure of the concrete at balanced-design ultimate moment. This, however, introduces an inconsistency, for the limit imposed on $p - p'$ precludes balanced design.

By Eq. 9, $q_b = 0.85(0.80)(87/137) = 0.432$; $q_{max} = 0.75(0.432) = 0.324$; $p_{max} = 0.324(5/50) = 0.0324$.

2. *Compute M_{u1}, M_{u2}, and C_{u2}*

Thus, $M_u = 690,000(12) = 8,280,000$ in·lb (935,474.4 N·m). Since two rows of tension bars are probably required, $d = 24 - 3.5 = 20.5$ in (520.7 mm). By Eq. 6, $M_{u1} = 0.90(14)(20.5)^2(5000)$

$\times$ (0.324)(0.809) = 6,940,000 in·lb (784,081.2 N·m); M_{u2} = 8,280,000 − 6,940,000 = 1,340,000 in·lb (151,393.2 N·m); C_{u2} = $M_{u2}/(d - d')$ = 1,340,000/(20.5 − 2.5) = 74,400 lb (330,931.2 N).

3.   *Compute the value of ϵ'_s under the balanced-design ultimate moment*

Compare this value with the strain at incipient yielding. By Eq. 3, c_b = $1.18q_bd/k_1$ = 1.18(0.432)(20.5)/0.80 = 13.1 in (332.74 mm); ϵ'_s/ϵ_c = (13.1 − 2.5)/13.1 = 0.809; ϵ'_s = 0.809(0.003) = 0.00243; ϵ_y = 50/29,000 = 0.0017 < ϵ'_s. The compression reinforcement will therefore yield before the concrete fails, and $f'_s = f_y$ may be used.

4.   *Alternatively, test the compression steel for yielding*

Apply

$$p - p' \geq \frac{0.85k_1f'_cd'(87,000)}{f_yd(87,000 - f_y)} \tag{12}$$

If this relation obtains, the compression steel will yield. The value of the right-hand member is 0.85(0.80)(5/50)(2.5/20.5)(87/37) = 0.0195. From the preceding calculations, $p - p'$ = 0.0324 > 0.0195. This is acceptable.

5.   *Determine the areas of reinforcement*

By Eq. 2, A_s = A'_s = $q_{max}bdf'_c/f_y$ = 0.324(14)(20.5)(5/50) = 9.30 in² (60.00 cm²); A'_s = $C_{u2}/(\phi f_y)$ = 74,400/[0.90(50,000)] = 1.65 in² (10.646 cm²); A_s = 9.30 + 1.65 = 10.95 in² (70.649 cm²).

6.   *Verify the solution*

Apply the following equations for the ultimate-moment capacity:

$$a = \frac{(A_s - A'_s)f_y}{0.85f'_cb} \tag{13}$$

So a = 9.30(50,000)/[0.85(5000)(14)] = 7.82 in (198.628 mm). Also,

$$M_u = \phi f_y\left[(A_s - A'_s)\left(d - \frac{a}{2}\right) + A'_s(d - d')\right] \tag{14}$$

So M_u = 0.90(50,000)(9.30 × 16.59 + 1.65 × 18) = 8,280,000 in·lb (935,474.4 N·m), as before. Therefore, the solution has been verified.

DESIGN OF WEB REINFORCEMENT

A 15-in (381-mm) wide 22.5-in (571.5-mm) effective-depth beam carries a uniform ultimate load of 10.2 kips/lin ft (148.86 kN/m). The beam is simply supported, and the clear distance between supports is 18 ft (5.5 m). Using f'_c = 3000 lb/in² (20,685 kPa) and f_y = 40,000 lb/in² (275,800 kPa), design web reinforcement in the form of vertical U stirrups for this beam.

Calculation Procedure:

1.   *Construct the shearing-stress diagram for half-span*

The ACI *Code* provides two alternative methods for computing the allowable shearing stress on an unreinforced web. The more precise method recognizes the contribution of both the shearing stress and flexural stress on a cross section in producing diagonal tension. The less precise and more conservative method restricts the shearing stress to a stipulated value that is independent of the flexural stress.

For simplicity, the latter method is adopted here. A section of the *Code* sets ϕ = 0.85 with respect to the design of web reinforcement. Let v_u = nominal ultimate shearing stress, lb/in² (kPa); v_c = shearing stress resisted by concrete, lb/in² (kPa); v'_u = shearing stress resisted by the web reinforcement, lb/in² (kPa); A_v = total cross-sectional area of stirrup, in² (cm²); V_u = ultimate vertical shear at section, lb (N); s = center-to-center spacing of stirrups, in (mm).

The shearing-stress diagram for half-span is shown in Fig. 111. Establish the region AF within

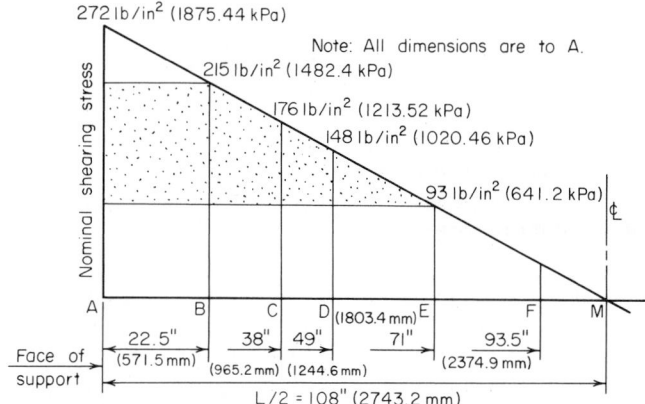

FIG. 111 Shearing-stress diagram.

which web reinforcement is required. The *Code* sets the allowable shearing stress in the concrete at

$$v_c = 2\phi(f_c')^{0.5} \tag{15}$$

The equation for nominal ultimate shearing stress is

$$v_u = \frac{V_u}{bd} \tag{16}$$

Then, $v_c = 2(0.85)(3000)^{0.5} = 93$ lb/in² (641.2 kPa).

At the face of the support, $V_u = 9(10,200) = 91,800$ lb (408,326.4 N); $v_u = 91,800/[15(22.5)]$ = 272 lb/in² (1875.44 kPa). The slope of the shearing-stress diagram = $-272/108 = -2.52$ lb/ (in²·in) (-0.684 kPa/mm). At distance d from the face of the support, $v_u = 272 - 22.5(2.52)$ = 215 lb/in² (1482.4 kPa); $v_u' = 215 - 93 = 122$ lb/in² (841.2 kPa).

Let E denote the section at which $v_u = v_c$. Then, $AE = (272 - 93)/2.52 = 71$ in (1803.4 mm). A section of the *Code* requires that web reinforcement be continued for a distance d beyond the section where $v_u = v_c$; $AF = 71 + 22.5 = 93.5$ in (2374.9 mm).

2. *Check the beam size for* Code *compliance*

Thus, $v_{u,\text{max}} = 10\phi(f_c')^{0.5} = 466 > 215$ lb/in² (1482.4 kPa). This is acceptable.

3. *Select the stirrup size*

Equate the spacing near the support to the minimum practical value, which is generally considered to be 4 in (101.6 mm). The equation for stirrup spacing is

$$s = \frac{\phi A_v f_y}{v_u' b} \tag{17}$$

Then $A_v = sv_u'b/(\phi f_y) = 4(122)(15)/[0.85(40,000)] = 0.215$ in² (1.3871 cm²). Since each stirrup is bent into the form of a U, the total cross-sectional area is twice that of a straight bar. Use no. 3 stirrups for which $A_v = 2(0.11) = 0.22$ in² (1.419 cm²).

4. *Establish the maximum allowable stirrup spacing*

Apply the criteria of the *Code*, or $s_{\text{max}} = d/4$ if $v_u > 6\phi(f_c')^{0.5}$. The right-hand member of this inequality has the value 279 lb/in² (1923.70 kPa), and this limit therefore does not apply. Then $s_{\text{max}} = d/2 = 11.25$ in (285.75 mm), or $s_{\text{max}} = A_v/(0.0015b) = 0.22/[0.0015(15)] = 9.8$ in (248.92 mm). The latter limit applies, and the stirrup spacing will therefore be restricted to 9 in (228.6 mm).

5. *Locate the beam sections at which the required stirrup spacing is 6 in (152.4 mm) and 9 in (228.6 mm)*

Use Eq. 17. Then $\phi A_v f_y/b = 0.85(0.22)(40,000)/15 = 499$ lb/in (87.38 kN/m). At C: $v'_u = 499/6 = 83$ lb/in^2 (572.3 kPa); $v_u = 83 + 93 = 176$ lb/in^2 (1213.52 kPa); $AC = (272 - 176)/2.52 = 38$ in (965.2 mm). At D: $v'_u = 499/9 = 55$ lb/in^2 (379.2 kPa); $v_u = 55 + 93 = 148$ lb/in^2 (1020.46 kPa); $AD = (272 - 148)/2.52 = 49$ in (1244.6 mm).

6. *Devise a stirrup spacing conforming to the computed results*

The following spacing, which requires 17 stirrups for each half of the span, is satisfactory and conforms with the foregoing results:

Quantity	Spacing, in (mm)	Total, in (mm)	Distance from last stirrup to face of support, in (mm)
1	2 (50.8)	2 (50.8)	2 (50.8)
9	4 (101.6)	36 (914.4)	38 (965.2)
2	6 (152.4)	12 (304.8)	50 (1270)
5	9 (228.6)	45 (1143)	95 (2413)

DETERMINATION OF BOND STRESS

A beam of 4000-lb/in^2 (27,580-kPa) concrete has an effective depth of 15 in (381 mm) and is reinforced with four no. 7 bars. Determine the ultimate bond stress at a section where the ultimate shear is 72 kips (320.3 kN). Compare this with the allowable stress.

Calculation Procedure:

1. *Determine the ultimate shear flow* h_u

The adhesion of the concrete and steel must be sufficiently strong to resist the horizontal shear flow. Let u_u = ultimate bond stress, lb/in^2 (kPa); V_u = ultimate vertical shear, lb (N); Σo = sum of perimeters of reinforcing bars, in (mm). Then the ultimate shear flow at any plane between the neutral axis and the reinforcing steel is $h_u = V_u/(d - a/2)$.

In conformity with the notational system of the working-stress method, the distance $d - a/2$ is designated as jd. Dividing the shear flow by the area of contact in a unit length and introducing the capacity-reduction factor yield

$$u_u = \frac{V_u}{\phi \Sigma o jd} \tag{18}$$

A section of the ACI *Code* sets $\phi = 0.85$ with respect to bond, and j is usually assigned the approximate value of 0.875 when this equation is used.

2. *Calculate the bond stress*

Thus, $\Sigma o = 11.0$ in (279.4 mm), from the ACI *Handbook*. Then $u_u = 72,000/[0.85(11.0)(0.875) \times (15)] = 587$ lb/in^2 (4047.4 kPa).

The allowable stress is given in the *Code* as

$$u_{u,\text{allow}} = \frac{9.5(f'_c)^{0.5}}{D} \tag{19}$$

but not above 800 lb/in^2 (5516 kPa). Thus, $u_{u,\text{allow}} = 9.5(4,000)^{0.5}/0.875 = 687$ lb/in^2 (4736.9 kPa).

DESIGN OF INTERIOR SPAN OF A ONE-WAY SLAB

A floor slab that is continuous over several spans carries a live load of 120 lb/ft^2 (5745 N/m^2) and a dead load of 40 lb/ft^2 (1915 N/m^2), exclusive of its own weight. The clear spans are 16 ft (4.9

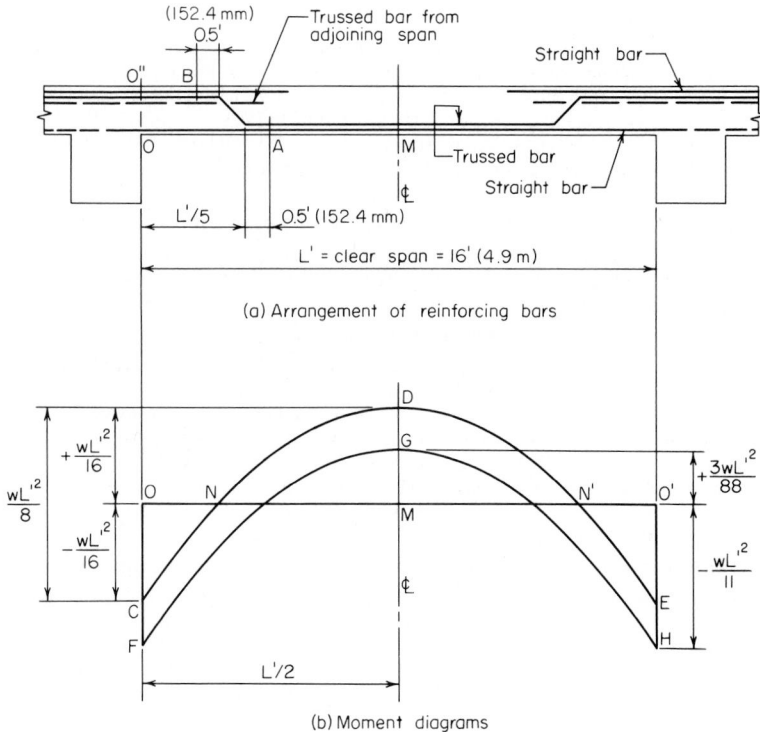

(a) Arrangement of reinforcing bars

(b) Moment diagrams

FIG. 112

m). Design the interior span, using $f'_c = 3000$ lb/in² (20,685 kPa) and $f_y = 50,000$ lb/in² (344,750 kPa).

Calculation Procedure:

1. Find the minimum thickness of the slab as governed by the Code

Refer to Fig. 112. The maximum potential positive or negative moment may be found by applying the type of loading that will induce the critical moment and then evaluating this moment. However, such an analysis is time-consuming. Hence, it is wise to apply the moment equations recommended in the ACI *Code* whenever the span and loading conditions satisfy the requirements given there. The slab is designed by considering a 12-in (304.8-mm) strip as an individual beam, making $b = 12$ in (304.8 mm).

Assuming that $L = 17$ ft (5.2 m), we know the minimum thickness of the slab is $t_{min} = L/35 = 17(12)/35 = 5.8$ in (147.32 mm).

2. Assuming a slab thickness, compute the ultimate load on the member

Tentatively assume $t = 6$ in (152.4 mm). Then the beam weight $= (6/12)(150$ lb/ft³ $= 75$ lb/lin ft (1094.5 N/m). Also, $w_u = 1.5(40 + 75) + 1.8(120) = 390$ lb/lin ft (5691.6 N/m).

3. Compute the shearing stress associated with the assumed beam size

From the *Code* for an interior span, $V_u = \frac{1}{2}w_u L' = \frac{1}{2}(390)(16) = 3120$ lb (13,877.8 N); $d = 6 - 1 = 5$ in (127 mm); $v_u = 3120/[12(5)] = 52$ lb/in² (358.54 kPa); $v_c = 93$ lb/in² (641.2 kPa). This is acceptable.

4. Compute the two critical moments

Apply the appropriate moment equations. Compare the computed moments with the moment capacity of the assumed beam size to ascertain whether the size is adequate. Thus, $M_{u,neg} = (\frac{1}{11})w_uL'^2 = (\frac{1}{11})(390)(16)^2(12) = 108,900$ in·lb (12,305.5 N·m), where the value 12 converts the dimension to inches. Then $M_{u,pos} = \frac{1}{16}w_uL'^2 = 74,900$ in·lb (8462.2 N·m). By Eq. 10, $q_{max} = 0.6375(0.85)(87/137) = 0.344$. By Eq. 6, $M_{u,allow} = 0.90(12)(5)^2(3000)(0.344)(0.797) = 222,000$ in·lb (25,081.5 N·m). This is acceptable. The slab thickness will therefore be made 6 in (152.4 mm).

5. Compute the area of reinforcement associated with each critical moment

By Eq. 7, $bdf_c = 12(5)(2.55) = 153.0$ kips (680.54 kN); then $2bf_cM_{u,neg}/\phi = 2(12)(2.55)(108.9)/0.90 = 7405$ kips^2 (146,505.7 kN^2), $A_{s,neg} = [153.0 - (153.0^2 - 7405)^{0.5}]/50 = 0.530$ in^2 (3.4196 cm^2). Similarly, $A_{s,pos} = 0.353$ in^2 (2.278 cm^2).

6. Select the reinforcing bars, and locate the bend points

For positive reinforcement, use no. 4 trussed bars 13 in (330.2 mm) on centers, alternating with no. 4 straight bars 13 in (330.2 mm) on centers, thus obtaining $A_s = 0.362$ in^2 (2.336 cm^2).

For negative reinforcement, supplement the trussed bars over the support with no. 4 straight bars 13 in (330.2 mm) on centers, thus obtaining $A_s = 0.543$ in^2 (3.502 cm^2).

The trussed bars are usually bent upward at the fifth points, as shown in Fig. 112a. The reinforcement satisfies a section of the ACI *Code* which requires that "at least . . . one-fourth the positive moment reinforcement in continuous beams shall extend along the same face of the beam into the support at least 6 in (152.4 mm)."

7. Investigate the adequacy of the reinforcement beyond the bend points

In accordance with the *Code*, $A_{min} = A_t = 0.0020bt = 0.0020(12)(6) = 0.144$ in^2 (0.929 cm^2).

A section of the *Code* requires that reinforcing bars be extended beyond the point at which they become superfluous with respect to flexure a distance equal to the effective depth or 12 bar diameters, whichever is greater. In the present instance, extension $= 12(0.5) = 6$ in (152.4 mm). Therefore, the trussed bars in effect terminate as positive reinforcement at section A (Fig. 112). Then $L'/5 = 3.2$ ft (0.98 m); $AM = 8 - 3.2 - 0.5 = 4.3$ ft (1.31 m).

The conditions immediately to the left of A are $M_u = M_{u,pos} - \frac{1}{2}w_u(AM)^2 = 74,900 - \frac{1}{2}(390)(4.3)^2(12) = 31,630$ in·lb (3573.56 N·m); $A_{s,pos} = 0.181$ in^2 (1.168 cm^2); $q = 0.181(50)/[12(5)(3)] = 0.0503$. By Eq. 5, $M_{u,allow} = 0.90(0.181)(50,000)(5)(0.970) = 39,500$ in·lb (4462.7 N·m). This is acceptable.

Alternatively, Eq. 11 may be applied to obtain the following conservative approximation: $M_{u,allow} = 74,900(0.181)/0.353 = 38,400$ in·lb (4338.43 N·m).

The trussed bars in effect terminate as negative reinforcement at B, where $O''B = 3.2 - 0.33 - 0.5 = 2.37$ ft (72.23 m). The conditions immediately to the right of B are $|M_u| = M_{u,neg} - 12(3120 \times 2.37 - \frac{1}{2} \times 390 \times 2.37^2) = 33,300$ in·lb (3762.23 N·m). Then $A_{s,neg} = 0.362$ in^2 (2.336 cm^2). As a conservative approximation, $M_{u,allow} = 108,900(0.362)/0.530 = 74,400$ in·lb (8405.71 N·m). This is acceptable.

8. Locate the point at which the straight bars at the top may be discontinued

9. Investigate the bond stresses

In accordance with Eq. 19, $u_{u,allow} = 800$ lb/in^2 (5516 kPa).

If *CDE* in Fig. 112b represents the true moment diagram, the bottom bars are subjected to bending stress in the interval NN'. Manifestly, the maximum bond stress along the bottom occurs at these boundary points (points of contraflexure), where the shear is relatively high and the straight bars alone are present. Thus $MN = 0.354L'$; V_u at N/V_u at support $= 0.354L'/(0.5L') = 0.71$; V_u at $N = 0.71(3120) = 2215$ lb (9852.3 N). By Eq. 18, $u_u = V_u/(\phi\Sigma ojd) = 2215/[0.85(1.45)(0.875)(5)] = 411$ lb/in^2 (2833.8 kPa). This is acceptable. It is apparent that the maximum bond stress in the top bars has a smaller value.

ANALYSIS OF A TWO-WAY SLAB BY THE YIELD-LINE THEORY

The slab in Fig. 113a is simply supported along all four edges and is isotropically reinforced. It supports a uniformly distributed ultimate load of w_u lb/ft^2 (kPa). Calculate the ultimate unit moment m_u for which the slab must be designed.

Calculation Procedure:

1. Draw line GH perpendicular to AE at E; express distances b and c in terms of a

Consider a slab to be reinforced in orthogonal directions. If the reinforcement in one direction is identical with that in the other direction, the slab is said to be *isotropically reinforced;* if the

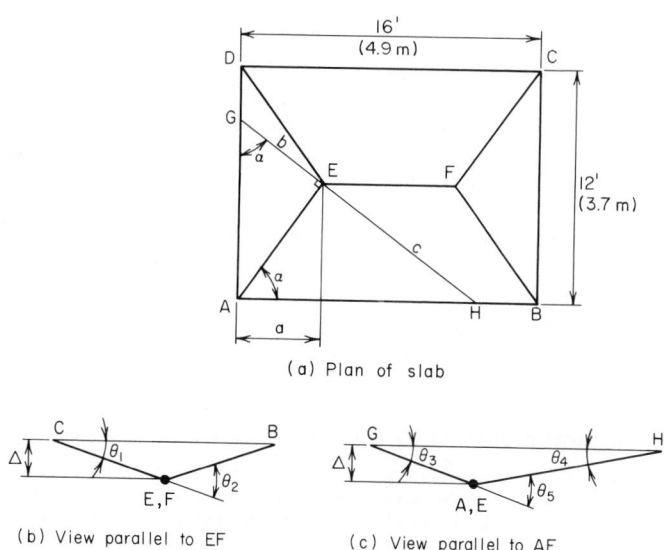

(a) Plan of slab

(b) View parallel to EF

(c) View parallel to AE

FIG. 113 Analysis of two-way slab by mechanism method.

reinforcements differ, the slab is described as *orthogonally anisotropic.* In the former case, the capacity of the slab is identical in all directions; in the latter case, the capacity has a unique value in every direction. In this instance, assume that the slab size is excessive with respect to balanced design, the result being that the failure of the slab will be characterized by yielding of the steel.

In a steel beam, a plastic hinge forms at a *section;* in a slab, a plastic hinge is assumed to form along a *straight line,* termed a *yield line.* It is plausible to assume that by virtue of symmetry of loading and support conditions the slab in Fig. 113a will fail by the formation of a central yield line *EF* and diagonal yield lines such as *AE,* the ultimate moment at these lines being positive. The ultimate *unit* moment m_u is the moment acting on a unit length.

Although it is possible to derive equations that give the location of the yield lines, this procedure is not feasible because the resulting equations would be unduly cumbersome. The procedure followed in practice is to assign a group of values to the distance a and to determine the corresponding values of m_u. The true value of m_u is the highest one obtained. Either the static or mechanism method of analysis may be applied; the latter will be applied here.

Expressing the distances b and c in terms of a gives $\tan \alpha = 6/a = AE/b = c/(AE)$; $b = aAE/6$; $c = 6AE/a$.

2. Find the rotation of the plastic hinges

Allow line *EF* to undergo a virtual displacement Δ after the collapse load is reached. During the virtual displacement, the portions of the slab bounded by the yield lines and the supports rotate as planes. Refer to Fig. 113b and c: $\theta_1 = \Delta/6$; $\theta_2 = 2\theta_1 = \Delta/3 = 0.333\Delta$; $\theta_3 = \Delta/b$; $\theta_4 = \Delta/c$; $\theta_5 = \Delta(1/b + 1/c) = [\Delta/(AE)](6/a + a/6)$.

3. Select a trial value of a, and evaluate the distances and angles

Using $a = 4.5$ ft (1.37 m) as the trial value, we find $AE = (a^2 + 6^2)^{0.5} = 7.5$ ft (2.28 m); $b = 5.63$ ft (1.716 m); $c = 10$ ft (3.0 m); $\theta_5 = (\Delta/7.5)(6/4.5 + 4.5/6) = 0.278\Delta$.

4. Develop an equation for the external work W_E performed by the uniform load on a surface that rotates about a horizontal axis

In Fig. 114, consider that the surface ABC rotates about axis AB through an angle θ while carrying a uniform load of w lb/ft² (kPa). For the elemental area dA, the deflection, total load, and external work are $\delta = x\theta$; $dW = w\,dA$; $dW_E = \delta\,dW = x\theta w\,dA$. The total work for the surface is $W_E = w\theta \int x\,dA$, or

$$W_E = w\theta Q \tag{20}$$

where Q = static moment of total area, with respect to the axis of rotation.

5. Evaluate the external and internal work for the slab

Using the assumed value, we see $a = 4.5$ ft (1.37 m), $EF = 16 - 9 = 7$ ft (2.1 m). The external work for the two triangles is $2w_u(\Delta/4.5)(\frac{1}{3})(12)(4.5)^2 = 18w_u\Delta$. The external work for the two trapezoids is $2w_u(\Delta/6)(\frac{1}{6})(16 + 2 \times 7)(6)^2 = 60w_u\Delta$. Then $W_E = w_u\Delta(18 + 60) = 78w_u\Delta$; $W_I = m_u(7\theta_2 + 4 \times 7.5\theta_5) = 10.67m_u\Delta$.

6. Find the value of m_u corresponding to the assumed value of a

Equate the external and internal work to find this value of m_u. Thus, $10.67m_u\Delta = 78w_u\Delta$; $m_u = 7.31w_u$.

7. Determine the highest value of m_u

Assign other trial values to a, and find the corresponding values of m_u. Continue this procedure until the highest value of m_u is obtained. This is the true value of the ultimate unit moment.

Design of Flexural Members by the Working-Stress Method

As demonstrated earlier, the analysis or design of a composite beam by the working-stress method is most readily performed by transforming the given beam to an equivalent homogeneous beam. In the case of a reinforced-concrete member, the transformation is made by replacing the reinforcing steel with a strip of concrete having an area nA_s and located at the same distance from the neutral axis as the steel. This substitute concrete is assumed capable of sustaining tensile stresses.

The following symbols, shown in Fig. 115, are to be added to the notational system given earlier: kd = distance from extreme compression fiber to neutral axis, in (mm); jd = distance

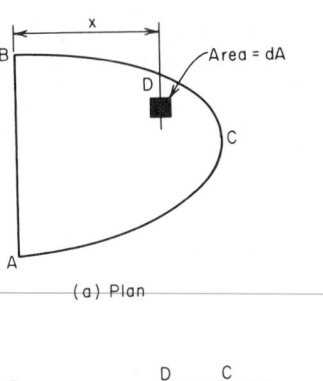

(a) Plan

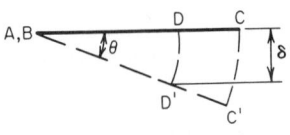

(b) Elevation

FIG. 114

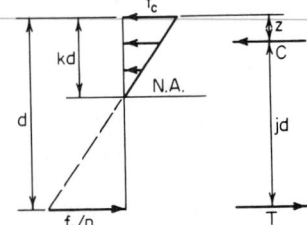

FIG. 115 Stress and resultant forces.

between action lines of C and T, in (mm); z = distance from extreme compression fiber to action line of C, in (mm).

The basic equations for the working-stress design of a rectangular beam reinforced solely in tension are

$$k = \frac{f_c}{f_c + f_s/n} \tag{21}$$

$$j = 1 - \frac{k}{3} \tag{22}$$

$$M = Cjd = \tfrac{1}{2}f_ckjbd^2 \tag{23}$$

$$M = \tfrac{1}{6}f_ck(3 - k)bd^2 \tag{24}$$

$$M = Tjd = f_sA_sjd \tag{25}$$

$$M = f_spjbd^2 \tag{26}$$

$$M = \frac{f_sk^2(3 - k)bd^2}{6n(1 - k)} \tag{27}$$

$$p = \frac{f_ck}{2f_s} \tag{28}$$

$$p = \frac{k^2}{2n(1 - k)} \tag{29}$$

$$k = [2pn + (pn)^2]^{0.5} - pn \tag{30}$$

For a given set of values of f_c, f_s, and n, M is directly proportional to the beam property bd^2. Let K denote the constant of proportionality. Then

$$M = Kbd^2 \tag{31}$$

where

$$K = \tfrac{1}{2}f_ckj = f_spj \tag{32}$$

The allowable flexural stress in the concrete and the value of n, which are functions of the ultimate strength f'_c, are given in the ACI *Code*, as is the allowable flexural stress in the steel. In all instances in the following procedures, the assumption is that the reinforcement is intermediate-grade steel having an allowable stress of 20,000 lb/in² (137,900 kPa).

Consider that the load on a beam is gradually increased until a limiting stress is induced. A beam that is so proportioned that the steel and concrete simultaneously attain their limiting stress is said to be in *balanced design*. For each set of values of f'_c and f_s, there is a corresponding set of values of K, k, j, and p associated with balanced design. These values are recorded in Table 6.

TABLE 6   Values of Design Parameters at Balanced Design

f'_c and n	f_c	f_s	K	k	j	p
2500 10	1125	20,000	178	0.360	0.880	0.0101
3000 9	1350	20,000	223	0.378	0.874	0.0128
4000 8	1800	20,000	324	0.419	0.860	0.0188
5000 7	2250	20,000	423	0.441	0.853	0.0248

In Fig. 116, *AB* represents the stress line of the transformed section for a beam in balanced design. If the area of reinforcement is increased while the width and depth remain constant, the neutral axis is depressed to O', and $A'O'B$ represents the stress line under the allowable load. But if the width is increased while the depth and area of reinforcement remain constant, the neutral axis is elevated to O'', and $AO''B'$ represents the stress line under the allowable load. This analysis leads to these conclusions: If the reinforcement is in excess of that needed for balanced design, the concrete is the first material to reach its limiting stress under a gradually increasing load. If the beam size is in excess of that needed for balanced design, the steel is the first material to reach its limiting stress.

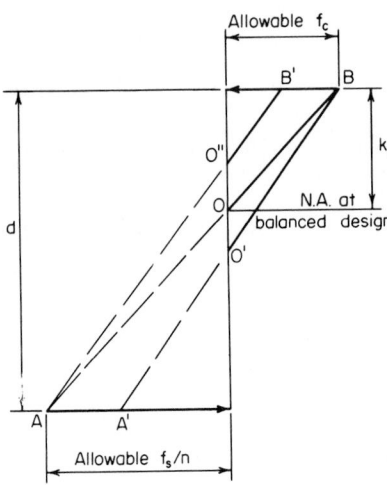

FIG. 116   Stress diagrams.

STRESSES IN A RECTANGULAR BEAM

A beam of 2500-lb/in^2 (17,237.5-kPa) concrete has a width of 12 in (304.8 mm) and an effective depth of 19.5 in (495.3 mm). It is reinforced with one no. 9 and two no. 7 bars. Determine the flexural stresses caused by a bending moment of 62 ft·kips (84.1 kN·m) (*a*) without applying the basic equations of reinforced-concrete beam design; (*b*) by applying the basic equations.

Calculation Procedure:

1.   Record the pertinent beam data

Thus $f'_c = 2500$ lb/in^2 (17,237.5 kPa); $\therefore n = 10$; $A_s = 2.20$ in^2 (14.194 cm^2); $nA_s = 22.0$ in^2 (141.94 cm^2). Then $M = 62,000(12) = 744,000$ in·lb (84,057.1 N·m).

2.   Transform the given section to an equivalent homogeneous section, as in Fig. 117b

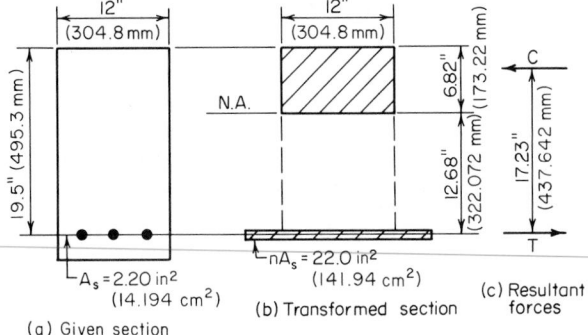

FIG. 117

3.   Locate the neutral axis of the member

The neutral axis coincides with the centroidal axis of the transformed section. To locate the neutral axis, set the static moment of the transformed area with respect to its centroidal axis equal to zero: $12(kd)^2/2 - 22.0(19.5 - kd) = 0$; $kd = 6.82$; $d - kd = 12.68$ in (322.072 mm).

4. Calculate the moment of inertia of the transformed section

Then evaluate the flexural stresses by applying the stress equation: $I = (\frac{1}{3})(12)(6.82)^3 + 22.0(12.68)^2 = 4806$ in^4 (200,040.6 cm^4); $f_c = Mkd/I = 744,000(6.82)/4806 = 1060$ lb/in^2 (7308.7 kPa); $f_s = 10(744,000)(12.68)/4806 = 19,600$ lb/in^2

5. Alternatively, evaluate the stresses by computing the resultant forces C and T

Thus $jd = 19.5 - 6.82/3 = 17.23$ in (437.642 mm); $C = T = M/jd = 744,000/17.23 = 43,200$ lb (192,153.6 N). But $C = \frac{1}{2}f_c(6.82)12$; $\therefore f_c = 1060$ lb/in^2 (7308.7 kPa); and $T = 2.20f_s$; $\therefore f_s = 19,600$ lb/in^2 (135,142 kPa). This concludes part a of the solution. The next step constitutes the solution to part b.

6. Compute pn and then apply the basic equations in the proper sequence

Thus $p = A_s/(bd) = 2.20/[12(19.5)] = 0.00940$; $pn = 0.0940$. Then by Eq. 30, $k = [0.188 + (0.094)^2]^{0.5} - 0.094 = 0.350$. By Eq. 22, $j = 1 - 0.350/3 = 0.883$. By Eq. 23, $f_c = 2M/(kjbd^2) = 2(744,000)/[0.350(0.883)(12)(19.5)^2] = 1060$ lb/in^2 (7308.7 kPa). By Eq. 25, $f_s = M/(A_s jd) = 744,000/[2.20(0.883)(19.5)] = 19,600$ lb/in^2 (135,142 kPa).

CAPACITY OF A RECTANGULAR BEAM

The beam in Fig. 118a is made of 2500-lb/in^2 (17,237.5-kPa) concrete. Determine the flexural capacity of the member (a) without applying the basic equations of reinforced-concrete beam design; (b) by applying the basic equations.

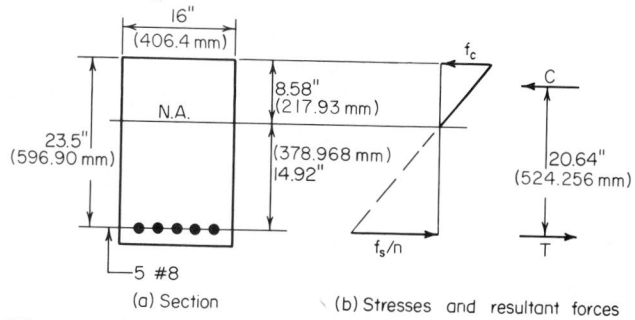

(a) Section   (b) Stresses and resultant forces

FIG. 118

Calculation Procedure:

1. Record the pertinent beam data

Thus, $f'_c = 2500$ lb/in^2 (17,237.5 kPa); $\therefore f_{c,\text{allow}} = 1125$ lb/in^2 (7756.9 kPa); $n = 10$; $A_s = 3.95$ in^2 (25.485 cm^2); $nA_s = 39.5$ in^2 (254.85 cm^2).

2. Locate the centroidal axis of the transformed section

Thus, $16(kd)^2/2 - 39.5(23.5 - kd) = 0$; $kd = 8.58$ in (217.93 mm); $d - kd = 14.92$ in (378.968 mm).

3. Ascertain which of the two allowable stresses governs the capacity of the member

For this purpose, assume that $f_c = 1125$ lb/in^2 (7756.9 kPa). By proportion, $f_s = 10(1125)(14.92/8.58) = 19,560$ lb/in^2 (134,866 kPa) $< 20,000$ lb/in^2 (137,900 kPa). Therefore, concrete stress governs.

4. Calculate the allowable bending moment

Thus, $jd = 23.5 - 8.58/3 = 20.64$ in (524.256 mm); $M = Cjd = \frac{1}{2}(1125)(16)(8.58)(20.64) = 1,594,000$ in·lb (180,090.1 N·m); or $M = Tjd = 3.95(19,560)(20.64) = 1,594,000$ in·lb (180,090.1 N·m). This concludes part a of the solution. The next step comprises part b.

5. Compute p and compare with p_b to identify the controlling stress

Thus, from Table 6, $p_b = 0.0101$; then $p = A_s/(bd) = 3.95/[16(23.5)] = 0.0105 > p_b$. Therefore, concrete stress governs.

Applying the basic equations in the proper sequence yields $pn = 0.1050$; by Eq. 30, $k = [0.210 + 0.105^2]^{0.5} - 0.105 = 0.365$; by Eq. 24, $M = (\%)(1125)(0.365)(2.635)(16)(23.5)^2 = 1,593,000$ in·lb (179,977.1 N·m). This agrees closely with the previously computed value of M.

DESIGN OF REINFORCEMENT IN A RECTANGULAR BEAM OF GIVEN SIZE

A rectangular beam of 4000-lb/in^2 (27,580-kPa) concrete has a width of 14 in (355.6 mm) and an effective depth of 23.5 in (596.9 mm). Determine the area of reinforcement if the beam is to resist a bending moment of (a) 220 ft·kips (298.3 kN·m); (b) 200 ft·kips (271.2 kN·m).

Calculation Procedure:

1. Calculate the moment capacity of this member at balanced design

Record the following values: $f_{c,allow} = 1800$ lb/in^2 (12,411 kPa); $n = 8$. From Table 6, $j_b = 0.860$; $K_b = 324$ lb/in^2 (2234.0 kPa); $M_b = K_b b d^2 = 324(14)(23.5)^2 = 2,505,000$ in·lb (283,014.9 N·m).

2. Determine which material will be stressed to capacity under the stipulated moment

For part a, $M = 220,000(12) = 2,640,000$ in·lb (3,579,840 N·m) $> M_b$. This result signifies that the beam size is deficient with respect to balanced design, and the concrete will therefore be stressed to capacity.

3. Apply the basic equations in proper sequence to obtain A_s

By Eq. 24, $k(3 - k) = 6M/(f_c bd^2) = 6(2,640,000)/[1800(14)(23.5)^2] = 1.138$; $k = 0.446$. By Eq. 29, $p = k^2/[2n(1 - k)] = 0.446^2/[16(0.554)] = 0.0224$; $A_s = pbd = 0.0224(14)(23.5) = 7.37$ in^2 (47.551 cm^2).

4. Verify the result by evaluating the flexural capacity of the member

For part b, compute A_s by the exact method and then describe the approximate method used in practice.

5. Determine which material will be stressed to capacity under the stipulated moment

Here $M = 200,000(12) = 2,400,000$ in·lb (3,254,400 N·m) $< M_b$. This result signifies that the beam size is excessive with respect to balanced design, and the steel will therefore be stressed to capacity.

6. Apply the basic equations in proper sequence to obtain A_s

By using Eq. 27, $k^2(3 - k)/(1 - k) = 6nM/(f_s bd^2) = 6(8)(2,400,000)/[20,000(14)(23.5)^2] = 0.7448$; $k = 0.411$. By Eq. 22, $j = 1 - 0.411/3 = 0.863$. By Eq. 25, $A_s = M/(f_s jd) = 2,400,000/[20,000(0.863)(23.5)] = 5.92$ in^2 (38.196 cm^2).

7. Verify the result by evaluating the flexural capacity of this member

The value of j obtained in step 6 differs negligibly from the value $j_b = 0.860$. Consequently, in those instances where the beam size is only moderately excessive with respect to balanced design, the practice is to consider that $j = j_b$ and to solve Eq. 25 directly on this basis. This practice is conservative, and it obviates the need for solving a cubic equation, thus saving time.

DESIGN OF A RECTANGULAR BEAM

A beam on a simple span of 13 ft (3.9 m) is to carry a uniformly distributed load, exclusive of its own weight, of 3600 lb/lin ft (52,538.0 N/m) and a concentrated load of 17,000 lb (75,616 N) applied at midspan. Design the section, using $f'_c = 3000$ lb/in^2 (20,685 kPa).

Calculation Procedure:

1. Record the basic values associated with balanced design

There are two methods of allowing for the beam weight: (a) to determine the bending moment with an estimated beam weight included; (b) to determine the beam size required to resist the external loads alone and then increase the size slightly. The latter method is used here.

From Table 6, $K_b = 223$ lb/in^2 (1537.6 kPa); $p_b = 0.0128$; $j_b = 0.874$.

2. Calculate the maximum moment caused by the external loads

Thus, the maximum moment $M_e = \frac{1}{4}PL + \frac{1}{8}wL^2 = \frac{1}{4}(17,000)(13)(12) + \frac{1}{8}(3600)(13)^2(12) = 1,576,000$ in·lb (178,056.4 N·m).

3. Establish a trial beam size

Thus, $bd^2 = M/K_b = 1,576,000/223 = 7067$ in^3 (115,828.1 cm^3). Setting $b = (\frac{2}{3})d$, we find $b = 14.7$ in (373.38 mm), $d = 22.0$ in (558.8 mm). Try $b = 15$ in (381 mm) and $d = 22.5$ in (571.5 mm), producing an overall depth of 25 in (635 mm) if the reinforcing bars may be placed in one row.

4. Calculate the maximum bending moment with the beam weight included; determine whether the trial section is adequate

Thus, beam weight $= 15(25)(150)/144 = 391$ lb/lin ft (5706.2 N/m); $M_w = (\frac{1}{8})(391)(13)^2(12) = 99,000$ in·lb (11,185.0 N·m); $M = 1,576,000 + 99,000 = 1,675,000$ in·lb (189,241.5 N·m); $M_b = K_b bd^2 = 223(15)(22.5)^2 = 1,693,000$ in·lb (191,275.1 N·m). The trial section is therefore satisfactory because it has adequate capacity.

5. Design the reinforcement

Since the beam size is slightly excessive with respect to balanced design, the steel will be stressed to capacity under the design load. Equation 25 is therefore suitable for this calculation. Thus, $A_s = M/(f_s jd) = 1,675,000/[20,000(0.874)(22.5)] = 4.26$ in^2 (27.485 cm^2).

An alternative method of calculating A_s is to apply the value of p_b while setting the beam width equal to the dimension actually required to produce balanced design. Thus, $A_s = 0.0128(15)(1675)(22.5)/1693 = 4.27$ in^2 (27.550 cm^2).

Use one no. 10 and three no. 9 bars, for which $A_s = 4.27$ in^2 (27.550 cm^2) and $b_{min} = 12.0$ in (304.8 mm).

6. Summarize the design

Thus, beam size is 15 × 25 in (381 × 635 mm); reinforcement is with one no. 10 and three no. 9 bars.

DESIGN OF WEB REINFORCEMENT

A beam 14 in (355.6 mm) wide with an 18.5-in (469.9-mm) effective depth carries a uniform load of 3.8 kips/lin ft (55.46 N/m) and a concentrated midspan load of 2 kips (8.896 kN). The beam is simply supported, and the clear distance between supports is 13 ft (3.9 m). Using $f'_c = 3000$ lb/in^2 (20,685 kPa) and an allowable stress f_v in the stirrups of 20,000 lb/in^2 (137,900 kPa), design web reinforcement in the form of vertical U stirrups.

Calculation Procedure:

1. Construct the shearing-stress diagram for half-span

The design of web reinforcement by the working-stress method parallels the design by the ultimate-strength method, given earlier. Let v = nominal shearing stress, lb/in^2 (kPa); v_c = shearing stress resisted by concrete; v' = shearing stress resisted by web reinforcement.

The ACI *Code* provides two alternative methods of computing the shearing stress that may be resisted by the concrete. The simpler method is used here. This sets

$$v_c = 1.1(f'_c)^{0.5} \tag{33}$$

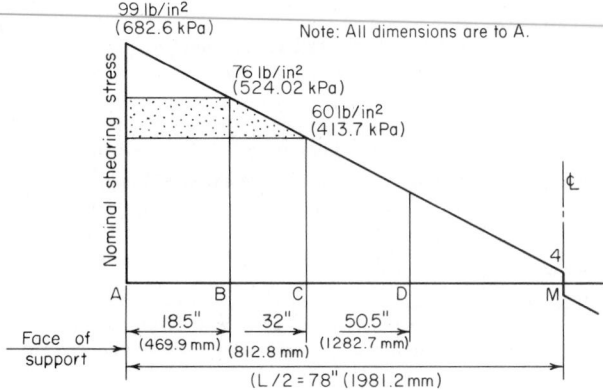

FIG. 119   Shearing-stress diagram.

The equation for nominal shearing stress is

$$v = \frac{V}{bd} \tag{34}$$

The shearing-stress diagram for a half-span is shown in Fig. 119. Establish the region AD within which web reinforcement is required. Thus, $v_c = 1.1(3000)^{0.5} = 60$ lb/in^2 (413.7 kPa). At the face of the support, $V = 6.5(3800) + 1000 = 25,700$ lb (114,313.6 N); $v = 25,700/[14(18.5)] = 99$ lb/in^2 (682.6 kPa).

At midspan, $V = 1000$ lb (4448 N); $v = 4$ lb/in^2 (27.6 kPa); slope of diagram $= -(99 - 4)/78 = -1.22$ lb/(in$^2\cdot$in) (-0.331 kPa/mm). At distance d from the face of the support, $v = 99 - 18.5(1.22) = 76$ lb/in^2 (524.02 kPa); $v' = 76 - 60 = 16$ lb/in^2 (110.3 kPa); $AC = (99 - 60)/1.22 = 32$ in (812.8 mm); $AD = AC + d = 32 + 18.5 = 50.5$ in (1282.7 mm).

2.   Check the beam size for compliance with the Code

Thus, $v_{\max} = 5(f_c')^{0.5} = 274$ lb/in^2 (1889.23 kPa) > 76 lb/in^2 (524.02 kPa). This is acceptable.

3.   Select the stirrup size

Use the method given earlier in the ultimate-strength calculation procedure to select the stirrup size, establish the maximum allowable spacing, and devise a satisfactory spacing.

CAPACITY OF A T BEAM

Determine the flexural capacity of the T beam in Fig. 120a, using $f_c' = 3000$ lb/in^2 (20,685 kPa).

Calculation Procedure:

1.   Record the pertinent beam values

The neutral axis of a T beam often falls within the web. However, to simplify the analysis, the resisting moment developed by the concrete lying between the neutral axis and the flange is usually disregarded. Let A_f denote the flange area. The pertinent beam values are $f_{c,\text{allow}} = 1350$ lb/in^2 (9308.3 kPa); $n = 9$; $k_b = 0.378$; $nA_s = 9(4.00) = 36.0$ in^2 (232.3 cm^2).

2.   Tentatively assume that the neutral axis lies in the web

Locate this axis by taking static moments with respect to the top line. Thus $A_f = 5(16) = 80$ in^2 (516.2 cm^2); $kd = [80(2.5) + 36.0(21.5)]/(80 + 36.0) = 8.40$ in (213.36 mm).

3.   Identify the controlling stress

Thus $k = 8.40/21.5 = 0.391 > k_b$; therefore, concrete stress governs.

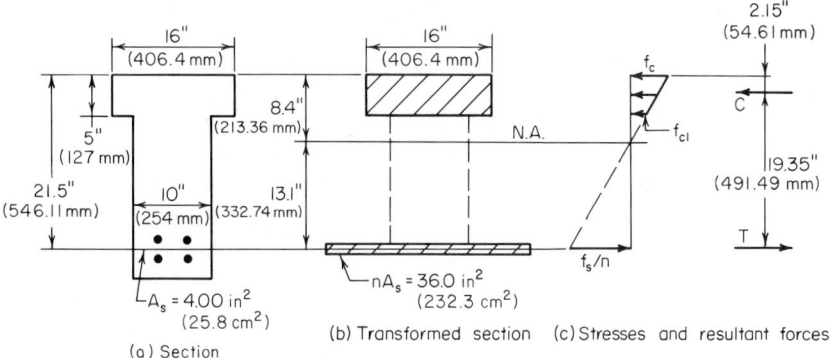

FIG. 120

(a) Section

(b) Transformed section   (c) Stresses and resultant forces

4.   *Calculate the allowable bending moment*

Using Fig. 120c, we see $f_{c1} = 1350(3.40)/8.40 = 546$ lb/in² (3764.7 kPa); $C = \frac{1}{2}(80)(1350 + 546) = 75,800$ lb (337,158.4 N). The action line of this resultant force lies at the centroidal axis of the stress trapezoid. Thus, $z = (\frac{5}{3})(1350 + 2 \times 546)/(1350 + 546) = 2.15$ in (54.61 mm); or $z = (\frac{5}{3})(8.40 + 2 \times 3.40)/(8.40 + 3.40) = 2.15$ in (54.61 mm); $M = Cjd = 75,800(19.35) = 1,467,000$ in·lb (165,741 N·m).

5.   *Alternatively, calculate the allowable bending moment by assuming that the flange extends to the neutral axis*

Then apply the necessary correction. Let C_1 = resultant compressive force if the flange extended to the neutral axis, lb (N); C_2 = resultant compressive force in the imaginary extension of the flange, lb (N). Then $C_1 = \frac{1}{2}(1350)(16)(8.40) = 90,720$ lb (403,522.6 N); $C_2 = 90,720(3.40/8.40)^2 = 14,860$ lb (66,097.3 N); $M = 90,720(21.5 - 8.40/3) - 14,860(21.5 - 5 - 3.40/3) = 1,468,000$ in·lb (165,854.7 N·m).

DESIGN OF A T BEAM HAVING CONCRETE STRESSED TO CAPACITY

A concrete girder of 2500-lb/in² (17,237.5-kPa) concrete has a simple span of 22 ft (6.7 m) and is built integrally with a 5-in (127-mm) slab. The girders are spaced 8 ft (2.4 m) on centers; the overall depth is restricted to 20 in (508 mm) by headroom requirements. The member carries a load of 4200 lb/lin ft (61,294.4 N/m), exclusive of the weight of its web. Design the section, using tension reinformcement only.

Calculation Procedure:

1.   *Establish a tentative width of web*

Since the girder is built integrally with the slab that it supports, the girder and slab constitute a structural entity in the form of a T beam. The effective flange width is established by applying the criteria given in the ACI *Code*, and the bending stress in the flange is assumed to be uniform across a line parallel to the neutral axis. Let A_f = area of flange in² (cm²); b = width of flange, in (mm); b' = width of web, in (mm); t = thickness of flange, in (mm); s = center-to-center spacing of girders.

To establish a tentative width of web, try $b' = 14$ in (355.6 mm). Then the weight of web = $14(15)(150)/144 = 219$, say 220 lb/lin ft (3210.7 N/m); $w = 4200 + 220 = 4420$ lb/lin ft (64,505.0 N/m).

Since two rows of bars are probably required, $d = 20 - 3.5 = 16.5$ in (419.1 mm). The critical shear value is $V = w(0.5L - d) = 4420(11 - 1.4) = 42,430$ lb (188,728.7 N); $v = V/b'd = 42,430/[14(16.5)] = 184$ lb/in² (1268.7 kPa). From the *Code*, $v_{max} = 5(f'_c)^{0.5} = 250$ lb/in² (1723.8 kPa). This is acceptable.

Upon designing the reinforcement, consider whether it is possible to reduce the width of the web.

2. Establish the effective width of the flange according to the Code

Thus, $\frac{1}{4}L = \frac{1}{4}(22)(12) = 66$ in (1676.4 mm); $16t + b' = 16(5) + 14 = 94$ in (2387.6 mm); $s = 8(12) = 96$ in (2438.4 mm); therefore $b = 66$ in (1676.4 mm).

3. Compute the moment capacity of the member at balanced design

Compare the result with the moment in the present instance to identify the controlling stres. With Fig. 120 as a guide, $k_b d = 0.360(16.5) = 5.94$ in (150.876 mm); $A_f = 5(66) = 330$ in^2 (2129.2 cm^2); $f_{c1} = 1125(0.94)/5.94 = 178$ lb/in^2 (1227.3 kPa); $C_b = T_b = \frac{1}{2}(330)(1125 + 178) = 215,000$ lb (956,320 N); $z_b = (\frac{3}{6})(5.94 + 2 \times 0.94)/(5.94 + 0.94) = 1.89$ in (48.0 mm); $jd = 14.61$ in (371.094 mm); $M_b = 215,000(14.61) = 3,141,000$ in·lb (354,870.2 N·m); $M = (\frac{1}{8})(4420)(22)^2(12) = 3,209,000$ in·lb (362,552.8 N·m).

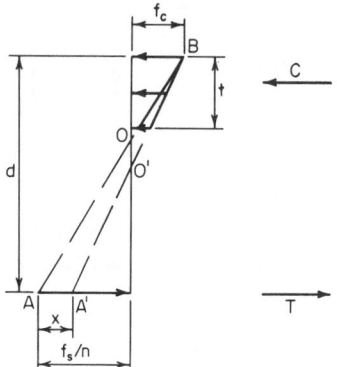

FIG. 121 Stress diagram for T beam.

The beam size is slightly deficient with respect to balanced design, and the concrete will therefore be stressed to capacity under the stipulated load. In Fig. 121, let AOB represent the stress line associated with balanced design and $A'O'B$ represent the stress line in the present instance. (The magnitude of AA' is exaggerated for clarity.)

4. Develop suitable equations for the beam

Refer to Fig. 121:

$$T = T_b + \frac{bt^2 x}{2d} \tag{35}$$

where T and T_b = tensile force in present instance and at balanced design, respectively. And

$$M = M_b + \frac{bt^2(3d - 2t)x}{6d} \tag{36}$$

5. Apply the equations from step 4

Thus, $M - M_b = 68,000$ in·lb (7682.6 N·m). By Eq. 36, $x = 68,000(6)(16.5)/[66(25)(49.5 - 10)] = 103$ lb/in^2 (710.2 kPa); $f_s = 20,000 - 10(103) = 18,970$ lb/in^2 (130,798.2 kPa). By Eq. 35, $T = 215,000 + 66(25)(103)/33 = 220,200$ lb (979,449.6 N).

6. Design the reinforcement; establish the web width

Thus $A_s = 220,200/18,970 = 11.61$ in^2 (74.908 cm^2). Use five no. 11 and three no. 10 bars, placed in two rows. Then $A_s = 11.61$ in^2 (74.908 cm^2); $b'_{min} = 14.0$ in (355.6 mm). It is therefore necessary to maintain the 14-in (355.6-mm) width.

7. Summarize the design

Width of web: 14 in (355.6 mm); reinforcement: five no. 11 and three no. 10 bars.

8. Verify the design by computing the capacity of the member

Thus $nA_s = 116.1$ in^2 (749.08 cm^2); $kd = [330(2.5) + 116.1(16.5)]/(330 + 116.1) = 6.14$ in (155.956 mm); $k = 6.14/16.5 = 0.372 > k_b$; therefore, concrete is stressed to capacity. Then $f_s = 10(1125)(10.36)/6.14 = 18,980$ lb/in^2 (130,867.1 kPa); $z = (\frac{3}{6})(6.14 + 2 \times 1.14)/(6.14 + 1.14) = 1.93$ in (49.022 mm); $jd = 14.57$ in (370.078 mm); $M_{allow} = 11.61(18,980)(14.57) = 3,210,000$ in·lb (362,665.8 N·m). This is acceptable.

DESIGN OF A T BEAM HAVING STEEL STRESSED TO CAPACITY

Assume that the girder in the previous calculation procedure carries a total load, including the weight of the web, of 4100 lb/lin ft (59,835.0 N/m). Compute the area of reinforcement.

Calculation Procedure:

1.   Identify the controlling stress

Thus, $M = (\frac{1}{8})(4100)(22)^2(12) = 2,977,000$ in·lb (336,341.5 N·m). From the previous calculation procedure, $M_b = 3,141,000$ in·lb (354,870.2 N·m). Since $M_b > M$, the beam size is slightly excessive with respect to balanced design, and the steel will therefore be stressed to capacity under the stipulated load.

2.   Compute the area of reinforcement

As an approximation, this area may be found by applying the value of jd associated with balanced design, although it is actually slightly larger. From the previous calculation procedure, $jd = 14.61$ in (371.094 mm). Then $A_s = 2,977,000/[20,000(14.61)] = 10.19$ in^2 (65.746 cm^2).

3.   Verify the design by computing the member capacity

Thus, $nA_s = 101.9$ in^2 (657.46 cm^2); $kd = (330 \times 2.5 + 101.9 \times 16.5)/(330 + 101.9) = 5.80$ in (147.32 mm); $z = (\frac{1}{3})(5.80 + 2 \times 0.80)/(5.80 + 0.80) = 1.87$ in (47.498 mm); $jd = 14.63$ in (371.602 mm); $M_{\text{allow}} = 10.19(20,000)(14.63) = 2,982,000$ in·lb (336,906.4 N·m). This is acceptable.

REINFORCEMENT FOR DOUBLY REINFORCED RECTANGULAR BEAM

A beam of 4000-lb/in^2 (27,580-kPa) concrete that will carry a bending moment of 230 ft·kips (311.9 kN·m) is restricted to a 15-in (381-mm) width and a 24-in (609.6-mm) total depth. Design the reinforcement.

Calculation Procedure:

1.   Record the pertinent beam data

In Fig. 122, where the imposed moment is substantially in excess of that corresponding to balanced design, it is necessary to reinforce the member in compression as well as tension. The loss in concrete area caused by the presence of the compression reinforcement may be disregarded.

Since plastic flow generates a transfer of compressive stress from the concrete to the steel, the ACI *Code* provides that "in doubly reinforced beams and slabs, an effective modular ratio of $2n$ shall be used to transform the compression reinforcement and compute its stress, which shall not be taken as greater than the allowable tensile stress." This procedure is tantamount to considering that the true stress in the compression reinforcement is twice the value obtained by assuming a linear stress distribution.

Let A_s = area of tension reinforcement, in^2 (cm^2); A'_s = area of compression reinforcement, in^2 (cm^2); f_s = stress in tension reinforcement, lb/in^2 (kPa); f'_s = stress in compression reinforcement, lb/in^2 (kPa); C' = resultant force in compression reinforcement, lb (N); M_1 = moment capacity of member if reinforced solely in tension to produce balanced design; M_2 = incremental moment capacity resulting from use of compression reinforcement.

The data recorded for the beam are $f_c = 1800$ lb/in^2 (12.411 kPa); $n = 8$; $K_b = 324$ lb/in^2 (2234.0 kPa); $k_b = 0.419$; $j_b = 0.860$; $M = 230,000(12) = 2,760,000$ in·lb (311,824.8 N·m).

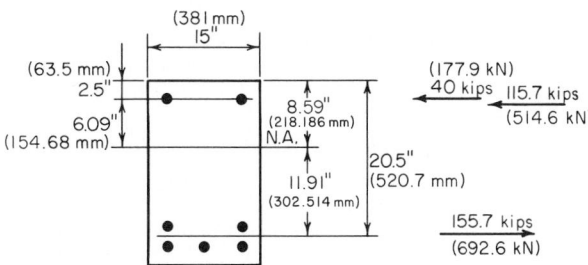

FIG. 122   Doubly reinforced beam.

2. *Ascertain whether one row of tension bars will suffice*

Assume tentatively that the presence of the compression reinforcement does not appreciably alter the value of j. Then $jd = 0.860(21.5) = 18.49$ in (469.646 mm); $A_s = M/(f_s jd) = 2,760,000/[20,000(18.49)] = 7.46$ in^2 (48.132 cm^2). This area of steel cannot be accommodated in the 15-in (381-mm) beam width, and two rows of bars are therefore required.

3. *Evaluate the moments M_1 and M_2*

Thus, $d = 24 - 3.5 = 20.5$ in (520.7 mm); $M_1 = K_b bd^2 = 324(15)(20.5)^2 = 2,040,000$ in·lb (230,479.2 N·m); $M_2 = 2,760,000 - 2,040,000 = 720,000$ in·lb (81,345.6 N·m).

4. *Compute the forces in the reinforcing steel*

For convenience, assume that the neutral axis occupies the same position as it would in the absence of compression reinforcement. For M_1, arm $= j_b d = 0.860(20.5) = 17.63$ in (447.802 mm); for M_2, arm $= 20.5 - 2.5 = 18.0$ in (457.2 mm); $T = 2,040,000/17.63 + 720,000/18.0 = 155,700$ lb (692,553.6 N); $C' = 40,000$ lb (177,920 N).

5. *Compute the areas of reinforcement and select the bars*

Thus $A_s = T/f_s = 155,700/20,000 = 7.79$ in^2 (50.261 cm^2); $kd = 0.419(20.5) = 8.59$ in (218.186 mm); $d - kd = 11.91$ in (302.514 mm). By proportion, $f'_s = 2(20,000)(6.09)/11.91 = 20,500$ lb/in^2 (141,347.5 kPa); therefore, set $f'_s = 20,000$ lb/in^2 (137,900 kPa). Then, $A'_s = C'/f'_s = 40,000/20,000 = 2.00$ in^2 (12.904 cm^2). Thus tension steel: five no. 11 bars, $A_s = 7.80$ in^2 (50.326 cm^2); compression steel: two no. 9 bars, $A_s = 2.00$ in^2 (12.904 cm^2).

DEFLECTION OF A CONTINUOUS BEAM

The continuous beam in Fig. 123a and b carries a total load of 3.3 kips/lin ft (48.16 kN/m). When it is considered as a T beam, the member has an effective flange width of 68 in (1727.2 mm). Determine the deflection of the beam upon application of full live load, using $f'_c = 2500$ lb/in^2 (17,237.5 kPa) and $f_y = 40,000$ lb/in^2 (275,800 kPa).

Calculation Procedure:

1. *Record the areas of reinforcement*

At support: $A_s = 4.43$ in^2 (28.582 cm^2) (top); $A'_s = 1.58$ in^2 (10.194 cm^2) (bottom). At center: $A_s = 3.16$ in^2 (20.388 cm^2) (bottom).

2. *Construct the bending-moment diagram*

Apply the ACI equation for maximum midspan moment. Refer to Fig. 123c: $M_1 = (\frac{1}{8})wL'^2 = (\frac{1}{8})3.3(22)^2 = 200$ ft·kips (271.2 kN·m); $M_2 = (\frac{1}{16})wL'^2 = 100$ ft·kips (135.6 kN·m); $M_3 = 100$ ft·kips (135.6 kN·m).

3. *Determine upon what area the moment of inertia should be based*

Apply the criterion set forth in the ACI *Code* to determine whether the moment of inertia is to be based on the transformed gross section or the transformed cracked section. At the support $pf_y = 4.43(40,000)/[14(20.5)] = 617 > 500$. Therefore, use the cracked section.

4. *Determine the moment of inertia of the transformed cracked section at the support*

Refer to Fig. 123d: $nA_s = 10(4.43) = 44.3$ in^2 (285.82 cm^2); $(n - 1)A'_s = 9(1.58) = 14.2$ in^2 (91.62 cm^2). The static moment with respect to the neutral axis is $Q = -\frac{1}{2}(14y^2) + 44.3(20.5 - y) - 14.2(y - 2.5) = 0$; $y = 8.16$ in (207.264 mm). The moment of inertia with respect to the neutral axis is $I_1 = (\frac{1}{3})14(8.16)^3 + 14.2(8.16 - 2.5)^2 + 44.3(20.5 - 8.16)^2 = 9737$ in^4 (40.53 dm^4).

5. *Calculate the moment of inertia of the transformed cracked section at the center*

Referring to Fig. 123e and assuming tentatively that the neutral axis falls within the flange, we see $nA_s = 10(3.16) = 31.6$ in^2 (203.88 cm^2). The static moment with respect to the neutral axis is $Q = \frac{1}{2}(68y^2) - 31.6(20.5 - y) = 0$; $y = 3.92$ in (99.568 mm). The neutral axis therefore falls within the flange, as assumed. The moment of inertia with respect to the neutral axis is $I_2 = (\frac{1}{3})68(3.92)^3 + 31.6(20.5 - 3.92)^2 = 10,052$ in^4 (41.840 dm^4).

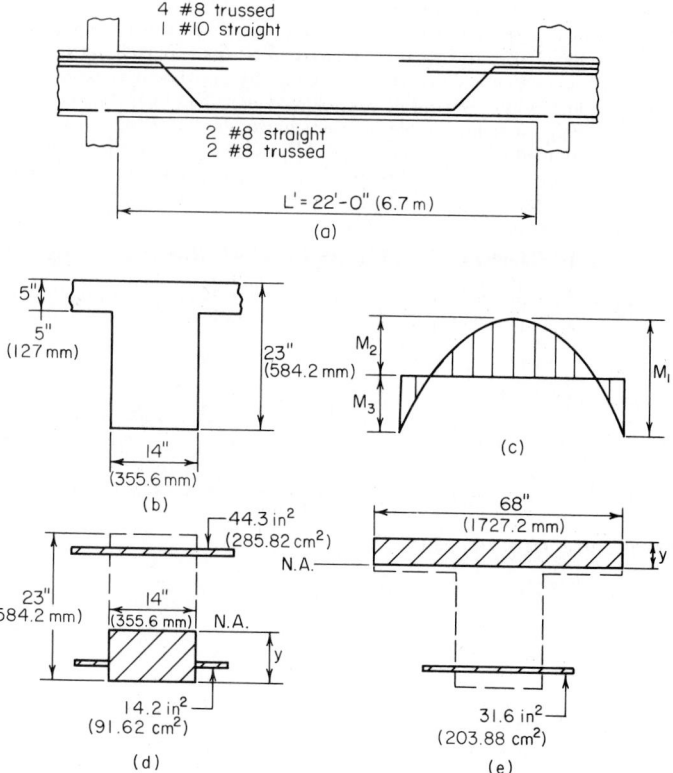

4 #8 trussed
I #10 straight

2 #8 straight
2 #8 trussed

L' = 22'-0" (6.7 m)

(a)

5"

5"
(127 mm)

23"
(584.2 mm)

14"
(355.6 mm)

(b)

M_2

M_1

M_3

(c)

44.3 in²
(285.82 cm²)

N.A.

23"
(584.2 mm)

14"
(355.6 mm)

N.A.

y

14.2 in²
(91.62 cm²)

(d)

68"
(1727.2 mm)

N.A.

y

31.6 in²
(203.88 cm²)

(e)

FIG. 123

6. Calculate the deflection at midspan

Use the equation

$$\Delta = \frac{L'^2}{EI}\left(\frac{5M_1}{48} - \frac{M_3}{8}\right) \tag{37}$$

where I = average moment of intertia, in⁴ (dm⁴). Thus, $I = \frac{1}{2}(9737 + 10,052) = 9895$ in⁴ (41.186 dm⁴); $E = 145^{1.5} \times 33(f'_c)^{0.5} = 57,600 (2500)^{0.5} = 2,880,000$ lb/in² (19,857.6 MPa). Then $\Delta = [22^2 \times 1728/(2880 \times 9895)](5 \times 200/48 - 100/8) = 0.244$ in (6.198 mm).

Where the deflection under sustained loading is to be evaluated, it is necessary to apply the factors recorded in the ACI *Code*.

Design of Compression Members by Ultimate-Strength Method

The notational system is P_u = ultimate axial compressive load on member, lb (N); P_b = ultimate axial compressive load at balanced design, lb (N); P_0 = allowable ultimate axial compressive load in absence of bending moment, lb (N); M_u = ultimate bending moment in member, lb·in (N·m); M_b = ultimate bending moment at balanced design; d' = distance from exterior surface to centroidal axis of adjacent row of steel bars, in (mm); t = overall depth of rectangular section or diameter of circular section, in (mm).

A compression member is said to be *spirally reinforced* if the longitudinal reinforcement is held in position by spiral hooping and *tied* if this reinforcement is held by means of intermittent lateral ties.

The presence of a bending moment in a compression member reduces the ultimate axial load that the member may carry. In compliance with the ACI *Code*, it is necessary to design for a minimum bending moment equal to that caused by an eccentricity of 0.05*t* for spirally reinforced members and 0.10*t* for tied members. Thus, every compression member that is designed by the ultimate-strength method must be treated as a beam column. This type of member is considered to be in balanced design if failure would be characterized by the simultaneous crushing of the concrete, which is assumed to occur when $\epsilon_c = 0.003$, and incipient yielding of the tension steel, which occurs when $f_s = f_y$. The ACI *Code* set $\phi = 0.75$ for spirally reinforced members and $\phi = 0.70$ for tied members.

ANALYSIS OF A RECTANGULAR MEMBER BY INTERACTION DIAGRAM

A short tied member having the cross section shown in Fig. 124*a* is to resist an axial load and a bending moment that induces compression at *A* and tension at *B*. The member is made of 3000-

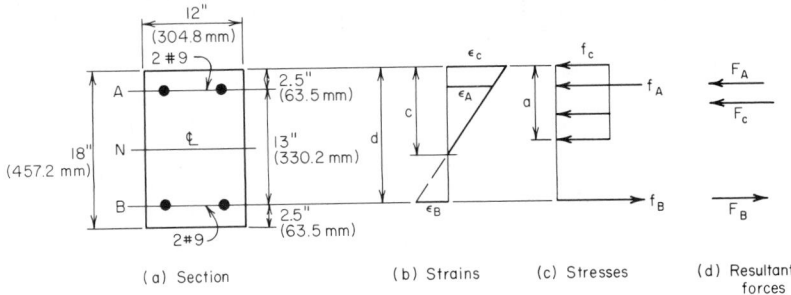

(a) Section   (b) Strains   (c) Stresses   (d) Resultant forces

FIG. 124

lb/in² (20,685-kPa) concrete, and the steel has a yield point of 40,000 lb/in² (275,800 kPa). By starting with $c = 8$ in (203.2 mm) and assigning progressively higher values to c, construct the interaction diagram for this member.

Calculation Procedure:

1.   Compute the value of c associated with balanced design

An *interaction diagram*, as the term is used here, is one in which every point on the curve represents a set of simultaneous values of the ultimate moment and allowable ultimate axial load. Let ϵ_A and ϵ_B = strain of reinforcement at *A* and *B*, respectively; ϵ_c = strain of extreme fiber of concrete; f_A and f_B = stress in reinforcement at *A* and *B*, respectively, lb/in² (kPa); F_A and F_B = resultant force in reinforcement at *A* and *B*, respectively; F_c = resultant force in concrete, lb (N).

Compression will be considered positive and tension negative. For simplicity, disregard the slight reduction in concrete area caused by the steel at *A*.

Referring to Fig. 124*b*, compute the value of c associated with balanced design. Computing P_b and M_b yields $c_b/d = 0.003/(0.003 + f_y/E_s) = 87,000/(87,000 + f_y)$; $c_b = 10.62$ in (269.748 mm). Then $\epsilon_A/\epsilon_B = (10.62 - 2.5)/(15.5 - 10.62) > 1$; therefore, $f_A = f_y$; $a_b = 0.85(10.62) = 9.03$ in (229.362 mm); $F_c = 0.85(3000)(12a_b) = 276,300$ lb (1,228,982.4 N); $F_A = 40,000(2.00) = 80,000$ lb ((355,840 N); $F_B = -80,000$ lb ($-355,840$ N); $P_b = 0.70(276,300) = 193,400$ lb (860,243.2 N). Also,

$$M_b = 0.70 \left[\frac{F_c(t - a)}{2} + \frac{(F_A - F_B)(t - 2d')}{2} \right] \tag{38}$$

Thus, $M_b = 0.70[276,300(18 - 9.03)/2 + 160,000(6.5)] = 1,596,000$ in·lb (180,316.1 N·m).

When $c > c_b$, the member fails by crushing of the concrete; when $c < c_b$, it fails by yielding of the reinforcement at line *B*.

2. Compute the value of c associated with incipient yielding of the compression steel

Compute the corresponding values of P_u and M_u. Since ϵ_A and ϵ_B are numerically equal, the neutral axis lies at N. Thus, $c = 9$ in (228.6 mm); $a = 0.85(9) = 7.65$ in (194.31 mm); $F_c = 30,600(7.65) = 234,100$ lb (1,041,276.8 N); $F_A = 80,000$ lb (355,840 N); $F_B = -80,000$ lb ($-355,840$ N); $P_u = 0.70\ (234,100) = 163,900$ lb (729,027.2 N); $M_u = 0.70(234,100 \times 5.18 + 160,000 \times 6.5) = 1,577,000$ in·lb (178,169.5 N·m).

3. Compute the minimum value of c at which the entire concrete area is stressed to 0.85f'_c

Compute the corresponding values of P_u and M_u. Thus, $a = t = 18$ in (457.2 mm); $c = 18/0.85 = 21.18$ in (537.972 mm); $f_B = \epsilon_c E_s(c - d)/c = 87,000(21.18 - 15.5)/21.18 = 23,300$ lb/in^2 (160,653.5 kPa); $F_c = 30,600(18) = 550,800$ lb (2,449,958.4 N); $F_A = 80,000$ lb (355,840 N); $F_B = 46,600$ lb (207,276.8 N); $P_u = 0.70(550,800 + 80,000 + 46,600) = 474,200$ lb (2,109,241.6 N); $M_u = 0.70(80,000 - 46,600)6.5 = 152,000$ in·lb (17,192.9 N·m).

4. Compute the value of c at which M_u = 0; compute P_0

The bending moment vanishes when F_B reaches 80,000 lb (355,840 N). From the calculation in step 3, $f_B = 87,000(c - d)/c = 40,000$ lb/in^2 (275,800 kPa); therefore, $c = 28.7$ in (728.98 mm); $P_0 = 0.70(550,800 + 160,000) = 497,600$ lb (2,213,324.8 N).

5. Assign other values to c, and compute P_u and M_u

By assigning values to c ranging from 8 to 28.7 in (203. 2 to 728.98 mm), typical calculations are: when $c = 8$ in (203.2 mm), $f_B = -40,000$ lb/in^2 ($-275,800$ kPa); $f_A = 40,000(5.5/7.5) = 29,300$ lb/in^2 (202,023.5 kPa); $a = 6.8$ in (172.72 mm); $F_c = 30,600(6.8) = 208,100$ lb (925,628.8 N); $P_u = 0.70(208,100 + 58,600 - 80,000) = 130,700$ lb (581,353.6 N); $M_u = 0.70\ (208,100 \times 5.6 + 138,600 \times 6.5) = 1,446,000$ in·lb (163,369.1 N·m).

When $c = 10$ in (254 mm), $f_A = 40,000$ lb/in^2 (275,800 kPa); $f_B = -40,000$ lb/in^2 ($-275,800$ kPa); $a = 8.5$ in (215.9 mm); $F_c = 30,600(8.5) = 260,100$ lb (1,156,924.8 N); $P_u = 0.70(260,100) = 182,100$ lb (809,980 N); $M_u = 0.70(260,100 \times 4.75 + 160,000 \times 6.5) = 1,593,000$ in·lb (179,997.1 N·m).

When $c = 14$ in (355.6 mm), $f_B = 87,000(14 - 15.5)/14 = -9320$ lb/in^2 ($-64,261.4$ kPa); $a = 11.9$ in (302.26 mm); $F_c = 30,600(11.9) = 364,100$ lb (1,619,516.8 N); $P_u = 0.70(364,100 + 80,000 - 18,600) = 297,900$ lb (1,325,059.2 N); $M_u = 0.70(364,100 \times 3.05 + 98,600 \times 6.5) = 1,226,000$ in·lb (138,513.5 N·m).

6. Plot the points representing computed values of P_u and M_u in the interaction diagram

Figure 125 shows these points. Pass a smooth curve through these points. Note that when $P_u < P_b$, a reduction in M_u is accompanied by a reduction in the allowable load P_u.

AXIAL-LOAD CAPACITY OF RECTANGULAR MEMBER

The member analyzed in the previous calculation procedure is to carry an eccentric longitudinal load. Determine the allowable ultimate load if the eccentricity as measured from N is (a) 9.2 in (233.68 mm); (b) 6 in (152.4 mm).

Calculation Procedure:

1. Evaluate the eccentricity associated with balanced design

Let e denote the eccentricity of the load and e_b the eccentricity associated with balanced design. Then $M_u = P_u e$. In Fig. 125, draw an arbitrary radius vector OD; then $\tan \theta = ED/OE =$ eccentricity corresponding to point D.

Proceeding along the interaction diagram from A to C, we see that the value of c increases and the value of e decreases. Thus, c and e vary in the reverse manner. To evaluate the allowable loads, it is necessary to identify the portion of the interaction diagram to which each eccentricity applies.

From the computations of the previous calculation procedure, $e_b = M_b/P_b = 1,596,000/193,400 = 8.25$ in (209.55 mm). This result discloses that an eccentricity of 9.2 in (233.68 mm)

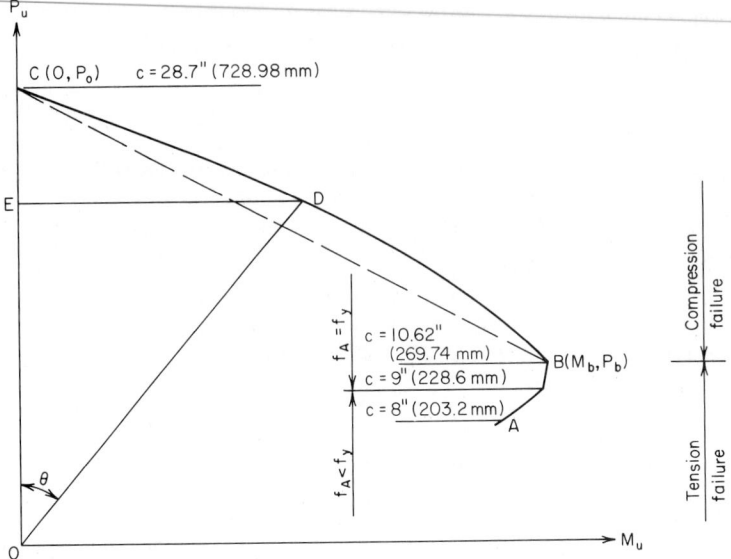

FIG. 125   Interaction diagram.

corresponds to a point on *AB* and an eccentricity of 6 in (152.4 mm) corresponds to a point on *BC*.

2. Evaluate P_u when e = 9.2 in (233.68 mm)

It was found that $c = 9$ in (228.6 mm) is a significant value. The corresponding value of e is $1,577,000/163,900 = 9.62$ in (244.348 mm). This result discloses that in the present instance $c > 9$ in (228.6 mm) and consequently $f_A = f_y$; $F_A = 80,000$ lb (355,840 N); $F_B = -80,000$ lb ($-355,840$ N); $F_c = 30,600a$; $P_u/0.70 = 30,600a$; $M_u/0.70 = 30,600a(18 - a)/2 + 160,000(6.5)$; $e = M_u/P_u = 9.2$ in (233.68 mm). Solving gives $a = 8.05$ in (204.47 mm), $P_u = 172,400$ lb (766,835.2 N).

3. Evaluate P_u when e = 6 in (152.4 mm)

To simplify this calculation, the ACI *Code* permits replacement of curve *BC* in the interaction diagram with a straight line through *B* and *C*. The equation of this line is

$$P_u = P_o - (P_o - P_b)\frac{M_u}{M_b} \tag{39}$$

By replacing M_u with $P_u e$, the following relation is obtained:

$$P_u = \frac{P_o}{1 + (P_o - P_b)e/M_b} \tag{39a}$$

In the present instance, $P_o = 497,600$ lb (2,213,324.8 N); $P_b = 193,400$ lb (860,243.2 N); $M_b = 1,596,000$ in·lb (180,316.1 N·m). Thus $P_u = 232,100$ lb (1,032,380 N).

ALLOWABLE ECCENTRICITY OF A MEMBER

The member analyzed in the previous two calculation procedures is to carry an ultimate longitudinal load of 150 kips (667.2 kN) that is eccentric with respect to axis *N*. Determine the maximum eccentricity with which the load may be applied.

Calculation Procedure:

1. Express P_u in terms of c, and solve for c

From the preceding calcluation procedures, it is seen that the value of c corresponding to the maximum eccentricity lies between 8 and 9 in (203.2 and 228.6 mm), and therefore $f_A < f_y$. Thus $f_B = -40,000$ lb/in^2 ($-275,800$ kPa); $f_A = 40,000(c - 2.5)/(15.5 - c)$; $F_c = 30,600(0.85c) = 26,000c$; $150,000 = 0.70\{26,000c + 80,000[(c - 2.5)/(15.5 - c) - 1]\}$; $c = 8.60$ in (218.44 mm).

2. Compute M_u and evaluate the eccentricity

Thus, $a = 7.31$ in (185.674 mm); $F_c = 223,700$ lb (995,017.6 N); $f_A = 35,360$ lb/in^2 (243,807.2 kPa); $M_u = 0.70(223,700 \times 5.35 + 150,700 \times 6.5) = 1,523,000$ in·lb (172,068.5 N·m); $e = M_u/P_u = 10.15$ in (257.81 mm).

Design of Compression Members by Working-Stress Method

The notational system is as follows: A_g = gross area of section, in^2 (cm^2); A_s = area of tension reinforcement, in^2 (cm^2); A_{st} = total area of longitudinal reinforcement, in^2 (cm^2); D = diameter of circular section, in (mm); $p_g = A_{st}/A_g$; P = axial load on member, lb (N); f_s = allowable stress in longitudinal reinforcement, lb/in^2 (kPa); $m = f_y/(0.85f'_c)$.

The working-stress method of designing a compression member is essentially an adaptation of the ultimate-strength method. The allowable ultimate loads and bending moments are reduced by applying an appropriate factor of safety, and certain simplifications in computing the ultimate values are introduced.

The allowable concentric load on a short spirally reinforced column is $P = A_g(0.25f'_c + f_sp_g)$, or

$$P = 0.25f'_cA_g + f_sA_{st} \tag{40}$$

where $f_s = 0.40f_y$, but not to exceed 30,000 lb/in^2 (206,850 kPa).

The allowable concentric load on a short tied column is $P = 0.85A_g(0.25f'_c + f_sp_g)$, or

$$P = 0.2125f'_cA_g + 0.85f_sA_{st} \tag{41}$$

A section of the ACI *Code* provides that p_g may range from 0.01 to 0.08. However, in the case of a circular column in which the bars are to be placed in a single circular row, the upper limit of p_g is often governed by clearance. This section of the *Code* also stipulates that the minimum bar size to be used is no. 5 and requires a minimum of six bars for a spirally reinforced column and four bars for a tied column.

DESIGN OF A SPIRALLY REINFORCED COLUMN

A short circular column, spirally reinforced, is to support a concentric load of 420 kips (1868.16 kN). Design the member, using $f'_c = 4000$ lb/in^2 (27,580 kPa) and $f_y = 50,000$ lb/in^2 (344,750 kPa).

Calculation Procedure:

1. Assume $p_g = 0.025$ and compute the diameter of the section

Thus, $0.25f'_c = 1000$ lb/in^2 (6895 kPa); $f_s = 20,000$ lb/in^2 (137,900 kPa). By Eq. 40, $A_g = 420/(1 + 20 \times 0.025) = 280$ in^2 (1806.6 cm^2). Then $D = (A_g/0.785)^{0.5} = 18.9$ in (130.32 mm). Set $D = 19$ in (131.01 mm), making $A_g = 283$ in^2 (1825.9 cm^2).

2. Select the reinforcing bars

The load carried by the concrete = 283 kips (1258.8 kN). The load carried by the steel = 420 − 283 = 137 kips (609.4 kN). Then the area of the steel is $A_{st} = 137/20 = 6.85$ in^2 (44.196 cm^2). Use seven no. 9 bars, each having an area of 1 in^2 (6.452 cm^2). Then $A_{st} = 7.00$ in^2 (45.164 cm^2). The *Reinforced Concrete Handbook* shows that a 19-in (482.6-mm) column can accommodate 11 no. 9 bars in a single row.

3. *Design the spiral reinforcement*

The portion of the column section bounded by the outer circumference of the spiral is termed the *core* of the section. Let A_c = core area, in^2 (cm^2); D_c = core diameter, in (mm); a_s = cross-sectional area of spiral wire, in^2 (cm^2); g = pitch of spiral, in (mm); p_s = ratio of volume of spiral reinforcement to volume of core.

The ACI *Code* requires 1.5-in (38.1-mm) insulation for the spiral, with g restricted to a maximum of $D_c/6$. Then $D_c = 19 - 3 = 16$ in (406.4 mm); $A_c = 201$ in^2 (1296.9 cm^2); $D_c/6 = 2.67$ in (67.818 mm). Use a 2.5-in (63.5-mm) spiral pitch. Taking a 1-in (25.4-mm) length of column,

$$p_s = \frac{\text{volume of spiral}}{\text{volume of core}} = \frac{a_s \pi D_c / g}{\pi D_c^2 / 4}$$

or

$$a_s = \frac{g D_c p_s}{4} \tag{42}$$

The required value of p_s as given by the ACI *Code* is

$$p_s = \frac{0.45(A_g/A_c - 1)f_c'}{f_y} \tag{43}$$

or $p_s = 0.45(283/201 - 1)4/50 = 0.0147$; $a_s = 2.5(16)(0.0147)/4 = 0.147$ in^2 (0.9484 cm^2). Use ½-in (12.7-mm) diameter wire with $a_s = 0.196$ in^2 (1.2646 cm^2).

4. *Summarize the design*

Thus: column size: 19-in (482.6-mm) diameter; longitudinal reinforcement: seven no. 9 bars; spiral reinforcement: ½-in (12.7-mm) diameter wire, 2.5-in (63.5-mm) pitch.

ANALYSIS OF A RECTANGULAR MEMBER BY INTERACTION DIAGRAM

A short tied member having the cross section shown in Fig. 126 is to resist an axial load and a bending moment that induces rotation about axis N. The member is made of 4000-lb/in^2 (27,580-kPa) concrete, and the steel has a yield point of 50,000 lb/in^2 (344,750 kPa). Construct the interaction diagram for this member.

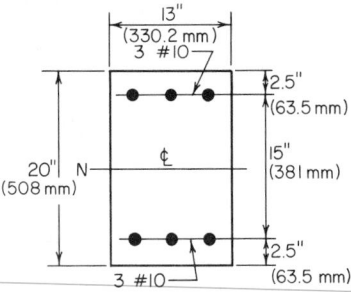

FIG. 126

Calculation Procedure:

1. *Compute P_a and M_f*

Consider a composite member of two materials having equal strength in tension and compression, the member being subjected to an axial load P and bending moment M that induce the allowable stress in one or both materials. Let P_a = allowable axial load in absence of bending moment, as computed by dividing the allowable ultimate load by a factor of safety; M_f = allowable bending moment in absence of axial load, as computed by dividing the allowable ultimate moment by a factor of safety.

Find the simultaneous allowable values of P and M by applying the interaction equation

$$\frac{P}{P_a} + \frac{M}{M_f} = 1 \tag{44}$$

Alternate forms of this equation are

$$M = M_f\left(1 - \frac{P}{P_a}\right) \qquad P = P_a\left(1 - \frac{M}{M_f}\right) \tag{44a}$$

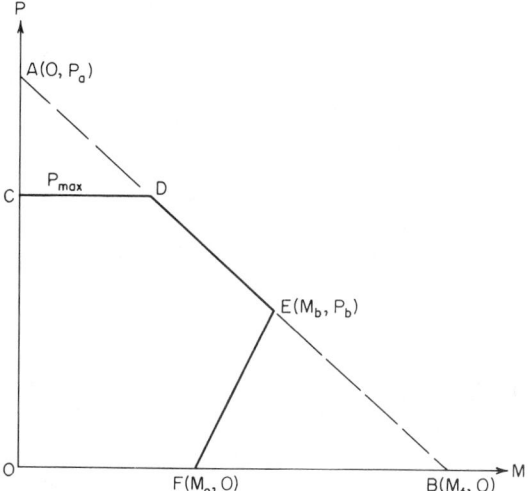

FIG. 127 Interaction diagram.

$$P = \frac{P_a M_f}{M_f + P_a M / P} \qquad (44b)$$

Equation 44 is represented by line AB in Fig. 127; it is also valid with respect to a reinforced-concrete member for a certain range of values of P and M. This equation is not applicable in the following instances: (a) If M is relatively small, Eq. 44 yields a value of P in excess of that given by Eq. 41. Therefore, the interaction diagram must contain line CD, which represents the maximum value of P.

(b) If M is relatively large, the section will crack, and the equal-strength assumption underlying Eq. 44 becomes untenable.

Let point E represent the set of values of P and M that will cause cracking in the extreme concrete fiber. And let P_b = axial load represented by point E; M_b = bending moment represented by point E; M_o = allowable bending moment in reinforced-concrete member in absence of axial load, as computed by dividing the allowable ultimate moment by a factor of safety. (M_o differs from M_f in that the former is based on a cracked section and the latter on an uncracked section. The subscript b as used by the ACI *Code* in the present instance does *not* refer to balanced design. However, its use illustrates the analogy with ultimate-strength analysis.) Let F denote the point representing M_o.

For simplicity, the interaction diagram is assumed to be linear between E and F. The interaction equation for a cracked section may therefore be expressed in any of the following forms:

$$M = M_o + \left(\frac{P}{P_b}\right)(M_b - M_o) \qquad P = P_b\left(\frac{M - M_o}{M_b - M_o}\right) \qquad (45a)$$

$$P = \frac{P_b M_o}{M_o - M_b + P_b M / P} \qquad (45b)$$

The ACI *Code* gives the following approximations: For spiral columns:

$$M_o = 0.12 A_{st} f_y D_s \qquad (46a)$$

where D_s = diameter of circle through center of longitudinal reinforcement. For symmetric tied columns:

$$M_o = 0.40 A_s f_y (d - d') \qquad (46b)$$

For unsymmetric tied columns:

$$M_o = 0.40A_s f_y jd \qquad (46c)$$

For symmetric spiral columns:

$$\frac{M_b}{P_b} = 0.43p_g m D_s + 0.14t \qquad (47a)$$

For symmetric tied columns:

$$\frac{M_b}{P_b} = d(0.67p_g m + 0.17) \qquad (47b)$$

For unsymmetric tied columns:

$$\frac{M_b}{P_b} = \frac{p'm(d - d') + 0.1d}{(p' - p)m + 0.6} \qquad (47c)$$

where p' = ratio of area of compression reinforcement to effective area of concrete. The value of P_a is taken as

$$P_a = 0.34f'_c A_g (1 + p_g m) \qquad (48)$$

The value of M_f is found by applying the section modulus of the transformed uncracked section, using a modular ratio of $2n$ to account for stress transfer between steel and concrete engendered by plastic flow. (If the steel area is multiplied by $2n - 1$, allowance is made for the reduction of the concrete area.)

Computing P_a and M_f yields $A_g = 260$ in^2 (1677.5 cm^2); $A_{st} = 7.62$ in^2 (49.164 cm^2); $p_g = 7.62/260 = 0.0293$; $m = 50/[0.85(4)] = 14.7$; $p_g m = 0.431$; $n = 8$; $P_a = 0.34(4)(260)(1.431) = 506$ kips (2250.7 kN).

The section modulus to be applied in evaluating M_f is found thus: $I = (\frac{1}{12})(13)(20)^3 + 7.62(15)(7.5)^2 = 15,100$ in^4 (62.85 dm^4); $S = I/c = 15,100/10 = 1510$ in^3 (24,748.9 cm^3); $M_f = Sf_c = 1510(1.8) = 2720$ in·kips (307.3 kN·m).

2. Compute P_b and M_b

By Eq. 47b, $M_b/P_b = 17.5(0.67 \times 0.431 + 0.17) = 8.03$ in (203.962 mm). By Eq. 44b, $P_b = P_a M_f/(M_f + 8.03P_a) = 506 \times 2720/(2720 + 8.03 \times 506) = 203$ kips (902.9 kN); $M_b = 8.03(203) = 1630$ in·kips (184.2 kN·m).

3. Compute M_o

By Eq. 46b, $M_o = 0.40(3.81)(50)(15) = 1140$ in·kips (128.8 kN·m).

4. Compute the limiting value of P

As established by Eq. 41, $P_{max} = 0.2125(4)(260) + 0.85(20)(7.62) = 351$ kips (1561.2 kN).

5. Construct the interaction diagram

The complete diagram is shown in Fig. 127.

AXIAL-LOAD CAPACITY OF A RECTANGULAR MEMBER

The member analyzed in the previous calculation procedure is to carry an eccentric longitudinal load. Determine the allowable load if the eccentricity as measured from N is (a) 10 in (254 mm); (b) 6 in (152.4 mm).

Calculation Procedure:

1. Evaluate P when $e = 10$ in (254 mm)

As the preceding calculations show, the eccentricity corresponding to point E in the interaction diagram is 8.03 in (203.962 mm). Consequently, an eccentricity of 10 in (254 mm) corresponds to a point on EF, and an eccentricity of 6 in (152.4 mm) corresponds to a point on ED.

By Eq. 45b, $P = 203(1140)/(1140 - 1630 + 203 \times 10) = 150$ kips (667.2 kN).

2.   *Evaluate P when e = 6 in (152.4 mm)*

By Eq. 44b, $P = 506(2720)/(2720 + 506 \times 6) = 239$ kips (1063.1 kN).

Design of Column Footings

A reinforced-concrete footing supporting a single column differs from the usual type of flexural member in the following respects: It is subjected to bending in all directions, the ratio of maximum vertical shear to maximum bending moment is very high, and it carries a heavy load concentrated within a small area. The consequences are as follows: The footing requires two-way reinforcement, its depth is determined by shearing rather than bending stress, the punching-shear stress below the column is usually more critical than the shearing stress that results from ordinary beam action, and the design of the reinforcement is controlled by the bond stress as well as the bending stress.

(a) Plan

(b) Elevation

FIG. 128

Since the footing weight and soil pressure are collinear, the former does not contribute to the vertical shear or bending moment. It is convenient to visualize the footing as being subjected to an upward load transmitted by the underlying soil and a downward reaction supplied by the column, this being, of course, an inversion of the true form of loading. The footing thus functions as an overhanging beam. The effective depth of footing is taken as the distance from the top surface to the center of the upper row of bars, the two rows being made identical to avoid confusion.

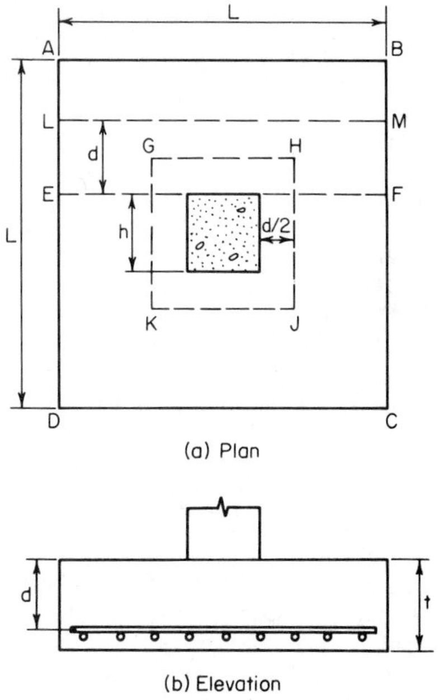

Refer to Fig. 128, which shows a square footing supporting a square, symmetrically located concrete column. Let P = column load, kips (kN); p = net soil pressure (that caused by the column load alone), lb/ft^2 (kPa); A = area of footing, ft^2 (m^2); L = side of footing, ft (m); h = side of column, in (mm); d = effective depth of footing, ft (m); t = thickness of footing, ft (m); f_b = bearing stress at interface of column, lb/in^2 (kPa); v_1 = nominal shearing stress under column, lb/in^2 (kPa); v_2 = nominal shearing stress caused by beam action, lb/in^2 (kPa); b_o = width of critical section for v_1, ft (m); V_1 and V_2 = vertical shear at critical section for stresses v_1 and v_2, respectively.

In accordance with the ACI *Code*, the critical section for v_1 is the surface *GHJK*, the sides of which lie at a distance $d/2$ from the column faces. The critical section for v_2 is plane *LM*, located at a distance d from the face of the column. The critical section for bending stress and bond stress is plane *EF* through the face of the column. In calculating v_2, f, and u, no allowance is made for the effects of the orthogonal reinforcement.

DESIGN OF AN ISOLATED SQUARE FOOTING

A 20-in (508-mm) square tied column reinforced with eight no. 9 bars carries a concentric load of 380 kips (1690.2 kN). Design a square footing by the working-stress method using these values: the allowable soil pressure is 7000 lb/ft^2 (335.2 kPa); $f_c' = 3000$ lb/in^2 (20,685 kPa); and $f_s = 20,000$ lb/in^2 (137,900 kPa).

Calculation Procedure:

1. Record the allowable shear, bond, and bearing stresses

From the ACI *Code* table, $v_1 = 110$ lb/in² (758.5 kPa); $v_2 = 60$ lb/in² (413.7 kPa); $f_b = 1125$ lb/in² (7756.9 kPa); $u = 4.8(f'_c)^{0.5}$/bar diameter = 264/bar diameter.

2. Check the bearing pressure on the footing

Thus, $f_b = 380/[20(20)] = 0.95$ kips/in² (7.258 MPa) < 1.125 kips/in² (7.7568 MPa). This is acceptable.

3. Establish the length of footing

For this purpose, assume the footing weight is 6 percent of the column load. Then $A = 1.06(380)/7 = 57.5$ ft² (5.34 m²). Make $L = 7$ ft 8 in = 7.67 ft (2.338 m); $A = 58.8$ ft² (5.46 m²).

4. Determine the effective depth as controlled by v_1

Apply

$$(4v_1 + p)d^2 + h(4v_1 + 2p)d = p(A - h^2) \qquad (49)$$

Verify the result after applying this equation. Thus $p = 380/58.8 = 6.46$ kips/ft² (0.309 MPa); $v_1 = 0.11(144) = 15.84$ kips/ft² (0.758 MPa); $69.8d^2 + 127.1d = 361.8$; $d = 1.54$ ft (0.469 m). Checking in Fig. 128, we see $GH = 1.67 + 1.54 = 3.21$ ft (0.978 m); $V_1 = 6.46(58.8 - 3.21^2) = 313$ kips (1392.2 kN); $v_1 = V_1/(b_o d) = 313/[4(3.21)(1.54)] = 15.83$ kips/ft² (0.758 MPa). This is acceptable.

5. Establish the thickness and true depth of footing

Compare the weight of the footing with the assumed weight. Allowing 3 in (76.2 mm) for insulation and assuming the use of no. 8 bars, we see that $t = d + 4.5$ in (114.3 mm). Then $t = 1.54(12) + 4.5 = 23.0$ in (584.2 mm). Make $t = 24$ in (609.6 mm); $d = 19.5$ in = 1.63 ft (0.496 m). The footing weight = 58.8(2)(0.150) = 17.64 kips (1384.082 kN). The assumed weight = 0.06(380) = 22.8 kips (101.41 kN). This is acceptable.

6. Check v_2

In Fig. 128, $AL = (7.67 - 1.67)/2 - 1.63 = 1.37$ ft (0.417 m); $V_2 = 380(1.37/7.67) = 67.9$ kips (302.02 kN); $v_2 = V_2/(Ld) = 67,900/[92(19.5)] = 38$ lb/in² (262.0 kPa) < 60 lb/in² (413.7 kPa). This is acceptable.

7. Design the reinforcement

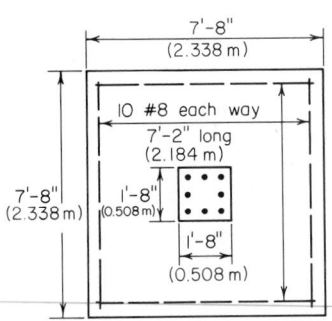

In Fig. 128, $EA = 3.00$ ft (0.914 m); $V_{EF} = 380(3.00/7.67) = 148.6$ kips (666.97 kN); $M_{EF} = 148.6(\frac{1}{2})(3.00)(12) = 2675$ in·kips (302.22 kN·m); $A_s = 2675/[20(0.874)(19.5)] = 7.85$ in² (50.648 cm²). Try 10 no. 8 bars each way. Then $A_s = 7.90$ in² (50.971 cm²); $\Sigma o = 31.4$ in (797.56 mm); $u = V_{EF}/\Sigma ojd = 148,600/[31.4(0.874)(19.5)] = 278$ lb/in² (1916.81 kPa); $u_{allow} = 264/1 = 264$ lb/in² (1820.3 kPa).

The bond stress at EF is slightly excessive. However, the ACI *Code*, in sections based on ultimate-strength considerations, permits disregarding the local bond stress if the average bond stress across the length of embedment is less than 80 percent of the allowable stress. Let L_e denote this length. Then $L_e = EA - 3 = 33$ in (838.2 mm); $0.80u_{allow} = 211$ lb/in² (1454.8 kPa); $u_{av} = A_s f_s/(L_e \Sigma o) = 0.79(20,000)/[33(3.1)] = 154$ lb/in² (1061.8 kPa). This is acceptable.

8. Design the dowels to comply with the Code

The function of the dowels is to transfer the compressive force in the column reinforcing bars to the

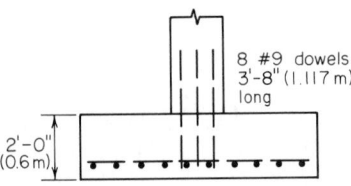

FIG. 129

footing. Since this is a tied column, assume the stress in the bars is $0.85(20,000) = 17,000$ lb/in^2 (117,215.0 kPa). Try eight no. 9 dowels with $f_y = 40,000$ lb/in^2 (275,800.0 kPa). Then $u = 264/(9/8) = 235$ lb/in^2 (1620.3 kPa); $L_e = 1.00(17,000)/[235(3.5)] = 20.7$ in (525.78 mm). Since the footing can provide a 21-in (533.4-mm) embedment length, the dowel selection is satisfactory. Also, the length of lap $= 20(9/8) = 22.5$ in (571.5 mm); length of dowels $= 20.7 + 22.5 = 43.2$, say 44 in (1117.6 mm). The footing is shown in Fig. 129.

COMBINED FOOTING DESIGN

An 18 -in (457.2-mm) square exterior column and a 20-in (508.0-mm) square interior column carry loads of 250 kips (1112 kN) and 370 kips (1645.8 kN), respectively. The column centers are 16 ft (4.9 m) apart, and the footing cannot project beyond the face of the exterior column. Design a combined rectangular footing by the working-stress method, using $f'_c = 3000$ lb/in^2 (20,685.0 kPa), $f_s = 20,000$ lb/in^2 (137,900.0 kPa), and an allowable soil pressure of 5000 lb/in^2 (239.4 kPa).

Calculation Procedure:

1. Establish the length of footing, applying the criterion of uniform soil pressure under total live and dead loads

In many instances, the exterior column of a building cannot be individually supported because the required footing would project beyond the property limits. It then becomes necessary to use a combined footing that supports the exterior column and the adjacent interior column, the footing being so proportioned that the soil pressure is approximately uniform.

The footing dimensions are shown in Fig. 130a, and the reinforcement is seen in Fig. 131. It is convenient to visualize the combined footing as being subjected to an upward load transmitted by the underlying soil and reactions supplied by the columns. The member thus functions as a beam that overhangs one support. However, since the footing is considerably wider than the columns, there is a transverse bending as well as longitudinal bending in the vicinity of the columns. For simplicity, assume that the transverse bending is confined to the regions bounded by planes AB and EF and by planes GH and NP, the distance m being $h/2$ or $d/2$, whichever is smaller.

In Fig. 130a, let Z denote the location of the resultant of the column loads. Then $x = 370(16)/(250 + 370) = 9.55$ ft (2.910 m). Since Z is to be the centroid of the footing, $L = 2(0.75 + 9.55) = 20.60$ ft (6.278 m). Set $L = 20$ ft 8 in (6.299 m), but use the value 20.60 ft (6.278 m) in the stress calculations.

2. Construct the shear and bending-moment diagrams

The net soil pressure per foot of length $= 620/20.60 = 30.1$ kips/lin ft (439.28 kN/m). Construct the diagrams as shown in Fig. 130.

3. Establish the footing thickness

Use

$$(Pv_2 + 0.17VL + Pp')d - 0.17Pd^2 = VLp' \tag{50}$$

where P = aggregate column load, kips (kN); V = maximum vertical shear at a column face, kips (kN); p' = gross soil pressure, kips/ft^2 (MPa).

Assume that the longitudinal steel is centered 3½ in (88.9 mm) from the face of the footing. Then $P = 620$ kips (2757.8 kN); $V = 229.2$ kips (1019.48 kN); $v_2 = 0.06(144) = 8.64$ kips/ft^2 (0.414 MPa); $9260d - 105.4d^2 = 23,608$; $d = 2.63$ ft (0.801 m); $t = 2.63 + 0.29 = 2.92$ ft. Set $t = 2$ ft 11 in (0.889 m); $d = 2$ ft 7½ in (0.800 m).

4. Compute the vertical shear at distance d from the column face

Establish the width of the footing. Thus $V = 229.2 - 2.63(30.1) = 150.0$ kips (667.2 kN); $v = V/(Wd)$, or $W = V/(vd) = 150/[8.64(2.63)] = 6.60$ ft (2.012 m). Set $W = 6$ ft 8 in (2.032 m).

5. Check the soil pressure

The footing weight $= 20.67(6.67)(2.92)(0.150) = 60.4$ kips (268.66 kN); $p' = (620 + 60.4)/[(20.67)(6.67)] = 4.94$ kips/ft^2 (0.236 MPa) < 5 kips/ft^2 (0.239 MPa). This is acceptable.

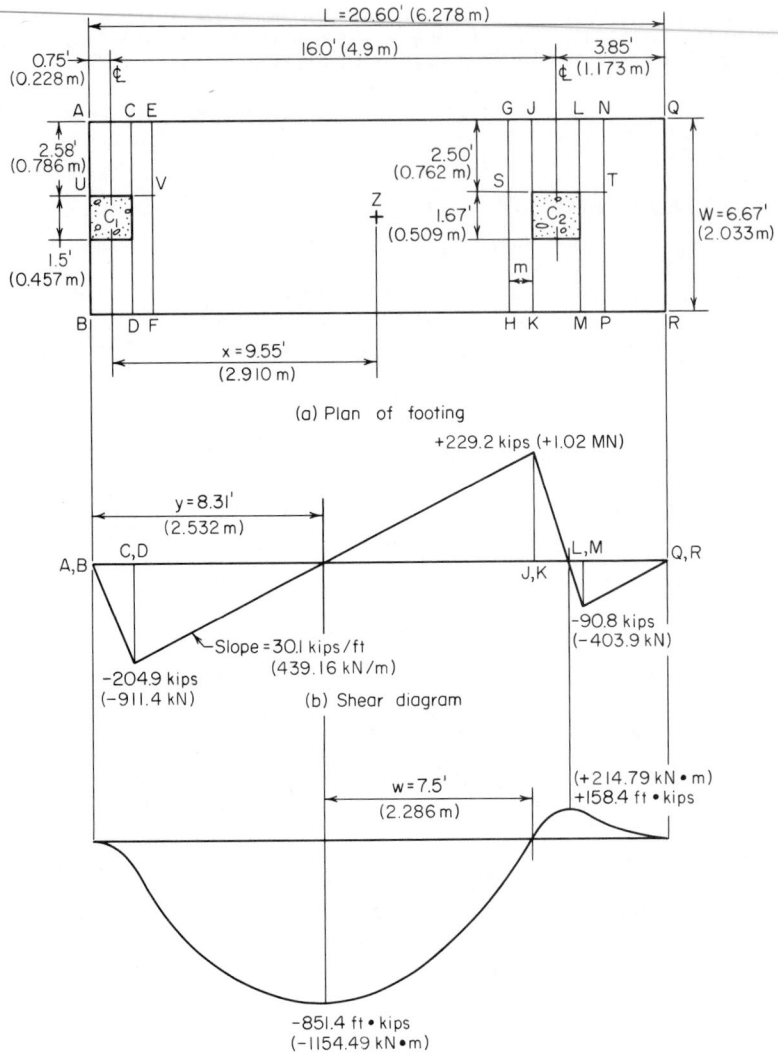

(a) Plan of footing

(b) Shear diagram

(c) Bending-moment diagram

FIG. 130

6. Check the punching shear

Thus, $p = 4.94 - 2.92(0.150) = 4.50$ kips/ft² (0.215 MPa). At $C1$: $b_o = 18 + 31.5 + 2 (18 + 15.8) = 117$ in (2971.8 mm); $V = 250 - 4.50(49.5)(33.8)/144 = 198$ kips (880.7 kN); $v_1 = 198,000/[117(31.5)] = 54$ lb/in² (372.3 kPa) < 110 lb/in² (758.5 kPa); this is acceptable.

At $C2$: $b_o = 4(20 + 31.5) = 206$ in (5232.4 mm); $V = 370 - 4.50(51.5)^2/144 = 287$ kips (1276.6 kN); $v_1 = 287,000/[206(31.5)] = 44$ lb/in² (303.4 kPa). This is acceptable.

7. Design the longitudinal reinforcement for negative moment

Thus, $M = 851,400$ ft·lb $= 10,217,000$ in·lb (1,154,316.6 N·m); $M_b = 223(80)(31.5)^2 = 17,700,000$ in·lb (1,999,746.0 N·m). Therefore, the steel is stressed to capacity, and $A_s =$

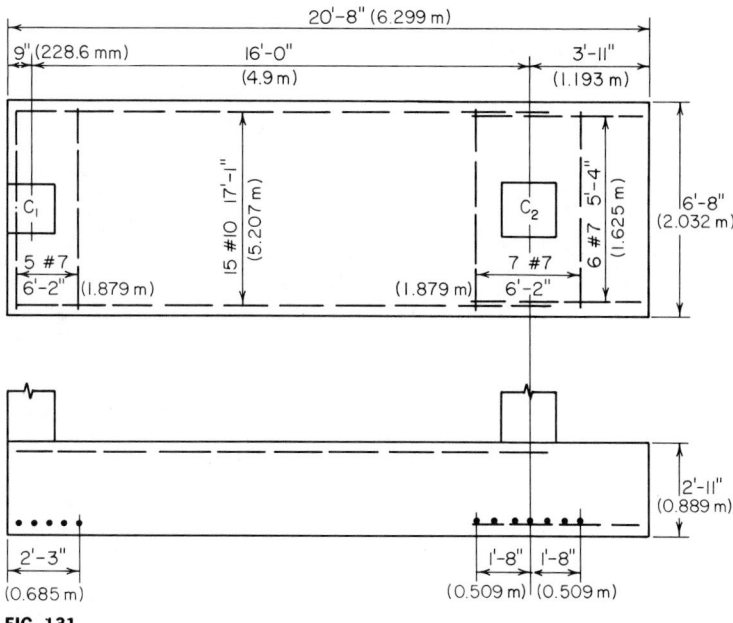

FIG. 131

$10,217,000/[20,000(0.874)(31.5)] = 18.6 \text{ in}^2 (120.01 \text{ cm}^2)$. Try 15 no. 10 bars with $A_s = 19.1 \text{ in}^2$ (123.2 cm^2); $\Sigma o = 59.9 \text{ in} (1521.46 \text{ mm})$.

The bond stress is maximum at the point of contraflexure, where $V = 15.81(30.1) - 250 = 225.9$ kips (1004.80 kN); $u = 225,900/[59.9(0.874)(31.5)] = 137 \text{ lb/in}^2 (944.6 \text{ kPa})$; $u_{allow} = 3.4(3000)^{0.5}/1.25 = 149 \text{ lb/in}^2 (1027.4 \text{ kPa})$. This is acceptable.

8. Design the longitudinal reinforcement for positive moment

For simplicity, design for the maximum moment rather than the moment at the face of the column. Then $A_s = 158,400(12)/[20,000(0.874)(31.5)] = 3.45 \text{ in}^2 (22.259 \text{ cm}^2)$. Try six no. 7 bars with $A_s = 3.60 \text{ in}^2 (23.227 \text{ cm}^2)$; $\Sigma o = 16.5 \text{ in} (419.10 \text{ mm})$. Take LM as the critical section for bond, and $u = 90,800/[16.5(0.874)(31.5)] = 200 \text{ lb/in}^2 (1379.0 \text{ kPa})$; $u_{allow} = 4.8(3000)^{0.5}/0.875 = 302 \text{ lb/in}^2 (2082.3 \text{ kPa})$. This is acceptable.

9. Design the transverse reinforcement under the interior column

For this purpose, consider member $GNPH$ as an independent isolated footing. Then $V_{ST} = 370(2.50/6.67) = 138.8$ kips (617.38 kN); $M_{ST} = \frac{1}{2}(138.8)(2.50)(12) = 2082 \text{ in}\cdot\text{kips} (235.22 \text{ kN}\cdot\text{m})$. Assume $d = 35 - 4.5 = 30.5 \text{ in} (774.7 \text{ mm})$; $A_s = 2,082,000/[20,000(0.874)(30.5)] = 3.91 \text{ in}^2 (25.227 \text{ cm}^2)$. Try seven no. 7 bars; $A_s = 4.20 \text{ in}^2 (270.098 \text{ cm}^2)$; $\Sigma o = 19.2 \text{ in} (487.68 \text{ mm})$; $u = 138,800/[19.2(0.874)(30.5)] = 271 \text{ lb/in}^2 (1868.5 \text{ kPa})$; $u_{allow} = 302 \text{ lb/in}^2 (2082.3 \text{ kPa})$. This is acceptable.

Since the critical section for shear falls outside the footing, shearing stress is not a criterion in this design.

10. Design the transverse reinforcement under the exterior column; disregard eccentricity

Thus, $V_{UV} = 250(2.58/6.67) = 96.8$ kips (430.57 kN); $M_{UV} = \frac{1}{2}(96.8)(2.58)(12) = 1498 \text{ in}\cdot\text{kips}$ (169.3 kN·m); $A_s = 2.72 \text{ in}^2 (17.549 \text{ cm}^2)$. Try five no. 7 bars; $A_s = 3.00 \text{ in}^2 (19.356 \text{ cm}^2)$; $\Sigma o = 13.7 \text{ in} (347.98 \text{ mm})$; $u = 96,800/[13.7 (0.874)(31.5)] = 257 \text{ lb/in}^2 (1772.0 \text{ kPa})$. This is acceptable.

Cantilever Retaining Walls

Retaining walls having a height ranging from 10 to 20 ft (3.0 to 6.1 m) are generally built as reinforced-concrete cantilever members. As shown in Fig. 132, a cantilever wall comprises a vertical stem to retain the soil, a horizontal base to support the stem, and in many instances a key that projects into the underlying soil to augment the resistance to sliding. Adequate drainage is an essential requirement, because the accumulation of water or ice behind the wall would greatly increase the horizontal thrust.

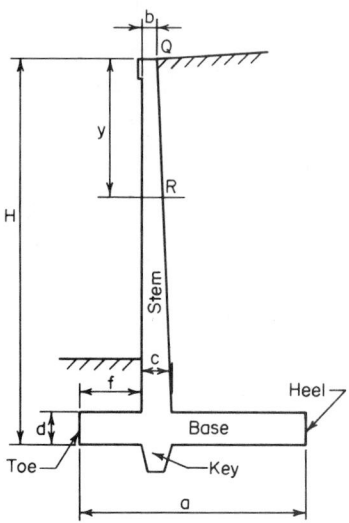

The calculation of earth thrust in this section is based on Rankine's theory, which is developed in a later calculation procedure. When a live load, termed a *surcharge*, is applied to the retained soil, it is convenient to replace this load with a hypothetical equivalent prism of earth. Referring to Fig. 132, consider a portion QR of the wall, R being at distance y below the top. Take the length of wall normal to the plane of the drawing as 1 ft (0.3 m). Let T = resultant earth thrust on QR; M = moment of this thrust with respect to R; h = height of equivalent earth prism that replaces surcharge; w = unit weight of earth; C_a = coefficient of active earth pressure; C_p = coefficient of passive earth pressure. Then

$$T = \tfrac{1}{2}C_a wy(y + 2h) \tag{51}$$

FIG. 132 Cantilever retaining wall.

$$M = (\tfrac{1}{6})C_a wy^2(y + 3h) \tag{52}$$

DESIGN OF A CANTILEVER RETAINING WALL

Applying the working-stress method, design a reinforced-concrete wall to retain an earth bank 14 ft (4.3 m) high. The top surface is horizontal and supports a surcharge of 500 lb/ft² (23.9 kPa). The soil weighs 130 lb/ft³ (20.42 kN/m³), and its angle of internal friction is 35°; the coefficient of friction of soil and concrete is 0.5. The allowable soil pressure is 4000 lb/ft² (191.5 kPa); $f'_c = 3000$ lb/in² (20,685 kPa) and $f_y = 40,000$ lb/in² (275,800 kPa). The base of the structure must be set 4 ft (1.2 m) below ground level to clear the frost line.

Calculation Procedure:

1. Secure a trial section of the wall

Apply these relations: $a = 0.60H$; $b \geq 8$ in (203.2 mm); $c = d = b + 0.045h$; $f = a/3 - c/2$.

The trial section is shown in Fig. 133a, and the reinforcement is shown in Fig. 134. As the calculation will show, it is necessary to provide a key to develop the required resistance to sliding. The sides of the key are sloped to ensure that the surrounding soil will remain undisturbed during excavation.

2. Analyze the trial section for stability

The requirements are that there be a factor of safety (FS) against sliding and overturning of at least 1.5 and that the soil pressure have a value lying between 0 and 4000 lb/ft² (0 and 191.5 kPa). Using the equation developed later in this handbook gives h = surcharge/soil weight = 500/130 = 3.85 ft (1.173 m); sin 35° = 0.574; tan 35° = 0.700; $C_a = 0.271$; $C_p = 3.69$; $C_a w = 35.2$ lb/ft³ (5.53 kN/m³); $C_p w = 480$ lb/ft³ (75.40 kN/m³); $T_{AB} = \tfrac{1}{2}(35.2)18(18 + 2 \times 3.85) = 8140$ lb (36,206.7 N); $M_{AB} = (\tfrac{1}{6})35.2(18)^2(18 + 3 \times 3.85) = 56,200$ ft·lb (76,207.2 N·m).

The critical condition with respect to stability is that in which the surcharge extends to G. The moments of the stabilizing forces with respect to the toe are computed in Table 7. In Fig. 133c, $x = 81,030/21,180 = 3.83$ ft (1.167 m); $e = 5.50 - 3.83 = 1.67$ ft (0.509 m). The fact that the

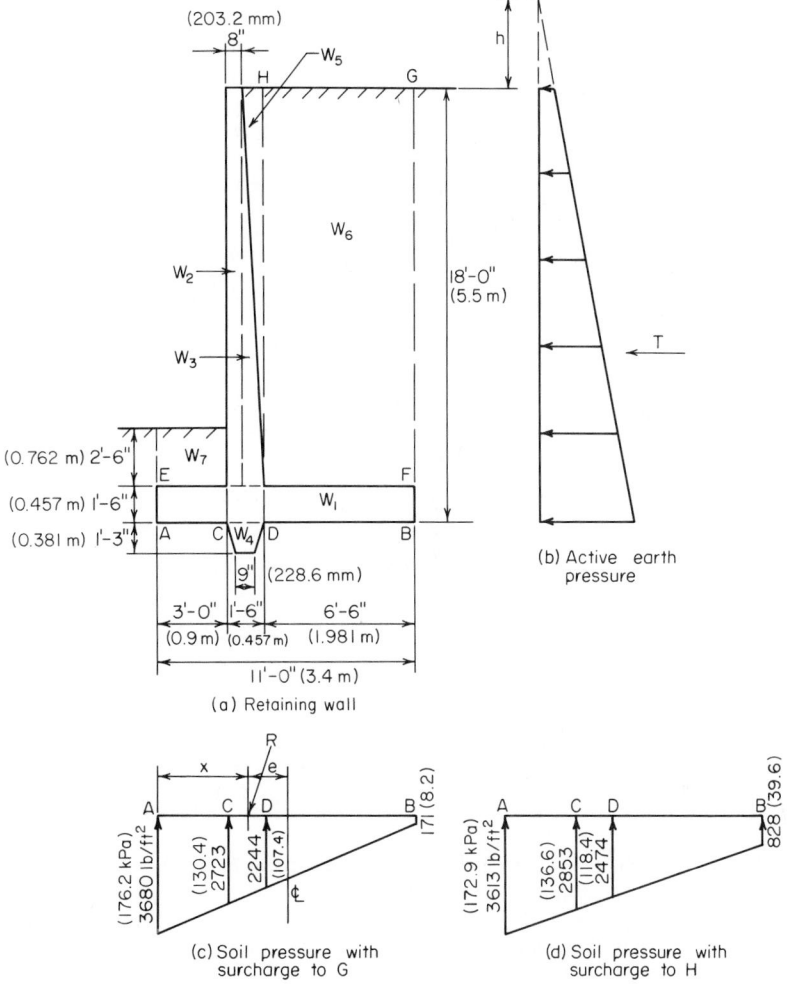

(a) Retaining wall

(b) Active earth pressure

(c) Soil pressure with surcharge to G

(d) Soil pressure with surcharge to H

FIG. 133

resultant strikes the base within the middle third attests to the absence of uplift. By $f = (P/A)(1 \pm 6e_x/d_x \pm 6e_y/d_y)$, $p_a = (21{,}180/11)(1 + 6 \times 1.67/11) = 3680$ lb/ft² (176.2 kPa); $p_b = (21{,}180/11)(1 - 6 \times 1.67/11) = 171$ lb/ft² (8.2 kPa). Check: $x = (11/3)(3680 + 2 \times 171)/(3680 + 171) = 3.83$ ft (1.167 m), as before. Also, $p_c = 2723$ lb/ft² (130.4 kPa); $p_d = 2244$ lb/ft² (107.4 kPa); FS against overturning = 137,230/56,200 = 2.44. This is acceptable.

Lateral displacement of the wall produces sliding of earth on earth to the left of C and of concrete on earth to the right of C. In calculating the passive pressure, the layer of earth lying above the base is disregarded, since its effectiveness is unknown. The resistance to sliding is as follows: friction, A to C (Fig. 133): ½(3680 + 2723)(3)(0.700) = 6720 lb (29,890.6 N); friction, C to B: ½(2723 + 171)(8)(0.5) = 5790 lb (25,753.9 N); passive earth pressure: ½(480)(2.75)² = 1820 lb (8095.4 N). The total resistance to sliding is the sum of these three items, or 14,330 lb (63,739.8 N). Thus, the FS against sliding is 14,330/8140 = 1.76. This is acceptable because it exceeds 1.5. Hence the trial section is adequate with respect to stability.

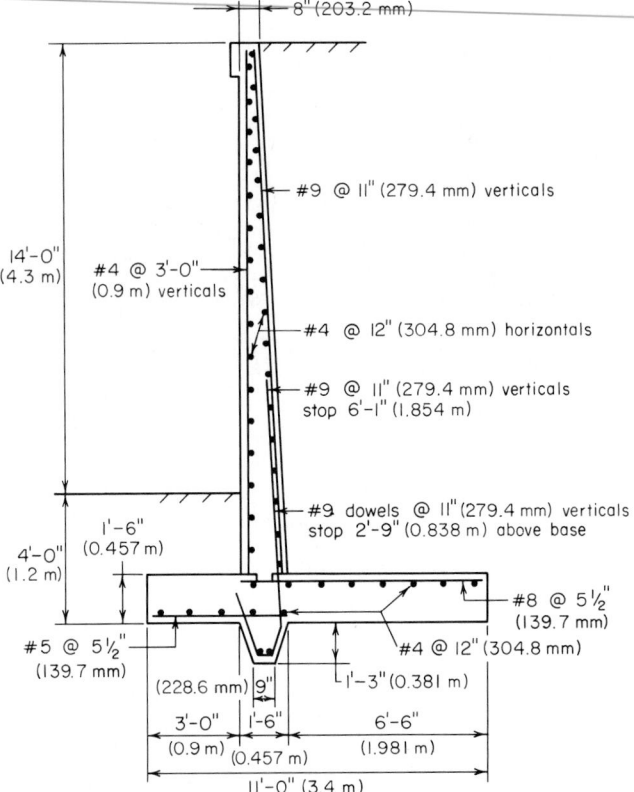

FIG. 134

3. Calculate the soil pressures when the surcharge extends to H

Thus $W_s = 500(6.5) = 3250$ lb (14,456 N); $\Sigma W = 21,180 + 3250 = 24,430$ lb (108,664.6 N); $M_a = 81,030 + 3250(7.75) = 106,220$ ft·lb (144,034.3 N·m); $x = 106,220/24,430 = 4.35$ ft (1.326 m); $e = 1.15$ ft (0.351 m); $p_a = 3613$ lb/ft² (173 kPa); $p_b = 828$ lb/ft² (39.6 kPa); $p_c = 2853$ lb/ft² (136.6 kPa); $p_d = 2474$ lb/ft² (118.5 kPa).

TABLE 7 Stability of Retaining Wall

Force, lb (N)		Arm, ft (m)	Moment, ft·lb (N·m)
W_1 1.5(11)(150)	= 2,480 (11,031.0)	5.50 (1.676)	13,640 (18,495.8)
W_2 0.67(16.5)(150)	= 1,650 (7,339.2)	3.33 (1.015)	5,500 (7,458.0)
W_3 0.5(0.83)(16.5)(150)	= 1,030 (4,581.4)	3.95 (1.204)	4,070 (5,518.9)
W_4 1.25(1.13)(150)	= 210 (934.1)	3.75 (1.143)	790 (1,071.2)
W_5 0.5(0.83)(16.5)(130)	= 890 (3,958.7)	4.23 (1.289)	3,760 (5,098.6)
W_6 6.5(16.5)(130)	= 13,940 (62,005.1)	7.75 (2.362)	108,000 (146,448.0)
W_7 2.5(3)(130)	= 980 (4,359.1)	1.50 (0.457)	1,470 (1993.3)
Total	21,180 (94,208.6)	. . . .	137,230 (186,083.8)
Overturning moment		. . . .	56,200 (76,207.2)
Net moment about A		. . . .	81,030 (109,876.6)

4. *Design the stem*

At the base of the stem, $y = 16.5$ ft (5.03 m) and $d = 18 - 3.5 = 14.5$ in (368.30 mm); $T_{EF} = 7030$ lb (31,269.4 N); $M_{EF} = 538,000$ in·lb (60,783.24 N·m). The allowable shear at a distance d above the base is $V_{\text{allow}} = vbd = 60(12)(14.5) = 10,440$ lb (46,437.1 N). This is acceptable. Also, $M_b = 223(12)(14.5)^2 = 563,000$ in·lb (63,607.74 N·m); therefore, the steel is stressed to capacity, and $A_s = 538,000/[20,000(0.874)(14.5)] = 2.12$ in^2 (13.678 cm^2). Use no. 9 bars 5½ in (139.70 mm) on centers. Thus, $A_s = 2.18$ in^2 (14.065 cm^2); $\Sigma o = 7.7$ in (195.58/mm); $u = 7030/[7.7(0.874)(14.5)] = 72$ lb/in^2 (496.5 kPa); $u_{\text{allow}} = 235$ lb/in^2 (1620.3 kPa). This is acceptable.

Alternate bars will be discontinued at the point where they become superfluous. As the following calculations demonstrate, the theoretical cutoff point lies at $y = 11$ ft 7 in (3.531 m), where $M = 218,400$ in·lb (24,674.8 N·m); $d = 4.5 + 10(11.58/16.5) = 11.52$ in (292.608 mm); $A_s = 218,400/[20,000 \ (0.874)(11.52)] = 1.08$ in^2 (6.968 cm^2). This is acceptable. Also, $T = 3930$ lb (17,480.6 N); $u = 101$ lb/in^2 (696.4 kPa). This is acceptable. From the ACI *Code*, anchorage $= 12(9/8) = 13.5$ in (342.9 mm).

The alternate bars will therefore be terminated at 6 ft 1 in (1.854 m) above the top of the base. The *Code* requires that special precautions be taken where more than half the bars are spliced at a point of maximum stress. To circumvent this requirement, the short bars can be extended into the footing; therefore only the long bars require splicing. For the dowels, $u_{\text{allow}} = 0.75(235) = 176$ lb/in^2 (1213.5 kPa); length of lap $= 1.00(20,000)/[176(3.5)] = 33$ in (838.2 mm).

5. *Design the heel*

Let V and M denote the shear and bending moment, respectively, at section D. Case 1: surcharge extending to G—downward pressure $p = 16.5(130) + 1.5(150) = 2370$ lb/ft^2 (113.5 kPa); $V = 6.5[2370 - \frac{1}{2}(2244 + 171)] = 7560$ lb (33,626.9 N); $M = 12(6.5)^2[\frac{1}{2} \times 2370 - \frac{1}{6}(2244 + 2 \times 171)] = 383,000$ in·lb (43,271.3 N·m).

Case 2: surcharge extending to H—$p = 2370 + 500 = 2870$ lb/ft^2 (137.4 kPa); $V = 6.5[2870 - \frac{1}{2}(2474 + 828)] = 7920$ lb (35,228.1 N) $< V_{\text{allow}}$; $M = 12(6.5)^2[\frac{1}{2} \times 2870 - \frac{1}{6}(2474 + 2 \times 828)] = 379,000$ in·lb (42,819.4 N·m); $A_s = 2.12(383/538) = 1.51$ in^2 (9.742 cm^2).

To maintain uniform bar spacing throughout the member, use no. 8 bars 5½ in (139.7 mm) on centers. In the heel, tension occurs at the *top* of the slab, and $A_s = 1.72$ in^2 (11.097 cm^2); $\Sigma o = 6.9$ in (175.26 mm); $u = 91$ lb/in^2 (627.4 kPa); $u_{\text{allow}} = 186$ lb/in^2 (1282.5 kPa). This is acceptable.

6. *Design the toe*

For this purpose, assume the absence of backfill on the toe, but disregard the minor modification in the soil pressure that results. Let V and M denote the shear and bending moment, respectively, at section C (Fig. 133). The downward pressure $p = 1.5(150) = 225$ lb/ft^2 (10.8 kPa).

Case 1: surcharge extending to G (Fig. 133)—$V = 3[\frac{1}{2}(3680 + 2723) - 225] = 8930$ lb (39,720.6 N); $M = 12(3)^2[(\frac{1}{6})(2723 + 2 \times 3680) - \frac{1}{2}(225)] = 169,300$ in·lb (19,127.5 N·m).

Case 2: surcharge extending to H (Fig. 133)—$V = 9020$ lb (40,121.0 N) $< V_{\text{allow}}$; $M = 169,300$ in·lb (19,127.5 N·m); $A_s = 2.12(169,300/538,000) = 0.67$ in^2 (4.323 cm^2). Use no. 5 bars 5½ in (139.7 mm) on centers. Then $A_s = 0.68$ in^2 (4.387 cm^2); $\Sigma o = 4.3$ in (109.22 mm); $u = 166$ lb/in^2 (1144.4 kPa); $u_{\text{allow}} = 422$ lb/in^2 (2909.7 kPa). This is acceptable.

The stresses in the key are not amenable to precise evaluation. Reinforcement is achieved by extending the dowels and short bars into the key and bending them.

In addition to the foregoing reinforcement, no. 4 bars are supplied to act as temperature reinforcement and spacers for the main bars, as shown in Fig. 134.

Prestressed Concrete

Prestressed-concrete construction is designed to enhance the suitability of concrete as a structural material by inducing prestresses opposite in character to the stresses resulting from gravity loads. These prestresses are created by the use of steel wires or strands, called *tendons*, that are incorporated in the member and subjected to externally applied tensile forces. This prestressing of the steel may be performed either before or after pouring of the concrete. Thus, two methods of prestressing a concrete beam are available: *pretensioning* and *posttensioning*.

In pretensioning, the tendons are prestressed to the required amount by means of hydraulic jacks, their ends are tied to fixed abutments, and the concrete is poured around the tendons. When hardening of the concrete has advanced to the required state, the tendons are released. The tendons now tend to contract longitudinally to their original length and to expand laterally to their original diameter, both these tendencies being opposed by the surrounding concrete. As a result of the longitudinal restraint, the concrete exerts a tensile force on the steel and the steel exerts a compressive force on the concrete. As a result of the lateral restraint, the tendons are deformed to a wedge shape across a relatively short distance at each end of the member. It is within this distance, termed the *transmission length*, that the steel becomes bonded to the concrete and the two materials exert their prestressing forces on each other. However, unless greater precision is warranted, it is assumed for simplicity that the prestressing forces act at the end sections.

The tendons may be placed either in a straight line or in a series of straight-line segments, being deflected at designated points by means of holding devices. In the latter case, prestressing forces between steel and concrete occur both at the ends and at these deflection points.

In posttensioning, the procedure usually consists of encasing the tendons in metal or rubber hoses, placing these in the forms, and then pouring the concrete. When the concrete has hardened, the tendons are tensioned and anchored to the ends of the concrete beam by means of devices called *end anchorages*. If the hoses are to remain in the member, the void within the hose is filled with grout. Posttensioning has two important advantages compared with pretensioning: It may be performed at the job site, and it permits the use of parabolic tendons.

The term *at transfer* refers to the instant at which the prestressing forces between steel and concrete are developed. (In posttensioning, where the tendons are anchored to the concrete one at a time, in reality these forces are developed in steps.) Assume for simplicity that the tendons are straight and that the resultant prestressing force in these tendons lies below the centroidal axis of the concrete section. At transfer, the member cambers (deflects upward), remaining in contact with the casting bed only at the ends. Thus, the concrete beam is compelled to resist the prestressing force and to support its own weight simultaneously.

At transfer, the prestressing force in the steel diminishes because the concrete contracts under the imposed load. The prestressing force continues to diminish as time elapses as a result of the relaxation of the steel and the shrinkage and plastic flow of the concrete subsequent to transfer. To be effective, prestressed-concrete construction therefore requires the use of high-tensile steel in order that the reduction in prestressing force may be small in relation to the initial force. In all instances, we assume that the ratio of final to initial prestressing force is 0.85. Moreover, to simplify the stress calculations, we also assume that the full initial prestressing force exists at transfer and that the entire reduction in this force occurs during some finite interval following transfer.

Therefore, two loading states must be considered in the design: the initial state, in which the concrete sustains the initial prestressing force and the beam weight; and the final state, in which the concrete sustains the final prestressing force, the beam weight, and all superimposed loads. Consequently, the design of a prestressed-concrete beam differs from that of a conventional type in that designers must consider two stresses at each point, the initial stress and the final stress, and these must fall between the allowable compressive and tensile stresses. A beam is said to be in *balanced design* if the critical initial and final stresses coincide precisely with the allowable stresses.

The term *prestress* designates the stress induced by the *initial* prestressing force. The terms *prestress shear* and *prestress moment* refer to the vertical shear and bending moment, respectively, that the initial prestressing force induces in the concrete at a given section.

The *eccentricity* of the prestressing force is the distance from the action line of this resultant force to the centroidal axis of the section. Assume that the tendons are subjected to a uniform prestress. The locus of the centroid of the steel area is termed the *trajectory* of the steel or of the prestressing force.

The sign convention is as follows: The eccentricity is positive if the action line of the prestressing force lies below the centroidal axis. The trajectory has a positive slope if it inclines downward to the right. A load is positive if it acts downward. The vertical shear at a given section is positive if the portion of the beam to the left of this section exerts an upward force on the concrete. A bending moment is positive if it induces compression above the centroidal axis and tension below it. A compressive stress is positive; a tensile stress, negative.

The notational system is as follows. *Cross-sectional properties:* A = gross area of section, in^2 (cm^2) A_s = area of prestressing steel, in^2 (cm^2); d = effective depth of section at ultimate strength, in (mm); h = total depth of section, in (mm); I = moment of inertia of gross area, in^4 (cm^4); y_b = distance from centroidal axis to bottom fiber, in (mm); S_b = section modulus with respect to bottom fiber $= I/y_b$, in^3 (cm^3); k_b = distance from centroidal axis to lower kern point, in (mm); k_t = distance from centroidal axis to upper kern point, in (mm). *Forces and moments:* F_i = initial prestressing force, lb (N); F_f = final prestressing force, lb (N); $\eta = F_f/F_i$; e = eccentricity of prestressing force, in (mm); e_{con} = eccentricity of prestressing force having concordant trajectory; θ = angle between trajectory (or tangent to trajectory) and horizontal line; m = slope of trajectory; w = vertical load exerted by curved tendons on concrete in unit distance; w_w = unit beam weight; w_s = unit superimposed load; w_{DL} = unit dead load; w_{LL} = unit live load; w_u = unit ultimate load; V_p = prestress shear; M_p = prestress moment; M_w = bending moment due to beam weight; M_s = bending moment due to superimposed load; C_u = resultant compressive force at ultimate load; T_u = resultant tensile force at ultimate load. *Stresses:* f_c' = ultimate compressive strength of concrete, lb/in^2 (kPa); f_{ci}' = compressive strength of concrete at transfer; f_s' = ultimate strength of prestressing steel; f_{su} = stress in prestressing steel at ultimate load; f_{bp} = stress in bottom fiber due to initial prestressing force; f_{bw} = bending stress in bottom fiber due to beam weight; f_{bs} = bending stress in bottom fiber due to superimposed loads; f_{bi} = stress in bottom fiber at initial state = $f_{bp} + f_{bw}$; f_{bf} = stress in bottom fiber at final state = $\eta f_{bp} + f_{bw} + f_{bs}$; f_{cai} = initial stress at centroidal axis. *Camber:* Δ_p = camber due to initial prestressing force, in (mm); Δ_w = camber due to beam weight; Δ_i = camber at initial state; Δ_f = camber at final state.

The symbols that refer to the bottom fiber are transformed to their counterparts for the top fiber by replacing the subscript b with t. For example, f_{ti} denotes the stress in the top fiber at the initial state.

DETERMINATION OF PRESTRESS SHEAR AND MOMENT

The beam in Fig. 135a is simply supported at its ends and prestressed with an initial force of 300 kips (1334.4 kN). At section C, the eccentricity of this force is 8 in (203.2 mm), and the slope of

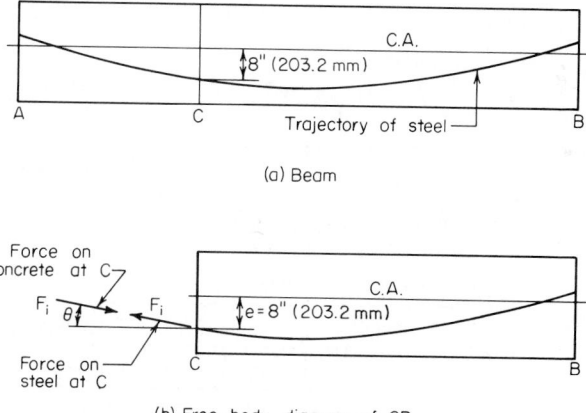

(a) Beam

(b) Free-body diagram of CB

FIG. 135

the trajectory is 0.014. (In the drawing, vertical distances are exaggerated in relation to horizontal distances.) Find the prestress shear and prestress moment at C.

Calculation Procedure:

1. Analyze the prestressing forces

If the composite concrete-and-steel member is regarded as a unit, the prestressing forces that the steel exerts on the concrete are purely internal. Therefore, if a beam is simply supported, the prestressing force alone does not induce any reactions at the supports.

Refer to Fig. 135b, and consider the forces acting on the beam segment *CB* solely as a result of F_i. The left portion of the beam exerts a tensile force F_t on the tendons. Since CB is in equilibrium, the left portion also induces compressive stresses on the concrete at C, these stresses having a resultant that is numerically equal to and collinear with F_i.

2. *Express the prestress shear and moment in terms of F_i*

Using the sign convention described, express the prestress shear and moment in terms of F_i and θ. (The latter is positive if the slope of the trajectory is positive.) Thus $V_p = -F_i \sin \theta$; $M_p = -F_i e \cos \theta$.

3. *Compute the prestress shear and moment*

Since θ is minuscule, apply these approximations: $\sin \theta = \tan \theta$, and $\cos \theta = 1$. Then

$$V_p = -F_i \tan \theta \tag{53}$$

Or, $V_p = -300{,}000(0.014) = -4200$ lb $(-18{,}681.6$ N).
   Also,

$$M_p = -F_i e \tag{54}$$

Or, $M_p = -300{,}000(8) = -2{,}400{,}000$ in$\cdot$lb $(-271{,}152$ N$\cdot$m).

STRESSES IN A BEAM WITH STRAIGHT TENDONS

A 12×18 in (304.8×457.2 mm) rectangular beam is subjected to an initial prestressing force of 230 kips (1023.0 kN) applied 3.3 in (83.82 mm) below the center. The beam is on a simple span of 30 ft (9.1 m) and carries a superimposed load of 840 lb/lin ft (12,258.9 N/m). Determine the initial and final stresses at the supports and at midspan. Construct diagrams to represent the initial and final stresses along the span.

Calculation Procedures:

1. *Compute the beam properties*

Thus, $A = 12(18) = 216$ in^2 (1393.6 cm^2); $S_b = S_t = (\frac{1}{6})(12)(18)^2 = 648$ in^3 (10,620.7 cm^3); $w_w = (216/144)(150) = 225$ lb/lin ft (3,283.6 N/m).

2. *Calculate the prestress in the top and bottom fibers*

Since the section is rectangular, apply $f_{bp} = (F_i/A)(1 + 6e/h) = (230{,}000/216)(1 + 6 \times 3.3/18) = +2236$ lb/in^2 ($+15{,}417.2$ kPa); $f_{tp} = (F_i/A)(1 - 6e/h) = -106$ lb/in^2 (-730.9 kPa).
   For convenience, record the stresses in Table 8 as they are obtained.

TABLE 8   Stresses in Prestressed-Concrete Beam

	At support		At midspan	
	Bottom fiber	Top fiber	Bottom fiber	Top fiber
(a) Initial prestress, lb/in^2 (kPa)	+2,236 (+15,417.2)	−106 (−730.9)	+2,236 (+15,417.2)	−106 (−730.9)
(b) Final prestress, lb/in^2 (kPa)	+1,901 (+13,107.4)	−90 (−620.6)	+1,901 (+13,107.4)	−90 (−620.6)
(c) Stress due to beam weight, lb/in^2 (kPa)	. . . . .	. . . . .	−469 (−3,233.8)	+469 (3,233.8)
(d) Stress due to superimposed load, lb/in^2 (kPa)	. . . . . .	. . . . . .	−1,750 (−12,066.3)	+1,750 (+12,066.3)
Initial stress: (a) + (c)	+2,236 (+15,417.2)	−106 (−730.9)	+1,767 (+12,183.5)	+363 (+2,502.9)
Final stress: (b) + (c) + (d)	+1,901 (+13,107.4)	−90 (−620.6)	−318 (−2,192.6)	+2,129 (+14,679.5)

3. Determine the stresses at midspan due to gravity loads

Thus $M_s = (\tfrac{1}{8})(840)(30)^2(12) = 1{,}134{,}000$ in·lb (128,119.32 N·m); $f_{ts} = -1{,}134{,}000/648 = -1750$ lb/in^2 (−12,066.3 kPa); $f_{ts} = +1750$ lb/in^2 (12,066.3 kPa). By proportion, $f_{bw} = -1750(225/840) = -469$; $f_{tw} = +469$ lb/in^2 (+3233.8 kPa).

4. Compute the initial and final stresses at the supports

Thus, $f_{bi} = +2236$ lb/in^2 (+15,417.2 kPa); $f_{ti} = -106$ lb/in^2 (−730.9 kPa); $f_{bf} = 0.85(2236) = +1901$ lb/in^2 (+13,107.4 kPa); $f_{tf} = 0.85(-106) = -90$ lb/in^2 (−620.6 kPa).

5. Determine the initial and final stresses at midspan

Thus $f_{bi} = +2236 - 469 = +1767$ lb/in^2 (+12,183.5 kPa); $f_{ti} = -106 + 469 = +363$ lb/in^2 (+2502.9 kPa); $f_{bf} = +1901 - 469 - 1750 = -318$ lb/in^2 (−2192.6 kPa); $f_{tf} = -90 + 469 + 1750 = +2129$ lb/in^2 (+14,679.5 kPa).

6. Construct the initial-stress diagram

In Fig. 136a, construct the initial-stress diagram $A_t A_b\,BC$ at the support and the initial-stress diagram $M_t M_b\,DE$ at midspan. Draw the parabolic arcs BD and CE. The stress diagram at an inter-

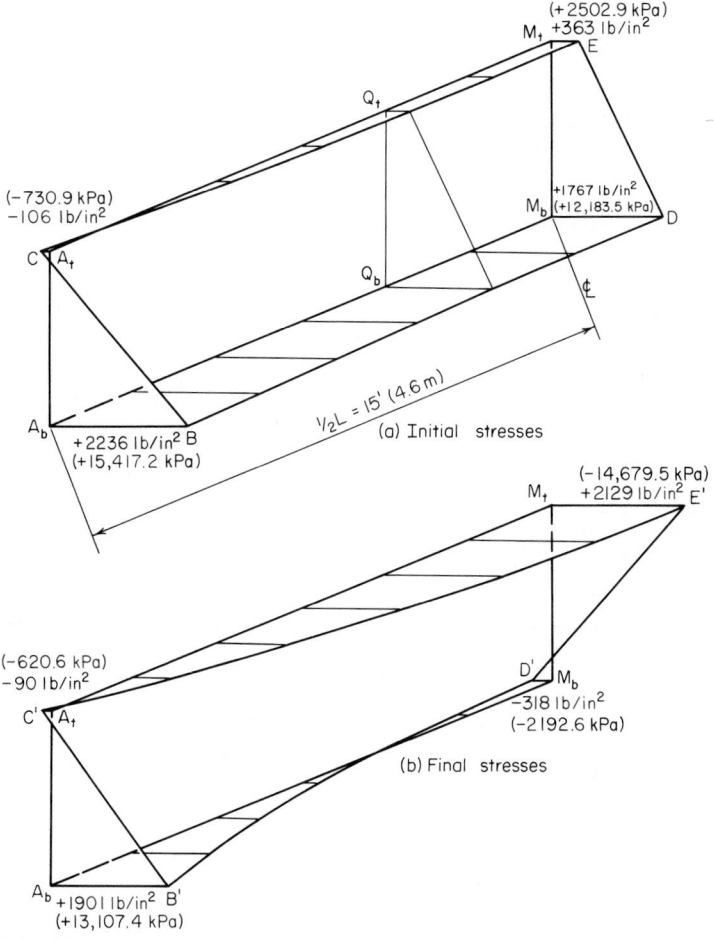

FIG. 136 Isometric stress diagrams for half-span.

mediate section Q is obtained by passing a plane normal to the longitudinal axis. The offset from a reference line through B to the arc BD represents the value of f_{bw} at that section.

7. Construct the final-stress diagram

Construct Fig. 136b in an analogous manner. The offset from a reference line through B' to the arc B'D' represents the value of $f_{bw} + f_{bs}$ at the given section.

8. Alternatively, construct composite stress diagrams for the top and bottom fibers

The diagram pertaining to the bottom fiber is shown in Fig. 137. The difference between the ordinates to DE and AB represents f_{bi}, and the difference between the ordinates to FG and AC represents f_{bf}.

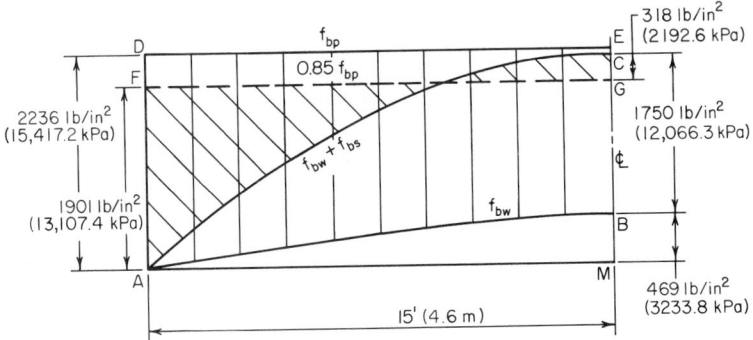

FIG. 137   Stresses in bottom fiber along half-span.

This procedure illustrates the following principles relevant to a beam with straight tendons carrying a uniform load: At transfer, the critical stresses occur at the supports; under full design load, the critical stresses occur at midspan if the allowable final stresses exceed η times the allowable initial stresses in absolute value.

The primary objective in prestressed-concrete design is to maximize the capacity of a given beam by maximizing the absolute values of the prestresses at the section having the greatest superimposed-load stresses. The three procedures that follow, when taken as a unit, illustrate the manner in which the allowable prestresses may be increased numerically by taking advantage of the beam-weight stresses, which are opposite in character to the prestresses. The next procedure will also demonstrate that when a beam is not in balanced design, there is a range of values of F_i that will enable the member to carry this maximum allowable load. In summary, the objective is to maximize the capacity of a given beam and to provide the minimum prestressing force associated with this capacity.

DETERMINATION OF CAPACITY AND PRESTRESSING FORCE FOR A BEAM WITH STRAIGHT TENDONS

An 8 × 10 in (203.2 × 254 mm) rectangular beam, simply supported on a 20-ft (6.1-m) span, is to be prestressed by means of straight tendons. The allowable stresses are: *initial*, + 2400 and −190 lb/in² (+16,548 and −1310.1 kPa); *final*, + 2250 and −425 lb/in² (+15,513.8 and −2930.3 kPa). Evaluate the allowable unit superimposed load, the maximum and minimum prestressing force associated with this load, and the corresponding eccentricities.

Calculation Procedure:

1. Compute the beam properties

Here $A = 80$ in² (516.16 cm²); $S = 133$ in² (858.1 cm²); $w_w = 83$ lb/lin ft (1211.3 N/m).

2. *Compute the stresses at midspan due to the beam weight*

Thus, $M_w = (\frac{1}{8})(83)(20)^2(12) = 49,800$ in·lb (5626.4 N·m); $f_{bw} = -49,800/133 = -374$ lb/in^2 (-2578.7 kPa); $f_{tw} = +374$ lb/in^2 (2578.7 kPa).

3. *Set the critical stresses equal to their allowable values to secure the allowable unit superimposed load*

Use Fig. 136 or 137 as a guide. At support: $f_{bi} = +2400$ lb/in^2 ($+16,548$ kPa); $f_{ti} = -190$ lb/in^2 (-1310.1 kPa); at midspan, $f_{bf} = 0.85(2400) - 374 + f_{bs} = -425$ lb/in^2 (-2930.4 kPa); $f_{tf} = 0.85(-190) + 374 + f_{ts} = +2250$ lb/in^2 ($+15,513.8$ kPa). Also, $f_{bs} = -2091$ lb/in^2 ($-14,417.4$ kPa); $f_{ts} = +2038$ lb/in^2 ($+14,052$ kPa).

Since the superimposed-load stresses at top and bottom will be numerically equal, the latter value governs the beam capacity. Or $w_s = w_w f_{ts}/f_{tw} = 83(2038/374) = 452$ lb/lin ft (6596.4 N/m).

4. *Find $F_{i,\max}$ and its eccentricity*

The value of w_s was found by setting the critical value of f_{ti} and of f_{tf} equal to their respective allowable values. However, since S_b is excessive for the load w_s, there is flexibility with respect to the stresses at the bottom. The designer may set the critical value of either f_{bi} or f_{bf} equal to its allowable value or produce some intermediate condition. As shown by the calculations in step 3, f_{bf} may vary within a range of $2091 - 2038 = 53$ lb/in^2 (365.4 kPa). Refer to Fig. 138, where the lines represent the stresses indicated.

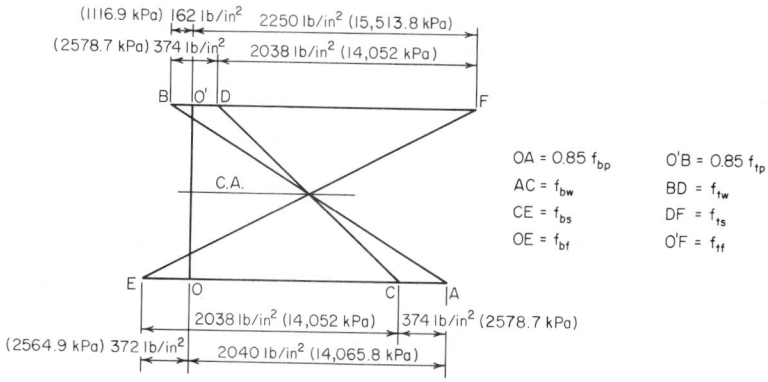

FIG. 138 Stresses at midspan under maximum prestressing force.

Points B and F are fixed, but points A and E may be placed anywhere within the 53-lb/in^2 (365.4-kPa) range. To maximize F_i, place A at its limiting position to the right; i.e., set the critical value of f_{bi} rather than that of f_{bf} equal to the allowable value. Then $f_{cai} = F_{i,\max}/A = \frac{1}{2}(2400 - 190) = +1105$ lb/in^2 ($+7619.0$ kPa); $F_{i,\max} = 1105(80) = 88,400$ lb (393,203.2 N); $f_{bp} = 1105 + 88,400e/133 = +2400$; $e = 1.95$ in (49.53 mm).

5. *Find $F_{i,\min}$ and its eccentricity*

For this purpose, place A at its limiting position to the left. Then $f_{bp} = 2,400 - (53/0.85) = +2338$ lb/in^2 ($+16,120.5$ kPa); $f_{cai} = +1074$ lb/in^2 ($+7405.2$ kPa); $F_{i,\min} = 85,920$ lb (382,172.2 N); $e = 1.96$ in (49.78 mm).

6. *Verify the value of $F_{i,\max}$ by checking the critical stresses*

At support: $f_{bi} = +2400$ lb/in^2 ($+16,548.0$ kPa); $f_{ti} = -190$ lb/in^2 (-1310.1 kPa). At midspan: $f_{bf} = +2040 - 374 - 2038 = -372$ lb/in^2 (-2564.9 kPa); $f_{tf} = -162 + 374 + 2038 = +2250$ lb/in^2 ($+15,513.8$ kPa).

7.   Verify the value of $F_{i,min}$ by checking the critical stresses

At support: $f_{bi} = + 2338$ lb/in^2 (16,120.5 kPa); $f_{ti} = -190$ lb/in^2 (-1310.1 kPa). At midspan: $f_{bf} = 0.85(2338) - 374 - 2038 = -425$ lb/in^2 (-2930.4 kPa); $f_{tf} = +2250$ lb/in^2 ($+15,513.8$ kPa).

BEAM WITH DEFLECTED TENDONS

The beam in the previous calculation procedure is to be prestressed by means of tendons that are deflected at the quarter points of the span, as shown in Fig. 139a. Evaluate the allowable unit superimposed load, the magnitude of the prestressing force, the eccentricity e_1 in the center interval, and the maximum and minimum allowable values of the eccentricity e_2 at the supports. What increase in capacity has been obtained by deflecting the tendons?

Calculation Procedure:

1.   Compute the beam-weight stresses at B

In the composite stress diagram, Fig. 139b, the difference between an ordinate to EFG and the corresponding ordinate to AHJ represents the value of f_{ti} at the given section. It is apparent that

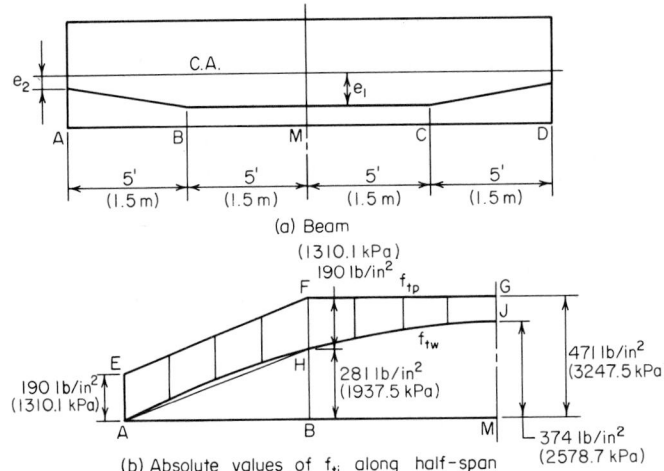

(a) Beam

(b) Absolute values of f_{ti} along half-span

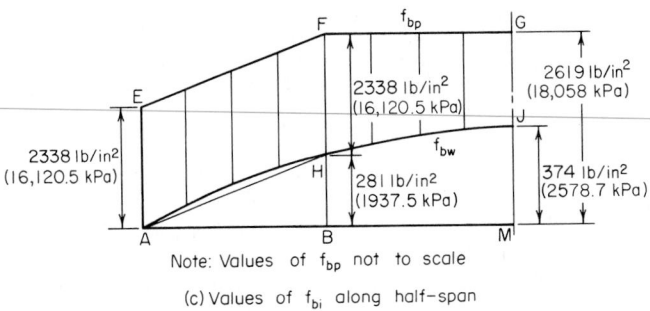

Note: Values of f_{bp} not to scale

(c) Values of f_{bi} along half-span

FIG. 139

if AE does not exceed HF, then f_{ti} does not exceed HF in absolute value anywhere along the span. Therefore, for the center interval BC, the critical stresses at transfer occur at the boundary sections B and C. Analogous observations apply to Fig. 139c.

Computing the beam-weight stresses at B yields $f_{bw} = (\frac{3}{4})(-374) = -281$ lb/in^2 (-1937.5 kPa); $f_{tw} = +281$ lb/in^2 ($+1937.5$ kPa).

2.   Tentatively set the critical stresses equal to their allowable values to secure the allowable unit superimposed load

Thus, at B: $f_{bi} = f_{bp} - 281 = +2400$; $f_{ti} = f_{tp} + 281 = -190$; $f_{bp} = +2681$ lb/in^2 ($+18,485.5$ kPa); $f_{tp} = -471$ lb/in^2 (-3247.5 kPa).

At M: $f_{bf} = 0.85(2681) - 374 + f_{bs} = -425$; $f_{tf} = 0.85(-471) + 374 + f_{ts} = +2250$; $f_{bs} = -2330$ lb/in^2 ($-16,065.4$ kPa); $f_{ts} = +2277$ lb/in^2 ($+15,699.9$ kPa). The latter value controls.

Also, $w_s = 83(2277/374) = 505$ lb/lin ft (7369.9 N/m); $505/452 = 1.12$. The capacity is increased 12 percent.

When the foregoing calculations are compared with those in the previous calculation procedure, the effect of deflecting the tendons is to permit an increase of 281 lb/in^2 (1937.5 kPa) in the absolute value of the prestress at top and bottom. The accompanying increase in f_{ts} is $0.85(281) = 239$ lb/in^2 (1647.9 kPa).

3.   Find the minimum prestressing force and the eccentricity e_1

Examination of Fig. 138 shows that f_{cai} is not affected by the form of trajectory used. Therefore, as in the previous calculation procedure, $F_i = 85,920$ lb (382,172.2 N); $f_{tp} = 1074 - 85,920e_1/133 = -471$; $e_1 = 2.39$ in (60.706 mm).

Although it is not required, the value of $f_{bp} = 1074 + 1074 - (-471) = +2619$ lb/in^2 ($+18,058$ kPa), or $f_{bp} = 2681 - 53/0.85 = +2619$ lb/in^2 ($+18,058$ kPa).

4.   Establish the allowable range of values of e_2

At the supports, the tendons may be placed an equal distance above or below the center. Then $e_{2,max} = 1.96$ in (23.44 mm); $e_{2,min} = -1.96$ in (-23.44 mm).

BEAM WITH CURVED TENDONS

The beam in the second previous calculation procedure is to be prestressed by tendons lying in a parabolic arc. Evaluate the allowable unit superimposed load, the magnitude of the prestressing force, the eccentricity of this force at midspan, and the increase in capacity accruing from the use of curved tendons.

Calculation Procedure:

1.   Tentatively set the initial and final stresses at midspan equal to their allowable values to secure the allowable unit superimposed load

Since the prestressing force has a parabolic trajectory, lines EFG in Fig. 139b and c will be parabolic in the present case. Therefore, it is possible to achieve the full allowable initial stresses at midspan. Thus, $f_{bi} = f_{bp} - 374 = +2400$; $f_{ti} = f_{tp} + 374 = -190$; $f_{bp} = +2774$ lb/in^2 ($+19,126.7$ kPa); $f_{tp} = -564$ lb/in^2 (-3888.8 kPa); $f_{bf} = 0.85(2774) - 374 + f_{bs} = -425$; $f_{tf} = 0.85(-564) + 374 + f_{ts} = +2250$; $f_{bs} = -2409$ lb/in^2 ($-16,610.1$ kPa); $f_{ts} = +2356$ lb/in^2 ($+16,244.6$ kPa). The latter value controls.

Also, $w_s = 83(2356/374) = 523$ lb/lin ft (7632.6 N/m); $523/452 = 1.16$. Thus the capacity is increased 16 percent.

When the foregoing calculations are compared with those in the earlier calculation procedure, the effect of using parabolic tendons is to permit an increase of 374 lb/in^2 (2578.7 kPa) in the absolute value of the prestress at top and bottom. The accompanying increase in f_{ts} is $0.85(374) = 318$ lb/in^2 (2192.6 kPa).

2.   Find the minimum prestressing force and its eccentricity at midspan

As before, $F_i = 85,920$ lb (382,172.2 N); $f_{tp} = 1074 - 85,920e/133 = -564$; $e = 2.54$ in (64.516 mm).

DETERMINATION OF SECTION MODULI

A beam having a cross-sectional area of 500 in² (3226 cm²) sustains a beam-weight moment equal to 3500 in·kips (395.4 kN·m) at midspan and a superimposed moment that varies parabolically from 9000 in·kips (1016.8 kN·m) at midspan to 0 at the supports. The allowable stresses are: initial, $+2400$ and -190 lb/in² ($+16,548$ and -1310.1 kPa); final, $+2250$ and -200 lb/in² ($+15,513.8$ and -1379 kPa). The member will be prestressed by tendons deflected at the quarter points. Determine the section moduli corresponding to balanced design, the magnitude of the prestressing force, and its eccentricity in the center interval. Assume that the calculated eccentricity is attainable (i.e., that the centroid of the tendons will fall within the confines of the section while satisfying insulation requirements).

Calculation Procedure:

1.   Equate the critical initial stresses, and the critical final stresses, to their allowable values

Let M_w and M_s denote the indicated moments at midspan; the corresponding moments at the quarter point are three-fourths as large. The critical initial stresses occur at the quarter point, while the critical final stresses occur at midspan. After equating the stresses to their allowable values, solve the resulting simultaneous equations to find the section moduli and prestresses. Thus: *stresses in bottom fiber*, $f_{bi} = f_{bp} - 0.75M_w/S_b = +2400$; $f_{bf} = 0.85f_{bp} - M_w/S_b - M_s/S_b = -200$. Solving gives $S_b = (M_s + 0.3625M_w)/2240 = 4584$ in³ (75,131.7 cm³) and $f_{bp} = +2973$ lb/in² ($+20,498.8$ kpa); *stresses in top fiber*, $f_{ti} = f_{tp} + 0.75(M_w/S_t) = -190$; $f_{tf} = 0.85f_{tp} + M_w/S_t + M_s/S_t = +2250$. Solving yields $S_t = (M_s + 0.3625M_w)/2412 = 4257$ in³ (69,772.2 cm³) and $f_{tp} = -807$ lb/in² (-5564.2 kPa).

2.   Evaluate F_i and e

In this instance, e denotes the eccentricity in the center interval. Thus $f_{bp} = F_i/A + F_ie/S_b = +2973$; $f_{tp} = F_i/A - F_ie/S_t = -807$; $F_i = (2973S_b - 807S_t)A/(S_b + S_t) = 576,500$ lb (2,564,272.0 N); $e = 2973S_b/F_i - S_b/A = 14.47$ in (367.538 mm).

3.   Alternatively, evaluate F_i by assigning an arbitrary depth to the member

Thus, set $h = 10$ in (254 mm); $y_b = S_th/(S_b + S_t) = 4.815$ in (122.301 mm); $f_{cai} = f_{bp} - (f_{bp} - f_{tp})y_b/h = 2973 - (2973 + 807)0.4815 = +1153$ lb/in² ($+7949.9$ kPa); $F_i = 1153(500) = 576,500$ lb (2,564,272.0 N).

EFFECT OF INCREASE IN BEAM SPAN

Consider that the span of the beam in the previous calculation procedure increases by 10 percent, thereby causing the midspan moment due to superimposed load to increase by 21 percent. Show that the member will be adequate with respect to flexure if all cross-sectional dimensions are increased by 7.2 percent. Compute the new eccentricity in the center interval, and compare this with the original value.

Calculation Procedure:

1.   Calculate the new section properties and bending moments

Thus $A = 500(1.072)^2 = 575$ in² (3709.9 cm²); $S_b = 4584(1.072)^3 = 5647$ in³ (92,554.3 cm³); $S_t = 4257(1.072)^3 = 5244$ in³ (85,949.2 cm³); $M_s = 9000(1.21) = 10,890$ in·kips (1230.4 kN·m); $M_w = 3500(1.072)^2(1.21) = 4867$ in·kips (549.9 kN·m).

2.   Compute the required section moduli, prestresses, prestressing force, and its eccentricity in the central interval, using the same sequence as in the previous calculation procedure

Thus $S_b = 5649$ in³ (92,587.1 cm³); $S_t = 5246$ in³ (85,981.9 cm³). Both these values are acceptable. Then $f_{bp} = +3046$ lb/in² ($+21,002.2$ kPa); $f_{tp} = -886$ lb/in² (-6108.9 kPa); $F_i = 662,800$ lb (2,948,134.4 N); $e = 16.13$ in (409.7 mm). The eccentricity has increased by 11.5 percent.

In practice, it would be more efficient to increase the vertical dimensions more than the horizontal dimensions. Nevertheless, as the span increases, the eccentricity increases more rapidly than the depth.

EFFECT OF BEAM OVERLOAD

The beam in the second previous calculation procedure is subjected to a 10 percent overload. How does the final stress in the bottom fiber compare with that corresponding to the design load?

Calculation Procedure:

1. Compute the value of f_{bs} under design load

Thus, $f_{bs} = -M_s/S_b = -9,000,000/4584 = -1963$ lb/in^2 ($-13,534.8$ kPa).

2. Compute the increment or f_{bs} caused by overload and the revised value of f_{bf}

Thus, $\Delta f_{bs} = 0.10(-1963) = -196$ lb/in^2 (-1351.4 kPa); $f_{bf} = -200 - 196 = -396$ lb/in^2 (-2730.4 kPa). Therefore, a 10 percent overload virtually doubles the tensile stress in the member.

PRESTRESSED-CONCRETE BEAM DESIGN GUIDES

On the basis of the previous calculation procedures, what conclusions may be drawn that will serve as guides in the design of prestressed-concrete beams?

Calculation Procedure:

1. Evaluate the results obtained with different forms of tendons

The capacity of a given member is increased by using deflected rather than straight tendons, and the capacity is maximized by using parabolic tendons. (However, in the case of a pretensioned beam, an economy analysis must also take into account the expense incurred in deflecting the tendons.)

2. Evaluate the prestressing force

For a given ratio of y_b/y_t, the prestressing force that is required to maximize the capacity of a member is a function of the cross-sectional area and the allowable stresses. It is independent of the form of the trajectory.

3. Determine the effect of section moduli

If the section moduli are in excess of the minimum required, the prestressing force is minimized by setting the critical values of f_{bf} and f_{ti} equal to their respective allowable values. In this manner, points A and B in Fig. 138 are placed at their limiting positions to the left.

4. Determine the most economical short-span section

For a short-span member, an I section is most economical because it yields the required section moduli with the minimum area. Moreover, since the required values of S_b and S_t differ, the area should be disposed unsymmetrically about middepth to secure these values.

5. Consider the calculated value of e

Since an increase in span causes a greater increase in the theoretical eccentricity than in the depth, the calculated value of e is not attainable in a long-span member because the centroid of the tendons would fall beyond the confines of the section. For this reason, long-span members are generally constructed as T sections. The extensive flange area elevates the centroidal axis, thus making it possible to secure a reasonably large eccentricity.

6. Evaluate the effect of overload

A relatively small overload induces a disproportionately large increase in the tensile stress in the beam and thus introduces the danger of cracking. Moreover, owing to the presence of many variable quantities, there is not a set relationship between the beam capacity at allowable final stress and the capacity at incipient cracking. It is therefore imperative that every prestressed-concrete

~~beam be subjected to an ultimate-strength analysis~~ to ensure that the beam provides an adequate factor of safety.

KERN DISTANCES

The beam in Fig. 140 has the following properties: $A = 850$ in^2 (5484.2 cm^2); $S_b = 11,400$ in^3 (186,846.0 cm^3); $S_t = 14,400$ in^3 (236,016.0 cm^3). A prestressing force of 630 kips (2802.2 kN) is applied with an eccentricity of 24 in (609.6 mm) at the section under investigation. Calculate f_{bp} and f_{tp} by expressing these stresses as functions of the kern distances of the section.

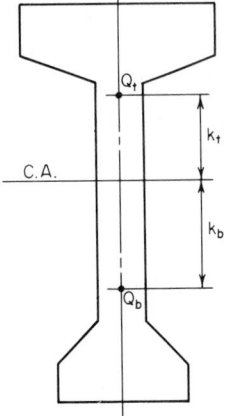

FIG. 140   Kern points.

Calculation Procedure:

1.   Consider the prestressing force to be applied at each kern point, and evaluate the kern distances

Let Q_b and Q_t denote the points at which a compressive force must be applied to induce a zero stress in the top and bottom fiber, respectively. These are referred to as the *kern points* of the section, and the distances k_b and k_t from the centroidal axis to these points are called the *kern distances*.

Consider the prestressing force to be applied at each kern point in turn. Set the stresses f_{tp} and f_{bp} equal to zero to evaluate the kern distances k_b and k_t, respectively. Thus $f_{tp} = F_i/A - F_i k_b/S_t = 0$, Eq. *a*; $f_{bp} = F_i/A - F_i k_t/S_b = 0$, Eq. *b*. Then

$$k_b = \frac{S_t}{a} \quad \text{and} \quad k_t = \frac{S_b}{A} \tag{55}$$

And, $k_b = 14,400/850 = 16.9$ in (429.26 mm); $k_t = 11,400/850 = 13.4$ in (340.36 mm).

2.   Express the stresses f_{bp} and f_{tp} associated with the actual eccentricity as functions of the kern distances

By combining the stress equations with Eqs. *a* and *b*, the following equations are obtained:

$$f_{bp} = \frac{F_i(k_t + e)}{S_b} \quad \text{and} \quad f_{tp} = \frac{F_i(k_b - e)}{S_t} \tag{56}$$

Substituting numerical values gives $f_{bp} = 630,000(13.4 + 24)/11,400 = +2067$ lb/in^2 ($+14,252.0$ kPa); $f_{tp} = 630,000(16.9 - 24)/14,400 = -311$ lb/in^2 (-2144.3 kPa).

3.   Alternatively, derive Eq. 56 by considering the increase in prestress caused by an increase in eccentricity

Thus, $\Delta f_{bp} = F_i \Delta e/S_b$; therefore, $f_{bp} = F_i(k_t + e)/S_b$.

MAGNEL DIAGRAM CONSTRUCTION

The data pertaining to a girder having curved tendons are $A = 500$ in^2 (3226.0 cm^2); $S_b = 5000$ in^3 (81,950 cm^3); $S_t = 5340$ in^3 (87,522.6 cm^3); $M_w = 3600$ in·kips (406.7 kN·m); $M_s = 9500$ in·kips (1073.3 kN·m). The allowable stresses are: *initial,* $+ 2400$ and -190 lb/in^2 ($+16,548$ and -1310.1 kPa); *final,* $+ 2250$ and $- 425$ lb/in^2 ($+15,513.8$ and -2930.4 kPa). (*a*) Construct the Magnel diagram for this member. (*b*) Determine the minimum prestressing force and its eccentricity by referring to the diagram. (*c*) Determine the prestressing force if the eccentricity is restricted to 18 in (457.2 mm).

Calculation Procedure:

1.   Set the initial stress in the bottom fiber at midspan equal to or less than its allowable value, and solve for the reciprocal of F_i

In this situation, the superimposed load is given, and the sole objective is to minimize the prestressing force. The Magnel diagram is extremely useful for this purpose because it brings into sharp focus the relationship between F_i and e. In this procedure, let f_{bi}, f_{bf} and so forth represent the *allowable* stresses.

Thus,

$$\frac{1}{F_i} \geq \frac{k_t + e}{M_w + f_{bi}S_b} \tag{57a}$$

2. Set the final stress in the bottom fiber at midspan equal to or algebraically greater than its allowable value, and solve for the reciprocal of F_i

Thus

$$\frac{1}{F_i} \leq \frac{\eta(k_t + e)}{M_w + M_s + f_{bf}S_b} \tag{57b}$$

3. Repeat the foregoing procedure with respect to the top fiber

Thus,

$$\frac{1}{F_i} \geq \frac{e - k_b}{M_w - f_{ti}S_t} \tag{57c}$$

and

$$\frac{1}{F_i} \leq \frac{\eta(e - k_b)}{M_w + M_s - f_{tf}S_t} \tag{57d}$$

4. Substitute numerical values, expressing F_i in thousands of kips

Thus, $1/F_i \geq (10 + e)/15.60$, Eq. a; $1/F_i \leq (10 + e)/12.91$, Eq. b; $1/F_i \geq (e - 10.68)/4.61$, Eq. c; $1/F_i \leq (e - 10.68)/1.28$, Eq. d.

5. Construct the Magnel diagram

In Fig. 141, consider the foregoing relationships as equalities, and plot the straight lines that represent them. Each point on these lines represents a set of values of $1/F_i$ and e at which the designated stress equals its allowable value.

When the section moduli are in excess of those corresponding to balanced design, as they are

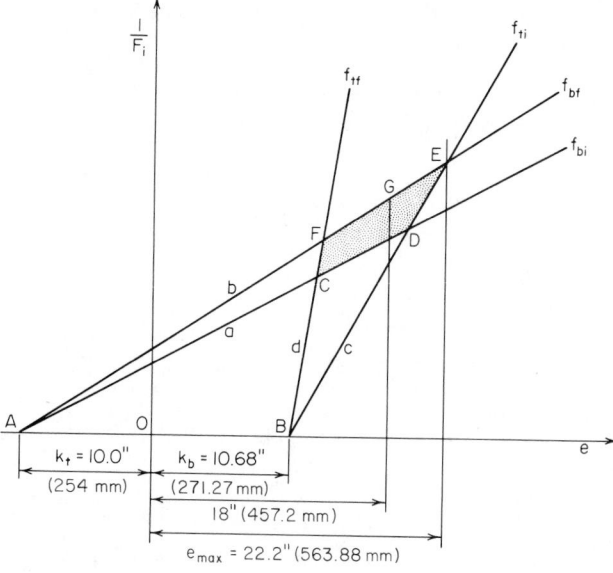

FIG. 141 Magnel diagram.

in the present instance, line b makes a greater angle with the e axis than does a, and line d makes a greater angle than does c. From the sense of each inequality, it follows that $1/F_i$ and e may have any set of values represented by a point within the quadrilateral $CDEF$ or on its circumference.

6.   To minimize F_i, determine the coordinates of point E at the intersection of lines b and c

Thus, $1/F_i = (10 + e)/12.91 = (e - 10.68)/4.61$; so $e = 22.2$ in (563.88 mm); $F_i = 401$ kips (1783.6 kN).

The Magnel diagram confirms the third design guide presented earlier in the section.

7.   For the case where e is restricted to 18 in (457.2 mm), minimize F_i by determining the ordinate of point G on line b

Thus, in Fig. 141, $1/F_i = (10 + 18)/12.91$; $F_i = 461$ kips (2050.5 kN).

The Magnel diagram may be applied to a beam having deflected tendons by substituting for M_w in Eqs. 57a and 57c the beam-weight moment at the deflection point.

CAMBER OF A BEAM AT TRANSFER

The following pertain to a simply supported prismatic beam: $L = 36$ ft (11.0 m); $I = 40,000$ in^4 (166.49 dm^4); $f'_{ci} = 4000$ lb/in^2 (27,580.0 kPa); $w_w = 340$ lb/lin ft (4961.9 N/m); $F_i = 430$ kips (1912.6 kN); $e = 8.8$ in (223.5 mm) at midspan. Calculate the camber of the member at transfer under each of these conditions: (a) the tendons are straight across the entire span; (b) the tendons are deflected at the third points, and the eccentricity at the supports is zero; (c) the tendons are curved parabolically, and the eccentricity at the supports is zero.

Calculation Procedure:

1.   Evaluate E_c at transfer, using the ACI Code

Review the moment-area method of calculating beam deflections, which is summarized earlier. Consider an upward displacement (camber) as positive, and let the symbols Δ_p, Δ_w, and Δ_i, defined earlier, refer to the camber at midspan.

Thus, using the ACI *Code*, $E_c = (145)^{1.5}(33)(4000)^{0.5} = 3,644,000$ lb/in^2 (25,125.4 MPa).

2.   Construct the prestress-moment diagrams associated with the three cases described

See Fig. 142. By symmetry, the elastic curve corresponding to F_i is horizontal at midspan. Consequently, Δ_p equals the deviation of the elastic curve at the support from the tangent to this curve at midspan.

3.   Using the literal values shown in Fig. 142, develop an equation for Δ_p by evaluating the tangential deviation; substitute numerical values

Thus, case a:

$$\Delta_p = \frac{ML^2}{8E_cI} \tag{58}$$

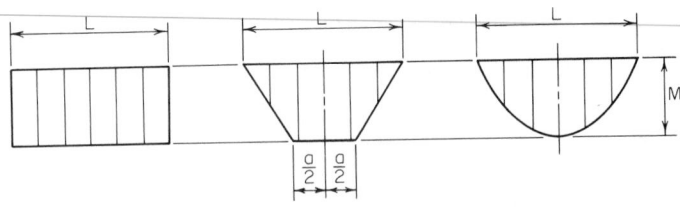

(a) Straight tendons   (b) Deflected tendons   (c) Parabolic tendons

FIG. 142   Prestress-moment diagrams.

or $\Delta_p = 430{,}000(8.8)(36)^2(144)/[8(3{,}644{,}000)(40{,}000)] = 0.61$ in (15.494 mm). For case *b*:

$$\Delta_p = \frac{M(2L^2 + 2La - a^2)}{24E_cI} \qquad (59)$$

or $\Delta_p = 0.52$ in (13.208 mm). For case *c*:

$$\Delta_p = \frac{5ML^2}{48E_cI} \qquad (60)$$

or $\Delta_p = 0.51$ in (12.954 mm).

4. Compute Δ_w

Thus, $\Delta_w = -5w_w L^4/(384E_cI) = -0.09$ in (-2.286 mm).

5. Combine the foregoing results to obtain Δ_i

Thus: case *a*, $\Delta_i = 0.61 - 0.09 = 0.52$ in (13.208 mm); case *b*, $\Delta_i = 0.52 - 0.09 = 0.43$ in (10.922 mm); case *c*, $\Delta_i = 0.51 - 0.09 = 0.42$ in (10.688 mm).

DESIGN OF A DOUBLE-T ROOF BEAM

The beam in Fig. 143 was selected for use on a simple span of 40 ft (12.2 m) to carry the following loads: roofing, 12 lb/ft^2 (574.5 N/m^2) snow, 40 lb/ft^2 (1915.1 N/m^2); total, 52 lb/ft^2 (2489.6 N/

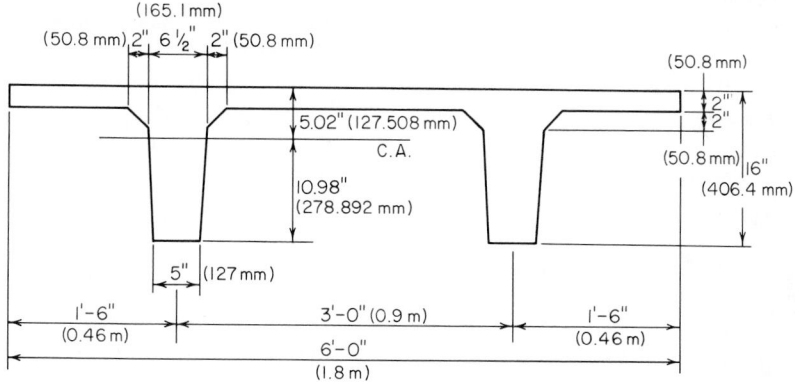

FIG. 143   Double-T roof beam.

m^2). The member will be pretensioned with straight seven-wire strands, $\%_6$ in (11.11 mm) diameter, having an area of 0.1089 in^2 (0.70262 cm^2) each and an ultimate strength of 248,000 lb/in^2 (1,709,960.0 kPa). The concrete strengths are $f'_c = 5000$ lb/in^2 (34,475.0 kPa) and $f'_{ci} = 4000$ lb/in^2 (27,580.0 kPa). The allowable stresses are: *initial*, $+2400$ and -190 lb/in^2 ($+16{,}548.0$ and -1310.1 kPa); *final*, $+2250$ and -425 lb/in^2 ($+15{,}513.8$ and -2930.4 kPa). Investigate the adequacy of this section, and design the tendons. Compute the camber of the beam after the concrete has hardened and all dead loads are present. For this calculation, assume that the final value of E_c is one-third of that at transfer.

Calculation Procedure:

1. Compute the properties of the cross section

Let f_{bf} and f_{tf} denote the respective stresses at *midspan* and f_{bi} and f_{ti} denote the respective stresses *at the support*. Previous calculation procedures demonstrated that where the section moduli are excessive, the minimum prestressing force is obtained by setting f_{bf} and f_{ti} equal to their allowable values.

Thus $A = 316$ in^2 (2038.8 cm^2); $I = 7240$ in^4 (30.14 dm^4); $y_b = 10.98$ in (278.892 mm); $y_t = 5.02$ in (127.508 mm); $S_b = 659$ in^3 (10,801.0 cm^3); $S_t = 1442$ in^3 (23,614 cm^3); $w_w = (316/144)150 = 329$ lb/lin ft (4801.4 N/m).

2.   *Calculate the total midspan moment due to gravity loads and the corresponding stresses*

Thus $w_s = 52(6) = 312$ lb/lin ft (4553.3 N/m); $w_w = 329$ lb/lin ft (4801.4 N/m); and $M_w + M_s = (\frac{1}{8})(641)(40^2)(12) = 1,538,000$ in·lb (173,763.2 N·m); $f_{bw} + f_{bs} = -1,538,000/659 = -2334$ lb/in^2 ($-16,092.9$ kPa); $f_{tw} + f_{ts} = +1,538,000/1442 = +1067$ lb/in^2 ($+7357.0$ kPa).

3.   *Determine whether the section moduli are excessive*

Do this by setting f_{bf} and f_{ti} equal to their allowable values and computing the corresponding values of f_{bi} and f_{tf}. Thus, $f_{bf} = 0.85f_{bp} - 2334 = -425$; therefore, $f_{bp} = +2246$ lb/in^2 ($+15,486.2$ kPa); $f_{ti} = f_{tp} = -190$ lb/in^2 (-1310.1 kPa); $f_{bi} = f_{bp} = +2246 < 2400$ lb/in^2 ($+16,548.0$ kPa). This is acceptable. Also, $f_{tf} = 0.85(-190) + 1067 = +905 < 2250$ lb/in^2 ($+15,513.8$ kPa); this is acceptable. The section moduli are therefore excessive.

4.   *Find the minimum prestressing force and its eccentricity*

Refer to Fig. 144. Thus, $f_{bp} = +2246$ lb/in^2 ($+15,486.2$ kPa); $f_{tp} = -190$ lb/in^2 (-1310.1 kPa); slope of $AB = 2246 - (-190)/16 = 152.3$ lb/(in^2·in) (41.33 MPa/m); $F_i/A = CD = 2246 - 10.98(152.3) = 574$ lb/in^2 (3957.7 kPa); $F_i = 574(316) = 181,400$ lb (806,867.2 N); slope of $AB = F_ie/I = 152.3$; $e = 152.3(7240)/181,400 = 6.07$ in (154.178 mm).

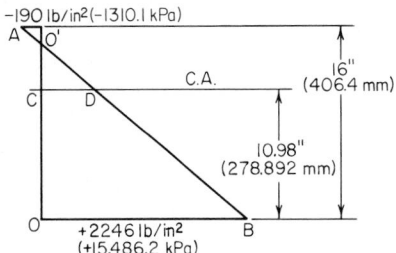

−190 lb/in²(−1310.1 kPa)

C.A.

16" (406.4 mm)

10.98" (278.892 mm)

+2246 lb/in² (+15,486.2 kPa)

FIG. 144 Prestress diagram.

5.   *Determine the number of strands required, and establish their disposition*

In accordance with the ACI *Code*, allowable initial force per strand = 0.1089(0.70)(248,000) = 18,900 lb (84,067.2 N); number required = 181,400/18,900 = 9.6. Therefore, use 10 strands (5 in each web) stressed to 18,140 lb (80,686.7 N) each.

Referring to the ACI *Code* for the minimum clear distance between the strands, we find the allowable center-to-center spacing = 4(⅞₆) = 1¾ in (44.45 mm). Use a 2-in (50.8-mm) spacing. In Fig. 145, locate the centroid of the steel, or $y = (2 \times 2 + 1 \times 4)/5 = 1.60$ in (40.64 mm); $v = 10.98 - 6.07 - 1.60 = 3.31$ in (84.074 mm); set $v = 3\frac{5}{16}$ in (84.138 mm).

6.   *Calculate the allowable ultimate moment of the member in accordance with the ACI Code*

Thus, $A_s = 10(0.1089) = 1.089$ in^2 (7.0262 cm^2); $d = y_t + e = 5.02 + 6.07 = 11.09$ in (281.686 mm); $p = A_s/(bd) = 1.089/[72(11.09)] = 0.00137$.

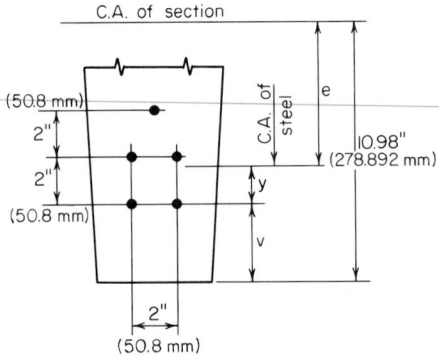

C.A. of section

(50.8 mm) 2"

2"

(50.8 mm)

C.A. of steel

e

10.98" (278.892 mm)

y

v

2"

(50.8 mm)

FIG. 145 Location of tendons.

Compute the steel stress and resultant tensile force at ultimate load:

$$f_{su} = f_s'\left(1 - \frac{0.5\,pf_s'}{f_c'}\right) \tag{61}$$

Or, $f_{su} = 248{,}000(1 - 0.5 \times 0.00137 \times 248{,}000/5000) = 240{,}00$ lb/in^2 (1,654,800 kPa); $T_u = A_s f_{su} = 1.089(240{,}000) = 261{,}400$ lb (1,162,707.2 N).

Compute the depth of the compression block. This depth, a, is found from $C_u = 0.85(5000)(72a) = 261{,}400$ lb (1,162,707.2 N); $a = 0.854$ in (21.6916 mm); $jd = d - a/2 = 10.66$ in (270.764 mm); $M_u = \phi T_u jd = 0.90(261{,}400)(10.66) = 2{,}500{,}000$ in·lb (282,450.0 N·m).

Calculate the steel index to ascertain that it is below the limit imposed by the ACI *Code*, or $q = pf_{su}/f_c' = 0.00137\,(240{,}000)/5000 = 0.0658 < 0.30$. This is acceptable.

7. Calculate the required ultimate-moment capacity as given by the ACI Code

Thus, $w_{DL} = 329 + 12(6) = 401$ lb/lin ft (5852.2 N/m); $w_{LL} = 40(6) = 240$ lb/lin ft (3502.5 N/m); $w_u = 1.5w_{DL} + 1.8w_{LL} = 1034$ lb/lin ft (15,090.1 N/m); M_u required = $(\frac{1}{8})(1034)(40)^2(12) = 2{,}480{,}000 < 2{,}500{,}000$ in·lb (282,450.0 N·m). The member is therefore adequate with respect to its ultimate-moment capacity.

8. Calculate the maximum and minimum area of web reinforcement in the manner prescribed in the ACI Code

Since the maximum shearing stress does not vary linearly with the applied load, the shear analysis is performed at ultimate-load conditions. Let A_v = area of web reinforcement placed perpendicular to the longitudinal axis; V_c' = ultimate-shear capacity of concrete; V_p' = vertical component of F_f at the given section; V_u' = ultimate shear at given section; s = center-to-center spacing of stirrups; f_{pc}' = stress due to F_f, evaluated at the centroidal axis, or at the junction of the web and flange when the centroidal axis lies in the flange.

Calculate the ultimate shear at the critical section, which lies at a distance $d/2$ from the face of the support. Then distance from midspan to the critical section = $\frac{1}{2}(L - d) = 19.54$ ft (5.955 m); $V_u' = 1034(19.54) = 20{,}200$ lb (89,849.6 N).

Evaluate V_c' by solving the following equations and selecting the smaller value:

$$V_{ci}' = 1.7b'd(f_c')^{0.5} \tag{62}$$

where d = effective depth, in (mm); b' = width of web at centroidal axis, in (mm); $b' = 2(5 + 1.5 \times 10.98/12) = 12.74$ in (323.596 mm); $V_{ci}' = 1.7(12.74)(11.09)(5000)^{0.5} = 17{,}000$ lb (75,616.0 N). Also,

$$V_{cw}' = b'd(3.5f_c'^{0.5} + 0.3f_{pc}') + V_p' \tag{63}$$

where d = effective depth or 80 percent of the overall depth, whichever is greater, in (mm). Thus, $d = 0.80(16) = 12.8$ in (325.12 mm); $V_p' = 0$. From step 4, $f_{pc}' = 0.85(574) = +488$ lb/in^2 (3364.8 kPa); $V_{cw}' = 12.74(12.8)(3.5 \times 5000^{0.5} + 0.3 \times 488) = 64{,}300$ lb (286,006.4 N); therefore, $V_c' = 17{,}000$ lb (75,616.0 N).

Calculate the maximum web-reinforcement area by applying the following equation:

$$A_v = \frac{s(V_u' - \phi V_c')}{\phi d f_y} \tag{64}$$

where d = effective depth at section of maximum moment, in (mm). Use $f_y = 40{,}000$ lb/in^2 (275,800.0 kPa), and set $s = 12$ in (304.8 mm). Then $A_v = 12(20{,}200 - 0.85 \times 17{,}000)/[0.85(11.09)(40{,}000)] = 0.184$ in^2/ft (3.8949 cm^2/m). This is the area required at the ends.

Calculate the minimum web-reinforcement area by applying

$$A_v = \frac{A_s f_s'}{80 f_y}\frac{s}{(b'd)^{0.5}} \tag{65}$$

or $A_v = (1.089/80)(248{,}000/40{,}000)12/(12.74 \times 11.09)^{0.5} = 0.085$ in^2/ft (1.7993 cm^2/m).

9. Calculate the camber under full dead load

From the previous procedure, $E_c = (\frac{1}{3})(3.644)(10)^6 = 1.215 \times 10^6$ lb/in^2 (8.377 × 10^6 kPa); $E_c I = 1.215(10)^6(7240) = 8.8 \times 10^9$ lb·in^2 (25.25 × 10^6 N·m^2); $\Delta_{DL} = -5(401)(40)^4(1728)/$

$[384(8.8)(10)^9] = -2.62$ in $(-66.548$ mm$)$. By Eq. 58, $\Delta_p = 0.85(181,400)(6.07)(40)^2(144)/$ $[8(8.8)(10)^9] = 3.06$ in $(77.724$ mm$)$; $\Delta = 3.06 - 2.62 = 0.44$ in $(11.176$ mm$)$.

DESIGN OF A POSTTENSIONED GIRDER

The girder in Fig. 146 has been selected for use on a 90-ft (27.4-m) simple span to carry the following superimposed loads: dead load, 1160 lb/lin ft (16,928.9 N/m), live load, 1000 lb/lin ft

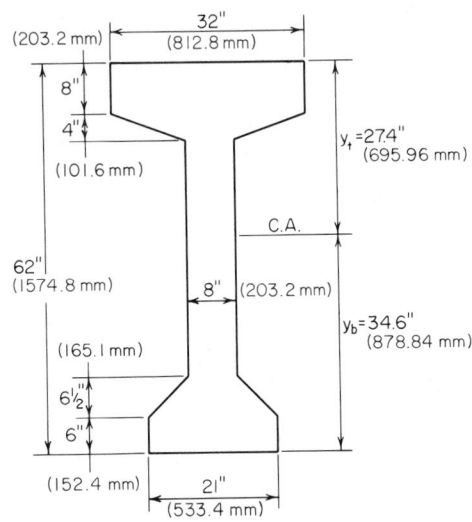

FIG. 146

(14,593.9 N/m). The girder will be posttensioned with Freyssinet cables. The concrete strengths are $f_c' = 5000$ lb/in^2 (34,475 kPa) and $f_{ci}' = 4000$ lb/in^2 (27,580 kPa). The allowable stresses are: *initial*, $+2400$ and -190 lb/in^2 ($+16,548$ and -1310.1 kPa); *final*, $+2250$ and -425 lb/in^2 ($+15,513.8$ and -2930.4 kPa). Complete the design of this member, and calculate the camber at transfer.

Calculation Procedure:

1.   *Compute the properties of the cross section*

Since the tendons will be curved, the initial stresses at midspan may be equated to the allowable values. The properties of the cross section are $A = 856$ in^2 (5522.9 cm^2); $I = 394,800$ in^4 (1643 dm^4); $y_b = 34.6$ in (878.84 mm); $y_t = 27.4$ in (695.96 mm); $S_b = 11,410$ in^3 (187,010 cm^3); $S_t = 14,410$ in^3 (236,180 cm^3); $w_w = 892$ lb/lin ft (13,017.8 N/m).

2.   *Calculate the stresses at midspan caused by gravity loads*

Thus $f_{bw} = -950$ lb/in^2 (-6550.3 kPa); $f_{bs} = -2300$ lb/in^2 ($-15,858.5$ kPa); $f_{tw} = +752$ lb/in^2 ($+5185.0$ kPa); $f_{ts} = +1820$ lb/in^2 ($+12,548.9$ kPa).

3.   *Test the section adequacy*

To do this, equate f_{bf} and f_{ti} to their allowable values and compute the corresponding values of f_{bi} and f_{tf}. Thus $f_{bf} = 0.85 f_{bp} - 950 - 2300 = -425$; $f_{ti} = f_{tp} + 752 = -190$; therefore, $f_{bp} = +3324$ lb/in^2 ($+22,919.0$ kPa) and $f_{tp} = -942$ lb/in^2 (-6495.1 kPa); $f_{bi} = +3324 - 950 = +2374 < 2400$ lb/in^2 (16,548.0 kPa). This is acceptable. And $f_{tf} = 0.85(-942) + 752 + 1820 = +1771 < 2250$ lb/in^2 (15,513.8 kPa). This is acceptable. The section is therefore adequate.

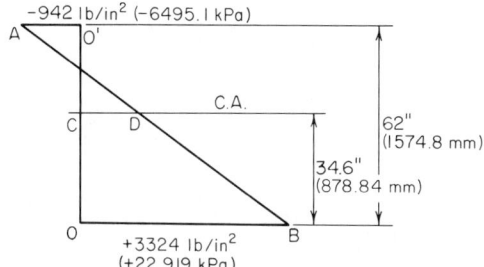

FIG. 147   Prestress diagram.

4.   Find the minimum prestressing force and its eccentricity at midspan

Do this by applying the prestresses found in step 3. Refer to Fig. 147. Slope of AB = [3324 − (−942)]/62 = 68.8 lb/(in^2·in) (18.68 kPa/mm); F_i/A = CD = 3324 − 34.6(68.8) = 944 lb/in^2 (6508.9 kPa); F_i = 944(856) = 808,100 lb (3,594,428.8 N); slope of AB = F_ie/I = 68.8; e = 68.8(394,800)/808,100 = 33.6 in (853.44 mm). Since y_b = 34.6 in (878.84 mm), this eccentricity is excessive.

5.   Select the maximum feasible eccentricity; determine the minimum prestressing force associated with this value

Try e = 34.6 − 3.0 = 31.6 in (802.64 mm). To obtain the minimum value of F_i, equate f_{bf} to its allowable value. Check the remaining stresses. As before, f_{bp} = +3324 lb/in^2 (+22,919 kPa). But f_{bp} = $F_i/856$ + 31.6F_i/11,410 = +3324; therefore F_i = 844,000 lb (3754.1 kN). Also, f_{tp} = −865 lb/in^2 (−5964.2 kPa); f_{bi} = +2374 lb/in^2 (+16,368.7 kPa); f_{ti} = − 113 lb/in^2 (−779.1 kPa); f_{tf} = +1837 lb/in^2 (+12,666.1 kPa).

6.   Design the tendons, and establish their pattern at midspan

Refer to a table of the properties of Freyssinet cables, and select 12/0.276 cables. The designation indicates that each cable consists of 12 wires of 0.276-in (7.0104-mm) diameter. The ultimate strength is 236,000 lb/in^2 (1,627,220 kPa). Then A_s = 0.723 in^2 (4.6648 cm^2) per cable. Outside diameter of cable = 1⅝ in (41.27 mm). Recommended final prestress = 93,000 lb (413,664 N) per cable; initial prestress = 93,000/0.85 = 109,400 lb (486,611.2 N) per cable. Therefore, use eight cables at an initial prestress of 105,500 lb (469,264.0 N) each.

A section of the ACI *Code* requires a minimum cover of 1½ in (38.1 mm) and another section permits the ducts to be bundled at the center. Try the tendon pattern shown in Fig. 148. Thus, y = [6(2.5) + 2(4.5)]/8 = 3.0 in (76.2 mm). This is acceptable.

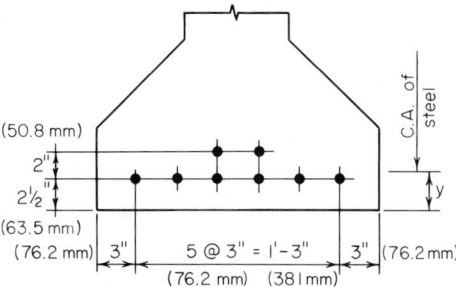

FIG. 148   Location of tendons at midspan.

7.   Establish the trajectory of the prestressing force

Construct stress diagrams to represent the initial and final stresses in the bottom and top fibers along the entire span.

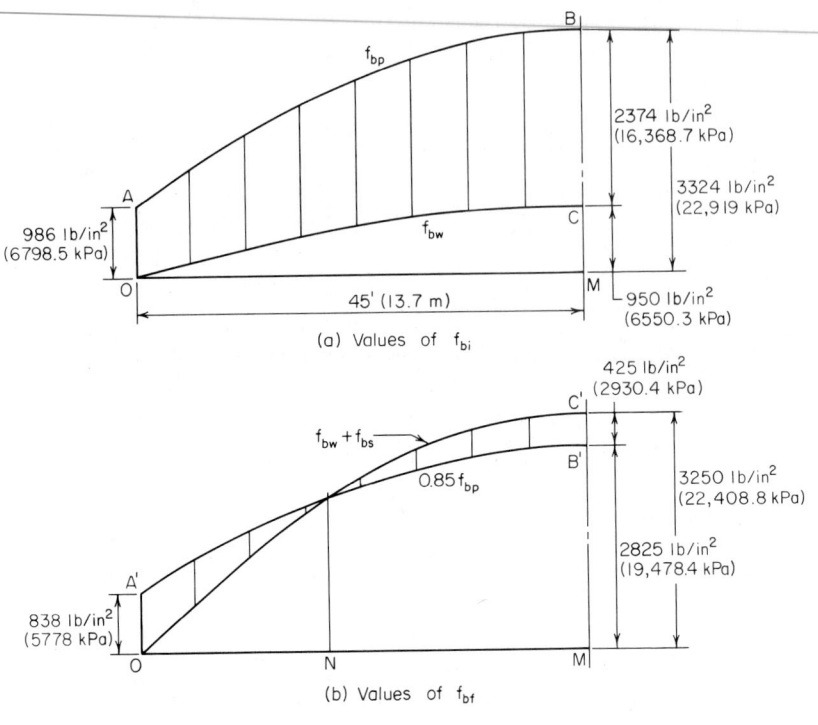

986 lb/in² (6798.5 kPa)

2374 lb/in² (16,368.7 kPa)

3324 lb/in² (22,919 kPa)

45' (13.7 m)

950 lb/in² (6550.3 kPa)

(a) Values of f_{bi}

425 lb/in² (2930.4 kPa)

3250 lb/in² (22,408.8 kPa)

2825 lb/in² (19,478.4 kPa)

838 lb/in² (5778 kPa)

(b) Values of f_{bf}

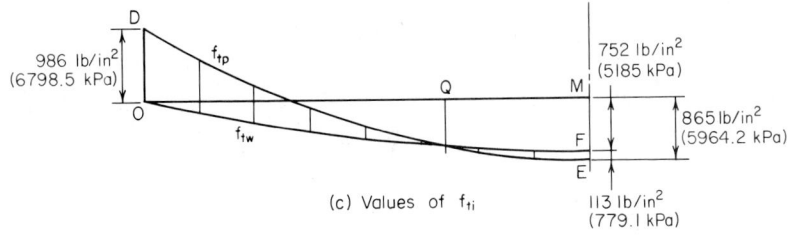

986 lb/in² (6798.5 kPa)

752 lb/in² (5185 kPa)

865 lb/in² (5964.2 kPa)

113 lb/in² (779.1 kPa)

(c) Values of f_{ti}

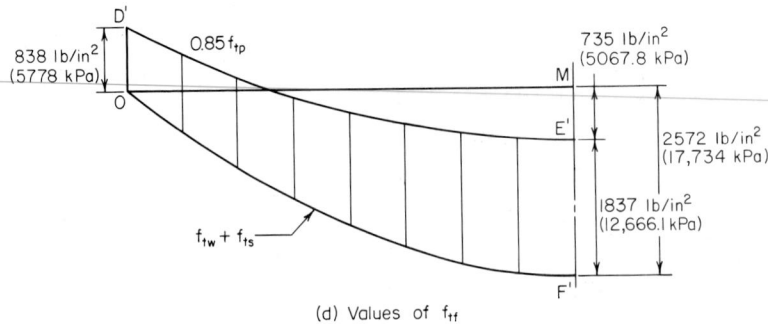

838 lb/in² (5778 kPa)

735 lb/in² (5067.8 kPa)

2572 lb/in² (17,734 kPa)

1837 lb/in² (12,666.1 kPa)

(d) Values of f_{tf}

FIG. 149

1.184

For convenience, set $e = 0$ at the supports. The prestress at the ends is therefore $f_{bp} = f_{tp} = 844,000/856 = +986$ lb/in² ($+6798.5$ kPa). Since e varies parabolically from maximum at mid-span to zero at the supports, it follows that the prestresses also vary parabolically.

In Fig. 149a, draw the parabolic arc AB with summit at B to represent the absolute value of f_{bp}. Draw the parabolic arc OC in the position shown to represent f_{bw}. The vertical distance between the arcs at a given section represents the value of f_{bi}; this value is maximum at midspan.

In Fig. 149b, draw $A'B'$ to represent the absolute value of the final prestress; draw OC' to represent the absolute value of $f_{bw} + f_{bs}$. The vertical distance between the arcs represents the value of f_{bf}. This stress is compressive in the interval ON and tensile in the interval NM.

Construct Fig. 149c and d in an analogous manner. The stress f_{ti} is compressive in the interval OQ.

8. Calculate the allowable ultimate moment of the member

The midspan section is critical in this respect. Thus, $d = 62 - 3 = 59.0$ in (1498.6 mm); $A_s = 8(0.723) = 5.784$ in² (37.3184 cm²); $p = A_s/(bd) = 5.784/[32(59.0)] = 0.00306$.

Apply Eq. 61, or $f_{su} = 236,000(1 - 0.5 \times 0.00306 \times 236,000/5000) = 219,000$ lb/in² (1,510,005.0 kPa). Also, $T_u = A_s f_{su} = 5.784(219,000) = 1,267,000$ lb (5,635,616.0 N). The concrete area under stress $= 1,267,000/[0.85(5,000)] = 298$ in² (1922.7 cm²). This is the shaded area in Fig. 150, as the following calculation proves: $32(9.53) - 4.59(1.53) = 305 - 7 = 298$ in² (1922.7 cm²).

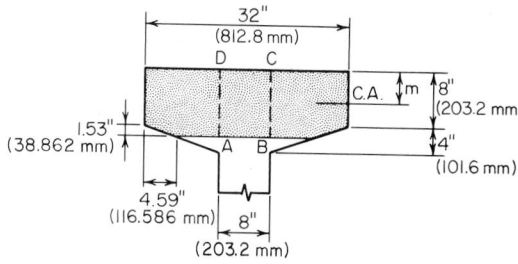

FIG. 150 Concrete area under stress at ultimate load.

Locate the centroidal axis of the stressed area, or $m = [305(4.77) - 7(9.53 - 0.51)]/298 = 4.67$ in (118.618 mm); $M_u = \phi T_u jd = 0.90(1,267,000)(59.0 - 4.67) = 61,950,000$ in·lb (6,999,111.0 N·m).

Calculate the steel index to ascertain that it is below the limit imposed by the ACI *Code*. Refer to Fig. 150. Or, area of $ABCD = 8(9.53) = 76.24$ in² (491.900 cm²). The steel area A_{sr} that is required to balance the force on this web strip is $A_{sr} = 5.784(76.24)/298 = 1.48$ in² (9.549 cm²); $q = A_{sr}f_{su}/(b'\,df_c') = 1.48(219,000)/[8(59.0)(5000)] = 0.137 < 0.30$. This is acceptable.

9. Calculate the required ultimate-moment capacity as given by the ACI Code

Thus, $w_u = 1.5(892 + 1160) + 1.8(1000) = 4878$ lb/lin ft (71,189.0 N/m); M_u required $= (\frac{1}{8})(4878)(90)^2(12) = 59,270,000$ in·lb (6,696,324.6 N·m). This is acceptable. The member is therefore adequate with respect to its ultimate-moment capacity.

10. Design the web reinforcement

Follow the procedure given in step 8 of the previous calculation procedure.

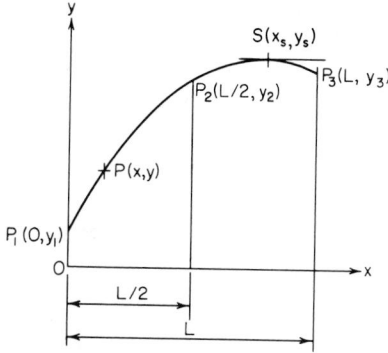

FIG. 151 Parabolic arc.

11.   *Design the end block*

This is usually done by applying isobar charts to evaluate the tensile stresses caused by the concentrated prestressing forces. Refer to Winter et al.—*Design of Concrete Structures*, McGraw-Hill.

12.   *Compute the camber at transfer*

Referring to earlier procedures in this section, we see that $E_cI = 3.644(10)^6(394,800) = 1.44 \times 10^{12}$ lb·in^2 $(4.132 \times 10^9$ N·m$^2)$. Also, $\Delta_w = -5(892)(90)^4(1728)/[384(1.44)(10)^{12}] = -0.91$ in $(-23.11$ mm). Apply Eq. 60, or $\Delta_p = 5(844,000)(31.6)(90)^2(144)/[48(1.44)(10)^{12}] = 2.25$ in $(57.15$ mm); $\Delta_i = 2.25 - 0.91 = 1.34$ in $(34.036$ mm).

PROPERTIES OF A PARABOLIC ARC

Figure 151 shows the literal values of the coordinates at the ends and at the center of the parabolic arc $P_1P_2P_3$. Develop equations for y, dy/dx, and d^2y/dx^2 at an arbitrary point P. Find the slope of the arc at P_1 and P_3 and the coordinates of the summit S. (This information is required for the analysis of beams having parabolic trajectories.)

Calculation Procedure:

1.   *Select a slope for the arc*

Let m denote the slope of the arc.

2.   *Present the results*

The equations are

$$y = 2(y_1 - 2y_2 + y_3)\left(\frac{x}{L}\right)^2 - (3y_1 - 4y_2 + y_3)\frac{x}{L} + y_1 \tag{66}$$

$$m = \frac{dy}{dx} = 4(y_1 - 2y_2 + y_3)\left(\frac{x}{L^2}\right) - \frac{3y_1 - 4y_2 + y_3}{L} \tag{67}$$

$$\frac{dm}{dx} = \frac{d^2y}{dx^2} = \frac{4}{L^2}(y_1 - 2y_2 + y_3) \tag{68}$$

$$m_1 = \frac{-(3y_1 - 4y_2 + y_3)}{L} \tag{69a}$$

$$m_3 = \frac{y_1 - 4y_2 + 3y_3}{L} \tag{69b}$$

$$x_s = \frac{(L/4)(3y_1 - 4y_2 + y_3)}{y_1 - 2y_2 + y_3} \tag{70a}$$

$$y_s = \frac{-(1/8)(3y_1 - 4y_2 + y_3)^2}{y_1 - 2y_2 + y_3} + y_1 \tag{70b}$$

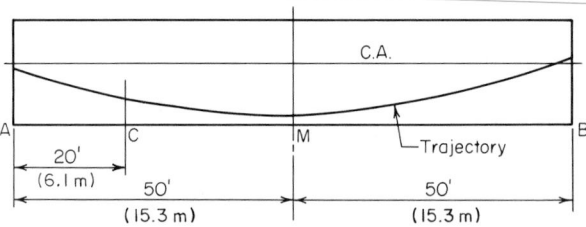

FIG. 152

ALTERNATIVE METHODS OF ANALYZING A BEAM WITH PARABOLIC TRAJECTORY

The beam in Fig. 152 is subjected to an initial prestressing force of 860 kips (3825.3 kN) on a parabolic trajectory. The eccentricities at the left end, midspan, and right end, respectively, are $e_a = 1$ in (25.4 mm); $e_m = 30$ in (762.0 mm); $e_b = -3$ in (-76.2 mm). Evaluate the prestress shear and prestress moment at section C (a) by applying the properties of the trajectory at C; (b) by considering the prestressing action of the steel on the concrete in the interval AC.

Calculation Procedure:

1. Compute the eccentricity and slope of the trajectory at C

Use Eqs. 66 and 67. Let m denote the slope of the trajectory. This is positive if the trajectory slopes downward to the right. Thus $e_a - 2e_m + e_b = 1 - 60 - 3 = -62$ in (1574.8 mm); $3e_a - 4e_m + e_b = 3 - 120 - 3 = -120$ in (-3048 mm); $e_c = 2(-62)(20/100)^2 + 120(20/100) + 1 = 20.04$ in (509.016 mm); $m_c = 4(-62/12)(20/100^2) - (-120/12 \times 100) = 0.0587$.

2. Compute the prestress shear and moment at C

Thus $V_{pc} = -m_c F_i = -0.0587(860,000) = -50,480$ lb ($-224,535.0$ N); $M_{pc} = -F_i e = -860,000(20.04) = -17,230,000$ in·lb ($-1,946,645.4$ N·m). This concludes the solution to part a.

3. Evaluate the vertical component w of the radial force on the concrete in a unit longitudinal distance

An alternative approach to this problem is to analyze the forces that the tendons exert on the concrete in the interval AC, namely, the prestressing force transmitted at the end and the radial forces resulting from curvature of the tendons.

Consider the component w to be positive if directed downward. In Fig. 153, $V_{pr} - V_{pq} = -F_i(m_r - m_q)$; therefore, $\Delta V_p/\Delta x = -F_i \Delta m/\Delta x$. Apply Eq. 68: $dV_p/dx = -F_i\, dm/dx = -(4F_i/L^2)(e_a - 2e_m + e_b)$; but $dV_p/dx = -w$. Therefore,

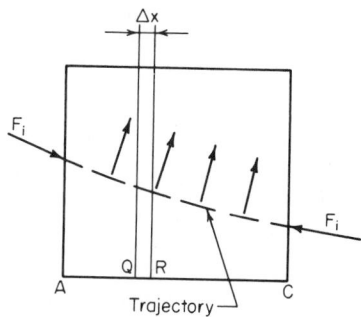

$$w = F_i \frac{dm}{dx} = \left(\frac{4F_i}{L^2}\right)(e_a - 2e_m + e_b) \quad (71)$$

This result discloses that when the trajectory is parabolic, w is uniform across the span. The radial forces are always directed toward the center of curvature, since the tensile forces applied at their ends tend to straighten the tendons. In the present instance, $w = (4F_i/100^2)(-62/12) = -0.002067F_i$ lb/lin ft ($-0.00678F_i$ N/m).

FIG. 153 Free-body diagram of concrete.

4. Find the prestress shear at C

By Eq. 69a, $m_a = -[-120/(100 \times 12)] = 0.1$; $V_{pa} = -0.1F_i$; $V_{pc} = V_{pa} - 20w = F_i(-0.1 + 20 \times 0.002067) = -0.0587F_i = -50,480$ lb ($-224,535.0$ N).

5. Find the prestress moment at C

Thus, $M_{pc} = M_{pa} + V_{pa}(240) - 20w(120) = F_i(-1 - 0.1 \times 240 + 20 \times 0.002067 \times 120) = -20.04F_i = -17,230,000$ in·lb (1,946,645.4 N·m).

PRESTRESS MOMENTS IN A CONTINUOUS BEAM

The continuous prismatic beam in Fig. 154 has a prestressing force of 96 kips (427.0 kN) on a parabolic trajectory. The eccentricities are $e_a = -0.40$ in (-10.16 mm); $e_d = +0.60$ in (15.24 mm); $e_b = -1.20$ in (-30.48 mm); $e_e = +0.64$ in (16.256 mm); $e_c = -0.60$ in (-15.24 mm). Construct the prestress-moment diagram for this member, indicating all significant values.

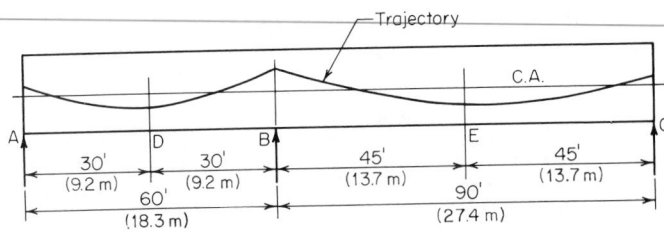

FIG. 154

Calculation Procedure:

1. Find the value of $wL^2/4$ for each span by applying Eq. 71

Refer to Fig. 155. Since members AB and BC are constrained to undergo an identical rotation at B, there exists at this section a bending moment M_{kb} in addition to that resulting from the eccentricity of F_i. The moment M_{kb} induces reactions at the supports. Thus, at every section of the beam there is a moment caused by continuity of the member as well as the moment $-F_i e$. The moment M_{kb} is termed the *continuity moment*; its numerical value is directly proportional to the distance from the given section to the end support.

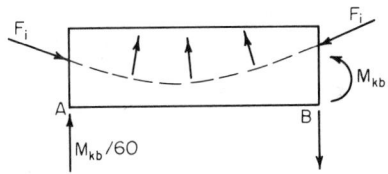

FIG. 155   Free-body diagram of concrete.

The continuity moment may be evaluated by adopting the second method of solution in the previous calculation procedure, since this renders the continuous member amenable to analysis by the theorem of three moments or moment distribution.

Determine $wL^2/4$ for each span: span AB, $w_1 L_1^2/4 = F_i(-0.40 - 1.20 - 1.20) = -2.80F_i$ in·lb $(-0.3163F_i$ N·m); span BC, $w_2 L_2^2/4 = F_i(-1.20 - 1.28 - 0.60) = -3.08F_i$ in·lb $(-0.3479F_i$ N·m).

2. Determine the true prestress moment at B in terms of F_i

Apply the theorem of three moments; by subtraction, find M_{kb}. Thus, $M_{pa}L_1 + 2M_{pb}(L_1 + L_2) + M_{pc}L_2 = -w_1 L_1^3/4 - w_2 L_2^3/4$. Substitute the value of L_1 and L_2, in feet (meters), and divide each term by F_i, or $0.40(60) + (2M_{pb} \times 150)/F_i + 0.60(90) = 2.80(60) + 3.08(90)$. Solving gives $M_{pb} = 1.224F_i$ in·lb $(0.1383F_i$ N·m). Also, $M_{kb} = M_{pb} - (-F_i e_b) = F_i(1.224 - 1.20) = 0.024F_i$. Thus, the continuity moment at B is positive.

3. Evaluate the prestress moment at the supports and at midspan

Using foot-pounds (newton-meters) in the moment evaluation yields $M_{pa} = 0.40(96,000)/12 = 3200$ ft·lb $(4339.2$ N·m); $M_{pb} = 1.224(96,000)/12 = 9792$ ft·lb $(13,278$ N·m); $M_{pc} = 0.60(96,000)/12 = 4800$ ft·lb $(6508.0$ N·m); $M_{pd} = -F_i e_d + M_{kd} = F_i(-0.60 + \frac{1}{2} \times 0.024)/12 = -4704$ ft·lb $(-6378$ N·m); $M_{pe} = F_i(-0.64 + \frac{1}{2} \times 0.024)/12 = -5024$ ft·lb $(-6812$ N·m).

4. Construct the prestress-moment diagram

Figure 156 shows this diagram. Apply Eq. 70 to locate and evaluate the maximum negative moments. Thus, $AF = 25.6$ ft $(7.80$ m); $BG = 49.6$ ft $(15.12$ m); $M_{pf} = -4947$ ft·lb $(-6708$ N·m); $M_{pg} = -5151$ ft·lb $(-6985$ N·m).

PRINCIPLE OF LINEAR TRANSFORMATION

For the beam in Fig. 154, consider that the parabolic trajectory of the prestressing force is displaced thus: e_a and e_c are held constant as e_b is changed to -2.0 in $(-50.80$ mm), the eccentricity at any intermediate section being decreased algebraically by an amount directly proportional to the distance from that section to A or C. Construct the prestress-moment diagram.

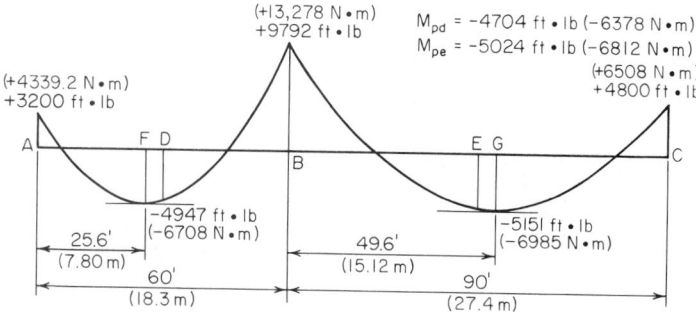

FIG. 156 Prestress-moment diagram.

Calculation Procedure:

1. Compute the revised eccentricities

The modification described is termed a *linear transformation* of the trajectory. Two methods are presented. Steps 1 through 4 comprise method 1; the remaining steps comprise method 2.

The revised eccentricities are $e_a = -0.40$ in (-10.16 mm); $e_d = +0.20$ in (5.08 mm); $e_b = -2.00$ in (-50.8 mm); $e_e = +0.24$ in (6.096 mm); $e_c = -0.60$ in (-15.24 mm).

2. Find the value of $wL^2/4$ for each span

Apply Eq. 71: span AB, $w_1L_1^2/4 = F_i(-0.40 - 0.40 - 2.00) = -2.80F_i$; span BC, $w_2L_2^2/4 = F_i(-2.00 - 0.48 - 0.60) = -3.08F_i$.

These results are identical with those obtained in the previous calculation procedure. The change in e_b is balanced by an equal change in $2e_d$ and $2e_e$.

3. Determine the true prestress moment at B by applying the theorem of three moments; then find M_{kb}

Refer to step 2 in the previous calculation procedure. Since the linear transformation of the trajectory has not affected the value of w_1 and w_2, the value of M_{pb} remains constant. Thus, $M_{kb} = M_{pb} - (-F_i e_b) = F_i(1.224 - 2.0) = -0.776F_i$.

4. Evaluate the prestress moment at midspan

Thus, $M_{pd} = -F_i e_d + M_{kd} = F_i(-0.20 - \frac{1}{2} \times 0.776)/12 = -4704$ ft·lb (-6378.6 N·m); $M_{pe} = F_i(-0.24 - \frac{1}{2} \times 0.776)/12 = -5024$ ft·lb (-6812.5 N·m).

These results are identical with those in the previous calculation procedure. The change in the eccentricity moment is balanced by an accompanying change in the continuity moment. Since three points determine a parabolic arc, the prestress moment diagram coincides with that in Fig. 156. This constitutes the solution by method 1.

5. Evaluate the prestress moments

Do this by replacing the prestressing system with two hypothetical systems that jointly induce eccentricity moments identical with those of the true system.

Let e denote the original eccentricity of the prestressing force at a given section and Δe the change in eccentricity that results from the linear transformation. The final eccentricity moment is $-F_i(e + \Delta e) = -(F_i e + F_i \Delta e)$.

Consider the beam as subjected to two prestressing forces of 96 kips (427.0 kN) each. One has the parabolic trajectory described in the previous calculation procedure; the other has the linear trajectory shown in Fig. 157, where $e_a = 0$, $e_b = -0.80$ in (-20.32 mm), and $e_c = 0$. Under the latter prestressing system, the tendons exert three forces on the concrete—one at each end and one at the deflection point above the interior support caused by the change in direction of the prestressing force.

The horizontal component of the prestressing force is considered equal to the force itself; it therefore follows that the force acting at the deflection point has no horizontal component.

Since the three forces that the tendons exert on the concrete are applied directly at the sup-

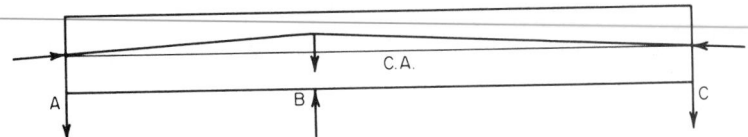

FIG. 157 Hypothetical prestressing system and forces exerted on concrete.

ports, their vertical components do not induce bending. Similarly, since the forces at A and C are applied at the centroidal axis, their horizontal components do not induce bending. Consequently, the prestressing system having the trajectory shown in Fig. 157 does not cause any prestress moments whatsoever. The prestress moments for the beam in the present instance are therefore identical with those for the beam in the previous calculation procedure.

The second method of analysis is preferable to the first because it is general. The first method demonstrates the equality of prestress moments before and after the linear transformation where the trajectory is parabolic; the second method demonstrates this equality without regard to the form of trajectory.

In this calculation procedure, the extremely important *principle of linear transformation* for a two-span continuous beam was developed. This principle states: The prestress moments remain constant when the trajectory of the prestressing force is transformed linearly. The principle is frequently applied in plotting a trial trajectory for a continuous beam.

Two points warrant emphasis. First, in a linear transformation, the eccentricities at the end supports remain constant. Second, the hypothetical prestressing systems introduced in step 5 are equivalent to the true system solely with respect to bending stresses; the axial stress F_i/A under the hypothetical systems is double that under the true system.

CONCORDANT TRAJECTORY OF A BEAM

Referring to the beam in the second previous calculation procedure, transform the trajectory linearly to obtain a concordant trajectory.

Calculation Procedure:

1. *Calculate the eccentricities of the concordant trajectory*

Two principles apply here. First, in a continuous beam, the prestress moment M_p consists of two elements, a moment $-F_i e$ due to eccentricity and a moment M_k due to continuity. The continuity moment varies linearly from zero at the ends to its maximum numerical value at the interior support. Second, in a linear transformation, the change in $-F_i e$ is offset by a compensatory change in M_k, with the result that M_p remains constant.

It is possible to transform a given trajectory linearly to obtain a new trajectory having the characteristic that $M_k = 0$ along the entire span, and therefore $M_p = -F_i e$. The latter is termed a *concordant trajectory*. Since M_p retains its original value, the concordant trajectory corresponding to a given trajectory is found simply by equating the final eccentricity to $-M_p/F_i$.

Refer to Fig. 154, and calculate the eccentricities of the concordant trajectory. As before, $e_a = -0.40$ in $(-10.16$ mm$)$ and $e_c = -0.60$ in $(-15.24$ mm$)$. Then $e_d = 4704(12)/96,000 = +0.588$ in $(+14.9352$ mm$)$; $e_b = -9792(12)/96,000 = -1.224$ in $(-31.0896$ mm$)$; $e_e = 5024(12)/96,000 = +0.628$ in $(15.9512$ mm$)$.

2. *Analyze the eccentricities*

All eccentricities have thus been altered by an amount directly proportional to the distance from the adjacent end support to the given section, and the trajectory has undergone a linear transformation. The advantage accruing from plotting a concordant trajectory is shown in the next calculation procedure.

DESIGN OF TRAJECTORY TO OBTAIN ASSIGNED PRESTRESS MOMENTS

The prestress moments shown in Fig. 156 are to be obtained by applying an initial prestressing force of 72 kips $(320.3$ kN$)$ with an eccentricity of -2 in $(-50.8$ mm$)$ at B. Design the trajectory.

Calculation Procedure:

1. *Plot a concordant trajectory*

Set $e = M_p/F_i$, or $e_a = -3200(12)/72,000 = -0.533$ in $(-13.5382$ mm); $e_d = +0.784$ in $(19.9136$ mm); $e_b = -1.632$ in $(-41.4528$ mm); $e_e = +0.837$ in $(21.2598$ mm); $e_c = -0.800$ in $(-20.32$ mm).

2. *Set e_b = desired eccentricity, and transform the trajectory linearly*

Thus, $e_a = -0.533$ in $(-13.5382$ mm); $e_c = -0.800$ in $(-20.32$ mm); $e_d = +0.784 - \frac{1}{2}(2.000 - 1.632) = +0.600$ in $(+15.24$ mm); $e_e = +0.837 - 0.184 = +0.653$ in $(+16.5862$ mm).

EFFECT OF VARYING ECCENTRICITY AT END SUPPORT

For the beam in Fig. 154, consider that the parabolic trajectory in span AB is displaced thus: e_b is held constant as e_a is changed to -0.72 in $(-18.288$ mm), the eccentricity at every intermediate section being decreased algebraically by an amount directly proportional to the distance from that section to B. Compute the prestress moment at the supports and at midspan caused by a prestressing force of 96 kips (427.0 kN).

Calculation Procedure:

1. *Apply the revised value of e_a; repeat the calculations of the earlier procedure*

Thus, $M_{pa} = 5760$ ft·lb (7810.6 N·m); $M_{pd} = -3680$ ft·lb (-4990.1 N·m); $M_{pb} = 9280$ ft·lb $(12,583.7$ N·m); $M_{pe} = -5280$ ft·lb $(-7159.7$ N·m); $M_{pc} = 4800$ ft·lb (6508.8 N·m).

The change in prestress moment caused by the displacement of the trajectory varies linearly across each span. Figure 158 compares the original and revised moments along span AB. This constitutes method 1.

2. *Replace the prestressing system with two hypothetical systems that jointly induce eccentricity moments identical with those of the true system*

This constitutes method 2. For this purpose, consider the beam to be subjected to two prestressing forces of 96 kips (427.0 kN) each. One has the parabolic trajectory described in the earlier procedure; the other has a trajectory that is linear in each span, the eccentricities being $e_a = -0.72 - (-0.40) = -0.32$ in $(-8.128$ mm), $e_b = 0$, and $e_c = 0$.

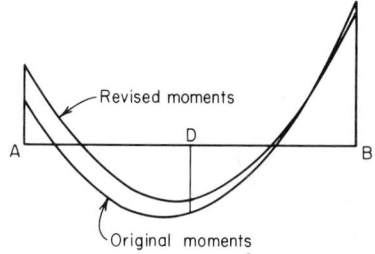

FIG. 158   Prestress-moment diagrams.

3. *Evaluate the prestress moments induced by the hypothetical system having the linear trajectory*

The tendons exert a force on the concrete at A, B, and C, but only the force at A causes bending moment.

Thus, $M_{pa} = -F_i e_a = -96,000(-0.32)/12 = 2560$ ft·lb (3471.4 N·m). Also, $M_{pa}L_1 + 2M_{pb}(L_1 + L_2) + M_{pc}L_2 = 0$. But $M_{pc} = 0$; therefore, $M_{pb} = -512$ ft·lb $(-694.3$ N·m); $M_{pd} = \frac{1}{2}(2560 - 512) = 1024$ ft·lb (1388.5 N·m); $M_{pe} = \frac{1}{2}(-512) = -256$ ft·lb $(-347.1$ N·m).

4. *Find the true prestress moments by superposing the two hypothetical systems*

Thus $M_{pa} = 3200 + 2560 = 5760$ ft·lb (7810.6 N·m); $M_{pd} = -4704 + 1024 = -3680$ ft·lb $(-4990.1$ N·m); $M_{pb} = 9792 - 512 = 9280$ ft·lb (12,583.7 N·m); $M_{pe} = -5024 - 256 = -5280$ ft·lb $(-7159.7$ N·m); $M_{pc} = 4800$ ft·lb (6508.8 N·m).

DESIGN OF TRAJECTORY FOR A TWO-SPAN CONTINUOUS BEAM

A T beam that is continuous across two spans of 120 ft (36.6 m) each is to carry a uniformly distributed live load of 880 lb/lin ft (12,842.6 N/m). The cross section has these properties: $A = 1440$ in^2 (9290.8 cm^2); $I = 752,000$ in^4 (3130.05 dm^4); $y_b = 50.6$ in (1285.24 mm); $y_t = 23.4$ in

(594.36 mm). The allowable stresses are: *initial*, $+2400$ and -60 lb/in^2 ($+16{,}548.0$ and -413.7 kPa); *final*, $+2250$ and -60 lb/in^2 ($+15{,}513.8$ and 413.7 kPa). Assume that the minimum possible distance from the extremity of the section to the centroidal axis of the prestressing steel is 9 in (228.6 mm). Determine the magnitude of the prestressing force, and design the parabolic trajectory (*a*) using solely prestressed reinforcement; (*b*) using a combination of prestressed and non-prestressed reinforcement.

Calculation Procedure:

1. Compute the section moduli, kern distances, and beam weight

For part *a*, an exact design method consists of these steps: First, write equations for the prestress moment, beam-weight moment, maximum and minimum potential superimposed-load moment, expressing each moment in terms of the distance from a given section to the adjacent exterior support. Second, apply these equations to identify the sections at which the initial and final stresses are critical. Third, design the prestressing system to restrict the critical stresses to their allowable range. Whereas the exact method is not laborious when applied to a prismatic beam carrying uniform loads, this procedure adopts the conventional, simplified method for illustrative purposes. This consists of dividing each span into a suitable number of intervals and analyzing the stresses at each boundary section.

For simplicity, set the eccentricity at the ends equal to zero. The trajectory will be symmetric about the interior support, and the vertical component *w* of the force exerted by the tendons on the concrete in a unit longitudinal distance will be uniform across the entire length of member. Therefore, the prestress-moment diagram has the same form as the bending-moment diagram of a nonprestressed prismatic beam continuous over two equal spans and subjected to a uniform load across its entire length. It follows as a corollary that the prestress moments at the boundary sections previously referred to have specific *relative* values, although their absolute values are functions of the prestressing force and its trajectory.

The following steps constitute a methodical procedure: Evaluate the relative prestress moments, and select a trajectory having ordinates directly proportional to these moments. The trajectory thus fashioned is concordant. Compute the prestressing force required to restrict the stresses to the allowable range. Then transform the concordant trajectory linearly to secure one that lies entirely within the confines of the section. Although the number of satisfactory concordant trajectories is infinite, the one to be selected is that which requires the minimum prestressing force. Therefore, the selection of the trajectory and the calculation of F_i are blended into one operation.

Divide the left span into five intervals, as shown in Fig. 159. (The greater the number of intervals chosen, the more reliable are the results.)

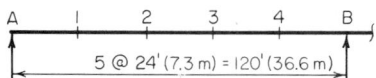

FIG. 159   Division of span into intervals.

Computing the moduli, kern distances, and beam weight gives $S_b = 14{,}860$ in^3 ($243{,}555.4$ cm^3); $S_t = 32{,}140$ in^3 ($526{,}774.6$ cm^3); $k_b = 22.32$ in (566.928 mm); $k_t = 10.32$ in (262.128 mm); $w_w = 1500$ lb/lin ft (21,890.9 N/m).

2. Record the bending-moment coefficients C_1, C_2, and C_3

Use Table 9 to record these coefficients at the boundary sections. The subscripts refer to these conditions of loading: 1, load on entire left span and none on right span; 2, load on entire right span and none on left span; 3, load on entire length of beam.

To obtain these coefficients, refer to the AISC *Manual*, case 29, which represents condition 1. Thus, $R_1 = (\tfrac{7}{16})wL$; $R_3 = -(\tfrac{1}{16})wL$. At section 3, for example, $M_1 = (\tfrac{7}{16})wL(0.6L) - \tfrac{1}{2}w(0.6L)^2 = [7(0.6) - 8(0.36)]wL^2/16 = 0.0825wL^2$; $C_1 = M_1/(wL^2) = +0.0825$.

To obtain condition 2, interchange R_1 and R_3. At section 3, $M_2 = -(\tfrac{1}{16})wL(0.6L) = -0.0375wL^2$; $C_2 = -0.0375$; $C_3 = C_1 + C_2 = +0.0825 - 0.0375 = +0.0450$.

These moment coefficients may be applied without appreciable error to find the maximum

TABLE 9   Calculations for Two-Span Beam: Part a

Section	1	2	3	4	B
1 C_1	+0.0675	+0.0950	+0.0825	+0.0300	−0.0625
2 C_2	−0.0125	−0.0250	−0.0375	−0.0500	−0.0625
3 C_3	+0.0550	+0.0700	+0.0450	−0.0200	−0.1250
4 f_{bw}, lb/in² (kPa)	−959 (−6,611)	−1,221 (−8,418)	−785 (−5,412)	+349 (+2406)	+2,180 (+15,029)
5 f_{bs1}, lb/in² (kPa)	−691 (−4,764)	−972 (−6,701)	−844 (−5,819)	−307 (−2,116)	+640 (+4,412)
6 f_{bs2}, lb/in² (kPa)	+128 (+882)	+256 (+1,765)	+384 (+2,647)	+512 (+3,530)	+640 (+4,412)
7 f_{tw}, lb/in² (kPa)	+444 (+3,060)	+565 (+3895)	+363 (+2,503)	−161 (−1,110)	−1,008 (−6,949)
8 f_{ts1}, lb/in² (kPa)	+319 (+2,199)	+450 (+3,102)	+390 (+2689)	+142 (+979)	−296 (−2,041)
9 f_{ts2}, lb/in² (kPa)	−59 (−407)	−118 (−813)	−177 (−1,220)	−237 (−1,634)	−296 (−2,041)
10 e_{con}, in (mm)	+17.19 (+436.6)	+21.87 (+555.5)	+14.06 (+357.1)	−6.25 (−158.8)	−39.05 (−991.9)
11 f_{bp}, lb/in² (kPa)	+2,148 (+14,808)	+2,513 (+17,325)	+1,903 (+13,119)	+318 (+2,192)	−2,243 (−15,463)
12 f_{tp}, lb/in² (kPa)	+185 (+128)	+16 (+110)	+298 (+2,054)	+1,031 (+7,108)	+2,215 (+15,270)
13 $0.85f_{bp}$, lb/in² (kPa)	+1,826 (+12,588)	+2,136 (+14,726)	+1,618 (+11,154)	+270 (+1,861)	−1,906 (−13,140)
14 $0.85f_{tp}$, lb/in² (kPa)	+157 (+1,082)	+14 (+97)	+253 (+1,744)	+876 (+6,039)	+1,883 (+12,981)

At midspan: $C_3 = +0.0625$ and $e_{con} = +19.53$ in (496.1 mm)

and minimum potential live-load bending moments at the respective sections. The values of C_3 also represent the relative eccentricities of a concordant trajectory.

Since the gravity loads induce the maximum positive moment at section 2 and the maximum negative moment at section B, the prestressing force and its trajectory will be designed to satisfy the stress requirements at these two sections. (However, the stresses at all boundary sections will be checked.) The Magnel diagram for section 2 is similar to that in Fig. 141, but that for section B is much different.

3. Compute the value of C_3 at midspan

Thus, $C_3 = +0.0625$.

4. Apply the moment coefficients to find the gravity-load stresses

Record the results in Table 9. Thus $M_w = C_3(1500)(120)^2(12) = 259,200,000C_3$ in·lb (29.3C_3 kN·m); $f_{bw} = -259,200,000C_3/14,860 = -17,440C_3$; $f_{bs1} = -10,230C_1$; $f_{bs2} = -10,230C_2$; $f_{tw} = 8065C_3$; $f_{ts1} = 4731C_1$; $f_{ts2} = 4731C_2$.

Since S_t far exceeds S_b, it is manifest that the prestressing force must be designed to confine the bottom-fiber stresses to the allowable range.

5. Consider that a concordant trajectory has been plotted; express the eccentricity at section B relative to that at section 2

Thus, $e_b/e_2 = -0.1250/+0.0700 = -1.786$; therefore, $e_b = -1.786e_2$.

6. Determine the allowable range of values of f_{bp} at sections 2 and B

Refer to Fig. 160. At section 2, $f_{bp} \leq +3621$ lb/in² (+24,966.8 kPa), Eq. a; $0.85f_{bp} \geq 1221 + 972 - 60$; therefore, $f_{bp} \geq +2509$ lb/in² (+17,299.5 kPa), Eq b. At section B, $f_{bp} \geq -2240$

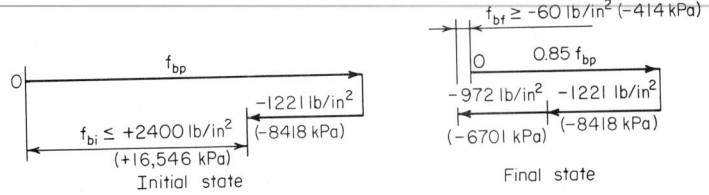

(a) Limiting values of f_{bp} at section 2

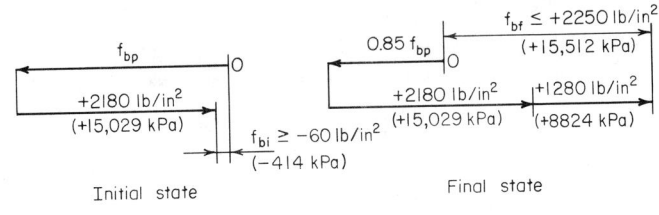

(b) Limiting values of f_{bp} at section B

FIG. 160

lb/in^2 ($-15{,}444.8$ kPa), Eq. c; $0.85f_{bp} \leq -(2180 + 1280) + 2250$; $f_{bp} \leq -1424$ lb/in^2 (-9818.5 kPa), Eq. d.

7. Substitute numerical values in Eq. 56, expressing e_b in terms of e_2

The values obtained are $1/F_i \geq (k_t + e_2)/(3621S_b)$, Eq. a'; $1/F_i \leq (k_t + e_2)/(2509S_b)$, Eq. b'; $1/F_i \geq (1.786e_2 - k_t)/(2240S_b)$, Eq. c'; $1/F_i \leq (1.786e_2 - k_t)/(1424S_b)$, Eq. d'.

8. Obtain the composite Magnel diagram

Considering the relations in step 7 as equalities, plot the straight lines representing them to obtain the composite Magnel diagram in Fig. 161. The slopes of the lines have these relative values: $m_a = 1/3621$; $m_b = 1/2509$; $m_c = 1.786/2240 = 1/1254$; $m_d = 1.786/1424 = 1/797$. The shaded area bounded by these lines represents the region of permissible sets of values of e_2 and $1/F_i$.

9. Calculate the minimum allowable value of F_i and the corresponding value of e_2

In the composite Magnel diagram, this set of values is represented by point A. Therefore, consider Eqs. b' and c' as equalities, and solve for the unknowns. Or, $(10.32 + e_2)/2509 = (1.786e_2 - 10.32)/2240$; solving gives $e_2 = 21.87$ in (555.5 mm) and $F_i = 1{,}160{,}000$ lb (5,159,680.0 N).

10. Plot the concordant trajectory

Do this by applying the values of C_3 appearing in Table 9; for example, $e_1 = +21.87(0.0550)/0.0700 = +17.19$ in (436.626 mm). At midspan, $e_m = +21.87(0.0625)/0.0700 = +19.53$ in (496.062 mm).

Record the eccentricities on line 10 of the table. It is apparent that this concordant trajectory is satisfactory in the respect that it may be linearly transformed to one falling within the confines of the section; this is proved in step 14.

11. Apply Eq. 56 to find f_{bp} and f_{tp}

Record the results in Table 9. For example, at section 1, $f_{bp} = 1{,}160{,}000(10.32 + 17.19)/14{,}860 = +2148$ lb/in^2 ($+14{,}810.5$ kPa); $f_{tp} = 1{,}160{,}000(22.32 - 17.19)/32{,}140 = +185$ lb/in^2 ($+1275.6$ kPa).

12. Multiply the values of f_{bp} and f_{tp} by 0.85, and record the results

These results appear in Table 9.

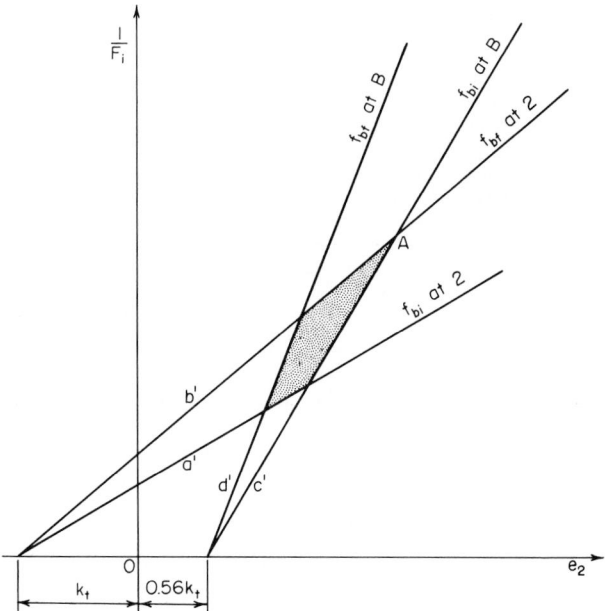

FIG. 161 Composite Magnel diagram.

13. *Investigate the stresses at every boundary section*

In calculating the final stresses, apply the live-load stress that produces a more critical condition. Thus, at section 1, $f_{bi} = -959 + 2148 = +1189$ lb/in^2 ($+8198.2$ kPa); $f_{bf} = -959 - 691 + 1826 = + 176$ lb/in^2 ($+1213.6$ kPa); $f_{ti} = + 444 + 185 = + 629$ lb/in^2 ($+4337.0$ kPa); $f_{tf} = + 444 + 319 + 157 = +920$ lb/in^2 ($+6343.4$ kPa). At section 2: $f_{bi} = -1221 + 2513 = +1292$ lb/in^2 ($+8908.3$ kPa); $f_{bf} = -1221 - 972 + 2136 = -57$ lb/in^2 (-393.0 kPa); $f_{ti} = + 565 + 16 = + 581$ lb/in^2 ($+4006.0$ kPa); $f_{tf} = +565 + 450 + 14 = + 1029$ lb/in^2 ($+7095.0$ kPa). At section 3: $f_{bi} = -785 + 1903 = +1118$ lb/in^2 ($+7706.8$ kPa); $f_{bf} = -785 - 844 + 1618 = - 11$ lb/in^2 (-75.8 kPa); $f_{ti} = + 363 + 298 = + 661$ lb/in^2 ($+4558.0$ kPa); $f_{tf} = +363 + 390 + 253 = + 1006$ lb/in^2 ($+6936.4$ kPa). At section 4: $f_{bi} = +349 + 318 = +667$ lb/in^2 ($+4599.0$ kPa); $f_{bf} = +349 - 307 + 270 = +312$ lb/in^2 ($+2151.2$ kPa); or $f_{bf} = +349 + 512 + 270 = +1131$ lb/in^2 (7798.2 kPa); $f_{ti} = -161 + 1031 = + 870$ lb/in^2 ($+5998.7$ kPa); $f_{tf} = -161 - 237 + 876 = +478$ lb/in^2 ($+3295.8$ kPa), or $f_{tf} = -161 + 142 + 876 = +857$ lb/in^2 ($+5909.0$ kPa). At section B: $f_{bi} = +2180 - 2243 = -63$ lb/in^2 (-434.4 kPa); $f_{bf} = +2180 + 1280 - 1906 = +1554$ lb/in^2 ($+10{,}714.8$ kPa); $f_{ti} = -1008 + 2215 = +1207$ lb/in^2 ($+8322.3$ kPa); $f_{tf} = -1008 - 592 + 1883 = +283$ lb/in^2 ($+1951.3$ kPa).

In all instances, the stresses lie within the allowable range.

14. *Establish the true trajectory by means of a linear transformation*

The imposed limits are $e_{max} = y_b - 9 = 41.6$ in (1056.6 mm), $e_{min} = -(y_t - 9) = -14.4$ in (-365.76 mm).

Any trajectory that falls between these limits and that is obtained by linearly transforming the concordant trajectory is satisfactory. Set $e_b = -14$ in (-355.6 mm), and compute the eccentricity at midspan and the maximum eccentricity.

Thus, $e_m = +19.53 + \frac{1}{2}(39.05 - 14) = +32.06$ in (814.324 mm). By Eq. 70b, $e_s = -(\frac{1}{8})(-4 \times 32.06 - 14)^2/(-2 \times 32.06 - 14) = +32.4$ in ($+823.0$ mm) < 41.6 in (1056.6 mm). This is acceptable. This constitutes the solution to part a of the procedure. Steps 15 through 20 constitute the solution to part b.

15. Assign eccentricities to the true trajectory, and check the maximum eccentricity

The preceding calculation shows that the maximum eccentricity is considerably below the upper limit set by the beam dimensions. Refer to Fig. 161. If the restrictions imposed by line c' are removed, e_2 may be increased to the value corresponding to a maximum eccentricity of 41.6 in (1056.6 mm), and the value of F_i is thereby reduced. This revised set of values will cause an excessive initial tensile stress at B, but the condition can be remedied by supplying nonprestressed reinforcement over the interior support. Since the excess tension induced by F_i extends across a comparatively short distance, the savings accruing from the reduction in prestressing force will more than offset the cost of the added reinforcement.

Assigning the following eccentricities to the true trajectory and checking the maximum eccentricity by applying Eq. 70b, we get $e_a = 0$; $e_m = +41$ in (1041.4 mm); $e_b = -14$ in (-355.6 mm); $e_s = -(\frac{1}{8})(-4 \times 41 - 14)^2/(-2 \times 41 - 14) = +41.3$ in (1049.02 mm). This is acceptable.

16. To analyze the stresses, obtain a hypothetical concordant trajectory by linearly transforming the true trajectory.

Let y denote the upward displacement at B. Apply the coefficients C_3 to find the eccentricities of the hypothetical trajectory. Thus, $e_m/e_b = (41 - \frac{1}{2}y)/(-14 - y) = +0.0625/-0.1250$; $y = 34$ in (863.6 mm); $e_a = 0$; $e_m = +24$ in (609.6 mm); $e_b = -48$ in (-1219.2 mm); $e_1 = -48$ ($+0.0550)/-0.1250 = +21.12$ in (536.448 mm); $e_2 = +26.88$ in (682.752 mm); $e_3 = +17.28$ in (438.912 mm); $e_4 = -7.68$ in (-195.072 mm).

17. Evaluate F_i by substituting in relation (b') of step 7

Thus, $F_i = 2509(14,860)/(10.32 + 26.88) = 1,000,000$ lb (4448 kN). Hence, the introduction of nonprestressed reinforcement served to reduce the prestressing force by 14 percent.

18. Calculate the prestresses at every boundary section; then find the stresses at transfer and under design load

Record the results in Table 10. (At sections 1 through 4, the final stresses were determined by applying the values on lines 5 and 8 in Table 9. The slight discrepancy between the final stress at 2 and the allowable value of -60 lb/in^2 (-413.7 kPa) arises from the degree of precision in the calculations.)

With the exception of f_{bt} at B, all stresses at the boundary sections lie within the allowable range.

TABLE 10   Calculations for Two-Span Continuous Beam: Part b

Section	1	2	3	4	B
e_{con}, in (mm)	+21.12	+26.88	+17.28	-7.68	-48.00
	(536.4)	(+682.8)	(+438.9)	(-195.1)	(-1,219.2)
f_{bp}, lb/in^2 (kPa)	+2,116	+2,503	+1,857	+178	-2,535
	(+14,588)	(+17,256)	(+12,802)	(+1,227)	(-17,476)
f_{tp}, lb/in^2 (kPa)	+37	-142	+157	+933	+2188
	(+255)	(-979)	(+1,082)	(+6,660)	(+15,084)
$0.85f_{bp}$, lb/in^2 (kPa)	+1,799	+2,128	+1,578	+151	-2,155
	(+12,402)	(+14,670)	(+10,879)	(+1,041)	(-14,857)
$0.85f_{tp}$, lb/in^2 (kPa)	+31	-121	+133	+793	+1,860
	(+214)	(-834)	(+917)	(+5,467)	(+12,823)
f_{bt}, lb/in^2 (kPa)	+1,157	+1,282	+1,072	+527	-355
	(+7,976)	(+8,838)	(+7,390)	(+3,633)	(-2,447)
f_{bf}, lb/in^2 (kPa)	+149	-65	-51	+193	+1,305
	(+1,027)	(-448)	(-352)	(+1,331)	(+8,997)
f_{ti}, lb/in^2 (kPa)	+481	+423	+520	+772	+1,180
	(+3,316)	(+2,916)	(+3,585)	(+5,322)	(+8,135)
f_{tf}, lb/in^2 (kPa)	+794	+894	+886	+774	+260
	(+5,474)	(+6,163)	(+6108)	(+5,336)	(+1,792)

19. Locate the section at which $f_{bi} = -60$ lb/in² (-413.7 kPa)

Since f_{bp} and f_{bw} vary parabolically across the span, their sum f_{bi} also varies in this manner. Let x denote the distance from the interior support to a given section. Apply Eq. 66 to find the equation for f_{bi}, using the initial-stress values at sections B, 3, and 1. Or, $-355 - 2 \times 1072 + 1157 = -1342$ (-9253.1 kPa); $3(-355) - 4(1072) + 1157 = -4196$ ($-28,931.4$ kPa); $f_{bi} = -2684(x/96)^2 + 4196x/96 - 355$. When $f_{bi} = -60$ (-413.7), $x = 7.08$ ft (2.15 m). The tensile stress at transfer is therefore excessive in an interval of only 14.16 ft (4.32 m).

20. Design the nonprestressed reinforcement over the interior support

As in the preceding procedures, the member must be investigated for ultimate-strength capacity. The calculation pertaining to any quantity that varies parabolically across the span may be readily checked by verifying that the values at uniformly spaced sections have equal "second differences." For example, with respect to the values of f_{bi} recorded in Table 10, the verification is:

$$
\begin{array}{ccccc}
+1157 & +1282 & +1072 & +527 & -355 \\
 & -125 & +210 & +545 & +882 \\
 & +335 & +335 & +337 &
\end{array}
$$

The values on the second and third lines represent the differences between successive values on the preceding line.

REACTIONS FOR A CONTINUOUS BEAM

With reference to the beam in the previous calculation procedure, compute the reactions at the supports caused by the initial prestressing force designed in part a.

Calculation Procedure:

1. Determine what causes the reactions at the supports

As shown in Fig. 155, the reactions at the supports result from the continuity at B, and $R_a = M_{kb}/L$.

2. Compute the continuity moment at B; then find the reactions

Thus, $M_p = -F_ie + M_k = -F_ie_{con}$; $M_k = F_i(e - e_{con}) = 1160(-14 + 39.05) = 29,060$ in·kips (3283 kN·m). $R_a = 29,060/[120(12)] = 20.2$ kips (89.8 kN); $R_B = -40.4$ kips (-179.8 kN).

Design of Highway Bridges

Where a bridge is supported by steel trusses, the stresses in the truss members are determined by applying the rules formulated in the truss calculation procedures given earlier in this handbook.

The following procedures show the design of a highway bridge supported by concrete or steel girders. Except for the deviations indicated, the *Standard Specifications for Highway Bridges*, published by the American Association of State Highway and Transportation Officials (AASHTO), are applied.

The AASHTO *Specification* recognizes two forms of truck loading: the H loading, and the HS loading. Both are illustrated in the *Specification*. For a bridge of relatively long span, it is necessary to consider the possibility that several trucks will be present simultaneously. To approximate this condition, the AASHTO *Specification* offers various lane loadings, and it requires that the bridge be designed for the lane loading if this yields greater bending moments and shears than does the corresponding truck loading.

In designing the bridge members, it is necessary to modify the wheel loads to allow for the effects of dynamic loading and the lateral distribution of loads resulting from the rigidity of the floor slab.

The basic notational system is: DF = factor for lateral distribution of wheel loads; IF = impact factor; P = resultant of group of concentrated loads.

The term *live load* as used in the following material refers to the wheel load after correction for distribution but before correction for impact.

DESIGN OF A T-BEAM BRIDGE

A highway bridge consisting of a concrete slab and concrete girders is to be designed for these conditions: loading, HS20-44; clear width, 28 ft (8.5 m); effective span, 54 ft (16.5 m); concrete strength, 3000 lb/in² (20,685 kPa); reinforcement, intermediate grade. The slab and girders will be poured monolithically, and the slab will include a ¾ in (19.05 mm) wearing surface. In addition, the design is to make an allowance of 15 lb/ft² (718 N/m²) for future paving. Design the slab and the cross section of the interior girders.

Calculation Procedure:

1.   *Record the allowable stresses and modular ratio given in the AASHTO Specification*

Refer to Fig. 162, which shows the spacing of the girders and the dimensions of the members. The sizes were obtained by a trial-and-error method. Values from the *Specification* are: $n = 10$

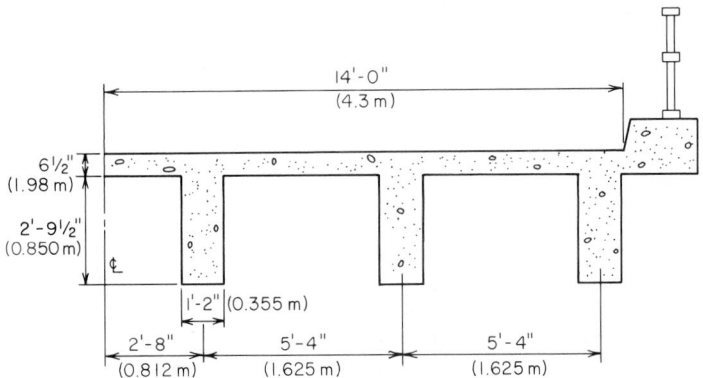

FIG. 162   Transverse section of T-beam bridge.

in stress calculations; $f_c = 0.4f'_c = 1200$ lb/in² (8274 kPa); for beams with web reinforcement, $v_{max} = 0.075f'_c = 225$ lb/in² (1551.4 kPa); $f_s = 20,000$ lb/in² (137.9 MPa); $u = 0.10f'_c = 300$ lb/in² (2068.5 kPa).

2.   *Compute the design coefficients associated with balanced design*

Thus, $k = 1200/(1200 + 2000) = 0.375$, using Eq. 21. Using Eq. 22, $j = 1 - 0.125 = 0.875$. By Eq. 32, $K = \frac{1}{2}(1200)(0.375)(0.875) = 197$ lb/in² (1358.3 kPa).

3.   *Establish the wheel loads and critical spacing associated with the designated vehicular loading*

As shown in the AASHTO *Specification*, the wheel-load system comprises two loads of 16 kips (71.2 kN) each and one load of 4 kips (17.8 kN). Since the girders are simply supported, an axle spacing of 14 ft (4.3 m) will induce the maximum shear and bending moment in these members.

4.   *Verify that the slab size is adequate and design the reinforcement*

The AASHTO *Specification* does not present moment coefficients for the design of continuous members. The positive and negative reinforcement will be made identical, using straight bars for both. Apply a coefficient of $\frac{1}{10}$ in computing the dead-load moment. The *Specification* provides that the span length S of a slab continuous over more than two supports be taken as the clear distance between supports.

In computing the effective depth, disregard the wearing surface, assume the use of No. 6 bars, and allow 1 in (25.4 mm) for insulation, as required by AASHTO. Then, $d = 6.5 - 0.75 - 1.0 - 0.38 = 4.37$ in (110.998 mm); $w_{DL} = (6.5/12)(150) + 15 = 96$ lb/lin ft (1401 N/m); $M_{DL} = (\frac{1}{10})w_{DL}S^2 = (\frac{1}{10})(96)(4.17)^2 = 167$ ft·lb (226 N·m); $M_{LL} = 0.8(S + 2)P_{20}/32$, by AASHTO, or $M_{LL} = 0.8(6.17)(16,000)/32 = 2467$ ft·lb (3345 N·m). Also by AASHTO, IF = 0.30; $M_{total} =$

$12(167 + 1.30 \times 2467) = 40{,}500$ in·lb (4.6 kN·m). The moment corresponding to balanced design is $M_b = K_b bd^2 = 197(12)(4.37)^2 = 45{,}100$ in·lb (5.1 kN·m). The concrete section is therefore excessive, but a 6-in (152.4-mm) slab would be inadequate. The steel is stressed to capacity at design load. Or, $A_s = 40{,}500/(20{,}000 \times 0.875 \times 4.37) = 0.53$ in^2 (3.4 cm^2). Use No. 6 bars 10 in (254 mm) on centers, top and bottom.

The transverse reinforcement resists the tension caused by thermal effects and by load distribution. By AASHTO, $A_t = 0.67(0.53) = 0.36$ in^2 (2.3 cm^2). Use five No. 5 bars in each panel, for which $A_t = 1.55/4.17 = 0.37$ in^2 (2.4 cm^2).

5. *Calculate the maximum live-load bending moment in the interior girder caused by the moving-load group*

The method of positioning the loads to evaluate this moment is described in an earlier calculation procedure in this handbook. The resultant, Fig. 163, has this location: $d = [16(14) + 4(28)]/(16$

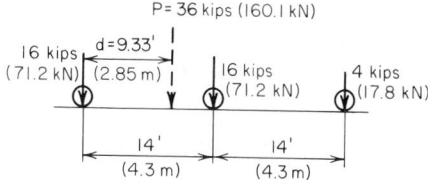

P = 36 kips (160.1 kN)

16 kips
(71.2 kN)   d = 9.33'
(2.85 m)   16 kips
(71.2 kN)   4 kips
(17.8 kN)

14'
(4.3 m)   14'
(4.3 m)

FIG. 163   Load group and its resultant.

$+ 16 + 4) = 9.33$ ft (2.85 m). Place the loads in the position shown in Fig. 164a. The maximum live-load bending moment occurs under the center load.

The AASHTO prescribes a distribution factor of $S/6$ in the present instance, where S denotes the spacing of girders. However, a factor of $S/5$ will be applied here. Then DF $= 5.33/5 = 1.066$; $16 \times 1.066 = 17.06$ kips (75.9 kN); $4 \times 1.066 = 4.26$ kips (18.9 kN); $P = 2(17.06) + 4.26 = 38.38$ kips (170.7 kN); $R_L = 38.38(29.33)/54 = 20.85$ kips (92.7 kN). The maximum live-load moment is $M_{LL} = 20.85(29.34) - 17.06(14) = 372.8$ ft·kips (505 kN·m).

6. *Calculate the maximum live-load shear in the interior girder caused by the moving-load group*

Place the loads in the position shown in Fig. 164b. Do not apply lateral distribution to the load at the support. Then, $V_{LL} = 16 + 17.06(40/54) + 4.26(26/54) = 30.69$ kips (136.5 kN).

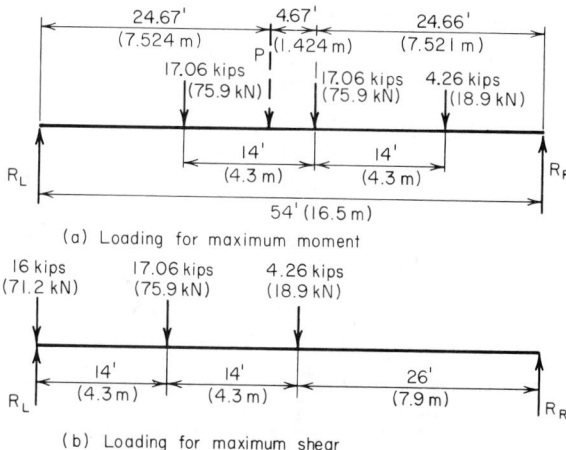

24.67'
(7.524 m)   4.67'
(1.424 m)   24.66'
(7.521 m)

P
17.06 kips
(75.9 kN)   17.06 kips
(75.9 kN)   4.26 kips
(18.9 kN)

R_L 14'
(4.3 m)   14'
(4.3 m) R_R

54' (16.5 m)

(a) Loading for maximum moment

16 kips
(71.2 kN)   17.06 kips
(75.9 kN)   4.26 kips
(18.9 kN)

R_L 14'
(4.3 m)   14'
(4.3 m)   26'
(7.9 m) R_R

(b) Loading for maximum shear

FIG. 164

7. Verify that the size of the girder is adequate and design the reinforcement

Thus, $w_{DL} = 5.33(96) + 14(33.5/144)(150) = 1000$ lb/lin ft (14.6 kN/m); $V_{DL} = 27$ kips (120.1 kN); $M_{DL} = (\frac{1}{8})(1)(54)^2 = 364.5$ ft·kips (494 kN·m). By AASHTO, IF $= 50/(54 + 125) = 0.28$; $V_{total} = 27 + 1.28(30.69) = 66.28$ kips (294.8 kN); $M_{total} = 12(364.5 + 1.28 \times 372.8) = 10,100$ in·kips (1141 N·m).

In establishing the effective depth of the girder, assume that No. 4 stirrups will be supplied and that the main reinforcement will consist of three rows of No. 11 bars. AASHTO requires 1½-in (38.1-mm) insulation for the stirrups and a clear distance of 1 in (25.4 mm) between rows of bars. However, 2 in (50.8 mm) of insulation will be provided in this instance, and the center-to-center spacing of rows will be taken as 2.5 times the bar diameter. Then, $d = 5.75 + 33.5 - 2 - 0.5 - 1.375(0.5 + 2.5) = 32.62$ in (828.548 mm); $v = V/b'jd = 66,280/(14 \times 0.875 \times 32.62) = 166 < 225$ lb/in² (1144.6 < 1551.4 kPa). This is acceptable.

Compute the moment capacity of the girder at balanced design. Since the concrete is poured monolithically, the girder and slab function as a T beam. Refer to Fig. 120 and its calculation procedure.

Thus, $k_b d = 0.375(32.62) = 12.23$ in (310.642 mm); $12.23 - 5.75 = 6.48$ in (164.592 mm). At balanced design, $f_{c1} = 1200(6.48/12.23) = 636$ lb/in² (4835.2 kPa). The effective flange width of the T beam as governed by AASHTO is 64 in (1625.6 mm); and $C_b = 5.75(64)(\frac{1}{2})(1.200 + 0.636) = 338$ kips (1503 kN); $jd = 32.62 - (5.75/3)(1200 + 2 \times 636)/(1200 + 636) = 30.04$ in (763.016 mm); $M_b = 338(30.04) = 10,150$ in·kips (1146 kN·m). The concrete section is therefore slightly excessive, and the steel is stressed to capacity, or $A_s = 10,100/20(30.04) = 16.8$ in² (108.4 cm²). Use 11 no. 11 bars, arranged in three rows.

AASHTO requires that the girders be tied together by diaphragms to obtain lateral rigidity of the structure.

COMPOSITE STEEL-AND-CONCRETE BRIDGE

The bridge shown in cross section in Fig. 165 is to carry an HS20-44 loading on an effective span of 74 ft 6 in (22.7 m). The structure will be unshored during construction. The concrete strength

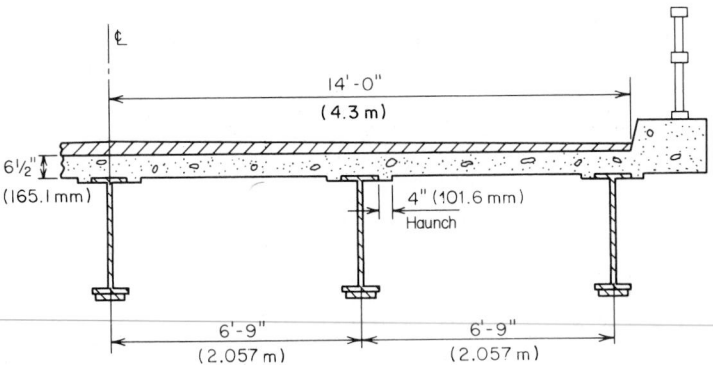

FIG. 165 Transverse section of composite bridge.

is 3000 lb/in² (20,685 kPa), and the entire slab is considered structurally effective; the allowable bending stress in the steel is 18,000 lb/in² (124.1 MPa). The dead load carried by the composite section is 250 lb/lin ft (3648 N/m). Preliminary design calculations indicate that the interior girder is to consist of W36 × 150 and a cover plate 10 × 1½ in (254 × 38.1 mm) welded to the bottom flange. Determine whether the trial section is adequate and complete the design.

Calculation Procedure:

1. Record the relevant properties of the W36 × 150

The design of a composite bridge consisting of a concrete slab and steel girders is governed by specific articles in the AASHTO *Specification.*

Composite behavior of the steel and concrete is achieved by adequately bonding the materials to function as a flexural unit. Loads that are present before the concrete has hardened are supported by the steel member alone; loads that are applied after hardening are supported by the composite member. Thus, the steel alone supports the concrete slab, and the steel and concrete jointly support the wearing surface.

Plastic flow of the concrete under sustained load generates a transfer of compressive stress from the concrete to the steel. Consequently, the stresses in the composite member caused by dead load are analyzed by using a modular ratio three times the value that applies for transient loads.

If a wide-flange shape is used without a cover plate, the neutral axis of the composite section is substantially above the center of the steel, and the stress in the top steel fiber is therefore far below that in the bottom fiber. Use of a cover plate depresses the neutral axis, reduces the disparity between these stresses, and thereby results in a more economical section. Let y' = distance from neutral axis of member to given point, in absolute value; $\bar{y}$ = distance from centroidal axis of WF shape to neutral axis of member. The subscripts b, ts, and tc refer to the bottom of member, top of steel, and top of concrete, respectively. The superscripts c and n refer to the composite and noncomposite member, respectively.

The relevant properties of the W36 × 150 are A = 44.16 in² (284.920 cm²); I = 9012 in⁴ (37.511 dm⁴); d = 35.84 in (910.336 mm); S = 503 in³ (8244.2 cm³); flange thickness = 1 in (25.4 mm), approximately.

2. Compute the section moduli of the noncomposite section where the cover plate is present

To do this, compute the static moment and moment of inertia of the section with respect to the center of the W shape; record the results in Table 11. Refer to Fig. 166: $\bar{y}$ = −280/59.16 = −4.73 in (−120.142 mm); y'_b = 19.42 − 4.73 = 14.69 in (373.126 mm); y'_{ts} = 17.92 + 4.73 = 22.65 in (575.31 mm). By the moment-of-inertia equation, I = 5228 + 9012 − 59.16(4.73)² = 12,916 in⁴ (53.76 dm⁴); S_b = 879 in³ (14,406.8 cm³); S_{ts} = 570 in³ (9342.3 cm³).

3. Transform the composite section, with cover plate included, to an equivalent homogeneous section of steel; compute the section moduli

In accordance with AASHTO, the effective flange width is 12(6.5) = 78 in (1981.2 mm). Using the method of an earlier calculation procedure, we see that when n = 30, $\bar{y}$ = 78/76.06 = 1.03

TABLE 11   Calculations for Girder with Cover Plate

	A	y	Ay	Ay^2	I_o
Noncomposite:					
W36 × 150	44.16	0	0	0	9,012
Cover plate	15.00	−18.67	−280	5,228	0
Total	59.16	. . . . . .	−280	5,228	9,012
Composite, n = 30:					
Steel (total)	59.16	. . . . . .	−280	5,228	9,012
Slab	16.90	21.17	358	7,574	60
Total	76.06	. . . . . .	78	12,802	9,072
Composite, n = 10:					
Steel (total)	59.16	. . . . . .	−280	5,228	9,012
Slab	50.70	21.17	1,073	22,722	179
Total	109.86	. . . . . .	793	27,950	9,191

in (26.162 mm); $y_b' = 19.42 + 1.03 = 20.45$ in (519.43 mm); $y_{ts}' = 17.92 - 1.03 = 16.89$ in (429.006 mm); $y_{tc}' = 16.89 + 6.50 = 23.39$ in (594.106 mm); $I = 12,802 + 9072 - 76.06(1.03)^2 = 21,793$ in^4 (90.709 dm^4); $S_b = 1066$ in^3 (17,471.7 cm^3); $S_{ts} = 1,290$ in^3 (21,143.1 cm^3); $S_{tc} = 932$ in^3 (15,275.5 cm^3).

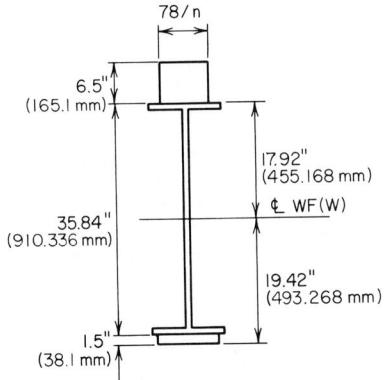

When $n = 10$: $y = 7.22$ in (183.388 mm); $y_b' = 26.64$ in (676.66 mm); $y_{ts}' = 10.70$ in (271.78 mm); $y_{tc}' = 17.20$ in (436.88 mm); $I = 27,950 + 9191 - 109.86(7.22)^2 = 31,414$ in^4 (130.7545 dm^4); $S_b = 1179$ in^3 (19,320.3 cm^3); $S_{ts} = 2936$ in^3 (48,121.0 cm^3); $S_{tc} = 1826$ in^3 (29,928.1 cm^3).

4. Transform the composite section, exclusive of the cover plate, to an equivalent homogeneous section of steel, and compute the values shown below

Thus, when $n = 30$, $y_b' = 23.78$ in (604.012 mm); $y_{ts}' = 12.06$ in (306.324 mm); $I = 14,549$ in^4 (60.557 dm^4); $S_b = 612$ in^3 (10,030.7 cm^3). When $n = 10$, $y_b' = 29.23$ in (742.442 mm); $y_{ts}' = 6.61$ in (167.894 mm); $I = 19,779$ in^4 (82.326 dm^4); $S_b = 677$ in^3 (11,096.0 cm^3).

FIG. 166 Transformed section.

5. Compute the dead load carried by the noncomposite member

Thus,

	lb/lin ft	N/m
Beam	150	2189.1
Cover plate	51	744.3
Slab: 0.54(6.75)(150)	547	7982.8
Haunch: 0.67(0.083)(150)	8	116.8
Diaphragms (approximate)	12	175.1
Shear connectors (approximate)	6	87.6
Total	774, say 780	11,383.2

6. Compute the maximum dead-load moments

Thus, $M_{DL}^c = (\frac{1}{8})(0.250)(74.5)^2(12) = 2080$ in·kips (235.00 kN·m); $M_{DL}^n = (\frac{1}{8})(0.780)(74.5)^2(12) = 6490$ in·kips (733.24 kN·m).

7. Compute the maximum live-load moment, with impact included

In accordance with the AASHTO, the distribution factor is DF $= 6.75/5.5 = 1.23$; IF $= 50/(74.5 + 125) = 0.251$, and $16(1.23)(1.251) = 24.62$ kips (109.510 kN); $4(1.23)(1.251) = 6.15$ kips (270.355 kN); $P_{LL+I} = 2(24.62) + 6.15 = 55.39$ kips (246.375 kN). Refer to Fig. 164a as a guide. Then, $M_{LL+I} = 12[(55.39 \times 39.58 \times 39.58/74.5) - 24.62(14)] = 9840$ in·kips (1111.7 kN·m).

For convenience, the foregoing results are summarized here:

	M, in·kips (kN·m)	S_b, in^3 (cm^3)	S_{ts}, in^3 (cm^3)	S_{tc}, in^3 (cm^3)
Noncomposite	6,490 (733.2)	879 (14,406.8)	570 (9,342.3)	
Composite, dead loads	2,080 (235.0)	1,066 (17,471.7)	1,290 (21,143.1)	932 (15,275.5)
Composite, moving loads	9,840 (1,111.7)	1,179 (19,323.8)	2,936 (48,121.0)	1,826 (29,928.1)

8. Compute the critical stresses in the member

To simplify the calculations, consider the sections of maximum live-load and dead-load stresses to be coincident. Then $f_b = 6490/879 + 2080/1066 + 9840/1179 = 17.68$ kips/in^2 (121.9 MPa); $f_{ts} = 6490/570 + 2080/1290 + 9840/2936 = 16.35$ kips/in^2 (112.7 MPa); $f_{tc} = 2080/(30 \times 932) + 9840/(10 \times 1826) = 0.61$ kips/in^2 (4.21 MPa). The section is therefore satisfactory.

9. Determine the theoretical length of cover plate

Let K denote the theoretical cutoff point at the left end. Let L_c = length of cover plate exclusive of the development length; b = distance from left support to K; $m = L_c/L$; d = distance from heavier exterior load to action line of resultant, as shown in Fig. 163; $r = 2d/L$.

From these definitions, $b = (L - L_c)/2 = L(1 - m)/2; m = 1 - b/(0.5L)$. The maximum moment at K due to live load and impact is

$$M_{LL+I} = \frac{(P_{LL+I}L)(1 - r + rm - m^2)}{4} \tag{72}$$

The diagram of dead-load moment is a parabola having its summit at midspan.

To locate K, equate the bottom-fiber stress immediately to the left of K, where the cover plate is inoperative, to its allowable value. Or, $(P_{LL+I}L)/4 = 55.39(74.5)(12)/4 = 12{,}380$ in·kips (1398.7 kN·m); $d = 9.33$ ft (2.844 m); $r = 18.67/74.5 = 0.251$; $6490(1 - m^2)/503 + 2080(1 - m^2)/612 + 12{,}380(0.749 + 0.251m - m^2)/677 = 18$ kips/in^2 (124.1 MPa); $m = 0.659$; $L_c = 0.659(74.5) = 49.10$ ft (14.97 m).

The plate must be extended toward each support and welded to the W shape to develop its strength.

10. Verify the result obtained in step 9

Thus, $b = \frac{1}{2}(74.5 - 49.10) = 12.70$ ft (3.871 m). At K: $M_{DL}^n = 12(\frac{1}{2} \times 74.5 \times 0.780 \times 12.70 - \frac{1}{2} \times 0.780 \times 12.70^2) = 3672$ in·kips (414.86 kN·m); $M_{DL}^c = 3672(250/780) = 1177$ in·kips (132.98 kN·m). The maximum moment at K due to the moving-load system occurs when the heavier exterior load lies directly at this section. Also $M_{LL+I} = 55.39(74.5 - 12.70 - 9.33)(12.70)(12)/74.5 = 5945$ in·kips (671.7 kN·m); $f_b = 3672/503 + 1177/612 + 5945/677 = 18.0$ kips/in^2 (124.11 MPa). This is acceptable.

11. Compute V_{DL} and V_{LL+I} at the support and at K

At the support $V_{DL}^c = \frac{1}{2}(0.250 \times 74.5) = 9.31$ kips (41.411 kN); IF = 0.251.

Consider that the load at the support is not subject to distribution. By applying the necessary correction, the following is obtained: $V_{LL+I} = 55.39(74.5 - 9.33)/74.5 - 16(1.251)(0.23) = 43.85$ kips (195.045 kN). At K: $V_{DL}^c = 9.31 - 12.70(0.250) = 6.13$ kips (27.266 kN); IF = $50/(61.8 + 125) = 0.268$; $P_{LL+I} = 36(1.268)(1.23) = 56.15$ kips (249.755 kN); $V_{LL+I} = 56.15(74.5 - 12.70 - 9.33)/74.5 = 39.55$ kips (175.918 kN).

12. Select the shear connectors, and determine the allowable pitch p at the support and immediately to the right of K

Assume use of $\frac{3}{4}$-in (19.1-mm) studs, 4 in (101.6 mm) high, with four studs in each transverse row, as shown in Fig. 167. The capacity of a connector as established by AASHTO is $110d^2(f_c')^{0.5} = 110 \times 0.75^2(3000)^{0.5} = 3390$ lb (15,078.7 N). The capacity of a row of connectors = $4(3390) = 13{,}560$ lb (60,314.9 N).

The shear flow at the bottom of the slab is found by applying $q = VQ/I$, or $q_{DL}^c = 9310(16.90)(12.06 + 3.25)/14{,}549 = 166$ lb/lin in (29,071.0 N/m); $q_{LL+I} = 43{,}850(50.70)(6.61 + 3.25)/19{,}779 = 1108$ lb/in^2 (7639.7 kPa); $p = 13{,}560/(166 + 1108) = 10.6$ in (269.24 mm).

Directly to the right of K: $q_{DL}^c = 6130(16.90)(16.89 + 3.25)/21{,}793 = 96$ lb/lin in (16,812.2 N/m); $q_{LL+I} = 39{,}550(50.70)(10.70 + 3.25)/31{,}414 = 890$ lb/lin in (155,862.9 N/m); $p = 13{,}560/(96 + 890) = 13.8$ in (350.52 mm).

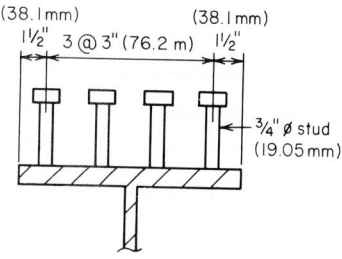

(38.1mm)   (38.1mm)
$1\frac{1}{2}''$ 3 @ 3" (76.2 m) $1\frac{1}{2}''$

$\frac{3}{4}''$ ⌀ stud (19.05 mm)

FIG. 167 Shear connectors.

~~It is necessary to determine the allowable pitch at other sections and to devise a suitable spac-~~ ing of connectors for the entire span.

13.   Design the weld connecting the cover plate to the W shape

The calculations for shear flow are similar to those in step 12. The live-load deflection of an unshored girder is generally far below the limit imposed by AASHTO. However, where an investigation is warranted, the deflection at midspan may be calculated by assuming, for simplicity, that the position of loads for maximum deflection coincides with the position for maximum moment. The theorem of reciprocal deflections, presented in an earlier calculation procedure, may conveniently be applied in calculating this deflection. The girders are usually tied together by diaphragms at the ends and at third points to obtain lateral rigidity of the structure.

Fluid Mechanics

Hydrostatics

The notational system used in hydrostatics is as follows: W = weight of floating body, lb (N); V = volume of displaced liquid, ft^3 (m^3); w = specific weight of liquid, lb/ft^3 (N/m^3); for water w = 62.4 lb/ft^3 (9802 N/m^3), unless another value is specified.

BUOYANCY AND FLOTATION

A timber member 12 ft (3.65 m) long with a cross-sectional area of 90 in^2 (580.7 cm^2) will be used as a buoy in saltwater. What volume of concrete must be fastened to one end so that 2 ft (60.96 cm) of the member will be above the surface? Use these specific weights: timber = 38 lb/ft^3 (5969 N/m^3); saltwater = 64 lb/ft^3 (10,053 N/m^3); concrete = 145 lb/ft^3 (22,777 N/m^3).

Calculation Procedure:

1.   Express the weight of the body and the volume of the displaced liquid in terms of the volume of concrete required

Archimedes' principle states that a body immersed in a liquid is subjected to a vertical buoyant force equal to the weight of the displaced liquid. In accordance with the equations of equilibrium, the buoyant force on a floating body equals the weight of the body. Therefore,

$$W = Vw \tag{73}$$

Let x denote the volume of concrete. Then $W = (90/144)(12)(38) + 145x = 285 + 145x$; $V = (90/144)(12 - 2) + x = 6.25 + x$.

2.   Substitute in Eq. 73 and solve for x

Thus, $285 + 145x = (6.25 + x)64$; $x = 1.42$ ft^3 (0.0402 m^3).

HYDROSTATIC FORCE ON A PLANE SURFACE

In Fig. 168, AB is the side of a vessel containing water, and CDE is a gate located in this plane. Find the magnitude and location of the resultant thrust of the water on the gate when the liquid surface is 2 ft (60.96 cm) above the apex.

Calculation Procedure:

1.   State the equations for the resultant magnitude and position

In Fig. 168, FH denotes the centroidal axis of area CDE that is parallel to the liquid surface, and G denotes the point of application of the resultant force. Point G is termed the *pressure center*.

Let A = area of given surface, ft^2 (cm^2); P = hydrostatic force on given surface, lb (N); p_m = mean pressure on surface, lb/ft^2 (kPa); y_{CA} and y_{PC} = vertical distance from centroidal axis and pressure center, respectively, to liquid surface, ft (m); z_{CA} and z_{PC} = distance along plane of given surface from the centroidal axis and pressure center, respectively, to line of intersection of this plane and the liquid surface, ft (m); I_{CA} = moment of inertia of area with respect to its centroidal axis, ft^4 (m^4).

Consider an elemental surface of area dA at a vertical distance y below the liquid surface. The hydrostatic force dP on this element is normal to the surface and has the magnitude

$$dP = wy\,dA \tag{74}$$

By applying Eq. 74, develop the following equations for the magnitude and position of the resultant force on the entire surface:

$$P = wy_{CA}A \tag{75}$$

$$z_{PC} = \frac{I_{CA}}{Az_{CA}} + z_{CA} \tag{76}$$

2. Compute the required values, and solve the equations in step 1

Thus $A = \frac{1}{2}(5)(6) = 15$ ft^2 (1.39 m^2); $y_{CA} = 2 + 4 \sin 60° = 5.464$ ft (166.543 cm); $z_{CA} = 2$ csc $60° + 4 = 6.309$ ft (192.3 cm); $I_{CA}/A = (bd^3/36)/(bd/2) = d^2/18 = 2$ ft^2 (0.186 m^2); $P = 62.4(5.464)(15) = 5114$ lb (22,747 N); $z_{PC} = 2/6.309 + 6.309 = 6.626$ ft (201.960 cm); $y_{PC} = 6.626 \sin 60° = 5.738$ ft (174.894 cm). By symmetry, the pressure center lies on the centroidal axis through C.

An alternative equation for P is

$$P = p_m A \tag{77}$$

Equation 75 shows that the mean pressure occurs at the centroid of the area. The above two steps constitute method 1 for solving this problem. The next three steps constitute method 2.

3. Now construct the pressure "prism" associated with the area

In Fig. 169, construct the pressure prism associated with area CDE. The pressures are as follows: at apex, $p = 2w$; at base, $p = (2 + 6 \sin 60°)w = 7.196w$.

The force P equals the volume of this prism, and its action line lies on the centroidal plane parallel to the base. For convenience, resolve this prism into a triangular prism and rectangular pyramid, as shown.

4. Determine P by computing the volume of the pressure prism

Thus, $P = Aw[2 + \frac{2}{3}(5.196)] = Aw(2 + 3.464) = 15(62.4)(5.464) = 5114$ lb (22,747 N).

5. Find the location of the resultant thrust

Compute the distance h from the top line to the centroidal plane. Then find y_{PC}. Or, $h = [2(\frac{2}{3})(6) + 3.464(\frac{3}{4})(6)]/5.464 = 4.317$ ft (131.582 cm); $y_{PC} = 2 + 4.317 \sin 60° = 5.738$ ft (174.894 cm).

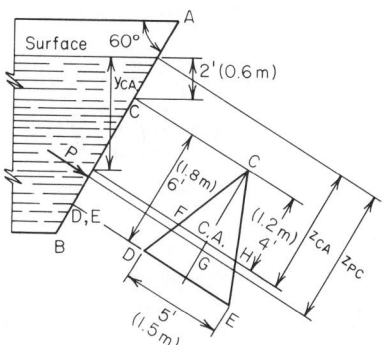

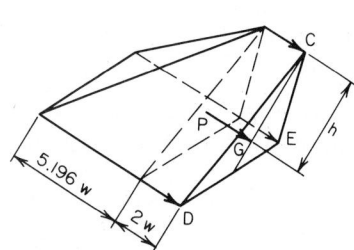

FIG. 168 Hydrostatic thrust on plane surface.   **FIG. 169** Pressure prism.

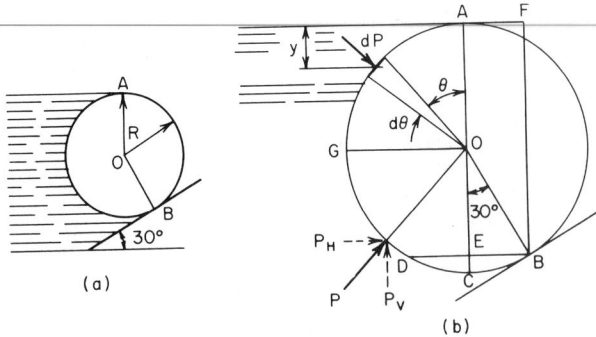

FIG. 170

HYDROSTATIC FORCE ON A CURVED SURFACE

The cylinder in Fig. 170a rests on an inclined plane and is immersed in liquid up to its top, as shown. Find the hydrostatic force on a 1-ft (30.48-cm) length of cylinder in terms of w and the radius R; locate the pressure center.

Calculation Procedure:

1. Evaluate the horizontal and vertical component of the force dP on an elemental surface having a central angle dθ

Refer to Fig. 170b. Adopt this sign convention: A horizontal force is positive if directed to the right; a vertical force is positive if directed upward. The first three steps constitute method 1.

Evaluating dP yields $dP_H = wR^2(\sin\theta - \sin\theta\cos\theta)\,d\theta$; $dP_V = wR^2(-\cos\theta + \cos^2\theta)\,d\theta$.

2. Integrate these equations to obtain the resultant forces P$_H$ and P$_V$; then find P

Here, $P_H = wR^2(-\cos\theta + \frac{1}{2}\cos^2\theta)]_0^{7\pi/6} = wR^2[-(-0.866-1) + \frac{1}{2}(0.75-1)] = 1.741wR^2$, to right; $P_V = wR^2(-\sin\theta + \frac{1}{2}\theta + \frac{1}{4}\sin 2\theta)]_0^{7\pi/6} = wR^2(0.5 + 1.833 + 0.217) = 2.550wR^2$, upward; $P = wR^2(1.741^2 + 2.550^2)^{0.5} = 3.087wR^2$.

3. Determine the value of θ at the pressure center

Since each elemental force dP passes through the center of the cylinder, the resultant force P also passes through the center. Thus, $\tan(180° - \theta_{PC}) = P_H/P_V = 1.741/2.550$; $\theta_{PC} = 145°41'$.

4. Evaluate P$_H$ and P$_V$

Apply these principles: P_H = force on an imaginary surface obtained by projecting the wetted surface on a vertical plane; $P_V = \pm$ weight of real or imaginary liquid lying between the wetted surface and the liquid surface. Use the plus sign if the *real* liquid lies below the wetted surface and the minus sign if it lies above this surface.

Then P_H = force, to right, on AC + force, to left, on EC = force, to right, on AE; $AE = 1.866R$; $p_m = 0.933\,wR$; $P_H = 0.933\,wR(1.866R) = 1.741\,wR^2$; P_V = weight of imaginary liquid above GCB − weight of real liquid above GA = weight of imaginary liquid in cylindrical sector $AOBG$ and in prismoid $AOBF$. Volume of sector $AOBG = [(7\pi/6)/(2\pi)](\pi R^2) = 1.833R^2$; volume of prismoid $AOBF = \frac{1}{2}(0.5R)(R + 1.866R) = 0.717R^2$; $P_V = wR^2(1.833 + 0.717) = 2.550wR^2$.

STABILITY OF A VESSEL

The boat in Fig. 171 is initially floating upright in freshwater. The total weight of the boat and cargo is 182 long tons (1813 kN); the center of gravity lies on the longitudinal (i.e., the fore-and-aft) axis of the boat and 8.6 ft (262.13 cm) above the bottom. A wind causes the boat to list through an angle of 6° while the cargo remains stationary relative to the boat. Compute the righting or upsetting moment (*a*) without applying any set equation; (*b*) by applying the equation for meta-centric height.

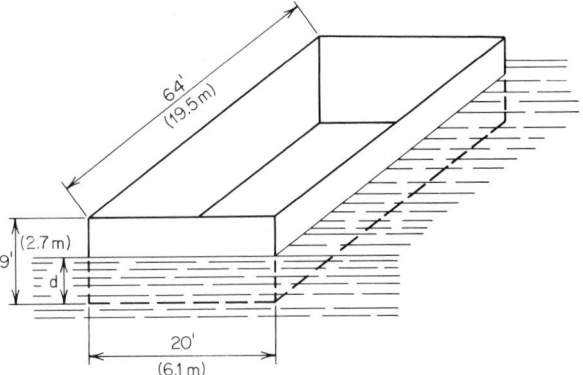

FIG. 171

Calculation Procedure:

1. Compute the displacement volume and draft when the boat is upright

The buoyant force passes through the center of gravity of the displaced liquid; this point is termed the *center of buoyancy*. Figure 172 shows the cross section of a boat rotated through an angle ϕ. The center of buoyancy for the upright position is B; B' is the center of buoyancy for the position shown, and G is the center of gravity of the boat and cargo.

In the position indicated in Fig. 172, the weight W and buoyant force R constitute a couple that tends to restore the boat to its upright position when the disturbing force is removed; their moment is therefore termed *righting*. When these forces constitute a couple that increases the rotation, their moment is said to be *upsetting*. The wedges OAC and $OA'C'$ are termed the *wedge of emersion* and *wedge of immersion*, respectively. Let h = horizontal displacement of center of buoyancy caused by rotation; h' = horizontal distance between centroids of wedge of emersion and wedge of immersion; V' = volume of wedge of emersion (or immersion). Then

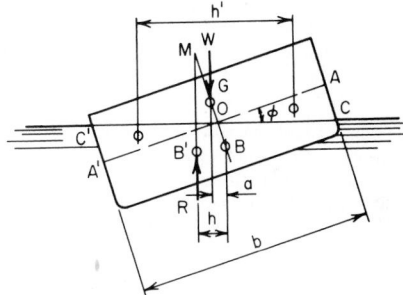

FIG. 172 Location of resultant forces on inclined vessel.

$$h = \frac{V'h'}{V} \qquad (78)$$

The displacement volume and the draft when the boat is upright are $W = 182(2240) = 407,700$ lb (1813 N); $V = W/w = 407,700/62.4 = 6530$ ft³ (184.93 m³); $d = 6530/[64(20)] = 5.10$ ft (155.448 cm).

2. Find h, using Eq. 78

Since ϕ is relatively small, apply this approximation: $h' = 2b/3 = 2(20)/3 = 13.33$ ft (406.298 cm), $h = \frac{1}{2}(10)(10 \tan 6°)(13.33)/[5.10(20)] = 0.687$ ft (20.940 cm).

3. Compute the horizontal distance a (Fig. 172)

Thus, $BG = 8.6 - \frac{1}{2}(5.10) = 6.05$ ft (184.404 cm); $a = 6.05 \sin 6° = 0.632$ ft (19.263 cm).

4. Compute the moment of the vertical forces

Thus, $M = W(h - a) = 407,700(0.055) = 22,400$ ft·lb (30,374.4 N·m). Since $h > a$, the moment is righting. This constitutes the solution to part a. The remainder of this procedure is concerned with part b.

In Fig. 172, let M denote the point of intersection of the vertical line through B' and the line

~~BG prolonged. Then~~ M ~~is termed the~~ *metacenter* associated with this position, and the distance GM is called the *metacentric height*. Also BG is positive if G is above B, and GM is positive if M is above G. Thus, the moment of vertical forces is righting or upsetting depending on whether the metacentric height is positive or negative, respectively.

5.   *Find the lever arm of the vertical forces*

Use the relation for metacentric height:

$$GM = \frac{I_{WL}}{V \cos \phi} - BG \tag{79}$$

where I_{WL} = moment of inertia of original waterline section about axis through O. Or, $I_{WL} = (\frac{1}{12})(64)(20)^3 = 42,670$ ft^4 (368.3 m^4); $GM = 42,670/6530 \cos 6° - 6.05 = 0.52$ ft (15.850 cm); $h - a = 0.52 \sin 6° = 0.054$ ft (1.646 cm), which agrees closely with the previous result.

Mechanics of Incompressible Fluids

The notational system is a = acceleration; A = area of stream cross section; C = discharge coefficient; D = diameter of pipe or depth of liquid in open channel; F = force; g = gravitational acceleration; H = total head, or total specific energy; h_F = loss of head between two sections caused by friction; h_L = total loss of head between two sections; h_V = difference in velocity heads at two sections if no losses occur; L = length of stream between two sections; M = mass of body; N_R = Reynolds number; p = pressure; Q = volumetric rate of flow, or discharge; s = hydraulic gradient = $-dH/dL$; T = torque; V = velocity; w = specific weight; z = elevation above datum plane; ρ = density (mass per unit volume); μ = dynamic (or absolute) viscosity; ν = kinematic viscosity = μ/ρ; τ = shearing stress. The units used for each symbol are given in the calculation procedure where the symbol is used.

If the discharge of a flowing stream of liquid remains constant, the flow is termed *steady*. Let subscripts 1 and 2 refer to cross sections of the stream, 1 being the upstream section. From the definition of steady flow,

$$Q = A_1 V_1 = A_2 V_2 = \text{constant} \tag{80}$$

This is termed the *equation of continuity*. Where no statement is made to the contrary, it is understood that the flow is steady.

Conditions at two sections may be compared by applying the following equation, which is a mathematical statement of Bernoulli's theorem:

$$\frac{V_1^2}{2g} + \frac{p_1}{w} + z_1 = \frac{V_2^2}{2g} + \frac{p_2}{w} + z_2 + h_L \tag{81}$$

The terms on each side of this equation represent, in their order of appearance, the *velocity head*, *pressure head*, and *potential head* of the liquid. Alternatively, they may be considered to represent forms of specific energy, namely, kinetic, pressure, and potential energy.

The force causing a change in velocity is evaluated by applying the basic equation

$$F = Ma \tag{82}$$

Consider that liquid flows from section 1 to section 2 in a time interval t. At any instant, the volume of liquid bounded by these sections is Qt. The force required to change the velocity of this body of liquid from V_1 to V_2 is found from: $M = Qwt/g$; $a = (V_2 - V_1)/t$. Substituting in Eq. (82) gives $F = Qw(V_2 - V_1)/g$, or

$$F = \frac{A_1 V_1 w(V_2 - V_1)}{g} = \frac{A_2 V_2 w(V_2 - V_1)}{g} \tag{83}$$

VISCOSITY OF FLUID

Two horizontal circular plates 9 in (228.6 mm) in diameter are separated by an oil film 0.08 in (2.032 mm) thick. A torque of 0.25 ft·lb (0.339 N·m) applied to the upper plate causes that plate to rotate at a constant angular velocity of 4 revolutions per second (r/s) relative to the lower plate. Compute the dynamic viscosity of the oil.

Calculation Procedure:

1. Develop equations for the force and torque

Consider that the fluid film in Fig. 173a is in motion and that a fluid particle at boundary A has a velocity dV relative to a particle at B. The shearing stress in the fluid is

$$\tau = \mu \frac{dV}{dx} \tag{84}$$

Figure 173b shows a cross section of the oil film, the shaded portion being an elemental surface. Let m = thickness of film; R = radius of plates; ω = angular velocity of one plate relative

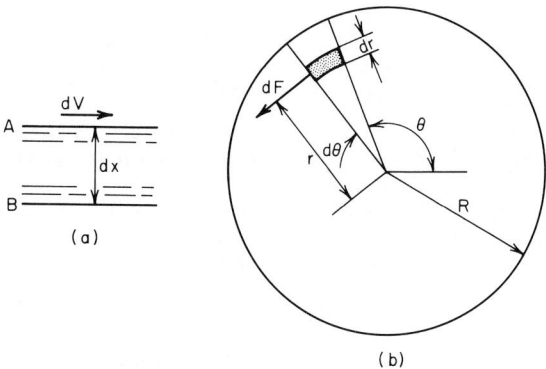

FIG. 173

to the other; dA = area of elemental surface; dF = shearing force on elemental surface; dT = torque of dF with respect to the axis through the center of the plate.

Applying Eq. 84, develop these equations: $dF = 2\pi\omega\mu r^2 \, dr \, d\theta/m$; $dT = r \, dF = 2\pi\omega\mu r^2 \, dr \, d\theta/m$.

2. Integrate the foregoing equation to obtain the resulting torque; solve for μ

Thus,

$$\mu = \frac{Tm}{\pi^2\omega R^4} \tag{85}$$

$T = 0.25$ ft·lb (0.339 N·m); $m = 0.08$ in (2.032 mm); $\omega = 4 \ r/s$; $R = 4.5$ in (114.3 mm); $\mu = 0.25(0.08)(12)^3/[\pi^2(4)(4.5)^4] = 0.00214$ lb·s/ft^2 (0.1025 N·s/m^2).

APPLICATION OF BERNOULLI'S THEOREM

A steel pipe is discharging 10 ft^3/s (283.1 L/s) of water. At section 1, the pipe diameter is 12 in (304.8 mm), the pressure is 18 lb/in^2 (124.11 kPa), and the elevation is 140 ft (42.67 m). At section 2, farther downstream, the pipe diameter is 8 in (203.2 mm), and the elevation is 106 ft (32.31 m). If there is a head loss of 9 ft (2.74 m) between these sections due to pipe friction, what is the pressure at section 2?

Calculation Procedure:

1. Tabulate the given data

Thus $D_1 = 12$ in (304.8 mm); $D_2 = 8$ in (203.2 mm); $p_1 = 18$ lb/in^2 (124.11 kPa); $p_2 = ?$; $z_1 = 140$ ft (42.67 m); $z_2 = 106$ ft (32.31 m).

2. Compute the velocity at each section

Applying Eq. 80 gives $V_1 = 10/0.785 = 12.7$ ft/s (387.10 cm/s); $V_2 = 10/0.349 = 28.7$ ft/s (874.78 cm/s).

3. Compute p_2 by applying Eq. 81

Thus, $(p_2 - p_1)/w = (V_1^2 - V_2^2)/(2g) + z_1 - z_2 - h_F = (12.7^2 - 28.7^2)/64.4 + 140 - 106 - 9 = 14.7$ ft (448.06 cm); $p_2 = 14.7(62.4)/144 + 18 = 24.4$ lb/in^2 (168.24 kPa).

FLOW THROUGH A VENTURI METER

A venturi meter of 3-in (76.2-mm) throat diameter is inserted in a 6-in (152.4-mm) diameter pipe conveying fuel oil having a specific gravity of 0.94. The pressure at the throat is 10 lb/in^2 (68.95 kPa), and that at an upstream section 6 in (152.4 mm) higher than the throat is 14.2 lb/in^2 (97.91 kPa). If the discharge coefficient of the meter is 0.97, compute the flow rate in gallons per minute (liters per second).

Calculation Procedure:

1. Record the given data, assigning the subscript 1 to the upstream section and 2 to the throat

The loss of head between two sections can be taken into account by introducing a *discharge coefficient* C. This coefficient represents the ratio between the actual discharge Q and the discharge Q_i that would occur in the absence of any losses. Then $Q = CQ_i$, or $(V_2^2 - V_1^2)/(2g) = C^2 h_V$.

Record the given data: $D_1 = 6$ in (152.4 mm); $p_1 = 14.2$ lb/in^2 (97.91 kPa); $z_1 = 6$ in (152.4 mm); $D_2 = 3$ in (76.2 mm); $p_2 = 10$ lb/in^2 (68.95 kPa); $z_2 = 0$; $C = 0.97$.

2. Express V_1 in terms of V_2 and develop velocity and flow relations

Thus,

$$V_2 = C \left[\frac{2gh_V}{1 - (A_2/A_1)^2} \right]^{0.5} \tag{86a}$$

Also

$$Q = CA_2 \left[\frac{2gh_V}{1 - (A_2/A_1)^2} \right]^{0.5} \tag{86b}$$

If V_1 is negligible, these relations reduce to

$$V_2 = C(2gh_V)^{0.5} \tag{87a}$$

and

$$Q = CA_2(2gh_V)^{0.5} \tag{87b}$$

3. Compute h_V by applying Eq. 81

Thus, $h_V = (p_1 - p_2)/w + z_1 - z_2 = 4.2(144)/[0.94(62.4)] + 0.5 = 10.8$ ft (3.29 m).

4. Compute Q by applying Eq. 86b

Thus, $(A_2/A_1)^2 = (D_2/D_1)^4 = \frac{1}{16}$; $A_2 = 0.0491$ ft^2 (0.00456 m^2); and $Q = 0.97(0.0491)[64.4 \times 10.8/(1 - \frac{1}{16})]^{0.5} = 1.30$ ft^3/s or, by using the conversion factor of 1 ft^3/s = 449 gal/min (28.32 L/s), the flow rate is $1.30(449) = 584$ gal/min (36.84 L/s).

FLOW THROUGH AN ORIFICE

Compute the discharge through a 3-in (76.2-mm) diameter square-edged orifice if the water on the upstream side stands 4 ft 8 in (1.422 m) above the center of the orifice.

Calculation Procedure:

1. Determine the discharge coefficient

For simplicity, the flow through a square-edged orifice discharging to the atmosphere is generally computed by equating the area of the stream to the area of the opening and then setting the discharge coefficient $C = 0.60$ to allow for contraction of the issuing stream. (The area of the issuing stream is about 0.62 times that of the opening.)

2. Compute the flow rate

Since the velocity of approach is negligible, use Eq. 87b. Or, $Q = 0.60(0.0491)(64.4 \times 4.67)^{0.5}$ $= 0.511 \text{ ft}^3/\text{s}$ (14.4675 L/s).

FLOW THROUGH THE SUCTION PIPE OF A DRAINAGE PUMP

Water is being evacuated from a sump through the suction pipe shown in Fig. 174. The entrance-end diameter of the pipe is 3 ft (91.44 cm); the exit-end diameter, 1.75 ft (53.34 cm). The exit pressure is 12.9 in (32.77 cm) of mercury vacuum. The head loss at the entry is one-fifteenth of the velocity head at that point, and the head loss in the pipe due to friction is one-tenth of the velocity head at the exit. Compute the discharge flow rate.

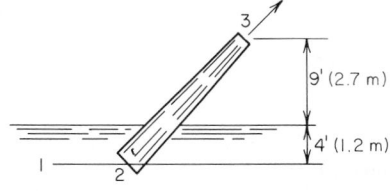

FIG. 174

Calculation Procedure:

1. Convert the pressure head to feet of water

The discharge may be found by comparing the conditions at an upstream point 1, where the velocity is negligible, with the conditions at point 3 (Fig. 174). Select the elevation of point 1 as the datum.

Converting the pressure head at point 3 to feet of water and using the specific gravity of mercury as 13.6, we have $p_3/w = -(12.9/12)13.6 = -14.6 \text{ ft}$ (−4.45 m).

2. Express the velocity head at 2 in terms of that at 3

By the equation of continuity, $V_2 = A_3V_3/A_2 = (1.75/3)^2V_3 = 0.34V_3$.

3. Evaluate V_3 by applying Eq. 81; then determine Q

Thus, $V_1^2/(2g) + p_1/w + z_1 = V_3^2/(2g) + p_3/w + z_3 + (1/15)V_2^2/(2g) + (1/10)V_3^2/(2g)$, or $0 + 4 + 0 = V_3^2/(2g) - 14.6 + 13 + [V_3^2/(2g)](1/15 \times 0.34^2 + 1/10)$; $V_3 = 18.0 \text{ ft/s}$ (548.64 cm/s); then $Q_3 = A_3V_3 = 0.785(1.75)^2(18.0) = 43.3 \text{ ft}^3/\text{s}$ (1225.92 L/s).

POWER OF A FLOWING LIQUID

A pump is discharging 8 ft³/s (226.5 L/s) of water. Gages attached immediately upstream and downstream of the pump indicate a pressure differential of 36 lb/in² (248.2 kPa). If the pump efficiency is 85 percent, what is the horsepower output and input?

Calculation Procedure:

1. Evaluate the increase in head of the liquid

Power is the rate of performing work, or the amount of work performed in a unit time. If the fluid flows with a specific energy H, the total energy of the fluid discharged in a unit time is

~~QwH.~~ This expression thus represents the work that the flowing fluid can perform in a unit time and therefore the power associated with this discharge. Since 1 hp = 550 ft·lb/s,

$$1 \text{ hp} = \frac{QwH}{550} \tag{88}$$

In this situation, the power developed by the pump is desired. Therefore, H must be equated to the specific energy added by the pump.

To evaluate the increase in head, consider the differences of the two sections being considered. Since both sections have the same velocity and elevation, only their pressure heads differ. Thus, $p_2/w - p_1/w = 36(144)/62.4 = 83.1$ ft (2532.89 cm).

2. Compute the horsepower output and input

Thus, $\text{hp}_{\text{out}} = 8(62.4)(83.1)/550 = 75.4$ hp; $\text{hp}_{\text{in}} = 75.4/0.85 = 88.7$ hp.

DISCHARGE OVER A SHARP-EDGED WEIR

Compute the discharge over a sharp-edged rectangular weir 4 ft (121.9 cm) high and 10 ft (304.8 cm) long, with two end contractions, if the water in the canal behind the weir is 4 ft 9 in (144.78 cm) high. Disregard the velocity of approach.

Calculation Procedure:

1. Adopt a standard relation for this weir

The discharge over a sharp-edged rectangular weir without end contractions in which the velocity of approach is negligible is given by the Francis formula as

$$Q = 3.33bh^{1.5} \tag{89a}$$

where b = length of crest and h = head on weir, i.e., the difference between the elevation of the crest and that of the water surface upstream of the weir.

2. Modify the Francis equation for end contractions

With two end contractions, the discharge of the weir is

$$Q = 3.33(b - 0.2h)h^{1.5} \tag{89b}$$

Substituting the given values yields $Q = 3.33(10 - 0.2 \times 0.75)0.75^{1.5} = 21.3$ ft^3/s (603.05 L/s).

LAMINAR FLOW IN A PIPE

A tank containing crude oil discharges 340 gal/min (21.4 L/s) through a steel pipe 220 ft (67.1 m) long and 8 in (203.2 mm) in diameter. The kinematic viscosity of the oil is 0.002 ft^2/s (1.858 cm^2/s). Compute the difference in elevation between the liquid surface in the tank and the pipe outlet.

Calculation Procedure:

1. Identify the type of flow in the pipe

To investigate the discharge in a pipe, it is necessary to distinguish between two types of fluid flow—*laminar* and *turbulent*. Laminar (or *viscous*) flow is characterized by the telescopic sliding of one circular layer of fluid past the adjacent layer, each fluid particle traversing a straight line. The velocity of the fluid flow varies parabolically from zero at the pipe wall to its maximum value at the pipe center, where it equals twice the mean velocity.

Turbulent flow is characterized by the formation of eddy currents, with each fluid particle traversing a sinuous path.

In any pipe the type of flow is ascertained by applying a dimensionless index termed the *Reynolds number,* defined as

$$N_R = \frac{DV}{\nu} \tag{90}$$

Flow is considered laminar if $N_R < 2100$ and turbulent if $N_R > 3000$.

In laminar flow the head loss due to friction is

$$h_F = \frac{32L\nu V}{gD^2} \tag{91a}$$

or

$$h_F = \left(\frac{64}{N_R}\right)\left(\frac{L}{D}\right)\left(\frac{V^2}{2g}\right) \tag{91b}$$

Let 1 denote a point on the liquid surface and 2 a point at the pipe outlet. The elevation of 2 will be taken as datum.

To identify the type of flow, compute N_R. Thus, $D = 8$ in (203.2 mm); $L = 220$ ft (6705.6 cm); $\nu = 0.002$ ft^2/s (1.858 cm^2/s); $Q = 340/449 = 0.757$, converting from gallons per minute to cubic feet per second. Then $V = Q/A = 0.757/0.349 = 2.17$ ft/s (66.142 cm/s). And $N_R = 0.667(2.17)/0.002 = 724$. Therefore, the flow is laminar because N_R is less than 2100.

2. *Express all losses in terms of the velocity head*

By Eq. 91b, $h_F = (64/724)(220/0.667)V^2/(2g) = 29.2V^2/(2g)$. Where $L/D > 500$, the following may be regarded as negligible in comparison with the loss due to friction: loss at pipe entrance, losses at elbows, velocity head at the discharge, etc. In this instance, include the secondary items. The loss at the pipe entrance is $h_E = 0.5V^2/(2g)$. The total loss is $h_L = 29.7V^2/(2g)$.

3. *Find the elevation of 1 by applying Eq. 81*

Thus, $z_1 = V_2^2/(2g) + h_L = 30.7V_2^2/(2g) = 30.7(2.17)^2/64.4 = 2.24$ ft (68.275 cm).

TURBULENT FLOW IN PIPE—APPLICATION OF DARCY-WEISBACH FORMULA

Water is pumped at the rate of 3 ft^3/s (85.0 L/s) through an 8-in (203.2-mm) fairly smooth pipe 2600 ft (792.48 m) long to a reservoir where the water surface is 180 ft (50.86 m) higher than the pump. Determine the gage pressure at the pump discharge.

Calculation Procedure:

1. *Compute h_F*

Turbulent flow in a pipe flowing full may be investigated by applying the Darcy-Weisbach formula for friction head

$$h_F = \frac{fLV^2}{2gD} \tag{92}$$

where f is a friction factor. However, since the friction head does not vary precisely in the manner implied by this equation, f is dependent on D and V, as well as the degree of roughness of the pipe. Values of f associated with a given set of values of the independent quantities may be obtained from Fig. 175.

Accurate equations for h_F are the following:
Extremely smooth pipes:

$$h_F = \frac{0.30LV^{1.75}}{1000D^{1.25}} \tag{93a}$$

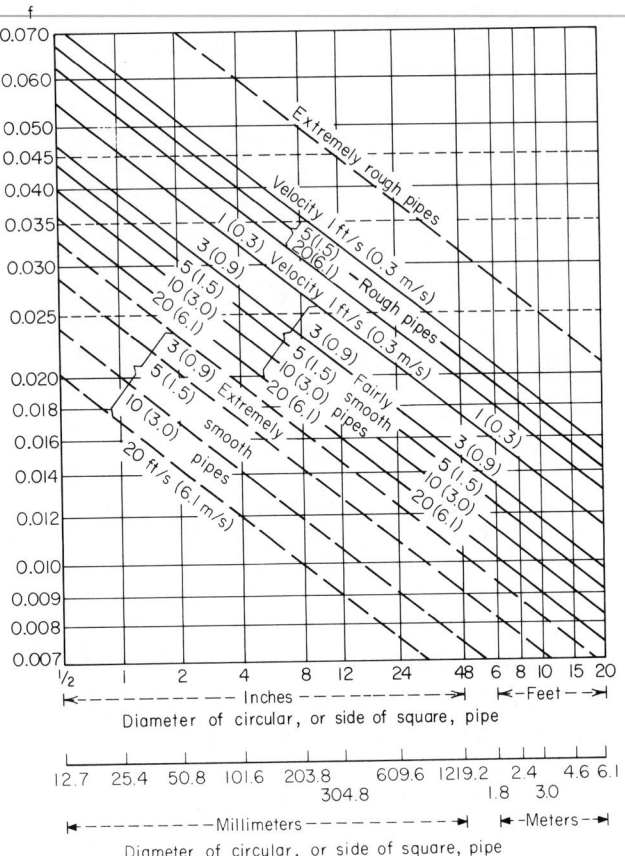

FIG. 175 Flow of water in pipes. *(From E. W. Schoder and F. M. Dawson, Hydraulics, McGraw-Hill Book Company, New York, 1934. By permission of the publishers.)*

Fairly smooth pipes:

$$h_F = \frac{0.38 L V^{1.86}}{1000 D^{1.25}} \tag{93b}$$

Rough pipes:

$$h_F = \frac{0.50 L V^{1.95}}{1000 D^{1.25}} \tag{93c}$$

Extremely rough pipes:

$$h_F = \frac{0.69 L V^{2}}{1000 D^{1.25}} \tag{93d}$$

Using Eq. 93b gives $V = Q/A = 3/0.349 = 8.60$ ft/s (262.128 cm/s); $h_F = 0.38(2.6)(8.60)^{1.86}/0.667^{1.25} = 89.7$ ft (27.34 m).

2.   *Alternatively, determine h_F using Eq. 92*

First obtain the appropriate f value from Fig. 175, or $f = 0.020$ for this pipe. Then $h_F = 0.020(2{,}600/0.667)(8.60^2/64.4) = 89.6$ ft (27.31 m).

3.   *Compute the pressure at the pump discharge*

Use Eq. 81. Since $L/D > 500$, ignore the secondary items. Then $p_1/w = z_2 + h_F = 180 + 89.6 = 269.6$ ft (82.17 m); $p_1 = 269.6(62.4)/144 = 117$ lb/in^2 (806.7 kPa).

DETERMINATION OF FLOW IN A PIPE

Two reservoirs are connected by a 7000-ft (2133.6-m) fairly smooth cast-iron pipe 10 in (254.0 mm) in diameter. The difference in elevation of the water surfaces is 90 ft (27.4 m). Compute the discharge to the lower reservoir.

Calculation Procedure:

1.   *Determine the fluid velocity and flow rate*

Since the secondary items are negligible, the entire head loss of 90 ft (27.4 m) results from friction. Using Eq. 93b and solving for V, we have $90 = 0.38(7)V^{1.86}/0.833^{1.25}$; $V = 5.87$ ft/s (178.918 cm/s). Then $Q = VA = 5.87(0.545) = 3.20$ ft^3/s (90.599 L/s).

2.   *Alternatively, assume a value of f and compute V*

Referring to Fig. 175, select a value for f. Then compute V by applying Eq. 92. Next, compare the value of f corresponding to this result with the assumed value of f. If the two values differ appreciably, assume a new value of f and repeat the computation. Continue this process until the assumed and actual values of f agree closely.

PIPE-SIZE SELECTION BY THE MANNING FORMULA

A cast-iron pipe is to convey water at 3.3 ft^3/s (93.430 L/s) on a grade of 0.001. Applying the Manning formula with $n = 0.013$, determine the required size of pipe.

Calculation Procedure:

Compute the pipe diameter

The Manning formula, which is suitable for both open and closed conduits, is

$$V = \frac{1.486R^{2/3}s^{1/2}}{n} \tag{94}$$

where n = roughness coefficient; R = hydraulic radius = ratio of cross-sectional area of pipe to the wetted perimeter of the pipe; s = hydraulic gradient = dH/dL. If the flow is uniform, i.e., the area and therefore the velocity are constant along the stream, then the loss of head equals the drop in elevation, and the grade of the conduit is s.

For a circular pipe flowing full, Eq. 94 becomes

$$D = \left(\frac{2.159Qn}{s^{1/2}}\right)^{3/8} \tag{94a}$$

Substituting numerical values gives $D = (2.159 \times 3.3 \times 0.013/0.001^{1/2})^{3/8} = 1.50$ ft (45.72 cm). Therefore, use an 18-in (457.2-mm) diameter pipe.

LOSS OF HEAD CAUSED BY SUDDEN ENLARGEMENT OF PIPE

Water flows through a pipe at 4 ft^3/s (113.249 L/s). Compute the loss of head resulting from a change in pipe size if (a) the pipe diameter increases abruptly from 6 to 10 in (152.4 to 254.0 mm); (b) the pipe diameter increases abruptly from 6 to 8 in (152.4 to 203.2 mm) at one section and then from 8 to 10 in (203.2 to 254.0 mm) at a section farther downstream.

Calculation Procedure:

1. Evaluate the pressure-head differential required to decelerate the liquid

Where there is an abrupt increase in pipe size, the liquid must be decelerated upon entering the larger pipe, since the fluid velocity varies inversely with area. Let subscript 1 refer to a section immediately downstream of the enlargement, where the higher velocity prevails, and let subscript 2 refer to a section farther downstream, where deceleration has been completed. Disregard the frictional loss.

Using Eq. 83 we see $p_2/w = p_1/w + (V_1V_2 - V_2^2)/g$.

2. Combine the result of step 1 with Eq. 81

The result is Borda's formula for the head loss h_E caused by sudden enlargement of the pipe cross section:

$$h_E = \frac{(V_1 - V_2)^2}{2g} \tag{95}$$

As this investigation shows, only part of the drop in velocity head is accounted for by a gain in pressure head. The remaining head h_E is dissipated through the formation of eddy currents at the entrance to the larger pipe.

3. Compute the velocity in each pipe

Thus

Pipe diam, in (mm)	Pipe area, ft^2 (m^2)	Fluid velocity, ft/s (cm/s)
6 (152.4)	0.196 (0.0182)	20.4 (621.79)
8 (203.2)	0.349 (0.0324)	11.5 (350.52)
10 (254.0)	0.545 (0.0506)	7.3 (222.50)

4. Find the head loss for part a

Thus, $h_E = (20.4 - 7.3)^2/64.4 = 2.66$ ft (81.077 cm).

5. Find the head loss for part b

Thus, $h_E = [(20.4 - 11.5)^2 + (11.5 - 7.3)^2]/64.4 = 1.50$ ft (45.72 cm). Comparison of these results indicates that the eddy-current loss is attenuated if the increase in pipe size occurs in steps.

DISCHARGE OF LOOPING PIPES

A pipe carrying 12.5 ft^3/s (353.90 L/s) of water branches into three pipes of the following diameters and lengths; $D_1 = 6$ in (152.4 mm); $L_1 = 1000$ ft (304.8 m); $D_2 = 8$ in (203.2 mm); $L_2 = 1300$ ft (396.2 m); $D_3 = 10$ in (254.0 mm); $L_3 = 1200$ ft (365.8 m). These pipes rejoin at their downstream ends. Compute the discharge in the three pipes, considering each as fairly smooth.

Calculation Procedure:

1. Express Q as a function of D and L

Since all fluid particles have the same energy at the juncture point, irrespective of the loops they traversed, the head losses in the three loops are equal. The flow thus divides itself in a manner that produces equal values of h_F in the loops.

Transforming Eq. 93b,

$$Q = \frac{kD^{2.67}}{L^{0.538}} \tag{96}$$

where k is a constant.

2. Establish the relative values of the discharges; then determine the actual values

Thus, $Q_2/Q_1 = (8/6)^{2.67}/1.3^{0.538} = 1.87$; $Q_3/Q_1 = (10/6)^{2.67}/1.2^{0.538} = 3.55$. Then $Q_1 + Q_2 + Q_3 = Q_1(1 + 1.87 + 3.55) = 12.5$ ft³/s (353.90 L/s). Solving gives $Q_1 = 1.95$ ft³/s (55.209 L/s): $Q_2 = 3.64$ ft³/s (103.056 L/s); $Q_3 = 6.91$ ft³/s (195.637 L/s).

FLUID FLOW IN BRANCHING PIPES

The pipes AM, MB, and MC in Fig. 176 have the diameters and lengths indicated. Compute the water flow in each pipe if the pipes are considered rough.

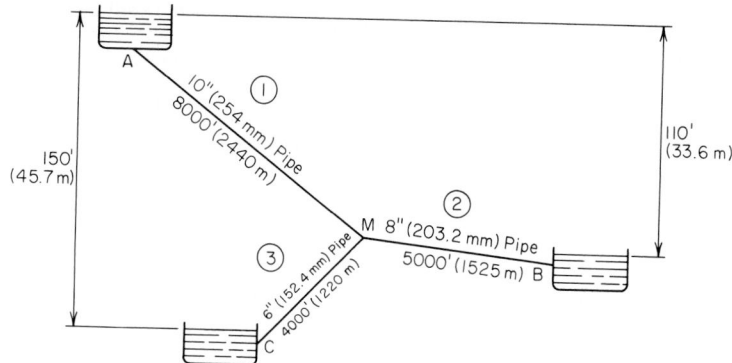

FIG. 176 Branching pipes.

Calculation Procedure:

1. Write the basic equations governing the discharges

Let subscripts 1, 2 and 3 refer to AM, MB, and MC, respectively. Then $h_{F1} + h_{F2} = 110$; $h_{F1} + h_{F3} = 150$, Eq. a; $Q_1 = Q_2 + Q_3$, Eq. b.

2. Transform Eq. 93c

The transformed equation is

$$Q = 38.7D^{2.64}\left(\frac{h_F}{L}\right)^{0.513} \tag{97}$$

3. Assume a trial value for h_{F1} and find the discharge; test the result

Use Eqs. a and 97 to find the discharges. Test the results for compliance with Eq. b. If we assume $h_{F1} = 70$ ft (21.3 m), then $h_{F2} = 40$ ft (12.2 m) and $h_{F3} = 80$ ft (24.4 m); $Q_1 = 38.7(0.833)^{2.64}(70/8000)^{0.513} = 2.10$ ft³/s (59.455 L/s). Similarly, $Q_2 = 1.12$ ft³/s (31.710 L/s) and $Q_3 = 0.83$ ft³/s (23.499 L/s); $Q_2 + Q_3 = 1.95 < Q_1$. The assumed value of h_{F1} is excessive.

4. Make another assumption for h_{F1} and the corresponding revisions

Assume $h_{F1} = 66$ ft (20.1 m). Then $Q_1 = 2.10(66/70)^{0.513} = 2.04$ ft³/s (57.757 L/s). Similarly, $Q_2 = 1.18$ ft³/s (33.408 L/s); $Q_3 = 0.85$ ft³/s (24.065 L/s). $Q_2 + Q_3 = 2.03$ ft³/s (57.736 L/s). These results may be accepted as sufficiently precise.

UNIFORM FLOW IN OPEN CHANNEL—DETERMINATION OF SLOPE

It is necessary to convey 1200 ft³/s (33,974.6 L/s) of water from a dam to a power plant in a canal of rectangular cross section, 24 ft (7.3 m) wide and 10 ft (3.0 m) deep, having a roughness coefficient of 0.016. The canal is to flow full. Compute the required slope of the canal in feet per mile (meters per kilometer).

Calculation Procedure:

1.   Apply Eq. 94

Thus, $A = 24(10) = 240 \text{ ft}^2$ (22.3 m^2); wetted perimeter = WP = $24 + 2(10) = 44 \text{ ft}$ (13.4 m); $R = 240/44 = 5.45 \text{ ft}$ (1.661 m); $V = 1200/240 = 5 \text{ ft/s}$ (152.4 cm/s); $s = [nV/(1.486R^{2/3})]^2$ $= [0.016 \times 5/(1.486 \times 5.45^{2/3})]^2 = 0.000302$; slope = $0.000302(5280 \text{ ft/mi}) = 1.59 \text{ ft/mi}$ (0.302 m/km).

REQUIRED DEPTH OF CANAL FOR SPECIFIED FLUID FLOW RATE

A trapezoidal canal is to carry water at 800 ft^3/s (22,649.7 L/s). The grade of the canal is 0.0004; the bottom width is 25 ft (7.6 m); the slope of the sides is 1½ horizontal to 1 vertical; the roughness coefficient is 0.014. Compute the required depth of the canal, to the nearest tenth of a foot.

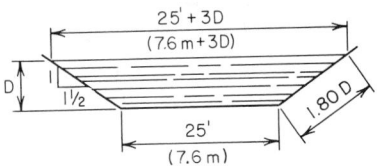

FIG. 177

Calculation Procedure:

1.   Transform Eq. 94 and compute $AR^{2/3}$

Thus, $AR^{2/3} = nQ/(1.486s^{1/2})$, Eq. 94b. Or, $AR^{2/3} = 0.014(800)/[1.486(0.0004)^{1/2}] = 377$.

2.   Express the area and wetted perimeter in terms of D (Fig. 177)

Side of canal = $D(1^2 + 1.5^2)^{0.5} = 1.80D$. $A = D(25 + 1.5D)$; WP = $25 + 360D$.

3.   Assume the trial values of D until Eq. 94b is satisfied

Thus, assume $D = 5 \text{ ft}$ (152.4 cm); $A = 162.5 \text{ ft}^2$ (15.10 m^2); WP = 43 ft (1310.6 cm); $R = 3.78$ ft (115.2 cm); $AR^{2/3} = 394$. The assumed value of D is therefore excessive because the computed $AR^{2/3}$ is greater than the value computed in step 1.

Next, assume a lower value for D, or $D = 4.9 \text{ ft}$ (149.35 cm); $A = 158.5 \text{ ft}^2$ (14.72 m^2); WP = 42.64 ft (1299.7 cm); $R = 3.72 \text{ ft}$ (113.386 cm); $AR^{2/3} = 381$. This is acceptable. Therefore, $D = 4.9 \text{ ft}$ (149.35 cm).

ALTERNATE STAGES OF FLOW; CRITICAL DEPTH

A rectangular channel 20 ft (609.6 cm) wide is to discharge 500 ft^3/s (14,156.1 L/s) of water having a specific energy of 4.5 ft·lb/lb (1.37 J/N). (a) Using $n = 0.013$, compute the required slope of the channel. (b) Compute the maximum potential discharge associated with the specific energy of 4.5 ft·lb/lb (1.37 J/N). (c) Compute the minimum specific energy required to maintain a flow of 500 ft^3/s (14,156.1 L/s).

Calculation Procedure:

1.   Evaluate the specific energy of an elemental mass of liquid at a distance z above the channel bottom

To analyze the discharge conditions at a given section in a channel, it is advantageous to evaluate the specific energy (or head) by taking the elevation of the bottom of the channel *at the given section* as datum. Assume a uniform velocity across the section, and let D = depth of flow, ft (cm); H_e = specific energy as computed in the prescribed manner; Q_u = discharge through a unit width of channel, ft^3/(s·ft) [L/(s·cm)].

Evaluating the specific energy of an elemental mass of liquid at a given distance z above the channel bottom, we get

$$H_e = \frac{Q_u^2}{2gD^2} + D \tag{98}$$

Thus, H_e is constant across the entire section. Moreover, if the flow is uniform, as it is here, H_e is constant along the entire stream.

2. *Apply the given values and solve for D*

Thus, H_e = 4.5 ft·lb/lb (1.37 J/N); Q_u = 500/20 = 25 ft³/(s·ft) [2323 L/(s·m)]. Rearrange Eq. 98 to obtain

$$D^2(H_e - D) = \frac{Q_u^2}{2g} \tag{98a}$$

Or, $D^2(4.5 - D)$ = $25^2/64.4$ = 9.705. This cubic equation has two positive roots, D = 1.95 ft (59.436 cm) and D = 3.84 ft (117.043 cm). There are therefore two stages of flow that accommodate the required discharge with the given energy. [The third root of the equation is D = -1.29 ft (-39.319 cm), an impossible condition.]

3. *Compute the slope associated with the computed depths*

Using Eq. 94, at the lower stage we have D = 1.95 ft (59.436 cm); A = 20(1.95) = 39.0 ft² (36,231.0 cm²); WP = 20 + 2(1.95) = 23.9 ft (728.47 cm); R = 39.0/23.9 = 1.63 ft (49.682 cm); V = 25/1.95 = 12.8 ft/s (390.14 cm/s); s = $[nV/(1.486R^{2/3})]^2$ = (0.013 × 12.8/1.486 × $1.63^{2/3})^2$ = 0.00654.

At the upper stage D = 3.84 ft (117.043 cm); A = 20(3.84) = 76.8 ft² (71,347.2 cm²); WP = 20 + 2(3.84) = 27.68 ft (843.686 cm); R = 76.8/27.68 = 2.77 ft (84.430 cm); V = 25/3.84 = 6.51 ft/s (198.4 cm/s); s = $[0.013 × 6.51/(1.486 × 2.77^{2/3})]^2$ = 0.000834. This constitutes the solution to part a.

4. *Plot the D-Q_u curve*

For part b, consider H_e as remaining constant at 4.5 ft·lb/lb (1.37 J/N) while Q_u varies. Plot the D-Q_u curve as shown in Fig. 178a. The depth that provides the maximum potential discharge is called the *critical depth* with respect to the given specific energy.

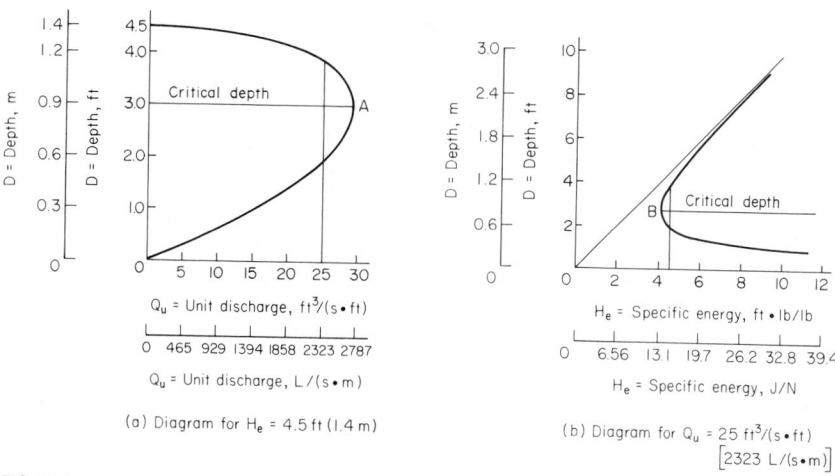

(a) Diagram for H_e = 4.5 ft (1.4 m)

(b) Diagram for Q_u = 25 ft³/(s·ft) $[2323 \text{ L}/(s \cdot m)]$

FIG. 178

5. *Differentiate Eq. 98 to find the critical depth; then evaluate $Q_{u,\text{max}}$*

Differentiating Eq. 98 and setting dQ_u/dD = 0 yield

$$\text{Critical depth } D_c = \tfrac{2}{3}H_e \tag{99}$$

Or, D_c = ⅔(4.5) = 3.0 ft (91.44 cm); $Q_{u,\text{max}}$ = $[64.4(4.5 × 3.0^2 - 3.0^2)]^{0.5}$ = 29.5 ft³/(s·ft) [2741 L/(s·m)]; Q_{max} = 29.5(20) = 590 ft³/s (16,704.2 L/s). This constitutes the solution to part b.

6.   *Plot the D-H$_e$ curve*

For part c, consider Q_u as remaining constant at 25 ft^3/(s·ft) [2323 L/(s·m)] while H_c varies. Plot the D-H_e curve as shown in Fig. 178b. (This curve is asymptotic with the straight lines $D = H_e$ and $D = 0$.) The depth at which the specific energy is minimum is called the *critical depth* with respect to the given unit discharge.

7.   *Differentiate Eq. 98 to find the critical depth; then evaluate H$_{e,min}$*

Differentiating gives

$$D_c = \left(\frac{Q_u^2}{g}\right)^{1/3} \tag{100}$$

Then $D_c = (25^2/32.2)^{1/3} = 2.69$ ft (81.991 cm). Then $H_{e,min} = 25^2/[64.4(2.69)^2] + 2.69 = 4.03$ ft·lb/lb (1.229 J/N).

The values of D as computed in part a coincide with those obtained by referring to the two graphs in Fig. 178. The equations derived in this procedure are valid solely for rectangular channels, but analogous equations pertaining to other channel profiles may be derived in a similar manner.

DETERMINATION OF HYDRAULIC JUMP

Water flows over a 100-ft (30.5-m) long dam at 7500 ft^3/s (212,400 L/s). The depth of tailwater on the level apron is 9 ft (2.7 m). Determine the depth of flow immediately upstream of the hydraulic jump.

Calculation Procedure:

1.   *Find the difference in hydrostatic forces per unit width of channel required to decelerate the liquid*

Refer to Fig. 179. *Hydraulic jump* designates an abrupt transition from lower-stage to upper-stage flow caused by a sharp decrease in slope, sudden increase in roughness, encroachment of

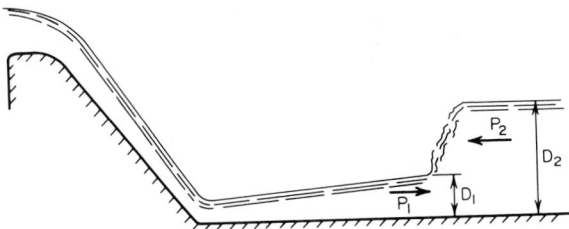

FIG. 179   Hydraulic jump on apron of dam.

backwater, or some other factor. The deceleration of liquid requires an increase in hydrostatic pressure, but only part of the drop in velocity head is accounted for by a gain in pressure head. The excess head is dissipated in the formation of a turbulent standing wave. Thus, the phenomenon of hydraulic jump resembles the behavior of liquid in a pipe at a sudden enlargement, as analyzed in an earlier calculation procedure.

Let D_1 and D_2 denote the depth of flow immediately upstream and downstream of the jump, respectively. Then $D_1 < D_c < D_2$. Refer to Fig. 178b. Since the hydraulic jump is accompanied by a considerable drop in energy, the point on the D-H_e diagram that represents D_2 lies both above and to the left of that representing D_1. Therefore, the upstream depth is less than the depth that would exist in the absence of any loss.

Using literal values, apply Eq. 83 to find the difference in hydrostatic forces per unit width of channel that is required to decelerate the liquid. Solve the resulting equation for D_1:

$$D_1 = -\frac{D_2}{2} + \left(\frac{2V_2^2 D_2}{g} + \frac{D_2^2}{4}\right)^{0.5} \tag{101}$$

2. *Substitute numerical values in Eq. 101*

Thus, $Q_u = 7500/100 = 75$ ft^3/(s·ft) [6969 L/(s·m)]; $V_2 = 75/9 = 8.33$ ft/s (2.538 m/s); $D_1 = -\% + (2 \times 8.33^2 \times 9/32.2 + 9^2/4)^{0.5} = 3.18$ ft (0.969 m).

RATE OF CHANGE OF DEPTH IN NONUNIFORM FLOW

The unit discharge in a rectangular channel is 28 ft^3/(s·ft) [2602 L/(s·m)]. The energy gradient is 0.0004, and the grade of the channel bed is 0.0010. Determine the rate at which the depth of flow is changing in the downstream direction (i.e., the grade of the liquid surface with respect to the channel bed) at a section where the depth is 3.2 ft (0.97 m).

Calculation Procedure:

1. *Express H as a function of D*

Let H = total specific energy at a given section as evaluated by selecting a fixed horizontal reference plane; L = distance measured in downstream direction; z = elevation of given section with respect to datum plane; s_b = grade of channel bed = $-dz/dL$; s_e = energy gradient = $-dH/dL$.

Express H as a function of D by annexing the potential-energy term to Eq. 98. Thus,

$$H = \frac{Q_u^2}{2gD^2} + D + z \tag{102}$$

2. *Differentiate this equation with respect to L to obtain the rate of change of D; substitute numerical values*

Differentiating gives

$$\frac{dD}{dL} = \frac{s_b - s_e}{1 - Q_u^2/(gD^3)} \tag{103a}$$

or in accordance with Eq. 100,

$$\frac{dD}{dL} = \frac{s_b - s_e}{1 - D_c^3/D^3} \tag{103b}$$

Substituting yields $Q_u^2/(gD^3) = 28^2/(32.2 \times 3.2^3) = 0.743$; $dD/dL = (0.0010 - 0.0004)/(1 - 0.743) = 0.00233$ ft/ft (0.00233 m/m). The depth is increasing in the downstream direction, and the water is therefore being decelerated.

As Eq. 103b reveals, the relationship between the actual depth at a given section and the critical depth serves as a criterion in ascertaining whether the depth is increasing or decreasing.

DISCHARGE BETWEEN COMMUNICATING VESSELS

In Fig. 180, liquid is flowing from tank A to tank B through an orifice near the bottom. The area of the liquid surface is 200 ft^2 (18.58 m^2) in A and 150 ft^2 (13.93 m^2) in B. Initially, the difference in water levels is 14 ft (4.3 m), and the discharge is 2 ft^3/s (56.6 L/s). Assuming that the discharge coefficient remains constant, compute the time required for the water level in tank A to drop 1.8 ft (0.54 m).

Calculation Procedure:

1. *By expressing the change in h during an elemental time interval, develop the time-interval equation*

Let A_a and A_b denote the area of the liquid surface in tanks A and B, respectively; let subscripts 1 and 2 refer to the beginning and end, respectively, of a time interval t. Then

$$t = \frac{2A_a A_b(h_1 - [h_1 h_2]^{0.5})}{Q_1(A_a + A_b)} \tag{104}$$

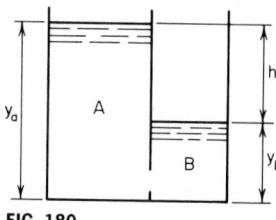

FIG. 180

2. Find the value of h when y_a diminishes by 1.8 ft (0.54 m)

Thus, $\Delta y_b = (-A_a/A_b)(\Delta y_a) = -(200/150)(-1.8) = 2.4$ ft (0.73 m); $\Delta h = \Delta y_a - \Delta y_b = -1.8 - 2.4 = -4.2$ ft (−1.28 m); $h_1 = 14$ ft (4.3 m); $h_2 = 14 - 4.2 = 9.8$ ft (2.99 m).

3. Substitute numerical values in Eq. 104 and solve for t

Thus, $t = 2(200)(150)[14 - (14 \times 9.8)^{0.5}]/[2(200 + 150)] = 196$ s $= 3.27$ min.

VARIATION IN HEAD ON A WEIR WITHOUT INFLOW TO THE RESERVOIR

Water flows over a weir of 60-ft (18.3-m) length from a reservoir having a surface area of 50 acres (202,350 m²). If the inflow to the reservoir ceases when the head on the weir is 2 ft (0.6 m), what will the head be at the expiration of 1 h? Consider that the instantaneous discharge is given by Eq. 89a.

Calculation Procedure:

1. Develop the time-interval equation

Let A = surface area of reservoir and C = numerical constant in discharge equation; and subscripts 1 and 2 refer to the beginning and end, respectively, of a time interval t. By expressing the change in head during an elemental time interval,

$$t = \frac{2A}{Cb(1/h_2^{0.5} - 1/h_1^{0.5})} \tag{105}$$

2. Substitute numerical values in Eq. 105; solve for h_2

Thus, $A = 50(43,560) = 2,178,000$ ft² (202,336.2 m²); $t = 3600$ s; solving gives $h_2 = 1.32$ ft (0.402 m).

VARIATION IN HEAD ON A WEIR WITH INFLOW TO THE RESERVOIR

Water flows over an 80-ft (24.4-m) long weir from a reservoir having a surface area of 6,000,000 ft² (557,400.0 m²) while the rate of inflow to the reservoir remains constant at 2175 ft³/s (61,578.9 L/s). How long will it take for the head on the weir to increase from zero to 95 percent of its maximum value? Consider that the instantaneous rate of flow over the weir is $3.4bh^{1.5}$.

Calculation Procedure:

1. Compute the maximum head on the weir by equating outflow to inflow

The water in the reservoir reaches its maximum height when equilibrium is achieved, i.e., when the rate of outflow equals the rate of inflow. Let Q_i = rate of inflow; Q_o = rate of outflow at a given instant; t = time elapsed since the start of the outflow.

Equating outflow to inflow yields $3.4(80h_{max}^{1.5}) = 2175$; $h_{max} = 4.0$ ft (1.2 m); $0.95h_{max} = 3.8$ ft (1.16 m).

2. Using literal values, determine the time interval dt during which the water level rises a distance dh

Thus, with C = numerical constant in the discharge equation,

$$dt = \frac{A}{Q_i - Cbh^{1.5}} dh \tag{106}$$

The right side of this equation is not amenable to direct integration. Consequently, the only feasible way of computing the time is to perform an approximate integration.

3. Obtain the approximate value of the required time

Select suitable increments of h, calculate the corresponding increments of t, and total the latter to obtain an approximate value of the required time. In calculating Q_o, apply the mean value of h associated with each increment.

The precision inherent in the result thus obtained depends on the judgment used in selecting the increments of h, and a clear visualization of the relationship between h and t is essential. Let $m = dt/dh = A/(Q_i - Cbh^{1.5})$. The m-h curve is shown in Fig. 181a. Then, $t = \int m\ dh$ = area between the m-h curve and h axis.

This area is approximated by summing the areas of the rectangles as indicated in Fig. 181, the length of each rectangle being equal to the value of m at the center of the interval. Note that as h increases, the increments Δh should be made progressively smaller to minimize the error introduced in the procedure.

Select the increments shown in Table 12, and perform the indicated calculations. The symbols h_b and h_m denote the values of h at the beginning and center, respectively, of an interval. The following calculations for the third interval illustrate the method: $h_m = \frac{1}{2}(2.0 + 2.8) = 2.4$ ft $(0.73$ m$)$; $m = 6{,}000{,}000/(2175 - 3.4 \times 80 \times 2.4^{1.5}) = 5160$ s/ft $(16{,}929.1$ s/m$)$; $\Delta t = m\ \Delta h = 5160(0.8) = 4130$ s. From Table 12, the required time is $t = 24{,}090$ s $= 6$ h 41.5 min.

The t-h curve is shown in Fig. 181b. The time required for the water to reach its maximum height is difficult to evaluate with precision because m becomes infinitely large as h approaches $h_{\max}$; that is, the water level rises at an imperceptible rate as it nears its limiting position.

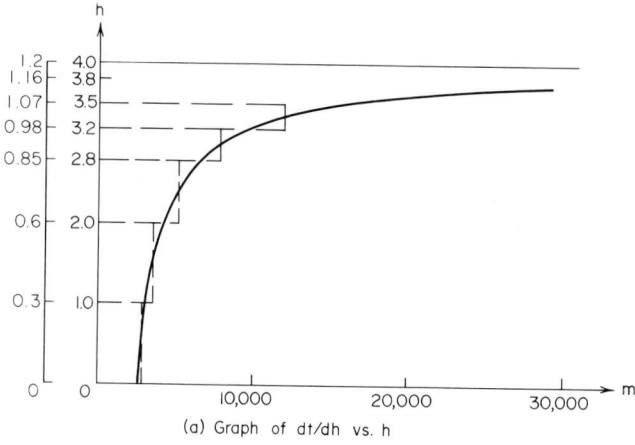

(a) Graph of dt/dh vs. h

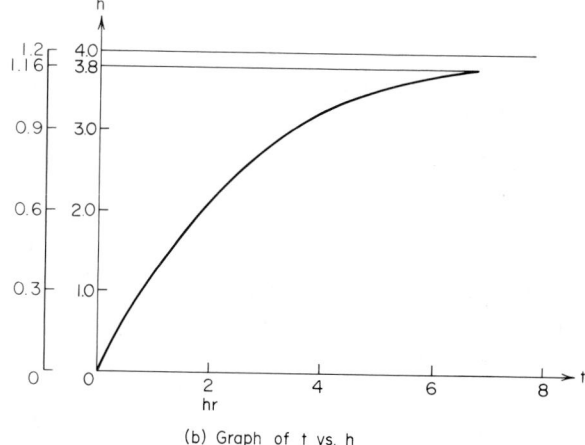

(b) Graph of t vs. h

FIG. 181

TABLE 12 — Approximate Integration of Eq. 106

Δh, ft (m)	h_b, ft (m)	h_m, ft (m)	m, s/ft (s/m)	Δt, s
1.0 (0.30)	0 (0.00)	0.5 (0.15)	2,890 (9,633.3)	2,890
1.0 (0.30)	1.0 (0.30)	1.5 (0.46)	3,580 (11,933.3)	3,580
0.8 (0.24)	2.0 (0.61)	2.4 (0.73)	5,160 (17,308.3)	4,130
0.4 (0.12)	2.8 (0.85)	3.0 (0.91)	7,870 (26,250.0)	3,150
0.3 (0.09)	3.2 (0.98)	3.35 (1.02)	11,830 (39,444.4)	3,550
0.2 (0.06)	3.5 (1.07)	3.6 (1.10)	18,930 (63,166.7)	3,790
0.1 (0.03)	3.7 (1.13)	3.75 (1.14)	30,000 (100,000)	3,000
Total				24,090

DIMENSIONAL ANALYSIS METHODS

The velocity of a raindrop in still air is known or assumed to be a function of these quantities: gravitational acceleration, drop diameter, dynamic viscosity of the air, and the density of both the water and the air. Develop the dimensionless parameters associated with this phenomenon.

Calculation Procedure:

1. Using a generalized notational system, record the units in which the six quantities of this situation are expressed

Dimensional analysis is an important tool both in theoretical investigations and in experimental work because it clairfies the relationships intrinsic in a given situation.

A quantity that appears in every dimensionless parameter is termed repeating; a quantity that appears in only one parameter is termed *nonrepeating*. Since the engineer is usually more accustomed to dealing with units of force rather than of mass, the force-length-time system of units is applied here. Let F, L, and T denote units of force, length, and time, respectively.

By using this generalized notational system, it is convenient to write the appropriate USCS units and then replace these with the general units. For example, with respect to acceleration: USCS units, ft/s²; general units, L/T^2 or LT^{-2}. Similarly, with respect to density (w/g): USCS units, $(\text{lb/ft}^3)/(\text{ft/s}^2)$; general units, FL^{-3}/LT^{-2} or $FL^{-4}T^2$.

The results are shown in the following table.

Quantity	Units
V = velocity of raindrop	LT^{-1}
g = gravitational acceleration	LT^{-2}
D = diameter of drop	L
μ_a = air viscosity	$FL^{-2}T$
ρ_w = water density	$FL^{-4}T^2$
ρ_a = air density	$FL^{-4}T^2$

2. Compute the number of dimensionless parameters present

This phenomenon contains six physical quantities and three units. Therefore, as a consequence of Buckingham's pi theorem, the number of dimensionless parameters is $6 - 3 = 3$.

3. Select the repeating quantities

The number of repeating quantities must equal the number of units (three here). These quantities should be independent, and they should collectively contain all the units present. The quantities g, D, and μ_a satisfy both requirements and therefore are selected as the repeating quantities.

4. Select the dependent variable V as the first nonrepeating quantity

Then write $\pi_1 = g^x D^y \mu_a^z V$, Eq. *a*, where π_1 is a dimensionless parameter and *x*, *y*, and *z* are unknown exponents that may be evaluated by experiment.

5. Transform Eq. a to a dimensional equation

Do this by replacing each quantity with the units in which it is expressed. Then perform the necessary expansions and multiplications. Or, $F^0 L^0 T^0 = (LT^{-2})^x L^y \times (FL^{-2}T)^z LT^{-1}$, $F^0 L^0 T^0 = F^z L^{x+y-2z+1} T^{-2x+z-1}$, Eq. *b*.

Every equation must be dimensionally homogeneous; i.e., the units on one side of the equation must be consistent with those on the other side. Therefore, the exponent of a unit on one side of Eq. *b* must equal the exponent of that unit on the other side.

6. Evaluate the exponents x, y, and z

Do this by applying the principle of dimensional homogeneity to Eq. *b*. Thus, $0 = z$; $0 = x + y - 2z + 1$; $0 = -2x + z - 1$. Solving these simultaneous equations yields $x = -\frac{1}{2}$; $y = -\frac{1}{2}$; $z = 0$.

7. Substitute these values in Eq. a

Thus, $\pi_1 = g^{-1/2} D^{-1/2} V$, or $\pi_1 = V/(gD)^{1/2}$.

8. Follow the same procedure for the remaining nonrepeating quantities

Select ρ_w and ρ_a in turn as the nonrepeating quantities. Follow the same procedure as before to obtain the following dimensionless parameters: $\pi_2 = \rho_w (gD^3)^{1/2}/\mu_a$, and $\pi_3 = \rho_a (gD^3)^{1/2}/\mu_a = (gD^3)^{1/2}/\nu_a$, where $\nu_a =$ kinematic viscosity of air.

HYDRAULIC SIMILARITY AND CONSTRUCTION OF MODELS

A dam discharges 36,000 ft³/s (1,019,236.7 L/s) of water, and a hydraulic jump occurs on the apron. The power loss resulting from this jump is to be determined by constructing a geometrically similar model having a scale of 1:12. (*a*) Determine the required discharge in the model. (*b*) Determine the power loss on the dam if the power loss on the model is found to be 0.18 hp (0.134 kW).

Calculation Procedure:

1. Determine the value of Q_m

Two systems are termed similar if their corresponding variables have a constant ratio. A hydraulic model and its prototype must possess three forms of similarity: geometric, or similarity of shape; kinematic, or similarity of motion; and dynamic, or similarity of forces.

In the present instance, the ratio associated with the geometric similarity is given, i.e., the ratio of a linear dimension in the model to the corresponding linear dimension in the prototype. Let r_g denote this ratio, and let subscripts *m* and *p* refer to the model and prototype, respectively.

Apply Eq. 89*a* to evaluate Q_m. Or $Q = C_1 bh^{1.5}$, where C_1 is a constant. Then $Q_m/Q_p = (b_m/b_p)(h_m/h_p)^{1.5}$. But $b_m/b_p = h_m/h_p = r_g$; therefore, $Q_m/Q_p = r_g^{2.5} = (\frac{1}{12})^{2.5} = 1/499$; $Q_m = 36,000/499 = 72$ ft³/s (2038.5 L/s).

2. Evaluate the power loss on the dam

Apply Eq. 88 to evaluate the power loss on the dam. Thus, hp $= C_2 Qh$, where C_2 is a constant. Then $hp_p/hp_m = (Q_p/Q_m)(h_p/h_m)$. But $Q_p/Q_m = (1/r_g)^{2.5}$, and $h_p/h_m = 1/r_g$; therefore, $hp_p/hp_m = (1/r_g)^{3.5} = 12^{3.5} = 5990$. Hence, $hp_p = 5990(0.18) = 1078$ hp (803.86 kW).

Surveying and Route Design

PLOTTING A CLOSED TRAVERSE

Complete the following table for a closed traverse.

Course	Bearing	Length, ft (m)
a	N32°27′E	110.8 (33.77)
b		83.6 (25.48)
c	S8°51′W	126.9 (38.68)
d	S73°31′W	
e	N18°44′W	90.2 (27.49)

Calculation Procedure:

1. Draw the known courses; then form a closed traverse

Refer to Fig. 182a. A line PQ is described by expressing its length L and its bearing α with respect to a reference meridian NS. For a closed traverse, such as *abcde* in Fig. 182b, the algebraic sum

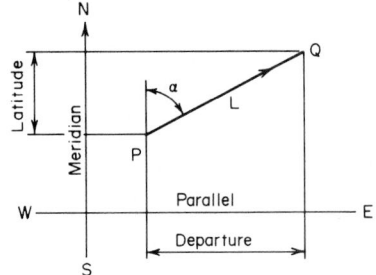
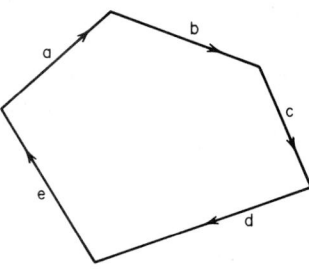

(a) Latitude and departure   (b) Closure of traverse

FIG. 182

of the latitudes and the algebraic sum of the departures must equal zero. A positive latitude corresponds to a northerly bearing, and a positive departure corresponds to an easterly bearing.

In Fig. 183, draw the known courses a, c, and e. Then introduce the hypothetical course f to form a closed traverse.

2. Calculate the latitude and departure of the courses

Use these relations:

$$\text{Latitude} = L \cos \alpha \tag{107}$$

$$\text{Departure} = L \sin \alpha \tag{108}$$

Computing the results for courses a, c, e, and f, we have the values shown in the following table.

Course	Latitude, ft (m)	Departure, ft (m)
a	+93.5	+59.5
c	−125.4	−19.5
e	+85.4	−29.0
Total	+53.5 (+16.306)	+11.0 (+3.35)
f	−53.5 (−16.306)	−11.0 (−3.35)

3. Find the length and bearing of f

Thus, $\tan \alpha_f = 11.0/53.5$; therefore, the bearing of f = S11°37′W; length of f = $53.5/\cos \alpha_f$ = 54.6 ft (16.64 m).

4. *Complete the layout*

Complete Fig. 183 by drawing line d through the upper end of f with the specified bearing and by drawing a circular arc centered at the lower end of f having a radius equal to the length of b.

5. *Find the length of d and the bearing of b*

Solve the triangle fdb to find the length of d and the bearing of b. Thus, $B = 73°31' - 11°37' = 61°54'$. By the law of sines, sin $F = f \sin B/b = 54.6 \sin 61°54'/83.6$; $F = 35°11'$; $D = 180° - (61°54' + 35°11') = 82°55'$; $d = b \sin D/\sin B = 83.6 \sin 82°55'/\sin 61°54' = 94.0$ ft $(28.65$ m$)$; $\alpha_b = 180° - (73°31' + 35°11') = 71°18'$. The bearing of $b = $ S71°18'E.

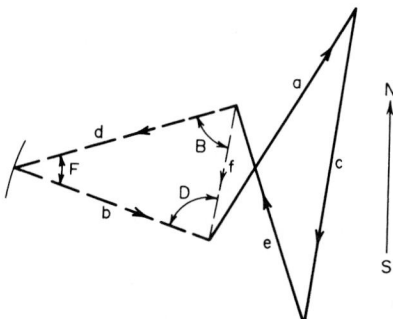

FIG. 183

AREA OF TRACT WITH RECTILINEAR BOUNDARIES

The balanced latitudes and departures of a closed transit-and-tape traverse are recorded in the table below. Compute the area of the tract by the DMD method.

Course	Latitude, ft (m)	Departure, ft (m)
AB	−132.3 (−40.33)	−135.6 (−41.33)
BC	+9.6 (2.93)	−77.5 (−23.62)
CD	+97.9 (29.84)	−198.5 (−60.50)
DE	+161.9 (49.35)	+143.6 (43.77)
EF	−35.3 (−10.76)	+246.7 (75.19)
FA	−101.8 (−31.03)	+21.3 (6.49)

Calculation Procedure:

1. *Plot the tract*

Refer to Fig. 184. The sum of m_1 and m_2 is termed the *double meridian distance* (DMD) of course AB. Let D denote the departure of a course. Then

$$\text{DMD}_n = \text{DMD}_{n-1} + D_{n-1} + D_n \tag{109}$$

where the subscripts refer to two successive courses.

The area of trapezoid $ABba$, which will be termed the *projection area* of AB, equals half the product of the DMD and latitude of the course. A projection area may be either positive or negative.

Plot the tract in Fig. 185. Since D is the most westerly point, pass the reference meridian through D, thus causing all DMDs to be positive.

2. *Establish the DMD of each course by successive applications of Eq. 109*

Thus, $\text{DMD}_{DE} = 143.6$ ft (43.77 m); $\text{DMD}_{EF} = 143.6 + 143.6 + 246.7 = 533.9$ ft (162.73 m); $\text{DMD}_{FA} = 533.9 + 246.7 + 21.3 = 801.9$ ft (244.42 m); $\text{DMD}_{AB} = 801.9 + 21.3 - 135.6 = 687.6$ ft (209.58 m); $\text{DMD}_{BC} = 687.6 - 135.6 - 77.5 = 474.5$ ft (144.62 m); $\text{DMD}_{CD} = 474.5 - 77.5 - 198.5 = 198.5$ ft (60.50 m). This is acceptable.

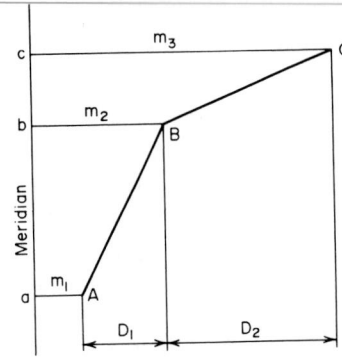

FIG. 184   Double meridian distance.

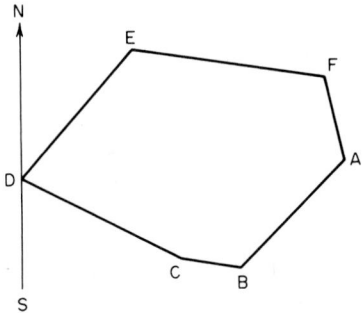

FIG. 185

3.   Calculate the area of the tract

Use the following theorem: The area of a tract is numerically equal to the aggregate of the projection areas of its courses. The results of this calculation are

Course	Latitude	×	DMD	=	2 × Projection area
AB	−132.3		687.6		−90,970
BC	+9.6		474.5		+4,555
CD	+97.9		198.5		+19,433
DE	+161.9		143.6		+23,249
EF	−35.3		533.9		−18,847
FA	−101.8		801.9		−81,634
Total					−144,214

$$\text{Area} = \tfrac{1}{2}(144{,}214) = 72{,}107 \text{ ft}^2 \ (6698.74 \text{ m}^2)$$

PARTITION OF A TRACT

The tract in the previous calculation procedure is to be divided into two parts by a line through E, the part to the west of this line having an area of 30,700 ft^2 (2852.03 m^2). Locate the dividing line.

Calculation Procedure:

1.   Ascertain the location of the dividing line EG

This procedure requires the solution of an oblique triangle. Refer to Fig. 186. It will be necessary to apply the following equations, which may be readily developed by drawing the altitude BD:

$$\text{Area} = \tfrac{1}{2} \, bc \sin A \tag{110}$$

$$\tan C = \frac{c \sin A}{b - c \cos A} \tag{111}$$

In Fig. 187, let EG represent the dividing line of this tract. By scaling the dimensions and making preliminary calculations or by using a planimeter, ascertain that G lies between B and C.

2.   Establish the properties of the hypothetical course EC

By balancing the latitudes and departures of DEC, latitude of $EC = -(+161.9 + 97.9) = -259.8$ ft (-79.18 m); departure of $EC = -(+143.6 - 198.5) = +54.9$ ft ($+16.73$ m); length

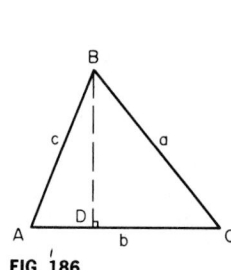

FIG. 186

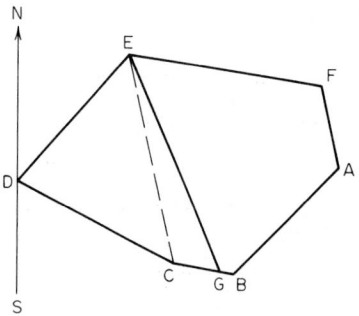

FIG. 187 Partition of tract.

of $EC = (259.8^2 + 54.9^2)^{0.5} = 265.5$ ft (80.92 m). Then $\text{DMD}_{DE} = 143.6$ ft (43.77 m); DMD_{EC} $= 143.6 + 143.6 + 54.9 = 342.1$ ft (104.27 m); $\text{DMD}_{CD} = 342.1 + 54.9 - 198.5 = 198.5$ ft (60.50 m). This is acceptable.

3. Determine angle GCE by finding the bearings of courses EC and BC

Thus $\tan \alpha_{EC} = 54.9/259.8$; bearing of $EC = S11°55.9'E$; $\tan \alpha_{BC} = 77.5/9.6$; bearing of $BC =$ N82°56.3'W; angle $GCE = 180° - (82°56.3' - 11°55.9') = 108°59.6'$.

4. Determine the area of triangle GCE

Calculate the area of triangle DEC; then find the area of triangle GCE by subtraction. Thus

Course	Latitude	×	DMD	=	2 × Projection area
CD	+97.9		198.5		+19,433
DE	+161.9		143.6		+23,249
EC	−259.8		342.1		−88,878
Total					−46,196

So the area of $DEC = \frac{1}{2}(46,196) = 23,098$ ft^2 (2145.8 m^2); area of $GCE = 30,700 - 23,098 = 7602$ ft^2 (706.22 m^2).

5. Solve triangle GCE completely

Apply Eqs. 110 and 111. To ensure correct substitution, identify the corresponding elements, making A the known angle GCE and c the known side EC. Thus

Fig. 186	Fig. 187	Known values	Calculated values
A	GCE	108°59.6'	
B	CEG		11°21.6'
C	EGC		59°38.8'
a	EG		291.0 ft (88.70 m)
b	GC		60.6 ft (18.47 m)
c	EC	265.5 ft (80.92 m)	

By Eq. 110, $7602 = \frac{1}{2}GC(265.5 \sin 108°59.6')$; solving gives $GC = 60.6$ ft (18.47 m). By Eq. 111, $\tan EGC = 265.5 \sin 108°59.6'/(60.6 - 265.5 \cos 108°59.6')$; $EGC = 59°38.8'$. By the law of sines, $EG/\sin GCE = EC/\sin EGC$; $EG = 291.0$ ft (88.70 m); $CEG = 180° - (108°59.6' + 59°38.8') = 11°21.6'$.

6. Find the bearing of course EG

Thus, $\alpha_{EG} = \alpha_{EC} + CEG = 11°55.9' + 11°21.6' = 23°17.5'$; bearing of $EG = S23°17.5'E$.

The surveyor requires the length and bearing of EG to establish this line of demarcation. She or he is able to check the accuracy of both the fieldwork and the office calculations by ascertaining that the point G established in the field falls on BC and that the measured length of GC agrees with the computed value.

AREA OF TRACT WITH MEANDERING BOUNDARY: OFFSETS AT IRREGULAR INTERVALS

The offsets below were taken from stations on a traverse line to a meandering stream, all data being in feet. What is the encompassed area?

Station	0 + 00	0 + 25	0 + 60	0 + 75	1 + 10
Offset	29.8	64.6	93.2	58.1	28.5

Calculation Procedure:

1. Assume a rectilinear boundary between successive offsets; develop area equations

Refer to Fig. 188. When a tract has a meandering boundary, this boundary is approximated by measuring the perpendicular offsets of the boundary from a straight line AB. Let d_r denote the distance along the traverse line between the first and the rth offset, and let $h_1, h_2, \ldots, h_n$ denote the offsets.

Developing the area equations yields

$$\text{Area} = \tfrac{1}{2}[d_2(h_1 - h_3) + d_3(h_2 - h_4) + \cdots + d_{n-1}(h_{n-2} - h_n) + d_n(h_{n-1} + h_n)] \quad (112)$$

Or,

$$\text{Area} = \tfrac{1}{2}[h_1 d_2 + h_2 d_3 + h_3(d_4 - d_2) + h_4(d_5 - d_3) + \cdots + h_n(d_n - d_{n-1})] \quad (113)$$

2. Determine the area, using Eq. 112

Thus, area $= \tfrac{1}{2}[25(29.8 - 93.2) + 60(64.6 - 58.1) + 75(93.2 - 28.5) + 110(58.1 + 28.5)] = 6590$ ft^2 (612.2 m^2).

3. Determine the area, using Eq. 113

Thus, area $= \tfrac{1}{2}[29.8 \times 25 + 64.6 \times 60 + 93.2(75 - 25) + 58.1(110 - 60) + 28.5(110 - 75)] = 6590$ ft^2 (612.2 m^2). Hence, both equations yield the same result. The second equation has a distinct advantage over the first because it has only positive terms.

DIFFERENTIAL LEVELING PROCEDURE

Complete the following level notes, and show an arithmetic check.

Point	BS, ft (m)	HI	FS ft (m)	Elevation, ft (m)
BM42	2.076 (0.63)	...	......	180.482 (55.01)
TP1	3.408 (1.04)	...	8.723 (2.66)	
TP2	1.987 (0.61)	...	9.826 (2.99)	
TP3	2.538 (0.77)	...	10.466 (3.19)	
TP4	2.754 (0.84)	...	8.270 (2.52)	
BM43	...	...	11.070 (3.37)	

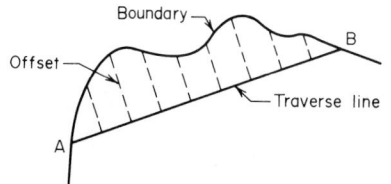

FIG. 188   Tract with irregular boundary.

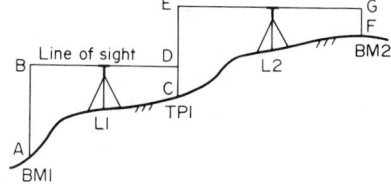

FIG. 189   Differential leveling.

Calculation Procedure:

1.   *Obtain the elevation for each point*

Differential leveling is used to ascertain the difference in elevation between two successive benchmarks by finding the elevations of several convenient intermediate points, called *turning points* (TP). In Fig. 189, consider that the instrument is set up at $L1$ and C is selected as a turning point. The rod reading AB represents the backsight (BS) of BM1, and rod reading CD represents the foresight (FS) of TP1. The elevation of BD represents the height of instrument (HI). The instrument is then set up at $L2$, and rod readings CE and FG are taken. Let a and b designate two successive turning points. Then

$$\text{Elevation}_a + \text{BS}_a = \text{HI} \tag{114}$$

$$\text{HI} - \text{FS}_b = \text{elevation}_b \tag{115}$$

Therefore,

$$\text{Elevation BM2} - \text{elevation BM1} = \Sigma\text{BS} - \Sigma\text{FS} \tag{116}$$

Apply Eqs. 114 and 115 successively to obtain the elevations recorded in the accompanying table.

Point	BS, ft (m)	HI, ft (m)	FS, ft (m)	Elevation, ft (m)
BM42	2.076 (0.63)	182.558 (55.64)	. . . . . .	180.482 (55.01)
TP1	3.408 (1.04)	177.243 (54.02)	8.723 (2.66)	173.835 (52.98)
TP2	1.987 (0.61)	169.404 (51.63)	9.826 (2.99)	167.417 (51.03)
TP3	2.538 (0.77)	161.476 (49.22)	10.466 (3.19)	158.938 (48.44)
TP4	2.754 (0.84)	155.960 (47.54)	8.270 (2.52)	153.206 (46.70)
BM43	. . . . . .	. . . . . .	11.070 (3.37)	144.890 (44.16)
Total	12.763 (3.89)	. . . . . .	48.355 (14.73)	

2.   *Verify the result by summing the backsights and foresights*

Substitute the results in Eq. 116: $144.890 - 180.482 = 12.763 - 48.355 = -35.592$.

STADIA SURVEYING

The following stadia readings were taken with the instrument at a station of elevation 483.2 ft (147.28 m), the height of instrument being 5 ft (1.5 m). The stadia interval factor is 100, and the value of C is negligible. Compute the horizontal distances and elevations.

Point	Rod intercept, ft (m)	Vertical angle
1	5.46 (1.664)	$+2°40'$ on 8 ft (2.4 m)
2	6.24 (1.902)	$+3°12'$ on 3 ft (0.9 m)
3	4.83 (1.472)	$-1°52'$ on 4 ft (1.2 m)

Calculation Procedure:

1. State the equations used in stadia surveying

Refer to Fig. 190 for the notational system pertaining to stadia surveying. The transit is set up over a reference point O, the rod is held at a control point N, and the telescope is sighted at a point Q on the rod; P and R represent the apparent locations of the stadia hairs on the rod.

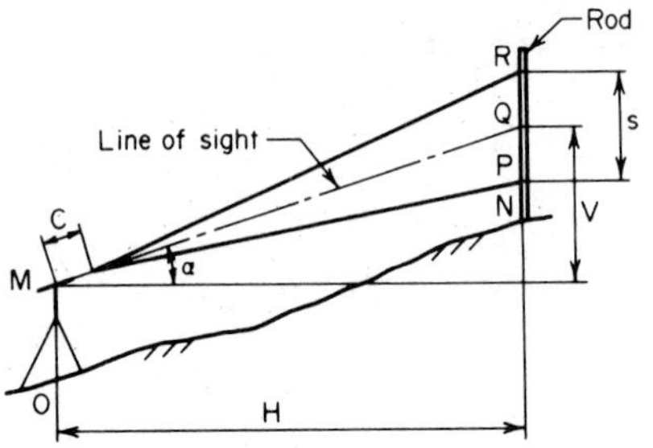

The first column in these notes presents the rod intercepts s, and the second column presents the vertical angle α and the distance NQ. Then

FIG. 190 Stadia surveying.

$$H = Ks \cos^2\alpha + C \cos \alpha \qquad (117)$$

$$V = \tfrac{1}{2} Ks \sin 2\alpha + C \sin \alpha \qquad (118)$$

Elevation of N = elevation of O

$$+ OM + V - NQ \qquad (119)$$

where K = stadia interval factor; C = distance from center of instrument to principal focus.

2. Substitute numerical values in the above equations

The results obtained are shown:

Point	H, ft (m)	V, ft (m)	Elevation, ft (m)
1	544.8 (166.06)	25.4 (7.74)	505.6 (154.11)
2	622.0 (189.59)	34.8 (10.61)	520.0 (158.50)
3	482.5 (147.07)	−15.7 (−4.79)	468.5 (142.80)

VOLUME OF EARTHWORK

Figure 191a and b represent two highway cross sections 100 ft (30.5 m) apart. Compute the volume of earthwork to be excavated, in cubic yards (cubic meters). Apply both the average-end-area method and the prismoidal method.

Calculation Procedure:

1. Resolve each section into an isosceles trapezoid and a triangle; record the relevant dimensions

Let A_1 and A_2 denote the areas of the end sections, L the intervening distance, and V the volume of earthwork to be excavated or filled.

Method 1: The average-end-area method equates the average area to the mean of the two end areas. Then

$$V = \frac{L(A_1 + A_2)}{2} \qquad (120)$$

Figure 191c shows the first section resolved into an isosceles trapezoid and a triangle, along with the relevant dimensions.

2. Compute the end areas, and apply Eq. 120

Thus: $A_1 = [24(40 + 64) + (32 - 24)64]/2 = 1504 \text{ ft}^2 (139.72 \text{ m}^2)$; $A_2 = [36(40 + 76) + (40 - 36)76]/2 = 2240 \text{ ft}^2 (208.10 \text{ m}^2)$; $V = 100(1504 + 2240)/[2(27)] = 6933 \text{ yd}^3 (5301.0 \text{ m}^3)$.

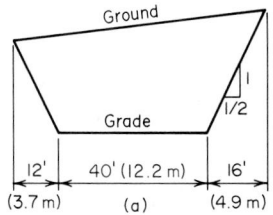

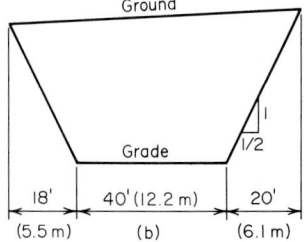

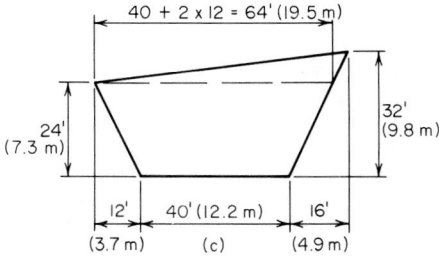

FIG. 191

3. Apply the prismoidal method

Method 2: The prismoidal method postulates that the earthwork between the stations is a prismoid (a polyhedron having its vertices in two parallel planes). The volume of a prismoid is

$$V = \frac{L(A_1 + 4A_m + A_2)}{6} \tag{121}$$

where A_m = area of center section.

Compute A_m. Note that each coordinate of the center section of a prismoid is the arithmetic mean of the corresponding coordinates of the end sections. Thus, $A_m = [30(40 + 70) + (36 - 30)70]/2 = 1860$ ft^2 (172.8 m^2).

4. Compute the volume of earthwork

Using Eq. 121 gives $V = 100(1504 + 4 \times 1860 + 2240)/[6(27)] = 6904$ yd^3 (5278.8 m^3).

DETERMINATION OF AZIMUTH OF A STAR BY FIELD ASTRONOMY

An observation of the sun was made at a latitude of 41°20′N. The altitude of the center of the sun, after correction for refraction and parallax, was 46°48′. By consulting a solar ephemeris, it was found that the declination of the sun at the instant of observation was 7°58′N. What was the azimuth of the sun?

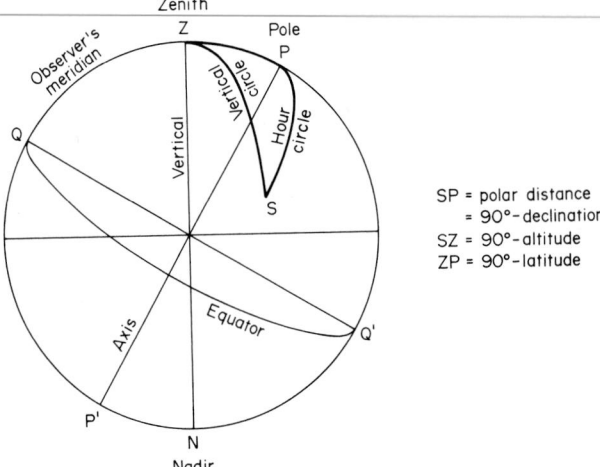

FIG. 192   The celestial sphere.

SP = polar distance
   = 90°-declination
SZ = 90°-altitude
ZP = 90°-latitude

Calculation Procedures:

1. *Calculate the azimuth of the body*

Refer to Fig. 192. The *celestial sphere* is an imaginary sphere on the surface of which the celestial bodies are assumed to be located; this sphere is of infinite radius and has the earth as its center. The *celestial equator*, or *equinoctial*, is the great circle along which the earth's equatorial plane intersects the celestial sphere. The *celestial axis* is the prolongation of the earth's axis of rotation. The *celestial poles* are the points at which the celestial axis pierces the celestial sphere. An *hour circle*, or a *meridian*, is a great circle that passes through the celestial poles.

The *zenith* and *nadir* of an observer are the points at which the vertical (plumb) line at the observer's site pierces the celestial sphere, the former being visible and the latter invisible to the observer. A *vertical circle* is a great circle that passes through the observer's zenith and nadir. The *observer's meridian* is the meridian that passes through the observer's zenith and nadir; it is both a meridian and a vertical circle.

In Fig. 192, P is the celestial pole, S is the apparent position of a star on the celestial sphere, and Z is the observer's zenith.

The coordinates of a celestial body *relative to the observer* are the *azimuth*, which is the angular distance from the observer's meridian to the vertical circle through the body as measured along the observer's horizon in a clockwise direction; and the *altitude*, which is the angular distance of the body from the observer's horizon as measured along a vertical circle.

The *absolute* coordinates of a celestial body are the *right ascension*, which is the angular distance between the vernal equinox and the hour circle through the body as measured along the celestial equator; and the *declination*, which is the angular distance of the body from the celestial equator as measured along an hour circle.

The relative coordinates of a body at a given instant are obtained by observation; the absolute coordinates are obtained by consulting an almanac of astronomical data. The latitude of the observer's site equals the angular distance of the observer's zenith from the celestial equator as measured along the meridian. In the astronomical triangle PZS in Fig. 192, the arcs represent the indicated coordinates, and angle Z represents the azimuth of the body as measured from the north. Calculating the azimuth of the body yields

$$\tan^2 \tfrac{1}{2} Z = \frac{\sin (S - L) \sin (S - h)}{\cos S \cos (S - p)}$$

(122)

where L = latitude of site; h = altitude of star; p = polar distance = $90°$ − declination; S = ½$(L + h + p)$; $L = 41°20'$; $h = 46°48'$; $p = 90° − 7°58' = 82°02'$; $S = ½(L + h + p) = 85°05'$; $S − L = 43°45'$; $S − h = 38°17'$; $S − p = 3°03'$.
Then

log sin 43°45′ =		9.839800
log sin 38°17′ =		9.792077
		9.631877
log cos 85°05′ =	8.933015	
log cos 3°03′ =	9.999384	8.932399
2 log tan ½Z =		0.699478
log tan ½Z =		0.349739
½Z = 65°55′03.5″	Z = 131°50′07″	

2. *Verify the solution by calculating Z in an alternative manner*

To do this, introduce an auxiliary angle M, defined by

$$\cos^2 M = \frac{\cos p}{\sin h \sin L} \tag{123}$$

Then

$$\cos (180° − Z) = \tan h \tan L \sin^2 M \tag{124}$$

Then

log cos 82°02′ =		9.141754
log sin 46°48′ =	9.862709	
log sin 41°20′ =	9.819832	9.682541
2 log cos M =		9.459213
log cos M =		9.729607
log sin M =		9.926276
2 log sin M =		9.852552
log tan 46°48′ =		0.027305
log tan 41°20′ =		9.944262
log cos (180° − Z) =		9.824119
Z = 131°50′07″, as before		

TIME OF CULMINATION OF A STAR

Determine the Eastern Standard Time (75th meridian time) of the upper culmination of Polaris at a site having a longitude 81°W of Greenwich. Reference to an almanac shows that the Greenwich Civil Time (GCT) of upper culmination for the date of observation is $3^h20^m05^s$.

Calculation Procedure:

1. *Convert the longitudes to the hour-minute-second system*

The rotation of the earth causes a star to appear to describe a circle on the celestial sphere centered at the celestial axis. The star is said to be *at culmination* or *transit* when it appears to cross the observer's meridian.

In Fig. 193, P and M represent the position of Polaris and the mean sun, respectively, when Polaris is at the Greenwich meridian, and P' and M' represent the position of these bodies when Polaris is at the observer's meridian. The distances h and h' represent, respectively, the time of culmination of Polaris at Greenwich and at the observer's site, measured from local noon. Since the apparent velocity of the mean sun is less than that of the stars, h' is less than h, the difference being approximately 10 s/h of longitude.

By converting the longitudes, 360° corresponds to 24 h; therefore, 15° corresponds to 1 h. Longitude of site = 81° = 5.4^h = $5^h24^m00^s$; standard longitude = 75° = 5^h.

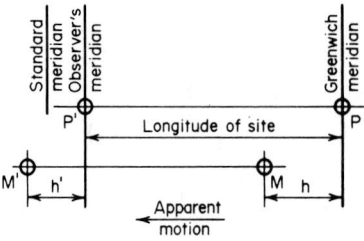

FIG. 193   Culmination of Polaris.

2. Calculate the time of upper culmination at the site

Correct this result to Eastern Standard Time. Since the standard meridian is east of the observer's meridian, the standard time is greater. Thus

GCT of upper culmination at Greenwich	$3^h20^m05^s$
Correction for longitude, 5.4 × 10 s	54^s
Local civil time of upper culmination at site	$3^h19^m11^s$
Correction to standard meridian	24^m00^s
EST of upper culmination at site	$3^h43^m11^s$a.m.

PLOTTING A CIRCULAR CURVE

A horizontal circular curve having an intersection angle of 28° is to have a radius of 1200 ft (365.7 m). The point of curve is at station 82 + 30. (a) Determine the tangent distance, long chord, middle ordinate, and external distance. (b) Determine all the data necessary to stake the curve if the *chord* distance between successive stations is to be 100 ft (30.5 m). (c) Calculate all the data necessary to stake the curve if the *arc* distance between successive stations is to be 100 ft (30.5 m).

Calculation Procedure:

1. Determine the geometric properties of the curve

Part a. Refer to Fig. 194: A is termed the *point of curve* (PC), B is the *point of tangent* (PT), and V the *point of intersection* (PI), or vertex. The notational system is Δ = intersection angle = angle between back and forward tangents = central angle AOB; R = radius of curve; T = tangent distance = AV = VB; C = long chord = AB; M = middle ordinate = DC; E = external distance = CV.

From the geometric relationships,

$$T = R \tan \tfrac{1}{2} \Delta \tag{125}$$
$$T = 1200(0.2493) = 299.2 \text{ ft (91.20 m)}$$

Also

$$C = 2R \sin \tfrac{1}{2} \Delta \tag{126}$$
$$C = 2(1200)(0.2419) = 580.6 \text{ ft (176.97 m)}$$

And,

$$M = R(1 - \cos \tfrac{1}{2} \Delta) \tag{127}$$
$$M = 1200(1 - 0.9703) = 35.6 \text{ ft (10.85 m)}$$

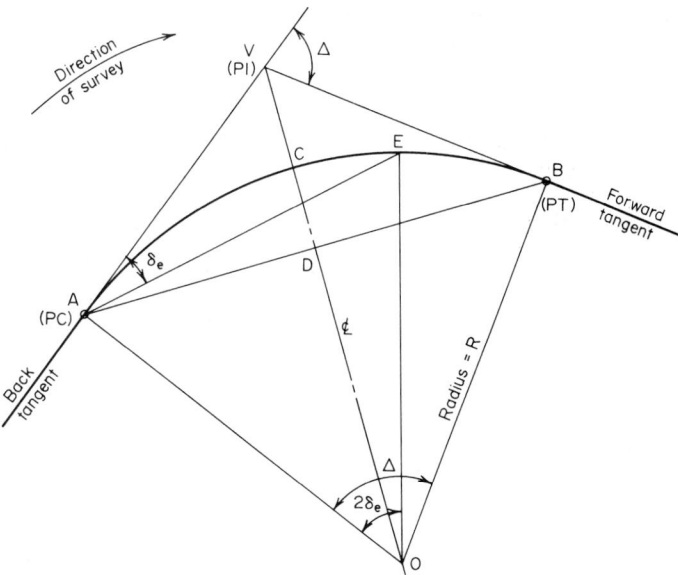

FIG. 194 Circular curve.

Lastly,

$$E = R \tan \tfrac{1}{2} \Delta \tan \tfrac{1}{4} \Delta \tag{128}$$
$$E = 1200(0.2493)(0.1228) = 36.7 \text{ ft } (11.19 \text{ m})$$

2. Verify the results in step 1

Use the pythagorean theorem on triangle ADV. Or, $AD = \tfrac{1}{2}(580.6) = 290.3$ ft (88.48 m); $DV = 35.6 + 36.7 = 72.3$ ft (22.04 m); then $290.3^2 + 72.3^2 = 89,500 \text{ ft}^2$ (8314.6 m²), to the nearest hundred; $299.2^2 = 89,500 \text{ ft}^2$ (8314.6 m²); this is acceptable.

3. Calculate the degree of curve D

Part b. In Fig. 194, let E represent a station along the curve. Angle VAE is termed the *deflection angle* δ_e of this station; it is equal to one-half the central angle AOE. In the field, the curve is staked by setting up the transit at the PC and then locating each station by means of its deflection angle and its chord distance from the preceding station.

Calculate the *degree of curve D*. This is the central angle formed by the radii to two successive stations or, what is the same in this instance, the central angle subtended by a *chord* of 100 ft (30.5 m). Then

$$\sin \tfrac{1}{2} D = \frac{50}{R} \tag{129}$$

So $\tfrac{1}{2}D = \arcsin 50/1200 = \arcsin 0.04167$; $\tfrac{1}{2}D = 2°23.3'$; $D = 4°46.6'$.

4. Determine the station at the PT

Number of stations on the curve $= 28°/4°46.6' = 5.862$; station of PT $= (82 + 30) + (5 + 86.2) = 88 + 16.2$.

5. Calculate the deflection angle of station 83 and the difference between the deflection angles of station 88 and the PT

For simplicity, assume that central angles are directly proportional to their corresponding chord lengths; the resulting error is negligible. Then $\delta_{83} = 0.70(2°23.3') = 1°40.3'$; $\delta_{PT} - \delta_{88} = 0.162(2°23.3') = 0°23.2'$.

6. *Calculate the deflection angle of each station*

Do this by adding ½D to that of the preceding station. Record the results thus:

Station	Deflection angle
82 + 30	0
83	1°40.3′
84	4°03.6′
85	6°26.9′
86	8°50.2′
87	11°13.5′
88	13°36.8′
88 + 16.2	14°00′

7. *Calculate the degree of curve in the present instance*

Part c. Since the subtended central angle is directly proportional to its arc length, $D/100 = 360/(2\pi R)$; therefore,

$$D = 18{,}000/\pi R = 5729.58/R \text{ degrees} \tag{130}$$

Then, $D = 5729.58/1200 = 4.7747° = 4°46.5′$.

8. *Repeat the calculations in steps 4, 5, and 6*

INTERSECTION OF CIRCULAR CURVE AND STRAIGHT LINE

In Fig. 195, MN represents a straight railroad spur that intersects the curved higway route AB. Distances on the route are measured along the arc. Applying the recorded data, determine the station of the intersection point P.

Calculation Procedure:

1. *Apply trigonometric relationships to determine three elements in triangle ONP*

Draw line OP. The problem resolves itself into the calculation of the central angle AOP, and this may be readily found by solving the oblique triangle ONP. Applying trigonometric relationships gives $AV = T = 800 \tan 54° = 1101.1$ ft (335.62 m); $AM = 1101.1 - 220 = 881.1$ ft (268.56 m); $AN = AM \tan 28° = 468.5$ ft (142.80 m); $ON = 800 - 468.5 = 331.5$ ft (101.04 m); $OP = 800$ ft (243.84 m); $ONP = 90° + 28° = 118°$.

2. *Establish the station of P*

Solve triangle ONP to find the central angle; then calculate arc AP and establish the station of P. By the law of sines, $\sin OPN = \sin ONP(ON)/OP$; therefore, $OPN = 21°27.7′$; $AOP = 180° - (118° + 21°27.7′) = 40°32.3′$; arc $AP = 2\pi(800)(40°32.3′)/360° = 566.0$ ft (172.52 m); station of $P = (22 + 00) + (5 + 66) = 27 + 66$.

REALIGNMENT OF CIRCULAR CURVE BY DISPLACEMENT OF FORWARD TANGENT

In Fig. 196, the horizontal circular curve AB has a radius of 720 ft (219.5 m) and an intersection angle of 126°. The curve is to be realigned by rotating the forward tangent through an angle of 22° to the new position $V'B$ while maintaining the PT at B. Compute the radius, and locate the PC of the new curve.

Calculation Procedure:

1. *Find the tangent distance of the new curve*

Solve triangle $BV'V$ to find the tangent distance of the new curve and the location of V'. Thus, $\Delta' = 126° - 22° = 104°$; $VB = 720 \tan 63° = 1413.1$ ft (430.71 m). By the law of sines, $V'B$

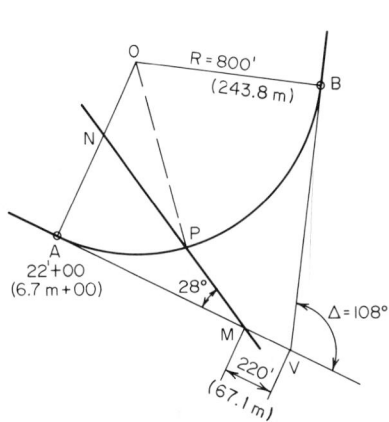

FIG. 195 Intersection of curve and straight line.

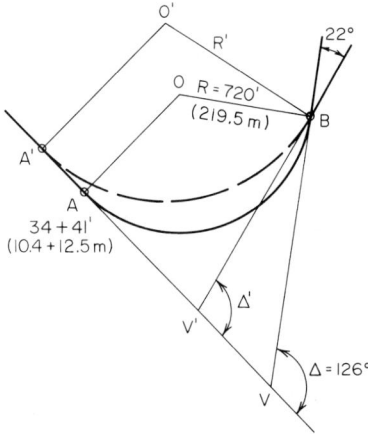

FIG. 196 Displacement of forward tangent.

$= 1413.1 \sin 126°/\sin 104° = 1178.2$ ft (359.12 m); $V'V = 1413.1 \sin 22°/\sin 104° = 545.6$ ft (166.30 m).

2. Compute the radius R'

By Eq. 125, $R' = 1178.2 \cot 52° = 920.5$ ft (280.57 m).

3. Determine the station of A'

Thus, $AV = VB = 1413.1$ ft (403.71 m); $A'V' = V'B = 1178.2$ ft (359.12 m); $A'A = A'V' + V'V - AV = 310.7$ ft (94.70 m); station of new PC $= (34 + 41) - (3 + 10.7) = 31 + 30.3$.

4. Verify the foregoing results

Draw the long chords AB and $A'B$. Then apply the computed value of R' to solve triangle $BA'A$ and thereby find $A'A$. By Eq. 126, $A'B = 2R' \sin \frac{1}{2}\Delta' = 1450.7$ ft (442.17 m); $AA'B = \frac{1}{2}\Delta' = 52°$; $A'AB = 180° - \frac{1}{2}\Delta = 117°$; $ABA' = 180° - (52° + 117°) = 11°$. By the law of sines, $A'A = 1450.7 \sin 11°/\sin 117° = 310.7$ ft (94.70 m). This is acceptable.

CHARACTERISTICS OF A COMPOUND CURVE

The tangents to a horizontal curve intersect at an angle of 68°22′. To fit the curve to the terrain, it is necessary to use a compound curve having tangent lengths of 955 ft (291.1 m) and 800 ft (243.8 m), as shown in Fig. 197. The minimum allowable radius is 1000 ft (304.8 m). Compute the larger radius and the two central angles.

Calculation Procedure:

1. Calculate the latitudes and departures of the known sides

A *compound curve* is a curve that comprises two successive circular arcs of unequal radii that are tangent at their point of intersection, the centers of the arcs lying on the same side of their common tangent. (Where the centers lie on opposite sides of this tangent, the curve is termed a *reversed curve*). In Fig. 197, C is the point of intersection of the arcs, and DE is the common tangent.

This situation is analyzed without applying any set equation to illustrate the general method of solution for compound and reversed curves. There are two unknown quantities: the radius R_1 and a central angle. (Since $\Delta_1 + \Delta_2 = \Delta$, either central angle may be considered the unknown.)

If the polygon $AVBO_2O_1$ is visualized as a closed traverse, the latitudes and departures of its sides are calculated, and the sum of the latitudes and sum of the departures are equated to zero,

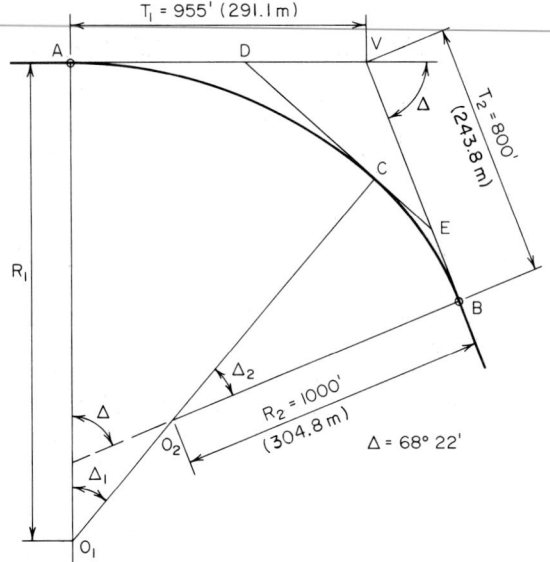

FIG. 197   Compound curve.

two simultaneous equations containing these two unknowns are obtained. For convenience, select O_1A as the reference meridian. Then

Side	Length, ft (m)	Bearing	Latitude	Departure
AV	955 (291.1)	90°	0	+955.00
VB	800 (243.8)	21°38′	−743.65	+294.93
BO_2	1000 (304.8)	68°22′	−368.67	−929.56
Total	. . .	. . .	−1112.32	+320.37

2.   Express the latitudes and departures of the unknown sides in terms of R_1 and Δ_1

Thus, for side O_2O_1: length = $R_1 - 1000$; bearing = Δ_1; latitude = $-(R_1 - 1000) \cos \Delta_1$; departure = $-(R_1 - 1000) \sin \Delta_1$.

Also, for side O_1A: length = R_1; bearing = 0; latitude = R_1; departure = 0.

3.   Equate the sum of the latitudes and sum of the departures to zero; express Δ_1 as a function of R_1

Thus, $\Sigma \text{lat} = R_1 - (R_1 - 1000) \cos \Delta_1 - 1112.32 = 0$; $\cos \Delta_1 = (R_1 - 1112.32)/(R_1 - 1000)$, or $1 - \cos \Delta_1 = 112.32/(R_1 - 1000)$, Eq. a. Also, $\Sigma \text{dep} = -(R_1 - 1000) \sin \Delta_1 + 320.37 = 0$; $\sin \Delta_1 = 320.37/(R_1 - 1000)$, Eq. b.

4.   Divide Eq. a by Eq. b, and determine the central angles

Thus, $(1 - \cos \Delta_1)/\sin \Delta_1 = \tan \frac{1}{2}\Delta_1 = 112.32/320.37$; $\frac{1}{2}\Delta_1 = 19°19′13″$; $\Delta_1 = 38°38′26″$; $\Delta_2 = 68°22′ - \Delta_1 = 29°43′34″$.

5.   Substitute the value of Δ_1 in Eq. b to find R_1

Thus, $R_1 = 1513.06$ ft (461.181 m).

6. *Verify the foregoing results by analyzing triangle DEV*

Thus, $AD = R_1 \tan \frac{1}{2}\Delta_1 = 530.46$ ft (161.684 m); $DV = 955 - 530.46 = 424.54$ ft (129.400 m); $EB = R_2 \tan \frac{1}{2}\Delta_2 = 265.40$ ft (80.894 m); $VE = 800 - 265.40 = 534.60$ ft (162.946 m); $DE = 530.46 + 265.40 = 795.86$ ft (242.578 m). By the law of cosines, $\cos \Delta = -(DV^2 + VE^2 - DE^2)/[2(DV)(VE)]$; $\Delta = 68°22'$. This is correct.

ANALYSIS OF A HIGHWAY TRANSITION SPIRAL

A horizontal circular curve for a highway is to be designed with transition spirals. The PI is at station 34 + 93.81, and the intersection angle is 52°48′. In accordance with the governing design criteria, the spirals are to be 350 ft (106.7 m) long and the degree of curve of the circular curve is to be 6° (arc definition). The approach spiral will be staked by setting the transit at the TS and locating 10 stations on the spiral by means of their deflection angles from the main tangent. Compute all data needed for staking the approach spiral. Also, compute the long tangent, short tangent, and external distance.

Calculation Procedure:

1. *Calculate the basic values*

In the design of a road, a spiral is interposed between a straight-line segment and a circular curve to effect a gradual transition from rectilinear to circular motion, and vice versa. The type of spiral most frequently used is the clothoid, which has the property that the curvature at a given point is directly proportional to the distance from the start of the curve to the given point, measured along the curve.

Refer to Fig. 198. The key points are identified by the following notational system: PI = point of intersection of main tangents; TS = point of intersection of main tangent and approach spiral (tangent-to-spiral point); SC = point of intersection of approach spiral and circular curve (spiral-to-curve point); CS = point of intersection of circular curve and departure spiral (curve-to-spiral point); ST = point of intersection of departure spiral and main tangent (spiral-to-tangent point); PC and PT = point at which tangents to the circular curve prolonged are parallel to the main tangents (also referred to as the *offsets*). Distances are designated in the following manner: L_s = length of spiral from TS to SC; L = length of spiral from TS to given point on spiral; R_c radius of circular curve; R = radius of curvature at given point on spiral; T_s = length of main tangent from TS to PI; E_s = external distance, i.e., distance from PI to midpoint of circular curve.

In addition, there is a long tangent (LT), short tangent (ST), and long chord (LC), as indicated with respect to the departure spiral.

Place the origin of coordinates at the TS and the x axis on the main tangent. Then x_c and y_c = coordinates of SC; k and p = abscissa and ordinate, respectively, of PC. The coordinates of the SC and PC are useful as parameters in the calculation of required distances. The distance p is termed the *throw*, or *shift*, of the curve; it represents the displacement of the circular curve from the main tangent resulting from interposition of the spiral.

The basic angles are Δ = angle between main tangents, or intersection angle; Δ_c = angle between radii at SC and CS, or central angle of circular curve; θ_s = angle between radii of spiral at TS and SC, or central angle of entire spiral; D_c = degree of curve of circular curve; D = degree of curve at given point on spiral; δ_s = deflection angle of SC from main tangent, with transit at TS; δ = deflection angle of given point on spiral from main tangent, with transit at TS.

Although extensive tables of spiral values have been compiled, this example is solved without recourse to these tables in order to illuminate the relatively simple mathematical relationships that inhere in the clothoid. Consider that a vehicle starts at the TS and traverses the approach spiral at constant speed. The degree of curve, which is zero at the TS, increases at a uniform rate to become D_c at the SC. The basic equations are

$$\theta_s = \frac{L_s D_c}{200} = \frac{L_s}{2R_c} \tag{131}$$

$$x_c = L_s \left(1 - \frac{\theta_s^2}{10}\right) \tag{132}$$

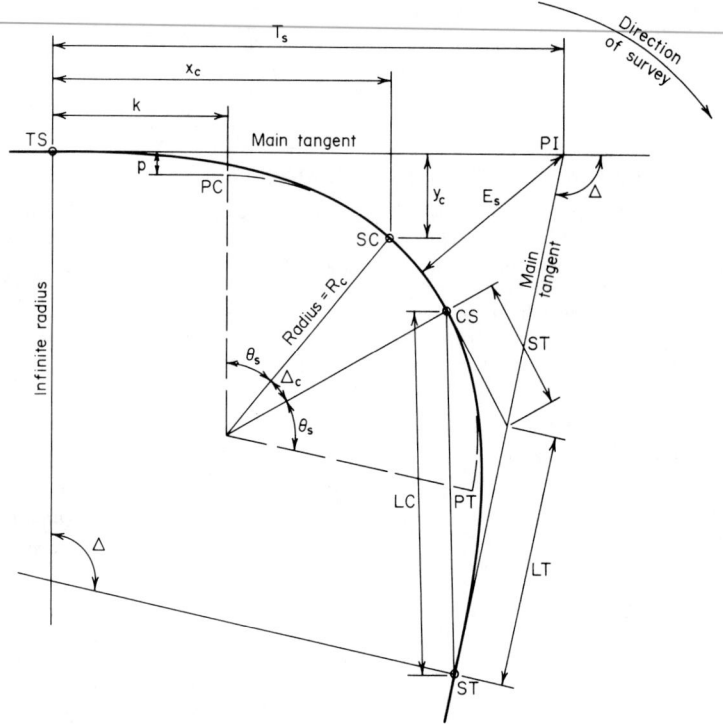

FIG. 198 Notational system for transition spirals.

$$y_c = L_s \left(\frac{\theta_s}{3} - \frac{\theta_s^3}{42} \right) \tag{133}$$

$$k = x_c - R_c \sin \theta_s \tag{134}$$

$$p = y_c - R_c(1 - \cos \theta_s) \tag{135}$$

$$\delta_s = \frac{L_s}{6R_c} = \frac{\theta_s}{3} \tag{136}$$

$$y = \left(\frac{L}{L_s} \right)^3 y_c \tag{137}$$

$$\delta = \left(\frac{L}{L_s} \right)^2 \delta_s \tag{138}$$

$$T_s = (R_c + p) \tan \tfrac{1}{2}\Delta + k \tag{139}$$

$$E_s = (R_c + p)(\sec \tfrac{1}{2}\Delta - 1) + p \tag{140}$$

$$LT = x_c - y_c \cot \theta_s \tag{141}$$

$$ST = y_c \csc \theta_s \tag{142}$$

Even though several of the foregoing equations are actually approximations, their use is valid when the value of D_c is relatively small.

Calculating the basic values yields $\Delta = 52°48'$; $L_s = 350$ ft (106.7 m); $D_c = 6°$; $\theta_s = L_s D_c/200 = 350(6)/200 = 10.5° = 10°30'$, or $\theta_s = 10.5(0.017453) = 0.18326$ rad; $\Delta_c = 52°48' - 2(10°30') = 31°48'$; $D_c = 6(0.017453) = 0.10472$ rad; $R_c = 100/D_c = 954.93$ ft (291.063 m); $x_c = 350(1 - 0.18326^2/10) = 348.83$ ft (106.323 m); $y_c = 350(0.18326/3 - 0.18326^2/42) = 21.33$ ft (6.501 m); $k = 348.83 - 954.93 \sin 10°30' = 174.80$ ft (53.279 m); $p = 21.33 - 954.93(1 - \cos 10°30') = 5.34$ ft (1.628 m).

2. Locate the TS and SC

Thus, $T_s = (954.93 + 5.34) \tan 26°24' + 174.80 = 651.47$; station of TS = $(34 + 93.81) - (6 + 51.47) = 28 + 42.34$; station of SC = $(28 + 42.34) + (3 + 50.00) = 31 + 92.34$.

3. Calculate the deflection angles

Thus, $\delta_s = 10°30'/3 = 3°30' = 3.5°$. Apply Eq. 138 to find the deflection angles at the intermediate stations. For example, for point 7, $\delta = 0.7^2(3.5°) = 1.715° = 1°42.9'$.

TABLE 13 Deflection Angles on Approach Spiral

Point	Station	Deflection angle
TS	28 + 42.34	0
1	77.34	0°02.1'
2	29 + 12.34	0°08.4'
3	47.34	0°18.9'
4	82.34	0°33.6'
5	30 + 17.34	0°52.5'
6	52.34	1°15.6'
7	87.34	1°42.9'
8	31 + 22.34	2°14.4'
9	57.34	2°50.1'
SC	92.34	3°30.0'

Record the results in Table 13. The chord lengths between successive stations differ from the corresponding arc lengths by negligible amounts, and therefore each chord length may be taken as 35.00 ft (1066.8 cm).

4. Compute the LT, ST, and E_s

Thus, LT = $348.83 - 21.33 \cot 10°30' = 233.75$ ft (7124.7 cm); ST = $21.33 \csc 10°30' = 117.04$ ft (3567.4 cm); $E_s = (954.93 + 5.34)(\sec 26°24' - 1) + 5.34 = 117.14$ ft (3570.4 cm).

5. Verify the last three calculations by substituting in the following test equation

Thus

$$\frac{\text{ST} + R_c \tan \tfrac{1}{2}\Delta_c}{\cos \tfrac{1}{2}\Delta} = \frac{T_s - \text{LT}}{\cos \tfrac{1}{2}\Delta_c} = \frac{E_s - R_c(\sec \tfrac{1}{2}\Delta_c - 1)}{\sin \theta_s} \tag{143}$$

TRANSITION SPIRAL: TRANSIT AT INTERMEDIATE STATION

Referring to the transition spiral in the previous calculation procedure, assume that lack of visibility from the TS makes these setups necessary: Points 4, 5, 6, and 7 will be located with the transit at point 3; points 8 and 9 and the SC will be located with the transit at point 7. Compute the orientation and deflection angles.

Calculation Procedure:

1. Consider that the transit is set up at point 3 and a backsight is taken to the TS; find the orientation angle

In Fig. 199, assume that the spiral has been staked up to P with the transit set up at the TS and that the remainder of the spiral is to be staked with the transit set up at P. Deflection angles are measured from the tangent through P (the *local* tangent). The instrument is oriented by backsighting to a preceding point B and then turning the angle δ_b. The orientation angle to B and deflection angle to a subsequent point F are

$$\delta_b = (2L_p + L_b)(L_p - L_b)\frac{\theta_s}{3L_s^2} \tag{144}$$

$$\delta_f = (2L_p + L_f)(L_f - L_p)\frac{\theta_s}{3L_s^2} \tag{145}$$

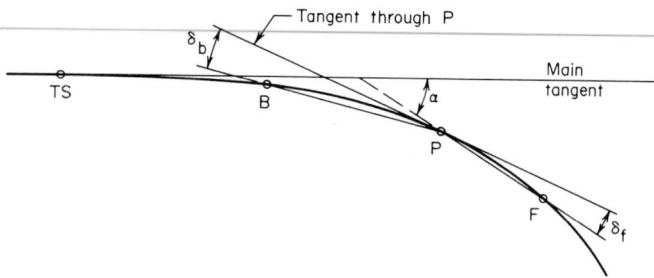

FIG. 199   Deflection angles from local tangent to spiral.

If B, P, and F are points obtained by dividing the spiral into an integral number of arcs, these equations may be converted to these more suitable forms:

$$\delta_b = (2n_p + n_b)(n_p - n_b)\frac{\theta_s}{3n_s^2} \tag{144a}$$

$$\delta_f = (2n_p + n_f)(n_f - n_p)\frac{\theta_s}{3n_s^2} \tag{145a}$$

where n denotes the number of arcs to the designated point.

Applying Eq. 144a to find the orientation angle and using data from the previous calculation procedure, we find $\theta_s = 10.5°$, $\theta_s/(3n_s^2) = 10.5°/[3(10^2)] = 0.035° = 2.1'$; $n_b = 0$; $n_p = 3$; $\delta_b = 6(3)(2.1') = 0°37.8'$.

2.   Find the deflection angles from the tangent through point 3

Thus, by Eq. 145a: point 4, $\delta = (6 + 4)(2.1') = 0°21'$; point 5, $\delta = (6 + 5)(2)(2.1') = 0°46.2'$; point 6, $\delta = (6 + 6)(3)(2.1') = 1°15.6'$; point 7, $\delta = (6 + 7)(4)(2.1') = 1°49.2'$.

3.   Consider that the transit is set up at point 7 and a backsight is taken to point 3; compute the orientation angle

Thus $n_b = 3$; $n_p = 7$; $\delta_b = (14 + 3)(4)(2.1') = 2°22.8'$.

4.   Compute the deflection angles from the tangent through point 7

Thus point 8, $\delta = (14 + 8)(2.1') = 0°46.2'$; point 9, $\delta = (14 + 9)(2)(2.1') = 1°36.6'$; SC, $\delta = (14 + 10)(3)(2.1') = 2°31.2'$.

5.   Test the results obtained

In Fig. 199, extend chord PF to its intersection with the main tangent, and let α denote the angle between these lines. Then

$$\alpha = (n_f^2 + n_f n_p + n_p^2)\frac{\theta_s}{3n_s^2} \tag{146}$$

This result should equal the sum of the angles applied in staking the curve from the TS to F. This procedure will be shown with respect to point 9.

For point 9, let P and F refer to points 7 and 9, respectively. Then $\alpha = (9^2 + 9 \times 7 + 7^2)(2.1') = 6°45.3'$. Summing the angles leading from the TS to point 9, we get

Deflection angle from main tangent to point 3	0°18.9'
Orientation angle at point 3	0°37.8'
Deflection angle from local tangent to point 7	1°49.2'
Orientation angle at point 7	2°22.8'
Deflection angle from local tangent to point 9	1°36.6'
Total	6°45.3'

This test may be applied to each deflection angle beyond point 3.

PLOTTING A PARABOLIC ARC

A grade of −4.6 percent is followed by a grade of +1.8 percent, the grades intersecting at station 54 + 20 of elevation 296.30 ft (90.312 m). The change in grade is restricted to 2 percent in 100 ft (30.5 m). Compute the elevation of every 50-ft (15.24-m) station on the parabolic curve, and locate the sag (lowest point of the curve). Apply both the average-grade method and the tangent-offset method.

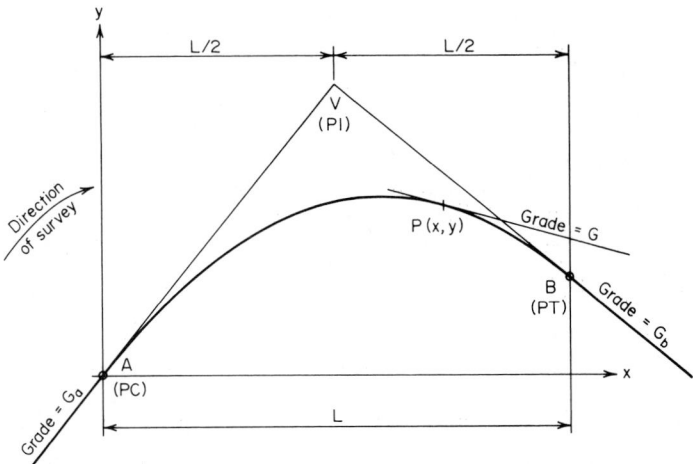

FIG. 200   Parabolic arc.

Calculation Procedure:

1. Compute the required length of curve

Using the notation in Figs. 200 and 201, we have G_a = −4.6 percent; G_b = +1.8 percent; r = rate of change in grade = 0.02 percent per foot; $L = (G_b − G_a)/r = [1.8 − (−4.6)]/0.02 = 320$ ft (97.5 m).

2. Locate the PC and PT

The station of the PC = station of the PI − $L/2$ = (54 + 20) − (1 + 60) = 52 + 60; station of PT = (54 + 20) + (1 + 60) = 55 + 80; elevation of PC = elevation of PI − $G_a L/2$ = 296.30 + 0.046(160) = 303.66 ft (92.556 m); elevation of PT = 296.30 + 0.018(160) = 299.18 ft (91.190 m).

3. Use the average-grade method to find the elevation of each station

Calculate the grade at the given station; calculate the average grade between the PI and that station, and multiply the average grade by the horizontal distance to find the ordinate. Equations used in analyzing a parabolic arc are

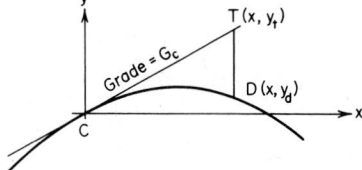

FIG. 201   Tangent offset.

$$y = \frac{rx^2}{2} + G_a x \tag{147}$$

$$G = rx + G_a \tag{148}$$

$$y = (G_a + G)\frac{x}{2} \tag{149}$$

$$DT = -\frac{rx^2}{2} \tag{150a}$$

$$DT = (G_c - G_d)\frac{x}{2} \tag{150b}$$

where DT = distance in Fig. 201.

Applying Eq. 148 with respect to station 53 + 00 yields x = 40 ft (12.2 m); G = 0.0002(40) − 0.046 = −0.038; G_{av} = (−0.046 − 0.038)/2 = −0.042; y = −0.042(40) = −1.68 ft (−51.206 cm); elevation = 303.66 − 1.68 = 301.98 ft (9204.350 cm). Perform these calculations for each station, and record the results in tabular form as shown:

Station	x, ft (m)	G	G_{av}	y, ft (m)	Elevation, ft (m)
52 + 60	0 (0)	−0.046	−0.046	0 (0)	303.66 (92.56)
53 + 00	40 (12.2)	−0.038	−0.042	−1.68 (−0.51)	301.98 (92.04)
53 + 50	90 (27.4)	−0.028	−0.037	−3.33 (−1.01)	300.33 (91.54)
54 + 00	140 (42.7)	−0.018	−0.032	−4.48 (−1.37)	299.18 (91.19)
54 + 50	190 (57.9)	−0.008	−0.027	−5.13 (−1.56)	298.53 (90.99)
55 + 00	240 (73.2)	+0.002	−0.022	−5.28 (−1.61)	298.38 (90.95)
55 + 50	290 (88.4)	+0.012	−0.017	−4.93 (−1.50)	298.73 (91.05)
55 + 80	320 (97.5)	+0.018	−0.014	−4.48 (−1.37)	299.18 (91.19)

4. *Verify the foregoing results*

Apply the principle that for a uniform horizontal spacing the "second differences" between the ordinate are equal. The results are shown:

Calculation of Differences

Elevations, ft (m)	First differences, ft (m)	Second differences, ft (m)
301.98 (92.04)		
	1.65 (0.5029)	
300.33 (91.54)		0.50 (0.1524)
	1.15 (0.3505)	
299.18 (91.19)		0.50 (0.1525)
	0.65 (0.1981)	
298.53 (90.99)		0.50 (0.1524)
	0.15 (0.0457)	
298.38 (90.95)		0.50 (0.1524)
	−0.35 (0.10668)	
298.73 (91.05)		

5. *Apply the tangent-offset method to find the elevation of each station*

Since this method is based on Eq. 147, substitute directly in that equation. For the present case, the equation becomes $y = rx^2/2 + G_a x = 0.0001x^2 - 0.046x$. Record the calculations for y in tabular form. The results, as shown, agree with those obtained by the average-grade method.

Tangent-Offset Method

Station	x, ft (m)	$0.0001x^2$, ft (m)	$0.046x$, ft (m)	y, ft (m)
52 + 60	0 (0)	0 (0)	0 (0)	0 (0)
53 + 00	40 (12.19)	0.16 (0.05)	1.84 (0.56)	−1.68 (−0.51)
53 + 50	90 (27.43)	0.81 (0.25)	4.14 (1.26)	−3.33 (−1.01)
54 + 00	140 (42.67)	1.96 (0.60)	6.44 (1.96)	−4.48 (−1.37)
54 + 50	190 (57.91)	3.61 (1.10)	8.74 (2.66)	−5.13 (−1.56)
55 + 00	240 (73.15)	5.76 (1.76)	11.04 (3.36)	−5.28 (−1.61)
55 + 50	290 (88.39)	8.41 (2.56)	13.34 (4.07)	−4.93 (−1.50)
55 + 80	320 (97.54)	10.24 (3.12)	14.72 (4.49)	−4.48 (−1.37)

6. Locate the sag S

Since the grade is zero at this point, Eq. 148 yields $G_s = rx_s + G_a = 0$; therefore $x_s = -G_a/r$ $= -(-0.046/0.0002) = 230$ ft (70.1 m); station of sag $= (52 + 60) + (2 + 30) = 54 + 90$; $G_{av} = \frac{1}{2} G_a = -0.023$; $y_s = -0.023(230) = -5.29$ ft (1.61 m); elevation of sag $= 303.66 - 5.29 = 298.37$ ft (90.943 m).

7. Verify the location of the sag

Do this by ascertaining that the offsets of the PC and PT from the tangent through S, which is horizontal, satisfy the tangent-offset principle. From the preceding results, tangent offset of PC $= 5.29$ ft (1.612 m); tangent offset of PT $= 5.29 - 4.48 = 0.81$ ft (0.247 m); distance to PC $= 230$ ft (70.1 m); distance to PT $= 320 - 230 = 90$ ft (27.4 m); $5.29/0.81 = 6.53$; $230^2/90^2 = 6.53$. Therefore, the results are verified.

LOCATION OF A SINGLE STATION ON A PARABOLIC ARC

The PC of a vertical parabolic curve is at station $22 + 00$ of elevation 165.30, and the grade at the PC is $+3.2$ percent. The elevation of the station $24 + 00$ is 168.90 ft (51.481 m). What is the elevation of station $25 + 50$?

Calculation Procedure:

1. Compute the offset of P_1 from the tangent through the PC

Refer to Fig. 202. The tangent-offset principle offers the simplest method of solution. Thus $x_1 = 200$ ft (61.0 m); $y_1 = 168.90 - 165.30 = 3.60$ ft (1.097 m); $Q_1T_1 = 200(0.032) = 6.40$ ft (1.951 m); $P_1T_1 = 6.40 - 3.60 = 2.80$ ft (0.853 m).

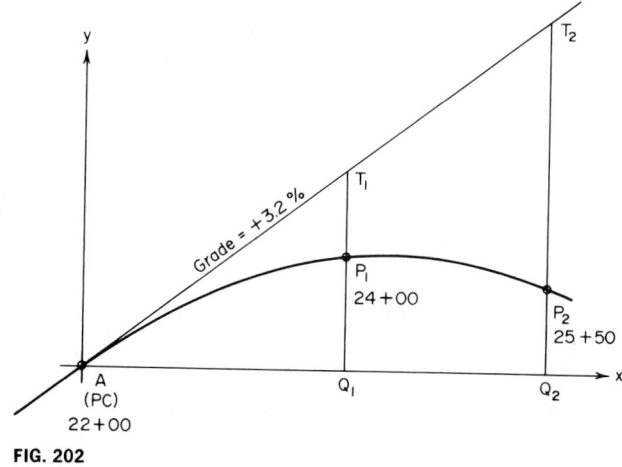

FIG. 202

2. Compute the offset of P_2 from the tangent through the PC; find the elevation of P_2

Thus $x_2 = 350$ ft (106.7 m); $P_2T_2/(P_1T_1) = x_2^2/x_1^2$; $P_2T_2 = 2.80(^{350}\!/_{200})^2 = 8.575$ ft (2.6137 m); $Q_2T_2 = 350(0.032) = 11.2$ ft (3.41 m); $Q_2P_2 = 11.2 - 8.575 = 2.625$ ft (0.8001 m); elevation of $P_2 = 165.30 + 2.625 = 167.925$ ft (51.184 m).

LOCATION OF A SUMMIT

An approach grade of $+1.5$ percent intersects a grade of -2.5 percent at station $29 + 00$ of elevation 226.30 ft (68.976 m). The connecting parabolic curve is to be 800 ft (243.8 m) long. Locate the summit.

Calculation Procedure:

1. Locate the PC

Draw a freehand sketch of the curve, and record all values in the sketch as they are obtained. Thus, station of PC = station of PI − $L/2$ = 25 + 00; elevation of PC = 226.30 − 400(0.015) = 220.30 ft (67.147 m).

2. Calculate the rate of change in grade; locate the summit

Apply the average-grade method to locate the summit. Thus, $r = (-2.5 - 1.5)/800 = -0.005$ percent per foot.

Place the origin of coordinates at the PC. By Eq. 148, $x_s = -G_a/r = 1.5/0.005 = 300$ ft (91.44 m). From the PC to the summit, $G_{av} = \frac{1}{2}G_a = 0.75$ percent. Then $y_s = 300(0.0075) = 2.25$ ft (0.686 m); station of summit = (25 + 00) + (3 + 00) = 28 + 00; elevation of summit = 220.30 + 2.25 = 222.55 ft (67.833 m). The summit can also be located by the tangent-offset method.

PARABOLIC CURVE TO CONTAIN A GIVEN POINT

A grade of −1.6 percent is followed by a grade of +3.8 percent, the grades intersecting at station 42 + 00 of elevation 210.00 ft (64.008 m). The parabolic curve connecting these grades is to pass through station 42 + 60 of elevation 213.70 ft (65.136 m). Compute the required length of curve.

Calculation Procedure:

1. Compute the tangent offsets; establish the horizontal location of P in terms of L

Refer to Fig. 203, where P denotes the specified point. The given data enable computation of the tangent offsets CP and DP, thus giving a relationship between the horizontal distances from A to P and from P to B. Since the distance from the centerline of curve to P is known, the length of curve may readily be found.

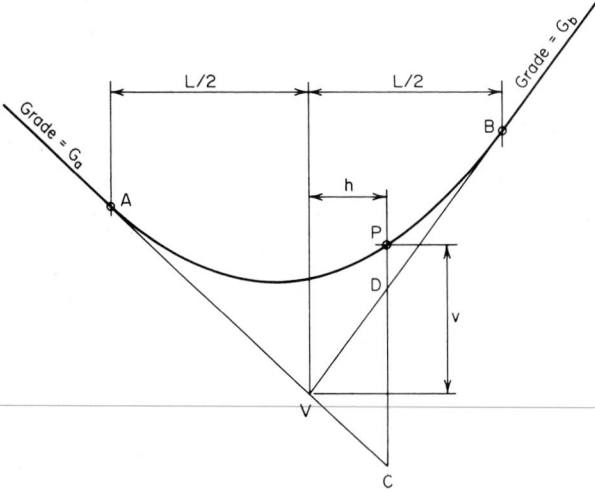

FIG. 203

Computing the tangent offsets gives $CP = v - G_a h$ and $DP = v - G_b h$; but $CP/DP = (L/2 + h)^2/(L/2 - h)^2 = (L + 2h)^2/(L - 2h)^2$; therefore,

$$\frac{L + 2h}{L - 2h} = \left(\frac{v - G_a h}{v - G_b h}\right)^{1/2} \tag{151}$$

2. *Substitute numerical values and solve for L*

Thus, $G_a = -1.6$ percent; $G_b = +3.8$ percent; $h = 60$ ft (18.3 m); $v = 3.70$ ft (1.128 m); then $(L + 120)/(L - 120) = [(3.7 \times 0.016 \times 60)/(3.7 - 0.038 \times 60)]^{1/2} = 1.81$; so $L = 416$ ft (126.8 m).

3. *Verify the solution*

There are many ways of verifying the solution. The simplest way is to compare the offsets of P and B from a tangent through A. By Eq. 150b, offset of B from tangent through $A = 208(0.016 + 0.038) = 11.232$ ft (3.4235 m). From the preceding calculations, offset of P from tangent through $A = 4.66$ ft (1.4203 m); $4.66/11.232 = 0.415$; $(208 + 60)^2/(416)^2 = 0.415$. This is acceptable.

SIGHT DISTANCE ON A VERTICAL CURVE

A vertical summit curve has tangent grades of $+2.6$ and -1.5 percent. Determine the minimum length of curve that is needed to provide a sight distance of 450 ft (137.2 m) to an object 4 in (101.6 mm) in height. Assume that the eye of the motorist is 4.5 ft (1.37 m) above the roadway.

Calculation Procedure:

1. *State the equation for minimum length when S < L*

The vertical curvature of a road must be limited to ensure adequate visibility across the summit. Consequently, the distance across which a given change in grade may be effected is subject to a lower limit imposed by the criterion of sight distance.

Let S denote the required sight distance and L the minimum length of curve. In Fig. 204, let E denote the position of the motorist's eye and P the top of an object. Assume that the curve has the maximum allowable curvature, so that the distance from E to P equals S.

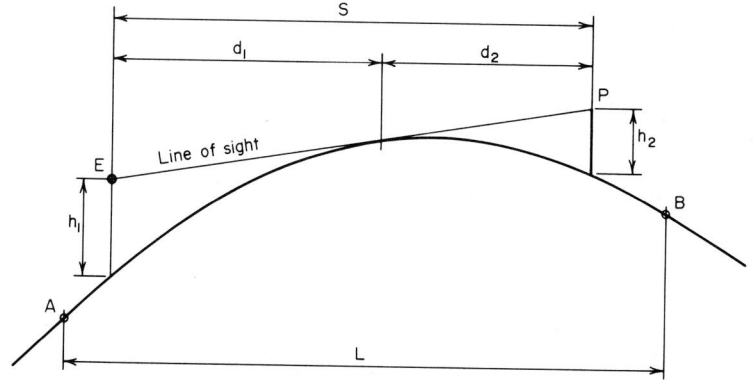

FIG. 204   Visibility on vertical summit curve.

Applying Eq. 150a gives

$$L = \frac{AS^2}{100[(2h_1)^{1/2} + (2h_2)^{1/2})]^2} \qquad (152)$$

whee $A = G_a - G_b$, in percent.

2. *State the equation for L when S > L*

Thus,

$$L = 2S - \frac{200(h_1^{1/2} + h_2^{1/2})^2}{A} \qquad (153)$$

3. Assuming, tentatively, that S < L, compute L

Thus h_1 = 4.5 ft (1.37 m); h_2 = 4 in = 0.33 ft (0.1 m); A = 2.6 + 1.5 = 4.1 percent; L = $4.1(450)^2/[100(9^{1/2} + 0.67^{1/2})^2]$ = 570 ft (173.7 m). Therefore, the assumption that S < L is valid because 450 < 570.

MINE SURVEYING: GRADE OF DRIFT

A vein of ore has a strike of S38°20′E and a northeasterly dip of 33°14′. What is the grade of a drift having a bearing of S43°10′E?

Calculation Procedure:

1. Express β as a function of α and θ

A vein of ore is generally assumed to have plane faces. The *strike*, or *trend*, of the vein is the bearing of any horizontal line in a face, and the *dip* is the angle of inclination of its face. A *drift* is a slightly sloping passage that follows the vein. Any line in a plane perpendicular to the horizontal is a *dip line*. The dip line is the steepest line in a plane, and the dip of the plane equals the angle of inclination of this line. With reference to the inclined plane ABCD in Fig. 205a, let α

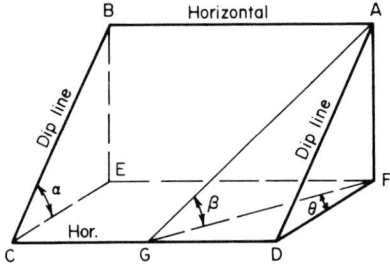
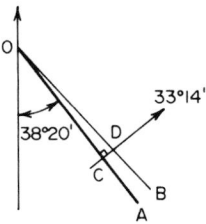

(a) Isometric view of inclined plane   (b) Strike-and-dip diagram (plan)

FIG. 205

= dip of plane; β = angle of inclination of arbitrary line AG; θ = angle between horizontal projections of AG and dip line.

By expressing β as a function of α and θ: tan β = AF/GF; tan α = AF/DF; tan β/tan α = DF/GF = cos θ.

$$\tan \beta = \tan \alpha \cos \theta \qquad (154)$$

2. Find the grade of the drift

Apply Eq. 154. In Fig. 205b, OA is a horizontal line in the vein, OB is the horizontal projection of the drift, and the arrow indicates the direction of dip. Then angle COD = 43°10′ − 38°20′ = 4°50′; θ = angle CDO = 90° − 4°50′ = 85°10′; tan β = tan 33°14′ cos 85°10′ = 0.0552; grade of drift = 5.52 percent.

3. Alternatively, solve without the use of Eq. 154

In Fig. 205b, set OD = 100 ft (30.5 m); let D′ denote the point on the face of the vein vertically below D. Then CD = 100 sin 4°50′ = 8.426 ft (2.5682 m); drop in elevation from O to D′ = drop in elevation from C to D′ = 8.426 tan 33°14′ = 5.52 ft (1.682 m). Therefore, grade = 5.52 percent.

DETERMINING STRIKE AND DIP FROM TWO APPARENT DIPS

Three points on the hanging wall (upper face) of a vein of ore have been located by vertical boreholes. These points, designated P, Q, and R, have these relative positions: P is 142 ft (43.3 m) above Q and 130 ft (39.6 m) above R; horizontal projection of PQ, length = 180 ft (54.9 m) and bearing = S55°32'W; horizontal projection of PR, length = 220 ft (67.1 m) and bearing = N19°26'W. Determine the strike and dip of the vein by both graphical construction and trigonometric calculations.

Calculation Procedure:

1. Plot the given data for the graphical procedure

In Fig. 206a, draw lines PQ and PR in plan in accordance with the given data for their horizontal projections. The angle of inclination of any line other than a dip line is an *apparent dip* of the vein.

2. Draw the elevations

In Fig. 206b and c, draw elevations normal to PQ and PR, respectively, locating the points in accordance with the given differences in elevation. Find the points S and T lying on PQ and PR, respectively, at an arbitrary distance v below P.

3. Draw the representation of the strike of the vein

Locate points S and T in Fig. 206a, and connect them with a straight line. This line is horizontal, and its bearing ϕ, therefore, represents the strike of the vein.

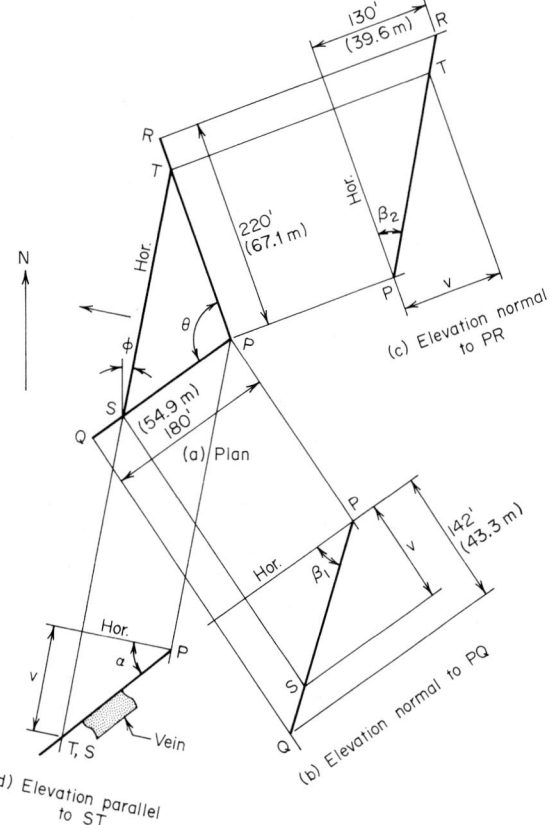

FIG. 206 Determination of strike and dip by orthographic projections.

4. Draw an edge view of the vein

In Fig. 206d, draw an elevation parallel to ST. Since this is an edge view of one line in the face, it is an edge view of the vein itself; it therefore represents the dip α of the vein in its true magnitude.

5. Determine the strike and dip

Scale angles ϕ and α, respectively. In Fig. 206a, the direction of dip is represented by the arrow perpendicular to ST.

6. Draw the dip line for the trigonometric solution

In Fig. 207, draw an isometric view of triangle PST, and draw the dip line PW. Its angle of inclination α equals the dip of the vein. Let O denote the point on a vertical line through P at the same elevation as S and T. Let β_1 and β_2 denote the angle of inclination of PS and PT, respectively; and let θ = angle SOT, θ_1 = angle SOW, θ_2 = angle TOW, m = $\tan \beta_2/\tan \beta_1$.

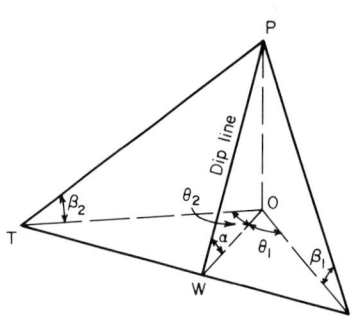

FIG. 207

7. Express θ_1 in terms of the known angles β_1, β_2, and θ

Then substitute numerical values to find the strike ϕ of the vein. Thus

$$\tan \theta_1 = \frac{m - \cos \theta}{\sin \theta} \qquad (155)$$

For this vein, m = $\tan \beta_2/\tan \beta_1$ = $130(180)/[220(142)]$ = 0.749040; θ = $180°$ − $(55°32' + 19°26')$ = $105°02'$. Substituting gives $\tan \theta_1$ = $(0.749040 + 0.259381)/0.965775$; θ_1 = $46°14'15''$; ϕ = $55°32' + 46°14'15'' − 90°$ = $11°46'15''$; strike of vein = N11°46'15''E.

8. Compute the dip of the vein

Use Eq. 154, considering PS as the line of known inclination. Thus, $\tan \alpha$ = $\tan \beta_1/\cos \theta_1$; α = $48°45'25''$.

9. Verify these results

Apply Eq. 154, considering PT as the line of known inclination. Thus θ_2 = $\theta − \theta_1$ = $105°02' − 46°14'15''$ = $58°47'45''$; $\tan \alpha$ = $\tan \beta_2/\cos \theta_2$; α = $48°45'25''$. This value agrees with the earlier computed value.

DETERMINATION OF STRIKE, DIP, AND THICKNESS FROM TWO SKEW BOREHOLES

In Fig. 208a, A and B represent points on the earth's surface through which skew boreholes were sunk to penetrate a vein of ore. Point B is 110 ft (33.5 m) due south of A. The data for these boreholes are as follows. *Borehole through A:* surface elevation = 870 ft (265.2 m); inclination = 49°; bearing of horizontal projection = N58°30′E. The hanging wall and footwall (lower face of vein) were struck at distances of 55 ft (16.8 m) and 205 ft (62.5 m), respectively, measured along the borehole. *Borehole through B:* surface elevation = 842 ft (256.6 m); inclination = 73°; bearing of horizontal projection = S44°50′E. The hanging wall and footwall were struck at distances of 98 ft (29.9 m) and 182 ft (55.5 m), respectively, measured along the borehole. Determine the strike, dip, and thickness of the vein by both graphical construction and trigonometric calculations.

Calculation Procedure:

1. Draw horizontal projections of the boreholes

Since the vein is assumed to have uniform thickness, the hanging wall and footwall are parallel. Two straight lines determine a plane. In the present instance, two points on the hanging wall and

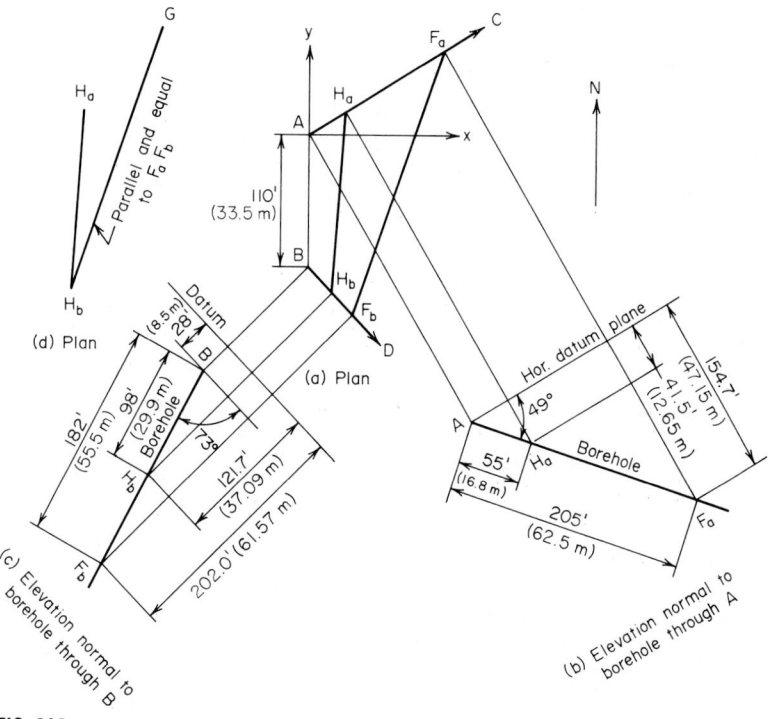

FIG. 208

two points on the footwall are given, enabling one line to be drawn in each of two parallel planes. These planes may be located by using these principles:

a. Consider a plane P and line L parallel to each other. If through any point on P a line is drawn parallel to L, this line lies in plane P.

b. Lines that are parallel and equal in length appear to be parallel and equal in length in all orthographic views.

These principles afford a means of locating a second line in the hanging wall or footwall.

Applying the specified bearings, draw the horizontal projections AC and BD of the boreholes in Fig. 208a.

2. *Locate the points of intersection with the hanging wall and footwall in elevation*

In Fig. 208b, draw an elevation normal to the borehole through A; locate the points of intersection H_a and F_a with the hanging wall and footwall, respectively. Select the horizontal plane through A as datum.

3. *Repeat the foregoing construction with respect to borehole through B*

This construction is shown in Fig. 208c.

4. *Locate the points of intersection in plan*

In Fig. 208a, locate the points of intersection. Draw lines H_aH_b and F_aF_b. The former lies in the hanging wall and the latter in the footwall. To avoid crowding, reproduce the plan of line H_aH_b in Fig. 208d.

5. *Draw the plan of a line H_bG that is parallel and equal in length to F_aF_b*

Do this by applying the second principle given above. In accordance with principle a, H_bG lies in the hanging wall, and this plane is therefore determined. The ensuing construction parallels that in the previous calculation procedure.

6.   Establish a system of rectangular coordinate axes

Use A as the origin (Fig. 208a). Make x the east-west axis, y the north-south axis, and z the vertical axis.

7.   Apply the given data to obtain the coordinates of the intersection points and point G

For example, with respect to F_a, $y = 205 \cos 49° \cos 58°30'$. The coordinates of G are obtained by adding to the coordinates of H_b the differences between the coordinates of F_a and F_b. The results are shown:

Point	x, ft (m)	y, ft (m)	z, ft (m)
H_a	30.8 (9.39)	18.9 (5.76)	-41.5 (-12.65)
H_b	20.2 (6.16)	-130.3 (-39.72)	-121.7 (-37.09)
F_a	114.7 (34.96)	70.3 (21.43)	-154.7 (-47.15)
F_b	37.5 (11.43)	-147.7 (-45.02)	-202.0 (-61.57)
G	97.4 (29.69)	87.7 (26.73)	-74.4 (-22.68)

8.   For convenience, reproduce the plan of the intersection points, and G

This is shown in Fig. 209a.

9.   Locate the point S at the same elevation as G

In Fig. 209b, draw an elevation normal to H_aH_b, and locate the point S on this line at the same elevation as G.

10.   Establish the strike of the plane

Locate S in Fig. 209a, and draw the horizontal line SG. Since both S and G lie on the hanging wall, the strike of this plane is now established.

11.   Complete the graphical solution

In Fig. 209c, draw an elevation parallel to SG. The line through H_a and H_b and that through F_a and F_b should be parallel to each other. This drawing is an edge view of the vein, and it presents the dip α and thickness t in their true magnitude. The graphical solution is now completed.

12.   Reproduce the plan view

For convenience, reproduce the plan of H_a, H_b, and G in Fig. 209d. Draw the horizontal projection of the dip line, and label the angles as indicated.

13.   Compute the lengths of lines H_aH_b and H_aG

Compute these lengths as projected on each coordinate axis and as projected on a horizontal plane. Use absolute values. Thus, line H_aH_b: $L_x = 30.8 - 20.2 = 10.6$ ft (3.23 m); $L_y = 18.9 - (-130.3) = 149.2$ ft (45.48 m); $L_z = -41.5 - (-121.7) = 80.2$ ft (24.44 m); $L_{\text{hor}} = (10.6^2 + 149.2^2)^{0.5} = 149.6$ ft (45.60 m). Line H_aG: $L_x = 97.4 - 30.8 = 66.6$ ft (20.30 m); $L_y = 87.7 - 18.9 = 68.8$ ft (20.97 m); $L_z = -41.5 - (-74.4) = 32.9$ ft (10.03 m); $L_{\text{hor}} = (66.6^2 + 68.8^2)^{0.5} = 95.8$ ft (29.20 m).

14.   Compute the bearing and inclination of lines H_aH_b and H_aG

Let ϕ_1 = bearing of H_aH_b; ϕ_2 = bearing of H_aG; β_1 = angle of inclination of H_aH_b; β_2 = angle of inclination of H_aG. Then $\tan \phi_1 = 10.6/149.2$; ϕ_1 = S4°04'W; $\tan \phi_2 = 66.6/68.8$; ϕ_2 = N44°04'E; $\tan \beta_1 = 80.2/149.6$; $\tan \beta_2 = 32.9/95.8$.

15.   Compute angle θ shown in Fig. 209d; determine the strike of the vein, using Eq. 155

Thus, $\theta = 180° + \phi_1 - \phi_2 = 140°00'$; $m = \tan \beta_2/\tan \beta_1 = 0.6406$; $\tan \theta_1 = (m - \cos 140°00')/\sin 140°00'$; $\theta_1 = 65°26'$; $\theta_2 = 74°34'$. The bearing of the horizontal projection of the dip line $= \theta_1 - \phi_1 = $ S61°22'E; therefore, the strike of the vein = N28°38'E.

16.   Compute the dip α of the vein

By Eq. 154, $\tan \alpha = \tan \beta_1/\cos \theta_1$; $\alpha = 52°12'$; or $\tan \alpha = \tan \beta_2/\cos \theta_2$; $\alpha = 52°14'$. This slight discrepancy between the two computed values falls within the tolerance of these calculations. Use the average value $\alpha = 52°13'$.

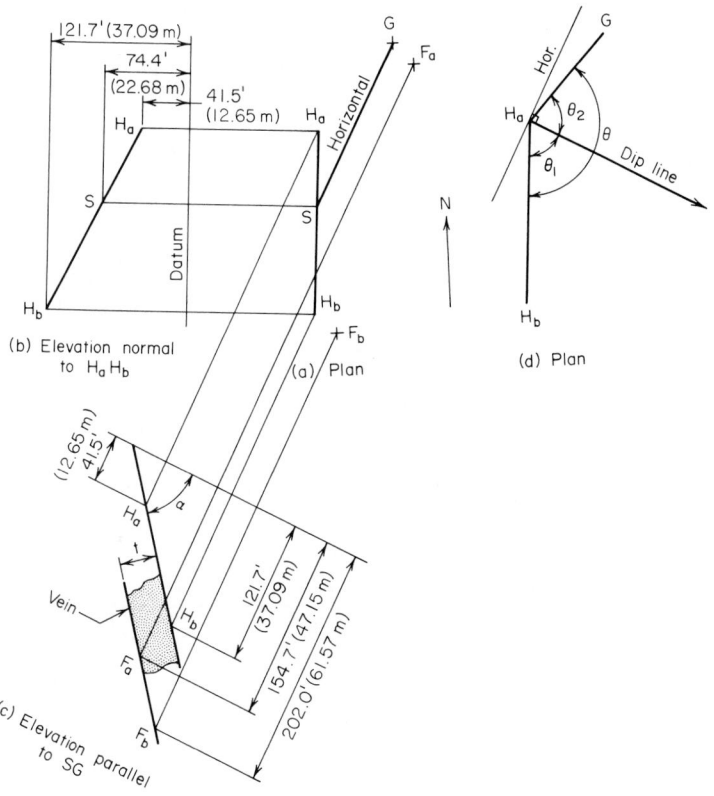

FIG. 209

17.   *Establish the relationship between the true thickness t of a vein and its apparent thickness t′ as measured along a skew borehole*

Refer to Figs. 208 and 209. Let δ = angle of inclination of borehole; γ = angle in plan between downward-sloping segments of borehole and dip line of vein. Then

$$t = t'(\cos \alpha \sin \delta - \sin \alpha \cos \delta \cos \gamma) \tag{156}$$

18.   *Find the true thickness, using Eq. 156*

Thus, borehole through A: $\delta = 49°$; $\gamma = 180° - (58°30' + 61°22') = 60°08'$; $t' = 205 - 55 = 150$ ft (45.7 m); $t = 150(\cos 52°13' \sin 49° - \sin 52°13' \cos 49° \cos 60°08') = 30.6$ ft (9.3 m). For the borehole through B: $\delta = 73°$; $\gamma = 61°22' - 44°50' = 16°32'$; $t' = 182 - 98 = 84$ ft (25.6 m); $t = 84(\cos 52°13' \sin 73° - \sin 52°13' \cos 73° \cos 16°32') = 30.6$ ft (9.3 m). This agrees with the value previously computed.

Aerial Photogrammetry

FLYING HEIGHT REQUIRED TO YIELD A GIVEN SCALE

At what altitude above sea level must an aircraft fly to obtain vertical photography having an average scale of 1 cm = 120 m if the camera lens has a focal length of 152 mm and the average elevation of the terrain to be surveyed is 290 m?

Calculation Procedure:

1.   *Write the equation for the scale of a vertical photograph*

In aerial photogrammetry, the term *photograph* generally refers to the positive photograph, and the plane of this photograph is considered to lie on the object side of the lens. A photograph is said to be *vertical* if the optical axis of the lens is in a vertical position at the instant of exposure. Since the plane of the photograph is normal to the optical axis, this plane is horizontal.

In Fig. 210a, point L is the *front nodal point* of the lens; a ray of light directed at this point leaves the lens without undergoing a change in direction. The point o at which the optical axis intersects the plane of the photograph is called the *principal point*. The distance from the ground to the camera may be considered infinite in relation to the dimensions of the lens, and so the distance Lo is equal to the focal length of the lens. The aircraft is assumed to be moving in a horizontal straight line, termed the *line of flight*, and the elevation of L above the horizontal

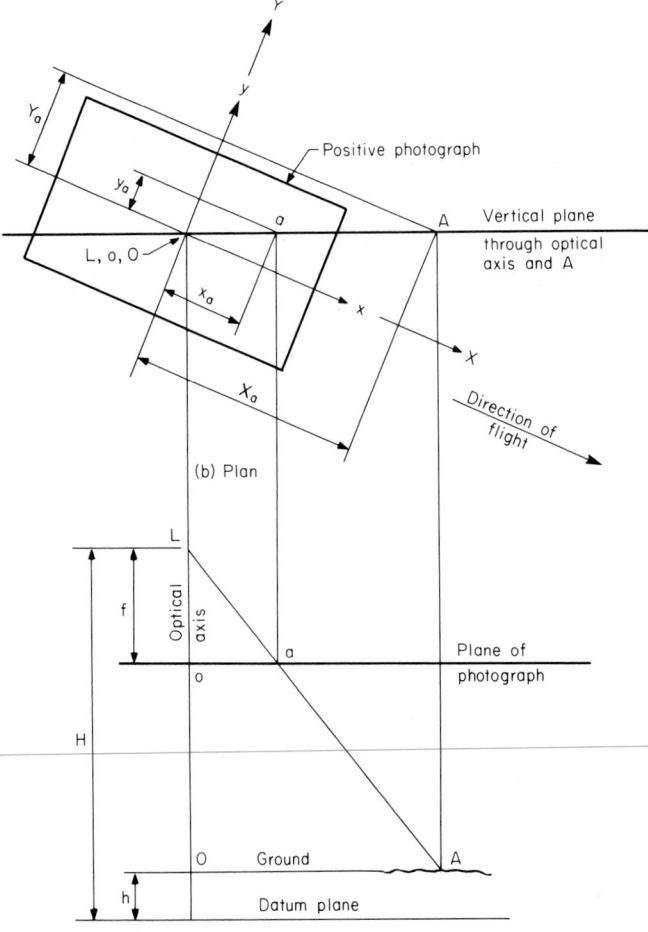

(b) Plan

(a) Elevation normal to vertical plane through
optical axis and point A

FIG. 210

datum plane is called the *flying height*. The position of L in space at the instant of exposure is called the *exposure station*. Where the area to be surveyed is relatively small, the curvature of the earth may be disregarded.

Since the plane of the photograph is horizontal, Fig. 210*b* is a view normal to this plane and so presents all distances in this plane in their true magnitude. In the photograph, the origin of coordinates is placed at o. The x axis is placed parallel to the line of flight, with x values increasing in the direction of flight, and the y axis is placed normal to the x axis.

In Fig. 210, A is a point on the ground, a is the image of A on the photograph, and O is a point at the same elevation as A that lies on the prolongation of Lo. Thus, o is the image of O. The *scale* of a photograph, expressed as a fraction, is the ratio of a distance in the photograph to the corresponding distance along the ground. In this case, the ratio is 1 cm/120 m = 0.01 m/120 m = 1/12,000.

Let H = flying height; h = elevation of A above datum; f = focal length; S = scale of photograph, expressed as a fraction. From Fig. 210, $S = oa/OA$, and by similar triangles $S = f/(H - h)$.

2.   *Solve this equation for the flying height*

Take sea level as datum. From the foregoing equation, with the meter as the unit of length, $H = h + f/S = 290 + 0.152/(1/12,000) = 290 + 0.152(12,000) = 2114$ m. This is the required elevation of L above sea level.

DETERMINING GROUND DISTANCE BY VERTICAL PHOTOGRAPH

Two points A and B are located on the ground at elevations of 250 and 190 m, respectively, above sea level. The images of A and B on a vertical aerial photograph are a and b, respectively. After correction for film shrinkage and lens distortion, the coordinates of a and b in the photograph are $x_a = -73.91$ mm, $y_a = +44.78$ mm, $x_b = +84.30$ mm, and $y_b = -21.65$ mm, where the subscript identifies the point. The focal length is 209.6 mm, and the flying height is 2540 m above sea level. Determine the distance between A and B as measured along the ground.

Calculation Procedure:

1.   *Determine the relationship between coordinates in the photograph and those in the datum plane*

Refer to Fig. 210, and let X and Y denote coordinate axes that are vertically below the x and y axes, respectively, and in the datum plane. Omitting the subscript, we have $x/X = y/Y = oa/OA = S = f/(H - h)$, giving $X = x(H - h)/f$ and $Y = y(H - h)/f$.

2.   *Compute the coordinates of A and B in the datum plane*

For A, $H - h = 2540 - 250 = 2290$ m. Substituting gives $X_A = (-0.07391)(2290)/0.2096 = -807.5$ m and $Y_A = (+0.04478)(2290)/0.2096 = +489.2$ m. For B, $H - h = 2540 - 190 = 2350$ m. Then $X_B = (+0.08430)(2350)/0.2096 = +945.2$ m, and $Y_B = (-0.02165)(2350)/0.2096 = -242.7$ m.

3.   *Compute the required distance*

Let $\Delta X = X_A - X_B$, $\Delta Y = Y_A - Y_B$, and AB = distance between A and B as measured along the ground. Disregarding the difference in elevation of the two points, we have $(AB)^2 = (\Delta X)^2 + (\Delta Y)^2$. Then $\Delta X = -1752.7$ m, $\Delta Y = 731.9$ m, and $(AB)^2 = (1752.7)^2 + (731.9)^2$, or $AB = 1899$ m.

DETERMINING THE HEIGHT OF A STRUCTURE BY VERTICAL PHOTOGRAPH

In Fig. 211, points A and B are located at the top and bottom, respectively, and on the vertical centerline of a tower. These points have images a and b, respectively, on a vertical aerial photograph having a scale of 1:10,800 with reference to the ground, which is approximately level. In the photograph, $oa = 76.61$ mm and $ob = 71.68$ mm. The focal length is 210.1 mm. Find the height of the tower.

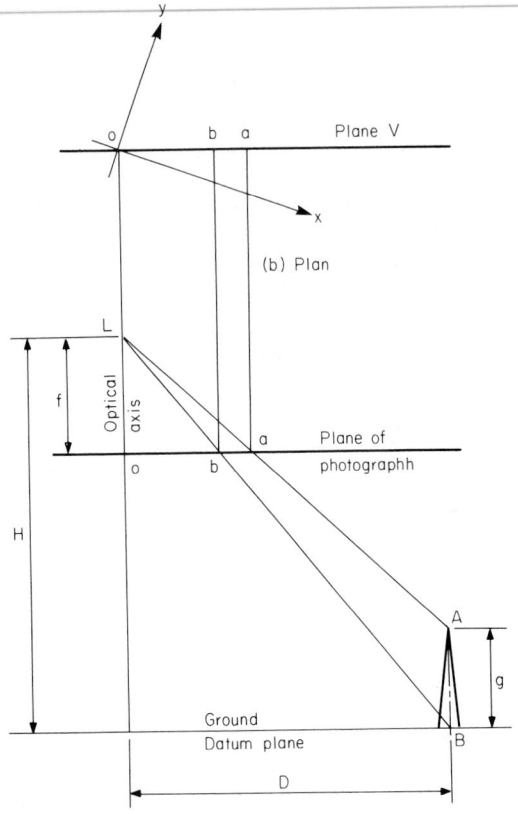

(a) Elevation normal to vertical plane through
optical axis and center of tower (plane V)

FIG. 211

Calculation Procedure:

1. Compute the flying height with reference to the ground

Take the ground as datum. Then scale $S = f/H$, or $H = f/S = 0.2101/(1/10,800) = 0.2101(10,800) = 2269$ m.

2. Establish the relationship between height of tower and distances in the photograph

Let g = height of tower. In Fig. 211, $oa/D = f/(H - g)$ and $ob/D = f/H$. Thus, $oa/ob = H/(H - g)$. Solving gives $g = H(1 - ob/oa)$.

3. Compute the height of tower

Substituting in the foregoing equation yields $g = 2269(1 - 71.68/76.61) = 146$ m.

Related Calculations: Let A denote a point at an elevation h above the datum, let B denote a point that lies vertically below A and in the datum plane, and let a and b denote the images of A and B, respectively. As Fig. 211 shows, a and b lie on a straight line that passes through o, which is called a *radial line*. The distance $d = ba$ is the displacement of the image of A resulting from its elevation above the datum, and it is termed the *relief displacement* of A. Thus, the relief displacement of a point is radially outward if that point lies above datum and radially inward if it lies below datum. From above, $ob/oa = (H - h)/H$, where H = flying height above datum. Then $d = oa - ob = (oa)h/H$.

DETERMINING GROUND DISTANCE BY TILTED PHOTOGRAPH

Two points A and B are located on the ground at elevations of 180 and 130 m, respectively, above sea level. Points A and B have images a and b, respectively, on an aerial photograph, and the coordinates of the images are $x_a = +40.63$ mm, $y_a = -73.72$ mm, $x_b = -78.74$ mm, and $y_b = +20.32$ mm. The focal length is 153.6 mm, and the flying height is 2360 m above sea level. By use of ground control points, it was established that the photograph has a tilt of 2°54′ and a swing of 162°. Determine the distance between A and B.

Calculation Procedure:

1. *Compute the transformed coordinates of the images*

Refer to Fig. 212, where L again denotes the front nodal point of the lens and o denotes the principal point. A photograph is said to be *tilted*, or *near vertical*, if by inadvertence the optical

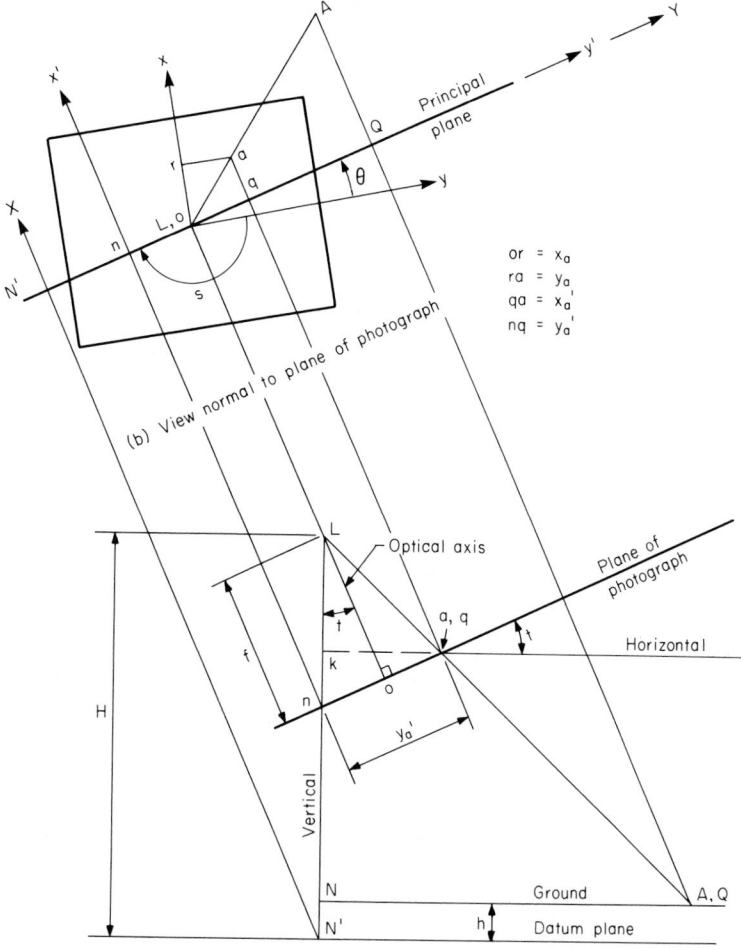

$$or = x_a$$
$$ra = y_a$$
$$qa = x_a'$$
$$nq = y_a'$$

(b) View normal to plane of photograph

(a) Elevation normal to principal plane

FIG. 212

axis of the lens is displaced slightly from the vertical at the time of exposure. The *tilt t* is the angle between the optical axis and the vertical. The *principal plane* is the vertical plane through the optical axis. Since the plane of the photograph is normal to the optical axis, it is normal to the principal plane. Therefore, Fig. 212a is an edge view of the plane of the photograph. Moreover, the angle between the plane of the photograph and the horizontal equals the tilt. In Fig. 212, A is a point on the ground and a is its image. Line AQ is normal to the principal plane, Q lies in that plane, and q is the image of Q.

Consider the vertical line through L. The points n and N at which this line intersects the plane of the photograph and the ground are called the *nadir point* and *ground nadir point*, respectively. The line of intersection of the principal plane and the plane of the photograph, which is line *no* prolonged, is termed the *principal line*. Now consider the vertical plane through o parallel to the line of flight. In the photograph, the x axis is placed on the line at which this vertical plane intersects the plane of the photograph, with x values increasing in the direction of flight. The y axis is normal to the x axis, and the origin lies at o. The *swing s* is the angle in the plane of the photograph, measured in a clockwise direction, between the positive side of the y axis and the radial line extending from o to n.

Transform the x and y axes in this manner: First, rotate the axes in a counterclockwise direction until the y axis lies on the principal line with its positive side on the upward side of the photograph; then displace the origin from o to n. Let x' and y' denote, respectively, the axes to which the x and y axes have been transformed. The x' axis is horizontal. Let θ denote the angle through which the axes are rotated in the first step of the transformation. From Fig. 212b, $\theta = 180° - s$.

The transformed coordinates of a point in the plane of the photograph are $x' = x \cos \theta + y \sin \theta$; $y' = -x \sin \theta + y \cos \theta + f \tan t$. In this case, $t = 2°54'$ and $\theta = 180° - 162° = 18°$. Then $x'_a = +40.63 \cos 18° - 73.72 \sin 18° = +15.86$ mm; $y'_a = -(+40.63) \sin 18° + (-73.72) \cos 18° + 153.6 \tan 2°54' = -74.89$ mm. Similarly, $x'_b = -78.74 \cos 18° + 20.32 \sin 18° = -68.61$ mm; $y'_b = -(-78.74) \sin 18° + 20.32 \cos 18° + 153.6 \tan 2°54' = +51.44$ mm.

2.   *Write the equations of the datum-plane coordinates*

Let X and Y denote coordinate axes that lie in the datum plane and in the same vertical planes as the x' and y' axes, respectively, as shown in Fig. 212. Draw the horizontal line kq in the principal plane. Then $kq = y'_a \cos t$ and $Lk = f \sec t - y'_a \sin t$. From Fig. 212b, $QA/qa = LQ/Lq$. From Fig. 212a, $LQ/Lq = LN/Lk = (H - h)/(f \sec t - y'_a \sin t)$. Setting $QA = X_A$, we have $qa = x'_a$, and omitting subscripts gives $X = x'(H - h)/(f \sec t - y' \sin t)$, Eq. a. From Fig. 212a, $NQ/kq = LN/Lk = (H - h)/(f \sec t - y'_a \sin t)$. Setting $NQ = Y_A$ and omitting subscripts, we get $Y = y'[(H - h)/(f \sec t - y' \sin t)] \cos t$, Eq. b.

3.   *Compute the datum-plane coordinates*

First compute $f \sec t = 153.6 \sec 2°54' = 153.8$ mm. For A, $H - h = 2360 - 180 = 2180$ m, and $f \sec t - y' \sin t = 153.8 - (-74.89) \sin 2°54' = 157.6$ mm. Then $(H - h)/(f \sec t - y' \sin t) = 2180/0.1576 = 13,830$. By Eq. a, $X_A = (+0.01586)(13,830) = +219.3$ m. By Eq. b, $Y_A = (-0.07489)(13,830) \cos 2°54' = -1034.4$ m.

Similarly, for B, $H - h = 2360 - 130 = 2230$ m and $f \sec t - y' \sin t = 153.8 - 51.44 \sin 2°54' = 151.2$ mm. Then $(H - h)/(f \sec t - y' \sin t) = 2230/0.1512 = 14,750$. By Eq. a, $X_B = (-0.06861)(14,750) = -1012.0$ m. By Eq. b, $Y_B = (+0.05144)(14,750) \cos 2°54' = +757.8$ m.

4.   *Compute the required distance*

Disregarding the difference in elevation of the two points and proceeding as in the second previous calculation procedure, we have $\Delta X = +219.3 - (-1012.0) = +1231.3$ m, and $\Delta Y = -1034.4 - 757.8 = -1792.2$ m. Then $(AB)^2 = (1231.3)^2 + (1792.2)^2$, or $AB = 2174$ m.

Related Calculations: The X and Y coordinates found in step 3 can be verified by assuming that these values are correct, calculating the corresponding x' and y' coordinates, and comparing the results with the values in step 2. The procedure is as follows. In Fig. 212a, let v_A = angle NLQ. Then $\tan v_A = NQ/LN = Y_A/(H - h)$. Also, angle $oLq = v_A - t$. Now, $x'_a/X_A = Lq/LQ = f \sec (v_A - t)/[(H - h) \sec v_A]$. Rearranging and omitting subscripts, we get $x' = Xf \cos v_A/[(H - h) \cos (v_A - t)]$, Eq. c. Similarly, $y'_a = no + oq = f \tan t + f \tan (v_A - t)$. Omitting the subscript gives $y' = f[\tan t + \tan (v_A - t)]$, Eq. d.

As an illustration, consider point A in the present calculation procedure, which has the computed coordinates $X_A = +219.3$ m and $Y_A = -1034.4$ m. Then $\tan v_A = -1034.4/2180 = -0.4745$. Thus, $v_A = -25°23'$ and $v_A - t = -25°23' - 2°54' = -28°17'$. By Eq. c, $x' = (+219.3)(0.1536)(0.9035)/(2180)(0.8806) = +0.01585$ m $= +15.85$ mm. Applying Eq. d with $t = 2°54'$ gives $y' = 153.6(0.0507 - 0.5381) = -74.86$ mm. If we allow for roundoff effects, these values agree with those in step 1.

The following equation, which contains the four coordinates x', y', X, and Y, can be applied to test these values for consistency:

$$\frac{f^2 + (y' - f \tan t)^2}{x'^2} = \frac{(H - h)^2 + Y^2}{X^2}$$

DETERMINING ELEVATION OF A POINT BY OVERLAPPING VERTICAL PHOTOGRAPHS

Two overlapping vertical photographs contain point P and a control point C that lies 284 m above sea level. The air base is 768 m, and the focal length is 152.6 mm. The micrometer readings on a parallax bar are 15.41 mm for P and 11.37 mm for C. By measuring the displacement of the initial principal point and obtaining its micrometer reading, it was established that the parallax of a point equals its micrometer reading plus 76.54 mm. Find the elevation of P.

Calculation Procedure:

1. Establish the relationship between elevation and parallax

Two successive photographs are said to overlap if a certain amount of terrain appears in both. The ratio of the area that is common to the two photographs to the total area appearing in one photograph is called the *overlap*. (In practice, this value is usually about 60 percent.) The distance between two successive exposure stations is termed the *air base*. If a point on the ground appears in both photographs, its image undergoes a displacement from the first photograph to the second, and this displacement is known as the *parallax* of the point. This quantity is evaluated by using the micrometer of a *parallax bar* and then increasing or decreasing the micrometer reading by some constant.

Assume that there is no change in the direction of flight. As stated, the x axis in the photograph is parallel to the line of flight, with x values increasing in the direction of flight. Refer to Fig. 213, where photographs 1 and 2 are two successive photographs and the subscripts correspond to the photograph numbers. Let A denote a point in the overlapping terrain, and let a denote its image, with the proper subscript. Figure 213c discloses that $y_{1a} = y_{2a}$; thus, parallax occurs solely in the direction of flight. Let p = parallax and B = air base. Then $p = x_{1a} - x_{2a} = o_1m_1 - o_2m_2 = o_1m_1 + m_2o_2$. Thus, $m_1m_2 = B - p$. By proportion, $(B - p)/B = (H - h - f)/(H - h)$, giving $p/B = f/(H - h)$, or $p = Bf/(H - h)$, Eq. a. Thus, the parallax of a point is inversely proportional to the vertical projection of its distance from the front nodal point of the lens.

2. Determine the flying height

From the given data, $B = 768$ m and $f = 152.6$ mm. Take sea level as datum. For the control point, $h = 284$ m and $p = 11.37 + 76.54 = 87.91$ mm. From Eq. a, $H = h + Bf/p$, or $H = 284 + 768(0.1526)/0.08791 = 1617$ m.

3. Compute the elevation of P

For this point, $p = 15.41 + 76.54 = 91.95$ mm. From Eq. a, $h = H - Bf/p$, or $h = 1617 - 768(0.1526)/0.09195 = 342$ m above sea level.

DETERMINING AIR BASE OF OVERLAPPING VERTICAL PHOTOGRAPHS BY USE OF TWO CONTROL POINTS

The air base of two successive vertical photographs is to be found by using two control points, R and S, that lie in the overlapping area. The images of R and S are r and s, respectively. The following data were all obtained by measurement: The length of the straight line RS is 2073 m.

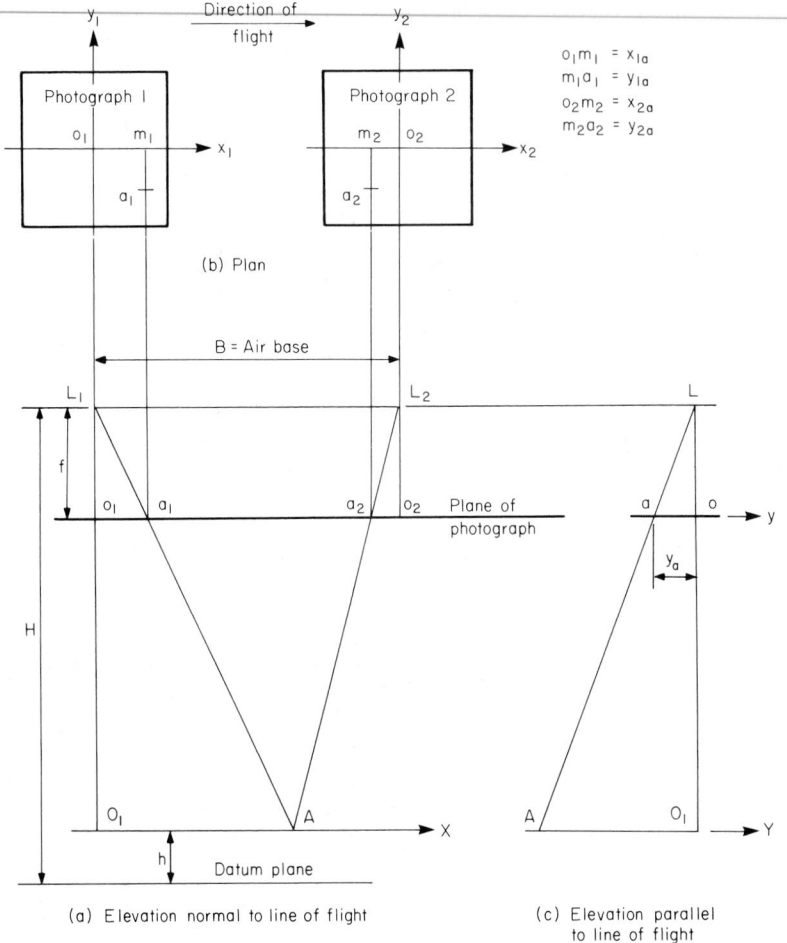

(b) Plan

B = Air base

(a) Elevation normal to line of flight

(c) Elevation parallel to line of flight

FIG. 213

The parallax of R is 92.03 mm, and that of S is 91.85 mm. The coordinates of the images in the left photograph are $x_r = +86.46$ mm, $y_r = -54.32$ mm, $x_s = +29.41$ mm, and $y_s = +56.93$ mm. Compute the air base.

Calculation Procedure:

1. *Express the ground coordinates of the endpoints in terms of the air base*

Refer to Fig. 213, and let X and Y denote coordinate axes that lie vertically below the x_1 and y_1 axes, respectively, and at the same elevation as A. Thus, O_1 is the origin of this system of coordinates. With reference to point A, by proportion, $X_A/x_{1a} = Y_A/y_{1a} = (H - h)/f$. From the previous calculation procedure, $(H - h)/f = B/p$. Omitting the subscript 1, we have $X_A = (x_a/p)B$ and $Y_A = (y_a/p)B$. Then $X_R = (+86.46/92.03)B = +0.9395B$; $Y_R = (-54.32/92.03)B = -0.5902B$; $X_S = (+29.41/91.85)B = +0.3202B$; $Y_S = (+56.93/91.85)B = +0.6198B$.

2. *Express the distance between the control points in terms of the air base; solve the resulting equation*

Disregarding the difference in elevation of the two points, we have $(RS)^2 = (X_R - X_S)^2 + (Y_R - Y_S)^2$. Now, $X_R - X_S = +0.6193B$ and $Y_R - Y_S = -1.2100B$. Then $2073^2 = [(0.6193)^2 + (1.2100)^2]B^2$, or $B = 1525$ m.

DETERMINING SCALE OF OBLIQUE PHOTOGRAPH

In a high-oblique aerial photograph, the distance between the apparent horizon and the principal point as measured along the principal line is 86.85 mm. The flying height is 2925 m above sea level, and the focal length is 152.7 mm. What is the scale of this photograph along a line that is normal to the principal line and at a distance of 20 mm above the principal point as measured along the principal line?

Calculation Procedure:

1. *Locate the true horizon in the photograph*

Refer to Fig. 214. An *oblique* aerial photograph is one that is taken with the optical axis intentionally displaced from the vertical, and a *high-oblique* photograph is one in which this displace-

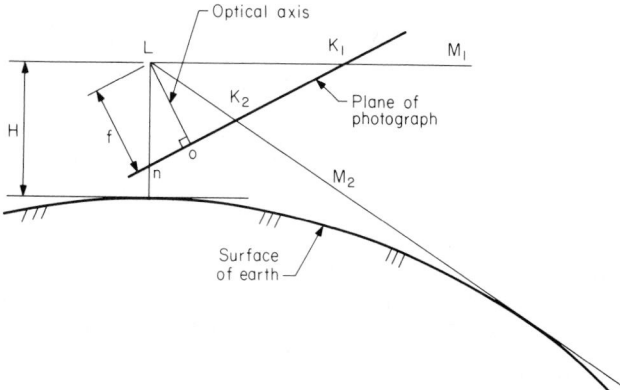

FIG. 214   Elevation normal to principal plane.

ment is sufficiently large to bring the earth's surface into view. By definition, the principal plane is the vertical plane that contains the optical axis, and the principal line is the line of intersection of this vertical plane and the plane of the photograph.

Assume that the terrain is truly level. The *apparent horizon* is the slightly curved boundary line in the photograph between earth and sky. Consider a conical surface that has its vertex at the front nodal point L and that is tangent to the spherical surface of the earth. If atmospheric refraction were absent, the apparent horizon would be the arc along which this conical surface intersected the plane of the photograph. The *true horizon* is the straight line along which the horizontal plane through L intersects the plane of the photograph; it is normal to the principal line. In Fig. 214, M_1 and M_2 are lines in the principal plane that pass through L; line M_1 is horizontal, and M_2 is tangent to the earth's surface. Points K_1 and K_2 are the points at which M_1 and M_2, respectively, intersect the plane of the photograph; these points lie on the principal line. Point K_1 lies on the true horizon; if atmospheric refraction is tentatively disregarded, K_2 lies on the apparent horizon.

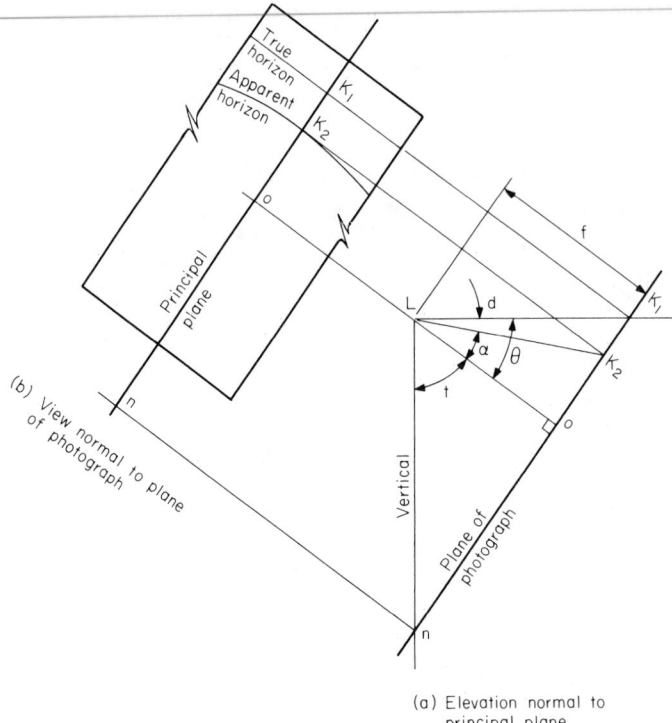

(a) Elevation normal to
principal plane

FIG. 215

Refer to Fig. 215a. The principal plane contains the *angle of dip d*, which is angle K_2LK_1; the *apparent depression angle* α, which is angle oLK_2; the (true) *depression angle* θ, which is angle oLK_1. Then $\theta = d + \alpha$. Let $H =$ flying height above sea level in meters, and $d' =$ angle of dip in minutes. Then $d' = 1.775 \sqrt{H}$, Eq. *a*. This relationship is based on the mean radius of the earth, and it includes allowance for atmospheric refraction. From Fig. 215a, $\tan \alpha = oK_2/f$, Eq. *b*. Then $d' = 1.775 \sqrt{2925} = 96.0'$, or $d = 1°36'$. Also, $\tan \alpha = 86.85/152.7 = 0.5688$, giving $\alpha = 29°38'$. Thus, $\theta = 1°36' + 29°38' = 31°14'$. From Fig. 215a, $oK_1 = f \tan \theta$, or $oK_1 = 152.7(0.6064) = 92.60$ mm. This dimension serves to establish the true horizon.

2.   Write the equation for the scale of a constant-scale line

Since the optical axis is inclined, the scale S of the photograph is constant only along a line that is normal to the principal line, and so such a line is called a *constant-scale line*. As we shall find, every constant-scale line has a unique value of S.

Refer to Fig. 216, where A is a point on the ground and a is its image. Line AQ is normal to the principal plane, Q lies in that plane, and q is the image of Q. Line Rq is a horizontal line in the principal plane. If the terrain is truly level and curvature of the earth may be disregarded, the vertical projection of the distance from A to L is H. Let $e =$ distance in photograph from true horizon to line qa. Along this line, $S = qa/QA = Lq/LQ = LR/LN$. But $LR = e \cos \theta$ and $LN = H$. Thus, $S = (e \cos \theta)/H$, Eq. *c*.

3.   Compute the scale along the specified constant-scale line

From above, $\theta = 31°14'$ and $oK_1 = 92.60$ mm. Then $e = 92.60 - 20 = 72.60$ mm. By Eq. *c*, $S = (0.07260)(0.8551)/2925 = 1/47,120$.

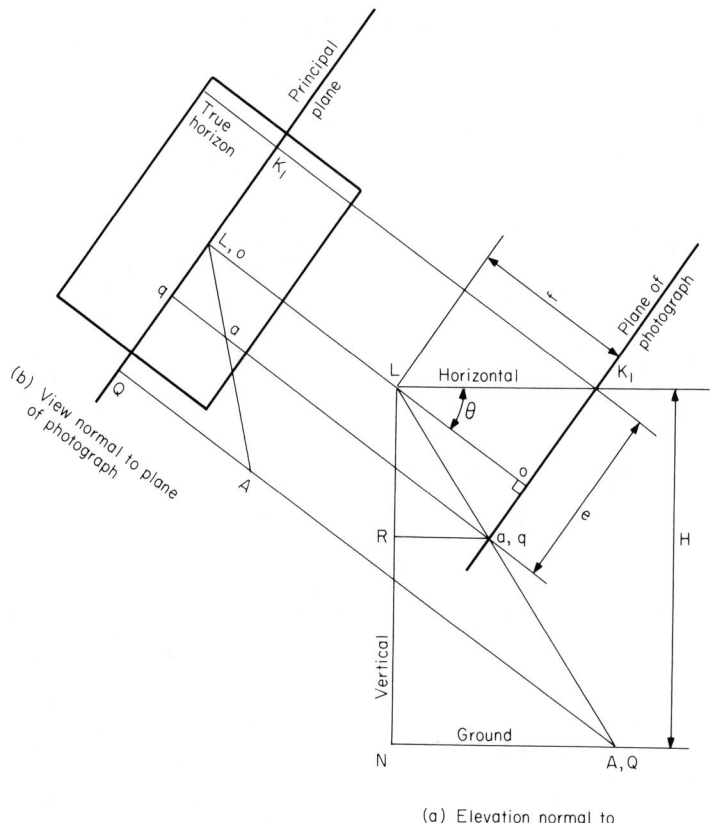

(a) Elevation normal to
principal plane

FIG. 216

Soil Mechanics

The basic notational system used is c = unit cohesion; s = specific gravity; V = volume; W = total weight; w = specific weight; ϕ = angle of internal friction; τ = shearing stress; σ = normal stress.

COMPOSITION OF SOIL

A specimen of moist soil weighing 122 g has an apparent specific gravity of 1.82. The specific gravity of the solids is 2.53. After the specimen is oven-dried, the weight is 104 g. Compute the void ratio, porosity, moisture content, and degree of saturation of the original mass.

Calculation Procedure:

1. *Compute the weight of moisture, volume of mass, and volume of each ingredient*

In a three-phase soil mass, the voids, or pores, between the solid particles are occupied by moisture and air. A mass that contains moisture but not air is termed *fully saturated;* this constitutes a two-phase system. The term *apparent specific gravity* denotes the specific gravity of the mass.

Let the subscripts s, w, and a refer to the solids, moisture, and air, respectively. Where a subscript is omitted, the reference is to the entire mass. Also, let e = void ratio = $(V_w + V_a)/V_s$; n = porosity = $(V_w + V_a)/V$; MC = moisture content = W_w/W_s; S = degree of saturation = $V_w/(V_w + V_a)$.

Refer to Fig. 217. A horizontal line represents volume, a vertical line represents specific gravity, and the area of a rectangle represents the weight of the respective ingredient in grams.

Computing weight and volume gives W = 122 g; W_s = 104 g; W_w = 122 − 104 = 18 g; V = 122/1.82 = 67.0 cm³; V_s = 104/2.53 = 41.1 cm³; V_w = 18.0 cm³; V_a = 67.0 − (41.1 + 18.0) = 7.9 cm³.

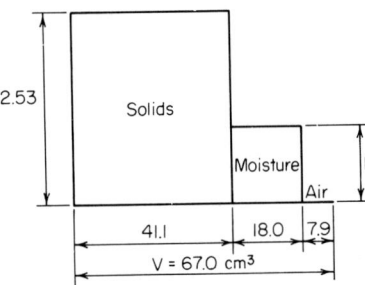

FIG. 217   Soil ingredients.

2. Compute the properties of the original mass

Thus, e = 100(18.0 + 7.9)/41.1 = 63.0 percent; n = 100(18.0 + 7.9)/67.0 = 38.7 percent; MC = 100(18)/104 = 17.3 percent; S = 100(18.0)/(18.0 + 7.9) + 69.5 percent. The factor of 100 is used to convert to percentage.

Soil composition is important from an environmental standpoint. Ever since the passage of the Environmental Protection Agency (EPA) Superfund Program by Congress, greater attention has been paid to soil composition by cities, states, and the federal government.

The major concern of regulators is with soil contaminated by industrial waste and trash. Liquid wastes can pollute soil and streams. Solid waste can produce noxious odors in the atmosphere. Some solid wastes are transported to "safe" sites for burning, where they may pollute the local atmosphere. Superfund money pays for the removal and burning of solid wastes.

A tax on chemicals provides the money for Superfund operations. Public and civic reaction to Superfund activities is most positive. Thus, quick removal of leaking drums of dangerous materials by federal agencies has done much to reduce soil contamination. Further, the Superfund Program has alerted industry to the dangers and effects of careless disposal of undesirable materials.

There are some 1200 dump sites on the Superfund Program agenda requiring cleanup. The work required at some sites ranges from excavation of buried waste to its eventual disposal by incineration. Portable and mobile incinerators are being used for wastes that do not pollute the air. Before any incineration can take place—either in fixed or mobile incinerators—careful analysis of the effluent from the incinerator must be made. For all these reasons, soil composition is extremely important in engineering studies.

SPECIFIC WEIGHT OF SOIL MASS

A specimen of sand has a porosity of 35 percent, and the specific gravity of the solids is 2.70. Compute the specific weight of this soil in pounds per cubic foot (kilograms per cubic meter) in the saturated and in the submerged state.

Calculation Procedure:

1. Compute the weight of the mass in each state

Set V = 1 cm³. The (apparent) weight of the mass when submerged equals the true weight less the buoyant force of the water. Thus, $V_w + V_a = nV$ = 0.35 cm³; V_s = 0.65 cm³. In the saturated state, W = 2.70(0.65) + 0.35 = 2.105 g. In the submerged state, W = 2.105 − 1 = 1.105 g; or W = (2.70 − 1)0.65 = 1.105 g.

2. Find the weight of the soil

Multiply the foregoing values by 62.4 to find the specific weight of the soil in pounds per cubic foot. Thus: saturated, w = 131.4 lb/ft³ (2104.82 kg/m³); submerged, w = 69.0 lb/ft³ (1105.27 kg/m³).

ANALYSIS OF QUICKSAND CONDITIONS

Soil having a void ratio of 1.05 contains particles having a specific gravity of 2.72. Compute the hydraulic gradient that will produce a quicksand condition.

Calculation Procedure:

1. *Compute the minimum gradient causing quicksand*

As water percolates through soil, the head that induces flow diminishes in the direction of flow as a result of friction and viscous drag. The drop in head in a unit distance is termed the *hydraulic gradient*. A quicksand condition exists when water that is flowing upward has a sufficient momentum to float the soil particles.

Let i denote the hydraulic gradient in the vertical direction and i_c the minimum gradient that causes quicksand. Equate the buoyant force on a soil mass to the submerged weight of the mass to find i_c. Or

$$i_c = \frac{s_s - 1}{1 + e} \tag{157}$$

For this situation, $i_c = (2.72 - 1)/(1 + 1.05) = 0.84$.

MEASUREMENT OF PERMEABILITY BY FALLING-HEAD PERMEAMETER

A specimen of soil is placed in a falling-head permeameter. The specimen has a cross-sectional area of 66 cm² and a height of 8 cm; the standpipe has a cross-sectional area of 0.48 cm². The head on the specimen drops from 62 to 40 cm in 1 h 18 min. Determine the coefficient of permeability of the soil, in centimeters per minute.

Calculation Procedure:

1. *Using literal values, equate the instantaneous discharge in the specimen to that in the standpipe*

The velocity at which water flows through a soil is a function of the *coefficient of permeability*, or *hydraulic conductivity*, of the soil. By Darcy's law of laminar flow,

$$v = ki \tag{158}$$

where i = hydraulic gradient, k = coefficient of permeability, v = velocity.

In a falling-head permeameter, water is allowed to flow vertically from a standpipe through a soil specimen. Since the water is not replenished, the water level in the standpipe drops as flow continues, and the velocity is therefore variable. Let A = cross-sectional area of soil specimen; a = cross-sectional area of standpipe; h = head on specimen at given instant; h_1 and h_2 = head at beginning and end, respectively, of time interval T; L = height of soil specimen; Q = discharge at a given instant.

Using literal values, we have $Q = Aki = -a \, dh/dt$.

2. *Evaluate k*

Since the head h is dissipated in flow through the soil, $i = h/L$. By substituting and rearranging, $(Ak/L)dT = -a \, dh/h$; integrating gives $AkT/L = a \ln (h_1/h_2)$, where ln denotes the natural logarithm. Then

$$k = \frac{aL}{AT} \ln \frac{h_1}{h_2} \tag{159}$$

Substituting gives $k = (0.48 \times 8/66 \times 78) \ln (62/40) = 0.000326$ cm/min.

CONSTRUCTION OF FLOW NET

State the Laplace equation as applied to two-dimensional flow of moisture through a soil mass, and list three methods of constructing a flow net that are based on this equation.

Calculation Procedure:

1.   Plot flow lines and equipotential lines

The path traversed by a water particle flowing through a soil mass is termed a *flow line, stream-line*, or *path of percolation*. A line that connects points in the soil mass at which the head on the water has some assigned value is termed an *equipotential line*. A diagram consisting of flow lines and equipotential lines is called a *flow net*.

In Fig. 218a, where water flows under a dam under a head H, lines AB and CD are flow lines and EF and GH are equipotential lines.

2.   Discuss the relationship of flow and equipotential lines

Since a water particle flowing from one equipotential line to another of smaller head will traverse the shortest path, it follows that flow lines and equipotential lines intersect at right angles, thus forming a system of orthogonal curves. In a flow net, the equipotential lines should be so spaced that the difference in head between successive lines is a constant, and the flow lines should be so spaced that the discharge through the space between successive lines is a constant. A flow net constructed in compliance with these rules illustrates the basic characteristics of the flow. For example, a close spacing of equipotential lines signifies a rapid loss of head in that region.

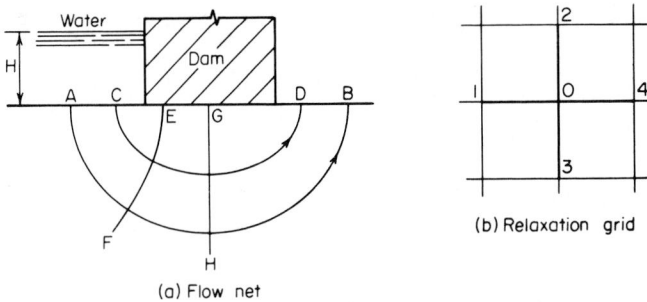

(a) Flow net   (b) Relaxation grid

FIG. 218

3.   Write the velocity equation

Let h denote the head on the water at a given point. Equation 158 can be written as

$$v = - k \frac{dh}{dL} \tag{158a}$$

where dL denotes an elemental distance along the flow line.

4.   State the particular form of the general Laplace equation

Let x and z denote a horizontal and vertical coordinate axis, respectively. By investigating the two-dimensional flow through an elemental rectangular prism of homogeneous, isentropic soil, and combining Eq. 158a with the equation of continuity, the particular form of the general Laplace equation

$$\frac{\partial^2 h}{\partial x^2} + \frac{\partial^2 h}{\partial z^2} = 0 \tag{160}$$

is obtained.

This equation is analogous to the equation for the flow of an electric current through a conducting sheet of uniform thickness and the equation of the trajectory of principal stress. (This is a curve that is tangent to the direction of a principal stress at each point along the curve. Refer to earlier calculation procedures for a discussion of principal stresses.)

The seepage of moisture through soil may be investigated by analogy with either the flow of an electric current or the stresses in a body. In the latter method, it is merely necessary to load a body in a manner that produces identical boundary conditions and then to ascertain the directions of the principal stresses.

5. *Apply the principal-stress analogy*

Refer to Fig. 218a. Consider the surface directly below the dam to be subjected to a uniform pressure. Principal-stress trajectories may be readily constructed by applying the principles of elasticity. In the flow net, flow lines correspond to the minor-stress trajectories and equipotential lines correspond to the major-stress trajectories. In this case, the flow lines are ellipses having their foci at the edges of the base of the dam, and the equipotential lines are hyperbolas.

A flow net may also be constructed by an approximate, trial-and-error procedure based on the method of relaxation. Consider that the area through which discharge occurs is covered with a grid of squares, a part of which is shown in Fig. 218b. If it is assumed that the hydraulic gradient is constant within each square, Eq. 160 leads to

$$h_1 + h_2 + h_3 + h_4 - 4h_0 = 0 \qquad (161)$$

Trial values are assigned to each node in the grid, and the values are adjusted until a consistent set of values is obtained. With the approximate head at each node thus established, it becomes a simple matter to draw equipotential lines. The flow lines are then drawn normal thereto.

SOIL PRESSURE CAUSED BY POINT LOAD

A concentrated vertical load of 6 kips (26.7 kN) is applied at the ground surface. Compute the vertical pressure caused by this load at a point 3.5 ft (1.07 m) below the surface and 4 ft (1.2 m) from the action line of the force.

Calculation Procedure:

1. *Sketch the load conditions*

Figure 219 shows the load conditions. In Fig. 219, O denotes the point at which the load is applied, and A denotes the point under consideration. Let R denote the length of OA and r and z denote the length of OA as projected on a horizontal and vertical plane, respectively.

2. *Determine the vertical stress σ_z at A*

Apply the Boussinesq equation:

$$\sigma_z = \frac{3Pz^3}{2\pi R^5} \qquad (162)$$

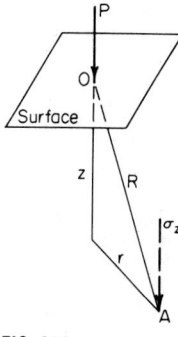

FIG. 219

Thus, with $P = 6000$ lb (26,688.0 N), $r = 4$ ft (1.2 m), $z = 3.5$ ft (1.07 m), $R = (4^2 + 3.5^2)^{0.5}$ = 5.32 ft (1.621 m); then $\sigma_z = 3(6000)(3.5)^3/[2\pi(5.32)^5] = 28.8$ lb/ft^2 (1.38 kPa).

Although the Boussinesq equation is derived by assuming an idealized homogeneous mass, its results agree reasonably well with those obtained experimentally.

VERTICAL FORCE ON RECTANGULAR AREA CAUSED BY POINT LOAD

A concentrated vertical load of 20 kips (89.0 kN) is applied at the ground surface. Determine the resultant vertical force caused by this load on a rectangular area 3 × 5 ft (91.4 × 152.4 cm) that lies 2 ft (61.0 cm) below the surface and has one vertex on the action line of the applied force.

Calculation Procedure:

1. *State the equation for the total force*

Refer to Fig. 220a, where A and B denote the dimensions of the rectangle, H its distance from the surface, and F is the resultant vertical force. Establish rectangular coordinate axes along the

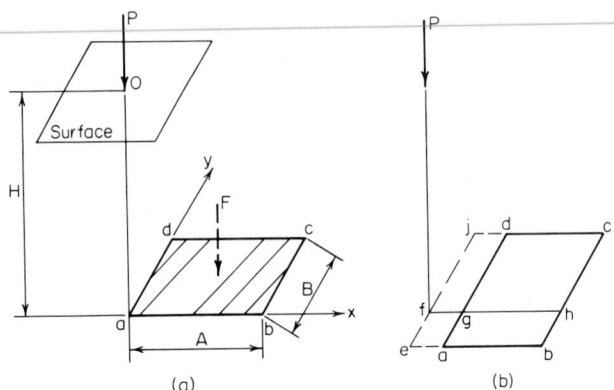

FIG. 220

sides of the rectangle, as shown. Let $C = A^2 + H^2$, $D = B^2 + H^2$, $E = A^2 + B^2 + H^2$, $\theta = \sin^{-1} H(E/CD)^{0.5}$ deg.

The force dF on an elemental area dA is given by the Boussinesq equation as $dF = [3Pz^3/(2\pi R^5)] \, dA$, where $z = H$ and $R = (H^2 + x^2 + y^2)^{0.5}$. Integrate this equation to obtain an equation for the total force F. Set $dA = dx \, dy$; then

$$\frac{F}{P} = 0.25 - \frac{\theta}{360°} + \frac{ABH}{2\pi E^{0.5}}\left(\frac{1}{C} + \frac{1}{D}\right) \tag{163}$$

2. Substitute numerical values and solve for F

Thus, $A = 3$ ft (91.4 cm); $B = 5$ ft (152.4 cm); $H = 2$ ft (61.0 cm); $C = 13$; $D = 29$; $E = 38$; $\theta = \sin^{-1} 0.6350 = 39.4°$; $F/P = 0.25 - 0.109 + 0.086 = 0.227$; $F = 20(0.227) = 4.54$ kips (20.194 kN).

The resultant force on an area such as *abcd* (Fig. 220*b*) may be found by expressing the area in this manner: *abcd* = *ebhf* − *eagf* + *fhcj* − *fgdj*. The forces on the areas on the right side of this equation are superimposed to find the force on *abcd*. Various diagrams and charts have been devised to expedite the calculation of vertical soil pressure.

VERTICAL PRESSURE CAUSED BY RECTANGULAR LOADING

A rectangular concrete footing 6 × 8 ft (182.9 × 243.8 cm) carries a total load of 180 kips (800.6 kN), which may be considered to be uniformly distributed. Determine the vertical pressure caused by this load at a point 7 ft (213.4 cm) below the center of the footing.

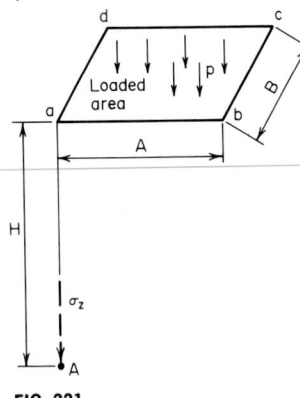

FIG. 221

Calculation Procedure:

1. State the equation for σ_z

Referring to Fig. 221, let p denote the uniform pressure on the rectangle *abcd* and σ_z the resulting vertical pressure at a point A directly below a vertex of the rectangle. Then

$$\frac{\sigma_z}{p} = 0.25 - \frac{\theta}{360} + \frac{ABH}{2\pi E^{0.5}}\left(\frac{1}{C} + \frac{1}{D}\right) \tag{164}$$

2. Substitute given values and solve for σ_z

Resolve the given rectangle into four rectangles having a vertex above the given point. Then $p = 180,000/[6(8)] = 3750$ lb/ft² (179.6 kPa). With $A = 3$ ft (91.4 cm); $B = 4$ ft (121.9 cm); $H = 7$ ft (213.4 cm); $C = 58$; $D = 65$; $E = 74$; $\theta = \sin^{-1} 0.9807 = 78.7°$; $\sigma_z/p = 4(0.25 - 0.218 + 0.051) = 0.332$; $\sigma_z = 3750(0.332) = 1245$ lb/ft² (59.6 kPa).

APPRAISAL OF SHEARING CAPACITY OF SOIL BY UNCONFINED COMPRESSION TEST

In an unconfined compression test on a soil sample, it was found that when the axial stress reached 2040 lb/ft^2 (97.7 kPa), the soil ruptured along a plane making an angle of 56° with the horizontal. Find the cohesion and angle of internal friction of this soil by constructing Mohr's circle.

Calculation Procedure:

1. *Construct Mohr's circle in Fig. 222b*

Failure of a soil mass is characterized by the sliding of one part past the other; the failure is therefore one of shear. Resistance to sliding occurs from two sources: cohesion of the soil and friction.

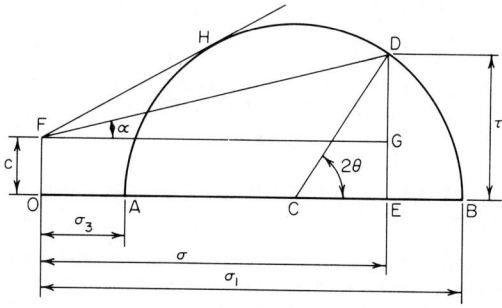

(a) Mohr's diagram for triaxial-stress condition

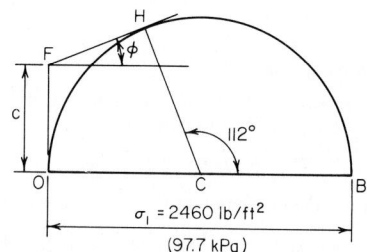

(b) Mohr's diagram for unconfined compression test

FIG. 222

Consider that the shearing stress at a given point exceeds the cohesive strength. It is usually assumed that the soil has mobilized its maximum potential cohesive resistance plus whatever frictional resistance is needed to prevent failure. The mass therefore remains in equilibrium if the ratio of the computed frictional stress to the normal stress is below the coefficient of internal friction of the soil.

Consider a soil prism in a state of triaxial stress. Let Q denote a point in this prism and P a plane through Q. Let c = unit cohesive strength of soil; σ = normal stress at Q on plane P; σ_1 and σ_3 = maximum and minimum normal stress at Q, respectively; τ = shearing stress at Q, on plane P; θ = angle between P and the plane on which σ_1 occurs; ϕ = angle of internal friction of the soil.

For an explanation of Mohr's circle of stress, refer to an earlier calculation procedure; then refer to Fig. 222a. The shearing stress ED on plane P may be resolved into the cohesive stress EG and the frictional stress GD. Therefore, $\tau = c + \sigma \tan \alpha$. The maximum value of α associated with point Q is found by drawing the tangent FH.

Assume that failure impends at Q. Two conclusions may be drawn: The angle between FH and the base line OAB equals ϕ, and the angle between the plane of impending rupture and the plane on which σ_1 occurs equals one-half angle BCH. (A soil mass that is on the verge of failure is said to be in *limit equilibrium*.)

In an unconfined compression test, the specimen is subjected to a vertical load without being restrained horizontally. Therefore, σ_1 occurs on a horizontal plane.

Constructing Mohr's circle in Fig. 222b, apply these values: $\sigma_1 = 2040$ lb/ft² (97.7 kPa); $\sigma_3 = 0$; angle $BCH = 2(56°) = 112°$.

2. Construct a tangent to the circle

Draw a line through H tangent to the circle. Let F denote the point of intersection of the tangent and the vertical line through O.

3. Measure OF and the angle of inclination of the tangent

The results are $OF = c = 688$ lb/ft² (32.9 kPa); $\phi = 22°$.

In general, in an unconfined compression test,

$$c = \tfrac{1}{2}\sigma_1 \cot \theta' \qquad \phi = 2\theta' - 90° \qquad (165)$$

where θ' denotes the angle between the plane of failure and the plane on which σ_1 occurs. In the special case where frictional resistance is negligible, $\phi = 0$; $c = \tfrac{1}{2}\sigma_1$.

APPRAISAL OF SHEARING CAPACITY OF SOIL BY TRIAXIAL COMPRESSION TEST

Two samples of a soil were subjected to triaxial compression tests, and it was found that failure occurred under the following principal stresses: sample 1, $\sigma_1 = 6960$ lb/ft² (333.2 kPa) and $\sigma_3 = 2000$ lb/ft² (95.7 kPa); sample 2, $\sigma_1 = 9320$ lb/ft² (446.2 kPa) and $\sigma_3 = 3000$ lb/ft² (143.6 kPa). Find the cohesion and angle of internal friction of this soil, both trigonometrically and graphically.

Calculation Procedure:

1. State the equation for the angle φ

Trigonometric method: Let S and D denote the sum and difference, respectively, of the stresses σ_1 and σ_3. By referring to Fig. 222a, develop this equation:

$$D - S \sin \phi = 2c \cos \phi \qquad (166)$$

Since the right-hand member represents a constant that is characteristic of the soil, $D_1 - S_1 \sin \phi = D_2 - S_2 \sin \phi$, or

$$\sin \phi = \frac{D_2 - D_1}{S_2 - S_1} \qquad (167)$$

where the subscripts correspond to the sample numbers.

2. Evaluate φ and c

By Eq. 167, $S_1 = 8960$ lb/ft² (429.0 kPa); $D_1 = 4960$ lb/ft² (237.5 kPa); $S_2 = 12,320$ lb/ft² (589.9 kPa); $D_2 = 6320$ lb/ft² (302.6 kPa); $\sin \phi = (6320 - 4960)/(12,320 - 8960)$; $\phi = 23°53'$. Evaluating c, using Eq. 166, gives $c = \tfrac{1}{2}(D \sec \phi - S \tan \phi) = 729$ lb/ft² (34.9 kPa).

3. For the graphical solution, use the Mohr's circle

Draw the Mohr's circle associated with each set of principal stresses, as shown in Fig. 223.

4. Draw the envelope; measure its angle of inclination

Draw the envelope (common tangent) FHH', and measure OF and the angle of inclination of the envelope. In practice, three of four samples should be tested and the average value of ϕ and c determined.

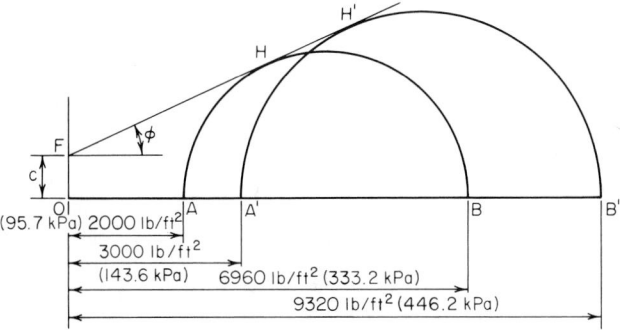

FIG. 223 Composite Mohr's diagram for triaxial compression tests.

EARTH THRUST ON RETAINING WALL CALCULATED BY RANKINE'S THEORY

A retaining wall supports sand weighing 100 lb/ft^3 (15.71 kN/m^3) and having an angle of internal friction of 34°. The back of the wall is vertical, and the surface of the backfill is inclined at an angle of 15° with the horizontal. Applying Rankine's theory, calculate the active earth pressure on the wall at a point 12 ft (3.7 m) below the top.

Calculation Procedure:

1. *Construct the Mohr's circle associated wtih the soil prism*

Rankine's theory of earth pressure applies to a uniform mass of dry cohesionless soil. This theory considers the state of stress at the instant of impending failure caused by a slight yielding of the wall. Let h = vertical distance from soil surface to a given point, ft (m); p = resultant pressure on a vertical plane at the given point, lb/ft^2 (kPa); ϕ = ratio of shearing stress to normal stress on given plane; θ = angle of inclination of earth surface. The quantity o may also be defined as the tangent of the angle between the resultant stress on a plane and a line normal to this plane; it is accordingly termed the *obliquity* of the resultant stress.

Consider the elemental soil prism *abcd* in Fig. 224*a*, where faces *ab* and *dc* are parallel to the surface of the backfill and faces *bc* and *ad* are vertical. The resultant pressure p_v on *ab* is vertical,

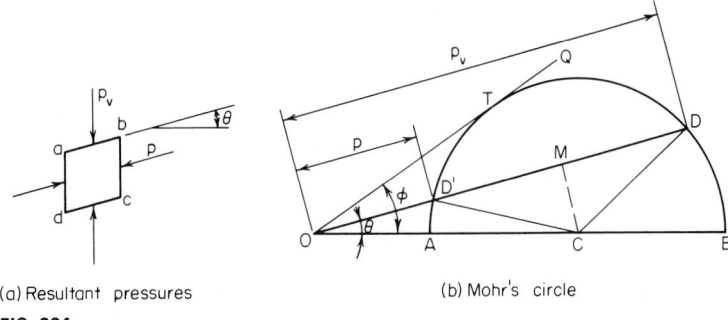

(a) Resultant pressures

(b) Mohr's circle

FIG. 224

and p is parallel to the surface. Thus, the resultant stresses on *ab* and *bc* have the same obliquity, namely, tan θ. (Stresses having equal obliquities are called *conjugate* stresses.) Since failure impends, there is a particular plane for which the obliquity is tan ϕ.

In Fig. 224*b*, construct Mohr's circle associated with this soil prism. Using a suitable scale, draw line *OD*, making an angle θ with the base line, where *OD* represents p_v. Draw line *OQ*, making an angle ϕ with the base line. Draw a circle that has its center *C* on the base line, passes through *D*, and is tangent to *OQ*. Line *OD'* represents p. Draw *CM* perpendicular to *OD*.

2. Using the Mohr's circle, state the equation for p

Thus,

$$p = \frac{[\cos \theta - (\cos^2 \theta - \cos^2 \phi)^{0.5}] \, wh}{\cos \theta + (\cos^2 \theta - \cos^2 \phi)^{0.5}} \tag{168}$$

By substituting, $w = 100$ lb/ft^3 (15.71 kN/m^3); $h = 12$ ft (3.7 m); $\theta = 15°$; $\phi = 34°$; $p = 0.321(100)(12) = 385$ lb/ft^2 (18.4 kPa).

The lateral pressure that accompanies a slight displacement of the wall *away from* the retained soil is termed *active pressure;* that which accompanies a slight displacement of the wall *toward* the retained soil is termed *passive pressure.* By an analogous procedure, the passive pressure is

$$p = \frac{[\cos \theta + (\cos^2 \theta - \cos^2 \phi)^{0.5}] wh}{\cos \theta - (\cos^2 \theta - \cos^2 \phi)^{0.5}} \tag{169}$$

The equations of active and passive pressure are often written as

$$p_a = C_a wh \qquad p_p = C_p wh \tag{170}$$

where the subscripts identify the type of pressure and C_a and C_p are the coefficients appearing in Eqs. 168 and 169, respectively.

In the special case where $\theta = 0$, these coefficients reduce to

$$C_a = \frac{1 - \sin \phi}{1 + \sin\phi} = \tan^2 (45° - \tfrac{1}{2}\phi) \tag{171}$$

$$C_p = \frac{1 + \sin \phi}{1 - \sin \phi} = \tan^2 (45° + \tfrac{1}{2}\phi) \tag{172}$$

The planes of failure make an angle of $45° + \tfrac{1}{2}\phi$ with the principal planes.

EARTH THRUST ON RETAINING WALL CALCULATED BY COULOMB'S THEORY

A retaining wall 20 ft (6.1 m) high supports sand weighing 100 lb/ft^3 (15.71 kN/m^3) and having an angle of internal friction of 34°. The back of the wall makes an angle of 8° with the vertical; the surface of the backfill makes an angle of 9° with the horizontal. The angle of friction between the sand and wall is 20°. Applying Coulomb's theory, calculate the total thrust of the earth on a 1-ft (30.5-cm) length of the wall.

Calculation Procedure:

1. Determine the resultant pressure P of the wall

Refer to Fig. 225a. Coulomb's theory postulates that as the wall yields slightly, the soil tends to rupture along some plane BC through the heel.

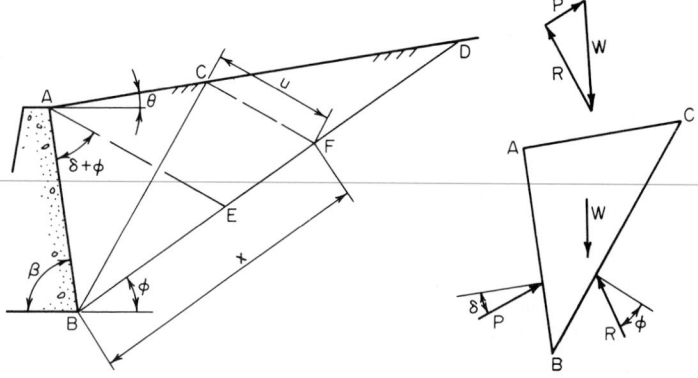

(a) Location of plane of failure

(b) Free-body diagram of sliding wedge

FIG. 225

Let δ denote the angle of friction between the soil and wall. As shown in Fig. 225b, the wedge ABC is held in equilibrium by three forces: the weight W of the wedge, the resultant pressure R of the soil beyond the plane of failure, and the resultant pressure P of the wall, which is equal and opposite to the thrust exerted by the earth on the wall. The forces R and P have the directions indicated in Fig. 225b. By selecting a trial wedge and computing its weight, the value of P may be found by drawing the force polygon. The problem is to identify the wedge that yields the maximum value of P.

In Fig. 225a, perform this construction: Draw a line through B at an angle ϕ with the horizontal, intersecting the surface at D. Draw line AE, making an angle $\delta + \phi$ with the back of the wall; this line makes an angle $\beta - \delta$ with BD. Through an arbitrary point C on the surface, draw CF parallel to AE. Triangle BCF is similar to the triangle of forces in Fig. 225b. Then $P = Wu/x$, where $W = w(\text{area } ABC)$.

2. Set dP/dx = 0 and state Rebhann's theorem

This theorem states: The wedge that exerts the maximum thrust on the wall is that for which triangles ABC and BCF have equal areas.

3. Considering BC as the true plane of failure, develop equations for x^2, u, and P

Thus,

$$x^2 = BE(BD) \tag{173}$$

$$u = \frac{AE(BD)}{x + BD} \tag{174}$$

$$P = \tfrac{1}{2}wu^2 \sin (\beta - \delta) \tag{175}$$

4. Evaluate P, using the foregoing equations

Thus, $\phi = 34°$; $\delta = 20°$; $\theta = 9°$; $\beta = 82°$; $\angle ABD = 64°$; $\angle BAE = 54°$; $\angle AEB = 62°$; $\angle BAD = 91°$; $\angle ADB = 25°$; $AB = 20 \csc 82° = 20.2$ ft (6.16 m). In triangle ABD: $BD = AB \sin 91°/\sin 25° = 47.8$ ft (14.57 m). In triangle ABE: $BE = AB \sin 54°/\sin 62° = 18.5$ ft (5.64 m); $AE = AB \sin 64°/\sin 62° = 20.6$ ft (6.28 m); $x^2 = 18.5(47.8)$; $x = 29.7$ ft (9.05 m); $u = 20.6(47.8)/(29.7 + 47.8) = 12.7$ ft (3.87 m); $P = \tfrac{1}{2}(100)(12.7)^2 \sin 62°$; $P = 7120$ lb/ft (103,909 N/m) of wall.

5. Alternatively, determine u graphically

Do this by drawing Fig. 225a to a suitable scale.

Many situations do not lend themselves to analysis by Rebhann's theorem. For instance, the backfill may be nonhomogeneous, the earth surface may not be a plane, a surcharge may be applied over part of the surface, etc. In these situations, graphical analysis gives the simplest solution. Select a trial wedge, compute its weight and the surcharge it carries, and find P by constructing the force polygon a shown in Fig. 225b. After several trial wedges have been investigated, the maximum value of P will become apparent.

If the backfill is cohesive, the active pressure on the retaining wall is reduced. However, in view of the difficulty of appraising the cohesive capacity of a disturbed soil, most designers prefer to disregard cohesion.

EARTH THRUST ON TIMBERED TRENCH CALCULATED BY GENERAL WEDGE THEORY

A timbered trench of 12-ft (3.7-m) depth retains a cohesionless soil having a horizontal surface. The soil weighs 100 lb/ft^3 (15.71 kN/m^3), its angle of internal friction is 26°30′, and the angle of friction between the soil and timber is 12°. Applying Terzaghi's general wedge theory, compute the total thrust of the soil on a 1-ft (30.5-cm) length of trench. Assume that the resultant acts at middepth.

Calculation Procedure:

1. Start the graphical construction

Refer to Fig. 226. The soil behind a timbered trench and that behind a cantilever retaining wall tend to fail by dissimilar modes, for in the former case the soil is restrained against horizontal

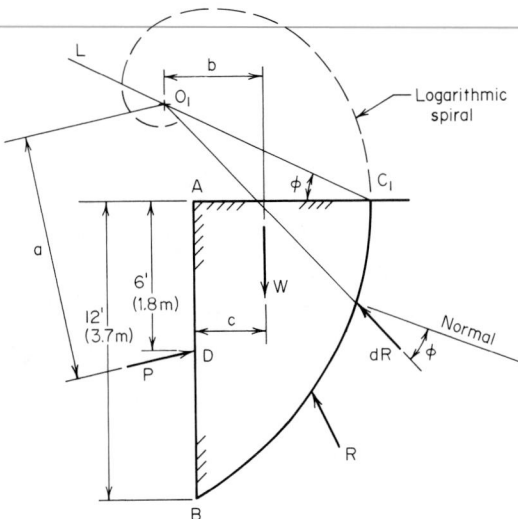

FIG. 226   General wedge theory applied to timbered trench.

movement at the surface by bracing across the trench. Consequently, the soil behind a trench tends to fail along a curved surface that passes through the base and is vertical at its intersection with the ground surface. At impending failure, the resultant force dR acting on any elemental area on the failure surface makes an angle ϕ with the normal to this surface.

The general wedge theory formulated by Terzaghi postulates that the arc of failure is a logarithmic spiral. Let v_o denote a reference radius vector and v denote the radius vector to a given point on the spiral. The equation of the curve is

$$r = r_o e^{\alpha \tan \phi} \tag{176}$$

where r_o = length of v_o; r = length of v; α = angle between v_o and v; e = base of natural logarithms = 2.718. . . .

The property of this curve that commends it for use in this analysis is that at every point the radius vector and the normal to the curve make an angle ϕ with each other. Therefore, if the failure line is defined by Eq. 176, the action line of the resultant force dR at any point is a radius vector or, in other words, the action line passes through the center of the spiral. Consequently, the action line of the total resultant force R also passes through the center.

The pressure distribution on the wall departs radically from a hydrostatic one, and the resultant thrust P is applied at a point considerably above the lower third point of the wall. Terzaghi recommends setting the ratio BD/AB at between 0.5 and 0.6.

Perform the following construction: Using a suitable scale, draw line AB to represent the side of the trench, and draw a line to represent the ground surface. At middepth, draw the action line of P at an angle of 12° with the horizontal.

On a sheet of transparent paper, draw the logarithmic spiral representing Eq. 176, setting ϕ = 26°30′ and assigning any convenient value to r_o. Designate the center of the spiral as O.

Select a point C_1 on the ground surface, and draw a line L through C_1 at an angle ϕ with the horizontal. Superimpose the drawing containing the spiral on the main drawing, orienting it in such a manner that O lies on L and the spiral passes through B and C_1. On the main drawing, indicate the position of the center of the spiral, and designate this point as O_1. Line AC_1 is normal to the spiral at C_1 because it makes an angle ϕ with the radius vector, and the spiral is therefore vertical at C_1.

2.   *Compute the total weight W of the soil above the failure line*

Draw the action line of W by applying these approximations:

$$\text{Area of wedge} = \tfrac{2}{3}(AB)AC_1 \qquad c = 0.4AC_1 \tag{177}$$

Scale the lever arms a and b.

3.   *Evaluate P by taking moments with respect to O_1*

Since R passes through this point,

$$P = \frac{bW}{a} \tag{178}$$

4.   *Select a second point C_2 on the ground surface; repeat the foregoing procedure*

5.   *Continue this process until the maximum value of P is obtained*

After investigating this problem intensively, Peckworth concluded that the distance AC to the true failure line varies between $0.4h$ and $0.5h$, where h is the depth of the trench. It is therefore advisable to select some point that lies within this range as the first trial position of C.

THRUST ON A BULKHEAD

The retaining structure in Fig. 227a supports earth that weighs 114 lb/ft³ (17.91 kN/m³) in the dry state, is 42 percent porous, and has an angle of internal friction of 34° in both the dry and submerged state. The backfill carries a surcharge of 320 lb/ft² (15.3 kPa). Applying Rankine's theory, compute the total pressure on this structure between A and C.

Calculation Procedure:

1.   *Compute the specific weight of the soil in the submerged state*

The lateral pressure of the soil below the water level consists of two elements: the pressure exerted by the solids and that exerted by the water. The first element is evaluated by applying the appropriate equation with w equal to the weight of the soil in the submerged state. The second element is assumed to be the full hydrostatic pressure, as though the solids were not present. Since there is water on both sides of the structure, the hydrostatic pressures balance one another and may therefore be disregarded.

In calculating the forces on a bulkhead, it is assumed that the pressure distribution is hydro-

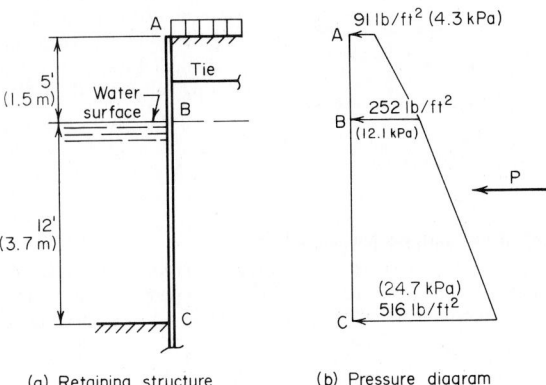

(a) Retaining structure   (b) Pressure diagram

FIG. 227

static (i.e., that the pressure varies linearly with the depth), although this is not strictly true with regard to a flexible wall.

Computing the specific weight of the soil in the submerged state gives $w = 114 - (1 - 0.42)62.4 = 77.8$ lb/ft³ (12.22 kN/m³).

2. Compute the vertical pressure at A, B, and C caused by the surcharge and weight of solids

Thus, $p_A = 320$ lb/ft² (15.3 kPa); $p_B = 320 + 5(114) = 890$ lb/ft² (42.6 kPa); $p_C = 890 + 12(77.8) = 1824$ lb/ft² (87.3 kPa).

3. Compute the Rankine coefficient of active earth pressure

Determine the lateral pressure at A, B and C. Since the surface is horizontal, Eq. 171 applies, with $\phi = 34°$. Refer to Fig. 227b. Then $C_a = \tan^2 (45° - 17°) = 0.283$; $p_A = 0.283(320) = 91$ lb/ft² (4.3 kPa); $p_B = 252$ lb/ft² (12.1 kPa); $p_C = 516$ lb/ft² (24.7 kPa).

4. Compute the total thrust between A and C

Thus, $P = \frac{1}{2}(5)(91 + 252) + \frac{1}{2}(12)(252 + 516) = 5466$ lb (24,312.7 N).

CANTILEVER BULKHEAD ANALYSIS

Sheet piling is to function as a cantilever retaining wall 5 ft (1.5 m) high. The soil weighs 110 lb/ft³ (17.28 kN/m³) and its angle of internal friction is 32°; the backfill has a horizontal surface. Determine the required depth of penetration of the bulkhead.

Calculation Procedure:

1. Take moments with respect to C to obtain an equation for the minimum value of d

Refer to Fig. 228a, and consider a 1-ft (30.5-cm) length of wall. Assume that the pressure distribution is hydrostatic, and apply Rankine's theory.

The wall pivots about some point Z near the bottom. Consequently, passive earth pressure is mobilized to the left of the wall betwen B and Z and to the right of the wall between Z and C.

Let P = resultant active pressure on wall; R_1 and R_2 = resultant passive pressure above and below center of rotation, respectively.

The position of Z may be found by applying statics. But to simplify the calculations, these assumptions are made: The active pressure extends from A to C; the passive pressure to the left of the wall extends from B to C; and R_2 acts at C. Figure 228b illustrates these assumptions.

By taking moments with respect to C and substituting values for R_1 and R_2,

(a) Cantilever   (b) Assumed pressures and
    bulkhead   resultant forces

FIG. 228

$$d = \frac{h}{(C_p/C_a)^{1/3} - 1} \qquad (179)$$

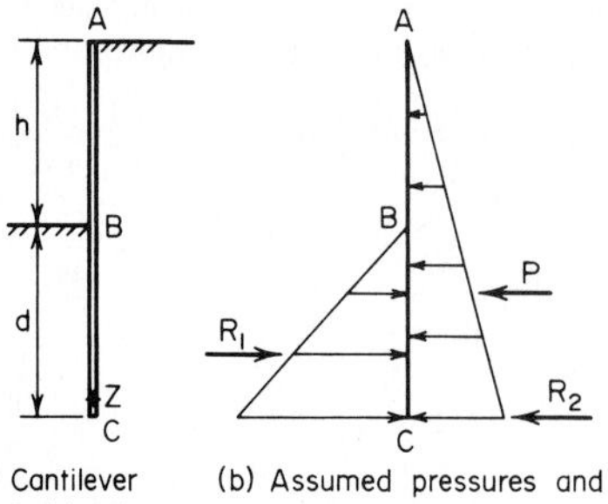

2. Substitute numerical values and solve for d

Thus, $45° + \frac{1}{2}\phi = 61°$; $45° - \frac{1}{2}\phi = 29°$. By Eqs. 171 and 172, $C_p/C_a = (\tan 61°/\tan 29°)^2 = 10.6$; $d = 5/[(10.6)^{1/3} - 1] = 4.2$ ft (1.3 m). Add 20 percent of the computed value to provide a factor of safety and to allow for the development of R_2. Thus, penetration $= 4.2(1.2) = 5.0$ ft (1.5 m).

ANCHORED BULKHEAD ANALYSIS

Sheet piling is to function as a retaining wall 20 ft (6.1 m) high, anchored by tie rods placed 3 ft (0.9 m) from the top at an 8-ft (2.4-m) spacing. The soil weighs 110 lb/ft³ (17.28 kN/m³), and its angle of internal friction is 32°. The backfill has a horizontal surface and carries a surcharge of 200 lb/ft² (9.58 kPa). Applying the equivalent-beam method, determine the depth of penetration to secure a fixed earth support, the tension in the tie rod, and the maximum bending moment in the piling.

Calculation Procedure:

1.   *Locate C and construct the net-pressure diagram for AC*

Refer to Fig. 229a. The depth of penetration is readily calculated if stability is the sole criterion. However, when the depth is increased beyond this minimum value, the tension in the rod and

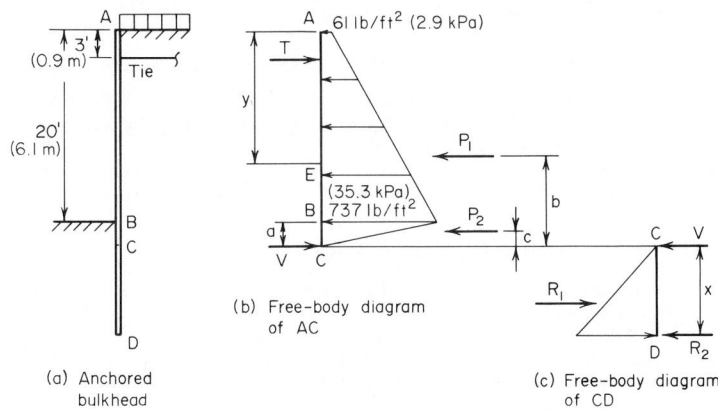

(a) Anchored
bulkhead

(b) Free-body diagram
of AC

(c) Free-body diagram
of CD

FIG. 229

the bending moment in the piling are reduced; the net result is a saving in material despite the increased length.

Investigation of this problem discloses that the most economical depth of penetration is that for which the tangent to the elastic curve at the lower end passes through the anchorage point. If this point is considered as remaining stationary, this condition can be described as one in which the elastic curve is vertical at D, the surrounding soil acting as a fixed support. Whereas an equation can be derived for the depth associated with this condition, such an equation is too cumbersome for rapid solution.

When the elastic curve is vertical at D, the lower point of contraflexure lies close to the point where the net pressure (the difference between active pressure to the right and passive pressure to the left of the wall) is zero. By assuming that the point of contraflexure and the point of zero pressure are in fact coincident, this problem is transformed to one that is statically determinate. The method of analysis based on this assumption is termed the *equivalent-beam* method.

When the piling is driven to a depth greater than the minimum needed for stability, it deflects in such a manner as to mobilize passive pressure to the right of the wall at its lower end. However, the same simplifying assumption concerning the pressure distribution as made in the previous calculation procedure is made here.

Let C denote the point of zero pressure. Consider a 1-ft (30.5-cm) length of wall, and let T = reaction at anchorage point and V = shear at C.

Locate C and construct the net-pressure diagram for AC as shown in Fig. 229b. Thus, w = 110 lb/ft³ (17.28 kN/m³) and $\phi = 32°$. Then $C_a = \tan^2(45° - 16°) = 0.307$; $C_p = \tan^2(45° + 16°) = 3.26$; $C_p - C_a = 2.953$; $p_A = 0.307(200) = 61$ lb/ft² (2.9 kPa); $p_B = 61 + 0.307(20)(110) = 737$ lb/ft² (35.3 kPa); $a = 737/[2.953(110)] = 2.27$ ft (0.69 m).

2.   *Calculate the resultant forces P₁ and P₂*

Thus, $P_1 = \frac{1}{2}(20)(61 + 737) = 7980$ lb (35,495.0 N); $P_2 = \frac{1}{2}(2.27)(737) = 836$ lb (3718.5 N); $P_1 + P_2 = 8816$ lb (39,213.6 N).

3. *Equate the bending moment at C to zero to find T, V, and the tension in the tie rod*

Thus $b = 2.27 + (\frac{2}{3})(737 + 2 \times 61)/(737 + 61) = 9.45$ ft (2.880 m); $c = \frac{2}{3}(2.27) = 1.51$ ft (0.460 m); $\Sigma M_C = 19.27T - 9.45(7980) - 1.51(836) = 0$; $T = 3980$ lb (17,703.0 N); $V = 8816 - 3980 = 4836$ lb (21,510.5 N). The tension in the rod $= 3980(8) = 31,840$ lb (141,624.3 N).

4. *Construct the net-pressure diagram for CD*

Refer to Fig. 229c and calculate the distance x. (For convenience, Fig. 229c is drawn to a different scale from that of Fig. 229b.) Thus $p_D = 2953(110x) = 324.8x$; $R_1 = \frac{1}{2}(324.8x^2) = 162.4x^2$; $\Sigma M_D = R_1x/3 - Vx = 0$; $R_1 = 3V$; $162.4x^2 = 3(4836)$; $x = 9.45$ ft (2.880 m).

5. *Establish the depth of penetration*

To provide a factor of safety and to compensate for the slight inaccuracies inherent in this method of analysis, increase the computed depth by about 20 percent. Thus, penetration $= 1.20(a + x) = 14$ ft (4.3 m).

6. *Locate the point of zero shear; calculate the piling maximum bending moment*

Refer to Fig. 229b. Locate the point E of zero shear. Thus $p_E = 61 + 0.307(110y) = 61 + 33.77y$; $\frac{1}{2}y(p_A + p_E) = T$; or $\frac{1}{2}y(122 + 33.77y) = 3980$; $y = 13.6$ ft (4.1 m), and $p_E = 520$ lb/ft^2 (24.9 kPa); $M_{max} = M_E = 3980[10.6 - (13.6/3)(520 + 2 \times 61)/(520 + 61)] = 22,300$ ft·lb/ft (99.190 N·m/m) of piling. Since the tie rods provide intermittent rather than continuous support, the piling sustains biaxial bending stresses.

STABILITY OF SLOPE BY METHOD OF SLICES

Investigate the stability of the slope in Fig. 230 by the method of slices (also known as the Swedish method). The properties of the upper and lower soil strata, designated as A and B, respectively, are A—$w = 110$ lb/ft^3 (17.28 kN/m^3); $c = 0$; $\phi = 28°$; B—$w = 122$ lb/ft^3 (19.16 kN/m^3); $c = 650$ lb/ft^2 (31.1 kPa); $\phi = 10°$. Stratum A is 36 ft (10.9 m) deep. A surcharge of 8000 lb/lin ft (116,751.2 N/m) is applied 20 ft (6.1 m) from the edge.

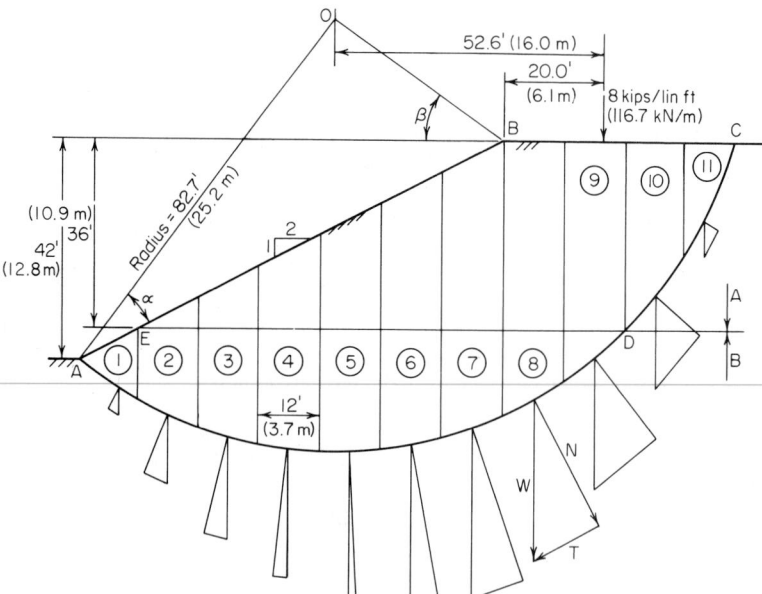

FIG. 230 Analysis of stability of slope by slices.

Calculation Procedure:

1. Locate the center of the trial arc of failure passing through the toe

It is assumed that failure of an embankment occurs along a circular arc, the prism of soil above the failure line tending to rotate about an axis through the center of the arc. However, there is no direct method of identifying the arc along which failure is most likely to occur, and it is necessary to resort to a cut-and-try procedure.

Consider a soil mass having a thickness of 1 ft (30.5 cm) normal to the plane of the drawing; let O denote the center of a trial arc of failure that passes through the toe. For a given inclination of embankment, Fellenius recommends certain values of α and β in locating the first trial arc.

Locate O by setting $\alpha = 25°$ and $\beta = 35°$.

2. Draw the arc AC and the boundary line ED of the two strata

3. Compute the length of arc AD

Scale the radius of the arc and the central angle AOD, and compute the length of the arc AD. Thus, radius = 82.7 ft (25.2 m); arc AD = 120 ft (36.6 m).

4. Determine the distance horizontally from O to the applied load

Scale the horizontal distance from O to the applied load. This distance is 52.6 ft (16.0 m).

5. Divide the soil mass into vertical strips

Starting at the toe, divide the soil mass above AC into vertical strips of 12-ft (3.7-m) width, and number the strips. For simplicity, consider that D lies on the boundary line between strips 9 and 10, although this is not strictly true.

6. Determine the volume and weight of soil in each strip

By scaling the dimensions or using a planimeter, determine the volume of soil in each strip; then compute the weight of soil. For instance, for strip 5: volume of soil A = 252 ft³ (7.13 m³); volume of soil B = 278 ft³ (7.87 m³); weight of soil = 252(110) + 278(122) = 61,600 lb (273,996.8 N). Record the results in Table 14.

7. Draw a vector below each strip

This vector represents the weight of the soil in the strip. (Theoretically, this vector should lie on the vertical line through the center of gravity of the soil, but such refinement is not warranted in this analysis. For the interior strips, place each vector on the vertical centerline.)

TABLE 14 Stability Analysis of Slope

Strip	Weight, kips (kN)	Normal component, kips (kN)	Tangential component, kips (kN)
1	10.3 (45.81)	8.9 (39.59)	−5.2 (−23.13)
2	28.1 (124.99)	26.0 (115.65)	−10.7 (−47.59)
3	41.9 (186.37)	40.6 (180.59)	−10.4 (−46.26)
4	53.0 (235.74)	52.7 (234.41)	−5.5 (−24.46)
5	61.6 (274.00)	61.5 (273.55)	2.6 (11.56)
6	67.7 (301.13)	66.5 (295.79)	12.8 (56.93)
7	71.0 (315.81)	67.0 (298.02)	23.4 (104.08)
8	67.1 (298.46)	58.8 (261.54)	32.4 (144.12)
9	54.8 (243.75)	43.0 (191.26)	34.0 (151.23)
10	38.3 (170.36)	24.9 (110.76)	29.1 (129.44)
11	14.3 (63.61)	7.0 (31.14)	12.5 (55.60)
Total, 1 to 9		425.0 (1890.40)	
Total, 10 and 11		31.9 (141.89)	
Grand total		456.9 (2032.29)	115.0 (511.52)

8.   Resolve the soil weights vectorially into components normal and tangential to the circular arc

9.   Scale the normal and tangential vectors; record the results in Table 14

10.   Total the normal forces acting on soils A and B; total the tangential forces

Failure of the embankment along arc AC would be characterized by the clockwise rotation of the soil prism above this arc about an axis through O, this rotation being induced by the unbalanced tangential force along the arc and by the external load. Therefore, consider a tangential force as positive if its moment with respect to an axis through O is clockwise and negative if this moment is counterclockwise. In the method of slices, it is assumed that the lateral forces on each soil strip approximately balance each other.

11.   Evaluate the moment tending to cause rotation about O

In the absence of external loads,

$$DM = r\Sigma T \qquad (180)$$

wehre DM = disturbing moment; r = radius of arc; ΣT = algebraic sum of tangential forces.
     In the present instance, DM = 82.7(115) + 52.6(8) = 9930 ft·kips (13,465.1 kN·m).

12.   Sum the frictional and cohesive forces to find the maximum potential resistance to rotation; determine the stabilizing moment

In general,

$$F = \Sigma N \tan \phi \qquad C = cL \qquad (181)$$

$$SM = r(F + C) \qquad (182)$$

where F = frictional force; C = cohesive force; ΣN = sum of normal forces; L = length of arc along which cohesion exists; SM = stabilizing moment.
     In the present instance, F = 425 tan 10° + 31.9 tan 28° = 91.9 kips (408.77 kN); C = 0.65(120) = 78.0; total of $F + C$ = 169.9 kips (755.72 kN); SM = 82.7(169.9) = 14,050 ft·kips (19,051.8 kN·m).

13.   Compute the factor of safety against failure

The factor of safety is FS = SM/DM = 14,050/9930 = 1.41.

14.   Select another trial arc of failure; repeat the foregoing procedure

15.   Continue this process until the minimum value of FS is obtained

The minimum allowable factor of safety is generally regarded as 1.5.

STABILITY OF SLOPE BY ϕ-CIRCLE METHOD

Investigate the stability of the slope in Fig. 231 by the ϕ-circle method. The properties of the soil are $w = 120$ lb/ft³ (18.85 kN/m³); $c = 550$ lb/ft² (26.3 kPa); $\phi = 4°$.

Calculation Procedure:

1.   Locate the first trial position

The ϕ-circle method of analysis formulated by Krey is useful where standard conditions are encountered. In contrast to the assumption concerning the stabilizing forces stated earlier, the ϕ-circle method assumes that the soil has mobilized its maximum potential *frictional* resistance plus whatever cohesive resistance is needed to prevent failure. A comparison of the maximum available cohesion with the required cohesion serves as an index of the stability of the embankment.
     In Fig. 231, O is the center of an assumed arc of failure AC. Let W = weight of soil mass above arc AC; R = resultant of all normal and frictional forces existing along arc AC; C = resultant cohesive force developed; L_a = length of arc AC; L_c = length of chord AC. The soil above the arc is in equilibrium under the forces W, R, and C. Since W is known in magnitude and direction, the magnitude of C may be readily found if the directions of R and C are determined.

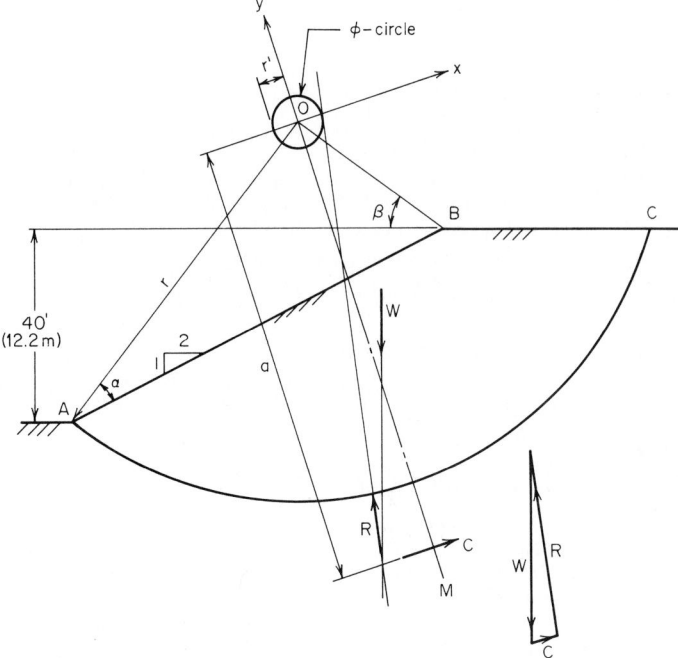

FIG. 231   Analysis of stability of slope by ϕ-circle method.

Locate the first trial position of O by setting $\alpha = 25°$ and $\beta = 35°$.

2.   Draw the arc AC and the radius OM bisecting this arc

3.   Establish rectangular coordinate axes at O, making OM the y axis

4.   Obtain the needed basic data

Scale the drawing or make the necessary calculations. Thus, $r = 78.8$ ft (24.02 m); $L_a = 154.6$ ft (47.12 m); $L_c = 131.0$ ft (39.93 m); area above arc $= 4050$ ft² (376.2 m²); $W = 4050(120) = 486,000$ lb (2,161,728 N); horizontal distance from A to centroid of area $= 66.7$ ft (20.33 m).

5.   Draw the vector representing W

Since the soil is homogeneous, this vector passes through the centroid of the area.

6.   State the equation for C; locate its action line

Thus,

$$C = C_x = cL_c \tag{183}$$

The action line of C is parallel to the x axis. Determine the distance a by taking moments about O. Thus $M = aC = acL_c$,

$$a = \left(\frac{L_a}{L_c}\right)r \tag{184}$$

Or $a = (154.6/131.0)78.8 = 93.0$ ft (28.35 m). Draw the action line of C.

7.   Locate the action line of R

For this purpose, consider the resultant force dR acting on an elemental area. Its action line is inclined at an angle ϕ with the radius at that point, and therefore the perpendicular distance r'

~~from *O* to this action line is~~

$$r' = r \sin \phi \tag{185}$$

Thus, r' is a constant for the arc AC. It follows that regardless of the position of dR along this arc, its action line is tangent to a circle centered at O and having a radius r'; this is called the ϕ *circle*, or *friction circle*. It is plausible to conclude that the action line of the total resultant is also tangent to this circle.

Draw a line tangent to the ϕ circle and passing through the point of intersection of the action lines of W and C. This is the action line of R. (The moment of R about O is counterclockwise, since its frictional component opposes clockwise rotation of the soil mass.)

8. *Using a suitable scale, determine the magnitude of C*

Draw the triangle of forces; obtain the magnitude of C by scaling. Thus, $C = 67,000$ lb (298,016 N).

9. *Calculate the maximum potential cohesion*

Apply Eq. 183, equating c to the unit cohesive capacity of the soil. Thus, $C_{\max} = 550(131) = 72,000$ lb (320,256 N). This result indicates a relatively low factor of safety. Other arcs of failure should be investigated in the same manner.

ANALYSIS OF FOOTING STABILITY BY TERZAGHI'S FORMULA

A wall footing carrying a load of 58 kips/lin ft (846.4 kN/m) rests on the surface of a soil having these properties: $w = 105$ lb/ft^3 (16.49 kN/m^3); $c = 1200$ lb/ft^2 (57.46 kPa); $\phi = 15°$. Applying Terzaghi's formula, determine the minimum width of footing required to ensure stability, and compute the soil pressure associated with this width.

Calculation Procedure:

1. *Equate the total active and passive pressures and state the equation defining conditions at impending failure*

While several methods of analyzing the soil conditions under a footing have been formulated, the one proposed by Terzaghi is gaining wide acceptance.

The soil underlying a footing tends to rupture along a curved surface, but the Terzaghi method postulates that this surface may be approximated by straight-line segments without introducing any significant error. Thus, in Fig. 232, the soil prism OAB tends to heave by sliding downward along OA under active pressure and sliding upward along AB against passive pressure. As stated earlier, these planes of failure make an angle of $\alpha = 45° + \frac{1}{2}\phi$ with the principal planes.

Let $b = $ width of footing; $h = $ distance from ground surface to bottom of footing; $p = $ soil

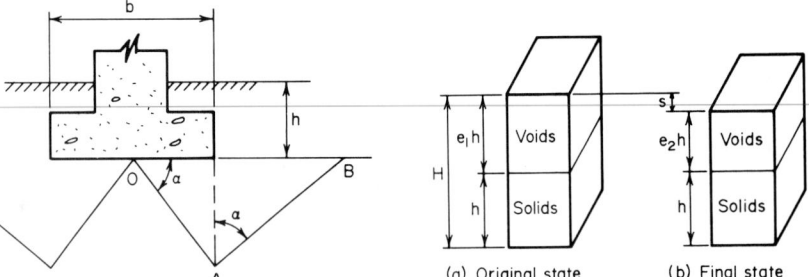

FIG. 232 Failure of soil under footing in accordance with Terzaghi's assumption.

FIG. 233

CIVIL ENGINEERING   **1.285**

pressure directly below footing. By equating the total active and passive pressures, state the following equation defining the conditions at impending failure:

$$p = wh \tan^4 \alpha + \frac{wb(\tan^5 \alpha - \tan \alpha)}{4} + 2c\,(\tan \alpha + \tan^3 \alpha) \qquad (186)$$

2. Substitute numerical values; solve for b; evaluate p

Thus, $h = 0$; $p = 58/b$; $\phi = 15°$; $\alpha = 45° + 7°30' = 52°30'$; $58/b = 0.105b(3.759 - 1.303)/4 + 2(1.2)(1.303 + 2.213)$; $b = 6.55$ ft (1.996 m); $p = 58{,}000/6.55 = 8850$ lb/ft² (423.7 kPa).

SOIL CONSOLIDATION AND CHANGE IN VOID RATIO

In a laboratory test, a load was applied to a soil specimen having a height of 30 in (762.0 mm) and a void ratio of 96.0 percent. What was the void ratio when the load settled ½ in (12.7 mm)?

Calculation Procedure:

1. Construct a diagram representing the volumetric composition of the soil in the original and final states

According to the Terzaghi theory of consolidation, the compression of a soil mass under an increase in pressure results primarily from the expulsion of water from the pores. At the instant the load is applied, it is supported entirely by the water, and the hydraulic gradient thus established induces flow. However, the flow in turn causes a continuous transfer of load from the water to the solids.

Equilibrium is ultimately attained when the load is carried entirely by the solids, and the expulsion of the water then ceases. The time rate of expulsion, and therefore of consolidation, is a function of the permeability of the soil, the number of drainage faces, etc. Let H = original height of soil stratum; s = settlement; e_1 = original void ratio; e_2 = final void ratio. Using the given data, construct the diagram in Fig. 233, representing the volumetric composition of the soil in the original and final states.

2. State the equation relating the four defined quantities

Thus,

$$s = \frac{H(e_1 - e_2)}{1 + e_1} \qquad (187)$$

3. Solve for e_2

Thus: $H = 30$ in (762.0 mm); $s = 0.50$ in (12.7 mm); $e_1 = 0.960$; $e_2 = 92.7$ percent.

COMPRESSION INDEX AND VOID RATIO OF A SOIL

A soil specimen under a pressure of 1200 lb/ft² (57.46 kPa) is found to have a void ratio of 103 percent. If the compression index is 0.178, what will be the void ratio when the pressure is increased to 5000 lb/ft² (239.40 kPa)?

Calculation Procedure:

1. Define the compression index

By testing a soil specimen in a consolidometer, it is possible to determine the void ratio associated with a given compressive stress. When the sets of values thus obtained are plotted on semilogarithmic scales (void ratio vs. logarithm of stress), the resulting diagram is curved initially but becomes virtually a straight line beyond a specific point. The slope of this line is termed the *compression index*.

2. Compute the soil void ratio

Let C_c = compression index; e_1 and e_2 = original and final void ratio, respectively; σ_1 and σ_2 = original and final normal stress, respectively.

Write the equation of the straight-line portion of the diagram:

$$e_1 - e_2 = C_c \log \frac{\sigma_2}{\sigma_1} \qquad (188)$$

Substituting and solving give $1.03 - e_2 = 0.178 \log (5000/1200)$; $e_2 = 92.0$ percent. Note that the logarithm is taken to the base 10.

Landfills—where municipal and industrial wastes are discarded—are subject to soil consolidation because of gradual contraction of the components. To hasten this contraction and reduce the space needed for trash, some communities are mining established landfills.

When a trash landfill is mined, more than half of the contents may be combustible in an incinerator. Useful electric power can be produced by burning the recovered trash. In one landfill, some 54 percent of the trash is burned, 36 percent is recycled to cover new trash, and 10 percent is returned to the landfill as unrecoverable. The overall effect is to obtain useful power from the mined trash while reducing the volume of the trash by 75 percent. This allows more new trash to be stored at the landfill without increasing the area required.

Mining of landfills also saves closing costs, which can run into millions of dollars for even the smallest landfill. Current EPA regulations require a landfill to be monitored for environmental risks for 30 years after closing. Mining the landfill eliminates the need for closure while producing moneymaking power and reducing the storage volume needed for a specific amount of trash. So before soil consolidation tests are made for a landfill, plans for its possible mining should be reviewed.

SETTLEMENT OF FOOTING

An 8-ft (2.4-m) square footing carries a load of 150 kips (667.2 kN) that may be considered uniformly distributed, and it is supported by the soil strata shown in Fig. 234. The silty clay has a compression index of 0.274; its void ratio prior to application of the load is 84 percent. Applying the unit weights indicated in Fig. 234, calculate the settlement of the footing caused by consolidation of the silty clay.

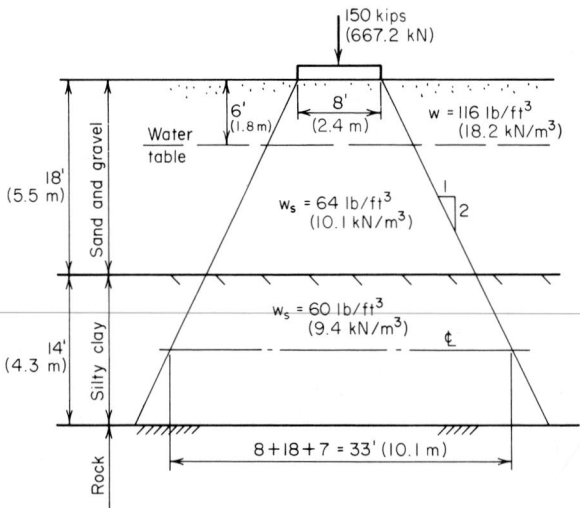

FIG. 234 Settlement of footing.

Calculation Procedure:

1. Compute the vertical stress at middepth before and after application of the load

To simplify the calculations, assume that the load is transmitted through a truncated pyramid having side slopes of 2 to 1 and that the stress is uniform across a horizontal plane. Take the stress at middepth of the silty-clay stratum as the average for that stratum.

Compute the vertical stress σ_1 and σ_2 at middepth before and after application of the load, respectively. Thus: $\sigma_1 = 6(116) + 12(64) + 7(60) = 1884$ lb/ft^2 (90.21 kPa); $\sigma_2 = 1884 + 150,000/33^2 = 2022$ lb/ft^2 (96.81 kPa).

2. Compute the footing settlement

Combine Eqs. 187 and 188 to obtain

$$s = \frac{HC_c \log (\sigma_2/\sigma_1)}{1 + e_1} \tag{189}$$

Solving gives $s = 14(0.274) \log (2022/1884)/(1 + 0.84) = 0.064$ ft $= 0.77$ in (19.558 mm).

DETERMINATION OF FOOTING SIZE BY HOUSEL'S METHOD

A square footing is to transmit a load of 80 kips (355.8 kN) to a cohesive soil, the settlement being restricted to ⅝ in (15.9 mm). Two test footings were loaded at the site until the settlement reached this value. The results were

Footing size	Load, lb (N)
1 ft 6 in × 2 ft (45.72 × 60.96 cm)	14,200 (63,161.6)
3 ft × 3 ft (91.44 × 91.44 cm)	34,500 (153,456.0)

Applying Housel's method, determine the size of the footing in plan.

Calculation Procedure:

1. Determine the values of p and s corresponding to the allowable settlement

Housel considers that the ability of a cohesive soil to support a footing stems from two sources: bearing strength and shearing strength. This concept is embodied in

$$W = Ap + Ps \tag{190}$$

where W = total load; A = area of contact surface; P = perimeter of contact surface; p = bearing stress directly below footing; s = shearing stress along perimeter.

Applying the given data for the test footings gives: footing 1, $A = 3$ ft^2 (2787 cm^2), $P = 7$ ft (2.1 m); footing 2, $A = 9$ ft^2 (8361 cm^2), $P = 12$ ft (3.7 m). Then $3p + 7s = 14,200$; $9p + 12s = 34,500$; $p = 2630$ lb/ft^2 (125.9 kPa); $s = 900$ lb/lin ft (13,134.5 N/m).

2. Compute the size of the footing to carry the specified load

Let x denote the side of the footing. Then, $2630x^2 + 900(4x) = 80,000$; $x = 4.9$ ft (1.5 m). Make the footing 5 ft (1.524 m) square.

APPLICATION OF PILE-DRIVING FORMULA

A 16 × 16 in (406.4 × 406.4 mm) pile of 3000-lb/in^2 (20,685-kPa) concrete, 45 ft (13.7 m) long, is reinforced with eight no. 7 bars. The pile is driven by a double-acting steam hammer. The weight of the ram is 4600 lb (20,460.8 N), and the energy delivered is 17,000 ft·lb (23,052 J) per blow. The average penetration caused by the final blows is 0.42 in (10.668 mm). Compute the bearing capacity of the pile by applying Redtenbacker's formula and using a factor of safety of 3.

Calculation Procedure:

1.   Find the weight of the pile and the area of the transformed section

The work performed in driving a pile into the soil is a function of the reaction of the soil on the pile and the properties of the pile. Therefore, the soil reaction may be evaluated if the work performed by the hammer is known. Let A = cross-sectional area of pile; E = modulus of elasticity; h = height of fall of ram; L = length of pile; P = allowable load on pile; R = reaction of soil on pile; s = penetration per blow; W = weight of falling ram; w = weight of pile.

Redtenbacker developed the following equation by taking these quantities into consideration: the work performed by the soil in bringing the pile to rest; the work performed in compressing the pile; and the energy delivered to the pile:

$$Rs + \frac{R^2L}{2AE} = \frac{W^2h}{W + w} \tag{191}$$

Finding the weight of the pile and the area of the transformed section, we get w = 16(16) (0.150)(45)/144 = 12 kips (53.4 kN). The area of a no. 7 bar = 0.60 in^2 (3.871 cm^2); n = 9; A = 16(16) + 8(9 − 1)0.60 = 294 in^2 (1896.9 cm^2).

2.   Apply Eq. 191 to find R; evaluate P

Thus, s = 0.42 in (10.668 mm); L = 540 in (13,716 mm); E_c = 3160 kips/in^2 (21,788.2 MPa.); W = 4.6 kips (20.46 kN); Wh = 17 ft·kips = 204 in·kips (23,052 J). Substituting gives 0.42R + 540R^2/[2(294)(3160)] = 4.6(204)/(4.6 + 12); R = 84.8 kips (377.19 kN); P = $R/3$ = 28.3 kips (125.88 kN).

CAPACITY OF A GROUP OF FRICTION PILES

A structure is to be supported by 12 friction piles of 10-in (254-mm) diameter. These will be arranged in four rows of three piles each at a spacing of 3 ft (91.44 cm) in both directions. A test pile is found to have an allowable load of 32 kips (142.3 kN). Determine the load that may be carried by this pile group.

Calculation Procedure:

1.   State a suitable equation for the load

When friction piles are compactly spaced, the area of soil that is needed to support an individual pile overlaps that needed to support the adjacent ones. Consequently, the capacity of the group is less than the capacity obtained by aggregating the capacities of the individual piles. Let P = capacity of group; P_i = capacity of single pile; m = number of rows; n = number of piles per row; d = pile diameter; s = center-to-center spacing of piles; θ = tan^{-1} d/s deg. A suitable

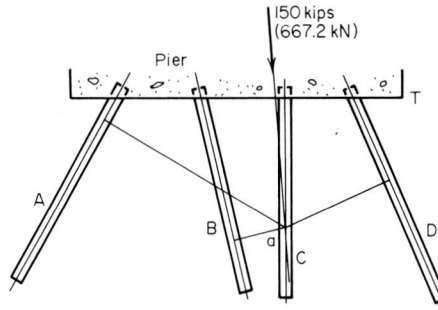

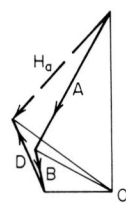

(a) Pile group and load

(b) Force polygon
for H_a

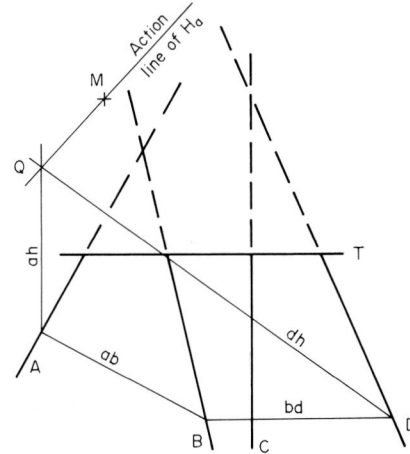

(c) Construction to locate action line of H_a

FIG. 235

equation using these variables is the Converse-Labarre equation

$$\frac{P}{P_i} = mn - \left(\frac{\theta}{90°}\right)[m(n-1) + n(m-1)]$$ (192)

2. *Compute the load*

Thus P_i = 32 kips (142.3 kN); m = 4; n = 3; d = 10 in (254 mm); s = 36 in (914.4 mm); θ = $\tan^{-1} 10/36$ = 15.5°. Then $P/32$ = 12 − (15.5/90)(4 × 2 + 3 × 3); P = 290 kips (1289.9 kN).

LOAD DISTRIBUTION AMONG HINGED BATTER PILES

Figure 235*a* shows the relative positions of four steel bearing piles that carry the indicated load. The piles, which may be considered as hinged at top and bottom, have identical cross sections and the following relative effective lengths: *A*, 1.0; *B*, 0.95; *C*, 0.93; *D*, 1.05. Outline a graphical procedure for determining the load transmitted to each pile.

Calculation Procedure:

1.   Subject the structure to a load for purposes of analysis

Steel and timber piles may be considered to be connected to the concrete pier by frictionless hinges, and bearing piles that extend a relatively short distance into compact soil may be considered to be hinge-supported by the soil.

Since four unknown quantities are present, the structure is statically indeterminate. A solution to this problem therefore requires an analysis of the deformation of the structure.

As the load is applied, the pier, assumed to be infinitely rigid, rotates to some new position. This displacement causes each pile to rotate about its base and to undergo an axial strain. The contraction or elongation of each pile is directly proportional to the perpendicular distance p from the axis of rotation to the longitudinal axis of that pile. Let P denote the load induced in the pile. Then $P = \Delta L\, AE/L$. Since ΔL is proportional to p and AE is constant for the group, this equation may be transformed to

$$P = \frac{kp}{L} \tag{193}$$

where k is a constant of proportionality.

If the center of rotation is established, the pile loads may therefore be found by scaling the p distances. Westergaard devised a simple graphical method of locating the center of rotation. This method entails the construction of string polygons, described in the first calculation procedure of this handbook.

In Fig. 235a select any convenient point a on the action line of the load. Consider the structure to be subjected to a load H_a that causes the pier to rotate about a as a center. The object is to locate the action line of this hypothetical load.

It is often desirable to visualize that a load is applied to a body at a point that in reality lies outside the body. This condition becomes possible if the designer annexes to the body an infinitely rigid arm containing the given point. Since this arm does not deform, the stresses and strains in the body proper are not modified.

2.   Scale the perpendicular distance from a to the longitudinal axis of each pile; divide this distance by the relative length of the pile

In accordance with Eq. 193, the quotient represents the relative magnitude of the load induced in the pile by the load H_a. If rotation is assumed to be counterclockwise, piles A and B are in compression and D is in tension.

3.   Using a suitable scale, construct the force polygon

This polygon is shown in Fig. 235b. Construct this polygon by applying the results obtained in step 2. This force polygon yields the direction of the action line of H_a.

4.   In Fig. 235b, select a convenient pole O and draw rays to the force polygon

5.   Construct the string polygon shown in Fig. 235c

The action line of H_a passes through the intersection point Q of rays ah and dh, and its direction appears in Fig. 235b. Draw this line.

6.   Select a second point on the action line of the load

Choose point b on the action line of the 150-kip (667.2-kN) load, and consider the structure to be subjected to a load H_b that causes the pier to rotate about b as center.

7.   Locate the action line of H_b

Repeat the foregoing procedure to locate the action line of H_b in Fig. 235c. (The construction has been omitted for clarity.) Study of the diagram shows that the action lines of H_a and H_b intersect at M.

8.   Test the accuracy of the construction

Select a third point c on the action line of the 150-kip (667.2-kN) load, and locate the action line of the hypothetical load H_c causing rotation about c. It is found that H_c also passes through M. In summary, these hypothetical loads causing rotation about specific points on the action line of the true load are all concurrent.

Thus, M is the center of rotation of the pier under the 150-kip (667.2-kN) load. This conclusion stems from the following analysis: Load H_a applied at M causes zero deflection at a. Therefore, in accordance with Maxwell's theorem of reciprocal deflections, if the true load is applied at a, it will cause zero deflection at M in the direction of H_a. Similarly, if the true load is applied at b, it will cause zero deflection at M in the direction of H_b. Thus, M remains stationary under the 150-kip (667.2-kN) load; that is, M is the center of rotation of the pier.

9.   Scale the perpendicular distance from M to the longitudinal axis of each pile

Divide this distance by the relative length of the pile. The quotient represents the relative magnitude of the load induced in the pile by the 150-kip (667.2-kN) load.

10.   Using a suitable scale, construct the force polygon by applying the results from step 9

If the work is accurate, the resultant of these relative loads is parallel to the true load.

11.   Scale the resultant; compute the factor needed to correct this value to 150 kips (667.2 kN)

12.   Multiply each relative pile load by this correction factor to obtain the true load induced in the pile

LOAD DISTRIBUTION AMONG PILES WITH FIXED BASES

Assume that the piles in Fig. 235a penetrate a considerable distance into a compact soil and may therefore be regarded as restrained against rotation at a certain level. Outline a procedure for determining the axial load and bending moment induced in each pile.

Calculation Procedure:

1.   State the equation for the length of a dummy pile

Since the Westergaard construction presented in the previous calculation procedure applies solely to hinged piles, the group of piles now being considered is not directly amenable to analysis by this method.

As shown in Fig. 236a, the pile AB functions in the dual capacity of a column and cantilever beam. In Fig. 236b, let A' denote the position of A following application of the load. If secondary effects are disregarded, the axial force P transmitted to this pile is a function of Δ_y, and the transverse force S is a function of Δ_x.

Consider that the fixed support at B is replaced with a hinged support and a pile AC of identical cross section is added perpendicular to AB, as shown in Fig. 236b. If pile AC deforms an amount Δ_x under an axial force S, the forces transmitted by the pier at each point of support are not affected by this modification of supports. The added pile is called a *dummy* pile. Thus, the given pile group may be replaced with an equivalent group consisting solely of hinged piles.

Stating the equation for the length L' of the dummy pile, equate the displacement Δ_x in the equivalent pile group to that in the actual group. Or, $\Delta_x = SL'/AE = SL^3/3EI$.

$$L' = \frac{AL^3}{3I} \tag{194}$$

2.   Replace all fixed supports in the given pile group with hinged supports

Add the dummy piles. Compute the lengths of these piles by applying Eq. 194.

3.   Determine the axial forces induced in the equivalent pile group

Using the given load, apply Westergaard's construction, as described in the previous calculation procedure.

4.   Remove the dummy piles; restore the fixed supports

Compute the bending moments at these supports by applying the equation $M = SL$.

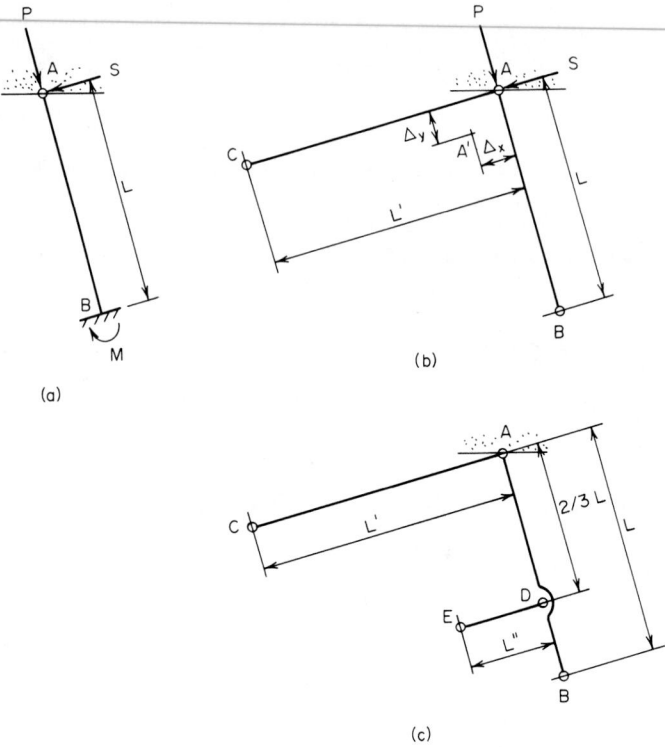

FIG. 236   Real and dummy piles.

LOAD DISTRIBUTION AMONG PILES FIXED AT TOP AND BOTTOM

Assume that the piles in Fig. 235a may be regarded as having fixed supports both at the pier and at their bases. Outline a procedure for determining the axial load and bending moment induced in each pile.

Calculation Procedure:

1.   *Describe how dummy piles may be used*

A pile made of reinforced concrete and built integrally with the pier is restrained against rotation relative to the pier. As shown in Fig. 236c, the fixed supports of pile *AB* may be replaced with hinges provided that dummy piles *AC* and *DE* are added, the latter being connected to the pier by means of a rigid arm through *D*.

2.   *Compute the lengths of the dummy piles*

If *D* is placed at the lower third point as indicated, the lengths to be assigned to the dummy piles are

$$L' = \frac{AL^3}{3I} \quad \text{and} \quad L'' = \frac{AL^3}{9I} \tag{195}$$

Replace the given group of piles with its equivalent group, and follow the method of solution in the previous calculation procedure.

SECTION 2

ARCHITECTURAL ENGINEERING

MAX KURTZ, P.E.
CONSULTING ENGINEER

GREGORY T. HICKS, R.A.
GREGORY T. HICKS AND ASSOCIATES, ARCHITECTS

STRUCTURAL DESIGN 2.2
   Design of an Eyebar 2.3
   Analysis of a Steel Hanger 2.4
   Analysis of a Gusset Plate 2.5
   Design of a Semirigid Connection 2.6
   Riveted Moment Connection 2.7
   Design of a Welded Flexible Beam Connection 2.9
   Design of a Welded Seated Beam Connection 2.10
   Design of a Welded Moment Connection 2.12
   Rectangular Knee of Rigid Bent 2.13
   Curved Knee of Rigid Bent 2.14
   Base Plate for Steel Column Carrying Axial Load 2.15
   Base for Steel Column with End Moment 2.15
   Grillage Support for Column 2.16
   Wind-Stress Analysis by Portal Method 2.19
   Wind-Stress Analysis by Cantilever Method 2.21
   Wind-Stress Analysis by Slope-Deflection Method 2.23
   Wind Drift of a Building 2.25
   Reduction in Wind Drift by Using Diagonal Bracing 2.26
   Light-Gage Steel Beam with Unstiffened Flange 2.27
   Light-Gage Steel Beam with Stiffened Compression Flange . . . 2.28
   Steel Beam Encased in Concrete 2.30
   Composite Steel-and-Concrete Beam 2.32
   Design of a Concrete Joist in a Ribbed Floor 2.34

Design of a Stair Slab 2.35
Free Vibratory Motion of a Rigid Bent 2.37
PLUMBING AND DRAINAGE 2.37
Determination of Plumbing-System Pipe Sizes 2.37
Design of Roof and Yard Rainwater Drainage Systems 2.42
Sizing Cold- and Hot-Water-Supply Piping 2.44
Sprinkler-System Selection and Design 2.52
Sizing Gas Piping for Heating and Cooking 2.55
Swimming Pool Selection, Sizing, and Servicing 2.58
HEATING, VENTILATING, AND AIR CONDITIONING 2.61
Building or Structure Heat-Loss Determination 2.61
Heating-System Selection and Analysis 2.62
Required Capacity of a Unit Heater 2.65
Steam Consumption of Heating Apparatus 2.70
Selection of Air Heating Coils 2.72
Radiant-Heating-Panel Choice and Sizing 2.75
Snow-Melting Heating-Panel Choice and Sizing 2.78
Heat Recovery from Lighting Systems for Space Heating 2.79
Air-Conditioning-System Heat-Load Determination—General Method 2.80
Air-Conditioning-System Heat-Load Determination—Numerical Computation . .   2.84
Air-Conditioning-System Cooling-Coil Selection 2.88
Mixing of Two Airstreams 2.93
Selection of an Air-Conditioning System for a Known Load 2.94
Sizing Low-Velocity Air-Conditioning-System Ducts—Equal-Friction Method . .   2.96
Sizing Low-Velocity Air-Conditioning Ducts—Static-Regain Method 2.102
Humidifier Selection for Desired Atmospheric Conditions 2.105
Use of the Psychrometric Chart in Air-Conditioning Calculations 2.110
Designing High-Velocity Air-Conditioning Ducts 2.112
Air-Conditioning-System Outlet- and Return-Grille Selection 2.114
Selecting Roof Ventilators for Building 2.118
Vibration-Isolator Selection for an Air Conditioner 2.120
Selection of Noise-Reduction Materials 2.122
SOLAR ENERGY 2.126
Designing a Flat-Plate Solar-Energy Heating and Cooling System 2.126
Determination of Solar Insolation on Solar Collectors under Differing Conditions   2.131
Sizing Collectors for Solar-Energy Heating Systems 2.134
F Chart Method for Determining Useful Energy Delivery in Solar Heating . . .   2.135
Domestic Hot-Water-Heater Collector Selection 2.141
Passive Solar-Heating System Design 2.145

Structural Design

REFERENCES: Crawley and Dillon—*Steel Buildings: Analysis and Design*, Wiley; Callender—*Time-Saver Standards for Architectural Design Data*, McGraw-Hill; Tutt and Adler—*VNR Metric Handbook of Architectural Standards*, Van Nostrand; Knowles—*Composite Steel and Concrete Construction*, Halsted Press; Parker and Hauf—*Simplified Design of Structural Wood*, Wiley; Timber Engineering Company—*Timber Design and Construction Handbook*, McGraw-Hill; Shuttleworth—*Mechanical and Electrical Systems for Construction*, McGraw-Hill; Aluminum Association—*Aluminum Construction Manual*; American Concrete Institute—*Building Code Requirements for Reinforced Concrete*; American Institute of Steel Construction—*Manual of Steel Construction*; American Iron and Steel Institute—*Light Gage Cold-Formed Steel Design Manual*; National Forests Products Association—*National Design Specification for Stress-Grade Lumber and Its Fastenings*; Beddle, et al.—*Structural Steel Design*, Ronald; Ferguson—*Reinforced Concrete Fundamentals*, Wiley; Gaylord and Gaylord—*Design of Steel Structures*, McGraw-Hill; Grinter—*Design of Modern Steel Structures*, Macmillan; Hoadley—*Essentials of Structural Design*, Wiley; Leontovich—*Frames and Arches*, McGraw-Hill; Lothers—*Advanced Design in Structural Steel*, Prentice-Hall; Norris, et al.—*Structural Design for Dynamic Loads*, McGraw-Hill; Salvadori and Levy—*Structural Design in Architecture*, Prentice-Hall;

Sanks—*Statistically Indeterminate Structural Analysis*, Ronald; Sutherland and Bowman—*Structural Theory*, Wiley; Timoshenko and Young—*Vibration Problems in Engineering*, Van Nostrand; Viest, Fountain, and Singleton—*Composite Construction in Steel and Concrete*, McGraw-Hill; Weidlinger—*Aluminum in Modern Architecture, Vol.* II, Reinhold; Williams and Harris—*Structural Design in Metals*, Ronald; Winter, et al.—*Design of Concrete Structures*, McGraw-Hill.

The basic principles of structural design are developed and applied in Sec. 1, Civil Engineering.

In the following Calculation Procedures, structural steel members are designed in accordance with the *Specification for the Design, Fabrication and Erection of Structural Steel for Buildings* of the American Institute of Steel Construction. In the absence of any statement to the contrary, it is to be understood that the structural-steel members are made of ASTM A36 steel, which has a yield-point stress of 36,000 lb/in^2 (248.2 MPa).

Reinforced-concrete members are designed in accordance with the specification *Building Code Requirements for Reinforced Concrete* of the American Concrete Institute.

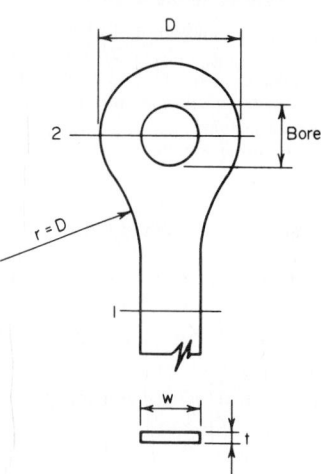

FIG. 1 Eyebar hanger.

DESIGN OF AN EYEBAR

A hanger is to carry a load of 175 kips (778.4 kN). Design an eyebar of A440 steel.

Calculation Procedure:

1. *Record the yield-point stresses of the steel*

Refer to Fig. 1 for the notational system. Let subscripts 1 and 2 refer to cross sections through the body of the bar and through the center of the pin hole, respectively.

Eyebars are generally flame-cut from plates of high-strength steel. The design provisions of the AISC *Specification* reflect the results of extensive testing of such members. A section of the *Specification* permits a tensile stress of $0.60f_y$ at 1 and $0.45f_y$ at 2, where f_y denotes the yield-point stress.

From the AISC *Manual* for A440 steel:

If $t \leq 0.75$ in (19.1 mm), $f_y = 50$ kips/in^2 (344.7 MPa).
If $0.75 < t \leq 1.5$ in (38 mm), $f_y = 46$ kips/in^2 (317.1 MPa).
If $1.5 < t \leq 4$ in (102 mm), $f_y = 42$ kips/in^2 (289.5 MPa).

2. *Design the body of the member, using a trial thickness*

The *Specification* restricts the ratio w/t to a value of 8. Compute the capacity P of a ¾ in (19.1-mm) eyebar of maximum width. Thus $w = 8(¾) = 6$ in (152 mm); $f = 0.6(50) = 30$ kips/in^2 (206.8 MPa); $P = 6(0.75)30 = 135$ kips (600.5 kN). This is not acceptable because the desired capacity is 175 kips (778.4 kN). Hence, the required thickness exceeds the trial value of ¾ in (19.1 mm). With t greater than ¾ in (19.1 mm), the allowable stress at 1 is $0.60f_y$, or $0.60(46$ kips/$in^2)$ = 27.6 kips/in^2 (190.3 MPa); say 27.5 kips/in^2 (189.6 MPa) for design use. At 2 the allowable stress is $0.45(46) = 20.7$ kips/in^2 (142.7 MPa), say 20.5 kips/in^2 (141.3 MPa) for design purposes.

To determine the required area at 1, use the relation $A_1 = P/f$, where f = allowable stress as computed above. Thus, $A_1 = 175/27.5 = 6.36$ in^2 (4103 mm²). Use a plate 6½ × 1 in (165 × 25.4 mm) in which $A_1 = 6.5$ in^2 (4192 mm²).

3. *Design the section through the pin hole*

The AISC *Specification* limits the pin diameter to a minimum value of $7w/8$. Select a pin diameter of 6 in (152 mm). The bore will then be 6 6/32 in (153 mm) diameter. The net width required will be $P/(ft) = 175/[20.5(1.0)] = 8.54$ in (217 mm); $D_{min} = 6.03 + 8.54 = 14.57$ in (370 mm). Set $D = 14¾$ in (375 mm), $A_2 = 1.0(14.75 − 6.03) = 8.72$ in^2 (5626 mm²); $A_2/A_1 = 1.34$. This result is satisfactory, because the ratio of A_2/A_1 must lie between 1.33 and 1.50.

4. Determine the transition radius r

In accordance with the *Specification*, set $r = D = 14\frac{3}{4}$ in (374.7 mm).

ANALYSIS OF A STEEL HANGER

A $12 \times \frac{1}{2}$ in (305×12.7 mm) steel plate is to support a tensile load applied 2.2 in (55.9 mm) from its center. Determine the ultimate load.

Calculation Procedure:

1. Determine the distance x

The plastic analysis of steel structures is developed in Sec. 1 of this handbook. Figure 2a is the load diagram, and Fig. 2b is the stress diagram at plastification. The latter may be replaced for convenience with the stress diagram in Fig. 2c, where $T_1 = C$; P_u = ultimate load; e = eccentricity; M_u = ultimate moment = $P_u e$; f_y = yield-point stress; d = depth of section; t = thickness of section.

By using Fig. 2c,

$$P_u = T_2 = f_y t(d - 2x) \tag{1}$$

Also, $T_1 = f_y tx$, and $M_u = P_u e = T_1(d - x)$, so

$$x = \frac{d}{2} + e - \left[\left(\frac{d}{2} + e \right)^2 - ed \right]^{0.5} \tag{2}$$

Or, $x = 6 + 2.2 - [(6 + 2.2)^2 - 2.2 \times 12]^{0.5} = 1.81$ in (45.9 mm).

2. Find P_u

By Eq. 1, $P_u = 36,000(0.50)(12 - 3.62) = 151,000$ lb (671.6 kN).

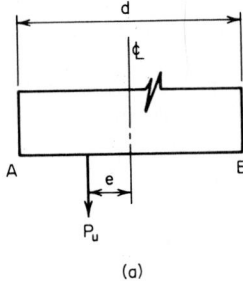

(a)

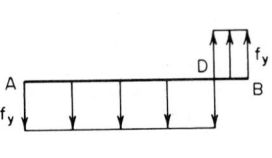

(b)

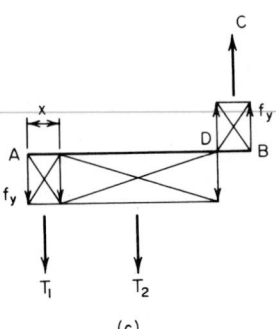

(c)

FIG. 2

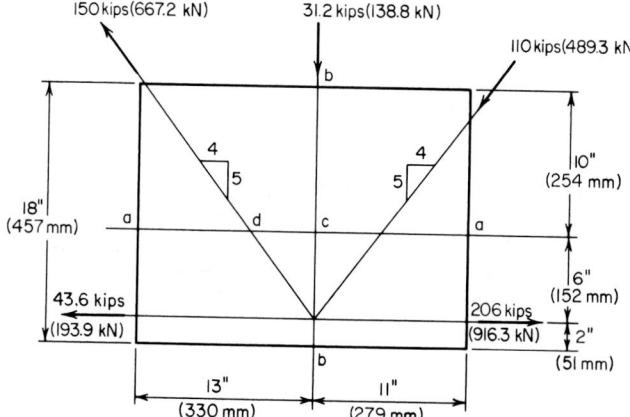

FIG. 3 Gusset plate.

ANALYSIS OF A GUSSET PLATE

The gusset plate in Fig. 3 is ½ in (12.7 mm) thick and connects three web members to the bottom chord of a truss. The plate is subjected to the indicated ultimate forces, and transfer of these forces from the web members to the plate is completed at section a-a. Investigate the adequacy of this plate. Use 18,000 lb/in² (124.1 MPa) as the yield-point stress in shear, and disregard interaction of direct stress and shearing stress in computing the ultimate-load and ultimate-moment capacity.

Calculation Procedure:

1. Resolve the diagonal forces into their horizontal and vertical components

Let H_u and V_u denote the ultimate shearing force on a horizontal and vertical plane, respectively. Resolving the diagonal forces into their horizontal and vertical components gives $(4^2 + 5^2)^{0.5} = 6.40$. Horizontal components: $150(4/6.40) = 93.7$ kips (416.8 kN); $110(4/6.40) = 68.7$ kips (305.6 kN). Vertical components: $150(5/6.40) = 117.1$ kips (520.9 kN); $110(5/6.40) = 85.9$ kips (382.1 kN).

2. Check the force system for equilibrium

Thus, $\Sigma F_H = 206.0 - 43.6 - 93.7 - 68.7 = 0$; this is satisfactory, as is $\Sigma F_V = 117.1 - 85.9 - 31.2 = 0$.

3. Compare the ultimate shear at section a-a with the allowable value

Thus, $H_u = 206.0 - 43.6 = 162.4$ kips (722.4 kN). To compute $H_{u,\text{allow}}$, assume that the shearing stress is equal to the yield-point stress across the entire section. Then $H_{u,\text{allow}} = 24(0.5)(18) = 216$ kips (960.8 kN). This is satisfactory.

4. Compare the ultimate shear at section b-b with the allowable value

Thus, $V_u = 117.1$ kips (520.9 kN); $V_{u,\text{allow}} = 18(0.5)(18) = 162$ kips (720.6 kN). This is satisfactory.

5. Compare the ultimate moment at section a-a with the plastic moment

Thus, $cd = 4(6)/5 = 4.8$ in (122 mm); $M_u = 4.8(117.1 + 85.9) = 974$ in·kips (110.1 kN·m). Or, $M_u = 6(206 - 43.6) = 974$ in·kips (110.1 kN·m). To find the plastic moment M_p, use the relation $M_p = f_y bd^2/4$, or $M_p = 36(0.5)(24)^2/4 = 2592$ in·kips (292.9 kN·m). This is satisfactory.

6. Compare the ultimate direct force at section b-b with the allowable value

Thus, $P_u = 93.7 + 43.6 = 137.3$ kips (610.7 kN); or $P_u = 206.0 - 68.7 = 137.3$ kips (610.7 kN); $e = 9 - 2 = 7$ in (177.8 mm). By Eq. 2, $x = 9 + 7 - [(9 + 7)^2 - 7 \times 18]^{0.5} = 4.6$ in

(116.8 mm). By Eq. 1, $P_{u,allow} = 36{,}000(0.5)(18 - 9.2) = 158.4$ kips (704.6 kN). This is satisfactory.

On horizontal sections above a-a, the forces in the web members have not been completely transferred to the gusset plate, but the eccentricities are greater than those at a-a. Therefore, the calculations in step 5 should be repeated with reference to one or two sections above a-a before any conclusion concerning the adequacy of the plate is drawn.

DESIGN OF A SEMIRIGID CONNECTION

A W14 × 38 beam is to be connected to the flange of a column by a semirigid connection that transmits a shear of 25 kips (111.2 kN) and a moment of 315 in · kips (35.6 kN · m). Design the connection for the moment, using A141 shop rivets and A325 field bolts of ⅞-in (22.2-mm) diameter.

Calculation Procedure:

1. *Record the relevant properties of the W14 × 38*

A semirigid connection is one that offers only partial restraint against rotation. For a relatively small moment, a connection of the type shown in Fig. 4a will be adequate. In designing this type of connection, it is assumed for simplicity that the moment is resisted entirely by the flanges; and the force in each flange is found by dividing the moment by the beam depth.

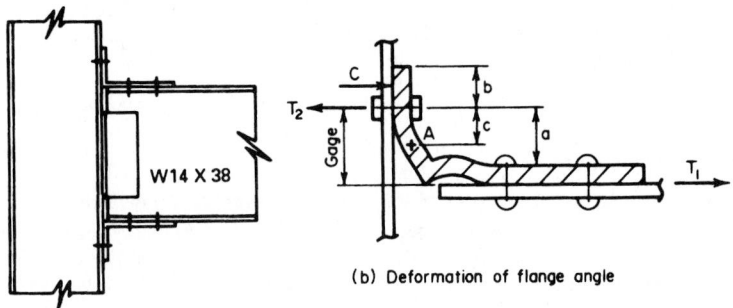

(a) Semirigid connection

(b) Deformation of flange angle

FIG. 4   (*a*) Semirigid connection; (*b*) deformation of flange angle.

Figure 4b indicates the assumed deformation of the upper angle, A being the point of contraflexure in the vertical leg. Since the true stress distribution cannot be readily ascertained, it is necessary to make simplifying assumptions. The following equations evolve from a conservative analysis of the member: $c = 0.6a$; $T_2 = T_1(1 + 3a/4b)$.

Study shows that use of an angle having two rows of bolts in the vertical leg would be unsatisfactory because the bolts in the outer row would remain inactive until those in the inner row yielded. If the two rows of bolts are required, the flange should be connected by means of a tee rather than an angle.

The following notational system will be used with reference to the beam dimensions: b = flange width; d = beam depth; t_f = flange thickness; t_w = web thickness.

Record the relevant properties of the W14 × 38; $d = 14.12$ in (359 mm); $t_f = 0.513$ in (13 mm). (Obtain these properties from a table of structural-shape data.)

2. *Establish the capacity of the shop rivets and field bolts used in transmitting the moment*

From the AISC *Specification*, the rivet capacity in single shear = 0.6013(15) = 9.02 kips (40.1 kN); rivet capacity in bearing = 0.875(0.513)(48.5) = 21.77 kips (96.8 kN); bolt capacity in tension = 0.6013(40) = 24.05 kips (106.9 kN).

3. *Determine the number of rivets required in each beam flange*

Thus, T_1 = moment/d = 315/14.12 = 22.31 kips (99.7 kN); number of rivets = T_1/rivet capacity in single shear = 22.31/9.02 = 2.5; use four rivets, the next highest even number.

4. *Assuming tentatively that one row of field bolts will suffice, design the flange angle*

Try an angle 8 × 4 × ¾ in (203 × 102 × 19 mm), 8 in (203 mm) long, having a standard gage of 2½ in (63.5 mm) in the vertical leg. Compute the maximum bending moment M in this leg. Thus, c = 0.6(2.5 − 0.75) = 1.05 in (26.7 mm); M = T_1c = 23.43 in·kips (2.65 kN·m). Then apply the relation $f = M/S$ to find the flexural stress. Or, f = 23.43/[(⅛)(8)(0.75)²] = 31.24 kips/in² (215.4 MPa).

Since the cross section is rectangular, the allowable stress is 27 kips/in² (186.1 MPa), as given by the AISC *Specification*. (The justification for allowing a higher flexural stress in a member of rectangular cross section as compared with a wide-flange member is presented in Sec. 1.)

Try a ⅞-in (22-mm) angle, with c = 0.975 in (24.8 mm); M = 21.75 in·kips (2.46 kN·m); f = 21.75/(⅛)(8)(0.875)² = 21.3 kips/in² (146.8 MPa). This is an acceptable stress.

5. *Check the adequacy of the two field bolts in each angle*

Thus, T_2 = 22.31[1 + 3 × 1.625/(4 × 1.5)] = 40.44 kips (179.9 kN); the capacity of two bolts = 2(24.05) = 48.10 kips (213.9 kN). Hence the bolts are acceptable because their capacity exceeds the load.

6. *Summarize the design*

Use angles 8 × 4 × ⅞ in (203 × 102 × 19 mm), 8 in (203 mm) long. In each angle, use four rivets for the beam connection and two bolts for the column connection. For transmitting the shear, the standard web connection for a 14-in (356-mm) beam shown in the AISC *Manual* is satisfactory.

RIVETED MOMENT CONNECTION

A W18 × 60 beam frames to the flange of a column and transmits a shear of 40 kips (177.9 kN) and a moment of 2500 in·kips (282.5 kN · m). Design the connection, using ⅞-in (22-mm) diameter rivets of A141 steel for both the shop and field connections.

Calculation Procedure:

1. *Record the relevant properties of the W18 × 60*

The connection is shown in Fig. 5a. Referring to the row of rivets in Fig. 5b, consider that there are n rivets having a uniform spacing p. The moment of inertia and section modulus of this rivet group with respect to its horizontal centroidal axis are

$$I = p^2 n \times \frac{n^2 - 1}{12} \qquad S = \frac{pn(n + 1)}{6} \qquad (3)$$

Record the properties of the W18 × 60: d = 18.25 in (463.6 mm); b = 7.558 in (192 mm); k = 1.18 in (30.0 mm); t_f = 0.695 in (17.7 mm); t_w = 0.416 in (10.6 mm).

2. *Establish the capacity of a rivet*

Thus: single shear, 9.02 kips (40.1 kN); double shear, 18.04 kips (80.2 kN); bearing on beam web, 0.875(0.416)(48.5) = 17.65 kips (78.5 kN).

3. *Determine the number of rivets required on line 1 as governed by the rivet capacity*

Try 15 rivets having the indicated disposition. Apply Eq. 3 with n = 17; then make the necessary correction. Thus, I = 9(17)(17² − 1)/12 − 2(9)² = 3510 in² (22,645 cm²); S = 3510/24 = 146.3 in (3716 mm).

Let F denote the force on a rivet, and let the subscripts x and y denote the horizontal and vertical components, respectively. Thus, F_x = M/S = 2500/146.3 = 17.09 kips (76.0 kN); F_y = 40/15 = 2.67 kips (11.9 kN); F = (17.09² + 2.67²)$^{0.5}$ = 17.30 < 17.65. Therefore, this is acceptable.

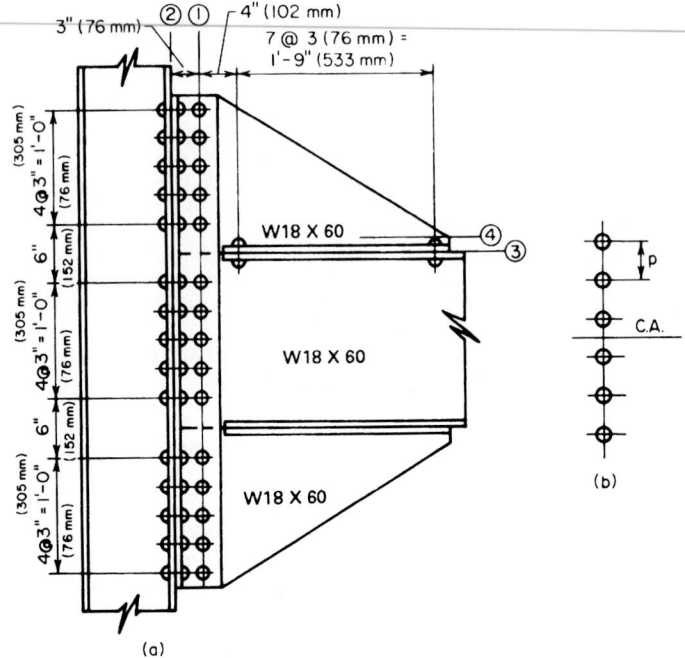

FIG. 5   Riveted moment connection.

4.   *Compute the stresses in the web plate at line 1*

The plate is considered continuous; the rivet holes are assumed to be 1 in (25.4 mm) in diameter for the reasons explained earlier.

The total depth of the plate is 51 in (1295.4 mm); the area and moment of inertia of the net section are $A_n = 0.416(51 - 15 \times 1) = 14.98$ in^2 (96.6 cm^2) and $I_n = (1/12)(0.416)(51)^3 - 1.0(0.416)(3510) = 3138$ in^4 (130,603.6 cm^4).

Apply the general shear equation. Since the section is rectangular, the maximum shearing stress is $v = 1.5V/A_n = 1.5(40)/14.98 = 4.0$ kips/in^2 (27.6 MPa). The AISC *Specification* gives an allowable stress of 14.5 kips/in^2 (99.9 MPa).

The maximum flexural stress is $f = Mc/I_n = 2500(25.5)/3138 = 20.3 < 27$ kips/in^2 (186.1 MPa). This is acceptable. The use of 15 rivets is therefore satisfactory.

5.   *Compute the stresses in the rivets on line 2*

The center of rotation of the angles cannot be readily located because it depends on the amount of initial tension to which the rivets are subjected. For a conservative approximation, assume that the center of rotation of the angles coincides with the horizontal centroidal axis of the rivet group. The forces are $F_x = 2500/[2(146.3)] = 8.54$ kips (37.9 kN); $F_y = 40/30 = 1.33$ kips (5.9 kN). The corresponding stresses in tension and shear are $s_t = F_x/A = 8.54/0.6013 = 14.20$ kips/in^2 (97.9 MPa); $s_s = F_y/A = 1.33/0.6013 = 2.21$ kips/in^2 (15.2 MPa). The *Specification* gives $s_{t,\text{allow}} = 28 - 1.6(2.21) > 20$ kips/in^2 (137.9 kPa). This is acceptable.

6.   *Select the size of the connection angles*

The angles are designed by assuming a uniform bending stress across a distance equal to the spacing p of the rivets; the maximum stress is found by applying the tensile force on the extreme rivet.

Try 4 × 4 × ¾ in (102 × 102 × 19 mm) angles, with a standard gage of 2½ in (63.5 mm) in the outstanding legs. Assuming the point of contraflexure to have the location specified in the previous calculation procedure, we get $c = 0.6(2.5 - 0.75) = 1.05$ in (26.7 mm); $M = 8.54(1.05)$ $= 8.97$ in·kips (1.0 kN·m); $f = 8.97/[(\frac{1}{6})(3)(0.75)^2] = 31.9 > 27$ kips/in² (186.1 MPa). Use 5 $\times 5 \times \frac{7}{8}$ in (127 $\times$ 127 $\times$ 22 mm) angles, with a 2½-in (63.5-mm) gage in the outstanding legs.

7. Determine the number of rivets required on line 3

The forces on the rivets above this line are shown in Fig. 6a. The resultant forces are $H = 64.11$ kips (285.2 kN); $V = 13.35$ kips (59.4 kN). Let M_3 denote the moment of H with respect to line 3. Then $a = \frac{1}{2}(24 - 18.25) = 2.88$ in (73.2 mm); $M_3 = 633.3$ in·kips (71.6 kN·m).

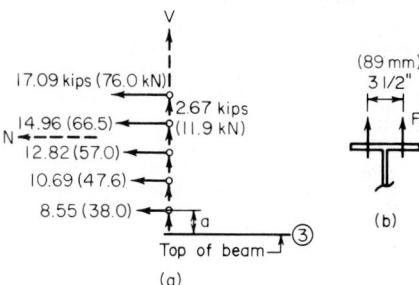

FIG. 6

With reference to Fig. 6b, the tensile force F_y in the rivet is usually limited by the bending capacity of the beam flange. As shown in the AISC *Manual*, the standard gage in the W18 × 60 is 3½ in (88.9 mm). Assume that the point of contraflexure in the beam flange lies midway between the center of the rivet and the face of the web. Referring to Fig. 4b, we have $c = \frac{1}{2}(1.75 - 0.416/2) = 0.771$ in (19.6 mm); $M_{allow} = fS = 27(\frac{1}{6})(3)(0.695)^2 = 6.52$ in·kips (0.74 kN·m). If the compressive force C is disregarded, $F_{y,allow} = 6.52/0.771 = 8.46$ kips (37.6 kN).

Try 16 rivets. The moment on the rivet group is $M = 633.3 - 13.35(14.5) = 440$ in·kips (49.7 kN·m). By Eq. 3, $S = 2(3)(8)(9)/6 = 72$ in (1829 mm). Also, $F_y = 440/72 + 13.35/16 = 6.94 < 8.46$ kips (37.6 kN). This is acceptable. (The value of F_y corresponding to 14 rivets is excessive.)

The rivet stresses are $s_t = 6.94/0.6013 = 11.54$ kips/in² (79.6 MPa); $s_s = 64.11/[16(0.6013)] = 6.67$ kips/in² (45.9 MPa). From the *Specification*, $s_{t,allow} = 28 - 1.6(6.67) = 17.33$ kips/in² (119.5 MPa). This is acceptable. The use of 16 rivets is therefore satisfactory.

8. Compute the stresses in the bracket at the toe of the fillet (line 4)

Since these stresses are seldom critical, take the length of the bracket as 24 in (609.6 mm) and disregard the eccentricity of V. Then $M = 633.3 - 64.11(1.18) = 558$ in·kips (63.1 kN·m); $f = 558/[(\frac{1}{6})(0.416)(24)^2] + 13.35/[0.416(24)] = 15.31$ kips/in² (105.5 MPa). This is acceptable. Also, $v = 1.5(64.11)/[0.416(24)] = 9.63$ kips/in² (66.4 MPa). This is also acceptable.

DESIGN OF A WELDED FLEXIBLE BEAM CONNECTION

A W18 × 64 beam is to be connected to the flange of its supporting column by means of a welded framed connection, using E60 electrodes. Design a connection to transmit a reaction of 40 kips (177.9 kN). The AISC table of welded connections may be applied in selecting the connection, but the design must be verified by computing the stresses.

Calculation Procedure:

1. Record the pertinent properties of the beam

It is necessary to investigate both the stresses in the weld and the shearing stress in the beam induced by the connection. The framing angles must fit between the fillets of the beam. Record the properties: $T = 15\frac{5}{8}$ in (390.5 mm); $t_w = 0.403$ in (10.2 mm).

2. Select the most economical connection from the AISC Manual

The most economical connection is: angles 3 × 3 × ⁵⁄₁₆ in (76 × 76 × 7.9 mm), 12 in (305 mm) long; weld size ³⁄₁₆ in (4.8 mm) for connection to beam web, ¼ in (6.4 mm) for connection to the supporting member.

According to the AISC table, weld A has a capacity of 40.3 kips (179.3 kN), and weld B has a capacity of 42.8 kips (190.4 kN). The minimum web thickness required is 0.25 in (6.4 mm). The connection is shown in Fig. 7a.

3. Compute the unit force in the shop weld

The shop weld connects the angles to the beam web. Refer to Sec. 1 for two calculation procedures for analyzing welded connections.

The weld for one angle is shown in Fig. 7b. The allowable force, as given in Sec. 1, is $m = 2(2.5)(1.25)/[2(2.5) + 12] = 0.37$ in (9.4 mm); $P = 20,000$ lb (88.9 kN); $M = 20,000(3 - 0.37) = 52,600$ in·lb (5942.7 N·m); $I_x = (\frac{1}{12})(12)^3 + 2(2.5)(6)^2 = 324$ in³ (5309.4 cm³); $I_y = 12(0.37)^2 + 2(\frac{1}{12})(2.5)^3 + 2(2.5)(0.88)^2 = 8$ in³ (131.1 cm³); $J = 324 + 8 = 332$ in³ (5440.5 cm³); $f_x = My/J = 52,600(6)/332 = 951$ lb/lin in (166.5 N/mm); $f_y = Mx/J = 52,600(2.5)(0.37)/332 = 337$ lb/lin in (59.0 N/mm); $f_y = 20,000/(2 \times 2.5 + 12) = 1176$ lb/lin in (205.9 N/mm); $F_x = 951$ lb/lin in (166.5 N/mm); $F_y = 337 + 1176 = 1513$ lb/lin in (265.0 N/mm); $F = (951^2 + 1513^2)^{0.5} = 1787 < 1800$, which is acceptable.

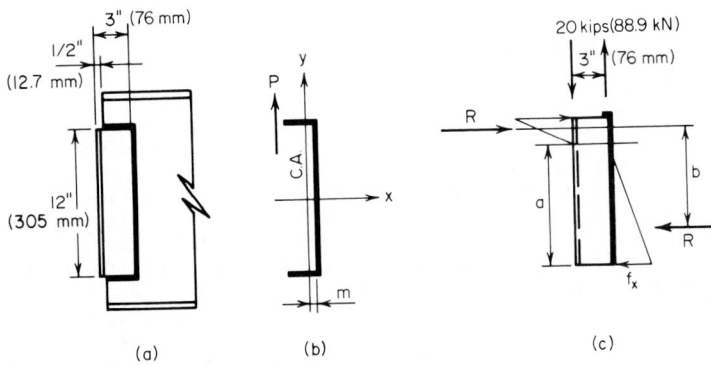

FIG. 7   Welded flexible beam connection.

4. Compute the shearing stress in the web

The allowable stress given in the AISC *Manual* is 14,500 lb/in² (99.9 MPa). The two angles transmit a unit shearing force of 3574 lb/lin in (0.64 kN/mm) to the web. The shearing stress is $v = 3574/0.403 = 8870$ lb/in² (61.1 MPa), which is acceptable.

5. Compute the unit force in the field weld

The field weld connects the angles to the supporting member. As a result of the 3-in (76.2-mm) eccentricity on the outstanding legs, the angles tend to rotate about a neutral axis located near the top, bearing against the beam web above this axis and pulling away from the web below this axis. Assume that the distance from the top of the angle to the neutral axis is one-sixth of the length of the angle. The resultant forces are shown in Fig. 7c. Then $a = (\frac{5}{6})12 = 10$ in (254 mm); $b = (\frac{5}{6})12 = 8$ in (203 mm); $B = 20,000(3)/8 = 7500$ lb (33.4 kN); $f_x = 2R/a = 1500$ lb/lin in (262.7 N/mm); $f_y = 20,000/12 = 1667$ lb/lin in (291.9 N/mm); $F = (1500^2 + 1667^2)^{0.5} = 2240 < 2400$ lb/lin in (420.3 N/mm), which is acceptable. The weld is returned a distance of ½ in (12.7 mm) across the top of the angle, as shown in the AISC *Manual*.

DESIGN OF A WELDED SEATED BEAM CONNECTION

A W27 × 94 beam with a reaction of 77 kips (342.5 kN) is to be supported on a seat. Design a welded connection, using E60 electrodes.

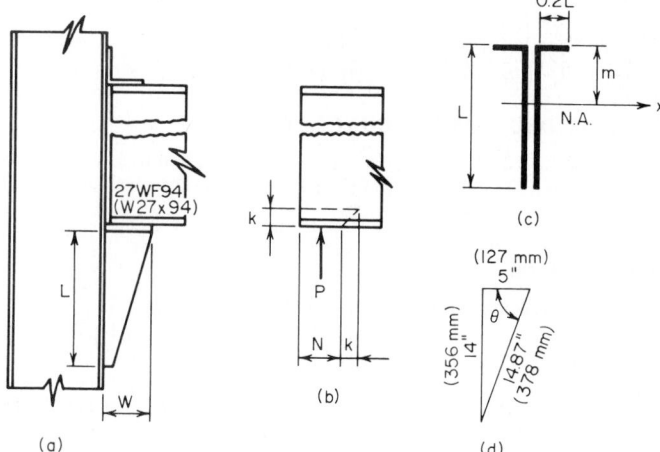

FIG. 8 Welded seated beam connection.

Calculation Procedure:

1. *Record the relevant properties of the beam*

Refer to the AISC *Manual*. The connection will consist of a horizontal seat plate and a stiffener plate below the seat, as shown in Fig. 8a. Record the relevant properties of the W27 × 94: k = 1.44 in (36.6 mm); b = 9.99 in (253.7 mm); t_f = 0.747 in (19.0 mm); t_w = 0.490 in (12.4 mm).

2. *Compute the effective length of bearing*

Equate the compressive stress at the toe of the fillet to its allowable value of 27 kips/in² (186.1 MPa) as given in the AISC *Manual*. Assume that the reaction distributes itself through the web at an angle of 45°. Refer to Fig. 8b. Then $N = P/27t_w - k$, or $N = 77/27(0.490) - 1.44 = 4.38$ in (111.3 mm).

3. *Design the seat plate*

As shown in the AISC *Manual*, the beam is set back about ½ in (12.7 mm) from the face of the support. Make $W = 5$ in (127.0 mm). The minimum allowable distance from the edge of the seat plate to the edge of the flange equals the weld size plus 5⁄16 in (7.8 mm). Make the seat plate 12 in (304.8 mm) long; its thickness will be made the same as that of the stiffener.

4. *Design the weld connecting the stiffener plate to the support*

The stresses in this weld are not amenable to precise analysis. The stiffener rotates about a neutral axis, bearing against the support below this axis and pulling away from the support above this axis. Assume for simplicity that the neutral axis coincides with the centroidal axis of the weld group; the maximum weld stress occurs at the top. A weld length of 0.2L is supplied under the seat plate on each side of the stiffener. Refer to Fig. 8c.

Compute the distance e from the face of the support to the center of the bearing, measuring N from the edge of the seat. Thus, $e = W - N/2 = 5 - 4.38/2 = 2.81$ in (71.4 mm); $P = 77$ kips (342.5 kN); $M = 77(2.81) = 216.4$ in·kips (24.5 kN·m); $m = 0.417L$; $I_x = 0.25L^3$; $f_1 = Mc/I_x = 216.4(0.417L)/0.25L^3 = 361.0/L^2$ kips/lin in; $f_2 = P/A = 77/2.4L = 32.08/L$ kips/lin in. Use a 5⁄16-in (7.9-mm) weld, which has a capacity of 3 kips/lin in (525.4 N/mm). Then $F^2 = f_1^2 + f_2^2 = 130,300/L^4 + 1029/L^2 \leq 3^2$. This equation is satisfied by $L = 14$ in (355.6 mm).

5. *Determine the thickness of the stiffener plate*

Assume this plate is triangular (Fig. 8d). The critical section for bending is assumed to coincide with the throat of the plate, and the maximum bending stress may be obtained by applying $f = (P/tW \sin^2 \theta)(1 + 6e'/W)$, where e' = distance from center of seat to center of bearing.

Using an allowable stress of 22,000 lb/in^2 (151.7 MPa), we have $e' = e - 2.5 = 0.31$ in (7.9 mm); $t = \{77/[22 \times 5(14/14.87)^2]\}(1 + 6 \times 0.31/5) = 1.08$ in (27.4 mm).

Use a 1⅛-in (28.6-mm) stiffener plate. The shearing stress in the plate caused by the weld is $v = 2(3000)/1.125 = 5330 < 14,500$ lb/in^2 (99.9 MPa), which is acceptable.

DESIGN OF A WELDED MOMENT CONNECTION

A W16 × 40 beam frames to the flange of a W12 × 72 column and transmits a shear of 42 kips (186.8 kN) and a moment of 1520 in · kips (171.1 kN · m). Design a welded connection, using E60 electrodes.

Calculation Procedure:

1.   *Record the relevant properties of the two sections*

In designing a welded moment connection, it is assumed for simplicity that the beam flanges alone resist the bending moment. Consequently, the beam transmits three forces to the column: the tensile force in the top flange, the compressive force in the bottom flange, and the vertical load. Although the connection is designed ostensibly on an elastic design basis, it is necessary to consider its behavior at ultimate load, since a plastic hinge would form at this joint. The connection is shown in Fig. 9.

Record the relevant properties of the sections: for the W16 × 40, $d = 16.00$ in (406.4 mm); $b = 7.00$ in (177.8 mm); $t_f = 0.503$ in (12.8 mm); $t_w = 0.307$ in (7.8 mm); $A_f = 7.00(0.503) = 3.52$ in^2 (22.7 cm^2). For the W12 × 72, $k = 1.25$ in (31.8 mm); $t_f = 0.671$ in (17.04 mm); $t_w = 0.403$ in (10.2 mm).

2.   *Investigate the need for column stiffeners; design the stiffeners if they are needed*

The forces in the beam flanges introduce two potential modes of failure: crippling of the column web caused by the compressive force, and fracture of the weld transmitting the tensile force as a result of the bending of the column flange. The AISC *Specification* establishes the criteria for ascertaining whether column stiffeners are required. The first criterion is obtained by equating the compressive stress in the column web at the toe of the fillet to the yield-point stress f_y; the second criterion was obtained empirically. At the ultimate load, the capacity of the unreinforced web = $(0.503 + 5 \times 1.25)0.430f_y = 2.904f_y$; capacity of beam flange = $3.52f_y$; $0.4(A_f)^{0.5} = 0.4(3.52)^{0.5} = 0.750 > 0.671$ in (17.04 mm).

Stiffeners are therefore required opposite both flanges of the beam. The required area is $A_{st} = 3.52 - 2.904 = 0.616$ in^2 (3.97 cm^2). Make the stiffener plates 3½ in (88.9 mm) wide to match the beam flange. From the AISC, $t_{min} = 3.5/8.5 = 0.41$ in (10.4 mm). Use two 3½ × ½ in (88.9 × 12.7 mm) stiffener plates opposite both beam flanges.

3.   *Design the connection plate for the top flange*

Compute the flange force by applying the total depth of the beam. Thus, $F = 1520/16.00 = 95$ kips (422.6 kN); $A = 95/22 = 4.32$ in^2 (27.87 cm^2).

Since the beam flange is 7 in (177.8 mm) wide, use a plate 5 in (127 mm) wide and ⅞ in (22.2 mm) thick, for which $A = 4.38$ in^2 (28.26 cm^2). This plate is butt-welded to the column flange and fillet-welded to the beam flange. In accordance with the AISC *Specification*, the minimum weld size is ⁵⁄₁₆ in (7.94 mm) and the maximum size is ¹³⁄₁₆ in (20.6 mm). Use a ⅝-in (15.9-mm) weld, which has a capacity of 6000 lb/lin in (1051 N/mm). Then, length of weld = 95/6 = 15.8

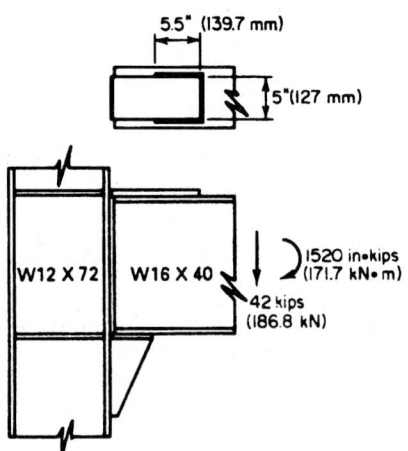

5.5" (139.7 mm)

5"(127 mm)

W12 X 72   W16 X 40

1520 in•kips (171.7 kN•m)

42 kips (186.8 kN)

FIG. 9   Welded moment connection.

in (401.3 mm), say 16 in (406.4 mm). To ensure that yielding of the joint at ultimate load will occur in the plate rather than in the weld, the top plate is left unwelded for a distance approximately equal to its width, as shown in Fig. 9.

4. Design the seat

The connection plate for the bottom flange requires the same area and length of weld as does the plate for the top flange. The stiffener plate and its connecting weld are designed in the same manner as in the previous calculation procedure.

RECTANGULAR KNEE OF RIGID BENT

Figure 10*a* is the elevation of the knee of a rigid bent. Design the knee to transmit an ultimate moment of 8100 in·kips (914.5 kN·m).

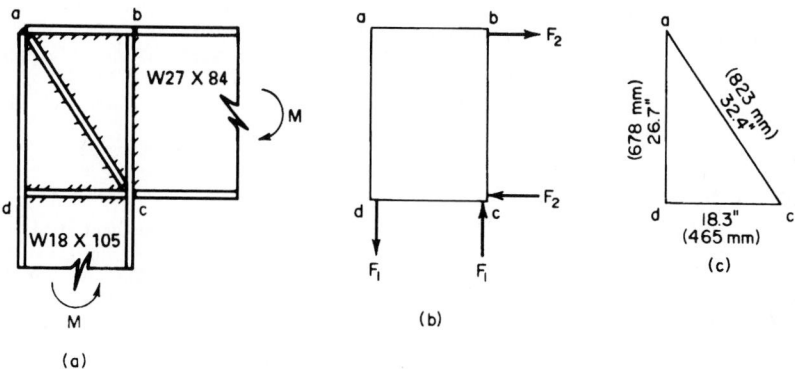

FIG. 10 Rectangular knee.

Calculation Procedure:

1. Record the relevant properties of the two sections

Refer to the AISC *Specification* and *Manual*. It is assumed that the moment in each member is resisted entirely by the flanges and that the distance between the resultant flange forces is 0.95 times the depth of the member.

Record the properties of the members: for the W18 × 105, d = 18.32 in (465.3 mm); b_f = 11.79 in (299.5 mm); t_f = 0.911 in (23.1 mm); t_w = 0.554 in (14.1 mm); k = 1.625 in (41.3 mm). For the W27 × 84, d = 26.69 in (677.9 mm); b_f = 9.96 in (253 mm); t_f = 0.636 in (16.2 mm); t_w = 0.463 in (11.8 mm).

2. Compute F_1

Thus, $F_1 = M_u/(0.95d) = 8100/[0.95(18.32)] = 465$ kips (2068.3 kN).

3. Determine whether web stiffeners are needed to transmit F_1

The shearing stress is assumed to vary linearly from zero at *a* to its maximum value at *d*. The allowable average shearing stress is taken as $f_y/(3)^{0.5}$, where f_y denotes the yield-point stress. The capacity of the web = $0.554(26.69)(36/3^{0.5}) = 307$ kips (1365.5 kN). Therefore, use diagonal web stiffeners.

4. Design the web stiffeners

Referring to Fig. 10*c*, we see that $ac = (18.3^2 + 26.7^2)^{0.5} = 32.4$ in (823 mm). The force in the stiffeners = $(465 - 307)32.4/26.7 = 192$ kips (854.0 kN). (The same result is obtained by computing F_2 and considering the capacity of the web across *ab*.) Then, $A_{st} = 192/36 = 5.33$ in^2 (34.39 cm^2). Use two plates 4 × ¾ in (101.6 × 19.1 mm).

5. Design the welds, using E60 electrodes

The AISC *Specification* stipulates that the weld capacity at ultimate load is 1.67 times the capacity at the working load. Consequently, the ultimate-load capacity is 1000 lb/lin in (175 N/mm) times the number of sixteenths in the weld size. The welds are generally designed to develop the full moment capacity of each member. Refer to the AISC *Specification*.

Weld at ab. This weld transmits the force in the flange of the 27-in (685.8-mm) member to the web of the 18-in (457.2-mm) member. Then $F = 9.96(0.636)(36) = 228$ kips (1014.1 kN), weld force = $228/[2(d - 2t_f)] = 228/[2(18.32 - 1.82)] = 6.91$ kips/lin in (1210.1 N/mm). Use a $\frac{7}{16}$-in (11.1-mm) weld.

Weld at bc. Use a full-penetration butt weld.

Weld at ac. Use the minimum size of $\frac{1}{4}$ in (6.4 mm). The required total length of weld is $L = 192/4 = 48$ in (1219.2 mm).

Weld at dc. Let F_3 denote that part of F_2 that is transmitted to the web of the 18-in (457.2-mm) member through bearing, and let F_4 denote the remainder of F_2. Force F_3 distributes itself through the 18-in (457.2-mm) member at 45° angles, and the maximum compressive stress occurs at the toe of the fillet. Find F_3 by equating this stress to 36 kips/in² (248.2 MPa); or $F_3 = 36(0.554)(0.636 + 2 \times 1.625) = 78$ kips (346.9 kN). To evaluate F_4, apply the moment capacity of the 27-in (685.8-mm) member. Or $F_4 = 228 - 78 = 150$ kips (667.2 kN).

The minimum weld size of $\frac{1}{4}$ in (6.4 mm) is inadequate. Use a $\frac{5}{16}$-in (7.9-mm) weld. The required total length is $L = 150/5 = 30$ in (762.0 mm).

CURVED KNEE OF RIGID BENT

In Fig. 11 the rafter and column are both W21 × 82, and the ultimate moment at the two sections of tangency—p and q—is 6600 in · kips (745.7 kN · m). The section of contraflexure in each member lies 84 in (2133.6 mm) from the section of tangency. Design the knee.

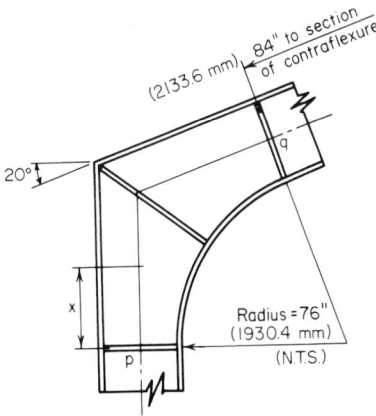

FIG. 11 Curved knee.

Calculation Procedure:

1. Record the relevant properties of the members

Refer to the Commentary in the AISC *Manual*. The notational system is the same as that used in the *Manual*, plus a = distance from section of contraflexure to section of tangency; b = member flange width; x = distance from section of tangency to given section; M = ultimate moment at given section; M_p = plastic-moment capacity of knee at the given section.

Assume that the moment gradient dM/dx remains constant across the knee. The web thickness of the knee is made equal to that of the main material. The flange thickness of the knee, however, must exceed that of the main material, for this reason: As x increases, both M and M_p increase, but the former increases at a faster rate when x is small. The critical section occurs where $dM/dx = dM_p/dx$.

An exact solution to this problem is possible, but the resulting equation is rather cumbersome. An approximate solution is given in the AISC *Manual*.

Record the relevant properties of the W21 × 82: $d = 20.86$ in (529.8 mm); $b = 8.96$ in (227.6 mm); $t_f = 0.795$ in (20.2 mm); $t_w = 0.499$ in (12.7 mm).

2. Design the cross section of the knee, assuming tentatively that flexure is the sole criterion

Use a trial thickness of $\frac{1}{2}$ in (12.7 mm) for the web plate and a 9-in (228.6-mm) width for the flange plate. Then $a = 84$ in (2133.6 mm); $n = a/d = 84/20.86 = 4.03$. From the AISC *Manual*,

$m = 0.14 \pm$; $t' = t(1 + m) = 0.795(1.14) = 0.906$ in (23.0 mm). Make the flange plate 1 in (25.4 mm) thick.

3. Design the stiffeners; investigate the knee for compliance with the AISC Commentary

From the Commentary, *item 5:* Provide stiffener plates at the sections of tangency and at the center of the knee. Make the stiffener plates $4 \times \frac{7}{8}$ in (102 × 22 mm), one on each side of the web.

Item 3: Thus, $\phi = \frac{1}{2}(90° - 20°) = 35°$; $\phi = 35/57.3 = 0.611$ rad; $L = R\phi = 76(0.611) = 46.4$ in (1178.6 mm); or $L = \pi R(70°/360°) = 46.4$ in (1178.6 mm); $L_{cr} = 6b = 6(9) = 54$ in (1373 mm), which is acceptable.

Item 4: Thus, $b/t' = 9$; $2R/b = 152/9 = 16.9$, which is acceptable.

BASE PLATE FOR STEEL COLUMN CARRYING AXIAL LOAD

A W14 × 53 column carries a load of 240 kips (1067.5 kN) and is supported by a footing made of 3000-lb/in² (20,682-kPa) concrete. Design the column base plate.

Calculation Procedure:

1. Compute the required area of the base plate; establish the plate dimensions

Refer to the base-plate diagram in the AISC *Manual.* The column load is assumed to be uniformly distributed within the indicated rectangle, and the footing reaction is assumed to be uniformly distributed across the base plate. The required thickness of the plate is established by computing the bending moment at the circumference of the indicated rectangle. Let f = maximum bending stress in plate; p = bearing stress; t = thickness of plate.

The ACI *Code* permits a bearing stress of 750 lb/in² (5170.5 kPa) if the entire concrete area is loaded and 1125 lb/in² (7755.8 kPa) if one-third of this area is loaded. Applying the 750-lb/in² (5170.5-kPa) value, we get plate area = load, lb/750 = 240,000/750 = 320 in² (2064.5 cm²).

The dimensions of the W14 × 53 are $d = 13.94$ in (354.3 mm); $b = 8.06$ in (204.7 mm); $0.95d = 13.24$ in (335.3 mm); $0.80b = 6.45$ in (163.8 mm). For economy, the projections m and n should be approximately equal. Set $B = 15$ in (381 mm) and $C = 22$ in (558.8 mm); then, area = 15(22) = 330 in² (2129 cm²); $p = 240,000/330 = 727$ lb/in² (5011.9 kPa).

2. Compute the required thickness of the base plate

Thus, $m = \frac{1}{2}(22 - 13.24) = 4.38$ in (111.3 mm), which governs. Also, $n = \frac{1}{2}(15 - 6.45) = 4.28$ in (108.7 mm).

The AISC *Specification* permits a bending stress of 27,000 lb/in² (186.1 MPa) in a rectangular plate. The maximum bending stress is $f = M/S = 3pm^2/t^2$; $t = m(3p/f)^{0.5} = 4.38(3 \times 727/27,000)^{0.5} = 1.24$ in (31.5 mm).

3. Summarize the design

Thus, $B = 15$ in (381 mm); $C = 22$ in (558.8 mm); $t = 1\frac{1}{4}$ in (31.8 mm).

BASE FOR STEEL COLUMN WITH END MOMENT

A steel column of 14-in (355.6-mm) depth transmits to its footing an axial load of 30 kips (133.4 kN) and a moment of 1100 in·kips (124.3 kN·m) in the plane of its web. Design the base, using A307 anchor bolts and 3000-lb/in² (20.7-MPa) concrete.

Calculation Procedure:

1. Record the allowable stresses and modular ratio

Refer to Fig. 12. If the moment is sufficiently large, it causes uplift at one end of the plate and thereby induces tension in the anchor bolt at that end. A rigorous analysis of the stresses in a column base transmitting a moment is not possible. For simplicity, compute the stresses across a horizontal plane through the base plate by treating this as the cross section of a reinforced-concrete beam, the anchor bolt on the tension side acting as the reinforcing steel. The effects of initial tension in the bolts are disregarded.

The anchor bolts are usually placed 2½ (63.5 mm) or 3 in (76.2 mm) from the column flange. Using a plate of 26-in (660-mm) depth as shown in Fig. 12a, let A_s = anchor-bolt cross-sectional area; B = base-plate width; C = resultant compressive force on base plate; T = tensile force in anchor bolt; f_s = stress in anchor bolt; p = maximum bearing stress; p' = bearing stress at column face; t = base-plate thickness.

Recording the allowable stresses and modular ratio by using the ACI *Code*, we get p = 750 lb/in² (5170 kPa) and n = 9. From the AISC *Specification*, f_s = 14,000 lb/in² (96.5 MPa); the allowable bending stress in the plate is 27,000 lb/in² (186.1 MPa).

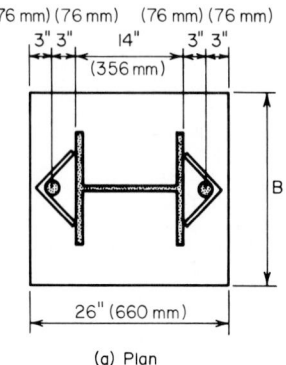

(76 mm) (76 mm) (76 mm) (76 mm)
3" 3"   14"   3" 3"
(356 mm)

B

26" (660 mm)

(a) Plan

2. Construct the stress and force diagrams

These are shown in Fig. 13. Then f_s/n = 14/9 = 1.555 kips/in² (10.7 MPa); kd = 23(0.750/2.305) = 7.48 in (190.0 mm); jd = 23 − 7.48/3 = 20.51 in (521.0 mm).

3. Design the base plate

Thus, C = ½(7.48)(0.750B) = 2.805B. Take moments with respect to the anchor bolt, or ΣM = 30(10) + 1100 − 2.805B(20.51) = 0; B = 24.3 in (617.2 mm).

Assume that the critical bending stress in the base plate occurs at the face of the column. Compute the bending moment at the face for a 1-in (25.4-mm) width of plate. Referring to Fig. 13c, we have p' = 0.750(1.48/7.48) = 0.148 kips/in² (1020.3 kPa); M = (6²/6)(0.148 + 2 × 0.750) = 9.89 in·kips (1.12 kN·m); t^2 = 6M/27 = 2.20 in² (14.19 cm²); t = 1.48 in (37.6 mm). Make the base plate 25 in (635 mm) wide and 1½ in (38.1 mm) thick.

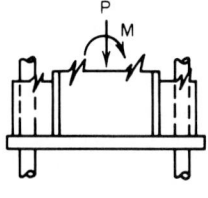

P
M

(b) Elevation

4. Design the anchor bolts

From the calculation in step 3, C = 2.805B = 2.805(24.3) = 68.2 kips (303.4 kN); T = 68.2 − 30 = 38.2 kips (169.9 kN); A_s = 38.2/14 = 2.73 in² (17.61 cm²). Refer to the AISC

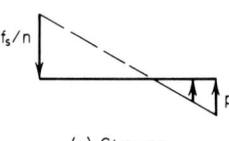

f_s/n

p'

(c) Stresses

FIG. 12 Anchor-bolt details. (a) Plan; (b) elevation; (c) stresses.

Manual. Use 2¼-in (57.2-mm) anchor bolts, one on each side of the flange. Then A_s = 3.02 in² (19.48 cm²).

5. Design the anchorage for the bolts

The bolts are held by angles welded to the column flange, as shown in Fig. 12 and in the AISC *Manual.* Use ½-in (12.7-mm) angles 12 in (304.8 mm) long. Each line of weld resists a force of ½T. Refer to Fig. 13d and compute the unit force F at the extremity of the weld. Thus, M = 19.1(3) = 57.3 in·kips (6.47 kN·m); S_x = (⅙)(12)² = 24 in² (154.8 cm²); F_x = 57.3/24 = 2.39 kips/lin in (0.43 kN/mm); F_y = 19.1/12 = 1.59 kips/lin in (0.29 kN/mm); F = (2.39² + 1.59²)⁰·⁵ = 2.87 kips/lin in (0.52 kN/mm). Use a ⁵⁄₁₆-in (4.8-mm) fillet weld of E60 electrodes, which has a capacity of 3 kips/lin in (0.54 kN/mm).

GRILLAGE SUPPORT FOR COLUMN

A steel column in the form of a W14 × 320 reinforced with two 20 × 1½ in (508 × 38.1 mm) cover plates carries a load of 2790 kips (12,410 kN). Design the grillage under this column, using

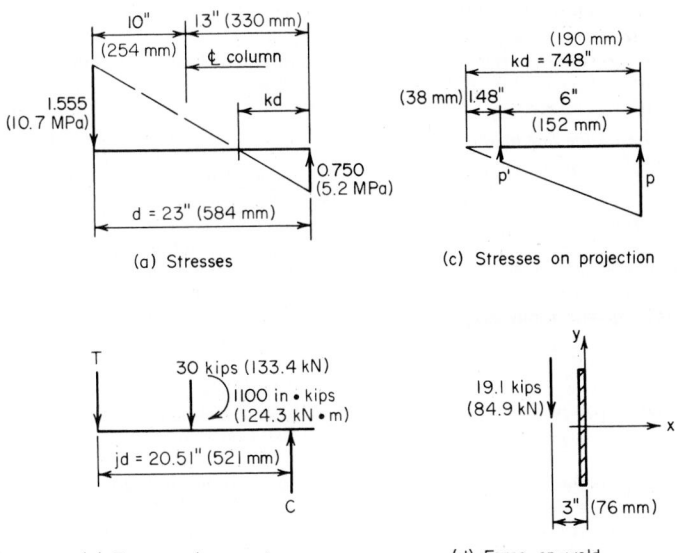

FIG. 13   (*a*) Stresses; (*b*) forces and moment; (*c*) stresses on projection; (*d*) force on weld.

an allowable bearing stress of 750 lb/in² (5170.5 kPa) on the concrete. The space between the beams will be filled with concrete.

Calculation Procedure:

1. *Establish the dimensions of the grillage*

Refer to Fig. 14. A load of this magnitude cannot be transmitted from the column to its footing through the medium of a base plate alone. It is therefore necessary to interpose steel beams between the base plate and the footing; these may be arranged in one tier or in two orthogonal tiers. Integrity of each tier is achieved by tying the beams together by pipe separators. This type

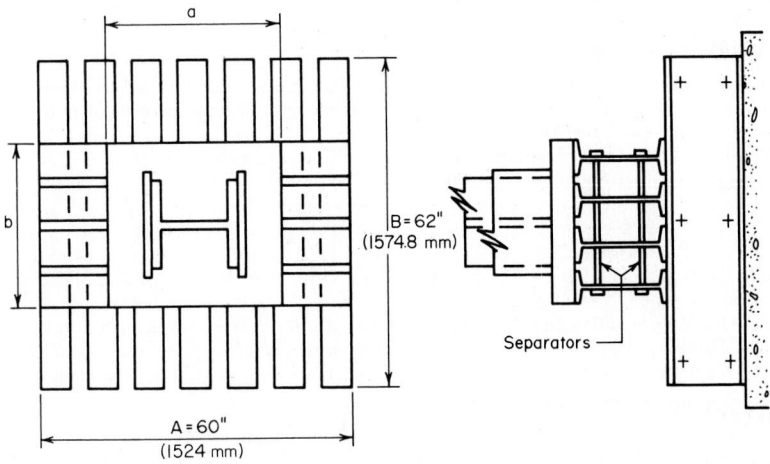

FIG. 14   Grillage under column.

of column support is termed a *grillage*. In designing the grillage, it is assumed that bearing pressures are uniform across each surface under consideration.

The area of grillage required = load, kips/allowable stress, kips/in^2 = 2790/0.750 = 3720 in^2 (23,994 cm^2). Set A = 60 in (1524 mm) and B = 62 in (1574.8 mm), giving an area of 3720 in^2 (23,994 cm^2), as required.

2. Design the upper-tier beams

There are three criteria: bending stress, shearing stress, and compressive stress in the web at the toe of the fillet. The concrete between the beams supplies lateral restraint, and the allowable bending stress is therefore 24 kips/in^2 (165.5 MPa).

Since the web stresses are important criteria, a grillage is generally constructed of S shapes rather than wide-flange beams to take advantage of the thick webs of S shapes. The design of the beams requires the concurrent determination of the length a of the base plate. Let f = bending stress; f_b = compressive stress in web at fillet toe; v = shearing stress; P = load carried by single beam; S = section modulus of single beam; k = distance from outer surface of beam to toe of fillet; t_w = web thickness of beam; a_1 = length of plate as governed by flexure; a_2 = length of plate as governed by compressive stress in web.

Select a beam size on the basis of stresses f and f_b, and then investigate v. The maximum bending moment occurs at the center of the span; its value is M = $P(A - a)/8$ = fS; therefore, a_1 = $A - 8fS/P$.

At the toe of the fillet, the load P is distributed across a distance $a + 2k$. Then f_b = $P/(a + 2k)t_w$; therefore, a_2 = $P/f_b t_w - 2k$. Try four beams; then P = 2790/4 = 697.5 kips (3102.5 kN); f = 24 kips/in^2 (165.5 MPa); f_b = 27 kips/in^2 (186.1 MPa). Upon substitution, the foregoing equations reduce to a_1 = 60 − 0.275S; a_2 = 25.8/t_w − 2k.

Select the trial beam sizes shown in the accompanying table, and calculate the corresponding values of a_1 and a_2.

Size	S, in^3 (cm^3)	t_w, in (mm)	k, in (mm)	a_1, in (mm)	a_2, in (mm)
S18 × 54.7	88.4 (1448.6)	0.460 (11.68)	1.375 (34.93)	35.7 (906.8)	53.3 (1353.8)
S18 × 70	101.9 (1669.8)	0.711 (18.06)	1.375 (34.93)	32.0 (812.8)	33.6 (853.4)
S20 × 65.4	116.9 (1915.7)	0.500 (12.7)	1.563 (39.70)	27.9 (708.7)	48.5 (1231.9)
S20 × 75	126.3 (2069.7)	0.641 (16.28)	1.563 (39.70)	25.3 (642.6)	37.1 (942.3)

Try S18 × 70, with a = 34 in (863.6 mm). The flange width is 6.25 in (158.8 mm). The maximum vertical shear occurs at the edge of the plate; its magnitude is V = $P(A - a)/(2A)$ = 697.5(60 − 34)/[2(60)] = 151.1 kips (672.1 kN); v = 151.1/[18(0.711)] = 11.8 < 14.5 kips/in^2 (99.9 MPa), which is acceptable.

3. Design the base plate

Refer to the second previous calculation procedure. To permit the deposition of concrete, allow a minimum space of 2 in (50.8 mm) between the beam flanges. The minimum value of b is therefore b = 4(6.25) + 3(2) = 31 in (787.4 mm).

The dimensions of the effective bearing area under the column are 0.95(16.81 + 2 × 1.5) = 18.82 in (478.0 mm); 0.80(20) = 16 in (406.4 mm). The projections of the plate are (34 − 18.82)/2 = 7.59 in (192.8 mm); (31 − 16)/2 = 7.5 in (190.5 mm).

Therefore, keep b = 31 in (787 mm), because this results in a well-proportioned plate. The pressure under the plate = 2790/[34(31)] = 2.65 kips/in^2 (18.3 MPa). For a 1-in (25.4-mm) width of plate, M = ½(2.65)(7.59)^2 = 76.33 in·kips (8.6 kN·m); S = M/f = 76.33/27 = 2.827 in^3 (46.33 cm^3); t = (6S)$^{0.5}$ = 4.12 in (104.6 mm).

Plate thicknesses within this range vary by ⅛-in (3.2-mm) increments, as stated in the AISC *Manual*. However, a section of the AISC *Specification* requires that plates over 4 in (102 mm) thick be planed at all bearing surfaces. Set t = 4½ in (114.3 mm) to allow for the planing.

4. Design the beams at the lower tier

Try seven beams. Thus, P = 2790/7 = 398.6 kips (1772.9 kN); M = 398.6(62 − 31)/8 = 1545 in·kips (174.6 kN·m); S_3 = 1545/24 = 64.4 in^3 (1055.3 cm^3).

Try S15 × 50. Then $S = 64.2$ in³ (1052.1 cm³); $t_w = 0.550$ in (14.0 mm); $k = 1.25$ in (31.8 mm); $b = 5.64$ in (143.3 mm). The space between flanges is $[60 - 7 × 5.64]/6 = 3.42$ in (86.9 mm). This result is satisfactory. Then $f_b = 398.6/[0.550(31 + 2 × 1.25)] = 21.6 < 27$ kips/in² (186.1 MPa), which is satisfactory; $V = 398.6(62 - 31)/[2(62)] = 99.7$ kips (443.5 kN); $v = 99.7/[15(0.550)] = 12.1 < 14.5$, which is satisfactory.

5.   *Summarize the design*

Thus: $A = 60$ in (1524 mm); $B = 62$ in (1574.8 mm); base plate is 31 × 34 × 4½ in (787.4 × 863.6 × 114.3 mm); upper-tier steel, four beams S18 × 70; lower-tier steel, seven beams 15150.0.

WIND-STRESS ANALYSIS BY PORTAL METHOD

The bent in Fig. 15 resists the indicated wind loads. Applying the portal method of analysis, calculate all shears, end moments, and axial forces.

Calculation Procedure:

1.   *Compute the shear factor for each column*

The portal method is an approximate and relatively simple method of wind-stress analysis that is frequently applied to regular bents of moderate height. It considers the bent to be composed of a group of individual portals and makes the following assumptions. (1) The wind load is distributed among the aisles of the bent in direct proportion to their relative widths. (2) The point of contraflexure in each member lies at its center.

Because of the first assumption, the shear in a given column is directly proportional to the average width of the adjacent aisles. (An alternative form of the portal method assumes that the wind load is distributed uniformly among the aisles, irrespective of their relative widths.)

In this analysis, we consider the *end moments* of a member, i.e., the moments exerted at the ends of the member by the joints. The sign conventions used are as follows. An end moment is positive if it is clockwise. The shear is positive if the lateral forces exerted on the member by the joints constitute a couple having a counterclockwise moment. An axial force is positive if it is tensile.

Figure 16a and b represents a beam and column, respectively, having positive end moments and positive shear. By applying the second assumption, $M_a = M_b = M$, Eq. *a*; $V = 2M/L$, or $M = VL/2$, Eq. *b*; $H = 2M/L$, or $M = HL/2$, Eq. *c*. In Fig. 15, the calculated data for each member are recorded in the order indicated.

The shear factor equals the ratio of the average width of the adjacent aisles to the total width. Or, line A, $15/75 = 0.20$; line B, $(15 + 12)/75 = 0.36$; line C, $(12 + 10.5)/75 = 0.30$; line D, $10.5/75 = 0.14$. For convenience, record these values in Fig. 15.

2.   *Compute the shear in each column*

For instance, column A-2-3, $H = -3900(0.20) = -780$ lb (-3.5 kN); column C-1-2, $H = -(3900 + 7500)0.30 = -3420$ lb (-15.2 kN).

3.   *Compute the end moments of each column*

Apply Eq. *c*. For instance, column A-2-3, $M = ½(-780)15 = -5850$ ft·lb (-7932.6 N·m); column D-0-1, $M = ½(-2751)18 = -24,759$ ft·lb ($-33,573.2$ N·m).

4.   *Compute the end moments of each beam*

Do this by equating the algebraic sum of end moments at each joint to zero. For instance, at line 3: $M_{AB} = 5850$ ft·lb (7932.6 N·m); $M_{BC} = -5850 + 10,530 = 4680$ ft·lb (6346.1 N·m); $M_{CD} = -4680 + 8775 = 4095$ ft·lb (5552.8 N·m). At line 2: $M_{AB} = 5850 + 17,100 = 22,950$ ft·lb (31,120.2 N·m); $M_{BC} = -22,950 + 30,780 + 10,530 = 18,360$ ft·lb (24,896.0 N·m).

5.   *Compute the shear in each beam*

Do this by applying Eq. *b*. For instance, beam B-2-C, $V = 2(18,360)/24 = 1530$ lb (6.8 kN).

6.   *Compute the axial force in each member*

Do this by drawing free-body diagrams of the joints and applying the equations of equilibrium. It is found that the axial forces in the interior columns are zero. This condition stems from the

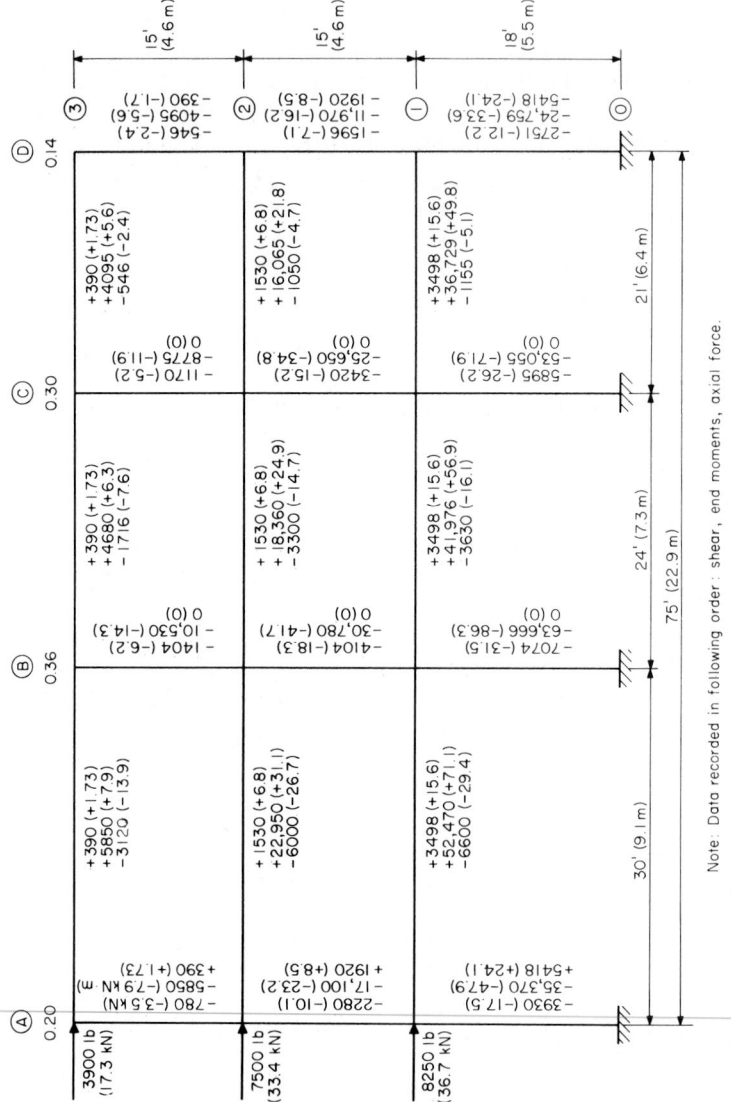

FIG. 15 Wind-stress analysis by portal method.

Note: Data recorded in following order: shear, end moments, axial force.

first assumption underlying the portal method and the fact that each interior column functions as both the leeward column of one portal and the windward column of the adjacent portal.

The absence of axial forces in the interior columns in turn results in the equality of the shear in the beams at each tier. Thus, the calculations associated with the portal method of analysis are completely self-checking.

WIND-STRESS ANALYSIS BY CANTILEVER METHOD

For the bent in Fig. 17, calculate all shears, end moments, and axial forces induced by the wind loads by applying the cantilever method of wind-stress analysis. For this purpose, assume that the columns have equal cross-sectional areas.

Calculation Procedure:

1. Compute the shear and moment on the bent at midheight of each horizontal row of columns

The cantilever method, which is somewhat more rational than the portal method, considers that the bent behaves as a vertical cantilever. Consequently, the direct stress in a column is directly proportional to the distance from the column to the centroid of the combined column area. As in the portal method, the assumption is made that the point of contraflexure in each member lies at its center. Refer to the previous calculation procedure for the sign convention.

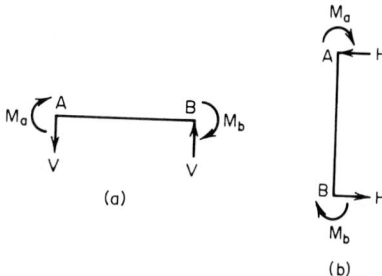

FIG. 16

Computing the shear and moment on the bent at midheight, we have the following. Upper row: $H = 3900$ lb (17.3 kN); $M = 3900(7.5) = 29,250$ ft·lb (39,663.0 N·m). Center row: $H = 3900 + 7500 = 11,400$ lb (50.7 kN); $M = 3900(22.5) + 7500(7.5) = 144,000$ ft·lb (195.3 kN·m). Lower row: $H = 11,400 + 8250 = 19,650$ lb (87.5 kN); $M = 3900(39) + 7500(24) + 8250(9) = 406,400$ ft·lb (551.1 kN·m), or $M = 144,000 + 11,400(16.5) + 8250(9) = 406,400$ ft·lb (551.1 kN·m), as before.

2. Locate the centroidal axis of the combined column area, and compute the moment of inertia of the area with respect to this axis

Take the area of one column as a unit. Then $x = (30 + 54 + 75)/4 = 39.75$ ft (12.12 m); $I = 39.75^2 + 9.75^2 + 14.25^2 + 35.25^2 = 3121$ ft^2 (289.95 m^2).

3. Compute the axial force in each column

Use the equation $f = My/I$. The y/I values are

	A	B	C	D
y	39.75	9.75	−14.25	−35.25
y/I	0.01274	0.00312	−0.00457	−0.01129

Then column A-2-3, $P = 29,250(0.01274) = 373$ kips (1659 kN); column B-0-1, $P = 406,400(0.00312) = 1268$ kips (5640 kN).

4. Compute the shear in each beam by analyzing each joint as a free body

Thus, beam A-3-B, $V = 373$ lb (1659 N); beam B-3-C, $V = 373 + 91 = 464$ lb (2.1 kN); beam C-3-D, $V = 464 - 134 = 330$ lb (1468 N); beam A-2-B, $V = 1835 - 373 = 1462$ lb (6.5 kN); beam B-2-C, $V = 1462 + 449 - 91 = 1820$ lb (8.1 kN).

5. Compute the end moments of each beam

Apply Eq. *b* of the previous calculation procedure. Or for beam A-3-B, $M = \frac{1}{2}(373)(30) = 5595$ ft·lb (7586.8 N·m).

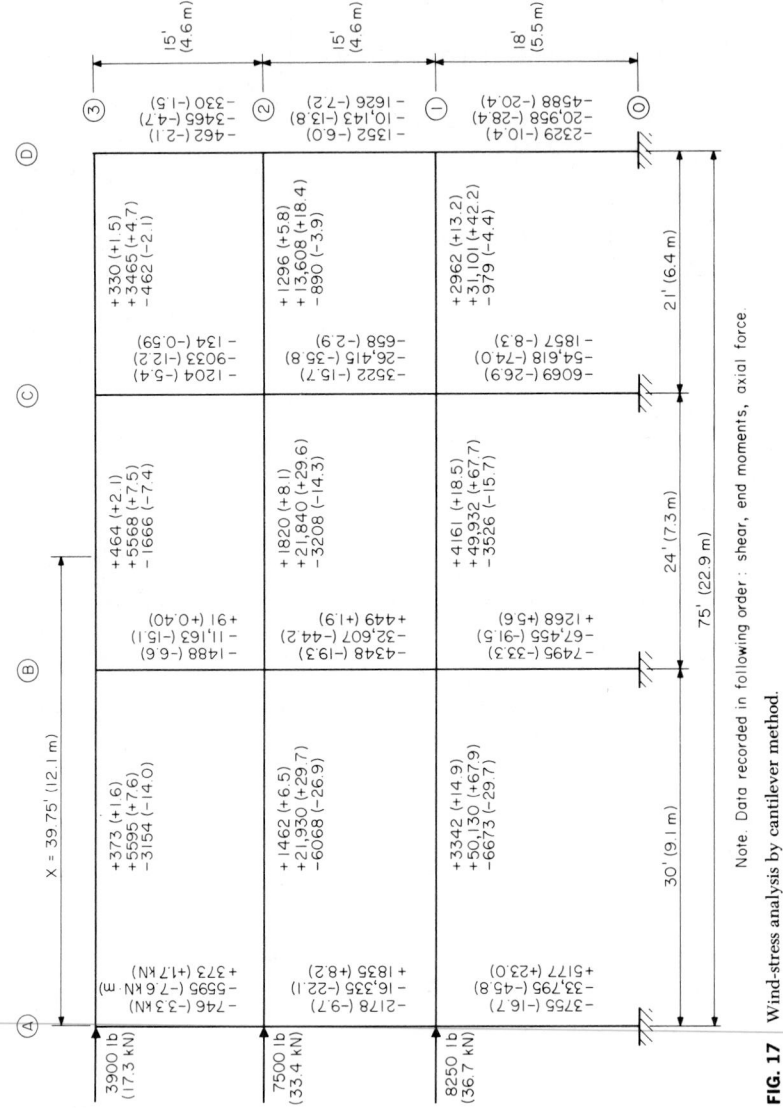

FIG. 17 Wind-stress analysis by cantilever method.

Note: Data recorded in following order: shear, end moments, axial force

2.22

6.   Compute the end moments of each column

Do this by equating the algebraic sum of the end moments at each joint to zero.

7.   Compute the shear in each column

Apply Eq. c of the previous calculation procedure. The sum of the shears in each horizontal row of columns should equal the wind load above that plane. For instance, for the center row, $\Sigma H = -(2178 + 4348 + 3522 + 1352) = -11,400$ lb $(-50.7$ kN), which is correct.

8.   Compute the axial force in each beam by analyzing each joint as a free body

Thus, beam A-3-B, $P = -3900 + 746 = -3154$ lb $(-14.0$ kN); beam B-3-C, $P = -3154 + 1488 = -1666$ lb $(-7.4$ kN).

WIND-STRESS ANALYSIS BY SLOPE-DEFLECTION METHOD

Analyze the bent in Fig. 18a by the slope-deflection method. The moment of inertia of each member is shown in the drawing.

Calculation Procedure:

1.   Compute the end rotations caused by the applied moments and forces; superpose the rotation caused by the transverse displacement

This method of analysis has not been applied extensively in the past because the arithmetic calculations involved become voluminous where the bent contains many joints. However, the increasing use of computers in structural design is overcoming this obstacle and stimulating a renewed interest in the method.

Figure 19 is the elastic curve of a member subjected to moments and transverse forces applied solely at its ends. The sign convention is as follows: an end moment is positive if it is clockwise;

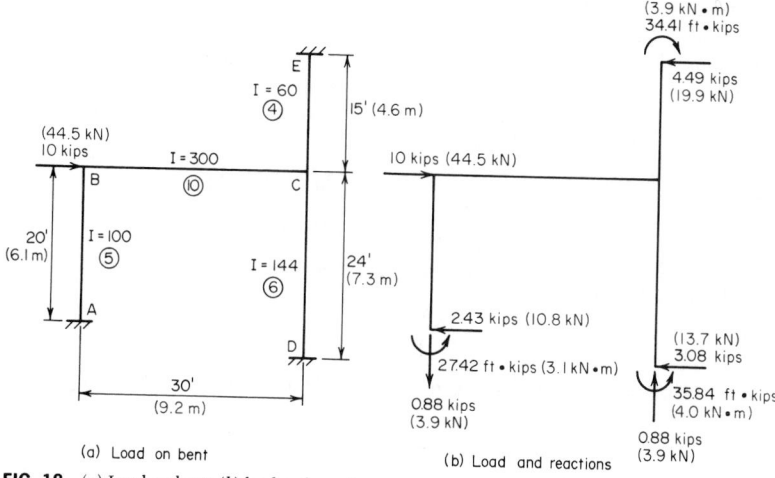

FIG. 18   (a) Load on bent; (b) load and reactions.

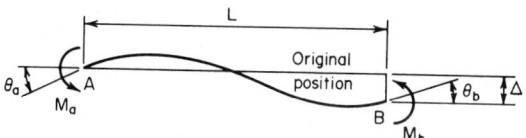

FIG. 19   Elastic curve of beam.

an angular displacement is positive if the rotation is clockwise; the transverse displacement Δ is positive if it rotates the member in a clockwise direction.

Computing the end rotations gives $\theta_a = (L/6EI)(2M_a - M_b) + \Delta/L$; $\theta_b = (L/6EI)(-M_a + 2M_b) + \Delta/L$. These results may be obtained by applying the moment-area method or unit-load method given in Sec. 1.

2. Solve the foregoing equations for the end moments

Thus,

$$M_a = \left(\frac{2EI}{L}\right)\left(2\theta_a + \theta_b - \frac{3\Delta}{L}\right) \qquad M_b = \left(\frac{2EI}{L}\right)\left(\theta_a + 2\theta_b - \frac{3\Delta}{L}\right) \tag{4}$$

These are the basic slope-deflection equations.

3. Compute the value of I/L for each member of the bent

Let K denote this value, which represents the relative stiffness of the member. Thus $K_{ab} = 100/20 = 5$; $K_{cd} = 144/24 = 6$; $K_{bc} = 300/30 = 10$; $K_{ce} = 60/15 = 4$. These values are recorded in circles in Fig. 18.

4. Apply Eq. 4 to each joint in turn

When the wind load is applied, the bent will deform until the horizontal reactions at the supports total 10 kips (44.5 kN). It is evident, therefore, that the end moments of a member are functions of the *relative* rather than the absolute stiffness of that member. Therefore, in writing the moment equations, the coefficient $2EI/L$ may be replaced with I/L; to view this in another manner, $E = \frac{1}{2}$.

Disregard the deformation associated with axial forces in the members, and assume that joints B and C remain in a horizontal line. The symbol M_{ab} denotes the moment exerted on member AB at joint A. Thus $M_{ab} = 5(\theta_b - 3\Delta/20) = 5\theta_b - 0.75\Delta$; $M_{dc} = 6(\theta_c - 3\Delta/24) = 6\theta_c - 0.75\Delta$; $M_{ec} = 4(\theta_c + 3\Delta/15) = 4\theta_c + 0.80\Delta$; $M_{ba} = 5(2\theta_b - 3\Delta/20) = 10\theta_b - 0.75\Delta$; $M_{cd} = 6(2\theta_c - 3\Delta/24) = 12\theta_c - 0.75\Delta$; $M_{ce} = 4(2\theta_c + 3\Delta/15) = 8\theta_c + 0.80\Delta$; $M_{bc} = 10(2\theta_b + \theta_c) = 20\theta_b + 10\theta_c$; $M_{cb} = 10(\theta_b + 2\theta_c) = 10\theta_b + 20\theta_c$.

5. Write the equations of equilibrium for the joints and for the bent

Thus, joint B, $M_{ba} + M_{bc} = 0$, Eq. a; joint C, $M_{cb} + M_{cd} + M_{ce} = 0$, Eq. b. Let H denote the horizontal reaction at a given support. Consider a horizontal force positive if directed toward the right. Then $H_a + H_d + H_e + 10 = 0$, Eq. c.

6. Express the horizontal reactions in terms of the end moments

Rewrite Eq. c. Or, $(M_{ab} + M_{ba})/20 + (M_{dc} + M_{cd})/24 - (M_{ec} + M_{ce})/15 + 10 = 0$, or $6M_{ab} + 6M_{ba} + 5M_{dc} + 5M_{cd} - 8M_{ec} - 8M_{ce} = -1200$, Eq. c'.

7. Rewrite Eqs. a, b, and c' by replacing the end moments with the expressions obtained in step 4

Thus, $30\theta_b + 10\theta_c - 0.75\Delta = 0$, Eq. A; $10\theta_b + 40\theta_c + 0.05\Delta = 0$, Eq. B; $900\theta_b - 60\theta_c - 29.30\Delta = -1200$, Eq. C.

8. Solve the simultaneous equations in step 7 to obtain the relative values of θ_b, θ_c, and Δ

Thus $\theta_b = 1.244$; $\theta_c = -0.367$; $\Delta = 44.85$.

9. Apply the results in step 8 to evaluate the end moments

The values, in foot-kips, are: $M_{ab} = -27.42$ (-37.18 kN·m); $M_{dc} = -35.84$ (-48.6 kN·m); $M_{ec} = 34.41$ (46.66 kN·m); $M_{ba} = -21.20$ (-28.75 kN·m); $M_{cd} = -38.04$ (-51.58 kN·m); $M_{ce} = 32.94$ (44.67 kN·m); $M_{bc} = 21.21$ (28.76 kN·m); $M_{cb} = 5.10$ (6.92 kN·m).

10. Compute the shear in each member by analyzing the member as a free body

The shear is positive if the transverse forces exert a counterclockwise moment. Thus $H_{ab} = (M_{ab} + M_{ba})/20 = -2.43$ kips (-10.8 kN); $H_{cd} = -3.08$ kips (-13.7 kN); $H_{ce} = 4.49$ kips (19.9 kN); $V_{bc} = 0.88$ kip (3.9 kN).

11.   *Compute the axial force in AB and BC*

Thus $P_{ab} = 0.88$ kip (3.91 kN); $P_{bc} = -7.57$ kips (-33.7 kN). The axial forces in EC and CD are found by equating the elongation of one to the contraction of the other.

12.   *Check the bent for equilibrium*

The forces and moments acting on the structure are shown in Fig. 18*b*. The three equations of equilibrium are satisfied.

WIND DRIFT OF A BUILDING

Figure 20*a* is the partial elevation of the steel framing of a skyscraper. The wind shear directly above line 11 is 40 kips (177.9 kN), and the wind force applied at lines 11 and 12 is 4 kips (17.8 kN) each. The members represented by solid lines have the moments of inertia shown in Table 1, and the structure is to be analyzed for wind stress by the portal method. Compute the wind drift for the bent bounded by lines 11 and 12; that is, find the horizontal displacement of the joints on line 11 relative to those on line 12 as a result of wind.

Calculation Procedure:

1.   *Using the portal method of wind-stress analysis, compute the shear in each column caused by the unit loads*

Apply the unit-load method presented in Sec. 1. For this purpose, consider that unit horizontal loads are applied to the structure in the manner shown in Fig. 20*b*.

The results obtained in steps 1, 2, and 3 below are recorded in Fig. 20*b*. To apply the portal method of wind-stress analysis, see the fourteenth calculation procedure in this section.

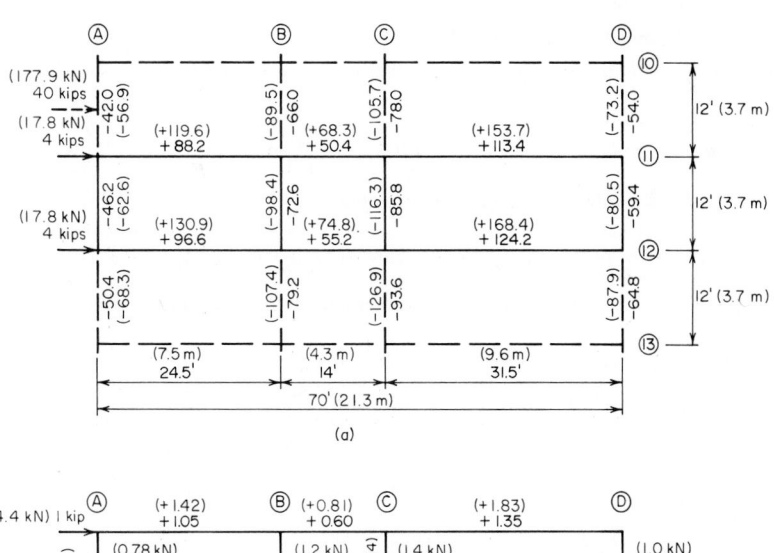

FIG. 20

TABLE 1   Calculation of Wind Drift

Member	I, in⁴ (cm⁴)	L, ft (m)	M_e, ft-kips (kN·m)	m_e, ft-kips (kN·m)	$M_e m_e L/I$
A-11-12	1,500 (62,430)	12 (3.66)	46.2 (62.6)	1.05 (1.42)	0.39
B-11-12	1,460 (60,765)	12 (3.66)	72.6 (98.5)	1.65 (2.24)	0.98
C-11-12	1,800 (74,916)	12 (3.66)	85.8 (116.3)	1.95 (2.64)	1.12
D-11-12	2,000 (83,240)	12 (3.66)	59.4 (80.6)	1.35 (1.83)	0.48
A-11-B	660 (27,469)	24.5 (7.47)	88.2 (119.6)	1.05 (1.42)	3.44
B-11-C	300 (12,486)	14 (4.27)	50.4 (68.3)	0.60 (0.81)	1.41
C-11-D	1,400 (58,268)	31.5 (9.60)	113.4 (153.8)	1.35 (1.83)	3.44
A-12-B	750 (31,213)	24.5 (7.47)	96.6 (130.9)	1.05 (1.42)	3.31
B-12-C	400 (16,648)	14 (4.27)	55.2 (74.9)	0.60 (0.81)	1.16
C-12-D	1,500 (62,430)	31.5 (9.60)	124.2 (168.4)	1.35 (1.83)	3.52
Total	. . . . . .	. . . . . .	. . . . . .	. . . . . .	19.25

2.   Compute the end moments of each column caused by the unit loads

3.   Equate the algebraic sum of end moments at each joint to zero; from this find the end moments of the beams caused by the unit loads

4.   Find the end moments of each column

Multiply the results obtained in step 2 by the wind shear in each panel to find the end moments of each column in Fig. 20a. For instance, the end moments of column C-11-12 are $-1.95(44) = -85.8$ ft·kips (-116.3 kN·m). Record the result in Fig. 20a.

5.   Find the end moments of the beams caused by the true loads

Equate the algebraic sum of end moments at each joint to zero to find the end moments of the beams caused by the true loads.

6.   State the equation for wind drift

In Fig. 21, M_e and m_e denote the end moments caused by the true load and unit load, respectively. Then the

$$\text{Wind drift } \Delta = \frac{\Sigma\, M_e m_e L}{3EI} \tag{5}$$

7.   Compute the wind drift by completing Table 1

In recording end moments, algebraic signs may be disregarded because the product $M_e m_e$ is always positive. Taking the total of the last column in Table 1, we find $\Delta = 19.25(12)^3/[3(29)(10)^3]$ $= 0.382$ in (9.7 mm). For dimensional homogeneity, the left side of Eq. 5 must be multiplied by 1 kip (4.45 kN). The product represents the external work performed by the unit loads.

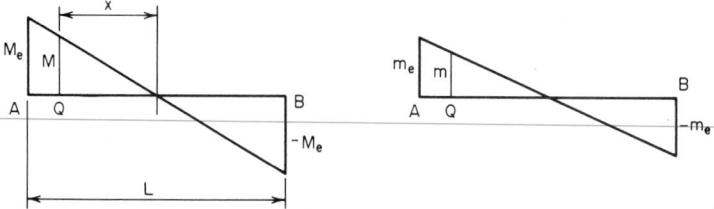

FIG. 21   Bending-moment diagrams.

REDUCTION IN WIND DRIFT BY USING DIAGONAL BRACING

With reference to the previous calculation procedure, assume that the wind drift of the bent is to be restricted to 0.20 in (5.1 mm) by introducing diagonal bracing between lines B and C. Design the bracing, using the gross area of the member.

Calculation Procedure:

1. State the change in length of the brace

The bent will be reinforced against lateral deflection by a pair of diagonal cross braces, each brace being assumed to act solely as a tension member. Select the lightest single-angle member that will satisfy the stiffness requirements; then compute the wind drift of the reinforced bent.

Assume that the bent in Fig. 22 is deformed in such a manner that B is displaced a horizontal distance Δ relative to D. Let A = cross-sectional area of member CB; P = axial force in CB; P_h = horizontal component of P; δL = change in length of CB. From the geometry of Fig. 22, δL = $\Delta \cos \theta = a\Delta/L$ approximately.

2. Express P_h in terms of Δ

Thus, $P = aAE\Delta/L^2$; $P_h = P \cos \theta = Pa/L$; then

$$P_h = \frac{a^2 AE\Delta}{L^3} \qquad (6)$$

3. Select a trial size for the diagonal bracing; compute the tensile capacity

A section of the AISC *Specification* limits the slenderness ratio for bracing members in tension to 300, and another section provides an allowable stress of 22 kips/in^2 (151.7 MPa). Thus, $L^2 = 14^2$

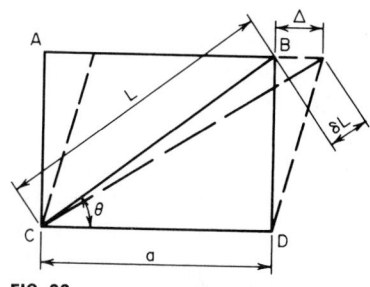

FIG. 22

+ 12^2 = 340 ft^2 (31.6 m^2); L = 18.4 ft (5.61 m); r_{min} = (18.4 × 12)/300 = 0.74 in (18.8 mm).

Try a 4 × 4 × ¼ in (101.6 × 101.6 × 6.35 mm) angle; r = 0.79 in (20.1 mm); A = 1.94 in^2 (12.52 cm^2); P_{max} = 1.94(22) = 42.7 kips (189.9 kN).

4. Compute the wind drift if the assumed size of bracing is used

By Eq. 6, P_h = {196/[(340)(18.4)(12)]} 1.94(29)(10)$^3\Delta$ = 147Δ kips (653.9Δ N). The wind shear resisted by the columns of the bent is reduced by P_h, and the wind drift is reduced proportionately.

From the previous calculation procedure, the following values are obtained: without diagonal bracing, Δ = 0.382 in (9.7 mm); with diagonal bracing, Δ = 0.382(44 − P_h)/44 = 0.382 − 1.28Δ. Solving gives Δ = 0.168 < 0.20 in (5.1 mm), which is acceptable.

5. Check the axial force in the brace

Thus, P_h = 147(0.168) = 24.7 kips (109.9 kN); $P = P_h L/a$ = 24.7(18.4)/14 = 32.5 < 42.7 kips (189.9 kN), which is satisfactory. Therefore, the assumed size of the member is satisfactory.

LIGHT-GAGE STEEL BEAM WITH UNSTIFFENED FLANGE

A beam of light-gage cold-formed steel consists of two 7 × 1½ in (177.8 × 38.1 mm) by no. 12 gage channels connected back to back to form an I section. The beam is simply supported on a 16-ft (4.88-m) span, has continuous lateral support, and carries a total dead load of 50 lb/lin ft (730 N/m). The live-load deflection is restricted to 1/360 of the span. If the yield-point stress f_y is 33,000 lb/in^2 (227.5 MPa), compute the allowable unit live load for this member.

Calculation Procedure:

1. Record the relevant properties of the section

Apply the AISI *Specification for the Design of Light Gage Cold-Formed Steel Structural Members*. This is given in the AISI publication *Light Gage Cold-Formed Steel Design Manual*. Use the same notational system, except denote the flat width of an element by g rather than w.

The publication mentioned above provides a basic design stress of 20,000 lb/in^2 (137.9 MPa) for this grade of steel. However, since the compression flange of the given member is unstiffened in accordance with the definition in one section of the publication, it may be necessary to reduce the allowable compressive stress. A table in the *Manual* gives the dimensions, design properties, and allowable stress of each section, but the allowable stress will be computed independently in this calculation procedure.

Let V = maximum vertical shear; M = maximum bending moment; w = unit load; f_b = basic design stress; f_c = allowable bending stress in compression; v = shearing stress; Δ = maximum deflection.

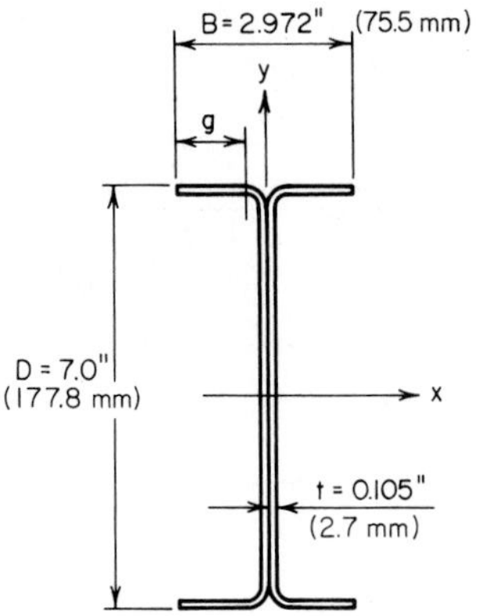

Record the relevant properties of the section as shown in Fig. 23: I_x = 12.4 in⁴ (516.1 cm⁴); S_x = 3.54 in³ (58.0 cm³); R = ³⁄₁₆ in (4.8 mm).

2. Compute f_c

Thus, $g = B/2 - t - R$ = 1.1935 in (30.3 mm); g/t = 1.1935/0.105 = 11.4. From the *Manual*, the allowable stress corresponding to this ratio is $f_c = 1.667f_b - 8640 - [(f_b - 12,950)g/t]/15 = 1.667(20,000) - 8640 - (20,000 - 12,950)11.4/15$ = 19,340 lb/in² (133.3 MPa).

3. Compute the allowable unit live load if flexure is the sole criterion

Thus $M = f_cS_x$ = 19,340(3.54)/12 = 5700 ft·lb (7729.2 N·m); $w = 8M/L^2$ = 8(5700)/16² = 178 lb/lin ft (2.6 kN/m); w_{LL} = 178 − 50 = 128 lb/lin ft (1.87 kN/m).

4. Investigate the deflection under the computed live load

D = 7.0" (177.8 mm)

t = 0.105" (2.7 mm)

B = 2.972" (75.5 mm)

FIG. 23

Using E = 29,500,000 lb/in² (203,373 MPa) as given in the AISI *Manual*, we have $\Delta_{LL} = 5w_{LL}L^4/(384EI_x)$ = 5(128)(16)⁴(12)³/[384(29.5)(10)⁶12.4] = 0.516 in (13.1 mm); $\Delta_{LL,allow}$ = 16(12)/360 = 0.533 in (13.5 mm), which is satisfactory.

5. Investigate the shearing stress under the computed total load

Refer to the AISI *Specification*. For the individual channel, $h = D - 2t$ = 6.79 in (172.5 mm); h/t = 64.7; 64,000,000/64.7² > ⅔f_b; therefore, v_{allow} = 13,330 lb/in² (91.9 MPa); the web area = 0.105(6.79) = 0.713 in² (4.6 cm²); V = ¼(178)16 = 712 lb (3.2 kN); v = 712/0.713 < v_{allow}, which is satisfactory. The allowable unit live load is therefore 128 lb/lin ft (1.87 kN/m).

LIGHT-GAGE STEEL BEAM WITH STIFFENED COMPRESSION FLANGE

A beam of light-gage cold-formed steel has a hat cross section 8 × 12 in (203.2 × 304.8 mm) of no. 12 gage, as shown in Fig. 24. The beam is simply supported on a span of 13 ft (3.96 m). If the yield-point stress is 33,000 lb/in² (227.5 MPa), compute the allowable unit load for this member and the corresponding deflection.

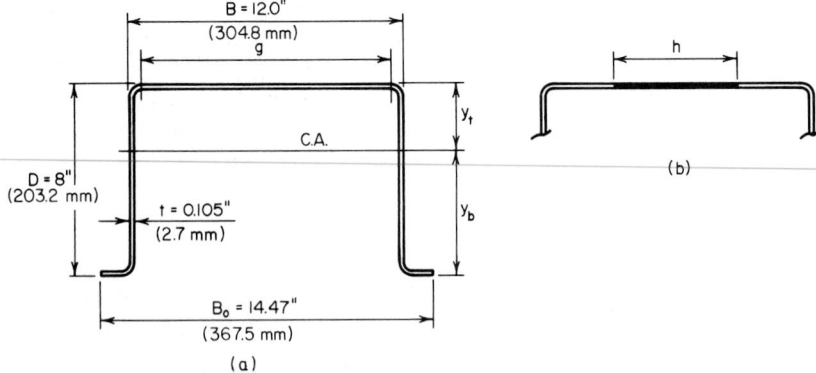

B = 12.0" (304.8 mm)

g

C.A.

y_t

(b)

h

D = 8" (203.2 mm)

t = 0.105" (2.7 mm)

y_b

B_0 = 14.47" (367.5 mm)

(a)

FIG. 24

Calculation Procedure:

1. Record the relevant properties of the entire cross-sectional area

Refer to the AISI *Specification* and *Manual*. The allowable load is considered to be the ultimate load that the member will carry divided by a load factor of 1.65. At ultimate load, the bending stress varies considerably across the compression flange. To surmount the difficulty that this condition introduces, the AISI *Specification* permits the designer to assume that the stress is uniform across an *effective flange width* to be established in the prescribed manner. The investigation is complicated by the fact that the effective flange width and the bending stress in compression are interdependent quantities, for the following reason. The effective width depends on the compressive stress; the compressive stress, which is less than the basic design stress, depends on the location of the neutral axis; the location of the neutral axis, in turn, depends on the effective width.

The beam deflection is also calculated by establishing an effective flange width. However, since the beam capacity is governed by stresses at the ultimate load and the beam deflection is governed by stresses at working load, the effective widths associated with these two quantities are unequal.

A table in the AISI *Manual* contains two design values that afford a direct solution to this problem. However, the values are computed independently here to demonstrate how they are obtained. The notational system presented in the previous calculation procedure is used, as well as A' = area of cross section exclusive of compression flange; H = static moment of cross-sectional area with respect to top of section; y_b and y_t = distance from centroidal axis of cross section to bottom and top of section, respectively.

We use the AISI *Manual* to determine the relevant properties of the entire cross-sectional area, as shown in Fig. 24: $A = 3.13$ in^2 (20.2 cm^2); $y_b = 5.23$ in (132.8 mm); $I_x = 26.8$ in^4 (1115.5 cm^4); $R = \%_6$ in (4.8 mm).

2. Establish the value of f_c for load determination

Use the relation $(8040t^2/f_c^{0.5})\{1 - 2010/[(f_c^{0.5}g)/t]\} = (H/D)(f_c + f_b)/f_c - A'$. Substituting gives $g = B - 2(t + R) = 12.0 - 2(0.105 + 0.1875) = 11.415$ in (289.9 mm); $g/t = 108.7$; $gt = 1.20$ in^2 (7.74 cm^2); $A = 3.13 - 1.20 = 1.93$ in^2 (12.45 cm^2); $y_t = 8.0 - 5.23 = 2.77$ in (70.36 cm); $H = 3.13(2.77) = 8.670$ in^3 (142.1 cm^3). The foregoing equation then reduces to $(88.64/f_c^{0.5})(1 - 18.49/f_c^{0.5}) = 1.084(f_c + 20,000)/f_c - 1.93$. By successive approximations, $f_c = 14,800$ lb/in^2 (102.0 MPa).

3. Compute the corresponding effective flange width for load determination in accordance with the AISI **Manual**

Thus, $b = (8040t/f_c^{0.5})\{1 - 2010/[(f_c^{0.5}g)/t]\} = (8040 \times 0.105/14,800^{0.5})[1 - 2010/(14,800^{0.5} \times 108.7)] = 5.885$ in (149.5 mm).

4. Locate the centroidal axis of the cross section having this effective width; check the value of f_c

Refer to Fig. 24b. Thus $h = g - b = 11.415 - 5.885 = 5.530$ in (140.5 mm); $ht = 0.581$ in^2 (3.75 cm^2); $A = 3.13 - 0.581 = 2.549$ in^2 (16.45 cm^2); $H = 8.670$ in^3 (142.1 cm^3); $y_t = 8.670/2.549 = 3.40$ in (86.4 mm); $y_b = 4.60$ in (116.8 mm); $f_c = y_t f_b/y_b = 3.40(20,000)/4.60 = 14,800$ lb/in^2 (102.0 MPa), which is satisfactory.

5. Compute the allowable load

The moment of inertia of the net section may be found by applying the value of the gross section and making the necessary corrections. Applying $S_x = I_x/y_b$, we get $I_x = 26.8 + 3.13(3.40 - 2.77)^2 - 0.581(3.40 - 0.053)^2 = 21.53$ in^4 (896.15 cm^4). Then $S_x = 21.53/4.60 = 4.68$ in^3 (76.69 cm^3). This value agrees with that recorded in the AISI *Manual*.

Then $M = f_b S_x = 20,000(4.68)/12 = 7800$ ft·lb (10,576 N·m); $w = 8M/L^2 = 8(7800)/13^2 = 369$ lb/lin ft (5.39 kN/m).

6. Establish the value of f_c for deflection determination

Apply $(10,320t^2/f_c^{0.5})[1 - 2580/(f_c^{0.5}g/t)] = (H/D)(f_c + f_b)/f_c - A'$, or $(113.8/f_c^{0.5}) \times (1 - 23.74/f_c^{0.5}) = 1.084(f_c + 20,000)/f_c - 1.93$. By successive approximation, $f_c = 13,300$ lb/in^2 (91.7 MPa).

7.   *Compute the corresponding effective flange width for deflection determination*

Thus, $b = (10,320t/f_c^{0.5})[1 - 2580/(f_c^{0.5}g/t)] = (10,320 \times 0.105/13,300^{0.5})[1 - 2580/(13,300^{0.5} \times 108.7)] = 7.462$ in (189.5 mm).

8.   *Locate the centroidal axis of the cross section having this effective width; check the value of f_c*

Thus $h = 11.415 - 7.462 = 3.953$ in (100.4 mm); $ht = 0.415$ in^2 (2.68 cm^2); $A = 3.13 - 0.415 = 2.715$ in^2 (17.52 cm^2); $H = 8.670$ in^3 (142.1 cm^3); $y_t = 8.670/2.715 = 3.19$ in (81.0 mm); $y_b = 4.81$ in (122.2 mm); $f_c = (3.19/4.81)20,000 = 13,300$ lb/in^2 (91.7 MPa), which is satisfactory.

9.   *Compute the deflection*

For the net section, $I_x = 26.8 + 3.13(3.19 - 2.77)^2 - 0.415(3.19 - 0.053)^2 = 23.3$ in^4 (969.8 cm^4). This value agrees with that tabulated in the AISI *Manual*. The deflection is $\Delta = 5wL^4/(384EI_x) = 5(369)(13)^4(12)^3/[384(29.5)(10)^6 23.3] = 0.345$ in (8.8 mm).

STEEL BEAM ENCASED IN CONCRETE

A concrete floor slab is to be supported by steel beams spaced 10 ft (3.05 m) on centers and having a span of 28 ft 6 in (8.69 m). The beams will be encased in concrete with a minimum cover of 2 in (50.8 mm) all around; they will remain unshored during construction. The slab has been designed as 4½ in (114.3 mm) thick, with $f_c' = 3000$ lb/in^2 (20.7 MPa). The loading includes the following: live load, 120 lb/ft^2 (5.75 kPa); finished floor and ceiling, 25 lb/ft^2 (1.2 kPa). The steel beams have been tentatively designed as W16 × 40. Review the design.

Calculation Procedure:

1.   *Record the relevant properties of the section and the allowable flexural stresses*

In accordance with the AISC *Specification*, the member may be designed as a composite steel-and-concrete beam, reliance being placed on the natural bond of the two materials to obtain composite action. Refer to Sec. 1 for the design of a composite bridge member. In the design of a composite building member, the effects of plastic flow are usually disregarded. Since the slab is poured monolithically, the composite member is considered continuous. Apply the following equations in computing bending moments in the composite beams: at midspan, $M = (\frac{1}{20})wL^2$; at support, $M = -(\frac{1}{12})wL^2$.

The subscripts c, ts, and bs refer to the extreme fiber of concrete, top of steel, and bottom of steel, respectively. The superscripts c and n refer to the composite and noncomposite sections, respectively.

Record the properties of the W16 × 40: $A = 11.77$ in^2 (75.94 cm^2); $d = 16.00$ in (406.4 mm); $I = 515.5$ in^4 (21.457 cm^4); $S = 64.4$ in^3 (1055.3 cm^3); flange width = 7 in (177.8 mm). By the AISC *Specification*, $f_s = 24,000$ lb/in^2 (165.5 MPa). By the ACI *Code*, $n = 9$ and $f_c = 1350$ lb/in^2 (9306.9 kPa).

2.   *Transform the composite section in the region of positive moment to an equivalent section of steel; compute the section moduli*

Refer to Fig. 25a and the AISC *Specification*. Use the gross concrete area. Then the effective flange width $= \frac{1}{4}L = \frac{1}{4}(28.5)12 = 85.5$ in (2172 mm); spacing of beams = 120 in (3048 mm); $16t + 11 = 16(4.5) + 11 = 83$ in (2108 mm); this governs. Transformed width = 83/9 = 9.22 in (234.2 mm).

Assume that the neutral axis lies within the flange, and take static moments with respect to this axis; or $\frac{1}{2}(9.22y^2) - 11.77(10 - y) = 0$; $y = 3.93$ in (99.8 mm).

Compute the moment of inertia. Slab: $(\frac{1}{3})9.22(3.93)^3 = 187$ in^4 (7783.5 cm^4). Beam: $515.5 + 11.77 \times (10 - 3.93)^2 = 949$ in^4 (39,500.4 cm^4); $I = 187 + 949 = 1136$ in^4 (47,283.9 cm^4); $S_c = 1136/3.93 = 289.1$ in^3 (4737.5 cm^3); $S_{bs} = 1136/14.07 = 80.7$ in^3 (1322.4 cm^3).

3.   *Transform the composite section in the region of negative moment to an equivalent section of steel; compute the section moduli*

Referring to Fig. 25b, we see that the transformed width = 11/9 = 1.22 in (31.0 mm). Take static moments with respect to the neutral axis. Or, $11.77(10 - y) - \frac{1}{2}(1.22y^2) = 0$; $y = 7.26$ in

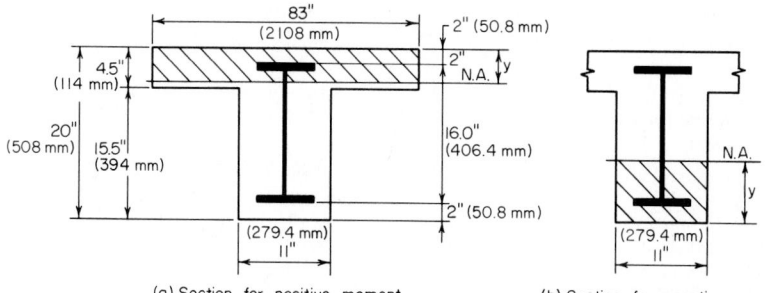

FIG. 25 Steel beam encased in concrete. (*a*) Section for positive moment; (*b*) section for negative moment.

(184.4 mm). Compute the moment of inertia. Thus, slab: $(\frac{1}{8})1.22(7.26)^3 = 155.6$ in^4 (6476.6 cm^4). Beam: $515.5 + 11.77(10 - 7.26)^2 = 603.9$ in^4 (25,136.2 cm^4); $I = 155.6 + 603.9 = 759.5$ in^4 (31,612.8 cm^4). Then $S_c = 759.5/7.26 = 104.6$ in^3 (1714.1 cm^3); $S_{ts} = 759.5/10.74 = 70.7$ in^3 (1158.6 cm^3).

4. *Compute the bending stresses at midspan*

The loads carried by the noncomposite member are: slab, $(4.5)150(10)/12 = 563$ lb/lin ft (8.22 kN/m); stem, $11(15.5)150/144 = 178$ lb/lin ft (2.6 kN/m); steel, 40 lb/lin ft (0.58 kN/m); total $= 563 + 178 + 40 = 781$ lb/lin ft (11.4 kN/m). The load carried by the composite member $= 145(10) = 1450$ lb/lin ft (21.2 kN/m). Then $M^n = (\frac{1}{8})781(28.5)^2 12 = 951,500$ in·lb (107.5 kN·m); $M^c = (\frac{1}{20})1450(28.5)^2 12 = 706,600$ in·lb (79.8 kN·m); $f_c = 706,600/[289.1(9)] = 272$ lb/in^2 (1875 kPa), which is acceptable. Also, $f_{bs} = (951,500/64.4) + (706,600/80.7) = 23,530$ lb/in^2 (162.2 MPa), which is acceptable.

5. *Compute the bending stresses at the support*

Thus, $M^c = 706,600(\frac{20}{12}) = 1,177,700$ in·lb (132.9 kN·m); $f_c = 1,177,700/[104.6(9)] = 1251$ lb/in^2 (8.62 MPa), which is satisfactory. Also, $f_{ts} = 1,177,700/70.7 = 16,600$ lb/in^2 (114.9 MPa), which is acceptable. The design is therefore satisfactory with respect to flexure.

6. *Investigate the composite member with respect to horizontal shear in the concrete at the section of contraflexure*

Assume that this section lies at a distance of 0.2L from the support. The shear at this section is $V^c = 1450(0.3)(28.5) = 12,400$ lb (55.2 kN).

Refer to Sec. 1. Where the bending moment is positive, the critical plane for horizontal shear is considered to be the surface *abcd* in Fig. 26*a*. For simplicity, however, compute the shear flow at the neutral axis. Apply the relation $q = VQ/I$, where $Q = \frac{1}{2}(9.22)(3.93)^2 = 71.20$ in^3 (1166.8 cm^3) and $q = 12,400(71.20)/1136 = 777$ lb/lin in (136 N/mm).

Resistance to shear flow is provided by the bond between the steel and concrete along *bc* and by the pure-shear strength of the concrete along *ab* and *cd*. (The term *pure shear* is used to

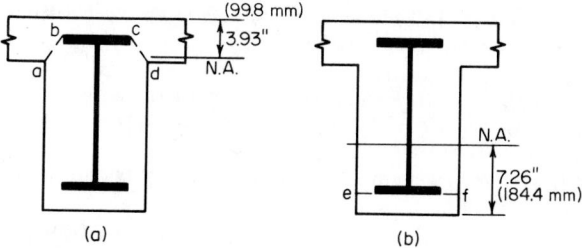

FIG. 26 Critical planes for horizontal shear.

distinguish this from shear that is used as a measure of diagonal tension.) The allowable stresses in bond and pure shear are usually taken as $0.03f'_c$ and $0.12f'_c$, respectively. Thus $bc = 7$ in (177.8 mm); $ab = (2.5^2 + 2^2)^{0.5} = 3.2$ in (81.3 mm); $q_{allow} = 7(90) + 2(3.2)360 = 2934$ lb/lin in (419 N/mm), which is satisfactory.

7. Investigate the composite member with respect to horizontal shear in the concrete at the support

The critical plane for horizontal shear is ef in Fig. 26b. Thus $V^c = 1450(0.5)28.5 = 20,660$ lb (91.9 kN); $Q = 1.22(2)(7.26 - 1) = 15.27$ in^3 (250.2 cm^3); $q = 20,660(15.27)/759.5 = 415$ lb/lin in (72.7 N/mm); $q_{allow} = 7(90) + 2(2)360 = 2070$ lb/lin in (363 N/mm), which is satisfactory.

Mechanical shear connectors are not required to obtain composite action, but the beam is wrapped with wire mesh.

COMPOSITE STEEL-AND-CONCRETE BEAM

A concrete floor slab is to be supported by steel beams spaced 11 ft (3.35 m) on centers and having a span of 36 ft (10.97 m). The beams will be supplied with shear connectors to obtain composite action of the steel and concrete. The slab will be 5 in (127 mm) thick and made of 3000-lb/in^2 (20.7-MPa) concrete. Loading includes the following: live load, 200 lb/ft^2 (9.58 kPa); finished floor, ceiling, and partition, 30 lb/ft^2 (1.44 kPa). In addition, each girder will carry a dead load of 10 kips (44.5 kN) applied as a concentrated load at midspan prior to hardening of the concrete. Conditions at the job site preclude the use of temporary shoring. Design the interior girders, limiting the overall depth of steel to 20 in (508 mm), if possible.

Calculation Procedure:

1. Compute the unit loads w_1, w_2, and w_3

Refer to the AISC *Specification* and *Manual*. Although ostensibly we apply the elastic-stress method, the design of a composite steel-and-concrete beam in reality is based on the ultimate-strength behavior of the member. Loads that are present before the concrete has hardened are supported by the steel member alone; loads that are present after the concrete has hardened are considered to be supported by the composite member, regardless of whether these loads originated before or after hardening. The effects of plastic flow are disregarded.

The subscripts 1, 2, and 3 refer, respectively, to dead loads applied before hardening of the concrete, dead loads applied after hardening of the concrete, and live loads. The subscripts b, ts, and tc refer to the bottom of the member, top of the steel, and top of the concrete, respectively. The superscripts c and n refer to the composite and noncomposite member, respectively.

We compute the unit loads for a slab weight of 63 lb/lin ft (0.92 kN/m) and an assumed steel weight of 80 lb/lin ft (1167.5 N/m): $w_1 = 63(11) + 80 = 773$ lb/lin ft (11.3 kN/m); $w_2 = 30(11) = 330$ lb/lin ft (4.8 kN/m); $w_3 = 200(11) = 2200$ lb/lin ft (32.1 kN/m).

2. Compute all bending moments required in the design

Thus, $M_1 = 12[(\frac{1}{8})0.773(36)^2 + \frac{1}{4}(10)36] = 2583$ in·kips (291.8 kN·m); $M_2 = (\frac{1}{8})0.330(36)^2 12 = 642$ in·kips (72.5 kN·m). $M_3 = (\frac{1}{8})2.200(36)^2 12 = 4277$ in·kips (483.2 kN·m); $M^c = 2583 + 642 + 4277 = 7502$ in·kips (847.6 kN·m); $M^n = 2583$ in·kips (291.8 kN·m); $M_{DL} = 2583 + 642 = 3225$ in·kips (364.4 kN·m); $M_{LL} = 4277$ in·kips (483.2 kN·m).

3. Compute the required section moduli with respect to the steel, using an allowable bending stress of 24 kips/in^2 (165.5 MPa)

In the composite member, the maximum steel stress occurs at the bottom; in the noncomposite member, it occurs at the top of the steel if a bottom-flange cover plate is used.

Thus, composite section, $S_b = 7502/24 = 312.6$ in^3 (5122.6 cm^3); noncomposite section, $S_{ts} = 2583/24 = 107.6$ in^3 (1763.3 cm^3).

4. Select a trial section by tentatively assuming that the composite-design tables in the AISC Manual are applicable

The *Manual* shows that a composite section consisting of a 5-in (127-mm) concrete slab, a W18 × 55 steel beam, and a cover plate having an area of 9 in^2 (58.1 cm^2) provides $S_b = 317.5$ in^3 (5202.9 cm^3). The noncomposite section provides $S_{ts} = 113.7$ in^3 (1863.2 cm^3).

Since unshored construction is to be used, the section must conform with the *Manual* equation $1.35 + 0.35 M_{LL}/M_{DL} = 1.35 + 0.35(4277/3225) = 1.81$. And $S_b^c/S_b^n = 317.5/213.6 = 1.49$, which is satisfactory.

The flange width of the W18 × 55 is 7.53 in (191.3 mm). The minimum allowable distance between the edge of the cover plate and the edge of the beam flange equals the size of the fillet weld plus $\frac{5}{16}$ in (7.9 mm). Use a 9 × 1 in (229 × 25 mm) plate. The steel section therefore coincides with that presented in the AISC *Manual*, which has a cover plate thickness t_p of 1 in (25.4 mm). The trial section is therefore W18 × 55; cover plate is 9 × 1 in (229 × 25 mm).

5. Check the trial section

The AISC composite-design tables are constructed by assuming that the effective flange width of the member equals 16 times the slab thickness plus the flange width of the steel. In the present instance, the effective flange width, as governed by the AISC, is $\frac{1}{4}L = \frac{1}{4}(36)12 = 108$ in (2743 mm); spacing of beams = 132 in (3353 mm); $16t + 7.53 = 16(5) + 7.53 = 87.53$ in (2223.3 mm), which governs.

The cross section properties in the AISC table may be applied. The moment of inertia refers to an equivalent section obtained by transforming the concrete to steel. Refer to Sec. 1. Thus $y_{tc} = 5 + 18.12 + 1 - 16.50 = 7.62$ in (194 mm); $S_{tc} = I/y_{tc} = 5242/7.62 = 687.9$ in³ (11,272.7 cm³). From the ACI *Code*, $f_c = 1350$ lb/in² (9.31 MPa) and $n = 9$. Then $f_c = M^c/(nS_{tc}) = 7,502,000/[9(687.9)] = 1210$ lb/in² (8.34 MPa), which is satisfactory.

6. Record the relevant properties of the W18 × 55

Thus, $A = 16.19$ in² (104.5 cm²); $d = 18.12$ in (460 mm); $I = 890$ in⁴ (37,044.6 cm⁴); $S = 98.2$ in³ (1609 cm³); flange thickness = 0.630 in (16.0 mm).

7. Compute the section moduli of the composite section where the cover plate is absent

To locate the neutral axis, take static moments with respect to the center of the steel. Thus, transformed flange width = 87.53/9 = 9.726 in (247.0 mm). Further,

Element	A, in² (cm²)	y, in (mm)	Ay, in³ (cm³)	Ay², in⁴ (cm⁴)	I_o, in⁴ (cm⁴)
W18 × 55	16.19 (104.5)	0 (0)	0 (0)	0 (0)	890 (37,044.6)
Slab	48.63 (313.7)	11.56 (294)	562.2 (9,212.8)	6,499 (270,509)	101 (4,203.9)
Total	64.82 (418.2)	. . . . . .	562.2 (9,212.8)	6,499 (270,509)	991 (41,248.5)

Then $\bar{y} = 562.2/64.82 = 8.67$ in (220 mm); $I = 6499 + 991 - 64.82(8.67)^2 = 2618$ in⁴ (108,969.4 cm⁴); $y_b = 9.06 + 8.67 = 17.73$ in (450 mm); $y_{tc} = 9.06 + 5 - 8.67 = 5.39$ in (136.9 mm); $S_b = 2618/17.73 = 147.7$ in³ (2420 cm³); $S_{tc} = 2618/5.39 = 485.7$ in³ (7959 cm³).

8. Verify the value of S_b

Apply the value of the K factor in the AISC table. This factor is defined by $K^2 = 1 - S_b$ without plate/S_b with plate. The S_b value without the plate = $317.5(1 - 0.73^2) = 148$ in³ (2425 cm³), which is satisfactory.

9. Establish the theoretical length of the cover plate

In Fig. 27, let C denote the section at which the cover plate becomes superfluous with respect to flexure. Then, for the composite section, $w = 0.773 + 0.330 + 2.200 = 3.303$ kips/lin ft (48.2 kN/m); $P = 10$ kips (44.5 kN); $M_m = 7502$ in·kips (847.6 kN·m); $R_a = 64.45$ kips (286.7 kN). The allowable values of M_c are, for concrete, $M_c = 485.7(9)1.35/12 = 491.8$ ft·kips (666.9 kN·m) and, for steel, $M_c = 147.7(24)/12 = 295.4$ ft·kips (400.6 kN·m), which governs. Then $R_a x - \frac{1}{2}wx^2 = 295.4$; $x = 5.30$ ft (1.62 m). The theoretical length = $36 - 2(5.30) = 25.40$ ft (7.74 m).

For the noncomposite section, investigate the stresses at the section C previously located. Thus: $w = 0.773$ kips/lin ft (11.3 kN/m); $P = 10$ kips (44.5 kN); $R_a = 18.91$ kips (84.1 kN); $M_c = 18.91(5.30) - \frac{1}{2}(0.773) \times 5.30^2 = 89.4$ ft·kips (121.2 kN·m); $f_b = f_{ts} = 89.4(12)/98.2 = 10.9$ kips/in² (75.1 MPa), which is satisfactory.

10. *Determine the axial force F in the cover plate at its end by computing the mean bending stress*

Thus $f_{mean} = My_{mean}/I = 295.4(12)(16.50 - 0.50)/5242 = 10.82$ kips/in^2 (74.6 MPa); $F = Af_{mean} = 9(10.82) = 97.4$ kips (433.2 kN). Alternatively, calculate F by applying the factor $12Q/I$ recorded in the AISC table. Thus, $F = 12QM/I = 0.33(295.4) = 97.5$ kips (433.7 kN).

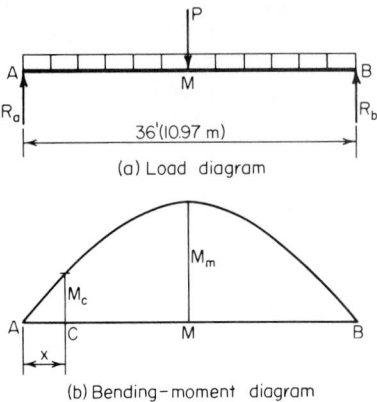

FIG. 27 (*a*) Load diagram; (*b*) bending-moment diagram.

11. *Design the weld required to develop the cover plate at each end*

Use fillet welds of E60 electrodes, placed along the sides but not along the end of the plate. The AISC *Specification* requires a minimum weld of ⁵⁄₁₆ in (7.9 mm) for a 1-in (25.4-mm) plate; the capacity of this weld is 3000 lb/lin in (525 N/mm). Then, length = 97,400/3000 = 32.5 in (826 mm). However, the AISC requires that the plate be extended 18 in (457 mm) beyond the theoretical cutoff point, thus providing 36 in (914 mm) of weld at each end.

12. *Design the intermittent weld*

The vertical shear at C is $V_c = R_a - 5.30w = 64.45 - 5.30(3.303) = 46.94$ kips (208.8 kN); $q = VQ/I = 46,940(0.33)/12 = 1290$ lb/lin in (225.9 N/mm). The AISC calls for a minimum weld length of 1½ in (3.81 mm). Let s denote the center-to-center spacing. Then $s = 2(1.5)3000/1290 = 7.0$ in (177.8 mm). The AISC imposes an upper limit of 24 times the thickness of the thinner part joined, or 12 in (304.8 mm). Thus, $s_{max} = 24(0.63) > 12$ in (304.8 mm). Use a 7-in (177.8-mm) spacing at the ends and increase the spacing as the shear diminishes.

13. *Design the shear connectors*

Use ¾-in (19.1-mm) studs, 3 in (76.2 mm) high. The design of the connectors is governed by the AISC *Specification*. The capacity of the stud = 11.5 kips (51.2 kN). From the AISC table, $V_h = 453.4$ kips (2016.7 kN). Total number of studs required = 2(453.4)/11.5 = 80. These are to be equally spaced.

DESIGN OF A CONCRETE JOIST IN A RIBBED FLOOR

The concrete floor of a building will be constructed by using removable steel pans to form a one-way ribbed slab. The loads are: live load, 80 lb/ft^2 (3.83 kPa); allowance for movable partitions, 20 lb/ft^2 (0.96 kPa); plastered ceiling, 10 lb/ft^2 (0.48 kPa); wood floor with sleepers in cinder-concrete fill, 15 lb/ft^2 (0.72 kPa). The joists will have a clear span of 17 ft (5.2 m) and be continuous over several spans. Design the interior joist by the ultimate-strength method, using $f'_c = 3000$ lb/in^2 (20.7 MPa) and $f_y = 40,000$ lb/in^2 (275.8 MPa).

Calculation Procedure:

1. *Compute the ultimate load carried by the joist*

A one-way ribbed floor consists of a concrete slab supported by closely spaced members called *ribs*, or *joists*. The joists in turn are supported by steel or concrete girders that frame to columns. Manufacturers' engineering data present the dimensions of steel-pan forms that are available and the average weight of floor corresponding to each form.

Try the cross section shown in Fig. 28, which has an average weight of 54 lb/ft^2 (2.59 kPa). Although the forms are tapered to facilitate removal, assume for design purposes that the joist has

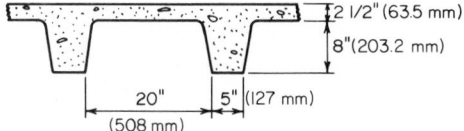

FIG. 28   Ribbed floor.

a constant width of 5 in (127 mm). The design of a ribbed floor is governed by the ACI *Code*. The ultimate-strength design of reinforced-concrete members is covered in Sec. 1.

Referring to the ACI *Code*, compute the ultimate load carried by the joist. Or, w_u = 2.08[1.5(54 + 20 + 10 + 15) + 1.8(80)] = 608 lb/lin ft (8.9 kN/m).

2.   *Determine whether the joist is adequate with respect to shear*

Since the joist is too narrow to permit the use of stirrups, the shearing stress must be limited to the value given in the ACI *Code*. Or, v_c = 1.1(2ϕ)(f_c')$^{0.5}$ = 1.1(2)(0.85)(3000)$^{0.5}$ = 102 lb/in^2 (703.2 kPa).

Assume that the reinforcement will consist of no. 4 bars. With ¾ in (19.1 mm) for fireproofing, as required by the ACI *Code*, d = 8 + 2.5 − 1.0 = 9.5 in (241.3 mm). The vertical shear at a distance d from the face of the support is V_u = (8.5 − 0.79)608 = 4690 lb (20.9 kN).

The critical shearing stress computed as required by the ACI *Code* is v_u = V_u/(bd) = 4690/ [5(9.5)] = 99 lb/in^2 (682.6 kPa) < v_c, which is satisfactory.

3.   *Compute the ultimate moments to be resisted by the joist*

Do this by applying the moment equations given in the ACI *Code*. Or, $M_{u,pos}$ = (¹⁄₁₆)608(17)^{2}12 = 132,000 in·lb (14.9 kN·m); $M_{u,neg}$ = (¹⁄₁₁)608(17)^{2}12 = 192,000 in·lb (21.7 kN·m).

Where the bending moment is positive, the fibers above the neutral axis are in compression, and the joist and tributary slab function in combination to form a T beam. Where the bending moment is negative, the joist functions alone.

4.   *Determine whether the joist is capable of resisting the negative moment*

Use the equation q_{max} = 0.6375$k_1$87,000/(87,000 + f_y), or q_{max} = 0.6375(0.85)87,000/127,000 = 0.371. By Eq. 6 of Sec. 1, M_u = $\phi bd^2 f_c'q$(1 − 0.59q), or M_u = 0.90(5)9.5^2 × (3000)0.371(0.781) = 353,000 in·lb (39.9 kN·m), which is satisfactory.

5.   *Compute the area of negative reinforcement*

Use Eq. 7 of Sec. 1. Or, f_c = 0.85(3) = 2.55 kips/in^2 (17.6 MPa); bdf_c = 5(9.5)2.55 = 121.1; 2$bf_c M_u$/ϕ = 2(5)2.55(192)/0.90 = 5440; A_s = [121.1 − (121.1^2 − 5440)$^{0.5}$]/40 = 0.63 in^2 (4.06 cm^2).

6.   *Compute the area of positive reinforcement*

Since the stress block lies wholly within the flange, apply Eq. 7 of Sec. 1, with b = 25 in (635 mm). Or, bdf_c = 605.6; 2$bf_c M_u$/ϕ = 18,700; A_s = [605.6 − (605.6^2 − 18,700)$^{0.5}$]/40 = 0.39 in^2 (2.52 cm^2).

7.   *Select the reinforcing bars and locate the bend points*

For positive reinforcement, use two no. 4 bars, one straight and one trussed, to obtain A_s = 0.40 in^2 (2.58 cm^2). For negative reinforcement, supplement the two trussed bars over the support with one straight no. 5 bar to obtain A_s = 0.71 in^2 (4.58 cm^2).

To locate the bend points of the trussed bars and to investigate the bond stress, follow the method given in Sec. 1.

DESIGN OF A STAIR SLAB

The concrete stair shown in elevation in Fig. 29*a*, which has been proportioned in conformity with the requirements of the local building code, is to carry a live load of 100 lb/ft^2 (4.79 kPa). The slab will be poured independently of the supporting members. Design the slab by the working-stress method, using f_c' = 3000 lb/in^2 (20.7 MPa) and f_s = 20,000 lb/in^2 (137.9 MPa).

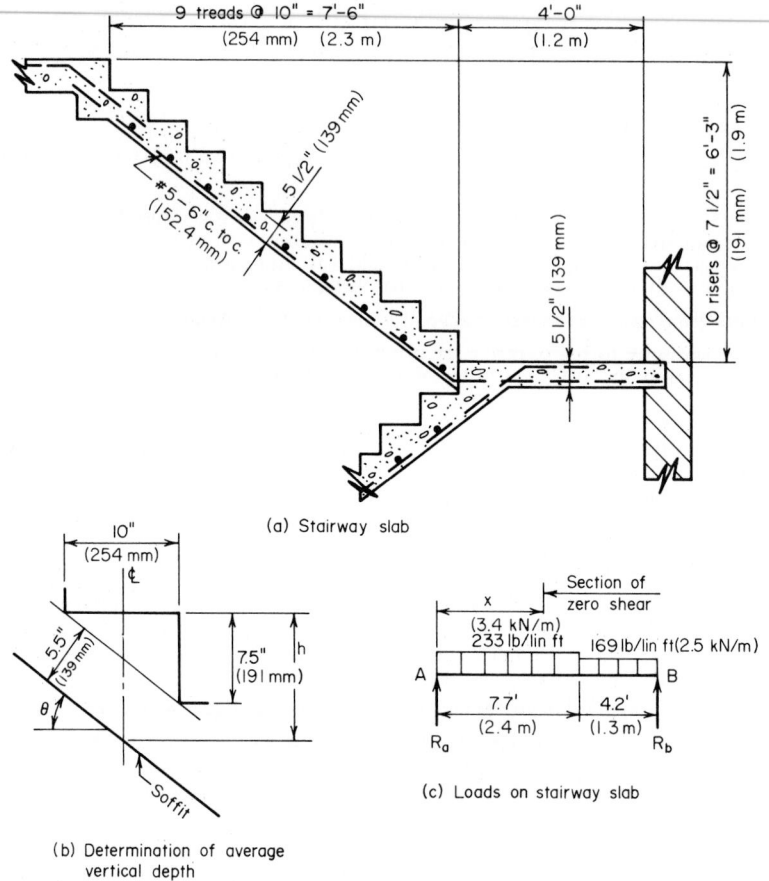

FIG. 29   (*a*) Stairway slab; (*b*) determination of average vertical depth; (*c*) loads on stairway slab.

Calculation Procedure:

1. *Compute the unit loads*

The working-stress method of designing reinforced-concrete members is presented in Sec. 1. The slab is designed as a simply supported beam having a span equal to the horizontal distance between the center of supports. For convenience, consider a strip of slab having a width of 1 ft (0.3 m).

Assume that the slab will be 5.5 in (139 mm) thick, the thickness of the stairway slab being measured normal to the soffit. Compute the average vertical depth in Fig. 29*b*. Thus sec θ = 1.25; h = 5.5(1.25) + 3.75 = 10.63 in (270.0 mm). For the stairway, w = 100 + 10.63(150)/12 = 233 lb/lin ft (3.4 kN/m); for the landing, w = 100 + 5.5(150)/12 = 169 lb/lin ft (2.5 kN/m).

2. *Compute the maximum bending moment in the slab*

Construct the load diagram shown in Fig. 29*c*, adding about 5 in (127 mm) to the clear span to obtain the effective span. Thus R_a = [169(4.2)2.1 + 233(7.7)8.05]/11.9 = 1339 lb (5.95 kN); x = 1339/233 = 5.75 ft (1.75 m); M_{max} = ½(1339)5.75(12) = 46,200 in·lb (5.2 kN·m).

3. *Design the reinforcement*

Refer to Table 6 in Sec. 1 to obtain the following values: $K_b = 223$ lb/in^2 (1.5 MPa); $j = 0.874$. Assume an effective depth of 4.5 in (114.3 mm). By Eq. 31, the moment capacity of the member at balanced design is $M_b = K_b bd^2 = 223(12)4.5^2 = 54,190$ in·lb (6.1 kN·m). The steel is therefore stressed to capacity. (Upon investigation, a 5-in (127-mm) slab is found to be inadequate.) By Eq. 25, $A_s = M/(f_s jd) = 46,200/[20,000(0.874)4.5] = 0.587$ in^2 (3.79 cm^2).

Use no. 5 bars, 6 in (152.4 mm) on centers, to obtain $A_s = 0.62$ in^2 (4.0 cm^2). In addition, place one no. 5 bar transversely under each tread to assist in distributing the load and to serve as temperature reinforcement. Since the slab is poured independently of the supporting members, it is necessary to furnish dowels at the construction joints.

FREE VIBRATORY MOTION OF A RIGID BENT

The bent in Fig. 30 is subjected to a horizontal load P applied suddenly at the top. Using literal values, determine the frequency of vibration of the bent. Make these simplifying assumptions: The girder is infinitely rigid; the columns have negligible mass; damping forces are absent.

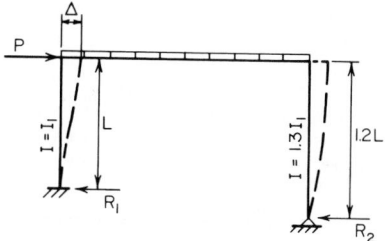

FIG. 30   Vibrating bent.

Calculation Procedure:

1. *Compute the spring constant*

The amplitude (maximum horizontal displacement of the bent from its position of static equilibrium) is a function of the energy imparted to the bent by the applied load. The frequency of vibration is independent of this energy. To determine the frequency, it is necessary to find the *spring constant* of the vibrating system. This is the static force that is required at the top to cause a horizontal displacement of one unit. Since the girder is considered to be infinitely rigid, the elastic curves of the columns are vertical at the top. Let f = frequency; k = spring constant; M = total mass of girder and bodies supported by girder.

Using cases 22 and 23 in the AISC *Manual*, we see that when $\Delta = 1$, the horizontal reactions are $R_1 = 12EI_1/L^3$; $R_2 = 3E(1.3I_1)/(1.2L)^3 = 2.26EI_1/L^3$; $k = R_1 + R_2 = 14.26EI_1/L^3$.

2. *Compute the frequency of vibration*

Use the equation $f = (1/2\pi)(k/M)^{0.5} = (1/2\pi)(14.26EI_1/ML^3)^{0.5} = 0.601(EI_1/ML^3)^{0.5}$ Hz.

Plumbing and Drainage

REFERENCES: Church—*Practical Plumbing Design Guide*, McGraw-Hill; Ripka—*Plumbing Installation and Design*, American Technical Society; Galeno—*Plumbing Estimating Handbook*, Van Nostrand; Page and Nation—*Estimator's Piping Manhour Manual*, Gulf Publishing; Nielson—*Standard Plumbing Engineering Design*, McGraw-Hill; Blenderman—*Design of Plumbing and Drainage Systems*, Industrial Press; D'Arcangelo—*Mathematics for Plumbers and Pipefitters*, Delmar; Miller and Gallina—*Estimating and Cost Control in Plumbing Design*, Van Nostrand; Manas—*National Plumbing Code Handbook*, McGraw-Hill; Babbitt—*Plumbing*, McGraw-Hill; American Standards Institute—*National Plumbing Code*; National Bureau of Standards—*Water Distributing Systems for Buildings*; National Association of Plumbing Contractors—*Water Supply Piping for the Plumbing System*; Copper and Brass Research Association—*Brass Pipe Handbook for Plumbing Installations* and *Copper Tube Handbook on Plumbing and Heating*.

DETERMINATION OF PLUMBING-SYSTEM PIPE SIZES

A two-story industrial plant has the following plumbing fixtures: first floor—six wall-lip urinals, three valve-operated water closets, three large-size lavatories, and six showers, each with a separate head; second floor—three wall-lip urinals, three valve-operated water closets, three large-size lavatories, and three showers, each with a separate head. Size the waste and vent stacks and the building house drain for this system. Use the *National Plumbing Code (NPC)* as the governing code for the plant locality. The branch piping and house drain will be pitched ¼ in (6.4 mm) per ft (m) of length.

Calculation Procedure:

1. *Select the upper-floor branch layout*

Sketch the layout of the proposed plumbing system, beginning with the upper, or second, floor. Figure 1 shows a typical plumbing-system sketch. Assume in this plant that the second-floor urinals, water closets, and lavatories are served by one branch drain and the showers by another branch. Both branch drains discharge into a vertical soil stack.

2. *Compute the upper-floor branch fixture units*

List each plumbing fixture as in Table 1.

Obtain the data for each numbered column of Table 1 in the following manner. (1) List the number of the floor being studied and number of each branch drain from the system sketch. Since it was decided to use two branch drains, number them accordingly. (2) List the name of each fixture that will be used. (3) List the number of each type of fixture that will be used. (4) Obtain from the *National Plumbing Code*, or Table 2, the number of *fixture units per fixture*, i.e., the average discharge, during use, of an arbitrarily selected fixture, such as a lavatory or toilet. Once this value is established in a plumbing code, the discharge rates of other types of fixtures are stated in terms of the basic unit. Plumbing codes adopted by various localities usually list the fixture units they recommend in a tabulation similar to Table 2. (5) Multiply the number of fixtures,

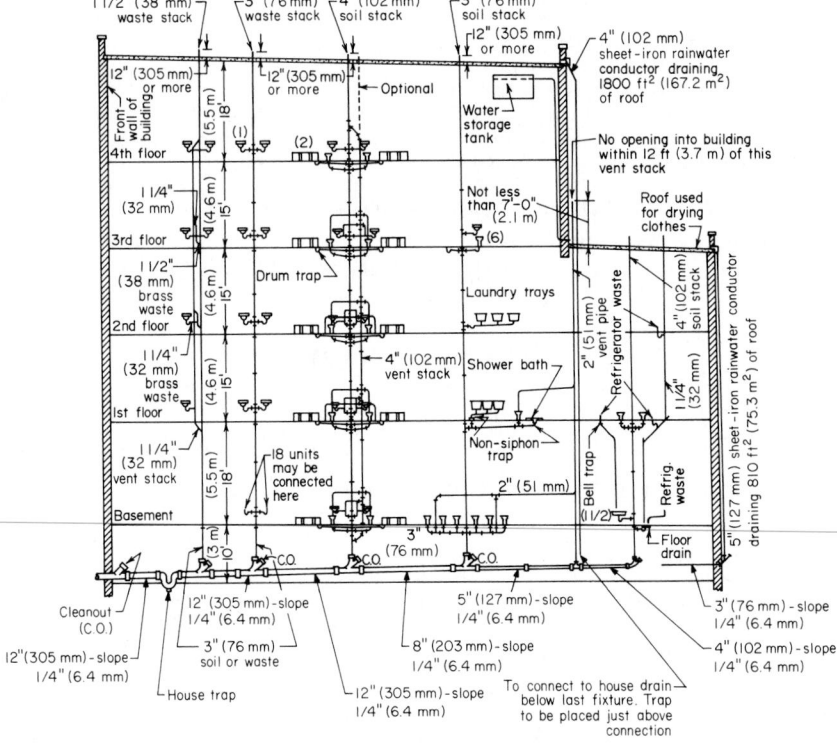

FIG. 1 Typical plumbing layout diagram for a multistory building.

TABLE 1   Floor-Fixture Analysis

(1) Floor	(2) Fixture name	(3) No. of fixtures	(4) Fixture units per fixture	(5) Total no. of fixture units
Floor 2	Urinals, wall-lip	3	4	12
Branch drain 1	Water closets, valve- operated	3	8	24
	Lavatories, large-size	3	2	6
Total	. . .	9	. . .	42
Branch drain 2	Showers	3	3	9
Total	. . .	3	. . .	9
Floor 1	Urinals, wall-lip	6	4	24
	Water closets, valve- operated	3	8	24
Branch drain 3	Lavatories, large-size	3	2	6
Total	. . .	12	. . .	54
Branch drain 4	Showers	6	3	18
Total	. . .	6	. . .	18

column 3, by the fixture units, column 4, to obtain the result in column 5. Thus, for the urinals, (3 urinals)(4 fixture units per urinal fixture) = 12 fixture units. Find the sum of the fixture units for each branch.

3. Size the upper-floor branch pipes

Refer to the *National Plumbing Code*, or Table 3, for the number of fixture units each branch can have connected to it. Thus, Table 3 shows that a 4-in (102-mm) branch pipe must be used for branch drain 1 because no more than 20 fixture units can be connected to the next smaller, or 3-in (76-mm) pipe. Hence, branch drain 1 will use a 4-in (102-mm) pipe because it serves 42 fixture units, step 2.

Branch drain 2 serves 9 fixture units, step 2. Hence, a 2½-in (64-mm) branch pipe will be suitable because it can serve 12 fixture units or less (Table 3).

4. Size the upper-floor stack

The two horizontal branch drains are sloped toward a vertical *stack* pipe that conducts the waste and water from the upper floors to the sewer. Use Table 3 to size the stack, which is three stories high, including the basement. The total number of second-floor fixture units the stack must serve is 42 + 9 = 51. Hence, for a 4-in (102-mm) stack, Table 3 must be used.

5. Size the upper-story vent pipe

Each branch drain on the upper floor must be vented. However, the stack can be extended upward and each branch vent connected to it, if desired. Use the *NPC*, or Table 4, to determine the vent size.

TABLE 2   Fixture Units per Fixture or Group°

Fixture type	Fixture-unit value as load factors	Minimum size of trap, in (mm)
One bathroom group consisting of water closet, lavatory, and bathtub or shower stall	Tank water closet, 6; flush-valve water closet, 8	
Bathtub† (with or without overhead shower)	2	1½ (38)
Bathtub†	3	2 (51)
Bidet	3	Nominal, 1½ (38)
Combination sink and tray	3	1½ (38)
Combination sink and tray with food-disposal unit	4	Separate traps, 1½ (38)
Dental unit or cuspidor	1	1¼ (32)
Dental lavatory	1	1¼ (32)
Drinking fountain	½	1 (25)
Dishwasher, domestic	2	1½ (38)
Floor drains‡	1	2 (51)
Kitchen sink, domestic	2	1½ (38)
Kitchen sink, domestic, with food-waste grinder	3	1½ (38)
Lavatory§	1	Small PO, 1¼ (32)
Lavatory§	2	Large PO, 1½ (38)
Lavatory, barber, beauty parlor	2	1½ (38)
Lavatory, surgeon's	2	1½ (38)
Laundry tray (one or two compartments)	2	1½ (38)
Shower stall, domestic	2	2 (51)
Showers (group) per head	3	
Sinks:		
Surgeon's	3	1½ (38)
Flushing rim (with valve)	8	3 (76)
Service (trap standard)	3	3 (76)
Service (P trap)	2	2 (51)
Pot, scullery, etc.	4	1½ (38)
Urinal, pedestal, siphon jet, blowout	8	Nominal, 3 (76)
Urinal, wall lip	4	1½ (38)
Urinal, stall, washout	4	2 (51)
Urinal trough [each 2-ft (0.61-m) section]	2	1½ (38)
Wash sink (circular or multiple) each set of faucets	2	Nominal, 1½ (38)
Water closet, tank-operated	2	Nominal, 3 (76)
Water closet, valve-operated	8	3 (76)

°From *National Plumbing Code*.
†A shower head over a bathtub does not increase the fixture value.
‡Size of floor drain shall be determined by the area of surface water to be drained.
§Lavatories with 1¼- (32-mm) or 1½-in (38-mm) trap have the same load value; larger PO (plumbing orifice) plugs have greater flow rate.

As a guide, the diameter of a branch vent or vent stack is one-half or more of the branch or stack it serves, but not less than 1¼ in (32 mm). Thus branch drain 1 would have a 4/2 = 2-in (51-mm) vent, whereas branch drain 2 would have a 2½/2 = 1¼-in (32-mm) vent.

6. *Select the lower-floor branch layout*

Assume that the six urinals, three water closets, and three lavatories are served by one branch drain and the six showers by another. Indicate these on the system sketch. Further, arrange both branch drains so that they discharge into the vertical stack serving the second floor.

TABLE 3   Horizontal Fixture Branches and Stacks°

Diameter of pipe, in (mm)	Maximum number of fixture units that may be connected to			
	Any horizontal† fixture branch	One stack of three stories in height or three intervals	More than three stories in height	
			Total for stack	Total at one story or branch interval
1¼ (32)	1	2	2	1
1½ (38)	3	4	8	2
2 (51)	6	10	24	6
2½ (64)	12	20	42	9
3 (76)	20‡	30§	60§	16‡
4 (102)	160	240	500	90
5 (127)	360	540	1100	200
6 (152)	620	960	1900	350

°From *National Plumbing Code.*
†Does not include branches of the building drain.
‡Not over two water closets.
§Not over six water closets.

7.   *Compute the lower-floor branch fixture units*

Use the same procedure as in step 2, listing the fixtures and their respective fixture units in the lower part of Table 1.

8.   *Size the lower-floor branch pipes*

By Table 3, branch drain 3 must be 4 in (102 mm) because it serves a total of 54 fixture units. Branch 4 must be 3 in (76 mm) because it serves a total of 18 fixture units.

9.   *Size the lower-floor stack*

The lower-floor stack serves both the upper- and lower-floor branch drains, or a total of 42 + 9 + 54 + 18 = 123 fixture units. Table 3 shows that a 4-in (102-mm) stack will be satisfactory.

10.   *Size the lower-floor vents*

By the one-half rule of step 5, the vent for branch drain 3 must be 2 in (51 mm), whereas that for branch drain 4 must be 1½ in (38 mm).

TABLE 4   Sizes of Building Drains and Sewers°

Diameter of pipe, in (mm)	Maximum number of fixture units that may be connected to any portion† of the building drain or the building sewer			
	Fall per foot (meter)			
	⅟₁₆ in (1.6 mm)	⅛ in (3.2 mm)	¼ in (6.4 mm)	½ in (12.7 mm)
2 (51)	. . .	. . .	21	26
2½ (64)	. . .	. . .	24	31
3 (76)	. . .	20‡	27‡	36‡
4 (102)	. . .	180	216	250
5 (127)	. . .	390	480	575
6 (152)	. . .	700	840	1000

°From *National Plumbing Code.*
†Includes branches of the building drain.
‡Not over two water closets.

11. *Size the building drain*

The building drain serves all the fixtures installed in the building and slopes down toward the city sewer. Hence, the total number of fixture units it serves is 42 + 9 + 54 + 18 = 123. This is the same as the vertical stack. Table 4 shows that a 4-in (102-mm) drain that is sloped ¼-in/ft (21 mm/m) will serve 216 fixture units. Thus, a 4-in (102-mm) drain will be satisfactory. The house trap that is installed in the building drain should also be a 4-in (102-mm) unit.

Related Calculations: Where a local plumbing code exists, use it instead of the *NPC*. If no local code exists, follow the *NPC* for all classes of buildings. Use the general method given here to size the various pipes in the system. Select piping materials (cast iron, copper, clay, steel, brass, wrought iron, lead, etc.) in accordance with the local or *NPC* recommendations. Where the house drain is below the level of the public sewer line, it is often arranged to discharge into a suitably sized *sump pit*. Sewage is discharged from the sump pit to the public sewer by a pneumatic ejector or motor-driven pump.

DESIGN OF ROOF AND YARD RAINWATER DRAINAGE SYSTEMS

An industrial plant is 300 ft (91.4 m) long and 100 ft (30.5 m) wide. The roof of the building is flat except for a 50-ft (15.2-m) long, 100-ft (30.5-m) wide, 80-ft (24.4-m) high machinery room at one end of the roof. Size the leaders and horizontal drains for this roof for a maximum rainfall of 4 in/h (102 mm/h). What size storm drain is needed if the drain is sloped ¼ in/ft (2.1 cm/m) of length?

Calculation Procedure:

1. *Sketch the building roof*

Figure 2 shows the building roof and machinery room roof. Indicate on the sketch the major dimensions of the roof and machinery room.

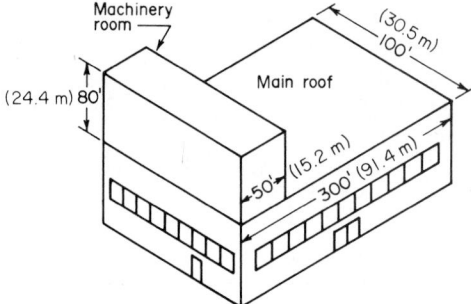

FIG. 2   Building roof areas.

2. *Compute the roof area to be drained*

Two roof areas must be drained, the machinery-room roof and the main roof. The respective areas are: machinery room roof area = 50 × 100 = 5000 ft² (464.5 m²); main roof area = 250 × 100 = 25,000 ft² (2322.5 m²).

The wall of the machinery room facing the main roof will also collect rain to some extent. This must be taken into consideration when the roof leaders are sized. Do this by computing the area of the wall facing the main roof and adding one-half this area to the main roof area. Thus, wall area = 80 × 100 = 8000 ft² (743.2 m²). Adding half this area to the main roof area gives 25,000 + 8000/2 = 29,000 ft² (2694 m²).

3. *Select the leader size for each roof*

Decide whether the small roof area, i.e., the machinery room roof, will be drained by separate leaders to the ground or to the main roof area. If the small roof area is drained separately, treat it as a building unto itself. Where the small roof drains onto the main roof, add the two roof areas to determine the leader size.

By treating the two roofs as separate units, Table 5 shows that a 5-in (127-mm) leader is needed for the 5000-ft² (464.5-m²) machinery room roof. This same table shows that an 8-in (203-mm) leader is needed for the 29,000-ft² (2694-m²) main roof, including the machinery room wall.

4. *Size the storm drain for each roof*

The lower portion of Table 5 shows that a 6-in (152-mm) storm drain is needed for the 5000-ft² (464.5-m²) roof. A 10-in (254-mm) storm drain (Table 5) is needed for the 29,000-ft² (2694-m²) main roof.

When any storm drain is connected to a building sanitary drain or storm sewer, a trap should be used at the inlet to the sanitary drain or storm sewer. The trap prevents sewer gases entering the storm leaders.

Related Calculations: Size roof leaders in strict accordance with the *National Plumbing Code (NPC)* or the local applicable code. Undersized roof leaders are dangerous because they can cause water buildup on a roof, leading to excessive roof loads. Where gutters are used on a building, size them in accordance with Table 6.

Where a roof leader discharges into a sanitary drain, convert the roof area to equivalent fixture units to determine the load on the sanitary drain. To convert roof area to fixture units, take the first 1000 ft² (92.9 m²) of roof area as equivalent to 256 fixture units when designing for a maximum rainfall of 4 in/h (102 mm/h). Where the total roof area exceeds 1000 ft² (92.9 m²), divide

TABLE 5 Sizes of Vertical Leaders and Horizontal Storm Drains°

Vertical leaders	
Size of leader or conductor,† in (mm)	Maximum projected roof area, ft² (m²)
2 (51)	720 (66.9)
2½ (64)	1,300 (120.8)
3 (76)	2,200 (204.4)
4 (102)	4,600 (427.3)
5 (127)	8,650 (803.6)
6 (152)	13,500 (1,254.2)
8 (203)	29,000 (2,694)

Horizontal storm drains			
Diameter of drain, in (mm)	Maximum projected roof area for drains of various slopes, ft² (m²)		
	⅛-in (3.2-mm) slope	¼-in (6.4-mm) slope	½-in (12.7-mm) slope
3 (76)	822 (76.4)	1,160 (107.8)	1,644 (152.7)
4 (102)	1,880 (174.4)	2,650 (246.2)	3,760 (349.3)
5 (127)	3,340 (310.3)	4,720 (438.5)	6,680 (620.6)
6 (152)	5,350 (497.0)	7,550 (701.4)	10,700 (994.0)
8 (203)	11,500 (1,068.4)	16,300 (1,514.3)	23,000 (2,136.7)
10 (254)	20,700 (1,923.0)	29,200 (2,712.7)	41,400 (3,846.1)
12 (305)	33,300 (3,093.6)	47,000 (4,366.3)	66,600 (6,187.1)
15 (381)	59,500 (5,527.6)	84,000 (7,803.6)	119,000 (11,055.1)

°From *National Plumbing Code.*
†The equivalent diameter of square or rectangular leader may be taken as the diameter of that circle that may be inscribed within the cross-sectional area of the leader.

TABLE 6   Size of Gutters°

Diameter of gutter,† in (mm)	Maximum projected roof area for gutters of various slopes, ft² (m²)			
	$\frac{1}{16}$-in (1.6-mm) slope	$\frac{1}{8}$-in (3.2-mm) slope	$\frac{1}{4}$-in (6.4-mm) slope	$\frac{1}{2}$-in (12.7-mm) slope
3 (76)	170 (15.8)	240 (22.3)	340 (31.6)	480 (44.6)
4 (102)	360 (33.4)	510 (47.4)	720 (66.9)	1,020 (94.8)
5 (127)	625 (58.1)	880 (81.8)	1,250 (116.1)	1,770 (164.4)
6 (152)	960 (89.2)	1,360 (126.3)	1,920 (178.4)	2,770 (257.3)
7 (178)	1,380 (128.2)	1,950 (181.2)	2,760 (256.4)	3,900 (362.3)
8 (203)	1,990 (184.9)	2,800 (260.1)	3,980 (369.7)	5,600 (520.2)
10 (254)	3,600 (334.4)	5,100 (473.8)	7,200 (668.9)	10,000 (929.0)

°From *National Plumbing Code*.
†Gutters other than semicircular may be used provided they have an equivalent cross-sectional area.

the remaining roof area by 3.9 ft² (0.36 m²) per fixture unit to determine the fixture load for the remaining area.

Thus, the machinery room roof in the above plant is equivalent to 256 + 4000/3.9 = 1281 fixture units. The main roof and machinery room wall are equivalent to 256 + 28,000/3.9 = 7436 fixture units. These roofs, if taken together, would place a total load of 1281 + 7436 = 8717 fixture units on a sanitary drain.

Where the rainfall differs from 4 in/h (102 mm/h), compute the load on the drain in the same way as described above. Choose the drain size from the appropriate table. Then multiply the drain size by actual maximum rainfall, in (mm)/4. If the drain size obtained is nonstandard, as will often be the case, use the next *larger* standard drain size. Thus, with a 6-in (152-mm) rainfall and a 5-in (127-mm) leader based on the 4-in (102-mm) rainfall tables, leader size = (5)(6/4) = 7.5 in (191 mm). Since this is not a standard size, use the next larger size, or 8 in (203 mm). Roof areas should be drained as quickly as possible to prevent excessive structural stresses caused by water accumulations.

To compute the required size of drains for paved areas, yards, courts, and courtyards, use the same procedure and tables as for roofs. Where the rainfall differs from 4 in (102 mm), apply the conversion ratio discussed in the previous paragraph. Note that the flow capacity of floor and roof drains must equal, or exceed, the flow capacity of the leader to which either unit is connected.

SIZING COLD- AND HOT-WATER-SUPPLY PIPING

An industrial building has the following plumbing fixtures: 2 showers, 200 private lavatories, 200 service sinks, 20 public lavatories, 1 dishwasher, 25 flush-valve water closets, and 20 stall urinals. Size the cold- and hot-water piping for these fixtures, using an upfeed system. The highest fixture is 50 ft (15.2 m) above the water main. The minimum water pressure available in the water main is 60 lb/in² (413.6 kPa); the pressure loss in the water meter is 8.3 lb/in² (57.2 kPa).

Calculation Procedure:

1. *Sketch the proposed piping system*

Draw a single-line diagram of the proposed cold- and hot-water piping. Thus, Fig. 3a shows the proposed basement layout of the water piping, and Fig. 3b shows two of the risers used in this industrial plant. Indicate on each branch line the "weight" in *fixture units* of fixtures served and the required water flow. Table 7 shows the rate of flow and required pressure during flow to different types of fixtures.

2. *Compute the demand weight of the fixtures*

List the fixtures as in Table 8. Next to the name and number of each fixture, list the demand weight for cold or hot water, or both, from Table 9. Note that when a fixture has both a cold-water and hot-water supply, only three-fourths of the fixture weight listed in Table 9 is used for

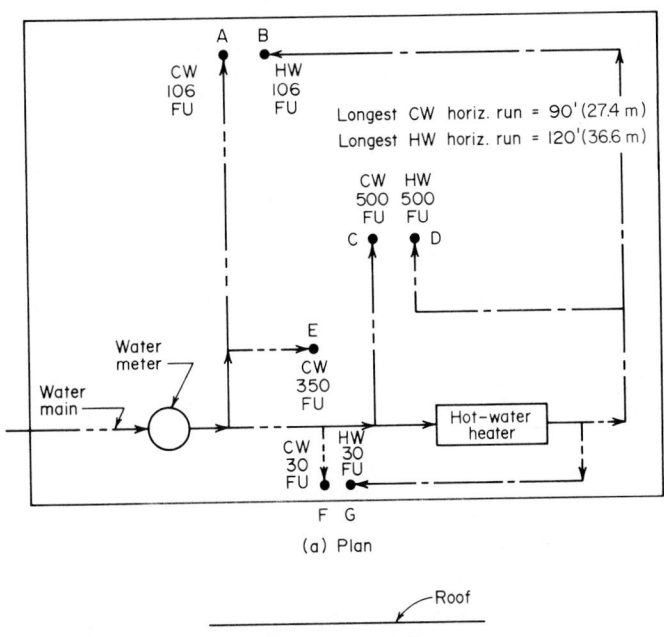

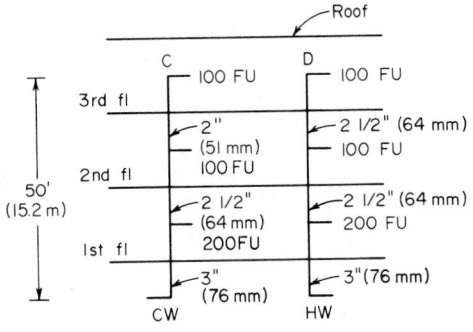

FIG. 3   (a) Plan of industrial-plant water piping; (b) elevation of building water-supply risers.

each cold-water and each hot-water outlet. Thus, with a total demand weight of 1 for a private lavatory, the cold-water demand weight is $0.75(1) = 0.75$ fixture unit, and the hot-water demand weight is $0.75(1) = 0.75$ fixture unit.

Find the product of the number of each type of fixture and the demand weight per fixture for cold and hot water; enter the result in the last two columns of Table 8. The sum of the cold- and hot-water-fixture demand weights, 986 and 636 fixture units, respectively, gives the total demand weight for the building, in fixture units, except for the dishwasher.

3. Compute the building water demand

Using Fig. 4a, enter at the bottom with the number of fixture units and project vertically upward to the curve. At the left read the demand—210 gal/min (13.3 L/s) of cold water and 160 gal/min (10.1 L/s) of hot water, excluding the dishwasher.

Table 9 shows that a dishwasher serving 500 people in an industrial plant requires 250 gal/h (0.26 L/s) with a demand factor of 0.40. This is equivalent to a demand of (demand, gal/h) (demand factor), or $(250)(0.40) = 100$ gal/h (0.11 L/s) or 100 gal/h/(60 min/h) = 1.66 gal/min

TABLE 7 Rate of Flow and Required Pressure during Flow for Different Fixtures°

Fixture	Flow pressure† lb/in²	Flow pressure† kPa	Flow rate gal/min	Flow rate L/s
Ordinary basin faucet	8	55.2	3.0	0.19
Self-closing basin faucet	12	82.7	2.5	0.16
Sink faucet, ⅜ in (9.5 mm)	10	68.9	4.5	0.28
Sink faucet, ½ in (12.7 mm)	5	34.5	4.5	0.28
Bathtub faucet	5	34.5	6.0	0.38
Laundry-tub cock, ½ in (12.7 mm)	5	34.5	5.0	0.32
Shower	12	82.7	5.0	0.32
Ball cock for closet	15	103.4	3.0	0.19
Flush valve for closet	10–20	68.9–137.9	15–40‡	0.95–2.5
Flush valve for urinal	15	103.4	15.0	0.95
Garden hose, 50 ft (15.2 m) and sill cock	30	206.8	5.0	0.32

°From *National Plumbing Code.*
†Flow pressure is the pressure in the pipe at the entrance to the particular fixture considered.
‡Wide range due to variation in design and type of flush-valve closets.

TABLE 8 Fixture Demand Weight

Fixture name	No. of fixtures	Demand weight per fixture in fixture units Cold water	Demand weight per fixture in fixture units Hot water	Total fixture demand weight in fixture units Cold water	Total fixture demand weight in fixture units Hot water
Shower	2	3	3	6	6
Lavatory, private	200	0.75	0.75	150	150
Lavatory, public	20	1.5	1.5	30	30
Sink, service	200	2.25	2.25	450	450
Dishwasher	1	. . . .	°	. . .	. . .
Water closet, flush-valve	25	10	. . .	250	. . .
Urinal, stall	20	5	. . .	100	. . . .
Total	. . .	. . . .	. . .	986	636

°Not given in *National Plumbing Code* tabulation.

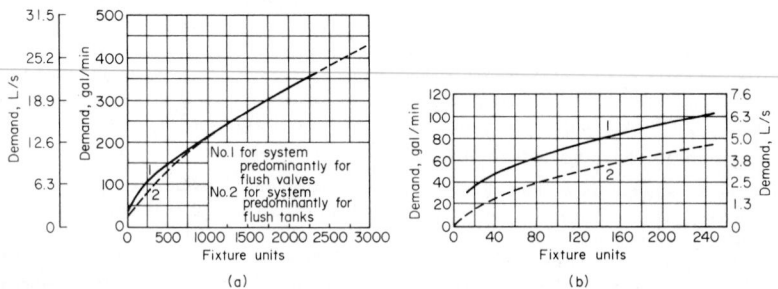

FIG. 4 (*a*) Domestic water demand for various fixtures; (*b*) enlargement of low-demand portion of (*a*).

TABLE 9 Demand Weight of Fixtures in Fixture Units°

Fixture or group†	Occupancy	Type of supply control	Weight in fixture units‡
Water closet	Public	Flush valve	10
Water closet	Public	Flush tank	5
Pedestal urinal	Public	Flush valve	10
Stall or wall urinal	Public	Flush valve	5
Stall or wall urinal	Public	Flush tank	3
Lavatory	Public	Faucet	2
Bathtub	Public	Faucet	4
Shower head	Public	Mixing valve	4
Service sink	Office, etc.	Faucet	3
Kitchen sink	Hotel or restaurant	Faucet	4
Water closet	Private	Flush valve	6
Water closet	Private	Flush tank	3
Lavatory	Private	Faucet	1
Bathtub	Private	Faucet	2
Shower head	Private	Mixing valve	2
Bathroom group	Private	Flush valve for closet	8
Bathroom group	Private	Flush tank for closet	6
Separate shower	Private	Mixing valve	2
Kitchen sink	Private	Faucet	2
Laundry trays (one to three)	Private	Faucet	3
Combination fixture	Private	Faucet	3

°From *National Plumbing Code.* For supply outlets likely to impose continuous demands, estimate continuous supply separately and add to total demand for fixtures.

†For fixtures not listed, weights may be assumed by comparing the fixture to a listed one using water in similar quantities and at similar rates.

‡The given weights are for total demand. For fixtures with both hot- and cold-water supplies, the weights for maximum separate demands may be taken as three-fourths the listed demand for supply.

(0.10 L/s), say 2.0 gal/min (0.13 L/s). Hence, the total hot-water demand is $160 + 2 = 162$ gal/min (10.2 L/s). The total building water demand is therefore $210 + 162 = 372$ gal/min (23.5 L/s).

4. *Compute the allowable piping pressure drop*

The minimum inlet water pressure generally recommended for a plumbing fixture is 8 lb/in^2 (55.2 kPa), although some authorities use a lower limit of 5 lb/in^2 (34.5 kPa). Flushometers normally require an inlet pressure of 15 lb/in^2 (103.4 kPa). Table 7 lists the usual inlet pressure and flow rates required for various plumbing fixtures.

Assume a 15-lb/in^2 (103.4-kPa) inlet pressure at the highest fixture. This fixture is 50 ft (15.2 m) above the water main (Fig. 3). To convert elevation in feet to pressure in pounds per square inch, multiply by 0.434, or (50 ft)(0.434) = 21.7 lb/in^2 (149.6 kPa). Last, the pressure loss in the water meter is 8.3 lb/in^2 (57.2 kPa), as given in the problem statement. Thus, the pressure loss in this or any other water-supply system, not considering piping friction loss, is fixture inlet pressure, lb/in^2, + vertical elevation loss, lb/in^2, + water-meter pressure loss, lb/in^2 = $15 + 21.7 + 8.3$ = 45 lb/in^2 (310.2 kPa). Hence, the pressure available to overcome the piping frictional resistance = $60 - 45 = 15$ lb/in^2 (103.4 kPa).

Note: The pressure loss in water meters of various sizes can be obtained from manufacturers' engineering data, or Fig. 5, for disk-type meters.

5. *Compute the allowable friction loss in the piping*

Figure 3a shows that the longest horizontal run of pipe is $90 + 50 = 140$ ft (42.7 m). Allowing 50 percent of the straight run for the equivalent length of valves and fittings in the longest run and riser gives the total equivalent length of cold-water piping as $140 + 0.50 = 210$ ft (64.0 m).

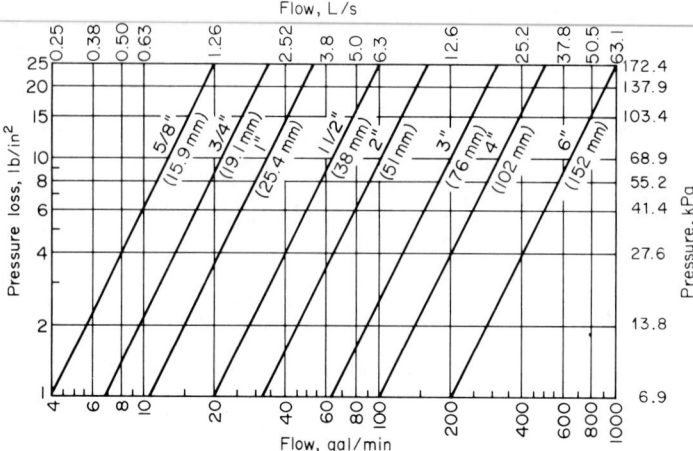

FIG. 5   Pressure loss in disk-type water meters.

Compute the allowable friction loss per 100 ft (30.5 m) of cold-water pipe from $F = 100$ (pressure available to overcome piping frictional resistance, lb/in²)/equivalent length of cold-water piping, ft. Or, $F = 100(15)/210 = 7.14$ lb/in² per 100 ft (1.62 kPa/m); use 7.0 lb/in² per 100 ft (1.58 kPa/m) for design purposes.

By the same procedure for the hot-water pipe, $F = 100(15)/255 = 5.88$ lb/in² per 100 ft (1.33 kPa/m); use 5.75 lb/in² per 100 ft (1.30 kPa/m). Reducing the design pressure loss for the cold- and hot-water piping design pressure loss to the next lower convenient pressure is done only to save time. If desired, the actual computed value can be used. *Never* round off to the next higher convenient pressure loss because this can lead to undersized pipes and reduced flow from the fixture.

6. Size the water main

Step 3 shows that the total building water demand is 372 gal/min (23.5 L/s). Using the cold-water friction loss of 7.0 lb/in² per 100 ft (1.58 kPa/m), enter Fig. 6 at the bottom at 7.0 and project vertically upward to 372 gal/min (23.5 L/s). Read the main size as 4 in (102 mm). This size would be run to the water heater (Fig. 3) unless the run were extremely long. With a long run, the main size would be reduced after each branch takeoff to the risers to reduce the cost of the piping.

7. Compute the water flow in each riser

List the risers in Fig. 3 as shown in Table 10. Next to the letter identifying a riser, list the water it handles (hot or cold), the number of fixture units served by the riser, and the flow. Find the flow by entering Fig. 4 with the number of fixture units served by the riser and projecting up to the flush-valve curve. Read the gallons per minute (liters per second) at the left of Fig. 4.

8. Choose the riser size

Enter the pressure loss, lb/in² per 100 ft (kPa per 30.5 m), found in step 5 next to each riser (Table 10). Using Fig. 4 and the appropriate pressure loss, size each riser and enter the chosen size in Table 10. Thus, riser A conveys 70 gal/min (4.4 L/s) with a pressure loss of 7.0 lb/in² per 100 ft (1.58 kPa/m). Figure 4 shows that a 2-in (50.8-mm) riser is suitable. When Fig. 4 indicates a pipe size that is between two standard pipe sizes, use the next *larger* pipe size.

9. Choose the fixture supply-pipe size

Use Table 11 as a guide for choosing the fixture supply-pipe size. Note that these tabulated sizes are the minimum recommended. Where the supply-pipe run is more than 3 ft (1 m), or where more than one fixture is served, use a larger size.

10. Select the hot-water-heater capacity

Table 12 shows that the demand factor for a hot-water heater in an industrial plant is 0.40 times the hourly hot-water demand. Step 3 shows that the total hot-water demand is 162 gal/min (10.2

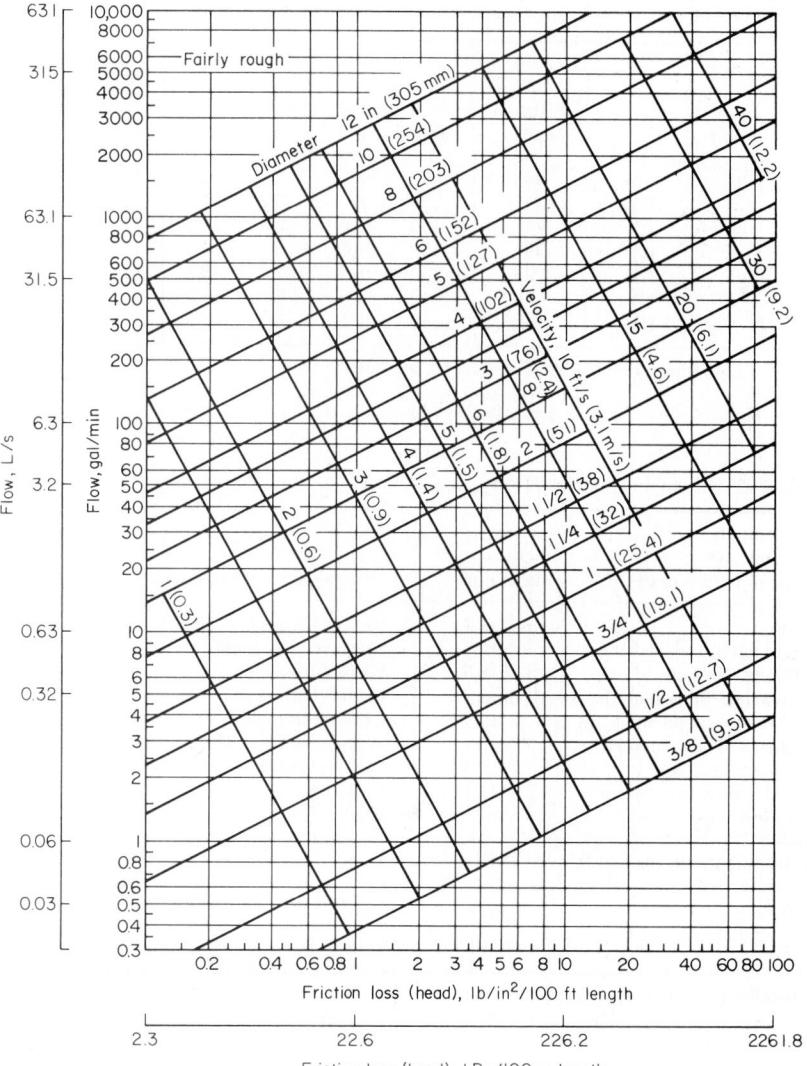

FIG. 6 Chart for selecting water-pipe size for various flow rates.

L/s), or 162(60) = 9720 gal/h (10.2 L/s). Therefore, this hot-water heater must have a heating coil capable of heating at least 0.4(9720) = 3888 gal/h, say 3900 gal/h (4.1 L/s).

The storage capacity should equal the product of hourly water demand and the storage factor from Table 12. Thus, storage capacity = 9720(1.0) = 9720 gal (36,790 L). Table 13 shows the usual hot-water temperature used for various services in different types of structures.

Related Calculations: Size the risers serving each floor, using the same procedure as in steps 6 and 7. Thus, risers C and D are each 3 in (76 mm) up to the first-floor branch. Between this and the second-floor branch, a 2½-in (64-mm) riser is needed. Between the second and third floors, a 2-in (51-mm) cold-water riser and a 2½-in (64-mm) hot-water riser are needed.

In a *downfeed* water-supply system, an elevated roof tank generally supplies cold water to the fixtures. To provide a 15-lb/in² (103.4-kPa) inlet pressure to the highest fixtures, the bottom

TABLE 10 Riser Sizing Calculations

Riser	Type°	Fixture units	Rate gal/min	Rate L/s	Pressure loss lb/in²	Pressure loss kPa	Riser size in	Riser size mm
A	CW	106	70	4.4	7.0	48.3	2	51
B	HW	106	70	4.4	5.75	39.6	2½	64
C	CW	500	150	9.5	7.0	48.3	3	76
D	HW	500	150	9.5	5.75	39.6	3	76
E	CW	350	130	8.2	7.0	48.3	2½	64
F	CW	30	40	2.5	7.0	48.3	2	51
G	HW	30	40	2.5	5.75	39.6	2	51

°CW stands for cold water; HW stands for hot water.

TABLE 11 Minimum Sizes for Fixture-Supply Pipes°

Type of fixture or device	Pipe size in	Pipe size mm	Type of fixture or device	Pipe size in	Pipe size mm
Bathtubs	½	12.7	Shower (single head)	½	12.7
Combination sink and tray	½	12.7	Sinks (service, slop)	½	12.7
Drinking fountain	⅜	9.5	Sinks, flushing rim	¾	19.1
Dishwasher (domestic)	½	12.7	Urinal (flush tank)	½	12.7
Kitchen sink, residential	½	12.7	Urinal (direct flush valve)	¾	19.1
Kitchen sink, commercial	¾	19.1	Water closet (tank type)	⅜	9.5
Lavatory	⅜	9.5	Water closet (flush valve type)	1	25.4
Laundry tray, one, two or three compartments	½	12.7	Hose bibs	½	12.7
			Wall hydrant	½	12.7

°From *National Plumbing Code*.

TABLE 12 Hot-Water Demand per Fixture for Various Building Types°

Type of fixture	Apartment house	Hospital	Hotel	Industrial plant	Office building
Basins, private lavatories	2 (7.6)	2 (7.6)	2 (7.6)	2 (7.6)	2 (7.6)
Basins, public lavatories	4 (15.1)	6 (22.7)	8 (30.3)	12 (45.4)	6 (22.7)
Showers	75 (283.9)	75 (283.9)	75 (283.9)	225 (851.6)	—
Slop sinks	20 (75.7)	20 (75.7)	30 (113.6)	20 (75.7)	15 (56.8)
Dishwashers (per 500 people)	250 (946.3)	250 (946.3)	250 (946.3)	250 (946.3)	250 (946.3)
Pantry sinks	5 (18.9)	10 (37.9)	10 (37.9)	—	—
Demand factor	0.30 (1.1)	0.35 (1.32)	0.25 (0.95)	0.40 (1.51)	0.30 (1.1)
Storage factor	1.25 (4.7)	0.60 (2.27)	0.80 (3.0)	1.00 (3.79)	2.00 (7.57)

°Based on average conditions for types of buildings listed, gallons of water per hour (liters per hour) per fixture at 140°F (60°C).

TABLE 13 Hot-Water Temperatures for Various Services, °F (°C)

Cafeterias (serving areas)	130	(54)
Lavatories and showers	130	(54)
Slop sinks (floor cleaning)	150	(65.6)
Slop sinks (other cleaning)	130	(54)
Cafeteria kitchens	130 + steam	(54 + steam)

of the tank must be $(15 \text{ lb/in}^2)(2.31 \text{ ft}\cdot\text{in}^2/\text{lb of water}) = 34.6$ ft (10.6 m) above the fixture inlet. Where this height cannot be obtained because the building design prohibits it, tank-type fixtures requiring only a 3 lb/in^2 (20.7 kPa) or $(3 \text{ lb/in}^2)(2.31) = 6.93$-ft (2.1-m) elevation at the fixture inlet may be used on the upper floors. Valve-type fixtures are used on the lower floors where the tank elevation provides the required 15-lb/in^2 (103.4-kPa) inlet pressure.

To design a downfeed system: (*a*) Compute the pressure available at the highest fixture resulting from the tank elevation from lb/in^2 = 0.434 (tank elevation above inlet to highest fixture, ft) (9.8 kPa/m). (*b*) Subtract the required inlet pressure to the highest fixture from the pressure obtained in *a*. (*c*) Compute the pressure available to overcome the friction in 100 ft (30.5 m) of piping, using the method of step 5 of the upfeed design procedure and substituting the value found in item *b*. (*d*) Size the main from the tank so it is large enough to provide the needed flow to all the upper- and lower-floor fixtures. (*e*) Note that the pressure in each supply main increases as the distance from the tank bottom becomes greater. Thus, the hydraulic pressure increases 0.43 lb/in$^2\cdot$ft) (9.8 kPa/m) of distance from the tank bottom. Usual design practice allows a 15-lb/in^2 (103.4-kPa) drop through the fittings and valves in the main. The remaining pressure produced by the tank elevation is then available for overcoming pipe friction.

Note that both cold and hot water can be supplied from separate overhead tanks. However, hot water is usually supplied from the building basement by a pump. In exceptionally high buildings, water tanks may be located on several intermediate floors as well as the roof. Hot-water heaters may also be located on intermediate floors, although the usual location is in the basement.

In a *zoned* system, one water tank and one set of hot-water heaters serve several floors or one or more wings of a building. The piping in each zone is designed as described above, using the appropriate method for an upfeed or downfeed system.

To provide hot water as soon as possible after a fixture is opened, the water may be continuously recirculated to the fixtures (Fig. 7). Recirculation is used with both upfeed and downfeed systems. To determine the required hot-water temperature in a system, use Table 13, which shows the usual hot-water temperatures used for various services in buildings of different types. Hot-water piping is generally insulated to reduce heat losses.

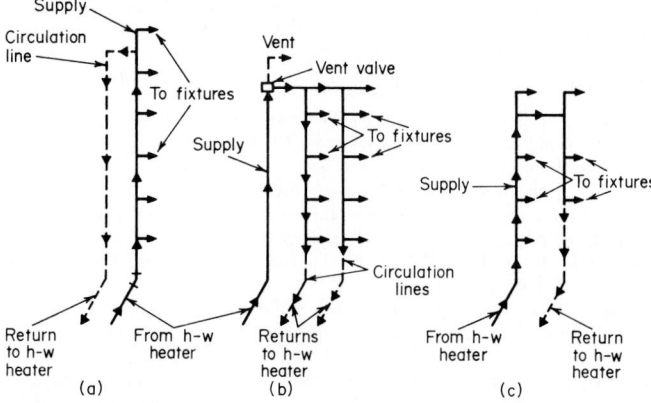

FIG. 7 Hot-water piping systems.

SPRINKLER-SYSTEM SELECTION AND DESIGN

Select and design a sprinkler system for the warehouse building shown in Fig. 8. The materials stored in this warehouse are not flammable. The warehouse is built of fire-resistive materials.

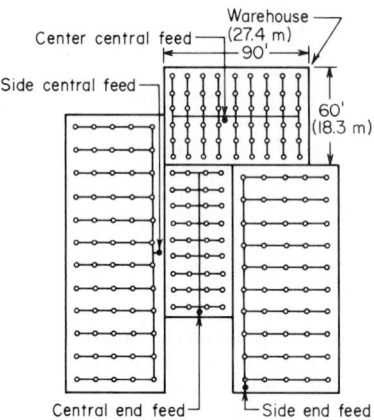

FIG. 8 Typical arrangement of sprinkler piping.

Calculation Procedure:

1. *Determine the type of occupancy of the building*

The classifications of occupancy used by the National Board of Fire Underwriters (NBFU) are (1) light hazard, such as apartment houses, asylums, clubhouses, colleges, churches, dormitories, hospitals, hotels, libraries, museums, office buildings, and schools; (2) ordinary hazard, such as mercantile buildings, warehouses, manufacturing plants, and occupancies not classed as light or extra hazardous; (3) extra-hazard occupancies, i.e., those buildings or portions of buildings where the inspection agency having jurisdiction determines that the hazard is severe.

Since this is a warehouse used to store non-flammable materials, it can be tentatively classed as an ordinary-hazard occupancy.

2. *Compute the number of sprinkler heads required*

Consult the local fire-prevention code and the fire underwriters regarding the type, size, and materials required for sprinkler systems. Typical codes recommend that each sprinkler head in an ordinary-hazard fire-resistive building protect not more than 100 ft² (9.29 m²) and that the sprinkler branch pipes, and the sprinklers themselves, be not more than 12 ft (3.7 m) apart, center to center.

The area of the warehouse floor is 90(60) = 5400 ft² (501.7 m²). With each sprinkler protecting 100 ft² (9.29 m²) of area, the number of sprinkler heads required is 5400 ft²/100 ft² (501.7 m²/9.29 m²) per sprinkler = 54 heads.

3. *Sketch the sprinkler layout*

If the warehouse has a centrally located support column or piping cluster, a center central feed pipe (Fig. 8) can be used. Assuming the sprinkler branch pipes are spaced on 10-ft (3.05-m) centers, sketch the branches and heads as shown in Fig. 8. Use a small circle to indicate each sprinkler head.

Space the end sprinkler heads and branch pipes away from the walls by an amount equal to one-half the center-to-center distance between branch pipes. Thus, the end sprinkler heads and branch pipes will be 10/2 = 5 ft (1.5 m) from the walls.

4. *Size the branch and main sprinkler pipes*

Use the local code, Fig. 9, or Table 14. Table 14 shows that a 1¼-in (32-mm) branch line will be suitable for three sprinklers in an ordinary-hazard occupancy such as this warehouse. Hence, each branch line having three sprinklers will be this size.

The horizontal overhead main supplying the branches will progressively decrease in diameter as it runs farther from the vertical center central feed and serves fewer sprinklers. To the right of the vertical feed, the horizontal main serves 30 sprinklers. Table 14 shows that a 3-in (76-mm) pipe can serve up to 40 sprinklers. Hence, this size will be used because the next smaller size, 2½ in (64 mm), can serve only 20 sprinklers.

Since the first branch has six sprinklers, a 3-in (76-mm) pipe is still needed for the main because Table 14 shows that a 2½-in (64-mm) pipe can serve only 20, or fewer, sprinklers. However, beyond the second branch, the diameter of the horizontal main can be reduced to 2½ in (64 mm) because the number of sprinklers served is 30 − 12 = 18. Beyond the fourth branch, the main size can be reduced to 2 in (51 mm) because only six sprinklers are served. Size the left-hand horizontal main in the same way.

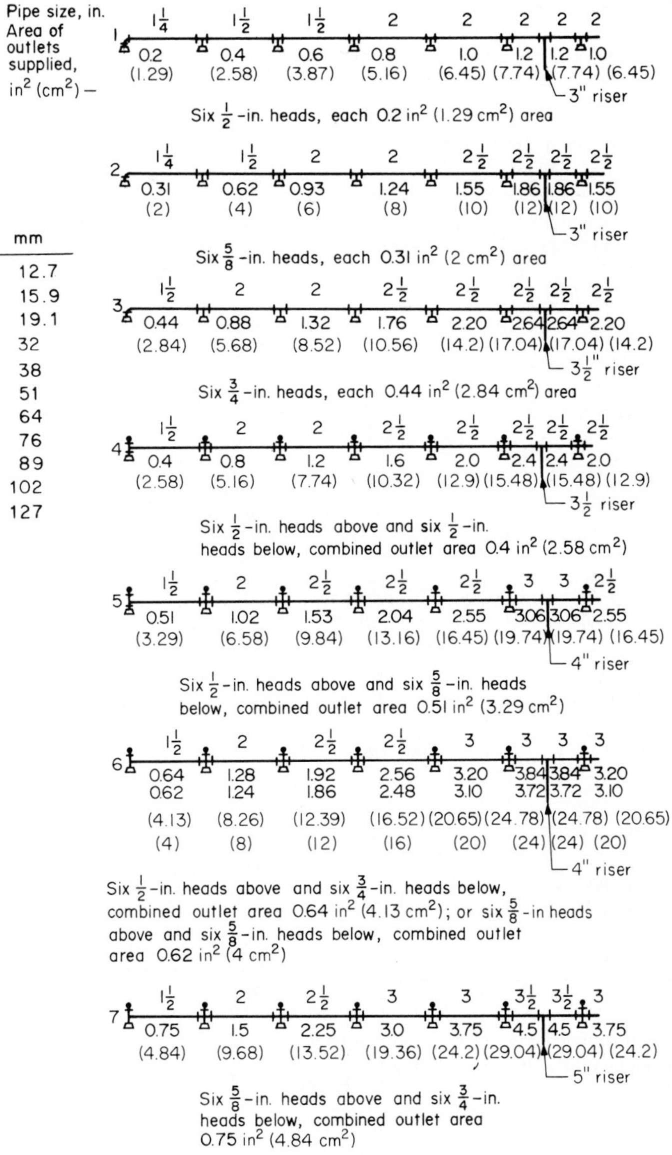

Pipe size, in.
Area of outlets supplied, in² (cm²) —

Row 1:
Pipe size, in.: 1¼ 1½ 1½ 2 2 2 2 2
0.2 0.4 0.6 0.8 1.0 1.2 1.2 1.0
(1.29) (2.58) (3.87) (5.16) (6.45) (7.74) (7.74) (6.45)
— 3″ riser
Six ½-in. heads, each 0.2 in² (1.29 cm²) area

Row 2:
1¼ 1½ 2 2 2½ 2½ 2½ 2½
0.31 0.62 0.93 1.24 1.55 1.86 1.86 1.55
(2) (4) (6) (8) (10) (12) (12) (10)
— 3″ riser
Six ⅝-in. heads, each 0.31 in² (2 cm²) area

Row 3:
1½ 2 2 2½ 2½ 2½ 2½ 2½
0.44 0.88 1.32 1.76 2.20 2.64 2.64 2.20
(2.84) (5.68) (8.52) (10.56) (14.2) (17.04) (17.04) (14.2)
— 3½″ riser
Six ¾-in. heads, each 0.44 in² (2.84 cm²) area

Row 4:
1½ 2 2 2½ 2½ 2½ 2½ 2½
0.4 0.8 1.2 1.6 2.0 2.4 2.4 2.0
(2.58) (5.16) (7.74) (10.32) (12.9) (15.48) (15.48) (12.9)
— 3½″ riser
Six ½-in. heads above and six ½-in. heads below, combined outlet area 0.4 in² (2.58 cm²)

Row 5:
1½ 2 2½ 2½ 2½ 3 3 2½
0.51 1.02 1.53 2.04 2.55 3.06 3.06 2.55
(3.29) (6.58) (9.84) (13.16) (16.45) (19.74) (19.74) (16.45)
— 4″ riser
Six ½-in. heads above and six ⅝-in. heads below, combined outlet area 0.51 in² (3.29 cm²)

Row 6:
1½ 2 2½ 2½ 3 3 3 3
0.64 1.28 1.92 2.56 3.20 3.84 3.84 3.20
0.62 1.24 1.86 2.48 3.10 3.72 3.72 3.10
(4.13) (8.26) (12.39) (16.52) (20.65) (24.78) (24.78) (20.65)
(4) (8) (12) (16) (20) (24) (24) (20)
— 4″ riser
Six ½-in. heads above and six ¾-in. heads below, combined outlet area 0.64 in² (4.13 cm²); or six ⅝-in heads above and six ⅝-in. heads below, combined outlet area 0.62 in² (4 cm²)

Row 7:
1½ 2 2½ 3 3 3½ 3½ 3
0.75 1.5 2.25 3.0 3.75 4.5 4.5 3.75
(4.84) (9.68) (13.52) (19.36) (24.2) (29.04) (29.04) (24.2)
— 5″ riser
Six ⅝-in. heads above and six ¾-in. heads below, combined outlet area 0.75 in² (4.84 cm²)

FIG. 9 Sprinkler pipe sizes.

in	mm
1/2	12.7
5/8	15.9
3/4	19.1
1-1/4	32
1-1/2	38
2	51
2-1/2	64
3	76
3-1/2	89
4	102
5	127

The vertical center central feed pipe serves 54 sprinklers. Hence, a 3½-in (89-mm) pipe must be used, according to Table 14.

5. *Choose the primary and secondary water supply*

Usual codes require that each sprinkler system have two water supplies. The primary supply should be automatic and must have sufficient capacity and pressure to serve the system. Local codes usually specify the minimum pressure and capacity acceptable for sprinklers serving various occupancies.

The secondary supply is often a motor-driven, automatically controlled fire pump supplied from a water main or taking its suction under pressure from a storage system having sufficient capacity to meet the water requirements of the structure protected.

For light-hazard occupancy, the pump should have a capacity of at least 250 gal/min (15.8 L/s); when the pump supplies both sprinklers and hydrants, the capacity should be at least 500 gal/min (31.5 L/s). Where the occupancy is classed as an ordinary hazard, as this warehouse is, the capacity of the pump should be at least 500 gal/min (31.6 L/s) or 750 gal/min (47.3 L/s), depending on whether hydrants are supplied in addition to sprinklers. For extra-hazard occupancy, consult the underwriter and local fire-protection authorities.

Related Calculations: For fire-resistive construction and light-hazard occupancy, the area protected by each sprinkler should not exceed 196 ft^2 (18.2 m^2), and the center-to-center distance of the sprinkler pipes and sprinklers themselves should not exceed 14 ft (4.3 m). For extra-hazard occupancy, the area protected by each sprinkler should not exceed 90 ft^2 (8.4 m^2); the distance between pipes and between sprinklers should not be more than 10 ft (3.05 m). Local fire-protection codes and underwriters' requirements cover other types of construction, including mill, semimill, open-joist, and joist-type with a sheathed or plastered ceiling.

For protection of structures against exposure to fires, outside sprinklers may be used. They can be arranged to protect cornices, windows, side walls, ridge poles, mansard roofs, etc. They are also governed by underwriters' requirements. Figure 9 shows the pipe sizes used for sprinklers protecting outside areas of buildings, including cornices, windows, side walls, etc.

TABLE 14 Pipe-Size Schedule for Typical Sprinkler Installations

Occupancy and pipe size, in (mm)	No. of sprinklers
Light hazard	
1 (25)	2
1¼ (32)	3
1½ (38)	5
2 (51)	10
2½ (64)	40
3 (76)	No limit
Ordinary hazard	
1 (25)	2
1¼ (32)	3
1½ (38)	5
2 (51)	10
2½ (64)	20
3 (76)	40
3½ (89)	65
4 (102)	100
5 (127)	160
6 (152)	250
Extra hazard	
1 (25)	1
1¼ (32)	2
1½ (38)	5
2 (51)	8
2½ (64)	15
3 (76)	27
3½ (89)	40
4 (102)	55
5 (127)	90
6 (152)	150

Four common types of automatic sprinkler systems are in use today: wet pipe, dry pipe, preaction, and deluge. The type of system used depends on a number of factors, including occupancy classification, local code requirements, and the requirements of the building fire underwriters. Since the requirements vary from one area to another, no attempt is made here to list those of each locality or underwriter. The Standards of the National Board of Fire Underwriters, as recommended by the National Fire Protection Association, are excerpted instead because they are so widely used that they are applicable for the majority of buildings. In general, the type of sprinkler chosen does not change the design procedure given above. Figure 10 shows a typical layout of the water-supply piping for an industrial-plant sprinkler system. Figure 11 shows how sprinklers are positioned with respect to a building ceiling.

Use the same general design procedure presented here for sprinklers in other types of buildings—hotels, office buildings, schools, churches, dormitories, colleges, museums, libraries, clubhouses, hospitals, and asylums.

Note: Do not finalize a sprinkler system design until after it is approved by local fire authorities and the fire underwriters insuring the building.

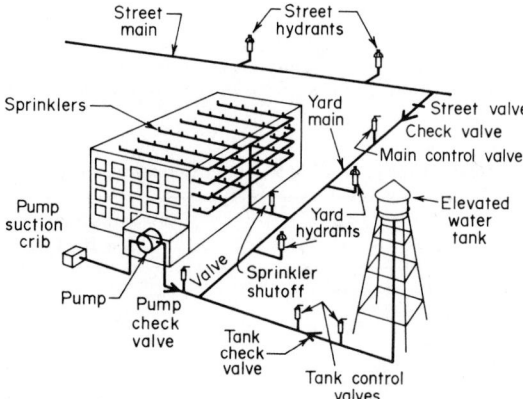

FIG. 10 Water-supply piping for sprinklers.

SIZING GAS PIPING FOR HEATING AND COOKING

An industrial building has two 8-gal/min (0.5-L/s) water heaters and ten ranges, each of which has four top burners and one oven burner. What maximum gas consumption must be provided for if carbureted water gas is used as the fuel? Determine the pressure in the longest run of gas pipe in this building if the total equivalent length of pipe in the longest run is 150 ft (45.7 m) and the specific gravity of the gas is 0.60 relative to air. What would the pressure loss of a 0.35-gravity gas be?

Calculation Procedure:

1. Compute the heat input to the appliances

Table 15 lists the typical heat input to various gas-burning appliances. By using the tabulated data for the 8-gal/min (0.5-L/s) water heaters and the four-burner stove, the maximum heat input = 2(300,000) + 10(62,500) = 1,225,000 Btu/h (359.0 kW). The gas-supply pipe must handle sufficient gas to supply this heat input because all burners might be operated simultaneously.

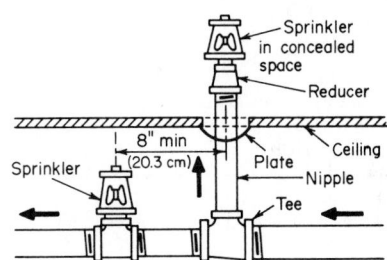

FIG. 11 Sprinkler positioning with respect to a building ceiling.

TABLE 15 Heat Input to Common Appliances

Unit	Approximate input, Btu/h (kW)
Water heater, side-arm or circulating type	25,000 (7.32)
Water heater, automatic instantaneous:	
4 gal/min (0.25 L/s)	150,000 (43.9)
6 gal/min (0.38 L/s)	225,000 (65.9)
8 gal/min (0.5 L/s)	300,000 (87.9)
Refrigerator	2,500 (0.73)
Ranges, domestic:	
Four top burners, one oven burner	62,500 (18.3)
Four top burners, two oven burners	82,500 (24.2)

2. *Compute the required gas-flow rate*

Table 16 shows that the heating value of carbureted water gas is 508 Btu/ft^3 (18,928 kJ/m^3). Using a value of 500 Btu/ft^3 (18,629 kJ/m^3) to provide a modest safety factor, we find gas flow required, ft^3 = maximum heat input required, Btu/h/fuel heating value, Btu/ft^3 = 1,225,000/500 = 2450 ft^3/h (69.4 m^3/h).

TABLE 16   Typical Heating Values of Commercial Gases

Gas	Net heating value, Btu/ft^3 (kJ/m^3)
Natural gas (Los Angeles)	971 (36,178.4)
Natural gas (Pittsburgh)	1,021 (38,041.3)
Coke-oven gas	514 (19,151.1)
Carbureted water gas	508 (18,927.5)
Commercial propane	2,371 (88,340.9)
Commercial butane	2,977 (110,919.8)

3. *Compute the pressure loss in the gas pipe*

The longest equivalent run is 150 ft (45.7 m). Gas flows through the pipe at the rate of 2450 ft^3/h (69.4 m^3/h). Enter Table 17 at this flow rate, or at the next *larger* tabulated flow rate. and project horizontally to the first pressure drop listed, or 3.5 in/100 ft (89 mm/30 m) in a 2-in (51-mm) pipe.

The pressure loss listed in Table 17 is for 100 ft (30.5 m) of pipe if the gas has a specific gravity of 0.6 in relation to air. To find the pressure loss in 150 ft (45.7 m) of pipe, use the relation actual pressure loss, in of water = (tabulated pressure loss, in per 100 ft)(actual pipe length, ft/tabulated pipe length, ft) = 3.5(150/100) = 5.25 in (13.3 mm) of water. Since the actual flow rate is less than 3000 ft^3/h (84.9 m^3/h), the actual pressure drop will be less than computed.

TABLE 17   Capacities of Gas Pipes [Losses of pressure are shown in inches of water per 100 ft (millimeters per 30.4 m) of pipe, due to the flow of gas with a specific gravity of 0.6 with respect to air]°

Rate of flow, ft^3/h (m^3/h)	Size of pipe, in (mm)			
	1¼ (32)	1½ (38)	2 (51)	2½ (64)
1000 (28.3)	. . . . . .	1.6 (40.6)	0.80 (20.3)	0.15 (3.8)
1500 (42.5)	8.8 (223.5)	3.6 (91.4)	0.86 (21.8)	0.33 (8.4)
2000 (56.6)	. . . . . .	6.3 (160.0)	1.50 (38.1)	0.60 (15.2)
3000 (84.9)	. . . . . .	. . . . . .	3.5 (88.9)	0.94 (23.9)
5000 (141.5)	. . . . . .	. . . . . .	8.8 (223.5)	3.6 (91.4)

Rate of flow, ft^3/h (m^3/h)	Size of pipe, in (mm)			
	3 (76)	3½ (89)	4 (102)	5 (127)
1000 (28.3)	0.05 (1.3)	. . . . . .	. . . . . .	. . . . . .
1500 (42.5)	0.11 (2.8)	0.05 (1.3)	. . . . . .	. . . . . .
2000 (56.6)	0.18 (4.6)	0.08 (2.0)	0.05 (1.3)	. . . . . .
3000 (84.9)	0.44 (11.2)	0.19 (4.8)	0.10 (2.5)	. . . . . .
5000 (141.5)	1.2 (30.5)	0.54 (13.7)	0.34 (8.6)	0.08 (2.0)

°To determine head losses for other lengths of pipe, multiply the head losses in this table by the length of the pipe and divide by 100 ft. (30.4 m). For head losses due to flow of gases with specific gravity other than 0.6, use the figures given in Table 18.

4. *Compute the pressure loss of the lighter gas*

To correct for a gas of a different specific gravity, multiply the actual gas flow by the appropriate factor from Table 18. Thus, equivalent flow rate for this plant when a gas of 0.5 gravity is flowing is 2450(0.77) = 1882 ft³/h (53.5 m³/h); say 2000 ft³/h (56.6 m³/h).

Entering Table 17 shows that a flow of 2000 ft³/h (56.6 m³/h) will have a pressure loss of 6.3 in (160 mm) of water in 100 ft (30.5 m) of 1½-in (38-mm) pipe. In 150 ft (45.7 m) of 1½-in (38-mm) pipe, the pressure loss will be 6.3(150/100) = 9.45 in (240 mm) of water. Increasing the pipe size to 2 in (51 mm) would reduce the pressure loss to 2.25 in (57.2 mm) of water.

Related Calculations: When gas flows upward in a vertical pipe to serve upper floors, there is a *gain* in the gas pressure if the gas is lighter than air. Table 19 shows the gain in gas pressure per 100 ft (30.5 m) of rise in a vertical pipe for gases of various specific gravities. This pressure gain must be recognized when piping systems are designed.

As with piping and fixtures for plumbing systems, gas piping and fixtures are subject to code regulations in most cities and towns. Natural and manufactured gases are widely used in stoves, water heaters, and space heaters of many designs. Since gas can form explosive mixtures when mixed with air, gas piping must be absolutely tight and free of leaks at all times. Usual codes cover every phase of gas-piping size, installation, and testing. The local code governing a particular building should be carefully followed during design and installation.

For gas supply, the usual practice is for the public-service gas company to run its pipes into the building cellar, terminating with a brass shutoff valve and gas meter inside the cellar wall. From this point, the plumbing contractor or gas-pipe fitter runs lines through the building to the various fixture outlets. When the pressure of the gas supplied by the public-service company is too high for the devices in the building, a pressure-reducing valve can be installed near the point where the line enters the building. The valve is usually supplied by the gas company.

Besides municipal codes governing the design and installation of gas piping and devices, the gas company serving the area will usually have a number of regulations that must be followed. In general, gas piping should be run in such a manner that it is unnecessary to locate the meter near a boiler, under a window or steps, or in any other area where it may be easily damaged. Where multiple-meter installations are used, the piping must be plainly marked by means of a metal tag showing which part of the building is served by the particular pipe. When two or more meters are used in a building to supply separate consumers, there should be no interconnection on the outlet side of the meters.

Materials used for gas piping include black iron, steel, and wrought iron. Copper tubing is also finding some use, and the values listed in Table 17 apply to it as well as schedule 40 (standard weight) pipe made of the materials listed above. Use the procedure given here to size gas pipes for industrial, commercial, and residential installations.

TABLE 18 Factors by Which Flows in Table 17 Must be Multiplied for Gases of Other Specific Gravity°

Specific gravity of gas	0.35	0.40	0.45	0.50	0.55	0.60	0.65	0.70
Factor	0.77	0.82	0.87	0.91	0.96	1.00	1.04	1.08

TABLE 19 Changes in Pressure in Gas Pipes

Gain in pressure per 100 ft (30.4 m) of rise in vertical pipe, in of water (mm of water)	0.96 (24.4)	0.89 (22.6)	0.81 (20.6)	0.74 (18.8)	0.66 (16.8)	0.59 (15.0)	0.52 (13.2)	0.44 (11.2)
Specific gravity of gas compared to air	0.35	0.40	0.45	0.50	0.55	0.60	0.65	0.70

SWIMMING POOL SELECTION, SIZING, AND SERVICING

Choose a swimming pool to serve 140 bathers with facilities for diving and swimming contests. Size the pumps for the pool. Select a suitable water-treatment system and the inlet and outlet pipe sizes. Determine the size of heater required for the pool, should heating of the water be required.

Calculation Procedure:

1. Compute the swimming-pool area required

Usual swimming pools are sized in accordance with the recommendations of the Joint Committee on Swimming Pools, which uses 25 ft² (2.3 m²) per bather as a desirable pool area. With 140 bathers, the recommended area = 25(140) = 3500 ft² (325.2 m²).

2. Choose the pool dimensions

Use Table 20 as a guide to usual pool dimensions. This tabulation shows that a 105-ft (32-m) long by 35-ft (10.7-m) wide pool is suitable for 147 bathers. Since the next smaller pool will handle only 108 bathers, the larger pool must be used.

To provide for swimming contests, lanes at least 7 ft (2.1 m) wide are required. Thus, this pool could have 35 ft/7 ft = 5 lanes for swimming contests. If more lanes are desired, the pool width must be increased, if there is sufficient space. Also, consideration of the pool length is required if swimming meets covering a specified distance are required. Assume that the 105 × 35 ft (32 × 10.7 m) pool chosen earlier is suitable with respect to contests and space.

To provide for diving contests, a depth of more than 9 ft (2.7 m) is recommended at the deep end of the pool. Table 20 shows that this pool has an actual maximum depth of 10 ft (3.05 m), which makes it better suited for diving contests. Some swimming specialists recommend a depth of at least 10 ft (3.05 m) for diving contests. Assume, therefore, that 10 ft (3.05 m) is acceptable for this pool.

The pool will have a capacity of 155,600 gal (588,946 L) of water (Table 20). If installed indoors, the pool would probably be faced with tile or glazed brick. An outdoor pool of this size is usually constructed of concrete, and the walls are a smooth finish.

3. Determine the pump capacity required

To keep the pool water as pure as possible, three *turnovers* (i.e., the number of times the water in the pool is changed each day) are generally used. This means that the water will be changed once each 24 h/3 changes = 8 h. The water is changed by recirculating it through filters, a chlorinator, strainer, and heater. Thus, the pump must handle water at the rate of pool capacity, gal/8 h = 155,600/8 = 19,450 gal/h (73,626 L/h), or 19,450/60 min/h = 324.1 gal/min (1266.9 L/min); say 325 gal/min (1230.3 L/min).

4. Choose the pump discharge head

Motor-driven centrifugal pumps find almost universal application for swimming pools. Reciprocating pumps are seldom suitable because they produce pulsations in the delivery pipe and pool filters. Either single- or double-suction single-stage centrifugal pumps can be used. The double-suction design is usually preferred because the balanced impeller causes less wear.

The discharge head that a swimming pool circulating pump must develop is a function of the resistance of the piping, fittings, heater, and filters. Of these four, the heater and filters produce the largest head loss.

The usual swimming pool heater causes a head loss of up to 10 ft (3.05 m) of water. Sand filters cause a head loss of about 50 ft (15.2 m) of water, whereas diatomaceous earth filters cause a head loss of about 90 ft (27.4 m) of water. To choose the pump discharge head, find the sum of the pump suction lift, piping and fitting head loss, and heater and filter head loss. Add a 10 percent allowance for overload. The result is the required pump discharge head in feet of water.

Most pools are equipped with two identical circulating pumps. The spare pump ensures constant operation of the pool should one pump fail. Also, the spare pump permits regular maintenance of the other pump.

5. Compute the quantity of makeup water required

Swimmers splash water over the gutter line of the pool. This water is drained away to the sewer in some pools; in others the water is treated and returned to the pool for reuse. Gutter drains are usually spaced at 15-ft (4.6-m) intervals.

Since the pool waterline is level with the gutter, every swimmer who enters the pool displaces

TABLE 20 Dimensions of Official Swimming Pools

Pool capacity, gal (L)	Bathing load, persons	Bathing capacity per day[a]	A	B	C	D	X	Y	Z	L	W
			Dimensions, ft (m)								
55,000 (208,197)	48	418	8 (2.4)	9 (2.7)	5 (1.5)	3.25 (1.0)	15 (4.6)	20 (6.1)	25 (7.6)	60 (18.3)	20 (6.1)
80,800 (305,860)	75	607	8 (2.4)	9.5 (2.9)	5 (1.5)	3.25 (1.0)	15 (4.6)	20 (6.1)	40 (12.2)	75 (22.9)	25 (7.6)
120,000 (454,248)	108	900	8 (2.4)	9.5 (2.9)	5 (1.5)	3.25 (1.0)	18 (5.5)	25 (7.6)	47 (14.3)	90 (27.4)	30 (9.1)
155,600 (589,008)	147	1,170	8 (2.4)	10 (3.0)	5 (1.5)	3.25 (1.0)	18 (5.5)	25 (7.6)	62 (18.9)	105 (32.0)	35 (10.7)
207,600 (785,849)	192	1,555	8 (2.4)	10 (3.0)	5 (1.5)	3.25 (1.0)	20 (6.1)	30 (9.1)	70 (21.3)	120 (36.6)	40 (12.2)
254,000 (961,492)	243	1,905	8 (2.4)	10 (3.0)	5 (1.5)	3.25 (1.0)	20 (6.1)	30 (9.1)	85 (25.9)	135 (41.2)	45 (13.7)
306,000 (1,158,332)	300	2,300	8 (2.4)	10 (3.0)	5 (1.5)	3.25 (1.0)	20 (6.1)	30 (9.1)	100 (30.5)	150 (45.7)	50 (15.2)
422,400 (1,598,953)	432	3,170	8 (2.4)	10 (3.0)	5 (1.5)	3.25 (1.0)	20 (6.1)	30 (9.1)	130 (39.6)	180 (54.9)	60 (18.3)
558,000 (2,112,253)	590	4,180	8 (2.4)	10 (3.0)	5 (1.5)	3.25 (1.0)	20 (6.1)	30 (9.1)	160 (48.8)	210 (64.0)	70 (21.3)

[a] Based on 8-h turnover.

some water, which enters the gutter and is drained away. This drainage must be made up by the pool recirculating system.

The water displaced by a swimmer is approximately equal to his or her weight. Assuming each swimmer weighs 160 lb (72.7 kg) this weight of water will be displaced into the gutter. Since 1 gal (3.79 L) of water weighs 8.33 lb (3.75 kg), each swimmer will displace 160 lb/(8.33 lb/gal) = 19.2 gal (72.7 L). With a maximum of 140 swimmers in the pool, the total quantity of water displaced is (140)(19.2) = 2695 gal/h (10,201.7 L/h), or 2695/(60 min/h) = 44.9 gal/min (169.9 L/min), say 45 gal/min (170.3 L/min).

Thus, to keep this pool operating, the water-supply system must be capable of delivering at least 45 gal/min (170.3 L/min). This quantity of water can come from a city water system, a well, or recirculation of the gutter water after purification.

6. Compute the required filter-bed area

Two types of filters are used in swimming pools: sand and diatomaceous earth. In either type, a flow rate of 2 to 4 gal/(min·ft^2) [81.5 to 162.5 L/(min·m^2)] is generally used. The lower flow rates, 2 to 2.5 gal/(min·ft^2) [81.5 to 101 L/(min·m^2)] are usually preferred. Assuming that a flow rate of 2.5 gal/(min·ft^2) [101 L/(min·m^2)] is used and that 325 gal/min (1230.3 L/min) flows through the filters, as computed in step 3, the filter-bed area required is 325 gal/min/[2.5 gal/(min·ft^2)] = 130 ft^2 (12.1 m^2).

Two filters are generally used in swimming pools to ensure continuity of service and back-washing of one filter while the other is in use. The required area of 130 ft^2 (12.1 m^2) could then be divided between the two filters. Some pools use three or more filters. Regardless of how many filters are used, the required area can be evenly divided among them.

7. Choose the number of water inlets and outlets for the pool

The pool inlets supply the recirculation water required. Usual practice rates each inlet for a flow of 10 to 20 gal/min (37.9 to 75.7 L/min). Given a 10-gal/min (37.9-L/min) flow for this pool, the number of inlets required is 325 gal/min/10 gal/min per inlet = 32.5, say 32.

Locate the inlets around the periphery of the pool and on each end. Space the inlets so that they provide an even distribution of the water. In general, inlets should not be located more than 30 ft (10 m) apart.

Size the pool drain to release the water in the pool within the desired time interval, usually 4 to 12 h. Since there is no harm in emptying a pool quickly—if the sewer into which the pool discharges has sufficient capacity—size the discharge line liberally. Thus, a 12-in (305-mm) discharge line can handle about 2000 gal/min (7570.8 L/min) when a swimming pool is drained.

8. Compute the quantity of disinfectant required

Chlorine, bromine, and ozone are some of the disinfectants used in swimming pools. Chlorine is probably the most popular. It is used in quantities sufficient to maintain 0.5 ppm chlorine in the water. Since this pool contains 155,600 gal (589,008 L) of water that is recirculated three times per day, the quantity of chlorine disinfectant that must be added each day is (155,600 gal)(3 changes per day)(8.33 lb/gal) [0.5 lb of chlorine per 10^6 lb (454,545.5 kg) of water] = 1.95 lb (0.89 kg) of chlorine per day. The required chlorine can be pumped into the pool inlet water or fed from cylinders.

9. Size the water heater for the pool

The usual swimming pool heater has a heating capacity, in gallons per hour, which is 10 times the gallons-per-minute rating of the circulating pump. Since the circulating pump for this pool is rated at 325 gal/min (1230.6 L/min), the heater should have a capacity of 10(325) = 3250 gal/h (12,306.6 L/h).

Since the entering water temperature may be as low as 40°F (4.4°C) in the winter, the heater should be chosen for this entering temperature. The outlet temperature of the water should be at least 80°F (26.7°C). To heat the entire contents of the pool from 40°F (4.4°C) to about 70°F (21°C), at least 48 h is generally allowed. Instantaneous hot-water heaters are usually chosen for swimming pool service.

10. Select the backwash sump pump

When the filter backwash flow cannot be discharged directly to a sewer, the usual practice is to pipe the backwash to a sump in the pool machinery room. The accumulated backwash is then pumped to the sewer by a sump pump mounted in the sump.

The sump should be large enough to store sufficient backwash to prevent overflowing. Assuming that either of the 130-ft² (12.1-m²) filter beds is backwashed with a flow of 12.5 gal/(min·ft²) [509.3 L/(min·m²)] of filter-bed area, the quantity of water entering the sump will be (12.5)(130) = 1725 gal/min. If there is room for a 5-ft deep, 8-ft wide, and 5-ft long (1.5-m deep, 2.4-m wide, and 1.5-m long) sump, its capacity will be 5 × 8 × 5 × 7.5 gal/ft³ = 1500 gal (5678 L). The difference, of 1625 − 1500 = 125 gal/min (473.2 L/min), must be discharged by the pump to prevent overflow of the sump. A 150-gal/min (567.8-L/min) sump pump should probably be chosen to provide a margin of safety. Further, it is usual practice to install duplicate sump pumps to ensure pool operation in the event one pump fails. Where water is collected from other drains and discharged to the sump, the pump capacity may have to be increased accordingly.

Related Calculations: Use the general procedure given here to choose swimming pools and their related equipment for schools, recreation centers, hotels, motels, cities, towns, etc. Wherever possible, follow the recommendations of local codes and of the Joint Committee on Swimming Pools.

Heating, Ventilating, and Air Conditioning

REFERENCES: Edwards—*Automatic Controls for Heating and Air Conditioning*, McGraw-Hill; Bowyer—*Central Heating*, David and Charles (England); Kut—*Heating and Hot Water Services in Buildings*, Pergamon; Stoecker—*Using SI Units in Heating, Air Conditioning and Refrigeration*, Business News; Oliker—*Cogeneration District Heating Applications*, ASME; Rizzi—*Design and Estimating for Heating, Ventilating and Air Conditioning*, Van Nostrand Reinhold; Edwards—*Handbook of Geothermal Energy*, Gulf Publishing; Cheremisinoff—*Cooling Towers: Selection, Design and Practice*, Ann Arbor Science; Dubin and Long—*Energy Conservation Standards for Building Design, Construction and Operation*, McGraw-Hill; Stoecker and Jones—*Refrigeration and Air Conditioning*, McGraw-Hill; Carrier Air Conditioning Company—*Handbook of Air Conditioning System Design*, McGraw-Hill; American Society of Heating, Refrigerating, and Air Conditioning Engineers—*Guide and Data Book*, Fundamentals and Equipment and *Guide and Data Book, Applications*; Buffalo Forge Company—*Fan Engineering*; Strock and Koral—*Handbook of Heating, Air Conditioning, and Ventilation*, Industrial Press; Carrier, Cherne, Grant, and Roberts—*Modern Air Conditioning, Heating and Ventilating*, Pitman; Emerick—*Heating Design and Practice*, McGraw-Hill; Holmes—*Air Conditioning in Summer and Winter*, McGraw-Hill; The Trane Company—*Air-Conditioning Manual*.

BUILDING OR STRUCTURE HEAT-LOSS DETERMINATION

An industrial building has 8-in (203.2 mm) thick uninsulated brick walls, a 2-in (50.8-mm) thick concrete uninsulated roof, and a concrete floor. The building is 150 ft (45.7 m) long, 75 ft (22.9 m) wide, and 15 ft (4.6 m) high. Each long wall contains eight 5 × 10 ft (1.5 × 3 m) double-glass windows, and each short wall contains two 4 × 8 ft (1.2 × 2.4 m) double-glass doors. What is the heat loss of this building per hour if the required indoor temperature is 70°F (21.1°C) and the design outside temperature is 0°F (−17.8°C)? How much will the heat loss increase if infiltration causes two air changes per hour?

Calculation Procedure:

1. *Compute the heat loss through the glass*

The usual heat-loss computation begins with the glass areas of a building. Hence, this procedure is followed here. However, a heat-loss computation can be started with any part of the building, provided each part of the structure is eventually considered.

To compute the heat loss through a building surface, use the general relation $H_L = UA\,\Delta t$, where H_L = heat loss, Btu/h, through the surface; U = overall coefficient of heat transmission for the material, Btu/(h·°F·ft²); A = area of heat-transmission surface, ft²; Δt = temperature difference, $F = t_i - t_o$, where t_i = inside temperature, °F; t_o = outside temperature, °F. Find U from Table 1 for the material in question.

This building has sixteen 5 × 10 ft (1.5 × 3 m) double-glass windows and four 4 × 8 ft (1.2 × 2.4 m) double-glass doors. Hence, the total glass area = 16 × 5 × 10 + 4 × 4 × 8 = 928 ft² (86.2 m²). The value of U for double glass is, from Table 1, 0.45. Thus, $H_L = UA\,\Delta t = 0.45(928)(70 - 0) = 29,200$ Btu/h (8639.8 W).

TABLE 1 Typical Overall Coefficients of Heat Transmission, Btu/(ft^2·h·°F) [W/(m^2·K)]

Building surface	Type	Type of insulation
Walls	8-in (203.2-mm) brick	0.50 (2.84)
Roof	2-in (50.8-mm) concrete	0.82 (4.66)
Windows	Single glass	1.13 (6.42)
	Double glass	0.45 (2.56)

2. Compute the heat loss through the building walls

Use the same relation as in step 1, substituting the wall heat-transfer coefficient and wall area. Thus, $U = 0.50$, and $A = 2 \times 150 \times 15 + 2 \times 75 \times 15 - 928 = 5822$ ft^2 (540.9 m^2). Then $H_L = UA\,\Delta t = 0.50(5,822)(70 - 0) = 204,000$ Btu/h (59,746.5 W).

3. Compute the heat loss through the building roof

Use the same relation as in step 1, substituting the roof heat-transfer coefficent and roof area. Thus, $U = 0.82$ and $A = 150 \times 75 = 11,250$ ft^2 (1045.1 m^2). Then $H_L = UA\,\Delta t = 0.82 \times (11,250)(70 - 0) = 646,000$ Btu/h (189,197.3 W).

4. Compute the total heat loss of the building

The total heat loss of a building is the sum of the individual heat losses of the walls, glass areas, roof, and floor. In large buildings the heat loss through concrete floors is usually negligible and can be ignored. Hence, the total heat loss of this building caused by transmission through the building surfaces is $H_T = 29,200 + 204,000 + 646,000 = 879,200$ Btu/h (257,495.7 W).

5. Compute the infiltration heat loss

The building volume is $150 \times 75 \times 15 = 168,500$ ft^3 (4773.1 m^3). With two air changes per hour, the volume of infiltration air that must be heated is $2 \times 168,500 = 337,000$ ft^3/h (9546.1 m^3/h). The heat that must be supplied to raise the temperature of this air is $H_i = $ (ft^3/h)(Δt)/55 $= (337,000)(70 - 0)/55 = 429,000$ Btu/h (125,643.4 W). Thus, the total heat loss of this building, including infiltration, is $H_T = 879,200 + 429,000 = 1,308,200$ Btu/h (383.1 kW).

Related Calculations: Determine the design outdoor temperature for a given locality from Baumeister—*Standard Handbook for Mechanical Engineers* or the ASHRAE *Guide and Data Book*, published by the American Society of Heating, Refrigerating and Air-Conditioning Engineers. Both these works are also suitable sources of comprehensive listings of U values for various materials and types of building construction. Since the winter-design outdoor temperature is usually for nighttime conditions, no credit is taken for heat given off by machinery, lights, people, etc., unless the structure will always operate on a 24-h basis. The safest design ignores these heat sources because the machinery in the building may be removed, the operating cycle changed, or the heat sources eliminated in some other way. However, where an internal heat source of any kind will be a permanent part of a building, simply subtract the hourly heat release from the total building heat loss. The result is the net heat loss of the building and is used in choosing the heating equipment for the building.

Most heat-loss calculations for large structures are made on a form available from heat equipment manufacturers. Such a form helps organize the calculations. The steps followed, however, are identical to those given above. Another advantage of the calculation form is that it helps the designer remember the various items—walls, roof, glass, infiltration, etc.—to consider.

When some areas in a structure will be kept at a lower temperature than others, compute the heat loss from one area to another in the manner shown above. Substitute the lower indoor temperature for the outdoor design temperature. For areas exposed to prevailing winds, some manufacturers recommend increasing the computed heat loss by 10 percent. Thus, if the north wall of a building were exposed to the prevailing winds and its heat loss is 50,000 Btu/h (14,655.0 W), this heat loss would be increased to $1.1(50,000) = 55,000$ Btu/h (16,120.5 W).

HEATING-SYSTEM SELECTION AND ANALYSIS

Choose a heating system suitable for an industrial plant consisting of a production area 150 ft (45.7 m) long and 75 ft (22.9 m) wide and an office area 75 ft (22.9 m) wide and 60 ft (18.3 m) long. The heat loss from the production area is 1,396,000 Btu/h (409.2 kW); the heat loss from the office area is 560,000 Btu/h (164.1 kW). Indoor design temperature for both areas is 70°F (21.1°C); outdoor design temperature is 0°F (−17.8°C). What will the fuel consumption of the chosen heating system be if the annual degree-days for the area in which the plant is located is 6000? Compare the annual fuel consumption of gas, oil, and coal.

Calculation Procedure:

1. *Choose the type of heating system to use*

Table 2 lists the various types of heating systems used today and typical applications. Study of this tabulation shows that steam unit heaters would probably be best for the production area because it is relatively large and open. Either a forced-warm-air or a two-pipe steam heating system could be used for the office area. Since the production area will use steam unit heaters, a two-pipe steam heating system would probably be best for the office area. Since steam unit heaters are almost universally two-pipe, the same method of supply and return is best chosen for the office system.

Note that Table 2 lists six different types of two-pipe steam heating systems. Choice of a particular type of two-pipe system depends on a number of factors, including economics, steam pressure required for nonheating (i.e., process) services in the building, type and pressure rating of boiler used, etc. Where high-pressure steam—30 to 150 lb/in² (gage) (206.8 to 1034.1 kPa), or higher—is used for process, the two-pipe system fitted with pressure-reducing valves between the process and heating mains is often an economical choice.

Hot-water heating is unsuitable for this building because unit heaters are required for the production area. An inlet air temperature less than 30°F (−1.1°C) is generally not recommended for hot-water unit heaters. Since the inlet-air temperature can be as low as 0°F (−17.8°C) in this plant, hot-water unit heaters could not be used.

2. *Compute the annual fuel consumption of the system*

Use the degree-day method to compute the annual fuel consumption. To apply the degree-day method, substitute the appropriate values in $F_g = DU_gRC$, for gas heating; $F_o = DU_oH_T C/1000$, for oil; $F_c = DU_cH_TC/1000$, for coal, where F = fuel consumption, the type of fuel being

TABLE 2   Typical Applications of Heating Systems

System type	Fuel°	Typical applications
Gravity warm air	G, O, C	Small residences, wooden or masonry
Forced warm air	G, O, C	Small and large residences, wooden or masonry; small and medium-sized industrial plants, offices
Steam heating:		
One-pipe	G, O, C	Small residences, wooden or masonry
Two-pipe†	G, O, C	Small and large residences, wooden or masonry; small and large industrial plants, offices. High-pressure systems [30 to 150 lb/in² (gage) (206.8 to 1034.1 kPa)] may be used in large industrial buildings having unit heaters or fan units. Unit heaters are used for large, open areas
Hot water:		
Gravity	G, O, C	Small residences, wooden or masonry
Forced‡	G, O, C	Small and large residences; small and large industrial plants
Radiant	G, O, C	Small and large residences and plants
Electric	Electricity	Small residences and plants

°G—gas; O—oil; C—coal.
†May be low-pressure, two-pipe vapor; two-pipe vacuum; two-pipe subatmospheric; two-pipe orifice; high-pressure.
‡May be one-pipe; two-pipe.

identified by the subscript, g for gas, o for oil, c for coal, with the unit of consumption being the therm for gas, gal for oil, and lb for coal; D = degree-days during the heating season; U = unit fuel consumption, the unit again being identified by the subscript; R = ft^2 EDR (equivalent direct radiation) in the heating system; C = a correction factor for the outdoor design temperatures. Values of U and C are given in Table 3. Note the fuel heating values on which this table is based. For other heating values, see Related Calculations, below.

For gas heating, $F_g = DU_gRC$, therms, where 1 therm = 100,000 Btu (105,500 kJ). To select the correct value for U_g, compute the ft^2 EDR (see the next calculation procedure) from $H_T/240$, or ft^2 (m^2) EDR = 1,956,000/240 = 8150 (757.1). Hence, U_g = 0.00116 from Table 3. From the same table, C = 1.00 for an outdoor design temperature of 0°F (-17.8°C). Hence, F_g = (6000)(0.00116)(8150)(1.00) = 56,800 therms. Assuming gas having a heating value h_v of 1000 Btu/ft^3 (37,266 kJ/m^3) is burned in this heating system, the annual gas consumption is $F_g \times 10^5/h_v$. Or, 56,800 $\times$ 10^5/1000 = 5,680,000 ft^3 (160,801 m^3).

Using Table 3 for oil heating and assuming a heating plant efficiency of 70 percent, U_o = 0.00437 gal per 1000 Btu/h [0.01568 L/(MJ·h)] heat loss, and C = 1.00 for the design conditions. Then $F_o = DU_oH_TC/1000$ = (6000)(0.00437)(1,956,000)(1.00)/1000 = 51,400 gal (194,570 L).

Using Table 3 for coal heating and assuming a heating-plant efficiency of 70 percent, we get U_c = 0.0507 lb coal per 1000 Btu/h [0.2163 kg/(MJ·h)] heat loss, and C = 1.00 for the design conditions. Then $F_c = DU_cH_TC/1000$ = (6000)(0.0507)(1,956,000)(1.00)/1000 = 595,000 lb, or 297 tons of 2000 lb (302 t) each.

Related Calculations: When the outdoor-design temperature is above or below 0°F (-18°C), a correction factor must be applied as shown in the equations given in step 2. The appropriate correction factor is given in Table 3. Fuel-consumption values listed in Table 3 are based on an indoor-design temperature of 70°F (21°C) and an outdoor-design temperature of 0°F (-18°C).

TABLE 3   Factors for Estimating Fuel Consumption*

Fuel	Consumption per degree-day	Heating conditions	
		Steam	Hot water
Gas	Therms [100,000 Btu (105,500 kJ)]	0.00127 [300-ft^2 (28-m^2) EDR] 0.00121 [300–700 ft^2 (28–65 m^2) EDR] 0.00116 [700-ft^2 (65-m^2) EDR]	0.000743 [500-ft^2 (46-m^2) EDR] 0.000709 [500–1000 ft^2 (46–92 m^2) EDR] 0.000675 [1200-ft^2 (111-m^2) EDR]
		Heating-plant efficiency	
Oil [heating value = 141,000 Btu/gal (39,297 MJ/L)]	Gal per 1000-Btu/h heat loss [L/(MJ·h)]	70% 0.00437 (0.01568)	80% 0.00383 (0.01374)
Coal [heating value = 12,000 Btu/lb (27,912 kJ/kg)]	Lb per 1000-Btu/h heat loss [kg/(MJ·h)]	70% 0.0507 (0.02163)	80% 0.0444 (0.01893)

Outside-design temperature correction factor

Outside design temperature, °F (°C)	-20 (-28.9)	-10 (-23.3)	0 (-17.8)	+10 (-12.2)	+20 (-6.7)
Correction factor	0.778	0.875	1.000	1.167	1.400

*ASHRAE *Guide and Data Book* with permission.

The oil consumption values are based on oil having a heating value of 141,000 Btu/gal (39,296.7 MJ/L). Where oil having a different heating value is burned, multiply the fuel consumption value selected by 141,000/(heating value, Btu/gal, of oil burned).

The coal consumption values are based on coal having a heating value of 12,000 Btu/lb (27,912 kJ/kg). Where coal having a different heating value is burned, multiply the fuel consumption value selected by 12,000/(heating value, Btu/lb, of coal burned).

For example, if the oil burned in the heating system in step 2 has a heating value of 138,000 Btu/gal (38,460.6 MJ/L), U_o = 0.00437(141,000/138,000) = 0.00446. And if the coal has a heating value of 13,500 Btu/lb (31,401 kJ/kg), U_c = 0.0507(12,000/13,500) = 0.0451.

Steam consumption of a heating system can also be computed by the degree-day method. Thus, the weight of steam W required for the degree-day period—from a day to an entire heating season—is $W = 24H_T D/1000$, where all the symbols are as given earlier. Thus, for the industrial building analyzed above, W = 24(1,956,000) × (6000)/1000 = 281,500,000 lb (127,954,546 kg) of steam per heating season. The denominator in this equation is based on low-pressure steam having an enthalpy of vaporization of approximately 1000 Btu/lb (2326 kJ/kg). Where high-pressure steam is used and the enthalpy of vaporization is lower, substitute the actual enthalpy in the denominator to obtain more accurate results.

Steam consumption for building heating purposes can also be given in pounds (kilograms) of steam per 1000 ft³ (28.3 m³) of building space per degree-day and of steam per 1000 ft² (92.9 m²) of EDR per degree-day. On the building-volume basis, steam consumption in the United States can range from a low of 0.130 (0.06) to a high of 2.07 lb (0.94 kg) of steam per 1000 ft³/degree-day (28.3 m³/degree-day), depending on the building type (apartment house, bank, church, department store, garage, hotel, or office building) and the building location (southwest or far north). Steam consumption can range from a low of 21 lb (10.9 kg) per 1000 ft² (92.9 m²) EDR per degree-day to a high of 120 lb (54.6 kg) per 1000 ft² (92.9 m²) EDR per degree-day, depending on the building type and location. The ASHRAE *Guide and Data Book* lists typical steam consumption values for buildings of various types in different locations.

REQUIRED CAPACITY OF A UNIT HEATER

An industrial building is 150 ft (45.7 m) long, 75 ft (22.9 m) wide, and 30 ft (9.1 m) high. The heat loss from the building is 350,000 Btu/h (102.6 kW). Choose suitable unit heaters for this building if 5-lb/in² (gage) (30.5-kPa) steam and 200°F (93.3°C) hot water are available to supply heat. Air enters the unit heater at 0°F (−18°C); an indoor temperature of 70°F (21°C) is desired in the building. What capacity unit heaters are needed if 20,000 ft³/min (566.2 m³/min) is exhausted from the building?

Calculation Procedure:

1. *Compute the total heat loss of the building*

The heat loss through the building walls and roof is given as 350,000 Btu/h (102.6 kW). However, there is an additional heat loss caused by infiltration of outside air into the building. Compute this loss as follows.

Find the cubic content of the building from volume, ft³ = LWH, where L = building length, ft; W = building width, ft; H = building height, ft. Thus, volume = (150)(75)(30) = 337,500 ft³ (9554.6 m³).

Determine the heat loss caused by infiltration by estimating the number of air changes per hour caused by leakage of air into and out of the building. For the usual industrial building, one to two air changes per hour are produced by infiltration. At one air change per hour, the quantity of infiltration air that must be heated from the outside to the inside temperature = building volume = 337,500 ft³/h (9554.6 m³/h). Had two air changes per hour been assumed, the quantity of infiltration air that must be heated = 2 × building volume.

Compute the heat required to raise the temperature of the infiltration air from the outside to the inside temperature from $H_i = (\text{ft}^3/\text{h})(\Delta t)/55$, where H_i = heat required to raise the temperature of the air, Btu/h, through Δt, where $\Delta t = t_i - t_o$, and t_i = inside temperature, °F; t_o = outside temperature, °F. For this building, H_i = 337,500(70 − 0)/55 = 429,000 Btu/h (125.7 kW). Hence the total heat loss of this building H_t = 350,000 + 429,000 Btu/h (228.3 kW), without exhaust ventilation.

2. *Determine the extra heat load caused by exhausting air*

When air is exhausted from a building, an equivalent amount of air must be supplied by infiltration or ventilation. In either case, an amount of air equal to that exhausted must be heated from the outside temperature to room temperature.

With an exhaust rate of 10,000 ft^3/min = 60(10,000) = 600,000 ft^3/h (19,986 m^3/h), the heat required is H_e = (ft^3/h)(Δt)/55, where H_e = heat required to raise the temperature of the air that replaces the exhaust air, Btu/h. Or, H_e = 600,000(70 − 0)/55 = 764,000 Btu/h (223.9 kW).

To determine the total heat loss from a building when both infiltration and exhaust occur, add the larger of the two heat requirements—infiltration or exhaust—to the heat loss caused by transmission through the building walls and roof. Since, in this building, $H_e > H_i$, H_t = 350,000 + 764,000 = 1,114,000 Btu/h (326.5 kW).

3. *Choose the location and number of unit heaters*

This building is narrow; i.e., it is half as wide as it is long. In such a building, three vertical-discharge unit heaters (Fig. 1) will provide good distribution of the heated air. With three heaters, the capacity of each should be 764,000/3 = 254,667 Btu/h (74.6 kW) without ventilation and 1,114,000/3 = 371,333 Btu/h (108.8 kW) with ventilation. Once the capacity of each unit heater is chosen, the spread diameter of the heated air discharged by the heater can be checked to determine whether it is sufficient to provide the desired comfort.

4. *Select the capacity of the unit heaters*

Use the engineering data published by a unit-heater manufacturer, such as Table 4, to determine the final air temperature, Btu delivered per hour, cubic feet of air handled, and the quantity of condensate formed. Thus, Table 4 shows that vertical-discharge model D unit heater delivers 277,900 Btu/h (81.5 kW) when the entering air is at 0°F (−17.8°C) and the heating steam is at a pressure of 5 lb/in^2 (gage) (34.5 kPa). This heater discharges 3400 ft^3/min (96.3 m^3/min) of

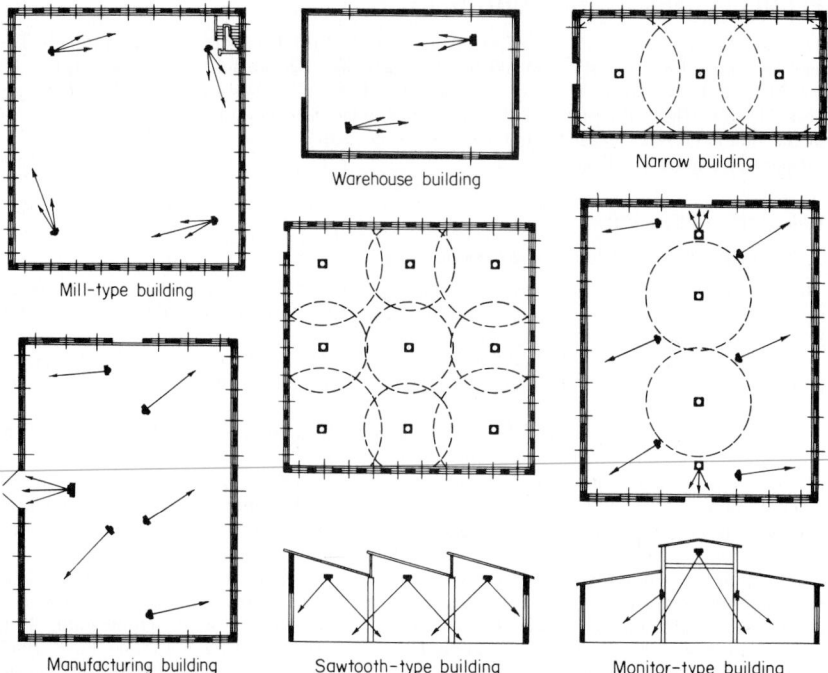

FIG. 1 Recommended arrangements of unit heaters in various types of buildings. (*Modine Manufacturing Company.*)

TABLE 4 Typical Vertical-Delivery Unit-Heater Capacities°

Model	Mtg. ht., ft (m)	Spread diam., ft (m)	Motor speed, r/min	ft³/min† (m³/ min)	0°F (−17.8°C) entering air			50°F (10°C) entering air		
					Btu/h (kW)	Final temp., °F (°C)	Cond., lb/h (kg/h)	Btu/h (kW)	Final temp., °F (°C)	Conden- sate, lb/ h (kg/h)
D	14 (4.3)	48, 54 (14.8, 16.5)	1,135	3,400 (96)	277,900 (81.5)	76 (24.4)	290 (161)	208,900 (61.2)	111 (43.9)	218 (99)
E	15 (4.6)	56, 62 (17.3, 18.9)	1,135	4,920 (139.3)	388,400 (113.8)	73 (22.8)	404 (183)	292,000 (85.6)	109 (42.8)	304 (151)

Mounting height correction factors

Steam press., lb/in² (gage) (kPa)		Water temp., °F (°C)	Normal room temp., °F (°C)		
			60 (15.6)	70 (21.1)	80 (26.7)
. . .	. . .	210 (98.9)	1.05	1.10	1.20
0–5	(0–34.4)	220 (104.4)	1.00	1.07	1.14
6–15	(41.4–103.4)	. . . . . .	0.88	0.94	1.00
16–30 (110.3–206.8)		. . . . . .	0.77	0.81	0.86
31–50 (213.7–344.7)		. . . . . .	0.70	0.73	0.77

°5-lb/in² (gage) (34. 5-kPa) steam supply.
†ft³/min capacity at *final* air temperature. For horizontal discharge unit heaters, the ft³/min capacity is usually stated at the *entering* air temperature.

heated air at 76°F (24.4°C) and forms 290 lb/h (131.8 kg/h) of condensate. The capacity table for a horizontal-discharge unit heater is similar to Table 4. When the building is ventilated, a model E unit heater delivering 388,400 Btu/h (113.8 kW) could be used with entering air at 0°F (−17.8°C). This heater, as Table 4 shows, delivers 4920 ft³/min (139.3 m³/min) of air at 73°F (22.8°C) and forms 404 lb (183.6 kg) of condensate.

5. *Check the spread diameter produced by the heater*

Table 4 shows the different spread diameters, i.e., diameter of the heated-air blast at the floor level for different mounting heights of the unit heater. Thus, at a 14-ft (4.3-m) mounting level above the floor, model D will produce a spread diameter of 48 ft (14.6 m) or 54 ft (16.5 m), depending on the type of outlet cone used. These spread diameters are based on 2-lb/in² (gage) (13.8-kPa) steam and 60°F (15.6°C) room temperature. For 5-lb/in² (gage) (34.5-kPa) steam and 70°F (21.1°C) room temperature, multiply the tabulated spread diameter and mounting height by the correction factor shown at the bottom of Table 4. Thus, for model D, spread diameter = 1.07(48) = 51.4 ft (15.7 m) and 1.07(54) = 57.8 ft (17.6 m), whereas the mounting height could be 1.07(14) = 14.98 ft (4.6 m).

Find the spread diameter for model E in the same way, or, 56 ft (17.1 m) and 62 ft (18.9 m) at a 15-ft (4.6-m) height with 2-lb/in² (gage) (13.8-kPa) steam. With 5-lb/in² (gage) (34.5-kPa) steam, the spread diameters are 1.07(56) = 59.9 ft (18.3 m) and 1.07(62) = 66.4 ft (20.2 m), whereas the mounting height could be 1.07(15) = 16 ft (4.9 m).

6. *Compute the hot-water-heater capacity required*

Study several manufacturers' engineering data to determine the capacity of a suitable hot-water unit heater. This study will show that hot-water unit heaters are generally not available for inlet-air temperatures less than 30°F (−1.1°C). Since the inlet-air temperature in this building is 0°F (−17.8°C), a hot-water unit heater would be unsuitable; hence, it cannot be used. The *minimum* outlet-air temperature often recommended for unit heaters is 95°F (35.0°C).

Related Calculations: Use the same general method to choose horizontal-delivery unit heaters. The heated air delivered by these units travels horizontally or can be deflected down toward the floor. Tables in manufacturers' engineering data list the heat-throw distance for horizontal-delivery unit heaters.

Standard ratings of steam unit heaters are given for 2-lb/in² (gage) (13.8-kPa) steam and 60°F (15.6°C) entering air; hot-water unit heaters for 180°F (82.2°C) entering water and 60°F (15.6°C) entering air. For other steam pressures, water temperatures, or entering air temperatures, *divide* the Btu per hour at stated conditions by the appropriate correction factor from Table 5 to obtain the required rating at *standard* conditions. Thus, a steam unit heater rated at 18,700 Btu/h (5.5 kW) at 20 lb/in² (gage) (137.9 kPa) and 70°F (21.1°C) entering air has a standard rating of 18,700/1.178 = 15,900 Btu/h (4.7 kW), closely, at 2 lb/in² (gage) (13.8 kPa) and 60°F (15.6°C) entering air temperature, using the correction factor from Table 5. Conversely, a steam unit heater rated at 228,000 Btu/h (66.8 kW) at 2 lb/in² (gage) (13.8 kPa) and 60°F (15.6°C) will deliver 228,000(1.421) = 324,000 Btu/h (94.9 kW) at 50 lb/in² (gage) (344.7 kPa) and 70°F (21.1°C) entering air. Electric and gas-fired unit heaters are also rated on the basis of 60°F (15.6°C) entering air.

When a unit heater is supplied both outside air and recirculated air from within the building, use the temperature of the combined airstreams as the inlet air temperature. Thus, with 1000 ft³/min (28.3 m³/min) of 10°F (−12.2°C) outside air and 4000 ft³/min (113.2 m³/min) of 70°F (21.1°C) recirculated air, the temperature of the air entering the heater is $t_e = (t_{oa}cfm_{oa} + t_r cfm_r)/cfm_t$, where t_e = temperature of air entering heater, °F; t_{oa} = temperature of outside air, °F; cfm_{oa} = quantity (cubic *feet* per *minute*) of outside air entering the heater, ft³/min; t_r = temperature of room air entering heater, °F; cfm_r = quantity of room air entering heater, ft³/min; cfm_t = total cfm entering heater = $cfm_{oa} + cfm_r$. Thus, $t_e = (10 \times 1000 + 70 \times 4000)/5000 = 58°F$ (14.4°C). This relation is valid for both steam and hot-water unit heaters.

To find the approximate outlet air temperature of a unit heater when the capacity, cfm, and entering air temperature are known, use the relation $t_o = t_i + (460 + t_i)/[575 \text{ ft}^3 \cdot \text{h}/(\text{min} \cdot \text{Btu})] - 1$, where t_o = unit-heater outlet-air temperature, °F; t_i = temperature of air entering the heater, °F; cfm = quantity of air passing through the heater, ft³/min; Btu/h = rated capacity of the unit heater. Thus, for a heater rated at 73,500 Btu/h (21.5 kW), 1530 ft³/min (43.4 m³/min), with 50°F (10.0°C) entering air temperature, $t_o = 50 + (460 + 50)/(575 \times 1530/73,500) - 1 = 94.1°F$ (34.5°C). The unit-heater capacity used in this equation should be the capacity at standard conditions: 2 lb/in² (gage) (13.8 kPa) steam or 180°F (82.2°C) water and 60°F (15.6°C) entering air temperature. Results obtained with this equation are only approximate. In any event, the outlet air temperature should never be less than the room air temperature.

TABLE 5   Unit-Heater Conversion Factors

	Entering air temperature, °F (°C)			
	40 (4.4)	50 (10.0)	60 (15.6)	70 (21.1)
Steam pressure, lb/in² (gage) (kPa)	Horizontal-delivery steam heaters			
2 (13.8)	1.153	1.076	1.000	0.927
5 (34.5)	1.209	1.131	1.055	0.981
10 (68.9)	1.288	1.209	1.132	1.057
20 (137.9)	1.413	1.333	1.254	1.178
40 (275.8)	1.593	1.510	1.430	1.351
50 (344.7)	1.664	1.582	1.500	1.421
Entering water temperature, °F (°C)	Horizontal-delivery hot-water heaters			
150 (65.6)	0.911	0.790	0.676	0.568
160 (71.1)	1.027	0.900	0.783	0.670
170 (76.7)	1.142	1.012	0.890	0.773
180 (82.2)	1.262	1.127	1.000	0.880
190 (87.8)	1.384	1.245	1.115	0.990
200 (93.3)	1.503	1.365	1.231	1.102

TABLE 6 Unit-Heater Outlet Velocity and Blow Distance

Unit-heater type	Outlet velocity, ft/min (m/min)	Blow distance, ft (m)
Centrifugal fan	1500–2500 (457–762)	20–200 (6–61)
Horizontal propeller fan	400–1000 (122–305)	30–100 (9–30)
Vertical propeller fan	1200–2200 (366–671)	70 (21)

For the actual outlet temperature of a specific unit heater, refer to the manufacturers' engineering data.

The air discharged by a unit heater should be at a temperature greater than the room temperature because air in motion tends to chill the occupants of a room. Choose the outlet temperature by referring to the heated-air velocity. Thus, with a velocity of 20 ft/min (6.1 m/min), the air temperature should be at least 76°F (24.4°C) at the heater outlet. As the air velocity increases, higher air temperatures are required.

The outlet air velocity and distance of blow of typical unit heaters are shown in Table 6.

When a unit heater of any type discharges against an external resistance, such as a duct or grille, its heating capacity and air capacity, as compared to standard conditions, are reduced. However, the final air temperature usually increases a few degrees over that at standard conditions.

To convert the rated output of any steam unit heater to ft^2 (m^2) of equivalent direct radiation [abbreviated ft^2 EDR (m^2 EDR)], divide the unit-heater rated capacity in Btu/h (W) by 240 Btu/(h·ft^2 EDR) (70.3 W/m^2 EDR). Thus, a unit heater rated at 240,000 Btu/h (70.34 W) has a heat output of 240,000/240 = 1000 ft^2 EDR (1000 m^2 EDR). For hot-water unit heaters, use the conversion factor of 150 Btu/(h·ft^2 EDR) (43.9 W/m^2 EDR).

To determine the rate of condensate formation in a steam unit heater, divide the rated output in Btu/h (W) by an enthalpy of 930 Btu/lb (2163.2 kJ/kg) of steam. Most unit-heater rating tables list the rate of condensate formation for each heater. Table 7 shows typical pipe sizes recommended for various condensate loads of steam unit heaters. Thus, with 1000 lb/h (454.6 kg/h) of condensate and 30-lb/in^2 (gage) (206.8-kPa) steam supply to the unit heater, the supply main should be 2½-in (64-mm) nominal diameter and the return main should be 1¼-in (32-mm) nominal diameter.

Figure 1 shows how unit heaters of any type should be located in buildings of various types. The diagrams are also useful in determining the approximate number of heaters needed once the heat loss is known. Locate unit heaters so that the following general conditions prevail, if possible.

Unit heater type	Desirable conditions
Horizontal delivery	Discharge should wipe exposed walls at an angle of about 30°. With multiple units, the airstreams should support each other.
Vertical delivery	With only vertical units, the airstream should blanket exposed walls with warm air.

The unit-heater arrangements shown in Fig. 1 illustrate a number of important principles.[1] The basic principle of unit-heater location is shown in the *mill-type* building. Here the heated-air flow from each unit heater supports the air flow from the other unit heaters and tends to set up a general circulation of air in the space heated. In the *warehouse* building arrangement, max-

[1]Modine Manufacturing Company.

TABLE 7 Typical Steam Unit-Heater Pipe Diameters,° in (mm)

| Condensate, lb/h (kg/h) | Steam-supply pressure, lb/in² (gage) (kPa) | | | |
| | 5 (34.5) | | 30 (206.8) | |
	Supply	Return†	Supply	Return
100 (45)	2 (51)	1 (25)	1¼ (32)	¾ (19)
400 (180)	3 (76)	2 (51)	2 (51)	1 (25)
800 (360)	4 (102)	2½ (64)	2½ (64)	1¼ (32)
1000 (450)	5 (127)	2½ (64)	2½ (64)	1¼ (32)

°Modine Manufacturing Company.
†Gravity return.

imum area coverage is obtained with a minimum number of units. The *narrow* building uses vertical-discharge unit heaters that blanket the building walls with warmed air.

In the *manufacturing* building, circular air movement is sacrificed to offset a large roof heat loss and to permit short runouts from a single steam main. Note how a long-throw unit heater blankets a frequently used doorway.

Vertical-discharge unit heaters are used in the medium-height *sawtooth-type* building shown in Fig. 1. The *monitor-type*-building installation combines both horizontal and vertical unit heaters. Horizontal-discharge unit heaters are located in the low-ceiling areas and vertical-discharge units in the high-ceiling areas above the craneway. Much of the data in this procedure were supplied by Modine Manufacturing Company.

STEAM CONSUMPTION OF HEATING APPARATUS

Determine the probable steam consumption of the non-space-heating equipment in a building equipped with the following: three bain-maries, each 100 ft² (9.3 m²), two 50-gal (189.3-L) coffee urns, one jet-type dishwasher, one plate warmer having a 60-ft³ (1.7-m³) volume, two steam tables, each having an area of 50 ft² (4.7 m²) and one water still having a capacity of 75 gal/h (283.9 L/h). The available steam pressure is 40 lb/in² (gage) (275.8 kPa); the kitchen equipment will operate at 20 lb/in² (gage) (137.9 kPa) and the water still at 40 lb/in² (gage) (275.8 kPa).

Calculation Procedure:

1. Determine steam consumption of the equipment

The general procedure in determining heating-equipment steam consumption is to obtain engineering data from the manufacturer of the unit. When these data are unavailable, Table 8 will provide enough information for a reasonably accurate first approximation. Hence, this tabulation is used here to show how the data are applied.

Since the supply steam pressure is 40 lb/in² (gage) (275.8 kPa), a pressure-reducing valve will have to be used between the steam main and the kitchen supply main. The capacity of this valve depends on the steam consumption of the equipment. Hence, the valve capacity cannot be determined until the equipment steam consumption is known.

During equipment operation the steam consumption is different from the consumption during startup. Since the operating consumption must be known before the starting consumption can be computed, the former is determined first.

Using data from Table 8, we see the three 100-ft² (9.3-m²) bain-maries will require (3 lb/h)(3 units)(100 ft² each) = 900 lb/h (409.1 kg/h) of steam. The two 50-gal (189.3-L) coffee urns require (2.75 lb/h)(2 units)(50 gal per unit) = 275 lb/h (125 kg/h), using the average steam consumption. One jet-type dishwasher will require 60 lb/h (27.3 kg/h) of steam. A 60-ft³ (1.7-m³) plate warmer will require (60 ft³/(1 unit) [(30 lb/h)/20-ft³ unit] = 90 lb/h (40.9 kg/h). Two 50-ft² (4.7-m²) steam tables require (1.5 lb/h)(2 units) (50 ft² per unit) = 150 lb/h. The 75-gal/h

TABLE 8   Typical Steam Consumption of Heating Equipment

Equipment	Steam, lb/h at 20–40 lb/in^2 (gage) (kg/h at 138–276 kPa)
Bain-marie, per ft^2 (m^2) of surface	3.0 (14.5)
Coffee urns, per gal (L)	2.5–3.0 (0.30–0.36)
Dishwashers, jet-type	60 (27.0)
Plate warmer, per 20 ft^3 (0.57 m^3)	30 (13.5)
Soup or stock kettle, 60 gal (0.23 L); 40 gal (0.15 L)	60; 45 (27.0; 20.3)
Steam table, per ft^2 (m^2) of surface	1.5 (7.3)
Vegetable steamer, per compartment 5-lb (2.3-kg) press	30 (13.5)
Water still, per gal (L) capacity per h	9 (1.1)

(283.9 L/h) water still will consume (9 lb/h)(1 unit)(75 gal/h) = 675 lb/h (306.8 kg/h). Hence, the total operating consumption of 20-lb/in^2 (gage) (137.9 kPa) steam is 900 + 275 + 60 + 90 + 150 = 1475 lb/h (670.5 kg/h). The water still will consume 675 lb/h (306.8 kg/h) of 40-lb/in^2 (gage) (275.8-kPa) steam.

Since the 20-lb/in^2 (gage) (137.9-kPa) steam must pass through the pressure-reducing valve before entering the 20-lb/in^2 (gage) (137.9-kPa) main, the required *operating* capacity of this valve is 1475 lb/h (670.5 kg/h). However, the total steam consumption during operation, without an allowance for condensation in the pipelines, is 1475 + 675 = 2150 lb/h (977.3 kg/h).

2. Compute the system condensation losses

Condensation losses can range from 25 to 50 percent of the steam supplied, depending on the type of insulation used on the piping, the ambient temperature in the locality of the pipe, and the degree of superheat, if any, in the steam. Since the majority of the steam used in this building is 20-lb/in^2 (gage) (137.9-kPa) steam reduced in pressure from 40 lb/in^2 (gage) (275.8 kPa) and there will be a small amount of superheating during pressure reduction, a 25 percent allowance for pipe condensation is probably adequate. Hence, the total operating steam consumption = (1.25)(2150) = 2688 lb/h (1221.8 kg/h).

3. Compute the startup steam consumption

During equipment startup there is additional condensation caused by the cold metal and, possibly, some cold products in the equipment. Therefore, the startup steam consumption is different from the operating consumption.

One rule of thumb estimates the startup steam consumption as two times the operating consumption. Thus, by this rule of thumb, startup steam consumption = 2(2688) = 5376 lb/h (2443.6 kg/h). Note that this consumption rate is of relatively short duration because the metal parts are warmed rapidly. However, the pressure-reducing valve must be sized for this flow rate unless slower warming is acceptable.

The actual rate of condensate formation can be computed if the weight of the equipment, the specific heat of the materials of construction, and initial and final temperatures of the equipment are known. Use the relation steam condensation, lb/h = $60\,Ws(\Delta t)/h_{fg}T$, where W = weight of equipment and piping being heated, lb; s = specific heat of the equipment and piping, Btu/(lb·°F); Δt = temperature rise of the equipment from the cold to the hot state, °F; h_{fg} = latent heat of vaporization of the heating steam, Btu/lb; T = heating period, min. In SI units, condensation is found from the same relationship, except W is in kg; s is in kJ/(kg·°C); t = temperature change, °C; h_{fg} = latent heat of vaporization, kJ/kg; other variables the same.

This relation assumes that the final temperature of the equipment approximately equals the temperature of the heating steam. Where the specific heat of the equipment is different from that of the piping, solve for the steam condensation rate of each unit and sum the results. Where products in the equipment must be heated, use the same relation but substitute the product weight and specific heat.

Related Calculations: Use the general method given here to compute the steam consumption of any type of industrial equipment for which the unit steam consumption is known or can be determined.

SELECTION OF AIR HEATING COILS

Select a steam heating coil to heat 80,000 ft^3/m (2264.8 m^3/min) of outside air from 10°F (−12.2°C) to 150° (65.6°C) for steam at 15 lb/in^2 (abs) (103.4 kPa). The heated air will be used for factory space heating. Illustrate how a steam coil is piped. Show the steps for choosing a hot-water heating coil.

Calculation Procedure:

1. Compute the required face area of the coil

If the coil-face air velocity is not given, a suitable air velocity must be chosen. In usual air-conditioning and heating practice, the air velocity across the face of the coil can range from 300 to 1000 ft/min (91.4 to 305 m/min) with 500, 800, and 1000 ft/min (152.4, 243.8, and 305 m/min) being common choices. The higher velocities—up to 1000 ft/min (305 m/min)—are used for industrial installations where noise is not a critical factor. Assume a coil face velocity of 800 ft/min (243.8 m/min) for this installation.

TABLE 9 Air-Volume Conversion Factors[°][†]

Air temp., °F (°C)	Factor	Air temp., °F (°C)	Factor
0 (−17.18)	1.152	90 (32.2)	0.964
10 (−12.2)	1.128	100 (37.8)	0.946
20 (−6.7)	1.104	110 (43.3)	0.930
30 (−1.1)	1.082	120 (48.9)	0.914
40 (4.4)	1.060	130 (54.4)	0.898
50 (10.0)	1.039	140 (60.0)	0.883
60 (15.6)	1.019	150 (65.6)	0.869
70 (21.1)	1.000	160 (71.1)	0.855
80 (26.7)	0.981	170 (76.7)	0.841

[°]McQuay, Inc.
[†]The air volume in heating and air-conditioning applications is usually measured in ft^3/m at 70°F (21.1°C) and 29.92 in (759.97 mm) Hg. When the actual air volume is at another temperature, multiply the ft^3/m or velocity by the factor given for the temperature of the air. The result is the ft^3/mim or velocity at standard conditions and is the values used when entering heater capacity tables and the friction chart, Fig. 2.

Compute the required face area from $A_c = cfm/V_a$, where A_c = required coil face area, ft^2: cfm = quantity of air to be heated by the coil, ft^3/min, at 70°F (21.1°C) and 29.92 in (759.97 mm) Hg; V_a = air velocity through the coil, ft/min. To correct the air quantity to standard conditions, when the air is being delivered at nonstandard conditions, multiply the flow in at the other temperature by the appropriate factor from Table 9. Thus, with the incoming air at 10°F (−12.2°C), ft^3/m at 70°F (21.1°C) and 29.92 in (759.97 mm) Hg = (1.128)(80,000) = 90,400 ft^3/m (2559.2 m^3/min). Hence, A_c = 90,400/800 = 112.8 ft^2 (10.5 m^2).

2. Compute the coil outlet temperature

The capacity, final temperature, and condensate formation rate for steam heating coils for air-conditioning and heating systems are usually based on steam supplied at 5 lb/in^2 (abs) (34.5 kPa) and inlet air at 0°F (−17.8°C). At other steam pressures a correction factor must be applied to the tabulated outlet temperature for 5-lb/in^2 (abs) (34.5-kPa) coils with 0°F (−17.8°C) inlet air.

Table 10 shows an excerpt from a typical coil-rating table and excerpts from coil correction-factor tables. To use such a tabulation for a coil supplied steam at 5 lb/in^2 (abs) (34.5 kPa), enter at the air inlet temperature and coil face velocity. Find the final air temperature equal to, or higher than, the required final air temperature. Opposite this read the number of rows of tubes required. Thus, in a 5-lb/in^2 (abs) (34.5-kPa) coil with 0°F (−17.8°C) inlet air, 800-ft/min (243.8-m/min) face velocity, and a 165°F (73.9°C) final air temperature, Table 10 shows that five rows of tubes would be required. This table shows that the coil forms condensate at the rate of 149.8 lb/(h·ft^2) [732.9 kg/(h·m^2)] of net fin area when the final air temperature is 166°F (74.4°C). The coil thus chosen is the first-trial coil, which must be checked against the actual steam conditions as described below.

When a coil is supplied steam at a pressure different from 5 lb/in^2 (abs) (34.5 kPa), multiply the final air temperature given in the 5-lb/in^2 (abs) (34.5-kPa) table for 0°F (−17.8°C), at the

TABLE 10 Steam-Heating-Coil Final Temperatures and Condensate-Formation Rate

Inlet air temp., °F (°C)	Rows of tubes	Face velocity, 800 ft/min (243.8 m/min)	
		Final air temp., °F, (°C)	Condensate, lb/h (kg/h)
0 (−17.8)	2	51 (10.6)	83.7 (37.7)
0 (−17.8)	3	124 (51.1)	111.7 (50.3)
0 (−17.8)	4	148 (64.4)	132.9 (59.8)
0 (−17.8)	5	166 (74.4)	149.8 (67.4)

Temperature-rise correction factor

Actual inlet air temp., °F (°C)	Steam pressure, lb/in² (abs) (kPa)		
	10 (68.9)	15 (103.4)	20 (137.9)
0 (−17.8)	1.054	1.100	1.139
10 (−12.2)	1.010	1.056	1.095
20 (−6.7)	0.966	1.011	1.051

Condensate correction factors

0 (−17.8)	1.063	1.117	1.165
10 (−12.2)	1.019	1.072	1.120
20 (−6.7)	0.974	1.027	1.075

face velocity being used, by the correction factor given in Table 10 for the actual steam pressure and actual inlet air temperature. Thus, for this coil, which is supplied steam at 15 lb/in² (abs) (103.4 kPa) and has an inlet air temperature of 10°F (−12.2°C), the temperature correction factor from Table 10 is 1.056. Add the product of the correction factor and the tabulated final air temperature to the inlet air temperature to obtain the actual final air temperature. Several trials may be necessary before the desired outlet temperature is obtained.

The desired final air temperature for this coil is 150°F (65.6°C). Using the 124°F (51.1°C) final air temperature from Table 10 as the first-trial valve, we get the actual final air temperature = (1.056)(124) + 10 = 141°F (60.6°C). This is too low. Trying the next higher final air temperature gives the actual final air temperature = (1.056)(148) + 10 = 166.5°F (74.7°C). This is higher than required, but the steam supply can be reduced to produce the desired final air temperature. Thus, the coil will be four rows of tubes deep, as Table 10 shows. Hence, the five rows of coils originally indicated will not be needed. Instead, four rows will suffice.

3. Compute the quantity of condensate produced by the coil

Use the same general procedure as in step 2, or actual condensate formed, lb/(h·ft²) = [lb/(h·ft²)] [condensate from 5-lb/in² (abs), 0°F (34.5-kPa, −17.8°C) table] (correction factor from Table 10); or for this coil, (132.9)(1.072) = 142.47 lb/(h·ft²) [690.1 kg/(h·m²)]. Since the coil has a net fin face area of 112.8 ft² (10.5 m²), the total actual condensate formed = (112.8)ft²(142.47) = 16,100 lb/h (7318.2 kg/h).

4. Determine the coil friction loss

Most manufacturers publish a chart or table of coil friction losses in coils having various face velocities and tube rows. Thus, Fig. 2 shows that a coil having a face velocity of 800 ft/min (243.8 m/min) and four rows of tube has a friction loss of 0.45 in (11.4 mm) of water. Figure 2 is a typical friction-loss chart and can be safely used for all routine preliminary coil selections. How-

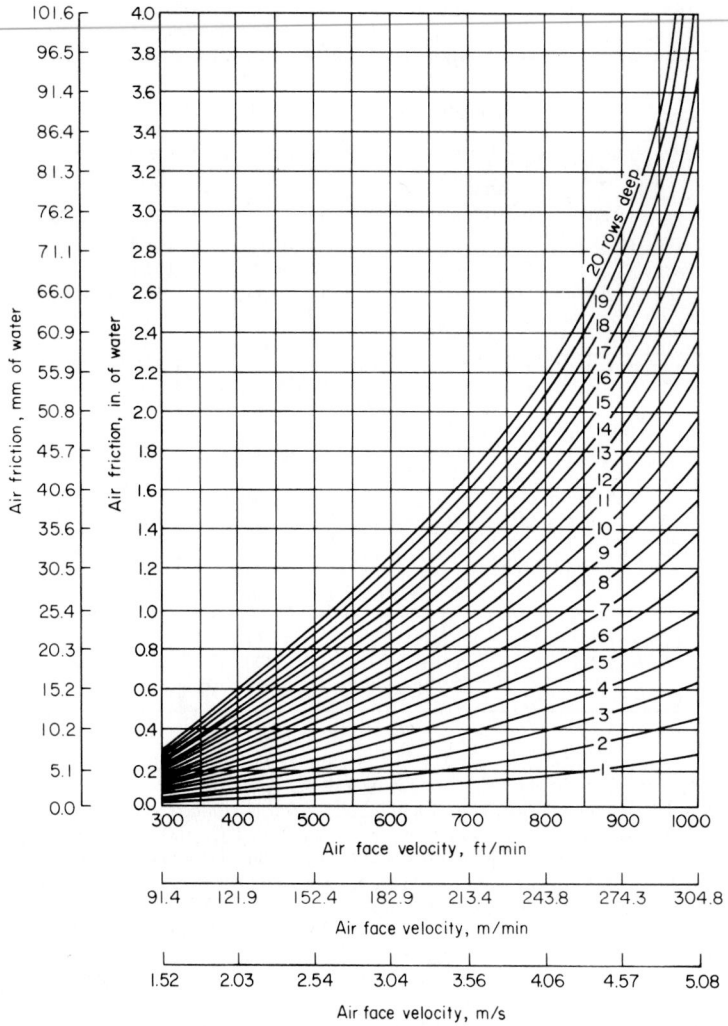

FIG. 2 Heating-coil air-friction chart for air at standard conditions of 70°F (21.1°C) and 29.92 in (76 cm) *(McQuay, Inc.)*

ever, when the final choice of a heating coil is made, use the friction chart or table prepared by the manufacturer of the coil chosen.

5. *Determine the coil dimensions*

Refer to the manufacturer's engineering data for the dimensions of the coil chosen. Each manufacturer has certain special construction features. Hence, there will be some variation in dimensions from one manufacturer to another.

6. *Indicate how the coil will be piped*

The ASHRAE *Guide and Data Book* shows piping arrangements for low- and high-pressure steam heating coils as recommended by various coil manufacturers. Follow the recommendations of the manufacturer whose coil is actually used when the final selection is made.

Related Calculations: Typical variables met in heating-coil selection are the *face velocity*, which varies from 300 to 1000 ft/min (91.4 to 304.8 m/min), with the higher velocities being used for industrial applications, the lower velocities for nonindustrial applications; the *final air temperature*, which ranges between 50 and 300°F (10 and 148.9°C), the lower temperatures being used for ventilation, the higher ones for heating; *steam pressures*, which vary from 2 to 150 lb/in^2 (gage) (13.8 to 1034.1 kPa), with the lower pressures—2 to 15 lb/in^2 (gage) (13.8 to 103.4 kPa)—being the most popular for heating.

Hot-water heating coils are also used for air heating. The general selection procedure is: (*a*) Compute the heating capacity, Btu/h, required from (1.08)(temperature rise of air, °F)(cfm heated).(*b*) Compute the coil face area required, ft^2, from cfm/face velocity, ft/min. Assume a suitable face velocity using the guide given above. (*c*) Compute the logarithmic mean-effective-temperature difference across the coil, using the method given elsewhere in this handbook. (*d*) Compute the required hot-water flow rate, gallons per minute (gal/min), from Btu/h heating capacity/(500)(temperature drop of water, °F). The usual temperature drop of the hot water during passage through the heating coil is 20°F (11.1°C), with water supplied at 150 to 225°F (65.6 to 107.2°C). (*e*) Determine the tube water velocity, ft/s, from (8.33)(gal/min)/(384)(number of tubes in heating coil). The number of tubes in the coil is obtained by making a preliminary selection of the coil, using heating-capacity tables similar to Table 10. The usual hot-water heating coil has a water velocity between 2 and 6 ft /s (0.6 and 1.8 m/s). (*f*) Compute the number of tube rows required from Btu/h heating capacity/(face area ft^2)(logarithmic mean-effective-temperature difference from step 3)(*K* factor from the manufacturer's engineering data). (*g*) Compute the coil air resistance or friction loss, using the manufacturer's chart or table. The usual friction loss ranges from 0.375 to 0.675 in (9.5 to 17.2 mm) of water for commercial applications to about 1.0 in (25 mm) of water for industrial installations.

RADIANT-HEATING-PANEL CHOICE AND SIZING

One room of a building has a heat loss of 13,900 Btu/h (4074.1 W). Choose and size a radiant heating panel suitable for this room. Illustrate the trial method of panel choice. The floor of the room is made of wooden blocks.

Calculation Procedure:

1. Choose the type and location of the heating coil

Compute the heat loss for a given room or building, using the method given earlier in this section. Once the heat loss is known, choose the type of heating panel to use—ceiling or floor. In some rooms or buildings, a combination of floor and ceiling panels may prove more effective than either type used alone. Wall panels are also used but not as extensively as floor and ceiling panels.

In general, ceiling panels are embedded in concrete or plaster, as are wall panels. Floor panels are almost always embedded in concrete. Hence, use of another type of floor—block, tile, wood, or metal—may rule out the use of floor panels. Since this room has a wooden-block floor, a ceiling panel will be chosen.

2. Size the heating panel

Table 11 shows the maximum Btu/h (W) heat output of ⅜-in (9.5-mm) copper-tube ceiling panels. The ⅜-in (9.5-mm) size is popular; however, other sizes—½-, ⅝-, and ⅞-in (12.7-, 15.9-, and 22.2-mm) diameter—are also used, depending on the heat load served. Using ⅜-in (9.5-mm) tubing on 6-in (152.4-mm) centers embedded in a plaster ceiling will provide a heat output of 60 Btu/(h·ft^2) (189.1 W/m^2) of tubing, as shown in Table 11.

To obtain the area of the heated panel, A ft^2, use the relation $A = H_L/P$, where H_L = room or building heat loss, Btu/h; P = panel maximum heat output, Btu/(h·ft^2). For this room, $A = 13,900/60 = 232$ ft^2 (21.6 m^2).

3. Determine the total length of tubing required

Use the appropriate tube-length factor from Table 11, or (2.0)(232) = 464 lin ft (141.4 m) of ⅜-in (9.5-mm) copper tubing.

4. Find the maximum panel tube length

To stay within the commercial limits of smaller hot-water circulating pumps, the maximum tube lengths per panel circuit given in Table 11 are generally used. This tabulation shows that the

TABLE 11 Heating Panel Characteristics

Maximum Btu/(h·ft²) (W/m²) of ⅜-in (9.5-mm) copper tube embedded in ceiling plaster		Maximum Btu/(h·ft²) (W/m²) of copper tube embedded in concrete floor	
	Approx.		*Approx.*
4½ in (114.3 mm) center to center	75 (236.4)	½ in (12.7 mm) 9 in (228.6 mm) center to center	50 (157.6)
6 in (152.4 mm) center to center	60 (189.1)	¾ in (19.1 mm) 9 in (228.6 mm) center to center	50 (157.6)
9 in (228.6 mm) center to center	45 (141.8)	¾ in (19.1 mm) 12 in (304.8 mm) center to center	50 (157.6)
		1 in (25.4 mm) 12 in (304.8 mm) center to center	50 (157.6)

Total length of tube required, ft (m)†

2.7 (0.82)	Where 4½-in (114.3-mm) centers are required
2 (0.61)	Where 6-in (152.4-mm) centers are required
1.3 (0.40)	Where 9-in (228.6-mm) centers are required
1 (0.30)	Where 12-in (304.8-mm) centers are required

Maximum panel unit tube length‡

Ceilings				Floors			
Nominal size, in (mm)	Centers, in (mm)	Btu/(h·ft) (W/m) of tube	Ft (m)	Nominal size, in (mm)	Centers, in (mm)	Btu/(h·ft) (W/m) of tube	Ft (m)
⅜ (9.5)	4½ (114.3)	27 (25.9)	175 (47.9)	½ (12.7)	9 (228.6)	38 (36.5)	220 (67.1)
⅜ (9.5)	6 (152.4)	30 (28.9)	165 (50.3)	¾ (19.1)	9 (228.6)	38 (36.5)	400 (121.9)
⅜ (9.5)	9 (228.6)	34 (32.7)	150 (45.7)	¾ (19.1)	12 (304.8)	50 (48.1)	350 (106.7)
				1 (25.4)	12 (304.8)	50 (48.1)	550 (167.7)

Number of panel circuits of maximum length§

Mains	Ceiling	Floor		
Diam., in (mm)	⅜ in—4½, 6, 9 center-to-center (9.53 mm—114.3, 152.4, 228.6 center-to-center)	½ in—9 center-to-center (12.7 mm—228.6 center-to-center)	¾ in— 9, 12 center-to-center (19.1 mm—228.6, 304.8 center-to-center)	1 in—12 center-to-center (25.4 mm—304.8 center-to-center)
2 (50.8)	27	16	8	5
1½ (38.1)	12	7	4	2
1¼ (31.6)	8	5	2	1
1 (25.4)	4	3	1	1
¾ (19.1)	2	1		

†To arrive at the required lin ft of tube per panel, multiply the ft² of heated panel by the factors given.
‡To keep within the commercial limits of the smaller pumps, as an example the above maximum tube lengths per panel circuit are suggested. These lengths alone will require not more than about a 4-ft (1.2-m) head. This does not include loss in mains.
§Use the information given in the section on maximum panel unit tube length, as given above, which can be supplied, allowing about 0.5 ft (150 mm) head required per 100 ft (30 m) of main (supply and return).

maximum panel unit tube length for ⅜-in (9.5-mm) tubing on 6-in (152.4-mm) centers is 165 ft (50.3 m). Such a length will not require a pump head of more than 4 ft (1.22 m), excluding the head loss in the mains.

5. Determine the number of panels required

Find the number of panels required by dividing the linear tubing length needed by the maximum unit tube length, or 464/165 = 2.81. Use three panels, the next larger whole number, because partial panels cannot be used. To conserve tubing and reduce the first cost of the installation, three panels, each having 155 lin ft (47.2 m) of piping, would be used. Note that the tubing length chosen for the actual panel must be *less than* the maximum length listed in Table 11.

6. Find the required piping main size required

Determine the number of panels required in the remainder of the building. Use Table 11 to select the proper main size for a pressure loss of about 0.5 ft/100 ft (150 mm/30 m) of the main, including the supply and return lines. Thus, if 12 ceiling panels of maximum length were used in the building, a 1½-in (38.1-mm) main would be used.

7. Use the trial method to choose the main size

If the size of the main for the panels required cannot be found in Table 11, compute the total Btu required for the panels from Table 11. Then, by trial, find the size of the main from Table 11 that will deliver approximately, and preferably somewhat more, than this total Btu/h requirement.

For instance, suppose an industrial building requires the following panel circuits:

No. of panel circuits	Tube size and location	Btu (kJ) required from Table 11 (number of circuits × Btu/ft of tube × maximum panel tube length)
4	⅜-in (9.5-mm) tubes on 4½-in (114.3-mm) centers, ceiling	4 × 27 × 175 = 18,900 (19,940)
1	½-in (12.7-mm) tubes on 9-in (228.6-mm) centers, floor	1 × 38 × 220 = 8,360 (8,820)
1	1-in (25.4-mm) tubes on 12-in (304.8-mm) centers, floor	1 × 50 × 550 = 27,500 (29,013)
3	¾-in (19.1-mm) tubes on 9-in (228.6-mm) centers, floor	3 × 38 × 400 = 45,600 (48,108)
9		Total Btu/h (kJ/h) required = 100,360 (105,881)

Trial 1: Assume 100 ft of 1½-in (30 m of 38-mm) main is used for seven floor circuits that are each 220 ft (67.1 m) long and made of ½-in (12.7-mm) tubing on 9-in (228.6-mm) centers. From Table 11 the output of these seven floor circuits is 7 circuits × 38 Btu/(h·ft) of tube (36.5 W/m) × 220 ft (67.1 m) of tubing = 58,520 Btu/h (17.2 kW). Since this output is considerably less than the required output of 100,360 Btu/h (29.4 kW), a 1½-in (38.1-mm) main is not large enough.

Trial 2: Assume 100 ft of 2-in (30 m of 50.8-mm) main for 16 circuits that are each 220 ft (67.1 mm) long and are made of ½-in (12.7-mm) tubing on 9-in (228.6-mm) centers. Then, as in trial 1, 16 × 38 × 220 = 133,760 Btu/h (39.2 kW) delivered. Hence, a 2-in (50.8-mm) main is suitable for the nine panels listed above.

Related Calculations: Use the same general method given here for heating panels embedded in the concrete floor of a building. The liquid used in most panel heating systems is water at about 130°F (54.4°C). This warm water produces a panel temperature of about 85°F (29.4°C) in floors and about 115°F (46.1°C) in ceilings. The maximum water temperature at the boiler is seldom allowed to exceed 150°F (65.6°C).

When the first-floor ceiling of a multistory building is not insulated, the floor above a ceiling panel develops about 17 Btu/(ft²·h) (53.6 W/m²) from the heated panel below. If this type of construction is used, the radiation into the room above can be deducted from the heat loss computed for that room. It is essential, however, to calculate only the heat output in the floor area directly above the heated panel.

Standard references, such as the ASHRAE *Guide and Data Book*, present heat-release data for heating panels in both graphical and tabular form. Data obtained from charts are used in the same way as described above for the tabular data. Tubing data for radiant heating are available from Anaconda American Brass Company.

SNOW-MELTING HEATING-PANEL CHOICE AND SIZING

Choose and size a snow-melting panel to melt a maximum snowfall of 3 in/h (76.2 mm/h) in a parking lot that has an area of 1000 ft² (92.9 m²). Heat losses downward, at the edges, and back of the slab are about 25 percent of the heat supplied; also, there is an atmospheric evaporation loss of 15 percent of the heat supplied. The usual temperature during snowfalls in the locality of the parking lot is 32°F (0°C).

Calculation Procedure:

1. *Compute the hourly snowfall weight rate*

The density of the snow varies from about 3 lb/ft³ at 5°F (48 kg/m³ at −15°C) to about 7.8 lb/ft³ at 34°F (124.9 kg/m³ at 1.1°C). Given a density of 7.3 lb/ft³ (116.9 kg/m³) for this installation, the hourly snowfall weight rate per ft² is (area, ft)(depth, ft)(density) = (1.0)(3/12)(7.3) = 1.83 lb/(h·ft²) [8.9 kg/(h·m²)]. In this computation, the rate of fall of 3 in/h (76.2 mm/h) is converted to a depth in ft by dividing by 12 in/ft.

2. *Compute the heat required for snow melting*

The heat of fusion of melting snow is 144 Btu/lb (334.9 kJ/kg). Since the snow accumulates at the rate of 1.83 lb/(h·ft²) [8.9 kg/(h·m²)], the amount of heat that must be supplied to melt the snow is (1.83)(144) = 264 Btu/(ft²·h) (832.1 W/m²).

Of the heat supplied, the percent lost is 25 + 15 = 40 percent as given. Of this total loss, 25 percent is lost downward and 15 percent is lost to the atmosphere. Hence, the total heat that must be supplied is (1.0 + 0.40)(264) = 370 Btu/(h·ft²) (1166 W/m²). With an area of 1000 ft² (92.9 m²) to be heated, the panel system must supply (1000)(370) = 370,000 Btu/h (108.6 kW).

3. *Determine the length of pipe or tubing required*

Consult the ASHRAE *Guide and Data Book* or manufacturer's engineering data to find the heat output per ft of tubing or pipe length. Suppose the heat output is 50 Btu/(h·ft) (48.1 W/m) of tube. Then the length of tubing required is 370,000/50 = 7400 ft (2256 m).

Some manufacturers rate their pipe or tubing on the basis of the rainfall equivalent of the snowfall and the wind velocity across the heated surface. Where this method is used, compute the heat required to melt the snow as the equivalent amount of heat to vaporize the water. This is Q_e = 1074(0.002V + 0.055)(0.185 − v_a), where Q_e = heat required to vaporize the water, Btu/(h·ft²); V = wind velocity over the heated surface, mi/h; v_a = vapor pressure of the atmospheric air, in Hg.

4. *Determine the quantity of heating liquid required*

Use the relation gpm = 0.125 H_t/dc Δt, for ethylene glycol, the most commonly used heating liquid. In this relation, H_t = total heat required for snow melting, Btu/h; d = density of the heating liquid, lb/ft³; c = specific heat of the heating liquid, Btu/(lb·°F); Δt = temperature loss of the heating liquid during passage through the heating coil, usually taken as 15 to 20 °F (−9.4 to −6.7°C).

Assuming that a 60 percent ethylene glycol solution is used for heating, we have d = 68.6 lb/ft³ (1098.3 kg/m³); c = 0.75 Btu/(lb·°F) [3140 J/(kg·K)]. Since the piping must supply 370,000 Btu/h (108.5 kW), gpm = (0.125)(370,000)/(68.6)(0.75)(20) = 45 gal/min (170.3 L/min) when the temperature loss of the heating liquid is 20°F (−6.7°C).

5. *Size the heater for the system*

The heater must provide at least 370,000 Btu/h (108.5 kW) to the ethylene glycol. If the heater has an overall efficiency of 60 percent, then the required heat input to deliver 370,000 Btu/h (108.5 kW) is 370,000/0.60 = 617,000 Btu/h (180.6 kW).

To avoid a long warmup time at the start of a snowfall, the usual practice is to operate the system for several hours prior to an expected snowfall. The heating liquid temperature during warmup is kept at about 100°F (37.8°C) and the pump is operated at half-speed.

Related Calculations: Use this general method to size snow-melting systems for sidewalks, driveways, loading docks, parking lots, storage yards, roads, and similar areas. To prevent an excessive warmup load on the system, provide for prestorm operation. Without prestorm operation, the load on the heater can be twice the normal hourly load.

Copper tubing and steel pipe are the most commonly used heating elements. For properties of tubing and piping important in snow-melting calculations, see Baumeister—*Standard Handbook for Mechanical Engineers.*

HEAT RECOVERY FROM LIGHTING SYSTEMS FOR SPACE HEATING

Determine the quantity of heat obtainable from 30 water-cooled fluorescent luminaires rated at 200 W each if the entering water temperature is 70°F (21.1°C) and the water flow rate is 1.0 gal/min (3.8 L/min). How much heat can be recovered from 10 air-cooled fluorescent luminaires rated at 100 W each?

Calculation Procedure:

1. *Compute the water-cooled luminaire heat recovery*

Luminaire manufacturers publish heat-recovery data in chart form (Fig. 3). This chart shows that with a flow rate of 0.5 gal/min (1.9 L/min) and a 70°F (21.1°C) entering water temperature, the heat recovery from a luminaire is 74 percent of the total input to the fixture.

For a group of lighting fixtures, total input, W = (number of fixtures)(rating per fixture, W). Or, for this installation, input, W = (30)(200) = 6000 W. Since 74 percent of this input is recoverable by the cooling water, recovered input = 0.74(6000) = 4400 W.

To convert incandescent lighting watts to Btu/h, multiply by 3.4. Where fluorescent lights are used, apply a factor of 1.25 to include the heat gain in the lamp ballast. Thus, the heat available for recovery from these fluorescent lamps is (3.4)(1.25)(4440) = 18,900 Btu/h (5.5 kW).

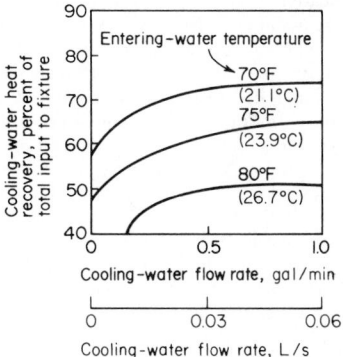

FIG. 3 Heat recovery in water-cooled lighting fixtures.

2. *Compute the temperature rise of the water*

Find the temperature rise of the water from Δt = (Btu/h)/500gpm, where Δt = temperature rise of the water, F; Btu/h = heat available; gpm = water flow rate through the luminaire. Or, Δt = 18,900/500(1.0) = 37.8°F (21.0°C). A temperature rise of this magnitude is seldom used in practice. However, this calculation shows the large amount of heat recoverable with water-cooled lighting fixtures.

3. *Compute the heat recoverable with air cooling*

In the usual air-cooled luminaire, 50 to 70 percent of the input energy is recoverable. Assuming a 60 percent recovery, with a total input of (10)(100) = 1000 W, we see that the energy recoverable is 0.60(1000) = 600 W. Converting to the heat recoverable gives (3.4)(1.25)(600) = 2545 Btu/h (745.9 W).

Related Calculations: Heat recovery from lighting fixtures is receiving increasing attention for many different structures because substantial fuel savings are possible. The water or air heated by the lighting is used to heat the supply or return air supplied to the conditioned space. Where the heat recovered must be rejected, as in the summer, either a cooling tower (for water) or an air-cooled condenser (for air) may be used.

Other popular sources of heat are refrigeration condensers and electric motors. In many installations the heat is absorbed from the condenser or motors, or both, by air.

AIR-CONDITIONING-SYSTEM HEAT-LOAD DETERMINATION—GENERAL METHOD

Show how to compute the total heat load for an air-conditioned industrial building fitted with windows having shades, internal heat loads from people and machines, and heat transmission gains through the walls, roof, and floor. Use the ASHRAE *Guide and Data Book* as a data source.

Calculation Procedure:[1]

1. *Determine the design outdoor and indoor conditions*

Refer to the ASHRAE *Guide and Data Book* (called *Guide*) for the state and city in which the building is located. Read from the *Guide* table for the appropriate city the design outdoor dry-bulb and wet-bulb temperatures. At the same time, determine from the *Guide* the indoor design conditions—temperature and relative humidity for the type of application being considered. The *Guide* lists a variety of typical applications such as apartment houses, motels, hotels, industrial plants, etc. It also lists the average summer wind velocity for a variety of locations. Where the exact location of a plant is not tabulated in the *Guide*, consult the nearest local branch of the weather bureau for information on the usual summer outdoor high and low dry- and wet-bulb temperatures, relative humidity, and velocity.

2. *Compute the sunlight heat gain*

The sunlight heat gain results from the solar radiation through the glass in the building's windows and the materials of construction in certain of the building's walls. If the glass or wall of a building is shaded by an adjacent solid structure, the sunlight heat gain for that glass and wall is usually neglected. The same is true for the glass and wall of the building facing the north.

Compute the glass sunlight heat gain, Btu/h, from (glass area, ft^2)(equivalent temperature difference from the appropriate *Guide* table)(factor for shades, if any are used). The equivalent temperature difference is based on the time of day and orientation of the glass with respect to the points of the compass. A latitude correction factor may have to be applied if the building is located in a tropical area. Use the equivalent temperature difference for the time of day on which the heat-load estimate is based. Several times may be chosen to determine at which time the greatest heat gain occurs. Where shades are used in the building, choose a suitable shade factor from the appropriate *Guide* table and insert it in the equation above.

Compute the sunlight heat gain, Btu/h, for the appropriate walls and the roof from (wall area, ft^2)(equivalent temperature difference from *Guide* for walls) [coefficient of heat transmission for the wall, Btu/(h·ft^2·°F)]. For the roof, find the heat gain, Btu/h, from (roof area, ft^2)(equivalent temperature difference from the appropriate *Guide* table for roofs) [coefficient of heat transmission for the roof, Btu/(h·ft^2·°F)].

3. *Compute the transmission heat gain*

All the glass in the building windows is subject to transmission of heat from the outside to the inside as a result of the temperature difference between the outdoor and indoor dry-bulb temperatures. This transmission gain is commonly called the *all-glass gain*. Find the all-glass transmission heat gain, Btu/h, from (total window-glass area, ft^2)(outdoor design dry-bulb temperature, °F—indoor design dry-bulb temperature,°F)[coefficient of heat transmission of the glass, Btu/(h·ft^2·°F), from the appropriate *Guide* table].

Compute the heat transmission, Btu/h, through the shaded walls, if any, from (total shaded wall area, ft^2)(equivalent temperature difference for shaded wall from the appropriate *Guide*

---

[1]SI units are given in later numerical procedures.

table, °F) [coefficient of heat transmission of the wall material, Btu/(h·ft²·°F), from the appropriate *Guide* table)].

Where the building has a machinery room or utility room that is not air-conditioned and is next to a conditioned space, and the temperature in the utility room is higher than in the conditioned space, find the heat gain, Btu/h, from (area of utility or machine room partition, ft²)(utility or machine room dry-bulb temperature, °F − conditioned-space dry-bulb temperature, °F)[coefficient of heat transmission of the utility or machine room partition, Btu/(h·ft²·°F), from the appropriate *Guide* table].

For buildings having a floor contacting the earth, or over unventilated and unheated basements, there is generally *no* heat gain through the floor because the ground is usually at a lower temperature than the floor. Where the floor is above the ground and in contact with the outside air, find the heat gain, Btu/h, through the floor from (floor area, ft²)(design outside dry-bulb temperature, °F—design inside dry-bulb temperature, °F) [coefficient of heat transmission of the floor material, Btu/(h·ft²·°F), from the appropriate *Guide* table]. When a machine room or utility room is below the floor, use the same relation but substitute the machine room or utility room dry-bulb temperature for the design outside dry-bulb temperature.

For floors above the ground, some designers reduce the difference between the design outdoor and indoor dry-bulb temperatures by 5°F (−15°C); other designers use the shaded-wall equivalent temperature difference from the *Guide*. Either method, or that given above, will provide safe results.

4.   *Compute the infiltration heat gain*

Use the relation infiltration heat gain, Btu/h = (window crack length, ft) [window infiltration, ft³/(ft·min) from the appropriate *Guide* table] (design outside dry-bulb temperature, °F − design inside dry-bulb temperature, °F)(1.08). Three aspects of this computation require explanation.

The window crack length used is usually one-half the total crack length in all the windows. Infiltration through cracks is caused by the wind acting on the building. Since the wind cannot act on all sides of a building at once, one-half the total crack length is generally used (but never less than one-half) in computing the infiltration heat gain. Note that the crack length varies with different types of windows. Thus the *Guide* gives, for metal sash, crack length = total perimeter of the movable section. For double-hung windows, the crack length = three times the width plus twice the height.

The window infiltration rate, ft³/(min·ft) of crack, is given in the *Guide* for various wind velocities. Some designers use the infiltration rate for a wind velocity of 10 mi/h (16.1 km/h); others use 5 mi/h (8.1 km/h). The factor 1.08 converts the computed infiltration to Btu/h (×0.32 = W).

5.   *Compute the outside-air bypass heat load*

Some outside air may be needed in the conditioned space to ventilate fumes, odors, and other undesirables in the conditioned space. This ventilation air imposes a cooling or dehumidifying load on the air conditioner because the heat or moisture, or both, must be removed from the ventilation air. Most air conditioners are arranged to permit some outside air to bypass the cooling coils. The bypassed outdoor air becomes a load within the conditioned space similar to infiltration air.

Determine heat load, Btu/h, of the outside air bypassing the air conditioner from (cfm of ventilation air)(design outdoor dry-bulb-temperature, °F − design indoor dry-bulb temperature, °F)(air-conditioner bypass factor)(1.08).

Find the ventilation-air quantity by multiplying the number of people in the conditioned space by the cfm per person recommended by the *Guide*. The cfm (m³/min) per person can range from a minimum of 5 (0.14) to a high of 50 (1.42) where heavy smoking is anticipated. If industrial processes within the conditioned space require ventilation, the air may be supplied by increasing the outside air flow, by a local exhaust system at the process, or by a combination of both. Regardless of the method used, outside air must be introduced to make up for the air exhausted from the conditioned space. The sum of the air required for people and processes is the total ventilation-air quantity.

Until the air conditioner is chosen, its bypass factor is unknown. However, to solve for the outside-air bypass heat load, a bypass factor must be applied. Table 12 shows typical bypass factors for various applications.

TABLE 12 Typical Bypass Factors°

a. For various applications

Coil bypass factor†	Type of application	Example
0.30–0.50	A *small* total load or a load that is somewhat larger with a low sensible heat factor (high latent load)	Residence
0.20–0.30	Typical comfort application with a *relatively small* total load or a low sensible heat factor with a somewhat larger load	Residence; small retail shop; factory
0.10–0.20	Typical comfort application	Department store; bank; factory
0.05–0.10	Applications with high internal sensible loads or requiring a large amount of outdoor air for ventilation	Department store; restaurant; factory
0–0.10	All outdoor air applications	Hospital; operating room; factory

b. For finned coils

	Without sprays		With sprays†	
	8 fins/in (25.4 mm)	14 fins/in (25.4 mm)	8 fins/in (25.4 mm)	14 fins/in (25.4 mm)
Depth of coils, rows	Velocity, ft/min (m/min)			
	300–700 (91.4–213.4)	300–700 (91.4–213.4)	300–700 (91.4–213.4)	300–700 (91.4–213.4)
2	0.42–0.55	0.22–0.38		
3	0.27–0.40	0.10–0.23		
4	0.19–0.30	0.50–0.14	0.12–0.22	0.03–0.10
5	0.12–0.23	0.02–0.09	0.08–0.14	0.01–0.08
6	0.08–0.18	0.01–0.06	0.06–0.11	0.01–0.05
8	0.03–0.08	. . .	0.02–0.05	

°Carrier Air Conditioning Company.
†The bypass factor with spray coils is decreased because spray provides more surface for contacting the air.

6. *Compute the heat load from internal heat sources*

Within an air-conditioned space, heat is given off by people, lights, appliances, machines, pipes, etc. Find the sensible heat, Btu/h, given off by people by taking the product (number of people in the air-conditioned space)(sensible heat release per person, Btu/h, from the appropriate *Guide* table). The sensible heat release per person varies with the activity of each person (seated, at rest, doing heavy work) and the room dry-bulb temperature. Thus, at 80°F (26.7°C), a person doing heavy work in a factory will give off 465 Btu/h (136.3 W) sensible heat; seated at rest in a theater at 80°F (26.7°C), a person will give off 195 Btu/h (57.2 W), sensible heat.

Find the heat, Btu/h, given off by electric lights from (wattage rating of all installed lights) (3.4). Where the installed lighting capacity is expressed in kilowatts, use the factor 3413 instead of 3.4.

For electric motors, find the heat, Btu/h, given off from (total installed motor hp)(2546)/motor efficiency expressed as a decimal. The usual efficiency assumed for electric motors is 85 percent.

Many other sensible-heat-generating devices may be used in an air-conditioned space. These devices include restaurant, beauty shop, hospital, gas-burning, and kitchen appliances. The *Guide* lists the heat given off by a variety of devices, as well as pipes, tanks, pumps, etc.

7. Compute the room sensible heat

Find the sum of the sensible heat gains computed in steps 2 (sunlight heat gain), 3 (transmission heat gain), 4 (infiltration heat gain), 5 (outside air heat gain), 6 (internal heat sources). This sum is the room sensible-heat subtotal.

A further sensible heat gain may result from supply-duct heat gain, supply-duct-leakage loss, and air-conditioning-fan horsepower. To the sum of these losses, a safety factor is usually added in the form of a percentage, since all the losses are also generally expressed as a percentage. The *Guide* provides means to estimate each loss and the safety factor. Assuming the sum of the losses and safety factor is x percent, the room sensible heat load, Btu/h, is $(1 + 0.01 \times x)$(room sensible heat subtotal, Btu/h).

8. Compute the room latent heat load

The room latent heat load results from the moisture entering the room with the infiltration air and bypass ventilation air, the moisture given off by room occupants, and any other moisture source such as open steam kettles, sterilizers, etc.

Find the infiltration-air latent heat load, Btu/h, from (cfm infiltration)(moisture content of outside air design at design outdoor conditions, g/lb − moisture content of the conditioned air at the design indoor conditions, g/lb)(0.68). Use a similar relation for the bypass ventilation air, or Btu/h latent heat = (cfm ventilation)(moisture content of the outside air at design outdoor conditions, g/lb − moisture content of conditioned air at design indoor conditions, g/lb)(bypass factor)(0.68).

Find the latent heat gain from the room occupants, Btu/h, from (number of occupants in the conditioned space)(latent heat gain, Btu/h per person, from the appropriate *Guide* table). Be sure to choose the latent heat gain that applies to the activity *and* conditioned-room dry-bulb temperature.

Nonhooded restaurant, hospital, laboratory, and similar equipment produces both a sensible and latent heat load in the conditioned space. Consult the *Guide* for the latent heat load for each type of unit in the space. Find the latent heat load of these units, Btu/h, from (number of units of each type)(latent heat load, Btu/h, per unit).

Take the sum of the latent heat loads for infiltration, ventilation bypass air, people, and devices. This sum is the room latent-heat subtotal, if water-vapor transmission through the building surfaces is neglected.

Water vapor flows through building structures, resulting in a latent heat load whenever a vapor-pressure difference exists across a structure. The latent heat load from this source is usually insignificant in comfort applications and need be considered only in low or high dew-point applications. Compute the latent heat gain from this source, using the appropriate *Guide* table, and add it to the room latent-heat subtotal.

Factors for the supply-duct leakage loss and for a safety margin are usually applied to the above sum. When all the latent heat subtotals are summed, the result is the room latent-heat total.

9. Compute the outside-air heat

Air brought in for space ventilation imposes a sensible and latent heat load on the air-conditioning apparatus. Compute the sensible heat load, Btu/h, from (cfm outside air)(design outdoor dry-bulb temperature − design indoor dry-bulb temperature)(1 − bypass factor)(1.08). Compute the latent heat load, Btu/h, from (cfm outside air) (design outdoor moisture content, g/lb − design indoor moisture content, g/lb)(1 − bypass factor)(0.68). Apply percentage factors for return duct heat and leakage gain, pump horsepower, dehumidifier, and piping loss; see the *Guide* for typical values.

10. Compute the grand-total heat and refrigeration tonnage

Take the sum of the room total heat and outside-air heat. The result is the grand-total heat load of the space, Btu/h.

Compute the refrigeration load, tons, from (grand-total heat, Btu/h)/[12,000 Btu/(h·ton) (3577 W) of refrigeration]. A refrigeration system having the next higher standard rating is generally chosen.

11.   *Compute the sensible heat factor and the apparatus dew point*

For any air-conditioning system, sensible heat factor = (room sensible heat, Btu/h)/(room total heat, Btu/h). Using a psychrometric chart and the known room conditions, we find the apparatus dew point. An alternative, and quicker, way to find the apparatus dew point is to use the Carrier *Handbook of Air Conditioning System Design* tables.

12.   *Compute the quantity of dehumidified air required*

Determine the dehumidified air temperature rise, °F, from (1 − bypass factor)(design indoor temperature, °F − apparatus dew point, °F). Compute the dehumidified air quantity, ft^3/min, from (room sensible heat, Btu/h)/1.08 (dehumidified air temperature rise, °F).

Related Calculations: The general procedure given above is valid for all types of air-conditioned spaces—offices, industrial plants, residences, hotels, apartment houses, motels, etc. Use the ASHRAE *Guide and Data Book* or the Carrier *Handbook of Air Conditioning System Design* as a source of data for the various calculations. Application of this method to an actual building is shown in the next calculation procedure.

In actual design work, a calculations form incorporating the calculations shown above is generally used. Such forms are obtainable from equipment manufacturers. Since the usual form does not provide any explanation of the calculations, the present calculation procedure is a useful guide to using the form. Refer to the Carrier *Handbook of Air Conditioning System Design* for one such form.

In using SI units in air-conditioning design calculations, the same general steps are followed as given above. Numerical usage of SI is shown in the next calculation procedure.

AIR-CONDITIONING-SYSTEM HEAT-LOAD DETERMINATION—NUMERICAL COMPUTATION

Determine the required capacity of an air-conditioning system to serve the industrial building shown in Fig. 4. The outside walls are 8-in (203.2-mm) brick with an interior finish of ⅜-in (9.5-mm) gypsum lath plastered and furred. A 6-in (152.4-mm) plain poured-concrete partition separates the machinery room from the conditioned space. The roof is 6-in (152.4-mm) concrete covered with ½-in (12.7-mm) thick insulating board, and the floor is 2-in (50.8-mm) concrete. The windows are double-hung, metal-frame locked units with light-colored shades three-quarters drawn. Internal heat loads are: 100 people doing light assembly work; twenty-five 1-hp (746-W) motors running continuously at full load; 20,000 W of light kept on at all times. The building is located in Port Arthur, Texas, at about 30° north latitude. The desired indoor design conditions are 80°F (26.7°C) dry bulb, 67°F (19.4°C) wet bulb, and 51 percent relative humidity. Air-conditioning equipment will be located in the machinery room. Use the general method given in the previous calculation procedure.

Calculation Procedure:

1.   *Determine the design outdoor and indoor conditions*

The ASHRAE *Guide and Data Book* lists the design dry-bulb temperature in common use for Port Arthur, Texas, as 95°F (35°C) and the design wet-bulb temperature as 79°F (26.1°C). Design indoor temperature and humidity conditions are given; if they were not given, the recommended conditions given in the *Guide* for an industrial building housing light assembly work would be used.

2.   *Compute the sunlight heat gain*

The east, south, and west windows and walls of the building are subject to sunlight heat gains. North-facing walls are neglected because the sunlight heat gain is usually less than the transmission heat gain. Reference to the *Guide* table for sunlight radiation through glass shows that the largest amount of heat radiation occurs through the east and west walls. The maximum radiation is 181 Btu/(h·ft^2) (570.5 W/m^2) of glass area at 8 a.m. for the east wall and the same for the west wall at 4 p.m. Radiation through the glass in the south wall never reaches this magnitude. Hence, only the east or west wall need be considered. Since the west wall has 22 windows compared with 20 in the east wall, the west-wall sunlight heat gain will be used because it has a *larger* heat gain. (If both walls had an equal number of windows, either wall could be used.) When the window

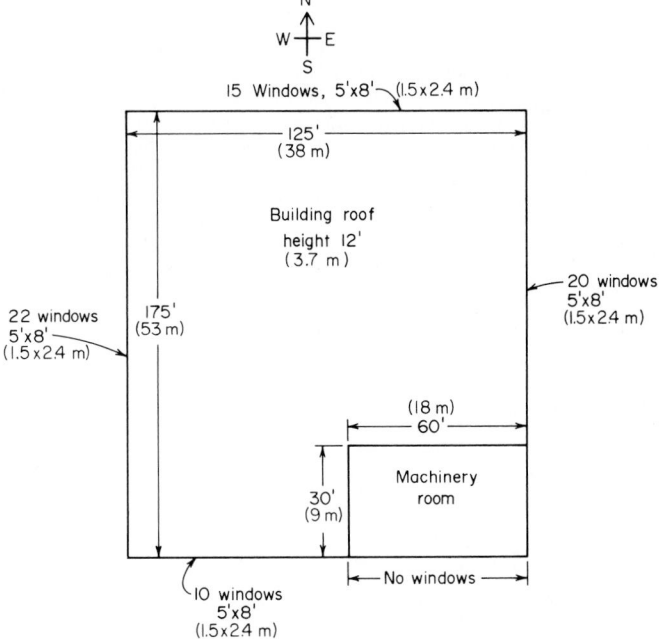

N wall area = 125 (12) − 15 (40) = 900 ft²(83.6 m²)

S wall area = 125 (12) − 5x8x10 = 1100 ft²(102.2 m²)

E wall area = (175)(12) − 5x8x20 = 1300 ft²(120.8 m²)

W wall area = (175)(12) − 5x8x22 = 1220 ft²(113.3 m²)

Roof area = (175)(125) = 21,875 ft²(2032.3 m²)

Glass area = (20x40) + (10x40) + 22(40) + 15(40) = 2680 ft²(248.9 m²)

Partition area = 90x12 = 1080 ft²(100.3 m²)

FIG. 4 Industrial building layout.

shades are normally three-quarters drawn, a value 0.6 times the tabulated sunlight radiation can be used. Hence, the west-glass sunlight heat gain = (22 windows)(5 × 8 ft each)(181)(0.6) = 95,600 Btu/h (28,020.4 W).

For the same time of day, 4 p.m., the *Guide* table shows that the east-glass radiation is 0 Btu/(h·ft²) and the south glass is 2 Btu/(h·ft²) (6.3 W/m²). Hence, the south-glass sunlight heat gain = (10 windows) (5 × 8 ft each)(2)(0.6) = 480 Btu/h (140.7 W).

The same three walls, and the roof, are also subject to sunlight heat gains. Reference to the *Guide* shows that with 8-in (203-mm) walls the temperature difference resulting from sunlight heat gains is 15°F (8.3°C) for south walls and 20°F (−6.7°C) for east and west walls. At 4 p.m. the roof temperature difference is given as 40°F (4.4°C). Hence, sunlight gain, south wall = (wall area, ft²)(temperature difference, °F)[wall coefficient of heat transfer, Btu/(h·ft²·°F)], or (1100)(15)(0.30) = 4950 Btu/h (1450.9 W). Likewise, the east-wall sunlight heat gain = (1300)(20)(0.30) = 7800 Btu/h (2286.2 W); the west-wall sunlight heat gain = (1220)(20)(0.30) = 7,320 Btu/h (2145.5 W); the roof sunlight heat gain = (21,875)(40)(0.33) = 289,000 Btu/h (84.7 kW). Note that the wall and roof coefficients of heat transfer are obtained from the appropriate *Guide* table.

The sum of the sunlight heat gains gives the total sunlight gain, or 405,150 Btu/h (118.7 kW).

3. Compute the glass transmission heat gain

All the glass in the building is subject to a transmission heat gain. Find the all-glass transmission heat gain, Btu/h, from (total glass area, ft²)(outdoor design dry-bulb temperature, °F − indoor design dry-bulb temperature, °F)[coefficient of heat transmission of glass, Btu/(h·ft²·°F), from *Guide*], or (2680)(95 − 80)(1.13) = 45,400 Btu/h (13.3 kW).

The transmission heat gain of the south, east, and west walls can be neglected because the sunlight heat gain is greater. Hence, only the north-wall transmission heat gain need be computed. For unshaded walls, the transmission heat gain, Btu/h, is (wall area, ft²)(design outdoor dry-bulb temperature, °F − design indoor dry-bulb temperature, °F)[coefficient of heat transmission, Btu/(h·ft²·°F)] = (900)(95 − 80)(0.30) = 4050 Btu/h (1187.1 W).

The heat gain from the ground can be neglected because the ground is usually at a lower temperature than the floor. Thus, the total transmission heat gain is the sum of the individual gains, or 66,530 Btu/h (19.5 kW).

4. Compute the infiltration heat gain

The total crack length for double-hung windows is 3 × width + 2 × height, or (67 windows)[(3 × 5) + (2 × 8)] = 2077 ft (633.1 m). By using one-half the total length, or 2077/2 = 1039 ft (316.7 m), and a wind velocity of 10 mi/h (16.1 km/h), the leakage cfm is (crack length, ft)(leakage per ft of crack) = (1039)(0.75) = 770 ft³/min, or 60(779) = 46,740 ft³/h (1323 m³/h).

The heat gain due to infiltration through the window cracks is (leakage, ft³/min) (design outdoor dry-bulb temperature, °F − design indoor dry-bulb temperature, °F)(1.08), or (779)(95 − 80)(1.08) = 12,610 Btu/h (3.7 kW).

5. Compute the outside-air bypass heat load

For factories, the *Guide* recommends a ventilation air quantity of 10 ft³/min (0.28 m³/min) per person. Local codes may require a larger quantity; hence the codes should be checked before a final choice is made of the ventilation-air quantity used per person. Since there are 100 people in this factory, the required ventilation quantity is 100(10) = 100 ft³/min (28 m³/min). Next, the bypass factor for the air-conditioning equipment must be chosen.

Table 12 shows that the usual factory air-conditioning equipment has a bypass factor ranging between 0.10 and 0.20. Assume a value of 0.10 for this installation.

The heat load, Btu/h, of the outside air bypassing the air conditioner is (cfm of ventilation air)(design outdoor dry-bulb temperature, °F − design indoor dry-bulb temperature, °F)(air-conditioner bypass factor)(1.08). Hence, (1000)(95 − 80)(0.10)(1.08) = 1620 Btu/h (474.8 W).

6. Compute the heat load from internal heat sources

The internal heat sources in this building are people, lights, and motors. Compute the sensible heat load of the people from Btu/h = (number of people in the air-conditioned space)(sensible heat release per person, Btu/h, from the appropriate *Guide* table). Thus, for this building with an 80°F (26.7°C) indoor dry-bulb temperature and 100 occupants doing light assembly work, the heat load produced by people = (100)(210) = 21,000 Btu/h (6.2 kW).

The motor heat load, Btu/h, is (motor hp)(2546)/motor efficiency. Given an 85 percent motor efficiency, the motor heat load = (25)(2546)/0.85 = 75,000 Btu/h (21.9 kW). Thus, the total internal heat load = 21,000 + 68,000 + 75,000 = 164,000 Btu/h (48.1 kW).

7. Compute the room sensible heat

Find the sum of the sensible heat gains computed in steps 2, 3, 4, 5, and 6. Thus, sensible heat load = 405,150 + 66,530 + 12,610 + 1620 + 164,000 = 649,910 Btu/h (190.5 kW), say 650,000 Btu/h (190.5 kW). Using an assumed safety factor of 5 percent to cover the various losses that may be encountered in the system, we find room sensible heat = (1.05)(650,000) = 682,500 Btu/h (200.0 kW).

8. Compute the room latent-heat load

The room latent load results from the moisture entering the air-conditioned space with the infiltration and bypass air, moisture given off by room occupants, and any other moisture sources.

Find the infiltration heat load, Btu/h, from (cfm infiltration)(moisture content of outside air at design outdoor conditions, g/lb − moisture content of the conditioned air at the design indoor conditions, g/lb)(0.68). Using a psychrometric chart or the *Guide* thermodynamic tables, we get the infiltration latent-heat load = (779)(124 − 78)(0.68) = 24,400 Btu/h (7.2 kW).

Using a similar relation for the ventilation air gives Btu/h latent heat = (cfm ventilation air)(moisture content of outside air at design outdoor conditions, g/lb − moisture content of the conditioned air at the design indoor conditions, g/lb)(bypass factor)(0.68). Or, (1000)(124 − 78)(0.10)(0.68) = 3130 Btu/h (917.4 W).

The latent heat gain from room occupants is Btu/h = (number of occupants in the conditioned space)(latent heat gain, Btu/h per person, from the appropriate *Guide* table). Or, (100)(450) = 45,000 Btu/h (13.2 kW).

Find the latent heat gain subtotal by taking the sum of the above heat gains, or 24,400 + 3130 + 45,000 = 72,530 Btu/h (21.2 kW). Using an allowance of 5 percent for supply-duct leakage loss and a safety margin gives the latent heat gain = (1.05)(72,530) = 76,157 Btu/h (22.3 kW).

9. *Compute the outside heat*

Compute the sensible heat load of the outside ventilation air from Btu/h = (cfm outside ventilation air)(design outdoor dry-bulb temperature − design indoor dry-bulb temperature)(1 − bypass factor)(1.08). For this system, with 1000 ft³/min (28.3 m³/min) outside air ventilation, sensible heat = (1000)(95 − 80)(1 − 0.10)(1.08) = 14,600 Btu/h (4.3 kW).

Compute the latent-heat load of the outside ventilation air from (cfm outside air) × (design outdoor moisture content, g/lb − design indoor moisture content, g/lb)(1 − bypass factor)(0.68). Using the moisture content from step 8 gives (1000)(124 − 78)(1 − 0.1) × (0.68) = 28,200 Btu/h (8.3 kW).

10. *Compute the grand-total heat and refrigeration tonnage*

Take the sum of the room total heat and the outside-air sensible and latent heat. The result is the grand-total heat load of the space, Btu/h (W or kW).

The room total heat = room sensible-heat total + room latent-heat total = 682,500 + 76,157 = 768,657 Btu/h (225.3 kW). Then the grand-total heat = 768,657 + 14,600 + 28,200 = 811,457 Btu/h, say 811,500 Btu/h (237.9 kW).

Compute the refrigeration load, tons, from (grand-total heat, Btu/h)/(12,000 Btu/h per ton of refrigeration), or 811,500/12,000 = 67.6 tons (237.7 kW), say, 70 tons (246.2 kW).

The quantity of cooling water required for the refrigeration-system condenser is Q gal/min = 30 × tons of refrigeration/condenser water-temperature rise, °F. Assuming a 75°F (23.9°C) entering water temperature and 95°F (35°C) leaving water temperature, which are typical values for air-conditioning practice Q = 30(70)/95 − 75 = 105 gal/min (397.4 L/min).

11. *Compute the sensible heat factor and apparatus dew point*

For any air-conditioning system, sensible heat factor = (room sensible heat, Btu/h)/(room total heat, Btu/h) = 682,500/768,657 = 0.888.

The *Guide* or the Carrier *Handbook of Air Conditioning Design* gives an apparatus dew point of 58°F (14.4°C), closely.

12. *Compute the quantity of dehumidified air required*

Determine the dehumidified air temperature rise first from F = (1 − bypass factor) (design indoor temperature, °F − apparatus dew point, °F), or F = (1 − 0.1)(80 − 58) = 19.8°F (−6.8°C).

Next, compute the dehumidified air quantity, ft³/min from (room sensible heat, Btu/h)/1.08(dehumidified air temperature rise, °F), or (682,500)/1.08(10.8) = 34,400 ft³/min (973.9 m³/min).

Related Calculations: Use this general procedure for any type of air-conditioned building or space—industrial, office, hotel, motel, apartment house, residence, laboratories, school, etc. Use the ASHRAE *Guide and Data Book* or the Carrier *Handbook of Air Conditioning System Design* as a source of data for the various calculations. In comparing the various values from the *Guide* used in this procedure, note that there may be slight changes in certain tabulated values from one edition of the *Guide* to the next. Hence, the values shown may differ slightly from those in the current edition. This should not cause concern, because the procedure is the same regardless of the values used.

AIR-CONDITIONING-SYSTEM COOLING-COIL SELECTION

Select an air-conditioning cooling coil to cool 15,000 ft^3/min (424.7 m^3/min) of air from 85°F (29.4°C) dry bulb, 67°F (19.4°C) wet bulb, 57°F (13.9°C) dew point, 38 percent relative humidity, to a dry-bulb temperature of 65°F (18.3°C) with cooling water at 50°F (10°C). Suppose the air were cooled below the dew point of the entering air. How would the calculation procedure differ?

Calculation Procedure

1. Compute the weight of air to be cooled

From a psychrometric chart (Fig. 5) find the specific volume of the entering air as 13.75 ft^3/lb (0.86 m^3/kg). To convert the air flow in ft^3/min to lb/h, use the relation lb/h = 60 cfm/v_s, where cfm = air flow, ft^3/min; v_s = specific volume of the entering air, ft^3/lb. Hence for this cooling coil, lb/h = 60(15,000)/13.75 = 65,500 lb/h (29,773 kg/h).

Where a cooling coil is rated by the manufacturer for air at 70°F (21.1°C) dry bulb and 50 percent relative humidity, as is often done, use the relation lb/h = 4.45cfm. Thus if this coil were rated for air at 70°F (21.1°C) dry bulb, the weight of air to be cooled would be lb/h = 4.45(15,000) = 66,800 (30,364 kg/h).

Since this air quantity is somewhat greater than when the entering air specific volume is used, and since cooling coils are often rated on the basis of 70°F (21.1°C) dry-bulb air, this quantity, 66,800 lb/h (30,364 kg/h), will be used. The procedure is the same in either case. [*Note:* 4.45 = 60/135 ft^3/lb, the specific volume of air at 70°F (21.1°C) and 50 percent relative humidity.]

2. Compute the quantity of heat to be removed

Use the relation $H_r = ws\Delta t$, where H_r = heat to be removed from the air, Btu/h; w = weight of air cooled, lb/h; s = specific heat of air = 0.24 Btu/(lb·°F); Δt = temperature drop of the air = entering dry-bulb temperature, °F − leaving dry-bulb temperature, °F. For this coil, H_r = (66,800)(0.24)(85 − 65) = 321,000 Btu/h (94.1 kW).

3. Compute the quantity of cooling water required

The quantity of cooling water required in gpm = $H_r/500 \, \Delta t_w$, where Δt_w = leaving water temperature, °F − entering water temperature, °F. Since the leaving water temperature is not known, a value must be assumed. The usual temperature rise of water during passage through an air-conditioning cooling coil is 4 to 12°F (2.2 to 6.7°C). Assuming a 10°F (5.6°C) rise, which is a typical value, gpm required = 321,000/500(10) = 64.2 gal/min (4.1 L/s).

4. Determine the logarithmic mean temperature difference

Use Fig. 4 in the heat transfer section of this handbook to determine the logarithmic mean temperature difference for the cooling coil. In this chart, greatest terminal difference = entering air temperature, °F − leaving water temperature, °F = 85 − 60 = 25°F (13.9°C), and least terminal temperature difference = leaving air temperature, °F − entering water temperature, °F = 65 − 50 = 15°F (8.3°C). Entering Fig. 4 at these two temperature values gives a logarithmic mean temperature difference (LMTD) of 19.5°F (10.8°C).

5. Compute the coil core face area

The coil core face area is the area exposed to the air flow; it does not include the area of the mounting flanges. Compute the coil core face area from $A_c = cfm/V_a$, where V_a = air velocity through the coil, ft/min. The usual air velocity through the coil, often termed face velocity, ranges from 300 to 800 ft/min (91.4 to 243.8 m/min) although special designs may use velocities down to 200 ft/min (60.9 m/min) or up to 1200 ft/min (365.8 m/min). Assuming a face velocity of 500 ft/min (152.4 m/min) gives A_c = 15,000/500 = 30 ft^2 (2.79 m^2).

6. Select the cooling coil for the load

Using the engineering data provided by the manufacturer whose coil is to be used, choose the coil. Table 13 summarizes typical engineering data provided by a coil manufacturer. This table shows that two 15.4-ft^2 1.43-m^2 coils placed side by side will provide a total face area of 2 × 15.4 = 30.8 ft^2. Hence, the actual air velocity through the coil is $V_a = cfm/A_c$ = 15,000/30.8 = 487 ft/min (148.4 m/min).

TABLE 13 Typical Cooling-Coil Characteristics

Face area, ft^2 (m^2)	14.0 (1.3)	15.4 (1.43)	17.9 (1.66)
Tube length, ft ·in (m)	5–6 (1.68)	6–0 (1.83)	7–0 (2.13)
Water velocity, ft/min (m/ min)	3.59 gal/min (4.14 L/min)	3.59 gal/min (4.14 L/min)	3.59 gal/min (4.14 L/min)

Coil heat-transfer factors, k = Btu/(h·ft^2·°F) LMTD row [W/(m^2·K)]

Water velocity, ft/min (m/min)	Air velocity, ft/min (m/min)		
	400 (121.9)	500 (152.4)	550 (167.6)
113 (34.4)	154 (46.9)	162 (49.4)	170 (51.8)
115 (35.1)	158 (48.2)	167 (50.9)	175 (53.3)
117 (35.7)	161 (49.1)	172 (52.4)	176 (53.6)

Coil water pressure drop, ft (m) of water per row

Tube length, ft · in (m)	Water velocity, ft/min (m/min)		
	90 (27.4)	120 (36.6)	150 (45.7)
5-6 (1.68)	0.16 (0.04)	0.26 (0.08)	0.38 (0.12)
6-0 (1.83)	0.18 (0.05)	0.29 (0.09)	0.42 (0.13)
7-0 (2.13)	0.21 (0.06)	0.32 (0.10)	0.46 (0.14)

Header water pressure drop, ft (m) of water

Coil type	Water velocity, ft/min (m/min)		
	90 (27.4)	120 (36.6)	150 (45.7)
A	0.26 (0.08)	0.48 (0.15)	0.72 (0.22)
B	0.34 (0.10)	0.62 (0.19)	0.92 (0.28)

7. *Compute the water velocity in the coil*

Table 13 shows that the coil water velocity, ft/min = 3.59*gpm*. Since the required water flow is, from step 3, 64.2 gal/min (243 L/min), the water flow for each unit will be half this, or 64.2/2 = 32.1 gal/min (121.5 L/min). Hence the water velocity = 3.59(32.1) = 115.2 ft/min (35.1 m/ min).

8. *Determine the coil heat-transfer factors*

Table 13 lists typical heat-transfer factors for various water and air velocities. Interpolating between 400- and 500-ft/min (121.9- and 152.4-m/min) air velocities at a water velocity of 115 ft/min (35.1 m/min) gives a heat transfer factor of k = 165 Btu/(h·ft^2·°F) [936.9 W/(m^2·K)] LMTD row. The increase in velocity from 15 to 15.2 ft/min (4.63 m/min) is so small that it can be ignored. If the actual velocity is midway, or more, between the two tabulated velocities, interpolate vertically also.

9. *Compute the number of tube rows required*

Use the relation number of tube rows = H_r/(LMTD)$(A_c)(k)$. Thus, number of rows = 321,000/ (19.5)(30.8)(165) = 3.24, or four rows, the next larger *even* number.

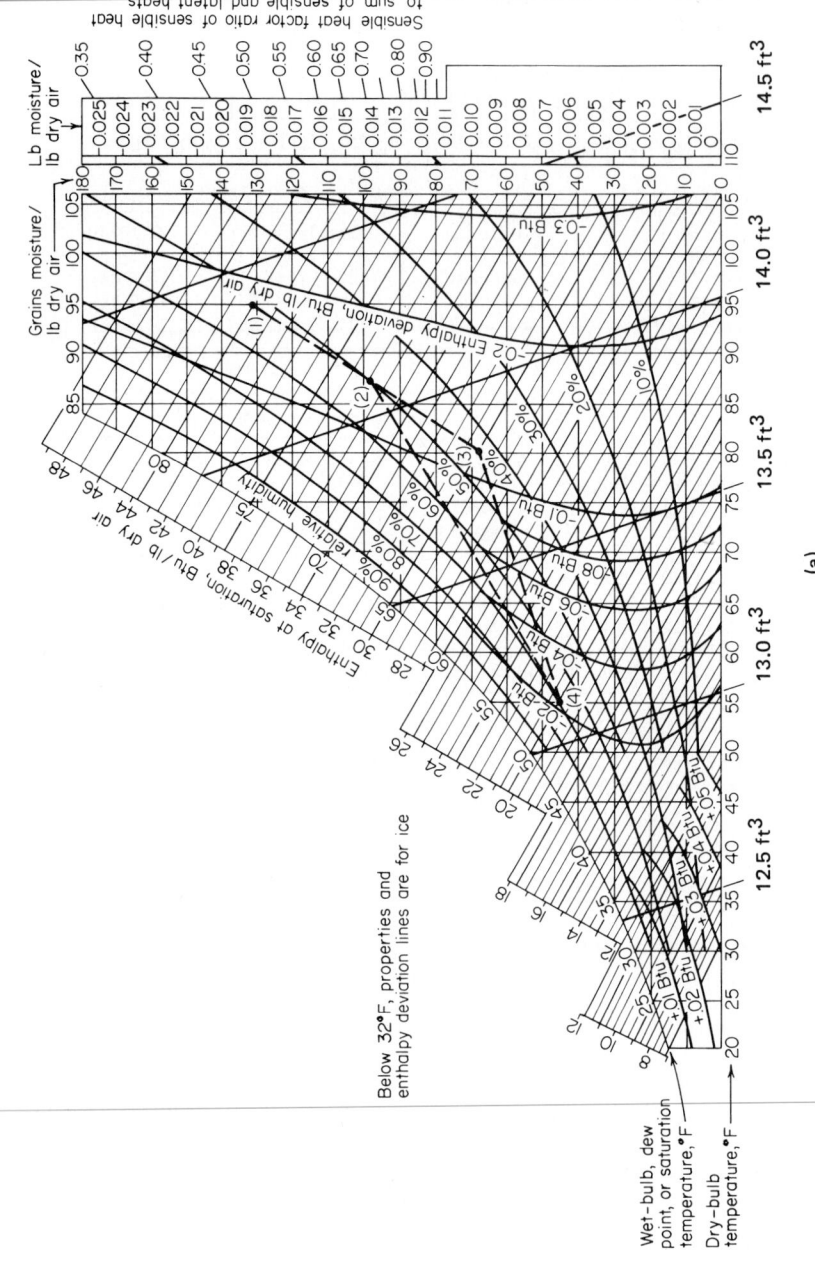

FIG. 5 Psychrometric charts for normal temperatures. (*a*) USCS version. (*Carrier Air Conditioning Company.*)

(a)

2.90

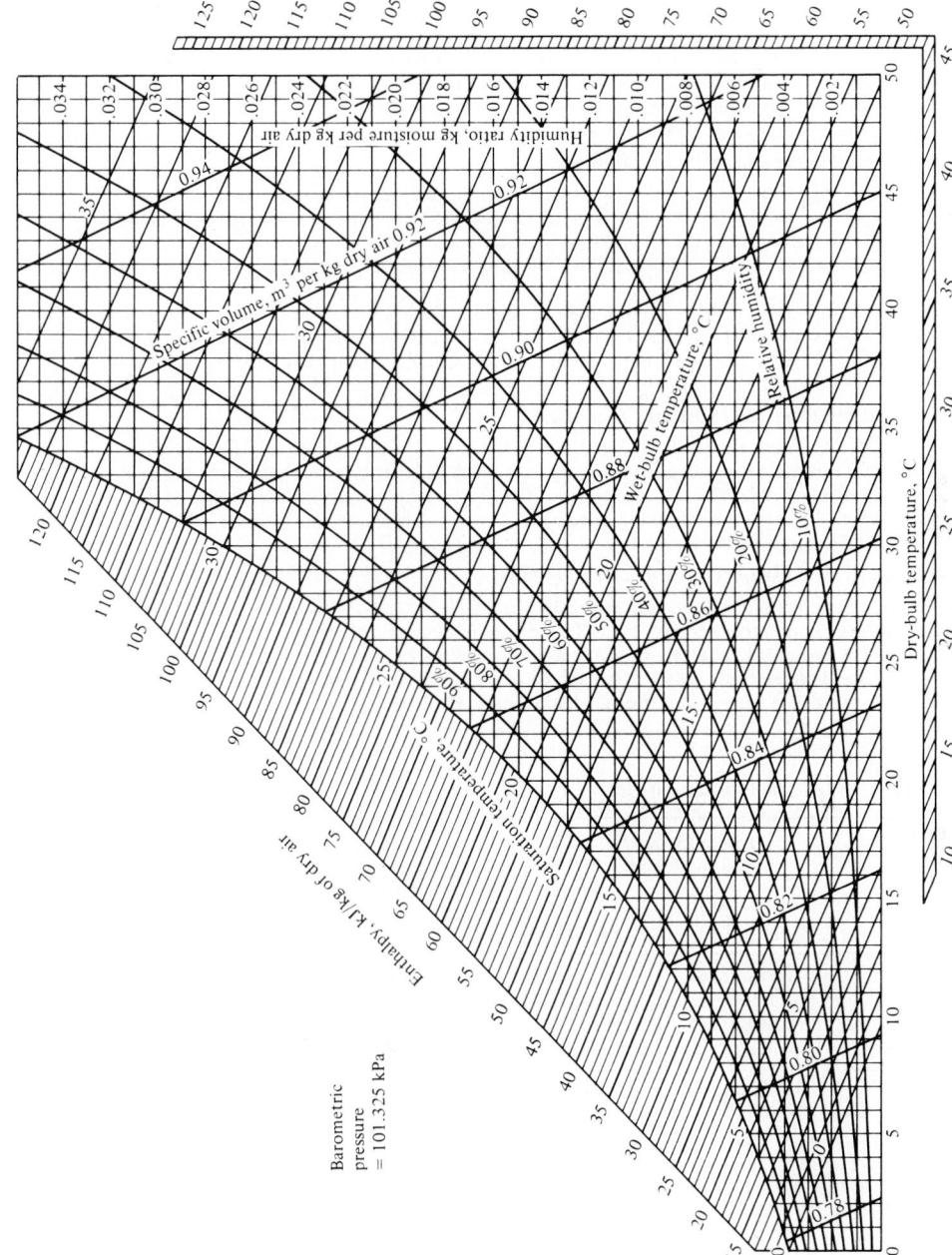

FIG. 5 (b) SI version. (*Stoecker and Jones*— *Refrigeration and Air Conditioning, McGraw-Hill.*)

Water cooling coils for air-conditioning service are usually built in units having two, four, six, or eight rows of coils. If the above calculation indicates that an odd number of coils should be used (i.e., the result was 3.24 rows), use the next smaller or larger *even* number of rows after increasing or decreasing the air and water velocity. Thus, to decrease the air velocity, use a coil having a larger face area. Recompute the air velocity and water velocity; find the new heat-transfer factor and the required number of rows. Continue doing this until a suitable number of rows is obtained. Usually only one recalculation is necessary.

10. Determine the coil water-pressure drop

Table 13 shows the water-pressure drop, ft (m) of water, for various tube lengths and water velocities. Interpolating between 90 and 120 ft/min (27.4 and 36.6 m/min) for a 6-ft (1.83-m) long tube gives a pressure drop of 0.27 ft (0.08 m) of water per row at a water velocity of 115.2 ft/min (35.1 m/min). Since the coil has four rows, total tube pressure drop = 4(0.27) = 1.08 ft (0.33 m) of water.

There is also a water pressure drop in the coil headers. Table 13 lists typical values. Interpolating between 90 and 120 ft/min (27.4 and 36.6 m/min) for a B-type coil gives a header pressure loss of 0.57 ft (0.17 m) of water at a water velocity of 115.2 ft/min (35.1 m/min). Hence, the total pressure loss in the coil is 1.08 + 0.57 = 1.65 ft (0.50 m) of water = coil loss + header loss.

11. Determine the coil resistance to air flow

Table 14 lists the resistance of coils having two to six rows of tubes and various air velocities. Interpolating for four tube rows gives a resistance of 0.225 in H_2O for an air velocity of 487 ft/min (148.4 m/min). The increase in resistance with a wet tube surface is, from Table 14, 28 percent at a 500-ft/min (152.4-m/min) air velocity. This occurs when the air is cooled below the entering air dew point and is discussed in step 13.

12. Check the coil selection in a coil-rating table

Many manufacturers publish precomputed coil-rating tables as part of their engineering data. Table 14 shows a portion of one such table. This tabulation shows that with an air velocity of 500 ft/min (152.4 m/min), four tube rows, an entering-air temperature of 85°F (29.4°C), and an entering water temperature of 50°F (10.0°C), the cooling coil has a cooling capacity of 10,300 Btu/(h·ft²) (32,497 W/m²), and a final air temperature of 65°F (18.3°C). Since the actual air

TABLE 14   Typical Cooling-Coil Resistance Characteristics

No. of tube rows	Air-flow resistance, in H_2O for 70°F air (mm H_2O for 21.1°C air)		
	Air-face velocity, ft/min (m/min)		
	400 (121.9)	500 (152.4)	600 (182.9)
2	0.081 (2.06)	0.122 (3.10)	0.164 (4.17)
4	0.162 (4.11)	0.234 (5.94)	0.318 (8.08)
6	0.234 (5.94)	0.344 (8.74)	0.472 (11.99)
8	0.312 (7.92)	0.454 (11.53)	0.622 (15.80)

Resistance increase due to wet tube surface, percent		
32	28	24

Coil cooling capacity, Btu/(h·ft²) (W/m²) face area and final air temperature, °F (°C)

Air velocity, ft/min (m/min)	No. of tube rows	Entering air temperature, °F (°C)	Entering water temperature, °F (°C)	
			45 (7.2)	50 (10)
500 (152.4)	4	85 (29.4)	11,900 (36,545) 63 (17.2)	10,300 (32,497) 65 (18.3)

velocity of 487 ft/min (148.4 m/min) is close to 500 ft/min (152.4 m/min), the tabulated cooling capacity closely approximates the actual cooling capacity. Hence the required heat-transfer area is $A_c = H_r/10{,}300 = 321{,}000/10{,}300 = 31.1$ ft^2 (2.98 m^2). This agrees closely with the area of 30.8 ft^2 (2.86 m^2) found in step 6.

In actual practice, designers use a coil cooling capacity table whenever it is available. However, the procedure given in steps 1 through 11 is also used when an exact analysis of a coil is desired or when a capacity table is not available.

13. Compute the heat removal for cooling below the dew point

When the temperature of the air leaving the cooling coil is lower than the dew point of the entering air, H_r = (weight of air cooled, lb)(total heat of entering air at its wet-bulb temperature, Btu/lb − total heat of the leaving air at its wet-bulb tempeature, Btu/lb). Once H_r is known, follow all the steps given above except that (a) a correction must be applied in step 11 for a wet tube surface. Obtain the appropriate correction factor from the manufacturer's engineering data, and apply it to the air-flow-resistance data, for the coil selected. (b) Also, the usual coil-rating table presents only the sensible-heat capacity of the coil. Where the ratio of sensible heat removed to latent heat removed is more than 2:1, the usual coil-rating table can be used. If the ratio is less than 2:1, use the procedure in steps 1 through 13.

Related Calculations: Use the method given here in steps 1 through 11 for any finned-type cooling coil mounted perpendicular to the air flow and having water as the cooling medium where the final air temperature leaving the cooling coil is *higher than* the dew point of the entering air. Follow step 13 for cooling below the dew point of the entering air.

Cooling and dehumidifying coils used in air-conditioning systems generally serve the following ranges of variables: (1) dry-bulb temperature of the entering air is 60 to 100 °F (15.6 to 37.8°C); wet-bulb temperature of entering air is 50 to 80°F (10 to 26.7°C); (2) coil core face velocity can range from 200 to 1200 ft/min (60.9 to 365.8 m/min) with 500 to 800 ft/min (152.4 to 243.8 m/min) being the most common velocity for comfort cooling applications; (3) entering water-temperature ranges from 40 to 65°F (4.4 to 18.3°C); (4) the water-temperature rise ranges from 4 to 12°F (2.2 to 6.7°C) during passage through the coil; (5) the water velocity ranges from 2 to 6 ft/s (0.61 to 1.83 m/s).

To choose an air-cooling coil using a direct-expansion refrigerant, follow the manufacturer's engineering data. Since most of the procedures are empirical, it is difficult to generalize about which procedure to use. However, the usual range of the volatile refrigerant temperature at the coil suction outlet is 25 to 55°F (−3.9 to 12.8°C). Where chilled water is circulated through the coil, the usual quantity range is 2 to 6 gal/(m·ton) (8.4 to 25 L/t).

MIXING OF TWO AIRSTREAMS

An air-conditioning system is designed to deliver 100,000 ft^3/min (2831 m^3/min) of air to a conditioned space. Of this total, 90,000 ft^3/min (2548 m^3/min) is recirculated indoor air at 72°F (22.2°C) and 40 percent relative humidity; 10,000 ft^3/min (283.1 m^3/min) is outdoor air at 0°F (−17.8°C). What are the enthalpy, temperature, moisture content, and relative humidity of the resulting air mixture? If air enters the room from the outlet grille at 60°F (15.6°C) after leaving the apparatus at a 50°F (10°C) dew point and the return air is at 75°F (23.9°C), what proportion of conditioned air and bypassed return air must be used to produce the desired outlet temperature at the grille?

Calculation Procedure:

1. Determine the proportions of each airstream

Use the relations $p_r = r/t$ and $p_0 = o/t$, where p_r = percent recirculated room air, expressed as a decimal; r = recirculated air quantity, ft^3/min; t = total air quantity, ft^3/min; p_0 = percent outside air, expressed as a decimal; o = outside air quantity, ft^3/min. For this system, $p_r = 90{,}000/100{,}000 = 0.90$, or 90 percent; $p_0 = 10{,}000/100{,}000 = 0.10$, or 10 percent. (The computation in SI units is identical.)

2. Determine the enthalpy of each airstream

Use a psychrometric chart or table to find the enthalpy of the recirculated indoor air as 24.6 Btu/lb (57.2 kJ/kg).

The enthalpy of the outdoor air is 0.0 Btu/lb (0.0 kJ/kg) because in considering heating or humidifying processes in winter it is always safest to assume that the outdoor air is completely dry. This condition represents the greatest heating and humidifying load because the enthalpy and the water-vapor content of the air are at a minimum when the air is considered dry at the outdoor temperature.

3.   Determine the moisture content of each airstream

The moisture content of the indoor air at a 72°F (22.2°C) dry-bulb temperature and 40 percent relative humidity is, from the psychrometric chart (Fig. 5), 47.2 gr/lb (6796.8 mg/kg). From a psychrometric table the moisture content of the 0°F (-17.8°C) outdoor air, which is assumed to be completely dry, is 0.0 gr/lb (0.0 mg/kg).

4.   Compute the enthalpy of the air mixture

Use the relation $h_m = (oh_0 + rh_r)/t$, where h_m = enthalpy of mixture, Btu/lb; h_0 = enthalpy of the outside air, Btu/lb; h_r = enthalpy of the recirculated room air, Btu/lb; other symbols as before. Hence, $h_m = (10{,}000 \times 0 + 90{,}000 \times 24.6)/100{,}000 = 22.15$ Btu/lb (51.5 kJ/kg).

5.   Compute the temperature of the air mixture

Use a similar relation to that in step 4, substituting the air temperature for the enthalpy. Or, $t_m = (ot_0 + rt_r)/t$, where t_m = mixture temperature, °F; t_0 = temperature of outdoor air, °F; t_r = temperature of recirculated room air, °F; other symbols as before. Hence, $t_m = (10{,}000 \times 0 + 90{,}000 \times 72)/100{,}000 = 64.9$°F (18.3°C).

6.   Compute the moisture content of the air mixture

Use a similar relation to that in step 4, substituting the moisture content for the enthalpy. Or, $g_m = (og_0 + rg_r)/t$, where g_m = gr of moisture per lb of mixture; g_0 = gr of moisture per lb of outdoor air; g_r = gr of moisture per lb of recirculated room air; other symbols as before. Thus, $g_m = (10{,}000 \times 0 + 90{,}000 \times 47.2)/100{,}000 = 42.5$ gr/lb (6120 mg/kg).

7.   Determine the relative humidity of the mixture

Enter the psychrometric chart at the temperature of the mixture, 64.9°F (18.3°C), and the moisture content, 42.5 gr/lb (6120 mg/kg). At the intersection of the two lines, find the relative humidity of the mixture as 47 percent relative humidity.

8.   Determine the required air proportions

Set up an equation in which x = proportion of conditioned air required to produce the desired outlet temperature at the grille and y = the proportion of bypassed air required. The air quantities will also be proportional to the dry-bulb temperatures of each airstream. Since the dew point of the air leaving an air-conditioning apparatus = dry-bulb temperature of the air, $50x + 75y = 60(x + y)$, or $15y = 10x$. Also, the sum of the two airstreams $x + y = 1$. Substituting and solving for x and y, we get x = 60 percent; y = 40 percent. Multiplying the actual air quantity supplied to the room by the percentage representing the proportion of each airstream will give the actual cfm required for supply and bypass air.

Related Calculations: Use this general procedure to determine the properties of any air mixture in which two airstreams are mixed without compression, expansion, or other processes involving a marked changed in the pressure or volume of either or both airstreams.

SELECTION OF AN AIR-CONDITIONING SYSTEM FOR A KNOWN LOAD

Choose the type of air-conditioning system to use for comfort conditioning of a factory having a heat gain that varies from 500,000 to 750,000 Btu/h (146.6 to 219.8 kW) depending on the outdoor temperature and the conditions inside the building. Indicate why the chosen system is preferred.

Calculation Procedure:

1.   Review the types of air-conditioning systems available

Table 15 summarizes the various types of air-conditioning systems *commonly used* for different applications. Economics and special design objectives dictate the final choice and modifications of the systems listed. Where higher-quality air conditioning is desired (often at a higher cost),

TABLE 15 Systems and Applications°

Applications	Systems†	Applications	Systems†
Single-purpose occupancies		Multipurpose occupancies	
Residential:		Office buildings	2e 2i 2j
Medium	1c	Hotels, dormitories	1e 1f 2j 2k
Large	1d 1e 2i	Motels	1e
Restaurants:		Apartment buildings	1f 2j 2k
Medium	1d 1h	Hospitals	1f 2h 2j
Large	1d 2f 2g 2h 2i	Schools and colleges	1f 2e 2f 2g 2h
Variety and specialty shops	1d	Museums	2h 2i
Bowling alleys	1d 2f	Libraries:	
Radio and TV studios:		Standard	2h 2f 2h 2i
Small	1d 2f 2h 2i	Rare books	2h
Large	1d 2f 2h 2i	Department stores	1d 2f
Country clubs	1d 2f 2h 2i	Shopping centers	1d 2f 2i
Funeral homes	1d 2i	Laboratories:	
Beauty salons	1c 1d	Small	1d 2e 2h 2i
Barber shops	1c 1d	Large building	2e 2g 2j
Churches	1d 2f 2i	Marine	2g 2j
Theaters	2f		
Auditoriums	2f		
Dance and roller skating			
pavillions	1d 2e 2f		
Factories (comfort)	1d 2f 2h		

°Carrier Air Conditioning—*Handbook of Air Conditioning Systems Design*, McGraw-Hill.
†The systems in the table are:

1. Individual room or zone unit systems
   a. DX self-contained
   b. All-water
   c. Room DX self-contained 0.5 to 2 tons (1.76 to 7.0 kW)
   d. Zone DX self-contained 2 tons and over (7.0 kW and over)
   e. All-water room fan-coil recirculating air
   f. All-water room fan-coil with outdoor air
2. Central station apparatus systems
   a. All-air
   b. Air-water
   c. All-air, single airstream
   d. Air-water, primary air systems
   e. All-air, single airstream, variable volume
   f. All-air, single airstream, bypass
   g. All-air, single airstream, reheat at terminal
   h. All-air, single airstream, reheat zone in duct
   i. All-air, single airstream, multizone single duct
   j. Air-water primary air systems, secondary water H-V H-P induction
   k. Air-water primary air systems, room fan-coil with outside air

    Systems listed for a particular application are the systems most commonly used. Economics and design objectives dictate the choice and deviations of systems listed above, other systems as listed in note 2, and some entirely new systems.
    Several systems are used in many of these applications when higher-quality air conditioning is desired (often at higher expense). They are dual-duct, dual-conduit, three-pipe induction and fan-coil, four-pipe induction and fan-coil, and panel-air.

    certain other systems may be considered. These are dual-duct, dual-conduit, three-pipe induction and fan-coil, four-pipe induction and fan-coil, and panel-air systems.
    Study of Table 15 shows that four main types of air-conditioning systems are popular: direct-expansion (termed DX), all-water, all-air, and air-water. These classifications indicate the methods

used to obtain the final within-the-space cooling and heating. The air surrounding the occupant is the end medium that is conditioned.

2.   *Select the type of air-conditioning system to use*

Table 15 indicates that direct-expansion and all-air air-conditioning systems are *commonly used* for factory comfort conditioning. The load in the factory being considered may range from 500,000 to 750,000 Btu/h (146.6 to 219.8 kW). This is the equivalent of a maximum cooling load of 750,000/12,000 Btu/(h·ton) of refrigeration = 62.5 tons (219.8 kW) of refrigeration.

Where a building has a varying heat load, bypass control wherein neutral air is recirculated from the conditioned space while the amount of cooling air is reduced is often used. With this arrangement, the full quantity of supply air is introduced to the cooled area at all times during system operation.

Self-contained direct-expansion systems can serve large factory spaces. Their choice over an all-air bypass system is largely a matter of economics and design objectives.

Where reheat is required, this may be provided by a reheater in a zone duct. Reheat control maintains the desired dry-bulb temperature within a space by replacing any decrease in sensible loads by an artificial heat load. Bypass control maintains the desired dry-bulb temperature within the space by modulating the amount of air to be cooled. Since the bypass all-air system is probably less costly for this building, it will be the first choice. A complete economic analysis would be necessary before this conclusion could be accepted as fully valid.

Related Calculations: Use this general method to make a preliminary choice of the 11 different types of air-conditioning systems for the 31 applications listed in Table 15. Where additional analytical data for comparison of systems are required, consult Carrier Air Conditioning Company's *Handbook of Air Conditioning System Design* or the ASHRAE *Guide and Data Book.*

SIZING LOW-VELOCITY AIR-CONDITIONING-SYSTEM DUCTS—EQUAL-FRICTION METHOD

An industrial air-conditioning system requires 36,000 ft³/min (1019.2 m³/min) of air. This low-velocity system will be fitted with enough air outlets to distribute the air uniformly throughout the conditioned space. The required operating pressure for each duct outlet is 0.20 in (5.1 mm) wg. Determine the duct sizes required for this system by using the equal-friction method of design. What is the required fan static discharge pressure?

Calculation Procedure:

1.   *Sketch the duct system*

The required air quantity, 36,000 ft³/min (1019.2 m³/min), must be distributed in approximately equal quantities to the various areas in the building. Sketch the proposed duct layout as shown in Fig. 6. Locate air outlets as shown to provide air to each area in the building.

Determine the required capacity of each air outlet from air quantity required, ft³/min per number of outlets, or outlet capacity = 36,000/18 = 2000 ft³/min (56.6 m³/min) per outlet. This is within the usual range of many commercially available air outlets. Where the required capacity per outlet is extremely large, say 10,000 ft³/min (283.1 m³/min), or extremely small, say 5 ft³/min (0.14 m³/min), change the number of outlets shown on the duct sketch to obtain an air quantity within the usual capacity range of commercially available outlets. Relocate each outlet so it serves approximately the same amount of floor area as each of the other outlets in the system. Thus, the duct sketch serves as a trial-and-error analysis of the outlet location and capacity.

Where a building area requires a specific amount of air, select one or more outlets to supply this air. Size the remaining outlets by the method described above, after subtracting the quantity of air supplied through the outlets already chosen.

2.   *Determine the required outlet operating pressure*

Consult the manufacturer's engineering data for the required operating pressure of each outlet. Where possible, try to use the same type of outlets throughout the system. This will reduce the initial investment. Assume that the required outlet operating pressure is 0.20-in (5.1-mm) wg for each outlet in this system.

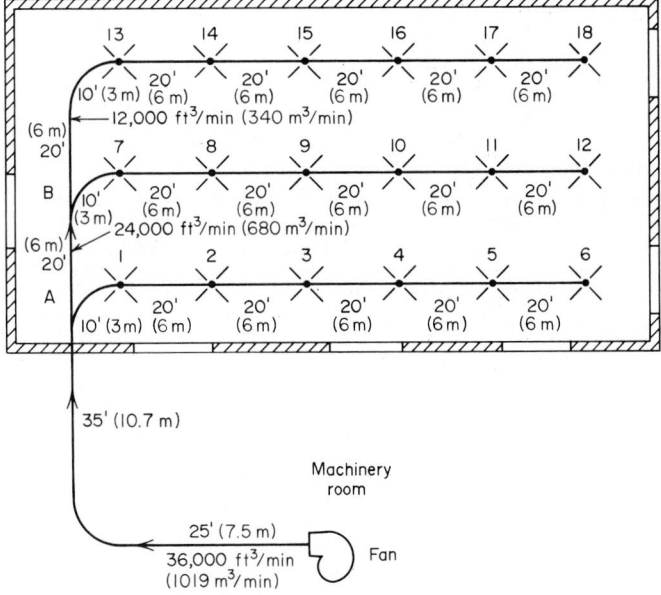

FIG. 6 Duct-system layout.

3. *Choose the air velocity for the main duct*

Use Table 16 to determine a suitable air velocity for the main duct of this system. Table 16 shows that an air velocity up to 2500 ft/min (762 m/min) can be used for main ducts where noise is the controlling factor; 3000 ft/min (914.4 m/min) where duct friction is the controlling factor. A velocity of 2500 ft/min (762 m/min) will be used for the main duct in this installation.

TABLE 16 Recommended Maximum Duct Velocities for Low-Velocity Systems, ft/min (m/min)°

Application	Controlling factor—noise generation, main ducts	Controlling factor—duct friction			
		Main ducts		Branch ducts	
		Supply	Return	Supply	Return
Residences	600 (183)	1000 (300)	800 (244)	600 (183)	600 (183)
Apartments, hotel bedrooms, hospital bedrooms	1000 (300)	1500 (457)	1300 (396)	1200 (366)	1000 (300)
Private offices, directors rooms, libraries	1200 (366)	2000 (610)	1500 (457)	1600 (488)	1200 (366)
Theaters, auditoriums	800 (244)	1300 (396)	1100 (335)	1000 (300)	800 (244)
General offices, high-class restaurants, high-class stores, banks	1500 (457)	2000 (610)	1500 (457)	1600 (488)	1200 (366)
Average stores, cafeterias	1800 (549)	2000 (610)	1500 (459)	1600 (488)	1200 (366)
Industrial	2500 (762)	3000 (914)	1800 (549)	2200 (671)	1500 (457)

°Carrier Air Conditioning Company.

4. *Determine the dimensions of the main duct*

The required duct area A ft^2 = (ft^3/min)/(ft/min) = 36,000/2500 = 14.4 ft^2 (1.34 m^2). A nearly square duct, i.e., a duct 46 × 45 in (117 × 114 cm), has an area of 14.38 ft^2 (1.34 m^2) and is a good first choice for this system because it closely approximates the outlet size of a standard centrifugal fan. Where possible, use a square main duct to simplify fan connections. Thus a 46 × 46 in (117 × 117 cm) duct might be the final choice for this system.

5. *Determine the main-duct friction loss*

Convert the duct area to the equivalent diameter in inches d, using $d = 2(144A/\pi)^{0.5} = 2(144 \times 14.4/\pi)^{0.5} = 51.5$ in (130.8 cm).

Enter Fig. 7 at 36,000 ft^3/min (1019.2 m^3/min) and project horizontally to a round-duct diameter of 51.5 in (130.8 cm). At the top of Fig. 7 read the friction loss as 0.13 in (3.3 mm) wg per 100 ft (30 m) of equivalent duct length.

6. *Size the branch ducts*

For many common air-conditioning systems the equal-friction method is used to size the ducts. In this method the supply, exhaust, and return-air ducts are sized so they have the same friction loss per foot of length for the entire system. The equal-friction method is superior to the velocity-reduction method of duct sizing because the former requires less balancing for symmetrical layouts.

The usual procedure in the equal-friction method is to select an initial air velocity in the main duct near the fan, using the sound level as the limiting factor. With this initial velocity and the design air flow rate, the required duct diameter is found, as in steps 4 and 5, above. Once the duct diameter is known, the friction loss is found from Fig. 7, as in step 5. This same friction loss is then maintained throughout the system, and the equivalent round-duct diameter is chosen from Fig. 7.

To expedite equal-friction calculations, Table 17 is often used instead of the friction chart. (It is valid for SI units also.) This provides the same duct sizes. Duct areas are determined from Table 17, and the area found is converted to a round-, rectangular-, or square-duct size suitable for the installation. This procedure of duct sizing automatically reduces the air velocity in the direction of air flow. Hence, the equal-friction method will be used for this system.

Compute the duct areas, using Table 17. Tabulate the results, using the duct run having the highest resistance. The friction loss through all elbows and fittings in the section must be included. The total friction loss in the duct having the highest resistance is the loss the fan must overcome.

Inspection of the duct layout (Fig. 6) shows that the duct run from the fan to outlet 18 probably has the highest resistance because it is the longest run. Tabulate the results as shown.

(1) Duct section	(2) Air quantity, ft^3/min (m^3/min)	(3) ft^3/min (m^3/min) capacity, percent	(4) Duct, percent	(5) Area, ft^2 (m^2)	(6) Duct size, in (cm)
Fan to A	36,000 (1,019.2)	100	100	14.4 (1.34)	46 × 45 (117 × 114)
A–B	24,000 (679.4)	67	73.5	10.6 (0.98)	39 × 39 (99 × 99)
B–13	12,000 (339.7)	33	41.0	5.9 (0.55)	30 × 29 (76 × 74)
13–14	10,000 (283.1)	28	35.5	5.1 (0.47)	27 × 27 (69 × 69)
14–15	8,000 (226.5)	22	29.5	4.3 (0.40)	25 × 25 (64 × 64)
15–16	6,000 (169.9)	17	24.0	3.5 (0.33)	23 × 22 (58 × 56)
16–17	4,000 (113.2)	11	17.5	2.5 (0.23)	20 × 18 (51 × 46)
17–18	2,000 (56.6)	6	10.5	1.5 (0.14)	15 × 15 (38 × 38)

The values in this tabulation are found as follows. Column 1 lists the longest duct run in the system. In column 2, the air leaving the outlets in branch A, or (6 outlets) (2000 ft^3/min per outlet) = 12,000 ft^3/min (339.7 m^3/min), is subtracted from the quantity of air, 36,000 ft^3/min (1019.2

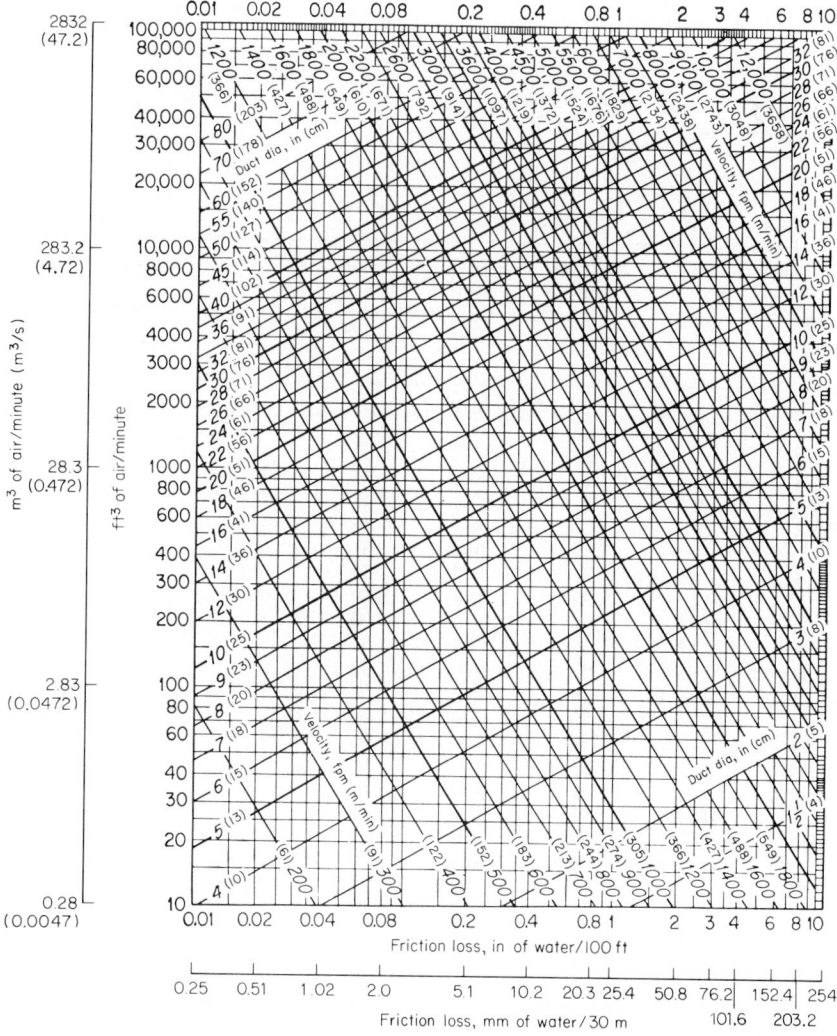

Friction loss for usual air conditions. This chart applies to smooth round galvanized iron ducts. See table below for corrections to apply when using other pipe.

| Type of pipe | Degree of roughness | Velocity | | Roughness factor (use as multiplier) |
		ft/min	m/min	
Concrete	Medium rough	1000–2000	300–610	1.4
Riveted steel	Very rough	1000–2000	300–610	1.9
Tubing	Very smooth	1000–2000	300–610	0.9

FIG. 7 Friction loss in round ducts.

TABLE 17   Percentage of Section Area in Branches for Maintaining Equal Friction°

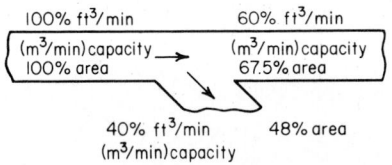

ft³/min (m³/min) capacity, %	Duct area,† %	ft³/min (m³/min) capacity, %	Duct area, %	ft³/min (m³/min) capacity, %	Duct area, %	ft³/min (m³/min) capacity, %	Duct area, %
1	2.0	26	33.5	51	59.0	76	81.0
2	3.5	27	34.5	52	60.0	77	82.0
3	5.5	28	35.5	53	61.0	78	83.0
4	7.0	29	36.5	54	62.0	79	84.0
5	9.0	30	37.5	55	63.0	80	84.5
6	10.5	31	39.0	56	64.0	81	85.5
7	11.5	32	40.0	57	65.0	82	86.0
8	13.0	33	41.0	58	65.5	83	87.0
9	14.5	34	42.0	59	66.5	84	87.5
10	16.5	35	43.0	60	67.5	85	88.5
11	17.5	36	44.0	61	68.0	86	89.5
12	18.5	37	45.0	62	69.0	87	90.0
13	19.5	38	46.0	63	70.0	88	90.5
14	20.5	39	47.0	64	71.0	89	91.5
15	21.5	40	48.0	65	71.5	90	92.0
16	23.0	41	49.0	66	72.5	91	93.0
17	24.0	42	50.0	67	73.5	92	94.0
18	25.0	43	51.0	68	74.5	93	94.5
19	26.0	44	52.0	69	75.5	94	95.0
20	27.0	45	53.0	70	76.5	95	96.0
21	28.0	46	54.0	71	77.0	96	96.5
22	29.5	47	55.0	72	78.0	97	97.5
23	30.5	48	56.0	73	79.0	98	98.0
24	31.5	59	57.0	74	80.0	99	99.0
25	32.5	50	58.0	75	80.5	100	100.0

°Carrier Air Conditioning Company.
†The same duct area percentage applies when flow is measured in m³/min or m³/s.

m³/min), discharged by the fan to give the air quantity flowing from *A-B.* A similar procedure is followed for each successive duct and air quantity.

Column 3 is found by dividing the air quantity in each branch listed in columns 1 and 2 by 36,000, the total air flow, and multiplying the result by 100. Thus, for run *B*-13, column 3 = 12,000 (100)/36,000 = 33 percent.

Column 4 values are found from Table 17. Enter that table with the cfm capacity from column 3 and read the duct area, percent. Thus, for branch 13-14 with 28 percent cfm capacity, the duct area from Table 17 is 35.5 percent. Determine the duct area, column 6, by taking the product, line by line, of column 4 and the main duct area. Thus, for branch 13-14, duct area = (0.355)(14.4) = 5.1 ft² (0.47 m²). Convert the duct area to a nearly square, or a square, duct by finding two dimensions that will produce the desired area.

Duct sections A through 6 and B through 12 have the same dimensions as the corresponding duct sections B through 18.

7. Find the total duct friction loss

Examination of the duct sketch (Fig. 6) indicates that the duct run from the fan to outlet 18 has the highest resistance. Compute the total duct run length and the equivalent length of the two elbows in the run thus as shown.

(1) Duct section	(2) System part	(3) Length ft	(3) Length m	(4) Elbow equivalent length ft	(4) Elbow equivalent length m
Fan to A	Duct	60	18.3		
	Elbow		. . .	30	9.1
A–B	Duct	20	6.1		
B–13	Duct	30	9.1		
	Elbow		. . .	15	4.6
13–14	Duct	20	6.1		
14–15	Duct	20	6.1		
15–16	Duct	20	6.1		
16–17	Duct	20	6.1		
17–18	Duct	20	6.1		
Total		210	64.0	45	13.7

Note several facts about this calculation. The duct lengths, column 3, are determined from the system sketch, Fig. 6. The equivalent length of the duct elbows, column 4, is determined from the *Guide* or Carrier *Handbook of Air Conditioning Design*. The total equivalent duct length = column 3 + column 4 = 210 + 45 = 255 ft (77.7 m).

8. Compute the duct friction loss

Use the general relation $h_T = Lf$, where h_T = total friction loss in duct, in wg; L = total equivalent duct length, ft; f = friction loss for the system, in wg per 100 ft (30 m). With the friction loss of 0.13 in (3.3 mm) wg per 100 ft (30 m), as determined in step 5, $h_r = (229/100)(0.13) = 0.2977$ in (7.6 mm) wg; say 0.30 in (7.6 mm) wg.

9. Determine the required fan static discharge pressure

The total static pressure required at the fan discharge = outlet operating pressure + duct loss − velocity regain between first and last sections of the duct, all expressed in in wg. The first two variables in this relation are already known. Hence, only the velocity regain need be computed.

The velocity v ft/min of air in any duct is v = cfm/duct area. For duct section A, v = 36,000/14.4 = 2500 ft/min (762 m/min); for the last duct section, 17-18, v = 2000/1.5 = 1333 ft/min (406.3 m/min).

When the fan discharge velocity is higher than the duct velocity in an air-conditioning system, use this relation to compute the static pressure regain $R = 0.75[(v_f/4000)^2 - (v_d/4000)^2]$, where R = regain, in wg; v_f = fan outlet velocity, ft/min; v_d = duct velocity, ft/min. Thus, for this system, $R = 0.75[(2500/4000)^2 - (1333/4000)^2] = 0.21$ in (5.3 mm) wg.

With the regain known, compute the total static pressure required as 0.20 + 0.30 − 0.21 = 0.29 in (7.4 mm) wg. A fan having a static discharge pressure of at least 0.30 in (7.6 mm) wg would probably be chosen for this system.

If the fan outlet velocity exceeded the air velocity in duct section A, the air velocity in this section would be used instead of the air velocity in the last duct section. Thus, in this circumstance, the last section becomes the duct connected to the fan outlet.

Related Calculations: Where the velocity in the fan outlet duct is *higher* than the fan outlet velocity, use the relation $l = 1.1[(v_d/4000)^2 - (v_f/4000)^2]$, where l = loss, in wg. This loss is the additional static pressure required of the fan. Hence, this loss must be *added* to the outlet operating pressure and the duct loss to determine the total static pressure required at the fan discharge.

The equal-friction method does not satisfy the design criteria of uniform static pressure at all branches and air terminals. To obtain the proper air quantity at the beginning of each branch, it is necessary to include a splitter damper to regulate the flow to the branch. It may also be necessary to have a control device (vanes, volume damper, or adjustable-terminal volume control) to regulate the flow at each terminal for proper air distribution.

The *velocity-reduction method* of duct design is not too popular because it requires a broad background of duct-design experience and knowledge to be within reasonable accuracy. It should be used for only the simplest layouts. Splitters and dampers should be included for balancing purposes.

To apply the velocity-reduction method: (1) Select a starting velocity at the fan discharge. (2) Make arbitrary reductions in velocity down the duct run. The starting velocity should not exceed the values in Table 16. Obtain the equivalent round-duct diameter from Fig. 7. Compute the required duct area from the round-duct diameter, and from this the duct dimensions, as shown in steps 4 and 5 above. (3) Determine the required fan static discharge pressure for the supply by using the longest run of duct, including all elbows and fittings. Note, however, that the longest run is not necessarily the run with the greatest friction loss, as shorter runs may have more elbows, fittings, and restrictions.

The equal-friction and velocity-reduction methods of air-conditioning system duct design are applicable only to low-velocity systems, i.e., systems in which the maximum air velocity is 3000 ft/min (914.4 m/min), or less. The methods presented in this calculation procedure are those used by the Carrier Air Conditioning Company at the time of this writing.

SIZING LOW-VELOCITY AIR-CONDITIONING DUCTS—STATIC-REGAIN METHOD

Using the same data as in the previous calculation procedure, an air velocity of 2500 ft/min (762.0 m/min) in the main duct section, an unvaned elbow radius of $R/D = 1.25$, and an operating pressure of 0.20 in (2.5 mm) wg for each outlet, size the system ducts, using the static-regain method of design for low-velocity systems.

Calculation Procedure:

1. Compute the fan outlet duct size

The fan outlet duct, also called the main duct section, will have an air velocity of 2500 ft/min (762.0 m/min). Hence, the required duct area is $A = 36,000/2500 = 14.4$ ft^2 (1.34 m^2). This corresponds to a round-duct diameter of $d = 2(144A/\pi)^{0.5} = 2(144 \times 14.4/\pi)^{0.5} = 51.5$ in (130.8 cm). A nearly square duct, i.e., a duct 46 × 45 in (116.8 × 114.3 cm), has an area of 14.38 ft^2 (1.34 m^2) and is a good first choice for this system because it closely approximates the outlet size of a standard centrifugal fan.

Where possible, use a square main duct to simplify fan connections. Thus, a 46 × 46 in (116.8 × 116.8 cm) duct might be the final choice of this system.

2. Compute the main-duct friction loss

Using Fig. 7, find the main-duct friction loss as 0.13 in (3.3 mm) wg per 100 ft (30 m) of equivalent duct length for a flow of 36,000 ft^3/min (1019.2 m^3/min) and a diameter of 51.5 in (130.8 cm).

3. Determine the friction loss up to the first branch duct

The length of the main duct between the fan and the first branch is 25 + 35 = 60 ft (18.3 m). The equivalent length of the elbow is, from the *Guide* or Carrier *Handbook of Air Conditioning Design*, 26 ft (7.9 m). Hence, the total equivalent length = 60 + 30 = 90 ft (27.4 m). The friction loss is then $h_T = L_f = (90/100)(0.13) = 0.117$ in (2.97 mm) wg.

4. *Size the longest duct run*

The longest duct run is from *A* to outlet 18 (Fig. 6). Size the duct using the following tabulation, preparing it as described below.

(1) Section number	(2) Air flow, ft³/min (m³/min)	(3) Equivalent length, ft (m)	(4) L/Q ratio	(5) Velocity, ft/ min (m/min)	(6) Duct area, ft² (m²) (2)/(5)	(7) Duct size, in (cm)
Fan to *A*	36,000 (1,019.2)	86 (26.2)	. . .	2,500 (762.0)	14.4 (1.34)	46 × 45 (116.8 × 114.3)
A–B	24,000 (679.4)	20 (6.1)	0.034	2,410 (734.6)	9.95 (0.92)	38 × 38 (96.5 × 96.5)
B–13	12,000 (339.7)	26° (7.9)°	0.088	2,200 (670.6)	5.45 (0.51)	28 × 28 (71.1 × 71.1)
13–14	10,000 (283.1)	20 (6.1)	0.072	2,040 (621.8)	4.90 (0.46)	27 × 27 (68.6 × 68.6)
14–15	8,000 (226.5)	20 (6.1)	0.083	1,850 (563.9)	4.33 (0.40)	25 × 25 (63.5 × 63.5)
15–16	6,000 (169.9)	20 (6.1)	0.098	1,700 (518.2)	3.53 (0.33)	24 × 23 (60.9 × 58.4)
16–17	4,000 (113.2)	20 (6.1)	0.130	1,520 (463.3)	2.63 (0.24)	20 × 19 (50.8 × 48.3)
17–18	2,000 (56.6)	20 (6.1)	0.195	1,300 (396.2)	1.54 (0.14)	15 × 15 (38.1 × 38.1)
B–7	12,000 (339.7)	25° (7.6)°	. . .	. . .	. . .	28 × 28 (71.1 × 71.1)
7–8	10,000 (283.1)	20 (6.1)	. . .	. . .	. . .	27 × 27 (68.6 × 68.6)
8–9	8,000 (226.5)	20 (6.1)	. . .	. . .	. . .	25 × 25 (63.5 × 63.5)
9–10	6,000 (169.9)	20 (6.1)	. . .	. . .	. . .	25 × 23 (60.9 × 58.4)
10–11	4,000 (113.2)	20 (6.1)	. . .	. . .	. . .	20 × 19 (50.8 × 48.3)
11–12	2,000 (56.1)	20 (6.1)	. . .	. . .	. . .	15 × 15 (38.1 × 38.1)
A–1	12,000 (339.7)	25° (7.6)	. . .	. . .	. . .	28 × 28 (71.1 × 71.1)
1–2	10,000 (283.1)	20 (6.1)	. . .	. . .	. . .	27 × 27 (68.6 × 68.6)
2–3	8,000 (226.5)	20 (6.1)	. . .	. . .	. . .	25 × 25 (63.5 × 64.5)
3–4	6,000 (169.9)	20 (6.1)	. . .	. . .	. . .	24 × 23 (60.9 × 58.4)
4–5	4,000 (113.2)	20 (6.1)	. . .	. . .	. . .	20 × 19 (50.8 × 48.3)
5–6	2,000 (56.1)	20 (6.1)	. . .	. . .	. . .	15 × 15 (38.1 × 38.1)

°See text.

List in column 1 the various duct sections in the longest duct run, as shown in Fig. 6. In column 2 list the air quantity flowing through each duct section. Tabulate in column 3 the equivalent length of each duct. Where a fitting is in the duct section, as in *B*-13, assume a duct size and compute the equivalent length using the *Guide* or Carrier fitting table. When the duct section does not have a fitting, as with section 13-14, the equivalent length equals the distance between the centerlines of two adjacent outlets.

Next, determine the *L/Q* ratio for each duct section, using Fig. 8. Enter Fig. 8 at the air quantity in the duct and project vertically upward to the curve representing the equivalent length of the duct. At the left read the *L/Q* ratio for this section of the duct. Thus, for duct section 13-14, Q = 10,000 ft³/min (283.1 m³/min), and L = 20 ft (6.1 m). Entering the chart as detailed above shows that L/Q = 0.72. Proceed in this manner, determining the *L/Q* ratio for each section of the duct in the longest duct run.

Determine the velocity of the air in the duct by using Fig. 9. Enter Fig. 9 at the *L/Q* ratio for the duct section, say 0.072 for section 13-14. Find the intersection of the *L/Q* curve with the velocity curve for the preceding duct section: 2200 ft/min (670.6 m/min) for section 13-14. At the bottom of Fig. 9 read the velocity in the duct section, i.e., after the previous outlet and in the duct section under consideration. Enter this velocity in column 5. Proceed in this manner, determining the velocity in each section of the duct in the longest duct run.

Determine the required duct area from column 2/column 5, and insert the result in column 6. Find the duct size, column 7, by converting the required duct area to a square- or rectangular-duct dimension. Thus a 27 × 17 in (68.6 × 43.2 cm) square duct has a cross-sectional area slightly greater than 4.90 ft² (0.46 m²).

Air quantity Q after takeoff, 100 m³/min

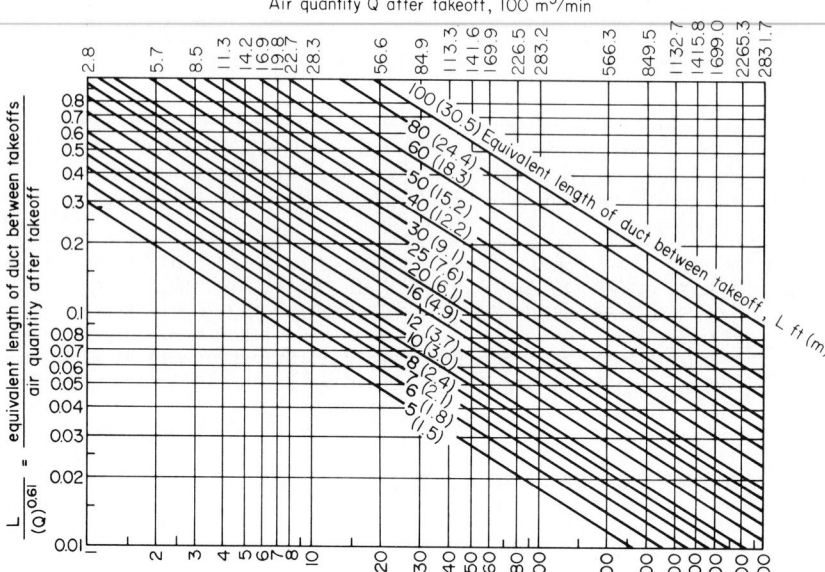

FIG. 8 L/Q ratio for air ducts. *(Carrier Air Conditioning Company.)*

5. *Determine the sizes of the other ducts in the system*

Since the ducts in runs *A* and *B* are symmetric with the duct containing the outlets in the longest run, they can be given the same size when the same quantity of air flows through them. Thus, duct section 7–8 is sized the same as section 13-14 because the same quantity of air, 10,000 ft³/min (4.72 m³/s), is flowing through both sections.

Where the duct section contains a fitting, as *B*-7 and *A*-1, assume a duct size and find the equivalent length, using the *Guide* or Carrier fitting table. These sections are marked with an asterisk.

6. *Determine the required fan discharge pressure*

The total pressure required at the fan discharge equals the sum of the friction loss in the main duct plus the terminal operating pressure. Hence the required fan static discharge pressure = 0.117 + 0.20 = 0.317 in (8.1 mm) wg.

Related Calculations: The basic principle of the static-regain method is to size a duct run so that the increase in static pressure (regain due to the reduction in velocity) at each branch or air terminal just offsets the friction loss in the succeeding section of duct. The static pressure is then the same before each terminal and at each branch.

As a *general* guide to the results obtained with the static-regain and equal-friction duct-design methods, the following should be helpful:

	Static regain	Equal friction
Main-duct sizes	Same	Same
Branch-duct sizes	Larger	Smaller
Sheet-metal weight	Greater	Less
Fan horsepower	Less	Greater
Balancing time	Less	Greater
Operating costs	Less	Greater

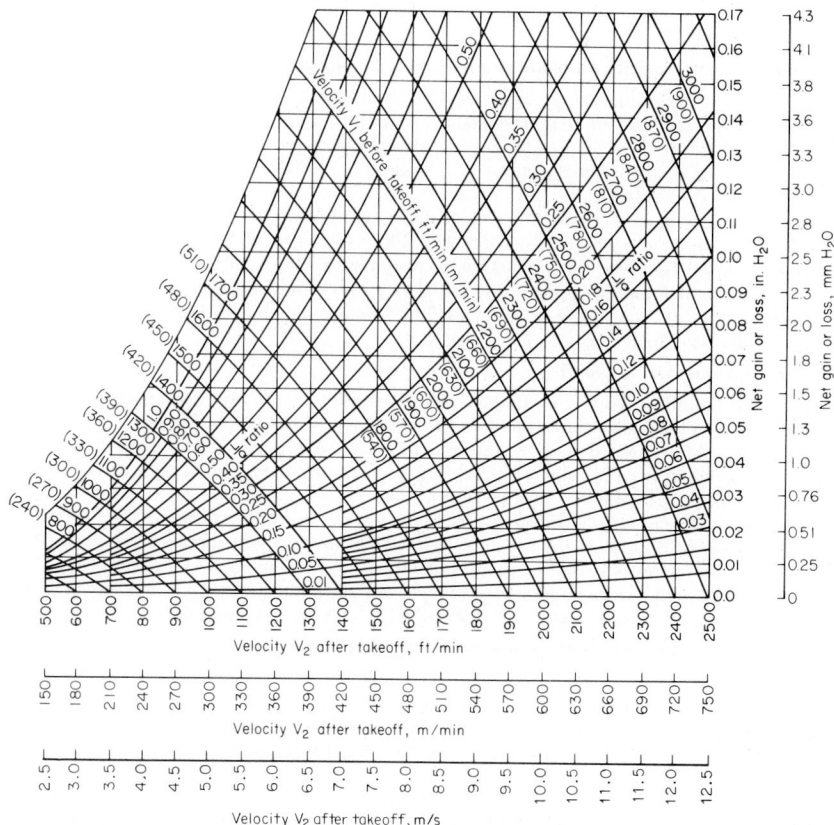

FIG. 9 Low-velocity static regain in air ducts. *(Carrier Air Conditioning Company.)*

Note that these tabulated results are *general* and may not apply to every system. The method presented in this calculation procedure is that used by the Carrier Air Conditioning Company at the time of this writing.

HUMIDIFIER SELECTION FOR DESIRED ATMOSPHERIC CONDITIONS

A paper mill has a storeroom with a volume of 500,000 ft³ (14,155 m³). The lowest recorded outdoor temperature in the mill locality is 0°F (−17.8°C). What capacity humidifier is required for this storeroom if a 70°F (21°C) dry-bulb temperature and a 65 percent relative humidity are required in it? Moisture absorption by the paper products in the room is estimated to be 450 lb/h (204.6 kg/h). The storeroom ventilating system produces three air changes per hour. What capacity humidifier is required if the room temperature is maintained at 60°F (15.5°C) and 65 percent relative humidity? The products release 400 lb/h (181.8 kg/h) of moisture. Steam at 25 lb/in² (gage) (172.4 kPa) is available for humidification. The outdoor air has a relative humidity of 50 percent and a minimum temperature of 5°F (−15°C).

Calculation Procedure:

1. Determine the outdoor design temperature

In choosing a humidifier, the usual procedure is to add 10°F (5.6°C) to the minimum outdoor recorded temperature because this temperature level seldom lasts more than a few hours. The

result is the design outdoor temperature. Thus, for this mill, design outdoor temperature = 0 +
~~10 = 10°F (−12°C).~~

2.   *Compute the weight of moisture required for humidification*

Enter Table 18 at an outdoor temperature of 10°F (−12.2°C), and project across to the desired
relative humidity, 65 percent. Read the quantity of steam required as 1.330 lb/h (0.6 kg/h) per
1000 ft³ (28.3 m³) of room volume for two air changes per hour. Since this room has three air
changes per hour, the quantity of moisture required is (3/2)(1330) = 1.995 lb/h (0.91 kg/h) per
1000 ft³ (28.3 m³) of volume.

The amount of moisture in the form of steam required for this storeroom = (room volume,
ft³/1000)(lb/h of steam per 1000 ft³) = (500,000/1000)(1.995) = 997.5 lb (453.4 kg) for humidification of the air. However, the products in the storeroom absorb 450 lb/h (202.5 kg/h) of moisture. Hence, the total moisture quantity required = moisture for air humidification + moisture
absorbed by products = 997.5 + 450.0 = 1447.5 lb/h (657.9 kg/h), say 1450 lb/h (659.1 kg/h)
for humidifier sizing purposes.

3.   *Select a suitable humidifier*

Table 19 lists typical capacities for humidifiers having orifices of various sizes and different steam
pressures. Study of Table 19 shows that one 1¼-in (32-mm) orifice humidifier and two ⅜-in (9.5-
cm) orifice humidifiers will discharge 1130 + (2)(174) = 1478 lb/h (671.8 kg/h) of steam when
the steam supply pressure is 25 lb/in² (gage) (172.4 kPa). Since the required capacity is 1450 lb/
h (659.1 kg/h), these humidifiers may be acceptable.

Large-capacity steam humidifiers usually must depend on existing ducts or large floor-type
unit heaters for distribution of the moisture. When such means of distribution are not available,
choose a larger number of smaller-capacity humidifiers and arrange them as shown in Fig. 10c.
Thus, if ⅜-in (9.5-mm) orifice humidifiers were selected, the number required would be (moisture
needed, lb/h)/(humidifier capacity, lb/h) = 1450/174 = 8.33, or 9 humidifiers.

4.   *Choose a humidifier for the other operating conditions*

Where the desired room temperature is different from 70°F (21°C), use Table 20 instead of Table
18. Enter Table 20 at the desired room temperature, 60°F (15.6°C), and read the moisture content of saturated air at this temperature, as 5.795 gr/ft³ (13.26 mL/dm³). The outdoor air at 5 +
10 = 15°F (9.4°C) contains, as Table 20 shows, 0.984 gr/ft³ (2.25 mL/dm³) of moisture when
fully saturated.

Find the moisture content of the air at the room and the outdoor conditions from moisture
content, gr/ft³ = (relative humidity of the air, expressed as a decimal) (moisture content of saturated air, gr/ft³). For the 60°F (15.6°C), 65 percent relative humidity room air, moisture content

TABLE 18   Steam Required for Humidification at 70°F (21°C)° †

Outdoor temp		Relative humidity desired indoors, percent									
		40		50		60		65		70	
°F	°C	lb	kg	lb	kg	lb	kg	lb	kg	lb	kg
50	10.0	0.045	0.02	0.271	0.12	0.501	0.23	0.616	0.28	0.731	0.33
40	4.4	0.307	0.14	0.537	0.24	0.767	0.35	0.882	0.40	1.000	0.45
30	1.1	0.503	0.23	0.734	0.33	0.964	0.43	1.079	0.49	1.194	0.54
20	−6.7	0.654	0.29	0.883	0.40	1.115	0.50	1.230	0.55	1.345	0.61
10	−12.2	0.754	0.34	0.985	0.44	1.215	0.55	1.330	0.60	1.445	0.65
0	−17.8	0.819	0.37	1.049	0.47	1.279	0.58	1.394	0.63	1.509	0.68
−10	−23.3	0.860	0.39	1.090	0.49	1.320	0.59	1.435	0.65	1.550	0.70
−20	−28.9	0.885	0.30	1.115	0.50	1.345	0.61	1.460	0.66	1.575	0.71

°Armstrong Machine Works.

†Pounds (kilograms) of steam per hour required per 1000 ft³ (28.3 m³) of space to secure desired indoor relative
humidity at 70°F (21°C), with various outdoor temperatures. Assuming two air changes per hour and outdoor
relative humidity of 75 percent.

TABLE 19 Humidifier Capacities*

Steam pressure, lb/in² (gage) (kPa)	Orifice size, in (mm)			
	7/16 (1.11)	3/8 (9.5)	1¼ (31.8)	1¾₄ (28.2)
5 (34.5)	100	. . . .	340	42
10 (68.9)	140	. . . .	610	
15 (103.4)	170	138	810	74
20 (137.9)	. . .	158	980	80
25 (172.4)	. . .	174	1130	90
30 (206.8)	. . .	190	1280	100

*Armstrong Machine Works.
†Continuous discharge capacity with steam pressures as indicated. No allowance for pressure drop after solenoid valve opens.

= (0.65)(5.795) = 3.77 gr/ft³ (8.63 mL/dm³). For the 15°F (−9.4°C) 50 percent relative humidity outdoor air, moisture content = (0.50)(0.984) = 0.492 gr/ft³ (1.13 mL/dm³). Thus, the humidifier must add the difference or 3.77 = 0.492 = 3.278 gr/ft³ (7.5 mL/dm³).

This storeroom has a volume of 500,000 ft³ (14,155 m³) and three air changes per hour. Thus, the weight of moisture that must be added per hour is (number of air changes per hour)(volume, ft³)(gr/ft³ of air)/7000 gr/lb or, for this storeroom, (3)(500,000)(3.278)/7000 = 701 lb/h (0.09 kg/s) excluding the product load. Since the product load is 400 lb/h (0.05 kg/s), the total humidification load is 701 + 400 = 1101 lb/h (0.14 kg/s). Choose the humidifiers for these conditions in the same way as described in step 4.

Related Calculations: Use the method given here to choose a humidifier for any normal industrial or comfort application. Table 21 summarizes typical recommended humidities and temperatures for a variety of industrial operations. The relative humidity maintained in industrial plants is extremely important because it can control the moisture content of hygroscopic materials.

TABLE 20 Moisture Content of Saturated Air

°F	°C	Grains of water	
		Per ft³	Per m³
15	−9.4	0.984	34.8
20	−6.7	1.242	43.9
40	4.4	2.863	101.1
50	10.0	4.106	145.0
60	15.6	5.795	204.7
70	21.1	8.055	284.5

Where the number of hourly air changes is not specified, assume two air changes, except in cotton mills where three or four may be necessary. If the plant ventilating system provides more than two air changes per hour, use the actual number of changes in computing the required humidifier capacity.

Many types of manufactured goods and raw materials absorb or release moisture during processing and storage. Since product quality usually depends directly on the moisture content, carefully controlled humidity will often reduce the number of rejects. The room humidifier must supply sufficient moisture for humidification of the air, plus any moisture absorbed by the products or materials in the room. Where these products or materials continuously release moisture to the atmosphere in the room, the quantity released can be subtracted from the moisture required

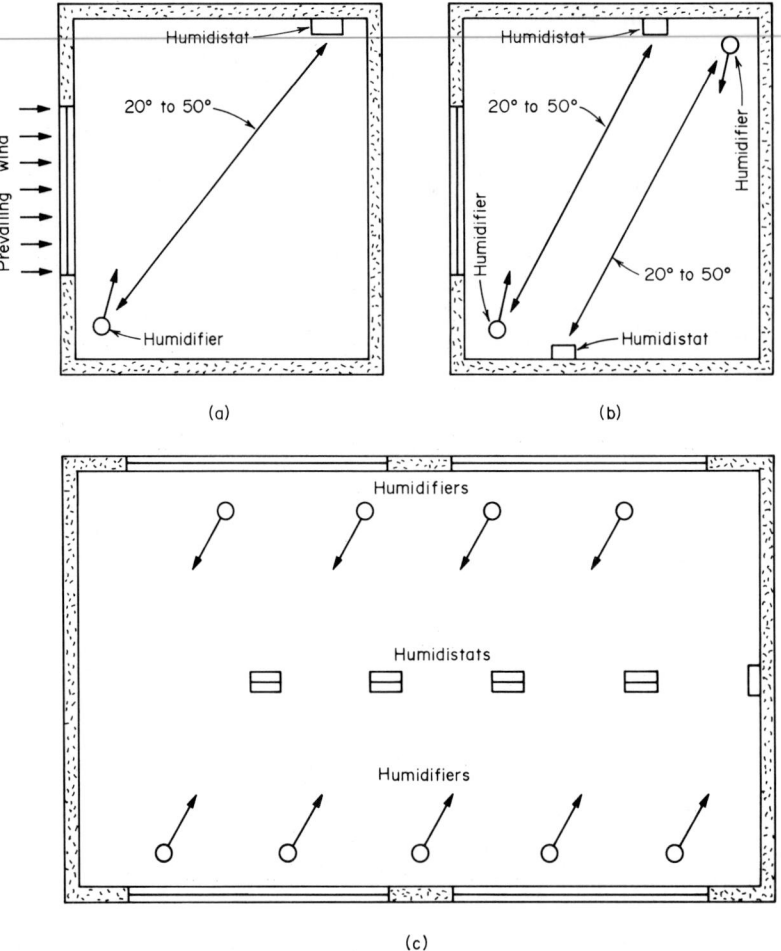

FIG. 10   Location of (a) a single humidifier, (b) two humidifiers, (c) multiple humidifiers.

for humidification. However, this condition can seldom be relied on. The usual procedure then is to select the humidifier on the basis of the moisture required for humidification of the air. The humidistat controls the operation of the humidifier, shutting it off when the products release enough moisture to supply the room requirements.

Correct locations for one or more humidifiers are shown in Fig. 10. Proper location of humidifiers is necessary if the design is to take advantage of the prevailing wind in the plant locality. Also, correct location provides a uniform, continuous circulation of air throughout the humidified area.

When only one humidifier is used, it is placed near the prevailing wind wall and arranged to discharge parallel to the wall exposed to the prevailing wind, Fig. 10a. Two humidifiers, Fig. 10b, are generally located in opposite corners of the manufacturing space and their discharges are used to produce a rotary air motion. Installations using more than two humidifiers generally have a slightly greater number of humidifiers on the windward wall to take advantage of the natural air drift from one side of the room to the other.

Pipe spray humidifiers are as shown in Fig. 11 unless the manufacturer advises otherwise. Size the return lines as shown in Table 22.

TABLE 21   Recommended Industrial Humidities and Temperatures

Industry	Degrees °F	°C	Relative humidity, %
Ceramics:			
Drying refractory shapes	110–150	43–65	50–60
Molding room	80	26	60
Confectionery:			
Chocolate covering	62–65	17–18	50–55
Hard-candy making	70–80	21–27	30–50
Electrical:			
Manufacture of cotton-covered wire, storage, general	60–80	21–27	60–70
Food storage:			
Apple	31–34	−0.5–1.1	75–85
Citrus fruit	32	0	80
Grain	60	16	30–45
Meat ripening	40	4	80
Paper products:			
Binding	70	21	45
Folding	77	25	65
Printing	75	24	60–78
Storage	75–80	24–27	40–60
Textile:			
Cotton carding	75–80	24–27	50–55
Cotton spinning	60–80	16–27	50–70
Rayon spinning, throwing	70	21	85
Silk processing	75–80	24–27	60–70
Wool spinning, weaving	75–80	24–27	55–60
Miscellaneous:			
Laboratory, analytical	60–70	16–21	60–70
Munitions, fuse loading	70	21	55
Cigar and cigarette making	70–75	21–24	55–65

Humidistats to start and stop the flow of moisture into the room may be either electrically or air (hygrostat) operated, according to the type of activities in the space. Where electric switches and circuits might cause a fire hazard, use an air-operated hygrostat instead of a humidistat. Locate either type of control to one side of the humidifying moisture stream, 20 to 50 ft (6.1 to 15.2 m) away.

TABLE 22   Steam- and Return-Pipe Sizes, in (mm)

Steam or condensate flow, lb/h (kg/h)	Steam pressure, lb/in² (gage) (kPa)				Length of return pipe, ft (m)	
	5 (34.5)	10 (68.9)	50 (344.7)	100 (689.4)	100 (30)	200 (60)
100 (45)	1½ (38.1)	1¼ (32)	1 (25)	1 (25)	1 (25)	1 (25)
200 (90)	2 (51)	2 (51)	1¼ (32)	1¼ (32)	1¼ (32)	1¼ (32)
400 (180)	3 (76)	2½ (64)	2 (51)	1½ (38.1)	1½ (38)	2 (51)
500 (225)	3 (76)	2½ (64)	2 (51)	2 (51)	2 (51)	2 (51)
1000 (450)	3½ (88.9)	3 (76)	2½ (64)	2½ (64)	2½ (64)	2½ (64)
2000 (900)	5 (127)	4 (102)	3 (76)	3 (76)	3 (76)	3 (76)
4000 (1800)	6 (152)	5 (127)	4 (102)	4 (102)	4 (102)	4 (102)

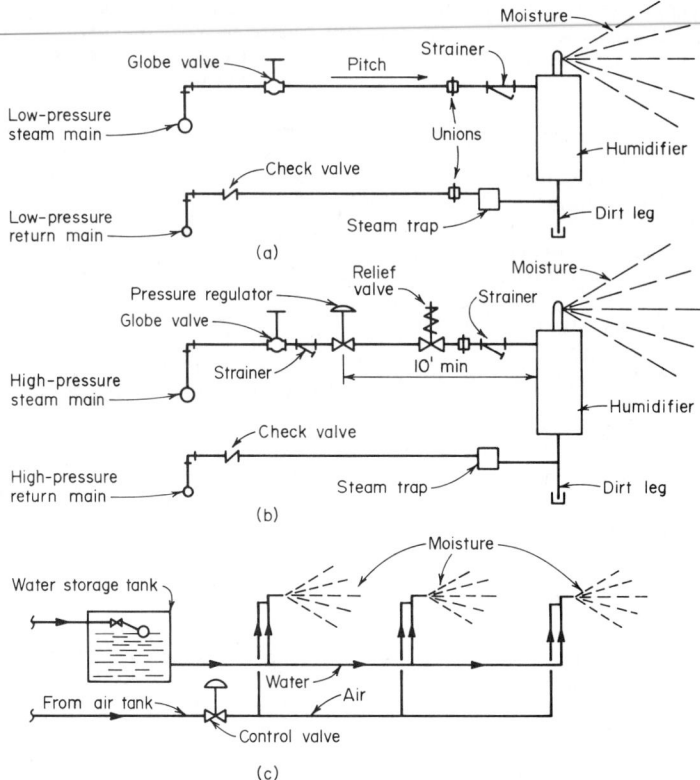

FIG. 11   Piping for spray-type humidifiers. (*a*) Low-pressure steam; (*b*) high-pressure steam; (*c*) water spray.

USE OF THE PSYCHROMETRIC CHART IN AIR-CONDITIONING CALCULATIONS

Determine the properties of air at 80°F (26.7°C) dry-bulb (db) temperature and 65°F (18.3°C) wet-bulb (wb) temperature, using the psychrometric chart. Determine the same properties of air if the wet-bulb temperature is 75°F (23.9°C) and the dew-point temperature is 67°F (19.4°C). Show on the psychrometric chart an air-conditioning process in which outside air at 95°F (35°C) db and 80°F (26.7°C) wb is mixed with return air from the room at 80°F (26.7°C) db and 65°F (18.3°C) wb. Air leaves the conditioning apparatus at 55°F (12.8°C) db and 50°F (10°C) wb.

Calculation Procedure:

1. *Determine the relative humidity of the air*

Using Fig. 5, enter the bottom of the chart at the first dry-bulb temperature, 80°F (26.7°C), and project vertically upward until the slanting 65°F (18.3°C) wet-bulb temperature line is intersected. At the intersection, or *state point*, read the relative humidity as 45 percent on the sloping curve. Note that the number representing the wet-bulb temperature appears on the saturation, or 100 percent relative humidity, curve and that the wet-bulb temperature line is a straight line sloping downward from left to right. The relative humidity curves slope upward from left to right and have the percentage of relative humidity marked on them.

When the wet-bulb and dew-point temperatures are given, enter the psychrometric chart at the wet-bulb temperature, 75°F (23.9°C) on the saturated curve. From here project downward along the wet-bulb temperature line until the horizontal line representing the dew-point temperature, 67°F (19.4°C), is intersected. At the intersection, or state point, read the dry-bulb temperature as 94.7°F (34.8°C) on the bottom scale of the chart. Read the relative humidity at the intersection as 40.05 percent because the intersection is very close to the 40 percent relative humidity curve.

2. Determine the moisture content of the air

Read the moisture content of the air in grains on the right-hand scale by projecting horizontally from the intersection, or state point. Thus, for the first condition of 80°F (26.7°C) dry bulb and 65°F (18.3°C) wet bulb, projection to the right-hand scale gives a moisture content of 68.5 gr/lb (9.9 gr/kg) of dry air.

For the second condition, 75°F wet bulb and 67°F (19.4°C) dew point, projection to the right-hand scale gives a moisture content of 99.2 gr/lb (142.9 gr/kg).

3. Determine the dew point of the air

This applies to the first condition only because the dew point is known for the second condition. From the intersection of the dry-bulb temperature, 80°F (26.7°C), and the wet-bulb temperature, 65°F (18.3°C), that is, the state point, project horizontally to the left to read the dew point on the horizontal intersection with the saturation curve as 56.8°F (13.8°C). Note that the temperatures plotted along the saturation curve correspond to both the wet-bulb and dew-point temperatures.

4. Determine the enthalpy of the air

Find the enthalpy (also called *total heat*) by reading the value on the sloping line on the central scale above the saturation curve at the state point for the air. Thus, for the first condition, 80°F (26.7°C) dry bulb and 65°F (18.3°C) wet bulb, the enthalpy is 30 Btu/lb (69.8 kJ/kg). The enthalpy value on the psychrometric chart includes the heat of 1 lb (0.45 kg) of dry air and the heat of the moisture in the air, in this case, 68.5 gr (98.6 gr) of water vapor.

For the second condition, 75°F (23.9°C) wet bulb and 67°F (19.4°C) dew point, read the enthalpy as 38.5 Btu/lb (89.6 kJ/kg) at the state point.

5. Determine the specific volume of the air

The specific volume lines slope downward from left to right from the saturation curve to the horizontal dry-bulb temperature. Values of specific volume increase by 0.5 ft³/lb (0.03 m³/kg) between each line.

For the first condition, 80°F (26.7°C) dry-bulb and 65°F (18.3°C) wet-bulb temperature, the stage point lies just to the right of the 13.8 line, giving a specific volume of 13.81 ft³/lb (0.86 m³/kg). For the second conditon, 75°F (23.9°C) wet-bulb and 67°F (19.4°C) dew-point temperatures, the specific volume, read in the same way, is 14.28 ft³/lb (0.89 m³/kg).

The weight of the air-vapor mixture can be found from 1.000 + 68.5 gr/lb of air/(7000 gr/lb) = 1.0098 lb (0.46 kg) for the first condition and 1.000 + 99.2/7000 = 1.0142 lb (0.46 kg). In both these calculations the 1000 lb (454.6 kg) represents the weight of the *dry* air and 68.5 gr (4.4 gr) and 99.2 gr (6.43 gr) represent the weight of the moisture for each condition.

6. Determine the vapor pressure of the moisture in the air

Read the vapor pressure by projecting horizontally from the state point to the extreme left-hand scale. Thus, for the first condition the pressure of the water vapor is 0.228 lb/in² (1.57 kPa). For the second condition the pressure of the water is 0.328 lb/in² (2.26 kPa).

7. Plot the air-conditioning process on the psychrometric chart

Air-conditioning processes are conveniently represented on the psychrometric chart. To represent any process, locate the various state points on the chart and convert the points by means of lines representing the process.

Thus, for the air-conditioning process being considered here, start with the outside air at 95°F (35°C) db and 80°F (26.7°C) wb, and plot point 1 (Fig. 5) at the intersection of the two temperature lines. Next, plot point 3, the return air from the room at 80°F (26.7°C) db and 65°F (18.3°C) wb. Point 2 is obtained by computing the final temperature of two airstreams that are mixed, using the method of the calculation procedure given earlier in this section. Plot point 4, using the given leaving temperatures for the apparatus, 55°F (12.8°C) db and 50°F (10°C) wb.

The process in this system is as follows: Air is supplied to the conditioned space along line 4-3. During passage along this line on the chart, the air absorbs heat and moisture from the room. While passing from point 3 to 2, the air absorbs additional heat and moisture while mixing with the warmer outside air. From point 1 to 2, the outside air is cooled while it is mixed with the indoor air. At point 2, the air enters the conditioning apparatus, is cooled, and has its moisture content reduced.

Related Calculations: Use the psychrometric chart for all applied air-conditioning problems where graphic representation of the state of the air or a process will save time. At any given state point of air, the relative humidity in percent can be computed from [partial pressure of the water vapor at the dew-point temperature, lb/in^2 (abs) + partial pressure of the water vapor at saturation corresponding to the dry-bulb temperature of the air, lb/in^2 (abs)](100). Determine the partial pressures from a table of air properties or from the steam tables.

In an *air washer* the temperature of the entering air is reduced. Well-designed air washers produce a leaving-air dry-bulb temperature that equals the wet-bulb and dew-point temperatures of the leaving air. The humidifier portion of an air-conditioning apparatus adds moisture to the air while the dehumidifier removes moisture from the air. In an ideal air washer, adiabatic cooling is assumed to occur.

By using the methods of step 7, any basic air-conditioning process can be plotted on the psychrometric chart. Once a process is plotted, the state points for the air are easily determined from the psychrometric chart.

When you make air-conditioning computations, keep these facts in mind: (1) The total enthalpy, sometimes termed *total heat*, varies with the wet-bulb temperature of the air. (2) The sensible heat of air depends on the wet-bulb temperature of the air; the enthalpy of vaporization, also called the *latent heat*, depends on the dew-point temperature of the air; the dry-bulb, wet-bulb, and dew-point temperatures of air are the same for a saturated mixture. (3) The dew-point temperature of air is fixed by the amount of moisture present in the air.

DESIGNING HIGH-VELOCITY AIR-CONDITIONING DUCTS

Design a high-velocity air-distribution system for the duct arrangement shown in Fig. 12 if the required total air flow is 5000 ft^3/min (2.36 m^3/s).

Calculation Procedure:

1.   *Determine the main-duct friction loss*

Many high-velocity air-conditioning systems are designed for a main-header velocity of 4000 ft/min (1219.2 m/min) and a friction loss of 1.0 in H$_2$O per 100 ft (0.08 cm/m) of equivalent duct length. The fan usually discharges into a combined air-diffuser noise-attenuator in which the static

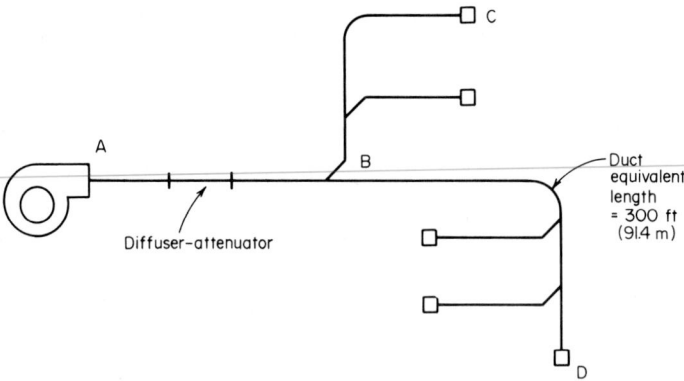

FIG. 12   High-velocity air-duct-system layout.

pressure of the air increases. This pressure increase must be considered in the choice of the fan-outlet static pressure, but the duct friction loss must be calculated first, as shown below.

Determine the main-duct friction loss by assuming a 1 in/100 ft (0.08 cm/m) static pressure loss for the main duct and a fan-outlet and main-duct velocity of 4000 ft/min (1219.2 m/min). Size the duct by using the equal-friction method. Thus, for the 300-ft (91.4-m) equivalent-length main duct in Fig. 12, the friction pressure loss will be (300 ft)(1.0 in/100 ft) = 3.0 in (76 mm) H_2O.

2.   *Compute the required fan-outlet pressure*

The total friction loss in the duct = duct friction, in H_2O + diffuser-attenuator static pressure, in H_2O. In typical installations the diffuser-attenuator static pressure varies from 0.3 to 0.5 in (7.6 to 9.7 mm) H_2O. This is the inlet pressure required to force air through the diffuser-attenuator with all outlets open. Using a value of 0.5 in (12.7 mm) gives the total friction loss in the duct = 3.0 + 0.5 = 3.5 in (8.9 mm).

At the fan outlet the required static pressure is less than the total friction loss in the main duct because there is static regain at each branch takeoff to the outlets. This static regain is produced by the reduction in velocity that occurs at each takeoff from the main duct. There is a recovery of static pressure (velocity regain) at the takeoff that offsets the friction loss in the succeeding duct section.

Assume that the velocity in branch C (Fig. 12) is 2000 ft/min (609.6 m/min). This is the usual maximum velocity in takeoffs to terminals. Then, the maximum static regain that could occur $R = (v_i/4005)^2 - (v_f/4000)^2$, where R = static regain, in of H_2O; v_i = initial velocity of the air, ft/min; v_f = final velocity of the air, ft/min. For this system with an initial velocity of 4000 ft/min (1219.2 m/min) and a final velocity of 2000 ft/min (609.6 m/min), $R = (4000/4005)^2 - (2000/4005)^2 = 0.75$.

The maximum static regain is seldom achieved. Actual static regains range from 0.5 to 0.8 of the maximum. With a value of 0.8, the actual static regain = 0.8(0.75) = 0.60 in (15.2 mm) H_2O. This static regain occurs at point B, the takeoff, and reduces the required fan discharge pressure to total friction loss in the duct − static region at first takeoff = 3.5 − 0.60 = 2.9 in (73.7 mm) H_2O. Thus, a fan developing a static discharge pressure of 3.0 in (76 mm) H_2O would probably be chosen for this system.

3.   *Find the branch-duct pressure loss*

To find the branch-duct pressure loss, find the pressure in the main duct at the takeoff point. Use the standard duct-friction chart (Fig. 7) to determine the pressure loss from the fan to the takeoff point. Subtract the sum of this loss and the diffuser-attenuator static pressure from the fan static discharge pressure. The result is the pressure available to force air through the branch duct. Size the branch duct by using the equal-friction method.

Related Calculations: Note that the design of a high-velocity duct system (i.e., a system design in which the air velocities and static pressures are higher than in conventional systems) is basically the same as for a low-velocity duct system designed for static regain. The air velocity is reduced at each takeoff to the riser and air terminals. Design of any high-velocity duct system involves a compromise between the reduced duct sizes (with a saving in materials, labor, and space costs) and higher fan horsepower.

Class II centrifugal fans (Table 23) are generally required for the higher static pressures used in high-velocity air-conditioning systems. Extra care must be taken in duct layout and construction. The high-velocity ducts are usually sealed to prevent air leakage that may cause objection-

TABLE 23 Classes of Construction for Centrifugal Fans

Class	Maximum total pressure, in H_2O (mm H_2O)
I	3¾ (95)—standard
II	6¾ (172)—standard
III	12¾ (324)—standard
IV	More than 12¾ (324)—recommended

able noise. Round ducts are preferred to rectangular ones because of the greater rigidity of the
round duct.

Use as many symmetric duct runs as possible in designing high-velocity duct systems. The
greater the system symmetry, the less time required for duct design, layout, balancing, construction, and installation.

The initial starting velocity used in the supply header depends on the number of hours of
operation. To achieve an economic balance between first cost and operating cost, lower air velocities in the header are recommended for 24-h operation, where space permits. Table 24 shows
typical air velocities used in high-velocity air-conditioning systems. Use this tabulation to select
suitable velocities for the main and branch ducts in high-velocity systems.

TABLE 24   Typical High-Velocity-System Air Velocities°

	Velocity, ft/min (m/min)
Header or main duct:	
12-h operation	3000–4000 (914.4–1219.2)
24-h operation	2000–3500 (609.6–1066.8)
Branch ducts:†	
90° conical tee	4000–5000 (1219.1–1524.0)
90° tee	3500–4000 (1066.8–1219.2)

°Carrier Air Conditioning Company.
†Branches are defined as a branch header or riser having four to five, or
more, takeoffs to terminals.

Carrier Air Conditioning Company recommends that the following factors be considered in
laying out header ductwork for high-velocity air-conditioning systems.

1. The design friction losses from the fan discharge to a point immediately upstream of the first
   riser takeoff from each branch header should be as nearly equal as possible.

2. To satisfy principle 1 above as applied to multiple headers leaving the fan, and to take maximum advantage of the allowable high velocity, adhere to the following basic rule whenever
   possible: Make as nearly equal as possible the ratio of the total equivalent length of each header
   run (fan discharge to the first riser takeoff) to the initial header diameter (L/D ratio). Thus,
   the longest header run should preferably have the highest air quantity so that the highest velocities can be used throughout.

3. Unless space conditions dictate otherwise, use a 90° tee or 90° conical tee for the takeoff from
   the header rather than a 45° tee. Fittings of 90° provide more uniform pressure drops to the
   branches throughout the system. Also, the first cost is lower.

AIR-CONDITIONING-SYSTEM OUTLET- AND RETURN-GRILLE SELECTION

Choose an air grille to deliver 425 ft^3/min (0.20 m^3/s) of air to a broadcast studio having a 12-ft
(3.7-m) ceiling height. The room is 10 ft (3.0 m) long and 10 ft (3.0 m) wide. Specify the temperature difference to use, the air velocity, grille static resistance, size, and face area.

Calculation Procedure:

1.   *Choose the outlet-grille velocity*

The air velocity specified for an outlet grille is a function of the type of room in which the grille
is used. Table 25 lists typical maximum outlet air velocities used in grilles serving various types
of rooms. Assuming a velocity of 350 ft/min (106.7 m/min) for the outlet grille in this broadcast
studio, compute the grille area required from $A = cfm/v$, where A = grille area, ft^2; cfm = air
flow through the grille, ft^3/min; v = air velocity, ft/min. Hence, $A = 425/350 = 1.214$ ft^2 (0.11
m^2).

TABLE 25   Typical Air Outlet Velocities

Type of room	Maximum velocity, ft/min (m/min)	
Broadcast studio	300–500	(91.4–152.4)
Apartments, private residences, churches, hotel	500–750	(152.4–228.6)
bedrooms, legitimate theatres, private offices	500–750	(152.4–228.6)
	500–750	(152.4–228.6)
	500–750	(152.4–228.6)
	500–750	(152.4–228.6)
	500–750	(152.4–228.6)
Movie theaters	1000	(304.8)
General offices	1200–1500	(365.8–457.2)
Stores—upper floors	1500	(365.8)
Stores—main floors	2000	(609.6)

2.   *Select the outlet-grille size*

Use the selected manufacturer's engineering data, such as that in Table 26. Examination of this table shows that there is no grille rated at 425 ft^3/min (0.20 m^3/s). Hence, the next larger capacity, 459 ft^3/min (0.22 m^3/s) must be used. This grille, as the third column from the right of Table 26 shows, is 24 in wide and 8 in high (60.9 cm wide and 20.3 cm high).

3.   *Choose the grille throw distance*

Throw is the horizontal distance the air will travel after leaving the grille. With a *fan spread*, the throw of this grille is 10 ft (3.0 m), Table 26. This throw is sufficient if the duct containing the grille is located at any point in the room, i.e., along one wall, in the center, etc. If desired, the grille can be adjusted to reduce the throw, but the throw cannot be increased beyond the distance tabulated. Hence, a fan-spread grille will be used.

4.   *Select the grille-mounting height*

The grille-mounting height is a function of several factors: the difference between the temperature of the entering air and the room air, the room-ceiling height, and the air *drop* (i.e., the distance the air falls from the time it passes through the outlet until it reaches the end of the throw).

By assuming a temperature difference of 20°F (11.1°C) between the entering air and the room air, Table 26 shows that the minimum ceiling height for this grille is 10 ft (3.0 m). Since the room is 12 ft (3.67 m) high, the grille can be mounted at any distance above the floor of 10 ft (3.0 m) or higher.

5.   *Determine the actual air velocity in the grille*

Table 26 shows that the actual air velocity in the grille is 375 ft/min (114.3 m/min). Table 25 shows that an air velocity of 300 to 500 ft/min (91.4 to 152.4 m/min) is suitable for broadcast studios. Hence, this grille is acceptable. If the actual velocity at the grille outlet were higher than that recommended in Table 25, a larger grille giving a velocity within the recommended range would have to be chosen.

6.   *Determine the grille static resistance*

Table 26 shows that the grille static resistance is 0.01 in (0.25 mm) H$_2$O. This is within the usual static resistance range of outlet grilles.

7.   *Determine the outlet-grille area*

Table 26 shows that the outlet grille has an area of 1.224 ft^2 (0.11 m^2), or 176 in^2 (1135.5 cm^2). This agrees well with the area computed in step 1, or 1.214 ft^2 (0.11 m^2).

8.   *Select the air-return grille*

Table 27 shows typical air velocities used for return grilles in various locations. By assuming that the air is returned through a wall louvre, a velocity of 500 ft/min (152.4 m/min) might be used. Hence, by using the equation of step 1, grille area $A = cfm/v = 425/500 = 0.85$ ft^2 (0.08 m^2).

TABLE 26 Air-Grille-Selection Table°

Air flow, ft³/min (m³/s)	Wall area per outlet, ft² (m²) Max.	Wall area per outlet, ft² (m²) Min.	Throw, ft (m)†	Min. ceiling height, ft (m) for temp. difference of 15°F (8.3°C)	20°F (11.1°C)	25°F (13.9°C)	Air velocity, ft/min (m/min)	Grille static resist., in H₂O (cm H₂O)	Outlet size, in (cm)	Grille face area ft² (m²)	in² (cm²)
306 (0.14)	25 (2.32)	8 (0.74)	S:11 (3.35) F:6 (1.83)	13 (3.96) 10 (3.05)	14 (4.27) 10 (3.05)	15 (4.57) 10 (3.05)	250 (76.2)	0.005 (0.01)	24 × 8 (60.9 × 20.3)	1.224 (0.11)	176 (1135.5)
459 (0.22)	57 (5.30)	17 (1.58)	S:20(6.1) F:10(3.0)	16 (4.88) 10 (3.05)	17 (5.18) 10 (3.05)	18 (5.49) 11 (3.35)	375 (114.3)	0.01 (0.03)	24 × 8 (60.9 × 20.3)	1.224 (0.11)	176 (1135.5)

°Waterloo Register.
†S—straight; F—fan spread.

TABLE 27   Lattice-Type Return-Grille Pressure Drop, in H$_2$O (mm H$_2$O)

Free area of grille, percent	Face velocity, ft/min (m/min)		Return-intake-air velocities[*]	
	400 (121.9)	600 (182.9)	Intake location	Velocity over gross area, ft/min (m/min)
50	0.04 (1.02)	0.09 (2.29)	Above occupied zone	800 and up (243.8 and up)
60	0.03 (0.76)	0.06 (1.52)	In occupied zone:	
70	0.02 (0.51)	0.05 (1.27)	Not near seats	600–800 (182.9–243.8)
80	0.01 (0.25)	0.03 (0.76)	Near seats	400–600 (121.9–182.9)
			Door or wall louvers	500–700 (152.4–213.4)
			Undercut door (through undercut area)	600 (182.9)

[*]ASHRAE *Guide.*

If a lattice-type return intake having a free area of 60 percent is used, Table 28 shows that the pressure drop during passage of the air through the grille is 0.04 in (1.02 mm) H$_2$O. Locate the return grille away from the supply grille to prevent short circuiting of the air and excessive noise. The pressure losses in Table 27 are typical for return grilles. Choice of the pressure drop to use is generally left with the system designer.

Related Calculations: Use this general method to choose outlet and return grilles for industrial, commercial, and domestic applications. Be certain not to exceed the tabulated velocities where noise is a factor in an installation. Excessive noise can lead to complaints from the room occupants.

The outlet-table excerpt presented here is typical of the table arrangement used by many manufacturers. Hence, the general procedure given for selecting an outlet is similar to that for any other manufacturer's outlet.

Many modern-design ceiling outlets are built so that the leaving air entrains some of the room air. The air being discharged by the outlet is termed *primary air*, and the room air is termed *secondary air.* The induction ratio R_i = (total air, ft^3/min)/(primary air, ft^3/min). Typical induction ratios run in the range of 30 percent.

For a given room, the total air in circulation, cfm = (outlet cfm)(induction ratio). Also, average room air velocity, fpm = 1.4 (total cfm in circulation)/area of wall, ft^2, opposite the outlet or outlets. The wall area in the last equation is the *clear* wall area. Any obstructions must be deducted. The multiplier 1.4 allows for blocking caused by the airstream. Where the room circulation factor K must be computed, use the relation K = (average room air velocity, ft/min)/ 1.4(induction ratio). The ideal room-air velocity for most applications is 25 ft/min (7.62 m/min). However, velocities up to 300 ft/min (91.4 m/min) are used in some factory air-conditioning applications.

TABLE 28   Approximate Pressure Drop for Lattice Return Intakes, in (mm) H$_2$O[*]

Percentage of free area	Face velocity, ft/min (m/min)			
	400 (121.9)	600 (182.9)	800 (243.8)	1000 (304.8)
50	0.06 (1.52)	0.13 (3.30)	0.22 (5.59)	0.35 (8.89)
60	0.04 (1.02)	0.09 (2.29)	0.16 (4.06)	0.24 (6.10)
70	0.03 (0.76)	0.07 (1.78)	0.12 (3.05)	0.18 (4.57)
80	0.02 (0.51)	0.05 (1.27)	0.09 (2.29)	0.14 (3.56)

[*]ASHRAE *Guide and Data Book.*

The types of outlets commonly used today are grille (perforated, fixed-bar, adjustable-bar), slotted, ejector, internal induction, pan, diffuser, and perforated ceiling. Choice of a given type depends on the room ceiling height, desired air-temperature difference, blow, drop, and spread, as well as other factors that are a function of the room, air quantity, and the activities in the room.

As a general guide to outlet selection, use the following pointers: (1) Choose the number of outlets for each room after considering the quantity of air required, throw or diffusion distance available, ceiling height, obstructions, etc. (2) Try to arrange the outlets symmetrically in the space available as shown by the room floor plan.

SELECTING ROOF VENTILATORS FOR BUILDINGS

A 10-bay building is 200 ft (60.9 m) long, 100 ft (30.5 m) wide, 50 ft (15.2 m) high to the top of the pitched roof, and 35 ft (10.7 m) high to the eaves. The building houses 15 turbine-driven generators and is classed as an engine room. Choose enough roof ventilators to produce a suitable number of air changes in the building. During reduced-load operating periods between 12 midnight and 7 a.m. on weekdays, and on weekends, only half the full-load air changes are required. The prevailing summer-wind velocity against the long side of the building is 10 mi/h (16.1 km/h). The total available open-window area on each long side is 300 ft^2 (27.9 m^2). The minimum difference between the outdoor and indoor temperatures will be 40°F (22.2°C).

Calculation Procedure:

1.   *Determine the cubic volume of the building*

To compute the cubic volume of a pitched-roof building, the usual procedure is to assume an average height from the eaves to the ridge. Since this building has a 15-ft (4.6-m) high ridge from the eaves, the average height = 15/2 = 7.5 ft (4.6 m). Since the height from the ground to the eaves is 35 ft (10.7 m), the building height to be used in the volume computation is 35 + 7.5 = 42.5 ft (13.9 m). Hence, the volume of the building, ft^3 = V = length × width × average height, all measured in ft = 200 × 100 × 42.5 = 850,000 ft^3 (24,064 m^3).

2.   *Determine the number of air changes required*

Table 29 shows that four to six air changes per hour are normally recommended for engine rooms. Using five air changes per hour will probably be satisfactory, and the roof ventilators will be chosen on this basis. During the early morning, and on weekends, 2.5 air changes will be satisfactory, since only half the normal number of air changes is needed during these periods.

3.   *Compute the required hourly air flow*

The required hourly air flow, ft^3/h = Q = (number of air changes per hour)(building volume, ft^3) = (5)(850,000) = 4,250,000 ft^3/h (120,318 m^3/h). During the early morning hours and on weekends when 2.5 air changes are used, Q = 2.5(850,000) = 2,125,000 ft^3/h (60,159 m^3/h).

TABLE 29   Number of Air Changes Required per Hour°

Auditoriums and assembly rooms	10–15	Libraries	3
Boiler rooms	10–15	Machine shops	6
Churches	10–15	Paint shops	10–15
Engine rooms	4–6	Paper mills	15–20
Factory buildings (general)	4	Pump rooms	8–10
Factory buildings (where excessive	15–20	Railroad shops	4
conditions of fumes, moisture, etc., are		Schools	10–12
present)		Textile mills (general)	4
Foundries	12	Textile mill dye houses	15–20
Garages	10–15	Theaters	5–8
General offices	3	Waiting rooms	4
Hotel dining rooms	4	Warehouses	4
Hotel kitchens	10–20	Wood-working shops	8
Laundries	15–25		

°DeBothezat Fans Division, AMETEK Inc.

4. Compute the air flow produced by natural ventilation

The ASHRAE *Guide and Data Book* lists the prevailing winter and summer wind velocities for a variety of locations. Usual practice, in designing natural-ventilation systems, is to use one-half the tabulated wind velocity for the season being considered. Since summer ventilation is usually of greater importance than winter ventilation, one-half the prevailing summer wind velocity is generally used in natural-ventilation calculations. As the prevailing summer wind velocity in this locality is 8 mi/h (12.9 km/h), a velocity of 8/2 = 4 mi/h (6.4 km/h) will be used to compute the air flow produced by the wind.

Use the relation $Q = VAE$ to find the air flow produced by the wind. In this relation, $Q =$ air flow produced by the wind, ft³/min; $V =$ design wind velocity, ft/min $= 88 \times$ mi/h; $A =$ free area of the air-inlet openings, ft²; $E =$ effectiveness of the air inlet openings—use 0.50 to 0.60 for openings perpendicular to the wind and 0.25 to 0.35 for diagonal winds.

Assuming $E = 0.50$, we get $Q = VAE = (4 \times 88)(300)(0.50) = 52,800$ ft³/min (1494.8 m³/min) or 60(52,800) = 3,168,000 ft³/h (89,686.1 m³/h). Step 3 shows that the required air flow is 4,250,000 ft³/h (120,318 m³/h) when all turbines are operating. Hence, the air flow produced by natural ventilation is inadequate for full-load operation. However, since the required flow of 2,125,000 ft³/h (60,159 m³/h) for the early morning hours and weekends is less than the natural-ventilation flow of 3,168,000 ft³/h (89,686.1 m³/h), natural ventilation may be acceptable during these periods.

5. Determine the number of stationary-type ventilators needed

A stationary-type roof ventilator (i.e., one that depends on the wind and air-temperature difference to produce the desired air movement) may be suitable for this application. If the stationary-type is not suitable, a powered-fan type of roof ventilator will be investigated and must be used. The Breidert-type ventilator is investigated here because the procedure is similar to that used for other stationary-type roof ventilators.

Stationary ventilators produce air flow out of a building by two means: suction caused by wind action across the ventilators and the stack effect caused by the temperature difference between the inside and outside air.

Figure 13 shows the air velocity produced in a stationary Breidert ventilator by winds of various velocities. Thus, with the average 5-mi/h (8.1-km/h) wind assumed earlier for this building, Fig. 13 shows that the air velocity through the ventilator produced by this wind velocity is 220 ft/min (67.1 m/min), closely.

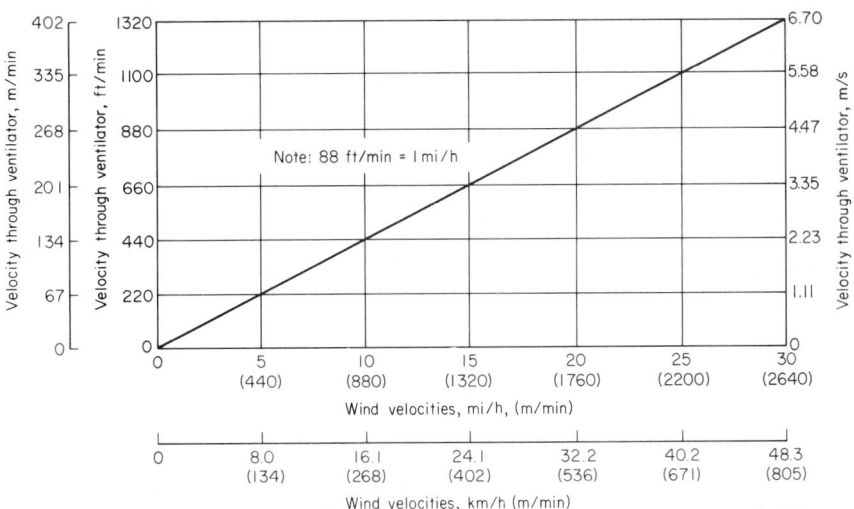

FIG. 13 Roof-ventilator air-exhaust capacity for various wind velocities. Add the extra velocity for temperature difference given in Table 30. *(G. C. Breidert Co.)*

Table 30 shows that a 1.0-ft^2 (0.093-m^2) ventilator installed on a 50-ft (15.2-m) high building having an air-temperature difference of 40°F (22.2°C) will produce, owing to the stack effect, an air-flow velocity of 482 ft/min (146.9 m/min). Hence, the total velocity through this ventilator resulting from the wind and stack action is 220 + 482 = 702 ft/min (213.9 m/min).

TABLE 30   Flow of Air in Natural-Draft Flues, ft^3/(min·ft^2) [m^3/(min·m^2)]°

Difference in temperature, °F (°C)	Height of flue, ft (m), same as height of room or building			
	30 (9.1)	40 (12.2)	50 (15.2)	60 (18.3)
10 (5.6)	188 (57.3)	217 (66.1)	242 (73.7)	264 (80.4)
20 (11.1)	265 (80.6)	306 (93.2)	342 (104.2)	373 (113.7)
30 (16.7)	325 (99.0)	375 (114.3)	419 (127.7)	461 (140.5)
40 (22.2)	374 (113.9)	431 (131.3)	482 (140.2)	529 (161.2)
50 (27.8)	419 (127.7)	484 (147.5)	541 (164.9)	594 (181.0)
60 (33.3)	460 (140.2)	532 (162.1)	595 (181.4)	650 (198.1)

°G. C. Breidert Co.

Since air flow, ft^3/min = (air velocity, ft/min)(area of ventilator opening, ft^2), an air flow of (702)(1.0) = 702 ft^3/min (19.9 m^3/min) will be produced by each square foot of ventilator-neck or inlet-duct area. Thus, to produce a flow of (4,250,000 ft^3/h)/(60 min/h) = 70,700 ft^3/min (2002 m^3/min) will require a ventilator area of 70,700/702 = 101 ft^2 (9.38 m^2). A 48-in (121.9-cm) Breidert ventilator has a neck area of 12.55 ft^2 (1.17 m^2). Hence, a 101/12.55 = 8.05, or eight ventilators will be required. Alternatively, the Breidert capacity table in the engineering data prepared by the manufacturer shows that a 48-in (121.9-cm) ventilator has a ventilating capacity of 8835 ft^3/min (250.1 m^3/min) when it is used for a 5-mi/h (8.1-km/h), 50-ft (15.2-m) high, 40°F (22.2°C) temperature-difference application. With this capacity, the number of ventilators required = 70,700/8835 = 8.02, say eight ventilators. These ventilators will be suitable for both full- and part-load operation.

6.   Determine the number of powered ventilators needed

Powered ventilators are equipped with single- or two-speed fans to produce a positive air flow independent of wind velocity and stack effect. For this reason, some engineers prefer powered ventilators where it is essential that air movement out of the building be maintained at all times.

Two-speed powered ventilators are usually designed so that the reduced-speed rpm is approximately one-half the full-speed rpm. The air flow at half-speed is about one-half that at full speed.

Checking the capacity table of a typical powered-ventilator manufacturer shows that ventilator capacities range from about 2100 ft^3/min (59.5 m^3/min) for a 21-in (53.3-cm) diameter unit at a ⅛-in (3.18-mm) static pressure difference to about 24,000 ft^3/min (679 m^3/min) for a 36-in (91.4-cm) diameter ventilator at the same pressure difference. With a 27-in (68.6-cm) diameter powered ventilator which has a capacity of 14,900 ft^3/min (421.8 m^3/min), the number required is [70,700 ft^3/min (2002 m^3/m)]/[14,900 ft^3/min (422 m^3/min) per unit] = 4.76, or 5.

7.   Choose the type of ventilator to use

Either a stationary or powered ventilator might be chosen for this application. Since a large amount of heat is generated in an engine room, the powered ventilator would probably be a better choice because there would be less chance of overheating during periods of little or no wind.

Related Calculations: Use the general method given here to choose stationary or powered ventilators for any of the 25 applications listed in Table 29. Usual practice is to locate one ventilator in each bay or sawtooth of a building.

With the greater interdiction of smoking in public places (factories, offices, hotels, restaurants, schools, etc.), special exhaust fans—often termed "smoke eaters"—are being installed. These high-velocity fans draw smoke-laden air from a designated smoking area and exhaust it to the atmosphere or to treatment devices.

Local building codes govern smoking in structures of various types, so the engineer must consult the local code before choosing the type of exhaust fan to use for a specific building. Many cities now prohibit all smoking inside a building. In such cities special exhaust fans are not needed to handle cigarette, cigar, or pipe smoke.

TABLE 31 Suggested Isolation Efficiencies

Equipment	Installed efficiency, %
Absorption units	95
Steam generators	95
Centrifugal compressors	98
Reciprocating compressors:	
Up to 15 hp (11.2 kW)	85
20–60 hp (14.9–44.8 kW)	90
75–150 hp (56.0–111.9 kW)	95
Packaged air conditioners	90
Centrifugal fans:	
80 r/min and above; all diameters	90–95
350–800 r/min; all diameters	70–90
200–350 r/min; 48-in (121.9-cm) diameter or smaller	°
200–350 r/min; 54-in (137.2-cm) diameter or larger	70–80
Centrifugal pumps	95
Cooling towers	85
Condensers	80
Fan coil units	80
Piping	95

°Installed for noise isolation only.

Restaurants, bowling alleys, billiard rooms, taverns, and similar gathering places still permit some indoor smoking. Most cities have not prohibited smoking in such establishments because it is a part of the ambiance of these places. Some cities, however, recommend—or require—designated smoking areas, particularly in restaurants. It is here that the engineer must select and specify a suitable exhaust fan to rid the area of tobacco smoke.

Follow the same procedure given above to choose the fan or fans. Be certain to use the required number of air changes specified by any local building code. While an excessive number of air changes will increase the winter heating load, many engineers overdesign to be certain they meet clean air requirements. Tobacco smoke must be handled decisively so that all patrons of an establishment are comfortable.

VIBRATION-ISOLATOR SELECTION FOR AN AIR CONDITIONER

Choose a vibration isolator for a packaged air conditioner operating at 1800 r/min. What minimum mounting deflection is required if the air conditioner is mounted on a basement floor? On an upper-story floor made of light concrete?

Calculation Procedure:

1. *Determine the suggested isolation efficiency*

Table 31 lists the suggested isolation efficiency for various components used in air-conditioning and refrigeration systems. This tabulation shows that the suggested isolation efficiency for a packaged air conditioner is 90 percent. This means that the vibration isolator or mounting should absorb 90 percent, or more, of the vibration caused by the machine. At this efficiency only 10 percent of the machine vibration would be transmitted to the supporting structure.

2. *Determine the static deflection caused by the vibration*

Use Fig. 14 to find the static deflection caused by the vibration. Enter at the bottom of Fig. 14 at 1800 r/min the disturbing frequency, and project vertically upward to the 90 percent efficiency curve. At the left read the static deflection as 0.11 in (2.79 mm).

3. *Select the type of vibration isolator to use*

Project to the right from the intersection with the efficiency curve (Fig. 14) to read the type of isolator to use. Thus, neoprene pads or neoprene-in-shear mounts will safely absorb up to 0.25-in (6.35-mm) static deflection. Hence, either type of isolator mounting could be used.

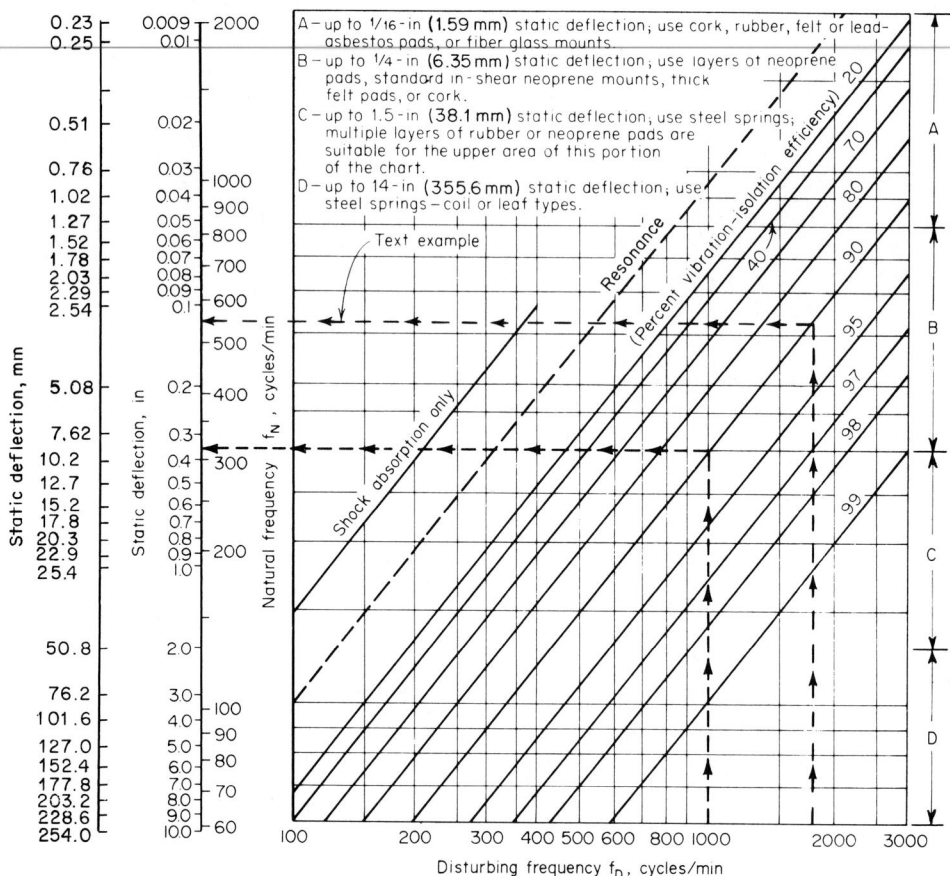

FIG. 14   Vibration-isolator deflection for various disturbing frequencies. *(Power.)*

4.   Check the isolator selection

Use Table 32 to check the theoretical isolation efficiency of the mounting chosen. Enter at the top at the rpm of the machine, and project vertically downward until an efficiency equal to, or greater than, that desired is intersected.

For this machine operating at 1800 r/min, single-deflection rubber mountings have an efficiency of 94 percent. Since neoprene is also called synthetic rubber, the isolator choice is acceptable because it yields a higher efficiency than required.

5.   Determine the minimum mounting deflection required

Table 33 lists the minimum mounting deflection required at various operating speeds for machines installed on various types of floors. Thus, at 1800 r/min, machines mounted on a basement floor must have isolator mountings that will absorb deflections up to 0.10 in (2.54 mm). Since the neoprene mountings chosen in step 3 will absorb up to 0.25 in (6.35 mm) of deflection, they will be acceptable for use on a basement-mounted machine.

For mounting on a light-concrete upper-story floor, Table 33 shows that the mounting must be able to absorb a deflection of 0.80 in (20.3 mm) for machines operating at 1800 r/min. Since the neoprene isolators can absorb only 0.25 in (6.35 mm), another type of mounting is needed if the machine is installed on an upper floor. Figure 14 shows that steel springs will absorb up to 1.5 in (38.1 mm) of static deflection. Hence, this type of mounting would be used for machines installed on upper floors of the building.

TABLE 32 Theoretical Vibration-Isolation Efficiencies[a]

Isolation material	Average static deflection, in (mm)		Average natural frequency	Efficiencies, percent									
				350 r/min	500 r/min	600 r/min	800 r/min	1000 r/min	1200 r/min	1500 r/min	1800 r/min	3000 r/min	3600 r/min
2-in (50.8-mm) thick standard-density cork	0.08	(2.03)	By test 1420	...	...	...	...	...	...	...	...	72	82
Type W waffle pad	Curvature corrected, 0.035	(0.89)	1000	...	...	...	...	...	...	20	55	87	92
Two layers of W waffle pad	Curvature corrected, 0.070	(1.78)	710	...	...	...	...	...	46	71	82	93	96
Single-deflection rubber mountings	0.20	(5.08)	420	...	...	...	62	79	86	91	94	98	99
Double-deflection rubber mountings	0.40	(10.16)	300	...	44	67	84	90	93	96	97	99	Almost perfect
Standard spring mountings	1.00	(25.4)	188	70	85	89	94	96	97	98	99	Almost perfect	Almost perfect
Double-deflection rubber and spring mountings	1.40	(35.6)	160	75	89	93	96	97	98	99	Almost perfect	Almost perfect	Almost perfect

[a] *Power* magazine.

TABLE 33 Minimum Mounting Deflections

Operating speed, r/ min	Basement— negligible floor deflection, in (mm)	Rigid concrete floor, in (mm)	Upper story— light-concrete floor, in (mm)	Wood floor, in (mm)
300	1.50 (38.1)	3.00 (76.2)	3.50 (88.9)	4.00 (101.6)
500	0.63 (16.0)	1.25 (31.8)	1.65 (41.9)	1.95 (49.5)
800	0.25 (6.35)	0.60 (15.2)	1.00 (25.4)	1.25 (31.6)
1200	0.20 (5.08)	0.45 (11.4)	0.80 (20.3)	1.00 (25.4)
1800	0.10 (2.54)	0.35 (8.9)	0.80 (20.3)	1.00 (25.4)
3600	0.03 (0.76)	0.20 (5.08)	0.80 (20.3)	1.00 (25.4)
7200	0.03 (0.76)	0.20 (5.08)	0.80 (20.3)	1.00 (25.4)

TABLE 34 Power and Intensity of Noise Sources

Sound source	Power range, W	Decibel range $(10^{-13}$ W)
Ram jet	100,000.0	180
Turbojet with 7000-lb (31,136-N) thrust	10,000.0	170
	1,000.0	160
Four-propeller airliner	100.0	150
75-piece orchestra, pipe organ; small aircraft engine	10.0	140
Chipping hammer	1.0	130
Piano, blaring radio	0.1	120
Centrifugal ventilating fan at 13,000 ft³/min (6.14 m³/s)	0.01	110
Automobile on roadway; vane-axial ventilating fan	0.001	100
Subway car, air drill	0.0001	90
Conversational voice; traffic on street corner	0.000 01	80
Street noise, average radio	0.000 001	70
Typical office	0.000 000 1	60
	0.000 000 01	50
Very soft whisper	0.000 000 001	40

Reference values relate decibel scales

db scale	Definition	Reference quantity
Sound-power level	$PWL = 10 \log \dfrac{W}{W \text{ re}}$	$W \text{ re} = 10^{-13}$ W
Sound-intensity level	$IL = 10 \log \dfrac{I}{I \text{ re}}$	$I \text{ re} = 10^{12}$ W/m² $= 10^{-16}$ W/cm²
Sound-pressure level	$SPL = 10 \log \dfrac{P^2}{P^2 \text{ re}} = 20 \log \dfrac{P}{P \text{ re}}$	$P \text{ re} = 0.000,02$ N/m² $= 0.0002$ μbar $= 0.0002$ dyn/cm²

° *Power* magazine.

Related Calculations: Use this general procedure for engines, compressors, turbines, pumps, fans, and similar rotating and reciprocating equipment. Note that the suggested isolation efficiencies in Table 31 are for air-conditioning equipment located in critical areas of buildings, such as offices, hospitals, etc. In noncritical areas, such as basements or warehouses, an isolation efficiency of 70 percent may be acceptable. Note that the efficiencies given in Table 31 are useful as general guides for all types of rotating machinery.

Although much emphasis is placed on atmospheric, soil, and water pollution, greater attention is being placed today on audio pollution than in the past. Audio pollution is the discomfort in human beings produced by excessive or high-pitch noise. One good example is the sound produced by jet aircraft during takeoff and landing.

Audio pollution can be injurious when it is part of the regular workplace environment. At home, audio pollution can interfere with one's life-style, making both indoor and outdoor activities unpleasant. For these reasons, regulatory agencies are taking stronger steps to curb audio pollution.

Control of audio pollution almost always reverts to engineering design. For this reason, engineers will be more concerned with the noise their designs produce because it is they who have more control of it than any others in the design, manufacture, and use of a product.

SELECTION OF NOISE-REDUCTION MATERIALS

A concrete-walled test laboratory is 25 ft (7.62 m) long, 20 ft (6.1 m) wide, and 10 ft (3.05 m) high. The laboratory is used for testing chipping hammers. What noise reduction can be achieved in this laboratory by lining it with acoustic materials?

Calculation Procedure:

1. Determine the noise level of devices in the room

Table 34 shows that a chipping hammer produces noise in the 130-dB range. Hence, the noise level of this room can be assumed to be 130 dB. This is rated as deafening by various authorities. Therefore, some kind of sound-absorption material is needed in this room if the uninsulated walls do not absorb enough sound.

TABLE 35 Noise-Reduction Coefficients°

Type	Material	Noise-reduction coefficient range [¾ in (19.1 mm) thick]
1	Regularly perforated cellulose-fiber tile	0.65–0.85
2	Randomly perforated cellulose-fiber tile	0.60–0.75
3	Textured, perforated, fissured, cellulose tile	0.50–0.70
4	Cellulose-fiber lay-in panels	0.50–0.60†
5	Perforated mineral-fiber tile	0.65–0.85
6	Fissured mineral-fiber tile	0.65–0.80
7	Textured, perforated or smooth mineral-fiber tile	0.65–0.85
8	Membrane-faced mineral-fiber tile	0.30–0.90
9	Mineral-fiber lay-in panels	0.20–0.90
10	Perforated-metal pans with mineral-fiber pads	0.60–0.80†
11	Perforated-metal lay-in panels with mineral-fiber pads	0.75–0.85
12	Mineral-fiber tile—fire-resistive assemblies	0.55–0.90
13	Mineral-fiber lay-in panels—fire-resistive units	0.65–0.75‡
14	Perforated-asbestos panels with mineral-fiber pads	0.65–0.75§
15	Sound-absorbent duct lining	0.65–0.75†
16	Special acoustic panels and materials	

°Acoustical and Insulating Materials Association.
†Noise-reduction coefficient 1 in (25.4 mm) thick.
‡Noise-reduction coefficient ⅝ in (15.9 mm) thick.
§Noise-reduction coefficient ¹⁵⁄₁₆ in (23.8 mm) thick.

2.   *Compute the total sound absorption of the room*

The *sound-absorption coefficient* of bare concrete is 0.1. This means that 10 percent of the sound produced in the room is absorbed by the bare concrete walls.

To find the total sound absorption by the walls and ceiling, find the product of the total area exposed to the sound and the sound absorption coefficient of the material. Thus, concrete area, excluding the floor but including the ceiling = two walls (25 ft long × 10 ft high) + two walls (20 ft wide × 10 ft high) + one ceiling (25 ft long × 20 ft wide) = 1400 ft² (130.1 m²). Then the total sound absorption = (1400)(0.1) = 140.

3.   *Compute the total sound absorption with acoustical materials*

Table 35 lists the sound- or noise-reduction coefficients for various acoustic materials. Assume that the four walls and ceiling are insulated with membrane-faced mineral-fiber tile having a sound absorption coefficient of 0.90, from Table 35.

Then, by the procedure of step 2, total noise reduction = (1400 ft²)(0.90) = 1260 ft² (117.1 m²).

4.   *Compute the noise reduction resulting from insulation use*

Use the relation noise reduction, dB = 10 log (total absorption *after* treatment/total absorption *before* treatment) = 10 log (1260/140) = 9.54 dB. Thus, the sound level in the room would be reduced to 130.00 − 9.54 = 120.46 dB. This is a reduction of 9.54/130 = 0.0733, or 7.33 percent. To obtain a further reduction of the noise in this room, the floor could be insulated or, preferably, the noise-producing device could be redesigned to give off less noise.

Related Calculations: Use this general procedure to determine the effectiveness of acoustic materials used in any room in a building, on a ship, in an airplane, etc.

Solar Energy

REFERENCES: ASHRAE—*Handbook of Fundamentals;* Balcomb, et al.—*Passive Solar Design Handbook, Vol. 2: Passive Solar Design Analysis,* DOE/CS-0127-2; Mazria—*The Passive Solar Energy Book,* Rodale Press; Strock and Koral—*Handbook of Air Conditioning, Heating and Ventilating,* Industrial Press; Kreider and Kreith—*Solar Heating and Cooling,* McGraw-Hill; Balcomb—*Passive Solar Design Handbook,* Solar Energy Information; Greeley—*Solar Heating and Cooling of Buildings,* Ann Arbor Science; Howell—*Thermal*

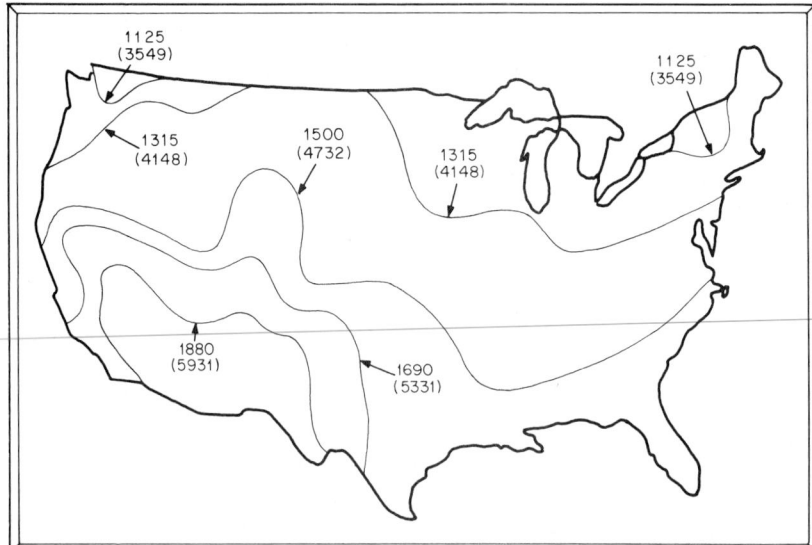

FIG. 15 Average amount of solar energy available, in Btu/(day · ft²) (W/m²), for different parts of the United States. (*Power.*)

Energy Systems: Analysis and Design, McGraw-Hill; Edwards—*Solar Collector Design,* Franklin Institute Press; Kreider and Kreith—*Solar Energy Handbook,* McGraw-Hill; Chauliaguet—*Solar Energy in Buildings,* Wiley.

DESIGNING A FLAT-PLATE SOLAR-ENERGY HEATING AND COOLING SYSTEM

Give general design guidelines for the planning of a solar-energy heating and cooling system for an industrial building in the Jacksonville, Florida, area to use solar energy for space heating and cooling and water heating. Outline the key factors considered in the design so they may be applied to solar-energy heating and cooling systems in other situations. Give sources of pertinent design data, where applicable.

Calculation Procedure:

1.   *Determine the average annual amount of solar energy available at the site*

Figure 15 shows the average amount of solar energy available, in Btu/(day·ft²) (W/m²) of panel area, in various parts of the United States. How much energy is collected depends on the solar panel efficiency and the characteristics of the storage and end-use systems.

Tables available from the National Weather Service and the American Society of Heating, Refrigerating and Air Conditioning Engineers (ASHRAE) chart the monthly solar-radiation impact for different locations and solar insolation [total radiation from the sun received by a surface, measured in Btu/(h·ft²) (W/m²); insolation is the sum of the direct, diffuse, and reflected radiation] for key hours of a day each month.

Estimate from these data the amount of solar radiation likely to reach the surface of a solar collector over 1 yr. Thus, for this industrial building in Jacksonville, Florida, Fig. 15 shows that the average amount of solar energy available is 1500 Btu/(day·ft²) (4.732 W/m²).

When you make this estimate, keep in mind that on a clear, sunny day direct radiation accounts for 90 percent of the insolation. On a hazy day only diffuse radiation may be available for collection, and it may not be enough to power the solar heating and cooling system. As a guide,

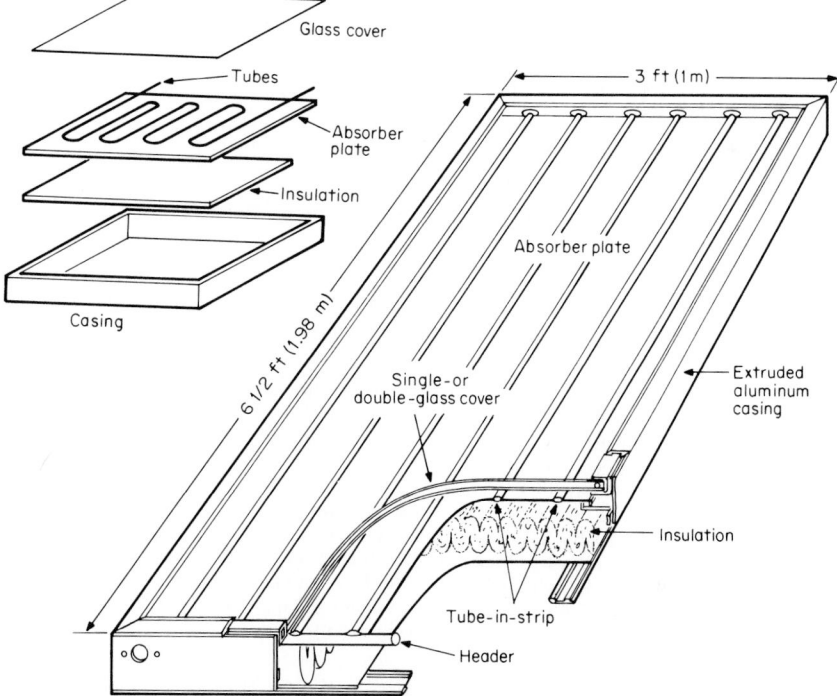

FIG. 16   Construction details of flat-plate solar collectors. *(Power.)*

the water temperatures required for solar heating and cooling systems are:

Space heating	Up to 170° F (76.7°C)
Space cooling with absorption air conditioning	From 200 to 240°F (93.3 to 115.6°C)
Domestic hot water	140°F (60°C)

2. *Choose collector type for the system*

There are two basic types of solar collectors: flat-plate and concentrating types. At present the concentrating type of collector is not generally cost-competitive with the flat-plate collector for normal space heating and cooling applications. It will probably find its greatest use for high-temperature heating of process liquids, space cooling, and generation of electricity. Since process heating applications are not the subject of this calculation procedure, concentrating collectors are discussed separately in another calculation procedure.

Flat-plate collectors find their widest use for building heating, domestic water heating, and similar applications. Since space heating and cooling are the objective of the system being considered here, a flat-plate collector system will be a tentative choice until it is proved suitable or unsuitable for the system. Figure 16 shows the construction details of typical flat-plate collectors.

3. *Determine the collector orientation*

Flat-plate collectors should face south for maximum exposure and should be tilted so the sun's rays are normal to the plane of the plate cover. Figure 17 shows the optimum tilt angle for the plate for various insolation requirements at different latitudes.

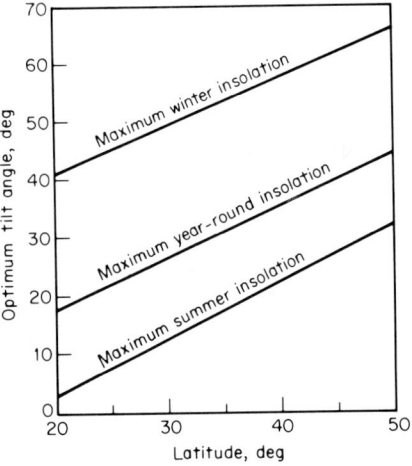

FIG. 17 Spacing of solar flat-plate collectors to avoid shadowing. *(Power.)*

Since Jacksonville, Florida, is approximately at latitude 30°, the tilt of the plate for maximum year-round insolation should be 25° from Fig. 17. As a general rule for heating with maximum winter insolation, the tilt angle should be 15° plus the angle of latitude at the site; for cooling, the tilt angle equals the latitude (in the south, this should be the latitude minus 10° for cooling); for hot water, the angle of tilt equals the latitude plus 5°. For combined systems, such as heating, cooling, and hot water, the tilt for the dominant service should prevail. Alternatively, the tilt for maximum year-round insolation can be used, as was done above.

When collector banks are set in back of one another in a sawtooth arrangement, low winter sun can cause shading of one collector by another. This can cause a loss in capacity unless the units are carefully spaced. Table 36 shows the minimum spacing to use between collector rows, based on the latitude of the installation and the collector tilt.

4. *Sketch the system layout*

Figure 18 shows the key components of a solar system using flat-plate collectors to capture solar radiation. The arrangement provides for heating, cooling, and hot-water production in this indus-

TABLE 36 Spacing to Avoid Shadowing, ft (m)°

Collector angle, deg	Latitude of installation, deg			
	30	35	40	45
30	9.9 (3.0)	11.1 (3.4)	12.6 (3.8)	13.7 (4.2)
45	10.6 (3.2)	12.2 (3.7)	14.4 (4.4)	16.0 (4.9)
60	10.7 (3.3)	12.6 (3.8)	15.4 (4.7)	17.2 (5.2)

° *Power* magazine.

trial building with sunlight supplying about 60 percent of the energy needed to meet these loads—a typical percentage for solar systems.

For this layout, water circulating in the rooftop collector modules is heated to 160°F (71.1°C) to 215°F (101.7°C). The total collector area is 10,000 ft² (920 m²). Excess heated hot water not needed for space heating or cooling or for domestic water is directed to four 6000-gal (22,740-L) tanks for short-term energy storage. Conventional heating equipment provides the hot water needed for heating and cooling during excessive periods of cloudy weather. During a period of 3 h around noon on a clear day, the heat output of the collectors is about 2 million Btu/h (586 kW), with an efficiency of about 50 percent at these conditions.

For this industrial building solar-energy system, a lithium-bromide absorption air-conditioning unit (a frequent choice for solar-heated systems) develops 100 tons (351.7 kW) of refrigeration for cooling with a coefficient of performance of 0.71 by using heated water from the solar collectors. Maximum heat input required by this absorption unit is 1.7 million Btu/h (491.8 kW) with a hot-water flow of 240 gal/min (909.6 L/min). Variable-speed pumps and servo-actuated valves control the water flow rates and route the hot-water flow from the solar collectors along several paths—to the best exchanger for heating or cooling of the building, to the absorption unit for cooling of the building, to the storage tanks for use as domestic hot water, or to short-term storage before other usage. The storage tanks hold enough hot water to power the absorption unit for several hours or to provide heating for up to 2 days.

Another—and more usual—type of solar-energy system is shown in Fig. 19. In it a flat-plate collector absorbs heat in a water/antifreeze solution that is pumped to a pair of heat exchangers.

From unit no. 1 hot water is pumped to a space-heating coil located in the duct work of the hot-air heating system. Solar-heated antifreeze solution pumped to unit no. 2 heats the hot water for domestic service. Excess heated water is diverted to fill an 8000-gal (30,320-L) storage tank. This heated water is used during periods of heavy cloud cover when the solar heating system cannot operate as effectively.

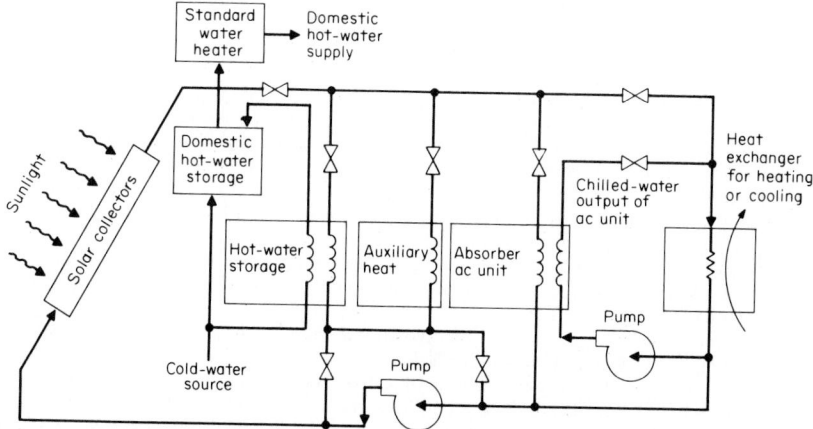

FIG. 18 Key components of a solar-energy system using flat-plate collectors. *(Power.)*

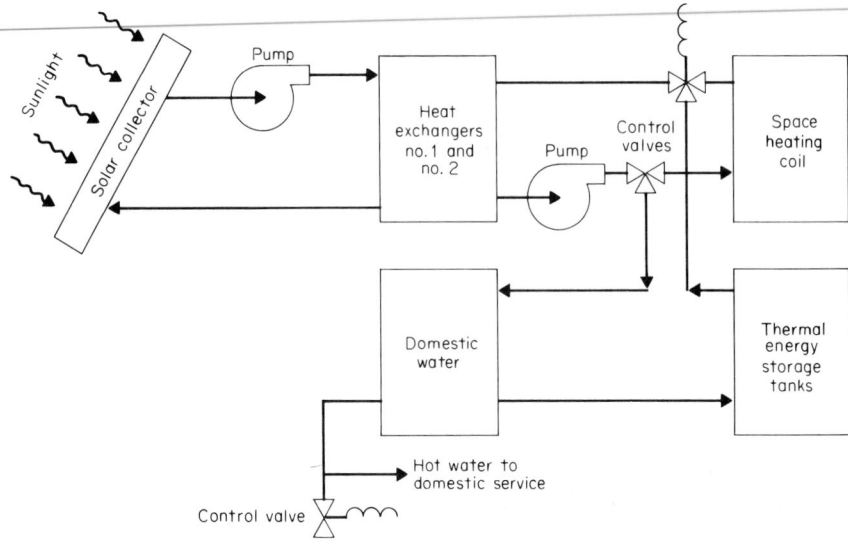

FIG. 19   Solar-energy system using flat-plate collectors and an antifreeze solution in a pair of heat exchangers. *(Power.)*

5.   *Give details of other techniques for solar heating*

Wet collectors having water running down the surface of a tilted absorber plate and collected in a gutter at the bottom are possible. While these "trickle-down" collectors are cheap, their efficiency is impaired by heat losses from evaporation and condensation.

Air systems using rocks or gravel to store heat instead of a liquid find use in residential and commercial applications. The air to be heated is circulated via ducts to the solar collector consisting of rocks, gravel, or a flat-plate collector. From here other ducts deliver the heated air to the area to be heated.

In an air system using rocks or gravel, more space is needed for storage of the solid media, compared to a liquid. Further, the ductwork is more cumbersome and occupies more space than the piping for liquid heat-transfer media. And air systems are generally not suitable for comfort cooling or liquid heating, such as domestic hot water.

Eutectic salts can be used to increase the storage capacity of air systems while reducing the volume required for storage space. But these salts are expensive, corrosive, and toxic, and they become less effective with repeated use. Where it is desired to store thermal energy at temperatures above 200°F (93.3°C), pressurized storage tanks are attractive.

Solar "heat wheels" can be used in the basic solar heating and cooling system in the intake and return passages of the solar system. The wheels permit the transfer of thermal energy from the return to the intake side of the system and offer a means of controlling humidity.

For solar cooling, high-performance flat-plate collectors or concentrators are needed to generate the 200 to 240°F (93.3 to 115.6°C) temperatures necessary for an absorption-chiller input. These chillers use either lithium bromide or ammonia with hot water to form an absorbent/refrigerant solution. Chiller operation is conventional.

Solar collectors can be used as a heat source for heat-pump systems in which the pump transfers heat to a storage tank. The hot water in the tank can then be used for heating, while the heat pump supplies cooling.

In summary, solar energy is a particularly valuable source of heat to augment conventional space-heating and cooling systems and for heating liquids. The practical aspects of system operation can be troublesome—corrosion, deterioration, freezing, condensation, leaks—but these problems can be surmounted. Solar energy is not "free" because a relatively high initial investment for equipment must be paid off over a long period. And the equipment requires some fossil-fuel energy to fabricate.

TABLE 37 Solar-Energy Design Selection Summary

Energy collection device	Typical heat-transfer applications	Typical uses of collected heat
Flat-plate tubed collector	Liquid heating	Space heating or cooling; water heating
Wet collectors	Liquid heating	Water or liquid heating
Rock or gravel collectors	Air heating	Space heating
Flat-plate air heaters	Air heating	Space heating or cooling
Eutectic salts	Air heating	Space heating
Evacuated-tube collectors	Liquid heating to higher temperature levels	Space cooling and heating; process heating
Concentrating collectors	Steam generation	Electric-power generation
Pressurized storage tank	Storage of thermal energy above 200°F (93.3°C)	Industrial processes
Heat wheels	Heat transfer	Humidity control

But even with these slight disadvantages, the more solar energy that can be put to work, the longer the supply of fossil fuels will last. And recent studies show that solar energy will become more cost-competitive as the price of fossil fuels continues to rise.

6. *Give design guides for typical solar systems*

To ensure the best performance from any solar system, keep these pointers in mind:

a. For space heating, size the solar collector to have an area of 25 to 50 percent of the building's floor area, depending on geographic location, amount of insulation, and ratio of wall to glass area in the building design.

b. For space cooling, allow 250 to 330 ft^2 (23.3 to 30.7 m^2) of collector surface for every ton of absorption air conditioning, depending on unit efficiency and solar intensity in the area. Insulate piping and vessels adequately to provide fluid temperatures of 200 to 240°F (93.3 to 115.6°C).

c. Size water storage tanks to hold between 1 and 2 gal/ft^2 (3.8 to 7.6 L/m^2) of collector surface area.

d. In large collector installations, gang collectors in series rather than parallel. Use the lowest fluid temperature suitable for the heating or cooling requirements.

e. Insulate piping and collector surfaces to reduce heat losses. Use an overall heat-transfer coefficient of less than 0.04 Btu/(h·ft^2·°F) [0.23 W/(m^2·K)] for piping and collectors.

f. Avoid water velocities of greater than 4 ft/s (1.2 m/s) in the collector tubes, or else efficiency may suffer.

g. Size pumps handling antifreeze solutions to carry the additional load caused by the higher viscosity of the solution.

Related Calculations: The general guidelines given here are valid for solar heating and cooling systems for a variety of applications (domestic, commercial, and industrial), for space heating and cooling, and for process heating and cooling, as either the primary or supplemental heat source. Further, note that solar energy is not limited to semitropical areas. There are numerous successful applications of solar heating in northern areas which are often considered to be "cold." And with the growing energy consciousness in all fields, there will be greater utilization of solar energy to conserve fossil-fuel use.

Energy experts in many different fields believe that solar-energy use is here to stay. Since there seems to be little chance of fossil-fuel price reductions (only increases), more and more energy users will be looking to solar heat sources to provide some of or all their energy needs. For example, Wagner College in Staten Island, New York, installed, at this writing, 11,100 ft^2 (1032.3 m^2) of evacuated-tube solar panels on the roof of their single-level parking structure. These panels provide heating, cooling, and domestic hot water for two of the buildings on the campus. Energy output of these evacuated-tube collectors is some 3 billion Btu (3.2 × 10^9 kJ), producing a fuel-cost savings of $25,000 during the first year of installation. The use of evacuated-tube collectors is planned in much the same way as detailed above. Other applications of such collectors include soft-drink bottling plants, nursing homes, schools, etc. More applications will be found as fossil-fuel price increases make solar energy more competitive in the years to come. Table 37 gives a summary of solar-energy collector choices for quick preliminary use.

Data in this procedure are drawn from an article in *Power* magazine prepared by members of the magazine's editorial staff and from Owens-Illinois, Inc.

DETERMINATION OF SOLAR INSOLATION ON SOLAR COLLECTORS UNDER DIFFERING CONDITIONS

A south-facing solar collector will be installed on a building in Glasgow, Montana, at latitude 48°13′N. What is the clear-day solar insolation on this panel at 10 a.m. on January 21 if the collector tilt angle is 48°? What is the daily surface total insolation for January 21 at this angle of collector tilt? Compute the solar insolation at 10:30 a.m. on January 21. What is the actual daily solar insolation for this collector? Calculate the effect on the clear-day daily solar insolation if the collector tilt angle is changed to 74°.

Calculation Procedure:

1. *Determine the insolation for the collector at the specified location*

The latitude of Glasgow, Montana, is 48°13′N. Since the minutes are less than 30, or one-half of a degree, the ASHRAE clear-day insolation table for 48° north latitude can be used. Entering Table 38 (which is an excerpt of the ASHRAE table) for 10 a.m. on January 21, we find the clear-day solar insolation on a south-facing collector with a 48° tilt is 206 Btu/(h·ft²) (649.7 W/m²). The daily clear-day surface total for January 21 is, from the same table, 1478 Btu/(day·ft²) (4661.6 W/m²) for a 48° collector tilt angle.

2. *Find the insolation for the time between tabulated values*

The ASHRAE tables plot the clear-day insolation at hourly intervals between 8 a.m. and 4 p.m. For other times, use a linear interpolation. Thus, for 10:30 a.m., interpolate in Table 38 between 10:00 and 11:00 a.m. values. Or, $(249 - 206)/2 + 206 = 227.5$ Btu/(h·ft²) (717.5 W/m²), where the 249 and 206 are the insolation values at 11 and 10 a.m., respectively. Note that the difference can be either added to or subtracted from the lower, or higher, clear-day insolation value, respectively.

3. *Find the actual solar insolation for the collector*

ASHRAE tables plot the clear-day solar insolation for particular latitudes. Dust, clouds, and water vapor will usually reduce the clear-day solar insolation to a value less than that listed.

To find the actual solar insolation at any location, use the relation $i_A = pi_T$, where i_A = actual solar insolation, Btu/(h·ft²) (W/m²); p = percentage of clear-day insolation at the location, expressed as a decimal; i_T = ASHRAE-tabulated clear-day solar insolation, Btu/(h·ft²) (W/m²). The value of $p = 0.3 + 0.65(S/100)$, where S = average sunshine for the locality, percent, from an ASHRAE or government map of the sunshine for each month of the year. For January, in Glasgow, Montana, the average sunshine is 50 percent. Hence, $p = 0.30 + 0.65(50/100) = 0.625$. Then $i_A = 0.625(1478) = 923.75$, say 923.5 Btu/(day·ft²) (2913.7 W/m²), by using the value found in step 1 of this procedure for the daily clear-day solar insolation for January 21.

4. *Determine the effect of a changed tilt angle for the collector*

Most south-facing solar collectors are tilted at an angle approximately that of the latitude of the location plus 15°. But if construction or other characteristics of the site prevent this tilt angle, the effect can be computed by using the ASHRAE tables and a linear interpolation.

Thus, for this 48°N location, with an actual tilt angle of 48°, a collector tilt angle of 74° will produce a clear-day solar insolation of $i_T = 1578[(74 - 68)/(90 - 68)](1578 - 1478) = 1551.0$ Btu/(day·ft²) (4894.4 W/m²), by the ASHRAE tables. In the above relation, the insolation values are for solar collector tilt angles of 68° and 90°, respectively, with the higher insolation value for the smaller angle. Note that the insolation (heat absorbed) is greater at 74° than at 48° tilt angle.

Related Calculations: This procedure demonstrates the flexibility and utility of the ASHRAE clear-day solar insolation tables. Using straight-line interpolation, the designer can obtain a number of intermediate clear-day values, including solar insolation at times other than

TABLE 38 Solar Position and Insolation Values for 48°N Latitude[°]

| Date | Solar time | | Solar position | | Btu·h/ft² (W/m²) total insolation on surfaces | | | | | | | |
	a.m.	p.m.	Alt.	Azm.	Normal	Horiz.	38	48	58	68	90
							South-facing surface angle with horizontal				
Jan 21	8	4	3.5	54.6	37 (116.6)	4 (12.6)	17 (53.6)	19 (59.9)	21 (66.2)	22 (69.4)	22 (69.4)
	9	3	11.0	42.6	185 (583.2)	46 (145.0)	120 (378.3)	132 (416.1)	140 (441.4)	145 (457.1)	139 (438.2)
	10	2	16.9	29.4	239 (753.4)	83 (261.7)	190 (598.9)	206 (649.4)	216 (680.9)	220 (693.6)	206 (649.4)
	11	1	20.7	15.1	261 (822.8)	107 (337.3)	231 (728.2)	249 (784.9)	260 (819.7)	263 (829.1)	243 (766.1)
	12		22.0	0.0	267 (841.7)	115 (362.5)	245 (772.4)	264 (832.3)	275 (866.9)	278 (876.4)	255 (803.9)
Surface daily totals					1710 (5390.7)	596 (1878.9)	1360 (4287.4)	1478 (4659.4)	1550 (4886.4)	1578 (4974.6)	1478 (4659.4)

[°] From ASHRAE, excerpted; used with permission from ASHRAE. Metrication supplied by handbook editor.

those listed, insolation at collector tilt angles different from those listed, insolation on both normal (vertical) and horizontal planes, and surface daily total insolation. The calculations are simple, provided the designer carefully observes the direction of change in the tabulated values and uses the latitude table for the collector location. Where an exact-latitude table is not available, the designer can interpolate in a linear fashion between latitude values less than and greater than the location latitude.

Remember that the ASHRAE tables give clear-day insolation values. To determine the actual solar insolation, the clear-day values must be corrected for dust, water vapor, and clouds, as shown above. This correction usually reduces the amount of insolation, requiring a larger collector area to produce the required heating or cooling. ASHRAE also publishes tables of the average percentage of sunshine for use in the relation for determining the actual solar insolation for a given location.

SIZING COLLECTORS FOR SOLAR-ENERGY HEATING SYSTEMS

Select the required collector area for a solar-energy heating system which is to supply 70 percent of the heat for a commercial building situated in Grand Forks, Minnesota, if the computed heat loss is 100,000 Btu/h (29.3 kW), the design indoor temperature is 70°F (21.1°C), the collector efficiency is given as 38 percent by the manufacturer, and collector tilt and orientation are adjustable for maximum solar-energy receipt.

Calculation Procedure:

1.   *Determine the heating load for the structure*

The first step in sizing a solar collector is to compute the heating load for the structure. This is done by using the methods given for other procedures in this handbook in Sec. 2, under Heating, Ventilating and Air Conditioning, and in Sec. 4, under Electric Comfort Heating. Use of these procedures would give the hourly heating load—in this instance, it is 100,000 Btu/h (29.3 kW).

2.   *Compute the energy insolation for the solar collector*

To determine the insolation received by the collector, the orientation and tilt angle of the collector must be known. Since the collector can be oriented and tilted for maximum results, the collector will be oriented directly south for maximum insolation. Further, the tilt will be that of the latitude

TABLE 39   Solar Energy Available for Heating

Month	Mean sunshine, percent	Total insolation Btu/ $(ft^2 \cdot day)$	Total insolation W/ $(m^2 \cdot day)$	Efficiency of collector, percent	Energy available from collector[*] Btu/ $(ft^2 \cdot day)$	Energy available from collector[*] W/ $(m^2 \cdot day)$	Heating-season energy available Btu/ $(month \cdot ft^2)$	Heating-season energy available W/ $(month \cdot m^2)$
Jan.	49	1,478	4,663.1	38	275.2	868.3	8,531.2	26,915.9
Feb.	54	1,972	6,221.7	38	404.7	1,276.8	11,331.6	35,751.2
Mar.	55	2,228	7,029.3	38	465.7	1,469.3	14,436.7	45,547.8
Apr.	57	2,266	7,149.2	38	490.8	1,548.5	14,724.0	46,454.2
May	60	2,234	7,048.3	38	509.4	1,607.2	15,791.4	49,821.9
June	64	2,204	6,953.6	38	NA[†]	NA	NA	NA
July	72	2,200	6,941.0	38	NA[†]	NA	NA	NA
Aug.	69	2,200	6,941.0	38	NA[†]	NA	NA	NA
Sept.	60	2,118	6,682.3	38	482.9	1,523.6	14,487.0	45,706.5
Oct.	54	1,860	5,868.3	38	381.7	1,204.3	11,832.7	37,332.2
Nov.	40	1,448	4,568.4	38	220.1	694.4	6,603.0	20,832.5
Dec.	40	1,250	3,943.8	38	190.0	599.5	5,890.0	18,582.9
							Total = 103,627.6 [326.9 kW/ $(month \cdot m^2)$]	

[*] Values in this column = (mean sunshine, %)[total insolation, Btu/(ft² · day)](efficiency of collector, %).

[†] Not applicable because not part of the heating season; June, July, and August are ignored in the calculation.

of Grand Forks, Minnesota, or 48°, since this produces the maximum performance for any solar collector.

Next, use tabulations of mean percentage of possible sunshine and solar position and insolation for the latitude of the installation. Such tabulations are available in ASHRAE publications and in similar reference works. List, for each month of the year, the mean percentage of possible sunshine and the insolation in Btu/(day·ft²) (W/m²), as in Table 39.

Using a heating season of September through May, we find the total solar energy available from the collector for these months is 103,627.6 Btu/ft² (326.9 kW/m²), found by taking the heat energy per month (= mean sunshine, percent)[total insolation, Btu/(ft²·day)](collector efficiency, percent) and summing each month's total. Heat available during the off season can be used for heating water for use in the building hot-water system.

3. Find the annual heating season heat load

Since the heat loss is 100,000 Btu/h (29.3 kW), the total heat load during the 9-month heating season from September through May, or 273 days, is H_a = (24 h)(273 days)(100,000) = 655,200,000 Btu (687.9 MJ).

4. Determine the collector area required

The calculation in step 2 shows that the total solar energy available during the heating season is 103,627.6 Btu/ft² (326.9 kW/m²). Then the collector area required is A ft² (m²) = H_a/S_a, where S_a = total solar energy available during the heating season, Btu/ft². Or A = 655,200,000/103,627.6 = 6322.64 ft² (587.4 m²) if the solar panel is to supply all the heat for the building. However, only 70 percent of the heat required by the building is to be supplied by solar energy. Hence, the required solar panel area = 0.7(6322.6) = 4425.8 ft² (411.2 m²).

With the above data, a collector of 4500 ft² (418 m²) would be chosen for this installation. This choice agrees well with the precomputed collector sizes published by the U.S. Department of Energy for various parts of the United States.

Related Calculations: The procedure shown here is valid for any type of solar collector—flat-plate, concentrating, or nonconcentrating. The two variables which must be determined for any installation are the annual heat loss for the structure and the annual heat flow available from the solar collector. Once these are known, the collector area is easily determined.

The major difficulty in sizing solar collectors for either comfort heating or water heating lies in determining the heat output of the collector. Factors such as collector tilt angle, orientation, and efficiency must be carefully evaluated before the collector final choice is made. And of these three factors, collector efficiency is probably the most important in the final choice of a collector.

F CHART METHOD FOR DETERMINING USEFUL ENERGY DELIVERY IN SOLAR HEATING

Determine the annual heating energy delivery of a solar space-heating system using a double-glazed flat-plate collector if the building is located in Bismarck, North Dakota, and the following specifications apply:

Building
Location: 47°N latitude
Space-heating load: 15,000 Btu/(°F·day) [8.5 kW/(m²·K·day)]

Solar System
Collector loss coefficient: $F_R U_C$ = 0.80 Btu/(h·ft²·°F) [4.5 W/(m²·K)]
Collector optical efficiency (average): $F_R(\overline{\tau\alpha})$ = 0.70
Collector tilt: $\beta = L + 15° = 62°$
Collector area: A_c = 600 ft² (55.7 m²)
Collector fluid flow rate: $\dot{m}_c/A_c$ = 11.4 lb/(h·ft²)(water) [0.0155 kg/(s·m²)]
Collector fluid heat capacity–specific gravity product: c_{pc} = 0.9 Btu/(lb·°F) [3.8 kJ/(kg·K)] (antifreeze)
Storage capacity: 1.85 gal/ft²(water) (75.4 L/m²)
Storage fluid flow rate: $\dot{m}_s/A_c$ = 20 lb/(h·ft²)(water) [0.027 kg/(s·m²)]
Storage fluid heat capacity: c_{ps} = 1 Btu/(lb·°F)(water) [4.2 kJ/(kg·K)]
Heat-exchanger effectiveness: 0.75

Climatic Data
Climatic data from the NWS are tabulated in Table 40

TABLE 40   Climatic and Solar Data for Bismarck, North Dakota

Month	Average ambient temperature		Heating, degree-days		Horizontal solar radiation, langleys/day
	°F	°C	°F	°C	
Jan.	8.2	−13.2	1761	978	157
Feb.	13.5	−10.3	1442	801	250
Mar.	25.1	−3.8	1237	687	356
Apr.	43.0	6.1	660	367	447
May	54.4	12.4	339	188	550
June	63.8	17.7	122	68	590
July	70.8	21.5	18	10	617
Aug.	69.2	20.7	35	19	516
Sept.	57.5	14.2	252	140	390
Oct.	46.8	8.2	564	313	272
Nov.	28.9	−1.7	1083	602	161
Dec.	15.6	−9.1	1531	850	124

Calculation Procedure:

1.   Determine the solar parameter P_s

The F chart is a common calculation procedure used in the United States to ascertain the useful energy delivery of active solar heating systems. The F chart applies only to the specific system designs of the type shown in Fig. 20 for liquid systems and Fig. 21 for air systems. Both systems find wide use today.

The F chart method consists of several empirical equations expressing the monthly solar heating fraction f_s as a function of dimensionless groups which relate system properties and weather data for a month to the monthly heating requirement. The several dimensionless parameters are grouped into two dimensionless groups called the *solar parameter* P_s and the *loss parameter* P_L.

The solar parameter P_s is the ratio of monthly solar energy absorbed by the collector divided

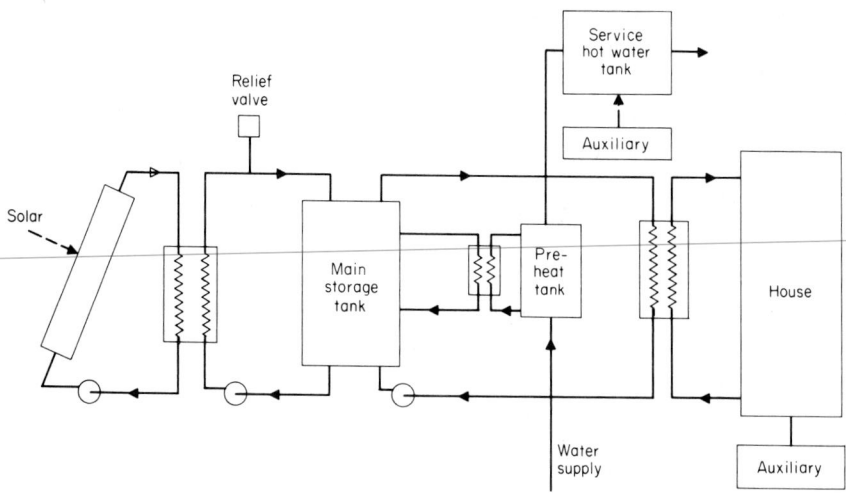

FIG. 20   Liquid-based solar space- and water-heating system. *(DOE/CS-0011.)*

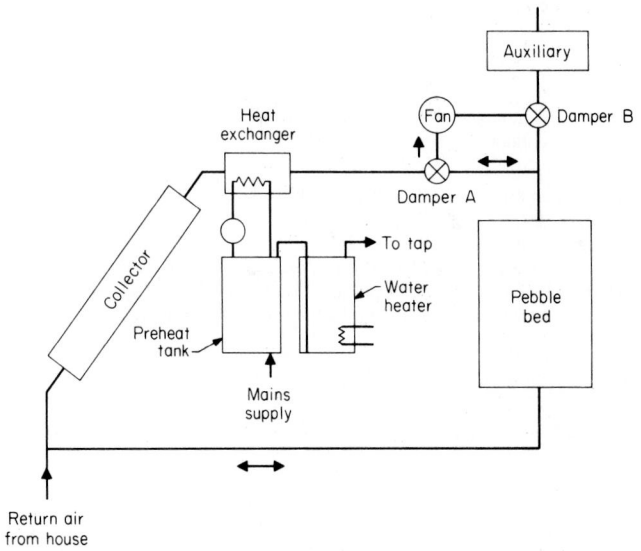

FIG. 21 Air-based solar space- and water-heating system. *(DOE/CS-0011.)*

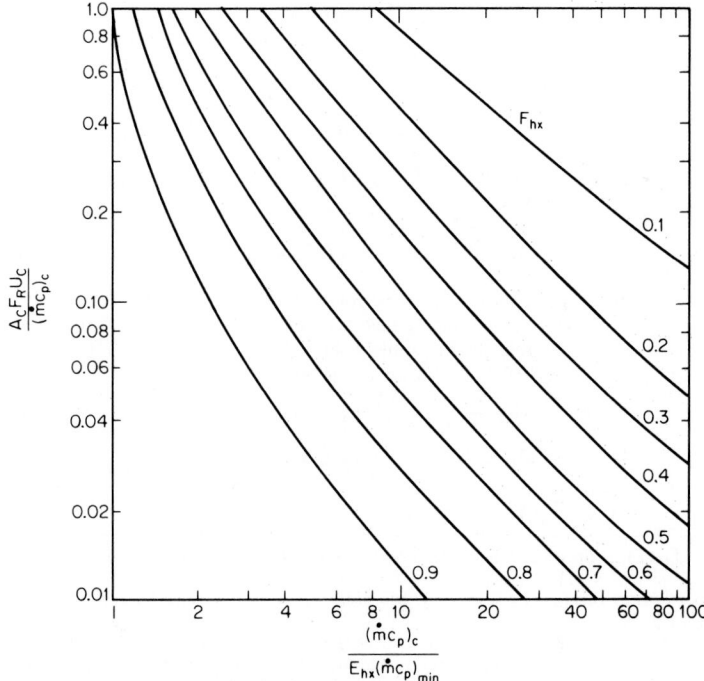

FIG. 22 Heat-exchanger penalty factor F_{hx}. When no exchanger is present, $F = 1$. *(Kreider—The Solar Heating Design Process, McGraw-Hill.)*

by the monthly heating load, or

$$P_s = \frac{K_{ldhx}F_{hx}(F_R\overline{\tau\alpha})\overline{I}_c N}{L}$$

where F_{hx} = heat-exchanger penalty factor (see Fig. 22)

$F_R\tau\alpha$ = average collector optical efficiency = 0.95 × collector efficiency curve intercept $F_R(\tau\alpha)_n$

$\overline{I}_c$ = monthly average insolation on collector surface from a listing of monthly solar and climatic data

N = number of days in a month

L = monthly heating load, *net of any passive system delivery* as calculated by the P chart, solar load ratio (SLR), or any other suitable method, Btu/month

K_{ldhx} = load-heat-exchanger correction factor for liquid systems, Table 41

(The P chart method and the SLR method are both explained in Related Calculations below.)

The value of P_s is found for each month of the year by substituting appropriate unit values in the above equation and tabulating the results (Table 42).

2. Determine the loss parameter P_L

The loss parameter P_L is related to the long-term energy losses from the collector divided by the monthly heating load:

$$P_L = (K_{stor}K_{flow}K_{DHW}) \frac{F_{hx}(F_R U_c)(T_r - \overline{T}_a) \Delta t}{L}$$

where $F_R U_c$ = magnitude of collector efficiency curve slope (can be modified to include piping and duct losses)

TABLE 41 F Chart K Factors

Correction factor	Air or liquid system	Correction factor	Validity range for factor
K_{flow}	A°	$\{[2 \text{ ft}^3/(\min\cdot\text{ft}_c^2)]/\text{actual flow}\}^{-0.28}$ $\{[0.61 \text{ m}^3/(\min\cdot\text{m}^2)]/\text{actual flow}\}^{-0.28}$	1–4 ft³/(min·ft²_c) [0.03 to 0.11 m³/(min·m²)]
	L	Small effect included in F_R and F_{hx} only	
K_{stor}	A†	$[(0.82 \text{ ft}^3/\text{ft}_c^2)/\text{actual volume}]^{0.30}$ $[(0.25 \text{ m}^3/\text{m}^2)/\text{actual volume}]^{0.30}$	0.4–3.3 ft³/ft²_c (0.012 to 0.10 m³/m²)
	L	$[(1.85 \text{ gal}/\text{ft}_c^2)/\text{actual volume}]^{0.25}$ $[(0.0754 \text{ L/m}^2)/\text{actual volume}]^{0.25}$	0.9–7.4 gal/ft²_c (36.7 to 301.5 L/m²)
K_{DHW}‡	L	$\dfrac{(1.18T_{w,o} + 3.86T_{w,i} - 2.32\overline{T}_a - 66.2)}{212 - \overline{T}_a}$	
K_{ldhx}§	L	$0.39 + 0.65 \exp[-0.139UA/E_L(\dot{m}c_p)_{air}]$	$0.5 < \dfrac{E_L(\dot{m}c_p)_{air}}{UA} < 50$
	A	NA	

°User must also include the effect of flow rate in F_R, i.e., in $F_R(\overline{\tau\alpha})$ and $F_R U_c$. Refer to manufacturer's data for this.

†For air systems using latent-heat storage, see J. J. Jurinak and S. I. Abdel-Khalik, *Energy*, vol. 4, p. 503 (1979) for the expression for K_{stor}.

‡Only applies for liquid storage DHW systems (air collectors can be used, however); $T_{w,o}$ = hot-water supply temperature, $T_{w,i}$ = cold-water supply temperature to water heater, both °F. Applies only to a specific water use schedule; predictions of performance for other schedules will have reduced accuracy.

§UA = unit building heat load, Btu/(h·°F); E_L = load-heat-exchanger effectiveness; $(\dot{m}c_p)_{air}$ = load-heat-exchanger air capacitance rate, Btu/(h·°F) = density × 60 × cubic feet per minute × 0.24.

Note: Metrication supplied by handbook editor.

TABLE 42 *F* Chart Summary

Month	Collector-plane radiation Btu/(day·ft²)	W/m²	Monthly energy demand L million Btu	MJ/month	P_L^a	P_s^a	f_s	Monthly delivery million Btu	MJ/month
Jan.	1506	4749.9	26.41	27.86	2.68	0.72	0.46	12.15	12.82
Feb.	1784	5626.7	21.63	22.82	2.88	0.95	0.60	12.98	13.69
Mar.	1795	5661.4	18.55	19.57	3.50	1.22	0.73	13.54	14.28
Apr.	1616	5096.9	9.90	10.44	5.75	2.00	0.94	9.31	9.82
May	1606	5065.3	5.08	5.36	10.78	>3.00	1.00	5.08	5.36
June	1571	4954.9	1.83	1.93	>20.00	>3.00	1.00	1.83	1.93
July	1710	5393.3	0.27	0.28	>20.00	>3.00	1.00	0.27	0.28
Aug.	1712	5399.6	0.52	0.55	>20.00	>3.00	1.00	0.52	0.55
Sept.	1721	5428.0	3.78	3.99	13.76	>3.00	1.00	3.78	3.99
Oct.	1722	5431.2	8.46	8.93	6.79	2.58	1.00	8.46	8.93
Nov.	1379	4349.4	16.24	17.13	3.79	1.04	0.61	9.91	10.46
Dec.	1270	4005.6	22.96	24.22	2.98	0.70	0.43	9.87	10.41
Annual			135.66	143.1			0.65	87.70	92.52

$^a P_s > 3.0$ or $P_L > 20.0$ implies $P_s = 3.0$ and $P_L = 20$, i.e., $f_s = 1.0$. The annual solar fraction $\bar{f}_s$ is 87.70/135.66, or 6 percent.

$$\overline{T}_a = \text{monthly average ambient temperature, °F (°C)}^1$$
$$\Delta t = \text{number of hours per month} = 24N$$
$$T_r = \text{reference temperature} = 212°F \ (100°C)$$
$$K_{stor} = \text{storage volume correction factor, Table 41}$$
$$K_{flow} = \text{collector flow rate correction factor, Table 41}$$
$$K_{DHW} = \text{conversion factor for parameter } P_L \text{ when } only \text{ a water heating system is to be studied, Table 41}$$

3. *Determine the monthly solar fraction*

The monthly solar fraction f_s depends only on these two parameters, P_s and P_L. For liquid heating systems using solar energy as their heat source, the monthly solar fraction is given by

$$f_s = 1.029P_s - 0.065P_L - 0.245P_s^2 + 0.0018P_L^2 + 0.0215P_s^3$$

if $P_s > P_L/12$ (if not, $f_s = 0$).

For air-based solar heating systems, the monthly solar heating fraction is given by

$$f_s = 1.040P_s - 0.065P_L - 0.159P_s^2 + 0.00187P_L^2 + 0.0095P_s^3$$

if $P_s > 0.07P_L$ (if not, $f_s = 0$).

Flow rate, storage, load heat-exchanger, and domestic hot-water correction factors for use in the equations for P_s and P_L, in steps 1 and 2 above, are given in Table 41 and Fig. 23.

When you use the *F* chart method of calculation for any system, follow this order: collector insolation, collector properties, monthly heat loads, monthly ambient temperatures, and monthly values of P_s and P_L. Once the parameter values are known, the monthly solar fraction and monthly energy delivery are readily calculated, as shown in Table 42.

The total of all monthly energy deliveries is the total annual useful energy produced by the solar system. And the total annual useful energy delivered divided by the total annual load is the annual solar load fraction.

---

[1]Note that $T_a \neq 65° -$ (degree heating days/N), contrary to statements of many U.S. government contractors and in many government reports. The equality is only valid for those months when $T_a < 65°F$ every day, i.e., only 3 or 4 months of the year at the most. The errors propagated through the solar industry by assuming the equality to be true are too many to count.

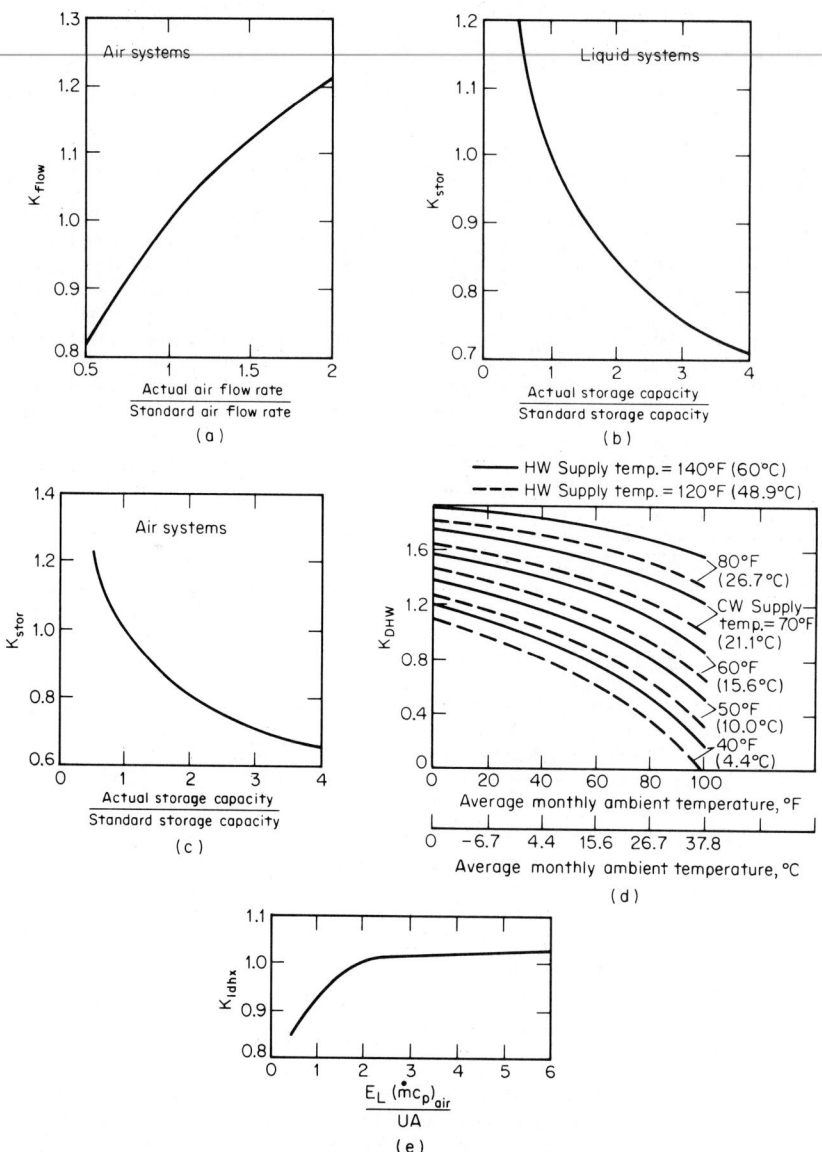

FIG. 23 *F* chart correction factors. (*a*) K_{flow} {standard value is 2 standard ft²/(min · ft²) [0.01 m³/(s · m²)]} (*b*), (*c*) K_{stor} [standard liquid value is 1.85 gal/ft² (75.4 L/m²); standard rock value is 0.82 ft³/ft² (0.025 m³/m²)]. (*d*) K_{DHW}. (*e*) K_{ldhx} (standard value of abscissa is 2.0). (*U.S. Dept of Housing and Urban Development and Kreider—The Solar Heating Design Process, McGraw-Hill.*)

4. Compute the monthly energy delivery

Set up a tabulation such as that in Table 42. Using weather data for Bismarck, North Dakota, list the collector-plane radiation, Btu/(day · ft²), monthly energy demand [= space-heating load, Btu/(°F · day)](degree days for the month, from weather data), P_L computed from the relation given, P_s computed from the relation given, f_s computed from the appropriate relation (water or air) given earlier, and the monthly delivery found from f_s (monthly energy demand).

Related Calculations: In applying the F chart method, it is important to use a consistent area basis for calculating the efficiency curve information and the solar and loss parameters, P_s and P_L. The early National Bureau of Standards (NBS) test procedures based the collector efficiency on net glazing area. A more recent and more widely used test procedure developed by ASHRAE (93.77) uses the gross-area basis. The gross area is the area of the glazing plus the area of opaque weatherstripping, seals, and supports. Hence, when ASHRAE test data are used, the solar and loss parameters must be based on gross area. The efficiency curve basis and F chart basis must be consistent for proper results.

The F chart method can be used for a number of other solar-heating calculations, including performance of an associated heat-pump backup system, collectors connected in series, etc. For specific steps in these specialized calculation procedures, see Kreider—*The Solar Heating Design Process*, McGraw-Hill. The calculation procedure given here is based on the Kreider book, with numbers and SI units being added to the steps in the calculation by the editor of this handbook.

The P chart mentioned as part of the P_s calculation is a trademark of the Solar Energy Design Corporation, POB 67, Fort Collins, Colorado 80521. Developed by Arney, Seward, and Kreider for passive predictions of solar performance, the P chart uses only the building heat load in Btu/(°F·day). The P chart will specify the solar fraction and optimum size of three passive systems.

The monthly solar load ratio is an empirical method of estimating monthly solar and auxiliary energy requirements for passive solar systems. For more data on both the P chart and SLR methods, see the Kreider work mentioned above.

DOMESTIC HOT-WATER-HEATER COLLECTOR SELECTION

Select the area for a solar collector to provide hot water for a family of six people in a residential building in Northport, New York, when the desired water outlet temperature is 140°F (60°C) and the water inlet temperature is 50°F (10°C). A pumped-liquid type of domestic hot-water (DHW) system (Fig. 24) is used. Compare the collector area required for 60, 80, and 90 percent of the DHW heading load.

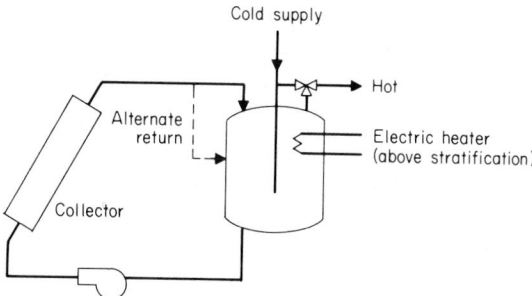

FIG. 24 Direct-heating pump circulation solar water heater. *(DOE/CS-0011.)*

Calculation Procedure:

1. Find the daily DHW heating load

A typical family in the United States uses about 20 gal (9.1 L) of hot water per person per day. Hence, a family of six will use a total of 6(20) = 120 gal (54.6 L) per day. Since water has a specific heat of unity (1.0) and weighs 8.34 lb/gal (1.0 kg/dm³), the daily DHW heating load is L = (120 gal/day)(8.34 lb/gal) [1.0 Btu/(lb·°F)](140 − 50) = 90,072 Btu/day (95,026 kJ/day). This is the 100 percent heating load.

2. Determine the average solar insolation for the collector

Use the month of January because this usually gives the minimum solar insolation during the year, providing the maximum collector area. Using Fig. 25 for eastern Long Island, where Northport is located, we find the solar insolation H = 580 Btu/(ft²·day) [1829.3 W/(m²·day)] on a horizontal surface. (The horizontal surface insolation is often used in DHW design because it provides conservative results.)

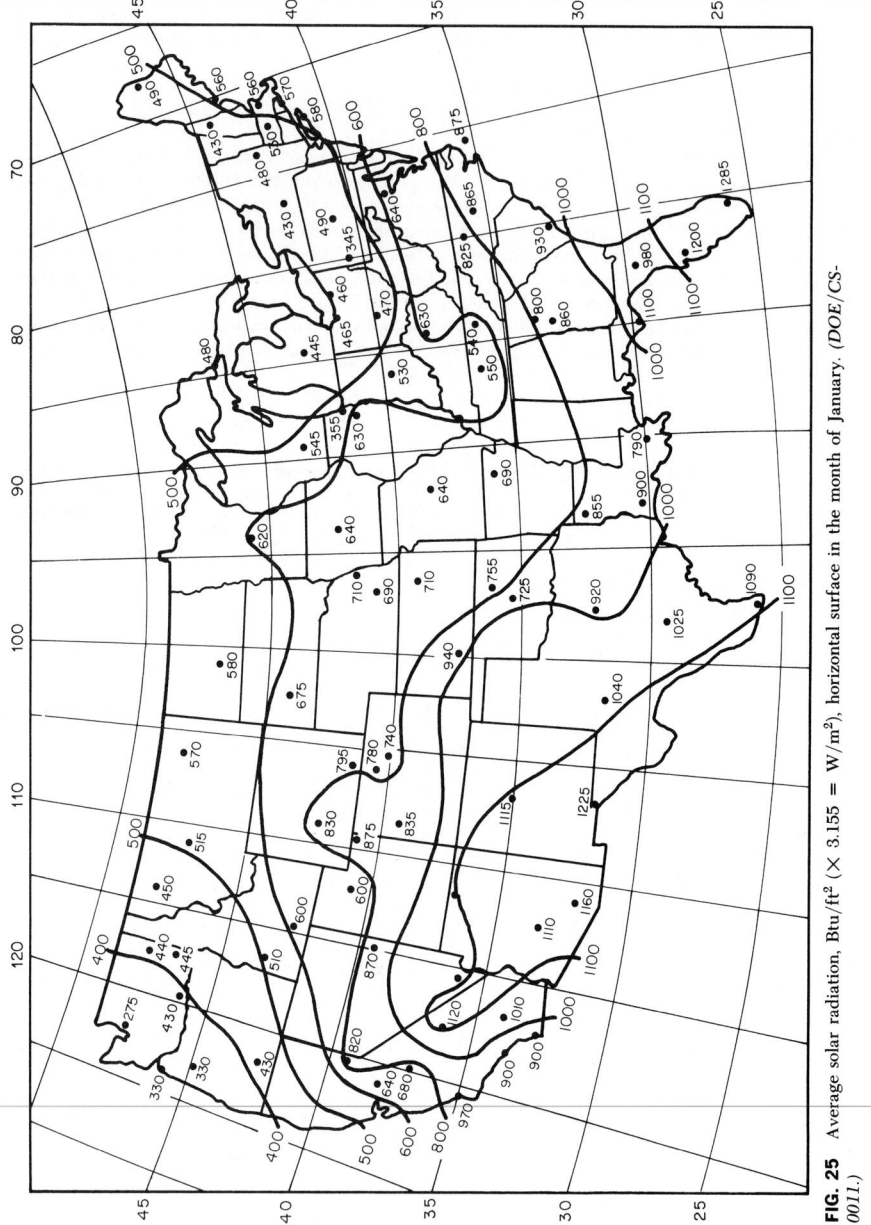

FIG. 25 Average solar radiation, Btu/ft² (× 3.155 = W/m²), horizontal surface in the month of January. (*DOE/CS-0011.*)

2.142

3. Find the HA/L ratio for this installation

Use Fig. 26 to find the HA/L ratio, where A = collector area, ft² (m²). Enter Fig. 26 on the left at the fraction F of the annual load supplied by solar energy; project to the right to the tinted area and then vertically downward to the HA/L ratio. Thus, for F = 60 percent, HA/L = 1.0; F = 80 percent, HA/L = 1.5; for F = 90 percent, HA/L = 2.0.

4. Compute the solar collector area required

Use the relation HA/L = 1 for the 60 percent fraction. Or, HA/L = 1, $A = L/H$ = 0.6(90,072)/ 580 = 93.18 ft² (8.7 m²). For HA/L = 1.5 for the 80 percent fraction, A = 0.8(90,072)/580 = 124.24 ft² (11.5 m²). And for the 90 percent fraction, A = 0.9(90,072)/580 = 139.77 ft² (12.98 m²).

Comparing these areas shows that the 80 percent factor area is 33 percent larger than the 60 percent factor area, while the 90 percent factor area is 50 percent larger. To evaluate the impact of the increased area, the added cost of the larger collector must be compared with the fuel that will be saved by reducing the heat input needed for DHW heating.

Related Calculations: Figures 25 and 26 are based on computer calculations for 11 different locations for the month of January in the United States ranging from Boulder, Colorado, to Boston; New York; Manhattan, Kansas; Gainesville, Florida; Santa Maria, California; St. Cloud, Minnesota; Washington; Albuquerque, New Mexico; Madison, Wisconsin; Oak Ridge, Tennessee. The separate curve above the shaded band in Fig. 26 is the result for Seattle, which is distinctly different from other parts of the country. Hot-water loads used in the computer computations range from 50 gal/day (189.2 L/day) to 2000 gal/day (7500 L/day). The sizing curves in Fig. 26 are approximate and should not be expected to yield results closer than 10 percent of the actual value.

Remember that the service hot-water load is nearly constant throughout the year while the solar energy collected varies from season to season. A hot-water system sized for January, such as that in Fig. 26, with collectors tilted at the latitude angle, will deliver high-temperature water and may even cause boiling in the summer. But a system sized to meet the load in July will not provide all the heat needed in winter. Orientation of the collector can partially overcome the month-to-month fluctuations in radiation and temperature.

Solar-energy water heaters cost from $300 for a roof-mounted collector to over $2000 for a collector mounted on a stand adjacent to the house. The latter are nonfreeze type collectors fitted with a draindown valve, 50 ft² (4.7 m²) of collector surface area, an 80-gal (302.8-L) water tank,

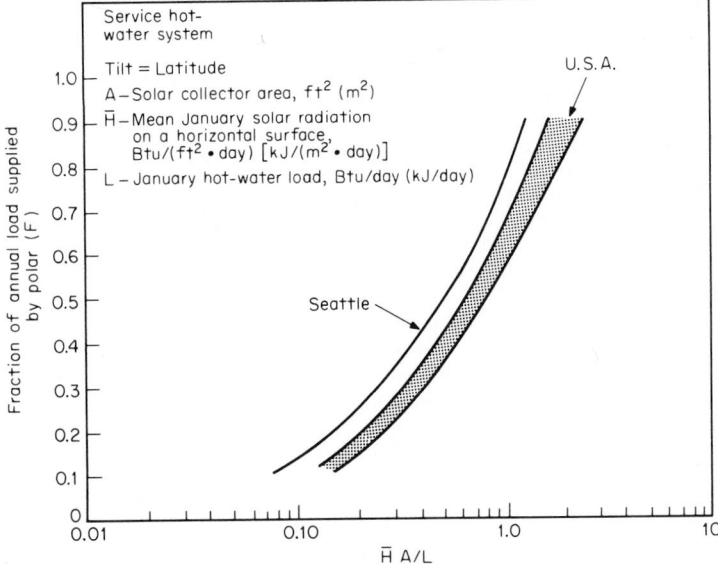

FIG. 26 Fraction of annual load supplied by a solar hot-water heating system as a function of January conditions. *(DOE/CS-0011.)*

and the needed pumps and controls. Estimates of the time to recover the investment in such a system can range from as little as 3 to as long as 8 yr, depending on the cost of the fuel saved. The charts used in this procedure were originally published in DOE/CS-0011, *Introduction to Solar Heating and Cooling—Design and Sizing,* available from the National Technical Information Service, Springfield, Virginia 22161.

DOE/CS-0011 notes that a typical family of four persons requires, in the United States, about 80 gal (302.8 L) of hot water per day. At a customary supply temperature of 140°F (60°C), the amount of heat required if the cold inlet water is at 60°F (15.5°C) is about 50,000 Btu/day (52,750 kJ/day). Further, there is a wide variation in the solar availability from region to region and from season to season in a particular location. There are also the short-term radiation fluctuations owing to cloudiness and the day-night cycle.

Seasonal variations in solar availability result in a 200 to 400 percent difference in the solar heat supply to a hot-water system. In the winter, for example, an average recovery of 40 percent of 1200 Btu/ft^2 (3785 W/m^2) of solar energy of sloping surface would require approximately 100 ft^2 (9.3 m^2) of collector for the 50,000-Btu (52,750-kJ) average daily requirement. Such a design would provide essentially all the hot-water needs on an average winter day, but would fall short on days of less than average sunshine. By contrast, a 50 percent recovery of an average summer radiant supply of 2000 Btu/ft^2 (6308 W/m^2) would involve the need for only 50 ft^2 (4.6 m^2) of collector to satisfy the average hot-water requirements.

If a 50-ft^2 (4.6-m^2) solar collector were installed, it could supply the major part of, or perhaps nearly all, the summer hot-water requirements, but it could supply less than half the winter needs. And if a 100-ft^2 (9.3-m^2) solar collector were used so that winter needs could be more nearly met, the system would be oversized for summer operation and excess solar heat would be wasted. In such circumstances, if an aqueous collection medium were used, boiling in the system would occur and collector or storage venting of steam would have to be provided.

The more important disadvantage of the oversized solar collector (for summer operation) is the economic penalty associated with investment in a collector that is not fully utilized. Although the cost of the 100-ft^2 (9.3-m^2) solar collector system would not be double that of the 50-ft^2 (4.6-m^2) unit, its annual useful heat delivery would be considerably less than double. It would, of course, deliver about twice as much heat in the winter season, when nearly all the heat could be used. But in the other seasons, particularly in summer, heat overflow would occur. The net effect of these factors is a lower economic return, per unit of investment, by the larger system. Stated another way, more Btu (kJ) per dollar of investment (hence cheaper solar heat) can be delivered by the smaller system.

If it is sized on average daily radiation in the sunniest months, the solar collector will be slightly oversized and a small amount of heat will be wasted on days of maximum solar input. On partly cloudy days during the warm season, some auxiliary heat must be provided. In the month of lowest average solar energy delivery, typically one-half to one-third as much solar-heated water can be supplied as during the warm season. Thus, fuel requirements for increasing the temperature of solar-heated water to the desired (thermostated) level could involve one-half to two-thirds of the total energy needed for hot-water heating in a midwinter month.

One disadvantage of solar DHW heating systems is the possibility of the water in the collector and associated pipe freezing during unexpectedly cold weather. Since they were introduced on a wide scale, thousands of solar DHW systems have suffered freeze damage, even in relatively warm areas of the world. Such damage is both costly and wasteful of energy.

Three ways are used to prevent freeze damage in solar DHW systems:

1. Pump circulation of warm water through the collector and piping during the night hours reduces the savings produced by the solar DHW heating system because the energy required to run the pump must be deducted from the fuel savings resulting from use of the solar panels.

2. Use an automatic drain-down valve or mechanism to empty the system of water during freezing weather. Since the onset of a freeze can be sudden, such systems must be automatic if they are to protect the collector and piping while the occupants of the building are away. Unfortunately, there is no 100 percent reliable drain-down valve or mechanism. A number of "fail-safe" systems have frozen during unusually sharp or sudden cold spells. Research is still being conducted to find the completely reliable drain-down device.

3. Indirect solar DHW systems use a nonfreeze fluid in the collector and piping to prevent freeze damage. The nonfreeze fluid passes through a heat exchanger wherein it gives up most of its heat to the potable water for the DHW system. To date, the indirect system gives the greatest

protection against freezing. Although there is a higher initial cost for an indirect system, the positive freeze protection is felt to justify this additional investment.

There are various sizing rules for solar DHW heating systems. Summarized below are those given by Kreider and Kreith—*Solar Heating and Cooling,* Hemisphere and McGraw-Hill:

Collector area: 1 ft^2/(gal·day) [0.025 m^2/(L·day)]; DHW storage tank capacity: 1.5 to 2 gal/ft^2 (61.1 to 81.5 L/m^2) of collector area; collector water flow rate: 0.025 gal/(min·ft^2) [0.000017 m^3/(s·m^2)]; indirect system storage flow rate: 0.03 to 0.04 gal/(min·ft^2) [0.0002 to 0.00027 m^3/(s·m^2)] of collector area; indirect system heat-exchanger area of 0.05 to 0.1 ft^2/collector ft^2 (0.005 to 0.009 m^2/collector m^2); collector tilt; latitude ±5°; indirect system expansion-tank volume: 12 percent of collector fluid loop; controller turnon ΔT: 15 to 20°F (27 to 36°C); controller turnoff: 3 to 5°F (5.4 to 9°C); system operating pressure: provide 3 lb/in^2 (20.7 kPa) at topmost collector manifold; storage-tank insulation: R-25 to R-30; mixing-valve set point: 120 to 140°F (48.8 to 59.9°C); pipe diameter: to maintain fluid velocity below 6 ft/s (1.83 m/s) and above 2 ft/s (0.61 m/s).

Most domestic solar hot-water heaters are installed to reduce fuel cost. Typically, domestic hot water is heated in an oil-burning boiler or heater. A solar collector reduces the amount of oil needed to heat water, thereby reducing fuel cost. Simple economic studies will show how long it will take to recover the cost of the collector, given the estimated fuel saving.

A welcome added benefit obtained when using a solar collector to heat domestic water is the reduced atmospheric pollution because less fuel is burned to heat the water. All combustion produces carbon dioxide, which is believed to contribute to atmospheric pollution and the possibility of global warming. Reducing the amount of fuel burned to heat domestic water cuts the amount of carbon dioxide emitted to the atmosphere.

Although the reduced carbon dioxide emission is difficult to evaluate on an economic basis, it is a positive factor to be considered in choosing a hot-water heating system. With greater emphasis on environmentally desirable design, solar heating of domestic hot water will receive more attention in the future.

PASSIVE SOLAR HEATING SYSTEM DESIGN

A south-facing passive solar collector will be designed for a one-story residence in Denver, Colorado. Determine the area of collector required to maintain an average inside temperature of 70°F (21°C) on a normal clear winter day for a corner room 15 ft (4.6 m) wide, 14 ft (4.3 m) deep, and 8 ft (2.4 m) high. The collector is located on the 15-ft (4.6-m) wide wall facing south, and the 14-ft (4.3-m) sidewall contains a 12-ft^2 (1.11-m^2) window. The remaining two walls adjoin heated space and so do not transfer heat. Find the volume and surface area of thermal storage material needed to prevent an unsuitable daytime temperature increase and to store the solar gain for nighttime heating. Estimate the passive solar-heating contribution for an average heating season.

Calculation Procedure:

1. *Compute the heat loss*

The surface areas and the coefficients of heat transmission of collector, windows, doors, walls, and roofs must be known to calculate the conductive heat losses of a space. The collector area can be estimated for purposes of heat-loss calculations from Table 43.

Table 43 lists ranges of the estimated ratio of collector area to floor area, g, of a space for latitudes 36°N to 48°N based on 4°F (2.2°C) intervals of average January temperature and on various types of passive solar collectors. Average January temperatures can be selected from government weather data. Denver has an average January temperature of 32°F (0°C). Choosing a direct-gain system for this installation, read down to the horizontal line for $t_o = 32°F$ (0°C), and then read right to the column for a direct-gain system. To find the estimated ratio of collector area to floor area, use a linear interpolation. Thus, for Denver, which is located at approximately 40°N, interpolate between 48°N and 36°N values. Or, $(0.24 - 0.20)/12 \times (40 - 36) + 0.20 = 0.21$, where 0.24 and 0.20 are the ratios at 48°N and 36°N, respectively; 12 is a constant derived from $48 - 36$; and 40 is the latitude for which a ratio is sought.

TABLE 43 Estimated Ratio of Collector Area to Floor Area, g
= $h_L(65 - t_o)/i_T$ for 36 to 48° North Latitude°

Average January temperature T_o, °F (°C)	Direct gain g	Water wall g	Masonry wall g
20 (−6.7)	0.27–0.32	0.54–0.64	0.69–0.81
24 (−4.4)	0.25–0.29	0.49–0.58	0.63–0.74
28 (−2.2)	0.22–0.27	0.44–0.52	0.56–0.67
32 (0)	0.20–0.24	0.39–0.47	0.50–0.60
36 (+2.2)	0.17–0.21	0.35–0.41	0.44–0.53
40 (+4.4)	0.15–0.18	0.30–0.35	0.38–0.45
44 (+6.7)	0.13–0.15	0.25–0.30	0.32–0.38

For SI temperatures, use the relation $g = h_L(18.33 - t_o)/i_T$.
° Based on a heat loss of 8 Btu/(day·ft²·°F) [0.58 W/(m²·K)].

Next, find the collector area by using the relation $A_C = (g)(A_F)$, where A_C = collector area, ft² (m²); g = ratio of collector area to floor area, expressed as a decimal; and A_F = floor area, ft² (m²). Therefore, $A_C = (0.21)(210) = 44$ ft² (4.1 m²).

To compute the conductive heat loss through a surface, use the general relation $H_C = UA \Delta t$, where H_C = conductive heat loss, Btu/h (W); U = overall coefficient of heat transmission of the surface, Btu/(h·ft²·°F) [W/(m²·K)]; A = area of heat transmission surface, ft² (m²); and Δt = temperature difference, °F = 65 − t_o (°C = 18.33 − t_o), where t_o = average monthly temperature, °F (°C). The U values of materials can be found in ASHRAE and architectural handbooks.

Since a direct-gain system was selected, the total area of glazing is the sum of the collector and noncollector glazing, 44 ft² (4.1 m²) + 12 ft² (1.1 m²) = 56 ft² (5.2 m²). Double glazing is recommended in all passive solar designs and is found to have a U value of 0.42 Btu/(h·ft²·°F) [2.38 W/(m²·K)] in winter. Thus, the conductive heat loss through the glazing is $H_C = UA \Delta t$ = (0.42)(56)(65 − 32) = 776 Btu/h (227.4 W).

The area of opaque wall surface subject to heat loss can be estimated by multiplying the wall height by the total wall length and then subtracting the estimated glazed areas from the total exterior wall area. Thus, the opaque wall area of this space is (8)(15 + 14) − 56 = 176 ft² (16.3 m²). Use the same general relation as above, substituting the U value and area of the wall. Thus, U = 0.045 Btu/(h·ft²·°F) [0.26 W/(m²·K)], and A = 176 ft² (16.3 m²). Then $H_C = UA \Delta t$ = (0.045)(176)(65 − 32) = 261 Btu/h (76.5 W).

To determine the conductive heat loss of the roof, use the same general relation as above, substituting the U value and area of the roof. Thus, U = 0.029 Btu/(h·ft²·°F) [0.16 W/(m²·K)] and A = 210 ft² (19.5 m²). Then $H_C = UA \Delta t$ = (0.029)(210)(65 − 32) = 201 Btu/h (58.9 W).

To calculate infiltration heat loss, use the relation $H_i = Vn \Delta t/55$, where V = volume of heated space, ft³ (m³); n = number of air changes per hour, selected from Table 44. The volume for this space is V = (15)(14)(8) = 1680 ft³ (47.6 m³). Entering Table 44 at the left for the

TABLE 44 Air Changes per Hour for Well-Insulated Spaces°

Description of space	Number of air changes per hour n
No windows or exterior doors	0.33
Windows or exterior doors on one side	0.67
Windows or exterior doors on two sides	1.0
Windows or exterior doors on three sides	1.33

° These figures are based on spaces with weatherstripped doors and windows or spaces with storm windows or doors. If the space does not have these features, increase the value listed for n by 50%.

physical description of the space, read to the right for n, the number of air changes per hour. This space has windows on two walls, so $n = 1$. Thus, $H_i = (1680)(1)(65 - 32)/55 = 1008$ Btu/h (295.4 W).

The total heat loss of the space is the sum of the individual heat losses of glass, wall, roof, and infiltration. Therefore, the total heat loss for this space is $H_T = 776 + 261 + 201 + 1008 = 2246$ Btu/h (658.3 W). Convert the total hourly heat loss to daily heat loss, using the relation $H_D = 24H_T$, where H_D = total heat loss per day, Btu/day (W). Thus, $H_D = 24(2246) = 53,904$ Btu/day (658.3 W).

2. Determine the daily insolation transmitted through the collector

Use government data or ASHRAE clear-day insolation tables. The latitude of Denver is 39°50′N. Since the minutes are greater than 30, or one-half of a degree, the ASHRAE table for 40°N is used. The collector is oriented due south. Hence, the average daily insolation transmitted through vertical south-facing single glazing for a clear day in January is $i_T = 1626$ Btu/ft^2 (5132 W/m^2), or double the half-day total given in the ASHRAE table. Since double glazing is used, correct the insolation transmitted through single glazing by a factor of 0.875. Thus, $i_T = (1626)(0.875) = 1423$ Btu/(day·ft^2) (4490 W/m^2) of collector.

3. Compute the area of unshaded collector required

Determine the area of unshaded collector needed to heat this space on an average clear day in January. An average clear day is chosen because sizing the collector for extreme or cloudy conditions would cause space overheating on clear days. January is used because it generally has the highest heating load of all the months.

To compute the collector area, use the relation $A_C = H_D/(E)(i_T)$, where E = a rule of thumb for the energy absorptance efficiency of the passive solar-heating system used, expressed as a decimal. Enter Table 45 for a direct gain system to find $E = 0.91$. Therefore, $A_C = 53,904/(0.91)(1423) = 42$ ft^2 (3.9 m^2).

TABLE 45 Energy Absorptance Efficiency of Passive Solar-Heating Systems

System	Efficiency E
Direct gain	0.91
Water thermal storage wall or roof pond	0.46
Masonry thermal storage wall	0.36
Attached greenhouse	0.18

If the area of unshaded collector computed in this step varies by more than 10 percent from the area of the collector estimated for heat-loss calculations in step 1, the heat loss should be recomputed with the new areas of collector and opaque wall. In this example, the computed and estimated collector areas are within 10 percent of each other, making a second computation of the collector area unnecessary.

4. Compute the insolation stored for nighttime heating

To compute the insolation to be stored for nighttime heating, the total daily insolation must be determined. Use the relation $i_D = (A_C)(i_T)(E)$, where i_D = total daily insolation collected, Btu (J). Therefore, $i_D = (42)(1423)(0.91) = 54,387$ Btu (57.4 kJ).

Typically 35 percent of the total space heat gain is used to offset daytime heat losses, requiring 65 percent to be stored for nighttime heating. Therefore, $i_S = (0.65)i_D$, where i_S = insolation stored, Btu (J). Thus, $i_S = (0.65)(54,387) = 35,352$ Btu (37.3 kJ). This step is not required for the design of thermal storage wall systems since the storage system is integrated within the collector.

5. Compute the volume of thermal storage material required

For a direct-gain system, use the formula $V_M = i_S/(d)(c_p)(\Delta t_S)(C_S)$, where V_M = volume of thermal storage material, ft^3 (m^3); d = density of storage material, lb/ft^3 (kg/m^3); c_p = specific heat of the material, Btu/(lb·°F) [kJ/(kg·K)]; Δt_S = temperature increase of the material, °F

TABLE 46   Properties of Thermal Storage Materials

Material	Density d		Specific heat c_p		Heat capacity	
	lb/ft^3	kg/m^3	Btu/(lb·°F)	kJ/(kg·K)	Btu/(ft^3·°F)	kJ/(m^3·K)
Water	62.4	999.0	1.00	4184.0	62.40	4180.8
Rock	153	2449.5	0.22	920.5	33.66	2255.2
Concrete	144	2305.4	0.22	920.5	31.68	2122.6
Brick	123	1969.2	0.22	920.5	27.06	1813.0
Adobe	108	1729.1	0.24	1004.2	25.92	1736.6
Oak	48	768.5	0.57	2384.9	27.36	1833.1
Pine	31	496.3	0.67	2803.3	20.77	1391.6

(°C); and, C_S = fraction of insolation absorbed by the material due to color, expressed as a decimal.

Select concrete as the thermal storage material. Entering Table 46, we find the density and specific heat of concrete to be 144 lb/ft^3 (2306.7 kg/m^3) and 0.22 Btu/(lb·°F) [0.921 kJ/(kg·K)], respectively.

A suitable temperature increase of the storage material in a direct-gain system is Δt_S = +15°F (+8.3°C). A range of +10 to +20°F (+5.6 to +11.1°C) can be used with smaller increases being more suitable. Select from Table 47. In this space, thermal energy will be stored in floors and walls, resulting in a weighted average of C_S = 0.60. Thus, V_M = 35,352/ (144)(0.22)(15)(0.60) = 124 ft^3 (3.5 m^3).

TABLE 47   Insolation Absorption Factors for Thermal Storage Material Based on Color

Color/Material	Factor C_S
Black, matte	0.95
Dark blue	0.91
Slate, dark gray	0.89
Dark green	0.88
Brown	0.79
Gray	0.75
Quarry tile	0.69
Red brick	0.68
Red clay tile	0.64
Concrete	0.60
Wood	0.60
Dark red	0.57
Limestone, dark	0.50
Limestone, light	0.35
Yellow	0.33
White	0.18

As a rule of thumb for thermal storage wall systems, provide a minimum of 1 ft^3 (0.30 m^3) of dark-colored thermal storage material per square foot (meter) of collector for masonry walls or 0.5 ft^3 (0.15 m^3) of water per square foot (meter) of collector for a water wall. This will provide enough thermal storage material to maintain the inside space temperature fluctuation within 15°F (8.33°C).

6.   Determine the surface area of storage material for a direct-gain space

In a direct-gain system, the insolation must be spread over the surface area of the storage material to prevent overheating. Generally, the larger the surface area of material, the lower the inside temperature fluctuation, and thus the space is more comfortable. To determine this area, enter Fig. 27 at the lower axis to select an acceptable space temperature fluctuation. Project vertically to the curve, and read left to the A_S/A_C ratio. This is the ratio of thermal storage material surface to collector area, where A_S = surface of storage material receiving direct, diffused or reflected insolation, ft^2 (m^2). In this example, 15°F (8.33°C) is selected, requiring A_S/A_C = 6.8. Thus, A_S = (6.8)(42) = 286 ft^2 (26.5 m^2).

This step is not required for the design of thermal storage wall systems in which A_S = A_C.

7.   Determine the average daily inside temperatures

To verify that the collector and thermal storage material are correctly sized, the average inside temperature must be determined. Use $t_I = t_o + 5 + (i_D)(65 - t_o)/H_D$, where t_I = average daily inside temperature, °F (°C), and 5°F (2.8°C) is an assumed inside temperature increase owing to internal heat generation such as lights, equipment, and people. Thus, t_I = 32 + 5 + (54,387)(65 − 32)/53,904 = 70.3°F (21.27°C).

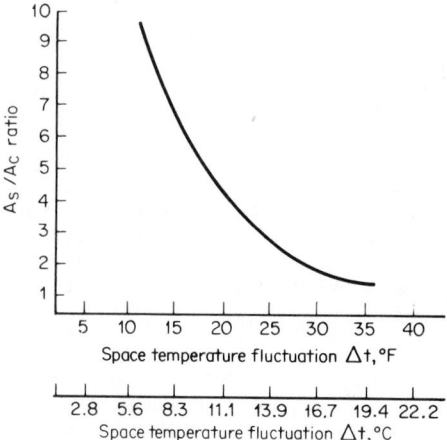

FIG. 27 Ratio of mass surface area to collector area. (*Based on data in DOE/CS-0127-2* Passive Solar Design Handbook, *vol. 2.*)

To determine the average daily low and high temperatures, use $t_L = t_I - \Delta t/2.5$, and $t_H = t_I + \Delta t/1.67$, where t_L = minimum average space temperature, °F (°C); t_H = maximum average space temperature, °F (°C); and Δt = inside space temperature fluctuation used in step 6. Thus, $t_L = 70.3 - 15/2.5 = 64.3°F$ (17.9°C), and $t_H = 70.3 + 15/1.67 = 79.3°F$ (26.27°C).

8. *Estimate the passive solar-heating contribution*

To estimate the passive solar-heating contribution (SHC) for an average month, use $SHC_M = 100(i_D)(p)/H_D$, where SHC_M = solar heating contribution of the total monthly space-heating needs, percent, and p = an insolation factor based on the percentage of clear days, expressed as a decimal. The value of $p = 0.30 + 0.65(S/100)$, where S = average sunshine for the month, percent, from an ASHRAE or government map of sunshine for each month. The average January sunshine for Denver is 67 percent. Hence, $p = 0.30 + 0.65(67/100) = 0.74$. Thus for this room in January, $SHC_M = 100(54,387)(0.74)/53,904 = 74.7$ percent of the total average space-heating needs are provided by the passive solar-heating system.

To estimate the average annual solar-heating contribution for a building, repeat steps 1, 2, and 7 for each space for each month of the heating season. Use the collector area computed in step 3 for an average clear day in January to determine i_D for each month unless part of the collector is shaded (in which case, determine the unshaded area and use that figure). Use $SHC_A = 100\Sigma(i_D)(p)(D)/\Sigma(H_D)(D)$, where SHC_A = annual passive solar-heating contribution, percent, and D = number of days of the month. The summation of the heat gains for each space for each month of the heating season is divided by the summation of the heat losses for each space for each month.

Related Calculations: These design procedures are suitable for buildings with skin-dominated heat loads such as heat losses through walls, roofs, perimeters, and infiltration. They are not applicable to buildings which have high internal heat loads or buildings which are so deep that it is difficult to collect solar heat. Therefore, these procedures generally should be limited to small and medium-size buildings with good solar access.

These procedures use an average clear-day method as a basis for sizing a passive solar-heating system. Average monthly and yearly data also are used. If the actual weather conditions vary substantially from the average, the performance of the system will vary. For instance, if a winter day is unseasonably warm, the passive solar-heating system will collect more heat than is required to offset the heat loss on that day, possibly causing space overheating. Since passive solar-heating systems rely on natural phenomena, temperature fluctuation and variability in performance are inherent in the system. Adjustable shading, reflectors, movable insulation, venting mechanisms, and backup heating systems are often used to stabilize system performance.

Since passive systems collect, store, and distribute heat through natural physical means, the system must be integrated with the architectural design. The actual efficiency of the system is highly variable and dependent on this integration within the architectural design. Efficiency ratings given in this procedure are rules of thumb. Detailed analyses of many variables and how they affect system performance can be found in DOE/CS-0127-2 and 3, *Passive Solar Design Handbook*, volumes 2 and 3, available from the National Technical Information Service, Springfield, Virginia, 22161. *The Passive Solar Energy Book,* by Edward Mazria, available from Rodale Press, Emmaus, PA, examines various architectural concepts and how they can be utilized to maximize system performance.

If thermal collection and storage to provide heating on cloudy days is desired, the collector area can be oversized by 10 percent. This necessitates the oversizing of the thermal storage material to store 75 percent of the total daily heat gain rather than 65 percent, as used in step 4. Oversizing the system will increase the average inside temperature. Step 7 should be used to verify that this higher average temperature is acceptable. Oversizing the system for cloudy-day storage is not recommended for excessively hazy or cloudy climates. Cloudy climates do not have enough clear days in a row to accumulate reserve heat for cloudy-day heating. This increased collector area may increase heat load in these climates. Cloudy-day storage should be considered only for climates with a ratio of several clear days to each cloudy day.

Passive solar-heating systems may overheat buildings if insolation reaches the collector during seasons when heating loads are low or nonexistent. Shading devices are recommended in passive solar-heated buildings to control unwanted heat. Shading devices should allow low-angle winter insolation to penetrate the collector but block higher-angle summer insolation. The shading device should allow enough insolation to penetrate the collector to heat the building during the lower-heating-load seasons of autumn and spring without overheating spaces. If shading devices are used, the area of unshaded collector must be calculated for each month to determine i_D. Methods to calculate the area of unshaded collector can be found in *The Passive Solar Energy Book* and in *Solar Control and Shading Devices,* by V. and A. Olgyay, available from Princeton University Press.

Passive solar-heating systems should be considered only for tightly constructed, well-insulated buildings. The cost of a passive system is generally higher than that of insulating and weatherstripping a building. A building that has a relatively small heat load will require a smaller collection and storage system and so will have a lower construction cost. The cost effectiveness of a passive solar-heating system is inversely related to the heat losses of the building. Systems which have a smaller ratio of collector area to floor area are generally more efficient.

Significant decreases in the size of the collector can be achieved by placing movable insulation over the collector at night. This is especially recommended for extremely cold climates in more northern latitudes. If night insulation is used, calculate heat loss for the uninsulated collector for 8 h with the daytime average temperature and for the insulated collector for 16 h with the nighttime average temperature.

Table 43 is based on a heat loss of 8 Btu/(day·ft^2) of floor area per °F [W/(m^2·K)]. Total building heat loss will increase with the increase in the ratio of collector to floor area because of the larger areas of glazing. However, it is assumed that this increase in heat loss will be offset by providing higher insulation values in noncollector surfaces. The tabulated values correspond to a residence with a compact plan, 8-ft-high ceilings, R-30 roof insulation, R-19 wall insulation, R-10 perimeter insulation, double glazing, and one air change per hour. It is provided for estimating purposes only. If the structure under consideration differs, the ratio of collector area to floor area, g, can be estimated for heat-loss calculations by using $g = h(65 - t_o)/i_T$, where h_L = estimated heat loss, Btu/(day·ft^2·°F) [W/(m^2·K)].

Passive solar heating is nonpolluting and is environmentally attractive. Other than the pollution (air, stream, and soil) possibly created in manufacturing the components of a passive solar heating system, this method of space heating is highly desirable from an environmental standpoint.

Solar heating does not produce carbon dioxide, as does the combustion of coal, gas, oil, and wood. Thus, there is no accumulation of carbon dioxide in the atmosphere from solar heating. It is the accumulated carbon dioxide in the earth's atmosphere that traps heat from the sun's rays and earth reradiation that leads to global warming.

Computer models of the earth's atmosphere and the warming that might be caused by excessive accumulation of carbon dioxide show that steps must be taken to control pollution. Although there is some disagreement about the true effect of carbon dioxide on global warming, most scientists believe that efforts to reduce carbon dioxide emissions are worthwhile. Both a

United Nations scientific panel and research groups associated with the National Academy of Sciences recommend careful study and tracking of the possibility of global warming.

For these reasons solar heating will receive more attention from designers. With more attention being paid to the environment, solar heating offers a nonpolluting alternative that can easily be incorporated in the design of most buildings.

SECTION 3

MECHANICAL ENGINEERING

JOSEPH LETO, P.E.
CONSULTING ENGINEER

GERALD M. EISENBERG
PROJECT ENGINEERING ADMINISTRATOR
AMERICAN SOCIETY OF MECHANICAL ENGINEERS

STEPHEN M. EBER, P.E.
EBASCO SERVICES, INC.

JEROME F. MUELLER, P.E.
MUELLER ENGINEERING CORP.

TYLER G. HICKS, P.E.
INTERNATIONAL ENGINEERING ASSOCIATES

EDGAR J. KATES, P.E.
CONSULTING ENGINEER

B.G.A. SKROTZKI, P.E.
POWER MAGAZINE

RAYMOND J. ROARK
PROFESSOR, UNIVERSITY OF WISCONSIN

S.W. SPIELVOGEL
PIPING ENGINEERING CONSULTANT

RUFUS OLDENBURGER
PROFESSOR, PURDUE UNIVERSITY

LYMAN F. SCHEEL
CONSULTING ENGINEER

MACHINE DESIGN AND ANALYSIS 3.7
    Energy Stored in a Rotating Flywheel 3.7
    Shaft Torque, Horsepower, and Driver Efficiency 3.8
    Pulley and Gear Loads on Shafts 3.9
    Shaft Reactions and Bending Moments 3.9
    Solid and Hollow Shafts in Torsion 3.10
    Solid Shafts in Bending and Torsion 3.11
    Equivalent Bending Moment and Ideal Torque for a Shaft 3.13
    Torsional Deflection of Solid and Hollow Shafts 3.13
    Deflection of a Shaft Carrying Concentrated and Uniform Loads . . 3.14
    Selection of Keys for Machine Shafts 3.15
    Selecting a Leather Belt for Power Transmission 3.16
    Selecting a Rubber Belt for Power Transmission 3.17
    Selecting a V Belt for Power Transmission 3.19
    Selecting Multiple V Belts for Power Transmission 3.22
    Selection of a Wire-Rope Drive 3.23
    Speeds of Gears and Gear Trains 3.24
    Selection of Gear Size and Type 3.25
    Gear Selection for Light Loads 3.29
    Selection of Gear Dimensions 3.31
    Horsepower Rating of Gears 3.32
    Moment of Inertia of a Gear Drive 3.33
    Bearing Loads in Geared Drives 3.34
    Force Ratio of Geared Drives 3.35
    Determination of Gear Bore Diameter 3.36
    Transmission Gear Ratio for a Geared Drive 3.37
    Epicyclic Gear Train Speeds 3.37
    Planetary-Gear-System Speed Ratio 3.38
    Selection of a Rigid Flange-Type Shaft Coupling 3.39
    Selection of a Flexible Coupling for a Shaft 3.42
    Selection of a Shaft Coupling for Torque and Thrust Loads . . . 3.44
    High-Speed Power-Coupling Characteristics 3.44
    Selection of Roller and Inverted-Tooth (Silent) Chain Drives . . 3.48
    Cam Clutch Selection and Analysis 3.50
    Timing-Belt Drive Selection and Analysis 3.51
    Geared Speed Reducer Selection and Application 3.54
    Power Transmission for a Variable-Speed Drive 3.56
    Bearing-Type Selection for a Known Load 3.56
    Shaft Bearing Length and Heat Generation 3.62
    Roller-Bearing Operating-Life Analysis 3.63
    Roller-Bearing Capacity Requirements 3.63
    Radial Load Rating for Rolling Bearings 3.64
    Rolling-Bearing Capacity and Reliability 3.65
    Porous-Metal Bearing Capacity and Friction 3.65
    Hydrostatic Thrust Bearing Analysis 3.67
    Hydrostatic Journal Bearing Analysis 3.69
    Hydrostatic Multidirection Bearing Analysis 3.72
    Load Capacity of Gas Bearings 3.75
    Spring Selection for a Known Load and Deflection 3.77
    Spring Wire Length and Weight 3.78
    Helical Compression and Tension Spring Analysis 3.78
    Selection of Helical Compression and Tension Springs 3.79
    Sizing Helical Springs for Optimum Dimensions and Weight . . . 3.80
    Selection of Square- and Rectangular-Wire Helical Springs . . . 3.81
    Curved Spring Design Analysis 3.82
    Round- and Square-Wire Helical Torsion-Spring Selection . . . 3.84
    Torsion-Bar Spring Analysis 3.86
    Multirate Helical Spring Analysis 3.87

Belleville Spring Analysis for Smallest Diameter 3.88
Ring-Spring Design Analysis 3.89
Liquid-Spring Selection 3.91
Selection of Air-Snubber Dashpot Dimensions 3.92
Design Analysis of Flat Reinforced-Plastic Springs 3.96
Life of Cyclically Loaded Mechanical Springs 3.98
Shock-Mount Deflection and Spring Rate 3.99
Clutch Selection for Shaft Drive 3.100
Brake Selection for a Known Load 3.102
Mechanical Brake Surface Area and Cooling Time 3.103
Involute Spline Size for Known Load 3.106
Friction Damping for Shaft Vibration 3.108
Designing Parts for Expected Life 3.110
Wear Life of Rolling Surfaces 3.112
Factor of Safety and Allowable Stress in Design 3.113
Rupture Factor and Allowable Stress in Design 3.116
Force and Shrink Fit Stress, Interference, and Torque 3.117
Hydraulic System Pump and Driver Selection 3.117
Selecting Bolt Diameter for Bolted Pressurized Joint 3.121
Determining Required Tightening Torque for a Bolted Joint 3.124
Selecting Safe Stress and Materials for Plastic Gears 3.125
Total Driving and Slip Torque for External-Spring Clutches 3.128
Design Methods for Noncircular Shafts 3.130
Hydraulic Piston Acceleration, Deceleration, Force, Flow, and Size
    Determination 3.138
Computation of Revolute Robot Proportions and Limit Stops 3.140
Hydropneumatic Accumulator Design for High Force Levels 3.144
Membrane Vibration 3.145
Power Savings Achievable in Industrial Hydraulic Systems 3.146
Sizing Dowel Pins 3.147
METALWORKING 3.148
Total Element Time and Total Operation Time 3.149
Cutting Speeds for Various Materials 3.149
Depth of Cut and Cutting Time for a Keyway 3.150
Milling-Machine Table Feed and Cutter Approach 3.151
Dimensions of Tapers and Dovetails 3.151
Angle and Length of Cut from Given Dimensions 3.152
Tool Feed Rate and Cutting Time 3.152
True Unit Time, Minimum Lot Size, and Tool-Change Time 3.153
Time Required for Turning Operations 3.153
Time and Power to Drill, Bore, Countersink, and Ream 3.154
Time Required for Facing Operations 3.155
Threading and Tapping Time 3.156
Turret-Lathe Power Input 3.157
Time to Cut a Thread on an Engine Lathe 3.158
Time to Tap with a Drilling Machine 3.158
Milling Cutting Speed, Time, Feed, Teeth Number, and Horsepower 3.159
Gang-, Multiple-, and Form-Milling Cutting Time 3.160
Shaper and Planer Cutting Speed, Strokes, Cycle Time, Power 3.161
Grinding Feed and Work Time 3.162
Broaching Time and Production Rate 3.163
Hobbing, Splining, and Serrating Time 3.163
Time to Saw Metal with Power and Band Saws 3.164
Oxyacetylene Cutting Time and Gas Consumption 3.164
Comparison of Oxyacetylene and Electric-Arc Welding 3.165
Presswork Force for Shearing and Bending 3.167
Mechanical-Press Midstroke Capacity 3.167
Stripping Springs for Pressworking Metals 3.167

Blanking, Drawing, and Necking Metals. 3.168
Metal Plating Time and Weight 3.169
Shrink- and Expansion-Fit Analyses 3.170
Press-Fit Force, Stress, and Slippage Torque 3.170
Learning-Curve Analysis and Construction 3.173
Learning-Curve Evaluation of Manufacturing Time 3.174
Determining Brinell Hardness 3.176
Economical Cutting Speeds and Production Rates 3.176
Optimum Lot Size in Manufacturing 3.177
Precision Dimensions at Various Temperatures 3.178
Horsepower Required for Metalworking 3.179
Cutting Speed for Lowest-Cost Machining 3.181
Reorder Quantity for Out-of-Stock Parts 3.181
Savings with More Machinable Materials 3.182
Time Required for Thread Milling 3.182
Drill Penetration Rate and Centerless Grinder Feed Rate 3.183
Bending, Dimpling, and Drawing Metal Parts 3.183
Blank Diameters for Round Shells 3.185
Breakeven Considerations in Manufacturing Operations 3.187
COMBUSTION 3.189
Combustion of Coal Fuel in a Furnace 3.189
Percent Excess Air while Burning Coal 3.192
Combustion of Fuel Oil in a Furnace 3.193
Combustion of Natural Gas in a Furnace 3.194
Combustion of Wood Fuel in a Furnace 3.199
Molal Method of Combustion Analysis 3.201
Final Combustion Products Temperature Estimate 3.203
Steam Boiler Heat Balance Determination 3.204
Gas Turbine Combustion Chamber Inlet Air Temperature 3.206
POWER GENERATION 3.208
Steam Mollier Diagram and Steam Table Use 3.209
Interpolation of Steam Table Values 3.211
Constant-Pressure Steam Process 3.214
Constant-Volume Steam Process 3.215
Constant-Temperature Steam Process 3.218
Constant-Entropy Steam Process 3.219
Irreversible Adiabatic Expansion of Steam 3.221
Irreversible Adiabatic Steam Compression 3.222
Throttling Processes for Steam and Water 3.223
Reversible Heating Process for Steam 3.225
Bleed-Steam Regenerative Cycle Layout and *T-S* Plot 3.227
Bleed Regenerative Steam Cycle Analysis 3.231
Reheat-Steam Cycle Performance 3.233
Mechanical-Drive Steam-Turbine Power-Output Analysis 3.236
Condensing Steam-Turbine Power-Output Analysis 3.238
Steam-Turbine Regenerative-Cycle Performance 3.240
Reheat-Regenerative Steam-Turbine Heat Rates 3.243
Steam Turbine–Gas Turbine Cycle Analysis 3.245
Steam-Condenser Performance Analysis 3.246
Steam-Condenser Air Leakage 3.249
Steam-Condenser Selection 3.250
Air-Ejector Analysis and Selection 3.252
Surface-Condenser Circulating-Water Pressure Loss 3.254
Surface-Condenser Weight Analysis 3.254
Barometric-Condenser Analysis and Selection 3.255
Cooling-Pond Size for a Known Heat Load 3.257
Direct-Contact Feedwater Heater Analysis 3.258
Closed Feedwater Heater Analysis and Selection 3.259

Power-Plant Heater Extraction-Cycle Analysis 3.264
Steam Boiler, Economizer, and Air-Heater Efficiency 3.268
Fire-Tube Boiler Analysis and Selection 3.270
Safety-Valve Steam-Flow Capacity 3.271
Safety-Valve Selection for a Watertube Steam Boiler 3.272
Steam-Quality Determination with a Throttling Calorimeter 3.277
Steam Pressure Drop in a Boiler Superheater 3.277
Selection of a Steam Boiler for a Given Load 3.278
Selecting Boiler Forced- and Induced-Draft Fans 3.281
Power-Plant Fan Selection from Capacity Tables 3.283
Analysis of Boiler Air Ducts and Gas Uptakes 3.285
Determination of the Most Economical Fan Control 3.290
Smokestack Height and Diameter Determination 3.292
Power-Plant Coal-Dryer Analysis 3.293
Coal Storage Capacity of Piles and Bunkers 3.295
Properties of a Mixture of Gases 3.295
Steam Injection in Air Supply 3.296
Regenerative-Cycle Gas-Turbine Analysis 3.297
Extraction Turbine in kW Output 3.298
INTERNAL-COMBUSTION ENGINES 3.301
Diesel Generating Unit Efficiency 3.301
Engine Displacement, Mean Effective Pressure, and Efficiency 3.302
Engine Mean Effective Pressure and Horsepower 3.302
Selection of an Industrial Internal-Combustion Engine 3.303
Engine Output at High Temperatures and High Altitudes 3.304
Indicator Use on Internal-Combustion Engines 3.305
Engine Piston Speed, Torque, Displacement, and Compression Ratio 3.306
Internal-Combustion Engine Cooling-Water Requirements 3.306
Design of a Vent System for an Engine Room 3.311
Design of a Bypass Cooling System for an Engine 3.312
Hot-Water Heat-Recovery System Analysis 3.317
Diesel Fuel Storage Capacity and Cost 3.318
Power Input to Cooling-Water and Lube-Oil Pumps 3.319
Lube-Oil Cooler Selection and Oil Consumption 3.320
Quantity of Solids Entering an Internal-Combustion Engine 3.321
Internal-Combustion Engine Performance Factors 3.321
AIR AND GAS COMPRESSORS AND VACUUM SYSTEMS 3.325
Compressor Selection for Compressed-Air Systems 3.325
Sizing Compressed-Air-System Components 3.329
Compressed-Air Receiver Size and Pump-Up Time 3.331
Vacuum-System Pump-Down Time 3.332
Vacuum-Pump Selection for High-Vacuum Systems 3.334
Vacuum-System Pumping Speed and Pipe Size 3.337
MATERIALS HANDLING 3.338
Bulk Material Elevator and Conveyor Selection 3.338
Screw Conveyor Power Input and Capacity 3.342
Design and Layout of Pneumatic Conveying Systems 3.344
PUMPS AND PUMPING SYSTEMS 3.352
Similarity or Affinity Laws for Centrifugal Pumps 3.352
Similarity or Affinity Laws in Centrifugal Pump Selection 3.353
Specific Speed Considerations in Centrifugal Pump Selection 3.353
Selecting the Best Operating Speed for a Centrifugal Pump 3.355
Total Head on a Pump Handling Vapor-Free Liquid 3.357
Pump Selection for any Pumping System 3.362
Analysis of Pump and System Characteristic Curves 3.368
Net Positive Suction Head for Hot-Liquid Pumps 3.374
Condensate Pump Selection for a Steam Power Plant 3.374
Minimum Safe Flow for a Centrifugal Pump 3.378

Selecting a Centrifugal Pump to Handle a Viscous Liquid 3.379
Pump Shaft Deflection and Critical Speed 3.381
Effect of Liquid Viscosity on Regenerative-Pump Performance 3.382
Effect of Liquid Viscosity on Reciprocating-Pump Performance 3.383
Effect of Viscosity and Dissolved Gas on Rotary Pumps 3.384
Selection of Materials for Pump Parts 3.386
Sizing a Hydropneumatic Storage Tank 3.386
Using Centrifugal Pumps as Hydraulic Turbines 3.387
Sizing Centrifugal-Pump Impellers for Safety Service 3.392
Pump Choice to Reduce Energy Consumption and Loss 3.394
PIPING AND FLUID FLOW 3.396
Pipe-Wall Thickness and Schedule Number 3.396
Pipe-Wall Thickness Determination by Piping Code Formula 3.397
Determining the Pressure Loss in Steam Piping 3.399
Piping Warm-Up Condensate Load 3.403
Steam Trap Selection for Industrial Applications 3.404
Selecting Heat Insulation for High-Temperature Piping 3.410
Orifice Meter Selection for a Steam Pipe 3.411
Selection of a Pressure-Regulating Valve for Steam Service 3.412
Hydraulic Radius and Liquid Velocity in Water Pipes 3.414
Friction-Head Loss in Water Piping of Various Materials 3.415
Chart and Tabular Determination of Friction Head 3.416
Relative Carrying Capacity of Pipes 3.420
Pressure-Reducing Valve Selection for Water Piping 3.420
Sizing a Water Meter 3.422
Equivalent Length of a Complex Series Pipeline 3.422
Equivalent Length of a Parallel Piping System 3.423
Maximum Allowable Height for a Liquid Siphon 3.424
Water-Hammer Effects in Liquid Pipelines 3.425
Specific Gravity and Viscosity of Liquids 3.425
Pressure Loss in Piping Having Laminar Flow 3.426
Determining the Pressure Loss in Oil Pipes 3.427
Flow Rate and Pressure Loss in Compressed-Air and Gas Piping 3.433
Flow Rate and Pressure Loss in Gas Pipelines 3.434
Selecting Hangers for Pipes at Elevated Temperatures 3.434
Hanger Spacing and Pipe Slope for an Allowable Stress 3.442
Effect of Cold Spring on Pipe Anchor Forces and Stresses 3.442
Reacting Forces and Bending Stress in Single-Plane Pipe Bend 3.443
Reacting Forces and Bending Stress in a Two-Plane Pipe Bend 3.446
Reacting Forces and Bending Stress in a Three-Plane Pipe Bend 3.451
Anchor Force, Stress, and Deflection of Expansion Bends 3.453
Slip-Type Expansion Joint Selection and Application 3.455
Corrugated Expansion Joint Selection and Application 3.457
Design of Steam Transmission Piping 3.460
Steam Desuperheater Analysis 3.468
Steam Accumulator Selection and Sizing 3.469
Selecting Plastic Piping for Industrial Use 3.471
Friction Loss in Pipes Handling Solids in Suspension 3.472
Desuperheater Water Spray Quantity 3.473
HEAT TRANSFER AND HEAT EXCHANGERS 3.474
Selecting Type of Heat Exchanger for a Specific Application 3.475
Shell-and-Tube Heat Exchanger Size 3.475
Heat Exchanger Actual Temperature Difference 3.480
Fouling Factors in Heat-Exchanger Sizing and Selection 3.482
Heat Transfer in Barometric and Jet Condensers 3.483
Selection of a Finned-Tube Heat Exchanger 3.484
Spiral-Type Heating Coil Selection 3.486
Sizing Electric Heaters for Industrial Use 3.488

Economizer Heat Transfer Coefficient 3.489
Boiler Tube Steam-Generating Capacity 3.490
REFRIGERATION 3.491
Refrigeration System Selection 3.491
Selection of a Refrigeration Unit for Product Cooling 3.493
Energy Required for Steam-Jet Refrigeration 3.498
Refrigeration Compressor Cycle Analysis 3.499
Reciprocating Refrigeration Compressor Selection 3.503
Centrifugal Refrigeration Machine Load Analysis 3.505
Heat Pump Cycle Analysis and Comparison 3.506
ENERGY CONSERVATION 3.509
Choice of Wind-Energy Conversion System 3.509
Fuel Savings Using High-Temperature Hot-Water Heating 3.516
Fuel Savings Produced by Heat Recovery 3.519
Cost of Heat Loss for Uninsulated Pipes 3.521
Heat-Rate Improvement Using Turbine-Driven Boiler Fans 3.522
Cost Separation of Steam and Electricity in a Cogeneration Power Plant Using
   the Energy Equivalence Method 3.526
Cogeneration Fuel Cost Allocation Based on an Established Electricity Cost . . 3.529
Boiler Fuel Conversion from Oil or Gas to Coal 3.533
Return on Investment for Energy-Saving Projects 3.536
Energy Savings from Reduced Boiler Scale 3.536
Ground Area and Unloading Capacity Required for Coal Burning 3.537
Heat Recovery from Boiler Blowdown Systems 3.538
Boiler Blowdown Percentage 3.540
Air-Cooled Heat Exchanger: Preliminary Selection 3.540
Fuel Savings Produced by Direct Digital Control of the Power-Generation
   Process 3.543
Small Hydro Power Considerations and Analysis 3.546
Sizing Flash Tanks to Conserve Energy 3.548
Flash Tank Output 3.550
Determining Waste-Heat Boiler Fuel Savings 3.551
Heat Exchangers: Quick Design and Evaluation 3.552
Figuring Flue-Gas Reynolds Number by Shortcuts 3.560

Machine Design and Analysis

REFERENCES: Deutschman—*Machine Design: Theory and Practice*, Macmillan; Johnson—*Mechanical Design Synthesis*, Kreiger; Stephenson and Collander—*Engineering Design*, Wiley-Interscience; Creamer—*Machine Design*, Addison-Wesley; Artobolevskii—*Mechanisms in Modern Engineering Design*, MIR Publishers (Moscow); Sandor and Erdman—*Advanced Mechanism Design: Analysis and Synthesis*, Prentice-Hall; Dhillon—*Reliability Engineering in Systems Design and Operation*, VNR; Chakraborty and Dhande—*Kinematics and Geometry of Planar and Spatial Cam Mechanisms*, Wiley; Lynwander—*Gear Drive Systems: Design and Application*, Dekker; Schwartz—*Composite Materials Handbook*, McGraw-Hill; Taraman—*CAD/CAM: Meeting Today's Productivity Challenge*, SME; Shtipelman—*Design and Manufacture of Hypoid Gears*, Wiley; Roark—*Formulas for Stress and Strain*, McGraw-Hill; Church—*Mechanical Vibrations*, Wiley; *Machinery's Handbook*, Industrial Press; Johnson—*Optimum Design of Mechanical Elements*, Wiley; Slaymaker—*Mechanical Design Analysis*, Wiley; Chironis—*Spring Design and Application*, McGraw-Hill; Spotts—*Design of Machine Elements*, Prentice-Hall; AGMA *Standards Books*, American Gear Manufacturers Association; Doughtie and Vallance—*Design of Machine Members*, McGraw-Hill; Buckingham—*Manual of Gear Design*, Industrial Press; Fuller—*Theory and Practice of Lubrication for Engineers*, Wiley; Dudley—*Gear Handbook*, McGraw-Hill; Churchman—*Prediction and Optimal Decision*, Wiley; Crandall—*Engineering Analysis*, McGraw-Hill; Ver Planck and Téare—*Engineering Analysis*, Wiley; Wahl—*Mechanical Springs*, McGraw-Hill; Haberman—*Engineering Systems Analysis*, Merrill; Shigley—*Mechanical Engineering Design*, McGraw-Hill; Ryder—*Creative Engineering Analysis*, Prentice-Hall; Baumeister and Marks—*Standard Handbook for Mechanical Engineers*, McGraw-Hill; Church—*Kinematics of Machines*, Wiley; Carmichael—*Kent's Mechanical Engineers' Handbook*, Wiley; Faires—*Design of Machine Elements*, Macmillan; Black—*Machine Design*, McGraw-Hill; Maleev—*Machine Design*, International; Bradford and Eaton—*Machine Design*, Wiley; Dudley—*Practical Gear Design*, McGraw-Hill; Shigley—*Simulation of Mechanical Systems*, McGraw-Hill.

ENERGY STORED IN A ROTATING FLYWHEEL

A 48-in (121.9-cm) diameter spoked steel flywheel having a 12-in wide $\times$ 10-in (30.5-cm $\times$ 25.4-cm) deep rim rotates at 200 r/min. How long a cut can be stamped in a 1-in (2.5-cm) thick aluminum plate if the stamping energy is obtained from this flywheel? The ultimate shearing strength of the aluminum is 40,000 lb/in² (275,789.9 kPa).

Calculation Procedure:

1.   *Determine the kinetic energy of the flywheel*

In routine design calculations, the weight of a spoked or disk flywheel is assumed to be concentrated in the rim of the flywheel. The weight of the spokes or disk is neglected. In computing the kinetic energy of the flywheel, the weight of a rectangular, square, or circular rim is assumed to be concentrated at the horizontal centerline. Thus, for this rectangular rim, the weight is concentrated at a radius of $48/2 - 10/2 = 19$ in (48.3 cm) from the centerline of the shaft to which the flywheel is attached.

Then the kinetic energy $K = Wv^2/(2g)$, where K = kinetic energy of the rotating shaft, ft·lb; W = flywheel weight of flywheel rim, lb; v = velocity of flywheel at the horizontal centerline of the rim, ft/s. The velocity of a rotating rim is $v = 2\pi RD/60$, where $\pi = 3.1416$; R = rotational speed, r/min; D = distance of the rim horizontal centerline from the center of rotation, ft. For this flywheel, $v = 2\pi(200)(19/12)/60 = 33.2$ ft/s (10.1 m/s).

The rim of the flywheel has a volume of (rim height, in)(rim width, in)(rim circumference measured at the horizontal centerline, in), or $(10)(12)(2\pi)(19) = 14,350$ in³ (235,154.4 cm³). Since machine steel weighs 0.28 lb/in³ (7.75 g/cm³), the weight of the flywheel rim is $(14,350)(0.28) = 4010$ lb (1818.9 kg). Then $K = (4010)(33.2)^2/[2(32.2)] = 68,700$ ft·lb (93,144.7 N·m).

2.   *Compute the dimensions of the hole that can be stamped*

A stamping operation is a shearing process. The area sheared is the product of the plate thickness and the length of the cut. Each square inch of the sheared area offers a resistance equal to the ultimate shearing strength of the material punched.

During stamping, the force exerted by the stamp varies from a maximum F lb at the point of contact to 0 lb when the stamp emerges from the metal. Thus, the average force during stamping is $(F + 0)/2 = F/2$. The work done is the product of $F/2$ and the distance through which this force moves, or the plate thickness t in. Therefore, the maximum length that can be stamped is that which occurs when the full kinetic energy of the flywheel is converted to stamping work.

With a 1-in (2.5-cm) thick aluminum plate, the work done is W ft·lb = (force, lb)(distance, ft). The work done when all the flywheel kinetic energy is used is $W = K$. Substituting the kinetic energy from step 1 gives $W = K = 68,700$ ft·lb (93,144.7 N·m) $= (F/2)(1/12)$; and solving for the force yields $F = 1,650,000$ lb (7,339,566.3 N).

The force F also equals the product of the plate area sheared and the ultimate shearing strength of the material stamped. Thus, $F = lts_u$, where l = length of cut, in; t = plate thickness, in; s_u = ultimate shearing strength of the material. Substituting the known values and solving for l, we get $l = 1,650,000/[(1)(40,000)] = 41.25$ in (104.8 cm).

Related Calculations: The length of cut computed above can be distributed in any form—square, rectangular, circular, or irregular. This method is suitable for computing the energy stored in a flywheel used for any purpose. Use the general procedure in step 2 for computing the principal dimension in blanking, punching, piercing, trimming, bending, forming, drawing, or coining.

SHAFT TORQUE, HORSEPOWER, AND DRIVER EFFICIENCY

A 4-in (10.2-cm) diameter shaft is driven at 3600 r/min by a 400-hp (298.3-kW) motor. The shaft drives a 48-in (121.9-cm) diameter chain sprocket having an output efficiency of 85 percent. Determine the torque in the shaft, the output force on the sprocket, and the power delivered by the sprocket.

Calculation Procedure:

1. *Compute the torque developed in the shaft*

For any shaft driven by any driver, the torque developed is T lb·in $= 63,000hp/R$, where $hp =$ horsepower delivered to, or by, the shaft; $R =$ shaft rotative speed, r/min. Thus, the torque developed by this shaft is $T = (63,000)(400)/3600 = 7000$ lb·in (790.9 N·m).

2. *Compute the sprocket output force*

The force developed at the output surface, tooth, or other part of a rotating member is given by $F = T/r$, where $F =$ force developed, lb; $r =$ radius arm of the force, in. In this drive the radius is $48/2 = 24$ in (61 cm). Hence, $F = 7000/24 = 291$ lb (1294.4 N).

3. *Compute the power delivered by the sprocket*

The work input to this shaft is 400 hp (298.3 kW). But the work output is less than the input because the efficiency is less than 100 percent. Since efficiency $=$ work output, hp/work input, hp, the work output, hp $=$ (work input, hp)(efficiency), or output hp $= (400)(0.85) = 340$ hp (253.5 kW).

Related Calculations: Use this procedure for any shaft driven by any driver—electric motor, steam turbine, internal-combustion engine, gas turbine, belt, chain, sprocket, etc. When computing the radius of toothed or geared members, use the pitch-circle or pitch-line radius.

PULLEY AND GEAR LOADS ON SHAFTS

A 500-r/min shaft is fitted with a 30-in (76.2-cm) diameter pulley weighing 250 lb (113.4 kg). This pulley delivers 35 hp (26.1 kW) to a load. The shaft is also fitted with a 24-in (61.0-cm) pitch-diameter gear weighing 200 lb (90.7 kg). This gear delivers 25 hp (18.6 kW) to a load. Determine the concentrated loads produced on the shaft by the pulley and the gear.

Calculation Procedure:

1. *Determine the pulley concentrated load*

The largest concentrated load caused by the pulley occurs when the belt load acts vertically downward. Then the total pulley concentrated load is the sum of the belt load and pulley weight.

For a pulley in which the tension of the tight side of the belt is twice the tension in the slack side of the belt, the maximum belt load is $F_p = 3T/r$, where $F_p =$ tension force, lb, produced by the belt load; $T =$ torque acting on the pulley, lb·in; $r =$ pulley radius, in. The torque acting on a pulley is found from $T = 63,000hp/R$, where $hp =$ horsepower delivered by pulley; $R =$ revolutions per minute (rpm) of shaft.

For this pulley, $T = 63,000(35)/500 = 4410$ lb·in (498.3 N·m). Hence, the total pulley concentrated load $= 882 + 250 = 1132$ lb (5035.1 N).

2. *Determine the gear concentrated load*

With a gear, the turning force acts only on the teeth engaged with the meshing gear. Hence, there is no slack force as in a belt. Therefore, $F_g = T/r$, where $F_g =$ gear tooth-thrust force, lb; $r =$ gear pitch radius, in: other symbols as before. The torque acting on the gear is found in the same way as for the pulley.

Thus, $T = 63,000(25)/500 = 3145$ lb·in (355.3 N·m). Then $F_g = 3145/12 = 263$ lb (1169.9 N). Hence, the total gear concentrated load is $263 + 200 = 463$ lb (2059.5 N).

Related Calculations: Use this procedure to determine the concentrated load produced by any type of gear (spur, herringbone, worm, etc.), pulley (flat, V, or chain belt), sprocket, or their driving member. When the power transmission belt or chain leaves the belt or sprocket at an angle other than the vertical, take the vertical component of the pulley force and add it to the pulley weight to determine the concentrated load.

SHAFT REACTIONS AND BENDING MOMENTS

A 30-ft (9.1-m) long steel shaft weighing 150 lb/ft (223.2 kg/m) of length has a 500-lb (2224.1-N) concentrated gear load 10 ft (3.0 m) from the left end of the shaft and a 2000-lb (8896.4-N) concentrated pulley load 15 ft (4.6 m) from the right end of the shaft. Determine the end reactions and the maximum bending moment in this shaft.

Calculation Procedure:

1. *Draw a sketch of the shaft*

Figure 1a shows a sketch of the shaft. Label the left- and right-hand reactions L_R and R_R, respectively.

2. *Compute the shaft end reactions*

Take moments about R_R to determine the magnitude of L_R. Since the shaft has a uniform weight per foot of length, assume that the total weight of the shaft is concentrated at its midpoint. Then $30L_R - 500(20) - 150(30)(15) - 2000(15) = 0$; $L_R = 3583.33$ lb (15,939.4 N). Take moments about L_R to determine R_R. Or, $30R_R - 500(10) - 150(30)(15) - 2000(15) = 0$; $R_R = 3416.67$ lb (15,198.1 N). Alternatively, the first reaction found could be subtracted from the sum of the vertical loads, or $500 + 30 \times 150 + 2000 - 3583.33 = 3416.67$ lb (15,198.1 N). However, taking moments about each support permits checking the results, because the sum of the reactions should equal the sum of the vertical loads, including the weight of the shaft.

3. *Compute the maximum bending moment*

The maximum bending moment in a shaft occurs where the shear is zero. Find the vertical shear at each point of applied load or reaction by taking the algebraic sum of the vertical forces to the left and right of the load. Use a plus sign for upward forces and a minus sign for downward forces.

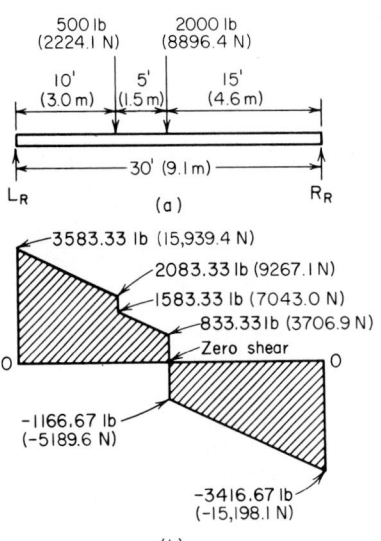

FIG. 1 Shaft bending-moment diagram.

Designate each shear force by V with a subscript number showing its location, in feet (meters) along the shaft from the left end. Use L and R to indicate whether the shear is to the left or right of the load. The shear at the left-hand reaction is $V_{LR} = +3583.33$ lb ($+15{,}939.5$ N); $V_{10L} = 3583.33 - 10 \times 150 = 2083.33$ lb (9267.1 N), where the product $10 \times 150 =$ the weight of the shaft from the point V_{LR} to the 500-lb (2224.1-N) load. At this load, $V_{10R} = 2083.33 - 500 = 1583.33$ lb (7043.0 N). To the right of the 500-lb (2224.1-N) load, at the 2000-lb (8896.4-N) load, $V_{20L} = 1583.33 - 5 \times 150 = 833.33$ lb (3706.9 N). To the right of the 2000-lb (8896.4-N) load, $V_{20R} = 833.33 - 2000 = -1166.67$ lb (-5189.6 N). At the left of V_R, $V_{30L} = -1166.67 - 15 \times 150 = -3416.67$ lb ($-15{,}198.1$ N). At the right hand end of the shaft $V_{30R} = -3416.67 + 3416.67 = 0$.

Draw the shear diagram (Fig. 1b). This diagram shows that zero shear occurs at a point 15 ft (4.6 m) from the left-hand reaction. Hence, the maximum bending moment M_m on this shaft is $M_m = 3583.33(15) - 500(5) - 150(15)(7.5) = 34{,}340$ lb·ft (46,558.8 N·m).

Related Calculations: Use this procedure for shafts of any metal—steel, bronze, aluminum, plastic, etc.—if the shaft is of uniform cross section. For nonuniform shafts, use the procedures discussed later in this section.

SOLID AND HOLLOW SHAFTS IN TORSION

A solid steel shaft will transmit 500 hp (372.8 kW) at 3600 r/min. What diameter shaft is required if the allowable stress in the shaft is 12,500 lb/in^2 (86,187.5 kPa)? What diameter hollow shaft is needed to transmit the same power if the inside diameter of the shaft is 1.0 in (2.5 cm)?

Calculation Procedure:

1. *Compute the torque in the solid shaft*

For any solid shaft, the torque T, lb·in $= 63,000 hp/R$, where $R =$ shaft rpm. Thus, $T = 63,000(500)/3600 = 8750$ lb·in (988.6 N·m).

2. *Compute the required shaft diameter*

For any solid shaft, the required diameter d, in $= 1.72 \, (T/s)^{1/3}$, where $s =$ allowable stress in shaft, lb/in². Thus, for this shaft, $d = 1.72(8750/12,500)^{1/3} = 1.526$ in (3.9 cm).

3. *Analyze the hollow shaft*

The usual practice is to size hollow shafts such that the ratio q of the inside diameter d_i into the outside diameter d_o in is 1:2 to 1:3 or some intermediate value. With a q in this range the shaft will have sufficient thickness to prevent failure in service.

Assume $q = d_i/d_o = 1/2$. Then with $d_i = 1.0$ in (2.5 cm), $d_o = d_i/q$, or $d_o = 1.0/0.5 = 2.0$ in (5.1 cm). With $q = 1/3$, $d_o = 1.0/0.33 = 3.0$ in (7.6 cm).

4. *Compute the stress in each hollow shaft*

For the hollow shaft $s = 5.1 T/d_o^3(1 - q^4)$, where the symbols are as defined above. Thus, for the 2-in (5.1-cm) outside-diameter shaft, $s = 5.1(8750)/[8(1 - 0.0625)] = 5950$ lb/in² (41,023.8 kPa).

By inspection, the stress in the 3-in (7.6-cm) outside-diameter shaft will be lower because the torque is constant. Thus, $s = 5.1 \, (8750)/[27(1 - 0.0123)] = 1672$ lb/in² (11,528.0 kPa).

5. *Choose the outside diameter of the hollow shaft*

Use a trial-and-error procedure to choose the hollow shaft's outside diameter. Since the stress in the 2-in (5.1-cm) outside-diameter shaft, 5950 lb/in² (41,023.8 kPa), is less than half the allowable stress of 12,500 lb/in² (86,187.5 kPa), select a smaller outside diameter and compute the stress while holding the inside diameter constant.

Thus, with a 1.5-in (3.8-cm) shaft and the same inside diameter, $s = 5.1(8750)/[3.38(1 - 0.197)] = 16,430$ lb/in² (113,284.9 kPa). This exceeds the allowable stress.

Try a larger outside diameter, 1.75 in (4.4 cm), to find the effect on the stress. Or $s = 5.1 \, (8750)/[5.35(1 - 0.107)] = 9350$ lb/in² (64,468.3 kPa). This is lower than the allowable stress.

Since a 1.5-in (3.8-cm) shaft has a 16,430-lb/in² (113,284.9-kPa) stress and a 1.75-in (4.4-cm) shaft has a 9350-lb/in² (64,468.3-kPa) stress, a shaft of intermediate size will have a stress approaching 12,500 lb/in² (86,187.5 kPa). Trying 1.625 in (4.1 cm) gives $s = 5.1(8750)/[4.4(1 - 0.143)] = 11,820$ lb/in² (81,489.9 kPa). This is within 680 lb/in² (4688.6 kPa) of the allowable stress and is close enough for usual design calculations.

Related Calculations: Use this procedure to find the diameter of any solid or hollow shaft acted on only by torsional stress. Where bending and torsion occur, use the next calculation procedure. Find the allowable torsional stress for various materials in Baumeister and Marks—*Standard Handbook for Mechanical Engineers*.

SOLID SHAFTS IN BENDING AND TORSION

A 30-ft (9.1-m) long solid shaft weighing 150 lb/ft (223.2 kg/m) is fitted with a pulley and a gear as shown in Fig. 2. The gear delivers 100 hp (74.6 kW) to the shaft while driving the shaft at 500 r/min. Determine the required diameter of the shaft if the allowable stress is 10,000 lb/in² (68,947.6 kPa).

Calculation Procedure:

1. *Compute the pulley and gear concentrated loads*

Using the method of the previous calculation procedure, we get $T = 63,000 hp/R = 63,000(100)/500 = 12,600$ lb·in (1423.6 N·m). Assuming that the maximum tension of the tight side of the belt is twice the tension of the slack side, we see the maximum belt load is $R_P = 3T/r = 3(12,600)/24 = 1575$ lb (7005.9 N). Hence, the total pulley concentrated load = belt load + pulley weight $= 1575 + 750 = 2325$ lb (10,342.1 N).

The gear concentrated load is found from $F_g = T/r$, where the torque is the same as computed for the pulley, or $F_g = 12,600/9 = 1400$ lb (6227.5 N). Hence, the total gear concentrated load is $1400 + 75 = 1475$ lb (6561.1 N).

Draw a sketch of the shaft showing the two concentrated loads in position (Fig. 2).

2.   *Compute the end reactions of the shaft*

Take moments about R_R to determine L_R, using the method of the previous calculation procedures. Thus, $L_R(30) - 2325(25) - 1475(8) - 150(30)(15) = 0$; $L_R = 4580$ lb (20,372.9 N). Taking moments about L_R to determine R_R yields $R_R(30) - 1475(22) - 2325(5) - 150(30)(15) = 0$; $R_R = 3720$ lb (16,547.4 N). Check by taking the sum of the upward forces: $4580 + 3720 = 8300$ lb (36,920.2 N) = sum of the downward forces, or $2325 + 1475 + 4500 = 8300$ lb (36,920.2 N).

3.   *Compute the vertical shear acting on the shaft*

Using the method of the previous calculation procedures, we find $V_{LR} = 4580$ lb (20,372.9 N); $V_{5L} = 4580 - 5(150) = 3830$ lb (17,036.7 N); $V_{5R} = 3830 - 2325 = 1505$ lb (6694.6 N); $V_{22L} = 1505 - 17(150) = -1045$ lb (−4648.4 N); $V_{22R} = -1045 - 1475 = -2520$ lb (−11,209.5 N); $V_{30L} = -2520 - 8(150) = -3720$ lb (−16,547.4 N); $V_{30R} = -3720 + 3720 = 0$.

4.   *Find the maximum bending moment on the shaft*

Draw the shear diagram shown in Fig. 2. Determine the point of zero shear by scaling it from the shear diagram or setting up an equation thus: positive shear $- x(150$ lb/ft$) = 0$, where the positive shear is the last recorded plus value, V_{5R} in this shaft, and $x =$ distance from V_{5R} where the shear is zero. Substituting values gives $1505 - 150x = 0$; $x = 10.03$ ft (3.1 m). Then $M_m = 4580(15.03) - 2325(10.03) - (150)(5 + 10.03)[(5 + 10.03)/2] = 28,575$ lb (127,108.3 N).

5.   *Determine the required shaft diameter*

Use the method of maximum shear theory to size the shaft. Determine the equivalent torque T_e from $T_e = (M_m^2 + T^2)^{0.5}$, where M_m is the maximum bending moment, lb·ft, acting on the shaft and T is the maximum torque acting on the shaft. For this shaft, $T_e = [28,575^2 + (12,600/12)^2]^{0.5} = 28,600$ lb·ft (38,776.4 N·m), where the torque in pound-inches is divided by 12 to convert it to pound-feet. To convert T_e to $T_{e'}$ lb·in, multiply by 12.

Once the equivalent torque is known, the shaft diameter d in is computed from $d = 1.72(T_{e'}/s)^{1/3}$, where $s =$ allowable stress in the shaft. For this shaft, $d = 1.72(28,500)(12)/(10,000)^{1/3} = 5.59$ in (14.2 cm). Use a 6.0-in (15.2-cm) diameter shaft.

Related Calculations: Use this procedure for any solid shaft of uniform cross section made of metal—steel, aluminum, bronze, brass, etc. The equation used in step 4 to determine the location of zero shear is based on a strength-of-materials principle: When zero shear occurs between two concentrated loads, find its location by dividing the last *positive* shear by the uniform load. If desired, the maximum principal stress theory can be used to combine the bending and torsional stresses in a shaft. The results obtained approximate those of the maximum shear theory.

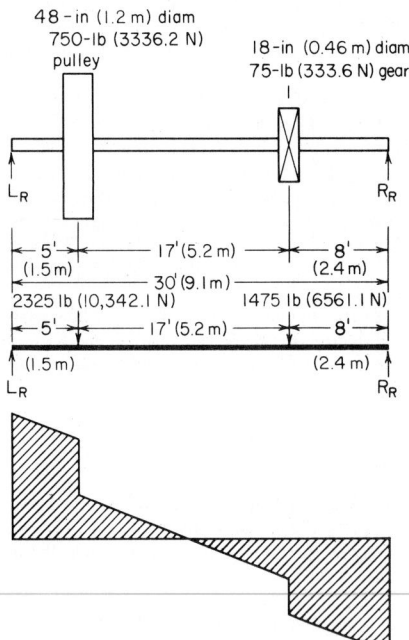

48-in (1.2 m) diam
750-lb (3336.2 N) pulley

18-in (0.46 m) diam
75-lb (333.6 N) gear

L_R R_R

5' (1.5 m)   17' (5.2 m)   8' (2.4 m)
30' (9.1 m)
2325 lb (10,342.1 N)   1475 lb (6561.1 N)

5' (1.5 m)   17' (5.2 m)   8' (2.4 m)
L_R R_R

FIG. 2 Solid-shaft bending moments.

EQUIVALENT BENDING MOMENT AND IDEAL TORQUE FOR A SHAFT

A 2-in (5.1-cm) diameter solid steel shaft has a maximum bending moment of 6000 lb·in (677.9 N·m) and an applied torque of 3000 lb·in (339.0 N·m). Is this shaft safe if the maximum allowable bending stress is 10,000 lb/in² (68,947.6 kPa)? What is the ideal torque for this shaft?

Calculation Procedure:

1. Compute the equivalent bending moment

The equivalent bending moment M_e lb·in for a solid shaft is $M_e = 0.5[M + (M^2 + T^2)^{0.5}]$, where M = maximum bending moment acting on the shaft, lb·in; T = maximum torque acting on the shaft, lb·in. For this shaft, $M_e = 0.5[6000 + (6000^2 + 3000^2)^{0.5}] = 6355$ lb·in (718.0 N·m).

2. Compute the stress in the shaft

Use the flexure relation $s = Mc/I$, where s = stress developed in the shaft, lb/in²; $M = M_e$ for a shaft; I = section moment of inertia of the shaft about the neutral axis; in⁴; c = distance from shaft neutral axis to outside fibers, in. For a circular shaft, $I = \pi d^4/64 = \pi(2)^4/64 = 0.785$ in⁴ (32.7 cm⁴); $c = d/2 = 2/2 = 1.0$. Then $s = Mc/I = (6355)(1.0)/0.785 = 8100$ lb/in² (55,849.5 kPa). Thus, the actual bending stress is 1900 lb/in² (13,100.5 kPa) less than the maximum allowable bending stress. Therefore the shaft is safe. Alternatively, compute the maximum equivalent bending moment from $M_e = sI/c = (10,000)(0.785)/1.0 = 7850$ lb·in (886.9 N·m). This is 7850 − 6355 = 1495 lb·in (168.9 N·m) greater than the actual equivalent bending moment. Hence, the shaft is safe.

3. Compute the ideal torque for the shaft

The ideal torque T_i lb·in for a shaft is $T_i = M + (M^2 + T^2)^{0.5}$, where M and T are the bending and torsional moments, respectively, acting on the shaft, lb·in. For this shaft, $T_i = 6000 + (6000^2 + 3000^2)^{0.5} = 12,710$ lb·in (1436.0 N·m).

 Related Calculations: Use this procedure for any shaft of uniform cross section made of metal—steel, aluminum, bronze, brass, etc.

TORSIONAL DEFLECTION OF SOLID AND HOLLOW SHAFTS

What diameter solid steel shaft should be used for a 500-hp (372.8-kW) 250-r/min application if the allowable torsional deflection is 1°, the maximum allowable stress is 10,000 lb/in² (68,947.6 kPa), and the modulus of rigidity is 13 × 10⁶ lb/in² (89.6 × 10⁶ kPa)? What diameter hollow steel shaft should be used if the ratio of the inside diameter to the outside diameter is 1:3, the allowable deflection is 1°, the allowable stress is 10,000 lb/in² (68,947.6 kPa), and the modulus of rigidity is 13 × 10⁶ lb/in² (89.6 × 10⁶ kPa)? What shaft has the greatest weight?

Calculation Procedure:

1. Determine the torque acting on the shaft

For any shaft, $T = 63,000hp/R$; or for this shaft, $T = 63,000(500)/250 = 126,000$ lb·in (14,236.1 N·m).

2. Compute the required diameter of the solid shaft

For a solid metal shaft, $d = (584Tl/G\alpha)^{1/3}$, where l = shaft length expressed as a number of shaft diameters, in; G = modulus of rigidity, lb/in²; α = angle of torsion deflection, deg.

 Usual specifications for noncritical applications of shafts require that the torsional deflection not exceed 1° in a shaft having a length equal to 20 diameters. Using this length gives $d = [584 \times 126,000 \times 20/(13 \times 10^6 \times 1.0)]^{1/3} = 4.84$ in (12.3 cm). Use a 5-in (12.7-cm) diameter shaft.

3. Compute the outside diameter of the hollow shaft

Assume that the shaft has a length equal to 20 diameters. Then for a hollow shaft $d = [584Tl/G\alpha(1 - q^4)]^{1/3}$, where $q = d_i/d_o$; d_i = inside diameter of the shaft, in; d_o = outside diameter of the shaft, in. For this shaft, $d = \{584 \times 126,000 \times 20/(13 \times 10^6 \times 1.0)[1 - (1/3)^4]\}^{1/3} =$

4.86 in (12.3 cm). Use a 5-in (12.7-cm) outside-diameter shaft. The inside diameter would be 5.0/3 = 1.667 in (4.2 cm).

4. Compare the weight of the shafts

Steel weighs approximately 480 lb/ft³ (7688.9 kg/m³). To find the weight of each shaft, compute its volume in cubic feet and multiply it by 480. Thus, for the 5-in (12.7-cm) diameter solid shaft, weight = $(\pi\, 5^2/4)(5 \times 20)(480)/1728 = 540$ lb (244.9 kg). The 5-in (12.7-cm) outside-diameter hollow shaft weighs $(\pi\, 5^2/4 - \pi 1.667^2/4)(5 \times 20)(480)/1728 = 242$ lb (109.8 kg). Thus, the hollow shaft weighs less than half the solid shaft. However, it would probably be more expensive to manufacture because drilling the central hole could be costly.

Related Calculations: Use this procedure to determine the steady-load torsional deflection of any shaft of uniform cross section made of any metal—steel, bronze, brass, aluminum, Monel, etc. The assumed torsional deflection of 1° for a shaft that is 20 times as long as the shaft diameter is typical for routine applications. Special shafts may be designed for considerably less torsional deflection.

DEFLECTION OF A SHAFT CARRYING CONCENTRATED AND UNIFORM LOADS

A 2-in (5.1-cm) diameter steel shaft is 6 ft (1.8 m) long between bearing centers and turns at 500 r/min. The shaft carries a 600-lb (2668.9-N) concentrated gear load 3 ft (0.9 m) from the left-hand center. Determine the deflection of the shaft if the modulus of elasticity E of the steel is 30 $\times$ 10⁶ lb/in² (206.8 $\times$ 10⁹ Pa). What would the shaft deflection be if the load were 2 ft (0.6 m) for the left-hand bearing? The shaft weighs 10 lb/ft (14.9 kg/m).

Calculation Procedure:

1. Compute the deflection caused by the concentrated load

When a beam carries both a concentrated and a uniformly distributed load, compute the deflection for each load separately and find the sum. This sum is the total deflection caused by the two loads.

For a beam carrying a concentrated load, the deflection Δ in $= Wl^3/48EI$, where $W =$ concentrated load, lb; $l =$ length of beam, in; $E =$ modulus of elasticity, lb/in²; $I =$ moment of inertia of shaft cross section, in⁴. For a circular shaft, $I = \pi d^4/64 = \pi(2)^4/64 = 0.7854$ in⁴ (32.7 cm⁴). Then $\Delta = 600(72)^3/[48(30)(10^6)(0.7854)] = 0.198$ in (5.03 mm). The deflection per foot of shaft length is $\Delta_f = 0.198/6 = 0.033$ in/ft (2.75 mm/m) for the concentrated load.

2. Compute the deflection due to shaft weight

For a shaft of uniform weight, $\Delta = 5wl^3/384EI$, where $w =$ total distributed load $=$ weight of shaft, lb. Thus, $\Delta = 5(60)(72)^3/[384(30 \times 10^6)(0.7854)] = 0.0129$ in (0.328 mm). The deflection per foot of shaft length is $\Delta_f = 0.0129/6 = 0.00214$ in/ft (0.178 mm/m).

3. Determine the total deflection of the shaft

The total deflection of the shaft is the sum of the deflections caused by the concentrated and uniform loads, or $\Delta_t = 0.198 + 0.0129 = 0.2109$ in (5.36 mm). The total deflection per foot of length is $0.033 + 0.00214 = 0.03514$ in/ft (2.93 mm/m).

Usual design practice limits the transverse deflection of a shaft of any diameter to 0.01 in/ft (0.83 mm/m) of shaft length. The deflection of this shaft is 3½ times this limit. Therefore, the shaft diameter must be increased if this limit is not to be exceeded.

Using a 3-in (7.6-cm) diameter shaft weighing 25 lb/ft (37.2 kg/m) and computing the deflection in the same way, we find the total transverse deflection is 0.0453 in (1.15 mm), and the total deflection per foot of shaft length is 0.00755 in/ft (0.629 mm/m). This is within the desired limits. By reducing the assumed shaft diameter in ⅛-in (0.32-cm) increments and computing the deflection per foot of length, a deflection closer to the limit can be obtained.

4. Compute the total deflection for the noncentral load

For a noncentral load, $\Delta = (Wc'/3EIl)[(cl/3 + cc'/3)^3]^{0.5}$, where $c =$ distance of concentrated load from left-hand bearing, in; $c' =$ distance of concentrated load from right-hand bearing, in.

Thus $c + c' = 1$, and for this shaft $c = 24$ in (61.0 cm) and $c' = 48$ in (121.9 cm). Then $\Delta = [600 \times 48/(3 \times 30 \times 10^6 \times 0.7854 \times 72)][(24 \times 72/x + 24 \times 48/3)^3]^{0.5} = 0.169$ in (4.29 mm).

The deflection caused by the weight of the shaft is the same as computed in step 2, or 0.0129 in (0.328 mm). Hence, the total shaft deflection is $0.169 + 0.0129 = 0.1819$ in (4.62 mm). The deflection per foot of shaft length is $0.1819/6 = 0.0303$ in (2.53 mm/m). Again, this exceeds 0.01 in/ft (0.833 mm/m).

Using a 3-in (7.6-cm) diameter shaft as in step 3 shows that the deflection can be reduced to within the desired limits.

Related Calculations: Use this procedure for any metal shaft—aluminum, brass, bronze, etc.—that is uniformly loaded or carries a concentrated load.

SELECTION OF KEYS FOR MACHINE SHAFTS

Select a key for a 4-in (10.2-cm) diameter shaft transmitting 1000 hp (745.7 kW) at 1000 r/min. The allowable shear stress in the key is 15,000 lb/in^2 (103,425.0 kPa), and the allowable compressive stress is 30,000 lb/in^2 (206,850.0 kPa). What type of key should be used if the allowable shear stress is 5000 lb/in^2 (34,475.0 kPa) and the allowable compressive stress is 20,000 lb/in^2 (137,900.0 kPa)?

Calculation Procedure:

1. Compute the torque acting on the shaft

The torque acting on the shaft is $T = 63,000 hp/R$, or $T = 63,000(1000/1000) = 63,000$ lb·in (7118.0 N·m).

2. Determine the shear force acting on the key

The shear force F_s lb acting on a key is $F_s = T/r$, where $T = $ torque acting on shaft, lb·in; $r = $ radius of shaft, in. Thus, $T = 63,000/2 = 31,500$ lb (140,118.9 N).

3. Select the type of key to use

When a key is designed so that its allowable shear stress is approximately one-half its allowable compressive stress, a square key (i.e., a key having its height equal to its width) is generally chosen. For other values of the stress ratio, a flat key is generally used.

Determine the dimensions of the key from Baumeister and Marks—*Standard Handbook for Mechanical Engineers*. This handbook shows that a 4-in (10.2-cm) diameter shaft should have a square key 1 in wide × 1 in (2.5 cm × 2.5 cm) high.

4. Determine the required length of the key

The length of a 1-in (2.5-cm) key based on the allowable shear stress is $l = 2F_s/(w_k s_s)$, where $w_k = $ width of key, in. Thus, $l = 31,500/[(1)(15,000)] = 2.1$ in (5.3 cm), say 2⅛ in (5.4 cm).

5. Check key length for the compressive load

The length of a 1-in (2.5-cm) key based on the allowable compressive stress is $l = 2F_s/(t s_c)$, where $t = $ key thickness, in; $s_c = $ allowable compressive stress, lb/in^2. Thus, $l = 2(13,500)/[(1)(30,000)] = 2.1$ in (5.3 cm). This agrees with the key length based on the allowable shear stress. The key length found in steps 4 and 5 should agree if the key is square in cross section.

6. Determine the key size for other stress values

When the allowable shear stress does not equal one-half the allowable compressive stress for a shaft key, a flat key is generally used. A flat key has a width greater than its height.

Find the recommended dimensions for a flat key from Baumeister and Marks—*Standard Handbook for Mechanical Engineers*. This handbook shows that a 4-in (10.2-cm) diameter shaft will use a 1-in (2.5-cm) wide by ¾-in (1.9-cm) thick flat key.

The length of the key based on the allowable shear stress is $l = F_s/(w_k s_s) = 31,500/[(l)(5000)] = 6.31$ in (16.0 cm). Use a 6⁵⁄₁₆-in (16.0-cm) long key.

Checking the key length based on the allowable compressive stress yields $l = 2F_s/(t s_c) = 2(31,500)/[(0.75)(20,000)] = 4.2$ in (10.7 cm). Use the longer length, 6⁵⁄₁₆ in (16.0 cm), because the shorter key would be overloaded in compression.

Related Calculations: Use this procedure for shafts and keys made of any metal (steel, bronze, brass, stainless steel, etc.). The dimensions of shaft keys can also be found in ANSI Standard B17f, Woodruff Keys, Keyslots and Cutters. Woodruff keys are used only for light-torque applications.

SELECTING A LEATHER BELT FOR POWER TRANSMISSION

Choose a leather belt to transmit 50 hp (37.3 kW) from a 1750-r/min squirrel-cage compensator-starting motor through a 12-in (30.5-cm) diameter pulley in an oily atmosphere. What belt width is needed with a 50-hp (37.3-kW) internal-combustion engine fitted with a 1750-r/min 12-in (30.5-cm) diameter pulley operating in an oily atmosphere?

Calculation Procedure:

1. Determine the belt speed

The speed of a belt S is found from $S = \pi RD$, where R = rpm of driving or driven pulley; D = diameter, ft, of driving or driven pulley. Thus, for this belt, $S = \pi(1750)(12/12) = 5500$ ft/min (27.9 m/s).

2. Determine the belt thickness needed

Use the National Industrial Leather Association recommendations. Enter Table 1 at the bottom at a belt speed of 5500 ft/min (27.9 m/s), i.e., between 4000 and 6000 ft/min (20.3 and 30.5 m/s); and project horizontally to the next smaller pulley diameter than that actually used. Thus, by entering at the line marked 4000–6000 ft/min (20.3–30.5 m/s) and projecting to the 10-in (25.4-cm) minimum diameter pulley, since a 12-in (30.5-cm) pulley is used, we see that a 23/64-in (0.91-cm) thick double-ply heavy belt should be used. Read the belt thickness and type at the top of the column in which the next smaller pulley diameter appears.

3. Determine the belt capacity factors

Enter the body of Table 1 at a belt speed of 5500 ft/min (27.9 m/s), i.e., between 4000 and 6000 ft/min (20.3 and 30.5 m/s); then project to the double-ply heavy column. Interpolating by eye gives a belt capacity factor of $K_c = 14.8$.

4. Determine the belt correction factors

Table 2 lists motor, pulley diameter, and operating correction factors, respectively. Thus, from Table 2, the motor correction factor $M = 1.5$ for a squirrel-cage compensator-starting motor.

TABLE 1   Leather-Belt Capacity Factors

Belt speed		Double ply	
		20/64 in (7.9 mm)	23/64 in (9.1 mm)
ft/min	m/s	Medium	Heavy
4000	20.3	10.9	12.6
5000	25.4	12.5	14.3
6000	30.5	13.2	15.2

Minimum pulley diameters

Up to 2500	Up to 12.7	5 in°	12.7 cm°	8 in°	20.3 cm°
2500–4000	12.7–20.3	6°	15.2°	9°	22.9°
4000–6000	20.3–30.5	7°	17.8°	10°	25.4°

°For belts 8 in (20.3 cm) and over, add 2 in (5.1 cm) to pulley diameter.

TABLE 2 Leather-Belt Correction Factors

	Correction factor
Characteristics or condition of motor and starter:	
Squirrel-cage, compensator-starting motor	$M = 1.5$
Squirrel-cage, line-starting	$M = 2.0$
Slip-ring, high starting torque	$M = 2.5$
Diameter of small pulley, in (cm):	
4 and under (10.2 and under)	$P = 0.5$
4.5 to 8 (11.4 to 20.3)	$P = 0.6$
9 to 12 (22.9 to 30.5)	$P = 0.7$
13 to 16 (33.0 to 40.6)	$P = 0.8$
17 to 30 (43.2 to 76.2)	$P = 0.9$
Over 30 (over 76.2)	$P = 1.0$
Operating conditions:	
Oily, wet, or dusty atmosphere	$F = 1.35$
Vertical drives	$F = 1.2$
Jerky loads	$F = 1.2$
Shock and reversing loads	$F = 1.4$

Also from Table 2, the smaller pulley diameter correction factor $P = 0.7$; and $F = 1.35$ for an oily atmosphere.

5. *Compute the required belt width*

The required belt width, in, is $W = hpMF/(K_cP)$, where $hp = $ horsepower transmitted by the belt; the other factors are as given above. For this belt, then, $W = (50)(1.5)(1.35)/[14.8(0.7)] = 9.7$ in (24.6 cm). Thus, a 10-in (25.4-cm) wide belt would be used because belts are commercially available in 1-in (2.5-cm) increments.

6. *Determine the belt width for the engine drive*

For a double-ply belt driven by a driver other than an electric motor, $W = 2750hp/dR$, where $d = $ driving pulley diameter, in; $R = $ driving pulley, r/min. Thus, $W = 2750(50)/[(12)(1750)] = 6.54$ in (16.6 cm). Hence, a 7-in (17.8-cm) wide belt would be used.

For a single-ply belt the above equation becomes $W = 1925\, hp/dR$.

    Related Calculations: Note that the relations in steps 1, 5, and 6 can be solved for any unknown variable when the other factors in the equations are known. Where the hp rating of a belt material is available from the manufacturer's catalog or other published data, find the required width from $W = hp_bF/K_cP$, where $hp_b = $ hp rating of the belt material, as stated by the manufacturer; other symbols as before. To find the tension T_b lb in a belt, solve $T_b = 33{,}000hp/S$ where $S = $ belt speed, ft/min. The tension per inch of belt width is $T_{bi} = T_b/W$. Where the belt speed exceeds 6000 ft/min (30.5 m/s), consult the manufacturer.

SELECTING A RUBBER BELT FOR POWER TRANSMISSION

Choose a rubber belt to transmit 15 hp (11.2 kW) from a 7-in (17.8-cm) diameter pulley driven by a shunt-wound dc motor. The pulley speed is 1300 r/min, and the belt drives an electric generator. The arrangement of the drive is such that the arc of contact of the belt on the pulley is 220°.

Calculation Procedure:

1. *Determine the belt service factor*

The belt *service factor* allows for the typical conditions met in the use of a belt with a given driver and driven machine or device. Table 3 lists typical service factors S_f used by the B. F.

TABLE 3   Service Factor S

Application	Squirrel-cage ac motor		Wound rotor ac motor (slip ring)	Single-phase capacitor motor	Dc shunt-wound motor	Diesel engine, four or more cylinders, above 700 r/min
	Normal torque, line start	High torque				
Agitators	1.0–1.2	1.2–1.4	1.2			
Compressors	1.2–1.4	. . . . . .	1.4	1.2	1.2	1.2
Belt conveyors (ore, coal, sand)	. . . . . .	1.4	. . . . . .	. .	1.2	
Screw conveyors	. . . . . .	1.8	. . . . . .	. .	1.6	
Crushing machinery	. . . . . .	1.6	1.6	. .	. . . . . .	1.4–1.6
Fans, centrifugal	1.2	. . . . . .	1.4	. .	1.4	1.4
Fans, propeller	1.4	2.0	1.6	. .	1.6	1.6
Generators and exciters	1.2	. . . . . .	. . . . . .	. .	1.2	2.0
Line shafts	1.4	. . . . . .	1.4	1.4	1.4	1.6
Machine tools	1.0–1.2	. . . . . .	1.2–1.4	1.0	1.0–1.2	
Pumps, centrifugal	1.2	1.4	1.4	1.2	1.2	
Pumps, reciprocating	1.2–1.4	. . . . . .	1.4–1.6	. .	. . . . . .	1.8–2.0

Goodrich Company. Entering Table 3 at the type of driver, a shunt-wound dc motor, and projecting downward to the driven machine, an electric generator, shows that $S_f = 1.2$.

2.   Determine the arc-of-contact factor

A rubber belt can contact a pulley in a range from about 140 to 220°. Since the hp capacity ratings for belts are based on an arc of contact of 180°, a correction factor must be applied for other arcs of contact.

Table 4 lists the arc-of-contact correction factor C_c. Thus, for an arc of contact of 220°, $C_c = 1.12$.

3.   Compute the belt speed

The belt speed is $S = \pi RD$, where S = belt speed, ft/min; R = pulley rpm; D = pulley diameter, ft. For this pulley, $S = \pi(1300)(7/12) = 2380$ ft/min (12.1 m/s).

4.   Choose the minimum pulley diameter and belt ply

Table 5 lists minimum recommended pulley diameters, belt material, and number of plies for various belt speeds. Choose the pulley diameter and number of plies for the next higher belt speed when the computed belt speed falls between two tabulated values. Thus, for a belt speed of 2380 ft/min (12.1 m/s), use a 7-in (17.8-cm) diameter pulley as listed under 2500 ft/min (12.7 m/s). The corresponding material specifications are found in the left-hand column and are four plies, 32-oz (0.9-kg) fabric.

5.   Determine the belt power rating

Enter Table 6 at 32 oz (0.9 kg) four-ply material specifications, and project horizontally to the belt speed. This occurs between the tabulated speeds of 2000 and 2500 ft/min (10.2 and 12.7 m/s). Interpolating, we find $[(2500 - 2380)/(2500 - 2000)](4.4 - 3.6) = 0.192$. Hence, the power rating of the belt hp_{bi} is $4.400 - 0.192 = 4.208$ hp/in (1.2 kW/cm) of width.

TABLE 4   Arc of Contact Factor K—Rubber Belts

Arc of contact, °	140	160	180	200	220
Factor K	0.82	0.93	1.00	1.06	1.12

TABLE 5   Minimum Pulley Diameters, in (cm)—Rubber Belts of 32-oz (0.9-kg) Hard Fabric

Ply	Belt speed, ft/min (m/s)			
	2000 (149.4)	2500 (186.4)	3000 (223.7)	4000 (298.3)
3	4 (10.2)	4 (10.2)	4 (10.2)	4 (10.2)
4	5 (12.7)	6 (15.2)	6 (15.2)	7 (17.8)
5	8 (20.3)	8 (20.3)	9 (22.9)	10 (25.4)
6	11 (27.9)	11 (27.9)	12 (30.5)	13 (33.0)
7	15 (38.1)	15 (38.1)	16 (40.6)	17 (43.2)
8	18 (45.7)	19 (48.3)	20 (50.8)	21 (53.3)
9	22 (55.9)	23 (58.4)	24 (61.0)	25 (63.5)
10	26 (66.0)	27 (68.6)	28 (71.1)	29 (73.7)

6. *Determine the required belt width*

The required belt width $W = hpS_f/(hp_{in}C_c)$, or $W = (15)(1.2)/[(4.208)(1.12)] = 3.82$ in (9.7 cm). Use a 4-in (10.2-cm) wide belt.

   Related Calculations: Use this procedure for rubber-belt drives of all types. For additional service factors, consult the engineering data published by B. F. Goodrich Company, The Goodyear Tire and Rubber Company, United States Rubber Company, etc.

SELECTING A V BELT FOR POWER TRANSMISSION

Choose a V belt to drive a 0.75-hp (559.3-W) stoker at about 900 r/min from a 1750-r/min motor. The stoker is fitted with a 3-in (7.6-cm) diameter sheave and the motor with a 6-in (15.2-cm) diameter sheave. The distance between the sheave shaft centerlines is 18 in (45.7 cm). The stoker handles soft coal free of hard lumps.

Calculation Procedure:

1. *Determine the design horsepower for the belt*

V-belt manufacturers publish service factors for belts used in various applications. Table 7 shows that a stoker is classed as heavy service and has a service factor of 1.4 to 1.6. By using the lower value, because the stoker handles soft coal free of hard lumps, the design horsepower for the belt

TABLE 6   Power Ratings of Rubber Belts [32-oz (0.9-kg) Hard Fabric]

(Hp = hp/in of belt width for 180° wrap)
(Power = kW/cm of belt width for 180° wrap)

Ply	Belt speed, ft/min (m/s)			
	2000 (10.2)	2500 (12.7)	3000 (15.2)	4000 (20.3)
3	2.9 (0.85)	3.5 (1.03)	4.1 (1.20)	5.1 (1.50)
4	3.9 (1.14)	4.7 (1.38)	5.5 (1.61)	6.8 (2.00)
5	4.9 (1.44)	5.9 (1.73)	6.9 (2.03)	8.5 (2.50)
6	5.9 (1.73)	7.1 (2.08)	8.3 (2.44)	10.2 (2.99)
7	6.9 (2.03)	8.3 (2.44)	9.7 (2.85)	11.9 (3.49)
8	7.9 (2.32)	9.5 (2.79)	11.1 (3.26)	13.6 (3.99)
9	8.9 (2.61)	10.6 (3.11)	12.4 (3.64)	15.3 (4.49)
10	9.8 (2.88)	11.7 (3.43)	13.7 (4.02)	17.0 (4.99)

TABLE 7 Service Factors for V-Belt Drives

Typical machines	Type of service	Service factors
Domestic washing machines, domestic ironers, advertising display fixtures, small fans and blowers	Light	1.0–1.2
Fans and blowers (heavy rotors), centrifugal pumps, oil burners, home workshop machines	Medium	1.2–1.4
Stokers, reciprocating pumps and compressors, refrigerators, drill presses, grinders, lathes, meat slicers, machines for industrial use	Heavy	1.4–1.6

is found by taking the product of the rated horsepower of the device driven by the belt and the service factor, or (0.75 hp)(1.4 service factor) = 1.05 hp (783.0 W). The belt must be capable of transmitting this, or a greater, horsepower.

2. Determine the belt speed and arc of contact

The belt speed $S = \pi RD$, where R = sheave rpm; D = sheave pitch diameter, ft = (sheave outside diameter, in $- 2X)/12$, where $2X$ = sheave dimension from Table 8. Before solving this equation, an assumption about the cross-sectional width of the belt must be made because $2X$ varies from 0.10 to 0.30 in (2.5 to 7.6 mm), and the exact cross section of the belt that will be used is not yet known. A value of $X = 0.15$ in (3.8 mm) is usually a safe assumption. It corresponds to a $3L$ belt cross section. Using $X = 0.15$ and the diameter and speed of the larger sheave, we see $S = \pi(1750)(6.0 - 0.15)/12 = 2675$ ft/min (13.6 m/s).

Compute the belt arc of contact from arc of contact, degrees = $180 - [60(d_1 - d_s)/l]$, where d_1 = large sheave nominal diameter, in; d_s = small sheave nominal diameter, in; l = distance between shaft centers, in. For this drive, arc = $180 - [60(6 - 3)/18] = 170°$. An arc-of-contact correction factor must be applied in computing the belt power capacity. Read this correction

TABLE 8 Sheave Dimensions—Light-Duty V Belt

Belt cross section	Sheave effective OD		Groove angle, deg	W		D		2X	
	in	cm		in	mm	in	mm	in	mm
2L	Under 1.5	Under 3.8	32	0.240	6.10	0.250	6.4	0.10	2.5
	1.5–1.99	3.8–5.05	34	0.243	6.17				
	2.0–2.5	5.08–6.4	36	0.246	6.25				
	Over 2.5	Over 6.4	38	0.250	6.35				
3L	Under 2.2	Under 5.6	32	0.360	9.14	0.406	10.3	0.15	3.8
	2.2–3.19	5.6–8.10	34	0.364	9.25				
	3.2–4.2	8.13–10.7	36	0.368	9.35				
	Over 4.20	Over 10.7	38	0.372	9.45				
4L	Under 2.65	Under 6.7	30	0.485	12.32	0.490	12.4	0.20	5.1
	2.65–3.24	6.7–8.23	32	0.490	12.45				
	3.25–5.65	8.26–14.4	34	0.494	12.55				
	Over 5.65	Over 14.4	38	0.504	12.80				

TABLE 9   Correction Factors for Arc of Contact—V-Belt Drives

Arc of contact, deg	Correction factor		Arc of contact, deg	Correction factor	
	V to V	V to flat°		V to V	V to flat°
180	1.00	0.75	130	0.86	0.86
170	0.98	0.77	120	0.82	0.82
160	0.95	0.80	110	0.78	0.78
150	0.92	0.82	100	0.74	0.74
140	0.89	0.84	90	0.69	0.69

° A V-to-flat drive has a small sheave and a larger-diameter flat pulley.

factor from Table 9 as $C_c = 0.98$ for a V-sheave to V-sheave drive and a 170° arc of contact. *Note:* If desired, the pitch diameters can be used in the above relation in place of the nominal diameters.

3.   *Select the belt to be used*

The $2X$ value used in step 3 corresponds to a $3L$ cross section belt. Check the power capacity of this belt by entering Table 10 at a belt speed of 2800 ft/min (14.2 m/s), the next larger tabulated speed, and projecting across to the appropriate small-sheave diameter—3 in (7.6 cm) or larger. Read the belt horsepower rating as 0.87 hp (648.8 W). This is considerably less than the required capacity of 1.05 hp (783.0 W) computed in step 1. Therefore, the $3L$ belt is unsatisfactory.

Try a $4L$ belt, Table 11, following the same procedure. A $4L$ belt with a 3-in (7.6-cm) diameter small sheave has a rating of 1.16 hp (865.0 W). Correct this for the actual arc of contact by multiplying by C_c, or (1.16)(0.98) = 1.137 (847.9 W). Thus, the belt is suitable for the design hp value of 1.05 hp (783.0 W).

As a final check, compute the actual belt speed using the actual $2X$ value from Table 8. Thus, for a $4L$ belt on a 6-in (15.2-cm) sheave, $2X = 0.20$, and $S = \pi(1750)(6 - 0.20)/12 = 2660$ ft/ min (13.5 m/s). Hence, use of 2800 ft/min (14.2 m/s) in selecting the belt was a safe assumption. Note that the difference between the belt speed based on the assumed value of $2X$, 2675 ft/min (13.6 m/s), and the actual belt speed, 2660 ft/min (13.5 m/s), is about 0.5 percent. This is negligible.

Related Calculations: Use this procedure when choosing a single V belt for a drive. Where multiple belts are used, follow the steps given in the next calculation procedure. The data presented for single V belts is abstracted from *Standards for Light-duty or Fractional-horsepower V-Belts*, published by the Rubber Manufacturers Association.

TABLE 10   Power Ratings of $3L$ Cross Section V Belts

(Based on 180° arc of contact on small sheave)

Belt speed		Effective OD of small sheave							
		1½ in (3.8 cm)		2 in (5.1 cm)		2½ in (6.4 cm)		3 in (7.6 cm) or more	
ft/min	m/s	hp	W	hp	W	hp	W	hp	W
2200	11.2	0.12	89.5	0.44	328.1	0.64	477.2	0.77	574.2
2400	12.2	0.10	74.6	0.45	335.6	0.66	492.2	0.81	604.0
2600	13.2	0.07	52.2	0.46	343.0	0.69	514.5	0.84	626.4
2800	14.2	0.04	29.8	0.46	343.0	0.70	522.0	0.87	648.8
3000	15.2	0.01	7.5	0.45	335.6	0.72	536.9	0.89	663.7

TABLE 11 Power Ratings of 4L Cross Section V Belts

(Based on 180° arc of contact on small sheave)

Belt speed		Effective OD of small sheave							
		2½ in (6.4 cm)		3 in (7.6 cm)		3½ in (8.9 cm)		4 in (10.2 cm) or more	
ft/min	m/s	hp	W	hp	W	hp	W	hp	W
2200	11.2	0.67	499.6	1.08	805.4	1.37	1021.6	1.58	1178.2
2400	12.2	0.68	507.1	1.12	835.2	1.43	1066.4	1.66	1237.9
2600	13.2	0.66	492.2	1.16	865.0	1.50	1118.5	1.75	1305.0
2800	14.2	0.65	484.7	1.18	879.9	1.54	1148.4	1.81	1349.7
3000	15.2	0.63	469.8	1.19	887.4	1.58	1178.2	1.87	1394.5

SELECTING MULTIPLE V BELTS FOR POWER TRANSMISSION

Choose the type and number of V belts needed to drive an air compressor from a 5-hp (3.7-kW) wound-rotor ac motor when the motor speed is 1800 r/min and the compressor speed is 600 r/min. The pitch diameter of the large sheave is 20 in (50.8 cm); and the distance between shaft centers is 36.0 in (91.4 cm).

Calculation Procedure:

1. *Choose the V-belt section*

Determine the design horsepower of the drive by finding the product of the service factor and the rated horsepower. Use Table 3 to find the service factor. The value of this factor is 1.4 for a compressor driven by a wound-rotor ac motor. Thus, the design horsepower = (5.0 hp)(1.4 service factor) = 7.0 hp (5.2 kW).

Enter Fig. 3 at 7.0 hp (5.2 kW), and project up to the small sheave speed, 600 r/min. Read the belt cross section as type B.

2. *Determine the small-sheave pitch diameter*

Use the speed ratio of the shafts to determine the diameter of the small sheave. The speed ratio of the shafts is the ratio of the speed of the high-speed shaft to that of the low-speed shaft, or $1800/600 = 3.0$. The sheave pitch diameters have the same ratio, or $20/PD_s = 3$; $PD_s = 20/3 = 6.67$ in (16.9 cm).

3. *Compute the belt speed*

The belt speed is $S = \pi RD$, where R = small-sheave rpm; D = small-sheave pitch diameter, ft. Thus, $S = \pi(1800)(6.67/12) = 3140$ ft/min (16.0 m/s).

4. *Determine the belt horsepower rating*

A tabulation of allowable belt horsepower ratings is used to determine the rating of a specific belt. To enter this table, the belt speed and the small-sheave equivalent diameter must be known.

Find the equivalent diameter d_e of the small sheave by taking the product of the small-sheave pitch diameter and the diameter factor, Table 12. Thus, for a speed range of 3.0 the small-diameter factor = 1.14, from Table 12. Hence, $d_e = (6.67)(1.14) = 7.6$ in (19.3 cm).

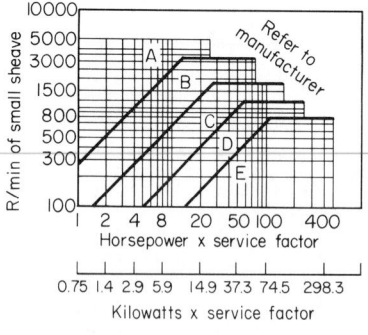

FIG. 3 V-belt cross section for required hp rating.

Enter Table 12A at a belt speed of 3200 ft/ min (16.3 m/s) and d_e = 7.6 in (19.3 cm). In the last column read the belt horsepower rating as 6.5 hp (4.85 kW). This rating must be corrected for the arc of contact and the belt length.

The arc of contact = $180 - [60(d_l - d_s)/l]$, where d_l and d_s = large- and small-sheave pitch diameters, respectively, in; l = distance between sheave shaft centers, in. Thus, arc of contact = $180 - [60(20 - 6.67)/36]$ = 157.8°. Using Table 4 and interpolating, we find the arc-of-contact correction factor C_c = 0.94.

Compute the belt pitch length from $L = 2l + 1.57(d_l + d_s) + (d_l - d_s)^2/(4l)$, where all the symbols are as given earlier. Thus, $L = 2(36) + 1.57(20 + 6.67) + (20 - 6.67)^2/[4(36)]$ = 115.1 in (292.4 cm).

Enter Table 13 by interpolating between the standard belt lengths of 105 and 120 in (266.7 and 304.8 cm), and find the length correction factor for a B cross section belt as 1.06.

TABLE 12 Small-Diameter Factors—Multiple V Belts

Speed ratio range	Small-diameter factor
1.000–1.019	1.00
1.020–1.032	1.01
1.033–1.055	1.02
1.056–1.081	1.03
1.082–1.109	1.04
1.110–1.142	1.05
1.143–1.178	1.06
1.179–1.222	1.07
1.223–1.274	1.08
1.275–1.340	1.09
1.341–1.429	1.10
1.430–1.562	1.11
1.563–1.814	1.12
1.815–2.948	1.13
2.949 and over	1.14

Find the product of the rated horsepower of the belt and the two correction factors—arc of contact and belt length, or (6.5)(0.94)(1.06) = 6.47 hp (4.82 kW).

5. *Choose the number of belts required*

The design horsepower, step 1, is 7.0 hp (5.2 kW). Thus, 7.0 design hp/6.47 rated belt hp = 1.08 belts; use two belts. Choose the next *larger* number of belts whenever a fractional number is indicated.

Related Calculations: The tables and data used here are based on engineering information which is available from and updated by the Mechanical Power Transmission Association and the Rubber Manufacturers Association. Similar engineering data are published by the various V-belt manufacturers. Data presented here may be used when manufacturer's engineering data are not available.

SELECTION OF A WIRE-ROPE DRIVE

Choose a wire-rope drive for a 3000-lb (1360.8-kg) traction-type freight elevator designed to lift freight or passengers totaling 4000 lb (1814.4 kg). The vertical lift of the elevator is 500 ft (152.4 m), and the rope velocity is 750 ft/min (3.8 m/s). The traction-type elevator sheaves are designed

TABLE 12A Power Ratings for Premium-Quality B-Section V Belts

Belt speed		6.6 in (16.8 cm)		7.0+ in (17.8+ cm)	
ft/min	m/s	hp	kW	hp	kW
3000	15.2	5.90	4.40	6.26	4.67
3200	16.3	6.12	4.56	6.50	4.85
3400	17.3	6.31	4.71	6.73	5.02

Equivalent diameter d_e

TABLE 13 Length Correction Factors—Multiple V Belts

Standard length designation		Belt cross section			
in	cm	A	B	C	D
96	243.8	1.08	. . .	0.92	
105	266.7	1.10	1.04	0.94	
120	304.8	1.13	1.07	0.97	0.86
136	345.4	. . .	1.09	0.99	

to accelerate the car to full speed in 60 ft (18.3 m) when it starts from a stopped position. A 48-in (1.2-m) diameter sheave is used for the elevator.

Calculation Procedure:

1. Select the number of hoisting ropes to use

The number of ropes required for an elevator is usually fixed by state or city laws. Check the local ordinances before choosing the number of ropes. Usual laws require at least four ropes for a freight elevator. Assume four ropes are used for this elevator.

2. Select the rope size and strength

Standard "blue-center" steel hoisting rope is a popular choice, as is "plow-steel" and "mild plow-steel" rope. Assume that four $\%_6$-in (14.3-mm) six-strand 19-wires-per-strand blue-center steel ropes will be suitable for this car. The 6 × 19 rope is commonly used for freight and passenger elevators. Once the rope size is chosen, its strength can be checked against the actual load. The breaking strength of $\%_6$ in (14.3 mm), 6 × 19 blue-center steel rope is 13.5 tons (12.2 t), and its weight is 0.51 lb/ft (0.76 kg/m). These values are tabulated in Baumeister and Marks—*Standard Handbook for Mechanical Engineers* and in rope manufacturers' engineering data.

3. Compute the total load on each rope

The weight of the car and its contents is 3000 + 4000 = 7000 lb (3175.1 kg). With four ropes, the load per rope is 7000/[4(2000 lb·ton)] = 0.875 ton (0.794 t).

With a 500-ft (152.4-m) lift, the length of each rope would be equal to the lift height. Hence, with a rope weight of 0.51 lb/ft (0.76 kg/m), the total weight of the rope = (0.51)(500)/2000 = 0.127 ton (0.115 t).

Acceleration of the car from the stopped condition places an extra load on the rope. The rate of acceleration of the car is found from $a = v^2/(2d)$, where a = car acceleration, ft/s²; v = final velocity of the car, ft/s; d = distance through which the acceleration occurs, ft. For this car, a = $(750/60)^2/[2(60)]$ = 1.3 ft/s² (39.6 cm/s²). The value 60 in the numerator of the above relation converts from feet per minute to feet per second.

The rope load caused by acceleration of the car is $L_r = Wa/$(number of ropes)(2000 lb/ton) [$g = 32.2$ ft/s² (9.8 m/s²)], where L_r = rope load, tons; W = weight of car and load, lb. Thus, $L_r = (7000)(1.3)/[(4)(2000)(32.2)]$ = 0.03351 ton (0.03040 t) per rope.

The rope load caused by acceleration of the rope is $L_r = Wa/32.2$, where W = weight of rope, tons. Or, $L_r = (0.127)(1.3)/32.2$ = 0.0512 ton (0.0464 t).

When the rope bends around the sheave, another load is produced. This bending load is, in pounds, $F_b = AE_r d_w/d_s$, where A = rope area, in²; E_r = modulus of elasticity of the whole rope = 12 × 10⁶ lb/in² (82.7 × 10⁶ kPa) for steel rope; d_w = rope diameter, in; d_s = sheave diameter, in. Thus, for this rope, $F_b = (0.0338)(12 × 10^6)(0.120/48)$ = 1014 lb, or 0.507 ton (0.460 t).

The total load on the rope is the sum of the individual loads, or 0.875 + 0.127 + 0.0351 + 0.507 + 0.051 = 1.545 tons (1.4 t). Since the rope has a breaking strength of 13.5 tons (12.2 t), the factor of safety FS = breaking strength, tons/rope load, tons = 13.5/1.545 = 8.74. The usual minimum acceptable FS for elevator ropes is 8.0. Hence, this rope is satisfactory.

Related Calculations: Use this general procedure when choosing wire-rope drivers for mine hoists, inclined-shaft hoists, cranes, derricks, car pullers, dredges, well drilling, etc. When standard hoisting rope is chosen, which is the type most commonly used, the sheave diameter should not be less than $30d_w$; the recommended diameter is $45d_w$. For *haulage rope* use $42d_w$ and $72d_w$, respectively; for special flexible *hoisting rope*, use $18d_w$ and $27d_w$ sheaves.

SPEEDS OF GEARS AND GEAR TRAINS

A gear having 60 teeth is driven by a 12-tooth gear turning at 800 r/min. What is the speed of the driven gear? What would be the speed of the driven gear if a 24-tooth idler gear were placed between the driving and driven gear? What would be the speed of the driven gear if two 24-tooth idlers were used? What is the direction of rotation of the driven gear when one and two idlers are used? A 24-tooth driving gear turning at 600 r/min meshes with a 48-tooth compound gear. The second gear of the compound gear has 72 teeth and drives a 96-tooth gear. What are the speed and direction of rotation of the 96-tooth gear?

Calculation Procedure:

1. *Compute the speed of the driven gear*

For any two meshing gears, the speed ratio $R_D/R_d = N_d/N_D$, where R_D = rpm of driving gear; R_d = rpm of driven gear; N_d = number of teeth in driven gear; N_D = number of teeth in driving gear. By substituting the given values, $R_D/R_d = N_d/N_D$, or $800/R_d = 60/12$; $R_d = 160$ r/min.

2. *Determine the effect of one idler gear*

An idler gear has *no* effect on the speed of the driving or driven gear. Thus, the speed of each gear would remain the same, regardless of the number of teeth in the idler gear. An idler gear is generally used to reduce the required diameter of the driving and driven gears on two widely separated shafts.

3. *Determine the effect of two idler gears*

The effect of more than one idler is the same as that of a single idler—i.e., the speed of the driving and driven gears remains the same, regardless of the number of idlers used.

4. *Determine the direction of rotation of the gears*

Where an odd number of gears are used in a gear train, the first and last gears turn in the *same* direction. Thus, with one idler, one driver, and one driven gear, the driver and driven gear turn in the *same* direction because there are three gears (i.e., an odd number) in the gear train.

Where an even number of gears is used in a gear train, the first and last gears turn in the *opposite* direction. Thus, with two idlers, one driver, and one driven gear, the driver and driven gear turn in the *opposite* direction because there are four gears (i.e., an even number) in the gear train.

5. *Determine the compound-gear output speed*

A compound gear has two gears keyed to the same shaft. One of the gears is driven by another gear; the second gear of the compound set drives another gear. In a compound gear train, the product of the number of teeth of the driving gears and the rpm of the first driver equals the product of the number of teeth of the driven gears and the rpm of the last driven gear.

In this gearset, the first driver has 24 teeth and the second driver has 72 teeth. The rpm of the first driver is 600. The driven gears have 48 and 96 teeth, respectively. Speed of the final gear is unknown. Applying the above rule gives $(24)(72)(600)2(48)(96)(R_d)$; $R_d = 215$ r/min.

Apply the rule in step 4 to determine the direction of rotation of the final gear. Since the gearset has an even number of gears, four, the final gear revolves in the opposite direction from the first driving gear.

Related Calculations: Use the general procedure given here for gears and gear trains having spur, bevel, helical, spiral, worm, or hypoid gears. Be certain to determine the correct number of teeth and the gear rpm before substituting values in the given equations.

SELECTION OF GEAR SIZE AND TYPE

Select the type and size of gears to use for a 100-ft^3/min (0.047-m^3/s) reciprocating air compressor driven by a 50-hp (37.3-kW) electric motor. The compressor and motor shafts are on parallel axes 21 in (53.3 cm) apart. The motor shaft turns at 1800 r/min while the compressor shaft turns at 300 r/min. Is the distance between the shafts sufficient for the gears chosen?

Calculation Procedure:

1. *Choose the type of gears to use*

Table 14 lists the kinds of gears in common use for shafts having parallel, intersecting, and non-intersecting axes. Thus, Table 14 shows that for shafts having parallel axes, spur or helical, external or internal, gears are commonly chosen. Since external gears are simpler to apply than internal gears, the external type is chosen wherever possible. Internal gears are the planetary type and are popular for applications where limited space is available. Space is not a consideration in this application; hence, an external spur gearset will be used.

Table 15 lists factors to consider in selecting gears by the characteristics of the application. As

TABLE 14 Types of Gears in Common Use°

Parallel axes	Intersecting axes	Nonintersecting parallel axes
Spur, external	Straight bevel	Crossed helical
Spur, internal	Zerol† bevel	Single-enveloping worm
Helical, external	Spiral bevel	Double-enveloping worm
Helical, internal	Face gear	Hypoid

°From Darle W. Dudley—*Practical Gear Design*, McGraw-Hill, 1954.
†Registered trademark of the Gleason Works.

TABLE 15 Gear Drive Selection by Application Characteristics°

Characteristic	Type of gearbox	Kind of teeth	Range of use
High power	Simple, branched, or epicyclic	Helical	Up to 40,000 hp (29,828 kW) per single mesh; over 60,000 hp (44,742 kW) in MDT designs; up to 40,000 hp (29,282 kW) in epicyclic units
	Simple, branched, or epicyclic	Spur	Up to 4000 hp (2983 kW) per single mesh; up to 10,000 hp (7457 kW) in an epicyclic
	Simple	Spiral bevel	Up to 15,000 hp (11,186 kW) per single mesh
		Zerol bevel	Up to 1000 hp (745.7 kW) per single mesh
High efficiency	Simple	Spur, helical or bevel	Over 99 percent efficiency in the most favorable cases—98 percent efficiency is typical
Light weight	Epicyclic	Spur or helical	Outstanding in airplane and helicopter drives
	Branched-MDT	Helical	Very good in marine main reductions
	Differential	Spur or helical	Outstanding in high-torque-actuating devices
		Bevel	Automobiles, trucks, and instruments
Compact	Epicyclic	Spur or helical	Good in aircraft nacelles
	Simple	Worm-gear	Good in high-ratio industrial speed reducers
	Simple	Spiroid	Good in tools and other applications
	Simple	Hypoid	Good in auto and truck rear ends plus other applications
	Simple	Worm-gear	Widely used in machine-tool index drives
	Simple	Hypoid	Used in certain index drives for machine tools
Precision	Simple or branched	Helical	A favorite for high-speed, high-accuracy power gears
	Simple	Spur	Widely used in radar pedestal gearing, gun control drives, navigation instruments, and many other applications
	Simple	Spiroid	Used where precision and adjustable backlash are needed

°*Mechanical Engineering*, November 1965.

3.26

TABLE 16   Gear Drive Selection for the Convenience of the User°

Consideration	Kind of teeth	Typical applications	Comments
Cost	Spur	Toys, clocks, instruments, industrial drives, machine tools, transmissions, military equipment, household applications, rocket boosters	Very widely used in all manner of applications where power and speed requirements are not too great—parts are often mass-produced at very low cost per part
Ease of use	Spur or helical	Change gears in machine tools, vehicle transmissions where gear shifting occurs	Ease of changing gears to change ratio is important
	Worm-gear	Speed reducers	High ratio drive obtained with only two gear parts
Simplicity	Crossed helical	Light power drives	No critical positioning required in a right-angle drive
	Face gear	Small power drives	Simple and easy to position for a right-angle drive
	Helical	Marine main drive units for ships, generator drives in power plants	Helical teeth with good accuracy and a design that provides good axial overlap mesh very smoothly
	Spiral bevel	Main drive units for aircraft, ships, and many other applications	Helical type of tooth action in a right-angle power drive
Noise	Hypoid	Automotive rear axle	Helical type of tooth provides high overlap
	Worm-gear	Small power drives in marine, industrial, and household appliance applications	Overlapping, multiple tooth contacts
	Spiroid-Gear	Portable tools, home appliances	Overlapping, multiple tooth contacts

° *Mechanical Engineering*, November 1965.

with Table 14, the data in Table 15 indicate that spur gears are suitable for this drive. Table 16, based on the convenience of the user, also indicates that spur gears are suitable.

2. *Compute the pitch diameter of each gear*

The distance between the driving and driven shafts is 21 in (53.3 cm). This distance is approximately equal to the sum of the driving gear pitch radius r_D in and the driven gear pitch radius r_d in. Or $d_D + r_d = 21$ in (53.3 cm).

In this installation the driving gear is mounted on the motor shaft and turns at 1800 r/min. The driven gear is mounted on the compressor shaft and turns at 300 r/min. Thus, the speed ratio of the gears (R_D, driver rpm/R_d, driven rpm) $= 1800/300 = 6$. For a spur gear, $R_D/R_d = r_d/r_D$, or $6 = r_d/r_D$, and $r_d = 6r_D$. Hence, substituting in $r_D + r_d = 21$, $r_D + 6r_D = 21$; $r_D = 3$ in (7.6 cm). Then $3 + r_d = 21$, $r_d = 18$ in (45.7 cm). The respective pitch diameters of the gears are $d_D = 2 \times 3 = 6.0$ in (15.2 cm); $d_d = 2 \times 18 = 36.0$ in (91.4 cm).

3. *Determine the number of teeth in each gear*

The number of teeth in a spur gearset, N_D and N_d, can be approximated from the ratio $R_D/R_d = N_d/N_D$, or $1800/300 = N_d/N_D$; $N_d = 6N_D$. Hence, the driven gear will have approximately six times as many teeth as the driving gear.

As a trial, assume that $N_d = 72$ teeth; then $N_D = N_d/6 = 72/6 = 12$ teeth. This assumption must now be checked to determine whether the gears will give the desired output speed. Since

TABLE 17 Gear Drive Selection by Arrangement of Driving and Driven Equipment[*]

Kind of teeth	Axes	Gearbox type	Type of tooth contact	Generic family
Spur[†]	Parallel	Simple (pinion and gear), epicyclic (planetary, star, solar), branched systems, idler for reverse	Line	Coplanar
Helical[†] (single or double helical, herringbone)	Parallel	Simple, epicyclic, branched	Overlapping line	Coplanar
Bevel	Right-angle or angular, but intersecting	Simple, epicyclic, branched	(Straight) line, (Zerol)[‡] line, (spiral) overlapping	Coplanar
Worm	Right-angle, nonintersecting	Simple	(Cylindrical) overlapping line, (double-enveloping)[§] overlapping line	Nonplanar
Crossed helical	Right-angle or skew, nonintersecting	Simple	Point	Nonplanar
Face gear	Right-angle, intersecting	Simple	Line (overlapping if helical)	Coplanar
Hypoid	Right-angle or angular, nonintersecting	Simple	Overlapping line	Nonplanar
Spiroid, helicon, planoid	Right-angle, nonintersecting	Simple	Overlapping line	Nonplanar

[*]*Mechanical Engineering*, November 1965.
[†]These kinds of teeth are often used to change rotary motion to linear motion by use of a pinion and rack.
[‡]Zerol is a registered trademark of the Gleason Works, Rochester, New York.
[§]The most widely used double-enveloping worm gear is the cone-drive type.

$R_D/R_d = N_d/N_D$, or $1800/300 = 72/12$; $6 = 6$. Thus, the gears will provide the desired speed change.

The distance between the shafts is 21 in (53.3 cm) $= r_D + r_d$. This means that there is no clearance when the gears are meshed. Since all gears require some clearance, the shafts will have to be moved apart slightly to provide this clearance. If the shafts cannot be moved apart, the gear diameter must be reduced. In this installation, however, the electric-motor driver can probably be moved a fraction of an inch to provide the desired clearance.

4. *Choose the final gear size*

Refer to a catalog of stock gears. From this catalog choose a driving and a driven gear having the required number of teeth and the required pitch diameter. If gears of the exact size required are not available, pick the nearest suitable stock sizes.

Check the speed ratio, using the procedure in step 3. As a general rule, stock gears having a slightly different number of teeth or a somewhat smaller or larger pitch diameter will provide nearly the desired speed ratio. When suitable stock gears are not available in one catalog, refer to one or more other catalogs. If suitable stock gears are still not available, and if the speed ratio is a critical factor in the selection of the gear, custom-sized gears may have to be manufactured.

Related Calculations: Use this general procedure to choose gear drives employing any of the 12 types of gears listed in Table 14. Table 17 lists typical gear selections based on the arrangement of the driving and driven equipment. These tables are the work of Darle W. Dudley.

GEAR SELECTION FOR LIGHT LOADS

Detail a generalized gear-selection procedure useful for spur, rack, spiral miter, miter, bevel, helical, and worm gears. Assume that the drive horsepower and speed ratio are known.

Calculation Procedure:

1. *Choose the type of gear to use*

Use Table 14 of the previous calculation procedure as a general guide to the type of gear to use. Make a tentative choice of the gear type.

2. *Select the pitch diameter of the pinion and gear*

Compute the pitch diameter of the pinion from $d_p = 2c/(R + 2)$, where $d_p =$ pitch diameter, in, of the pinion, which is the *smaller* of the two gears in mesh; $c =$ center distance between the gear shafts, in; $R =$ gear ratio = larger rpm, number of teeth, or pitch diameter + smaller rpm, smaller number of teeth, or smaller pitch diameter.

Compute the pitch diameter of the gear, which is the *larger* of the two gears in mesh, from $d_g = d_pR$.

3. *Determine the diametral pitch of the drive*

Tables 18 to 21 show typical diametral pitches for various horsepower ratings and gear materials. Enter the appropriate table at the horsepower that will be transmitted, and select the diametral pitch of the pinion.

4. *Choose the gears to use*

Enter a manufacturer's engineering tabulation of gear properties, and select the pinion and gear for the horsepower and rpm of the drive. Note that the rated horsepower of the pinion and the gear must equal, or exceed, the rated horsepower of the drive at this specified input and output rpm.

5. *Compute the actual center distance*

Find half the sum of the pitch diameter of the pinion and the pitch diameter of the gear. This is the actual center-to-center distance of the drive. Compare this value with the available space. If the actual center distance exceeds the allowable distance, try to rearrange the drive or select another type of gear and pinion.

TABLE 18 Spur-Gear Pitch Selection Guide[*]

(20° pressure angle)

Gear diametral pitch		Pinion		Gear	
in	cm	hp	W	hp	W
20	50.8	0.04–1.69	29.8–1,260	0.13–0.96	96.9–715.9
16	40.6	0.09–2.46	67.1–1,834	0.22–1.61	164.1–1,200
12	30.5	0.24–5.04	179.0–3,758	0.43–3.16	320.8–2,356
10	25.4	0.46–6.92	343.0–5,160	0.70–5.12	522.0–3,818
8	20.3	0.88–10.69	656.2–7,972	1.11–7.87	827.7–5,869
6	15.2	1.84–16.63	1,372–12,401	2.28–12.39	1,700–9,239
5	12.7	3.04–24.15	2,267–18,009	3.75–17.19	2,796–12,819
4	10.2	5.29–34.83	3,945–25,973	6.36–25.17	4,743–18,769
3	7.6	13.57–70.46	10,119–52,542	15.86–51.91	11,831–38,709

[*] Morse Chain Company.

TABLE 19 Miter and Bevel-Gear Pitch Selection Guide[*]

(20° pressure angle)

Gear diametral pitch		Hardened gear		Unhardened gear	
in	cm	hp	kW	hp	kW
Steel spiral miter					
18	45.7	0.07–0.70	0.053–0.522	0.04–0.42	0.030–0.313
12	30.5	0.15–1.96	0.112–1.462	0.09–1.17	0.067–0.873
10	25.4	0.50–4.53	0.373–3.378	0.30–2.70	0.224–2.013
8	20.3	1.56–7.15	1.163–5.331	0.93–4.26	0.694–3.177
7	17.8	1.93–9.30	1.439–6.935	1.15–5.54	0.858–4.131
Steel miter					
20	50.8	. . . .	. . . .	0.01–0.12	0.008–0.090
16	40.6	0.07–0.73	0.053–0.544	0.02–0.72	0.015–0.537
14	35.6	. . . .	. . . .	0.04–0.37	0.030–0.276
12	30.5	0.14–2.96	0.104–2.207	0.07–1.77	0.052–1.320
10	25.4	0.39–3.47	0.291–2.588	0.23–2.07	0.172–1.544

Steel and cast-iron bevel gears

Ratio	hp	W
1.5:1	0.04–2.34	29.8–1744.9
2:1	0.01–12.09	7.5–9015.5
3:1	0.04–8.32	29.8–6204.2
4:1	0.05–10.60	37.3–7904.4
6:1	0.07–2.16	52.2–1610.7

[*] Morse Chain Company.

6. *Check the drive speed ratio*

Find the actual speed ratio by dividing the number of teeth in the gear by the number of teeth in the pinion. Compare the actual ratio with the desired ratio. If there is a major difference, change the number of teeth in the pinion or gear or both.

Related Calculations: Use this general procedure to select gear drives for loads up to the ratings shown in the accompanying tables. For larger loads, use the procedures given elsewhere in this section.

SELECTION OF GEAR DIMENSIONS

A mild-steel 20-tooth 20° full-depth-type spur-gear pinion turning at 900 r/min must transmit 50 hp (37.3 kW) to a 300-r/min mild-steel gear. Select the number of gear teeth, diametral pitch of the gear, width of the gear face, the distance between the shaft centers, and the dimensions of the gear teeth. The allowable stress in the gear teeth is 800 lb/in² (55,160 kPa).

Calculation Procedure:

1. *Compute the number of teeth on the gear*

For any gearset, $R_D/R_d = N_d/N_D$, where R_D = rpm of driver; R_d = rpm of driven gear; N_d = number of teeth on the driven gear; N_D = number of teeth on driving gear. Thus, $900/300 = N_d/20$; $N_d = 60$ teeth.

2. *Compute the diametral pitch of the gear*

The diametral pitch of the gear must be the same as the diametral pitch of the pinion if the gears are to run together. If the diametral pitch of the pinion is known, assume that the diametral pitch of the gear equals that of the pinion.

When the diametral pitch of the pinion is not known, use a modification of the Lewis formula, shown in the next calculation procedure, to compute the diametral pitch. Thus, $P = (\pi\, SaYv/33,000\ hp)^{0.5}$, where all the symbols are as in the next calculation procedure, except that $a = 4$ for machined gears. Obtain $Y = 0.421$ for 60 teeth in a 20° full-depth gear from Baumeister and

TABLE 20   Helical-Gear Pitch Selection Guide[*]

Gear diametral pitch		Hardened-steel gear	
in	cm	hp	W
20	50.8	0.04–1.80	29.8–1,342.3
16	40.6	0.08–2.97	59.7–2,214.7
12	30.5	0.22–5.87	164.1–4,377.3
10	25.4	0.37–8.29	275.9–6,181.9
8	20.3	†0.66–11.71	492.2–8,732.1
		‡0.49–9.07	365.4–6,763.5
6	15.2	§1.44–19.15	1,073.8–14,280.2
		†1.15–15.91	857.6–11,864.1

[*] Morse Chain Company.
† 1-in (2.5-cm) face.
‡ ¾-in (1.9-cm) face.
§ 1½-in (3.8-cm) face.

TABLE 21   Worm-Gear Pitch Selection Guide[*]

Gear diametral pitch		Bronze gears					
		Single		Double		Quadruple	
in	cm	hp	W	hp	W	hp	W
12	30.5	0.04–0.64	29.8–477.2	0.05–1.21	37.3–902.3	0.05–3.11	37.3–2319
10	25.4	0.06–0.97	44.7–723.3	0.08–2.49	59.7–1856	0.13–4.73	96.9–3527
8	20.3	0.11–1.51	82.0–1126	0.15–3.95	111.9–2946	0.08–7.69	59.7–5734
				Triple			
5	12.7	0.51–4.61	380.3–3437	1.10–10.53	820.3–7852		
4	10.2	0.66–6.74	492.2–5026				

[*] Morse Chain Company.

Marks—*Standard Handbook for Mechanical Engineers.* Assume that v = pitch-line velocity = 1200 ft/min (6.1 m/s). This is a typical reasonable value for v. Then $P = [\pi \times 8000 \times 4 \times 0.421 \times 1200/(33,000 \times 50)]^{0.5} = 5.56$, say 6, because diametral pitch is expressed as a whole number whenever possible.

3. Compute the gear face width

Spur gears often have a face width equal to about four times the circular pitch of the gear. Circular pitch $p_c = \pi/P = \pi/6 = 0.524$. Hence, the face width of the gear = $4 \times 0.524 = 2.095$ in, say 2⅛ in (5.4 cm).

4. Determine the distance between the shaft centers

Find the exact shaft centerline distance from $d_c = (N_p + N_g)/2P$), where N_p = number of teeth on pinion gear; N_g = number of teeth on gear. Thus, $d_c = (20 + 60)/[2(6)] = 6.66$ in (16.9 cm).

5. Compute the dimensions of the gear teeth

Use AGMA *Standards*, Dudley—*Gear Handbook*, or the engineering tables published by gear manufacturers. Each of these sources provides a list of factors by which either the circular or diametral pitch can be multiplied to obtain the various dimensions of the teeth in a gear or pinion. Thus, for a 20° full-depth spur gear, using the circular pitch of 0.524 computed in step 3, we have the following:

		Factor		Circular pitch		Dimension, in (mm)
Addendum	=	0.3183	×	0.524	=	0.1668 (4.2)
Dedendum	=	0.3683	×	0.524	=	0.1930 (4.9)
Working depth	=	0.6366	×	0.524	=	0.3336 (8.5)
Whole depth	=	0.6866	×	0.524	=	0.3598 (9.1)
Clearance	=	0.05	×	0.524	=	0.0262 (0.67)
Tooth thickness	=	0.50	×	0.524	=	0.262 (6.7)
Width of space	=	0.52	×	0.524	=	0.2725 (6.9)
Backlash = width of space − tooth thickness	=			0.2725 − 0.262	=	0.0105 (0.27)

The dimensions of the pinion teeth are the same as those of the gear teeth.

Related Calculations: Use this general procedure to select the dimensions of helical, herringbone, spiral, and worm gears. Refer to the AGMA *Standards* for suitable factors and typical allowable working stresses for each type of gear and gear material.

HORSEPOWER RATING OF GEARS

What are the strength horsepower rating, durability horsepower rating, and service horsepower rating of a 600-r/min 36-tooth 1.75-in (4.4-cm) face-width 14.5° full-depth 6-in (15.2-cm) pitch-diameter pinion driving a 150-tooth 1.75-in (4.4-cm) face width 14.5° full-depth 25-in (63.5-cm) pitch-diameter gear if the pinion is made of SAE 1040 steel 245 BHN and the gear is made of cast steel 0.35/0.45 carbon 210 BHN when the gearset operates under intermittent heavy shock loads for 3 h/day under fair lubrication conditions? The pinion is driven by an electric motor.

Calculation Procedure:

1. Compute the strength horsepower, using the Lewis formula

The widely used Lewis formula gives the strength horsepower, $hp_s = SYFK_vv/(33,000P)$, where S = allowable working stress of gear material, lb/in^2; Y = tooth form factor (also called the Lewis factor); F = face width, in; K_v = dynamic load factor = $600/(600 + v)$ for metal gears, $0.25 + 150/(200 + v)$ for nonmetallic gears; v = pitchline velocity, ft/min = (pinion pitch diameter, in)(pinion rpm)(0.262); P = diametral pitch, in = number of teeth/pitch diameter, in. Obtain values of S and Y from tables in Baumeister and Marks—*Standard Handbook for Mechanical*

Engineers, or AGMA *Standards Books,* or gear manufacturers' engineering data. Compute the strength horsepower for the pinion and gear separately.

Using one of the above references for the pinion, we find $S = 25,000$ lb/in^2 (172,368.9 kPa) and $Y = 0.298$. The pitchline velocity for the metal pinion is $v = (6.0)(600)(0.262) = 944$ ft/min (4.8 m/s). Then $K_v = 600/(600 + 944) = 0.388$. The diametral pitch of the pinion is $P = N_p/d_p$, where $N_p =$ number of teeth on pinion; $d_p =$ diametral pitch of pinion, in. Or $P = 36/6 = 6$.

Substituting the above values in the Lewis formula gives $hp_s = (25,000)(0.298)(1.75)(0.388)(944)/[(33,000)(6)] = 24.117$ hp (17.98 kW) for the pinion.

Using the Lewis formula and the same procedure for the 150-tooth gear, $hp_s = (20,000)(0.374)(1.75)(0.388)(944)/[(33,000)(6)] = 24.2$ hp (18.05 kW). Thus, the strength horsepower of the gear is greater than that of the pinion.

2. Compute the durability horsepower

The durability horsepower of spur gears is found from $hp_d = F_iK_rD_oC_r$ for 20° pressure-angle full-depth or stub teeth. For 14.5° full-depth teeth, multiply hp_d by 0.75. In this relation, $F_i =$ face-width and built-in factor from AGMA *Standards;* $K_r =$ factor for tooth form, materials, and ratio of gear to pinion from AGMA *Standards;* $D_o = (d_p^2R_p/158,000)(1 - v^{0.5}/84)$, where $d_p =$ pinion pitch diameter, in; $R_p =$ pinion rpm; $v =$ pinion pitchline velocity, ft/min, as computed in step 1; $C_r =$ factor to correct for increased stress at the start of single-tooth contact as given by AGMA *Standards.*

Using appropriate values from these standards for low-speed gears of double speed reductions yields $hp_d = (0.75)(1.46)(387)(0.0865)(1.0) = 36.6$ hp (27.3 kW).

3. Compute the gearset service rating

Determine, by inspection, which is the lowest computed value for the gearset—the strength or durability horsepower. Thus, step 1 shows that the strength horsepower $hp_s = 24.12$ hp (18.0 kW) of the pinion is the lowest computed value. Use this lowest value in computing the gear-train service rating.

Using the AGMA *Standards,* determine the service factors for this installation. The load service factor for heavy shock loads and 3 h/day intermittent operation with an electric-motor drive is 1.5 from the *Standards.* The lubrication factor for a drive operating under fair conditions is, from the *Standards,* 1.25. To find the service rating, divide the lowest computed horsepower by the product of the load and lubrication factors; or, service rating = $24.12/(1.5)(1.25) = 7.35$ hp (9.6 kW).

Were this gearset operated only occasionally (0.5 h or less per day), the service rating could be determined by using the lower of the two computed strength horsepowers, in this case 24.12 hp (18.0 kW). Apply only the load service factor, or 1.25 for occasional heavy shock loads. Thus, the service rating for these conditions = $24.12/1.25 = 19.30$ hp (14.4 kW).

Related Calculations: Similar AGMA gear construction-material, tooth-form, face-width, tooth-stress, service, and lubrication tables are available for rating helical, double-helical, herringbone, worm, straight-bevel, spiral-bevel, and Zerol gears. Follow the general procedure given here. Be certain, however, to use the applicable values from the appropriate AGMA tables. In general, choose suitable stock gears first; then check the horsepower rating as detailed above.

MOMENT OF INERTIA OF A GEAR DRIVE

A 12-in (30.5-cm) outside-diameter 36-tooth steel pinion gear having a 3-in (7.6-cm) face width is mounted on a 2-in (5.1-cm) diameter 36-in (91.4-cm) long steel shaft turning at 600 r/min. The pinion drives a 200-r/min 36-in (91.4-cm) outside-diameter 108-tooth steel gear mounted on a 12-in (30.5-cm) long 2-in (5.1-cm) diameter steel shaft that is solidly connected to a 24-in (61.0-cm) long 4-in (10.2-cm) diameter shaft. What is the moment of inertia of the high-speed and low-speed assemblies of this gearset?

Calculation Procedure:

1. Compute the moment of inertia of each gear

The moment of inertia of a cylindrical body about its longitudinal axis is $I_i = WR^2$, where $I_i =$ moment of inertia of a cylindrical body, in^4/in of length; $W =$ weight of cylindrical material, lb/in^3; R = radius of cylinder to its outside surface, in. For a steel shaft or gear, this relation can be simplified to $I_t = D^4/35.997$, where D = shaft or gear diameter, in. When you are computing I for a gear, treat it as a solid blank of material. This is a safe assumption.

Thus, for the 12-in (30.5-cm) diameter pinion, $I = 12^4/35.997 = 576.05$ in^4/in (9439.8 cm^4/cm) of length. Since the gear has a 3-in (7.6-cm) face width, the moment of inertia for the total length is $I_t = (3.0)(576.05) = 1728.15$ in^4 (71,931.0 cm^4).

For the 36-in (91.4-cm) gear, $I_i = 36^4/35.997 = 46,659.7$ in^4/in (764,615.5 cm^4/cm) of length. With a 3-in (7.6-cm) face width, $I_t = (3.0)(46,659.7) = 139,979.1$ in^4 (5,826,370.0 cm^4).

2. Compute the moment of inertia of each shaft

Follow the same procedure as in step 1. Thus for the 36-in (91.4-cm) long 2-in (5.1-cm) diameter pinion shaft, $I_t = (2^4/35.997)(36) = 16.0$ in^4 (666.0 cm^4).

For the 12-in (30.5-cm) long 2-in (5.1-cm) diameter portion of the gear shaft, $I_t = (2^4/35.997)(12) = 5.33$ in^4 (221.9 cm^4). For the 24-in (61.0-cm) long 4-in (10.2-cm) diameter portion of the gear shaft, $I_t = (4^4/35.997)(24) = 170.69$ in^4 (7104.7 cm^4). The total moment of inertia of the gear shaft equals the sum of the individual moments, or $I_t = 5.33 + 170.69 = 176.02$ in^4 (7326.5 cm^4).

3. Compute the high-speed-assembly moment of inertia

The effective moment of inertia at the high-speed assembly input $= I_{thi} = I_{th} + I_{tl}/(R_h/R_l)^2$, where I_{th} = moment of inertia of high-speed assembly, in^4; I_{tl} = moment of inertia of low-speed assembly, in^4; R_h = high speed, r/min; R_l = low speed, r/min. To find I_{th} and I_{tl}, take the sum of the shaft and gear moments of inertia for the high- and low-speed assemblies, respectively. Or, $I_{th} = 16.0 + 1728.5 = 1744.15$ in^4 (72,597.0 cm^4); $I_{tl} = 176.02 + 139,979.1 = 140,155.1$ in^4 (5,833,695.7 cm^4).

Then $I_{thi} = 1744.15 + 140,155.1/(600/200)^2 = 17,324.2$ in^4 (721,087.6 cm^4).

4. Compute the low-speed-assembly moment of inertia

The effective moment of inertia at the low-speed assembly output is $I_{tlo} = I_{tl} + I_{th}(R_h/R_l)^2 = 140,155.1 + (1744.15)(600/200)^2 = 155,852.5$ in^4 (6,487,070.8 cm^4).

Note that $I_{thi} \neq I_{tlo}$. One value is approximately nine times that of the other. Thus, in stating the moment of inertia of a gear drive, be certain to specify whether the given value applies to the high- or low-speed assembly.

Related Calculations: Use this procedure for shafts and gears made of any metal—aluminum, brass, bronze, chromium, copper, cast iron, magnesium, nickel, tungsten, etc. Compute WR^2 for steel, and multiply the result by the weight of shaft material, lb/in^3/0.283.

BEARING LOADS IN GEARED DRIVES

A geared drive transmits a torque of 48,000 lb·in (5423.3 N·m). Determine the resulting bearing load in the drive shaft if a 12-in (30.5-cm) pitch-radius spur gear having a 20° pressure angle is used. A helical gear having a 20° pressure angle and a 14.5° spiral angle transmits a torque of 48,000 lb·in (5423.2 N·m). Determine the bearing load it produces if the pitch radius is 12 in (30.5 cm). Determine the bearing load in a straight bevel gear having the same proportions as the helical gear above, except that the pitch cone angle is 14.5°. A worm having an efficiency of 70 percent and a 30° helix angle drives a gear having a 20° normal pressure angle. Determine the bearing load when the torque is 48,000 lb·in (5423.3 N·m) and the worm pitch radius is 12 in (30.5 cm).

Calculation Procedure:

1. Compute the spur-gear bearing load

The tangential force acting on a spur-gear tooth is $F_t = T/r$, where F_t = tangential force, lb; T = torque, lb·in; r = pitch radius, in. For this gear, $F_t = T/r = 48,000/12 = 4000$ lb (17,792.9 N). This force is tangent to the pitch-diameter circle of the gear.

The separating force acting on a spur-gear tooth perpendicular to the tangential force is $F_s = F_t \tan \alpha$, where α = pressure angle, degrees. For this gear, $F_s = (4000)(0.364) = 1456$ lb (6476.6 N).

Find the resultant force R_f lb from $R_f = (F_t^2 + F_s^2)^{0.5} = (4000^2 + 1456^2)^{0.5} = 4260$ lb (18,949.4 N). This is the bearing load produced by the gear.

2. Compute the helical-gear load

The tangential force acting on a helical gear is $F_t = T/r = 48,000/12 = 4000$ lb (17,792.9 N). The separating force, acting perpendicular to the tangential force, is $F_s = F_t \tan \alpha/\cos \beta$, where β = the spiral angle. For this gear, $F_s = (4000)(0.364)/0.986 = 1503$ lb (6685.7 N). The resultant bearing load, which is a side thrust, is $R_f = (4000^2 + 1503^2)^{0.5} = 4380$ lb (19,483.2 N).

Helical gears produce an end thrust as well as the side thrust just computed. This end thrust is given by $F_e = F_t \tan \beta$, or $F_e = (4000)(0.259) = 1036$ lb (4608.4 N). The end thrust of the driving helical gear is equal and opposite to the end thrust of the driven helical gear when the teeth are of the opposite hand in each gear.

3. Compute the bevel-gear load

The tangential force acting on a bevel gear is $F_t = T/r = 48,000/12 = 4000$ lb (17,792.9 N). The separating force is $F_s = F_t \tan \alpha \cos \theta$, where θ = pitch cone angle. For this gear, $F_s = (4000)(0.364)(0.968) = 1410$ lb (6272.0 N).

Bevel gears produce an end thrust similar to helical gears. This end thrust is $F_e = F_t \tan \alpha \sin \theta$, or $F_e = (4000)(0.364)(0.25) = 364$ lb (1619.2 N). The side thrust in a bevel gear is $F_t = 4000$ lb (17,792.9 N) and acts tangent to the pitch-diameter circle. The resultant is an end thrust produced by F_s and F_e, or $R_f = (F_s^2 + F_e^2)^{0.5} = (1410^2 + 364^2)^{0.5} = 1458$ lb (6485.5 N). In a bevel-gear drive, F_t is common to both gears, F_s becomes F_e on the mating gear, and F_e becomes F_s on the mating gear.

4. Compute the worm-gear bearing load

The worm tangential force $F_t = T/r = 48,000/12 = 4000$ lb (17,792.9 N). The separating force is $F_s = F_t E \tan \alpha/\sin \phi$, where E = worm efficiency expressed as a decimal; ϕ = worm helix or lead angle. Thus, $F_s = (4999)(0.70)(0.364)/0.50 = 2040$ lb (9074.4 N).

The worm end thrust force is $F_e = F_t E \cot \phi = (4000)(0.70)(1.732) = 4850$ lb (21,573.9 N). This end thrust acts perpendicular to the separating force. Thus the resultant bearing load $R_f = (F_s^2 + F_e^2)^{0.5} = (2040^2 + 4850^2)^{0.5} = 5260$ lb (23,397.6 N).

Forces developed by the gear are equal and opposite to those developed by the worm tangential force if cancelled by the gear tangential force.

Related Calculations: Use these procedures to compute the bearing loads in any type of geared drive—open, closed, or semiclosed—serving any type of load. Computation of the bearing load is a necessary step in bearing selection.

FORCE RATIO OF GEARED DRIVES

A geared hoist will lift a maximum load of 1000 lb (4448.2 N). The hoist is estimated to have friction and mechanical losses of 5 percent of the maximum load. How much force is required to lift the maximum load if the drum on which the lifting cable reels is 10 in (25.4 cm) in diameter and the driving gear is 50 in (127.0 cm) in diameter? If the load is raised at a velocity of 100 ft/min (0.5 m/s), what is the hp output? What is the driving-gear tooth load if the gear turns at 191 r/min? A 15-in (38.1-cm) triple-reduction hoist has three driving gears with 48-, 42-, and 36-in (121.9-, 106.7-, and 91.4-cm) diameters, respectively, and two pinions of 12- and 10-in (30.5- and 25.4-cm) diameter. What force is required to lift a 1000-lb (4448.2-N) load if friction and mechanical losses are 10 percent?

Calculation Procedure:

1. Compute the total load on the hoist

The friction and mechanical losses *increase* the maximum load on the drum. Thus, the total load on the drum = maximum lifting load, lb + friction and mechanical losses, lb = 1000 + 1000(0.05) = 1050 lb (4670.6 N).

2. Compute the required lifting force

Find the lifting force from $L/D_g = F/d_d$, where L = total load on hoist, lb; D_g = diameter of driving gear, in; F = lifting force required, lb; d_d = diameter of lifting drum, in. For this hoist, $1050/50 = F/10$; $F = 210$ lb (934.1 N).

3. *Compute the horsepower input*

Find the horsepower input from $hp = Lv/33{,}000$, where v = load velocity, ft/min. Thus, $hp = (1050)(100)/33{,}000 = 3.19$ hp (2.4 kW).

Where the mechanical losses are not added to the load before the horsepower is computed, use the equation $hp = Lv/(1.00 - \text{losses})(33{,}000)$. Thus, $hp = (1000)(100)/(1 - 0.05)(33{,}000) = 3.19$ hp (2.4 kW), as before.

4. *Compute the driving-gear tooth load*

Assume that the entire load is carried by one tooth. Then the tooth load L_t lb = $33{,}000\,hp/v_g$, where v_g = peripheral velocity of the driving gear, ft/min. With a diameter of 50 in (127.0 cm) and a speed of 191 r/min, $v_g = \pi D_g R/12$, where R = gear rpm. Or, $v_g = \pi(50)(191)/12 = 2500$ ft/min (12.7 m/s). Then $L_t = (33{,}000)(3.19)/2500 = 42.1$ lb (187.3 N). This is a nominal tooth-load value.

5. *Compute the triple-reduction hoisting force*

Use the equation from step 2, but substitute the product of the three driving-gear diameters for D_g and the three driven-gear diameters for d_d. The total load $= 1000 + 0.10(1000) = 1100$ lb (4893.0 N). Then $L/D_g = F/d_d$, or $1100/(48 \times 42 \times 36) = F/(15 \times 12 \times 10)$; $F = 27.2$ lb (121.0 N). Thus, the triple-reduction hoist reduces the required lifting force to about one-tenth that required by a double-reduction hoist (step 2).

Related Calculations: Use this procedure for geared hoists of all types. Where desired, the number of gear teeth can be substituted for the driving- and driven-gear diameters in the force equation in step 2.

DETERMINATION OF GEAR BORE DIAMETER

Two helical gears transmit 500 hp (372.9 kW) at 3600 r/min. What should the bore diameter of each gear be if the allowable stress in the gear shafts is 12,500 lb/in² (86,187.5 kPa)? How should the gears be fastened to the shafts? The shafts are solid in cross section.

Calculation Procedure:

1. *Compute the required hub bore diameter*

The hub bore diameter must at least equal the outside diameter of the shaft, unless the gear is press- or shrink-fitted on the shaft. Regardless of how the gear is attached to the shaft, the shaft must be large enough to transmit the rated torque at the allowable stress.

Use the method of step 2 of "Solid and Hollow Shafts in Torsion" in this section to compute the required shaft diameter, after finding the torque by using the method of step 1 in the same procedure. Thus, $T = 63{,}000\,hp/R = (63{,}000)(500)/3600 = 8750$ lb·in (988.6 N·m). Then $d = 1.72(T/s)^{1/3} = 1.72(8750/12{,}500)^{1/3} = 1.526$ in (3.9 cm).

2. *Determine how the gear should be fastened to the shaft*

First decide whether the gears are to be permanently fastened or removable. This decision is usually based on the need for gear removal for maintenance or replacement. Removable gears can be fastened by a key, setscrew, spline, pin, clamp, or a taper and screw. Large gears transmitting 100 hp (74.6 kW) or more are usually fitted with a key for easy removal. See "Selection of Keys for Machine Shafts" in this section for the steps in choosing a key.

Permanently fastened gears can be shrunk, pressed, cemented, or riveted to the shaft. Shrink-fit gears generally transmit more torque before slippage occurs than do press-fit gears. With either type of fastening, interference is necessary; i.e., the gear bore is made smaller than the shaft outside diameter.

Baumeister and Marks—*Standard Handbook for Mechanical Engineers* shows that press- or shrink-fit gears on shafts of 1.19- to 1.58-in (3.0- to 4.0-cm) diameter should have an interference ranging from 0.3 to 4.0 thousandths of an inch (0.8 to 10.2 thousandths of a centimeter) on the diameter, depending on the class of fit desired.

Related Calculations: Use this general procedure for any type of gear—spur, helical, herringbone, worm, etc. Never reduce the shaft diameter below that required by the stress equation, step 1. Thus, if interference is provided by the shaft diameter, *increase* the diameter; do not reduce it.

TRANSMISSION GEAR RATIO FOR A GEARED DRIVE

A four-wheel vehicle must develop a drawbar pull of 17,500 lb (77,843.9 N). The engine, which develops 500 hp (372.8 kW) and drives through a gear transmission a 34-tooth spiral bevel pinion gear which meshes with a spiral bevel gear having 51 teeth. This gear is keyed to the drive shaft of the 48-in (121.9-cm) diameter rear wheels of the vehicle. What transmission gear ratio should be used if the engine develops maximum torque at 1500 r/min? Select the axle diameter for an allowable torsional stress of 12,500 lb/in^2 (86,187.5 kPa). The efficiency of the bevel-gear differential is 80 percent.

Calculation Procedure:

1. *Compute the torque developed at the wheel*

The wheel torque = (drawbar pull, lb)(moment arm, ft), where the moment arm = wheel radius, ft. For this vehicle having a wheel radius of 24 in (61.0 cm), or 24/12 = 2 ft (0.6 m), the wheel torque = (17,500)(2) = 35,000 lb·ft (47,453.6 N·m).

2. *Compute the torque developed by the engine*

The engine torque $T = 5250\, hp/R$, or $T = (5250)(500)/1500 = 1750$ lb·ft (2372.7 N·m), where R = rpm.

3. *Compute the differential speed ratio*

The differential speed ratio = $N_g/N_p = 51/34 = 1.5$, where N_g = number of gear teeth; N_p = number of pinion teeth.

4. *Compute the transmission gear ratio*

For any transmission gear, its ratio = (output torque, lb·ft)/[(input torque, lb·ft)(differential speed ratio)(differential efficiency)], or transmission gear ratio = $35,000/[(1750)(1.5)(0.80)]$ = 16.67. Thus, a transmission with a 16.67 ratio will give the desired output torque at the rated enging speed.

5. *Determine the required shaft diameter*

Use the relation $d = 1.72(T/s)^{1/3}$ from the previous calculation procedure to determine the axle diameter. Since the axle is transmitting a total torque of 35,000 lb·ft (47,453.6 N·m), each of the two rear wheels develops a torque for 35,000/2 = 17,500 lb·ft (23,726.8 N·m), and $d = 1.72(17,500/12,500)^{1/3} = 1.92$ in (4.9 cm).

Related Calculations: Use this general procedure for any type of differential—worm gear, herringbone gear, helical gear, or spiral gear—connected to any type of differential. The output torque can be developed through a wheel, propeller, impeller, or any other device. Note that although this vehicle has two rear wheels, the total drawbar pull is developed by *both* wheels. Either wheel delivers *half* the drawbar pull. If the total output torque were developed by only one wheel, its shaft diameter would be $d = 1.72(35,000/12,500)^{1/3} = 2.42$ in (6.1 cm).

EPICYCLIC GEAR TRAIN SPEEDS

Figure 4 shows several typical arrangements of epicyclic gear trains. The number of teeth and the rpm of the driving arm are indicated in each diagram. Determine the driven-member rpm for each set of gears.

Calculation Procedure:

1. *Compute the spur-gear speed*

For a gear arranged as in Fig. 4a, $R_d = R_D(1 + N_s/N_d)$, where R_d = driven-member rpm; R_D = driving-member rpm; N_s = number of teeth on the stationary gear; N_d = number of teeth on the driven gear. Given the values given for this gear and since the arm is the driving member, $R_d = 40(1 + 84/21)$; $R_d = 200$ r/min.

Note how the driven-gear speed is attained. During one planetary rotation around the stationary gear, the driven gear will rotate axially on its shaft. The number of times the driven gear rotates on its shaft $= N_s/N_d = 84/21 = 4$ times per planetary rotation about the stationary gear.

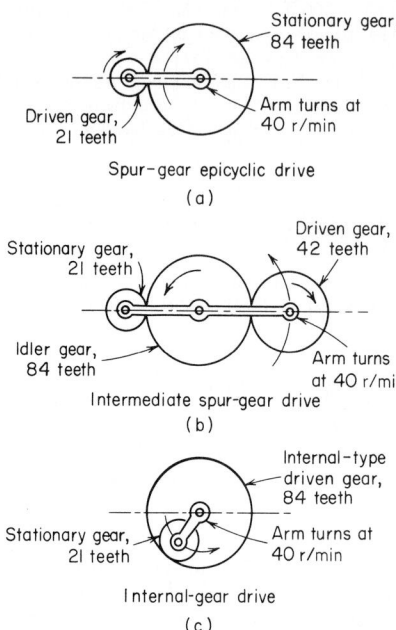

Stationary gear,
84 teeth

Arm turns at
40 r/min

Driven gear,
21 teeth

Spur–gear epicyclic drive

(a)

Stationary gear,
21 teeth

Driven gear,
42 teeth

Idler gear,
84 teeth

Arm turns
at 40 r/min

Intermediate spur-gear drive

(b)

Internal–type
driven gear,
84 teeth

Stationary gear,
21 teeth

Arm turns at
40 r/min

Internal-gear drive

(c)

FIG. 4   Epicyclic gear trains.

While rotating on its shaft, the driven gear makes a planetary rotation around the fixed gear. So while rotating axially on its shaft four times, the driven gear makes one additional planetary rotation about the stationary gear. Its total axial and planetary rotation is $4 + 1 = 5$ r/min per rpm of the arm. Thus, the gear ratio $G_r = R_D/R_d = 40/200 = 1{:}5$.

2. Compute the idler-gear train speed

The idler gear, Fig. 4b, turns on its shaft while the arm rotates. Movement of the idler gear causes rotation of the driven gear. For an epicyclic gear train of this type, $R_d = R_D(1 - N_s/N_d)$, where the symbols are as defined in step 1. Thus, $R_d = 40(1 - 21/42) = 20$ r/min.

3. Compute the internal gear drive speed

The arm of the internal gear drive, Fig. 4c, turns and carries the stationary gear with it. For a gear train of this type, $R_d = R_D(1 - N_s/N_d)$, or $R_d = 40(1 - 21/84) = 30$ r/min.

Where the internal gear is the driving gear that turns the arm, making the arm the driven member, the velocity equation becomes $r_d = R_D N_D/(N_D + N_s)$, where R_D = driving-member rpm; N_D = number of teeth on the driving member.

Related Calculations: The arm was the driving member for each of the gear trains considered here. However, any gear can be made the driving member if desired. Use the same relations as given above, but substitute the gear rpm for R_D. Thus, a variety of epicyclic gear problems can be solved by using these relations. Where unusual epicyclic gear configurations are encountered, refer to Dudley—*Gear Handbook* for a tabular procedure for determining the gear ratio.

PLANETARY-GEAR-SYSTEM SPEED RATIO

Figure 5 shows several arrangements of important planetary-gear systems using internal ring gears, planet gears, sun gears, and one or more carrier arms. Determine the output rpm for each set of gears.

Calculation Procedure:

1. Determine the planetary-gear output speed

For the planetary-gear drive, Fig. 5a, the gear ratio $G_r = (1 + N_4 N_2/N_3 N_1)/(1 - N_4 N_2/N_5 N_1)$, where $N_1, N_2, \ldots, N_5$ = number of teeth, respectively, on each of gears 1, 2, . . . , 5. Also, for any gearset, the gear ratio G_r = input rpm/output rpm, or G_r = driver rpm/driven rpm.

With ring gear 2 fixed and ring gear 5 the output gear, Fig. 5a, and the number of teeth shown, $G_r = \{1 + (33)(74)/[(9)(32)]\}/\{1 - (33)(74)/[(175)(32)]\} = -541.667$. The minus sign indicates that the output shaft revolves in a direction *opposite* to the input shaft. Thus, with an input speed of 5000 r/min, G_r = input rpm/output rpm; output rpm = input rpm/G_r, or output rpm $= 5000/541.667 = 9.24$ r/min.

2. *Determine the coupled planetary drive output speed*

The drive, Fig. 5b, has the coupled ring gear 2, the sun gear 3, the coupled planet carriers C and C', and the fixed ring gear 4. The gear ratio is $G_r = (1 - N_2N_4/N_1N_3)$, where the symbols are the same as before. Find the output speed for any given number of teeth by first solving for G_r and then solving G_r = input rpm/output rpm.

With the number of teeth shown, $G_r = 1 - (75)(75)/[(32)(12)] = -13.65$. Then output rpm = input rpm/G_r = 1200/13.65 = 87.9 r/min.

Two other arrangements of coupled planetary drives are shown in Fig. 5c and d. Compute the output speed in the same manner as described above.

3. *Determine the fixed-differential output speed*

Figure 5e and f shows two typical fixed-differential planetary drives. Compute the output speed in the same manner as step 2.

4. *Determine the triple planetary output speed*

Figure 5g shows three typical triple planetary drives. Compute the output speed in the same manner as step 2.

5. *Determine the output speed of other drives*

Figure 5h, i, j, k, and l shows the gear ratio and arrangement for the following drives: compound spur-bevel gear, plancentric, wobble gear, double eccentric, and Humpage's bevel gears. Compute the output speed for each in the same manner as step 2.

Related Calculations: Planetary and sun-gear calculations are simple once the gear ratio is determined. The gears illustrated here[1] comprise an important group in the planetary and sun-gear field. For other gear arrangements, consult Dudley—*Gear Handbook*.

SELECTION OF A RIGID FLANGE-TYPE SHAFT COUPLING

Choose a steel flange-type coupling to transmit a torque of 15,000 lb·in (1694.4 N·m) between two 2½-in (6.4-cm) diameter steel shafts. The load is uniform and free of shocks. Determine how many bolts are needed in the coupling if the allowable bolt shear stress is 3000 lb/in² (20,685.0 kPa). How thick must the coupling flange be, and how long should the coupling hub be if the allowable stress in bearing for the hub is 20,000 lb/in² (137,900.0 kPa) and in shear 6000 lb/in² (41,370.0 kPa)? The allowable shear stress in the key is 12,000 lb/in² (82,740.0 kPa). There is no thrust force acting on the coupling.

Calculation Procedure:

1. *Choose the diameter of the coupling bolt circle*

Assume a bolt-circle diameter for the coupling. As a first choice, assume the bolt-circle diameter is three times the shaft diameter, or $3 \times 2.5 = 7.5$ in (19.1 cm). This is a reasonable first assumption for most commercially available couplings.

2. *Compute the shear force acting at the bolt circle*

The shear force F_s lb acting at the bolt-circle radius r_b in is $F_s = T/r_b$, where T = torque on shaft, lb·in. Or, $F_s = 15,000/(7.5/2) = 4000$ lb (17,792.9 N).

3. *Determine the number of coupling bolts needed*

When the allowable shear stress in the bolts is known, compute the number of bolts N required from $N = 8F_s/(\pi d^2 s_s)$, where d = diameter of each coupling bolt, in; s_s = allowable shear stress in coupling bolts, lb/in².

The usual bolt diameter in flanged, rigid couplings ranges from ¼ to 2 in (0.6 to 5.1 cm), depending on the torque transmitted. Assuming that ½-in (1.3-cm) diameter bolts are used in this coupling, we see that $N = 8(4000)/[\pi(0.5)^2(3000)] = 13.58$, say 14 bolts.

Most flanged, rigid couplings have two to eight bolts, depending on the torque transmitted. A coupling having 14 bolts would be a poor design. To reduce the number of bolts, assume a larger

---

[1]John H. Glover, "Planetary Gear Systems," *Product Engineering*, Jan. 6, 1964.

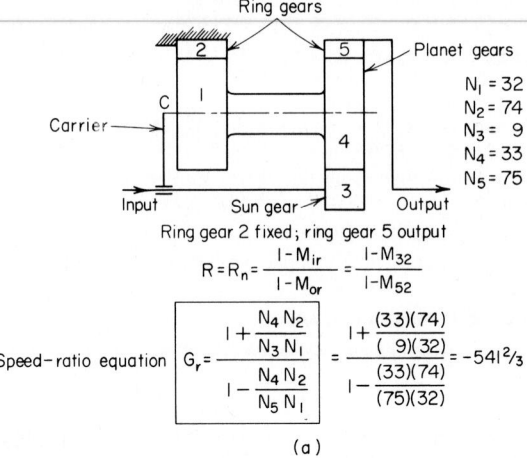

Ring gears

Planet gears

$N_1 = 32$
$N_2 = 74$
$N_3 = 9$
$N_4 = 33$
$N_5 = 75$

Carrier

C

Input   Sun gear   Output

Ring gear 2 fixed; ring gear 5 output

$$R = R_n = \frac{1 - M_{ir}}{1 - M_{or}} = \frac{1 - M_{32}}{1 - M_{52}}$$

Speed-ratio equation

$$G_r = \frac{1 + \dfrac{N_4 N_2}{N_3 N_1}}{1 - \dfrac{N_4 N_2}{N_5 N_1}} = \frac{1 + \dfrac{(33)(74)}{(9)(32)}}{1 - \dfrac{(33)(74)}{(75)(32)}} = -541\,^2/_3$$

(a)

Coupled planetary drives

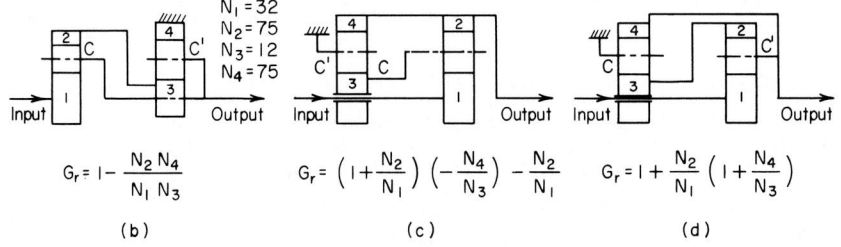

$N_1 = 32$
$N_2 = 75$
$N_3 = 12$
$N_4 = 75$

Input   Output

$$G_r = 1 - \frac{N_2 N_4}{N_1 N_3}$$

(b)

Input   Output

$$G_r = \left(1 + \frac{N_2}{N_1}\right)\left(-\frac{N_4}{N_3}\right) - \frac{N_2}{N_1}$$

(c)

Input   Output

$$G_r = 1 + \frac{N_2}{N_1}\left(1 + \frac{N_4}{N_3}\right)$$

(d)

Fixed-differential drives

Output is difference between speeds of two parts leading to high reduction ratios

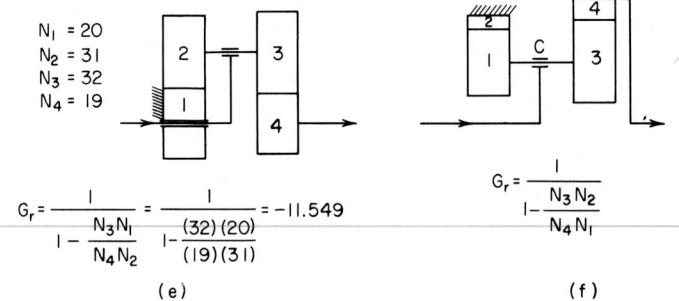

$N_1 = 20$
$N_2 = 31$
$N_3 = 32$
$N_4 = 19$

$$G_r = \frac{1}{1 - \dfrac{N_3 N_1}{N_4 N_2}} = \frac{1}{1 - \dfrac{(32)(20)}{(19)(31)}} = -11.549$$

(e)

$$G_r = \frac{1}{1 - \dfrac{N_3 N_2}{N_4 N_1}}$$

(f)

FIG. 5 Planetary gear systems. *(Product Engineering.)*

Triple planetary drives

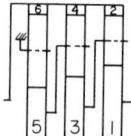

Input to gear 1, ouput from gear 6

$$G_r = \left(1 + \frac{N_2}{N_1}\right)\left[\left(1 + \frac{N_4}{N_3}\right)\left(-\frac{N_6}{N_5}\right) - \frac{N_4}{N_3}\right] - \frac{N_2}{N_1}$$

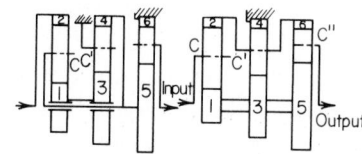

$$G_r = \left[1 + \frac{N_1}{N_2}\left(1 + \frac{N_4}{N_3}\right)\right]\left(1 + \frac{N_6}{N_5}\right)$$

$$G_r = \left[1 + \frac{N_4/N_3}{1 + (N_2/N_1)}\right] \Big/ \left[1 + \frac{N_4/N_3}{1 + (N_6/N_5)}\right]$$

(g)

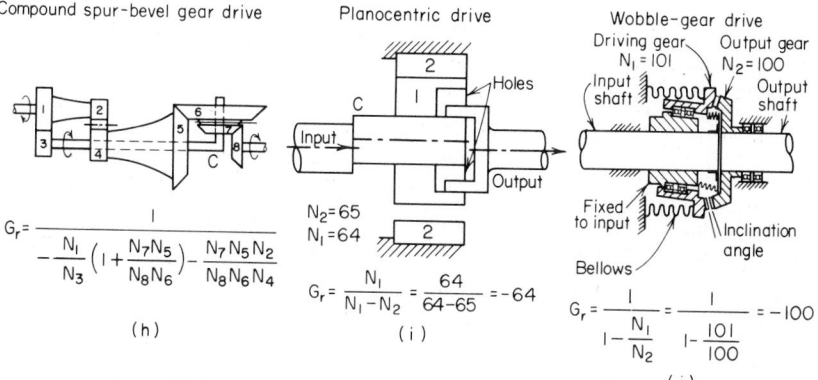

Compound spur-bevel gear drive

$$G_r = \cfrac{1}{-\cfrac{N_1}{N_3}\left(1 + \cfrac{N_7 N_5}{N_8 N_6}\right) - \cfrac{N_7 N_5 N_2}{N_8 N_6 N_4}}$$

(h)

Planocentric drive

Holes

Input

Output

$N_2 = 65$
$N_1 = 64$

$$G_r = \frac{N_1}{N_1 - N_2} = \frac{64}{64 - 65} = -64$$

(i)

Wobble-gear drive

Driving gear $N_1 = 101$ Output gear $N_2 = 100$

Input shaft   Output shaft

Fixed to input   Inclination angle

Bellows

$$G_r = \cfrac{1}{1 - \cfrac{N_1}{N_2}} = \cfrac{1}{1 - \cfrac{101}{100}} = -100$$

(j)

Double-eccentric drives

Two arrangements. Input is through double-throw crank (carrier). Gear 1 fixed to frame.

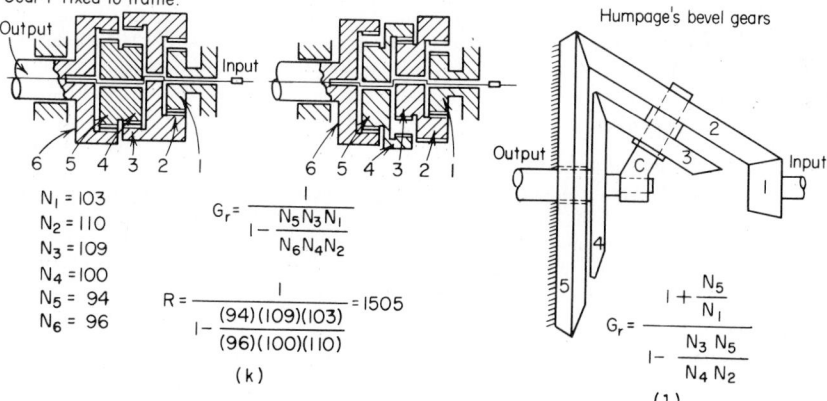

Output   Input   Output

6 5 4 3 2 1   6 5 4 3 2 1

$N_1 = 103$
$N_2 = 110$
$N_3 = 109$
$N_4 = 100$
$N_5 = 94$
$N_6 = 96$

$$G_r = \cfrac{1}{1 - \cfrac{N_5 N_3 N_1}{N_6 N_4 N_2}}$$

$$R = \cfrac{1}{1 - \cfrac{(94)(109)(103)}{(96)(100)(110)}} = 1505$$

(k)

Humpage's bevel gears

Output   Input

$$G_r = \cfrac{1 + \cfrac{N_5}{N_1}}{1 - \cfrac{N_3 N_5}{N_4 N_2}}$$

(l)

diameter, say 0.75 in (1.9 cm). Then $N = 8(4000)/[\pi(0.75)^2(3000)] = 6.03$, say eight bolts, because an odd number of bolts are seldom used in flanged couplings.

Determine the shear stress in the bolts by solving the above equation for $s_s = 8F_s/(\pi d^2 N) = 8(4000)/[\pi(0.75)^2(8)] = 2265$ lb/in^2 (15,617.2 kPa). Thus, the bolts are not overstressed, because the allowable stress is 3000 lb/in^2 (20,685.0 kPa).

4. Compute the coupling flange thickness required

The flange thickness t in for an allowable bearing stress s_b lb/in^2 is $t = 2F_s/(Nds_b) = 2(4000)/[(8)(0.75)(20,000)] = 0.0666$ in (0.169 cm). This thickness is much less than the usual thickness used for flanged couplings manufactured for off-the-shelf use.

5. Determine the hub length required

The hub length is a function of the key length required. Assuming a ¾-in (1.9-cm) square key, compute the hub length l in from $l = 2F_{ss}/(t_k s_t)$, where F_{ss} = force acting at shaft outer surface, lb; t_k = key thickness, in. The force $F_{ss} = T/r_h$, where r_h = inside radius of hub, in = shaft radius = 2.5/2 = 1.25 in (3.2 cm) for this shaft. Then $F_{ss} = 15,000/1.25 = 12,000$ lb·in (1355.8 N·m). Then $l = 2(12,000)/[(0.75)(20,000)] = 1.6$ in (4.1 cm).

When the allowable design stress for bearing, 20,000 lb/in^2 (137,895.1 kPa) here, is less than half the allowable design stress for shear, 12,000 lb/in^2 (82,740.0 kPa) here, the longest key length is obtained when the bearing stress is used. Thus, it is not necessary to compute the thickness needed to resist the shear stress for this coupling. If it is necessary to compute this thickness, find the force acting at the surface of the coupling hub from $F_h = T/r_h$, where r_h = hub radius, in. Then $t_s = F_h/\pi d_h s_s$, where d_h = hub diameter, in; s_s = allowable hub shear stress, lb/in^2.

Related Calculations: Couplings offered as standard parts by manufacturers are usually of sufficient thickness to prevent fatigue failure.

Since each half of the coupling transmits the total torque acting, the length of the key must be the same in each coupling half. The hub diameter of the coupling is usually 2 to 2.5 times the shaft diameter, and the coupling lip is generally made the same thickness as the coupling flange. The procedure given here can be used for couplings made of any metallic material.

SELECTION OF A FLEXIBLE COUPLING FOR A SHAFT

Choose a stock flexible coupling to transmit 15 hp (11.2 kW) from a 1000-r/min four-cylinder gasoline engine to a dewatering pump turning at the same rpm. The pump runs 8 h/day and is an uneven load because debris may enter the pump. The pump and motor shafts are each 1.0 in (2.5 cm) in diameter. Maximum misalignment of the shafts will not exceed 0.5°. There is no thrust force acting on the coupling, but the end float or play may reach ⅟₁₆ in (0.2 cm).

Calculation Procedure:

1. Choose the type of coupling to use

Consult Table 22 or the engineering data published by several coupling manufacturers. Make a tentative choice from Table 22 of the type of coupling to use, based on the maximum misalignment expected and the tabulated end-float capacity of the coupling. Thus, a roller-chain-type coupling (one in which the two flanges are connected by a double roller chain) will be chosen

TABLE 22 Allowable Flexible Coupling Misalignment

Coupling type	Angular misalignment	Parallel misalignment		End float	
		USCS	SI	USCS	SI
Plastic chain	Up to 1.0°	0.005 in	0.1 mm	⅟₁₆ in	2 mm
Roller chain	Up to 0.5°	2% of chain pitch	2% of chain pitch	⅟₁₆ in	2 mm
Silent chain	Up to 0.5°	2% of chain pitch	2% of chain pitch	¼ to ¾ in	0.6 to 1.9 cm
Neoprene biscuit	Up to 5.0°	0.01 to 0.05 in	0.3 to 1.3 mm	Up to ½ in	Up to 1.3 cm
Radial	Up to 0.5°	0.01 to 0.02 in	0.3 to 0.5 mm	Up to ⅟₁₆ in	Up to 2 mm

TABLE 23 Flexible Coupling Service Factors°

Type of drive			
Engine,† less than six cylinders	Engine, six cylinders or more	Electric motor; steam turbine	Type of load
2.0	1.5	1.0	Even load, 8 h/day; nonreversing, low starting torque
2.5	2.0	1.5	Uneven load, 8 h/day; moderate shock or torque, nonreversing
3.0	2.5	2.0	Heavy shock load, 8 h/day; reversing under full load, high starting torque

°Morse Chain Company.
†Gasoline or diesel.

from Table 22 for this drive because it can accommodate 0.5° of misalignment and an end float of up to $\frac{1}{16}$ in (0.2 cm).

2. Choose a suitable service factor

Table 23 lists typical service factors for roller-chain-type flexible couplings. Thus, for a four-cylinder gasoline engine driving an uneven load, the service factor SF = 2.5.

3. Apply the service factor chosen

Multiply the horsepower or torque to be transmitted by the service factor to obtain the coupling design horsepower or torque. Or, coupling design $hp = (15)(2.5) = 37.5$ hp (28.0 kW).

4. Select the coupling to use

Refer to the coupling design horsepower rating table in the manufacturer's engineering data. Enter the table at the shaft rpm, and project to a design horsepower slightly greater than the value computed in step 3. Thus, in Table 24 a typical rating tabulation shows that a coupling design horsepower rating of 38.3 hp (28.6 kW) is the next higher value above 37.5 hp (28.0 kW).

5. Determine whether the coupling bore is suitable

Table 24 shows that a coupling suitable for 38.3 hp (28.6 kW) will have a maximum bore diameter up to 1.75 in (4.4 cm) and a minimum bore diameter of 0.625 in (1.6 cm). Since the engine and pump shafts are each 1.0 in (2.5 cm) in diameter, the coupling is suitable.

The usual engineering data available from manufacturers include the stock keyway sizes, coupling weight, and principal dimensions of the coupling. Check the overall dimensions of the coupling to determine whether the coupling will fit the available space. Where the coupling bore diameter is too small to fit the shaft, choose the next larger coupling. If the dimensions of the coupling make it unsuitable for the available space, choose a different type or make a coupling.

TABLE 24 Flexible Coupling hp Ratings°

r/min			Bore diameter, in (cm)	
800	1000	1200	Maximum	Minimum
16.7	19.9	23.2	1.25 (3.18)	0.5 (1.27)
32.0	38.3	44.5	1.75 (4.44)	0.625 (1.59)
75.9	90.7	105.0	2.25 (5.72)	0.75 (1.91)

°Morse Chain Company.

Related Calculations: Use the general procedure given here to select any type of flexible coupling using flanges, springs, roller chain, preloaded biscuits, etc., to transmit torque. Be certain to apply the service factor recommended by the manufacturer. Note that biscuit-type couplings are rated in hp/100 r/min. Thus, a biscuit-type coupling rated at 1.60 hp/100 r/min (1.2 kW/100 r/min) and a maximum allowable speed of 4800 r/min could transmit a maximum of (1.80 hp)(4800/100) = 76.8 hp (57.3 kW).

SELECTION OF A SHAFT COUPLING FOR TORQUE AND THRUST LOADS

Select a shaft coupling to transmit 500 hp (372.9 kW) and a thrust of 12,500 lb (55,602.8 N) at 100 r/min from a six-cylinder diesel engine. The load is an even one, free of shock.

Calculation Procedure:

1. *Compute the torque acting on the coupling*

Use the relation $T = 5252hp/R$ to determine the torque, where T = torque acting on coupling, lb·ft; hp = horsepower transmitted by the coupling; R = shaft rotative speed, r/min. For this coupling, $T = (5252)(500)/100 = 26,260$ lb·ft (35,603.8 N·m).

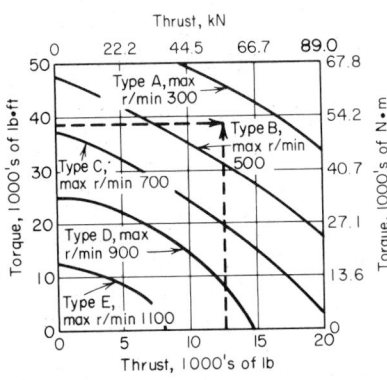

FIG. 6   Shaft-coupling characteristics.

2. *Find the service torque*

Multiply the torque T by the appropriate service factor from Table 23. This table shows that a service factor of 1.5 is suitable for an even load, free of shock. Thus, the service torque = (26,260 lb·ft)(1.5) = 39,390 lb·ft, say 39,500 lb·ft (53,554.8 N·m).

3. *Choose a suitable coupling*

Enter Fig. 6 at the torque on the left, and project horizontally to the right. Using the known thrust, 12,500 lb (55,602.8 N), enter Fig. 6 at the bottom and project vertically upward until the torque line is intersected. Choose the coupling model represented by the next higher curve. This shows that a type A coupling having a maximum allowable speed of 300 r/min will be suitable. If the plotted maximum rpm is lower than the actual rpm of the coupling, use the next plotted coupling type rated for the actual, or a higher, rpm.

In choosing a specific coupling, use the manufacturer's engineering data. This will resemble Fig. 6 or will be a tabulation of the ranges plotted.

Related Calculations: Use this procedure to select couplings for industrial and marine drives where both torque and thrust must be accommodated. See the Marine Engineering section of this handbook for an accurate way to compute the thrust produced on a coupling by a marine propeller. Always check to see that the coupling bore is large enough to accommodate the connected shafts. Where the bore is too small, use the next larger coupling.

HIGH-SPEED POWER-COUPLING CHARACTERISTICS

Select the type of power coupling to transmit 50 hp (37.3 kW) at 200 r/min if the angular misalignment varies from a minimum of 0 to a maximum of 45°. Determine the effect of angular misalignment on the shaft position, speed, and acceleration at angular misalignments of 30 and 45°

Calculation Procedure:

1. *Determine the type of coupling to use*

Table 25, developed by N. B. Rothfuss, lists the operating characteristics of eight types of high-speed couplings. Study of this table shows that a universal joint is the only type of coupling among

TABLE 25 Operating Characteristics of Couplings[a]

	Contoured diaphragm	Axial spring	Laminated disk	Universal joint	Ball-race	Gear	Chain	Elastomeric
Speed range, r/min	0–60,000	0–8,000	0–20,000	0–8,000	0–8,000	0–25,000	0–6,300	0–6,000
Power range, kW/100 r/min	1–500	1–9,000	1–100	1–100	1–100	1–2,000	1–200	0–400
Angular misalignment, degrees	0–8.0	0–2.0	0–1.5	0–45	0–40	0–3	0–2	0–4
Parallel misalignment, mm	0–2.5	0–2.5	0–2.5	None	None	0–2.5	0–2.5	0–2.5
Axial movement, cm	0–0.5	0–2.5	0–0.5	None	None	0–5.1	0–2.5	0–0.8
Ambient temperature, °C	900	Varies	900	Varies	Varies	Varies	Varies	Varies
Ambient pressure, kPa	Sea level to zero	Varies	Sea level to zero	Varies	Varies	Varies	Varies	Varies

[a] *Product Engineering.*

TABLE 26 Functional Characteristics of Couplings°

	Contoured diaphragm	Axial spring	Laminated disk	Universal joint	Ball-race	Gear	Chain	Elastomeric
No lubrication	✓	⋯	✓	⋯	⋯	⋯	⋯	✓
No backlash	✓	✓	✓	†	†	⋯	⋯	✓
Constant velocity ratio	✓	⋯	⋯	†	✓ †	†	†	✓
Containment	✓	⋯	†	⋯	✓	✓		
Angular only	✓	⋯	✓	✓	✓	✓	✓	
Axial and angular	✓	⋯	✓	⋯	✓	✓	✓	✓
Axial and parallel	✓	✓	✓	⋯	⋯	✓	✓	✓
Axial, angular, and parallel	✓	✓	✓	⋯	⋯	✓	✓	✓
High temperature	✓	⋯	✓	✓				
High altitude	✓	⋯	✓	✓	✓	✓	✓	✓
High torsional spring rate	✓	⋯	✓	✓	✓	✓	✓	✓
Low bending moment	✓	✓	✓	⋯	⋯	⋯	⋯	
No relative movement	✓	⋯	⋯	⋯	⋯	⋯	⋯	

° *Product Engineering.*
† Zero backlash and containment can be obtained by special design.
‡ Constant velocity ratio at small angles can be closely approximated.

TABLE 27   Universal Joint Output Variations°

Misalignment angle, deg	Maximum position error	Maximum speed error, percent	Ratio A/ω^2
5	0°06'34"	0.382	0.011747
10	0°26'18"	1.543	0.030626
15	0°59'36"	3.526	0.069409
20	1°46'54"	6.418	0.124966
25	2°48'42"	10.338	0.198965
30	4°06'42"	15.470	0.294571
35	5°42'20"	22.077	0.417232
40	7°36'43"	30.541	0.576215
45	9°52'26"	41.421	0.787200

°Caused by misalignment of the shaft. Table from *Machine Design.*

those listed that can handle an angular misalignment of 45°. Further study shows that a universal coupling has a suitable speed and hp range for the load being considered. The other items tabulated are not factors in this application. Therefore, a universal coupling will be suitable. Table 26 compares the functional characteristics of the couplings. Data shown support the choice of the universal joint.

2. *Determine the shaft position error*

Table 27, developed by David A. Lee, shows the output variations caused by misalignment between the shafts. Thus, at 30° angular misalignment, the position error is 4°06'42". This means that the output shaft position shifts from −4°06'42" to +4°06'42" twice each revolution. At a 45° misalignment the position error, Table 27, is 9°52'26". The shift in position is similar to that occurring at 30° angular misalignment.

3. *Compute the output-shaft speed variation*

Table 27 shows that at 30° angular misalignment the output-shaft speed variation is ±15.47 percent. Thus, the output-shaft speed varies between 200(1.00 ± 0.1547) = 169.06 and 230.94 r/min. This speed variation occurs *twice* per revolution.

For a 45° angular misalignment the speed variation, determined in the same way, is 117.16 to 282.84 r/min. This speed variation also occurs twice per revolution.

4. *Determine output-shaft acceleration*

Table 27 lists the ratio of maximum output-shaft acceleration A to the square of the input speed, ω^2, expressed in radians. To convert r/min to rad/s, use $rps = 0.1047$ r/min $= 0.1047(200) = 20.94$ rad/s.

For 30° angular misalignment, from Table 27 $A/\omega^2 = 0.294571$. Thus, $A = \omega^2(0.294571) = (20.94)^2(0.294571) = 129.6$ rad/s². This means that a constant input speed of 200 r/min produces an output acceleration ranging from −129.6 to +129.6 rad/s², and back, at a frequency of 2(200 r/min) = 400 cycles/min.

At a 45° angular misalignment, the acceleration range of the output shaft, determined in the same way, is −346 to +346 rad/s². Thus, the acceleration range at the larger shaft angle misalignment is 2.67 times that at the smaller, 30°, misalignment.

Related Calculations: Table 25 is useful for choosing any of seven other types of high-speed couplings. The eight couplings listed in this table are popular for high-horsepower applications. All are classed as rigid types, as distinguished from entirely flexible connectors such as flexible cables.

Values listed in Table 25 are nominal ones that may be exceeded by special designs. These values are guideposts rather than fixed; in borderline cases, consult the manufacturer's engineering data. Table 26 compares the functional characteristics of the couplings and is useful to the designer who is seeking a unit with specific operating characteristics. Note that the values in Table 25 are maximum and not additive. In other words, a coupling *cannot* be operated at the maximum angular and parallel misalignment and at the maximum horsepower and speed simultaneously—although in some cases the combination of maximum angular misalignment, maximum horsepower, and maximum speed would be acceptable. Where shock loads are anticipated, apply a suitable correction factor, as given in earlier calculation procedures, to the horsepower to be transmitted before entering Table 25.

SELECTION OF ROLLER AND INVERTED-TOOTH (SILENT) CHAIN DRIVES

Choose a roller chain and the sprockets to transmit 6 hp (4.5 kW) from an electric motor to a propeller fan. The speed of the motor shaft is 1800 r/min and of the driven shaft 900 r/min. How long will the chain be if the centerline distance between the shafts is 30 in (76.2 cm)?

Calculation Procedure:

1. *Determine, and apply, the load service factor*

Consult the manufacturer's engineering data for the appropriate load service factor. Table 28 shows several typical load ratings (smooth, moderate shock, heavy shock) for various types of driven devices. Use the load rating and the type of drive to determine the service factor. Thus, a propeller fan is rated as a heavy shock load. For this type of load and an electric-motor drive, the load service factor is 1.5, from Table 28.

Apply the load service factor by taking the product of it and the horsepower transmitted, or (1.5)(6 hp) = 9.0 hp (6.7 kW). The roller chain and sprockets must have enough strength to transmit this horsepower.

2. *Choose the chain and number of teeth in the small sprocket*

Using the manufacturer's engineering data, enter the horsepower rating table at the small-sprocket rpm and project to a horsepower value equal to, or slightly greater than, the required

TABLE 28   Roller Chain Loads and Service Factors*

Load rating	
Driven device	Type of load
Agitators (paddle or propeller)	Smooth
Brick and clay machinery	Heavy shock
Compressors (centrifugal and rotary)	Moderate shock
Conveyors (belt)	Smooth
Crushing machinery	Heavy shock
Fans (centrifugal)	Moderate shock
Fans (propeller)	Heavy shock
Generators and exciters	Moderate shock
Laundry machinery	Moderate shock
Mills	Heavy shock
Pumps (centrifugal, rotary)	Moderate shock
Textile machinery	Smooth

	Service factor		
	Internal-combustion engine		
Type of load	Hydraulic drive	Mechanical drive	Electric motor or turbine
Smooth	1.0	1.2	1.0
Moderate shock	1.2	1.4	1.3
Heavy shock	1.4	1.7	1.5

*Excerpted from Morse Chain Company data.

TABLE 29   Roller Chain Power Rating°

[*Single-strand, ⅝-in (1.6-cm) pitch roller chain*]

No. of teeth in small sprocket	Small sprocket rpm					
	1500		1800		2100	
	hp	kW	hp	kW	hp	kW
14	10.7	7.98	8.01	5.97	6.34	4.73
15	11.9	8.87	8.89	6.63	7.03	5.24
16	13.1	9.77	9.79	7.30	7.74	5.77
17	14.3	10.7	10.7	7.98		

°Excerpted from Morse Chain Company data.

rating. At this horsepower rating, read the number of teeth in the small sprocket, which is also listed in the table. Thus, in Table 29, which is an excerpt from a typical horsepower rating tabulation, 9.0 hp (6.7 kW) is not listed at a speed of 1800 r/min. However, the next higher horsepower rating, 9.79 hp (7.3 kW), will be satisfactory. The table shows that at this power rating, 16 teeth are used in the small sprocket.

This sprocket is a good choice because most manufacturers recommend that at least 16 teeth be used in the smaller sprocket, except at low speeds (100 to 500 r/min).

3. Determine the chain pitch and number of strands

Each horsepower rating table is prepared for a given chain pitch, number of chain strands, and various types of lubrication. Thus, Table 29 is for standard single-strand ⅝-in (1.6-cm) pitch roller chain. The 9.79-hp (7.3-kW) rating at 1800 r/min for this chain is with type III lubrication—oil bath or oil slinger—with the oil level maintained in the chain casing at a predetermined height. See the manufacturer's engineering data for the other types of lubrication (manual, drip, and oil stream) requirements.

4. Compute the drive speed ratio

For a roller chain drive, the speed ratio $S_r = R_h/R_l$, where R_h = rpm of high-speed shaft; R_l = rpm of low-speed shaft. For this drive, $S_r = 1800/900 = 2$.

5. Determine the number of teeth in the large sprocket

To find the number of teeth in the large sprocket, multiply the number of teeth in the small sprocket, found in step 2, by the speed ratio, found in step 4. Thus, the number of teeth in the large sprocket = (16)(2) = 32.

6. Select the sprockets

Refer to the manufacturer's engineering data for the dimensions of the available sprockets. Thus, one manufacturer supplies the following sprockets for ⅝-in (1.6-cm) pitch single-strand roller chain: 16 teeth, OD = 3.517 in (8.9 cm), bore = ⅝ in (1.6 cm); 32 teeth, OD = 6.721 in (17.1 cm), bore = ⅝ or ¾ in (1.6 or 1.9 cm). When choosing a sprocket, be certain to refer to data for the size and type of chain selected in step 3, because each sprocket is made for a specific type of chain. Choose the type of hub—setscrew, keyed, or taper-lock bushing—based on the torque that must be transmitted by the drive. See earlier calculation procedures in this section for data on key selection.

7. Determine the length of the chain

Compute the chain length in pitches L_p from $L_p = 2C + (S/2) + K/C$, where C = shaft center distance, in/chain pitch, in; S = sum of the number of teeth in the small and large sprocket; K = a constant from Table 30, obtained by entering this table with the value D = number of teeth in large sprocket − number of teeth in small sprocket. For this drive, $C = 30/0.625 = 48$; $S = 16 + 32 = 48$; $D = 32 − 16 = 16$; $K = 6.48$ from Table 30. Then, $L_p = 2(48) + 48/2 + 6.48/48 = 120.135$ pitches. However, a chain cannot contain a fractional pitch; therefore, use the next higher number of pitches, or $L_p = 121$ pitches.

Convert the length in pitches to length in inches, L_i, by taking the product of the chain pitch p in and L_p. Or $L_i = L_p p = (121)(0.625) = 75.625$ in (192.1 cm).

Related Calculations: At low-speed ratios, large-diameter sprockets can be used to reduce the roller-chain pull and bearing loads. At high-speed ratios, the number of teeth in the high-speed sprocket may have to be kept as small as possible to reduce the chain pull and bearing loads. The Morse Chain Company states: Ratios over 7:1 are generally not recommended for single-width roller chain drives. Very slow-speed drives (10 to 100 r/min) are often practical with as few as 9 or 10 teeth in the small sprocket, allowing ratios up to 12:1. In all cases where ratios exceed 5:1, the designer should consider the possibility of using compound drives to obtain maximum service life.

TABLE 30 Roller Chain Length Factors[*]

D	K	D	K
1	0.03	11	3.06
2	0.10	12	3.65
3	0.23	13	4.28
4	0.41	14	4.96
5	0.63	15	5.70
6	0.91	16	6.48
7	1.24	17	7.32
8	1.62	18	8.21
9	2.05	19	9.14
10	2.53	20	10.13

[*] Excerpted from Morse Chain Company data.

When you select standard inverted-tooth (silent) chain and high-velocity inverted-tooth silent-chain drives, follow the same general procedures as given above, except for the following changes.

Standard inverted-tooth silent chain: (a) Use a minimum of 17 teeth, and an odd number of teeth on one sprocket, where possible. This increases the chain life. (b) To achieve minimum noise, select sprockets having 23 or more teeth. (c) Use the proper service factor for the load, as given in the manufacturer's engineering data. (d) Where a long or fixed-center drive is necessary, use a sprocket or shoe idler where the largest amount of slack occurs. (e) Do not use an idler to reduce the chain wrap on small-diameter sprockets. (f) Check to see that the small-diameter sprocket bore will fit the high-speed shaft. Where the high-speed shaft diameter exceeds the maximum bore available for the chosen smaller sprocket, increase the number of teeth in the sprocket or choose the next larger chain pitch. (This general procedure also applies to roller chain sprockets.) (g) Compute the chain design horsepower from (drive hp)(chain service factor). (h) Select the chain pitch, number of teeth in the small sprocket, and chain *width* from the manufacturer's rating table. Thus, if the chain design horsepower = 36 hp (26.8 kW) and the chain is rated at 4 hp/in (1.2 kW/cm) of width, the required chain width = 36 hp/(4 hp/in) = 9 in (22.9 cm).

High-velocity inverted-tooth silent chain: (a) Use a minimum of 25 teeth and an odd number of teeth on one sprocket, where possible. This increases the chain life. (b) To achieve minimum noise, select sprockets with 27 or more teeth. (c) Use a larger service factor than the manufacturer's engineering data recommends, if trouble-free drives are desired. (d) Use a wider chain than needed, if an increased chain life is wanted. Note that the chain width is computed in the same way as described in item *h* above. (e) If a longer center distance between the drive shafts is desired, select a larger chain pitch [usual pitches are ¾, 1, 1½, or 2 in (1.9, 2.5, 3.8, or 5.1 cm)]. (f) Provide a means to adjust the centerline distance between the shafts. Such an adjustment *must* be provided in vertical drives. (g) Try to use an even number of pitches in the chain to avoid an offset link.

CAM CLUTCH SELECTION AND ANALYSIS

Choose a cam-type clutch to drive a centrifugal pump. The clutch must transmit 125 hp (93.2 kW) at 1800 r/min to the pump, which starts and stops 40 times per hour throughout its 12-h/day, 360-day/year operating period. The life of the pump will be 10 years.

Calculation Procedure:

1. *Compute the maximum torque acting on the clutch*

Compute the torque acting on the clutch from $T = 5252hp/R$, where the symbols are the same as in the previous calculation procedure. Thus, for this clutch, $T = 5252 \times 125/1800 = 365$ lb·ft (494.9 N·m).

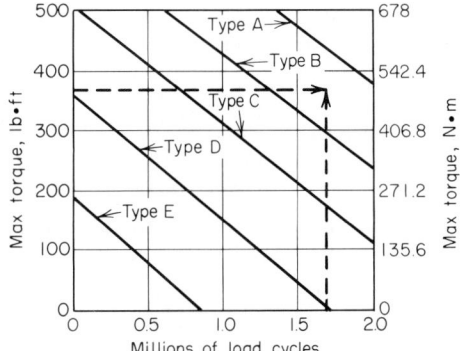

FIG. 7 Cam-type-clutch selection chart.

2. *Analyze the torque acting on the clutch*

For installations free of shock loads during starting and stopping, the running torque is the maximum torque that acts on the clutch. But if there is a shock load during starting or stopping, or at other times, the shock torque must be added to the running torque to determine the total torque acting. Compute the shock torque using the relation in step 1 and the actual hp and speed developed by the shock load.

3. *Compute the total number of load applications*

With 40 starts and stops (cycles) per hour, a 12-h day, and 360 operating days per year, the number of cycles per year is (40 cycles/h)(12 h/day)(360 days/year) = 172,800. In 10 years, the clutch will undergo (172,800 cycles/year)(10 years) = 1,728,000 cycles.

4. *Choose the clutch size*

Enter Fig. 7 at the maximum torque, 365 lb·ft (494.9 N·m), on the left, and the number of load cycles, 1,728,000, on the bottom. Project horizontally and vertically until the point of intersection is reached. Select the clutch represented by the next higher curve. Thus a type A clutch would be used for this load. (Note that the clutch capacity could be tabulated instead of plotted, but the results would be the same.)

5. *Check the clutch dimensions*

Determine whether the clutch bore will accommodate the shafts. If the clutch bore is too small, choose the next larger clutch size. Also check to see whether the clutch will fit into the available space.

Related Calculations: Use this general procedure to select cam-type clutches for business machines, compressors, conveyors, cranes, food processing, helicopters, fans, aircraft, printing machinery, pumps, punch presses, speed reducers, looms, grinders, etc. When choosing a specific clutch, use the manufacturer's engineering data to select the clutch size.

TIMING-BELT DRIVE SELECTION AND ANALYSIS

Choose a toothed timing belt to transmit 20 hp (14.9 kW) from an electric motor to a rotary mixer for liquids. The motor shaft turns at 1750 r/min and the mixer shaft is to turn at 600 ± 20 r/min. This drive will operate 12 h/day, 7 days/week. Determine the type of timing belt to use and the driving and driven pulley diameters if the shaft centerline distance is about 27 in (68.6 cm).

Calculation Procedure:

1. *Choose the service factor for the drive*

Timing-belt manufacturers publish service factors in their engineering data based on the type of prime mover, the type of driven machine (compressor, mixer, pump, etc.), type of drive (speedup), and drive conditions (continuous operation, use of an idler, etc.).

TABLE 31   Typical Timing-Belt Service Factors°

Type of drive	Type of load	Service factor
Electric motors, hydraulic motors, internal-combustion engines, line shafts	Shock-free	2.0
	Shocks	2.5
	Continuous operation or idler use	2.7
	Speed-up	3.0

°Use only for preliminary selection of belt. From Morse Chain Company data.

Usual service factors for any type of driver range from 1.3 to 2.5 for various types of driven machines. Correction factors for speed-up drives range from 0 to 0.40; the specific value chosen is *added* to the machine-drive correction factor. Drive conditions, such as 24-h continuous operation or the use of an idler pulley on the drive, cause an additional 0.2 to be added to the correction factor. Seasonal or intermittent operation *reduces* the machine-drive factor by 0.2.

Look up the service factor in Table 31, if the manufacturer's engineering data are not readily available. Table 31 gives safe data for usual timing-belt applications and is suitable for preliminary selection of belts. Where a final choice is being made, use the manufacturer's engineering data.

For a liquid mixer shock-free load, use a service factor of 2.0 from Table 31, since there are no other features which would require a larger value.

2.   *Compute the design horsepower for the belt*

The design horsepower $hp_d = hp_l \times SF$, where hp_l = load horsepower; SF = service factor. Thus, for this drive, $hp_d = (20)(2) = 40$ hp (29.8 kW).

3.   *Compute the drive speed ratio*

The drive speed ratio $S_r = R_h/R_l$, where R_h = rpm of high-speed shaft; R_l = rpm of low-speed shaft. For this drive $S_r = 1750/600 = 2.92{:}1$, the rated rpm. If the driven-pulley speed falls 10 r/min, $S_r = 1750/580 = 3.02{:}1$. Thus, the speed ratio may vary between 2.92 and 3.02.

4.   *Choose the timing-belt pitch*

Enter Table 32, or the manufacturer's engineering data, at the design horsepower and project to the driver rpm. Where the exact value of the design horsepower is not tabulated, use the next higher tabulated value. Thus, for this 1750-r/min drive having a design horsepower of 40 (29.8 kW), Table 32 shows that a ⅞-in (2.2-cm) pitch belt is required. This value is found by entering Table 32 at the next higher design horsepower, 50 (37.3 kW), and projecting to the 1750-r/min column. If 40 hp (29.8 kW) were tabulated, the table would be entered at this value.

5.   *Choose the number of teeth for the high-speed sprocket*

Enter Table 33, or the manufacturer's engineering data, at the timing-belt pitch and project across to the rpm of the high-speed shaft. Opposite this value read the minimum number of sprocket

TABLE 32   Typical Timing-Belt Pitch°

Design power		Speed of high-speed shaft, r/min					
		3500		1750		1160	
hp	kW	in	cm	in	cm	in	cm
25	18.6	½, ⅞	1.3, 2.2	½, ⅞	1.3, 2.2	⅞	2.2
50	37.3	½, ⅞	1.3, 2.2	⅞	2.2	⅞, 1¼	2.2, 3.2
60 and up	44.7 and up	⅞	2.2	⅞	2.2	⅞, 1¼	2.2, 3.2

°Morse Chain Company.

TABLE 33 Minimum Number of Sprocket Teeth°

Belt pitch		High-speed shaft, r/min	Minimum sprocket pitch distance		No. of teeth
in	cm		in	cm	
½	1.3	3500	3.501	8.9	20
		1750	3.183	8.1	18
		1160	2.865	7.3	16
⅞	2.2	3500	7.241	18.4	26
		1750	6.685	17.0	24
		1160	6.127	15.6	22
1¼	3.2	3500	10.345	26.3	26
		1750	9.549	24.3	24
		1160	8.753	22.2	22

°Morse Chain Company.

teeth. Thus, for a 1750-r/min ⅞-in (2.2-cm) pitch timing belt, Table 32 shows that the high-speed sprocket should have no less than 24 teeth nor a pitch diameter less than 6.685 in (17.0 cm). (If a smaller diameter sprocket were used, the belt service life would be reduced.)

6. Select a suitable timing belt

Enter Table 34, or the manufacturer's engineering data, at either the exact speed ratio, if tabulated, or the nearest value to the speed-ratio range. For this drive, having a ratio of 2.92:3.02, the nearest value in Table 34 is 3.00. This table shows that with a 24-tooth driver and a 72-tooth driven sprocket, a center distance of 27.17 in (69.0 cm) is obtainable. Since a center distance of about 27 in (68.6 cm) is desired, this belt is acceptable.

Where an exact center distance is specified, several different sprocket combinations may have to be tried before a belt having a suitable center distance is obtained.

7. Determine the required belt width

Each center distance listed in Table 34 corresponds to a specific pitch and type of belt construction. The belt construction is often termed XL, L, H, XH, and XXH. Thus, the belt chosen in step 6 is an XH construction.

Refer now to Table 35 or the manufacturer's engineering data. Table 35 shows that a 2-in (5.1-cm) wide belt will transmit 38 hp (28.3 kW) at 1750 r/min. This is too low, because the design horsepower rating of the belt is 40 hp (29.8 kW). A 3-in (7.6-cm) wide belt will transmit 60 hp (44.7 kW). Therefore, a 3-in (7.6-cm) belt should be used because it can safely transmit the required horsepower.

If five, or less, teeth are in mesh when a timing belt is installed, the width of the belt must be

TABLE 34 Timing-Belt Center Distances°

Speed ratio	No. of sprocket teeth		Center distance					
	Driver	Driven	XH		XH		XH	
			in	cm	in	cm	in	cm
2.80	30	84	22.81	57.94	30.11	76.48	37.30	94.74
3.00	24	72	27.17	69.01	34.34	87.22	41.46	105.3
3.20	30	96	19.19	48.74	26.84	68.17	34.19	86.84

°Morse Chain Company.

TABLE 35 Belt Power Rating°

[⅞ in (2.2 cm) pitch XH]

No. of teeth in high-speed sprocket	Belt width		Sprocket rpm					
			1700		1750		2000	
	in	cm	hp	kW	hp	kW	hp	kW
24	2	5.1	37	27.6	38	28.3	43	32.1
	3	7.6	59	44.0	60	44.7	67	50.0
	4	10.2	83	61.9	85	63.4	95	70.8

°Morse Chain Company.

increased to ensure sufficient load-carrying ability. To determine the required belt width to carry the load, divide the belt width by the appropriate factor given below.

Teeth in mesh	5	4	3	2
Factor	0.80	0.60	0.40	0.20

Thus, a 3-in (7.6-cm) belt with four teeth in mesh would have to be widened to 3/0.60 = 5.0 in (12.7 cm) to carry the desired load.

Related Calculations: Use this procedure to select timing belts for any of these drives: agitators, mixers, centrifuges, compressors, conveyors, fans, blowers, generators (electric), exciters, hammer mills, hoists, elevators, laundry machinery, line shafts, machine tools, paper-manufacturing machinery, printing machinery, pumps, sawmills, textile machinery, woodworking tools, etc. For exact selection of a specific make of belt, consult the manufacturer's tabulated or plotted engineering data.

GEARED SPEED REDUCER SELECTION AND APPLICATION

Select a speed reducer to lift a sluice gate weighing 200 lb (889.6 N) through a distance of 6 ft (1.8 m) in 5 s or less. The door must be opened and closed 12 times per hour. The drive for the door lifter is a 1150-r/min electric motor that operates 10 h/day.

Calculation Procedure:

1. *Choose the type of speed reducer to use*

There are many types of speed reducers available for industrial drives. Thus, a roller chain with different size sprockets, a V-belt drive, or a timing-belt drive might be considered for a speed-reduction application because all will reduce the speed of a driven shaft. Where a load is to be raised, often geared speed reducers are selected because they provide a positive drive without slippage. Also, modern geared drives are compact, efficient units that are easily connected to an electric motor. For these reasons, a right-angle worm-gear speed reducer will be tentatively chosen for this drive. If upon investigation this type of drive proves unsuitable, another type will be chosen.

2. *Determine the torque that the speed reducer must develop*

A convenient way to lift a sluice door is by means of a roller chain attached to a bracket on the door and driven by a sprocket keyed to the speed reducer output shaft. As a trial, assume that a 12-in (30.5-cm) diameter sprocket is used.

The torque T lb·in developed by sprocket $= T = Wr$, where $W =$ weight lifted, lb; $r =$ sprocket radius, in. For this sprocket, by assuming that the starting friction in the sluice-door guides produces an additional load of 50 lb (222.4 N), $T = (200 + 50)(6) = 1500$ lb·in (169.5 N·m).

3.   Compute the required rpm of the output shaft

The door must be lifted 6 ft (1.8 m) in 5 s. This is a speed of (6 ft × 60 s/min)/5 s = 72 ft/min (0.4 m/s). The circumference of the sprocket is $\pi d = \pi(1.0) = 3.142$ ft (1.0 m). To lift the door at a speed of 72 ft/min (0.4 m/s), the output shaft must turn at a speed of (ft/min)/(ft/r) = 72/3.142 = 22.9 r/min. Since a slight increase in the speed of the door is not objectionable, assume that the output shaft turns at 23 r/min.

4.   Apply the drive service factor

The AGMA *Standard Practice for Single and Double Reduction Cylindrical Worm and Helical Worm Speed Reducers* lists service factors for geared speed reducers driven by electric motors and internal-combustion engines. These factors range from a low of 0.80 for an electric motor driving a machine producing a uniform load for occasional 0.5-h service to a high of 2.25 for a single-cylinder internal-combustion engine driving a heavy shock load 24 h/day. The service factor for this drive, assuming a heavy shock load during opening and closing of the sluice gate, would be 1.50 for 10-h/day operation. Thus, the drive must develop a torque of at least (load torque, lb·in)(service factor) = (1500)(1.5) = 2250 lb·in (254.2 N·m).

5.   Choose the speed reducer

Refer to Table 36 or the manufacturer's engineering data. Table 36 shows that a single-reduction worm-gear speed reducer having an input of 1.24 hp (924.7 W) will develop 2300 lb·in (254.2 N·m) of torque at 23 r/min. This is an acceptable speed reducer because the required ouput torque is 2250 lb·in (254.2 N·m) at 23 r/min. Also, the allowable overhung load, 1367 lb (6080.7 N), is adequate for the sluice-gate weight. A 1.5-hp (1118.5-W) motor would be chosen for this drive.

Related Calculations: Use this general procedure to select geared speed reducers (single- or double-reduction worm gears, single-reduction helical gears, gear motors, and miter boxes) for machinery drives of all types, including pumps, loaders, stokers, welding positioners, fans, blowers, and machine tools. The starting friction load, applied to the drive considered in this procedure, is typical for applications where a heavy friction load is likely to occur. In rotating machinery of many types, the starting friction load is usually nil, except where the drive is connected to a loaded member, such as a conveyor belt. Where a clutch disconnects the driver from the load, there is negligible starting friction.

Well-designed geared speed reducers generally will not run at temperatures higher than 100°F (55.6°C) *above* the prevailing ambient temperature, measured in the lubricant sump. At higher operating temperatures the lubricant may break down, leading to excessive wear. Fan-cooled speed reducers can carry heavier loads than noncooled reducers without overheating.

TABLE 36   Speed Reducer Torque Ratings°

(Single-reduction worm gear)

Input power at 1150 r/min		Drive output				
			Torque		OHL†	
hp	kW	r/min	lb·in	N·m	lb	N
1.54	1.15	28.7	2416	273.0	1367	6080.7
1.24	0.92	23.0	2300	259.9	1367	6080.7
0.93	0.69	19.2	1970	222.6	1367	6080.7

°Extracted from Morse Chain Company data.
†Allowable overhung load on drive.

POWER TRANSMISSION FOR A VARIABLE-SPEED DRIVE

Choose the power-transmission system for a three-wheeled contractor's vehicle designed to carry a load of 1000 lb (4448.2 N) at a speed of 8 mi/h (3.6 m/s) over rough terrain. The vehicle tires will be 16 in (40.6 cm) in diameter, and the engine driving the vehicle will operate continuously. The empty vehicle weighs 600 lb (2668.9 N), and the engine being considered has a maximum speed of 4200 r/min.

Calculation Procedure:

1. Compute the horsepower required to drive the vehicle

Compute the required driving horsepower from $hp = 1.25\ Wmph/1750$, where W = total weight of *loaded* vehicle, lb (N); mph = maximum loaded vehicle speed, mi/h (km/h). Thus, for this vehicle, $hp = 1.25(1000 + 600)(8)/1750 = 9.15$ hp (6.8 kW).

2. Determine the maximum vehicle wheel speed

Compute the maximum wheel rpm from rpm_w = (maximum vehicle speed, mi/h) $\times$ (5280 ft/mi)/15.72 (tire rolling diameter, in). Or, $rpm_w = (8)(5280)/[(15.72)(16)] = 167.8$ r/min.

3. Select the power transmission for the vehicle

Refer to engineering data published by drive manufacturers. Choose a drive suitable for the anticipated load. The load on a typical contractor's vehicle is one of sudden starts and stops. Also, the drive must be capable of transmitting the required horsepower. A 10-hp (7.5-kW) drive would be chosen for this vehicle.

Small vehicles are often belt-driven by means of an infinitely variable transmission. Such a drive, having an overdrive or speed-increase ratio of 1:1.5 or 1:1, would be suitable for this vehicle. From the manufacturer's engineering data, a drive having an input rating of 10 hp (7.5 kW) will be suitable for momentary overloads of up to 25 percent. The operating temperature of any part of the drive should never exceed 250°F (121.1°C). For best results, the drive should be operated at temperatures well below this limit.

4. Compute the required output-shaft speed reduction

To obtain the maximum power output from the engine, the engine should operate at its maximum rpm when the vehicle is traveling at its highest speed. This prevents lugging of the engine at lower speeds.

The transmission transmits power from the engine to the driving axle. Usually, however, the transmission cannot provide the needed speed reduction between the engine and the axle. Therefore, a speed-reduction gear is needed between the transmission and the axle. The transmission chosen for this drive could provide a 1:1 or a 1:1.5 speed ratio. Assume that the 1:1.5 speed ratio is chosen to provide higher speeds at the maximum vehicle load. Then the speed reduction required = (maximum engine speed, r/min)(transmission ratio)/(maximum wheel rpm) = $(4200)(1.5)/167.8 = 37.6$.

Check the manufacturer's engineering data for the ratios of available geared speed reducers. Thus, a study of one manufacturer's data shows that a speed-reduction ratio of 38 is available by using a single-reduction worm-gear drive. This drive would be suitable if it were rated at 10 hp (7.5 kW) or higher. Check to see that the gear has a suitable horsepower rating before making the final selection.

Related Calculations: Use the general procedure given here to choose power transmissions for small-vehicle compressors, hoists, lawn mowers, machine tools, conveyors, pumps, snow sleds, and similar equipment. For nonvehicle drives, substitute the maximum rpm of the driven machine for the maximum wheel velocity in steps 2, 3, and 4.

BEARING-TYPE SELECTION FOR A KNOWN LOAD

Choose a suitable bearing for a 3-in (7.6-cm) diameter 100-r/min shaft carrying a total radial load of 12,000 lb (53,379 N). A reasonable degree of shaft misalignment must be allowed by the bearing. Quiet operation of the shaft is desired. Lubrication will be intermittent.

Calculation Procedure:

1. *Analyze the desired characteristics of the bearing*

Two major types of bearings are available to the designer, *rolling* and *sliding*. Rolling bearings are of two types, *ball* and *roller*. Sliding bearings are also of two types, *journal* for radial loads and *thrust* for axial loads only or for combined axial and radial loads. Table 37 shows the principal characteristics of rolling and sliding bearings. Based on the data in Table 37, a sliding bearing would be suitable for this application because it has a *fair* misalignment tolerance and a *quiet* noise level. Both factors are key considerations in the bearing choice.

TABLE 37   Key Characteristics of Rolling and Sliding Bearings°

	Rolling	Sliding
Life	Limited by fatigue properties of bearing metal	Unlimited, except for cyclic loading
Load:		
Unidirectional	Excellent	Good
Cyclic	Good	Good
Starting	Excellent	Poor
Unbalance	Excellent	Good
Shock	Good	Fair
Emergency	Fair	Fair
Speed limited by:	Centrifugal loading and material surface speeds	Turbulence and temperature rise
Starting friction	Good	Poor
Cost	Intermediate, but standardized, varying little with quantity	Very low in simple types or in mass production
Space requirements (radial bearing):		
Radial dimension	Large	Small
Axial dimension	⅛ to ½ shaft diameter	¼ to 2 times shaft diameter
Misalignment tolerance	Poor in ball bearings except where designed for at sacrifice of load capacity; good in spherical roller bearings; poor in cylindrical roller bearings	Fair
Noise	May be noisy, depending on quality and resonance of mounting	Quiet
Damping	Poor	Good
Low-temperature starting	Good	Poor
High-temperature operation	Limited by lubricant	Limited by lubricant
Type of lubricant	Oil or grease	Oil, water, other liquids, grease, dry lubricants, air, or gas
Lubrication, quantity required	Very small, except where large amounts of heat must be removed	Large, except in low-speed boundary-lubrication types
Type of failure	Limited operation may continue after fatigue failure but not after lubricant failure	Often permits limited emergency operation after failure
Ease of replacement	Function of type of installation; usually shaft need not be replaced	Function of design and installation; split bearings used in large machines

°*Product Engineering.*

TABLE 38 Materials for Sleeve Bearings[a]

[*Cost figures are for a 1-in (2.54-cm) sleeve bearing ordered in quantity*]

	Maximum load		Maximum speed		PV limit		Maximum operating temperature		Cost, $
	lb/in²	kPa	ft/min	m/s	(lb/in²)(ft/min)	kPa·m/s	°F	°C	
Porous bronze	4,000	27,579.0	1,500	7.6	50,000	1,751.3	150	65.6	0.11
Porous iron	8,000	55,158.1	800	4.1	50,000	1,751.3	150	65.6	0.09
Teflon fabric	60,000	413,685.4	50	0.3	25,000	875.6	500	260.0	0.04
Phenolic	6,000	41,368.5	2,500	12.7	15,000	525.4	200	93.3	0.05
Wood	6,000	41,368.5	2,000	10.2	15,000	525.4	150	65.6	0.40
Carbon-graphite	600	4,136.9	2,500	12.7	10,000	350.3	750	398.9	0.39
Reinforced Teflon	2,500	17,236.9	2,500	12.7	10,000	350.3	500	260.0	0.45
Nylon	1,000	6,894.8	1,000	5.1	3,000	105.1	200	93.3	0.04
Delrin	1,000	6,894.8	1,000	5.1	3,000	105.1	180	82.2	0.03
Lexan	1,000	6,894.8	1,000	5.1	3,000	105.1	220	104.4	0.05
Teflon	500	3,447.4	100	0.5	1,000	35.0	500	260.0	1.00

[a] *Product Engineering.*

2. Choose the bearing materials

Table 38 shows that a porous-bronze bearing, suitable for intermittent lubrication, can carry a maximum pressure load of 4000 lb/in² (27,580.0 kPa) at a maximum shaft speed of 1500 ft/min (7.62 m/s). By using the relation $l = L/(Pd)$, where l = bearing length, in, L = load, lb, d = shaft diameter, in, the required length of this sleeve bearing is $l = L/(Pd)$ = 12,000/[(4000)(3)] = 1 in (2.5 cm).

Compute the shaft surface speed V ft/min from $V = \pi dR/12$, where d = shaft diameter, in; R = shaft rpm. Thus, $V = \pi(3)(100)/12 = 78.4$ ft/min (0.4 m/s).

With the shaft speed known, the PV, or pressure-velocity, value of the bearing can be computed. For this bearing, with an operating pressure of 4000 lb/in² (27,580.0 kPa), PV = 4000 × 78.4 = 313,600 (lb/in²) (ft/min) (10,984.3 kPa·m/s). This is considerably in excess of the PV limit of 50,000 (lb/in²) (ft/min) (1751.3 kPa·m/s) listed in Table 38. To come within the recommended PV limit, the operating pressure of the bearing must be reduced.

Assume an operating pressure of 600 lb/in² (4137.0 kPa). Then $l = L/(Pd)$ = 12,000/[(600)(3)] = 6.67 in (16.9 cm), say 7 in (17.8 cm). The PV value of the bearing then is (600)(78.4) = 47,000 (lb/in²) (ft/min) (1646.3 kPa·m/s). This is a satisfactory value for a porous-bronze bearing because the recommended limit is 50,000 (lb/in²) (ft/min) (1751.3 kPa·m/s).

3. Check the selected bearing size

The sliding bearing chosen will have a diameter somewhat in excess of 3 in (7.6 cm) and a length of 7 in (17.8 cm). If this length is too great to fit in the allowable space, another bearing material will have to be studied, by using the same procedure. Figure 8 shows the space occupied by rolling and sliding bearings of various types.

Table 39 shows the load-carrying capacity and maximum operating temperatures for oil-film journal sliding bearings that are regularly lubricated. These bearings are termed *full film* because they receive a supply of lubricant at regular intervals. Surface speeds of 20,000 to 25,000 ft/min (101.6 to 127.0 m/s) are common for industrial machines fitted with these bearings. This corresponds closely to the surface speed for ball and roller bearings.

4. Evaluate oil-film bearings

Oil-film sliding bearings are chosen by the method of the next calculation procedure. The bearing size is made large enough that the maximum operating temperature listed in Table 39 is not exceeded. Table 40 lists typical design load limits for oil-film bearings in various services. Figure 9 shows the typical temperature limits for rolling and sliding bearings made of various materials.

5. Evaluate rolling bearings

Rolling bearings have lower starting friction (coefficient of friction f = 0.002 to 0.005) than sliding bearings (f = 0.15 to 0.30). Thus, the rolling bearing is preferred for applications requiring low starting torque [integral-horsepower electric motors up to 500 hp (372.9 kW), jet engines, etc.]. By pumping oil into a sliding bearing, its starting coefficient of friction can be reduced to nearly zero. This arrangement is used in large electric generators and certain mill machines.

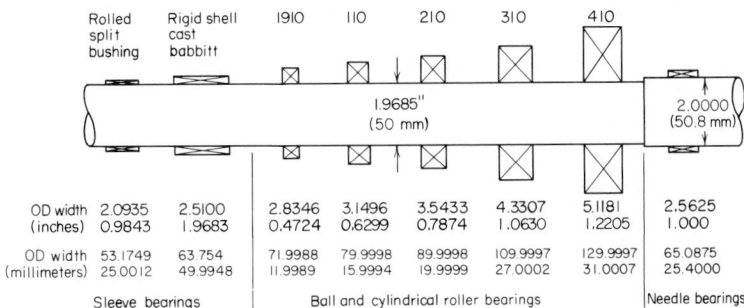

FIG. 8 Relative space requirements of sleeve and rolling-element bearings to carry the same diameter shaft. *(Product Engineering.)*

TABLE 39 Oil-Film Journal Bearing Characteristics[°]

Bearing material	Load-carrying capacity		Maximum operating temperature	
	lb/in^2	kPa	°F	°C
Tin-based babbitt	800–1,500	5,516.0–10,342.5	300	148.9
Lead-based babbitt	800–1,200	5,516.0–8,274.0	300	148.9
Alkali-hardened steel	1,200–1,500	8,274.0–10,342.5	500	260.0
Cadmium base	1,500–2,000	10,342.5–13,789.5	500	260.0
Copper-lead	1,500–2,500	10,342.5–17,236.9	350	176.7
Tin bronze	4,000	27,580.0	500+	260.0+
Lead bronze	3,000–4,000	20,685.0–27,580.0	450	232.2
Aluminum alloy	4,000	27,580.0	250	121.1
Silver (overplated)	4,000	27,580.0	500	260.0
Three-component bearings babbitt-surfaced	2,000–4,000	13,790.0–27,580.0	225–300	107.2–148.9

[°] *Product Engineering.*

TABLE 40 Typical Design Load Limits for Oil-Film Bearings[°]

Bearing	Maximum load on projected area	
	lb/in^2	kPa
Electric motors	200	1,379.0
Steam turbines	300	2,068.4
Automotive engines:		
Main bearings	3,500	24,131.6
Connecting rods	5,000	34,473.8
Diesel engines:		
Main bearings	3,000	20,684.7
Connecting rods	4,500	31,026.4
Railroad car axles	350	2,413.2
Steel mill roll necks:		
Steady	2,000	13,789.5
Peak	5,000	24,473.8

[°] *Product Engineering.*

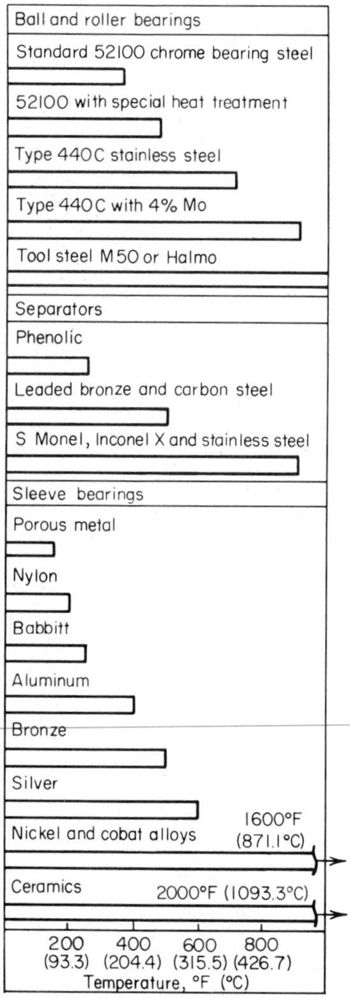

FIG. 9 Bearing temperature limits. (*Product Engineering.*)

TABLE 41 Relative Load Capacity, Cost, and Size of Rolling Bearings°

Bearing type (for 50-mm bore)	Radial capacity	Axial capacity	Cost	Outer diameter	Width
Ball bearings:					
Deep groove (Conrad)	1.0	1.0	1.0	1.0	1.0
Filling notch	1.2	Low	1.2	1.0	1.0
Double row	1.5	1.1	2.2	1.0	1.6
Angular contact	1.1	1.9	1.6	1.0	1.0
Duplex	1.8	1.9	2.0	1.0	2.0
Self-aligning	0.7	0.2	1.3	1.0	1.0
Ball thrust	0	0.9	0.8	0.7	0.8
Roller bearings:					
Cylindrical	1.6	. . .	1.9	1.0	1.0
Tapered	1.3	0.8	0.9	1.0	1.0
Spherical	3.0	1.1	5.0	1.0	1.5
Needle	1.0	0	0.3	0.5	1.6
Flat thrust	0	4.0	3.8	0.8	0.9

° *Product Engineering.*

The running friction of rolling bearings is in the range of $f = 0.001$ to 0.002. For oil-film sliding bearings, $f = 0.002$ to 0.005.

Rolling bearings are more susceptible to dirt than are sliding bearings. Also, rolling bearings are inherently noisy. Oil-film bearings are relatively quiet, but they may allow higher amplitudes of shaft vibration.

Table 41 compares the size, load capacity, and cost of rolling bearings of various types. Briefly, ball bearings and roller bearings may be compared thus: ball bearings (*a*) run at higher speeds without undue heating, (*b*) cost less per pound of load-carrying capacity for light loads, (*c*) have friction torque at light loads, (*d*) are available in a wider variety of sizes, (*e*) can be made in smaller sizes, and (*f*) have seals and shields for easy lubrication. Roller bearings (*a*) can carry heavier loads, (*b*) are less expensive for larger sizes and heavier loads, (*c*) are more satisfactory under shock and impact loading, and (*d*) may have lower friction at heavy loads. Table 42 shows the speed limit, termed the *dR limit* (equals bearing shaft bore *d* in mm multiplied by the shaft rpm *R*), for ball and roller bearings. Speeds higher than those shown in Table 42 may lead to early bearing failures. Since the *dR* limit is proportional to the shaft surface speed, the *dR* value gives an approximate measure of the bearing power loss and temperature rise.

Related Calculations: Use this general procedure to select shaft bearings for any type of regular service conditions. For unusual service (i.e., excessively high or low operating temperatures, large loads, etc.) consult the specific selection procedures given elsewhere in this section.

TABLE 42 Speed Limits for Ball and Roller Bearings°

Lubrication	DN limit, mm × r/min
Oil:	
Conventional bearing designs	300,000–350,000
Special finishes and separators	1,000,000–1,500,000
Grease:	
Conventional bearing designs	250,000–300,000
Silicone grease	150,000–200,000
Special finishes and separators high-speed greases	500,000–600,000

° *Product Engineering.*

Note that the PV value of a sliding bearing can also be expressed as $PV = L/(dl) \times \pi dR/12$ $= \pi LR/(12l)$. The bearing load and shaft speed are usually fixed by other requirements of a design. Where the PV equation is solved for the bearing length l and the bearing is too long to fit the available space, select a bearing material having a higher allowable PV value.

SHAFT BEARING LENGTH AND HEAT GENERATION

How long should a sleeve-type bearing be if the combined weight of the shaft and gear tooth load acting on the bearing is 2000 lb (8896 N)? The shaft is 1 in (2.5 cm) in diameter and is oil-lubricated. What is the rate of heat generation in the bearing when the shaft turns at 60 r/min? How much above an ambient room temperature of 70°F (21.1°C) will the temperature of the bearing rise during operation in still air? In moving air?

Calculation Procedure:

1. Compute the required length of the bearing

The required length l of a sleeve bearing carrying a load of L lb is $l = L/(Pd)$, where $l =$ bearing length, in; $L =$ bearing load, lb; $P =$ bearing reaction force, lb; $P =$ allowable mean bearing pressure, lb/in² [ranges from 25 to 2500 lb/in² (172.4 to 17,237.5 kPa) for normal service and up to 8000 lb/in² (55,160.0 kPa) for severe service], on the projected bearing area, in² $= ld$; $d =$ shaft diameter, in. Thus, for this bearing, assuming an allowable mean bearing pressure of 400 lb/in² (2758.0 kPa), $l = L/(Pd) = 2000/[(400)(1)] = 5$ in (12.7 cm).

2. Compute the rate of bearing heat generation

The rate of heat generation in a plain sleeve bearing is given by $h = fLdR/3000$, where $h =$ rate of heat generation in the bearing, Btu/min; $f =$ bearing coefficient of friction for the lubricant used; $R =$ shaft rpm; other symbols as in step 1.

The coefficient of friction for oil-lubricated bearings ranges from 0.005 to 0.030, depending on the lubricant viscosity, shaft rpm, and mean bearing pressure. Given a value of $f = 0.020$, $h = (0.020)(2000)(1)(60)/3000 = 0.8$ Btu/min, or $H = 0.8(60 \text{ min/h}) = 48.0$ Btu/h (14.1 kW).

3. Compute the bearing wall area

The wall area A of a small sleeve-type bearing, such as a pillow block fitted with a bushing, is $A = (10 \text{ to } 15)dl/144$, where $A =$ bearing wall area, ft²; other symbols as before. For larger bearing pedestals fitted with a cast-iron or steel bearing shell, the factor in this equation varies from 18 to 25.

Since this is a small bearing having a 1-in (2.5-cm) diameter shaft, the first equation with a factor of 15 to give a larger wall area can be used. The value of 15 was chosen to ensure adequate radiating surface. Where space or weight is a factor, the value of 10 might be chosen. Intermediate values might be chosen for other conditions. Substituting yields $A = (15)(1)(5)/144 = 0.521$ ft² (0.048 m²).

4. Determine the bearing temperature rise

In *still air*, a bearing will dissipate $H = 2.2A(t_w - t_a)$ Btu/h, where $t_w =$ bearing wall temperature, °F; t_a ambient air temperature, °F; other symbols as before. Since H and A are known, the temperature rise can be found by solving for $t_w - t_a = H/(2.2A) = 48.0/[(2.2)(0.521)] = 41.9$°F (23.3°C), and $t_w = 41.9 + 70 = 111.9$°F (44.4°C). This is a low enough temperature for safe operation of the bearing. The maximum allowable bearing operating temperature for sleeve bearings using normal lubricants is usually assumed to be 200°F (93.3°C). To reduce the operating temperature of a sleeve bearing, the bearing wall area must be increased, the shaft speed decreased, or the bearing load reduced.

In *moving air*, the heat dissipation from a sleeve-type bearing is $H = 6.5A(t_w - t_a)$. Solving for the temperature rise as before, we get $t_w - t_a = H/(6.5A) = 48.0/[(6.5)(0.521)] = 14.2$°F (7.9°C), and $t_w = 14.2 + 70 = 84.2$°F (29.0°C). This is a moderate operating temperature that could be safely tolerated by any of the popular bearing materials.

Related Calculations: Use this procedure to analyze sleeve-type bearings used for industrial line shafts, marine propeller shafts, conveyor shafts, etc. Where the ambient temperature varies during bearing operation, use the highest ambient temperature expected, in computing the bearing operating temperature.

ROLLER-BEARING OPERATING-LIFE ANALYSIS

A machine must have a shaft of about 5.5 in (14.0 cm) in diameter. Choose a roller bearing for this 5.5-in (14.0-cm) diameter shaft that turns at 1000 r/min while carrying a radial load of 20,000 lb (88,964 N). What is the expected life of this bearing?

Calculation Procedure:

1.   *Determine the bearing life in revolutions*

The operating life of rolling-type bearings is often stated in millions of revolutions. Find this life from $R_L = (C/L)^{10/3}$, where R_L = bearing operating life, millions of revolutions; C = dynamic capacity of the bearing, lb; L = applied radial load on bearing, lb.

Obtain the dynamic capacity of the bearing being considered by consulting the manufacturer's engineering data. Usual values of dynamic capacity range between 2500 lb (11,120.6 N) and 750,000 lb (3,338,166.5 N), depending on the bearing design, type, and bore. For a typical 5.5118-in (14.0-cm) bore roller bearing, C = 92,400 lb (411,015.7 N).

With C known, compute $R_L = (C/L)^{10/3} = (92,400/20,000)^{10/3} = 162 \times 10^6$.

2.   *Determine the bearing life*

The minimum life of a bearing in millions of revolutions, R_L, is related to its life in hours, h, by the expression $R_L = 60Rh/10^6$, where R = shaft speed, r/min. Solving gives $h = 10^6 R_L/(60R)$ = $(10^6)(162)/[(60)(1000)]$ = 2700-h minimum life.

Related Calculations: This procedure is useful for those situations where a bearing must fit a previously determined shaft diameter or fit in a restricted space. In these circumstances, the bearing size cannot be varied appreciably, and the machine designer is interested in knowing the minimum probable life that a given size of bearing will have. Use this procedure whenever the bearing size is approximately predetermined by the installation conditions in motors, pumps, engines, portable tools, etc. Obtain the dynamic capacity of any bearing under consideration from the manufacturer's engineering data.

ROLLER-BEARING CAPACITY REQUIREMENTS

A machine must be fitted with a roller bearing that will operate at least 30,000 h without failure. Select a suitable bearing for this machine in which the shaft operates at 3600 r/min and carries a radial load of 5000 lb (22,241.1 N).

Calculation Procedure:

1.   *Determine the bearing life in revolutions*

Use the relation $R_L = 60Rh/10^6$, where the symbols are the same as in the previous calculation procedure. Thus, $R_L = 60(3600)(30,000)/10^6$; R_L = 6480 million revolutions (Mr).

2.   *Determine the required dynamic capacity of the bearing*

Use the relation $R_L = (C/L)^{10/3}$, where the symbols are the same as in the previous calculation procedure. So $C = L(R_L)^{3/10} = (5000)(6480)^{3/10} = 69,200$ lb (307,187.0 N).

Choose a bearing of suitable bore having a dynamic capacity of 69,200 lb (307,187.0 N) or more. Thus, a typical 5.9055-in (15.0-cm) bore roller bearing has a dynamic capacity of 72,400 lb (322,051.3 N). It is common practice to undercut the shaft to suit the bearing bore, if such a reduction in the shaft does not weaken the shaft. Use the manufacturer's engineering data in choosing the actual bearing to be used.

Related Calculations: This procedure shows a situation in which the life of the bearing is of greater importance than its size. Such a situation is common when the reliability of a machine is a key factor in its design. A dynamic rating of a given amount, say 72,400 lb (322,051.3 N), means that if in a large group of bearings of this size each bearing has a 72,400-lb (322,051.3-N) load applied to it, 90 percent of the bearings in the group will complete, or exceed, 10^6 r before the first evidence of fatigue occurs. This average life of the bearing is the number of revolutions that 50 percent of the bearings will complete, or exceed, before the first evidence of fatigue develops. The average life is about 3.5 times the minimum life.

Use this procedure to choose bearings for motors, engines, turbines, portable tools, etc. Where extreme reliability is required, some designers choose a bearing having a much larger dynamic capacity than calculations show is required.

RADIAL LOAD RATING FOR ROLLING BEARINGS

A mounted rolling bearing is fitted to a shaft driven by a 4-in (10.2-cm) wide double-ply leather belt. The shaft is subjected to moderate shock loads about one-third of the time while operating at 300 r/min. An operating life of 40,000 h is required of the bearing. What is the required radial capacity of the bearing? The bearing has a normal rated life of 15,000 h at 500 r/min. The weight of the pulley and shaft is 145 lb (644.9 N).

Calculation Procedure:

1.   *Determine the bearing operating factors*

To determine the required radial capacity of a rolling bearing, a series of operating factors must be applied to the radial load acting on the shaft: life factor f_L, operating factor f_O, belt tension factor f_B, and speed factor f_S. Obtain each of the four factors from the manufacturer's engineering data because there may be a slight variation in the factor value between different bearing makers. Where a given factor does not apply to the bearing being considered, omit it from the calculation.

2.   *Determine the bearing life factor*

Rolling bearings are normally rated for a certain life, expressed in hours. If a different life for the bearing is required, a life factor must be applied. The bearing being considered here has a normal rated life of 15,000 h. The manufacturer's engineering data show that for a mounted bearing which must have a life of 40,000 h, a life factor $f_L = 1.340$ should be used. For this particular make of bearing, f_L varies from 0.360 at a 500-h to 1.700 at a 100,000-h life for mounted units. At 15,000 h, $f_L = 1.000$.

3.   *Determine the bearing operating factor*

A rolling-bearing operating factor is used to show the effect of peak and shock loads on the bearing. Usual operating factors vary from 1.00 for steady loads with any amount of overload to 2.00 for bearings with heavy shock loads throughout their operating period. For this bearing with moderate shock loads about one-third of the time, $f_O = 1.32$.

A combined operating factor, obtained by taking the product of two applicable factors, is used in some circumstances. Thus, when the load is an oscillating type, an additional factor of 1.25 must be applied. This type of load occurs in certain linkages and pumps. When the outer race of the bearing revolves, as in sheaves, truck wheels, or gyrating loads, an additional factor of 1.2 is used. To find the combined operating factor, first find the normal operating factor, as described earlier. Then take the product of the normal and the additional operating factors. The result is the combined operating factor.

4.   *Determine the bearing belt-tension factor*

When the bearing is used on a belt-driven shaft, a belt-tension factor must be applied. Usual values of this factor range from 1.0 for a chain drive to 2.30 for a single-ply leather belt. For a double-ply leather belt, $f_B = 2.0$.

5.   *Determine the bearing speed factor*

Rolling bearings are rated at various speeds. When the shaft operates at a speed different from the rated speed, a speed factor f_S must be applied. Since this shaft operates at 300 r/min while the rated speed of the bearing is 500 r/min, $f_S = 0.860$, from the manufacturer's engineering data. For a 500-r/min bearing, f_S varies from 0.245 at 5 r/min to 1.87 at 4000 r/min.

6.   *Determine the radial load on the bearing*

The radial load produced by a leather belt can vary from 130 lb/in (227.7 N/cm) of width for normal-tension belts to 450 lb/in (788.1 N/cm) of width for very tight belts. Assuming normal tension, the radial load for a double-ply leather belt is, from engineering data, 180 lb/in (315.2 N/cm) of width. Since this belt is 4 in (10.2 cm) wide, the radial belt load = 4(180) = 720 lb (3202.7 N). The total radial load R_T is the sum of the belt, shaft, and pulley loads, or $R_T = 720 + 145 = 865$ lb (3847.7 N).

7. *Compute the required radial capacity of the bearing*

The required radial capacity of a bearing $R_C = R_T f_L f_O f_B f_S = (865)(1.340)(1.32)(2.0)(0.86) = 2630$ lb (11,698.8 N).

8. *Select a suitable bearing*

Enter the manufacturer's engineering data at the shaft rpm (300 r/min for this shaft) and project to a bearing radial capacity equal to, or slightly greater than, the computed required radial capacity. Thus, one make of bearing, suitable for 2- and 2⅜-in (5.1- and 5.6-cm) diameter shafts, has a radial capacity of 2710 at 300 r/min. This is close enough for general selection purposes.

    Related Calculations: Use this general procedure for any type of rolling bearing. When comparing different makes of rolling bearings, be sure to convert them to the same life expectancies before making the comparison. Use the life-factor table presented in engineering handbooks or a manufacturer's engineering data for each bearing to convert the bearings being considered to equal lives.

ROLLING-BEARING CAPACITY AND RELIABILITY

What is the required basic load rating of a ball bearing having an equivalent radial load of 3000 lb (13,344.7 N) if the bearing must have a life of 400×10^6 r at a reliability of 0.92? The ratio of the average life to the rating life of the bearing is 5.0. Show how the required basic load rating is determined for a roller bearing.

Calculation Procedure:

1. *Compute the required basic load rating*

Use the Weibull two-parameter equation, which for a life ratio of 5 is $L_e/L_B = (1.898/R_L^{0.333})(\ln 1/R_e)^{0.285}$, where L_e = equivalent radial load on the bearing, lb; L_B = the required basic load rating of the bearing, lb, to give the desired reliability at the stated life; R_L = bearing operating life, Mr; ln = natural or Naperian logarithm to the base e; R_e = required reliability, expressed as a decimal.

    Substituting gives $3000/L_B = (1.898/400^{0.333})(\ln 1/0.92)^{0.285}$; $L_B = 23,425$ lb (104,199.6 N). Thus, a bearing having a basic load rating of at least 23,425 lb (104,199.6 N) would provide the desired reliability. Select a bearing having a load rating equal to, or slightly in excess of, this value. Use the manufacturer's engineering data as the source of load-rating data.

2. *Compute the roller-bearing basic load rating*

Use the following form of the Weibull equation for roller bearings: $L_e/L_B = (1.780/R_L^{0.30})(\ln 1/R_e)^{0.257}$. Substitute values and solve for L_B, as in step 1.

    Related Calculations: Use the Weibull equation as given here when computing bearing life in the range of 0.9 and higher. The ratio of average life/rating life = 5 is usual for commercially available bearings.[1]

POROUS-METAL BEARING CAPACITY AND FRICTION

Determine the load capacity ψ and coefficient of friction of a porous-metal bearing for a 1-in (2.5-cm) diameter shaft, 1-in (2.5-cm) bearing length, 0.2-in (0.5-cm) thick bearing, 0.001-in (0.003-cm) radial clearance, metal permeability $\phi = 5 \times 10^{-10}$ in (1.3×10^{-9} cm), shaft speed = 1500 r/min, eccentricity ratio $\epsilon = 0.8$, and an SAE-30 mineral-oil lubricant with a viscosity of 6×10^{-6} lb·s/in² (4.1×10^{-6} N·s/cm²).

Calculation Procedure:

1. *Sketch the bearing and shaft*

Figure 10 shows the bearing and shaft with the various known dimensions indicated by the identifying symbols given above.

---

[1]C. Mischke, "Bearing Reliability and Capacity," *Machine Design,* Sept. 30, 1965.

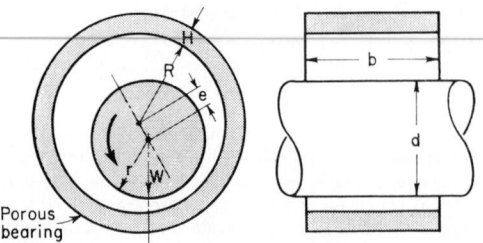

FIG. 10 Typical porous-metal bearing. *(Product Engineering.)*

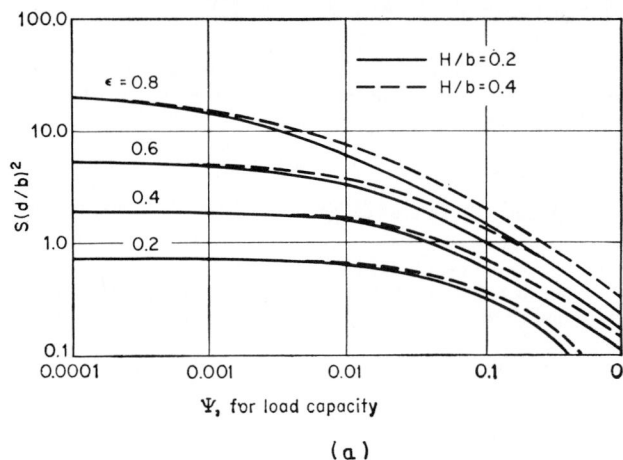

(a)

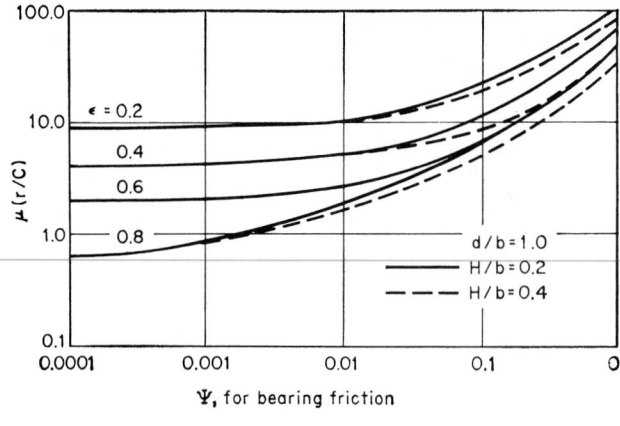

(b)

FIG. 11 (*a*) Bearing load capacity factors; (*b*) bearing friction factors. *(Product Engineering.)*

2. *Compute the load capacity factor*

The load capacity factor $\psi = \phi H/C^3$, where ϕ = metal permeability, in; H = bearing thickness, in; C = radial clearance of bearing, $R_b - r = 0.001$ in (0.003 cm) for this bearing. Hence, $\psi = (5 \times 10^{-10})(0.2)/(0.001)^3 = 0.10$.

3. *Compute the bearing thickness-length ratio*

The thickness-length ratio = $H/b = 0.2/1.0 = 0.2$.

4. *Determine the $S(d/b)^2$ value for the bearing*

In the $S(d/b)^2$ value for the bearing, S = the Summerfeld number for the bearing; the other values are as shown in Fig. 10.

Using the ψ, ϵ, and H/b values, enter Fig. 11a, and read $S(b/d)^2 = 1.4$ for an eccentricity ratio of $\epsilon = e/C = 0.8$. Substitute in this equation the known values for b and d and solve for S, or $S(1.0/1.0)^2 = 1.4$; $S = 1.4$.

5. *Compute the bearing load capacity*

Find the bearing load capacity from $S = (L/R_i \eta b)(C/r)^2$, where η = lubricant viscosity, lb·s/in^2; r = shaft radius, in; R_i = shaft velocity, in·s. Solving gives $L = (SR_i \eta b)/(C/r)^2 = (1.4)(78.5)(6 \times 10^{-6})(1.0)/(0.001/0.5)^2 = 164.7$ lb (732.6 N). (The shaft rotative velocity must be expressed in in/s in this equation because the lubricant viscosity is given in lb·s/in^2.)

6. *Determine the bearing coefficient of friction*

Enter Fig. 11b with the known values of ψ, ϵ, and H/b and read $u(r/C) = 7.4$. Substitute in this equation the known values for r and C, and solve for the bearing coefficient of friction μ, or $\mu = 7.4C/r = (7.4)(0.001)/0.5 = 0.0148$.

Related Calculations: Porous-metal bearings are similar to conventional sliding-journal bearings except that the pores contain an additional supply of lubricant to replace that which may be lost during operation. The porous-metal bearing is useful in assemblies where there is not enough room for a conventional lubrication system or where there is a need for improved lubrication during the starting and stopping of a machine. The permeability of the finished porous metal greatly influences the ability of a lubricant to work its way through the pores. Porous-metal bearings are used in railroad axle supports, water pumps, generators, machine tools, and other equipment. Use the procedure given here when choosing porous-metal bearings for any of these applications. The method given here was developed by Professor W. T. Rouleau, Carnegie Institute of Technology, and C. A. Rhodes, Senior Research Engineer, Jet Propulsion Laboratory, California Institute of Technology.

HYDROSTATIC THRUST BEARING ANALYSIS

An oil-lubricated hydrostatic thrust bearing must support a load of 107,700 lb (479,073.5 N). This vertical bearing has an outside diameter of 16 in (40.6 cm) and a recess diameter of 10 in (25.4 cm). What oil pressure and flow rate are required to maintain a 0.006-in (0.15-mm) lubricant film thickness with an SAE-20 oil having an absolute viscosity of $\eta = 42.4 \times 10^{-7}$ lb/(s·in^2) [2.9 $\times$ 10^{-6} N/(s·cm^2)] if the shaft turns at 750 r/min? What are the pumping loss and the viscous friction loss? What is the optimum lubricant-film thickness?

Calculation Procedure:

1. *Determine the required lubricant-supply pressure*

The design equations and methods developed at Franklin Institute by Dudley Fuller, Professor of Mechanical Engineering, Columbia University, are applicable to vertical hydrostatic bearings, Fig. 12, using oil, grease, or gas lubrication. By substituting the appropriate value for the lubricant viscosity, the same set of design equations can be used for any of the lubricants listed above. These equations are accurate, simple, and reliable; they are therefore used here.

Solve Fuller's applied load equation, $L = (p_i \pi/2)\{r^2 - r_i^2/[\ln(r/r_i)]\}$, for the lubricant-supply inlet pressure. In this equation, L = applied load on the bearing, lb; p_i = lubricant-supply inlet pressure, lb/in^2, r = shaft radius, in; r_i = recess or step radius, in; $\ln$ = natural or Naperian logarithm to the base e. Solving gives $p_i = 2L/\pi\{r^2 - r_i^2/[\ln(r/r_i)]\} = 2(107,700)/\pi [8^2 - 5^2/ (\ln 8/5)]$, or $p_i = 825$ lb/in^2 (5688.4 kPa).

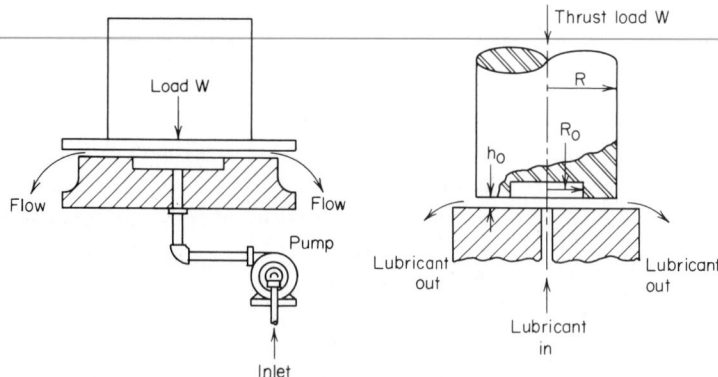

FIG. 12   Hydrostatic thrust bearings. *(Product Engineering.)*

2.   *Compute the required lubricant flow rate*

By Fuller's flow-rate equation, $Q = p_i \pi h^3 / [6\eta \ln (r/r_i)]$, where Q = lubricant flow rate, in³/s; h = lubricant-film thickness, in; η = lubricant absolute viscosity, lb·s/in²; other symbols as in step 1. Thus, with $h = 0.006$ in, $Q = (825\ \pi)(0.006)^3 / [6(42.4 \times 10^{-6})(0.470)] = 46.85$ in³/s (767.7 cm³/s).

3.   *Compute the pumping loss*

The pumping loss results from the work necessary to force the lubricant radially outward through the film space, or $H_p = Q(p_i - p_o)$, where H_p = power required to pump the lubricant = pumping loss, in·lb/s; p_o = lubricant outlet pressure, lb/in²; other symbols as in step 1. For circular thrust bearings it can be assumed that the lubricant outlet pressure p_0 is negligible, or p_0 = 0. Then H_p = 46.85(825 − 0) = 38,680 in·lb/s (4370.3 N·m/s) = 38,680/[550 ft·lb/(min· hp)](12 in/ft) = 5.86 hp (4.4 kW).

4.   *Compute the viscous friction loss*

The viscous-friction-loss equation developed by Fuller is $H_f = [(R^2 \eta/(58.05h)](r^4 - r_0^4)$, where H_f = viscous friction loss, in·lb/s; R = shaft rpm; other symbols as in step 1. Thus, H_f = $\{(750)^2(42.4 \times 10^{-7})/[(58.05)(0.006)]\}(8^4 - 5^4)$, or H_f = 23,770 in·lb/s (2685.6 N·m/s) = 23,770/[(550)(12)] = 3.60 hp (2.7 kW).

5.   *Compute the optimum lubricant-film thickness*

The film thickness that will produce a minimum combination of pumping loss and friction loss can be evaluated by determining the minimum point of the curve representing the sum of the respective energy losses (pumping and viscous friction) when plotted against film thickness.

With the shaft speed, lubricant viscosity, and bearing dimensions constant at the values given in the problem statement, the viscous-friction loss becomes $H_f = 0.0216/h$ for this bearing. Substitute various values for h ranging between 0.001 and 0.010 in (0.0254 and 0.254 mm) (the usual film thickness range), and solve for H_f. Plot the results as shown in Fig. 13.

Combine the lubricant-flow and pumping-loss equations to express H_p in terms of the lubricant-film thickness, or $H_p = (1000h)^3/36.85$, for this bearing with a pump having an efficiency of 100 percent. For a pump with a 50 percent efficiency, this equation becomes $H_p = (1000h)^3/18.42$. Substitute various values of h ranging between 0.001 and 0.010, and plot the results as in Fig. 13 for pumps with 100 and 50 percent efficiencies, respectively. Figure 13 shows that for a 100 percent efficient pump, the minimum total energy loss occurs at a film thickness of 0.004 in (0.102 mm). For 50 percent efficiency, the minimum total energy loss occurs at 0.0035-in (0.09-mm) film thickness, Fig. 13.

Related Calculations: Similar equations developed by Fuller can be used to analyze hydrostatic thrust bearings of other configurations. Figures 14 and 15 show the equations for modified square bearings and circular-sector bearings. To apply these equations, use the same general pro-

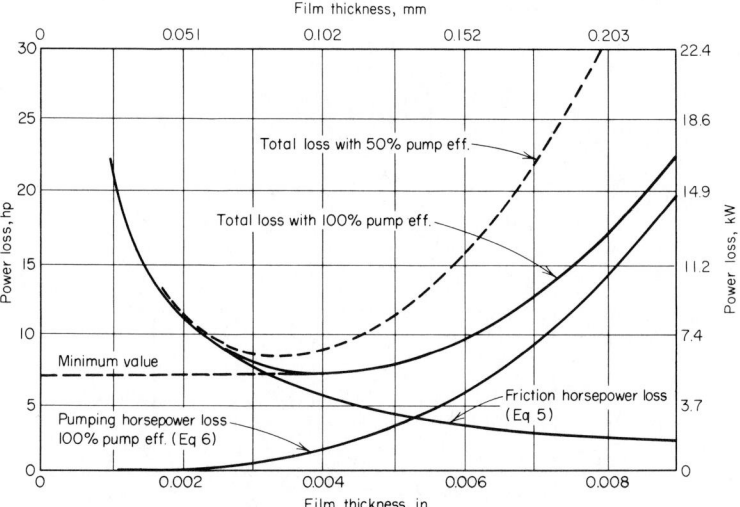

FIG. 13 Oil-firm thickness for minimum power loss in a hydrostatic thrust bearing. *(Product Engineering.)*

cedures shown above. Note, however, that each equation uses a factor K obtained from the respective design chart.

Also note that a hydrostatic bearing uses an externally fed pressurized fluid to keep two bearing surfaces *completely* separated. Compared with hydrodynamic bearings, in which the pressure is self-induced by the rotation of the shaft, hydrostatic bearings have (1) lower friction, (2) higher load-carrying capacity, (3) a lubricant-film thickness insensitive to shaft speed, (4) a higher spring constant, which leads to a self-centering effect, and (5) a relatively thick lubricant film permitting cooler operation at high shaft speeds. Hydrostatic bearings are used in rolling mills, instruments, machine tools, radar, telescopes, and other applications.

HYDROSTATIC JOURNAL BEARING ANALYSIS

A 4.000-in (10.160-cm) metal shaft rests in a journal bearing having an internal diameter of 4.012 in (10.190 cm). The lubricant is SAE-30 oil at 100°F (37.8°C) having a viscosity of 152×10^{-7} reyn. This lubricant is supplied under pressure through a groove at the lowest point in the bearing. The length of the bearing is 6 in (15.2 cm), the length of the groove is 3 in (7.6 cm), and the load on the bearing is 3600 lb (16,013.6 N). What lubricant-inlet pressure and flow rate are required to raise the shaft 0.002 in (0.051 mm) and 0.004 in (0.102 mm)?

Calculation Procedure:

1. *Determine the radial clearance and clearance modulus*

The design equations and methods developed at Franklin Institute by Dudley Fuller, Professor of Mechanical Engineering, Columbia University, are applicable to hydrostatic journal bearings using oil, grease, or gas lubrication. These equations are accurate, simple, and reliable; therefore they are used here.

By Fuller's method, the radial clearance c, in $= r_b - r_s$, Fig. 16, where r_b = bearing internal radius, in; r_s = shaft radius, in. Or, $c = (4.012/2) - (4.000/2) = 0.006$ in (0.152 mm).

Next, compute the clearance modulus m from $m = c/r_s = 0.006/2 = 0.003$ in/in (0.003 cm/cm). Typical values of m range from 0.005 to 0.003 in/in (0.005 to 0.003 cm/cm) for hydrostatic journal bearings.

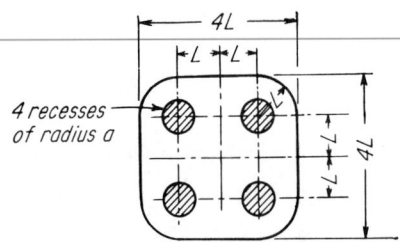

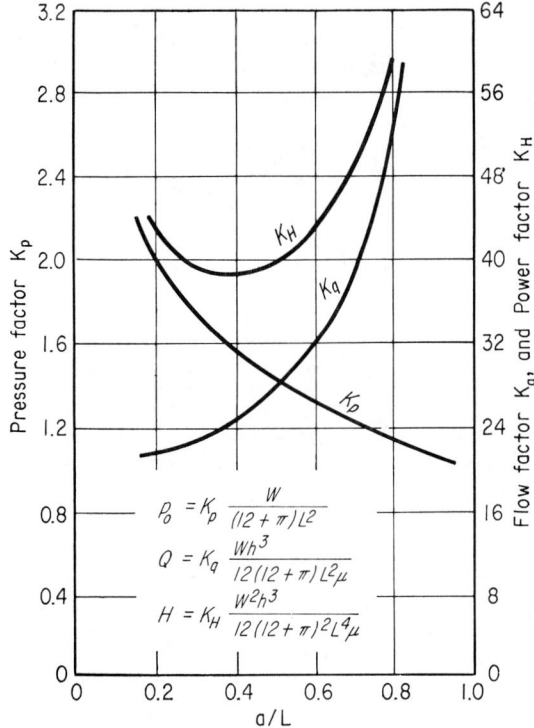

FIG. 14 Constants and equations for modified square hydrostatic bearings. (*Product Engineering.*)

2. Compute the shaft eccentricity in the clearance space

The numerical parameter used to describe the eccentricity of the shaft in the bearing clearance space is the ratio $\epsilon = 1 - h/(mr)$, where h = shaft clearance, in, during operation. With a clearance of $h = 0.002$ in (0.051 mm), $\epsilon = 1 - [0.002/(0.003 \times 2)] = 0.667$. With a clearance of $h = 0.004$ in (0.102 mm), $\epsilon = 1 - [0.004/(0.003 \times 2)] = 0.333$.

3. Compute the eccentricity constants

The eccentricity constant $A_k = 12[2 - \epsilon/(1 - \epsilon)^2] = 12[2 - 0.667/(1 - 0.667)^2] = 144.6$. A second eccentricity constant B is given by $B_k = 12\{\epsilon(4 - \epsilon^2)/[2(1 - \epsilon^2)^2] + 2 + \epsilon^2/(1 - \epsilon^2)^{2.5} \times \arctan[1 + \epsilon/(1 - \epsilon^2)^{0.5}]\}$. Since this relation is awkward to handle, Fig. 17 was developed by Fuller. From Fig. 17, $B_k = 183$.

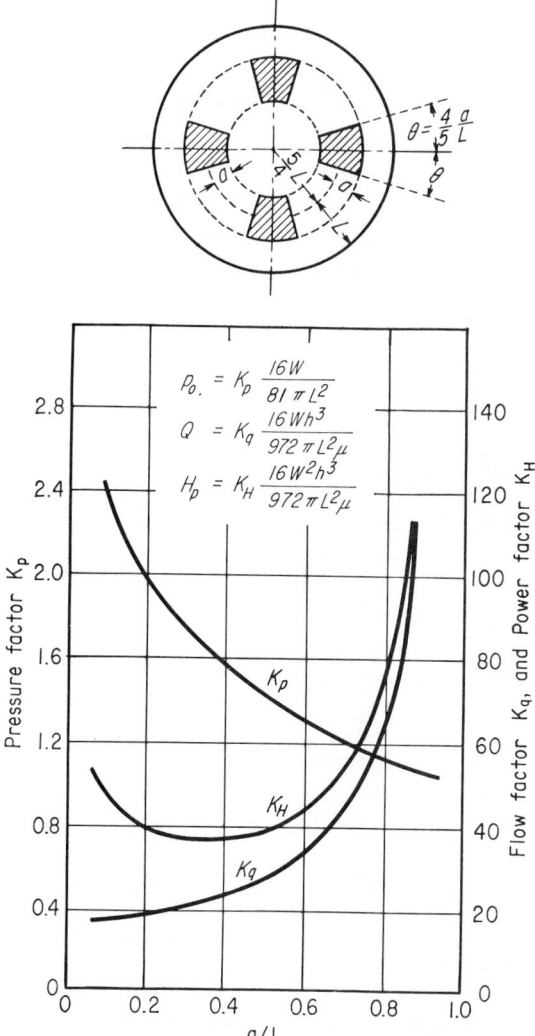

FIG. 15 Constants and equations for circular-sector hydrostatic bearings. *(Product Engineering.)*

4. Compute the required lubricant flow rate

The lubricant flow rate is found from Q in^3/s $= 2Lm^3r_s/(\eta A_k)$, where L = load acting on shaft, lb; η = lubricant viscosity, reyns. For the bearing with $h = 0.002$ in (0.051 mm), $Q = 2(3600)(0.003)^3(2)/[(152 \times 10^{-7})(144.6)] = 0.177$ in^3/s (2.9 cm^3/s), or 0.0465 gal/min (2.9 mL/s).

5. Compute the required lubricant-inlet pressure

The lubricant-inlet pressure is found from $p_i = \eta QB/(2bm^3r_s^2)$, where p_i = lubricant inlet pressure, lb/in^2; other symbols as before. Thus, $p_i = (152 \times 10^{-7})(0.177)(183)/[(2)(3)(0.003)^3(2)^2] = 759$ lb/in^2 (5233.3 kPa).

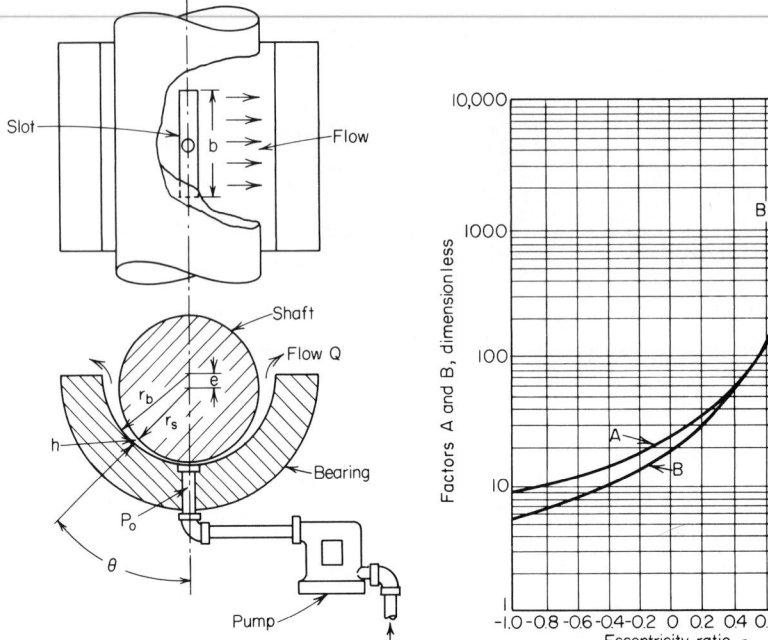

FIG. 16 Hydrostatic journal bearing. *(Product Engineering.)*

FIG. 17 Constants for hydrostatic journal bearing oil flow and load capacity. *(Product Engineering.)*

6.   Analyze the larger-clearance bearing

Use the same procedure as in steps 1 through 5. Then $\epsilon = 0.333$; $A = 45.0$; $B = 42.0$; $Q = 0.560$ in³/s (9.2 cm³/s) = 0.1472 gal/min (9.2 mL/s); $p_i = 551$ lb/in² (3799.1 kPa).

Related Calculations: Note that the closer the shaft is to the center of the bearing, the smaller the lubricant pressure required and the larger the oil flow. If the larger flow requirements can be met, the design with the thicker oil film is usually preferred, because it has a greater ability to absorb shock loads and tolerate thermal change.

Use the general design procedure given here for any applications where a hydrostatic journal bearing is applicable and there is no thrust load.

HYDROSTATIC MULTIDIRECTION BEARING ANALYSIS

Determine the lubricant pressure and flow requirements for the multidirection hydrostatic bearing shown in Fig. 18 if the vertical coplanar forces acting on the plate are 164,000 lb (729,508.4 N) upward and downward, respectively. The lubricant viscosity $\eta = 393 \times 10^{-7}$ reyn film thickness = 0.005 in (0.127 mm), $L = 7$ in (17.8 cm), $a = 3.5$ in (8.9 cm). What would be the effect of decreasing the film thickness h on one side of the plate by 0.001-in (0.025-mm) increments from 0.005 to 0.002 in (0.127 to 0.051 mm)? What is the bearing stiffness?

Calculation Procedure:

1.   Compute the required lubricant-inlet pressure

By Fuller's method, Fig. 19 shows that the bearing has four pressure pads to support the plate loads. Figure 19 also shows the required pressure, flow, and power equations, and the appropriate constants for these equations. The inlet-pressure equation is $p_i = K_p L_s/(16L^2)$, where $p_i =$

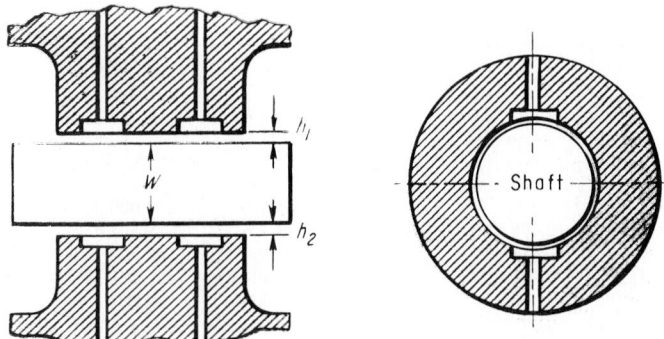

FIG. 18 Double-acting hydrostatic thrust bearing. *(Product Engineering.)*

required lubricant inlet pressure, lb/in²; K_p = pressure constant from Fig. 19; L_s = plate load, lb; L = bearing length, in.

Find K_p from Fig. 19 after setting up the ratio a/L = 3.5/7.0 = 0.5, where a = one-half the pad length, in. Then K_p = 1.4. Hence p_i = (1.4)(164,000)/[(16)(7)²] = 293 lb/in² (2020.2 kPa).

2. *Compute the required lubricant flow rate*

From Fig. 19, the required lubricant flow rate $Q = K_q L_s h^3/(192 L^2 \eta)$, where K_q = flow constant; h = lubricant-film thickness, in; other symbols as before. For a/L = 0.5, K_q = 36. Then Q = (36)(164,000)(0.005)³/[(192)(7)²(393 × 10⁻⁷)] = 1.99 in³/s (32.6 cm³/s). This can be rounded off to 2.0 in³/s (32.8 cm³/s) for usual design calculations.

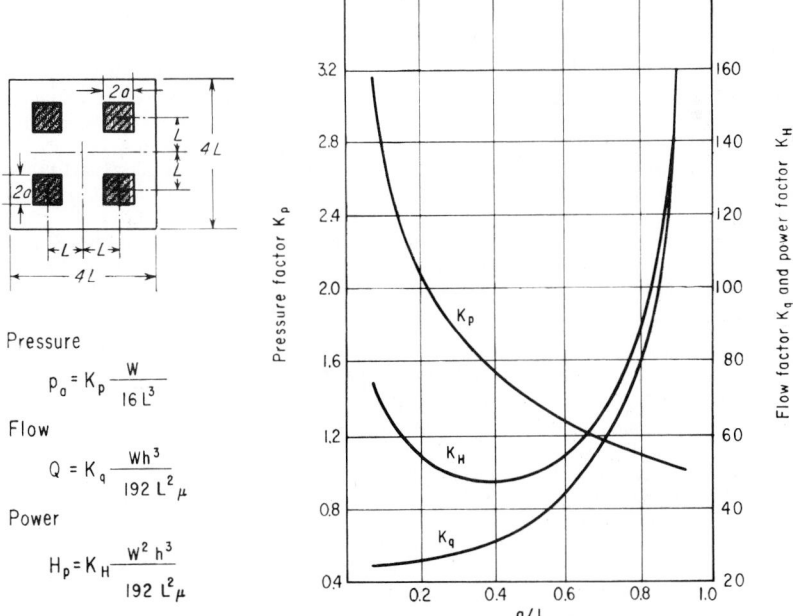

FIG. 19 Dimensions and equations for thrust-bearing design. *(Product Engineering.)*

TABLE 43 Load Capacity of Dual-Direction Bearing

Film thickness, in (cm)		Inlet pressure, lb/in² (kPa)		Load, lb (N)		Load capacity, lb (N)
h_2	h_1	p_{i2}	p_{i1}	L_{s2}	L_{s1}	$L_{s2} - L_{s1}$
0.005	0.005	293	293	164,000	164,000	0
(0.127)	(0.127)	(2,020.2)	(2,020.0)	(729,508.4)	(729,508.4)	(0.0)
0.004	0.006	571	170	320,000	95,000	225,000
(0.102)	(0.152)	(3,937.0)	(1,172.2)	(1,423,431.0)	(422,581.1)	(1,000,850.0)
0.003	0.007	1,360	106	760,000	59,700	700,300
(0.076)	(0.178)	(9,377.2)	(730.9)	(3,380,648.7)	(265,558.9)	(3,115,089.9)
0.002	0.008	4,570	71	2,560,000	40,000	2,520,000
(0.051)	(0.203)	(31,510.2)	(489.5)	(11,387,448.3)	(177,928.9)	(11,209,519.4)

3. Compute the pressure and load for other plate clearances

The sum of the plate lubricant-film thicknesses, $h_1 + h_2$, Fig. 18, is a constant. For this bearing, $h_1 + h_2 = 0.005 + 0.005 = 0.010$ in (0.254 mm). With no load on the plate, if oil is pumped into both bearing faces at the rate of 2.0 in³/s (32.8 cm³/s), the maximum recess pressure will be 293 lb/in² (2020.2 kPa). The force developed on each face will be 164,000 lb (729,508.4 N). Since the lower face is pushed up with this force and the top face is pushed down with the same force, the net result is zero.

With a downward external load imposed on the plate such that the lower film thickness h_2 is reduced to 0.004 in (0.102 mm), the upper film thickness will become 0.006 in (0.152 mm), since $h_1 + h_2 = 0.010$ in (0.254 mm) = a constant for this bearing. If the lubricant flow rate is held constant at 2.0 in³/s (32.8 cm³/s), then $K_q = 36$ from Fig. 19, since $a/L = 0.5$. With these constants, the load equation becomes $L_s = 0.0205/h^3$, and the inlet-pressure equation becomes $p_i = L_s/560$.

Using these equations, compute the upper and lower loads and inlet pressures for h_2 and h_1 ranging from 0.004 to 0.002 and 0.006 to 0.008 in (0.102 to 0.051 and 0.152 to 0.203 mm), respectively. Tabulate the results as shown in Table 43. Note that the allowable load is computed by using the respective film thickness for the lower and upper parts of the plate. The same is true of the lubricant-inlet pressure, except that the corresponding load is used instead.

Thus, for $h_2 = 0.004$ in (0.102 mm), $L_{s2} = 0.0205/(0.004)^3 = 320,000$ lb (1,423,431.0 N). Then $p_{i2} = 320,000/560 = 571$ lb/in² (3937.0 kPa). For $h_k = 0.006$ in (0.152 mm), $L_{s1} = 0.0205/(0.006)^3 = 95,000$ lb (422,581.1 N). Then $p_{i1} = 95,000/560 = 169.6$, say 170, lb/in² (1172.2 kPa).

The load difference $L_{s2} - L_{s1}$ = the bearing load capacity. For the film thicknesses considered above, $L_{s2} - L_{s1} = 320,000 - 95,000 = 225,000$ lb (1,000,850.0 N). Load capacities for various other film thicknesses are also shown in Table 43.

4. Determine the bearing stiffness

Plot the net load capacity of this bearing vs. the lower film thickness, Fig. 20. A tangent to the curve at any point indicates the stiffness of this bearing. Draw a tangent through the origin where $h_2 = h_1 = 0.005$ in (0.127 mm). The slope of this tangent = vertical value/horizontal value = $(725,000 - 0)/(0.005 - 0.002) = 241,000,000$ lb/in (42,205,571 N/m) = the bearing stiffness. This means that a load of 241,000,000 lb (1,072,021,502 N) would be required to displace the plate 1.0 in (2.5 cm). Since the plate cannot move this far, a load of 241,000 lb (1,072,021.5 N) would move the plate 0.001 in (0.025 mm).

If the lubricant flow rate to each face of the bearing were doubled, to $Q = 4.0$ in³/s (65.5 cm³/s), the stiffness of the bearing would increase to 333,000,000 lb/in (58,317,241 N/m), as shown in Fig. 20. This means that an additional load of 333,000 lb (1,481,257.9 N) would displace the plate 0.001 in (0.025 mm). The stiffness of a hydrostatic bearing can be controlled by suitable design, and a wide range of stiffness values can be designed into the bearing system.

Related Calculations: Hydrostatic bearings of the design shown here are useful for a variety

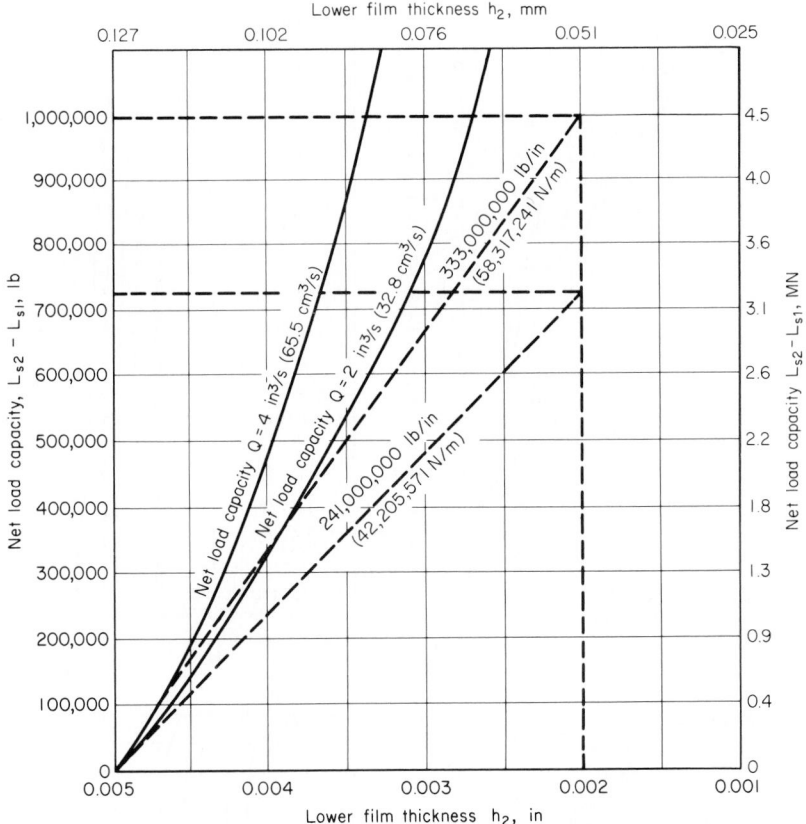

FIG. 20 Net load capacity for a double-direction thrust bearing. *(Product Engineering.)*

of applications. Journal bearings for multidirectional loads are analyzed in a manner similar to that described here.

LOAD CAPACITY OF GAS BEARINGS

Determine the load capacity and bearing stiffness of a hydrostatic air bearing, using 70°F (21.1°C) air if the bearing orifice radius = 0.0087 in (0.0221 cm), the radial clearance h = 0.0015 in (0.038 mm), the bearing diameter d = 3.00 in (7.6 cm), the bearing length L = 3 in (7.6 cm), the total number of air orifices N = 8, the ambient air pressure p_a = 14.7 lb/in^2 (abs) (101.4 kPa), the air supply pressure p_s = 15 lb/in^2 (gage) = 29.7 lb/in^2 (abs) (204.8 kPa), the (gas constant)(total temperature) = RT = 1.322 × 10^8 in^2/s^2 (8.5 × 10^8 cm^2/s^2), ϵ = the eccentricity ratio = 0.30, air viscosity η = 2.82 × 10^{-9} lb·s/in^2 (1.94 × 10^{-9} N·s/cm^2), the orifice coefficient α = 0.63, and the shaft speed ω = 2100 rad/s = 20,000 r/min.

Calculation Procedure:

1. Compute the bearing factors Λ and Λ$_T$

For a hydrostatic gas bearing, $\Lambda = (6\eta\omega/p_a)(d/zh)^2 = [(6)(2.82 \times 10^{-9})(2100)/14.7][3/(2 \times 0.0015)]^2 = 2.41$.

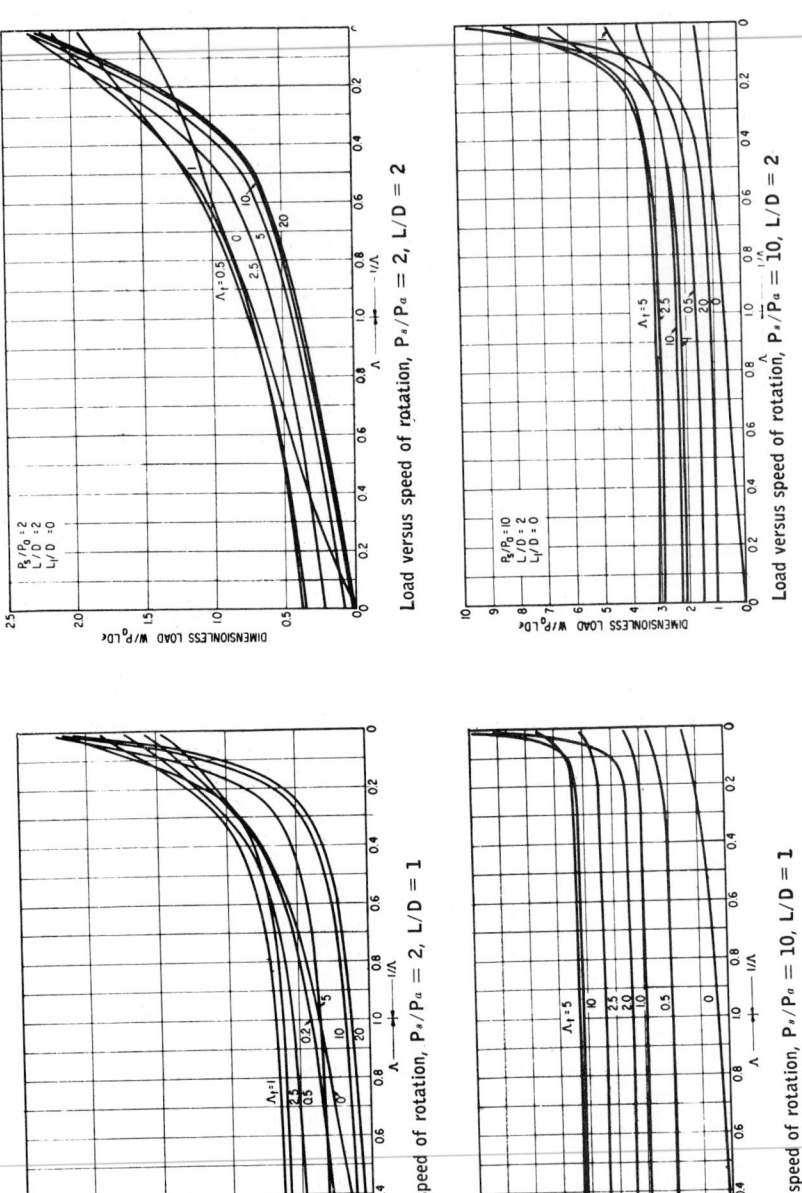

FIG. 21 Gas-bearing constants. (*Product Engineering.*)

Also, $\Lambda_T = 6\eta Na^2\alpha(RT)^{0.5}/p_ah^3 = (6)(2.82 \times 10^{-9})(8)(0.0087)^2(0.63)(1.322 \times 10^8)^2/$ $[(14.7)(0.0015)^3] = 1.5$.

2. Determine the dimensionless load

Since $L/d = 3/3 = 1$, and $p_s/p_a = 29.7/14.7 \approx 2$, use the first chart[1] in Fig. 21. Before entering the chart, compute $1/\Lambda = 1/2.41 = 0.415$. Then, from the chart, the dimensionless load = 0.92 $= L_d$.

3. Compute the bearing load capacity

The bearing load capacity $L_s = L_d p_a LD\epsilon = (0.92)(14.7)(3.00)(3.00)(0.3) = 36.5$ lb (162.4 N). If there were no shaft rotation, $\Lambda = 0$, $\Lambda_T = 1.5$, and $L_d = 0.65$, from the same chart. Then $L_s = L_d p_a LD\epsilon = (0.65)(14.7)(3.00)(3.00)(0.3) = 25.8$ lb (114.8 N). Thus, rotation of the shaft increases the load-carrying ability of the bearing by $[(36.5 - 25.8)/25.8](100) = 41.5$ percent.

4. Compute the bearing stiffness

For a hydrostatic gas bearing, the bearing stiffness $B_s = L_s/(h\epsilon) = 36.5/[(0.0015)(0.3)] = 81,200$ lb/in (142,203.0 N/cm).

Related Calculations: Use this procedure for the selection of gas bearings where the four charts presented here are applicable. The data summarized in these charts result from computer solutions of the complex equations for "hybrid" gas bearings. The work was done by Mechanical Technology Inc., headed by Beno Sternlicht.

SPRING SELECTION FOR A KNOWN LOAD AND DEFLECTION

Give the steps in choosing a spring for a known load and an allowable deflection. Show how the type and size of spring are determined.

Calculation Procedure:

1. Determine the load that must be handled

A spring may be required to absorb the force produced by a falling load or the recoil of a mass, to mitigate a mechanical shock load, to apply a force or torque, to isolate vibration, to support moving masses, or to indicate or control a load or torque. Analyze the load to determine the magnitude of the force that is acting and the distance through which it acts.

Once the magnitude of the force is known, determine how it might be absorbed—by compression or extension (tension) of a spring. In some applications, either compression or extension of the spring is acceptable.

2. Determine the distance through which the load acts

The load member usually moves when it applies a force to the spring. This movement can be in a vertical, horizontal, or angular direction, or it may be a rotation. With the first type of movement, a *compression*, or *tension*, spring is generally chosen. With a torsional movement, a *torsion-type* spring is usually selected. Note that the movement in either case may be negligible (i.e., the spring applies a large restraining force), or the movement may be large, with the spring exerting only a nominal force compared with the load.

3. Make a tentative choice of spring type

Refer to Table 44, entering at the type of load. Based on the information known about the load, make a tentative choice of the type of spring to use.

4. Compute the spring size and stress

Use the methods given in the following calculation procedures to determine the spring dimensions, stress, and deflection.

5. Check the suitability of the spring

Determine (a) whether the spring will fit in the allowable space, (b) the probable spring life, (c) the spring cost, and (d) the spring reliability. Based on these findings, use the spring chosen, if it

[1]"Gas Bearings," *Product Engineering*, July 8, 1963.

TABLE 44   Metal Spring Selection Guide

Type of load	Suitable spring type	Relative magnitude of load on spring	Deflection absorbed
Compression	Helical	Small to large	Small to large
	Leaf	Large	Moderate
	Flat	Small to large	Small to large
	Belleville	Small to large	Moderate
	Ring	Large	Small
Tension	Helical	Small to large	Small to large
	Leaf	Large	Moderate
	Flat	Small to large	Small to large
Torsion	Helical torsion	Small to large	Small to large
	Spiral	Moderate	Moderate
	Torsion bar	Large	Small

is satisfactory. If the spring is unsatisfactory, choose another type of spring from Table 44 and repeat the study.

SPRING WIRE LENGTH AND WEIGHT

How long a wire is needed to make a helical spring having a mean coil diameter of 0.820 in (20.8 mm) if there are five coils in the spring? What will this spring weigh if it is made of oil-tempered spring steel 0.055 in (1.40 mm) in diameter?

Calculation Procedure:

1.   Compute the spring wire length

Find the spring length from $l = \pi n d_m$, where l = wire length, in; n = number of coils in the spring, in. Thus, for this spring, $l = \pi(5)(0.820) = 12.9$ in (32.8 cm).

2.   Compute the weight of the spring

Find the spring weight from $w = 0.224 l d^2$, where w = spring weight, lb; d = spring wire diameter, in. For this spring, $w = 0.224(12.9)(0.055)^2 = 0.0087$ lb (0.0387 kg).

   Related Calculations: The weight equation in step 2 is valid for springs made of oil-tempered steel, chrome vanadium steel, silica-manganese steel, and silicon-chromium steel. For stainless steels, use a constant of 0.228, in place of 0.224, in the equation. The relation given in this procedure is valid for any spring having a continuous coil—helical, spiral, etc. Where a number of springs are to be made, simply multiply the length and weight of each by the number to be made to determine the total wire length required and the weight of the wire.

HELICAL COMPRESSION AND TENSION SPRING ANALYSIS

Determine the dimensions of a helical compression spring to carry a 5000-lb (22,241.1-N) load if it is made of hard-drawn steel wire having an allowable shear stress of 65,000 lb/in^2 (448,175.0 kPa). The spring must fit in a 2-in (5.1-cm) diameter hole. What is the deflection of the spring? The spring operates at atmospheric temperature, and the shear modulus of elasticity is 5×10^6 lb/in^2 (34.5 × 10^9 Pa).

Calculation Procedure:

1.   Choose the tentative dimensions of the spring

Since the spring must fit inside a 2-in (5.1-cm) diameter hole, the mean diameter of the coil should not exceed about 1.75 in (4.5 cm). Use this as a trial mean diameter, and compute the wire diameter from $d = [8 L d_m k/(\pi s_s)]^{1/3}$, where d = spring wire diameter, in; L = load on spring, lb; d_m = mean diameter of coil, in; k = spring curvature correction factor = $(4c - 1)/(4c - 4)$ + $0.615/c$ for heavily coiled springs, where $c = 2r_m/d = d_m/d$; s_s = allowable shear stress material, lb/in^2. For lightly coiled springs, $k = 1.0$. Thus, $d = [8 \times 5000 \times 1.75 \times 1.0/(\pi \times 65,000)]^{1/3} = 0.70$ in (1.8 cm). So the outside diameter d_o of the spring will be $d_o = d_m +$

$2(d/2) = d_m + d = 1.75 + 0.70 = 2.45$ in (6.2 cm). But the spring must fit a 2-in (5.1-cm) diameter hole. Hence, a smaller value of d_m must be tried.

Using $d_m = 1.5$ in (3.8 cm) and following the same procedure, we find $d = [8 \times 5000 \times 1.50 \times 1.0/(\pi \times 65,000)]^{1/3} = 0.665$ in (1.7 cm). Then $d_o = 1.5 + 0.665 = 2.165$ in (5.5 cm), which is still too large.

Using $d_m = 1.25$ in (3.2 cm), we get $d = [8 \times 1.25 \times 1.0/(\pi \times 65,000)]^{1/3} = 0.625$ in (1.6 cm). Then $d_o = 1.25 + 0.625 = 1.875$ in (4.8 cm). Since this is nearly 2 in (5.1 cm), the spring probably will be acceptable. However, the value of k should be checked.

Thus, $c = 2r_m/d = 2(1.25/2)/0.625 = 2.0$. Note that $r_m = d_m/2 = 1.25/2$ in this calculation for the value of c. Then $k = [(4 \times 2) - 1]/[(4 \times 2) - 4] + 0.615/2 = 2.0575$. Hence, the assumed value of $k = 1.0$ was inaccurate for this spring. Recalculating, $d = [8 \times 5000 \times 1.25 \times 2.0575/(\pi \times 65,000)]^{1/3} = 0.796$ in (2.0 cm). Now, $1.25 + 0.796 = 2.046$ in (5.2 cm), which is still too large.

Using $d_m = 1.20$ in (3.1 cm) and assuming $k = 2.0575$, then $d = [8 \times 5000 \times 1.20 \times 2.0575/(\pi \times 65,000)]^{1/3} = 0.785$ in (2.0 cm) and $d_o = 1.20 + 0.785 = 1.985$ in (5.0 cm). Checking the value of k gives $c = 1.20/0.785 = 1.529$ and $k = [(4 \times 1.529) - 1]/[(4 \times 1.529) - 4] + 0.615/1.529 = 2.820$. Recalculating again, $d = [8 \times 5000 \times 1.20 \times 2.820/(\pi \times 65,000)]^{1/3} = 0.872$ in (2.2 cm). Then $d_o = 1.20 + 0.872 = 2.072$ in (5.3 cm), which is worse than when $d_m = 1.25$ in (3.2 cm) was used.

It is now obvious that a practical trade-off must be utilized so that a spring can be designed to carry a 5000-lb (22,241.1-N) load and fit in a 2-in (5.1-cm) diameter hole, d_h, with suitable clearance. Such a trade-off is to use hard-drawn steel wire of greater strength at higher cost. This trade-off would not be necessary if d_h could be increased to, say, 2.13 in (5.4 cm).

Using $d_m = 1.25$ in (3.2 cm) and a clearance of, say, 0.08 in (0.20 cm), then $d = d_h - d_m - 0.08 = 2.00 - 1.25 - 0.08 = 0.67$ ion (1.7 cm). Also, $c = d_m/d = 1.25/0.67 = 1.866$ and $k = [(4 \times 1.866) - 1]/[(4 \times 1.866) - 4] + 0.615/1.866 = 2.196$. The new allowable shear stress can now be found from $s_s = 8Ld_mk/\pi d^3 = 8 \times 5000 \times 1.25 \times 2.196/(\pi \times 0.67^3) = 116,206$ lb/in^2 (801,212.1 kPa). Hard-drawn spring wire (ASTM A-227-47) is available with $s_s = 117,000$ lb/in^2 (806,686.6 kPa) and $G = 11.5 \times 10^6$ lb/in^2 (79.29 $\times$ 10^9 Pa).

Hence, $d = [8 \times 5000 \times 1.25 \times 2.196/(\pi \times 117,000)]^{1/3} = 0.668$ in (1.7 cm). Thus, $d_o = 1.25 + 0.668 = 1.918$ in (4.9 cm). Now, the spring is acceptable because further recalculations would show that d_o will remain less than 2 in (5.1 cm), regardless.

2. Compute the spring deflection

The deflection of a helical compression spring is given by $f = 64nr_m^3L/(d^4G) = 4\pi nr_m^2 s_s/(dGk)$, where f = spring deflection, in; n = number of coils in this spring; r_m = mean radius of spring coil, in; L, d, G, s_s, and k are as determined for the acceptable spring.

Assuming $n = 10$ coils, we find $f = 64 \times 10 \times (1.25/2)^3 \times 5000/(0.668^4 \times 11.5 \times 10^6) = 0.341$ in (0.9 cm). Or, $f = 4\pi \times 10 \times (1.25/2)^2 \times 117,000/(0.668 \times 11.5 \times 10^6 \times 2.196) = 0.340$ (0.9 cm), a close agreement.

The number of coils n assumed for this spring is based on past experience with similar springs. However, where past experience does not exist, several trial values of n can be used until a spring of suitable deflection and length is obtained.

Related Calculations: Use this general procedure to analyze helical coil compression or tension springs. As a general guide, the outside diameter of a spring of this type is taken as (0.96)(hole diameter). The active solid height of a compression-type spring, i.e., the height of the spring when fully closed by the load, usually is nd, or (0.9) (final height when compressed by the design load).

SELECTION OF HELICAL COMPRESSION AND TENSION SPRINGS

Choose a helical compression spring to carry a 90-lb (400.3-N) load with a stress of 50,000 lb/in^2 (344,750.0 kPa) and a deflection of about 2.0 in (5.1 cm). The spring should fit in a 3.375-in (8.6-cm) diameter hole. The spring operates at about 70°F (21.1°C). How many coils will the spring have? What will the free length of the spring be?

Calculation Procedure:

1. Determine the spring outside diameter

Using the usual relation between spring outside diameter and hole diameter, we get $d_o = 0.96d_h$, where d_h = hole diameter, in. Thus, $d_o = 0.96(3.375) = 3.24$ in, say 3.25 in (8.3 cm).

TABLE 45   Load and Spring Rates for Helical Compression and Tension Springs°

Spring wire diameter		Outside diameter of spring coil					
		in	cm	in	cm	in	cm
in	cm	3	7.6	3.25	8.3	3.5	8.9
0.207	0.5258	113†	502.6	104	462.6	97.2	432.4
		121	211.9	93.6	163.9	74.1	129.8
0.250	0.6350	198	880.7	183	814.0	170	756.2
		270	472.8	208	364.3	163	285.5
0.283	0.7188	285	1267.7	263	1169.9	247	1098.7
		460	805.6	352	616.4	276	483.4

°After H. F. Ross, "Application of Tables for Helical Compression and Extension Spring Design," *Transactions ASME*, vol. 69, p. 727.
†First figure given is loads in lb at 100,000-lb/in² (in N at 689,500-kPa) stress. Second figure is spring rate in lb/in (N/cm) per coil, $G = 11.5 \times 10^6$ lb/in² (79.3×10^9 Pa).

2.   *Determine the required wire diameter*

The equations in the previous calculation procedure can be used to determine the required wire diameter, if desired. However, the usual practice is to select the wire diameter by using precomputed tabulations of spring properties, charts of spring properties, or a special slide rule available from some spring manufacturers. The tabular solution will be used here because it is one of the most popular methods.

Table 45 shows typical loads and spring rates for springs of various outside diameters and wire diameters based on a corrected shear stress of 100,000 lb/in² (689,500.0 kPa) and a shear modulus of $G = 11.5 \times 10^6$ lb/in² (79.3×10^9 Pa).

Before Table 45 can be used, the actual load must be corrected for the tabulated stress. Do this by taking the product of (actual load, lb)(table stress, lb/in²)/(allowable spring stress, lb/in²). For this spring, tabular load, lb = (90)(100,000/50,000) = 180 lb (800.7 N). This means that a 90-lb (400.3-N) load at a 50,000-lb/in² (344,750.0-kPa) stress corresponds to a 180-lb (800.7-N) load at 100,000-lb/in² (689,500-kPa) stress.

Enter Table 45 at the spring outside diameter, 3.25 in (8.3 cm), and project vertically downward in this column until a load of approximately 180 lb (800.7 N) is intersected. At the left read the wire diameter. Thus, with a 3.25-in (8.3-cm) outside diameter and 183-lb (814.0-N) load, the required wire diameter is 0.250 in (0.635 cm).

3.   *Determine the number of coils required*

The allowable spring deflection is 2.0 in (5.1 cm), and the spring rate per single coil, Table 45, is 208 lb/in (364.3 N/cm) at a tabular stress of 100,000 lb/in² (689,500 kPa). We use the relation, deflection f, in = load, lb/desired spring rate, lb/in, S_R; or, 2.0 = 90/S_R; S_R = 90/20 = 45 lb/in (78.8 N/cm).

4.   *Compute the number of coils in the spring*

The number of active coils in a spring is n = (tabular spring rate, lb/in)/(desired spring rate, lb/in). For this spring, n = 208/45 = 4.62, say 5 coils.

5.   *Determine the spring free length*

Find the approximate length of the spring in its free, expanded condition from l in = $(n + i)d + f$, where l = approximate free length of spring, in; i = number of inactive coils in the spring; other symbols as before. Assuming two inactive coils for this spring, we get l = (5 + 2)(0.25) + 2 = 3.75 in (9.5 cm).

Related Calculations: Similar design tables are available for torsion springs, spiral springs, coned-disk (Belleville) springs, ring springs, and rubber springs. These design tables can be found in engineering handbooks and in spring manufacturers' engineering data. Likewise, spring design charts are available from many of these same sources. Spring design slide rules are generally available free of charge to design engineers from spring manufacturers.

SIZING HELICAL SPRINGS FOR OPTIMUM DIMENSIONS AND WEIGHT

Determine the dimensions of a helical spring having the minimum material volume if the initial, suddenly applied, load on the spring is 15 lb (66.7 N), the mean coil diameter is 1.02 in (2.6 cm), the spring stroke is 1.16 in (2.9 cm), the final spring stress is 100,000 lb/in² (689,500 kPa), and the spring modulus of torsion is 11.5×10^6 lb/in² (79.3×10^9 Pa).

Calculation Procedure:

1. Compute the minimum spring volume

Use the relation $v_m = 8fLG/s_f^2$, where v_m = minimum volume of spring, in³; f = spring stroke, in³; L = initial load on spring, lb; G = modulus of torsion of spring material, lb/in²; s_f = final stress in spring, lb/in². For this spring, $v_m = 8(1.16)(15)(11.5 \times 10^6)/(100,000)^2 = 0.16$ in³ (2.6 cm³). Note: $s_f = 2s_s$, where s_s = shear stress due to a static, or gradually applied, load.

2. Compute the required spring wire diameter

Find the wire diameter from $d = [16Ld_m/(\pi s_f)]^{1/3}$, where d = wire diameter in; d_m = mean diameter of spring, in; other symbols as before. For this spring, $d = [16 \times 15 \times 1.02/(\pi \times 100,000)]^{1/3} = 0.092$ in (2.3 mm).

3. Find the number of active coils in the spring

Use the relation $n = 4v_m/(\pi^2 d^2 d_m)$, where n = number of active coils; other symbols as before. Thus, $n = 4(0.16)/[\pi^2(0.092)^2(1.02)] = 7.5$ coils.

4. Determine the active solid height of the spring

The solid height $H_s = (n + 1)d$, in, or $H_s = (7.5 + 1)(0.092) = 0.782$ in (2.0 cm). For a practical design, allow 10 percent clearance between the solid height and the minimum compressed height H_c. Thus, $H_c = 1.1H_s = 1.1(0.782) = 0.860$ in (2.2 cm). The assembled height $H_a = H_c + f = 0.860 + 1.16 = 2.020$ in (5.13 cm).

5   Compute the spring load-deflection rate

The load-deflection rate $R = Gd^4/(8d_m^3 n)$, where R = load-deflection rate, lb/in; other symbols as before. Thus, $R = (11.5 \times 10^6)(0.092)^4/[8(1.02)^3(7.5)] = 12.9$ lb/in (2259.1 N/m).

The initial deflection of the spring is $f_i = L/R$ in, or $f_i = 15/12.9 = 1.163$ in (3.0 cm). Since the free height of a spring $H_f = H_a + f_i$, the free height of this spring is $H_f = 2.020 + 1.163 = 3.183$ in (8.1 cm).

Related Calculations: The above procedure for determining the minimum spring volume can be used to find the minimum spring weight by relating the spring weight W lb to the density of the spring material ρ lb/in³ in the following manner: For the required initial load L_1 lb, $W_{min} = \rho(8fL_1G/s_f^2)$. For the required energy capacity E in·lb, $W_{min} = \rho(4ED/s_f^2)$. For the required final load L_2 lb, $W_{min} = \rho(2f_2L_2G/s_f^2)$.

The above procedure assumes the spring ends are open and not ground. For other types of end conditions, the minimum spring volume will be greater by the following amount: For squared (closed) ends, $v_m = 0.5\pi^2 d^2 d_m$. For ground ends, $v_m = 0.25\pi^2 d^2 d_m$. The methods presented here were developed by Henry Swieskowski and reported in *Product Engineering*.

SELECTION OF SQUARE- AND RECTANGULAR-WIRE HELICAL SPRINGS

Choose a square-wire spring to support a load of 500 lb (2224.1 N) with a deflection of not more than 1.0 in (2.5 cm). The spring must fit in a 4.25-in (10.8-cm) diameter hole. The modulus of rigidity for the spring material is $G = 11.5 \times 10^6$ lb/in² (79.3×10^9 Pa). What is the shear stress in the spring? Determine the corrected shear stress for this spring.

Calculation Procedure:

1. Determine the spring dimensions

Assume that a 4-in (10.2-cm) diameter square-bar spring is used. Such a spring will fit the 4.25-in (10.8-cm) hole with a small amount of room to spare.

As a trial, assume that the width of the spring wire = 0.5 in (1.3 cm) = a. Since the spring is square, the height of the spring wire = 0.5 in (1.3 cm) = b.

With a 4-in (10.2-cm) outside diameter and a spring wire width of 0.5 in (1.3 cm), the mean radius of the spring coil r_m = 1.75 in (4.4 cm). This is the radius from the center of the spring to the center of the spring wire coil.

2.   Compute the spring deflection

The deflection of a square-wire tension spring is $f = 45 L r_m^3 n/(Ga^4)$, where f = spring deflection, in; L = load on spring, lb; n = number of coils in spring; other symbols as before. To solve this equation, the number of coils must be known. Assume, as a trial value, five coils. Then f = $45(500)(1.75)^3(5)/[(11.5 \times 10^6)(0.5)^4]$ = 0.838 in (2.1 cm). Since a deflection of not more than 1.0 in (2.5 cm) is permitted, this spring is probably acceptable.

3.   Compute the shear stress in the spring

Find the shear stress in a square-bar spring from $S_s = 4.8 L r_m/a^3$, where S_s = spring shear stress, lb/in²; other symbols as before. For this spring, S_s = $(4.8)(500)(1.75)/(0.5)^3$ = 33,600 lb/in² (231,663.8 kPa). This is within the allowable limits for usual spring steel.

4.   Determine the corrected shear stress

Find the shear stress in a square-bar spring from $S_s = 4.8 L r_m/a^3$, where s_s = spring correction factor $k = 1 + 1.2/c + 0.56/c^2 + 0.5/c^3$, where $c = 2r_m/a$. For this spring, $c = (2 \times 1.75)/0.5$ = 7.0. Then $k = 1 + 1.2/7 + 0.56/7^2 + 0.5/7^3$ = 1.184. Hence, the corrected shear stress is $S_s' = k s_s$, or $S_s' = (1.184)(33,600)$ = 39,800 lb/in² (274,411.3 kPa). This is still within the limits for usual spring steel.

Related Calculations: Use a similar procedure to select rectangular-wire springs. Once the dimensions are selected, compute the spring deflection from $f = 19.6 L r_m^3 n/[Gb^3(a - 0.566)]$, where all the symbols are as given earlier in this calculation procedure. Compute the uncorrected shear stress from $S_s = L r_m(3a + 1.8b)/(a^2 b^2)$. To correct the stress, use the Liesecke correction factor given in Wahl—*Mechanical Springs.* For most selection purposes, the uncorrected stress is satisfactory.

CURVED SPRING DESIGN ANALYSIS

Find the maximum load P, maximum deflection F, and spring constant C for the curved rectangular wire spring shown in Fig. 22 if the spring variables expressed in metric units are E = 14,500 kg/mm², S_b = 55 kg/mm², b = 1.20 mm, h = 0.30 mm, r_1 = 0.65 mm, r_2 = 1.75 mm, L = 9.7 mm, u_1 = 1.7 mm, and u_2 = 5.6 mm.

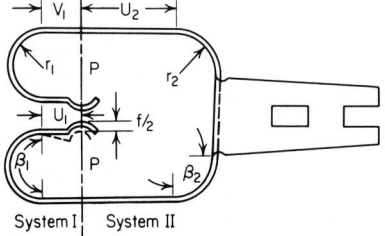

FIG. 22   Typical curved spring. *(Product Engineering.)*

Calculation Procedure:

1.   Divide the spring into analyzable components

Using Fig. 23, developed by J. Palm and K. Thomas of West Germany, as a guide, divide the spring to be analyzed into two or more analyzable components, Fig. 22. Thus, the given spring can be divided into two springs—a type D (Fig. 23), called system I, and a type A (Fig. 23), called system II.

2.   Compute the spring force

The spring force $P = P_I = P_{II}$. Since $(u_2 + r_2) > (u_1 + r_1)$, the spring in system II exerts a larger force. From Fig. 23 for $\beta = 90°$, $P = S\sigma_{max}/(u_2 + r_2)$, where S = section modulus, mm³, of the spring wire. Since $S = bh^2/6$ for a rectangle, $P = bh^2\sigma_{max}/[6(u_2 + r_2)]$ where b = spring wire width, mm; h = spring wire height, mm; σ_{max} = maximum bending stress in the spring, kg/mm³; other symbols as given in Fig. 22. Then $P = (1.20)(0.30)^2 \times (55)/[6(5.6 + 1.75)]$ = 0.135 kg.

3.   Compute the spring deflection

The total deflection of the springs is $F = 2F_I + F_{II}$, where F = spring deflection, mm, and the subscripts refer to each spring system. Taking the sum of the deflections as given in Fig. 23, we get $F = [2P/(3EI)][2K_1 r_1^3(m_1 + \beta_1/2)^2 + (v_1 - u_1)^3 + K_2 r_2^3(m_2 + \beta_2)^3]$, where E = Young's modulus, kg/mm²; I = spring wire moment of inertia, mm⁴; K = correction factor for the spring

Spring type	Spring deflection	Spring force and bending stresses
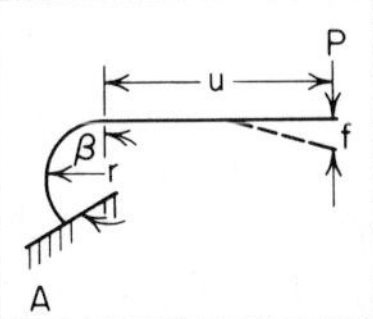 **A**	$F_1 = \dfrac{KPr^3}{3EI}(m+\beta)^3$ where $\alpha = \beta$ for finding K	When $\alpha = 0°$ to $90°$ When $\alpha = 90°$ to $180°$ $P = \dfrac{S\sigma}{u+\sin\beta}$ $P = \dfrac{S\sigma}{u+r}$ $\sigma = \dfrac{Pr(m+\sin\beta)}{S}$ $\sigma = \dfrac{Pr(m+1)}{S}$
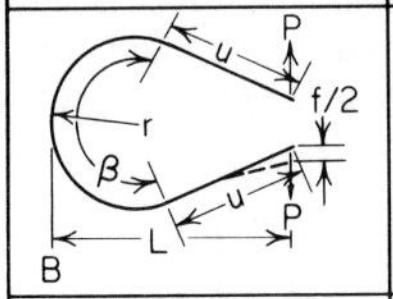 **B**	$F_2 = \dfrac{2KPr^3}{3EI}\left(m+\dfrac{\beta}{2}\right)^3$ where $\alpha = \dfrac{\beta}{2}$ for finding K	$P = \dfrac{S\sigma}{L}$
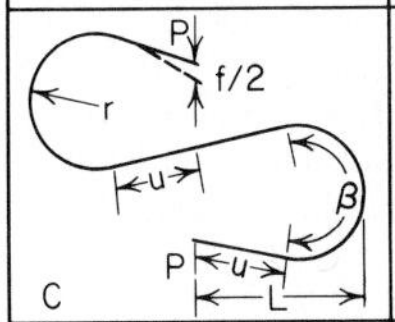 **C**	$F_3 = 2F_2 = \dfrac{4KPr^3}{3EI}\left(m+\dfrac{\beta}{2}\right)^3$ where $\alpha = \dfrac{\beta}{2}$ for finding K	$\sigma = \dfrac{PL}{S}$
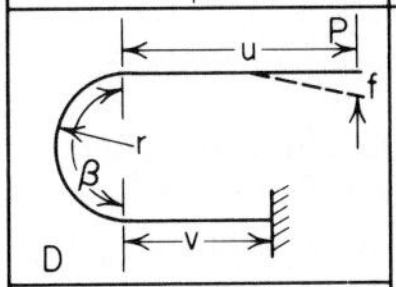 **D** **E**	$F_4 = F_5 = \dfrac{P}{3EI}\left[2Kr^3\left(m+\dfrac{\beta}{2}\right)^3 + (v-u)^3\right]$ where $\alpha = \dfrac{\beta}{2}$ for finding K	$P = \dfrac{S\sigma}{\lambda} = \dfrac{P\lambda}{S}$ <table><tr><td>First condition</td><td>Second condition</td><td>λ</td></tr><tr><td>u ≥ v</td><td>- - -</td><td>u + r</td></tr><tr><td>u < v</td><td>(u− v) < (u+r)</td><td>u + r</td></tr><tr><td>u < v</td><td>(v−u) > (u+r)</td><td>v − u</td></tr><tr><td>u = 0</td><td>v ≤ r</td><td>r</td></tr><tr><td>u = 0</td><td>v > r</td><td>v</td></tr></table>

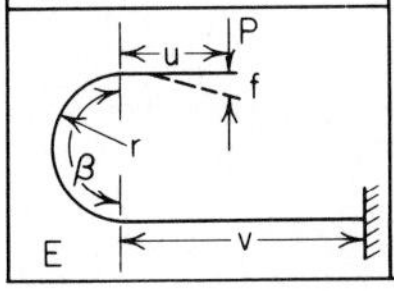

FIG. 23 Deflection, force, and stress relations for curved springs. *(Product Engineering.)*

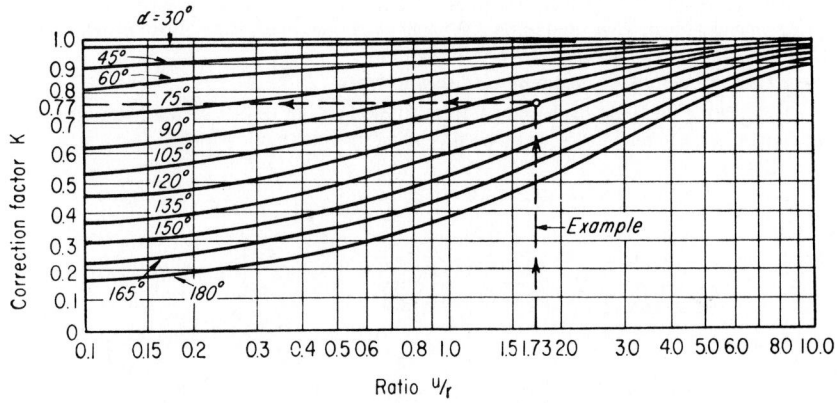

FIG. 24 Correction factors for curved springs. *(Product Engineering.)*

from Fig. 24, where the subscripts refer to the radius being considered in the relation u/r; m = u/r; β = angle of spring curvature, rad. Where the subscripts 1 and 2 are used in this equation, they refer to the respective radius identified by this subscript. Since $I = bh^3/12$ for a rectangle, or $I = (1.20)(0.30)^3/12 = 0.0027$ mm^4, $F = \{[2(0.135)/[(3)(14,500)(0.0027)]\}$ $[2(0.92)(0.65)(2.62 + 1.57)^3 + 0 + 0.94(1.75)^3(3.2 + 1.57)^3] = 1.34$ mm.

4. Compute the spring constant

The spring constant $C = P/F = 0.135 = 0.135/1.34 = 0.101$ kg/mm.

   Related Calculations: The relations given here can also be used for round-wire springs. For accurate results, h/r for flat springs and d_o/r for round-wire springs should be less than 0.6. The various symbols used in this calculation procedure are defined in the text and illustrations. Since the equations given here analyze the springs and do not contain any empirical constants, the equations can be used, as presented, for both metric and English units. Where a round spring is analyzed, $h = b = d_o$, where d_o = spring outside diameter, mm or in.

ROUND- AND SQUARE-WIRE HELICAL TORSION-SPRING SELECTION

Choose a round-music-wire torsion spring to handle a moment load of 15.0 lb·in (1.7 N·m) through a deflection angle of 250°. The mean diameter of the spring should be about 1.0 in (2.5 cm) to satisfy the space requirements of the design. Determine the required diameter of the spring wire, the stress in the wire, and the number of turns required in the spring. What is the maximum moment and angular deflection the spring can handle? What is the maximum moment and deflection without permanent set?

Calculation Procedure:

1. Select a suitable wire diameter

To reduce the manufacturing cost of a spring, a wire of standard diameter should be used, whenever possible, for the spring, Fig. 25. Usual torsion-spring wire diameters and the side of square-wire springs range from 0.02 to 0.60 in (0.05 to 1.52 cm), depending on the moment the spring must carry and the angular deflection.

   Assume a wire diameter of 0.10 in (0.25 cm) and a bending stress of 150,000 lb/in^2 (1.03 × 10^9 Pa) as trial values for this spring. [Typical round-wire and square-wire torsion-spring bending stresses range from 100,000 to 200,000 lb/in^2 (689.5 × 10^6 to 1.38 × 10^9 Pa), depending on the material used in the spring.]

   Compute the twisting moment corresponding to the assumed stress from $M_i = \pi d^3 S_b/32$, where M_i = twisting moment load, lb·in; d = spring wire diameter, in; S_b = bending stress in spring, lb/in^2. Thus, $M_i = \pi(0.10)^3(150,000)/32 = 14.7$ lb·in (1.66 N·m). This is very close to the actual moment load of 15.0 lb·in (1.7 N·m). Therefore, the assumed spring diameter and bending stress are acceptable, thus far.

2. Compute the actual spring stress

Use the following relation to find the actual bending stress S_b lb/in^2 in the spring: S_b = (actual spring moment lb·in/computed spring moment, lb·in)(assumed stress, lb/in^2); S_b = (15.0/14.7)(150,000) = 153,000 lb/in^2 (1.05 × 10^9 Pa).

3. Check the actual vs. recommended spring stress

Enter Fig. 26 at the wire diameter of 0.10 in (0.25 cm), and project vertically upward to the music-wire curve to read the recommended bending stress for music wire as 159,000 lb/in^2 (1.10

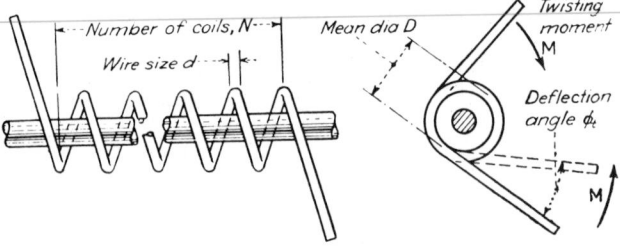

FIG. 25   Typical torsion spring. (*Product Engineering.*)

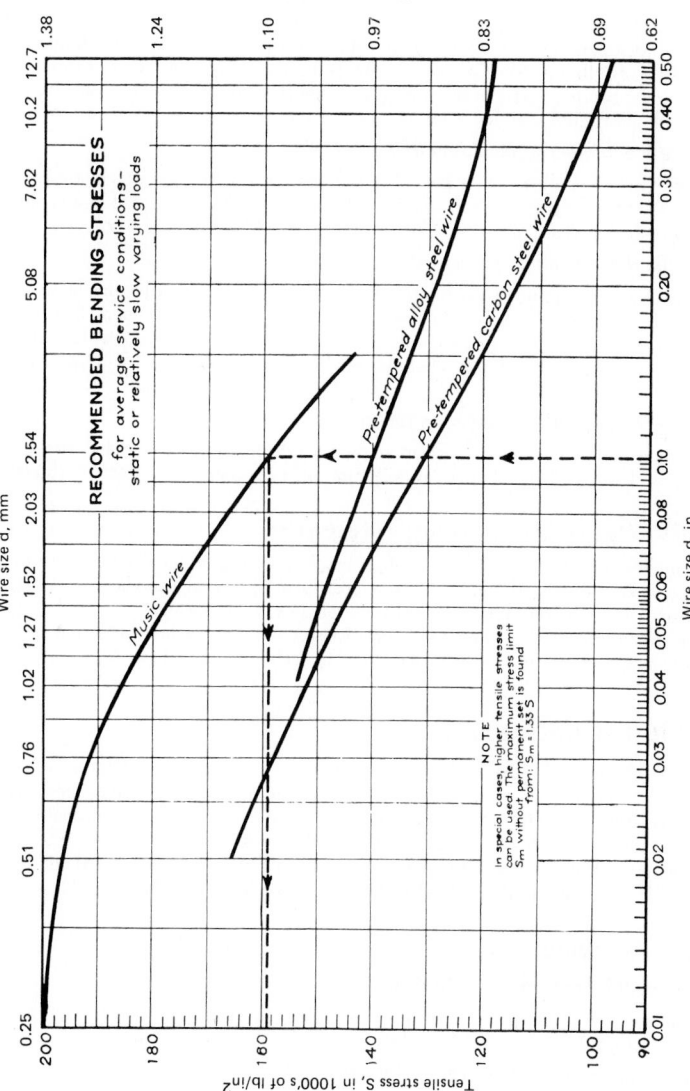

FIG. 26 Recommended bending stresses for torsion springs. (*Product Engineering.*)

3.85

$\times 10^9$ Pa). Since the actual stress, 153,000 lb/in^2 (1.05 $\times$ 10^9 Pa), is less than but reasonably close to the recommended stress, the selected wire diameter is acceptable for the planned load on the spring. This chart and calculation procedure were developed by H. F. Ross and reported in *Product Engineering*.

4.   Determine the angular deflection per spring coil

Compute the angular deflection per coil from $\phi = 360S_b d_m / (Ed)$, where ϕ = angular deflection per spring coil, degrees; d_m = mean diameter of spring, in; E = Young's modulus for spring material = 30 $\times$ 10^6 lb/in^2 (206.9 $\times$ 10^9 Pa) for spring steel; other symbols as before. Thus, by using the *assumed* bending stress in the spring, $\phi = 360(150,000)(1.0)/[(30 \times 10^6)(0.1)] = 18°$. This value is the maximum safe deflection per coil for the spring.

5.   Compute the number of coils required

The number of coils n required in a helical torsion spring is $n = \phi_t$(assumed stress, lb/in^2)/ϕ (actual stress, lb/in^2), where ϕ_t = total angular deflection of spring, degrees; ϕ = maximum safe deflection per coil, degrees. Thus, $n = 250(150,000)/[(18)(153,000)] = 13.6$ coils; use 14 coils.

6.   Determine the maximum moment the spring can handle

On the basis of the maximum recommended stress, the moment can be increased to M_i = [(maximum recommended stress, lb/in^2)/(assumed stress, lb/in^2)](actual moment, lb·in). Read the maximum recommended stress from Fig. 26 as 159,000 lb/in^2 (1.10 $\times$ 10^9 Pa) for 0.1-in (0.25-cm) diameter music wire, as in step 3. Thus, $M_i = (159,000/150,000)(14.7) = 15.6$ lb·in (1.8 N·m).

7.   Compute the maximum angular deflection

The maximum angular deflection per coil is ϕ = [(maximum recommended stress, lb/in^2)/(assumed stress, lb/in^2)](computed angular deflection per coil, degrees) = 159,000/150,000 $\times$ 18 = 19.1° per coil.

8.   Determine the special-case moment and deflection

The maximum moment M_{max} and deflection ϕ_{max}, without permanent set, can be one-third greater than in steps 6 and 7, or $M_{max} = 15.6(1.33) = 20.8$ lb·in (2.4 N·m), and $\phi_{max} = 19.1(1.33) = 25.5°$ per coil. These maximum values allow for overloads on the spring.

    Related Calculations: Use the same procedure for square-wire helical torsion springs, but substitute the length of the side of the square for d in each equation where d appears.

TORSION-BAR SPRING ANALYSIS

What must the diameter of a torsion bar be if it is to have a spring rate of 2400 lb·in/rad (271.2 N·m/rad) and a total angle of twist of 0.20 rad? The bar is made of 302 stainless steel, which has a proportional limit in tension of 35,000 lb/in^2 (241.3 $\times$ 10^6 Pa), and G = torsional modulus of elasticity = 10^7 lb/in^2 (68.95 $\times$ 10^9 Pa). The length of the torsion bar is 26.0 in (66.0 cm), and it is solid throughout. What size square torsion bar would be required? What size equilateral triangular section would be required? What is the energy storage of each bar form?

Calculation Procedure:

1.   Determine the proportional limit in shear

For stainless steel, the proportional limit S_s lb/in^2 in shear is 0.55 times that in tension, or S_s = 0.55(35,000) = 19,250 lb/in^2 (132.7 $\times$ 10^6 Pa).

2.   Compute the required diameter of the bar

Use the relation $d = 2S_s l/(G\theta)$, where d = torsion-bar diameter, in; l = torsion-bar length, in; θ = total angle of twist of torsion bar, rad; other symbols as before. Thus, $d = 2(19,250)(26.0)/[10^7(0.20)] = 0.50$ in (1.3 cm).

3.   Compute the square-bar size

Use the relation $d = 1.482S_s l/(G\theta)$, where d = side of the square bar, in. Thus $d = 1.482(19,250)(26.0)/[10^7(0.2)] = 0.371$ in (0.9 cm).

4. *Compute the triangular-bar size*

Use the relation $d = 2.31 S_s l/(G\theta)$, where $d =$ side of the triangular bar, in. Thus, $d = 2.31(19,250)(26.0)/[10^7(0.2)] = 0.578$ in (1.5 cm).

5. *Compute the energy storage of each bar*

For a solid circular torsion spring, the energy storage $e = S_s^2/(4G)$, where $e =$ energy storage in the bar, in·lb/in^3. Thus, $e = (19,250)^2/[4(10^7)] = 9.25$ in·lb/in^3 (6.4 N·cm/cm^3).

For a square bar, $e = S_s^2/(6.48G)$, where the symbols are the same as before, or, $e = (19,250)^2/[6.48(10^7)] = 5.71$ in·lb/in^3 (3.9 N·cm/cm^3).

For a triangular bar, $e = S_s^2/(7.5G)$, where the symbols are the same as before. Or, $e = (19,250)^2/[7.5(10^7)] = 4.94$ in·lb/in^3 (3.4 N·cm/cm^3).

Related Calculations: Use this procedure for torsion-bar springs made of any metal. The energy-storage capacity of various springs in terms of the spring weight is as follows:

	Energy storage of spring	
Type of spring	in·lb/lb	N·m/kg
Leaf	300–450	74.7–112.1
Helical round-		
wire coil	700–1100	174.4–274.0
Torsion-bar	1000–1500	249.1–373.6
Volute	500–1000	124.5–249.1
Rubber in shear	2000–4000	498.2–996.4

The analyses in this calculation procedure are based on the work of Donald Bastow and D. A. Derse and are reported in *Product Engineering*.

MULTIRATE HELICAL SPRING ANALYSIS

Determine the required spring rates, number of coils, coil clearances, and free length of two helical coil springs if spring 1 has preload of 1.2 lb (5.3 N) and spring 2 has a preload of 19.1 lb (85.0 N) in a double preload mechanism. The rod is to deflect 0.46 in (1.2 cm) before building up to the preload of 19.1 lb (85.0 N). Total deflection is to be 3.0 in (7.6 cm) with a load of 78 lb (347.0 N). The mean spring diameter $d_m = 1.29$ in (3.28 cm) for both springs; the wire diameter is $d = 0.148$ in (3.76 mm) for spring 1; $d = 0.156$ in (3.96 mm) for spring 2; $G = 11.5 \times 10^6$ lb/in^2 (79.3 × 10^9 Pa) for both springs.

Calculation Procedure:

1. *Determine the spring rate for each spring*

The spring rate, lb/in, is $R_s = $ (preload spring 2, lb − preload spring 1, lb)/deflection, in, before full preload. Thus, for spring 1, $R_{s1} = (19.1 - 1.2)/0.46 = 38.9$ lb/in (68.1 N/cm).

For the combination of the two springs $R_{st} = (78 - 19.1)/(3.0 - 0.46) = 23.1$ lb/in (40.5 N/cm).

For spring 2, $R_{s2} = R_{s1}R_{st}/(R_{s1} - R_{st})$, where the symbols are the same as before. Or, $R_{s2} = (38.9)(23.1)/(38.0 - 23.1) = 56.9$ lb/in (99.6 N/cm).

2. *Check the spring rate against the spring deflection*

The deflection, in, is $f = L/R$, where $L =$ load on the spring, lb; $R =$ spring rate, lb/in. Thus for spring 1, $f_1 = (78 - 1.2)/38.9 = 1.97$ in (5.0 cm). For spring 2, $f_2 = L_2/R_{s2} = (78 - 19.1)/56.9 = 1.03$ in (2.6 cm). For the two springs, $F_t = f_1 + f_2 = 1.97 + 1.03 = 3.00$ in (7.6 cm). This agrees with the allowable deflection of 3 in (7.6 cm) at the full load of 78 lb (347.0 N). Therefore, the computed spring rates and preloads are acceptable.

3. Compute the number of coils for each spring

The number of coils $n = Gd^4/(8d_m^3 R)$, where the symbols are as defined before. Thus, $n_1 = (11.5 \times 10^6)(0.148)^4/[8(1.29)^3(38.9)] = 8.25$ coils. And $n_2 = (11.5 \times 10^6)(0.156)^4/[8(1.29)^3(56.9)] = 7$ coils.

4. Compute the solid height of each spring

Allowing one inactive coil for each end of each spring, so that the ends may be squared and ground, we find the solid height $h_s = d(\text{number of coils} + 2)$. Or $h_{s1} = (0.148)(8.25 + 2) = 1.517$ in (3.85 cm). And $h_{s2} = (0.1567)(7 + 2) + 1.404$ in (3.57 cm).

5. Determine the coil clearances

Assume a coil clearance of 3 times the spring wire diameter. Then the coil clearance c, in, for each spring is $c_1 = (3)(0.148) = 0.444$ in (1.128 cm) and $c_2 = (3)(0.156) = 0.468$ in (1.189 cm).

6. Compute the free length of each spring

The free length of a helical spring $= l_f = \text{solid height} + \text{coil clearance} + \text{deflection} + [\text{preload}, \text{lb}/(\text{spring rate, lb/in})]$. For spring 1, $l_{f1} = 1.517 + 0.444 + 1.970 + (1.2/38.9) = 3.962$ in (10.06 cm). For spring 2, $l_{f2} = 1.404 + 0.468 + 1.030 + (19.1/56.9) = 3.235$ in (8.22 cm).

Related Calculations: Use this procedure for springs made of any metal. This analysis is based on the work of K. A. Flesher, as reported in *Product Engineering*.

BELLEVILLE SPRING ANALYSIS FOR SMALLEST DIAMETER

What are the minimum outside radius r_o and thickness t for a steel Belleville spring that carries a load of 1000 lb (4448.2 N) at a maximum compressive stress of 200,000 lb/in^2 (1.38×10^9 Pa) when compressed flat?

Calculation Procedure:

1. Determine the spring radius ratio and the height-thickness ratio

The radius ratio $r_r = r_o/r_i$ for a Belleville spring, where $r_o =$ outside radius of spring, in; $r_i =$ inside radius of spring, in = radius of hole in spring, in. Table 46 summarizes recommended values for the radius ratio for various values of the height-thickness ratio to produce the smallest diameter spring. In general, an r_r value of 1.75 usually produces a spring of suitably small size. When $r_i = 1.75$, Table 46 shows that the height-thickness ratio h/t with both values expressed in inches is 1.5. Assume that these two values are valid, and proceed with the calculation.

2. Determine the spring outside radius

Table 47 shows the stress constant $r_o s_c/L^{0.5}$, where $s_c =$ maximum compressive stress on the top surface at the inner edge, lb/in^2. Fig. 27; $L =$ total axial load on spring, lb. For $h/t = 1.5$ and $r_r = 1.75$, the stress constant $r_o s_c/L^{0.5} = -19,050$. Solving gives $r_o = -19,050L^{0.5}/s_c$. By substituting the given values, $r_o = 19,050(1000)^{0.5}/ -200,000 = 3.01$ in (7.65 cm). The negative sign is used for the spring stress because it is a compressive stress.

TABLE 46   Design Constants for Belleville Springs[°]

h/t	r_o/r_i
1.00	1.25
1.25	1.50
1.50	1.75
1.75	2.00
2.00	2.50

[°]*Product Engineering.*

TABLE 47   Stress Constants for Belleville Springs[°]

	$r_o/r_i = 1.75$	
h/t	K	$r_o s_c/L^{0.5}$
1.00	-3.2455	$-13,460$
1.25	-4.3734	$-16,220$
1.50	-5.6279	$-19,050$
1.75	-7.0090	$-21,970$

[°]*Product Engineering.*

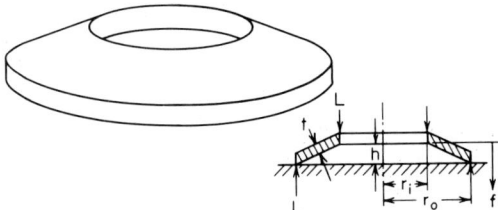

FIG. 27   Belleville spring: appearance and dimensions. *(Product Engineering.)*

3.   Determine the radius of the hole in the spring

For this Belleville spring, $r_i = r_o/r_r = 3.01/1.75 = 1.72$ in (4.37 cm).

4.   Compute the spring thickness

The thickness of a Belleville spring is given by $t = [s_c r_o^2/(KE)]^{0.5}$, where $K = $ a stress constant from Table 47; $E = $ modulus of elasticity of the spring material, lb/in²; other symbols as before. Thus, with $E = 30 \times 10^6$ (206.9 × 10⁹ Pa), $t = [-200,000 \times 3.01^2/(-5.6279 \times 30 \times 10^6)]^{0.5}$ = 0.1037 in (2.63 mm).

5.   Compute the spring height

Since $h/t = 1.5$ for this spring, $h = 1.5(0.1037) = 0.156$ in (3.96 mm).

Related Calculations: Professor M. F. Spotts developed the analytical procedure and data presented here. His studies show that space is usually the limiting factor in spring selection, and the designer generally must determine the minimum permissible outside diameter of the spring to carry a given load at a specified stress. Further, the ratio of the outside to the inside diameter for the smallest spring is about 1.75, assuming that the load spring is compressed nearly flat, which is the usual design assumption. A value of h/t of 1.5 is recommended for most spring applications. Belleville springs are used in disk brakes, the preloading of bolted assemblies, ball bearings, etc. The analysis presented here is useful for all usual applications of Belleville springs.

RING-SPRING DESIGN ANALYSIS

Determine the major dimensions of a ring spring made of material having an allowable stress of 175,000 lb/in² (1.21 × 10⁹ Pa), $E = 29 \times 10^6$ lb/in² (199.9 × 10⁹ Pa), a coefficient of friction of 0.12, an inside diameter of 7.0 in (17.8 cm), an outside diameter of 9.0 in (22.9 cm) or less, a taper angle of 14°, an axial load of 56 tons (50.8 t), and a deflection of not more than 8.0 in (20.3 cm).

Calculation Procedure:

1.   Determine the inner-ring dimensions

For the usual ring spring, the ring height h is 15 percent of the allowable outside diameter, or $(0.15)(9.0) = 1.35$ in (3.4 cm), Fig. 28. The axial gap between the rings g is usually 25 percent of the ring height.

Compute the area of the internal ring from $A_i = L/(\pi K_c s_i)$, where $A_i = $ area of internal ring, in²; $L = $ axial load on spring, lb; $K_c = $ spring constant from Fig. 29; $s_i = $ allowable stress in the inner ring of the spring, lb/in². With a coefficient of friction $\mu = 0.12$ and a taper angle of 14°, $K_c = 0.38$. Then $A_i = 56 \times 2000/[\pi(0.38)(175,000)] = 0.537$ in² (3.47 cm²).

The width w_i of the inner ring is $w_i = [A_i - (h_i^2 \tan\theta)/4]/h_i$, where $\tan\theta = $ tangent of taper angle; $h_i = $ height of inner ring, in. Thus, $w_i = [0.537 - (1.35^2 \tan 14°)/4]/1.35 = 0.314$ (7.98 mm).

Use a trial-and-error process to determine the dimensions of the outer ring. Do this by assuming a cross-sectional area for the outer ring; then compute whether the outside diameter and stress meet the specifications for the spring.

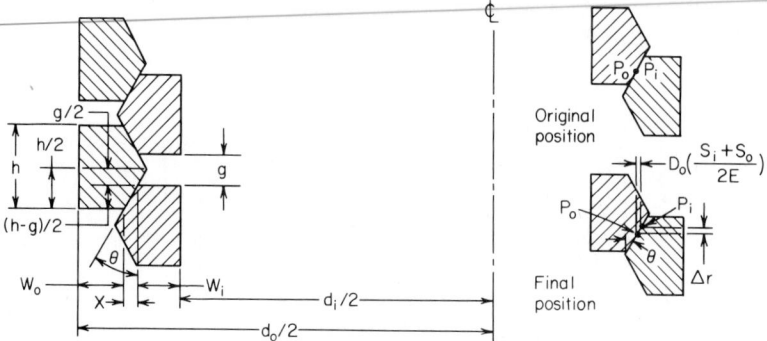

FIG. 28   Ring-spring positions and dimensions. *(Product Engineering.)*

Assume that $A_o = 0.609$ in² (3.93 cm²). Then $s_o = L/(\pi A_o K_c)$, where s_o = stress in outer ring, lb/in²; other symbols as before. So $s_o = 56 \times 2000/[\pi(0.609)(0.38)] = 154,200$ lb/in² (1.06 × 10⁹ Pa). This stress is within the allowable limits.

In the usual ring spring, $h_o = h_i = 1.35$ in (3.4 cm) for this spring. Then, by using a relation similar to that for the inner ring, $w_o = [A_o - (h_o^2 \tan\theta)/4]/h_o = [0.609 - (1.35^2 \tan 14°)/4]/1.35 = 0.366$ in (9.3 mm).

Find the outside diameter of the ring from $d_o = d_i + 2w_i + 2w_o + (h - g)\tan\theta$, where d_o = outside diameter of outer ring, in; d_i = inside diameter of inner ring, in; g = axial gap of rings, in = 25 percent of ring height for this spring, or 0.25(1.35) = 0.3375 in (8.57 mm). Hence, $d_o = 7.0 + 2(0.314) + 2(0.366) + (1.35 - 0.3375)\tan 14° = 8.613$ in (21.9 cm). This is close enough to the maximum allowable outside diameter of 9 in (22.9 cm) to be acceptable. Were the value of d_o unacceptable, another value of A_o would be assumed and the calculation repeated until the stress and d_o values were acceptable.

2.   *Compute the number of rings required*

Find the axial deflection per ring f, in, from $f = d_a[(s_i + s_o)/(2E)]\cot\theta$, where d_a = mean diameter of the spring, in; E = modulus of elasticity of the spring material, lb/in². Compute $d_a = [(d_o - 2w_o) + (d_i + 2w_i)]/2 = [(8.613 - 2 \times 0.366) + (7.0 + 2 \times 0.314)]/2 = 7.755$ in (19.7 cm). Then $f = 7.755[(175,000 + 154,200)/(2 \times 29 \times 10^6)]\cot 14° = 0.176$ in (4.47 mm). Since the axial deflection must not exceed 8 in (20.3 cm), the number of rings required = axial deflection, in/deflection per ring, in = 8.0/0.176 = 45.5, or 46 rings. Figure 30 shows the spring dimensions.

Related Calculations: Ring springs are suitable for pipe-vibration isolation, shock absorbers, plows, trench diggers, railroad couplers, etc. The recommended approximate proportions of ring springs are as follows: (1) Compressed height should be at least 4 times the deflection of the spring. (2) Ring height should be 15 to 20 percent of the ring outside diameter. (3) Spring outside diameter and height are usually as large as space permits.(4) Thin ring sections are preferred to thick ones. (5) Ring taper should be 1:4. (6) Coefficient of friction for ring springs varies from 0.10 to 0.18. (7) Allowable spring stresses are 160,000 lb/in² (1.10 × 10⁹ Pa) for nonmachined steel, 200,000 lb/in² (1.38 × 10⁹ Pa) for machined steel. For vibratory loads, the allowable stress is about one-half these values. (8) Load capacities

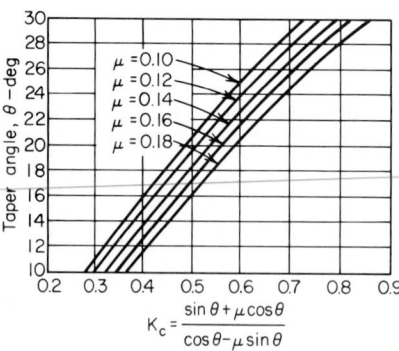

FIG. 29   Ring-spring compression constant in terms of the taper angle for various values of the coefficient of friction. *(Product Engineering.)*

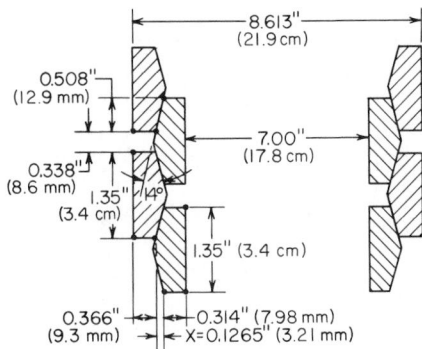

FIG. 30 Dimensions of a typical ring spring. (*Product Engineering.*)

of ring springs vary between 2 and 150 tons (1.8 and 136.1 t). (9) Spring deflections vary between 1 in (2.5 cm) and 1 ft (0.3 m). (10) The equations given above can be used for spring design or for analysis of an existing spring.

The design method given here was developed by Tyler G. Hicks and reported in *Product Engineering.*

LIQUID-SPRING SELECTION

Select a liquid spring to absorb a 50,000-lb (222,411.1-N) load with a 5-in (12.7-cm) stroke. The rod diameter is 1 in (2.5 cm). What is the probable temperature rise per stroke? Compare this spring with metal-coil, Belleville, and ring springs.

Calculation Procedure:

1. Compute the liquid volume required

Assume that the final pressure of the compressed liquid is 50,000 lb/in^2 (344,750 kPa) and that the liquid is compressed 18 percent on application of full load on the spring. This means that 82 percent (100 − 18) of the original volume remains after application of the load.

Compute the liquid volume required from $v = \pi S d^2/(4c)$, where v = liquid volume required, in^3; S = stroke length, in; d = rod diameter, in; c = liquid compressibility, expressed as a decimal. Thus $v = \pi(5)(1)^2/[4(0.18)] = 21.8$ in^3 (357.2 cm^3).

2. Determine the cylinder length

In a liquid spring, the cylinder inside diameter d_i is usually greater than that of the rod. Assuming an inside diameter of 1.8 in (4.6 cm) for the cylinder, we find length $= 4v/(\pi d_i^2)$, where d_i = cylinder inside diameter, in; other symbols as before. For this cylinder, length $= 4(21.8)/[\pi(1.8)^2]$ = 8.56, say 8.6, in (21.8 cm).

3. Determine the cylinder dimensions

With a 1.8-in (4.6-cm) inside diameter, a 3-in (7.6-cm) outside diameter will be required, based on the usual cylinder proportions. Allowing 3.4 in (8.6 cm) for the cylinder ends and seals and 5 in (12.7 cm) for the stroke, we find that the total length of the cylinder will be 8.6 + 3.4 + 5.0 = 17.0 in (43.2 cm).

4. Compute the cylinder temperature rise

Assume that the average friction load is 10 percent of the load on the spring, or $0.1 \times 50,000 = 5000$ lb (22,241.1 N). A friction load of 10 percent is typical for liquid springs.

The energy absorbed per stroke of the spring is $e = Fl$, where e = energy absorbed, ft·lb; F = friction force, lb; l = stroke length, ft. For this spring, $e = 5000(5/12) = 2085$ ft·lb (2826.9 N·m). Since 778.2 ft·lb = 1 Btu = 1.1 kJ, $e = 2085/778.2 = 2.68$ Btu (2827.6 J).

TABLE 48   Performance of Four Typical Spring Types°

	Coil	Nested Belleville washers	Tapered rings	Liquid
Useful range:				
Low load	1 oz (28.3 g)	20 lb (9.1 kg)	2 tons (1.8 t)	100 lb (45.4 kg)
High load	10 ton (9.1 t)	100 ton (90.7 t)	150 ton (136.1 t)	200 ton (181.4 t)
Force vs. deflection	Low to high	High	High	Medium to high
Stroke	Short to long	Short	Short to medium	Short to long
Damping ability	Low	Low	Low	Low to high
Relative cost	Low	Low	Medium	High

° *Product Engineering.*
Note: An example: For 50,000-lb (222,411.1-N) load, 5-in (12.7-cm) stroke:

Size	Length, in (cm)	68 (172.7)	37 (94.0)	24 (61.0)	17 (43.2)
	Diameter, in (cm)	11.5 (29.5)	8 (20.3)	5 (12.7)	3 (7.6)

An assembly of the dimensions computed in step 3 will weigh about 35 lb (15.9 kg) and will have an average overall specific heat of 0.15 Btu/(lb·°F) [628.0 J/(kg·°C)]. Hence, the temperature rise per stroke will be: Btu of heat generated per stroke/[(specific heat)(cylinder weight, lb)] = 0.51°F (0.28°C) per stroke. A temperature rise of this magnitude is easily dissipated by the external surfaces of the cylinder. But a smaller liquid spring under rapidly fluctuating loads may have an excessive temperature rise. Each spring must be analyzed separately.

5.   Compare the various types of springs

By using previously presented calculation procedures, Table 48 can be constructed. This table and the spring analysis given above are based on the work of Lloyd M. Polentz, Consulting Engineer. The tabulation shows that the liquid spring is the shortest and has the smallest diameter for the load in question. Figure 31 shows typical liquid springs; Fig. 32 shows the compressibility of liquids used in various liquid springs.

   Related Calculations: Use the method given here to select liquid springs for applications in any of a variety of services where a large load must be absorbed. The seals at the cylinder ends must be absolutely tight. Liquid springs are best applied in atmospheres where the temperature variation is minimal.

SELECTION OF AIR-SNUBBER DASHPOT DIMENSIONS

Determine the required orifice area, peak actuator pressure, peak negative acceleration, and the time required for the stroke of a 3-in^3 (49.2-cm^3) capacity air snubber if the total load mass M = 0.1 lbs·s^2/in (17.9 g·s^2/cm); the snubber pressure P_i = 100 lb/in^2 (689.5 kPa); piston area A_p = 3 in^2 (19.4 cm^2); initial snubber active length S = 1.0 in (2.5 cm); initial piston velocity v_i = 100 in/s (254 cm/s); piston velocity at the end of travel v = 29 in/s (73.7 cm/s); constant external force on snubber F = 150 lb (667.2 N); initial gas temperature T_i = 530 R (294.1 K); gas constant R = [639.6 in·lb/(lb·°R)] (air); C_D = orifice discharge coefficient = 0.9 dimensionless.

Calculation Procedure:

1.   Compute the snubber dimensionless parameters

The first dimensionless parameter K_E = stored energy/kinetic energy = $P_i V_i/(Mv_i^2)$ = $(100)(3)/[(0.1)(100)^2]$ = 0.3. The next parameter K_F = constant external force/initial pressure force = $F/$

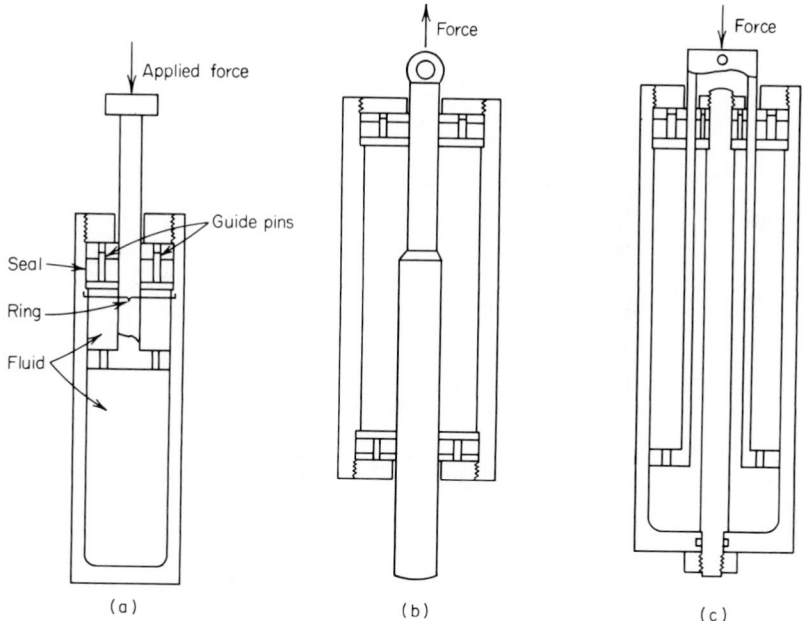

FIG. 31 Typical liquid springs. (a) General design; (b) tension type; (c) long-stroke type. (*Product Engineering.*)

$(P_i A_p) = 150/[(100)(3)] = 0.5$. The third parameter K_v = piston velocity at end of stroke/initial piston velocity = $v/v_i = 20/100 = 0.20$.

2. Determine the actual value of the orifice parameter

The parameter K_w = initial orifice flow/initial displacement flow = $w_i/(\rho A_p v_i)$, where ρ = gas density, lb/in³. Figure 33 gives values of K_w for $K_F = 0$ and $K_F = 1.0$. However, K_F for this snubber = 0.5. Therefore, it is necessary to interpolate between the charts for $K_F = 0$ and $K_F = 1.0$.

Interpolate by constructing a chart, Fig. 34, using values of K_w read from each chart in Fig. 33. Thus, when $K_F = 1$, $K_v = 0.2$, $K_E = 0.3$, $K_w = 0.295$. After the curve is constructed, read $K_w = 0.375$ for $K_f = 0.5$.

3. Compute the true flow through the orifice

The true initial flow rate w_i, lb/s = $K_w[P_i/(RT_i)]A_p v_i$, where all the symbols are as defined earlier. Thus, $w_i = (0.375)\{100/[(639.6)(530)]\}(3)(100) = 0.0332$ lb/s (15.1 g/s).

4. Compute the required orifice area

Use the equation $A_o = w_i/P_i C_D\{(kg/RT_i)[2/(k+1)](k+1)/(k-1)\}^{0.5}$, where $k = 1.4$; $g =$

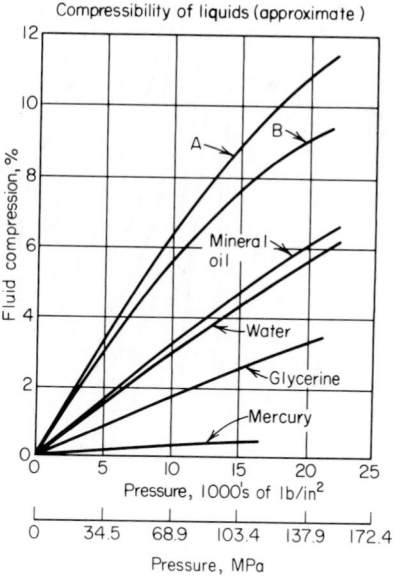

FIG. 32 Common fluids for liquid springs are Dow-Corning type F-4029, curve A, and type 200, curve B. (*Product Engineering.*)

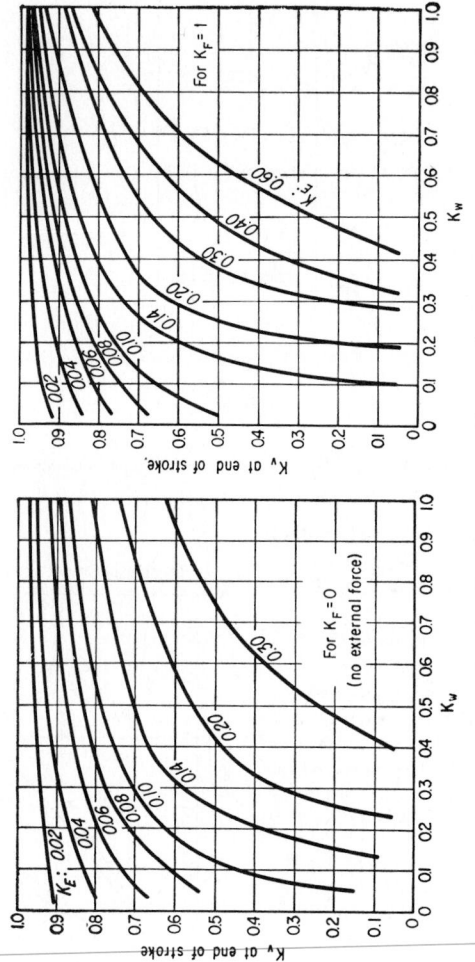

FIG. 33 Impact velocity vs. orifice flow (dimensionless) for air snubber. (*Product Engineering.*)

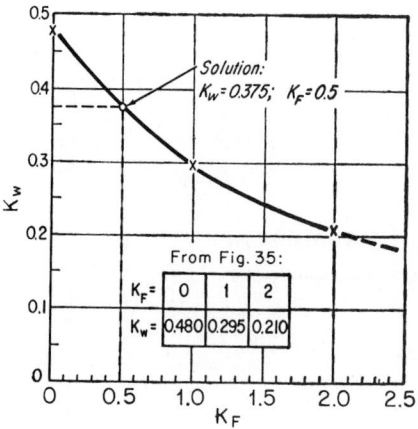

FIG. 34 Cross plot for an air snubber. *(Product Engineering.)*

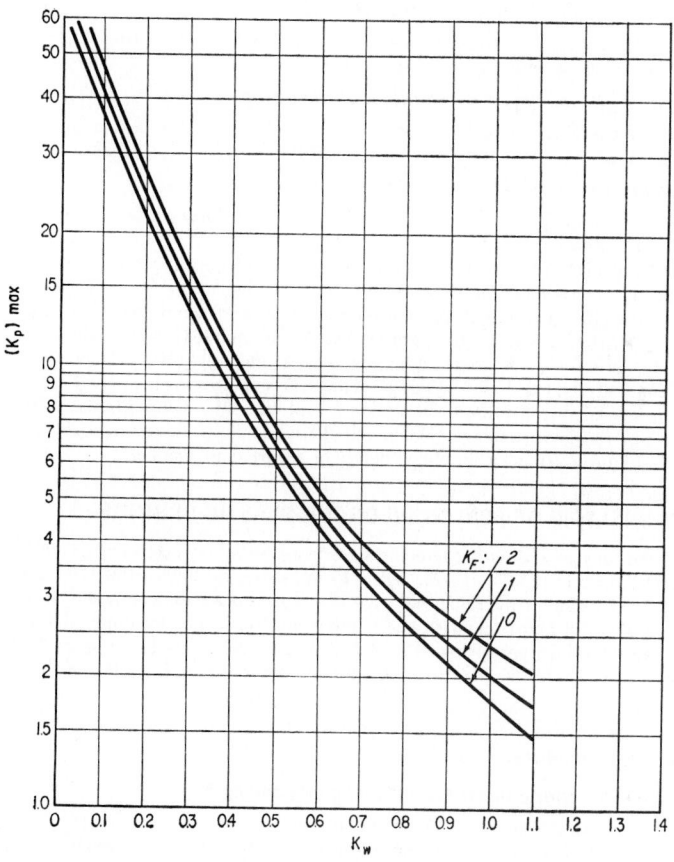

FIG. 35 Maximum pressure at impact (dimensionless). *(Product Engineering.)*

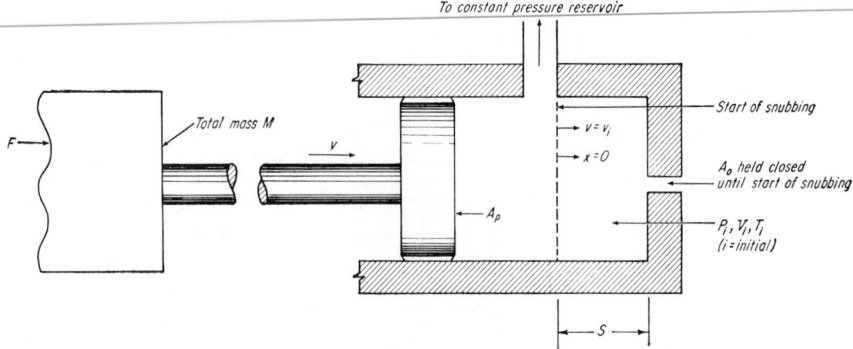

FIG. 36   Simplified air-snubber design. *(Product Engineering.)*

32.2 ft/s^2 (9.8 m/s^2); other symbols as before. Thus, $A_o = 0.0332/[100(0.9)]\{(1.4 \times 32.2/639.6 \times 530)[(2/(1.4 + 1)](1.4 + 1)/(1.4 - 1)]\}^{0.5} = 0.016$ in^2 (0.103 cm^2).

5.   Determine the maximum pressure at the end of the stroke

Read, from Fig. 35, $K_{p,max} = 10.2$ for $K_F = 0.5$, $K_w = 0.375$. Then the true P_{max} at end of stroke $= K_p P_i = 10.2(100) = 1020$ lb/in^2 (7032.9 kPa).

6.   Determine the maximum acceleration of the piston

For an air snubber, $K_{a,max} = K_F - K_p = 0.5 - 10.2 = -9.7$. Also, the maximum acceleration $a_{max} = K_a P_i A_p/M = (-9.7)(100)(3.0)/0.1 = -29,100$ in/s^2 (-739.1 m/s^2).

7.   Determine the approximate travel time of the piston

The travel time for the piston is $t = K_t S/v_i$, where t = travel time, s. Or, $t = 0.95 \times (1.0)/100 = 0.0095$ s, assuming $K_t = 0.95$.

   Related Calculations: The equations in this procedure were developed by Tom Carey and T. T. Hadeler, and are based on these assumptions: (1) They apply only to a piston-orifice-type dashpot; (2) the piston is firmly stopped at the end of the stroke and does not rebound, oscillate, or bounce; (3) friction is zero; (4) the external force is constant; (5) $k = 1.4$, which means that the equations are valid for air, hydrogen, nitrogen, oxygen, and any other gas having a specific-heat ratio of about 1.4; (6) the contained air or gas is ideal; (7) compression is adiabatic; (8) flow through the bleed orifice is critical (a valid assumption except when the dashpot initial pressure is atmospheric, as in screendoor snubbers). When actual friction exists, there is a slight increase in the value of K_t. Figure 36 shows a simplified design of a typical air snubber.

DESIGN ANALYSIS OF FLAT REINFORCED-PLASTIC SPRINGS

A large shaker unit in a vibrating screen system is supported on a series of six steel leaf springs, each a cantilever 6 in (15.2 cm) wide by 0.125 in (3.18 mm) thick. How thick should a single epoxy-glass leaf spring of the same width be if it is to replace the composite steel spring? The cantilever is 30 in (76.2 cm) long with a 24-in (60.9-cm) free length; maximum deflection = 0.375 in (9.53 mm); axial load per spring = 2500 lb (11,120.6 N); safety factor = 8; $E = 4.5 \times 10^6$ lb/in^2 (31.0 $\times$ 10^9 Pa) for the plastic spring; ultimate flexure strength = 100,000 lb/in^2 (689,475.7 kPa).

Calculation Procedure:

1.   Compute the spring thickness for minimum bending stress

The equation for the thickness giving the minimum bending stress is $t = [4LI^2/(wE)]^{1/3}$, where t = spring thickness, in; L = axial load on spring, lb; l = spring free length, in; w = spring width, in; E = modulus of elasticity of spring material, lb/in^2. For this spring, $t = [4 \times 2500 \times 24^2/(6 \times 4.5 \times 10^6)]^{1/3} = 0.598$, say 0.6 in (1.5 cm).

2. Determine the total combined stress in the beam

The maximum combined stress $s_B = (3Et/l^2 + 6L/wt^2)f$, where f = spring deflection, in; other symbols as before. Thus, $s_B = \{e \times 4.5 \times 10^6 \times 0.6/[24^2 + 6 \times 2500/(6 \times 0.6^2)]\} \times 0.375 = 7875$ lb/in^2 (54,290 kPa).

The total stress in the spring $s_T = s_B + L/A$, where A = spring cross-sectional area. Thus, $s_T = 7875 + 2500/(6 \times 0.6) = 8570$ lb/in^2 (59,082 kPa). The allowable stress = ultimate flexure strength, lb/in^2/factor of safety = 100,000/8 = 12,500 lb/in^2 (86,175.5 kPa). Since $s_T < 12,500$ lb/in^2 (86,175.5 kPa), the dimensions of the spring are satisfactory.

3. Check the critical buckling stress of the spring

To prevent buckling of the spring, the following must hold: $L/A \leq \pi^2 Et^2/36^2 = s_{CR}$, or $2500/(6 \times 0.6) \leq \pi^2 \times 4.5 \times 10^6 \times 0.6^2/(3 \times 24^2) = 694$ lb/in^2 (4784 kPa) < 9240 lb/in^2 (63,701 kPa). Hence, the spring dimensions are satisfactory.

4. Determine whether the computed thickness gives adequate stiffness

For a plastic spring to have a stiffness equal to a steel spring having n leaves, the plastic spring should have a thickness of $t = t_s(nE_s/E)^{1/3}$, where t_s and E_s refer to the thickness, in, and modulus of elasticity, lb/in^2, respectively, of the steel spring; other symbols as before. Thus, $t = 0.125[6 \times 30 \times 10^6/(4.5 \times 10^6)]^{1/3} = 0.43$ in (1.1 cm). Since $t = 0.60$ in (1.5 cm), as computed in step 1, the plastic spring is slightly too stiff.

5. Check the thickness required for equivalent thickness

Using the equations for s_B and s_T from step 2, and the equation for s_{CR} from step 3, compute the respective stresses for values of t less than, and greater than, 0.43. Thus

t		s_B		Q/A		s_T		s_{CR}	
in	cm	lb/in^2	kPa	lb/in^2	kPa	lb/in^2	kPa	lb/in^2	kPa
0.500	1.270	8,143	56,144.0	833	5,743.5	8,976	61,889.5	6,400	44,128.0
0.600	1.524	7,875	54,296.2	695	4,792.0	8,570	59,090.2	9,240	63,709.8
0.625	1.588	7,900	54,468.6	666	4,592.1	8,566	59,062.6	10,000	68,947.6

Plot the results as in Fig. 37. This plot clearly shows that $t = 0.43$ in (1.09 cm) gives $s_T < 12,500$ lb/in2 (86,175 kPa), and $Q/A < s_{CR}$. Hence, this thickness is satisfactory.

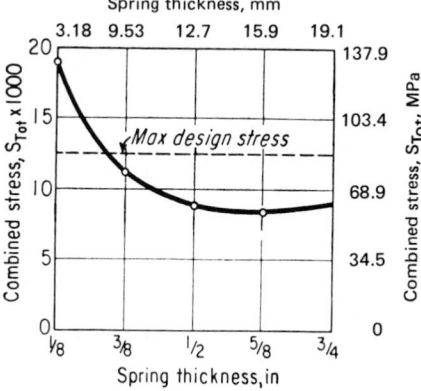

FIG. 37 Combined stress in a plastic spring. (*Product Engineering.*)

6. *Determine whether a thinner spring can be used*

A thinner spring will save money. From Fig. 37, $t = 0.375$ in (9.53 mm) gives an s_T value well below the maximum design stress, and the actual spring stress is one-third the critical buckling stress. If tests on a 0.375-in (9.53-mm) thick spring show no serious disruption of harmonic operation, then specify the thinner material to lower the cost. Otherwise, use the thicker 0.43-in (1.09-cm) spring.

   Related Calculations: Use this procedure for unidirectional, cross-plied, or isotropic-ply plastic springs. Obtain the allowable stress for the spring from the plastic manufacturer. The method given here is the work of L. A. Heggernes, reported in *Product Engineering.*

LIFE OF CYCLICALLY LOADED MECHANICAL SPRINGS

What is the probable life in cycles of a Belleville spring under a bending load if it is made of carbon steel having a Rockwell hardness of C48?

Calculation Procedure:

1. *Determine the spring material tensile strength*

Enter Fig. 38 at the Rockwell hardness C48, and project vertically upward to read the tensile strength of the carbon-steel spring material as 235,00 lb/in² (1620 MPa).

2. *Compute the actual stress in the spring*

Using the spring dimensions and the equations presented in the Belleville spring calculation procedure, compute the actual stress in the spring. For the spring in question, the actual stress is found to be 150,000 lb/in² (1034 MPa). This is $150,000(100)/235,000 = 63.8$, say 64, percent of the spring material tensile strength.

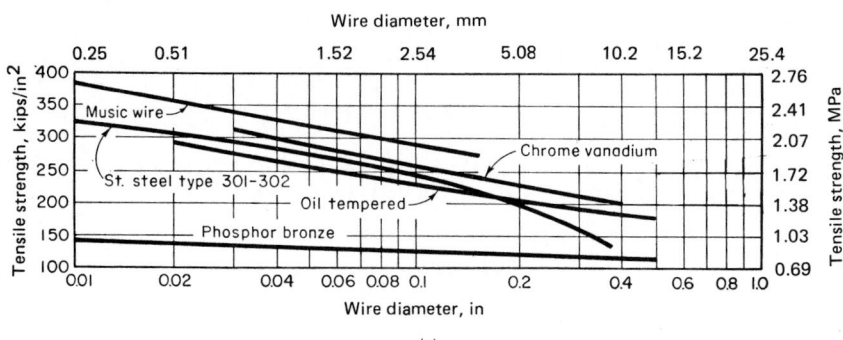

(a)

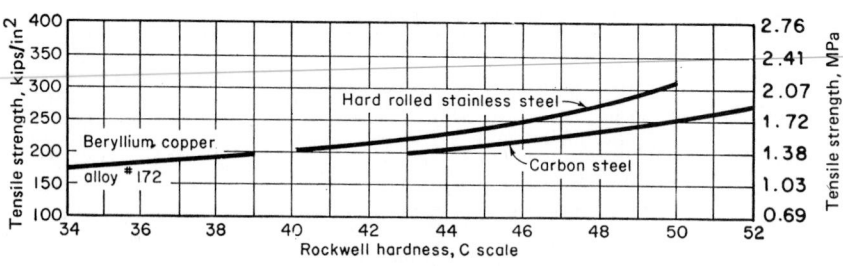

(b)

FIG. 38   (*a*) Tensile strength of spring wire; (*b*) tensile strength of spring strip. (*Product Engineering.*)

3. *Estimate the spring cycle life*

Enter the upper part of Table 49 for springs in bending. This tabulation shows that at a stress of 65 percent of the tensile strength, the spring will have a life between 10,000 and 100,000 stress cycles. Actual test of the spring caused failure at about 100,000 cycles.

Related Calculations: Use this procedure for helical torsion springs, cantilever springs, wave washers, flat springs, motor springs, helical compression and extension springs, torsion bars, and Belleville springs. Be sure to enter the proper portion of Table 49 when finding the approximate number of repetitive stress cycles. The method presented here is the work of George W. Kuasz and William R. Johnson and is reported in *Product Engineering*.

SHOCK-MOUNT DEFLECTION AND SPRING RATE

Determine the maximum probable acceleration, the shock isolator deflection, and the isolator spring rate for a 25-lb (11.3-kg) piece of electronic equipment which drops from a 24-in (61.0-cm) high tailgate of a truck onto a concrete road. The product lands on one corner point and should be considered as rigid steel for analysis purposes. In its carton, the load will be supported by 16 shock isolators.

Calculation Procedure:

1. *Compute the acceleration of the load*

Use the relation $g = (72/t)(h)^{0.5}$, where g = load acceleration in g [1 g = 32.2 ft/s^2 (9.8 m/s^2) at sea level]; t = shock-rise time, ms, from Table 50; h = drop height, in. From Table 50, $t = 2$ ms for rigid steel making point contact with concrete. Then $g = 72/2 \times (24)^{0.5} = 176.5g$.

2. *Compute the isolator deflection*

Use the relation $d = 2h/(g - 1)$, where d = isolator deflection, in; other symbols as before. For this load, $d = 2 \times 24/(176.5 - 1) = 0.273$ in (6.93 mm).

3. *Compute the required specific spring rate for the isolator*

Use the relation $K = g/d$, where D = isolator specific spring rate, lb/(in·lb). Thus $K = 176.5/0.273 = 646$ lb/(in·lb) [254.3 N/(N·cm)].

TABLE 49   Design Stresses for Springs°

No. of repetitive stress cycles	Maximum design stress (percent of the tensile strength shown in charts)
Design stress for springs in bending†	
10,000	80
	65‡
100,000	53
1,000,000	50
10,000,000	48
Design stress for springs in torsion§	
10,000	45
	35‡
100,000	35
1,000,000	33
10,000,000	30

° *Product Engineering.*
†For example, helical torsion springs, Bellevilles, cantilever springs, wave washers, flat springs, and motor springs.
‡For stainless-steel and phosphor-bronze materials. Tests show that such materials have low yield points.
§For example, helical compression springs, helical extension springs, and torsion bars.

TABLE 50   Typical Value for Shock-Time Rise

Condition	Shock-time rise, ms	
	Flat face	Point
Rigid steel against concrete	1	2
Rigid steel against wood or mastic	2–3	5–6
Steel or aluminum against compact earth	2–4	6–8
Steel or aluminum against sand	5–6	15
Product case against mud	15	20
Product case against 1-in (2.5-cm) felt	20	30

Note: Mass of struck surface is assumed to be at least 10 times the striking mass. Point contact with spherical radius of 1 in (2.5 cm).

4. *Determine the required spring rate per isolator*

With n shock isolators, the required spring rate, lb/in per isolator is $k = KW/n$, where $W =$ weight of part, lb; $n =$ number of isolators used in the carton. Thus, $k = (646)(25)/16 = 1020$ lb/in (182.2 kg/cm).

Related Calculations: Some of the largest stock loads encountered by equipment occur during transportation. Thus, vertical accelerations on the body of a 2-ton (1.8-t) truck traveling at 30 mi/h (13.4 m/s) on good pavement range from 1 to 2 g, with a rise time of 10 to 15 ms. Higher speeds, rougher roads, stiffer truck springs, and careless driving all decrease the rise time and thus double or triple the acceleration loads.

The highest acceleration forces in railroad freight cars occur during humping, when the impact loads on a product container may range from 4.5 to 28 g.

For most components that are sensitive to shock, suppliers include maximum safe acceleration loads in engineering data. Maximum allowable loads on vacuum tubes are 2 to 5 g; relays may withstand higher accelerations, depending on the type and direction of the acceleration. Transistors have low mass and good rigidity and are highly resistant to shock when properly supported. Ball-bearing races may be indented by the balls; sleeve bearings are usually much more resistant to shock.

The function of a shock mount is to provide enough protection to avoid damage under expected conditions. But overdesign can be costly, both in the design of the product and in the shock-mount components. Underdesign can lead to failures of the shock mount in service and possible damage to the product. Therefore, careful design of shock mounts is important. The method presented here is the work of Raymond T. Magner, reported in *Product Engineering*.

CLUTCH SELECTION FOR SHAFT DRIVE

Choose a clutch to connect a 50-hp (37.3-kW) internal-combustion engine to a 300-r/min single-acting reciprocating pump. Determine the general dimensions of the clutch.

TABLE 51   Clutch Characteristics

Type of clutch	Typical applications[*]
Friction:	
Cone	Varying loads; 0 to 200 hp (0 to 149.1 kW); losing popularity for many applications, particularly in the higher hp ranges
Disk or plate	Varying loads; 0 to 500 hp (0 to 372.9 kW); widely used; more popular than the cone clutch
Rim:	
Band	Varying loads; 0 to 100 hp (0 to 74.6 kW); not too widely used
Overrunning	Constant or moderately varying loads; 0 to 200 hp (0 to 149.1 kW); engages in one direction; freewheels in the opposite direction
Centrifugal	Constant loads; 0 to 50 hp (0 to 37.3 kW)
Inflatable	Varying loads; 0 to 5000 hp (0 to 3728.5 kW); compressed air inflates clutch; have 360° friction surface
Magnetic	Varying loads; 0 to 10,000 hp (0 to 7457.0 kW); high speeds; also used where disk clutch would be overloaded
Positive-engagement	Nonslip operation; low-speed (10 to 150 r/min) engagement; has sudden starting action
Fluid	Large, varying loads; 0 to 10,000 hp (0 to 7457.0 kW); variable-speed output; can produce a desired slip
Electromagnetic	Large, varying loads; 0 to 10,000 hp (0 to 7457.0 kW); variable-speed output; characteristics similar to fluid clutches

[*]Clutch capacity depends on the design, materials of constructions, type of load, shaft speed, and operating conditions. The applications and capacity ranges given here are typical but should not be taken as the only uses for which the listed clutches are suitable.

TABLE 52 Clutch Service Factors

Type of service	Service factor
Driver:	
Electric motor:	
Steady load	1.0
Fluctuating load	1.5
Gas engine:	
Single cylinder	1.5
Multiple cylinder	1.0
Diesel engine:	
High-speed	1.5
Large, slow-speed	2.0
Driven machine:	
Generator:	
Steady load	1.0
Fluctuating load	1.5
Blower	1.0
Compressor, depending on number of cylinders	2.0–2.5
Pumps:	
Centrifugal	1.0
Reciprocating, single-acting	2.0
Reciprocating, double-acting	1.5
Lineshaft	1.5
Woodworking machinery	1.75
Hoists, elevators, cranes, and shovels	2.0
Hammer mills, ball mills, and crushers	2.0
Brick machinery	3.0
Rock crushers	3.0

Calculation Procedure:

1. *Choose the type of clutch for the load*

Table 51 shows typical applications for the major types of clutches. Where economy is the prime consideration, a positive-engagement or a cone-type friction clutch would be chosen. Since a reciprocating pump runs at a slightly varying speed, a centrifugal clutch is not suitable. For greater dependability, a disk or plate friction clutch is more desirable than a cone clutch. Assume that dependability is more important than economy, and choose a disk-type friction clutch.

2. *Determine the required clutch torque starting capacity*

A clutch must start its load from a stopped condition. Under these circumstances the instantaneous torque may be two, three, or four times the running torque. Therefore, the usual clutch is chosen so it has a torque capacity of at least twice the running torque. For internal-combustion engine drives, a starting torque of three to four times the running torque is generally used. Assume 3.5 times is used for this engine and pump combination. This is termed the *clutch starting factor*.

Since $T = 63,000hp/R$, where T = torque, lb·in; hp = horsepower transmitted; R = shaft rpm; $T = 63,000(50)/300 = 10,500$ lb·in (1186.3 N·m). This is the required starting torque capacity of the clutch.

3. *Determine the total required clutch torque capacity*

In addition to the clutch starting factor, a service factor is also usually applied. Table 52 lists typical clutch service factors. This tabulation shows that the service factor for a single-reciprocating pump is 2.0. Hence, the total required clutch torque capacity = required starting torque capacity × service factor = $10,500 × 2.0 = 21,000$-lb·in (2372.7-N·m) torque capacity.

4. Choose a suitable clutch for the load

Consult a manufacturer's engineering data sheet listing clutch torque capacities for clutches of the type chosen in step 1 of this procedure. Choose a clutch having a rated torque equal to or greater than that computed in step 3. Table 53 shows a portion of a typical engineering data sheet. A size 6 clutch would be chosen for this drive.

TABLE 53   Clutch Ratings

Clutch number	Torque rating		Power (100 r/min)	
	lb·in	N·m	hp	kW
1	2,040	230.5	3	2.2
2	4,290	484.7	6	4.5
3	8,150	920.8	12	8.9
4	13,300	1,502.7	21	15.7
5	19,700	2,225.8	31	23.1
6	35,200	3,977.1	55	41.0
7	44,000	4,971.3	69	51.5

Related Calculations: Use the general method given here to select clutches for industrial, commercial, marine, automotive, tractor, and similar applications. Note that engineering data sheets often list the clutch rating in terms of torque, lb·in, and hp/(100 r/min).

Friction clutches depend, for their load-carrying ability, on the friction and pressure between two mating surfaces. Usual coefficients of friction for friction clutches range between 0.15 and 0.50 for dry surfaces, 0.05 and 0.30 for greasy surfaces, and 0.05 and 0.25 for lubricated surfaces. The allowable pressure between the surfaces ranges from a low of 8 lb/in² (55.2 kPa) to a high of 300 lb/in² (2068.5 kPa).

BRAKE SELECTION FOR A KNOWN LOAD

Choose a suitable brake to stop a 50-hp (37.3-kW) motor automatically when power is cut off. The motor must be brought to rest within 40 s after power is shut off. The load inertia, including the brake rotating member, will be about 200 lb·ft² (82.7 N·m²); the shaft being braked turns at 1800 r/min. How many revolutions will the shaft turn before stopping? How much heat must the brake dissipate? The brake operates once per minute.

Calculation Procedure:

1. Choose the type of brake to use

Table 54 shows that a shoe-type electric brake is probably the best choice for stopping a load when the braking force must be applied automatically. The only other possible choice—the eddy-current brake—is generally used for larger loads than this brake will handle.

2. Compute the average brake torque required to stop the load

Use the relation $T_a = Wk^2 n/(308t)$, where T_a = average torque required to stop the load, lb·ft; Wk^2 = load inertia, including brake rotating member, lb·ft², n = shaft speed prior to braking, r/min; t = required or desired stopping time, s. For this brake, $T_a = (200)(1800)/[308(40)] = 29.2$ lb·ft, or 351 lb·in (39.7 N·m).

3. Apply a service factor to the average torque

A service factor varying from 1.0 to 4.0 is usually applied to the average torque to ensure that the brake is of sufficient size for the load. Applying a service factor of 1.5 for this brake yields the required capacity = 1.5(351) = 526 in·lb (59.4 N·m).

4. Choose the brake size

Use an engineering data sheet from the selected manufacturer to choose the brake size. Thus, one manufacturer's data show that a 16-in (40.6-cm) diameter brake will adequately handle the load.

5. Compute the revolutions prior to stopping

Use the relation $R_s = tn/120$, where R = number of revolutions prior to stopping; other symbols as before. Thus, $R_s = (40)(1800)/120 = 600$ r.

6. Compute the heat the brake must dissipate

Use the relation $H = 1.7 \, FWk^2(n/100)^2$, where H = heat generated at friction surfaces, ft·lb/min; F = number of duty cycles per minute; other symbols as before. Thus, $H = 1.7(1)(200)(1800/100)^2 = 110,200$ ft·lb/min (2490.2 N·m/s).

TABLE 54 Mechanical and Electrical Brake Characteristics

Type of brake	Typical characteristics
Block	Wooden or cast-iron shoe bearing on iron or steel wheel; double blocks prevent bending of shaft; used where economy is prime consideration; leverage 5:1
Band	Asbestos fabric bearing on metal wheels; fabric may be reinforced with copper wire and impregnated with asphalt; bands are faced with wooden blocks; used where economy is a major consideration; leverage 10:1
Cone	Friction surface attached to metal cone; popular for cranes; coefficient of friction = 0.08 to 0.10; useful for intermittent braking applications
Disk	Have one or more flat braking surfaces; effective for large loads; continuous application
Internal-shoe	Popular for vehicles where shaft rotation occurs in both directions; self-energizing, i.e., friction makes shoe follow rotating brake drum; capable of large braking power
Eddy-current	Used for flywheels requiring quick braking and where large kinetic energy of rotating masses precludes use of block brakes because of excessive heating
Electric, shoe-type	Used where automatic application of brake is required as soon as power is turned off; spring-activated brake shoes apply the braking action
Electric, friction-disk type	Best for duty cycles requiring a number of stops and starts per minute; may have one or multiple disks

7. *Determine whether the brake temperature will rise*

From the manufacturer's data sheet, find the heat dissipation capacity of the brake while operating and while at rest. For a 16-in (40.6-cm) shoe-type brake, one manufacturer gives an operating heat dissipation H_o = 150,000 ft·lb/min (3389.5 N·m/s) and an at-rest heat dissipation of H_v = 35,000 ft·lb/min (790.9 N·m/s).

Apply the cycle time for the event; i.e., the brake operates for 40 s, or 40/60 of the time, and is at rest for 20 s, or 20/60 of the time. Hence, the heat dissipation of the brake is (150,000)(40/60) + (35,000)(20/60) = 111,680 ft·lb/min (2523.6 N·m/s). Since the heat dissipation, 111,680 ft·lb/min (2523.6 N·m/s), exceeds the heat generated, 110,200 ft·lb/min (2490.2 N·m/s), the temperature of the brake will remain constant. If the heat generated exceeded the heat dissipated, the brake temperature would rise constantly during the operation.

Brake temperatures higher than 250°F (121.1°C) can reduce brake life. In the 250 to 300°F (121.1 to 148.9°C) range, periodic replacement of the brake friction surfaces may be necessary. Above 300°F (148.9°C), forced-air cooling of the brake is usually necessary.

Related Calculations: Because electric brakes are finding wider industrial use, Tables 55 and 56, summarizing their performance characteristics and ratings, are presented here for easy reference.

The coefficient of friction for brakes must be carefully chosen; otherwise, the brake may "grab," i.e., attempt to stop the load instantly instead of slowly. Usual values for the coefficient of friction range between 0.08 and 0.50.

The methods given above can be used to analyze brakes applied to hoists, elevators, vehicles, etc. Where Wk^2 is not given, estimate it, using the moving parts of the brake and load as a guide to the relative magnitude of load inertia. The method presented is the work of Joseph F. Peck, reported in *Product Engineering*.

MECHANICAL BRAKE SURFACE AREA AND COOLING TIME

How much radiating surface must a brake drum have if it absorbs 20 hp (14.9 kW), operates for half the use cycle, and cannot have a temperature rise greater than 300°F (166.7°C)? How long will it take this brake to cool to a room temperature of 75°F (23.9°C) if the brake drum is made of cast iron and weighs 100 lb (45.4 kg)?

TABLE 55 Performance Characteristics of Electric Brakes

Brake type	Operational mode		Design characteristics					Brake functions performed					On-off duty-cycling capability
	On-off	Continuous	Torque adjustment	Torque-control range	Wear adjustment	Residual drag	Heat dissipation	Instant stop	Cushioned stop	Retard (drag)	Hold	Failsafe brake	
Magnetic particle	Yes	Yes	Electrical	Wide	Nonwearing	High	Limited	No	Yes	Yes	Yes	No	Limited by heat-dissipation capability to low-inertia loads
Eddy-current, air-cooled	Yes	Yes	Electrical	Wide	Nonwearing	Moderate to low	Good	No	Yes	Yes	No	No	Limited to long time cycles
Eddy-current, water-cooled	Yes	Yes	Electrical	Wide	Nonwearing	High to moderate	Excellent	No	Yes	Yes	No	No	Limited to long time cycles
Single-disk friction, electrically actuated	Yes	Yes	Electrical	Wide	Self-compensating	None	Excellent	Yes	Yes	Yes	Yes	No	Excellent—up to several hundred stops per minute
Multidisk friction, electrically actuated, direct-acting	Yes	No	Electrical	Moderate		Low	Limited	Yes	Yes	No	Yes	No	Same as comparable size electric motor: 12 stops per minute (maximum)
Multidisk friction, electrically actuated, indirect-acting	Yes	No	Mechanical	Limited	Mechanical			Yes	Semisoft	No	Yes	No	Same as comparable size electric motor: 12 stops per minute (maximum)
Multidisk friction, spring-actuated	Yes	No	Mechanical	Limited	Mechanical	Low	Limited	Yes	Semisoft	No	Yes	Yes	Same as comparable size electric motor: 12 stops per minute (maximum)
Shoe brake, spring-actuated	Yes	No	Mechanical	Limited	Mechanical	None	Good	Yes	Semisoft	No	Yes	Yes	Generally not over 3 stops per minute without derating

TABLE 56 Representative Range of Ratings and Dimensions for Electric Brakes

Brake type	hp (W)	Torque, maximum, lb·ft (N·m)	Shaft speed, maximum, r/min	Diameter, in (cm)	Length, in (cm)	Inertia of rotating member, lb·ft² (N·m²)
Magnetic particle brakes	1/50–25 (14.9–18,643)	0.6–150 (0.8–203.4)	1,000–2,000	2–10 (5.1–25.4)	2–6 (5.1–15.2)	1.5×10^{-4}–0.27 (6.2×10^{-5}–0.11)
Eddy-current brakes:						
Air cooled	3/4–75 (559.3–55,928)	5–1,740 (6.8–2,359)	2,000–900	6½–24¾ (16.5–62.9)	9½–43½ (24.1–110.5)	0.12–100 (0.05–41.3)
Water-cooled	40–800 (29,828–596,560)	130–4,600 (176.3–6,237)	1,800–1,200	14¾–36½ (37.5–92.7)	18½–43 (47.0–109.2)	8.5–725 (3.5–299.6)
Friction disk brakes:						
Single-disk, electrically actuated	1/50–200 (14.9–149,140)	0.17–700 (0.23–949.1)	10,000–1,800	1½–15¼ (3.8–38.7)	1½–4½ (3.8–11.4)	0.000125–3 (0.000052–1.2)
Multiple-disk, electrically actuated	¼–2,000 (186.4–1,491,400)	3–15,000 (4.1–20,337)	5,000–750	2¼–21 (5.7–53.3)	2–8 (5.1–20.3)	Up to 90 (Up to 37.2)
Multiple-disk, spring actuated	¼–2,000 (186.4–1,491,400)	4–7,500 (5.4–10,169)	5,000–1,200	4–29 (10.2–73.7)	2½–16½ (6.4–41.9)	
Shoe brakes, spring-actuated	1–2,500 (745.7–1,864,250)	3–10,000 (4.1–13,558)	10,000–1,200	2–28 (5.1–71.1)	4½–12 (11.4–30.5)	0.023–485 (0.010–200.4)

Calculation Procedure:

1. Compute the required radiating area of the brake

Use the relation $A = 42.4hpF/K$, where A = required brake radiating area, in^2; hp = power absorbed by the brakes; F = brake load factor = operating portion of use cycle; K = constant = Ct_r, where C = radiating factor from Table 57, t_r = brake temperature rise, °F. For this brake, assuming a full 300°F (166.7°C) temperature rise and using data from Table 57, we get

$$A = 42.4(20)(0.5)/[(0.00083)(300)] = 1702 \text{ in}^2$$
$$(10,980.6 \text{ cm}^2).$$

TABLE 57 Brake Radiating Factors

Temperature rise of brake		Radiating factor C
°F	°C	
100	55.6	0.00060
200	111.1	0.00075
300	166.7	0.00083
400	222.2	0.00090

2. Compute the brake cooling time

Use the relation $t = (cW \ln t_r)/(K_c A)$, where t = brake cooling time, min; c = specific heat of brake-drum material, Btu/(lb·°F); W = weight of brake drum, lb; t_r = drum temperature rise, °F; ln = log to base e = 2.71828; K_c = a constant varying from 0.4 to 0.8; other symbols as before. Using K_c = 0.4, c = 0.13, t = $(0.13 \times 100 \ln 300)/[(0.4)(1702)]$ = 0.1088 min.

Related Calculations: Use this procedure for friction brakes used to stop loads that are lifted or lowered, as in cranes, moving vehicles, rotating cylinders, and similar loads.

INVOLUTE SPLINE SIZE FOR KNOWN LOAD

Choose the type and size of involute spline to transmit a torque of 10,000 lb·in (1129.8 N·m) from an electric motor to a centrifugal pump. What are the required face width and number of teeth for this spline?

Calculation Procedure:

1. Select the type of spline to use

Involute splines are usually chosen for industrial drives because this type transmits more torque for its size than a parallel-side spline does. The involute spline has almost no speed limitation, being used at speeds of 10,000 r/min and higher. Further, an involute spline can be cut and measured by the same machines that cut and measure gear teeth. A spline, however, differs from a gear in that the spline has no rolling action and all teeth are in contact at once.

Involute splines may be either *flexible* or *fixed*. Flexible splines allow some rocking motion; and under torque, the teeth slip axially to accommodate axial expansion or runout. Fixed splines allow no relative or rocking motion between the internal and external teeth. The fixed-type spline can be either shrink-fitted or loosely fitted together. For a centrifugal-pump drive, the flexible-type spline is generally preferred. Therefore, a flexible involute spline will be chosen for this drive. A standard commercial grade will be acceptable.

2. Determine the pitch diameter of the spline

Enter Fig. 39 at a torque of 10,000 lb·in (1129.8 N·m), and project vertically upward to the curve marked *Commercial flexible*. From the intersection with this curve, project horizontally to the left to find the required spline pitch diameter as 3.75 in (9.5 cm). This is also the required outside diameter of a keyed shaft to transmit the same torque.

3. Determine the maximum effective face width

Enter Fig. 40 at the pitch diameter of 3.75 in (9.5 cm), and project horizontally to the curve marked *For flexible splines*. From the intersection, project vertically downward to read the maximum effective face width as 1.75 in (4.4 cm).

4. Choose the number of teeth for the spline

Table 58 lists the recommended minimum number of teeth for an involute spline. Cost and manufacturing considerations determine the number of teeth to use, because the number of teeth

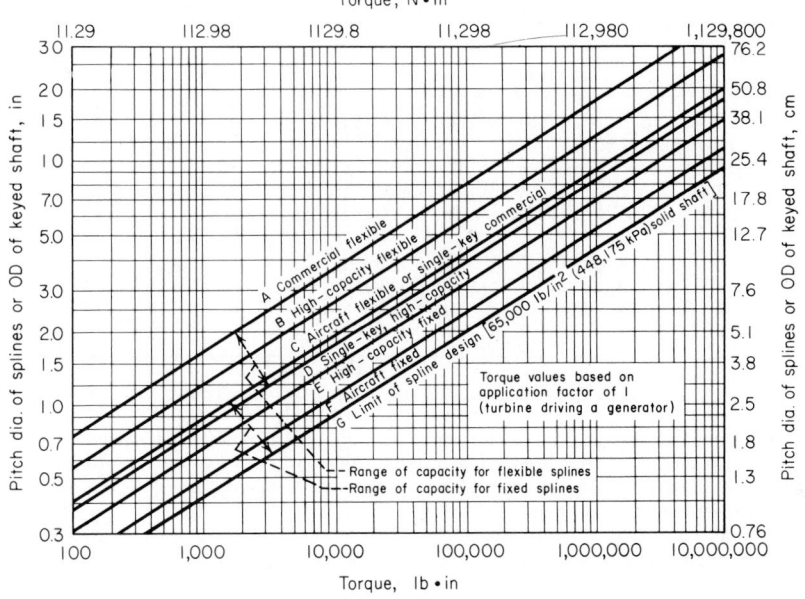

FIG. 39 Spline size based on diameter-torque relationships. *(Product Engineering.)*

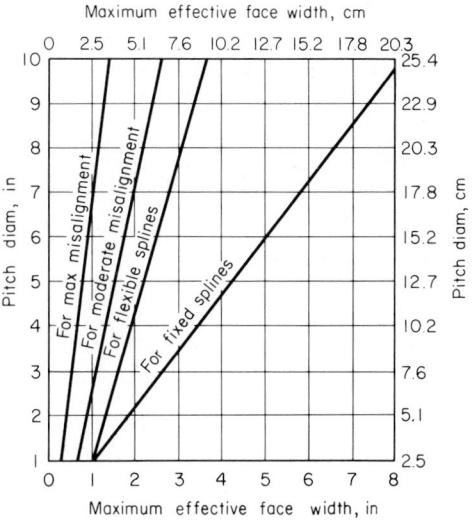

FIG. 40 Face width of splines for various applications. *(Product Engineering.)*

TABLE 58   Recommended Minimum Numbers of Spline Teeth°

Pitch diameter		Broaching			Shaping				Shaving or grinding	
		Angle: 30°	20° or 14½°		30°	25°	25°	14½°	25%	20°
in	cm	Depth: 50%	50%	30%	50%	70%	75%	30%	70%	75%
0.5	1.3	6	10	10	12	18				
0.8	2.0	8	12	10	14	20	22	16		
1.0	2.5	8	12	10	16	20	24	16		
2.0	5.1	8	12	12	20	20	24	24		
4.0	10.2	10	16	16	24	24	32	24	48	48
8.0	20.3	20	20	24	32	32	40	32	56	56
12.0	30.5	30	30	36	36	36	48	36	60	60

° *Product Engineering.*

chosen has no effect on tooth stress. An even number of teeth should be used whenever possible. When a large number of teeth are used on a spline, the root diameter of the external member is greater, tool design is easier, and lubrication is improved. Generally, however, the cost of the spline increases with a larger number of teeth.

For industrial drives, where the spline cost is usually more important than the weight of the spline or the space it occupies, a tooth with a 20° pressure angle is generally chosen. The nominal tooth depth, compared with gear teeth, is 75 percent. Using these data and a pitch diameter of 3.75 in (9.5 cm), as determined in step 2, shows that 32 teeth should be used.

Related Calculations: Involute splines for use in aircraft applications generally have a 30° pressure angle and 50 percent depth. In automotive service, shaved splines having the same proportions as the industrial splines mentioned above are often used. Rolled splines having 30 or 40° pressure angles and 50 and 40 percent depth, respectively, are also used. ANSA standards covering involute and straight-sided splines are available.

The method presented here is the work of Darle W. Dudley, reported in *Product Engineering.*

FRICTION DAMPING FOR SHAFT VIBRATION

Design a *dry-friction* (also termed *coulomb friction*) sleeve for a shaft transmitting power to an air compressor. The shaft has an outside diameter of 7.5 in (19.1 cm) and a length of 8 ft (2.4 m), and it drives the compressor as shown in Fig. 41. The angular value of torsional vibration should be limited to 10 percent of the steady displacement caused by the mean torque in the shaft. The compressor torque is 800,000 lb·in (90,387.8 N·m).

Calculation Procedure:

1. *Compute the required damping ratio*

To apply the friction-damping technique to a shaft, a sleeve (Fig. 42) is added which is attached to the shaft at one end, A. The sleeve is extended along the shaft and makes contact with some point on the shaft through the disk. This disk may be welded to or tightly pressed on the shaft and snugly fits into the sleeve.

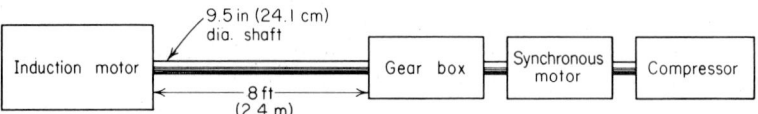

FIG. 41   Transmission system designed for friction damping. (*Product Engineering.*)

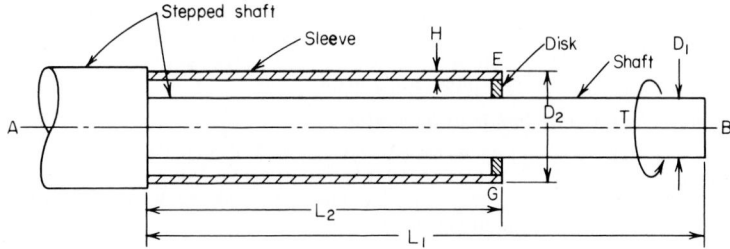

FIG. 42 Thin sleeve added to rotating shaft reduces torsional vibrations. *(Product Engineering.)*

In most dry-friction damping, about 3 percent of the damping takes place per cycle. If the forcing torque were reduced for one cycle, the strain energy would drop to 97 percent of its maximum value and the angular displacement θ of the shaft would drop to 0.97θ. Hence, the forcing torque must be such as to increase the angular displacement by an amount, or (in the absence of damping) $\Delta\theta = 0.03\theta$ per cycle $= 0.015\theta$ for a half-cycle.

Compute the damping ratio for the system from $R = 1 - \Delta\theta/\theta$, where $R =$ damping ratio. Thus, $R = 1 - 0.015\theta/(0.1\theta) = 0.85$. The value 0.1θ is used in the denominator because the design requires that $\Delta\theta$ be limited to 10 percent of the steady displacement θ, which results from the mean torque in the shaft.

2. Determine the shaft damping/critical damping value

With $R = 0.85$, enter Fig. 43, and find $m = 5.2$, where $m = D_1/(8HC^3) =$ ratio of the torsional stiffness of the shaft to that of the sleeve, a dimensional constant; $D_1 =$ shaft diameter, in; $H =$ thickness of the sleeve wall, in; $C = D_2/D_1$, where $D_2 =$ outside diameter of sleeve, in. Thus, damping/critical damping value $= 0.026$, or 2.6 percent, from Fig. 43, assuming $L_1/L_2 = 1.0$.

3. Select the sleeve outside diameter

Since $m = D_1/(8HC^3) = 5.2$, and $D_1 = 7.5$ in (19.1 cm), $HC^3 = 0.1802$, by substitution of the value of D_1 and m in this relation.

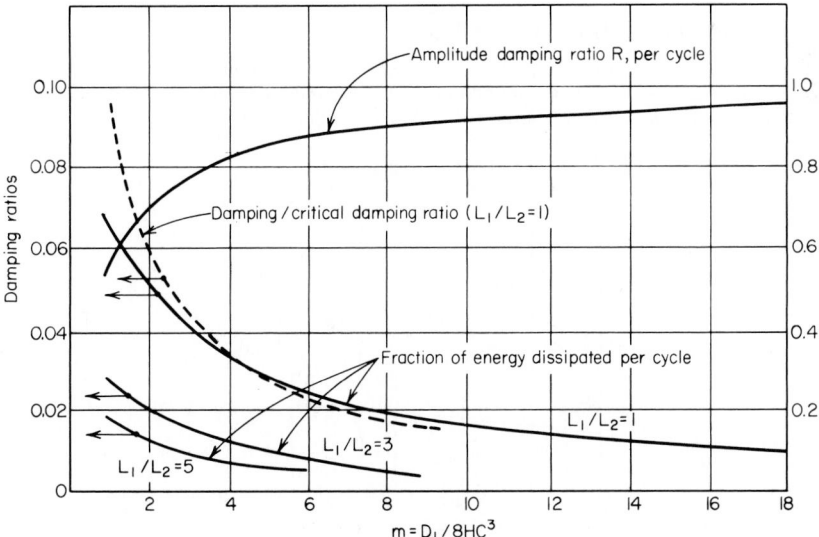

FIG. 43 Design chart for friction damping. *(Product Engineering.)*

Choose how HC^3 is to be made up. Assuming $D_2 = 2D_1$, $C = D_2/D_1 = 2.0$. Since $HC^3 = 0.1802$, $H = -0.1802/2^3 = 0.0225$ in (0.572 mm). This provides a sleeve thickness of about 24 gage. The sleeve will weigh only about 2.7 percent of the shaft weight. Thus, a 10:1 reduction in the vibration is obtained with very little extra weight.

4. *Compute the resisting torque of the system*

The ratio of resisting frictional torque applied by the sleeve T_r, lb·in to the applied torque on the shaft T lb·in is $T_r/T = 1[1 - (1 - 1/ar)^{0.5}]$, where $a = L_1/L_2$; $r = 1 + m$. Or, $T_r/T = 1\{1 - [1/(1 \times 6.2)]^{0.5}\} = 0.09$. Since the compressor torque $T = 800,000$ lb·in (90,387.8 N·m), $T_r = 0.09(800,000) = 72,000$ lb·in (8134.9 N·m).

5. *Compute the friction force on the sleeve*

In step 4 the sleeve diameter D_2 was chosen as $2D_1$. Since $D_1 = 7.5$ in (19.1 cm), $2D_1 = 15$ in (38.10 cm). The frictional torque acts, through the disk, over the circumference of the inner surface of the sleeve. The diameter of the inner surface of the sleeve is $15.0 - 2(0.0225) = 14.955$ in (38.09 cm), using the sleeve thickness obtained in step 3. The circumference of the inner surface of the sleeve is $14.955\pi = 47.0$ in (119.4 cm). Hence, the friction force acting on the sleeve $F_f = T_r/\text{circumference}$, in $= 72,000/47.0 = 1532$ lb (6814.7 N).

6. *Determine the disk normal force*

Assume that the disk has a coefficient of friction of 0.6. Then the normal force acting on the sleeve is $F_n = F_f/f$, where $f = $ coefficient of friction, or $F_n = 1532/0.6 = 2550$ lb (11,343.0 N).

Related Calculations: Dry-friction damping can be applied to industrial machines of many types, military equipment (submarines, missiles, aircraft), internal-combustion engines, and similar machinery. Vibration amplitudes in a shaft become a problem when the shaft length-to-thickness ratio L_1/D_1 becomes large. Although the shaft diameter can be increased to reduce the ratio, this adds to the weight and cost of the machine.

Here are several useful design pointers: (1) If weight is a primary objective, make the damping-sleeve diameter as large as possible. (2) If weight is not important, use a sleeve diameter only slightly larger than the shaft diameter. (3) Sleeve length can vary from 0.1 to 1.0 shaft length. With short sleeves, be sure the sleeve has sufficient rigidity and stiffness. (4) Reduce the sleeve wall thickness at the end of the sleeve in contact with the disk so that the contact pressure will not induce large stresses in the shaft. The method presented here was developed by Burt Zimmerman and reported in *Product Engineering*.

DESIGNING PARTS FOR EXPECTED LIFE

A machined and ground rod has an ultimate strength of $s_u = 90,000$ lb/in² (620,350 kPa) and a yield strength $s_y = 60,000$ lb/in² (413,700 kPa). It is grooved by grinding and has a stress concentration factor of $K_f = 1.5$. The expected loading in bending is 10,000 to 60,000 lb/in² (68,950

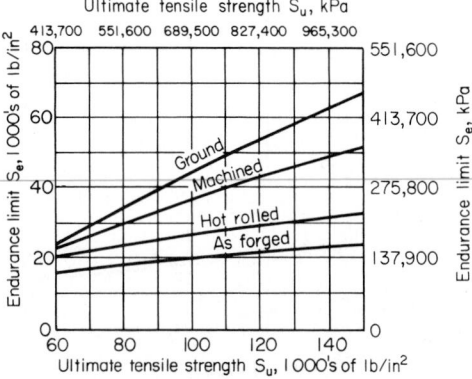

FIG. 44 Relationship between endurance limit and ultimate tensile strength. *(Product Engineering.)*

to 413,700 kPa) for 0.5 percent of the time; 20,000 to 50,000 lb/in^2 (137,900 to 344,750 kPa) for 9.5 percent of the time; 20,000 to 45,000 lb/in^2 (137,900 to 310,275 kPa) for 20 percent of the time; 30,000 to 40,000 lb/in^2 (206,850 to 275,800 kPa) for 30 percent of the time; 30,000 to 35,000 lb/in^2 (206,850 to 241,325 kPa) for 40 percent of the time. What is the expected fatigue life of this part in cycles?

Calculation Procedure:

1. *Determine the material endurance limit*

For s_u = 90,000 lb/in^2 (620,550 kPa), the endurance limit of the material s_e = 40,000 lb/in^2 (275,800 kPa), closely, from Fig. 44.

2. *Compute the equivalent completely reversed stress*

The largest equivalent completely reversed stress for each load in bending is $s_F = (s_e/s_y)s_a + K_f s_a$ lb/in^2; where s_a = average or steady stress, lb/in^2; other symbols as before. Since s_e/s_y = 40,000/60,000 = ⅔ and $K_f = 1.5$, $s_v = \frac{2}{3}s_a + 1.5s_a$. Then

$$s_{v1} = (\tfrac{2}{3})(35{,}000) + (1.5)(25{,}000) = 60{,}830 \text{ lb/in}^2 \ (419{,}422.9 \text{ kPa})$$

$$s_{v2} = (\tfrac{2}{3})(35{,}000) + (1.5)(15{,}000) = 45{,}830 \text{ lb/in}^2 \ (315{,}997.9 \text{ kPa})$$

$$s_{v3} = (\tfrac{2}{3})(32{,}500) + (1.5)(12{,}500) = 40{,}420 \text{ lb/in}^2 \ (278{,}695.9 \text{ kPa})$$

$$s_{v4} = (\tfrac{2}{3})(35{,}000) + (1.5)(5000) = 30{,}830 \text{ lb/in}^2 \ (212{,}572.9 \text{ kPa})$$

$$s_{v5} = (\tfrac{2}{3})(32{,}500) + (1.5)(2500) = 25{,}420 \text{ lb/in}^2 \ (175{,}270.9 \text{ kPa})$$

3. *Compute the fatigue life for the initial stress*

The initial stress is s_{v1}; the fatigue life at this stress is, in cycles, $N_1 = 1000(s_u/s_{vi})^{3/\log 1(s_u/s_e)}$ where s_{vi} = equivalent completely reversed stress, lb/in^2; other symbols as before. By taking the first value of $s_{vi} = s_{v1}$ = 60,830 lb/in^2 (419,422.9 kPa), $N_1 = 1000(90{,}000/60{,}830)^{3/\log 2.25}$ = 28,100 cycles.

4. *Compute the exponent for the fatigue-life equation*

The exponent for the fatigue-life equation is $2.55/\log(s_u/s_e) = 2.55/\log(90{,}000/40{,}000)$ = 7.2406.

5. *Compute the factors for the fatigue-life equation*

The factors needed for the fatigue-life equation are s_{v1}/s_{vi}, $(s_{vi}/s_{v1})^{2.55/\log(s_u/s_e)}$, and $\alpha_i(s_{vi}/s_{v1})^{2.55/\log(s_u/s_e)}$. In these factors, the value of $s_{vi} = s_{v2}, s_{v3}$, and so forth, as summarized in Table 59. The value α_i = percent-time duration of a stress, expressed as a decimal. The numerical values computed are summarized in Table 59.

6. *Compute the part fatigue life in cycles*

The part fatigue life, in cycles, is $N = N_1/\alpha_1 + \alpha_2[1/(s_{v1}/s_{v2})]^{2.55\log(s_u/s_e)} + 3[1/(s_{v1}/s_{v3})]^{2.55\log(s_u/s_e)} + \ldots$ for each bending load. In this equation, $\alpha_1, \alpha_2, \ldots$ = percent-time

TABLE 59 Values for Cycles-to-Failure Analysis

s_{vi}					
lb/in^2	kPa	s_{v1}/s_{vi}	$(s_{vi}/s_{v1})^{7.2406}$	i	$i(s_{vi}/s_{v1})^{7.2406}$
60,830	419,422.9	1.00	1.00	0.005	0.00500
45,830	315,997.9	1.3273	0.12873	0.095	0.01223
40,420	278,695.9	1.5051	0.05179	0.200	0.01036
30,830	212,572.9	1.9730	0.007296	0.300	0.00219
25,420	175,270.9	2.3934	0.001802	0.400	0.00072
					0.03050

duration of a stress, expressed as a decimal, the subscript referring to the stress mentioned in step 2, above.

Since Table 59 summarizes the denominator of the fatigue-life equation, $N = N_1/0.03050 = 28,100/0.03050 = 922,000$ cycles.

Related Calculations: Data on the endurance limit, yield point, and ultimate strength of ferrous materials are tabulated in Baumeister and Marks—*Standard Handbook for Mechanical Engineers*. The equations presented in this calculation procedure hold for both simple and complex loading. These equations can be used for analysis of an existing part or for design of a part to fail after a selected number of cycles. The latter procedure is sometimes used for components in an assembly in which the principal part has an accurately known life.

The method presented here is the work of Professor M. F. Spotts, reported in *Product Engineering*.

WEAR LIFE OF ROLLING SURFACES

Determine the maximum allowable bearing load in various bearing materials to avoid pronounced wear before 40×10^6 stress cycles if the roller bearing made of these materials has these dimensions: outside diameter = 4.3307 in (11.0 cm), bore = 2.3622 in (6.0 cm), width = 0.866 in (2.2

TABLE 60   Load-Stress K Factors for Various Materials[°]

Roller 1		Roller 2	K of roller 1 at number of cycles			
Material	Hardness	Material	10^6	10^7	4×10^7	10^8
Gray cast iron	130–180 BHN		4,000	2,000	1,300	
GM Meehanite	190–240 BHN	Same as roller 1	4,000	2,500	1,950	
Nodular cast iron	207–241 BHN		10,000	5,600	. . . .	3,400
Gray cast iron	270–290 BHN		7,500	5,300	4,200	
Gray cast iron, phosphate-coated	140–160 BHN		2,600	1,400	1,000	
	160–190 BHN		3,200	1,900	1,300	
	270–290 BHN		5,500	4,000	3,100	
SAE 1020 steel, phosphate-coated	130–150 BHN		4,500	2,700	1,700	
SAE 4150 steel, chromium-plated	270–300 BHN	Carbon tool steel	13,500	11,000	. . . .	9,000
SAE 6150 steel	270–300 BHN		2,600	1,300		
SAE 1020 steel, induction-hardened	45–55 RC	60–62 Re	21,000	14,500	. . . .	10,000
SAE 1340 steel, case-hardened	50–58 RC		26,000	20,000	. . . .	15,000
Phosphor bronze	67–77 BHN		3,600	1,600	1,000	
Yellow brass	Drawn		5,600	3,000	2,000	
	Extruded		4,500	2,400	1,700	
Zinc diecasting	. . . .		1,100	500	320	
Laminated-graphitized phenolic	. . . .		1,700	1,300	1,000	
Cast aluminum SAE 39	60–65 BHN	Gray cast iron 340–360 BHN	1,200	500	300	

[°] *Product Engineering*. Based on data presented by W. C. Cram at a University of Michigan symposium on surface damage.

cm), inner-race radius $r_2 = 1.439$ in (3.655 cm), roller diameter $= 0.468$ in (1.19 cm), roller width $f = 0.468$ in (1.19 cm), number of rollers $n = 16$. Materials being considered are gray cast iron, Meehanite; and hardened-steel rollers on cast-iron races, on heat-treated cast-iron races, on heat-treated and medium-steel races, and on carburized low-carbon steel races.

Calculation Procedure:

1. Determine the load-stress factor for each material

Table 60 lists the load-stress factor K for various materials at varying load cycles. Thus for gray cast iron, $K = 1300$ at 40×10^6 cycles.

2. Compute the maximum allowable bearing load

Use the relation $F = nfK/[5(1/r_1 + 1/r_2)]$ where all symbols are defined in the problem statement above. Thus, $F = (16)(0.468)(1300)/[5(1/0.234 + 1/1.439)] = 391$ lb (1739.3 N) for gray cast iron.

For the other materials, by using the appropriate K value from Table 60 and the above procedure, the allowable load is as follows:

Bearing materials	Allowable load, lb (N)
Meehanite rollers and races	587 (2,611.1)
Hardened-steel rollers, cast-iron races	300 (1,334.5)
Hardened-steel rollers, heat-treated cast-iron races	933 (4,150.2)
Hardened-steel rollers, heat-treated medium-steel races	2,700 (12,010.2)
Hardened-steel rollers, carburized low-carbon steel races	3,912 (17,401.4)

Related Calculations: The same, or similar, procedures can be used for computing the wear life of gears, cams, bearings, clutches, chains, and other devices having rolling surfaces. Thus, in a joint composed of a pin having radius r_1 in a hole of radius r_2, $F = fK/(1/r_1 - 1/r_2)$. This relation also applies to a roller chain. For a cam, $F = fK \cos \infty/(1/\rho - 1/r)$, where $\infty =$ cam pressure angle; $\rho =$ cam radius of contact, in; $r =$ radius of contact, in.

The method presented here is the work of Professor Donald J. Myatt, reported in *Product Engineering*.

FACTOR OF SAFETY AND ALLOWABLE STRESS IN DESIGN

Determine the cross section dimension b of a uniform square bar of structural steel having a tensile yield strength $s_y = 33,000$ lb/in^2 (227,535.0 kPa), if the bar carries a center load of 1000 lb (4448.2 N) as a beam with a span of 24 in (61.0 cm) with simply supported ends. Use both the allowable-stress and ultimate-strength methods to determine the required dimension.

Calculation Procedure:

1. Design first on the basis of allowable stress

From Table 61, in the section on buildings, the working or allowable stress $s_w = 0.6s_y = 0.6(33,000) = 19,800$ lb/in^2 (136,521.0 kPa).

2. Compute the maximum bending moment

For a central load and simple supports, the maximum bending moment is, from earlier sections of this handbook, $M_m = (W/2)(L/2)$, where $W =$ load on beam, lb; $L =$ length of beam, in. Thus, $M_m = (1000/2)(24/2) = 6000$ in·lb (677.9 N·m).

3. Compute the required cross section dimension

For a beam of square cross section, $I/c = [b(b)^3/12]/(b/2) = b^3/6$. Also, $s_w = M_m c/I$. Substituting and solving for b gives $b^3 = (6000)(6)/19,800$; $b = 1.22$ in (3.1 cm).

TABLE 61 Illustrative Allowable Stresses and Factors of Safety*

Application	Materials	Allowable stress s_w	Approximate factor of safety
Buildings and other structures	Structural steel	Direct tension $s_w = 0.6s_y$; $0.45s_y$ on net section at pin holes; bending $s_w = 0.6s_y$; shear $s_w = 0.45s_y$ on rivets, $0.40s_y$ on girder webs; bearing $s_w = 1.35s_y$ on rivets in double shear; $1s_y$ in single shear	1.70 for beams; 1.85 for continuous frames
	Structural aluminum 6061-T6, 6062-T6 $s_u = 38,000$ lb/in² (262,010 kPa), $s_y = 35,000$ lb/in² (241,325 kPa)	Direct tension $s_w = 17,000$ lb/in² (117,215 kPa); bending structural shapes, $s_w = 17,000$ lb/in² (117,215 kPa); rectangular sections $s_w = 23,000$ lb/in² (158,585 kPa); shear $s_w = 10,000$ lb/in² (68,950 kPa) on cold-driven rivets, $11,000$ lb/in² (75,845 kPa) on girder webs; bearing $s_w = 30,000$ lb/in² (206,850 kPa) on rivets	1.8 for beams; 2 for columns
	Reinforced concrete	Bending, compression in concrete, $s_w = 0.45s_c'$; bending, tension in steel, $s_w = 0.40s_y$; bending, tension in plain concrete footings, $s_w = 0.03s_c'$; shear, on concrete in unreinforced web, $s_w = 0.03s_c'$; compression in concrete column, $s_w = 0.225s_c'$	6 in general
	Wood	Bending, $s_w = \frac{9}{16}s_r'$; long compression, $s_w = \frac{1}{6}s_r'$; transverse-compression, $s_w = \frac{1}{6}$ elastic limit (400 for Douglas fir); shear parallel to grain $s_w = 120$ for Douglas fir	
Bridges	Structural metals and reinforced concrete	s_w about $0.9s_w$ for buildings	2.05
	Wood	s_w same as for buildings	6

Application	Material		n
	Steel (shafts, etc.)	Steady tension, compression or bending, $s_w = s_y/n$; pure shear, $s_w = s_y/2n$; tension s_t plus shear s_s; $s_t^2 + 4s_s^2 \leqq s_y/n$; alternating stress s_a plus mean stress s_m: point representing an alternating stress $nk_f s_a$ and a mean stress ns_m must lie below the Goodman diagram, Fig. 1	n usually between 1.5 and 2
Machinery	Steel (SAE 1095), leaf springs, thickness = t in $t > 0 < 0.10$	Static loading $s_w = 230,000 - 1,000,000t$ lb/in^2 (1,585,850 - 6,895,000t kPa); variable loading, 10^7 cycles, $s_w = 200,000 - 800,000t$ lb/in^2 (1,379,000 - 5,516,000t kPa); dynamic loading, 10^7 cycles, $s_w = 155,000 - 600,000t$ lb/in^2 (1,068,725 - 4,137,000t kPa)	
	Steel (wire, ASTM-A228, helical springs)	$s_w = 100,000$ lb/in^2 (689,500 kPa) [for $d = 0.2$ in (0.5 cm); 10^7 cycles repeated stress]	
	Carbon steel	Membrane stress, $s_w = 0.211s_u$; membrane plus discontinuity stresses, $s_w = 0.9s_y$ or $0.6s_u$	5
Pressure vessels (unfired)	Alloy steels	Membrane stress, $s_w = 0.25s_u$; membrane plus discontinuity stresses, $s_w = 0.95s_y$ or $0.6s_u$	4
	Cast iron	Membrane stress, $s_w = 0.1s_u$; bending stress $s_w = 0.15s_u$	10, 6.67
	Nonferrous metals	Same rule as alloy steels	4
Airplanes	Aluminum alloy and steel	Ultimate strength design	1.5 against ultimate 1 against yield
	Wood	Ultimate strength design	

*From Roark—*Formulas for Stress and Strain,* 4th ed., McGraw-Hill.
Note: s_w = allowable or working stress; s_u = ultimate tensile strength; s_y = tensile yield strength; s' = modulus of rupture in cross-bending of rectangular bar; s'_s = ultimate shear strength; s'_c = ultimate compressive strength; s_c = endurance limit or endurance strength for specified life; n = dividing factor applied to s_u, s_y, or s_c to obtain s_w.

4.   *Design the beam on the basis of ultimate strength and a factor of safety*

The safety factor from Table 61 is 1.70 for beams. This safety factor is applied by designing the beam to fail just under a load of 1.7(1000 lb) = 1700 lb (7562.0 N). Thus, to generalize, the design failure load = (factor of safety)(load on part, lb) = W_u.

For structural steel and other materials capable of fully plastic behavior, a simple beam or other member will collapse when the maximum moment equals the *plastic moment* M_p, which is developed when the stress throughout the section becomes equal to s_y. Hence, $M_p = s_y z$, where Z = plastic modulus = arithmetical sum of the static moments of the upper and lower parts of the cross section about the horizontal axis that divides the area in half. For a square with its edges horizontal and vertical, as assumed here, $Z = (\frac{1}{2}b^2)(\frac{1}{4}b) + (\frac{1}{2}b^2)(\frac{1}{4}b) = \frac{1}{4}b^3$. Hence, $M_p = 33,000(\frac{1}{4}b^3)$.

Set M_p = the ultimate bending moment, or $(W_u/2)(L/2)$, where W_u = design failure load = (factor of safety)(load on part, lb) = (1.70)(1000) = 1700 lb (7562.0 N) for this beam. Thus, $M_p = 33,000(\frac{1}{4}b^3) = (1700/2)(24/2)$; $b = 1.07$ in (2.7 cm).

By using the allowable stress, $b = 1.22$ in (3.1 cm), as compared with 1.07 in (2.7 cm). The difference in results for the two methods (about 12 percent) can be traced to the ratio of Z to I/c, called the *shape factor*. For the case under consideration, this ratio is $(\frac{1}{4}b^3)/(\frac{1}{6}d^3) = 1.5$.

5.   *Determine the beam size for a vertical diagonal*

Here $Z = 0.2357b^3$, $I/c = 0.1178b^3$, and the shape factor = $Z/(I/c) = 0.2357/0.1178 = 2.0$.

Designing by allowable stress, using steps 1, 2, and 3, gives $(0.1178b^3)(19,800) = 6000$; $b = 1.37$ in (3.5 cm). Designing by ultimate strength gives $(0.2357b^3)(33,000) = 10,200$; $b = 1.095$ in (2.8 cm).

This computation shows that a more economical design is generally obtained by ultimate-strength design, with the advantage becoming greater as the shape factor becomes larger.

Related Calculations: For conventional structural sections, such as I beams, the shape factor is not much greater than 1, and the two methods yield about the same result for statically determinate problems, such as this one. If the problem involves a statically indeterminate beam or a rigid frame, the advantage of the ultimate-strength method becomes more apparent because it takes account of the fact that collapse cannot occur, as a rule, until the plastic moment is developed at each of two or more sections.

The method given here is the work of Professor Raymond J. Roark, reported in *Product Engineering*.

RUPTURE FACTOR AND ALLOWABLE STRESS IN DESIGN

Determine the proper thickness t of a circular plate 40 in (101.6 cm) in diameter if the edge of the plate is simply supported and the plate carries a uniformly distributed pressure of 200 lb/in² (1379.0 kPa). The plate is made of cast iron having an ultimate strength of 50,000 lb/in² (344,750.0 kPa).

Calculation Procedure:

1.   *Design on the basis of allowable stress*

From Table 61 of the previous calculation procedure, note in the section for pressure vessels that the value of the working or allowable stress s_w lb/in² for tension due to bending is 0.15 s_u, where s_u = ultimate tensile strength of the material, lb/in². Thus, $s_w = 0.15(50,000) = 7500$ lb/in² (51,712.5 kPa).

2.   *Compute the required plate thickness*

The maximum stress for a simply supported plate is, from Roark—*Formulas for Stress and Strain*, $s_{max} = (3W/8\pi t^2)(3 + v)$, where W = total load on plate, lb; t = plate thickness, in; v = Poisson's ratio. For this plate, $W = (200 \text{ lb/in}^2)(\pi)(20)^2 = 251,330$ lb (1,117,971.6 N). Assuming $v = 0.3$ and solving for t, we find $7500 = [(3)(251,330)/8\pi t^2](3 + 0.3)$; $t = 3.63$ in (9.2 cm).

3.   *Design on the basis of ultimate strength*

For ultimate-strength design, from Table 61 of the previous calculation procedure, use the value of 6.67 for the factor of safety. Design the plate to break, theoretically, under a load. Hence, the breaking load would be (factor of safety) W, or 6.67(251,330) = 1,667,000 lb (7,415,186.1 N).

4. Apply the rupture factor to the design

With a brittle material like cast iron, the concept of the plastic moment does not apply. Use instead the *rupture factor*, which is the ratio of the calculated maximum tensile stress at rupture to the ultimate tensile strength, both expressed in lb/in^2. Whereas the rupture factor must be determined experimentally, a number of typical values are given in Table 62 for a variety of cases.

For case 2, which corresponds to this problem, the rupture factor for cast iron is $R_i = 1.9$. Using this in the same equation for s_w as in step 1, $s_w = 1.9(50,000) = 95,000 \ lb/in^2$ (655,025.0 kPa). Then, by using the procedure and equation in step 2 but substituting the breaking load for the plate, $95,000 = [(3)(1,677,000)/8\pi t^2] (3 + 0.3)$; $t = 2.63$ in (6.7 cm).

Related Calculations: The reliability of the solution using the ultimate-strength design technique depends on the accuracy of the rupture factor. When experimental or tabulated values of R_i are not available, the modulus of rupture may be used in place of $R_i s_u$. Typical values of Poisson's ratio v for various materials are given in Table 63.

The method presented here is the work of Professor Raymond J. Roark, reported in *Product Engineering*.

FORCE AND SHRINK FIT STRESS, INTERFERENCE, AND TORQUE

A 0.5-in (1.3-cm) thick steel band having a modulus of elasticity of $E = 30 \times 10^6 \ lb/in^2$ (206.8 $\times 10^9$ Pa) is to be forced on a 4-in (10.2-cm) diameter steel shaft. The maximum allowable stress in the band is 24,000 lb/in^2 (165,480.0 kPa). What interference should be used between the band and the shaft? How much torque can the fit develop if the band is 3 in (7.6 cm) long and the coefficient of friction is 0.20?

Calculation Procedure:

1. Compute the required interference

Use the relation $i = sd/E$, where i = the required interference to produce the maximum allowable stress in the band, in; s = stress in band or hub, lb/in^2; d = shaft diameter, in; E = modulus of elasticity of band or hub, lb/in^2. For this fit, $i = (24,000)(4.0)/(30 \times 10^6) = 0.0032$ in (0.081 mm).

2. Compute the torque the fit will develop

Use the relation $T = Eitl\pi f$, where T = fit torque, lb·in; t = band or hub thickness, in; l = band or hub length, in; f = coefficient of friction between the materials. For this joint, $T = 30 \times 10^6 \times 0.0032 \times 0.5 \times 3.0 \times \pi \times 0.20 = 90,432$ lb·in (10,217.4 N·m).

Related Calculations: Use this general procedure for either shrink or press fits. The axial force required for a press fit of two members made of the same material is F_a = axial force for the press fit, lb; p_c = radial pressure between the two members, $lb/in^2 = iE(d_c^2 - d_i^2)(d_o^2 - d_c^2)/zd_c^3(d_o^2 - d_i^2)$, where d_o = outside diameter of the external member, in; d_c = nominal diameter of the contact surfaces, in; d_i = inside diameter of the inner member, in.

HYDRAULIC SYSTEM PUMP AND DRIVER SELECTION

Choose the pump and the driver horsepower for a rubber-tired tractor bulldozer having four-wheel drive. The hydraulic system must propel the vehicle, operate the dozer, and drive the winch. Each main wheel will be driven by a hydraulic motor at a maximum wheel speed of 59.2 r/min and a maximum torque of 30,000 lb·in (3389.5 N·m). The wheel speed at maximum torque will be 29.6 r/min; maximum torque at low speed will be 74,500 lb·in (8417.4 N·m). The tractor speed must be adjustable in two ways: for overall forward and reverse motion and for turning, where the outside wheels turn at a faster rate than do the inside ones. Other operating details are given in the appropriate design steps below.

Calculation Procedure:

1. Determine the propulsion requirements of the system

Usual output requirements include speed, torque, force, and power for each function of the system, through the full capacity range.

TABLE 62 Values of the Rupture Factor for Brittle Materials°

Form of member and manner of loading	Rupture factor; ratio of computed maximum stress at rupture to ultimate tensile strength		Ratio of computed maximum stress at rupture to modulus of rupture in bending or torsion	
	Cast iron	Plaster	Cast iron	Plaster
1. Rectangular beam, end support, center loading, $l/d = 8$ or more	1.70	1.60	1	1
2. Solid circular plate, edge support, uniform loading, $a/t = 10$ or more	1.9	1.71	...	1.07
3. Solid circular plate edge support, uniform loading on concentric circular area	$2.4-0.5(r_o/a)^{1/6}$	$2.2-0.5(r_o/a)^{1/6}$	$1.40-0.3(r_o/a)^{1/6}$	$1.4-0.3(r_o/a)^{1/6}$

°From Roark—*Formulas for Stress and Strain*, 4th ed., McGraw-Hill.

TABLE 63 Poisson's Ratio for Various Materials

Material	Poisson's ratio
Aluminum:	
Cast	0.330
Wrought	0.330
Brass, cast, 66% Cu, 34% Zn	0.350
Bronze, cast, 85% Cu, 7.2% Zn, 6.4% Sn	0.358
Cast iron	0.260
Copper, pure	0.337
Phosphor bronze, cast, 92.5% Cu, 7.0% Sn, 0.5% Ph	0.380
Steel:	
Soft	0.300
1% C	0.287
Cast	0.280
Tin, cast, pure	0.330
Wrought iron	0.280
Nickel	0.310
Zinc	0.210

First analyze the *propel* power requirements. For any propel condition, $hp = Tn/63,000$, where hp = horsepower required; T = torque, lb·in, at n r/min. Thus, at maximum speed, $hp = (30,000)(59.2)/63,000 = 28.2$ hp (21.0 kW). At maximum torque, $hp = 74,500 \times 29.6/63,000 = 35.0$ (26.1 kW); at maximum speed and maximum torque, $hp = (74,500)(59.2)/63,000 = 70.0$ (52.2 kW).

The drive arrangement for a bulldozer generally uses hydraulic motors geared down to wheel speed. Choose a 3000-r/min step-variable type of motor for each wheel of the vehicle. Then each motor will operate at either of two displacements. At maximum vehicle loads, the higher displacement is used to provide maximum torque at low speed; at light loads, where a higher speed is desired, the lower displacement, producing reduced torque, is used.

Determine from a manufacturer's engineering data the motor specifications. For each of these motors the specifications might be: maximum displacement, 2.1 in³/r (34.4 cm³/r); rated pressure, 6000 lb/in² (41,370.0 kPa); rated speed, 3000 r/min; power output at rated speed and pressure, 90.5 hp (67.5 kW); torque at rated pressure, 1900 lb·in (214.7 N·m).

The gear reduction ratio GR between each motor and wheel = (output torque required, lb·in)/(input torque, lb·in, × gear reduction efficiency). Assuming a 92 percent gear reduction efficiency, a typical value, we find GR = 74,500/(1900 × 0.92) = 42.6:1. Hence, the maximum motor speed = wheel speed × GR = 59.2 × 42.6 + 2520 r/min. At full torque the motor speed is, by the same relation, 29.6 × 42.6 = 1260 r/min.

The required oil flow for the four motors is, at 1260 r/min, in³/r × 4 motors × (r/min)/(231 in³/gal) = 2.1 × 4 × 1260/231 = 45.8 gal/min (2.9 L/s). With a 10 percent leakage allowance, the required flow = 50 gal/min (3.2 L/s), closely, or 50/4 = 12.5 gal/min (0.8 L/s) per motor.

As computed above, the power output per motor is 35 hp (26.1 kW). Thus, the four motors will have a total output of 4(35) = 140 hp (104.4 kW).

2. *Determine the linear auxiliary power requirements*

The dozer uses a linear power output. Two hydraulic cylinders each furnish a maximum force of 10,000 lb (44,482.2 N) to the dozer at a maximum speed of 10 in/s (25.4 cm/s). Assuming that the maximum operating pressure of the system is 3500 lb/in² (24,132.5 kPa), we see that the piston area required per cylinder is: force developed, lb/operating pressure, lb/in² = 10,000/3500 = 2.86 in² (18.5 cm²), or about a 2-in (5.1-cm) cylinder bore. With a 2-in (5.1-cm) bore, the operating pressure could be reduced in the inverse ratio of the piston areas. Or, $2.86/(2^2\pi/4) = p/3500$, where p = cylinder operating pressure, lb/in². Hence, $p = 3180$ lb/in², say 3200 lb/in² (22,064.0 kPa).

By using a 2-in (5.1-cm) bore cylinder, the required oil flow, gal, to each cylinder = (cylinder volume, in³)(stroke length, in)/(231 in³/gal) = $(2^2\pi/4)(10)/231 = 0.1355$ gal/s, or 0.1355 × (60 s/min) = 8.15 gal/min (0.5 L/s), or 16.3 gal/min (1.0 L/s) for two cylinders. The power input to the two cylinders is hp = 16.3(3200)/1714 = 30.4 hp (22.7 kW).

3. Determine rotary auxiliary power requirements

The winch will be turned by one hydraulic motor. This winch must exert a maximum line pull of 20,000 lb (88,964.4 N) at a maximum linear speed of 280 ft/min (1.4 m/s) with a maximum drum torque of 200,000 lb·in (22,597.0 N·m) at a drum speed of 53.5 r/min.

Compute the drum hp from $hp = Tn/63{,}000$, where the symbols are the same as in step 1. Or, hp = (200,000)(53.5)/63,000 = 170 hp (126.8 kW).

Choose a hydraulic motor having these specifications: displacement = 6 in³/r (98.3 cm³/r); rated pressure = 6000 lb/in² (41,370.0 kPa); rated speed = 2500 r/min; output torque at rated pressure = 5500 lb·in (621.4 N·m); power output at rated speed and pressure = 218 hp (162.6 kW). This power output rating is somewhat greater than the computed rating, but it allows some overloading.

The gear reduction ratio GR between the hydraulic motor and winch drum, based on the maximum motor torque, is GR = (output torque required, lb·in)/(torque at rated pressure, lb·in, × reduction gear efficiency) = 20,000/(5500 × 0.92) = 39.5:1. Hence, by using this ratio, the maximum motor speed = 53.5 × 39.5 = 2110 r/min. Oil flow rate to the motor = in³/r × (r/min)/231 = 6 × 2110/231 = 54.8 gal/min (3.5 L/s), without leakage. With 5 percent leakage, flow rate = 1.05(54.8) = 57.2 gal/min (3.6 L/s).

4. Categorize the required power outputs

List the required outputs and the type of motion required—rotary or linear. Thus: propel = rotary; dozer = linear; winch = rotary.

5. Determine the total number of simultaneous functions

There are two simultaneous functions: (*a*) propel motors and dozer cylinders; (*b*) propel motors at slow speed and drive winch.

For function *a*, maximum oil flow = 50 + 16.3 = 66.3 gal/min (4.2 L/s); maximum propel motor pressure = 6000 lb/in² (41,370.0 kPa); maximum dozer cylinder pressure = 3200 lb/in² (22,064.0 kPa). Data for function *a* came from previous steps in this calculation procedure.

For function *b*, the maximum oil flow need not be computed because it will be less than for function *a*.

6. Determine the number of series nonsimultaneous functions

These are the dozer, propel, and winch functions.

7. Determine the number of parallel simultaneous functions

These are the propel and dozer functions.

8. Establish function priority

The propel and dozer functions have priority over the winch function.

9. Size the piping and values

Table 64 lists the normal functions required in this machine and the type of valve that would be chosen for each function. Each valve incorporates additional functions: The step variable selector valve has a built-in check valve; the propel directional valve and winch directional valve have built-in relief valves and motor overload valves; the dozer directional valve has a built-in relief valve and a fourth position called *float*. In the float position, all ports are interconnected, allowing the dozer blade to move up or down as the ground contour varies.

10. Determine the simultaneous power requirements

These are: Horsepower for propel and dozer = (gpm)(pressure, lb/in²)/1714 for the propel and dozer functions, or (50)(6000)/1714 + (16.3)(3200)/1714 = 205.4 hp (153.2 kW). Winch horsepower, by the same relation, is (57.2)(6000)/1714 = 200 hp (149.1 kW). Since the propel-dozer functions do not operate at the same time as the winch, the prime mover power need be only 205.4 hp (153.2 kW).

11. Plan the specific circuit layouts

To provide independent simultaneous flow to each of the four propel motors, plus the dozer cylinders, choose two split-flow piston-type pumps having independent outlet ports. Split the dis-

TABLE 64   Hydraulic-System Valving and Piping

Valving	
Function	Type of valve
Step variable selector	Three-way, two-position
Propel directional	Four-way three-position, tandem-center
Winch directional	Four-way, three-position, tandem-center
Dozer directional	Four-way, four-position

Piping			
Branch of circuit	Propel motor	Dozer cylinder	Winch motor
Maximum flow, gal/min (L/s)	12.5 (0.8)	16.3 (1.0)	57.2 (3.6)
Maximum pressure, lb/in^2 (kPa)	6000 (41,370)	3200 (22,064)	6000 (41,370)
Tube size, in (cm)	¾ (1.9)	¾ (1.9)	1½ (3.8)
Tube material, ASTM	4130	4130	4130
Tube wall, in (mm)	0.120 (3.05)	0.109 (2.77)	0.250 (6.35)

charge of each pump into three independent flows. Two pumps rated at $66.3/2 = 33.15$ gal/min (2.1 L/s) each at 6000 lb/in^2 (41,370.0 kPa) will provide the needed oil.

When the vehicle is steered, additional flow is required by the outside wheels. Design the circuit so oil will flow from three pump pistons to each wheel motor. Four pistons of one split-flow pump are connected through check valves to all four motors. With this arrangement, oil will flow to the motors with the least resistance.

To make use of all or part of the oil from the propel-dozer circuits for the winch circuit, the outlet series ports of the propel and dozer valves are connected into the winch circuit, since the winch circuit is inoperative only when both the propel *and* the dozer are operating. When only the propel function is in operation, the winch is able to operate slowly but at full torque.

12.   *Investigate adjustment of the winch gear ratio*

As computed in step 3, the winch gear ratio is based on torque. Now, because a known gpm (gallons per minute) is available for the winch motor from the propel and dozer circuits when these are not in use, the gear ratio can be based on the motor speed resulting from the available gpm.

Flow from the propel and dozer circuit = 66.3 gal/min (4.2 L/s); winch motor speed = 2450 r/min; required winch drum speed = 53.5 r/min. Thus, $GR = 2450/53.5 = 45.8:1$.

With the proposed circuit, the winch gear reduction should be increased from 39.5:1 to 45.8:1. The winch circuit pressure can be reduced to $(39.5/45.8)(6000) = 5180$ lb/in^2 (35,716.1 kPa). The required size of the winch oil tubing can be reduced to 0.219 in (5.6 mm).

13.   *Select the prime mover hp*

Using a mechanical efficiency of 89 percent, we see that the prime mover for the pumps should be rated at $205.4/0.89 = 230$ hp (171.5 kW). The prime mover chosen for vehicles of this type is usually a gasoline or diesel engine.

Related Calculations: The method presented here is also valid for fixed equipment using a hydraulic system, such as presses, punches, and balers. Other applications for which the method can be used include aircraft, marine, and on-highway vehicles. Use the method presented in an earlier section of this handbook to determine the required size of the connecting tubing.

The procedure presented above is the work of Wes Master, reported in *Product Engineering*.

SELECTING BOLT DIAMETER FOR BOLTED PRESSURIZED JOINT

Select a suitable bolt diameter for the typical bolted joint in Fig. 45 when the joint is used on a pressurized cylinder having a flanged head clamped to the body of the cylinder by eight equally spaced bolts. The vessel internal pressure, which may be produced by hydraulic fluid or steam,

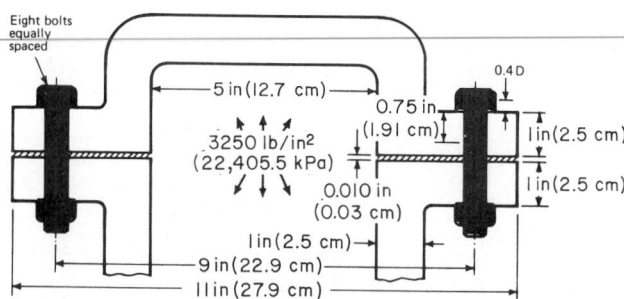

FIG. 45   Typical bolted joint analyzed in the calculation procedure. *(Product Engineering.)*

varies from 0 to 3250 lb/in^2 (0 to 22,405.5 kPa). What clamping force must be applied by each bolt to ensure that no leakage will occur? Check the selected bolt size to ensure long fatigue life under static and fluctuating loading.

Calculation Procedure:

1. Determine the axially pressurized area of the cylinder

The internal diameter of the cylinder is, as shown, 5 in (12.7 cm). The axially pressurized area of the cylinder head is $A = \pi D^2/4 = 5^2/4 = 19.63$ in^2 (126.62 cm^2).

2. Find the applied working load on each bolt

Since eight bolts are specified for this flanged joint, each bolt will carry one-eighth of the total load found from $F_A = PA/8$, where F_A = axial applied working load, lb; P = maximum pressure in vessel, lb/in^2; A = axially pressurized area, in^2. Or, $F_A = 3250(19.63)/8 = 7947.7$ lb, say 8000 lb (3633.4 kg) for calculation purposes.

3. Compute the bolt load produced by torquing

An air-stall power wrench will be used to tighten (torque) the nuts on these bolts. Such a wrench has a torque tightening factor C_T of 2.5 maximum, as shown in Table 65. Using $C_T = 2.5$, we find the load on each bolt P_Y, lb, causing a yield stress is $P_Y = C_T F_A = 2.5(8000) = 20,000$ lb (9090.9 kg).

4. Find the ultimate tensile strength P_U for the bolt

With the yield strength typically equal to 80 percent of the ultimate tensile strength of the bolt, $P_U = P_Y/0.80 = 20,000/0.80 = 25,000$ lb (11,363.6 kg).

TABLE 65   Torque Tightening Factor°

Method	C_T
Electronic bolt-torquing systems	1.0 to 1.5
Torque or power wrench with direct torque control	1.6 to 1.8
Power wrench by elongation measurement of calibrated bolts with the original clamped part	1.4 to 1.6
Power wrench using air-stall principle	1.7 to 2.5

°*Product Engineering.*

5. Determine the required bolt area and diameter

Select a grade 8 bolt having an ultimate strength of 150,000 lb/in^2 (1034.1 MPa). The nominal bolt area must then be $A_b = P_U/U_s = 25,000/150,000 = 0.1667$ in^2 (1.08 cm^2). Then the bolt diameter is $D_b = (4A_b/\pi)^{0.5} = [4(0.1667)/\pi]^{0.5} = 0.4607$ in (1.17 cm).

The closest standard bolt size is 0.5 in (1.27 cm). Choose 0.5-13NC bolts, keeping in mind that coarse-thread bolts generally have stronger threads.

6. Find the spring rate of the bolt chosen

The general equation for the spring rate K, lb/in, for a part under tension loading is $K = AE/$

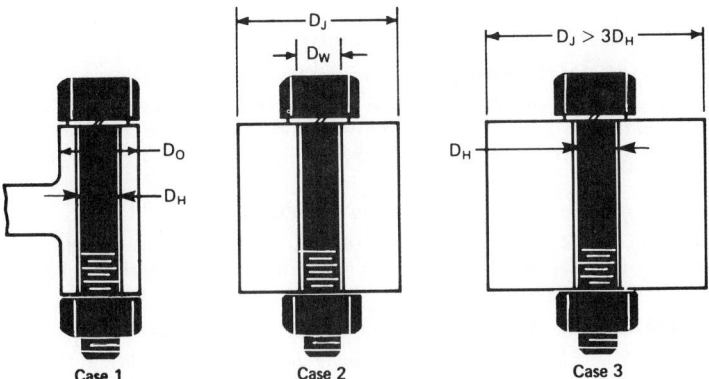

FIG. 46   Joint size influences the general equations for spring rate of the assembly. *(Product Engineering.)*

L, where A = part cross-sectional area, in^2; E = Young's modulus, lb/in^2; L = length of section, in. To find the spring rate of a part with different cross-sectional areas, as a bolt has, add the reciprocal of the spring rate of each section, or $1/K_{total} = 1/K_1 + 1/K_2 + \cdots$.

For a bolt, which consists of three parts (head, unthreaded portion, and nut), the spring rate determined by G. H. Junker, a consultant with Unbrako-SPS European Division, is given by $1/K_B = (1/E)[(0.4D/A_1) + (L_1/A_1) + (L_T/A_M) + (0.4D_M/A_M)]$, where D = nominal bolt diameter, in; A_1 = cross-sectional area of unthreaded portion of bolt, in^2; A_m = cross-sectional area, in^2, of minor threaded diameter D_M; L_1 = length of unthreaded portion, 0.75 in (1.91 cm) for this bolt; L = length of threaded portion being clamped, in. Values of $0.4D$ and $0.4D_M$ pertain to the elastic deformation in the head and nut areas, respectively, and were derived in tests in Germany.

For a 0.5-in (1.27-cm) diameter bolt, the cross-sectional area A_1 of the unthreaded portion of the body is $A_1 = \pi(0.5)^2/4 = 0.1964$ in^2 (1.27 cm^2).

The minor thread diameter of the 0.5-13NC thread bolt is, from a table of thread dimensions, 0.4056 in (1.03 cm). Thus, $A_M = \pi(0.4056)^2/4 = 0.1292$ in^2 (0.83 cm^2).

Substituting the appropriate values in the spring-rate equation for the bolt gives $1/K_B = (1/30 \times 10^6)[0.4 \times 0.5/0.1964 + 0.75/0.1964 + (2.010 - 0.75)/0.1292 + 0.4 \times 0.4056/0.1292]$; $K_B = 1.893 \times 10^6$ lb/in (338,752 kg/cm).

7. *Compute the spring rate of the joint*

Calculations of the spring rate of the joint can be simplified by assuming that the bolt head and nut, when compressing the joint as the bolt is tightened, will cause a stress distribution in the shape of a hollow cylinder, with most of the joint compression occurring in the vicinity under the bolt head and nut.

To calculate an equivalent area for the joint A_J in^2 for use in the spring-rate equation for the joint, the designer has a choice of one of three equations:

Case 1—When most of the outside diameter of the joint is equal to or smaller than the bolt-head diameter, as when parts of a bushing are clamped, Fig. 46, then $A_J = (\pi/4)(D_0^2 - D_H^2)$, where D_0 = outside diameter of the joint or bushing, in; D_H = diameter of bolt hole, in.

Case 2—When the outer diameter of the joint D_J in is greater than the effective bolt-head diameter or washer D_W in, but less than $3D_H$, then $A_J = (\pi/4)(D_W^2 - D_N^2) + (\pi/8)(D_J/D_W - 1)(D_W L_J/5 + L_J^2/100)$.

Case 3—When the joint diameter D_J in is equal to or greater than $3D_H$, then $A_J = (\pi/4)(D_W + L_J/10 - D_H^2)$.

For this bolted joint, case 2 applies. Assuming that the bearing diameter of the head or nut is 0.75 in (1.9 cm), then $A_J = (\pi/4)(0.75^2 - 0.5^2) + (\pi/8)(2/0.75 - 1)[0.75(2.010)/5 + (2.010)^2/$

$100] = 0.4692 \text{ in}^2 (3.03 \text{ cm}^2)$. Inserting this value of A_J in the spring-rate equation of step 6 gives $K_J = 0.4692(30 \times 10^6)/2.010 = 7.003 \times 10^6 \text{ lb/in} (1.253 \times 10^6 \text{ kg/cm})$.

8.   *Find the portion of the working load that unloads the clamped joint*

The loading constant C_L considers the bolt and joint elasticity and is given by $C_L = K_B/(K_B + K_J)$. For this joint, $C_L = 1.893/(1.893 + 7.003) = 0.2128$. Then F_p, the portion of the working load F_A that unloads the clamped joint, is $F_P = F_A(1 - C_L) = 8000(1 - 0.2128) = 6298 \text{ lb}$ (2862.7 kg).

9.   *Determine the loss of clamping force due to embedding*

Some embedding occurs after a bolt is tightened at its assembly. A recommended value is 10 percent. Or specific values can be obtained from tests. Thus, the loss of clamping force F_Z, lb = $0.10 F_A = 0.10(8000) = 800 \text{ lb} (363.6 \text{ kg})$.

10.   *Find the clamping force required for the joint*

Since the working load in the vessel fluctuates from 0 to 8000 lb (0 to 3636.4 kg), the minimum required clamping force $F_K = 0$. Now the maximum required clamping force F_M, lb, can be found from $F_M = C_T(F_P + F_K + F_Z)$, since all the variables are known. Or, $F_M = 2.5(6298 + 0 + 800) = 17,745 \text{ lb} (8065.9 \text{ kg})$.

Since the bolt has a yield strength P_Y of 20,000 lb (9090.9 kg), it has sufficient strength for static loading. For dynamic loading, however, the additional loading is $F_S = F_A C_L$, where $F_S =$ portion of working load that additionally loads the bolt, lb. Or, $F_S = 8000(0.2128) = 1702 \text{ lb}$ (773.6 kg).

11.   *Check the endurance limit of the selected bolt*

The endurance limit S_E, lb, should not be exceeded for long-life operation of the joint. Exact values of S_E can be obtained from the bolt manufacturer or computed from standard endurance-limit equations. For grade 8 bolts, $S_E = 4600 D^{1.59}$, or $S_E = 4600(0.5)^{1.59} = 1527 \text{ lb} (694.1 \text{ kg})$.

The value for F_S that was computed for the 0.5-in (1.27-cm) bolt, 1702 lb (773.6 kg), should not have exceeded 1527 lb (694.1 kg). Since it did, a bolt with a larger diameter must now be selected. Checking a 0.625-in (1.59-cm) bolt (0.625-11), we find the endurance limit becomes $S_E = 4600(0.625)^{1.59} = 2179 \text{ lb} (990.5 \text{ kg})$.

Thus, although the 0.5-in (1.27-cm) bolt would have sufficed for static loading, it would not have provided the necessary endurance strength for long-life operation. So the 0.625-in (1.59-cm) bolt is selected, and a new clamping force $F_M = 16,970 \text{ lb} (7713.6 \text{ kg})$ is calculated, by using the same series of steps detailed above.

Related Calculations: The procedure given here can be used to analyze bolted joints used in pressure vessels in many different applications—power plants, hydraulic systems, aircraft, marine equipment, structures, and piping systems. New thread forms permit longer fatigue life for bolted joints. For this reason, the bolted joint is becoming more popular than ever. Further, electronic bolt-tightening equipment allows a bolt to be tightened precisely up to its yield point safely. This gets the most out of a particular bolt, resulting in product and assembly cost savings.

The key to economical bolted joints is finding the minimum bolt size and clamping force to provide the needed seal. Further, the bolt size chosen and clamping force used must be correct. But some designers avoid the computations for these factors because they involve bolt loading, elasticity of the bolt and joint, reduction in preload resulting from embedding of the bolt, and the method used to tighten the bolt. The procedure given here, developed by G. H. Junker, a consultant with Unbrako-SPS European Division, simplifies the computations. Data given here were presented in *Product Engineering* magazine, edited by Frank Yeaple, M.E.

DETERMINING REQUIRED TIGHTENING TORQUE FOR A BOLTED JOINT

Determine the bolt-tightening torque required for the 0.625-in 11NC (1.59-cm) bolt analyzed in the previous calculation procedure. The bolt must provide a 17,000-lb (7727.3-kg) preload. The coefficient of friction between threads f_T is assumed to be 0.12, while the coefficient of friction between the nut and the washer f_N is assumed to be 0.14.

Calculation Procedure:

1. Determine the dimensions of the bolt

From tables of thread dimensions, the minor thread diameter $D_M = 0.5135$ in (1.3 cm), and the pitch diameter of the bolt threads $D_P = 0.5660$ in (1.44 cm). The mean bearing diameter of the nut is $D_N = 1.25D$, where D = nominal bolt diameter, in. Or, $D_N = 1.25(0.625) = 0.781$ in (1.98 cm).

For the thread coefficient of friction, $\tan \phi = 0.12$; $\phi = 6.84°$. The helix angle of the thread, $\beta = 2.93°$, is found from $\sin \beta = 1/\pi (11)(0.5660) = 0.05112$.

2. Find the dimensionless thread angle factor C_A

The thread angle factor $C_A = \tan(\beta + \phi)/\cos \alpha = \tan(2.36 + 6.84)/\cos 30° = 0.2$.

3. Compute the tightening torque applied

The tightening torque applied to the nut or bolt head, in·lb, is $T = F_M(D_N f_N + C_A D_P)/2$. Substituting gives $T = 17,000[0.781(0.14) + 0.2(0.566)]/2 = 1886$ in·lb = 157 ft·lb (212.7 J).

4. Find the combined stress induced in the bolt

The combined stress S_C in a bolt in a bolted joint is $S_C = T\{0.89 + 1.66[1 + (5.2C_A D_P/D_M)^2]^{0.5}\}/D_M^2(f_N D_N + C_A D_P)$. Substituting, we find $S_C = 1886\{0.89 + 1.66[1 + (5.2 \times 0.2 \times 0.566/0.5135)^2]^{0.5}\}/\{0.5135^2[0.14(0.781) + 0.2(0.566)]\} = 109,766$ lb/in² (756,812.4 kPa).

Generally, S_C should be kept within 68 percent of the ultimate strength S_U. Thus, $S_U = 1.47S_C$. Substituting for the above bolt gives $S_U = 1.47(109,766) = 161,356$ lb/in² (1.1 MPa).

Related Calculations: This procedure can be used for determining the bolt-tightening torque for any application in which a bolted joint is used. Numerous applications are listed in the previous calculation procedure. The equations and approach given here are those of Bernie J. Cobb, Mechanical Engineer, Missile Research and Development Command, Redstone Arsenal, as reported in *Product Engineering*, edited by Frank Yeaple.

SELECTING SAFE STRESS AND MATERIALS FOR PLASTIC GEARS

Determine the safe stress, velocity, and material for a plastic spur gear to transmit 0.125 hp (0.09 W) at 350 r/min 8 h/day under a steady load. Number of teeth in the gear = 75; diametral pitch = 32; pressure angle = 20°; pitch diameter = 2.34375 in (5.95 cm); face width = 0.375 in (0.95 cm).

Calculation Procedure:

1. Compute the velocity at the gear pitch circle

Use the relation $V = rpm(D_p)\pi/12$, where V = velocity at pitch-circle diameter, ft/min; rpm = gear speed, r/min; D_P = pitch diameter, in. Solving yields $V = 350(2.34375)\pi/12 = 215$ ft/min (2.15 m/s).

2. Find the safe stress for the gear

Use the relation $S_S = 55(600 + V)PC_S H_P/(FYV)$, where S_S = safe stress on the gear, lb/in²; P = diametral pitch, in; C_S = service factor from Table 66; H_P = horsepower transmitted by the

TABLE 66 Service Factor C_S for Horsepower Equations[°]

Type of load	8–10 h/day	24 h/day	Intermittent, 3 h/day	Occasional, 0.5 h/day
Steady	1.00	1.25	0.80	0.50
Light shock	1.25	1.50	1.00	0.80
Medium shock	1.50	1.75	1.25	1.00
Heavy shock	1.75	2.00	1.50	1.25

[°] *Product Engineering.*

TABLE 67 Tooth-Form Factor Y for Horsepower Equations°

Number of teeth	14½° Involute or cycloidal	20° Full-depth involute	20° Stub-tooth involute	20° Internal full depth	
				Pinion	Gear
50	0.352	0.408	0.474	0.437	0.613
75	0.364	0.434	0.496	0.452	0.581
100	0.371	0.446	0.506	0.462	0.581
150	0.377	0.459	0.518	0.468	0.565
300	0.383	0.471	0.534	0.478	0.534
Rack	0.390	0.484	0.550	. . . .	. . . .

°*Product Engineering.*

gear; F = face width of the gear, in; Y = tooth form factor, Table 67. Substituting, we find S_S = 55(600 + 215)(32)(1.0)(0.125)/[0.375(0.434)(215)] = 5124 lb/in² (35,561 kPa).

3. *Select the gear material*

Enter Table 68 at the safe stress, 5124 lb/in² (35,561 kPa), and choose either nylon or polycarbonate gear material because the safe stress falls within the allowed range, 600 lb/in² (41,364 kPa), for these two materials. If a glass-reinforced gear is to be used, any of a number of materials listed in Table 68 would be suitable.

Related Calculations: Two other v equations are used in the analysis of plastic gears. For helical gears: $H_P = S_S FYV/[423(78 + V^{0.5})P_N C_S]$. For straight bevel gears, $H_P = S_S FYV(C - F)^4/[55(600 + V)PCC_S]$, where all the symbols are as given earlier; C = pitch-cone diameter, in; P_N = normal diametral pitch.

In growing numbers of fractional-horsepower applications up to 1.5 hp (1.12 kW), gears molded of plastics are being chosen. Typical products are portable power tools, home appliances, instrumentation, and various automotive components.

The reasons for this trend are many. Besides offering the lowest initial cost, plastic gears can be molded as one piece to include other functional parts such as cams, ratchets, lugs, and other gears without need for additional assembly or finishing operations. Moreover, plastic gears are lighter and quieter than metal gears, and are self-lubricating, corrosion-resistant, and relatively free from maintenance. Also, they can be molded inexpensively in colors for coding during assembly or just for looks.

Improved molding techniques achieve high accuracy. Also new special gear-tooth forms improve strength and wear. "We now can hold tooth-to-tooth composite error to within 0.0005 in (0.00127 cm)," says Samuel Pierson, president of ABA Tool & Die Co. It is made possible by electric-discharge machining of the metal molds and computerized analysis of the effects of moisture absorption and other factors on size change of the plastic gears.

TABLE 68 Safe Stress Values for Horsepower Equations°

Plastic	Unfilled		Glass-reinforced	
	lb/in²	kPa	lb/in²	kPa
ABS	3,000	20,682	6,000	41,364
Acetal	5,000	34,470	7,000	48,258
Nylon	6,000	41,364	12,000	82,728
Polycarbonate	6,000	41,364	9,000	62,046
Polyester	3,500	24,129	8,000	55,152
Polyurethane	2,500	17,235	. . . . .	. . . . .

°*Product Engineering.*

"Furthermore, we now are recommending special tooth forms we developed to utilize full-fillet root radii for increased fatigue strength and tip relief for more uniform motion when teeth flex under load. Usually, it is too expensive for designers of machined metal gears to deviate from standard AGMA tooth forms. But with molded plastic gears, deviations add little to the cost," reports Mr. Pierson in *Product Engineering* magazine.

Plastic gears, however, are weaker than metal gears, have a relatively high rate of thermal expansion, and are temperature-limited.

If performance cannot be achieved solely with change in tooth form, try fillers that stabilize the molded part, boost load-carrying abilities, and improve self-lubricating and wear characteristics. Popular fillers include glass, polytetrafluoroethylene (PTFE), molybdenum disulfide, and silicones.

In short hairlike fibers, miniscule beads, or fine-milled powder, glass can markedly increase the tensile strength of a gear and reduce thermal expansion to as little as one-third the original value. Molybdenum disulfide, PTFE, and silicones, as built-in lubricants, reduce wear. Plastic formulations containing both glass and lubricant are becoming increasingly popular to combine strength and lubricity.

Six common plastics for molded gears are nylon, acetal, ABS, polycarbonate, polyester, and polyurethane.

The most popular gear plastic is still nylon. This workhorse has good strength, high abrasion resistance, and a low coefficient of friction. Furthermore, it is self-lubricating and unaffected by most industrial chemicals. Numerous manufacturers make nylon resins.

Nylon's main drawback is its tendency to absorb moisture. This is accompanied by an increase in the gear dimensions and toughness. The effects must be predicted accurately. Also, nylon is harder to mold than acetal—one of its main competitors.

All plastics have higher coefficients of thermal expansion than metals do. The coefficient for steel between 0 and 30°C is 1.23×10^5 per degree, whereas for nylon it lies between 7 and 10 $\times 10^{-5}$.

The grade usually preferred is nylon 6/6. Frequently it is filled with about 25 percent short glass fibers and a small amount of lubricant fillers. Do not hesitate to ask the injection molder to custom-blend resins and fillers.

The combination of a nylon gear running against an acetal gear results in a lower coefficient of friction than either a nylon against nylon or an acetal against acetal can. A good idea is to intersperse nylon with acetal in gear trains. Acetal is less expensive than nylon and is generally easier to mold. Strength is lower, however.

Two types of acetals are popular for gears: acetal homopolymer, which was the original acetal developed by DuPont, available under the tradename of Delrin; and acetal copolymer, manufactured by Celanese under the tradename Celcon. Other examples are glass-filled acetals (Fulton 404, from LNP Corp.).

All acetals are easily processed and have good natural lubricity, creep resistance, chemical resistance, and dimensional stability. But they have a high rate of mold shrinkage.

The main attraction of ABS plastic is its low cost, probably the lowest of the six classes of plastics used for gears. Some ABS is translucent and has a high gloss surface. It also offers ease of processibility, toughness, and rigidity. Two typical tradenames of ABS are Cycolac (Borg Warner), and Kralastic (Uniroyal).

The polycarbonates have high impact strength, high resistance to creep, a useful temperature range of -60 to $240°F$ (-51 to $115.4°C$), low water absorption and thermal expansion rates, and ease and accuracy in molding.

Because of a rather high coefficient of friction, polycarbonate formulations are available to boost flexible strength of the gear teeth.

Recently, thermoplastic polyesters have been available for high-performance injection-molded parts. Some polyesters are filled with reinforcing glass fibers for gear applications. The adhesion between the polyester matrix and the glass fibers results in a substantial increase in strength and produces a rigid material that is creep-resistant at elevated temperature [$330°F(148.7°C)$].

The glossy surface of some polyesters seems to improve lubricity against other thermoplastics and metals. It withstands most organic solvents and chemicals at room temperature and has long-term resistance to gasoline, motor oil, and transmission fluids up to $140°F$ ($60°C$).

The polyurethanes are elastomeric resins sought for noise dampening or shock absorption, as in gears for bedroom clocks or sprockets for snowmobiles. Many proprietary polyurethane versions are available, including Cyanaprene (American Cyanamide), Estane (B. F. Goodrich), Texin (Mobay Chemical), and Voranol (Dow Chemical).

TOTAL DRIVING AND SLIP TORQUE FOR EXTERNAL-SPRING CLUTCHES

Determine the capacity of a light, steel, external-spring clutch in which the spring is made of 0.05-in (0.13-cm) diameter round wire. The spring rides on 1-in (2.54-cm) diameter shafts. The inner diameter of the spring at rest is 0.98 in (2.49 cm). There are 16 coils, 8 per shaft, in the spring. The coefficient of friction of the steel $\mu = 0.10$, and the modulus of elasticity $E = 30 \times 10^6$ lb/in^2 (208.2 $\times 10^6$ kPa). What are the transmitted and slip torques for this overrunning clutch?

Calculation Procedure:

1. Find the transmitted torque for this clutch

Use the relation $T_t = [\pi(Ed^4i)/32D^4](e^{2\pi\mu N} - 1)$, where T_t = transmitted torque, lb·in; d = diameter of spring wire, in; i = diametral interference (the difference between the spring helix diameter and shaft diameter), in; D = drive shaft diameter, in; μ = coefficient of friction; N = number of spring coils per shaft; other symbols as before. Substituting gives $T_t = [\pi(30 \times 10^6)(0.05)^4(0.02)/32(1)^4](e^{2\pi(0.1)(8)} - 1) = (0.368)(152.4 - 1) = 55.7$ lb·in (6.29 N·m).

2. Compute the maximum slip torque for the clutch

Use the relation $T_s = [\pi(Ed^4i)/32D^4][1/(e^{2\pi\mu N} - 1)]$, where the symbols are as defined earlier and T_s = slip torque, lb·in. Substituting, we find $T_s = 0.368(1/52.4 - 1) = -0.366$ lb·in (-0.04 N·m). This shows that $T_s \simeq -M$, where M = spring moment, lb·in. It also demonstrates that although the capacity of a spring clutch depends on the spring moment, such a clutch can be made relatively insensitive to changes in the coefficient of friction or number of coils. These are the factors appearing in the exponent of the above equations.

For many years, spring clutches have been almost ignored in the clutch clan, even in the relatively small and specialized family of overrunning clutches. A number of popular design texts omit the spring clutch entirely, so the key design formulas are hard to come by. Yet most spring or overrunning clutches function in two modes: either they are locked and driving the output (the usual mode), or they are overrunning, with the input stopped and the output continuing to rotate undriven.

The spring clutch may well be the simplest overrunning clutch. It has a helical spring, wound from rectangular steel stock and coiled inside the shaft bores, bridging the gap between them (Fig. 47a). The helix diameter at either end of the spring is slightly larger than that of the shaft in which it nestles, so the steel coils remain in continuous contact with the shaft. (This internal design differs from most other versions of the spring clutch, in which the spring is wrapped around the outsides of two butting, coaxial shafts.)

When the drive shaft rotates in the same sense as the twist of the helix, that is, when the rotation tends to unwind the internal spring, then the coils try to expand, gripping both shaft bores tightly and pushing the output.

When the relative rotation of the input shaft reverses, however, when the motion tends to wind the internal spring tighter, as it would if the input stopped or slowed, leaving the output to spin under its own momentum, then the helix contracts, loosening its grip and slipping freely inside the bores—*freewheeling*.

The spring inevitably drags a bit in the overrunning mode; it could not tighten its grip again if it did not. But the residual friction is relatively small and well controlled. The consequent low wear rate and high reliability are the internal spring's principal advantage. And the internal-spring design is virtually immune to centrifugal effects, another critical consideration for high-speed use, unlike the external-spring version, which can loosen its grip at high speeds.

The wrapped-spring clutch, Fig. 47b, does lend itself to a number of sophisticated applications, despite this limitation. It can serve as an overrunning clutch, although its operation is the mirror image of that of the internal spring. The wrapped spring loosens and freewheels when driven by an unwinding torque; and when subjected to a winding torque, it squeezes down on the shafts like a child's "Chinese Handcuffs" toy, pulling the output along after the input.

By bending the spring's input end outward (Fig. 47b), clutch makers can fashion an on/off

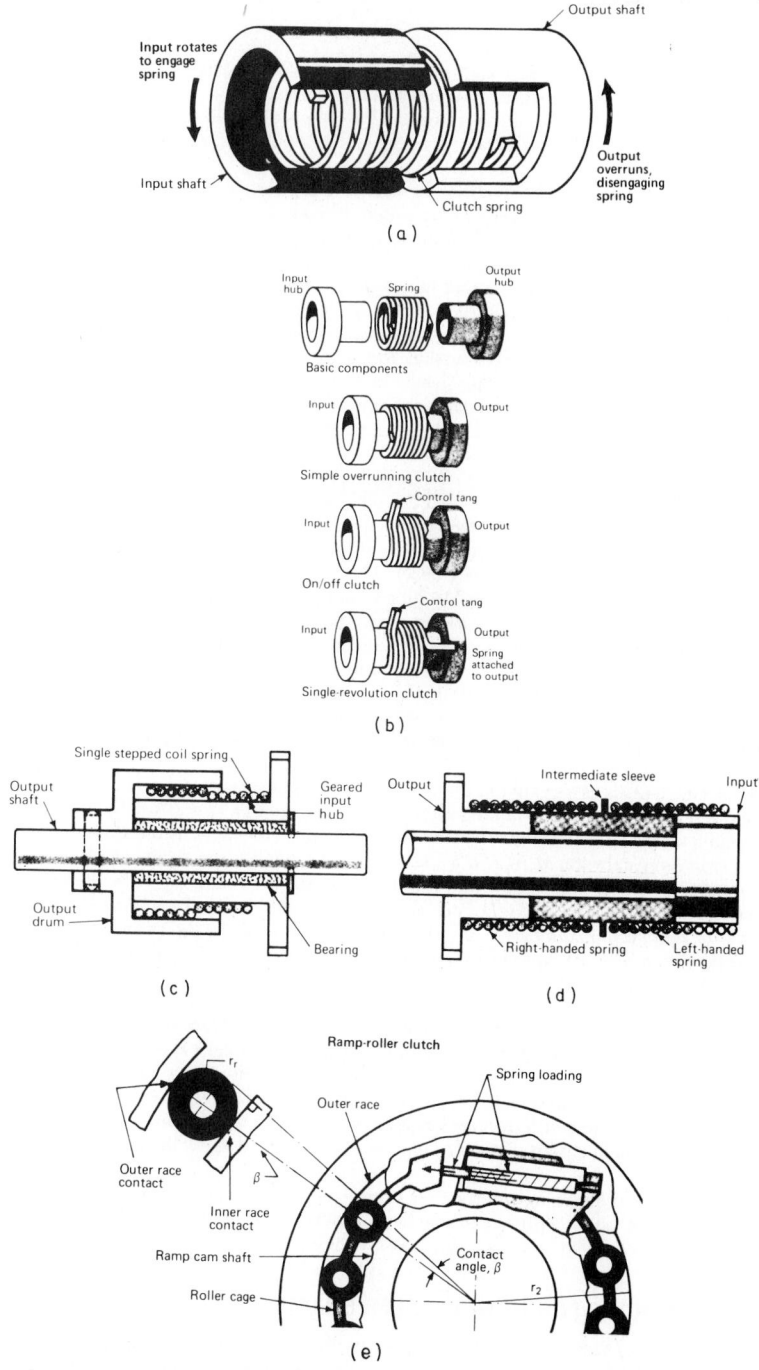

FIG. 47 (*a*) Internal-spring overrunning clutch. (*b*) Wrapped-spring clutches. (*c*) Two-way slip clutch. (*d*) Mated right- and left-handed spring slip clutch. (*e*) Ramp-roller clutch. (*Design Engineering.*)

clutch. When this tang is released, the spring wraps down on the shafts for a positive drive; when the tang is engaged again, the spring unwinds and releases the output.

In a variation, the other end of the spring is permanently attached to the output shaft, giving single-revolution control. When the tang is engaged, the input clutches out and the spring stops the output (Fig. 47b).

The spring clutch also can be modified to work as a slip clutch or drag brake. Such a device is, in a sense, the opposite of an overrunning clutch: It works by beefing up residual drag so that the external spring can continue to transmit torque, even when it is being "unwound" in the overrunning direction.

By varying the spring dimensions and material, this drag can be precisely controlled to produce an extremely useful slip clutch. Indeed, spring clutches have been modified to give a predetermined slip for either direction of rotation. In one version, Fig. 47c, a stepped helix creates a dual external/internal spring. Another version, Fig. 47d, attaches a right-handed spring to a left-handed spring through an intermediate sleeve. In each case, one spring component slips with a controlled drag on one shaft while its partner clamps down, Chinese-handcuff-like, on the other shaft. If the input rotation is reversed, the two springs shift roles. Spring clutches are available from many sources.

Two factors control any application of wrapped spring clutches, whether overrunning or slip: the torque capacity in the driving or locked mode (where the spring grips the shaft) and the frictional drag torque capacity in the overrunning or slip direction. For overrunning applications, one would obviously want high driving torque and low drag torque. In a slip clutch, one would want to adjust the drag torque to the specific job. Happily, each factor can be calculated, by using equations derived by Joseph Kaplan of Machine Components.

The total driving torque T_t is the product of two factors: an exponential function of the coefficient of friction and number of coils wrapping each shaft; and the spring's moment M, a function of shaft diameter and the spring's dimensions and modulus of elasticity.

The drag or slip torque capacity T_s is the product of the same spring moment M and an inverse exponential function of the number of coils and the coefficient of friction. Since this exponential factor generally approaches -1, the approximation of $T_s = -M$ is good enough for most applications.

These clutch-analysis equations, using the symbols given earlier, are useful: *For spring clutches:* The maximum torque in the locked or drive direction is $T_t = M(e^{2\mu\pi N} - 1)$ lb·in. The overrunning slip or friction torque is $T_s = M(e^{-2\mu\pi N} - 1)$. Where M is the spring moment for circular cross-sectional wire, $M = \pi E d^4 \delta/(32D^2)$. For rectangular cross-sectional wire, $M = Ebt^3\delta/(6D^2)$. Large drums or high speeds may add a centrifugal correction $\Delta\delta$ to the diametral interference δ; $\Delta\delta = 2\rho D^5\omega^2/ed^2$; $\Delta\delta = 1.5\rho D^5\omega^2/et^2$. *For sprag clutches:* $T_t = (r_1 nLK \tan \alpha)/(1/r_1) + 1/r_s$. *For ramp-roller clutches:* $T_t = nr_r r_2 LK \tan (\beta/2)$. Table 69 shows the results of a study by Sikorsky Aircraft of 10 types of overrunning clutches. Figures of merit were developed for each.

Additional symbols in these equations are $e = $ Euler's number $\simeq 2.7183$; $K = $ hertzian stress constant factor from Table 70; $L = $ length of roller or sprag, in; $M = $ spring moment, lb·in; $N = $ number of spring coils per shaft; $n = $ number of sprags or rollers; $r_1 = $ radius of inner race, in; $r_2 = $ radius of outer race, in $r_r = $ radius of roller, in; $r_s = $ radius of sprag at contact of inner race, in; $T_t = $ transmitted torque, lb·in; $t = $ thinness of rectangular wire, in; $\alpha = $ sprag gripping angle, degrees; $\beta = $ ramp-race contact angle; $\delta = $ diametral interference (difference between spring helix diameter and shaft diameter), in; $\rho = $ coil material density, lb/in^3; $\omega = $ shaft angular velocity, rad/s.

This procedure, containing information from a study conducted by Sikorsky Aircraft, was reported in *Design Engineering* magazine in an article by Doug McCormick, Associate Editor, and uses equations derived by Joseph Kaplan of Machine Components.

DESIGN METHODS FOR NONCIRCULAR SHAFTS

Find the maximum shear stress and the angular twist per unit length produced by a torque of 20,000 lb·in (2258.0 N·m) imposed on a double-milled steel shaft with a four-splined hollow core, Fig. 48. The shaft outer diameter is 2 in (5.1 cm); the inner diameter is 1 in (2.5 cm). Modulus of rigidity of the shaft is $G = 12 \times 10^6$ lb/in^2 (83.3 $\times 10^6$ kPa). The flat surfaces of the outer, milled

TABLE 69 Sikorsky's Ratings of 10 Clutches[*]

	Weight	Cost	Reliability	Maintainability	Centrifugal effects	Vibration	Transient torque	Thermal effects	Startup control	Failure modes	Operation limits	Unusual testing	Multi-engine	Bearing brinelling	Figure of merit
Spring clutch	11.1	9.4	22.7	8.2	7.0	3.5	3.0	2.0	2.3	2.6	3.9	3.0	1.6	1.0	81.3
Sprag, type A	12.3	10.9	18.4	7.4	6.8	4.7	2.8	1.3	3.2	1.5	3.3	3.0	1.5	0.3	77.4
Sprag, type B	12.2	11.0	20.5	7.7	2.5	4.5	3.0	1.3	3.2	2.5	3.3	3.0	1.5	0.3	76.5
Ramp roller (RR)	10.8	8.9	21.3	7.6	6.5	4.5	2.8	1.3	3.2	2.4	3.3	3.0	1.4	0.7	77.7
Actuated RR	9.2	7.7	17.1	8.1	7.0	4.5	2.8	0.7	3.1	3.0	3.7	3.0	2.0	0.7	72.6
Ball bearing RR	11.9	10.9	13.9	5.3	1.0	5.2	2.8	1.1	1.6	2.4	3.3	3.0	1.4	0.3	64.1
Roller gear RR	10.4	9.9	14.5	6.7	1.0	5.0	2.8	1.1	3.2	2.3	3.3	3.0	1.3	0.7	65.2
Positive ratchet	10.4	8.2	18.1	8.7	6.8	5.2	2.8	2.0	3.6	1.8	3.9	3.0	1.8	0.7	76.0
Face ratchet	10.8	8.8	16.4	8.4	1.8	4.9	2.8	0.7	3.8	1.8	3.7	2.0	1.8	0.7	68.4
Link ratchet	11.3	12.3	19.0	8.8	4.5	5.9	2.8	2.0	3.8	2.8	3.3	2.0	1.8	0.3	80.6
Of a possible	13	13	30	9	7	6	3	2	4	3	4	3	2	1	100

[*] *Design Engineering.*

TABLE 70   *K* Factors for Various Materials

Material°	K at number of cycles		
	10^6	10^7	10^8
Gray cast-iron rolls (130–180 BHN)	4,000	2,000	1,300†
GM Meehanite cast-iron rolls (190–240 BHN)	4,000	2,500	1,950†
Gray cast-iron rolls with hardened-steel rolls:			
Cast iron, 140–160 BHN	2,600	1,400	1,000†
Cast iron, 160–190 BHN	3,200	1,900	1,300†
Cast iron, 270–290 BHN	5,500	4,000	3,100†
Medium-carbon-steel rolls with hardened-steel rolls (medium-carbon steel, 270–300 BHN)	14,000	11,000	9,000
Carburized-carbon-steel rolls with hardened-steel rolls (carburized-carbon steel, 55–58 R_c)	24,000	18,000	15,000‡

°Lubricated with straight mineral oil.
†At 4×10^7 cycles.
‡At 5×10^8 cycles.

shaft are cut down 0.1 in (0.25 cm), while the inner contour has a radius r_i = 0.5 in (1.27 cm) with four splines of half-width A = 0.1 in (0.25 cm) and height B = 0.1 in (0.25 cm).

Calculation Procedure:

1. *Find the torsional-stiffness and shear-stress factors for the outer shaft*

The proportionate mill height H/R for the outer shaft is, from Fig. 49 and the given data, H/R = 0.1/1 = 0.1. Entering Fig. 49 at H/R = 0.1, project to the torsional-stiffness factor (V) curve for shaft type 1, and read V = 0.71. Likewise, from Fig. 49, f = shear-stress factor = 0.82.

2. *Find the torsional-stiffness factor for the inner contour*

The spline dimensions, from Fig. 50, are A/B = 0.1/0.1 = 1.0, and B/R = 0.1/0.5 = 0.2. Entering Fig. 50 at B/R = 0.2, project upward to the A/B = 1.0 curve, and read the torsional-stiffness factor V = 0.85.

3. *Compute the angular twist of the shaft*

The composite torsional stiffness is defined as $V_t R^4$, where V_t = composite-shaft stiffness factor and R = the internal diameter. For a composite shaft, $V_t R^4 = \Sigma V_i r_i^4 = (0.71)(1^4) - (0.85)(0.5)^4$

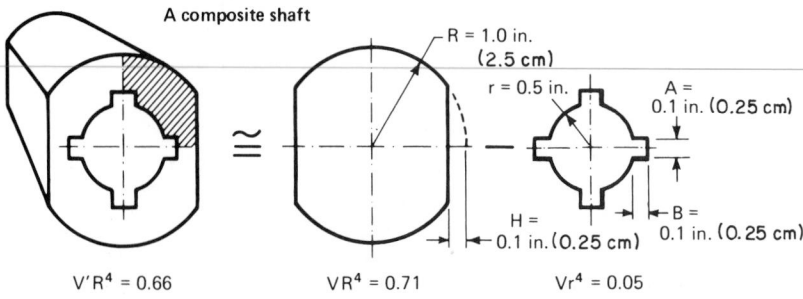

FIG. 48   Typical composite shaft. *(Design Engineering.)*

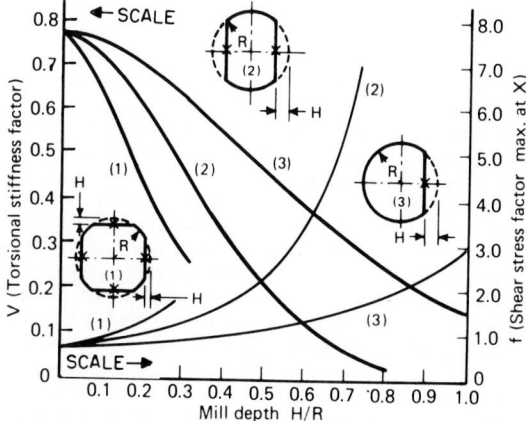

FIG. 49   Milled shafts. *(Design Engineering.)*

= 0.66 for this shaft. (The first term in this summation is the outer shaft; the second is the inner cross section.)

To find the angular twist θ of this composite shaft, substitute in the relation $T = 2G\theta V_t R^4$, where T = torque acting on the shaft, lb·in; other symbols as defined earlier. Substituting, we find $20,000 = 2(12 \times 10^6)\theta(0.66)$; $\theta = 0.0012$ rad/in (0.000472 rad/cm).

4.   *Find the maximum shear stress in the shaft*

For a *solid* double-milled shaft, the maximum shear stress, S_s, lb/in² $= Tf/R^3$, where the symbols are as given earlier. Substituting for this shaft, assuming for now that the shaft is solid, we get $S_s = Tf/R^2 = (20,000)(0.82)/(1^2) = 16,400$ lb/in² (113,816 kPa), where the value of f is from step 1.

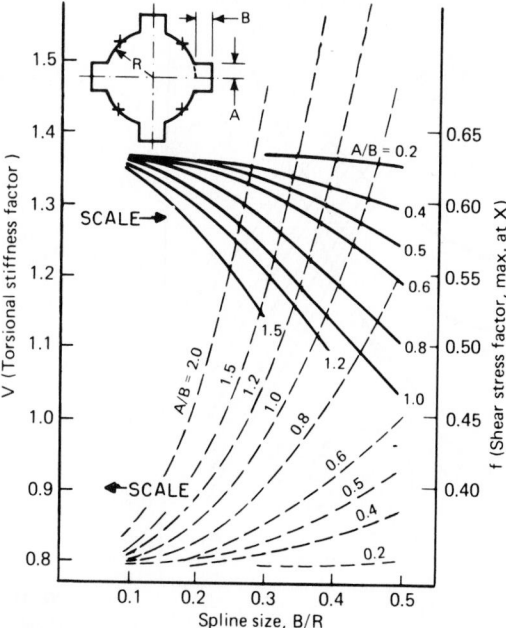

FIG. 50   Four-spline shaft. *(Design Engineering.)*

Because the shaft is hollow, however, the solid-shaft maximum shear stress must be multiplied by the ratio $V/V_t = 0.71/0.66 = 1.076$, where the value of V is from step 1 and the value of V_t is from step 3.

The hollow-shaft maximum shear stress is $S_s' = S_s(V/V_t) = 16,400(1.076) = 17,646$ lb/in² (122,993 kPa).

Related Calculations: By using the charts (Figs. 49 through 58) and equations presented here, quantitative and performance factors can be calculated to well within 5 percent for a variety of widely used shafts. Thus, designers and engineers now have a solid analytical basis for choosing shafts, instead of having to rely on rules of thumb, which can lead to application problems.

Although design engineers are familiar with torsion and shear stress analyses of uniform circular shafts, usable solutions for even the most common noncircular shafts are often not only unfamiliar, but also unavailable. As a circular bar is twisted, each infinitesimal cross section rotates about the bar's longitudinal axis: plane cross sections remain plane, and the radii within each cross section remain straight. If the shaft cross section deviates even slightly from a circle, however, the situation changes radically and calculations bog down in complicated mathematics.

The solution for the circular cross section is straightforward: The shear stress at any point is proportional to the point's distance from the bar's axis; at each point, there are two equal stress vectors perpendicular to the radius through the point, one stress vector lying in the plane of the cross section and the other parallel to the bar's axis. The maximum stress is tangent to the shaft's outer surface. At the same time, the shaft's torsional stiffness is a function of its material, angle of twist, and the polar moment of inertia of the cross section.

The stress and torque relations can be summarized as $\theta = T/(JG)$, or $T = G\theta J$, and $S_s = TR/J$ or $S_s = G\theta R$, where J = polar moment of inertia of a circular cross section ($= \pi R^4/2$); other symbols are as defined earlier.

If the shaft is splined, keyed, milled, or pinned, then its cross sections do not remain plane in torsion, but warp into three-dimensional surfaces. Radii do not remain straight, the distribution of shear stress is no longer linear, and the directions of shear stress are no longer perpendicular to the radius.

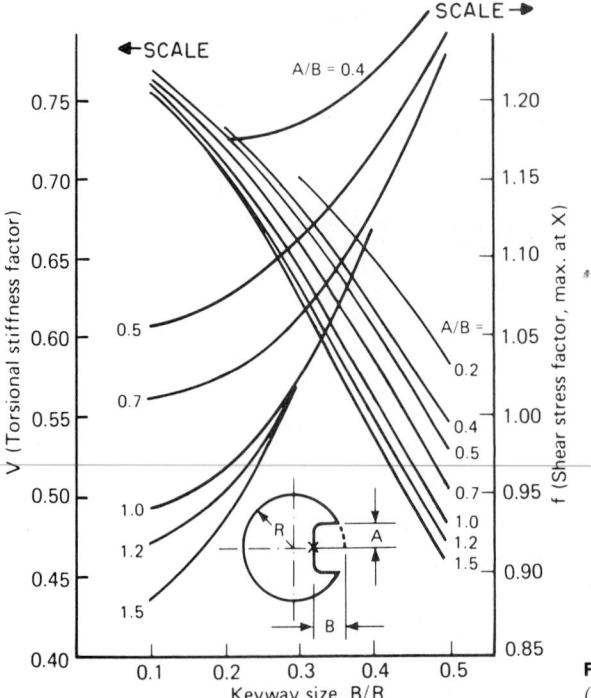

FIG. 51 Single-keyway shaft. (*Design Engineering.*)

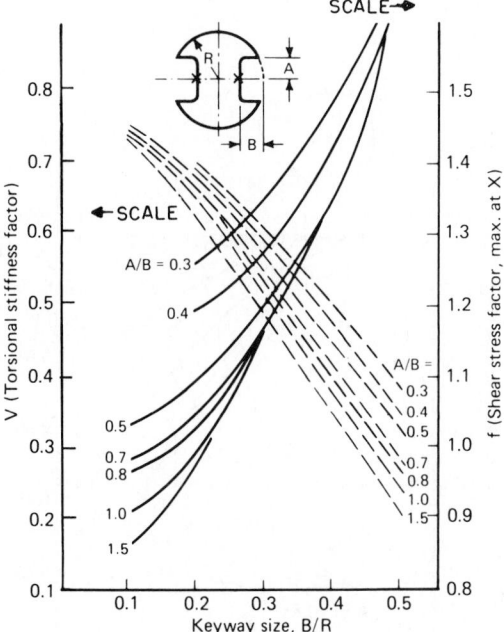

FIG. 52 Two-keyway shaft. *(Design Engineering.)*

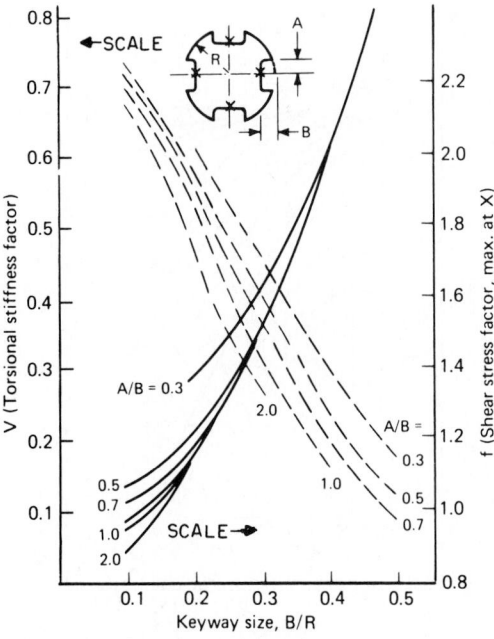

FIG. 53 Four-keyway shaft. *(Design Engineering.)*

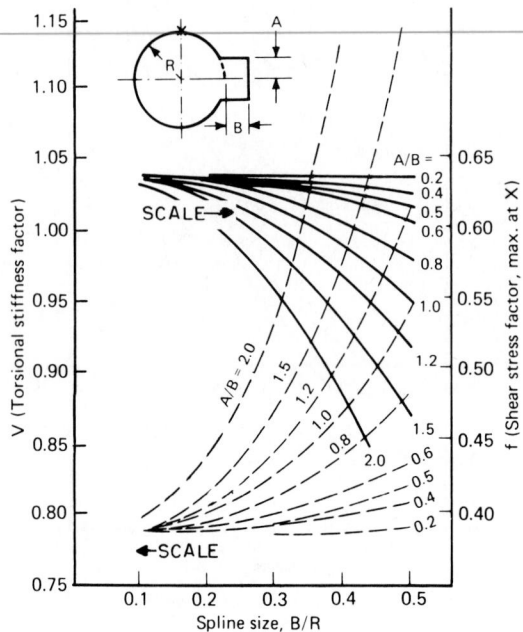

FIG. 54 Single-spline shaft. (*Design Engineering.*)

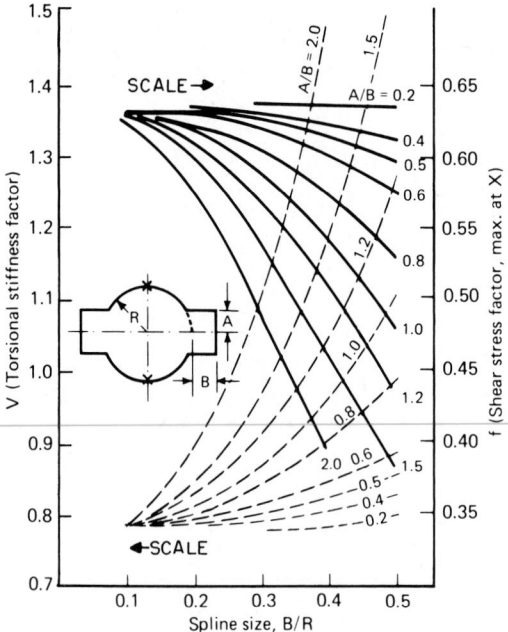

FIG. 55 Two-spline shaft. (*Design Engineering.*)

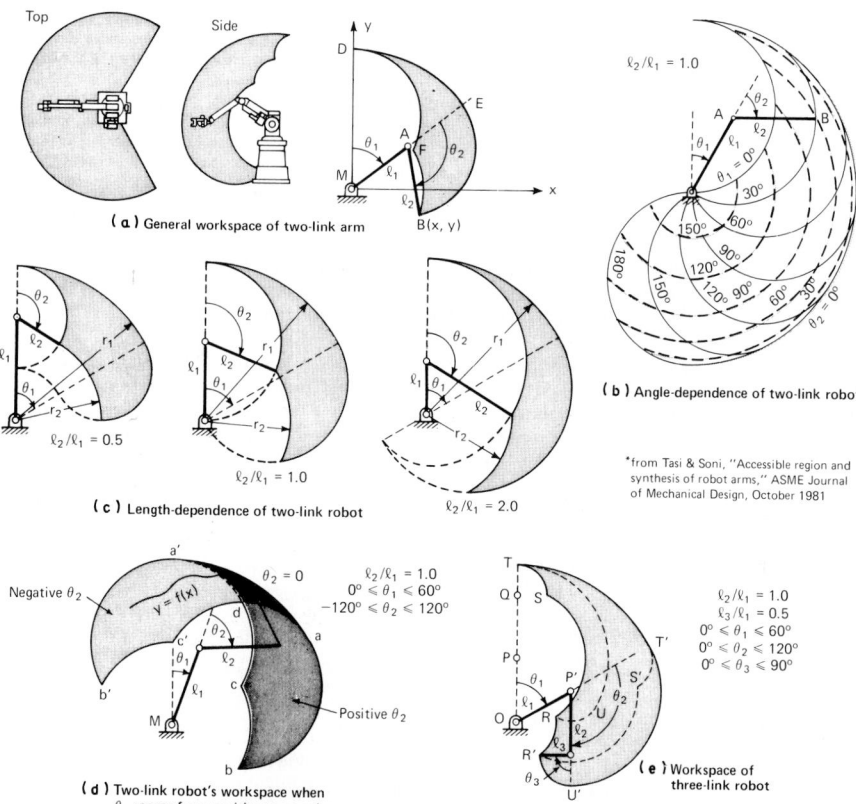

Top   Side

(a) General workspace of two-link arm

$\ell_2/\ell_1 = 0.5$

$\ell_2/\ell_1 = 1.0$

$\ell_2/\ell_1 = 2.0$

(c) Length-dependence of two-link robot

$\ell_2/\ell_1 = 1.0$

(b) Angle-dependence of two-link robot

*from Tasi & Soni, "Accessible region and synthesis of robot arms," ASME Journal of Mechanical Design, October 1981

Negative θ_2

$y = f(x)$

$\theta_2 = 0$ $\ell_2/\ell_1 = 1.0$
$0° \leqslant \theta_1 \leqslant 60°$
$-120° \leqslant \theta_2 \leqslant 120°$

Positive θ_2

(d) Two-link robot's workspace when θ_2 ranges from positive to negative

$\ell_2/\ell_1 = 1.0$
$\ell_3/\ell_1 = 0.5$
$0° \leqslant \theta_1 \leqslant 60°$
$0° \leqslant \theta_2 \leqslant 120°$
$0° \leqslant \theta_3 \leqslant 90°$

(e) Workspace of three-link robot

FIG. 59   Revolute robots are common in industrial applications. The robot's angular limits and the relative length of its limbs determine the size and shape of the workspace of the robot. *(Tasi and Soni, ASME Journal of Mechanical Design and Design Engineering.)*

2. Define the shape and area of the robot's workspace

Angular travel limitations are particularly important on robots whose major joints are powered by linear actuators, generally hydraulic cylinders. Figure 59b shows how maximum and minimum values for θ_1 and θ_2 affect the workspace envelope of planar projection of a common 3R robot.

Other robots—notably those powered by rotary actuators or motor-reducer sets—may be double-jointed at the elbow; θ may be either negative or positive. These robots produce the reflected workspace cross sections shown in Fig. 59d.

The relative lengths of the upper arm and forearm also strongly influence the shape of the two-link robot's workspace, Fig. 59c. Tsai and Soni's calculations show that, for a given total reach $L = l_1 + l_2$, the area bounded by the four arcs—the workspace—is greatest when $l_1/l_2 = 1.0$.

Last, the shape and area of the workspace depend on the ratio l_2/l_1, on $\theta_{2,max}$, and on the difference $(\theta_{1,max} - \theta_{1,min})$. And given a constant rate of change for θ_2, Tsai and Soni found that the arm can cover the most ground when the elbow is bent 90°.

3. Specify parameters for a two-link robot able to reach any collection of points

To reach any collection of points (x_i, y_i) in the cross-sectional plane, Tsai and Soni transform the equations for f_1 and f_2 into a convenient procedure by turning f_1 and f_2 around to give equations for the angles θ_{1i} and $\theta 2_i$ needed to reach each of the points (x_i, y_i). These equations are:

$$\theta_{1i} = \cos^{-1}\left[\frac{y_i - y_0}{\sqrt{(x_i - x_0)^2 + (y_i - y_0)^2}}\right] - \cos^{-1}\left[\frac{(x_i - x_0)^2 + (y_i - y_0)^2 + l_1^2 - l_2^2}{2l_1\sqrt{(x_i - x_0)^2 + (y_i - y_0)^2}}\right]$$

$$\theta_{2i} = \cos^{-1}\left[\frac{(x_i - x_0)^2 + (y_i - y_0)^2 - l_1^2 + l_2^2}{2l_1 l_2}\right]$$

Whereas the original equations assumed that the robot's shoulder is located at (0,0), these equations allow for a center of rotation (x_0, y_0) anywhere in the plane.

Using the above equations, the designer then does the following: (a) She or he finds $x_{\min}$, $y_{\min}$, $x_{\max}$, and $y_{\max}$ among all the values (x_i, y_i). (b) If the location of the shoulder of the robot is constrained, the designer assigns the proper values (x_0, y_0) to the center of rotation. If there are no constraints, the designer assumes arbitrary values; the optimum position for the shoulder may be determined later.

(c) The designer finds the maximum necessary reach L from among all $L_i = [(x_i - x_0)^2 + (y_i - y_0)^2]^{0.5}$. Then set $l_1 = l_2 = L/2$. (d) Compute θ_{1i} and θ_{2i} from the equations above for every point (x_i, y_i). Then find the maximum and minimum values for both angles.

(e) Compute the area A_2 of the accessible region from $A = F(\theta_{1,\max} - \theta_{1,\min})(l_1 + l_2)^2$, where $F = (l_2/l_1)(\cos\theta_{2,\min} - \cos\theta_{2,\max})/[1 + (l_2/l_1)^2]$. (f) Use a grid search method, repeating steps b through e to find the optimum values for (x_0, y_0), the point at which A is at a minimum.

As Tsai and Soni note, this procedure can be computerized. The end result by either manual or computer computation is a set of optimum values for x_0, y_0, l_1, l_2, $\theta_{1,\max}$, $\theta_{1,\min}$, $\theta_{2,\max}$, and $\theta_{2,\min}$.

4. List the steps for three-link robot design

In practice, the pitch link of a robot's wrist extends the mechanism to produce a three-link 4R robot—equivalent to a 3R robot in the cross-sectional plane, Fig. 59. This additional link changes the shape and size of the workspace; it is generally short, and the additions are often minor.

Find the shape of the workspace thus: (a) Fix the first link at $\theta_{1,\min}$ and treat the links l_2 and l_3 (that is, PQ and QT) as a two-link robot to determine their accessible region $RSTU$. (b) Rotate the workspace $RSTU$ through the whole permissible angle $\theta_{1,\max} - \theta_{1,\min}$. The region swept out is the workspace.

The third link increases the workspace and permits the designer to specify the attitude of the last link and the "precision points" through which the arm's endpoint must pass.

Besides specifying a set of points (x_i, y_i), the designer may specify for each point a unit vector $\mathbf{e}_i$. In operation, the end link QT will point along $\mathbf{e}_i$. Thus, the designer specifies the location of two points: the endpoint T and the base of the third link Q, Fig. 59e.

Designing such a three-link device is quite similar to designing a two-link version. The designer must add three steps at the start of the design sequence: (a) Select an appropriate length l_3 for the third link. (b) Specify a unit vector $\mathbf{e}_i = e_{xi}\mathbf{i} + e_{yi}\mathbf{j}$, for each prescribed accessible point (x_i, y_i). (c) From these, specify a series of precision points (x_i', y_i') for the endpoint Q of the two-link arm $l_1 + l_2$; $x_i' = x_i - e_{xi} l_3$, $y_i' = y_i - e_{xi} l_3$.

The designer then creates a linkage that is able to reach all precision points (x_i', y_i'), using the steps outlined for a two-link robot. Tsai and Soni also synthesize five-bar mechanisms to generate prescribed coupler curves. They also show how to design equivalent single- and dual-cam mechanisms for producing the same motion.

Related Calculations: The robot is becoming more popular every year for a variety of industrial activities such as machining, welding, assembly, painting, stamping, soldering, cutting, grinding, etc. Kenichi Ohmae, a director of McKinsey and Company, refers to robots as "steel-collar workers." Outside of the industrial field robots are finding other widespread applications. Thus, on the space shuttle *Columbia* a 45-ft (13.7-m) robot arm hauled a 65,000-lb (29,545-kg) satellite out of earth orbit. Weighing only 905 lb (362 kg), the arm has a payload capacity 70 times its own weight. In the medical field, robots are helping disabled people and others who are incapacitated to lead more normal lives. Newer robots are being fitted with vision devices enabling them to distinguish between large and small parts. Designers look forward to the day when vision can be added to medical robots to further expand the life of people having physical disabilities.

Joseph Engelberger, pioneer roboticist, classifies robots into several different categories. Chief among these are as follows: (1) A cartesian robot must move its entire mass linearly during any x

The principal robot bodies – 3 degrees of freedom (with maximum workspaces)

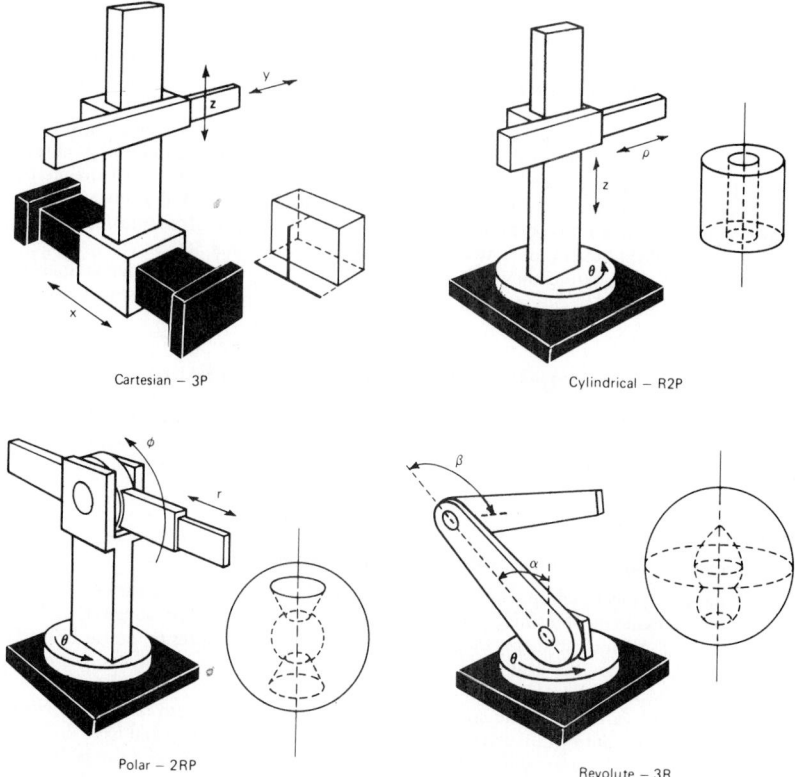

Cartesian – 3P

Cylindrical – R2P

Polar – 2RP

Revolute – 3R

FIG. 60 Types of robot bodies. *(Design Engineering, after Engelberger.)*

axis translation; this robot is well adapted for dealing with wide flat sheets as in painting and welding. The cartesian robot might be an inefficient choice for jobs needing many fast left-and-right moves. (2) Spherical-body robots might be best suited for loading machine tools. (3) Likewise, cylindrical robots are adapted to loading machine tools. (4) Revolute robots find a wide variety of applications in industry. Figure 60 shows a number of different robot bodies.

In the human body we get 7 degrees of freedom from just three joints. Most robots get only 6 degrees of freedom from six joints. This comparison gives one an appreciation of the construction of the human body compared to that of a robot. Nevertheless, robots are replacing humans in a variety of activities, saving labor and money for the organization using them.

This calculation procedure provides the designer with a number of equations for designing industrial, medical, and other robots. In designing a robot the designer must be careful not to use a robot which is too complex for the activity performed. Where simple operations are performed, such as painting, loading, and unloading, usually a simple one-directional robot will be satisfactory. Using more expensive multidirectional robots will only increase the cost of performing the operation and reduce the savings which might otherwise be possible.

Ohmae cities four ways in which robots are important in industry: (1) They reduce labor costs in industries which have a large labor component as part of their total costs. (2) Robots are easier to schedule in times of recession than are human beings. In many plants robots will reduce the breakeven point and are easier to "lay off" than human beings. (3) Robots make it easier for a small firm to enter precision manufacturing businesses. (4) Robots allow location of a plant to be made independent of the skilled-labor supply. For these reasons, there is a growing interest in the use of robots in a variety of industries.

A valuable reference for designers is Joseph Engelberger's book *Robotics in Practice*, published by Amacon, New York. This pioneer roboticist covers many topics important to the modern designer.

At the time of this writing, the robot population of the United States was increasing at the rate of 150 robots per month. The overhead cost of a robot in the automotive industry is currently under $5 per hour, compared to about $14 per hour for hourly employees. Robot maintenance cost is about 50 cents per hour of operation, while the operating labor cost of a robot is about 40 cents per hour. Downtime for robots is less than 2 percent, according to *Mechanical Engineering* magazine of the ASME. Mean time between failures for robots is about 500 h.

The procedure given here is the work of Y. C. Tsai and A. H. Soni of Oklahoma State University, as reported in *Design Engineering* magazine in an article by Doug McCormick, Associate Editor.

HYDROPNEUMATIC ACCUMULATOR DESIGN FOR HIGH FORCE LEVELS

Design a hydropneumatic spring to absorb the mechanical shock created by a 300-lb (136.4-kg) load traveling at a velocity of 20 ft/s (6.1 m/s). Space available to stop the load is limited to 4 in (10.2 cm).

Calculation Procedure:

1.   *Determine the kinetic energy which the spring must absorb*

Figure 61 shows a typical hydropneumatic accumulator which functions as a spring. The spring is a closed system made up of a single-acting cylinder (or sometimes a rotary actuator) and a gas-filled accumulator. As the load drives the piston, fluid (usually oil) compresses the gas in the flexible rubber bladder. Once the load is removed, either partially or completely, the gas pressure drives the piston back for the return cycle.

The flow-control valve limits the speed of the compression and return strokes. In custom-designed springs, flow-control valves are often combinations of check valves and fixed or variable orifices. Depending on the orientation of the check valve, the compression speed can be high with low return speed, or vice versa. Within the pressure limits of the components, speed and stroke length can be varied by changing the accumulator precharge. Higher precharge pressure gives shorter strokes, slower compression speed, and faster return speed.

The kinetic energy that must be absorbed by the spring is given by $E_k = 12WV^2/2g$, where E_k = kinetic energy that must be absorbed, in·lb/ W = weight of load, lb; V = load velocity, ft/s; g = acceleration due to gravity, 32.2 ft/s². From the given data, $E_k = 12(300)(20)^2/2(32.2) = 22,360$ in·lb (2526.3 N·m).

2.   *Find the final pressure of the gas in the accumulator*

To find the final pressure of the gas in the accumulator, first we must assume an accumulator size and pressure rating. Then we check the pressure developed and the piston stroke. If they are

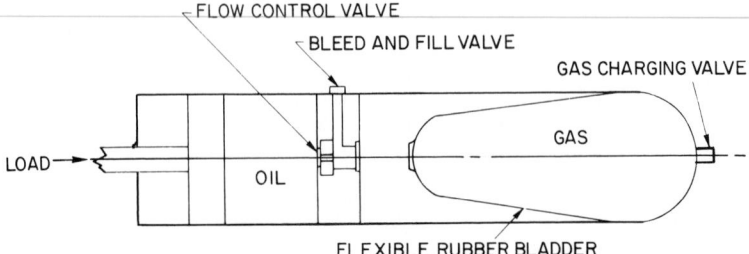

FIG. 61   Typical hydropneumatic accumulator. *(Machine Design.)*

within the allowable limits for the application, the assumptions were correct. If the limits are exceeded, we must make new assumptions and check the values again until a suitable design is obtained.

For this application, based on the machine layout, assume that a 2.5-in (6.35-cm) cylinder with a 60-in^3 (983.2-cm^3) accumulator is chosen and that both are rated at 2000 lb/in^2 (13,788 kPa) with a 1000-lb/in^2 (abs) (6894-kPa) precharge. Check that the final loaded pressure and volume are suitable for the load.

The final load pressure p_2 lb/in^2 (abs) (kPa) is found from $p_2^{(n-1)/n} = p_1^{(n-1)/n}\{[E_k(n-1)/(p_1v_1)] + 1\}$, where p_1 = precharge pressure of the accumulator, lb/in^2 (abs) (kPa); n = the polytropic gas constant = 1.4 for nitrogen, a popular charging gas; v_1 = accumulator capacity, in^3 (cm^3). Substituting gives $p_2^{(1.4-1)/1.4} = 1000^{(1.4-1)/1.4}\{[22,360(1.4-1)/(1000 \times 60)] + 1\} = 1626$ lb/in^2 (abs) (11,213.1 kPa). Since this is within the 2000-lb/in^2 (abs) limit selected, the accumulator is acceptable from a pressure standpoint.

3. Determine the final volume of the accumulator

Use the relation $v_2 = v_1(p_1/p_2)^{1/n}$, where v_2 = final volume of the accumulator, in^3; v_1 = initial volume of the accumulator, in^3; other symbols as before. Substituting, we get $v_2 = 60(1000/1626)^{1/1.4} = 42.40$ in^3 (694.8 cm^3).

4. Compute the piston stroke under load

Use the relation $L = 4(v_1 - v_2)/\pi D^2$, where L = length of stroke under load, in; D = piston diameter, in. Substituting yields $L = 4(60 - 42.40)/(\pi \times 2.5^2)$ 3.58 in (9.1 cm). Since this is within the allowable travel of 4 in (10 cm), the system is acceptable.

Related Calculations: Hydropneumatic accumulators have long been used as shock dampers and pulsation attenuators in hydraulic lines. But only recently have they been used as mechanical shock absorbers, or springs.

Current applications include shock absorption and seat-suspension systems for earth-moving and agricultural machinery, resetting mechanisms for plows, mill-roll loading, and rock-crusher loading. Potential applications include hydraulic hammers and shake tables.

In these relatively high-force applications, hydropneumatic springs have several advantages over mechanical springs. First, they are smaller and lighter, which can help reduce system costs. Second, they are not limited by metal fatigue, as mechanical springs are. Of course, their life is not infinite, for it is limited by wear of rod and piston seals.

Finally, hydropneumatic springs offer the inherent ability to control load speeds. With an orifice check valve or flow-control valve between actuator and accumulator, cam speed can be varied as needed.

One reason why these springs are not more widely used is that they are not packaged as off-the-shelf items. In the few cases where packages exist, they are often intended for other uses. Thus, package dimensions may not be those needed for spring applications, and off-the-shelf springs may not have all the special system parameters needed. But it is not hard to select individual off-the-shelf accumulators and actuators for a custom-designed system. The procedure given here is an easy method for calculating needed accumulator pressures and volumes. It is the work of Zeke Zahid, Vice President and General Manager, Greer Olaer Products Division, Greer Hydraulics, Inc., as reported in *Machine Design*.

MEMBRANE VIBRATION

A pressure-measuring device is to be constructed of a 0.005-in (0.0127-cm) thick alloy steel circular membrane stretched over a chamber opening, as shown in Fig. 62. The membrane is subjected to a uniform tension of 2000 lb (8900 N) and then secured in position over a 6-in (15.24-cm) diameter opening. The steel has a modulus of elasticity of 30,000,000 lb/in^2 (210.3 GPa) and weighs 0.3 lb/in^3 (1.1 N/cm^3). Vibration of the membrane due to pressure in the chamber is to be picked up by a strain gage mechanism; in order to calibrate the device, it is required to determine the fundamental mode of vibration of the membrane.

Calculation Procedure:

1. Compute the weight of the membrane per unit area

Weight of the membrane per unit area, $w = w_u \times t$, where the weight per unit volume, w_u = 0.3 lb/in^3 (1.1 N/cm^3); membrane thickness, t = 0.005 in (0.0127 cm). Hence, $w = 0.3 \times 0.005 = 0.0015$ lb/in^2 (0.014 N/cm^2).

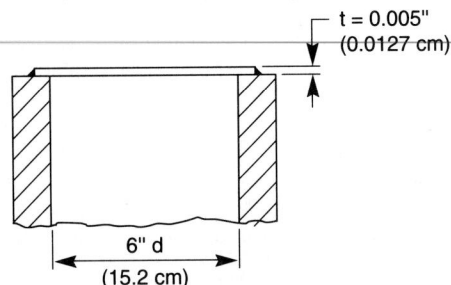

FIG. 62   Membrane for pressure-measuring device.

2.   *Compute the uniform tension per unit length of the membrane boundary*

Uniform tension per unit length of the membrane boundary, $S = F/L$, where the uniformly applied tensile force, $F = 2000$ lb (8900 N); length of the membrane boundary, $L = d = 6$ in (15.24 cm). Thus, $S = 2000/6 = 333$ lb/in (584 N/cm).

3.   *Compute the area of the membrane*

The area of the membrane, $A = \pi d^2/4 = \pi(6)^2/4 = 28.27$ in^2 (182.4 cm^2).

4.   *Compute the frequency of the fundamental mode of vibration in the membrane*

From *Marks' Standard Handbook for Mechanical Engineers*, 9th edition, McGraw-Hill, Inc., the frequency of the fundamental mode of vibration of the membrane, $f = (\alpha/2\pi)[(gS)/(wA)]^{1/2}$, where the membrane shape constant for a circle, $\alpha = 4.261$; gravitational acceleration, $g = 32.17 \times 12 = 386$ in/s^2 (980 cm/s^2); other values as before. Then, $f = (4.261/2\pi)$ [(386 $\times$ 333)/(0.0015 $\times$ 28.27)]$^{1/2} = 1181$ Hz.

Related Calculations: To determine the value for S in step 2 involves a philosophy similar to that for the hoop stress formula for thin-wall cylinders, i.e., the uniform tension per unit length of the membrane boundary depends on tensile forces created by uniformly stretching the membrane in all directions. Therefore, for symmetrical shapes other than a circle, such as those presented in *Marks' M.E. Handbook*, the value for L in the equation for S as given in this procedure is the length of the longest line of symmetry of the geometric shape of the membrane. The shape constant and other variable values change accordingly.

POWER SAVINGS ACHIEVABLE IN INDUSTRIAL HYDRAULIC SYSTEMS

An industrial hydraulic system can be designed with three different types of controls. At a flow rate of 100 gal/min (6.31 L/s), the pressure drop across the controls is as follows: Control A, 500 lb/in^2 (3447 kPa); control B, 1000 lb/in^2 (6894 kPa); control C, 2000 lb/in^2 (13,788 kPa). Determine the power loss and the cost of this loss for each control if the cost of electricity is 15 cents per kilowatthour. How much more can be spent on a control if it operates 3000 h/year?

Calculation Procedure:

1.   *Compute the horsepower lost in each control*

The horsepower lost during pressure drop through a hydraulic control is given by $hp = 5.82(10^{-4})Q \, \Delta P$, where $Q =$ flow rate through the control, gal/min; $\Delta P =$ pressure loss through the control. Substituting for each control and using the letter subscript to identify it, we find $hp_A = 5.82(10^{-4})(100)(500) = 29.1$ hp (21.7 kW); $hp_B = 5.82(10^{-4})(100)(1000) = 5.82$ hp (43.4 kW); $hp_c = 5.82(10^{-4})(100)(2000) = 116.4$ hp (86.8 kW).

2.   *Find the cost of the pressure loss in each control*

The cost in dollars per hour wasted $w = $ kW($/kWh) $ = hp(0.746)($/kWh). Substituting and using a subscript to identify each control, we get $w_A = 21.7(\$0.15) = \$3.26$; $w_B = 43.4(\$0.15) = \$6.51$; $w_C = 86.8(\$0.15) = \$13.02$.

The annual loss for each control with 3000-h operation is $w_{A,an} = 3000(\$3.26) = \$9780$; $w_{B,an} = 3000(\$6.51) = \$19{,}530$; $w_{C,an} = 3000(\$13.02) = \$39{,}060$.

3. Determine the additional amount that can be spent on a control

Take one of the controls as the base or governing control, and use it as the guide to the allowable extra cost. Using control C as the base, we can see that it causes an annual loss of \$39,060. Hence, we could spend up to \$39,060 for a more expensive control which would provide the desired function with a smaller pressure (and hence, money) loss.

The time required to recover the extra money spent for a more efficient control can be computed easily from (\$39,060 − loss with new control, \$), where the losses are expressed in dollars per year.

Thus, if a new control costs \$2500 and control C costs \$1000, while the new control reduced the annual loss to \$20,060, the time to recover the extra cost of the new control would be (\$2500 − \$1000)/(\$39,060 − \$20,060) = 0.08 year, or less than 1 month. This simple application shows the importance of careful selection of energy control devices.

And once the new control is installed, it will save \$39,060 − \$20,060 = \$19,000 per year, assuming its maintenance cost equals that of the control it replaces.

Related Calculations: This approach to hydraulic system savings can be applied to systems serving industrial plants, aircraft, ships, mobile equipment, power plants, and commercial installations. Further, the approach is valid for any type of hydraulic system using oil, water, air, or synthetic materials as the fluid.

With greater emphasis in all industries on energy conservation, more attention is being paid to reducing unnecessary pressure losses in hydraulic systems. Dual-pressure pumps are finding wider use today because they offer an economical way to provide needed pressures at lower cost. Thus, the alternative control considered above might be a dual-pressure pump, instead of a throttling valve.

Other ways that pressure (and energy) losses are reduced is by using accumulators, shutting off the pump between cycles, modular hydraulic valve assemblies, variable-displacement pumps, electronic controls, and shock absorbers. Data in this procedure are from *Product Engineering* magazine, edited by Frank Yeaple.

SIZING DOWEL PINS

A dowel pin shown in Fig. 63a is used to resist a moment created by a force of 110 lb (489 N) acting through a distance of 6 in (15.24 cm) on an outer mating part, the hub, that is tightly fitted on a cylindrical internal part, the shaft, which has a radius of 0.7 in (1.78 cm). Another dowel pin, the loose-fitting clevis pin shown in Fig. 63b, is intended to support a force of 550 lb (2450 N). The pin length subjected to compressive loading is 0.625 in (1.59 cm) and the distance between points of support for bending is 0.9375 in (2.38 cm). The joint is expected to oscillate. Allowable stresses are: 11,000 lb/in^2 (75.84 MPa) shear; 7000 lb/in^2 (48.26 MPa) bending; 2000 lb/in^2 (13.79 MPa) compression. Find the required dowel pin diameters.

Calculation Procedure:

1. Determine the pin diameter for the shear example

Since the mating parts are tightly fitted, check only for shear. Thus, use the relation $d_s = [2PL/(\pi r s_s)]^{1/2}$, where d_s = minimum pin diameter, in (m); P = applied force, lb (N); L = lever arm, in (m); r = shaft radius, in (m); s_s = allowable shear stress, lb/in^2 (Pa). Hence, $d_s = [2 \times 110 \times 6/(\pi \times 0.7 \times 11{,}000)]^{1/2} = 0.234$ in (0.59 cm). The dowel pin diameter should be no larger than $0.3D$, where $D = 2r$, the diameter of the smallest part, the shaft in this case, mating with the dowel pin. If the pin must be larger than $0.3D$, two dowel pins should be used, one on either side of the load. The dowel pin should be located no closer than $1.5D$ from the end of the hub.

2. Calculate the pin diameter for the bending example

The oscillating clevis pin is loosely fitted, hence, it is necessary to check for stresses in shear, s_s; bending, s_b; compression, s_c. The minimum pin diameters required are: to resist shear, d_s

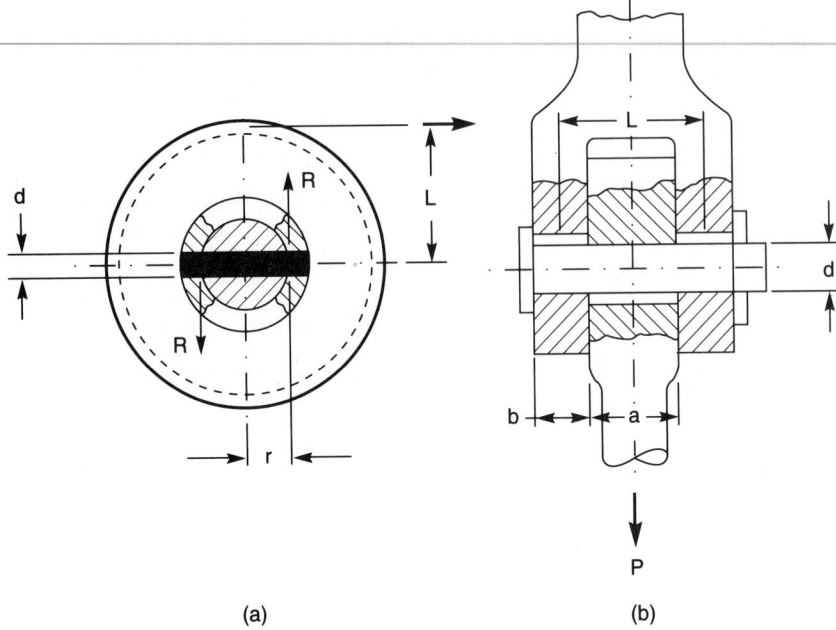

(a)   (b)

FIG. 63 (*a*) Dowel pin shear example. (*b*) Dowel pin bending example. (*Machine Design*).

$= [2P/(\pi \times s_s)]^{1/2} = [2 \times 550/(\pi \times 11,000)]^{1/2} = 0.178$ in (0.45 cm); to resist bending, $d_b = [(P/2)(L/2)/(0.1 \times s_b)]^{1/2} = [(550/2)(0.9375/2)(0.1 \times 7,100)]^{1/3} = 0.566 = (1.44$ cm); to resist compressive loads $d_c = P/(a \times s_c) = 550/(0.625 \times 2,000) = 0.440$ in (1.12 cm). The largest of these pin diameters $d_b = 0.566$ in (1.44 cm) is the pin diameter selected.

Related Calculations: Where the pin is stronger than the mating parts, or where its primary function is alignment or centering, dowel pins can be sized by these rules of thumb: for a pin stressed in shear, pin diameter should be $0.2D$ to $0.3D$. If the pin is stressed longitudinally, as in bending, its diameter should be $0.5D$ when $D \leqq 0.3125$ in (0.79 cm), or $0.4D$ if D is larger.

To locate nests of small parts such as gage plates, pin diameters from 0.125 in (0.32 cm) to 0.1875 in (0.48 cm) are acceptable. For locating dies, pin diameter should never be less than 0.25 in (0.64 cm). In general, pin diameter should be the same as that of the screws used to fasten the work. Within each plate or part to be doweled, the length of the dowel pin should be $1.5D$ to $2D$.

This procedure is based on an article by Federico Strasser, *Machine Design* magazine, November 14, 1983.

Metalworking

REFERENCES: Blazynski—*Metal Forming: Tool Profiles and Flow*, Halsted Press; Lippmann—*Engineering Plasticity: Theory of Metal Forming Processes*, Springer-Verlag; Ross—*Handbook of Metal Treatments and Testing*, Tavistock (England); Le Grand—*American Machinist's Handbook*, McGraw-Hill; Boston—*Metal Processing*, Wiley; Nordhoff—*Machine-Shop Estimating*, McGraw-Hill; *Machinery's Handbook*, Industrial Press; *Welding Handbook*, American Welding Society; ASTME—*Tool Engineer's Handbook*, McGraw-Hill; *Procedure Handbook of Arc Welding Design and Practice*, The Lincoln Electric Company; Black—*Theory of Metal Cutting*, McGraw-Hill; Doyle—*Manufacturing Processes and Materials for Engineers*, Prentice-Hall; Brierly and Siekmann—*Machining Principles and Cost Control*, McGraw-Hill;

Reason—*The Measurement of Surface Texture*, Cleaver-Hume; Bolz—*Production Processes: Their Influence on Design*, Penton; Harris—*A Handbook of Woodcutting*, HMSO, London; *Application Data, Cemented Carbides, Cemented Oxides*, Metallurgical Products Department, General Electric Company; Wood—*Final Report on Advanced Theoretical Formability Manufacturing Technology*, LTV, Inc., and USAF; Maynard—*Handbook of Business Administration*, McGraw-Hill; Niedzwiedzki—*Manual of Machinability and Tool Evaluation*, Huebner Publications, Cleveland; Hendriksen—*Chipbreakers*, The National Machine Tool Builders Association; ASTME—*Fundamentals of Tool Design*, Prentice-Hall; Crane—*Plastic Working of Metals and Power Press Operations*, Wiley; Jones—*Die Design and Die Making Practice*, Industrial Press; DeGarmo—*Materials and Processes in Manufacturing*, Macmillan; Jevons—*The Metallurgy of Deep Drawing and Pressing*, Wiley; Stanley—*Punches and Dies*, McGraw-Hill.

TOTAL ELEMENT TIME AND TOTAL OPERATION TIME

The observed times for a turret-lathe operation are as follows: (1) material to bar stop, 0.0012 h; (2) index turret, 0.0010 h; (3) point material, 0.0005 h; (4) index turret, 0.0012 h; (5) turn 0.300-in (0.8-cm) diameter part, 0.0075 h; (6) clear hexagonal turret, 0.0009 h; (7) advance cross-slide tool, 0.0008 h; (8) cutoff part, 0.0030 h; (9) aside with part, 0.0005 h. What is the total element time? What is the total operation time if 450 parts are processed? Pointing of the material was later found unnecessary. What effect does this have on the element and operation total time?

Calculation Procedure:

1. *Compute the total element time*

Compute the total element time by finding the sum of each of the observed times in the operation, or sum steps 1 through 9: 0.0012 + 0.0010 + 0.0005 + 0.0012 + 0.0075 + 0.0009 + 0.0008 + 0.0030 + 0.0005 = 0.0166 h = 0.0166 (60 min/h) = 0.996 minute per element.

2. *Compute the total operation time*

The total operation time = (element time, h)(number of parts processed). Or, (0.0166)(450) = 7.47 h.

3. *Compute the time savings on deletion of one step*

When one step is deleted, two or more times are usually saved. These times are the machine preparation and machine working times. In this process, they are steps 2 and 3. Subtract the sum of these times from the total element time, or 0.0166 − (0.0010 + 0.0005) = 0.0151 h. Thus, the total element time decreases by 0.0015 h. The total operation time will now be (0.0151)(450) = 6.795 h, or a reduction of (0.0015)(450) = 0.6750 h. Checking shows 7.470 − 6.795 = 0.675 h.

   Related Calculations: Use this procedure for any multiple-step metalworking operation in which one or more parts are processed. These processes may be turning, boring, facing, threading, tapping, drilling, milling, profiling, shaping, grinding, broaching, hobbing, cutting, etc. The time elements used may be from observed or historical data.

   Recent introduction of international quality-control specifications by the International Organization for Standardization (ISO) will require greater accuracy in all manufacturing calculations. The best-known set of specifications at this time is ISO 9000 covering quality standards and management procedures. All engineers and designers everywhere should familiarize themselves with ISO 9000 and related requirements so that their products have the highest quality standards. Only then will their designs survive in the competitive world of international commerce and trading.

CUTTING SPEEDS FOR VARIOUS MATERIALS

What spindle rpm is needed to produce a cutting speed of 150 ft/min (0.8 m/s) on a 2-in (5.1-cm) diameter bar? What is the cutting speed of a tool passing through 2.5-in (6.4-cm) diameter material at 200 r/min? Compare the required rpm of a turret-lathe cutter with the available spindle speeds.

Calculation Procedure:

1. Compute the required spindle rpm

In a rotating tool, the spindle rpm $R = 12C/\pi d$, where C = cutting speed, ft/min; d = work diameter, in. For this machine, $R = 12(150)/\pi(2) = 286$ r/min.

2. Compute the tool cutting speed

For a rotating tool, $C = R\pi d/12$. Thus, for this tool, $C = (200)(\pi)(2.5)/12 = 131$ ft/min (0.7 m/s).

The cutting-speed equation is sometimes simplified to $C = Rd/4$. Using this equation for the above machine, we see $C = 200(2.5)/4 = 125$ ft/min (0.6 m/s). In general, it is wiser to use the exact equation.

3. Compare the required rpm with the available rpm

Consult the machine nameplate, *American Machinist's Handbook,* or a manufacturer's catalog to determine the available spindle rpm for a given machine. Thus, one Warner and Swasey turret lathe has a spindle speed of 282 compared with the 286 r/min required in step 1. The part could be cut at this lower spindle speed, but the time required would be slightly greater because the available spindle speed is $286 - 282 = 4$ r/min less than the computed spindle speed.

When preparing job-time estimates, be certain to use the available spindle speed, because this is frequently less than the computed spindle speed. As a result, the actual cutting time will be longer when the available spindle speed is lower.

Related Calculations: Use this procedure for a cutting tool having a rotating cutter, such as a lathe, boring mill, automatic screw machine, etc. Tables of cutting speeds for various materials (metals, plastics, etc.) are available in the *American Machinist's Handbook,* as are tables of spindle rpm and cutting speed.

DEPTH OF CUT AND CUTTING TIME FOR A KEYWAY

What depth of cut is needed for a ¾-in (1.9-cm) wide keyway in a 3-in (7.6-cm) diameter shaft? The keyway length is 2 in (5.1 cm). How long will it take to mill this keyway with a 24-tooth cutter turning at 130 r/min if the feed is 0.005 per tooth?

Calculation Procedure:

1. Sketch the shaft and keyway

Figure 1 shows the shaft and keyway. Note that the depth of cut D in $= W/2 + A$, where W = keyway width, in; A = distance from the key horizontal centerline to the top of the shaft, in.

2. Compute the distance from the centerline to the shaft top

For a machined keyway, $A = [d - (d^2 - W^2)^{0.5}]/2$, where d = shaft diameter, in. With the given dimensions, $A = [3 - (3^2 - 0.75^2)^{0.5}]/2 = 0.045$ in (1.1 mm).

3. Compute the depth of cut for the keyway

The depth of cut $D = W/2 + A = 0.75/2 + 0.045 = 0.420$ in (1.1 cm).

4. Compute the keyway cutting time

For a single milling cutter, cutting time, min = length of cut, in/[(feed per tooth) × (number of teeth on cutter)(cutter rpm)]. Thus, for this keyway, cutting time $= 2.0/[(0.005)(24)(130)] = 0.128$ min.

Related Calculations: Use this procedure for square or rectangular keyways. For Woodruff key-seat milling, use the same cutting-time equation as in step 4. A Woodruff key seat is almost a semicircle, being one-half the width of the key *less than* a semicircle. Thus, a ⁹⁄₁₆-in (1.4-cm) deep Woodruff key seat containing a ⅜-in (1.0-cm) wide key will be (⅜)/2 = ³⁄₁₆ in (0.5 cm) less than a semicircle. The key seat would be cut with a cutter having a radius of ⁹⁄₁₆ + ³⁄₁₆ = ¹²⁄₁₆, or ¾ in (1.9 cm).

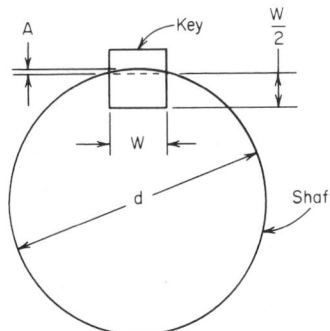

FIG. 1 Keyway dimensions.

MILLING-MACHINE TABLE FEED AND CUTTER APPROACH

A 12-tooth milling cutter turns at 400 r/min and has a feed of 0.006 per tooth per revolution. What table feed is needed? If this cutter is 8 in (20.3 cm) in diameter and is facing a 2-in (5.1-cm) wide part, determine the cutter approach.

Calculation Procedure:

1. *Compute the required table feed*

For a milling machine, the table feed F_T in/min $= f_t n R$, where $f_t =$ feed per tooth per revolution; $n =$ number of teeth in cutter; $R =$ cutter rpm. For this cutter, $F_T = (0.006) \times (12)(400) = 28.8$ in/min (1.2 cm/s).

2. *Compute the cutter approach*

The approach of a milling cutter A_c in $= 0.5D_c - 0.5(D_c^2 - w^2)^{0.5}$, where $D_c =$ cutter diameter, in; $w =$ width of face of cut, in. For this cutter, $A_c = 0.5(8) - 0.5(8^2 - 2^2)^{0.5} = 0.53$ in (1.3 cm).

  Related Calculations: Use this procedure for any milling cutter whose dimensions and speed are known. These cutters can be used for metals, plastics, and other nonmetallic materials.

DIMENSIONS OF TAPERS AND DOVETAILS

What are the taper per foot (TPF) and taper per inch (TPI) of an 18-in (45.7-cm) long part having a large diameter d_l of 3 in (7.6 cm) and a small diameter d_s of 1.5 in (3.8 cm)? What is the length of a part with the same large and small diameters as the above part if the TPF is 3 in/ft (25 cm/m)? Determine the dimensions of the dovetail in Fig. 2 if $B = 2.15$ in (5.15 cm), $C = 0.60$ in (1.5 cm), and $a = 30°$. A ⅜-in (1.0-cm) diameter plug is used to measure the dovetail.

Calculation Procedure:

1. *Compute the taper of the part*

For a round part TPF in/ft $= 12(d_l - d_s)/L$, where $L =$ length of part, in; other symbols as defined above. Thus for this part, TPF $= 12(3.0 - 1.5)/18 = 1$ in/ft (8.3 cm/m). And TPI in/in $= (d_l - d_s)/L_2$, or $(3.0 - 1.5)/18 = 0.0833$ in/in (0.0833 cm/cm).

  The taper of round parts may also be expressed as the angle measured from the shaft centerline, that is, one-half the included angle between the tapered surfaces of the shaft.

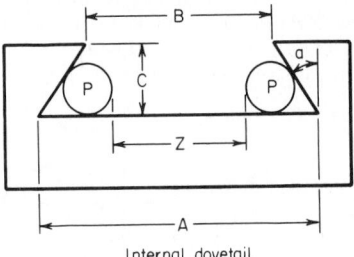

Internal dovetail

FIG. 2 Dovetail dimensions.

2. *Compute the length of the tapered part*

Converting the first equation of step 1 gives $L = 12(d_l - d_s)/\text{TPF}$. Or, $L = 12(3.0 - 1.5)/3.0 = 6$ in (15.2 cm).

3. *Compute the dimensions of the dovetail*

For external and internal dovetails, Fig. 2, with all dimensions except the angles in inches, $A = B + CF = I + HF$; $B = A - CF = G - HF$; $E = P \cot (90 + a/2) + P$; $D = P \cot (90 - a/2) + P$; $F = 2 \tan a$; $Z = A - D$. Note that P = diameter of plug used to measure the dovetail, in.

With the given dimensions, $A = B + CF$, or $A = 2.15 + (0.60)(2 \times 0.577) = 2.84$ in (7.2 cm). Since the plug P is ⅜ in (1.0 cm) in diameter, $D = P \cot (90 - a/2) + P = 0.375 \cot (90 - \tfrac{30}{4}) + 0.375 = 1.025$ in (2.6 cm). Then $Z = A - D = 2.840 - 1.025 = 1.815$ in (4.6 cm). Also $E = P \cot (90 + a/2) + P = 0.375 \cot (90 + \tfrac{30}{4}) + 0.375 = 0.591$ in (1.5 cm).

With flat-cornered dovetails, as at I and G, and $H = \tfrac{1}{8}$ in (0.3 cm), $A = I + HF$. Solving for I, we get $I = A - HF = 2.84 - (0.125)(2 \times 0.577) = 2.696$ in (6.8 cm). Then $G = B + HF = 2.15 + (0.125)(2 \times 0.577) = 2.294$ in (5.8 cm).

Related Calculations: Use this procedure for tapers and dovetails in any metallic and nonmetallic material. When a large number of tapers and dovetails must be computed, use the appropriate tables in the *American Machinist's Handbook.*

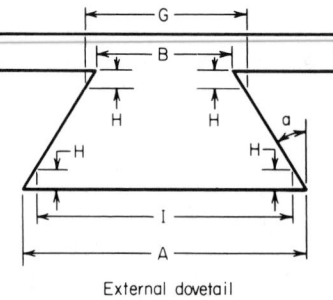

FIG. 2   *(Continued)*

External dovetail

ANGLE AND LENGTH OF CUT FROM GIVEN DIMENSIONS

At what angle must a cutting tool be set to cut the part in Fig. 3? How long is the cut in this part?

Calculation Procedure:

1. *Compute the angle of the cut*

Use trigonometry to compute the angle of the cut. Thus, $\tan a$ = opposite side/adjacent side = $(8 - 5)/6 = 0.5$. From a table of trigonometric functions, a = cutting angle = 26° 34′, closely.

2. *Compute the length of the cut*

Use trigonometry to compute the length of cut. Thus, $\sin a$ = opposite side/hypotenuse, or 0.4472 = $(8 - 5)/\text{hypotenuse}$; length of cut = length of hypotenuse = $3/0.4472 = 6.7$ in (17.0 cm).

Related Calculations: Use this general procedure to compute the angle and length of cut for any metallic or nonmetallic part.

TOOL FEED RATE AND CUTTING TIME

A part 3.0 in (7.6 cm) long is turned at 100 r/min. What is the feed rate if the cutting time is 1.5 min? How long will it take to cut a 7.0-in (17.8-cm) long part turning at 350 r/min if the feed is 0.020 in/r (0.51 mm/r)? How long will it take to drill a 5-in (12.7-cm) deep hole with a drill speed of 1000 r/min and a feed of 0.0025 in/r (0.06 mm/r)?

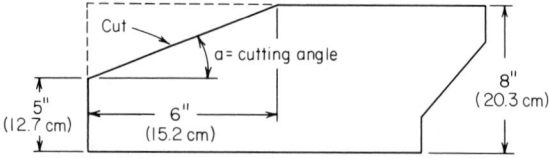

FIG. 3   Length of cut of a part.

Calculation Procedure:

1. Compute the tool feed rate

For a tool cutting a rotating part, $f = L/(Rt)$, where t = cutting time, min. For this part, $f = 3.0/[(100)(1.5)] = 0.02$ in/r (0.51 mm/r).

2. Compute the cutting time for the part

Transpose the equation in step 1 to yield $t = L/(Rf)$, or $t = 7.0/[(350)(0.020)] = 1.0$ min.

3. Compute the drilling time for the part

Drilling time is computed using the equation of step 2, or $t = 5.0/[(1000)(0.0025)] = 2.0$ min.

    Related Calculations: Use this procedure to compute the tool feed, cutting time, and drilling time in any metallic or nonmetallic material. Where many computations must be made, use the feed-rate and cutting-time tables in the *American Machinist's Handbook*.

TRUE UNIT TIME, MINIMUM LOT SIZE, AND TOOL-CHANGE TIME

What is the machine unit time to work 25 parts if the setup time is 75 min and the unit standard time is 5.0 min? If one machine tool has a setup standard time of 9 min and a unit standard time of 5.0 min, how many pieces must be handled if a machine with a setup standard of 60 min and a unit standard time of 2.0 min is to be more economical? Determine the minimum lot size for an operation requiring 3 h to set up if the unit standard time is 2.0 min and the maximum increase in the unit standard may not exceed 15 percent. Find the unit time to change a lathe cutting tool if the operator takes 5 min to change the tool and the tool cuts 1.0 min/cycle and has a life of 3 h.

Calculation Procedure:

1. Compute the true unit time

The true unit time for a machine $T_u = S_u/N + U_s$, where S_u = setup time, min; N = number of pieces in lot; U_s = unit standard time, min. For this machine, $T_u = 75/75 + 5.0 = 6.0$ min.

2. Determine the most economical machine

Call one machine X, the other Y. Then (unit standard time of X, min)(number of pieces) + (setup time of X, min) = (unit standard time of Y, min)(number of pieces) + (setup time of Y, min). For these two machines, since the number of pieces Z is unknown, $5.0Z + 9 = 2.0Z + 60$. So $Z = 17$ pieces. Thus, machine Y will be more economical when 17 or more pieces are made.

3. Compute the minimum lot size

The minimum lot size $M = S_u/(U_s K)$, where K = allowable increase in unit-standard time, percent. For this run, $M = (3 \times 60)/[(2.0)(0.15)] = 600$ pieces.

4. Compute the unit tool-changing time

The unit tool-changing time U_t to change from dull to sharp tools is $U_t = T_c C_t/l$, where T_c = total time to change tool, min; C_t = time tool is in use during cutting cycle, min; l = life of tool, min. For this lathe, $U_t = (5)(1)/[(3)(60)] = 0.0278$ min.

    Related Calculations: Use these general procedures to find true unit time, the most economical machine, minimum lot size, and unit tool-changing time for any type of machine tool—drill, lathe, milling machine, hobs, shapers, thread chasers, etc.

TIME REQUIRED FOR TURNING OPERATIONS

Determine the time to turn a 3-in (7.6-cm) diameter brass bar down to a 2½-in (6.4-cm) diameter with a spindle speed of 200 r/min and a feed of 0.020 in (0.51 mm) per revolution if the length of cut is 4 in (10.2 cm). Show how the turning-time relation can be used for relief turning, pointing of bars, internal and external chamfering, hollow mill work, knurling, and forming operations.

Calculation Procedure:

1. Compute the turning time

For a turning operation, the time to turn T_t min $= L/(fR)$, where L = length of cut, in; f = feed, in/r; R = work rpm. For this part, T_t = 4/[(0.02)(200)] = 1.00 min.

2. Develop the turning relation for other operations

For *relief turning* use the same relation as in step 1. Length of cut is the length of the relief, Fig. 4. A small amount of time is also required to hand-feed the tool to the minor diameter of the relief. This time is best obtained by observation of the operation.

The time required to *point a bar*, called *pointing*, is computed by using the relation in step 1. The length of cut is the distance from the end of the bar to the end of the tapered point, measured parallel to the axis of the bar, Fig. 4.

Use the relation in step 1 to compute the time to cut an internal or external chamfer. The length of cut of a chamfer is the horizontal distance L, Fig. 4.

A hollow mill reduces the external diameter of a part. The cutting time is computed by using the relation in step 1. The length of cut is shown in Fig. 4.

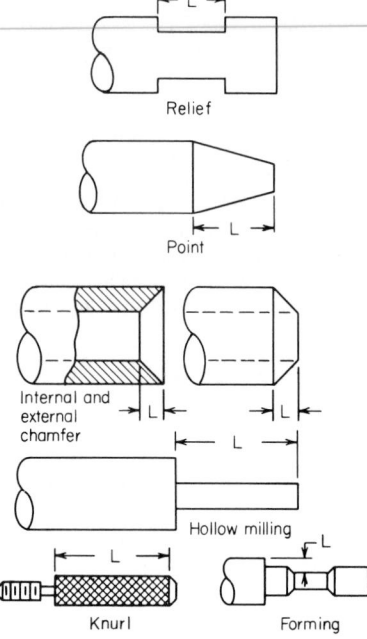

FIG. 4   Turning operations.

Compute the time to knurl, using the relation in step 1. The length of cut is shown in Fig. 4.
Compute the time for forming, using the relation in step 1. Length of cut is shown in Fig. 4.

TIME AND POWER TO DRILL, BORE, COUNTERSINK, AND REAM

Determine the time and power required to drill a 3-in (7.6-cm) deep hole in an aluminum casting if a ¾-in (1.9-cm) diameter drill turning at 1000 r/min is used and the feed is 0.030 in (0.8 mm) per revolution. Show how the drilling relations can be used for boring, countersinking, and reaming. How long will it take to drill a hole through a 6-in (15.2-cm) thick piece of steel if the cone height of the drill is 0.5 in (1.3 cm), the feed is 0.002 in/r (0.05 mm/r), and the drill speed is 100 r/min?

Calculation Procedure:

1. Compute the time required for drilling

The time required to drill T_d min = L/fR, where L = depth of hole = length of cut, in. In most drilling calculations, the height of the drill cone (point) is ignored. (Where the cone height is used, follow the procedure in step 4.) For this hole, T_d = 3/[(0.030) × (1000)] = 0.10 min.

2. Compute the power required to drill the hole

The power required to drill, in hp, is $hp = 1.3LfCK$, where C = cutting speed, ft/min, sometimes termed surface feet per minute $sfpm = \pi DR/12$; K = power constant from Table 1. For an aluminum casting, K = 3. Then hp = (1.3)(3)(0.030)(π × 0.75 × 1000/12)(3) = 66.0 hp (49.2 kW). The factor 1.3 is used to account for dull tools and for overcoming friction in the machine.

TABLE 1   Power Constants for Machining

Material	Power constant
Carbon steel C1010 to C1025	6
Manganese steel T1330 to T1350	9
Nickel steel 2015 to 2320	7
Molybdenum	9
Chromium	10
Stainless steels	11
Cast iron:	
Soft	3
Medium	3
Hard	4
Aluminum alloys:	
Castings	3
Bar	4
Copper	4
Brass (except manganese)	4
Monel metal	10
Magnesium alloys	3
Malleable iron:	
Soft	3
Medium	4
Hard	5

3.   *Adapt the drill relations to other operations*

The time and power required for boring are found from the two relations given above. The length of the cut = length of the bore. Also use these relations for undercutting, sometimes called *internal relieving* and for counterboring. These same relations are also valid for countersinking, center drilling, start or spot drilling, and reaming. In reaming, the length of cut is the total depth of the hole reamed.

4.   *Compute the time for drilling a deep hole*

With parts having a depth of 6 in (15.2 cm) or more, compute the drilling time from $T_d = (L + h)/(fR)$, where h = cone height, in. For this hole, $T_d = (6 + 0.5)/[(0.002)(100)] = 32.25$ min. This compares with $T_d = L/fR = 6/[(0.002)(100)] = 30$ min when the height of the drill cone is ignored.

TIME REQUIRED FOR FACING OPERATIONS

How long will it take to face a part on a lathe if the length of cut is 4 in (10.2 cm), the feed is 0.020 in/r (0.51 mm/r) and the spindle speed is 50 r/min? Determine the facing time if the same part is faced by an eight-tooth milling cutter turning at 1000 r/min and having a feed of 0.005 in (0.13 mm) per tooth per revolution. What table feed is required if the cutter is turning at 50 r/min? What is the feed per tooth with a table feed of 4.0 in/min (1.7 mm/s)? What added table travel is needed when a 4-in (10.2-cm) diameter cutter is cutting a 4-in (10.2-cm) wide piece of work?

Calculation Procedure:

1.   *Compute the lathe facing time*

For lathe facing, the time to face T_f min $= L/(fR)$, where the symbols are the same as given for previous calculation procedures in this section. For this part, $T_f = 4/[(0.02)(50)] = 4.0$ min.

2.   Compute the facing time using a milling cutter

With a milling cutter, $T_f = L/(f_t nR)$, where f_t = feed per tooth, in/r; n = number of teeth on cutter; other symbols as before. For this part, $T_f = 4/[(0.005)(8) \times (1000)] = 0.10$ min.

3.   Compute the required table feed

In a milling machine, the table feed F_t in/min = $f_t nR$. For this machine, $F_t = (0.005) \times (8)(50) = 2.0$ in/min (0.85 mm/s).

4.   Compute the feed per tooth

For a milling machine, the feed per tooth, in/r, $f_t = F_t/Rn$. In this machine, $f_t = 4.0/[(50)(8)] = 0.01$ in/r (0.25 mm/r).

5.   Compute the added table travel

In face milling, the added table travel A_t in = $0.5[D_c - (D_c^2 - W^2)^{0.5}]$, where the symbols are the same as given earlier. For this cutter and work, $A_t = 0.5[4 - (4^2 - 4^2)^{0.5}] = 2.0$ in (5.1 cm).

THREADING AND TAPPING TIME

How long will it take to cut a 4-in (10.2-cm) long thread at 100 r/min if the rod will have 12 threads per inch and a button die is used? The die is backed off at 200 r/min. What would the threading time be if a self-opening die were used instead of a button die? What will the threading time be for a single-pointed threading tool if the part being threaded is aluminum and the back-off speed is twice the threading speed? The rod is 1 in (2.5 cm) in diameter. How long will it take to tap a 2-in (5.1-cm) deep hole with a 1-14 solid tap turning at 100 r/min? How long will it take to mill-thread a 1-in (2.5-cm) diameter bolt having 15 threads per inch 3 in (7.6 cm) long if a 4-in (10.2-cm) diameter 20-flute thread-milling hob turning at 80 r/min with a 0.003 in (0.08-mm) feed is used?

Calculation Procedure:

1.   Compute the button-die threading time

For a multiple-pointed tool, the time to thread $T_t = Ln_t/R$, where L = length of cut = length of thread measured parallel to thread longitudinal axis, in; n_t = number of threads per inch. For this button die, $T_t = (4)(12)/100 = 0.48$ min. This is the time required to cut the thread.

Compute the back-off time B min from $B = Ln_t/R_B$, where R_B = back-off rpm, or $B = (4)(12)/200 = 0.24$ min. Hence, the total time to cut and back off = $T_t + B = 0.48 + 0.24 = 0.72$ min.

2.   Compute the self-opening die threading time

With a self-opening die, the die opens automatically when it reaches the end of the cut thread and is withdrawn instantly. Therefore, the back-off time is negligible. Hence the time to thread = $T_t = Ln_t/R = (4)(12)/100 = 0.48$ min. One cut is usually sufficient to make a suitable thread.

3.   Compute the single-pointed tool cutting time

With a single-pointed tool, more than one cut is usually necessary. Table 2 lists the number of cuts needed with a single-pointed tool working on various materials. The maximum cutting speed for threading and tapping is also listed.

Table 2 shows that four cuts are needed for an aluminum rod when a single-pointed tool is used. Before computing the cutting time, compute the cutting speed to determine whether it is within the recommended range given in Table 2. From a previous calculation procedure, $C = R\pi d/12$, or $C = (100)(\pi)(1.0)/12 = 26.2$ ft/min (13.3 cm/s). Since this is less than the maximum recommended speed of 30 r/min, Table 2, the work speed is acceptable.

Compute the time to thread from $T_t = Ln_t c/R$, where c = number of cuts to thread, from Table 2. For this part, $T_t = (4)(12)(4)/100 = 1.92$ min.

If the tool is backed off at twice the threading speed, and the back-off time $B = Ln_t c/R_B$, $B = (4)(12)(4)/200 = 0.96$ min. Hence, the total time to thread and back off = $T_t + B = 1.92 + 0.96 = 2.88$ min. In some shapes, a single-pointed tool may not be backed off; the tool may instead be repositioned. The time required for this approximates the back-off time.

TABLE 2 Number of Cuts and Cutting Speed for Dies and Taps

	No. of cuts°	Cutting speed†	
		ft/min	m/s
Aluminum	4	30	0.15
Brass (commercial)	3	30	0.15
Brass (naval)	4	30	0.15
Bronze (ordinary)	5	30	0.15
Bronze (hard)	7	20	0.10
Copper	5	20	0.10
Drill rod	8	10	0.05
Magnesium	4	30	0.15
Monel (bar)	8	10	0.05
Steel (mild)	5	20	0.10
Steel (medium)	7	10	0.05
Steel (hard)	8	10	0.05
Steel (stainless)	8	10	0.05

°Single-pointed threading tool; maximum spindle speed, 250 r/min.
†Maximum recommended speed for single- and multiple-pointed tools; maximum spindle speed for multiple-pointed tools = 150 r/min for dies and taps.

4. *Compute the tapping time*

The time to tap T_t min $= Ln_t/R$. With a solid tap, the tool is backed out at twice the tapping speed. With a collapsing tap, the tap is withdrawn almost instantly without reversing the machine or tap.

For this hole, $T_t = (2)(14)/100 = 0.28$ min. The back-off time $B = Ln_t/R_B = (2)(14)/200 = 0.14$ min. Hence, the total time to tap and back off $= T_t + B = 0.28 + 0.14 = 0.42$ min.

The maximum spindle speed for tapping should not exceed 250 r/min. Use the cutting-speed values given in Table 2 in computing the desirable speed for various materials.

5. *Compute the thread-milling time*

The time for thread milling is $T_t = L/(fnR)$, where L = length of cut, in = circumference of work, in; f = feed per flute, in; n = number of flutes on hob; R = hob rpm. For this bolt, $T_t = 3.1416/[(0.003)(20)(80)] = 0.655$ min.

Note that neither the length of the threaded portion nor the number of threads per inch enters into the calculation. The thread hob covers the entire length of the threaded portion and completes the threading in one revolution of the work head.

TURRET-LATHE POWER INPUT

How much power is required to drive a turret lathe making a ½-in (1.3-cm) deep cut in cast iron if the feed is 0.015 in/r (0.38 mm/r), the part is 2.0 in (5.1 cm) in diameter, and its speed is 382 r/min? How many 1.5-in (3.8-cm) long parts can be cut from a 10-ft (3.0-m) long bar if a ¼-in (6.4-mm) cutoff tool is used? Allow for end squaring.

Calculation Procedure:

1. *Compute the surface speed of the part*

The cutting, or surface, speed, as given in a previous calculation procedure, is $C = R\pi d/12$, or $C = (382)(\pi)(2.0)/12 = 200$ ft/min (1.0 m/s).

2. *Compute the power input required*

For a turret lathe, the hp input hp = 1.33$DfCK$, where D = cut depth, in; f = feed, in/r; K = material constant from Table 3. For cast iron, K = 3.0. Then hp = (1.33)(0.5)(0.015)(200)(3.0) = 5.98, say 6.0 hp (4.5 kW).

3. *Compute the number of parts that can be cut*

Allow 2 in (5.1 cm) on the bar end for checking and ½ in (1.3 cm) on the opposite end for squaring. With an original length of 10 ft = 120 in (304.8 cm), this leaves 120 − 2.5 = 117.5 in (298.5 cm) for cutting.

Each part cut will be 1.5 in (3.8 cm) long + 0.25 in (6.4 mm) for the cutoff, or 1.75 in (4.4 cm) of stock. Hence, the number of pieces which can be cut = 117.5/1.75 = 67.1, or 67 pieces.

Related Calculations: Use this procedure to find the turret-lathe power input for any of the materials, and similar materials, listed in Table 3. The parts cutoff computation can be used for any material—metallic or nonmetallic. Be sure to allow for the width of the cutoff tool.

TABLE 3   Turret-Lathe Power Constant

Material	Constant K
Bronze	3
Cast iron	3
SAE steels:	
1020	6
1045	8
3250	9
4150	9
4615	6
X1315	6
Straight tubing	6
Steel castings and forgings	9
Heat-treated steels:	
4150	10
52100	10

TIME TO CUT A THREAD ON AN ENGINE LATHE

How long will it take an engine lathe to cut an acme thread having a length of 5 in (12.7 cm), a major diameter of 2 in (5.1 cm), four threads per inch (1.575 threads per centimeter), a depth of 0.1350 in (3.4 mm), a cutting speed of 70 ft/min (0.4 m/s), and a depth of cut of 0.005 in (0.1 mm) per pass if the material cut is medium steel? How many passes of the tool are required?

Calculation Procedure:

1. *Compute the cutting time*

For an acme, square, or worm thread cut on an engine lathe, the total cutting time T_t min, excluding the tool positioning time, is found from $T_t = Ld_tDn_t/(4Cd_c)$, where L = thread length, in, measured parallel to the thread longitudinal axis; d_t = thread major diameter, in; D = depth of thread, in; n_t = number of threads per inch; C = cutting speed, ft/min; d_c = depth of cut per pass, in.

For this acme thread, T_t = (5)(2)(0.1350)(4)/[4(70)(0.005)] = 3.85 min. To this must be added the time required to position the tool for each pass. This equation is also valid for SI units.

2. *Compute the number of tool passes required*

The depth of cut per pass is 0.005 in (0.1 mm). A total depth of 0.1350 in (3.4 mm) must be cut. Therefore, the number of passes required = total depth cut, in/depth of cut per pass, in = 0.1350/0.005 = 27 passes.

Related Calculations: Use this procedure for threads cut in ferrous and nonferrous metals. Table 4 shows typical cutting speeds.

TIME TO TAP WITH A DRILLING MACHINE

How long will it take to tap a 4-in (10.2-cm) deep hole with a 1½-in (3.8-cm) diameter tap having six threads per inch (2.36 threads per centimeter) if the tap turns at 75 r/min?

TABLE 4   Thread Cutting Speeds

Material	Cutting speed	
	ft/min	m/s
Soft nonferrous metals	250	1.25
Mild steel	100	0.50
Medium steel	75	0.38
Hard steel	50	0.25

Calculation Procedure:

1. Compute the tap surface speed

By the method of a previous calculation procedure, $C = R\pi d/12 = (75)(\pi)(1.5)/12 = 29.5$ ft/min (0.15 m/s).

2. Compute the time to tap the hole and withdraw the tool

For tapping with a drilling machine, $T_t = Dn_tD_c\pi/(8\ C)$, where D = depth of cut = depth of hole tapped, in; n_t = number of threads per inch; D_c = cutter diameter, in = tap diameter, in. For this hole, $T_t = (4)(6)(1.5)\pi/[8(29.5)] = 0.48$ min, which is the time required to tap and withdraw the tool.

Related Calculations: Use this procedure for tapping ferrous and nonferrous metals on a drill press. The recommended tap surface speed for various metals is: aluminum, soft brass, ordinary bronze, soft cast iron, and magnesium: 30 ft/min (0.15 m/s); naval brass, hard bronze, medium cast iron, copper and mild steel: 20 ft/min (0.10 m/s); hard cast iron, medium steels, and hard stainless steel: 10 ft/min (0.05 m/s).

MILLING CUTTING SPEED, TIME, FEED, TEETH NUMBER, AND HORSEPOWER

What is the cutting speed of a 12-in (30.5-cm) diameter milling cutter turning at 190 r/min? How many teeth are needed in the cutter at this speed if the feed is 0.010 in (0.3 mm) per tooth, the depth of cut is 0.075 in; (1.9 mm), the length of cut is 5 in (12.7 cm), the power available at the cutter is 14 hp (10.4 kW), and the mill is cutting hard malleable iron? How long will it take the mill to make this cut? What is the maximum feed rate that can be used? What is the power input to the cutter if a 20-hp (14.9-kW) machine is used?

Calculation Procedure:

1. Compute the cutter cutting speed

For a milling cutter, use the simplified relation $C = Rd/4$, where the symbols are as given earlier in this section. Or, $C = (190)(12)/4 = 570$ ft/min (2.9 m/s).

2. Compute the number of cutter teeth required

For a carbide cutter, $n = K_mhp_c/(Df_tLR)$, where n = number of teeth on cutter; K_m = machinability constant or K factor from Table 5; hp_c = horsepower available at the milling cutter; D = depth of cut, in, f_t = cutter feed, inches per tooth; L = length of cut, in; R = cutter rpm.

Table 5 shows that $K_m = 0.90$ for malleable iron. Then $n = (0.90)(14)(0.075)(0.01) \times (5)(190)] = 17.68$, say 18 teeth. For general-purpose use, the Metal Cutting Institute recommends that $n = 1.5$(cutter diameter, in) for cutters having a diameter of more than 3 in (7.6 cm). For this cutter, $n = 1.5(12) = 18$ teeth. This agrees with the number of teeth computed with the cutter equation.

3. Compute the milling time

For a milling machine, the time to cut T_t min $= L/(f_tnR)$, where L = length of cut, in; f_t = feed per tooth, inches per tooth per revolution; n = number of teeth on the cutter; R = cutter rpm. Thus, the time to cut is $T_t = 5/(0.01)(18)(190) = 0.146$ min.

4. Compute the maximum feed rate

For a milling machine, the maximum feed rate f_m in/min $= K_mhp_c/(DL)$, where L = length of cut; other symbols are the same as in step 2. Thus, $f_m = (0.90)(14)/[(0.075)(5)] = (33.6)$ in/min (1.4 cm/s).

5. Compute the power input to the machine

The power available at the cutter is 14 hp (10.4 kW). The power required $hp_c = DLnRf_t/K_m$, where all symbols are as given above. Thus, $hp_c = (0.075)(5)(18)(190)(0.01)/0.90 = 14.25$ hp

TABLE 5   Machinability Constant K_m

Aluminum	2.28
Brass, soft	2.00
Bronze, hard	1.40
Bronze, very hard	0.65
Cast iron, soft	1.35
Cast iron, hard	0.85
Cast iron, chilled	0.65
Cast magnesium	2.50
Malleable iron	0.90
Steel, soft	0.85
Steel, medium	0.65
Steel, hard	0.48
Steel:	
100 Brinell	0.80
150 Brinell	0.70
200 Brinell	0.65
250 Brinell	0.60
300 Brinell	0.55
400 Brinell	0.50

TABLE 6   Typical Milling-Machine Efficiencies

Rated power of machine		Overall efficiency, percent
hp	kW	
3	2.2	40
5	3.7	48
7.5	5.6	52
10	7.5	52
15	11.2	52
20	14.9	60
25	18.6	65
30	22.4	70
40	29.8	75
50	27.3	80

(10.6 kW). This is slightly more than the available horsepower.

Milling machines have overall efficiencies ranging from a low of 40 percent to a high of 80 percent, Table 6. Assume a machine efficiency of 65 percent. Then the required power input is $14.25/0.65 = 21.9$ hp (16.3 kW). Therefore, a 20- or 25-hp (14.9- or 18.6-kW) machine will be satisfactory, depending on its actual operating efficiency.

Related Calculations: After selecting a feed rate, check it against the suggested feed per tooth for milling various materials given in the *American Machinist's Handbook*. Use the method of a previous calculation procedure in this section to determine the cutter approach. With the approach known, the maximum chip thickness, in = (cutter approach, in)(table advance per tooth, in)/(cutter radius, in). Also, the feed per tooth, in = (feed rate, in/min)/[(cutter rpm)(number of teeth on cutter)].

GANG-, MULTIPLE-, AND FORM-MILLING CUTTING TIME

How long will it take to gang mill a part if three cutters are used with a spindle speed of 70 r/min and there are 12 teeth on the smallest cutter, a feed of 0.015 in/r (0.4 mm/r) and a length of cut of 8 in (20.3 cm)? What will be the unit time to multiple mill four keyways if each of the four cutters has 20 teeth, the feed is 0.008 in (0.2 mm) per tooth, spindle speed is 150 r/min, and the keyway length is 3 in (7.6 cm)? Show how the cutting time for form milling is computed, and how the cutter diameter for straddle milling is computed.

Calculation Procedure

1. Compute the gang-milling cutting time

For any gang-milling operation, from the dimensions of the smallest cutter, the time to cut $T_t = L/f_t nR$, where L = length of cut, in; f_t = feed per tooth, in/r; n = number of teeth on cutter; R = spindle rpm. For this part, $T_t = 8/[(0.015)(12)(70)] = 0.635$ min.

Note that in all gang-milling cutting-time calculations, the number of teeth and feed of the *smallest* cutter are used.

2. Compute the multiple-milling cutting time

In multiple milling, the cutting time $T_t = L/(f_t nR_m)$, where n = number of milling cutter used. In multiple milling, the cutting time is termed the unit time. For this machine, $T_t = 3/[(0.008)(20)(150)(4)] = 0.0303$ min.

3. *Show how form milling time is computed*

Form-milling cutters are used on surfaces that are neither flat nor square. The cutters used for form milling resemble other milling cutters. The cutting time is therefore computed from $T_t = L/(f_t nR)$, where all symbols are the same in step 1.

4. *Show how the cutter diameter is computed for straddle milling*

In straddle milling, the cutter diameter must be large enough to permit the work to pass under the cutter arbor. The minimum-diameter cutter to straddle mill a part = (diameter of arbor, in) + 2 (face of cut, in + 0.25). The 0.25 in (6.4 mm) is the allowance for clearance of the arbor.

   Related Calculations: Use the equation of step 1 to compute the cutting time for metal slitting, screw slotting, angle milling, T-slot milling, Woodruff key-seat milling, and profiling and routing of parts. In T-slot milling, two steps are required—milling of the vertical member and milling of the horizontal member. Compute the milling time of each; the sum of the two is the total milling time.

SHAPER AND PLANER CUTTING SPEED, STROKES, CYCLE TIME, POWER

What is the cutting speed of a shaper making 54-strokes/min if the stroke length is 6 in (15.2 cm)? How many strokes per minute should the ram of a shaper make if it is shaping a 12-in (30.5-cm) long aluminum bar at a cutting speed of 200 ft/min? How long will it take to make a cut across a 12-in (30.5-cm) face of a cast-iron plate if the feed is 0.050 in (1.3 mm) per stroke and the ram makes 50 strokes/min? What is the cycle time of a planer if its return speed is 200 ft/min (1.0 m/s), the acceleration-deceleration constant is 0.05, and the cutting speed is 100 ft/min (0.5 m/s)? What is the planer power input if the depth of cut is ⅛ in (3.2 mm) and the feed is ¹⁄₁₆ in (1.6 mm) per stroke?

Calculation Procedure:

1. *Compute the shaper cutting speed*

For a shaper, the cutting speed, ft/min, is $C = SL/6$, where S = strokes/min; L = length of stroke, in; where the cutting-stroke time = return-stroke time. Thus, for this shaper, $C = (54)(6)/6 = 54$ ft/min (0.3 m/s).

2. *Compute the shaper stroke rate*

Transpose the equation of step 1 to $S = 6C/L$. Then $S = 6(200)/12 = 100$ strokes/min.

3. *Compute the shaper cutting time*

For a shaper the cutting time, min, is $T_t = L/(fS)$, where L = length of cut, in; f = feed, in/stroke; S = strokes/min. Thus, for this shaper, $T_t = 12/[(0.05)(50)] = 4.8$ min. Multiply T_t by the number of strokes needed; the result is the total cutting time, min.

TABLE 7 Power Factors for Planers°

Depth of cut		Feed		
in	cm	¹⁄₃₂ in (0.8 mm) per stroke	¹⁄₁₆ in (1.6 mm) per stroke	⅛ in (3.2 mm) per stroke
⅛	0.3	0.0115	0.0235	0.047
¼	0.6	0.023	0.047	0.094
⅜	1.0	0.035	0.070	0.141
½	1.3	0.047	0.094	0.189
⅝	1.6	0.063	0.118	0.236
¾	1.9	0.080	0.142	0.284
⅞	2.2	0.087	0.165	0.331
1	2.5	0.094	0.189	0.378

°Excerpted from the Cincinnati Planer Company and *American Machinist's Handbook.*

4. Compute the planer cycle time

The cycle time for a planer, min, $= (L/C) + (L/R_c) + k$, where R_c = cutter return speed, ft/min; k = acceleration-deceleration constant. Since the cutting speed is 100 ft/min (0.5 m/s) and the return speed is 200 ft/min (1.0 m/s), the cycle time $= (12/100) + (12/200) + 0.05 = 0.23$ min.

5. Compute the power input to the planer

Table 7 lists typical power factors for planers planing cast iron and steel. To find the power required, multiply the power factor by the cutter speed, ft/min. For the planer in step 3 with a cutting speed of 100 ft/min (0.5 m/s) and a power factor of 0.0235 for a ⅛-in (3.2-mm) deep cut and a 1⁄16-in (1.6-mm) feed, $hp_{input} = (0.0235)(100) = 2.35$ hp (1.8 kW).

For steel up to 40 points carbon, multiply the above result by 2; for steel above 40 points carbon, multiply by 2.25.

Related Calculations: Where a shaper has a cutting stroke time that does not equal the return-stroke time, compute its cutting speed from $C = SL/(12)$(cutting-stroke time, min/sum of cutting- and return-stroke time, min). Thus, if the shaper in step 1 has a cutting-stroke time of 0.8 min and a return-stroke time of 0.4 min, $C = (54)(6)/[(12) \times (0.8/1.2)] = 40.5$ ft/min (0.2 m/s).

GRINDING FEED AND WORK TIME

What is the feed of a centerless grinding operation if the regulating wheel is 8 in (20.3 cm) in diameter and turns at 100 r/min at an angle of inclination of 5°? How long will it take to rough grind on an external cylindrical grinder a brass shaft that is 3.0 in (7.6 cm) in diameter and 12 in (30.5 cm) long, if the feed is 0.003 in (0.076 mm), the spindle speed is 20 r/min, the grinding-wheel width is 3 in (7.6 cm) and the diameter is 8 in (20.3 cm), and the total stock on the part is 0.015 in (0.38 mm)? How long would it take to make a finishing cut on this grinder with a feed of 0.001 in (0.025 mm), stock of 0.010 in (0.25 mm), and a cutting speed of 100 ft/min (0.5 m/s)?

Calculation Procedure:

1. Compute the feed rate for centerless grinding

In centerless grinding, the feed, in/min, $f = \pi dR \sin \infty$, where $\pi = 3.1416$; d = diameter of the regulating wheel, in; R = regulating wheel rpm; ∞ = angle of inclination of the regulating wheel. For this grinder, $f = \pi(8)(100)(\sin 5°) = 219$ in/min (9.3 cm/s). Centerless grinders will grind as many as 50,000 1-in (2.5-cm) parts per hour.

2. Compute the rough-grinding time

The rough-grinding time T_t min $= Lt_s d/(2WfC)$, where L = length of ground part, in; t_s = total stock on part, in; W = width of grinding-wheel face, in; C = cutting speed, ft/min.

Compute the cutting speed first because it is not known. By the method of previous calculation procedures, $C = \pi dR/12 = \pi(8)(20)/12 = 42$ ft/min (0.2 m/s). Then $T_t = (12)(0.015)(3)/[2(3)(0.003)(42)] = 0.714$ min.

3. Compute the finish-grinding time

For finish grinding, use the same equation as in step 2, except that the factor 2 is omitted from the denominator. Thus, $T_t = Lt_s d/(WfC)$, or $T_t = (12)(0.010)(3)/[(3)(0.001)(100)] = 1.2$ min.

Related Calculations: Use the same equations as in steps 1 and 4 for internal cylindrical grinding. In surface grinding, about 250 in²/min (26.9 cm²/s) can be ground 0.001 in (0.03 mm) deep if the material is hard. For soft materials, about 1000 in² (107.5 cm²) and 0.001 in (0.03 mm) deep can be ground per minute.

In honing cast iron, the average stock removal is 0.006 to 0.008 in/(ft·min) [0.008 to 0.011 mm/(m·s)]. With hard steel or chrome plate, the rate of honing averages 0.003 to 0.004 in/ft·min [0.004 to 0.006 mm/(m·s)].

BROACHING TIME AND PRODUCTION RATE

How long will it take to broach a medium-steel part if the cutting speed is 20 ft/min (0.1 m/s), the return speed is 100 ft/min (0.5 m/s), and the stroke length is 36 in (91.4 cm)? What will the production rate be if starting and stopping occupy 2 s and loading 5 s with an efficiency of 85 percent?

Calculation Procedure:

1. Compute the broaching time

The broaching time T_t min $= (L/C) + (L/R_c)$, where L = length of stroke, ft; C = cutting speed, ft/min; R_c = return speed, ft/min; for this work, $T_t = (3/20) + (3/100) = 0.18$ min.

2. Compute the production rate

In a complete cycle of the broaching machine there are three steps: broaching; starting and stopping; and loading. The cycle time, at 100 percent efficiency, is the sum of these three steps, or $0.18 \times 60 + 2 + 5 = 17.8$ s, where the factor 60 converts 0.18 min to seconds. At 85 percent efficiency, the cycle time is greater, or $17.8/0.85 = 20.9$ s. Since there are 3600 s in 1 h, production rate $= 3600/20.9 = 172$ pieces per hour.

HOBBING, SPLINING, AND SERRATING TIME

How long will it take to hob a 36-tooth 12-pitch brass spur gear having a tooth length of 1.5 in (3.8 cm) by using a 2.75-in (7.0-cm) hob? The whole depth of the gear tooth is 0.1789 in (4.5 mm). How many teeth should the hob have? Hob feed is 0.084 in/r (2.1 mm/r). What would be the cutting time for a 47° helical gear? How long will it take to spline-hob a brass shaft which is 2.0 in (5.1 cm) in diameter, has 12 splines, each 10 in (25.4 cm) long, if the hob diameter is 3.0 in (7.6 cm), cutter feed is 0.050 in (1.3 mm), cutter speed is 120 r/min, and spline depth is 0.15 in (3.8 mm)? How long will it take to hob 48 serrations on a 2-in (5.1-cm) diameter brass shaft if each serration is 2 in (5.1 cm) long, the 18-flute hob is 2.5 in (6.4 cm) in diameter, the approach is 0.3 in (7.6 mm), the feed per flute is 0.008 in (0.2 mm), and the hob speed is 250 r/min?

Calculation Procedure:

1. Compute the hob approach

The hob approach $A_c = \sqrt{d_g(D_c - d_g)}$, where d_g = whole depth of gear tooth, in; D_c = hob diameter, in. For this hob, $A_c = \sqrt{0.1798(2.7500 - 0.798)} = 0.68$ in (1.7 cm).

TABLE 8   Gear-Hobbing Cutting Speeds

| Gear material | Spur gears | | Helical gears[*] | |
| | Cutting speed | | | |
	ft/min	m/s	Angle,°	Percentage of feed to use
Brass	150	0.8	0–36	100
Fiber	150	0.8	36–48	80
Cast iron (soft)	100	0.5	48–60	67
Steel (mild)	100	0.5	60–70	50
Steel (medium)	75	0.4	70–90	33
Steel (hard)	50	0.3		

[*] Reduce feed by percentage shown when helical gears are cut.

2. *Determine the cutting speed of the hob*

Table 8 shows that a cutting speed of $C = 150$ ft/min (0.8 m/s) is generally used for brass gears. With a 2.75-in (7.0-cm) diameter hob, this corresponds to a hob rpm of $R = 12C/(\pi D_c) = (12)(150)/[\pi(2.75)] = 208$ r/min.

3. *Compute the hobbing time*

The time to hob a spur gear T_t min $= N(L + A_c)/fR$, where $N =$ number of teeth in gear to be cut; $L =$ length of a tooth in the gear, in; $A_c =$ hob approach, in; $f =$ hob feed, in/r; $R =$ hob rpm. For this spur gear, $T_t = (36)(1.5 + 0.68)/[(0.084)(208)] = 4.49$ min.

4. *Compute the cutting time for a helical gear*

Table 8 shows that the feed for a 47° helical gear should be 80 percent of that for a spur gear. By the relation in step 3, $T_t = (36)(1.5 + 0.68)/[(0.80)(0.084)(208)] = 5.61$ min.

5. *Compute the time to spline hob*

Use the same procedure as for hobbing. Thus, $A_c = \sqrt{d_g(D_c - d_g)} = \sqrt{0.15(3.0 - 0.15)} = 0.654$ in (1.7 cm). Then $T_t = N(L + A)/fR$, where $N =$ number of splines; $L =$ length of spline, in; other symbols as before. For this shaft, $T_t = (12)(10 + 0.654)/[(0.05)(120)] = 21.3$ min.

6. *Compute the time to serrate*

The time to hob serrations T_t min $= N(L + A)/(fnR)$, where $N =$ number of serrations; $L =$ length of serration, in; $n =$ number of flutes on hob; other symbols as before. For this shaft, $T_t = (48)(2 + 0.30)/[(0.008)(18)(250)] = 3.07$ min.

TIME TO SAW METAL WITH POWER AND BAND SAWS

How long will it take to saw a rectangular piece of alloy-plate aluminum 6 in (15.2 cm) wide and 2 in (5.1 cm) thick if the length of cut is 6 in (15.2 cm), the power hacksaw makes 120 strokes/min, and the average feed per stroke is 0.0040 in (0.1 mm)? What would the sawing time be if a band saw with a 200-ft/min (1.0-m/s) cutting speed, 16 teeth per inch (6.3 teeth per centimeter), and a 0.0003-in (0.008-mm) feed per tooth is used?

Calculation Procedure:

1. *Compute the sawing time for a power saw*

For a power saw with positive feed, the time to saw T_t min $= L/(Sf)$, where $L =$ length of cut, in; $S =$ strokes/min of saw blade; $f =$ feed per stroke, in. In this saw, $T_t = (6)/[(120)(0.0040)] = 12.5$ min.

2. *Compute the band-saw cutting time*

For a band saw, the sawing time T_t min $= L/(12Cnf)$, where $L =$ length of cut, in; $C =$ cutting speed, ft/min; $n =$ number of saw teeth per inch; $f =$ feed, inches per tooth. With this band saw, $T_t = (6)/[(12)(200)(16)(0.0003)] = 0.521$ min.

   Related Calculations: When nested round, square, or rectangular bars are to be cut, use the greatest *width* of the nested bars as the length of cut in either of the above equations.

OXYACETYLENE CUTTING TIME AND GAS CONSUMPTION

How long will it take to make a 96-in (243.8-cm) long cut in a 1-in (2.5-cm) thick steel plate by hand and by machine? What will the oxygen and acetylene consumption be for each cutting method?

Calculation Procedure:

1. *Compute the cutting time*

For any flame cutting, the cutting time T_t min $= L/C$, where $L =$ length of cut, in; $C =$ cutting speed, in/min, from Table 9. With manual cutting, $T_t = 96/8 = 12$ min, using the lower manual cutting speed given in Table 9. At the higher manual cutting speed, $T_t = 96/12 = 8$ min. With machine cutting, $T_t = 96/14 = 6.86$ min, by using the lower machine cutting speed in Table 9. At the higher machine cutting speed, $T_t = 96/18 = 5.34$ min.

2. Compute the gas consumption

From Table 9 the oxygen consumption is 130 to 200 ft^3/h (1023 to 1573 cm^3/s). Thus, actual consumption, ft^3 = (cutting time, min/60) (consumption, ft^3/h) = (12/60)(130) = 26 ft^3 (0.7 m^3) at the minimum cutting speed and minimum oxygen consumption. For this same speed with maximum oxygen consumption, actual ft^3 used = (12/60)(200) = 40 ft^3 (1.1 m^3).

Compute the acetylene consumption in the same manner, or (12/60)(13) = 2.6 ft^3 (0.07 m^3), and (12/60)(16) = 3.2 ft^3 (0.09 m^3). Use the same procedure to compute the acetylene and oxygen consumption at the higher cutting speeds.

Related Calculations: Use the procedure given here for computing the cutting time and gas consumption when steel, wrought iron, or cast iron is cut. Thicknesses ranging up to 5 ft (1.5 m) are economically cut by an oxyacetylene torch. Alloying elements in steel may require preheating of the metal to permit cutting. To compute the gas required per lineal foot, divide the actual consumption for the length cut, in inches, by 12.

COMPARISON OF OXYACETYLENE AND ELECTRIC-ARC WELDING

Determine the time required to weld a 4-ft (1.2-m) long seam in a ⅜-in (9.5-mm) plate by the oxyacetylene and electric-arc methods. How much oxygen and acetylene are required? What weight of electrode will be used? What is the electric-power consumption? Assume that one weld bead is run in the joint.

Calculation Procedure:

1. Compute the welding time

For any welding operation, the time required to weld T_t min = L/C, where L = length of weld, in; C = welding speed, in/min. When oxyacetylene welding is used, $T_t = 48/1.0 = 48$ min, when a welding speed of 1.0 in/min (0.4 mm/s) is used. With electric-arc welding, $T_t = 48/18 = 2.66$ min when the welding speed = 18 in/min (7.6 mm/s) per bead. For plate thicknesses under 1 in (2.5 cm), typical welding speeds are in the range of 1 to 2 in/min (0.4 to 0.8 mm/s) for oxyacetylene and 18 in/min (7.6 mm/s) for electric-arc welding. For thicker plates, consult *The Welding Handbook*, American Welding Society.

2. Compute the gas consumption

Gas consumption for oxyacetylene welding is given in cubic feet per foot of weld. Using values from *The Welding Handbook*, or a similar reference, we see that oxygen consumption = (ft^3 O_2 per ft of weld) (length of weld, ft); acetylene consumption = (ft^3 acetylene per ft of weld) (length of weld, ft). For this weld, with only one bead, oxygen consumption = (10.0)(4) = 40 ft^3 (1.1 m^3); acetylene consumption = (9.0)(4) = 36 ft^3 (1.0 m^3).

3. Compute the weight of electrode required

The Welding Handbook tabulates the weight of electrode for various types of welds—square grooves, 90° grooves, etc., per foot of weld. Then the electrode weight required, lb = (rod consumption, lb/ft) (weld length, ft).

For oxyacetylene welding, the electrode weight required, from data in *The Welding Handbook*, is (0.597)(4) = 2.388 lb (1.1 kg). For electric-arc welding, weight = (0.18)(4) = 0.72 lb (0.3 kg).

4. Compute the electric-power consumption

In electric-arc welding the power consumption is $kW = (V)(A)/(1000)$ (efficiency). *The Welding Handbook* shows that for a ⅜-in (9.5-mm) thick plate, $V = 40$, $A = 450$ A, efficiency = 60 percent. Then power consumption = (40)(450)/[(1000)(0.60)] = 30 kW. For this press, $F = (8)(0.5)(16.0) = 64$ tons (58.1 t).

Related Calculations: Where more than one pass or bead is required, multiply the time for one bead by the number of beads deposited. If only 50 percent penetration is required for the bead, the welding speed will be twice that where full penetration is required.

TABLE 9 Oxyacetylene Cutting Speed and Gas Consumption

Metal thickness		Speed				Gas consumption			
		Manual		Machine		Oxygen		Acetylene	
in	cm	in/min	mm/s	in/min	mm/s	ft³/h	cm³/s	ft³/h	cm³/s
0.25	0.6	16–18	6.8–7.6	20–26	8.5–11.0	50–90	393.3–707.9	8–11	62.9–86.5
0.50	1.3	12–15	5.1–6.4	17–22	7.2–9.3	90–125	707.9–983.2	10–13	78.7–102.3
1	2.5	8–12	3.4–5.1	14–18	5.9–7.6	130–200	1023–1573	13–16	102.3–125.9
2	5.1	5–7	2.1–3.0	10–13	4.2–5.5	200–300	1573–2360	16–20	125.9–157.3
4	10.2	4–5	1.7–2.1	7–9	3.0–3.8	300–400	2360–3146	21–26	165.2–204.5
6	15.2	3–4	1.3–1.7	5–7	2.1–3.0	400–500	3146–3933	26–32	204.5–251.7
8	20.3	3–6	1.3–2.5	4–6	1.7–2.5	500–650	3933–5113	28–35	220.2–275.3
10	25.4	2–3	0.8–1.3	3–4	1.3–1.7	700–1000	5506–7860	30–38	236.0–298.9
12	30.5	2.5–3.5	1.1–1.5	3–4	1.3–1.7	720–880	5663–6922	42–52	330.4–409.0

PRESSWORK FORCE FOR SHEARING AND BENDING

What is the press force to shear an 8-in (20.3-cm) long 0.5-in (1.3-cm) thick piece of annealed bronze having a shear strength of 16.0 tons/in² (2.24 t/cm²)? What is the stripping load? Determine the force required to produce a U bend in this piece of bronze if the unsupported length is 4 in (10.2 cm), the bend length is 6 in (15.2 cm), and the ultimate tensile strength is 32.0 tons/in² (4.50 t/cm²).

Calculation Procedure:

1. Compute the required shearing force

For any metal in which a straight cut is made, the required shearing force, tons $= F = Lts$, where L = length of cut, in; t = metal thickness, in; s = shear strength of metal being cut, tons/in². Where round, elliptical, or other shaped holes are being cut, substitute the sum of the circumferences of all the holes for L in this equation.

2. Compute the stripping load

For the typical press, the stripping load is 3.5 percent of the required shearing force, or (0.035)(64) = 2.24 tons (2.0 t).

3. Compute the required bending force

When U bends or channels are pressed in a metal, $F = 2Lt^2s_t/W$, where s_t = ultimate tensile strength of the metal, tons/in²; W = width of unsupported metal, in = distance between the vertical members of a channel or U bend, measured to the *outside* surfaces, in. For this U bend, $F = 2(6)(0.5)^2(32)/4 = 24$ tons (21.8 t).

    Related Calculations: Right-angle edge bends require a bending force of $F = Lt^2s_t/(2W)$, while free V bends with a centrally located load require a bending force of $F = Lt^2s_t/W$. All symbols are as given in steps 1 and 2.

MECHANICAL-PRESS MIDSTROKE CAPACITY

Determine the maximum permissible midstroke capacity of single- and twin-driven 2-in (5.1-cm) diameter crankshaft presses if the stroke of the slide is 12 in (30.5 cm) for each.

Calculation Procedure:

1. Compute the single-driven press capacity

For a single-driven crankshaft press with a heat-treated 0.35 to 0.45 percent carbon-steel crankshaft having a shear strength of 6 tons/in² (0.84 t/cm²), the maximum permissible midstroke capacity F tons $= 2.4d^3/S$, where d = shaft diameter at main bearing, in; S = stroke length, in; or $F = (2.4)(2)^3/12 = 1.6$ tons (1.5 t).

2. Compute the twin-driven press capacity

Twin-driven presses with main (bull) gears on each end of the crankshaft have a maximum permissible midstroke capacity of $F = 3.6d^3/S$, when the shaft shearing strength is 9 tons/in². For this press, $F = 3.6(2)^3/12 = 2.4$ tons (2.2 t).

    Related Calculations: Use the equation in step 2 to compute the maximum permissible midstroke capacity of all wide (right-to-left) double-crank presses. Since gear eccentric presses are built in competition with crankshaft presses, their midstroke pressure capacity is within the same limits as in crankshaft presses. The diameters of the fixed pins on which the gear eccentrics revolve are usually made the same as the crankshaft in crankshaft presses of the same rated capacity.

STRIPPING SPRINGS FOR PRESSWORKING METALS

Determine the force required to strip the work from a punch if the length of cut is 5.85 in (14.9 cm) and the stock is 0.25 in (0.6 cm) thick. How many springs are needed for the punch if the force per inch deflection of the spring is 100 lb (175.1 N/cm)?

Calculation Procedure:

1. Compute the required stripping force

The required stripping force F_p lb needed to strip the work from a punch is $F_p = Lt/0.00117$, where L = length of cut, in; t = thickness of stock cut, in. For this punch, $F_p = (5.85)(0.25)/0.00117 = 1250$ lb (5560.3 N).

2. Compute the number of springs required

Only the first ⅛-in (0.3-cm) deflection of the spring can be used in the computation of the stripping force produced by the spring. Thus, for this punch, number of springs required = stripping force, lb/force, lb, to produce ⅛-in (0.3-cm) deflection of the spring, or $1250/100 = 12.5$ springs. Since a fractional number of springs cannot be used, 13 springs would be selected.

Related Calculations: In high-speed presses, the springs should not be deflected more than 25 percent of their free length. For heavy, slow-speed presses, the total deflection should not exceed 37.5 percent of the free length of the spring. The stripping force for aluminum alloys is generally taken as one-eighth the maximum blanking pressure.

BLANKING, DRAWING, AND NECKING METALS

What is the maximum blanking force for an aluminum part if the length of the cut is 30 in (76.2 cm), the metal is 0.125 in (0.3 cm) thick, and the yield strength is 2.5 tons/in² (0.35 t/cm²)? How much force is required to draw a 12-in (30.5-cm) diameter 0.25-in (0.6-cm) thick stainless steel shell if the yield strength is 15 tons/in² (2.1 t/cm²)? What force is required to neck a 0.125-in (0.3-cm) thick aluminum shell from a 3- to a 2-in (7.6- to 5.1-cm) diameter if the necking angle is 30° and the ultimate compressive strength of the material is 14 tons/in² (1.97 t/cm²)?

Calculation Procedure:

1. Compute the maximum blanking force

The maximum blanking force for any metal is given by $F = Lts$, where F = blanking force, tons; L = length of cut, in (= circumference of part, in); t = metal thickness, in; s = yield strength of metal, tons/in². For this part, $F = (30)(0.125)(2.5) = 0.375$ tons (0.34 t).

2. Compute the maximum drawing force

Use the same equation as in step 1, substituting the drawing-edge length or perimeter (circumference of part) for L. Thus, $F = (12\pi)(0.25)(15) = 141.5$ tons (128.4 t).

3. Compute the required necking force

The force required to neck a shell is $F = ts_c(d_1 - d_s)/\cos$ (necking angle), where F = necking force, tons; t = shell thickness, in; s_c = ultimate compressive strength of the material, tons/in²;

TABLE 10   Metal Yield Strength

Metal	Yield strength	
	tons/in²	t/cm²
Aluminum, 2S annealed	2.5	0.35
Aluminum, 24S heat-treated	23.0	2.23
Low brass, ¼ hard	24.5	3.46
Yellow brass, annealed	10.0	1.41
Cold-rolled steel, ¼ hard	16.0	2.25
Stainless steel, 18-8	15.0	2.11

Note: As a general rule, the necking angle should not exceed 35°.

d_1 = large diameter of shell, i.e., the diameter *before* necking, in; d_s = small diameter of shell, i.e., the diameter *after* necking, in. For this shell, F = $(0.125)(14)(3.0 - 2.0)/\cos 30°$ = 2.02 tons (1.8 t).

Related Calculations: Table 10 presents typical yield strengths of various metals which are blanked or drawn in metalworking operations. Use the given strength as shown above.

METAL PLATING TIME AND WEIGHT

How long will it take to electroplate a 0.004-in (0.1-mm) thick zinc coating on a metal plate if a current density of 25 A/ft^2 (269.1 A/m^2) is used at an 80 percent plating efficiency? How much zinc is required to produce a 0.001-in (0.03-mm) thick coating on an area of 60 ft^2 (5.6 m^2)?

Calculation Procedure:

1. *Compute the metal plating time*

The plating time T_p min = 60 $An/(A_a e)$, where A = A/ft^2 required to deposit 0.001 in (0.03 mm) of metal at 100 percent cathode efficiency; n = number of thousandths of inch actually deposited; A_a = current actually supplied, A/ft^2; e = plating efficiency, expressed as a decimal. Table 11 gives typical values of A for various metals used in electroplating. For plating zinc, from the value in Table 11, T_p = $60(14.3)(4)/[(25)(0.80)]$ = 171.5 min, or 171.5/60 = 2.86 h.

2. *Compute the weight of metal required*

The plating metal weight = (area plated, in^2) (plating thickness, in) (plating metal density, lb/in^3). For this plating job, given the density of zinc from Table 11, the plating metal weight = (60 × 144)(0.004)(0.258) = 8.91 lb (4.0 kg) of zinc. In this calculation the value 144 is used to convert 60 ft^2 to square inches.

Related Calculations: The efficiency of finishing cathodes is high, ranging from 80 to nearly 100 percent. Where the actual efficiency is unknown, assume a value of 80 percent and the results obtained will be safe for most situations.

TABLE 11   Electroplating Current and Metal Weight

Metal	Time to deposit, Ah		Metal density	
	0.001 in/ft^2 at 100% efficiency	0.01 mm/m^2 at 100% efficiency	lb/in^3	g/cm^3
Antimony, Sb	10.40	0.038	0.241	6.671
Cadmium, Cd	9.73	0.036	0.312	8.636
Chromium, Cr	51.80	0.189	0.256	7.086
Cobalt(ous), Co	19.00	0.069	0.322	8.913
Copper(ous), Cu	8.89	0.033	0.322	8.913
Copper(ic), Cu	17.80	0.065	0.322	8.913
Gold(ous), Au	6.20	0.023	0.697	19.29
Gold(ic), Au	18.60	0.068	0.697	19.29
Nickel, Ni	19.00	0.069	0.322	8.913
Platinum	27.80	0.102	0.775	21.45
Silver, Ag	6.20	0.023	0.380	10.52
Tin(ous), Sn	7.80	0.029	0.264	7.307
Tin(ic), Sn	15.60	0.057	0.264	7.307
Zinc, Zn	14.30	0.052	0.258	7.141

SHRINK- AND EXPANSION-FIT ANALYSES

To what temperature must an SAE 1010 steel ring 24 in (61.0 cm) in inside diameter be raised above a 68°F (20°C) room temperature to expand it 0.004 in (0.10 mm) if the linear coefficient of expansion of the steel is 0.0000068 in/(in·°F)[0.000012 cm/(cm·°C)]? To what temperature must a 2-in (5.08-cm) diameter SAE steel shaft be reduced to fit it into a 1.997-in (5.07-cm) diameter hole for an expansion fit? What cooling medium should be used?

Calculation Procedure:

1. Compute the required shrink-fit temperature rise

The temperature needed to expand a metal ring a given amount before making a shrink fit is given by $T = E/(Kd)$, where T = temperature rise *above* room temperature, °F; K = linear coefficient of expansion of the metal ring, in/(in·°F); d = ring internal diameter, in. For this ring, $T = 0.004/[(0.0000068)(24)] = 21.5°F$ (11.9°C). With a room temperature of 68°F (20.0°C), the final temperature of the ring must be $68 + 21.5 = 89.5°F$ (31.9°C) or higher.

2. Compute the temperature for an expansion fit

Nitrogen, air, and oxygen in liquid form have a low boiling point, as does dry ice (solid carbon dioxide). Nitrogen and dry ice are considered the safest cooling media for expansion fits because both are relatively inert. Liquid nitrogen boils at $-320.4°F$ ($-195.8°C$) and dry ice at $-109.3°F$ ($-78.5°C$). At $-320°F$ ($-195.6°C$) liquid nitrogen will reduce the diameter of metal parts by the amount shown in Table 12. Dry ice will reduce the diameter by about one-third the values listed in Table 12.

With liquid nitrogen, the diameter of a 2-in (5.1-cm) round shaft will be reduced by $(2.0)(0.0022) = 0.0044$ in (0.11 mm), given the value for SAE steels from Table 12. Thus, the diameter of the shaft at $-320.4°F$ ($-195.8°C$) will be $2.000 - 0.0044 = 1.9956$ in (5.069 cm). Since the hole is 1.997 in (5.072 cm) in diameter, the liquid nitrogen will reduce the shaft size sufficiently.

If dry ice were used, the shaft diameter would be reduced $0.0044/3 = 0.00146$ in (0.037 mm), giving a final shaft diameter of $2.00000 - 0.00146 = 1.99854$ in (5.076 cm). This is too large to fit into a 1.997-in (5.072-cm) hole. Thus, dry ice is unsuitable as a cooling medium.

PRESS-FIT FORCE, STRESS, AND SLIPPAGE TORQUE

What force is required to press a 4-in (10.2-cm) outside-diameter cast-iron hub on a 2-in (5.1-cm) outside-diameter steel shaft if the allowance is 0.001-in interference per inch (0.001 cm/cm) of

TABLE 12   Metal Shrinkage with Nitrogen Cooling

Metal	Shrinkage, in/in (cm/cm) of shaft diameter
Magnesium alloys	0.0046
Aluminum alloys	0.0042
Copper alloys	0.0033
Cr-Ni alloys (18-8 to 18-12)	0.0029
Monel metals	0.0023
SAE steels	0.0022
Cr steels (5 to 27% Cr)	0.0019
Cast iron (not alloyed)	0.0017

shaft diameter, the length of fit is 6 in (15.2 cm), and the coefficient of friction is 0.15? What is the maximum tensile stress at the hub bore? What torque is required to produce complete slippage of the hub on the shaft?

Calculation Procedure:

1. Determine the unit press-fit pressure

Figure 5 shows that with an allowance of 0.001 in interference per inch (0.001 cm/cm) of shaft diameter and a shaft-to-hub diameter ratio of $2/4 = 0.5$, the unit press-fit pressure between the hub and the shaft is $p = 6800$ lb/in^2 (46,886.0 kPa).

2. Compute the press-fit force

The press-fit force F tons $= \pi f p d L / 2000$, where $f =$ coefficient of friction between hub and shaft; $p =$ unit press-fit pressure, lb/in^2; $d =$ shaft diameter, in; $L =$ length of fit, in. For this press fit, $F = (\pi)(0.15)(6800)(2.0)(6)/2000 = 19.25$ tons (17.4 t).

3. Determine the hub bore stress

Use Fig. 6 to determine the hub bore stress. Enter the bottom of Fig. 6 at 0.0010 in (0.0010 cm) interference allowance per in of shaft diameter and project vertically to $d/D = 0.5$. At the left read the hub stress as 11,600 lb/in^2 (79,982 kPa).

4. Compute the slippage torque

The torque, in·lb, required to produce complete slippage of a press fit is $T = 0.5\,\pi f p L d^2$, or $T = 0.5(3.1416)(0.15)(6800)(6)(2)^2 = 38,450$ in·lb (4344.1 N·m).

   Related Calculations: Figure 7 shows the press-fit pressures existing with a steel hub on a steel shaft. The three charts presented in this calculation procedure are useful for many different press fits, including those using a hollow shaft having an internal diameter less than 25 percent of the external diameter and for all solid steel shafts.

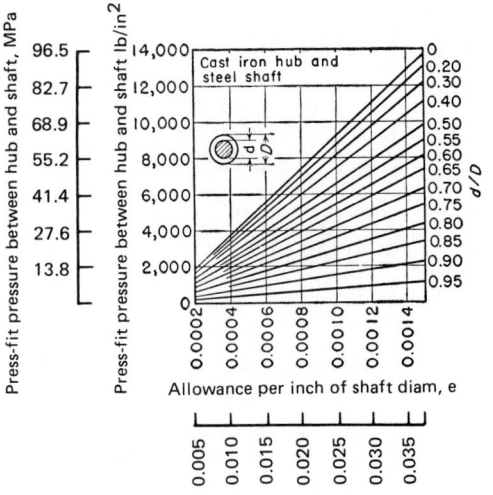

FIG. 5 Press-fit pressures between steel hub and shaft.

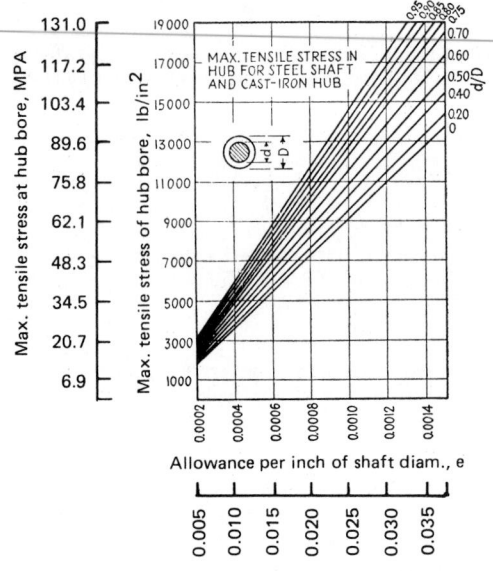

FIG. 6 Variation in tensile stress in cast-iron hub in press-fit allowance.

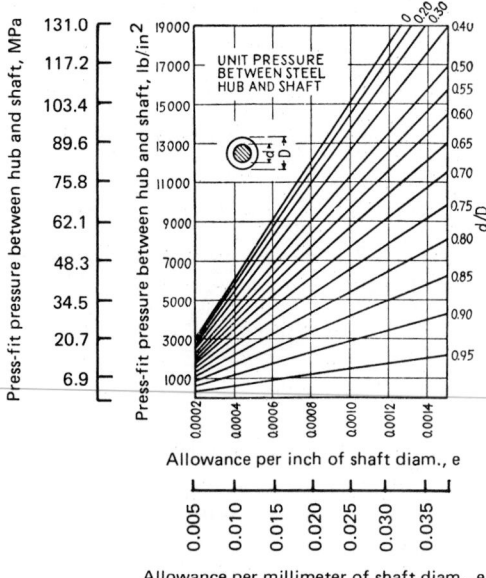

FIG. 7 Press-fit pressures between cast-iron hub and shaft.

LEARNING-CURVE ANALYSIS AND CONSTRUCTION

A short-run metalworking job requires five operators. The longest individual learning time for the new task is 3 days; 2 days are allowed for group familiarization with the task. If the normal output is 1000 units per 8-hr day, determine the daily allowance per operator when the standard for 100 percent performance is 0.8 worker-hour per 100 units produced.

Calculation Procedure:

1. Plot the learning curve

A learning curve shows the improvement that occurs with repetition of a task. Figure 8 is a typical learning curve with the learning period, days, plotted against the percent of methods time measurement (MTM) determined normal task. The shape of the curve, once determined for a given operation, does not change. The horizontal scale division is, however, changed to suit the minimum learning period for 100 percent performance. Thus, for a 3-day learning period the horizontal scale becomes 3 days. The coordinate at each of these three points (i.e., days) becomes the minimum expected task for each day. Performance above these tasks rates a bonus. The base of 60 percent of normal performance for the first day of learning for all jobs is attainable and meets management's minimum requirements.

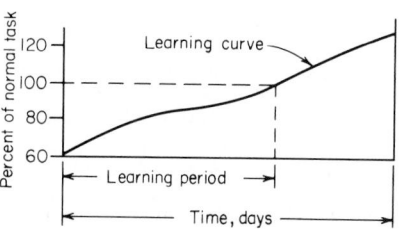

FIG. 8   Typical learning curve for a metalworking task.

2. Determine the learning period to allow

(a) Find the learning time, by test, for each work station in the group. (b) Select the longest individual learning time—in this instance, 3 days. (c) Add a group familiarization allowance when the group exceeds three operators—2 days here. (d) Find the sum of $b + \bar{c}$, or $3 + 2 = 5$ days. This is the learning period to allow.

3. Find the task for each day

Divide the horizontal learning-period axis into five parts, one part for each of the 5 learning-period days allowed. Draw an ordinate for each day, and read the percentage task for that day at the intersection with the learning curve, or: 60.0; 70.5; 75.5; 80.0; 87 percent for days 1, 2, 3, 4, and 5, respectively.

4. Compute the daily task and daily allowance

With a normal (100 percent) task of 1000 units for an 8-hr day, set up a table (like Table 13) of daily tasks and time allowance during the learning period. Begin with a column listing the number of learning days. In the next column, list the percentage learning performance read from Fig. 8. Find the daily task in units, column 3 of Table 13, by taking the product of 1000 units and the percentage learning performance, column 2, expressed as a decimal. Last, compute the daily

TABLE 13   Learning-Curve Analysis

Learning days	Percentage of learning performance	Daily task, units	Daily allowance per operator, in units
1	60.0	600	40% × 8 = 3.20
2	70.5	705	29.5% × 8 = 2.36
3	75.5	755	24.5% × 8 = 1.96
4	80.0	800	20.0% × 8 = 1.60
5	87.0	870	13.0% × 8 = 1.04
6	100.0	1000	0% × 8 = 0

allowance per operator by finding the product of 8 h and the difference between 1.00 and the percentage of learning performance; i.e., for day 1: $(1.00 - 0.60)(8) = 3.2$ h. Tabulate the results in the fourth column of Table 13.

5. Compute the incentive pay for the group

In this plant the incentive pay is found by taking the product of the production in units and the standard set for 100 percent performance, or 0.80 worker-hour per 100 units produced. Thus, production of 600 units on day 1 will earn $(600/100)(0.80) = 4.8$-h pay for each group. Add to this the learning allowance of 3.2 h for day 1, and each group has earned $4.8 + 3.2 = 8$-h pay for 8-h work.

If the group produced 700 units during day 1, it would earn $(700/100)(0.80) = 5.6$-h pay at this standard. With the learning allowance of 3.2 h, the daily earnings would be $5.6 + 3.2 = 8.8$-h pay for 8-h work. This is exactly what is desired. The operator is rewarded for learning quickly.

Related Calculations: Select the length of the learning period for any new short-run task by conferring with representatives of the manufacturing, industrial engineering, and industrial relations departments. A simple operation that will be performed 1000 to 2000 times in an 8-h period would require a 3-day learning period. This is considered the minimum time for bringing such an operation up to normal speed. This is also true if a small group (three or less operators) perform equally simple operations. With larger groups (four or more operators), both simple and complex operations require an additional allowance for operators to adjust themselves to each other. Two days is a justified allowance for up to 15 operators learning to cooperate with one another under incentive conditions.

To prepare a plant-wide learning curve, keep records of the learning rates for a number of short-run tasks. Combine these data to prepare a typical learning curve for a particular plant. The method developed here was first described in *Factory*, now *Modern Manufacturing*, magazine.

LEARNING-CURVE EVALUATION OF MANUFACTURING TIME

A metalworking process requires 1.00 h for manufacture of the first unit of a production run. If the operator has an improvement or learning rate of 90 percent, determine the time required to manufacture the 2d, 4th, 8th, and 16th units. What is the cumulative average unit time for the 16th unit? If 100 units are manufactured, what is the cumulative average time for the 100th unit? What is the unit manufacturing time for the 100th item?

Calculation Procedure:

1. Compute the unit time for the production cycle

The learning curve relates the production time to the number of units produced. When the number of units produced doubles, the time required to produce the unit representing the doubled quantity is: (Learning rate, percent)(time, h or min, to produce the unit representing one-half the doubled quantity). Or, for the production line being considered here:

Unit number	Production time, h
1	1.00
2	$0.90(1.00) = 0.900$
4	$0.90(0.90) = 0.810$
8	$0.90(0.81) = 0.729$
16	$0.90(0.729) = 0.656$

2. Compute the cumulative average unit time

The cumulative average unit time for any unit in a production run = (Σ unit time for each item in the run)/(number of items in the run). Thus, computing the time for items 1 through 16 as shown in step 1, and taking the sum, we get the cumulative average unit time = 12.044 h/16 units = 0.752 h.

TABLE 14   Learning-Curve Factors

	Learning rate, percent		
	85	90	95
No. of units	Time or cost, percent of unit 1	Time or cost, percent of unit 1	Time or cost, percent of unit 1
1	1.000	1.000	1.000
2	0.850	0.900	0.950
4	0.723	0.810	0.903
8	0.614	0.729	0.857
16	0.522	0.656	0.815
32	0.444	0.591	0.774
64	0.377	0.531	0.735
100	0.340	0.497	0.711

Learning-curve slopes

Learning rate, percent	Curve slope
70	−0.514
75	−0.415
80	−0.322
85	−0.234
90	−0.152
95	−0.074

3.   Compute the cumulative average time for the 100th unit

Set up a ratio of the learning factor for the 100th unit/learning factor for the 16th unit, and multiply the ratio by the cumulative average 16th unit time. Or, from the factors in Table 14, $(0.497/0.656)(0.752 \text{ h}) = 0.570 \text{ h}$.

4.   Compute the unit time for the 100th unit

Using the factor for the 90 percent learning curve in Table 14, the unit time for the 100th unit made $= (1.00 \text{ h})(0.497) = 0.497 \text{ h}$.

Related Calculations: When using learning curves, be extremely careful to distinguish between *unit time* and *cumulative average unit time*. The unit time is the time required to make a particular unit in a production run, say the 10th, 16th, etc. Thus, a unit time of 0.5 h for the 16th unit in a production run means that the time required to make the 16th unit is 0.5 h. The 15th unit will require *more* time to make it; the 17th unit will require *less* time.

The cumulative average unit time is the *average* time to manufacture a given number of identical items. To obtain the cumulative average unit time for any given number of items, take the sum of the time required for each item up to and including that item and divide the sum by the number of items.

Either the *unit time or cumulative average unit time* can be used in manufacturing time or cost estimates, as long as the estimator knows which time value is being used. Failure to recognize the respective time values can result in serious errors.

A learning curve plotted on log-log coordinates is a straight line, Fig. 9. The slope of typical learn-

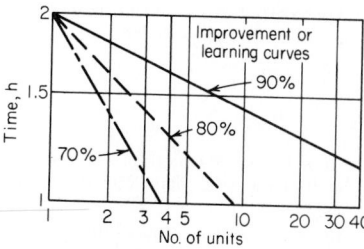

FIG. 9   Learning curves plotted on log-log scale.

ing curves is listed in Table 14. Since a learning curve slopes downward—i.e., the unit manufacturing time decreases as more units are produced—the slope is expressed as a negative value.

Typical improvement or learning rates are: machining, drilling, etc., 90 to 95 percent; short-cycle bench assembly, 85 to 90 percent; equipment maintenance, 75 to 80 percent; electronics assembly and welding, 80 to 90 percent; general assembly, 70 to 80 percent. When an operation consists of several tasks having different learning rates, compute the overall learning rate for the task by taking the sum of the product of each learning rate (LR) and the percentage of the total task it represents. Thus, with $LR_1 = 0.90$ for 60 percent of the total task; $LR_2 = 0.80$ for 20 percent of the total task; $LR_3 = 0.70$ for 10 percent of the total task, the overall learning rate LR $= 0.90(0.60) + 0.80(0.20) + 0.70(0.10) = 0.77$.

Note that in machine-paced operations—i.e., those in which the speed of the machine controls the operator's activities—there is less chance for the operator to learn. Hence, the learning rate will be higher—90 to 95 percent—than in worker-paced operations that have learning rates of 70 to 80 percent. When learning or improvement ceases, the operator has reached the level-off point, and the task cannot be performed any more rapidly. The ratio set up in step 3 can use any two items in a production run, provided that the cumulative average time for the smaller item is multiplied by the ratio.

DETERMINING BRINELL HARDNESS

A 3000-kg load is put on a 10-mm diameter ball to determine the Brinell hardness of a steel. The ball produces a 4-mm-diameter indentation in 30 s. What is the Brinell hardness of the steel?

Calculation Procedure:

1. Determine the Brinell hardness by using an exact equation

The standard equation for determining the Brinell hardness is BHN $= F/(\pi d_1/2)(d_1 - \sqrt{d_1^2 - d_s^2})$, where F = force on ball, kg; d_1 = ball diameter, mm; d_s = indentation diameter, mm. For this test, BHN $= 3000/(\pi \times 10/2)(10 - \sqrt{10^2 - 4^2}) = 229$.

2. Compute the Brinell hardness by using an approximate equation

One useful approximate equation for Brinell hardness is BHN $= (4F/\pi d_s^2) - 10$. For this test, BHN $= (4 \times 3000/\pi \times 4^2) - 10 = 228.5$. This compares favorably with the exact formula. For Brinell hardness exceeding 200, the approximate equation gives results that are less than 0.1 percent in error.

Related Calculations: Use this procedure for iron, steel, brass, bronze, and other hard or soft metals. A 500-kg test load is used for soft metals (brass, bronze, etc.). For Brinell hardness above 500, use a tungsten-carbide ball. The metal tested should be at least 10 times as thick as the indentation depth and wide enough so that no metal flows toward the edges of the specimen. The metal surface must be clean and free of defects.

ECONOMICAL CUTTING SPEEDS AND PRODUCTION RATES

A cutting tool used to cut beryllium costs $6 with its shank and can be reground for reuse five times. The average tool-changing time is 5 min. What is the most economical cutting speed if the machine labor rate is $3 per hour and the overhead is 200 percent? What is the cutting speed for the maximum production rate? The cost of regrinding the tool is 35 cents per edge.

Calculation Procedure:

1. Determine the tool cost factor

The cost factor $T_c + Y/X$ for a tool is composed of T_c = time to change tool, min; Y = tool cost per cutting edge, including prorated initial cost plus reconditioning costs, cents; X = machining rate, including labor and overhead, cents/min.

For this tool, $T_c = 5$ min. The tool can be reground five times after its original use, giving a total of $5 + 1 = 6$ cutting edges (five regrindings + the original edge) during its life. Since the tool costs $6 new, the prorated cost per edge = $6/6 edges = $1, or 100 cents. The regrinding cost = 35 cents per edge; thus $Y = 100 + 35 = 135$ cents per edge.

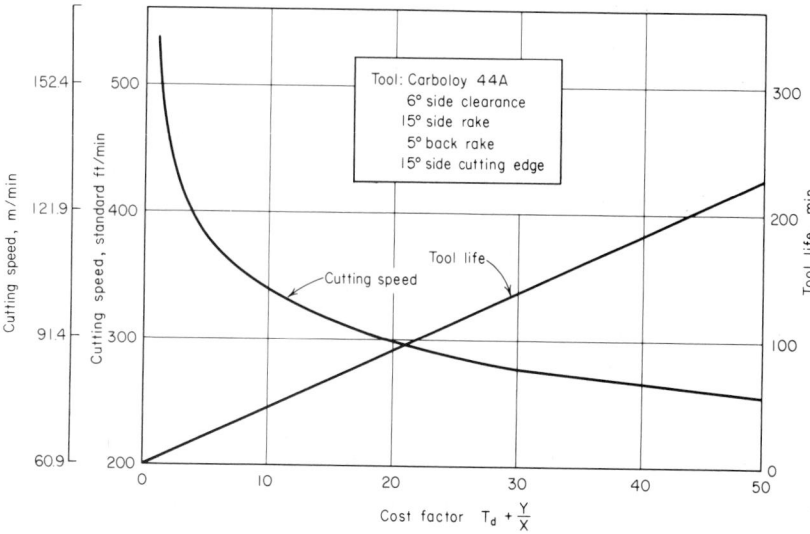

FIG. 10 Optimum cutting-speed chart. (*American Machinist.*)

With a machine labor rate of $3 per hour and an overhead factor of 200 percent, or 2.00($3) = $6, the value of X = machining rate = $3 + $6 = $9, or 900 cents per hour, or 900/60 = 15 cents per minute. Then, the cost factor $T_c + Y/X = 5 + 135/15 = 14$.

2. Determine the cutting speed for minimum cost

Enter Fig. 10 at a cost factor of 14 and project vertically upward until the cutting-speed curve is intersected. At the left, read the cutting speed for minimum tool cost as 320 surface ft/min.

3. Determine the probable tool life

Project upward from the cost factor of 14 in Fig. 10 to the tool-life curve. At the right, read the tool life as 66 min.

4. Determine the speed and life for the maximum production rate

Substitute the value of T_c for the cost factor $T_c + Y/X$ on the horizontal scale of Fig. 10. As before, read the cutting speed and tool life at the intersection with the respective curves. The plotted values apply when the chip-removal suction devices will operate efficiently at the cutting speeds indicated by the curves. Thus, with $T_c = 5$, the cutting speed is 370 surface ft/min, and the tool life is 30 min.

Related Calculations: Figure 10, and similar optimum cutting-speed charts, is plotted for a specific land wear—in this case 0.010 in (0.03 cm). For a land wear of 0.015 in (0.04 cm), multiply the cutting speeds obtained from Fig. 10 by 1.13. However, a land wear of 0.015 in (0.04 cm) is not recommended because the wear rates are accelerated. If Carboloy 883 tools are used in place of the 44A grade plotted in Fig. 10, multiply the cutting speeds obtained from this chart by 1.12 for a 0.010-in (0.03-cm) wear land or 1.26 for a 0.015-in (0.04-cm) wear land.

Charts similar to Fig. 10 for other tool materials can be obtained from tool manufacturers. Do not use Fig. 10 for any tool material other than Carboloy 44A. The method presented here is the work of D. R. Walker and J. Gubas, as reported in *American Machinist.*

OPTIMUM LOT SIZE IN MANUFACTURING

A manufacturing plant has a demand for 900 of its products per month on which the setup cost is $10. The cost of each unit is $5; the annual inventory charge is 12 percent/year of the average

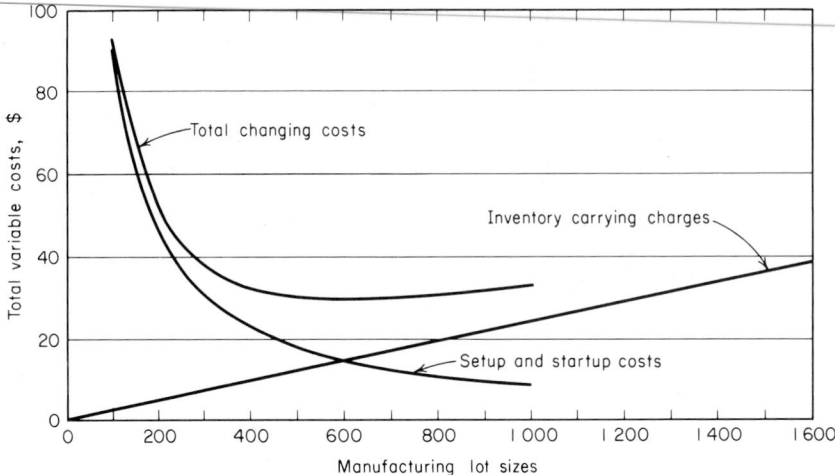

FIG. 11   Changing costs associated with manufacturing lot sizes. *(American Machinist.)*

dollar value held in stock; the period for which the demand has occurred is 1/12 year. What is optimum manufacturing lot size? Plot a cost chart for this plant.

Calculation Procedure:

1.   *Determine the optimum manufacturing lot size*

Optimum manufacturing lot size can be found from: {2(demand, units, per period)(cost per setup, $)/[(demand period, fraction of a year)($ cost per unit)(annual inventory charge, percent of average $ value held in stock)]}$^{0.5}$. For this run, optimum lot size = {2(900)($10)/ (1/12)($5)(0.12)]}$^{0.5}$ = 600 units.

2.   *Plot a cost chart for this plant*

Figure 11 shows a typical cost chart. Plot each curve using production runs of 100, 200, 300, 400, ... , 1600 units. The values for each curve are determined from: inventory carrying charges = (number of units in run)($ cost per unit)(annual inventory charge, percent)(demand period)/2; setup and startup costs = (demand during period, units)($ cost of setup)/(number of units in run); total changing costs = inventory carrying charges + setup and startup costs.

  The units for these equations are the same as given in step 1. Note that the total-changing-costs curve is a minimum at the point where the inventory-carrying-charges curve and setup-and-startup-costs curve intersect. Also, the two latter curves intersect at the optimum manufacturing lot size—600 units, as computed in step 1.

  Related Calculations: Economical lot-size relations are readily adaptable to machine-shop computations. With only slight changes, the same principles can be applied to determination of optimum-quantity purchases. The procedure described here is the work of I. Heitner, as reported in *American Machinist*.

PRECISION DIMENSIONS AT VARIOUS TEMPERATURES

A magnesium workpiece with a dimension of 12.5000 to 12.4996 in (31.750 to 31.749 cm) is at a temperature of 85°F (29.4°C) after machining. The steel gage with which the dimensions of the workpiece will be checked is at 75°F (23.9°C). The workpiece must be gaged immediately to determine whether further grinding is necessary. Tolerance on the work is ±0.0002 in (0.005 mm). What should the dimensions of the workpiece be if there is not enough time available to allow the gage and workpiece temperatures to equalize? The standard reference temperature is 68°F (20.0°C).

Calculation Procedure:

1. Compute the actual work dimensions

The temperature of the workpiece is 85°F (29.4°C), or $85 - 68 = 17°F$ (9.4°C) above the standard reference of 68°F (20.0°C). Since the actual temperature of the workpiece is greater than the standard temperature, the dimensions of the workpiece will be larger than at the standard temperature because the part expands as its temperature increases.

To find the amount by which the workpiece will be oversize at the actual temperature, multiply the nominal dimension of the workpiece, 12.5 in (31.75 cm), by the coefficient of linear expansion of the material and by the difference between the actual and standard temperatures. For this magnesium workpiece, the average oversize amount at the actual temperature, 85°F (29.4°C), is $(12.5)(14.4 \times 10^{-6})(85 - 68) = 0.003060$ in (0.078 mm).

2. Compute the actual gage dimension

Compute the actual gage dimension in a similar way, using the same dimension, 12.5 in (31.75 cm), but the coefficient of linear expansion of the gage material, steel, and the gage temperature, 75°F (23.9°C). Or, $(12.5)(6.4 \times 10^{-6})(75 - 68) = 0.000560$ in (0.014 mm).

3. Compute the workpiece dimension as a check

The workpiece dimension, corrected for tolerance, plus the difference between the oversize amounts computed in steps 1 and 2 is the dimension to which the part should be checked at the existing shop and gage temperature.

Applying the tolerance, ±0.0002 in (±0.005 mm), to the drawing dimension, 12.5000 − 12.4996, gives a drawing dimension of 12.4998 ± 0.0002 in (31.7495 ± 0.0005 cm). Adding the difference between oversize dimensions, $0.003060 - 0.000560 = 0.002500$ in (0.0635 mm), to 12.4998 ± 0.0002 in (31.7495 ± 0.0005 cm) gives a checking dimension of 12.5023 ± 0.0002 in (31.7558 ± 0.0005 cm). If personnel check the workpiece at this dimension, they will have full confidence that it will be the right size.

Related Calculations: This procedure can be used for any metal—bronze, aluminum, cast iron, etc.—for which the coefficient of linear expansion is known. Obtain the coefficient from Baumeister and Marks—*Standard Handbook for Mechanical Engineers*, or a similar reference. When a workpiece is at a temperature less than the National Bureau of Standards standard of 68°F (20.0°C), the part contracts instead of expanding. The dimension change computed in step 1 is then negative. This is also true of the gage, if it is at a temperature of less than 68°F (20.0°C). Note that the tolerance is constant regardless of the actual temperature of the part.

The procedure given here is the work of H. K. Eitelman, as reported in *American Machinist*.

HORSEPOWER REQUIRED FOR METALWORKING

What is the input horsepower required for machining, on a geared-head lathe, a 4-in (10.2-cm) diameter piece of AISI 4140 steel having a hardness of 260 BHN if the depth of cut is 0.25 in (0.6 cm), the cutting speed is 300 ft/min (1.5 m/s), and the feed per revolution is 0.025 in (0.6 mm)?

Calculation Procedure:

1. Determine the metal removal rate

Compute the metal removal rate (MRR) in^3/min from $MRR = 12fDC$, where f = tool feed rate, in/r; D = depth of cut, in; C = cutting speed, ft/min. For this workpiece, $MRR = 12(0.025)(0.25)(300) = 22.5$ in^3/min (6.1 cm^3/s).

2. Determine the unit horsepower required

Table 15 lists the average unit horsepower required for cutting various metals. The unit horsepower hp_u is the power required to remove 1 in^3 (1 cm^3) of metal per minute at 100 percent efficiency of the machine. Table 15 shows that AISI 4130 to 4345 of 250 to 300 BHN, the range into which AISI 4140 260 BHN falls, has a unit hp of 0.70 (8.5 unit kW).

TABLE 15 Average Unit hp (kW) Factors for Ferrous Metals and Alloys[°]

Material classification	Brinell hardness number		
	201–250	251–300	301–350
AISI 3160–3450	0.62 (7.52)	0.75 (9.1)	0.87 (10.6)
AISI 4130–4345	0.58 (7.04)	0.70 (8.5)	0.83 (10.1)
AISI 4615–4820	0.58 (7.04)	0.70 (8.5)	0.83 (10.1)

[°]General Electric Company.

The unit horsepower must be corrected for feed. From Fig. 12 and a feed of 0.025 in/r (0.6 mm/r), the correction factor is found to be 0.90. Thus, the true unit horsepower = (0.70)(0.90) = 0.63 hp/(in^3·min) [28.7 W/(cm^3·min)].

3. Compute the horsepower required at the cutter

The horsepower required at the cutter hp_c = (hp_u)(MRR), or hp_c = (0.63)(22.5) = 14.18 hp (10.6 kW).

4. Compute the motor horsepower required

The power required at the cutter is the input necessary after allowing for losses in gears, bearings, and other parts of the drive. Table 16 lists typical overall machine-tool efficiencies. A gear-head lathe has an efficiency of 70 percent. Thus, hp_m = hp_c/e, where e = machine-tool efficiency, expressed as a decimal. Or, hp_m = 14.18/0.70 = 20.25 hp (15.1 kW). A 20-hp (14.9-kW) motor would be satisfactory for this machine.

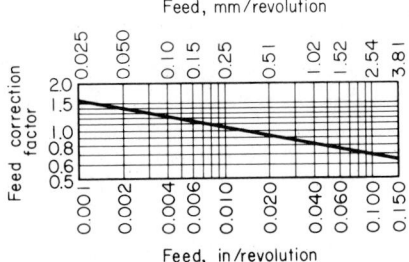

FIG. 12 Feed correction factors based on normal tool geometries. *(General Electric Co.)*

TABLE 16 Efficiencies of Metalworking Machines[°]

Typical overall machine-tool efficiency values (except milling machines), percent	Typical overall efficiencies for milling machines		
	Rated power of machine		Overall efficiency, percent
	hp	kW	
Direct spindle drive, 90	3	2.2	40
	5	3.7	48
One-belt drive, 85	7.5	5.6	52
Two-belt drive, 70	10	7.5	52
	15	11.2	52
Geared head, 70	20	14.9	60
	25	18.6	65
	30	22.4	70
	40	29.8	75
	50	37.3	80

[°]General Electric Company.

Related Calculations: Use this procedure for single or multiple tools. When more than one tool is working at the same time, compute hp_c for each tool and add the individual values to find the total hp_c. Divide the total hp_c by the machine efficiency to determine the required motor horsepower hp_m. This procedure makes ample allowance for dulling of the tools.

Compute metal removal rates for other operations as follows. *Face milling:* MRR = WDF_T, where W = width of cut, in; D = depth of cut, in; F_T = table feed, in/min. *Slot milling:* MRR = WDF_T, where all symbols are as before. *Planing or shaping:* MRR = $DfLS$, where D = depth of cut, in; f = feed, in per stroke or revolution; L = length of workpiece, in; S = strokes/min. *Multiple tools:* MRR = $(d_1^2 - d_s^2)\pi fR/4$, where d_1 = original diameter of workpiece, in, *before* cutting; d_s = workpiece diameter *after* cutting, in; R = rpm of workpiece; other symbols as before.

The procedure given here is the work of Robert G. Brierley and H. J. Siekmann as reported in *Machining Principles and Cost Control.*

CUTTING SPEED FOR LOWEST-COST MACHINING

What is the optimal cutting speed for a part if the maximum feed for which an acceptable finish is obtained at 169 r/min of the workpiece is 0.011 in/r (0.3 mm/r) when the cost of labor and overhead is $0.24 per hour, the number of pieces produced per tool change is 15, the cost per tool change is $0.62, and the length of cut is 8 in (20.3 cm)?

Calculation Procedure:

1. Compute the optimization factor

When the lowest-cost machining speed for an operation is determined, one popular procedure is to choose any speed and feed at which the operation meets the finish requirements. If desired, the speed and feed at which the operation is now running might be chosen. By keeping the speed constant, the feed is increased to the maximum value for which the finish is acceptable. This is the optimal value for the feed and is called the *optimal feed*. The number of pieces produced under these conditions is measured between tool changes. Then the optimal cutting speed is computed from: optimal cutting speed, r/min = (chosen speed, r/min)(optimization factor).

For any operation, the optimization factor = {(labor and overhead cost, $/min)(number of pieces per tool change)(length of cut, in)/[(3)(cost per tool change, $)(chosen speed, r/min)(optimal feed, in/r)]}$^{-4}$. Substitute the given values. Thus, the optimization factor = $\{(0.24)(15)(8)/[(3)(0.62)(169)(0.011)]\}^{-4}$ = 1.7.

2. Compute the optimal cutting speed

From the relation given in step 1, optimal cutting speed = (chosen speed, r/min)(optimization factor) = (169)(1.7) = 287 r/min.

Related Calculations: The relation given in step 1 for the optimization factor is valid for carbide tools. It can be modified to apply to high-speed tools by changing the fourth root to an eight root and changing the 3 in the denominator to 7.

REORDER QUANTITY FOR OUT-OF-STOCK PARTS

A metalworking process uses 10 parts during the lead time. How many parts should be reordered if an out-of-stock situation can be accepted for 10 percent of the time? For 35 percent of the time?

Calculation Procedure:

1. Determine the out-of-stock factor

Table 17 lists out-of-stock factors for various times during which a part might be out of stock. Thus, the acceptable out-of-stock factor for 10 percent is 1.29, and for 35 percent it is 0.39.

2. Compute the reorder quantity

For any manufacturing process, reorder quantity = (out-of-stock factor)(usage during lead time)$^{0.5}$ + (usage during lead time). Thus, the reorder point for this process with an acceptable out-of-stock factor for 10 percent is $(1.29)(10)^{0.5} + (10) = 14.08$ parts, say 15 parts. With 35 percent, reorder point $= (0.29)(10)^{0.5} + 10 = 11.23$, or 12 parts.

Related Calculations: Use this procedure for any types of parts ordered from either an internal or external source. In general, reducing the allowable stock-out time will increase the time during which a process using the parts can operate.

SAVINGS WITH MORE MACHINABLE MATERIALS

What are the gross and net savings made with a more machinable material that reduces the production time by 36 s per part when 800 lb (362.9 kg) of steel is required for 1000 parts and the total machine operating cost is $6 per hour? The more machinable material costs 4 cents per pound (8.8 cents per kilogram) more than the less machinable material, and 5000 parts are produced per day.

Calculation Procedure:

1. Compute the gross savings possible

The gross saving possible in a machining operation when a more machinable material is used is: gross saving, cents/lb = (machining time saved with new material, s per piece)(total cost of operating machine, cents/h)/[(3.6)(weight of material to make 1000 pieces, lb)]. For this operation, the gross saving $= (36)(600)/[(3.6)(800)] = 7.5$ cents per pound (16.5 cents per kilogram). With a production rate of 5000 parts per day, the gross saving is (5000 parts)(800 lb/1000 parts)(7.5 cents/lb) = 30,000 cents, or $300.

2. Compute the net savings possible

The more machinable materials cost 4 cents more per pound than the less machinable material. Hence, the net saving is $(7.5 - 4.0)$ = 3.5 cents per pound (7.7 cents per kilogram), or (5000 parts)(800 lb/1000 parts)(3.5 cents/lb) = 14,000 cents, or $140.

Related Calculations: Use this general procedure for parts made of any material—steel, brass, bronze, aluminum, plastic, etc.

TABLE 17 Out-of-Stock Factors°

Acceptable percentage of stock-outs	Out-of-stock factor
50	0.00
45	0.13
40	0.26
35	0.39
25	0.68
15	1.04
10	1.29
5	1.65
4	1.76
3.5	1.82
3	1.89
2	2.06
1	2.33
0	4.0

°Nyles V. Reinfeld in *American Machinist*.

TIME REQUIRED FOR THREAD MILLING

How long will it take to thread-mill a 2⅞-in (7.3-cm) diameter hard steel bolt with a 2½-in (6.4-cm) diameter 18-flute hob?

Calculation Procedure:

1. Determine the cutting speed and feed of the hob

Table 18 lists typical cutting speeds and feeds for various materials. For hard steel, the usual cutting speed in thread milling is 50 ft/min (0.3 m/s), and the feed per flute is 0.002 in (0.05 mm).

2. Compute the time required for thread milling

The time required for thread milling, T_t min $= \pi d/(fnR)$, where d = work diameter, in; f = feed, in per flute; n = number of flutes on hob; R = hob rpm.

From a previous calculation procedure, $R = 12C/(\pi d)$, where C = hob cutting speed, ft/min; d = hob diameter, in. For this hob, $R = (12)(50)/[\pi(2.5)] = 76.4$ r/min. Then $T_t = \pi(2⅞)/[(0.002)(18)(76.4)] = 3.29$ min.

TABLE 18 Thread-Milling Speeds and Feeds

Material threaded	Speed		Feed per flute	
	ft/min	m/s	in	mm
Aluminum	500	2.5	0.0015	0.038
Brass	250	1.3	0.0015	0.038
Mild steel	100	0.5	0.0020	0.051
Medium steel	75	0.4	0.0020	0.051
Hard steel	50	0.3	0.0020	0.051

Related Calculations: Use this procedure for any metallic or nonmetallic material—aluminum, brass, mild steel, medium steel, hard steel, plastics, etc.

DRILL PENETRATION RATE AND CENTERLESS GRINDER FEED RATE

What is the drill penetration rate when a drill turns at 1000 r/min and has a feed of 0.006 in/r (0.15 mm/r)? What is the feed rate of a centerless grinder having a 12-in (30.5-cm) diameter regulating wheel running at 60 r/min if the angle of inclination between the regulating and grinding wheel is 5°?

Calculation Procedure:

1. Compute the rate of drill penetration

The rate of drill penetration P in/min $= fR$, where f = drill feed, in/r; R = drill rpm. For this drill, $P = (0.006)(1000) = 6.0$ in/min (2.5 mm/s).

2. Compute the grinder feed rate

The work feed f in/min in a centerless grinder is $f = \pi dR \sin a$, where d = regulating-wheel diameter, in; R = regulating-wheel rpm; a = angle of inclination between the regulating and grinding wheel. For this grinder, $f = \pi(12)(60)(\sin 5°) = 197.6$ in/min (8.4 cm/s).

BENDING, DIMPLING, AND DRAWING METAL PARTS

What is the minimum bend radius R in for 0.02-gage Vascojet 1000 metal if it is bent transversely to an angle of 130°? What is the minimum radius R in of a bend in 0.040-gage Rene 41 metal bent longitudinally at an angle of 52° at room temperature? Determine the maximum length of dimple flange H in for AM-350 metal at 500°F (260°C) when the bend angle is 42° and the edge radius R is 0.250 in (6.4 mm). Find the maximum blank diameter and maximum cup depth for drawing Rene 41 metal at 400°F (204°C) when using a die diameter of 10 in (25.4 cm) and 0.063-gage material. Figure 13a, b, and c shows the anticipated manufacturing conditions.

Calculation Procedure:

1. Compute the minimum bend radius

Table 19 shows that the critical bend angle (i.e., maximum bend angle α without breakage) for Vascojet 1000 metal is 118°. Hence, the required bend angle is greater than the critical bend angle. Therefore, the required bend limit equals the critical bend limit, and $R/T = 1.30$, from Table 19. Hence, the minimum radius $R_m = (R/T)(T) = (1.30)(0.02) = 0.026$ in (0.66 mm).

With Rene 41 metal, bent longitudinally at room temperature, the critical bend angle is 122°, from Table 19. Since the required bend angle of 52° is less than critical, find the R/T value in the right-hand portion of Table 19. When the actual bend angle is between two tabulated angles, interpolate thus:

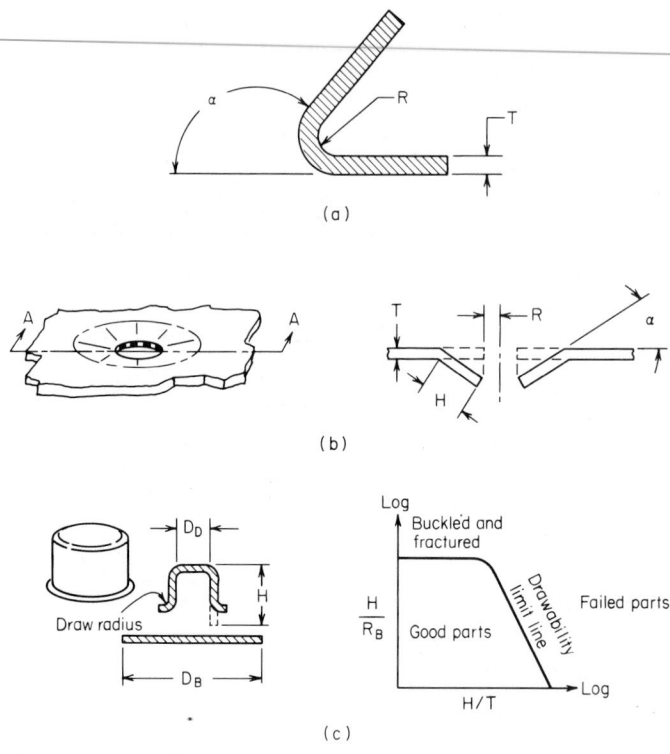

FIG. 13 (*a*) Brake-bent part shape and parameters; (*b*) ram-coin dimpling setup; (*c*) drawing setup. (*American Machinist.*)

TABLE 19 Brake-Bend Parts Parameters°

| Material | L/T | F | ∞ | R/T | \multicolumn{7}{c}{R/T for angles ∞ below critical} |
					30	45	60	75	90	105	120
Titanium (13V-11Cr-3Al)	L/T	RT	105	2.40	0.34	0.68	1.16	1.80	2.25	2.40	
Vascojet 1000	L/T	RT	118	1.30	0.18	0.38	0.64	0.92	1.13	1.26	1.30
USS 12 MoV	L/T	RT	119	1.20	0.16	0.34	0.60	0.84	1.04	1.16	1.20
17-7PH	L/T	RT	122	0.80	0.10	0.22	0.37	0.54	0.66	0.75	0.79
AM-350	L/T	RT	122	0.80	0.10	0.22	0.37	0.54	0.66	0.75	0.79
PH 15-7 Mo	L/T	RT	121	0.86	0.11	0.23	0.42	0.60	0.72	0.80	0.84
A-286	L/T	RT	124	0.66	0.07	0.15	0.29	0.43	0.54	0.62	0.65
Hastelloy X	L/T	RT	120	1.00	0.12	0.26	0.47	0.67	0.84	0.95	1.00
Inconel X	L/T	RT	124	0.64	0.06	0.14	0.28	0.41	0.52	0.60	0.63
Rene 41	L	RT	122	0.80	0.10	0.22	0.37	0.54	0.66	0.75	0.79
Rene 41	T	RT	113	1.64	0.28	0.53	0.84	1.16	1.44	1.58	1.64
J-1570	L	RT	124	0.68	0.08	0.16	0.30	0.45	0.56	0.64	0.67

° *American Machinist*, LTV, Inc.; USAF.
Note: L/T = grain direction, where L = longitudinal and T = transverse; F = bending temperature; ∞ = critical bend; R/T = critical bend limits.

Angle,°	R/T value
60	0.37
52	
45	0.22

$[(52 - 45)/(60 - 45)](0.37 - 0.22) = 0.07$. Then R/T for $52° = 0.22 + 0.07 = 0.29$. With R/T known for $52°$, compute the minimum radius $R_m = (R/T)(T) = (0.29)(0.040) = 0.0116$ in (0.2946 mm).

2. Determine the dimple-flange length

Table 20 shows typical dimpling limits to avoid radial splitting at the edge of the hole of various modern materials. With a bend angle between the tabulated angles, interpolate thus:

Angle, °	H/R value
45	1.10
42	
40	1.43

$[(42 - 40)/(45 - 40)](1.10 - 1.43) = -0.132$, and $H/R = 1.43 + (-0.132) = 1.298$ at $42°$. Then the maximum dimple-flange length $H_m = (H/R)(R) = (1.298)(0.250) = 0.325$ in (8.255 mm).

3. Determine the maximum blank diameter

Table 21 lists the drawing limits for flat-bottom cups made of various modern materials. For this cup, $D_D/T = 10/0.063 = 158.6$, say 159. The corresponding D_B/T and H/D_D ratios are not tabulated. Therefore, interpolate between D_D/T values of 150 and 200. Thus, for Rene 41:

D_D/T	D_B/D_D
200	1.52
159	
150	1.73

$[(159 - 150)/(200 - 150)](1.52 - 1.73) = -0.0378$, and $D_B/D_D = 1.73 + (-0.0378) = 1.692$, when $D_D/T = 159$. Then, the *maximum* value of $D_{Bm} = (D_B/D_D)(D_D) = (1.692)(10) = 16.92$ in (43.0 cm).

Interpolating as above yields $H/D_D = 0.48$ when $D_D/T = 159$. Then the maximum height $H_m = (H/D_D)(D) = (0.48)(10) = 4.8$ in (12.2 cm).

Related Calculations: The procedures given here are typical of those used for the newer "exotic" metals developed for use in aerospace, cryogenic, and similar advanced technologies. The three tables presented here were developed by LTV, Inc., for the U.S. Air Force, and reported in *American Machinist*.

BLANK DIAMETERS FOR ROUND SHELLS

What blank diameter D in is required for the round shells in Fig. 14a and b if $d = 12$ in (30.5 cm), $d_1 = 12$ in (30.5 cm), $d_2 = 14$ in (35.6 cm), and $h = 14$ in (35.6 cm)?

TABLE 20 Dimpling Limits to Avoid Radial Splitting at Hole Edge[a]

		Dimpling limit H/R				
		Standard, for various bend angles a; above and below standard bend angle				
Material	Temperature, °F (°C)	30°	35°	40°	45°	50°
2024-T3	70 (21.1)	2.15	1.60	1.20	0.93	0.80
Ti-8-1-1	70 (21.1)	1.88	1.42	1.08	0.82	0.70
TZM Moly	70 (21.1)	1.98	1.50	1.12	0.87	0.73
Cb-752	70 (21.1)	2.28	1.70	1.30	0.98	0.83
PH 15-7 Mo	500 (260.0)	2.43	1.84	1.40	1.07	0.90
AM-350	500 (260.0)	2.46	1.87	1.43	1.10	0.93
Ti-8-1-1	1200 (648.9)	2.30	1.72	1.30	1.00	0.85
Ti-13-11-3	1200 (648.9)	2.58	1.95	1.48	1.15	0.95

[a] *American Machinist*, LTV, Inc.; USAF.

TABLE 21 Drawing Limits for Flat-Bottom Cups[a]

		Die to blank diameter ratios D_B/D_D; cup-depth ratios H/D_D							
		For various D_D/T ratios							
Material	Temperature ratio, °F (°C)	25	50	100	150	200	250	300	400
Am-350	500 (260.0) D_B/D_D	2.22	2.18	2.00	1.71	1.54	1.42	1.40	1.30
	H/D_D	0.97	0.95	0.75	0.50	0.37	0.30	0.26	0.20
A-286	1000 (537.8) D_B/D_D	2.22	2.46	2.16	1.85	1.64	1.49	1.41	1.36
	H/D_D	1.00	1.21	0.87	0.57	0.42	0.34	0.28	0.22
Rene 41	400 (204.4) D_B/D_D	2.22	2.22	1.92	1.73	1.52	1.48	1.42	1.33
	H/D_D	0.97	0.97	0.73	0.51	0.37	0.31	0.27	0.21
L-605	500 (260.0) D_B/D_D	2.22	2.29	2.00	1.68	1.54	1.45	1.44	1.38
	H/D_D	0.97	1.05	0.74	0.47	0.35	0.30	0.26	0.21
T1-13-11-3	1200 (648.9) D_B/D	2.38	2.53	2.28	1.92	1.67	1.58	1.45	1.44
	H/D_D	1.15	1.34	0.94	0.60	0.44	0.35	0.30	0.24
Tungsten	600 (315.6) D_B/D_D	2.08	2.11	1.98	1.66	1.53	1.46	1.38	1.34
	H/D_D	0.83	0.87	0.69	0.45	0.35	0.29	0.24	0.20

[a] *American Machinist*, LTV, Inc.; USAF.

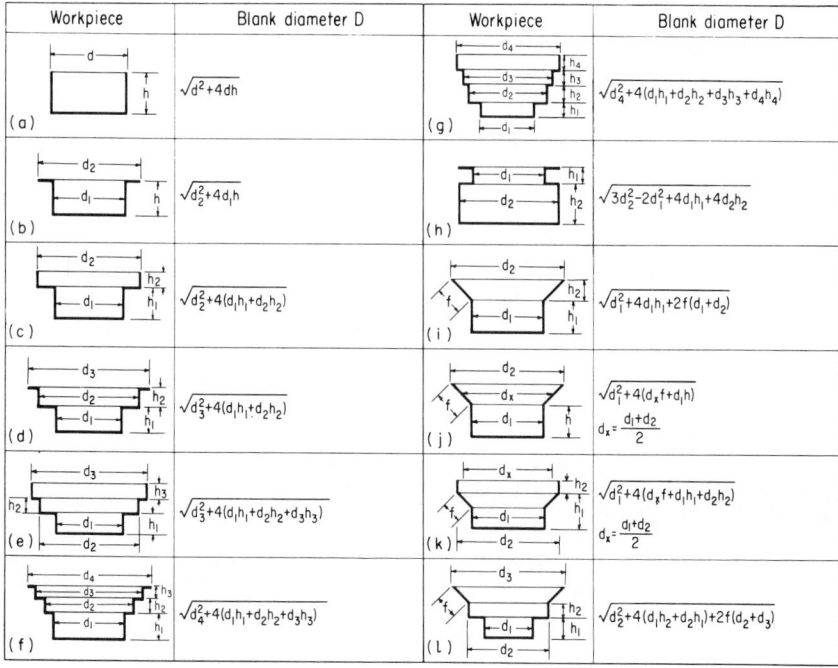

FIG. 14 Blank diameters for round shells. *(American Machinist.)*

Calculation Procedure:

1. *Compute the plain-cup blank diameter*

Figure 14a shows the plain cup. Compute the required blank diameter from $D = (d^2 + 4dh)^{0.5}$ $= (12^2 + 4 \times 12 \times 4)^{0.5} = 18.33$ in (46.6 cm).

2. *Compute the flanged-cup blank diameter*

Figure 14b shows the flanged cup. Compute the required blank diameter from $D = (d_2^2 + fd_1h)^{0.5}$ $= (14^2 + 4 \times 12 \times 4)^{0.5} = 19.7$ in (49.3 cm).

Related Calculations: Figure 14 gives the equations for computing 12 different round-shell blank diameters. Use the same general procedures as in steps 1 and 2 above. These equations were derived by Ferene Kuchta, Mechanical Engineer, J. Wiss & Sons Co., and reported in *American Machinist.*

BREAKEVEN CONSIDERATIONS IN MANUFACTURING OPERATIONS

A manufacturing plant has the net sales and fixed and variable expenses shown in Table 22. What is the breakeven point for this plant in units and sales? Plot a conventional and an alternative breakeven chart for this plant.

Calculation Procedure:

1. *Compute the breakeven units*

Use the relation BE_u = fixed expenses, $/[(sales income per unit, $ − variable costs per unit, $)]$, where BE_u = breakeven in units. By substituting, BE_u = $400,000/($20 − $12) = 50,000 units.

TABLE 22 Manufacturing Business Income and Expenses

Condensed Income Statement
For year ending Dec. 31, 19—

Net sales (60,000 units @ $20 per unit)	. . . .		$1,200,000
	Variable	*Fixed*	
Less costs and expenses:			
Direct material	$195,000	. . .	
Direct labor	215,000	. . .	
Manufacturing expenses	100,000	$200,000	
Selling expenses	50,000	150,000	
General and administrative expenses	160,000	50,000	
Total	720,000	400,000	1,120,000
Net profit before federal income taxes	. . . .		80,000

2. *Compute the breakeven sales*

Two methods can be used to compute the breakeven sales. With the breakeven units known from step 1, $BE_s = BE_u$(unit sales price, $), where BE_s = breakeven sales, $. By substituting, $BE_s = (50,000)(\$20) = \$1,000,000$.

Alternatively, compute the profit-volume (PV) ratio: $PV = (\text{sales, \$} - \text{variable costs, \$})/\text{sales,}$ $\$ = (\$1,200,000 - \$720,000)/\$1,200,000 = 0.40$. Then BE_s = fixed costs, $/PV = \$400,000/0.40 = \$1,000,000$. This is identical to the breakeven sales computed in the previous paragraph.

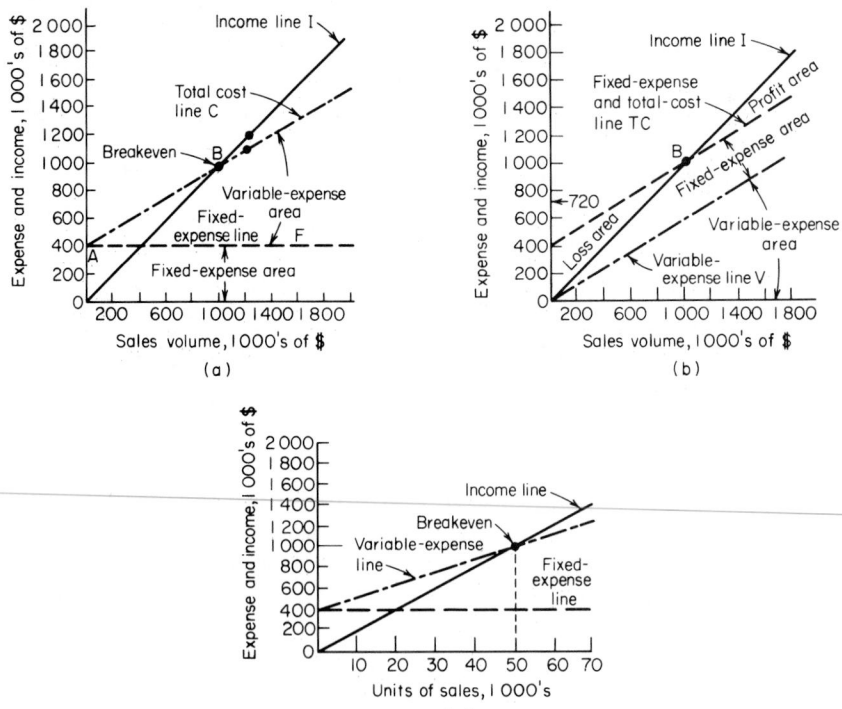

FIG. 15 Three forms of the breakeven chart as used in metalworking activities.

3. *Draw the conventional and alternative breakeven charts*

Figure 15a shows the conventional breakeven chart for this plant. Construct this chart by drawing the horizontal line *F* for the fixed expenses, the solid sloping line *I* for the income or sales, and the dotted sloping line *C* for the total costs. Note that the vertical axis is for the expenses and income and the horizontal axis is for the sales, all measured in monetary units.

The breakeven point is at the intersection of the income and total-cost curves, point *B*, Fig. 15a. Projecting vertically downward shows that point *B* corresponds to a sales volume of $1,000,000, as computed in step 2.

Alternative breakeven charts are shown in Fig. 15b and c. Both charts are constructed in a manner similar to Fig. 15a.

Related Calculations: Breakeven computations are valuable tools for analyzing any manufacturing operation. The concepts are also applicable to other business activities. Thus, typical PV values for various types of businesses are:

Business	Typical activity	Typical PV
Consumer appliances	Fully automated; high-volume output	0.15–0.25
Standard centrifugal pumps	Batch output in large volume	0.20–0.30
Acid-handling centrifugal pumps	Batch output in small volume	0.25–0.35
Standard prototype, one-of-a-kind	Ships, machine tools	0.30–0.40
Special-design one-of-a-kind prototype	Buildings, factories	0.35–0.50

Combustion

REFERENCES: Chigier—*Energy, Combustion and Environment*, McGraw-Hill; Lewis and Von Elbe—*Combustion, Flames and Explosion of Gases*, Pergamon Press; Zung—*Evaporation-Combustion of Fuel*, American Chemical Society; Johnson and Auth—*Fuels and Combustion Handbook*, McGraw-Hill; Babcock & Wilcox Company—*Steam: Its Generation and Use*; Combustion Engineering Corporation—*Combustion Engineering*; Gaffert—*Steam Power Stations*, McGraw-Hill; Skrotzki and Vopat—*Applied Energy Conversion*, McGraw-Hill; Popovich-Hering—*Fuels and Lubricants*, Wiley; ASME—*Power Test Code for Steam Boilers*; Moore—*Coal*, Wiley; Moore—*Liquid Fuels*, The Technical Press, Ltd., London; American Gas Association—*Combustion*; Dunstan—*Science of Petroleum*, Oxford, London; Trinks—*Industrial Furnaces*, Wiley; Perry—*Chemical Engineers Handbook*, McGraw-Hill.

COMBUSTION OF COAL FUEL IN A FURNACE

A coal has the following ultimate analysis (or percent by weight): C = 0.8339; H_2 = 0.0456; O_2 = 0.0505; N_2 = 0.0103; S = 0.0064; ash = 0.0533; total = 1.000 lb (0.45 kg). This coal is burned in a steam-boiler furnace. Determine the weight of air required for theoretically perfect combustion, the weight of gas formed per pound (kilogram) of coal burned, and the volume of flue gas, at the boiler exit temperature of 600°F (316°C) per pound (kilogram) of coal burned; air required with 20 percent excess air, and the volume of gas formed with this excess; the CO_2 percentage in the flue gas on a dry and wet basis.

Calculation Procedure:

1.   *Compute the weight of oxygen required per pound of coal*

To find the weight of oxygen required for theoretically perfect combustion of coal, set up the following tabulation, based on the ultimate analysis of the coal:

Element	$\times$	Molecular-weight ratio	=	lb (kg) O_2 required
C; 0.8339	$\times$	32/12	=	2.2237 (1.009)
H_2; 0.0456	$\times$	16/2	=	0.3648 (0.165)
O_2; 0.0505; decreases external O_2 required			=	−0.0505 (0.023)
N_2; 0.0103 is inert in combustion and is ignored				
S; 0.0064	$\times$	32/32	=	0.0064 (0.003)
Ash 0.0533 is inert in combustion and is ignored				
Total 1.0000 lb (kg) external O_2 per lb (kg) fuel			=	2.5444 (1.154)

Note that of the total oxygen needed for combustion, 0.0505 lb (0.023 kg), is furnished by the fuel itself and is assumed to reduce the total external oxygen required by the amount of oxygen present in the fuel. The molecular-weight ratio is obtained from the equation for the chemical reaction of the element with oxygen in combustion. Thus, for carbon $C + O_2 \rightarrow CO_2$, or $12 + 32 = 44$, where 12 and 32 are the molecular weights of C and O_2, respectively.

2.   *Compute the weight of air required for perfect combustion*

Air at sea level is a mechanical mixture of various gases, principally 23.2 percent oxygen and 76.8 percent nitrogen by weight. The nitrogen associated with the 2.5444 lb (1.154 kg) of oxygen required per pound (kilogram) of coal burned in this furnace is the product of the ratio of the nitrogen and oxygen weights in the air and 2.5444, or $(2.5444)(0.768/0.232) = 8.4228$ lb (3.820 kg). Then the weight of air required for perfect combustion of 1 lb (0.45 kg) of coal = sum of nitrogen and oxygen required $= 8.4228 + 2.5444 = 10.9672$ lb (4.975 kg) of air per pound (kilogram) of coal burned.

3.   *Compute the weight of the products of combustion*

Find the products of combustion by addition:

Fuel constituents	+	Oxygen	$\rightarrow$	Products of combustion	lb	kg
C; 0.8339	+	2.2237	$\rightarrow$	CO_2 =	3.0576	1.387
H; 0.0456	+	0.3648	$\rightarrow$	H_2O =	0.4104	0.186
O_2; 0.0505; this is *not* a product of combustion						
N_2; 0.0103; inert but passes through furnace				=	0.0103	0.005
S; 0.0064	+	0.0064	$\rightarrow$	SO_2 =	0.0128	0.006
Outside nitrogen from step 2				$= N_2$ =	8.4228	3.820
lb (kg) of flue gas per lb (kg) of coal burned				=	11.9139	5.404

4.   *Convert the flue-gas weight to volume*

Use Avogadro's law, which states that under the same conditions of pressure and temperature, 1 mol (the molecular weight of a gas expressed in lb) of any gas will occupy the same volume.

At 14.7 lb/in^2 (abs) (101.3 kPa) and 32°F (0°C), 1 mol of any gas occupies 359 ft^3 (10.2 m^3). The volume per pound of any gas at these conditions can be found by dividing 359 by the molecular weight of the gas and correcting for the gas temperature by multiplying the volume by the ratio of the absolute flue-gas temperature and the atmospheric temperature. To change the weight analysis (step 3) of the products of combustion to volumetric analysis, set up the calculation thus:

Products	Weight		Molecular weight	Temperature correction	Volume at	
	lb	kg			600°F, ft^3	316°C, m^3
CO_2	3.0576	1.3869	44	(359/44)(3.0576)(2.15) =	53.6	1.518
H_2O	0.4104	0.1862	18	(359/18)(0.4104)(2.15) =	17.6	0.498
Total N_2	8.4331	3.8252	28	(359/28)(8.4331)(2.15) =	232.5	6.584
SO_2	0.0128	0.0058	64	(359/64)(0.0128)(2.15) =	0.15	0.004
ft^3 (m^3) of flue gas per lb (kg) of coal burned				=	303.85	8.604

In this calculation, the temperature correction factor 2.15 = absolute flue-gas temperature, °R/ absolute atmospheric temperature, °R = (600 + 460)/(32 + 460). The total weight of N_2 in the flue gas is the sum of the N_2 in the combustion air and the fuel, or 8.4228 + 0.0103 = 8.4331 lb (3.8252 kg). The value is used in computing the flue-gas volume.

5. Compute the CO_2 content of the flue gas

The volume of CO_2 in the products of combustion at 600°F (316°C) is 53.6 ft^3 (1.518 m^3), as computed in step 4; and the total volume of the combustion products is 303.85 ft^3 (8.604 m^3). Therefore, the percent CO_2 on a wet basis (i.e., including the moisture in the combustion products) = ft^3 CO_2/total ft^3 = 53.6/303.85 = 0.1764, or 17.64 percent.

The percent CO_2 on a dry, or Orsat, basis is found in the same manner except that the weight of H_2O in the products of combustion, 17.6 lb (7.83 kg) from step 4, is subtracted from the total gas weight. Or, percent CO_2, dry, or Orsat basis = (53.6)/(303.85 − 17.6) = 0.1872, or 18.72 percent.

6. Compute the air required with the stated excess flow

With 20 percent excess air, the air flow required = (0.20 + 1.00)(air flow with no excess) = 1.20 (10.9672) = 13.1606 lb (5.970 kg) of air per pound (kilogram) of coal burned. The air flow with no excess is obtained from step 2.

7. Compute the weight of the products of combustion

The excess air passes through the furnace without taking part in the combustion and increases the weight of the products of combustion per pound (kilogram) of coal burned. Therefore, the weight of the products of combustion is the sum of the weight of the combustion products without the excess air and the product of (percent excess air)(air for perfect combustion, lb); or, given the weights from steps 3 and 2, respectively, = 11.9139 + (0.20)(10.9672) = 14.1073 lb (6.399 kg) of gas per pound (kilogram) of coal burned with 20 percent excess air.

8. Compute the volume of the combustion products and the percent CO_2

The volume of the excess air in the products of combustion is obtained by converting from the weight analysis to the volumetric analysis and correcting for temperature as in step 4, using the air weight from step 2 for perfect combustion and the excess-air percentage, or (10.9672)(0.20)(359/28.95)(2.15) = 58.5 ft^3 (1.656 m^3). In this calculation the value 28.95 is the molecular weight of air. The total volume of the products of combustion is the sum of the column for perfect combustion, step 4, and the excess-air volume, above, or 303.85 + 58.5 = 362.35 ft^3 (10.261 m^3).

By using the procedure in step 5, the percent CO_2, wet basis = 53.6/362.35 = 14.8 percent. The percent CO_2, dry basis = 53.8/(362.35 − 17.6) = 15.6 percent.

Related Calculations: Use the method given here when making combustion calculations for any type of coal—bituminous, semibituminous, lignite, anthracite, cannel, or coking—from any coal field in the world used in any type of furnace—boiler, heater, process, or waste-heat. When the air used for combustion contains moisture, as is usually true, this moisture is added to the combustion-formed moisture appearing in the products of combustion. Thus, for 80°F (26.7°C) air of 60 percent relative humidity, the moisture content is 0.013 lb/lb (0.006 kg/kg) of dry air. This amount appears in the products of combustion for each pound of air used and is a commonly assumed standard in combustion calculations.

Fossil-fuel-fired power plants release sulfur emissions to the atmosphere. In turn, this produces sulfates, which are the key ingredient in acid rain. The federal Clean Air Act regulates sulfur dioxide emissions from power plants. Electric utilities which burn high-sulfur coal are thought to produce some 35 percent of atmospheric emissions of sulfur dioxide in the United States.

Sulfur dioxide emissions by power plants have declined some 30 percent since passage of the Clean Air Act in 1970, and a notable decline in acid rain has been noted at a number of test sites. In 1990 the Acid Rain Control Program was created by amendments to the Clean Air Act. This program further reduces the allowable sulfur dioxide emissions from power plants, steel mills, and other industrial facilities.

The same act requires reduction in nitrogen oxide emissions from power plants and industrial facilities, so designers must keep this requirement in mind when designing new and replacement facilities of all types which use fossil fuels.

PERCENT EXCESS AIR WHILE BURNING COAL

A certain coal has the following composition by weight percentages: carbon 75.09, nitrogen 1.56, ash 3.38, hydrogen 5.72, oxygen 13.82, sulfur 0.43. When burned in an actual furnace, measurements showed that there was 8.93 percent combustible in the ash pit refuse and the following Orsat analysis in percentages was obtained: carbon dioxide 14.2, oxygen 4.7, carbon monoxide 0.3. If it can be assumed that there was no combustible in the flue gas other that the carbon monoxide reported, calculate the percentage of excess air used.

1.   *Compute the amount of theoretical air required per lb$_m$ (kg) of coal*

Theoretical air required per pound (kilogram) of coal, $w_{ta} = 11.5C' + 34.5[H_2' - O_2'/8)] + 4.32S'$, where C', H_2', O_2', and S' represent the percentages by weight, expressed as decimal fractions, of carbon, hydrogen, oxygen, and sulfur, respectively. Thus, $w_{ta} = 11.5(0.7509) + 34.5[0.0572 - (0.1382/8)] + 4.32(0.0043) = 10.03$ lb (4.55 kg) of air per lb (kg) of coal. The ash and nitrogen are inert and do not burn.

2.   *Compute the correction factor for combustible in the ash*

The correction factor for combustible in the ash, $C_1 = (w_f C_f - w_r C_r)/(w_f \times 100)$, where the amount of fuel, $w_f = 1$ lb (0.45 kg) of coal; percent by weight, expressed as a decimal fraction, of carbon in the coal, $C_f = 75.09$; percent by weight of the ash and refuse in the coal, $w_r = 0.0338$; percent by weight of combustible in the ash, $C_r = 8.93$. Hence, $C_1 = [(1 \times 75.09) - (0.0338 \times 8.93)]/(1 \times 100) = 0.748$.

3.   *Compute the amount of dry flue gas produced per lb (kg) of coal*

The lb (kg) of dry flue gas per lb (kg) of coal, $w_{dg} = C_1(4CO_2 + O_2 + 704)/[3(CO_2 + CO)]$, where the Orsat analysis percentages are for carbon dioxide, $CO_2 = 14.2$; oxygen, $O_2 = 4.7$; carbon monoxide, $CO = 0.3$. Hence, $w_{dg} = 0.748 \times [(4 \times 14.2) + 4.7 + 704)]/[3(14.2 + 0.3)] = 13.16$ lb/lb (5.97 kg/kg).

4.   *Compute the amount of dry air supplied per lb (kg) of coal*

The lb (kg) of dry air supplied per lb (kg) of coal, $w_{da} = w_{dg} - C_1 + 8[H_2' - (O_2'/8)] - (N_2'/N)$, where the percentage by weight of nitrogen in the fuel, $N_2' = 1.56$, and "atmospheric nitrogen" in the supply air, $N_2 = 0.768$; other values are as given or calculated. Then, $w_{da} = 13.16 - 0.748 + 8[0.0572 - (0.1382/8)] - (0.0156/0.768) = 12.65$ lb/lb (5.74 kg/kg).

5.   *Compute the percent of excess air used*

Percent excess air $= (w_{da} - w_{ta})/w_{ta} = (12.65 - 10.03)/10.03 = 0.261$, or 26.1 percent.

Related Calculations: The percentage by weight of nitrogen in "atmospheric air" in step 4 appears in *Principles of Engineering Thermodynamics*, 2nd edition, by Kiefer et al., John Wiley & Sons, Inc.

COMBUSTION OF FUEL OIL IN A FURNACE

A fuel oil has the following ultimate analysis: C = 0.8543; H_2 = 0.1131; O_2 = 0.0270; N_2 = 0.0022; S = 0.0034; total = 1.0000. This fuel oil is burned in a steam-boiler furnace. Determine the weight of air required for theoretically perfect combustion, the weight of gas formed per pound (kilogram) of oil burned, and the volume of flue gas, at the boiler exit temperature of 600°F (316°C), per pound (kilogram) of oil burned; the air required with 20 percent excess air, and the volume of gas formed with this excess; the CO_2 percentage in the flue gas on a dry and wet basis.

Calculation Procedure:

1. Compute the weight of oxygen required per pound (kilogram) of oil

The same general steps as given in the previous calculation procedure will be followed. Consult that procedure for a complete explanation of each step.

Using the molecular weight of each element, we find

Element	×	Molecular-weight ratio	=	lb (kg) O_2 required
C; 0.8543	×	32/12	=	2.2781 (1.025)
H_2; 0.1131	×	16/2	=	0.9048 (0.407)
O_2; 0.0270; decreases external O_2 required			=	−0.0270 (−0.012)
N_2; 0.0022 is inert in combustion and is ignored				
S; 0.0034	×	32/32	=	0.0034 (0.002)
Total 1.0000				
lb (kg) external O_2 per lb (kg) fuel			=	3.1593 (1.422)

2. Compute the weight of air required for perfect combustion

The weight of nitrogen associated with the required oxygen = (3.1593)(0.768/0.232) = 10.458 lb (4.706 kg). The weight of air required = 10.4583 + 3.1593 = 13.6176 lb/lb (6.128 kg/kg) of oil burned.

3. Compute the weight of the products of combustion

As before,

Fuel constituents	+	Oxygen	=	Products of combustion
C; 0.8543 + 2.2781	=	3.1324	=	CO_2
H_2; 0.1131 + 0.9148	=	1.0179	=	H_2O
O_2; 0.270; *not* a product of combustion				
N_2; 0.0022; inert but passes through furnace	=	0.0022	=	N_2
S; 0.0034 + 0.0034	=	0.0068	=	SO_2
Outside N_2 from Step 2	=	10.458	=	N_2
lb (kg) of flue gas per lb (kg) of oil burned	=	14.6173 (6.578)		

4.   *Convert the flue-gas weight to volume*

As before,

Products	Weight lb	Weight kg	Molecular weight	Temperature correction		Volume at 600°F, ft³	Volume at 316°C, m³
CO_2	3.1324	1.4238	44	(359/44)(3.1324)(2.15)	=	55.0	1.557
H_2O	1.0179	0.4626	18	(359/18)(1.0179)(2.15)	=	43.5	1.231
N_2 (total)	10.460	4.7545	28	(359/28)(10.460)(2.15)	=	288.5	8.167
SO_2	0.0068	0.0031	64	(359/64)(0.0068)(2.15)	=	0.82	0.023
ft³ (m³) of flue gas per lb (kg) of oil burned					=	387.82	10.978

In this calculation, the temperature correction factor 2.15 = absolute flue-gas temperature, °R/absolute atmospheric temperature, °R = $(600 + 460)/(32 + 460)$. The total weight of N_2 in the flue gas is the sum of the N_2 in the combustion air and the fuel, or 10.4580 + 0.0022 = 10.4602 lb (4.707 kg).

5.   *Compute the CO_2 content of the flue gas*

CO_2, wet basis, = 55.0/387.82 = 0.142, or 14.2 percent. CO_2, dry basis, = 55.0/(387.2 − 43.5) = 0.160, or 16.0 percent.

6.   *Compute the air required with stated excess flow*

The pounds (kilograms) of air per pound (kilogram) of oil with 20 percent excess air = (1.20)(13.6176) = 16.3411 lb (7.353 kg) of air per pound (kilogram) of oil burned.

7.   *Compute the weight of the products of combustion*

The weight of the products of combustion = product weight for perfect combustion, lb + (percent excess air)(air for perfect combustion, lb) = 14.6173 + (0.20)(13.6176) = 17.3408 lb (7.803 kilogram) of flue gas per pound (kilogram) of oil burned with 20 percent excess air.

8.   *Compute the volume of the combustion products and the percent CO_2*

The volume of excess air in the products of combustion is found by converting from the weight to the volumetric analysis and correcting for temperature as in step 4, using the air weight from step 2 for perfect combustion and the excess-air percentage, or (13.6176)(0.20)(359/28.95)(2.15) = 72.7 ft³ (2.058 m³). Add this to the volume of the products of combustion found in step 4, or 387.82 + 72.70 = 460.52 ft³ (13.037 m³).

By using the procedure in step 5, the percent CO_2, wet basis = 55.0/460.52 = 0.1192, or 11.92 percent. The percent CO_2, dry basis = 55.0/(460.52 − 43.5) = 0.1318, or 13.18 percent.

Related Calculations: Use the method given here when making combustion calculations for any type of fuel oil—paraffin-base, asphalt-base, Bunker C, no. 2, 3, 4, or 5—from any source, domestic or foreign, in any type of furnace—boiler, heater, process, or waste-heat. When the air used for combustion contains moisture, as is usually true, this moisture is added to the combustion-formed moisture appearing in the products of combustion. Thus, for 80°F (26.7°C) air of 60 percent relative humidity, the moisture content is 0.013 lb/lb (0.006 kg/kg) of dry air. This amount appears in the products of combustion for each pound (kilogram) of air used and is a commonly assumed standard in combustion calculations.

COMBUSTION OF NATURAL GAS IN A FURNACE

A natural gas has the following volumetric analysis at 60°F (15.5°C): CO_2 = 0.004; CH_4 = 0.921; C_2H_6 = 0.041; N_2 = 0.034; total = 1.000. This natural gas is burned in a steam-boiler furnace. Determine the weight of air required for theoretically perfect combustion, the weight of gas formed per pound of natural gas burned, and the volume of the flue gas, at the boiler exit temperature of 650°F (343°C), per pound (kilogram) of natural gas burned; air required with 20 percent excess air, and the volume of gas formed with this excess; CO_2 percentage in the flue gas on a dry and wet basis.

Calculation Procedure:

1. *Compute the weight of oxygen required per pound of gas*

The same general steps as given in the previous calculation procedures will be followed, except that they will be altered to make allowances for the differences between natural gas and coal.

The composition of the gas is given on a volumetric basis, which is the usual way of expressing a fuel-gas analysis. To use the volumetric-analysis data in combustion calculations, they must be converted to a weight basis. This is done by dividing the weight of each component by the total weight of the gas. A volume of 1 ft^3 (1 m^3) of the gas is used for this computation. Find the weight of each component and the total weight of 1 ft^3 (1 m^3) as follows, using the properties of the combustion elements and compounds given in Table 1:

Component	Percent by volume	Density lb/ft^3	kg/m^3	Component weight = column 1 × column 2 lb/ft^3	kg/m^3
CO_2	0.004	0.1161	1.859	0.0004644	0.007
CH_4	0.921	0.0423	0.677	0.0389583	0.624
C_2H_6	0.041	0.0792	1.268	0.0032472	0.052
N_2	0.034	0.0739	0.094	0.0025026	0.040
Total	1.000			0.0451725	0.723

$$\text{Percent } CO_2 = 0.0004644/0.0451725 = 0.01026, \text{ or } 1.03 \text{ percent}$$
$$\text{Percent } CH_4 \text{ by weight} = 0.0389583/0.0451725 = 0.8625 \text{ or } 86.25 \text{ percent}$$
$$\text{Percent } C_2H_6 \text{ by weight} = 0.0032472/0.0451725 = 0.0718, \text{ or } 7.18 \text{ percent}$$
$$\text{Percent } N_2 \text{ by weight} = 0.0025026/0.0451725 = 0.0554, \text{ or } 5.54 \text{ percent}$$

The sum of the weight percentages = $1.03 + 86.25 + 7.18 + 5.54 = 100.00$. This sum checks the accuracy of the weight calculation, because the sum of the weights of the component parts should equal 100 percent.

Next, find the oxygen required for combustion. Since both the CO_2 and N_2 are inert, they do not take part in the combustion; they pass through the furnace unchanged. Using the molecular weights of the remaining components in the gas and the weight percentages, we have

Compound	×	Molecular-weight ratio	=	lb (kg) O_2 required
CH_4; 0.8625	×	64/16	=	3.4500 (1.553)
C_2H_6; 0.0718	×	112/30	=	0.2920 (0.131)
lb (kg) external O_2 required per lb (kg) fuel			=	3.7420 (1.684)

In this calculation, the molecular-weight ratio is obtained from the equation for the combustion chemical reaction, or $CH_4 + 2O_2 = CO_2 + 2H_2O$, that is, $16 + 64 = 44 + 36$, and $C_2H_6 + \frac{7}{2}O_2 = 2CO_2 + 3H_2O$, that is, $30 + 112 = 88 + 54$. See Table 2 from these and other useful chemical reactions in combustion.

2. *Compute the weight of air required for perfect combustion*

The weight of nitrogen associated with the required oxygen = $(3.742)(0.768/0.232) = 12.39$ lb (5.576 kg). The weight of air required = $12.39 + 3.742 = 16.132$ lb/lb (7.259 kg/kg) of gas burned.

TABLE 1 Properties of Combustion Elements°

| Element or compound | Formula | Molecular weight | At 14.7 lb/in² (abs) (101.3 kPa) 60°F (15.6°C) | | Nature | | Heat value, Btu (kJ) | | |
			Weight, lb/ft³ (kg/m³)	Volume, ft³/lb (m³/kg)	Gas or solid	Combustible	Per lb (kg)	Per ft³ (m³) at 14.7 lb/in² (abs) (101.3 kPa), 60°F (15.6°C)	Per mole
Carbon	C	12	...	...	S	Yes	14,540 (33,820)	...	174,500
Hydrogen	H_2	2.02†	0.0053 (0.0849)	188 (11.74)	G	Yes	61,000 (141,886)	325 (12,109)	123,100
Sulfur	S	32	...	...	S	Yes	4,050 (9,420)	...	129,600
Carbon monoxide	CO	28	0.0739 (1.183)	13.54 (0.85)	G	Yes	4,380 (10,187)	323 (12,035)	122,400
Methane	CH_4	16	0.0423 (0.677)	23.69 (1.48)	G	Yes	24,000 (55,824)	1,012 (37,706)	384,000
Acetylene	C_2H_2	26	0.0686 (1.098)	14.58 (0.91)	G	Yes	21,500 (50,009)	1,483 (55,255)	562,000
Ethylene	C_2H_4	28	0.0739 (1.183)	13.54 (0.85)	G	Yes	22,200 (51,637)	1,641 (61,141)	622,400
Ethane	C_2H_6	30	0.0792 (1.268)	12.63 (0.79)	G	Yes	22,300 (51,870)	1,762 (65,650)	668,300
Oxygen	O_2	32	0.0844 (1.351)	11.84 (0.74)	G				
Nitrogen	N_2	28	0.0739 (1.183)	13.52 (0.84)	G				
Air‡	...	29	0.0765 (1.225)	13.07 (0.82)	G				
Carbon dioxide	CO_2	44	0.1161 (1.859)	8.61 (0.54)	G				
Water	H_2O	18	0.0475 (0.760)	21.06 (1.31)	G				

°P. W. Swain and L. N. Rowley, "Library of Practical Power Engineering" (collection of articles published in *Power*).
†For most practical purposes, the value of 2 is sufficient.
‡The molecular weight of 29 is merely the weighted average of the molecular weight of the constituents.

TABLE 2 Chemical Reactions

Combustible substance	Reaction	Mols	lb (kg)°
Carbon to carbon monoxide	$C + \frac{1}{2}O_2 = CO$	$1 + \frac{1}{2} = 1$	$12 + 16 = 28$
Carbon to carbon dioxide	$C + O_2 = CO_2$	$1 + 1 = 1$	$12 + 16 = 28$
Carbon monoxide to carbon dioxide	$CO + \frac{1}{2}O_2 = CO_2$	$1 + \frac{1}{2} = 1$	$28 + 16 = 44$
Hydrogen	$H_2 + \frac{1}{2}O_2 = H_2O$	$1 + \frac{1}{2} = 1$	$2 + 16 = 18$
Sulfur to sulfur dioxide	$S + O_2 = SO_2$	$1 + 1 = 1$	$32 + 32 = 64$
Sulfur to sulfur trioxide	$S + \frac{3}{2}O_2 = SO_3$	$1 + \frac{3}{2} = 1$	$32 + 48 = 80$
Methane	$CH_4 + 2O_2 = CO_2 + 2H_2O$	$1 + 2 = 1 + 2$	$16 + 64 = 44 + 36$
Ethane	$C_2H_6 + \frac{7}{2}O_2 = 2CO_2 + 3H_2O$	$1 + \frac{7}{2} = 2 + 3$	$30 + 112 = 88 + 54$
Propane	$C_3H_8 + 5O_2 = 3CO_2 + 4H_2O$	$1 + 5 = 3 + 4$	$44 + 160 = 132 + 72$
Butane	$C_4H_{10} + \frac{13}{2}O_2 = 4CO_2 + 5H_2O$	$1 + \frac{13}{2} = 4 + 5$	$58 + 208 = 176 + 90$
Acetylene	$C_2H_2 + \frac{5}{2}O_2 = 2CO_2 + H_2O$	$1 + \frac{5}{2} = 2 + 2$	$26 + 80 = 88 + 18$
Ethylene	$C_2H_4 + 3O_2 = 2CO_2 + 2H_2O$	$1 + 3 = 2 + 2$	$28 + 96 = 88 + 36$

°Substitute the molecular weights in the reaction equation to secure lb (kg). The lb (kg) on each side of the equation must balance.

3.197

3. Compute the weight of the products of combustion

Fuel constituents	+	Oxygen	=	Products of combustion	
				lb	kg
CO_2; 0.0103; inert but passes through the furnace			=	0.010300	0.005
CH_4; 0.8625	+	3.45	=	4.312500	1.941
C_2H_6; 0.003247	+	0.2920	=	0.032447	0.015
N_2; 0.0554; inert but passes through the furnace			=	0.055400	0.025
Outside N_2 from step 2			=	12.390000	5.576
lb (kg) of flue gas per lb (kg) of natural gas burned			=	16.800347	7.562

4. Convert the flue-gas weight to volume

The products of complete combustion of any fuel that does not contain sulfur are CO_2, H_2O, and N_2. Using the combustion equation in step 1, compute the products of combustion thus: $CH_4 + 2O_2 = CO_2 + H_2O$; $16 + 64 = 44 + 36$; or the CH_4 burns to CO_2 in the ratio of 1 part CH_4 to 44/16 parts CO_2. Since, from step 1, there is 0.03896 lb CH_4 per ft³ (0.624 kg/m³) of natural gas, this forms $(0.03896)(44/16) = 0.1069$ lb (0.048 kg) of CO_2. Likewise, for C_2H_6, $(0.003247)(88/30) = 0.00952$ lb (0.004 kg). The total CO_2 in the combustion products $= 0.00464 + 0.1069 + 0.00952 = 0.11688$ lb (0.053 kg), where the first quantity is the CO_2 in the fuel.

Using a similar procedure for the H_2O formed in the products of combustion by CH_4, we find $(0.03896)(36/16) = 0.0875$ lb (0.039 kg). For C_2H_6, $(0.003247)(54/30) = 0.005816$ lb (0.003 kg). The total H_2O in the combustion products $= 0.0875 + 0.005816 = 0.093316$ lb (0.042 kg).

Step 2 shows that 12.39 lb (5.58 kg) of N_2 is required per lb (kg) of fuel. Since 1 ft³ (0.028 m³) of the fuel weighs 0.04517 lb (0.02 kg), the volume of gas which weighs 1 lb (2.2 kg) is $1/0.04517 = 22.1$ ft³ (0.626 m³). Therefore, the weight of N_2 per ft³ of fuel burned $= 12.39/22.1 = 0.560$ lb (0.252 kg). This, plus the weight of N_2 in the fuel, step 1, is $0.560 + 0.0025 = 0.5625$ lb (0.253 kg) of N_2 in the products of combustion.

Next, find the total weight of the products of combustion by taking the sum of the CO_2, H_2O, and N_2 weights, or $0.11688 + 0.09332 + 0.5625 = 0.7727$ lb (0.35 kg). Now convert each weight to ft³ at 650°F (343°C), the temperature of the combustion products, or:

Products	Weight		Molecular weight	Temperature correction		Volume at	
	lb	kg				650°F, ft³	343°C, m³
CO_2	0.11688	0.05302	44	$(379/44)(0.11688)(2.255)$	=	2.265	0.0641
H_2O	0.09332	0.04233	18	$(379/18)(0.09332)(2.255)$	=	4.425	0.1252
N_2 (total)	0.5625	0.25515	28	$(379/28)(0.5625)(2.255)$	=	17.190	0.4866
ft³ (m³) of flue gas per ft³ (m³) of natural-gas fuel					=	23.880	0.6759

In this calculation, the value of 379 is used in the molecular-weight ratio because at 60°F (15.6°C) and 14.7 lb/in² (abs) (101.3 kPa), the volume of 1 lb (0.45 kg) of any gas $= 379$/gas molecular weight. The fuel gas used is initially at 60°F (15.6°C) and 14.7 lb/in² (abs) (101.3 kPa). The ratio $2.255 = (650 + 460)/(32 + 460)$.

5. Compute the CO_2 content of the flue gas

CO_2, wet basis $= 2.265/23.88 = 0.947$, or 9.47 percent. CO_2, dry basis $= 2.265/(23.88 - 4.425) = 0.1164$, or 11.64 percent.

6.   Compute the air required with the stated excess flow

With 20 percent excess air, $(1.20)(16.132) = 19.3584$ lb of air per lb (8.71 kg/kg) of natural gas, or $19.3584/22.1 = 0.875$ lb of air per ft^3 (13.9 kg/m^3) of natural gas. See step 4 for an explanation of the value 22.1.

7.   Compute the weight of the products of combustion

Weight of the products of combustion = product weight for perfect combustion, lb + (percent excess air) (air for perfect combustion, lb) = $16.80 + (0.20)(16.132) = 20.03$ lb (9.01 kg).

8.   Compute the volume of the combustion products and the percent CO_2

The volume of excess air in the products of combustion is found by converting from the weight to the volumetric analysis and correcting for temperature as in step 4, using the air weight from step 2 for perfect combustion and the excess-air percentage, or $(16.132/22.1)(0.20)(379/28.95)(2.255) = 4.31$ ft^3 (0.122 m^3). Add this to the volume of the products of combustion found in step 4, or $23.88 + 4.31 = 28.19$ ft^3 (0.798 m^3).

By the procedure in step 5, the percent CO_2, wet basis = $2.265/28.19 = 0.0804$, or 8.04 percent. The percent CO_2, dry basis = $2.265/(28.19 - 4.425) = 0.0953$, or 9.53 percent.

Related Calculations: Use the method given here when making combustion calculations for any type of gas used as a fuel—natural gas, blast-furnace gas, coke-oven gas, producer gas, water gas, sewer gas—from any source, domestic or foreign, in any type of furnace—boiler, heater, process, or waste-heat. When the air used for combustion contains moisture, as is usually true, this moisture is added to the combustion-formed moisture appearing in the products of combustion. Thus, for 80°F (26.7°C) air of 60 percent relative humidity, the moisture content is 0.013 lb/lb (0.006 kg/kg) of dry air. This amount appears in the products of combustion for each pound of air used and is a commonly assumed standard in combustion calculations.

COMBUSTION OF WOOD FUEL IN A FURNACE

The weight analysis of a yellow-pine wood fuel is: C = 0.490; H_2 = 0.074; O_2 = 0.406; N_2 = 0.030. Determine the weight of oxygen and air required with perfect combustion and with 20 percent excess air. Find the weight and volume of the products of combustion under the same conditions, and the wet and dry CO_2. The flue-gas temperature is 600°F (316°C). The air supplied for combustion has a moisture content of 0.013 lb/lb (0.006 kg/kg) of dry air.

Calculation Procedure:

1.   Compute the weight of oxygen required per pound of wood

The same general steps as given in earlier calculation procedures will be followed; consult them for a complete explanation of each step. Using the molecular weight of each element, we have

Element	×	Molecular-weight ratio	=	lb (kg) O_2 required	
C; 0.490	×	32/12	=	1.307	(0.588)
H_2; 0.074	×	16/2	=	0.592	(0.266)
O_2; 0.406; decreases external O_2 required			=	−0.406	(−0.183)
N_2; 0.030 inert in combustion					
Total 1.000					
lb (kg) external O_2 per lb (kg) fuel			=	1.493	(0.671)

2.   Compute the weight of air required for complete combustion

The weight of nitrogen associated with the required oxygen = $(1.493)(0.768/0.232) = 4.95$ lb (2.228 kg). The weight of air required = $4.95 + 1.493 = 6.443$ lb/lb (2.899 kg/kg) of wood burned, if the air is dry. But the air contains 0.013 lb of moisture per lb (0.006 kg/kg) of air. Hence, the total weight of the air = $6.443 + (0.013)(6.443) = 6.527$ lb (2.937 kg).

3. Compute the weight of the products of combustion

Use the following relation:

Fuel constituents	+	Oxygen	=	Products of combustion, lb (kg)
C; 0.490	+	1.307	=	1.797 (0.809) = CO_2
H_2; 0.074	+	0.592	=	0.666 (0.300) = H_2O
O_2; not a product of combustion				
N_2; inert but passes through the furnace			=	0.030 (0.014) = N_2
Outside N_2 from step 2			=	4.950 (2.228) = N_2
Outside moisture from step 2			=	0.237 (0.107)
lb (kg) of flue gas per lb (kg) of wood burned			=	7.680 (3.458)

4. Convert the flue-gas weight to volume

Use, as before, the following tabulation:

Products	Weight lb	Weight kg	Molecular weight	Temperature correction	Volume at 600°F, ft³	316°C, m³
CO_2	1.797	0.809	44	(359/44)(1.797)(2.15) =	31.5	0.892
H_2O (fuel)	0.666	0.300	18	(359/18)(0.666)(2.15) =	28.6	0.810
N_2 (total)	4.980	2.241	28	(359/28)(4.980)(2.15) =	137.2	3.884
H_2O (outside air)	0.837	0.377	18	(359/18)(0.837)(2.15) =	35.9	10.16
Cu ft (m³) of flue gas per lb (kg) of oil					233.2	6.602

In this calculation the temperature correction factor 2.15 = (absolute flue-gas temperature, °R)/ (absolute atmospheric temperature, °R) = (600 + 460)/(32 + 460). The total weight of N_2 is the sum of the N_2 in the combustion air and the fuel.

5. Compute the CO_2 content of the flue gas

The CO_2, wet basis = 31.5/233.2 = 0.135, or 13.5 percent. The CO_2, dry basis = 31.5/(233.2 − 28.6 − 35.9) = 0.187, or 18.7 percent.

6. Compute the air required with the stated excess flow

With 20 percent excess air, (1.20)(6.527) = 7.832 lb (3.524 kg) of air per lb (kg) of wood burned.

7. Compute the weight of the products of combustion

The weight of the products of combustion = product weight for perfect combustion, lb + (percent excess air)(air for perfect combustion, lb) = 8.280 + (0.20)(6.527) = 9.585 lb (4.313 kg) of flue gas per lb (kg) of wood burned with 20 percent excess air.

8. Compute the volume of the combustion products and the percent CO_2

The volume of the excess air in the products of combustion is found by converting from the weight to the volumetric analysis and correcting for temperature as in step 4, using the air weight from step 2 for perfect combustion and the excess-air percentage, or (6.527)(0.20)(359/28.95)(2.15) = 34.8 ft³ (0.985 m³). Add this to the volume of the products of combustion found in step 4, or 233.2 + 34.8 = 268.0 ft³ (7.587 m³).

By using the procedure in step 5, the percent CO_2, wet basis = 31.5/268 = 0.1174, or 11.74 percent. The percent CO_2, dry basis = 31.5/(268 − 28.6 − 35.9 − 0.20 × 0.837) = 0.155, or 15.5 percent. In the dry-basis calculation, the factor (0.20)(0.837) is the outside moisture in the excess air.

Related Calculations: Use the method given here when making combustion calculations for any type of wood or woodlike fuel—spruce, cypress, maple, oak, sawdust, wood shavings, tanbark, bagesse, peat, charcoal, redwood, hemlock, fir, ash, birch, cottonwood, elm, hickory, walnut, chopped trimmings, hogged fuel, straw, corn, cottonseed hulls, city refuse—in any type of furnace—boiler, heating, process, or waste-heat. Most of these fuels contain a small amount of ash—usually less than 1 percent. This was ignored in this calculation procedure because it does not take part in the combustion.

MOLAL METHOD OF COMBUSTION ANALYSIS

A coal fuel has this ultimate analysis: $C = 0.8339$; $H_2 = 0.0456$; $O_2 = 0.0505$; $N_2 = 0.0103$; S $= 0.0064$; ash $= 0.0533$; total $= 1.000$. This coal is completely burned in a boiler furnace. Using the molal method, determine the weight of air required per lb (kg) of coal with complete combustion. How much air is needed with 25 percent excess air? What is the weight of the combustion products with 25 percent excess air? The combustion air contains 0.013 lb of moisture per lb (0.006 kg/kg) of air.

Calculation Procedure:

1. *Convert the ultimate analysis to moles*

A mole of any substance is an amount of the substance having a weight equal to the molecular weight of the substance. Thus, 1 mol of carbon is 12 lb (5.4 kg) of carbon, because the molecular weight of carbon is 12. To convert an ultimate analysis of a fuel to moles, assume that 100 lb (45 kg) of the fuel is being considered. Set up a tabulation thus:

Ultimate analysis, %	Weight		Molecular weight	Moles per 100-lb (45-kg) fuel
	lb	kg		
C = 0.8339	83.39	37.526	12	6.940
H_2 = 0.0456	4.56	2.052	2	2.280
O_2 = 0.0505	5.05	2.678	32	0.158
N_2 = 0.0103	1.03	0.464	28	0.037
S = 0.0064	0.64	0.288	32	
Ash = 0.0533	5.33	2.399	Inert	
Total	100.00	45.407	. . .	9.435

2. *Compute the mols of oxygen for complete combustion*

From Table 2, the burning of carbon to carbon dioxide requires 1 mol of carbon and 1 mol of oxygen, yielding 1 mol of CO_2. Using the molal equations in Table 2 for the other elements in the fuel, set up a tabulation thus, entering the product of columns 2 and 3 in column 4:

(1) Element	(2) Moles per 100-lb (45-kg) fuel	(3) Moles O_2 per 100-lb (45-kg) fuel	(4) Total moles O_2
C	6.940	1.00	6.940
H_2	2.280	0.5	1.140
O_2	0.158	Reduces O_2 required	−0.158
N_2	0.037	Inert in combustion	
S	0.020	1.00	0.020
Total moles of O_2 required	. . .	. . .	7.942

TABLE 3   Molal Conversion Factors

Element or compound	Mol/mol of combustible for complete combustion; no excess air					
	For combustion			Combustion products		
	O_2	N_2	Air	CO_2	H_2O	N_2
Carbon,° C	1.0	3.76	4.76	1.0	...	3.76
Hydrogen, H_2	0.5	0.188	2.38	...	1.0	1.88
Oxygen, O_2						
Nitrogen, N_2						
Carbon monoxide, CO	0.5	1.88	2.38	1.0	...	1.88
Carbon dioxide, CO_2						
Sulfur,° S	1.0	3.76	4.76	1.0	...	3.76
Methane, CH_4	2.0	7.53	0.53	1.0	2.0	7.53
Ethane, C_2H_6	3.5	13.18	16.68	2.0	3.0	13.18

°In molal calculations, carbon and sulfur are considered as gases.

3. Compute the moles of air for complete combustion

Set up a similar tabulation for air, thus:

(1) Element	(2) Moles per 100-lb (45-kg) fuel	(3) Moles air per 100-lb (45-kg) fuel	(4) Total moles air
C	6.940	4.76	33.050
H_2	2.280	2.38	5.430
O_2	0.158	Reduces O_2 required	−0.752
N_2	0.037	Inert in combustion	
S	0.020	4.76	0.095
Total moles of air required		...	37.823

In this tabulation, the factors in column 3 are constants used for computing the total moles of air required for complete combustion of each of the fuel elements listed. These factors are given in the Babcock & Wilcox Company—*Steam: Its Generation and Use* and similar treatises on fuels and their combustion. A tabulation of these factors is given in Table 3.

An alternative, and simpler, way of computing the moles of air required is to convert the required O_2 to the corresponding N_2 and find the sum of the O_2 and N_2. Or, $3.76O_2 = N_2$; $N_2 + O_2$ = moles of air required. The factor 3.76 converts the required O_2 to the corresponding N_2. These two relations were used to convert the 0.158 mol of O_2 in the above tabulation to moles of air.

Using the same relations and the moles of O_2 required from step 2, we get $(3.76)(7.942) = 29.861$ mol of N_2. Then $29.861 + 7.942 = 37.803$ mol of air, which agrees closely with the 37.823 mol computed in the tabulation. The difference of 0.02 mol is traceable to roundings.

4. Compute the air required with the stated excess air

With 25 percent excess, the air required for combustion = $(125/100)(37.823) = 47.24$ mol.

5. Compute the mols of combustion products

Using data from Table 3, and recalling that the products of combustion of a sulfur-containing fuel are CO_2, H_2O, and SO_2, and that N_2 and excess O_2 pass through the furnace, set up a tabulation thus:

(1) Moles per 100-lb (45-kg) fuel	(2) Mol/mol of combustible	(3) Moles of combustion products per 100-lb (45-kg) of fuel
CO_2; 6.940	1	6.940
H_2O; 2.280 + (47.24)(0.021 + 0.158)	. . .	3.430
SO_2; 0.020	1	0.202
N_2; (47.24)(0.79)	. . .	37.320
Excess O_2; (1.25)(7.942) − 7.942	. . .	1.986

Total moles, wet combustion products = 49.878
Total moles, dry combustion products = 49.878 −3.232
= 46.646

In this calculation, the total moles of CO_2 is obtained from step 2. The moles of H_2 in 100 lb (45 kg) of the fuel, 2.280, is assumed to form H_2O. In addition, the air from step 4, 47.24 mol, contains 0.013 lb of moisture per lb (0.006 kg/kg) of air. This moisture is converted to moles by dividing the molecular weight of air, 28.95, by the molecular weight of water, 18, and multiplying the result by the moisture content of the air, or $(28.95/18)(0.013) = 0.0209$, say 0.021 mol of water per mol of air. The product of this and the moles of air gives the total moles of moisture (water) in the combustion products per 100 lb (45 kg) of fuel fired. To this is added the moles of O_2, 0.158, per 100 lb (45 kg) of fuel, because this oxygen is assumed to unite with hydrogen in the air to form water. The nitrogen in the products of combustion is that portion of the moles of air required, 47.24 mol from step 4, times the proportion of N_2 in the air, or 0.79. The excess O_2 passes through the furnace and adds to the combustion products and is computed as shown in the tabulation. Subtracting the total moisture, 3430 mol, from the total (or wet) combustion products gives the moles of dry combustion products.

Related Calculations: Use this method for molal combustion calculations for all types of fuels—solid, liquid, and gaseous—burned in any type of furnace—boiler, heater, process, or waste-heat. Select the correct factors from Table 3.

FINAL COMBUSTION PRODUCTS TEMPERATURE ESTIMATE

Pure carbon is burned to carbon dioxide at constant pressure in an insulated chamber. An excess air quantity of 20 percent is used and the carbon and the air are both initially at 77°F (25°C). Assume that the reaction goes to completion and that there is no dissociation. Calculate the final product's temperature using the following constants: Heating value of carbon, 14,087 Btu/lb (32.74×10^3 kJ/kg); constant-pressure specific heat of oxygen, nitrogen, and carbon dioxide are 0.240 Btu/lb$_m$ (0.558 kJ/kg), 0.285 Btu/lb$_m$ (0.662 kJ/kg), and 0.300 Btu/lb (0.697 kJ/kg), respectively.

Calculation Procedure:

1. Establish the chemical equation for complete combustion with 100 percent air

With 100 percent air: $C + O_2 + 3.78N_2 \rightarrow CO_2 + 3.78N_2$, where approximate molecular weights are: for carbon, $MC = 12$; oxygen, $MO_2 = 32$; nitrogen, $MN_2 = 28$; carbon dioxide, $MCO_2 = 44$. See the Related Calculations of this procedure for a general description of the 3.78 coefficient for N_2.

2. Establish the chemical equation for complete combustion with 20 percent excess air

With 20 percent excess air: $C + 1.2 O_2 + (1.2 \times 3.78)N_2 \rightarrow CO_2 + 0.2 O_2 + (1.2 \times 3.78)N_2$.

3. Compute the relative weights of the reactants and products of the combustion process

Relative weight = moles × molecular weight. Coefficients of the chemical equation in step 2 represent the number of moles of each component. Hence, for the reactants, the relative weights are: for C = 1 × MC = 1 × 12 = 12; O_2 = 1.2 × MO_2 = 1.2 × 32 = 38.4; N_2 = (1.2 × 3.78)MN_2 = (1.2 × 3.78 × 28) = 127. For the products, relative weights are: for CO_2 = 1 × MCO_2 = 1 × 44 = 44; O_2 = 0.2 × MO_2 = 0.2 × 32 = 6.4; N_2 = 127, unchanged. It should be noted that the total relative weight of the reactants equal that of the products at 177.4.

4. *Compute the relative weights of the products of combustion on the basis of a per unit relative weight of carbon*

Since the relative weight of carbon, C = 12 in step 3; hence, on the basis of a per unit relative weight of carbon, the corresponding relative weights of the products are: for carbon dioxide, wCO_2 = MCO_2/12 = 44/12 = 3.667; oxygen, wO_2 = MO_2/12 = 6.4/12 = 0.533; nitrogen, wN_2 = MN_2/12 = 127/12 = 10.58.

5. *Compute the final product's temperature*

Since the combustion chamber is insulated, the combustion process is considered adiabatic. Hence, on the basis of a per unit mass of carbon, the heating value (HV) of the carbon = the corresponding heat content of the products. Thus, relative to a temperature base of 77°F (25°C), 1 × HVC = [(wCO_2 × c_pCO_2) + (wO_2 × c_pO_2) + (wH_2 × c_pN_2)](t_2 − 77), where the heating value of carbon, HVC = 14,087 Btu/lb$_m$ (32.74 × 10^3 kJ/kg); the constant-pressure specific heat of carbon dioxide, oxygen, and nitrogen are c_pCO_2 = 0.300 Btu/lb (0.697 kJ/kg), c_pO_2 = 0.240 Btu/lb (0.558 kJ/kg), and c_pN_2 = 0.285 Btu/lb (0.662 kJ/kg), respectively; final product temperature is t_2; other values as before. Then, 1 × 14,087 = [(3.667 × 0.30) + (0.533 × 0.24) + (10.58 × 0.285)(t_2 − 77)]. Solving, t_2 = 3320 + 77 = 3397°F (1869°C).

   Related Calculations: In the above procedure it is assumed that the carbon is burned in dry air. Also, the nitrogen coefficient of 3.78 used in the chemical equation in step 1 is based on a theoretical composition of dry air as 79.1 percent nitrogen and 20.9 percent oxygen by volume, so that 79.1/20.9 = 3.78. For a more detailed description of this coefficient see the Related Calculations under the procedure for "Gas Turbine Combustion Chamber Inlet Air Temperature" in this subsection.

STEAM BOILER HEAT BALANCE DETERMINATION

A steam generator having a maximum rated capacity of 60,000 lb/h (27,000 kg/h) is operating at 45,340 lb/h (20,403 kg/h), delivering 125-lb/in² (gage) 400°F (862-kPa, 204°C) steam with a feedwater temperature of 181°F (82.8°C). At this generating rate, the boiler requires 4370 lb/h (1967 kg/h) of West Virginia bituminous coal having a heating value of 13,850 Btu/lb (32,215 kJ/kg) on a dry basis. The ultimate fuel analysis is: C = 0.7757; H_2 = 0.0507; O_2 = 0.0519; N_2 = 0.0120; S = 0.0270; ash = 0.0827; total = 1.0000. The coal contains 1.61 percent moisture. The boiler-room intake air and the fuel temperature = 79°F (26.1°C) dry bulb, 71°F (21.7°C) wet bulb. The flue-gas temperature is 500°F (260°C), and the analysis of the flue gas shows these percentages: CO_2 = 12.8; CO = 0.4; O_2 = 6.1; N_2 = 80.7; total = 100.0. Measured ash and refuse = 9.42 percent of dry coal; combustible in ash and refuse = 32.3 percent. Compute a heat balance for this boiler based on these test data. The boiler has four water-cooled furnace walls.

Calculation Procedure:

1. *Determine the heat input to the boiler*

In a boiler heat balance the input is usually stated in Btu per pound of fuel as fired. Therefore, input = heating value of fuel = 13,850 Btu/lb (32,215 kJ/kg).

2. *Compute the output of the boiler*

The output of any boiler = Btu/lb (kJ/kg) of fuel + the losses. In this step the first portion of the output, Btu/lb (kJ/kg) of fuel will be computed. The losses will be computed in step 3.

   First find W_s lb of steam produced per lb of fuel fired. Since 45,340 lb/h (20,403 kg/h) of steam is produced when 4370 lb/h (1967 kg/h) of fuel is fired, W_s = 45,340/4370 = 10.34 lb of steam per lb (4.65 kg/kg) of fuel.

   Once W_s is known, the output h_1 Btu/lb of fuel can be found from h_1 = W_s(h_s − h_w), where h_s = enthalpy of steam leaving the superheater, or boiler if a superheater is not used; h_w = enthalpy of feedwater, Btu/lb. For this boiler with steam at 125 lb/in² (gage) [= 139.7 lb/in² (abs)] and 400°F (930 kPa, 204°C), h_s = 1221.2 Btu/lb (2841 kJ/kg), and h_w = 180.92 Btu/lb (420.8 kJ/kg), from the steam tables. Then h_1 = 10.34(1221.2 − 180.92) = 10,766.5 Btu/lb (25,043 kJ/kg) of coal.

3. Compute the dry flue-gas loss

For any boiler, the dry flue-gas loss h_2 Btu/lb (kJ/kg) of fuel is given by $h_2 = 0.24W_g \times (T_g - T_a)$, where W_g = lb of dry flue gas per lb of fuel; T_g = flue-gas exit temperature, °F; T_a = intake-air temperature, °F.

Before W_g can be found, however, it must be determined whether any excess air is passing through the boiler. Compute the excess air, if any, from excess air, percent = 100 (O_2 − ½CO)/ [0.264N_2 − (O_2 − ½CO)], where the symbols refer to the elements in the flue-gas analysis. Substituting values from the flue-gas analysis gives excess air = 100(6.1 − 0.2)/[0.264 × 80.7 − (6.1 − 0.2)] = 38.4 percent.

Using the method given in earlier calculation procedures, find the air required for complete combustion as 10.557 lb/lb (4.751 kg/kg) of coal. With 38.4 percent excess air, the additional air required = (10.557)(0.384) = 4.053 lb/lb (1.82 kg/kg) of fuel.

From the same computation in which the air required for complete combustion was determined, the lb of *dry* flue gas per lb of fuel = 11.018 (4.958 kg/kg). Then, the total flue gas at 38.4 percent excess air = 11.018 + 4.053 = 15.071 lb/lb (6.782 kg/kg) of fuel.

With a flue-gas temperature of 500°F (260°C), and an intake-air temperature of 79°F (26.1°C), h_2 = 0.24(15.071)(500 − 70) = 1524 Btu/lb (3545 kJ/kg) of fuel.

4. Compute the loss due to evaporation of hydrogen-formed water

Hydrogen in the fuel is burned in forming H_2O. This water is evaporated by heat in the fuel, and less heat is available for producing steam. This loss is h_3 Btu/lb of fuel = $9H(1089 - T_f + 0.46T_g)$, where H = percent H_2 in the fuel ÷ 100; T_f = temperature of fuel *before* combustion, °F; other symbols as before. For this fuel with 5.07 percent H_2, h_3 = 9(5.07/100)(1089 − 79 + 0.46 × 500) = 565.8 Btu/lb (1316 kJ/kg) of fuel.

5. Compute the loss from evaporation of fuel moisture

This loss is h_4 Btu/lb of fuel = $W_{mf}(1089 - T_f + 0.46T_g)$, where W_{mf} = lb of moisture per lb of fuel; other symbols as before. Since the fuel contains 1.61 percent moisture, in terms of *dry* coal this is (1.61)/(100 − 1.61) = 0.0164, or 1.64 percent. Then h_4 = (1.64/100)(1089 − 79 + 0.46 × 500) = 20.34 Btu/lb (47.3 kJ/kg) of fuel.

6. Compute the loss from moisture in the air

This loss is h_5 Btu/lb of fuel = $0.46W_{ma}(T_g - T_a)$, W_{ma} = (lb of water per lb of dry air)(lb air supplied per lb fuel). From a psychrometric chart, the weight of moisture per lb of air at a 79°F (26.1°C) dry-bulb and 71°F (21.7°C) wet-bulb temperature is 0.014 (0.006 kg). The combustion calculation, step 3, shows that the total air required with 38.4 percent excess air = 10.557 + 4.053 = 14.61 lb of air per lb (6.575 kg/kg) of fuel. Then, W_{ma} = (0.014)(14.61) = 0.2045 lb of moisture per lb (0.092 kg/kg) of air. And h_5 = (0.46)(0.2045)(500 − 79) = 39.6 Btu/lb (92.1 kJ/kg) of fuel.

7. Compute the loss from incomplete combustion of C to CO_2 in the stack

This loss is h_6 Btu/lb of fuel = [CO/(CO + CO_2)](C)(10,190), where CO and CO_2 are the percent by volume of these compounds in the flue gas by Orsat analysis; C = lb carbon per lb of coal. With the given flue-gas analysis and the coal ultimate analysis, h_6 = 0.4/(0.4 + 12.8)[(77.57)/(100)](10,190) = 239.5 Btu/lb (557 kJ/kg) of fuel.

8. Compute the loss due to unconsumed carbon in the refuse

This loss is h_7 Btu/lb of fuel = W_c(14,150), where W_c = lb of unconsumed carbon in refuse per lb of fuel fired. With an ash and refuse of 9.42 percent of the dry coal and combustible in the ash and refuse of 32.3 percent, h_7 = (9.42/100)(32.3/100)(14,150) = 430.2 Btu/lb (1006 kJ/kg) of fuel.

9. Find the radiation loss in the boiler furnace

Use the American Boiler and Affiliated Industries (ABAI) chart, or the manufacturer's engineering data to approximate the radiation loss in the boiler. Either source will show that the radiation loss is 1.09 percent of the gross heat input. Since the gross heat input is 13,850 Btu/lb (32,215 kJ/kg) of fuel, the radiation loss = (13,850)(1.09/100) = 151.0 Btu/lb (351.2 kJ/kg) of fuel.

10. *Summarize the losses; find the unaccounted-for loss*

Set up a tabulation thus, entering the various losses computed earlier:

Item	Btu/lb fuel	kJ/kg fuel	Percent
1. Input	13,850.0	32,215.4	100.0
2. Output	10,770.0	25,051	77.75
Losses:			
3. Flue gas	1,524.0	3,545	11.00
4. Hydrogen	565.8	1,315	4.09
5. Water-fuel	20.3	47.2	0.15
6. Water-air	39.6	92.1	0.29
7. CO	239.5	557	1.73
8. Carbon-ash	430.2	1,001	3.11
9. Radiation	151.0	351.2	1.09
10. Unaccounted	109.6	254.9	0.79
Total	13,850.0	32,214.4	100.00

The unaccounted-for loss is found by summing all the other losses, 3 through 9, and subtracting from 100.00.

Related Calculations: Use this method to compute the heat balance for any type of boiler—watertube or firetube—in any kind of service—power, process, or heating—using any kind of fuel—coal, oil, gas, wood, or refuse. Note that step 3 shows how to compute excess air from an Orsat flue-gas analysis.

More stringent environmental laws are requiring larger investments in steam-boiler pollution-control equipment throughout the world. To control sulfur emissions, expensive scrubbers are required on large boilers. Without such scrubbers the sulfur emissions can lead to acid rain, smog, and reduced visibility in the area of the plant and downwind from it.

With the increased number of free-trade agreements between adjacent countries, cross-border pollution is receiving greater attention. The reason for this increased attention is because not all countries have the same environmental control requirements. When a country with less stringent requirements pollutes an adjacent country having more stringent pollution regulations, both political and regulatory problems can arise.

For example, two adjacent countries are currently discussing pollution problems of a cross-border type. One country's standard for particulate emissions is 10 times weaker than the adjacent country's, while its sulfur dioxide limit is 8 times weaker. With such a wide divergence in pollution requirements, cross-border flows of pollutants can be especially vexing.

All boiler-plant designers must keep up to date on the latest pollution regulations. Today there are some 90,000 environmental regulations at the federal, state, and local levels, and more than 40 percent of these regulations will change during the next 12 months. To stay in compliance with such a large number of regulations requires constant attention to those regulations applicable to boiler plants.

GAS TURBINE COMBUSTION CHAMBER INLET AIR TEMPERATURE

A gas turbine combustion chamber is well insulated so that heat losses to the atmosphere are negligible. Octane, C_8H_{18}, is to be used as the fuel and 400 percent of the stoichiometric air quantity is to be supplied. The air first passes through a regenerative heater and the air supply temperature at the combustion chamber inlet is to be set so that the exit temperature of the combustion gases is 1600°F (871°C). (See Fig. 16.) Fuel supply temperature is 77°F (25°C) and its heating value is to be taken as 19,000 Btu/lb$_m$ (44,190 kJ/kg) relative to a base of 77°F (15°C).

The air may be treated in calculations as a perfect gas with a constant-pressure specific heat of 0.24 Btu/(lb · °F) [1.005 kJ/(kg · °C)]. The products of combustion have an enthalpy of 15,400 Btu/lb · mol) [33,950 Btu/(kg · mol)] at 1600°F (871°C) and an enthalpy of 3750 Btu/(lb · mol) [8270 Btu/(kg · mol)] at 77°F (24°C). Determine, assuming complete combustion and neglecting dissociation, the required air temperature at the inlet of the combustion chamber.

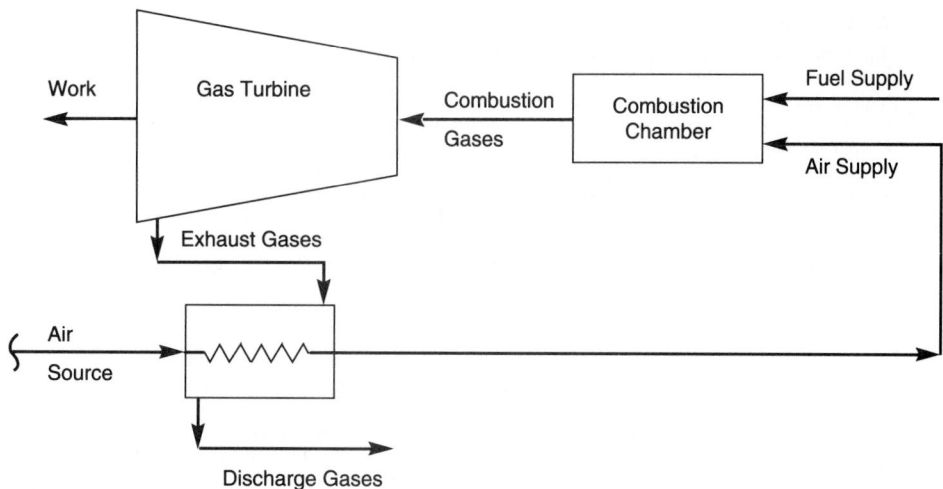

FIG. 16 Gas turbine flow diagram.

Calculation Procedure:

1. Find the amount of oxygen required for complete combustion of the fuel

Eight atoms of carbon in C_8 combine with 8 molecules of oxygen, O_2, and produce 8 molecules of carbon dioxide, $8CO_2$. Similarly, 9 molecules of hydrogen, H_2, in H_{18} combine with 9 atoms of oxygen, O, or 4.5 molecules of oxygen, to form 9 molecules of water, $9H_2O$. Thus, 100 percent, or the stoichiometric, air quantity required for complete combustion of a mole of fuel, C_8H_{18}, is proportional to $8 + 12.5$ moles of oxygen, O_2.

2. Establish the chemical equation for complete combustion with 100 percent air

With 100 percent air: $C_8H_{18} + 12.5\,O_2 + (3.784 \times 12.5)N_2 \rightarrow 8CO_2 + 9H_2O + 47.3N_2$, where 3.784 is a derived volumetric ratio of atmospheric nitrogen, (N_2), to oxygen, O_2, in dry air. The (N_2) includes small amounts of inert and inactive gases. See Related Calculations of this procedure.

3. Establish the chemical equation for complete combustion with 400 percent of the stoichiometric air quantity, or 300 percent excess air

With 400 percent air: $C_8H_{18} + 50\,O_2 + (4 \times 47.3)N_2 \rightarrow 8CO_2 + 9H_2O + 189.2N_2 + (3 \times 12.5)O_2$.

4. Compute the molecular weights of the components in the combustion process

Molecular weight of $C_8H_{18} = [(12 \times 8) + (1 \times 18)] = 114$; $O_2 = 16 \times 2 = 32$; $N_2 = 14 \times 2 = 28$; $CO_2 = [(12 \times 1) + (16 \times 2)] = 44$; $H_2O = [(1 \times 2) + (16 \times 1)] = 18$.

5. Compute the relative weights of the reactants and products of the combustion process

Relative weight = moles × molecular weight. Coefficients of the chemical equation in step 3 represent the number of moles of each component. Hence, for the reactants, the relative weights are: $C_8H_{18} = 1 \times 114 = 114$; $O_2 = 50 \times 32 = 1600$; $N_2 = 189.2 \times 28 = 5298$. Total relative weight of the reactants is 7012. For the products, the relative weights are: $CO_2 = 8 \times 44 = 352$; $H_2O = 9 \times 18 = 162$; $N_2 = 189.2 \times 28 = 5298$; $O_2 = 37.5 \times 32 = 1200$. Total relative weight of the products is 7012, also.

6. Compute the enthalpy of the products of the combustion process

Enthalpy of the products of combustion, $h_p = m_p(h_{1600} - h_{77})$, where m_p = number of moles of the products; h_{1600} = enthalpy of the products at 1600°F (871°C); h_{77} = enthalpy of the products at 77°F (25°C). Thus, $h_p = (8 + 9 + 189.2 + 37.5)(15,400 - 3750) = 2,839,100$ Btu [6,259,100 Btu (SI)].

7. *Compute the air supply temperature at the combustion chamber inlet*

Since the combustion process is adiabatic, the enthalpy of the reactants $h_r = h_p$, where $h_r =$ (relative weight of the fuel $\times$ its heating value) + [relative weight of the air $\times$ its specific heat $\times$ (air supply temperature − air source temperature)]. Therefore, $h_r = (114 \times 19{,}100) +$ [$(1600 + 5298) \times 0.24 \times (t_a - 77)$] + 2,839,100 Btu [6,259,100 Btu (SI)]. Solving for the air supply temperature, $t_a = [(2{,}839{,}100 - 2{,}177{,}400/1655.5] + 77 = 477°F$ (247°C).

Related Calculations: This procedure, appropriately modified, may be used to deal with similar questions involving such things as other fuels, different amounts of excess air, and variations in the condition(s) being sought under certain given circumstances.

The coefficient, (?) = 3.784 in step 2, is used to indicate that for each unit of volume of oxygen, O_2, 12.5 in this case, there will be 3.784 units of nitrogen, N_2. This equates to an approximate composition of air as 20.9 percent oxygen and 79.1 percent "atmospheric nitrogen," (N_2). In turn, this creates a paradox, because page 200 of *Principles of Engineering Thermodynamics*, by Kiefer, et al., John Wiley & Sons, Inc., states air to be 20.99 percent oxygen and 79.01 percent atmospheric nitrogen, where the ratio $(N_2)/O_2 = (?) = 79.01/20.99 = 3.764$.

Also, page 35 of *Applied Energy Conversion*, by Skrotski and Vopat, McGraw-Hill, Inc., indicates an assumed air analysis of 79 percent nitrogen and 21 percent oxygen, where (?) = 3.762. On that basis, a formula is presented for the amount of dry air chemically necessary for complete combustion of a fuel consisting of atoms of carbon, hydrogen, and sulfur, or C, H, and S, respectively. That formula is: $W_a = 11.5C + 34.5[H - (0/8)] + 4.32S$, lb air/lb fuel (kg air/kg fuel).

The following derivation for the value of (?) should clear up the paradox and show that either 3.784 or 3.78 is a sound assumption which seems to be wrong, but in reality is not. In the above equation for W_a, the carbon, hydrogen, or sulfur coefficient, $C_x = (MO_2/DO_2)M_x$, where MO_2 is the molecular weight of oxygen, O_2; DO_2 is the decimal fraction for the percent, by weight, of oxygen, O_2, in dry air containing "atmospheric nitrogen," (N_2), and small amounts of inert and inactive gases: M_x is the formula weight of the combustible element in the fuel, as indicated by its relative amount as a reactant in the combustion equation. The alternate evaluation of C_x is obtained from stoichiometric chemical equations for burning the combustible elements of the fuel, i.e., $C + O_2 + (?)N_2 \rightarrow CO_2 + (?)N_2$; $2H_2 + O_2 + (?)N_2 \rightarrow 2H_2O + (?)N_2$; $S + O_2 + (?)N_2 \rightarrow SO_2 + (?)N_2$. Evidently, $C_x = [MO_2 + (? \times MN_2)]/M_x$, where MN_2 is the molecular weight of nitrogen, N_2, and the other items are as before.

Equating the two expressions, $C_x = [MO_2 + (? \times MN_2)]/M_x = (MO_2/DO_2)M_x$, reveals that the M_x terms cancel out, indicating that the formula weight(s) of combustible components is irrelevant in solving for (?). Then, (?) = $(1 - DO_2)[MO_2/(MN_2 \times DO_2)]$. From the abovementioned book by Kiefer, et al., $DO_2 = 0.23188$. From *Marks' Standard Handbook for Mechanical Engineers*, McGraw-Hill, Inc., $MO_2 = 31.9988$ and $MN_2 = 28.0134$. Thus, (?) = $(1 - 0.23188)[31.9988/(28.0134 \times 0.23188)] = 3.7838$. This demonstrates that the use of (?) = 3.784, or 3.78, is justified for combustion equations.

By using either of the two evaluation equations for C_x, and with accurate values for M_x, i.e., $M_C = 12.0111$; $M_H = 2 \times 2 \times 1.00797 = 4.0319$; $M_S = 32.064$, from *Marks' M.E. Handbook*, the more precise values for C_C, C_H, and C_S are found out to be 11.489, 34.227, and 4.304, respectively. However, the actual C_x values, 11.5, 34.5, and 4.32, used in the formula for W_a are both brief for simplicity and rounded up to be on the safe side.

Power Generation

REFERENCES: El-Wakil—*Powerplant Technology*, McGraw-Hill; Goss—*Factors Affecting Power Plant Waste Heat Utilization*, Pergamon Press; Polimeros—*Energy Cogeneration Handbook*, Industrial Press; Yu—*Electric Power System Dynamics*, Academic Press; Hagel—*Alternative Energy Strategies*, Praeger; Aschner—*Planning Fundamentals of Thermal Power Plants*, Israel University Press; Komanoff—*Power Plant Cost Escalation*, VNR; Seeley—*Elements of Thermal Technology*, Dekker; Hunt—*Handbook of Energy Technology*, VNR; Blair, Cassel, and Edelstein—*Geothermal Energy: Investment Decisions and Commercial Development*, Wiley; Goodman and Love—*Geothermal Energy Projects: Planning and Management*, Pergamon Press; Edgerton—*Available Energy and Environmental Economics*, Heath; Meyers—*Handbook of Energy Technology and Economics*, Wiley; Babcock & Wilcox Company—*Steam: Its Generation and Use*; Combustion Engineering Corporation—*Combustion Engineering*; Skrotzki and Vopat—*Power Station Engineering and Economy,* McGraw-Hill; Heat Enchange Institute—*Steam Surface Condenser Standards;* Gaffert—*Steam Power Stations,* McGraw-Hill; ASME—*Test Code for Steam Generating Units;* Potter—*Steam Power Plants,* Ronald; Smith and Stinson—*Fuels and Combustion,* McGraw-Hill; Zerban and Nye—*Steam Power Plants,* International Textbook; Sorenson—*Gas Turbines,* Ronald; Salisbury—*Steam Turbines and Their Cycles,* Wiley; Dusinberre—*Gas Turbine Power,* International Textbook; Zemansky—*Heat and Thermodynamics,* McGraw-Hill; Jakob—*Heat Transfer,* Wiley; McAdams—*Heat Transmission,* McGraw-Hill; Buffalo Forge Company—*Fan Engineering;* Church—*Steam Turbines,* McGraw-Hill; Bleeder Heater Manufacturers Association, Inc.—*Standards;* Tubular Exchanger Manufacturers Association—*Standards.*

STEAM MOLLIER DIAGRAM AND STEAM TABLE USE

(1) Determine from the Mollier diagram for steam (*a*) the enthalpy of 100 lb/in² (abs) (689.5-kPa) saturated steam, (*b*) the enthalpy of 10-lb/in² (abs) (68.9-kPa) steam containing 40 percent moisture, (*c*) the enthalpy of 100-lb/in² (abs) (689.5-kPa) steam at 600°F (315.6°C). (2) Determine from the steam tables (*a*) the enthalpy, specific volume, and entropy of steam at 145.3 lb/in² (gage) (1001.8 kPa); (*b*) the enthalpy and specific volume of superheated steam at 1100 lb/in²

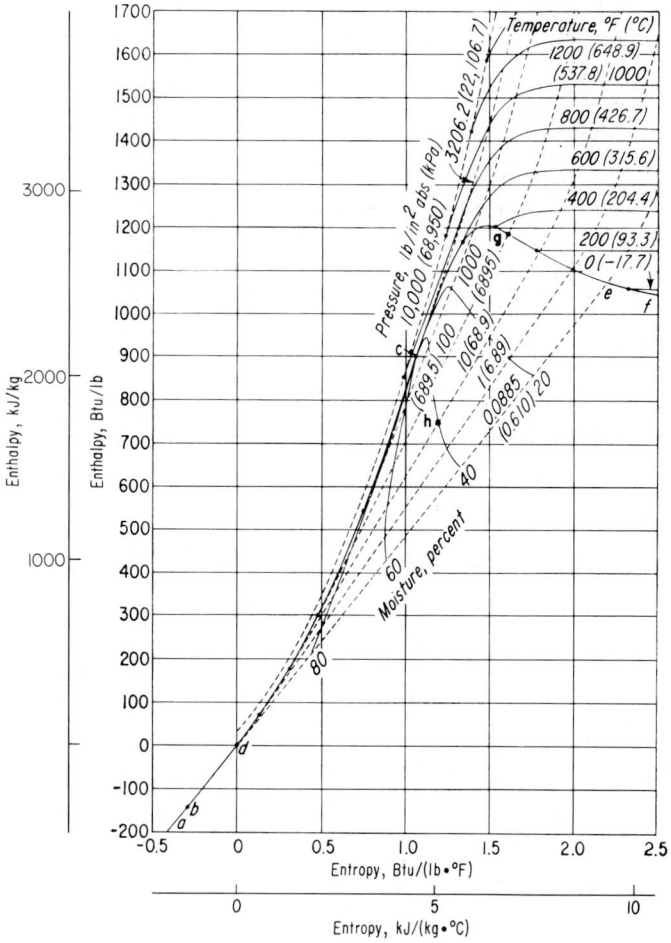

FIG. 1 Simplified Mollier diagram for steam.

(abs) (7584.2 kPa) and 600°F (315.6°C); (c) the enthalpy and specific volume of high-pressure steam at 7500 lb/in² (abs) (51,710.7 kPa) and 1200°F (648.9°C); (d) the enthalpy, specific volume, and entropy of 10-lb/in² (abs) (68.9-kPa) steam containing 40 percent moisture.

Calculation Procedure:

1. Use the pressure and saturation (or moisture) lines to find enthalpy

(a) Enter the Mollier diagram by finding the 100-lb/in² (abs) (689.5-kPa) pressure line, Fig. 1. In the Mollier diagram for steam, the pressure lines slope upward to the right from the lower left-hand corner. For saturated steam, the enthalpy is read at the intersection of the pressure line with the saturation curve *cef*, Fig. 1.

Thus, project along the 100-lb/in² (abs) (689.5-kPa) pressure curve, Fig. 1, until it intersects the saturation curve, point *g*. From here project horizontally to the left-hand scale of Fig. 1 and read the enthalpy of 100-lb/in² (abs) (689.5-kPa) saturated steam as 1187 Btu/lb (2761.0 kJ/kg). (The Mollier diagram in Fig. 1 has fewer grid divisions than large-scale diagrams to permit easier location of the major elements of the diagram.)

(b) On a Mollier diagram, the enthalpy of wet steam is found at the intersection of the saturation pressure line with the percentage-of-moisture curve corresponding to the amount of moisture in the steam. In a Mollier diagram for steam, the moisture curves slope downward to the right from the saturated liquid line *cd*, Fig. 1.

To find the enthalpy of 10-lb/in² (abs) (68.9-kPa) steam containing 40 percent moisture, project along the 10-lb/in² (abs) (68.9-kPa) saturation pressure line until the 40 percent moisture curve is intersected, Fig. 1. From here project horizontally to the left-hand scale and read the enthalpy of 10-lb/in² (abs) (68.9-kPa) wet steam containing 40 percent moisture as 750 Btu/lb (1744.5 kJ/kg).

2. Find the steam properties from the steam tables

(a) Steam tables normally list absolute pressures or temperature in degrees Fahrenheit as one of their arguments. Therefore, when the steam pressure is given in terms of a gage reading, it must be converted to an absolute pressure before the table can be entered. To convert gage pressure to absolute pressure, add 14.7 to the gage pressure, or $p_a = p_g + 14.7$. In this instance, $p_a = 145.3 + 14.7 = 160.0$ lb/in² (abs) (1103.2 kPa). Once the absolute pressure is known, enter the saturation pressure table of the steam table at this value, and project horizontally to the desired values. For 160-lb/in² (abs) (1103.2-kPa) steam, using the ASME or Keenan and Keyes—*Thermodynamic Properties of Steam*, we see that the enthalpy of evaporation $h_{fg} = 859.2$ Btu/lb (1998.5 kJ/kg), and the enthalpy of saturated vapor $h_g = 1195.1$ Btu/lb (2779.8 kJ/kg), read from the respective columns of the steam tables. The specific volume v_g of the saturated vapor of 160-lb/in² (abs) (1103.2-kPa) steam is, from the tables, 2.834 ft³/lb (0.18 m³/kg), and the entropy s_g is 1.5640 Btu/(lb·°F) [6.55 kJ/(kg·°C)].

(b) Every steam table contains a separate tabulation of properties of superheated steam. To enter the superheated steam table, two arguments are needed—the absolute pressure and the temperature of the steam. To determine the properties of 1100-lb/in² (abs) (7584.5-kPa) 600°F (315.6°C) steam, enter the superheated steam table at the given absolute pressure and project horizontally from this absolute pressure [1100 lb/in² (abs) or 7584.5 kPa] to the column corresponding to the superheated temperature (600°F or 315.6°C) to read the enthalpy of the superheated vapor as $h = 1236.7$ Btu/lb (2876.6 kJ/kg) and the specific volume of the superheated vapor $v = 0.4532$ ft³/lb (0.03 m³/kg).

(c) For high-pressure steam use the ASME—*Steam Table*, entering it in the same manner as the superheated steam table. Thus, for 7500-lb/in² (abs) (51,712.5 kPa) 1200°F (648.9°C) steam, the enthalpy of the superheated vapor is 1474.9 Btu/lb (3430.6 kJ/kg), and the specific volume of the superheated vapor is 0.1060 ft³/lb (0.0066 m³/kg).

(d) To determine the enthalpy, specific volume, and the entropy of wet steam having *y* percent moisture by using steam tables instead of the Mollier diagram, apply these relations: $h = h_g - yh_{fg}/100$; $v = v_g - yv_{fg}/100$; $s = s_g - ys_{fg}/100$, where *y* = percentage of moisture expressed as a whole number. For 10-lb/in² (abs) (68.9-kPa) steam containing 40 percent moisture, obtain the needed values—h_g, h_{fg}, v_g, v_{fg}, s_g, and s_{fg}—from the saturation-pressure steam table and substitute in the above relations. Thus,

$$h = 1143.3 - \frac{40(982.1)}{100} = 750.5 \text{ Btu/lb } (1745.7 \text{ kJ/kg})$$

$$v = 38.42 - \frac{40(38.40)}{100} = 23.06 \text{ ft}^3/\text{lb } (1.44 \text{ m}^3/\text{kg})$$

$$s = 1.7876 - \frac{40(1.5041)}{100} = 1.1860 \text{ Btu/(lb} \cdot {}^\circ\text{F)} [4.97 \text{ kJ/(kg} \cdot {}^\circ\text{C)}]$$

Note that Keenan and Keyes, in *Thermodynamic Properties of Steam*, do not tabulate v_{fg}. Therefore, this value must be obtained by subtraction of the tabulated values, or $v_{fg} = v_g - v_f$. The value v_{fg} thus obtained is used in the relation for the volume of the wet steam. For 10-lb/in² (abs) (68.9-kPa) steam containing 40 percent moisture, $v_g = 38.42$ ft³/lb (2.398 m³/kg) and $v_f = 0.017$ ft³/lb (0.0011 m³/kg). Then $v_{fg} = 38.42 - 0.017 = 28.403$ ft³/lb (1.773 m³/kg).

In some instances, the quality of steam may be given instead of its moisture content in percentage. The quality of steam is the percentage of vapor in the mixture. In the above calculation, the quality of the steam is 60 percent because 40 percent is moisture. Thus, quality $= 1 - m$, where $m =$ percentage of moisture, expressed as a decimal.

INTERPOLATION OF STEAM TABLE VALUES

(1) Determine the enthalpy, specific volume, entropy, and temperature of saturated steam at 151 lb/in² (abs) (1041.1 kPa). (2) Determine the enthalpy, specific volume, entropy, and pressure of saturated steam at 261°F (127.2°C). (3) Find the pressure of steam at 1000°F (537.8°C) if its specific volume is 2.6150 ft³/lb (0.16 m³/kg). (4) Calculate the enthalpy, specific volume, and entropy of 300-lb/in² (abs) (2068.5-kPa) steam at 567.22°F (297.3°C).

Calculation Procedure:

1. *Use the saturation-pressure table*

Study of the saturation-pressure table shows that there is no pressure value for 151 lb/in² (abs) (1041.1 kPa) listed. So it will be necessary to interpolate between the next higher and next lower tabulated pressure values. In this instance, these values are 152 and 150 lb/in² (abs) (1048.0 and 1034.3 kPa), respectively. The pressure for which properties are being found [151 lb/in² (abs) or 1041.1 kPa] is called the *intermediate pressure*. At 152 lb/in² (abs) (1048.0 kPa), $h_g = 1194.3$ Btu/lb (2777.5 kJ/kg); $v_g = 2.977$ ft³/lb (0.19 m³/kg); $s_g = 1.5683$ Btu/(lb·°F) [6.67 kJ/(kg·°C)]; $t = 359.46$°F (181.9°C). At 150 lb/in² (abs) (1034.3 kPa), $h_g = 1194.1$ Btu/lb (2777.5 kJ/kg); $v_g = 3.015$ ft³/lb (0.19 m³/kg); $s_g = 1.5694$ Btu/(lb·°F) [6.57 kJ/(kg·°C)]; $t = 358.42$°F (181.3°C).

For the enthalpy, note that as the pressure increases, so does h_g. Therefore, the enthalpy at 151 lb/in² (abs) (1041.1 kPa), the intermediate pressure, will equal the enthalpy at 150 lb/in² (abs) (1034.3 kPa) (the lower pressure used in the interpolation) plus the proportional change (difference between the intermediate pressure and the lower pressure) for a 1-lb/in² (abs) (6.9-kPa) pressure increase. Or, at any higher pressure, $h_{gi} = h_{gl} + [(p_i - p_l)/(p_h - p_l)](h_h - h_l)$, where $h_{gi} =$ enthalpy at the intermediate pressure; $h_{gl} =$ enthalpy at the lower pressure used in the interpolation; $h_h =$ enthalpy at the higher pressure used in the interpolation; $p_i =$ intermediate pressure; p_h and $p_l =$ higher and lower pressures, respectively, used in the interpolation. Thus, from the enthalpy values obtained from the steam table for 150 and 152 lb/in² (abs) (1034.3 and 1048.0 kPa), $h_{gi} = 1194.1 + [(151 - 150)/(152 - 150)](1194.3 - 1194.1) = 1194.2$ Btu/lb (2777.7 kJ/kg) at 151 lb/in² (abs) (1041.1 kPa) saturated.

Next study the steam table to determine the direction of change of specific volume between the lower and higher pressures. This study shows that the specific volume decreases as the pressure increases. Therefore, the specific volume at 151 lb/in² (abs) (1041.1 kPa) (the intermediate pressure) will equal the specific volume at 150 lb/in² (abs) (1034.3 kPa) (the lower pressure used in the interpolation) minus the proportional change (difference between the intermediate pressure and the lower interpolating pressure) for a 1-lb/in² (abs) pressure increase. Or, at any pressure,

$v_{gi} = v_{gl} - [(p_i - p_l)/(p_h - p_l)](v_l - v_h)$, where the subscripts are the same as above and $v =$ specific volume at the respective pressure. With the volume values obtained from steam tables for 150 and 152 lb/in² (abs) (1034.3 and 1048.0 kPa), $v_{gi} = 3.015 - [(151 - 150)/(152 - 150)](3.015 - 2.977) = 2.996$ ft³/lb (0.19 m³/kg) and 151 lb/in² (abs) (1041.1 kPa) saturated.

Study of the steam table for the direction of entropy change shows that entropy, like specific volume, decreases as the pressure increases. Therefore, the entropy at 151 lb/in² (abs) (1041.1 kPa) (the intermediate pressure) will equal the entropy at 150 lb/in² (abs) (1034.3 kPa) (the lower pressure used in the interpolation) minus the proportional change (difference between the intermediate pressure and the lower interpolating pressure) for a 1-lb/in² (abs) (6.9-kPa) pressure increase. Or, at any higher pressure, $s_{gi} = s_{gl} - [(p_i - p_l)/(p_h - p_l)](s_l - s_h) = 1.5164 - [(151 - 150)/(152 - 150)](1.5694 - 1.5683) = 1.56885$ Btu/(lb·°F) [6.6 kJ/(kg·°C)] at 151 lb/in² (abs) (1041.1 kPa) saturated.

Study of the steam table for the direction of temperature change shows that the saturation temperature, like enthalpy, increases as the pressure increases. Therefore, the temperature at 151 lb/in² (abs) (1041.1 kPa) (the intermediate pressure) will equal the temperature at 150 lb/in² (abs) (1034.3 kPa) (the lower pressure used in the interpolation) plus the proportional change (difference between the intermediate pressure and the lower interpolating pressure) for a 1-lb/in² (abs) (6.9-kPa) increase. Or, at any higher pressure, $t_{gi} = t_{gl} + [(p_i - p_l)/(p_h - p_l)](t_h - t_l) = 358.42 + [(151 - 150)/(152 - 150)](359.46 - 358.42) = 358.94$°F (181.6°C) at 151 lb/in² (abs) (1041.1 kPa) saturated.

2. *Use the saturation-temperature steam table*

Study of the saturation-temperature table shows that there is no temperature value of 261°F (127.2°C) listed. Therefore, it will be necessary to interpolate between the next higher and next lower tabulated values. In this instance these values are 262 and 260°F (127.8 and 126.7°C), respectively. The temperature for which properties are being found (261°F or 127.2°C) is called the intermediate temperature.

Temperature		h_g		v_g		s_g		p_g	
°F	°C	Btu/lb	kJ/kg	ft³/lb	m³/kg	Btu/ (lb·°F)	kJ/(kg· °C)	lb/in² (abs)	kPa
262	127.8	1168.0	2716.8	11.396	0.71	1.6833	7.05	36.646	252.7
260	126.7	1167.3	2715.1	11.763	0.73	1.6860	7.06	35.429	244.3

For enthalpy, note that as the temperature increases, so does h_g. Therefore, the enthalpy at 261°F (127.2°C) (the intermediate temperature) will equal the enthalpy at 260°F (126.7°C) (the lower temperature used in the interpolation) plus the proportional change (difference between the intermediate temperature and the lower temperature) for a 1°F (0.6°C) temperature increase. Or, at any higher temperature, $h_{gi} = h_{gl} + [(t_i - t_l)/(t_h - t_l)](h_h - h_l)$, where $h_{gl} =$ enthalpy at the lower temperature used in the interpolation; $h_h =$ enthalpy at the higher temperature used in the interpolation; $t_i =$ intermediate temperature; t_h and $t_l =$ higher and lower temperatures, respectively, used in the interpolation. Thus, from the enthalpy values obtained from the steam table for 260 and 262°F (126.7 and 127.8°C), $h_{gi} = 1167.3 + [(261 - 260)/(262 - 260)](1168.0 - 1167.3) = 1167.65$ Btu/lb (2716.0 kJ/kg) at 261°F (127.2°C).

Next, study the steam table to determine the direction of change of specific volume between the lower and higher temperatures. This study shows that the specific volume decreases as the pressure increases. Therefore, the specific volume at 261°F (127.2°C) (the intermediate temperature) will equal the specific volume at 260°F (126.7°C) (the lower temperature used in the interpolation) minus the proportional change (difference between the intermediate temperature and the lower interpolating temperature) for a 1°F (0.6°C) temperature increase. Or, at any higher temperature, $v_{gi} = v_{gl} - [(t_i - t_l)/(t_h - t_l)](v_l - v_h) = 11.763 - [(261 - 260)/(262 - 260)](11.763 - 11.396) = 11.5795$ ft³/lb (0.7 m³/kg) at 261°F (127.2°C) saturated.

Study of the steam table for the direction of entropy change shows that entropy, like specific volume, decreases as the temperature increases. Therefore, the entropy at 261°F (127.2°C) (the intermediate temperature) will equal the entropy at 260°F (126.7°C) (the lower temperature used in the interpolation) minus the proportional change (difference between the intermediate temperature and the lower temperature) for a 1°F (0.6°C) temperature increase. Or, at any higher temperature, $s_{gi} = s_{gl} - [(t_i - t_l)/(t_h - t_l)](s_l - s_h) = 1.6860 - [(261 - 260)/(262 - 260)](1.6860 - 1.6833) = 1.68465$ Btu/(lb·°F) [7.1 kJ/(kg·°C)] at 261°F (127.2°C).

Study of the steam table for the direction of pressure change shows that the saturation pressure, like enthalpy, increases as the temperature increases. Therefore, the pressure at 261°F (127.2°C) (the intermediate temperature) will equal the pressure at 260°F (126.7°C) (the lower temperature used in the interpolation) plus the proportional change (difference between the intermediate temperature and the lower interpolating temperature) for a 1°F (0.6°C) temperature increase. Or, at any higher temperature, $p_{gi} = p_{gl} + [(t_i - t_l)/(t_h - t_l)](p_h - p_l) = 35.429 + [(261 - 260)(262 - 260)](36.646 - 35.429) = 36.0375$ lb/in^2 (abs) (248.5 kPa) at 261°F (127.2°C) saturated.

3. Use the superheated steam table

Choose the superheated steam table for steam at 1000°F (537.9°C) and 2.6150 ft^3/lb (0.16 m^3/kg) because the highest temperature at which saturated steam can exist is 705.4°F (374.1°C). This is also the highest temperature tabulated in some saturated-temperature tables. Therefore, the steam is superheated when at a temperature of 1000°F (537.9°C).

Look down the 1000°F (537.9°C) columns in the superheated steam table until a specific volume value of 2.6150 (0.16) is found. This occurs between 325 lb/in^2 (abs) (2240.9 kPa, $v = 2.636$ or 0.16) and 330 lb/in^2 (abs) (2275.4 kPa, $v = 2.596$ or 0.16). Since there is no volume value exactly equal to 2.6150 tabulated, it will be necessary to interpolate. List the values from the steam table thus:

\multicolumn{2}{c}{p}		\multicolumn{2}{c}{t}		\multicolumn{2}{c}{v}	
lb/in^2 (abs)	kPa	°F	°C	ft^3/lb	m^3/kg
325	2240.9	1000	537.9	2.636	0.16
330	2275.4	1000	537.9	2.596	0.16

Note that as the pressure rises, at constant temperature, the volume decreases. Therefore, the intermediate (or unknown) pressure is found by subtracting from the higher interpolating pressure [330 lb/in^2 (abs) or 2275.4 kPa in this instance] the product of the proportional change in the specific volume and the difference in the pressures used for the interpolation. Or, $p_{gi} = p_h - [(v_i - v_h)/(v_l - v_h)](p_h - p_l)$, where the subscripts h, l, and i refer to the high, low, and intermediate (or unknown) pressures, respectively. In this instance, $p_{gi} = 330 - [(2.615 - 2.596)/(2.636 - 2.596)](330 - 325) = 327.62$ lb/in^2 (abs) (2,258.9 kPa) at 1000°F (537.9 kPa) and a specific volume of 2.6150 ft^3/lb (0.16 m^3/kg).

4. Use the superheated steam table

When a steam pressure and temperature are given, determine, before performing any interpolation, the state of the steam. Do this by entering the saturation-pressure table at the given pressure and noting the saturation temperature. If the given temperature exceeds the saturation temperature, the steam is superheated. In this instance, the saturation-pressure table shows that at 300 lb/in^2 (abs) (2068.5 kPa) the saturation temperature is 417.33°F (214.1°C). Since the given temperature of the steam is 567.22°F (297.3°C), the steam is superheated because its actual temperature is greater than the saturation temperature.

Enter the superheated steam table at 300 lb/in^2 (abs) (2068.5 kPa), and find the next temperature lower than 567.22°F (297.3°C); this is 560°F (293.3°C). Also find the next higher temperature; this is 580°F (304.4°C). Tabulate the enthalpy, specific volume, and entropy for each temperature thus:

t		h		v		s	
°F	°C	Btu/lb	kJ/kg	ft³/lb	m³/kg	Btu/(lb·°F)	kJ/(kg·°C)
560	293.3	1292.5	3006.4	1.9218	0.12	1.6054	6.72
580	304.4	1303.7	3032.4	1.9594	0.12	1.6163	6.77

Use the same procedures for each property—enthalpy, specific volume, and entropy—as given in step 2 above; but change the sign between the lower volume and entropy and the proportional factor (temperature in this instance), because for superheated steam the volume and entropy increase as the steam temperature increases. Thus

$$h_{gi} = 1292.5 + \frac{567.22 - 560}{580 - 560}(1303.7 - 1292.5) = 1269.6 \text{ Btu/lb (3015.9 kJ/kg)}$$

$$v_{gi} = 1.9128 + \frac{567.22 - 560}{580 - 560}(1.9594 - 1.9128) = 1.9296 \text{ ft}^3/\text{lb (0.12 m}^3/\text{kg)}$$

$$s_{gi} = 1.6054 + \frac{567.22 - 560}{580 - 560}(1.6163 - 1.6054) = 1.6093 \text{ Btu/(lb·°F) [6.7 kJ/(kg·°C)]}$$

Note: Also observe the direction of change of a property *before* interpolating. Use a *plus* or *minus* sign between the higher interpolating value and the proportional change depending on whether the tabulated value increases (+) or decreases (−).

CONSTANT-PRESSURE STEAM PROCESS

Three pounds of wet steam, containing 15 percent moisture and initially at a pressure of 400 lb/in² (abs) (2758.0 kPa), expands at constant pressure ($P = C$) to 600°F (315.6°C). Determine the initial temperature T_1, enthalpy H_1, internal energy E_1, volume V_1, entropy S_1, final entropy H_2, internal energy E_2, volume V_2, entropy S_2, heat added to the steam Q_1, work output W_2, change in initial energy ΔE, change in specific volume ΔV, change in entropy ΔS.

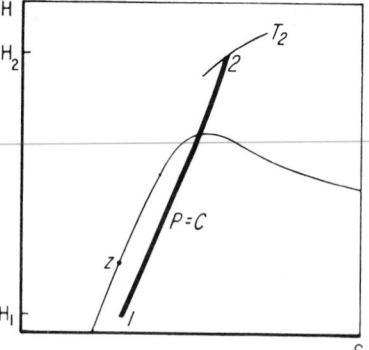
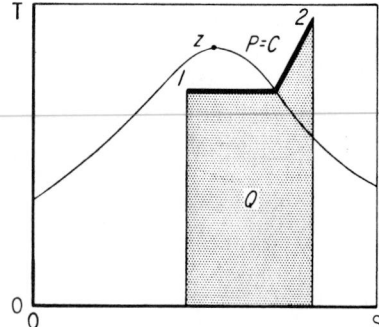

FIG. 2   Constant-pressure process.

Calculation Procedure:

1. Determine the initial steam temperature from the steam tables

Enter the saturation-pressure table at 400 lb/in^2 (abs) (2758.0 kPa), and read the saturation temperature as 444.59°F (229.2°C).

2. Correct the saturation values for the moisture of the steam in the initial state

Sketch the process on a pressure-volume (P-V), Mollier (H-S), or temperature-entropy (T-S) diagram, Fig. 2. In state 1, y = moisture content = 15 percent. Using the appropriate values from the saturation-pressure steam table for 400 lb/in^2 (abs) (2758.0 kPa), correct them for a moisture content of 15 percent:

$$H_1 = h_g - yh_{fg} = 1204.5 - 0.15(780.5) = 1087.4 \text{ Btu/lb (2529.3 kJ/kg)}$$

$$E_1 = u_g - yu_{fg} = 1118.5 - 0.15(695.9) = 1015.1 \text{ Btu/lb (2361.1 kJ/kg)}$$

$$V_1 = v_g - yv_{fg} = 1.1613 - 0.15(1.1420) = 0.990 \text{ ft}^3\text{/lb (0.06 m}^3\text{/kg)}$$

$$S_1 = s_g - ys_{fg} = 1.4844 - 0.15(0.8630) = 1.2945 \text{ Btu/(lb·°F) [5.4 kJ/(kg·°C)]}$$

3. Determine the steam properties in the final state

Since this is a constant-pressure process, the pressure in state 2 is 400 lb/in^2 (abs) (2758.0 kPa), the same as state 1. The final temperature is given as 600°F (315.6°C). This is greater than the saturation temperature of 444.59°F (229.2°C). Hence, the steam is superheated when in state 2. Use the superheated steam tables, entering at 400 lb/in^2 (abs) (2758.8 kPa) and 600°F (315.6°C). At this condition, H_2 = 1306.9 Btu/lb (3039.8 kJ/kg); V_2 = 1.477 ft^3/lb (0.09 m^3/kg). Then E_2 = $h_{2g} - P_2V_2/J$ = 1306.9 − 400(144)(1.477)/778 = 1197.5 Btu/lb (2785.4 kJ/kg). In this equation, the constant 144 converts pounds per square inch to pounds per square foot, absolute, and J = mechanical equivalent of heat = 778 ft·lb/Btu (1 N·m/J). From the steam tables, S_2 = 1.5894 Btu/(lb·°F) [6.7 kJ/(kg·°C)].

4. Compute the process inputs, outputs, and changes

$W_2 = (P_1/J)(V_2 - V_1)m = [400(144)/778](1.4770 - 0.9900)(3) = 108.1$ Btu (114.1 kJ). In this equation, m = weight of steam used in the process = 3 lb (1.4 kg). Then

$$Q_1 = (H_2 - H_1)m = (1306.9 - 1087.4)(3) = 658.5 \text{ Btu (694.4 kJ)}$$

$$\Delta E = (E_2 - E_1)m = (1197.5 - 1014.1)(3) = 550.2 \text{ Btu (580.2 kJ)}$$

$$\Delta V = (V_2 - V_1)m = (1.4770 - 0.9900)(3) = 1.461 \text{ ft}^3 \text{ (0.041 m}^3\text{)}$$

$$\Delta S = (S_2 - S_1)m = (1.5894 - 1.2945)(3) = 0.8847 \text{ Btu/°F (1.680 kJ/°C)}$$

5. Check the computations

The work output W_2 should equal the change in internal energy plus the heat input, or $W_2 = E_1 - E_2 + Q_1 = -550.2 + 658.5 = 108.3$ Btu (114.3 kJ). This value very nearly equals the computed value of W_2 = 108.1 Btu (114.1 kJ) and is close enough for all normal engineering computations. The difference can be traced to calculator input errors. In computing the work output, the internal-energy change has a negative sign because there is a decrease in E during the process.

   Related Calculations: Use this procedure for all constant-pressure steam processes.

CONSTANT-VOLUME STEAM PROCESS

Five pounds (2.3 kg) of wet steam initially at 120 lb/in^2 (abs) (827.4 kPa) with 30 percent moisture is heated at constant volume (V = C) to a final temperature of 1000°F (537.8°C). Determine the initial temperature T_1, enthalpy H_1, internal energy E_1, volume V_1, final pressure P_2, enthalpy H_2, internal energy E_2, volume V_2, heat added Q_1, work output W, change in internal energy ΔE, volume ΔV, and entropy ΔS.

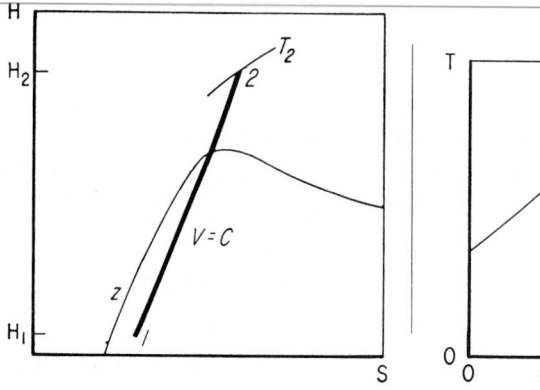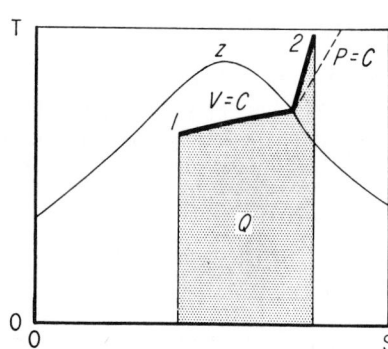

FIG. 3   Constant-volume process.

Calculation Procedure:

1.   Determine the initial steam temperature from the steam tables

Enter the saturation-pressure table at 120 lb/in² (abs) (827.4 kPa), the initial pressure, and read the saturation temperature $T_1 = 341.25°F$ (171.8°C).

2.   Correct the saturation values for the moisture in the steam in the initial state

Sketch the process on P-V, H-S, or T-S diagrams, Fig. 3. Using the appropriate values from the saturation-pressure table for 120 lb/in² (abs) (827.4 kPa), correct them for a moisture content of 30 percent:

$$H_1 = h_g - y h_{fg} = 1190.4 - 0.3(877.9) = 927.0 \text{ Btu/lb (2156.2 kJ/kg)}$$

$$E_1 = u_g - y u_{fg} = 1107.6 - 0.3(795.6) = 868.9 \text{ Btu/lb (2021.1 kJ/kg)}$$

$$V_1 = v_g - y v_{fg} = 3.7280 - 0.3(3.7101) = 2.6150 \text{ ft}^3/\text{lb (0.16 m}^3/\text{kg)}$$

$$S_1 = s_g - y s_{fg} = 1.5878 - 0.3(1.0962) = 1.2589 \text{ Btu/(lb·°F) [5.3 kJ/(kg·°C)]}$$

3.   Determine the steam volume in the final state

We are given $T_2 = 1000°F$ (537.8°C). Since this is a constant-volume process, $V_2 = V_1 = 2.6150$ ft³/lb (0.16 m³/kg). The total volume of the vapor equals the product of the specific volume and the number of pounds of vapor used in the process, or total volume = 2.6150(5) = 13.075 ft³ (0.37 m³).

4.   Determine the final steam pressure

The final steam temperature (1000°F or 537.8°C) and the final steam volume (2.6150 ft³/lb or 0.16 m³/kg) are known. To determine the final steam pressure, find in the steam tables the state corresponding to the above temperature and specific volume. Since a temperature of 1000°F (537.8°C) is higher than any saturation temperature (705.4°F or 374.1°C is the highest saturation temperature for saturated steam), the steam in state 2 must be superheated. Therefore, the superheated steam tables must be used to determine P_2.

Enter the 1000°F (537.8°C) column in the steam table, and look for a superheated-vapor specific volume of 2.6150 ft³/lb (0.16 m³/kg). At a pressure of 325 lb/in² (abs) (2240.9 kPa),

$$v = 2.636 \text{ ft}^3/\text{lb (0.16 m}^3/\text{kg)}$$

$$h = 1542.5 \text{ Btu/lb (3587.9 kJ/kg)}$$

$$s = 1.7863 \text{ Btu/(lb·°F) [7.48 kJ/(kg·°C)]}$$

and at a pressure of 330 lb/in^2 (abs) (2275.4 kPa)

$$v = 2.596 \text{ ft}^3/\text{lb} \ (0.16 \text{ m}^3/\text{kg})$$

$$h = 1524.4 \text{ Btu/lb} \ (3545.8 \text{ kJ/kg})$$

$$s = 1.7845 \text{ Btu}/(\text{lb} \cdot {}^\circ\text{F}) \ [7.47 \text{ kJ}/(\text{kg} \cdot {}^\circ\text{C})]$$

Thus, 2.6150 lies between 325 and 330 lb/in^2 (abs) (2240.9 and 2275.4 kPa). To determine the pressure corresponding to the final volume, it is necessary to interpolate between the specific-volume values, or $P_2 = 330 - [(2.615 - 2.596)/(2.636 - 2.596)](330 - 325) = 327.62$ lb/in^2 (abs) (2258.9 kPa). In this equation, the volume values correspond to the upper [330 lb/in^2 (abs) or 2275.4 kPa], lower [325 lb/in^2 (abs) or 2240.9 kPa], and unknown pressures.

5. *Determine the final enthalpy, entropy, and internal energy*

The final enthalpy can be interpolated in the same manner, using the enthalpy at each volume instead of the pressure. Thus $H_2 = 1524.5 - [(2.615 - 2.596)/(2.636 - 2.596)](1524.5 - 1524.4) = 1524.45$ Btu/lb (3545.8 kJ/kg). Since the difference in enthalpy between the two pressures is only 0.1 Btu/lb (0.23 kJ/kg) (=1524.5 - 1524.4), the enthalpy at 327.62 lb/in^2 (abs) could have been assumed equal to the enthalpy at the lower pressure [325 lb/in^2 (abs) or 2240.9 kPa], or 1524.4 Btu/lb (3545.8 kJ/kg), and the error would have been only 0.05 Btu/lb (0.12 kJ/kg), which is negligible. However, where the enthalpy values vary by more than 1.0 Btu/lb (2.3 kJ/kg), interpolate as shown, if accurate results are desired.

Find S_2 by interpolating between pressures, or

$$S = 1.7863 - \frac{327.62 - 325}{330 - 325}(1.7863 - 1.7845) = 1.7854 \text{ Btu}/(\text{lb} \cdot {}^\circ\text{F}) \ [7.5 \text{ kJ}/(\text{kg} \cdot {}^\circ\text{C})]$$

$$E_2 = H_2 - \frac{P_2 V_2}{J} = 1524.4 - \frac{327.62(144)(2.615)}{778} = 1365.9 \text{ Btu/lb} \ (3177.1 \text{ kJ/kg})$$

6. *Compute the changes resulting from the process*

Here $Q_1 = (E_2 - E_1)m = (1365.9 - 868.9)(5) = 2485$ Btu (2621.8 kJ); $\Delta S = (S_2 - S_1)m = (1.7854 - 1.2589)(5) = 2.6325$ Btu/°F (5.0 kJ/°C).

By definition, $W = 0$; $\Delta V = 0$; $\Delta E = Q_1$. Note that the curvatures of the constant-volume line on the *T-S* chart, Fig. 3, are different from the constant-pressure line, Fig. 2. Adding heat Q_1 to a constant-volume process affects only the internal energy. The total entropy change must take into account the total steam mass $m = 5$ lb (2.3 kg).

Related Calculations: Use this general procedure for all constant-volume steam processes.

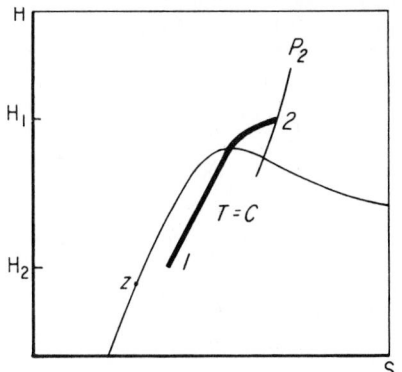
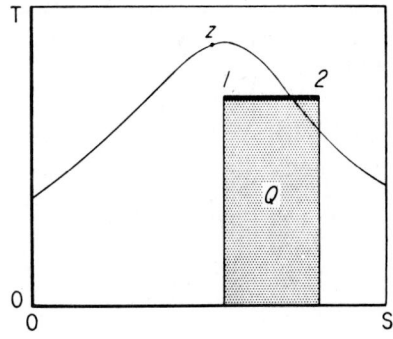

FIG. 4   Constant-temperature process.

CONSTANT-TEMPERATURE STEAM PROCESS

Six pounds (2.7 kg) of wet steam initially at 1200 lb/in² (abs) (8274.0 kPa) and 50 percent moisture expands at constant temperature ($T = C$) to 300 lb/in² (abs) (2068.5 kPa). Determine the initial temperature T_1, enthalpy H_1, internal energy E_1, specific volume V_1, entropy S_1, final temperature T_2, enthalpy H_2, internal energy E_2, volume V_2, entropy S_2, heat added Q_1, work output W_2, change in internal energy ΔE, volume ΔV, and entropy ΔS.

Calculation Procedure:

1. *Determine the initial steam temperature from the steam tables*

Enter the saturation-pressure table at 1200 lb/in² (abs) (8274.0 kPa), and read the saturation temperature $T_1 = 567.22°F$ (297.3°C).

2. *Correct the saturation values for the moisture in the steam in the initial state*

Sketch the process on *P-V*, *H-S*, or *T-S* diagrams, Fig. 4. Using the appropriate values from the saturation-pressure table for 1200 lb/in² (abs) (8274.0 kPa), correct them for the moisture content of 50 percent:

$$H_1 = h_g - y_1 h_{fg} = 1183.4 - 0.5(611.7) = 877.5 \text{ Btu/lb (2041.1 kJ/kg)}$$

$$E_1 = u_g - y_1 u_{fg} = 1103.0 - 0.5(536.3) = 834.8 \text{ Btu/lb (1941.7 kJ/kg)}$$

$$V_1 = v_g - y_1 v_{fg} = 0.3619 - 0.5(0.3396) = 0.19 \text{ ft}^3/\text{lb (0.012 m}^3/\text{kg)}$$

$$S_1 = s_g - y_1 s_{fg} = 1.3667 - 0.5(0.5956) = 1.0689 \text{ Btu/(lb·°F) [4.5 kJ/(kg·°C)]}$$

3. *Determine the steam properties in the final state*

Since this is a constant-temperature process, $T_2 = T_1 = 567.22°F$ (297.3°C); $P_2 = 300$ lb/in² (abs) (2068.5 kPa), given. The saturation temperature of 300 lb/in² (abs) (2068.5 kPa) is 417.33°F (214.1°C). Therefore, the steam is superheated in the final state because 567.22°F (297.3°C) > 417.33°F (214.1°C), the saturation temperature.

To determine the final enthalpy, entropy, and specific volume, it is necessary to interpolate between the known final temperature and the nearest tabulated temperatures greater and less than the final temperature.

	v		h		s	
	ft³/lb	m³/kg	Btu/lb	kJ/kg	Btu/(lb·°F)	kJ/(kg·°C)
At $T = 560°F$ (293.3°C)	1.9128	0.12	1292.5	3006.4	1.6054	6.72
At $T = 580°F$ (304.4°C)	1.9594	0.12	1303.7	3032.4	1.6163	6.76

Then

$$H_2 = 1292.5 + \frac{567.22 - 560}{580 - 560}(1303.7 - 1292.5) = 1296.5 \text{ Btu/lb (3015.7 kJ/kg)}$$

$$S_2 = 1.6054 + \frac{567.22 - 560}{580 - 560}(1.6163 - 1.6054) = 1.6093 \text{ Btu/(lb·°F) [6.7 kJ/(kg·°C)]}$$

$$V_2 = 1.9128 + \frac{567.22 - 560}{580 - 560}(1.9594 - 1.9128) = 1.9296 \text{ ft}^3/\text{lb (0.12 m}^3/\text{kg)}$$

$$E_2 = H_2 - \frac{P_2 V_2}{J} = 1296.5 - \frac{300(144)(1.9296)}{778} = 1109.3 \text{ Btu/lb (2580.2 kJ/kg)}$$

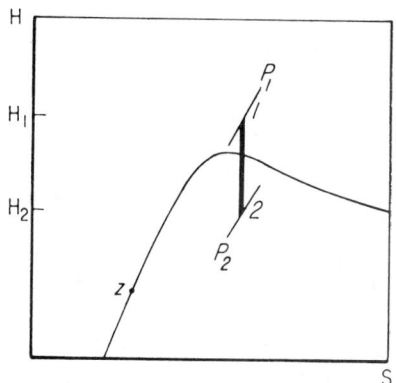 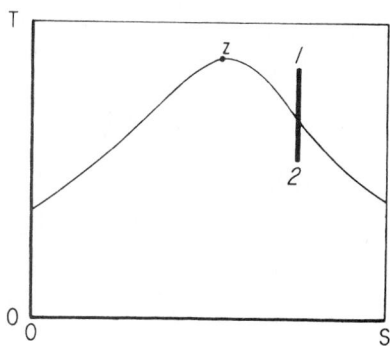

FIG. 5   Constant-entropy process.

4.   *Compute the process changes*

Here $Q_1 = T(S_2 - S_1)m$, where T_1 = absolute initial temperature, °R. So $Q_1 = (567.22 + 460)(1.6093 - 1.0689)(6) = 3330$ Btu (3513.3 kJ). Then

$$\Delta E = E_2 - E_1 = 1109.3 - 834.8 = 274.5 \text{ Btu/lb (638.5 kJ/kg)}$$

$$\Delta H = H_2 - H_1 = 1296.5 - 877.5 = 419.0 \text{ Btu/lb (974.6 kJ/kg)}$$

$$W_2 = (Q_1 - \Delta E)m = (555 - 274.5)(6) = 1.683 \text{ Btu (1.8 kJ)}$$

$$\Delta S = S_2 - S_1 = 1.6093 - 1.0689 = 0.5404 \text{ Btu/(lb·°F) [2.3 kJ/(kg·°C)]}$$

$$\Delta V = V_2 - V_1 = 1.9296 - 0.1921 = 1.7375 \text{ ft}^3/\text{lb (0.11 m}^3/\text{kg)}$$

Related Calculations: Use this procedure for any constant-temperature steam process.

CONSTANT-ENTROPY STEAM PROCESS

Ten pounds (4.5 kg) of steam expands under two conditions—nonflow and steady flow—at constant entropy ($S = C$) from an initial pressure of 2000 lb/in^2 (abs) (13,790.0 kPa) and a temperature of 800°F (426.7°C) to a final pressure of 2 lb/in^2 (abs) (13.8 kPa). In the steady-flow process, assume that the initial kinetic energy E_{k1} = the final kinetic energy E_{k2}. Determine the initial enthalpy H_1, internal energy E_1, volume V_1, entropy S_1, final temperature T_2, percentage of moisture y, enthalpy H_2, internal energy E_2, volume V_2, entropy S_2, change in internal energy ΔE, enthalpy ΔH, entropy ΔS, volume ΔV, heat added Q_1, and work output W_2.

Calculation Procedure:

1.   *Determine the initial enthalpy, volume, and entropy from the steam tables*

Enter the superheated-vapor table at 2000 lb/in^2 (abs) (13,790.0 kPa) and 800°F (427.6°C), and read $H_1 = 1335.5$ Btu/lb (3106.4 kJ/kg); $V_1 = 0.3074$ ft^3/lb (0.019 m^3/kg); $S_1 = 1.4576$ Btu/(lb·°F) [6.1 kJ/(kg·°C)].

2.   *Compute the initial energy*

$$E_1 = H_1 - \frac{P_1 V_1}{J} = 1335.5 - \frac{2000(144)(0.3074)}{778} = 1221.6 \text{ Btu/lb (2841.1 kJ/kg)}$$

3. Determine the vapor properties on the final state

Sketch the process on *P-V*, *H-S*, or *T-S* diagrams, Fig. 5. Note that the expanded steam is wet in the final state because the 2-lb/in² (abs) (13.8-kPa) pressure line is under the saturation curve on the *H-S* and *T-S* diagrams. Therefore, the vapor properties in the final state must be corrected for the moisture content. Read, from the saturation-pressure steam table, the liquid and vapor properties at 2 lb/in² (abs) (13.8 kPa). Tabulate these properties thus:

$$s_f = 0.1749 \text{ Btu/(lb·°F) [0.73 kJ/(kg·°C)]} \qquad s_{fg} = 1.7451 \text{ Btu/(lb·°F) [7.31 kJ/(kg·C)]}$$

$$h_f = 93.99 \text{ Btu/lb (218.6 kJ/kg)} \qquad h_{fg} = 1022.2 \text{ Btu/lb (2377.6 kJ/kg)}$$

$$u_f = 93.98 \text{ Btu/lb (218.6 kJ/kg)} \qquad u_{fg} = 957.9 \text{ Btu/lb (2228.1 kJ/kg)}$$

$$v_f = 0.016 \text{ ft}^3/\text{lb (0.0010 m}^3/\text{kg)} \qquad v_{fg} = 173.71 \text{ ft}^3/\text{lb (10.8 m}^3/\text{kg)}$$

$$s_g = 1.9200 \text{ Btu/(lb·°F) [8.04 kJ/(kg·C)} \qquad h_g = 1116.3 \text{ Btu/lb (2596.5 kJ/kg)}$$

$$u_g = 1051.9 \text{ Btu/lb (2446.7 kJ/kg)} \qquad v_g = 173.73 \text{ ft}^3/\text{lb (10.8 m}^3/\text{kg)}$$

Since this is a constant-entropy process, $S_2 = S_1 = s_g - y_2 s_{fg}$. Solve for y_2, the percentage of moisture in the final state. Or, $y_2 = (s_g - S_1)/s_{fg} = (1.9200 - 1.4576)/1.7451 = 0.265$, or, 26.5 percent. Then

$$H_2 = h_g - y_2 h_{fg} = 1116.2 - 0.265(1022.2) = 845.3 \text{ Btu/lb (1966.2 kJ/kg)}$$

$$E_2 = u_g - y_2 u_{fg} = 1051.9 - 0.265(957.9) = 798.0 \text{ Btu/lb (1856.1 kJ/kg)}$$

$$V_2 = v_g - y_2 v_{fg} = 173.73 - 0.265(173.71) = 127.7 \text{ ft}^3/\text{lb (8.0 m}^3/\text{kg)}$$

4. Compute the changes resulting from the process

The total change in properties is for 10 lb (4.5 kg) of steam, the quantity used in this process. Thus,

$$\Delta E = (E_1 - E_2)m = (1221.6 - 798.0)(10) = 4236 \text{ Btu (4469.2 kJ)}$$

$$\Delta H = (H_1 - H_2)m = (1335.5 - 845.3)(10) = 4902 \text{ Btu (5171.9 kJ)}$$

$$\Delta S = (S_1 - S_2)m = (1.4576 - 1.4576)(10) = 0 \text{ Btu/°F (0 kJ/°C)}$$

$$\Delta V = (V_1 - V_2)m = (0.3074 - 127.7)(10) = -1274 \text{ ft}^3 \ (-36.1 \text{ m}^3)$$

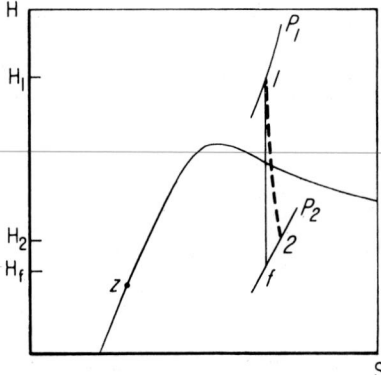
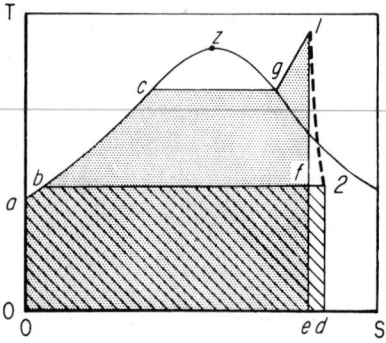

FIG. 6   Irreversible adiabatic process.

So $Q_1 = 0$ Btu. (By definition, there is no transfer of heat in a constant-entropy process.) Nonflow $W_2 = \Delta E = 4236$ Btu (4469.2 kJ). Steady flow $W_2 = \Delta H = 4902$ Btu (5171.9 kJ).

Note: In a constant-entropy process, the nonflow work depends on the change in internal energy. The steady-flow work depends on the change in enthalpy and is larger than the nonflow work by the amount of the change in the flow work.

IRREVERSIBLE ADIABATIC EXPANSION OF STEAM

Ten pounds (4.5 kg) of steam undergoes a steady-flow expansion from an initial pressure of 2000 lb/in^2 (abs) (13,790.0 kPa) and a temperature 800°F (426.7°C) to a final pressure of 2 lb/in^2 (abs) (13.9 kPa) at an expansion efficiency of 75 percent. In this steady flow, assume $E_{k1} = E_{k2}$. Determine ΔE, ΔH, ΔS, ΔV, Q, and W_2.

Calculation Procedure:

1. *Determine the initial vapor properties from the steam tables*

Enter the superheated-vapor tables at 2000 lb/in^2 (abs) (13,790.0 kPa) and 800°F (426.7°C), and read $H_1 = 1335.5$ Btu/lb (3106.4 kJ/kg); $V_1 = 0.3074$ ft^3/lb (0.019 m^3/kg); $E_1 = 1221.6$ Btu/lb (2840.7 kJ/kg); $S_1 = 1.4576$ Btu/(lb·°F) [6.1 kJ/(kg·°C)].

2. *Determine the vapor properties in the final state*

Sketch the process on P-V, H-S, or T-S diagram, Fig. 6. Note that the expanded steam is wet in the final state because the 2-lb/in^2 (abs) (13.9-kPa) pressure line is under the saturation curve on the H-S and T-S diagrams. Therefore, the vapor properties in the final state must be corrected for the moisture content. However, the actual final enthalpy cannot be determined until after the expansion efficiency $[H_1 - H_2(H_1 - H_{2s})]$ is evaluated.

To determine the final enthalpy H_2, another enthalpy H_{2s} must be computed by assuming a constant-entropy expansion to 2 lb/in^2 (abs) (13.8 kPa) and a temperature of 126.08°F (52.3°C). Enthalpy H_{2s} will then correspond to a constant-entropy expansion into the wet region, and the percentage of moisture will correspond to the final state. This percentage is determined by finding the ratio of $s_g - S_1$ to s_{fg}, or $y_{2s} = s_g - S_1/s_{fg} = 1.9200 - 1.4576/1.7451 = 0.265$, where s_g and s_{fg} are entropies at 2 lb/in^2 (abs) (13.8 kPa). Then $H_{2s} = h_g - y_{2s}h_{fg} = 1116.2 - 0.265(1022.2) = 845.3$ Btu/lb (1966.2 kJ/kg). In this relation, h_g and h_{fg} are enthalpies at 2 lb/in^2 (abs) (13.8 kPa).

The expansion efficiency, given as 0.75, is $H_1 - H_2/(H_1 - H_{2s}) = $ actual work/ideal work $= 0.75 = 1335.5 - H_2/(1335.5 - 845.3)$. Solve for $H_2 = 967.9$ Btu/lb (2251.3 kJ/kg).

Next, read from the saturation-pressure steam table the liquid and vapor properties at 2 lb/in^2 (abs) (13.8 kPa). Tabulate these properties thus:

$$h_f = 93.99 \text{ Btu/lb (218.6 kJ/kg)}$$

$$h_{fg} = 1022.2 \text{ Btu/lb (2377.6 kJ/kg)}$$

$$h_g = 1116.2 \text{ Btu/lb (2596.3 kJ/kg)}$$

$$s_f = 0.1749 \text{ Btu/(lb·°F) [0.73 kJ/(kg·°C)]}$$

$$s_{fg} = 1.7451 \text{ Btu/(lb·°F) [7.31 kJ/(kg·°C)]}$$

$$s_g = 1.9200 \text{ Btu/(lb·°F) [8.04 kJ/(kg·°C)]}$$

$$u_f = 93.98 \text{ Btu/lb (218.60 kJ/kg)}$$

$$u_{fg} = 957.9 \text{ Btu/lb (2228.1 kJ/kg)}$$

$$u_g = 1051.9 \text{ Btu/lb (2446.7 kJ/kg)}$$

$$v_f = 0.016 \text{ ft}^3\text{/lb (0.0010 m}^3\text{/kg)}$$

$$v_{fg} = 173.71 \text{ ft}^3\text{/lb (10.84 m}^3\text{/kg)}$$

$$v_g = 173.73 \text{ ft}^3\text{/lb (10.85 m}^3\text{/kg)}$$

Since the actual final enthalpy H_2 is different from H_{2s}, the final actual moisture y_2 must be computed by using H_2. Or, $y_2 = h_g - H_2/h_{fg} = 1116.1 - 967.9/1022.2 = 0.1451$. Then

$$E_2 = u_g - y_2 u_{fg} = 1051.9 - 0.1451(957.9) = 912.9 \text{ Btu/lb (2123.4 kJ/kg)}$$

$$V_2 = v_g - y_2 v_{fg} = 173.73 - 0.1451(173.71) = 148.5 \text{ ft}^3/\text{lb (9.3 m}^3/\text{kg)}$$

$$S_2 = s_g - y_2 s_{fg} = 1.9200 - 0.1451(1.7451) = 1.6668 \text{ Btu/(lb·°F) [7.0 kJ/kg·°C)]}$$

3. Compute the changes resulting from the process

The total change in properties is for 10 lb (4.5 kg) of steam, the quantity used in this process. Thus

$$\Delta E = (E_1 - E_2)m = (1221.6 - 912.9)(10) = 3087 \text{ Btu (3257.0 kJ)}$$

$$\Delta H = (H_1 - H_2)m = (1335.5 - 967.9)(10) = 3676 \text{ Btu (3878.4 kJ)}$$

$$\Delta S = (S_2 - S_1)m = (1.6668 - 1.4576)(10) = 2.092 \text{ Btu/°F (4.0 kJ/°C)}$$

$$\Delta V = (V_2 - V_1)m = (148.5 - 0.3074)(10) = 1482 \text{ ft}^3 \text{ (42.0 m}^3)$$

So $Q = 0$; by definition, $W_2 = \Delta H = 3676$ Btu (3878.4 kJ) for the steady-flow process.

IRREVERSIBLE ADIABATIC STEAM COMPRESSION

Two pounds (0.9 kg) of saturated steam at 120 lb/in^2 (abs) (827.4 kPa) with 80 percent quality undergoes nonflow adiabatic compression to a final pressure of 1700 lb/in^2 (abs) (11,721.5 kPa) at 75 percent compression efficiency. Determine the final steam temperature T_2, change in internal energy ΔE, change in entropy ΔS, work input W, and heat input Q.

Calculation Procedure:

1. Determine the vapor properties in the initial state

From the saturation-pressure steam tables, $T_1 = 341.25°F$ (171.8°C) at a pressure of 120 lb/in^2 (abs) (827.4 kPa) saturated. With $x_1 = 0.8$, $E_1 = u_f + x_1 u_{fg} = 312.05 + 0.8(795.6) = 948.5$ Btu/lb (2206.5 kJ/kg), from internal-energy values from the steam tables. The initial entropy is $S_1 = s_f + x_1 s_{fg} = 0.4916 + 0.8(1.0962) = 1.3686$ Btu/(lb·°F) [5.73 kJ/(kg·°C)].

2. Determine the vapor properties in the final state

Sketch a T-S diagram of the process, Fig. 7. Assume a constant-entropy compression from the initial to the final state. Then $S_{2s} = S_1 = 1.3686$ Btu/(lb·°F) [5.7 kJ/(kg·°C)].

The final pressure, 1700 lb/in^2 (abs) (11,721.5 kPa), is known, as is the final entropy, 1.3686 Btu/(lb·°F) [5.7 kJ/(kg·°C)] with constant-entropy expansion. The T-S diagram (Fig. 7) shows that the steam is superheated in the final state. Enter the superheated steam table at 1700 lb/in^2 (abs) (11,721.5 kPa), project across to an entropy of 1.3686, and read the final steam temperature as 650°F (343.3°C). (In most cases, the final entropy would not exactly equal a tabulated value, and it would be necessary to interpolate between tabulated entropy values to determine the intermediate pressure value.)

From the same table, at 1700 lb/in^2 (abs) (11.721.5 kPa) and 650°F (343.3°C), $H_{2s} = 1214.4$ Btu/lb (2827.4 kJ/kg); $V_{2s} = 0.2755$ ft^3/lb (0.017 m^3/lb). Then $E_{2s} = H_{2s} - P_2 V_{2s}/J = 1214.4 - 1700(144)(0.2755)/788 = 1127.8$ Btu/lb (2623.3 kJ/kg). Since E_1 and E_{2s} are known, the ideal work W can be computed. Or, $W = E_{2s} - E_1 = 1127.8 - 948.5 = 179.3$ Btu/lb (417.1 kJ/kg).

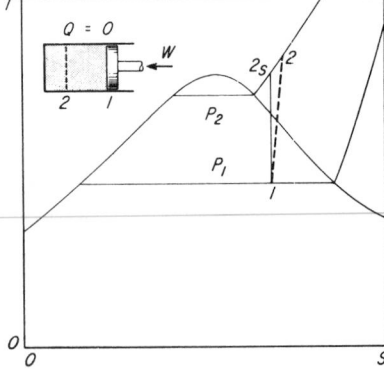

FIG. 7 Irreversible adiabatic compression process.

3. Compute the vapor properties of the actual compression

Since the compression efficiency is known, the actual final internal energy can be found from compression efficiency = ideal W/actual W =

$E_{2s} - E_1/(E_2 - E_1)$, or $0.75 = 1127.8 - 948.5/(E_2 - 948.5)$; $E_2 = 1187.6$ Btu/lb (2762.4 kJ/kg). Then $E = (E_2 - E_1)m = (1187.6 - 948.5)(2) = 478.2$ Btu (504.5 kJ) for 2 lb (0.9 kg) of steam. The actual work input $W = \Delta E = 478.2$ Btu (504.5 kJ). By definition, $Q = 0$.

Last, the actual final temperature and entropy must be computed. The final actual internal energy $E_2 = (1187.6$ Btu/lb (2762.4 kJ/kg) is known. Also, the T-S diagram shows that the steam is superheated. However, the superheated steam tables do not list the internal energy of the steam. Therefore, it is necessary to assume a final temperature for the steam and then compute its internal energy. The computed value is compared with the known internal energy, and the next assumption is adjusted as necessary. Thus, assume a final temperature of 720°F (382.2°C). This assumption is higher than the ideal final temperature of 650°F (343.3°C) because the T-S diagram shows that the actual final temperature is higher than the ideal final temperature. Using values from the superheated steam table for 1700 lb/in² (abs) (11,721.5 kPa) and 720°F (382.2°C), we find

$$E = H - \frac{PV}{J} = 1288.4 - \frac{1700(144)(0.3283)}{778} = 1185.1 \text{ Btu/lb (2756.5 kJ/kg)}$$

This value is less than the actual internal energy of 1187.6 Btu/lb (2762.4 kJ/kg). Therefore, the actual temperature must be higher than 720°F (382.2°C), since the internal energy increases with temperature. To obtain a higher value for the internal energy to permit interpolation between the lower, actual, and higher values, assume a higher final temperature—in this case, the next temperature listed in the steam table, or 740°F (393.3°C). Then, for 1700 lb/in² (abs) (11,721.5 kPa) and 740°F (393.3°C),

$$E = 1305.8 - \frac{1700(144)(0.3410)}{778} = 1198.5 \text{ Btu/lb (2757.7 kJ/kg)}$$

This value is greater than the actual internal energy of 1187.6 Btu/lb (2762.4 kJ/kg). Therefore, the actual final temperature of the steam lies somewhere between 720 and 740°F (382.2 and 393.3°C). Interpolate between the known internal energies to determine the final steam temperature and final entropy. Or,

$$T_2 = 720 + \frac{1178.6 - 1185.1}{1198.5 - 1185.1}(740 - 720) = 723.7°F \text{ (384.3°C)}$$

$$S_2 = 1.4333 + \frac{1187.6 - 1185.1}{1198.5 - 1185.1}(1.4480 - 1.4333) = 1.4360 \text{ Btu/(lb·°F) [6.0 kJ/(kg·°C)]}$$

$$\Delta S = (S_2 - S_1)m = (1.4360 - 1.3686)(2) = 0.1348 \text{ Btu/°F (0.26 kJ/°C)}$$

Note that the final actual steam temperature is 73.7°F (40.9°C) higher than that (650°F or 343.3°C) for the ideal compression.

Related Calculations: Use this procedure for any irreversible adiabatic steam process.

THROTTLING PROCESSES FOR STEAM AND WATER

A throttling process begins at 500 lb/in² (abs) (3447.5 kPa) and ends at 14.7 lb/in² (abs) (101.4 kPa) with (1) steam at 500 lb/in² (abs) (3447.5 kPa) and 500°F (260.0°C); (2) steam at 500 lb/in² (abs) (3447.5 kPa) and 4 percent moisture; (3) steam at 500 lb/in² (abs) (3447.5 kPa) with 50 percent moisture; and (4) saturated water at 500 lb/in² (abs) (3447.5 kPa). Determine the final enthalpy H_2, temperature T_2, and moisture content y_2 for each process.

Calculation Procedure:

1. Compute the final-state conditions of the superheated steam

From the superheated steam table for 500 lb/in² (abs) (3447.5 kPa) and 500°F (260.0°C), $H_1 = 1231.3$ Btu/lb (2864.0 kJ/kg). By definition of a throttling process, $H_1 = H_2 = 1231.3$ Btu/lb (2864.0 kJ/kg). Sketch the T-S diagram for a throttling process, Fig 8a.

To determine the final temperature, enter the superheated steam table at 14.7 lb/in² (abs)

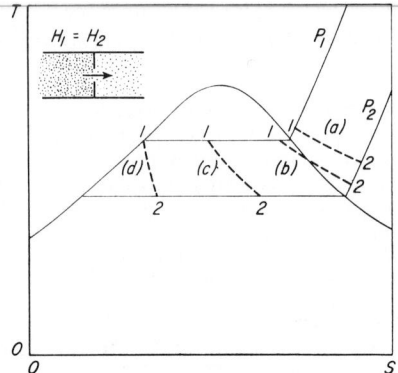

FIG. 8   Throttling process for steam.

(101.4 kPa), the final pressure, and project across to an enthalpy value equal to or less than the known enthalpy, 1231.3 Btu/lb (2864.0 kJ/kg). (The superheated steam table is used because the T-S diagram, Fig. 8, shows that the steam is superheated in the final state.) At 14.7 lb/in² (abs) (101.4 kPa) there is no tabulated enthalpy value that exactly equals 1231.3 Btu/lb (2864.0 kJ/kg). The next lower value is 1230 Btu/lb (2861.0 kJ/kg) at $T = 380°F$ (193.3°C). The next higher value at 14.7 lb/in² (abs) (101.4 kPa) is 1239.9 Btu/lb (2884.0 kJ/kg) at $T = 400°F$ (204.4°C). Interpolate between these enthalpy values to find the final steam temperature:

$$T_2 = 380 + \frac{1231.3 - 1230.5}{1239.9 - 1230.5}(400 - 380) = 381.7°F \ (194.3°C)$$

The steam does not contain any moisture in the final state because it is superheated.

2.   *Compute the final-state conditions of the slightly wet steam*

Determine the enthalpy of 500-lb/in² (abs) (3447.5-kPa) saturated steam from the saturation-pressure steam table:

$$h_g = 1204.4 \text{ Btu/lb } (2801.4 \text{ kJ/kg}) \qquad h_{fg} = 755.0 \text{ Btu/lb } (1756.1 \text{ kJ/kg})$$

Correct the enthalpy for moisture:

$$H_1 = h_g - y_1 h_{fg} = 1204.4 - 0.04(755.0) = 1174.2 \text{ Btu/lb } (2731.2 \text{ kJ/kg})$$

Then, by definition, $H_2 = H_1 = 1174.2$ Btu/lb (2731.2 kJ/kg).

Determine the final condition of the throttled steam (wet, saturated, or superheated) by studying the T-S diagram. If a diagram were not drawn, you would enter the saturation-pressure steam table at 14.7 lb/in² (abs) (101.4 kPa), the final pressure, and check the tabulated h_g. If the tabulated h_g were greater than H_1, the throttled steam would be superheated. If the tabulated h_g were less than H_1, the throttled steam would be saturated. Examination of the saturation-pressure steam table shows that the throttled steam is superheated because $H_1 > h_g$.

Next, enter the superheated steam table to find an enthalpy value of H_1 at 14.7 lb/in² (abs) (101.4 kPa). There is no value equal to 1174.2 Btu/lb (2731.2 kJ/kg). The next lower value is 1173.8 Btu/lb (2730.3 kJ/kg) at $T = 260°F$ (126.7°C). The next higher value at 14.7 lb/in² (abs) (101.4 kPa) is 1183.3 Btu/lb (2752.4 kJ/kg) at $T = 280°F$ (137.8°C). Interpolate between these enthalpy values to find the final steam temperature:

$$T_2 = 260 + \frac{1174.2 - 1173.8}{1183.3 - 1173.8}(280 - 260) = 260.8°F \ (127.1°C)$$

This is higher than the temperature of saturated steam at 14.7 lb/in² (abs) (101.4 kPa)—212°F (100°C)—giving further proof that the throttled steam is superheated. The throttled steam, therefore, does not contain any moisture.

3. *Compute the final-state conditions of the very wet steam*

Determine the enthalpy of 500-lb/in^2 (abs) (3447.5-kPa) saturated steam from the saturation-pressure steam table. Or, h_g = 1204.4 Btu/lb (2801.4 kJ/kg); h_{fg} = 755.0 Btu/lb (1756.1 kJ/kg). Correct the enthalpy for moisture:

$$H_1 = H_2 = h_g - y_1 h_{fg} = 1204.4 - 0.5(755.0) = 826.9 \text{ Btu/lb } (1923.4 \text{ kJ/kg})$$

Then, by definition, $H_2 = H_1$ = 826.9 Btu/lb (1923.4 kJ/kg).

Compare the final enthalpy, H_2 = 826.9 Btu/lb (1923.4 kJ/kg), with the enthalpy of saturated steam at 14.7 lb/in^2 (abs) (101.4 kPa), or 1150.4 Btu/lb (2675.8 kJ/kg). Since the final enthalpy is less than the enthalpy of saturated steam at the same pressure, the throttled steam is wet. Since $H_1 = h_g - y_2 h_{fg}$, $y_2 = (h_g - H_1)/h_{fg}$. With a final pressure of 14.7 lb/in^2 (abs) (101.4 kPa), use h_g and h_{fg} values at this pressure. Or,

$$y_2 = \frac{1150.4 - 826.9}{970.3} = 0.3335, \text{ or } 33.35\%$$

The final temperature of the steam T_2 is the same as the saturation temperature at the final pressure of 14.7 lb/in^2 (abs) (101.4 kPa), or T_2 = 212°F (100°C).

4. *Compute the final-state conditions of saturated water*

Determine the enthalpy of 500-lb/in^2 (abs) (3447.5-kPa) saturated water from the saturation-pressure steam table at 500 lb/in^2 (abs) (3447.5 kPa); $H_1 = h_f$ = 449.4 Btu/lb (1045.3 kJ/kg) = H_2, by definition. The *T-S* diagram, Fig. 8, shows that the throttled water contains some steam vapor. Or, comparing the final enthalpy of 449.4 Btu/lb (1045.3 kJ/kg) with the enthalpy of saturated liquid at the final pressure, 14.7 lb/in^2 (abs) (101.4 kPa), 180.07 Btu/lb (418.8 kJ/kg), shows that the liquid contains some vapor in the final state because its enthalpy is greater.

Since $H_1 = H_2 = h_g - y_2 h_{fg}$, $y_2 = (h_g - H_1)/h_{fg}$. Using enthalpies at 14.7 lb/in^2 (abs) (101.4 kPa) of h_g = 1150.4 Btu/lb (2675.8 kJ/kg) and h_{fg} = 970.3 Btu/lb (2256.9 kJ/kg) from the saturation-pressure steam table, we get y_2 = 1150.4 − 449.4/970.3 = 0.723. The final temperature of the steam is the same as the saturation temperature at the final pressure of 14.7 lb/in^2 (abs) (101.4 kPa), or T_2 = 212°F (100°C).

Note: Calculation 2 shows that when you start with slightly wet steam, it can be throttled (expanded) through a large enough pressure range to produce superheated steam. This procedure is often used in a throttling calorimeter to determine the initial quality of the steam in a pipe. When very wet steam is throttled, calculation 3, the net effect may be to produce drier steam at a lower pressure. Throttling saturated water, calculation 4, can produce partial or complete flashing of the water to steam. All these processes find many applications in power-generation and process-steam plants.

REVERSIBLE HEATING PROCESS FOR STEAM

Subcooled water at 1500 lb/in^2 (abs) (10,342.5 kPa) and 140°F (60.0°C), state 1, Fig. 9, is heated at constant pressure to state 4, superheated steam at 1500 lb/in^2 (abs) (10,342.5 kPa) and 1000°F (537.8°C). Find the heat added (1) to raise the compressed liquid to saturation temperature, (2) to vaporize the saturated liquid to saturated steam, (3) to superheat the steam to 1000°F (537.8°C), and (4) Q_1, ΔV, and ΔS from state 1 to state 4.

Calculation Procedure:

1. *Sketch the T-S diagram for this process*

Figure 9 is typical of a steam boiler and superheater. Feedwater fed to a boiler is usually subcooled liquid. If the feedwater pressure is relatively high, subcooling must be taken into account, if accurate results are desired. Some authorities recommend that at pressures below 400 lb/in^2 (abs)

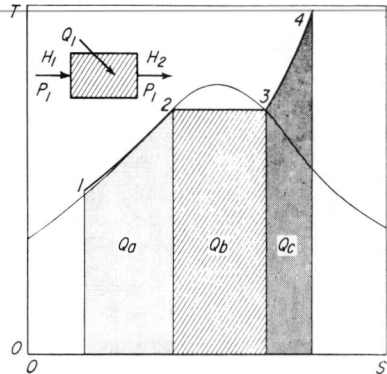

FIG. 9 Reversible heating process.

(2758.0 kPa) subcooling be ignored and values from the saturated-steam table be used. This means that the enthalpies and other properties listed in the steam table corresponding to the actual water temperature are sufficiently accurate. But above 400 lb/in^2 (abs) (2758.0 kPa), the compressed-liquid table should be used.

2. *Determine the initial properties of the liquid*

In the saturation-temperature steam table read, at 140°F (60.0°C), h_f = 107.89 Btu/lb (251.0 kJ/kg); p_f = 2.889 lb/in^2 (19.9 kPa); v_f = 0.01629 ft^3/lb (0.0010 m^3/kg); s_f = 0.1984 Btu/(lb·°F) [0.83 kJ/(kg·°C)].

Next, the enthalpy, volume, and entropy of the water at 1500 lb/in^2 (abs) (10,342.5 kPa) and 140°F (60.0°C) must be found. Since the water is at a much higher pressure than that corresponding to its temperature [1500 versus 2.889 lb/in^2 (abs)], the compressed-liquid portion of the steam table must be used. This table shows that three desired properties are plotted for 32, 100, and 200°F (0.0, 37.8, and 93.3°C) and higher temperatures. However, 140°F (60.0°C) is not included. Therefore, it is necessary to interpolate between 100 and 200°F (37.8 and 93.3°C). Thus, at 1500 lb/in^2 (abs) (10,342.5 kPa) in the compressed-liquid table:

Property	Temperature		Interpolation
	100°F (37.8°C)	200°F (93.3°C)	
$h - h_f$, Btu/lb (kJ/kg)	+3.99 (+9.28)	+3.36 (+7.82)	+3.74 (+8.70)
$(v - v_f)10^5$, ft^3/lb (m^3/kg)	−7.5 (−0.47)	−8.1 (−0.51)	−7.7 (−0.48)
$(s - s_f)10^3$, Btu/(lb·°F) [kJ/(kg·°C)]	−0.86 (−3.60)	−1.79 (−7.49)	−1.23 (−5.15)

Each property is interpolated in the following way:

$$h - h_f = 3.99 - \frac{3.99 - 3.36}{200 - 100}(140 - 100) = 3.99 - 0.25$$

$$= 3.74 \text{ Btu/lb (8.70 kJ/kg)}$$

$$(v - v_f)10^5 = -7.5 - \frac{8.1 - 7.5}{200 - 100}(140 - 100) = -7.5 - 0.24$$

$$= -7.74 \text{ ft}^3\text{/lb } (-0.48 \text{ m}^3\text{/kg})$$

$$(s - s_f)10^3 = -0.86 - \frac{1.79 - 0.86}{200 - 100}(140 - 100) = -0.86 - 0.37$$

$$= -1.23 \text{ Btu/(lb·°F) } [-5.15 \text{ kJ/(kg·°C)}]$$

These interpolated values must now be used to correct the saturation data at 140°F (60.0°C) to the actual subcooled state 1 properties. Thus, at 1500 lb/in^2 (abs) (10,342.5 kPa) and 140°F (60.0°C),

$$H_1 = h_f + \text{interpolated } h = 107.89 + 3.74 = 111.63 \text{ Btu/lb (259.7 kJ/kg)}$$

$$V_1 = v_f - \frac{\text{interpolated } v}{10^5} = 0.01629 - \frac{7.74}{10^5} = 0.01621 \text{ ft}^3/\text{lb (0.0010 m}^3/\text{kg)}$$

$$S_1 = s_f - \frac{\text{interpolated } s}{10^3} = 0.1984 - \frac{1.23}{10^3} = 0.1972 \text{ Btu/(lb·°F) [0.83 kJ/(kg·°C)]}$$

3. Compute the heat added to raise the compressed liquid to the saturation temperature

From the saturation-pressure steam table for 1500 lb/in² (abs) (10,342.5 kPa), the enthalpy of the saturated liquid $H_2 = 611.6$ Btu/lb (1422.6 kJ/kg). The heat added Q_a to raise the compressed liquid to the saturation temperature is $Q_a = H_2 - H_1 = 611.6 - 111.6 = 500$ Btu/lb (1163.0 kJ/kg).

4. Compute the heat added to vaporize the saturated liquid

Read from the saturation-pressure steam table the enthalpy of saturated vapor at 1500 lb/in² (abs) (10,342.5 kPa), $H_3 = 1167.9$ Btu/lb (2716.5 kJ/kg). Then the heat added to vaporize the saturated water $Q_b = H_3 - H_2 = 1167.9 - 611.6 = 556.3$ Btu/lb (1294.0 kJ/kg).

5. Compute the heat added to superheat the steam

Find in the superheated steam table for 1500 lb/in² (abs) (10,342.5 kPa) and 1000°F (537.8°C) the properties of the superheated steam: $H_4 = 1490.1$ Btu/lb (3466.0 kJ/kg); $V_4 = 0.5390$ ft³/lb (0.034 m³/kg); $S_4 = 1.6001$ Btu/(lb·°F) [6.7 kJ/(kg·°C)]. Then the heat added to superheat the saturated steam $Q_4 = H_4 - H_3 = 1490.1 - 1167.9 = 322.2$ Btu/lb (749.4 kJ/kg).

6. Determine the property changes during the process

$$Q_1 = Q_a + Q_b + Q_c = H_4 - H_1 = 1490.1 - 111.6 = 1378.5 \text{ Btu/lb (3206.4 kJ/kg)}$$

$$\Delta V = V_4 - V_1 = 0.5390 - 0.01621 = 0.5228 \text{ ft}^3/\text{lb (0.033 m}^3/\text{kg)}$$

$$\Delta S = S_4 - S_1 = 1.6001 - 0.1972 = 1.4029 \text{ Btu/(lb·°F) [5.9 kJ/(kg·°C)]}$$

BLEED-STEAM REGENERATIVE CYCLE LAYOUT AND T-S PLOT

Sketch the cycle layout, T-S diagram, and energy-flow chart for a regenerative bleed-steam turbine plant having three feedwater heaters and four feed pumps. Write the equations for the work-output available energy and the energy rejected to the condenser.

Calculation Procedure:

1. Sketch the cycle layout

Figure 10 shows a typical practical regenerative cycle having three feedwater heaters and four feedwater pumps. Number each point where steam enters and leaves the turbine and where steam enters or leaves the condenser and boiler. Also number the points in the feedwater cycle where feedwater enters and leaves a heater. Indicate the heater steam flow by m with a subscript corresponding to the heater number. Use W_p and a suitable subscript to indicate the pump work for each feed pump, except the last, which is labeled W_{pF}. The heat input to the steam generator is Q_a; the work output of the steam turbine is W_e; the heat rejected by the condenser is Q_r.

2. Sketch the T-S diagram for the cycle

To analyze any steam cycle, trace the flow of 1 lb (0.5 kg) of steam through the system. Thus, in this cycle, 1 lb (0.5 kg) of steam leaves the steam generator at point 2 and flows to the turbine.

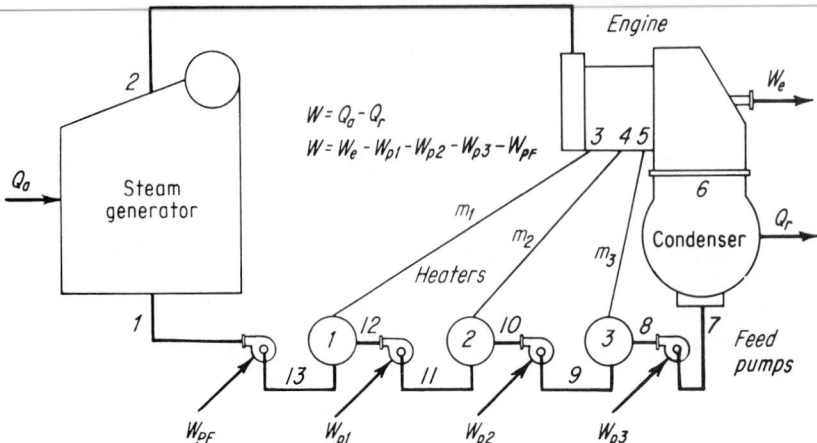

FIG. 10   Regenerative steam cycle uses bleed steam.

From state 2 to 3, 1 lb (0.5 kg) of steam expands at constant entropy (assumed) through the turbine, producing work output $W_1 = H_2 - H_3$, represented by area 1-a-2-3 on the T-S diagram, Fig. 11a. At point 3, some steam is bled from the turbine to heat the feedwater passing through heater 1. The quantity of steam bled, m_1 lb is less than the 1 lb (0.5 kg) flowing between points 2 and 3. Plot states 2 and 3 on the T-S diagram, Fig. 11a.

From point 3 to 4, the quantity of steam flowing through the turbine is $1 - m_1$ lb. This steam produces work output $W_2 = H_3 - H_4$. Plot point 4 on the T-S diagram. Then, area 1-3-4-12 represents the work output W_2, Fig. 11a.

At point 4, steam is bled to heater 2. The weight of this steam is m_2 lb. From point 4, the steam continues to flow through the turbine to point 5, Fig. 11a. The weight of the steam flowing between points 4 and 5 is $1 - m_1 - m_2$ lb. Plot point 5 on the T-S diagram, Fig. 11a. The work output between points 4 and 5, $W_3 = H_4 - H_5$, is represented by area 4-5-10-11 on the T-S diagram.

At point 5, steam is bled to heater 3. The weight of this bleed steam is m_3 lb. From point 5, steam continues to flow through the turbine to exhaust at point 6, Fig. 11a. The weight of steam flowing between points 5 and 6 is $1 - m_1 - m_2 - m_3$ lb. Plot point 6 on the T-S diagram, Fig. 11a.

The work output between points 5 and 6 is $W_4 = H_5 - H_6$, represented by area 5-6-7-9 on the T-S diagram, Fig. 11a. Area Q_r represents the heat given up by 1 lb (0.5 kg) of exhaust steam. Similarly, the area marked Q_a represents the heat absorbed by 1 lb (0.5 kg) of water in the steam generator.

3.   *Alter the T-S diagram to show actual cycle conditions*

As plotted in Fig. 11a, Q_a is true for this cycle since 1 lb (0.5 kg) of water flows through the steam generator and the first section of the turbine. But Q_r is much too large; only $1 - m_1 - m_2 - m_3$ lb of steam flows through the condenser. Likewise, the net areas for W_2, W_3, and W_4, Fig. 11a, are all too large, because less than 1 lb (0.5 kg) of steam flows through the respective turbine sections. The area for W_1, however, is true.

A true *proportionate-area* diagram can be plotted by applying the factors for actual flow, as in Fig. 11b. Here W_2, outlined by the heavy lines, equals the similarly labeled area in Fig. 11a, multiplied by $1 - m_1$. The states marked 11' and 12', Fig. 11b, are not true state points because of the ratioing factor applied to the area for W_2. The true state points 11 and 12 of the liquid before and after heater pump 3 stay as shown in Fig. 11a.

Apply $1 - m_1 - m_2$ to W_3 of Fig. 11a to obtain the proportionate area of Fig. 11b; to obtain W_4, multiply by $1 - m_1 - m_2 - m_3$. Multiplying by this factor also gives Q_r. Then all the areas in Fig. 11b will be in proper proportion for 1 lb (0.5 kg) of steam entering the turbine throttle but less in other parts of the cycle.

In Fig. 11b, the work can be measured by the difference of the area Q_a and the area Q_r. There is no simple net area left, because the areas coincide on only two sides. But the area enclosed by

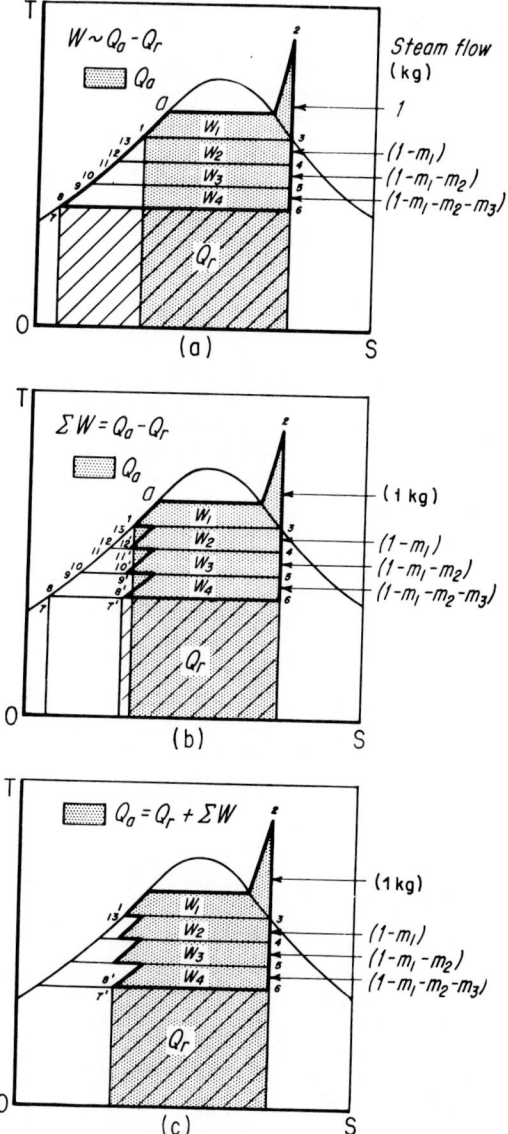

FIG. 11   (a) T-S chart for the bleed-steam regenerative cycle in Fig. 10; (b) actual fluid flow in the cycle; (c) alternative plot of (b).

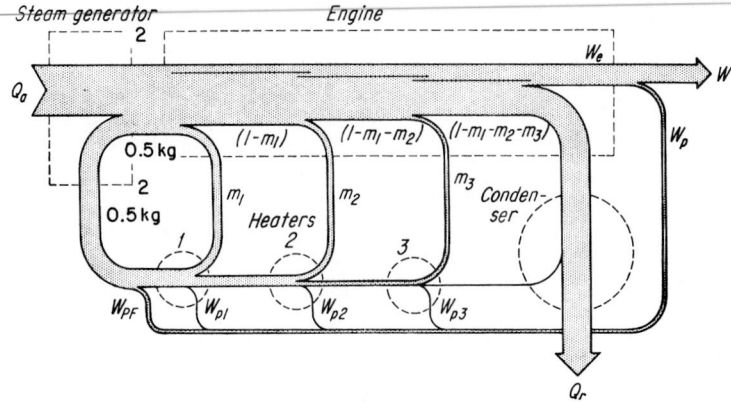

FIG. 12   Energy-flow chart of cycle in Fig. 10.

the heavy lines *is* the total net work W for the cycle, equal to the sum of the work produced in the various sections of the turbine, Fig. 11b. Then Q_a is the alternate area $Q_r + W_1 + W_2 + W_3 + W_4$, as shaded in Fig. 11$c$.

The sawtooth appearance of the liquid-heating line shows that as the number of heaters in the cycle increases, the heating line approaches a line of constant entropy. The best number of heaters for a given cycle depends on the steam state of the turbine inlet. Many medium-pressure and medium-temperature cycles use five to six heaters. High-pressure and high-temperature cycles use as many as nine heaters.

4.   *Draw the energy-flow chart*

Choose a suitable scale for the heat content of 1 lb (0.5 kg) of steam leaving the steam generator. A typical scale is 0.375 in per 1000 Btu/lb (0.41 cm per 1000 kJ/kg). Plot the heat content of 1 lb (0.5 kg) of steam vertically on line 2-2, Fig. 12. Using the same scale, plot the heat content in energy streams m_1, m_2, m_3, W_e, W, W_p, W_{pF}, and so forth. In some cases, as W_{p1}, W_{p2}, and so forth, the energy stream may be so small that it is impossible to plot it to scale. In these instances, a single thin line is used. The completed diagram, Fig. 12, provides a useful concept of the distribution of the energy in the cycle.

Related Calculations: The procedure given here can be used for all regenerative cycles, provided that the equations are altered to allow for more, or fewer, heaters and pumps. The following calculation procedure shows the application of this method to an actual regenerative cycle.

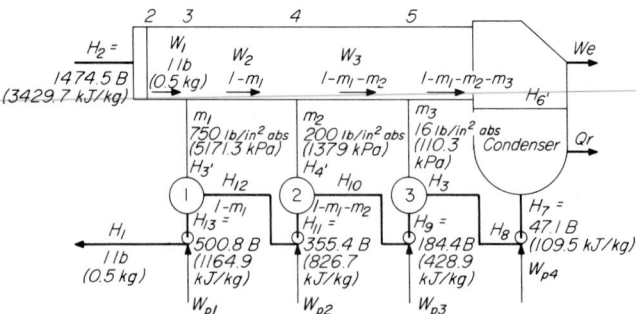

FIG. 13   Bleed-regenerative steam cycle.

BLEED REGENERATIVE STEAM CYCLE ANALYSIS

Analyze the bleed regenerative cycle shown in Fig. 13, determining the heat balance for each heater, plant thermal efficiency, turbine or engine thermal efficiency, plant heat rate, turbine or engine heat rate, and turbine or engine steam rate. Throttle steam pressure is 2000 lb/in² (abs) (13,790.0 kPa) at 1000°F (537.8°C); steam-generator efficiency = 0.88; station auxiliary steam consumption (excluding pump work) = 6 percent of the turbine or engine output; engine efficiency of each turbine or engine section = 0.80; turbine or engine cycle has three feedwater heaters and bleed-steam pressures as shown in Fig. 13; exhaust pressure to condenser is 1 inHg (3.4 kPa) absolute.

Calculation Procedure:

1. Determine the enthalpy of the steam at the inlet of each heater and the condenser

From a superheated-steam table, find the throttle enthalpy H_2 = 1474.5 Btu/lb (3429.7 kJ/kg) at 2000 lb/in² (abs) (13,790.0 kPa) and 1000°F (537.8°C). Next find the throttle entropy S_2 = 1.5603 Btu/(lb·°F) [6.5 kJ/(kg·°C)], at the same conditions in the superheated-steam table.

Plot the throttle steam conditions on a Mollier chart, Fig. 14. Assume that the steam expands from the throttle conditions at constant entropy = constant S to the inlet of the first feedwater heater, 1, Fig. 13. Plot this constant S expansion by drawing the straight vertical line 2-3 on the Mollier chart, Fig. 14, between the throttle condition and the heater inlet pressure of 750 lb/in² (abs) (5171.3 kPa).

Read on the Mollier chart H_3 = 1346.7 Btu/lb (3132.4 kJ/kg). Since the engine or turbine efficiency e_e = $H_2 - H_3/(H_2 - H_3)$ = 0.8 = 1474.5 - $H_3/(1474.5 - 1346.7)$; H_3 = actual enthalpy of the steam at the inlet to heater 1 = 1474.5 - 0.8(1474.5 - 1346.7) = 1372.2 Btu/lb (3191.7 kJ/kg). Plot this enthalpy point on the 750-lb/in² (abs) (5171.3-kPa) pressure line of the Mollier chart, Fig. 14. Read the entropy at the heater inlet from the Mollier chart as $S_{3'}$ =

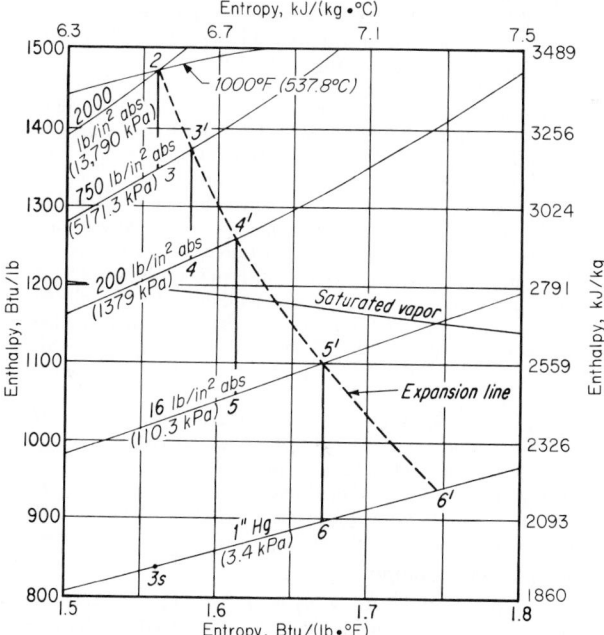

FIG. 14 Mollier-chart plot of the cycle in Fig. 13.

1.5819 Btu/(lb·°F) [6.6 kJ/(kg·°C)] at 750 lb/in² (abs) (5171.3 kPa) and 1372.2 Btu/lb (3191.7 kJ/kg).

Assume constant-S expansion from $H_{3'}$ to H_4 at 200 lb/in² (abs) (1379.0 kPa), the inlet pressure for feedwater heater 2. Draw the vertical straight line 3'-4 on the Mollier chart, Fig. 14. By using a procedure similar to that for heater 1, $H_{4'} = H_{3'} - e_e(H_{3'} - H_4) = 1372.2 - 0.8(1372.2 - 1230.0) = 1258.4$ Btu/lb (2927.0 kJ/kg). This is the actual enthalpy of the steam at the inlet to heater 2. Plot this enthalpy on the 200-lb/in² (abs) (1379.0-kPa) pressure line of the Mollier chart, and find $S_{4'} = 1.613$ Btu/(lb·°F) [6.8 kJ/(kg·°C)], Fig. 14.

Using the same procedure with constant-S expansion from $H_{4'}$, we find $H_5 = 1059.5$ Btu/lb (2464.4 kJ/kg) at 16 lb/in² (abs) (110.3 kPa), the inlet pressure to heater 3. Next find $H_{5'} = H_{4'} - e_e(H_{4'} - H_5) = 1258.4 - 0.8(1258.4 - 1059.5) = 1099.2$ Btu/lb (2556.7 kJ/kg). From the Mollier chart find $S_{5'} = 1.671$ Btu/(lb·°F) [7.0 kJ/(kg·°C)], Fig. 14.

Using the same procedure with constant-S expansion from $H_{5'}$ to H_6, find $H_6 = 898.2$ Btu/lb (2089.2 kJ/kg) at 1 inHg absolute (3.4 kPa), the condenser inlet pressure. Then $H_{6'} = H_{5'} - e_e(H_{5'} - H_6) = 1099.2 - 0.8(1099.2 - 898.2) = 938.4$ Btu/lb (2182.7 kJ/kg), the actual enthalpy of the steam at the condenser inlet. Find, on the Mollier chart, the moisture in the turbine exhaust = 15.1 percent.

2. Determine the overall engine efficiency

Overall engine efficiency e_e is higher than the engine-section efficiency because there is partial available-energy recovery between sections. Constant-S expansion from the throttle to the 1-inHg absolute (3.4-kPa) exhaust gives H_{3S}, Fig. 14, as 838.3 Btu/lb (1949.4 kJ/kg), assuming that all the steam flows to the condenser. Then, overall $e_e = H_2 - H_{6'}/(H_2 - H_{3S}) = 1474.5 - 938.4/1474.5 - 838.3 = 0.8425$, or 84.25 percent, compared with 0.8 or 80 percent, for individual engine sections.

3. Compute the bleed-steam flow to each feedwater heater

For each heater, energy in = energy out. Also, the heated condensate leaving each heater is a saturated liquid at the heater bleed-steam pressure. To simplify this calculation, assume negligible steam pressure drop between the turbine bleed point and the heater inlet. This assumption is permissible when the distance between the heater and bleed point is small. Determine the pump work by using the chart accompanying the compressed-liquid table in Keenan and Keyes—*Thermodynamic Properties of Steam*, or the ASME—*Steam Tables*.

For heater 1, energy in = energy out, or $H_{3'}m_1 + H_{12}(1 - m_1) = H_{13}$, where m = bleed-stream flow to the feedwater heater, lb/lb of throttle steam flow. (The subscript refers to the heater under consideration.) Then, $H_{3'}m_1 + (H_{11} + W_{p2})(1 - m_1) = H_{13}$, where W_{p2} = work done by pump 2, Fig. 13, in Btu/lb per pound of throttle flow. Then $1372.2m_1 + (355.4 + 1.7)(1 - m_1) = 500.8$; $m_1 = 0.1416$ lb/lb (0.064 kg/kg) throttle flow; $H_1 = H_{13} + W_{p1} = 500.8 + 4.7 = 505.5$ Btu/lb (1175.8 kJ/kg), where W_{p1} = work done by pump 1, Fig. 13. For each pump, find the work from the chart accompanying the compressed-liquid table in Keenan and Keyes—*Steam Tables* by entering the chart at the heater inlet pressure and projecting vertically at constant entropy to the heater outlet pressure, which equals the next heater inlet pressure. Read the enthalpy values at the respective pressures, and subtract the smaller from the larger to obtain the pump work during passage of the feedwater through the pump from the lower to the higher pressure. Thus, $W_{p2} = 1.7 - 0.0 = 1.7$ Btu/lb (4.0 kJ/kg), from enthalpy values for 200 lb/in² (abs) (1379.0 kPa) and 750 lb/in² (abs) (5171.3 kPa), the heater inlet and discharge pressures, respectively.

For heater 2, energy in = energy out, or $H_{4'}m_2 + H_{10}(1 - m_1 - m_2) = H_{11}(1 - m_1)H_{4'}m_2 + (H_9 + W_{p3})(1 - m_1 - m_2) = H_{11}(1 - m_1)1258.4m_2 + (184.4 + 0.5)(0.8584 - m_2) = 355.4(0.8584)m_2 = 0.1365$ lb/lb (0.0619 kg/kg) throttle flow.

For heater 3, energy in = energy out, or $H_{5'}m_3 + H_8(1 - m_1 - m_2 - m_3) = H_9(1 - m_1 - m_2)H_{5'}m_3 + (H_7 + W_{p4})(1 - m_1 - m_2 - m_3) = H_9(1 - m_1 - m_2)1099.2m_3 + (47.1 + 0.1)(0.7210 - m_3) = 184.4(0.7219)m_3 = 0.0942$ lb/lb (0.0427 kg/kg) throttle flow.

4. Compute the turbine work output

The work output per section W Btu is $W_1 = H_2 - H_{3'} = 1474.5 - 1372.1 = 102.3$ Btu (107.9 kJ), from the previously computed enthalpy values. Also $W_2 = (H_{3'} - H_{4'})(1 - m_1) = (1372.2 - 1258.4)(1 - 0.1416) = 97.7$ Btu (103.1 kJ); $W_3 = (H_{4'} - H_{5'})(1 - m_1 - m_2) = (1258.4 - 1099.2)(1 - 0.1416 - 0.1365) = 115.0$ Btu (121.3 kJ); $W_4 = (H_{5'} - H_{6'})(1 - m_1 - m_2 - m_3)$

$= (1099.2 - 938.4)(1 - 0.1416 - 0.1365 - 0.0942) = 100.9$ Btu (106.5 kJ). The total work output of the turbine $= W_e = \Sigma W = 102.3 + 97.7 + 115.0 + 100.9 = 415.9$ Btu (438.8 kJ). The total $W_p = \Sigma W_p = W_{p1} + W_{p2} + W_{p3} + W_{p4} = 4.7 + 1.7 + 0.5 + 0.1 = 7.0$ Btu (7.4 kJ).

Since the station auxiliaries consume 6 percent of W_e, the auxiliary consumption $= 0.6(415.9) = 25.0$ Btu (26.4 kJ). Then, net station work $w = 415.9 - 7.0 - 25.0 = 383.9$ Btu (405.0 kJ).

5. *Check the turbine work output*

The heat added to the cycle Q_a Btu/lb $= H_2 - H_1 = 1474.5 - 505.5 = 969.0$ Btu (1022.3 kJ). The heat rejected from the cycle Q_r Btu/lb $= (H_{6'} - H_7)(1 - m_1 - m_2 - m_3) = (938.4 - 47.1)(0.6277) = 559.5$ Btu (590.3 kJ). Then $W_e - W_p = Q_a - Q_r = 969.0 - 559.5 = 409.5$ Btu (432.0 kJ).

Compare this with $W_e - W_p$ computed earlier, or $415.9 - 7.0 = 408.9$ Btu (431.4 kJ), or a difference of $409.5 - 408.9 = 0.6$ Btu (0.63 kJ). This is an accurate check; the difference of 0.6 Btu (0.63 kJ) comes from errors in Mollier chart and calculator readings. Assume 408.9 Btu (431.4 kJ) is correct because it is the lower of the two values.

6. *Compute the plant and turbine efficiencies*

Plant energy input $= Q_a/e_b$, where $e_b =$ boiler efficiency. Then plant energy input $= 969.0/0.88 = 1101.0$ Btu (1161.6 kJ). Plant thermal efficiency $= W/(Q_a/e_b) = 383.9 = 1101.0 = 0.3486$. Turbine thermal efficiency $= W_e/Q_a = 415.9/969.0 = 0.4292$. Plant heat rate $= 3413/0.3486 = 9970$ Btu/kWh (10,329.0 kJ/kWh), where 3413 $=$ Btu/kWh. Turbine heat rate $= 3413/0.4292 = 7950$ Btu/kWh (8387.7 kJ/kWh). Turbine throttle steam rate $=$ (turbine heat rate)/$(H_2 - H_1) = 7950/(1474.5 - 505.5) = 8.21$ lb/kWh (3.7 kg/kWh).

Related Calculations: By using the procedures given, the following values can be computed for any actual steam cycle: engine or turbine efficiency e_e; steam enthalpy at the main-condenser inlet; bleed-steam flow to a feedwater heater; turbine or engine work output per section; total turbine or engine work output; station auxiliary power consumption; net station work output; plant energy input; plant thermal efficiency; turbine or engine thermal efficiency; plant heat rate; turbine or engine heat rate; turbine throttle heat rate. To compute any of these values, use the equations given and insert the applicable variables.

REHEAT-STEAM CYCLE PERFORMANCE

A reheat-steam cycle has a 2000 lb/in² (abs) (13,790-kPa) throttle pressure at the turbine inlet and a 400-lb/in² (abs) (2758-kPa) reheat pressure. The throttle and reheat temperature of the steam is 1000°F (537.8°C); condenser pressure is 1 inHg absolute (3.4 kPa); engine efficiency of the high-pressure and low-pressure turbines is 80 percent. Find the cycle thermal efficiency.

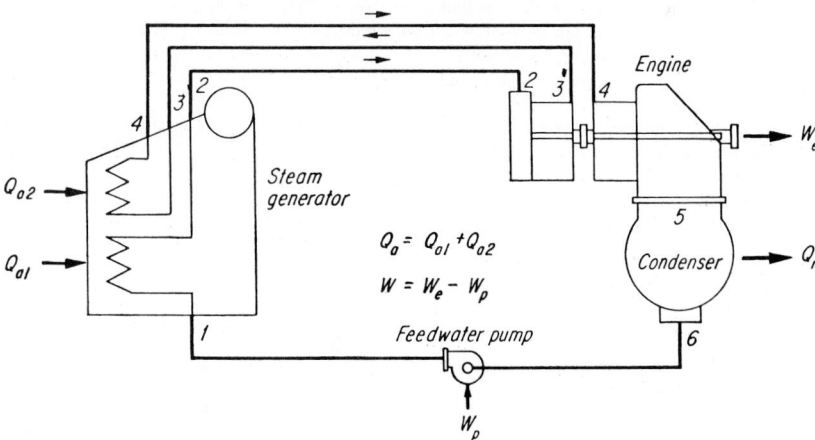

FIG. 15 Typical steam reheat cycle.

Calculation Procedure:

1. Sketch the cycle layout and cycle T-S diagram

Figures 15 and 16 show the cycle layout and T-S diagram with each important point numbered. Use a cycle layout and T-S diagram for every calculation of this type because it reduces the possibility of errors.

2. Determine the throttle-steam properties from the steam tables

Use the superheated steam tables, entering at 2000 lb/in² (abs) (13,790 kPa) and 1000°F (537.8°C) to find throttle-steam properties. Applying the symbols of the T-S diagram in Fig. 16, we get $H_2 = 1474.5$ Btu/lb (3429.7 kJ/kg); $S_2 = 1.5603$ Btu/(lb·°F) [6.5 kJ/(kg·°C)].

3. Find the reheat-steam enthalpy

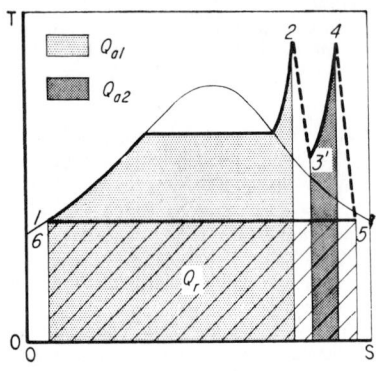

Assume a constant-entropy expansion of the steam from 2000 to 400 lb/in² (13,790 to 2758 kPa). Trace this expansion on a Mollier (H-S) chart, Fig. 1, where a constant-entropy process is a vertical line between the initial [2000 lb/in² (abs) or 13,790 kPa] and reheat [400 lb/in² (abs) or 2758 kPa] pressures. Read on the Mollier chart $H_3 = 1276.8$ Btu/lb (2969.8 kJ/kg) at 400 lb/in² (abs) (2758 kPa).

4. Compute the actual reheat properties

The ideal enthalpy drop, throttle to reheat = $H_2 - H_3 = 1474.5 - 1276.8 = 197.7$ Btu/lb (459.9 kJ/kg). The actual enthalpy drop = (ideal drop)(turbine efficiency) = $H_2 - H_{3'} = 197.5(0.8) = 158.2$ Btu/lb (368.0 kJ/kg) = W_{e1} = work output in the high-pressure section of the turbine.

FIG. 16 Irreversible expansion in reheat cycle.

Once W_{e1} is known, $H_{3'}$ can be computed from $H_{3'} = H_2 - W_{e1} = 1474.5 - 158.2 = 1316.3$ Btu/lb (3061.7 kJ/kg).

The steam now returns to the boiler and leaves at condition 4, where $P_4 = 400$ lb/in² (abs) (2758 kPa); $T_4 = 1000$°F (537.8°C); $S_4 = 1.7623$ Btu/(lb·°F) [7.4 kJ/(kg·°C)]; $H_4 = 1522.4$ Btu/lb (3541.1 kJ/kg) from the superheated-steam table.

5. Compute the exhaust-steam properties

Use the Mollier chart and an assumed constant-entropy expansion to 1 inHg (3.4 kPa) absolute to determine the ideal exhaust enthalpy, or $H_5 = 947.4$ Btu/lb (2203.7 kJ/kg). The ideal work of the low-pressure section of the turbine is then $H_4 - H_5 = 1522.4 - 947.4 = 575.0$ Btu/lb (1338 kJ/kg). The actual work output of the low-pressure section of the turbine is $W_{e2} = H_4 - H_{5'} = 575.0(0.8) = 460.8$ Btu/lb (1071.1 kJ/kg).

Once W_{e2} is known, $H_{5'}$ can be computed from $H_{5'} = H_4 - W_{e2} = 1522.4 - 460.0 = 1062.4$ Btu/lb (2471.1 kJ/kg).

The enthalpy of the saturated liquid at the condenser pressure is found in the saturation-pressure steam table at 1 inHg absolute (3.4 kPa) = $H_6 = 47.1$ Btu/lb (109.5 kJ/kg).

The pump work W_p from the compressed-liquid table diagram in the stream tables is $W_p = 5.5$ Btu/lb (12.8 kJ/kg). Then the enthalpy of the water entering the boiler $H_1 = H_6 + W_p = 47.1 + 5.5 = 52.6$ Btu/lb (122.3 kJ/kg).

6. Compute the cycle thermal efficiency

For any reheat cycle,

$$e = \text{cycle thermal efficiency}$$

$$= \frac{(H_2 - H_{3'}) + (H_4 - H_{5'}) - W_p}{(H_2 - H_1) + (H_4 - H_{3'})} = \frac{(1474.5 - 1316.3) + (1522.4 - 1062.4) - 5.5}{(1474.5 - 52.6) + (1522.4 - 1316.3)}$$

$$= 0.3766, \text{ or } 37.66 \text{ percent}$$

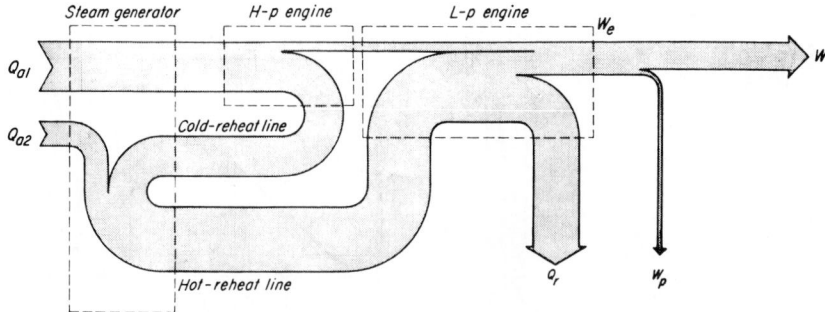

FIG. 17   Energy-flow diagram for reheat cycle in Fig. 15.

Figure 17 is an energy-flow diagram for the reheat cycle analyzed here. This diagram shows that the fuel burned in the steam generator to produce energy flow Q_{a1} is the largest part of the total energy input. The cold-reheat line carries the major share of energy leaving the high-pressure turbine.

Related Calculations: Reheat-regenerative cycles are used in some large power plants. Figure 18 shows a typical layout for such a cycle having three stages of feedwater heating and one stage of reheating. The heat balance for this cycle is computed as shown above, with the bleed-flow terms m computed by setting up an energy balance around each heater, as in earlier calculation procedures.

By using a T-S diagram, Fig. 19, the cycle thermal efficiency is

$$e = \frac{W}{Q_a} = \frac{Q_a - Q_r}{Q_a} = 1 - \frac{Q_r}{Q_{a1} + Q_{a2}}$$

Based on 1 lb (0.5 kg) of working fluid entering the steam generator and turbine throttle,

$$Q_r = (1 - m_1 - m_2 - m_3)(H_7 - H_8)$$

$$Q_{a1} = (H_2 - H_1)$$

$$Q_{a2} = (1 - m_1)(H_4 - H_3)$$

Figure 20 shows the energy-flow chart for this cycle.

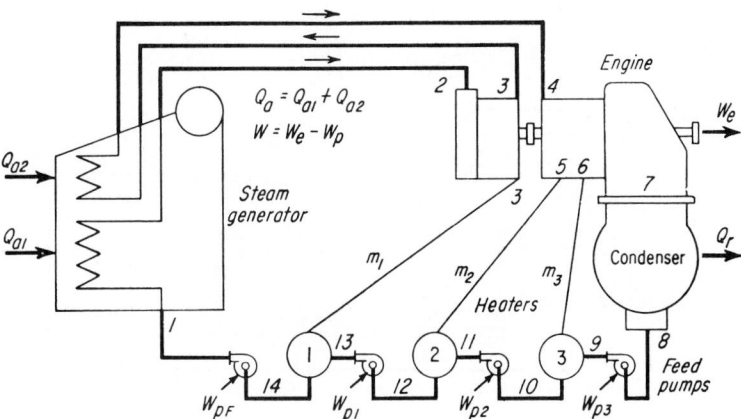

FIG. 18   Combined reheat and bleed-regenerative cycle.

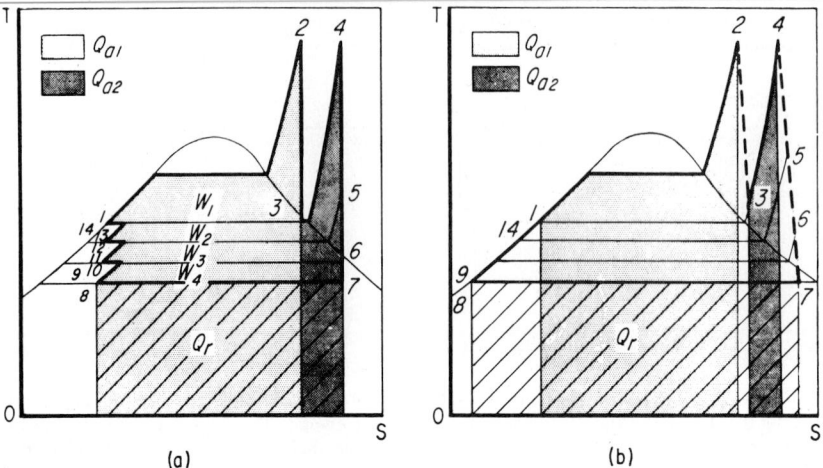

FIG. 19 (a) *T-S* diagram for ideal reheat-regenerative-bleed cycle; (b) *T-S* diagram for actual cycle.

Some high-pressure plants use two stages of reheating, Fig. 21, to raise the cycle efficiency. With two stages of reheating, the maximum number generally used, and values from Fig. 21,

$$ e = \frac{(H_2 - H_3) + (H_4 - H_5) + (H_6 - H_7) - W_p}{(H_2 - H_1) + (H_4 - H_3) + (H_6 - H_5)} $$

MECHANICAL-DRIVE STEAM-TURBINE POWER-OUTPUT ANALYSIS

Show the effect of turbine engine efficiency on the condition lines of a turbine having engine efficiencies of 100 (isentropic expansion), 75, 50, 25, and 0 percent. How much of the available energy is converted to useful work for each engine efficiency? Sketch the effect of different steam inlet pressures on the condition line of a single-nozzle turbine at various loads. What is the available energy, Btu/lb of steam, in a noncondensing steam turbine having an inlet pressure of 1000 lb/in² (abs) (6895 kPa) and an exhaust pressure of 100 lb/in² (gage) (689.5 kPa)? How much work

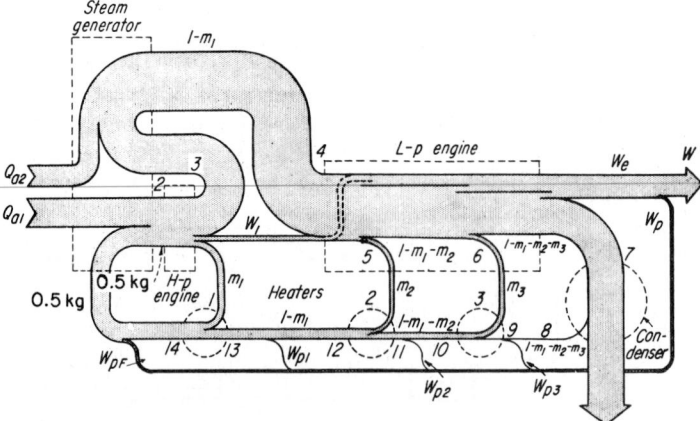

FIG. 20 Energy flow of cycle in Fig. 18.

will this turbine perform if the steam flow rate to it is 1000 lb/s (453.6 kg/s) and the engine efficiency is 40 percent?

Calculation Procedure:

1. Sketch the condition lines on the Mollier chart

Draw on the Mollier chart for steam initial- and exhaust-pressure lines, Fig. 22, and the initial-temperature line. For an isentropic expansion, the entropy is constant during the expansion, and the engine efficiency = 100 percent. The expansion or condition line is a vertical trace from h_1 on the initial-pressure line to h_{2s} on the exhaust-pressure line. Draw this line as shown in Fig. 22.

For zero percent engine efficiency, the other extreme in the efficiency range, $h_1 = h_2$ and the condition line is a horizontal line. Draw this line as shown in Fig. 22.

Between 0 and 100 percent engine efficiency, the condition lines become more nearly vertical as the engine efficiency approaches 100 percent, or an isentropic expansion. Draw the condition lines for 25, 50, and 75 percent efficiency, as shown in Fig. 22.

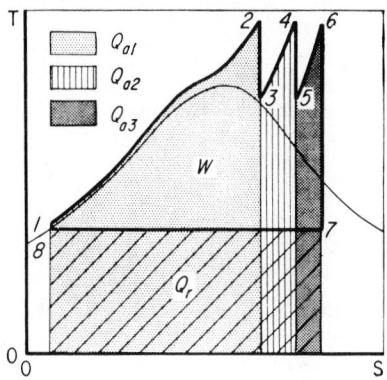

FIG. 21   *T-S* diagram for multiple reheat stages.

For the isentropic expansion, the available energy = $h_1 - h_{2s}$, Btu/lb of steam. This is the energy that an ideal turbine would make available.

For actual turbines, the enthalpy at the exhaust pressure $h_2 = h_1 -$ (available energy)(engine efficiency)/100, where available energy = $h_1 - h_{2s}$ for an ideal turbine working between the same initial and exhaust pressures. Thus, the available energy converted to useful work for any engine efficiency = (ideal available energy, Btu/lb)(engine efficiency, percent)/100. Using this relation, the available energy at each of the given engine efficiencies is found by substituting the ideal available energy and the actual engine efficiency.

2. Sketch the condition lines for various throttle pressures

Draw the throttle- and exhaust-pressure lines on the Mollier chart, Fig. 23. Since the inlet control valve throttles the steam flow as the load on the turbine decreases, the pressure of the steam

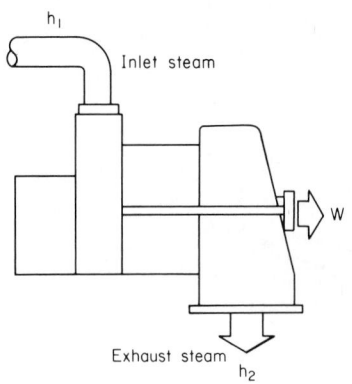

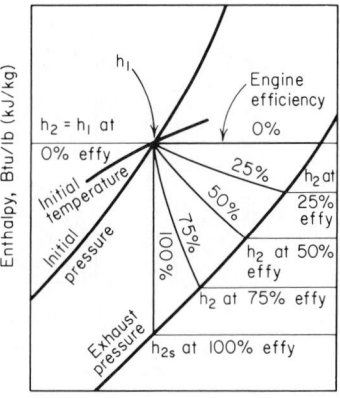

Entropy, Btu/(lb • °F) $\left[\text{kJ/(kg • °C)}\right]$

FIG. 22   Mollier chart of turbine condition lines.

entering the turbine nozzle is lower at reduced loads. Show this throttling effect by indicating the lower inlet pressure lines, Fig. 23, for the reduced loads. Note that the lowest inlet pressure occurs at the minimum plotted load—25 percent of full load—and the maximum inlet pressure at 125 percent of full load. As the turbine inlet steam pressure decreases, so does the available energy, because the exhaust enthalpy rises with decreasing load.

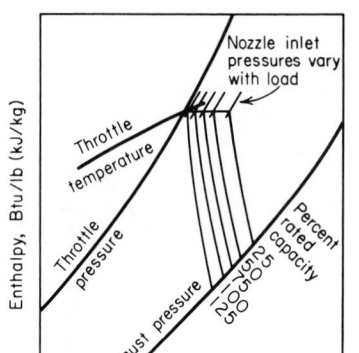

Entropy, Btu/(lb•°F) $\left[\text{kJ/(kg •°C)}\right]$

FIG. 23 Turbine condition line shifts as the inlet steam pressure varies.

3. *Compute the turbine available energy and power output*

Use a noncondensing-turbine performance chart, Fig. 24, to determine the available energy. Enter the bottom of the chart at 1000 lb/in² (abs) (6895 kPa) and project vertically upward until the 100-lb/in² (gage) (689.5-kPa) exhaust-pressure curve is intersected. At the left, read the available energy as 205 Btu/lb (476.8 kJ/kg) of steam.

With the available energy, flow rate, and engine efficiency known, the work output = (available energy, Btu/lb)(flow rate, lb/s)(engine efficiency/100)/[550 ft·lb/(s·hp)]. [*Note:* 550 ft·lb/(s·hp) = 1 N·m/(W·s).] For this turbine, work output = (205 Btu/lb)(1000 lb/s)(40/100)/550 = 149 hp (111.1 kW).

Related Calculations: Use the steps given here to analyze single-stage noncondensing mechanical-drive turbines for stationary, portable, or marine applications. Performance curves such as Fig. 24 are available from turbine manufacturers. Single-stage noncondensing turbines are used for feed-pump, draft-fan, and auxiliary-generator drive.

CONDENSING STEAM-TURBINE POWER-OUTPUT ANALYSIS

What is the available energy in steam supplied to a 5000-kW turbine if the inlet steam conditions are 1000 lb/in² (abs) (6895 kPa) and 800°F (426.7°C) and the turbine exhausts at 1 inHg absolute (3.4 kPa)? Determine the theoretical and actual heat rate of this turbine if its engine efficiency is 74 percent. What are the full-load output and steam rate of the turbine?

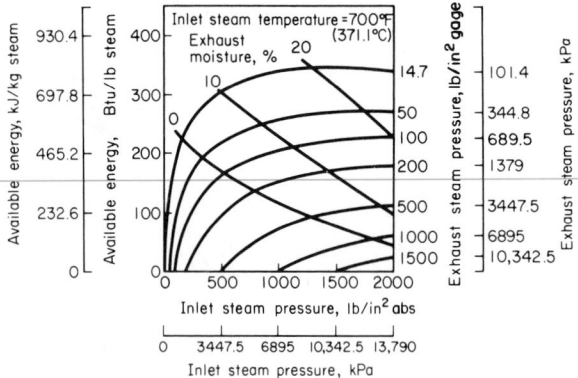

FIG. 24 Available energy in turbine depends on the initial steam state and the exhaust pressure.

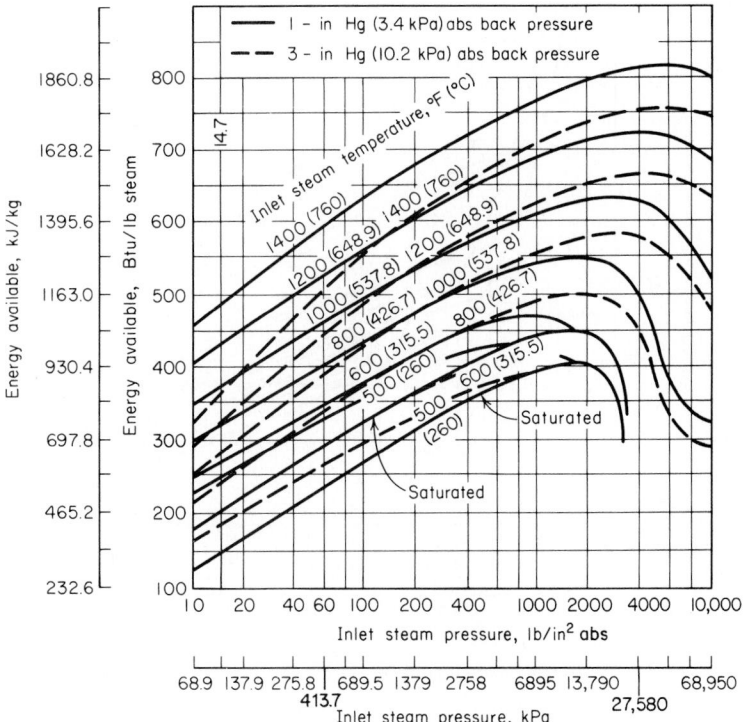

FIG. 25 Available energy for typical condensing turbines.

1. *Determine the available energy in the steam*

Enter Fig. 25 at the bottom at 1000-lb/in² (abs) (6895.0-kPa) inlet pressure, and project vertically upward to the 800°F (426.7°C) 1-in (3.4-kPa) exhaust-pressure curve. At the left, read the available energy as 545 Btu/lb (1267.7 kJ/kg) of steam.

2. *Determine the heat rate of the turbine*

Enter Fig. 26 at an initial steam temperature of 800°F (426.7°C), and project vertically upward to the 1000-lb/in² (abs) (6895.0-kPa) 1-in (3.4-kPa) curve. At the left, read the theoretical heat rate as 8400 Btu/kWh (8862.5 kJ/kWh).

When the theoretical heat rate is known, the actual heat rate is found from: actual heat rate HR, Btu/kWh = (theoretical heat rate, Btu/kWh)/(engine efficiency). Or, actual HR = 8400/0.74 = 11,350 Btu/kWh (11,974.9 kJ/kWh).

3. *Compute the full-load output and steam rate*

The energy converted to work, Btu/lb of steam = (available energy, Btu/lb of steam)(engine efficiency) = (545)(0.74) = 403 Btu/lb of steam (937.4 kJ/kg).

For any prime mover driving a generator, the full-load output, Btu = (generator kW rating)(3413 Btu/kWh) = (5000)(3413) = 17,060,000 Btu/h (4999.8 kJ/s).

The steam flow = (full-load output, Btu/h)/(work output, Btu/lb) = 17,060,000/403 = 42,300 lb/h (19,035 kg/h) of steam. Then the full-load steam rate of the turbine, lb/kWh = (steam flow, lb/h)/(kW output at full load) = 42,300/5000 = 8.46 lb/kWh (3.8 kg/kWh).

Related Calculations: Use this general procedure to determine the available energy, theoretical and actual heat rates, and full-load output and steam rate for any stationary, marine, or portable condensing steam turbine operating within the ranges of Figs. 25 and 26. If the actual

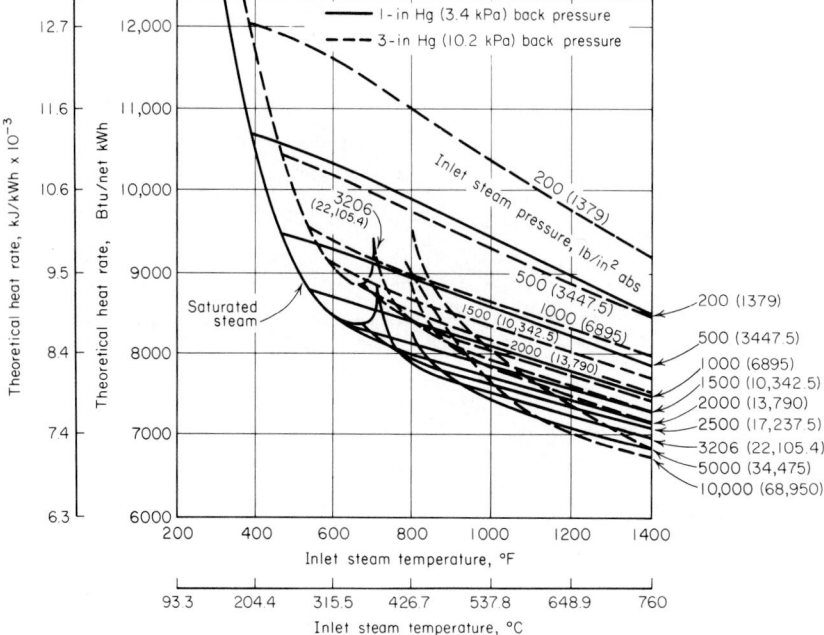

FIG. 26 Theoretical heat rate for condensing turbines.

performance curves are available, use them instead of Figs. 25 and 26. The curves given here are suitable for all preliminary estimates for condensing turbines operating with exhaust pressures of 1 or 3 inHg absolute (3.4 or 10.2 kPa). Many modern turbines operate under these conditions.

STEAM-TURBINE REGENERATIVE-CYCLE PERFORMANCE

When throttle steam is at 1000 lb/in² (abs) (6895 kPa) and 800°F (426.7°C) and the exhaust pressure is 1 inHg (3.4 kPa) absolute, a 5000-kW condensing turbine has an actual heat rate of 11,350 Btu/kWh (11,974.9 kJ/kWh). Three feedwater heaters are added to the cycle, Fig. 27, to heat the feedwater to 70 percent of the maximum possible enthalpy rise. What is the actual heat rate of the turbine? If 10 heaters instead of 3 were used and the water enthalpy were raised to 90 percent of the maximum possible rise in these 10 heaters, would the reduction in the actual heat rate be appreciable?

Calculation Procedure:

1. Determine the actual enthalpy rise of the feedwater

Enter Fig. 28 at the throttle pressure of 1000 lb/in² (abs) (6895 kPa), and project vertically upward to the 1-inHg (3.4-kPa) absolute back-pressure curve. At the left, read the maximum possible feedwater enthalpy rise as 495 Btu/lb (1151.4 kJ/kg). Since the actual rise is limited to 70 percent of the maximum possible rise by the conditions of the design, the actual enthalpy rise = (495)(0.70) = 346.5 Btu/lb (805.9 kJ/kg).

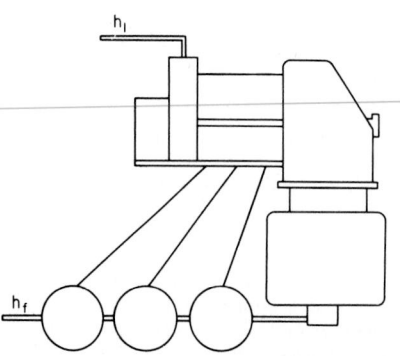

FIG. 27 Regenerative feedwater heating.

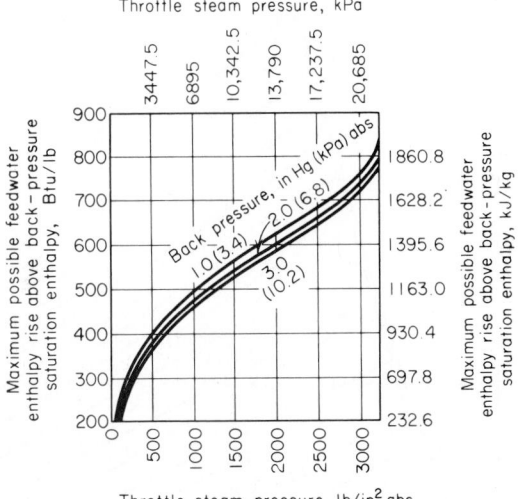

FIG. 28   Feedwater enthalpy rise.

2.   *Determine the heat-rate and heater-number correction factors*

Find the theoretical reduction in straight-condensing (no regenerative heaters) heat rates from Fig. 29. Enter the bottom of Fig. 29 at the inlet steam temperature, 800°F (426.7°C), and project vertically upward to the 1000-lb/in² (abs) (6895-kPa) 1-inHg (3.4-kPa) back-pressure curve. At the left, read the reduction in straight-condensing heat rate as 14.8 percent.

Next, enter Fig. 30 at the bottom at 70 percent of maximum possible rise in feedwater enthalpy, and project vertically to the three-heater curve. At the left, read the reduction in straight-condensing heat rate for the number of heaters and actual enthalpy rise as 0.71.

3.   *Apply the heat-rate and heater-number correction factors*

Full-load regenerative-cycle heat rate, Btu/kWh = (straight-condensing heat rate, Btu/kWh) [1 − (heat-rate correction factor)(heater-number correction factor)] = (13,350)[1 − (0.148)(0.71)] = 10,160 Btu/kWh (10,719.4 kJ/kWh).

4.   *Find and apply the correction factors for the larger number of heaters*

Enter Fig. 30 at 90 percent of the maximum possible enthalpy rise, and project vertically to the 10-heater curve. At the left, read the heat-rate reduction for the number of heaters and actual enthalpy rise as 0.89.

Using the heat-rate correction factor from step 2 and 0.89, found above, we see that the full-load 10-heater regenerative-cycle heat rate = (11,350)[1 − (0.148)(0.89)] = 9850 Btu/kWh (10,392.3 kJ/kWh), by using the same procedure as in step 3. Thus, adding 10 − 3 = 7 heaters reduces the heat rate by 10,160 − 9850 = 310 Btu/kWh (327.1 kJ/kWh). This is a reduction of 3.05 percent.

To determine whether this reduction in heat rate is appreciable, the carrying charges on the extra heaters, piping, and pumps must be compared with the reduction in annual fuel costs resulting from the lower heat rate. If the fuel saving is greater than the carrying charges, the larger number of heaters can usually be justified. In this case, tripling the number of heaters would probably increase the carrying charges to a level exceeding the fuel savings. Therefore, the reduction in heat rate is probably not appreciable.

Related Calculations: Use the procedure given here to compute the actual heat rate of steam-turbine regenerative cycles for stationary, marine, and portable installations. Where necessary, use the steps of the previous procedure to compute the actual heat rate of a straight-condensing cycle before applying the present procedure. The performance curves given here are suitable for first approximations in situations where actual performance curves are unavailable.

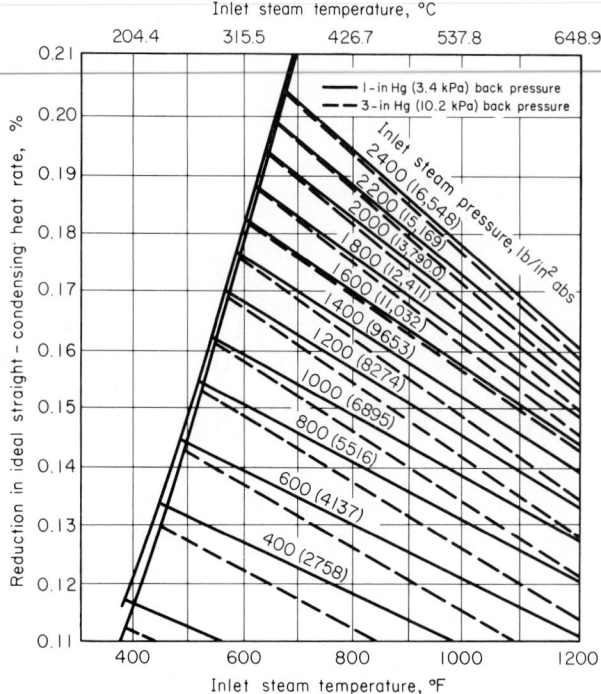

FIG. 29 Reduction in straight-condensing heat rate obtained by regenerative heating.

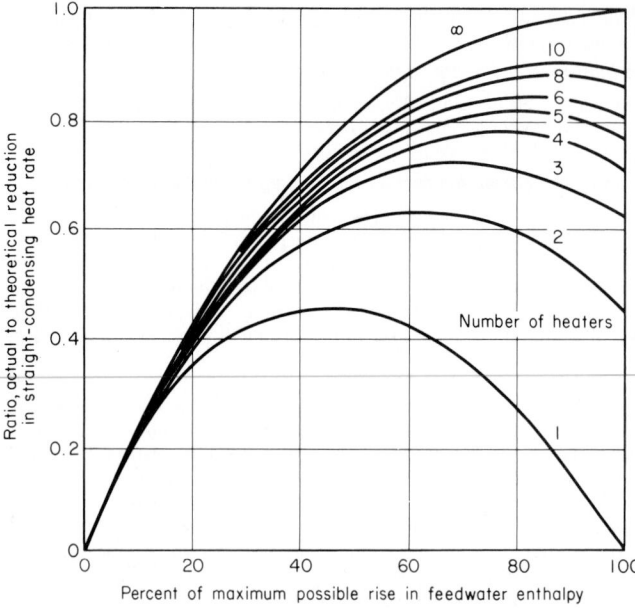

FIG. 30 Maximum possible rise in feedwater enthalpy varies with the number of heaters used.

3.242

REHEAT-REGENERATIVE STEAM-TURBINE HEAT RATES

What are the net and gross heat rates of a 300-kW reheat turbine having an initial steam pressure of 3500 lb/in² (gage) (24,132.5 kPa) with initial and reheat steam temperatures of 1000°F (537.8°C) with 1.5 inHg (5.1 kPa) absolute back pressure and six stages of regenerative feedwater heating? Compare this heat rate with that of 3500 lb/in² (gage) (24,132.5 kPa) 600-mW cross-compound four-flow turbine with 3600/1800 r/min shafts at a 300-mW load.

Calculation Procedure:

1. *Determine the reheat-regenerative heat rate*

Enter Fig. 31 at 3500-lb/in² (gage) (24,132.5-kPa) initial steam pressure, and project vertically to the 300-mW capacity net-heat-rate curve. At the left, read the net heat rate as 7680 Btu/kWh (8102.6 kJ/kWh). On the same vertical line, read the gross heat rate as 7350 Btu/kWh (7754.7 kJ/kWh). The gross heat rate is computed by using the generator-terminal output; the net heat rate is computed after the feedwater-pump energy input is deducted from the generator output.

2. *Determine the cross-compound turbine heat rate*

Enter Fig. 32 at 350 mW at the bottom, and project vertically upward to 1.5-inHg (5.1-kPa) exhaust pressure midway between the 1- and 2-inHg (3.4-and 6.8-kPa) curves. At the left, read the net heat rate as 7880 Btu/kWh (8313.8 kJ/kWh). Thus, the reheat-regenerative unit has a lower net heat rate. Even at full rated load of the cross-compound turbine, its heat rate is higher than the reheat unit.

   Related Calculations: Use this general procedure for comparing stationary and marine high-pressure steam turbines. The curves given here are typical of those supplied by turbine manufacturers for their turbines.

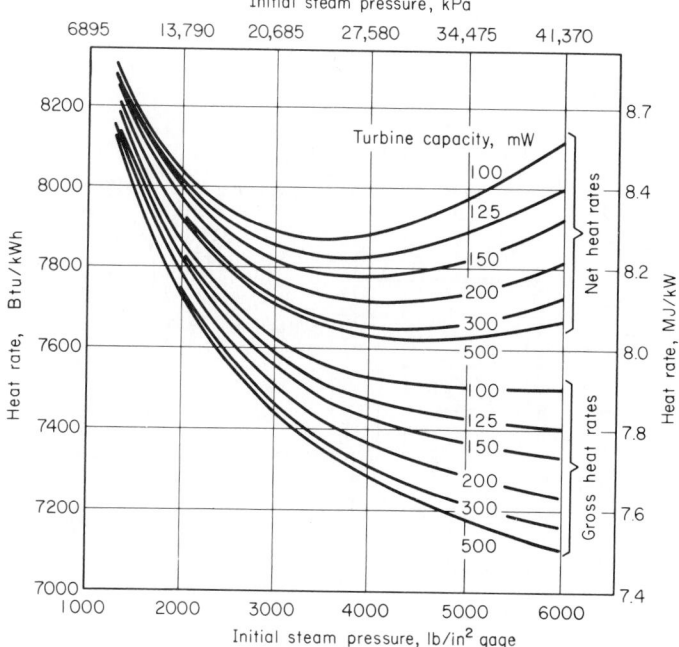

FIG. 31 Full-load heat rates for steam turbines with six feedwater heaters, 1000°F/1000°F (538°C/538°C) steam, 1.5-in (38.1-mm) Hg (abs) exhaust pressure.

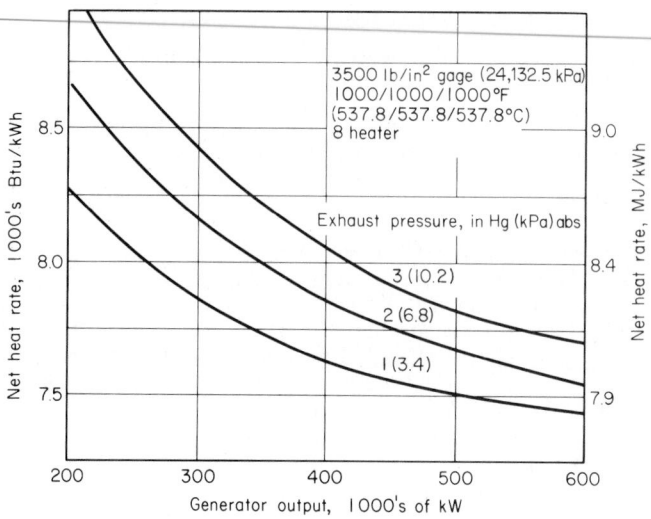

FIG. 32 Heat rate of a cross-compound four-flow steam turbine with 3600/1800-r/min shafts.

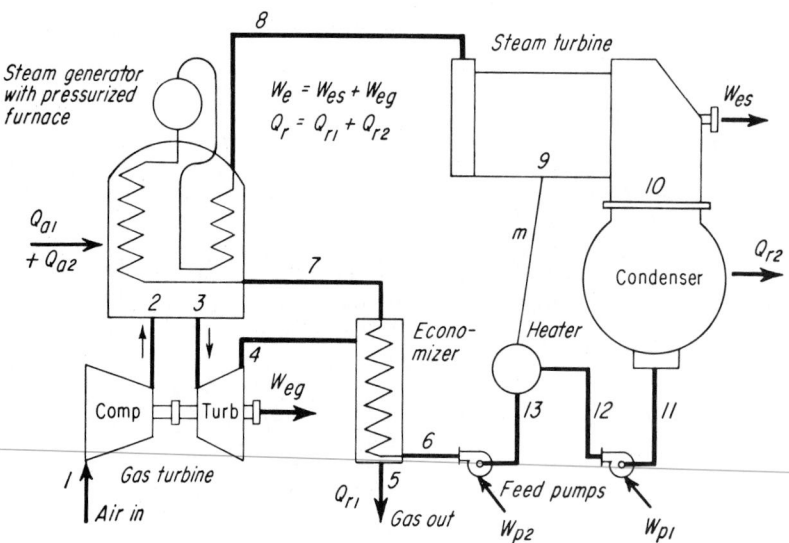

FIG. 33 Combined gas turbine–steam turbine cycle.

STEAM TURBINE–GAS TURBINE CYCLE ANALYSIS

Sketch the cycle layout, T-S diagram, and energy-flow chart for a combined steam turbine–gas turbine cycle having one stage of regenerative feedwater heating and one stage of economizer feedwater heating. Compute the thermal efficiency and heat rate of the combined cycle.

Calculation Procedure:

1. Sketch the cycle layout

Figure 33 shows the cycle. Since the gas-turbine exhaust-gas temperature is usually higher than the bleed-steam temperature, the economizer is placed after the regenerative feedwater heater. The feedwater will be progressively heated to a higher temperature during passage through the regenerative heater and the gas-turbine economizer. The cycle shown here is only one of many possible combinations of a steam plant and a gas turbine.

2. Sketch the T-S diagram

Figure 34 shows the T-S diagram for the combined gas turbine–steam turbine cycle. There is irreversible heat transfer Q_T from the gas-turbine exhaust to the feedwater in the economizer, which helps reduce the required energy input Q_{a2}.

3. Sketch the energy-flow chart

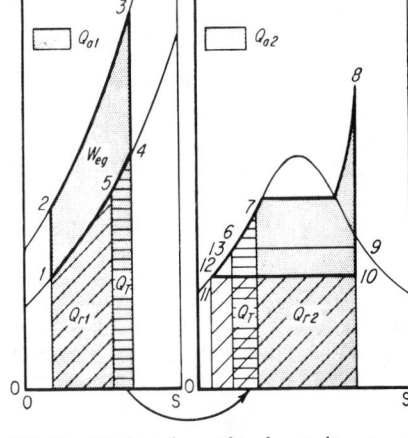

FIG. 34 T-S charts for combined gas turbine–steam turbine cycle have irreversible heat transfer Q from gas-turbine exhaust to the feedwater.

Choose a suitable scale for the energy input, and proportion the energy flow to each of the other portions of the cycle. Use a single line when the flow is too small to plot to scale. Figure 35 shows the energy-flow chart.

4. Determine the thermal efficiency of the cycle

Since $e = W/Q_a$, $e = Q_a - Q_r/Q_a = 1 - [Q_{r1} + Q_{r2}/(Q_{a1} + Q_{a2})]$, given the notation in Figs. 33, 34, and 35.

The relative weight of the gas w_g to 1 lb (0.5 kg) of water must be computed by taking an energy balance about the economizer. Or, $H_7 - H_6 = w_g(H_4 - H_5)$. Using the actual values for the enthalpies, solve this equation for w_g.

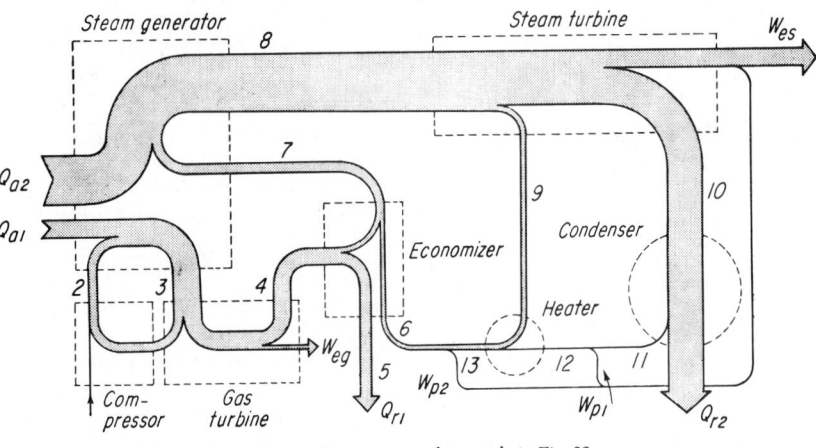

FIG. 35   Energy-flow chart of the gas turbine–steam turbine cycle in Fig. 33.

With w_g known, the other factors in the efficiency computation are

$$Q_{r1} = w_g(H_5 - H_1)$$

$$Q_{r2} = (1 - m)(H_{10} - H_{11})$$

$$Q_{a1} = w_g(H_3 - H_2)$$

$$Q_{a2} = H_8 - H_7$$

The bleed-steam flow m is calculated from an energy balance about the feedwater heater. Note that the units for the above equations can be any of those normally used in steam- and gas-turbine analyses.

STEAM-CONDENSER PERFORMANCE ANALYSIS

(a) Find the required tube surface area for a shell-and-tube type of condenser serving a steam turbine when the quantity of steam condensed S is 25,000 lb/h (3.1 kg/s); condenser back pressure = 2 inHg absolute (6.8 kPa); steam temperature t_s = 101.1°F (38.4°C); inlet water temperature t_1 = 80°F (26.7°C); tube length per pass L = 14 ft (4.3 m); water velocity V = 6.5 ft/s (2.0 m/ s); number of passes = 2; tube size and gage: ¾ in (1.9 cm), no. 18 BWG; cleanliness factor = 0.80. (b) Compute the required area and cooling-water flow rate for the same conditions as (a) except that cooling water enters at 85°F (29.4°C). (c) If the steam flow through the condenser in (a) decreases to 15,000 lb/h (1.9 kg/s), what will be the absolute steam pressure in the condenser shell?

Calculation Procedure:

1. *Sketch the condenser, showing flow conditions*

(a) Figure 36 shows the condenser and the flow conditions prevailing.

2. *Determine the condenser heat-transfer coefficient*

Use standard condenser-tube engineering data available from the manufacturer or Heat Exchange Institute. Table 1 and Fig. 37 show typical condenser-tube data used in condenser selection. These data are based on a minimum water velocity of 3 ft/s (0.9 m/s) through the condenser tubes, a minimum absolute pressure of 0.7 inHg (2.4 kPa) in the condenser shell, and a minimum Δt terminal temperature difference $t_s - t_2$ of 5°F (2.8°C). These conditions are typical for power-plant surface condensers.

Enter Fig. 37 at the bottom at the given water velocity, 6.5 ft/s (2.0 m/s), and project vertically upward until the ¾-in (1.9-cm) OD tube curve is intersected. From this point, project horizontally to the left to read the heat-transfer coefficient U = 690 Btu/(ft²·°F) [14,104.8 kJ/(m²· °C)] LMTD (log mean temperature difference). Also read from Fig. 37 the temperature correction factor for an inlet-water temperature of 80°F (26.7°C) by entering at the bottom at 80°F (26.7°C) and projecting vertically upward to the temperature-correction curve. From the intersection with this curve, project to the right to read the correction as 1.04. Correct U for temperature and cleanliness by multiplying the value obtained from the chart by the correction factors, or U = 690(1.04)(0.80) = 574 Btu/(ft²·h·°F) [11,733.6 kJ/(m²·h· °C)] LMTD.

3. *Compute the tube constant*

Read from Table 1, for two passes through ¾-in (1.9-cm) OD 18 BWG tubes, k = a constant = 0.377. Then kL/V = 0.377(14)/6.5 = 0.812.

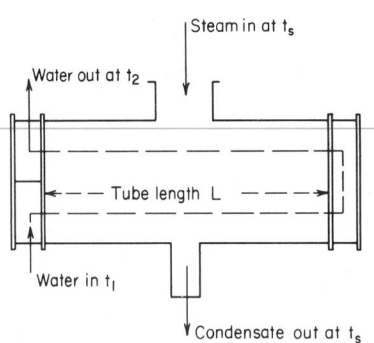

FIG. 36 Temperatures governing condenser performance.

TABLE 1   Standard Condenser Tube Data

Tube OD, in (cm)	Tube gage BWG	Tube ID, in (cm)	Surface area, ft²/ft (m²/m)		Velocity, ft/s for 1 gal/min (m/s for 1 L/min)	Value of k for number of tube passes		
			Outside	Inside		One	Two	Three
¾ (1.9)	18	0.652 (1.656)	0.1963 (0.0598)	0.1706 (0.0520)	0.9611 (0.0774)	0.188	0.377	0.565
	16	0.620 (1.575)	0.1963 (0.0598)	0.1613 (0.0492)	1.063 (0.0856)	0.208	0.417	0.625
	14	0.584 (1.483)	0.1963 (0.0598)	0.1528 (0.0466)	1.198 (0.0965)	0.235	0.470	0.705

4. *Compute the outlet-water temperature*

The equation for outlet-water temperature is $t_2 = t_s - (t_s - t_1/e^x)$, where $x = (kL/V)(U/500)$, or $x = 0.812(574/500) = 0.932$. Then $e^x = 2.7183^{0.932} = 2.54$. With this value known, $t_2 = 101.1 - (101.1 - 80/2.54) = 92.8°F$ (33.8°C). Check to see that $\Delta t(t_s - t_2)$ is less than the minimum 5°F (2.8°C) terminal difference. Or, $101.1 - 92.8 = 8.3°F$ (4.6°C), which is greater than 5°F (2.8°C).

5. *Compute the required tube surface area*

The required cooling-water flow, gal/min $= 950S/[500(t_2 - t_1)] = 950(25,000)/[500(92.8 - 80)] = 3700$ gal/min (233.4 L/s). This equation assumes that 950 Btu is to be removed from each

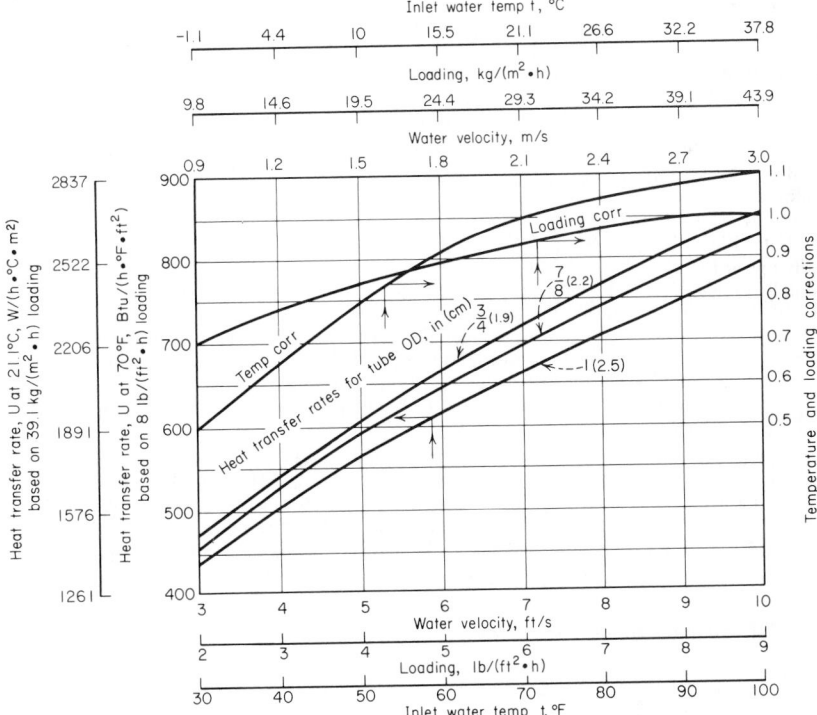

FIG. 37   Heat-transfer and correction curves for calculating surface-condenser performances.

pound (2209.7 kJ/kg) of steam condensed. When a different quantity of heat must be removed, use the actual quantity in place of the 950 in this equation.

With the tube constant kL/V and cooling-water flow rate known, the required area is computed from $A = (kL/V)(gpm) = (0.812)(3700) = 3000$ ft^2 (278.7 m^2).

Since the value of U was not corrected for condenser loading, it is necessary to check whether such a correction is needed. Condenser loading $= S/A = 25,000/3000 = 8.33$ lb/ft^2 (40.7 kg/m^2). Figure 37 shows that no correction (correction factor = 1.0) is necessary for loadings greater than 8.0 lb/ft^2 (39.1 kg/m^2). Therefore, the loading for this condenser is satisfactory without correction.

This step concludes the general calculation procedure for a surface condenser serving any steam turbine. The next procedure shows the method to follow when a higher cooling-water inlet temperature prevails.

6. Compute the cooling-water outlet temperature

(b) Higher cooling water temperature. From Fig. 37 for 85°F (29.4°C) cooling-water inlet temperature and a 0.80 cleanliness factor, $U = 690(1.06)(0.80) = 585$ Btu/(ft$^2 \cdot$h$\cdot$°F) [3.3 kJ/(m$^2 \cdot$°C$\cdot$s)] LMTD.

Given data from Table 1, the tube constant $kL/V = 0.377(14)/6.5 = 0.812$. Then $x = (kL/V)(U/500) = 0.812(585/500) = 0.950$. Using this exponent, we get $e^x = 2.8183^{0.950} = 2.586$. The cooling-water outlet temperature is then $t_2 = t_s - (t_s - t_1/e^x) = 101.1 - (101.1 - 85)/2.586 = 94.9$°F (34.9°C). Check to see that $\Delta t(t_s - t_2)$ is greater than the minimum 5°F (2.8°C) terminal temperature difference. Or, $101.1 - 94.9 = 6.5$°F (3.6°C), which is greater than 5°F (2.8°C).

7. Compute the water flow rate, required area, and loading

The required cooling-water flow, gal/min $= 950S/[500(t_2 - t_1)] = 950(25,000)/[500(94.9 - 85)] = 4800$ gal/min (302.8 L/s).

With the tube constant kL/V and cooling-water flow rate known, the required area is computed from $A = (kL/V)(gpm) = 0.812(4800) = 3900$ ft^2 (362.3 m^2). Then loading $= S/A = 25,000/3900 = 6.4$ lb/ft^2 (31.2 kg/m^2).

Since the loading is less than 8 lb/ft^2 (39.1 kg/m^2), refer to Fig. 37 to obtain the loading correction factor. Enter at the bottom at 6.4 lb/ft^2 (31.2 kg/m^2), and project vertically to the loading curve. At the right, read the loading correction factor as 0.95. Now the value of U already computed must be corrected, and all dependent quantities recalculated.

8. Recalculate the condenser proportions

First, correct U for loading. Or, $U = 585(0.95) = 555$. Then $x = 0.812(555/500) = 0.90$; $e^x = 2.7183^{0.90} = 2.46$; $t_2 = 101.1 - (101.1 - 85/2.46) = 94.6$°F (34.8°C). Check $\Delta t = t_s - t_2 = 101.1 - 94.6 = 6.5$°F (3.6°C), which is greater than 5°F (2.8°C). The cooling-water flow rate, gal/min $= 950 (25,000)/[500(94.6 - 85)] = 4950$ gal/min (312.3 L/s). Then $A = 0.812(4950) = 4020$ ft^2 (373.5 m^2), and loading $= 25,000/4020 = 6.23$ lb/ft^2 (30.4 kg/m^2).

Check the correction factor for this loading in Fig. 37. The correction factor is 0.94, compared with 0.95 for the first calculation. Since the value of U would be changed only about 1 percent by using the lower factor, the calculations need not be revised further. Where U would change by a larger amount—say 5 percent or more—it would be necessary to repeat the procedure just detailed, applying the new correction factor.

Note that the 5°F (2.8°C) increase in cooling-water temperature (from 80 to 85°F or 26.7 to 29.4°C) requires an additional 1020 ft^2 (94.8 m^2) of condenser surface and 125 gal/min (7.9 L/s) of cooling-water flow to maintain the same back pressure. These increments will vary, depending on the temperature level at which the increase occurs. The effect of reduced steam flow on the steam pressure in the condenser shell will not be computed because the recalculation above is the last step in part (b) of this procedure.

(c) Reduced steam flow to condenser.

9. Determine the condenser loading

From procedure (a) above, the cooling-water flow $= 3700$ gal/min (233.4 L/s); condenser surface $A = 3000$ ft^2 (278.7 m^2). Then, with a 15,000-lb/h (1.9-kg/s) steam flow, loading $= S/A = 15,000/3000 = 5$ lb/ft^2 (24.4 kg/m^2).

10. Compute the heat-transfer coefficient

Correct the previous heat-transfer rate $U = 690$ Btu/(ft$^2 \cdot$ h $\cdot$ °F) [3.9 kJ/(m$^2 \cdot$ °C $\cdot$ s)] LMTD for temperature, cleanliness, and loading. Or, $U = 690(1.04)(0.80)(0.89) = 511$ Btu/(ft$^2 \cdot$ h $\cdot$ °F) [2.9 kJ/(m$^2 \cdot$ °C $\cdot$ s)] LMTD, given the correction factors from Fig. 37.

11. Compute the final steam temperature

As before, $x = (kL/V)(U/500) = (0.377)(14/6.5)(511/500) = 0.830$. Then $\Delta t = t_2 - t_1 = 950S/(500gpm) = 950(15,000)/[500(3700)] = 7.7$°F (4.3°C). With $t_1 = 80$°F (26.7°C), $t_2 = \Delta t + t_1 = 7.7 + 80 = 87.7$°F (30.9°C). Since $t_2 = t_s - (t_s - t_1)/e^x$, $e^x = t_s - t_1/(t_s - t_2)$, or $2.7183^{0.830} = t_s - 80/(t_s - 87.7)$. Solve for t_s; or, $t_s = 201.1 - 80/1.294 = 93.6$°F (34.2°C).

At a saturation temperature of 93.6°F (34.2°C), the steam table (saturation temperature) shows that the steam pressure in the condenser shell is 1.59 inHg (5.4 kPa).

Check the Δt terminal temperature difference. Or, $\Delta t = t_s - t_2 = 93.6 - 87.7 = 5.9$°F (3.3°C). Since the terminal temperature difference is greater than 5°F (2.8°C), the calculated performance can be realized.

Related Calculations: The procedures and data given here can be used to compute the required cooling-water flow, cooling-water temperature rise, quantity of steam condensed by a given cooling-water flow rate and temperature rise, required condenser surface area, tube length per pass, water velocity, steam temperature in condenser, cleanliness factor, and heat-transfer rate. Whereas Fig. 37 is suitable for all usual condenser calculations for the ranges given, check the Heat Exchange Institute for any new curves that might have been made available before you make the final selection of very large condensers (more than 100,000 lb/h or 12.6 kg/s of steam flow).

Note: The design water temperature used for condensers is either the average summer water temperature or the average annual water temperature, depending on which is higher. The design steam load is the maximum steam flow expected at the full-load rating of the turbine or engine. Usual shell-and-tube condensers have tubes that vary in length from about 8 ft (2.4 m) in the smallest sizes to about 40 ft (12.2 m) or more in the largest sizes. Each square foot of tube surface will condense 7 to 20 lb/h (0.88 to 2.5 g/s) of steam with a cooling-water circulating rate of 0.1 to 0.25 gal/(lb · min) [0.014 to 0.035 L/(kg · s)] of steam condensed. The method presented here is the work of Glenn C. Boyer.

STEAM-CONDENSER AIR LEAKAGE

The air leakage into a condenser is estimated to be 12 ft^3/min (0.34 m^3/min) of 70°F (21°C) air at 14.7 lb/in^2 (101 kPa). At the air outlet connection on the condenser, the temperature is 84°F (29°C) and the total (mixture) pressure is 1.80 inHg absolute (6.1 kPa). Determine the quantity of steam, lb$_m$/h (kg/h), lost from the condenser.

Calculation Procedure:

1. Compute the mass rate of flow per hour of the air leakage

The mass rate of flow per hour of the estimated dry air leakage into the condenser, $w_a = pV/R_aT$, where the air pressure, $p = 14.7 \times 144$ lb$_f$/ft^2 (101 kPa); volumetric flow rate, $V = 12 \times 60 = 720$ ft^3/h (20.4 m^3/h); gas constant for air, $R_a = 53.34$ ft · lb/(lb · °R) [287(m · N/kg · K)]; air temperature, $T = 70 + 460 = 530$°R (294 K). Then, $w_a = (14.7 \times 144)(720)/(53.34 \times 530) = 53.9$ lb/h (24.4 kg/h).

2. Determine the partial pressure of the air in the mixture

The partial pressure of the air in the mixture of air and steam, $p_a = p_m - p_v$, where the mixture pressure, $p_m = 1.80 \times 0.491 = 0.884$ lb/in^2 (6.09 kPa); partial vapor pressure, $p_v = 0.577$ lb/in^2 (3.98 kPa), as found in the Steam Tables mentioned under Related Calculations of this procedure. Then, $p_a = 0.884 - 0.577 = 0.307$ lb/in^2 (2.1 kPa).

3. Compute the humidity ratio of the mixture

The humidity ratio of the mixture, $w_v = R_a p_v/(R_v p_a)$, where the gas constant for steam vapor,

TABLE 2 Typical Design Conditions for Steam Condensers

Cooling water temperature		Steam pressure		Temperature difference $t_s - t_1$	
°F	°C	inHg	kPa	°F	°C
70	21.1	1.5–2.0	5.1–6.8	21.7–31.1	12.1–17.3
75	23.9	2.0–2.5	6.8–8.5	26.1–33.7	14.5–18.7
80	26.7	2.0–4.0	6.8–13.5	21.1–45.4	11.7–25.2

$R_v = 85.8$ ft · lb/(lb$_m$ · °R) [462(J/kg · K)], as found in a reference mentioned under Related Calculations of this procedure. Then, $w_v = 53.34 \times 0.577/(85.8 \times 0.307) = 1.17$ lb vapor/lb dry air (0.53 kg/kg).

4. Compute the rate of steam lost from the condenser

Steam is lost from the condenser at the rate of $w_h = w_v \times w_a = 1.17 \times 53.9 = 63.1$ lb/h (28.6 kg/h).

Related Calculations: The partial vapor pressure in step 2 was found at 84°F (29°C) under Table 1, Saturation: Temperatures of *Thermodynamic Properties of Water Including Vapor, Liquid, and Solid Phases*, 1969, Keenan, et al., John Wiley & Sons, Inc. Use later versions of such tables whenever available, as necessary. The gas constant for water vapor in step 3 was obtained from *Principles of Engineering Thermodynamics*, 2d edition, by Kiefer, et al., John Wiley & Sons, Inc.

STEAM-CONDENSER SELECTION

Select a condenser for a steam turbine exhausting 150,000 lb/h (18.9 kg/s) of steam at 2 inHg absolute (6.8 kPa) with a cooling-water inlet temperature of 75°F (23.9°C). Assume a 0.85 condition factor, ⅞-in (2.2-cm) no. 18 BWG tubes, and an 8-ft/s (2.4-m/s) water velocity. The water supply is restricted. Obtain condenser constants from the Heat Exchange Institute, *Steam Surface Condenser Standards*.

Calculation Procedure:

1. Select the $t_s - t_1$ temperature difference

Table 2 shows customary design conditions for steam condensers. With an inlet-water temperature of 75°F (23.9°C) and an exhaust steam pressure of 2.0 inHg absolute (6.8 kPa), the customary temperature difference $t_s - t_1 = 26.1$°F (14.5°C). With a sufficient water supply and a siphonic circuitry, $(t_2 - t_1)/(t_s - t_1)$ is usually between 0.5 and 0.55. For a restricted water supply or high frictional resistance and static head, the value of this factor ranges from 0.55 to 0.75.

2. Compute the LMTD across the condenser

With 75°F (23.9°C) inlet water, $t_s - t_1 = 101.14 - 75 = 26.14$°F (14.5°C), given the steam temperature in the saturation-pressure table. Once $t_s - t_1$ is known, it is necessary to assume a value for the ratio $(t_2 - t_1)/(t_s - t_1)$. As a trial, assume 0.60, since the water supply is restricted. Then $(t_2 - t_1)/(t_s - t_1) = 0.60 = (t_2 - t_1)/26.14$. Solving, we get $t_2 - t_1 = 15.68$°F (8.7°C). The difference between the steam temperature t_s and the outlet temperature t_2 is then $t_s - t_2 = 26.14 - 15.68 = 10.46$°F (5.8°C). Checking, we find $t_2 = t_1 + (t_2 - t_1) = 75 + 15.68 = 90.68$°F (50.38°C); $t_s - t_2 = 101.14 - 90.68 = 10.46$°F (5.8°C). This value is greater than the required minimum value of 5°F (2.8°C) for $t_s - t_2$. The assumed ratio 0.60 is therefore satisfactory.

Were $t_s - t_2$ less than 5°F (2.8°C), another ratio value would be assumed and the difference computed again. You would continue doing this until a value of $t_s - t_2$ greater than 5°F (2.8°C) were obtained. Then LMTD $= (t_2 - t_1)/\ln[(t_s - t_1)/(t_s - t_2)]$; LMTD $= 15.68/\ln(26.1/10.46) = 17.18$°F (9.5°C).

3.   *Determine the heat-transfer coefficient*

From the Heat Exchange Institute or manufacturer's data U is 740 Btu/(ft^2·h·°F) [4.2 kJ/(m^2· °C·s)] LMTD for a water velocity of 8 ft/s (2.4 m/s). If these data are not available, Fig. 37 can be used with complete safety for all preliminary selections.

Now U must be corrected for the inlet-water temperature, 75°F (23.9°C), and the condition factor, 0.85, which is a term used in place of the correction factor by some authorities. From Fig. 37, the correction for 75°F (23.9°C) inlet water = 1.04. Then actual U = 740(1.04)(0.85) = 655 Btu/(ft^2·h·°F) [3.7 kJ/(m^2·°C·s)] LMTD.

4.   *Compute the steam condensation rate*

The heat-transfer rate per square foot of condenser surface with a 17.18°F (9.5°C) LMTD is U(LMTD) = 655(17.18) = 11,252.9 Btu/(ft^2·h) [35.5 kJ/(m^2·s)].

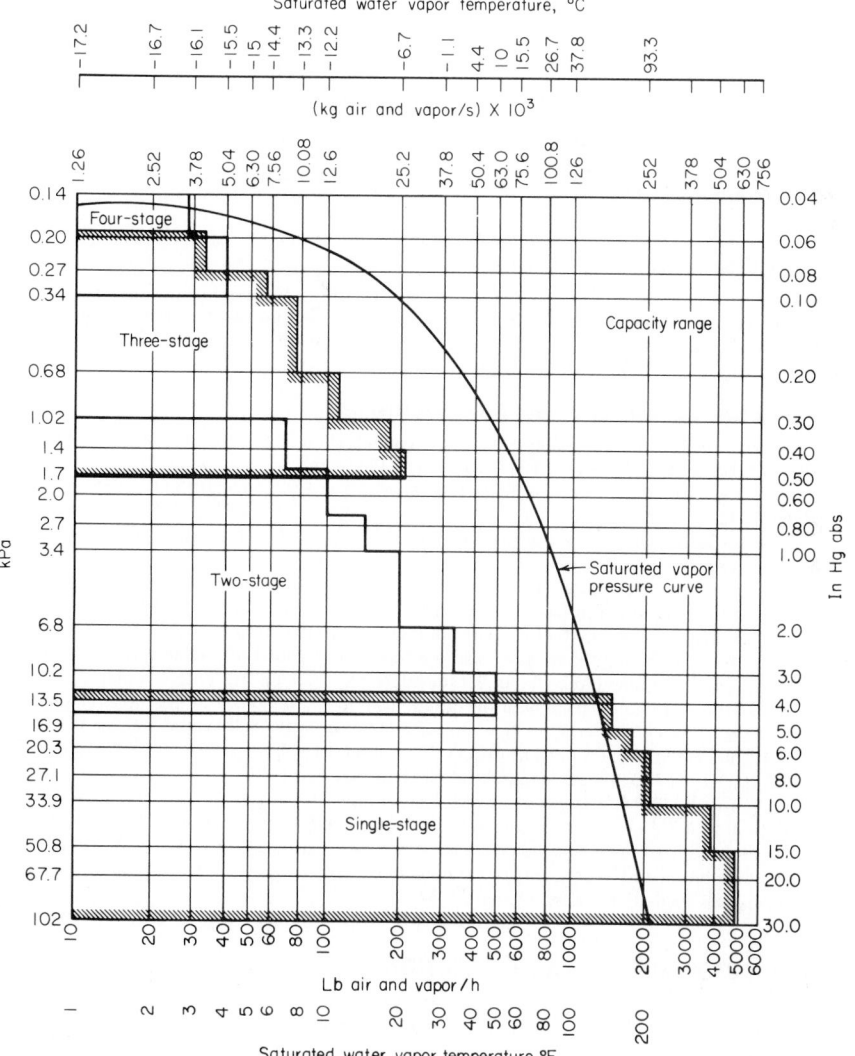

FIG. 38   Steam-ejector capacity-range chart.

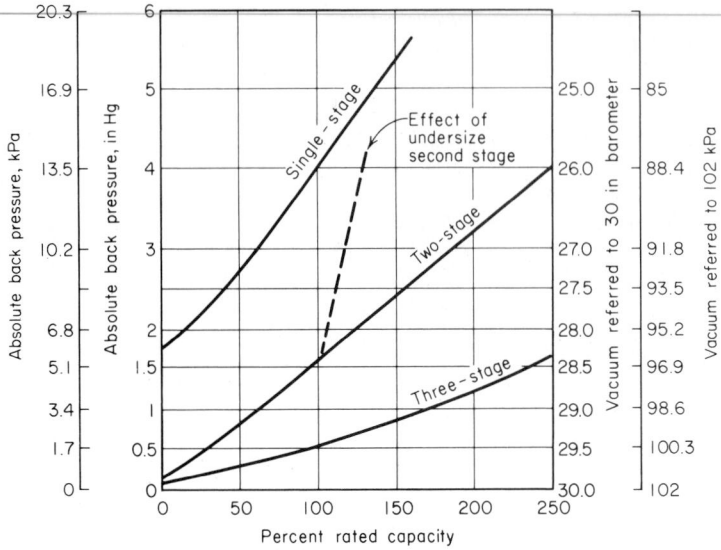

FIG. 39   Steam-jet ejector characteristics.

Condensers serving steam turbines are assumed, for design purposes, to remove 950 Btu/lb (2209.7 kJ/kg) of steam condensed. Therefore, the steam condensation rate for any condenser is [Btu/(ft$^2 \cdot$h)]/950, or 1252.9/950 = 11.25 lb/(ft$^2 \cdot$h) [15.3 g/(m$^2 \cdot$s)].

5.   Compute the required surface area and water flow

The required surface area = steam flow (lb/h)/[condensation rate, lb/(ft$^2 \cdot$h)], or with a 150,000-lb/h (18.9-kg/s) flow, 150,000/11.25 = 13,320 ft^2 (1237.4 m^2).

The water flow rate, gal/min = 950S/[500($t_2 - t_1$)] = 950(150,000)/[500(15.68)] = 18,200 gal/min (1148.1 L/s).

Related Calculations: See the previous calculation procedure for steps in determining the water-pressure loss through a surface condenser.

To choose a surface condenser for a steam engine, use the same procedures as given above, except that the heat removed from the exhaust steam is 1000 Btu/lb (2326.9 kJ/kg). Use a condition (cleanliness) factor of 0.65 for steam engines because the oil in the exhaust steam fouls the condenser tubes, reducing the rate of heat transfer. The condition (cleanliness) factor for steam turbines is usually assumed to be 0.8 to 0.9 for relatively clean, oil-free cooling water.

At loads greater than 50 percent of the design load, $t_s - t_1$ follows a straight-line relationship. Thus, in the above condenser, $t_s - t_1 = 26.14°$F (14.5°C) at the full load of 150,000 lb/h (18.9 kg/s). If the load falls to 60 percent (90,000 lb/h or 11.3 kg/s), then $t_s - t_1 = 26.14(0.60) = 15.7°$F (8.7°C). At 120 percent load (180,000 lb/h or 22.7 kg/s), $t_s - t_1 = 26.14(1.20) = 31.4°$F (17.4°C). This straight-line law is valid with constant inlet-water temperature and cooling-water flow rate. It is useful in analyzing condenser operating conditions at other than full load.

Single- or multiple-pass surface condensers may be used in power services. When a liberal supply of water is available, the single-pass condenser is often chosen. With a limited water supply, a two-pass condenser is often chosen.

AIR-EJECTOR ANALYSIS AND SELECTION

Choose a steam-jet air ejector for a condenser serving a 250,000-lb/h (31.5-kg/s) steam turbine exhausting at 2 inHg absolute (6.8 kPa). Determine the number of stages to use, the approximate steam consumption, and the quantity of air and vapor mixture the ejector will handle.

Calculation Procedure:

1. Select the number of stages for the ejector

Use Fig. 38 as a preliminary guide to the number of stages required in the ejector. Enter at 2-inHg absolute (6.8-kPa) condenser pressure, and project horizontally to the stage area. This shows that a two-stage ejector will probably be satisfactory.

Check the number of stages above against the probable overload range of the prime mover by using Fig. 39. Enter at 2-inHg absolute (6.8-kPa) condenser pressure, and project to the two-stage curve. This curve shows that a two-stage ejector can readily handle a 25-percent overload of the prime mover. Also, the two-stage curve shows that this ejector could handle up to 50 percent overload with an increase in the condenser absolute pressure of only 0.4 inHg (1.4 kPa). This is shown by the pressure, 2.4 inHg absolute (8.1 kPa), at which the two-stage curve crosses the 150 percent overload ordinate (Fig. 39).

2. Determine the ejector operating conditions

Use the Heat Exchange Institute or manufacturer's data. Table 3 excerpts data from the Heat Exchange Institute for condensers in the range considered in this procedure.

Study of Table 3 shows that a two-stage condensing ejector unit serving a 250,000-lb/h (31.5-kg/s) steam turbine will require 450 lb/h (56.7 g/s) of 300-lb/in^2 (gage) (2068.5-kPa) steam. Also, the ejector will handle 7.5 ft^3/min (0.2 m^3/min) of free, dry air, or 33.75 lb/h (4.5 g/s) of air. It will remove up to 112.5 lb/h (14.2 g/s) of an air-vapor mixture.

The actual air leakage into a condenser varies with the absolute pressure in the condenser, the tightness of the joints, and the conditions of the tubes. Some authorities cite a maximum leakage of about 250-lb/h (31.5-g/s) steam flow. At 400,000 lb/h (50.4 kg/s), the leakage is 160 lb/h (20.2 g/s); at 250,000 lb/h (31.5 kg/s), it is 130 lb/h (16.4 g/s) of air-vapor mixture. A condenser in good condition will usually have less leakage.

For an installation in which the manufacturer supplies data on the probable air leakage, use a psychrometric chart to determine the weight of water vapor contained in the air. Thus, at 2 inHg absolute (6.8 kPa) and 80°F (26.7°C), each pound of air will carry with it 0.68 lb (0.68 kg/kg) of water vapor. In a surface condenser into which 20 lb (9.1 kg) of air leaks, the ejector must handle 20 + 20(0.68) = 33.6 lb/h (4.2 g/s) of air-vapor mixture. Table 3 shows that this ejector can readily handle this quantity of air-vapor mixture.

Related Calculations: When you choose an air ejector for steam-engine service, double the Heat Exchange Institute steam-consumption estimates. For most low-pressure power-plant service, a two-stage ejector with inter- and aftercondensers is satisfactory, although some steam engines operating at higher absolute exhaust pressures require only a single-stage ejector. Twin-element ejectors have two sets of stages; one set serves as a spare and may also be used for capacity regulation in stationary and marine service. The capacity of an ejector is constant for a given steam pressure and suction pressure. Raising the steam pressure will not increase the ejector capacity.

TABLE 3 Air-Ejector Capacities for Surface Condensers for Steam Turbines°

Steam load, lb/h (kg/s)	Free, dry air at 70°F (21.1°C), ft^3/min (cm^3/s)	Air, lb/h (g/s)	Air-vapor mixture at 30 percent dry air, lb/h (g/s)	Steam consumption at 300 lb/in^2 (gage) (2068.5 kPa), lb/h (g/s)
100,001– 250,000 (12.6– 31.5)	7.5 (3539.6)	33.75 (4.3)	112.5 (14.2)	450 (56.7)

°Two-stage condensing ejector unit.

SURFACE-CONDENSER CIRCULATING-WATER PRESSURE LOSS

Determine the circulating-water pressure loss in a two-pass condenser having 12,000 ft^2 (1114.8 m^2) of condensing surface, a circulating-water flow rate of 10,000 gal/min (630.8 L/s), ¾-in (1.9-cm) no. 16 BWG tubes, a water flow rate of 7 ft/s (2.1 m/s), external friction of 20 ft of water (59.8 kPa), and a 10-ft-of-water (29.9-kPa) siphonic effect on the circulating-water discharge.

Calculation Procedure:

1. *Determine the water flow rate per tube*

Use a tabulation of condenser-tube engineering data available from the manufacturer or the Heat Exchange Institute, or complete the water flow rate from the physical dimensions of the tube thus: ¾ in (1.9 cm) no. 16 BWG tube ID = 0.620 in (1.6 cm) from a tabulation of condenser-tube data, such as Table 1. Assume a water velocity of 1 ft/s (0.3 m/s). Then a 1-ft (0.3-m) length of the tube will contain $(12)(0.620)^2\pi/4 = 3.62$ in^3 (59.3 cm^3) of water. This quantity of water will flow through the tube for each foot of length per second of water velocity [194.6 cm^3/(m·s)]. The flow per minute will be 3.62 (60 s/min) = 217.2 in^3/min (3559.3 cm^3/min). Since 1 U.S. gal = 231 in^3 (3.8 L), the gal/min flow at a 1 ft/s (0.3 m/s) velocity = 217.2/231 = 0.94 gal/min (0.059 L/s).

With an actual velocity of 7 ft/s (2.1 m/s), the water flow rate per tube is 7(0.94) = 6.58 gal/min (0.42 L/s).

2. *Determine the number of tubes and length of water travel*

Since the water flow rate through the condenser is 10,000 gal/min (630.8 L/s) and each tube conveys 6.58 gal/min (0.42 L/s), the number of tubes = 10,000/6.58 = 1520 tubes per pass.

Next, the total length of water travel for a condenser having A ft^2 of condensing surface is computed from A(number of tubes)(outside area per linear foot, ft^2). The outside area of each tube can be obtained from a table of tube properties, such as Table 1, or computed from (OD, in)$(\pi)(12)/144$, or $(0.75)(\pi)(12)/144 = 0.196$ ft^2/lin ft (0.06 m^2/m). Then, total length of travel = 12,000/[(1520)(0.196)] = 40.2 ft (12.3 m). Since the condenser has two passes, the length of tube per pass = 40.2/2 = 20.1 ft (6.1 m). Since each pass has an equal number of tubes and there are two passes, the total number of tubes in the condenser = 2 passes (1520 tubes per pass) = 3040 tubes.

3. *Compute the friction loss in the system*

Use the Heat Exchange Institute or manufacturer's curves to find the friction loss per foot of condenser tube. At 7 ft/s (2.1 m/s), the Heat Exchange Institute curve shows the head loss is 0.4 ft of head per foot (3.9 kPa/m) of travel for ¾-in (1.9-cm) no. 16 BWG tubes. With a total length of 40.2 ft (12.3 m), the tube head loss is 0.4(40.2) = 16.1 ft (48.1 kPa).

Use the Heat Exchange Institute or manufacturer's curves to find the head loss through the condenser water boxes. From the first reference, for a velocity of 7 ft/s (2.1 m/s), head loss = 1.4 ft (4.2 kPa) of water for a single-pass condenser. Since this is a two-pass condenser, the total water-box head loss = 2(1.4) = 2.8 ft (8.4 kPa).

The total condenser friction loss is then the sum of the tube and water-box losses, or 16.1 + 2.8 = 18.9 ft (56.5 kPa) of water. With an external friction loss of 20 ft (59.8 kPa) in the circulating-water piping, the total loss in the system, without siphonic assistance, is 18.9 + 20 = 38.9 ft (116.3 kPa). Since there is 10 ft (29.9 kPa) of siphonic assistance, the total friction loss in the system with siphonic assistance is 38.9 − 10 = 28.9 ft (86.3 kPa). In choosing a pump to serve this system, the frictional resistance of 28.9 ft (86.3 kPa) would be rounded to 30 ft (89.7 kPa), and any factor of safety added to this value of head loss.

Note: The most economical cooling-water velocity in condenser tubes is 6 to 7 ft/s (1.8 to 2.1 m/s); a velocity greater than 8 ft/s (2.4 m/s) should not be used, unless warranted by special conditions.

SURFACE-CONDENSER WEIGHT ANALYSIS

A turbine exhaust nozzle can support a weight of 100,000 lb (444,822.2 N). Determine what portion of the total weight of a surface condenser must be supported by the foundation if the weight of the condenser is 275,000 lb (1,223,261.1 N), the tubes and water boxes have a capacity of 8000 gal (30,280.0 L), and the steam space has a capacity of 30,000 gal (113,550.0 L) of water.

Calculation Procedure:

1. Compute the maximum weight of the condenser

The maximum weight on a condenser foundation occurs when the shell, tubes, and water boxes are full of water. This condition could prevail during accidental flooding of the steam space or during tests for tube leaks when the steam space is purposefully flooded. In either circumstance, the condenser foundation and spring supports, if used, must be able to carry the load imposed on them. To compute this load, find the sum of the individual weights:

Condenser weight, dry	275,000 lb (1,223,261.1 N)
Water in tubes and boxes = (8.33 lb/gal)(8000 gal)	66,640 lb (296,429.5 N)
Water in steam space = (8.33)(30,000)	249,900 lb (1,111,610.7 N)
Maximum weight when full of water	592,540 lb (2,631,301.2 N)

2. Compute the foundation load

The turbine nozzle can support 100,000 lb (444,822.2 N). Therefore, the foundation must support $591,540 - 100,000 = 491,540$ lb (2,186,479.0 N). For foundation design purposes this would be rounded to 495,000 lb (2,201,869.9 N).

Related Calculations: When you design a condenser foundation, do the following: (1) Leave enough room at one end to permit withdrawal of faulty tubes and insertion of new tubes. Since some tubes may exceed 40 ft (12.3 m) in length, careful planning is needed to provide sufficient installation space. During the design of a power plant, a template representing the tube length is useful for checking the tube clearance on a scale plan and side view of the condenser installation. When there is insufficient room for tube removal with one shape of condenser, try another shape with shorter tubes.

(2) Provide enough headroom under the condenser to produce the required submergence on the condensate-pump impeller. Most condensate pumps require at least 3-ft (0.9-m) submergence. If necessary, the condensate pump can be installed in a pit under the condenser, but this should be avoided if possible.

BAROMETRIC-CONDENSER ANALYSIS AND SELECTION

Select a countercurrent barometric condenser to serve a steam turbine exhausting 25,000 lb/h (3.1 kg/s) of steam at 5 inHg absolute (16.9 kPa). Determine the quantity of cooling water required if the water inlet temperature is 50°F (10.0°C). What is the required dry-air capacity of the ejector? What is the required pump head if the static head is 40 ft (119.6 kPa) and the pipe friction is 15 ft of water (44.8 kPa)?

Calculation Procedure:

1. Find the steam properties from the steam tables

At 5 inHg absolute (16.9 kPa), $h_g = 1119.4$ Btu/lb (2603.7 kJ/kg), from the saturation-pressure table. If the condensing water were to condense the steam without subcooling the condensate, the final temperature of the condensate, from the steam tables, would be 133.76°F (56.5°C), corresponding to the saturation temperature. However, subcooling almost always occurs, and the usual practice in selecting a countercurrent barometric condenser is to assume the final condensate temperature t_c will be 5°F (2.8°C) below the saturation temperature corresponding to the absolute pressure in the condenser. Given a 5°F (2.8°C) difference, $t_c = 133.76 - 5 = 128.76$°F (53.7°C). Interpolating in the saturation-temperature steam table, we find the enthalpy of the condensate h_f at 128.76°F (53.7°C) is 96.6 Btu/lb (224.8 kJ/kg).

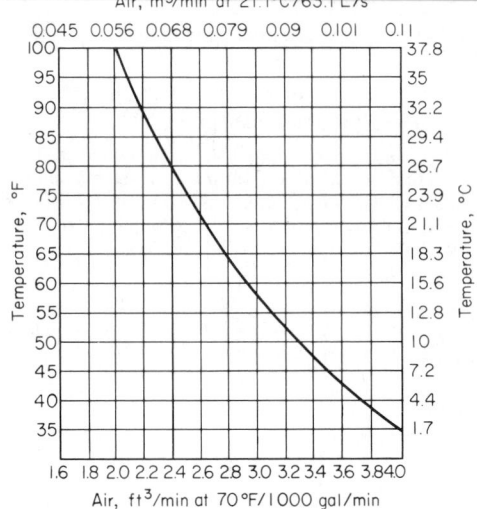

FIG. 40   Allowance for air in condenser injection water.

2.   *Compute the quantity of condensing water required*

In any countercurrent barometric condenser, the quantity of cooling water Q lb/h required is $Q = W(h_g - h_t)/(t_c - t_1)$, where W = weight of steam condensed, lb/h; t_1 = cooling-water inlet temperature, °F. Then $Q = 25{,}000(1119.4 - 96.66)/(128.76 - 50) = 325{,}000$ lb/h (40.9 kg/s). By converting to gallons per minute, $Q = 325{,}000/500 = 650$ gal/min (41.0 L/s).

3.   *Determine the required ejector dry-air capacity*

Use the Heat Exchanger Institute or a manufacturer's tabulation of free, dry-air leakage and the allowance for air in the cooling water to determine the required dry-air capacity. Thus, from Table 4, the free, dry-air leakage for a barometric condenser serving a turbine is 3.0 ft³/min (0.08 m³/min) of air and vapor. The allowance for air in the 50°F (10.0°C) cooling water is 3.3 ft³/min (0.09 m³/min) of air at 70°F (21.1°C) per 1000 gal/min (63.1 L/s) of cooling water, Fig. 40. The total dry-air leakage is the sum, or $3.0 + 3.3 = 6.3$ ft³/min (0.18 m³/min). Thus, the ejector must be capable of handling at least 6.3 ft³/min (0.18 m³/min) of dry air to serve this barometric condenser at its rated load of 25,000 lb/h (3.1 kg/s) of steam.

Where the condenser will operate at a lower vacuum (i.e., a higher absolute pressure), overloads up to 50 percent may be met. To provide adequate dry-air handling capacity at this overload with the same cooling-water inlet temperature, find the free, dry-air leakage at the higher condensing rate from Table 4 and add this to the previously found allowance for air in the cooling water. Or, $4.5 + 3.3 = 7.8$ ft³/min (0.22 m³/min). An ejector capable of handling up to 10 ft³/min (0.32 m³/min) would be a wise choice for this countercurrent barometric condenser.

4.   *Determine the pump head required*

Since a countercurrent barometric condenser operates at pressures below atmospheric, it assists the cooling-water pump by "sucking" the water into the condenser. The maximum assist that can be assumed is $0.75V$, where V = design vacuum, inHg.

In this condenser with a 26-in (88.0-kPa) vacuum, the maximum assist is $0.75(26) = 19.5$ inHg (66.0 kPa). Converting to feet of water, using 1.0 inHg = 1.134 ft (3.4 kPa) of water, we find $19.5(1.134) = 22.1$ ft (66.1 kPa) of water. The total head on the pump is then the sum of the static and friction heads less $0.75V$, expressed in feet of water. Or, the total head on the pump = $40 + 15 - 22.1 = 32.9$ ft (98.4 kPa). A pump with a total head of at least 35 ft (104.6 kPa) of water would be chosen for this condenser. Where corrosion or partial clogging of the piping is

TABLE 4 Free, Dry-Air Leakage

[*ft³/min (m³/s) at 70°F or 21.1°C air and vapor mixture, 7½° below vacuum temperature or 4.2° for Celsius*]

Maximum steam condensation		Barometric and low-level jet condensers								
		Serving turbines				Serving engines			
lb/h	kg/s	ft³/min	m³/s	lb/h	g/s	ft³/min	m³/s	lb/h	g/s
75,000–150,000	9.4–18.9	6.5	0.0031	97.5	12.3	13.0	0.0061	195.0	24.6
150,001–250,000	18.9–31.5	8.5	0.0040	127.5	16.1				
250,001–350,000	31.5–44.1	10.0	0.0047	150.0	18.9				

expected, a pump with a total head of 50 ft (149.4 kPa) would probably be chosen to ensure sufficient head even though the piping is partially clogged.

Related Calculations: (1) When a condenser serving a steam engine is being chosen, use the appropriate dry-air leakage value from Table 4. (2) For ejector-jet barometric condensers, assume the final condensate temperature t_c as 10 to 20°F (5.6 to 11.1°C) below the saturation temperature corresponding to the absolute pressure in the condenser. This type of condenser does not use an ejector, but it requires 25 to 50 percent more cooling water than the countercurrent barometric condenser for the same vacuum. (3) The total pump head for an ejector-jet barometric condenser is the sum of the static and friction heads plus 10 ft (29.9 kPa). The additional positive head is required to overcome the pressure loss in spray nozzles.

COOLING-POND SIZE FOR A KNOWN HEAT LOAD

How many spray nozzles and what surface area are needed to cool 10,000 gal/min (630.8 L/s) of water from 120 to 90°F (48.9 to 32.2°C) in a spray-type cooling pond if the average wet-bulb temperature is 650°F (15.6°C)? What would the approximate dimensions of the cooling pond be? Determine the total pumping head if the static head is 10 ft (29.9 kPa), the pipe friction is 35 ft of water (104.6 kPa), and the nozzle pressure is 8 lb/in² (55.2 kPa).

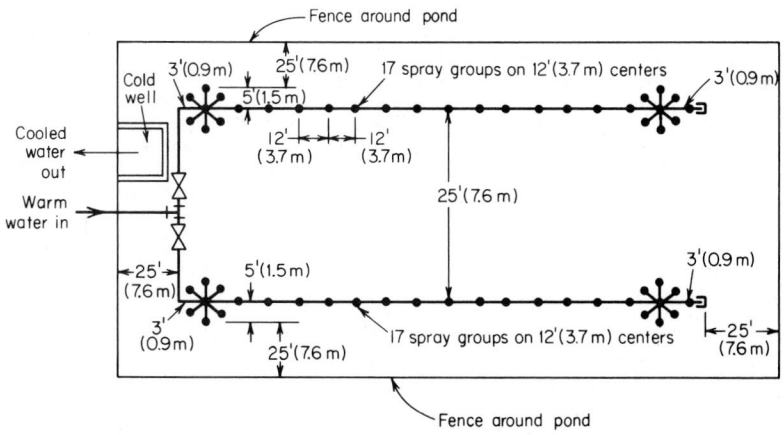

FIG. 41 Spray-pond nozzle and piping layout.

Calculation Procedure:

1.   Compute the number of nozzles required

Assume a water flow of 50 gal/min (3.2 L/s) per nozzle; this is a typical flow rate for usual cooling-pond nozzles. Then the number of nozzles required = (10,000 gal/min)/(50 gal/min per nozzle) = 200 nozzles. If 6 nozzles are used in each spray group, a series of crossed arms, with each arm containing one or more nozzles, then 200 nozzles/6 nozzles per spray group = 33⅓ spray groups will be needed. Since a partial spray group is seldom used, 34 spray groups would be chosen.

2.   Determine the surface area required

Usual design practice is to provide 1 ft² (0.09 m²) of pond area per 250 lb (113.4 kg) of water cooled for water quantities exceeding 1000 gal/min (63.1 L/s). Thus, in this pond, the weight of water cooled = (10,000 gal/min)(8.33 lb/gal)(60 min/h) = 4,998,000, say 5,000,000 lb/h (630.0 kg/s). Then, the area required, given 1 ft² of pond area per 250 lb of water (0.82 m² per 1000 kg) cooled = 5,000,000/250 = 20,000 ft² (1858.0 m²).

As a cross-check, use another commonly accepted area value: 125 Btu/(ft²·°F) [2555.2 kJ/(m²·°C)] is the difference between the air wet-bulb temperature and the warm entering-water temperature. This is the equivalent of (120 − 60)(125) = 7500 Btu/ft² (85,174 kJ/m²) in this spray pond, because the air wet-bulb temperature is 60°F (15.6°C) and the warm-water temperature is 120°F (48.9°C). The heat removed from the water is (lb/h of water)(temperature decrease, °F)(specific heat of water) = (5,000,000)(120 − 90)(1.0) = 150,000,000 Btu/h (43,960.7 kW). Then, area required = (heat removed, Btu/h)/(heat removal, Btu/ft²) = 150,000,000/7500 = 20,000 ft² (1858.0 m²). This checks the previously obtained area value.

3.   Determine the spray-pond dimensions

Spray groups on the same header or pipe main are usually arranged on about 12-ft (3.7-m) centers with the headers or pipe mains spaced on about 25-ft (7.6-m) centers, Fig. 41. Assume that 34 spray groups are used, instead of the required 33⅓, to provide an equal number of groups in two headers and a small extra capacity.

Sketch the spray pond and headers, Fig. 41. This shows that the length of each header will be about 204 ft (62.2 m) because there are seventeen 12-ft (3.7-m) spaces between spray groups in each header. Allowing 3 ft (0.9 m) at each end of a header for fittings and clean-outs gives an overall header length of 210 ft (64.0 m). The distance between headers is 25 ft (7.6 m). Allow 25 ft (7.6 m) between the outer sprays and the edge of the pond. This gives an overall width of 85 ft (25.9 m) for the pond, if we assume the width of each arm in a spray group is 10 ft (3.0 m). The overall length will then be 210 + 25 + 25 = 260 ft (79.2 m). A cold well for the pump suction and suitable valving for control of the incoming water must be provided, as shown in Fig. 41. The water depth in the pond should be 2 to 3 ft (0.6 to 0.9 m).

4.   Compute the total pumping head

The total head, ft of water = static head + friction head + required nozzle head = 10 + 35 + 80(0.434) = 48.5 ft (145.0 kPa) of water. A pump having a total head of at least 50 ft (15.2 m) of water would be chosen for this spray pond. If future expansion of the pond is anticipated, compute the probable total head required at a future date, and choose a pump to deliver that head. Until the pond is expanded, the pump would operate with a throttled discharge. Normal nozzle inlet pressures range from about 6 to 10 lb/in² (41.4 to 69.0 kPa). Higher pressures should not be used, because there will be excessive spray loss and rapid wear of the nozzles.

Related Calculations: Unsprayed cooling ponds cool 4 to 6 lb (1.8 to 2.7 kg) of water from 100 to 70°F/ft² (598.0 to 418.6°C/m²) of water surface. An alternative design rule is to assume that the pond will dissipate 3.5 Btu/(ft²·h) (11.0 W/m²) water surface per degree difference between the wet-bulb temperature of the air and the entering warm water.

DIRECT-CONTACT FEEDWATER HEATER ANALYSIS

Determine the outlet temperature of water leaving a direct-contact open-type feedwater heater if 250,000 lb/h (31.5 kg/s) of water enters the heater at 100°F (37.8°C). Exhaust steam at 10.3 lb/in² (gage) (71.0 kPa) saturated flows to the heater at the rate of 25,000 lb/h (31.5 kg/s). What saving is obtained by using this heater if the boiler pressure is 250 lb/in² (abs) (1723.8 kPa)? Determine the approximate volume of the heater if a 2-min storage capacity is provided in it.

Calculation Procedure:

1. Compute the water outlet temperature

Assume the heater is 90 percent efficient. Then $t_o = t_i w_w + 0.9 w_s h_g / (w_w + 0.9 w_s)$, where $t_o =$ outlet water temperature, °F; $t_i =$ inlet water temperature, °F; $w_w =$ weight of water flowing through heater, lb/h; 0.9 = heater efficiency, expressed as a decimal; $w_s =$ weight of steam flowing to the heater, lb/h; $h_g =$ enthalpy of the steam flowing to the heater, Btu/lb.

For saturated steam at 10.3 lb/in² (gage) (71.0 kPa), or 10.3 + 14.7 = 25 lb/in² (abs) (172.4 kPa), $h_g = 1160.6$ Btu/lb (2599.6 kJ/kg), from the saturation pressure steam tables. Then

$$t_o = \frac{100(250,000) + 0.9(25,000)(1160.6)}{250,000 + 0.9(25,000)} = 187.5°\text{F } (86.4°\text{C})$$

2. Compute the savings obtained by feed heating

The percentage of saving, expressed as a decimal, obtained by heating feedwater is $(h_o - h_i)/(h_b - h_i)$ where h_o and $h_i =$ enthalpy of the water leaving and entering the heater, respectively, Btu/lb; $h_b =$ enthalpy of the steam at the boiler operating pressure, Btu/lb. For this plant, from the steam tables, $h_o - h_i/(h_b - h_i) = 155.44 - 67.97/(1201.1 - 67.97) = 0.077$, or 7.7 percent.

A popular rule of thumb states that for every 11°F (6.1°C) rise in feedwater temperature in a heater, there is approximately a 1 percent saving in the fuel that would otherwise be used to heat the feedwater. Checking the above calculation with this rule of thumb shows reasonably good agreement.

3. Determine the heater volume

With a capacity of W lb/h of water, the volume of a direct-contact or open-type heater can be approximated from $v = W/10,000$, where $v =$ heater internal volume, ft³. For this heater, $v = 250,000/10,000 = 25$ ft³ (0.71 m³).

Related Calculations: Most direct-contact or open feedwater heaters store in 2-min supply of feedwater when the boiler load is constant, and the feedwater supply is all makeup. With little or no makeup, the heater volume is chosen so that there is enough capacity to store 5 to 30 min feedwater for the boiler.

CLOSED FEEDWATER HEATER ANALYSIS AND SELECTION

Analyze and select a closed feedwater heater for the third stage of a regenerative steam-turbine cycle in which the feedwater flow rate is 37,640 lb/h (4.7 kg/s), the desired temperature rise of the water during flow through the heater is 80°F (44.4°C) (from 238 to 318°F or, 114.4 to 158.9°C), bleed heating steam is at 100 lb/in² (abs) (689.5 kPa) and 460°F (237.8°C), drains leave the heater at the saturation temperature corresponding to the heating steam pressure [100 lb/in² (abs) or 689.5 kPa], and ⅝-in (1.6-cm) OD admiralty metal tubes with a maximum length of 6 ft (1.8 m) are used. Use the *Standards of the Bleeder Heater Manufacturers Association, Inc.*, when analyzing the heater.

Calculation Procedure:

1. Determine the LMTD across heater

When heat-transfer rates in feedwater heaters are computed, the average film temperature of the feedwater is used. In computing this, the *Standards of the Bleeder Heater Manufacturers Association* specify that the *saturation temperature* of the heating steam be used. At 100 lb/in² (abs) (689.5 kPa), $t_s = 327.81°$F (164.3°C). Then

$$\text{LMTD} = t_m = \frac{(t_s - t_i) - (t_s - t_o)}{\ln [t_s - t_i/(t_s - t_o)]}$$

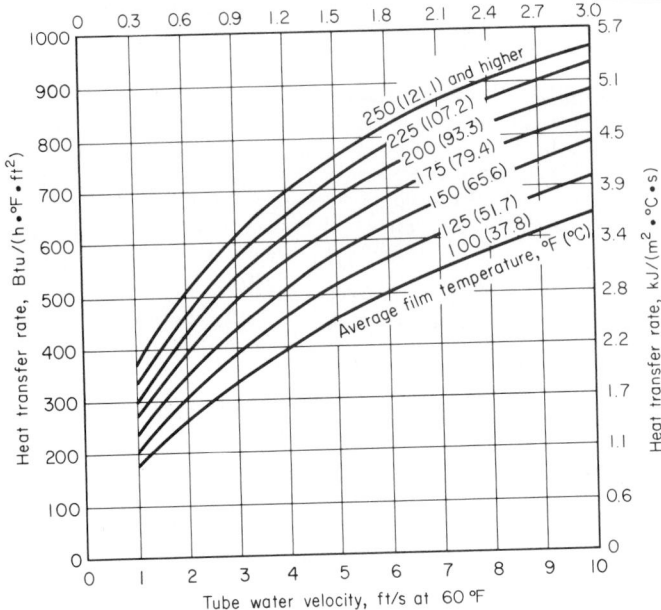

FIG. 42   Heat-transfer rates for closed feedwater heaters. *(Standards of Bleeder Heater Manufacturers Association, Inc.)*

where the symbols are as defined in the previous calculation procedure. Thus,

$$t_m = \frac{(327.81 - 238) - (327.81 - 318)}{\ln [327.81 - 238/(327.81 - 318)]}$$

$$= 36.5°F \ (20.3°C)$$

The average film temperature t_f for any closed heater is then

$$t_f = t_s - 0.8t_m$$

$$= 327.81 - 29.2 = 298.6°F \ (148.1°C)$$

2.   *Determine the overall heat-transfer rate*

Assume a feedwater velocity of 8 ft/s (2.4 m/s) for this heater. This velocity value is typical for smaller heaters handling less than 100,000-lb/h (12.6-kg/s) feedwater flow. Enter Fig. 42 at 8 ft/s (2.4 m/s) on the lower horizontal scale, and project vertically upward to the 250°F (121.1°C) average film temperature curve. This curve is used even though $t_f = 298.6°F$ (148.1°C), because the standards recommend that heat-transfer rates higher than those for a 250°F (121.1°C) film temperature not be used. So, from the 8-ft/s (2.4-m/s) intersection with the 250°F (121.1°C) curve in Fig. 42, project to the left to read $U =$ the overall heat-transfer rate = 910 Btu/(ft^2·°F·h) [5.2 kJ/(m^2·°C·s)].

Next, check Table 5 for the correction factor for U. Assume that no. 18 BWG ⅝-in (1.6-cm) OD arsenical copper tubes are used in this exchanger. Then the correction factor from Table 5 is 1.00, and $U_{corr} = 910(1.00) = 910$. If no. 9 BWG tubes are chosen, $U_{corr} = 910(0.85) = 773.5$ Btu/(ft^2·°F·h) [4.4 kJ/(m^2·°C·s)], given the correction factor from Table 5 for arsenical copper tubes.

TABLE 5   Multipliers for Base Heat-Transfer Rates

[*For tube OD ⅝ to 1 in (1.6 to 2.5 cm) inclusive*]

BWG	As-Cu	Adm	90/10 Cu-Ni	80/20 Cu-Ni	70/30 Cu-Ni	Monel
			Tube material			
18	1.00	1.00	0.97	0.95	0.92	0.89
17	1.00	1.00	0.94	0.91	0.87	0.85
16	1.00	1.00	0.91	0.88	0.84	0.82
15	1.00	0.99	0.89	0.86	0.82	0.79
14	1.00	0.96	0.85	0.82	0.77	0.75
13	0.98	0.93	0.81	0.78	0.73	0.70
12	0.95	0.90	0.77	0.73	0.68	0.65
11	0.92	0.87	0.74	0.70	0.65	0.62
10	0.89	0.83	0.69	0.66	0.60	0.58
9	0.85	0.80	0.65	0.62	0.56	0.54

3. Compute the amount of heat transferred by the heater

The enthalpy of the entering feedwater at 238°F (114.4°C) is, from the saturation-temperature steam table, $h_{fi} = 206.32$ Btu/lb (479.9 kJ/kg). The enthalpy of the leaving feedwater at 318°F (158.9°C) is, from the same table, $h_{fo} = 288.20$ Btu/lb (670.4 kJ/kg). Then the heater transferred H_t Btu/h is $H_t = w_w(h_{fo} - h_{fi})$, where w_w = feedwater flow rate, lb/h. Or, $H_t = 37,640(288.20 - 206.32) = 3,080,000$ Btu/h (902.7 kW).

4. Compute the surface area required in the exchanger

The surface area required A ft^2 = H_t/Ut_m. Then $A = 3,080,000/[(910)(36.5)] = 92.7$ ft^2 (8.6 m^2).

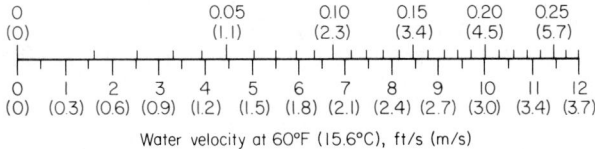

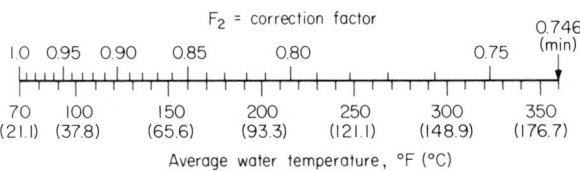

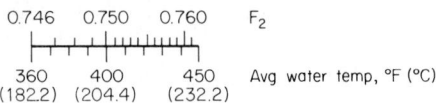

FIG. 43   Correction factors for closed feedwater heaters. (*Standards of Bleeder Heater Manufactuers Association, Inc.*)

5. Determine the number of tubes per pass

Assume the heater has only one pass, and compute the number of tubes required. Once the number of tubes is known, a decision can be made about the number of passes required. In a closed heater, number of tubes $= w_w$ (passes) (ft^3/s per tube)/[v(ft^2 per tube open area)], where $w_w =$ lb/h of feedwater passing through heater; $v =$ feedwater velocity in tubes, ft/s.

Since the feedwater enters the heater at 238°F (114.4°C) and leaves at 318°F (158.9°C), its specific volume at 278°F (136.7°C), midway between t_i and t_o, can be considered the average specific volume of the feedwater in the heater. From the saturation-pressure steam table, $v_f =$ 0.01691 ft^3/lb (0.0011 m^3/kg) at 278°F (136.7°C). Convert this to cubic feet per second per tube by dividing this specific volume by 3600 (number of seconds in 1 h) and multiplying by the pounds per hour of feedwater per tube. Or, ft^3/s per tube $= (0.01691/3600)$(lb/h per tube).

Since no. 18 BWG ⅝-in (1.6-cm) OD tubes are being used, ID $= 0.625 - 2$(thickness) $= 0.625 - 2(0.049) = 0.527$ in (1.3 cm). Then, open area per tube, ft$^2 = (\pi d^2/4)/144 = 0.7854(0.527)^2/144 = 0.001525$ ft^2 (0.00014 m^2) per tube. Alternatively, this area could be obtained from a table of tube properties.

With these data, compute the total number of tubes from number of tubes $= [(37,640)(1)(0.01681/3600)]/[(8)(0.001525)] = 14.49$ tubes.

6. Compute the required tube length

Assume that 14 tubes are used, since the number required is less than 14.5. Then, tube length l, ft $= A/$(number of tubes per pass)(passes)(area per ft of tube). Or, tube length for 1 pass $= 92.7/[(14)(1)(0.1636)] = 40.6$ ft (12.4 m). The area per ft of tube length is obtained from a table of tube properties or computed from 12π(OD)/144 $= 12\pi(0.625)/155 = 0.1636$ ft^2 (0.015 m^2).

7. Compute the actual number of passes and the actual tube length

Since the tubes in this heater cannot exceed 6 ft (1.8 m) in length, the number of passes required $=$ (length for one pass, ft)/(maximum allowable tube length, ft) $= 40.6/6 = 6.77$ passes. Since a fractional number of passes cannot be used and an even number of passes permit a more convenient layout of the heater, choose eight passes.

From the same equation for tube length as in step 6, $l =$ tube length $= 92.7/[(14)(8)(0.1636)] = 5.06$ ft (1.5 m).

8. Determine the feedwater pressure drop through heater

In any closed feedwater heater, the pressure loss Δp lb/in^2 is $\Delta p = F_1 F_2 (L + 5.5D)N/D^{1.24}$, where $\Delta p =$ pressure drop in the feedwater passing through the heater, lb/in^2; F_1 and $F_2 =$ correction factors from Fig. 43; $L =$ total lin ft of tubing divided by the number of tube holes in one tube sheet; $D =$ tube ID; $N =$ number of passes. In finding F_2, the average water temperature is taken as $t_s - t_m$.

For this heater, using correction factors from Fig. 43,

$$\Delta p = (0.136)(0.761)\left[\frac{5.06(8)(14)}{(8)(14)} + 5.5(0.527)\right]\frac{8}{0.527^{1.24}}$$

$$= 14.6 \text{ lb/in}^2 \text{ (100.7 kPa)}$$

9. Find the heater shell outside diameter

The total number of tubes in the heater $=$ (number of passes)(tubes per pass) $= 8(14) = 112$ tubes. Assume that there is ⅜-in (1.0-cm) clearance between each tube and the tube alongside, above, or below it. Then the pitch or center-to-center distance between tubes $=$ pitch $+$ tube OD $= $ ⅜ $+$ ⅝ $= 1$ in (2.5 cm).

The number of tubes per ft^2 of tube sheet $= 166/$(pitch)^2, or $166/1^2 = 166$ tubes per ft^2 (1786.8 per m^2). Since the heater has 112 tubes, the area of the tube sheet $= 112/166 = 0.675$ ft^2, or 97 in^2 (625.8 cm^2).

The inside diameter of the heater shell $=$ (tube sheet area, in^2/0.7854)$^{0.5} = (97/0.7854)^{0.5} = 11.1$ in (28.2 cm). With a 0.25-in (0.6-cm) thick shell, the heater shell OD $= 11.1 + 2(0.25) = 11.6$ in (29.5 cm).

10. *Compute the quantity of heating steam required*

Steam enters the heater at 100 lb/in² (abs) (689.5 kPa) and 460°F (237.8°C). The enthalpy at this pressure and temperature is, from the superheated steam table, h_g = 1258.8 Btu/lb (2928.0 kJ/kg). The steam condenses in the heater, leaving as condensate at the saturation temperature corresponding to 100 lb/in² (abs) (689.5 kPa), or 327.81°F (164.3°C). The enthalpy of the saturated liquid at this temperature is, from the steam tables, h_f = 298.4 Btu/lb (694.1 kJ/kg).

The heater steam consumption for any closed-type feedwater heater is W, lb/h = $w_w(\Delta t)(h_g - h_f)$, where Δt = temperature rise of feedwater in heater, °F; c = specific heat of feedwater,

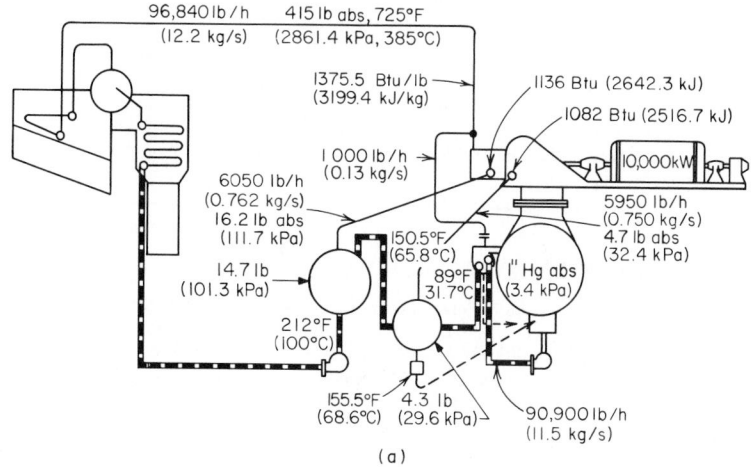

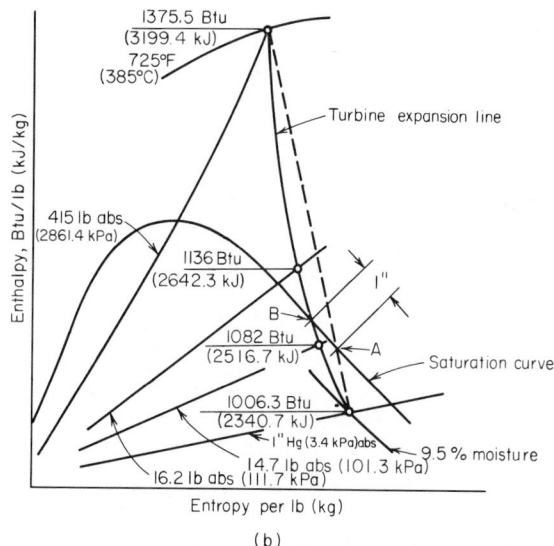

FIG. 44 (*a*) Two stages of feedwater heating in a steam plant; (*b*) Mollier chart of the cycle in (*a*).

Btu/(lb·°F). Assume $c = 1.00$ for the temperature range in this heater, and $W = (37,640)(318 - 238)(1.00)/(1258.8 - 298.4) = 3140$ lb/h (0.40 kg/s).

Related Calculations: The procedure used here can be applied to closed feedwater heaters in stationary and marine service. A similar procedure is used for selecting hot-water heaters for buildings, marine, and portable service. Various authorities recommend the following terminal difference (heater condensate temperature minus the outlet feedwater temperature) for closed feedwater heaters:

Feedwater outlet temperature		Terminal difference	
°F	°C	°F	°C
86 to 230	30.0 to 110.0	5	2.8
230 to 300	110.0 to 148.9	10	5.6
300 to 400	148.9 to 204.4	15	8.3
400 to 525	204.4 to 273.9	20	11.1

POWER-PLANT HEATER EXTRACTION-CYCLE ANALYSIS

A steam power plant operates at a boiler-drum pressure of 460 lb/in² (abs) (3171.7 kPa), a turbine throttle pressure of 415 lb/in² (abs) (2861.4 kPa) and 725°F (385.0°C), and a turbine capacity of 10,000 kW (or 13,410 hp). The Rankine-cycle efficiency ratio (including generator losses) is: full load, 75.3 percent; three-quarters load, 74.75 percent; half load, 71.75 percent. The turbine exhaust pressure is 1 inHg absolute (3.4 kPa); steam flow to the steam-jet air ejector is 1000 lb/h (0.13 kg/s). Analyze this cycle to determine the possible gains from two stages of extraction for feedwater heating, with the first stage a closed heater and the second stage a direct-contact or mixing heater. Use engineering-office methods in analyzing the cycle.

Calculation Procedure:

1. Sketch the power-plant cycle

Figure 44a shows the plant with one closed heater and one direct-contact heater. Values marked on Fig. 44a will be computed as part of this calculation procedure. Enter each value on the diagram as soon as it is computed.

2. Compute the throttle flow without feedwater heating extraction

Use the superheated steam tables to find the throttle enthalpy $h_f = 1375.5$ Btu/lb (3199.4 kJ/kg) at 415 lb/in² (abs) (2861.4 kPa) and 725°F (385.0°C).

Assume an irreversible adiabatic expansion between throttle conditions and the exhaust pressure of 1 inHg (3.4 kPa). Compute the final enthalpy H_{2s} by the same method used in earlier calculation procedures by finding y_{2s}, the percentage of moisture at the exhaust conditions with 1-inHg absolute (3.4-kPa) exhaust pressure. Do this by setting up the ratio $y_{2s} = (s_y - S_1)/s_{fg}$, where s_g and s_{fg} are entropies at the exhaust pressure; S_1 is entropy at throttle conditions. From the steam tables, $y_{2s} = 2.0387 - 1.6468/1.9473 = 0.201$. Then $H_{2s} = h_g - y_{2s}h_{fg}$, where h_g and h_{fg} are enthalpies at 1 inHg absolute (3.4 kPa). Substitute values from the steam table for 1 inHg absolute (3.4 kPa); or, $H_{2s} = 1096.3 - 0.201(1049.2) = 885.3$ Btu/lb (2059.2 kJ/kg).

The available energy in this irreversible adiabatic expansion is the difference between the throttle and exhaust conditions, or $1375.5 - 885.3 = 490.2$ Btu/lb (1140.2 kJ/kg). The work at full load on the turbine is: (Rankine-cycle efficiency)(adiabatic available energy) = $(0.753)(490.2) = 369.1$ Btu/lb (858.5 kJ/kg). Enthalpy at the exhaust of the actual turbine = throttle enthalpy minus full-load actual work, or $1375.5 - 369.1 = 1006.4$ Btu/lb (2340.9 kJ/kg). Use the Mollier chart to find, at 1.0 inHg absolute (3.4 kPa) and 1006.4 Btu/lb (2340.9 kJ/kg), that the exhaust steam contains 9.5 percent moisture.

Now the turbine steam rate SR = 3413(actual work output, Btu). Or, SR = 3413/369.1 =

9.25 lb/kWh (4.2 kg/kWh). With the steam rate known, the nonextraction throttle flow is (SR)(kW output) = 9.25(10,000) = 92,500 lb/h (11.7 kg/s).

3. *Determine the heater extraction pressures*

With steam extraction from the turbine for feedwater heating, the steam flow to the main condenser will be reduced, even with added throttle flow to compensate for extraction.

Assume that the final feedwater temperature will be 212°F (100.0°C) and that the heating range for each heater is equal. Both assumptions represent typical practice for a moderate-pressure cycle of the type being considered.

Feedwater leaving the condenser hotwell at 1 inHg absolute (3.4 kPa) is at 79.03°F (26.1°C). This feedwater is pumped through the air-ejector intercondensers and aftercondensers, where the condensate temperature will usually rise 5 to 15°F (2.8 to 8.3°C), depending on the turbine load. Assume that there is a 10°F (5.6°C) rise in condensate temperature from 79 to 89°F (26.1 to 31.7°C). Then the temperature range for the two heaters is 212 − 89 = 123°F (68.3°C). The temperature rise per heater is 123/2 = 61.5°F (34.2°C), since there are two heaters and each will have the same temperature rise. Since water enters the first-stage closed heater at 89°F (31.7°C), the exit temperature from this heater is 89 + 61.5 = 150.5°F (65.8°C).

The second-stage heater is a direct-contact unit operating at 14.7 lb/in² (abs) (101.4 kPa), because this is the saturation pressure at an outlet temperature of 212°F (100.0°C). Assume a 10 percent pressure drop between the turbine and heater steam inlet. This is a typical pressure loss for an extraction heater. Extraction pressure for the second-stage heater is then 1.1(14.7) = 16.2 lb/in² (abs) (111.7 kPa).

Assume a 5°F (2.8°C) terminal difference for the first-stage heater. This is a typical terminal difference, as explained in an earlier calculation procedure. The saturated steam temperature in the heater equals the condensate temperature = 150.5°F (65.8°C) exit temperature + 5°F (2.8°C) terminal difference = 155.5°F (68.6°C). From the saturation-temperature steam table, the pressure at 155.5°F (68.6°C) is 4.3 lb/in² (abs) (29.6 kPa). With a 10 percent pressure loss, the extraction pressure = 1.1(4.3) = 4.73 lb/in² (abs) (32.6 kPa).

4. *Determine the extraction enthalpies*

To establish the enthalpy of the extracted steam at each stage, the actual turbine-expansion line must be plotted. Two points—the throttle inlet conditions and the exhaust conditions—are known. Plot these on a Mollier chart, Fig. 44b. Connect these two points by a dashed straight line, Fig. 44b.

Next, measure along the saturation curve 1 in (2.5 cm) from the intersection point A back toward the enthalpy coordinate, and locate point B. Now draw a gradually sloping line from the throttle conditions to point B; from B increase the slope to the exhaust conditions. The enthalpy of the steam at each extraction point is read where the lines of constant pressure cross the expansion line. Thus, for the second-stage direct-contact heater where p = 16.2 lb/in² (abs) (111.7 kPa), h_g = 1136 Btu/lb (2642.3 kJ/kg). For the first-stage closed heater where p = 4.7 lb/in² (abs) (32.4 kPa), h_g = 1082 Btu/lb (2516.7 kJ/kg).

When the actual expansion curve is plotted, a steeper slope is used between the throttle superheat conditions and the saturation curve of the Mollier chart, because the turbine stages using superheated steam (stages above the saturation curve) are more efficient than stages using wet steam (stages below the saturation curve).

5. *Compute the extraction steam flow*

To determine the extraction flow rates, two assumptions must be made—condenser steam flow rate and first-stage closed-heater extraction flow rate. The complete cycle will be analyzed, and the assumption checked. If the assumptions are incorrect, new values will be assumed, and the cycle analyzed again.

Assume that the condenser steam flow from the turbine is 84,000 lb/h (10.6 kg/s) when it is operating with extraction. Note that this value is less than the nonextraction flow of 92,500 lb/h (11.7 kg/s). The reason is that extraction of steam will reduce flow to the condenser because the steam is bled from the turbine after passage through the throttle but before the condenser inlet.

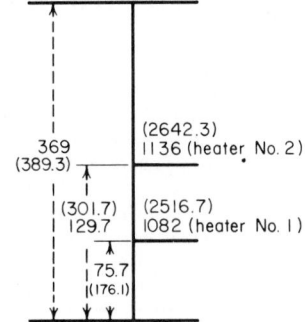

Enthalpy at throttle 1375.3 Btu/lb (3198.9 kJ/kg)

369
(389.3)

(2642.3)
1136 (heater No. 2)

(301.7)
129.7

(2516.7)
1082 (heater No. 1)

75.7
(176.1)

Enthalpy at exhaust 1006.3 Btu/lb (2340.7 kJ/kg)

FIG. 45   Diagram of turbine-expansion line.

Then, for the first-stage closed heater, condensate flow is as follows:

From condenser	84,000 lb/h (10.6 kg/s) assumed
From steam-jet ejector	1,000 lb/h (0.13 kg/s)
From first-stage heater	5,900 lb/h (0.74 kg/s) assumed
Total	90,900 lb/h (11.5 kg/s)

The value of 5900 lb/h (0.74 kg/s) of condensate from the first-stage heater is the second assumption made. Since it will be checked later, an error in the assumption can be detected.

Assume a 2 percent heat radiation loss between the turbine and heater. This is a typical loss. Then

Steam enthalpy at heater = 1082(0.98)	= 1060.4 Btu/lb (2465.5 kJ/kg)
Enthalpy of condensate at 155.5°F (68.6°C)	= −123.4 Btu/lb (−287.0 kJ/kg)
Heat given up per lb (kg) of steam condensed	= 937.0 Btu/lb (2179.5 kJ/kg)
Enthalpy of feedwater at 150.5°F (65.8°C)	= 118.3 Btu/lb (275.2 kJ/kg)
Enthalpy of feedwater to heater at 89°F (31.7°C)	= −57.0 Btu/lb (−132.6 kJ/kg)
Heat absorbed by feedwater	= 61.3 Btu/lb (142.6 kJ/kg)

Required extraction = (total condensate flow, lb/h) [(heat absorbed by feedwater, Btu/lb)/ (heat given up per lb of steam condensed, Btu/lb)], or required extraction = (90,900)(61.3/ 937) = 5950 lb/h (0.75 kg/s)

Compare the required extraction, 5950 lb/h (0.75 kg/s), with the assumed extraction, 5900 lb/h (0.74 kg/s). The difference is only 50 lb/h (0.006 kg/s), which is less than 1 percent. Therefore, the assumed flow rate is satisfactory, because estimates within 1 percent are considered sufficiently accurate for all routine analyses.

For the second-stage direct-contact heater, condensate flow, lb/h, is as follows:

From the first-stage heater	90,900 lb/h (11.5 kg/s)
Steam enthalpy at heater = 1135(0.98)	= 1112.3 Btu/lb (2587.2 kJ/kg)
Enthalpy of condensate at 212°F (100.0°C)	= −180.0 Btu/lb (−418.7 kJ/kg)
Heat given up per lb of steam condensed	= 932.3 Btu/lb (2168.5 kJ/kg)
Enthalpy of feedwater at 212°F (100.0°C)	= 180.0 Btu/lb (418.7 kJ/kg)
Enthalpy of feedwater at 150.5°F (65.8°C)	= 118.3 Btu/lb (275.2 kJ/kg)
Heat absorbed by feedwater	= 61.7 Btu/lb (143.5 kJ/kg)

The required extraction, calculated in the same way as for the first-stage heater, is (90,900)(61.7/932.2) = 6050 lb/h (0.8 kg/s).

The computed extraction flow for the second-stage heater is not compared with an assumed value because an assumption was not necessary.

6. Compute the actual condenser steam flow

Sketch a vertical line diagram, Fig. 45, showing the enthalpies at the throttle, heaters, and exhaust. From this diagram, the work lost by the extracted steam can be computed. As Fig. 45 shows, the total enthalpy drop from the throttle to the exhaust is 369 Btu/lb (389.3 kJ/kg). Each pound of extracted steam from the first- and second-stage bleed points causes a work loss of 75.7 Btu/lb (176.1 kJ/kg) and 129.7 Btu/lb (301.7 kJ/kg), respectively. To carry the same load, 10,000 kW, with extraction, it will be necessary to supply the following additional compensation steam to the turbine throttle: (heater flow, lb/h)(work loss, Btu/h)/(total work, Btu/h). Then

	lb/h	kg/s
First-stage closed heater:		
(5950)(75.7/369)	1220	0.15
Second-stage direct-contact heater:		
(6050)(129.7/369)	2120	0.27
Total additional throttle flow to compensate for extraction	3340	0.42

Check the assumed condenser flow using nonextraction throttle flow + additional throttle flow − heater extraction = condenser flow. Set up a tabulation of the flows as follows:

Flow	lb/h	kg/s
Throttle; nonextraction	92,500	11.65
Added flow (compensation)	3,340	0.42
Throttle; extraction	95,840	12.07
Extraction (5950 + 6050)	−12,000	−1.51
Condenser flow	83,840	10.56

Compare this actual flow, 83,840 lb/h (10.6 kg/s), with the assumed flow, 84,000 lb/h (10.6 kg/s). The difference, 160 lb/h (0.02 kg/s), is less than 1 percent. Since an accuracy within 1 percent is sufficient for all normal power-plant calculations, it is not necessary to recompute the cycle. Had the difference been greater than 1 percent, a new condenser flow would be assumed and the cycle recomputed. Follow this procedure until a difference of less than 1 percent is obtained.

7. Determine the economy of the extraction cycle

For a nonextraction cycle operating in the same pressure range,

	Btu/lb	kJ/kg
Enthalpy of throttle steam	1375.3	3198.9
Enthalpy of condensate at 79°F		
(26.1°C)	−47.0	−109.3
Heat supplied by boiler	1328.3	3089.6

Heat chargeable to turbine = (throttle flow + air-ejector flow)(heat supplied by boiler)/(kW output of turbine) = (92,500 + 1000)(1328.3)/10,000 = 12,410 Btu/kWh (13,093.2 kJ/kWh), which is the actual heat rate HR of the nonextraction cycle.

For the extraction cycle using two heaters,

	Btu/lb	kJ/kg
Enthalpy of throttle steam	1375.3	3198.9
Enthalpy of feedwater leaving second heater	−180.0	−418.7
Heat supplied by boiler	1195.3	2780.3

As before, heat chargeable to turbine = (95,840 + 1000)(1195.3)/10,000 = 11,580 Btu/kWh (12,217.5 kJ/kWh). Therefore, the improvement = (nonextraction HR − extraction HR)/nonextraction HR = (12,410 − 11,580)/12,410 = 0.0662, or 6.62 percent.

Related Calculations: (1) To determine the percent improvement in a steam cycle resulting from additional feedwater heaters in the cycle, use the same procedure as given above for three, four, five, six, or more heaters. Plot the percent improvement vs. number of stages of extraction, Fig. 46, to observe the effect of additional heaters. A plot of this type shows the decreasing gains made by additional heaters. Eventually the gains become so small that the added expenditure for an additional heater cannot be justified.

(2) Many simple marine steam plants use only two stages of feedwater heating. To analyze such a cycle, use the procedure given, substituting the hp output for the kW output of the turbine.

(3) Where a marine plant has more than two stages of feedwater heating, follow the procedure given in (1) above.

STEAM BOILER, ECONOMIZER, AND AIR-HEATER EFFICIENCY

Determine the overall efficiency of a steam boiler generating 56,000 lb/h (7.1 kg/s) of 600 lb/in² (abs) (4137.0 kPa) 800°F (426.7°C) steam. The boiler is continuously blown down at the rate of 2500 lb/h (0.31 kg/s). Feedwater enters the economizer at 300°F (148.9°C). The furnace burns 5958 lb/h (0.75 kg/s) of 13,100-Btu/lb (30,470.6-kJ/kg) HHV (higher heating value) coal having an ultimate analysis of 68.5 percent C, 5 percent H₂, 8.9 percent O₂, 1.2 percent N₂, 3.2 percent S, 8.7 percent ash, and 4.5 percent moisture. Air enters the boiler at 63°F (17.2°C) dry-bulb and 56°F (13.3°C) wet-bulb temperature, with 56 gr of vapor per lb (123.5 gr/kg) of dry air. Carbon in the fuel refuse is 7 percent, refuse is 0.093 lb/lb (0.2 kg/kg) of fuel. Feedwater leaves the economizer at 370°F (187.8°C). Flue gas enters the economizer at 850°F (454.4°C) and has an analysis of 15.8 percent CO₂, 2.8 percent O₂, and 81.4 percent N₂. Air enters the air heater at 63°F (17.2°C) with 56 gr/lb (123.5 gr/kg) of dry air; air leaves the heater at 480°F (248.9°C). Gas enters the air heater at 570°F (298.9°C), and 14 percent of the air to the furnace comes from the mill fan. Determine the steam generator overall efficiency, economizer efficiency, and air-heater efficiency. Figure 47 shows the steam generator and the flow factors that must be considered.

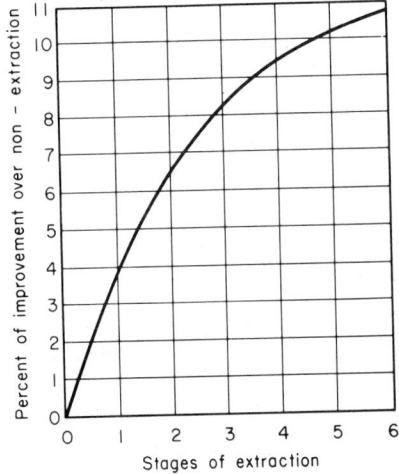

FIG. 46 Percentage of improvement in turbine heat rate vs. stages of extraction.

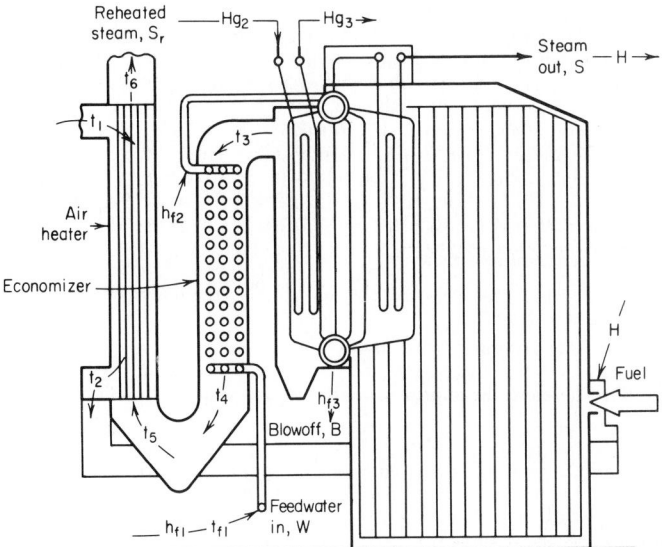

FIG. 47 Points in a steam generator where temperatures and enthalpies are measured in determining the boiler efficiency.

Calculation Procedure:

1. Determine the boiler output

The boiler output $= S(h_g - h_{f1}) + S_r(h_{g3} - h_{g2}) + B(h_{f3} - h_{f1})$, where S = steam generated, lb/h; h_g = enthalpy of the generated steam, Btu/lb; h_{f1} = enthalpy of inlet feedwater; S_r = reheated steam flow, lb/h (if any); h_{g3} = outlet enthalpy of reheated steam; h_{g2} = inlet enthalpy of reheated steam; B = blowoff, lb/h; h_{f3} = blowoff enthalpy, where all enthalpies are in Btu/lb. Using the appropriate steam table and deleting the reheat factor because there is no reheat, we get boiler output $= 56,000(1407.7 - 269.6) + 2500(471.6 - 269.6) = 64,238,600$ Btu/h (18,826.5 kW).

2. Compute the heat input to the boiler

The boiler input $= FH$, where F = fuel input, lb/h (as fired); H = higher heating value, Btu/lb (as fired). Or, boiler input $= 5958(13,100) = 78,049,800$ Btu/h (22,874.1 kW).

3. Compute the boiler efficiency

The boiler efficiency $=$ (output, Btu/h)/(input, Btu/h) $= 64,238,600/78,049,800 = 0.822$, or 82.2 percent.

4. Determine the heat absorbed by the economizer

The heat absorbed by the economizer, Btu/h $= w_w(h_{f2} - h_{f1})$, where w_w = feedwater flow, lb/h; h_{f2} and h_{f1} = enthalpies of feedwater leaving and entering the economizer, respectively, Btu/lb. For this economizer, with the feedwater leaving the economizer at 370°F (187.8°C) and entering at 300°F (148.9°C), heat absorbed $= (56,000 + 2500)(342.79 - 269.59) = 4,283,000$ Btu/h (1255.2 kW). Note that the total feedwater flow w_w is the sum of the steam generated and the continuous blowdown rate.

5. Compute the heat available to the economizer

The heat available to the economizer, Btu/h $= H_g F$, where H_g = heat available in flue gas, Btu/lb of fuel = heat available in dry gas + heat available in flue-gas vapor, Btu/lb of fuel = $(t_3 - t_{f1})(0.24G) + (t_3 - t_{f1})(0.46)\{M_f + 8.94H_2 + M_a[G - C_b - N_2 - 7.94(H_2 - O_2/8)]\}$, where

$G = \{[11CO_2 + 8O_2 + 7(N_2 + CO)]/[3(CO_2 + CO)]\}(C_b + S/2.67) + S/1.60$; $M_f =$ lb of moisture per lb fuel burned; $M_a =$ lb of moisture per lb of dry air to furnace; $C_b =$ lb of carbon burned per lb of fuel burned $= C - RC_r$; $C_r =$ lb of combustible per lb of refuse; $R =$ lb of refuse per lb of fuel; H_2, N_2, C, O_2, $S =$ lb of each element per lb of fuel, as fired; CO_2, CO, O_2, $N_2 =$ percentage parts of volumetric analysis of dry combustion gas entering the economizer. Substituting gives $C_b = 0.685 - (0.093)(0.07) = 0.678$ lb/lb (0.678 kg/kg) fuel; $G = [11(0.158) + 8(0.028) + 7(0.814)]/[3(0.158)] \times (0.678 + 0.032/2.67) + 0.032/1.60$; $G = 11.18$ lb/lb (11.18 kg/kg) fuel. $H_g = (800 - 300)(0.24) \times (11.18) + (800 - 300)(0.46)\{0.045 + (8.9)(0.05) + 56/7000[11.18 - 0.678 - 0.012 - 7.94 \times (0.05 - 0.089/8)]\}$; $H_g = 1473$ Btu/lb (3426.2 kJ/kg) fuel. Heat available $= H_gF = (1473)(5958) = 8,770,000$ Btu/h (2570.2 kW).

6. Compute the economizer efficiency

The economizer efficiency $=$ (heat absorbed, Btu/h)/(heat available, Btu/h) $= 4,283,000/8,770,000 = 0.488$, or 48.8 percent.

7. Compute the heat absorbed by air heater

The heat absorbed by the air heater, Btu/lb of fuel, $= A_h(t_2 - t_1)(0.24 + 0.46M_a)$, where $A_h =$ air flow through heater, lb/lb fuel $= A - A_m$; $A =$ total air to furnace, lb/lb fuel $= G - C_b - N_2 - 7.94(H_2 - O_2/8)$; $G =$ similar to economizer but based on gas at the furnace exit; $A_m =$ external air supplied by the mill fan or other source, lb/lb of fuel. Substituting shows $G = [11(0.16) + 8(0.26) + 7(0.184)]/[3(0.16)](0.678 + 0.032/2.67) + 0.032/1.60$; $G = 11.03$ lb/lb (11.03 kg/kg) fuel; $A = 11.03 - 0.69 - 0.012 - 7.94(0.05 - 0.089/8)$; $A = 10.02$ lb/lb (10.02 kg/kg) fuel. Heat absorbed $= (1 - 0.15)(10.02)(480 - 63)(0.24 + 56/7000) = 865.5$ Btu/lb (2013.2 kJ/kg) fuel.

8. Compute the heat available to the air heater

The heat available to the air heater, Btu/h $= (t_5 - t_1)0.24G + (t_5 - t_1)0.46(M_f + 8.94H_2 + M_aA)$. In this relation, all symbols are the same as for the economizer except that G and A are based on the gas entering the heater. Substituting gives $G = [11(0.15) + 8(0.036) + 7(0.814)]/[3(0.15)](0.678 + 0.032/2.67) + 0.032/1.60$; $G = 11.72$ lb/lb (11.72 kg/kg) fuel. And $A = 11.72 - 0.69 - 0.012 - 7.94(0.05 - 0.089/8) = 10.71$ lb/lb (10.71 kg/kg) fuel. Heat available $= (570 - 3)(0.24)(11.72) + (570 - 63)(0.46)[0.045 + 8.94(0.05) + 56/7000(10.71)] = 1561$ Btu/lb (3630.9 kJ/kg).

9. Compute the air-heater efficiency

The air-heater efficiency $=$ (heat absorbed, Btu/lb fuel)/(heat available, Btu/lb fuel) $= 865.5/1561 = 0.554$, or 55.4 percent.

Related Calculations: The above procedure is valid for all types of steam generators, regardless of the kind of fuel used. Where oil or gas is the fuel, alter the combustion calculations to reflect the differences between the fuels. Further, this procedure is also valid for marine and portable boilers.

FIRE-TUBE BOILER ANALYSIS AND SELECTION

Determine the heating surface in an 84-in (213.4-cm) diameter fire-tube boiler 18 ft (5.5 m) long having 84 tubes of 4-in (10.2-cm) ID if 25 percent of the upper shell ends are heat-insulated. How much steam is generated if the boiler evaporates 34.5 lb/h of water per 12 ft² [3.9 g/(m²·s)] of heating surface? How much heat is added by the boiler if it operates at 200 lb/in² (abs) (1379.0 kPa) with 200°F (93.3°C) feedwater? What is the factor of evaporation for this boiler? How much hp is developed by the boiler if 7,000,000 Btu/h (2051.4 kW) is delivered to the water?

Calculation Procedure:

1. Compute the shell area exposed to furnace gas

Shell area $= \pi DL(1 - 0.25)$, where $D =$ boiler diameter, ft; $L =$ shell length, ft; $1 - 0.25$ is the portion of the shell in contact with the furnace gas. Then shell area $= \pi(84/12)(18)(0.75) = 297$ ft² (27.0 m²).

2.   Compute the tube area exposed to furnace gas

Tube area $= \pi dLN$, where $d =$ tube ID, ft; $L =$ tube length, ft; $N =$ number of tubes in boiler. Substituting gives tube area $= \pi(4/12)(18)(84) = 1583$ ft^2 (147.1 m^2).

3.   Compute the head area exposed to furnace gas

The area exposed to furnace gas is twice (since there are *two* heads) the exposed head area minus twice the area occupied by the tubes. The exposed head area is (total area)$(1 -$ portion covered by insulation, expressed as a decimal). Substituting, we get $2\pi D^2/4 - (2)(84)\pi d^2/4 = 2\pi/4(84/12)^2(0.75) - (2)(84)\pi(4/12)^2/4 =$ head area $= 43.1$ ft^2 (4.0 m^2).

4.   Find the total heating surface

The total heating surface of any fire-tube boiler is the sum of the shell, tube, and head areas, or $297.0 + 1583 + 43.1 = 1923$ ft^2 (178.7 m^2), total heating surface.

5.   Compute the quantity of steam generated

Since the boiler evaporates 34.5 lb/h of water per 12 ft^2 [3.9 g/(m$^2\cdot$s)] of heating surface, the quantity of steam generated $= 34.5$ (total heating surface, ft^2)$/12 = 34.5(1923.1)/12 = 5200$ lb/h (0.66 kg/s).

Note: Evaporation of 34.5 lb/h (0.0043 kg/s) from and at 212°F (100.0°C) is the definition of the now-discarded term *boiler horsepower*. However, this term is still met in some engineering examinations and is used by some manufacturers when comparing the performance of boilers. A term used in lieu of boiler horsepower, with the same definition, is *equivalent evaporation*. Both terms are falling into disuse, but they are included here because they still find some use today.

6.   Determine the heat added by the boiler

Heat added, Btu/lb of steam $= h_g - h_{f1}$; from steam table values $1198.4 - 167.99 = 1030.41$ Btu/lb (2396.7 kJ/kg). An alternative way of computing heat added is $h_g -$ (feedwater temperature, °F, $- 32$), where 32 is the freezing temperature of water on the Fahrenheit scale. By this method, heat added $= 1198.4 - (200 - 32) = 1030.4$ Btu/lb (2396.7 kJ/kg). Thus, both methods give the same results in this case. In general, however, use of steam table values is preferred.

7.   Compute the factor of evaporation

The factor of evaporation is used to convert from the actual to the equivalent evaporation, defined earlier. Or, factor of evaporation $=$ (heat added by boiler, Btu/lb)$/970.3$, where 970.3 Btu/lb (2256.9 kJ/kg) is the heat added to develop 1 boiler hp (bhp) (0.75 kW). Thus, the factor of evaporation for this boiler $= 1030.4/970.3 = 1.066$.

8.   Compute the boiler hp output

Boiler hp $=$ (actual evaporation, lb/h) (factor of evaporation)$/34.5$. In this relation, the actual evaporation must be computed first. Since the furnace delivers 7,000,000 Btu/h (2051.5 kW) to the boiler water and the water absorbs 1030.4 Btu/lb (2396.7 kJ/kg) to produce 200-lb/in^2 (abs) (1379.0-kPa) steam with 200°F (93.3°C) feedwater, the steam generated, lb/h $=$ (total heat delivered, Btu/h)/(heat absorbed, Btu/lb) $= 7,000,000/1030.4 = 6670$ lb/h (0.85 kg/s). Then boiler hp $= (6760)(1.066)/34.5 = 209$ hp (155.9 kW).

The rated hp output of horizontal fire-tube boilers with separate supporting walls is based on 12 ft^2 (1.1 m^2) of heating surface per boiler hp. Thus, the rated hp of this boiler $= 1923.1/12 = 160$ hp (119.3 kW). When producing 209 hp (155.9 kW), the boiler is operating at 209/160, or 1.305 times its normal rating, or $(100)(1.305) = 130.5$ percent of normal rating.

Note: Today most boiler manufacturers rate their boilers in terms of pounds per hour of steam generated at a stated pressure. Use this measure of boiler output whenever possible. Inclusion of the term *boiler hp* in this handbook does not indicate that the editor favors or recommends its use. Instead, the term was included to make the handbook as helpful as possible to users who might encounter the term in their work.

SAFETY-VALVE STEAM-FLOW CAPACITY

How much saturated steam at 150 lb/in^2 (abs) (1034.3 kPa) can a 2.5-in (6.4-cm) diameter safety valve having a 0.25-in (0.6-cm) lift pass if the discharge coefficient of the valve c_d is 0.75? What is the capacity of the same valve if the steam is superheated 100°F (55.6°C) above its saturation temperature?

Calculation Procedure:

1. Determine the area of the valve annulus

Annulus area, in^2 = A = πDL, where D = valve diameter, in; L = valve lift, in. Annulus area = $\pi(2.5)(0.25)$ = 1.966 in^2 (12.7 cm^2).

2. Compute the ideal flow for this safety valve

Ideal flow F_i lb/s for any safety valve handling saturated steam is F_i = $p_s^{0.97} A/60$, where p_s = saturated-steam pressure, lb/in^2 (abs). For this valve, F_i = $(150)^{0.97}$ $(1.966)/60$ = 4.24 lb/s (1.9 kg/s).

3. Compute the actual flow through the valve

Actual flow F_a = $F_i c_d$ = $(4.24)(0.75)$ = 3.18 lb/s (1.4 kg/s) = $(3.18)(3600$ s/h) = 11,448 lb/h (1.44 kg/s).

4. Determine the superheated-steam flow rate

The ideal superheated-steam flow F_{is} lb/s is F_{is} = $p_s^{0.97} A/[60(1 + 0.0065t_s)]$, where t_s = superheated temperature, above saturation temperature, °F. Then F_{is} = $(150)^{0.97}(1.966)/[60(1 + 0.0065 \times 100)]$ = 3.96 lb/s (1.8 kg/s). The actual flow is F_{as} = $F_{is} c_d$ = $(3.96)(0.75)$ = 2.97 lb/s (1.4 kg/s) = $(2.97)(3600)$ = 10,700 lb/h (1.4 kg/s).

Related Calculations: Use this procedure for safety valves serving any type of stationary or marine boiler.

SAFETY-VALVE SELECTION FOR A WATERTUBE STEAM BOILER

Select a safety valve for a watertube steam boiler having a maximum rating of 100,000 lb/h (12.6 kg/s) at 800 lb/in^2 (abs) (5516.0 kPa) and 900°F (482.2°C). Determine the valve diameter, size of boiler connection for the valve, opening pressure, closing pressure, type of connection, and valve material. The boiler is oil-fired and has a total heating surface of 9200 ft^2 (854.7 m^2) of which 1000 ft^2 (92.9 m^2) is in waterwall surface. Use the ASME *Boiler and Pressure Vessel Code* rules when selecting the valve. Sketch the escape-pipe arrangement for the safety valve.

Calculation Procedure:

1. Determine the minimum valve relieving capacity

Refer to the latest edition of the *Code* for the relieving-capacity rules. Recent editions of the *Code* require that the safety valve have a *minimum* relieving capacity based on the pounds of steam generated per hour per square foot of boiler heating surface and waterwall heating surface. In the edition of the *Code* used in preparing this handbook, the relieving requirement for oil-fired boilers was 10 lb/(ft$^2 \cdot$h) of steam [13.6 g/(m$^2 \cdot$s)] of boiler heating surface, and 16 lb/(ft$^2 \cdot$h) of steam [21.9 g/(m$^2 \cdot$s)] of waterwall surface. Thus, the minimum safety-valve relieving capacity for this boiler, based on total heating surface, would be $(8200)(10) + (1000)(16)$ = 92,000 lb/h (11.6 kg/s). In this equation, 1000 ft^2 (92.9 m^2) of waterwall surface is deducted from the total heating surface of 9200 ft^2 (854.7 m^2) to obtain the boiler heating surface of 8200 ft^2 (761.8 m^2).

The minimum relieving capacity based on total heating surface is 92,000 lb/h (11.6 kg/s); the maximum rated capacity of the boiler is 100,000 lb/h (12.6 kg/s). Since the *Code* also requires that "the safety valve or valves will discharge all the steam that can be generated by the boiler," the minimum relieving capacity must be 100,000 lb/h (12.6 kg/s), because this is the maximum capacity of the boiler and it exceeds the valve capacity based on the heating-surface calculation. If the valve capacity based on the heating-surface steam generation were larger than the stated maximum capacity of the boiler, the *Code* heating-surface valve capacity would be used in safety-valve selection.

2. Determine the number of safety valves needed

Study the latest edition of the *Code* to determine the requirements for the number of safety valves. The edition of the *Code* used here requires that "each boiler shall have at least one safety valve and if it [the boiler] has more than 500 ft^2 (46.5 m^2) of water heating surface, it shall have two or more safety valves." Thus, at least two safety valves are needed for this boiler. The *Code* further specifies, in the edition used, that "when two or more safety valves are used on a boiler, they may be mounted either separately or as twin valves made by placing individual valves on Y bases, or duplex valves having two valves in the same body casing. Twin valves made by placing individual valves on Y bases, or duplex valves having two valves in the same body, shall be of equal sizes." Also, "when not more than two valves of different sizes are mounted singly, the relieving capacity of the smaller valve shall not be less than 50 percent of that of the larger valve."

Assume that two equal-size valves mounted on a Y base will be used on the steam drum of this boiler. Two or more equal-size valves are usually chosen for the steam drum of a watertube boiler.

Since this boiler handles superheated steam, check the *Code* requirements regarding super-heaters. The *Code* states that "every attached superheater shall have one or more safety valves near the outlet." Also, "the discharge capacity of the safety valve, or valves, on an attached super-heater may be included in determining the number and size of the safety valves for the boiler, provided there are no intervening valves between the superheater safety valve and the boiler, and provided the discharge capacity of the safety valve, or valves, on the boiler, as distinct from the superheater, is at least 75 percent of the aggregate valve capacity required."

Since the safety valves used must handle 100,000 lb/h (12.6 kg/s), and one or more super-heater safety valves are required by the *Code*, assume that the two steam-drum valves will handle, in accordance with the above requirement, 80,000 lb/h (10.1 kg/s). Assume that one superheater safety valve will be used. Its capacity must then be at least $100,000 - 80,000 = 20,000$ lb/h (2.5 kg/s). (Use as few superheater safety valves as possible, because this simplifies the installation and reduces cost.) With this arrangement, each steam-drum valve must handle $80,000/2 = 40,000$ lb/h (5.0 kg/s) of steam, since there are two safety valves on the steam drum.

3. Determine the valve pressure settings

Consult the *Code*. It requires that "one or more safety valves on the boiler proper shall be set at or below the maximum allowable working pressure." For modern boilers, the maximum allowable working pressure is usually 1.5, or more, times the rated operating pressure in the lower [under 1000 lb/in^2 (abs) or 6895.0 kPa] pressure ranges. To prevent unnecessary operation of the safety valve and to reduce steam losses, the lowest safety-valve setting is usually about 5 percent higher than the boiler operating pressure. For this boiler, the lowest pressure setting would be $800 + 800(0.05) = 840$ lb/in^2 (abs) (5791.8 kPa). Round this to 850 lb/in^2 (abs) (5860.8 kPa, or 6.25 percent) for ease of selection from the usual safety-valve rating tables. The usual safety-valve pressure setting is between 5 and 10 percent higher than the rated operating pressure of the boiler.

Boilers fitted with superheaters usually have the superheater safety valve set at a lower pres-sure than the steam-drum safety valve. This arrangement ensures that the superheater safety valve opens first when overpressure occurs. This provides steam flow through the superheater tubes at all times, preventing tube burnout. Therefore, the superheater safety valve in this boiler will be set to open at 850 lb/in^2 (abs) (5860.8 kPa), the lowest opening pressure for the safety valves chosen. The steam-drum safety valves will be set to open at a higher pressure. As decided earlier, the superheater safety valve will have a capacity of 20,000 lb/h (2.5 kg/s).

Between the steam drum and the superheater safety valve, there is a pressure loss that varies from one boiler to another. The boiler manufacturer supplies a performance chart showing the drum outlet pressure for various percentages of the maximum continuous steaming capacity of the boiler. This chart also shows the superheater outlet pressure for the same capacities. The dif-ference between the drum and superheater outlet pressure for any given load is the superheater pressure loss. Obtain this pressure loss from the performance chart.

Assume, for this boiler, that the superheater pressure loss, plus any pressure losses in the nonre-turn valve and dry pipe, at maximum rating, is 60 lb/in^2 (abs) (413.7 kPa). The steam-drum operating pressure will then be superheater outlet pressure + superheater pressure loss = $800 + 60 = 860$ lb/in^2 (abs) (5929.7 kPa). As with the superheater safety valve, the steam-drum safety valve is usually set to open at about 5 percent above the drum operating pressure at maximum steam output. For this boiler then, the drum safety-valve set pressure = ~~860 + 860(0.05) = 903~~ lb/in^2 (abs) (6226.2 kPa). Round this to 900 lb/in^2 (abs) (6205.5 kPa) to simplify valve selection.

Some designers add the drum safety-valve blowdown or blowback pressure (difference between the valve opening and closing pressures, lb/in^2) to the total obtained above to find the drum operating pressure. However, the 5 percent allowance used above is sufficient to allow for the blowdown in boilers operating at less than 1000 lb/in^2 (abs) (6895.0 kPa). At pressures of 1000 lb/in^2 (abs) (6895.0 kPa) and higher, add the drum safety-valve blowdown *and* the 5 percent allowance to the superheater outlet pressure and pressure loss to find the drum pressure.

4. Determine the required valve orifice discharge area

Refer to a safety-valve manufacturer's engineering data listing valve capacities at various working pressures. For the two steam-drum valves, enter the table at 900 lb/in^2 (abs) (6205.5 kPa), and project horizontally until a capacity of 40,000 lb/h (5.0 kg/s), or more, is intersected. Here is an excerpt from a typical manufacturer's capacity table for safety valves handling *saturated steam:*

Set pressure		Orifice area					
lb/in^2 (abs)	kPa	0.994 in^2	6.41 cm^2	1.431 in^2	9.23 cm^2	2.545 in^2	16.42 cm^2
890	6,136.6	41,750	5.26	60,000	7.56	107,200	13.5
900	6,205.5	42,200	5.32	60,900	7.67	108,000	13.6
910	6,274.5	42,700	5.38	61,600	7.76	109,300	13.8

Thus, at 900 lb/in^2 (abs) (6205.5 kPa) a valve with an orifice area of 0.994 in^2 (6.4 cm^2) will have a capacity of 42,200 lb/h (5.3 kg/s) of saturated steam. This is 5.5 percent greater than the required capacity of 40,000 lb/h (5.0 kg/s) for each steam-drum valve. However, the usual selection cannot be made at exactly the desired capacity. Provided that the valve chosen has a greater steam relieving capacity than required, there is no danger of overpressure in the steam drum. Be careful to note that safety valves for saturated steam are chosen for the steam drum because superheating of the steam does not occur in the steam drum.

The superheater safety valve must handle 20,000 lb/h (2.5 kg/s) of 850 lb/in^2 (abs) (5860.8-kPa) steam at 900°F (482.2°C). Safety valves handling superheated steam have a smaller capacity than when handling saturated steam. To obtain the capacity of a safety valve handling super-heated steam, the saturated steam capacity is multiplied by a correction factor that is less than 1.00. An alternative procedure is to divide the required superheated-steam capacity by the same correction factor to obtain the saturated-steam capacity of the valve. The latter procedure will be used here because it is more direct.

Obtain the correction factor from the safety-valve manufacturer's engineering data by entering at the steam pressure and projecting to the steam temperature, as shown below.

Set pressure		Steam temperature	
lb/in^2 (abs)	kPa	880°F (471.1°C)	900°F (482.2°C)
800	5516.0	0.80	0.80
850	6205.5	0.81	0.80
900	5860.8	0.81	0.80

Thus, at 850 lb/in^2 (abs) (5860.8 kPa) and 900°F (482.2°C), the correction factor is 0.80. The required saturated steam capacity then is 20,000/0.80 = 25,000 lb/h (3.1 kg/s).

Refer to the manufacturer's saturated-steam capacity table as before, and at 850 lb/in^2 (abs) (5860.8 kPa) find the closest capacity as 31,500 lb/h (4.0 kg/s) for a 0.785-in^2 (5.1-cm^2) orifice. As with the steam-drum valves, the actual capacity of the safety valve is somewhat greater than

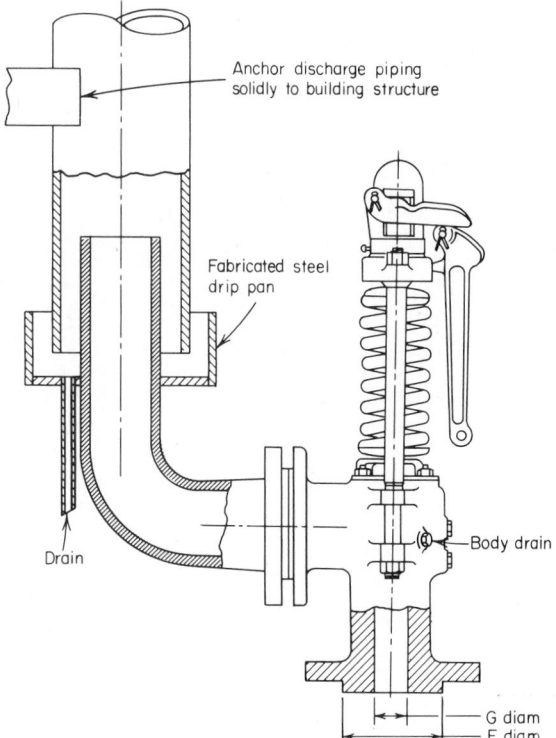

FIG. 48 Typical boiler safety-valve discharge elbow and drip-pan connection. *(Industrial Valve and Instrument Division of Dresser Industries Inc.)*

the required capacity. In general, it is difficult to find a valve with exactly the required steam relieving capacity.

5. *Determine the valve nominal size and construction details*

Turn to the data section of the safety-valve engineering manual to find the valve construction features. For the steam-drum valves having 0.994-in^2 (6.4-cm^2) orifice areas, the engineering data show, for 900-lb/in^2 (abs) (6205.5-kPa) service, each valve is 1½-in (3.8-cm) unit rated for temperatures up to 1050°F (565.6°C). The inlet is a 900-lb/in^2 (6205.5-kPa) 1½-in (3.8-cm) flanged connection, and the outlet is a 150-lb/in^2 (1034.3-kPa) 3-in (7.6-cm) flanged connection. Materials used in the valve include: body, cast carbon steel; disk seat, stainless steel AISI 321. The overall height is 27⅞ in (70.8 cm); dismantled height is 32¾ in (83.2 cm).

Similar data for the superheated steam valve show, for a maximum pressure of 900 lb/in^2 (abs) (6205.5 kPa), that it is a 1½-in (3.8-cm) unit rated for temperatures up to 1000°F (537.8°C). The inlet is a 900-lb/in^2 (6205.5-kPa) 1½-in (3.8-cm) flanged connection, and the outlet is a 150-lb/in^2 (1034.3-kPa) 3-in (7.6-cm) flanged connection. Materials used in the valve include: body, cast alloy steel, ASTM 217-WC6; spindle, stainless steel; spring, alloy steel; disk seat, stainless steel. Overall height is 21⅜ in (54.3 cm); dismantled height is 25¼ in (64.1 cm). Checking the *Code* shows that "every safety valve used on a superheater discharging superheated steam at a temperature over 450°F (232.2°C) shall have a casing, including the base, body, bonnet and spindle, of steel, steel alloy, or equivalent heat-resisting material. The valve shall have a flanged inlet connection."

Thus, the superheater valve selected is satisfactory.

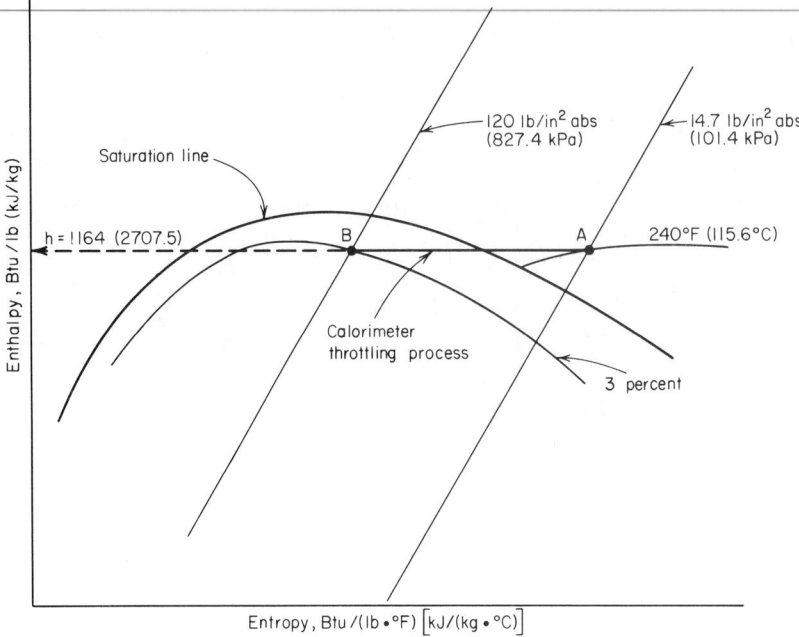

FIG. 49   Mollier-diagram plot of a throttling-calorimeter process.

6.   *Compute the steam-drum connection size*

The *Code* requires that "when a boiler is fitted with two or more safety valves on one connection, this connection to the boiler shall have a cross-sectional area not less than the combined areas of inlet connections of all safety valves with which it connects."

The inlet area for each valve $= \pi D^2/4 = \pi(1.5)^2/4 = 1.77$ in^2 (11.4 cm^2). For two valves, the total inlet area $= 2(1.77) = 3.54$ in^2 (22.8 cm^2). The required minimum diameter of the boiler connection is $d = 2(A/\pi)^{0.5}$, where $A =$ inlet area. Or, $d = 2(3.54/\pi)^{0.5} = 2.12$ in (5.4 cm). Select a $2\frac{1}{2} \times 1\frac{1}{2} \times 1\frac{1}{2}$ in (6.4 $\times$ 3.8 $\times$ 3.8 cm). Y for the two steam-drum valves and a $2\frac{1}{2}$-in (6.4-cm) steam-drum outlet connection.

7.   *Compute the safety-valve closing pressure*

The *Code* requires safety valves "to close after blowing down not more than 4 percent of the set pressure." For the steam-drum valves the closing pressure will be $900 - (900)(0.04) = 865$ lb/in^2 (abs) (5964.2 kPa). The superheater safety valve will close at $850 - (850)(0.04) = 816$ lb/in^2 (abs) (5626.3 kPa).

8.   *Sketch the discharge elbow and drip pan*

Figure 48 shows a typical discharge elbow and drip-pan connection. Fit all boiler safety valves with escape pipes to carry the steam out of the building and away from personnel. Extend the escape pipe to at least 6 ft (1.8 m) above the roof of the building. Use an escape pipe having a diameter equal to the valve outlet size. When the escape pipe is more than 12 ft (3.7 m) long, some authorities recommend increasing the escape-pipe diameter by $\frac{1}{2}$ in (1.3 cm) for each additional 12-ft (3.7-m) length. Excessive escape-pipe length without an increase in diameter can cause a backpressure on the safety valve because of flow friction. The safety valve may then chatter excessively.

Support the escape pipe independently of the safety valve. Fit a drain to the valve body and drip pan as shown in Fig. 48. This prevents freezing of the condensate and also eliminates the possibility of condensate in the escape pipe raising the valve opening pressure. When a muffler is fitted to the escape pipe, the inlet diameter of the muffler should be the same as, or larger than, the escape-pipe diameter. The outlet area should be greater than the inlet area of the muffler.

Related Calculations: Compute the safety-valve size for fire-tube boilers in the same way as described above, except that the *Code* gives a tabulation of the required area for safety-valve boiler connections based on boiler operating pressure and heating surface. Thus, with an operating pressure of 200 lb/in^2 (gage) (1379.0 kPa) and 1800 ft^2 (167.2 m^2) of heating surface, the *Code* table shows that the safety-valve connection should have an area of at least 9.148 in^2 (59.0 cm^2). A 3½-in (8.9-cm) connection would provide this area; or two smaller connections could be used provided that the sum of their areas exceeded 9.148 in^2 (59.0 cm^2).

Note: Be sure to select safety valves approved for use under the *Code* or local law governing boilers in the area in which the boiler will be used. Choice of an unapproved valve can lead to its rejection by the bureau or other agency controlling boiler installation and operation.

STEAM-QUALITY DETERMINATION WITH A THROTTLING CALORIMETER

Steam leaves an industrial boiler at 120 lb/in^2 (abs) (827.4 kPa) and 341.25°F (171.8°C). A portion of the steam is passed through a throttling calorimeter and is exhausted to the atmosphere when the barometric pressure is 14.7 lb/in^2 (abs) (101.4 kPa). How much moisture does the steam leaving the boiler contain if the temperature of the steam at the calorimeter is 240°F (115.6°C)?

Calculation Procedure:

1. Plot the throttling process on the Mollier diagram

Begin with the endpoint, 14.7 lb/in^2 (abs) (101.4 kPa) and 240°F (115.6°C). Plot this point on the Mollier diagram as point *A*, Fig. 49. Note that this point is in the superheat region of the Mollier diagram, because steam at 14.7 lb/in^2 (abs) (101.4 kPa) has a temperature of 212°F (100.0°C), whereas the steam in this calorimeter has a temperature of 240°F (115.6°C). The enthalpy of the calorimeter steam is, from the Mollier diagram, 1164 Btu/lb (2707.5 kJ/kg).

2. Trace the throttling process on the Mollier diagram

In a throttling process, the steam expands at constant enthalpy. Draw a straight, horizontal line from point *A* to the left on the Mollier diagram until the 120-lb/in^2 (abs) (827.4-kPa) pressure curve is intersected, point *B*, Fig. 49. Read the moisture content of the steam as 3 percent where the 1164-Btu/lb (2707.5-kJ/kg) horizontal trace *AB*, the 120-lb/in^2 (abs) (827.4-kPa) pressure line, and the 3 percent moisture line intersect.

Related Calculations: A throttling calorimeter *must* produce superheated steam at the existing atmospheric pressure if the moisture content of the supply steam is to be found. Where the throttling calorimeter cannot produce superheated steam at atmospheric pressure, connect the calorimeter outlet to an area at a pressure less than atmospheric. Expand the steam from the source, and read the temperature at the calorimeter. If the steam temperature is greater than that corresponding to the absolute pressure of the vacuum area—for example, a temperature greater than 133.76°F (56.5°C) in an area of 5 inHg (16.9 kPa) absolute pressure—follow the same procedure as given above. Point *A* would then be in the below-atmospheric area of the Mollier diagram. Trace to the left to the origin pressure, and read the moisture content as before.

STEAM PRESSURE DROP IN A BOILER SUPERHEATER

What is the pressure loss in a boiler superheater handling w_s = 200,000 lb/h (25.2 kg/s) of saturated steam at 500 lb/in^2 (abs) (3447.5 kPa) if the desired outlet temperature is 750°F (398.9°C)? The steam free-flow area through the superheater tubes A_s ft^2 is 0.500, friction factor f is 0.025, tube ID is 2.125 in (5.4 cm), developed length l of a tube in one circuit is 150 in (381.0 cm), and the tube bend factor B_f is 12.0.

Calculation Procedure:

1. Determine the initial conditions of the steam

To compute the pressure loss in a superheater, the initial specific volume of the steam v_g and the mass-flow ratio w_s/A_s must be known. From the steam table, v_g = 0.9278 ft^3/lb (0.058 m^3/kg) at 500 lb/in² (abs) (3447.5 kPa) saturated. The mass-flow ratio $w_s/A_s = 200,000/0.500 = 400,000$.

2.   *Compute the superheater entrance and exit pressure loss*

Entrance and exit pressure loss p_E lb/in² $= v_f/8(0.00001w_s/A_s) = 0.9278/8[(0.00001) \times (400,000)]^2 = 1.856$ lb/in² (12.8 kPa).

3.   *Compute the pressure loss in the straight tubes*

Straight-tube pressure loss p_s lb/in² $= v_f lf/ID(0.00001w_s/A_s)^2 = 0.9278(150)(0.025)/2.125[(0.00001)(400,000)]^2 = 26.2$ lb/in² (abs) (180.6 kPa).

4.   *Compute the pressure loss in the superheater bends*

Bend pressure loss $p_b = 0.0833B_f(0.00001w_s/A_s)^2 = 0.0833(12.0)[(0.00001)(400,000)]^2 = 16.0$ lb/in² (110.3 kPa).

5.   *Compute the total pressure loss*

The total pressure loss in any superheater is the sum of the entrance, straight-tube, bend, and exit-pressure losses. These losses were computed in steps 2, 3, and 4 above. Therefore, total pressure loss $p_t = 1.856 + 26.2 + 16.0 = 44.056$ lb/in² (303.8 kPa).

Note: Data for superheater pressure-loss calculations are best obtained from the boiler manufacturer. Several manufacturers have useful publications discussing superheater pressure losses. These are listed in the references at the beginning of this section.

SELECTION OF A STEAM BOILER FOR A GIVEN LOAD

Choose a steam boiler, or boilers, to deliver up to 250,000 lb/h (31.5 kg/s) of superheated steam at 800 lb/in² (abs) (5516 kPa) and 900°F (482.2°C). Determine the type or types of boilers to use, the capacity, type of firing, feedwater-quality requirements, and best fuel if coal, oil, and gas are all available. The normal continuous steam requirement is 200,000 lb/h (25.2 kg/s).

Calculation Procedure:

1.   *Select type of steam generator*

Use Fig. 50 as a guide to the usual types of steam generators chosen for various capacities and different pressure and temperature conditions. Enter Fig. 50 at the left at 800 lb/in² (abs) (5516 kPa), and project horizontally to the right, along *AB*, until the 250,000-lb/h (31.5-kg/s) capacity ordinate *BC* is intersected. At *B*, the operating point of this boiler, Fig. 50 shows that a watertube boiler should be used.

Boiler units presently available can deliver steam at the desired temperature of 900°F (482.2°C). The required capacity of 250,000 lb/h (31.5 kg/s) is beyond the range of *packaged watertube boilers*—defined by the American Boiler Manufacturer Association as "a boiler equipped and shipped complete with fuel-burning equipment, mechanical-draft equipment, automatic controls, and accessories."

Shop-assembled boilers are larger units, where all assembly is handled in the builder's plant but with some leeway in the selection of controls and auxiliaries. The current maximum capacity of shop-assembled boilers is about 100,000 lb/h (12.6 kg/s). Thus, a standard-design, larger-capacity boiler is required.

Study manufacturers' engineering data to determine which types of watertube boilers are available for the required capacity, pressure, and temperature. This study reveals that, for this installation, a standard, field-assembled, welded-steel-cased, bent-tube, single-steam-drum boiler with a completely water-cooled furnace would be suitable. This type of boiler is usually fitted with an air heater, and an economizer might also be used. The induced- and forced-draft fans are not integral with the boiler. Capacities of this type of boiler usually available range from 50,000 to 350,000 lb/h (6.3 to 44.1 kg/s); pressure from 160 to 1050 lb/in² (1103.2 to 7239.8 kPa); steam temperature from saturation to 950°F (510.0°C); fuels—pulverized coal, oil, gas, or a combination; controls—manual to completely automatic; efficiency—to 90 percent.

2.   *Determine the number of boilers required*

The normal continuous steam requirement is 200,000 lb/h (25.2 kg/s). If a 250,000-lb/h (31.5-kg/s) boiler were chosen to meet the maximum required output, the boiler would normally oper-

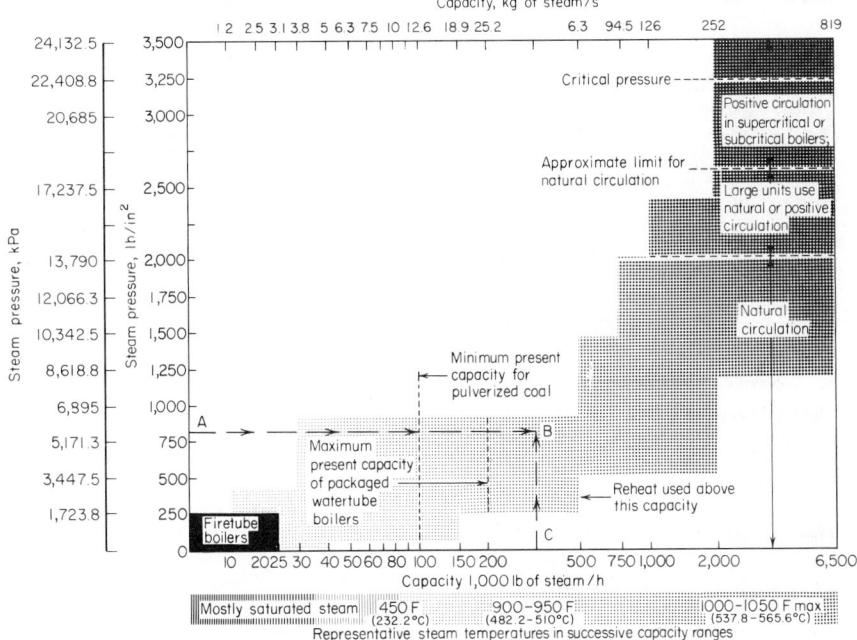

FIG. 50 Typical pressure and capacity relationships for steam generators. *(Power.)*

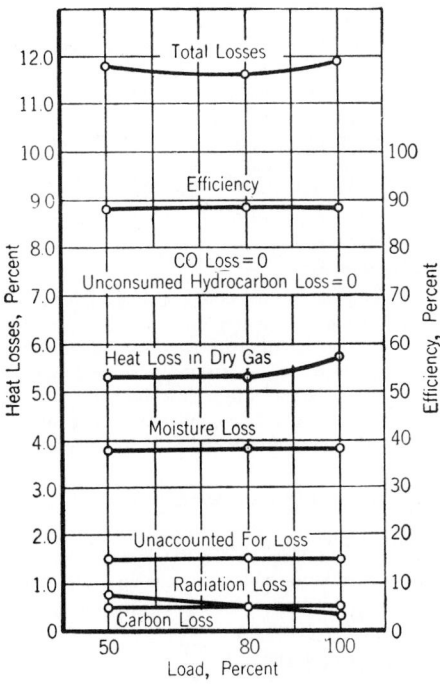

FIG. 51 Typical watertube steam-generator losses and efficiency.

3.279

ate at 200,000/250,000, or 80 percent capacity. Obtain the performance chart, Fig. 51, from the manufacturer and study it. This chart shows that at 80 percent load, the boiler efficiency is about equal to that at 100 percent load. Thus, there will not be any significant efficiency loss when the unit is operated at its normal continuous output. The total losses in the boiler are lower at 80 percent load than at full (100 percent) load.

Since there is not a large efficiency decrease at the normal continuous load, and since there are not other factors that require or make more than one boiler desirable, a single boiler unit would be most suitable for this installation. One boiler is more desirable than two or more because installation of a single unit is simpler and maintenance costs are lower. However, where the load fluctuates widely and two or more boilers could best serve the steam demand, the savings in installation and maintenance costs would be insignificant compared with the extra cost of operating a relatively large boiler installed in place of two or more smaller boilers. Therefore, each installation must be carefully analyzed and a decision made on the basis of the existing conditions.

3.   *Determine the required boiler capacity*

The stated steam load is 250,000 lb/h (31.5 kg/s) at maximum demand. Study the installation to determine whether the steam demand will increase in the future. Try to determine the rate of increase in the steam demand; for example, installation of several steam-using process units each year during the next few years will increase the steam demand by a predictable amount every year. By using these data, the rate of growth and total steam demand can be estimated for each year. Where the growth will exceed the allowable overload capacity of the boiler—which can vary from 0 to 50 percent of the full-load rating, depending on the type of unit chosen—consider installing a larger-capacity boiler now to meet future load growth. Where the future load is unpredictable, or where no load growth is anticipated, a unit sized to meet today's load would be satisfactory. If this situation existed in this plant, a 250,000-lb/h unit (31.5-kg/s) would be chosen for the load. Any small temporary overloads could be handled by operating the boiler at a higher output for short periods.

Alternatively, assume that a load of 25,000 lb/h (3.1 kg/s) will be added to the maximum demand on this boiler each year for the next 5 years. This means that in 5 years the maximum demand will be $250,000 + 25,000(5) = 375,000$ lb/h (47.2 kg/s). This is an overload of $(375,000 - 250,000)/250,000 = 0.50$, or 50 percent. It is unlikely that the boiler could carry a continuous overload of 50 percent. Therefore it might be wise to install a 375,000-lb/h (47.2-kg/s) boiler to meet present and future demands. Base this decision on the accuracy of the future-demand prediction and the economic advantages or disadvantages of investing more money now for a demand that will not occur until some future date. Refer to the section on engineering economics for procedures to follow in economics calculations of this type.

Thus, with no increase in the future load, a 250,000-lb/h (31.5-kg/s) unit would be chosen. With the load increase specified, a 375,000-lb/h (47.2-kg/s) unit would be the choice, if there were no major economic disadvantages.

4.   *Choose the type of fuel to use*

Watertube boilers of the type being considered will economically burn the three fuels available— coal, oil, or gas—either singly or in combination. In the design considered here, the furnace water-cooled surfaces and boiler surfaces are integral parts of each other. For this reason the boiler is well suited for pulverized-coal firing in the 50,000- to 300,000-lb/h (6.3- to 37.8-kg/s) capacity range. Thus, if a 250,000-lb/h (31.5-kg/s) unit were chosen, it could be fired by pulverized coal. With a larger unit of 375,000 lb/h (47.2 kg/s), pulverized-coal, oil, or gas firing might be used. Use an economic comparison to determine which fuel would give the lowest overall operating cost for the life of the boiler.

5.   *Determine the feedwater-quality requirements*

Watertube boilers of all types require careful control of feedwater quality to prevent scale and sludge deposits in tubes and drums. Corrosion of the interior boiler surfaces must be controlled. Where all condensate is returned to the boiler, the makeup water must be treated to prevent the conditions just cited. Therefore, a comprehensive water-treating system must be planned for, particularly if the raw-water supply is poor.

6.   *Estimate the boiler space requirements*

The space occupied by steam-generating units is an important consideration in plants in municipal areas and where power-plant buildings are presently crowded by existing equipment. The manufacturer's engineering data for this boiler show that for pulverized-coal firing, the hopper-type furnace bottom is best. The data also show that the smallest boiler with a hopper bottom occupies a space 21 ft (6.4 m) wide, 31 ft (9.4 m) high, and 14 ft (4.3 m) front to rear. The largest boiler occupies a space 21 ft (6.4 m) wide, 55 ft (16.8 m) high, and 36 ft (11.0 m) front to rear. Check these dimensions against the available space to determine whether the chosen boiler can be installed without major structural changes. The steel walls permit outdoor or indoor installation with top or bottom support of the boiler optional in either method of installation.

Related Calculations: Use this general procedure to select boilers for industrial, central-station, process, and marine applications.

Where a boiler is to burn hazardous industrial waste as a fuel, the designer must carefully observe two waste laws: the 1980 Superfund law and the 1976 Resource Conservation and Recovery Act. These laws regulate the firing of hazardous wastes in boilers to control air pollution and explosion dangers.

Since hazardous wastes from industrial operations can vary in composition, it is important that the designer know what variables might be met during actual firing. Without correct analysis of the wastes, air pollution can become a severe problem in the plant locale.

The Environmental Protection Agency (EPA) and state regulatory agencies should be carefully consulted before any final design decisions are made for new and expanded boiler plants. While the firing of hazardous wastes can be a convenient way to dispose of them, the potential impact on the environment must be considered before any design is finalized.

SELECTING BOILER FORCED- AND INDUCED-DRAFT FANS

Combustion calculations show that an oil-fired watertube boiler requires 200,000 lb/h (25.2 kg/s) of air for combustion at maximum load. Select forced- and induced-draft fans for this boiler if the average temperature of the inlet air is 75°F (23.9°C) and the average temperature of the combustion gas leaving the air heater is 350°F (176.7°C) with an ambient barometric pressure of 29.9 inHg (101.0 kPa). Pressure losses on the air-inlet side are as follows, in inH$_2$O: air heater, 1.5 (0.37 kPa); air-supply ducts, 0.75 (0.19 kPa); boiler windbox, 1.75 (0.44 kPa); burners, 1.25 (0.31 kPa). Draft losses in the boiler and related equipment are as follows, in inH$_2$O: furnace pressure, 0.20 (0.05 kPa); boiler, 3.0 (0.75 kPa); superheater, 1.0 (0.25 kPa); economizer, 1.50 (0.37 kPa); air heater, 2.00 (0.50 kPa); uptake ducts and dampers, 1.25 (0.31 kPa). Determine the fan discharge pressure and horsepower input. The boiler burns 18,000 lb/h (2.3 kg/s) of oil at full load.

Calculation Procedure:

1. *Compute the quantity of air required for combustion*

The combustion calculations show that 200,000 lb/h (25.2 kg/s) of air is theoretically required for combustion in this boiler. To this theoretical requirement must be added allowances for excess air at the burner and leakage out of the air heater and furnace. Allow 25 percent excess air for this boiler. The exact allowance for a given installation depends on the type of fuel burned. However, a 25 percent excess-air allowance is an average used by power-plant designers for coal, oil, and gas firing. With this allowance, the required excess air = 200,000(0.25) = 50,000 lb/h (6.3 kg/s).

Air-heater air leakage varies from about 1 to 2 percent of the theoretically required airflow. Using 2 percent, we see the air-heater leakage allowance = 200,000(0.02) = 4000 lb/h (0.5 kg/s).

Furnace air leakage ranges from 5 to 10 percent of the theoretically required airflow. With 7.5 percent, the furnace leakage allowance = 200,000(0.075) = 15,000 lb/h (1.9 kg/s).

The total airflow required is the sum of the theoretical requirement, excess air, and leakage. Or, 200,000 + 50,000 + 4000 + 15,000 = 269,000 lb/h (33.9 kg/s). The forced-draft fan must supply at least this quantity of air to the boiler. Usual practice is to allow a 10 to 20 percent safety factor for fan capacity to ensure an adequate air supply at all operating conditions. This factor of safety is applied to the total airflow required. Using a 10 percent factor of safety, we see that fan capacity = 269,000 + 269,000(0.1) = 295,900 lb/h (37.3 kg/s). Round this to 296,000-lb/h (37.3-kg/s) fan capacity.

2.   *Express the required airflow in cubic feet per minute*

Convert the required flow in pounds per hour to cubic feet per minute. To do this, apply a factor of safety to the ambient air temperature to ensure an adequate air supply during times of high ambient temperature. At such times, the density of the air is lower, and the fan discharges less air to the boiler. The usual practice is to apply a factor of safety of 20 to 25 percent to the known ambient air temperature. Using 20 percent, we see the ambient temperature for fan selection = $75 + 75(0.20) = 90°F$ (32.2°C). The density of air at 90°F (32.2°C) is 0.0717 lb/ft^3 (1.15 kg/m^3), found in Baumeister and Marks—*Standard Handbook for Mechanical Engineers*. Converting gives ft^3/min = (lb/h)/(60 lb/ft^3) = 296,000/60(0.0717) = 69,400 ft^3/min (32.8 m^3/s). This is the minimum capacity the forced-draft fan may have.

3.   *Determine the forced-draft discharge pressure*

The total resistance between the forced-draft fan outlet and furnace is the sum of the losses in the air heater, air-supply ducts, boiler windbox, and burners. For this boiler, the total resistance, inH$_2$O = 1.5 + 0.75 + 1.75 + 1.25 = 5.25 inH$_2$O (1.3 kPa). Apply a 15 to 30 percent factor of safety to the required discharge pressure to ensure adequate airflow at all times. Or, fan discharge pressure, with a 20 percent factor of safety = 5.25 + 5.25(0.20) = 6.30 inH$_2$O (1.6 kPa). The fan must therefore deliver at least 69,400 ft^3/min (32.8 m^3/s) at 6.30 inH$_2$O (1.6 kPa).

4.   *Compute the power required to drive the forced-draft fan*

The air horsepower for any fan = $0.0001753 H_f C$, where H_f = total head developed by fan, inH$_2$O; C = airflow, ft^3/min. For this fan, air hp = 0.0001753(6.3)(69,400) = 76.5 hp (57.0 kW). Assume or obtain the fan and fan-driver efficiencies at the rated capacity (69,400 ft^3/min, or 32.8 m^3/s) and pressure (6.30 inH$_2$O, or 1.6 kPa). With a fan efficiency of 75 percent and assuming the fan is driven by an electric motor having an efficiency of 90 percent, we find the overall efficiency of the fan-motor combination is (0.75)(0.90) = 0.675, or 67.5 percent. Then the motor horsepower required = air horsepower/overall efficiency = 76.5/0.675 = 113.2 hp (84.4 kW). A 125-hp (93.2-kW) motor would be chosen because it is the nearest, next larger unit readily available. Usual practice is to choose a *larger* driver capacity when the computed capacity is lower than a standard capacity. The next larger standard capacity is generally chosen, except for extremely large fans where a special motor may be ordered.

5.   *Compute the quantity of flue gas handled*

The quantity of gas reaching the induced-draft fan is the sum of the actual air required for combustion from step 1, air leakage in the boiler and furnace, and the weight of fuel burned. With an air leakage of 10 percent in the boiler and furnace (this is a typical leakage factor applied in practice), the gas flow is as follows:

	lb/h	kg/s
Actual airflow required	296,000	37.3
Air leakage in boiler and furnace	29,600	3.7
Weight of oil burned	18,000	2.3
Total	343,600	43.3

Determine from combustion calculations for the boiler the density of the flue gas. Assume that the combustion calculations for this boiler show that the flue-gas density is 0.045 lb/ft^3 (0.72 kg/m^3) at the exit-gas temperature. To determine the exit-gas temperature, apply a 10 percent factor of safety to the given exit temperature, 350°F (176.6°C). Hence, exit-gas temperature = 350 + 350(0.10) = 385°F (196.1°C). Then flue-gas flow, ft^3/min = (flue-gas flow, lb/h)/(60)(flue-gas density, lb/ft^3) = 343,600/[(60)(0.045)] = 127,000 ft^3/min (59.9 m^3/s). Apply a 10 to 25 percent factor of safety to the flue-gas quantity to allow for increased gas flow. With a 20 percent factor of safety, the actual flue-gas flow the fan must handle = 127,000 + 127,000(0.20) = 152,400 ft^3/min (71.8 m^3/s), say 152,500 ft^3/min (71.9 m^3/s) for fan-selection purposes.

6.   Compute the induced-draft fan discharge pressure

Find the sum of the draft losses from the burner outlet to the induced-draft fan inlet. These losses are as follows for this boiler:

	inH$_2$O	kPa
Furnace draft loss	0.20	0.05
Boiler draft loss	3.00	0.75
Superheater draft loss	1.00	0.25
Economizer draft loss	1.50	0.37
Air heater draft loss	2.00	0.50
Uptake ducts and damper draft loss	1.25	0.31
Total draft loss	8.95	2.23

Allow a 10 to 25 percent factor of safety to ensure adequate pressure during all boiler loads and furnace conditions. With a 20 percent factor of safety for this fan, the total actual pressure loss $= 8.95 + 8.95(0.20) = 10.74$ inH$_2$O (2.7 kPa). Round this to 11.0 inH$_2$O (2.7 kPa) for fan-selection purposes.

7.   Compute the power required to drive the induced-draft fan

As with the forced-draft fan, air horsepower $= 0.0001753 H_f C = 0.0001753(11.0) \times (127,000) = 245$ hp (182.7 kW). If the combined efficiency of the fan and its driver, assumed to be an electric motor, is 68 percent, the motor horsepower required $= 245/0.68 = 360.5$ hp (268.8 kW). A 375-hp (279.6-kW) motor would be chosen for the fan driver.

8.   Choose the fans from a manufacturer's engineering data

Use the next calculation procedure to select the fans from the engineering data of an acceptable manufacturer. For larger boiler units, the forced-draft fan is usually a backward-curved blade centrifugal-type unit. Where two fans are chosen to operate in parallel, the pressure curve of each fan should decrease at the same rate near shutoff so that the fans divide the load equally. Be certain that forced-draft fans are heavy-duty units designed for continuous operation with well-balanced rotors. Choose high-efficiency units with self-limiting power characteristics to prevent overloading the driving motor. Airflow is usually controlled by dampers on the fan discharge.

Induced-draft fans handle hot, dusty combustion products. For this reason, extreme care must be taken to choose units specifically designed for induced-draft service. The usual choice for large boilers is a centrifugal-type unit with forward- or backward-curved, or flat blades, depending on the type of gas handled. Flat blades are popular when the flue gas contains large quantities of dust. Fan bearings are generally water-cooled.

Related Calculations: Use the procedure given above for the selection of draft fans for all types of boilers—fire-tube, packaged, portable, marine, and stationary. Obtain draft losses from the boiler manufacturer. Compute duct pressure losses by using the methods given in later procedures in this handbook.

POWER-PLANT FAN SELECTION FROM CAPACITY TABLES

Choose a forced-draft fan to handle 69,400 ft^3/min (32.8 m^3/s) of 90°F (32.2°C) air at 6.30-inH$_2$O (1.6-kPa) static pressure and an induced-draft fan to handle 152,500 ft^3/min (72.0 m^3/s) of 385°F (196.1°C) gas at 11.0-inH$_2$O (2.7-kPa) static pressure. The boiler that these fans serve is installed at an elevation of 5000 ft (1524 m) above sea level. Use commercially available capacity tables for making the fan choice. The flue-gas density is 0.045 lb/ft^3 (0.72 kg/m^3) at 385°F (196.1°C).

Calculation Procedure:

1.   Compute the correction factors for the forced-draft fan

Commercial fan-capacity tables are based on fans handling standard air at 70°F (21.1°C) at a barometric pressure of 29.92 inHg (101.0 kPa) and having a density of 0.075 lb/ft^3 (1.2 kg/m^3).

TABLE 6 Fan Correction Factors

Temperature		Correction factor	Altitude		Correction factor
°F	°C		ft	m	
80	26.7	1.009	4500	1371.6	1.086
90	32.2	1.018	5000	1524.0	1.095
100	37.8	1.028	5500	1676.4	1.106
375	190.6	1.255			
400	204.4	1.273			
450	232.2	1.310			

Where different conditions exist, the fan flow rate must be corrected for temperature and altitude.

Obtain the engineering data for commercially available forced-draft fans, and turn to the temperature and altitude correction-factor tables. Pick the appropriate correction factors from these tables for the prevailing temperature and altitude of the installation. Thus, in Table 6, select the correction factors for 90°F (32.2°C) air and 5000-ft (1524.0-m) altitude. These correction factors are $C_T = 1.018$ for 90°F (32.2°C) air and $C_A = 1.095$ for 5000-ft (1524.0-m) altitude.

Find the composite correction factor (CCF) by taking the product of the temperature and altitude correction factors. Or, CCF = (1.018)(1.095) = 1.1147. Now divide the given cubic feet per minute (cfm) by the correction factor to find the capacity-table cfm. Or, capacity-table cfm = 69,400/1.147 = 62,250 ft³/min (29.4 m³/s).

2. Choose the fan size from the capacity table

Turn to the fan-capacity table in the engineering data, and look for a fan delivering 62,250 ft³/min (29.4 m³/s) at 6.3-inH$_2$O (1.6-kPa) static pressure. Inspection of the table shows that the capacities are tabulated for 6.0- and 6.5-inH$_2$O (1.5- and 1.6-kPa) static pressure. There is no tabulation for 6.3-inH$_2$O (1.57-kPa) static pressure.

Enter the table at the nearest capacity to that required, 62,250 ft³/min (29.4 m³/s), as shown in Table 7. This table, excerpted with permission from the American Standard Inc. engineering data, shows that the nearest capacity of this particular type of fan is 62,595 ft³/min (29.5 m³/s). The difference, or 62,595 − 62,250 = 345 ft³/min (0.16 m³/s), is only 345/62,250 = 0.0055, or 0.55 percent. This is a negligible difference, and the 62,595-ft³/min (29.5-m³/s) fan is well suited for its intended use. The extra static pressure of 6.5 − 6.3 = 0.2 inH$_2$O (0.05 kPa) is desirable in a forced-draft fan because furnace or duct resistance may increase during the life of the boiler. Also, the extra static pressure is so small that it will not markedly increase the fan power consumption.

3. Compute the fan speed and power input

Multiply the capacity-table rpm and brake horsepower (bhp) by the composite factor to determine the actual rpm and bhp. Thus, with data from Table 7, the actual rpm = (1096)(1.1147) = 1221.7 r/min. Actual bhp = (99.08)(1.1147) = 110.5 bhp (82.4 kW). This is the horsepower input required to drive the fan and is close to the 113.2 hp (84.4 kW) computed in the previous calculation procedure. The actual motor horsepower would be the same in each case because a

TABLE 7 Typical Fan Capacities

Capacity		Outlet velocity		Outlet velocity pressure		Ratings at 6.5-inH$_2$O (1.6-kPa) static pressure		
ft³/min	m³/s	ft/min	m/s	inH$_2$O	kPa	r/min	bhp	kW
61,204	28.9	4400	22.4	1.210	0.3011	1083	95.45	71.2
62,595	29.5	4500	22.9	1.266	0.3150	1096	99.08	73.9
63,975	30.2	4600	23.4	1.323	0.3212	1109	103.0	76.8

standard-size motor would be chosen. The difference of $113.2 - 110.5 = 2.7$ hp (2.0 kW) results from the assumed efficiencies that depart from the actual values. Also, a sea-level altitude was assumed in the previous calculation procedure. However, the two methods used show how accurately fan capacity and horsepower input can be estimated by judicious evaluation of variables.

4. *Compute the correction factors for the induced-draft fan*

The flue-gas density is 0.045 lb/ft³ (0.72 kg/m³) at 385°F (196.1°C). Interpolate in the temperature correction-factor table because a value of 385°F (196.1°C) is not tabulated. Find the correction factor for 385°F (196.1°C) thus: [(Actual temperature − lower temperature)/(higher temperature − lower temperature)] × (higher temperature correction factor − lower temperature correction factor) + lower temperature correction factor. Or, [(385 − 375)/(400 − 375)](1.273 − 1.255) + 1.255 = 1.262.

The altitude correction factor is 1.095 for an elevation of 5000 ft (1524.0 m), as shown in Table 6.

As for the forced-draft fan, CCF $= C_T C_A = (1.262)(1.095) = 1.3819$. Use the CCF to find the capacity-table cfm in the same manner as for the forced-draft fan. Or, capacity-table cfm = (given cfm)/CCF = 152,500/1.3819 = 110,355 ft³/min (52.1 m³/s).

5. *Choose the fan size from the capacity table*

Check the capacity table to be sure that it lists fans suitable for induced-draft (elevated-temperature) service. Turn to the 11-inH₂O (2.7-kPa) static-pressure capacity table, and find a capacity equal to 110,355 ft³/min (52.1 m³/s). In the engineering data used for this fan, the nearest capacity at 11-inH₂O (2.7-kPa) static pressure is 110,467 ft³/min (52.1 m³/s), with an outlet velocity of 4400 ft/min (22.4 m/s), an outlet velocity pressure of 1.210 inH₂O (0.30 kPa), a speed of 1222 r/min, and an input horsepower of 255.5 bhp (190.5 kW). The tabulation of these quantities is of the same form as that given for the forced-draft fan, step 2. The selected capacity of 110,467 ft³/min (52.1 m³/s) is entirely satisfactory because it is only 110,467 − 110,355/110,355 = 0.00101, or 0.1 percent, higher than the desired capacity.

6. *Compute the fan speed and power input*

Multiply the capacity-table rpm and brake horsepower by the CCF to determine the actual rpm and brake horsepower. Thus, the actual rpm = (1222)(1.3819) = 1690 r/min. Actual brake horsepower = (255.5)(1.3819) = 353.5 bhp (263.6 kW). This is the horsepower input required to drive the fan and is close to the 360.5 hp (268.8 kW) computed in the previous calculation procedure. The actual motor horsepower would be the same in each case because a standard-size motor would be chosen. The difference in horsepower of 360.5 − 353.5 = 7.0 hp (5.2 kW) results from the same factors discussed in step 3.

Note: The static pressure is normally used in most fan-selection procedures because this pressure value is used in computing pressure and draft losses in boilers, economizers, air heaters, and ducts. In any fan system, the total air pressure = static pressure + velocity pressure. However, the velocity pressure at the fan discharge is not considered in draft calculations unless there are factors requiring its evaluation. These requirements are generally related to pressure losses in the fan-control devices.

Related Calculations: Use the fan-capacity table to obtain these additional details of the fan: outlet inside dimensions (length and width), fan-wheel diameter and circumference, fan maximum bhp, inlet area, fan-wheel peripheral velocity, NAFM fan class, and fan arrangement. Use the engineering data containing the fan-capacity table to find the fan dimensions, rotation and discharge designations, shipping weight, and, for some manufacturers, prices.

ANALYSIS OF BOILER AIR DUCTS AND GAS UPTAKES

Three oil-fired boilers are supplied air through the breeching shown in Fig. 52a. Each boiler will burn 13,600 lb/h (1.71 kg/s) of fuel oil at full load. The draft loss through each boiler is 8 inH₂O (2.0 kPa). Uptakes from the three boilers are connected as shown in Fig. 52b. Determine the draft loss through the entire system if a 50-ft (15.2-m) high metal stack is used and the gas temperature at the stack inlet is 400°F (204.4°C).

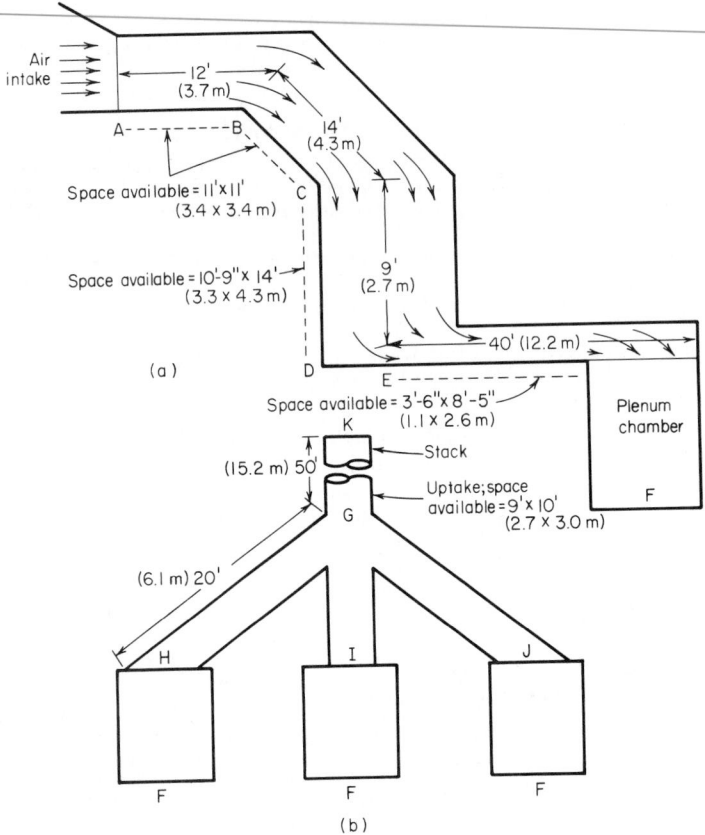

FIG. 52 (a) Boiler intake-air duct; (b) boiler uptake ducts.

Calculation Procedure:

1. Determine the airflow through the breeching

Compute the airflow required, cubic feet per pound of oil burned, using the methods given in earlier calculation procedures. For this installation, assume that the combustion calculation shows that 250 ft³/lb (15.6 m³/kg) of oil burned is required. Then the total airflow required = (number of boilers)(lb/h oil burned per boiler)(ft³/lb oil)/(60 min/h) = (3)(13,600)(250)/60 = 170,000 ft³/min (80.2 m³/s).

2. Select the dimensions for each length of breeching duct

With the airflow rate of 170,000 ft³/min (80.2 m³/s) known, the duct area can be determined by assuming an air velocity and computing the duct area A_d ft² from A_d = (airflow rate, ft³/min)/(air velocity, ft/min). Once the area is known, the duct can be sized to give this area. Thus, if 9 ft² (0.8 m²) is the required duct area, a duct 3 × 3 ft (0.9 × 0.9 m) or 2 × 4.5 ft (0.6 × 1.4 m) would provide the required area.

In the usual power plant, the room available for ducts limits the maximum allowable duct size. So the designer must try to fit a duct of the required area into the available space. This is done by changing the duct height and width until a duct of suitable area fitting the available space is found. If the duct area is reduced below that required, compute the actual air velocity to determine whether it exceeds recommended limits.

In this power plant, the space available in the open area between A and C, Fig. 52, is a square 11×11 ft (3.4×3.4 m). By allowing a 3-in (7.6-cm) clearance around the outside of the duct and using a square duct, its dimensions would be 10.5×10.5 ft (3.2×3.2 m), or a cross-sectional area of $(10.5)(10.5) = 110$ ft^2 (10.2 m^2), closely. With 170,000 ft^3/min (80.2 m^3/s) flowing through the duct, the air velocity v ft/min $=$ ft^3/min/$A_d = 170,000/110 = 1545$ ft/min (7.8 m/s). This is a satisfactory air velocity because the usual plant air system velocity is 1200 to 3600 ft/min (6.1 to 18.3 m/s).

Between C and D the open area in this power plant is 10 ft 9 in (3.3 m) by 14 ft (4.3 m). Using the same 3-in (7.6-cm) clearance all around the duct, we find the dimensions of the vertical duct CD are 10.25×13 ft (3.1×4.0 m), or a cross-sectional area of $10.25 \times 13 = 133$ ft^2 (12.5 m^2), closely. The air velocity in this section of the duct is $v = 170,000/133 = 1275$ ft/min (6.5 m/s). Since it is desirable to maintain, if possible, a constant velocity in all sections of the duct where space permits, the size of this duct might be changed so it equals that of AB, 10.5×10.5 ft (3.2×3.2 m). However, the installation costs would probably be high because the limited space available would require alteration of the power-plant structure. Also, the velocity is section CD is above the usual minimum value of 1200 ft/min (6.1 m/s). For these reasons, the duct will be installed in the 10.25×13 ft (3.1×4.0 m) size.

Between E and F the vertical distance available for installation of the duct is 3.5 ft (1.1 m), and the horizontal distance is 8.5 ft (2.6 m). Using the same 3-in (7.6-cm) clearance as before gives a 3×8 ft (0.9×2.4 m) duct size, or a cross-sectional area of $(3)(8) = 24$ ft^2 (2.2 m^2). At E the duct divides into three equal-size branches, one for each boiler, and the same area, 24 ft^2 (2.2 m^2), is available for each branch duct. The flow in any branch duct is then $170,000/3 = 56,700$ ft^3/min (26.8 m^3/s). The velocity in any of the three equal branches is $v = 56,700/24 = 2360$ ft/min (12.0 m/s). When a duct system has two or more equal-size branches, compute the pressure loss in one branch only because the losses in the other branches will be the same. The velocity in branch EF is acceptable because it is within the limits normally used in power-plant practice. At F the air enters a large plenum chamber, and its velocity becomes negligible because of the large flow area. The boiler forced-draft fan intakes are connected to the plenum chamber. Each of the three ducts feeds into the plenum chamber.

3. *Compute the pressure loss in each duct section*

Begin the pressure-loss calculations at the system inlet, point A, and work through each section to the stack outlet. This procedure reduces the possibility of error and permits easy review of the calculations for detection of errors. Assign letters to each point of the duct where a change in section dimensions or directions, or both, occurs. Use these letters:

Point A: Assume that 70°F (21.1°C) air having a density of 0.075 lb/ft^3 (1.2 kg/m^3) enters the system when the ambient barometric pressure is 29.92 inHg (101.3 kPa). Compute the velocity pressure at point A, in inH$_2$O, from $p_v = v^2/[3.06(10^4)(460 + t)]$, where $t =$ air temperature, °F. Since the velocity of the air at A is 1545 ft/min (11.7 m/s), $p_v = (1545)^2/[3.06(10^4)(530)] = 0.147$ inH$_2$O (0.037 kPa) at 70°F (21.1°C).

The entrance loss at A, where there is a sharp-edged duct, is $0.5p_v$, or $0.5(0.147) = 0.0735$ inH$_2$O (0.018 kPa). With a rounded inlet, the loss in velocity pressure would be negligible.

Section AB: There is a pressure loss due to duct friction between A and B, and B and C. Also, there is a bend loss at points B and C. Compute the duct friction first.

For any circular duct, the static pressure loss due to friction p_s inH$_2$O $= (0.03L/d^{1.24})(v/1000)^{1.84}$, where $L =$ duct length, ft; $d =$ duct diameter, in. To convert any rectangular or square duct with sides a and b ft high and wide, respectively, to an equivalent round duct of D-ft diameter, use the relation $D = 2ab/(a + b)$. For this duct, $d = 2(10.5)(10.5)/(10.5 + 10.5) = 10.5$ ft (3.2 m) $= 126$ in (320 cm) $= d$. Since this duct is 12 ft (3.7 m) long between A and B, $p_s = [0.03(12)/126^{1.24}](1.545/1000)^{1.84} = 0.002$ inH$_2$O (0.50 Pa).

Point B: The 45° bend at B has, from Baumeister and Marks—*Standard Handbook for Mechanical Engineers*, a pressure drop of 60 percent of the velocity head in the duct, or $(0.60)(0.147) = 0.088$ inH$_2$O (20.5 Pa) loss.

Section BC: Duct friction in the 14-ft (4.3-m) long downcomer BC is $p_s = [0.03(14)/126^{1.24}](1545/1000)^{1.84} = 0.0023$ inH$_2$O (0.56 Pa). Point C: The 45° bend at C has a velocity head loss of 60 percent of the velocity pressure. Determine the velocity pressure in this duct in the same manner as for point A, or $p_v = (1545)^2/[3.06(10^4)(530)] = 0.147$ inH$_2$O (36.1 Pa), since the velocity at points B and C is the same. Then the velocity head loss = $(0.60)(0.147) = 0.088$ inH$_2$O (21.9 Pa).

Section CD: The equivalent round-duct diameter is $D = (2)(10.25)(13)/(10.25 + 13) = 11.45$ ft (3.5 m) = 137.3 in (348.7 cm). Duct friction is then $p_s = [0.03(9)/137.3^{1.24}](1275/1000)^{1.84} = 0.000934$ inH$_2$O (0.23 Pa). Velocity pressure in the duct is $p_v = (1275)^2/[3.06(10^4)(530)] = 0.100$ inH$_2$O (24.9 Pa). Since there is no room for a transition piece—that is, a duct providing a gradual change in flow area between points C and D—the decrease in velocity pressure from 0.147 to 0.100 in (36.6 to 24.9 Pa), or $0.147 - 0.10 = 0.047$ inH$_2$O (11.7 Pa), is not converted to static pressure and is lost.

Point E: The pressure loss in the right-angle bend at E is, from Baumeister and Marks—*Standard Handbook for Mechanical Engineers*, 1.2 times the velocity head, or $(1.2)(0.1) = 0.12$ inH$_2$O (29.9 Pa). Also, since this is a sharp-edged elbow, there is an additional loss of 50 percent of the velocity head, or $(0.5)(0.10) = 0.05$ inH$_2$O (12.4 Pa).

The velocity pressure at point E is $p_v = (2360)^2/[3.06(10^4)(530)] = 0.343$ inH$_2$O (85.4 Pa).

Section EF: The equivalent round-duct diameter is $D = (2)(3)(8)/(3 + 8) = 4.36$ ft (1.3 m) = 52.4 in (133.1 cm). Duct friction $p_s = [0.03(40)/52.4^{1.24}](2360/1000)^{1.84} = 0.0247$ inH$_2$O (6.2 Pa).

Air entering the large plenum chamber at F loses all its velocity. There is no static-pressure regain; therefore, the velocity-head loss = $0.348 - 0.0 = 0.348$ inH$_2$O (86.6 Pa).

4.   *Compute the losses in the uptake and stack*

Convert the airflow of 250 ft^3/lb (15.6 m^3/kg) of fuel oil to pounds of air per pound of fuel oil by multiplying by the density, or $250(0.075) = 18.75$ lb of air per pound of oil. The flue gas will contain 18.75 lb of air + 1 lb of oil per pound of fuel burned, or $(18.75 + 1)/18.75 = 1.052$ times as much gas leaves the boiler as air enters; this can be termed the *flue-gas factor*.

Point G: The quantity of flue gas entering the stack from each boiler (corrected to a 400°F or 204.4°C outlet temperature) is, in °R, (cfm air to furnace)(stack, °R/air, °R)(flue-gas factor). Or stack flue-gas flow = $(56,700)[(400 + 460)/(70 + 460)](1.052) = 97,000$ ft^3/min (45.8 m^3/s) per boiler.

The total duct area available for the uptake leading to the stack is 9×10 ft (2.7×3.0 m) = 90 ft^2 (8.4 m^2), based on the clearance above the boilers. The flue-gas velocity for three boilers is $v = (3)(97,000)/90 = 3235$ ft/min (16.4 m/s). The velocity pressure in the uptake is $p = (3235)^2/[3.06(10^4)(460 + 400)] = 0.397$ inH$_2$O (98.8 Pa).

Point H: The flue-gas flow from all the boilers is divided equally between three ducts, HG, IG, JG, Fig. 52. It is desirable to maintain the same gas velocity in each duct and have this velocity equal to that in the uptake. The same velocity can be obtained in each duct by making each duct one-third the area of the uptake, or 90/3 = 30 ft^2 (2.8 m^2). Then $v = 97,000/30 = 3235$ ft/min (16.4 m/s) in each duct. Since the velocity in each duct equals the velocity in the uptake, the velocity pressure in each duct equals that in the uptake, or 0.397 inH$_2$O (98.8 Pa).

Ducts HG and JG have two 45° bends in them, or the equivalent of one 90° bend. The velocity-pressure loss in a 90° bend is 1.20 times the velocity head in the duct; or, for either HG or JG, $(1.20)(0.397) = 0.476$ inH$_2$O (118.5 Pa).

Section HG: The equivalent duct diameter for a 30-ft^2 (2.8-m^2) duct is $D = 2(30/\pi)^{0.5} = 6.19$ ft (1.9 m) = 74.2 in (188.4 cm). The duct friction in HG, which equals that in JG, is $p_s = [0.03(20)/74.2^{1.24}](530/860)(3235/1000)^{1.84} = 0.01536$ inH$_2$O (3.8 Pa), if we correct for the flue-gas temperature with the ratio $(70 + 460)/(400 + 460) = 530/860$.

Section GK: The stack joins the uptake at point G. Assume that this installation is designed for a stack-gas area of 500 lb of oil per square foot (2441.2 kg/m^2) of stack; or, for three boilers, stack area = $(3)(13,600$ lb/h oil)/500 = 81.5 ft^2 (7.6 m^2). The stack diameter will then be $D = 2(8.15/\pi)^{0.5} = 10.18$ ft (3.1 m) = 122 in (309.9 cm).

The gas velocity in the stack is $v = (3)(97,000)/81.5 = 3570$ ft/min (18.1 m/s). The friction in the stack is $p_s = [0.03(50)/122^{1.24}](3570/1000)^{1.84}(503/860) = 0.0194$ inH$_2$O (4.8 Pa).

5. *Compute the total losses in the system*

Tabulate the individual losses and find the sum as follows:

	inH$_2$O	kPa
Point A; entrance loss	0.0735	0.0183
Section AB; duct friction	0.0020	0.0005
Point B; bend loss	0.0880	0.0219
Section BC; duct friction	0.0023	0.0006
Point C; bend loss	0.0880	0.0219
Section CD; duct friction	0.0009	0.0002
Section CD; velocity-pressure loss	0.0470	0.0117
Point E; bend loss	0.1200	0.0299
Point E; sharp-edge loss	0.0500	0.0124
Section EF; duct friction	0.0247	0.0061
Section EF; plenum velocity-head loss	0.3480	0.0866
Boiler friction loss	8.0000	1.9907
Section HG; duct friction	0.0154	0.0038
Points H and G total bend loss	0.4760	0.1184
Section GK; stack friction	0.0194	0.0048
Total loss	9.3552	2.3279

The total loss computed here is the minimum static pressure that must be developed by the draft fans or blowers. This total static pressure can be divided between the forced- and induced-draft fans or confined solely to the forced-draft fans in plants not equipped with an induced-draft fan. If only a forced-draft fan is used, its static discharge pressure should be at least 20 percent greater than the losses, or $(1.2)(9.3552) = 11.21$ inH$_2$O (2.8 kPa) at a total airflow of 97,000 ft^3/min (45.8 m^3/s). If more than one forced-draft fan were used for each boiler, each fan would have a total static pressure of at least 11.21 inH$_2$O (2.8 kPa) and a capacity of less than 97,000 ft^3/min (45.8 m^3/s). In making the final selection of the fan, the static pressure would be rounded to 12 inH$_2$O (3.0 kPa).

Where dampers are used for combustion-air control, include the wide-open resistance of the dampers in computing the total losses in the system at full load on the boilers. Damper resistance values can be obtained from the damper manufacturer. Note that as the damper is closed to reduce the airflow at lower boiler loads, the resistance through the damper is increased. Check the fan head-capacity curve to determine whether the head developed by the fan at lower capacities is sufficient to overcome the greater damper resistance. Since the other losses in the system will decrease with smaller airflow, the fan static pressure is usually adequate.

Note: (1) Follow the notational system used here to avoid errors from plus and minus signs applied to atmospheric pressures and draft. Use of the plus and minus signs does not simplify the calculation and can be confusing.

(2) A few designers, reasoning that the pressure developed by a fan varies as the square of the air velocity, square the percentage safety-factor increase before multiplying by the static pressure. Thus, in the above forced-draft fan, the static discharge pressure with a 20 percent increase in pressure would be $(1.2)^2(9.3552) = 13.5$ inH$_2$O (3.4 kPa). This procedure provides a wider margin of safety but is not widely used.

(3) Large steam-generating units, some ship propulsion plants, and some packaged boilers use only forced-draft fans. Induced-draft fans are eliminated because there is a saving in the total fan hp required, there is no air infiltration into the boiler setting, and a slightly higher boiler efficiency can be obtained.

(4) The duct system analyzed here is typical of a study-type design where no refinements are used in bends, downcomers, and other parts of the system. This type of system was chosen for the analysis because it shows more clearly the various losses met in a typical duct installation. The system could be improved by using a bellmouthed intake at A, dividing vanes or splitters in the elbows, a transition in the downcomer, and a transition at F. None of these improvements would be expensive, and they would all reduce the static pressure required at the fan discharge.

(5) Do not subtract the stack draft from the static pressure the forced- or induced-draft fan must produce. Stack draft can vary considerably, depending on ambient temperature, wind velocity, and wind direction. Therefore, the usual procedure is to ignore any stack draft in fan-selection calculations because this is the safest procedure.

Related Calculations: The procedure given here can be used for all types of boilers fitted with air-supply ducts and uptake breechings—heating, power, process, marine, portable, and packaged.

DETERMINATION OF THE MOST ECONOMICAL FAN CONTROL

Determine the most economical fan control for a forced- or induced-draft fan designed to deliver 140,000 ft^3/min (66.1 m^3/s) at 14 inH$_2$O (3.5 kPa) at full load. Plot the power-consumption curve for each type of control device considered.

Calculation Procedure:

1.   *Determine the types of controls to consider*

There are five types of controls used for forced- and induced-draft fans: (*a*) a damper in the duct with constant-speed fan drive; (*b*) two-speed fan driver; (*c*) inlet vanes or inlet louvres with a

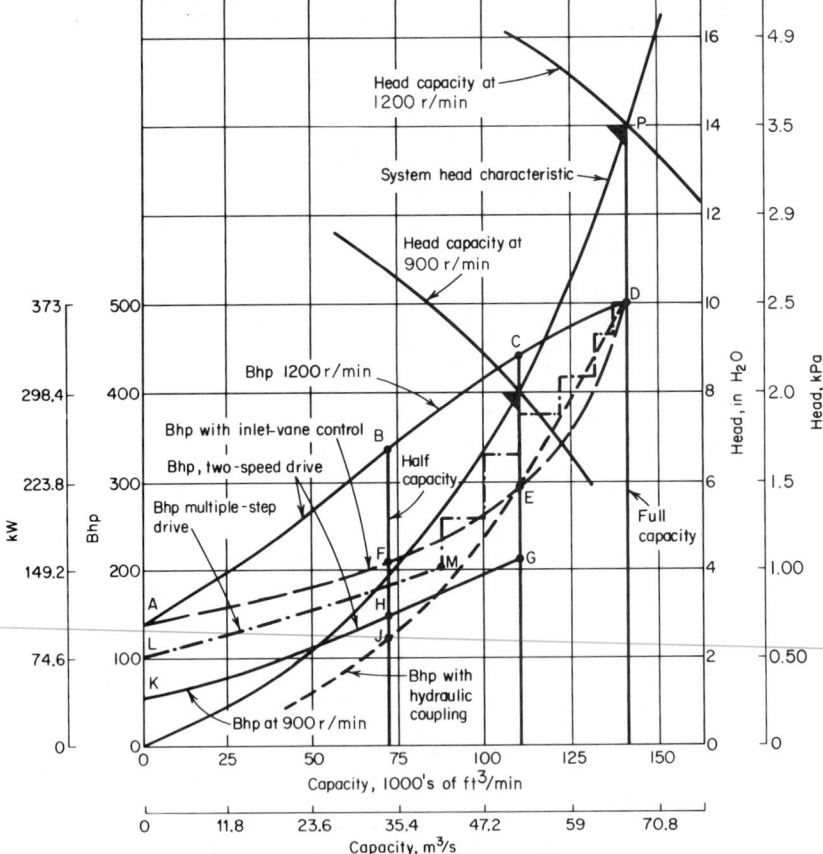

FIG. 53 Power requirements for a fan fitted with different types of controls. *(American Standard Inc.)*

constant-speed fan drive; (*d*) multiple-step variable-speed fan drive; and (*e*) hydraulic or electric coupling with constant-speed driver giving wide control over fan speed.

2.   *Evaluate each type of fan control*

Tabulate the selection factors influencing the control decision as follows, using the control letters in step 1:

Control type	Control cost	Required power input	Advantages (A), and disadvantages (D)
a	Low	High	(A) Simplicity; (D) high power input
b	Moderate	Moderate	(A) Lower input power; (D) higher cost
c	Low	Moderate	(A) Simplicity; (D) ID fan erosion
d	Moderate	Moderate	(D) Complex; also needs dampers
e	High	Low	(A) Simple; no dampers needed

3.   *Plot the control characteristics for the fans*

Draw the fan head-capacity curve for the airflow or gasflow range considered, Fig. 53. This plot shows the maximum capacity of 140,000 ft^3/min (66.1 m^3/s) and required static head of 14 inH$_2$O (3.5 kPa), point *P*.

Plot the power-input curve *ABCD* for a constant-speed motor or turbine drive with damper control—type *a*, listed above—after obtaining from the fan manufacturer, or damper builder, the input power required at various static pressures and capacities. Plotting these values gives curve *ABCD*. Fan speed is 1200 r/min.

Plot the power-input curve *GHK* for a two-speed drive, type *b*. This drive might be a motor with an additional winding, or it might be a second motor for use at reduced boiler capacities. With either arrangement, the fan speed at lower boiler capacities is 900 r/min.

Plot the power-input curve *AFED* for inlet-vane control on the forced-draft fan or inlet-louvre control on induced-draft fans. The data for plotting this curve can be obtained from the fan manufacturer.

Multiple-step variable-speed fan control, type *d*, is best applied with steam-turbine drives. In a plant with ac auxiliary motor drives, slip-ring motors with damper integration must be used between steps, making the installation expensive. Although dc motor drives would be less costly, few power plants other than marine propulsion plants have direct current available. And since marine units normally operate at full load 90 percent of the time or more, part-load operating economics are unimportant. If steam-turbine drive will be used for the fans, plot the power-input curve *LMD*, using data from the fan manufacturer.

A hydraulic coupling or electric magnetic coupling, type *e*, with a constant-speed motor drive would have the power-input curve *DEJ*.

Study of the power-input curves shows that the hydraulic and electric couplings have the smallest power input. Their first cost, however, is usually greater than any other types of power-saving devices. To determine the return on any extra investment in power-saving devices, an economic study, including a load-duration analysis of the boiler load, must be made.

4.   *Compare the return on the extra investment*

Compute and tabulate the total cost of each type of control system. Then determine the extra investment for each of the more costly control systems by subtracting the cost of type *a* from the cost of each of the other types. With the extra investment known, compute the lifetime savings in power input for each of the more efficient control methods. With the extra investment and savings resulting from it known, compute the percentage return on the extra investment. Tabulate the findings as in Table 8.

In Table 8, considering control type *c*, the extra cost of type *c* over type *b* = $75,000 − 50,000 = $25,000. The total power saving of $6500 is computed on the basis of the cost of energy in the plant for the life of the control. The return on the extra investment then = $6500/$25,000 = 0.26, or 26 percent. Type *e* control provides the highest percentage return on the extra invest-

TABLE 8   Fan Control Comparison

	Type of control used				
	a	b	c	d	e
Total cost, $	30,000	50,000	75,000	89,500	98,000
Extra cost, $	.....	20,000	25,000	14,500	8,500
Total power saving, $	.....	8,000	6,500	3,000	6,300
Return on extra investment, %	.....	40	26	20.7	74.2

ment. It would probably be chosen if the only measure of investment desirability is the return on the extra investment. However, if other criteria are used, such as a minimum rate of return on the extra investment, one of the other control types might be chosen. This is easily determined by studying the tabulation in conjunction with the investment requirement.

Related Calculations: The procedure used here can be applied to heating, power, marine, and portable boilers of all types. Follow the same steps given above, changing the values to suit the existing conditions. Work closely with the fan and drive manufacturer when analyzing drive power input and costs.

SMOKESTACK HEIGHT AND DIAMETER DETERMINATION

Determine the required height and diameter of a smokestack to produce 1.0-inH$_2$O (0.25-kPa) draft at sea level if the average air temperature is 60°F (15.6°C); barometric pressure is 29.92 inHg (101.3 kPa); the boiler flue gas enters the stack at 500°F (260.0°C); the flue-gas flow rate is 100 lb/s (45.4 kg/s); the flue-gas density is 0.045 lb/ft^3 (0.72 kg/m^3); and the flue-gas velocity is 30 ft/s (9.1 m/s). What diameter and height would be required for this stack if it were located 5000 ft (1524.0 m) above sea level?

Calculation Procedure:

1. Compute the required stack height

The required stack height S_h ft = $d_s/0.256pK$, where d_s = stack draft, inH$_2$O; p = barometric pressure, inHg; $K = 1/T_a - 1/T_g$, where T_a = air temperature, °R; T_g = average temperature of stack gas, °R. In applying this equation, the temperature of the gas at the stack outlet must be known to determine the average temperature of the gas in the stack. Since the outlet temperature cannot be measured until after the stack is in use, an assumed outlet temperature must be used for design calculations. The outlet temperature depends on the inlet temperature, ambient air temperature, and materials used in the stack construction. For usual smokestacks, the gas temperature will decrease 100 to 200°F (55.6 to 111.1°C) between the stack inlet and outlet. Using a 100°F (55.6°C) gas-temperature decrease for this stack, we get S_h = (1.0) + 0.256(29.92)(1/520 − 1/910) = 159 ft (48.5 m). Apply a 10 percent factor of safety. Then the stack height = (159)(1.10) = 175 ft (53.3 m).

2. Compute the required stack diameter

Stack diameter d_s ft is found from $d_s = 0.278(W_gT_g/Vd_gp)^{0.5}$, where W_g = flue-gas flow rate in stack, lb/s; V = flue-gas velocity in stack, ft/s; d_g = flue-gas density, lb/ft^3. For this stack, d_s = 0.278{(100)(910)/[(30)(0.045)(29.92)]}$^{0.5}$ = 13.2 ft (4.0 m), or 13 ft 3 in (4 m 4 cm), rounding to the nearest inch diameter.

  Note: Use this calculation procedure for any stack material—masonary, steel, brick, or plastic. Most boiler and stack manufacturers use charts based on the equations above to determine the economical height and diameter of a stack. Thus, the Babcock & Wilcox Company, New York, presents four charts for stack sizing, in *Steam: Its Generation and Use*. Combustion Engineering, Inc., also presents four charts for stack sizing, in *Combustion Engineering*. The equations used in the present calculation procedure are adequate for a quick, first approximation of stack height and diameter.

3. *Compute the required stack height and diameter at 5000-ft (1524.0-m) elevation*

Fuels require the same amount of oxygen for combustion regardless of the altitude at which they are burned. Therefore, this stack must provide the same draft as at sea level. But as the altitude above sea level increases, more air must be supplied to the fuel to sustain the same combustion rate, because air above sea level contains less oxygen per cubic foot than at sea level. To accommodate the larger air and flue-gas flow rate without an increase in the stack friction loss, the stack diameter must be increased.

To determine the required stack height S_e ft at an elevation above sea level, multiply the sea-level height S_h by the ratio of the sea-level and elevated-height barometric pressures, inHg. Since the barometric pressure at 5000 ft (1524.0 m) is 24.89 inHg (84.3 kPa) and the sea-level barometric pressure is 29.92 inHg (101.3 kPa), $S_e = (175)(29.92/24.89) = 210.2$ ft (64.1 m).

The stack diameter d_e ft at an elevation above sea level will vary as the 0.40 power of the ratio of the sea-level and altitude barometric pressures, or $d_e = d_s(p_e/p)^{0.4}$, where $p_e =$ barometric pressure of altitude, inHg. For this stack, $d_e = (13.2)(29.92/24.89)^{0.4} = 14.2$ ft (4.3 m), or 14 ft 3 in (4 m 34 cm).

Related Calculations: The procedure given here can be used for heating, power, marine, industrial, and residential smokestacks or chimneys, regardless of the materials used for construction. When designing smokestacks for use at altitudes above sea level, use step 3, or substitute the actual barometric pressure at the elevated location in the height and diameter equations of steps 1 and 2.

POWER-PLANT COAL-DRYER ANALYSIS

A power-plant coal dryer receives 180 tons/h (163.3 t/h) of wet coal containing 15 percent free moisture. The dryer is arranged to drain 6 percent of the moisture from the coal, and a moisture content of 1 percent is acceptable in the coal delivered to the power plant. Determine the volume and temperature of the drying gas required for the dryer, the total heat, grate area, and combustion-space volume needed. Ambient air temperature during drying is 70°F (21.1°C).

Calculation Procedure:

1. *Compute the quantity of moisture to be removed*

The total moisture in the coal = 15 percent. Of this, 6 percent is drained and 1 percent can remain in the coal. The amount of moisture to be removed is therefore $15 - 6 - 1 = 8$ percent. Since 180 tons (163.3 t) of coal are received per hour, the quantity of moisture to be removed per minute is $[180/(60 \text{ min/h})](2000 \text{ lb/ton})(0.08) = 480$ lb/min (3.6 kg/s).

2. *Compute the airflow required through the dryer*

Air enters the dryer at 70°F (21.1°C). Assume that evaporation of the moisture on the coal takes place at 125°F (51.7°C)—this is about midway in the usual evaporation temperature range of 110 to 145°F (43.3 to 62.8°C). Determine the moisture content of saturated air at each temperature, using the psychrometric chart for air. Thus, for saturated air at 70°F (21.1°C) dry-bulb temperature, the weight of the moisture it contains is w_m lb (kg) of water per pound (kilogram) of dry air = 0.0159 (0.00721), whereas at 125°F (51.7°C), $w_m = 0.09537$ lb of water per pound (0.04326 kg/kg) of dry air. The weight of water removed per pound of air passing through the dryer is the difference between the moisture content at the leaving temperature, 125°F (51.7°C), and the entering temperature, 70°F (21.1°C), or $0.09537 - 0.01590 = 0.07947$ lb of water per pound (0.03605 kg/kg) of dry air.

Since air at 70°F (21.1°C) has a density of 0.075 lb/ft³ (1.2 kg/m³), $1/0.075 = 13.3$ ft³ (0.4 m³) of air at 70°F (21.1°C) must be supplied to absorb 0.07947 lb of water per pound (0.03605 kg/kg) of dry air. With 480 lb/min (3.6 kg/s) of water to be evaporated in the dryer, each cubic foot of air will absorb $0.07947/13.3 = 0.005945$ lb (0.095 kg/m³) of moisture, and the total airflow must be $(480 \text{ lb/min})/(0.005945) = 80,800$ ft³/min (38.1 m³/s), given a dryer efficiency of 100 percent. However, the usual dryer efficiency is about 75 percent, not 100 percent. Therefore, the total actual airflow through the dryer should be $80,800/0.75 = 107,700$ ft³/min (50.8 m³/s).

Note: If desired, a table of moist air properties can be used instead of a psychrometric chart to determine the moisture content of the air at the dryer inlet and outlet conditions. The moisture content is read in the humidity ratio W_s column. See the ASHRAE—*Guide and Data Book* for such a tabulation of moist-air properties.

3. *Compute the required air temperature*

Assume that the heating air enters at a temperature t greater than 125°F (51.7°C). Set up a heat balance such that the heat given up by the air in cooling from t to 125°F (51.7°C) = the heat required to evaporate the water on the coal + the heat required to raise the temperature of the coal and water from ambient to the evaporation temperature + radiation losses.

The heat given up by the air, Btu = (cfm)(density of air, lb/ft³)[specific heat of air, Btu/(lb·°F)]$(t$ − evaporation temperature, °F). The heat required to evaporate the water, Btu = (weight of water, lb/min)$(h_{fg}$ at evaporation temperature). The heat required to raise the temperature of the coal and water from ambient to the evaporation temperature, Btu = (weight of coal, lb/min)(evaporation temperature − ambient temperature)[specific heat of coal, Btu/(lb·°F)] + (weight of water, lb/min)(evaporation temperature − ambient temperature)[specific heat of water, Btu/(lb·°F)]. The heat required to make up for radiation losses, Btu = {(area of dryer insulated surfaces, ft²)[heat-transfer coefficient, Btu/(ft²·°F·h)]$(t$ − ambient temperature) + (area of dryer uninsulated surfaces, ft²)[heat-transfer coefficient, Btu/(ft²·°F·h)]$(t$ − ambient temperature)}/60.

Compute the heat given up by the air, Btu, as $(107,700)(0.075)(0.24)(t - 70)$, where 0.075 is the air density and 0.24 is the specific heat of air.

Compute the heat required to evaporate the water, Btu, as $(480)(1022.9)$, where $1022.9 = h_{fg}$ at 125°F (51.7°C) from the steam tables.

Compute the heat required to raise the temperature of the coal and water from ambient to the evaporation temperature, Btu, as $(6000)(t - 70)(0.30) + (480)(t - 70)(1.0)$, where 0.30 is the specific heat of the coal and 1.0 is the specific heat of water.

Compute the heat required to make up the radiation losses, assuming 3000 ft² (278.7 m²) of insulated and 1500 ft² (139.4 m²) of uninsulated surface in the dryer, with coefficients of heat transfer of 0.35 and 3.0 for the insulated and uninsulated surfaces, respectively. Then radiation heat loss, Btu = $(3000)(0.35)(t - 70) + (1500)(3.0)(t - 70)$.

Set up the heat balance thus and solve for t: $(107,700)(0.075)(0.24)(t - 70) = (480)(1022.9) + (6000)(125 - 70)(0.30) + (480)(125 - 70)(1.0) + [(3000)(0.35)(t - 70) + (1500)(3.0)(t - 70)]/60$; so $t = 406°F$ (207.8°C). In this heat balance, the factor 60 is divided into the radiation heat loss to convert flow in Btu/h to Btu/min because all the other expressions are in Btu/min.

4. *Determine the total heat required by the dryer*

Using the equation of step 3 with $t = 406°F$ (207.8°C), we find the total heat = $(107,770)(0.075)(0.24)(406 - 70) = 651,000$ Btu/min, or $60(651,000) = 39,060,000$ Btu/h (11,439.7 kW).

5. *Compute the dryer-furnace grate area*

Assume that heat for the dryer is produced from coal having a lower heating value of 13,000 Btu/lb (30,238 kJ/kg) and that 40 lb/h of coal is burned per square foot [0.05 kg/(m²·s)] of grate area with a combustion efficiency of 70 percent.

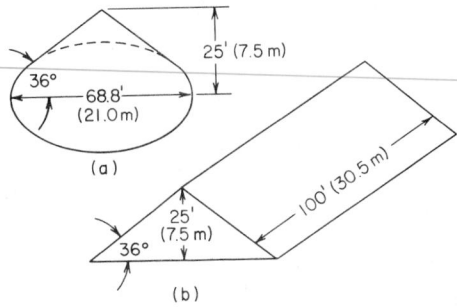

FIG. 54 (*a*) Conical coal pile; (*b*) triangular coal pile.

The rate of coal firing = (Btu/min to dryer)/(coal heating value, Btu/lb)(combustion efficiency) = 651,000/(13,000)(0.70) = 71.5 lb/min, or 60(71.5) = 4990 lb/h (0.63 kg/s). Grate area = 4990/40 = 124.75 ft^2, say 125 ft^2 (11.6 m^2).

6.   Compute the dryer-furnace volume

The usual heat-release rates for dryer furnaces are about 50,000 Btu/(h·ft^3) (517.5 kW/m^3) of furnace volume. For this furnace, which burns 4900 lb/h (0.63 kg/s) of 13,000-Btu/lb (30,238-kJ/kg) coal, the total heat released is 4990(13,000) = 64,870,000 Btu/h (18,998.8 kW). With an allowable heat release of 50,000 Btu/(h·ft^3) (517.1 kW/m^3), the required furnace volume = 64,870,000/50,000 = 1297.4 ft^3, say 1300 ft^3 (36.8 m^3).

Related Calculations: The general procedure given here can be used for any air-heated dryer used to dry moist materials. Thus, the procedure is applicable to chemical, soil, and fertilizer drying, as well as coal drying. In each case, the specific heat of the material dried must be used in place of the specific heat of coal given above.

COAL STORAGE CAPACITY OF PILES AND BUNKERS

Bituminous coal is stored in a 25-ft (7.5-m) high, 68.8-ft (21.0-m) diameter, circular-base conical pile. How many tons of coal does the pile contain if its base angle is 36°? How much bituminous coal is contained in a 25-ft (7.5-m) high rectangular pile 100 ft (30.5 m) long if the pile cross section is a triangle having a 36° base angle?

Calculation Procedure:

1.   Sketch the coal pile

Figure 54a and b shows the two coal piles. Indicate the pertinent dimensions—height, the diameter, length, and base angle—on each sketch.

2.   Compute the volume of the coal pile

Volume of a right circular cone, ft^3 = $\pi r^2 h/3$, where r = radius, ft; h = cone height, ft. Volume of a triangular pile = $bal/2$, where b = base length, ft; a = altitude, ft; l = length of pile, ft.

For this conical pile, volume = $\pi(3.4)^2(25)/3$ = 31,000 ft^3 (877.8 m^3). Since 50 lb of bituminous coal occupies about 1 ft^3 of volume (800.9 kg/m^3), the weight of coal in the conical pile = (31,000 ft^3)(50 lb/ft^3) = 1,550,000 lb, or (1,550,000 lb)/(2000 lb/ton) = 775 tons (703.1 t).

For the triangular pile, base length = $2h$/tan 36° = (2)(25)/0.727 = 68.8 ft (21.0 m). Then volume = (68.8)(25)(100)/2 = 86,000 ft^3 (2435.2 m^3). The weight of bituminous coal in the pile is, as for the conical pile, (86,000)(50) = 4,300,000 lb, or (4,300,000 lb)/(2000 lb/ton) = 2150 tons (1950.4 t).

Related Calculations: Use this general procedure to compute the weight of coal in piles of all shapes, and in bunkers, silos, bins, and similar storage compartments. The procedure can be used for other materials also—grain, sand, gravel, coke, etc. Be sure to use the correct density when converting the total storage volume to total weight. Refer to Baumeister and Marks—*Standard Handbook for Mechanical Engineers* for a comprehensive tabulation of the densities of various materials.

PROPERTIES OF A MIXTURE OF GASES

A 10-ft^3 (0.3-m^3) tank holds 1 lb (0.5 kg) of hydrogen (H$_2$), 2 lb (0.9 kg) of nitrogen (N$_2$), and 3 lb (1.4 kg) of carbon dioxide (CO$_2$) at 70°F (21.1°C). Find the specific volume, pressure, specific enthalpy, internal energy, and specific entropy of the individual gases and of the mixture and the mixture density. Use Avogadro's and Dalton's laws and Keenan and Kaye—*Thermodynamic Properties of Air, Products of Combustion and Component Gases*, Krieger, commonly termed the *Gas Tables*.

Calculation Procedure:

1.   Compute the specific volume of each gas

Using H, N, and C as subscripts for the respective gases, we see that the specific volume of any gas v ft^3/lb = total volume of tank, ft^3/weight of gas in tank, lb. Thus, v_H = 10/1 = 10 ft^3/lb

(0.6 m³/kg); v_N = 10/2 = 5 ft³/lb (0.3 m³/kg); v_C = 10/3 = 3.33 ft³/lb (0.2 m³/kg). Then the specific volume of the mixture of gases is v_t ft³/lb = total volume of gas in tank, ft³/sum of weight of individual gases, lb = 10/(1 + 2 + 3) = 1.667 ft³/lb (0.1 m³/kg).

2.   _Determine the absolute pressure of each gas_

Using $P = RTw/v_tM$, where P = absolute pressure of the gas, lb/ft² (abs); R = universal gas constant = 1545; T = absolute temperature of the gas, °R = °F + 459.9, usually taken as 460; w = weight of gas in the tank, lb; v_t = total volume of the gas in the tank, ft³; M = molecular weight of the gas. Thus, P_H = (1545)(70 + 460)(1.0)/[(10)(2.0)] = 40,530 lb/ft² (abs) (1940.6 kPa); P_N = (1545)(70 + 460)(2.0)/[(10)(28)] = 5850 lb/ft² (abs) (280.1 kPa); P_C = (1545)(70 + 460)(3.0)/[(10)(44)] = 5583 lb/ft² (abs) (267.3 kPa); $P_t = \Sigma P_H, P_N, P_C$ = 40,530 + 5850 + 5583 = 51,963 lb/ft² (abs) (2488.0 kPa).

3.   _Determine the specific enthalpy of each gas_

Refer to the _Gas Tables_, entering the left-hand column of the table at the absolute temperature, 530°R (294 K), for the gas being considered. Opposite the temperature, read the specific enthalpy in the h column. Thus, h_H = 1796.1 Btu/lb (4177.7 kJ/kg); h_N = 131.4 Btu/lb (305.6 kJ/kg); h_C = 90.17 Btu/lb (209.7 kJ/kg). The total enthalpy of the mixture of the gases is the sum of the products of the weight of each gas and its specific enthalpy, or (1)(1796.1) + (2)(131.4) + (3)(90.17) = 2329.4 Btu (2457.6 kJ) for the 6 lb (2.7 kg) or 10 ft³ (0.28 m³) of gas. The specific enthalpy of the mixture is the total enthalpy/gas weight, lb, or 2329.4/(1 + 2 + 3) = 388.2 Btu/lb (903.0 kJ/kg) of gas mixture.

4.   _Determine the internal energy of each gas_

Using the _Gas Tables_ as in step 3, we find E_H = 1260.0 Btu/lb (2930.8 kJ/kg); E_N = 93.8 Btu/lb (218.2 kJ/kg); E_C = 66.3 Btu/lb (154.2 kJ/kg). The total energy = (1)(1260.0) + (2)(93.8) + (3)(66.3) = 1646.5 Btu (1737.2 kJ). The specific enthalpy of the mixture = 1646.5/(1 + 2 + 3) = 274.4 Btu/lb (638.3 kJ/kg) of gas mixture.

5.   _Determine the specific entropy of each gas_

Using the _Gas Tables_ as in step 3, we get S_H = 15.52 Btu/(lb·°F) [65.0 kJ/(kg·°C)]; S_N = 1.558 Btu/(lb·°F) [4.7 kJ/(kg·°C)]. The entropy of the mixture = (1)(12.52) + (2)(1.558) + (3)(1.114) = 18.978 Btu/°F (34.2 kJ/°C). The specific entropy of the mixture = 18.978/(1 + 2 + 3) = 3.163 Btu/(lb·°F) [13.2 kJ/(kg·°C)] of the gas mixture.

6.   _Compute the density of the mixture_

For any gas, the total density d_t = sum of the densities of the individual gases. And since density of a gas = 1/specific volume, $d_t = 1/v_t = 1/v_H + 1/v_N + 1/v_C$ = 1/10 + 1/5 + 1/3.33 = 0.6 lb/ft³ (9.6 kg/m³) of mixture. This checks with step 1, where v_t = 1.667 ft³/lb (0.1 m³/kg), and is based on the principle that all gases occupy the same volume.

Related Calculations: Use this method for any gases stored in any type of container—steel, plastic, rubber, canvas, etc.—under any pressure from less than atmospheric to greater than atmospheric at any temperature.

STEAM INJECTION IN AIR SUPPLY

In a certain manufacturing process, a mixture of air and steam at a total mixture pressure of 300 lb/in² absolute (2068 kPa) and 400°F (204°C) is desired. The relative humidity of the mixture is to be 60 percent. For a required mixture flow rate of 500 lb/h (3.78 kg/s) determine (a) the volume flow rate of dry air in ft³/min (m³/s) of free air, where air is understood to be air at 14.7 lb/in² (101 kPa) and 70°F (21°C); and (b) the required rate of steam injection in lb/h (kg/s).

Calculation Procedure:

1.   _Determine the partial pressure of the vapor and that of the air_

From Table 1, Saturation: Temperatures of the Steam Tables mentioned under Related Calculations of this procedure, at 400°F (204°C) the steam saturation pressure, P_{vs} = 247.31 lb/in² (1705 kPa), by interpolation. Since the vapor pressure is approximately proportional to the grains of moisture in the mixture, the partial pressure of vapor in the mixture, $P_{vp} = \phi P_{vs} = 0.6 \times 247.31 = 148.4$ lb/in^2 absolute (1023 kPa), where ϕ is the relative humidity as a decimal. Then, the partial pressure of the air in the mixture, $P_a = P_m - P_{vs} = 300 - 148.4 = 151.6$ lb/in^2 absolute (1045 kPa), where P_m is the total mixture pressure.

2. Compute the density of air in the mixture

The air density, $\rho_a = P_a/(R_a T_a)$, where $P_a = 151.6 \times 144 = 21.83 \times 10^3$ lb/ft^2 (1045 kPa); the gas constant for air, $R_a = 53.3$ ft · lb/(lb · R) [287 J/(kg · K)]; absolute temperature of the air, $T_a = 400 + 460 = 860°$R (478 K). Then, $\rho_a = 21.83 \times 10^3/(53.3 \times 860) = 0.4762$ lb/ft^3 (7.63 kg/m^3).

3. Find the specific volume of the vapor in the mixture

From Table 3, Vapor of the Steam Tables, at 148.4 lb/in^2 absolute (1023 kPa) and 400°F (204°C), the specific volume of the vapor, $v_v = 3.261$ ft^3/lb (0.2036 m^3/kg), by interpolation.

4. Compute the density of the vapor and that of the mixture

The density of the vapor, $\rho_v = 1/v_v = 1/3.261 = 0.3066$ lb/ft^3 (4.91 kg/m^3). The density of the mixture, $\rho_m = \rho_a + \rho_v = 0.4762 + 0.3066 = 0.7828$ lb/ft^3 (12.54 kg/m^3).

5. Compute the amount of air in 500 lb/h (3.78 kg/s) of mixture

In 500 lb/h (3.78 kg/s) of mixture, w_m, the amount of air, $w_a = \rho_a \times w_m/\rho_m = 0.4762 \times 500/0.7828 = 304$ lb/h (2.30 kg/s).

6. Compute the flow rate of dry air

(a) The flow rate of dry air at 14.7 lb/in^2 (101 kPa) and 70°F (21°C), $V_a = w_a \times R_a \times T/P$, where the free air temperature, $T = 70 + 460 = 530°$R (294 K); free air pressure, $P = 14.7 \times 144 = 2.117 \times 10^3$ lb/ft^2 (101 kPa); other values as before. Hence, $V_a = 304 \times 53.3 \times 530/(2.117 \times 10^3) = 4060$ ft^3/h = 67.67 ft^3/min (1.92 × 10^{-3} m^3/s).

7. Compute the rate of steam injection

(b) The rate of steam injection, $w_s = w_s - w_a = 500 - 304 = 196$ lb/h (1.48 kg/s).

Related Calculations: The Steam Tables appear in *Thermodynamic Properties of Water Including Vapor, Liquid, and Solid Phases*, 1969, Keenan, et al., John Wiley & Sons, Inc. This procedure considers the air and steam as ideal gases which behave in accordance with the Gibbs-Dalton law of gas mixtures having complete homogeneous molecular dispersion and additive pressures. Also, calculations in steps 2 and 6 are based on Boyle's law and Charles' law, which relate pressure, volume, and temperature of a gas, or gas mixture. Clear and concise presentations of these and other significant definitions appear in *Thermodynamics and Heat Power*, 4th edition, by Irving Granet, Regents/Prentice-Hall, Englewood Cliffs, NJ 07632.

REGENERATIVE-CYCLE GAS-TURBINE ANALYSIS

What is the cycle air rate, lb/kWh, for a regenerative gas turbine having a pressure ratio of 5, an air inlet temperature of 60°F (15.6°C), a compressor discharge temperature of 1500°F (815.6°C), and performance in accordance with Fig. 55? Determine the cycle thermal efficiency and work ratio. What is the power output of a regenerative gas turbine if the work input to the compressor is 4400 hp (3281.1 kW)?

Calculation Procedure:

1. Determine the cycle rate

Use Fig. 55, entering at the pressure ratio of 5 in Fig. 55c and projecting to the 1500°F (815.6°C) curve. At the left, read the cycle air rate as 52 lb/kWh (23.6 kg/kWh).

2. Find the cycle thermal efficiency

Enter Fig. 55b at the pressure ratio of 5 and project vertically to the 1500°F (815.6°C) curve. At left, read the cycle thermal efficiency as 35 percent. Note that this point corresponds to the maximum efficiency obtainable from this cycle.

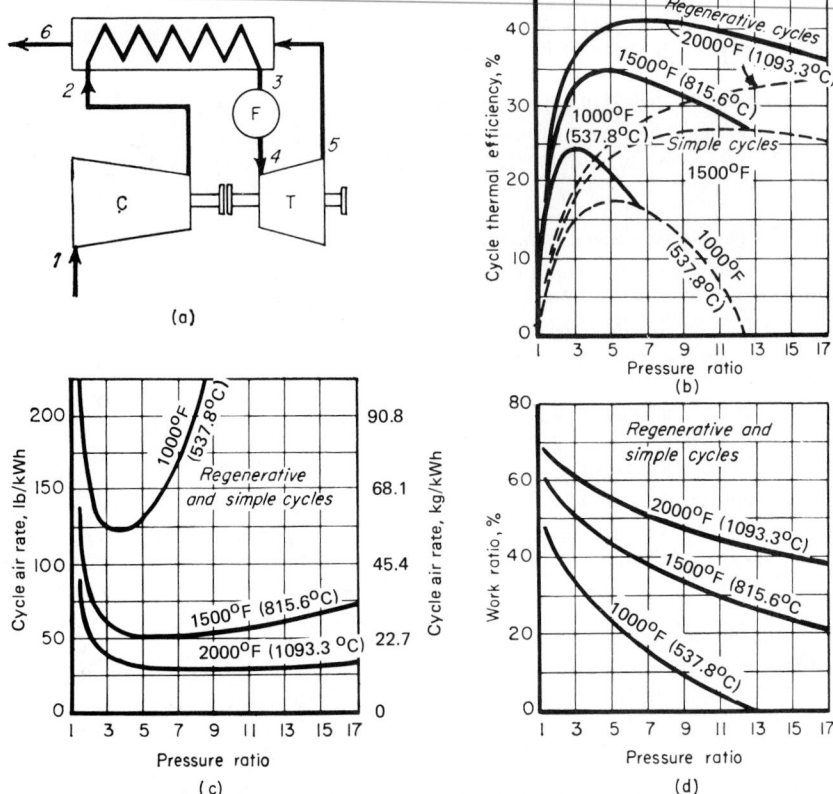

FIG. 55 (*a*) Schematic of regenerative gas turbine; (*b*), (*c*), and (*d*) gas-turbine performance based on a regenerator effectiveness of 70 percent, compressor and turbine efficiency of 85 percent; air inlet = 60°F (15.6°C); no pressure losses.

3. Find the cycle work ratio

Enter Fig. 55*d* at the pressure ratio of 5 and project vertically to the 1500°F (815.6°C) curve. At the left, read the work ratio as 44 percent.

4. Compute the turbine power output

For any gas turbine, the work ratio, percent = $100w_c/w_t$, where w_c = work input to the turbine, hp; w_t = work output of the turbine, hp. Substituting gives $44 = 100(4400)/w_t$; $w_t = 100(4400)/44 = 10,000$ hp (7457.0 kW).

   Related Calculations: Use this general procedure to analyze gas turbines for power-plant, marine, and portable applications. Where the operating conditions are different from those given here, use the manufacturer's engineering data for the turbine under consideration.

   Figure 56 shows the effect of turbine-inlet temperature, regenerator effectiveness, and compressor-inlet-air temperature on the performance of a modern gas turbine. Use these curves to analyze the cycles of gas turbines being considered for a particular application if the operating conditions are close to those plotted.

EXTRACTION TURBINE kW OUTPUT

An automatic extraction turbine operates with steam at 400 lb/in² absolute (2760 kPa), 700°F (371°C) at the throttle, its extraction pressure is 200 lb/in² (1380 kPa) and it exhausts at 110 lb/

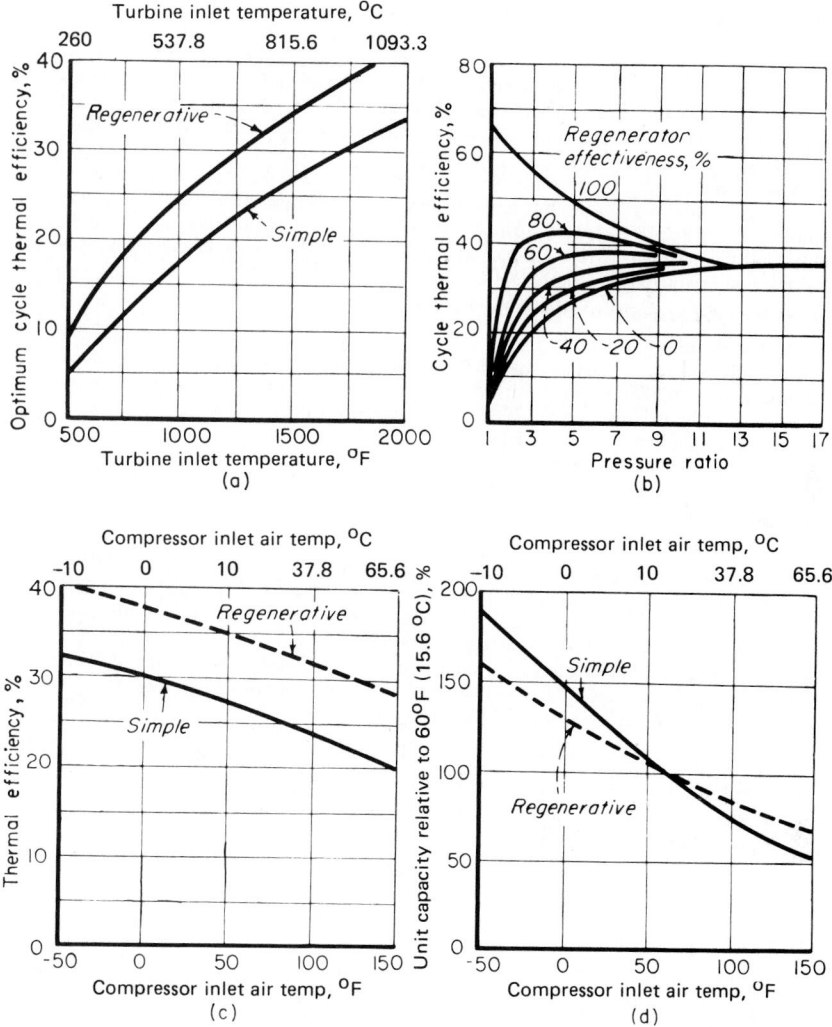

FIG. 56 (*a*) Effect of turbine-inlet on cycle performance; (*b*) effect of regenerator effectiveness; (*c*) effect of compressor inlet-air temperature; (*d*) effect of inlet-air temperature on turbine-cycle capacity. These curves are based on a turbine and compressor efficiency of 85 percent, a regenerator effectiveness of 70 percent, and a 1500°F (815.6°C) inlet-gas temperature.

in² absolute (760 kPa). At full load 80,000 lb/h (600 kg/s) is supplied to the throttle and 20,000 lb/h (150 kg/s) is extracted at the bleed point. What is the kW output?

Calculation Procedure:

1. Determine steam conditions at the throttle, bleed point, and exhaust

Steam flow through the turbine is indicated by "enter" at the throttle, "extract" at the bleed point, and "exit" at the exhaust, as shown in Fig. 57*a*. The steam process is considered to be at constant entropy, as shown by the vertical isentropic line in Fig. 57b. At the throttle, where the steam enters at the given pressure, p_1 = 400 lb/in² absolute (2760 kPa) and temperature, t_1 = 700°F (371°C), steam enthalpy, h_1 = 1362.7 Btu/lb (3167.6 kJ/kg) and its entropy, s_1 = 1.6398, as indicated by Table 3, Vapor of the Steam Tables mentioned under Related Calculations of this procedure. From the Mollier chart, a supplement to the Steam Tables, the following conditions are found along the vertical isentropic line where $s_1 = s_x = s_2$ = 1.6398 Btu/(lb · °F) (6.8655 kJ/kg · °C):

At the bleed point, where the given extraction pressure, p_x = 200 lb/in² (1380 kPa) and the entropy, s_x, is as mentioned above, the enthalpy, h_x = 1284 Btu/lb (2986 kJ/kg) and the temperature t_x = 528°F (276°C). At the exit, where the given exhaust pressure, p_2 = 110 lb/in² (760 kPa) and the entropy, s_2, is as mentioned above, the enthalpy, h_2 = 1225 Btu/lb (2849 kJ/kg) and the temperature, t_2 = 400°F (204°C).

2. Compute the total available energy to the turbine

Between the throttle and the bleed point the available energy to the turbine, $AE_1 = Q_1(h_1 - h_x)$, where the full load rate of steam flow, Q_1 = 80,000 lb/h (600 kg/s); other values are as before. Hence, AE_1 = 80,000 × (1362.7 − 1284) = 6.296 × 10⁶ Btu/h (1845 kJ/s). Between the bleed point and the exhaust the available energy to the turbine, $AE_2 = (Q_1 - Q_2)(h_x - h_2)$, where the extraction flow rate, Q_x = 20,000 lb$_m$/h (150 kg/s); other values as before. Then, AE_2 = (80,000 − 20,000)(1284 − 1225) = 3.54 × 10⁶ Btu/h (1037 kJ/s). Total available energy to the turbine, $AE = AE_1 + AE_2$ = 6.296 × 10⁶ + (3.54 × 10⁶) = 9.836 × 10⁶ Btu/h (172.8 × 10³ kJ/s).

3. Compute the turbine's kW output

The power available to the turbine to develop power at the shaft, in kilowatts, $kW = AE/(Btu/kW · h)$ = 9.836 × 10⁶/3412.7 = 2880 kW. However, the actual power developed at the shaft, $kW_a = kW × e$, where e is the mechanical efficiency of the turbine. Thus, for an efficiency, e = 0.90, then kW_a = 2880 × 0.90 = 2590 kW (2590 kJ/s).

Related Calculations: The Steam Tables appear in *Thermodynamic Properties of Water Including Vapor, Liquid, and Solid Phases*, 1969, Keenan, et al., John Wiley & Sons, Inc. Use later versions of such tables whenever available, as necessary.

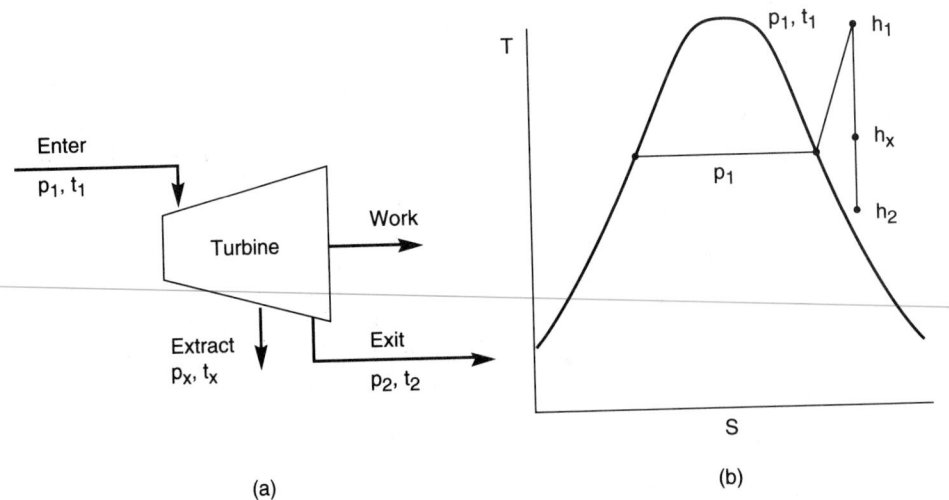

FIG. 57   (a) Turbine steam flow diagram. (b) Temperature-entropy schematic for steam flow.

Internal-Combustion Engines

REFERENCES: Benson—*Internal Combustion Engines*, Pergamon; Kates and Luck—*Diesel and High-Compression Gas Engines*, American Technical Society; Ranney—*Fuel Additives for Internal Combustion Engines*, Noyes; Blackman and Thomas—*Fuel Economy of the Gasoline Engine*, Halsted Press; Sitkei—*Heat Transfer and Thermal Loading in Internal Combustion Engines*, International Publications Services; Baxa—*Noise Control in Internal Combustion Engines*, Wiley; Diesel Engine Manufacturers Associations—*Standard Practices for Stationary Diesel Engines*; Lichty—*Internal-Combustion Engines*, McGraw-Hill; Allen—*The Modern Diesel*, Prentice-Hall; Maleev—*Internal-Combustion Engines*, McGraw-Hill; *Diesel Engineering Handbook*, Diesel Publications; Adams—*Elements of Diesel Engineering*, Henley; Severns and Degler—*Steam, Air and Gas Power*, Wiley; Ricardo—*The High-Speed Internal-Combustion Engine*, Blackie; Obert—*Internal Combustion Engines*, International Textbooks; Fors—*Practical Marine Diesel Engineering*, Simmons-Boardman.

DIESEL GENERATING UNIT EFFICIENCY

A 3000-kW diesel generating unit performs thus: fuel rate, 1.5 bbl (238.5 L) of 25° API fuel for a 900-kWh output; mechanical efficiency, 82.0 percent; generator efficiency, 92.0 percent. Compute engine fuel rate, engine-generator fuel rate, indicated thermal efficiency, overall thermal efficiency, brake thermal efficiency.

Calculation Procedure:

1. Compute the engine fuel rate

The fuel rate of an engine driving a generator is the weight of fuel, lb, used to generate 1 kWh at the generator input shaft. Since this engine burns 1.5 bbl (238.5 L) of fuel for 900 kW at the generator terminals, the total fuel consumption is (1.5 bbl)(42 gal/bbl) = 63 gal (238.5 L), at a generator efficiency of 92.0 percent.

To determine the weight of this oil, compute its specific gravity s from $s = 141.5/(131.5 + °API)$, where °API = API gravity of the fuel. Hence, $s = 141.5(131.5 + 25) = 0.904$. Since 1 gal (3.8 L) of water weighs 8.33 lb (3.8 kg) at 60°F (15.6°C), 1 gal (3.8 L) of this oil weighs $(0.904)(8.33) = 7.529$ lb (3.39 kg). The total weight of fuel used when burning 63 gal is (63 gal)(7.529 lb/gal) = 474.5 lb (213.5 kg).

The generator is 92 percent efficient. Hence, the engine actually delivers enough power to generate $900/0.92 = 977$ kWh at the generator terminals. Thus, the engine fuel rate = 474.5 lb fuel/977 kWh = 0.485 lb/kWh (0.218 kg/kWh).

2. Compute the engine-generator fuel rate

The engine-generator fuel rate takes these two units into consideration and is the weight of fuel required to generate 1 kWh at the generator terminals. Using the fuel-consumption data from step 1 and the given output of 900 kW, we see that engine-generator fuel rate = 474.5 lb fuel/900 kWh output = 0.527 lb/kWh (0.237 kg/kWh).

3. Compute the indicated thermal efficiency

Indicated thermal efficiency is the thermal efficiency based on the *indicated* horsepower of the engine. This is the horsepower developed in the engine cylinder. The engine fuel rate, computed in step 1, is the fuel consumed to produce the brake or shaft horsepower output, after friction losses are deducted. Since the mechanical efficiency of the engine is 82 percent, the fuel required to produce the indicated horsepower is 82 percent of that required for the brake horsepower, or $(0.82)(0.485) = 0.398$ lb/kWh (0.179 kg/kWh).

The indicated thermal efficiency of an internal-combustion engine driving a generator is $e_i = 3413/f_i(\text{HHV})$, where e_i = indicated thermal efficiency, expressed as a decimal; f_i = indicated fuel consumption, lb/kWh; HHV = higher heating value of the fuel, Btu/lb.

Compute the HHV for a diesel fuel from HHV = 17,680 + 60 × °API. For this fuel, HHV = 17,680 + 60(25) = 19,180 Btu/lb (44,612.7 kJ/kg).

With the HHV known, compute the indicated thermal efficiency from $e_i = 3,413/[(0.398)(19,180)] = 0.447$, or 44.7 percent.

4. Compute the overall thermal efficiency

The overall thermal efficiency e_o is computed from $e_o = 3413/f_o(\text{HHV})$, where f_o = overall fuel consumption, Btu/kWh; other symbols as before. Using the engine-generator fuel rate from step 2, which represents the overall fuel consumption $e_o = 3413/[(0.527)(19,180)] = 0.347$, or 34.7 percent.

5. Compute the brake thermal efficiency

The engine fuel rate, step 1, corresponds to the brake fuel rate f_b. Compute the brake thermal efficiency from $e_b = 3413/f_b(\text{HHV})$, where f_b = brake fuel rate, Btu/kWh; other symbols as before. For this engine-generator set, $e_b = 3413/[(0.485)(19,180)] = 0.367$, or 36.7 percent.

Related Calculations: Where the fuel consumption is given or computed in terms of lb/(hp·h), substitute the value of 2545 Btu/(hp·h) (1.0 kW/kWh) in place of the value 3413 Btu/kWh (3600.7 kJ/kWh) in the numerator of the e_i, e_o, and e_b equations. Compute the indicated, overall, and brake thermal efficiencies as before. Use the same procedure for gas and gasoline engines, except that the higher heating value of the gas or gasoline should be obtained from the supplier or by test.

ENGINE DISPLACEMENT, MEAN EFFECTIVE PRESSURE, AND EFFICIENCY

A 12×18 in (30.5×44.8 cm) four-cylinder four-stroke single-acting diesel engine is rated at 200 bhp (149.2 kW) at 260 r/min. Fuel consumption at rated load is 0.42 lb/(bhp·h) (0.25 kg/kWh). The higher heating value of the fuel is 18,920 Btu/lb (44,008 kJ/kg). What are the brake mean effective pressure, engine displacement in ft³/(min·bhp), and brake thermal efficiency?

Calculation Procedure:

1. Compute the brake mean effective pressure

Compute the brake mean effective pressure (bmep) for an internal-combustion engine from $bmep = 33,000\,bhp_n/LAn$, where $bmep$ = brake mean effective pressure, lb/in²; bhp_n = brake horsepower output delivered per cylinder, hp; L = piston stroke length, ft; a = piston area, in²; n = cycles per minute per cylinder = crankshaft rpm for a two-stroke cycle engine, and 0.5 the crankshaft rpm for a four-stroke cycle engine.

For this engine at its rated bhp, the output per cylinder is 200 bhp/4 cylinders = 50 bhp (37.3 kW). Then $bmep = 33,000(50)/[(18/12)(12)^2(\pi/4)(260/2)] = 74.8$ lb/in² (516.1 kPa). (The factor 12 in the denominator converts the stroke length from inches to feet.)

2. Compute the engine displacement

The total engine displacement V_d ft³ is given by $V_d = LAnN$, where A = piston area, ft²; N = number of cylinders in the engine; other symbols as before. For this engine, $V_d = (18/12)(12/12)^2(\pi/4)(260/2)(4) = 614$ ft³/min (17.4 m³/min). The displacement is in cubic feet per minute because the crankshaft speed is in r/min. The factor of 12 in the denominators converts the stroke and area to ft and ft², respectively. The displacement per bhp = (total displacement, ft³/min)/bhp output of engine = $614/200 = 3.07$ ft³/(min·bhp) (0.12 m³/kW).

3. Compute the brake thermal efficiency

The brake thermal efficiency e_b of an internal-combustion engine is given by $e_b = 2545/(sfc)(\text{HHV})$, where sfc = specific fuel consumption, lb/(bhp·h); HHV = higher heating value of fuel, Btu/lb. For this engine, $e_b = 2545/[(0.42)(18,920)] = 0.32$, or 32.0 percent.

Related Calculations: Use the same procedure for gas and gasoline engines. Obtain the higher heating value of the fuel from the supplier, a tabulation of fuel properties, or by test.

ENGINE MEAN EFFECTIVE PRESSURE AND HORSEPOWER

A 500-hp (373-kW) internal-combustion engine has a brake mean effective pressure of 80 lb/in² (551.5 kPa) at full load. What are the indicated mean effective pressure and friction mean effective pressure if the mechanical efficiency of the engine is 85 percent? What are the indicated horsepower and friction horsepower of the engine?

Calculation Procedure:

1. Determine the indicated mean effective pressure

Indicated mean effective pressure $imep$ lb/in^2 for an internal-combustion engine is found from $imep = bmep/e_m$, where $bmep$ = brake mean effective pressure, lb/in^2; e_m = mechanical efficiency, percent, expressed as a decimal. For this engine, $imep = 80/0.85 = 94.1$ lb/in^2 (659.3 kPa).

2. Compute the friction mean effective pressure

For an internal-combustion engine, the friction mean effective pressure $fmep$ lb/in^2 is found from $fmep = imep - bmep$, or $fmep = 94.1 - 80 = 14.1$ lb/in^2 (97.3 kPa).

3. Compute the indicated horsepower of the engine

For an internal-combustion engine, the mechanical efficiency $e_m = bhp/ihp$, where ihp = indicated horsepower. Thus, $ihp = bhp/e_m$, or $ihp = 500/0.85 = 588$ ihp (438.6 kW).

4. Compute the friction hp of the engine

For an internal-combustion engine, the friction horsepower is $fhp = ihp - bhp$. In this engine, $fhp = 588 - 500 = 88$ fhp (65.6 kW).

Related Calculations: Use a similar procedure to determine the *indicated engine efficiency* $e_{ei} = e_i/e$, where e = ideal cycle efficiency; *brake engine efficiency*, $e_{eb} = e_b e$; *combined engine efficiency* or *overall engine thermal efficiency* $e_{eo} = e_o/e$. Note that each of these three efficiencies is an *engine* efficiency and corresponds to an actual thermal efficiency, e_i, e_b, and e_o.

Engine efficiency $e_e = e_t/e$, where e_t = actual *engine* thermal efficiency. Where desired, the respective *actual* indicated brake, or overall, output can be susbstituted for e_i, e_b, and e_o in the numerator of the above equations if the ideal output is substituted in the denominator. The result will be the respective engine efficiency. Output can be expressed in Btu per unit time, or horsepower. Also, e_e = actual *mep*/ideal *mep*, and e_{ei} = *imep*/ideal *mep*; e_{eb} = *bmep*/ideal *mep*; e_{eo} = overall *mep*/ideal *mep*. Further, $e_b = e_m e_i$, and $bmep = e_m(imep)$. Where the actual heat supplied by the fuel, HHV Btu/lb, is known, compute $e_i e_b$ and e_o by the method given in the previous calculation procedure. The above relations apply to any reciprocating internal-combustion engine using any fuel.

SELECTION OF AN INDUSTRIAL INTERNAL-COMBUSTION ENGINE

Select an internal-combustion engine to drive a centrifugal pump handling 2000 gal/min (126.2 L/s) of water at a total head of 350 ft (106.7 m). The pump speed will be 1750 r/min, and it will run continuously. The engine and pump are located at sea level.

Calculation Procedure:

1. Compute the power input to the pump

The power required to pump water is $hp = 8.33GH/33,000e$, where G = water flow, gal/min; H = total head on the pump, ft of water; e = pump efficiency, expressed as a decimal. Typical centrifugal pumps have operating efficiencies ranging from 50 to 80 percent, depending on the pump design and condition and liquid handled. Assume that this pump has an efficiency of 70 percent. Then $hp = 8.33(2000)/(350)/[(33,000)(0.70)] = 252$ hp (187.9 kW). Thus, the internal-combustion engine must develop at least 252 hp (187.9 kW) to drive this pump.

2. Select the internal-combustion engine

Since the engine will run continuously, extreme care must be used in its selection. Refer to a tabulation of engine ratings, such as Table 1. This table shows that a diesel engine that delivers 275 continuous brake horsepower (205.2 kW) (the nearest tabulated rating equal to or greater than the required input) will be rated at 483 bhp (360.3 kW) at 1750 r/min.

The gasoline-engine rating data in Table 1 show that for continuous full load at a given speed, 80 percent of the tabulated power can be used. Thus, at 1750 r/min, the engine must be rated at $252/0.80 = 315$ bhp (234.9 kW). A 450-hp (335.7-kW) unit is the only one shown in Table 1 that would meet the needs. This is too large; refer to another builder's rating table to find an engine rated at 315 to 325 bhp (234.9 to 242.5 kW) at 1750 r/min.

TABLE 1　Internal-Combustion Engine Rating Table

Diesel engines						
Continuous bhp (kW) at given rpm				Rated bhp	No. of cylinders	Cooling°
1400	1600	1750	1800			
187 (139.5)	214 (159.6)	227 (169.3)	230 (171.6)	300 (223.8)	6	E
230 (171.6)	256 (190.0)	275 (205.2)	280 (208.9)	438 (326.7)	12	R
240 (179.0)	273 (203.7)	295 (220.0)	305 (227.5)	438 (326.7)	12	E

Gasoline engines†						
405 (302.1)	430 (320.8)	450 (335.7)	475 (354.4)	595 (438.9)	12	R

°E = heat-exchanger-cooled; R = radiator-cooled.
†Use 80 percent of tabulated power if engine is to run at continuous full load.

The unsuitable capacity range in the gasoline-engine section of Table 1 is a typical situation met in selecting equipment. More time is often spent in finding a suitable unit at an acceptable price than is spent computing the required power output.

Related Calculations: Use this procedure to select any type of reciprocating internal-combustion engine using oil, gasoline, liquefied-petroleum gas, or natural gas for fuel.

ENGINE OUTPUT AT HIGH TEMPERATURES AND HIGH ALTITUDES

An 800-hp (596.8-kW) diesel engine is operated 10,000 ft (3048 m) above sea level. What is its output at this elevation if the intake air is at 80°F (26.7°C)? What will the output at 10,000-ft (3048-m) altitude be if the intake air is at 110°F (43.4°C)? What would the output be if this engine were equipped with an exhaust turbine-driven blower?

Calculation Procedure:

1. *Compute the engine output at altitude*

Diesel engines are rated at sea level at atmospheric temperatures of not more than 90°F (32.3°C). The sea-level rating applies at altitudes up to 1500 ft (457.2 m). At higher altitudes, a correction factor for elevation must be applied. If the atmospheric temperature is higher than 90°F (32.2°C), a temperature correction must be applied.

Table 2 lists both altitude and temperature correction factors. For an 800-hp (596.8-kW) engine at 10,000 ft (3048 m) above sea level and 80°F (26.7°C) intake air, hp output = (sea-level hp) (altitude correction factor), or output = (800)(0.68) = 544 hp (405.8 kW).

TABLE 2　Correction Factors for Altitude and Temperature

Engine altitude		Engine type		Intake temperature		Correction factor
ft	m	Nonsupercharged	Supercharged	°F	°C	
7,000	2,134	0.780	0.820	90 or less	32.3 or less	1.000
8,000	2,438	0.745	0.790	95	35	0.986
9,000	2,743	0.712	0.765	100	37.8	0.974
10,000	3,048	0.680	0.740	105	40.6	0.962
12,000	3,658	0.612	0.685	110	43.3	0.950
				115	46.1	0.937
				120	48.9	0.925
				125	51.7	0.913
				130	54.4	0.900

TABLE 3 Atmospheric Pressure at Various Altitudes

Altitude		Pressure	
ft	m	inHg	mm
Sea Level		29.92	759.97
4,000	1,219	25.84	656.3
5,000	1,524	24.89	632.2
6,000	1,829	23.98	609.1
8,000	2,438	22.22	564.4
10,000	3,048	20.58	522.7
12,000	3,658	19.03	483.4

Note: A 500- to 1500-ft altitude is considered equivalent to sea level by the Diesel Engine Manufacturers Association if the atmospheric pressure is not less than 28.25 inHg (717.6 mmHg).

2. *Compute the engine output at the elevated temperature*

When the intake air is at a temperature greater than 90°F (32.3°C), a temperature correction factor must be applied. Then output = (sea-level hp)(altitude correction factor)(intake-air-temperature correction factor), or output = (800)(0.68)(0.95) = 516 hp (384.9 kW), with 110°F (43.3°C) intake air.

3. *Compute the output of a supercharged engine*

A different altitude correction factor is used for a supercharged engine, but the same temperature correction factor is applied. Table 2 lists the altitude correction factors for supercharged diesel engines. Thus, for this supercharged engine at 10,000-ft (3048-m) altitude with 80°F (26.7°C) intake air, output = (sea-level hp)(altitude correction factor) = (800)(0.74) = 592 hp (441.6 kW).

At 10,000-ft (3048-m) altitude with 110°F (43.3°C) inlet air, output = (sea-level hp)(altitude correction factor)(temperature correction factor) = (800)(0.74)(0.95) = 563 hp (420.1 kW).

Related Calculations: Use the same procedure for gasoline, gas, oil, and liquefied-petroleum gas engines. Where altitude correction factors are not available for the type of engine being used, other than a diesel, multiply the engine sea-level brake horsepower by the ratio of the altitude-level atmospheric pressure to the atmospheric pressure at sea level. Table 3 lists the atmospheric pressure at various altitudes.

An engine located below sea level can theoretically develop more power than at sea level because the intake air is denser. However, the greater potential output is generally ignored in engine-selection calculations.

INDICATOR USE ON INTERNAL-COMBUSTION ENGINES

An indicator card taken on an internal-combustion engine cylinder has an area of 5.3 in^2 (34.2 cm^2) and a length of 4.95 in (12.7 cm). What is the indicated mean effective pressure in this cylinder? What is the indicated horsepower of this four-cycle engine if it has eight 6-in (15.6-cm) diameter cylinders, an 18-in (45.7-cm) stroke, and operates at 300 r/min? The indicator spring scale is 100 lb/in (1.77 kg/mm).

Calculation Procedure:

1. *Compute the indicated mean effective pressure*

For any indicator card, *imep* = (card area, in^2) (indicator spring scale, lb)/(length of indicator card, in), where *imep* = indicated mean effective pressure, lb/in^2. Thus, for this engine, *imep* = (5.3)(100)/4.95 = 107 lb/in^2 (737.7 kPa).

2. *Compute the indicated horsepower*

For any reciprocating internal-combustion engine, $ihp = (imep)LAn/33,000$, where ihp = indicated horsepower per cylinder; L = piston stroke length, ft; A = piston area, in^2; n = number of cycles/min. Thus, for this four-cycle engine where n = 0.5 r/min, $ihp = (107)(18/12)(6)^2(\pi/4)(300/2)/33,000$ = 20.6 ihp (15.4 kW) per cylinder. Since the engine has eight cylinders, total ihp = (8 cylinders)(20.6 ihp per cylinder) = 164.8 ihp (122.9 kW).

Related Calculations: Use this procedure for any reciprocating internal-combustion engine using diesel oil, gasoline, kerosene, natural gas, liquefied-petroleum gas, or similar fuel.

ENGINE PISTON SPEED, TORQUE, DISPLACEMENT, AND COMPRESSION RATIO

What is the piston speed of an 18-in (45.7-cm) stroke 300 = r/min engine? How much torque will this engine deliver when its output is 800 hp (596.8 kW)? What are the displacement per cylinder and the total displacement if the engine has eight 12-in (30.5-cm) diameter cylinders? Determine the engine compression ratio if the volume of the combustion chamber is 9 percent of the piston displacement.

Calculation Procedure:

1. *Compute the engine piston speed*

For any reciprocating internal-combustion engine, piston speed = $fpm = 2L(rpm)$, where L = piston stroke length, ft; rpm = crankshaft rotative speed, r/min. Thus, for this engine, piston speed = $2(18/12)(300)$ = 9000 ft/min (2743.2 m/min).

2. *Determine the engine torque*

For any reciprocating internal-combustion engine, $T = 63,000(bhp)/rpm$, where T = torque developed, in·lb; bhp = engine brake horsepower output; rpm = crankshaft rotative speed, r/min. Or $T = 63,000(800)/300$ = 168,000 in·lb (18,981 N·m).

Where a prony brake is used to measure engine torque, apply this relation: $T = (F_b - F_o)r$, where F_b = brake scale force, lb, with engine operating; F_o = brake scale force with engine stopped and brake loose on flywheel; r = brake arm, in = distance from flywheel center to brake knife edge.

3. *Compute the displacement*

The displacement per cylinder d_c in^3 of any reciprocating internal-combustion engine is $d_c = L_iA_i$ where L_i = piston stroke, in; A_i = piston head area, in^2. For this engine, $d_c = (18)(12)^2(\pi/4)$ = 2035 in^3 (33,348 cm^3) per cylinder.

The total displacement of this eight-cylinder engine is therefore (8 cylinders)(2035 in^3 per cylinder) = 16,280 in^3 (266,781 cm^3).

4. *Compute the compression ratio*

For a reciprocating internal-combustion engine, the compression ratio $r_c = V_b/V_a$, where V_b = cylinder volume at the start of the compression stroke, in^3 or ft^3; V_a = combustion-space volume at the end of the compression stroke, in^3 or ft^3. When this relation is used, both volumes must be expressed in the same units.

In this engine, V_b = 2035 in^3 (33,348 cm^3); V_a = (0.09)(2035) = 183.15 in^3. Then r_c = 2035/183.15 = 11.1:1.

Related Calculations: Use these procedures for any reciprocating internal-combustion engine, regardless of the fuel burned.

INTERNAL-COMBUSTION ENGINE COOLING-WATER REQUIREMENTS

A 1000-bhp (746-kW) diesel engine has a specific fuel consumption of 0.360 lb/(bhp·h) (0.22 kg/kWh). Determine the cooling-water flow required if the higher heating value of the fuel is 10,350 Btu/lb (24,074 kJ/kg). The net heat rejection rates of various parts of the engine are, in percent: jacket water, 11.5; turbocharger, 2.0; lube oil, 3.8; aftercooling, 4.0; exhaust, 34.7; radiation, 7.5; How much 30 lb/in^2 (abs) (206.8 kPa) steam can be generated by the exhaust gas if this is a four-cycle engine? The engine operates at sea level.

Calculation Procedure:

1. Compute the engine heat balance

Determine the amount of heat used to generate 1 bhp·h (0.75 kWh) from: heat rate, Btu/(bhp·h) = (sfc)(HHV), where sfc = specific fuel consumption, lb/(bhp·h); HHV = higher heating value of fuel, Btu/lb. Or, heat rate = (0.36)(19,350) = 6967 Btu/(bhp·h) (2737.3 W/kWh).

Compute the heat balance of the engine by taking the product of the respective heat rejection percentages and the heat rate as follows:

			Btu/(bhp·h)	W/kWh
Jacket water	(0.115)(6967)	=	800	314.3
Turbocharger	(0.020)(6967)	=	139	54.6
Lube oil	(0.038)(6967)	=	264	103.7
Aftercooling	(0.040)(6967)	=	278	109.2
Exhaust	(0.347)(6967)	=	2420	880.1
Radiation	(0.075)(6967)	=	521	204.7
Total heat loss		=	4422	1666.6

Then the power output = 6967 − 4422 = 2545 Btu/(bhp·h) (999.9 W/kWh), or 2545/6967 = 0.365, or 36.5 percent. Note that the sum of the heat losses and power generated, expressed in percent, is 100.0.

2. Compute the jacket cooling-water flow rate

The jacket water cools the jackets and the turbocharger. Hence, the heat that must be absorbed by the jacket water is 800 + 139 = 939 Btu/(bhp·h) (369 W/kWh), using the heat rejection quantities computed in step 1. When the engine is developing its full rated output of 1000 bhp (746 kW), the jacket water must absorb [939 Btu/(bhp·h)](1000 bhp) = 939,000 Btu/h (275,221 W).

Apply a safety factor to allow for scaling of the heat-transfer surfaces and other unforeseen difficulties. Most designers use a 10 percent safety factor. Applying this value of the safety factor for this engine, we see the total jacket-water heat load = 939,000 + (0.10)(939,000) = 1,032,900 Btu/h (302.5 kW).

Find the required jacket-water flow from $G = H/500\Delta t$, where G = jacket-water flow, gal/min; H = heat absorbed by jacket water, Btu/h; Δt = temperature rise of the water during passage through the jackets, °F. The usual temperature rise of the jacket water during passage through a diesel engine is 10 to 20°F (5.6 to 11.1°C). Using 10°F for this engine, we find G = 1,032,900/[(500)(10)] = 206.58 gal/min (13.03 L/s), say 207 gal/min (13.06 L/s).

3. Determine the water quantity for radiator cooling

In the usual radiator cooling system for large engines, a portion of the cooling water is passed through a horizontal or vertical radiator. The remaining water is recirculated, after being tempered by the cooled water. Thus, the radiator must dissipate the jacket, turbocharger, and lube-oil cooler heat.

The lube oil gives off 264 Btu/(bhp·h) (103.8 W/kWh). With a 10 percent safety factor, the total heat flow is 264 + (0.10)(264) = 290.4 Btu/(bhp·h) (114.1 W/kWh). At the rated output of 1000 bhp (746 kW), the lube-oil heat load = [290.4 Btu/(bhp·h)](1000 bhp) = 290,400 Btu/h (85.1 kW). Hence, the total heat load on the radiator = jacket + lube-oil heat load = 1,032,900 + 290,400 = 1,323,300 Btu/h (387.8 kW).

Radiators (also called fan coolers) serving large internal-combustion engines are usually rated for a 35°F (19.4°C) temperature reduction of the water. To remove 1,323,300 Btu/h (387.8 kW) with a 35°F (19.4°C) temperature decrease will require a flow of $G = H/(500\Delta t)$ = 1,323,300/[(500)(35)] = 76.1 gal/min (4.8 L/s).

4. Determine the aftercooler cooling-water quantity

The aftercooler must dissipate 278 Btu/(bhp·h) (109.2 W/kWh). At an output of 1000 bhp (746 kW), the heat load = [278 Btu/(bhp·h)](1000 bhp) = 278,000 Btu/h (81.5 kW). In general,

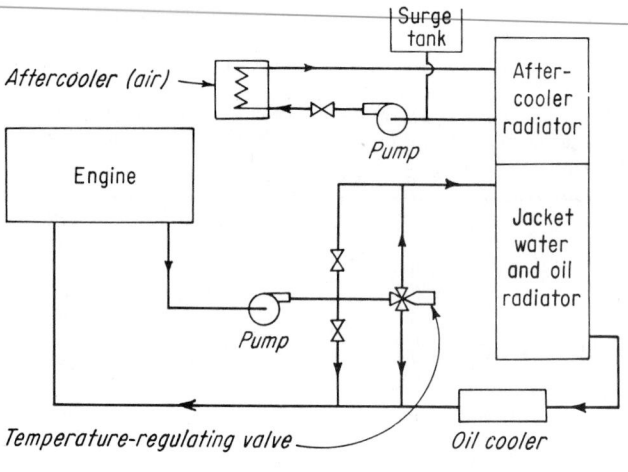

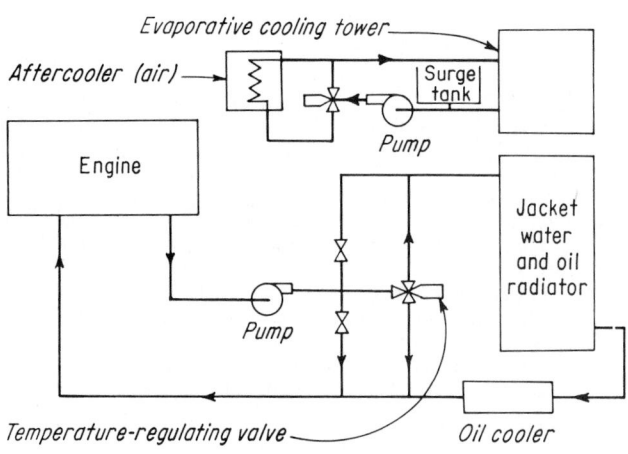

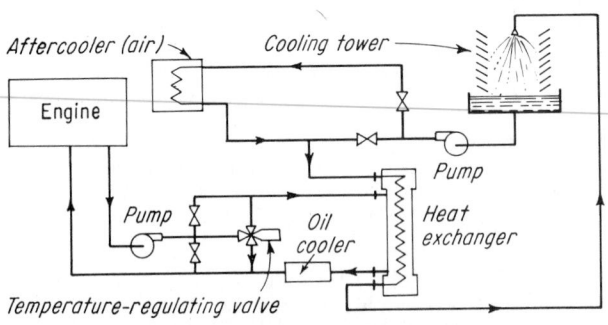

FIG. 1 Internal-combustion engine cooling systems: (*a*) radiator type; (*b*) evaporating cooling tower; (*c*) cooling tower. (*Power.*)

Bhp (kW)	Engine jackets			Oil cooler			Turbo aftercooler		
	Btu/h (kW)	Q_{JW} gal/min (L/s)	Q_R gal/min (L/s)	Btu/h (kW)	Q_R gal/min (L/s)	Q_O gal/min (L/s)	Btu/h (kW)	Q_A lb/s (kg/s)	Q_{AW} gal/min (L/s)
1000 (750)	1,032,000 (302.5)	207 (13.1)	—	290,000 (85)	75 (4.7)	140 (8.8)	278,000 (81.5)	3.3 (1.5)	110 (6.9)
	1,322,000 (387.5)	—	75 (4.7)						

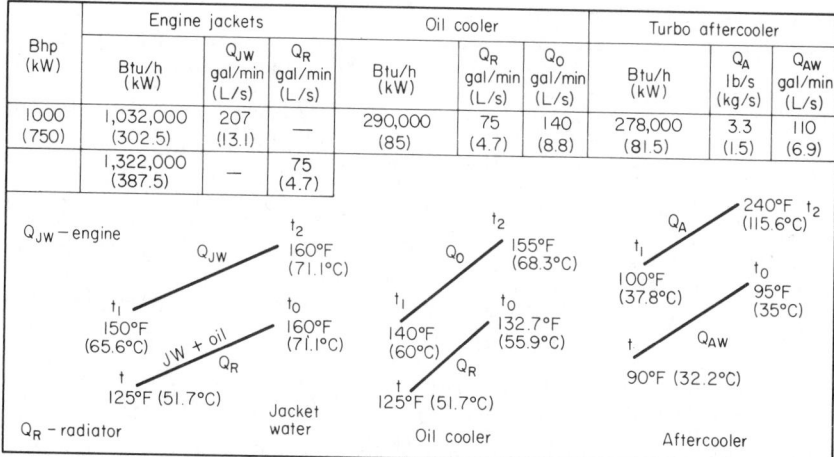

FIG. 2 Slant diagrams for internal-combustion engine heat exhangers. *(Power.)*

designers do not use a factor of safety for the aftercooler because there is less chance of fouling or other difficulties.

With a 5°F (2.8°C) temperature rise of the cooling water during passage through the aftercooler, the quantity of water required $G = H/(500\Delta t) = 278,000/[(500)(5)] = 111$ gal/min (7.0 L/s).

5. Compute the quantity of steam generated by the exhaust

Find the heat available in the exhaust by using $H_e = Wc\Delta t_e$, where H_e = heat available in the exhaust, Btu/h; W = exhaust-gas flow, lb/h; c = specific heat of the exhaust gas = 0.252 Btu/(lb·°F) (2.5 kJ/kg); Δt_e = exhaust-gas temperature at the boiler inlet, °F − exhaust-gas temperature at the boiler outlet,°F.

The exhaust-gas flow from a four-cycle turbocharged diesel is about 12.5 lb/(bhp·h) (7.5 kg/kWh). At full load this engine will exhaust [12.5 lb/(bhp·h)](1000 bhp) = 12,500 lb/h (5625 kg/h).

The temperature of the exhaust gas will be about 750°F (399°C) at the boiler inlet, whereas the temperature at the boiler outlet is generally held at 75°F (41.7°C) higher than the steam temperature to prevent condensation of the exhaust gas. Steam at 30 lb/in² (abs) (206.8 kPa) has a temperature of 250.33°F (121.3°C). Thus, the exhaust-gas outlet temperature from the boiler will be 250.33 + 75 = 325.33°F (162.9°C), say 325°F (162.8°C). Then H_e = (12,500)(0.252)(750 − 325) = 1,375,000 Btu/h (403.0 kW).

At 30 lb/in² (abs) (206.8 kPa),the enthalpy of vaporization of steam is 945.3 Btu/lb (2198.9 kJ/kg), found in the steam tables. Thus, the exhaust heat can generate 1,375,000/945.3 = 1415 lb/h (636.8 kg/h) if the boiler is 100 percent efficient. With a boiler efficiency of 85 percent, the steam generated = (1415 lb/h)(0.85) = 1220 lb/h (549.0 kg/h), or (1220 lb/h)/1000 bhp = 1.22 lb/(bhp·h) (0.74 kg/kWh).

Related Calculations: Use this procedure for any reciprocating internal-combustion engine burning gasoline, kerosene, natural gas, liquified-petroleum gas, or similar fuel. Figure 1 shows typical arrangements for a number of internal-combustion engine cooling systems.

When ethylene glycol or another antifreeze solution is used in the cooling system, alter the denominator of the flow equation to reflect the change in specific gravity and specific heat of the antifreeze solution, as compared with water. Thus, with a mixture of 50 percent glycol and 50 percent water, the flow equation in step 2 becomes $G = H/(436\Delta t)$. With other solutions, the numerical factor in the denominator will change. This factor = (weight of liquid, lb/gal)(60 min/h), and the factor converts a flow rate of lb/h to gal/min when divided into the lb/h flow rate. Slant diagrams, Fig. 2, are often useful for heat-exchanger analysis.

TABLE 4   Total Air Volume Needs[°]

Set kW	ft³/min (m³/min) for combustion	ft³/min (m³/min) for radiator	Maximum room temperature rise	
			Maximum ambient temperature of inlet air, °F (°C)	Room air rise, °F (°C)
20	130 (3.7)	3000 (84.9)	90 (32.2)	20 (11.1)
30	195 (5.5)	5000 (141.6)	95–105 (35–40.6)	15 (8.3)
40	260 (7.4)	5500 (155.7)	110–120 (43.3–48.9)	10 (5.6)
60	390 (1.0)	6000 (169.9)		

[°] *Power.*

Two-cycle engines may have a larger exhaust-gas flow than four-cycle engines because of the scavenging air. However, the exhaust temperature will usually be 50 to 100°F (27.7 to 55.6°C) lower, reducing the quantity of steam generated.

Where a dry exhaust manifold is used on an engine, the heat rejection to the cooling system is reduced by about 7.5 percent. Heat rejected to the aftercooler cooling water is about 3.5 percent of the total heat input to the engine. About 2.5 percent of the total heat input to the engine is rejected by the turbocharger jacket.

The jacket cooling water absorbs 11 to 14 percent of the total heat supplied. From 3 to 6 percent of the total heat supplied to the engine is rejected in the oil cooler.

The total heat supplied to an engine = (engine output, bhp)[heat rate, $Btu/(bhp \cdot h)$]. A jacket-water flow rate of 0.25 to 0.60 gal/(min·bhp) (0.02 to 0.05 kg/kW) is usually recommended. The normal jacket-water temperature rise is 10°F (5.6°C); with a jacket-water outlet temperature of 180°F (82.2°C) or higher, the temperature rise of the jacket water is usually held to 7°F (3.9°C) or less.

To keep the cooling-water system pressure loss within reasonable limits, some designers recommend a pipe velocity equal to the nominal pipe size used in the system, or 2 ft/s for 2-in pipe (0.6 m/s for 50.8-mm); 3 ft/s for 3-in pipe (0.9 m/s for 76.2-mm); etc. The maximum recommended velocity is 10 ft/s for 10 in (3.0 m/s for 254 mm) and larger pipes. Compute the actual pipe diameter from $d = (G/2.5v)^{0.5}$, where G = cooling-water flow, gal/min; v = water velocity, ft/s.

Air needed for a four-cycle high-output turbocharged diesel engine is about 3.5 ft³/(min·bhp) (0.13 m³/kW); 4.5 ft³/(min·bhp) (0.17 m³/kW) for two-cycle engines. Exhaust-gas flow is about 8.4 ft³/(min·bhp) (0.32 m³/kW) for a four-cycle diesel engine; 13 ft³/(min·bhp) (0.49 m³/kW) for two-cycle engines. Air velocity in the turbocharger blower piping should not exceed 3300 ft/min (1006 m/min); gas velocity in the exhaust system should not exceed 6000 ft/min (1828 m/

TABLE 5   Heat Radiated from Typical Internal-Combustion Units, Btu/min (W)[°]

	Cooling by radiator and fan		Cooling by radiator, fan, and city water	
Alternator, kW	40	60	40	60
Engine-alternator set, silencer, and 25 ft (7.6 m) of exhaust pipe, Btu/min (W)	1830 (8.94)	2625 (12.8)	1701 (8.3)	2500 (12.2)
Exhaust pipe beyond silencer:				
Length 5 ft (1.5 m)	24 (0.12)	35 (0.17)	20 (0.10)	22 (0.11)
Length 10 ft (3.0 m)	45 (0.22)	65 (0.32)	39 (0.19)	40 (0.20)
Length 15 ft (4.6 m)	65 (0.32)	89 (0.44)	57 (0.38)	55 (0.27)

[°] *Power.*

TABLE 6 Range of Discharge Temperature°

| Room fan discharge temperature range | | | Wind to water gage | | | |
| | | | Wind velocity | | Inlet pressure water gage | |
°F	°C	K	mph	km/h	in	mm
80–89	26.7–31.7	57	60	96.5	1.75	44.5
90–99	32.3–37.2	58	30	48.3	0.43	10.9
100–110	37.8–43.3	59				
111–120	43.9–48.9	60				
121–130	49.4–54.4	61				

° *Power.*

min). The exhaust-gas temperature should not be reduced below 275°F (135°C), to prevent condensation.

The method presented here is the work of W. M. Kauffman, reported in *Power*.

DESIGN OF A VENT SYSTEM FOR AN ENGINE ROOM

A radiator-cooled 60-kW internal-combustion engine generating set operates in an area where the maximum summer ambient temperature of the inlet air is 100°F (37.8°C). How much air does this engine need for combustion and for the radiator? What is the maximum permissible temperature rise of the room air? How much heat is radiated by the engine-alternator set if the exhaust pipe is 25 ft (7.6 m) long? What capacity exhaust fan is needed for this engine room if the engine room has two windows with an area of 30 ft² (2.8 m²) each, and the average height between the air inlet and outlet is 5 ft (1.5 m)? Determine the rate of heat dissipation by the windows. The engine is located at sea level.

Calculation Procedure:

1. *Determine engine air-volume needs*

Table 4 shows typical air-volume needs for internal-combustion engines installed indoors. Thus, a 60-kW set requires 390 ft³/min (11.0 m³/min) for combustion and 6000 ft³/min (169.9 m³/min) for the radiator. Note that in the smaller ratings, the combustion air needed is 6.5 ft³/(min · kW)(0.18 m³/kW), and the radiator air requirement is 150 ft³/(min·kW)(4.2 m³/kW).

2. *Determine maximum permissible air temperature rise*

Table 4 also shows that with an ambient temperature of 95 to 105°F (35 to 40.6°C), the maximum permissible room temperature rise is 15°F (8.3°C). When you determine this value, be certain to use the highest inlet air temperature expected in the engine locality.

TABLE 7 Air Density at Various Elevations°

| Elevation above sea level | | Multiplying factor, A | Approximate air density percent compared with sea level for same temperature |
ft	m		
4,000	1,219	1.158	86.4
5,000	1,524	1.202	83.2
6,000	1,829	1.247	80.2
7,000	2,134	1.296	77.2
10,000	3,048	1.454	68.8

° *Power.*

3. Determine the heat radiated by the engine

Table 5 shows the heat radiated by typical internal-combustion engine generating sets. Thus, a 60-kW radiator-and fan-cooled set radiates 2625 Btu/min (12.8 W) when the engine is fitted with a 25-ft (7.6-m) long exhaust pipe and a silencer.

4. Compute the airflow produced by the windows

The two windows can be used to ventilate the engine room. One window will serve as the air inlet; the other, as the air outlet. The area of the air outlet must at least equal the air-inlet area. Airflow will be produced by the stack effect resulting from the temperature difference between the inlet and outlet air.

The airflow C ft^3/min resulting from the stack effect is $C = 9.4A(h\Delta t_a)^{0.5}$, where A = free area of the air inlet, ft^2; h = height from the middle of the air-inlet opening to the middle of the air-outlet opening, ft; Δt_a = difference between the average indoor air temperature at point H and the temperature of the incoming air, °F. In this plant, the maximum permissible air temperature rise is 15°F (8.3°C), from step 2. With a 100°F (37.8°C) outdoor temperature, the maximum indoor temperature would be $100 + 15 = 115$°F (46.1°C). Assume that the difference between the temperature of the incoming and outgoing air is 15°F (8.3°C). Then $C = 9.4(30)(5 \times 15)^{0.5} = 2445$ ft^3/min (69.2 m^3/min).

5. Compute the cooling airflow required

This 60-kW internal-combustion engine generating set radiates 2625 Btu/min (12.8 W), step 3. Compute the cooling airflow required from $C = HK/\Delta t_a$, where C = cooling airflow required, ft^3/min; H = heat radiated by the engine, Btu/min; K = constant from Table 6; other symbols as before. Thus, for this engine with a fan discharge temperature of 111 to 120°F (43.9 to 48.9°C), Table 6, $K = 60$; $\Delta t_a = 15$°F (8.3°C) from step 4. Then $C = (2625)(60)/15 = 10,500$ ft^3/min (297.3 m^3/min).

The windows provide 2445 ft^3/min (69.2 m^3/min), step 4, and the engine radiator gives 6000 ft^3/min (169.9 m^3/min), step 1, or a total of $2445 + 6000 = 8445$ ft^3/min (239.1 m^3/min). Thus, $10,500 - 8445 = 2055$ ft^3/min (58.2 m^3/min) must be removed from the room. The usual method employed to remove the air is an exhaust fan. An exhaust fan with a capacity of 2100 ft^3/min (59.5 m^3/min) would be suitable for this engine room.

Related Calculations: Use this procedure for engines burning any type of fuel—diesel, gasoline, kerosene, or gas—in any type of enclosed room at sea level or elevations up to 1000 ft (304.8 m). Where windows or the fan outlet are fitted with louvers, screens, or intake filters, be certain to compute the net free area of the opening. When the radiator fan requires more air than is needed for cooling the room, an exhaust fan is unnecessary.

Be certain to select an exhaust fan with a sufficient discharge pressure to overcome the resistance of exhaust ducts and outlet louvers, if used. A propeller fan is usually chosen for exhaust service. In areas having high wind velocity, an axial-flow fan may be needed to overcome the pressure produced by the wind on the fan outlet.

Table 6 shows the pressure developed by various wind velocities. When the engine is located above sea level, use the multiplying factor in Table 7 to correct the computed air quantities for the lower air density.

An engine radiates 2 to 5 percent of its total heat input. The total heat input = (engine output, bhp) [heat rate, Btu/(bhp·h)]. Provide 12 to 20 air changes per hour for the engine room. The most effective ventilators are power-driven exhaust fans or roof ventilators. Where the heat load is high, 100 air changes per hour may be provided. Auxiliary-equipment rooms require 10 air changes per hour. Windows, louvers, or power-driven fans are used. A four-cycle engine requires 3 to 3.5 ft^3/min of air per bhp (0.11 to 0.13 m^3/kW); a two-cycle engine, 4 to 5 ft^3/(min · bhp) (0.15 to 0.19 m^3/kW).

The method presented here is the work of John P. Callaghan, reported in *Power*.

DESIGN OF A BYPASS COOLING SYSTEM FOR AN ENGINE

The internal-combustion engine in Fig. 3 is rated at 402 hp (300 kW) at 514 r/min and dissipates 3500 Btu/(bhp·h) (1375 W/kW) at full load to the cooling water from the power cylinders and water-cooled exhaust manifold. Determine the required cooling-water flow rate if there is a 10°F

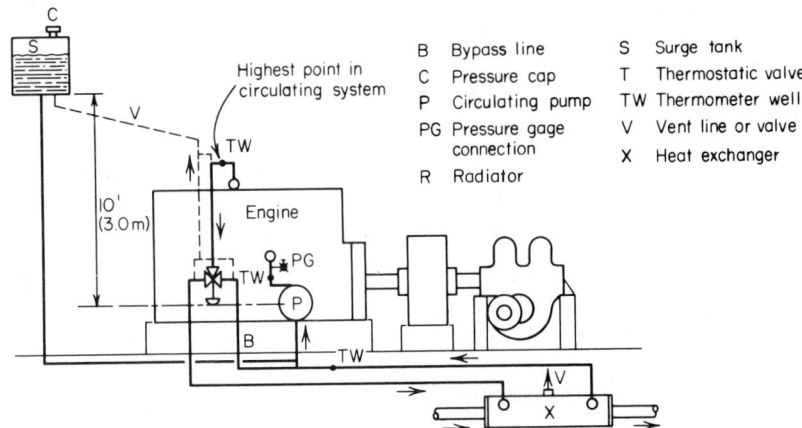

FIG. 3   Engine cooling-system hookup. *(Mechanical Engineering.)*

(5.6°C) temperature rise during passage of the water through the engine. Size the piping for the cooling system, using the head-loss data in Fig. 4, and the pump characteristic curve, Fig. 5. Choose a surge tank of suitable capacity. Determine the net positive suction head requirements for this engine. The total length of straight piping in the cooling system is 45 ft (13.7 m). The engine is located 500 ft (152.4 m) above sea level.

Calculation Procedure:

1.   *Compute the cooling-water quantity required*

The cooling-water quantity required is $G = H/(500\Delta t)$, where G = cooling-water flow, gal/min; H = heat absorbed by the jacket water, Btu/h = (maximum engine hp) [heat dissipated, Btu/ (bhp·h)]; Δt = temperature rise of the water during passage through the engine, °F. Thus, for this engine, $G = (402)(3500)/[500(10)] = 281$ gal/min (17.7 L/s).

2.   *Choose the cooling-system valve and pipe size*

Obtain the friction head-loss data for the engine, the heat exchanger, and the three-way valve from the manufacturers of the respective items. Most manufacturers have curves or tables available for easy use. Plot the head losses, as shown in Fig. 4, for the engine and heat exchanger.

   Before the three-way valve head loss can be plotted, a valve size must be chosen. Refer to a three-way valve capacity tabulation to determine a suitable valve size to handle a flow of 281 gal/ min (17.7 L/s). One such tabulation recommends a 3-in (76.2-mm) valve for a flow of 281 gal/ min (17.7 L/s). Obtain the head-loss data for the valve, and plot it as shown in Fig. 4.

   Next, assume a size for the cooling-water piping. Experience shows that a water velocity of 300 to 600 ft/min (91.4 to 182.9 m/min) is satisfactory for internal-combustion engine cooling systems. Using the Hydraulic Institute's *Pipe Friction Manual* or Cameron's *Hydraulic Data*, enter at 280 gal/min (17.6 L/s), the approximate flow, and choose a pipe size to give a velocity of 400 to 500 ft/min (121.9 to 152.4 m/min), i.e., midway in the recommended range.

   Alternatively, compute the approximate pipe diameter from $d = 4.95$ [gpm/velocity, ft/ min]$^{0.5}$. With a velocity of 450 ft/min (137.2 m/min), $d = 4.95(281/450)^{0.5} = 3.92$, say 4 in (101.6 mm). The *Pipe Friction Manual* shows that the water velocity will be 7.06 ft/s (2.2 m/s), or 423.6 ft/min (129.1 m/min), in a 4-in (101.6-mm) schedule 40 pipe. This is acceptable. Using a 3½-in (88.9-mm) pipe would increase the cost because the size is not readily available from pipe suppliers. A 3-in (76.2-mm) pipe would give a velocity of 720 ft/min (219.5 m/min), which is too high.

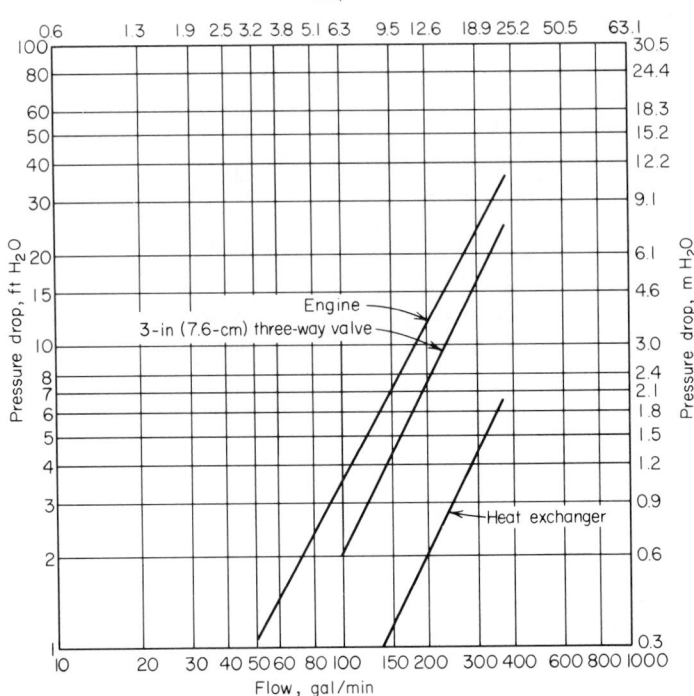

FIG. 4 Head-loss data for engine cooling-system components. *(Mechanical Engineering.)*

3. Compute the piping-system head loss

Examine Fig. 3, which shows the cooling system piping layout. Three flow conditions are possible: (*a*) all the jacket water passes through the heat exchanger, (*b*) a portion of the jacket water passes through the heat exchanger, and (*c*) none of the jacket water passes through the heat exchanger—instead, all the water passes through the bypass circuit. The greatest head loss usually occurs when the largest amount of water passes through the longest circuit (or flow condition *a*). Compute the head loss for this situation first.

Using the method given in the piping section of this handbook, compute the equivalent length of the cooling-system fitting and piping, as shown in Table 8. Once the equivalent length of the pipe and fittings is known, compute the head loss in the piping system, using the method given in the piping section of this handbook with a Hazen-Williams constant of *C* = 130 and a rounded-off flow rate of 300 gal/min (18.9 L/s). Summarize the results as shown in Table 8.

The total head loss is produced by the water flow through the piping, fittings, engine, three-way valve, and heat exchanger. Find the head loss for the last components in Fig. 4 for a flow of 300 gal/min (18.9 L/s). List the losses in Table 8, and find the sum of all the losses. Thus, the total circuit head loss is 57.61 ft (17.6 m) of water.

Compute the head loss for 0, 0.2, 0.4, 0.6, and 0.8 load on the engine, using the same procedure as in steps 1, 2, and 3 above. Plot on the pump characteristic curve, Fig. 5, the system head loss for each load. Draw a curve *A* through the points obtained, Fig. 5.

Compute the system head loss for condition *b* with half the jacket water [150 gal/min (9.5 L/s)] passing through the heat exchanger and half [150 gal/min (9.5 L/s)] through the bypass circuit. Make the same calculation for 0, 0.2, 0.4, 0.6, and 0.8 load on the engine. Plot the result as curve *B*, Fig. 5.

TABLE 8 Sample Calculation for Full Flow through Cooling Circuit°
(Fittings and Piping in Circuit)

Fitting or pipe	Number in circuit	Equivalent length of straight pipe	
		ft	m
3-in (76.2-mm) elbow	1	5.5	1.7
3 × 4 (76.2 × 101.6-mm) reducer	4	7.2	2.2
4-in (101.6-mm) elbow	7	50.4	15.4
4-in (101.6-mm) tee	1	23.0	7.0
3-in (76.2-mm) pipe	. . .	0.67	0.2
4-in (101.6-mm) pipe	. . .	45.0	13.7
Total equivalent length of pipe:			
3-in (76.2-mm) pipe, standard weight	. . .	13.37	4.1
4-in (101.6-mm) pipe, standard weight	. . .	118.4	36.1

Head loss calculation: Calculation for a flow rate of 300 gal/min (18.9 L/s) through circuit:
    Using the Hazen-Williams friction-loss equation with a *C* factor of 130 (surface roughness constant), with 300 gal/min (18.9 L/s) flowing through the pipe, the head loss per 100 ft (30.5 m) of pipe is 21.1 ft (6.4 m) and 5.64 ft (1.1 m) for the 3-in (76.2-mm) and 4-in (101.6-mm) pipes, respectively. Thus head loss in piping is†

$$3 \text{ in } \frac{21.1}{100} \times 13.37 = 2.83 \text{ ft } (0.86 \text{ m})$$

$$4 \text{ in } \frac{5.64}{100} \times 118.4 = 6.68 \text{ ft } (2.0 \text{ m})$$

	ft	m
From Fig. 5 the head loss is:		
Through engine	26.00	7.9
Through 3-in (76.2-mm) three-way valve	17.50	5.3
Through heat exchanger	4.6	1.4
Total circuit head loss	57.61	14.6

°*Mechanical Engineering.*
†Shaw and Loomis, *Cameron Hydraulic Data Book,* 12th ed., Ingersoll-Rand Company, 1951, p. 27.

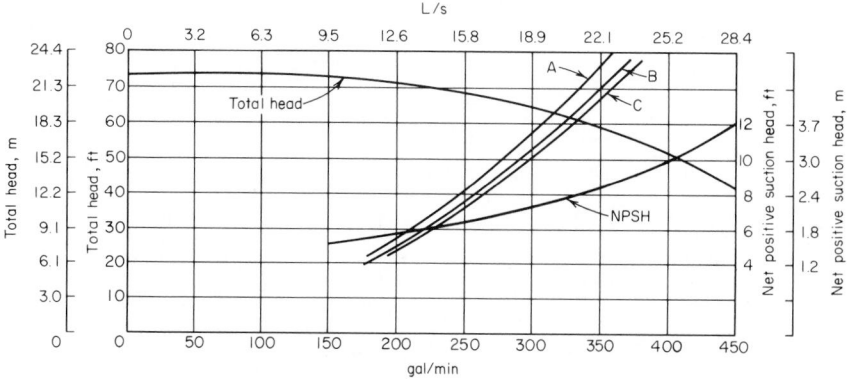

FIG. 5 Pump and system characteristics for engine cooling system. *(Mechanical Engineering.)*

Perform a similar calculation for condition c—full flow through the bypass circuit. Plot the results as curve C, Fig. 5.

4. Compute the actual cooling-water flow rate

Find the points of intersection of the pump total-head curve and the three system head-loss curves A, B, and C, Fig. 5. These intersections occur at 314, 325, and 330 gal/min (19.8, 20.5, and 20.8 L/s), respectively.

The initial design assumed a 10°F (5.6°C) temperature rise through the engine with a water flow rate of 281 gal/min (17.7 L/s). Rearranging the equation in step 1 gives $\Delta t = H/(400G)$. Substituting the flow rate for condition a gives an actual temperature rise of $\Delta t = (402)(3500)/[(500)(314)] = 8.97°F$ (4.98°C). If a 180°F (82.2°C) rated thermostatic element is used in the three-way valve, holding the outlet temperature t_o to 180°F (82.2°C), the inlet temperature t_i will be $\Delta t = t_o - t_i = 8.97$; $180 - t_i = 8.97$; $t_i = 171.03°F$ (77.2°C).

5. Determine the required surge-tank capacity

The surge tank in a cooling system provides storage space for the increase in volume of the coolant caused by thermal expansion. Compute this expansion from $E = 62.4g\Delta V$, where E = expansion, gal (L); g = number of gallons required to fill the cooling system; ΔV = specific volume, ft^3/lb (m^3/kg) of the coolant at the operating temperature − specific volume of the coolant, ft^3/lb (m^3/kg) at the filling temperature.

The cooling system for this engine must have a total capacity of 281 gal (1064 L), step 1. Round this to 300 gal (1136 L) for design purposes. The system operating temperature is 180°F (82.2°C), and the filling temperature is usually 60°F (15.6°C). Using the steam tables to find the specific volume of the water at these temperatures, we get $E = 62.4(300)(0.01651 - 0.01604) = 8.8$ gal (33.3 L).

Usual design practice is to provide two to three times the required expansion volume. Thus, a 25-gal (94.6-L) tank (nearly three times the required capacity) would be chosen. The extra volume provides for excess cooling water that might be needed to make up water lost through minor leaks in the system.

Locate the surge tank so that it is the highest point in the cooling system. Some engineers recommend that the bottom of the surge tank be at least 10 ft (3 m) above the pump centerline and connected as close as possible to the pump intake. A 1½- or 2-in (38.1- or 50.8-mm) pipe is usually large enough for connecting the surge tank to the system. The line should be sized so that the head loss of the vented fluid flowing back to the pump suction will be negligible.

6. Determine the pump net positive suction head

The pump characteristic curve, Fig. 5, shows the net positive suction head (NPSH) required by this pump. As the pump discharge rate increases, so does the NPSH. This is typical of a centrifugal pump.

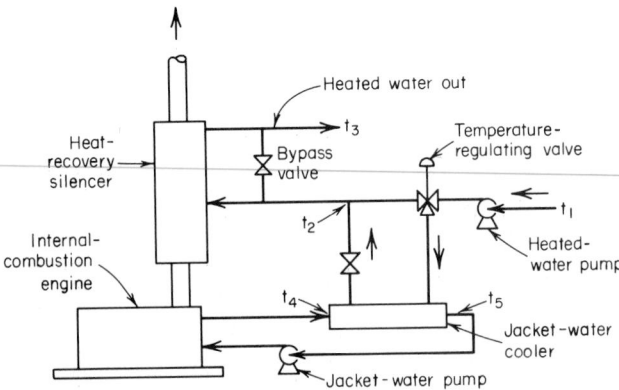

FIG. 6 Internal-combustion engine cooling system.

The greatest flow, 330 gal/min (20.8 L/s), occurs in this system when all the coolant is diverted through the bypass circuit, Figs. 4 and 5. At a 330-gal/min (20.8-L/s) flow rate through the system, the required NPSH for this pump is 8 ft (2.4 m), Fig. 5. This value is found at the intersection of the 330-gal/min (20.8-L/s) ordinate and the NPSH curve.

Compute the existing NPSH, ft (m), from NPSH $= H_s - H_f + 2.31(P_s - P_v)/s$, where H_s = height of minimum surge-tank liquid level above the pump centerline, ft (m); H_f = friction loss in the suction line from the surge-tank connection to the pump inlet flange, ft (m) of liquid; P_s = pressure in surge tank, or atmospheric pressure at the elevation of the installation, lb/in^2 (abs) (kPa); P_v = vapor pressure of the coolant at the pumping temperature, lb/in^2 (abs) (kPa); s = specific gravity of the coolant at the pumping temperature.

7. Determine the operating temperature with a closed surge tank

A pressure cap on the surge tank, or a radiator, will permit operation at temperatures above the atmospheric boiling point of the coolant. At a 500-ft (152.4-m) elevation, water boils at 210°F (98.9°C). Thus, without a closed surge tank fitted with a pressure cap, the maximum operating temperature of a water-cooled system would be about 200°F (93.3°C).

If a 7-lb/in^2 (gage) (48.3-kPa) pressure cap were used at the 500-ft (152.4-m) elevation, then the pressure in the vapor space of the surge tank could rise to $P_s = 14.4 + 7.0 = 21.4$ lb/in^2 (abs) (147.5 kPa). The steam tables show that water at this pressure boils at 232°F (111.1°C). Checking the NPSH at this pressure shows that NPSH $= (10 - 1.02) + 2.31(21.4 - 21.4)/0.0954 = 8.98$ ft (2.7 m). This is close to the required 8-ft (2.4-m) head. However, the engine could be safely operated at a slightly lower temperature, say 225°F (107.2°C).

8. Compute the pressure at the pump suction flange

The pressure at the pump suction flange P lb/in^2 (gage) $= 0.433s(H_s - H_f) = (0.433)(0.974)(10.00 - 1.02) = 3.79$ lb/in^2 (gage) (26.1 kPa).

A positive pressure at the pump suction is needed to prevent the entry of air along the shaft. To further ensure against air entry, a mechanical seal can be used on the pump shaft in place of packing.

Related Calculations: Use this general procedure in designing the cooling system for any type of reciprocating internal-combustion engine—gasoline, diesel, gas, etc. Where a coolant other than water is used, follow the same procedure but change the value of the constant in the denominator of the equation of step 1. Thus, for a mixture of 50 percent glycol and 50 percent water, the constant = 436, instead of 500.

The method presented here is the work of Duane E. Marquis, reported in *Mechanical Engineering*.

HOT-WATER HEAT-RECOVERY SYSTEM ANALYSIS

An internal-combustion engine fitted with a heat-recovery silencer and a jacket-water cooler is rated at 1000 bhp (746 kW). It exhausts 13.0 lb/(bhp·h) [5.9 kg/(bhp·h)] of exhaust gas at 700°F (371.1°C). To what temperature can hot water be heated when 500 gal/min (31.5 L/s) of jacket water is circulated through the hookup in Fig. 6 and 100 gal/min (6.3 L/s) of 60°F (15.6°C) water is heated? The jacket water enters the engine at 170°F (76.7°C) and leaves at 180°F (82.2°C).

Calculation Procedure:

1. Compute the exhaust heat recovered

Find the exhaust-heat recovered from $H_e = Wc\Delta t_e$, where the symbols are the same as in the previous calculation procedures. Since the final temperature of the exhaust gas is not given, a value must be assumed. Temperatures below 275°F (135°C) are undesirable because condensation of corrosive vapors in the silencer may occur. Assume that the exhaust-gas outlet temperature from the heat-recovery silencer is 300°F (148.9°C). Then $H_e = (1000)(13)(0.252)(700 - 300) = 1,310,000$ Btu/h (383.9 kW).

2. Compute the heated-water outlet temperature from the cooler

Using the temperature notation in Fig. 6, we see that the heated-water outlet temperature from the jacket-water cooler is $t_z = (w_z/w_1)(t_4 - t_5) + t_1$, where w_1 = heated-water flow, lb/h; w_z

= jacket-water flow, lb/h; the other symbols are indicated in Fig. 6. To convert gal/min of water flow to lb/h, multiply by 500. Thus, w_1 = (100 gal/min)(500) = 50,000 lb/h (22,500 kg/h), and w_2 = (500 gal/min)(500) = 250,000 lb/h (112,500 kg/h). Then t_z = (250,000/50,000)(180 − 170) + 60 = 110°F (43.3°C).

3. *Compute the heated-water outlet temperature from the silencer*

The silencer outlet temperature $t_3 = H_e/w_1 + t_z$, or t_3 = 1,310,000/50,000 + 110 = 136.2°F (57.9°C).

   Related Calculations: Use this method for any type of engine—diesel, gasoline, or gas—burning any type of fuel. Where desired, a simple heat balance can be set up between the heat-releasing and heat-absorbing sides of the system instead of using the equations given here. However, the equations are faster and more direct.

DIESEL FUEL STORAGE CAPACITY AND COST

A diesel power plant will have six 1000-hp (746-kW) engines and three 600-hp (448-kW) engines. The annual load factor is 85 percent and is nearly uniform throughout the year. What capacity day tanks should be used for these engines? If fuel is delivered every 7 days, what storage capacity is required? Two fuel supplies are available; a 24° API fuel at $0.0825 per gallon ($0.022 per liter) and a 28° API fuel at $0.0910 per gallon ($0.024 per liter). Which is the better buy?

Calculation Procedure:

1. *Compute the engine fuel consumption*

Assume, or obtain from the engine manufacturer, the specific fuel consumption of the engine. Typical modern diesel engines have a full-load heat rate of 6900 to 7500 Btu/(bhp·h) (2711 to 3375 W/kWh), or about 0.35 lb/(bhp·h) of fuel (0.21 kg/kWh). Using this value of fuel consumption for the nine engines in this plant, we see the hourly fuel consumption at 85 percent load factor will be (6 engines)(1000 hp)(0.35)(0.85) + (3 engines)(600 hp)(0.35)(0.85) = 2320 lb/h (1044 kg/h).

   Convert this consumption rate to gal/h by finding the specific gravity of the diesel oil. The specific gravity s = 141.5/(131.5 + °API). For the 24° API oil, s = 141.5/(131.5 + 24) = 0.910. Since water at 60°F (15.6°C) weighs 8.33 lb/gal (3.75 kg/L), the weight of this oil is (0.910)(8.33) = 7.578 lb/gal (3.41 kg/L). For the 28° API oil, s = 141.5/(131.5 + 28) = 0.887, and the weight of this oil is (0.887)(8.33) = 7.387 lb/gal (3.32 kg/L). Using the lighter oil, since this will give a larger gal/h consumption, we get the fuel rate = (2320 lb/h)/(7.387 lb/gal) = 315 gal/h (1192 L/h).

   The daily fuel consumption is then (24 h/day)(315 gal/h) = 7550 gal/day (28,577 L/day). In 7 days the engines will use (7 days)(7550 gal/day) = 52,900, say 53,000 gal (200,605 L).

2. *Select the tank capacity*

The actual fuel consumption is 53,000 gal (200,605 L) in 7 days. If fuel is delivered exactly on time every 7 days, a fuel-tank capacity of 53,000 gal (200,605 L) would be adequate. However, bad weather, transit failures, strikes, or other unpredictable incidents may delay delivery. Therefore, added capacity must be provided to prevent engine stoppage because of an inadequate fuel supply.

   Where sufficient space is available, and local regulations do not restrict the storage capacity installed, use double the required capacity. The reason is that the additional storage capacity is relatively cheap compared with the advantages gained. Where space or storage capacity is restricted, use 1½ times the required capacity.

   Assuming double capacity is used in this plant, the total storage capacity will be (2)(53,000) = 106,000 gal (401,210 L). At least two tanks should be used, to permit cleaning of one without interrupting engine operation.

   Consult the National Board of Fire Underwriters bulletin *Storage Tanks for Flammable Liquids* for rules governing tank materials, location, spacing, and fire-protection devices. Refer to a tank capacity table to determine the required tank diameter and length or height depending on whether the tank is horizontal or vertical. Thus, the Buffalo Tank Corporation *Handbook* shows that a 16.5-ft (5.0-m) diameter 33.5-ft (10.2-m) long horizontal tank will hold 53,600 gal (202,876

L) when full. Two tanks of this size would provide the desired capacity. Alternatively, a 35-ft (10.7-m) diameter 7.5-ft (2.3-m) high vertical tank will hold 54,000 gal (204,390 L) when full. Two tanks of this size would provide the desired capacity.

Where a tank capacity table is not available, compute the capacity of a cylindrical tank from capacity = $5.87D^2L$, where D = tank diameter, ft; L = tank length or height, ft. Consult the NBFU or the tank manufacturer for the required tank wall thickness and vent size.

3. Select the day-tank capacity

Day tanks supply filtered fuel to an engine. The day tank is usually located in the engine room and holds enough fuel for a 4- to 8-h operation of an engine at full load. Local laws, insurance requirements, or the NBFU may limit the quantity of oil that can be stored in the engine room or a day tank. One day tank is usually used for each engine.

Assume that a 4-h supply will be suitable for each engine. Then the day tank capacity for a 1000-hp (746-kW) engine = (1000 hp)[0.35 lb/(bhp·h) fuel] (4 h) = 1400 lb (630 kg), or 1400/ 7.387 = 189.6 gal (717.6 L), given the lighter-weight fuel, step 1. Thus, one 200-gal (757-L) day tank would be suitable for each of the 1000-hp (746-kW) engines.

For the 600-hp (448-kW) engines, the day-tank capacity should be.(600 hp)[0.35 lb/(bhp·h) fuel] (4 h) = 840 lb (378 kg), or 840/7.387 = 113.8 gal (430.7 L). Thus, one 125-gal (473-L) day tank would be suitable for each of the 600-hp (448-kW) engines.

4. Determine which is the better fuel buy

Compute the higher heating value HHV of each fuel from HHV = 17,645 + 54(°API), or for 24° fuel, HHV = 17,645 + 54(24) = 18,941 Btu/lb (44,057 kJ/kg). For the 28° fuel, HHV = 17,645 + 54(28) = 19,157 Btu/lb (44,559 kJ/kg).

Compare the two oils on the basis of cost per 10,000 Btu (10,550 kJ), because this is the usual way of stating the cost of a fuel. The weight of each oil was computed in step 1. Thus the 24° API oil weighs 7.578 lb/gal (0.90 kg/L), while the 28° API oil weighs 7.387 lb/gal (0.878 kg/L).

Then the cost per 10,000 Btu (10,550 kJ) = (cost, $/gal)/[(HHV, Btu/lb)/10,000](oil weight, lb/gal). For the 24° API oil, cost per 10,000 Btu (10,550 kJ) = $0.0825/[(18.941/10,000)(7.578)] = $0.00574, or 0.574 cent per 10,000 Btu (10,550 kJ). For the 28° API oil, cost per 10,000 Btu (10,550 kJ) = $0.0910/[(19,157/10,000)(7387)] = $0.00634, or 0.634 cent per 10,000 Btu (10,550 kJ). Thus, the 24° API is the better buy because it costs less per 10,000 Btu (10,550 kJ).

Related Calculations: Use this method for engines burning any liquid fuel. Be certain to check local laws and the latest NBFU recommendations before ordering fuel storage or day tanks.

Low-sulfur diesel amendments were added to the federal Clean Air Act in 1991. These amendments require diesel engines to use low-sulfur fuel to reduce atmospheric pollution. Reduction of fuel sulfur content will not require any change in engine operating procedures. If anything, the lower sulfur content will reduce engine maintenance requirements and costs.

The usual distillate fuel specification recommends a sulfur content of not more than 1.5 percent by weight, with 2 percent by weight considered satisfactory. Refineries are currently producing diesel fuel that meets federal low-sulfur requirements. While there is a slight additional cost for such fuel at the time of this writing, when the regulations went into effect, predictions are that the price of low-sulfur fuel will decline as more is manufactured.

Automobiles produce 50 percent of the air pollution throughout the developed world. The Ozone Transport Commission, set up by Congress as part of the 1990 Clear Air Act, is enforcing emission standards for new automobiles and trucks. To date, the cost of meeting such standards has been lower than anticipated. By the year 2003, all new automobiles will be pollution-free— if they comply with the requirements of the act. Stationary diesel plants using low-sulfur fuel will emit extremely little pollution.

POWER INPUT TO COOLING-WATER AND LUBE-OIL PUMPS

What is the required power input to a 200-gal/min (12.6-L/s) jacket-water pump if the total head on the pump is 75 ft (22.9 m) of water and the pump has an efficiency of 70 percent when it handles freshwater and saltwater? What capacity lube-oil pump is needed for a four-cycle 500-hp (373-kW) turbocharged diesel engine having oil-cooled pistons? What is the required power input to this pump if the discharge pressure is 80 lb/in^2 (551.5 kPa) and the efficiency of the pump is 68 percent?

Calculation Procedure:

1. Determine the power input to the jacket-water pump

The power input to jacket-water and raw-water pumps serving internal-combustion engines is often computed from the relation $hp = Gh/Ce$, where hp = hp input; G = water discharged by pump, gal/min; h = total head on pump, ft of water; C = constant = 3960 for freshwater having a density of 62.4 lb/ft^3 (999.0 kg/m^3); 3855 for saltwater having a density of 64 lb/ft^3 (1024.6 kg/m^3).

For this pump handling freshwater, $hp = (200)(75)/(3960)(0.70) = 5.42$ hp (4.0 kW). A 7.5-hp (5.6-kW) motor would probably be selected to handle the rated capacity plus any overloads.

For this pump handling saltwater, $hp = (200)(75)/[(3855)(0.70)] = 5.56$ hp (4.1 kW). A 7.5-hp (5.6-kW) motor would probably be selected to handle the rated capacity plus any overloads. Thus, the same motor could drive this pump whether it handles freshwater or saltwater.

2. Compute the lube-oil pump capacity

The lube-oil pump capacity required for a diesel engine is found from $G = H/200\Delta t$, where G = pump capacity, gal/min; H = heat rejected to the lube oil, Btu/(bhp·h); Δt = lube-oil temperature rise during passage through the engine, °F. Usual practice is to limit the temperature rise of the oil to a range of 20 to 25°F (11.1 to 13.9°C), with a maximum operating temperature of 160°F (71.1°C). The heat rejection to the lube oil can be obtained from the engine heat balance, the engine manufacturer, or *Standard Practices for Stationary Diesel Engines*, published by the Diesel Engine Manufacturers Association. With a maximum heat rejection rate of 500 Btu/(bhp·h) (196.4 W/kWh) from *Standard Practices* and an oil-temperature rise of 20°F (11.1°C), $G = [500 \text{ Btu/(bhp·h)}](1000 \text{ hp})/[(200)(20)] = 125$ gal/min (7.9 L/s).

By using the *lowest* temperature rise and the *highest* heat rejection rate, a safe pump capacity is obtained. Where the pump cost is a critical factor, use a higher temperature rise and a lower heat rejection rate. Thus, with a heat rejection rate of 300 Btu/(bhp·h) (117.9 W/kWh) from *Standard Practices*, the above pump would have a capacity of $G = (300)(1000)/[(200)(25)] = 60$ gal/min (3.8 L/s).

3. Compute the lube-oil pump power input

The power input to a separate oil pump serving a diesel engine is given by $hp = Gp/1720e$, where G = pump discharge rate, gal/min; p = pump discharge pressure, lb/in^2; e = pump efficiency. For this pump, $hp = (125)(80)/[(1720)(0.68)] = 8.56$ hp (6.4 kW). A 10-hp (7.5-kW) motor would be chosen to drive this pump.

With a capacity of 60 gal/min (3.8 L/s), the input is $hp = (60)(80)/[1720)(0.68)] = 4.1$ hp (3.1 kW). A 5-hp (3.7-kW) motor would be chosen to drive this pump.

Related Calculations: Use this method for any reciprocating diesel engine, two- or four-cycle. Lube-oil pump capacity is generally selected 10 to 15 percent oversize to allow for bearing wear in the engine and wear of the pump moving parts. Always check the selected capacity with the engine builder. Where a bypass-type lube-oil system is used, be sure to have a pump of sufficient capacity to handle *both* the engine and cooler oil flow.

Raw-water pumps are generally duplicates of the jacket-water pump, having the same capacity and head ratings. Then the raw-water pump can serve as a standby jacket-water pump, if necessary.

LUBE-OIL COOLER SELECTION AND OIL CONSUMPTION

A 500-hp (373-kW) internal-combustion engine rejects 300 to 600 Btu/(bhp·h) (118 to 236 W/kWh) to the lubricating oil. What capacity and type of lube-oil cooler should be used for this engine if 10 percent of the oil is bypassed? If this engine consumes 2 gal (7.6 L) of lube oil per 24 h at full load, determine its lube-oil consumption rate.

Calculation Procedure:

1. Determine the required lube-oil cooler capacity

Base the cooler capacity on the maximum heat rejection rate plus an allowance for overloads. The usual overload allowance is 10 percent of the full-load rating for periods of not more than 2 h in any 24 h period.

For this engine, the maximum output with a 10 percent overload is $500 + (0.10)(500) = 550$ hp (410 kW). Thus, the maximum heat rejection to the lube oil would be (550 hp)[600 Btu/(bhp·h)] = 330,000 Btu/h (96.7 kW).

2. Choose the type and capacity of lube-oil cooler

Choose a shell-and-tube type heat exchanger to serve this engine. Long experience with many types of internal-combustion engines shows that the shell-and-tube heat exchanger is well suited for lube-oil cooling.

Select a lube-oil cooler suitable for a heat-transfer load of 330,000 Btu/h (96.7 kW) at the prevailing cooling-water temperature difference, which is usually assumed to be 10°F (5.6°C). See previous calculation procedures for the steps in selecting a liquid cooler.

3. Determine the lube-oil consumption rate

The lube-oil consumption rate is normally expressed in terms of bhp·h/gal. Thus, if this engine operates for 24 h and consumes 2 gal (7.6 L) of oil, its lube-oil consumption rate = (24 h)(500 bhp)/2 gal = 6000 bhp·h/gal (1183 kWh/L).

Related Calculations: Use this procedure for any type of internal-combustion engine using any fuel.

QUANTITY OF SOLIDS ENTERING AN INTERNAL-COMBUSTION ENGINE

What weight of solids annually enters the cylinders of a 1000-hp (746-kW) internal-combustion engine if the engine operates 24 h/day, 300 days/year in an area having an average dust concentration of 1.6 gr per 1000 ft^3 of air (28.3 m^3)? The engine air rate (displacement) is 3.5 ft^3/(min·bhp) (0.13 m^3/kW). What would the dust load be reduced to if an air filter fitted to the engine removed 80 percent of the dust from the air?

Calculation Procedure:

1. Compute the quantity of air entering the engine

Since the engine is rated at 1000 hp (746 kW) and uses 3.5 ft^3/(min·bhp) [0.133 m^3/(min·kW)], the quantity of air used by the engine each minute is (1000 hp)[3.5 ft^3/(min·hp)] = 3500 ft^3/min (99.1 m^3/min).

2. Compute the quantity of dust entering the engine

Each 1000 ft^3 (28.3 m^3) of air entering the engine contains 1.6 gr (103.7 mg) of dust. Thus, during every minute of engine operation, the quantity of dust entering the engine is (3500/1000)(1.6) = 5.6 gr (362.8 mg). The hourly dust intake = (60 min/h)(5.6 gr/min) = 336 gr/h (21,772 mg/h).

During the year the engine operates 24 h/day for 300 days. Hence, the annual intake of dust is (24 h/day)(300 days/year)(336 gr/h) = 2,419,200 gr (156.8 kg). Since there is 7000 gr/lb, the weight of dust entering the engine per year = 2,419,200 gr/(7000 gr/lb) = 345.6 lb/year (155.5 kg/year).

3. Compute the filtered dust load

With the air filter removing 80 percent of the dust, the quantity of dust reaching the engine is $(1.00 - 0.80)(345.6 \text{ lb/year}) = 69.12$ lb/year (31.1 kg/year). This shows the effectiveness of an air filter in reducing the dust and dirt load on an engine.

Related Calculations: Use this general procedure to compute the dirt load on an engine from any external source.

INTERNAL-COMBUSTION ENGINE PERFORMANCE FACTORS

Discuss and illustrate the important factors in internal-combustion engine selection and performance. In this discussion, consider both large and small engines for a full range of usual applications.

Calculation Procedure:

1. Plot typical engine load characteristics

Figure 7 shows four typical load patterns for internal-combustion engines. A continuous load, Fig. 7a, is generally considered to be heavy-duty and is often met in engines driving pumps or electric generators.

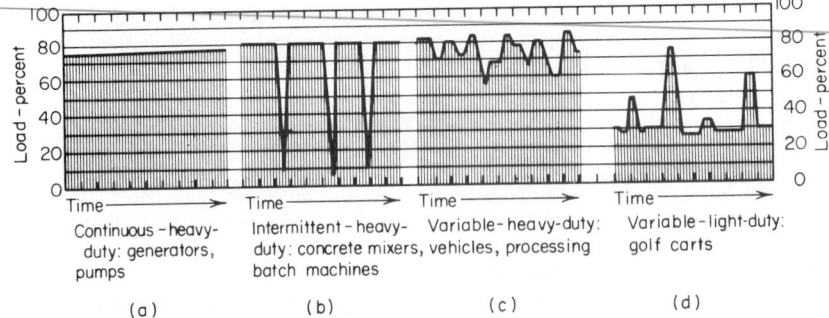

FIG. 7 Typical internal-combustion engine load cycles: (*a*) continuous, heavy-duty; (*b*) intermittent, heavy-duty; (*c*) variable, heavy-duty; (*d*) variable, light-duty. (*Product Engineering.*)

Intermittent heavy-duty loads, Fig. 7*b*, are often met in engines driving concrete mixers, batch machines, and similar loads. Variable heavy-duty loads, Fig. 7*c*, are encountered in large vehicles, process machinery, and similar applications. Variable light-duty loads, Fig. 7*d*, are met in small vehicles like golf carts, lawn mowers, chain saws, etc.

2. Compute the engine output torque

Use the relation $T = 5250 \text{ bhp}/(\text{r/min})$ to compute the output torque of an internal-combustion engine. In this relation, bhp = engine bhp being developed at a crankshaft speed having rotating speed of *rpm*.

3. Compute the hp output required

Knowing the type of load on the engine (generator, pump, mixer, saw blade, etc.), compute the power output required to drive the load at a constant speed. Where a speed variation is expected, as in variable-speed drives, compute the average power needed to accelerate the load between two desired speeds in a given time.

4. Choose the engine output speed

Internal-combustion engines are classified in three speed categories: high (1500 r/min or more), medium (750 to 1500 r/min), and low (less than 750 r/min).

Base the speed chosen on the application of the engine. A high-speed engine can be lighter and smaller for the same hp rating, and may cost less than a medium-speed or slow-speed engine serving the same load. But medium-speed and slow-speed engines, although larger, offer a higher torque output for the equivalent hp rating. Other advantages of these two speed ranges include longer service life and, in some instances, lower maintenance costs.

Usually an application will have its own requirements, such as allowable engine weight, available space, output torque, load speed, and type of service. These requirements will often indicate that a particular speed classification must be used. Where an application has no special speed requirements, the speed selection can be made on the basis of cost (initial, installation, maintenance, and operating cost), type of parts service available, and other local conditions.

5. Analyze the engine output torque required

In some installations, an engine with good lugging power is necessary, especially in tractors, harvesters, and hoists, where the load frequently increases above normal. For good lugging power, the engine should have the inherent characteristic of increasing torque with drooping speed. The engine can then resist the tendency for increased load to reduce the output speed, giving the engine good lugging qualities.

One way to increase the torque delivered to the load is to use a variable-ratio hydraulic transmission. The transmission will amplify the torque so that the engine will not be forced into the lugging range.

Other types of loads, such as generators, centrifugal pumps, air conditioners, and marine drives, may not require this lugging ability. So be certain to consult the engine power curves and torque characteristic curve to determine the speed at which the maximum torque is available.

TABLE 9 Comparison of Fuels for Internal-Combustion Engines[*]

	Storage life (quantities)		Consistency, Btu/ft³	Initial cost of engine, relative	Cost of fuel	Residue	Antiknock rating	Filtering necessary	Weight		Heat content			
	Small	Large							lb/gal	kg/L	Btu/vol	mJ/vol	Btu/lb	mJ/kg
Gasoline	Good	Poor (6 months)	Good	Low	High	High	Best is costly	Medium	6.000	0.714	123,039 Btu/gal	34,291 kJ/L	20,627	47.9
Diesel:														
No. 1	Good	Fair (1 year)	Good	High	Low	Low if properly filtered	. . .	High	6.850	0.815	135,800 Btu/gal	37,847 kJ/L	19,750	45.9
No. 2	Good	Fair (1 year)	Good	High	Low	Low if properly filtered	. . .	High	7.020	0.835	139,000 Btu/gal	38,739 kJ/L	19,786	46.0
Natural gas	Not necessary	Not necessary	Poor	Medium	Medium	Low	High	Very little	. . .	. . .	1,000 Btu/ft³	37,250 kJ/m³		
LPG:														
Propane	Good	Good	Poor	Medium	Medium	Low	Good	Very little	4.235	0.504	91,740 Btu/gal	25,568 kJ/L	21,308	49.6
Butane	Good	Good	Poor	Medium	Medium	Low	Good	Very little	4.873	0.580	103,830 Btu/gal	28,937 kJ/L	20,627	47.9

[*]Product Engineering.

TABLE 10 Performance Table for Small Internal-Combustion Engines [Less than 7 hp (5 kW)][°]

	Variety of models available	Typical weight lb/hp (kg/kW)	Operating speeds		Lugging ability	Torque output	Relative life expectancy, h	Relative cost	Fuel required	Shaft direction	Noise level	Starters	Integral optional Pto's	Ignition	Cost of operation	Variety of options and accessories
			Typical maximum	Typical efficient minimum												
Lightweight: 2-stroke	Narrow	2:1 (1.2:1)	3,600 (governed) to 7,500	2,000 to 3,000	Poor to fair	Fair	500	Lowest	Gasoline oil mixed	Vertical, horizontal, or universal	High	Rope, recoil, impulse	No	Magneto	High	Standard—extremely low custom—wide
4-stroke	Wide	6:1; 10:1 (3.6:1; 6.1:1)	4,000	2,000 to 2,400	Fair to good	Good	500	1 to 2	Gasoline (LPG)	Vertical or horizontal	Moderate	Rope, recoil, impulse, electric	Several	Magneto	Moderate	Standard—wide
Heavyweight: 4-stroke	Wide	11:1; 20:1 (6.6:1; 12.1:1)	4,000	1,600 to 1,800	Good to excellent	Good	7,500	2 to 4	Gasoline (LPG)	Vertical or horizontal	Moderate	Rope, recoil, impulse, crank, electric	No	Magneto, distributor	Moderate	Standard—moderately wide
Diesel	Narrow	35:1 (21.1:1)	2,400	1,500	Excellent	Good	25,000	4	Diesel	Horizontal	Moderate to high	Electric	No	Battery, distributor, glow plugs	Low	Narrow

[°] Product Engineering.

6. Evaluate the environmental conditions

Internal-combustion engines are required to operate under a variety of environmental conditions. The usual environmental conditions critical in engine selection are altitude, ambient temperature, dust or dirt, and special or abnormal service. Each of these, except the last, is considered in previous calculation procedures.

Special or abnormal service includes such applications as fire fighting, emergency flood pumps and generators, and hospital standby service. In these applications, an engine must start and pick up a full load without warmup.

7. Compare engine fuels

Table 9 compares four types of fuels and the internal-combustion engines using them. Note that where the cost of the fuel is high, the cost of the engine is low; where the cost of the fuel is low, the cost of the engine is high. This condition prevails for both large and small engines in any service.

8. Compare the performance of small engines

Table 10 compares the principal characteristics of small gasoline and diesel engines rated at 7 hp (5 kW) or less. Note that engine life expectancy can vary from 500 to 25,000 h. With modern, mass-produced small engines it is often just as cheap to use short-life replaceable two-stroke gasoline engines instead of a single long-life diesel engine. Thus, the choice of a small engine is often based on other considerations, such as ease and convenience of replacement, instead of just hours of life. Chances are, however, that most long-life applications of small engines will still require a long-life engine. But the alternatives must be considered in each case.

Related Calculations: Use the general data presented here for selecting internal-combustion engines having ratings up to 200 hp (150 kW). For larger engines, other factors such as weight, specific fuel consumption, lube-oil consumption, etc., become important considerations. The method given here is the work of Paul F. Jacobi, as reported in *Product Engineering*.

Air and Gas Compressors and Vacuum Systems

REFERENCES: Hawthorne—*Aerodynamics of Turbines and Compressors*, Princeton University Press; Tramm and Dean—*Centrifugal Compressor and Pump Stability, Stall, and Surge*, ASME; *Chemical Engineering* Magazine—*Fluid Movers: Pumps, Compressors, Fans and Blowers*, McGraw-Hill; Martini—*Practical Seal Design*, Dekker; Cheremisinoff and Gupta—*Handbook of Fluids in Motion*, Butterworths; Van Atta—*Vacuum Science and Engineering*, McGraw-Hill; Dushman—*Scientific Foundations of Vacuum Technique*, Wiley; Guthrie and Wakerling—*Vacuum Equipment and Techniques*, McGraw-Hill; Yarwood—*High Vacuum Techniques*, Wiley; Lewin—*Vacuum Science and Technology*, McGraw-Hill; Pirani and Yarwood—*Principles of Vacuum Engineering*, Reinhold; Reimann—*Vacuum Technique*, Chapman and Hall; Steinherz—*Handbook of High Vacuum Engineering*, Reinhold; Compressed Air and Gas Institute—*Compressed Air and Gas Handbook*; Ingersoll-Rand Company—*Compressed Air Data*.

COMPRESSOR SELECTION FOR COMPRESSED-AIR SYSTEMS

Determine the required capacity, discharge pressure, and type of compressor for an industrial-plant compressed-air system fitted with the tools listed in Table 1. The plant is located at sea level and operates 16 h/day.

Calculation Procedure:

1. Compute the required airflow rate

List all the tools and devices in the compressed-air system that will consume air, Table 1. Then obtain from Table 2 the probable air consumption, ft³/min, of each tool. Enter this value in column 1, Table 1. Next list the number of each type of tool that will be used in the system in column 2. Find the maximum probable air consumption of each tool by taking the product, line by line, of columns 1 and 2. Enter the result in column 3, Table 1, for each tool.

The air consumption values shown in column 3 represent the airflow rate required for continuous operation of each type and number of tools listed. However, few air tools operate continually. To provide for this situation, a load factor is generally used when an air compressor is selected.

TABLE 1 Typical Computation of Compressed-Air Requirements

Tool	(1) Air consumption		(2)	(3) Air required, (1) × (2)		(4)	(5) Probable air demand, (3) × (4)	
	ft³/min	m³/min	Number of tools	ft³/min	m³/min	Load factor	ft³/min	m³/min
Grinding wheel, 6 in (15.2 cm)	50	1.4	5	250	7.1	0.3	75	2.1
Rotary sander, 9-in (22.9-cm) pad	55	1.6	2	110	3.1	0.5	55	1.6
Chipping hammers, 13 lb (5.9 kg)	30	0.85	8	240	6.8	0.4	96	2.7
Nut setters, 5⁄16 in (0.79 cm)	20	0.57	10	200	5.7	0.6	120	3.4
Paint spray	10	0.28	1	10	0.28	0.1	1	0.03
Plug drills	40	1.1	3	120	3.4	0.2	24	0.68
Riveters, 18 lb (8.1 kg)	35	0.99	5	175	4.9	0.4	70	1.9
Steel drill, 7⁄8 in (2.2 cm), 25 lb (11.3 kg)	80	2.3	5	400	11.3	0.4	160	4.5
Total							601*	16.9*

*To this sum must be added allowance for future needs and expected leakage loss, if any.

TABLE 2 Approximate Air Needs of Pneumatic Tools

	ft³/min	m³/min
Grinders:		
6- and 8-in (15.2- and 20.3-cm) diameter wheels	50	1.4
2- and 2½-in (5.1- and 6.4-cm) diameter wheels	14–20	0.40–0.57
File and burr machines	18	0.51
Rotary sanders, 9-in (22.9-cm) diameter pads	55	1.56
Sand rammers and tampers:		
1 × 4 in (2.5 × 10.2 cm) cylinder	25	0.71
1¼ × 5 in (3.2 × 12.7 cm) cylinder	28	0.79
1½ × 6 in (3.8 × 15.2 cm) cylinder	39	1.1
Chipping hammers:		
10 to 13 lb (4.5 to 5.9 kg)	28–30	0.79–0.85
2 to 4 lb (0.9 to 1.8 kg)	12	0.34
Nut setters:		
To 5⁄16 in, 8 lb (0.79 cm, 3.6 kg)	20	0.57
½ to ¾ in, 18 lb (1.3 to 1.9 cm, 8.1 kg)	30	0.85
Paint spray	2–20	0.06–0.57
Plug drills	40–50	1.1–1.4
Riveters:		
3⁄32- to ⅛-in (0.24- to 0.32-cm) rivets	12	0.34
Larger, weighing 18 to 22 lb (8.1 to 9.9 kg)	35	0.99
Rivet busters	35–39	0.51–0.75
Steel drills, rotary motors:		
To ¼ in (0.64 cm) weighing 1¼ to 4 lb (0.56 to 1.8 kg)	18–20	0.57–1.1
¼ to ⅜ in (0.69 to 0.95 cm) weighing 6 to 8 lb (2.7 to 3.6 kg)	20–40	1.98
½ to ¾ in (1.27 to 1.91 cm) weighing 9 to 14 lb (4.1 to 6.3 kg)	70	2.27
⅞ to 1 in (2.2 to 2.5 cm) weighing 25 lb (11.25 kg)	80	1.1
Wood borers to 1-in (2.5 cm) diameter, weighing 14 lb (6.3 kg)	40	

2. Select the equipment load factor

The equipment load factor = (actual air consumption of the tool or device, ft^3/min)/(full-load continuous air consumption of the tool or device, ft^3/min). Load factors for compressed-air operated devices are usually less than 1.0.

Two variables are involved in the equipment load factor. The first is the *time factor*, or the percentage of the total time the tool or device actually uses compressed air. The second is the *work factor*, or percentage of maximum possible work output done by the tool. The load factor is the product of these two variables.

Determine the load factor for a given tool or device by consulting the manufacturer's engineering data, or by estimating the factor value by using previous experience as a guide. Enter the load factor in column 4, Table 1. The values shown represent typical load factors encountered in industrial plants.

3. Compute the actual air consumption

Take the product, line by line, of columns 3 and 4, Table 1. Enter the result, i.e., the probable air demand, in column 5, Table 1. Find the sum of the values in column 5, or 601 ft^3/min. This is the probable air demand of the system.

4. Apply allowances for leakage and future needs

Most compressed-air system designs allow for 10 percent of the required air to be lost through leaks in the piping, tools, hoses, etc. Whereas some designers claim that allowing for leakage is a poor design procedure, observation of many installations indicates that air leakage is a fact of life and must be considered when an actual system is designed.

With a 10 percent leakage factor, the required air capacity = 1.1(601) = 661 ft^3/min (18.7 m^3/min).

Future requirements are best estimated by predicting what types of tools and devices will probably be used. Once this is known, prepare a tabulation similar to Table 1, listing the predicted future tools and devices and their air needs. Assume that the future air needs, column 5, are 240 ft^3/min (6.8 m^3/min). Then the total required air capacity = 661 + 240 = 901 ft^3/min (25.5 m^3/min), say 900 ft^3/min (25.47 m^3/min) = present requirements + leakage allowance + predicted future needs, all expressed in ft^3/min.

5. Choose the compressor discharge pressure and capacity

In selecting the type of compressor to use, two factors are of key importance: discharge pressure required and capacity required.

Most air tools and devices are designed to operate at a pressure of 90 lb/in^2 (620 kPa) at the tool inlet. Hence, usual industrial compressors are rated for a discharge pressure of 100 lb/in^2 (689 kPa), the extra lb/in^2 providing for pressure loss in the piping between the compressor and the tools. Since none of the tools used in this plant are specialty items requiring higher than the normal pressure, a 100-lb/in^2 (689-kPa) discharge pressure will be chosen.

Where the future air demands are expected to occur fairly soon—within 2 to 3 years—the general practice is to choose a compressor having the capacity to satisfy present and future needs. Hence, in this case, a 900-ft^3/min (25.5-m^3/min) compressor would be chosen.

6. Compute the power required to compress the air

Table 3 shows the power required to compress air to various discharge pressures at different altitudes above sea level. Study of this table shows that at sea level a single-stage compressor requires 22.1 bhp/(100 ft^3/min) (5.8 kW/m^3) when the discharge pressure is 100 lb/in^2 (689 kPa). A two-stage compressor requires 19.1 bhp (14.2 kW) under the same conditions. This is a saving of 3.0 bhp/(100 ft^3/min) (0.79 kW/m^3). Hence, a two-stage compressor would probably be a better investment because this hp will be saved for the life of the compressor. The usual life of an air compressor is 20 years. Hence, by using a two-stage compressor, the approximate required bhp = (900/100)(19.1) = 171.9 bhp (128 kW), say 175 bhp (13.1 kW).

7. Choose the type of compressor to use

Reciprocating compressors find the widest use for stationary plant air supply. They may be single- or two-stage, air- or water-cooled. Here is a general guide to the types of reciprocating compressors that are satisfactory for various loads and service:

Single-stage air-cooled compressor up to 3 hp (2.2 kW), pressures to 150 lb/in^2 (1034 kPa), for light and intermittent running up to 1 h/day.

TABLE 3   Air Compressor Brake Horsepower (kW) Input[°]

Altitude, ft (m)	Single-stage discharge pressure, lb/in² (gage) (kPa)			Two-stage discharge pressure, lb/in² (gage) (kPa)		
	60 (414)	80 (552)	100 (689)	60 (414)	80 (552)	100 (689)
0 (0)	16.3 (12.2)	19.5 (14.6)	22.1 (16.5)	14.7 (10.9)	17.1 (12.8)	19.1 (14.3)
2000 (610)	15.9 (11.9)	18.9 (14.1)	21.3 (15.9)	14.3 (10.7)	16.5 (12.3)	18.4 (13.7)
4000 (1212)	15.4 (11.5)	18.2 (13.6)	20.6 (15.4)	13.8 (10.3)	15.8 (11.8)	17.7 (13.2)
6000 (1820)	15.0 (11.2)	17.6 (13.1)	20.0 (14.9)	13.3 (9.9)	15.2 (11.3)	17.0 (12.7)

[°]*Courtesy Ingersoll-Rand.* Values shown are the approximate bhp input required per 100 ft³/min (2.8 m³/min) of free air actually delivered. The bhp input can vary considerably with the type and size of compressor.

Two-stage air-cooled compressor up to 3 hp (2.2 kW), pressures to 150 lb/in² (1034 kPa), for 4 to 8 h/day running time.

Single-stage air-cooled compressor up to 15 hp (11.2 kW) for pressures to 80 lb/in² (552 kPa); above 80 lb/in² (552 kPa), use two-stage air-cooled compressor.

Single-stage horizontal double-acting water-cooled compressor for pressures to 100 lb/in² (689 kPa) horsepowers of 10 to 100 (7.5 to 75 kW), for 24 h/day or less operating time.

Two-stage, single-acting air-cooled compressor for 10 to 100 hp (7.5 to 75 kW), 5 to 10 h/day operation.

Two-stage double-acting water-cooled compressor for 100 hp (75 kW), or more, 24 h/day, or less operating time.

Using this general guide, choose a two-stage double-acting water-cooled reciprocating compressor, because more than 100-hp (75-kW) input is required and the compressor will operate 16 h/day.

Rotary compressors are not as widely used for industrial compressed-air systems as reciprocating compressors. The reason is that usual rotary compressors discharge at pressures under 100 lb/in² (68.9 kPa), unless they are multistage units.

Centrifugal compressors are generally used for large airflows—several thousand ft³/min or more. Hence, they usually find use for services requiring large air quantities, such as steel-mill blowing, copper conversion, etc. As a general rule, machines discharging at pressures of 35 lb/in² (241 kPa) or less are termed *blowers;* machines discharging at pressures greater than 35 lb/in² (241 kPa) are termed *compressors.*

Using these facts as a guide enables the designer to choose, as before, a two-stage double-acting water-cooled compressor for this application. Refer to the manufacturer's engineering data for the compressor dimensions and weight.

8. *Select the compressor drive*

Air compressors can be driven by electric motors, gasoline engines, diesel engines, gas turbines, or steam turbines. The most popular drive for reciprocating air compressors is the electric motor—either direct-connected or belt-connected. Where either dc or ac power supply is available, the usual choice is an electric-motor drive. However, special circumstances, such as the availability of low-cost fuel, may dictate another choice of drive for economic reasons. Assuming that there are no special economic reasons for choosing another type of drive, an electric motor would be chosen for this installation.

With an ac power supply, the squirrel-cage induction motor is generally chosen for belt-driven compressors. Synchronous motors are also used, particularly when power-factor correction is desired. Motor-driven air compressors generally operate at constant speed and are fitted with cylinder unloaders to vary the quantity of air delivered to the air receiver. A typical power input to a large reciprocating compressor is 22 hp (16.4 kW) per 100 ft³/min (2.8 m³/min) of free air compressed.

Air compressors are almost always rated in terms of *free air* capacity, i.e., air at the compressor intake location. Since the altitude, barometric pressure, and air temperature may vary at any

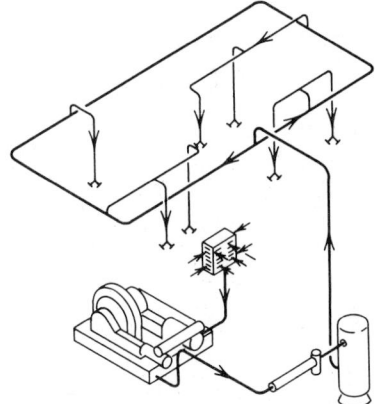

FIG. 1   Central system for compressed-air supply.

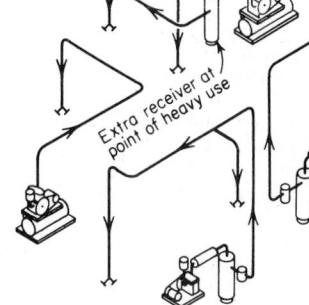

FIG. 2   Unit system for compressed-air supply.

locality, the term *free air* does not mean air under standard or uniform conditions. The displacement of an air compressor is the volume of air displaced per unit of time, usually stated in $ft^3/$min. In a multistage compressor, the displacement is that of the low-pressure cylinder only.

9.   *Choose the type of air distribution system*

Two types of air distribution systems are in use in industrial plants: *central* and *unit*. In a central system, Fig. 1, one or more large compressors centrally located in the plant supply compressed air to the areas needing it. The supply piping often runs in the form of a loop around the areas needing air.

A unit system, Fig. 2, has smaller compressors located in the areas where air is used. In the usual plant, each compressor serves only the area in which it is located. Emergency connections between the various areas may or may not be installed.

Central systems have been used for many years in large industrial plants and give excellent service Unit systems are used in both small and large plants but probably find more use in smaller plants today. With the large quantity of air required by this plant, a central system would probably be chosen, unless the air was needed at widely scattered locations in the plant, leading to excessive pressure losses in the distribution piping of a central system. In such a situation, a unit system with the capacity divided between compressors as necessary would be chosen.

Related Calculations: Where possible, choose a larger compressor than the calculations indicate is needed, because air use in industrial plants tends to increase. Avoid choosing a compressor having a free-air capacity less than one-third the required free-air capacity.

When choosing a water-cooled compressor instead of an air-cooled unit, remember that water cooling is more expensive than air cooling. However, the power input to water-cooled compressors is usually less than to air-cooled compressors of the same capacity. For either type of cooling, a two-stage compressor, with intercooling, is more economical when the compressor must operate 4 h or more in a 24-h period. Table 4 shows the typical cooling-water requirements of various types of water-cooled compressors.

When the inlet air temperature is above or below 60°F (15.6°C), the compressor delivery will vary. Table 5 shows the relative delivery of compressors handling air at various inlet temperatures.

SIZING COMPRESSED-AIR-SYSTEM COMPONENTS

What is the minimum capacity air receiver that should be used in a compressed-air system having a compressor displacing 800 $ft^3/$min (0.38 $m^3/$s) when the intake pressure is 14.7 lb/in^2 (abs) (101.4 kPa) and the discharge pressure is 120 lb/in^2 (abs) (827.4 kPa)? How long will it take for this compressor to pump up a 300-ft^3 (8.5-m^3) receiver from 80 to 120-lb/in^2 (551.6 to 827.4 kPa) if the average volumetric efficiency of the compressor is 68 percent? For how long can an 80-lb/in^2 (abs) (551.6-kPa) tool be operated from a 120-lb/in^2 (abs) (827.4-kPa), 300-ft^3 (8.5-m^3) receiver

TABLE 4   Cooling Water Recommended for Intercoolers, Cylinder Jackets, Aftercoolers

	Actual free air, gal/min per 100 ft³/min (L/s per 100 m³/s)
Intercooler separate	2.5–2.8 (334.2–374.3)
Intercooler and jackets in series	2.5–2.8 (334.2–374.3)
Aftercoolers:	
80 to 100 lb/in² (551.6 to 689.5 kPa), two-stage	1.25 (167.1)
80 to 100 lb/in² (551.6 to 689.5 kPa), single-stage	1.8 (240.6)
Two-stage jackets alone (both)	0.8 (106.9)
Single-stage jackets:	
40 lb/in² (275.8 kPa)	0.6 (80.2)
60 lb/in² (413.7 kPa)	0.8 (106.9)
80 lb/in² (551.6 kPa)	1.1 (147.0)
100 lb/in² (689.5 kPa)	1.3 (173.8)

if the tool uses 10 ft³/min (0.005 m³/s) of free air and the receiver pressure is allowed to fall to 85 lb/in² (abs) (586.1 kPa) when the atmospheric pressure is 14.7 lb/in² (abs) (101.4 kPa)? What diameter air piston is required to produce a 1000-lb (4448.2-N) force if the pressure of the air is 150 lb/in² (abs) (1034.3 kPa)?

Calculation Procedure:

1.   Compute the required volume of the air receiver

Use the relation $V_m = dp_1/p_2$, where V_m = minimum receiver volume needed, ft³; d = compressor displacement, ft³/min (use only the first-stage displacement for two-stage compressors); p_1 = compressor intake pressure, lb/in² (abs); p_2 = compressor discharge pressure, lb/in² (abs). Thus, for this compressor, $V_m = 800(14.7/120) = 97$ ft³ (2.7 m³). To provide a reserve capacity, a receiver having a volume of 150 or 200 ft³ (4.2 or 5.7 m³) would probably be chosen.

2.   Compute the receiver pump-up time

Use the relation $t = V(p_f - p_i)/(14.7de)$, where t = receiver pump-up time, min; p_f = final pressure, lb/in² (abs); p_i = initial receiver pressure, lb/in² (abs); d = compressor piston displacement, ft³/min; e = compressor volumetric efficiency, percent. Thus, $t = 300(120 - 80)/[14.7(800)(0.68)] = 1.5$ min. When the compressor discharge capacity is given in ft³/min of free air instead of in terms of piston displacement, drop the volumetric efficiency term from the above relation before computing the pump-up time.

TABLE 5   Effect of Initial Temperature on Delivery of Air Compressors
[*Based on a nominal intake temperature of 60°F (15.6°C)*]

Initial temperature				Relative delivery
°F	°R	°C	K	
40	500	4.4	277.4	1.040
50	510	10.0	283.0	1.020
60	520	15.6	288.6	1.000
70	530	21.1	294.1	0.980
80	540	26.7	299.7	0.961

TABLE 6   Air-Consumption Altitude Factors (100-lb/in² or 689.5-kPa air supply)

Altitude		Factor
ft	m	
6,000	1,828.8	1.224
8,000	2,438.3	1.310
10,000	3,048.0	1.404

Note: For pressure losses in compressed-air piping systems, see the index.

3. *Compute the air supply time*

Use the relation $t_s = V(p_{max} - p_{min})/(cp_a m)$, where t_s = time in minutes during which the receiver of volume V ft^3 will supply air from the receiver maximum pressure p_{max} lb/in^2 (abs) to the minimum pressure p_{min} lb/in^2 (abs); c = ft^3/min of free air required to operate the tool; p_a = atmospheric pressure, lb/in^2 (abs). Or, $t_s = 300(120 - 85)/[(10)(14.7)] = 7.15$ min.

Note that in this relation p_{min} is the minimum air pressure to operate the air tool. A higher minimum tank pressure was chosen here because this provides a safer estimate of the time duration for the supply of air. Had the tool operating pressure been chosen instead, the time available, by the same relation, would be $t_s = 81.5$ min.

This calculation shows that it is often wise to install an auxiliary receiver at a distance from the compressor but near the tools drawing large amounts of air. Use of such an auxiliary receiver, particularly near the end of a long distribution line, can often eliminate the need for purchasing another air compressor.

4. *Compute the required piston diameter*

Use the relation $A_p = F/p_m$, where A_p = required piston area to produce the desired force, in^2; F = force produced, lb; p_m = maximum air pressure available for the piston, lb/in^2 (abs). Or, $A_p = 1000/150 = 6.66$ in^2 (43.0 cm^2). The piston diameter d is $d = 2(A_p/\pi)^{0.5} = 2.91$ in (7.4 cm).

Related Calculations: The air consumption of power tools is normally expressed in ft^3/min of free air at sea level; the actual capacity of any type of air compressor is expressed in the same units. At locations above sea level, the quantity of free air required to operate an air tool increases because the atmospheric pressure is lower. To find the air consumption of an air tool at an altitude above sea level in terms of ft^3/min of free air at the elevation location, multiply the sea-level consumption by the appropriate factor from Table 6. Thus, a tool that consumes 10 ft^3/min (0.005 m^3/s) of free air at sea level will use 10 (1.310) = 13.1 ft^3/min (0.006 m^3/s) of 100 lb/in^2 (689.5-kPa) free air at an 8000-ft (2438.4-m) altitude.

COMPRESSED-AIR RECEIVER SIZE AND PUMP-UP TIME

What is the minimum size receiver that can be used in a compressed-air system having a compressor rated at 800 ft^3/min (0.4 m^3/s) of free air if the intake pressure is 14.7 lb/in^2 (abs) (101.4 kPa) and the discharge pressure is 120 lb/in^2 (abs) (827.4 kPa)? How long will it take the compressor to pump up the receiver from 60 lb/in^2 (abs) (413.7 kPa) to 120 lb/in^2 (abs) (827.4 kPa)? The compressor is a two-stage water-cooled unit. How much cooling water is required for the intercooler and jacket if they are piped in series and for the aftercooler?

Calculation Procedure

1. *Compute the required minimum receiver volume*

For any air compressor, the minimum receiver volume v_m ft^3 = Dp_i/p_d, where D = compressor displacement, ft^3/min free air (use only the first-stage displacement for multistage compressors); p_i = compressor inlet pressure, lb/in^2 (abs); p_d = compressor discharge pressure, lb/in^2 (abs). For this compressor, $v_m = (800)(14.7)/(120) = 98$ ft^3 (2.8 m^3). To provide a reserve supply of air, a receiver having a volume of 150 or 200 ft^3 (4.2 or 5.7 m^3) would probably be chosen. Be certain that the receiver chosen is a standard unit; otherwise, its cost may be excessive.

2. *Compute the pump-up time required*

Assume that a 150-ft^3 (4.2-m^3) receiver is chosen. Then, for any receiver, the pump-up time t min = $v_r(p_e - p_s)/De$, where v_r = receiver volume, ft^3; p_e = pressure at end of pump-up, lb/in^2 (abs); p_s = pressure at start of pump-up, lb/in^2 (abs); e = compressor volumetric efficiency, expressed as a decimal (0.50 to 0.75 for single-stage and 0.80 to 0.90 for multistage compressors). For this compressor, with a volumetric efficiency of 0.85, $t = (150)(120 - 60)/[(800)(0.85)] = 13.22$ min.

3. *Determine the quantity of cooling water required*

Use the Compressed Air and Gas Institute (CAGI) cooling-water recommendations given in the *Compressed Air and Gas Handbook*, or Baumeister and Marks—*Standard Handbook for*

Mechanical Engineers. For 80 to 125 lb/in^2 (gage) (551.6 to 861.9 kPa) discharge pressure with the intercooler and jacket in series, CAGI recommends a flow of 2.5 to 2.8 gal/min per 100 ft^3/min (334.2 to 374.3 L/s per 100 m^3/s) of free air. Using 2.5 gal/min (334.2 L/s), we see that the cooling water required for the intercooler and jackets = (2.5)(800/100) = 20.0 gal/min (2673.9 L/s). CAGI recommends 1.25 gal/min per 100 ft^3/min (167.1 L/s per 100 m^3/s) of free air for an aftercooler serving a two-stage 80 to 125 lb/in^2 (gage) (551.6- to 861.9-kPa) compressor, or (1.25)(800/100) = 10.0 gal/min (1377.3 L/s) for this compressor. Thus, the total quantity of cooling water required for this compressor is 20 + 10 = 30 gal/min (4010.9 L/s).

Related Calculations: Use this procedure for any type of air compressor serving an industrial, commercial, utility, or residential load of any capacity. Follow CAGI or the manufacturer's recommendations for cooling-water flow rate. When a compressor is located above or below sea level, multiply its rated free-air capacity by the appropriate altitude correction factor obtained from the CAGI—*Compressed Air and Gas Handbook* or Baumeister and Marks—*Standard Handbook for Mechanical Engineers.*

VACUUM-SYSTEM PUMP-DOWN TIME

An industrial vacuum system with a 200-ft^3 (5.7-m^3) receiver serving cleaning outlets is to operate to within 2.5 inHg (9.7 kPa) absolute of the barometer when the barometer is 29.8 inHg (115.1 kPa). How long will it take to evacuate the receiver to this pressure when a single-stage vacuum pump with a displacement of 60 ft^3/min (0.03 m^3/s) is used? The pump is rated to dead end at a 29.0-inHg (112.1-kPa) vacuum when the barometer is 30.0 inHg (115.9 kPa). The pump volumetric efficiency is shown in Fig. 3.

Calculation Procedure:

1. Compute the pump operating vacuum

The pump must operate to within 2.5 inHg (9.7 kPa) of the barometer, or a vacuum of 29.8 − 2.5 = 27.3 inHg (105.5 kPa).

2. Compute the quantity of free air removed from the receiver

Select a number of absolute pressures between 29.8 inHg (115.1 kPa), the actual barometric pressure, and the final receiver pressure, 2.5 inHg (9.7 kPa); and list them in the first column of a table such as Table 7. Assume equal pressure reductions—say 3 inHg (11.6 kPa)—for each step except the last few, where smaller reductions have been assumed to ensure greater accuracy.

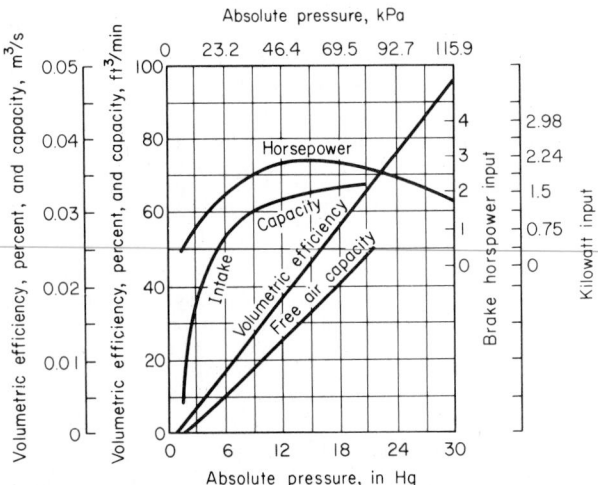

FIG. 3 Capacity, power-input, and efficiency curves for a typical reciprocating vacuum pump.

TABLE 7 Evacuation Time Calculations

Absolute pressure in receiver, inHg (kPa)	P_r/P_a	Quantity of free air, ft^3 (m^3)		Average volumetric efficiency, Fig. 2	Free-air capacity, ft^3/min (m^3/s)	Evacuation time, min
		In receiver	Removed			
29.8 (115.1)	1.000	200.0 (61.0)	0.0 (0.0)			
26.8 (103.5)	0.899	179.8 (54.8)	20.2 (6.2)	0.91	54.6 (0.026)	0.370
23.8 (92.0)	0.798	159.6 (48.6)	20.2 (6.2)	0.81	48.6 (0.023)	0.415
20.8 (80.4)	0.698	139.6 (42.6)	20.0 (6.1)	0.72	43.2 (0.020)	0.464
17.8 (68.8)	0.597	119.4 (36.4)	20.2 (6.2)	0.62	37.2 (0.018)	0.544

Total time required   9.019

Enter in the second column of Table 7 the ratio of the absolute pressure in the receiver to the atmospheric pressure, or P_r/P_a, both expressed in inHg. Thus, for the second step, $P_r/P_a = 26.8/29.8 = 0.899$.

The amount of air remaining in the receiver, measured at atmospheric conditions, is then the product of the receiver volume, 200 ft^3 (5.7 m^3), and the ratio of the pressures. Or, for the second pressure reduction, $200(0.899) = 179.8$ ft^3 (5.1 m^3). Enter the result in the third column of Table 7. This computation is a simple application of the gas laws with the receiver temperature assumed constant. Assumption of a constant air temperature is valid because, although the air temperature varies during pumping down, the overall effect is that of a constant temperature.

Find the quantity of air removed from the receiver by successive subtraction of the values in the third column. Thus, for the second pressure step, the air removed from the receiver $= 200.0 - 179.8 = 20.2$ ft^3 (0.6 m^3) and so on for the remaining steps. Enter the result of each subtraction in the fourth column of Table 7.

3. Compute the actual quantity of air handled by the pump

The volumetric efficiency of a vacuum pump varies during each pressure reduction. To simplify the pump-down time calculation, an average value for the volumetric efficiency can be used for each step in the receiver pressure reduction. Find the average volumetric efficiency for this vacuum pump from Fig. 3. Thus, for the pressure reduction from 29.8 to 26.8 inHg (115.1 to 103.5 kPa), the volumetric efficiency is found at $(29.8 + 26.8)/2 = 28.3$ inHg (109.3 kPa) to be 91 percent. Enter this value in the fifth column of Table 7. Follow the same procedure to find the remaining values, and enter them as shown.

The actual quantity of free air this vacuum pump can handle is numerically equal to the product of the volumetric efficiency, column 5, Table 7, and the pump piston displacement. Or, for the above pressure reduction, free-air capacity $= 0.91(60) = 54.6$ ft^3/min (0.026 m^3/s). Enter this result in column 6, Table 7.

4. Compute the pump-down time for each pressure reduction

The second line of Table 7 shows, in column 4, that at an absolute pressure of 26.8 inHg (103.5 kPa), 20.2 ft^3 (0.6 m^3) of free air is removed from the receiver. However, the vacuum pump can handle 54.6 ft^3/min (0.03 m^3/s), column 6. Since the time required to remove air from the receiver is (ft^3 removed)/(cylinder capacity, ft^3/min), the time required to remove 20.2 ft^3 (0.6 m^3) is $20.2/54.6$ ft$^3 = 0.370$ min.

Compute the required time for each pressure step in the same manner. The total pump-down time is then the sum of the individual times, or 9.019 min, column 7, Table 7. This result is suitable for all usual design purposes because it closely approximates the actual time required, and the errors involved are so slight as to be negligible. Leakage into industrial-plant vacuum systems often equals the volume handled by the vacuum pump.

5. Use the pump-down time for compressor selection

To choose an industrial vacuum pump using the pump-down procedure described in steps 1 to 4, (a) obtain the characteristics curves for several makes and capacities of vacuum pumps; (b) compute the pump-down time for each pump, using the procedure in steps 1 to 4; (c) compute the air inflow to the system, based on the free-air capacity of each outlet and the number of outlets in the system; (d) compute how long the pump must run to handle the air inflow; and (e) choose the pump having the shortest running time and smallest required power input.

Thus, with 10 vacuum outlets each having a free-air flow of 50 ft³/h (1.4 m³/s), the total air inflow is 10(50) = 500 ft³/h (14.2 m³/s). This means that a 200-ft³ (5.7-m³) receiver would be filled 500/200 = 2.5 times per hour. Since the pump discussed in steps 1 to 4 requires approximately 9 min to reduce the receiver pressure from atmospheric to 2.5 inHg absolute (9.7 kPa), its running time to serve these outlets would be 9(2.5) = 22.5 min, approximately. The power input to this vacuum pump, Fig. 3, ranges from a minimum of about 1 hp (0.7 kW) to a maximum of about 3 hp (2.2 kW).

If another pump could evacuate this receiver in 6 min and needed only 2.5 hp (1.9 kW) as the maximum power input, it might be a better choice, provided that its first cost were not several times that of the other pump. Use the methods of engineering economics to compare the economic merits of the two pumps.

Related Calculations: Note carefully that the procedure given here applies to industrial vacuum systems used for cleaning, maintenance, and similar purposes. The procedure should not be used for high-vacuum systems applied to production processes, experimental laboratories, etc. Use instead the method given in the next calculation procedure in this section.

To be certain that the correct pump-down time is obtained, many engineers include the volume of the system piping in the computation. This is done by computing the volume of all pipes in the system and adding the result to the receiver volume. This, in effect, increases the receiver volume that must be pumped down and gives a more accurate estimate of the probable pump-down time. Some engineers also add a leakage allowance of up to 100 percent of the sum of the receiver and piping volume. Thus, if the piping volume in the above system were 50 ft³ (1.4 m³), the total volume to be evacuated would be 2(200 + 50) = 500 ft³ (14.2 m³). The factor 2 in this expression was inserted to reflect the 100 percent leakage; i.e., the pump must handle the receiver and piping volume plus the leakage, or twice the sum of the receiver and piping volume.

Some industrial vacuum pumps are standard reciprocating air compressors run in the reverse of their normal direction after slight modification. The vacuum lines are connected to the receiver, from which the compressor takes it suction. After removing air from the receiver, the compressor discharges to the atmosphere.

VACUUM-PUMP SELECTION FOR HIGH-VACUUM SYSTEMS

Choose a mechanical vacuum pump for use in a laboratory fitted with a vacuum system having a total volume, including the piping, of 12,000 ft³ (339.8 m³). The operating pressure of the system is 0.10 torr (0.02 kPa), and the optimum pump-down time is 150 min. (*Note:* 1 torr = 1 mmHg = 0.2 kPa.)

Calculation Procedure:

1. Make a tentative choice of pump type

Mechanical vacuum pumps of the reciprocating type are well suited for system pressures in the 0.0001- to 760-torr (2×10^{-5} to 115.6-kPa) range. Hence, this type of pump will be considered first to see whether it meets the desired pump-down time.

2. Obtain the pump characteristic curves

Many manufacturers publish pump-down factor curves such as those in Fig. 4a and b. These curves are usually published as part of the engineering data for a given line of pumps. Obtain the curves from the manufacturers whose pumps are being considered.

3. Compute the pump-down time for the pumps being considered

Three reciprocating pumps can serve this system: a single-stage pump, a compound or two-stage pump, or a combination of a mechanical booster and a single-stage backing or roughing-down pump. Figure 4 gives the pump-down factor for each type of pump.

To use the pump-down factor, apply this relation: $t = VF/d$, where t = pump-down time, min; V = system volume, ft³; F = pump-down factor for the pump; d = pump displacement, ft³/min.

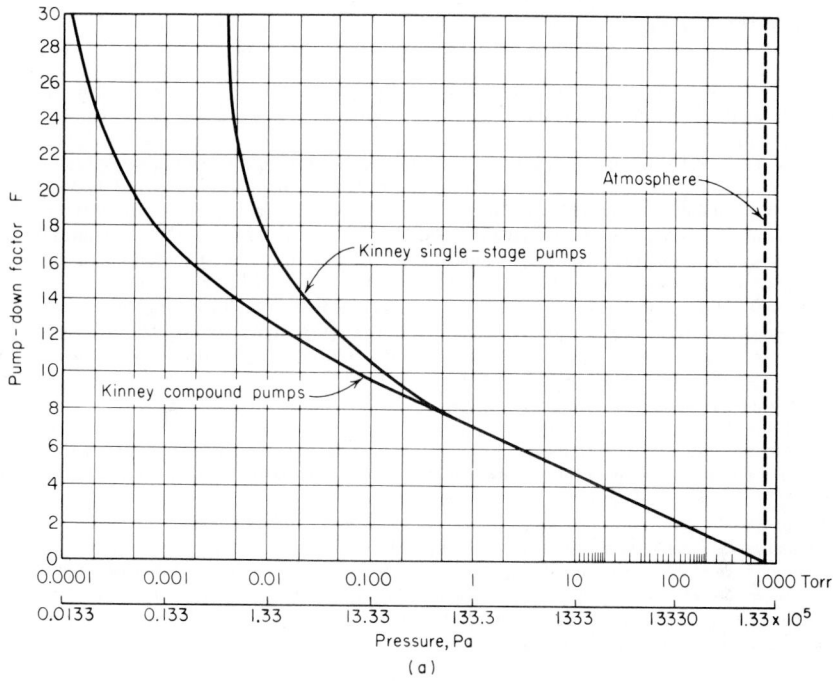

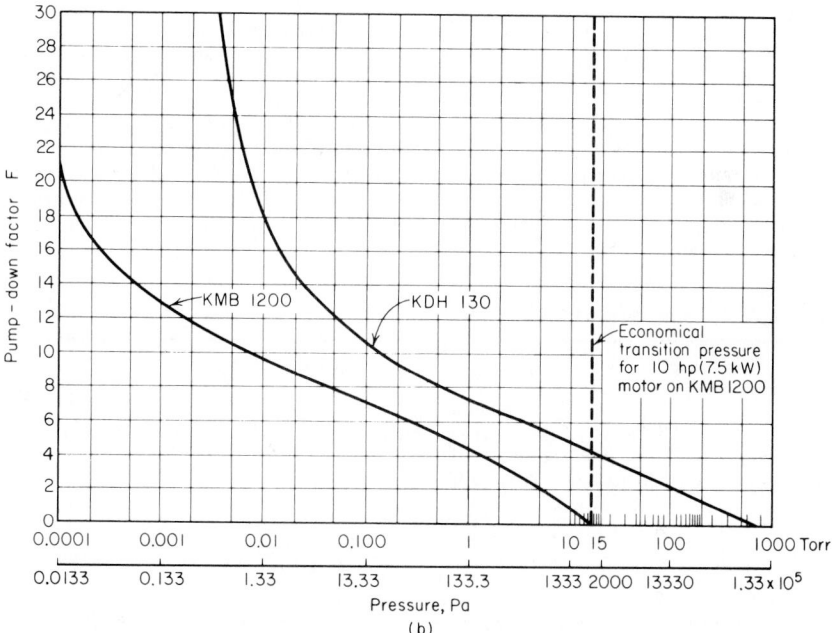

FIG. 4 (*a*) Pump-down factor for single-stage and compound vacuum pumps; (*b*) pump-down factor for mechanical booster and backing pump. (*After Kinney Vacuum Division, The New York Air Brake Company, and Van Atta.*)

Thus, for a single-stage pump, Fig. 4a shows that $F = 10.8$ for a pressure of 0.10 torr (1.5 kPa). Assuming a pump displacement of 1000 ft³/min (0.5 m³/s), $t = 12,000(10.8)/1000 = 129.6$ min, say 130 min.

For a compound pump, $F = 9.5$ from Fig. 4a. Hence, a compound pump having the same displacement, or 1000 ft³/min (0.5 m³/s), will require $t = 12,000(9.5)/1000 = 114.0$ min.

With a combination arrangement, the backing or roughing pump, a 130-ft³/min (0.06-m³/s) unit, reduces the system pressure from atmospheric, 760 torr (115.6 kPa), to the economical transition pressure, 15 torr (2.3 kPa), Fig. 4b. Then the single-stage mechanical booster pump, a 1200-ft³/min (0.6-m³/s) unit, takes over and in combination with the backing pump reduces the pressure to the desired level, or 0.10 torr (1.5 Pa). During this part of the cycle, the unit operates as a two-stage pump. Hence, the total pump-down time consists of the sum of the backing-pump and booster-pump times. The pump-down factors are, respectively, 4.2 for the backing pump at 15 torr (2.3 kPa) and 6.9 for the booster pump at 0.10 torr (1.5 Pa). Hence, the respective pump-down times are $t_1 = 12,000(4.2)/130 = 388$ min; $t_2 = 12,000(6.9)/1200 = 69$ min. The total time is thus $388 + 69 = 457$ min.

The pump-down time with the combination arrangment is greater than the optimum 150 min. Where a future lower operating pressure is anticipated, making the combination arrangement desirable, an additional large-capacity single-stage roughing pump can be used to assist the 130-ft³/min (0.06-m³/s) unit. This large-capacity unit is operated until the transition pressure is reached and roughing down is finished. The pump is then shut off, and the balance of the pumping down is carried on by the combination unit. This keeps the power consumption at a minimum.

Thus, if a 1200-ft³/min (0.06-m³/s) single-stage roughing pump were used to reduce the pressure to 15 torr (2.3 kPa), its pump-down time would be $t = 12,000(4.0)/1200 = 40$ min. The total pump-down time for the combination would then be $40 + 69 = 109$ min, using the time computed above for the two pumps in combination.

4.   *Apply the respective system factors*

Studies and experience show that the calculated pump-down time for a vacuum system must be corrected by an appropriate system factor. This factor makes allowance for the normal outgassing of surfaces exposed to atmospheric air. It also provides a basis for judging whether a system is pumping down normally or whether some problem exists that must be corrected. Table 8 lists typical system factors that have proved reliable in many tests. To use the system factor for any pump, apply it this way: $t_a = tS$, where t_a = actual pump-down time, min; t = computed pump-down time from step 3, min; S = system factor for the type of pump being considered.

Thus, by using the appropriate system factor for each pump, the actual pump-down time for the single-stage mechanical pump is $t_a = 130(1.5) = 195$ min. For the compound mechanical pump, $t_a = 114(1.25) = 142.5$ min. For the combination mechanical booster pump, $t_a = 190(1.35) = 147$ min.

TABLE 8   Recommended System Factors°

Pressure range		System factors		
torr	Pa	Single-stage mechanical pump	Compound mechanical pump	Mechanical booster pump°
760–20	115.6 kPa–3000	1.0	1.0	. . .
20–1	3000–150	1.1	1.1	1.15
1–0.5	150–76	1.25	1.25	1.15
0.5–0.1	76–15	1.5	1.25	1.35
0.1–0.02	15–3	. . .	1.25	1.35
0.02–0.001	3–0.15	. . .	. . .	2.0

°Based on bypass operation until the booster pump is put into operation. Larger system factors apply if rough pumping flow must pass through the idling mechanical booster. Any time needed for operating valves and getting the mechanical booster pump up to speed must also be added.
Source: From Van Atta—*Vacuum Science and Engineering*, McGraw-Hill.

5. *Choose the pump to use*

Based on the actual pump-down time, either the compound mechanical pump or the combination mechanical booster pump can be used. In the final choice of the pump, other factors should be taken into consideration—first cost, operating cost, maintenance cost, reliability, and probable future pressure requirements in the system. Where future lower pressure requirements are not expected, the compound mechanical pump would be a good choice. However, if lower operating pressures are anticipated in the future, the combination mechanical booster pump would probably be a better choice.

Van Atta[1] gives the following typical examples of pumps chosen for vacuum systems:

Pressure range, torr	Typical pump choice
Down to 50 (7.6 kPa)	Single-stage oil-sealed rotary; large water or vapor load may require use of refrigerated traps
0.05 to 0.01 (7.6 to 1.5 Pa)	Single-stage or compound oil-sealed pump plus refrigerated traps, particularly at the lower pressure limit
0.01 to 0.005 (1.5 to 0.76 Pa)	Compound oil-sealed plus refrigerated traps, or single-stage pumps backing diffusion pumps if a continuous large evolution of gas is expected
1 to 0.0001 (152.1 to 0.015 Pa)	Mechanical booster and backing pump combination with interstage refrigerated condenser and cooled vapor trap at the high-vacuum inlet for extreme freedom from vapor contamination
0.0005 and lower (0.076 Pa and lower)	Single-stage pumps backing diffusion pumps, with refrigerated traps on the high-vacuum side of the diffusion pumps and possibly between the single-stage and diffusion pumps if evolution of condensable vapor is expected

VACUUM-SYSTEM PUMPING SPEED AND PIPE SIZE

A laboratory vacuum system has a volume of 500 ft^3 (14.2 m^3). Leakage into the system is expected at the rate of 0.00035 ft^3/min (0.00001 m^3/min). What backing pump speed, i.e., displacement, should an oil-sealed vacuum pump serving this system have if the pump blocking pressure is 0.150 mmHg and the desired operating pressure is 0.0002 mmHg? What should the speed of the diffusion pump be? What pipe size is needed for the connecting pipe of the backing pump if it has a displacement or pumping speed of 380 ft^3/min (10.8 m^3/min) at 0.150 mmHg and a length of 15 ft (4.6 m)?

Calculation Procedure:

1. *Compute the required backing pump speed*

Use the relation $d_b = G/P_b$, where d_b = backing pump speed or pump displacement, ft^3/min; G = gas leakage or flow rate, mm · min/ft^3. To convert the gas or leakage flow rate to mm · min/ ft^3, multiply the ft^3/min by 760 mm, the standard atmospheric pressure, mmHg. Thus, $d_b = 760(0.00035)/0.150 = 1.775$ ft^3/min (0.05 m^3/min).

2. *Select the actual backing pump speed*

For practical purposes, since gas leakage and outgassing are impossible to calculate accurately, a backing pump speed or displacement of at least twice the computed value, or 2(1.775) = 3.550 ft^3/min (0.1 m^3/min), say 4 ft^3/min (0.11 m^3/min), would probably be used.

If this backing pump is to be used for pumping down the system, compute the pump-down

[1]C. M. Van Atta—*Vacuum Science and Engineering*, McGraw-Hill, New York, 1965.

time as shown in the previous calculation procedure. Should the pump-down time be excessive, increase the pump displacement until a suitable pump-down time is obtained.

3. *Compute the diffusion pump speed*

The diffusion pump reduces the system pressure from the blocking point, 0.150 mmHg, to the system operating pressure of 0.0002 mmHg. (*Note:* 1 torr = 1 mmHg.) Compute the diffusion pump speed from $d_d = G/P_d$, where d_d = diffusion pump speed, ft^3/min; P_d = diffusion-pump operating pressure, mmHg. Or, d_d = 760(0.00035)/0.0002 = 1330 ft^3/min (37.7 m^3/min). To allow for excessive leaks, outgassing, and manifold pressure loss, a 3000- or 4000-ft^3/min (84.9- or 113.2-m^3/min) diffusion pump would be chosen. To ensure reliability of service, two diffusion pumps would be chosen so that one could operate while the other was being overhauled.

4. *Compute the size of the connecting pipe*

In usual vacuum-pump practice, the pressure drop in pipes serving mechanical pumps is not allowed to exceed 20 percent of the inlet pressure prevailing under steady operating conditions. A correctly designed vacuum system, where this pressure loss is not exceeded, will have a pump-down time which closely approximates that obtained under ideal conditions.

Compute the pressure drop in the high-pressure region of vacuum pumps from $p_d = 1.9 d_b L/d^4$, where p_d = pipe pressure drop, μm; d_b = backing pump displacement or speed, ft^3/min; L = pipe length, ft; d = inside diameter of pipe, in. Since the pressure drop should not exceed 20 percent of the inlet or system operating pressure, the drop for a backing pump is based on its blocking pressure, or 0.150 mmHg, or 150 μm. Hence p_d = 0.20(150) = 30 μm. Then 30 = 1.9(380)(15)/d^4, and d = 4.35 in (110.5 mm). Use a 5-in (127.0-mm) diameter pipe.

In the low-pressure region, the diameter of the converting pipe should equal, or be larger than, the pump inlet connection. Whenever the size of a pump is increased, the diameter of the pipe should also be increased to conform with the above guide.

Related Calculations: Use the general procedures given here for laboratory- and production-type high-vacuum systems.

Materials Handling

REFERENCES: Apple—*Materials Handling Systems Design*, Wiley; Wasp—*Slurry Pipeline Transportation*, Trans Tech; Machinery Studies—*Materials Handling Equipment*, Business Trends; Bolz—*Materials Handling Handbook*, Wiley; Reisner and Eisenhart—*Bins and Bunkers for Handling Bulk Materials*, Trans Tech; *Chemical Engineering* Magazine—*Pneumatic Conveying of Bulk Materials*, McGraw-Hill; Hudson—*Conveyors*, Wiley; Buffalo Forge Company—*Fan Engineering*; Stanier—*Plant Engineering Handbook*, McGraw-Hill; Baumeister and Marks—*Standard Handbook for Mechanical Engineers*, McGraw-Hill.

BULK MATERIAL ELEVATOR AND CONVEYOR SELECTION

Choose a bucket elevator to handle 150 tons/h (136.1 t/h) of abrasive material weighing 50 lb/ft^3 (800.5 kg/m^3) through a vertical distance of 75 ft (22.9 m) at a speed of 100 ft/min (30.5 m/min). What hp input is required to drive the elevator? The bucket elevator discharges onto a horizontal conveyor which must transport the material 1400 ft (426.7 m). Choose the type of conveyor to use, and determine the required power input needed to drive it.

Calculation Procedure:

1. *Select the type of elevator to use*

Table 1 summarizes the various characteristics of bucket elevators used to transport bulk materials vertically. This table shows that a continuous bucket elevator would be a good choice, because it is a recommended type for abrasive materials. The second choice would be a pivoted bucket elevator. However, the continuous bucket type is popular and will be chosen for this application.

2. *Compute the elevator height*

To allow for satisfactory loading of the bulk material, the elevator length is usually increased by about 5 ft (1.5 m) more than the vertical lift. Hence, the elevator height = 75 + 5 = 80 ft (24.4 m).

TABLE 1 Bucket Elevators

	Centrifugal discharge	Perfect discharge	Continuous bucket	Gravity discharge	Pivoted bucket
Carrying paths	Vertical	Vertical to inclination 15° from vertical	Vertical to inclination 15° from vertical	Vertical and horizontal	Vertical and horizontal
Capacity range, tons/h (t/h), material weighing 50 lb/ft³ (800.5 kg/m³)	78 (70.8)	34 (30.8)	345 (312.9)	191 (173.3)	255 (231.3)
Speed range, ft/min (m/min)	306 (93.3)	120 (36.6)	100 (30.5)	100 (30.5)	80 (24.4)
Location of loading point	Boot	Boot	Boot	On lower horizontal run	On lower horizontal run
Location of discharge point	Over head wheel	Over head wheel	Over head wheel	On horizontal run	On horizontal run
Handling abrasive materials	Not preferred	Not preferred	Recommended	Not recommended	Recommended

Source: Link-Belt Div. of FMC Corp.

3.   Compute the required power input to the elevator

Use the relation $hp = 2CH/1000$, where C = elevator capacity, tons/h; H = elevator height, ft. Thus, for this elevator, $hp = 2(150)(80)/1000 = 24.0$ hp (17.9 kW).

The power input relation given above is valid for continuous-bucket, centrifugal-discharge, perfect-discharge, and super-capacity elevators. A 25-hp (18.7-kW) motor would probably be chosen for this elevator.

4.   Select the type of conveyor to use

Since the elevator discharges onto the conveyor, the capacity of the conveyor should be the same, per unit time, as the elevator. Table 2 lists the characteristics of various types of conveyors. Study of the tabulation shows that a belt conveyor would probably be best for this application, based on the speed, capacity, and type of material it can handle. Hence, it will be chosen for this installation.

5.   Compute the required power input to the conveyor

The power input to a conveyor is composed of two portions: the power required to move the empty belt conveyor and the power required to move the load horizontally.

Determine from Fig. 1 the power required to move the empty belt conveyor, after choosing the required belt width. Determine the belt width from Table 3.

Thus, for this conveyor, Table 3 shows that a belt width of 42 in (106.7 cm) is required to transport up to 150 tons/h (136.1 t/h) at a belt speed of 100 ft/min (30.5 m/min). [Note that the next *larger* capacity, 162 tons/h (146.9 t/h), is used when the exact capacity required is not tabulated.] Find the horsepower required to drive the empty belt by entering Fig. 1 at the belt distance between centers, 1400 ft (426.7 m), and projecting vertically upward to the belt width, 42 in (106.7 cm). At the left, read the required power input as 7.2 hp (5.4 kW).

Compute the power required to move the load horizontally from $hp = (C/100)(0.4 + 0.00345L)$, where L = distance between conveyor centers, ft; other symbols as before. For this conveyor, $hp = (150/100)(0.4 + 0.00325 \times 1400) = 6.83$ hp (5.1 kW). Hence, the total horsepower to drive this horizontal conveyor is $7.2 + 6.83 = 14.03$ hp (10.5 kW).

The total horsepower input to this conveyor installation is the sum of the elevator and conveyor belt horsepowers, or $14.03 + 24.0 = 38.03$ hp (28.4 kW).

TABLE 2 Conveyor Characteristics

	Belt conveyor	Apron conveyor	Flight conveyor	Drag chain	En masse conveyor	Screw conveyor	Vibratory conveyor
Carrying paths	Horizontal to 18°	Horizontal to 25°	Horizontal to 45°	Horizontal or slight incline, 10°	Horizontal to 90°	Horizontal to 15°; may be used up to 90° but capacity falls off rapidly	Horizontal or slight incline, 5° above or below horizontal
Capacity range, tons/h (t/h) material weighing 50 lb/ft³ (800.5 kg/m³)	2160 (1959.5)	100 (90.7)	360 (326.6)	20 (18.1)	100 (90.7)	150 (136.1)	100 (90.7)
Speed range, ft/min (m/min)	600 (182.9)	100 (30.5)	150 (45.7)	20 (6.1)	80 (24.4)	100 (30.5)	40 (12.2)
Location of loading point	Any point	Any point	Any point	Any point	On horizontal runs	Any point	Any point
Location of discharge point	Over end wheel and intermediate points by tripper or plow	Over end wheel	At end of trough and intermediate points by gates	At end of trough	Any point on horizontal runs by gate	At end of trough and intermediate points by gates	At end of trough
Handling abrasive materials	Recommended	Recommended	Not recommended	Recommended with special steels	Not recommended	Not preferred	Recommended

Source: Link-Belt Div. of FMC Corp.

3.340

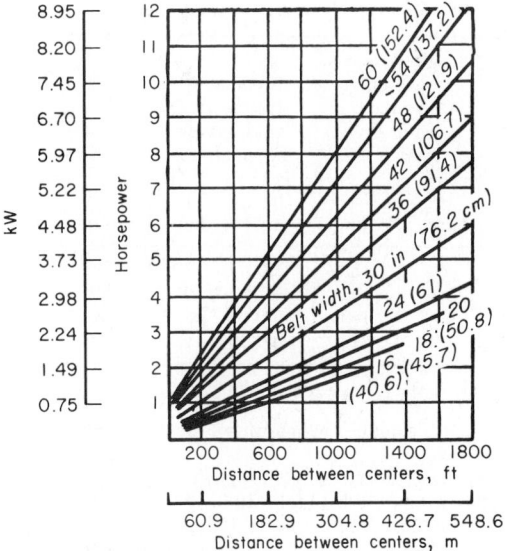

FIG. 1 Horsepower (kilowatts) required to move an empty conveyor belt at 100 ft/min (30.5 m/min).

Related Calculations: The procedure given here is valid for conveyors using rubber belts reinforced with cotton duck, open-mesh fabric, cords, or steel wires. It is also valid for stitched-canvas belts, balata belts, and flat-steel belts. The required horsepower input includes any power absorbed by idler pulleys.

Table 4 shows the minimum recommended belt widths for lumpy materials of various sizes. Maximum recommended belt speeds for various materials are shown in Table 5.

TABLE 3 Capacities of Troughed Rest [tons/h (t/h) with Belt Speed of 100 ft/min (30.5 m/min)]

Belt width, in (cm)	Weight of material, lb/ft^3 (kg/m^3)			
	30 (480.3)	50 (800.5)	100 (1601)	150 (2402)
30 (76.2)	47 (42.6)	79 (71.7)	158 (143.3)	237 (214.9)
36 (91.4)	69 (62.6)	114 (103.4)	228 (206.8)	342 (310.2)
42 (106.7)	97 (87.9)	162 (146.9)	324 (293.9)	486 (440.9)
48 (121.9)	130 (117.9)	215 (195.0)	430 (390.1)	645 (585.1)
60 (152.4)	207 (187.8)	345 (312.9)	690 (625.9)	1035 (938.9)

Source: United States Rubber Co.

TABLE 4 Minimum Belt Width for Lumps

Belt width, in (mm)	24 (609.6)	36 (914.4)	42 (1066.8)	48 (1219.2)
Sized materials, in (mm)	4½ (114.3)	8 (203.2)	10 (254)	12 (304.9)
Unsized material, in (mm)	8 (203.2)	14 (355.6)	20 (508)	35 (889)

TABLE 5   Maximum Belt Speeds for Various Materials

Width of belt		Light or free-flowing materials, grains dry sand, etc.		Moderately free-flowing sand, gravel, fine stone, etc.		Lump coal, coarse stone, crushed ore		Heavy sharp lumpy materials, heavy ores, lump coke	
in	mm	ft/min	m/min	ft/min	m/min	ft/min	m/min	ft/min	m/min
12–14	305–356	400	122	250	76	—	—	—	—
16–18	406–457	500	152	300	91	250	76	—	—
20–24	508–610	600	183	400	122	350	107	250	76
30–36	762–914	750	229	500	152	400	122	300	91

When a conveyor belt is equipped with a tripper, the belt must rise about 5 ft (1.5 m) above its horizontal plane of travel.

This rise must be included in the vertical-lift power input computation. When the tripper is driven by the belt, allow 1 hp (0.75 kW) for a 16-in (406.4-mm) belt, 3 hp (2.2 kW) for a 36-in (914.4-mm) belt, and 7 hp (5.2 kW) for a 60-in (1524-mm) belt. Where a rotary cleaning brush is driven by the conveyor shaft, allow about the same power input to the brush for belts of various widths.

SCREW CONVEYOR POWER INPUT AND CAPACITY

What is the required power input for a 100-ft (30.5-m) long screw conveyor handling dry coal ashes having a maximum density of 40 lb/ft^3 (640.4 kg/m^3) if the conveyor capacity is 30 tons/h (27.2 t/h)?

Calculation Procedure:

1.   *Select the conveyor diameter and speed*

Refer to a manufacturer's engineering data or Table 6 for a listing of recommended screw conveyor diameters and speeds for various types of materials. Dry coal ashes are commonly rated as group 3 materials, Table 7, i.e., materials with small mixed lumps with fines.

To determine a suitable screw diameter, assume two typical values and obtain the recommended rpm from the sources listed above or Table 6. Thus, the maximum rpm recommended for a 6-in (152.4-mm) screw when handling group 3 material is 90, as shown in Table 6; for a 20-in (508.0-mm) screw, 60 r/min. Assume a 6-in (152.4-mm) screw as a trial diameter.

TABLE 6   Screw Conveyor Capacities and Speeds

Material group	Maximum material density		Maximum r/min for diameters of:	
	lb/ft^3	kg/m^3	6 in (152 mm)	20 in (508 mm)
1	50	801	170	110
2	50	801	120	75
3	75	1201	90	60
4	100	1601	70	50
5	125	2001	30	25

TABLE 7 Material Factors for Screw Conveyors

Material group	Material type	Material factor
1	Lightweight: Barley, beans, flour, oats, pulverized coal, etc.	0.5
2	Fines and granular: Coal—slack or fines Sawdust, soda ash Flyash	0.9 0.7 0.4
3	Small lumps and fines: Ashes, dry alum Salt	4.0 1.4
4	Semiabrasives; small lumps: Phosphate, cement Clay, limestone Sugar, white lead	1.4 2.0 1.0
5	Abrasive lumps: Wet ashes Sewage sludge Flue dust	5.0 6.0 4.0

2. *Determine the material factor for the conveyor*

A material factor is used in the screw conveyor power input computation to allow for the character of the substance handled. Table 7 lists the material factor for dry ashes as $F = 4.0$. Standard references show that the average weight of dry coal ashes is 35 to 40 lb/ft³ (640.4 kg/m³).

3. *Determine the conveyor size factor*

A size factor that is a function of the conveyor diameter is also used in the power input computation. Table 8 shows that for a 6-in (152.4-mm) diameter conveyor the size factor $A = 54$.

4. *Compute the required power input to the conveyor*

Use the relation $hp = 10^{-6}(ALN + CWLF)$, where hp = hp input to the screw conveyor head shaft; A = size factor from step 3; L = conveyor length, ft; N = conveyor rpm; C = quantity of material handled, ft³/h; W = density of material, lb/ft³; F = material factor from step 2. For this conveyor, given the data listed above, $hp = 10^{-6}(54 \times 100 \times 60 + 1500 \times 40 \times 100 \times 4.0) = 24.3$ hp (18.1 kW). With a 90 percent motor efficiency, the required motor rating would be 24.3/0.90 = 27 hp (20.1 kW). A 30-hp (22.4-kW) motor would be chosen to drive this conveyor. Since this is not an excessive power input, the 6-in (152.4-mm) conveyor is suitable for this application.

If the calculation indicates that an excessively large power input, say 50 hp (37.3 kW) or more, is required, then the larger-diameter conveyor should be analyzed. In general, a higher initial investment in conveyor size that reduces the power input will be more than recovered by the savings in power costs.

Related Calculations: Use the procedure given here for screw or spiral conveyors and feeders handling any material that will flow. The usual screw or spiral conveyor is suitable for conveying materials for distances up to about 200 ft (60.9 m), although special designs can be built

TABLE 8 Screw Conveyor Size Factors

Conveyor diameter, in (mm)	6 (152.4)	9 (228.6)	10 (254)	12 (304.8)	16 (406.4)	18 (457.2)	20 (508)	24 (609.6)
Size factor	54	96	114	171	336	414	510	690

for greater distances. Conveyors of this type can be sloped upward to angles of 35° with the horizontal. However, the capacity of the conveyor decreases as the angle of inclination is increased. Thus the reduction in capacity at a 10° inclination is 10 percent over the horizontal capacity; at 35° the reduction is 78 percent.

The capacities of screw and spiral conveyors are generally stated in ft³/h (m³/h) of various classes of materials at the maximum recommended shaft rpm. As the size of the lumps in the material conveyed increases, the recommended shaft rpm decreases. The capacity of a screw or spiral conveyor at a lower speed is found from (capacity at given speed, ft³/h) [(lower speed, r/min)/(higher speed, r/min)]. Table 6 shows typical screw conveyor capacities at usual operating speeds.

Various types of screws are used for modern conveyors. These include short-pitch, variable-pitch, cut flights, ribbon, and paddle screws. The procedure given above also applies to these screws.

DESIGN AND LAYOUT OF PNEUMATIC CONVEYING SYSTEMS

A pneumatic conveying system for handling solids in an industrial exhaust installation contains two grinding-wheel booths and one lead each for a planer, sander, and circular saw. Determine the required duct sizes, resistance, and fan capacity for this pneumatic conveying system.

Calculation Procedure:

1.   *Sketch the proposed exhaust system*

Make a freehand sketch, Fig. 2, of the proposed system. Show the main and branch ducts and the booths and hoods. Indicate all major structural interferences, such as building columns, deep girders, beams, overhead conveyors, piping, etc. Draw the layout approximately to scale.

Mark on the sketch the length of each duct run. Avoid, if possible, vertical drops or rises in the main exhaust duct between the hoods and the fan. Do this by locating the main duct centerline 10 ft (3 m) or so above the finished floor.

Number each hood or booth, and give each duct run an identifying letter. Although it is not absolutely necessary, it is more convenient during the design process to have the hoods in numerical order and the duct runs in alphabetical order.

2.   *Determine the required air quantities and velocities*

Prepare a listing, columns 1 and 2, Table 9, of the booths, hoods, and duct runs. Enter the required air quantities and velocities for each booth or hood and duct in Table 9, columns 3 and 4. Select

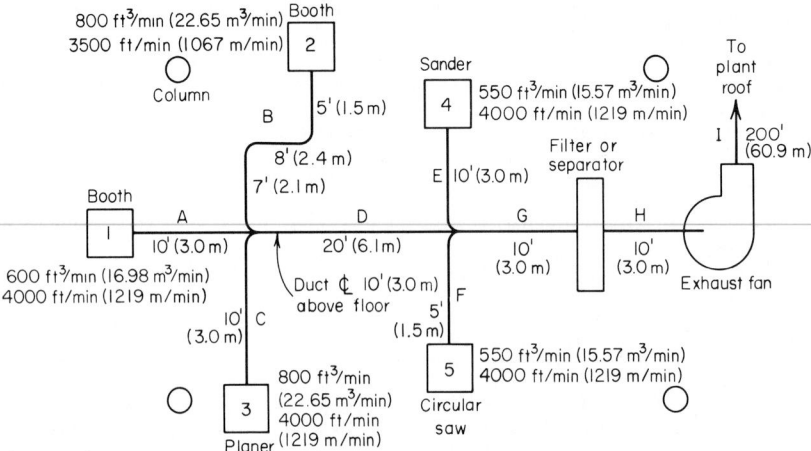

FIG. 2   Exhaust system layout.

the air quantities and velocities from the local code covering industrial exhaust systems, if such a code is available. If a code does not exist, use the ASHRAE *Guide* or Table 10.

Use extreme care in selecting the air quantities and velocities, because insufficient flow may cause dangerous atmospheric conditions. Harmful process wastes in the form of dust, gas, or moisture may injure plant personnel.

3. Size the main and branch ducts

Determine the required duct area by dividing the air quantity, ft^3/min (m^3/min), by the air velocity in the duct, or column 3/column 4, Table 9. Enter the result in column 5, Table 9.

Once the required duct area is known, find from Table 11 the nearest whole-number duct diameter corresponding to the required area. Avoid fractional diameters at this stage of the calculation, because ducts of these sizes are usually more expensive to fabricate. Later, if necessary, two or three duct sizes may be changed to fractional values. By selecting only whole-number diameters in the beginning, the cost of duct fabrication may be reduced somewhat. Enter the duct whole-number diameter in column 6, Table 9.

4. Compute the actual air velocity in the duct

Use Fig. 3 to determine the actual velocity in each duct. Enter the chart at the air quantity corresponding to that in the duct, and project vertically to the diameter curve representing the duct size. Read the actual velocity in the duct on the velocity scale, and enter the value in column 7 of Table 9.

The actual velocity in the duct should, in all cases, be equal to or greater than the design velocity shown in column 4, Table 9. If the actual velocity is less than the design velocity, decrease the duct diameter until the actual velocity is equal to or greater than the design velocity.

5. Compute the duct velocity pressure

With the actual velocity known, compute the corresponding velocity pressure in the duct from $h_v = (v/4005)^2$, where h_v = velocity pressure in the duct, inH$_2$O; v = air velocity in the duct, ft/min. Thus, for the duct run A in which the actual air velocity is 4300 ft/min (1310.6 m/min), $h_v = (4300/4005)^2 = 1.15$ in (29.2 mm) H$_2$O. Compute the actual velocity pressure in each duct run, and enter the result in column 8, Table 9.

6. Compute the equivalent length of each duct

Enter the total straight length of each duct, including any vertical drops, in column 9, Table 9. Use accurate lengths, because the system resistance is affected by the duct length.

Next list the equivalent length of each elbow in the duct runs in column 10, Table 9. For convenience, assume that the equivalent length of an elbow is 12 times the duct diameter in ft. Thus, an elbow in a 6-in (152.4-mm) diameter duct has an equivalent resistance of (6-in diameter/[(12 in/ft)(12)]) = 6 ft (1.83 m) of straight duct. When making this calculation, assume that all elbows have a radius equal to twice the diameter of the duct. Consider 45° bends as having the same resistance as 90° elbows. Note that branch ducts are usually arranged to enter the main duct at an angle of 45° or less. These assumptions are valid for all typical industrial exhaust systems and pneumatic conveying systems.

Find the total equivalent length of each duct by taking the sum of columns 9 and 10, Table 9, horizontally, for each duct run. Enter the result in column 11, Table 9.

7. Determine the actual friction in each duct

Using Fig. 3, determine the resistance, inH$_2$O (mmH$_2$O) per 100 ft (30.5 m) of each duct by entering with the air quantity and diameter of that duct. Enter the frictional resistance thus found in column 12, Table 9.

Compute actual friction in each duct by multiplying the friction per 100 ft (30.5 m) of duct, column 12, Table 9, by the total duct length, column 11 ÷ 100. Thus for duct run A, actual friction = 5.4(10/100) = 0.54 in (13.7 mm) H$_2$O. Compute the actual friction for the other duct runs in the same manner. Tabulate the results in column 13, Table 9.

8. Compute the hood entrance losses

Hoods are used in industrial exhaust systems to remove vapors, dust, fumes, and other undesirable airborne contaminants from the work area. The hood entrance loss, which depends upon the hood configuration, is usually expressed as a certain percentage of the velocity pressure in the branch

TABLE 9 Exhaust System Design Calculations

(1) Booth or hood	(2) Duct run	(3) ft³/min (m³/min) in duct	(4) Design velocity, ft/min (m/min)	(5) Duct area = column 3/column 4, ft² (m²)	(6) Duct diameter, in (mm)	(7) Actual velocity, ft/min (m/min)	(8) Actual velocity pressure, inH₂O (mmH₂O)	(9) Length of straight duct, ft (m)	(10) Equivalent length of elbows, ft (m)	(11) Total duct length = column 9 + column 10, ft (m)	(12) Friction per 100 ft (30 m) of duct, inH₂O (mmH₂O)	(13) Actual friction, inH₂O (mmH₂O)
1	A	600 (16.98)	4000 (1219)	0.150 (0.014)	5 (127)	4300 (1311)	1.15 (29.2)	10 (3.0)	0 (0)	10 (3.0)	5.4 (137.2)	0.54 (13.7)
2	B	800 (22.65)	3500 (1067)	0.228 (0.021)	6 (152)	4200 (1280)	1.0 (25.4)	20 (6.1)	18 (5.5)	38 (11.6)	4.0 (101.6)	1.57 (39.9)
3	C	800 (22.65)	4000 (1219)	0.200 (0.019)	6 (152)	4200 (1280)	1.0 (25.4)	10 (3.0)	6 (1.8)	16 (4.8)	4.0 (101.6)	0.64 (16.3)
	D	2200 (62.28)	4000 (1219)	0.550 (0.051)	10 (254)	4000 (1219)	1.0 (25.4)	20 (6.1)	0 (0)	20 (6.1)	2.1 (53.3)	0.42 (10.7)
4	E	550 (15.57)	4000 (1219)	0.137 (0.013)	5 (127)	4000 (1219)	1.0 (25.4)	10 (3.0)	5 (1.5)	15 (4.5)	4.6 (116.8)	0.69 (17.5)
5	F	550 (15.57)	4000 (1219)	0.137 (0.013)	5 (127)	4000 (1219)	1.0 (25.4)	5 (1.5)	5 (1.5)	10 (3.0)	4.6 (116.8)	0.46 (11.7)
	G	3300 (93.42)	4000 (1219)	0.825 (0.077)	12 (305)	4200 (1280)	1.0 (25.4)	10 (3.0)	0 (0)	10 (3.0)	1.9 (48.3)	0.19 (4.8)
	H	3300 (93.42)	3000 (914)	1.10 (0.102)	14 (356)	3000 (914)	0.55 (13.9)	10 (3.0)	14 (4.3)	24 (7.3)	0.84 (21.3)	0.20 (5.1)
	I	3300 (93.42)	2000 (610)	1.65 (0.153)	18 (457)	2000 (610)	0.25 (6.4)	200 (60.9)	0 (0)	200 (60.9)	0.25 (6.4)	0.50 (12.7)

TABLE 9 Exhaust System Design Calculations (*Continued*)

		System resistance	Hood number			
		1	2	3	4	5
Velocity pressure in hood branch, in (mm) H_2O		1.15 (29.2) 50	1.0 (25.4) 11	1.0 (25.4) 50	1.0 (25.4) 60	1.0 (25.4) 60
Entrance loss (% of velocity pressure)		(50)	(11)	(50)	(60)	(60)
Entrance loss, in (mm) H_2O		0.58 (14.6)	0.11 (2.8)	0.50 (12.7)	0.60 (15.2)	0.60 (15.2)
Branch and main duct resistances	A	0.54 (13.7)				
	B	......	1.57 (39.9)			
	C	......	......	0.64 (16.3)		
	D	0.42 (10.7)	0.42 (10.7)	0.42 (10.7)		
	E	......	......	......	0.69 (17.5)	
	F	......	......	......	......	0.46 (11.7)
	G	0.19 (4.8)	0.19 (4.8)	0.19 (4.8)	0.19 (4.8)	0.19 (4.8)
	H	0.20 (5.1)	0.20 (5.1)	0.20 (5.1)	0.20 (5.1)	0.20 (5.1)
	I	0.50 (12.7)	0.50 (12.7)	0.50 (12.7)	0.50 (12.7)	0.50 (12.7)
Collector or filter resistance, in (mm) H_2O		2.00 (50.8)	2.00 (50.8)	2.00 (50.8)	2.00 (50.8)	2.00 (50.8)
Total resistance in each branch, in (mm) H_2O		4.43 (112.4)	4.99 (126.8)	4.45 (113.1)	4.18 (106.1)	3.95 (100.3)

TABLE 10 Recommended Exhaust Air Quantities

Operation	ft³/min (m³/min)	Branch duct velocity, ft/ min (m/min)	Branch duct diameter, in (mm)
Sanding:			
Single drum, [10-in (25.4-cm) diameter]	400 (11.32)	4000 (1219)	4 (101.6)
Disk	550 (15.57)	4000 (1219)	5 (127)
Circular saws [16- to 24-in (40.6- to 60.9-cm) diameter]	450 (12.74)	4000 (1219)	4.5 (114.3)
Shoe machinery	550 (15.57)	4000 (1219)	5 (127)
Buffing and polishing wheels [16- to 24-in (40.6- to 60.9-cm) diameter]	600 (16.98)	4500 (1372)	5 (127)
Grinding wheels [16- to 20-in (40.6- to 50.8-cm) diameter]	600 (16.98)	4500 (1372)	5 (127)
Abrasive blast rooms	. . .	3500 (1067)	
Pharmaceuticals	. . .	3000 (1067)	

Conveying velocities

Material conveyed	Conveying velocity, ft/min (m/min)
Vapors, gases, fumes, fine dusts	1500 to 2000 (457 to 610)
Fine dry dusts	3000 (914)
Average industrial dusts	3500 (1067)
Coarse particles	3500 to 4500 (1067 to 1372)
Large particles, heavy loads, moist materials, pneumatic conveying	4500 and higher (1372 and higher)

TABLE 11 Duct Diameters and Areas

Diameter		Area	
in	mm	ft²	m²
4.0	102	0.0873	0.008
5.0	127	0.1364	0.013
6.0	152.4	0.1964	0.018
7.0	178	0.2673	0.025
8.0	203.2	0.3491	0.032
10.0	254	0.5454	0.051
12	305	0.7854	0.073
14	356	1.069	0.099
16	406.4	1.396	0.130
18	457.2	1.767	0.164
20	508	2.182	0.203
22	559	2.640	0.245
24	610	3.142	0.292

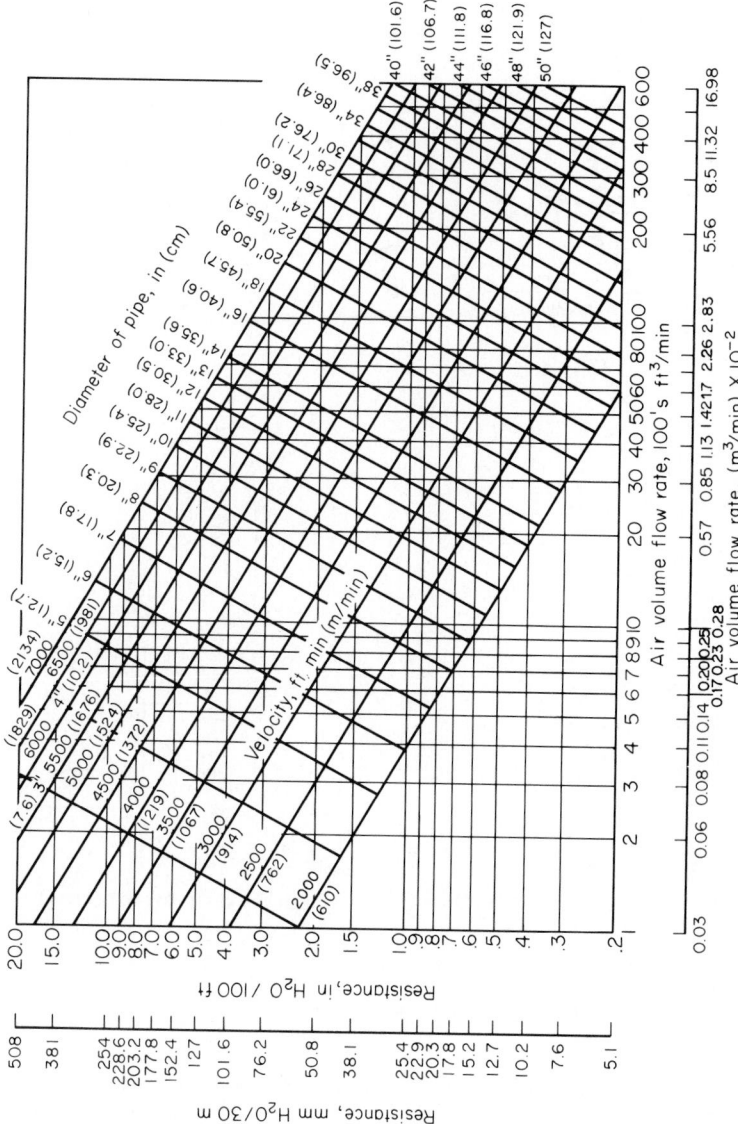

FIG. 3 Duct resistance chart. *(American Air Filter Co.)*

3.349

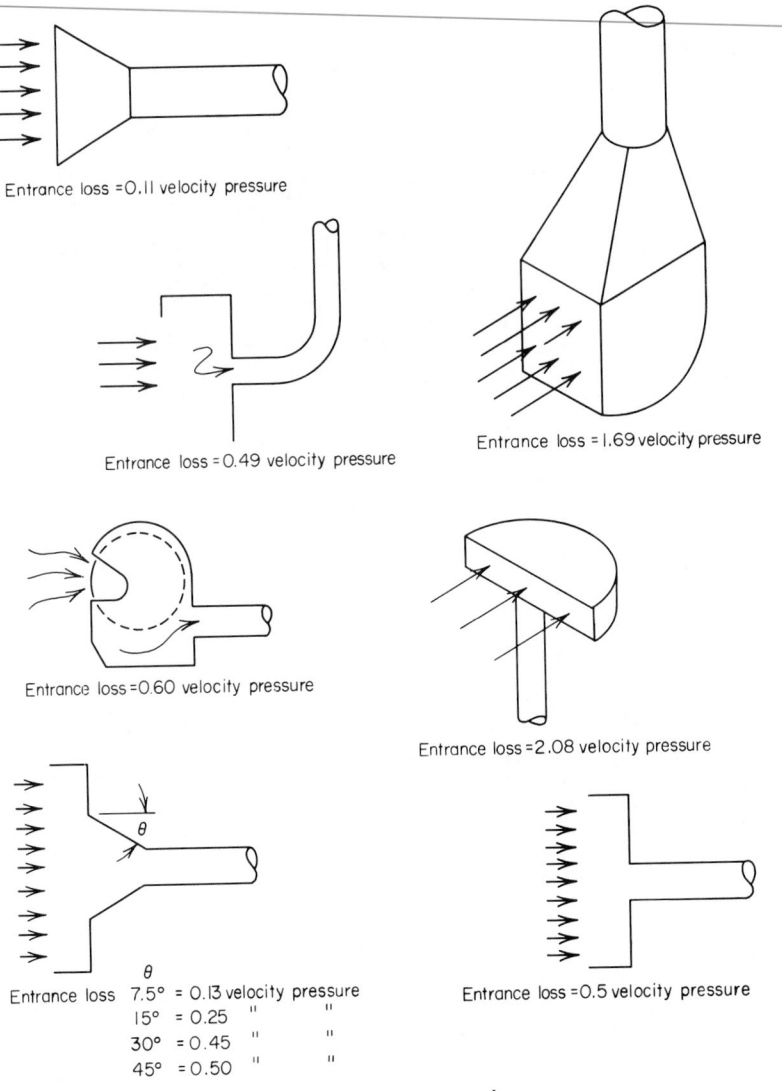

Entrance loss = 0.11 velocity pressure

Entrance loss = 0.49 velocity pressure

Entrance loss = 1.69 velocity pressure

Entrance loss = 0.60 velocity pressure

Entrance loss = 2.08 velocity pressure

Entrance loss 7.5° = 0.13 velocity pressure
            15° = 0.25   "   "
            30° = 0.45   "   "
            45° = 0.50   "   "

Entrance loss = 0.5 velocity pressure

FIG. 4 Entrance losses for various types of exhaust-system intakes.

duct connected to the hood, Fig. 4. Since the hood entrance loss usually accounts for a large portion of the branch resistance, the entrance loss chosen should always be on the safe side.

List the hood designation number under the "System Resistance" heading, as shown in Table 9. Under each hood designation number, list the velocity pressure in the branch connected to that hood. Obtain this value from column 8, Table 9. List under the velocity pressure, the hood entrance loss from Fig. 4 for the particular type of hood used in that duct run. Take the product of these two values, and enter the result under the hood number on the "entrance loss, inH₂O" line. Thus, for hood 1, entrance loss = 1.15(0.50) = 0.58 in (14.7 mm) H₂O. Follow the same procedure for the other hoods listed.

9. Find the resistance of each branch run

List the main and branch runs, A through F, Table 9. Trace out each main and branch run in Fig. 2, and enter the actual friction listed in column 3 of Table 9. Thus for booth 1, the main and branch runs consist of A, D, G, H, and I. Insert the actual friction, in (mm) H_2O, as shown in Table 9, or A = 9.54(242.3), D = 0.42(10.7), G = 0.19(4.8), H = 0.20(5.1), I = 0.50(12.7).

Determine the filter friction loss from the manufacturer's engineering data. It is common practice to design industrial exhaust systems on the basis of dirty filters or separators; i.e., the frictional resistance used in the design calculations is the resistance of a filter or separator containing the maximum amount of dust allowable under normal operating conditions. The frictional resistance of dirty filters can vary from 0.5 to 6 in (12.7 to 152.4 mm) H_2O or more. Assume that the frictional resistance of the filter used in this industrial exhaust system is 2.0 in (50.8 mm) H_2O.

Add the filter resistance to the main and branch duct resistance as shown in Table 9. Find the sum of each column in the table, as shown. This is the total resistance in each branch, in H_2O, Table 9.

10. Balance the exhaust system

Inspection of the lower part of Table 9 shows that the computed branch resistances are unequal. This condition is usually encountered during system design. To balance the system, certain duct sizes must be changed to produce equal resistance in all ducts. Or, if possible, certain ducts can be shortened. If duct shortening is not possible, as is often the case, an exhaust fan capable of operating against the largest resistance in a branch can be chosen. If this alternative is selected, special dampers must be fitted to the air inlets of the booths or ducts. For economical system operation, choose the balancing method that permits the exhaust fan to operate against the minimum resistance.

In the system being considered here, a fairly accurate balance can be obtained by decreasing the size of ducts E and F to 4.75 in (120.7 mm) and 4.375 in (111.1 mm), respectively. Duct B would be increased to 6.5 in (165.1 mm) in diameter.

11. Choose the exhaust fan capacity and static pressure

Find the required exhaust fan capacity in ft^3/min from the sum of the airflows in the ducts, A through H, column 3, Table 9, or 3300 ft^3/min (93.5 m^3/min). Choose a static pressure equal to or greater than the total resistance in the branch duct having the greatest resistance. Since this is slightly less than 4.5 in (114.3 mm) H_2O, a fan developing 4.5 in (114.3 mm) H_2O static pressure will be chosen. A 10 percent safety factor is usually applied to these values, giving a capacity of 3600 ft^3/min (101.9 m^3/min) and a static pressure of 5.0 in (127 mm) H_2O for this system.

12. Select the duct material and thickness

Galvanized sheet steel is popular for industrial exhaust systems, except where corrosive fumes and gases rule out galvanized material. Under these conditions, plastic, tile, stainless steel, or composition ducts may be substituted for galvanized ducts. Table 12 shows the recommended metal gage for galvanized ducts of various diameters. Do not use galvanized-steel ducts for gas temperatures higher than 400°F (204°C).

Hoods should be two gages heavier than the connected branch duct. Use supports not more than 12 ft (3.7 m) apart for horizontal ducts up to 8-in (203.2-mm) diameter. Supports can be spaced up to 20 ft (6.1 m) apart for larger ducts. Fit a duct cleanout opening every 10 ft (3 m). Where changes of diameter are made in the main duct, fit an eccentric taper with a length of at least 5 in (127 mm) for every 1-in (25.4-mm) change in diameter. The end of the

TABLE 12   Exhaust-System Duct Gages

Duct diameter, in (mm)	Metal gage
Up to 8 (203.2)	22
9 to 18 (228.6 to 457.2)	20
19 to 30 (482.6 to 762)	18
31 and larger (787.4 and larger)	16

main duct is usually extended 6 in (152.4 mm) beyond the last branch and closed with a removable cap. For additional data on industrial exhaust system design, see the newest issue of the ASHRAE Guide.

Related Calculations: Use this procedure for any type of industrial exhaust system, such as those serving metalworking, woodworking, plating, welding, paint spraying, barrel filling, foundry, crushing, tumbling, and similar operations. Consult the local code or ASHRAE *Guide* for specific airflow requirements for these and other industrial operations.

This design procedure is also valid, in general, for industrial pneumatic conveying systems. For several comprehensive, worked-out designs of pneumatic conveying systems, see Hudson—*Conveyors*, Wiley.

Pumps and Pumping Systems

REFERENCES: Karassik—*Pump Handbook*, McGraw-Hill; Warring—*Pumps—Selection, Systems, and Applications*, Trade and Technical Press (England); Crawford—*Marine and Offshore Pumping and Piping Systems*, Butterworth; *Europump Terminology: Glossary of Pump Applications in English, German, Italian, and Spanish*, International Ideas; Isman—*Fire Service Pumps and Hydraulics*, Delmar; Pollak—*Pump User's Handbook*, Gulf Publishing; Anderson—*Centrifugal Pumps*, Trade and Technical Press (England); Walker—*Pump Selection*, Ann Arbor Science Press; Bartlett—*Pumping Stations for Water and Sewage*, Halsted Press; Koutitas—*Elememts of Computational Hydraulics*, Chapman and Hall; Blevins—*Applied Fluid Dynamics Handbook*, VNR; Herbich—*Offshore Pipeline Design Elements*, Dekker; Zienkiewicz—*Numerical Methods in Offshore Engineering*, Wiley; The Hydraulic Institute—*Standards of the Hydraulic Institute*; Allis-Chambers Manufacturing Company—*Pic-A-Pump*; Hicks and Edwards—*Pump Application Engineering*, McGraw-Hill; Stepanoff—*Centrifugal and Axial Flow Pumps*, Wiley; Karassik and Carter—*Centrifugal Pumps*, McGraw-Hill; Allen—*Using Centrifugal Pumps*, Oxford; Buffalo Pumps—*Centrifugal Pump Applications Manual*, Kristal and Annett—*Pumps*, McGraw-Hill; Economy Pumps, Inc.—*Pump Data*; Molloy—*Pumps and Pumping*, Chemical Publishing; Moore et al.—*The Vertical Pump*, Johnston Pump Company; Karassik—*Engineers' Guide to Centrifugal Pumps*, McGraw-Hill; Kovats and Desmur—*Pompes, Ventilateurs, Compresseurs*, Dunod, Paris; Fuchslocher and Schulz—*Die Pumpen*, Springer-Verlag, Berlin; Pfleiderer—*Die Kreiselpumpen*, Springer-Verlag, Berlin.

SIMILARITY OR AFFINITY LAWS FOR CENTRIFUGAL PUMPS

A centrifugal pump designed for a 1800-r/min operation and a head of 200 ft (60.9 m) has a capacity of 3000 gal/min (189.3 L/s) with a power input of 175 hp (130.6 kW). What effect will a speed reduction to 1200 r/min have on the head, capacity, and power input of the pump? What will be the change in these variables if the impeller diameter is reduced from 12 to 10 in (304.8 to 254 mm) while the speed is held constant at 1800 r/min?

Calculation Procedure:

1. Compute the effect of a change in pump speed

For any centrifugal pump in which the effects of fluid viscosity are negligible, or are neglected, the similarity or affinity laws can be used to determine the effect of a speed, power, or head change. For a *constant impeller diameter*, the laws are $Q_1/Q_2 = N_1/N_2$; $H_1/H_2 = (N_1/N_2)^2$; $P_1/P_2 = (N_1/N_2)^3$. For a *constant speed*, $Q_1/Q_2 = D_1/D_2$; $H_1/H_2 = (D_1/D_2)^2$; $P_1/P_2 = (D_1/D_2)^3$. In both sets of laws, Q = capacity, gal/min; N = impeller rpm; D = impeller diameter, in; H = total head, ft of liquid; P = bhp input. The subscripts 1 and 2 refer to the initial and changed conditions, respectively.

For this pump, with a constant impeller diameter, $Q_1/Q_2 = N_1/N_2$; $3000/Q_2 = 1800/1200$; $Q_2 = 2000$ gal/min (126.2 L/s). And, $H_1/H_2 = (N_1/N_2)^2 = 200/H_2 = (1800/1200)^2$; $H_2 = 88.9$ ft (27.1 m). Also, $P_1/P_2 = (N_1/N_2)^3 = 175/P_2 = (1800/1200)^3$; $P_2 = 51.8$ bhp (38.6 kW).

2. Compute the effect of a change in impeller diameter

With the speed constant, use the second set of laws. Or, for this pump, $Q_1/Q_2 = D_1/D_2$; $3000/Q_2 = {}^{12}\!/_{10}$; $Q_2 = 2500$ gal/min (157.7 L/s). And $H_1/H_2 = (D_1/D_2)^2$; $200/H_2 = ({}^{12}\!/_{10})^2$; $H_2 = 138.8$ ft (42.3 m). Also, $P_1/P_2 = (D_1/D_2)^3$; $175/P_2 = ({}^{12}\!/_{10})^3$; $P_2 = 101.2$ bhp (75.5 kW).

Related Calculations: Use the similarity laws to extend or change the data obtained from centrifugal pump characteristic curves. These laws are also useful in field calculations when the pump head, capacity, speed, or impeller diameter is changed.

The similarity laws are most accurate when the efficiency of the pump remains nearly constant. Results obtained when the laws are applied to a pump having a constant impeller diameter are somewhat more accurate than for a pump at constant speed with a changed impeller diameter. The latter laws are more accurate when applied to pumps having a low specific speed.

If the similarity laws are applied to a pump whose impeller diameter is increased, be certain to consider the effect of the higher velocity in the pump suction line. Use the similarity laws for any liquid whose viscosity remains constant during passage through the pump. However, the accuracy of the similarity laws decreases as the liquid viscosity increases.

SIMILARITY OR AFFINITY LAWS IN CENTRIFUGAL PUMP SELECTION

A test-model pump delivers, at its best efficiency point, 500 gal/min (31.6 L/s) at a 350-ft (106.7-m) head with a required net positive suction head (NPSH) of 10 ft (3 m) a power input of 55 hp (41 kW) at 3500 r/min, when a 10.5-in (266.7-mm) diameter impeller is used. Determine the performance of the model at 1750 r/min. What is the performance of a full-scale prototype pump with a 20-in (50.4-cm) impeller operating at 1170 r/min? What are the specific speeds and the suction specific speeds of the test-model and prototype pumps?

Calculation Procedure:

1. Compute the pump performance at the new speed

The similarity or affinity laws can be stated in general terms, with subscripts p and m for prototype and model, respectively, as $Q_p = K_d^3 K_n Q_m$; $H_p = K_d^2 K_n^2 H_m$; $NPSH_p = K_d^2 K_n^2 NPSH_m$; $P_p = K_d^5 K_n^3 P_m$, where K_d = size factor = prototype dimension/model dimension. The usual dimension used for the size factor is the impeller diameter. Both dimensions should be in the same units of measure. Also, K_n = (prototype speed, r/min)/(model speed, r/min). Other symbols are the same as in the previous calculation procedure.

When the model speed is reduced from 3500 to 1750 r/min, the pump dimensions remain the same and $K_d = 1.0$; $K_n = 1750/3500 = 0.5$. Then $Q = (1.0)(0.5)(500) = 250$ r/min; $H = (1.0)^2(0.5)^2(350) = 87.5$ ft (26.7 m); NPSH $= (1.0)^2(0.5)^2(10) = 2.5$ ft (0.76 m); $P = (1.0)^5(0.5)^3(55) = 6.9$ hp (5.2 kW). In this computation, the subscripts were omitted from the equations because the same pump, the test model, was being considered.

2. Compute performance of the prototype pump

First, K_d and K_n must be found: $K_d = 20/10.5 = 1.905$; $K_n = 1170/3500 = 0.335$. Then $Q_p = (1.905)^3(0.335)(500) = 1158$ gal/min (73.1 L/s); $H_p = (1.905)^2(0.335)^2(350) = 142.5$ ft (43.4 m); $NPSH_p = (1.905)^2(0.335)^2(10) = 4.06$ ft (1.24 m); $P_p = (1.905)^5(0.335)^3(55) = 51.8$ hp (38.6 kW).

3. Compute the specific speed and suction specific speed

The specific speed or, as Horwitz[1] says, "more correctly, discharge specific speed," is $N_s = N(Q)^{0.5}/(H)^{0.75}$, while the suction specific speed $S = N(Q)^{0.5}/(NPSH)^{0.75}$, where all values are taken at the best efficiency point of the pump.

For the model, $N_s = 3500(500)^{0.5}/(350)^{0.75} = 965$; $S = 3500(500)^{0.5}/(10)^{0.75} = 13,900$. For the prototype, $N_s = 1170(1158)^{0.5}/(142.5)^{0.75} = 965$; $S = 1170(1156)^{0.5}/(4.06)^{0.75} = 13,900$. The specific speed and suction specific speed of the model and prototype are equal because these units are geometrically similar or homologous pumps and both speeds are mathematically derived from the similarity laws.

Related Calculations: Use the procedure given here for any type of centrifugal pump where the similarity laws apply. When the term *model* is used, it can apply to a production test pump or to a standard unit ready for installation. The procedure presented here is the work of R. P. Horwitz, as reported in *Power* magazine.[1]

SPECIFIC-SPEED CONSIDERATIONS IN CENTRIFUGAL PUMP SELECTION

What is the upper limit of specific speed and capacity of a 1750-r/min single-stage double-suction centrifugal pump having a shaft that passes through the impeller eye if it handles clear water at 85°F (29.4°C) at sea level at a total head of 280 ft (85.3 m) with a 10-ft (3-m) suction lift? What is the efficiency of the pump and its approximate impeller shape?

[1]R. P. Horwitz, "Affinity Laws and Specific Speed Can Simplify Centrifugal Pump Selection," *Power*, November 1964.

Calculation Procedure:

1. Determine the upper limit of specific speed

Use the Hydraulic Institute upper specific-speed curve, Fig. 1, for centrifugal pumps or a similar curve, Fig. 2, for mixed- and axial-flow pumps. Enter Fig. 1 at the bottom at 280-ft (85.3-m) total head, and project vertically upward until the 10-ft (3-m) suction-lift curve is intersected. From here, project horizontally to the right to read the specific speed $N_S = 2000$. Figure 2 is used in a similar manner.

2. Compute the maximum pump capacity

For any centrifugal, mixed- or axial-flow pump, $N_S = (gpm)^{0.5}(rpm)/H_t^{0.75}$, where H_t = total head on the pump, ft of liquid. Solving for the maximum capacity, we get $gpm = (N_S H_t^{0.75}/rpm)^2 = (2000 \times 280^{0.75}/1750)^2 = 6040$ gal/min (381.1 L/s).

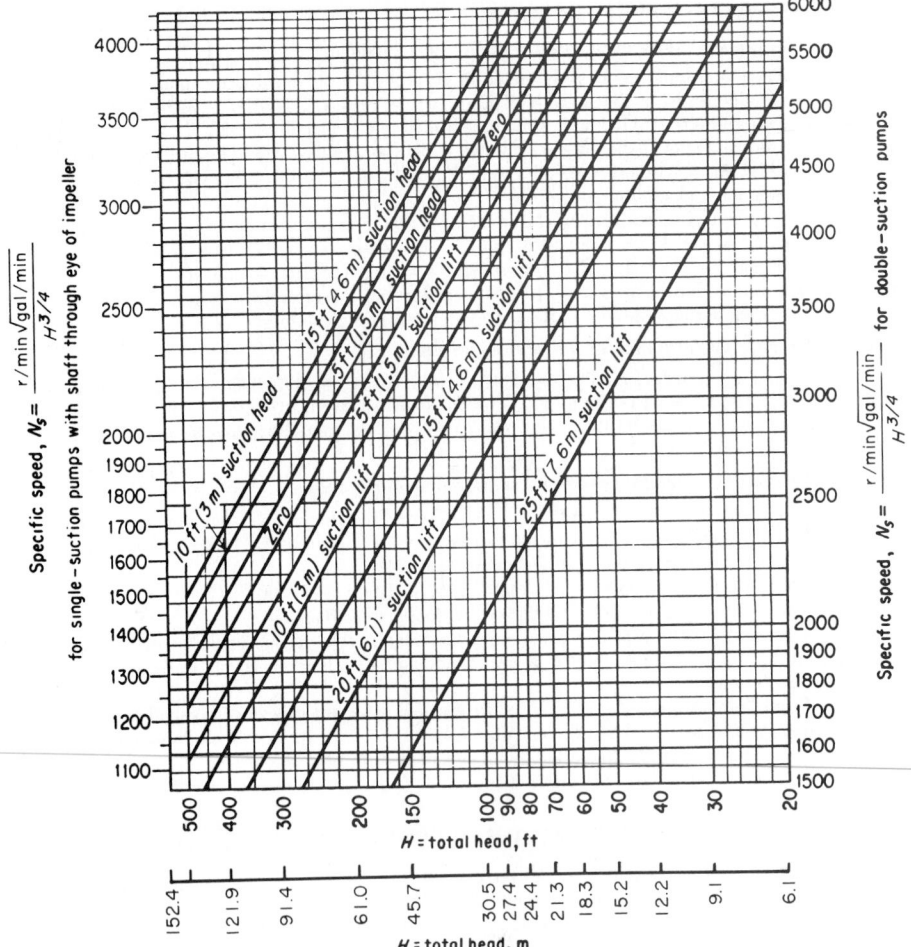

FIG. 1 Upper limits of specific speeds of single-stage, single- and double-suction centrifugal pumps handling clear water at 85°F (29.4°C) at sea level. (*Hydraulic Institute.*)

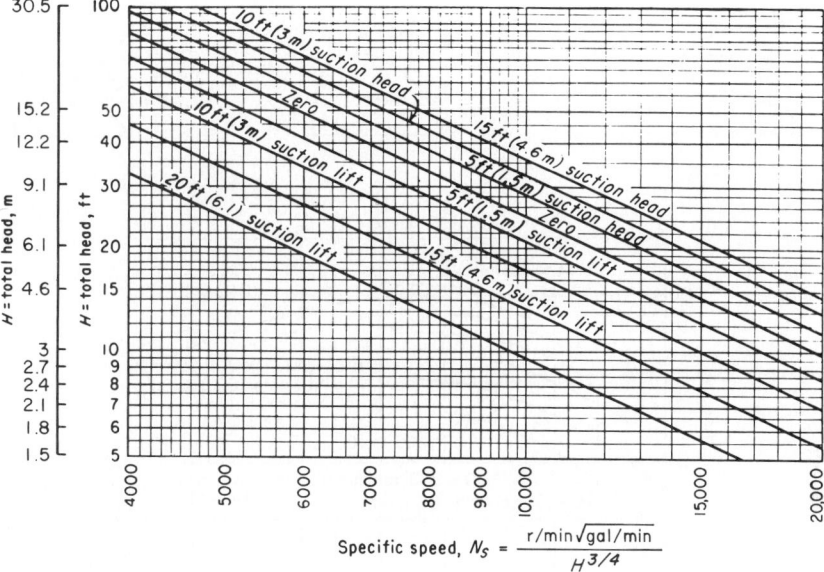

FIG. 2 Upper limits of specific speeds of single-suction mixed-flow and axial-flow pumps. *(Hydraulic Institute.)*

3. *Determine the pump efficiency and impeller shape*

Figure 3 shows the general relation between impeller shape, specific speed, pump capacity, efficiency, and characteristic curves. At $N_S = 2000$, efficiency = 87 percent. The impeller, as shown in Fig. 3, is moderately short and has a relatively large discharge area. A cross section of the impeller appears directly under the $N_S = 2000$ ordinate.

Related Calculations: Use the method given here for any type of pump whose variables are included in the Hydraulic Institute curves, Figs. 1 and 2, and in similar curves available from the same source. *Operating specific speed*, computed as above, is sometimes plotted on the performance curve of a centrifugal pump so that the characteristics of the unit can be better understood. *Type specific speed* is the operating specific speed giving maximum efficiency for a given pump and is a number used to identify a pump. Specific speed is important in cavitation and suction-lift studies. The Hydraulic Institute curves, Figs. 1 and 2, give upper limits of speed, head, capacity and suction lift for cavitation-free operation. When making actual pump analyses, be certain to use the curves (Figs. 1 and 2) in the latest edition of the *Standards of the Hydraulic Institute*.

SELECTING THE BEST OPERATING SPEED FOR A CENTRIFUGAL PUMP

A single-suction centrifugal pump is driven by a 60-Hz ac motor. The pump delivers 10,000 gal/min (630.9 L/s) of water at a 100-ft (30.5-m) head. The available net positive suction head = 32 ft (9.7 m) of water. What is the best operating speed for this pump if the pump operates at its best efficiency point?

Calculation Procedure:

1. *Determine the specific speed and suction specific speed*

Ac motors can operate at a variety of speeds, depending on the number of poles. Assume that the motor driving this pump might operate at 870, 1160, 1750, or 3500 r/min. Compute the specific

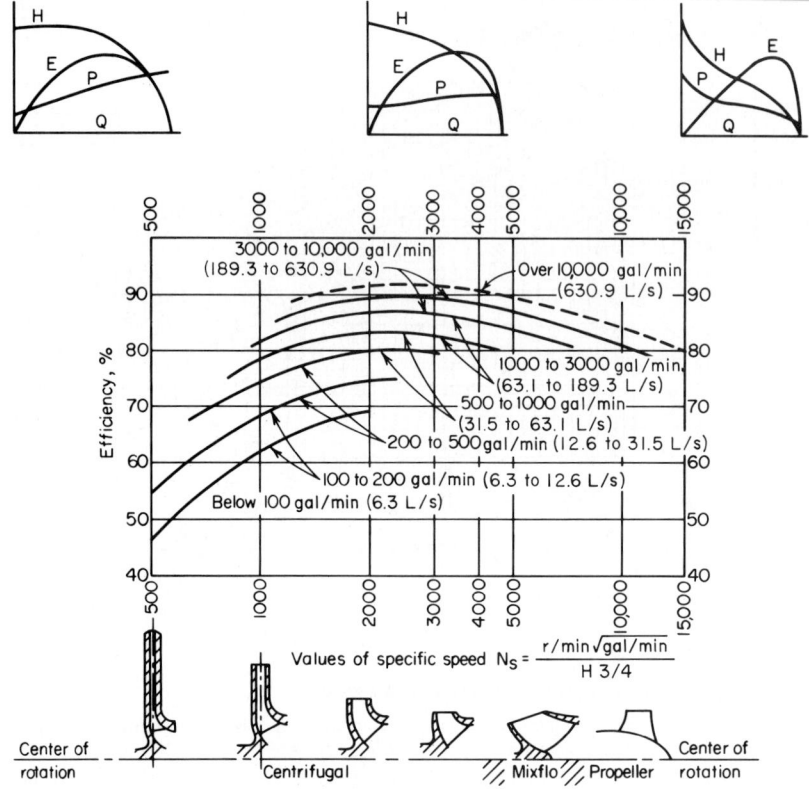

FIG. 3 Approximate relative impeller shapes and efficiency variations for various specific speeds of centrifugal pumps. *(Worthington Corporation.)*

speed $N_S = N(Q)^{0.5}/(H)^{0.75} = N(10{,}000)^{0.5}/(100)^{0.75} = 3.14N$ and the suction specific speed $S = N(Q)^{0.5}/(NPSH)^{0.75} = N(10{,}000)^{0.5}/(32)^{0.75} = 7.43N$ for each of the assumed speeds. Tabulate the results as follows:

Operating speed, r/min	Required specific speed	Required suction specific speed
870	2,740	6,460
1,160	3,640	8,620
1,750	5,500	13,000
3,500	11,000	26,000

2. *Choose the best speed for the pump*

Analyze the specific speed and suction specific speed at each of the various operating speeds, using the data in Tables 1 and 2. These tables show that at 870 and 1160 r/min, the suction specific-speed rating is poor. At 1750 r/min, the suction specific-speed rating is excellent, and a turbine or mixed-flow type pump will be suitable. Operation at 3500 r/min is unfeasible because a suction specific speed of 26,000 is beyond the range of conventional pumps.

TABLE 1   Pump Types Listed by Specific Speed°

Specific speed range	Type of pump
Below 2,000	Volute, diffuser
2,000–5,000	Turbine
4,000–10,000	Mixed-flow
9,000–15,000	Axial-flow

°Peerless Pump Division, FMC Corporation.

TABLE 2   Suction Specific-Speed Ratings°

Single-suction pump	Double-suction pump	Rating
Above 11,000	Above 14,000	Excellent
9,000–11,000	11,000–14,000	Good
7,000–9,000	9,000–11,000	Average
5,000–7,000	7,000–9,000	Poor
Below 5,000	Below 7,000	Very poor

°Peerless Pump Division, FMC Corporation.

Related Calculations: Use this procedure for any type of centrifugal pump handling water for plant services, cooling, process, fire protection, and similar requirements. This procedure is the work of R. P. Horwitz, Hydrodynamics Division, Peerless Pump, FMC Corporation, as reported in *Power* magazine.

TOTAL HEAD ON A PUMP HANDLING VAPOR-FREE LIQUID

Sketch three typical pump piping arrangements with static suction lift and submerged, free, and varying discharge head. Prepare similar sketches for the same pump with static suction head. Label the various heads. Compute the total head on each pump if the elevations are as shown in Fig. 4 and the pump discharges a maximum of 2000 gal/min (126.2 L/s) of water through 8-in (203.2-mm) schedule 40 pipe. What hp is required to drive the pump? A swing check valve is used on the pump suction line and a gate valve on the discharge line.

Calculation Procedure:

1. Sketch the possible piping arrangements

Figure 4 shows the six possible piping arrangements for the stated conditions of the installation. Label the total static head, i.e., the *vertical* distance from the surface of the source of the liquid supply to the free surface of the liquid in the discharge receiver, or to the point of free discharge from the discharge pipe. When both the suction and discharge surfaces are open to the atmosphere, the total static head equals the vertical difference in elevation. Use the free-surface elevations that cause the maximum suction lift and discharge head, i.e., the *lowest* possible level in the supply tank and the *highest* possible level in the discharge tank or pipe. When the supply source is *below* the pump centerline, the vertical distance is called the *static suction lift*; with the supply *above* the pump centerline, the vertical distance is called *static suction head*. With variable static suction head, use the lowest liquid level in the supply tank when computing total static head. Label the diagrams as shown in Fig. 4.

2. Compute the total static head on the pump

The total static head H_{ts} ft = static suction lift, h_{sl} ft + static discharge head h_{sd} ft, where the pump has a suction lift, s in Fig. 4a, b, and c. In these installations, $H_{ts} = 10 + 100 = 110$ ft (33.5 m). Note that the static discharge head is computed between the pump centerline and the water level with an underwater discharge, Fig. 4a; to the pipe outlet with a free discharge, Fig. 4b; and to the maximum water level in the discharge tank, Fig. 4c. When a pump is discharging into a closed compression tank, the total discharge head equals the static discharge head plus the head equivalent, ft of liquid, of the internal pressure in the tank, or 2.31 × tank pressure, lb/in².

Where the pump has a static suction head, as in Fig. 4d, e, and f, the total static head H_{ts} ft = h_{sd} − static suction head h_{sh} ft. In these installations, $H_t = 100 − 15 = 85$ ft (25.9 m).

The total static head, as computed above, refers to the head on the pump without liquid flow. To determine the total head on the pump, the friction losses in the piping system during liquid flow must be also determined.

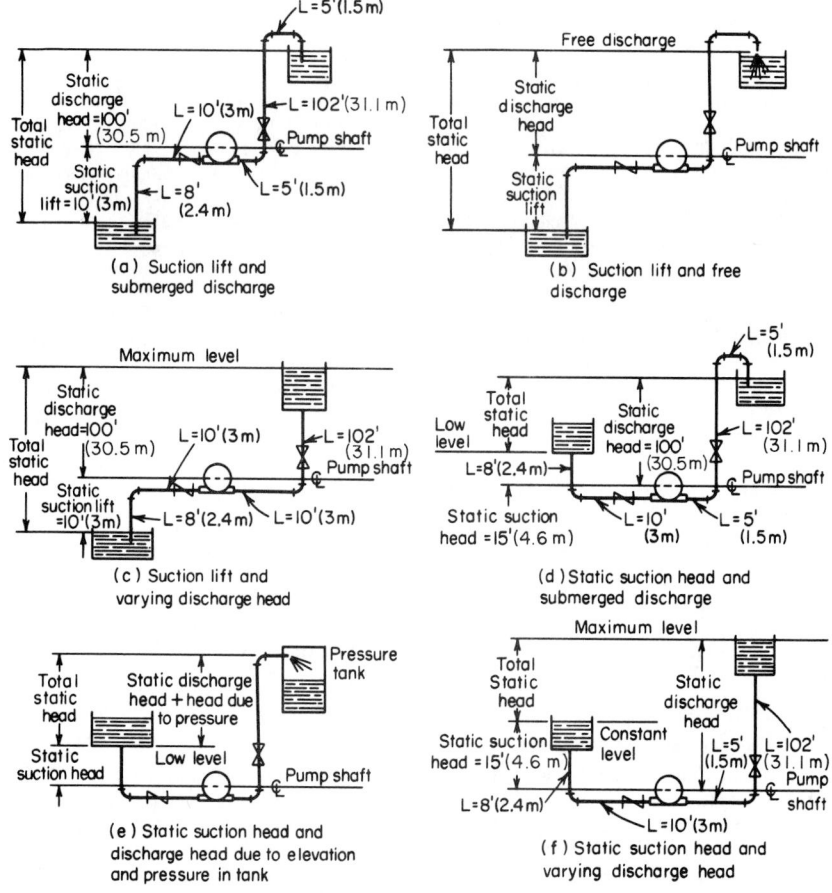

FIG. 4   Typical pump suction and discharge piping arrangements.

3.   *Compute the piping friction losses*

Mark the length of each piece of straight pipe on the piping drawing. Thus, in Fig. 4a, the total length of straight pipe L_t ft = 8 + 10 + 5 + 102 + 5 = 130 ft (39.6 m), if we start at the suction tank and add each length until the discharge tank is reached. To the total length of straight pipe must be added the *equivalent* length of the pipe fittings. In Fig. 4a there are four long-radius elbows, one swing check valve, and one globe valve. In addition, there is a minor head loss at the pipe inlet and at the pipe outlet.

The equivalent length of one 8-in (203.2-mm) long-radius elbow is 14 ft (4.3 m) of pipe, from Table 3. Since the pipe contains four elbows, the total equivalent length = 4(14) = 56 ft (17.1 m) of straight pipe. The open gate valve has an equivalent resistance of 4.5 ft (1.4 m); and the open swing check valve has an equivalent resistance of 53 ft (16.2 m).

The entrance loss h_e ft, assuming a basket-type strainer is used at the suction-pipe inlet, is h_e ft = $Kv^2/2g$, where K = a constant from Fig. 5; v = liquid velocity, ft/s; g = 32.2 ft/s^2 (980.67 cm/s^2). The exit loss occurs when the liquid passes through a sudden enlargement, as from a pipe to a tank. Where the area of the tank is large, causing a final velocity that is zero, $h_{ex} = v^2/2g$.

The velocity v ft/s in a pipe = $gpm/2.448d^2$. For this pipe, v = 2000/[(2.448)(7.98)^2] = 12.82 ft/s (3.91 m/s). Then h_e = 0.74(12.82)^2/[2(32.2)] = 1.89 ft (0.58 m), and h_{ex} = (12.82)^2/

TABLE 3 Resistance of Fittings and Valves (length of straight pipe giving equivalent resistance)

Pipe size		Standard ell		Medium-radius ell		Long-radius ell		45° Ell		Tee		Gate valve, open		Globe valve, open		Swing check, open	
in	mm	ft	m	ft	m	ft	m	ft	m	ft	m	ft	m	ft	m	ft	m
6	152.4	16	4.9	14	4.3	11	3.4	7.7	2.3	33	10.1	3.5	1.1	160	48.8	40	12.2
8	203.2	21	6.4	18	5.5	14	4.3	10	3.0	43	13.1	4.5	1.4	220	67.0	53	16.2
10	254.0	26	7.9	22	6.7	17	5.2	13	3.9	56	17.1	5.7	1.7	290	88.4	67	20.4
12	304.8	32	9.8	26	7.9	20	6.1	15	4.6	66	20.1	6.7	2.0	340	103.6	80	24.4

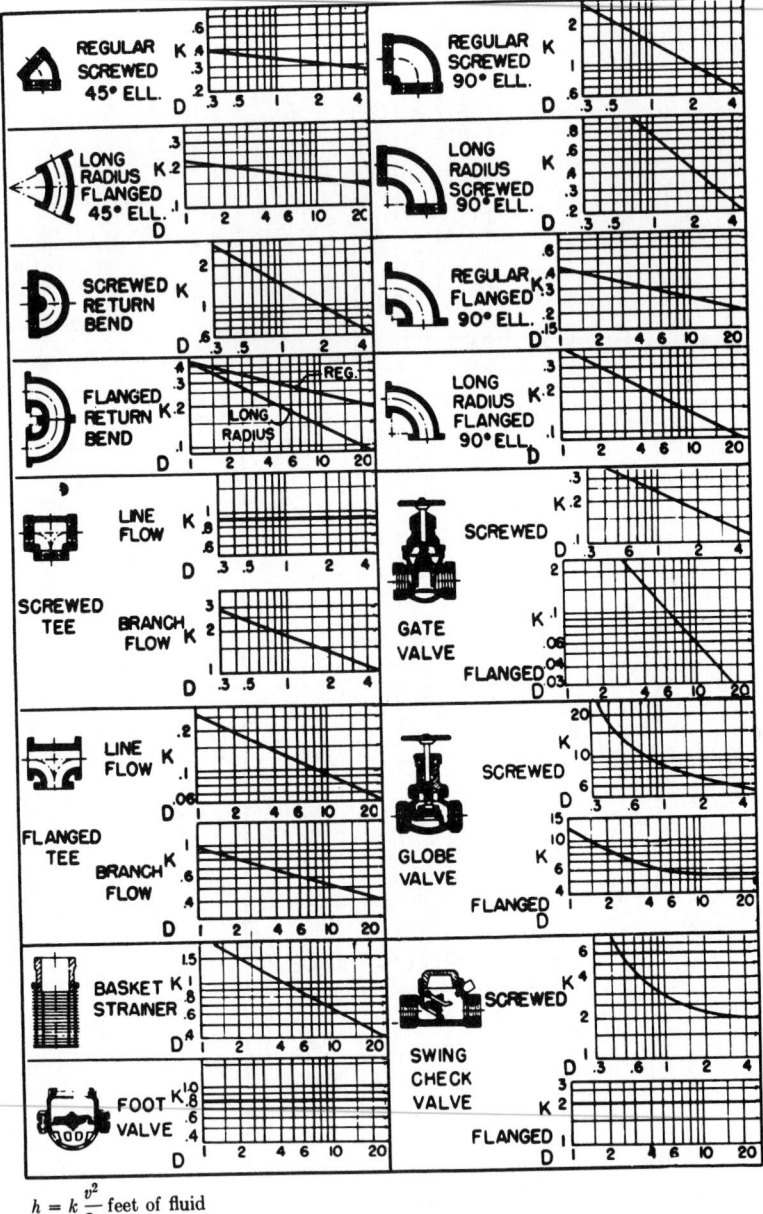

$$h = k \frac{v^2}{2g} \text{ feet of fluid}$$

FIG. 5 Resistance coefficients of pipe fittings. To convert to SI in the equation for h, v^2 would be measured in m/s and feet would be changed to meters. The following values would also be changed from inches to millimeters: 0.3 to 7.6, 0.5 to 12.7, 1 to 25.4, 2 to 50.8, 4 to 101.6, 6 to 152.4, 10 to 254, and 20 to 508. (*Hydraulic Institute.*)

TABLE 4 Pipe Friction Loss for Water (wrought-iron or steel schedule 40 pipe in good condition)

Diameter		Flow		Velocity		Velocity head		Friction loss per 100 ft (30.5 m) of pipe	
in	mm	gal/min	L/s	ft/s	m/s	ft water	m water	ft water	m water
6	152.4	1000	63.1	11.1	3.4	1.92	0.59	6.17	1.88
6	152.4	2000	126.2	22.2	6.8	7.67	2.3	23.8	7.25
6	152.4	4000	252.4	44.4	13.5	30.7	9.4	93.1	28.4
8	203.2	1000	63.1	6.41	1.9	0.639	0.195	1.56	0.475
8	203.2	2000	126.2	12.8	3.9	2.56	0.78	5.86	1.786
8	203.2	4000	252.4	25.7	7.8	10.2	3.1	22.6	6.888
10	254.0	1000	63.1	3.93	1.2	0.240	0.07	0.497	0.151
10	254.0	3000	189.3	11.8	3.6	2.16	0.658	4.00	1.219
10	254.0	5000	315.5	19.6	5.9	5.99	1.82	10.8	3.292

$[(2)(32.2)] = 2.56$ ft (0.78 m). Hence, the total length of the piping system in Fig. 4a is 130 + 56 + 4.5 + 53 + 1.89 + 2.56 = 247.95 ft (75.6 m), say 248 ft (75.6 m).

Use a suitable head-loss equation, or Table 4, to compute the head loss for the pipe and fittings. Enter Table 4 at an 8-in (203.2-mm) pipe size, and project horizontally across to 2000 gal/min (126.2 L/s) and read the head loss as 5.86 ft of water per 100 ft (1.8 m/30.5 m) of pipe.

The total length of pipe and fittings computed above is 248 ft (75.6 m). Then total friction-head loss with a 2000 gal/min (126.2-L/s) flow is H_f ft = (5.86)(248/100) = 14.53 ft (4.5 m).

4. Compute the total head on the pump

The total head on the pump $H_t = H_{ts} + H_f$. For the pump in Fig. 4a, $H_t = 110 + 14.53 = 124.53$ ft (37.95 m), say 125 ft (38.1 m). The total head on the pump in Fig. 4b and c would be the same. Some engineers term the total head on a pump the *total dynamic head* to distinguish between static head (no-flow vertical head) and operating head (rated flow through the pump).

The total head on the pumps in Fig. 4d, c, and f is computed in the same way as described above, except that the total static head is less because the pump has a static suction head. That is, the elevation of the liquid on the suction side reduces the total distance through which the pump must discharge liquid; thus the total static head is less. The static suction head is *subtracted* from the static discharge head to determine the total static head on the pump.

5. Compute the horsepower required to drive the pump

The brake horsepower input to a pump $bhp_i = (gpm)(H_t)(s)/3960e$, where s = specific gravity of the liquid handled; e = hydraulic efficiency of the pump, expressed as a decimal. The usual hydraulic efficiency of a centrifugal pump is 60 to 80 percent; reciprocating pumps, 55 to 90 percent; rotary pumps, 50 to 90 percent. For each class of pump, the hydraulic efficiency decreases as the liquid viscosity increases.

Assume that the hydraulic efficiency of the pump in this system is 70 percent and the specific gravity of the liquid handled is 1.0. Then $bhp_i = (2000)(127)(1.0)/(3960)(0.70) = 91.6$ hp (68.4 kW).

The theoretical or *hydraulic horsepower* $hp_h = (gpm)(H_t)(s)/3960$, or $hp_h = (2000) \times (127)(1.0)/3900 = 64.1$ hp (47.8 kW).

Related Calculations: Use this procedure for any liquid—water, oil, chemical, sludge, etc.—whose specific gravity is known. When liquids other than water are being pumped, the specific gravity and viscosity of the liquid, as discussed in later calculation procedures, must be taken into consideration. The procedure given here can be used for any class of pump—centrifugal, rotary, or reciprocating.

Note that Fig. 5 can be used to determine the equivalent length of a variety of pipe fittings. To use Fig. 5, simply substitute the appropriate K value in the relation $h = Kv^2/2g$, where h = equivalent length of straight pipe; other symbols as before.

PUMP SELECTION FOR ANY PUMPING SYSTEM

Give a step-by-step procedure for choosing the class, type, capacity, drive, and materials for a pump that will be used in an industrial pumping system.

Calculation Procedure:

1. Sketch the proposed piping layout

Use a single-line diagram, Fig. 6, of the piping system. Base the sketch on the actual job conditions. Show all the piping, fittings, valves, equipment, and other units in the system. Mark the *actual* and *equivalent* pipe length (see the previous calculation procedure) on the sketch. Be certain to include all vertical lifts, sharp bends, sudden enlargements, storage tanks, and similar equipment in the proposed system.

2. Determine the required capacity of the pump

The required capacity is the flow rate that must be handled in gal/min, million gal/day, ft^3/s, gal/h, bbl/day, lb/h, acre·ft/day, mil/h, or some similar measure. Obtain the required flow rate from the process conditions, for example, boiler feed rate, cooling-water flow rate, chemical feed

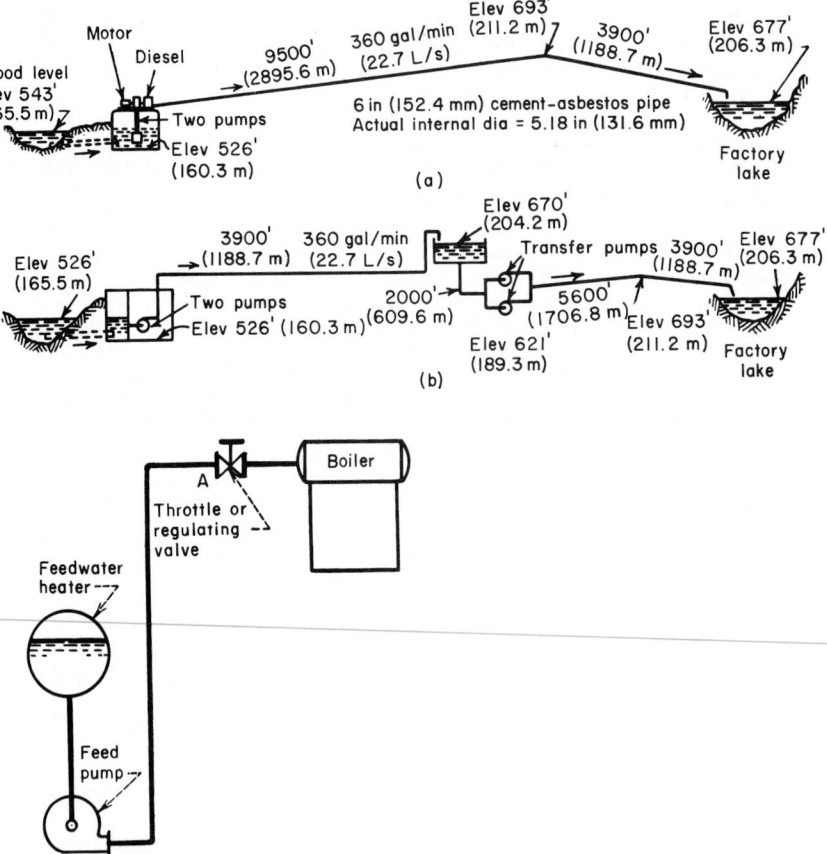

FIG. 6 (*a*) Single-line diagrams for an industrial pipeline; (*b*) single-line diagram of a boiler-feed system. (*Worthington Corporation.*)

rate, etc. The required flow rate for any process unit is usually given by the manufacturer or can be computed by using the calculation procedures given throughout this handbook.

Once the required flow rate is determined, apply a suitable factor of safety. The value of this factor of safety can vary from a low of 5 percent of the required flow to a high of 50 percent or more, depending on the application. Typical safety factors are in the 10 percent range. With flow rates up to 1000 gal/min (63.1 L/s), and in the selection of process pumps, it is common practice to round a computed required flow rate to the next highest round-number capacity. Thus, with a required flow rate of 450 gal/min (28.4 L/s) and a 10 percent safety factor, the flow of 450 + 0.10(450) = 495 gal/min (31.2 L/s) would be rounded to 500 gal/min (31.6 L/s) *before* the pump was selected. A pump of 500-gal/min (31.6-L/s), or larger, capacity would be selected.

3. *Compute the total head on the pump*

Use the steps given in the previous calculation procedure to compute the total head on the pump. Express the result in ft (m) of water—this is the most common way of expressing the head on a pump. Be certain to use the exact specific gravity of the liquid handled when expressing the head in ft (m) of water. A specific gravity less than 1.00 *reduces* the total head when expressed in ft (m) of water; whereas a specific gravity greater than 1.00 *increases* the total head when expressed in ft (m) of water. Note that variations in the suction and discharge conditions can affect the total head on the pump.

4. *Analyze the liquid conditions*

Obtain complete data on the liquid pumped. These data should include the name and chemical formula of the liquid, maximum and minimum pumping temperature, corresponding vapor pressure at these temperatures, specific gravity, viscosity at the pumping temperature, pH, flash point, ignition temperature, unusual characteristics (such as tendency to foam, curd, crystallize, become gelatinous or tacky), solids content, type of solids and their size, and variation in the chemical analysis of the liquid.

Enter the liquid conditions on a pump selection form like that in Fig. 7. Such forms are available from many pump manufacturers or can be prepared to meet special job conditions.

5. *Select the class and type of pump*

Three *classes* of pumps are used today—centrifugal, rotary, and reciprocating, Fig. 8. Note that these terms apply only to the mechanics of moving the liquid—not to the service for which the pump was designed. Each class of pump is further subdivided into a number of *types*, Fig. 8.

Use Table 5 as a general guide to the class and type of pump to be used. For example, when a large capacity at moderate pressure is required, Table 5 shows that a centrifugal pump would probably be best. Table 5 also shows the typical characteristics of various classes and types of pumps used in industrial process work.

Consider the liquid properties when choosing the class and type of pump, because exceptionally severe conditions may rule out one or another class of pump at the start. Thus, screw- and gear-type rotary pumps are suitable for handling viscous, nonabrasive liquid, Table 5. When an abrasive liquid must be handled, either another class of pump or another type of rotary pump must be used.

Also consider all the operating factors related to the particular pump. These factors include the type of service (continuous or intermittent), operating-speed preferences, future load expected and its effect on pump head and capacity, maintenance facilities available, possibility of parallel or series hookup, and other conditions peculiar to a given job.

Once the class and type of pump is selected, consult a rating table (Table 6) or rating chart, Fig. 9, to determine whether a suitable pump is available from the manufacturer whose unit will be used. When the hydraulic requirements fall between two standard pump models, it is usual practice to choose the next larger size of pump, unless there is some reason why an exact head and capacity are required for the unit. When one manufacturer does not have the desired unit, refer to the engineering data of other manufacturers. Also keep in mind that some pumps are custom-built for a given job when precise head and capacity requirements must be met.

Other pump data included in manufacturer's engineering information include characteristic curves for various diameter impellers in the same casing, Fig. 10, and variable-speed head-capacity curves for an impeller of given diameter, Fig. 11. Note that the required power input is given in Figs. 9 and 10 and may also be given in Fig. 11. Use of Table 6 is explained in the table.

Performance data for rotary pumps are given in several forms. Figure 12 shows a typical plot

Summary of Essential Data Required in Selection of Centrifugal Pumps

1. Number of Units Required

2. Nature of the Liquid to Be Pumped
   Is the liquid:
   a. Fresh or salt water, acid or alkali, oil, gasoline, slurry, or paper stock?
   b. Cold or hot and if hot, at what temperature? What is the vapor pressure of the liquid at the pumping temperature?
   c. What is its specific gravity?
   d. Is it viscous or nonviscous?
   e. Clear and free from suspended foreign matter or dirty and gritty? If the latter, what is the size and nature of the solids, and are they abrasive? If the liquid is of a pulpy nature, what is the consistency expressed either in percentage or in lb per cu ft of liquid? What is the suspended material?
   f. What is the chemical analysis, pH value, etc.? What are the expected variations of this analysis? If corrosive, what has been the past experience, both with successful materials and with unsatisfactory materials?

3. Capacity
   What is the required capacity as well as the minimum and maximum amount of liquid the pump will ever be called upon to deliver?

4. Suction Conditions
   Is there:
   a. A suction lift?
   b. Or a suction head?
   c. What are the length and diameter of the suction pipe?

5. Discharge Conditions
   a. What is the static head? Is it constant or variable?
   b. What is the friction head?
   c. What is the maximum discharge pressure against which the pump must deliver the liquid?

6. Total Head
   Variations in items 4 and 5 will cause variations in the total head.

7. Is the service continuous or intermittent?

8. Is the pump to be installed in a horizontal or vertical position? If the latter,
   a. In a wet pit?
   b. In a dry pit?

9. What type of power is available to drive the pump and what are the characteristics of this power?

10. What space, weight, or transportation limitations are involved?

11. Location of installation
    a. Geographical location
    b. Elevation above sea level
    c. Indoor or outdoor installation
    d. Range of ambient temperatures

12. Are there any special requirements or marked preferences with respect to the design, construction, or performance of the pump?

FIG. 7   Typical selection chart for centrifugal pumps. *(Worthington Corporation.)*

of the head and capacity ranges of different types of rotary pumps. Reciprocating-pump capacity data are often tabulated, as in Table 7.

6. *Evaluate the pump chosen for the installation*

Check the specific speed of a centrifugal pump, using the method given in an earlier calculation procedure. Once the specific speed is known, the impeller type and approximate operating efficiency can be found from Fig. 3.

Check the piping system, using the method of an earlier calculation procedure, to see whether the available net positive suction head equals, or is greater than, the required net positive suction head of the pump.

Class   Type

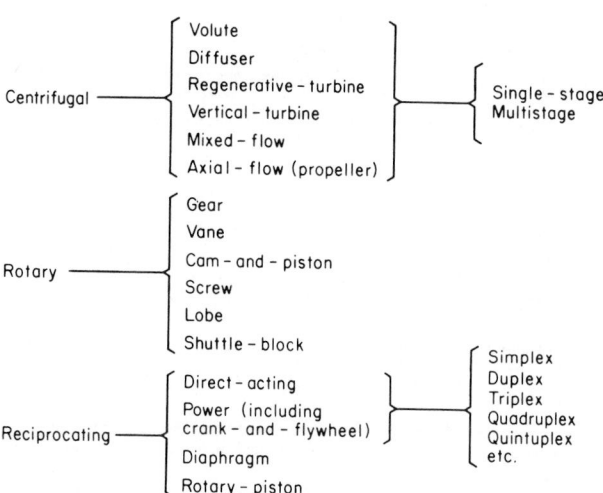

FIG. 8 Modern pump classes and types.

TABLE 5 Characteristics of Modern Pumps

	Centrifugal		Rotary	Reciprocating		
	Volute and diffuser	Axial flow	Screw and gear	Direct acting steam	Double acting power	Triplex
Discharge flow Usual maximum suction lift, ft (m)	Steady 15 (4.6)	Steady 15 (4.6)	Steady 22 (6.7)	Pulsating 22 (6.7)	Pulsating 22 (6.7)	Pulsating 22 (6.7)
Liquids handled	Clean, clear; dirty, abrasive; liquids with high solids content		Viscous; non-abrasive	Clean and clear		
Discharge pressure range	Low to high		Medium	Low to highest produced		
Usual capacity range	Small to largest available		Small to medium	Relatively small		
How increased head affects: Capacity Power input	Decrease Depends on specific speed		None Increase	Decrease Increase	None Increase	None Increase
How decreased head affects: Capacity Power input	Increase Depends on specific speed		None Decrease	Small increase Decrease	None Decrease	None Decrease

TABLE 6 Typical Centrifugal-Pump Rating Table

Size		Total head			
gal/min	L/s	20 ft, r/min—hp	6.1 m, r/min—kW	25 ft, r/min—hp	7.6 m, r/min—kW
3 CL:					
200	12.6	910—1.3	910–0.97	1010—1.6	1010—1.19
300	18.9	1000—1.9	1000–1.41	1100—2.4	1100—1.79
400	25.2	1200—3.1	1200—2.31	1230—3.7	1230—2.76
500	31.5	—	—	—	—
4 C:					
400	25.2	940—2.4	940—1.79	1040—3	1040—2.24
600	37.9	1080—4	1080—2.98	1170—4.6	1170—3.43
800	50.5	—	—	—	—

Example: 1080—4 indicates pump speed is 1080 r/min; actual input required to operate the pump is 4 hp (2.98 kW).
Source: Condensed from data of Goulds Pumps, Inc.; SI values added by handbook editor.

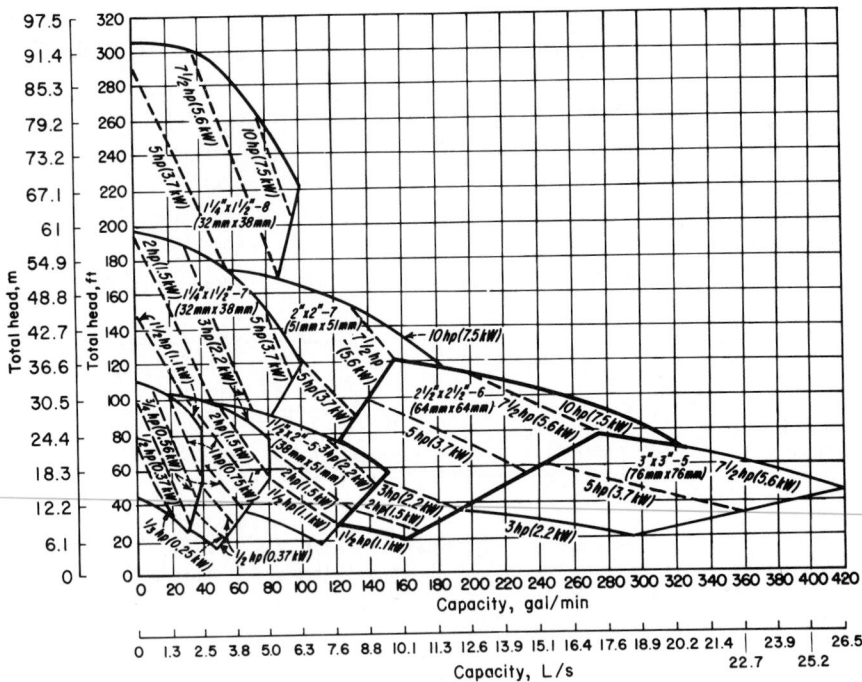

FIG. 9 Composite rating chart for a typical centrifugal pump. *(Goulds Pumps, Inc.)*

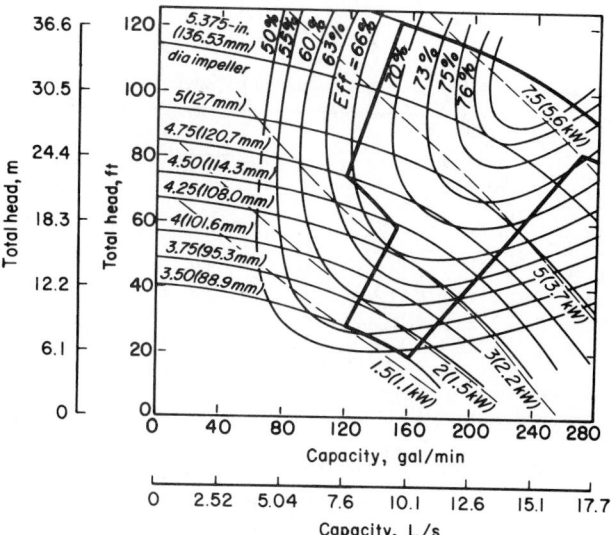

FIG. 10 Pump characteristics when impeller diameter is varied within the same casing.

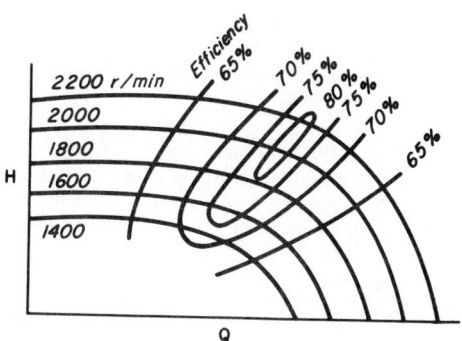

FIG. 11 Variable-speed head-capacity curves for a centrifugal pump.

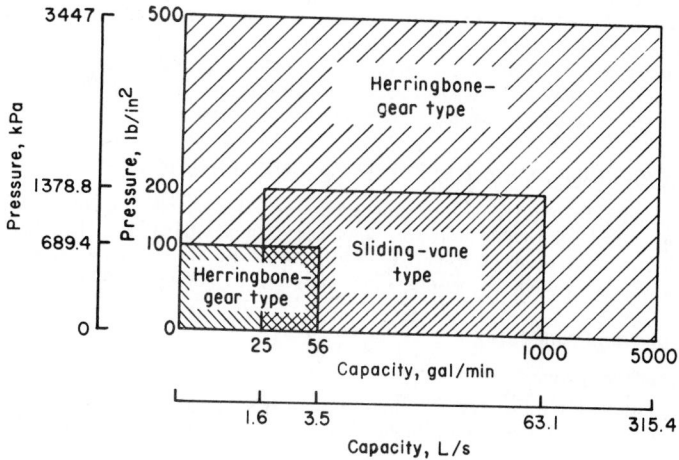

FIG. 12 Capacity ranges of some rotary pumps. *(Worthington Corporation.)*

TABLE 7 Capacities of Typical Horizontal Duplex Plunger Pumps

Size		Cold-water pressure service			
				Piston speed	
in	cm	gal/min	L/s	ft/min	m/min
6 × 3½ × 6	15.2 × 8.9 × 15.2	60	3.8	60	18.3
7½ × 4½ × 10	19.1 × 11.4 × 25.4	124	7.8	75	22.9
9 × 5 × 10	22.9 × 12.7 × 25.4	153	9.7	75	22.9
10 × 6 × 12	25.4 × 15.2 × 30.5	235	14.8	80	24.4
12 × 7 × 12	30.5 × 17.8 × 30.5	320	20.2	80	24.4

Size		Boiler-feed service					
				Boiler		Piston speed	
in	cm	gal/min	L/s	hp	kW	ft/min	m/min
6 × 3½ × 6	15.2 × 8.9 × 15.2	36	2.3	475	354.4	36	10.9
7½ × 4½ × 10	19.1 × 11.4 × 25.4	74	4.7	975	727.4	45	13.7
9 × 5 × 10	22.9 × 12.7 × 25.4	92	5.8	1210	902.7	45	13.7
10 × 6 × 12	25.4 × 15.2 × 30.5	141	8.9	1860	1387.6	48	14.6
12 × 7 × 12	30.5 × 17.8 × 30.5	192	12.1	2530	1887.4	48	14.6

Source: Courtesy of Worthington Corporation.

Determine whether a vertical or horizontal pump is more desirable. From the standpoint of floor space occupied, required NPSH, priming, and flexibility in changing the pump use, vertical pumps may be preferable to horizontal designs in some installations. But where headroom, corrosion, abrasion, and ease of maintenance are important factors, horizontal pumps may be preferable.

As a general guide, single-suction centrifugal pumps handle up to 50 gal/min (3.2 L/s) at total heads up to 50 ft (15.2 m); either single- or double-suction pumps are used for the flow rates to 1000 gal/min (63.1 L/s) and total heads to 300 ft (91.4 m); beyond these capacities and heads, double-suction or multistage pumps are generally used.

Mechanical seals are becoming more popular for all types of centrifugal pumps in a variety of services. Although they are more costly than packing, the mechanical seal reduces pump maintenance costs.

Related Calculations: Use the procedure given here to select any class of pump—centrifugal, rotary, or reciprocating—for any type of service—power plant, atomic energy, petroleum processing, chemical manufacture, paper mills, textile mills, rubber factories, food processing, water supply, sewage and sump service, air conditioning and heating, irrigation and flood control, mining and construction, marine services, industrial hydraulics, iron and steel manufacture.

ANALYSIS OF PUMP AND SYSTEM CHARACTERISTIC CURVES

Analyze a set of pump and system characteristic curves for the following conditions: friction losses without static head; friction losses with static head; pump without lift; system with little friction, much static head; system with gravity head; system with different pipe sizes; system with two discharge heads; system with diverted flow; and effect of pump wear on characteristic curve.

Calculation Procedure:

1. *Plot the system-friction curve*

Without static head, the system-friction curve passes through the origin (0,0), Fig. 13, because when no head is developed by the pump, flow through the piping is zero. For most piping systems, the friction-head loss varies as the square of the liquid flow rate in the system. Hence, a system-friction curve, also called a friction-head curve, is parabolic—the friction head increases as the flow rate or capacity of the system increases. Draw the curve as shown in Fig. 13.

2.   *Plot the piping system and system-head curve*

Figure 14a shows a typical piping system with a pump operating against a static discharge head. Indicate the total static head, Fig. 14b, by a dashed line—in this installation H_{ts} = 110 ft. Since static head is a physical dimension, it does not vary with flow rate and is a constant for all flow rates. Draw the dashed line parallel to the abscissa, Fig. 14b.

From the point of no flow—zero capacity—plot the friction-head loss at various flow rates—100, 200, 300 gal/min (6.3, 12.6, 18.9 L/s), etc. Determine the friction-head loss by computing it as shown in an earlier calculation procedure. Draw a curve through the points obtained. This is called the *system-head curve.*

Plot the pump head-capacity (*H-Q*) curve of the pump on Fig. 14b. The *H-Q* curve can be obtained from the pump manufacturer or from a tabulation of *H* and *Q* values for the

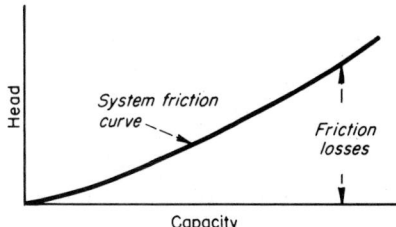

FIG. 13   Typical system-friction curve.

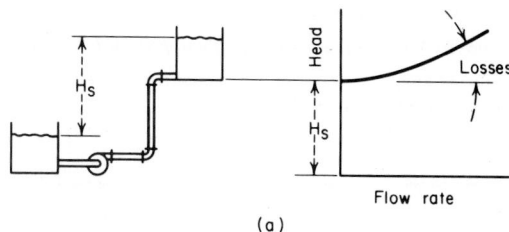

(a)

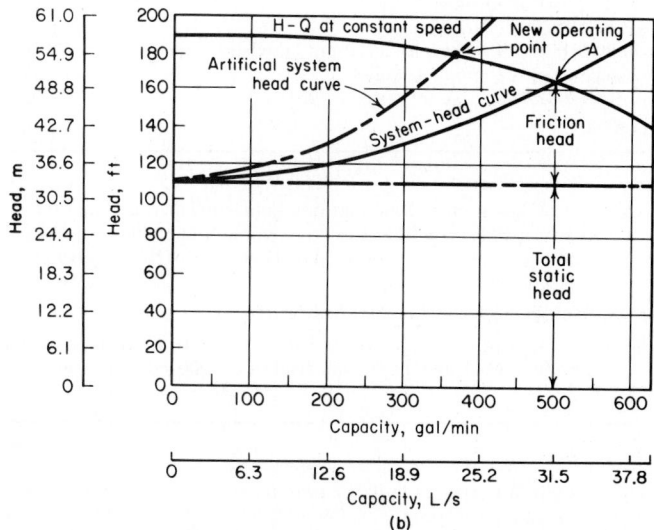

FIG. 14   (a) Significant friction loss and lift; (b) system-head curve superimposed on pump head-capacity curve. *(Peerless Pumps.)*

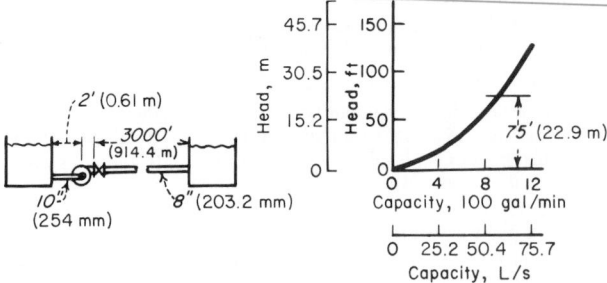

FIG. 15 No lift; all friction head. *(Peerless Pumps.)*

pump being considered. The point of intersection A between the H-Q and system-head curves is the operating point of the pump.

Changing the resistance of a given piping system by partially closing a valve or making some other change in the friction alters the position of the system-head curve and pump operating point. Compute the frictional resistance as before, and plot the artificial system-head curve as shown. Where this curve intersects the H-Q curve is the new operating point of the pump. System-head curves are valuable for analyzing the suitability of a given pump for a particular application.

3. Plot the no-lift system-head curve and compute the losses

With no static head or lift, the system-head curve passes through the origin (0,0), Fig. 15. For a flow of 900 gal/min (56.8 L/s) in this system, compute the friction loss as follows, using the Hydraulic Institute *Pipe Friction Manual* tables or the method of earlier calculation procedures:

	ft	m
Entrance loss from tank into 10-in (254-mm) suction pipe, $0.5v^2/2g$	0.10	0.03
Friction loss in 2 ft (0.61 m) of suction pipe	0.02	0.01
Loss in 10-in (254-mm) 90° elbow at pump	0.20	0.06
Friction loss in 3000 ft (914.4 m) of 8-in (203.2-mm) discharge pipe	74.50	22.71
Loss in fully open 8-in (203.2-mm) gate valve	0.12	0.04
Exit loss from 8-in (203.2-mm) pipe into tank, $v^2/2g$	0.52	0.16
Total friction loss	75.46	23.01

Compute the friction loss at other flow rates in a similar manner, and plot the system-head curve, Fig. 15. Note that if all losses in this system except the friction in the discharge pipe were ignored, the total head would not change appreciably. However, for the purposes of accuracy, all losses should always be computed.

4. Plot the low-friction, high-head system-head curve

The system-head curve for the vertical pump installation in Fig. 16 starts at the total static head, 15 ft (4.6 m), and zero flow. Compute the friction head for 15,000 gal/min as follows:

	ft	m
Friction in 20 ft (6.1 m) of 24-in (609.6-mm) pipe	0.40	0.12
Exit loss from 24-in (609.6-mm) pipe into tank, $v^2/2g$	1.60	0.49
Total friction loss	2.00	0.61

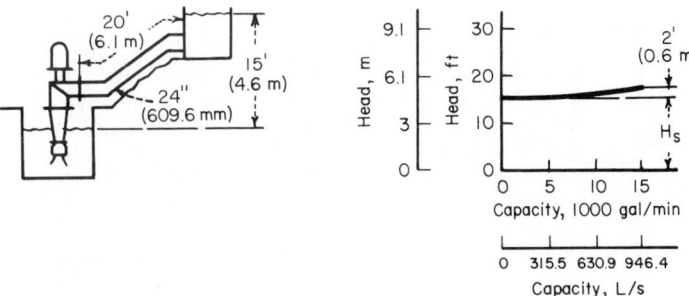

FIG. 16 Mostly lift; little friction head. *(Peerless Pumps.)*

Hence, almost 90 percent of the total head of 15 + 2 = 17 ft (5.2 m) at 15,000-gal/min (946.4-L/s) flow is static head. But neglect of the pipe friction and exit losses could cause appreciable error during selection of a pump for the job.

5. *Plot the gravity-head system-head curve*

In a system with gravity head (also called negative lift), fluid flow will continue until the system friction loss equals the available gravity head. In Fig. 17 the available gravity head is 50 ft (15.2 m). Flows up to 7200 gal/min (454.3 L/s) are obtained by gravity head alone. To obtain larger flow rates, a pump is needed to overcome the friction in the piping between the tanks. Compute the friction loss for several flow rates as follows:

	ft	m
At 5000 gal/min (315.5 L/s) friction loss in 1000 ft (305 m) of 16-in (406.4-mm) pipe	25	7.6
At 7200 gal/min (454.3 L/s), friction loss = available gravity head	50	15.2
At 13,000 gal/min (820.2 L/s), friction loss	150	45.7

Using these three flow rates, plot the system-head curve, Fig. 17.

6. *Plot the system-head curves for different pipe sizes*

When different diameter pipes are used, the friction loss vs. flow rate is plotted independently for the two pipe sizes. At a given flow rate, the total friction loss for the system is the sum of the loss for the two pipes. Thus, the combined system-head curve represents the sum of the static head and the friction losses for all portions of the pipe.

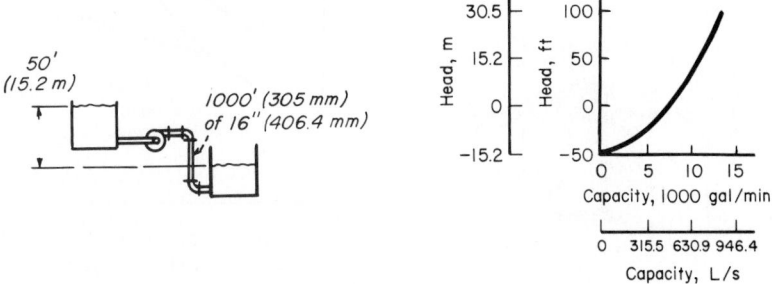

FIG. 17 Negative lift (gravity head). *(Peerless Pumps.)*

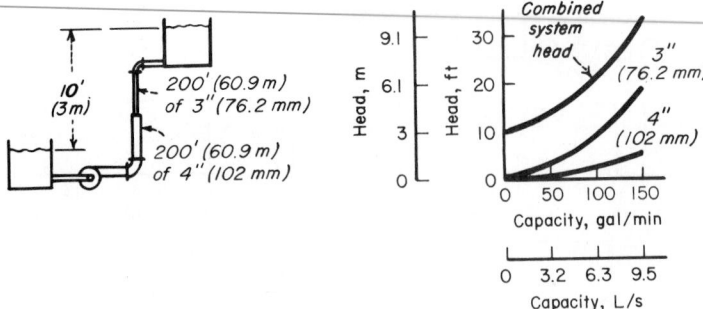

FIG. 18 System with two different pipe sizes. *(Peerless Pumps.)*

Figure 18 shows a system with two different pipe sizes. Compute the friction losses as follows:

	ft	m
At 150 gal/min (9.5 L/s), friction loss in 200 ft (60.9 m) of 4-in (102-mm) pipe	5	1.52
At 150 gal/min (9.5 L/s), friction loss in 200 ft (60.9 m) of 3-in (76.2-mm) pipe	19	5.79
Total static head for 3- (76.2-) and 4-in (102-mm) pipes	10	3.05
Total head at 150-gal/min (9.5-L/s) flow	34	10.36

Compute the total head at other flow rates, and then plot the system-head curve as shown in Fig. 18.

7. Plot the system-head curve for two discharge heads

Figure 19 shows a typical pumping system having two different discharge heads. Plot separate system-head curves when the discharge heads are different. Add the flow rates for the two pipes at the same head to find points on the combined system-head curve, Fig. 19. Thus,

	ft	m
At 550 gal/min (34.7 L/s), friction loss in 1000 ft (305 m) of 8-in (203.2-mm) pipe	10	3.05
At 1150 gal/min (72.6 L/s), friction	38	11.6
At 1150 gal/min (72.6 L/s), friction + lift in pipe 1	88	26.8
At 550 gal/min (34.7 L/s), friction + lift in pipe 2	88	26.8

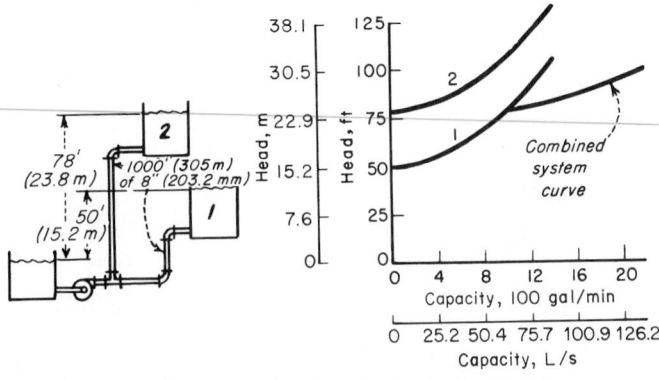

FIG. 19 System with two different discharge heads. *(Peerless Pumps.)*

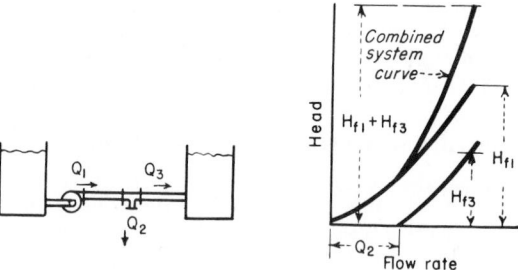

FIG. 20   Part of the fluid flow is diverted from the main pipe. *(Peerless Pumps.)*

The flow rate for the combined system at a head of 88 ft (26.8 m) is $1150 + 550 = 1700$ gal/min (107.3 L/s). To produce a flow of 1700 gal/min (107.3 L/s) through this system, a pump capable of developing an 88-ft (26.8-m) head is required.

8.   *Plot the system-head curve for diverted flow*

To analyze a system with diverted flow, assume that a constant quantity of liquid is tapped off at the intermediate point. Plot the friction loss vs. flow rate in the normal manner for pipe 1, Fig. 20. Move the curve for pipe 3 to the right at zero head by an amount equal to Q_2, since this represents the quantity passing through pipes 1 and 2 but not through pipe 3. Plot the combined system-head curve by adding, at a given flow rate, the head losses for pipes 1 and 3. With $Q = 300$ gal/min (18.9 L/s), pipe 1 = 500 ft (152.4 m) of 10-in (254-mm) pipe, and pipe 3 = 50 ft (15.2 m) of 6-in (152.4-mm) pipe.

	ft	m
At 1500 gal/min (94.6 L/s) through pipe 1, friction loss	11	3.35
Friction loss for pipe 3 (1500 − 300 = 1200 gal/min) (75.7 L/s)	8	2.44
Total friction loss at 1500-gal/min (94.6-L/s) delivery	19	5.79

9.   *Plot the effect of pump wear*

When a pump wears, there is a loss in capacity and efficiency. The amount of loss depends, however, on the shape of the system-head curve. For a centrifugal pump, Fig. 21, the capacity loss is greater for a given amount of wear if the system-head curve is flat, as compared with a steep system-head curve.

Determine the capacity loss for a worn pump by plotting its *H-Q* curve. Find this curve by testing the pump at different capacities and plotting the corresponding head. On the same chart, plot the *H-Q* curve for a new pump of the same size, Fig. 21. Plot the system-head curve, and determine the capacity loss as shown in Fig. 21.

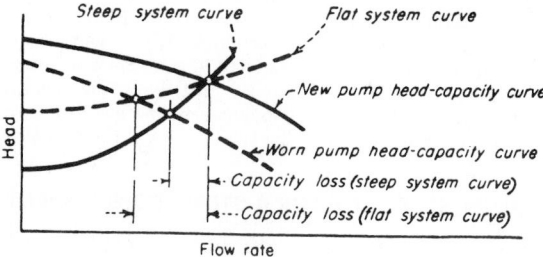

FIG. 21   Effect of pump wear on pump capacity. *(Peerless Pumps.)*

Related Calculations: Use the techniques given here for any type of pump—centrifugal, reciprocating, or rotary—handling any type of liquid—oil, water, chemicals, etc. The methods given here are the work of Melvin Mann, as reported in *Chemical Engineering*, and Peerless Pump Division of FMC Corp.

NET POSITIVE SUCTION HEAD FOR HOT-LIQUID PUMPS

What is the maximum capacity of a double-suction condensate pump operating at 1750 r/min if it handles 100°F (37.8°C) water from a hot well in a condenser having an absolute pressure of 2.0 in (50.8 mm) Hg if the pump centerline is 10 ft (30.5 m) below the hot-well liquid level and the friction-head loss in the suction piping and fitting is 5 ft (1.52 m) of water?

Calculation Procedure:

1. Compute the net positive suction head on the pump

The net positive suction head h_n on a pump when the liquid supply is *above* the pump inlet = pressure on liquid surface + static suction head − friction-head loss in suction piping and pump inlet − vapor pressure of the liquid, all expressed in ft absolute of liquid handled. When the liquid supply is *below* the pump centerline—i.e., there is a static suction lift—the vertical distance of the lift is *subtracted* from the pressure on the liquid surface instead of added as in the above relation.

The density of 100°F (37.8°C) water is 62.0 lb/ft³ (992.6 kg/m³), computed as shown in earlier calculation procedures in this handbook. The pressure on the liquid surface, in absolute ft of liquid = (2.0 inHg)(1.133)(62.4/62.0) = 2.24 ft (0.68 m). In this calculation, 1.133 = ft of 39.2°F (4°C) water = 1 inHg; 62.4 = lb/ft³ (999.0 kg/m³) of 39.2°F (4°C) water. The temperature of 39.2°F (4°C) is used because at this temperature water has its maximum density. Thus, to convert inHg to ft absolute of water, find the product of (inHg)(1.133)(water density at 39.2°F)/(water density at operating temperature). Express both density values in the same unit, usually lb/ft³.

The static suction head is a physical dimension that is measured in ft (m) of liquid at the operating temperature. In this installation, h_{sh} = 10 ft (3 m) absolute.

The friction-head loss is 5 ft (1.52 m) of water. When it is computed by using the methods of earlier calculation procedures, this head loss is in ft (m) of water at maximum density. To convert to ft absolute, multiply by the ratio of water densities at 39.2°F (4°C) and the operating temperature, or (5)(62.4/62.0) = 5.03 ft (1.53 m).

The vapor pressure of water at 100°F (37.8°C) is 0.949 lb/in² (abs) (6.5 kPa) from the steam tables. Convert any vapor pressure to ft absolute by finding the result of [vapor pressure, lb/in² (abs)] (144 in²/ft²)/liquid density at operating temperature, or (0.949)(144)/62.0 = 2.204 ft (0.67 m) absolute.

With all the heads known, the net positive suction head is h_n = 2.24 + 10 − 5.03 − 2.204 = 5.01 ft (1.53 m) absolute.

2. Determine the capacity of the condensate pump

Use the Hydraulic Institute curve, Fig. 22, to determine the maximum capacity of the pump. Enter at the left of Fig. 22 at a net positive suction head of 5.01 ft (1.53 m), and project horizontally to the right until the 3500-r/min curve is intersected. At the top, read the capacity as 278 gal/min (17.5 L/s).

Related Calculations: Use this procedure for any condensate or boiler-feed pump handling water at an elevated temperature. Consult the *Standards of the Hydraulic Institute* for capacity curves of pumps having different types of construction. In general, pump manufacturers who are members of the Hydraulic Institute rate their pumps in accordance with the *Standards,* and a pump chosen from a catalog capacity table or curve will deliver the stated capacity. A similar procedure is used for computing the capacity of pumps handling volatile petroleum liquids. When you use this procedure, be certain to refer to the latest edition of the *Standards.*

CONDENSATE PUMP SELECTION FOR A STEAM POWER PLANT

Select the capacity for a condensate pump serving a steam power plant having a 140,000 lb/h (63,000 kg/h) exhaust flow to a condenser that operates at an absolute pressure of 1.0 in (25.4

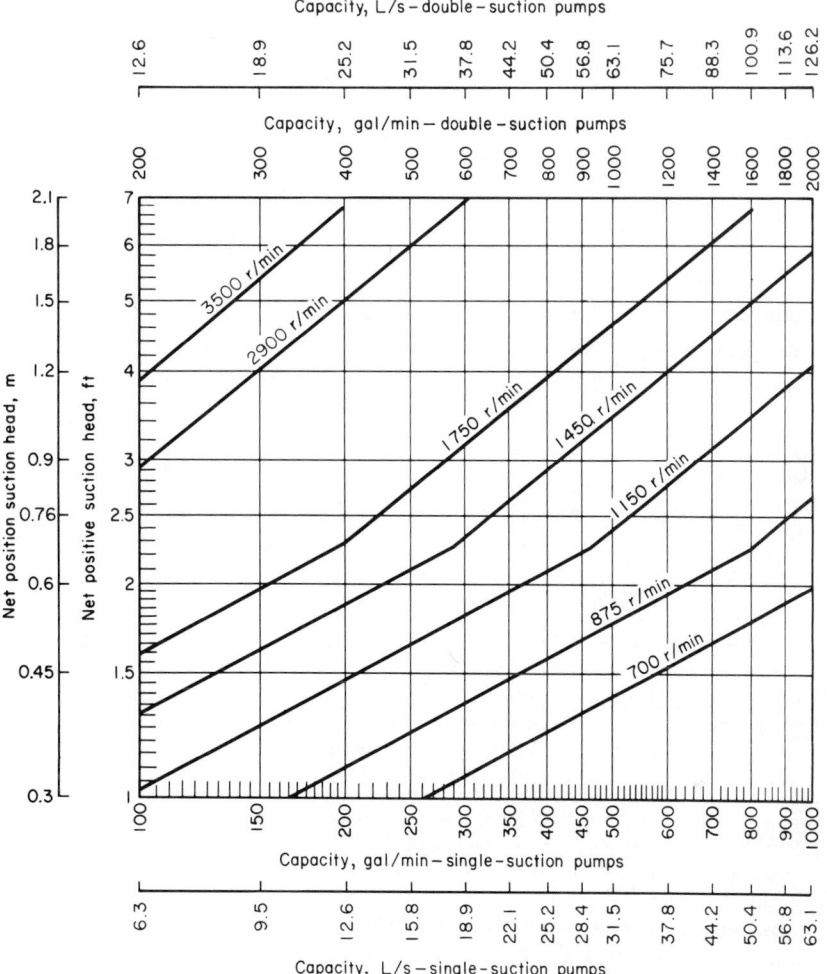

FIG. 22 Capacity and speed limitations of condensate pumps with the shaft through the impeller eye. (*Hydraulic Institute.*)

mm) Hg. The condensate pump discharges through 4-in (101.6-mm) schedule 40 pipe to an air-ejector condenser that has a frictional resistance of 8 ft (2.4 m) of water. From here, the condensate flows to and through a low-pressure heater that has a frictional resistance of 12 ft (3.7 m) of water and is vented to the atmosphere. The total equivalent length of the discharge piping, including all fittings and bends, is 400 ft (121.9 m), and the suction piping total equivalent length is 50 ft (15.2 m). The inlet of the low pressure heater is 75 ft (22.9 m) above the pump centerline, and the condenser hot-well water level is 10 ft (3 m) above the pump centerline. How much power is required to drive the pump if its efficiency is 70 percent?

Calculation Procedure:

1. Compute the static head on the pump

Sketch the piping system as shown in Fig. 23. Mark the static elevations and equivalent lengths as indicated.

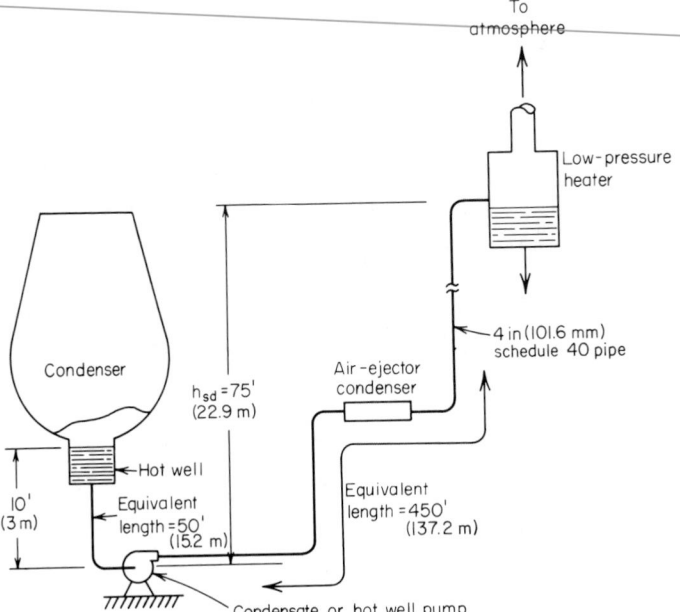

FIG. 23 Condensate pump serving a steam power plant.

The total head on the pump $H_t = H_{ts} + H_f$, where the symbols are the same as in earlier calculation procedures. The total static head $H_{ts} = h_{sd} - h_{sh}$. In this installation, $h_{sd} = 75$ ft (22.9 m). To make the calculation simpler, convert all the heads to absolute values. Since the heater is vented to the atmosphere, the pressure acting on the surface of the water in it = 14.7 lb/in² (abs) (101.3 kPa), or 34 ft (10.4 m) of water. The pressure acting on the condensate in the hot well is 1 in (25.4 mm) Hg = 1.133 ft (0.35 m) of water. [An absolute pressure of 1 in (25.4 mm) Hg = 1.133 ft (0.35 m) of water.] Thus, the absolute discharge static head = 75 + 34 = 109 ft (33.2 m), whereas the absolute suction head = 10 + 1.13 = 11.13 ft (3.39 m). Then $H_{ts} = h_{hd} - h_{sh}$ = 109.00 − 11.13 = 97.87 ft (29.8 m), say 98 ft (29.9 m) of water.

2. Compute the friction head in the piping system

The total friction head H_f = pipe friction + heater friction. The pipe friction loss is found first, as shown below. The heater friction loss, obtained from the manufacturer or engineering data, is then added to the pipe-friction loss. Both must be expressed in ft (m) of water.

To determine the pipe friction, use Fig. 24 of this section and Table 22 and Fig. 13 of the *Piping* section of this handbook in the following manner. Find the product of the liquid velocity, ft/s, and the pipe internal diameter, in, or vd. With an exhaust flow of 140,000 lb/h (63,636 kg/h) to the condenser, the condensate flow is the same, or 140,000 lb/h (63,636 kg/h) at a temperature of 79.03°F (21.6°C), corresponding to an absolute pressure in the condenser of 1 in (25.4 mm) Hg, obtained from the steam tables. The specific volume of the saturated liquid at this temperature and pressure is 0.01608 ft³/lb (0.001 m³/kg). Since 1 gal (0.26 L) of liquid occupies 0.13368 ft³ (0.004 m³), specific volume, gal/lb, is (0.01608/0.13368) = 0.1202 (1.01 L/kg). Therefore, a flow of 140,000 lb/h (63,636 kg/h) = a flow of (140,000)(0.1202) = 16,840 gal/h (63,739.4 L/h), or 16,840/60 = 281 gal/min (17.7 L/s). Then the liquid velocity $v = gpm/2.448d^2$ = 281/2.448(4.026)² = 7.1 ft/s (2.1 m/s), and the product vd = (7.1)(4.026) = 28.55.

Enter Fig. 24 at a temperature of 79°F (26.1°C), and project horizontally to the right to vd = 28.55 and then vertically upward to the water curve. From the intersection, project horizontally to the right to vd = 28.55 and then vertically upward to read R = 250,000. Using Table 22 and Fig. 13 of the *Piping* section and R = 250,000, find the friction factor f = 0.0185. Then the head loss due to pipe friction $H_f = (L/D)(v^2/2g)$

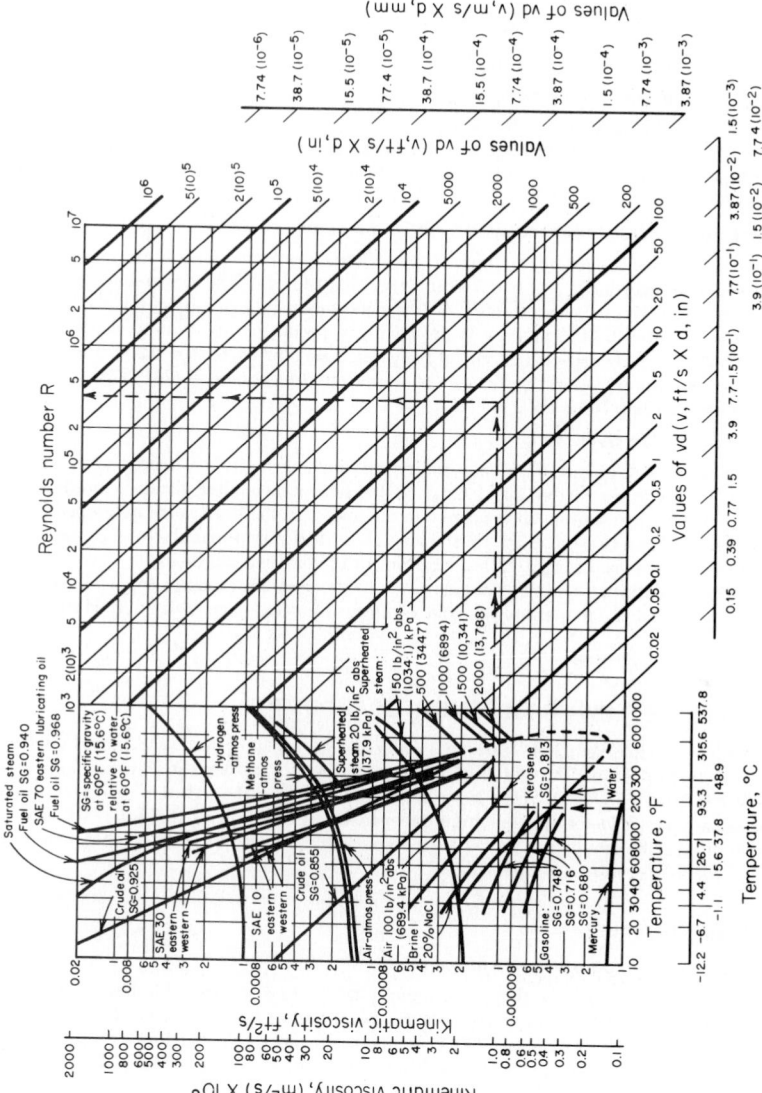

FIG. 24 Kinematic viscosity and Reynolds number chart. *(Hydraulic Institute.)*

$= 0.0185 \ (450/4.026/12)/[(7.1)^2/2(32.2)] = 19.18$ ft (5.9 m). In this computation, L = total equivalent length of the pipe, pipe fittings, and system valves, or 450 ft (137.2 m).

3. Compute the other head losses in the system

There are two other head losses in this piping system: the entrance loss at the square-edged hot-well pipe leading to the pump and the sudden enlargement in the low-pressure heater. The velocity head $v^2/2g = (7.1)^2/2(32.2) = 0.784$ ft (0.24 m). Using k values from Fig. 5 in this section, $h_e = kv^2/2g = (0.5)(0.784) = 0.392$ ft (0.12 m); $h_{ex} = v^2/2g = 0.784$ ft (0.24 m).

4. Find the total head on the pump

The total head on the pump $H_t = H_{ts} + H_f = 97.87 + 19.18 + 8 + 12 + 0.392 + 0.784 = 138.226$ ft (42.1 m), say 140 ft (42.7 m) of water. In this calculation, the 8- (2.4-m) and 12-ft (3.7-m) head losses are those occurring in the heaters. With a 25 percent safety factor, total head = $(1.25)(140) = 175$ ft (53.3 m).

5. Compute the horsepower required to drive the pump

The brake horsepower input $bhp_i = (gpm)(H_t)(s)/3960e$, where the symbols are the same as in earlier calculation procedures. At 1 in (25.4 mm) Hg, 1 lb (0.45 kg) of the condensate has a volume of 0.01608 ft^3 (0.000455 m^3). Since density = 1/specific volume, the density of the condensate = $1/0.01608 = 62.25$ ft^3/lb (3.89 m^3/kg). Water having a specific gravity of unity weighs 62.4 lb/ft^3 (999 kg/m^3). Hence, the specific gravity of the condensate is $62.25/62.4 = 0.997$. Then, assuming that the pump has an operating efficiency of 70 percent, we get $bhp_i = (281)(175) \times (0.997)/[3960(0.70)] = 17.7$ bhp (13.2 kW).

6. Select the condensate pump

Condensate or hot-well pumps are usually centrifugal units having two or more stages, with the stage inlets opposed to give better axial balance and to subject the sealing glands to positive internal pressure, thereby preventing air leakage into the pump. In the head range developed by this pump, 175 ft (53.3 m), two stages are satisfactory. Refer to a pump manufacturer's engineering data for specific stage head ranges. Either a turbine or motor drive can be used.

   Related Calculations: Use this procedure to choose condensate pumps for steam plants of any type—utility, industrial, marine, portable, heating, or process—and for combined steam-diesel plants.

MINIMUM SAFE FLOW FOR A CENTRIFUGAL PUMP

A centrifugal pump handles 220°F (104.4°C) water and has a shutoff head (with closed discharge valve) of 3200 ft (975.4 m). At shutoff, the pump efficiency is 17 percent and the input brake horsepower is 210 (156.7 kW). What is the minimum safe flow through this pump to prevent overheating at shutoff? Determine the minimum safe flow if the NPSH is 18.8 ft (5.7 m) of water and the liquid specific gravity is 0.995. If the pump contains 500 lb (225 kg) of water, determine the rate of the temperature rise at shutoff.

Calculation Procedure:

1. Compute the temperature rise in the pump

With the discharge valve closed, the power input to the pump is converted to heat in the casing and causes the liquid temperature to rise. The temperature rise $t = (1 - e) \times H_s/778e$, where t = temperature rise during shutoff, °F; e = pump efficiency, expressed as a decimal; H_s = shutoff head, ft. For this pump, $t = (1 - 0.17)(3200)/[778(0.17)] = 20.4$°F (36.7°C).

2. Compute the minimum safe liquid flow

For general-service pumps, the minimum safe flow M gal/min = 6.0(bhp input at shutoff)/t. Or, $M = 6.0(210)/20.4 = 62.7$ gal/min (3.96 L/s). This equation includes a 20 percent safety factor.

   Centrifugal boiler-feed pumps usually have a maximum allowable temperature rise of 15°F (27°C). The minimum allowable flow through the pump to prevent the water temperature from rising more than 15°F (27°C) is 30 gal/min (1.89 L/s) for each 100-bhp (74.6-kW) input at shutoff.

3. Compute the temperature rise for the operating NPSH

An NPSH of 18.8 ft (5.73 m) is equivalent to a pressure of $18.8(0.433)(0.995) = 7.78 \text{ lb/in}^2$ (abs) (53.6 kPa) at 220°F (104.4°C), where the factor 0.433 converts ft of water to lb/in^2. At 220°F (104.4°C), the vapor pressure of the water is 17.19 lb/in^2 (abs) (118.5 kPa), from the steam tables. Thus, the total vapor pressure the water can develop before flashing occurs = NPSH pressure + vapor pressure at operating temperature $= 7.78 + 17.19 = 24.97 \text{ lb/in}^2$ (abs) (172.1 kPa). Enter the steam tables at this pressure, and read the corresponding temperature as 240°F (115.6°C). The allowable temperature rise of the water is then $240 - 220 = 20°F$ (36.0°C). Using the safe-flow relation of step 2, we find the minimum safe flow is 62.9 gal/min (3.97 L/s).

4. Compute the rate of temperature rise

In any centrifugal pump, the rate of temperature rise t_r, °F/min $= 42.4$(bhp input at shutoff)/wc, where w = weight of liquid in the pump, lb; c = specific heat of the liquid in the pump, Btu/(lb·°F). For this pump containing 500 lb (225 kg) of water with a specific heat, $c = 1.0$, $t_r = 42.4(210)/[500(1.0)] = 17.8°F/\text{min}$ (32°C/min). This is a very rapid temperature rise and could lead to overheating in a few minutes.

Related Calculations: Use this procedure for any centrifugal pump handling any liquid in any service—power, process, marine, industrial, or commercial. Pump manufacturers can supply a temperature-rise curve for a given model pump if it is requested. This curve is superimposed on the pump characteristic curve and shows the temperature rise accompanying a specific flow through the pump.

SELECTING A CENTRIFUGAL PUMP TO HANDLE A VISCOUS LIQUID

Select a centrifugal pump to deliver 750 gal/min (47.3 L/s) of 1000-SSU oil at a total head of 100 ft (30.5 m). The oil has a specific gravity of 0.90 at the pumping temperature. Show how to plot the characteristic curves when the pump is handling the viscous liquid.

Calculation Procedure:

1. Determine the required correction factors

A centrifugal pump handling a viscous liquid usually must develop a greater capacity and head, and it requires a larger power input than the same pump handling water. With the water performance of the pump known—from either the pump characteristic curves or a tabulation of pump performance parameters—Fig. 25, prepared by the Hydraulic Institute, can be used to find suitable correction factors. Use this chart only within its scale limits; do not extrapolate. Do not use the chart for mixed-flow or axial-flow pumps or for pumps of special design. Use the chart only for pumps handling uniform liquids; slurries, gels, paper stock, etc., may cause incorrect results. In using the chart, the available net positive suction head is assumed adequate for the pump.

To use Fig. 25, enter at the bottom at the required capacity, 750 gal/min (47.3 L/s), and project vertically to intersect the 100-ft (30.5-m) head curve, the required head. From here project horizontally to the 1000-SSU viscosity curve, and then vertically upward to the correction-factor curves. Read $C_E = 0.635$; $C_Q = 0.95$; $C_H = 0.92$ for $1.0Q_{NW}$. The subscripts E, Q, and H refer to correction factors for efficiency, capacity, and head, respectively; and NW refers to the water capacity at a particular efficiency. At maximum efficiency, the water capacity is given as $1.0Q_{NW}$; other efficiencies, expressed by numbers equal to or less than unity, give different capacities.

2. Compute the water characteristics required

The water capacity required for the pump $Q_w = Q_v/C_Q$ where Q_v = viscous capacity, gal/min. For this pump, $Q_w = 750/0.95 = 790$ gal/min (49.8 L/s). Likewise, water head $H_w = H_v/C_H$, where H_v = viscous head. Or, $H_w = 100/0.92 = 108.8$ (33.2 m), say 109 ft (33.2 m) of water.

Choose a pump to deliver 790 gal/min (49.8 L/s) of water at 109-ft (33.2-m) head of water, and the required viscous head and capacity will be obtained. Pick the pump so that it is operating at or near its maximum efficiency on water. If the water efficiency $E_w = 81$ percent at 790 gal/min (49.8 L/s) for this pump, the efficiency when handling the viscous liquid $E_v = E_w C_E$. Or, $E_v = 0.81(0.635) = 0.515$, or 51.5 percent.

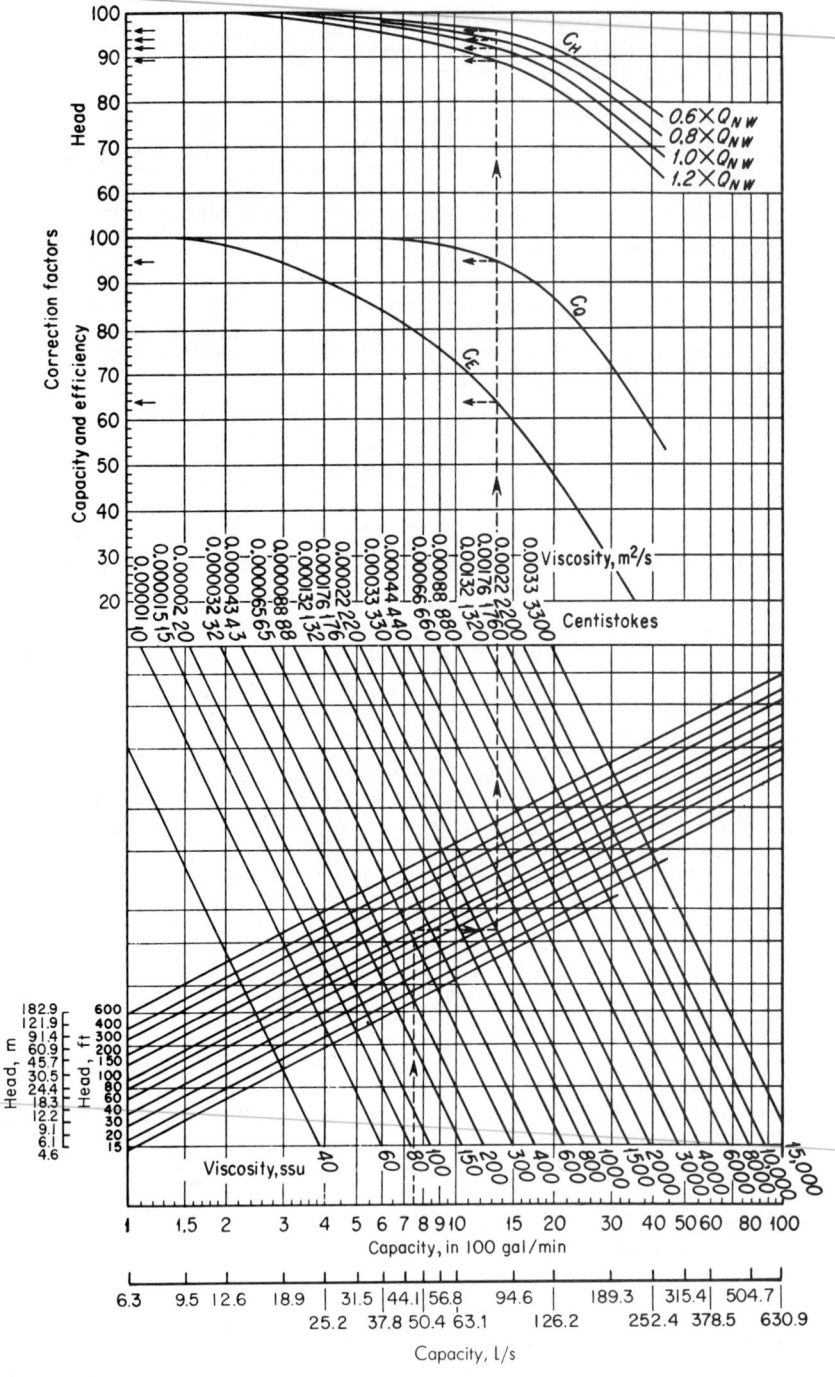

FIG. 25 Correction factors for viscous liquids handled by centrifugal pumps. *(Hydraulic Institute.)*

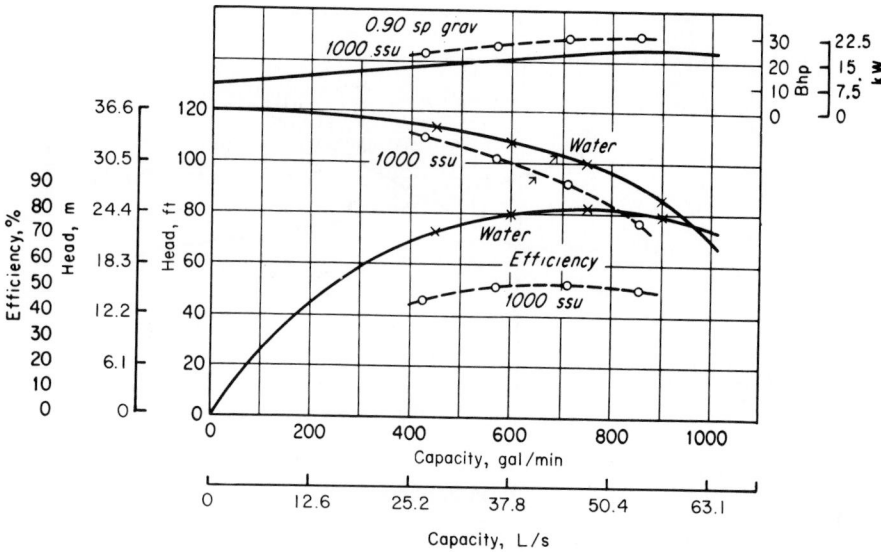

FIG. 26 Characteristics curves for water (solid line) and oil (dashed line). (*Hydraulic Institute.*)

The power input to the pump when handling viscous liquids is given by $P_v = Q_v H_v s / 3960 E_v$, where $s =$ specific gravity of the viscous liquid. For this pump, $P_v = (750) \times (100)(0.90)/[3960(0.515)] = 33.1$ hp (24.7 kW).

3. Plot the characteristic curves for viscous-liquid pumping

Follow these eight steps to plot the complete characteristic curves of a centrifugal pump handling a viscous liquid when the water characteristics are known: (*a*) Secure a complete set of characteristic curves (*H, Q, P, E*) for the pump to be used. (*b*) Locate the point of maximum efficiency for the pump when handling water. (*c*) Read the pump capacity, *Q* gal/min, at this point. (*d*) Compute the values of 0.6*Q*, 0.8*Q*, and 1.2*Q* at the maximum efficiency. (*e*) Using Fig. 25, determine the correction factors at the capacities in steps *c* and *d*. Where a multistage pump is being considered, use the head per stage (= total pump head, ft/number of stages), when entering Fig. 25. (*f*) Correct the head, capacity, and efficiency for each of the flow rates in *c* and *d*, using the correction factors from Fig. 25. (*g*) Plot the corrected head and efficiency against the corrected capacity, as in Fig. 26. (*h*) Compute the power input at each flow rate and plot. Draw smooth curves through the points obtained, Fig. 26.

Related Calculations: Use the method given here for any uniform viscous liquid—oil, gasoline, kerosene, mercury, etc—handled by a centrifugal pump. Be careful to use Fig. 25 only within its scale limits; *do not extrapolate.* The method presented here is that developed by the Hydraulic Institute. For new developments in the method, be certain to consult the latest edition of the Hydraulic Institute *Standards.*

PUMP SHAFT DEFLECTION AND CRITICAL SPEED

What are the shaft deflection and approximate first critical speed of a centrifugal pump if the total combined weight of the pump impellers is 23 lb (10.4 kg) and the pump manufacturer supplies the engineering data in Fig. 27?

Calculation Procedure:

1. Determine the deflection of the pump shaft

Use Fig. 27 to determine the shaft deflection. Note that this chart is valid for only one pump or series of pumps and must be obtained from the pump builder. Such a chart is difficult to prepare from test data without extensive test facilities.

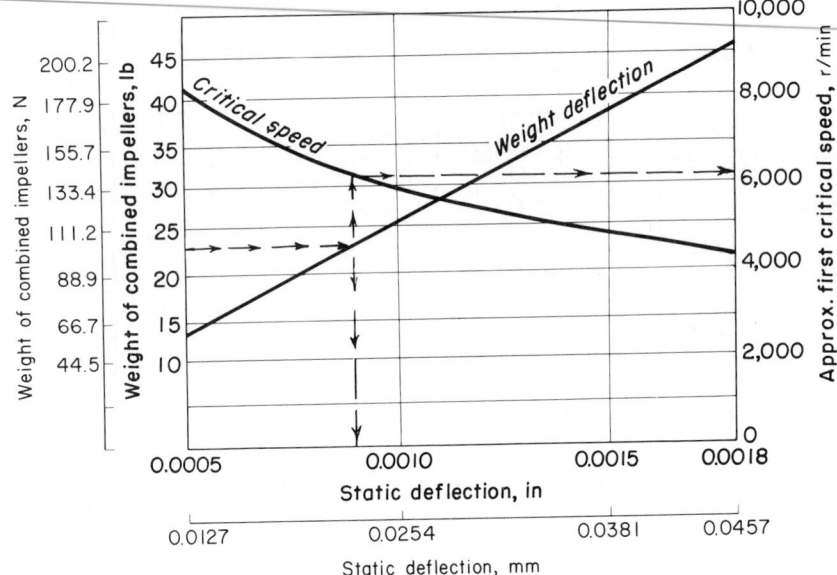

FIG. 27   Pump shaft deflection and critical speed. *(Goulds Pumps, Inc.)*

Enter Fig. 27 at the left at the total combined weight of the impellers, 23 lb (10.4 kg), and project horizontally to the right until the weight-deflection curve is intersected. From the intersection, project vertically downward to read the shaft deflection as 0.009 in (0.23 mm) at full speed.

2.   Determine the critical speed of the pump

From the intersection of the weight-deflection curve in Fig. 27 project vertically upward to the critical-speed curve. Project horizontally right from this intersection and read the first critical speed as 6200 r/min.

Related Calculations: Use this procedure for any class of pump—centrifugal, rotary, or reciprocating—for which the shaft-deflection and critical-speed curves are available. These pumps can be used for any purpose—process, power, marine, industrial, or commercial.

EFFECT OF LIQUID VISCOSITY ON REGENERATIVE-PUMP PERFORMANCE

A regenerative (turbine) pump has the water head-capacity and power-input characteristics shown in Fig. 28. Determine the head-capacity and power-input characteristics for four different viscosity oils to be handled by the pump—400, 600, 900, and 1000 SSU. What effect does increased viscosity have on the performance of the pump?

Calculation Procedure:

1.   Plot the water characteristics of the pump

Obtain a tabulation or plot of the water characteristics of the pump from the manufacturer or from their engineering data. With a tabulation of the characteristics, enter the various capacity and power points given, and draw a smooth curve through them, Fig. 28.

2.   Plot the viscous-liquid characteristics of the pump

The viscous-liquid characteristics of regenerative-type pumps are obtained by test of the actual unit. Hence, the only source of this information is the pump manufacturer. Obtain these characteristics from the pump manufacturer or their test data, and plot them on Fig. 28, as shown, for each oil or other liquid handled.

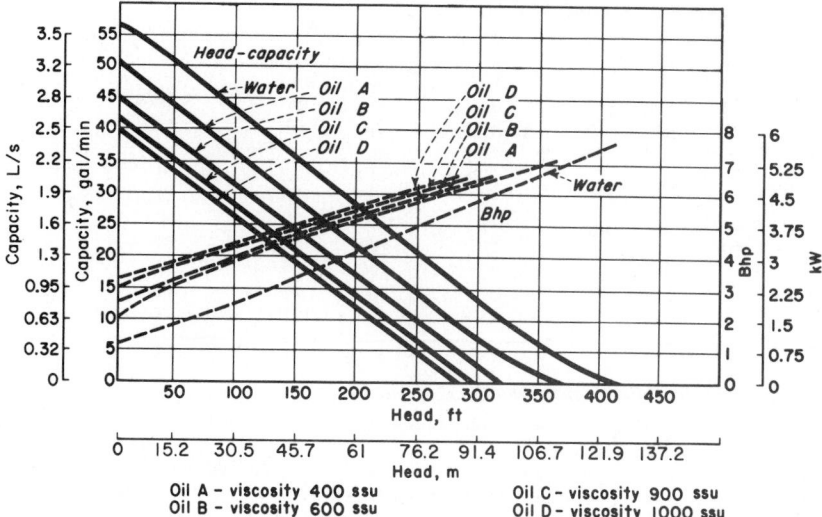

FIG. 28 Regenerative pump performance when handling water and oil. (*Aurora Pump Division, The New York Air Brake Company.*)

3. Evaluate the effect of viscosity on pump performance

Study Fig. 28 to determine the effect of increased liquid viscosity on the performance of the pump. Thus at a given head, say 100 ft (30.5 m), the capacity of the pump decreases as the liquid viscosity increases. At 100-ft (30.5-m) head, this pump has a water capacity of 43.5 gal/min (2.74 L/s), Fig. 28. The pump capacity for the various oils at 100-ft (30.5-m) head is 36 gal/min (2.27 L/s) for 400 SSU; 32 gal/min (2.02 L/s) for 600 SSU; 28 gal/min (1.77 L/s) for 900 SSU; and 26 gal/min (1.64 L/s) for 1000 SSU, respectively. There is a similar reduction in capacity of the pump at the other heads plotted in Fig. 28. Thus, as a general rule, the capacity of a regenerative pump decreases with an increase in liquid viscosity at constant head. Or conversely, at constant capacity, the head developed decreases as the liquid viscosity increases.

Plots of the power input to this pump show that the input power increases as the liquid viscosity increases.

Related Calculations: Use this procedure for a regenerative-type pump handling any liquid—water, oil, kerosene, gasoline, etc. A decrease in the viscosity of a liquid, as compared with the viscosity of water, will produce the opposite effect from that of increased viscosity.

EFFECT OF LIQUID VISCOSITY ON RECIPROCATING-PUMP PERFORMANCE

A direct-acting steam-driven reciprocating pump delivers 100 gal/min (6.31 L/s) of 70°F (21.1°C) water when operating at 50 strokes per minute. How much 2000-SSU crude oil will this pump deliver? How much 125°F (51.7°C) water will this pump deliver?

Calculation Procedure:

1. Determine the recommended change in pump performance

Reciprocating pumps of any type—direct-acting or power—having any number of liquid-handling cylinders—one to five or more—are usually rated for maximum delivery when handling 250-SSU liquids or 70°F (21.1°C) water. At higher liquid viscosities or water temperatures, the speed—strokes or rpm—is reduced. Table 8 shows typical recommended speed-correction factors for reciprocating pumps for various liquid viscosities and water temperatures. This table shows that with a liquid viscosity of 2000 SSU the pump speed should be reduced 20 percent. When 125°F (51.7°C) water is handled, the pump speed should be reduced 25 percent, as shown in Table 8.

TABLE 8 Speed-Correction Factors

Liquid viscosity, SSU	Speed reduction, %	Water temperature		Speed reduction, %
		°F	°C	
250	0	70	21.1	0
500	4	80	26.7	9
1000	11	100	37.8	18
2000	20	125	51.7	25
3000	26	150	65.6	29
4000	30	200	93.3	34
5000	35	250	121.1	38

2. *Compute the delivery of the pump*

The delivery capacity of any reciprocating pump is directly proportional to the number of strokes per minute it makes or to its rpm.

When 2000-SSU oil is used, the pump strokes per minute must be reduced 20 percent, or $(50)(0.20) = 10$ strokes/min. Hence, the pump speed will be $50 - 10 = 40$ strokes/min. Since the delivery is directly proportional to speed, the delivery of 2000-SSU oil = $(40/50)(100) = 80$ gal/min (5.1 L/s).

When handling 125°F (51.7°C) water, the pump strokes/min must be reduced 25 percent, or $(50)(0.5) = 12.5$ strokes/min. Hence, the pump speed will be $50.0 - 12.5 = 37.5$ strokes/min. Since the delivery is directly proportional to speed, the delivery of 125°F (51.7°C) water = $(37.5/50)(100) = 75$ gal/min (4.7 L/s).

Related Calculations: Use this procedure for any type of reciprocating pump handling liquids falling within the range of Table 8. Such liquids include oil, kerosene, gasoline, brine, water, etc.

EFFECT OF VISCOSITY AND DISSOLVED GAS ON ROTARY PUMPS

A rotary pump handles 8000-SSU liquid containing 5 percent entrained gas and 10 percent dissolved gas at a 20-in (508-mm) Hg pump inlet vacuum. The pump is rated at 1000 gal/min (63.1 L/s) when handling gas-free liquids at viscosities less than 600 SSU. What is the output of this pump without slip? With 10 percent slip?

TABLE 9 Rotary Pump Speed Reduction for Various Liquid Viscosities

Liquid viscosity, SSU	Speed reduction, percent of rated pump speed
600	2
800	6
1,000	10
1,500	12
2,000	14
4,000	20
6,000	30
8,000	40
10,000	50
20,000	55
30,000	57
40,000	60

Calculation Procedure:

1. *Compute the required speed reduction of the pump*

When the liquid viscosity exceeds 600 SSU, many pump manufacturers recommend that the speed of a rotary pump be reduced to permit operation without excessive noise or vibration. The speed reduction usually recommended is shown in Table 9.

With this pump handling 8000-SSU liquid, a speed reduction of 40 percent is necessary, as shown in Table 9. Since the capacity of a rotary pump varies directly with its speed, the output of this pump when handling 8000-SSU liquid = (1000 gal/min) × (1.0 − 0.40) = 600 gal/min (37.9 L/s).

2. *Compute the effect of gas on the pump output*

Entrained or dissolved gas reduces the output

TABLE 10 Effect of Entrained or Dissolved Gas on the Liquid Displacement of Rotary Pumps (liquid displacement: percent of displacement)

Vacuum at pump inlet, inHg (mmHg)	Gas entrainment					Gas solubility					Gas entrainment and gas solubility combined				
	1%	2%	3%	4%	5%	2%	4%	6%	8%	10%	1% 2%	2% 4%	3% 6%	4% 8%	5% 10%
5 (127)	99	97½	96½	95	93½	99½	99	98½	97	97½	98½	96½	96	92	91
10 (254)	98½	97¼	95½	94	92	99	97½	97	95	95	97½	95	90	90	88½
15 (381)	98	96½	94½	92½	90½	97	96	94	92	90½	96	93	89½	86½	83½
20 (508)	97½	94½	92	89	86½	96	92	89	86	83	94	88	83	78	74
25 (635)	94	89	84	79	75½	90	83	76½	71	66	85½	75½	68	61	55

For example, with 5 percent gas entrainment at 15 inHg (381 mmHg) vacuum, the liquid displacement will be 90½ percent of the pump displacement, neglecting slip, or with 10 percent dissolved gas liquid displacement will be 90½ percent of the pump displacement; and with 5 percent entrained gas combined with 10 percent dissolved gas, the liquid displacement will be 83½ percent of pump replacement.
Source: Courtesy of Kinney Mfg. Div., The New York Air Brake Co.

of a rotary pump, as shown in Table 10. The gas in the liquid expands when the inlet pressure of the pump is below atmospheric and the gas occupies part of the pump chamber, reducing the liquid capacity.

With a 20-in (508-mm) Hg inlet vacuum, 5 percent entrained gas, and 10 percent dissolved gas, Table 10 shows that the liquid displacement is 74 percent of the rated displacement. Thus, the output of the pump when handling this viscous, gas-containing liquid will be (600 gal/min) (0.74) = 444 gal/min (28.0 L/s) without slip.

3. Compute the effect of slip on the pump output

Slip reduces rotary-pump output in direct proportion to the slip. Thus, with 10 percent slip, the output of this pump = (444 gal/min)(1.0 − 0.10) = 369.6 gal/min (23.3 L/s).

Related Calculations: Use this procedure for any type of rotary pump—gear, lobe, screw, swinging-vane, sliding-vane, or shuttle-block, handling any clear, viscous liquid. Where the liquid is gas-free, apply only the viscosity correction. Where the liquid viscosity is less than 600 SSU but the liquid contains gas or air, apply the entrained or dissolved gas correction, or both corrections.

SELECTION OF MATERIALS FOR PUMP PARTS

Select suitable materials for the principal parts of a pump handling cold ethylene chloride. Use the Hydraulic Institute recommendations for materials of construction.

Calculation Procedure:

1. Determine which materials are suitable for this pump

Refer to the data section of the Hydraulic Institute *Standards*. This section contains a tabulation of hundreds of liquids and the pump construction materials that have been successfully used to handle each liquid.

The table shows that for cold ethylene chloride having a specific gravity of 1.28, an all-bronze pump is satisfactory. In lieu of an all-bronze pump, the principal parts of the pump—casing, impeller, cylinder, and shaft—can be made of one of the following materials: austenitic steels (low-carbon 18-8; 18-8/Mo; highly alloyed stainless); nickel-base alloys containing chromium, molybdenum, and other elements, and usually less than 20 percent iron; or nickel-copper alloy (Monel metal). The order of listing in the *Standards* does not necessarily indicate relative superiority, since certain factors predominating in one instance may be sufficiently overshadowed in others to reverse the arrangement.

2. Choose the most economical pump

Use the methods of earlier calculation procedures to select the most economical pump for the installation. Where the corrosion resistance of two or more pumps is equal, the standard pump, in this instance an all-bronze unit, will be the most economical.

Related Calculations: Use this procedure to select the materials of construction for any class of pump—centrifugal, rotary, or reciprocating—in any type of service—power, process, marine, or commercial. Be certain to use the latest edition of the Hydraulic Institute *Standards*, because the recommended materials may change from one edition to the next.

SIZING A HYDROPNEUMATIC STORAGE TANK

A 200-gal/min (12.6-L/s) water pump serves a pumping system. Determine the capacity required for a hydropneumatic tank to serve this system if the allowable high pressure in the tank and system is 60 lb/in² (gage) (413.6 kPa) and the allowable low pressure is 30 lb/in² (gage) (206.8 kPa). How many starts per hour will the pump make if the system draws 3000 gal/min (189.3 L/ s) from the tank?

Calculation Procedure:

1. Compute the required tank capacity

In the usual hydropneumatic system, a storage-tank capacity in gal of 10 times the pump capacity in gal/min is used, if this capacity produces a moderate running time for the pump. Thus, this system would have a tank capacity of (10)(200) = 2000 gal (7570.8 L).

2. *Compute the quantity of liquid withdrawn per cycle*

For any hydropneumatic tank the withdrawal, expressed as the number of gallons (liters) withdrawn per cycle, is given by $W = (v_L - v_H)/C$, where v_L = air volume in tank at the lower pressure, ft^3 (m^3); v_H = volume of air in tank at higher pressure, ft^3 (m^3); C = conversion factor to convert ft^3 (m^3) to gallons (liters), as given below.

Compute V_L and V_H using the gas law for v_H and either the gas law or the reserve percentage for v_L. Thus, for v_H, the gas law gives $v_H = p_L v_L / p_H$, where p_L = lower air pressure in tank, lb/in^2 (abs) (kPa); p_H = higher air pressure in tank lb/in^2 (abs) (kPa); other symbols as before.

In most hydropneumatic tanks a liquid reserve of 10 to 20 percent of the total tank volume is kept in the tank to prevent the tank from running dry and damaging the pump. Assuming a 10 percent reserve for this tank, $v_L = 0.1\ V$, where V = tank volume in ft^3 (m^3). Since a 2000-gal (7570-L) tank is being used, the volume of the tank is 2000/7.481 ft^3/gal = 267.3 ft^3 (7.6 m^3). With the 10 percent reserve at the 44.7 lb/in^2 (abs) (308.2-kPa) lower pressure, $v_L = 0.9\ (267.3)$ = 240.6 ft^3 (6.3 m^3), where $0.9 = V - 0.1\ V$.

At the higher pressure in the tank, 74.7 lb/in^2 (abs) (514.9 kPa), the volume of the air will be, from the gas law, $v_H = p_L v_L / p_H = 44.7\ (240.6)/74.7 = 143.9$ ft^3 (4.1 m^3). Hence, during withdrawal, the volume of liquid removed from the tank will be $W_g = (240.6 - 143.9)/0.1337 = 723.3$ gal (2738 L). In this relation the constant converts from cubic feet to gallons and is 0.1337. To convert from cubic meters to liters, use the constant 1000 in the denominator.

3. *Compute the pump running time*

The pump has a capacity of 200 gal/min (12.6 L/s). Therefore, it will take 723/200 = 3.6 min to replace the withdrawn liquid. To supply 3000 gal/h (11,355 L/h) to the system, the pump must start 3000/723 = 4.1, or 5 times per hour. This is acceptable because a system in which the pump starts six or fewer times per hour is generally thought satisfactory.

Where the pump capacity is insufficient to supply the system demand for short periods, use a smaller reserve. Compute the running time using the equations in steps 2 and 3. Where a larger reserve is used—say 20 percent—use the value 0.8 in the equations in step 2. For a 30 percent reserve, the value would be 0.70, and so on.

Related Calculations: Use this procedure for any liquid system having a hydropneumatic tank—well drinking water, marine, industrial, or process.

USING CENTRIFUGAL PUMPS AS HYDRAULIC TURBINES

Select a centrifugal pump to serve as a hydraulic turbine power source for a 1500-gal/min (5677.5-L/min) flow rate with 1290 ft (393.1 m) of head. The power application requires a 3600-r/min speed, the specific gravity of the liquid is 0.52, and the total available exhaust head is 20 ft (6.1 m). Analyze the cavitation potential and operating characteristics at an 80 percent flow rate.

Calculation Procedure:

1. *Choose the number of stages for the pump*

Search of typical centrifugal-pump data shows that a head of 1290 ft (393.1 m) is too large for a single-stage pump of conventional design. Hence, a two-stage pump will be the preliminary choice for this application. The two-stage pump chosen will have a design head of 645 ft (196.6 m) per stage.

2. *Compute the specific speed of the pump chosen*

Use the relation $N_s = $ pump $rpm(Q)^{0.5}/H^{0.75}$, where N_s = specific speed of the pump; rpm = r/min of pump shaft; Q = pump capacity or flow rate, gal/min; H = pump head per stage, ft. Substituting, we get $N_s = 3600(1500)^{0.5}/(645)^{0.75} = 1090$. Note that the specific speed value is the same regardless of the system of units used—USCS or SI.

3. *Convert turbine design conditions to pump design conditions*

To convert from turbine design conditions to pump design conditions, use the pump manufacturer's conversion factors that relate turbine best efficiency point (bep) performance with pump bep performance. Typically, as specific speed N_s varies from 500 to 2800, these bep factors generally vary as follows: the conversion factor for capacity (gal/min or L/min) C_Q, from 2.2 to 1.1;

the conversion factor for head (ft or m) C_H, from 2.2 to 1.1; the conversion factor for efficiency C_E, from 0.92 to 0.99. Applying these conversion factors to the turbine design conditions yields the pump design conditions sought.

At the specific speed for this pump, the values of these conversion factors are determined from the manufacturer to be $C_Q = 1.24$; $C_H = 1.42$; $C_E = 0.967$.

Given these conversion factors, the turbine design conditions can be converted to the pump design conditions thus: $Q_p = Q_t/C_Q$, where Q_p = pump capacity or flow rate, gal/min or L/min; Q_t = turbine capacity or flow rate in the same units; other symbols are as given earlier. Substituting gives $Q_p = 1500/1.24 = 1210$ gal/min (4580 L/min).

Likewise, the pump discharge head, in feet of liquid handled, is $H_p = H_t/C_H$. So $H_p = 645/1.42 = 454$ ft (138.4 m).

4.   Select a suitable pump for the operating conditions

Once the pump capacity, head, and rpm are known, a pump having its best bep at these conditions can be selected. Searching a set of pump characteristic curves and capacity tables shows that a two-stage 4-in (10-cm) unit with an efficiency of 77 percent would be suitable.

5.   Estimate the turbine horsepower developed

To predict the developed horsepower, convert the pump efficiency to turbine efficiency. Use the conversion factor developed above. Or, the turbine efficiency $E_t = E_pC_E = (0.77)(0.967) = 0.745$, or 74.5 percent.

With the turbine efficiency known, the output brake horsepower can be found from bhp = $Q_tH_tE_ts/3960$, where s = fluid specific gravity; other symbols as before. Substituting, we get bhp = $1500(1290)(0.745)(0.52)/3960 = 198$ hp (141 kW).

6.   Determine the cavitation potential of this pump

Just as pumping requires a minimum net positive suction head, turbine duty requires a net positive exhaust head. The relation between the total required exhaust head (TREH) and turbine head per stage is the cavitation constant σ_r = TREH/H. Figure 29 shows σ_r vs. N_s for hydraulic turbines. Although a pump used as a turbine will not have exactly the same relationship, this curve provides a good estimate of σ_r for turbine duty.

To prevent cavitation, the total available exhaust head (TAEH) must be greater than the TREH. In this installation, $N_s = 1090$ and TAEH = 20 ft (6.1 m). From Fig. 29, $\sigma_r = 0.028$ and TREH = $0.028(645) = 18.1$ ft (5.5 m). Because TAEH > TREH, there is enough exhaust head to prevent cavitation.

7.   Determine the turbine performance at 80 percent flow rate

In many cases, pump manufacturers treat conversion factors as proprietary information. When this occurs, the performance of the turbine under different operating conditions can be predicted from the general curves in Figs. 30 and 31.

At the 80 percent flow rate for the turbine, or 1200 gal/min (4542 L/min), the operating point is 80 percent of bep capacity. For a specific speed of 1090, as before, the percentages of bep head and efficiency are shown in Figs. 30 and 31: 79.5 percent of bep head and 91 percent of bep efficiency. To find the actual performance, multiply by the bep values. Or, $H_t = 0.795(1290) = 1025$ ft (393.1 m); $E_t = 0.91(74.5) = 67.8$ percent.

The bhp at the new operating condition is then bhp = $1200(1025)(0.678)(0.52)/3960 = 110$ hp (82.1 kW).

In a similar way, the constant-head curves in Figs. 32 and 33 predict turbine performance at different speeds. For example, speed is 80 percent of bep speed at 2880 r/min. For a specific speed of 1090, the percentages of bep capacity, efficiency, and power are 107 percent of the capacity, 94 percent of the efficiency, and 108 percent of the bhp. To get the actual performance, convert as before: $Q_t = 107(1500) = 1610$ gal/min (6094 L/min); $E_t = 0.94(74.5) = 70.0$ percent; bhp = $1.08(189) = 206$ hp (153.7 kW).

Note that the bhp in this last instance is higher than the bhp at the best efficiency point. Thus more horsepower can be obtained from a given unit by reducing the speed and increasing the flow rate. When the speed is fixed, more bhp cannot be obtained from the unit, but it may be possible to select a smaller pump for the same application.

Related Calculations: Use this general procedure for choosing a centrifugal pump to drive—as a hydraulic turbine—another pump, a fan, a generator, or a compressor, where high-pressure liquid is available as a source of power. Because pumps are designed as fluid movers,

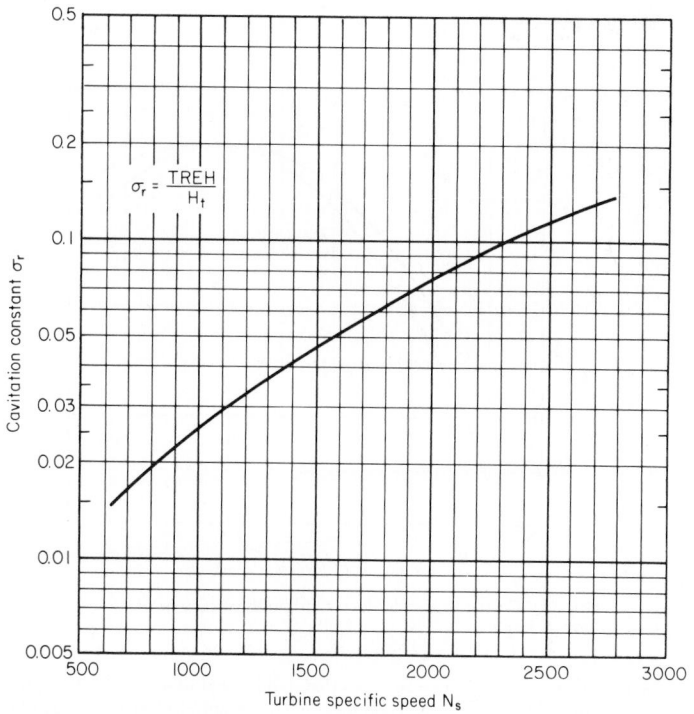

FIG. 29 Cavitation constant for hydraulic turbines. *(Chemical Engineering.)*

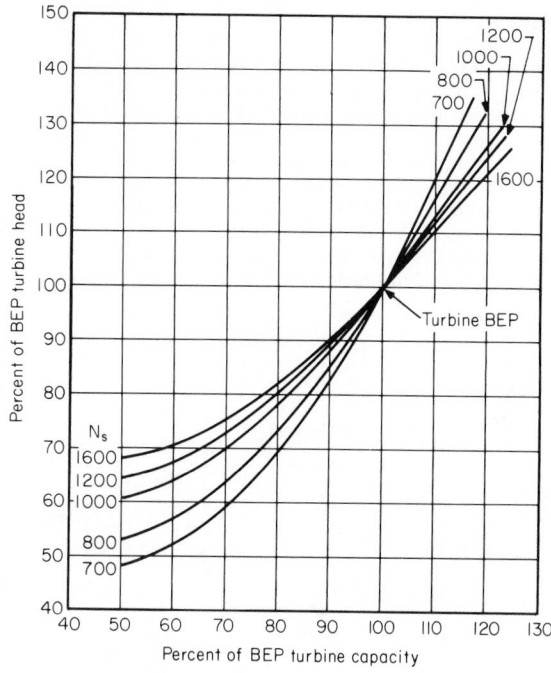

FIG. 30 Constant-speed curves for turbine duty. *(Chemical Engineering.)*

3.389

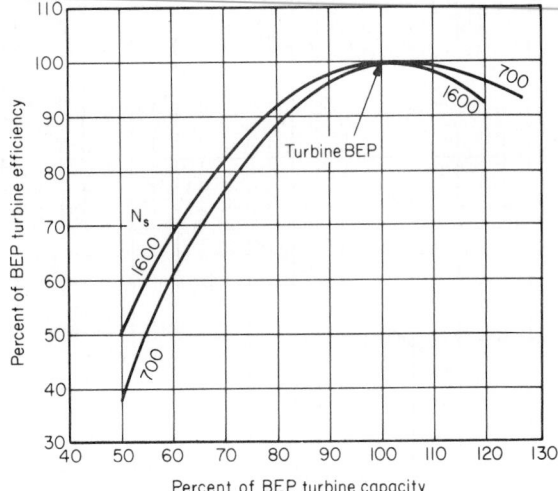

FIG. 31   Constant-speed curves for turbine duty. (*Chemical Engineering.*)

they may be less efficient as hydraulic turbines than equipment designed for that purpose. Steam turbines and electric motors are more economical where steam or electricity is available.

But using a pump as a turbine can pay off in remote locations where steam or electric power would require additional wiring or piping, in hazardous locations that require nonsparking equipment, where energy may be recovered from a stream that otherwise would be throttled, and when a radial-flow centrifugal pump is immediately available but a hydraulic turbine is not.

In the most common situation, there is a liquid stream with fixed head and flow rate and an application requiring a fixed rpm; these are the turbine design conditions. The objective is to pick a pump with a turbine bep at these conditions. With performance curves such as Fig. 34, turbine design conditions can be converted to pump design conditions. Then you select from a manufacturer's catalog a model that has its pump bep at those values.

The most common error in pump selection is using the turbine design conditions in choosing a pump from a catalog. Because catalog performance curves describe pump duty, not turbine duty, the result is an oversized unit that fails to work properly.

This procedure is the work of Fred Buse, Chief Engineer, Standard Pump Aldrich Division of Ingersoll-Rand Co., as reported in *Chemical Engineering* magazine.

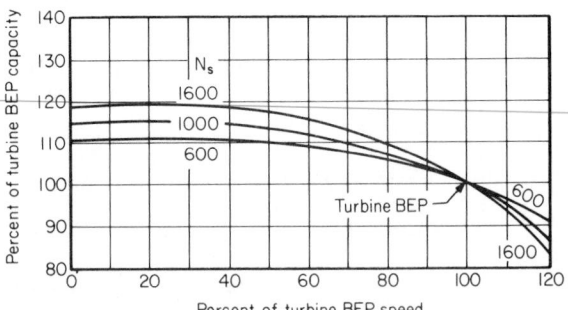

FIG. 32   Constant-head curves for turbine duty. (*Chemical Engineering.*)

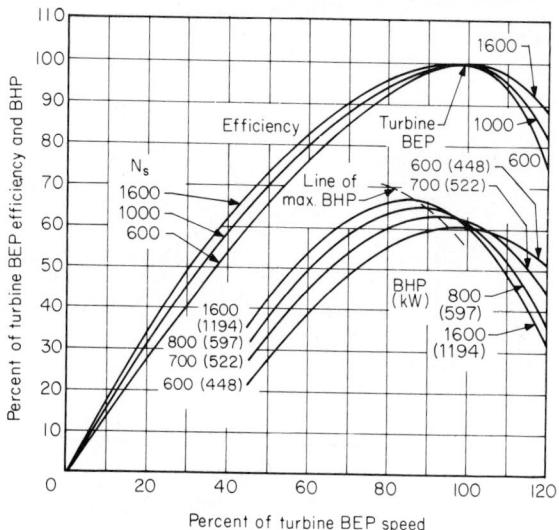

FIG. 33 Constant-head curves for turbine only. *(Chemical Engineering.)*

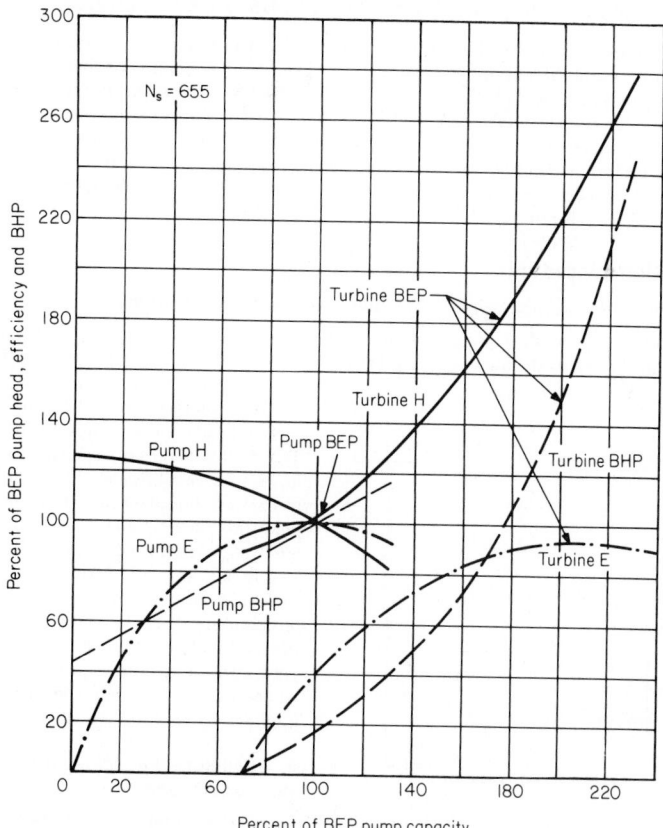

FIG. 34 Performance of a pump at constant speed in pump duty and turbine duty. *(Chemical Engineering.)*

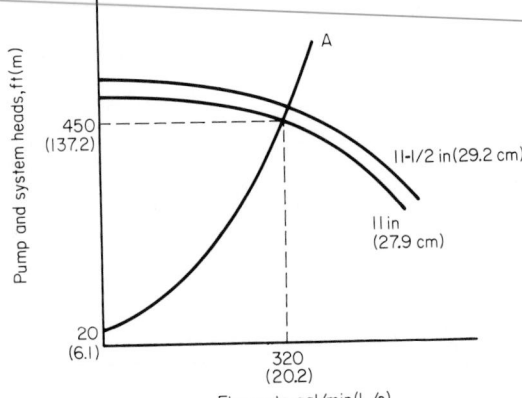

FIG. 35 System-head curves without recirculation flow. *(Chemical Engineering.)*

SIZING CENTRIFUGAL-PUMP IMPELLERS FOR SAFETY SERVICE

Determine the impeller size of a centrifugal pump that will provide a safe continuous-recirculation flow to prevent the pump from overheating at shutoff. The pump delivers 320 gal/min (20.2 L/s) at an operating head of 450 ft (137.2 m). The inlet water temperature is 220°F (104.4°C), and the system has an NPSH of 5 ft (1.5 m). Pump performance curves and the system-head characteristic curve for the discharge flow (without recirculation) are shown in Fig. 35, and the piping layout is shown in Fig. 36. The brake horsepower (bhp) for an 11-in (27.9-cm) and an 11.5-in (29.2-cm) impeller at shutoff is 53 and 60, respectively. Determine the permissible water temperature rise for this pump.

Calculation Procedure:

1. Compute the actual temperature rise of the water in the pump

Use the relation $P_0 = P_v + P_{NPSH}$, where P_0 = pressure corresponding to the actual liquid temperature in the pump during operation, lb/in² (abs) (kPa); P_v = vapor pressure in the pump at the inlet water temperature, lb/in² (abs) (kPa); P_{NPSH} = pressure created by the net positive suction head on the pumps, lb/in² (abs) (kPa). The head in feet (meters) must be converted to lb/in² (abs) (kPa) by the relation lb/in² (abs) = (NPSH, ft) (liquid density at the pumping temperature, lb/ft³)/(144 in²/ft²). Substituting yields $P_0 = 17.2$ lb/in² (abs) + 5(59.6)/144 = 19.3 lb/in² (abs) (133.1 kPa).

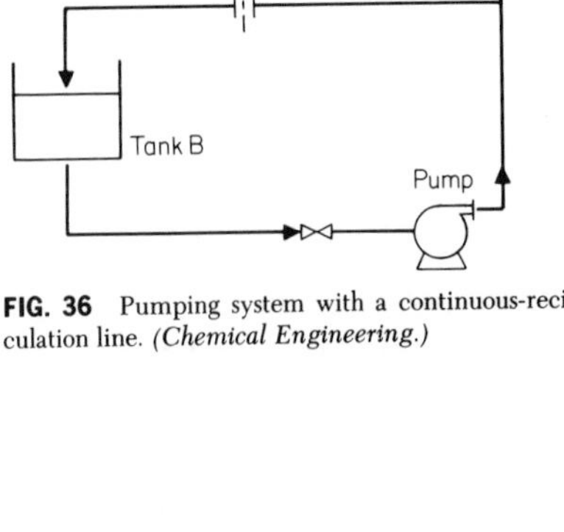

Using the steam tables, find the saturation temperature T_s corresponding to this absolute pressure as $T_s = 226.1$°F (107.8°C). Then the permissible temperature rise is $T_p = T_s - T_{op}$, where T_{op} = water temperature in the pump inlet. Or, $T_p = 226.1 - 220 = 6.1$°F (3.4°C).

2. Compute the recirculation flow rate at the shutoff head

From the pump characteristic curve with recirculation, Fig. 37, the continuous-recircu-

FIG. 36 Pumping system with a continuous-recirculation line. *(Chemical Engineering.)*

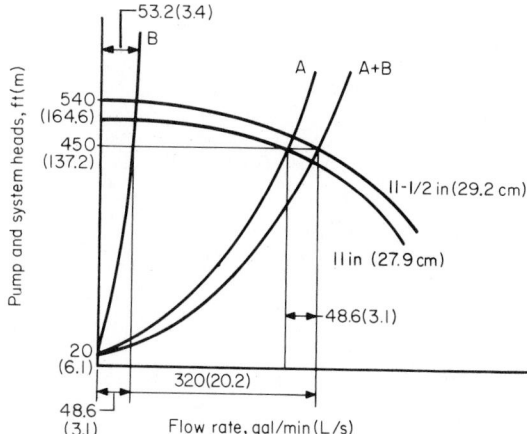

FIG. 37 System-head curves with recirculation flow. *(Chemical Engineering.)*

lation flow Q_B for an 11.5-in (29.2-cm) impeller at an operating head of 450 ft (137.2 m) is 48.6 gal/min (177.1 L/min). Find the continuous-recirculation flow at shutoff head H_s ft (m) of 540 ft (164.6 m) from $Q_s = Q_B(H_s/H_{op})^{0.5}$, where H_{op} = operating head, ft (m). Or, $Q_s = 48.6(540/450) = 53.2$ gal/min (201.4 L/min).

3. Find the minimum safe flow for this pump

The minimum safe flow, lb/h, is given by $w_{min} = 2545bhp/[C_pT_p + (1.285 \times 10^{-3})H_s]$, where C_p = specific head of the water; other symbols as before. Substituting, we find $w_{min} = 2545(60)/[1.0(6.1) + (1.285 \times 10^{-3})(540)] = 22{,}476$ lb/h (2.83 kg/s). Converting to gal/min yields $Q_{min} = w_{min}/[(ft^3/h)(gal/min)(lb/ft^3)]$ for the water flowing through the pump. Or, $Q_{min} = 22{,}476/[(8.021)(59.6)] = 47.1$ gal/min (178.3 L/min).

4. Compare the shutoff recirculation flow with the safe recirculation flow

Since the shutoff recirculation flow $Q_s = 53.2$ gal/min (201.4 L/min) is greater than $Q_{min} = 47.1$ gal/min (178.3 L/min), the 11.5-in (29.2-cm) impeller is adequate to provide safe continuous recirculation. An 11.25-in (28.6-cm) impeller would not be adequate because $Q_{min} = 45$ gal/min (170.3 L/min) and $Q_s = 25.6$ gal/min (96.9 L/min).

Related Calculations: Safety-service pumps are those used for standby service in a variety of industrial plants serving the chemical, petroleum, plastics, aircraft, auto, marine, manufacturing, and similar businesses. Such pumps may be used for fire protection, boiler feed, condenser cooling, and related tasks. In such systems the pump is usually oversized and has a recirculation loop piped in to prevent overheating by maintaining a minimum safe flow. Figure 35 shows a schematic of such a system. Recirculation is controlled by a properly sized orifice rather than by valves because an orifice is less expensive and highly reliable.

The general procedure for sizing centrifugal pumps for safety service, using the symbols given earlier, is this: (1) Select a pump that will deliver the desired flow Q_A, using the head-capacity characteristic curves of the pump and system. (2) Choose the next larger diameter pump impeller to maintain a discharge flow of Q_A to tank A, Fig. 35, and a recirculation flow Q_B to tank B, Fig. 35. (3) Compute the recirculation flow Q_s at the pump shutoff point from $Q_s = Q_B(H_s/H_{op})^{0.5}$. (4) Calculate the minimum safe flow Q_{min} for the pump with the larger impeller diameter. (5) Compare the recirculation flow Q_s at the pump shutoff point with the minimum safe flow Q_{min}. If $Q_s \geq Q_{min}$, the selection process has been completed. If $Q_s < Q_{min}$, choose the next larger size impeller and repeat steps 3, 4, and 5 above until the impeller size that will provide the minimum safe recirculation flow is determined.

This procedure is the work of Mileta Mikasinovic and Patrick C. Tung, design engineers, Ontario Hydro, as reported in *Chemical Engineering* magazine.

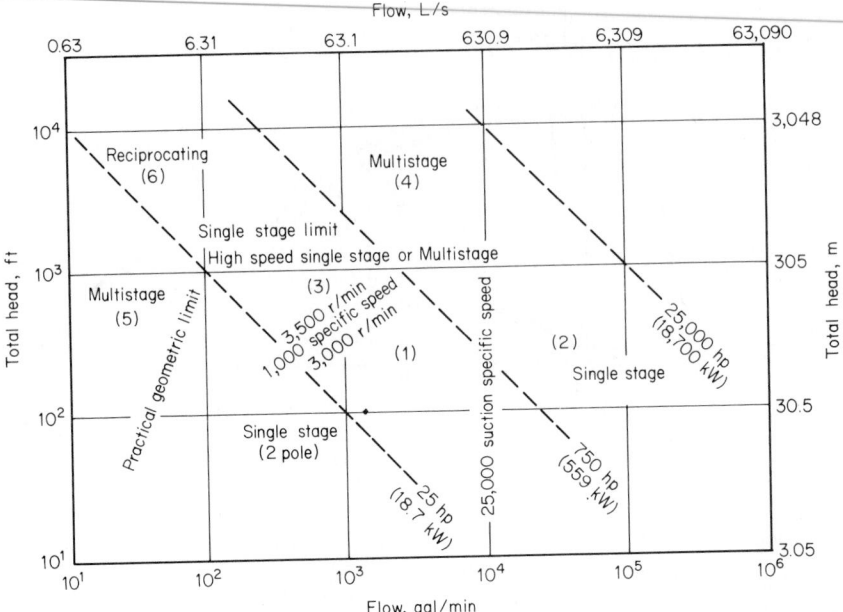

FIG. 38 Selection guide is based mainly on specific speed, which indicates impeller geometry. *(Chemical Engineering.)*

PUMP CHOICE TO REDUCE ENERGY CONSUMPTION AND LOSS

Choose an energy-efficient pump to handle 1000 gal/min (3800 L/min) of water at 60°F (15.6°C) at a total head of 150 ft (45.5 m). A readily commercially available pump is preferred for this application.

Calculation Procedure:

1. *Compute the pump horsepower required*

For any pump, $bhp_i = (gpm)(H_t)(s)/3960e$, where bhp_i = input brake (motor) horsepower to the pump; H_t = total head on the pump, ft; s = specific gravity of the liquid handled; e = hydraulic efficiency of the pump. For this application where s = 1.0 and a hydraulic efficiency of 70 percent can be safely assumed, $bhp_i = (1000)(150)(1)/(3960)(0.70) = 54.1$ bhp (40.3 kW).

2. *Choose the most energy-efficient pump*

Use Fig. 38, entering at the bottom at 1000 gal/min (3800 L/min) and projecting vertically upward to a total head of 150 ft (45.5 m). The resulting intersection is within area 1, showing from Table 11 that a single-stage 3500-r/min electric-motor-driven pump would be the most energy-efficient.

   Related Calculations: The procedure given here can be used for pumps in a variety of applications—chemical, petroleum, commercial, industrial, marine, aeronautical, air-conditioning, cooling-water, etc., where the capacity varies from 10 to 1,000,000 gal/min (38 to 3,800,000 L/min) and the head varies from 10 to 10,000 ft (3 to 3300 m). Figure 38 is based primarily on the characteristic of pump specific speed $N_s = NQ^2/H^{3/4}$, where N = pump rotating speed, r/min; Q = capacity, gal/min (L/min); H = total head, ft (m).

   When N_s is less than 1000, the operating efficiency of single-stage centrifugal pumps falls off dramatically; then either multistage or higher-speed pumps offer the best efficiency.

   Area 1 of Fig. 38 is the densest, crowded both with pumps operating at 1750 and 3500 r/min,

TABLE 11 Type of Pump for Highest Energy Efficiency°

Area 1: Single-stage, 3500 r/min
Area 2: Single-stage, 1750 r/min or lower
Area 3: Single-stage, above 3500 r/min, or multistage, 3500 r/min
Area 4: Multistage
Area 5: Multistage
Area 6: Reciprocating

°Includes ANSI B73.1 standards; see area number in Fig. 38.

because years ago 3500-r/min pumps were not thought to be as durable as 1750-r/min ones. Since the adoption of the AVS standard in 1960 (superseded by ANSI B73.1), pumps with stiffer shafts have been proved reliable.

Also responsible for many 1750-r/min pumps in area 1 has been the impression that the higher (3500-r/min) speed causes pumps to wear out faster. However, because impeller tip speed is the same at both 3500 and 1750 r/min [as, for example, a 6-in (15-cm) impeller at 3500 r/min and a 12-in (30-cm) one at 1750 r/min], so is the fluid velocity, and so should be the erosion of metal surface. Another reason for not limiting operating speed is that improved impeller inlet design allows operation at 3500 r/min to capacities of 5000 gal/min (19,000 L/min) and higher.

Choice of operating speed also may be indirectly limited by specifications pertaining to suction performance, such as that fixing the top suction specific speed S directly or indirectly by choice of the sigma constant or by reliance on Hydraulic Institute charts.

Values of S below 8000 to 10,000 have long been accepted for avoiding cavitation. However, since the development of the inducer, S values in the range of 20,000 to 25,000 have become commonplace, and values as high as 50,000 have become practical.

The sigma constant, which relates NPSH to total head, is little used today, and Hydraulic Institute charts (which are being revised) are conservative.

In light of today's designs and materials, past restrictions resulting from suction performance limitations should be reevaluated or eliminated entirely.

Even if the most efficient pump has been selected, there are a number of circumstances in which it may not operate at peak efficiency. Today's cost of energy has made these considerations more important.

A centrifugal pump, being a hydrodynamic machine, is designed for a single peak operating-point capacity and total head. Operation at other than this best efficiency point (bep) reduces efficiency. Specifications now should account for such factors as these:

1. A need for a larger number of smaller pumps. When a process operates over a wide range of capacities, as many do, pumps will often work at less than full capacity, hence at lower efficiency. This can be avoided by installing two or three pumps in parallel, in place of a single large one, so that one of the smaller pumps can handle the flow when operations are at a low rate.

2. Allowance for present capacity. Pump systems are frequently designed for full flow at some time in the future. Before this time arrives, the pumps will operate far from their best efficiency points. Even if this interim period lasts only 2 or 3 years, it may be more economical to install a smaller pump initially and to replace it later with a full-capacity one.

3. Inefficient impeller size. Some specifications call for pump impeller diameter to be no larger than 90 or 95 percent of the size that a pump could take, so as to provide reserve head. If this reserve is used only 5 percent of the time, all such pumps will be operating at less than full efficiency most of the time.

4. Advantages of allowing operation to the right of the best efficiency point. Some specifications, the result of such thinking as that which provides reserve head, prohibit the selection of pumps that would operate to the right of the best efficiency point. This eliminates half of the pumps that might be selected and results in oversized pumps operating at lower efficiency.

This procedure is the work of John H. Doolin, Director of Product Development, Worthington Pumps, Inc., as reported in *Chemical Engineering* magazine.

Piping and Fluid Flow

REFERENCES: Severud and Marr—*Elevated Temperature Piping Design*, ASME; Jeppson—*Analysis of Flow in Pipe Networks*, Butterworths/Ann Arbor Science Press; Sherwood and Whistance—*Piping Guide*, Syentek Books; Williams—*Pipelines and Permafrost*, Longmans; Watters—*Modern Analysis and Control of Unsteady Flow in Pipelines*, Butterworths/Ann Arbor Science Press; Lambert—*Pipeline Instrumentation and Controls Handbook*, Gulf Publishing; Marks—*Oceanic Pipeline Computations*, Penwell; Kentish—*Industrial Pipework*, McGraw-Hill (UK); Brebbia and Ferrante—*Computational Hydraulics*, Butterworths; King and Crocker—*Piping Handbook*, McGraw-Hill; ANSA—*Code for Pressure Piping* (commonly called the *Piping Code*); ASME—*Fluid Meters—Their Theory and Application*; King and Brater—*Handbook of Hydraulics*, McGraw-Hill; Ingersoll-Rand Company—*Cameron Hydraulic Data*; The Hydraulic Institute—*Standards of the Hydraulic Institute*; Baumeister and Marks—*Standard Handbook for Mechanical Engineers*, McGraw-Hill; Littleton—*Industrial Piping*, McGraw-Hill; The Hydraulic Institute—*Pipe Friction Manual*; Black, Sivalls, and Bryson—*Valve Sizing Book* and *Cv Book*; Fluid Controls Institute—*Recommended Voluntary Standard Formulas for Sizing Control Valves*; Bell—*Petroleum Transportation Handbook*, McGraw-Hill; Perry—*Chemical Engineers' Handbook*, McGraw-Hill; Spielvogel—*Piping Stress Calculations Simplified*, Spielvogel Publishing; Grinnell Company, Inc.—*Piping Design and Engineering*; M. W. Kellogg Co.—*Design of Piping Systems*, Wiley; National Valve and Manufacturing Co.—*Piping Catalog*; McClain—*Fluid Flow in Pipes*, Industrial Press; Tube Turns Division of Chemetron Corp.—*Piping Engineering*.

PIPE-WALL THICKNESS AND SCHEDULE NUMBER

Determine the minimum wall thickness t_m in (mm) and schedule number SN for a branch steam pipe operating at 900°F (482.2°C) if the internal steam pressure is 1000 lb/in² (abs) (6894 kPa). Use ANSA B31.1 *Code for Pressure Piping* and the ASME *Boiler and Pressure Vessel Code* valves and equations where they apply. Steam flow rate is 72,000 lb/h (32,400 kg/h).

Calculation Procedure:

1. *Determine the required pipe diameter*

When the length of pipe is not given or is as yet unknown, make a first approximation of the pipe diameter, using a suitable velocity for the fluid. Once the length of the pipe is known, the pressure loss can be determined. If the pressure loss exceeds a desirable value, the pipe diameter can be increased until the loss is within an acceptable range.

Compute the pipe cross-sectional area a in² (cm²) from $a = 2.4Wv/V$, where W = steam flow rate, lb/h (kg/h); v = specific volume of the steam, ft³/lb (m³/kg); V = steam velocity, ft/min (m/min). The only unknown in this equation, other than the pipe area, is the steam velocity V. Use Table 1 to find a suitable steam velocity for this branch line.

Table 1 shows that the recommended steam velocities for branch steam pipes range from 6000 to 15,000 ft/min (1828 to 4572 m/min). Assume that a velocity of 12,000 ft/min (3657.6 m/min) is used in this branch steam line. Then, by using the steam table to find the specific volume of steam at 900°F (482.2°C) and 1000 lb/in² (abs) (6894 kPa), $a = 2.4(72,000)(0.7604)/12,000 = 10.98$ in² (70.8 cm²). The inside diameter of the pipe is then $d = 2(a/\pi)^{0.5} = 2(10.98/\pi)^{0.5} = 3.74$ in (95.0 mm). Since pipe is not ordinarily made in this fractional internal diameter, round it to the next larger size, or 4-in (101.6-mm) inside diameter.

2. *Determine the pipe schedule number*

The ANSA *Code for Pressure Piping*, commonly called the *Piping Code*, defines schedule number as SN = 1000 P_i/S, where P_i = internal pipe pressure, lb/in² (gage); S = allowable stress in the pipe, lb/in², from *Piping Code*. Table 2 shows typical allowable stress values for pipe in power piping systems. For this pipe, assuming that seamless ferritic alloy steel (1% Cr, 0.55% Mo) pipe is used with the steam at 900°F (482°C), SN = (1000)(1014.7)/13,100 = 77.5. Since pipe is not ordinarily made in this schedule number, use the next *highest* readily available schedule number, or SN = 80. [Where large quantities of pipe are required, it is sometimes economically wise to order pipe of the exact SN required. This is not usually done for orders of less than 1000 ft (304.8 m) of pipe.]

3. *Determine the pipe-wall thickness*

Enter a tabulation of pipe properties, such as in Crocker and King—*Piping Handbook*, and find the wall thickness for 4-in (101.6-mm) SN 80 pipe as 0.337 in (8.56 mm).

TABLE 1   Recommended Fluid Velocities in Piping

	Velocity of fluid	
Service	ft/min	m/s
Boiler and turbine leads	6,000–12,000	30.5–60.9
Steam headers	6,000–8,000	30.5–40.6
Branch steam lines	6,000–15,000	30.5–76.2
Feedwater lines	250–850	1.3–4.3
Exhaust and low-pressure steam lines	6,000–15,000	30.5–76.2
Pump suction lines	100–300	0.51–1.52
Bleed steam lines	4,000–6,000	20.3–30.5
Service water mains	120–300	0.61–1.52
Vacuum steam lines	20,000–40,000	101.6–203.2
Steam superheater tubes	2,000–5,000	10.2–25.4
Compressed-air lines	1,500–2,000	7.6–10.2
Natural-gas lines (large cross-country)	100–150	0.51–0.76
Economizer tubes (water)	150–300	0.76–1.52
Crude-oil lines [6 to 30 in (152.4 to 762.0 mm)]	50–350	0.25–1.78

Related Calculations: Use the method given here for any type of pipe—steam, water, oil, gas, or air—in any service—power, refinery, process, commercial, etc. Refer to the proper section of B31.1 *Code for Pressure Piping* when computing the schedule number, because the allowable stress S varies for different types of service.

The *Piping Code* contains an equation for determining the minimum required pipe-wall thickness based on the pipe internal pressure, outside diameter, allowable stress, a temperature coefficient, and an allowance for threading, mechanical strength, and corrosion. This equation is seldom used in routine piping-system design. Instead, the schedule number as given here is preferred by most designers.

PIPE-WALL THICKNESS DETERMINATION BY PIPING CODE FORMULA

Use the ANSA B31.1 *Code for Pressure Piping* wall-thickness equation to determine the required wall thickness for an 8.625-in (219.1-mm) OD ferritic steel plain-end pipe if the pipe is used in 900°F (482°C) 900-lb/in^2 (gage) (6205-kPa) steam service.

Calculation Procedure:

1. *Determine the constants for the thickness equation*

Pipe-wall thickness to meet ANSA *Code* requirements for power service is computed from $t_m = \{DP/[2(S + YP)]\} + C$, where t_m = minimum wall thickness, in; D = outside diameter of pipe, in; P = internal pressure in pipe, lb/in^2 (gage); S = allowable stress in pipe material, lb/in^2; Y = temperature coefficient; C = end-condition factor, in.

Values of S, Y, and C are given in tables in the *Code for Pressure Piping* in the section on Power Piping. Using values from the latest edition of the *Code*, we get S = 12,500 lb/in^2 (86.2 MPa) for ferritic-steel pipe operating at 900°F (482°C); Y = 0.40 at the same temperature; C = 0.065 in (1.65 mm) for plain-end steel pipe.

2. *Compute the minimum wall thickness*

Substitute the given and *Code* values in the equation in step 1, or t_m = [(8.625)(900)]/[2(12,500 + 0.4 × 900)] + 0.065 = 0.367 in (9.32 mm).

Since pipe mills do not fabricate to precise wall thicknesses, a tolerance above or below the computed wall thickness is required. An allowance must be made in specifying the wall thickness found with this equation by *increasing* the thickness by 12½ percent. Thus, for this pipe, wall thickness = 0.367 + 0.125(0.367) = 0.413 in (10.5 mm).

Refer to the *Code* to find the schedule number of the pipe. Schedule 60 8-in (203-mm) pipe

TABLE 2 Allowable Stresses (S Values) for Alloy-Steel Pipe in Power Piping Systems°
(Abstracted from ASME Power Boiler Code and Code for Pressure Piping, ASA B31.1)

Material	ASTM specification	Grade or symbol	Minimum tensile strength		S values for metal temperatures not to exceed†					
			lb/in²	MPa	850°F	454°C	900°F	482°C	950°F	510°C
Seamless ferritic steels:										
Carbon-molybdenum	A335	P1	55,000	379.2	13,150	90.7	12,500	86.2	...	...
0.65 Cr, 0.55 Mo	A335	P2	55,000	379.2	13,150	90.7	12,500	86.2	10,000	68.9
1.00 Cr, 0.55 Mo	A335	P12	60,000	413.6	14,200	97.9	13,100	90.3	11,000	75.8

°Crocker and King—Piping Handbook.
†Where welded construction is used, consideration should be given to the possibility of graphite formation in carbon-molybdenum steel above 875°F (468°C) or in chromium-molybdenum steel containing less than 0.60 percent chromium above 975°F (523.9°C).

has a wall thickness of 0.406 in (10.31 mm), and schedule 80 pipe has a wall thickness of 0.500 in (12.7 mm). Since the required thickness of 0.413 in (10.5 mm) is greater than schedule 60 but less than schedule 80, the higher schedule number, 80, should be used.

3. Check the selected schedule number

From the previous calculation procedure, SN $= 1000 P_t / S$. For this pipe, SN $= 1000(900)/12,500$ $= 72$. Since piping is normally fabricated for schedule numbers 10, 20, 30, 40, 60, 80, 100, 120, 140, and 160, the next larger schedule number higher than 72, that is 80, will be used. This agrees with the schedule number found in step 2.

   Related Calculations: Use this method in conjunction with the appropriate *Code* equation to determine the wall thickness of pipe conveying air, gas, steam, oil, water, alcohol, or any other similar fluids in any type of service. Be certain to use the correct equation, which in some cases is simpler than that used here. Thus, for lead pipe, $t_m = Pd/2S$, where P = safe working pressure of the pipe, lb/in^2 (gage); d = inside diameter of pipe, in; other symbols as before.

   When a pipe will operate at a temperature between two tabulated *Code* values, find the allowable stress by interpolating between the tabulated temperature and stress values. Thus, for a pipe operating at 680°F (360°C), find the allowable stress at 650°F (343°C) [= 9500 lb/in^2 (65.5 MPa)] and 700°F (371°C) [= 9000 lb/in^2 (62.0 MPa)]. Interpolate thus: allowable stress at 680°F (360°C) = [(700°F − 680°F)/(700°F − 650°F)](9500 − 9000) + 9000 = 200 + 9000 = 9200 lb/in^2 (63.4 MPa). The same result can be obtained by interpolating downward from 9500 lb/in^2 (65.5 MPa), or allowable stress at 680°F (360°C) = 9500 − [(680 − 650)/(700 − 650)](9500 − 9000) = 9200 lb/in^2 (63.4 MPa).

DETERMINING THE PRESSURE LOSS IN STEAM PIPING

Use a suitable pressure-loss chart to determine the pressure loss in 510 ft (155.5 m) of 4-in (101.6-mm) flanged steel pipe containing two 90° elbows and four 45° bends. The schedule 40 piping conveys 13,000 lb/h (5850 kg/h) of 40-lb/in^2 (gage) (275.8-kPa) 350°F (177°C) superheated steam. List other methods of determining the pressure loss in steam piping.

Calculation Procedure:

1. Determine the equivalent length of the piping

The equivalent length of a pipe L_e ft = length of straight pipe, ft + equivalent length of fittings, ft. Using data from the Hydraulic Institute, Crocker and King—*Piping Handbook*, earlier sections of this handbook, or Fig. 1, find the equivalent length of a 90° 4-in (101.6-mm) elbow as 10 ft (3 m) of straight pipe. Likewise, the equivalent length of a 45° bend is 5 ft (1.5 m) of straight pipe. Substituting in the above relation and using the straight lengths and the number of fittings of each type, we get $L_e = 510 + (2)(10) + 4(5) = 550$ ft (167.6 m) of straight pipe.

2. Compute the pressure loss, using a suitable chart

Figure 2 presents a typical pressure-loss chart for steam piping. Enter the chart at the top left at the superheated steam temperature of 350°F (177°C), and project vertically downward until the 40-lb/in^2 (gage) (275.8-kPa) superheated steam pressure curve is intersected. From here, project horizontally to the right until the outer border of the chart is intersected. Next, project through the steam flow rate, 13,000 lb/h (5900 kg/h) on scale *B*, Fig. 2, to the pivot scale *C*. From this point, project through 4-in (101.6-mm) schedule 40 pipe on scale *D*, Fig. 2. Extend this line to intersect the pressure-drop scale, and read the pressure loss as 7.25 lb/in^2 (50 kPa) per 100 ft (30.4 m) of pipe.

   Since the equivalent length of this pipe is 550 ft (167.6 m), the total pressure loss in the pipe is $(550/100)(7.25) = 39.875$ lb/in^2 (274.9 kPa), say 40 lb/in^2 (275.8 kPa).

3. List the other methods of computing pressure loss

Numerous pressure-loss equations have been developed to compute the pressure drop in steam piping. Among the better known are those of Unwin, Fritzche, Spitzglass, Babcock, Gutermuth, and others. These equations are discussed in some detail in Crocker and King—*Piping Handbook* and in the engineering data published by valve and piping manufacturers.

   Most piping designers use a chart to determine the pressure loss in steam piping because a

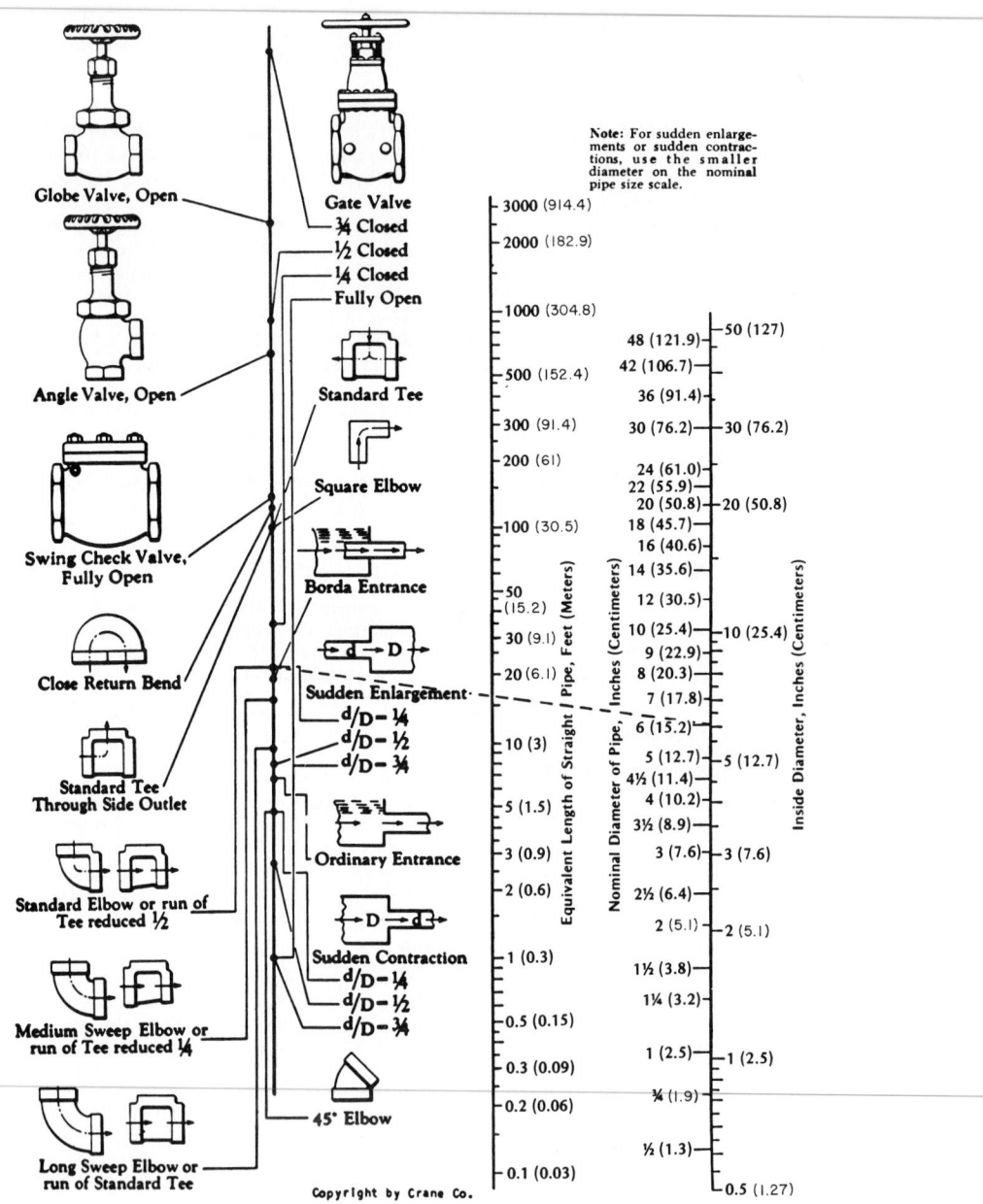

Note: For sudden enlargements or sudden contractions, use the smaller diameter on the nominal pipe size scale.

Globe Valve, Open

Angle Valve, Open

Swing Check Valve, Fully Open

Close Return Bend

Standard Tee Through Side Outlet

Standard Elbow or run of Tee reduced ½

Medium Sweep Elbow or run of Tee reduced ¼

Long Sweep Elbow or run of Standard Tee

Gate Valve
¾ Closed
½ Closed
¼ Closed
Fully Open

Standard Tee

Square Elbow

Borda Entrance

Sudden Enlargement
d/D – ¼
d/D – ½
d/D – ¾

Ordinary Entrance

Sudden Contraction
d/D – ¼
d/D – ½
d/D – ¾

45° Elbow

Copyright by Crane Co.

Equivalent Length of Straight Pipe, Feet (Meters)

3000 (914.4)
2000 (182.9)
1000 (304.8)
500 (152.4)
300 (91.4)
200 (61)
100 (30.5)
50 (15.2)
30 (9.1)
20 (6.1)
10 (3)
5 (1.5)
3 (0.9)
2 (0.6)
1 (0.3)
0.5 (0.15)
0.3 (0.09)
0.2 (0.06)
0.1 (0.03)

Nominal Diameter of Pipe, Inches (Centimeters)

48 (121.9)
42 (106.7)
36 (91.4)
30 (76.2)
24 (61.0)
22 (55.9)
20 (50.8)
18 (45.7)
16 (40.6)
14 (35.6)
12 (30.5)
10 (25.4)
9 (22.9)
8 (20.3)
7 (17.8)
6 (15.2)
5 (12.7)
4½ (11.4)
4 (10.2)
3½ (8.9)
3 (7.6)
2½ (6.4)
2 (5.1)
1½ (3.8)
1¼ (3.2)
1 (2.5)
¾ (1.9)
½ (1.3)

Inside Diameter, Inches (Centimeters)

50 (127)
30 (76.2)
20 (50.8)
10 (25.4)
5 (12.7)
3 (7.6)
2 (5.1)
1 (2.5)
0.5 (1.27)

FIG. 1 Equivalent length of pipe fittings and valves. *(Crane Company.)*

3.400

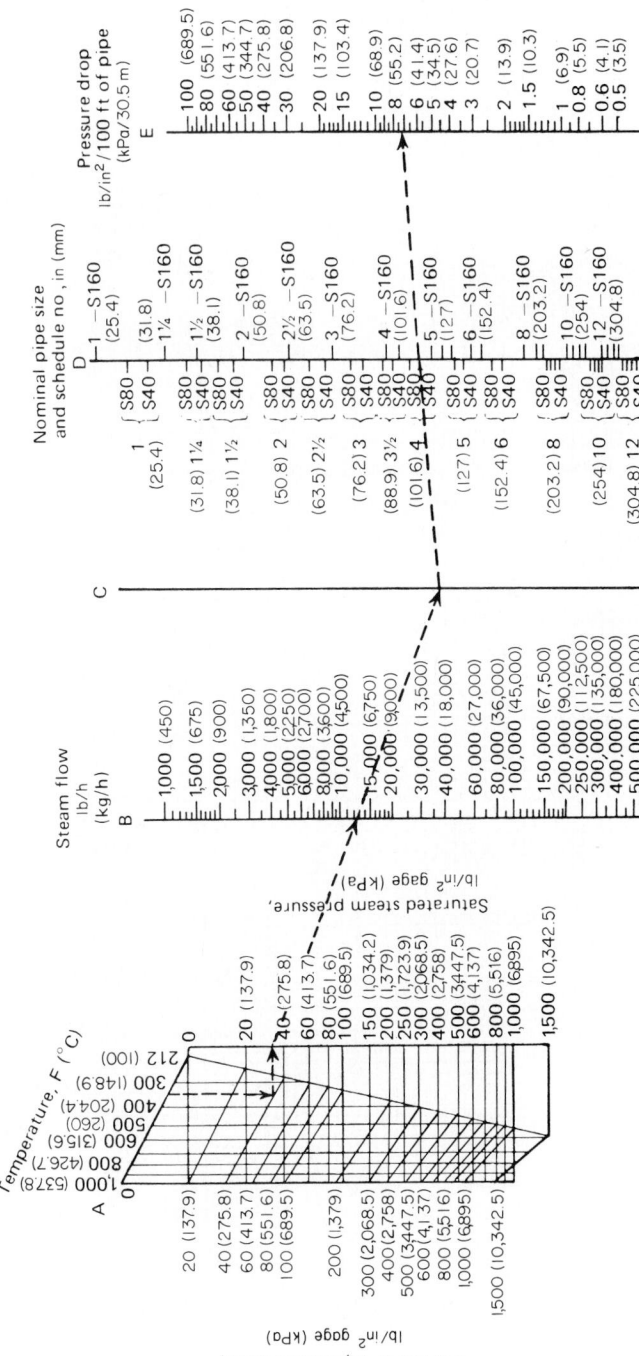

FIG. 2 Pressure loss in steam pipes based on the Fritzche formula. (*Power.*)

3.401

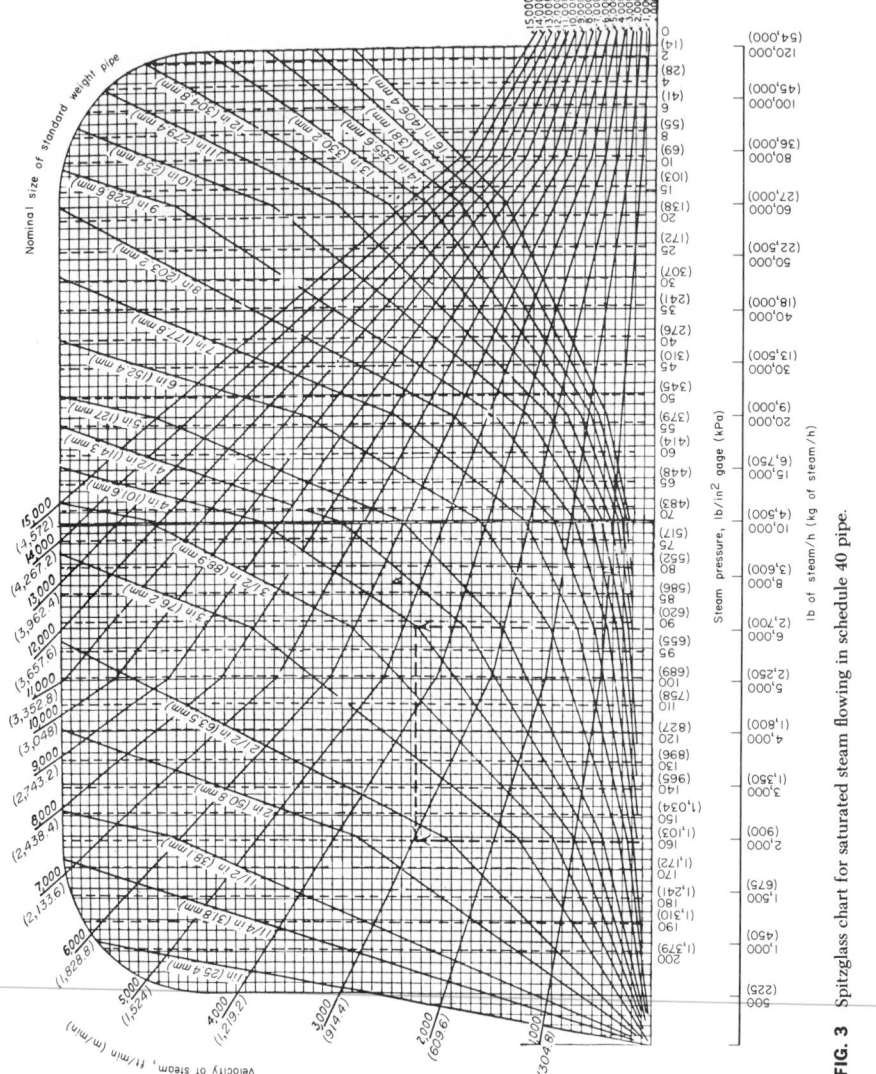

FIG. 3 Spitzglass chart for saturated steam flowing in schedule 40 pipe.

TABLE 3 Steam Velocities Used in Pipe Design

Steam condition	Steam pressure		Steam use	Steam velocity	
	lb/in²	kPa		ft/min	m/min
Saturated	0–15	0–103.4	Heating	4,000–6,000	1,219.2–1,828.8
Saturated	50–150	344.7–1,034.1	Process	6,000–10,000	1,828.8–3,048.0
Superheated	200 and higher	1,378.8 and higher	Boiler leads	10,000–15,000	3,048.0–4,572.0

chart saves time and reduces the effort involved. Further, the accuracy obtained is sufficient for all usual design practice.

Figure 3 is a popular flowchart for determining steam flow rate, pipe size, steam pressure, or steam velocity in a given pipe. Using this chart, the designer can determine any one of the four variables listed above when the other three are known. In solving a problem on the chart in Fig. 3, use the steam-quantity lines to intersect pipe sizes and the steam-pressure lines to intersect steam velocities. Here are two typical applications of this chart.

Example: What size schedule 40 pipe is needed to deliver 8000 lb/h (3600 kg/h) of 120-lb/in² (gage) (827.3-kPa) steam at a velocity of 5000 ft/min (1524 m/min)?

Solution: Enter Fig. 3 at the upper left at a velocity of 5000 ft/min (1524 m/min), and project along this velocity line until the 120-lb/in² (gage) (827.3-kPa) pressure line is intersected. From this intersection, project horizontally until the 8000-lb/h (3600-kg/h) vertical line is intersected. Read the *nearest* pipe size as 4 in (101.6 mm) on the *nearest* pipe-diameter curve.

Example: What is the steam velocity in a 6-in (152.4-mm) pipe delivering 20,000 lb/h (9000 kg/h) of steam at 85 lb/in² (gage) (586 kPa)?

Solution: Enter the bottom of the chart, Fig. 3, at the flow rate of 20,000 lb/h (9000 kg/h), and project vertically upward until the 6-in (152.4-mm) pipe curve is intersected. From this point, project horizontally to the 85-lb/in² (gage) (586-kPa) curve. At the intersection, read the velocity as 7350 ft/min (2240.3 m/min).

Table 3 shows typical steam velocities for various industrial and commercial applications. Use the given values as guides when sizing steam piping.

PIPING WARM-UP CONDENSATE LOAD

How much condensate is formed in 5 min during warm-up of 500 ft (152.4 m) of 6-in (152.4-mm) schedule 40 steel pipe conveying 215-lb/in² (abs) (1482.2-kPa) saturated steam if the pipe is insulated with 2 in (50.8 mm) of 85 percent magnesia and the minimum external temperature is 35°F (1.7°C)?

Calculation Procedure:

1. Compute the amount of condensate formed during pipe warm-up

For any pipe, the condensate formed during warm-up C_h lb/h = $60(W_p)(\Delta t)(s)/h_{fg}N$, where W_p = total weight of pipe, lb; Δt = difference between final and initial temperature of the pipe, °F; s = specific heat of pipe material, Btu/(lb·°F); h_{fg} = enthalpy of vaporization of the steam, Btu/lb; N = warm-up time, min.

A table of pipe properties shows that this pipe weighs 18.974 lb/ft (28.1 kg/m). The steam table shows that the temperature of 215-lb/in² (abs) (1482.2-kPa) saturated steam is 387.89°F (197.7°C), say 388°F (197.8°C); the enthalpy h_{fg} = 837.4 Btu/lb (1947.8 kJ/kg). The specific heat of steel pipe s = 0.144 Btu/(lb·°F) [0.6 kJ/(kg·°C)]. Then C_h = 60(500 × 18.974)(388 − 35)(0.114)/[(837.4)(5)] = 5470 lb/h (2461.5 kg/h).

2. Compute the radiation-loss condensate load

Condensate is also formed by radiation of heat from the pipe during warm-up and while the pipe is operating. The warm-up condensate load decreases as the radiation load increases, the peak occurring midway (2½ min in this case) through the warm-up period. For this reason, one-half the normal radiation load is added to the warm-up load. Where the radiation load is small, it is often disregarded. However, the load must be computed before its magnitude can be determined.

For any pipe, $C_r = (L)(A)(\Delta t)(H)/h_{fg}$, where L = length of pipe, ft; A = external area of pipe, ft² per ft of length; H = heat loss through bare pipe or pipe insulation, Btu/(ft²·h·°F), from the piping or insulation tables. This 6-in (152.4-mm) schedule 40 pipe has an external area A = 1.73 ft²/ft (0.53 m²/m) of length. The heat loss through 2 in (50.8 mm) of 85 percent magnesia, from insulation tables, is H = 0.286 Btu/(ft²·h·°F) [1.62 W/(m²·°C)]. Then C_r = (500) × (1.73)(388 − 35)(0.286)/837.4 = 104.2 lb/h (46.9 kg/h). Adding half the radiation load to the warm-up load gives 5470 + 52.1 = 5522.1 lb/h (2484.9 kg/h).

3. Apply a suitable safety factory to the condensate load

Trap manufacturers recommend a safety factor of 2 for traps installed between a boiler and the end of a steam main; traps at the end of a long steam main or ahead of pressure-regulating or shutoff valves usually have a safety factor of 3. With a safety factor of 3 for this pipe, the steam trap should have a capacity of at least 3(5522.1) = 16,566.3 lb/h (7454.8 kg/h), say 17,000 lb/h (7650.0 kg/h).

Related Calculations: Use this method to find the warm-up condensate load for any type of steam pipe—main or auxiliary— in power, process, heating, or vacuum service. The same method is applicable to other vapors that form condensate—Dowtherm, refinery vapors, process vapors, and others.

STEAM TRAP SELECTION FOR INDUSTRIAL APPLICATIONS

Select steam traps for the following four types of equipment: (1) the steam directly heats solid materials as in autoclaves, retorts, and sterilizers; (2) the steam indirectly heats a liquid through a metallic surface, as in heat exchangers and kettles, where the quantity of liquid heated is known and unknown; (3) the steam indirectly heats a solid through a metallic surface, as in dryers using cylinders or chambers and platen presses; and (4) the steam indirectly heats air through metallic surfaces, as in unit heaters, pipe coils, and radiators.

Calculation Procedure:

1. Determine the condensate load

The first step in selecting a steam trap for any type of equipment is determination of the condensate load. Use the following general procedure.

a. Solid materials in autoclaves, retorts, and sterilizers. How much condensate is formed when 2000 lb (900.0 kg) of solid material with a specific heat of 1.0 is processed in 15 min at 240°F (115.6°C) by 25-lb/in² (gage) (172.4-kPa) steam from an initial temperature of 60°F in an insulated steel retort?

For this type of equipment, use $C = WSP$, where C = condensate formed, lb/h; W = weight of material heated, lb; s = specific heat, Btu/(lb·°F); P = factor from Table 4. Thus, for this application, C = (2000)(1.0)(0.193) = 386 lb (173.7 kg) of condensate. Note that P is based on a temperature rise of 240 − 60 = 180°F (100°C) and a steam pressure of 25 lb/in² (gage) (172.4 kPa). For the retort, using the specific heat of steel from Table 5, C = (4000)(0.12)(0.193) = 92.6 lb of condensate, say 93 lb (41.9 kg). The total weight of condensate formed in 15 min is 386 + 93 = 479 lb (215.6 kg). In 1 h, 479(60/15) = 1916 lb (862.2 kg) of condensate is formed.

TABLE 4 Factors $P = (T − t)/L$ to Find Condensate Load

Pressure		Temperature		
lb/in² (abs)	kPa	160°F (71.1°C)	180°F (82.2°C)	200°F (93.3°C)
20	137.8	0.170	0.192	0.213
25	172.4	0.172	0.193	0.214
30	206.8	0.172	0.194	0.215

TABLE 5 Use These Specific Heats to Calculate Condensate Load

Solids	Btu/(lb·°F)	kJ/(kg·°C)	Liquids	Btu/(lb·°F)	kJ/(kg·°C)
Aluminum	0.23	0.96	Alcohol	0.65	2.7
Brass	0.10	0.42	Carbon tetrachloride	0.20	0.84
Copper	0.10	0.42	Gasoline	0.53	2.22
Glass	0.20	0.84	Glycerin	0.58	2.43
Iron	0.13	0.54	Kerosene	0.47	1.97
Steel	0.12	0.50	Oils	0.40–0.50	1.67–2.09

A safety factor must be applied to compensate for radiation and other losses. Typical safety factors used in selecting steam traps are as follows:

Steam mains and headers	2–3
Steam heating pipes	2–6
Purifiers and separators	2–3
Retorts for process	2–4
Unit heaters	3
Submerged pipe coils	2–4
Cylinder dryers	4–10

With a safety factor of 4 for this process retort, the trap capacity = (4)(1916) = 7664 lb/h (3449 kg/h), say 7700 lb/h (3465 kg/h).

b(1). Submerged heating surface and a known quantity of liquid. How much condensate forms in the jacket of a kettle when 500 gal (1892.5 L) of water is heated in 30 min from 72 to 212°F (22.2 to 100°C) with 50-lb/in² (gage) (344.7-kPa) steam?

For this type of equipment, $C = GwsP$, where G = gal of liquid heated; w = weight of liquid, lb/gal. Substitute the appropriate values as follows: C = (500)(8.33)(1.0) × (0.154) = 641 lb (288.5 kg), or (641)(60/30) = 1282 lb/h (621.9 kg/h). With a safety factor of 3, the trap capacity = (3)(1282) = 3846 lb/h (1731 kg/h), say 3900 lb/h (1755 kg/h).

b(2). Submerged heating surface and an unknown quantity of liquid. How much condensate is formed in a coil submerged in oil when the oil is heated as quickly as possible from 50 to 250°F (10 to 121°C) by 25-lb/in² (gage) (172.4-kPa) steam if the coil has an area of 50 ft² (4.66 m²) and the oil is free to circulate around the coil?

For this condition, $C = UAP$, where U = overall coefficient of heat transfer, Btu/(h·ft²·°F), from Table 6; A = area of heating surface, ft². With free convection and a condensing-vapor-to-liquid type of heat exchanger, U = 10 to 30. With an average value of U = 20, C = (20)(50)(0.214) = 214 lb/h (96.3 kg/h) of condensate. Choosing a safety factor 3 gives trap capacity = (3)(214) = 642 lb/h (289 kg/h), say 650 lb/h (292.5 kg/h).

b(3). Submerged surfaces having more area than needed to heat a specified quantity of liquid in a given time with condensate withdrawn as rapidly as formed. Use Table 7 instead of step b(1) or b(2). Find the condensation rate by multiplying the submerged area by the appropriate factor from Table 7. Use this method for heating water, chemical solutions, oils, and other liquids. Thus, with steam at 100 lb/in² (gage) (689.4 kPa) and a temperature of 338°F (170°C) and heating oil from 50 to 226°F (10 to 108°C) with a submerged surface having an area of 500 ft² (46.5 m²), the mean temperature difference = steam temperature minus the average liquid temperature = Mtd = 338 − (50 + 226/2) = 200°F (93.3°C). The factor from Table 7 for 100 lb/in² (gage) (689.4 kPa) steam and a 200°F (93.3°C) Mtd is 56.75. Thus, the condensation rate = (56.75)(500) = 28,375 lb/h (12,769 kg/h). With a safety factor of 2, the trap capacity = (2)(28,375) = 56,750 lb/h (25,538 kg/h).

c. Solids indirectly heated through a metallic surface. How much condensate is formed in a chamber dryer when 1000 lb (454 kg) of cereal is dried to 750 lb (338 kg) by 10-lb/in² (gage) (68.9-kPa) steam? The initial temperature of the cereal is 60°F (15.6°C), and the final temperature equals that of the steam.

TABLE 6 Ordinary Ranges of Overall Coefficients of Heat Transfer

Type of heat exchanger	State of controlling resistance		Typical fluid	Typical apparatus
	Free convection, U	Forced convection, U		
Liquid to liquid	25–60 [141.9–340.7]	150–300 [851.7–1703.4]	Water	Liquid-to-liquid heat exchangers
Liquid to liquid	5–10 [28.4–56.8]	20–50 [113.6–283.9]	Oil	
Liquid to gas°	1–3 [5.7–17.0]	2–10 [11.4–56.8]	—	Hot-water radiators
Liquid to boiling liquid	20–60 [113.6–340.7]	50–150 [283.9–851.7]	Water	Brine coolers
Liquid to boiling liquid	5–20 [28.4–113.6]	25–60 [141.9–340.7]	Oil	
Gas° to liquid	1–3 [5.7–17.0]	2–10 [11.4–56.8]	—	Air coolers, economizers
Gas° to gas	0.6–2 [3.4–11.4]	2–6 [11.4–34.1]	—	Steam superheaters
Gas° to boiling liquid	1–3 [5.7–17.0]	2–10 [11.4–56.8]	—	Steam boilers
Condensing vapor to liquid	50–200 [283.9–1136]	150–800 [851.7–4542.4]	Steam to water	Liquid heaters and condensers
Condensing vapor to liquid	10–30 [56.8–170.3]	20–60 [113.6–340.7]	Steam to oil	
Condensing vapor to liquid	40–80 [227.1–454.2]	60–150 [340.7–851.7]	Organic vapor to water	
Condensing vapor to liquid	—	15–300 [85.2–1703.4]	Steam-gas mixture	Steam pipes in air, air heaters
Condensing vapor to gas°	1–2 [5.7–11.4]	2–10 [11.4–56.8]	—	Scale-forming evaporators
Condensing vapor to boiling liquid	40–100 [227.1–567.8]	—	—	
Condensing vapor to boiling liquid	300–800 [1703.4–4542.4]	—	Steam to water	
Condensing vapor to boiling liquid	50–150 [283.9–851.7]	—	Steam to oil	

° At atmospheric pressure.

Note: $U = $ Btu/(h·ft²·°F) [W/(m²·°C)]. Under many conditions, either higher or lower values may be realized.

TABLE 7 Condensate Formed in Submerged Steel° Heating Elements, lb/(ft²·h) [kg/(m²·min)]

MTD†		Steam pressure				
°F	°C	75 lb/in² (abs) (517.1 kPa)	100 lb/in² (abs) (689.4 kPa)	150 lb/in² (abs) (1034.1 kPa)	Btu/(ft²·h)	kW/m²
175	97.2	44.3 (3.6)	45.4 (3.7)	46.7 (3.8)	40,000	126.2
200	111.1	54.8 (4.5)	56.8 (4.6)	58.3 (4.7)	50,000	157.7
250	138.9	90.0 (7.3)	93.1 (7.6)	95.7 (7.8)	82,000	258.6

°For copper, multiply table data by 2.0; for brass, by 1.6.
†Mean temperature difference, °F or °C, equals temperature of steam minus average liquid temperature. Heat-transfer data for calculating this table obtained from and used by permission of the American Radiator & Standard Sanitary Corp.

For this condition, $C = 970(W - D)/h_{fg} + WP$, where D = dry weight of the material, lb; h_{fg} = enthalpy of vaporization of the steam at the trap pressure, Btu/lb. From the steam tables and Table 4, $C = 970(1000 - 750)/952 + (1000)(0.189) = 443.5$ lb/h (199.6 kg/h) of condensate. With a safety factor of 4, the trap capacity = (4)(443.5) = 1774 lb/h (798.3 kg/h).

d. Indirect heating of air through a metallic surface. How much condensate is formed in a unit heater using 10-lb/in² (gage) (68.9-kPa) steam if the entering-air temperature is 30°F (−1.1°C) and the leaving-air temperature is 130°F (54.4°C)? Airflow is 10,000 ft³/min (281.1 m³/min).

Use Table 8, entering at a temperature difference of 100°F (37.8°C) and projecting to a steam pressure of 10 lb/in² (gage) (68.9 kPa). Read the condensate formed as 122 lb/h (54.9 kg/h) per 1000 ft³/min (28.3 m³/min). Since 10,000 ft³/min (283.1 m³/min) of air is being heated, the condensation rate = (10,000/1000)(122) = 1220 lb/h (549 kg/h). With a safety factor of 3, the trap capacity = (3)(1220) = 3660 lb/h (1647 kg/h), say 3700 lb/h (1665 kg/h).

Table 9 shows the condensate formed by radiation from bare iron and steel pipes in still air and with forced-air circulation. Thus, with a steam pressure of 100 lb/in² (gage) (689.4 kPa) and an initial air temperature of 75°F (23.9°C), 1.05 lb/h (0.47 kg/h) of condensate will be formed per ft² (0.09 m²) of heating surface in still air. With forced-air circulation, the condensate rate is (5)(1.05) = 5.25 (lb/(h·ft²) [25.4 kg/(h·m²)] of heating surface.

Unit heaters have a *standard rating* based on 2-lb/in² (gage) (13.8-kPa) steam with entering air at 60°F (15.6°C). If the steam pressure or air temperature is different from these standard conditions, multiply the heater Btu/h capacity rating by the appropriate correction factor form, Table 10. Thus, a heater rated at 10,000 Btu/h (2931 W) with 2-lb/in² (gage) (13.8-kPa) steam and 60°F (15.6°C) air would have an output of (1.290)(10,000) = 12,900 Btu/h (3781 W) with 40°F (4.4°C) inlet air and 10-lb/in² (gage) (68.9-kPa) steam. Trap manufacturers usually list heater Btu ratings and recommend trap model numbers and sizes in their trap engineering data. This allows easier selection of the correct trap.

TABLE 8 Steam Condensed by Air, lb/h at 1000 ft³/min (kg/h at 28.3 m³/min)°

Temperature difference		Pressure		
°F	°C	5 lb/in² (gage) (34.5 kPa)	10 lb/in² (gage) (68.9 kPa)	50 lb/in² (gage) (344.7 kPa)
50	27.8	61 (27.5)	61 (27.5)	63 (28.4)
100	55.6	120 (54.0)	122 (54.9)	126 (56.7)
150	83.3	180 (81.0)	183 (82.4)	189 (85.1)

°Based on 0.0192 Btu (0.02 kJ) absorbed per ft³ (0.028 m³) of saturated air per °F (0.556°C) at 32°F (0°C). For 0°F (−17.8°C), multiply by 1.1.

TABLE 9   Condensate Formed by Radiation from Bare Iron and Steel, $lb/(ft^2 \cdot h)$ $[kg/(m^2 \cdot h)]$

Air temperature		Steam pressure			
°F	°C	50 lb/in² (gage) (344.7 kPa)	75 lb/in² (gage) (517.1 kPa)	100 lb/in² (gage) (689.5 kPa)	150 lb/in² (gage) (1034 kPa)
65	18.3	0.82 (3.97)	1.00 (5.84)	1.08 (5.23)	1.32 (6.39)
70	21.2	0.80 (3.87)	0.98 (4.74)	1.06 (5.13)	1.21 (5.86)
75	23.9	0.77 (3.73)	0.88 (4.26)	1.05 (5.08)	1.19 (5.76)

*Based on still air; for forced-air circulation, multiply by 5.

2. *Select the trap size based on the load and steam pressure*

Obtain a chart or tabulation of trap capacities published by the manufacturer whose trap will be used. Figure 4 is a capacity chart for one type of bucket trap manufactured by Armstrong Machine Works. Table 11 shows typical capacities of impulse traps manufactured by the Yarway Company.

To select a trap from Fig. 4, when the condensation rate is uniform and the pressure across the trap is constant, enter at the left at the condensation rate, say 8000 lb/h (3600 kg/h) (as obtained from step 1). Project horizontally to the right to the vertical ordinate representing the pressure across the trap [$= \Delta p =$ steam-line pressure, lb/in² (gage) − return-line pressure with trap valve closed, lb/in² (gage)]. Assume $\Delta p = 20$ lb/in² (gage) (138 kPa) for this trap. The intersection of the horizontal 8000-lb/h (3600-kg/h) projection and the vertical 20-lb/in² (gage) (137.9-kPa) projection is on the sawtooth capacity curve for a trap having a $\frac{9}{16}$-in (14.3-mm) diameter orifice. If these projections intersected beneath this curve, a $\frac{9}{16}$-in (14.3-mm) orifice would still be used if the point were between the verticals for this size orifice.

The dashed lines extending downward from the sawtooth curves show the capacity of a trap at reduced Δp. Thus, the capacity of a trap with a $\frac{3}{8}$-in (9.53-mm) orifice at $\Delta p = 30$ lb/in² (gage) (207 kPa) is 6200 lb/h (2790 kg/h), read at the intersection of the 30-lb/in² (gage) (207-kPa) ordinate and the dashed curve extended from the $\frac{3}{8}$-in (9.53-mm) solid curve.

To select an impulse trap from Table 11, enter the table at the trap inlet pressure, say 125 lb/in² (gage) (862 kPa), and project to the desired capacity, say 8000 lb/h (3600 kg/h), determined from step 1. Table 11 shows that a 2-in (50.8-mm) trap having an 8530-lb/h (3839-kg/h) capacity must be used because the next smallest size has a capacity of 5165 lb/h (2324 kg/h). This capacity is less than that required.

Some trap manufacturers publish capacity tables relating various trap models to specific types of equipment. Such tables simplify trap selection, but the condensation rate must still be computed as given here.

Related Calculations: Use the procedure given here to determine the trap capacity required for any industrial, commercial, or domestic application including acid vats, air dryers, asphalt tanks, autoclaves, baths (dyeing), belt presses, bleach tanks, blenders, bottle washers, brewing kettles, cabinet dryers, calenders, can washers, candy kettles, chamber dryers, chambers (reaction), cheese kettles, coils (cooking, kettle, pipe, tank, tank-car), confectioners' kettles, continuous dryers,

TABLE 10   Unit-Heater Correction Factors

Steam pressure		Temperature of entering air		
lb/in² (gage)	kPa	20°F (−6.7°C)	40°F (4.4°C)	60°F (15.6°C)
5	34.5	1.370	1.206	1.050
10	68.9	1.460	1.290	1.131
15	103.4	1.525	1.335	1.194

Source: Yarway Corporation; SI values added by handbook editor.

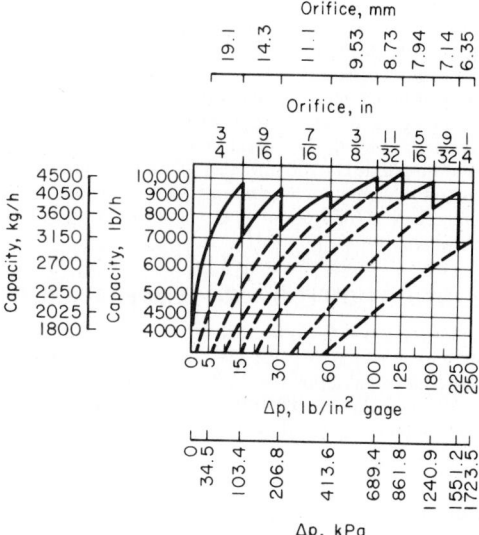

FIG. 4 Capacities of one type of bucket steam trap. *(Armstrong Machine Works.)*

conveyor dryers, cookers (nonpressure and pressure), cooking coils, cooking kettles, cooking tanks, cooking vats, cylinder dryers, cylinders (jacketed), double-drum dryers, drum dryers, drums (dyeing), dry cans, dry kilns, dryers (cabinet, chamber, continuous, conveyor, cylinder, drum, festoon, jacketed, linoleum, milk, paper, pulp, rotary, shelf, stretch, sugar, tray, tunnel), drying rolls, drying rooms, drying tables, dye vats, dyeing baths and drums, dryers (package), embossing-press platens, evaporators, feedwater heaters, festoon dryers, fin-type heaters, fourdriniers, fuel-oil pre-heaters, greenhouse coils, heaters (steam), heat exchangers, heating coils and kettles, hot-break tanks, hot plates, kettle coils, kettles (brewing, candy, cheese, confectioners', cooking, heating, process), kiers, kilns (dry), liquid heaters, mains (steam), milk-bottle washers, milk-can washers, milk dryers, mixers, molding-press platens, package dryers, paper dryers, percolators, phonograph-record press platens, pipe coils (still- and circulating-air), platens, plating tanks, plywood press platens, preheaters (fuel-oil), preheating tanks, press platens, pressure cookers, process kettles, pulp dryers, purifiers, reaction chambers, retorts, rotary dryers, steam mains (risers, separators), stocking boarders, storage-tank coils, storage water heaters, stretch dryers, sugar dryers, tank-car coils, tire-mold presses, tray dryers, tunnel dryers, unit heaters, vats, veneer press platens, vulcanizers, and water stills. Hospital equipment—such as autoclaves and sterilizers—can be analyzed in the same way, as can kitchen equipment—bain marie, compartment cooker, egg boiler,

TABLE 11 Capacities of Impulse Traps, lb/h (kg/h)
[*Maximum continuous discharge of condensate, based on condensate at 30°F (16.7°C) below steam temperature.*]

Pressure at trap inlet		Trap nominal size	
lb/in² (gage)	kPa	1.25 in (38.1 mm)	2.0 in (50.8 mm)
125	861.8	6165 (2774)	8530 (3839)
150	1034.1	6630 (2984)	9075 (4084)
200	1378.8	7410 (3335)	9950 (4478)

Source: Yarway Corporation.

~~kettles, steam table, and urns; and laundry equipment~~—blanket dryers, curtain dryers, flatwork ironers, presses (dry-cleaning, laundry), sock forms, starch cookers, tumblers, etc.

When using a trap capacity diagram or table, be sure to determine the basis on which it was prepared. Apply any necessary correction factors. Thus, *cold-water capacity ratings* must be corrected for traps operating at higher condensate temperatures. Correction factors are published in trap engineering data. The capacity of a trap is greater at condensate temperatures less than 212°F (100°C) because at or above this temperature condensate forms flash steam when it flows into a pipe or vessel at atmospheric [14.7 lb/in² (abs) (101.3 kPa)] pressure. At altitudes above sea level, condensate flashes into steam at a lower temperature, depending on the altitude.

The method presented here is the work of L. C. Campbell, Yarway Corporation, as reported in *Chemical Engineering*.

SELECTING HEAT INSULATION FOR HIGH-TEMPERATURE PIPING

Select the heat insulation for a 300-ft (91.4-m) long 10-in (254-mm) turbine lead operating at 570°F (299°C) for 8000 h/year in a 70°F (21.1°C) turbine room. How much heat is saved per year by this insulation? The boiler supplying the turbine has an efficiency of 80 percent when burning fuel having a heating value of 14,000 Btu/lb (32.6 MJ/kg). Fuel costs $6 per ton ($5.44 per metric ton). How much money is saved by the insulation each year? What is the efficiency of the insulation?

Calculation Procedure:

1. *Choose the type of insulation to use*

Refer to an insulation manufacturer's engineering data or Crocker and King—*Piping Handbook* for recommendations about a suitable insulation for a pipe operating in the 500 to 600°F (260 to 316°C) range. These references will show that calcium silicate is a popular insulation for this temperature range. Table 12 shows that a thickness of 3 in (76.2 mm) is usually recommended for 10-in (254-mm) pipe operating at 500 to 599°F (260 to 315°C).

2. *Determine heat loss through the insulation*

Refer to an insulation manufacturer's engineering data to find the heat loss through 3-in (76.2-mm) thick calcium silicate as 0.200 Btu/(h·ft²·°F) [1.14 W/(m²·°C)]. Since 10-in (254-mm) pipe has an area of 2.817 ft²/ft (0.86 m²/m) of length and since the temperature difference across the pipe is $570 - 70 = 500$°F (260°C), the heat loss per hour = $(0.200)(2.817)(500) = 281.7$ Btu/ (h·ft) (887.9 W/m²). The heat loss from bare 10-in (254-mm) pipe with a 500°F (260°C) temperature difference is, from an insulation manufacturer's engineering data, 4.640 Btu/(h·ft²·°F) [26.4 W/(m²·°C)], or $(4.64)(2.817)(500) = 6510$ Btu/(h·ft) (6.3 kW/m).

3. *Determine annual heat saving*

The heat saved = bare-pipe loss, Btu/h — insulated-pipe loss, Btu/h = $6510 - 281.7 = 6228.3$ Btu/(h·ft) (5989 W/m) of pipe. Since the pipe is 300 ft (91.4 m) long and operates 8000 h per year, the annual heat saving = $(300)(8000)(6228.3) = 14,940,000,000$ Btu/year (547.4 kW).

4. *Compute the money saved by the heat insulation*

The heat saved in fuel as fired = (annual heat saving, Btu/year)/(boiler efficiency) = $14,940,000,000/0.80 = 18,680,000,000$ Btu/year (5473 MW). Weight of fuel saved = (annual ~~heat saving, Btu/year)/(heating value of fuel, Btu/lb)(2000 lb/ton)~~ = 18,680,000,000/ $[(14,000)(2000)] = 667$ tons (605 t). At $6 per ton ~~($5.44 per metric ton), the monetary saving is~~ ($6)(667) = $4002 per year.

5. *Determine the insulation efficiency*

Insulation efficiency = (bare-pipe loss — insulated-pipe loss)/bare pipe loss, all expressed in Btu/ h, or bare-pipe loss = $(6510.0 - 281.7)/6510.0 = 0.957$, or 95.7 percent.

Related Calculations: Use this method for any type of insulation—magnesia, fiber-glass, asbestos, felt, diatomaceous, mineral wool, etc.—used for piping at elevated temperatures conveying steam, water, oil, gas, or other fluids or vapors. To coordinate and simplify calculations, become familiar with the insulation tables in a reliable engineering handbook or comprehensive insulation catalog. Such familiarity will simplify routine calculations.

TABLE 12   Recommended Insulation Thickness

Nominal pipe size		Pipe temperature					
		400–499°F	204–259°C	500–599°F	260–315°C		
in	mm	in	mm	in	mm		
6	152.4	2½°	63.5	2½	63.5		
8	203.2	2½	63.5	3	76.2		
10	254.0	2½	63.5	3	76.2		
12	304.8	3	76.2	3	76.2		
14 and over	355.6 and over	3	76.2	3½	88.9		

°Available in single- or double-layer insulation.

ORIFICE METER SELECTION FOR A STEAM PIPE

Steam is metered with an orifice meter in a 10-in (254-mm) boiler lead having an internal diameter of $d_p = 9.760$ in (247.9 mm). Determine the maximum rate of steam flow that can be measured with a steel orifice plate having a diameter of $d_0 = 5.855$ in (148.7 mm) at 70°F (21.1°C). The upstream pressure tap is $1D$ ahead of the orifice, and the downstream tap is $0.5D$ past the orifice. Steam pressure at the orifice inlet $p_p = 250$ lb/in² (gage) (1724 kPa), temperature is 640°F (338°C). A differential gage fitted across the orifice has a maximum range of 120 in (304.8 cm) of water. What is the steam flow rate when the observed differential pressure is 40 in (101.6 cm) of water? Use the ASME Research Committee on Fluid Meters method in analyzing the meter. Atmospheric pressure is 14.696 lb/in² (abs) (101.3 kPa).

Calculation Procedure:

1.   Determine the diameter ratio and steam density

For any orifice meter, diameter ratio = β = meter orifice diameter, in/pipe internal diameter, in = 5.855/9.760 = 0.5999.

Determine the density of the steam by entering the superheated steam table at 250 + 14.696 = 264.696 lb/in² (abs) (1824.8 kPa) and 640°F (338°C) and reading the specific volume as 2.387 ft³/lb (0.15 m³/kg). For steam, the density = 1/specific volume = $d_s = 1/2.387 = 0.4193$ lb/ft³ (6.7 kg/m³).

2.   Determine the steam viscosity and meter flow coefficient

From the ASME publication, *Fluid Meters—Their Theory and Application*, the steam viscosity gu_1 for a steam system operating at 640°F (338°C) is $gu_1 = 0.0000141$ in·lb/(°F·s·ft²) [0.000031 N·m/(°C·s·m²)].

Find the flow coefficient K from the same ASME source by entering the 10-in (254-mm) nominal pipe diameter table at $\beta = 0.5999$ and projecting to the appropriate Reynolds number column. Assume that the Reynolds number = 10^7, approximately, for the flow conditions in this pipe. Then $K = 0.6486$. Since the Reynolds number for steam pressures above 100 lb/in² (689.4 kPa) ranges from 10^6 to 10^7, this assumption is safe because the value of K does not vary appreciably in this Reynolds number range. Also, the Reynolds number cannot be computed yet because the flow rate is unknown. Therefore, assumption of the Reynolds number is necessary. The assumption will be checked later.

3.   Determine the expansion factor and the meter area factor

Since steam is a compressible fluid, the expansion factor Y_1 must be determined. For superheated steam, the ratio of the specific heat at constant pressure c_p to the specific heat at constant volume c_v is $k = c_p/c_v = 1.3$. Also, the ratio of the differential maximum pressure reading h_w, in of water, to the maximum pressure in the pipe, lb/in² (abs) = 120/246.7 = 0.454. From the expansion-factor curve in the ASME *Fluid Meters*, $Y_1 = 0.994$ for $\beta = 0.5999$ and the pressure ratio =

0.454. And, from the same reference, the meter area factor $F_a = 1.0084$ for a steel meter operating at 640°F (338°C).

4. Compute the rate of steam flow

For square-edged orifices, the flow rate, lb/s $= w = 0.0997F_aKd^2Y_1(h_wd_s)^{0.5} =$ $(0.0997)(1.0084)(0.6486)(5.855)^2(0.994)(120 \times 0.4188)^{0.5} = 15.75$ lb/s (7.1 kg/s).

5. Compute the Reynolds number for the actual flow rate

For any steam pipe, the Reynolds number $R = 48w/(d_p gu_1) = 48(15.75)/$ $[(3.1416)(0.760)(0.0000141)] = 1,750,000$.

6. Adjust the flow coefficient for the actual Reynolds number

In step 2, $R = 10^7$ was assumed and $K = 0.6486$. For $R = 1,750,000$, $K = 0.6489$, from ASME *Fluid Meters*, by interpolation. Then the actual flow rate $w_h =$ (computed flow rate)(ratio of flow coefficients based on assumed and actual Reynolds numbers) $= (15.75)(0.6489/0.6486)(3.600) = 56,700$ lb/h (25,515 kg/h), closely, where the value 3600 is a conversion factor for changing lb/s to lb/h.

7. Compute the flow rate for a specific differential gage deflection

For a 40-in (101.6-cm) H_2O deflection, F_a is unchanged and equals 1.0084. The expansion factor changes because $h_w/p_p = 40/264.7 = 0.151$. From the ASME *Fluid Meters*, $Y_1 = 0.998$. By assuming again that $R = 10^7$, $K = 0.6486$, as before, $w = (0.0997)$ $(1.0084)(0.6486)(5.855)^2(0.998)(40 \times 0.4188)^{0.5} = 9.132$ lb/s (4.1 kg/s). Computing the Reynolds number as before, gives $R = (40)(0.132)/[(3.1416)(0.76)(0.0000141)] = 1,014,000$. The value of K corresponding to this value, as before, is from ASME—*Fluid Meters: K* $= 0.6497$. Therefore, the flow rate for a 40 in (101.6 cm) H_2O reading, in lb/h $= w_h = (0.132)(0.6497/0.6486)(3600)$ $= 32,940$ lb/h (14,823 kg/h).

Related Calculations: Use these steps and the ASME *Fluid Meters* or comprehensive meter engineering tables giving similar data to select or check an orifice meter used in any type of steam pipe—main, auxiliary, process, industrial, marine, heating, or commercial, conveying wet, saturated, or superheated steam.

SELECTION OF A PRESSURE-REGULATING VALVE FOR STEAM SERVICE

Select a single-seat spring-loaded diaphragm-actuated pressure-reducing valve to deliver 350 lb/ h (158 kg/h) of steam at 50 lb/in² (gage) (344.7 kPa) when the initial pressure is 225 lb/in² (gage) (1551 kPa). Also select an integral pilot-controlled piston-operated single-seat pressure-regulating valve to deliver 30,000 lb/h (13,500 kg/h) of steam at 40 lb/in² (gage) (275.8 kPa) with an initial pressure of 225 lb/in² (gage) (1551 kPa) saturated. What size pipe must be used on the downstream side of the valve to produce a velocity of 10,000 ft/min (3048 m/min)? How large should the pressure-regulating valve be if the steam entering the valve is at 225 lb/in² (gage) (1551 kPa) and 600°F (316°C)?

Calculation Procedure:

1. Compute the maximum flow for the diaphragm-actuated valve

For best results in service, pressure-reducing valves are selected so that they operate 60 to 70 percent open at normal load. To obtain a valve sized for this opening, divide the desired delivery, lb/h, by 0.7 to obtain the maximum flow expected. For this valve then, the maximum flow $= 350/0.7 = 500$ lb/h (225 kg/h).

2. Select the diaphragm-actuated valve size

Using a manufacturer's engineering data for an acceptable valve, enter the appropriate valve capacity table at the valve inlet steam pressure, 225 lb/in² (gage) (1551 kPa), and project to a capacity of 500 lb/h (225 kg/h), as in Table 13. Read the valve size as ¾ in (19.1 mm) at the top of the capacity column.

3. Select the size of the pilot-controlled pressure-regulating valve

Enter the capacity table in the engineering data of an acceptable pilot-controlled pressure-regulating valve, similar to Table 14, at the required capacity, 30,000 lb/h (13,500 kg/h). Project

TABLE 13 Pressure-Reducing-Valve Capacity, lb/h (kg/h)

Inlet pressure		Valve size		
lb/in^2 (gage)	kPa	½ in (12.7 mm)	¾ in (19.1 mm)	1 in (25.4 mm)
200	1379	420 (189)	460 (207)	560 (252)
225	1551	450 (203)	500 (225)	600 (270)
250	1724	485 (218)	560 (252)	650 (293)

Source: Clark-Reliance Corporation.

across until the correct inlet steam pressure column, 225 lb/in^2 (gage) (1551 kPa), is intercepted, and read the required valve size as 4 in (101.6 mm).

Note that it is not necessary to compute the maximum capacity before entering the table, as in step 1, for the pressure-reducing valve. Also note that a capacity table such as Table 14 can be used only for valves conveying saturated steam, unless the table notes state that the values listed are valid for other steam conditions.

4. *Determine the size of the downstream pipe*

Enter Table 14 at the required capacity, 30,000 lb/h (13,500 kg/h); project across to the valve *outlet pressure*, 40 lb/in^2 (gage) (275.8 kPa); and read the required pipe size as 8 in (203.2 mm) for a velocity of 10,000 ft/min (3048 m/min). Thus, the pipe immediately downstream from the valve must be enlarged from the valve size, 4 in (101.6 mm), to the required pipe size, 8 in (203.2 mm), to obtain the desired steam velocity.

5. *Determine the size of the valve handling superheated steam*

To determine the correct size of a pilot-controlled pressure-regulating valve handling superheated steam, a correction must be applied. Either a factor or a tabulation of corrected pressures, Table 15, may be used. To use Table 15, enter at the valve inlet pressure, 225 lb/in^2 (gage) (1551.2 kPa), and project across to the total temperature, 600°F (316°C), to read the corrected pressure, 165 lb/in^2 (gage) (1137.5 kPa). Enter Table 14 at the *next highest* saturated steam pressure, 175 lb/in^2 (gage) (1206.6 kPa); project down to the required capacity, 30,000 lb/h (13,500 kg/h); and read the required valve size as 5 in (127 mm).

Related Calculations: To simplify pressure-reducing and pressure-regulating valve selection, become familiar with two or three acceptable valve manufacturers' engineering data. Use the procedures given in the engineering data or those given here to select valves for industrial, marine, utility, heating, process, laundry, kitchen, or hospital service with a saturated or superheated steam supply.

Do not oversize reducing or regulating valves. Oversizing causes chatter and excessive wear.

When an anticipated load on the downstream side will not develop for several months after installation of a valve, fit to the valve a reduced-area disk sized to handle the present load. When the load increases, install a full-size disk. Size the valve for the ultimate load, not the reduced load.

Where there is a wide variation in demand for steam at the reduced pressure, consider installing two regulators piped in parallel. Size the smaller regulator to handle light loads and the larger

TABLE 14 Pressure-Regulating-Valve Capacity

Steam capacity		Initial steam pressure, saturated			
lb/h	kg/h	40 lb/in^2 (gage) (276 kPa)	175 lb/in^2 (gage) (1206 kPa)	225 lb/in^2 (gage) (1551 kPa)	300 lb/in^2 (gage) (2068 kPa)
20,000	9,000	6° (152.4)	4 (101.6)	4 (101.6)	3 (76.2)
30,000	13,500	8 (203.2)	5 (127.0)	4 (101.6)	4 (101.6)
40,000	18,000	—	5 (127.0)	5 (127.0)	4 (101.6)

°Valve diameter measured in inches (millimeters).
Source: Clark-Reliance Corporation.

TABLE 15 Equivalent Saturated Steam Values for Superheated Steam at Various Pressures and Temperatures

Steam pressure		Steam temperature		Total temperature					
				500°F	600°F	700°F	260.0°C	315.6°C	371.1°C
lb/in² (gage)	kPa	°F	°C	Steam values, lb/in² (gage)			Steam values, kPa		
205	1413.3	389	198	171	149	133	1178.9	1027.2	916.9
225	1551.2	397	203	190	165	147	1309.9	1137.5	1013.4
265	1826.9	411	211	227	200	177	1564.9	1378.8	1220.2

Source: Clark-Reliance Corporation.

regulator to handle the difference between 60 percent of the light load and the maximum heavy load. Set the larger regulator to open when the minimum allowable reduced pressure is reached. Then both regulators will be open to handle the heavy load. Be certain to use the actual regulator inlet pressure and not the boiler pressure when sizing the valve if this is different from the inlet pressure. Data in this calculation procedure are based on valves built by the Clark-Reliance Corporation, Cleveland, Ohio.

Some valve manufacturers use the valve flow coefficient C_v for valve sizing. This coefficient is defined as the flow rate, lb/h, through a valve of given size when the pressure loss across the valve is 1 lb/in² (6.89 kPa). Tabulations like Tables 13 and 14 incorporate this flow coefficient and are somewhat easier to use. These tables make the necessary allowances for downstream pressures less than the critical pressure (= 0.55 × absolute upstream pressure, lb/in², for superheated steam and hydrocarbon vapors; and 0.58 × absolute upstream pressure, lb/in², for saturated steam). The accuracy of these tabulations equals that of valve sizes determined by using the flow coefficient.

HYDRAULIC RADIUS AND LIQUID VELOCITY IN WATER PIPES

What is the velocity of 1000 gal/min (63.1 L/s) of water flowing through a 10-in (254-mm) inside-diameter cast-iron water main? What is the hydraulic radius of this pipe when it is full of water? When the water depth is 8 in (203.2 mm)?

Calculation Procedure:

1. Compute the water velocity in the pipe

For any pipe conveying water, the liquid velocity is v ft/s = gal/min/(2.448d^2), where d = internal pipe diameter, in. For this pipe, v = 1000/[2.448(100)] = 4.08 ft/s (1.24 m/s), or (60)(4.08) = 244.8 ft/min (74.6 m/min).

2. Compute the hydraulic radius for a full pipe

For any pipe, the hydraulic radius is the ratio of the cross-sectional area of the pipe to the wetted perimeter, or $d/4$. For this pipe, when full of water, the hydraulic radius = 10/4 = 2.5.

3. Compute the hydraulic radius for a partially full pipe

Use the hydraulic radius tables in King and Brater—*Handbook of Hydraulics*, or compute the wetted perimeter by using the geometric properties of the pipe, as in step 2. From the King and Brater table, the hydraulic radius = Fd, where F = table factor for the ratio of the depth of water, in/diameter of channel, in = 8/10 = 0.8. For this ratio, F = 0.304. Then, hydraulic radius = (0.304)(10) = 3.04 in (77.2 mm).

Related Calculations: Use this method to determine the water velocity and hydraulic radius in any pipe conveying cold water—water supply, plumbing, process, drain, or sewer.

TABLE 16   Values of C in Hazen-Williams Formula

Type of pipe	$C°$	Type of pipe	$C°$
Cement-asbestos	140	Cast iron or wrought iron	100
Asphalt-lined iron or steel	140	Welded or seamless steel	100
Copper or brass	130	Concrete	100
Lead, tin, or glass	130	Corrugated steel	60
Wood stave	110		

°Values of C commonly used for design. The value of C for pipes made of corrosive materials decreases as the age of the pipe increases; the values given are those that apply at an age of 15 to 20 years. For example, the value of C for cast-iron pipes 30 in (762 mm) in diameter or greater at various ages is approximately as follows: new, 130; 5 years old, 120; 10 years old, 115; 20 years old, 100; 30 years old, 90; 40 years old, 80; and 50 years old, 75. The value of C for smaller-size pipes decreases at a more rapid rate.

FRICTION-HEAD LOSS IN WATER PIPING OF VARIOUS MATERIALS

Determine the friction-head loss in 2500 ft (762 m) of clean 10-in (254-mm) new tar-dipped cast-iron pipe when 2000 gal/min (126.2 L/s) of cold water is flowing. What is the friction-head loss 20 years later? Use the Hazen-Williams and Manning formulas, and compare the results.

Calculation Procedure:

1. *Compute the friction-head loss by the Hazen-Williams formula*

The Hazen-Williams formula is $h_f = [v/(1.318CR_h^{0.63})]^{1.85}$, where h_f = friction-head loss per ft of pipe, ft of water; v = water velocity, ft/s; C = a constant depending on the condition and kind of pipe; R_h = hydraulic radius of pipe, ft.

For a water pipe, v = gal/min/$(2.44d^2)$; for this pipe, $v = 2000/[2.448(10)^2] = 8.18$ ft/s (2.49 m/s). From Table 16 or Crocker and King—*Piping Handbook*, C for new pipe = 120; for 20-year-old pipe, C = 90; $R_h = d/4$ for a full-flow pipe = 10/4 = 2.5 in, or 2.5/12 = 0.208 ft (63.4 mm). Then $h_f = [8.18/(1.318 \times 120 \times 0.208^{0.63})]^{1.85} = 0.0263$ ft (8.0 mm) of water per ft (m) of pipe. For 2500 ft (762 m) of pipe, the total friction-head loss = 2500(0.0263) = 65.9 ft (20.1 m) of water for the new pipe.

For 20-year-old pipe and the same formula, except with C = 90, h_f = 0.0451 ft (13.8 mm) of water per ft (m) of pipe. For 2500 ft (762 m) of pipe, the total friction-head loss = 2500(0.0451) = 112.9 ft (34.4 m) of water. Thus, the friction-head loss nearly doubles [from 65.9 to 112.9 ft (20.1 to 34.4 m)] in 20 years. This shows that it is wise to design for future friction losses; otherwise, pumping equipment may become overloaded.

2. *Compute the friction-head loss from the Manning formula*

The Manning formula is $h_f = n^2v^2/2.208R_h^{4/3}$, where n = a constant depending on the condition and kind of pipe; other symbols as before.

Using n = 0.011 for new coated cast-iron pipe from Table 17 or Crocker and King—*Piping Handbook*, we find $h_f = (0.011)^2(8.18)^2/[2.208(0.208)^{4/3}] = 0.0295$ ft (8.9 mm) of water per ft (m) of pipe. For 2500 ft (762 m) of pipe, the total friction-head loss = 2500(0.0295) = 73.8 ft (22.5 m) of water, as compared with 65.9 ft (20.1 m) of water computed with the Hazen-Williams formula.

For coated cast-iron pipe in fair condition, n = 0.013, and h_f = 0.0411 ft (12.5 mm) of water. For 2500 ft (762 m) of pipe, the total friction-head loss = 2500(0.0411) = 102.8 ft (31.3 m) of water, as compared with 112.9 ft (34.4 m) of water computed with the Hazen-Williams formula. Thus, the Manning formula gives results higher than the Hazen-Williams in one case and lower in another. However, the differences in each case are not excessive; (73.8 − 65.9)/65.9 = 0.12, or 12 percent higher, and (112.9 − 102.8)/102.8 = 0.0983, or 9.83 percent lower. Both these differences are within the normal range of accuracy expected in pipe friction-head calculations.

TABLE 17   Roughness Coefficients (Manning's n) for Closed Conduits

	Manning's n	
Type of conduit	Good construction°	Fair construction°
Concrete pipe	0.013	0.015
Corrugated metal pipe or pipe arch, 2⅔ × ½ in (67.8 × 12.7 mm) corrugation, riveted:		
Plain	0.024	

Paved invert:				
Percent of circumference paved	25	50		
Depth of flow:				
Full	0.021	0.018		
0.8D	0.021	0.016		
0.6D	0.019	0.013		

Type of conduit	Good construction°	Fair construction°
Vitrified clay pipe	0.012	0.014
Cast-iron pipe, uncoated	0.013	
Steel pipe	0.011	
Brick	0.014	0.017
Monolithic concrete:		
Wood forms, rough	0.015	0.017
Wood forms, smooth	0.012	0.014
Steel forms	0.012	0.013
Cemented-rubble masonry walls:		
Concrete floor and top	0.017	0.022
Natural floor	0.019	0.025
Laminated treated wood	0.015	0.017
Vitrified-clay liner plates	0.015	

°For poor-quality construction, use larger values of n.

Related Calculations: The Hazen-Williams and Manning formulas are popular with many piping designers for computing pressure losses in cold-water piping. To simplify calculations, most designers use the precomputed tabulated solutions available in Crocker and King—*Piping Handbook*, King and Brater—*Handbook of Hydraulics*, and similar publications. In the rush of daily work these precomputed solutions are also preferred over the more complex Darcy-Weisbach equation used in conjunction with the friction factor f, the Reynolds number R, and the roughness-diameter ratio.

Use the method given here for sewer lines, water-supply pipes for commercial, industrial, or process plants, and all similar applications where cold water at temperatures of 33 to 90°F (0.6 to 32.2°C) flows through a pipe made of cast iron, riveted steel, welded steel, galvanized iron, brass, glass, wood-stove, concrete, vitrified, common clay, corrugated metal, unlined rock, or enameled steel. Thus, either of these formulas, used in conjunction with a suitable constant, gives the friction-head loss for a variety of piping materials. Suitable constants are given in Tables 16 and 17 and in the above references. For the Hazen-Williams formula, the constant C varies from about 70 to 140, while n in the Manning formula varies from about 0.017 for $C = 70$ to 0.010 for $C = 140$. Values obtained with these formulas have been used for years with satisfactory results. At present, the Manning formula appears the more popular.

CHART AND TABULAR DETERMINATION OF FRICTION HEAD

Figure 5 shows a process piping system supplying 1000 gal/min (63.1 L/s) of 70°F (21.1°C) water. Determine the total friction head, using published charts and pipe-friction tables. All the valves and fittings are flanged, and the piping is 10-in (254-mm) steel, schedule 40.

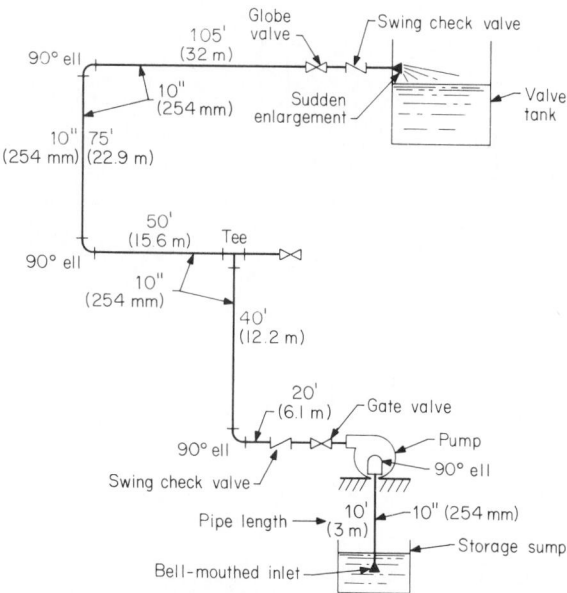

FIG. 5 Typical industrial piping system.

Calculation Procedure:

1. Determine the total length of the piping

Mark the length of each piping run on the drawing after scaling it or measuring it in the field. Determine the total length by adding the individual lengths, starting at the supply source of the liquid. In Fig. 5, beginning at the storage sump, the total length of piping = 10 + 20 + 40 + 50 + 75 + 105 = 300 ft (91.4 m). Note that the physical length of the fittings is included in the length of each run.

2. Compute the equivalent length of each fitting

The frictional resistance of pipe fittings (elbows, tees, etc.) and valves is greater than the actual length of each fitting. Therefore, the equivalent length of straight piping having a resistance equal to that of the fittings must be determined. This is done by finding the equivalent length of each fitting and taking the sum for all the fittings.

Use the equivalent length table in the pump section of this handbook or in Crocker and King— *Piping Handbook*, Baumeister and Marks—*Standard Handbook for Mechanical Engineers*, or *Standards of the Hydraulic Institute*. Equivalent length values will vary slightly from one reference to another.

Starting at the supply source, as in step 1, for 10-in (254-mm) flanged fittings throughout, we see the equivalent fitting lengths are: bell-mouth inlet, 2.9 ft (0.88 m); 90° ell at pump, 14 ft (4.3 m); gate valve, 3.2 ft (0.98 m); swing check valve, 120 ft (36.6 m); 90° ell, 14 ft (4.3 m); tee, 30 ft (9.1 m); 90° ell, 14 ft (4.3 m); 90° ell, 14 ft (4.3 m); globe valve, 310 ft (94.5 m); swing check valve, 120 ft (36.6 m); sudden enlargement = (liquid velocity, ft/s)^2/2g = $(4.07)^2/2(32.2)$ = 0.257 ft (0.08 m), where the terminal velocity is zero, as in the tank. Find the liquid velocity as shown in a previous calculation procedure in this section. The sum of the fitting equivalent lengths is 2.9 + 14 + 3.2 + 120 + 14 + 30 + 14 + 14 + 310 + 120 + 0.257 = 642.4 ft (159.8 m). Adding this to the straight length gives a total length of 642.4 + 300 = 942.4 ft (287.2 m).

3. Compute the friction-head loss by using a chart

Figure 6 is a popular friction-loss chart for fairly rough pipe, which is any ordinary pipe after a few years' use. Enter at the left at a flow of 1000 gal/min (63.1 L/s), and project to the right until

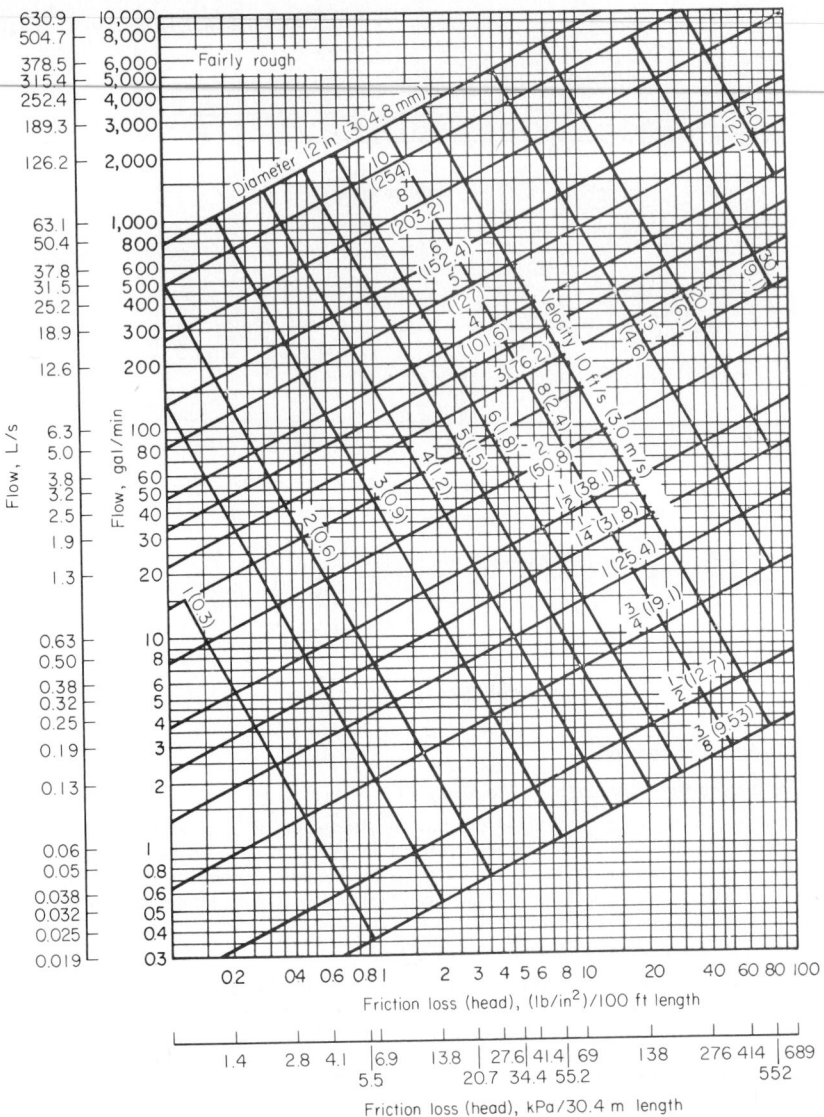

FIG. 6 Friction loss in water piping.

TABLE 18 Absolute Roughness Classification of Pipe Surfaces for Selection of Friction Factor f in Fig. 7

Commercial pipe surface (new)	Absolute roughness ϵ		Commercial pipe surface (new)	Absolute roughness ϵ	
	ft	mm		ft	mm
Glass, drawn brass, copper, lead	Smooth	Smooth	Cast iron	0.00085	0.26
Wrought iron, steel	0.00015	0.05	Wood stave	0.0006–0.003	0.18–0.91
Asphalted cast iron	0.0004	0.12	Concrete	0.001–0.01	0.30–3.05
Galvanized iron	0.0005	0.15	Riveted steel	0.003–0.03	0.91–9.14

the 10-in (254-mm) diameter curve is intersected. Read the friction-head loss at the top or bottom of the chart as 0.4 lb/in² (2.8 kPa), closely, per 100 ft (30.5 m) of pipe. Therefore, total friction-head loss = $(0.4)(942.4/100)$ = 3.77 lb/in² (26 kPa). Converting gives $(3.77)(2.31)$ = 8.71 ft (2.7 m) of water.

4. *Compute the friction-head loss from tabulated data*

Using the *Standards of the Hydraulic Institute* pipe-friction table, we find that the friction head h_f of water per 100 ft (30.5 m) of pipe = 0.500 ft (0.15 m). Hence, the total friction head = $(0.500)(942.4/100)$ = 4.71 ft (1.4 m) of water. The Institute recommends that 15 percent be added to the tabulated friction head, or $(1.15)(4.71)$ = 5.42 ft (1.66 m) of water.

Using the friction-head tables in Crocker and King—*Piping Handbook*, the friction head = 6.27 ft (1.9 m) per 1000 ft (304.8 m) of pipe with C = 130 for new, very smooth pipe. For this piping system, the friction-head loss = $(942.4/1000)(6.27)$ = 5.91 ft (1.8 m) of water.

5. *Use the Reynolds number method to determine the friction head*

In this method, the friction factor is determined by using the Reynolds number R and the relative roughness of the pipe ϵ/D, where ϵ = pipe roughness, ft, and D = pipe diameter, ft.

For any pipe, $R = Dv/\nu$, where v = liquid velocity, ft/s, and ν = kinematic viscosity, ft²/s. Using King and Brater—*Handbook of Hydraulics*, v = 4.07 ft/s (1.24 m/s), and ν = 0.00001059 ft²/s (0.00000098 m²/s) for water at 70°F (21.1°C). Then $R = (10/12)(4.07)/0.00001059$ = 320,500.

From Table 18 or the above reference, ϵ = 0.00015, and ϵ/D = 0.00015/(10/12) = 0.00018.

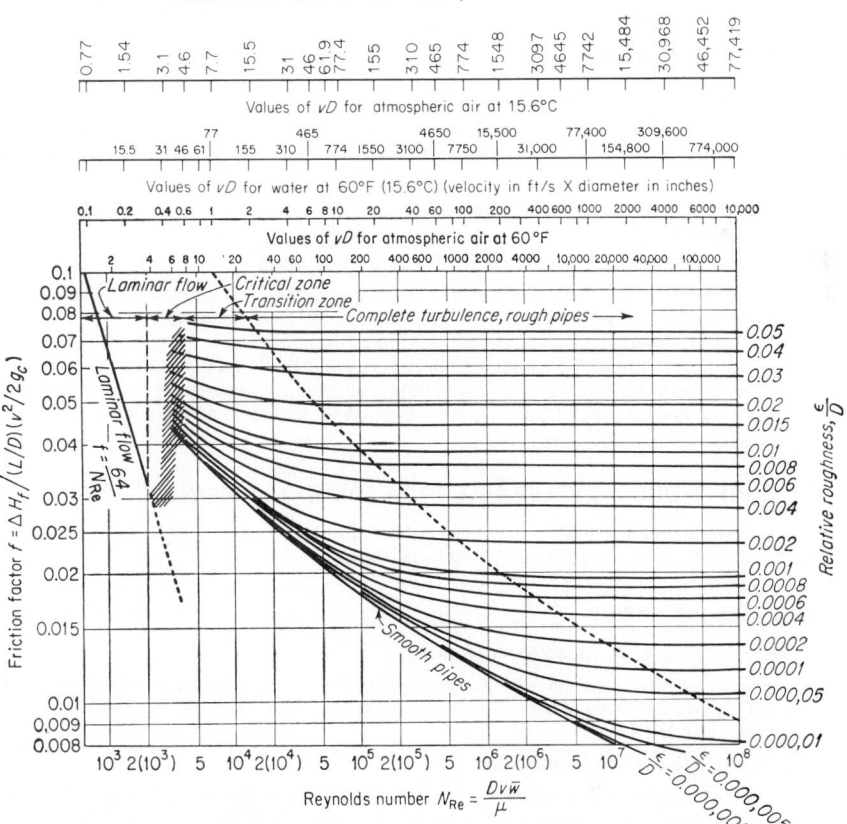

FIG. 7   Friction factors for laminar and turbulent flow.

From the Reynolds-number, relative-roughness, friction-factor curve in Fig. 7 or in Baumeister—*Standard Handbook for Mechanical Engineers*, the friction factor $f = 0.016$.

Apply the Darcy-Weisbach equation $h_f = f(l/D)(v^2/2g)$, where l = total pipe length, including the fittings' equivalent length, ft. Then $h_f = (0.016)(942.4/10/12)(4.07)^2/(2 \times 32.2) = 4.651$ ft (1.43 m) of water.

6. Compare the results obtained

Three different friction-head values were obtained: 8.71, 5.91, and 4.651 ft (2.7, 1.8, and 1.4 m) of water. The results show the variations that can be expected with the different methods. Actually, the Reynolds number method is probably the most accurate. As can be seen, the other two methods give safe results—i.e., the computed friction head is higher. The *Pipe Friction Manual*, published by the Hydraulic Institute, presents excellent simplified charts for use with the Reynolds number method.

Related Calculations: Use any of these methods to compute the friction-head loss for any type of pipe. The Reynolds number method is useful for a variety of liquids other than water—mercury, gasoline, brine, kerosene, crude oil, fuel oil, and lube oil. It can also be used for saturated and superheated steam, air, methane, and hydrogen.

RELATIVE CARRYING CAPACITY OF PIPES

What is the equivalent steam-carrying capacity of a 24-in (609.6-mm) inside-diameter pipe in terms of a 10-in (254-mm) inside-diameter pipe? What is the equivalent water-carrying capacity of a 23-in (584.2-mm) inside-diameter pipe in terms of a 13.25-in (336.6-mm) inside-diameter pipe?

Calculation Procedure:

1. Compute the relative carrying capacity of the steam pipes

For steam, air, or gas pipes, the number N of small pipes of inside diameter d_2 in equal to one pipe of larger inside diameter d_1 in is $N = (d_1^3\sqrt{d_2 + 3.6})/(d_2^3 + \sqrt{d_1 + 3.6})$. For this piping system, $N = (24^3 + \sqrt{10} + 3.6)/(10^3 + \sqrt{24} + 3.6) = 9.69$, say 9.7. Thus, a 24-in (609.6-mm) inside-diameter steam pipe has a carrying capacity equivalent to 9.7 pipes having a 10-in (254-mm) inside diameter.

2. Compute the relative carrying capacity of the water pipes

For water, $N = (d_2/d_1)^{2.5} = (23/13.25)^{2.5} = 3.97$. Thus, one 23-in (584-cm) inside-diameter pipe can carry as much water as 3.97 pipes of 13.25-in (336.6-mm) inside diameter.

Related Calculations: Crocker and King—*Piping Handbook* and certain piping catalogs (Crane, Walworth, National Valve and Manufacturing Company) contain tabulations of relative carrying capacities of pipes of various sizes. Most piping designers use these tables. However, the equations given here are useful for ranges not covered by the tables and when the tables are unavailable.

PRESSURE-REDUCING VALVE SELECTION FOR WATER PIPING

What size pressure-reducing valve should be used to deliver 1200 gal/h (1.26 L/s) of water at 40 lb/in^2 (275.8 kPa) if the inlet pressure is 140 lb/in^2 (965.2 kPa)?

Calculation Procedure:

1. Determine the valve capacity required

Pressure-reducing valves in water systems operate best when the nominal load is 60 to 70 percent of the maximum load. Using 60 percent, we see that the maximum load for this valve = 1200/0.6 = 2000 gal/h (2.1 L/s).

2. Determine the valve size required

Enter a valve capacity table in suitable valve engineering data at the valve inlet pressure, and project to the exact, or next higher, valve capacity. Thus, enter Table 19 at 140 lb/in^2 (965.2 kPa)

TABLE 19   Maximum Capacities of Water Pressure-Reducing Valves, gal/h (L/s)

Inlet pressure		Valve size		
lb/in² (gage)	kPa	¾ in (19.1 mm)	1 in (25.4 mm)	1¼ in (31.8 mm)
120	827.3	1550 (1.6)	2000 (2.1)	4500 (4.7)
140	965.2	1700 (1.8)	2200 (2.3)	5000 (5.3)
160	1103.0	1850 (1.9)	2400 (2.5)	5500 (5.8)

Source: Clark-Reliance Corporation.

and project to the next higher capacity, 2200 gal/h (2.3 L/s), since a capacity of 2000 gal/h (2.1 L/s) is not tabulated. Read at the top of the column the required valve size as 1 in (25.4 mm).

Some valve manufacturers present the capacity of their valves in graphical instead of tabular form. One popular chart, Fig. 8, is entered at the difference between the inlet and outlet pressures on the abscissa, or 140 − 40 = 100 lb/in² (689.4 kPa). Project vertically to the flow rate of 2000/60 = 33.3 gal/min (2.1 L/s). Read the valve size on the intersecting valve capacity curve, or on the next curve if there is no intersection with the curve. Figure 8 shows that a 1-in (25.4-mm) valve should be used. This agrees with the tabulated capacity.

Related Calculations: Use this method for pressure-reducing valves in any type of water piping—process, domestic, commercial—where the water temperature is 100°F (37.8°C) or less. Table 19 is from data prepared by the Clark-Reliance Corporation; Fig. 8 is from Foster Engineering Company data.

Some valve manufacturers use the valve flow coefficient C_v for valve sizing. This coefficient is defined as the flow rate, gal/min, through a valve of given size when the pressure loss across the valve is 1 lb/in² (6.9 kPa). Tabulations like Table 19 and flowcharts like Fig. 8 incorporate this flow coefficient and are somewhat easier to use. Their accuracy equals that of the flow coefficient method.

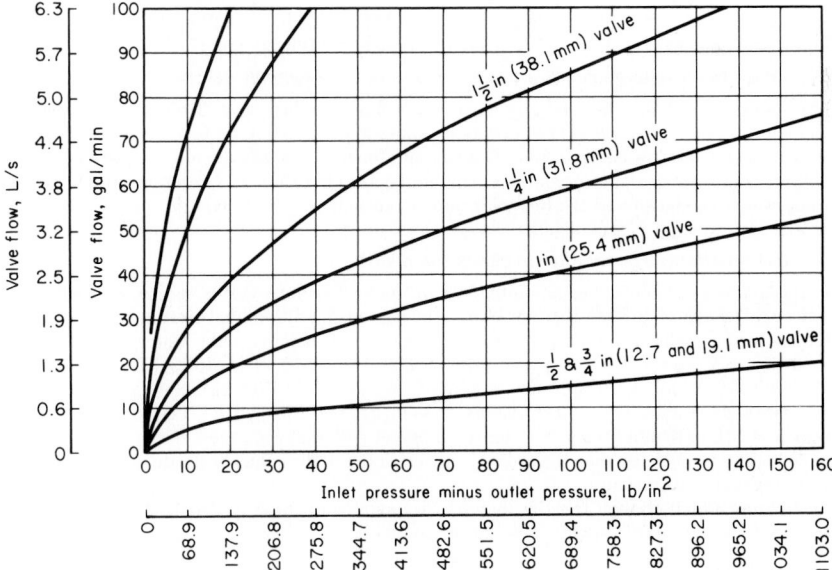

FIG. 8   Pressure-reducing valve flow capacity. *(Foster Engineering Company.)*

SIZING A WATER METER

A 6 × 4 in (152.4 × 101.6 mm) Venturi tube is used to measure water flow rate in a piping system. The dimensions of the meter are: inside pipe diameter d_p = 6.094 in (154.8 mm); throat diameter d = 4.023 in (102.2 mm). The differential pressure is measured with a mercury manometer having water on top of the mercury. The average manometer reading for 1 h is 10.1 in (256.5 mm) of mercury. The temperature of the water in the pipe is 41°F (5.0°C), and that of the room is 77°F (25°C). Determine the water flow rate in lb/h, gal/h, and gal/min. Use the ASME Research Committee on Fluid Meters method in analyzing the meter.

Calculation Procedure:

1. Convert the pressure reading to standard conditions

The ASME meter equation constant is based on a manometer liquid temperature of 68°F (20.0°C). Therefore, the water and mercury density at room temperature, 77°F (25°C), and the water density at 68°F (20.0°C), must be used to convert the manometer reading to standard conditions by the equation $h_w = h_m(m_d - w_d)/w_s$, where h_w = equivalent manometer reading, in (mm) H_2O at 68°F (20.0°C); h_m = manometer reading at room temperature, in mercury; m_d = mercury density at room temperature, lb/ft³; w_d = water density at room temperature, lb/ft³; w_s = water density at standard conditions, 68°F (20.0°C), lb/ft³. From density values from the ASME publication *Fluid Meters: Their Theory and Application*, h_w = 10.1(844.88 − 62.244)/62.316 = 126.8 in (322.1 cm) of water at 68°F (20.0°C).

2. Determine the throat-to-pipe diameter ratio

The throat-to-pipe diameter ratio β = 4.023/6.094 = 0.6602. Then $1/(1 - \beta^4)^{0.5} = 1/(1 - 0.6602^4)^{0.5}$ = 1.1111.

3. Assume a Reynolds number value, and compute the flow rate

The flow equation for a Venturi tube is w lb/h = $359.0(Cd^2/\sqrt{1 - \beta^4})(w_{dp}h_w)^{0.5}$, where C = meter discharge coefficient, expressed as a function of the Reynolds number; w_{dp} = density of the water at the pipe temperature, lb/ft³. With a Reynolds number greater than 250,000, C is a constant. As a first trial, assume $R > 250,000$ and C = 0.984 from *Fluid Meters*. Then w = 359.0(0.984)(4.023)²(1.1111)(62.426 × 126.8)^{0.5} = 565,020 lb/h (254,259 kg/h), or 565,020/8.33 lb/gal = 67,800 gal/h (71.3 L/s), or 67,800/60 min/h = 1129 gal/min (71.23 L/s).

4. Check the discharge coefficient by computing the Reynolds numbers

For a water pipe, $R = 48w_s/(\pi d_p g u)$, where w_s = flow rate, lb/s = $w/3600$; u = coefficient of absolute viscosity. Using *Fluid Meters* data for water at 41°F (5°C), we find R = 48(156.95)/[(π × 6.094)(0.001004)] = 391,900. Since C is constant for $R > 250,000$, use of C = 0.984 is correct, and no adjustment in the computations is necessary. Had the value of C been incorrect, another value would be chosen and the Reynolds number recomputed. Continue this procedure until a satisfactory value for C is obtained.

5. Use an alternative solution to check the results

Fluid Meters gives another equation for Venturi meter flow rate, that is, w lb/s = $0.525(Cd^2/\sqrt{1 - \beta^4})[w_{dp}(p_1 - p_2)]^{0.5}$, where $p_1 - p_2$ is the manometer differential pressure in lb/in². Using the conversion factor in *Fluid Meters* for converting in of mercury under water at 77°F (25°C) to lb/in² (kPa), we get $p_1 - p_2$ = (10.1)(0.4528) = 4.573 lb/in² (31.5 kPa). Then w = (0.525)(0.984)(4.023)²(1.1111)(62.426 × 4.573)^{0.5} = 156.9 lb/s (70.6 kg/s), or (156.9)(3600 s/h) = 564,900 lb/h (254,205 kg/h), or 564,900/8.33 lb/gal = 67,800 gal/h (71.3 L/s), or 67,800/60 min/h = 1129 gal/min (71.2 L/s). This result agrees with that computed in step 3 within 1 part in 5600. This is much less than the probable uncertainties in the values of the discharge coefficient and the differential pressure.

Related Calculations: Use this method for any Venturi tube serving cold-water piping in process, industrial, water-supply, domestic, or commercial service.

EQUIVALENT LENGTH OF A COMPLEX SERIES PIPELINE

Figure 9 shows a complex series pipeline made up of four lengths of different size pipe. Determine the equivalent length of this pipe if each size of pipe has the same friction factor.

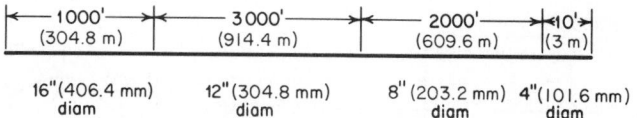

FIG. 9 Complex series pipeline.

Calculation Procedure:

1. *Select the pipe size for expressing the equivalent length*

The usual procedure in analyzing complex pipelines is to express the equivalent length in terms of the smallest, or next to smallest, diameter pipe. Choose the 8-in (203.2-mm) size as being suitable for expressing the equivalent length.

2. *Find the equivalent length of each pipe*

For any complex series pipeline having equal friction factors in all the pipes, L_e = equivalent length, ft, of a section of constant diameter = (actual length of section, ft) (inside diameter, in, of pipe used to express the equivalent length/inside diameter, in, of section under consideration)^5.

For the 16-in (406.4-mm) pipe, $L_e = (1000)(7.981/15.000)^5 = 42.6$ ft (12.9 m). The 12-in (304.8-mm) pipe is next; for it $L_e = (3000)(7.981/12.00)^5 = 390$ ft (118.9 m). For the 8-in (203.2-mm) pipe, the equivalent length = actual length = 2000 ft (609.6 m). For the 4-in (101.6-mm) pipe, $L_e = (10)(7.981/4.026)^5 = 306$ ft (93.3 m). Then the total equivalent length of 8-in (203.2-mm) pipe = sum of the equivalent lengths = $42.6 + 390 + 2000 + 306 = 2738.6$ ft (834.7 m); or, by rounding off, 2740 ft (835.2 m) of 8-in (203.2-mm) pipe will have a frictional resistance equal to the complex series pipeline shown in Fig. 9. To compute the actual frictional resistance, use the methods given in previous calculation procedures.

Related Calculations: Use this general procedure for any complex series pipeline conveying water, oil, gas, steam, etc. See Crocker and King—*Piping Handbook* for derivation of the flow equations. Use the tables in Crocker and King to simplify finding the fifth power of the inside diameter of a pipe. The method of the next calculation procedure can also be used if a given flow rate is assumed.

Choosing a flow rate of 1000 gal/min (63.1 L/s) and using the tables in the Hydraulic Institute *Pipe Friction Manual* give an equivalent length of 2770 ft (844.3 m) for the 8-in (203.2-mm) pipe. This compares favorably with the 2740 ft (835.2 m) computed above. The difference of 30 ft (9.1 m) is negligible and can be accounted for by calculator variations.

The equivalent length is found by summing the friction-head loss for 1000-gal/min (63.1-L/s) flow for each length of the four pipes—16, 12, 8, and 4 in (406, 305, 203, and 102 mm)—and dividing this by the friction-head loss for 1000 gal/min (63.1 L/s) flowing through an 8-in (203.2-mm) pipe. Be careful to observe the units in which the friction-head loss is stated, because errors are easy to make if the units are ignored.

EQUIVALENT LENGTH OF A PARALLEL PIPING SYSTEM

Figure 10 shows a parallel piping system used to supply water for industrial needs. Determine the equivalent length of a single pipe for this system. All pipes in the system are approximately horizontal.

Calculation Procedure:

1. *Assume a total head loss for the system*

To determine the equivalent length of a parallel piping system, assume a total head loss for the system. Since this head loss is assumed for computation purposes only, its value need not be exact or even approximate. Assume a total head loss of 50 ft of water for each pipe in this system.

2. *Compute the flow rate in each pipe in the system*

Assume that the roughness coefficient C in the Hazen-Williams formula is equal for each of the pipes in the system. This is a valid assumption. Using the assumed value of C, compute the flow rate in each pipe. To allow for possible tuberculation of the pipe, assume that $C = 100$.

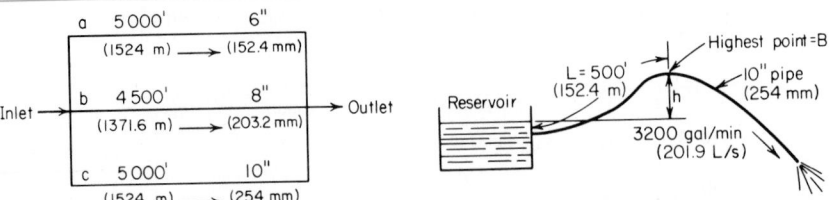

FIG. 10 Parallel piping system.

FIG. 11 Liquid siphon piping system.

The Hazen-Williams formula is given in a previous calculation procedure and can be used to solve for the flow rate in each pipe. A more rapid way to make the computation is to use the friction-loss tabulations for the Hazen-Williams formula in Crocker and King—*Piping Handbook*, the Hydraulic Institute—*Pipe Friction Manual*, or a similar set of tables.

Using such a set of tables, enter at the friction-head loss equal to 50 ft (15.2 m) per 5000 ft (1524 m) of pipe for the 6-in (152.4-mm) line. Find the corresponding flow rate Q gal/min. Using the Hydraulic Institute tables, $Q_a = 270$ gal/min (17.0 L/s); $Q_b = 580$ gal/min (36.6 L/s); $Q_c = 1000$ gal/min (63.1 L/s). Hence, the total flow = 270 + 580 + 1000 = 1850 gal/min (116.7 L/s).

3. Find the equivalent size and length of the pipe

Using the Hydraulic Institute tables again, look for a pipe having a 50-ft (15.2 m) head loss with a flow of 1850 gal/min (116.7 L/s). Any pipe having a discharge equal to the sum of the discharge rates for all the pipes, at the assumed friction head, is an equivalent pipe.

Interpolating friction-head values in the 14-in (355.6-mm) outside-diameter [13.126-in (333.4-mm) inside-diameter] table shows that 5970 ft (1820 m) of this pipe is equivalent to the system in Fig. 10. This equivalent size can be used in any calculations related to this system—selection of a pump, determination of head loss with longer or shorter mains, etc. If desired, another equivalent-size pipe could be found by entering a different pipe-size table. Thus, 5310 ft (1621.5 m) of 14-in (355.6-mm) pipe [12.814-in (326.5-mm) inside diameter] is also equivalent to this system.

Related Calculations: Use this procedure for any liquid—water, oil, gasoline, brine—flowing through a parallel piping system. The pipes are assumed to be full at all times.

MAXIMUM ALLOWABLE HEIGHT FOR A LIQUID SIPHON

What is the maximum height h ft (m), Fig. 11, that can be used for a siphon in a water system if the length of the pipe from the water source to its highest point is 500 ft (152.4 m), the water velocity is 13.0 ft/s (3.96 m/s), the pipe diameter is 10 in (254 mm), and the water temperature is 70°F (21.1°C) if 3200 gal/min (201.9 L/s) is flowing?

Calculation Procedure:

1. Compute the velocity of the water in the pipe

From an earlier calculation procedure, $v = gpm/(2.448d^2)$. With an internal diameter of 10.020 in (254.5 mm), $v = 3200/[(2.448)(10.02)^2] = 13.0$ ft/s (3.96 m/s).

2. Determine the vapor pressure of the water

Using a steam table, we see that the vapor pressure of water at 70°F (21.1°C) is $p_v = 0.3631$ lb/in² (abs) (2.5 kPa), or (0.3631) (144 in²/ft²) = 52.3 lb/ft² (2.5 kPa). The specific volume of water at 70°F (21.1°C) is, from a steam table, 0.01606 ft³/lb (0.001 m³/kg). Converting this to density at 70°F (21.1°C), density = 1/0.01606 = 62.2 lb/ft³ (995.8 kg/m³). The vapor pressure in ft of 70°F (21.1°C) water is then $f_v = (52.3$ lb/ft²)/(62.2 lb/ft³) = 0.84 ft (0.26 m) of water.

3. Compute or determine the friction-head loss and velocity head

From the reservoir to the highest point of the siphon, B, Fig. 11, the friction head in the pipe must be overcome. Use the Hazen-Williams or a similar formula to determine the friction head, as given in earlier calculation procedures or a pipe-friction table. From the Hydraulic Institute

Pipe Friction Manual, h_f = 4.59 ft per 100 ft (1.4 m per 30.5 m), or (500/100)(4.59) = 22.95 ft (7.0 m). From the same table, velocity head = 2.63 ft/s (0.8 m/s).

4. Determine the maximum height for the siphon

For a siphon handling water, the maximum allowable height h at sea level with an atmospheric pressure of 14.7 lb/in^2 (abs) (101.3 kPa) = [14.7 × (144 in^2/ft^2)/(density of water at operating temperature, lb/ft^3) − (vapor pressure of water at operating temperature, ft + 1.5 × velocity head, ft + friction head, ft)]. For this pipe, h = 14.7 × 144/62.2 − (0.84 + 1.5 × 2.63 + 22.95) = 11.32 ft (3.45 m). In actual practice, the value of h is taken as 0.75 to 0.8 the computed value. Using 0.75 gives h = (0.75)(11.32) = 8.5 ft (2.6 m).

Related Calculations: Use this procedure for any type of siphon conveying a liquid—water, oil, gasoline, brine, etc. Where the liquid has a specific gravity different from that of water, i.e., less than or greater than 1.0, proceed as above, expressing all heads in ft of liquid handled. Divide the resulting siphon height by the specific gravity of the liquid. At elevations above atmospheric, use the actual atmospheric pressure instead of 14.7 lb/in^2 (abs) (101.3 kPa).

WATER-HAMMER EFFECTS IN LIQUID PIPELINES

What is the maximum pressure developed in a 200-lb/in^2 (1378.8-kPa) water pipeline if a valve is closed nearly instantly or pumps discharging into the line are all stopped at the same instant? The pipe is 8-in (203.2-mm) schedule 40 steel, and the water flow rate is 2800 gal/min (176.7 L/s). What maximum pressure is developed if the valve closes in 5 s and the line is 5000 ft (1524 m) long?

Calculation Procedure:

1. Determine the velocity of the pressure wave

For any pipe, the velocity of the pressure wave during water hammer is found from v_w = 4720/$(1 + Kd/Et)^{0.5}$, where v_w = velocity of the pressure wave in the pipeline, ft/s; K = bulk modulus of the liquid in the pipeline = 300,000 for water; d = internal diameter of pipe, in; E = modulus of elasticity of pipe material, lb/in^2 = 30 × 10^6 lb/in^2 (206.8 GPa) for steel; t = pipe-wall thickness, in. For 8-in (203.2-mm) schedule 40 steel pipe and data from a table of pipe properties, v_w = 4720/[1 + 300,000 × 7.981/(30 × 10^6 × 0.322)]$^{0.5}$ = 4225.6 ft/s (1287.9 m/s).

2. Compute the pressure increase caused by water hammer

The pressure increase p_1 lb/in^2 due to water hammer = $v_w v$/[32.2(2.31)], where v = liquid velocity in the pipeline, ft/s; 32.2 = acceleration due to gravity, ft/s^2; 2.31 ft of water = 1-lb/in^2 (6.9-kPa) pressure.

For this pipe, v = 0.4085gpm/d^2 = 0.4085(2800)/(7.981)^2 = 18.0 ft/s (5.5 m/s). Then p_i = (4225.6)(18)/[32.2(2.31)] = 1022.56 lb/in^2 (7049.5 kPa). The maximum pressure developed in the pipe is then p_i + pipe operating pressure = 1022.56 + 200 = 1222.56 lb/in^2 (8428.3 kPa).

3. Compute the hammer pressure rise caused by valve closure

The hammer pressure rise caused by valve closure p_v lb/in^2 = $2p_iL/v_wT$, where L = pipeline length, ft; T = valve closing time, s. For this pipeline, p_v = 2(1022.56)(5000)/[(4225.6)(5)] = 484 lb/in^2 (3336.7 kPa). Thus, the maximum pressure in the pipe will be 484 + 200 = 648 lb/in^2 (4467.3 kPa).

Related Calculations: Use this procedure for any type of liquid—water, oil, etc.—in a pipeline subject to sudden closure of a valve or stoppage of a pump or pumps. The effects of water hammer can be reduced by relief valves, slow-closing check valves on pump discharge pipes, air chambers, air spill valves, and air injection into the pipeline.

SPECIFIC GRAVITY AND VISCOSITY OF LIQUIDS

An oil has a specific gravity of 0.8000 and a viscosity of 200 SSU (Saybolt Seconds Universal) at 60°F (15.6°C). Determine the API gravity and Bé gravity of this oil and its weight in lb/gal (kg/L). What is the kinematic viscosity in cSt? What is the absolute viscosity in cP?

Calculation Procedure:

1. Determine the API gravity of the liquid

For any oil at 60°F (15.6°C), its specific gravity S, in relation to water at 60°F (15.6°C), is $S = 141.5/(131.5 + °API)$; or $°API = (141.5 - 131.5S)/S$. For this oil, $°API = [141.5 - 131.5(0.80)]/0.80 = 45.4 °API$.

2. Determine the Bé gravity of the liquid

For any liquid lighter than water, $S = 140/(130 + Bé)$; or $Bé = (140 - 130S)/S$. For this oil, $Bé = [140 - 130(0.80)]/0.80 = 45 Bé$.

3. Compute the weight per gal of liquid

With a specific gravity of S, the weight of 1 ft³ of oil $= (S)$[weight of 1 ft³ (1 m³) of fresh water at 60°F (15.6°C)] $= (0.80)(62.4) = 49.92$ lb/ft³ (799.2 kg/m³). Since 1 gal (3.8 L) of liquid occupies 0.13368 ft³ the weight of this oil is $(49.92)(0.13368) = 6.66$ lb/gal (0.79 kg/L).

4. Compute the kinematic viscosity of the liquid

For any liquid having an SSU viscosity greater than 100 s, the kinematic viscosity $k = 0.220$ (SSU) $- 135/SSU$ cSt. For this oil, $k = 0.220(200) - 135/200 = 43.325$ cSt.

5. Convert the kinematic viscosity to absolute viscosity

For any liquid, the absolute viscosity, cP $=$ (kinematic viscosity, cSt)(density). Thus, for this oil, the absolute viscosity $= (43.325)(49.92) = 2163$ cP.

Related Calculations: For liquids *heavier* than water, $S = 145/(145 - Bé)$. When the SSU viscosity is between 32 and 99 SSU, $k = 0.226$ (SSU) $- 195/SSU$ cSt. Modern terminology for absolute viscosity is dynamic viscosity. Use these relations for any liquid—brine, gasoline, crude oil, kerosene, Bunker C, diesel oil, etc. Consult the *Pipe Friction Manual* and Crocker and King—*Piping Handbook* for tabulations of typical viscosities and specific gravities of various liquids.

PRESSURE LOSS IN PIPING HAVING LAMINAR FLOW

Fuel oil at 300°F (148.9°C) and having a specific gravity of 0.850 is pumped through a 30,000-ft (9144-m) long 24-in (609.6-mm) pipe at the rate of 500 gal/min (31.6 L/s). What is the pressure loss if the viscosity of the oil is 75 cP (0.075 Pa·s)?

Calculation Procedure:

1. Determine the type of flow that exists

Flow is laminar (also termed *viscous*) if the Reynolds number R for the liquid in the pipe is less than 1200. Tubulent flow exists if the Reynolds number is greater than 2500. Between these values is a zone in which either condition may exist, depending on the roughness of the pipe wall, entrance conditions, and other factors. Avoid sizing a pipe for flow in this critical zone because excessive pressure drops result without a corresponding increase in the pipe discharge.

Compute the Reynolds number from $R = 3.162G/kd$, where $G = $ flow rate gal/min (L/s); $k = $ kinematic viscosity of liquid, cSt $= $ viscosity z, cP/specific gravity of the liquid S; $d = $ inside diameter of pipe, in (cm). From a table of pipe properties, $d = 22.626$ in (574.7 mm). Also, $k = z/S = 75/0.85 = 88.2$ cSt. Then $R = 3162(500)/[88.2(22.626)] = 792$. Since $R < 1200$, laminar flow exists in this pipe.

2. Compute the pressure loss by using the Poiseuille formula

The Poiseuille formula gives the pressure drop p_d lb/in² (kPa) $= 2.73(10^{-4})luG/d^4$, where $l = $ total length of pipe, including equivalent length of fittings, ft; $u = $ absolute viscosity of liquid, cP (Pa·s); $G = $ flow rate, gal/min (L/s); $d = $ inside diameter of pipe, in (cm). For this pipe, $p_d = 2.73(10^{-4})(10,000)(75)(500)/262,078 = 1.17$ lb/in² (8.1 kPa).

Related Calculations: Use this procedure for any pipe in which there is laminar flow of the liquid. Other liquids for which this method can be used include water, molasses, gasoline, brine, kerosene, and mercury. Table 20 gives a quick summary of various ways in which the Reynolds number can be expressed. The symbols in Table 20, in the order of their appearance, are $D = $ inside diameter of pipe, ft (m); $v = $ liquid velocity, ft/s (m/s); $\rho = $ liquid density, lb/ft³ (kg/m³); $\mu = $ absolute viscosity of liquid, lb mass/(ft·s) [kg/(m·s)]; $d = $ inside diameter of pipe, in

TABLE 20   Reynolds Number

Reynolds number R	Numerator				Denominator	
	Coefficient	First symbol	Second symbol	Third symbol	Fourth symbol	Fifth symbol
Dvp/μ	. . .	ft	ft/s	lb/ft³	lb mass/(ft·s)	
$124dv\rho/z$	124	in	ft/s	lb/ft³	cP	
$50.7G\rho/dz$	50.7	gal/min	lb/ft³	. . .	in	cP
$6.32W/dz$	6.32	lb/h	. . .	. . .	in	cP
$35.5B\rho/dz$	35.5	bbl/h	lb/ft³	. . .	in	cP
$7742dv/k$	7,742	in	ft/s	. . .	. . .	cP
$3162G/dk$	3,162	gal/min	. . .	. . .	in	cP
$2214B/dk$	2,214	bbl/h	. . .	. . .	in	cP
$22,735q\rho/dz$	22,735	ft³/s	lb/ft³	. . .	in	cP
$378.9Q\rho/dz$	378.9	ft³/min	lb/ft³	. . .	in	cP

(cm). From a table of pipe properties, $d = 22.626$ in (574.7 mm). Also, $k = z/S$ liquid flow rate, lb/h (kg/h); B = liquid flow rate, bbl/h (L/s); k = kinematic viscosity of the liquid, cSt; q = liquid flow rate, ft³/s (m³/s); Q = liquid flow rate, ft³/min (m³/min). Use Table 20 to find the Reynolds number for any liquid flowing through a pipe.

DETERMINING THE PRESSURE LOSS IN OIL PIPES

What is the pressure drop in a 5000-ft (1524-m) long 6-in (152.4-mm) oil pipe conveying 500 bbl/h (22.1 L/s) of kerosene having a specific gravity of 0.813 at 65°F (18.3°C), which is the temperature of the liquid in the pipe? The pipe is schedule 40 steel.

Calculation Procedure:

1. Determine the kinematic viscosity of the oil

Use Fig. 12 and Table 21 or the Hydraulic Institute—*Pipe Friction Manual* kinematic viscosity and Reynolds number chart to determine the kinematic viscosity of the liquid. Enter Table 12 at kerosene, and find the coordinates as $X = 10.2$, $Y = 16.9$. Using these coordinates, enter Fig. 12 and find the absolute viscosity of kerosene at 65°F (18.3°C) as 2.4 cP. By the method of a previous calculation procedure, the kinematic viscosity = absolute viscosity, cP/specific gravity of the liquid = 2.4/0.813 = 2.95 cSt. This value agrees closely with that given in the *Pipe Friction Manual*.

2. Determine the Reynolds number of the liquid

The Reynolds number can be found from the *Pipe Friction Manual* chart mentioned in step 1 or computed from $R = 2214B/(dk) = 2214(500)/[(6.065)(2.95)] = 61,900$.

To use the *Pipe Friction Manual* chart, compute the velocity of the liquid in the pipe by converting the flow rate to ft³/s. Since there is 42 gal/bbl (0.16 L) and 1 gal (0.00379 L) = 0.13368 ft³ (0.00378 m³), 1 bbl = (42)(0.13368) = 5.6 ft³ (0.16 m³). With a flow rate of 500 bbl/h (79.5 m³/h) the equivalent flow = (500)(5.6) = 2800 ft³/h (79.3 m³/h), or 2800/3600 s/h = 0.778 ft³/s (0.02 m³/s). Since 6-in (152.4-mm), schedule 40 pipe has a cross-sectional area of 0.2006 ft² (0.02 m²) internally, the liquid velocity = 0.778/0.2006 = 3.88 ft/s (1.2 m/s). Then, the product (velocity, ft/s)(internal diameter, in) = (3.88)(6.065) = 23.75 ft/s. In the *Pipe Friction Manual*, project horizontally from the kerosene specific-gravity curve to the vd product of 23.75, and read the Reynolds number as 61,900, as before. In general, the Reynolds number can be found more quickly by computing it using the appropriate relation given in an earlier calculation procedure, unless the flow velocity is already known.

3. Determine the friction factor of this pipe

Enter Fig. 13 at the Reynolds number value of 61,900, and project to the curve 4 as indicated by Table 22. Read the friction factor as 0.0212 at the left. Alternatively, the *Pipe Friction Manual* friction-factor chart could be used, if desired.

VISCOSITIES

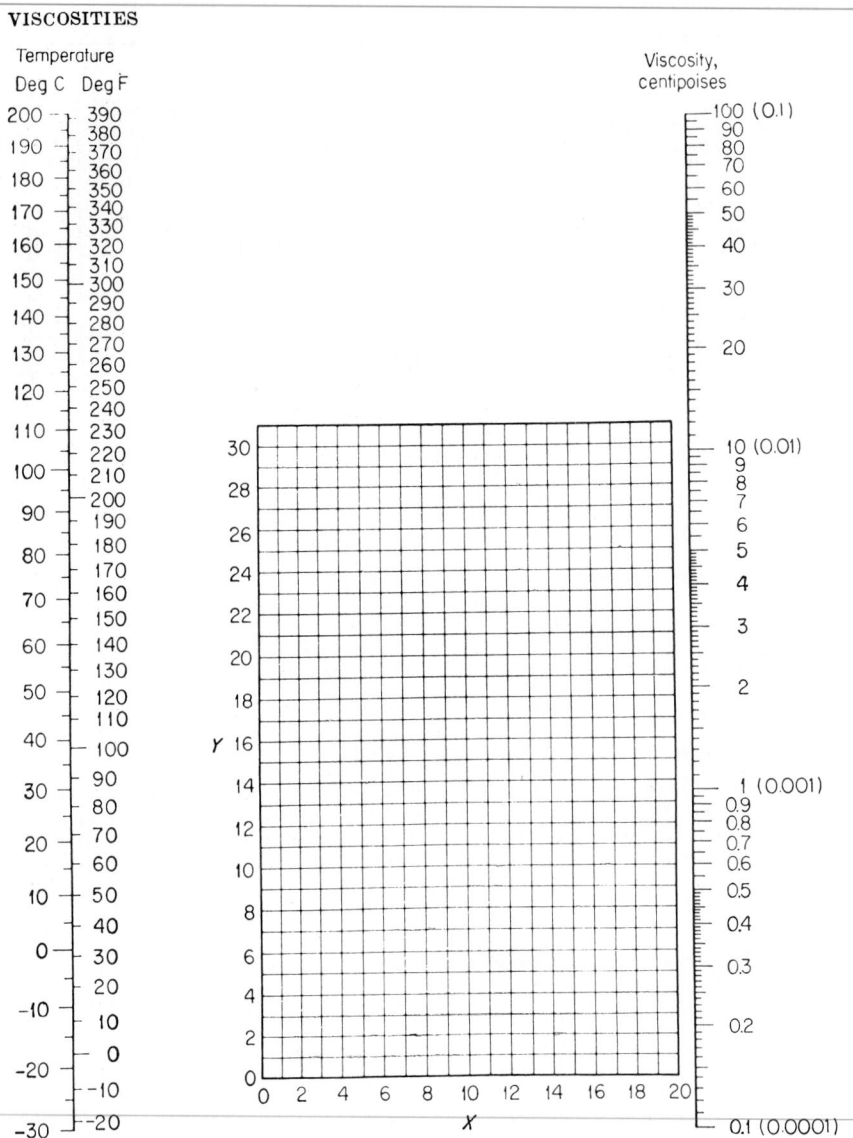

FIG. 12 Viscosities of liquids at 1 atm. For coordinates, see Table 21.

4. Compute the pressure loss in the pipe

Use the Fanning formula $p_d = 1.06(10^{-4})f\rho lB^2/d^5$. In this formula, ρ = density of the liquid, lb/ft³. For kerosene, ρ = (density of water, lb/ft³)(specific gravity of the kerosene) = $(62.4)(0.813) = 50.6$ lb/ft³ (810.1 kg/m³). Then $p_d = 1.06(10^{-4})(0.0212)(50.6)(5000)(500)^2/8206$ = 17.3 lb/in² (119.3 kPa).

Related Calculations: The Fanning formula is popular with oil-pipe designers and can be stated in various ways: (1) with velocity v ft/s, $p_d = 1.29(10^{-3})f\rho v^2 l/d$; (2) with velocity V ft/min, $p_d = 3.6(10^{-7})f\rho V^2 l/d$; (3) with flow rate in G gal/min, $p_d = 2.15(10^{-4})f\rho lG^2/d^2$; (4) with the flow rate in W lb/h, $p_d = 3.36(10^{-6})flW^2/d^5\rho$.

TABLE 21 Viscosities of Liquids
Coordinates for use with Fig. 12

No.	Liquid	X	Y	No.	Liquid	X	Y
1	Acetaldehyde	15.2	4.8	56	Freon-22	17.2	4.7
	Acetic acid:			57	Freon-13	12.5	11.4
2	100%	12.1	14.2		Glycerol:		
3	70%	9.5	17.0	58	100%	2.0	30.0
4	Acetic anhydride	12.7	12.8	59	50%	6.9	19.6
	Acetone:			60	Heptene	14.1	8.4
5	100%	14.5	7.2	61	Hexane	14.7	7.0
6	35%	7.9	15.0	62	Hydrochloric acid, 31.5%	13.0	16.6
7	Allyl alcohol	10.2	14.3	63	Isobutyl alcohol	7.1	18.0
	Ammonia:			64	Isobutyric acid	12.2	14.4
8	100%	12.6	2.0	65	Isopropyl alcohol	8.2	16.0
9	26%	10.1	13.9	66	Kerosene	10.2	16.9
10	Amyl acetate	11.8	12.5	67	Linseed oil, raw	7.5	27.2
11	Amyl alcohol	7.5	18.4	68	Mercury	18.4	16.4
12	Aniline	8.1	18.7		Methanol:		
13	Anisole	12.3	13.5	69	100%	12.4	10.5
14	Arsenic trichloride	13.9	14.5	70	90%	12.3	11.8
15	Benzene	12.5	10.9	71	40%	7.8	15.5
	Brine:			72	Methyl acetate	14.2	8.2
16	CaCl$_2$, 25%	6.6	15.9	73	Methyl chloride	15.0	3.8
17	NaCl, 25%	10.2	16.6	74	Methyl ethyl ketone	13.9	8.6
18	Bromine	14.2	13.2	75	Naphthalene	7.9	18.1
19	Bromotoluene	20.0	15.9		Nitric acid:		
20	Butyl acetate	12.3	11.0	76	95%	12.8	13.8
21	Butyl alcohol	8.6	17.2	77	60%	10.8	17.0
22	Butyric acid	12.1	15.3	78	Nitrobenzene	10.6	16.2
23	Carbon dioxide	11.6	0.3	79	Nitrotoluene	11.0	17.0
24	Carbon disulfide	16.1	7.5	80	Octane	13.7	10.0
25	Carbon tetrachloride	12.7	13.1	81	Octyl alcohol	6.6	21.1
26	Chlorobenzene	12.3	12.4	82	Pentachloroethane	10.9	17.3
27	Chloroform	14.4	10.2	83	Pentane	14.9	5.2
28	Chlorosulfonic acid	11.2	18.1	84	Phenol	6.9	20.8
	Chlorotoluene:			85	Phosphorus tribromide	13.8	16.7
29	Ortho	13.0	13.3	86	Phosphorus trichloride	16.2	10.9
30	Meta	13.3	12.5	87	Propionic acid	12.8	13.8
31	Para	13.3	12.5	88	Propyl alcohol	9.1	16.5
32	Cresol, meta	2.5	20.8	89	Propyl bromide	14.5	9.6
33	Cyclohexanol	2.9	24.3	90	Propyl chloride	14.4	7.5
34	Dibromoethane	12.7	15.8	91	Propyl iodide	14.1	11.6
35	Dichloroethane	13.2	12.2	92	Sodium	16.4	13.9
36	Dichloromethane	14.6	8.9	93	Sodium hydroxide, 50%	3.2	25.8
37	Diethyl oxalate	11.0	16.4	94	Stannic chloride	13.5	12.8
38	Dimethyl oxalate	12.3	15.8	95	Sulfur dioxide	15.2	7.1
39	Diphenyl	12.0	18.3		Sulfuric acid:		
40	Dipropyl oxalate	10.3	17.7	96	110%	7.2	27.4
41	Ethyl acetate	13.7	9.1	97	98%	7.0	24.8
	Ethyl alcohol:			98	60%	10.2	21.3
42	100%	10.5	13.8	99	Sulfuryl chloride	15.2	12.4
43	95%	9.8	14.3	100	Tetrachloroethane	11.9	15.7
44	40%	6.5	16.6	101	Tetrachloroethylene	14.2	12.7
45	Ethyl benzene	13.2	11.5	102	Titanium tetrachloride	14.4	12.3
46	Ethyl bromide	14.5	8.1	103	Toluene	13.7	10.4
47	Ethyl chloride	14.8	6.0	104	Trichloroethylene	14.8	10.5
48	Ethyl ether	14.5	5.3	105	Turpentine	11.5	14.9
49	Ethyl formate			106	Vinyl acetate		
50	Ethyl iodide	14.7	10.3	107	Water	10.2	13.0
51	Ethylene glycol	6.0	23.6		Xylene:		
52	Formic acid	10.7	15.8	108	Ortho	13.5	12.1
53	Freon-11	14.4	9.0	109	Meta	13.9	10.6
54	Freon-12	16.8	5.6	110	Para	13.9	10.9
55	Freon-21	15.7	7.5				

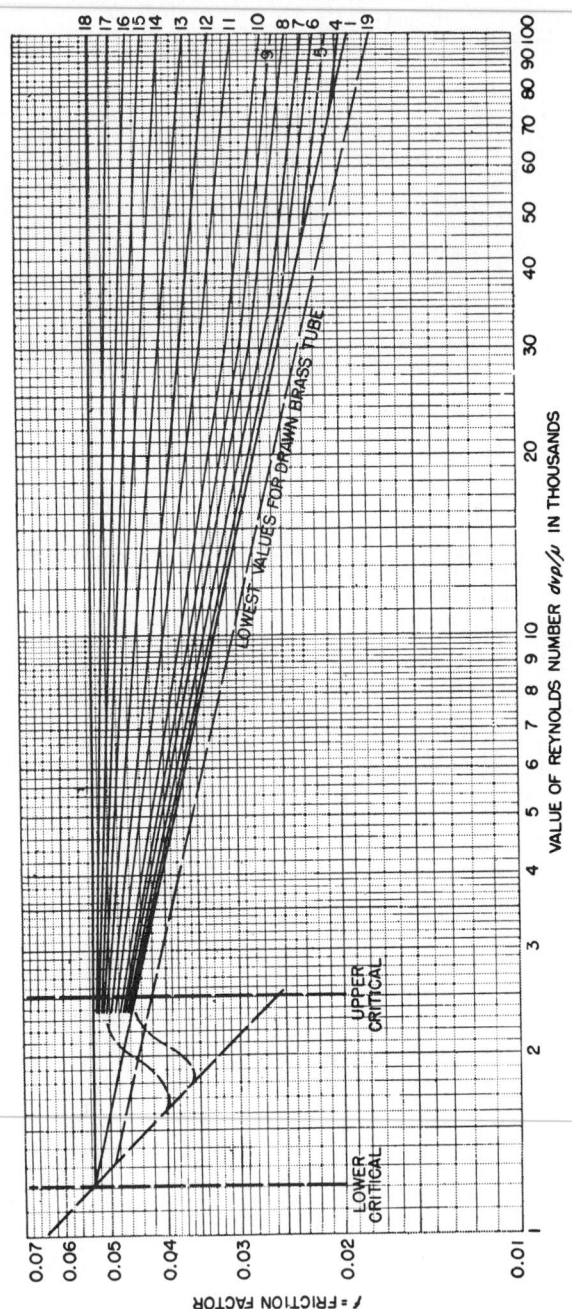

3.430

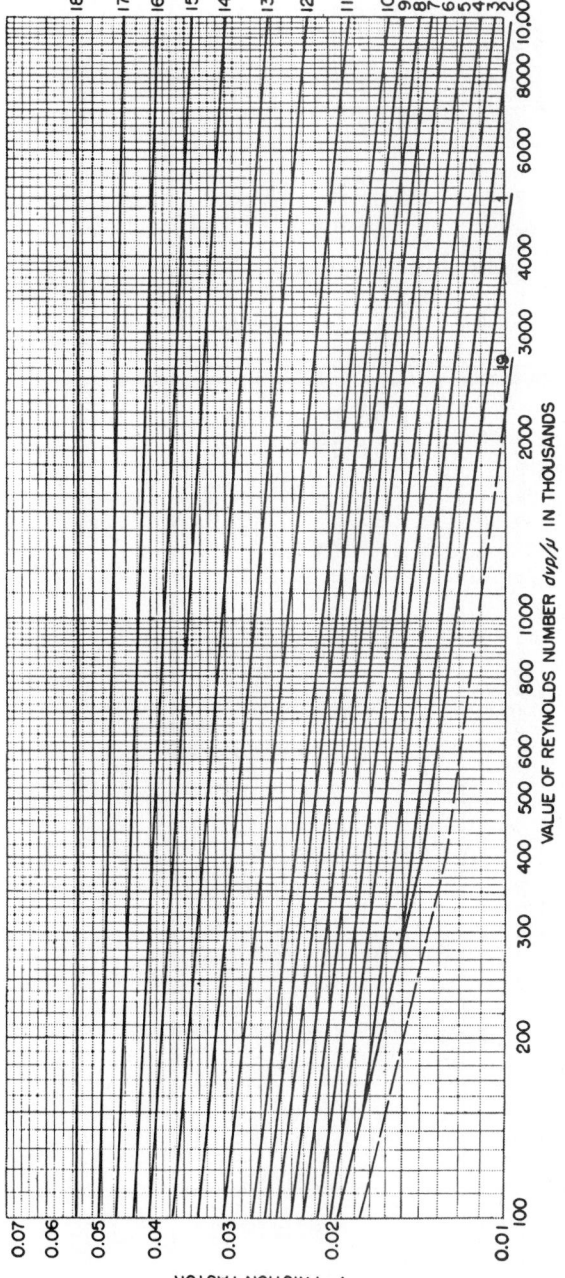

FIG. 13 Friction-factor curves. *(Mechanical Engineering.)*

3.431

TABLE 22 Data for Fig. 13

Percentage of roughness	For value of f see curve	Diameter (actual of drawn tubing, nominal of standard-weight pipe)											
		Drawn tubing, brass, tin, lead, glass		Clean steel, wrought iron		Clean, galvanized		Best cast iron		Average cast iron		Heavy riveted, spiral riveted	
		in	mm	in	mm	in	mm	in	mm	in	mm	in	mm
0.2	1	0.35 up	8.89 up	72	1829	—	—	—	—	—	—	—	—
1.35	4	—	—	6–12	152–305	10–24	254–610	20–48	508–1219	42–96	1067–2438	84–204	2134–5182
2.1	5	—	—	4–5	102–127	6–8	152–203	12–16	305–406	24–36	610–914	48–72	1219–1829
3.0	6	—	—	2–3	51–76	3–5	76–127	5–10	127–254	10–20	254–508	20–42	508–1067
3.8	7	—	—	1½	38	2½	64	3–4	76–102	6–8	152–203	16–18	406–457
4.8	8	—	—	1–1¼	25–32	1½–2	38–51	2–2½	51–64	4–5	102–127	10–14	254–356
6.0	9	—	—	¾	19	1¼	32	1½	38	3	76	8	203
7.2	10	—	—	½	13	1	25	1¼	32	—	—	5	127
10.5	11	—	—	⅜	9.5	¾	19	1	35	—	—	4	102
14.5	12	—	—	¼	6.4	½	13	—	—	—	—	3	76
24.0	14	0.125	3.18	—	—	⅜	9.5	—	—	—	—	—	—
31.5	16	—	—	—	—	¼	6.4	—	—	—	—	—	—
37.5	18	0.0625	1.588	—	—	⅛	3.2	—	—	—	—	—	—

Use this procedure for any petroleum product—crude oil, kerosene, benzene, gasoline, naphtha, fuel oil, Bunker C, diesel oil toluene, etc. The tables and charts presented here and in the *Pipe Friction Manual* save computation time.

FLOW RATE AND PRESSURE LOSS IN COMPRESSED-AIR AND GAS PIPING

Dry air at 80°F (26.7°C) and 150 lb/in² (abs) (1034 kPa) flows at the rate of 500 ft³/min (14.2 m³/min) through a 4-in (101.6-mm) schedule 40 pipe from the discharge of an air compressor. What are the flow rate in lb/h and the air velocity in ft/s? Using the Fanning formula, determine the pressure loss if the total equivalent length of the pipe is 500 ft (152.4 m).

Calculation Procedure:

1.   *Determine the density of the air or gas in the pipe*

For air or a gas, $pV = MRT$, where p = absolute pressure of the gas, lb/ft² (abs); V = volume of M lb of gas, ft³; M = weight of gas, lb; R = gas constant, ft·lb/(lb·°F); T = absolute temperature of the gas, °R. For this installation, using 1 ft³ of air, $M = pV/(RT)$, $M = (150)(144)/[(53.33)(80 + 459.7)] = 0.750$ lb/ft³ (12.0 kg/m³). The value of R in this equation was obtained from Table 23.

2.   *Compute the flow rate of the air or gas*

For air or a gas, the flow rate W_h lb/h = (60) (density, lb/ft³)(flow rate, ft³/min); or $W_h = (60)(0.750)(500) = 22,500$ lb/h (10,206 kg/h).

3.   *Compute the velocity of the air or gas in the pipe*

For any air or gas pipe, velocity of the moving fluid v ft/s $= 183.4\, W_h/3600\, d^2\rho$, where d = internal diameter of pipe, in; ρ = density of fluid, lb/ft³. For this system, $v = (183.4)(22,500)/[(3600)(4.026)^2(0.750)] = 94.3$ ft/s (28.7 m/s).

4.   *Compute the Reynolds number of the air or gas*

The viscosity of air at 80°F (26.7°C) is 0.0186 cP, obtained from Crocker and King—*Piping Handbook*, Perry et al.—*Chemical Engineers' Handbook*, or a similar reference. Then, by using the Reynolds number relation given in Table 20, $R = 6.32W/(dz) = (6.32)(22,500)/[(4.026)(0.0186)] = 1,899,000$.

TABLE 23   Gas Constants

Gas	R ft·lb/(lb·°F)	R J/(kg·K)	C for critical-velocity equation
Air	53.33	286.9	2870
Ammonia	89.42	481.1	2080
Carbon dioxide	34.87	187.6	3330
Carbon monoxide	55.14	296.7	2820
Ethane	50.82	273.4	
Ethylene	54.70	294.3	2480
Hydrogen	767.04	4126.9	750
Hydrogen sulfide	44.79	240.9	
Isobutane	25.79	138.8	
Methane	96.18	517.5	2030
Natural gas	—	—	2070–2670
Nitrogen	55.13	296.6	2800
n-butane	25.57	137.6	
Oxygen	48.24	259.5	2990
Propane	34.13	183.6	
Propylene	36.01	193.7	
Sulfur dioxide	23.53	126.6	3870

5. *Compute the pressure loss in the pipe*

Using Fig. 13 or the Hydraulic Institute *Pipe Friction Manual*, we get f = 0.0142 to 0.0162 for a 4-in (101.6-mm) schedule 40 pipe when the Reynolds number = 3,560,000. From the Fanning formula from an earlier calculation procedure and the higher value of f, p_d = $3.36(10^{-6})flW^2/d^5\rho$, or p_d = $3.36(10^{-6})(0.0162)(500)(22,500)^2/[(4.026)^5(0.750)]$ = 17.37 lb/in^2 (119.8 kPa).

Related Calculations: Use this procedure to compute the pressure loss, velocity, and flow rate in compressed-air and gas lines of any length. Gases for which this procedure can be used include ammonia, carbon dioxide, carbon monoxide, ethane, ethylene, hydrogen, hydrogen sulfide, isobutane, methane, nitrogen, *n*-butane, oxygen, propane, propylene, and sulfur dioxide.

Alternate relations for computing the velocity of air or gas in a pipe are v = $144W_s/a\rho$; v = $183.4W_s/d^2\rho$; v = $0.0509W_sv_g/d^2$, where W_s = flow rate, lb/s; a = cross-sectional area of pipe, in^2, v_g = specific volume of the air or gas at the operating pressure and temperature, ft^3/lb.

FLOW RATE AND PRESSURE LOSS IN GAS PIPELINES

Using the Weymouth formula, determine the flow rate in a 10-mi (16.1-km) long 4-in (101.6-mm) schedule 40 gas pipeline when the inlet pressure is 200 lb/in^2 (gage) (1378.8 kPa), the outlet pressure is 20 lb/in^2 (gage) (137.9 kPa), the gas has a specific gravity of 0.80, a temperature of 60°F (15.6°C), and the atmospheric pressure is 14.7 lb/in^2 (abs) (101.34 kPa).

Calculation Procedure:

1. *Compute the flow rate from the Weymouth formula*

The Weymouth formula for flow rate is Q = $28.05[(p_i^2 - p_0^2)d^{5.33}/sL]^{0.5}$, where p_i = inlet pressure, lb/in^2 (abs); p_0 = outlet pressure, lb/in^2 (abs); d = inside diameter of pipe, in; s = specific gravity of gas; L = length of pipeline, mi. For this pipe, Q = $28.05 \times [(214.7^2 - 34.7^2)4.026^{5.33}/0.8 \times 10]^{0.5}$ = 86,500 lb/h (38,925 kg/h).

2. *Determine if the acoustic velocity limits flow*

If the outlet pressure of a pipe is less than the critical pressure p_c lb/in^2 (abs), the flow rate in the pipe cannot exceed that obtained with a velocity equal to the critical or acoustic velocity, i.e., the velocity of sound in the gas. For any gas, p_c = $Q(T_i)^{0.5}/d^2C$, where T_i = inlet temperature, °R; C = a constant for the gas being considered.

Using C = 2070 from Table 23, or Crocker and King—*Piping Handbook*, p_c = $(86,500)(60 + 460)^{0.5}/[(4.026)^2(2070)]$ = 58.8 lb/in^2 (abs) (405.4 kPa). Since the outlet pressure p_0 = 34.7 lb/in^2 (abs) (239.2 kPa), the critical or acoustic velocity limits the flow in this pipe because $p_c > p_0$. When $p_c < p_0$, critical velocity does not limit the flow.

Related Calculations: Where a number of gas pipeline calculations must be made, use the tabulations in Crocker and King—*Piping Handbook* and Bell—*Petroleum Transportation Handbook*. These tabulations will save much time. Other useful formulas for gas flow include the Panhandle, Unwin, Fritsche, and rational. Results obtained with these formulas agree within satisfactory limits for normal engineering practice.

Where the outlet pressure is unknown, assume a value for it and compute the flow rate that will be obtained. If the computed flow is less than desired, check to see that the outlet pressure is less than the critical. If it is, increase the diameter of the pipe. Use this procedure for natural gas from any gas field, manufactured gas, or any other similar gas.

To find the volume of gas that can be stored per mile of pipe, solve V_m = $1.955p_md^2K$, where p_m = mean pressure in pipe, lb/in^2 (abs) $\approx (p_i + p_0)/2$; K = $(1/Z)^{0.5}$, where Z = supercompressibility factor of the gas, as given in Baumeister and Marks—*Standard Handbook for Mechanical Engineers* and Perry—*Chemical Engineer's Handbook*. For exact computation of p_m, use p_m = $(\frac{2}{3})(p_i + p_0 - p_ip_0/p_i + p_0)$.

SELECTING HANGERS FOR PIPES AT ELEVATED TEMPERATURES

Select the number, capacity, and types of pipe hanger needed to support the 6-in (152.5-mm) schedule 80 pipe in Fig. 14 when the installation temperature is 60°F (15.6°C) and the operating temperature is 700°F (371.1°C). The pipe is insulated with 85 percent magnesia weighing 11.4

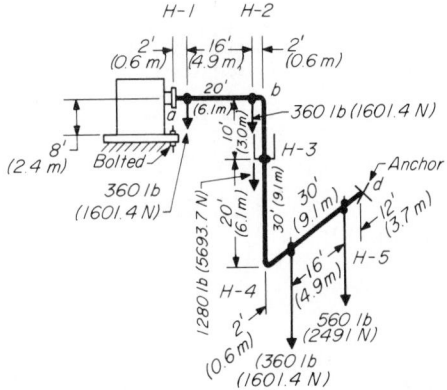

FIG. 14 Typical complex pipe operating at high temperature.

lb/ft (16.63 N/m). The pipe and unit served by the pipe have a coefficient of thermal expansion of 0.0575 in/ft (0.48 cm/m) between the 60°F (15.6°C) installation temperature and the 700°F (371.1°C) operating temperature.

Calculation Procedure:

1.   Draw a freehand sketch of the pipe expansion

Use Fig. 15 as a guide and sketch the expanded pipe, using a dashed line. The sketch need not be exactly to scale; if the proportions are accurate, satisfactory results will be obtained. The shapes shown in Fig. 15 cover the 11 most common situations met in practice.

2.   Tentatively locate the required hangers

Begin by locating hangers H-1 and H-5 close to the supply and using units, Fig. 14. Keeping a hanger close to each unit (boiler, turbine, pump, engine, etc.) prevents overloading the connection on the unit.

Space intermediate hangers H-2, H-3, and H-4 so that the recommended distances in Table 24 or hanger engineering data (e.g., Grinnell Corporation *Pipe Hanger Design and Engineering*) are not exceeded. Indicate the hangers on the piping drawing as shown in Fig. 14.

3.   Adjust the hanger locations to suit structural conditions

Study the building structural steel in the vicinity of the hanger locations, and adjust these locations so that each hanger can be attached to a support having adequate strength.

4.   Compute the load each hanger must support

From a table of pipe properties, such as in Crocker and King—*Piping Handbook*, find the weight of 6-in (152.4-mm) schedule 80 pipe as 28.6 lb/ft (41.7 N/m). The insulation weighs 11.4 lb/ft (16.6 N/m), giving a total weight of insulated pipe of 28.6 + 11.4 = 40.0 lb/ft (58.4 N/m).

Compute the load on the hangers supporting horizontal pipes by taking half the length of the pipe on each side of the hanger. Thus, for hanger H-1, there is (2 ft)(½) + (16 ft) × (½) = 9 ft (2.7 m) of horizontal pipe, Fig. 14, which it supports. Since this pipe weighs 40 lb/ft (58.4 N/m), the total load on hanger H-1 = (9 ft)(40 lb/ft) = 360 lb (1601.4 N). A similar analysis for hanger H-2 shows that it supports (8 + 1)(40) = 360 lb (1601.4 N).

Hanger H-3 supports the entire weight of the vertical pipe, 30 ft (9.14 m), plus 1 ft (0.3 m) at the top bend and 1 ft (0.3 m) at the bottom bend, or a total of 1 + 30 + 1 = 32 ft (9.75 m). The total load on hanger H-3 is therefore (32)(40) = 1280 lb (5693.7 N).

Hanger H-4 supports (1 + 8)(40) = 360 lb (1601.4 N), and hanger H-5 supports (8 + 6)(40) = 560 lb (2491 N).

As a check, compute the total weight of the pipe and compare it with the sum of the endpoint

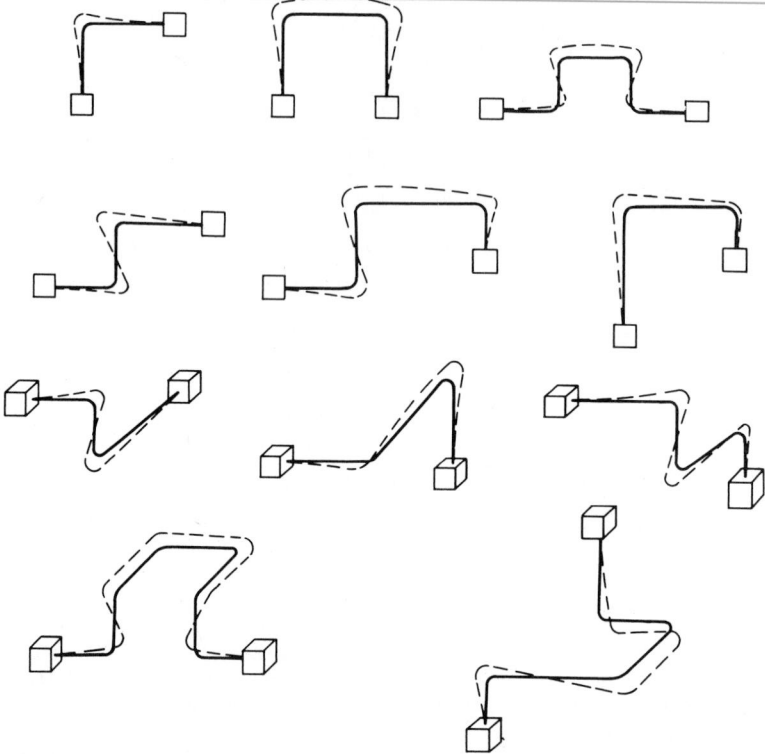

FIG. 15 Pipe shapes commonly used in power and process plants assume the approximate forms shown by the dotted lines when the pipe temperature rises. *(Power.)*

and hanger loads. Thus, there is 100 ft (30.5 m) of pipe weighing (80)(40) = 3200 lb (14.2 kN). The total load the hangers will support is 360 + 360 + 1280 + 360 + 560 = 2920 lb (12.9 kN). The first endpoint will support (1)(40) = 40 lb (177.9 N), and the anchor will support (6)(40) = 240 lb (1067 N). The total hanger and endpoint support = 2920 + 40 + 240 = 3200 lb (14.2 kN); therefore, the pipe weight = the hanger load.

5. *Sketch the shape of the hot pipe*

Use Fig. 15 as a guide, and draw a dotted outline of the approximate shape the pipe will take when hot. Start with the first corner point nearest the unit on the left, Fig. 16. This point will move away from the unit, as in Fig. 16. Do the same for the first corner point near the other unit served by the pipe and for intermediate corner points. Use arrows to indicate the probable direction of pipe movement at each corner. When sketching the shape of the hot pipe, remember that a straight pipe expanding against a piece of pipe at right angles to itself will bend the latter. The

TABLE 24 Maximum Recommended Spacing between Pipe Hangers

Nominal pipe size, in (mm)	4 (101.6)	5 (127)	6 (152.4)	8 (203.2)	10 (254)	12 (304.8)
Maximum span, ft (m)	14 (4.3)	16 (4.9)	17 (5.2)	19 (5.8)	22 (6.7)	23 (7.0)

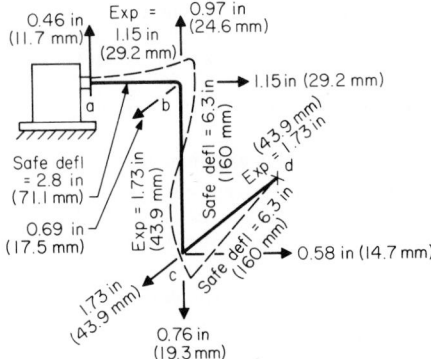

FIG. 16 Expansion of the various parts of the pipe shown in Fig. 14. (*Power*.)

TABLE 25 Deflection, in (mm), that Produces 14,000-lb/in² (96,530-kPa) Tensile Stress in Pipe Legs Acting as a Cantilever Beam, Load at Free End

Cantilever length, ft (m)	Nominal pipe size, in (mm)		
	4 (101.6)	6 (152.4)	8 (203.2)
5 (1.5)	0.26 (6.6)	0.17 (4.3)	0.13 (3.3)
10 (3.0)	1.03 (26.2)	0.70 (17.8)	0.54 (13.7)
15 (4.6)	2.32 (58.9)	1.58 (40.1)	1.21 (30.7)
20 (6.1)	4.12 (104.6)	2.80 (71.1)	2.15 (54.6)
25 (7.6)	6.44 (163.6)	4.38 (111.3)	3.35 (85.1)
30 (9.1)	9.26 (235.2)	6.30 (160.0)	4.83 (122.7)

distance that various lengths of pipe will bend while producing a tensile stress of 14,000 lb/in² (96.5 MPa) is given in Table 25. This stress is a typical allowable value for pipes in industrial systems.

6.   *Determine the thermal movement of units served by the pipe*

If either or both fixed units (boiler, turbine, etc.) operate at a temperature above or below atmospheric, determine the amount of movement at the flange of the unit to which the piping connects, using the thermal data in Table 26. Do this by applying the thermal expansion coefficient for the metal of which the unit is made. Determine the vertical and horizontal distance of the flange face from the point of no movement of the unit.

The point of no movement is the point or surface where the unit is fastened to *cold* structural steel or concrete.

The flange, point *a*, Fig. 16, is 8 ft (2.4 m) above the bolted end of the unit and directly in line with the bolt, Fig. 14. Since the bolt and flange are on a common vertical line, there will not be any *horizontal* movement of the flange because the bolt is the no-movement point of the unit.

Since the flange is 8 ft (2.4 m) away from the point of no movement, the amount that the flange will move = (distance away from the point of no movement, ft)(coefficient of ther-

TABLE 26 Thermal Expansion of Pipe, in/ft (mm/m) (Carbon and Carbon-Moly Steel and WI)

Operating temperature, °F (°C)	Installation temperature	
	32°F (0°C)	60°F (15.6°C)
600 (316)	0.050 (4.17)	0.0475 (3.96)
650 (343)	0.055 (4.58)	0.0525 (4.38)
700 (371)	0.060 (5.0)	0.0575 (4.79)
750 (399)	0.065 (5.42)	0.0624 (5.2)
800 (427)	0.070 (5.83)	0.0674 (5.62)

mal expansion, in/ft) = (8)(0.0575) = 0.46 in (11.7 mm) *away* (up) from the point of no movement. If the unit were operating at a temperature *less than* atmospheric, it would contract and the flange would move *toward* (down) the point of no movement. Mark the flange movement on the piping sketch, Fig. 16.

Anchor *d*, Fig. 16, does not move because it is attached to either cold structural steel or concrete.

7. *Compute the amount of expansion in each pipe leg*

Expansion of the pipe, in = (pipe length, ft)(coefficient of linear expansion, in/ft). For length *ab*, Fig. 14, the expansion = (20)(0.0575) = 1.15 in (29.2 mm); for *bc*, (30)(0.0575) = 1.73 in (43.9 mm); for *cd*, (30)(0.0575) = 1.73 in (43.9 mm). Mark the amount and direction of expansion on Fig. 16.

8. *Determine the allowable deflection for each pipe leg*

Enter Table 25 at the nominal pipe size and find the allowable deflection for a 14,000-lb/in^2 (96.5-MPa) tensile stress for each pipe leg. Thus, for *ab*, the allowable deflection = 2.80 in (71.1 mm) for a 20-ft (6.1-m) long leg; for *bc*, 6.30 in (160 mm) for a 30-ft (9.1-m) long leg; for *cd*, 6.30 in (160 mm) for a 30-ft (9.1-m) long leg. Mark these allowable deflections on Fig. 16, using dashed arrows.

9. *Compute the actual vertical and horizontal deflections*

Sketch the vertical deflection diagram, Fig. 17*a*, by drawing a triangle showing the total expansion in each direction in proportion to the length of the parts at right angles to the expansion. Thus,

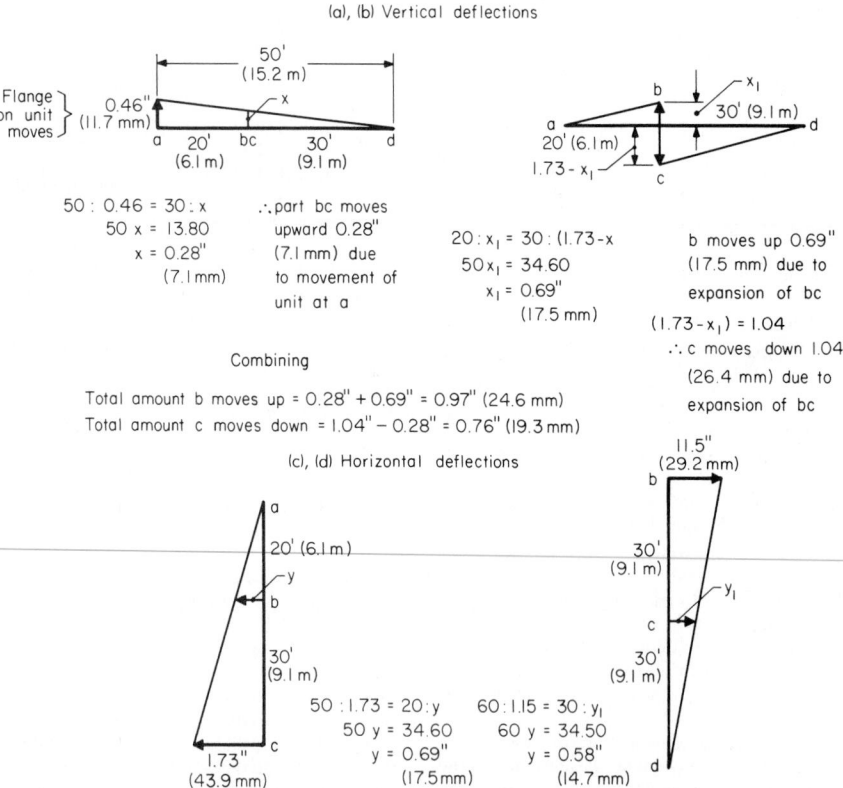

FIG. 17 (*a*), (*b*) Vertical deflection diagrams for the pipe in Fig. 14; (*c*), (*d*) horizontal deflection diagrams for the pipe in Fig. 14. (*Power.*)

the 0.46-in (11.7-mm) upward expansion at the flange, a, is at right angles to leg ab and is drawn as the altitude of the right triangle. Lay off 20 ft (6.1 m), ab, on the base of the triangle. Since bc is parallel to the direction of the flange movement, it is shown as a point, bc, on the base of this triangle. From point bc, lay off cd on the base of the triangle, Fig. 17a, since it is at right angles to the expansion of point a. Then, by similar triangles, $50:46 = 30:x$; $x = 0.28$ in (7.1 mm). Therefore, leg bc moves upward 0.28 in (7.1 mm) because of the flange movement at a.

Now draw the deflection diagram, Fig. 17b, showing the upward movement of leg ab and the downward movement of leg cd along the length of each leg, or 20 and 30 ft (6.1 and 9.1 m), respectively. Solve the similar triangles, or $20:x_1 = 30:(1.73 - x_1)$; $x_1 = 0.69$ in (17.5 mm). Therefore, point b moves *up* 0.69 in (17.5 mm) as a result of the expansion of leg bc. Then 1.73 $- x_1 = 1.73 - 0.69 = 1.04$ in (26.4 mm). Thus, point c moves *down* 1.04 in (26.4 mm) as a result of the expansion of bc. The total distance b moves up $= 0.28 + 0.69 = 0.97$ in (24.6 mm), whereas the total distance c moves down $= 1.04 - 0.28 = 0.76$ in (19.3 mm). Mark these actual deflections on Fig. 16.

Find the actual horizontal deflections in a similar fashion by constructing the triangle, Fig. 17c, formed by the vertical pipe bc and the horizontal pipe ab. Since point a does not move horizontally but point b does, lay off leg ab at right angles to the direction of movement, as shown. From point b lay off leg bc. Then, since leg bc expands 1.73 in (43.9 mm), lay this distance off perpendicular to ac, Fig. 17c. By similar triangles, $20 + 30:1.73 = 20:y$; $y = 0.69$ in (17.5 mm). Hence, point b deflects 0.69 in (17.5 mm) in the direction shown in Fig. 16.

Follow the same procedure for leg cd, constructing the triangle in Fig. 17d. Beginning with point b, lay off legs bc and cd. The altitude of this right triangle is then the distance point c moves when leg ab expands, or 1.15 in (29.2 mm). By similar triangles, $30 + 30:1.15 = 30:y_1$; $y_1 =$ deflection of point $c = 0.58$ in (14.7 mm).

10.   *Select the type of pipe hanger to use*

Figure 18 shows several popular types of pipe hangers, together with the movements that they are designed to absorb. For hangers H-1 and H-2, use type E, Fig. 18, because the pipe moves both vertically and horizontally at these points, as Fig. 17 shows. Use type F, Fig. 18, for hanger H-3, because riser bc moves both vertically and horizontally. Hangers H-4 and H-5 should be type E, because they must absorb both horizontal and vertical movements.

Once the hangers are selected from Fig. 18, refer to hanger engineering data for the exact design details of the hangers that will be selected. During the study of the data, look for other hangers that absorb the same movement or movements but may be more adaptable to the existing structural steel conditions.

11.   *Select the hanger-rod diameter for each hanger*

Use Table 27 to find the required hanger-rod diameter. Since the pipe operates at 700°F (371°C), select the maximum safe load from the 750°F (399°C) column. Tabulate the loads and diameters as follows:

Hanger	Load, lb (kN)	Rod diameter, in (mm)
H-1	360 (1.6)	⅜ (9.5)
H-2	360 (1.6)	⅜ (9.5)
H-3	1280 (5.7)	2-½ each (2–12.7)
H-4	360 (1.6)	⅜ (9.5)
H-5	560 (2.5)	½ (12.7)

Select standard springs for spring-loaded hangers from pipe-hanger engineering data. Springs are listed in the data on the basis of loading per inch of travel. For small movements [less than 1 in (25.4 mm)], it is generally desirable to select a lighter spring and precompress it at installation so that it has a light loading. Hanger movement will then load the spring to the desired value. This approach is desirable from another standpoint: any error in estimating hanger movement will not cause as large an unbalanced load on the pipe as would a heavier spring with a greater loading per inch of travel.

Related Calculations: Use this procedure for any type of pipe operating at elevated temperature—steam, oil, water, gas, etc.—serving a load in a power plant, process plant, ship, barge,

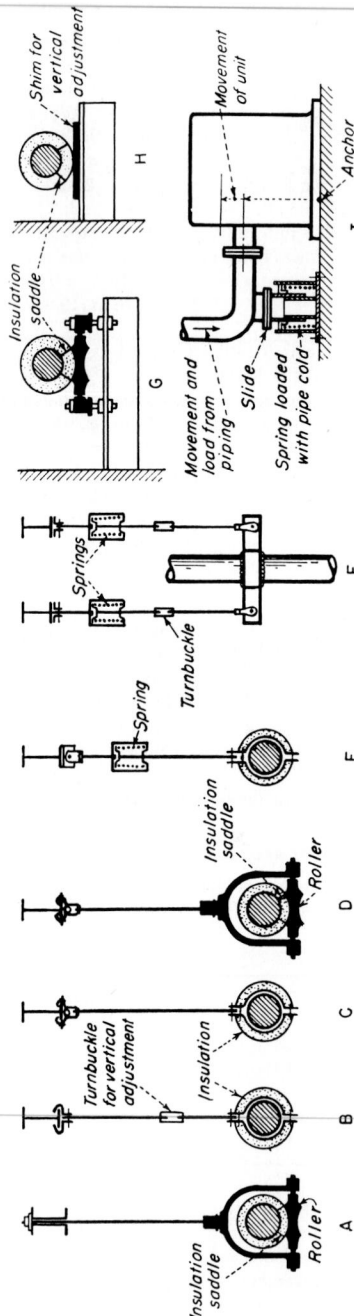

FIG. 18 Pipe hangers chosen depend on the movement expected. Hangers *A* and *B* are suitable for pipe movement in one horizontal direction. Hangers *C* and *D* permit pipe movement in two horizontal directions. Vertical and horizontal movement requires use of hangers such as *E* for horizontal pipes and *F* for vertical pipes. (*G*) Cantilever support; (*H*) sliding movement in two horizontal directions; (*I*) base elbow support.

TABLE 27   Hanger-Rod Load-Carrying Capacity (Hot-Rolled Steel Rod)

Nominal diameter of rod, in (mm)	Thread root area, in² (mm²)	Maximum safe load on rod, lb (kN), at rod temperature of:	
		450°F (232°C)	750°F (399°C)
⅜ (9.5)	0.068 (43.9)	610 (2.7)	510 (2.3)
½ (12.7)	0.126 (81.3)	1130 (5.0)	940 (4.2)
⅝ (15.9)	0.202 (130.3)	1810 (8.1)	1510 (6.7)
¾ (19.1)	0.302 (194.8)	2710 (12.1)	2260 (10.1)
⅞ (22.2)	0.419 (270.3)	3770 (16.8)	3150 (14.0)
1 (25.4)	0.552 (356.1)	4960 (22.1)	4150 (18.5)

aircraft, or other type of installation. In piping systems having very little or no increase in temperature during operation, the steps for computing the expansion can be eliminated. In this type of installation, the weight of the piping is the primary consideration in the choice of the hangers.

If desired, hanger loads can also be determined by taking moments about an arbitrarily selected axis on either side of the hanger. This method gives the same results as the procedure used above. The weight of bends is assumed to be concentrated at the center of gravity of each bend, whereas the weight of valves is assumed to be concentrated at the vertical centerline of the valve. Figure 19 shows typical moment arms, a and c, for valves and other fittings. The moment

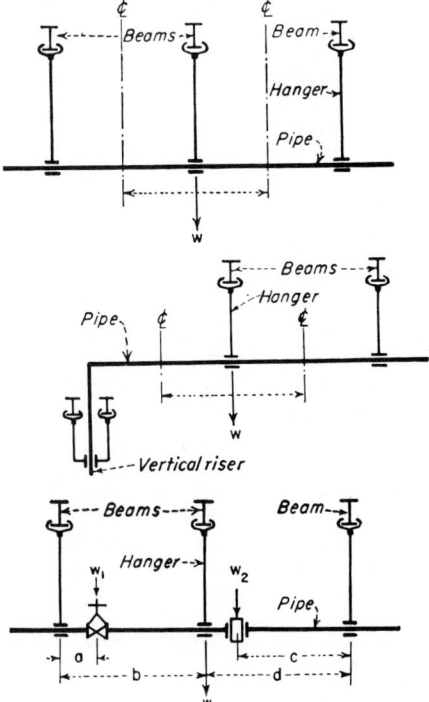

FIG. 19 Compute hanger loads of uniformly loaded pipes as shown. Use beam relations for concentrated loads. (*Power.*)

for W_1 about the hanger to the left of it is W_1a, and the moment for W_2 is W_2c about the hanger to the right of it. The weight of the pipe is assumed to be concentrated at a point midway between the hangers, and the moment is (weight of pipe, lb)(distance between hangers, ft/2). The method given here was developed by Frank Kamarck, Mechanical Engineer, and reported in *Power* magazine.

HANGER SPACING AND PIPE SLOPE FOR AN ALLOWABLE STRESS

An 8-in (203.2-mm) schedule 40 water pipe has an allowable bending stress of 10,000 lb/in^2 (68,950 kPa). What is the maximum allowable distance between hangers for this pipe? What slope will the allowable hanger span require to prevent pocketing of water in the pipe? Describe the method for computing hanger span and pipe slope for empty pipe. How are hanger distances computed when the pipe contains concentrated loads?

Calculation Procedure:

1. Compute the allowable span between hangers

For a pipe filled with water, $S = WL^2/8m$, where S = bending stress in pipe, lb/in^2; W = weight of pipe and water lb/lin in; L = maximum allowable distance between hangers, in; m = section modulus of pipe, in^3. By using a table of pipe properties, as in Crocker and King—*Piping Handbook*, $L = (8mS/W)^{0.5} = (8 \times 16.81 \times 10,000/4.18)^{0.5} = 568$ in, or $568/12 = 47.4$ ft (14.5 m).

2. Compute the pipe slope required by the span

To prevent pocketing of water or condensate at the low point in the pipe, the pipe must be pitched so that the outlet is lower than the lowest point in the span. When the pipe has no concentrated loads—such as valves, cross connections, or meters—the deflection of the pipe is y in = $22.5wl^4/(EI)$, where w = weight of the pipe and its contents, lb/ft; l = distance between hangers, ft; E = modulus of elasticity of pipe, lb/in^2 = 30×10^6 for steel; I = moment of inertia of the pipe, in^4. Substituting values gives $y = (22.5)(50.24)(47.4)^4/[(30 \times 10^6)(72.5)] = 2.61$ in (66.3 mm).

With the deflection y known, the pipe slope, expressed as 1 in (25.4 mm) per G ft of pipe length, is 1 in (25.4 mm) per G ft = $\frac{1}{4}y$, or $G = (47.4)/[(4)(2.61)] = 4.53$. Thus, a pipe slope of 1 in (25.4 mm) in 4.53 ft (1.38 m) is necessary to prevent pocketing of the water when the hanger span is 47.4 ft (14.5 m). With this slope, the outlet of the pipe would be $47.4/4.53 = 10.45$ in (265.4 mm) below the inlet.

3. Compute the empty-pipe hanger span and pipe slope

Use the same procedure as in steps 1 and 2, except that the empty weight of the pipe is substituted in the equations instead of the weight of the pipe when full of water. For pipes containing steam, gas, or vapor, compute the flowing-fluid weight and add it to the pipe weight. Follow the same procedure for insulated pipes, adding the insulation weight to the pipe weight.

4. Determine the hanger span and slope with concentrated loads

Hanger span and pipe slope can be computed from standard beam relations. However, most piping designers use the deflection chart and deflection factors for concentrated loads in Crocker and King—*Piping Handbook*. The chart and correction factors simplify the calculations considerably. The computation involves only simple multiplication and division.

Related Calculations: Use this procedure for piping in any type of installation—power, process, marine, industrial, or utility—for any type of liquid, vapor, or gas.

EFFECT OF COLD SPRING ON PIPE ANCHOR FORCES AND STRESSES

A carbon molybdenum pipe operates at 800°F (427°C) and has an anchor force of 5000 lb (22.2 kN) and a maximum bending stress s_b of 15,000 lb/in^2 (103.4 MPa) without cold spring. Compute the anchor force and bending stress in the hot and cold condition when the pipe is cold-sprung an amount equal to the expansion e and $0.5e$. The total expansion of the pipe is 24 in (609.6 mm).

Calculation Procedure:

1. *Compute the hot-condition force and stress*

The allowable cold-spring adjustment is expressed as a ratio $(e - 2S/3)/e$, where e = the total expansion of the pipe, in; S = cold-spring distance, in. This ratio is multiplied by the original anchor force and bending stress at the maximum operating temperature *without* cold spring to find the anchor force and bending stress *with* cold spring in the hot condition. If the ratio is less than $2/3$, the value of $2/3$ is used where maximum credit for cold spring is desired.

For this pipe, with maximum cold spring, the ratio = $(24 - 2 \times 24/3)/24 = 1/3$. Since this is less than $2/3$, use $2/3$. Then, the anchor force $F = (2/3)(5000) = 3333$ lb (14.8 kN), and the bending stress $s_b = (2/3)(15,000) = 10,000$ lb/in² (68.9 MPa).

With $S = 0.5e = (0.5)(24) = 12$ in (304.8 mm), the ratio = $(24 - 2 \times 12/3)/24 = 2/3$. Hence, $F = (2/3)(5000) = 3333$ lb (14.8 kN); $s_b = (2/3)(15,000) = 10,000$ lb/in² (68.9 MPa).

2. *Compute the cold-condition force and stress*

For the cold condition, the adjustment ratio = $-S/eM_R$, where M_R = modulus ratio for the pipe material = modulus of elasticity, lb/in², of the pipe material at the operating temperature, °F/ modulus of elasticity of the pipe material, lb/in², at 70°F (21.1°C). For this pipe, $M_R = 0.865$, from a table of pipe properties. The minus sign in the ratio indicates that the anchor force and stress are reversed in the cold condition as compared with the hot condition.

For this pipe, with maximum cold spring, the ratio = $-24/[(24)(0.865)] = -1.156$. Then the anchor force in the cold condition = $(-1.156)(15,000) = -5790$ lb (25.7 kN), and the bending stress = $(-1.156)(15,000) = -17,350$ lb/in² (119.6 MPa).

With $S = 0.5e = (0.5)(24) = 12$ in (304.8 mm), the ratio = $-12/[(24)(0.865)] = -0.578$. Then the anchor force in the cold condition $(-0.578)(5000) = -2895$ lb (12.9 kN), and the bending stress = $(-0.578)(15,000) = -8670$ lb/in² (59,771 kPa).

These calculations show that cold spring reduces the anchor force and bending stress when the pipe is in the hot condition, step 1. With a cold spring of one-half the pipe expansion, the anchor force and bending stress are reduced and reversed when in the cold condition, step 2. When the cold spring equals the expansion, the anchor force and bending stress increase in the cold condition, step 2.

Related Calculations: Use this procedure for a pipe conveying steam, oil, gas, water, and similar vapors, liquids, and gases.

REACTING FORCES AND BENDING STRESS IN SINGLE-PLANE PIPE BEND

Determine the horizontal and vertical reacting forces in the single-plane pipe bend of Fig. 20 if the pipe is 6-in (152.4-mm) schedule 40 carbon steel A106 seamless operating at 500°F (260°C). What is the maximum bending stress in the pipe and the resultant reacting or anchor force? Determine the maximum bending stress if a long-radius welded elbow is used at point C, Fig. 20. Use the tabular method of solution.

Calculation Procedure:

1. *Compute the horizontal reacting or anchor force*

Several methods are available for determining the reacting or anchor forces and maximum bending stress in a single-plane pipe bend. Crocker and King—*Piping Handbook* presents simplified, analytical, and graphical methods for computing forces and stresses in single- and multiplane piping systems. Another useful reference, *Design of Piping Systems,* written by members of the engineering departments of the M. W. Kellogg Company, presents both simplified and analytical methods and an excellent history and discussion of piping flexibility analysis. Probably the simplest method for routine piping flexibility analyses is that developed by the Grinnell Company, Inc., and S. W. Spielvogel. This method uses tabulated constants for specific pipe shapes in one, two, and three planes. It is satisfactory for the majority of piping problems met in normal engineering practice. To assist the practicing engineer, a number of Grinnell-Spielvogel tabulations for common pipe shapes are included here. For uncommon pipe shapes, refer to Grinnell Company—

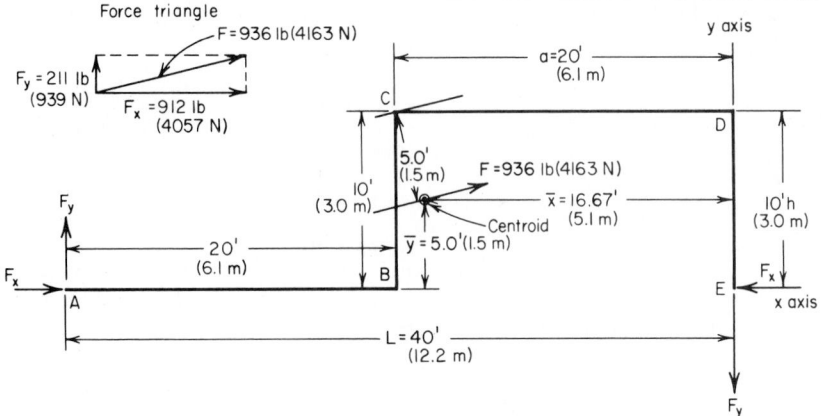

FIG. 20   U-shaped pipe with single tangent.

Piping Design and Engineering or to Spielvogel—*Piping Stress Calculations Simplified*. Both these references contain complete tabulations for a variety of pipe shapes.

To apply the Grinnell-Spielvogel solution procedure, compute the horizontal reaction force F_x lb from $F_x = k_x cI_p/L^2$, where k_x = a constant from Table 28 for the bend shape shown in Fig. 20; c = expansion factor = (pipe expansion, in/100 ft)$(EM_R/172{,}800)$, where E = modulus of elasticity of the pipe material being used, lb/in^2; M_R = modulus ratio = E at the operating temperature, F/E at 70°F (21.1°C) = 0.932; I_p = moment of inertia of pipe cross section, in^4; L = length of bend, ft, as shown in Fig. 20.

To enter Table 28 for the shape in Fig. 20, the values of L/a and L/h must be known, or $L/a = 40/20 = 2$; $L/h = 40/10 = 4$. Entering Table 28 at these values, read $k_x = 91$; $k_y = 21$; $k_b = 120$. From the Spielvogel c table or by computation, $c = 570$ for carbon-steel pipe operating at 500°F (260°C). From a table of pipe properties, $I_p = 28.14$ in^4 (1171.3 cm^4) for 6-in (152.4-mm) schedule 40 pipe. Then $F_x = (91)(570)(28.14)/(40)^2 = 912$ lb (4057 N).

2.   Compute the vertical reacting or anchor force

Use the same procedure as in step 1, except that the vertical reacting force F_y lb = $k_y cI_p/L^2$, or $F_y = (21)(570)(28.14)/(40)^2 = 211$ lb (939 N), by using the appropriate value from Table 28.

3.   Compute the resultant reacting or anchor force

The resultant reacting or anchor force F lb is found by drawing and solving the force triangle in Fig. 20. From the pythagorean theorem $F = (912^2 + 211^2)^{0.5} = 936$ lb (4163 N). Draw the force triangle to scale, as shown in Fig. 20.

4.   Compute the maximum bending stress in the pipe

The pipe bending stress s_b lb/in^2 is found in a similar manner from $s_b = k_b cD/L$, where k_b = bending-stress factor from Table 28; D = outside diameter of pipe, in. For 6-in (152.4-mm) schedule 40 pipe having an outside diameter of 6.625 in (168.3 mm), $s_b = (120)(570)(6.625)/40 = 11{,}330$ lb/in^2 (78.1 MPa).

5.   Determine the bending stress in the welded elbow

The tables presented here are accurate when all the turns in the piping system analyzed are miters or rigid fittings. When all the turns are welded elbows or bends, the anchor forces derived from Table 28 are accurate for practical systems. The actual forces will be somewhat smaller than the values obtained from Table 28. Stresses in the elbows or bends may, however, exceed the values computed from Table 28 if the stress intensification factor β for these curved sections is >1. If the proportion of the straight to curved pipe is large, use the following procedure to obtain a close approximation of the stress in the curved section:

Determine the value of β from a table of pipe properties. For a 6-in (152.4-mm) schedule 40

TABLE 28 U Shape with Single Tangent

Reacting Force $\qquad F_x = k_x \cdot c \cdot \dfrac{I_p}{L^2}$ lb (N)

Reacting Force $\qquad F_y = k_y \cdot c \cdot \dfrac{I_p}{L^2}$ lb (N)

Maximum Bending Stress $\quad s_B = k_b \cdot c \cdot \dfrac{D}{L}$ lb/in^2 (Pa)

I_p in in^4 (cm^4) L in ft (m) D in in (cm)

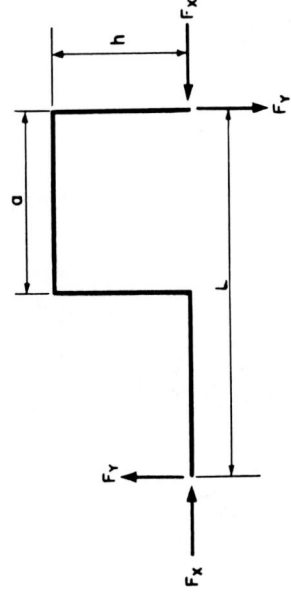

	L/a								
	1.5			**2**			**3**		
L/h	k_x	k_y	k_b	k_x	k_y	k_b	k_x	k_y	k_b
1.0	2.63 (0.0261)	0.75 (0.0074)	10.5 (8,690)	2.8 (0.0278)	1.41 (0.0140)	11.3 (9,350)	3.3 (0.0327)	2.3 (0.0228)	12.5 (10,300)
2.0	14.5 (0.1439)	3.4 (0.0337)	33.6 (27,800)	16 (0.1588)	5.8 (0.0575)	38 (31,400)	20 (0.1984)	8.4 (0.0833)	42 (34,700)
3.0	39 (0.3870)	7.7 (0.0764)	67 (55,400)	45 (0.4465)	12.4 (0.1230)	75 (62,100)	53 (0.5259)	16 (0.1588)	77 (63,700)
4.0	79 (0.7838)	13.5 (0.1339)	108 (89,400)	91 (0.9029)	21 (0.2084)	120 (99,300)	108 (1.072)	26 (0.2580)	124 (103,000)
5.0	139 (1.3792)	21.8 (0.2163)	159 (131,600)	156 (1.548)	31 (0.3076)	173 (143,000)	185 (1.836)	37 (0.3671)	174 (144,000)

long-radius welded elbow, $\beta = 2.22$. Therefore, the actual stress may exceed the table-computed stress, because $\beta > 1$.

Lay out the pipe bend to scale and compute the centroid of the bend by taking line moments about the x and y axes, Fig. 20.

	x axis			y axis	
AB	$(20')(0')$	$= \quad 0$	AB	$(20')(30')$	$= \quad 600$
BC	$(10')(5')$	$= \quad 50$	BC	$(10')(20')$	$= \quad 200$
CD	$(20')(10')$	$= 200$	CD	$(20')(10')$	$= \quad 200$
DE	$(10')(5')$	$= \quad 50$	DE	$(10')(0')$	$= \quad 0$
	60	300		60	1000

$$\bar{y} = 300/60 = 5 \text{ ft } (1.5 \text{ m}) \qquad \bar{x} = 1000/60 = 16.67 \text{ ft } (5.1 \text{ m})$$

In this calculation, the first value is the length of the pipe segment, and the second value is the distance of the center of gravity of the segment from the axis. For a straight section of pipe, the center of gravity is taken as the midpoint of the pipe section. The welded elbows are ignored in stress calculations based on table values.

Lay off $\bar{x}$ and $\bar{y}$ to scale, Fig. 20. Scale the distance to the tangent to the centerline of the long-radius elbow at C. This distance $d = 5.0$ ft (1.52 m). The moment at point C, $M_c = Fd$ lb·ft; or $m_c = 12Fd$ lb·in (1.36 N·m), and $m_c = 12(936)(5) = 56,160$ lb·in (6.34 kN·m), after force F is transposed from the force triangle to the centroid, Fig. 20.

The bending stress at any point in a pipe is $s_b = m\beta/S_m$, where $S_m =$ section modulus of the pipe cross section, in³. For 6-in (152.4-mm) schedule 40 pipe, $S_m = 8.50$ in³ (139.3 cm³), from a table of pipe properties. Then $s_b = (56,160)(2.22)/8.50 = 14,700$ lb/in² (101.3 MPa). This is somewhat greater than the 11,330 lb/in² (78.1 MPa) computed in step 4 but within the allowable stress of 15,000 lb/in² (103.4 MPa) for seamless carbon steel A106 pipe at 500°F (260°C).

By inspection of the scale drawing, Fig. 20, the stress in the long-radius elbows at B and D is less than at C because the moment arm at each of these points is less than at C.

Related Calculations: Tables 29, 30, and 31 present Grinnell-Spielvogel reaction and stress factors for three other single-plane bends—90° turn, U shape with equal tangents, and U shape with unequal legs. Use these tables and the factors in them in the same way as described above. Correct for curved elbows in the same manner. The tables can be used for piping conveying steam, water, gas, oil, and similar liquids, vapors, or gases. For bends of different shape, the analytical method must be used.

REACTING FORCES AND BENDING STRESS IN A TWO-PLANE PIPE BEND

Determine the horizontal reacting forces and bending and torsional stresses in the two-plane pipe bend shown in Table 32 if the dimensions of the bend are $L = 20$ ft (6.1 m); $h = 5$ ft (1.5 m); $a = 5$ ft (1.5 m); $b = 5$ ft (1.5 m). Use the tabular method of solution. The pipe is a 10-in (254-mm) carbon steel schedule 80 line operating at 750 lb/in² (gage) (5170.5 kPa) and 750°F (398.9°C). Determine the combined stress in the pipe.

Calculation Procedure:

1. *Compute the tabular factors for the pipe bend*

To apply the Grinnell-Spielvogel method to two-plane pipe bends, three tabular factors are required: L/a, a/b, and L/h. From the given values, $L/a = 20/5 = 4$; $a/b = 5/5 = 1$; $L/h = 20/5 = 4$.

TABLE 29 90° Turn

Reacting Force $\qquad F_x = k_x \cdot c \cdot \dfrac{I_p}{L^2}$ lb (N)

Reacting Force $\qquad F_y = k_y \cdot c \cdot \dfrac{I_p}{L^2}$ lb (N)

Maximum Bending Stress $\quad s_B = k_b \cdot c \cdot \dfrac{D}{L}$ lb/in^2 (Pa)

I_p in in^4 (cm^4) L in ft (m) D in in (cm)

L/h	k_x	k_y	k_b
1.0	12.0 (0.1191)	12.0 (0.1191)	36 (29,800)
2.0	54.0 (0.5358)	16.6 (0.1647)	102 (84,400)
3.0	150 (1.488)	23.5 (0.2332)	209 (173,000)
4.0	315 (3.125)	31.5 (0.3125)	349 (289,000)
5.0	570 (5.656)	39.5 (0.3919)	528 (437,000)

2. Determine the force and stress factors for the pipe

From Table 32, for the factors in step 1, $k_x = 21.3$; $k_b = 24.5$; $k_t = 7.40$.

3. Compute the horizontal reacting force of the bend

The horizontal reacting force $F_x = k_x c I_p / L^2$, where the symbols are the same as in the preceding calculation procedure, except for L. Substituting the values for 10-in (254-mm) carbon-steel schedule 80 pipe operating at 750°F (399°C), we get $F_x = (21.3)(874)(244.9)/(20)^2 = 11,380$ lb (52.6 kN).

4. Compute the bending stress in the pipe

The bending stress in the pipe is found from $s_b = k \, cD/L$, where the symbols are the same as in the previous calculation procedure. Substituting values gives $s_b = (24.5)(874)(10.75)/(20) = 11,510$ lb/in^2 (79.4 MPa). Table 32 shows that the maximum combined stress in the pipe occurs at the two upper bends, D.

5. Compute the torsional stress in the pipe

The torsional stress in the pipe is found from $s_t = k_t \, cD/L$, where the symbols are the same as in the previous calculation procedure. Substituting values yields $s_t = (7.40)(874)(10.75)/20 = 3475$ lb/in^2 (24 MPa). Table 32 shows that the maximum combined stress in the pipe occurs at the two upper bends, D.

6. Determine the combined stress in the pipe

For any multiplane piping system, the combined stress s_{co} lb/in^2 = $0.5\{s_l + s_c + [4s_t^2 + (s_l - s_c)^2]^{0.5}\}$. In this equation, $s_l = s_b + s_p$, where s_p = pressure due to internal pressure, lb/in^2 (kPa); s_c = circumferential or hoop stress, lb/in^2 (kPa); other symbols are as given earlier. Also, $s_p = pA_i/A_m$, where p = operating pressure, lb/in^2 (gage) (kPa); A_i = inside area of pipe cross section, in^2 (cm^2); A_m = metal area of pipe cross section, in^2 (cm^2). Likewise, $s_c = p(D - t)/2t$, where D = outside diameter of pipe, in; t = pipe-wall thickness, in (cm).

TABLE 30 U Shape with Equal Tangents

Reacting Force $\qquad F_x = k_x \cdot c \cdot \dfrac{I_P}{L^2}$ lb (N)

Maximum Bending Stress $\quad s_B = k_b \cdot c \cdot \dfrac{D}{L}$ lb/in² (Pa)

I_P in in⁴ (cm⁴) L in ft (m) D in in (cm)

L/h	L/a = 2		3		4		5		6	
	k_x	k_b	k_x	k_b	k_x	k_b	k_x	k_b	k_x	k_b
1.0	2.40 (0.0238)	7.20 (5,960)	2.46 (0.0244)	8.2 (6,780)	2.52 (0.0250)	8.82 (7,300)	2.58 (0.0256)	9.29 (7,690)	2.64 (0.0262)	9.69 (8,020)
2.0	12.00 (0.1191)	18.00 (14,900)	12.5 (0.1240)	21.8 (18,000)	13.24 (0.1314)	24.8 (20,500)	13.87 (0.1376)	27.1 (22,400)	14.4 (0.1429)	28.8 (23,800)
3.0	29.45 (0.2922)	29.45 (24,400)	31.2 (0.3096)	37.4 (30,900)	33.6 (0.3334)	43.7 (36,200)	35.8 (0.3552)	48.7 (40,300)	37.7 (0.371)	52.7 (43,600)
4.0	54.9 (0.5447)	41.1 (34,000)	58.5 (0.5804)	53.6 (44,300)	64.0 (0.6350)	64.0 (53,000)	69.1 (0.6856)	72.5 (60,000)	73.6 (0.7303)	79.7 (65,900)
5.0	88.2 (0.8751)	52.9 (43,800)	95.3 (0.9456)	70.8 (58,600)	104.6 (1.0378)	85.2 (70,500)	114.7 (1.1380)	97.8 (80,900)	122.5 (1.2154)	107.5 (88,900)

TABLE 31 U Shape with Unequal Legs

Reacting Force $\qquad F_x = k_x \cdot c \cdot \dfrac{I_P}{L^2}$ lb (N)

Reacting Force $\qquad F_y = k_y \cdot c \cdot \dfrac{I_P}{L^2}$ lb (N)

Maximum Bending Stress $\quad s_B = k_b \cdot c \cdot \dfrac{D}{L}$ lb/in² (Pa)

I_P in in⁴ (cm⁴) L in ft (m) D in in (cm)

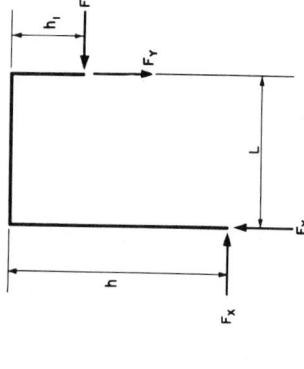

L/h	h/h_1								
	4/3			2			3		
	k_x	k_y	k_b	k_x	k_y	k_b	k_x	k_y	k_b
0.2	0.07 (0.0007)	0.6 (0.0059)	1.5 (1,240)	0.29 (0.0029)	1.8 (0.0179)	7 (5,790)	0.53 (0.0053)	3.4 (0.0337)	11 (9,100)
0.8	2.4 (0.0238)	0.9 (0.0089)	9.5 (7,860)	3.6 (0.0357)	2.5 (0.0248)	15 (12,400)	4.8 (0.0476)	4.4 (0.0436)	20 (16,500)
1.0	4.3 (0.0127)	1.2 (0.0119)	16 (13,200)	6.2 (0.0615)	3.0 (0.0298)	21 (17,400)	8 (0.0794)	4.9 (0.0486)	26 (21,500)
2.0	27 (0.2679)	2.3 (0.0228)	58 (48,000)	37 (0.3671)	6.0 (0.0595)	75 (62,100)	44 (0.4366)	8.0 (0.0883)	88 (72,800)
3.0	81 (0.8037)	3.8 (0.0377)	124 (102,600)	110 (1.0914)	10.0 (0.0992)	162 (134,000)	128 (1.2700)	15 (0.1488)	185 (153,000)

3.449

TABLE 32 Two-Plane U with Tangents

Reacting Force $\quad F_x = k_x \cdot c \cdot \dfrac{I_P}{L^2}$ lb (N)

Bending Stress $\quad s_B = k_B \cdot c \cdot \dfrac{D}{L}$ lb/in² (Pa)

Torsional Stress $\quad s_T = k_T \cdot c \cdot \dfrac{D}{L}$ lb/in² (Pa)

I_P in in⁴ (cm⁴) $\quad L$ in ft (m) $\quad D$ in in (cm)

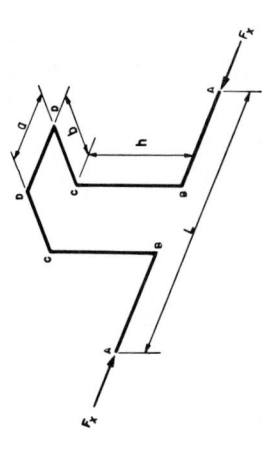

| | a/b | | | | | | | | | L/a = 4 | | |
| | 0.25 | | | 0.5 | | | 1 | | | 2 | | |
L/h	k_x	k_b	k_t	k_x	k_b	k_t	k_x	k_b	k_t	k_x	k_b	k_t
1	0.67 (0.0066)	D 3.20 (2,650)	…	1.22 (0.0121)	A 4.35 (3,600)	A 0.30 (250)	1.67 (0.0166)	A 5.2 (4,300)	A 0.15 (120)	2.0 (0.0198)	C 6.3 (5,200)	
2	1.35 (0.0134)	D 5.80 (6,800)	…	4.30 (0.0427)	D 9.96 (8,240)	D 2.45 (2,030)	6.96 (0.0691)	A 11.0 (9,100)	…	9.3 (0.0923)	C 15.0 (12,400)	
3	1.70 (0.0169)	D 7.00 (5,790)	…	6.23 (0.0618)	D 13.8 (11,400)	D 2.28 (1,890)	14.0 (0.1389)	D 16.5 (13,700)	6.55 (5,420)	21.2 (0.2103)	C 24.0 (19,900)	
4	1.88 (0.0187)	D 7.44 (6,150)	…	7.84 (0.0778)	D 16.9 (14,000)	D 2.09 (1,730)	21.3 (0.2113)	D 24.5 (20,300)	D 7.40 (6,120)	36.2 (0.3592)	C 30.0 (24,800)	
5	2.01 (0.0199)	D 7.75 (6,410)	…	8.94 (0.0887)	D 18.8 (15,600)	D 1.89 (1,560)	27.8 (0.2758)	31.4 (26,000)	D 7.75 (6,410)	52.6 (0.5219)	D 31.0 (25,600)	D 17.3 (14,300)

Computing stress values for this 10-in (254-mm) schedule 80 carbon-steel pipe operating at 750 lb/in² (gage) (5.2 MPa) and 750°F (399°C), and using values from a table of pipe properties, we get $s_p = (750)(71.8/18.92) = 2845$ lb/in² (19.6 MPa); $s_c = (750)(10.75 - 0.593)/2(0.593) = 6420$ lb/in² (44.3 MPa). Then $s_1 = s_b + s_p = 11,510 + 2845 = 14,355$ lb/in² (99 MPa), where s_b is from step 4. By substituting in the combined-stress equation, $s_{co} = 0.5\{14,355 + 6420 + [4 \times 3475^2 + (14,355 - 6420)^2]^{0.5}\} = 16,648$ lb/in² (114.8 MPa). This is higher than the stress allowed in carbon-steel pipe by the ANSA *Piping Code*, unless the pipe conforms to the special conditions of certain paragraphs of the *Code*.

Related Calculations: Use this procedure for piping conveying steam, water, gas, oil, and similar liquids, vapors, or gases. For bends of different shape, the analytical method is generally used.

REACTING FORCES AND BENDING STRESS IN A THREE-PLANE PIPE BEND

Determine the three reacting forces and moments and bending and torsional stresses in the three-plane pipe bend shown in Table 33 if the dimensions of the bend are $L_1 = 20$ ft (6.1 m), $L_2 = 10$ ft (3.0 m), $L_3 = 5$ ft (1.5 m). The pipe is 10-in (254-mm) carbon-steel schedule 80 operating at 750 lb/in² (gage) (5.2 MPa) and 750°F (399°C).

Calculation Procedure:

1. Compute the tabular factors for the pipe bend

From the Grinnell-Spielvogel method, the two tabular factors required are $m = L_1/L_3$ and $n = L_2/L_3$; or $m = 20/5 = 4$ and $n = 10/5 = 2$.

2. Determine the force and stress factors for the pipe

From Table 33, for the factors in step 1, $k_b = 8.0$; $k_t = 3.6$; $k_x = 1.48$; $k_y = 0.13$; $k_z = 0.80$; $k_{xy} = 1.3$; $k_{xz} = 1.2$; $k_{yz} = 0.51$.

3. Compute the longitudinal reaction force of the bend

The horizontal reacting force $F_x = k_x c I_p/L_3^2$, where the symbols are the same as in the preceding calculation procedure, except for L_3. By substituting values for 10-in (254-mm) carbon-steel schedule 80 pipe operating at 750°F (398.9°C), $F_x = (1.48)(874)(244.9)/(5)^2 = 12,680$ lb (56.4 kN).

4. Compute the vertical reacting force of the bend

The vertical reacting force $F_y = k_y c I_p/L_3^2$, where the symbols are the same as in the preceding calculation procedure, except for L_3. Substituting values for this pipe, we find $F_y = (0.13)(874)(244.9)/(5)^2 = 1115$ lb (4.96 kN).

5. Compute the horizontal reacting force of the bend

The horizontal reacting force $F_z = k_z c I_p/L_3^2$, where the symbols are the same as in the preceding calculation procedure, except for K_z and L_3. Substituting values for this pipe gives $F_z = (0.80)(874)(244.9)/(5)^2 = 6850$ lb (30.5 kN).

6. Compute the bending and torsional stresses in the pipe

The bending stress $s_b = k_b cD/L_3$, where the symbols are the same as in the preceding calculation procedure, except for L_3. Substituting values for this pipe gives $s_b = (8.0)(874)(10.75)/5 = 15,020$ lb/in² (103.6 MPa).

The torsional stress $s_t = k_t cD/L_3$, where the symbols are the same as in the preceding calculation procedure, except for L_3. By substituting values for this pipe, $s_t = (3.6)(874)(10.75)/5 = 6760$ lb/in² (46.6 MPa).

7. Compute the three reacting moments at the pipe end

For each bending moment M ft·lb $= kc I_p/L_3$, where the symbols are the same as given in the previous steps in this calculation procedure, except that k is the appropriate bending-moment factor.

For the xy moment $M_{xy} = (1.3)(874)(244.9)/5 = 55,700$ ft·lb (75.5 kN·m). For the xz moment $M_{xz} = (1.2)(874)(244.9)/5 = 51,400$ ft·lb (69.7 kN·m). For the yz moment $M_{yz} = (0.51)(874)(244.9)/5 = 22,250$ ft·lb (30.2 kN·m).

TABLE 33 Three-Dimensional 90° Turns

$L_1 \geq L_3$ $L_1 = m$ $\dfrac{L_2}{L_3} = n$

Bending Stress $s_B = k_b \cdot c \cdot \dfrac{D}{L}$ lb/in² (Pa)

Torsional Stress $s_T = k_t \cdot c \cdot \dfrac{D}{L_3}$ lb/in² (Pa)

Reacting Force $F_x = k_x \cdot c \cdot \dfrac{I_P}{L_3^2}$ lb (N)

Reacting Force $F_y = k_y \cdot c \cdot \dfrac{I_P}{L_3^2}$ lb (N)

Reacting Force $F_z = k_z \cdot c \cdot \dfrac{I_P}{L_3^2}$ lb (N)

Reacting Moment $M_{xy} = k_{xy} \cdot x \cdot \dfrac{I_P}{L_3}$ ft·lb (N·m)

Reacting Moment $M_{xx} = k_{xx} \cdot c \cdot \dfrac{I_P}{L_3}$ ft·lb (N·m)

Reacting Moment $M_{yz} = k_{yz} \cdot c \cdot \dfrac{I_P}{L_3}$ ft·lb (N·m)

I_p in in⁴ (cm⁴) L in ft (m) D in in (cm)

m = 3

n	k_b	k_t	k_x	k_y	k_z	k_{xy}	k_{xz}	k_{yz}
1	A 22.3 (18,500)	A 4.86 (4,020)	4.5 (0.045)	0.54 (0.0054)	1.50 (0.0148)	3.6 (0.0357)	1.62 (0.0161)	0.74 (0.0073)
2	D 9.3 (7,690)	D 0.15 (120)	1.4 (0.014)	0.22 (0.0022)	1.10 (0.0109)	1.1 (0.0109)	1.00 (0.0099)	0.71 (0.0070)
3	D 10.0 (8,270)	D 0.24 (199)	0.76 (0.007)	0.13 (0.0013)	1.08 (0.0107)	0.60 (0.0059)	0.74 (0.0073)	0.70 (0.0069)

m = 4

k_b	k_t	k_x	k_y	k_z	k_{xy}	k_{xz}	k_{yz}	n
A 24.0 (19,900)	A 6.0 (4,960)	4.80 (0.0476)	0.37 (0.0037)	1.10 (0.0110)	3.9 (0.0387)	2.0 (0.0199)	0.61 (0.0061)	1
A 8.0 (6,620)	A 3.6 (2,980)	1.48 (0.0147)	0.13 (0.0013)	0.80 (0.0079)	1.3 (0.0129)	1.2 (0.0119)	0.51 (0.0051)	2
D 7.26 (6,000)	D 0.10 (83)	0.76 (0.0075)	0.09 (0.0009)	0.65 (0.0064)	0.6 (0.0059)	0.88 (0.0088)	0.42 (0.0042)	3

Related Calculations: Use this procedure for piping conveying steam, water, gas, oil, and similar liquids, vapors, or gases. For bends of different shape, the analytical method must be used. Compute the combined stress in the same way as in step 6 of the previous calculation procedure. Table 33 shows that the maximum combined stress occurs at point A in this piping system.

ANCHOR FORCE, STRESS, AND DEFLECTION OF EXPANSION BENDS

Determine the deflection and anchor force in an 8-in (203.2-mm) schedule 40 double-offset expansion U bend having a radius of 64 in (1626 mm) if the bending stress is 10,000 lb/in² (68.9 MPa). What would the deflection and anchor force be with a bending stress of 15,000 lb/in² (103.4 MPa) if the bend tangents are guided and the pipe is carbon steel operating at 500°F (260°C)? With a bending stress of 8000 lb/in² (55.2 MPa)? Tabulate the deflection and anchor-force equations for the popular types of expansion bends when the expanding pipe is guided axially.

Calculation Procedure:

1. Compute the deflection of the pipe bend

For a double-offset expansion U bend, the deflection d in $= 0.728R^2K/D\beta$, where $R =$ bend radius, ft; $K =$ flexibility factor for curved pipe, from a table of pipe properties or from $K = (12\lambda^2 + 10)/(12\lambda^2 + 1)$, where $\lambda = 12tR/r^2$, where $t =$ pipe thickness, in, $r = (D - t)/.2$; $D =$ outside diameter of pipe, in; $\beta =$ stress coefficient for curved pipe from a table of pipe properties or from $\beta = (2K/3)[(6\lambda^2 + 5)/18]^{0.5}$ when $\lambda \le 1.47$; $\beta = (12\lambda^2 - 2)/(12\lambda^2 + 1)$ when $\lambda > 1.47$.

For this bend, $R = 64/12 = 5.33$ ft (1.62 m); $K = 1.49$ from a table of pipe properties or by computation; $D = 8.625$ in (219.1 mm) from a table of pipe properties; $\beta = 0.86$ from a table of pipe properties, or by computation. Then $d = (0.728)(5.33)^2(1.49)/[(8.625)(0.86)] = 4.15$ in (105.4 mm).

2. Compute the anchor force of the pipe bend

For a double-offset expansion U bend, the anchor force F_x lb $= 976 \, I_p/(RD\beta)$, where $I_p =$ moment of inertia of pipe cross section, in⁴. For this pipe, use values from a table of pipe properties; or computing the values, $F_x = (976)(72.5)/[(5.33)(8.625)(0.86)] = 1790$ lb (7.96 kN).

3. Compute the deflection and anchor force for a larger bending stress

With a larger bending stress—15,000 lb/in² (103.4 MPa) in this instance—and a greater deflection at the higher stress d_h, solve $d_h = (d)(\text{allowable stress, lb/in}^2)/(10,000M_R)$, or $d_h = (4.15)(15,000)/[10,000(0.932)] = 6.68$ in (169.7 mm). As in a preceding calculation, $M_R =$ modulus ratio $= 0.932$.

The anchor force at the larger bending stress is $F_h = F_x d_h M_R/d$, or $F_x = (1790)(6.68)(0.932)/4.15 = 2680$ lb (11,921 N).

4. Compute the deflection and anchor force for a smaller bending stress

Use the same equation as in step 3, except that the lower bending stress is substituted for the higher one. Or, $d_1 = (4.15)(8000)/[10,000(0.932)] = 3.56$ in (90.4 mm), and $F_1 = (1790)(3.56)(0.932)/4.15 = 1432$ lb (6370 N).

5. Tabulate the deflection and anchor-force equations

Do this as shown in the table on page 3.442.

Related Calculations: Use the procedures given here for piping conveying steam, water, oil, gas, air, and similar vapors, liquids, and gases. The value of E in step 5, 29×10^6 lb/in² (199.9 MPa), is satisfactory for pipes made of carbon steel, carbon moly steel, chromium moly steel, nickel steel, and chromium nickel steel. These materials are commonly used in piping systems requiring expansion bends.

Note that the equations in step 5 apply to pipe bends having guides to direct the axial expansion of the pipe. This is the usual arrangement used today because unguided bends require too much space. For design of unrestrained bends, multiply d by 1.5 to find the deflection at the higher stress, as in step 3. This factor, 1.5, is an approximation, but it is on the safe side in almost every case. The equations given in step 5 are presented in great detail in Grinnell—*Piping Design and Engineering* and Crocker and King—*Piping Handbook*.

Bend type	Deflection for 10,000 lb/in² (68.9-MPa) s_b		Anchor force for 10,000 lb/in² (68.9-MPa) s_b	
	$E = 29 \times 10^6$ lb/in²	$E = 199.9$ MPa	$E = 29 \times 10^6$ lb/in²	$E = 199.9$ MPa
Double-offset U	$d = 0.728R^2K/D\beta$	$d = 5.056 \times 10^{-3}R^2K/D\beta$	$F_x = 976\,I_p/RD\beta$	$F_x = 8070 \times I_p/RD\beta$
Expansion U bend (no tangents)	$d = 0.312R^2K/D\beta$	$d = 2.167 \times 10^{-3}R^2K/D\beta$	$F_x = 1667I_p/RD\beta$	$F_x = 13,780 \times I_p/RD\beta$
Expansion U bend [tangents = 2 ft (0.6 m)]	$d = [(0.312R^3 + 0.795R^2 + 0.624R)K + 0.132]/(R + 1)D\beta$	$d = [(2.167R^3 + 165R^2 + 3895R) \times 10^{-3}K + 24.7]/(R + 30)D\beta$	$F_x = 1667I_p/(R + 1)D\beta$	$F_x = 13,780I_p/(R + 30)D\beta$
Expansion U bend (tangents = R)	$d = (0.577 + 0.011)R^2/D\beta$	$d = (4.007K + 0.076)R^2/D\beta$	$F_x = 1111I_p/RD\beta$	$F_x = 9190I_p/RD\beta$
Expansion U bend (tangents = $2R$)	$d = (0.865K + 0.0662)R^2/D\beta$	$d = (6.00 \times 10^{-3}K + 0.459 \times 10^{-3})R^2/D\beta$	$F_x = 833I_p/RD\beta$	$F_x = 6890I_p/RD\beta$
Expansion U bend (tangents = $4R$)	$d = (1.465K + 0.353)R^2/D\beta$	$d = (10.17K + 2.45) \times 10^{-3} \times R^2/D\beta$	$F_x = 556I_p/RD\beta$	$F_x = 4600I_p/RD\beta$
Double-offset U bend	$d = 0.260R^2K/D\beta$	$d = 1.806 \times 10^{-3}R^2K/D\beta$	$F_x = 1209I_p/RD\beta$	$F_x = 9997I_p/RD\beta$
Single-offset quarter bend	$d = 0.0366R^2K/D\beta$	$d = 2.5 \times 10^{-4}R^2K/D\beta$	$F_x = 2763I_p/RD\beta$	$F_x = 22,850I_p/RD\beta$
Circle bend	$d = 0.312R^2K/D\beta$	$d = 2.167 \times 10^{-3}R^2K/D\beta$	$\begin{cases} F_y = 0.066F_x \\ F_x = 1667I_p/RD\beta \end{cases}$	$\begin{cases} F_y = 0.066F_x \\ F_x = 13,780 \times I_p/RD\beta \end{cases}$

3.454

SLIP-TYPE EXPANSION JOINT SELECTION AND APPLICATION

Select and size slip-type expansion joints for the 20-in (508-mm) carbon-steel schedule 40 pipeline in Fig. 21 if the pipe conveys 125-lb/in^2 (gage) (861.6-kPa) steam having a temperature of 380°F (193°C). The minimum temperature expected in the area where the pipe is installed is 0°F (-17.8°C). Determine the anchor loads that can be expected. The steam inlet to the pipe is at A; the outlet is at F.

Calculation Procedure:

1. Determine the expansion of each section of pipe

From Fig. 22, the expansion of steel pipe at 380°F (193°C) with a 0°F (-17.8°C) minimum temperature is 3.4 in (88.9 mm) per 100 ft (30.5 m) of pipe. Expansion of each section of pipe is then e in = (3.4)(pipe length, ft/ 100). For AB, e = (3.4)(140/100) = 4.76 in (120.9 mm); for BC, e = (3.4)(90/100) = 3.06 in (77.7 mm); for CD, e = (3.4)(220/100) =

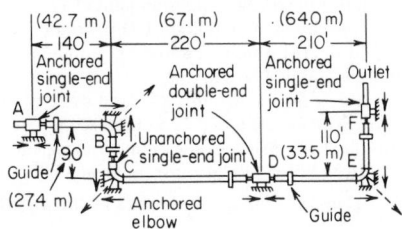

FIG. 21 Slip-type expansion joints in a piping system. *(Yarway Corporation.)*

7.48 in (190 mm); for DE, e = (3.4)(210/100) = 71.4 in (1813.6 mm); for EF, e = (3.4)(110/100) = 3.74 in (95 mm).

2. Select the type and the traverse of each expansion joint

The slip-type expansion joint at A will absorb expansion from only one direction—the right-hand side. This expansion will occur in pipe section AB and is 4.76 in (120.9 mm) from step 1. Therefore, a single-end slip-type expansion joint (one that absorbs expansion on only one side) can be used. The traverse—the amount of expansion a slip joint will absorb—is usually given in multiples of 4 in (101.6 mm), that is, 4, 8, and 12 in (101.6, 203.2, and 304.8 mm). Hence, an 8-in (203.2-mm) traverse slip-type single-end joint will be suitable at A because the expansion is 4.76 in (120.9 mm). A 4-in (101.6-mm) traverse joint would be unsatisfactory because it could not absorb at 4.76-in (120.9-mm) expansion.

The next joint, at C, must absorb the expansion in the vertical pipe BC. Since the elbow beneath the joint is anchored, an unanchored joint can be used. With pipe expansion in only one direction—from B to C—a single-end joint can be used. Since the expansion of section BC is 3.06

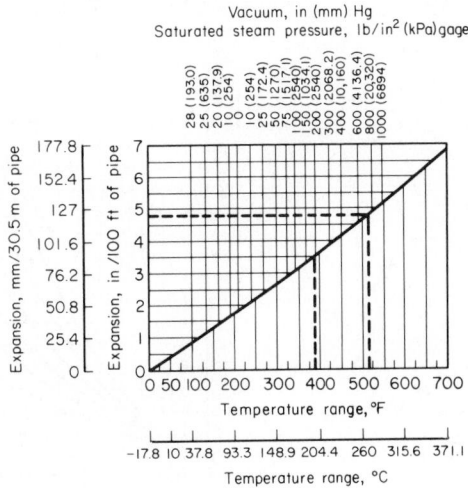

FIG. 22 Expansion of steel pipe.
(Yarway Corporation.)

in (77.7 mm), use a single-end 4-in (101.6-mm) traverse slip-type expansion joint, unanchored at C.

The expansion joint at *D* must absorb expansion from two directions—from *C* to *D* and from *E* to *D*. Therefore, a double-end joint (one that can absorb expansion on each end) must be used. The double-end joint must be anchored because the pipe expands *away* from the anchored elbow *C* in section *CD* and *away* from the anchored elbow *E* in section *DE*. In both instances the pipe expands *toward* the expansion joint at *D*.

The expansion in section *CD* is, from step 1, 7.48 in (190 mm), whereas the expansion in *DE* is 7.14 in (181.4 mm). Therefore, a double-end anchored joint with an 8-in (203.2-mm) traverse at *each* end will be suitable.

Since the pipe outlet is at *F* and there is no anchor in the pipe at *F*, the expansion joint at this point must be anchored. The pipe section between *E* and *F* will expand vertically upward into the joint for a distance of 3.74 in (95 mm), as computed in step 1. Therefore, a single-end anchored joint with a 4-in (101.6-mm) traverse will be suitable.

3. *Compute the anchor loads in the pipeline*

Use Fig. 23 to determine the anchor loads on intermediate and end anchors (those where the pipe makes a sharp change in direction). Enter Fig. 23 at the bottom at a pipe size of 20-in (508-mm) diameter, and project vertically upward to the dashed curve labeled *intermediate anchor—all pressures*. At the left read the anchor load at each intermediate anchor, *A*, *D*, and *F*, as 20,000 lb (88.9 kN). Note that the joint expansion load = joint contraction load = 20,000 lb (88.9 kN).

The end anchors, *B*, *C*, and *E*, have, from Fig. 23, a possible maximum load of 58,000 lb (258 kN), found by projecting vertically upward from the 20-in (508-mm) pipe size to 125-lb/in^2 (gage) (862-kPa) steam pressure, which lies midway between the 100- and 150-lb/in^2 (gage) (689.5- and 1034-kPa) curves. Indicate the possible maximum end-anchor loads by the solid arrows at each elbow, as shown in Fig. 21. The resultant *R* of the loads at any end anchor is found by the pythagorean theorem to be $R = (58,000^2 + 58,000^2)^{0.5} = 82,200$ lb (365.6 kN). Indicate the resultant by a dotted arrow, as shown in Fig. 21.

Contraction loads on the end anchors are in the reverse direction and consist only of friction. This friction load equals the joint expansion load, or 20,000 lb (88.9-kN). The resultant of the joint expansion loads is $(20,000^2 + 20,000^2)^{0.5} = 28,350$ lb (126.1 kN).

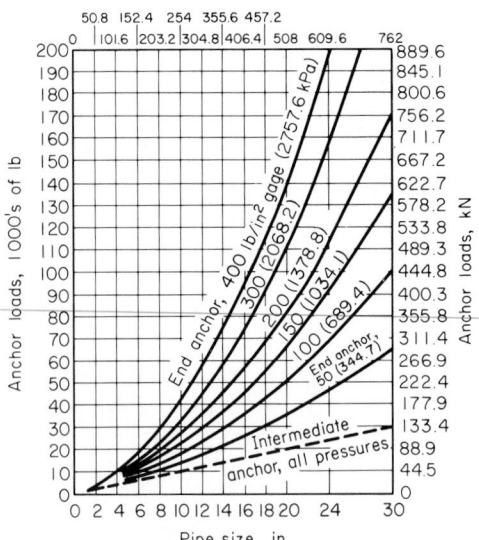

FIG. 23 End- and intermediate-anchor loads in piping systems. *(Yarway Corporation.)*

TABLE 34 Guide and Support Spacing

| Nominal pipe size, in (mm) | Distance between guide and joint, ft (m) | | Distance between guides, ft (m) |
| | Packing type | | |
	Gun	Gland	
18 (457)	24 (7.3)	11 (3.4)	100 (30.5)
20 (508)	25 (7.6)	12 (3.7)	105 (32)
24 (610)	26 (7.9)	12 (3.7)	110 (33.5)

Locate guides within 25 or 12 ft (7.62 or 3.66 m) of the expansion joint, depending on the type of packing used, Table 34. These guides should allow free axial movement of the pipe into and out of the joint with minimum friction.

Related Calculations: Use this procedure to choose slip-type expansion joints for pipes conveying steam, water, air, oil, gas, and similar vapors, liquids, and gases. In some instances, the gland friction and pressure thrust is used instead of Fig. 23 to determine anchor loads. With either method, the results are about the same.

CORRUGATED EXPANSION JOINT SELECTION AND APPLICATION

Select corrugated expansion joints for the 8-, 6-, and 4-in (203.2-, 152.4-, and 101.6-mm) carbon-steel pipeline in Fig. 24 if the steam pressure in the pipe is 75 lb/in^2 (gage) (517.1 kPa), the steam temperature is 340°F (171°C), and the installation temperature is 60°F (15.6°C).

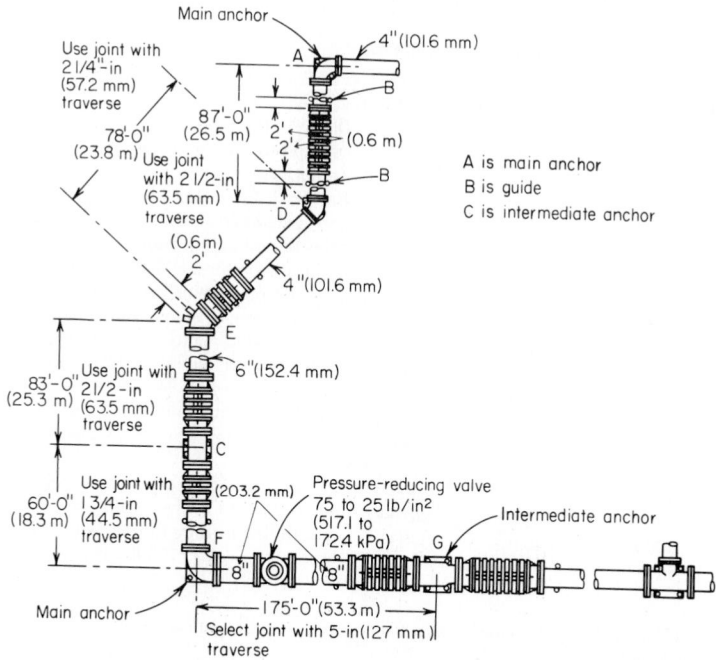

FIG. 24 Piping system fitted with expansion joints. *(Flexonics Division, Universal Oil Products Company.)*

Calculation Procedure:

1. Determine the expansion of each section of pipe

From a table of thermal expansion of pipe, the expansion of carbon-steel pipe at 340°F (171°C) is 2.717 in/100 ft (2.26 mm/30.5 m) from 0 to 340°F (−17.8 to 171°C). Between 0 and 60°F (−17.8 and 15.6°C) the expansion is 0.448 in/100 ft (50 mm/30.5 m). Hence, the expansion between 60 and 340°F (15.6 and 171°C) is $2.717 − 0.448 = 2.269$ in/100 ft (1.89 mm/m). This factor can now be applied to each length of pipe by finding the product of (pipe-section length, ft/100)(expansion, in/100 ft) = expansion of section, in = e.

For section AD, $e = (87/100)(2.269) = 1.97$ in (50 mm); for DE, $e = (78/100)(2.269) = 1.77$ in (45 mm); for EC, $e = (83/100)(2.269) = 1.88$ in (47.8 mm); for CF, $e = (60/100)(2.269) = 1.36$ in (34.5 mm); for FG, $e = (175/100)(2.269) = 3.97$ in (100.8 mm).

In selecting corrugated expansion joints, the usual practice is to increase the computed expansion by a suitable safety factor to allow for any inaccuracies in temperature measurement. By applying a 25 percent safety[1] factor: for AD, $e = (1.97)(1.25) = 2.46$ in (62.5 mm); for DE, $e = (1.77)(1.25) = 2.13$ in (54.1 mm); for EC, $e = (1.88)(1.25) = 2.35$ in (59.7 mm); for CF, $e = (1.36)(1.25) = 1.70$ (43.2 mm); for FG, $e = (3.97)(1.25) = 4.96$ in (126 mm).

2. Select the traverse for, and type of, each expansion joint

Obtain corrugated-expansion joint engineering data, and select a joint with the next largest traverse for each section of pipe. Thus, traverse $AD \geq 2\frac{1}{2}$ in (63.5 mm); traverse $DE \geq 2\frac{1}{4}$ in (57.2 mm); traverse $EC \geq 2\frac{1}{2}$ in (63.5 mm); traverse $CF \geq 1\frac{3}{4}$ in (44.5 mm); traverse $FG \geq 5.0$ in (127 mm).

Two types of expansion joints are commonly used: free-flexing and controlled-flexing. Free-flexing joints are generally used where the pressures in the pipeline are relatively low and the required motion is relatively small. Controlled-flexing expansion joints are generally used for higher pressures and larger motions. Both types of expansion joints are available in stainless steel in both single and dual units. For precise data on a given joint being considered, consult the expansion-joint manufacturer. Corrugated expansion joints are characterized by their freedom from any maintenance needs.

3. Compute the anchor loads in the pipeline

Main anchors are used between expansion joints, as at F and A, Fig. 24, and at turns such as at F and A. The force[2] a main anchor must absorb is given by F_i lb $= F_p + F_e$, where $F_p =$ pressure thrust in the pipe, lb $= pA$, where $p =$ pressure in pipe, lb/in² (gage); $A =$ effective internal cross-sectional area of expansion joint, in² (see Table 35 for cross-sectional areas of typical corrugated joints); $F_e =$ force required to compress the expansion joint, lb $= [300$ lb/in (52.5 N/mm)](joint inside diameter, in) for stainless-steel self-equalizing joints, and [200 lb/in (35 N/mm)] (joint inside diameter, in) for copper nonequalizing joints. Determining the main anchor force for the 8-in (203.2-mm) pipeline gives $F_i = (75)(85) + (300)(8) = 8775$ lb (39.0 kN). In this equation, the area of 85 in² (548.3 cm²) in the first term is obtained from Table 35.

The total force at a main anchor, as at A and F, Fig. 24, is the vector sum of the forces in each line leading to the anchor. Thus, at F, there is a force of 8775 lb (39.0 kN) in the 8-in

TABLE 35 Effective Area of Corrugated Expansion Joints

Joint inside diameter		Joint effective area	
in	mm	in²	cm²
6	152.4	51.0	329.0
8	203.2	85.0	548.4
10	254.0	120.0	774.2
12	304.8	174.0	1122.6
14	355.6	215.0	1387.1
16	406.4	270.0	1741.9
18	457.2	310.0	1999.9
20	508.0	390.0	2516.1
24	609.6	540.0	3483.9

[1]This value is for illustration purposes only. Contact the expansion-joint manufacturer for the exact value of the safety factor to use.

[2]This is an approximate method for finding the anchor force. For a specific make of expansion joint, consult the joint manufacturer.

(203.2-mm) line and a force of $F_i = (75)(51) + (300)(6) = 5625$ lb (25 kN) in the 6-in (152.4-mm) line connected to the elbow outlet. Since the elbow at F is a right angle, use the pythagorean theorem, or $R =$ resultant anchor force, lb $= (8775^2 + 5625^2)^{0.5} = 10,400$ lb (46.3 kN).

Where two lines containing corrugated expansion joints are connected by a bend of other than 90°, as at D and E, use a force triangle to determine the anchor force after computing F_i for each pipe. Thus, at E, F_i for the 6-in (152.4-mm) pipe $= 5625$ lb (25 kN), and $F_i = (75) \times (23.5) + (300)(4) = 2963$ lb (13.2 kN) for the 4-in (101.6-m) pipe. Draw the force triangle in Fig. 25 with the 6-in (152.4-mm) pipe F_i and the 4-in (101.6-mm) pipe F_i as two sides and the bend angle, 45°, as the included angle. Connect the third side, or resultant, to the ends of the force vectors, and scale the resultant as 4125 lb (18.4 kN), or compute the resultant from the law of cosines. Find the resultant force at D in a similar manner as 2963 lb (13.2 kN).

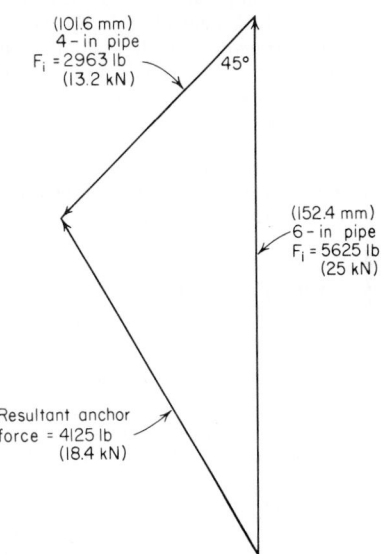

Intermediate anchors, as at C and G, must withstand only one force—the unbalanced (differential) spring force. With approximate force calculations,[1] starting at C, for a 6-in (152.4-mm) expansion joint, $F_e = (300)(6) = 1800$ lb (8 kN). At G, for an 8-in (203.2-mm) expansion joint, $F_e = (300)(8) = 2400$ lb (10.7 kN). Thus, the loads the intermediate anchors must withstand are considerably less than the main-anchor loads.

Provide the pipe guides at suitable locations in accordance with the joint manufacturer's recommendations and at suitable intervals on the pipeline to prevent any lateral and buckling forces on the joint and adjacent piping. Intermediate anchors between two joints in a straight run of pipe ensure that each joint will absorb its share of the total pipe motion. Slope the pipe in the direction of fluid flow to prevent condensate accumulation. Use enough pipe hangers to prevent sagging of the pipe.

FIG. 25 Force triangle for determining piping anchor force.

Related Calculations: Use this procedure to choose corrugated-type expansion joints for pipes conveying steam, water, air, oil, gas, and similar vapors, liquids, and gases. When choosing a specific make of corrugated expansion joint, use the manufacturer's engineering data, where available, to determine the maximum allowable traverse. One popular make has a maximum traverse of 7.5 in (190.5 mm) or a maximum allowable lateral motion of 1.104 in (28.0 mm) in its various joint sizes. The larger the lateral motion, the greater the number of corrugations required in the joint.

In some pipelines there is an appreciable pressure thrust caused by a change in direction of the pipe. This pressure or centrifugal thrust F_c is usually negligible, but the wise designer makes a practice of computing this thrust from $F_c = (2A\rho v^2/32.2) \times (\sin \theta/2)$ lb, where $A =$ inside area of pipe, ft²; $\rho =$ density of fluid or vapor, lb/ft³; $v =$ fluid or vapor velocity, ft/s; $\theta =$ change in direction of the pipeline.

The number of corrugations required in a joint varies with the expansion and lateral motion to be absorbed. A typical free-flexing joint can absorb 6.25 in (158.8 mm) of expansion and a variable amount of lateral motion, depending on joint size and operating condition. Free-flexing joints are commonly built in diameters up to 48 in (1219 mm), while controlled-flexing joints are commonly built in diameters up to 24 in (609.6 mm). For a more precise calculation procedure, consult the Flexonics Division, Universal Oil Products Company.

---

[1]Consult the expansion-joint manufacturer for an exact procedure for computing the anchor forces.

DESIGN OF STEAM TRANSMISSION PIPING

Design a steam transmission pipe to supply a load that is 1700 ft (518.2 m) from the power plant. The terrain permits a horizontal run between the power plant and the load. Maximum steam flow required by the load is 300,000 lb/h (135,000 kg/h), whereas the average steam flow required is estimated as 150,000 lb/h (67,500 kg/h). The maximum steam pressure at the load must not exceed 150 lb/in^2 (abs) (1034.1 kPa) saturated. Superheated steam at 450 lb/in^2 (abs) (3102.7 kPa) and 600°F (316°C) is available at the power plant. Two schemes are proposed for the line: (1) Reduce the steam pressure to 180 lb/in^2 (abs) (1240.9 kPa) at the line inlet, thus allowing a 180 − 150 = 30-lb/in^2 (206.8-kPa) loss in the 1700-ft (518.2-m) long line. This scheme is called the *nominal pressure-loss line*. (2) Admit high-pressure steam to the line and thereby allow the steam pressure to fall to a level slightly greater than 150 lb/in^2 (abs) (1034.1 kPa). Since 600°F (316°C) steam would probably cause expansion and heat-loss difficulties in the pipe, assume that the inlet temperature of the steam is reduced to 455°F (235°C) in a desuperheater in the power plant. There is a 10-lb/in^2 (68.9-kPa) pressure loss between the power plant and the line, reducing the line inlet pressure to 440 lb/in^2 (abs) (3033.4 kPa). Since the pressure can fall about 440 − 150 = 290 lb/in^2 (1999.3 kPa), this will be called the *maximum pressure-loss line*. During design, determine which line is the most economical.

Calculation Procedure:

1.   *Determine the required pipe diameter for each condition*

The average steam pressure in the nominal pressure-loss line is (inlet pressure + outlet pressure)/2 = (180 + 150)/2 = 165 lb/in^2 (abs) (1138 kPa). Use this average pressure to determine the pipe size, because the average pressure is more representative of actual conditions in the pipe. Assume that there will be a 5-lb/in^2 (34.5-kPa) pressure drop through any expansion bends and other fittings in the pipe. Then, the allowable friction-pressure drop = 30 − 5 = 25 lb/in^2 (172.4 kPa).

Use the Thomas saturated-steam formula to determine the required pipe diameter, or $d = (80,000W/Pv)^{0.5}$, where d = inside pipe diameter, in; W = weight of steam flowing, lb/min; P = average steam pressure, lb/in^2 (abs); v = steam velocity, ft/min. Assuming a steam velocity of 10,000 ft/min (3048 m/min), which is typical for a long steam transmission line, we get $d = [(80,000 \times 300,000/60)/(165 \times 10,000)]^{0.5}$ = 15.32 in (389.1 mm).

The inside diameter of a schedule 40 16-in (406-mm) outside-diameter pipe is, from a table of pipe properties, 15.00 in (381 mm). Assume that a 16-in (406-mm) pipe will be used if schedule 40 wall thickness is satisfactory for the nominal pressure-loss line. Note that the larger flow was used in computing the size of this line because a pipe satisfactory for the larger flow will be acceptable for the smaller flow.

The maximum pressure-loss line will have an average pressure that is a function of the inlet pressure at the pressure-reducing valve at the line outlet. Assume that there is a 10-lb/in^2 (68.9-kPa) drop through this reducing valve. Then steam will enter the valve at 150 + 10 = 160 lb/in^2 (abs) (1103 kPa), and the average line pressure = (440 + 160)/2 = 300 lb/in^2 (abs) (2068 kPa). Using a higher steam velocity [15,000 ft/min (4572 m/min)] for this maximum pressure-loss line than for the nominal pressure-loss line [10,000 ft/min (3048 m/min)], because there is a larger allowable pressure drop, compute the required inside diameter from the Thomas saturated-steam formula because the steam has a superheat of only 456.28 − 455.00 = 1.28°F (2.3°C). Or, $d = [(80,000 \times 300,000/60)/(300 \times 15,000)]^{0.5}$ = 9.44 in (239.8 mm). Since a 10-in (254-mm) schedule 40 pipe has an inside diameter of 10,020 in (254.5 mm), use this size for the maximum pressure-loss line.

2.   *Compute the required pipe-wall thickness*

As shown in an earlier calculation procedure, the schedule number SN = $1000P_i/S$. Assuming that seamless carbon-steel ASTM A53 grade A pipe is used for both lines, the *Piping Code* allows a stress of 12,000 lb/in^2 (82.7 MPa) for this material at 600°F (316°C). Then SN = (1000) × (435)/12,000 = 36.2; use schedule 40 pipe, the next largest schedule number for both lines. This computation verifies the assumption in step 1 of the suitability of schedule 40 for each line.

3. Check the pipeline for critical velocity

In a steam line, $p_c = W'/Cd^2$, where p_c = critical pressure in pipe, lb/in^2 (abs), W' = steam flow rate, lb/h; C = constant from Crocker and King—*Piping Handbook*; d = inside diameter of pipe, in.

When the pressure loss in a pipe exceeds 50 to 58 percent of the initial pressure, flow may be limited by the fluid velocity. The limiting velocity that occurs under these conditions is called the *critical velocity*, and the coexisting pipeline pressure, the *critical pressure*.

Critical velocity may limit flow in the 10-in (254-mm) maximum pressure-loss line because the terminal pressure of 150 lb/in^2 (abs) (1034 kPa) is less than 58 percent of 440 lb/in^2 (abs) (3033.4 kPa), the inlet pressure. Use the above equation to find the critical pressure. Or, p_c = $(300,000)/[(75.15)(10.02)^2]$ = 39.7 lb/in^2 (abs) (273.7 kPa), using the constant from the *Piping Handbook* after interpolating for the initial enthalpy of 1205.4 Btu/lb (2804 kJ/kg), which is obtained from steam-table values.

Critical velocity would limit flow if the pipeline terminal pressure were equal to, or less than, 39.7 lb/in^2 (abs) (273.7 kPa). Since the terminal pressure of 150 lb/in^2 (abs) (1034.1 kPa) is greater than 39.7 lb/in^2 (abs) (237.7 kPa), critical velocity does not limit the steam flow. With smaller flow rates, the critical pressure will be lower because the denominator in the equation remains constant for a given pipe. Hence, the 10-in (254-mm) line will readily transmit 300,000-lb/h (135,000-kg/h) and smaller flows.

If critical pressure existed in the pipeline, the diameter of the pipe might have to be increased to transmit the desired flow. The 16-in (406.4-mm) line does not have to be checked for critical pressure because its final pressure is more than 58 percent of the initial pressure.

4. Compute the heat loss for each line

Assume that 2-in (50.8-mm) thick 85 percent magnesia insulation is used on each line and that the lines will run above the ground in an area having a minimum temperature of 40°F (4.4°C). Set up a computation form as follows:

	16 (406.5)	10 (254.0)
Pipe size, in (mm)	16 (406.5)	10 (254.0)
Steam temperature, °F (°C)	373 (189)	455 (235)
Air temperature, °F (°C)	40 (4.4)	40 (4.4)
Temperature difference, °F (°C)	333 (184.6)	415 (184.6)
Insulation heat loss, Btu/(h·ft^2·°F)° [W/(m^2·°C)]	1.11 (6.3)	0.704 (3.99)
Heat loss, Btu/(h·1in ft) (W/m)	370 (356)	292 (281)
Heat loss, Btu/h (kW), for 1700 ft (518 m)	629,000 (184)	496,400 (145.6)
Total heat loss, Btu/h (kW), with a 25% safety factor	786,250 (230)	620,500 (182.0)
Heat loss, Btu/lb (W/kg) of steam, for 300,000-lb/h (135,000-kg/h) flow	2.62 (1.7)	2.07 (1.35)
Heat loss, Btu/lb (W/kg), for the average flow of 150,000 lb/h (67,500 kg/h)	5.24 (3.4)	4.14 (2.69)

°From table of pipe insulation, Ehret Magnesia Manufacturing Company.

In this form, the following computations were made for both pipes: heat loss, Btu/(h·lin ft) = [insulation heat loss, Btu/(h·ft^2·°F)] (temperature difference,°F); heat loss, Btu/h for 1700 ft (518.2 m) = [heat loss, Btu/(h·lin ft)] (1700); total heat loss, Btu/h, 25 percent safety factor = (heat loss, Btu/h) (1700 ft)(1.25); heat loss, Btu/lb steam = (total heat loss, Btu/h, with a 25 percent safety factor)/(300,000-lb steam).

5. Compute the leaving enthalpy of the steam in each line

Acceleration of steam in each line results from an enthalpy decrease of $h_a = (v_2^2 - v_1^2)/2g(778)$, where h_a = enthalpy decrease, Btu/lb; v_2 and v_1 = final and initial velocity of the steam, respectively, ft/s; $g = 32.2$ ft/s^2. The velocity at any point x in the pipe is found from the continuity equation $v_x = (W'v_g)/3600\, A_x$, where v_x = steam velocity, ft/s, when the steam volume is v_g ft³/ lb, and A_x is the cross-sectional area of pipe, ft², at the point being considered.

For the 16-in (406.4-mm) nominal pressure-loss line with a flow of 300,000 lb/h (135,000 kg/ h) at 180-lb/in² (abs) (1241-kPa) entering and 150-lb/in² (abs) (1034.1-kPa) leaving pressure, using steam and piping table values, $v_1 = (300,000)(2.53)/[(3600)(1.23)] = 171.5$ ft/s (52.3 m/ s); $v_2 = 300,000(3.015)/[(3600)(1.23)] = 205$ ft/s (62.5 m/s). Then $h_a = [(204.5)^2 - (171.5)^2]/ [(64.4)(778)] = 0.2504$ Btu/lb (0.58 kJ/kg), say 0.25 Btu/lb (0.58 kJ/kg).

By an identical calculation, $h_a = 3.7$ Btu/lb (8.6 kJ/kg) for the 10-in (254-mm) maximum pressure-loss line when the leaving steam is assumed to be 150 lb/in² (abs) (1034.1 kPa), saturated.

Enthalpy of the 180-lb/in² (abs) (1241-kPa) saturated steam entering the 16-in (406.4-mm) line is 1196.9 Btu/lb (2784 kJ/kg). Heat loss during 300,000-lb/h (135,000-kg/h) flow is 2.62 Btu/ lb (6.1 kJ/kg), as computed in step 4. The enthalpy drop of 0.25 Btu/lb (0.58 kJ/kg) accelerates the steam. Hence, the calculated leaving enthalpy is 1196.9 − (2.62 + 0.25) = 1194.03 Btu/lb (2777.3 kJ/kg). The enthalpy of the leaving steam at 150 lb/in² (abs) (1034.1 kPa) saturated is 1194.1 Btu/lb (2777.5 kJ/kg). To have saturated steam leave the line, 1194.10 − 1194.03, or 0.07 Btu/lb (0.16 kJ/kg), must be supplied to the steam. This heat will be obtained from the enthalpy of vaporization given off by condensation of some of the steam in the line.

Make a group of identical calculations for the 10-in (254-mm) maximum pressure-loss line. The enthalpy of 440-lb/in (abs) (3033.4-kPa) 455°F (235°C) entering steam is 1205.4 Btu/lb (2803.8 kJ/kg), found by interpolation in the steam tables. Heat loss during 300,000-lb/h (135,000-kg/h) flow is 2.07 Btu/lb (4.81 kJ/kg). An enthalpy drop of 3.7 Btu/lb (8.6 kJ/kg) accelerates the steam. Hence, the calculated leaving enthalpy = 1205.4 − (2.07 + 3.7) = 1199.63 Btu/lb (2790.3 kJ/kg).

The enthalpy of the leaving steam at 150-lb/in² (abs) (1034-kPa) saturated is 1194.1 Btu/lb (2777.5 kJ/kg). As a result, under maximum flow conditions, the steam will be superheated from the entering point to the leaving point of the line. The enthalpy difference of 5.53 Btu/lb = 1199.63 − 1194.10) (12.9 kJ/kg) produces this superheat. Because the steam is superheated throughout the line length, condensation of the steam will not occur during maximum flow conditions.

For most industrial applications, the steam leaving the line may be considered as saturated at the desired pressure. But for precise temperature regulation, some form of pressure-temperature control must be used at the end of long lines.

During average flow conditions of 150,000 lb/h (67,500 kg/h), the line heat loss is 4.14 Btu/ lb (9.6 kJ/kg), as computed in step 4. The enthalpy drop to accelerate the steam is 0.925 Btu/lb (2.2 kJ/kg). As in the case of maximum flow, the steam is superheated throughout the length of the 10-in (254-mm) maximum pressure-loss line because the calculated leaving enthalpy is 1205.40 − 5.07 = 1200.33 Btu/lb (2791.9 kJ/kg).

6. *Compute the quantity of condensate formed in each line*

For either line, the quantity of condensate formed, lb/h = $C = W'(h_g$ at leaving pressure − calculated leaving h_g)/outlet pressure h_{fg}.

Using computed values from step 5 and steam-table values, we see the 16-in (406.4-mm) line with 300,000 lb/h (135,000 kg/h) flowing forms $C = (300,000)(0.07)/863.6 = 24.35$, say 24.4 lb/h (10.9 kg/h) of condensate.

Condensation during an average flow of 150,000 lb/h (67,500 kg/h) is found in the same way. The enthalpy drop to accelerate the steam is neglected for average flow in normal pressure-loss lines because the value is generally small. For the 150,000-lb/h (67,500-kg/h) flow, the calculated leaving enthalpy = 1196.90 − 5.24 = 1191.66 Btu/lb (536.3 kg/h). Hence, $C = (150,000)(1194.10 − 1191.66)/863.6 = 424$ lb/h (190.8 kg/h) say 425 lb/h (191.3 kg/h).

The largest amount of condensate is formed during line warm-up. Condensate-removal equipment—traps and related piping—must be sized up on the basis of the warm-up not the average steam flow. Using a warm-up time of 30 min and the method of an earlier calculation procedure, we see the condensate formed in 16-in (406.4-mm) schedule 40 pipe weighing 83 lb/ft (122.8 kg/ m) is, with a 25 percent safety factor to account for radiation, $C = 1.25 \times (60)(83)(1700)(373 − 40)(0.12)/[(30)(850.8)] = 16,550$ lb/h (7448 kg/h). Thus, the trap or traps should have a capacity of about 17,000 lb/h (7650 kg/h) to remove the condensate during the 30-min warm-up period.

Condensate does not form in the 10-in (254-mm) maximum pressure-loss line during either maximum or average flow. Warm-up condensate for a 30-min warm-up period and a 25 percent safety factor is $C = 1.25(60)(40.5)(1700)(455 - 40)(0.12)/[(30)(770.0)] = 11,120$ lb/h (5004 kg/ h). Thus, the trap or traps should have a capacity of about 11,500 lb/h (5175 kg/h) to remove the condensate during the 30-min warm-up period.

In general, traps sized on a warm-up basis have adequate capacity for the condensate formed during the maximum and average flows. However, the condensate formed under all three conditions must be computed to determine the maximum rate of formation for trap and drain-line sizing.

7. Determine the number of plain U bends needed

A 1700-ft (518-m) long steel steam line operating at a temperature in the 400°F (204°C) range will expand nearly 50 in (1270 mm) during operation. This expansion must be absorbed in some way without damaging the pipe. There are four popular methods for absorbing expansion in long transmission lines: plain U bends, double-offset expansion U bends, slip or corrugated expansion joints, and welded-elbow expansion bends. Each of these will be investigated to determine which is the most economical.

Assume that the governing code for piping design in the locality in which the line will be installed requires that the combined stress resulting from bending and pressure S_{bp} not exceed three-fourths the sum of the allowable stress for the piping material at atmospheric temperature S_a and the allowable stress at the operating temperature S_o of the pipe. This is a common requirement. In equation form, $S_{bp} = 0.75(S_a + S_o)$, where each stress is in lb/in^2.

By using allowable stress values from the *Piping Code* or the local code for 16-in (406.4-mm) seamless carbon-steel ASTM A53 grade A pipe operating at 373°F (189°C), $S_{bp} = 0.75(12,000 + 12,000) = 18,000$ lb/in^2 (124.1 MPa).

Determine the longitudinal pressure stress P_L by dividing the end force due to internal pressure F_e lb by the cross-sectional area of the pipe wall a_m in^2, or $P_L = F_e/a_m$. In this equation, $F_e = pa$, where $p =$ pipe operating pressure, lb/in^2 (gage); $a =$ cross-sectional area of the pipe, in^2. Since the 16-in (406.4-mm) line operates at $180 - 14.7 = 165.3$ lb/in^2 (gage) (1139.6 kPa) and, from a table of pipe properties, $a = 176.7$ in^2 (1140 cm^2) and $a_m = 24.35$ in^2 (157.1 cm^2), $P_L = (165.3)(176.7)/24.35 = 1197$, say 1200 lb/in^2 (8.3 MPa). The allowable bending stress at 373°F (189°C), the pipe operating temperature, is then $S_{np} - P_L = 18,000 - 1200 = 16,800$ lb/in^2 (115.8 MPa).

Assume that the expansion U bend will have a radius of seven times the nominal pipe diameter, or $(7)(16$ in$) = 112$ in (284.5 cm). The allowable bending stress is 16,800 lb/in^2 (115.8 MPa). Full *Piping Code* allowable credit will be taken for cold spring; i.e., the pipe will be cut short by 50 percent or more of the computed expansion and sprung into position.

Referring to Crocker and King—*Piping Handbook*, or a similar tabulation of allowable U-bend overall lengths for various operating temperatures, and choosing the length for 400°F (204°C), we see that the next higher tabulated temperature greater than the 373°F (189.4°C) operating temperature, an allowable length of 157.0 ft (47.9 m) is obtained for the bend. Plot a curve of the allowable bend length vs. temperature at 200, 300, 400, and 500°F (93.3, 148.9, 204.4, and 260°C). From this curve, the allowable bend length at 373°F (189.4°C) is found to be 175 ft (53.3 m). This length is based on an allowable pipe stress of 12,000 lb/in^2 (82.7 MPa) and no cold spring. Since the allowable stress is 16,800 lb/in^2 (115.8 MPa) and maximum cold spring is used, permitting a length 1.5 times the tabulated length, the total allowable length per bend $= (175.0)(16,800/12,000)(1.5) = 367.5$ ft (112 m). With a total length of pipe between the power plant and a load of 1700 ft (518.2 m), the number of bends required $= 1700/367.5 = 4.64$ bends. Since only a whole number of bends can be used, the next larger whole number, or five bends, would be satisfactory for this 16-in (406.4-mm) line. Each bend would have an overall length, Fig. 26, of $1700/5 = 340$ ft (103.6 m).

Find the actual stress S_a in the pipe when five 340-ft (103.6-m) bends are used by setting up a proportion between the tabulated stress and bend length. Thus, the *Piping Handbook* chart is based on a stress of 12,000 lb/in^2 (82.7 MPa) without cold spring. For this stress, the maximum allowable bend length is 175 ft (53.3 m) found by graphical interpolation of the tabular values, as discussed above. When a 340-ft (103.6-m) bend with maximum cold spring is used, the pipe stress is such that the allowable bend length is $340/1.5 = 226.5$ ft (69 m). The actual stress in the pipe is therefore $S_a/12,000 = 226.5/175$, or $S_a = 15,520$ lb/in^2 (107 MPa). This compares favorably with the allowable stress of 16,800 lb/in^2 (115.8 MPa). The actual stress is less because the overall bend length was reduced.

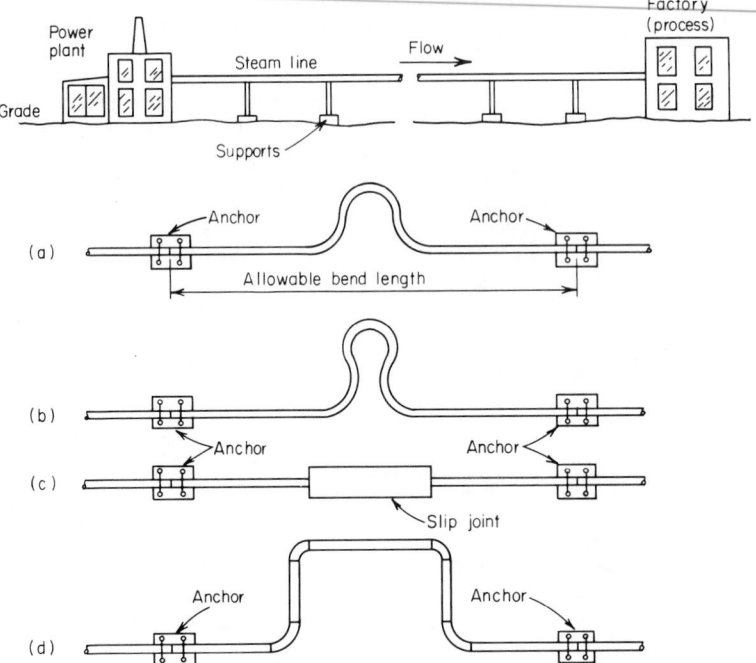

FIG. 26   Process steam line and different schemes for absorbing pipe thermal expansion.

Use the *Piping Handbook* or the method of an earlier calculation procedure to find the anchor reaction forces for these bends. Using the *Piping Handbook* method with graphical interpolation, the anchor reacting force for a 16-in (406.4-mm) schedule 80 bend having a radius of seven times the pipe diameter is 10,550 lb (46.9 kN) at 373°F (189.4°C), based on a 12,000-lb/in^2 (82.7-MPa) stress in the pipe. This tabular reaction must be corrected for the actual pipe stress and for schedule 40 pipe instead of schedule 80 pipe. Thus, the actual anchor reaction, lb = (tabular reaction, lb) [(actual stress, lb/in^2) (tabular stress, lb/in^2)] (moment of inertia, schedule 40 pipe, in^4/moment of inertia of schedule 80 pipe, in^4) = (10,550)(15,520/12,000)(731.9/1156.6) = 8650 lb (38.5 kN). With a reaction of this magnitude, each anchor would be designed to withstand a force of 10,000 lb (44.5 kN). Good design would locate the bends midway between the anchor points; that is, there would be an anchor at each end of each bend. Adjustment for cold spring is not necessary, because it has negligible effect on anchor forces.

Use the same procedure for the 10-in (254-mm) maximum pressure-loss line. If 100-in (254-cm) radius bends are used, seven are required. The bending stress is 14,700 lb/in^2 (65.4 kN), and the anchor force is 2935 lb (13.1 kN). Anchors designed to withstand 3000 lb (13.3 kN) would be used.

8.   *Determine the number of double-offset U bends needed*

By the same procedure and the *Piping Handbook* tabulation similar to that in step 7, the 16-in (406.4-mm) nominal pressure-loss line requires two 850-ft (259.1-m) long 112-in (284.5-cm) radius bends. Stress in the pipe is 15,610 lb/in^2 (107.6 MPa), and the anchor reaction is 4780 lb (21.3 kN).

The 10-in (254-mm) maximum pressure-loss line requires five 340-ft (103.6-m) long 70-in (177.8-cm) radius bends. Stress in the pipe is 12,980 lb/in^2 (89.5 MPa), and the anchor reaction is 2090 lb (9.3 kN).

Note that a smaller number of double-offset U bends are required—two rather than five for the 16-in (406.4-mm) pipe and five rather than seven for the 10-in (254-mm) pipe. This shows that double-offset U bends can absorb more expansion than plain U bends.

9. *Determine the number of expansion joints needed*

For any pipe, the total linear expansion e_t in at an elevated temperature above 32°F (0°C) is $e_t = (c_e)(\Delta t)(l)$, where c_e = coefficient of linear expansion, in/(ft·°F); Δt = operating temperature, °F − installation temperature, °F; l = length of straight pipe, ft. Using Crocker and King—*Piping Handbook* as the source for c_e for both lines, we see the expansion of the 373°F (189°C) 16-in (406.4-mm) line with a 40°F (4.4°C) installation temperature is $e_t = (12)(0.0000069)(373 − 40)(1700) = 46.8$ in (1189 mm). For the 10-in (254-mm) 455°F (235°C) line, $e_t = (12)(0.0000072)(455 − 40)(1700) = 61$ in (1549 mm). The factor 12 is used in each of these computations because Crocker and King give c_e in in/in; therefore, the pipe total length must be converted to inches by multiplying by 12.

Double-ended slip-type expansion joints that can absorb up to 24 in (609.6 mm) of expansion are available. Hence, the number of joints N needed for each line is: 16-in (406.4-mm) line, $N = 46.8/24$, or 2; 10-in (254-mm) line, $N = 61/24$, or 3.

The joints for each line would be installed midway between anchors, Fig. 26. Joints in both lines would be anchored to the ground or a supporting structure. Between the joints, the pipe must be adequately supported and free to move. Roller supports that guide and permit longitudinal movement are usually best for this service. Whereas roller-support friction varies, it is usually assumed to be about 100 lb (444.8 N) per support. At least six supports per 100 ft (30.5 m) are needed for the 16-in (406.4-mm) line and sever per 100 ft (30.5 m) for the 10-in (254-mm) line. Support friction and the number of rollers required are obtained from Crocker and King—*Piping Handbook* or piping engineering data.

The required anchor size and strength depend on the pipe diameter, steam pressure, slip-joint construction, and type of supports used. During expansion of the pipe, friction at the supports and in the joint packing sets up a force that must be absorbed by the anchor. Also, steam pressure in the joint tends to force it apart. The magnitude of these forces is easily computed. With the total force known, a satisfactory anchor can be designed. Slip-joint packing-gland friction varies with different manufacturers, type of joint, and packing used. Gland friction in one popular type of slip joint is about 2200 lb/in (385.3 N/mm) of pipe diameter. Assuming use of these joints in both lines, compute the anchor forces as follows:

	lb	kN
16-in (406.4-mm) nominal pressure-loss line		
Support friction = [(1700 ft)(6 supports per 100 ft)		
(100 lb per support)]/100	= 10,200	45.3
Gland friction = (2200 lb/in diameter)(16 in)	= 35,200	156.6
Pressure force = [165.3 lb/in² (gage)](176.7-in² pipe area)	= 29,200	129.9
Total force to be absorbed by anchor	= 74,600	331.8
10-in (254-mm) maximum pressure-loss line		
Support friction = [(1700)(7)(100)]/100	= 11,900	52.8
Gland friction = (2200)(10)	= 22,000	97.9
Pressure force = (425.3)(78.9)	= 33,600	149.5
Total force to be absorbed by anchor	= 67,500	300.2

Comparing these results shows that the 10-in (254-mm) line requires smaller anchors than does the 16-in (406.4-mm) line. However, the 16-in (406.4-mm) line requires only three anchors whereas the 10-in (254-mm) line needs four anchors. The total cost of anchors for both lines will be about equal because of the difference in size of the anchors.

The advantages of slip joints become apparent when the piping layout is studied. Only a minimum of pipe is needed because the pipe runs in a straight line between the point of supply and point of use. The amount of insulation is likewise a minimum.

Corrugated expansion joints could be used in place of slip-type joints. These would reduce the required anchor size somewhat because there would be no gland friction. The selection procedure resembles that given for slip-type joints.

10.   *Select welded-elbow expansion bends*

Use the graphical analysis in Crocker and King—*Piping Handbook* or in any welding fittings engineering data. Using either method shows that three bends of the most economical shape are suitable for the 16-in (406.4-mm) line and four for the 10-in (254-mm) line. The most economical bend is obtained when the bend width, divided by the distance between the anchor points, is 0.50. With these proportions, the longitudinal stress at the top and bottom of the bend is the same. Use of such bends, although desirable, is not always feasible, because existing piping or structures interfere.

When bend dimensions other than the most economical must be used, the maximum longitudinal stress occurs at the top of the bend when the width/anchor distance < 0.5. When this ratio is > 0.5, the maximum stress occurs at the bottom of the bend. Regardless of the bend type—plain U, double-offset U, or welded—the actual stress in the pipe should not exceed 40 percent of the tensile strength of the pipe material.

11.   *Determine the materials, quantities, and costs*

Set up tabulations showing the materials needed and their cost. Table 36 shows the materials required. Piping length is computed by using standard bend tables available in the cited references.

Table 37 shows the approximate material costs for each pipeline. The costs used in preparing this table were the most accurate available at the time of writing. However, the actual numerical values given in the table should not be used for similar design work because price changes may cause them to be incorrect. The important findings in such a tabulation are the differences in total cost. These differences will remain substantially constant even though prices change. Hence, if an $8000 difference exists between two sizes of pipe, this difference will not change appreciably with a moderate rise or fall in unit prices of materials.

Study of Table 37 shows that, in general, lines using double-offset U bends or welding elbows have the lowest material first cost. However, higher first costs do not rule out slip joints or plain U bends. Frequently, use of slip joints will eliminate offsets to clear existing buildings or piping because the pipe path is a straight line. Plain U bends have smaller overall heights than double-offset U bends. For this reason, the plain bend is often preferable where the pipe is run through congested areas of factories.

In some cases, past piping practice will govern line selection. For instance, in a factory that has made wide use of slip joints, the slightly higher cost of such a line might be overlooked. Preference might also be shown for plain U bends, double-offset U bends, or welded bends.

The values given in Table 37 do not include installation, annual operating costs, or depreciation. These have been omitted because accurate estimates are difficult to make unless actual conditions are known. Thus, installation costs may vary considerably according to who does the work. Annual costs are a function of the allowable depreciation, nature of process served, and location of the line. For a given transmission line of the type considered here, annual costs will usually be less for the smaller line.

The economic analysis, as made by the pipeline designer, should include all costs relative to the installation and operation of the line. The allowable cost of money and recommended depreciation period can be obtained from the accounting department.

TABLE 36   Summary of Material Requirements for Various Lines

Means used to absorb expansion	Number of anchors required		Approximate number of supports required		Approximate feet (meters) of pipe and insulation required	
	Pipe size, in (mm)					
	10 (254)	16 (406.4)	10 (254)	16 (406.4)	10 (254)	16 (406.4)
Plain U bends	9	5	127	120	2120 (646.2)	1970 (600.5)
Double-offset U bends	6	3	119	114	1985 (605.0)	1820 (554.7)
Slip joints	4	3	102	102	1700 (518.2)	1700 (518.2)
Welding elbows	5	4	106	106	1760 (536.4)	1760 (536.4)

TABLE 37 Approximate Material Costs for Various Lines

Pipe size, in (mm)

Means used to absorb expansion	Total material cost, $		Condensate removal equipment		Cost of anchors, $		Cost of supports, $		Cost of insulation, $		Cost of pipe and bends or joints, $	
	10 (254)	16 (406.4)	10 (254)	16 (406.4)	10 (254)	16 (406.4)	10 (254)	16 (406.4)	10 (254)	16 (406.4)	10 (254)	16 (406.4)
Plain U bends	26,500	51,350	2,000	3,000	1,000	1,800	1,800	2,400	3,700	5,650	18,000	38,500
Double-offset U bends	23,800	44,000	2,000	3,000	600	800	1,700	2,300	3,500	5,400	16,000	32,500
Slip joints	29,650	51,775	2,000	3,000	400	600	1,500	2,000	3,000	4,675	22,750	41,500
Welding elbows	23,800	43,975	2,000	3,000	400	800	1,800	2,300	3,600	5,375	16,000	32,500

12. *Select the most economical pipe size*

Table 37 shows that from the standpoint of first costs, the smaller line is more economical. This lower first cost is not, however, obtained without losing some large-line advantages.

Thus, steam leaves the 16-in (406.4-mm) line at 150 lb/in^2 (abs) (1034.1 kPa) saturated, the desired outlet condition. Special controls are unnecessary. With the 10-in (254-mm) line, the desired leaving conditions are not obtained. Slightly superheated steam leaves the line unless special controls are used. Where an exact leaving temperature is needed by the process served, a desuperheater at the end of the 10-in (254-mm) line will be needed. Neglecting this disadvantage, the 10-in (254-mm) line is more economical than is the 16-in (406.4-mm) line.

Besides lower first cost, the small line loses less heat to the atmosphere, has smaller anchor forces, and does not cause steam condensation during average flows. Lower heat losses and condensation reduce operating costs. Therefore, if special temperature controls are acceptable, the 10-in (254-mm) maximum pressure-loss line will be a more economical investment.

Such a conclusion neglects the possibility of future plant expansion. Where expansion is anticipated, installation of a small line now and another line later to handle increased steam requirements is uneconomical Instead, installation of a large nominal pressure-loss line now that can later be operated as a maximum pressure-loss line will be found more economical. Besides the advantage of a single line in crowded spaces, there is a reduction in installation and maintenance costs.

13. *Provide for condensate removal*

Fit a condensate drip line for every 100 ft (30.5 m) of pipe, regardless of size. Attach a trap of suitable capacity (see step 6) to each drip line. Pitch the steam-transmission pipe toward the trap, if possible. Where the condensate must flow *against* the steam, the steam-transmission pipe *must* be sloped in the direction of condensate flow. Every vertical rise of the main line must also be dripped. Where water is scarce, return the condensate to the boiler.

Related Calculations: Use this method to design long steam, gas, liquid, or vapor lines for factories, refineries, power plants, ships, process plants, steam heating systems, and similar installation. Follow the applicable piping code when designing the pipeline.

STEAM DESUPERHEATER ANALYSIS

A spray- or direct-contact-type desuperheater is to remove the superheat from 100,000 lb/h (45,000 kg/h) of 300-lb/in^2 (abs) (2068-kPa) 700°F (371°C) steam. Water at 200°F (93.3°C) is available for desuperheating. How much water must be furnished per hour to produce 30-lb/in^2 (abs) (206.8-kPa) saturated steam? How much steam leaves the desuperheater? If a shell-and-tube type of noncontact desuperheater is used, determine the required water flow rate if the overall coefficient of heat transfer U = 500 Btu/(h·ft^2·°F) [2.8 kW/(m^3·°C)]. How much tube area A is required? How much steam leaves the desuperheater? Assume that the desuperheating water is not allowed to vaporize in the desuperheater.

Calculation Procedure:

1. *Compute the heat absorbed by the water*

Water entering the desuperheater must be heated from the entering temperature, 200°F (93.3°C), to the saturation temperature of 300-lb/in^2 (abs) (2068-kPa) steam, or 417.3°F (214°C). Using the steam tables, we see the sensible heat that must be absorbed by the water = h_f at 417.3°F (214°C) − h_f at 200°F (93.3°C) = 393.81 − 167.99 = 255.81 Btu/lb (525.2 kJ/kg) of water used.

Once the desuperheating water is at 417.3°F (214°C), the saturation temperature of 300°F (148.9°C) steam, the water must be vaporized if additional heat is to be absorbed. From the steam tables, the enthalpy of vaporization at 300 lb/in^2 (abs) (2068 kPa) is h_{fg} = 809.0 Btu/lb (1881.7 kJ/kg). This is the amount of heat the water will absorb when vaporized from 417.3°F (214°C).

Superheated steam at 300 lb/in^2 (abs) (2068 kPa) and 700°F (371°C) has an enthalpy of h_g = 1368.3 Btu/lb (3182.7 kJ/kg), and the enthalpy of 300-lb/in^2 (abs) (2068-kPa) saturated steam is h_g = 1202.8 Btu/lb (2797.7 kJ/kg). Thus 1368.3 − 1202.8 = 165.5 Btu/lb (384.9 kJ/kg) must be absorbed by the water to desuperheat the steam from 700°F (371°C) to saturation at 300 lb/in^2 (abs) (2068 kPa).

2. *Compute the weight of water required for the spray*

The weight of water evaporated by 1 lb (0.45 kg) of steam while it is being desuperheated = heat absorbed by water, Btu/lb of steam/heat required to evaporate 1 lb (0.45 kg) of water entering the desuperheater at 200°F (93.3°C), Btu = 165.5/(225.81 + 809.0) = 0.16 lb (0.07 kg) of water. Since 100,000 lb/h (45,000 kg/h) of steam is being desuperheated, the water flow rate required = (0.16)(100,000) = 16,000 lb/h (7200 kg/h). Water for direct-contact desuperheating can be taken from the feedwater piping or from the boiler.

Note that 16,000 lb/h (7200 kg/h) of additional steam will leave the desuperheater because the superheated steam is not condensed while being desuperheated. Thus, the total flow from the desuperheater = 100,000 + 16,000 = 116,000 lb/h (52,200 kg/h).

3. *Compute the tube area required in the desuperheater*

The total heat transferred in the desuperheater, Btu/h = UAt_m, where t_m = logarithmic mean temperature difference across the heater. Using the method for computing the logarithmic temperature difference given elsewhere in this handbook, or a graphical solution as in Perry—*Chemical Engineers' Handbook*, we find t_m = 134°F (74.4°C) with desuperheating water entering at 200°F (93.3°C) and leaving at 430°F (221.1°C), a temperature about 13°F (7°C) higher than the leaving temperature of the saturated steam, 417.3°F (214°C). Steam enters the desuperheater at 700°F (371°C). Assumption of a leaving water temperature 10 to 15°F (5.6 to 8.3°C) higher than the steam temperature is usually made to ensure an adequate temperature difference so that the desired heat-transfer rate will be obtained. If the graphical solution is used, the greatest temperature difference then becomes 700 − 200 = 500°F (278°C), and the least temperature difference = 430 − 417.3 = 12.7°F (7°C).

Then the heat transferred = (500)(A)(134), whereas the heat given up by the steam is, from step 1, (100,000 lb/h)(165.5 Btu/lb) [(45,000 kg/h)(384.9 kJ/kg)]. Since the heat transferred = the heat absorbed, (500)(A)(134) = (100,000)(165.5); A = 247 ft² (22.9 m²), say 250 ft² (23.2 m²).

4. *Compute the required water flow*

Heat transferred to the water = (500)(247)(134) Btu/h (W). The temperature rise of the water during passage through the desuperheater = outlet temperature, °F − inlet temperature = outlet temperature, °F = 430 − 200 = 230°F (127.8°C). Since the specific heat of water = 1.0, closely, the heat absorbed by the water = (flow rate, lb/h)(230)(1.0). Then the heat transferred = heat absorbed, or (500)(247)(134) = (flow rate, lb/h)(230)(1.0); flow rate = 72,000 lb/h (32,400 kg/h). Since the water and steam do *not* mix, the steam output of the desuperheater = steam input = 100,000 lb/h (45,000 kg/h).

Only about 25 percent as much water, 16,000 lb/h (7200 kg/h), is required by the direct-contact desuperheater as compared with the indirect desuperheater. The indirect type of superheater requires more cooling water because the enthalpy of vaporization, nearly 1000 Btu/lb (2326 kJ/kg) of water, is not used to absorb heat. Some indirect-type desuperheaters are designed to permit the desuperheating water to vaporize. This steam is returned to the boiler. The water-consumption determination and the calculation procedure for this type are similar to the spray-type discussed earlier. Where the water does not vaporize, it must be kept at a high enough pressure to prevent vaporization.

Related Calculations: Use this method to analyze steam desuperheaters for any type of steam system—industrial, utility, heating, process, or commercial.

STEAM ACCUMULATOR SELECTION AND SIZING

Select and size a steam accumulator to deliver 10,000 lb/h (4500 kg/h) of 25-lb/in² (abs) (172.4-kPa) steam for peak loads in a steam system. Charging steam is available at 75 lb/in² (abs) (517.1 kPa). Room is available for an accumulator not more than 30 ft (9.1 m) long, 20 ft (6.1 m) wide, and 20 ft (6.1 m) high. How much steam is required for startup?

Calculation Procedure:

1. *Determine the required water capacity of the accumulator*

One lb (0.45 kg) of water stored in this accumulator at 75 lb/in² (abs) (517.1 kPa) has a saturated liquid enthalpy h_f = 277.43 Btu/lb (645.3 kJ/kg) from the steam tables; whereas for 1 lb (0.45

kg) of water at 25 lb/in² (abs) (172.4 kPa), h_f = 208.42 Btu/lb (484.8 kJ/kg). In an accumulator, the stored water flashes to steam when the pressure on the outlet is reduced. For this accumulator, when the pressure on the 75-lb/in² (abs) (517.1-kPa) water is reduced to 25 lb/in² (abs) (172.4 kPa) by a demand for steam, each pound of stored 75-lb/in² (abs) (517.1-kPa) water flashes to steam, releasing 277.43 − 208.42 = 69.01 Btu/lb (160.5 kJ/kg).

The enthalpy of vaporization of 25 lb/in² (abs) (172.4-kPa) steam is h_{fg} = 952.1 Btu/lb (2215 kJ/kg). Thus, 1 lb (0.45 kg) of 75-lb/in² (abs) (517.1-kPa) water will form 69.01/952.1 = 0.0725 lb (0.03 kg) of steam. To supply 10,000 lb/h (4500 kg/h) of steam, the accumulator must store 10,000/0.0725 = 138,000 lb/h (62,100 kg/h) of 75-lb/in² (abs) (517.1-kPa) water.

Saturated water at 75 lb/in² (abs) (517.1 kPa) has a specific volume of 0.01753 ft³/lb (0.001 m³/kg) from the steam tables. Since density = 1/specific volume, the density of 75-lb/in² (abs) (517.1-kPa) saturated water = 1/0.01753 = 57 lb/ft³ (912.6 kg/m³). The volume required in the accumulator to store 138,000 lb (62,100 kg) of 75-lb/in² (abs) (517.1-kPa) water = total weight, lb/density of water = 138,000/57 = 2420 ft³ (68.5 m³).

2. *Select the accumulator dimensions*

Many steam accumulators are cylindrical because this shape permits convenient manufacture. Other shapes—rectangular, cubic, etc.—may also be used. However, a cylindrical shape is assumed here because it is the most common.

The usual accumulator that serves as a reserve steam supply between a boiler and a load (often called a Ruths-type accumulator) can safely release steam at the rate of 0.3 [accumulator storage pressure, lb/in² (abs)] lb/ft² of water surface per hour [kg/(m²·h)]. Thus, this accumulator can release (0.3)(75) = 22.5 lb/(ft²·h) [112.5 kg/(m²·h)]. Since a release rate of 10,000 lb/h (4500 kg/h) is desired, the surface area required = 10,000/225 = 445 ft² (41.3 m²).

Space is available for a 30-ft (9.1-m) long accumulator. A cylindrical accumulator of this length would require a diameter of 445/30 = 14.82 ft (4.5 m), say 15 ft (4.6 m). When half full of water, the accumulator would have a surface area (30)(15) = 450 ft² (41.8 m²).

Once the accumulator dimensions are known, its storage capacity must be checked. The volume of a horizontal cylinder of d-ft diameter and l-ft length = $(\pi d^2/4)(l)$ = $(\pi \times 15^2/4)(3)$ = 5300 ft³ (150 m³). When half full, this accumulator could store 5300/2 = 2650 ft³ (75 m³). Since, from step 1, a capacity of 2420 ft³ (68.5 m³) is required, a 15 × 30 ft (4.6 × 9.1 m) accumulator is satisfactory. A water-level controller must be fitted to the accumulator to prevent filling beyond about the midpoint. In this accumulator, the water level could rise to about 60 percent, or (0.60)(15) = 9 ft (2.7 m), without seriously reducing the steam capacity. When an accumulator delivers steam from a more-than-half-full condition, its releasing capacity increases as the water level falls to the midpoint, where the release area is a maximum. Since most accumulators function for only short periods, say 5 or 10 min, it is more important that the vessel be capable of delivering the desired rate of flow than that it deliver the last pound of steam in its lb/h rating.

If the size of the accumulator computed as shown above is unsatisfactory from the standpoint of space, alter the dimensions and recompute the size.

3. *Compute the quantity of charging steam required*

To start an accumulator, it must first be partially filled with water and then charged with steam at the charging pressure. The usual procedure is to fill the accumulator from the plant feedwater system. Assume that the water used for this accumulator is at 14.7 lb/in² (abs) and 212°F (101.3 kPa and 100°C) and that the accumulator vessel is half-full at the start.

For any accumulator, the weight of charging steam required is found by solving the following heat-balance equation: (weight of starting water, lb)(h_f of starting water, Btu/lb) + (weight of charging steam, lb)(charging steam h_g, Btu/lb) = (weight of charging steam, lb + weight of starting water, lb)(h_f at charging pressure, Btu/lb). For this accumulator with a 75-lb/in² (abs) (517-kPa) charging pressure and 212°F (100°C) starting water, the first step is to compute the weight of water in the half-full accumulator. Since, from step 2, the accumulator must contain 2420 ft³ (68.5 m³) of water, this water has a total weight of (volume of water, ft³)/(specific volume of water, ft³/lb) = 2420/0.01672 = 144,600 lb (65,070 kg). However, the accumulator can actually store 2650 ft³ (75 m³) of water. Hence, the actual weight of water = 2650/0.1672 = 158,300 lb (71,235 kg). Then, with C = weight of charging steam, lb, (158,300)(180.07) + (C)(1181.9) = $(C + 158,300)$(277.43); C = 17,080 lb (7686 kg) of steam.

Once the accumulator is started up, less steam will be required. The exact amount is computed in the same manner, by using the steam and water conditions existing in the accumulator.

Related Calculations: Use this method to size an accumulator for any type of steam service—heating, industrial, process, utility. The operating pressure of the accumulator may be greater or less than atmospheric.

SELECTING PLASTIC PIPING FOR INDUSTRIAL USE

Select the material, schedule number, and support spacing for a 1-in (25.4-mm) nominal-diameter plastic pipe conveying ethyl alcohol liquid having a temperature of 75°F (23.9°C) and a pressure of 400 lb/in² (2758 kPa). What expansion must be anticipated if a 1000-ft (304.8-m) length of the pipe is installed at a temperature of 50°F (10°C)? How does the cost of this plastic pipe compare with galvanized-steel pipe of the same size and length?

Calculation Procedure:

1. Determine the required schedule number

Refer to Baumeister and Marks—*Standard Handbook for Mechanical Engineers* or a plastic pipe manufacturer's engineering data for the required schedule number. Table 38 shows typical pressure ratings for various sizes and schedule number polyvinyl chloride (PVC) (plastic) piping.

Table 38 shows that schedule 40 normal-impact grade 1-in (25.4-mm) pipe is unsuitable because its maximum operating pressure with fluid at 75°F (24°C) is 310 lb/in² (2.13 MPa). Plain-end 1-in (25.4-mm) schedule 80 pipe is, however, satisfactory because it can withstand pressures up to 435 lb/in² (2.99 MPa). Note that threaded schedule 80 pipe can withstand pressures only to 255 lb/in² (1757 kPa). Therefore, plain-end normal-impact grade pipe must be used for this installation. High-impact grade pipe, in general, has lower allowable pressure ratings at 75°F (24°C) because the additive used to increase the impact resistance lowers the tensile strength, temperature, and chemical resistance. Data shown in Table 38 are also presented in graphical form in some engineering data.

2. Select a suitable piping material

Refer to piping engineering data to determine the corrosion resistance of PVC to ethyl alcohol. A Grinnell Company data sheet rates PVC normal-impact and high-impact pipe as having excellent corrosion resistance to ethyl alcohol at 72 and 140°F (22.2 and 60°C). Therefore, PVC is a suitable piping material for this liquid at its operating temperature of 75°F (24°C).

3. Find the required support spacing

Use a tabulation or chart in the plastic-pipe engineering data to find the required support spacing for the pipe. Be sure to read the spacing under the correct schedule number. Thus, a Grinnell Company plastic-piping tabulation recommends a 5-ft 4-in (162.6-cm) spacing for schedule 80 1-in (25.4-mm) PVC pipe that weighs 0.382 lb/ft (0.57 kg/m) when empty. The pipe hangers should not clamp the pipe tightly; instead, free axial movement should be allowed.

4. Compute the expansion of the pipe

The temperature of the pipe rises from 50 to 75°F (10 to 24°C) when it is put in operation. This is a rise of $75 - 50 = 25°F$ (14°C). Table 39 shows the thermal expansion of various types of plastic piping.

TABLE 38 Maximum Operating Pressure, PVC Pipe [normal-impact grade, fluid temperature 75°F (23.9°C) or less]

Pipe size		Schedule 40, plain end		Schedule 80			
				Plain end		Threaded	
in	mm	lb/in²	MPa	lb/in²	MPa	lb/in²	MPa
½	12.7	410	2.83	575	3.96	330	2.28
¾	19.1	335	2.31	470	3.24	285	1.97
1	25.4	310	2.14	435	2.99	255	1.76
1½	38.1	230	1.59	325	2.24	205	1.41

TABLE 39   Thermal Expansion of Plastic Pipe

Piping material	Expansion	
	in/(ft·°F)	cm/(m·°C)
Butyrate	0.00118	0.018
Kralastic	0.00067	0.010
Polyethylene	0.00108	0.016
Polyvinyl chloride	0.00054	0.008
Saran	0.00126	0.019

The thermal expansion of any plastic pipe is found from $E_t = LC \, \Delta t$, where E_t = total expansion, in; L = pipe length, ft; C = coefficient of thermal expansion, in/(ft·°F), from Table 39, Δt = temperature change of the pipe, °F. For this pipe, $E_t = (1000)(0.00054)(25) = 13.5$ in (342.9 mm) when the temperature rises from 50 to 75°F (10 to 24°C).

5. Determine the relative cost of the pipe

Check the prices of galvanized-steel and PVC pipe as quoted by various suppliers. These quotations will permit easy comparison. In this case, the two materials will be approximately equal in per-foot cost.

Related Calculations: Use the method given here for selecting plastic pipe for any service—process, domestic, or commercial—conveying any fluid or gas. Note that the maximum operating pressure of plastic piping is normally taken as about 20 percent of the bursting pressure. The allowable operating pressure decreases with an increase in temperature. The maximum allowable operating temperature is usually 150°F (65.5°C). The pressure loss caused by pipe friction in plastic pipe is usually about one-half the pressure loss in galvanized-steel pipe of the same diameter. Pressure loss for plastic piping is computed in the same way as for steel piping.

FRICTION LOSS IN PIPES HANDLING SOLIDS IN SUSPENSION

What is the friction loss in 800 ft (243.8 m) of 6-in (152.4-mm) schedule 40 pipe when 400 gal/min (25.2 L/s) of sulfate paper stock is flowing? The consistency of the sulfate stock is 6 percent.

Calculation Procedure:

1. Determine the friction loss in the pipe

There are few general equations for friction loss in pipes conveying liquids having solids in suspension. Therefore, most practicing engineers use plots of friction loss available in engineering handbooks, *Cameron Hydraulic Data, Standards of the Hydraulic Institute,* and from pump engineering data. Figure 27 shows one set of typical friction-loss curves based on work done at the University of Maine on the data of Brecht and Heller of the Technical College, Darmstadt, Germany, and published by Goulds Pumps, Inc. There is a similar series of curves for commonly used pipe sizes from 2 through 36 in (50.8 through 914.4 mm).

Enter Fig. 27 at the pipe flow rate, 400 gal/min (25.2 L/s), and project vertically upward to the 6 percent consistency curve. From the intersection, project horizontally to the left to read the friction loss as 60 ft (18.3 m) of liquid per 100 ft (30.5 m) of pipe. Since this pipe is 800 ft (243.8 m) long the total friction-head loss in the pipe = (800/100)(60) = 480 ft (146.3 m) of liquid flowing.

2. Correct the friction loss for the liquid consistency

Friction-loss factors are usually plotted for one type of liquid, and correction factors are applied to determine the loss for similar, but different, liquids. Thus, with the Goulds charts, a factor of 0.9 is used for soda, sulfate, bleached sulfite, and reclaimed paper stocks. For ground wood, the factor is 1.40.

When the stock consistency is less than 1.5 percent, water-friction values are used. Below a consistency of 3 percent, the velocity of flow should not exceed 10 ft/s (3.05 m/s). For suspensions of 3 percent and above, limit the maximum velocity in the pipe to 8 ft/s (2.4 m/s).

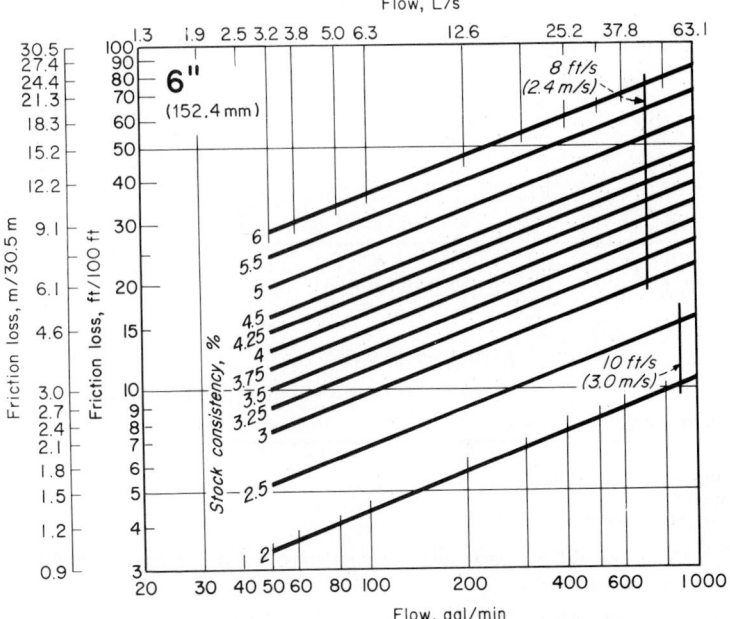

FIG. 27   Friction loss of paper stock in 4-in (101.6-mm) steel pipe. *(Goulds Pumps, Inc.)*

Since the liquid flowing in this pipe is sulfate stock, use the 0.9 correction factor, or the actual total friction head = (0.9)(480) = 432 ft (131.7 m) of sulfate liquid. Note that Fig. 27 shows that the liquid velocity is less than 8 ft/s (2.4 m/s).

Related Calculations: Use this procedure for soda, sulfate, bleached sulfite, and reclaimed and ground-wood paper stock. The values obtained are valid for both suction and discharge piping. The same general procedure can be used for sand mixtures, sewage, slurries, trash, sludge, and foods in suspension in a liquid.

DESUPERHEATER WATER SPRAY QUANTITY

A pressure- and temperature-reducing station in a steam line is operating under the following conditions: pressure and temperature ahead of the station are 1400 lb/in^2 absolute (5650 kPa), 950°F (510°C); the reduced temperature and pressure after the station are 600°F (315°C), 200 lb/in^2 absolute (1380 kPa). If 450,000 lb/h (3400 kg/s) of steam is required at 200 lb/in^2 (1380 kPa), how much water, which is available at 200 lb/in^2 absolute (1380 kPa) and 635.8°F (335.4°C), must be sprayed in at the superheater? See Fig. 28.

Calculation Procedure:

1. Determine the quantity of heat entering the desuperheater via the spray in terms of the amount of water

The quantity of heat entering the desuperheater via the spray, $Q = w \times h_f$, where the amount of water is w, lb/h (kg/s); from Table 2, Saturation Pressures, of the Steam Tables mentioned under Related Calculations of this procedure, heat content of water, saturated steam. At 200 lb/in^2 (1380 kPa), $h_f = 355.4$ Btu/lb$_m$ (826 kJ/kg). Thus, $Q = w \times 355.4$ lb/h (kg/s). It should be noted that the Steam Tables show the saturation temperature to be 381.79°F (194.3°C) at the given pressure, shown on Fig. 28. Obviously, the 635.8°F (335.4°C) given in the problem is not correct, because at that temperature there would either be superheated steam, vapor, or the water would have to be under a pressure of 2000 lb/in^2 (13.8 × 10^3 kPa).

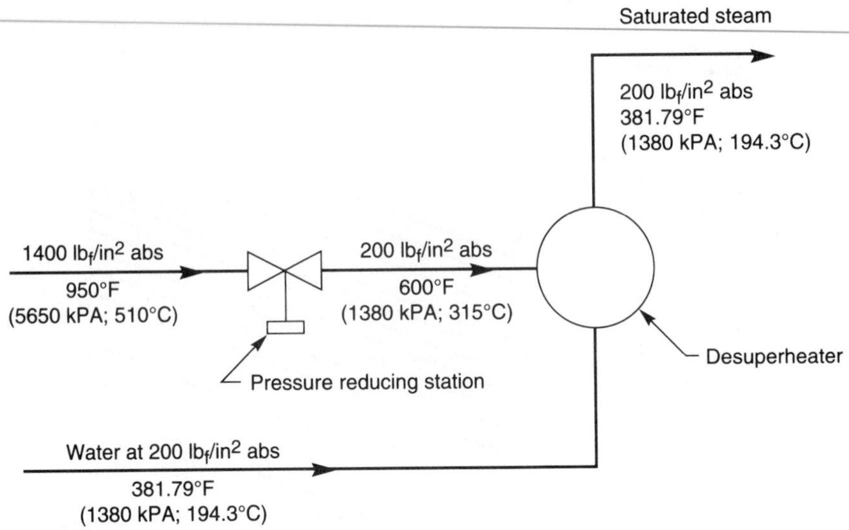

Saturated steam

200 lbf/in² abs
381.79°F
(1380 kPA; 194.3°C)

1400 lbf/in² abs
950°F
(5650 kPA; 510°C)

200 lbf/in² abs
600°F
(1380 kPA; 315°C)

Pressure reducing station

Desuperheater

Water at 200 lbf/in² abs
381.79°F
(1380 kPA; 194.3°C)

FIG. 28   Desuperheater fluid flow diagram.

2.   Find the enthalpy of the superheated steam entering the desuperheater and the enthalpy of the saturated steam leaving

From Table 3, Vapor, of the Steam Tables, superheated steam entering the desuperheater at 200 lb/in² (1380 kPa) and 600°F (315°C) has an enthalpy, h = 1322.1 Btu/lb (2075 kJ/kg). From Table 2, Saturation: Pressures, of the Steam Tables, saturated steam leaving the superheater at 200 lb/in² (1380 kPa) and 381.79°F (194.3°C) as saturated vapor has an enthalpy, h_g = 1198.4 Btu/lb$_m$ (2787 kJ/kg).

3.   Compute the amount of water which must be sprayed into the desuperheater

The amount of water which must be sprayed into the desuperheater, w_w lb/h (kg/s), can be found by the use of a heat balance equation where, as an adiabatic process, the amount of heat into the superheater equals the amount of heat out. Then, $w \times h_f + (450,000 - w)h = 450,000 \times h_g$. Or, $w \times 355.4 + (450,000 - w)(1322.1) = 450,000 \times 1198.4$. Solving, $w = 450,000 \times 123.7/966.7 = 57,580$ lb/h (435.3 kg/s).

Related Calculations: Strictly speaking, the given pressure and temperature conditions before the pressure-reducing station were irrelevant in the Calculation Procedure. Also, the incorrect given saturation temperature of 635.8°F (335.4°C) is an example of possible distractions which should be guarded against while solving such problems. The Steam Tables appear in *Thermodynamic Properties of Water Including, Liquid, and Solid Phases*, 1969, Keenan, et al., John Wiley & Sons, Inc. Use later versions of such tables whenever available, as necessary.

Heat Transfer and Heat Exchangers

REFERENCES: Goldstein—*Heat Transfer in Energy Conservation*, ASME; Isachenko—*Heat Transfer*, Mir (Moscow); Karlekar and Desmond—*Engineering Heat Transfer*, West; Kays and Crawford—*Convective Heat and Mass Transfer*, McGraw-Hill; Butterworth and Hewitt—*Two Phase Flow and Heat Transfer*, Oxford University Press; French—*Heat Transfer and Fluid Flow in Nuclear Systems*, Pergamon Press; Frost—*Heat Transfer at Low Temperatures*, Plenum Press; McAdams—*Heat Transmission*, McGraw-Hill; Kern—*Process Heat Transfer*, McGraw-Hill; General Electric Company—*Electric Heaters and Heating Devices*; Jakob—*Heat Transfer*, Wiley; Bosworth—*Heat Transfer Phenomena*, Wiley; Kays and London—*Compact Heat Exchangers*, McGraw-Hill; Kraus—*Cooling Electronic Equipment*, Prentice-Hall; Fraas and Ozisik—*Heat Exchanger Design*, Wiley; Heat Transfer Research, Inc.—*Design Manual*; API Standards—*Heat Exchangers for General Refinery Service*; Giedt—*Principles of Engineering Heat Transfer*, Van Nostrand; Eckert and Drake—*Heat and Mass Transfer*, McGraw-Hill; Schnieder—*Conduction Heat Transfer*, Addison-Wesley; Kreith—*Principles of Heat Transfer*, International Textbook; Perry—*Chemical Engineers' Handbook*, McGraw-Hill; Carslaw and Jaeger—*Conduction of Heat in Solids*, Oxford; Wilkes—*Heat Insulation*, Wiley.

SELECTING TYPE OF HEAT EXCHANGER FOR A SPECIFIC APPLICATION

Determine the type of heat exchanger to use for each of the following applications: (1) heating oil with steam; (2) cooling internal-combustion engine liquid coolant; (3) evaporating a hot liquid. For each heater chosen, specify the typical pressure range for which the heater is usually built and the typical range of the overall coefficient of heat transfer U.

Calculation Procedure:

1. Determine the heat-transfer process involved

In a heat exchanger, one or more of four processes may occur: heating, cooling, boiling, or condensing. Table 1 lists each of these four processes and shows the usual heat-transfer fluids involved. Thus, the heat exchangers being considered here involve (a) oil heater—heating—vapor-liquid; (b) internal-combustion engine coolant—cooling—gas-liquid; (c) hot-liquid evaporation—boiling—liquid-liquid.

2. Specify the heater action and the usual type selected

Using the same identifying letters for the heaters being selected, Table 1 shows the action and usual type of heater chosen. Thus,

	Action	Type
a.	Steam condensed; oil heated	Shell-and-tube
b.	Air heated; water cooled	Tubes in open air
c.	Waste liquid cooled; water boiled	Shell-and-tube

3. Specify the usual pressure range and typical U

Using the same identifying letters for the heaters being selected, Table 1 shows the action and usual type of heater chosen. Thus,

	Usual pressure range	Typical U range Btu/(h·°F·ft²)	Typical U range W/(m²·°C)
a.	0–500 lb/in² (abs) (0 to 3447 kPa)	20–60	113.6–340.7
b.	0–100 lb/in² (abs) (0 to 689.4 kPa)	2–10	11.4–56.8
c.	0–500 lb/in² (abs) (0 to 3447 kPa)	40–150	227.1–851.7

4. Select the heater for each service

Where the heat-transfer conditions are normal for the type of service met, the type of heater listed in step 2 can be safely used. When the heat-transfer conditions are unusual, a special type of heater may be needed. To select such a heater, study the data in Table 1 and make a tentative selection. Check the selection by using the methods given in the following calculation procedures in this section.

   Related Calculations: Use Table 1 as a general guide to heat-exchanger selection in any industry—petroleum, chemical, power, marine, textile, lumber, etc. Once the general type of heater and its typical U value are known, compute the required size, using the procedures given later in this section.

SHELL-AND-TUBE HEAT EXCHANGER SIZE

What is the required heat-transfer area for a parallel-flow shell-and-tube heat exchanger used to heat oil if the entering oil temperature is 60°F (15.6°C), the leaving oil temperature is 120°F (48.9°C), and the heating medium is steam at 200 lb/in² (abs) (1378.8 kPa)? There is no subcool-

TABLE 1 Heat-Exchanger Selection Guide°

	Heat-transfer fluids	Equipment	Action	Type†	Pressure range‡	Typical range of U§
	Liquid-liquid	Boiler-water blowdown exchanger	Blowdown cooled, feedwater heated	S	M, H	50–300 (0.28–1.7)
		Laundry-water heat reclaimer	Waste water cooled, feed heated	S	L	30–200 (0.17–1.1)
		Service-water heater	Waste liquid cooled, water heated	S	L, H	50–300 (0.28–1.7)
	Vapor-liquid	Bleeder heater	Steam condensed, feedwater heated	S	L, H	200–800 (1.1–4/5)
		Deaerating feed heater	Steam condensed, feedwater heated	M	L, M	DC
		Jet heater	Steam condensed, water heated	M	L	DC
		Process kettle	Steam condensed, liquid heated	S	L, M	100–500 (0.57–2.8)
		Oil heater	Steam condensed, oil heated	S	L, M	20–60 (0.11–0.34)
		Service-water heater	Steam condensed, water heated	S	L, M	200–800 (1.1–4.5)
Heating		Open flow-through heater	Steam condensed, water heated	M	L	DC
		Liquid-sodium steam superheater	Sodium cooled, steam superheated	S	M, H	50–200 (0.28–1.1)
	Gas-liquid	Waste-heat water heater	Waste gas cooled, water heated	T	L	2–10 (0.011–0.057)
		Boiler economizer	Flue gas cooled, feedwater heated	T	M, H	2–10 (0.011–0.057)
		Hot-water radiator	Water cooled, air heated	T	L	1–10 (0.0057–0.057)
	Gas-gas	Boiler air heater	Flue gas cooled, combustion air heated	T, R	L	2–10 (0.011–0.057)
		Gas-turbine regenerator	Flue gas cooled, combustion air heated	T	L	2–10 (0.011–0.057)
	Vapor-gas	Boiler superheater	Combustion gas cooled, steam superheated	T	M, H	2–20 (0.011–0.11)
		Steam pipe coils	Steam condensed, air heated	T	L, M	2–10 (0.011–0.057)
		Steam radiator	Steam condensed, air heated	T	L	2–10 (0.011–0.057)
Cooling	Liquid-liquid	Oil cooler	Water heated, oil cooled	S, D	L, M	20–200 (0.11–1.1)
		Water chiller	Refrigerant boiled, water cooled	S	L, M	30–151 (0.17–0.86)
		Brine cooler	Refrigerant boiled, brine cooled	S	L, M	30–150 (0.17–0.86)
		Transformer-oil cooler	Water heated, oil cooled	S	L, M	20–50 (0.11–0.88)
	Vapor-liquid	Boiler desuperheater	Boiler water heated, steam desuperheated	S, M	M, H	150–800 (0.85–4.5)

			Type of design[†]	[‡]	[§]	
Cooling	Gas-liquid	Compressor intercoolers and aftercoolers	Water heated, compressed air cooled	S	L, H	10–20	(0.057–0.11)
		Internal-combustion-engine radiator	Air heated, water cooled	T	L	2–10	(0.011–0.057)
		Generator hydrogen, air coolers	Water heated, hydrogen or air cooled	S	L	2–10	(0.011–0.057)
		Air-conditioning cooler	Water heated, air cooled	T	L	2–10	(0.011–0.057)
		Refrigeration heat exchanger	Brine heated, air cooled	T	L, M	2–10	(0.011–0.057)
	Vapor-gas	Refrigeration evaporator	Refrigerant boiled, air cooled	T	L, M	2–10	(0.011–0.057)
		Boiler desuperheater	Flue gas heated, steam desuperheated	T	M, H	2–8	(0.011–0.045)
	Liquid-liquid	Hot-liquid evaporator	Waste liquid cooled, water boiled	S	L, H	40–150	(0.23–0.85)
		Liquid-sodium steam generator	Sodium cooled, water boiled	S	M, H	500–1000	(2.8–5.7)
Boiling	Vapor-liquid	Evaporator (vacuum)	Steam condensed, water boiled	S	L	400–600	(2.3–3.4)
		Evaporator (high pressure)	Steam condensed, water boiled	S	L, M	400–600	(2.3–3.4)
		Mercury condenser-boiler	Mercury condensed, water boiled	S	M, H	500–700	(2.8–4.0)
	Gas-liquid	Waste-heat steam boiler	Flue gas cooled, water boiled	T	L, H	2–10	(0.011–0.057)
		Direct-fired steam boiler	Combustion gas cooled, water boiled	T	L, H	2–10	(0.011–0.057)
Condensing	Vapor-liquid	Refrigeration condenser	Water heated, refrigerant condensed	S, D	L, M	80–250	(0.45–1.4)
		Steam surface condenser	Water heated, steam condensed	S	L	300–800	(1.7–4.5)
		Steam mixing condenser	Water heated, steam condensed	M	L	DC	
		Intercondenser and aftercondenser	Condensate heated, steam condensed	S	L	15–300	(0.085–1.7)
	Vapor-gas	Air-cooled surface condenser	Air heated, steam condensed	T	L	2–16	(0.011–0.091)

Power.

†S—shell-and-tube exchanger; T—tubes in path of moving fluid, or exchanger open to surrounding air; R—regenerative plate-type or simple plate-type exchanger; D—double-tube exchanger; M—direct contact mixing exchanger.

‡L—highest pressure ranges from 0 to 100 lb/in² (abs) (0 to 689.4 kPa); M—highest pressure from 100 to 500 lb/in² (abs) (689.4 to 3447 kPa); H—500 lb/in² (abs) (3447 kPa) up.

§Values of U represent range of overall heat-transfer coefficients that might be expected in various exchangers. Coefficients are stated in Btu/(h·°F·ft²) [W/(m²·°C)] of heating surface. Total heat transferred in exchanger, in Btu/h, is obtained by multiplying a specific value of U for that type of exchanger by the surface and the log mean temperature difference. DC indicates direct exchange of heat.

ing of condensate in the heat exchanger. The overall coefficient of heat transfer $U = 25$ Btu/(h·°F·ft^2) [141.9 W/(m^2·°C)]. How much heating steam is required if the oil flow rate through the heater is 100 gal/min (6.3 L/s), the specific gravity of the oil is 0.9, and the specific heat of the oil is 0.5 Btu/(lb·°F) [2.84 W/(m^2·°C)]?

Calculation Procedure:

1. Compute the heat-transfer rate of the heater

With a flow rate of 100 gal/min (6.3 L/s) or (100 gal/min)(60 min/h) = 6000 gal/h (22,710 L/h), the weight flow rate of the oil, using the weight of water of specific gravity 1.0 as 8.33 lb/gal, is (6000 gal/h) (0.9 specific gravity)(8.33 lb/gal) = 45,000 lb/h (20,250 kg/h), closely.

Since the temperature of the oil rises $120 - 60 = 60°$F (33.3°C) during passage through the heat exchanger and the oil has a specific heat of 0.50, find the heat-transfer rate of the heater from the general relation $Q = wc\ \Delta t$, where Q = heat-transfer rate, Btu/h; w = oil flow rate, lb/h; c = specific heat of the oil, Btu/(lb·°F); Δt = temperature rise of the oil during passage through the heater. Thus, $Q = (45,000)(0.5)(60) = 1,350,000$ Btu/h (0.4 MW).

2. Compute the heater logarithmic mean temperature difference

The logarithmic mean temperature difference (LMTD) is found from LMTD = $(G - L)/\ln (G/L)$, where G = greater terminal temperature difference of the heater, °F; L = lower terminal temperature difference of the heater, °F; ln = logarithm to the base e. This relation is valid for heat exchangers in which the number of shell passes equals the number of tube passes.

In general, for parallel flow of the fluid streams, $G = T_1 - t_1$ and $L = T_2 - t_2$, where T_1 = heating fluid inlet temperature, °F; T_2 = heating fluid outlet temperature, °F; t_1 = heated fluid inlet temperature, °F; t_2 = heated fluid outlet temperature, °F. Figure 1 shows the maximum and minimum terminal temperature differences for various fluid flow paths.

For this parallel-flow exchanger, $G = T_1 - t_1 = 382 - 60 = 322°$F (179°C), where 382°F (194°C) = the temperature of 200-lb/in^2 (abs) (1379-kPa) saturated steam, from a table of steam properties. Also, $L = T_2 - t_2 = 382 - 120 = 262°$F (145.6°C), where the condensate temperature = the saturated steam temperature because there is no subcooling of the condensate. Then LMTD = $G - L/\ln(G/L) = (322 - 262)/\ln(322/262) = 290°$F (161°C).

3. Compute the required heat-transfer area

Use the relation $A = Q/U \times$ LMTD, where A = required heat-transfer area, ft^2; U = overall coefficient of heat transfer, Btu/(ft^2·h·°F). Thus, $A = 1,350,000/[(25)(290)] = 186.4$ ft^2 (17.3 m^2), say 200 ft^2 (18.6 m^2).

4. Compute the required quantity of heating system

The heat added to the oil = $Q = 1,350,000$ Btu/h, from step 1. The enthalpy of vaporization of 200-lb/in^2 (abs) (1379-kPa) saturated steam is, from the steam tables, 843.0 Btu/lb (1960.8 kJ/kg). Use the relation $W = Q/h_{fg}$, where W = flow rate of heating steam, lb/h; h_{fg} = enthalpy of vaporization of the heating steam, Btu/lb. Hence, $W = 1,350,000/843.0 = 1600$ lb/h (720 kg/h).

Related Calculations: Use this general procedure to find the heat-transfer area, fluid outlet temperature, and required heating-fluid flow rate when true parallel flow or counterflow of the fluids occurs in the heat exchanger. When such a true flow does *not* exist, use a suitable correction factor, as shown in the next calculation procedure.

The procedure described here can be used for heat exchangers in power plants, heating systems, marine propulsion, air-conditioning systems, etc. Any heating or cooling fluid—steam, gas, chilled water, etc.—can be used.

To select a heat exchanger by using the results of this calculation procedure, enter the engineering data tables available from manufacturers at the computed heat-transfer area. Read the heater dimensions directly from the table. Be sure to use the next *larger* heat-transfer area when the exact required area is not available.

When there is little movement of the fluid on either side of the heat-transfer area, such as occurs during heat transmission through a building wall, the arithmetic mean (average) temperature difference can be used instead of the LMTD. Use the LMTD when there is rapid movement of the fluids on either side of the heat-transfer area and a rapid change in temperature in one, or both, fluids. When one of the two fluids is partially, but not totally, evaporated or condensed, the

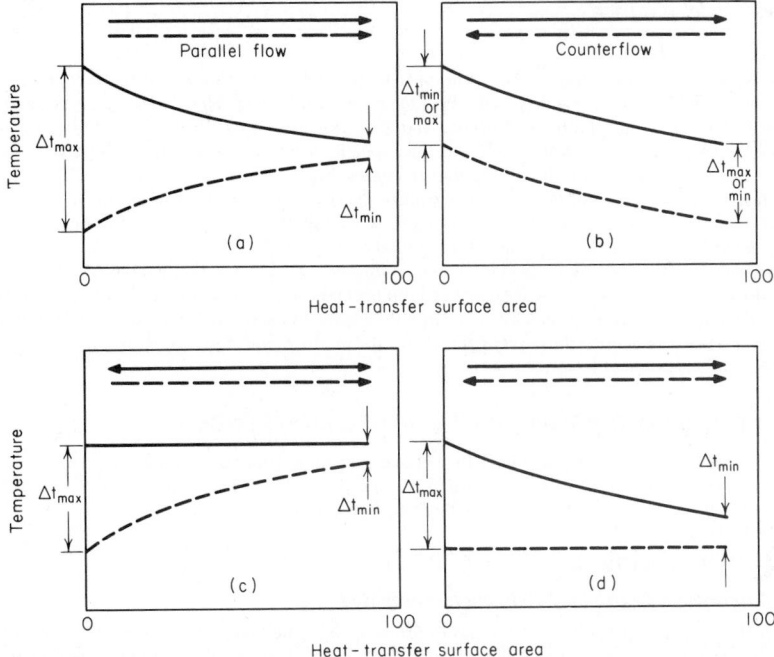

FIG. 1 Temperature relations in typical parallel-flow and counterflow heat exchangers.

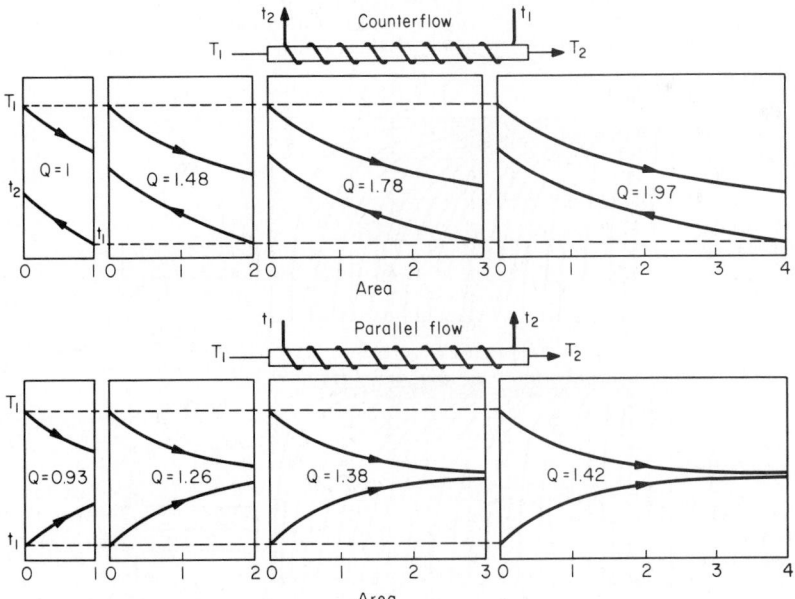

FIG. 2 For certain conditions, the area between the temperature curves measures the amount of heat being transferred.

true mean temperature difference is different from the arithmetic mean and the LMTD. Special methods, such as those presented in Perry—*Chemical Engineers' Handbook*, must be used to compute the actual temperature difference under these conditions.

When two liquids or gases with constant specific heats are exchanging heat in a heat exchanger, the area between their temperature curves, Fig. 2, is a measure of the total heat being transferred. Figure 2 shows how the temperature curves vary with the amount of heat-transfer area for counterflow and parallel-flow exchangers when the fluid inlet temperatures are kept constant. As Fig. 2 shows, the counterflow arrangement is superior.

If enough heating surface is provided, in a counterflow exchanger, the leaving cold-fluid temperature can be raised above the leaving hot-fluid temperature. This cannot be done in a parallel-flow exchanger, where the temperatures can only approach each other regardless of how much surface is used. The counterflow arrangement transfers more heat for given conditions and usually proves more economical to use.

HEAT-EXCHANGER ACTUAL TEMPERATURE DIFFERENCE

A counterflow shell-and-tube heat exchanger has one shell pass for the heating fluid and two shell passes for the fluid being heated. What is the actual LMTD for this exchanger if $T_1 = 300°F$ (148.9°C), $T_2 = 250°F$ (121°C), $t_1 = 100°F$ (37.8°C), and $t_2 = 230°F$ (110°C)?

Calculation Procedure:

1. *Determine how the LMTD should be computed*

When the numbers of shell and tube passes are unequal, true counterflow does not exist in the heat exchanger. To allow for this deviation from true counterflow, a correction factor must be applied to the logarithmic mean temperature difference (LMTD). Figure 3 gives the correction factor to use.

2. *Compute the variables for the correction factor*

The two variables that determine the correction factor are shown in Fig. 3 as $P = (t_2 - t_1)/(T_1 - t_1)$ and $R = (T_1 - T_2)/(t_2 - t_1)$. Thus, $P = (230 - 100)/(300 - 100) = 0.65$, and $R = (300 - 250)/(230 - 100) = 0.385$. From Fig. 3, the correction factor is $F = 0.90$ for these values of P and R.

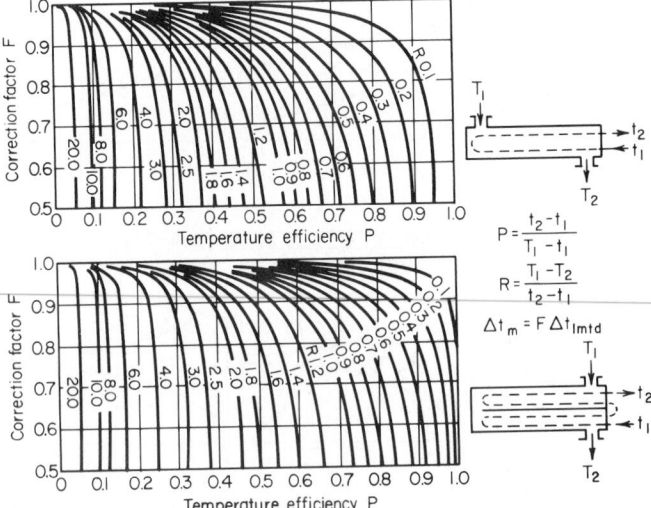

FIG. 3   Correction factors for LMTD when the heater flow path differs from true counterflow. (*Power.*)

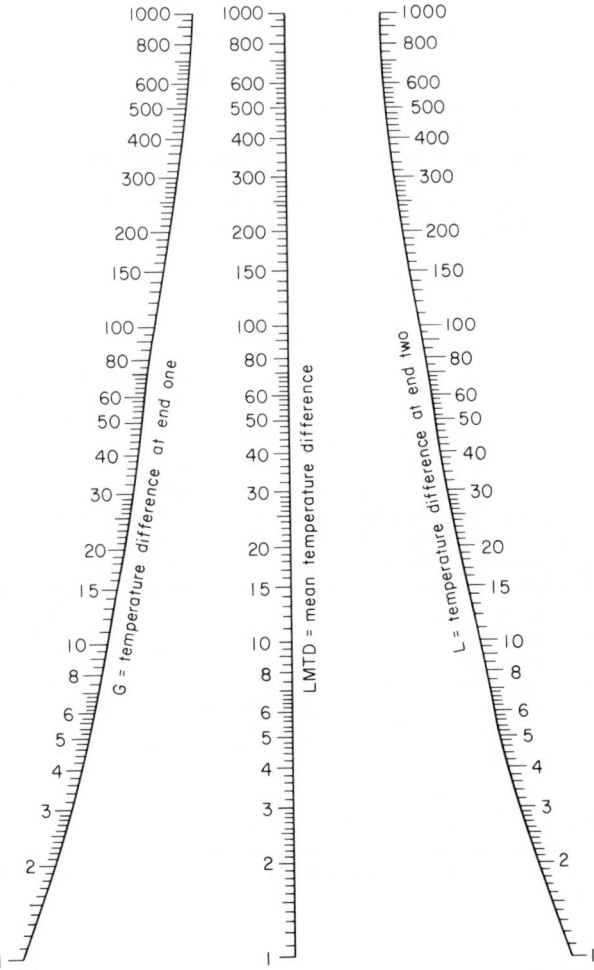

FIG. 4 Logarithmic mean temperature for a variety of heat-transfer applications.

3. *Compute the theoretical LMTD*

Use the relation LMTD = $(G - L)/\ln(G/L)$, where the symbols for counterflow heat exchange are $G = T_2 - t_1$; $L = T_1 - t_2$; ln = logarithm to the base e. All temperatures in this equation are expressed in °F. Thus, $G = 250 - 100 = 150$°F (83.3°C); $L = 300 - 230 = 70$°F (38.9°C). Then LMTD = $(150 - 70)/\ln(150/70) = 105$°F (58.3°C).

4. *Compute the actual LMTD for this exchanger*

The actual LMTD for this or any other heat exchanger is $\text{LMTD}_{\text{actual}} = F(\text{LMTD}_{\text{computed}}) = 0.9(105) = 94.5$°F (52.5°C). Use the actual LMTD to compute the required exchanger heat-transfer area.

   Related Calculations: Once the corrected LMTD is known, compute the required heat-exchanger size in the manner shown in the previous calculation procedure. The method given here is valid for both two- and four-pass shell-and-tube heat exchangers. Figure 4 simplifies the computation of the uncorrected LMTD for temperature differences ranging from 1 to 1000°F (−17 to 537.8°C). It gives LMTD with sufficient accuracy for all normal industrial and commercial heat-exchanger applications. Correction-factor charts for three shell passes, six or more tube passes, four shell passes, and eight or more tube passes are published in the *Standards of the Tubular Exchanger Manufacturers Association*.

FOULING FACTORS IN HEAT-EXCHANGER SIZING AND SELECTION

A heat exchanger having an overall coefficient of heat transfer of $U = 100$ Btu/(ft$^2 \cdot$ h $\cdot$ °F) [567.8 W/(m$^2 \cdot$ °C)] is used to cool lean oil. What effect will the tube fouling have on the value of U for this exchanger?

Calculation Procedure:

1. Determine the heat exchanger fouling factor

Use Table 2 to determine the fouling factor for this exchanger. Thus, the fouling factor for lean oil = 0.0020.

2. Determine the actual U for the heat exchanger

Enter Fig. 5 at the bottom with the clean heat-transfer coefficient of $U = 100$ Btu/(h $\cdot$ ft$^2 \cdot$ °F) [567.8 W/(m$^2 \cdot$ °C)] and project vertically upward to the 0.002 fouling-factor curve. From the intersection with this curve, project horizontally to the left to read the design or actual heat-transfer coefficient as $U_a = 78$ Btu/(h $\cdot$ ft$^2 \cdot$ °F) [442.9 W/(m$^2 \cdot$ °C)]. Thus, the fouling of the tubes causes a reduction of the U value of $100 - 78 = 22$ Btu/(h $\cdot$ ft$^2 \cdot$ °F) [124.9 W/(m$^2 \cdot$ °C)]. This means that the required heat transfer area must be increased by nearly 25 percent to compensate for the reduction in heat transfer caused by fouling.

   Related Calculations: Table 2 gives fouling factors for a wide variety of service conditions in applications of many types. Use these factors as described above; or add the fouling factor to the film resistance for the heat exchanger to obtain the total resistance to heat transfer. Then U = the reciprocal of the total resistance. Use the actual value U_a of the heat-transfer coefficient when sizing a heat exchanger. The method given here is that used by Condenser Service and Engineering Company, Inc.

TABLE 2   Heat-Exchanger Fouling Factors°

Fluid heated or cooled	Fouling factor
Fuel oil	0.0055
Lean oil	0.0020
Clean recirculated oil	0.0010
Quench oils	0.0042
Refrigerants (liquid)	0.0011
Gasoline	0.0006
Steam-clean and oil-free	0.0001
Refrigerant vapors	0.0023
Diesel exhaust	0.013
Compressed air	0.0022
Clean air	0.0011
Seawater under 130°F (54°C)	0.0006
Seawater over 130°F (54°C)	0.0011
City or well water under 130°F (54°C)	0.0011
City or well water over 130°F (54°C)	0.0021
Treated boiler feedwater under 130°F, 3 ft/s (54°C, 0.9 m/s)	0.0008
Treated boiler feedwater over 130°F, 3 ft/s (54°C, 0.9 m/s)	0.0009
Boiler blowdown	0.0022

°Condenser Service and Engineering Company, Inc.

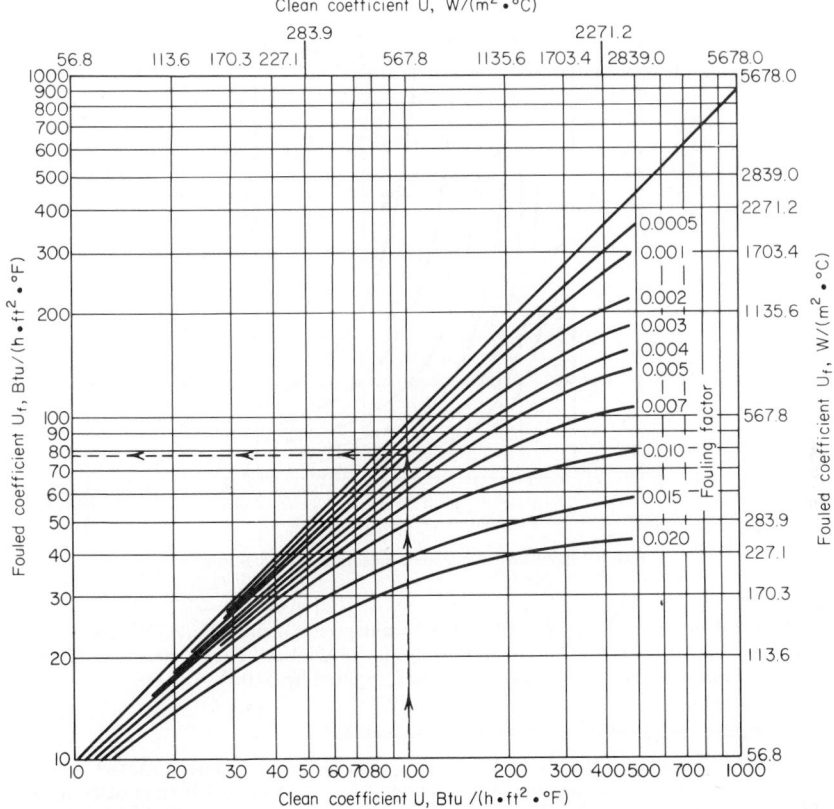

FIG. 5 Effect of heat-exchanger fouling on the overall coefficient of heat transfer. *(Condenser Service and Engineering Co., Inc.)*

HEAT TRANSFER IN BAROMETRIC AND JET CONDENSERS

A counterflow barometric condenser must maintain an exhaust pressure of 2 lb/in^2 (abs) (13.8 kPa) for an industrial process. What condensing-water flow rate is required with a cooling-water inlet temperature of 60°F (15.6°C); of 80°F (26.7°C)? How much air must be removed from this barometric condenser if the steam flow rate is 25,000 lb/h (11,250 kg/h); 250,000 lb/h (112,500 kg/h)?

Calculation Procedure:

1. *Compute the required unit cooling-water flow rate*

Use Fig. 6 as a quick guide to the required cooling-water flow rate for counterflow barometric condensers. Thus, entering the bottom of Fig. 6 at 2-lb/in^2 (abs) (13.8-kPa) exhaust pressure and projecting vertically upward to the 60°F (15.6°C) and 80°F (26.7°C) cooling-water inlet temperature curves show that the required flow rate is 52 gal/min (3.2 L/s) and 120 gal/min (7.6 L/s), respectively, per 1000 lb/h (450 kg/h) of steam condensed.

2. *Compute the total cooling-water flow rate required*

Use this relation: total cooling water required, gal/min = (unit cooling-water flow rate, gal/min per 1000 lb/h of steam condensed) (steam flow, lb/h)/1000. Or, total gpm = (52)(250,000/1000)

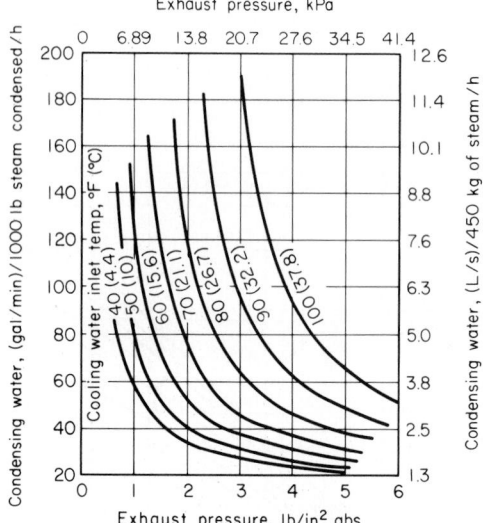

FIG. 6 Barometric condenser condensing-water flow rate.

= 13,000 gal/min (820.2 L/s) of 60°F (15.6°C) cooling water. For 80°F (26.7°C) cooling water, total gpm = (120)(250,000/1000) = 30,000 gal/min (1892.7 L/s). Thus, a 20°F (11.1°C) rise in the cooling-water temperature raises the flow rate required by 30,000 − 13,000 = 17,000 gal/min (1072.5 L/s).

3. Compute the quantity of air that must be handled

With a steam flow of 25,000 lb/h (11,250 kg/h) to a barometric condenser, manufacturers' engineering data show that the quantity of air entering with the steam is 3 ft³/min (0.08 m³/min); with a steam flow of 250,000 lb/h (112,500 kg/h), air enters at the rate of 10 ft³/min (0.28 m³/min). Hence, the quantity of air in the steam that must be handled by this condenser is 10 ft³/min (0.28 m³/min).

Air entering with the cooling water varies from about 2 ft³/min per 1000 gal/min of 100°F (0.06 m³/min per 3785 L/min of 37.8°C) water to 4 ft³/min per 1000 gal/min at 35°F (0.11 m³/min per 3785 L/min at 1.7°C). Using a value of 3 ft³/min (0.08 m³/min) for this condenser, we see the quantity of air that must be handled is (ft³/min per 1000 gal/min)(cooling-water flow rate, gal/min)/1000, or cfm of air = (3)(13,000/1000) = 39 ft³/min at 60°F (1.1 m³/min at 15.6°C). At 80°F (26.7°C) cfm = (3)(30,000/1000) = 90 ft³/min (2.6 m³/min).

Hence, the total air quantity that must be handled is 39 + 10 = 49 ft³/min (1.4 m³/min) with 60°F (15.6°C) cooling water, and 90 + 10 = 100 ft³/min (2.8 m³/min) with 80°F (26.7°C) cooling water. The air is usually removed from the barometric condenser by a two-stage air ejector.

Related Calculations: For help in specifying conditions for parallel-flow and counterflow barometric condensers, refer to *Standards of Heat Exchange Institute—Barometric and Low-Level Jet Condensers*. Whereas Fig. 6 can be used for a first approximation of the cooling water required for parallel-flow barometric condensers, the results obtained will not be as accurate as for counterflow condensers.

SELECTION OF A FINNED-TUBE HEAT EXCHANGER

Choose a finned-tube heat exchanger for a 1000-hp (746-kW) four-cycle turbocharged diesel engine having oil-cooled pistons and a cooled exhaust manifold. The heat exchanger will be used only for jacket-water cooling.

TABLE 3 Approximate Rates of Heat Rejection to Cooling Systems[*]

	Four-cycle engines			
Engine type	Normally aspirated, dry pistons, water-jacketed exhaust manifold, Btu/(bhp·hr) (kJ/kWh)	Normally aspirated, oil-cooled pistons, water-jacketed manifold, Btu/(bhp·h) (kJ/kWh)	Turbocharged, oil-cooled pistons, dry manifold, Btu/(bhp·h) (kJ/kWh)	Turbocharged, oil-cooled pistons, cooled manifold, Btu/(bhp·h) (kJ/kWh)
Jacket water	2200–2600 (12.5–14.8)	2000–2500 (11.3–14.2)	1450–1750 (8.2–9.9)	1800–2200 (10.2–12.5)
Lubricating oil	175–350 (1.0–2.0)	300–600 (1.7–3.4)	300–500 (1.7–2.8)	300–500 (1.7–2.8)
Raw water	2375–2950 (13.5–16.7)	2300–3100 (13.1–17.6)	1750–2250 (9.9–12.8)	2100–2700 (11.9–15.3)

	Two-cycle engines		
	Loop scavenging oil-cooled pistons, Btu/(bhp·h) (kJ/kWh)	Uniflow scavenging oil-cooled pistons	
Engine type		Opposed piston, Btu/(bhp·h) (kJ/kWh)	Valve in head, Btu/(bhp·h) (kJ/kWh)
Jacket water	1300–1900 (7.4–10.8)	1200–1600 (6.8–9.1)	1700–2100 (9.6–11.9)
Lubricating oil	500–700 (2.8–4.0)	900–1100 (5.1–6.2)	400–750 (2.3–4.3)
Raw water	1800–2600 (10.2–14.8)	2100–2700 (11.9–15.3)	2100–2850 (11.9–16.2)

[*]Diesel Engine Manufacturers Association; SI values added by handbook editor.

Calculation Procedure:

1. *Determine the heat-exchanger cooling load*

The Diesel Engine Manufacturers Association (DEMA) tabulation, Table 3, lists the heat rejection to the cooling system by various types of diesel engines. Table 3 shows that the heat rejection from the jacket water of a four-cycle turbocharged engine having oil-cooled pistons and a cooled manifold is 1800 to 2200 Btu/(bhp·h) (0.71 to 0.86 kW/kW). Using the higher value, we see the jacket-water heat rejection by this engine is (1000 bhp)[2200 Btu/(bhp·h)] = 2,200,000 Btu/h (644.8 kW).

2. *Determine the jacket-water temperature rise*

DEMA reports that a water temperature rise of 15 to 20°F (8.3 to 11.1°C) is common during passage of the cooling water through the engine. The maximum water discharge temperature reported by DEMA ranges from 140 to 180°F (60 to 82.2°C). Assume a 20°F (11.1°C) water temperature rise and a 160°F (71.1°C) water discharge temperature for this engine.

3. *Determine the air inlet and outlet temperatures*

Refer to weather data for the locality of the engine installation. Assume that the weather data for the locality of this engine show that the maximum dry-bulb temperature met in summer is 90°F (32.2°C). Use this as the air inlet temperature.

Before the required surface area can be determined, the air outlet temperature from the radiator must be known. This outlet temperature cannot be computed directly. Hence, it must be assumed and a trial calculation made. If the area obtained is too large, a higher outlet air temperature must be assumed and the calculation redone. Assume an outlet air temperature of 150°F (65.6°C).

4. Compute the LMTD for the radiator

The largest temperature difference for this exhanger is $160 - 90 = 70°F$ (38.9°C), and the smallest temperature difference is $150 - 140 = 10°F$ (5.6°C). In the smallest temperature difference expression, $140°F$ (77.8°C) = water discharge temperature from the engine − cooling-water temperature rise during passage through the engine, or $160 - 20 = 140°F$ (77.8°C). Then LMTD = $(70 - 10)/[\ln(70/10)] = 30°F$ (16.7°C). (Figure 4 could also be used to compute the LMTD).

5. Compute the required exchanger surface area

Use the relation $A = Q/U \times$ LMTD, where A = surface area required, ft²; Q = rate of heat transfer, Btu/h; U = overall coefficient of heat transfer, Btu/(h·ft²·°F). To solve this equation, U must be known.

Table 1 in the first calculation procedure in this section shows that U ranges from 2 to 10 Btu/(h·ft²·°F) [56.8 W/(m²·°C)] in the usual internal-combustion-engine finned-tube radiator. Using a value of 5 for U, we get $A = 2,200,000/[(5)(30)] = 14,650$ ft² (1361.0 m²).

6. Determine the length of finned tubing required

The total area of a finned tube is the sum of the tube and fin area per unit length. The tube area is a function of the tube diameter, whereas the finned area is a function of the number of fins per inch of tube length and the tube diameter.

Assume that 1-in (2.5-cm) tubes having 4 fins per inch (6.35 mm per fin) are used in this radiator. A tube manufacturer's engineering data show that a finned tube of these dimensions has 5.8 ft² of area per linear foot (1.8 m²/lin m) of tube.

To compute the linear feet L of finned tubing required, use the relation $L = A/(\text{ft}^2/\text{ft})$, or $L = 14,650/5.8 = 2530$ lin ft (771.1 m) of tubing.

7. Compute the number of individual tubes required

Assume a length for the radiator tubes. Typical lengths range between 4 and 20 ft (1.2 and 6.1 m), depending on the size of the radiator. With a length of 16 ft (4.9 m) per tube, the total number of tubes required = $2530/16 = 158$ tubes. This number is typical for finned-tube heat exchangers having large heat-transfer rates [more than 10^6 Btu/h (100 kW)].

8. Determine the fan horsepower required

The fan horsepower required can be computed by determining the quantity of air that must be moved through the heat exchanger, after assuming a resistance—say 1.0 in of water (0.025 Pa)—for the exchanger. However, the more common way of determining the fan horsepower is by referring to the manufacturer's engineering data.

Thus, one manufacturer recommends three 5-hp (3.7-kW) fans for this cooling load, and another recommends two 8-hp (5.9-kW) fans. Hence, about 16 hp (11.9 kW) is required for the radiator.

Related Calculations: The steps given here are suitable for the initial sizing of finned-tube heat exchangers for a variety of applications. For exact sizing, it may be necessary to apply a correction factor to the LMTD. These correction factors are published in Kern—*Process Heat Transfer*, McGraw-Hill, and McAdams—*Heat Transfer*, McGraw-Hill.

The method presented here can be used for finned-tube heat exchangers used for air heating or cooling, gas heating or cooling, and similar industrial and commercial applications.

SPIRAL-TYPE HEATING COIL SELECTION

How many feet of heating coil are required to heat 1000 gal/h (1.1 L/s) of 0.85-specific-gravity oil if the specific heat of the oil is 0.50 Btu/(lb·°F) [2.1 kJ/(kg·°C)], the heating medium is 65-lb/in² (gage) (448.2-kPa) steam, and the oil enters at 60°F (15.6°C) and leaves at 125°F (51.7°C)? There is no subcooling of the condensate.

Calculation Procedure:

1. Compute the LMTD for the heater

Steam at $65 + 14.7 = 79.7$ lb/in² (abs) (549.5 kPa) has a temperature of approximately 312°F (155.6°C), as given by the steam tables. Condensate at this pressure has the same approximate temperature. Hence, the entering and leaving temperatures of the heating fluid are approximately the same.

Oil enters the heater at 60°F (15.6°C) and leaves at 125°F (51.7°C). Therefore, the greater temperature difference G across the heater is $G = 312 - 60 = 252°F$ (140.0°C), and the lesser temperature difference L is $L = 312 - 125 = 187°F$ (103.9°C). Hence, the LMTD $= (G - L)/[\ln(G/L)]$, or $(252 - 187)/[\ln(252/187)] = 222°F$ (123.3°C). In this relation, $\ln =$ logarithm to the base e $= 2.7183$. (Figure 4 could also be used to determine the LMTD.)

2. Compute the heat required to raise the oil temperature

Water weighs 8.33 lb/gal (1.0 kg/L). Since this oil has a specific gravity of 0.85, it weighs $(8.33)(0.85) = 7.08$ lb/gal (0.85 kg/L). With 1000 gal/h (1.1 L/s) of oil to be heated, the weight of oil heated is (1000 gal/h)(7.08 lb/gal) = 7080 lb/h (0.89 kg/s). Since the oil has a specific heat of 0.5 Btu/(lb·°F) [2.1 kJ/(kg·°C)] and this oil is heated through a temperature range of $125 - 60 = 65°F$ (36.1°C), the quantity of heat Q required to raise the temperature of the oil is $Q =$ (7080 lb/h) [0.5 Btu/(lb·°F) (65°F)] = 230,000 Btu/h (67.4 kW).

3. Compute the heat-transfer area required

Use the relation $A = Q/(U \times LMTD)$, where $Q =$ heat-transfer rate, Btu/h; $U =$ overall coefficient of heat transfer, Btu/(h·ft²·°F). For heating oil to 125°F (51.7°C), the U value given in Table 1 is 20 to 60 Btu/(h·ft²·°F) [0.11 to 0.34 kW/(m²·°C)]. Using a value of $U = 30$ Btu/(h·ft²·°F) [0.17 kW/(m²·°C)] to produce a conservatively sized heater, we find $A = 230,000/[(30)(222)] = 33.4$ ft² (3.1 m²) of heating surface.

4. Choose the coil material for the heater

Spiral-type tank heating coils are usually made of steel because this material has a good corrosion resistance in oil. Hence, this coil will be assumed to be made of steel.

5. Compute the heating steam flow required

To determine the steam flow rate required, use the relation $S = Q/h_{fg}$, where $S =$ steam flow, lb/h; $h_{fg} =$ latent heat of vaporization of the heating steam, Btu/lb, from the steam tables; other symbols as before. Hence, $S = 230,000/901.1 = 256$ lb/h (0.03 kg/s), closely.

6. Compute the heating coil pipe diameter

Steam-heating coils submerged in the liquid being heated are usually chosen for a steam velocity of 4000 to 5000 ft/min (20.3 to 25.4 m/s). Compute the heating pipe cross-sectional area a in² from $a = 2.4Sv_g/V$, where $v_g =$ specific volume of the steam at the coil operating pressure, ft³/lb, from the steam tables; $V =$ steam velocity in the heating coil, ft/min; other symbols as before. With a steam velocity of 4000 ft/min (20.3 m/s), $a = 2.4(256)(5.47)/4000 = 0.838$ in² (5.4 cm²).

Refer to a tabulation of pipe properties. Such a tabulation shows that the internal transverse area of a schedule 40 1-in (2.5-cm) diameter nominal steel pipe is 0.863 in² (5.6 cm²). Hence, a 1-in (2.5-cm) pipe will be suitable for this heating coil.

7. Determine the length of coil required

A pipe property tabulation shows that 2.9 lin ft (0.9 m) of 1-in (2.5-cm) schedule 40 pipe has 1.0 ft² (0.09 m²) of external area. Hence, the total length of pipe required in this heating coil = (33.1 ft²)(2.9 ft/ft²) = 96 ft (29.3 m).

Related Calculations: Use this general procedure to find the area and length of spiral heating coil required to heat water, industrial solutions, oils, etc. This procedure also can be used to find the area and length of cooling coils used to cool brine, oils, alcohol, wine, etc. In every case, be certain to substitute the correct specific heat for the liquid being heated or cooled. For typical values of U, consult Perry—*Chemical Engineers' Handbook*, McGraw-Hill; McAdams—*Heat Transmission*, McGraw-Hill; or Kern—*Process Heat Transfer*, McGraw-Hill.

SIZING ELECTRIC HEATERS FOR INDUSTRIAL USE

Choose the heating capacity of an electric heater to heat a pot containing 600 lb (272.2 kg) of lead from the charging temperature of 70°F (21.1°C) to a temperature of 750°F (398.9°C) if 600 lb (272.2 kg) of the lead is to be melted and heated per hour. The pot is 30 in (76.2 cm) in diameter and 18 in (45.7 cm) deep.

Calculation Procedure:

1. Compute the heat needed to reach the melting point

When a solid is melted, first it must be raised from its ambient or room temperature to the melting temperature. The quantity of heat required is H = (weight of solid, lb)[specific heat of solid, Btu/(lb·°F)]$(t_m - t_i)$, where H = Btu required to raise the temperature of the solid, °F; t_i = room, charging, or initial temperature of the solid, °F; t_m = melting temperature of the solid, °F.

For this pot with lead having a melting temperature of 620°F (326.7°C) and an average specific heat of 0.031 Btu/(lb·°F) [0.13 kJ/(kg·°C)], H = (600)(0.031)(620 − 70) = 10,240 Btu/h (3.0 kW), or (10,240 Btu/h)/(3412 Btu/kWh) = 2.98 kWh.

2. Compute the heat required to melt the solid

The heat H_m Btu required to melt a solid is H_m = (weight of solid melted, lb)(heat of fusion of the solid, Btu/lb). Since the heat of fusion of lead is 10 Btu/lb (23.3 kJ/kg), H_m = (600)(10) = 6000 Btu/h, or 6000/3412 = 1.752 kWh.

3. Compute the heat required to reach the working temperature

Use the same relation as in step 1, except that the temperature range is expressed as $t_w - t_m$, where t_w = working temperature of the melted solid. Thus, for this pot, H = (600)(0.031)(750 − 620) = 2420 Btu/h (709.3 W), or 2420/3412 = 0.709 kWh.

4. Determine the heat loss from the pot

Use Fig. 7 to determine the heat loss from the pot. Enter at the bottom of Fig. 7 at 750°F (398.9°C), and project vertically upward to the 10-in (25.4-cm) diameter pot curve. At the left, read the heat loss at 7.3 kWh/h.

5. Compute the total heating capacity required

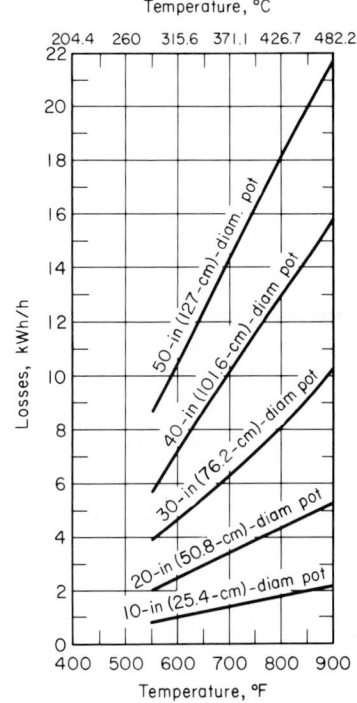

FIG. 7 Heat losses from melting pots. (General Electric Co.)

The total heating capacity required is the sum of the individual capacities, or 2.98 + 1.752 + 0.708 + 7.30 = 12.74 kWh. A 15-kW electric heater would be chosen because this is a standard size and it provides a moderate extra capacity for overloads.

Related Calculations: Use this general procedure to compute the capacity required for an electric heater used to melt a solid of any kind—lead, tin, type metal, solder, etc. When the substance being heated is a liquid—water, dye, paint, varnish, oil, etc.—use the relation H = (weight of liquid heated, lb) [specific heat of liquid, Btu/(lb·°F)] (temperature rise desired, °F), when the liquid is heated to approximately its boiling temperature, or a lower temperature.

For space heating of commercial and residential buildings, two methods used for computing

TABLE 4 Two Methods for Determining Wattage for Heating Buildings Electrically*

	W/ft^3 method	W/m^3 method
1. Interior rooms with no or little outside exposure	0.75 to 1.25	25.6 to 44.1
2. Average rooms with moderate windows and doors	1.25 to 1.75	44.1 to 61.8
3. Rooms with severe exposure and great window and door space	1.0 to 4.0	35.3 to 141.3
4. Isolated rooms, cabins, watchhouses, and similar buildings	3.0 to 6.0	105.9 to 211.9

	The "35" method
1. Volume in ft^3 for one air change × 0.35 =	0.01 W
2. Exposed net wall, roof, or ceiling and floor in ft^2 × 3.5 =	0.1 W
3. Area of exposed glass and doors in ft^2 × 35.0 =	1 W

*General Electric Company.

the approximate wattage required are the W/ft^3 and the "35" method. These are summarized in Table 4. In many cases, the results given by these methods agree closely with more involved calculations. When the desired room temperature is different from 70°F (21.1°C), increase or decrease the required kilowatt capacity proportionately, depending on whether the desired temperature is higher than or lower than 70°F (21.1°C).

For heating pipes with electric heaters, use a heater capacity of 0.8 W/ft^2 (8.6 W/m^2) of uninsulated exterior pipe surface per °F temperature difference between the pipe and the surrounding air. If the pipe is insulated with 1 in (2.5 cm) of insulation, use 30 percent of this value, or 0.24 (W/(ft^2·°F) [4.7 W/(m^2·°C)].

The types of electric heaters used today include immersion (for water, oil, plating, liquids, etc.), strip, cartridge, tubular, vane, fin, unit, and edgewound resistor heaters. These heaters are used in a wide variety of applications including liquid heating, gas and air heating, oven warming, deicing, humidifying, plastics heating, pipe heating, etc.

For pipe heating, a tubular heating element can be fastened to the bottom of the pipe and run parallel with it. For large-wattage applications, the heater can be spiraled around the pipe. For temperatures below 165°F (73.9°C), heating cable can be used. Electric heating is often used in place of steam tracing of outdoor pipes.

The procedure presented above is the work of General Electric Company.

ECONOMIZER HEAT TRANSFER COEFFICIENT

A 4530-ft^2 (421-m^2) heating surface counterflow economizer is used in conjunction with a 150,000-lb/h (68,040-kg/h) boiler. The inlet and outlet water temperatures are 210°F (99°C) and 310°F (154°C). The inlet and outlet gas temperatures are 640°F (338°C) and 375°F (191°C). Find the overall heat transfer coefficient in Btu/(h·ft^2·°F) [W/(m^2·°C)] [kJ/(h·m^2·°C)].

Calculation Procedure:

1. *Determine the enthalpy of water at the inlet and outlet temperatures*

From Table 1, Saturation: Temperatures, of the Steam Tables mentioned under Related Calculations of this procedure, for water at inlet temperature, $t_1 = 210$°F (99°C), the enthalpy, $h_1 = 178.14$ Btu/lb (414 kJ/kg), and at the outlet temperature, $t_2 = 310$°F (154°C), the enthalpy, $h_2 = 279.81$ Btu/lb$_m$ (651 kJ/kg).

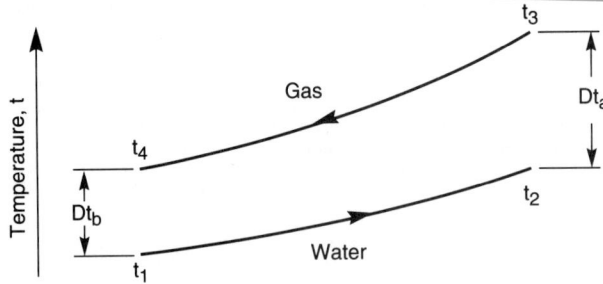

FIG. 8 Temperature vs surface area of economizer.

2. Compute the logarithmic mean temperature difference between the gas and water

As shown in Fig. 8, the temperature difference of the gas entering and the water leaving, Δt_a = $t_3 - t_2$ = 640 − 310 = 330°F (166°C) and for the gas leaving and the water entering, Δt_b = $t_4 - t_1$ = 375 − 210 = 165°F (74°C). Then, the logarithmic mean temperature difference, $\Delta t_m = (\Delta t_a - \Delta t_b)/[2.3 \times \log_{10} (\Delta t_a/\Delta t_b)] = (330 - 165)/[2.3 \times \log_{10} (330/165)] = 238°F$ (115°C).

3. Compute the economizer heat transfer coefficient

All the heat lost by the gas is considered to be transferred to the water, hence the heat lost by the gas, $Q = w(h_2 - h_1) = UA \, \Delta t_m$, where the water rate of flow, w = 150,000 lb/h (68,000 kg/h); U is the overall heat transfer coefficient; heating surface area, A = 4530 ft² (421 m²); other values as before. Then, 150,000 × (279.81 − 178.41) = U(4530)(238). Solving, U = [150,000 × (279.81 − 178.14)]/(4530 × 238) = 14.1 Btu/(h·ft²·°F) [80 W/(m²·°C)] [288 kJ/(h·m²·°C)].

Related Calculations: The Steam Tables appear in *Thermodynamic Properties of Water Including Vapor, Liquid, and Solid Phases*, 1969, Keenan, et al., John Wiley & Sons, Inc. Use later versions of such tables whenever available, as necessary.

BOILER TUBE STEAM-GENERATING CAPACITY

A counterflow bank of boiler tubes has a total area of 900 ft² (83.6 m²) and its overall coefficient of heat transfer is 13 Btu/(h·ft²·°F) [73.8 W/(m²·K). The boiler tubes generate steam at a pressure of 1000 lb/in² absolute (6900 kPa). The tube bank is heated by flue gas which enters at a temperature of 2000°F (1367 K) and at a rate of 450,000 lb/h (56.7 kg/s). Assume an average specific heat of 0.25 Btu/(lb·°F) [1.05 kJ/(kg·K)] for the gas and calculate the temperature of the gas that leaves the bank of boiler tubes. Also, calculate the rate at which the steam is being generated in the tube bank.

Calculation Procedure:

1. Find the temperature of steam at 1000 lb,/in² (6900 kPa)

From Table 2, Saturation: Pressures, of the Steam Tables mentioned under Related Calculations of this procedure, the saturation temperature of steam at 1000 lb/in² (6900 kPa), t_s = 544.6°F (558 K), a constant value as indicated in Fig. 9.

2. Determine the logarithmic mean temperature difference in terms of the flue-gas leaving temperature

The logarithmic mean temperature difference, $\Delta t_m = (\Delta t_1 - \Delta t_2)/\{2.3 \times \log_{10} [(t_1 - t_s)/(t_2 - t_s)]\}$, where Δt_1 = flue gas entering temperature − steam temperature = $(t_1 - t_s)$ = (2000 − 544.6); Δt_2 = flue-gas leaving temperature − steam temperature = $(t_2 - t_s)$ = $(t_2 - 544.6)$; $(\Delta t_1 - \Delta t_2)$ = [(2000 − 544.6) − $(t_2 - 544.6)$] = $(2000 - t_2)$; $[(t_1 - t_s)/(t_2 - t_s)]$ = [(2000 − 544.6)/$(t_2 - 544.6)$]. Hence, $\Delta t_m = [(2000 - t_2)/\{2.3 \times \log_{10} [(1455.4)/(t_2 - 544.6)]\}$.

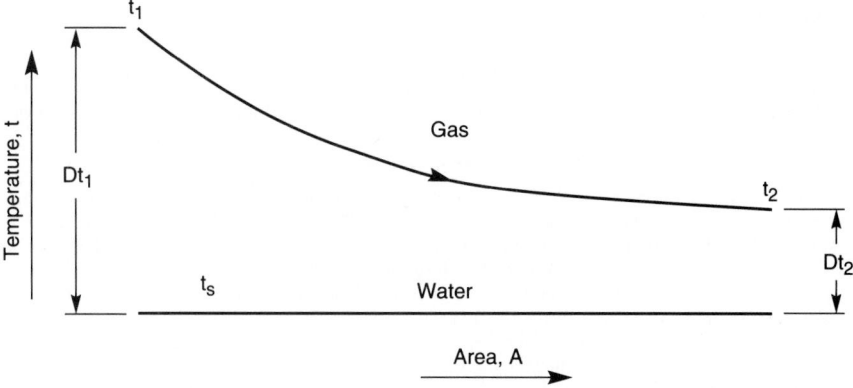

FIG. 9   Temperature vs surface area of boiler tubes.

3.   *Compute the flue-gas leaving temperature*

Heat transferred to the boiler water, $Q = w_g \times c_p \times (t_1 - t_2) = UA\Delta t_m$, where the flow rate of flue gas, $w_g = 450,000$ lb/h (56.7 kg/s); flue-gas average specific heat, $c_p = 0.25$ Btu/(lb/°F) [1.05 kg/(kg·K)]; overall coefficient of heat transfer of the boiler tubes, $U = 13$ Btu/(h·ft²·°F) [73.8 W/(m²·K)]; area of the boiler tubes exposed to heat, $A = 900$ ft² (83.6 m²); other values as before.

Then, $Q = 450,000 \times 0.25 \times (2000 - t_2) = 13 \times 900 \times [(2000 - t_2)/\{2.3 \times \log_{10}[(1455.4)/(t_2 - 544.6)]\}$. Or, $\log_{10}[(1455.4/(t_2 - 544.6)] = 13 \times 900/(2.3 \times 450,000 \times 0.25) = 0.0452$. The antilog of $0.0452 = 1.11$, hence, $[(1455.4/(t_2 - 544.7)] = 1.11$, and $t_2 = (1455.4/1.11) + 544.6 = 1850$°F (1280 K).

4.   *Find the heat of vaporization of the water*

From the Steam Tables, the heat of vaporization of the water at 1000 lb/in² (6900 kPa), $h_{fg} = 649.5$ Btu/lb (1511 kJ/kg).

5.   *Compute the steam-generating rate of the boiler tube bank*

Heat absorbed by the water = heat transferred by the flue gas, or $Q = w_s \times h_{fg} = w_g \times c_p \times (t_1 - t_2)$, where the mass of steam generated is w_s in lb/h (kg/s); other values as before. Then, $w_s \times 649.5 = 450,000 \times 0.25 \times (2000 - 1850) = 16.9 \times 10^6$ Btu/h (4950 kJ/s) (4953 kW). Thus, $w_s = 16.9 \times 10^6/649.5 = 26,000$ lb/h (200 kg/s).

   Related Calculations: The Steam Tables appear in *Thermodynamic Properties of Water Including Vapor, Liquid, and Solid Phases*, 1969, Keenan, et al., John Wiley & Sons, Inc. Use later versions of such tables whenever available, as required.

Refrigeration

REFERENCES: Trott—*Refrigeration and Air Conditioning*, McGraw-Hill; Hallowell—*Cold and Freezer Storage Manual*, AVI; Munton and Stott—*Refrigeration at Sea*, Applied Science (England); International Institute of Refrigeration—*Low Temperature and Electric Power*, Pergamon; Betts—*Refrigeration and Thermometry below One Kelvin*, Crane-Russak; Emerick—*Heating Design and Practice*, McGraw-Hill; Carrier Air Conditioning Company—*Handbook of Air Conditioning System Design*, McGraw-Hill; Severns and Fellows—*Air Conditioning and Refrigeration*, Wiley; ASHRAE—*Guide and Data Book: Fundamentals and Equipment*; ASHRAE—*Guide and Data Book: Applications*; Strock and Koral—*Handbook of Heating, Air Conditioning, and Ventilation*, Industrial Press; American Blower Corporation—*Air Conditioning and Engineering*; MacIntire-Hutchinson—*Refrigeration Engineering*, Wiley.

REFRIGERATION SYSTEM SELECTION

Choose a refrigeration system for a given load. Show the steps that the designer should follow in choosing a suitable refrigeration system for various types of loads.

Calculation Procedure:

1.   *Determine the refrigeration load*

Use the method given in the next calculation procedure. In any refrigeration plant, the total refrigeration load = heat gain from external sources, tons + product load, tons + sensible heat load, tons.

2.   *Choose the type of refrigeration system to use*

Table 1 shows the usual compressor choices for various refrigeration loads. Thus, reciprocating compressors find wide use for refrigeration loads up to 400 tons (362.9 t). Up to loads of about 5 tons (4.5 t), *unit systems* that combine the compressor, drive, evaporator, and condenser in a compact unit are popular. In some instances, larger-capacity unit systems may be available from certain manufacturers. Some large unit systems, called *central-station systems,* are built with capacities of 100 to 150 tons (90.7 to 136.1 t).

From 5- to 400-ton (4.5- to 362.9-t) capacity, *built-up central systems* are popular. In these systems, the manufacturer supplies the compressor, evaporator, and condenser as separate units. These are connected by suitable piping. The refrigeration equipment manufacturer may or may not supply the compressor driving unit. This driver may be an electric motor, steam turbine, internal-combustion engine, or some other type of prime mover.

For loads greater than 400 tons (362.9 t), the centrifugal refrigeration compressor is often chosen. Whereas this may be a built-up system, more and more manufacturers today supply completely fabricated systems containing all the needed components, including the controls, driver, etc.

Steam-jet refrigeration units find some application for loads of 50 tons (45.4 t) or more. The steam-jet refrigeration system is used for a large number of applications where steam is available.

TABLE 1   Typical Refrigeration System Choices°

System load		System type		
tons	t	Often used	Occasionally used	Rarely used
0–5	0–4.5	Unit system with reciprocating compressor	Central-station built-up system; reciprocating compressor	Central-station built-up units
5–25	4.5–22.7	Central-station built-up systems; reciprocating compressor	Central station built-up systems; reciprocating compressor	Absorption or adsorption units
25–50	22.7–45.4	Central-station built-up systems; reciprocating compressor	Central-station built-up systems; centrifugal compressor	Absorption units
50–400	45.4–362.9	Central-station built up systems; reciprocating compressor	Central-station built-up systems; steam-jet and centrifugal compressors	
400 and up	362.9 and up	Central system; centrifugal and/or absorption unit	Central-station built-up steam-jet unit	

°Adapted from ASHRAE data.

Typical applications include comfort air conditioning, industrial process cooling, and similar service. In recent years, some large office buildings have used steam-jet systems mounted in the building penthouse. These units provide the cooling needed for the building air-conditioning system.

Absorption refrigeration systems were once popular for a variety of cooling tasks in industry, food storage, etc. In recent years, the absorption system has found renewed use in medium- and large-size air-conditioning systems. The usual absorbent used today is lithium bromide; the refrigerant is ordinary tap water. Absorption refrigeration systems are popular in areas where fuel costs are low, electric rates are high, waste steam is available, low-pressure heating boilers are unused during the cooling season, or steam or gas utility companies desire to promote summer loads. Absorption refrigeration systems can be installed in almost any location in a building where the floor is of adequate strength and reasonably level. Absence of heavy moving parts practically eliminates vibration and reduces the noise level to a minimum.

Combination absorption-centrifugal refrigeration systems are well suited for many large-tonnage air-conditioning and industrial loads. These systems are extremely economical where medium- or high-pressure steam is used as the energy source.

Using the expected refrigeration load from step 1 and the data above, make a preliminary choice of the type of refrigeration systems to use. Remember that the necessity for part-load operation might change the preliminary choice of the system type.

3. Choose the system components

Manufacturers' engineering data generally list compatible components for a given capacity compressor. These components include the condenser, expansion valve, evaporator, receiver, cooling tower, etc. Later calculation procedures in this section give specific instructions for selecting these and other important components of the system. When a unit system is chosen, the important components are preselected by the manufacturer.

4. Have the system choice verified

Have the manufacturer whose equipment will be used verify the selection for the given load. This ensures a correct choice.

Related Calculations: Use this general procedure to select the type of refrigeration system serving air-conditioning, product-cooling, liquid-cooling, ice-making, and similar applications in stationary (land) and marine service.

SELECTION OF A REFRIGERATION UNIT FOR PRODUCT COOLING

What capacity and type of refrigeration system are needed for a walk-in cooler having inside dimensions of $8 \times 6 \times 10$ ft ($2.4 \times 1.8 \times 3.1$ m) if it is insulated with 4-in (10.2-cm) thick cork? The user estimates that a maximum of 400 lb (181.4 kg) of beef will be placed in the cooler daily, arriving at $70°F$ ($21.1°C$). The average hottest summer day in the cooler locality is, according to weather bureau records, $92°F$ ($33.3°C$). The meat is to be stored at $36°F$ ($2.2°C$). A ⅛-hp (0.12-kW) blower circulates air in the cooler. What refrigeration capacity is required for the same cooler, if the meat is stored at $-10°F$ ($-23.3°C$) and the cork insulation is 8 in (20.3 cm) thick? Two ⅛-hp (0.09-kW) blowers will be used in the cooler.

Calculation Procedure:

1. Compute the outside area of the cooler

The outside dimensions of this cooler are 9 ft (2.7 m) high, 7 ft (2.1 m) wide, and 11 ft (3.4 m) long, including the cork insulation and the supporting structure. Hence, the total outside area of the cooler, including the floor and roof, is $2(9 \times 7) + 2(9 \times 11) + 2(7 \times 11) = 478$ ft^2 (44.4 m^2).

2. Compute the heat gain and service load

There is a heat gain into the cooler through the insulated surfaces caused by the difference between the inside and outside temperatures. Also, there is a service load, that is, a heat gain caused by the opening and shutting of the cooler door. Since meat will be loaded only once a day, it is safe to assume that the service load is a normal one—i.e., the door will be opened less than 5 times per hour.

TABLE 2 Heat Leakage Factors[a]

	Btu and kJ per degree temperature difference per ft² (m²) of outside surface											
	Insulation thickness											
	in						cm					
	1	2	4	6	8	10	2.5	3.1	10.2	15.2	20.3	25.4
Heat leakage only	0.178	0.127	0.079	0.059	0.046	0.038	0.0010	0.0007	0.0005	0.0003	0.0003	0.0002
Heat leakage plus normal service load	0.216	0.163	0.110	0.090	0.077	0.069	0.0012	0.0009	0.0006	0.0005	0.0004	0.0004

[a]Brunner Manufacturing Company. SI values added by handbook editor.
Note: Light duty—multiply factor by 0.90; heavy duty—multiply factor by 1.10; single glass—multiply factor by 15; double glass—multiply factor by 6.5; triple glass—multiply factor by 5. If any wall or ceiling of a cooler is exposed to the sun, increase the temperature difference by 20°F (11.1°C) for that wall.

For product storage, cooling, heat, and service load, Btu/h = (total outside area of cooler, ft²)(maximum outside temperature, °F − minimum inside temperature, °F)(factor from Table 2), or (478)(92 − 36)(0.110) = 2944 Btu/h (0.86 kW).

3. Compute the product heat load

Use this relation: product heat load, Btu/h = (lb/h of product cooled)(temperature of product entering cooler, °F − temperature of product leaving cooler, °F)[specific heat of product, Btu/(lb·°F)]. For this cooler, given the specific heat from Table 3, the product heat load = (400 lb/24 h)(70 − 36)(0.8) = 453 Btu/h (132.8 W).

4. Compute the total heat load

The total heat load = sum of heat gain and service load + product heat load + supplementary heat load, Btu/h, or 2994 + 453 + 424 = 3821 Btu/h (1.1 kW).

5. Compute the refrigeration-system capacity required

In cooler operation, it is essential to ensure defrosting of the evaporator during the off cycle. To permit this defrosting, select a condensing unit to operate 18 h per 24-h day. With an 18-h operating time, the required condensing-unit capacity to handle the 24-h load is (24 h/operating time, h)(total heat load, Btu/h) = (24/18)(3821) = 5082 Btu/h (1.5 kW).

6. Select the refrigeration unit

Since the required capacity of this refrigeration system is between 0 and 5 tons (0 and 4.5 t), the previous calculation procedure indicates that a unit system with a reciprocating compressor is the most common type used. Referring to a manufacturer's engineering data shows that a 5000-Btu/h (1.46-kW) 0.5-hp (0.37-kW) air-cooled unit having a 20°F (−6.7°C) suction temperature is available. This unit will operate about 18.5 h/day to carry the actual heat load of 5082 Btu/h (1.48 kW) if the evaporator is chosen on a 16°F (36°F − 20°F) (8.9°C) temperature difference between the room and refrigerant.

The exact size of a condensing unit cannot be selected until a choice is made of the evaporating temperature, or suction pressure, at which the compressor is to work. In general, a difference of between 10 and 20°F (5.6 and 11.1°C) should be maintained between the product or room temperature and the evaporator temperature. Thus, the 16°F (8.9°C) temperature difference used above is within the normal working range.

A better plan for this product cooler would be to select a larger evaporator on a 10°F (5.6°C) temperature-difference basis. The running time of the condensing unit would be decreased because of the higher operating suction temperature, that is, 26 instead of 20°F (−3.3 instead of −6.7°C).

A standard refrigeration unit having the same characteristics as the unit described above, except that the evaporating temperature is 25°F (−3.9°C), has a capacity of 5550 Btu/h (1.6 kW) with refrigerant 12. The evaporating pressure is 24.6 lb/in² (169.6 kPa). The compressor is a two-cylinder unit and is belt-driven by an electric motor. A finned-tube air-cooled condenser is used. The receiver is mounted below the compressor, on the same frame. Refrigerant 12 (formerly called Freon-12) is most satisfactory for low-temperature systems.

7. Compute the required capacity at below-freezing temperature

The six steps above are for product storage at temperatures above freezing, i.e., above 32°F (0°C). For temperatures below 32°F (0°C), the same procedure is followed except that the product load is computed in three steps—cooling to 32°F (0°C), freezing, and cooling to the final temperature.

Thus, heat and service load = 478[92 − (−10)](0.085) = 4140 Btu/h (1.2 kW), given the heavy-duty factor from Table 2.

For cooling to 32°F (0°C), product load = (400/24)(70 − 32)(0.8) = 504 Btu/h (0.15 kW). For the freezing process, heat removal = (lb/h of product cooled)(latent heat or enthalpy of freezing, Btu/lb, from Table 3), or (400 lb/24 h)(98) = 1635 Btu/h (0.48 kW). To cool below freezing, heat removal = (lb/h of product cooled)(32°F − temperature of storage room, °F)[specific heat of product at temperature below freezing, Btu/(lb·°F), from Table 3], or (400/24)[32 − (−10)](0.404) = 282 Btu/h (0.08 kW). Then the total product load = 504 + 1635 + 282 = 2421 Btu/h (0.71 kW).

The supplementary load with two ⅛-hp (0.09-kW) blowers is (2)(⅛)(2545) = 635 Btu/h (0.19 kW).

The total load is thus 4140 + 2421 + 635 = 7196 Btu/h (2.1 kW). Assuming a 16-h/day

TABLE 3 Typical Specific and Latent Heats°

Article	Specific heat				Latent heat of freezing		Cold-storage temperature	
	Above freezing		Below freezing					
	Btu/(lb·°F)	kJ/(kg·°C)	Btu/(lb·°F)	kJ/(kg·°C)	Btu/(lb·°F)	kJ/(kg·°C)	°F	°C
Canned goods:								
Fruits	As fresh	As fresh	As fresh	As fresh	····	····	35–40	1.7–4.4
Meats	As fresh	As fresh	As fresh	As fresh	····	····	35–40	1.7–4.4
Sardines	0.760	3.18	0.410	1.72	101.0	234.9	35–40	1.7–4.4
Butter, eggs, etc.:								
Butter	0.302	1.26	0.238	1.00	18.4	42.8	18–20	−7.8–6.7
Cheese	0.480	2.01	0.305	1.28	50.5	117.5	34	1.1
Eggs	0.760	3.18	0.410	1.72	100.0	232.6	31	−0.6
Milk, ice cream	0.900	3.77	0.462	1.93	124.0	288.4	35	1.7
Flour, meal (wheat)	0.26–0.38	1.1–1.6	0.21–0.28	0.9–1.2	14.4–28.8	34–67	36–40	2.2–4.4
Vegetables:								
Asparagus	0.952	3.99	0.482	2.02	134.0	311.7	34–35	1.1–1.7
Cabbage	0.928	3.88	0.473	1.98	131.0	304.7	34–35	1.1–1.7
Carrots	0.864	3.62	0.449	1.88	119.5	278.0	34–35	1.1–1.7
Celery (edible portion)	0.952	3.99	0.482	2.02	135.0	314.0	34–35	1.1–1.7
Dried beans	0.300	1.26	0.237	0.99	18.0	41.9	32–45	0.0–7.2
Dried corn	0.284	1.19	0.231	0.97	15.1	35.1	35–45	1.7–7.2

Dried peas	0.276	1.16	0.224	0.94	13.7	31.9	35–45	1.7–7.2
Onions	0.900	3.77	0.462	1.93	126.0	293.1	36	2.2
Parsnips	0.864	3.62	0.449	1.88	119.5	278.0	34–35	1.1–1.7
Potatoes	0.792	3.32	0.422	1.97	106.5	247.7	36–40	2.2–4.4
Sauerkraut	0.912	3.82	0.467	1.45	128.0	297.7	35	1.7
Miscellaneous:								
Cigars, tobacco	…	…	…	…	…	…	35–42	1.7–5.6
Furs, woolens, etc.	…	…	…	…	…	…	35	1.7
Honey	0.344	1.44	0.254	1.06	25.9	60.2	36–40	2.2–4.4
Hops	…	…	…	…	…	…	32–40	0.0–4.4
Maple syrup	0.488	2.08	0.308	1.29	51.8	120.5	40–45	4.4–7.2
Maple sugar	0.240	1.00	0.215	0.98	7.2	16.7	40–45	4.4–7.2
Poultry, dressed and iced	0.790	3.31	0.421	1.76	105.0	244.7	28–30	−2.2–1.1
Poultry, dry-packed	0.720	3.01	0.395	1.65	93.5	217.5	26–28	−2.2–1.1
Poultry, scalded	0.800	3.35	0.425	1.65	108.0	251.2	20	−6.7
Game, frozen	0.680	2.85	0.380	1.59	86.5	201.2	15–28	−9.4–2.2
Poultry, frozen	0.680	2.85	0.380	1.59	86.5	201.2	15–28	−9.4–2.2
Nuts (dried)	0.21–0.29	0.9–1.2	0.20–0.24	0.8–1.0	4.3–14.4	10–34	35–40	1.7–4.4
Water	1.000	4.20	0.500	2.09	144.0	334.9	…	…
Meats:								
Fresh (typical only)	0.800	3.35	0.404	1.69	…	…	20–40	−6.7–4.4
Fruits:								
Fresh (typical only)	0.700	2.93	0.387	1.62	…	…	32–55	0.0–12.2

*Brunner Manufacturing Company; SI values added by handbook editor.

3.497

operating time for the refrigeration unit, the condensing capacity required is 7196(24/16) = 10,800 Btu/h (3.2 kW).

Choose an evaporator for a 10°F (5.6°C) temperature difference with a capacity of 10,800 Btu/h (3.2 kW) at a suction temperature of −20°F (−28.9°C). Checking a manufacturer's engineering data shows that a 3-hp (2.2-kW) air-cooled two-cylinder unit will be suitable.

Related Calculations: Use this general procedure to choose refrigeration units for stationary, mobile (truck), and marine applications of walk-in coolers, display cases, milk and bottle coolers, ice cream freezers and hardeners, air conditioning, etc. Note that one procedure is used for applications above 32°F (0°C) and another for applications below 32°F (0°C).

In general, choose a unit for a 10 to 20°F (5.6 to 11.1°C) difference between the product and evaporator temperatures. Thus, where a room is maintained at 40°F (4.4°C), choose a condensing unit capacity corresponding to above a 25°F (−3.9°C) evaporator temperature. If brine is to be cooled to 5°F (−15.0°C), select a condensing unit for about −10°F (−23.3°C). Where a high relative humidity is desired in a cold room, select cooling coils with a large surface area, so that the minimum operating differential temperature can be maintained between the room and the coil. The procedure and data given here were published by the Brunner Manufacturing Company based on ASHRAE data.

ENERGY REQUIRED FOR STEAM-JET REFRIGERATION

A steam-jet refrigeration system operates with an evaporator temperature of 45°F (7.2°C) and a chilled-water inlet temperature of 60°F (15.6°C). The condenser operating pressure is 1.135 lb/in^2 (abs) (7.8 kPa), and the steam-jet ejectors use 3.1 lb of boiler steam per pound (1.4 kg/kg) of vapor removed from the evaporator. How many pounds of boiler steam are required per ton of refrigeration produced? How much steam is required per hour for a 100-ton (90.6-t) capacity steam-jet refrigeration unit?

Calculation Procedure:

1. Determine the system pressures and enthalpies

Using the steam tables, find the following values. At 45°F (7.2°C), P = 0.1475 lb/in^2 (abs) (1.0 kPa); h_f = 13.06 Btu/lb (30.6 kJ/kg); h_{fg} = 1068.4 Btu/lb (2485.1 kJ/kg). At 60°F (15.6°C), h_f = 28.06 Btu/lb (65.3 kJ/kg). At 1.135 lb/in^2 (abs) (7.8 kPa), h_f = 73.95 Btu/lb (172.0 kJ/kg), where P = absolute pressure, lb/in^2 (abs); h_f = enthalpy of liquid, Btu/lb; h_{fg} = enthalpy of vaporization, Btu/lb.

2. Compute the chilled-water heat pickup

The chilled-water inlet temperature is 60°F (15.6°C), and the chilled-water outlet temperature is the same as the evaporator temperature, or 45°F (7.2°C), as shown in Fig. 1. Hence, the chilled-water heat pickup = enthalpy at 60°F (15.6°C) − enthalpy at 45°F (7.2°C), both expressed in Btu/lb. Or, heat pickup = 28.06 − 13.06 = 15.0 Btu/lb (34.9 kJ/kg).

3. Compute the required chilled-water flow rate

Since a ton of refrigeration corresponds to a heat removal rate of 12,000 Btu/h (3.5 kW), the chilled-water flow rate = (12,000 Btu/h)/(chilled-water heat pickup, Btu/lb) = 12,000/15 = 8000 lb/(h·ton) [1.0 kg/(s·t)].

4. Compute the quantity of chilled water that vaporizes

Figure 1 shows the three fluid cycles involved: (a) chilled-water flow from the evaporator to the cooling coils and back, (b) chilled-water vapor flow from the evaporator through the ejector to the condenser and back as makeup, and (c) boiler steam flow from the boiler to the ejector to the condenser and back to the boiler as condensate.

Base the calculations on 1 lb (0.5 kg) of chilled water flowing through the cooling coils. For the throttling process from 3 to 4 in the evapoator, Fig. 1, the enthalpy remains constant, but part of the chilled water vaporizes at the lower, or evaporator, pressure. Hence, H_3 = H_4 = h_f + xh_{fg}, where x = lb of vapor formed per lb of chilled water entering, or 28.06 = 13.06 + x(1068.4); x = 0.01405 lb of vapor per lb (0.0063 kg/kg) of chilled water entering. The quantity of chilled water remaining at 1 in the evaporator is 1.0 − 0.01405 = 0.98595 lb/lb (0.4436 kg/kg) of chilled water recirculating.

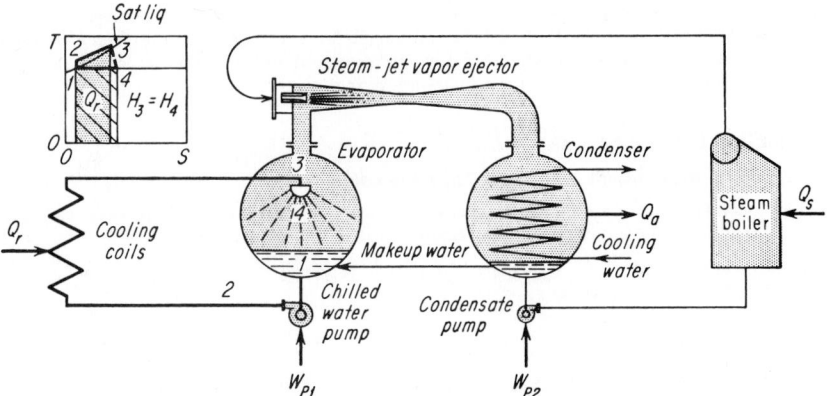

FIG. 1 Steam-jet refrigeration unit and T-S diagram of its operating cycle.

5. *Compute the quantity of makeup vaporized*

Some of the condensate in the condenser returns to the evaporator as makeup, Fig. 1. This makeup throttles into the evaporator and part of it evaporates. Hence, $H_m = h_f + x_m h_{fg}$, where H_m = enthalpy of condensate, Btu/lb; x_m = quantity of makeup vaporized, lb/lb of makeup water. Since the enthalpy of the condensate at the condenser pressure of 1.135 lb/in^2 (abs) (7.8 kPa) is 73.95 Btu/lb (172.0 kJ/kg), 73.95 = 13.06 + x_m(1068.4); x_m = 0.057 lb of makeup vaporized per lb (0.025 kg/kg) of makeup water entering the evaporator.

Makeup vapor simply recirculates between the evaporator and the condenser. So the total makeup water entering the evaporator must replace both the chilled-water vapor and the makeup vapor formed by the two throttling processes.

6. *Compute the makeup vapor and water quantities*

The lb of makeup vapor per lb of makeup water remaining in the evaporator = $x_m/(1.0 - x_m)$ = 0.0570/(1.0 − 0.0570) = 0.0604.

The total makeup water to the evaporator needed to replace the vapor = x(1 + lb of makeup vapor per lb of makeup water) = 0.01405(1 + 0.0604) = 0.01491 lb/lb (0.0067 kg/kg) of chilled water circulating. This is also the vapor removed from the evaporator by the ejector.

7. *Compute the total vapor removed from the evaporator*

The total vapor removed from the evaporator = [lb/(h·ton) chilled water] × (makeup water per lb of chilled water circulated) = (8000)(0.01491) = 119.3 lb/ton (54.1 kg/t) of refrigeration.

8. *Compute the boiler steam required*

The boiler steam required = (vapor removed from the evaporator, lb/ton of refrigeration)(steam-jet steam, lb/lb of vapor removed from the evaporator) = (119.3)(3.1) = 370 lb of boiler steam per ton of refrigeration (167.8 kg/t). For a 100-ton (90.6-t) machine, the boiler steam required = (100)(370) = 37,000 lb/h (4.7 kg/s).

Related Calculations: Use this general method for any steam-jet refrigeration system using water and steam to produce a low temperature for air conditioning, product cooling, manufacturing processes, or other applications. Note that any of the eight items computed can be found when the other variables are known.

REFRIGERATION COMPRESSOR CYCLE ANALYSIS

An ammonia refrigeration compressor takes its suction from the evaporator, Fig. 2a, at a temperature of −20°F (−28.9°C) and a quality of 95 percent. The compressor discharges at a pressure of 100 lb/in^2 (abs) (689.5 kPa). Liquid ammonia leaves the condenser at 50°F (10.0°C). Find the heat absorbed by the evaporator, the work input to the compressor, the heat rejected to the condenser, the coefficient of performance (COP) of the cycle, horsepower per ton of refrigeration, the quality of the refrigerant at state 2, quantity of refrigerant circulated per ton of refrigeration, required rate of condensing-water flow for a 100-ton (90.6-t) load, compressor displacement for a 100-ton (90.6-t) capacity. What cylinder dimensions are required for a 100-ton (90.6-t) capacity if the stroke = 1.3(cylinder bore) and the compressor makes 200 r/min?

Calculation Procedure:

1. Compute the enthalpy and entropy at cycle points

Assume a constant-entropy compression process for this cycle. This is the usual procedure in analyzing a refrigeration compressor whose actual performance is not known.

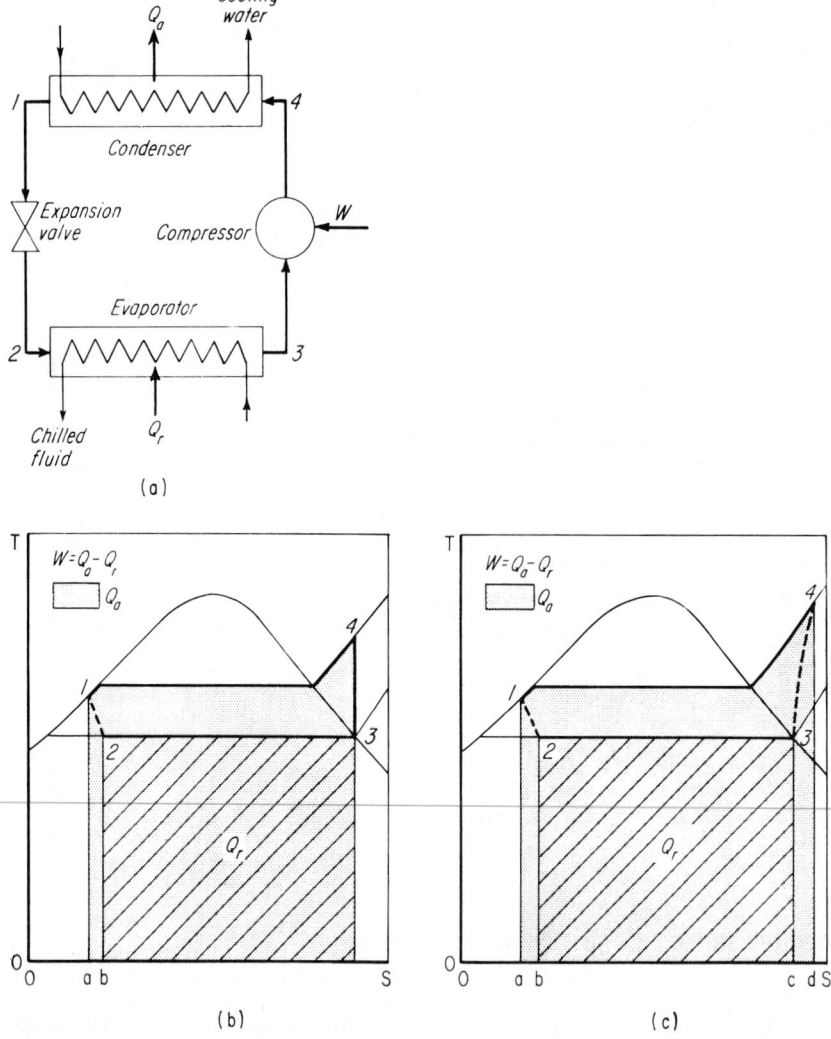

FIG. 2 (*a*) Components of a vapor refrigeration system; (*b*) ideal refrigeration cycle *T-S* diagram; (*c*) actual refrigeration cycle *T-S* diagram.

TABLE 4 Thermodynamic Properties of Ammonia°

Saturated ammonia

Temperature t, °F (°C)	Pressure p, lb/in² (abs) (kPa)	Volume, ft³/lb (m³/kg)		Enthalpy, Btu/lb (kJ/kg)			Entropy, Btu/(lb·°F) [kJ/(kg·°C)]	
		Liquid v_f	Vapor v_g	Liquid h_f	Evaporation h_{fg}	Vapor h_g	Liquid s_f	Vapor s_g
0 (−17.8)	30.42 (209.7)	0.0242 (0.00151)	9.116 (0.569)	42.9 (99.8)	568.9 (1323.3)	611.8 (1423.0)	0.0975 (0.408)	1.3352 (5.590)
20 (−6.7)	48.21 (332.4)	0.0247 (0.00154)	5.910 (0.369)	64.7 (150.5)	553.1 (1286.5)	617.8 (1437.0)	0.1437 (0.564)	1.2969 (5.430)
100 (37.8)	211.9 (1461.1)	0.0272 (0.00170)	1.419 (0.089)	155.2 (361.0)	477.8 (1111.4)	633.0 (1472.4)	0.3166 (1.326)	1.1705 (4.901)
120 (48.9)	286.4 (1974.7)	0.0284 (0.00177)	1.047 (0.065)	179.0 (416.4)	455.0 (1058.3)	634.0 (1474.7)	0.3576 (1.497)	1.1427 (4.787)

Superheated ammonia

Temperature, °F (°C)	50 lb/in² (abs) (344.8 kPa) [21.67°F] (−5.7°C) saturation			100 lb/in² (abs) (689.5 kPa) [56.05°F] (13.4°C) saturation			150 lb/in² (abs) (1034.3 kPa) [78.81°F] (26.0°C) saturation		
	v	h	s	v	h	s	v	h	s
100 (37.8)	6.843 (0.427)	663.7 (1543.8)	1.3816 (5.783)	3.304 (0.206)	655.2 (1524.0)	1.2891 (5.397)	2.118 (0.132)	645.9 (1502.4)	1.2289 (5.145)
120 (48.9)	7.117 (0.444)	674.7 (1569.4)	1.4009 (5.865)	3.454 (0.216)	667.3 (1552.1)	1.3104 (5.486)	2.228 (0.139)	659.4 (1533.98)	1.2526 (5.244)
140 (60.0)	7.387 (0.461)	685.7 (1594.9)	1.4195 (5.943)	3.600 (0.225)	679.2 (1579.8)	1.3305 (5.571)	2.334 (0.146)	672.3 (1563.8)	1.2745 (5.336)

Using Fig. 2b as a guide, we see that $H_3 = h_f + xh_{fg}$, where H_3 = enthalpy at point 3, Btu/lb; h_f = enthalpy of liquid ammonia, Btu/lb from Table 4; h_{fg} = enthalpy of evaporation, Btu/lb, from the same table; x = vapor quality, expressed as a decimal. Since point 3 represents the suction conditions of the compressor, $H_3 = 21.4 + 0.95(583.6) = 575.8$ Btu/lb (1339.3 kJ/kg).

The entropy at point 3 is $S_3 = s_f + xs_{fg}$, where the subscripts refer to the same fluid states as above and the S and s values are the entropy. Or, $S_3 = 0.0497 + 0.95(1.3277) = 1.3110$ Btu/(lb·°F) [5.465 kJ/(kg·°C)].

2.   *Compute the final cycle temperature and enthalpy*

The compressor discharges at 100 lb/in² (abs) (689.5 kPa) at an entropy of $S_4 = 1.3110$ Btu/(lb·°F) [5.465 kJ/(kg·°C)]. Inspection of the saturated ammonia properties, Table 4, shows that at 100 lb/in² (abs) (689.5 kPa) the entropy of saturated vapor is less than that computed. Hence, the vapor discharged by the compressor must be superheated.

Enter Table 4 at $S_4 = 1.3110$ Btu/(lb·°F) [5.465 kJ/(kg·°C)]. Inspection shows that the final cycle temperature T_4 lies between 120 and 130°F (48.9 and 54.4°C) because the actual entropy value lies between the entropy values for these two temperatures. Interpolating gives $T_4 = 130 - [(S_{130} - S_4)/(S_{130} - S_{120})] \times (130 - 120)$, where the subscripts refer to the respective temperatures. Or, $T_4 = 130 - [(1.3206 - 1.3110)/(1.3206 - 1.3104)](130 - 120) = 120.6°F$ (49.2°C).

Interpolating in a similar fashion for the final enthalpy, using the enthalpy at 130°F (54.4°C) as the base, we find $H_4 = 673.3 - [(1.3206 - 1.3110)/(1.3206 - 1.3104)](673.3 - 667.3) = 667.7$ Btu/lb (1553.1 kJ/kg).

3.   *Compute the heat absorbed by the evaporator*

The heat absorbed by the evaporator is $Q_r = H_3 - H_2$, where Q_r = Btu/lb of refrigerant. Or, for this system, $Q_r = 575.8 - 97.9 = 477.9$ Btu/lb (1111.6 kJ/kg).

4.   *Compute the work input to the compressor*

Find the work input to the compressor from $W = H_4 - H_3$, where W = work input, Btu/lb of refrigerant. Or, $W = 667.7 - 575.8 = 91.9$ Btu/lb (213.8 kJ/kg) of refrigerant circulated.

5.   *Compute heat rejected to the condenser*

The heat rejected to the condenser is $Q_a = H_4 - H_1$, where Q_a = heat rejection, Btu/lb of refrigerant. Or, $Q_a = 667.7 - 97.9 = 569.8$ Btu/lb (1325.4 kJ/kg) of refrigerant circulated.

6.   *Compute the coefficient of performance of the machine*

For any refrigerating machine, the coefficient of performance (COP) = Q_r/W, where the symbols are as defined earlier. Or COP = 477.9/91.9 = 5.20.

7.   *Compute the horsepower per ton for this system*

For any refrigerating system, the horsepower per ton $hp_t = 4.72/\text{COP}$. Or, for this system, $hp_t = 4.72/5.20 = 0.908$ hp/ton (0.68 kW).

8.   *Compute the refrigerant quality at the evaporator inlet*

At the evaporator inlet, or point 2, the quality of the refrigerant $x = (H_2 - h_f)/h_{fg}$, where the enthalpies are those at −20°F (−28.9°C), the evaporator operating temperature. Or, $x = (97.9 - 21.4)/583.6 = 0.1311$, or 13.11 percent quality.

9.   *Compute the quantity of refrigerant circulated per ton capacity*

Find the quantity of refrigerant circulated, lb/(min·ton) of refrigeration produced from $q_t = 200/Q_r$, or $q_t = 200/477.9 = 0.419$ lb/(min·ton) [(0.0035 kg/(s·t)] of refrigeration.

10.   *Compute the required rate of condensing-water flow*

The heat rejected to the condenser Q_a must be absorbed by the condenser cooling water. The quantity of water that must be circulated is $q_w = Q_a/\Delta t$, where q_w = weight of water circulated per lb of refrigerant; Δt = temperature rise of the cooling water during passage through the condenser, °F. Assuming a 20°F (11.1°C) temperature rise of the cooling water, we find $q_w = 569.8/20 = 28.49$, say 28.5 lb of water per lb (12.8 kg/kg) of refrigerant circulated.

Since 0.419 lb/(min·ton) [0.0035 kg/(t·s)] of ammonia must be circulated, step 9, at a load of 100 tons (90.7 t), the quantity of refrigerant circulated will be 100(0.419) = 41.9 lb/min (0.32 kg/s). The condenser cooling water required is then (28.5)(41.9) = 1191 lb/min (9.0 kg/s), or 1191/8.33 = 143.4 gal/min (9.1 L/s).

11. *Compute the compressor displacement*

Use the relation $V_d = q_t v_g T$, where V_d = required compressor displacement, ft^3/min; q_t = quantity of refrigerant circulated, lb/(ton·min); v_g = specific volume of suction gas, ft^3/lb; T = refrigeration capacity, tons. For a 100-ton (90.7-t) capacity with the suction gas at $-20°$F $(-28.9°$C), $V_d = (0.419)(14.68)(100) = 614$ ft^3/min (0.29 m^3/s), given the specific volume for $-20°$F $(-28.9°$C) suction gas from Table 4.

12. *Compute the compressor cylinder dimensions*

For any reciprocating refrigeration compressor, V_d = (shaft rpm)(piston displacement, ft^3/stroke) = v_d, or $614 = 200 v_d$; $v_d = 3.07$ ft^3 (0.087 m^3).

Also, $D = (V_d/0.785)^{1/3}$, where D = piston diameter, ft; r = ratio of stroke length to cylinder bore. Or, $D = [3.07/(0.785 \times 1.3)]^{1/3} = 1.447$ ft (0.44 m). Then $L = 1.3D = 1.3(1.447) = 1.88$ ft (0.57 m).

Related Calculations: Employ the method given here for any reciprocating compressor using any refrigerant. Note that where the volumetric efficiency E_V of a compressor is given, the actual volume of gas drawn into the cylinder, ft^3 = E_V × piston displacement, ft^3. When analyzing an actual compressor, be sure you use the enthalpies which actually prevail. Thus, the gas entering the compressor suction may be superheated instead of saturated, as assumed here.

RECIPROCATING REFRIGERATION COMPRESSOR SELECTION

Choose the compressor capacity and hp, and determine the heat rejection rate for a 36-ton (32.7-t) load, a 30°F ($-1.1°$C) evaporator temperature, a 20°F ($-6.7°$C) evaporator coil superheat, a suction-line pressure drop of 2 lb/in^2 (13.8 kPa), a condensing temperature of 105°F (40.6°C), a compressor speed of 1750 r/min, a subcooling of the refrigerant of 5°F (2.8°C) in the water-cooled condenser, and use of refrigerant 12. Determine the required condensing-water flow rate when the entering water temperature is 70°F (21.1°C). How many gal/min of chilled water can be handled if the water temperature is reduced 10°F (5.6°C) by the evaporator chiller?

Calculation Procedure:

1. *Compute the compressor suction temperature*

With refrigerant 12, a pressure change of 1 lb/in^2 (6.9 kPa) at 0°F ($-17.8°$C) is equivalent to a temperature change of 2°F (1.1°C); at 50°F (10.0°C), a 1-lb/in^2 (6.9-kPa) pressure change is equivalent to 1°F (0.6°C) temperature change. At the evaporator temperature of 30°F ($-1.1°$C), the temperature change is about 1.4°F·in^2/lb (0.11°C/kPa), obtained by interpolation between the ranges given above. Then, suction temperature, °F = evaporator temperature, °F $-$ (suction-line loss, °F·in^2/lb)(suction-line pressure drop, lb/in^2), or $30 - 1.4 \times 2 = 27.2$, say 27°F ($-2.8°$C).

2. *Compute the compressor equivalent capacity*

To compute the compressor equivalent capacity, two correction factors must be applied: the superheat correction factor and the subcooling correction factor. Both are given in the engineering data available from compressor manufacturers.

To apply correction-factor listings, such as those in Table 5, use the following as guides: (a)

TABLE 5 Open Compressor Ratings

| Suction temperature | | Condensing temperature, 105°F (40.6°C) | | | | | |
| | | Capacity | | Power input | | Heat rejection | |
°F	°C	tons	t	bhp	kW	tons	t
10	-12.2	26.2	23.8	41.3	30.8	34.9	31.7
20	-6.7	34.0	30.8	45.3	33.8	43.6	39.6
30	-1.1	43.0	39.0	48.6	36.2	53.2	48.3

TABLE 6 Rating Basis and Capacity Multipliers—Refrigerant 12 and Refrigerant 500°

Saturated suction temperature		Actual suction gas temperature to compressor			
°F	°C	30°F (−1.1°C)	40°F (4.4°C)	50°F (10.0°C)	60°F (15.6°C)
20	−6.7	0.969	0.978	0.987	0.996
30	−1.1	0.970	0.979	0.987	0.996
40	4.4	. . .	0.987	0.992	0.997

°Carrier Air Conditioning Company; SI values added by handbook editor.

Superheating of the suction gas can result from heat pickup by the gas outside the cooled space. Superheating increases the refrigeration compressor capacity 0.3 to 1.0 percent per 10°F (5.6°C) with refrigerant 12 or 500 if the heat absorbed represents useful refrigeration, such as coil super-heat, and not superheating from a liquid suction heat exchanger. (b) Subcooling increases the potential refrigeration effect by reducing the percentage of liquid flashed during expansion. For each °F of subcooling, the compressor capacity is increased about 0.5 percent owing to the increased refrigeration effect per pound of refrigerant flow.

Applying guide (a) to a 27°F (−2.8°C) suction, 20°F (−6.7°C) superheat, interpolate in Table 6 between the 40 and 50°F (4.4 and 10.0°C) actual suction-gas temperatures for a 30°F (−1.1°C) saturated suction temperature, because the actual suction temperature is 27 + 20 = 47°F (8.3°C) and the saturated suction temperature is given as 30°F (−1.1°C). Or, (0.987 − 0.979)[(47 − 40)/(50 − 40)] + 0.979 = 0.9846, say 0.985.

Applying guide (b), we see that subcooling = 5°F (2.8°C), as given. Then subcooling correction = 1 − 0.0005(15 − 5) = 0.95, where 0.005 = 0.5 percent, expressed as a decimal; 15°F (−9.4°C) = the liquid subcooling on which the compressor capacity is based. This value is given in the compressor rating, Table 6.

With the superheat and subcooling correction factors known, compute the compressor equivalent capacity from (load, tons)/[(superheat correction factor)(subcooling correction factor)], or 36/[(0.985)(0.95)] = 38.5 tons (34.9 t).

3. Select the compressor unit

Use Table 5. Choose an eight-cylinder compressor. Interpolate for a 27°F (−2.8°C) suction and 105°F (40.6°C) condensing temperature to find compressor capacity = 40.3 tons (36.6 t); power input = 47.6 bhp (35.5 kW); heat rejection = 50.3 tons (45.6 t).

4. Compute the required condensing-water flow rate

From step 3, the condensing temperature of the compressor chosen is 105°F (40.6°C). Assume a condenser-water outlet temperature of 95°F (35.0°C), a typical value. Then the required condenser-water flow rate, gal/min = 24 × condenser load/(condensing-water outlet temperature, °F − entering condenser-water temperature, °F). Or 24(50.3)/(95 − 70) = 48.4 gal/min (3.1 L/s). This is within the normal flow for water-cooled condensing units. Thus, city-water quantities range from 1 to 2 gal/(min·ton) [0.07 to 0.14 L/(s·t)]; cooling-tower quantities are usually chosen for 3 gal/(min·ton) [0.21 L/s·t)].

5. Compute the quantity of chilled water that can be handled

Use this relation: chilled water, gal/min = 24 × capacity, tons/chilled-water temperature range, or inlet − outlet temperature, °F. Since, from step 3, the compressor capacity is 40.3 tons (36.6 t) and the chilled-water temperature range is 10°F (5.6°C), *gpm* = 24(40.3)/10 = 96.7 gal/min (6.1 L/s).

The temperature of the chilled water leaving the evaporator chiller is selected so that it equals the inlet temperature required at the heat-load source. The required inlet temperature is a function of the type of heat exchanger, type of load, and similar factors.

Related Calculations: The standard operating conditions for an air-conditioning refrigeration system, as usually published by the manufacturer, are based on an entering saturated refrigerant vapor temperature of 40°F (4.4°C), an actual entering refrigerant vapor temperature of

55°F (12.8°C), a leaving saturated refrigerant vapor temperature of 105°F (40.6°C), and an ambient of 90°F (32.2°C) and no liquid subcooling.

The Air Conditioning and Refrigeration Institute (ARI) standards for a reciprocating compressor liquid-chilling package establish a standard rating condition for a water-cooled model of a leaving chilled-water temperature of 44°F (6.7°C), a chilled-water range of 10°F (5.6°C), a 0.0005 fouling factor in the cooler and the condenser, a leaving condenser-water temperature of 95°F (35.0°C), and a condenser-water temperature rise of 10°F (5.6°C). The standard rating conditions for a condenserless model are a leaving chilled-water temperature of 44°F (6.7°C), a chilled-water temperature range of 10°F (5.6°C), a 0.0005 fouling factor in the cooler, and a condensing temperature of 105 or 120°F (40.6 or 48.9°C).

Use these standard rating conditions to make comparisons between compressors. When catalog ratings of compressors of different manufacturers are compared, the rating conditions must be known, particularly the amount of subcooling and superheating needed to produce the capacities shown.

General guides for reciprocating compressors using refrigerants 12, 22, and 500 are as follows:

1. Lowering the evaporator temperature 10°F (5.6°C) from a base of 40 and 105°F (4.4 and 40.6°C) reduces the system (evaporator) capacity about 24 percent and at the same time increases the compressor hp/ton by about 18 percent.

2. Increasing the condensing temperature 15°F (8.3 °C) from a base of 40 and 105°F (4.4 and 40.6°C) reduces the capacity about 13 percent and at the same time increases the compressor hp/ton by about 27 percent.

3. In air-conditioning service at normal loads, a piping loss equivalent to approximately 2°F (1.1°C) is allowed in the suction piping and to 2°F (1.1°C) in the hot-gas discharge piping. Thus when an evaporator requires a refrigerant temperature of 42°F (5.6 °C) to handle a load, the compressor must be selected for a 40°F (4.4°C) suction temperature. Correspondingly, if the condenser requires 103°F (39.4°C) to reject the proper amount of heat, the compressor must be selected for a 103 + 2 = 105°F (40.6°C) condensing temperature.

4. Compressor manufacturers generally state the operating limits for each compressor in the capacity table describing it. These limits should not be exceeded.

5. To select a condenser to match a compressor, the heat rejection of the compressor must be known. For an open-type compressor, heat rejection, tons = 0.212(compressor power input, bhp) + tons refrigeration capacity of the compressor. For a gas-cooled hermetic-type compressor, heat rejection, tons = 0.285 (kW input to the compressor) + refrigeration capacity, tons. The selection procedure and other data given here were developed by the Carrier Air Conditioning Company.

Environmental restrictions on chlorofluorocarbons (CFC) require that they be phased out over a period of time. This will restrict the use of R-12 refrigerant, which is being replaced in automative applications by a non-CFC refrigerant named R-134a. Some changes may be required in the automative refrigerant system when R-134a is substituted for R-12.

In air conditioning systems for buildings, ships, and other similar installations, R-123 is being substituted for R-11 and R-12 refrigerants. Conversion of existing refrigeration systems to the new non-CFC refrigerants is considered essential. Replacement parts for existing CFC plants will gradually become scarcer, as will qualified repair personnel.

CENTRIFUGAL REFRIGERATION MACHINE LOAD ANALYSIS

Select a centrifugal refrigeration machine to cool 720 gal/min (45.4 L/s) of chilled water from an entering temperature of 60°F (15.6°C) to a leaving temperature of 45°F (7.2°C).

Calculation Procedure:

1. Compute the load on the machine

Use this relation: load, tons = $gpm \times \Delta t/24$, where gpm = quantity of chilled water cooled, gal/min; Δt = temperature reduction of the chilled water during passage through the evaporator chiller, °F. For this machine, load = 720(50 − 45)/24 = 450 tons (408.2 t).

2. Choose the compressor to use

Table 7 shows typical hermetic centrifugal refrigeration machine ratings. In a hermetic machine, the driver is built into the housing, completely isolating the refrigerant space from the atmosphere. An open machine has a shaft that projects outside the compressor housing. The shaft must

TABLE 7   Typical Hermetic Centrifugal Machine Ratings° [Refrigeration Capacity, tons (t)]

Leaving chilled-water temperature		Leaving condenser-water temperature					
°F	°C	85°F	29.4°C	90°F	32.2°C	95°F	35.0°C
44	6.7	442†	401.0†	435	394.6	424	384.6
45	7.2	450	408.2	441	400.1	430	390.1
46	7.8	457†	414.6†	447	405.5	435	394.6

°Carrier Air Conditioning Company.
†These ratings require less than 330-kW input. All ratings shown are based on a two-pass cooler using 380 to 1260 gal/min (24.0 to 79.5 L/s) and on a two-pass condenser using 430 to 1430 gal/min (27.1 to 90.2 L/s).

be fitted with a suitable seal to prevent refrigerant leakage. Open machines are available in capacities up to approximately 4500 tons (4085 t) at air-conditioning load temperatures. Hermetic machines are available in capacities up to approximately 2000-ton (1814-t) capacity.

Study of Table 7 shows that a 450-ton (408.5-t) unit is available with a leaving chilled-water temperature of 45°F (7.2°C) and a leaving condenser-water temperature of 85°F (29.4°C). If the condenser water were available at temperatures of 75°F (23.9°C) or lower, this machine would probably be chosen.

Related Calculations: The factors involved in the selection of a centrifugal machine are load; chilled-water, or brine quantity; temperature of the chilled water or brine; condensing medium (usually water) to be used; quantity of the condensing medium and its temperature; type and quantity of power available; fouling-factor allowance; amount of usable space available; and the nature of the load, whether variable or constant. The final selection is usually based on the least expensive combination of machine and heat rejection device, as well as a reasonable machine operating cost.

Brine cooling normally requires special selection of the machine by the manufacturer. As a general rule, multiple-machine applications are seldom made on normal air-conditioning loads less than about 400 tons (362.9 t).

The optimum machine selection involves matching the correct machine and cooling tower as well as the correct entering chilled-water temperature and temperature reduction. A selection of several machines and cooling towers often results in finding one combination having a minimum first cost. In many instances, it is possible to reduce the condenser-water quantity and increase the leaving condenser-water temperature, resulting in a smaller tower.

Centrifugal refrigeration machines are used for air-conditioning, process, marine, manufacturing, and many other cooling applications throughout industry.

HEAT PUMP CYCLE ANALYSIS AND COMPARISON

Determine the quantity of water required to supply heat to a heat pump that must deliver 70,000 Btu/h (20.5 kW) to a building. Refrigerant 12 is used; the temperature of the water in the heat sink is 50°F (10.0°C). Air must be delivered to the heating system at a temperature of 118°F (47.8°C).

Calculation Procedure:

1. *Determine the compressor suction temperature to use*

To produce sufficient heat transfer between the water and the evaporator, a temperature difference of at least 10°F (5.6°C) must exist. With a water temperature of 50°F (10.0°C), this means that a suction temperature of 40°F (4.4°C) might be satisfactory. A suction temperature of 40°F (4.4°C) corresponds to a suction pressure of 51.68 lb/in² (abs) (356.3 kPa), as a table of thermodynamic properties of refrigerant 12 shows.

Since water entering the evaporator heat exchanger cannot be reduced to 40°F (4.4°C), the refrigerant temperature, the actual outlet temperature must be either assumed or computed. Assume that the water leaves the evaporator heat exchanger at 44°F (6.7°C). Then, each pound of water passing through the evaporator yields 50°F − 44°F = 6 Btu (6.3 kJ). Since 1 gal of water weighs 8.33 lb (1 kg/L), the quantity of heat released by the water is (6 Btu/lb)(8.33 lb/gal) = 49.98 Btu/gal, say 50 Btu/gal (13.9 kJ/L).

As an alternative solution, assume a suction temperature of 35°F (1.7°C) and an evaporator exit temperature of 39°F (3.9°C). Then each pound of water will yield 50 − 39 = 11 Btu (11.6 kJ). This is equal to (11)(8.33) = 91.6 Btu/gal (25.5 kJ/L). This comparison indicates that for every °F the cooling range of the water heat sink is extended, an additional 8.33 Btu/gal (2.3 kJ/L) of water is obtained.

2. *Evaluate the effect of suction-temperature decrease*

As the compressor suction temperature is reduced, the specific volume of the suction gas increases. Thus the compressor must handle more gas to evaporate the same quantity of refrigerant. However, the displacement of the usual reciprocating compressor used in a heat-pump system cannot be varied easily, if at all, in some designs. Also, at the lower suction temperature, the enthalpy of vaporization of the refrigerant increases only slightly.

Study of a table of thermodynamic properties of refrigerant 12 shows that reducing the suction temperature from 40 to 35°F (4.4 to 1.7°C) increases the specific volume from 0.792 to 0.862 ft^3/lb (0.0224 to 0.0244 m^3/kg). The enthalpy of vaporization increases from 65.71 to 66.28 Btu/lb (152.8 to 154.2 kJ/kg), but the total enthalpy decreases from 82.71 to 82.16 Btu/lb (192.4 to 191.1 kJ/kg). Hence, the advisability of reducing the suction temperature must be carefully investigated before a final decision is made.

3. *Determine the required compressor discharge temperature*

Air must be delivered to the heating system at 118°F (47.8°C), according to the design requirements. To produce a satisfactory transfer of heat between the condenser and the air, a 10°F (5.6°C) temperature difference is necessary. Hence, the compressor discharge temperature must be at least 118 + 10 = 128°F (53.3°C).

Checking a table of thermodynamic properties of refrigerant 12 shows that a temperature of 128°F (53.3°C) corresponds to a discharge pressure of 190.1 lb/in^2 (abs) (1310.7 kPa). The table also shows that the enthalpy of the vapor at the 118°F (47.8°C) condensing temperature is 90.01 Btu/lb (209.4 kJ/kg), whereas the enthalpy of the liquid is 35.65 Btu/lb (82.9 kJ/kg).

With a suction temperature of 40°F (4.4°C), the enthalpy of the vapor is 82.71 Btu/lb (192.4 kJ/kg). Hence, the heat supplied by the evaporator is: enthalpy of vapor at 40°F (4.4°C) − enthalpy of liquid at 118°F = 82.71 − 35.65 = 47.06 Btu/lb (109.5 kJ/kg). This heat is abstracted from the water that is drawn from the heat sink.

The gas leaving the evaporator contains 82.71 Btu/lb (192.4 kJ/kg). When this gas enters the condenser, it contains 90.64 Btu/lb (210.8 kJ/kg). The difference, or 90.64 − 82.71 = 7.93 Btu/lb (18.4 kJ/kg), is added to the gas by the compressor and represents a portion of the work input to the compressor.

4. *Compute the evaporator and compressor heat contribution*

The total heat delivered to the air = evaporator heat + compressor heat = 47.06 + 7.93 = 54.99 Btu/lb (127.9 kJ/kg). Then the evaporator supplies 47.06/54.99 = 0.856, or 85.6 percent of the total heat, and the compressor supplies 7.93/54.99 = 0.144, or 14.4 percent of the total heat.

5. *Determine the actual evaporator and compressor heat contribution*

Since this heat pump is rated at 70,000 Btu/h (20.5 kW), the evaporator contributes 0.856 × 70,000 = 59,920 Btu/h (17.5 kW), and the compressor supplies 0.144(70,000) = 10,080 Btu/h (3.0 kW). As a check, 59,920 + 10,080 = 70,000 Btu/h (20.5 kW).

6. *Compute the sink-water flow rate required*

The evaporator obtains its heat, or 59,920 Btu/h (17.5 kW), from the sink water. Since, from step 1, each gallon of water delivers 50 Btu (52.8 kJ) at a 40°F (4.4°C) suction temperature, the flow rate required to contribute the evaporator heat is 59,920/50 = 1198.4 gal/h, or 1198.4/60 = 19.9 gal/min (1.3 L/s).

7. *Evaluate the lower suction temperature*

At 35°F (1.7°C), the evaporator will supply 82.16 − 35.65 = 46.51 Btu/lb (108.2 kJ/kg), by the same reasoning as in step 3. The balance, or 90.64 − 82.16 = 8.48 Btu/lb (19.7 kJ/kg), must be supplied by the compressor.

8. *Compute the required refrigerant gas flow*

At a 40°F (4.4°C) suction temperature, a table of thermodynamic properties of refrigerant 12 shows that the specific volume of the gas is 0.792 ft³/lb (0.050 m³/kg). Step 4 shows that the heat pump must deliver 54.99 Btu/lb (127.9 kJ/kg) of refrigerant to the air, or $54.99/0.792 = 69.4$ Btu/ft³ (2585.8 kJ/m³) of gas. With a total heat requirement of 70,000 Btu/h (20.5 kW), the compressor must handle $70,000/69.4 = 1010$ ft³/h (0.0079 m³/s).

As noted earlier, the cubic capacity of a reciprocating compressor is a fixed value at a given speed. Hence, a compressor chosen to handle this quantity of gas cannot handle a larger heat load.

At a 35°F (1.7°C) suction temperature, using the same procedure as above, we see that the required heat content of the gas is $54.99/0.862 = 63.7$ Btu/ft³ (2373.4 kJ/m³). The compressor capacity must be $70,000/63.7 = 1099$ ft³/h (0.0086 m³/s).

If the compressor were selected to handle 1010 ft³/h (0.0079 m³/s), then reducing the suction temperature to 35°F (1.7°C) would give a heat capacity of only $(1010)(63.7) = 64,400$ Btu/h (18.9 kW). This is inadequate because the system requires 70,000 Btu/h (20.5 kW).

9. *Compute the water flow rate at the lower suction temperature*

Step 6 shows the procedure for finding the sink-water flow rate required at a 40°F (4.4°C) suction temperature. Suppose, however, that the heat output of 64,400 Btu/h (18.9 kW) at the 35°F (1.7°C) suction temperature was acceptable. The evaporator portion of this load, by the method of step 4, is $(82.16 - 35.65)/54.99 = 0.847$, or 84.7 percent. Hence, the quantity of water required is $(64,400)(0.847)/[(91.6)(60)] = 9.92$ gal/min (0.63 L/s). In this equation, the value of 91.6 Btu/gal is obtained from step 1. The factor 60 converts hours to minutes. Thus, reducing the suction temperature from 40 to 35°F (4.4 to 1.7°C) just about halves the water quantity—from 19.9 to 9.92 gal/min (1.26 to 0.63 L/s).

10. *Compare the pumping power requirements*

The power input to the pump is hp $= 8.33(gpm)$(head, ft)/33,000(pump efficiency)(motor efficiency). If the total head on the pump is computed as being 40 ft (12.2 m), the efficiency of the pump is 60 percent, and the efficiency of the motor is 85 percent, then with a 40°F (4.4°C) suction temperature, the pump horsepower is $8.33(19.9)(40)/33,000(0.60)(0.85) = 0.394$, say 0.40 hp (0.30 kW).

At a 35°F (1.7°C) suction temperature with a flow rate of 9.92 gal/min (0.03 L/s) and all the other factors the same, hp $= 8.33(9.92)(40)/33,000(0.60)(0.85) = 0.1965$, say 0.20 hp (0.15 kW). Thus, the 35°F (1.7°C) suction temperature requires only half the pump hp that the 40°F (4.4°C) suction temperature requires.

11. *Compute the compressor power input and power cost*

At the 40°F (4.4°C) suction temperature, the compressor delivers 7.93 Btu/lb (18.4 kJ/kg) of refrigerant gas, step 3. Since the total weight of gas delivered by the compressor per hour is (70,000 Btu/h)/(54.99 Btu/lb) $= 1272$ lb/h (0.16 kg/s), the compressor's total heat contribution is (1272 lb/h)(7.93 Btu/lb) $= 10,100$ Btu/h (3.0 kW).

With a compressor-driving motor having an efficiency of 85 percent, the hourly motor input is equivalent to $10,100/0.85 = 11,880$ Btu/h (3.5 kW), or 11,880/2545 Btu/(hp·h) $= 4.66$ hp·h $= (4.66)[746$ Wh/(hp·h)$] = 3480$ Wh $= 3.48$ kWh. Also, the pump requires 0.4 hp·h, step 10, or $(0.4)(746) = 299$ Wh $= 0.299$ kWh. Hence, the total power consumption at a 40°F (4.4°C) suction temperature is $3.48 + 0.299 = 3.779$ kWh. At a power cost of 5 cents per kilowatthour, the energy cost is 3.779(5.0) $= 18.9$ cents per hour.

With a 35°F (1.7°C) suction temperature and using the lower heating capacity obtained with the smaller, fixed-capacity compressor, step 9, we see the weight of gas handled by the compressor will be (64,400 Btu/h)/(54.99 Btu/lb) $= 1172$ lb/h (0.15 kg/s). From step 7, the compressor must supply 8.48 Btu/lb (19.7 kJ/kg). Therefore, the compressor's total heat contribution is (1172 lb/h)(8.48 Btu/lb) $= 9950$ Btu/h (2.9 kW). With a motor efficiency of 85 percent, the hourly motor input is equivalent to $9950/0.85 = 11,700$ Btu/h (3.4 kW).

Using the same procedure as above, we see that the electric power input to the compressor will be 3.43 kWh, while the pump electric power input is 0.150 kWh. The total electric power input is 3.58 kWh at a cost of $(3.58)(5.0) = 17.9$ cents per hour. Hence, the hourly savings with the 35°F (1.7°C) suction temperature is $18.9 - 17.9 = 1.0$ cent. Note, however, that the heat output at the 35°F (1.7°C) suction temperature, 64,400 Btu/h (18.9 kW), is 5600 Btu/h (1.6 kW) less than at the 40°F (4.4°C) suction temperature. If the lower heat output were unacceptable, the higher suction temperature or a larger compressor would have to be used. Either alternative would increase the power cost.

Related Calculations: With a water sink as the heat source, the usual water consumption of a heat pump ranges from 1.1 to more than 4 gal/(min·ton) [0.06 to 0.23 L/(s·t)]. A consumption range this broad requires that the actual flow rate be computed because a guess could be considerably in error.

Either air or the earth may be used as a heat source instead of water. When the cooling load rather than the heating load establishes the basic equipment size, an ideal situation exists for the use of the heat pump with air as the heat source. This occurs in localities where the minimum outdoor temperature in the winter is 20°F (-6.7°C) or higher.

Ground coils can be bulky, costly, and troublesome. One study shows that the temperature difference between the evaporating refrigerant in a ground coil and the surrounding earth is about equal to the number of Btu/h that may be drawn from each linear foot of coil. Thus, with a temperature difference of 15°F (8.3°C) and a 70,000-Btu (73,853.9-kJ) load of which 85 percent is supplied by the coil, the length of coil needed is (70,000)(0.85)/15 = 3970 ft (1210.1 m).

The coefficient of performance of a heat pump = heat rejected by condenser, Btu/heat equivalent of the net work of compression, Btu. The usual single-stage air source heat pump has a coefficient of performance ranging between 2.25 and 3.0. The procedure and data presented were developed by Robert Henderson Emerick, P.E., Consulting Mechanical Engineer.

Energy Conservation

REFERENCES: Hunt—*Windpower*, Van Nostrand Reinhold; Burberry—*Building for Energy Conservation*, Halsted Press; Chiogioji—*Industrial Energy Conservation*, Dekker; Courtney—*Energy Conservation in the Built Environment*, Longman; Culp—*Principles of Energy Conversion*, McGraw-Hill; Dorf—*The Energy Factbook*, McGraw-Hill; Dubin—*Energy Conservation Standards*, McGraw-Hill; *Energy Conservation in the International Energy Agency*, OECD; Grant—*Energy Conservation in the Chemical & Process Industries*, Institute of Chemical Engineers, England; Helcke—*The Energy Saving Guide*, Commission of the European Communities; Jarmul—*The Architect's Guide to Energy Conservation*, McGraw-Hill; Kovah—*Thermal Energy Storage*, Pergamon; Meckler—*Energy Conservation in Buildings & Industrial Plants*, McGraw-Hill; Payne—*Energy Managers' Handbook*, Butterworth; Pindyck—*The Structure of World Energy Demands*, M.I.T. Press; Reay—*Industrial Energy Conservation*, Pergamon; Smith—*Industrial Energy Management for Cost Reduction*, Ann Arbor Science; Yaverbaum—*Energy Saving by Increasing Boiler Efficiency*, Noyes; Considine—*Energy Technology Handbook*, McGraw-Hill.

CHOICE OF WIND-ENERGY CONVERSION SYSTEM

Select a wind-energy conversion system to generate electric power at constant speed and constant frequency in a sea-level area where winds average 18 m/h (29 km/h), a cut-in speed of 8 mi/h (13 km/h) is sought, blades will be fully feathered (cut out) at wind speeds greater than 60 mi/h (100 km/h), and the system must withstand maximum wind velocities of 150 mi/h (240 km/h). Determine typical costs which might be expected. The maximum rotor diameter allowable for the site is 125 ft (38 m).

Calculation Procedure:

1. Determine the total available wind power

Figure 1 shows the total available power in a freely flowing windstream at sea level for various wind speeds and cross-sectional areas of windstream. Since the maximum blade diameter, given that a blade-type conversion device will be used, is 125 ft (38 m), the area of the windstream will be $A = \pi d^2/4 = \pi(125)^2/4 = 12,271.9$ ft² (1140.1 m²). Entering Fig. 1 at this area and projecting vertically to a wind speed of 18 mi/h (29 km/h), we see that the total available power is 200 kW.

2. Select a suitable wind machine

Typical modern wind machines are shown in Fig. 2. In any wind-energy conversion system there are three basic subsystems: the aerodynamic system, the mechanical transmission system (gears, shafts, bearings, etc.), and the electrical generating system. Figure 2 gives the taxonomy of the

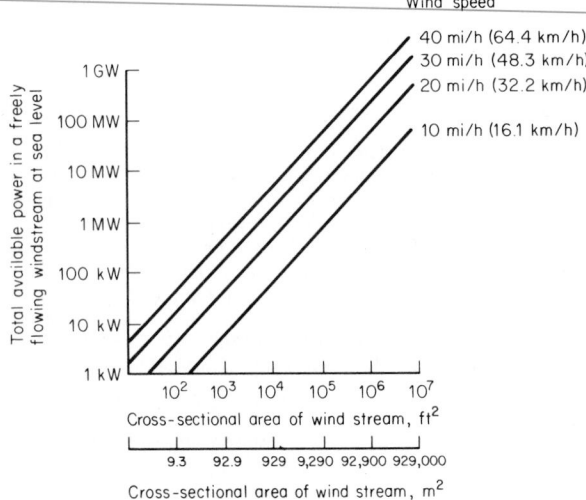

FIG. 1 The total available power in a freely flowing windstream at sea level versus the cross-sectional area of the windstream and the wind speed. *(Mechanical Engineering.)*

more practical versions of wind machines (the aerodynamic system) available today. "Almost any physical configuration which produces an asymmetric force in the wind can be made to rotate, translate, or oscillate—thereby generating power. The governing consideration is economic—how much power for how much size and cost," according to Fritz Hirschfeld, Member, ASME.

Continuing, Hirschfeld notes, "The power coefficient of an ideal wind machine rotor varies with the ratio of blade tip speed to free-flow windstream speed, and approaches the maximum of 0.59 when this ratio reaches a value of 5 or 6. Experimental evidence indicates that two-bladed rotors of good aerodynamic design—running at high rotational speeds where the ratio of the blade-tip-speed-to-free-flow-speed of the windstream is 5 or 6—will have power coefficients as high as 0.47. Figure 3 outlines the maximum power coefficients obtainable for several rotor designs. Figure 4 plots the typical performance curves of a number of different wind machines."

Choose a horizontal-axis double-bladed rotor wind machine for this application with a power coefficient C_p of 0.375. This type of wind machine is being chosen because (1) the power coefficient is relatively high (0.375), providing efficient conversion of the energy of the wind; (2) the allowable blade diameter, 125 ft (38 m), is suitable for the double-bladed design; (3) a double-bladed rotor will operate well in the average wind speed, 18 mi/h (29 km/h), prevailing in the installation area; and (4) the double-bladed rotor is well suited for the constant-speed constant-frequency (CSCF) system desired for this installation.

3. Compute the maximum electric power output of the wind machine

The power of the wind P_w is converted to mechanical power P_m by the wind machine. In any wind machine, $P_m = C_p P_w$, where C_p = power coefficient. The mechanical power is then converted to electric power by the generator. Since there is an applicable efficiency for each of the systems, that is, C_p for the aerodynamic system, η_m for the mechanical system (gears, usually), or η_g for the generator, the electric power generated is $P_e = P_w \eta_m \eta_g$.

In actual practice, the maximum electric power output in kilowatts of horizontal-axis bladed wind machines geared to a 70 percent efficiency electric generator can be quickly computed from $P_e = 0.38 d^2 V^3 / 10^6$, where d = blade diameter, ft (m); V = maximum wind velocity, ft/s (m/s). For this wind machine, $P_e = (0.38)(125)^2(26.4)^3/10^6 = 109.2$ kW. This result agrees closely with the actual machine on which the calculation procedure is based, which has a rated output of 100 kW.

Horizontal axis

Single-bladed

Double-bladed

Three-bladed

U.S. farm windmill
multibladed

Bicycle
multibladed

Upwind

Downwind

Enfield-Andeau

Sail wing

Multirotor

Counter-
rotating blades

Crosswind
Savonius

Crosswind
paddles

Diffuser

Concentrator

Unconfined
vortex

FIG. 2 Wind machines come in all shapes and sizes. Some of the more practical design categories are illustrated in this taxonomy. *(Mechanical Engineering.)*

4. *Determine the typical capital cost of this machine*

Figures 5 and 6 show typical capital costs for small conventional wind machines. Larger wind machines, such as the one being considered here, are estimated to have a cost of $150,000 for a 100-kW unit, or $1500 per kilowatt. Such costs may be safely used in first approximations with the base-year cost given in the illustration being suitably adjusted by a factor for inflation.

   Related Calculations: Use this general procedure to choose wind machines for other duties—pumping, battery charging, supplying power to utility lines, etc. Be certain to check with manufacturers to determine whether the calculated results agree with actual practice in the field. In general, good agreement will be found to exist.

   Wind power is a renewable, nonpolluting energy source in plentiful supply in certain parts of the world. For example, a recent report by the Union of Concerned Scientists points out that four states—Kansas, Nebraska, and North and South Dakota—have enough wind to gen-

Vertical axis

Primarily drag-type

Savonius   Multibladed Savonius   Plates   Shield   Cupped

Primarily lift-type

ϕ-Darrieus $\triangle$-Darrieus   Giromill   Turbine

Combinations

Savonius/ ϕ Darrieus   Split Savonius   Magnus   Airfoil

Others

Deflector   Sunlight   Venturi   Confined vortex

FIG. 2   (*Continued*)

erate—in theory—all the electricity needed in the United States today.

In addition, the agricultural resources of the midwestern part of the United States could provide crops and crop residues to be used as fuel in power plants. Likewise, logging and wood residues could be used to fire boilers to generate electricity. The resulting air pollution would be much less than that produced by coal-fired plants. Crop- and crop-residue-fired plants could eventually supply some 10 percent of the electrical energy needed by the midwestern states.

Current cost to produce a kilowatt of electricity using wind power is less than 6 cents. This compares favorably with the 4–6 cents cost per kilowatt for the typical coal-fired generating

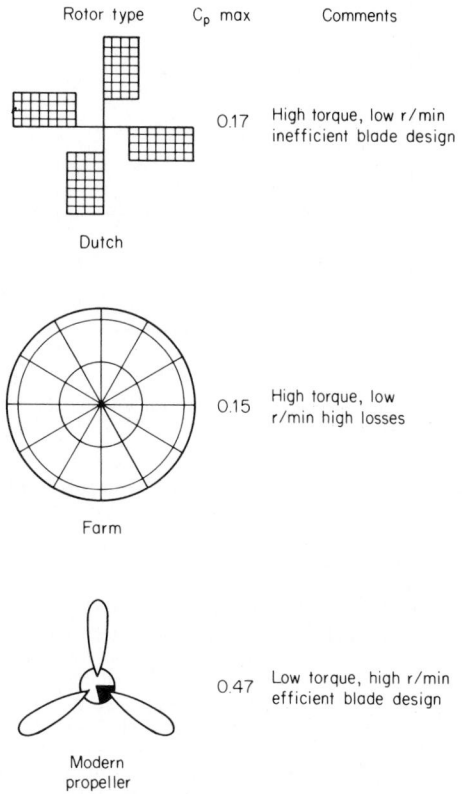

Rotor type	C_p max	Comments

Dutch — 0.17 — High torque, low r/min inefficient blade design

Farm — 0.15 — High torque, low r/min high losses

Modern propeller — 0.47 — Low torque, high r/min efficient blade design

FIG. 3 The maximum power coefficients for several types of rotor designs. *(Mechanical Engineering.)*

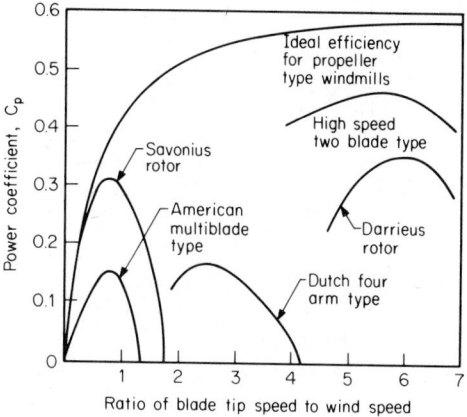

FIG. 4 Typical performance curves for different types of wind machines. *(Mechanical Engineering.)*

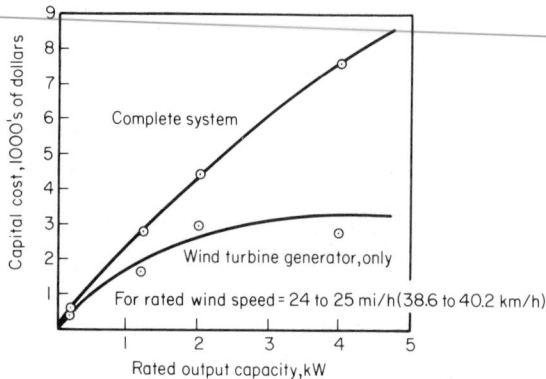

FIG. 5 Capital cost of small conventional wind machine. *(Mechanical Engineering.)*

station. When the relatively low cost of wind power is combined with its nonpolluting features, this form of electrical generation is extremely attractive to environmentally conscious engineers and scientists.

The advantages cited above apply equally well to many other nations throughout the world. In some areas wind power offers a simple, low-cost solution to energy needs without resorting to complex technical methods. Wind power does have a worldwide future.

The illustrations and much of the data in this procedure are the work of Fritz Hirschfeld, as reported in *Mechanical Engineering* magazine. Also reported in the magazine is a proposal by J. S. Goela of Physical Sciences, Inc., to use kites to extract energy from the wind.

Kites avoid the use of high-capital-cost components such as windmill towers and large rotors. Further, a kite can utilize the full available potential of the wind. As Fig. 7 shows, the earth's boundary layer extends up to 5000 ft (1500 m) above sea level. In this boundary layer, the average wind velocity increases while the air density decreases with altitude. Consequently, the total available wind power per unit area ($= \frac{1}{2}\rho V^3$) increases with altitude until at an altitude of 5000 ft (1500 m) a maximum is reached. The ratio of available wind power in New England at 5000 ft (1500 m) and 150 ft [many wind systems operate at an altitude of 150 ft (50 m) or less] is 25. This is a large factor which makes it very attractive to employ systems that use an energy extraction device located at an altitude of 5000 ft (1500 m). Even at an altitude of 1000 ft (300 m), this ratio is large, approximately equal to 10.

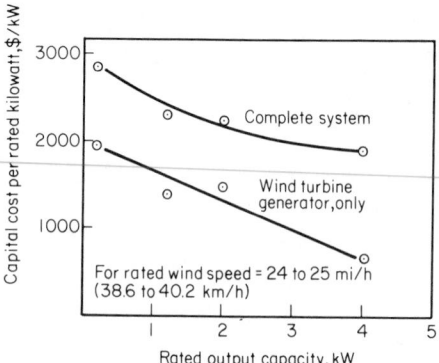

FIG. 6 Capital cost, per rated kilowatt, for small conventional wind machines. *(Mechanical Engineering.)*

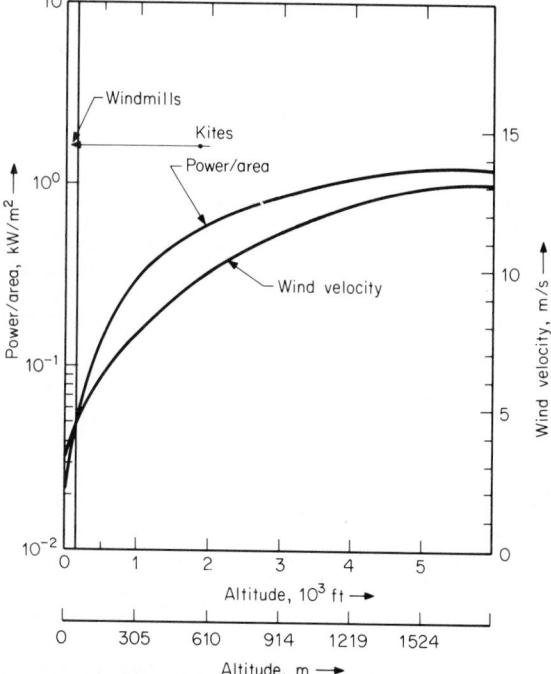

FIG. 7 Variation of mean annual free-air wind velocity and total available wind power per unit area ($= \frac{1}{2}pV^3$) with altitude in New England. (*Mechanical Engineering.*)

To understand how the proposed scheme will extract energy from the wind, consider the following: The motion of air generates a pull in the rope that holds the kite. This pull is a function of both the angle of attack of the kite and the kite area normal to the wind direction, and by varying any of these we can vary the pull on the rope. On the surface of the earth, this rope will be suitably connected to an energy system which will convert the variation in developed force on the rope to the rotational energy of a rotor.

Whenever a period of calm occurs, the kite will tend to lose its altitude. One solution to this difficulty is to fill the empty spaces in the kite with helium gas such that the upward pull from the helium gas will balance the downward gravity force due to the weight of the kite and its string. Another possibility is to tie a ballon to the kite.

A detailed theoretical analysis of the proposed scheme has been carried out. This analysis indicates that the proposed scheme is scalable, that the drag on the kite string is small in comparison to the pull in the string for large devices, and that approximately 0.38 kW of power theoretically can be obtained from a kite 1.2 yd² (1 m²) in area. In addition, there are no material or systems constraints which will prevent the kite from achieving an elevation of 5000 ft (1500 m).

Even though it is difficult to estimate the cost of wind power from the proposed scheme, a rough estimate indicates that for a 100-kW system, the capital cost per unit of energy from the proposed scheme will be less—approximately by a factor of 3—in comparison with the capital cost of one unit of energy produced from other 100-kW wind-energy systems.

The most important application of kite-based energy systems is in developing countries where these systems can be used to pump water from wells, grind grain, and generate electricity. A majority of developing countries do not have an adequate supply of indigenous oil and gas, nor can they afford to buy substantial quantities of fossil fuel at international prices. What these countries prefer is a system that could generate useful energy by using as inputs resources that are available within the country. The simple scheme proposed here is ideally suited for the needs of developing countries.

The kite-based system may also be economically attractive in comparison to a small windmill of less than 1 hp (750 W) which has been used in rural and farm areas in the western United States to pump water, generate electricity, and irrigate land. Another application of the proposed scheme is to generate auxiliary power in large sailboats, motorboats, and ships where conventional wind-energy schemes cannot be employed. The wind-energy system employing kites can also be used as a fuel saver in conjunction with already existing transmission lines.

A kite flying at an altitude of 5000 ft (1500 m) may present a hazard to low-flying airplanes. One way to avoid a collision with the kite or an entanglement with the kite lines is to enhance their visibility by providing flashing lights around the kite structure and along its retaining line. Another approach is to fly the kites at lower altitudes. For instance, even at an altitude of 1000 ft (300 m), the total available wind power is larger by a factor of 10 in comparison with that at 150 ft (50 m), Fig. 7.

Wind energy was welcomed by environmentalists when first introduced on a large-scale basis in the late 70s and early 80s. Today all environmental groups support wind energy—but with less enthusiasm than earlier. The reason for the loss of enthusiasm? Windmills are killing birds that fly into them.

Some of the birds killed by windmills are endangered species, so environmentalists seek a solution to prevent the killing of hundreds of birds a year that crash into the 17,000 100-ft (30.5-m) windmills erected in California where 80 percent of the world's wind power is produced.

A number of states are considering construction of windmill "farms" where thousands of power-generating windmills will be clustered together. Environmentalists are concerned that such farms will raise the death toll amongst birds. Wind turbine manufacturers estimate that from 2 to 6 birds are killed per year per 100 wind turbines.

FUEL SAVINGS USING HIGH-TEMPERATURE HOT-WATER HEATING

Determine the fuel savings possible by using high-temperature-water (HTW) heating instead of steam if 50,000 lb/h (6.3 kg/s) of steam at 150 lb/in^2 (gage) (1034 kPa) is to be produced for delivering heat to equipment 1000 ft (305 m) from the boiler. The saturation temperature of the steam is 360°F (182°C); specific volume = 2.75 ft^3/lb (0.017 m^3/kg); enthalpy of evaporation = 857 Btu/lb (1996.8 kJ/kg); enthalpy of saturated vapor = 1195.6 Btu/lb (2785.7 kJ/kg); ambient temperature = 70°F (21.1°C); steam velocity = 5000 ft/min (1524 m/s); density of water at 240°F (171.1°C) = 56 lb/ft^3 (896.6 kg/m^3).

Calculation Procedure:

1.   Compute the required pipe cross-sectional area

The required pipe cross-sectional area is A ft^2 (m^2) = Wv/V, where W = steam flow rate, lb/h (kg/h); V = specific volume of the steam, ft^3/lb (m^3/kg); V = steam velocity, ft/h (m/h). Or, A = 50,000(2.75)/[5000(60)] = 0.46 ft^2 or 66 in^2 (429 cm^2).

2.   Choose the size of the steam pipe

For a 5-lb/in^2 (34.4-kPa) pressure drop in the 1000-ft (305-m) pipeline, standard pressure-loss calculations given elsewhere in this handbook show that a 10-in (25.4-cm) diameter pipe would be suitable when used in conjunction with a 5-in (12.7-cm) condensate-return line. The 10-in (25.4-cm) line would have 2-in (5.1-cm) thick calcium silicate insulation, while the 5-in (12.7-cm) line would have 1-in (2.5-cm) thick insulation of the same material.

3.   Compute the heat losses in the two lines

Using the insulation heat-loss calculation methods given elsewhere in this handbook, we find the heat loss in the 10-in (25.4-cm) line is 183,200 Btu/h (53,678 W), while the heat loss in the 5-in (12.7-cm) condensate line is 78,100 Btu/h (22,883 W). Summing these, we see the total heat loss for the steam system is 261,300 Btu/h (76.6 kW).

4. Determine the amount of condensate formed

The amount of condensate formed w_c = (steam-line heat loss, Btu/h)/(enthalpy of vaporization, Btu/lb), or w_c = 183,200/857 = 214 lb/h (97.3 kg/h).

5. Compute the amount of heat delivered to the load

The amount of heat delivered to the load is H Btu/h = (steam flow rate, lb/h − condensate formation rate for heat loss, lb/h) (enthalpy of vaporization of the steam, Btu/lb). Or, for this steam system, H = (50,000 − 214)857 = 42,666,602 Btu/h (12,501 kW).

6. Determine the condensate flash-out losses

If the flash vapor is produced when the condensate is flashed out to atmospheric pressure in the return line and condensate receiver, the losses from flash-out will equal the enthalpy of the saturated water at 365°F (185°C) minus the enthalpy of the saturated water at 212°F (100°C), or 338.5 − 180 = 158.5 Btu/lb (369 kJ/kg).

To produce 857 Btu (904 J) of latent heat per pound of steam, the boiler must supply 1195.6 − 180 = 1015.6 Btu/lb (2366.3 kJ/kg), assuming that the condensate is returned to the boiler at 212°F (100°C). Hence, condensate losses from flash-out = (158.5/1015.6)100 = 15.6 percent. In addition, there is an approximate 5 percent loss due to leakage of steam and condensate, plus blowdown losses, which brings the total losses to 15.6 + 5.0 = 20.6, say 20 percent.

7. Compute the total boiler heat input required

With a condensate loss of 20 percent, as computed above, the amount of condensate returned to the boiler = 0.80(50,000 lb/h of steam) = 40,000 lb/h (5.04 kg/s). Hence, the enthalpy of the feedwater to the boiler, including makeup water, is 40,000 lb (18,181.8 kg) of condensate at 212°F (100°C) = 40,000(180 Btu/lb) = 7,200,000 Btu (7596 kJ); 10,000 lb (4545.4 kg) of makeup water at 50°F (10°C), 18 Btu/lb (41.9 kJ/kg), is 10,000(18) = 180,000 Btu (189,900 J); the sum = 7,380,000 Btu (7785 kJ). The boiler must therefore produce 50,000 × 1195.6 − 7,380,000 = 52,400,000 Btu/h (15,353.2 kW).

Assuming 75 percent boiler efficiency for this unit (a valid assumption for the usual steam heating boiler), we find the adjusted total amount of energy needed for steam heat = 52,400,000/0.75 = 69,867,000 Btu (73,710 kJ).

8. Compute the hourly water flow rate

To deliver 42,666,600 Btu/h (45,013.2 kJ) to the equipment, assume a 40°F (4.4°C) temperature drop between the supply and the return. Then the hourly flow rate = Btu/h heat required/temperature drop of the water, °F = 42,666,600/40 = 1,066,700 lb/h (134.3 kg/s).

9. Choose the size pipe to use for the supply and return

Assume a water flow velocity of 10 ft/s (3.05 m/s). Then the pipe area needed, from the relation in step 1 of this procedure, is 1,066,700/(3600)(10)(56) = 0.529 ft² (0.049 m²). This area requires a 10-in (25.4-cm) pipe.

10. Compute the heat loss in the piping

The supply and return lines would require 2000 ft (609.6 m) of 10-in (25.4-cm) pipe with 2-in (5.1-cm) thick calcium silicate insulation. If the supply temperature is 360°F (182.2°C) and the return temperature is 320°F (160°C), the mean temperature would be (360 + 320)/2 = 340°F (171.1°C). Using the insulation heat-loss calculation methods given elsewhere in this handbook, we see that the heat loss in the supply and return lines is 326,800 Btu/h (96.8 kW). Hence, the total amount of heat which must be supplied to the water is 326,800 + 42,666,600 = 42,993,400 Btu/h (12,597 kW) before allowance is made for the efficiency of the boiler.

11. Compare the steam and hot-water systems

A typical hot-water heating boiler for a system such as this will have an operating efficiency of 77 percent. Using this value, we find the heat which must be supplied by the fuel = 42,993,400/0.77 = 55,835,600 Btu/h (16,360 kW).

As computed earlier, the heat required by the steam system exceeds that required by the hot-water system by 69,867,000 − 55,835,600 = 14,031,400 Btu/h (4111 kW), or 20 percent. This means that the high-temperature hot-water system will use 20 percent less fuel than the steam system for this installation.

Related Calculations: Use this approach when comparing or designing HTW systems for airports, military installations, hospitals, shopping centers, multifamily dwellings, garden apartments, industrial plants, central heating for large districts, university campuses, chemical-process plants, and similar installations. High-temperature-water systems are those using water in the 250 to 420°F (121.1 to 215.5°C) range, corresponding to a steam pressure of 300 lb/in² (gage) (2068.5 kPa). Mechanical problems caused by high water pressures above 420°F (215.5°C) make this temperature the practical upper limit. HTW systems can produce fuel savings 20 percent greater than systems using steam.

Studies show that conversion from steam to HTW is attractive—particularly for systems rated at 20,000,000 Btu/h (5860 kW) or higher. At this rating the conversion cost can usually be paid off in about 2 years. Smaller HTW systems, from 5,000,000 to 15,000,000 Btu/h (1470 to 4395 kW), are only marginally more economical to operate than steam, but they are still favored because they provide much more accurate and uniform temperature control.

HTW systems can give fuel savings of 20 to 50 percent, compared to an equivalent steam heating system. For new installations, the total capital investment is about the same for both steam and HTW systems. However, the savings in fuel costs and maintenance make the payout period for a new HTW system shorter than for conversion of an existing steam system.

Many plants use their steam boilers for both process and space heating. Cascade (direct-contact) heaters can generate up to 350°F (176.7°C) water from 150-lb/in² (1034-kPa) steam [or 400°F (204.4°C) from a 250-lb/in² (1724-kPa) boiler]. This water temperature is adequate for the rolls, presses, extruders, evaporators, conveyors, and reactors used in many industrial plants. Steam-pressure reducing valves are not needed to maintain the different temperature levels required by each machine.

Plants having steam boilers can convert to HTW heating simply and quickly by installing direct-contact water heaters in, or adjacent to, the boiler room. Such heaters can also serve as heat

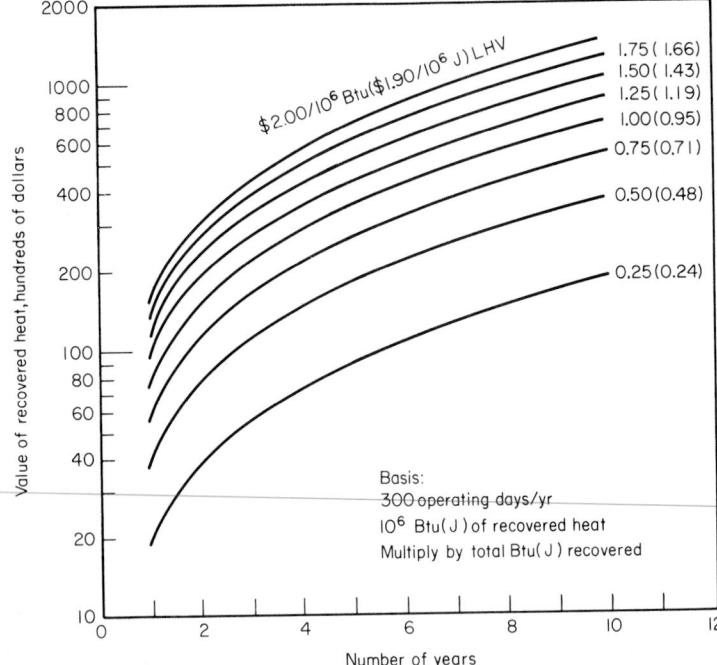

FIG. 8   Chart yields value of 1 million Btu of recovered heat. This value is based on the projected average costs for primary fuel. (*Chemical Engineering.*)

reservoirs, absorbing sudden peak loads and allowing the boilers to operate at fairly constant loads. HTW systems can easily supply water at elevated temperatures for process loads and water at lower temperatures for process loads and space-heating loads. Distribution efficiency of such systems approaches 95 percent overall.

For process applications requiring extremely close temperature control, the water circulating rate through a secondary loop can be designed to limit the difference between inlet and outlet temperatures to $\pm 2°F$ ($\pm 1.11°C$). The greater heat capacity of hot water over steam and the narrower pipelines required are other advantages. The usual HTW line need be only one or two sizes larger than the condensate line required in a steam system. The ratio of absolute heat-storing capacity is 42 to 1 in favor of HTW over steam. Where steam is needed in an all-HTW system, it can be obtained easily by flashing some of the water to steam. This calculation procedure is the work of William M. Teller, William Diskant, and Louis Malfitani, all of American Hydrotherm Corporation, as reported in *Chemical Engineering* magazine.

FUEL SAVINGS PRODUCED BY HEAT RECOVERY

Determine the primary-fuel saving which can be produced by heat recovery if 150 M Btu/h (158.3 MJ/h) in the form of 650-lb/in^2 (gage) (4481.1-kPa) steam superheated to 750°F (198.9°C) is recovered. The projected average primary-fuel cost (such as coal, gas, oil, etc.) over a 12-year evaluation period for this proposed heat recovery scheme is $0.75 per 10^6 Btu ($0.71 per million joules) lower heating value (LHV). Expected thermal efficiency of a conventional power boiler to produce steam at the equivalent pressure and temperature is 86 percent, based on the LHV of the fuel.

Calculation Procedure:

1. *Determine the value of the heat recovered during 1 year*

Enter Fig. 8 at the bottom at 1 year and project vertically to the curve marked $0.75 per 10^6 LHV. From the intersection with the curve, project to the left to read the value of the heat recovered as $5400 per year per MBtu/h recovered ($5094 per MJ).

2. *Find the total value of the recovered heat*

The total value of the recovered heat = (hourly value of the heat recovered, $/10^6 Btu)(heat recovered, 10^6 Btu/h)(life of scheme, years). For this scheme, total value of recovered heat = ($5400)(150 $\times$ 10^6 Btu/h)(12 years) = $9,720,000.

3. *Compute the total value of the recovered heat, taking the boiler efficiency into consideration*

Since the power boiler has an efficiency of 86 percent, the equivalent cost of the primary fuel would be $0.75/0.86 = $0.872 per 10^6 Btu ($0.823 per million joules). The total value of the recovered heat if bought as primary fuel would be $9,720,000($0.872/$0.75) = $11,301,119. This is nearly $1 million a year for the 12-year evaluation period—a significant amount of money in almost any business. Thus, for a plant producing 1000 tons/day (900 t/day) of a product, the heat recovery noted above will reduce the cost of the product by about $3.14 per ton, based on 258 working days per year.

Related Calculations: This general procedure can be used for any engineered installation where heat is available for recovery, such as power-generating plants, chemical-process plants, petroleum refineries, marine steam-propulsion plants, nuclear generating facilities, air-conditioning and refrigeration plants, building heating systems, etc. Further, the procedure can be used for these and any other heat-recovery projects where the cost of the primary fuel can be determined. Offsetting the value of any heat saving will be the cost of the equipment needed to effect this saving. Typical equipment used for heat savings include waste-heat boilers, insulation, heat pipes, incinerators, etc.

With the almost certain continuing rise in fuel costs, designers are seeking new and proven ways to recover heat. Ways which are both popular and effective include the following:

1. Converting recovered heat to high-pressure steam in the 600- to 1500-lb/in^2 (gage) (4137- to 10,343-kPa) range where the economic value of the steam is significantly higher than at lower pressures.

2. Superheating steam using elevated-temperature streams to both recover heat and add to the economic value of the steam.

3. Using waste heat to raise the temperature of incoming streams of water, air, raw materials, etc.

4. Recovering heat from circulating streams of liquids which might otherwise be wasted.

In evaluating any heat-recovery system, the following facts should be included in the calculation of the potential savings:

1. The economic value of the recovered heat should exceed the value of the primary energy required to produce the equivalent heat at the same temperature and/or pressure level. An efficiency factor must be applied to the primary fuel in determining its value compared to that obtained from heat recovery. This was done in the above calculation.

2. An economic evaluation of a heat-recovery system must be based on a projection of fuel costs over the average life of the heat-recovery equipment.

3. Environmental pollution restrictions must be kept in mind at all times because they may force the use of a more costly fuel.

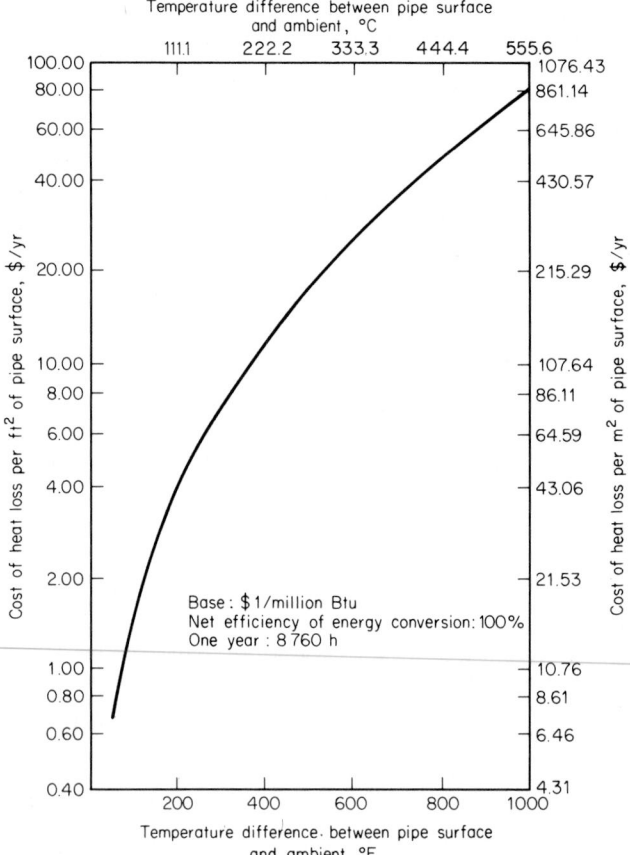

FIG. 9   Cost of heat loss when insulation is missing. *(Chemical Engineering.)*

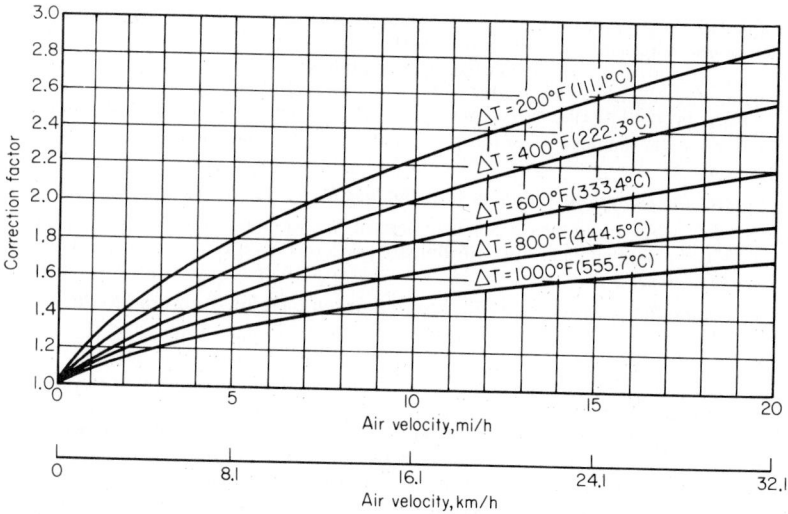

FIG. 10 Correction for wind velocity; ΔT is difference between pipe and ambient temperature. *(Chemical Engineering.)*

4. Many elevated-temperature process streams require cooling over a long temperature range. In such instances, the economic analysis should credit the heat-recovery installation with the savings that result from eliminating non-heat-recovery equipment that normally would have been provided. Also, if the heat-recovery equipment permits faster cooling of a stream and this time saving has an economic value, this value must be included in the study.

5. Where heat-recovery equipment reduces primary-fuel consumption, it is possible that plant operations can be continued with the use of such equipment whereas without the equipment the continued operation of a plant might not be possible.

The above calculations and comments on heat recovery are the work of J. P. Fanaritis and H. J. Streich, both of Struthers Wells Corp., as reported in *Chemical Engineering* magazine.

Where the primary-fuel cost exceeds or is different from the values plotted in Fig. 8, use the value of $1.00 per 10^6 Btu (J) (LHV) and multiply the result by the ratio of (actual cost, dollars per 10^6 Btu/$1) ($/J/$1). Thus, if the actual cost is $3 per 10^6 Btu (J), solve for $1 per 10^6 Btu (J) and multiply the result by 3. And if the actual cost were $0.80, the result would be multiplied by 0.8.

COST OF HEAT LOSS FOR UNINSULATED PIPES

What is the annual cost of the heat loss from 10 ft (3 m) of uninsulated 6-in (15-cm) diameter pipe conveying steam at 375°F (190.6°C) if the average ambient temperature is 75°F (23.9°C), the fuel cost is $1.44 per 10^6 Btu ($1.36 per MJ), and the net energy conversion efficiency is 72 percent? What effect would a wind velocity of 10 mi/h (16.1 km/h) have on the heat loss from the pipe?

Calculation Procedure:

1. *Determine the heat loss in still air per unit length of pipe*

The heat loss from any vessel or pipe depends on the temperature difference between the pipe or vessel and the medium in contact with the warmer surface, or ΔT. For this installation, $\Delta T = 375 - 75 = 300°F$ (166.7°C).

To determine the cost of the heat loss per unit length of pipe, enter Fig. 9 at the bottom at

300°F (166.7°C) and project vertically upward to the intersection with the curve; from the intersection project horizontally to the left to read the cost as $7.60 per year per square foot ($84.44 per year per square meter) of pipe surface. Convert this cost to the cost per linear foot (meter) of pipe by multiplying by 1.734 ft^2/ft, the surface area per foot of length of 6-in (15.2-cm) nominal pipe. Hence, ($7.60)(1.734) = $13.18 per foot per year ($43.24 per meter per year) for this pipe.

To compute the annual cost for this pipe, multiply by the total pipe length, or 10($13.18) = $131.80 per year.

2. *Correct the computed annual cost of the loss for net efficiency and fuel cost*

The net efficiency of energy conversion is 72 percent, and the annual fuel cost = (cost, $ per foot per year)(ft of pipe)(fuel cost, $ per 10^6 Btu per year)/(net energy conversion efficiency, %) = ($131.80)(1.44/0.72) = $263.60.

3. *Determine the annual cost of the heat loss when the pipe is exposed to the wind*

Enter Fig. 10 at the air velocity, 10 mi/h (16.1 km/h) and project upward to a ΔT of 300°F (166.7°C). At the left, read the wind correction factor as 2.14. Then the annual cost in a location having a prevailing wind = (annual cost for wind-free location)(air velocity correction factor) = ($263.60)(2.14) = $564.10.

Related Calculations: For piping whose surface has been left uninsulated or whose insulation has been so severely damaged as to be useless, the yearly (8760-h) cost of the heat loss per square foot can be found from Fig. 9. Costs so determined will generally be within the accuracy of data that can be obtained in the field, and are suitable for most applications without further refinement.

Figure 9 costs are based on a net efficiency of energy conversion of 100 percent, a fuel price of $1.00 per 10^6 Btu, and heat transfer to still air. Corrections are necessary to arrive at annual costs for other conditions:

Net efficiency of less than 100 percent: To correct for a different net efficiency, divide the cost derived from Fig. 9 by the actual net efficiency of energy conversion. (For intermediate-size boilers most common in usual plants, the net efficiency is typically 72 percent.)

Fuel cost other than $1.00 per 10^6 Btu: To correct for a different cost, multiply the cost derived from Fig. 9 by the actual cost of the fuel.

Outdoor installation exposed to moving air: To correct for variant wind conditions, multiply the cost obtained from Fig. 9 by a correction factor from Fig. 10 for the actual wind velocity. For most of the continental United States, 10 mi/h (16.1 km/h) represents a reasonable annual average.

The unit costs provided by Fig. 9 are average values for a range of pipe sizes in still air. Although there is some variation for different pipe sizes and ambient temperatures, the variations are small compared to the cost of heat loss caused by low air velocities.

The heat loss resulting from missing insulation is the difference between the heat loss from bare pipe and that through normal insulation. Insulation standards have changed in recent years because of varying fuel and insulation costs and other economic factors. And although the loss through insulation will vary with the standard adopted for a project, such variations will be negligible compared to the heat loss from bare pipe.

In general, corrections for precise ambient conditions, pipe size, and insulation thickness will not be justified because they go beyond the reasonable accuracy of field data.

The procedure and illustrations presented here are the work of Rene Cordero, piping and process mechanical equipment design engineer, Allied Chemical Corp., as reported in *Chemical Engineering* magazine. The data provided are valid for piping in chemical, petrochemical, factory, marine, power, and similar plants where it is desired to determine the cost of heat loss from uninsulated or partially insulated pipes.

HEAT-RATE IMPROVEMENT USING TURBINE-DRIVEN BOILER FANS

What is the net heat-rate improvement and net kilowatt gain in a steam power plant having a main generating unit rated at 870,000 kW at 2.5 in (6.35 cm) HgA, 0 percent makeup with motor-driven fans if turbine-driven fans are substituted? Plant data are as follows: (*a*) tandem-compound turbine, four-flow, 3600-r/min 33.5 in (85.1 cm) last-stage buckets with 264-ft^2 (24.5-m^2) total

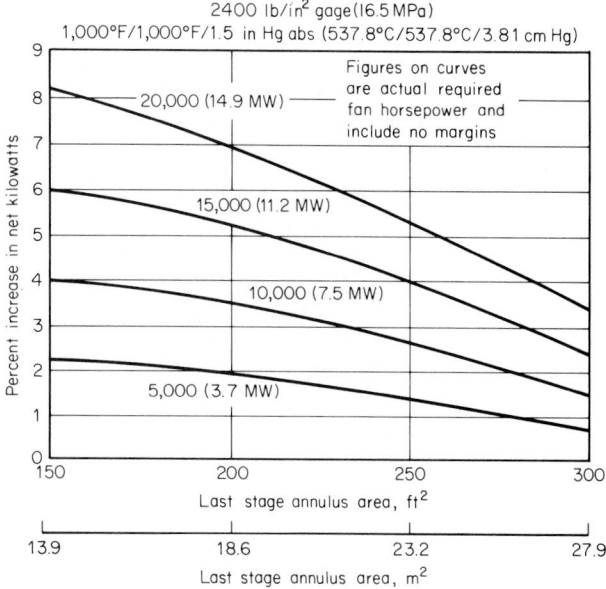

FIG. 11 Percentage increase in net kilowatts vs. last stage annulus area for 2400-lb/in² (gage) when turbine-driven fans are used as compared to motors. *(Combustion.)*

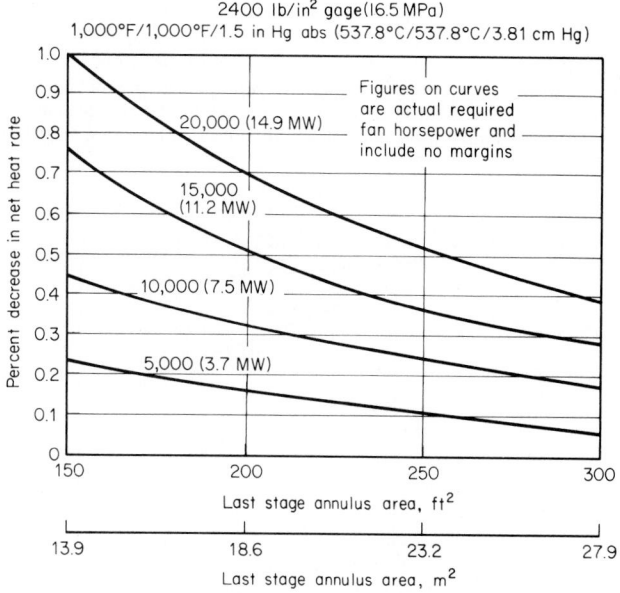

FIG. 12 Percentage decrease in net heat rate vs. last-stage annulus area for 2400-lb/in² (gage) when turbine-driven fans are used as compared to motors. *(Combustion.)*

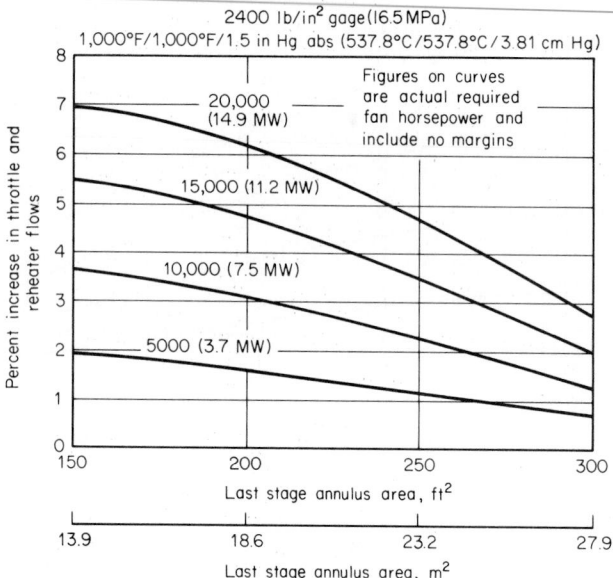

FIG. 13 Percentage increase in throttle and reheater flows vs. last-stage annulus area for 2400-lb/in² (gage) when turbine-driven fans are used as compared to motors. *(Combustion.)*

last-stage annulus area; (*b*) steam conditions 3500 lb/in² (gage) (24,133 kPa), 1000°F/1000°F (537.8°C/537.8°C); (*c*) with main-unit valves wide open, overpressure with motor-driven fans, generator output = 952,000 kW at 2.5 in (6.35 cm) HgA and 0 percent makeup; net heat rate = 7770 Btu/kWh (8197.4 kJ/kWh); (*d*) actual fan horsepower = 14,000(10,444 W) at valves wide open, overpressure with no flow or head margins; (*e*) motor efficiency = 93 percent; transmission efficiency = 98 percent; inlet-valve efficiency = 88 percent; total drive efficiency = 80 percent; difference between the example drive efficiency and base drive efficiency = 80 − 76.7 = 3.3 percent.

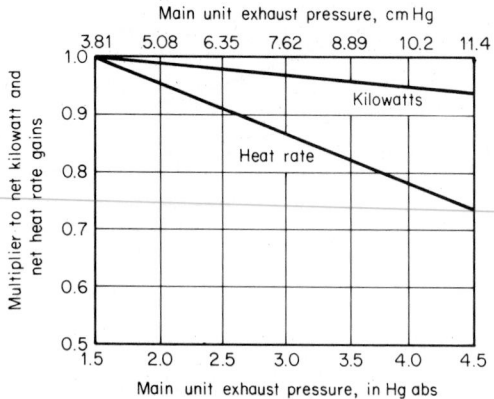

FIG. 14 Multiplier to net kilowatt and net heat-rate gains to correct for main-unit exhaust pressure higher than 1.5 inHg (38.1 mmHg). *(Combustion.)*

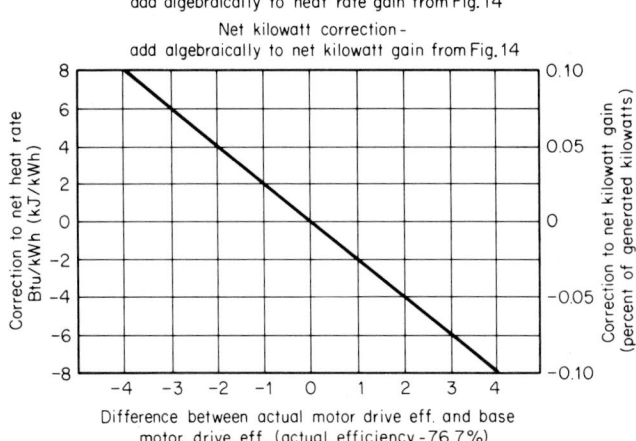

Net heat rate correction -
add algebraically to heat rate gain from Fig. 14

Net kilowatt correction -
add algebraically to net kilowatt gain from Fig. 14

FIG. 15 Corrections for differences in motor-drive system efficiency. *(Combustion.)*

Calculation Procedure:

1. Determine the percentage increase in net kilowatt output when turbine-driven fans are used

Enter Fig. 11 at 264-ft^2 (24.5-m^2) annulus area and 14,000 required fan horsepower, and read the increase as 3.6 percent. Hence, the net plant output increase = 34,272 kW (= 0.036 × 952,000).

2. Compute the net heat improvement

From Fig. 12, the net heat rate improvement = 0.31 percent. Or, 0.0031(7770) = 24 Btu (25.3 J).

3. Determine the increase in the throttle and reheater steam flow

From Fig. 13, the increase in the throttle and reheater flow is 3.1 percent. This is the additional boiler steam flow required for the turbine-driven fan cycle.

4. Compute the net kilowatt gain and the net heat-rate improvement

From Fig. 14 the multipliers for the 2.5 in (6.35 cm) HgA backpressure are 0.98 for net kilowatt gain and 0.91 for net heat rate. Hence, net kW gain = 34,272(0.98) = 33.587 kW, and net heat-rate improvement = 24 × 0.91 = 22.0 Btu (23.0 J).

5. Determine the overall cycle benefits

From Fig. 15 the correction for a drive efficiency of 80 percent compared to the base case of 76.7 percent is obtained. Enter the curve with 3.3 percent (= 80 − 76.7) and read −6.6-Btu (−6.96-J) correction on the net heat rate and −0.08 percent of generated kilowatts.

To determine the overall cycle benefits, add algebraically to the values obtained from step 4, or net kW gain = 33,587 + (−0.0008 × 952,000) = 32,825 kW; net heat-rate improvement = 22.1 + (−6.6) = 15.5 Btu (16.4 J).

Related Calculations: This calculation procedure can be used for any maximum-loaded main turbine in utility stations serving electric loads in metropolitan or rural areas. A maximum-loaded main turbine is one designed and sized for the maximum allowable steam flow through its last-stage annulus area.

Turbine-driven fans have been in operation in some plants for more than 10 years. Next to feed pumps, the boiler fans are the second largest consumer of auxiliary power in utility stations.

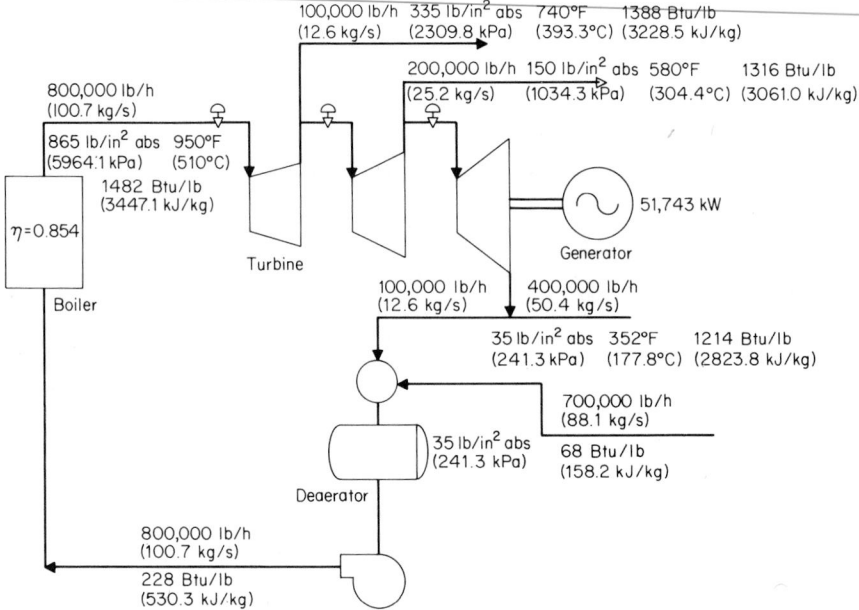

FIG. 16 Dual-purpose turbine heat balance. *(Combustion.)*

Current studies indicate that turbine-driven fans can be economic at 700 MW and above, and possibly as low as 500 MW. Although the turbine-driven fan system will have a higher initial capital cost when compared to a motor-driven fan system, the additional cost will be more than offset by the additional net output in kilowatts. In certain cases, economic studies may show that turbine drives for fans may be advantageous in constant-throttle-flow evaluations.

As power plants for utility use get larger, fan power required for boilers is increasing. Environmental factors such as use of SO_2 removal equipment are also increasing the required fan power. With these increased fan-power requirements, turbine drive will be the more economic arrangement for many large fossil plants. Further, these drives enable the plant designer to obtain a greater output from each unit of fuel input.

This calculation procedure is based on the work of E. L. Williamson, J. C. Black, A. F. Destribats, and W. N. Iuliano, all of Southern Services, Inc., and F. A. Reed, General Electric Company, as reported in *Combustion* magazine and in a paper presented before the American Power Conference, Chicago.

COST SEPARATION OF STEAM AND ELECTRICITY IN A COGENERATION POWER PLANT USING THE ENERGY EQUIVALENCE METHOD

Allocate—using the energy equivalence method—the steam and electricity costs in a power plant having a double automatic-extraction, noncondensing steam turbine for process steam and electric generation. Turbine throttle steam flow is 800,000 lb/h (100.7 kg/s) at 865 lb/in² (abs) (5964.1 kPa). Process steam is extracted from the turbine in the amounts of 100,000 lb/h (12.6 kg/s) at 335 lb/in² (abs) (2309.8 kPa) and 200,000 lb/h (25.2 kg/s) at 150 lb/in² (abs) (1034.3 kPa) and is delivered to process plants. A total of 500,000 lb/h (62.9 kg/s) is exhausted at 35 lb/in² (abs) (241.3 kPa) with 100,000 lb/h (12.6 kg/s) of this exhaust steam for deaerator heating in the cycle and 400,000 lb/h (50.4 kg/s) sent to process plants. The turbine has a gross electric output of 51,743 kW, and the heat balance for the dual-purpose turbine cycle is shown in Fig. 16. Efficiency of

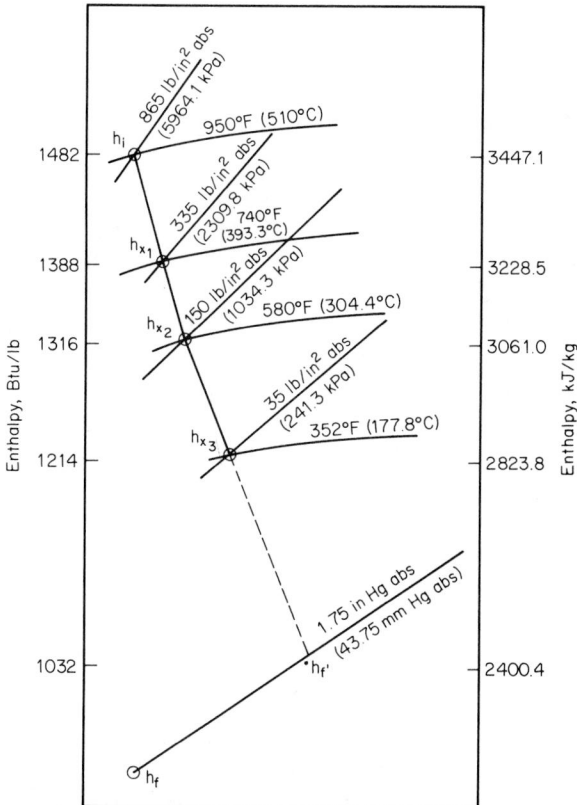

FIG. 17 Turbine expansion curve. *(Combustion.)*

the steam boiler is 85.4 percent, while the fuel is priced at \$0.50 per 10^6 Btu (\$0.47 per MJ). If a condensing turbine is used, an attainable backpressure is 1.75 inHg (abs) [43.75 mmHg (abs)], while the assumed turbine efficiency is 82 percent and the exhaust enthalpy is $h_{f'} = 1032$ Btu/lb (2400.4 MJ/kg). Figure 17 shows the expansion-state curve of the turbine on a Mollier diagram. Final feedwater enthalpy is 228 Btu/lb (530.3 mJ/kg). Allocate the fuel cost to each energy use by using the energy equivalence method.

Calculation Procedure:

1. Compute the hourly total fuel cost

The total fuel cost for this plant per hour is $C_f = (1/\text{boiler efficiency})(0.50/10^6)(\text{throttle steam}$ flow rate m_t, lb/h)$(h_t - h_{fw})$, where h_t = throttle enthalpy, Btu/lb, and h_{fw} = feedwater enthalpy, Btu/lb. Substituting gives $C_f = (1/0.854)(0.50/10^6)(800,000)(1482 - 228) = \$587.35$ per hour.

2. Compute the nonextraction ultimate electric output

The ultimate electric output is $E_u = m_t(h_i - h_{f'})/3413$, where m_t = total turbine inlet steam flow, lb/h; h_i = turbine initial enthalpy, Btu/lb; other symbols as before. Substituting, we find $E_u = (800,00)(1482 - 1032)/3413 = 105,480$ kW.

TABLE 1   Energy Equivalence Method of Fuel Cost Allocation[a]

Utility	Base	+	Unit cost heating steam	=	Total	Total fuel cost	Percent of total
Electricity 51,743 kW†	5.568		0.296		5.864 mi/kWh	$303.40	51.65
Steam @ 335 lb/in² (abs) (2309 kPa) 100,000 lb/h (45,000 kg/h)	0.5808		0.031		$0.6118 per 1000 lb (450 kg)	$ 61.18	10.41
Steam @ 150 lb/in² (abs) (1034 kPa) 200,000 lb/h (90,000 kg/h)	0.4633		0.0246		$0.4879 per 1000 lb (450 kg)	$ 97.57	16.62
Steam @ 35 lb/in² (abs) (241 kPa) 400,000 lb/h (180,000 kg/h)	0.2969		0.0158		0.3127 per 1000 lb (450 kg)	$125.20	21.32
					Total:	$587.35	100.00

[a] *Combustion* magazine.
†Net kW delivered to process plants should be delivered after deducting fixed mechanical and electrical losses of the alternator. Electricity unit cost charged to production would be slightly higher after this adjustment.

3.   Determine the actual electric output of the dual-purpose turbine

Use the relation $E_a = W(\text{actual})/3413 =$ the work done by the extraction steam between the throttle inlet and the extraction point, plus the work done by the nonextraction steam between the throttle and the exhaust. Or, from the turbine expansion curve in Fig. 17, $E_a = [100,000(1482 - 1388) + 200,000(1482 - 1316) + 500,000(1482 - 1214)]/3413 = 51,743$ kW.

4.   Compute the extraction steam kilowatt equivalence

Again from Fig. 17, $E_{x1} = (h_{x1} - h_f)/3413 = 100,000(1388 - 1032)/3413 = 10,432$ kW; $E_{x2} = 200,000(1316 - 1032)/3413 = 16,642$ kW; $E_{x3} = 500,000(1214 - 1032)/3413 = 26,663$ kW. Hence, the nonextraction turbine ultimate electric output $= E_a + E_{x1} + E_{x2} + E_{x3} = 51,743 + 10,432 + 16,642 + 26,663 = 105,480$ kW.

5.   Determine the base fuel cost of electricity and steam

The base fuel cost of electricity $= C_f/E_u$, or $587.35/105,480 = \$0.005568$ per kilowatthour, or 5.568 mil/kWh.

   Now the base fuel cost of the steam at the different pressures can be found from (kW equivalence)(base cost of electricity, mil)/(rate of steam use, lb/h). Thus, for the 335-lb/in² (abs) (2309.8-kPa) extraction steam used at the rate of 100,000 lb/h (12.6 kg/s), base fuel cost $=$ 10,432(5.568)/100,000 $= \$0.5808$ per 1000 lb ($0.2640 per 1000 kg). For the 150-lb/in² (abs) (1034.3-kPa) steam, base fuel cost $=$ 16,642(5.568)/200,000 $= \$0.4633$ per 1000 lb ($0.21059 per 1000 kg). And for the 35-lb/in² (abs) (241.3-kPa) steam, base fuel cost $=$ 26,663(5.568)/500,000 $= \$0.2969$ per 1000 lb ($0.13495 per 1000 kg). Since 100,000 lb/h (12.6 kg/s) of the 500,000-lb/h (62.9 kg/s) is used for deaerator heating, the cost of this heating steam $=$ (100,000/1000)($0.2969) $= \$29.69$ per hour.

   Since 100,000 lb (45,000 kg) of steam utilizes its energy for deaerator heating within the cycle, its equivalent electric output of 26,663/5 $=$ 5333 kW should be deducted from the 26,663-kW electric energy equivalency of the 35-lb/in² (abs) (241.3-kPa) steam. The remaining equivalent energy of 21,330 kW ($=$ 26,663 $-$ 5333) represents 35-lb/in² (abs) (241.3-kPa) extraction steam to be delivered to process plants.

6.   Determine the added unit fuel cost

The deaerator-steam fuel cost of $29.69 per hour would be shared by both process steam and electricity in terms of energy equivalency as 105,480 $-$ 5333 $=$ 100,147 kW. Using this output as the denominator, we see that the added unit fuel cost for electricity based on sharing the cost of this heat energy input to the deaerator is $26.69/100,147 = $0.000296 per kilowatthour, or 0.296 mil/kWh.

Likewise, added fuel cost of 335-lb/in² (abs) (2309-kPa) steam = 10,432(100,000)/0.296 = $0.031 per 1000 lb (450 kg); added fuel cost of 150-lb/in² (abs) (1034-kPa) steam = $0.0246 per 1000 lb (450 kg); added fuel cost of 35-lb/in² (abs) (241-kPa) steam = $0.0158 per 1000 lb (450 kg). The fuel-cost allocation of steam and electricity is summarized in Table 1.

Related Calculations: The energy equivalence method is based on the fact that the basic energy source for process steam and electricity is the heat from fuel (combustion or fission). The cost of the fuel must be charged to the process steam and electricity. Since the analysis does not distinguish between types of fuels or methods of heat release, this procedure can be used for coal, oil, gas, wood, peat, bagasse, etc. Also, the procedure can be used for steam generated by nuclear fission.

Cogeneration is suitable for a multitude of industries such as steel, textile, shipbuilding, aircraft, food, chemical, petrochemical, city and town district heating, etc. With the increasing cost of all types of fuel, cogeneration will become more popular than in the past. This calculation procedure is the work of Paul Leung of Bechtel Corporation, as reported at the 34th Annual Meeting of the American Power Conference and published in *Combustion* magazine. Since the procedure is based on thermodynamic and economic principles, it has wide applicability in a variety of industries. For a complete view of the allocation of costs in cogeneration plants, the reader should carefully study the Related Calculations in the next calculation procedure.

COGENERATION FUEL COST ALLOCATION BASED ON AN ESTABLISHED ELECTRICITY COST

A turbine of the single-purpose type, operating at initial steam conditions identical to those in the previous calculation procedure, and a condenser backpressure of 1.75 in (43.75 mm) Hg (abs), would have a turbine heat rate of 9000 Btu/kWh (9495 kJ/kWh). Compute the fuel cost allocation to that of steam by using the established-electricity-cost method.

Calculation Procedure:

1. Compute the unit cost of the electricity

The unit cost of the electricity is F_e = (fuel price, $)(turbine heat rate, Btu/kWh)/(boiler efficiency). For this plant, $F_e = (0.5/10^6)(9000)/(0.854) = 0.00527 per kilowatthour, or 5.27 mil/kWh.

2. Determine where the deaerator heating steam should be charged

The turbine heat rate of 9000 Btu/kWh is a reasonable and economically justifiable heat rate of a regenerative cycle with a certain degree of feedwater heating. Hence, in this case, the deaerator heating steam should not be charged to the electricity. Instead, this portion of the deaerator-heating-steam cost should be charged to the process steam.

3. Allocate the fuel cost to steam

The total fuel cost from previous calculation procedure is $587.35 per hour. The electricity cost allocation = (kW generated)(cost $/kWh) = (51,743)(0.00527) = $273. Hence, the fuel cost to the steam is $587.35 − $273.00 = $314.35.

4. Compute the power equivalence of the steam

From the previous calculation procedure, $E_x = E_{x1} + E_{x2} + E_{x3}$, where E_x = equivalent electric output of the extraction steam, kW; E_{x1}, . . . = equivalent electric output of the various extraction steam flows, kW. Hence, $\Sigma E_x = 10,432 + 16,642 + 26,663 = 53,737$ kW.

5. Determine the ratio of each extraction steam flow to the total extraction steam flow

The ratio for any flow is $E_x/\Sigma E_x$. Thus, $E_{x1}/\Sigma E_x = 10,432/53,737 = 0.194$; $E_{x2}/\Sigma E_x = 16,663/53,737 = 0.310$; and $E_{x3}/\Sigma E_x = 26,663/53,737 = 0.496$.

6. Compute the base unit fuel cost of steam

Use the relation $(E_x/\Sigma E_x)$(fuel cost to steam)$/m$, where m = (steam flow rate, lb/h)/1000. Hence, for 335-lb/in² (abs) (2309.8-kPa) steam, base unit fuel cost = (0.194)($314.35)/100 = $0.610 per

TABLE 2 Established-Electricity-Cost Method of Fuel-Cost Allocation[*]

Utility	Base	+	Unit cost heating steam	=	Total	Total fuel cost	Percent of total
Electricity 51,743 kW[†]	5.27		0		5.27 mil/kWh	$273.00	46.48
Steam @ 335 lb/in² (abs) (2309 kPa) 100,000 lb/h (45,000 kg/h)	0.610		0.067		$0.677 per 1000 lb (450 kg)	$ 67.70	11.53
Steam @ 150 lb/in² (abs) (1034 kPa) 200,000 lb/h (90,000 kg/h)	0.487		0.053		$0.541 per 1000 lb (450 kg)	$108.25	18.43
Steam @ 35 lb/in² (abs) (241 kPa) 400,000 lb/h (180,000 kg/h)	0.312		0.034		$0.346 per 1000 lb (450 kg)	$138.40	23.56
					Total:	$587.35	100.00

[*] *Combustion* magazine.
[†] Net kw delivered to process plants should be delivered after deducting fixed mechanical loss and electrical loss of the alternator. Electricity unit cost charged to production would be slightly higher after this adjustment.

1000 lb ($0.277 per 1000 kg); for 150-lb/in² (abs) (1034.3-kPa) steam, base unit fuel cost = $(0.310)($314.35)/200 = 0.487 per 1000 lb ($0.2213 per 1000 kg); for 35-lb/in² (abs) (241.3-kPa) steam, base unit fuel cost = $(0.496)($314.35)/500 = 0.312 per 1000 lb ($0.1418 per 1000 kg). Since the deaeration steam is at 35 lb/in² (abs) (241 kPa), the cost of this steam = $(100,000/1000)($0.312) = 31.20 per hour.

7. Determine the unit fuel cost from sharing the cost of the deaerator heating steam

If the 5333-kW power equivalence of the deaerator heating steam is deducted from the electric power equivalence of the extraction steam, the kilowatt equivalence of all steam to production centers becomes $53,737 - 5333 = 48,404$ kW. The unit fuel cost from sharing the cost of the deaerator heating steam is then ($31.20/h)/(48,404) = 0.000644 per kilowatthour, or 0.644 mil/kWh.

8. Compute the added fuel cost of steam at each pressure

The added fuel cost at each pressure is (kW output at that pressure/steam flow rate, lb/h)(0.644). Thus, added fuel cost of 335-lb/in² (abs) (2309.8-kPa) steam = $(10,431/100,000)(0.644) = 0.067 per 1000 lb ($0.03045 per 1000 kg); added fuel cost for 150-lb/in² (abs) (1034.3-kPa) steam = $0.053 per 1000 lb ($0.02409 per 1000 kg); added fuel cost for 35-lb/in² (abs) (241.3-kPa) steam = $0.034 per 1000 lb ($0.01545 per 1000 kg). Table 2 summarizes the fuel-cost allocation of steam and electricity by using this approach.

   Related Calculations: The established-electricity-cost method is based on the assumption (or existence) of a reasonable and economically justifiable heat rate of the cycle being considered or used. The cost of the fuel must be charged to the process steam and electricity. Since the analysis does not distinguish between types of fuels or methods of heat release, this procedure can be used for coal, oil, gas, wood, peat, bagasse, etc. Also, the procedure can be used for steam generated by nuclear fission.

   Cogeneration is suitable for a multitude of industries such as steel, textile, shipbuilding, aircraft, food, chemical, petrochemical, city and town district heating, etc. With the increasing cost of all types of fuels, cogeneration will become more popular than in the past.

   Other approaches to cost allocations for cogeneration include: (1) capital cost segregation, (2)

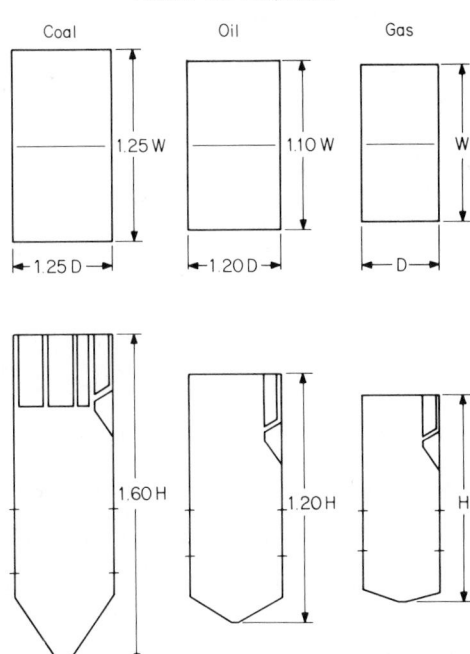

FIG. 18 Furnace size comparisons. *(Combustion.)*

capital cost allocation by cost separation of major functions, (3) cost separation of joint components, (4) capital cost allocation based on single-purpose electric generating plant capital cost, (5) unit cost based on fixed annual capacity factor, and (6) unit cost based on fixed peak demand. Each method has its advantages, depending on the particular design situation.

In the two examples given here (the present and previous calculation procedures), water return to the dual-purpose turbine cycle is assumed to be of condensate quality. Hence, no capital and operating costs of water have been included. In actual cases, a cost account should be set up based on the quantity of the returned condensate. Special charges would be necessary for the unreturned

TABLE 3 Coal Properties—Nominal 600-MW Unit[°]

Type of coal	Eastern bituminous	Midwestern bituminous	Subbituminous C	Texas lignite	Northern plains lignite
HHV, Btu/lb	12,000	10,000	8,400	7,300	6,800
(kJ/kg)	(27,912)	(23,260)	(19,538)	(16,980)	(15,817)
Moisture, %	6	12	27	32	37
lb $H_2O/10^6$ Btu	5	12	32	44	54
(kg $H_2O/10^6$ kJ)	(0.00002)	(0.00005)	(0.000014)	(0.000019)	(0.000023)
Fuel fired, lb/h	450,000	540,000	643,000	740,000	794,000
(kJ/h)	(202,500)	(243,000)	(289,350)	(333,000)	(357,300)

[°]*Combustion* magazine.

TABLE 4 Pulverizer Requirements—Nominal 600-MW Unit[*]

	Eastern bituminous	Midwestern bituminous	Subbituminous	Texas lignite	Northern plains lignite
Hardgrove grindability	55	56	43	48	35
No. required	6	6	6	6	7
Nominal capacity[†]	50 tons/h (50.8 t/h)	63 tons/h (64 t/h)	85 tons/h (86.4 t/h)	92 tons/h (93.5 t/h)	100 tons/h (101.6 t/h)
Primary air temperature for drying coal	525°F (274°C)	640°F (338°C)	725°F (385°C)	750°F (399°C)	750°F (399°C)

[*]*Combustion* magazine.
[†]Mill selection based on one full spare with remaining mills at 0.9 × new capacity.

portion of the water. Although the examples presented are for a fossil-fueled cycle, the methods are equally valid for a nuclear steam-turbine cycle. For a contrasting approach and for more data on where this procedure can be used, review the Related Calculations portion of the previous calculation procedure.

This calculation procedure is the work of Paul Leung of Bechtel Corporation, as reported at the 34th Annual Meeting of the American Power Conference and published in *Combustion* magazine. Since the procedure is based on thermodynamic and economic principles, it has wide applicability in a variety of industries.

With utility power plants—some 3500—reaching their 30th birthday within the next few years, designers are evaluating ways of repowering. When a plant is repowered, emissions are reduced, efficiency rises, as do reliability, output, and service life. So repowering has many attractions, including environmental benefits. More than 20 GW of capacity are estimated candidates for repowering.

Repowering replaces older facilities with new or different equipment. Several types of repowering are used today: (1) *Partial repowering*—which combines an existing plant system,

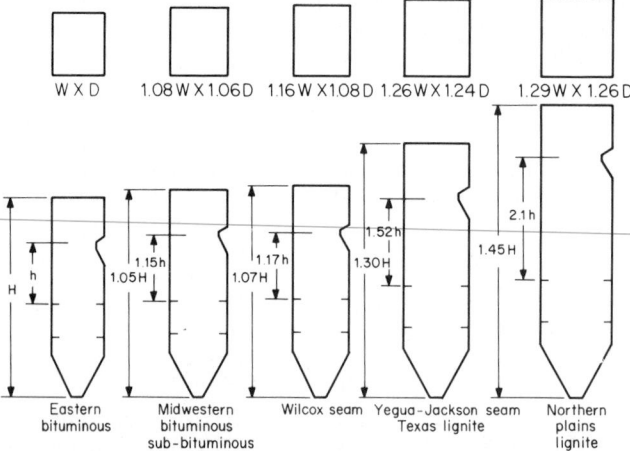

FIG. 19 Furnace sizes needed for various coals for efficient operation. (*Combustion.*)

infrastructure, and new equipment to provide increased output. *Example*: Combined-cycle repowering using a heat-recovery steam generator (HRSG) that recovers waste heat from the exhaust of a new gas-turbine/generator. *Example*: New gas-turbine/generator exhausts into existing boilers eliminating combustion-air-forced-draft needs while increasing the efficiency of the steam-generation cycle. Capital requirements are smaller than for an HRSG. This form of repowering is popular in Europe.

(2) *Station repowering*—reuses existing buildings, water-treatment systems, and fuel-handling system—but *not* the original steam cycle. New generating capability is installed to replace the existing steam plant—usually in the form of one or more gas turbines.

(3) *Site repowering*—uses an existing site but none of the equipment, such as boilers or turbines. Reusing an existing site eases permitting requirements, compared to developing a new site. To reduce overall project costs, it may be possible to reuse the infrastructure supporting the plant—such as power line and water- and fuel-delivery systems.

Specific methods for repowering include: (1) Combined-cycle repowering uses a new gas turbine and an HRSG to repower an existing facility by replacing or augmenting an existing boiler. (2) Gas turbines serving multiple-pressure HRSG provide power output and steam to existing steam-turbine generators. The gas-turbine power output goes directly to the utility's power lines. Natural gas fuels the gas turbine.

(3) Pressurized fluidized-bed boilers are installed in place of existing boilers. Hot gases from the new boiler are used to drive a gas-turbine generator to increase the overall plant output. (4) Hot windbox repowering (also called the turbocharged boiler) adds a gas turbine/generator to an existing plant. The high-temperature exhaust from the gas turbine is used as combustion air in the existing boiler. This eliminates—in most cases—the need for a forced-draft fan while the gas turbine is operating. Plants using this method of repowering, which is prevalent in Europe, boost efficiency by 10 to 15 percent and output by 20 to 33 percent.

Data presented here on repowering were reported by Steven Collins, assistant editor, in *Power* magazine.

BOILER FUEL CONVERSION FROM OIL OR GAS TO COAL

An industrial plant uses three 400,000-lb/h (50.4-kg/s) boilers fired by oil, a 600-MW generating unit, and two 400-MW units fired by oil. The high cost of oil, and the predictions that its cost will continue to rise in future years, led the plant owners to seek conversion of the boilers to coal firing. Outline the numerical and engineering design factors which must be considered in any such conversion.

Calculation Procedure:

1. *Evaluate the furnace size considerations*

The most important design consideration for a steam-generating unit is the fuel to be burned. Furnace size, fuel-burning and preparation equipment, heating-surface quantity and placement, heat-recovery equipment, and air-quality control devices are all fuel-dependent. Further, these items vary considerably among units, depending on the kind of fuel being used.

Figure 18 shows the difference in furnace size required between a coal-fired design boiler and an oil- or gas-fired design for the same steaming capacity in lb/h (kg/s). The major differences between coal firing and oil or natural-gas firing result from the solid form of coal prior to burning and the ash in the products of combustion. Oil produces only small amounts of ash; natural gas produces no ash. Coal must be stored, conveyed, and pulverized before being introduced into a furnace. Oil and gas require little preparation. For these reasons, a boiler designed to burn oil as its primary fuel makes a poor conversion candidate for coal firing.

2. *Evaluate the coal properties from various sources*

Table 3 shows coal properties from many parts of the United States. Note that the heating values range from 12,000 Btu/lb (27,960 kJ/kg) to 6800 Btu/lb (15,844 kJ/kg). For a 600-MW unit, the coal firing rates [450,000 to 794,000 lb/h (56.7 or 99.9 kg/s)] to yield comparable heat inputs provide an appreciation of the coal storage yard and handling requirements for the various coals. On an hourly usage ratio alone, the lower-heating-value coal requires 1.76 times more fuel to be handled.

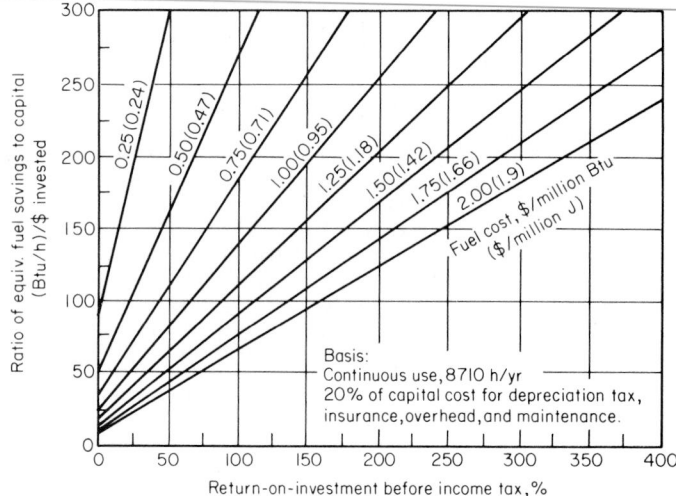

FIG. 20   ROI evaluation of energy-conservation projects. *(Chemical Engineering.)*

Pulverizer requirements are shown in Table 4 while furnace sizes needed for the various coals are shown in Fig. 19.

3.   *Evaluate conversion to coal fuel*

Most gas-fired boilers can readily be converted to oil at reasonable cost. Little or no derating (reduction of steam or electricity output) is normally required.

From an industrial or utility view, conversion of oil- or gas-fired boilers not initially designed to fire coal is totally impractical from an economic viewpoint. Further, the output of the boiler would be severely reduced.

For example, the overall plant site requirements for a typical station having a pair of 400-MW units designed to fire natural gas could be an area of 624,000 ft^2 (57,970 m^2). This area would be for turbine bays, steam generators, and cooling towers. (With a condenser, the area required would be less.)

To accommodate the same facilities for a coal-fired plant with two 400-MW units, the ground area required would be 20 times greater. The additional facilities required include coal storage yard, ash disposal area, gas-cleaning equipment (scrubbers and precipitators), railroad siding, etc.

A coal-fired furnace is nominally twice the size of a gas-fired furnace. For some units the coal-fired boiler requires 4 times the volume of a gas-fired unit. Severe deratings of 40 to 70 percent are usually required for oil- and/or gas-fired boilers not originally designed for coal firing when they are switched to coal fuel. Further, such boilers cannot be economically converted to coal unless they were originally designed to be.

As an example of the derating required, the 400,000-lb/h (50.4-kg/s) units considered here would have to be derated to 265,000 lb/h (33.4 kg/s) if converted to pulverized-coal firing. This is 66 percent of the original rating. If a spreader stoker were used to fire the boiler, the maximum capacity obtainable would be 200,000 lb/h (25.2 kg/s) of steam. Extensive physical alteration of the boiler would also be required. Thus, a spreader stoker would provide only 50 percent of the original steaming capacity.

Related Calculations: Conversion of boilers from oil and/or gas firing to coal firing requires substantial capital investment, lengthy outage of the unit while alterations are being made, and derating of the boiler to about half the designed capacity. For these reasons, most engineers do not believe that conversion of oil- and/or gas-fired boilers to coal firing is economically feasible.

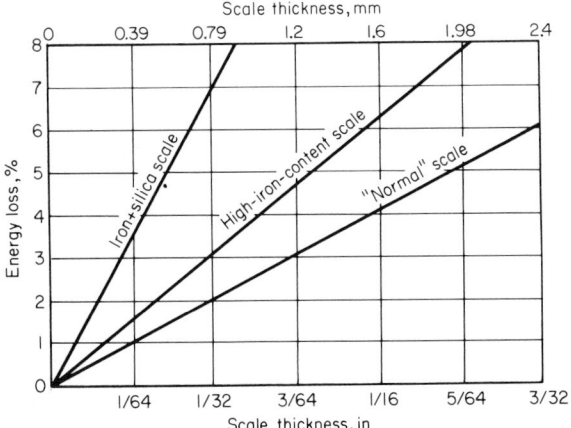

FIG. 21 Effects of scale on boiler operation. (*Chemical Engineering.*)

The types of boilers which are most readily convertible from oil or gas to coal are those which were originally designed to burn coal (termed *reconversion*). These are units which were mandated to convert to oil in the late 1960s because of environmental legislation.

Where the land originally used for coal storage was not sold or used for other purposes, the conversion problem is relatively minor. But if the land was sold or converted to other uses, there could be a difficult problem finding storage space for the coal.

Most of these units were designed to burn low-ash, low-moisture, high-heating-value, and high-ash-fusion coals. Fuels of this quality may no longer be available. Hence reconversion to coal firing may require significant downrating of the boiler.

Another important aspect of reconversion is the restoration of the coal storage, handling, and pulverizing equipment. This work will probably require considerable attention. Further, pulverizer capacity may not be sufficient, given the lower grade of fuel that would probably have to be burned.

This procedure is based on the work of C. L. Richards, Vice-President, Fossil Power Systems Engineering Research & Development, C-E Power Systems, Combustion Engineering, Inc., as reported in *Combustion* magazine.

To comply with environmental regulations, a number of coal-burning power plants have installed scrubbers ahead of the stack inlet to reduce sulfur dioxide emissions. Estimates show that some 22 tons/h of waste can be generated by scrubbers installed in the United States alone. This waste contains ash, limestone, and gypsum.

Research at Ohio State University is now directed at using scrubber waste to reclaim coal strip mines, fertilize farm soil by enriching it, and to create concretelike building materials.

When scrubber waste is used to treat soil from strip mines, the soil's acidity is reduced to a level where hardy grasses and alfalfa grow well. It is hoped that the barren sites of strip mines can be converted to useful fields using the scrubber waste.

Grasses grown on such reclaimed sites are safe for animals to eat. Water leached from treated sites meets Enviromental Protection Agency standards for agricultural use. Approval by EPA for use of scrubber waste at such sites is being sought.

Further experiments are being conducted on using the scrubber waste on acidic farmland. It will be used alone, or in combination with nutrient-rich sewer sludge. The third use for scrubber waste is as a sort of concrete for roads or the floors of feedlots.

Productive use of scrubber waste promises better control of the environment, reducing sulfur dioxide while recovering land that yielded the fuel that produced the CO_2.

RETURN ON INVESTMENT FOR ENERGY-SAVING PROJECTS

An industrial plant is considering an energy-saving installation of insulation which is projected to save 22.5 million Btu/h (65.9 MW) at a cost of $100,000. What is the return on investment (ROI) for this project if fuel is estimated to cost $1.25 per million Btu ($1.25 per 1,055,000 J), the plant is in continuous use, and 20 percent of the capital cost is for depreciation, tax, insurance, overhead, and maintenance?

Calculation Procedure:

1.   *Compute the savings-to-capital ratio*

The savings-to-capital ratio is S = (Btu/h saved)/(capital investment to achieve this saving). Or, for this installation, S = 22,500,000/$100,000 = 225 Btu/h per $1 invested (65.9 W per $1 invested).

2.   *Determine the return on investment*

Enter Fig. 20 at S = 225 Btu/h, and project horizontally to the right to the fuel-cost curve for $1.25 per 10^6 Btu. From the intersection project vertically downward to the ROI scale to read 225 percent.

Related Calculations: This procedure can be used for any type of energy-saving installation—be it new machinery, more effective insulation, alternative heat sources (such as solar, wood, coal, gas, etc.), or other investments which save energy. Further, the ROI concept is valid for industrial, commercial, and residential installations where profit is a prime factor in business decisions. Here is how the ROI is put to use in business decisions:

(1) The ROI for alternative energy-saving schemes is computed, as described above. (2) These values are compared, and the highest value is chosen for further study. (3) By using the firm's target ROI for new projects, the new energy-saving project ROI is compared with that of the target. (4) If the energy-saving project ROI exceeds tha target ROI, the project is attractive from an investment standpoint. But if the energy-saving ROI is less than the firm's target ROI, the investment is not attractive from a business standpoint. Note, however, that in times of acute fuel shortages, an energy-saving project with a low ROI might still be attractive if it saves fuel. Further, each firm's management chooses the target ROI to be used in evaluating new projects of various kinds.

ROI is one of the most popular measures or indices for judging the business attractiveness of a proposed investment. When expressed as a percentage, ROI is the annual rate of return on the original investment made for an energy-saving project, new product, new structure, etc.

This procedure is the work of Jack Robertson, Senior Technical Specialist, Dow Chemical U.S.A., as reported in *Chemical Engineering* magazine.

ENERGY SAVINGS FROM REDUCED BOILER SCALE

A boiler generates 16,700 lb/h (2.1 kg/s) at 100 percent rating with an efficiency of 75 percent. If 1/32 in (0.79 mm) of "normal" scale is allowed to form on the tubes, determine what savings can be made if 144,000-Btu/gal (40,133-MJ/m³) fuel oil costs $1 per gallon ($1 per 3.8 L) and the boiler uses 16.74 million Btu/h (4.9 MW) operating 8000 h/year.

Calculation Procedure:

1.   *Determine the annual energy usage*

Compute the annual energy usage from (million Btu/h) (hours of operation annually)/efficiency. For this boiler, annual energy usage = (16.74)(8000)/0.75 = 178,560 million Btu (188,380 kJ).

2. Find the energy loss caused by scale on the tubes

Enter Fig. 21 at the scale thickness, $\frac{1}{32}$ in (0.79 mm), and project vertically upward to the "normal" scale (salts of Ca and Mg) curve. At the left read the energy loss as 2 percent. Hence, the annual energy loss in heat units $= (178,560 \text{ million Btu/year})(0.02) = 3571$ million Btu/year (130.8 kW).

3. Compute the annual savings if the scale is removed

If the scale is removed, then the energy lost, computed in step 2, will be saved. Thus, the annual dollar savings after scale removal $=$ (heat loss in energy units) (fuel price, \$/gal)/(fuel heating value, Btu/gal). Or, savings $= (3751 \times 10^6)(\$1.00)/144,000 = \$26,049$.

Related Calculations: This approach can be used with any type of boiler—watertube, firetube, etc. The data are also applicable to tubed water heaters which are directly fired.

Note that when the scale is high in iron and silica that the energy loss is much greater. Thus, with scale of the same thickness [$\frac{1}{32}$ in (0.79 mm)], the energy loss for scale high in iron and silica is 7 percent, from Fig. 21. Then the annual loss $= 178,560(0.07) = 12,500$ million Btu/year (3.63 MW). Removing the scale and preventing its reformation will save, assuming the same heating value and cost for the fuel oil, $(12,500 \times 10^6 \text{ Btu/year}) (\$1.00)/144,000 = \$86,805$ per year.

While this calculation gives the energy savings from reduced boiler scale, the results also can be used to determine the amount that can be invested in a water-treatment system to prevent scale formation in a boiler, water heater, or other heat exchanger. Thus, the initial investment in treating equipment can at least equal the projected annual savings produced by the removal of scale.

This procedure is the work of Walter A. Hendrix and Guillermo H. Hoyos, Engineering Experiment Station, Georgia Institute of Technology, as reported in *Chemical Engineering* magazine.

GROUND AREA AND UNLOADING CAPACITY REQUIRED FOR COAL BURNING

An industrial plant is considering switching from oil to coal firing to reduce fuel costs. Determine the ground area required for 60 days' coal storage if the plant generates 100,000 lb/h (45,360 kg/h) of steam at a 60 percent winter load factor with a steam pressure of 150 lb/in^2 (gage) (1034 kPa), average boiler evaporation is 9.47 lb steam/lb coal (4.3 kg/kg), coal density $= 50$ lb/ft^3 (800 kg/m^3), boiler efficiency is 83 percent with an economizer, and the average storage pile height for the coal is 20 ft (6.096 m).

Calculation Procedure:

1. Determine the storage area required for the coal

The storage area, A ft^2, can be found from $A = 24WFN/EdH$, where $H =$ steam generation rate, lb/h; $F =$ load factor, expressed as a decimal; $N =$ number of days storage required; $E =$ average boiler evaporation rate, lb/h; $d =$ density of coal, lb/ft^3; $H =$ height of coal pile allowed, ft. Substituting yields $A = 24(100,000)(0.6)(60)/[(9.47)(50)(20)] = 9123$ ft^2 (847 m^2).

2. Find the maximum hourly burning rate of the boiler

The maximum hourly burning rate in tons per hour is given by $B = W/2000E$, where the symbols are as defined earlier. Substituting, we find $B = 100,000/2000(9.47) = 5.28$ tons/h (4.79 t/h). With 24-h use in any day, maximum daily use $= 24 \times 5.28 = 126.7$ tons/day (115 t/day).

3. Find the required unloading rate for this plant

As a general rule, the unloading rate should be about 9 times the maximum total plant burning rate. Higher labor and demurrage costs justify higher unloading rates and less manual supervision of coal handling. Find the unloading rate in tons per hour from $U = 9W/2000E$, where the symbols are as defined earlier. Substituting gives $U = 9(100,000)/2000(8.47) = 47.5$ tons/h (43.1 t/h).

Related Calculations: With the price of oil, gas, wood, and waste fuels rising to ever-higher levels, coal is being given serious consideration by industrial, central-station, commercial, and marine plants. Factors which must be included in any study of conversion to (or original use of) coal include coal delivery to the plant, storage before use, and delivery to the boiler.

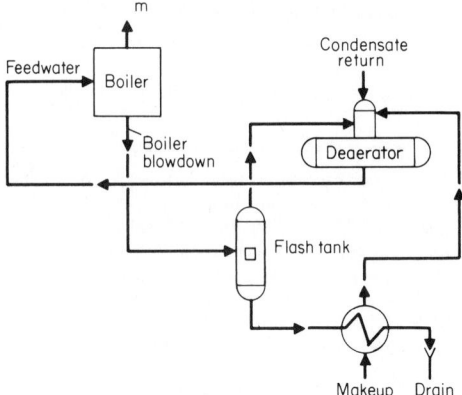

FIG. 22 Typical blowdown heat-recovery system. *(Combustion.)*

For land installations, coal is usually received in railroad hopper-bottom cars in net capacities ranging between 50 and 100 tons with 50- and 70-ton (45.4- and 63.5-t) capacity cars being most common.

Because cars require time for spotting and moving on the railroad siding, coal is actually delivered to storage for only a portion of the unloading time. Thawing of frozen coal and car shaking also tend to reduce the actual delivery. True unloading rate may be as low as 50 percent of the continuous-flow capacity of the handling system. Hence, the design coal-handling rate of the conveyor system serving the unloading station should be twice the desired unloading rate. So, for the installation considered in this procedure, the conveyor system should be designed to handle 2(47.5) = 95 tons/h (86.2 t/h). This will ensure that at least six rail cars of 60-ton (54.4-t) average capacity will be emptied in an 8-h shift, or about 360 tons/day (326.7 t/day).

With a maximum daily usage of 126.7 tons/day (115 t/day), as computed in step 2 above, the normal handling of coal, from rail car delivery during the day shift, will accumulate about 3 days' peak use during an 8-h shift. If larger than normal shipments arrive, the conveyor system can be operated more than 8 h/day to reduce demurrage charges.

This procedure is the work of E. R. Harris, Department Head, G. F. Connell, and F. Dengiz, all of the Environmental and Energy Systems, Argonaut Realty Division, General Motors Corporation, as reported in *Combustion* magazine.

HEAT RECOVERY FROM BOILER BLOWDOWN SYSTEMS

Determine the heat lost per day from sewering the blowdown from a 600-lb/in^2 (gage) (4137-kPa) boiler generating 1 million lb/day (18,939.4 kg/h) of steam at 80 percent efficiency. Compare this loss to the saving from heat recovery if the feedwater has 20 cycles of concentration (that is, 5 percent blowdown), ambient makeup water temperature is 70°F (21°C), flash tank operating pressure is 10 lb/in^2 (gage) (69 kPa) with 28 percent of the blowdown flashed, blowdown heat exchanger effluent temperature is 120°F (49°C), fuel cost is \$2 per 10^6 Btu [\$2 per (9.5)^6 J], and the piping is arranged as shown in Fig. 22.

Calculation Procedure:

1. *Compute the feedwater flow rate*

The feedwater flow rate, 10^6 lb/day = (steam generated, 10^6 lb/day)/(100 − blowdown percentage), or 10^6/(100 − 5) = 1.053 × 10^6 lb/day (0.48 × 10^6 kg/day).

2. *Find the steam-production equivalent of the blowdown flow*

The steam-production equivalent of the blowdown = feedwater flow rate − steam flow rate = 1.053 − 1.0 = 53,000 lb/day (24,090 kg/day).

3.   Compute the heat loss per lb of blowdown

The heat loss per lb (kg) of blowdown = saturation temperature of boiler water − ambient temperature of makeup water. Or, heat loss = (488 − 70) = 418 Btu/lb (973.9 kJ/kg).

4.   Find the total heat loss from sewering

When the blowdown is piped to a sewer (termed *sewering*), the heat in the blowdown stream is lost forever. With today's high cost of all fuels, the impact on plant economics can be significant. Thus, total heat loss from sewering = (heat loss per lb of blowdown) (blowdown flow rate, lb/ day) = 418 Btu/lb (53,000 lb/day) = 22.2 × 10⁶ Btu/day (23.4 × 10⁶ J/day).

5.   Determine the fuel-cost equivalent of the blowdown

The fuel-cost equivalent of the blowdown = (heat loss per day, 10⁶ Btu)(fuel cost, $ per 10⁶ Btu)/ (boiler efficiency, %), or (22.2)(2)/0.8 = $55.50 per day.

6.   Find the blowdown flow to the heat exchanger

With 28 percent of the blowdown flashed to steam, this means that 100 − 28 = 72 percent of the blowdown is available for use in the heat exchanger. Since the blowdown total flow rate is 53,000 lb/day (24,090 kg/day), the flow rate to the blowdown heat exchanger will be 0.72(53,000) = 38,160 lb/day (17,345 kg/day).

7.   Determine the daily heat loss to the sewer

As Fig. 22 shows, the blowdown water which is not flashed, flows through the heat exchanger to heat the incoming makeup water and then is discharged to the sewer. It is the heat in this sewer discharge which is to be computed here.

   With a heat-exchanger effluent temperature of 120°F (49°C) and a makeup water temperature of 70°F (21°C), the heat loss to the sewer is 120 − 70 = 50 Btu/lb (116.5 kJ/kg). And since the flow rate to the sewer is 38,160 lb/day (17,345 kg/day), the total heat loss to the sewer is 50(38,160) = 1.91 × 10⁶ Btu/day (2.02 × 10⁶ kJ/day).

8.   Compare the two systems in terms of heat recovered

The heat recovered − heat loss by sewering = heat loss with recovery = 22.2 × 10⁶ Btu/day − 1.91 × 10⁶ Btu/day = 20.3 × 10⁶ Btu/day (21.4 × 10⁶ J/day).

9.   Determine the percentage of the blowdown heat recovered and dollar savings

The percentage of heat recovered = (heat recovered, Btu/day)/(original loss, Btu/day) = (20.3/ 22.2)(100) = 91 percent. Since the cost of the lost heat was $55.50 per day without any heat recovery, the dollar savings will be 91 percent of this, or 0.91($55.50) = $50.51 per day, or $18,434.33 per year with 365 days of operation. And as fuel costs rise, which they are almost certain to do in future years, the annual saving will increase. Of course, the cost of the blowdown heat-recovery equipment must be offset against this saving. In general, the savings warrant the added investment for the extra equipment.

   Related Calculations: This procedure is valid for any type of steam-generating equipment for residential, commercial, industrial, central-station, or marine installations. (In the latter installation the "sewer" is the sea.) The typical range of blowdown heat recovery is in the 80 to 90 percent area. In view of the rapid rise in fuel prices, this range of heat recovery is significant. Hence, much wider use of blowdown heat recovery can be expected in all types of steam-generating plants.

TABLE 5   Heat-Transfer Coefficients for Air-Cooled Heat Exchangers°

Liquid cooled	Heat-transfer coefficient	
	Btu/(h·ft²·°F)	W/(m²·K)
Diesel oil	45–55	255.5–312.3
Kerosene	55–60	312.3–340.7
Heavy naphtha	60–65	340.7–369.1

° *Chemical Engineering.*

To reduce scale buildup in boilers, low cycles of boiler water concentration are preferred. This means that high blowdown rates will be used. To prevent wasting expensive heat present in the blowdown, heat-recovery equipment such as that discussed above is used. In industrial plants (which are subject to many sources of condensate contamination), cycles of concentration are seldom allowed to exceed 50 (2 percent blowdown). In the above application, the cycles of concentration = 20, or 5 percent blowdown.

To prevent boiler scale buildup, good pretreatment of the makeup is recommended. Typical current selections for pretreatment equipment, by using the boiler operating pressure as the main criterion, are thus:

Boiler pressure, lb/in^2 (kPa)	Pretreatment equipment
0–600 (4137)	Sodium zeolite softening
600–900 (4137–6205)	Hot line/demineralizers
Above 900 (6205)	Demineralizers

This procedure is the work of A. A. Askew, Betz Laboratories, Inc., as reported in *Combustion* magazine.

BOILER BLOWDOWN PERCENTAGE

The allowable concentration in a certain drum is 2000 ppm. Pure condensate is fed to the drum at the rate of 85,000 gal/h (89.4 L/s). Make-up, containing 50 grains (gr)/gal (856 mg/L) of sludge-producing impurities, is also delivered to the drum at the rate of 1500 gal/h (1.58 L/s). Calculate the blowdown as a percentage of the boiler steaming capacity.

Calculation Procedure:

1. *Compute the ppm of impurities per gallon of make-up water*

There are 58,410 gr/gal (10^6 mg/L). See the Related Calculations of this procedure for the basis of this factor. Parts per million of impurities, $ppm_i = [(gr/gal)_i/(gr/gal)_i/(gr/gal)] \times 10^6$, where $(gr/gal)_i$ = quality of, or impurities in, the make-up water, gr/gal (mg/L); (gr/gal) = grains/gal (mg/L). Hence, $ppm_i = (50/58,410) \times 10^6 = 50 \times 17.12 = 856$.

2. *Compute the blowdown rate*

To maintain impurities in the drum at a certain concentration, the parts fed to the drum = parts discharged by the blowdown. Thus, $ppm_i = (gal/h)_m = ppm_a \times (gal/h)_a$, where the subscripts stand for i = impurities; m = make-up; a = allowable; b = blowdown. Then, $856 \times 1500 = 2000 \times (gal/h)_b$. Solving: $(gal/h)_b = 856 \times 1500/2000 = 642$ (0.675 L/s).

3. *Compute the boiler steaming capacity*

The boiler steaming capacity $(gal/h)_s = (gal/h)_f + (gal/h)_m - (gal/h)_b$, where $(gal/h)_f$ = feedwater flow rate. Then, $(gal/h)_s = 85,000 + 1500 - 642 = 85,858$ (90.3 L/s).

4. *Compute the blowdown percentage*

Blowdown percentage = $[(gal/h)_b/(gal/h)_s] \times 100 = (642/85,858) \times 100 = 0.747$ percent.

Related Calculations: The gr/gal factor in step 1 is based on the density of impurities being considered as equal to the maximum density of clean fresh water, 8.3443 lb/gal (1.0 kg/L). Since 1 lb = 7000 gr, then $8.3443 \times 7000 = 58,410$ gr/gal (10^6 mg/L).

AIR-COOLED HEAT EXCHANGER: PRELIMINARY SELECTION

Kerosene flowing at a rate of 250,000 lb/h (31.5 kg/s) is to be cooled from 160°F (71°C) to 125°F (51.6°C), for a total heat duty of 4.55 million Btu/h (1.33 MW). How large an air cooler (sometimes called a *dry* heat exchanger) is needed for this service if the design dry-bulb temperature of the air is 95°F (35°C)?

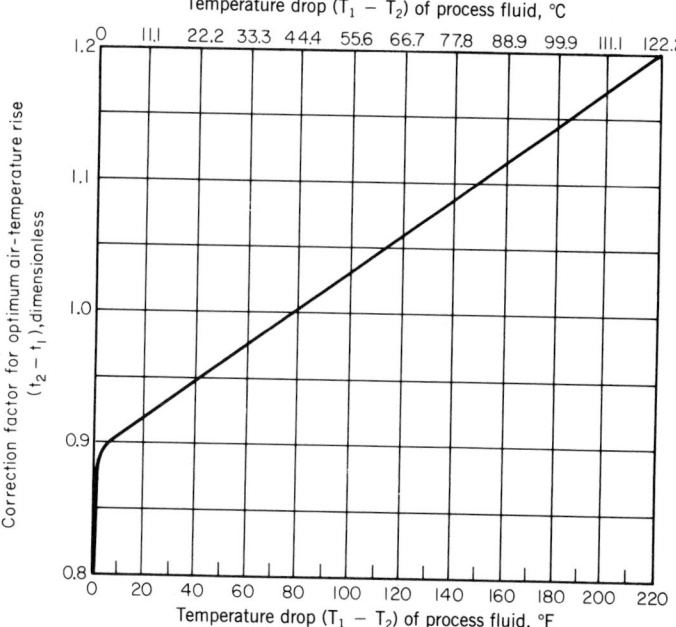

FIG. 23 Correction factors for estimated temperature rise. *(Chemical Engineering.)*

Calculation Procedure:

1. Determine the temperature rise of the air during passage through the cooler

From Table 5 estimate the overall heat-transfer coefficient for an air cooler handling kerosene at 55 Btu/(h·ft²·°F) [312.3 W/(m²·K)]. Then the air-temperature rise is $t_2 - t_1 = 0.005U\{[(T_1 + T_2)/2] - t_1\}$, where t_1 = inlet air temperature, °F or °C; t_2 = outlet air temperature, °F or °C; U = overall heat-transfer coefficient, Btu/(h·ft²·°F) [W/(m²·K)]. T_1 = cooled fluid inlet temperature, °F or °C; T_2 = cooled fluid outlet temperature, °F or °C. Substituting yields $t_2 - t_1 = 0.005(55)\{[(160 + 125)/2] - 95\} = 13.06$°F (7.2°C).

Next, from Fig. 23, the correction factor for a process-fluid temperature rise of $160 - 125 = 35$°F (19.4°C) is 0.94. So the corrected temperature rise $= f(t_2 - t_1) = 0.94(13.06) = 12.28$°F (6.8°C). Therefore, $t_2 = 95 + 12.28 = 107.28$°F (41.8°C).

2. Find the log mean temperature difference (LMTD) for the heat exchanger

Use the relation LMTD $= (\Delta t_2 - \Delta t_1)/\ln (\Delta t_2/\Delta t_1)$. Or, LMTD $= [(160 - 107.28) - (125 - 95)]/\ln [(160 - 107.28)/(125 - 95)] = 40.30$. This value of the LMTD must be corrected by using Fig. 24 for temperature efficiency P and a correlating factor R. Thus, $P = (t_2 - t_1)/(T_1 - t_1) = (107.28 - 95)/(160 - 95) = 0.189$. Also, $R = (T_1 - T_2)/(t_2 - t_1) = (160 - 125)/(107.28 - 95) = 2.85$. Then, from Fig. 24, LMTD correction factor = 0.95, and the corrected LMTD $= f(\text{LMTD}) = 0.95(40.30) = 38.29$°F (21.2°C).

3. Determine the hypothetical bare-tube area needed for the exchanger

Use the relation $A = Q/U\Delta T$, where A = hypothetical bare-tube area required, ft² (m²); Q = heat transferred, Btu/h (W); ΔT = effective temperature difference across the exchanger = corrected LMTD. Substituting gives $A = 4,550,000/55(38.29) = 2160$ ft² (200.7 m²).

4. Choose the cooler size and number of fans

Enter Table 6 with the required bare-tube area, and choose a 12-ft (3.6-m) wide cooler with either four rows of 40-ft (12-m) long tubes with two fans, for a total bare surface of 2284 ft² (205.6 m²),

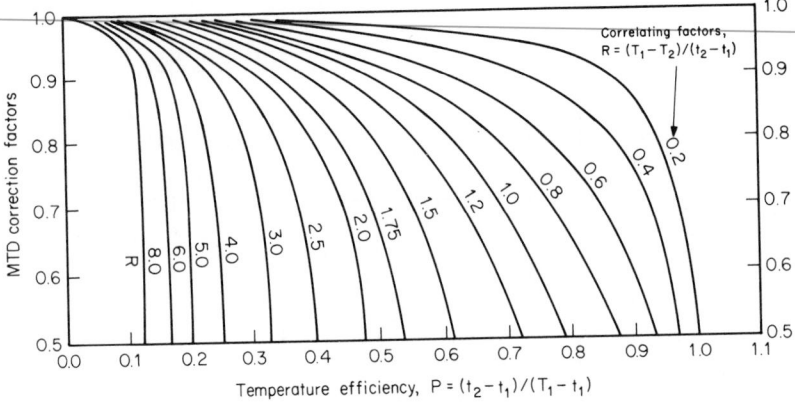

FIG. 24   MTD correction factors for one-pass crossflow with both shell side and tube side unmixed. *T* represents hot-fluid characteristics, and *t* represents cold-fluid characteristics. Subscripts 1 and 2 represent inlet and outlet, respectively. *(Chemical Engineering.)*

or five rows of 32-ft (9.6-m) long with two fans for 2288 ft² (205.5 m²) of surface. From Fig. 25, the fan horsepower for the cooler would be 1.56(2284/100) = 35.63 hp (25.6 kW).

   Related Calculations: Air coolers are widely used in industrial, commercial, and some residential applications because the fluid cooled is not exposed to the atmosphere, air is almost always available for cooling, and energy is saved because there is no evaporation loss of the fluid being cooled.

   Typical uses in these applications include process-fluid cooling, engine jacket-water cooling, air-conditioning condenser-water cooling, vapor cooling, etc. Today there are about seven leading design manufacturers of air coolers in the United States.

   The procedure given here depends on three key assumptions: (1) an overall heat-transfer coefficient is assumed, depending on the fluid cooled and its temperature range; (2) the air temperature rise $t_2 - t_1$ is calculated by an empirical formula; (3) bare tubes are assumed and fan horsepower (kW) is estimated on this basis to avoid the peculiarities of one fin type. By using the empirical formula given in step 1 of this procedure, the size air cooler obtained will be within 25 percent of optimum. This is adjusted for greater accuracy through use of the correction factor shown in Fig. 23.

   Since no existing computer program is capable of considering all variables in optimizing air coolers, the procedure given here is useful as a first trial in calculating an optimum design. The flow pattern and correction factors used for this estimating procedure are those for one-pass crossflow with both tube fluid and air unmixed as they flow through the exchanger.

TABLE 6   Typical Air-Cooled Heat-Exchanger Cooling Area, ft² (m²)°

Approximate cooler width		Tube length		Fans per unit	No. of 1-in (2.5-cm) tube rows in depth on 2⅜-in (6-cm) pitch	
ft	m	ft	m		4	5
12	3.66	32	9.8	2	1827 (169.7)	2288 (212.6)
		36	10.9	2	2056 (191.0)	2574 (239.1)
		40	12.2	2	2284 (212.2)	2861 (265.8)
14	4.27	14	4.3	1	931 (86.5)	1166 (108.3)
		16	4.9	1	1064 (98.8)	1333 (123.8)

°*Chemical Engineering.*

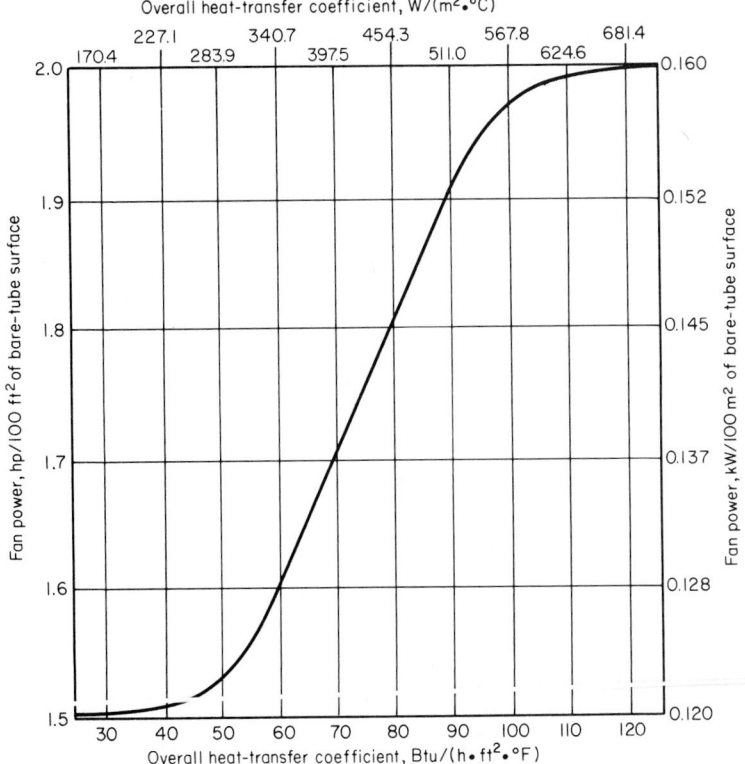

Overall heat-transfer coefficient, W/(m²•°C)

FIG. 25 Approximate fan power requirements for air coolers. *(Chemical Engineering.)*

Where additional correction factors are needed for different flow patterns across the exchanger, the designer should consult the standards of the Tubular Exchanger Manufacturers Association (TEMA). Similar data will be found in reference books on heat exchange.

The procedure given here is the work of Robert Brown, General Manager, Happy Division, Therma Technology, Inc., as reported in *Chemical Engineering* magazine. Note that the procedure given is for a preliminary selection. The final selection will usually be made in conjunction with advice and guidance from the manufacturer of the air cooler.

FUEL SAVINGS PRODUCED BY DIRECT DIGITAL CONTROL OF THE POWER-GENERATION PROCESS

A 200-MW steam-turbine generating unit supplied steam at 1000°F (538°C) and 2400 lb/in² (gage) (16,548 kPa) has an existing variability of 20°F (6.7°C) in the steam-temperature control. It is desired to reduce the variability of the steam temperature and thus allow a closer approach to the turbine design warrantee limits of 1050°F (566°C). A digital-control system will allow a 30°F (16.7°C) higher operating temperature at the turbine throttle. What will be the effect of this more precise temperature control on the efficiency and fuel cost of this unit? Fuel cost is $2.50 per 10^6 Btu ($2.38 per 10^6 J), and the plant heat rate is 9061 Btu/kWh (9559 kJ/kWh), with a turbine backpressure of 1 inHg (2.5 cmHg).

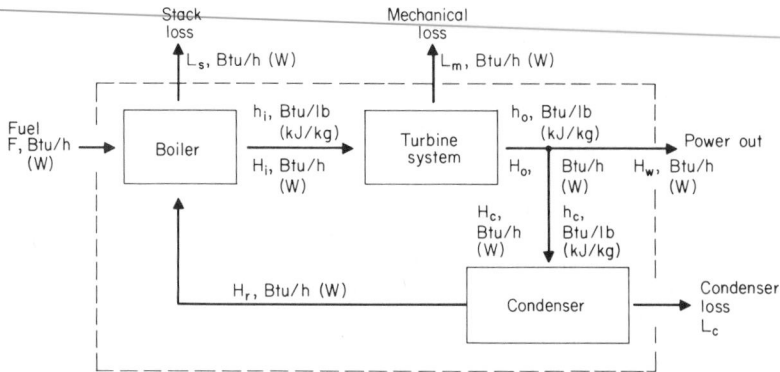

FIG. 26   Flow diagram showing input, output, and losses. *(Combustion.)*

Calculation Procedure:

1.   *Sketch the unit flow diagram; write the overall efficiency equation*

Figure 26 shows the flow diagram for this unit with the input, output, and losses indicated. The overall efficiency e of the system is determined by dividing the power output H_w, Btu/h (W), by the fuel input F, Btu/h (W).

2.   *Express the efficiency equation with the system losses shown*

The losses in a typical steam-turbine generating unit are the stack loss L_s, Btu/h (W); mechanical loss in turbine L_m, Btu/h (W); condenser loss L_c, Btu/h (W). The power output can now be expressed as $H_w = F - L_s - L_m - L_c$. Hence, $e = (F - L_s - L_m - L_c)/F$.

3.   *Write the loss relations for the unit*

The boiler efficiency and turbine efficiency can each be assumed to be about 90 percent. This is a safe assumption for such an installation. Then $L_s = 0.1F$; $L_m = 0.1H_c$; $H_i = 0.9F$; $H_o = 0.9H_i$, by using the symbols shown in Fig. 26.

4.   *Write the plant efficiency equation at the higher temperature*

With the temperature at the turbine inlet increased by 30°F (16.7°C), the condenser inlet enthalpy h_i' will change to 1480.9 Btu/lb (3444.6 kJ/kg), based on steam-table values. Setting up a ratio between the condenser inlet enthalpy after the throttle-temperature increase and before yields $1480.9/1023 = 1.0135h_i$. This ratio can be used for the other values, if we allow the prime symbol to indicate the values at the higher temperatures. Or, $H_i' = 1.0135H_i$; $L_m' = 1.0135H_i$; $L_c' = 1.0058L_c$. Then $e' = (F' - L_s' - L_m' - L_c')/F'$.

5.   *Compute the efficiency and heat-rate improvement*

The improvement in plant efficiency $\Delta e = (e' - e)/e = e'/e - 1$. Substituting values, we find $\Delta e = [0.8209 - (L_c/F)(1.0058)]/\{1.0135[0.81 - (L_c/F)] - 1\}$.

With a heat rate of 9061 Btu/kWh (9559 kJ/kWh), the overall efficiency $e = (3412 \text{ Btu/kWh}/9061)100 = 37.65$ percent. Substituting this value of e in the general efficiency relation given in step 2 above shows that $L_c/F = 0.4335$. Then, substituting this value in the Δe equation above, we find $\Delta e = 0.3849/0.3816 - 1 = 0.0086$, or 0.86 percent.

6.   *Convert the efficiency improvement into annual fuel-cost savings*

The annual fuel cost C can be computed from $C = 3.412(\text{fuel cost, \$ per } 10^6 \text{ Btu}) (\text{hours of operation per year}) (\text{plant MW capacity})/e$. Or, $C = 3.412 (2.5) (8760)(200)/0.3765 = \$39,693,386$ per year. Then, with an efficiency improvement of 0.86 percent, the annual fuel-cost saving $S = 0.0086(\$39,693,386) = \$341,362$. In 10 years, with no increase in fuel costs (a highly unlikely condition), the fuel-cost savings with more precise steam-temperature control would be nearly \$3.5 million.

TABLE 7 Basic Reasons for Using Automatic Control°

1. Increase in quantity or number of products (generation for fixed investment)
2. Improved product quality
3. Improved product uniformity (steam-temperature variability)
4. Savings in energy (improved efficiency or heat rate)
5. Raw-material savings (fixed savings)
6. Savings in plant equipment (more capacity from fixed investment)
7. Decrease in human drudgery (increased operator effectiveness)

° *Combusion* magazine.

TABLE 8 Functions of Direct Digital Controls°

1. Feedwater
2. Air flow and furnace draft
3. Fuel flow
4. RH and SH steam temperature
5. Primary air pressure and temperature
6. Minor loop control
   a. Cold-end metal temperature
   b. Cold-end metal temperature
   c. Turbine lube oil temperature
   d. Generator stator coolant to secondary coolant pressure differential
   c. Hydrogen temperature controls

° *Combustion* magazine.

TABLE 9 Evolution of Boiler Controls°

1. 1905–1920 Hand control with regulator assistance
2. 1920–1940 Analog boiler control systems acceptance
3. 1940–1950 Pneumatic direct connected analog systems
4. 1950–1960 Pneumatic transmitted analog systems
5. 1960–1970 Discrete component solid-state electric analog systems, burner control, and digital computers
6. 1970–1980 Integrated-circuit digital and analog systems

° *Combustion* magazine.

Related Calculations: Although this procedure is based on the use of digital-control systems, it is equally applicable for all forms of advanced control systems which can improve operator effectiveness and thereby save money in operating costs. Table 7 shows the basic reasons for using automatic control in both central-station and industrial power plants. With the ever-increasing energy costs forecast for the future, automatic control of power equipment will become of greater importance.

Table 8 shows the functions of direct digital controls (DDCs) for a variety of power-plant types. Performance improvements of up to 5:1 or more have been reported with DDC. With the expected life of today's plants at 40 years, the annual savings produced by DDC can have a significant impact on life-cycle costs. Table 9 shows the evolution of boiler controls—six phases of evolution. Central-station and large industrial plants today are in a distributed digital "revolution." Changes will occur in measurement, control, information, systems, and actuators. Clear benefits will be measured in terms of installed cost, ease of startup, reliability, control performance, and flexibility. This procedure is based on the work of M. A. Keys, Vice-President, Engineering, M. P. Lukas, Manager, Application Engineering, Bailey Controls Company, as reported in *Combustion* magazine.

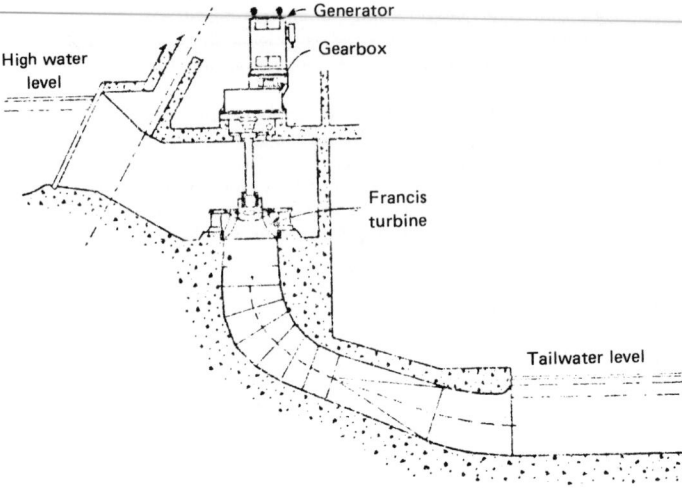

FIG. 27   Vertical Francis turbine in open pit was adapted to 8-m head in an existing Norwegian dam. *(Power.)*

SMALL HYDRO POWER CONSIDERATIONS AND ANALYSIS

A city is considering a small hydro power installation to save fossil fuel. To obtain the savings, the following steps will be taken: refurbish an existing dam, install new turbines, operate the generating plant. Outline the considerations a designer must weigh before undertaking the actual construction of such a plant.

Calculation Procedure:

1. *Analyze the available head*

Most small hydro power sites today will have a head of less than 50 ft (15.2 m) between the high-water level and tail-water level, Fig. 27. The power-generating capacity will usually be 25 MW or less.

2. *Relate absolute head to water flow rate*

Because heads across the turbine in small hydro installations are often low in magnitude, the tail-water level is important in assessing the possibilities of a given site. At high-water flows, tail-water levels are often high enough to reduce turbine output, Fig. 28a. At some sites, the available head at high flow is extremely low, Fig. 28b.

The actual power output from a hydro station is $P = HQwe/550$, where P = horsepower output; H = head across turbine, ft; Q = water flow rate, ft^3/s; w = weight of water, lb/ft^3; e = turbine efficiency. Substituting in this equation for the plant shown in Fig. 28b, for flow rates of 500 and 1500 m^3/s, we see that a tripling of the water flow rate increases the power output by only 38.7 percent, while the absolute head drops 53.8 percent (from 3.9 to 1.8 m). This is why the tail-water level is so important in small hydro installations.

Figure 28c shows how station costs can rise as head decreases. These costs were estimated by the Department of Energy (DOE) for a number of small hydro power installations. Figure 28d shows that station cost is more sensitive to head than to power capacity, according to DOE estimates. And the prohibitive costs for developing a completely new small hydro site mean that nearly all work will be at existing dams. Hence, any water exploitation for power must not encroach seriously on present customs, rights, and usages of the water. This holds for both upstream and downstream conditions.

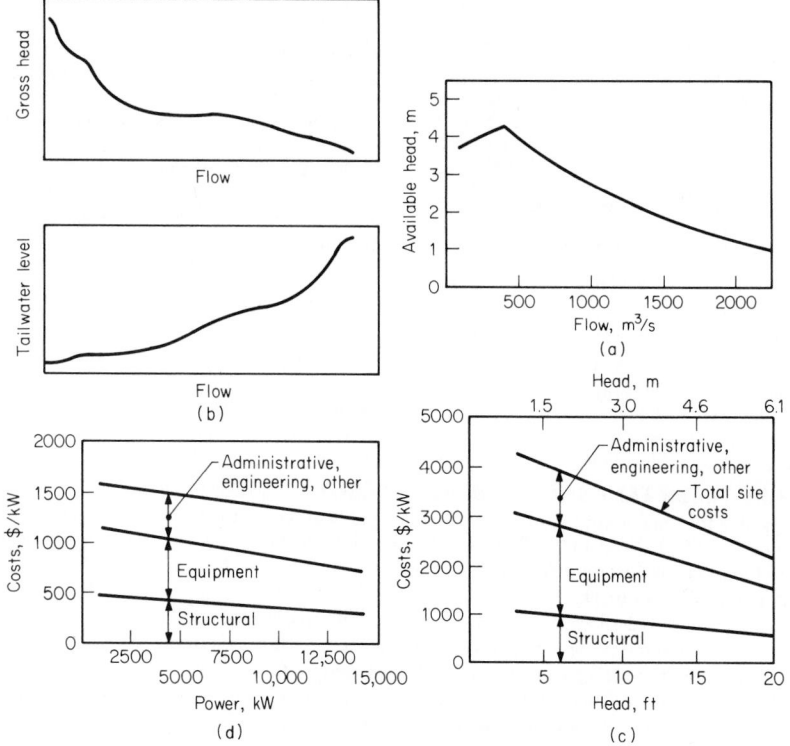

FIG. 28 (*a*) Rising tail-water level in small hydro projects can seriously curtail potential. (*b*) Anderson-Cottonwood dam head dwindles after a peak at low flow. (*c*) Low heads drive DOE estimates up. (*d*) Linear regression curves represent DOE estimates of costs of small sites. (*Power.*)

3. *Outline machinery choice considerations*

Small-turbine manufacturers, heeding the new needs, are producing a good range of semistandard designs that will match any site needs in regard to head, capacity, and excavation restrictions.

The Francis turbine, Fig. 27, is a good example of such designs. A horizontal-shaft Francis turbine may be a better choice for some small projects because of lower civil-engineering costs and compatibility with standard generators.

Efficiency of small turbines is a big factor in station design. The problem of full-load versus part-load efficiency, Fig. 29, must be considered. If several turbines can fit the site needs, then good part-load efficiency is possible by load sharing.

Fitting new machinery to an existing site requires ingenuity. If enough of the old powerhouse is left, the same setup for number and type of turbines might be used. In other installations the powerhouse may be absent, badly deteriorated, or totally unsuitable. Then river-flow studies should be made to determine which of the new semistandard machines will best fit the conditions.

Personnel costs are extremely important in small hydro projects. Probably very few small hydro projects centered on redevelopment of old sites can carry the burden of workers in constant attendance. Hence, personnel costs should be given close attention.

Tube and bulb turbines, with horizontal or nearly horizontal shafts, are one way to solve the problem of fitting turbines into a site without heavy excavation or civil engineering works. Several standard and semistandard models are available.

In low head work, the turbine is usually low-speed, far below the speed of small generators.

A speed-increasing gear box is therefore required. A simple helical-gear unit is satisfactory for vertical-shaft and horizontal-shaft turbines. Where a vertical turbine drives a horizontal generator, a right-angle box makes the turn in the power flow.

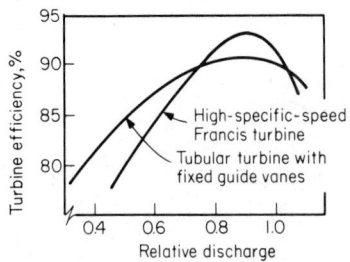

FIG. 29 Steep Francis-turbine efficiency falloff frequently makes multiple units advisable. *(Power.)*

Governing and control equipment is not a serious problem for small hydro plants.

Related Calculations: Most small hydro projects are justified on the basis of continuing inflation which will make the savings they produce more valuable as time passes. Although this practice is questioned by some people, the recent history of inflation seems to justify the approach.

As fossil-fuel prices increase, small hydro installations will become more feasible. However, the considerations mentioned in this procedure should be given full weight before proceeding with the final design of any plant. The data in this procedure were drawn from an ASME meeting on the subject with information from papers, panels, and discussion summarized by William O'Keefe, Senior Editor, *Power* magazine, in an article in that publication.

SIZING FLASH TANKS TO CONSERVE ENERGY

Determine the dimensions required for a commercial flash tank if the flash tank pressure is 5 lb/in^2 (gage) (34.5 kPa) and 14,060 lb/h (1.77 kg/s) of flash steam is available. Would the flash tank be of the centrifugal or top-inlet type?

Calculation Procedure:

Two major types of flash tanks are in use today: top-inlet and centrifugal-inlet tanks, as shown in Fig. 30. Tank and overall height and outside diameter are also shown in Fig. 30.

1. Determine the rating and type of flash tank required

Refer to Table 10. Locate the 5-lb/in^2 (gage) (34.5-kPa) flash tank pressure column, and project downward to the minimum value that exceeds 14,060 lb/h (1.77 kg/s). Note that a no. 5 centrifugal flash tank with a maximum rating of 20,000 lb/h (2.5 kg/s) of flash steam is appropriate, and no standard top-inlet type has sufficient capacity at this pressure for this flow rate.

2. Determine the dimensions of the tank

In Table 10 locate tank no. 5, and read the dimensions horizontally to the right. Hence, the dimensions required for the tank are 60-in (152.4-cm) OD, 78-in (198.1-cm) tank height, 88-in (223.5-cm) overall height, inlet pipe size of 6 in (15.2 cm), steam outlet pipe of 8 in (20.3 cm), and a water outlet pipe of 6 in (15.2 cm).

Related Calculations: Use this procedure for choosing a flash tank for a variety of applications—industrial power plants, central stations, marine steam plants, and nuclear stations. Flash tanks can conserve energy by recovering steam that might otherwise be wasted. This steam can be used for space heating, feedwater heating, industrial processes, etc. Condensate remaining after the flashing can be used as boiler feedwater because it is usually pure and contains valuable heat. Or the condensate may be used in an industrial process requiring pure water at an elevated temperature.

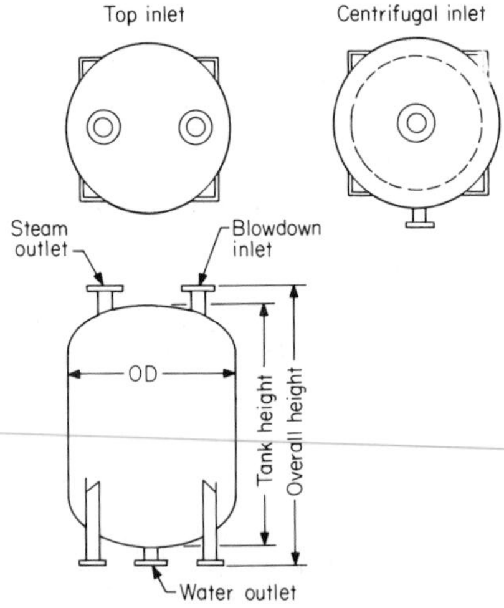

FIG. 30 Centrifugal and top-inlet flash-tank dimensions. *(Chemical Engineering.)*

TABLE 10 Maximum Ratings for Centrifugal and Top-Inlet Flash Tanks, 1000 lb/h (1000 kg/s)°

Tank no.	Flash-tank pressure, lb/in² (gage) (kPa)					
	1(6.9)	5 (34.5)	10 (69.0)	20 (138.0)	50 (345.0)	100 (690.0)
Centrifugal flash tanks						
4	6.0 (0.76)	7.1 (0.89)	8.8 (1.11)	12.0 (1.51)	21.0 (2.64)	34.0 (4.28)
5	16.0 (2.01)	20.0 (2.52)	24.0 (3.02)	32.0 (4.03)	58.0 (7.30)	100.0 (12.59)
6	27.0 (3.40)	34.0 (4.28)	42.0 (5.29)	58.0 (7.30)	105.0 (13.22)	180.0 (22.66)
Top-inlet flash tanks						
2	1.1 (0.14)	1.3 (0.16)	1.7 (0.21)	2.2 (0.28)	4.0 (0.50)	6.90 (0.87)
3	2.2 (0.28)	2.9 (0.37)	3.5 (0.44)	4.9 (0.62)	8.7 (1.10)	14.80 (1.86)
4	4.3 (9.54)	5.2 (0.65)	6.5 (0.82)	8.7 (1.10)	15.0 (1.89)	25.0 (3.15)

Dimensions of commercial flash tanks

Tank no.	Outside diameter		Tank height		Overall height		Inlet pipe		Outlet pipe			
									Steam		Water	
	in	cm	in	cm	in	cm	in	cm	in	cm	in	cm
Centrifugal flash tanks												
4	48	121.9	67	170.2	77	195.6	4	10.2	6	15.2	4	10.2
5	60	152.4	78	198.1	88	223.5	6	15.2	8	20.3	6	15.2
6	72	182.9	89	226.1	99	251.5	8	20.3	10	25.4	6	15.2
Top-inlet flash tanks												
2	24	60.9	56	142.2	65.5	166.4	3	7.6	3	7.6	1.5	3.8
3	36	91.4	62	157.5	71.5	181.6	4	10.2	4	10.2	2	5.1
4	48	121.9	67	170.2	76.5	194.3	6	15.2	6	15.2	4	10.2

°*Chemical Engineering magazine.*

Flashing steam can cause a violent eruption of the liquid from which the steam is formed. Hence, any flash tank must be large enough to act as a separator to remove entrained moisture from the steam. The dimensions given in Table 10 are for flash tanks of proven design. Hence, the values obtained from Table 10 are satisfactory for all normal design activities. The procedure given here is the work of T. R. MacMillan, as reported in *Chemical Engineering*.

FLASH TANK OUTPUT

A boiler operating with a drum pressure of 1400 lb/in² absolute (9650 kPa) delivers 200,000 lb (90,720 kg) of steam per hour and has a continuous blowdown of 2 percent of its output in order to keep the boiler water at proper dissolved solids. The water blowdown passes to a flash tank operating at slightly above atmospheric pressure in which part of the water flashes to steam, which in turn passes to an open feedwater heater. How much steam is flashed per hour?

Calculation Procedure:

1. Determine the amount of blowdown
Amount of blowdown $B = 0.02\ D$, where D is the steam delivery. Hence, $B = 0.02 \times 200,000 = 4000$ lb/h (30 kg/s).

2. Find the enthalpy of the blowdown-saturated liquid
Blowdown water leaves the boiler at point d in Fig. 31a as saturated liquid, poind d in Fig. 31b. Blowdown at a pressure of $p_d = 1400$ lb/in² (9650 kPa) has, from saturated steam tables mentioned under Related Calculations, an enthalpy $h_d = 598.7$ Btu/lb $= (1392$ kJ/kg).

3. Find the enthalpy of the blowdown fluid at the flash tank
The blowdown fluid is assumed to undergo an isenthalpic, or constant-enthalpy, throttling process from point d to point e on Fig. 31b where, at the flash tank, $h_e = h_d$, found above.

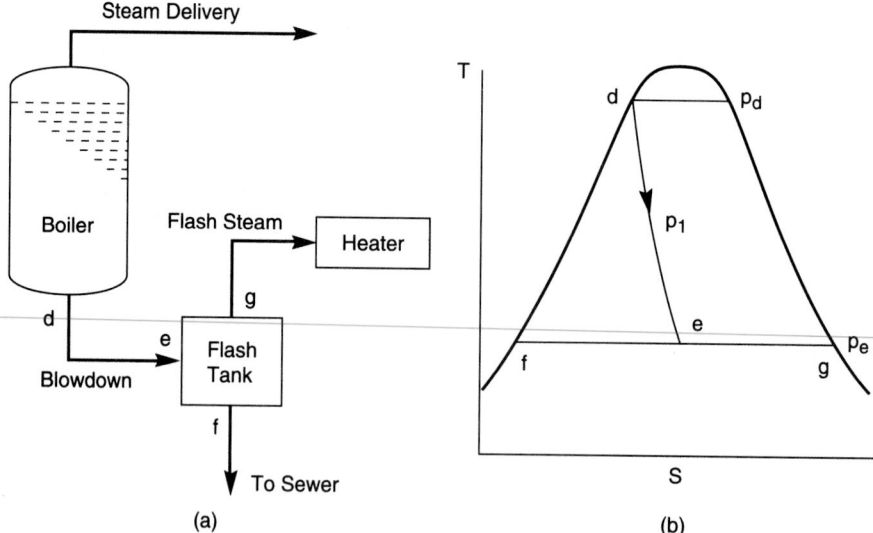

FIG. 31 (a) Boiler blowdown flow diagram. (b) Temperature-entropy schematic for blowdown.

4. Find the enthalpy of saturated liquid within the flash tank

From the saturated steam tables, at $p_e = 15 \text{ lb/in}^2$ (103 kPa), slightly above atmospheric pressure, the enthalpy of the saturated liquid at point f on Fig. 31b, $h_f = 181.1$ Btu/lb (421 kJ/kg).

5. Find the enthalpy of evaporation within the flash tank

From the saturated steam tables, the heat required to evaporate 1 lb (0.45 kg) of water under the pressure p_e within the flash tank is $h_g - h_f = h_{fg} = 969.7$ Btu/lb (2254 kJ/kg).

6. Calculate the amount of steam flashed per hour

The tank flashes steam at the rate of $F = B[(h_e - h_f)] = 4000[598.7 - 181.1)/969.7] = 1723$ lb/h (13 kg/s).

Related Calculations: Saturation steam tables appear in *Thermodynamic Properties of Water Including Vapor, Liquid, and Solid Phases*, 1969, John Wiley & Sons, Inc.

The equation for F in step 6 stems from the presumption of an adiabatic heat balance where $F \times h_{fg} = B(h_e - h_f)$.

DETERMINING WASTE-HEAT BOILER FUEL SAVINGS

An industrial plant has 3000 standard ft^3/min (1.42 m^3/day) of waste gas at 1500°F (816°C) available. How much steam can be generated by this waste gas if the waste-heat boiler has an efficiency of 85 percent, the specific heat of the gas is 0.0178 Btu/(standard ft$^3 \cdot$°F) (1.19 kJ/cm^2), the exit gas temperature is 400°F (204°C), and the enthalpy of vaporization of the steam to be generated is 970.3 Btu/lb (2256.9 kJ/kg)? What fuel savings will be obtained if the plant burns no. 6 fuel oil having a heating value of 140,000 Btu/gal (39,200 kJ/L) and a current cost of $1.00 per gallon ($1 per 3.785 L) and a future cost of $1.35 per gallon ($1.35 per 3.785 L)? The waste-heat boiler is expected to operate 24 h/day, 330 days/year. Efficiency of fuel boilers in this plant is 80 percent.

Calculation Procedure:

1. Compute the steam production rate from the waste heat

Use the relation $S = C_v V(T - t)60E/h_v$, where S = steam production rate, lb/h; C_v = specific heat of gas, Btu/(standard ft$^3 \cdot$°F); V = volumetric flow rate of waste gas, standard ft^3/min; T = waste-gas temperature at boiler exit, °F; E = waste-heat boiler efficiency, expressed as a decimal; h_v = heat of vaporization of the steam being generated by the waste gas, Btu/lb. Substituting gives $S = 0.0178(3000)(1500 - 400)60(0.85)/970.3 = 3087.7$ lb/h (1403.3 kg/h).

2. Find the present and future fuel savings potential

The cost equivalent C dollars per hour of the savings produced by using the waste-heat gas can be found from $C = Sh_v K/E_b$, where the symbols are as given earlier and K = fuel cost, $ per Btu as fired ($ per 1.055 kJ), E_b = efficiency of fuel-fired boilers in the plant. Substituting for the current fuel cost of $1 per gallon, we find $C = 3087.4(970.3)($1/140,000)/0.8 = 26.75. Since the waste-heat boiler will operate 24 h/day, the daily savings will be 24($26.75) = $642. With 330-days/year operation, the annual saving is (330 days)($642 per day) = $211,860. This saving could be used to finance the investment in the waste-heat boiler.

Where the exit gas temperature from the waste-heat boiler will be different from 400°F (204.4°C), adjust the steam output and dollar savings by using the difference in the equation in step 1.

Related Calculations: This procedure can be used for finding the savings possible from recovering heat from a variety of gas streams such as diesel-engine and gas-turbine exhausts, process-gas streams, refinery equipment exhausts, etc. To apply the procedure, several factors must be known or assumed: waste-heat boiler steam pressure, feedwater temperature, final exit gas temperature, heating value of fuel being saved, and operating efficiency of the waste-heat and fuel-fired boilers in the plant. Note that the exit gas temperature must be higher than the saturation temperature of the steam generated in the waste-heat boiler for heat transmission between the waste gas and the water in the boiler to occur.

As a guide, the exit gas temperature should be 100°F (51.1°C) above the steam temperature in the waste-heat boiler. For economic reasons, the temperature difference should be at least

150°F (76.6°C). Otherwise, the amount of heat transfer area required in the waste-heat boiler will make the investment uneconomical.

This procedure is the work of George V. Vosseller, P. E., Toltz, King, Durvall, Anderson and Associates, Inc., as reported in *Chemical Engineering* magazine.

HEAT EXCHANGERS: QUICK DESIGN AND EVALUATION

Find the required surface area and shell-side flow rate of a cross-flow heat exchanger with four single-pass tube rows being designed to meet the following conditions: tube mass flow rate $\dot{m}$ = 22,200 lb/h (10,070 kg/h); tube specific heat capacity, c = 0.20 Btu/lb·°F (0.84 kJ/kg·°C); shell specific heat capacity, C = 0.24 Btu/lb·°F (1.00 kJ/kg·°C); tube-side inlet fluid temperature, t_1 = 500°F (260°C); tube-side outlet fluid temperature, t_2 = 320°F (160°C); shell-side inlet liquid temperature, T_1 = 86°F (30°C); shell-side outlet liquid temperature, T_2 = 131°F (55°C); overall heat-transfer coefficient, U = 9.0 Btu/h·ft²·°F (51.1 W/m²·°C) (183.9 kJ/h·m²·°C).

Another unit, a 1-shell-pass and 2-tube-pass heat exchanger with a vertical shell-side baffle for divided flow, performs as follows: $\dot{m}$ = 33,500 lb/h (15,200 kg/h); c = 0.98 Btu/lb·°F (4.10 kJ/kg·°C); shell mass flow rate $\dot{M}$ = 50,000 lb/h (22,680 kg/h); C = 0.60 Btu/lb·°F (2.51 kJ/kg·°C); t_1 = 270°F (132.2°C); T_1 = 520°F (271.1°C); U = 12.6 Btu/h·ft²·°F (71.5 W/m²·°C) (257.4 kJ/h·m²·°C); heat exchanger surface area, A = 2200 ft² (204.4 m²). Evaluate the performance of this unit by finding its thermal effectiveness, its efficiency, the outlet temperature of the tube-side fluid, and the outlet temperature of the shell-side vapor.

Calculation Procedure:

1. *Compute the ratio of shell-side liquid to tube-side fluid temperature differences and the thermal effectiveness, or temperature efficiency, of the cross-flow unit*

The shell-to-tube ratio of temperature differences is found by $R = (T_1 - T_2)/(t_2 - t_1) = (86 - 131)/(320 - 500) = 0.25$. Thermal effectiveness, $P = (t_2 - t_1)/(T_1 - t_1) = (320 - 500)/(86 - 500) = 0.43$.

2. *Compute the shell-side vapor flow rate*

Shell-side vapor flow rate is $\dot{M} = \dot{m}c/RC = (22,200)(0.20)/[(0.25)(0.24)] = 74,000$ lb/h (33,570 kg/h).

3. *Determine the number of transfer units and heat exchanger efficiency*

The point where $R = 0.25$ and $P = 0.43$ on Fig. 40, shown amongst several figures appearing after step 7, corresponds to values for the number of transfer units, NTU = 0.62 and heat exchanger efficiency, $F = 0.99$. This value for F shows that the design has an efficiency close to that for a pure countercurrent configuration where $F = 1.00$.

4. *Compute the heat exchanger surface area*

To find the area use the formula $A = (\text{NTU})(\dot{m}c)/U = (0.62)(22,200)(0.20)/9.0 = 306$ ft² (28.4 m²).

5. *Compute R and NTU for the shell-and-tube unit*

Substitute appropriate values into the following equations, thus $R = \dot{m}c/\dot{M}C = (33,500)(0.98)/(50,000)(0.60) = 1.09$ and NTU $= UA/\dot{m}c = (12.6)(2200)/(33,500)(0.98) = 0.84$.

6. *Determine the thermal effectiveness and heat exchanger efficiency*

The point where curves for R and NTU intersect on Fig. 35 corresponds to a thermal effectiveness, $P = 0.42$ and an efficiency, $F = 0.89$. For the configuration of this unit, the value of F can be considered acceptable.

7. *Find the exit temperatures of the tube-side fluid and the shell-side vapor*

The tube-side fluid exit temperature, $t_2 = P(T_1 - t_1) + t_1 = 0.42(520 - 270) + 270 = 375°F$ (190.6°C), and the shell-side vapor exit temperature, $T_2 = T_1 - R(t_2 - t_1) = 520 - 1.09(375 - 270) = 405°F$ (297.2°C).

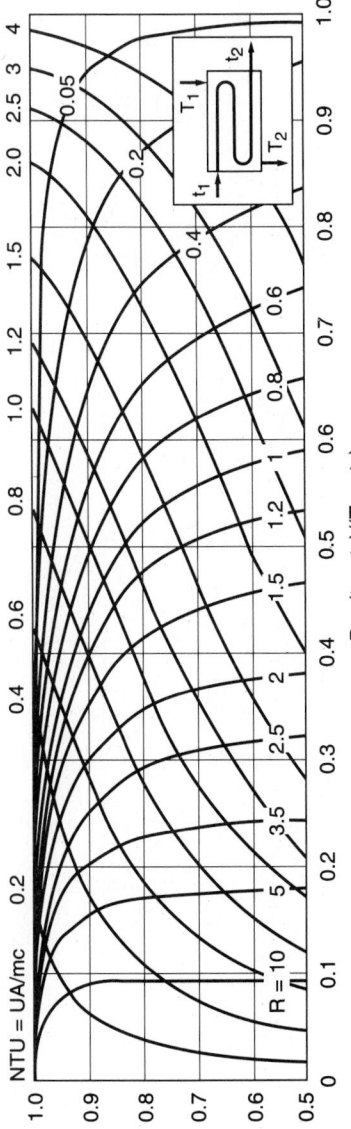

FIG. 32 Design and performance chart for a 1-shell-pass and 3-tube-pass exchanger. (*Chemical Engineering.*)

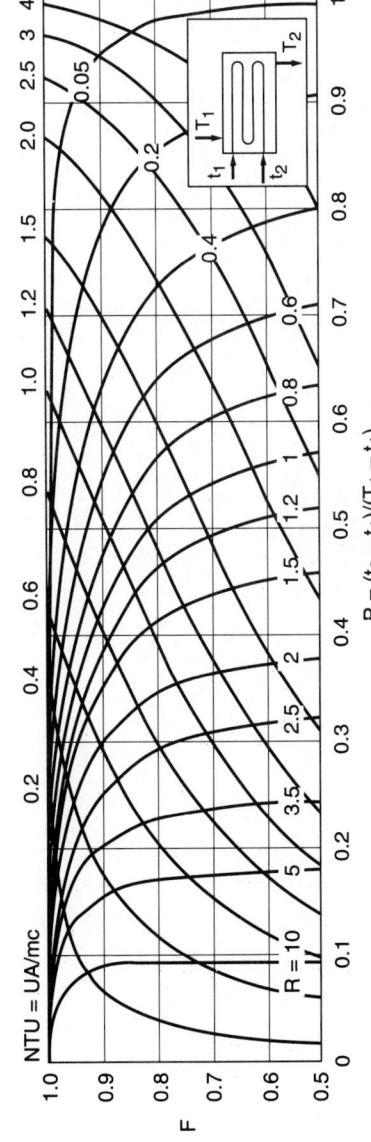

FIG. 33 Design and performance chart for a 1-shell-pass and 4-tube-pass exchanger. (*Chemical Engineering.*)

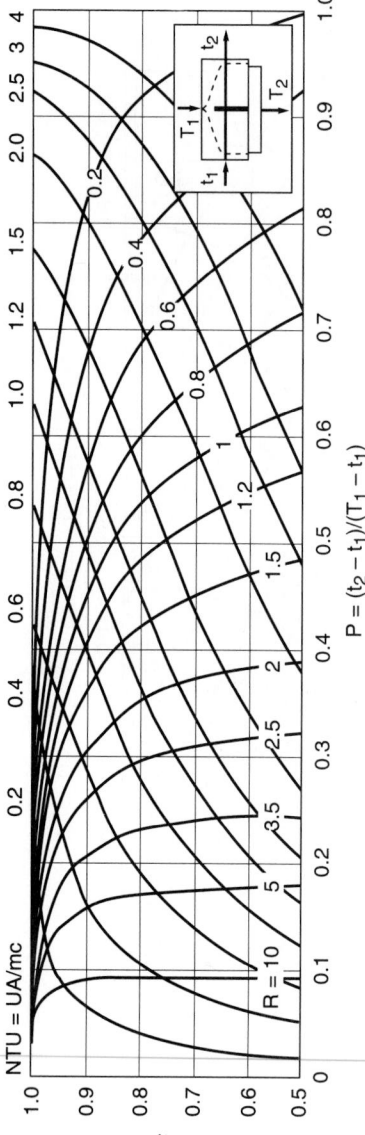

FIG. 34 Design and performance chart for a 1-shell-pass and 1-tube-pass exchanger with a vertical shell-side baffle. (*Chemical Engineering.*)

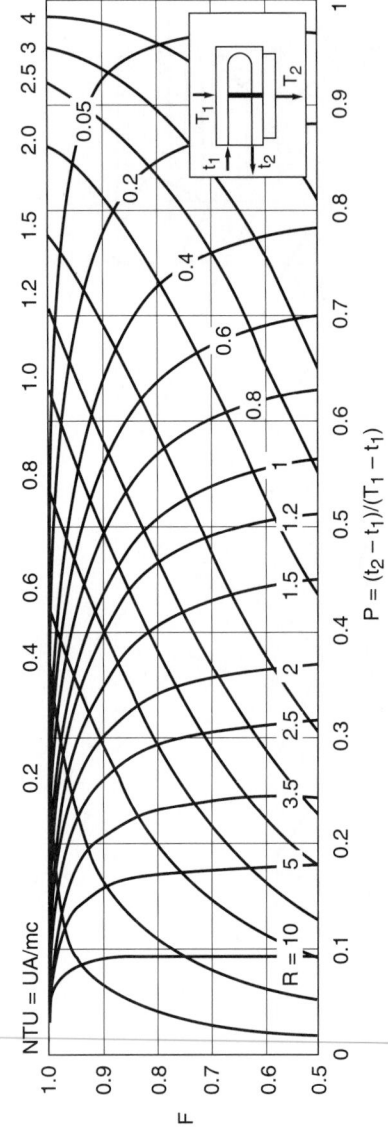

FIG. 35 Design and performance chart for a 1-shell-pass and 2-tube-pass exchanger with a vertical shell-side baffle. (*Chemical Engineering.*)

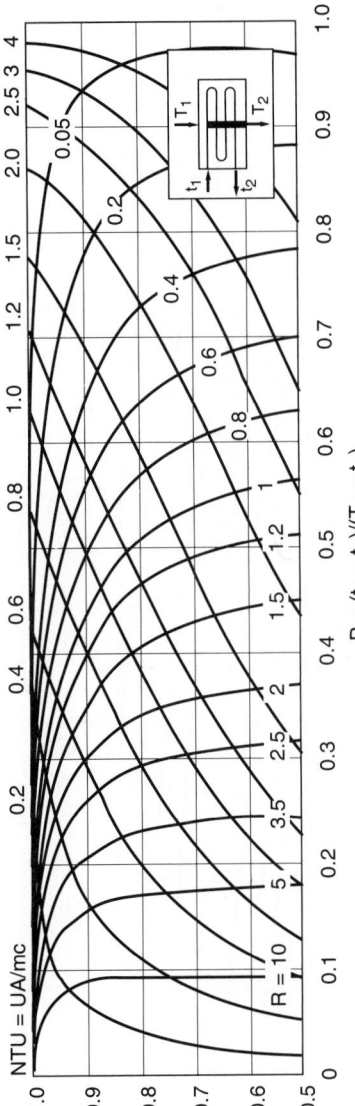

FIG. 36 Design and performance chart for a 1-shell-pass and 4-tube-pass exchanger with a vertical shell-side baffle. (*Chemical Engineering.*)

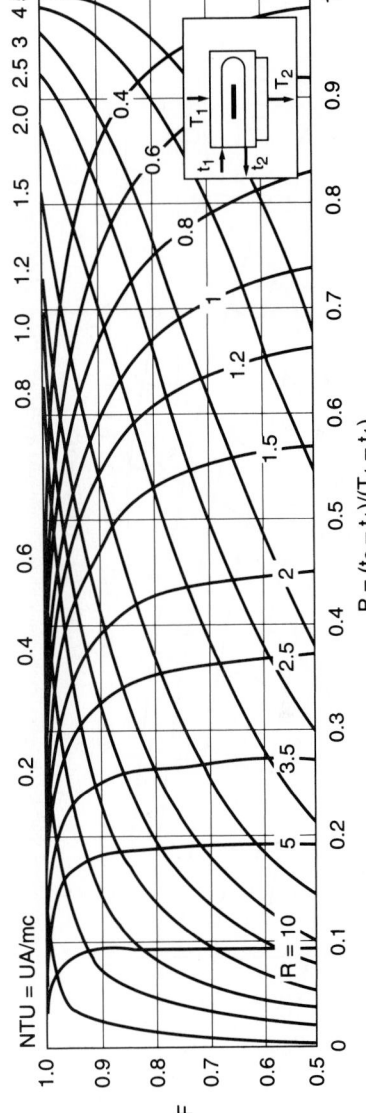

FIG. 37 Design and performance chart for a 1-shell-pass and 2-tube-pass exchanger with a horizontal shell-side baffle. (*Chemical Engineering.*)

3.555

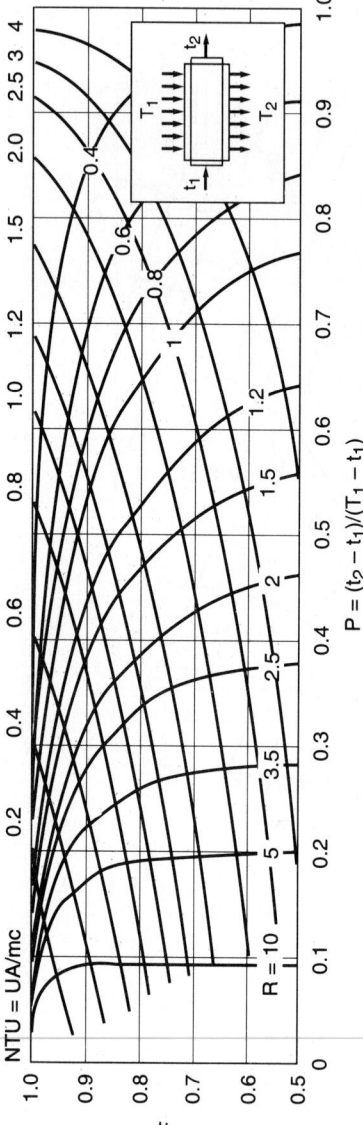

FIG. 38 Design and performance chart for cross-flow exchanger with two single-pass row tubes. (*Chemical Engineering.*)

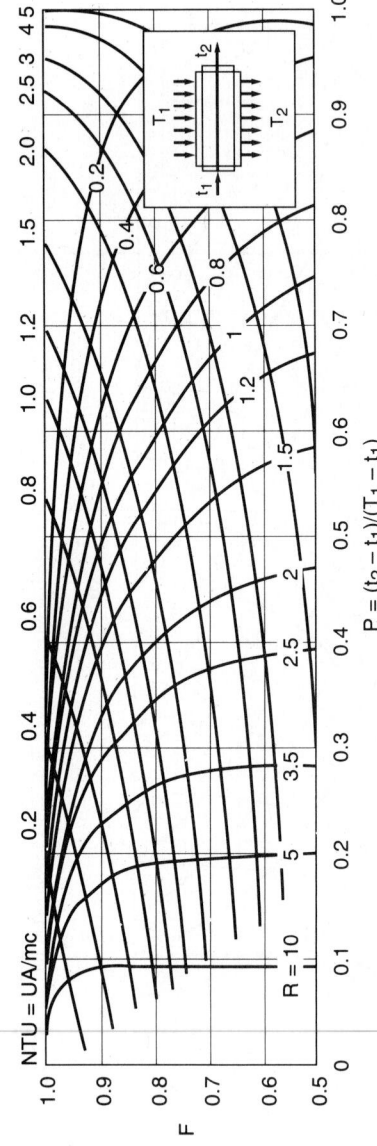

FIG. 39 Design and performance chart for a cross-flow exchanger with three single-pass tube rows. (*Chemical Engineering.*)

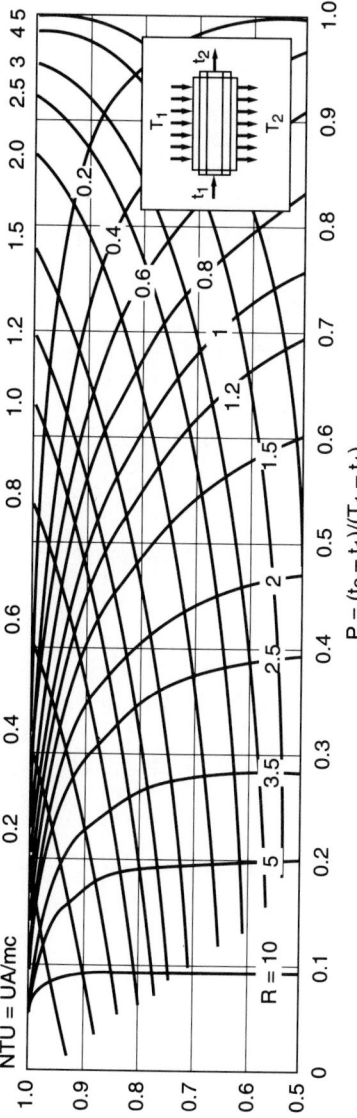

FIG. 40 Design and performance chart for cross-flow exchanger with four single-pass tube rows. (*Chemical Engineering.*)

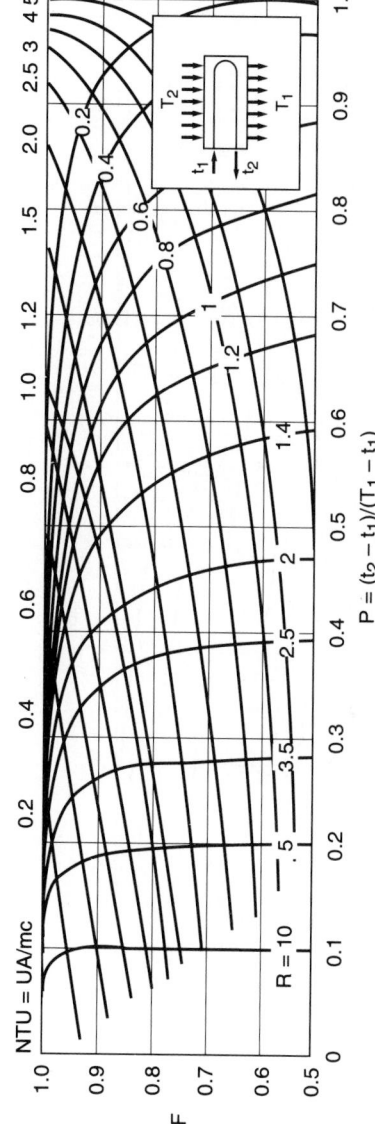

FIG. 41 Design and performance chart for a cross-flow exchanger with a 2-tube-pass. (*Chemical Engineering.*)

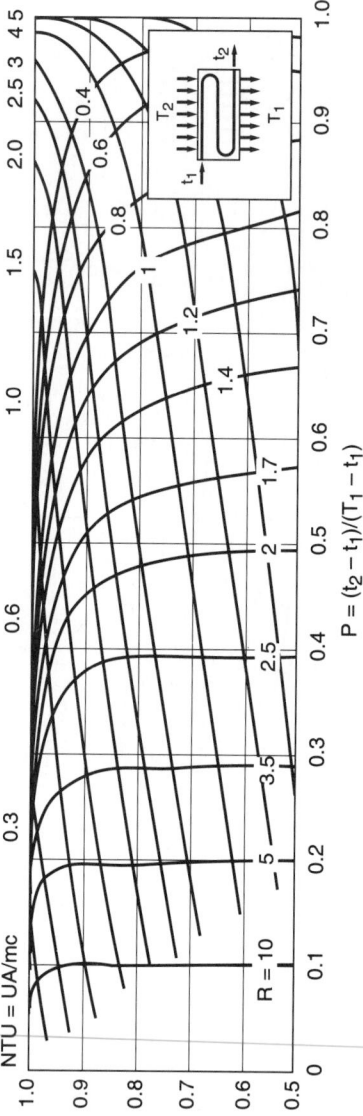

FIG. 42 Design and performance chart for a cross-flow exchanger with a 3-tube pass. (*Chemical Engineering.*)

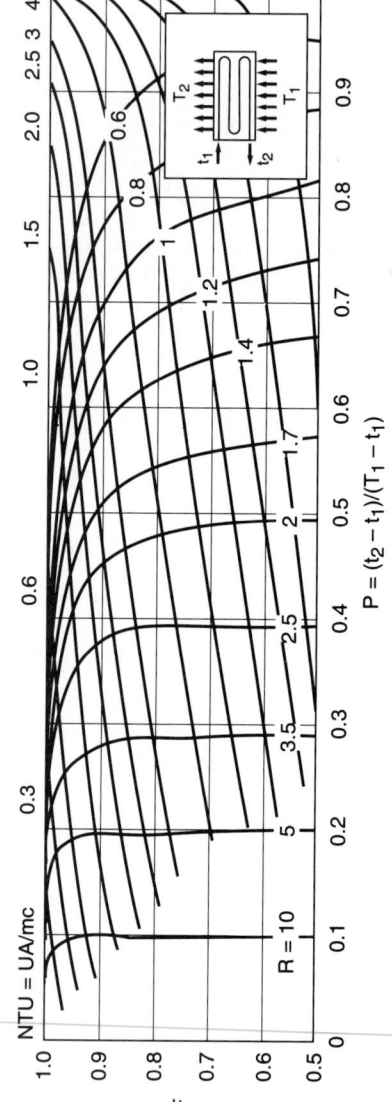

FIG. 43 Design and performance chart for a cross-flow exchanger with a 4-tube pass. (*Chemical Engineering.*)

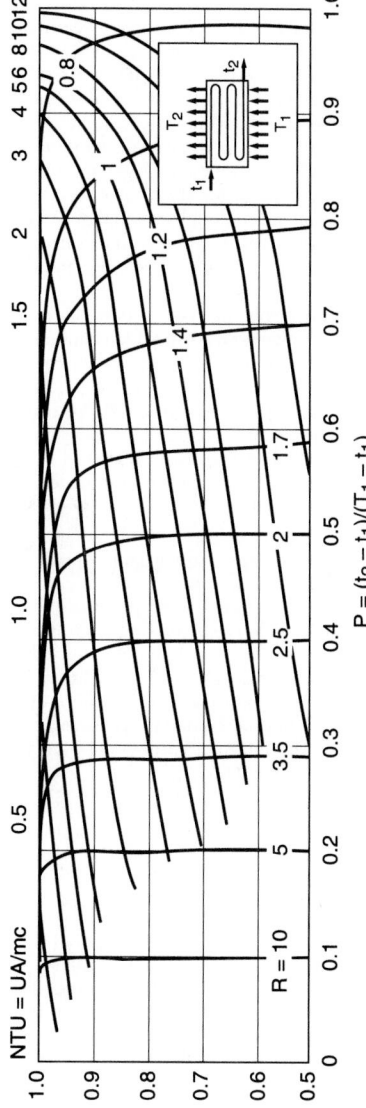

FIG. 44 Design and performance chart for a cross-flow exchanger with a 5-tube pass. (*Chemical Engineering.*)

Related Calculations: On the design and performance charts, Figs. 32 through 44, heat exchanger efficiency, F = true mean temperature difference/logarithmic temperature difference for countercurrent flow and relates the actual rate of heat transfer, Q, to the theoretical rate, $U \times A \times$ logarithmic temperature difference. True mean temperature differences has been solved analytically for each configuration. This, in conjunction with the new NTU curves, eliminates the need for trial-and-error calculations to determine the design and evaluate the performance of specific heat exchangers. By establishing desired conditions which in effect specify any two of the four parameters, the other two parameters may then be read directly from a chart and used to design a unit and/or to evaluate its performance.

The 1-shell-pass–3-tube-pass and 1-shell-pass–2-tube-pass heat exchangers represent conventional shell-and-tube units such as those for steam heating, heating of one process stream by cooling another, and condensation and cooling by a cooling-water utility. Shell-side pressure drops through divided-flow units are typically one-eighth of those through conventional shell-and-tube heat exchangers; hence divided flow units are recommended where low shell-side pressure drops are required.

Cross-flow heat exchangers differ from the conventional and divided-flow shell-and-tube units in that they have tube-bank arrangements over which another stream flows perpendicular to the tubes. Typical applications of cross-flow units include air cooling of overhead condensate streams and trim-product coolers.

This calculation procedure is based upon the work of Jeff Bowman, E.I. du Pont de Nemours Co., and Richard Turton, assistant professor of chemical engineering at West Virginia University, as reported in *Chemical Engineering* magazine. Note that final selection of a unit will usually be made in conjunction with advice and guidance from the manufacturer of the unit.

FIGURING FLUE-GAS REYNOLDS NUMBER BY SHORTCUTS

A low-sulfur No. 2 distillate fuel oil has a chemical composition of 87.4 percent carbon and 12.6 percent hydrogen by weight, ignoring sulfur. The fuel's higher heating value (HHV), or ΔH_{gross}, the standard (60°F) (15.5°C) heat of combustion (based on stoichiometric air usage), is 18,993 Btu/lb (44.148 kJ/kg) of the fuel. With a volumetric proportion of 79 percent atmospheric nitrogen, including rare gases, to 21 percent oxygen, the molar ratio of N_2 to O_2 in air is 3.76:1. What is the Reynolds number for the flow of the flue gas produced by that fuel if it is completely burned in 50 percent excess air at the rate of 25.3 lb/h (11.5 kg/h) and the flue gas leaves a 1-ft (0.3-m) diameter duct at 2000°F (1093°C)?

Calculation Procedure:

1. Compute the volume flow rate of the flue gas

Based on stoichiometric air usage, the standard (32°F, 1-atm) (0°C, 101.3-kPa) volume of flue gas (std ft³/lb) (std m³/kg) of fuel burned is $V_{std} = (\Delta H_{gross}/100)[1 + $ (percent excess air)/100 percent]. Then, $V_{std} = (18,993/100)[1 + (50/100)] = 285$ std ft³/lb (17.79 std m³/kg).

Adjust for temperature expansion by using the ideal-gas law to get the per lb (kg) of fuel actual amount of flue gas, $V' = V_{std}(460 + T)/(460 + 32)$, where T is the flue-gas temperature in °F. Thus, $V' = 285(460 + 2000)/(492) = 1425$ ft³/lb (89 m³/kg). The metric result can be verified as follows: $V_{std} = 17.79[273 + (2000 - 32)/1.8]/273 = 89$ m³/kg.

The flue-gas approximate flow rate, $V = V'W/3600$, where W is the given hourly burning rate of the fuel. Hence, $V = 1425 \times 25.3/3600 = 10.0$ ft³/s (0.28 m³/s).

2. Determine the viscosity of the flue gas

Boiler and incinerator flue gases are composed of several gases, hence a precise calculation of the Reynolds number can be cumbersome. By assuming that the flue gas behaves like nitrogen, it is possible to obtain fast and accurate preliminary approximations for both boilers and incinerators. Then, by means of the graph in Fig. 45, read the dynamic viscosity of nitrogen as μ' = 0.054 cp (54 $\times$ 10⁻⁶ Pa).

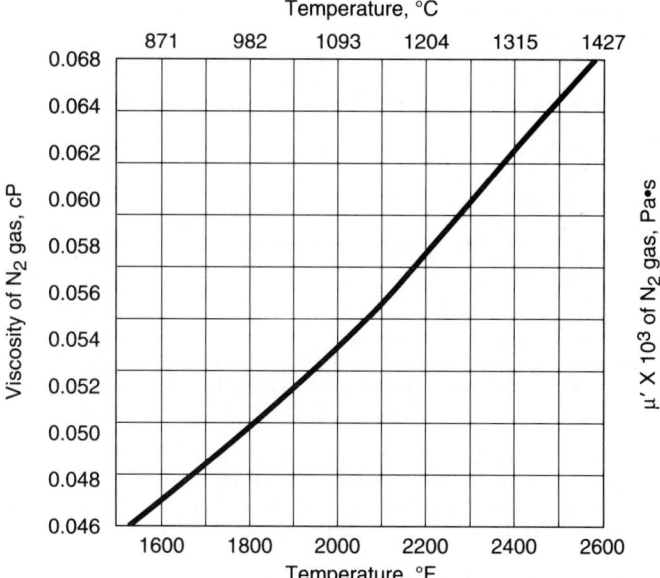

FIG. 45   Dynamic viscosity of nitrogen gas. *(Chemical Engineering.)*

3. *Compute the Reynolds number of the flue gas*

By algebraic manipulations, as mentioned under Related Calculations of this procedure, the shortcut formula for an estimate of the Reynolds number is found to be $R_e = 73,700/[D\mu'(460 + T)]$, where the given duct diameter, $D = 1.0$ ft (0.3 m) and the other values are as previously determined. Thus, $R_e = 72,700 \times 10.0[(1)(0.054)(460 + 2000)] = 5470$.

Related Calculations: The shortcut formula for the value of R_e is derived by algebraic manipulation of four expressions listed below. Symbols adequately defined previously are not redefined.

(1) Reynolds number $R_e = D\rho U/\mu$, where $D =$ diameter of a circular cross-section, or equivalent diameter of some other cross-section, ft (m); $\rho =$ density of the gas, lb/ft³ (kg/m³); $U =$ average linear velocity of the gas, ft/s (m/s); $\mu =$ viscosity of the gas, lb/(ft·s) (Pa·s).

(2) At high temperatures, the average densities and viscosities of flue gas closely approximate those of nitrogen alone. Thus, ρ and μ for nitrogen can be used to estimate R_e with no significant error. Hence, from the ideal-gas law an estimate of the density of nitrogen, $\rho = (28)(460 + 32)/[(359)(460 + T)]$, where 28 is the molecular weight of nitrogen; 359 is the volume, ft³, of 1 lb · mol of gas at 32°F (0°C) and at atmospheric pressure, 14.7 lb/in² (6.89 kPa). In SI units the factor is 22.41 m³/kg · mol.

(3) $U = 4V/\pi D^2$

(4) $\mu = \mu'/1488$

Turbulent flow in a boiler or incinerator assures adequate mixing and near-complete or complete combustion. Flow is turbulent when its Reynolds number is greater than 2000 or 3000, and is more turbulent when R_e is greater. This reduces the amount of excess air required for complete combustion and hence increases boiler and incinerator efficiency.

TABLE 11 Flue-gas components

Component	lb·mol/ lb$_m$ oil		Molar proportions		Molecular weight		Weight proportion
CO_2	0.0728/0.7763	=	0.094	×	44	=	4.14
H_2O	0.0630		0.081		18		1.46
O_2	0.0522		0.067		32		2.14
N_2	0.5883		0.758		28		21.22
Flue gas	0.7763		1.000		28.96		28.96

A review of the shortcut equations reveals that, for a given set of values for D, M, and T and that μ' depends on T, the degree of accuracy of the shortcut value of R_e is reflected by the precision of the value of V_{std} found in step 1 of this procedure. This can be done by checking the stoichiometry of the combustion process per lb (kg) of the fuel with the given percentages of C and H_2, as follows: lb·mol C/lb oil = 0.874/12 = 0.0728; lb·mol H_2/lb oil = 0.126/2 = 0.0630.

Then, 0.0728 C + 0.0630 H_2 + [1 + (percent excess air/100 percent](0.0728 + 0.0630/2) O_2 + [1 + (percent excess air/100 percent](3.76)(0.0728 + 0.0630/2) N_2 → 0.0728 CO_2 + 0.0630 H_2O + (percent excess air/100 percent)(0.1043) O_2 + [1 + (percent excess air/100 percent)](3.76)(0.1043) N_2. Table 11 indicates that the lb·mol/lb oil is 0.7763; hence V_{std} = 0.7763 × 359 = 279 std ft³/lb (17.42 std m³/kg) of oil. This shows the error in the shortcut estimate for V_{std} to be about 2 percent in this case.

Also, Table 11 shows that this flue gas has a composition of 9.4 percent CO_2; 8.1 percent H_2O; 6.7 percent O_2; 75.8 percent N_2; and has an average molecular weight of 28.96. By the method shown on page 3.279 in *Perry's Chemical Engineers' Handbook*, 6th edition, McGraw-Hill, this flue gas has a calculated mixture viscosity of μ' = 0.0536 cP (53.6 × 10⁻⁶ Pa). Using this value and the average molecular weight of 28.96 instead of 28 in the gas density formula in step 2, the R_e estimate would be 5700. This indicates the shortcut estimate of 5470 to be in error by about 4 percent in this case. There are other factors that could contribute to errors in the shortcut calculations. In practice, wood and municipal solid waste contain considerable amounts of moisture, which reduces their heating values that refer to dry conditions, only. The shortcut calculations are very accurate for fossil fuels, such as coal, fuel oil, and natural gas, and wood. They are useful for wastes or waste-fuel mixtures. Errors by shortcut calculations seldom exceed ±10 percent when excess air is less than 150 to 200 percent.

Though errors for fossil fuels with 100 percent or less excess air are generally 5 percent or less, there are factors that increase the error of the shortcut method: (1) High water content in the fuel or waste; (2) high halogen content; (3) excess air above 100 percent. However, the shortcut method can still be used to give a quick first approximation even when these factors are present.

This shortcut method is based on two articles written by Irwin Frankel of The Mitre Corp., Metrek Div., 1820 Dolley Madison Blvd., McLean, VA 22102. The articles, "Shortcut calculations for fluegas volume" and "Figure fluegas Reynolds number," appeared in the *Chemical Engineering* magazine issues of June 1, 1981, and August 24, 1981, respectively.

SECTION 4

ELECTRICAL ENGINEERING

ANDREW W. EDWARDS
POWER ENGINEER
WESTINGHOUSE ELECTRIC CORP.

HAROLD L. RORDEN
CONSULTING ENGINEER
AMERICAN ELECTRIC POWER SERVICE CORP.

FREDERICK W. SUHR
CONSULTING ENGINEER
GENERAL ELECTRIC CO.

Direct-Current Circuit Analysis 4.2
Kirchhoff's Laws for DC Circuit Analysis 4.5
Alternating-Current Circuit Analysis 4.7
Vector Algebra in AC Circuit Analysis 4.8
Lightning-Arrester Selection and Application 4.9
Direct-Current Generator Selection 4.12
Alternator Selection for a Known Load 4.13
Selecting Electric-Motor Starting and Speed Controls 4.14
Basic Short-Circuit Current Determination 4.17
Power-System Short-Circuit Current 4.18
Transformer Characteristics and Performance 4.20
Transformer Selection for an Industrial Load 4.21
System Power-Factor Analysis 4.24
Power-Factor Determination and Improvement 4.26
Wiring-Size Choice for Primary Distribution 4.27
Wire- and Cable-Size Determination for a Known Load 4.30
Preliminary Electric Load Estimating 4.33
Plant Power-Distribution-System Planning 4.37
Industrial-Battery Selection and Sizing 4.40

Electric-Motor Selection for a Known Load 4.44
Starting Time and Current for AC Electric Motors 4.47
Interior-Lighting-System Selection and Sizing 4.51
Outdoor-Lighting Selection and Sizing 4.59
Feeder Sizing for a Combination Electric Load 4.63
Sizing Residential-Service Demand Load 4.64
Electric Comfort Heating Load Determination 4.68
Electric-Motor Choice Based on Total Cost 4.79
Direct-Current Permanent-Magnet Motor Analysis 4.81
Electrical-Measurement Analysis of Permanent-Magnet Motors 4.85
Air Cooling of Electric-Motor Drives 4.87
Flywheel Selection for Electric-Motor Drives 4.87
Lightning Protection of Industrial Plants 4.90

REFERENCES: Roe—*Procedures of Industrial Electrical Design,* McGraw-Hill; Kuffel and Abdullah—*High Voltage Engineering,* Pergamon; Lazar—*Electrical Systems Analysis & Design,* McGraw-Hill; Snow—*Electrical Drafting & Design,* Prentice-Hall, IEEE—*Recommended Practice for Protection & Coordination of Industrial & Commercial Power Systems,* Wiley; Thumann—*Electrical Design, Safety & Energy Conservation,* Fairmont Press; Bell and Whitehead—*Basic Electrical Engineering & Instrumentation for Engineers,* Granada; Cooper and Fordham—*Electrical Safety Engineering,* Butterworths; Cruz—*System Sensitivity Analysis,* Academic Press; Slurzberg and Osterheld—*Essentials of Electricity-Electronics,* McGraw-Hill; Corcoran—*Basic Electrical Engineering,* Wiley; Oppenheimer and Borchers—*Direct and Alternating Currents,* McGraw-Hill; Chang—*Energy Conversion,* Prentice-Hall; Rosenblatt and Friedman—*Direct and Alternating Current Machinery,* McGraw-Hill; Pender—*Electrical Engineers Handbook,* Wiley; Fink and Carroll—*Standard Handbook for Electrical Engineers,* McGraw-Hill; Timbie-Willson—*Industrial Electricity,* Wiley; Abbott and Steka—*National Electrical Code Handbook,* McGraw-Hill; Richter—*Practical Electrical Wiring,* McGraw-Hill; Steka and Brandon—*NFPA Handbook of the National Electrical Code,* McGraw-Hill; Hubert—*Operational Electricity,* Wiley; Croft and Carr—*American Electrician's Handbook,* McGraw-Hill; Beeman—*Industrial Power Systems Handbook,* McGraw-Hill; Gibbs—*Transformer Principles and Practices,* McGraw-Hill; Libby—*Motor Selection and Application,* McGraw-Hill; U.S. Government Printing Office—*Electric Current Abroad,* IEEE—*Electric Systems for Commercial Buildings;* IEEE—*Electric Power Distribution for Industrial Plants;* Power Magazine Editors—*Industrial Electrical Systems,* McGraw-Hill; Dawes—*Electrical Engineering,* McGraw-Hill; NEMA—*Standards for Motors and Generators;* NEMA—*Standards for Industrial Control;* ANSA—*Induction Machines;* ANSA—*Synchronous Motors;* McPartland and Novak—*Electrical Systems Design,* McGraw-Hill; McPartland and Novak—*Electrical Design Details,* McGraw-Hill; McPartland and Novak—*Electrical Equipment Manual,* McGraw-Hill; McPartland and Novak—*Electrical Systems for Power and Light,* McGraw-Hill.

DIRECT-CURRENT CIRCUIT ANALYSIS

A direct-current (dc) circuit contains 15 resistors arranged as shown in Fig. 1. Compute the current flow through, and the voltage drop across, each resistor in this circuit.

Calculation Procedure:

1.   *Divide the circuit into sections and groups*

In analyzing a complex dc combination circuit, the simplest procedure is to divide the circuit into sections that can later be combined into single or multiple resistances in series. Once this is done, the circuit is much easier to analyze.

Mark on the circuit diagram the sections decided upon. Figure 1 shows that the three groups of series-parallel resistances can be divided into three sections. Indicate each section as shown.

Further, subdivide each section into parallel and series groups. Mark the groups as shown in Fig. 1.

2.   *Combine the group resistances to obtain a single section resistance*

Use the rules for parallel and series circuits to combine the resistance values in each group to obtain one single equivalent resistance for each section. Use R = group resistance, Ω, and r = resistor resistance, Ω.

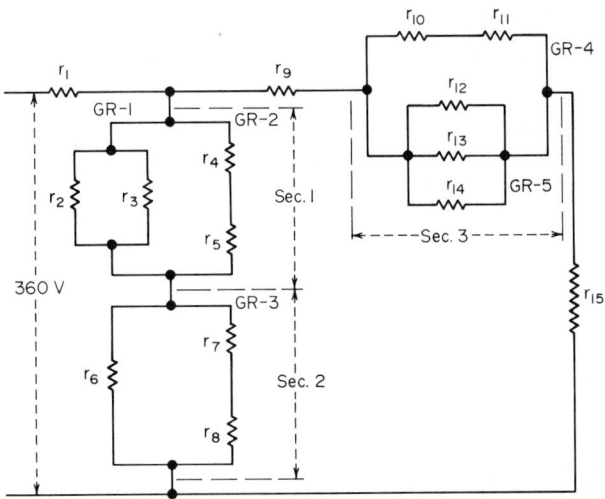

FIG. 1   Direct-current circuit containing numerous resistances (values in ohms): $r_1 = 20$; $r_2 = 1000$; $r_3 = 1500$; $r_4 = 200$; $r_5 = 400$; $r_6 = 600$; $r_7 = 400$; $r_8 = 200$; $r_9 = 20$; $r_{10} = 200$; $r_{11} = 40$; $r_{12} = 100$; $r_{13} = 200$; $r_{14} = 600$; $r_{15} = 52$.

For group 1, Fig. 1,

$$R_1 = \frac{1}{1/r_2 + 1/r_3} = \frac{1}{1/1000 + 1/1500} = 600\ \Omega$$

since r_2 and r_3 are resistances in parallel. For group 2: $R_2 = r_4 + r_5 = 200 + 400 = 600\ \Omega$, since r_4 and r_5 are resistances in series. For section 1, Fig. 1,

$$R_{S1} = \frac{1}{1/R_1 + 1/R_2} = \frac{1}{1/600 + 1/600} = 300\ \Omega$$

since R_1 and R_2 are two resistances in parallel that replace the former r_2, r_3, r_4, and r_5 resistances. R_{S1} is the equivalent single resistance for groups 1 and 2.

Follow the same procedure for group 3, Fig. 1. Thus, $R_3 = r_7 + r_8 = 400 + 200 = 600\ \Omega$. Then, since r_6 and group 3 are in parallel,

$$R_{S2} = \frac{1}{1/r_6 + 1/R_3} = \frac{1}{1/600 + 1/600} = 300\ \Omega$$

Follow the same procedure for groups 4 and 5, Fig. 1. Or, $R_4 = r_{10} + r_{11} = 200 + 40 = 240\ \Omega$.

$$R_5 = \frac{1}{1/r_{12} + 1/r_{13} + 1/r_{14}} = \frac{1}{1/100 + 1/200 + 1/600} = 60\ \Omega$$

Then, for section e,

$$R_{S3} = \frac{1}{1/R_4 + 1/R_5} = \frac{1}{1/240 + 1/60} = 48\ \Omega$$

3.   *Sketch the equivalent circuit*

Sections 1, 2, and 3 have been reduced to equivalent single-series resistances. Sketch them in place as shown in Fig. 2.

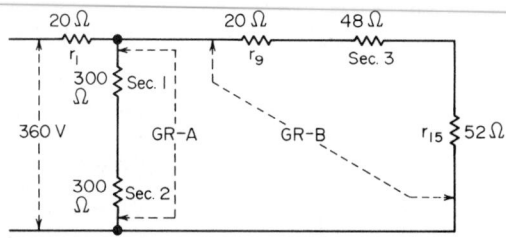

FIG. 2   Direct-current circuit of Fig. 1 reduced to two groups of resistances.

4. *Study the equivalent circuit for its makeup*

Study of Fig. 2 shows that the circuit now consists of a simple parallel-series circuit; i.e., group A is in parallel with group B while both are in series with r_1.

5. *Analyze the equivalent circuit to find the line resistance*

Using the rules for parallel and series circuits as in step 2, we see the resistance of group A is R_A = $R_{S1} + R_{S2} = 300 + 300 = 600 \ \Omega$. Likewise for group B, $R_B = R_9 + R_3 + r_{15} = 20 + 48 + 52 = 120 \ \Omega$.

The equivalent resistance of the parallel circuit consisting of groups A and B is then

$$R_{AB} = \frac{1}{1/R_A + 1/R_B} = \frac{1}{1/600 + 1/120} = 100 \ \Omega$$

Last, combine this series resistance with the one remaining series resistance r_1 to find the total resistance of the line, or $R_T = r_1 + R_{AB} = 20 + 100 = 120 \ \Omega$. Hence, the total equivalent resistance of this circuit is 120 Ω.

6. *Compute the line current*

Use the relation $I = E/R_T$, where I = line current, A; E = line voltage; R_T = line resistance, Ω. Or $I = 360/120 = 3$ A.

7. *Compute the equivalent-circuit current and voltage*

The voltage across any resistance is $e_r = I_r r_r$, where I and r are the current and resistance, respectively, for the resistance in question. Thus, $e_1 = I_T r_1 = 3 \times 20 = 60$ V.

Then, $E_A = E_B = E_T - e_1$ because the voltage acting on groups A and B is that potential remaining after the voltage drop across r_1. Or, $E_A = E_B = 360 - 60 = 300$ V.

The current in each group is $I = E/R$, or $I_A = E_A/R_A = 300/600 = 0.50$ A; $I_B = E_B/R_B = 300/120 = 2.50$ A. As a check, $I_T = I_A + I_B = 0.50 + 2.50 = 3.0$ A, as computed in step 6.

8. *Compute the section current and voltage*

Use the same reasoning as in step 7: $e_{S1} = I_A R_{S1} = 0.50 \times 300 = 150$ V; $e_{S2} = I_A R_{S2} = 0.50 \times 300 = 150$ V.

The voltage across group A equals the sum of the voltage drops across the resistances in this group, or $E_A = e_{S1} + e_{S2} = 150 + 150 = 300$ V. This checks with the computation in step 7.

Using the same procedure for group B gives $e_9 = I_B r_9 = 2.50 \times 20 = 50$ V; $e_{S3} = I_B R_{S3} = 2.50 \times 48 = 120$ V; $e_{15} = I_{S3} + e_{15} = 2.50 \times 52 = 130$ V. As a check, $E_B = e_9 + e_{S3} + e_{15} = 50 + 120 + 130 = 300$ V, as previously computed in step 7.

9. *Compute the original circuit current and voltage*

The current in each resistor, section, and group of the original circuit is given by the general relation $i = e/r$, where the appropriate voltage and resistance are substituted. Thus, for section 1 of the original circuit, $i_2 = e_{S1}/R_2 = 150/600 = 0.250$ A; $i_{r2} = e_{S1}/r_2 = 150/1000 = 0.150$ A; $i_{r3} = e_{S1}/r_3 = 150/1500 = 0.100$ A. Then $I_{S1} = i_2 + i_{r2} + i_{r3} = 0.250 + 0.150 + 0.100 = 0.500$ A, and $e_{r4} = i_2 r_4 = 0.250 \times 400 = 100$ V; $e_{S1} = e_{r4} + e_{r5} = 50 + 100 = 150$ V.

ELECTRICAL ENGINEERING **4.5**

Likewise, for the voltage across section 1, $e_{r4} = i_2r_4 = 0.250 \times 200 = 50$ V; $e_{r5} = i_2r_5 = 0.250 \times 400 = 100$ V; $e_{S1} = e_{r4} + e_{r5} = 50 + 100 = 150$ V.

For section 2 find the current in the same way as for section 1: $i_{r6} = e_{S2}/r_6 = 150/600 = 0.250$ A; $i_3 = e_{S2}/R_3 = 150/600 = 0.250$ A; $I_{S2} = i_{r6} + i_3 = 0.250 + 0.250 = 0.500$ A.

The voltage across section 2 is made up of $e_{r7} = i_3r_7 = 0.250 \times 400 = 100$ V; $e_{r8} = i_3r_8 = 0.250 \times 200 = 50$ V; $e_{S2} = e_{r7} + e_{r8} = 100 + 50 = 150$ V.

For section 3, $i_4 = e_{S3}/R_4 = 120/240 = 0.5$ A; $i_5 = e_{S3}/R_5 = 120/60 = 2.0$ A; $I_{S3} = i_4 + i_5 = 0.50 + 2.0 = 2.5$ A. Also; $e_{r10} = i_4r_{10} = 0.50 \times 200 = 100$ V; $e_{r11} = i_4r_{11} = 0.50 \times 40 = 20$ V; $e_{S3} = e_{10} + e_{11} = 100 + 20 = 120$ V.

For group 5, $i_{r12} = e_{S3}/r_{12} = 120/100 = 1.20$ A; $i_{r13} = e_{S3}/r_{13} = 120/200 = 0.60$ A; $i_{r14} = e_{S3}/r_{14} = 120/600 = 0.20$ A; $i_5 = i_{r12} + i_{r14} = 1.20 + 0.60 + 0.20 = 2.00$ A.

Related Calculations: Any reducible dc circuit (i.e., any circuit that can be reduced to one equivalent resistance with a single power source), no matter how complex, can be solved in a similar manner to that described above. When each circuit is solved separately and combined whenever possible, the current through each resistor and the voltage drop across it may be obtained. To analyze such circuits, perform the following steps in the order listed: (1) Combine the resistance values in each group to obtain one single equivalent resistance value for each section. (2) Combine the resistance values of all sections to obtain one single equivalent resistance value for the line. (3) Solve for the line current. (4) Find the current flowing in each resistor. (5) Find the voltage across each resistor. In the analysis of some circuits, steps 4 and 5 may have to be interchanged.

KIRCHHOFF'S LAWS FOR DC CIRCUIT ANALYSIS

Analyze the circuit of the previous calculation procedure by applying Kirchhoff's laws.

Calculation Procedure:

1. Label all circuit elements as to name and value

Figure 3a shows the circuit of Fig. 2 with its elements identified by symbol and value. The 360-V generator represents the circuit voltage source.

2. Label the current direction in each branch of the circuit

Draw an arrow alongside each branch of the circuit to indicate the direction of electron flow in the branch. Use judgment in assuming the probable direction of electron flow in each branch of the circuit (Fig. 3b).

3. Mark all circuit connecting points with a reference letter

Follow a clockwise direction in marking the connecting points with the reference letter (Fig. 3b).

4. Set up current equations at each junction of three or more circuit elements

When you set up the current junction equations, consider the currents entering the junction as algebraically *positive;* consider currents leaving the junction as algebraically *negative*. Thus, at junction C (Fig. 3b), $i_1 + i_2 - I_T = 0$; at junction F, $I_T - i_1 - i_2 = 0$.

In this case, except for the algebraic signs, the two equations are the same. The reason for this condition is that the number of *independent equations* that can be used is always one less than the total number of junctions in a circuit. (Kirchhoff's current law states: *The algebraic sum of all currents at any point in a circuit must be zero.*)

5. Set up voltage equations for each closed path in the circuit

Kirchhoff's voltage law states: *The algebraic sum of all the voltages in any closed path of a circuit must be zero.* So be sure to indicate the polarity of the voltages. When you set up the voltage-loop equations, follow these two rules:

1. The voltage of a power source is positive when the direction of the path being traced is from the negative to the positive terminal and negative when in the reverse direction.

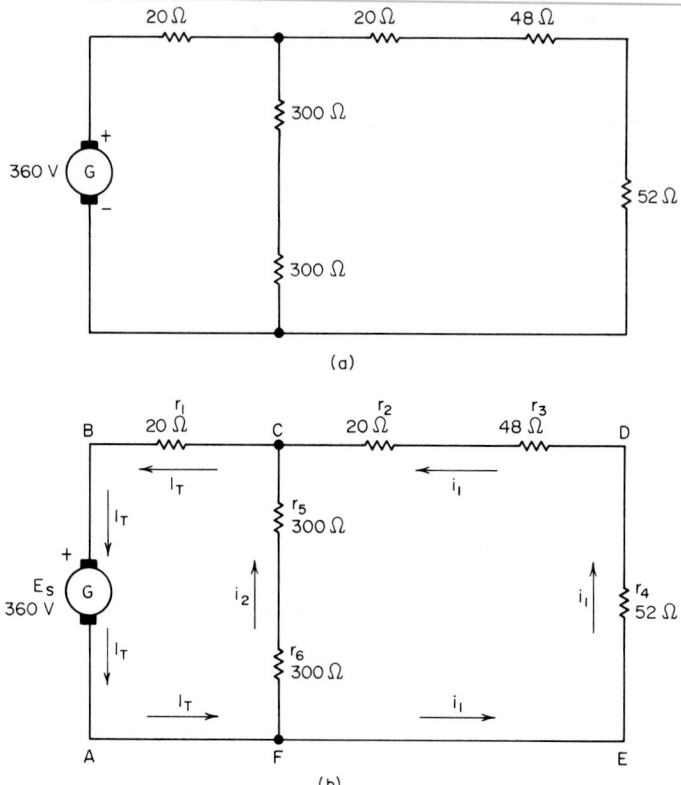

FIG. 3 (*a*) Typical dc circuit; (*b*) current and loop designations for analysis of a circuit by use of Kirchhoff's law.

2. The polarity of a voltage at a resistor depends on the direction of the electron flow through it. When the indicated direction of electron flow is opposite to the direction in which the voltage loop is being traced, the voltage at a resistor is negative. When the directions are both the same, the voltage is positive.

Using these rules for loop $ABCDEFA$ (Fig. 3*b*) gives $E_S - I_T r_1 - i_1 r_2 - i_1 r_3 - i_1 r_4 = 0$. By taking the path of this loop in the opposite direction as $AFEDCBA$, $i_1 r_4 + i_1 r_3 + i_1 r_2 + i_T r_1 - E_S = 0$.

Substitute known circuit element values in the first equation to obtain $360 - 20I_T - 120i_1 = 0$ for loop $ABCDEFA$.

Setting up the voltage equation for loop $ABCFA$ in a similar manner, we find $E_S - I_T r_1 - i_2 r_5 - i_2 r_6 = 0$. Substituting known circuit element values yields $360 - 20I_T - 600i_2 = 0$. Likewise for loop $FCDEF$, $i_2 r_6 + i_2 r_5 - i_1 r_2 - i_1 r_3 - i_1 r_4 = 0$, and $600i_2 - 120i_1 = 0$.

6.   *Solve the independent current and voltage equations simultaneously*

When a circuit has n basic unknowns, n equations that contain each unknown at least once are necessary for the solution of the problem. Since there are three basic unknowns in this circuit, namely, I_T, i_1, and i_2, only three equations are needed, if each equation contains each unknown at least once.

From one of the current equations, write one of the unknowns in terms of the other unknown or unknowns. Or, $i_1 + i_2 - I_T = 0$; hence, $I_T = i_1 + i_2$. Substituting $i_1 + i_2$ for I_T in the current equation for loop $AFEDCBA$ gives $360 - 20(i_1 + i_2) - 120i_1 = 0$, or $360 - 140i_1 - 20i_2 = 0$.

Solve this last equation for one of the unknowns in terms of the other unknown, or $i_1 = 2.57 - 0.143i_2$. Substitute this quantity in the current equation for loop $FCDEF$, or $600i_2 - 120(2.57 - 0.143i_2) = 0$; $i_2 = 0.5$ A. Substitute this value for i_2 in the first combined equation, or $360 - 140i_1 - 20i_2 = 0$, and solve for i_1, or $i_1 = 2.5$ A. Since i_1 and i_2 are known, the equation $I_T = i_1 + i_2$ can be solved, or $I_T = 2.5 + 0.5 = 3.0$ A.

7. Compute the unknown voltages, using Ohm's law

Thus, $e_{r1} = I_T r_1 = 3 \times 20 = 60$ V; $e_{r2} = i_1 r_2 = 2.5 \times 20 = 50$ V; $e_{r3} = i_1 r_3 = 2.5 \times 48 = 120$ V; $e_{r4} = i_1 r_4 = 2.5 \times 52 = 130$ V; $e_{r5} = i_2 r_5 = 0.5 \times 300 = 150$ V; $e_{r6} = i_2 r_6 = 0.5 \times 300 = 150$ V.

Related Calculations: The above calculation procedure is an application of Kirchhoff's laws to a *reducible circuit*. Use the same general method and steps for *irreducible* circuits.

Note that when a circuit has two power sources, such as often occurs in irreducible circuits, the directions of the currents are not known. Hence, current directions must be assumed. If an incorrect direction is chosen, it will show up in the solution as a negative current. However, the magnitude of the current, i.e., the numerical value, will be correct regardless of the direction chosen. Use the negative sign for the current in any remaining calculations. As a final check, substitute the values obtained in any unused current or voltage equations so that all unknown current values are used at least once.

ALTERNATING-CURRENT CIRCUIT ANALYSIS

A series-parallel alternating-current (ac) circuit is connected as in Fig. 4a. Find the current, power, apparent power, power factor, and phase angle of the line. Find the voltage and current of each part of the circuit. Draw a vector diagram for the circuit.

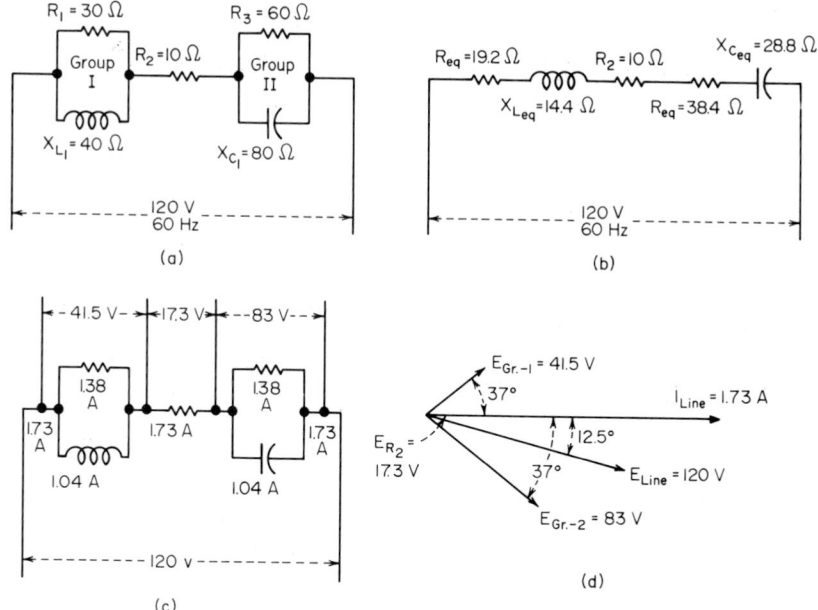

FIG. 4 (a) Typical ac circuit; (b) circuit converted to series elements; (c) voltages across the circuit elements; (d) circuit vector diagram.

Calculation Procedure:

1. Reduce each parallel group to a single value of resistance and reactance

The purpose of this reduction is to produce the same effect when the circuit elements are connected in series as the original parallel group. These values are called the *equivalent resistance* and *equivalent reactance*. Do this by (*a*) assigning a convenient assumed voltage to the group for finding the impedance and phase angle of the group, (*b*) finding the equivalent resistance, and (*c*) finding the equivalent reactance.

Assume a 120-V supply. Then, for group I, Fig. 4*a*: $I_R = E/R_1 = 120/30 = 4$ A; $I_{XL} = E/X_L = 120/40 = 3$ A; $I_{\text{line}} = (I_R^2 + I_{XL}^2)^{0.5} = (4^2 + 3^2)^{0.5} = 5$ A; $Z_{\text{eq}} = E/I_{\text{line}} = 120/5 = 24$ Ω, where $Z_{\text{eq}} =$ equivalent reactance of group I.

Find the power factor of group I from $\cos\theta = I_R/I_{\text{line}} = \tfrac{4}{5} = 0.80$; $\sin\theta = I_X/I_{\text{line}} = \tfrac{3}{5} = 0.60$. From a table of trigonometric functions, $\theta = 37°$ lagging.

Use the relation $R_{\text{eq}} = Z_{\text{eq}} \cos\theta$ to find the equivalent resistance, R_{eq} Ω. Or, $R_{\text{eq}} = 24 \times 0.80 = 19.2$ Ω. Likewise, the equivalent reactance X_{eq} is $X_{\text{eq}} = Z_{\text{eq}} \sin\theta = 24 \times 0.60 = 14.4$ Ω.

Using a similar procedure for group II, Fig. 4*a*, after assuming a 240-V supply, we find $I_{R3} = E/R_3 = 240/60 = 4$ A; $I_{XC} = E/X_C = 240/80 = 3$ A; $I_{\text{line}} = (I_{R3}^2 + I_{XC}^2)^{0.5} = 5$ A; $Z_{\text{eq}} = E/I_{\text{line}} = 240/5 = 48$ Ω; $\cos\theta = I_R/I_{\text{line}} = \tfrac{4}{5} = 0.80$; $\sin\theta = I_{XC}/I_{\text{line}} = \tfrac{3}{5} = 0.60$; $\theta = 37°$ leading; $R_{\text{eq}} = Z_{\text{eq}} \cos\theta = 48 \times 0.80 = 38.4$ Ω; $X_{\text{eq}} = Z_{\text{eq}} \sin\theta = 48 \times 0.60 = 28.8$ Ω.

2. Draw an equivalent series-circuit diagram

Figure 4*b* shows the equivalent series-circuit diagram in which the computed equivalent resistance and equivalent reactance replace parallel groups I and II.

3. Compute the desired quantities for the equivalent circuit

Use the relation $Z_{cct} = [(R_{\text{eq}} + R)^2 + (X_{Leq} - X_{ceq})^2]^{0.5}$, where $Z_{cct} =$ circuit impedance, Ω. Or, $Z_{cct} = [(19.2 + 38.4 + 10)^2 + (14.4 - 28.8)^2]^{0.5} = 69.1$ Ω. Then $I_{\text{line}} = E_{\text{line}}/Z_{cct} = 120/69.1 = 1.73$ A. The line power is $P_{cct} = I^2R_{cct}$ W, or $1.73^2 \times (19.2 + 38.4) = 203$ W. The apparent power of the circuit is $AP_{cct} = E_{\text{line}}I_{\text{line}} = 120 \times 1.73 = 208$ VA. The power factor of the circuit is $PF_{cct} = P_{cct}/AP_{cct} = 203/208 = 0.976$, or $\theta_{cct} = 12.5°$ leading.

4. Compute the voltage and current of the original circuit

The voltage at group I is $I_{\text{line}}Z_{\text{eq}} = 1.73 \times 24 = 41.5$ V; at R_2 the voltage is $I_2R_2 = 1.73 \times 10 = 17.3$ V. Also, the voltage at group II $= I_{II}Z_{\text{eq}} = 1.73 \times 48 = 83$ V.

The current through R_1 is $I_{R1} = E_{GRI}/R_1 = 41.5/30 = 1.38$ A; the current through X_{L1} is $I_{XL1} = E_{GRI}/X_{L1} = 41.5/40 = 1.04$ A. As a check, the line current to group I should be $I_{\text{line}} = (I_{R1}^2 + I_{XL1}^2)^{0.5} = (1.38^2 + 1.04^2)^{0.5} = 1.73$ A.

The current through group II $= I_{R2} = I_{\text{line}} = 1.73$ A. The current through R_3 is $I_{R3} = E_{GRII}/R_3 = 83/60 = 1.38$ A; also $I_{XC} = E_{GRII}/X_C = 83/80 = 1.04$ A. Figure 4*c* shows the voltage and current distribution in the original circuit, and Fig. 4*d* shows the vector diagram for the circuit. Note that the line voltage $E_{\text{line}} = 120$ V = the assumed voltage used in step 1.

Related Calculations: To analyze parallel-series ac circuits, use the following steps: (1) Find the impedance, current, power, $\cos\theta$, and $\sin\theta$ for each parallel branch, using the general relations given above. (2) Resolve each current into its in-phase (resistance) component and quadrature (90° angle) or reactance component, using the relations $I_R = I \cos\theta$ and $I_X = I \sin\theta$, where $I =$ branch current, $\theta =$ leading or lagging power factor angle. (3) Compute the line current by combining the in-phase (resistance) components and the quadrature (reactance) components of all the branch currents, using the general relation $I_{\text{line}} = [(I_1 \cos\theta_1 + I_2 \cos\theta_2, \ldots)^2 + (\pm I_1 \sin\theta_1 \pm I_2 \sin\theta_2, \ldots)^2]^{0.5}$. In the expression $\pm I_1 \sin\theta_1 \ldots$, use + for leading currents and − for lagging currents. When the sum of the second parentheses in the I_{line} expression is positive, this indicates that I_{line} is a leading current; a negative sum indicates a lagging line current. (4) Compute the impedance of the circuit. (5) Compute the power taken by the circuit. (6) Compute the VA of the circuit. (7) Compute the line power factor. Although this is the recommended solution sequence, it may have to be altered, depending on the characteristics of the circuit analyzed.

VECTOR ALGEBRA IN AC CIRCUIT ANALYSIS

Use vector algebra to analyze the circuit in Fig. 4. Find the impedance, phase angle, current, power factor, and power of the line, and the voltage and current of each part of the circuit.

Calculation Procedure:

1. Compute the impedance of each circuit group

For circuit groups with two components in parallel, use the relation $Z_T = Z_1 Z_2/(Z_1 + Z_2)$, where Z_T is the total impedance, Ω; Z_1 and Z_2 are the impedances, respectively, of the two components in parallel. For group I, using vector notation, $Z_{GI} = [(30 + j40)/(30 + j40)] \times [(30 - j40)/(30 - j40)] = 19.2 + j14.4 = 24\underline{/37°}\ \Omega$. For group II, $Z_{GII} = \{[60(-j80)]/(60 - j80)\}[(60 + j80)/(60 + j80)] = 38.4 - j28.8 = 48\underline{/-37°}\ \Omega$.

2. Draw the equivalent series circuit

Using the computed impedances of the parallel groups, draw the equivalent series circuit (Fig. 4b).

3. Compute the line impedance

The line impedance $Z_{\text{line}} = Z_{GI} + R_2 + Z_{GII}\ \Omega$. Or in words, the line impedance equals the sum of the parallel-circuit group impedances plus the line resistance. For the series circuit in Fig. 4b, $Z_{\text{line}} = 19.2 + j14.4 + 10 + 38.4 - j28.8 = 67.6 - j14.4 = 69\ \underline{/12°}\ \Omega$.

4. Compute the line current

Use the relation $I_{\text{line}} = E/Z_{\text{line}}\ \Omega$. Or, $I_{\text{line}} = 120/69 = 1.74$ A.

5. Determine the line power factor

Step 3 shows the line phase angle as 12°. Since $PF_{\text{line}} = \cos \theta_{\text{line}} = \cos 12°$, $PF_{\text{line}} = 0.978$.

6. Compute the line power

Use the relation $P_{\text{line}} = EIPF$ W, or $120 \times 1.74 \times 0.978 = 204$ W.

7. Compute the voltage and current in the circuit parts

For R_1, $E_{R1} = E_{XL1} = E_{G1} = I_{\text{line}} Z_{GI} = 1.74 \times 24 = 41.8$ V. Then $I_{R1} = E_{G1}/R_1 = 41.8/30 = 1.39$ A. Also, $I_{XL1} = E_{G1}/X_{L1} = 41.8/40 = 1.04$ A. This completes the voltage and current computations for group I.

For group II: $E_{GII} = I_{\text{line}}R_2 = 1.74 \times 10 = 17.4$ V; $I_{R2} = I_{\text{line}} = 1.74$ A; $E_{R3} = E_{XC1} = E_{GII} = I_{\text{line}}Z_{GII} = 1.74 \times 48 = 83.6$ V; $I_{R3} = E_{GII}/R_3 = 83.6/60 = 1.39$ A; $I_{XC1} = E_{GII}/X_{C1} = 83.6/80 = 1.04$ A.

Related Calculations: Vector algebra saves many steps in complex ac circuit problems. When polar notation is used, as in this calculation procedure, solutions are obtained more rapidly. Note that the circuit analyzed in this procedure is the circuit analyzed in the previous procedure.

LIGHTNING-ARRESTER SELECTION AND APPLICATION

Select lightning arresters for a three-phase 13.8-kV industrial plant electric system fitted with 4.16-kV rotating machines.

Calculation Procedure:

1. Select the distribution-system arrester voltage rating

Table 1 shows the typical voltage ratings of lightning arresters usually chosen for three-phase power systems. Thus, either a 15- or a 12-kV arrester may be used for this system, depending on how the system neutral is grounded. Where the type of grounding is not known or is yet to be chosen, the higher-rated arrester is a safer choice. Also, an effectively grounded neutral may, under fault conditions or other emergencies, leave a portion of the system ungrounded. For these reasons the higher-rated arrester is often preferred. Hence, a 15-kV arrester will be chosen for this distribution system.

2. Choose the type of arrester to use

Table 2 shows the typical arrester types used for various required voltage ratings. Thus, either a distribution- or station-type arrester can be used for a required voltage rating of 3 to 15 kV. Further study of Table 2 shows that protection of industrial-plant electric systems is usually accomplished by use of a station-type lightning arrester. The reason is that the value of the equipment and the importance of uninterrupted service warrant the use of station-type arresters

TABLE 1   Voltage Ratings of Arresters Usually Selected for Three-Phase Systems

Nominal system voltage, kV	Voltage rating of arrester, kV	
	System neutral ungrounded or resistance grounded	System neutral effectively grounded
0.120/0.208 Y	0.65	0.175
0.240	0.65	0.65
0.480	0.65	0.65
0.600	0.65	0.65
2.4	3	3
2.4/4.16Y	4.5° or 6	3,† 4.5,° or 6
4.16	4.5° or 6	4.5° or 6
4.8	6	4.5° or 6
6.9	7.5° or 9	6
12	15	12
7.2/12.47 Y	15	9† or 12
13.2 (or 13.8)	15	12
23	25	20
34.5	37	30
46	50	40
69	73	60
115	121	97
138	145	121

°The 4.5- and 7.5-kV arresters are available only in the station type.
†The use of these arresters requires an X_0/X_1 ratio less than that necessary to make the system "effectively grounded."
SOURCE: Beeman—*Industrial Power Systems Handbook*, McGraw-Hill.

TABLE 2   Lightning-Arrester Applications

Required arrester voltage rating, kV	Type of arrester used
3–15	Distribution or station
20–73	Line or station

Equipment protected	Typical arrester choice
Industrial plant	Station
Liquid-filled transformers and substations rated 1000 kVA and less; short cables between overhead lines and apparatus; small breakers; disconnects	Distribution or line
Rotating machines	Distribution

throughout the voltage range of the plant. Hence, this type of arrester will be chosen for this plant.

3. Choose the rotating-machinery arresters

Table 3 shows the voltage rating of lightning arresters used to protect three-phase ac rotating machines. Since it has already been decided (step 2) to use station-type arresters in this plant, the

TABLE 3 Protective Equipment for Three-Phase AC Rotating Machines

Machine voltage rating (phase-to-phase)	For installation at machine terminals or on machine bus						For installation 1500 to 2000 ft (457 to 610 m) out on directly connected exposed overhead lines		
	Protective capacitors			Station-type arresters			Distribution-type arresters		
	Voltage rating	Microfarads per pole	Single-pole units required	Voltage rating		Single-pole units required	Voltage rating		Single-pole units required
				Ungrounded or resistance-grounded system	Effectively grounded system		Ungrounded or resistance-grounded system	Effectively grounded system	
0–650	0–650	1.0	3°	650	650	3°	650	650	3
2,400	2,400	0.5	3°	3,000	3,000	3	3,000	3,000	3
4,160	4,160	0.5	3°	4,500	3,000†	3	6,000	3,000†	3
4,800	4,800	0.5	3	6,000	4,500	3	6,000	6,000	3
6,900	6,900	0.5	3	7,500	6,000	3	9,000	6,000	3
11,500	11,500	0.25	3 or 6‡	12,000	9,000	3	12,000	9,000	3
13,800	13,800	0.25	3 or 6‡	15,000	12,000	3	15,000	12,000	3

° A single three-pole unit is commonly used.

† The use of 3000-V arresters on a 4160-V system requires an X_0/X_1 ratio less than that necessary to make the system "effectively grounded."

‡ Use six capacitor units (0.5 μF per phase) where both of the following conditions apply: (1) Machine is directly connected to the exposed overhead lines, is connected through an autotransformer, or is connected through a Y-Y transformer with both Y's grounded. (2) Machine is ungrounded, is neutral-grounded through a resistance greater than 50 Ω, or is neutral-grounded through a reactance greater than 5 Ω (60-cycle basis). In all other cases three capacitor units (0.25 μF per phase) will suffice.

SOURCE: Beeman—*Industrial Power Systems Handbook*, McGraw-Hill.

choice of the rotating-machinery type of arrester is simplified. Thus, all that need be done is to choose the station-type arrester voltage rating.

Table 3 shows that either 4500- or 3000-V station-type arresters are suitable for 4160-V machines, depending on the type of system grounding. Once again, unless the system is known to be effectively grounded, the higher voltage rating is the safer choice. Thus, either the higher or lower voltage rating would be used, depending on the grounding method employed.

Related Calculations: Other ac equipment that can benefit from lightning-arrester protection includes metal-clad switchgear, generators, transformers, distribution lines, circuit breakers, overhead feeders, etc.

Direct-current motors and generators connected to exposed overhead lines are generally protected by capacitor-type dc arresters. These arresters can be installed at the machine terminals, on the bus, or at the station on each outgoing feeder. Mercury-arc rectifiers and their transformers can be protected by a set of station-type or distribution-type arresters on the supply side of the transformer. If the dc feeders are exposed, suitable dc arresters should also be used at the dc terminals of the rectifier, on the dc bus, or on the exposed dc feeders.

DIRECT-CURRENT GENERATOR SELECTION

Select an alternator for a new industrial plant having an expected demand for 8000 kW. The load served by the generator requires an input voltage of at least 230 V at all ratings up to full load.

Calculation Procedure:

1. Compute the generator amperage output

Use the relation $I = P/E$, where I = generator amperage output; P = generator power output, W; E = generator terminal voltage. For this generator, $I = 1,000,000/230 = 4350$ A.

2. Select the generator rating

Typical standard ratings for direct-current generators are 1000, 2000, 3000, 4000, 5000, 6000, . . . A. Since the required amperage output of this generator is between two standard ratings, select the next higher rating, or 5000 A. This provides extra capacity for a moderate growth in the generator load.

3. Select the type of generator to use

Two classes of dc generators are used today: separately excited and self-excited. The self-excited generator is, in general, more popular for power service. Three types of self-excited dc generators are used for power generation: shunt, series, and compound wound. The first two have *drooping* voltage characteristics, i.e., the generator output voltage decreases as the external load increases.

For general-purpose power-supply service, a cumulative-compound-type generator is preferred. This type of generator may be *flat-compounded* (Fig. 5) so that it produces the same voltage at full load as at no load, or it may be *overcompounded* so it produces a higher voltage at full load than at no load. Other wiring arrangements are *undercompounded* and *differential-compounded*. Since the load served by this generator requires an input voltage of at least 230 V, either a flat-compounded or an overcompounded generator can be used. To provide for the possibility of load growth, an overcompounded generator would be the best choice because it will provide the desired voltage at all loads within the generator rating.

4. Compute the generator efficiency

The manufacturer's engineering data list the losses in the generator. These losses are usually expressed in watts and are shunt-field loss = shunt-field current × generator voltage rating; armature loss = (shunt-field current + armature current)^2 × armature resistance, Ω; series-field loss = (series-field current + armature current)^2 × series field resistance, Ω; and the stray power loss of the generator.

Assume that the sum of all these losses for this generator is given as 80,000 W. Then, since generator efficiency = output, W/(output + losses, W), efficiency = $1,000,000/(1,000,000 + 80,000) = 0.925$, or 92.5 percent.

Related Calculations: Use this general method to choose dc generators for power, emergency marine, and similar applications. Be sure to relate the generator output-voltage character-

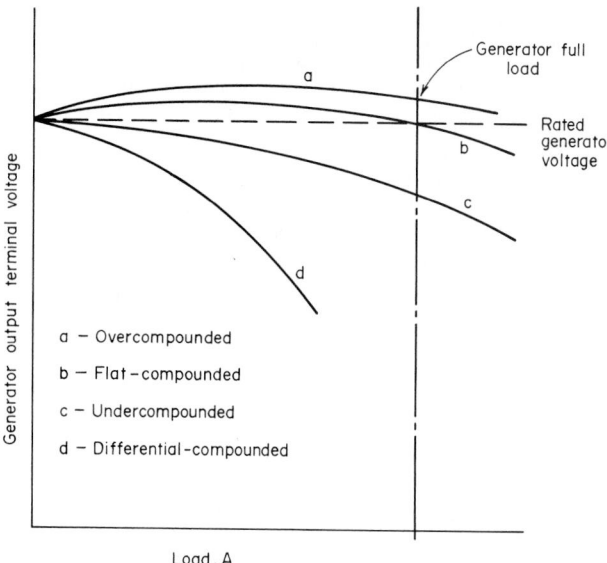

FIG. 5   Voltage characteristics of dc generators.

istic to the load being served. If this important aspect of dc generator choice is overlooked, the unit chosen may be unsuitable for the load. Also, be sure that the rpm of the generator selected can be obtained using a suitable prime mover.

ALTERNATOR SELECTION FOR A KNOWN LOAD

Select an alternator for a new industrial plant having an expected demand for 8000 kW at a power factor (pf) of 0.8. Continuous operation of the plant is expected, and interruptions of generating service must be avoided because they are costly. Hence, reserve capacity must be provided for forced outages and regular maintenance. The plant will have a continuously high load factor.

Calculation Procedure:

1.   *Select the alternator capacity*

In general, the cost per kilowatt of installed generating capacity decreases as the size of the unit increases. Also, the efficiency of steam-driven generating units increases as unit size increases. Operating labor costs are nearly proportional to the number of generating units installed. The efficiency of a generating unit is usually low at light loads and rises to a "best point" somewhere between 75 percent and full load. The exact shape of the efficiency curve of an alternator depends on the type of prime mover and, to a lesser degree, on the size, design, and other features of the alternator and prime mover.

These characteristics of generating units point toward supplying a load with one large alternator. Yet this is rarely possible because of load conditions.

Where continuity of operation is required or where interruptions are costly, reserve capacity is required, not only for forced outages but also for scheduled maintenance outages. It is usually more economical to provide reserve capacity for several smaller generating units than for one large unit. A typical plant might use three units, any two of which can carry the load. This arrangement also provides economically for load growth—adding a unit gives four machines with any three able to carry the full plant load. Where the continuity of service is an important consideration, check to see what standby service, if any, the local utility can provide and what it costs.

Using the above facts as a general guide, make a tentative choice of three units for this 8000-kW plant. The capacities of the three units chosen will be 6000, 4000, and 4000 kW to fulfill the requirement that any two units be able to carry the plant load. Thus, the capacity of the units, when operated in twos, is 6000 + 4000 = 10,000 kW or 4000 + 4000 = 8000 kW.

2. Study the plant load factor

The plant load factor also exerts a strong influence on alternator choice. A low load factor discourages the use of one or two large units because much of the operating time will be on the lower part of the efficiency curve. A special problem in some industries is the weekend load, which may require a special small unit. The load-duration curve provides a valuable insight for relating unit sizes to plant load conditions. Since this plant has a continuously high load factor, the three units will probably be suitable.

3. Select the alternator operating voltage

Figure 6 shows the typical standard voltage ratings for alternators of various capacities. Thus, 4000-kW alternators are rated between 4.16 and 6.9 kV. In the larger sizes—5000 kW and up—13.8 kV is a standard voltage.

Since any two of these alternators must operate in parallel, the same voltage should be used for each. In this plant the smaller alternators will use 6.8 kV because higher voltages are the trend today. Hence, the large generator should be chosen for the same voltage because it must operate with either smaller unit.

When choosing alternator voltage, keep in mind the trend toward generating at higher voltages (up to 13.8 kV) in industrial plants and distributing at that voltage to secondary substations. Where a later change from delta to Y connections to increase the alternator voltage is planned, be sure the alternator is now connected delta.

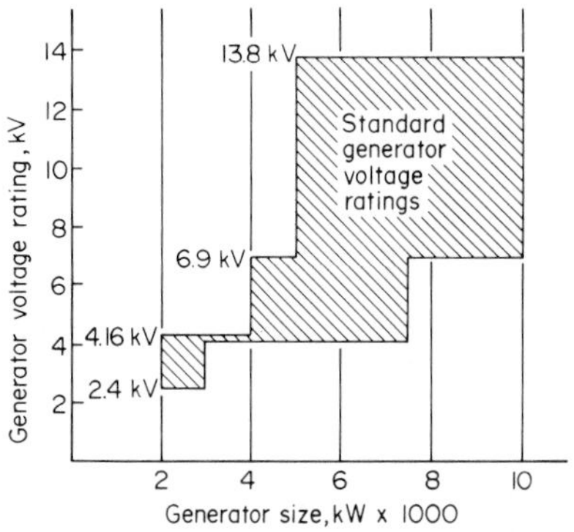

FIG. 6   Standard alternator voltage ratings.

4. Check the generator regulation

The standard alternator regulation of 40 to 45 percent at 0.8 pf will satisfy most industrial-plant needs. Where the plant has one load which is much larger than any other, such as the chipper motor in a paper mill, a different regulation from that cited above might be required. The exact regulation required varies with the motor rating and application.

5. Check the paralleling characteristics of the alternators

To parallel two alternators, the voltage, frequency, waveform, phasing, and phase rotation of both alternators must be the same. During paralleling, slight adjustments can be made in the alternator voltage and frequency. Check the specification sheet of each alternator to see that the requirements listed above are met.

Related Calculations: Use this general method to choose alternators for commercial, marine, portable, and similar applications. Be sure to use the manufacturer's specification sheet when considering a specific alternator.

SELECTING ELECTRIC-MOTOR STARTING AND SPEED CONTROLS

Choose a suitable starter and speed control for a 500-hp (372.8-kW) wound-rotor ac motor that must have a speed range of 2 to 1 with a capability for low-speed jogging. The motor is to operate at about 1800 r/min with current supplied at 4160 V, 60 Hz. An enclosed starter and a controller are desirable from the standpoint of protection. What is the actual motor speed if the motor has four poles and a slip of 3 percent?

Calculation Procedure:

1. Select the type of starter to use

Table 4 shows that a magnetic starter is suitable for wound-rotor motors in the 220- to 4500-V and 5 to 1000-hp (3.7 to 745.7-kW) range. Since the motor is in this voltage and horsepower range, a magnetic starter will probably be suitable. Also, the magnetic starter is available in an enclosed cabinet, making it suitable for this installation.

Table 5 shows that a motor starting torque of approximately 200 percent of the full-load motor torque and current is obtained on the first point of acceleration.

2. Compute the full-load speed of the motor

Use the relation $S = [(100 - s)/100]120f/n$, where S = motor full-load speed, r/min; s = slip, percent; f = frequency of supply current, Hz; n = number of poles in the motor. For this motor, $S = [(100 - 3)/100]120(60)/4 = 1750$ r/min.

3. Choose the type of speed control to use

Table 5 summarizes the various types of adjustable-speed drives available today. This listing shows that power-operated contactors used with wound-rotor motors will give a 3:1 speed range with

TABLE 4 Typical Alternating-Current Motor Starters°

Motor type	Starter type	Typical range		
		Voltage	hp	kW
Squirrel cage	Magnetic, full voltage	110–550	1.5–600	1.1–447
	With fusible or nonfusible disconnect or circuit breaker	208–550	2–200	1.5–149
	Reversible	110–550	1.5–200	1.1–149
	Manual, full voltage	110–550	1.5–7.5	1.1–5.6
	Manual, reduced voltage, autotransformer	220–2500	5–150	3.7–112
	Magnetic, reduced voltage, autotransformer	220–5000	5–1750	3.7–1305
	Magnetic, reduced voltage, resistor	220–550	5–600	3.7–447
Wound rotor	Magnetic, primary and secondary control	220–4500	5–1000	3.7–746
	Drums and resistors for secondary control	1000 max	5–750	3.7–559
Synchronous	Reduced voltage, magnetic	220–5000	25–3000	19–2237
	Reduced voltage, semimagnetic	220–2500	20–175	15–131
	Full voltage, magnetic	220–5000	25–3000	19–2237
High-capacity induction	Magnetic, full voltage	2300–4600	To 2250	To 1678
	Magnetic, reduced voltage	2300–4600	To 2250	To 1678
High-capacity synchronous	Magnetic, full voltage	2300–4600	To 2500	To 1864
	Magnetic, reduced voltage	2300–4600	To 2500	To 1864
High-capacity wound rotor	Magnetic, primary and secondary	2300–4600	To 2250	To 1678

°Based on Allis-Chalmers, General Electric, and Westinghouse units.

TABLE 5 Adjustable-Speed Drives

Drive features	Drive types						
	Constant-voltage dc	Adjust.-voltage dc motor-generator set	Adjust.-voltage rectifier	Eddy-current clutch	Wound-rotor ac, standard	Wound-rotor thyratron	Wound-rotor dc-motor set
Power units required	Rectifier, dc motor	Ac motor, dc generator, dc motor	Rectifier, reactor,[a] dc motor	Ac motor, eddy-current clutch	Ac motor	Ac motor, thyratrons	Ac motor, dc motor, rectifier
Normal speed range	4:1	8:1 c-t+[b] 4:1 c-hp[c]	8:1 c-t+ 4:1 c-hp[c]	34:1, 2 pole; 17:1, 4 pole	3:1	10:1[c]	3:1
Low speed for jogging	No[d]	Yes	Yes	Yes	Yes	Yes	Yes
Torque available	c-hp	c-t	c-t	c-t	c-t	c-t	c-t, c-hp
Speed regulation	10–15%	5% with regulator	5% with regulator	2% with regulator	Poor	±3%	5–7½%
Speed control	Field rheostat	Rheostats or pots	Rheostats or pots	Rheostats or pots	Steps, power contactors	Rheostats or pots	Rheostats or pots
Enclosures available	All	All	All	Open[e]	All	All	All
Braking:							
Regen	No	Yes	No	No	Yes	Yes	No
Dynamic	Yes	Yes	Yes	No[f]	Yes	Yes	Yes
Multiple operation	Yes	Yes	Yes	Yes	Yes	Yes	No
Parallel operation	Yes	Yes	Yes	Yes	No	Yes	Yes
Controlled acceleration, deceleration	Yes	Yes	Yes	Yes	No	Yes	No
Efficiency	80–85%	63–73%	70–80%	80–85%	80–85%	80–85%	80–85%
Top speed at maximum torque	83–87%	60–67%	60–70%	29%	29%	85–90%	73–78%
Rotor inertia[g]	100%[h]	100%	100%	75%	90%	90%	175%
Starting torque	200–300%	200–300%	200–300%	200–300%	200%	200–300%	200–300%
Number of comm. rings	1 comm.	2 comm.	1 comm.	None	1 set rings	1 set rings	1 comm., 1 set rings

[a]Used only in saturable-reactor designs.
[b]c-t—constant-torque; c-hp—constant horsepower.
[c]Units of 200:1 speed range are available.
[d]Low speed can be obtained by using armature resistance.
[e]Totally enclosed units must be water- or oil-cooled.
[f]Eddy-current brake may be integral with unit.
[g]Based on standard dc motor.
[h]Normally is a larger dc motor since it has slower base speed.

low-speed jogging. Since a 2:1 speed range is required, the proposed controller is suitable because it gives a wider speed range than needed.

Note from Table 5 that if a wider speed range were required, a thyratron control could produce a range up to 10:1 on a wound-rotor motor. Also, a wound-rotor dc motor set might be used too. In such an arrangement, an ac and dc motor are combined on the same shaft. The rotor current is converted to dc by external silicon rectifiers and fed back to the dc armature through the commutator.

Related Calculations: Use the two tables presented here to guide the selection of starters and controls for ac motors serving industrial, commercial, marine, portable, and residential applications.

To choose a dc motor starter, use Table 6 as a guide.

Speed controls for dc motors can be chosen by using Table 7 as a guide. Dc motors are finding increasing use in industry. They are also popular in marine service.

TABLE 6   Direct-Current Motor Starters

Type of starter	Typical uses
Across-the-line	Limited to motors of less than 2 hp (1.5 kW)
Reduced voltage, manual control (face-plate type)	Used for motors up to 50 hp (37.3 kW) where starting is infrequent
Reduced voltage, multiple switch	Motors of more than 50 hp (37.3 kW)
Reduced voltage, drum switch	Large motors; frequent starting and stopping
Reduce voltage, magnetic switch	Frequent starting and stopping; large motors

TABLE 7   Direct-Current Motor-Speed Controls

Type of motor	Speed characteristic	Type of control
Series-wound	Varying; wide-speed regulation	Armature shunt and series resistors
Shunt-wound	Constant at selected speed	Armature shunt and series resistors; field weakening; variable armature voltage
Compound-wound	Regulation about 25 percent	Armature shunt and series resistors; field weakening; variable armature voltage

BASIC SHORT-CIRCUIT CURRENT DETERMINATION

A 50-hp (37.4-kW) ac motor draws 63 A at full load. This 40-Ω apparent impedance motor is supplied from an "infinite" bus through a transformer with a rated output of 440 V, 200 A, and a 0.2-Ω impedance. Determine the short-circuit current flow if a fault occurs between the transformer and the motor. What will be the effect of using a 2000-A, 0.02-Ω impedance transformer for the same motor to provide for a load growth?

Calculation Procedure:

1. *Sketch the circuit hookup*

Figure 7 shows the typical circuit hookup for an installation of this type. The circuit breaker at point X must have a large enough rating to handle the short-circuit current.

2. *Compute the short-circuit current with the small transformer*

With a short circuit at F, the only impedance limiting the short-circuit current flow is the transformer impedance of 0.2 Ω, because the current will take the path of least resistance. The motor apparent impedance of 40 Ω is so much larger than the transformer impedance that the short-circuit current will rush out at F.

Compute the short-circuit current from $I_S = E/Z_t$, where I_S = short-circuit current, A; E = bus voltage rating, i.e., transformer-output rating, V; Z_t = transformer impedance, Ω. Thus, I_S = 440/0.2 = 2200 A. Hence, the circuit breaker must handle at least 2200 A to protect this circuit.

3. *Compute the short-circuit current with the large transformer*

Use the same relation as in step 2. Or, I_S = 440/0.02 = 22,000 A. Thus, the larger transformer, installed to handle the greater load, will require a circuit breaker with a much higher rating. Note

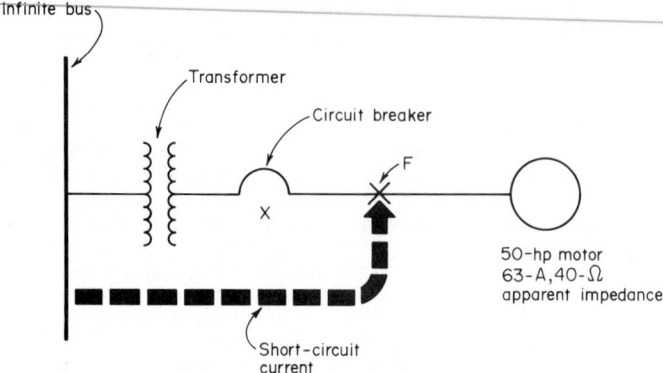

FIG. 7   Typical motor circuit with a step-down transformer and protective circuit breaker.

that the motor-load current will remain the same, yet the short-circuit current increases tenfold as the system load increases.

   Related Calculations: This simple short-circuit computation shows the basic procedure to use. As a circuit and its components become more complex, so do the short-circuit computations. Typical methods are shown in the following calculation procedure.

POWER-SYSTEM SHORT-CIRCUIT CURRENT

A three-phase power system has two generating stations that supply one substation. If the generator ratings, line voltages, and line reactances are as listed below, determine the short-circuit current when a fault occurs in the distribution line beyond the substation.

Unit	Rating, kVA	Line-to-line kV	Reactance, Ω
Generator x	80,000	13.8	2.3
Generator y	85,000	13.8	1.1
Line xz	80,000	13.8	0.65
Line yz	85,000	13.8	0.40
Line z to short	150,000	13.8	0.45

Calculation Procedure:

1.   *Express the system reactances on a per-unit basis*

Select a kva (kilovolt-ampere) base, such as 50,000, 100,000, 200,000 kVA, etc. Any easily manipulated value can be used. Once a kva base is chosen, compute the reactance per unit X_{pu} from $X_{pu} = $ (kva base)$X/(kv^2 x1000)$, where X = unit reactance, Ω; kv = line-to-line kV for the unit.

   Selecting a kva base of 100,000 kVA and using this relation for generator x, we find $X_{pu} = (100,000)(2.3)/(13.8)^2(1000) = 1.21$. Using the same kva base, compute the per-unit reactance for the other generator and each line. Draw a single-line diagram of the system. Mark the reactance values on the system single-line diagram, Fig. 8a.

2.   *Sketch the network representing this power system*

Figure 8b shows the network representing this system. It consists of a parallel network of two parts, each part having two X_{pu} reactances in series.

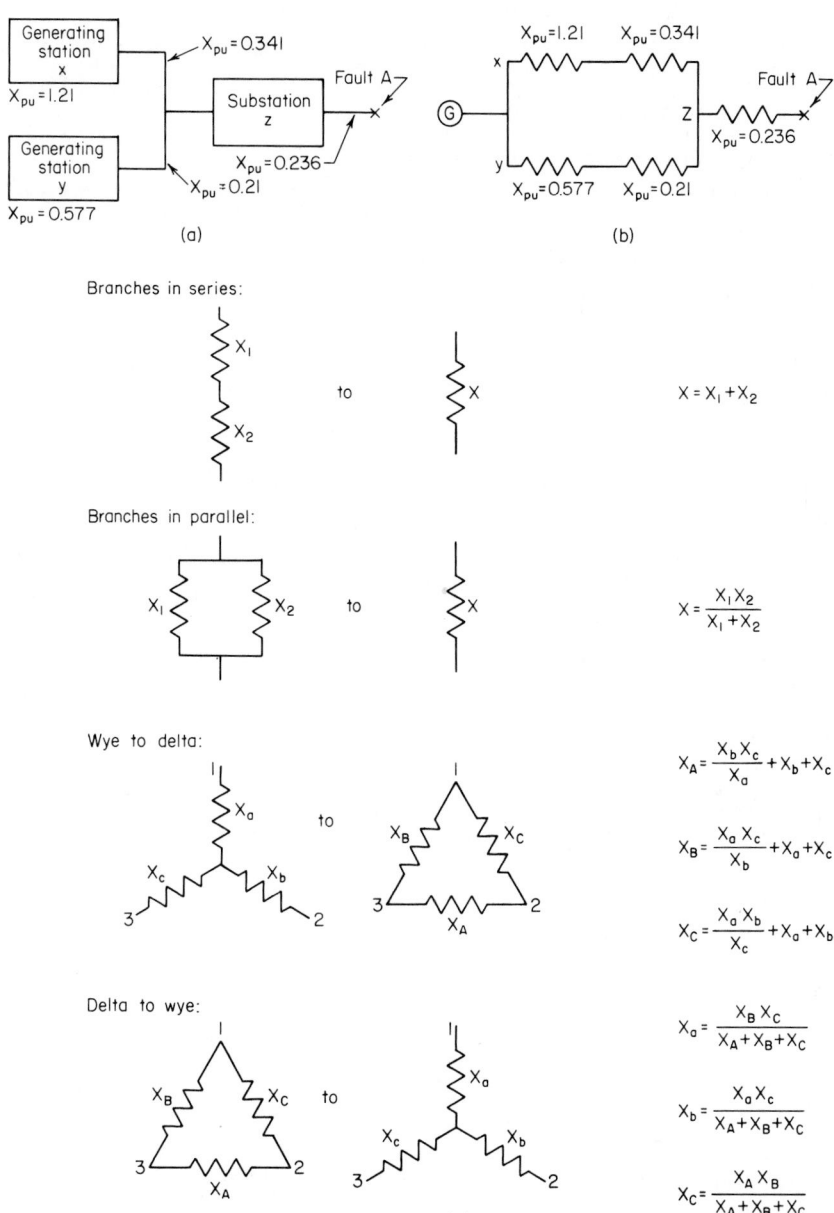

FIG. 8 (*a*) Electric power system with two generating stations and one substation; (*b*) circuit diagram of generating system in (*a*); (*c*) reactances for various circuit arrangements.

3. *Compute the equivalent reactance between the generator and the fault*

Use the relation $X_{eq} = X_x X_y/(X_x + X_y) + X_{3A}$, where X_{eq} = equivalent reactance of the network, per unit; X_x = total per unit reactance for leg xz; X_y = total per unit reactance for leg yz; X_{zA} = total per unit reactance for leg zA. Substituting gives $X_{eq} = [(1.21 + 0.341)(0.577 + 0.21)]/(1.21 + 0.341 + 0.577 + 0.21) + 0.236 = 0.759$ per unit.

4. *Compute the normal line current at the fault*

Use the relation I_n = kva base/$kv\sqrt{3}$ for the normal line current at the fault, where I_n = normal line current at the fault, A; other symbols as before. Thus, $I_f = 100,000/13.8\sqrt{3} = 4175$ A.

5. *Compute the current in each line at the fault*

Use the relation $I_f = I_n/X_{eq}$ for a three-phase short circuit, where I_f = current in each line at the fault, A; other symbols as before. Thus, $I_f = 4175/0.759 = 5500$ A.

6. *Compute the fault current in each line*

For generator x and transmission line xz, $I_x = I_f(X_{pux} + X_{puxz})/\Sigma X_{pu}$, where I_x = generator and transmission-line fault current, A; X_{pux} = generator reactance, per unit; X_{puxz} = transmission-line reactance, per unit; ΣX_{pu} = sum of the individual network reactances; other symbols as before. Substituting gives $I = 5500(1.21 + 0.341)/(1.21 + 0.341 + 0.577 + 0.21) = 3650$ A.

Using a similar relation for generator y and transmission-line yz, we find $I_y = 5500(0.577 + 0.21)/(1.21 + 0.341 + 0.577 + 0.21) = 1850$ A. As a check, the sum of the two generator and transmission-line currents should equal I_f, or $3650 + 1850 = 5500$ A = I_f.

Related Calculations: The general method presented here is valid for simple and complex short-circuit calculations. Summarized, this procedure is as follows. (1) Obtain equipment and line reactances from equipment characteristics and tabulated data. (2) Draw the system single-line diagram. (3) Convert reactances to per unit on a base kva. (4) Combine reactances to obtain a reactance diagram without series reactances. (5) Compute the fault current or kva.

To combine reactances, use the equations shown in Fig. 8c.

Last, Table 8 shows the short-circuit current theoretically possible at various operating voltages and kva ratings.

TABLE 8   Theoretical Short-Circuit Currents

Short-circuit energy, three-phase kVA	Short-circuit amperage per phase for various voltages			
	15,000 V	2,500 V	480 V	208 V
25,000	1,000	5,800	31,000	70,000
50,000	2,000	11,500	63,000	140,000
100,000	4,000	23,000	125,000	280,000

Note: These are calculated current values, discounting impedance that may be met in typical plant systems on secondaries of unit substations.

TRANSFORMER CHARACTERISTICS AND PERFORMANCE

A 60-Hz 1000-kVA three-winding transformer is rated at 4800-V primary voltage and 600- and 480-V secondary voltages. The transformer has 800 primary turns, and the rating of each secondary winding is 500 kVA. Compute the number of turns in each secondary winding, the rated primary current at unit power factor (pf), rated primary current at 0.8 pf, lagging current, rated current of the 600- and 480-V secondary windings, primary current when the rated current flows in the 480-V winding with pf = 1.0, and the rated current flow in the 600-V winding with a 0.8 lagging pf.

Calculation Procedure:

1. Compute the number of turns in the secondary windings

Use the relation $N_s = v_s N_p / v_p$, where N_s = number of turns in secondary winding; v_s = voltage rating of secondary winding; N_p = number of turns in the primary winding; v_p = voltage rating of the primary winding.

For the 600-V secondary winding, $N_s = 600(800)/4800 = 100$ turns. For the 480-V secondary winding, $N_s = 480(800)/4800 = 80$ turns.

2. Compute the rated primary current

Use the relation $I_p = va/v_p$, where I_p = primary current, A; va = transformer volt-ampere rating; v = transformer primary voltage rating. For this transformer at unity power factor, $I_p = 1,000,000/4800 = 208$ A.

At 0.8-pf lagging current, the rated primary current is the same as the rated primary current at unity power factor, or 208 A. The reason for this is that the power factor relates to the transformer or secondary load. The power factor of the load does not affect the rated primary-winding current.

3. Compute the secondary-winding current

Use the relation $I_s = va_s/v_s$, where I_s = secondary current flow, A; va_s = secondary volt-ampere rating; v_s = secondary voltage rating.

For the 600-V secondary winding, $I_s = 500,000/600 = 833$ A. For the 480-V secondary winding, $I_s = 500,000/480 = 1041$ A.

4. Compute the primary current at the given secondary loads

Use the relation $I_p = (va_{s1}/v_p)(\cos^{-1} pf_1) + (va_{s2}/v_p)(\cos^{-1} pf_2)$, where I_p = primary current, A; va_{s1} = volt-amperes of the first secondary coil; v_p = transformer primary voltage; $\cos^{-1} pf$ = angle, expressed in degrees, whose cosine = the power factor of the load on the first secondary coil; other symbols are the same, except that they refer to the other secondary coil of the transformer.

Thus, $I_p = (600 \times 833/4800)\underline{/-36.8°} + (480 \times 1041/4800) \underline{/0°} = 104.1 \cos 36.8°$ $- j104.1 \sin 36.8° + 104.1 \cos 0° + 104.1 \sin 0° = 83.4 - j62.4 + 0' + 104.1$, or $I_p = 187.5$ $- j62.4$. Converting this expression using vector algebra, we get $I_p = (187.5^2 - 62.4^2)^{0.5} = 197.8$ A.

Related Calculations: Use this general method to analyze transformers with one or more secondary coil windings used for power, distribution, residential, or commercial service.

TRANSFORMER SELECTION FOR AN INDUSTRIAL LOAD

Select a three-phase transformer for an industrial plant having an expected load of 300 kVA located 400 ft (121.9 m) from the transformer. The transformer will be located in an area of high humidity. What are the transformer voltage drop, copper loss, and core loss if the primary voltage is 4160 and the secondary 480 V with a power factor of 0.80? What sound level can be expected with this transformer?

Calculation Procedure:

1. Choose between an outdoor substation and a load center

Outdoor substations were once the popular way of supplying power to industrial plants. Today a load center is often used in place of the outdoor substation. In a load center, the primary power is brought close to the plant load instead of being ended at an outdoor substation.

Load centers are not always the most economical choice, however. Here are general guides for the choice of outdoor substation versus load-center distribution. With a 120/240-V single-phase load, a load center is usually best if the transformer-to-load distance is more than 160 ft (48.8 m) for a 25-kVA load, more than 90 ft (27.4 m) for a 75-kVA load, and more than 60 ft (18.3 m) for a 200-kVA load. For a 480-V three-phase distribution system, load centers are generally best for a 150-kVA load more than 400 ft (121.9 m) from the transformer; a 300-kVA load more than 300 ft (91.4 m); a 750-kVA load more than 150 ft (45.7 m) from the transformer.

Since this installation serves a 300-kVA three-phase load 400 ft (121.9 m) from the transformer, a load-center distribution system will probably be best, according to the above guide. Hence, this will be the tentative first choice for this plant.

2. Choose the type of transformer cooling

Basically, two types of transformer cooling are used today: liquid and air. The usual liquid coolants currently used are askarel, a synthetic nonflammable liquid, and mineral oil. The liquid-cooled transformer predominates in sizes over 500 kVA. By far the largest number of transformers in terms of kva capacity are oil-cooled.

Air-cooled transformers are termed *dry-type* units. Dry-type transformers can be sealed units, open units, forced-air-cooled, or self- and forced-air-cooled.

Table 9 shows the important factors that should be considered in choosing load-center transformers. Study of this table shows that in areas of high humidity an askarel-cooled transformer is generally chosen.

3. Specify the transformer variables

Table 10 lists many of the variables for load-center transformers. Thus, askarel-filled load-center transformers normally use class A insulation, have an average temperature rating of 55°C rise and 65°C hottest-spot rise. The weight of an askarel-filled unit is about 1.25 times the weight of an oil-filled transformer; the floor space required by the two is the same.

4. Determine the transformer electrical characteristics

Table 11 lists the electrical ratings of typical modern transformers. This table shows that a 300-kVA rating is a standard one. Hence, such a unit is readily available from manufacturers.

5. Determine the transformer losses

Table 12 shows typical losses for three-phase distribution transformers with ratings to 400 kVA. Thus, at full load on a transformer with a 0.80 power factor, the voltage drop through the trans-

TABLE 9   Transformer Coolant Characteristics°

	Dry	Askarel
Exposure to lightning		
Where transformers connect directly to lightning-exposed circuits with only usual lightning protection	No	Yes
If lightning exposure is negligible or if unit is suitably protected, if feeder cable is underground, or if overhead supply having usual lightning protection feeds into primary cable at least 1500 ft (457.2 m) long	Yes	Yes
Location and atmospheric conditions		
Relatively clean atmosphere: assembly plants, etc.	Yes	Yes
Bad atmosphere: foundries, cement, flour mills, etc.	No	Yes
Areas of high humidity or of possible flooding	No	Yes
Acid, oil, or corrosive vapors present	No	Yes
Units to be installed in hazardous locations (see National Electrical Code)	No	Yes
Transformers to be overhead on platforms or roof trusses, all other conditions being satisfactory	Yes	No

°In general, the sealed dry-type transformer can be used under all the given conditions that would rule out ventilated dry-type units.

TABLE 10   Comparison of Load-Center Unit Substation Transformer Sections

Type of transformer	Liquid-filled		Dry type		
	Oil	Askarel	Open ventilated	Sealed class B	Sealed class H
Impulse strength	100%	100%	50%°	50%°	50%°
Total loss at 75°	100%	100%	100%	100%	100%
Insulation	Class A	Class A	Class B	Class B	Class H
Temperature ratings:					
Average rise	55°C	55°C	80°C	120°C	120°C°
Hottest-spot rise	65°C	65°C	110°C	140°C	140°C°
Audio sound level	X dB	X dB	(X + 10 dB°)	(X + 10 dB°)	(X + 10 dB°)
Weights	100%	125%	80%	125%	125%
Dimensions:					
Floor space	100%	100%	100%	120%	120%
Height	100%	100%	90%	110%	110%
Normally available for application:					
Indoor or outdoor	Outdoor	All	Indoor only	All†	All†
Submersible	Submersible				
Fire and explosion-resistant	No	Yes	Yes	Yes (plus)	Yes (plus)
Maintenance required:					
Liquid	Normal	Infrequent	None	None	None
Internal cleaning	None	None	Frequent	None	None
External cleaning and painting expense	Normal	Normal	Subnormal	Minimum	Minimum
Special precautions before energizing either initially or after shutdown	None	None	Yes	None	None

°Not yet covered by industry standards.
†Applicable for all types of installation assuming no exposure to lightning or assuming adequate protection against impulse voltages can be provided.
SOURCE: Beeman—*Industrial Power Systems Handbook*, McGraw-Hill.

former is 2.83 percent. The copper loss is 0.92 percent, and the core loss is 0.80 percent. Note that Table 12 also shows the impedance for the transformer.

6.   *Determine the transformer and sound level*

Table 13 shows that the average factory has a sound level of 70 to 75 dB. Table 14 shows that a 300-kVA oil-immersed transformer has a sound level of 55 dB. Since the transformer sound level is less than the ambient sound level in the average factory, the transformer chosen is a suitable unit. As a general guide, choose a transformer having a decibel rating *lower* than the ambient sound level of the area in which the transformer will be installed.

    Related Calculations: Use this general procedure to choose transformers for industrial, commercial, and residential use. As a general guide, the following definitions are useful. *Power transformers* are used in generating plants to step up voltage and in substations to step down voltage. Power transformers are usually rated at 500 kVA and larger.

    Distribution transformers step voltage down to 600 or 480 V for industrial use or 240 and 120 V for residential or commercial use. *Instrument transformers*, classed *potential* or *current*, serve low-voltage meters and relays. *Specialty transformers* include units used to change the voltage for specific applications such as signs, arc lamps, bells, etc.

TABLE 11   Electrical Ratings of Transformers for Power and
Distribution Service°

kVa rating		Voltage ratings	
Single-phase	Three-phase	Primary	Secondary
1.5	5.0	120	120
2.5	7.5	240	240
3.0	10	480	480
5.0	15	600	600
7.5	25	2,400	2,400
10	50	2,500	4,160
15	75	4,160	6,900
25	100	4,330	11,500
37.5	150	4,800	13,800
50	200	6,900	23,000
75	300	11,500	34,500
100	450	13,800	46,000
150	600	25,000	69,000
200	750	34,500	
250	1,000	46,000	
300	1,500	69,000	
500	2,000	92,000	
1,000	2,500	115,000	
1,250	3,000	138,000	
1,500	5,000	161,000	

°Partial listing of commercially available units.

SYSTEM POWER-FACTOR ANALYSIS

A system has three types of loads: lighting, induction motors, and synchronous motors. Determine the system power factor for the following conditions: lighting load = 400 kVA at unity (1.0) power factor; induction-motor load = 700 kVA at 0.85 power factor, lagging; synchronous motor load = 300 kVA at 0.75 power factor, leading. What is the system power factor?

Calculation Procedure:

1.   *Determine the lighting-load kvar*

The 400-kVA lighting load has a unity (1.0) power factor. With a unity power factor, the kilovoltampere (kva) load = the kilowatt load. Also, since there is no reactive current when the power factor is unity, the lighting-load kvar = 0.

2.   *Compute the induction-motor kvar*

Find the induction-motor kilowatt load from $kw = kva(pf)$, where kva = induction-motor kva; pf = induction-motor power factor. Or, $kw = 700(0.85) = 595$ kW.
      Compute the induction-motor kvar from $kvar = (kva^2 - kw^2)^{0.5}$, or $(700^2 - 595^2)^{0.5} = 370$ kvar.

3.   *Compute the synchronous-motor kvar*

As in step 2, for the synchronous motor $kw = kva(pf) = 300(0.75) = 225$ kW. Then $kvar = (kva^2 - kw^2)^{0.5}$, as in step 2. Or $kvar = (300^2 - 225^2)^{0.5} = 198.5$ kvar.

4.   *Compute the system kw and kvar*

The kilowatt load is the sum of the individual kilowatt loads, or lighting kw + motor $kw = 400 + 595 + 225 = 1220$ kW. The kvar of the system is found in the same way, except that the

TABLE 12   Electrical Characteristics of Single-Phase and Three-Phase 60-Hz Distribution Transformers

Size, single-phase, kVA	Impedance, percent $Z/\phi°$	Percent voltage drop through transformer with full-load current			Cu loss, %	Core loss, %
		97% pf	80% pf	50% pf		
Single-phase transformers; voltage rating: 2400/4160 to 120/240 V						
3	$2.7/32.90°$	2.56	2.7	2.40	2.27	0.93
5	$2.7/38.40°$	2.46	2.7	2.51	2.12	0.72
7½	$2.7/43.5°$	2.35	2.66	2.58	1.96	0.64
10	$2.7/45.9°$	2.29	2.66	2.62	1.88	0.57
15	$2.8/51.6°$	2.21	2.71	2.76	1.74	0.51
25	$2.8/56°$	2.08	2.64	2.79	1.56	0.46
37½	$2.9/61.8°$	1.95	2.62	2.9	1.37	0.394
50	$2.9/64.7°$	1.84	2.56	2.89	1.24	0.372
75	$3.5/69.4°$	1.99	2.94	3.45	1.24	0.370
100	$3.5/69.9°$	1.96	2.93	3.45	1.20	0.370
Three-phase transformers; voltage rating: 2400/4160 to 240/480 V						
5	$4.2/41.3°$	3.73	4.19	4.00	3.12	1.28
10	$3.8/46.80°$	3.20	3.74	3.70	2.55	0.88
15	$3.75/53.1°$	2.91	3.60	3.72	2.2	0.74
25	$3.9/57.2°$	2.85	3.66	3.89	2.05	0.63
37½	$4.02/60°$	2.79	3.70	4.02	1.93	0.58
50	$4.03/61.8°$	2.72	3.66	4.02	1.82	0.51
75	$3.54/63.1°$	2.32	3.17	3.53	1.60	0.53
100	$3.7/65°$	2.31	3.26	3.68	1.56	0.57
200	$3.92/71.4°$	2.11	3.23	3.84	1.25	0.47
300	$3.61/75.25°$	1.74	2.83	3.46	0.92	0.80
400	$3.77/71.9°$	1.97	3.09	3.70	1.17	0.75

lighting-load kvar is zero and the leading (synchronous-motor) kvar offsets the lagging (induction motor) kvar, or 370.0 = 198.5 = 171.5 kvar.

The reason for taking the difference between the induction- and synchronous-motor kvar is because a synchronous motor can supply kvar to the system.

5. Compute the system kva and pf

Use the relation $kva = (kw^2 + kvar^2)^{0.5} = [(1220)^2 + (171.5)^2]^{0.5} = 1333$ kVA. Compute the system power factor from pf = system kw/system kva = 1220/1333 = 0.915 lagging. The power factor is termed *lagging* because the lagging or induction-motor kvar exceeds the leading or syn-chronous-motor kvar. Capacitors, synchronous motors, or synchronous generators could be used to improve the power factor of this system.

Related Calculations: Use this general method to analyze the power factor of any power system—industrial, commercial, or residential.

TABLE 13 Average Sound Levels for Various Occupancies

Occupancy	Decibel range
Apartments and hotels	35–45
Average factory	70–75
Classrooms and lecture rooms	35–50
Hospitals, auditoriums, and churches	35–40
Private offices and conference rooms	40–45
Offices:	
Small	53
Medium (3 to 10 desks)	58
Large	64
Factory	61
Stores:	
Average	45–55
Large (5 or more clerks)	61
Residence:	
Without radio	53
With radio, conversation	60
Radio, recording, and television	25–30
Theaters and music rooms	30–35
Street:	
Average	80

Note: Manufacturers now sound-rate dry-type transformers to meet or exceed NEMA audible-sound-level standards. Select a transformer with a decibel rating lower than the ambient sound level of the area in which it is to be installed.

TABLE 14 NEMA Audible Sound Levels

For dry-type general-purpose specialty transformers 600 V or less, single- or three-phase		Oil-immersed and dry-type self-cooled transformers, 15,000-V insulation class and below			
Transformer rating, kVA	Average sound level, dB	kVA	Oil immersed, dB	Dry type, dB Ventilated	Sealed
0–9	40	0–300	55	58	57
10–50	45	301–500	56	60	59
51–150	50	501–700	57	62	61
151–300	55	701–1000	58	64	63
		1001–1500	60	65	64
301–500	60	1501–2000	61	66	65
		2001–3000	63	68	66

POWER-FACTOR DETERMINATION AND IMPROVEMENT

Determine the power factor in a 440-V three-phase power system when the load draws 135 A 85 kW. If it is desired to improve the power factor of this circuit to 88 percent with capacitors, what capacitor rating is required?

Calculation Procedure:

1. Compute the circuit kva

For a three-phase ac circuit, use the relation $kva = \sqrt{3}va/1000$, where kva = kilovolt-amperes of the circuit; v = circuit voltage; a = circuit amperage. Or, $kva = \sqrt{3} \times (440)(135)/1000 = 103$ kVA.

2. Compute the circuit power factor

Use the relation power factor = $pf = kw/kva$, where kw = circuit kilowatt load. Or, $pf = 85/103 = 0.825$, or 82.5 percent.

3. Compute the circuit kvar

The total current in an ac circuit is usually made up of two components: power-producing current and magnetizing current. Other terms used for these currents are *working current* and *reactive current*, respectively. From a power standpoint, the terms used are *true power* in W or kW, and reactive power, in var or kvar. The abbreviation "var" stands for volt-ampere reactive.

In an ac circuit, $kvar = (kva^2 - kw^2)^{0.5}$. Thus, for this circuit, $kvar = (103^2 - 85^2)^{0.5} = 58.5$ kvar.

4. Compute the kvar at the new power factor

At the new power factor of 88 percent, the circuit kw is the same, or 85. However, the circuit kva will be kw/new pf, or $85/0.88 = 96.5$ kVA.

The new circuit $kvar = (kva^2 - kw^2)^{0.5} = (96.5^2 - 85^2)^{0.5} = 45.6$ kvar. Thus, the circuit provides 45.6 of the 58.5 kvar required by the load.

5. Compute the required capacitor kvar

The capacitor must provide the difference between the load and circuit kvar, or $58.5 - 45.6 = 12.9$ kvar, say 13 kvar. Thus to improve the power factor of this circuit from 82.5 to 88 percent, a 13-kvar capacitor is required.

Related Calculations: The method given above is useful for determining the capacitor, synchronous-motor, or synchronous-condenser rating required to produce a given power-factor increase. As a general rule, the synchronous condenser is usually too costly a device for power-factor improvement service in industrial plants. Hence, it is seldom used for this purpose in industrial plants.

WIRING-SIZE CHOICE FOR PRIMARY DISTRIBUTION

Determine the proper size of underground cable to use to supply a 1000-kW load at 80 percent power factor at a distance of 0.4 mi (643.7 m) with an allowable voltage drop of 5.0 percent. The receiving-end voltage is 2200 V, three-phase.

Calculation Procedure:

1. Compute the system kva-miles

Use the relation kva-miles = (load, kW/pf)(line length, mi), where pf = system power factor. For this system, kva-miles = $(1000/0.80)(0.4) = 500$ kVA·mi (804.7 MVA·m).

2. Choose the wire size to use

Table 15 shows that with an 80 percent power factor, 2400 V, three-phase line, a 4/0 cable will have a 1 percent voltage drop per 174.0 kVA-mi (235.3 MVA·m). However, the receiving voltage of this system is 2200 V. To correct for the difference between the actual and tabulated receiving voltage, multiply the tabulated kva-miles by the ratio (actual receiving voltage/tabulated receiving voltage)^2, or $(2200/2400)^2(174.0) = 146.2$ kVA·mi (159.3 MVA·m) per 1.0 percent voltage drop.

3. Compute the actual voltage drop

Use the relation actual voltage drop, percent = system kva-miles/kva-miles per 1 percent voltage drop, or $500/146.2 = 3.41$ percent actual voltage drop. Since the allowable voltage drop is 5 percent, this is within the desired range.

TABLE 15 Voltage Regulation of 5- and 10-kV Cables and Overhead-Line-Wire Primary Distribution Voltages, kVA·mi (kVA·km) per 1% Voltage Drop, Balanced Load°

Underground distribution, 5- and 10-kV single-conductor cables, annealed copper

Cable size	Voltage rating, kV	Power factor								
		4160 V, three-phase			2400 V, three-phase			2400 V, single-phase		
		kVA·mi (kVA·km) per 1% drop			kVA·mi (kVA·km) per 1% drop			kVA·mi (kVA·km) per 1% drop		
		97%	80%	50%	97%	80%	50%	97%	80%	50%
1/0	10	295.0 (474.7)	297.0 (477.9)	355.0 (571.2)	98.2 (158.0)	98.7 (158.8)	118.0 (189.9)	49.0 (78.9)	49.7 (79.9)	59.0 (94.9)
1/0	5	300.0 (482.8)	306.5 (493.3)	377.0 (606.6)	100.0 (160.9)	102.2 (164.5)	125.7 (202.3)	50.0 (80.5)	51.1 (82.2)	62.8 (101.1)
2/0	10	364.0 (585.8)	354.0 (569.7)	407.0 (654.9)	120.7 (194.2)	118.0 (189.9)	135.5 (218.1)	60.6 (97.5)	59.0 (94.9)	68.0 (109.4)
4/0	10	545.0 (877.1)	492.0 (791.8)	525.0 (844.7)	181.7 (292.4)	164.0 (263.9)	175.0 (281.6)	91.0 (146.5)	82.0 (131.9)	87.5 (140.8)
4/0	5	560.0 (901.2)	523.0 (841.7)	572.0 (920.3)	186.0 (299.3)	174.0 (280.0)	191.0 (307.4)	93.2 (149.9)	87.0 (140.0)	95.3 (153.4)
350 Mcm	5	852.0 (1371.2)	723.0 (1163.6)	728.0 (1171.4)	282.0 (453.8)	240.0 (386.2)	242.0 (389.5)	142.0 (228.5)	120.5 (193.9)	121.0 (194.7)

Overhead distribution copper-line wire

Wire size	Equivalent spacing, in (cm)	4160 V, three-phase — kVA·mi (kVA·km) per 1% drop			2400 V, three-phase — kVA·mi (kVA·km) per 1% drop			2400 V, single-phase — kVA·mi (kVA·km) per 1% drop		
1/0	28 (71.1)	248.0 (399.1)	207.0 (333.1)	206.0 (331.5)	82.6 (132.9)	69.0 (111.0)	68.7 (110.6)	41.3 (66.5)	34.5 (55.5)	34.4 (55.4)
2/0	28 (71.1)	298.0 (479.6)	236.0 (379.8)	225.0 (362.1)	99.3 (159.8)	78.7 (126.7)	75.0 (120.7)	49.6 (79.8)	39.4 (63.4)	37.5 (60.4)
4/0	28 (71.1)	415.0 (667.9)	295.0 (474.8)	260.0 (418.4)	138.3 (223.4)	98.3 (158.2)	86.6 (139.4)	69.2 (111.4)	49.2 (79.2)	43.3 (69.7)
350 Mcm	28 (71.1)	574.0 (923.8)	363.0 (584.2)	297.0 (477.9)	191.2 (307.7)	121.0 (194.7)	99.0 (159.3)	95.5 (153.7)	60.5 (97.4)	59.4 (95.6)
500 Mcm	28 (71.1)	700.0 (1126.5)	405.0 (651.8)	318.5 (512.6)	233.3 (375.5)	135.0 (217.3)	106.2 (170.9)	116.6 (187.6)	67.4 (108.5)	63.1 (101.5)

Wire size	Equivalent spacing, in (cm)	6900 V, three-phase — kVA·mi (kVA·km) per 1% drop			11,950 V, three-phase — kVA·mi (kVA·km) per 1% drop		
1/0	60 (152.4)	627 (1009.1)	537 (864.2)	519 (835.2)	1879 (3023.9)	1610 (2591.0)	1555 (2502.5)
2/0	60 (152.4)	739 (1189.3)	601 (967.2)	560 (901.2)	2217 (3567.9)	1803 (2901.6)	1680 (2703.7)
3/0	60 (152.4)	859 (1382.4)	673 (1083.1)	601 (967.2)	2576 (4145.7)	2019 (3249.3)	1803 (2901.6)
4/0	60 (152.4)	988 (1590.0)	739 (1189.3)	640 (1029.9)	2963 (4768.5)	2217 (3567.9)	1919 (3088.3)
250 Mcm	60 (152.4)	1097 (1765.5)	805 (1295.5)	677 (1089.5)	3291 (5296.4)	2415 (3886.6)	2032 (3270.2)

°For receiving voltages slightly different from the given values, multiply kva-miles in the tables by the square of the ratio of the new voltage to the voltage in the table. For example, for 4000 V multiply by $(4000/4160)^2 = 0.924$.

4.   Check the next smaller size cable

Since there is a difference of $5.00 - 3.41 = 1.59$ percent between the actual and the allowable voltage drop, the next smaller cable might be satisfactory. A smaller cable will cost less and is therefore worth checking.

Using the same procedure as in step 2, except that a 2/0 10-kV cable is used, we find that kva-mile per 1 percent voltage drop = $(2200/2400)^2(118.0) = 99.0$. Then, as in step 3, actual voltage drop = $500/99.0 = 5.06$ percent. Hence, the smaller cable is unsuitable because it produces a higher voltage drop than allowed.

Related Calculations: Table 15 can be used to determine (1) the voltage drop in an existing circuit when the load is known, (2) the proper size of a conductor to use to limit the voltage drop to a predetermined value, or (3) the proper voltage rating of a new line. The 50 percent power-factor column in Table 15 can be used to calculate the instantaneous drop in voltage caused by the starting of a large motor on the line. The average power factor of a motor is 50 percent when the motor is being started. Note that data are listed in Table 15 for both underground and overhead distribution systems.

WIRE- AND CABLE-SIZE DETERMINATION FOR A KNOWN LOAD

An electrical installation consists of 1 motor and 48 lights. What size wire and fuses are needed for the 50-hp (37.3-kW) squirrel-cage induction motor that is started at its rated three-phase 440-V 60-Hz current if the motor is 280 ft (85.3 m) from the power panel and the voltage drop must not exceed 1 percent of the supply voltage? The lighting load totals 10,000 W. This load is supplied by two 110-V circuits. The voltage drop in the lighting circuits must not exceed 1 percent. What size wire is needed if each branch circuit is 170 ft (51.8 m) long?

Calculation Procedure:

1.   Sketch the circuit layout

Figure 9 shows a single-line diagram of a typical light and power branch-circuit layout. Figure 10 shows the typical equipment used in such a circuit.

2.   Determine the motor full-load current

Use the *National Electrical Code (NEC)* table of motor full-load current or Table 16 to determine the full-load current of the motor. Thus, Table 16 shows that a 440-V 50-hp (37.3-kW) induction motor requires 63 A at full load.

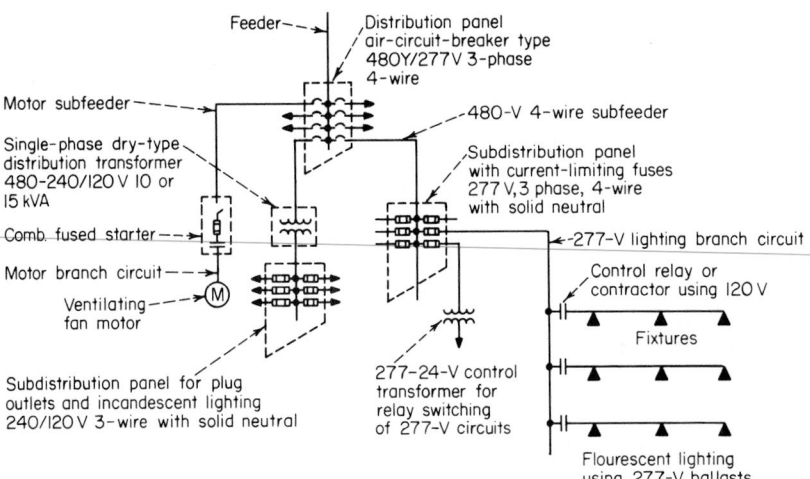

FIG. 9   Typical light and power branch-circuit layout served from a 480-Y/277-V system.

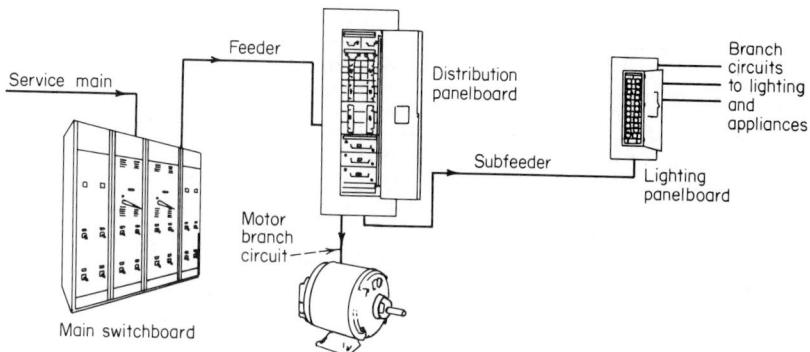

FIG. 10 Typical equipment used in motor and lighting circuits.

3. Determine the required carrying capacity of the motor circuit

The *NEC* recommends that individual branch circuits to motors have a current-carrying capacity at least 125 percent of the motor full-load running current. Thus, for this 50-hp (37.3-kW) motor, the current-carrying capacity of the branch circuit supplying the motor must be at least $1.25(63$ A$) = 78.7$ A, say 80 A.

4. Determine the required wire size

Table 17 lists the allowable wire lengths for a 1 percent, or less, voltage drop in 220- to 230-V and 440- to 460-V systems. To use Table 17 for 440- to 460-V three- or four-wire feeders, multiply the tabulated lengths by 2.

Thus, with an 80-A flow, a 2/0 wire will have a 1 percent voltage drop in a length of $2(151) = 302$ ft (92.0 m), using the data from Table 17. Since the motor is 280 ft (85.3 m) from the power panel, the voltage drop will be less 1 percent if a 2/0 wire is used.

5. Select the fuse capacity for the motor circuit

For motors of more than 1 hp the *NEC* allows an external overcurrent device actuated by the motor running current and set to open at not more than 125 percent of the motor full-load current for motors with a temperature rise not over 40°C, or 115 percent of the motor full-load current for all other motors.

Where fuses are used to protect a circuit, the fuses should have a rating of at least 300 percent the motor full-load current. Or, for this motor $3(63$ A$) = 189$ A. A 200-A fuse would be suitable.

Note that the fuse capacity cannot be arbitrarily increased without a study of the circuit and the devices in it. Thus, a 300-A fuse could not be used for this circuit because the excessive current might damage the motor before the fuse blew.

Were a circuit breaker used instead of fuses, the breaker would be rated at $1.25(63$ A$) = 78.7$, say 75 A. The circuit-breaker rating is reduced to the next lowest standard rating to provide greater protection for the motor. For motors over 1500 hp (1118.5 kW), an embedded temperature detector may be used to cause opening of the current supply when the motor temperature becomes excessive.

6. Compute the current flow in the lighting circuits

With two lighting circuits, the load in each circuit = total load, W per number of circuits, or $11,000/2 = 5500$ W per circuit.

Use the relation $I = W/v$ to determine the current flow in each lighting circuit. In this relation, I = current flow, A; w = lighting load of the circuit, W; v = circuit voltage. Thus, for each of these circuits, $I = 5500/110 = 50$ A.

7. Select the wire size for the lighting circuits

Table 17 shows the wire size for 115/230-V single-phase circuits having a 1 percent voltage drop for various lengths. This tabulation is also applicable to 110-V ac circuits.

TABLE 16 Electric-Motor Full-Load Current

Full-load current° direct-current motors

hp	kW	115 V	230 V	550 V
¼	0.19	3	1.5	
⅓	0.25	3.8	1.9	
½	0.37	5.4	2.7	
¾	0.56	7.4	3.7	1.6
1	0.75	9.6	4.8	2.0
1½	1.12	13.2	6.6	2.7
2	1.49	17	8.5	3.6
3	2.24	25	12.5	5.2
5	3.73	40	20	8.3
7½	5.59	58	29	12
10	7.46	76	38	16
15	11.2	112	56	23
20	14.9	148	74	31
25	18.6	184	92	38
30	22.4	220	110	46
40	29.8	292	146	61
50	37.3	360	180	75
60	44.7	430	215	90
75	55.9	536	268	111
100	74.6	. . .	355	148
125	93.2	. . .	443	184
150	111.9	. . .	534	220
200	149.1	. . .	712	295

Full-load current,† single-phase ac motors

hp	kW	115 V	230 V	440 V
⅙	0.12	4.4	2.2	
¼	0.19	5.8	2.9	
⅓	0.25	7.2	3.6	
½	0.37	9.8	4.9	
¾	0.56	13.8	6.9	
1	0.75	16	8	
1½	1.12	20	10	
2	1.49	24	12	
3	2.24	34	17	
5	3.73	56	28	
7½	5.59	80	40	21
10	7.46	100	50	26

Three-phase ac motors

hp	kW	Induction type, squirrel-cage and wound-rotor, A					Synchronous type, unity power factor, A			
		110 V	220 V	440 V	550 V	2300 V	220 V	440 V	550 V	2300 V
½	0.37	4	2	1	0.8					
¾	0.56	5.6	2.8	1.4	1.1					
1	0.75	7	3.5	1.8	1.4					
1½	1.12	10	5	2.5	2.0					
2	1.49	13	6.5	3.3	2.6					
3	2.24	. . .	9	4.5	4					

TABLE 16   Electric-Moter Fuel-Load Current (*Continued*)

		Three-phase ac motors							
		Induction type, squirrel-cage and wound-rotor, A				Synchronous type, unity power factor, A			
hp	kW	110 V	220 V	440 V	550 V	2300 V	220 V	440 V	550 V	2300 V
5	3.73	. . .	15	7.5	6					
7½	5.59	. . .	22	11	9					
10	7.46	. . .	27	14	11					
15	11.2	. . .	40	20	16					
20	14.9	. . .	52	26	21					
25	18.6	. . .	64	32	26	7	54	27	22	5.4
30	22.4	. . .	78	39	31	8.5	65	33	26	6.5
40	29.8	. . .	104	52	41	10.5	86	43	35	8
50	37.3	. . .	125	63	50	13	108	54	44	10
60	44.7	. . .	150	75	60	16	128	64	51	12
75	55.9	. . .	185	93	74	19	161	81	65	15
100	74.6	. . .	246	123	98	25	211	106	85	20
125	93.2	. . .	310	155	124	31	264	132	106	25
150	111.9	. . .	360	180	144	37	. . .	158	127	30
200	149.1	. . .	480	240	192	49	. . .	210	168	40

°These values of full-load current are for motors running at speeds usual for belted motors and motors with normal torque characteristics. Motors built for especially low speeds or high torques may require more running current, in which case the name-plate current rating should be used.

For full-load currents of 208- and 200-V motors, increase the corresponding 220-V motor full-load current by 6 and 10 percent, respectively.

†These values are for motors with usual speeds and torque characteristics. Name-plate current values should be used for motors with low speeds or high torques.

To obtain full-load currents of 208- and 200-V motors, increase 230-V current values by 10 and 15 percent, respectively.

Inspection of the tabulated values shows that a 1/0 wire is required for a 1 percent voltage drop in a 170-ft (51.8-m) long circuit. Hence, this size wire will be used for the lighting circuit.

Related Calculations: In planning electric circuits, the loading and lengths of feeders and runs between outlets must be related to the voltage drop and the need for spare capacity in the circuit for possible future increases in load. Each lamp, appliance, or other utilization device in the circuit is designed for best performance at a particular operating voltage. Although such devices will operate at voltages on either side of the design value, there generally will be adverse effects for operation at voltages lower than the specified value.

A 1 percent drop in voltage at an incandescent lamp produces about a 3 percent decrease in light output; a 10 percent voltage drop will decrease the output about 30 percent. In resistance-type heating devices, a voltage drop has a similar effect on the heat output. In motor-operated appliances, low voltage to the motor affects the starting and pull-out torque. Also, the current drawn from the line increases with the drop in voltage. Overheating of the windings may result.

The method presented here is valid for lighting, power, heating, and similar circuits. For comprehensive tabulations of wire sizes and allowable loads, consult the latest edition of the *National Electrical Code*. Where a local electrical code governs wiring and system design, be sure to consult it when selecting wire sizes.

PRELIMINARY ELECTRIC LOAD ESTIMATING

Estimate the lighting and power loads for an electronics factory that is 200 ft (61.0 m) long and 100 ft (30.5 m) wide. The fluorescent lights in this one-story plant will be located 12 ft (3.7 m) above the floor and will provide an illumination level of 50 fc (538.2 lx). What is the demand amperage if the power is supplied at 13,800 V?

TABLE 17 Average Circuit Lengths [ft (m)] for 1 Percent Voltage Drop

Single-phase ac loads 115/230 V, 60 Hz, 100% pf

Load, A	Wire size—Circular mils				Wire size—B & S or A.W.G.	
	500	400	300	250	2/0	1/0
40	1106 (337)	898 (274)	669 (204)	558 (170)	299 (91)	239 (73)
50	885 (270)	719 (219)	535 (163)	447 (136)	240 (73)	191 (58)
60	737 (225)	599 (183)	446 (136)	372 (113)	200 (61)	159 (48)

Three-phase delta ac loads 230 V, 60 Hz, 85% pf

Load, A	Wire size—Circular mils				Wire size—B & S or A.W.G.		
	500	400	300	250	3/0	2/0	1/0
70	406 (124)	357 (109)	303 (92)	271 (83)	208 (63)	173 (53)	145 (44)
80	355 (108)	312 (95)	265 (81)	238 (73)	182 (55)	151 (46)	127 (39)
90	316 (96)	278 (85)	235 (72)	211 (64)	162 (49)	134 (41)	113 (34)
100	284 (87)	250 (76)	212 (65)	190 (58)	146 (45)	121 (37)	101 (31)

Calculations based on copper resistance of 12.5 Ω per CM-ft at 50°C (122°F).
Reactance and impedance losses calculated for each wire.
Conductors closely grouped in metallic conduit.
Balanced three-wire loads: drop is 1.15 V for given length.
Two-wire, 230-V loads: drop is 2.3 V for given length.
For 208-V, four-wire Y feeders, multiply given length by 0.9.
For 230-V, single-phase feeders, multiply given length by 0.85.
For 460-V, three- or four-wire feeders, multiply given length by 2.
For aluminum wire, multiply given lengths by 0.7 or use length of copper wire which is two sizes smaller than the aluminum size under consideration.

Balanced lighting loads
Three- and four-wire, 115 V 1 percent drop from supply cabinet to first outlet supplying permanently connected appliance or fixture°

Maximum overcurrent circuit protection†	Intermittent loads			Continuous loads		
	100% F 2–3 C	80% F 4–6 C	70% F 7–9 C	100% F 2–3 C	80% F 4–6 C	70% F 7–9 C
15 A	15 A 1725 W	12 A 1380 W	10.5 A 1207 W	12 A 1380 W	9.6 A 1104 W	8.4 A 966 W
20 A	20 A 2300 W	16 A 1840 W	14 A 1610 W	16 A 1840 W	12.8 A 1472 W	11.2 A 1288 W

Loads and lengths in ft (m) for 1% drop on three- and four-wire 115-V circuits

Load, A	No. 10 wire	No. 12 wire	No. 14 wire
1	946 (288)	596 (182)	374 (114)
2	474 (144)	298 (91)	188 (57)
3	316 (96)	198 (60)	124 (38)
4	236 (72)	148 (45)	94 (28)
5	190 (58)	120 (36)	76 (23)
6	158 (48)	100 (30)	62 (19)
7	136 (41)	86 (26)	54 (16)

TABLE 17   Average Circuit Lengths [ft (m)] for 1 Percent Voltage Drop (*Continued*)

Load, A	No. 10 wire	No. 12 wire	No. 14 wire
8	118 (36)	74 (22)	46 (14)
9	106 (32)	66 (20)	42 (13)
10	94 (29)	60 (18)	38 (12)

°Calculations based on copper resistance of 13 Ω per CM-ft at 60°C (140°F).
For two-phase, three-wire circuits tapped off a three-phase, four-wire Y service, multiply given lengths by 0.67.
†A—amperes; W—watts; C—conduit conductor; F—fills.

Calculation Procedure:

1. *Compute the lighting power requirements*

For preliminary electric load estimates, tabulated average power demands are often sufficient. Thus, Table 18 lists the typical power requirements in terms of the lighting-demand factor for lights mounted at various heights above the floor. To use the values from this table, solve the relation $VA/ft^2 = $ (lighting-demand factor)(illumination level, fc). Thus, for this electronics factory using fluorescent lights mounted 12 ft (3.7 m) above the floor, $VA/ft^2 = (0.060)(50) = 3.0$ (32.3 VA/m^2). With a plant area of $200 \times 100 = 20,000$ ft^2 (1858.1 m^2), the total *estimated* power requirement for lighting is 3.0(20,000) = 60,000 VA.

TABLE 18   Power Requirements for Lighting°

Fixture height, ft (m)	Lighting demand factor†	
	Incandescent	Fluorescent
Less than 14 (4.3)	0.12	0.060
14–35 (4.3–10.7)	0.13	0.065
35–50 (10.7–15.2)	0.15	0.070

°Based on median-characteristic fixtures and a 75-ft (22.9-m) wide room, 200 ft (61 m) or more long.
†VA = (lighting demand factor)(intensity of illumination, ft) = *c*.
SOURCE: Beeman—*Industrial Power Systems Handbook*, McGraw-Hill.

2. *Compute the plant power demand*

Table 19 lists the combined power and lighting load densities for various industries. This table shows that the load density for electronic equipment manufacture is 10 VA/ft^2 (107.6 VA/m^2). Since the lighting demand was computed in step 1 as 3.0 VA/ft^2 (32.3 VA/m^2), the power demand will be the difference between 3.0 VA and the tabulated total of 10.0 VA/ft^2 (107.6 VA/m^2) for power and light, or $10.0 - 3.0 = 7.0$ VA/ft^2 (75.3 VA/m^2). With a total area of 20,000 ft^2 (1858.1 m^2), the power demand is 7.0(20,000) = 140,000 VA.

3. *Compute the total load demand*

The total load demand is the sum of the loads computed in steps 1 and 2, or the product of the floor area of the plant and the appropriate value from Table 19. Or $60,000 + 140,000 = 10(20,000) = 200,000$ VA.

Table 20 shows the typical voltampere demands for a variety of plants. Based on actual plants, this table shows both average values and the ranges encountered in the plants surveyed.

TABLE 19 Typical Power-Load Densities

	Demand for power and light	
Factory type	VA/ft²	VA/m²
Beet-sugar factory and refinery	19	204.5
Paper mills	14	150.7
Textile mills, engine building	12	129.2
Cigarette manufacturing	11	118.4
General manufacturing, chemicals, electronic equipment	10	107.6
Small-appliance manufacturing, machine repair shops	7.5	80.7
Lamp manufacturing	5	53.8
Small-device manufacturing	3.5	37.7

SOURCE: Beeman—*Industrial Power Systems Handbook*, McGraw-Hill.

TABLE 20 Recorded Load Data for a Group of Plants (Summary)

	Demand				Annual power				Annual kWh/ VA demand	
	VA/ft²		VA/m²		kWh/ft²		kWh/m²			
Type of plant	Average	Range	Average	Range	Average	Range	Average	Range	Average	Range
Chemical	10.0	6–13	107.6	65–140	33.7	14–54	362.7	151–581	3.3	2–4
Electronics	10.3	3–20	110.9	32–215	25.8	11–67	277.7	118–721	2.4	1–4
Foundry	9.9	. . .	106.6	. . .	32.4	. . .	348.8	. . .	3.3	. . .
Lamps:										
General	4.8	2–12	51.7	22–129	19.5	5–53	209.9	54–571	4.2	2–6
Wire works	8.7	6–13	93.6	65–140	30.0	13–64	322.9	140–689	2.8	2–5
Base works	4.0	. . .	43.1	. . .	25.0	. . .	269.1	. . .	6.0	. . .
Glass works	4.5	2–7	48.2	22–75	24.2	13–37	260.5	140–398	5.4	4–6
Porcelain	2.0	. . .	21.5	. . .	10.9	. . .	117.3	. . .	5.4	
Printing	3.0	. . .	32.3	. . .	8.0	. . .	86.1	. . .	3.0	. . .
Small appliances	7.4	2–13	39.7	22–140	20.5	4–46	220.7	43–495	2.9	1–7
Small device	3.6	2–7	38.8	22–75	9.5	3–27	102.3	32–291	2.4	1–5
General:										
Large (over 5000 kVA)	10.0	5–17	107.6	54–183	70.0	8–195	753.5	86–2099	7.0	1–19
Small (under 5000 kVA)	9.8	3–18	105.5	32–194	31.0	5–50	333.7	54–538	5.2	1–27

SOURCE: Beeman—*Industrial Power System Handbook*, McGraw-Hill.

4.36

4. *Compute the demand amperage*

The product of volts and amperes, voltamperes (va), is termed *apparent power* in an ac circuit. In a single-phase unity-power-factor circuit, va $= EI$. In a balanced three-wire three-phase delta or wye circuit, va $= 1.732EI$. Thus, with a power supply at 13,800 V, the demand amperage is $200,000 = 13,800I; I = 14.5$ A.

Related Calculations: Use this general method for preliminary estimates of the electric load of any type of industrial plant. For exact determination of the electrical demand, use the methods given in later calculation procedures in this section.

PLANT POWER-DISTRIBUTION-SYSTEM PLANNING

Select the power-distribution system for an industrial plant having motors ranging from ½ to 1000 hp (0.37 to 745.7 kW). The total demand of the plant is estimated to be 15,000 kVA. Choose the type of distribution system for use within this plant.

Calculation Procedure:

1. *Select the system voltage rating*

Table 21 lists the typical voltage ratings (classes) for industrial power equipment. Popular main generation and distribution voltages are 4160 and 13,800 V. Utilization voltage, i.e., the voltage

TABLE 21   Voltage Ratings for Power Equipment

Nominal system voltage class	Generator or transformer no-load rated voltage	Utilization equipment rated voltage
240	240	220
480°	480°	440°
600	600	550
2,400°	2,400°	2,300°
4,160°	4,160°	4,000°
6,900	6,900	5,600
13,800°	13,800°	13,200°

°The voltages marked by an asterisk are the preferred ratings for most applications because of the availability of equipment and overall sound-system engineering.

at which motors and similar large equipment is operated, is usually 2400 and/or 480 V. As a general guide, the higher the voltage in a given voltage class, the lower the overall system cost.

Figure 11 shows typical voltages used for various loads. Based on the installed cost of all the components comprising a distribution system, a 4160-V system has a lower first cost than a 13,800-V system for plants with a demand up to about 10,000 kVA. For plants in the range of 10,000- to 20,000-kVA capacity, the costs of the two systems are comparable. Above 20,000-kVA capacity, a 13,800-V system is usually preferred.

Since this plant has a 15,000-kVA demand, a 13,800-V distribution voltage will be chosen. This choice will permit easier future expansion without a higher initial investment. As a further verification of this voltage choice, Fig. 12 shows the approximate limitations on the amount of power that can be fed from a single

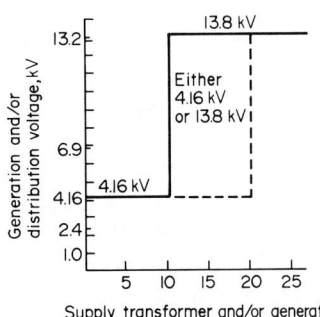

FIG. 11   Relationship of distribution voltage to load.

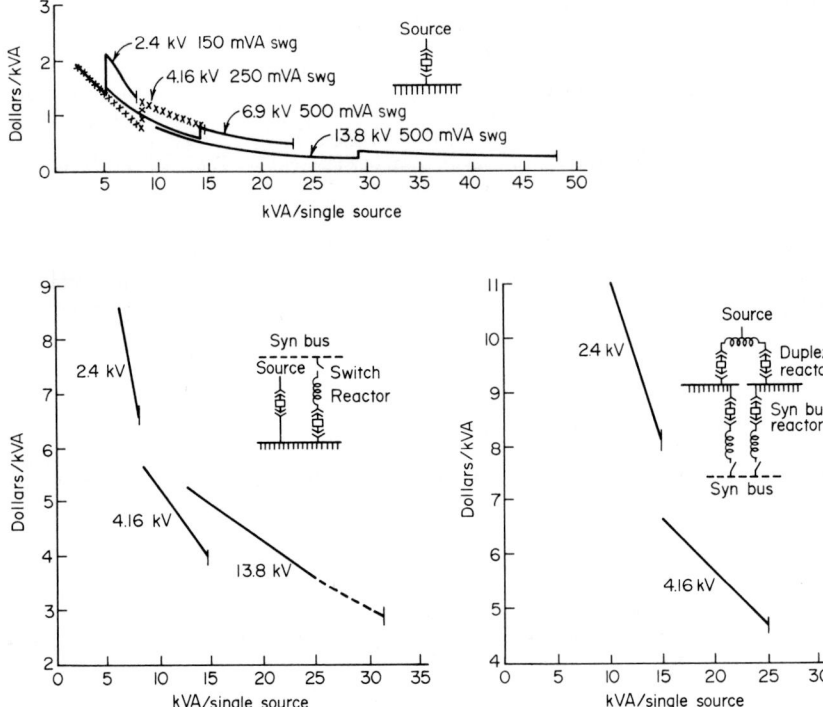

FIG. 12   Typical loads that can be fed from a single source at different voltages.

source at different voltages. These curves indicate that a 13,800-V distribution voltage can adequately supply the demand. The costs shown are relative; do not use the absolute values shown. Transformer costs are not shown because the cost of a transformer is about the same for secondary voltages of 2400, 4160, and 13,800 V.

2.   *Choose the motor operating voltage*

Figure 13 shows the recommended utilization voltage for motors rated at 10 to 10,000 hp (7.5 to 7457.0 kW) This chart shows that with a 13,800-V source a 2300-V motor utilization voltage is recommended for motors rated 150 to 3500 hp (111.9 to 2609.9 kW). Hence, this is an acceptable operating voltage for the larger motors in this plant. For motors less than 150 hp (111.9 kW), 440 V would be used, as Fig. 13 indicates. This voltage could be obtained by using step-down transformers.

In the layout of a new plant, it is usually best to consider first the selection of the system voltage from the standpoint of power distribution only. Too often there is a tendency to compromise between distribution voltage and motor voltage. The result is that the final voltage will be too low from the standpoint of good overall system design.

If a motor application occurs near the ends of the range for each voltage class in Fig. 13, review the motor application carefully. Other factors such as starting equipment, speed, type of motor, etc., may affect the voltage selection sufficiently to make a different voltage more desirable.

3.   *Select the power-distribution system*

The *load-center* power-distribution system (Fig. 14) is probably the most economical system for this plant. For new or rearranged plants, the load-center system of power distribution is the most flexible way to supply existing loads or to meet changing load demands with the lowest invest-

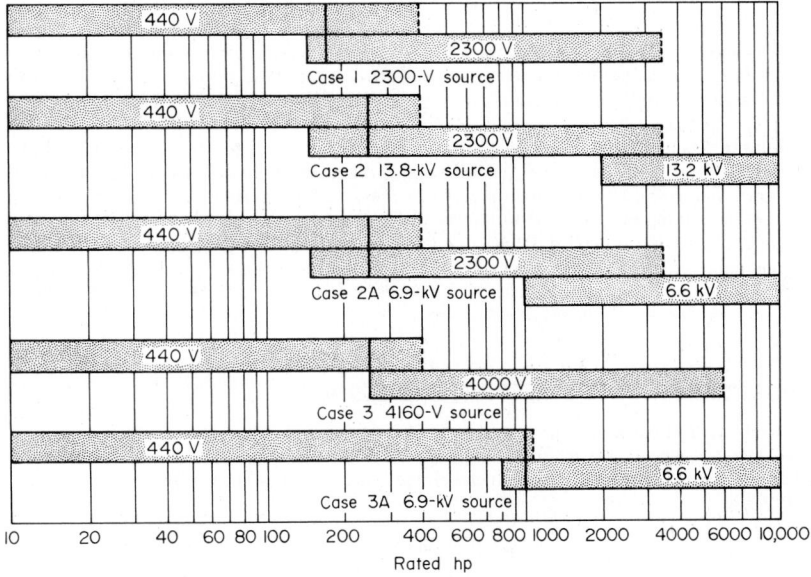

FIG. 13 Motor ratings recommended for various utilization voltages.

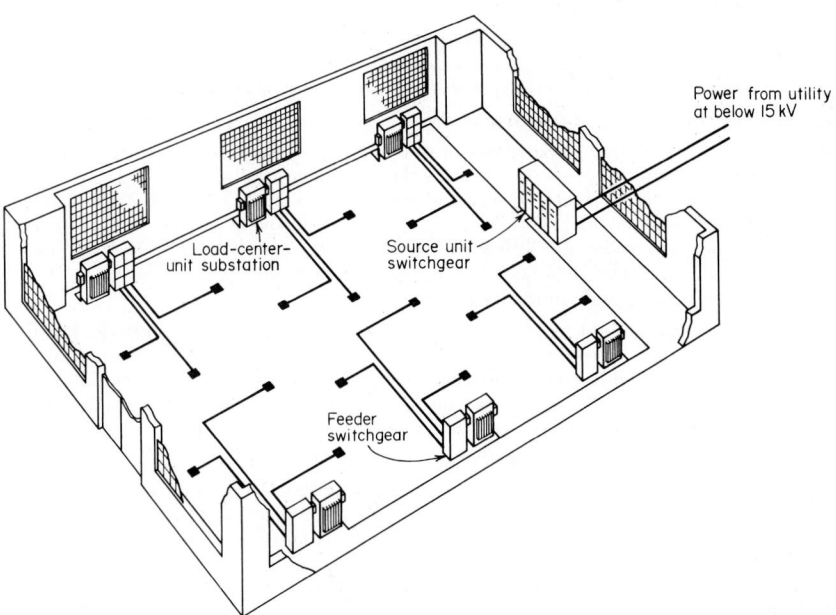

FIG. 14 Power-distribution layout with source unit, load-center substation, feeder switchgear, and cables.

ment. A unit substation can supply power to load centers at 13,800, 4160, 2400 V, or other voltages.

The fundamental approach is to use a relatively high voltage for transmitting power to the load centers where it is stepped down to the utilization voltage. The primary voltage, as shown in step 1, varies as a function of the total plant load, and the secondary voltage as a function of the motor size, step 2.

Related Calculations: The general method given here is valid for all types of industrial plants—chemical, petroleum, food, textile, metalworking, etc. Note that the problem of selecting the proper voltage for a plant's main power-distribution system is similar whether power is purchased from a utility or generated within the plant. Power may be generated in industrial plants at 13,800 V or below, as desired. Power purchased from a utility may be supplied at 13,800 V or at some higher or lower voltage, depending on the utility system and the industrial load requirements. The recommendations given here were developed by W. B. Wilson, General Electric Company, and reported in *Chemical Engineering*.

INDUSTRIAL-BATTERY SELECTION AND SIZING

Choose the type of battery to use to start a diesel-engine standby generating unit for an industrial plant. The starting motor is rated at 10 hp (7.5 kW). It is desired to keep the battery investment as low as possible. How long could the chosen battery supply a 24-V emergency lighting load of 4800 W?

Calculation Procedure:

1. *Compute the starting-motor amperage*

The starting motor is rated at 10-hp (7.5-kW) output, or (10 hp)(746 W/hp) = 7460 W = 7.46 kW. To compute the amperage, use the dc relation $P = IE$, where P = power, W; I = current flow, A; E = rated circuit voltage. Solving gives $I = P/E = 7460/24 = 310.8$, say 311 A.

2. *Choose the type of battery to use*

Table 22 lists the major characteristics, advantages, and disadvantages of rechargeable industrial batteries. Study of this tabulation shows that either a lead-acid or nickel-cadmium battery would be suitable for this application. Both types of batteries have a suitable voltage range.

3. *Compute the number of cells required*

When a battery is discharging, the voltage at its terminals decreases. To determine the number of cells required, divide the required voltage by the voltage during discharge.

Table 22 shows that the discharge voltage of a lead-acid battery is 2.1 to 1.46 V and that of a nickel-cadmium battery is 1.3 to 0.75 V. Using the average discharge voltage for each battery, we see the number of lead-acid cells required = 24 V/1.78 = 13.46, say 14 cells. The number of cells required for a nickel-cadmium battery = 24 V/1.03 = 23.3, say 24 cells.

4. *Check the relative costs of the batteries*

Table 22 shows that the lead-acid battery has the lowest first cost of all the batteries listed. A nickel-cadmium battery, according to Table 22, costs two to three times as much as a lead-acid battery of the same ampere-hour rating. Since both batteries must have the same rating, the nickel-cadmium battery will cost twice as much as the lead-acid battery. To keep the battery investment as low as possible, a lead-acid battery would therefore be chosen.

5. *Compute the battery operating time*

The battery must supply 311 A during the starting cycle. A 300-Ah lead-acid battery (i.e., a battery capable of supplying 300 A for 1 h) would be suitable for the engine starting load because it would have sufficient amperage capacity to supply the 311 A for the short starting cycle—usually less than 1 min.

The emergency-lighting load of 480 W requires $I = P/E = 4800/24 = 200$ A. With a 300-Ah battery, the operating time with this emergency load would be $T = \text{Ah}/\text{A} = 300/200 = 1.5$ h.

Related Calculations: Use this general method to choose rechargeable batteries for starting engines, powering industrial trucks, supplying emergency lighting, powering portable tools, furnishing inverter power supply, etc. Where an economic analysis of a battery installation must be made, use the methods of engineering economy to compare the various alternatives.

For industrial, automotive, and similar heavy-demand services, the output of a battery is often stated thus: 150 A for 4.1 min; 66 A for 20 min; 53 Ah for 20 h. The actual rating of this battery is 53 Ah for 20 h. Most manufacturers list three discharge rates for each battery. Figure 15 shows how the output current of three different nickel-cadmium plate designs compares with a typical lead-acid cell of the same ampere-hour capacity. Two types of emergency-power battery systems are shown in Fig. 16.

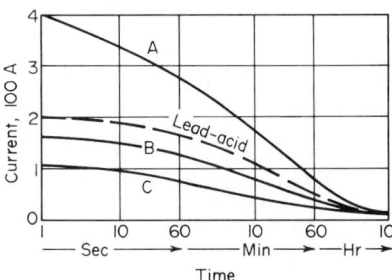

FIG. 15 Discharge rates for three nickel-cadmium plate designs *(A, B, C)* compared with lead-acid cells at the same ampere-hour capacity.

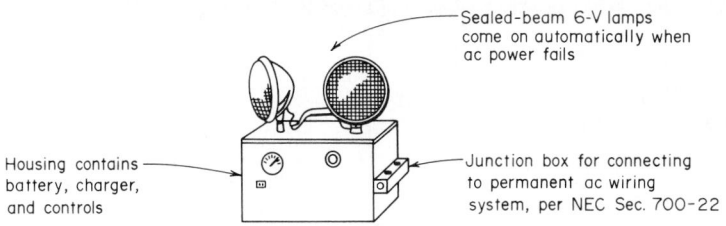

Typical Unit Emergency Light
for Use on Branch Circuits

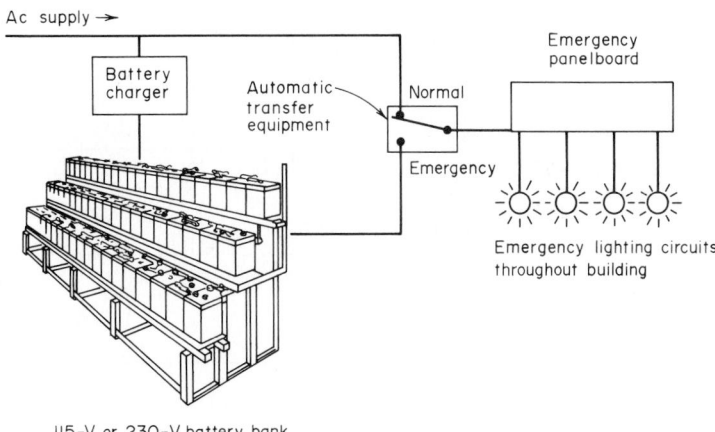

Basic Layout of Full-system
DC Emergency Lighting

FIG. 16 Two types of emergency power systems served by batteries.

TABLE 22 Battery Characteristics

Type of battery	Typical plant jobs	Typical battery voltage	Typical ampere-hour (6-h rating)	Typical life (good mtce)	Volts per cell
Lead-acid	Truck starting and small truck motive power.	12	160–300	1–4 (yr)	2.14 (initial); 2.1 to 1.46 under discharge
	Industrial truck motive power.	12–72	200–1000	5–10	
	Packaged emergency lighting.	6	5–50	4–6	
	Diesel standby generator starting.	12–32	120–300	1–3	
	Switchgear control source.	32–250	50–200+	6–12	
	Inverter continuous power source.	24–130	50–500+	6–14	
	Portable tools or instruments.	6–12 (sealed)	5–12	1–4	
Nickel-cadmium	Diesel standby generator starting.	12–32	40–190	15+	1.34 (initial); 1.3 to 0.75 under discharge
	Packaged emergency lighting.	6	5–50	15+	
	Switchgear control source.	32–250	10–200+	15+	
	Inverter continuous power source.	24–130	50–500+	15+	
	Portable tools or instruments.	6–12 (sealed)	To 15+	15+	
Nickel-iron	Industrial truck motive power.	24–36	340–1000	8–20	Same as Ni-Cd
Silver-zinc	Portable tools or instruments.	6–12	3–8+	1–2	1.86, 1.55 to 1.1
Silver-cadmium	Portable tools or instruments.	6–12	3–8+	2–3	1.34, 1.3 to 0.8

Composition	Typical advantages	Typical disadvantages
Lead and lead oxide electrodes and H_2SO_4 electrolyte	Lowest of all in first cost; long-term price picture relatively stable and competitive. Knowledge of maintenance requirements is widespread. Supplies high output under emergency overload conditions (at some sacrifice in overall life). Emergency replacements readily available everywhere. Easy to repair or rebuild damaged cells. Significant scrap value.	Deteriorates quickly under neglect or incompetent maintenance. Fumes during charging are corrosive, explosive, annoying—must be vented. Needs frequent addition of water. Heavy and bulky to handle (acid hazard). Freezes readily when discharged. Cells will not be uniform unless very close quality control is maintained in manufacture. Vibration shortens life—shakes loose electrode ingredients.
Nickel oxide and cadmium electrodes and KOH electrolyte	Withstands neglect or incompetent maintenance. Needs little water per year. Will not freeze. Substantially smaller and lighter than lead acid. Holds charge without attention for months or years. Very high short-time current capability (often enables use of smaller battery). Withstands vibration. Fumes do not corrode enclosure.	Up to two or three times as expensive as lead-acid for same ampere-hour rating. Requires over ⅛ more cells (and extra connections) for same load as lead-acid. Small number of market sources for industrial types (virtually all plates are imported—U.S. assembled). Multicell sealed batteries sometimes need individual cell charging. Charger characteristics are critical on sealed cells.
Nickel oxide, iron, KOH	Extremely long life. Needs little maintenance (has most advantages of Ni-Cd).	Requires most space of all. Nearly as heavy as lead-acid.
Silver oxide, zinc, KOH	Very high energy per unit volume.	May cost about 10 times that of lead-acid. Relatively short cycle life.
Silver oxide, cadmium, KOH	High energy per unit volume. Withstands frequent cycling.	Very expensive. Limited operating life.

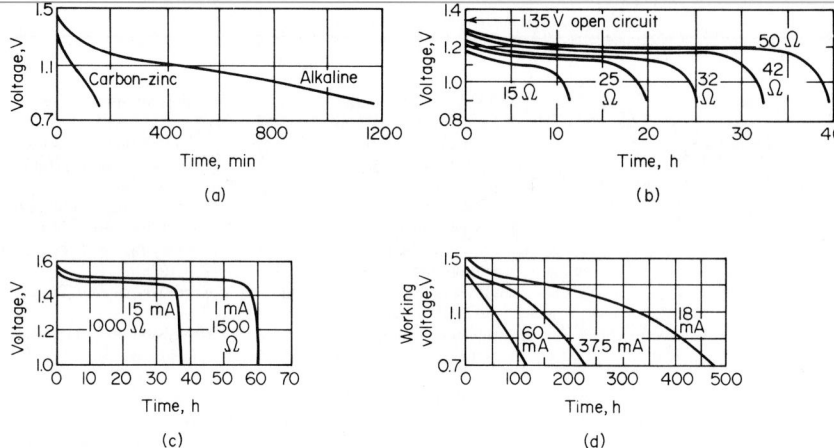

FIG. 17   Output voltage of small cells. (*a*) Alkaline-manganese-zinc; (*b*) mercury; (*c*) silver-zinc; (*d*) carbon-zinc.

Figure 17 shows how the output voltage of four different types (alkaline-manganese zinc, mercury, silver-zinc, and carbon-zinc) of small batteries varies with time. Small cells can be designed for optimum characteristics. These batteries find wide use in a variety of portable devices.

ELECTRIC-MOTOR SELECTION FOR A KNOWN LOAD

Select an electric motor to drive a 3600-r/min centrifugal pump rated at 3000 gal/min (189.3 L/s) of water at a 100-ft (30.5-m) total head.

Calculation Procedure:

1. Compute the required horsepower input

Use an hp input relation that is applicable to the driven unit. For a centrifugal pump the input hp is $hp_i = (gpm)(h)(s)/3960$ (eff), where gpm = quantity of fluid handled, gal/min; h = pump total head, ft of liquid handled; s = specific gravity of liquid handled; eff = efficiency of pump, expressed as a decimal.

Assume an efficiency for the driven device if it is not known or given. Thus, assume that this pump has an efficiency of 85 percent, a typical value for well-constructed, centrifugal pumps. Then $hp_i = (3000)(100)(1.0)/3960(0.85) = 89.1$ hp (66.4 kW).

2. Select the motor hp

Checking a tabulation of standard motor hp ratings, such as Table 23, shows that the next highest standard hp rating is 100 hp (74.6 kW). Hence, this rating will be tentatively chosen to drive the pump. (As a general guide, choose the next *larger* motor rating for usual industrial applications).

3. Select the motor electrical supply

In almost every situation, the electrical supply will be determined by the existing supply in the area or facility, except for small portable battery-powered motors. Thus, the availability of an ac supply will make the choice of an ac motor more likely. The same is true where the local electrical supply is direct current.

Assume that the electrical supply is 60-Hz alternating current. This immediately indicates that an ac motor of some type will be used to drive the pump unless there are special design requirements that dictate use of a dc motor. Since no such design requirements are stated, an ac motor will probably be acceptable.

TABLE 23   Standard Motor Power Ratings

hp	kW	hp	kW	hp	kW	hp	kW	hp	kW	hp	kW
⅛	0.09	2	1.49	30	22.4	200	149.1	700	522.0	2250	1667.8
⅙	0.12	3	2.24	40	29.8	250	186.4	800	596.6	2500	1864.2
¼	0.19	5	3.73	50	37.3	300	223.7	900	671.1	3000	2237.1
⅓	0.25	7.5	5.59	60	44.7	350	261.0	1000	745.7	3500	2609.9
½	0.37	10	7.46	75	55.9	400	298.3	1250	932.1	4000	2982.8
¾	0.56	15	11.2	100	74.6	450	335.6	1500	1118.5	4500	3355.6
1	0.75	20	14.9	125	93.2	500	372.8	1750	1305.0	5000	3728.5
1.5	1.12	25	18.6	150	111.9	600	447.4	2000	1491.4		

4.   *Determine the driven-machine torque requirements*

The motor hp rating alone is generally insufficient for selecting large motors. Besides the required output hp, two other factors must be known: the speed-torque characteristic of the motor and load and the moment of inertia, or WK^2, of the motor and load, where W = weight accelerated, lb; K = radius of gyration of the load, in.

Figure 18 shows that the starting torque of a typical centrifugal pump with its discharge valve closed is about 40 percent of the rated torque. Hence, the type of motor chosen should develop at least 40 percent of the rated torque on starting.

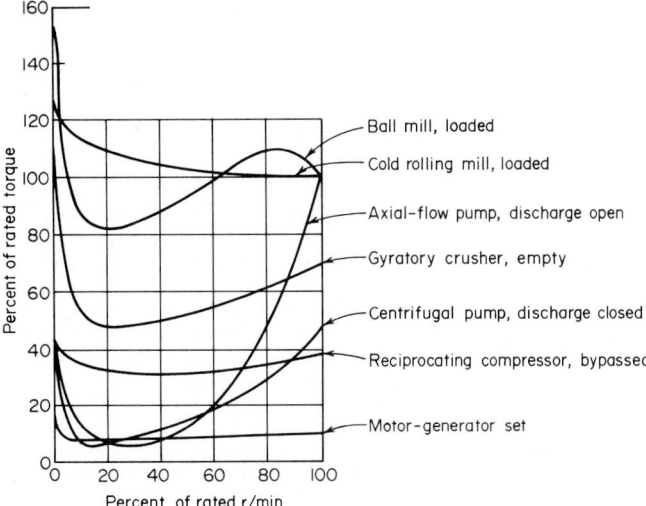

FIG. 18   Speed-torque characteristics for typical motor-driven loads. Flywheel effect and the time required for load acceleration are neglected here.

5.   *Choose the motor type for the driven load*

Figure 19 shows that induction motors are generally chosen for speeds above 500 r/min and power ratings from 0 to 1000 hp (0 to 745.7 kW), or more. The three-phase induction motor is also called a *squirrel-cage motor* because of the appearance of its rotor.

The starting torque of an induction motor depends on its NEMA classification. Usual starting torques range from about 60 to 150 percent of the full-load torque. Typical starting-torque ranges

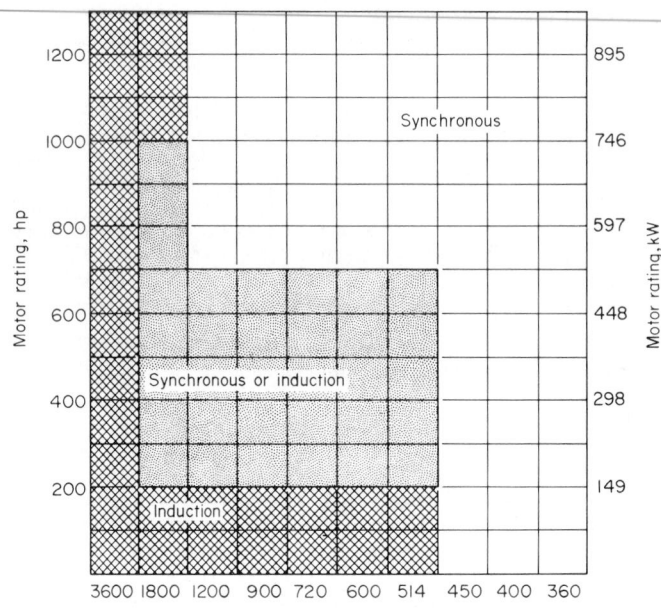

FIG. 19 Typical speed- and power-range applications for ac motors.

of ac motors are plotted in the *Standard Handbook for Electrical Engineers*. These plots show that any NEMA induction motor would be suitable for this pump drive because they all develop a starting torque greater than the driven machine's starting torque. Also, the WK^2 value of a centrifugal pump is characteristically low, so this factor can be neglected here. The WK^2 value of the load affects the length of the starting period, which in turn determines the heating of the motor. Thus, the motor selected to drive this pump will be a general-purpose 100-hp (74.6-kW) induction type without special controls.

Related Calculations: Use this general method to select ac and dc motors for all types of industrial and commercial drives—fans, compressors, crushers, motor-generator sets, blowers, etc. Figure 20 shows the speed-torque characteristics of the three popular types of dc motors—series, shunt, and compound—that might be used for any of these loads.

As a further guide to motor selection, Table 24 summarizes ac and dc characteristics and applications.

Where there is a choice between synchronous and induction motors, use the following rule of thumb as a guide to relative cost: *Synchronous motors are less expensive than induction motors if the motor-rated hp exceeds 1 hp · min/r.* This rule, however, overlooks the higher power factor and efficiency of the synchronous motor. These two factors are important in low-speed-motor choice.

As a general guide, the WK^2 of a driven unit is usually relatively high when the starting torque of the driven unit is high. Thus, the WK^2 and starting torque of crushers, ball mills, rolling mills, and similar equipment are usually high. Figure 18 shows typical starting torques for these units.

To pick a new NEMA *rerate motor* of greater hp that will fit the mounting pad of an old motor, use the following guides:

1. The first two digits of the frame number are four times the D dimension (Fig. 21) in in. If D = 9, the first two digits are $4 \times 9 = 36$.

2. The third digit of the frame number depends on the value of the F dimension in in. Select this digit from the 0 to 9 headings over the F dimensions in Table 25. For example, if $D = 9$ and

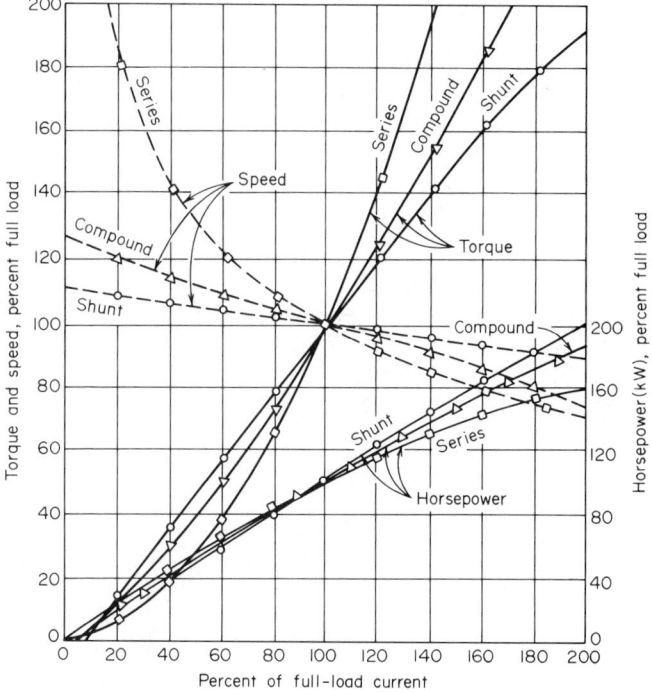

FIG. 20 Speed-torque characteristics of three types of dc motor.

$F = 7$, the third digit is 6, from Table 25. The motor frame number for the rerate motor is therefore 366.

3. The letter A is added after some frame numbers to designate an industrial dc motor or generator. The A denotes that certain detailed dimensions may differ from those of an ac motor or generator having the same frame number. Usually this is not a critical factor, and dimension variables are available in the manufacturer's engineering data.

When a new motor of given frame size, such as 366, is to be installed, the user may need the D and F dimensions to check the foundation requirements. Working with the frame number, say xyz, or 366, determine these values from $D = xy/4 = 36/4 = 9$.

From the tabulated data (Table 25), if z (the third digit) is 6, then $F = 7$ in (177.8 mm). With this information, any shimming or relocating of the mounting boltholes necessary for installing a new motor can be easily determined.

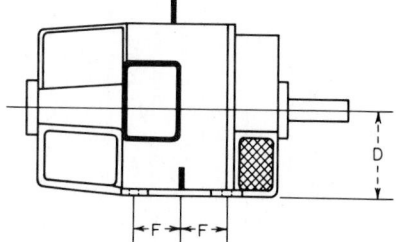

FIG. 21 Important dimensions of NEMA rerate motors.

STARTING TIME AND CURRENT FOR AC ELECTRIC MOTORS

A 300-hp (223.8-kW) 1176-r/min (1200-r/min synchronous speed) motor drives a directly connected reciprocating pump. The motor and pump have a WK^2 or inertia of 28,000 lb·ft² (11,571.1 N·m²). How long will it take to bring this load up to full speed if the starting torque is negligible?

TABLE 24 Summary of Motor Characteristics and Applications

Speed regulation	Speed control	Starting torque	Breakdown torque	Applications
		Polyphase motors		
General-purpose squirrel cage (design B):				
Drops about 3% for large to 5% for small sizes	None, except multispeed types, designed for two to four fixed speeds	100% for large, 275% for 1-hp (0.75-kW) four-pole unit	200% of full load	Constant-speed service where starting torque is not excessive. Fans, blowers, rotary compressors, and centrifugal pumps
High-torque squirrel cage (design C):				
Drops about 3% for large to 6% for small sizes	None, except multispeed types, designed for two to four fixed speeds	250% of full load for high-speed to 200% for low-speed designs	200% of full load	Constant-speed where fairly high starting torque is required infrequently with starting current about 550% of full load. Reciprocating pumps and compressors, crushers, etc.
High-slip squirrel cage (design D):				
Drops 10 to 15% from no load to full load	None, except multispeed types, designed for two to four fixed speeds	225 to 300% of full load, depending on speed with rotor resistance	200%. Will usually not stall until loaded to maximum torque, which occurs at standstill	Constant-speed and high starting torque, if starting is not too frequent, and for high-peak loads with or without flywheels. Punch presses, shears, elevators, etc.
Low-torque squirrel cage (design F):				
Drops about 3% for large to 5% for small sizes	None, except multispeed types, designed for two to four fixed speeds	50% of full load for high-speed to 90% for low-speed designs	135 to 170% of full load	Constant-speed service where starting duty is light. Fans, blowers, centrifugal pumps, and similar loads

	Speed regulation	Speed control	Starting torque	Starting current	Applications
Wound-rotor:	With rotor rings short-circuited drops about 3% for large to 5% for small sizes	Speed can be reduced to 50% by rotor resistance. Speed varies inversely as load	Up to 300% depending on external resistance in rotor circuit and how distributed	300% when rotor slip rings are short-circuited	Where high starting torque with low starting current or where limited speed control is required. Fans, centrifugal and plunger pumps, compressors, conveyors, hoists, cranes, etc.
Synchronous:	Constant	None, except special motors designed for two fixed speeds	40% for slow to 160% for medium speed 80% pf. Specials develop higher	Unity-pf motors, 170%; 80%-pf motors, 225%. Specials up to 300%	For constant-speed service, direct connection to slow-speed machines and where power-factor correction is required
Compound:	Drops 7 to 20% from no load to full load depending on amount of compounding	Any desired range, depending on design and type of control	Higher than for shunt, depending on amount of compounding	High. Limited by commutation, heating, and line capacity	Where high starting torque and fairly constant speed are required. Plunger pumps, punch presses, shears, bending rolls, geared elevators, conveyors, hoists
Split-phase:	Drops about 10% from no load to full load	None	75% for large to 175% for small sizes	150% for large to 200% for small sizes	Constant-speed service where starting is easy. Small fans, centrifugal pumps, and light-running machines, where polyphase is not available

TABLE 24 Summary of Motor Characteristics and Applications (*Continued*)

Speed regulation	Speed control	Starting torque	Breakdown torque	Applications
Capacitor: Drops about 5% for large to 10% for small sizes	None	150 to 350% of full load depending on design and size	150% for large to 200% for small sizes	Constant-speed service for any starting duty and quiet operation, where polyphase current cannot be used
Commutator: Drops about 5% for large to 10% for small sizes	Repulsion induction, none. Brush-shifting types, 4:1 at full full load	250% for large to 350% for small sizes	150% for large to 250% for small sizes	Constant-speed service for any starting duty where speed control is required and polyphase current cannot be used
		Dc and single-phase motors		
Series: Varies inversely as load. Races on light loads and full voltage	Zero to maximum depending on control and load	High. Varies as square of voltage. Limited by commutation, heating, capacity	High. Limited by commutation, heating, and line capacity	Where high starting torque is required and speed can be regulated. Traction, bridges, hoists, gates, car dumpers, car retarders
Shunt: Drops 3 to 5% from no load to full load	Any desired range depending on design type and type of system	Good. With constant field, varies directly as voltage applied to armature	High. Limited by commutation, heating, and line capacity	Where constant or adjustable speed is required and starting conditions are not severe. Fans, blowers, centrifugal pumps, conveyors, wood and metal-working elevators

TABLE 25   NEMA Motor Frame Numbers

| D | \multicolumn{10}{c}{F dimension, third digit in frame number} | Frame series |
	0	1	2	3	4	5	6	7	8	9	
4½	1¾	2	2¼	2½	2¾	3⅛	3½	4	4½	5	180
5	2	2¼	2½	2¾	3⅛	3½	4	4½	5	5½	200
5¼	2	2½	2½	2¾	3¼	3½	4	4½	5	5½	210
5½	2¼	2½	2¾	3⅛	3⅜	3¾	4½	5	5½	6¼	220
6¼	2½	2¾	3⅛	3½	4⅛	4½	5	5½	6¼	7	250
7	2¾	3⅛	3½	4	4¾	5	5½	6¼	7	8	280
8	3⅛	3½	4	4½	5¼	5½	6	7	8	9	320
9	3½	4	4½	5	5⅝	6⅛	7	8	9	10	360
10	4	4½	5	5½	6⅛	6⅞	8	9	10	11	400
11	4½	5	5½	6¼	7¼	8¼	9	10	11	12½	440
12½	5	5½	6¼	7	8	9	10	11	12½	14	500
14½	5½	6¼	7	8	9	10	11	12½	14	16	580
17	7	8	9	10	11	12½	14	16	18	20	680

What is the energy lost in overcoming the rotor and load inertia during starting? How much starting current will this motor draw if it is a 220-V three-phase NEMA type B motor?

Calculation Procedure:

1. Compute the full-load torque

Use the relation $T = 5252hp/rpm$, where T = full-load torque developed by the motor, lb·ft; hp = rated horsepower of the motor; rpm = motor operating speed, r/min. Thus, $T = 5252(300)/1176 = 1342$ lb·ft (1819.5 N·m). This relation assumes that full-load torque is developed during acceleration.

2. Compute the starting time

Use the relation $S = WK^2(rpm)/308T$, where S = starting time, s; other symbols as before. Thus, $S = 28,000(1176)/308(1342) = 79.5$ s. Hence, this motor will take approximately 1⅓ min to bring this load up to the operating speed.

3. Compute the energy required to overcome inertia

Use the relation $E = 2.31WK^2(rpm)^2 10^{-7}$, where E = energy required to overcome the starting inertia, kW·s; other symbols as before. Thus, $E = 2.31(28,000)(1176)^2 \times (10^{-7}) = 7600$ kW·s.

4. Compute the motor starting current

Use the relation $A = (\text{kVA/hp})(\text{motor hp})/\text{constant}$, where A = motor starting current, A; kVA/hp is the locked-rotor current per hp from Table 26; constant = value from Table 27 for the motor operating voltage. Since this is known to be a NEMA type B motor, $A = (3.15)(300)/0.381 = 2480$ A, by using the appropriate data from Tables 26 and 27. Note that the lower value of kVA/hp given in Table 26 was used in this relation. Either value can be used, depending on the anticipated current requirements—small or large.

   Related Calculations: Use this general method to determine the starting time and current of any ac induction or synchronous motor. Note that the motor is assumed to exert full-load torque during the acceleration. As a general guide, the starting current of a motor can be up to about 5.5 times the full-load operating current. The NEMA code-letter designation of a motor can be determined from the catalog description of the motor or from the motor nameplate.

INTERIOR-LIGHTING-SYSTEM SELECTION AND SIZING

Choose the type of lighting and number of fixtures required for a machine shop that is 80 ft (24.4 m) long and 48 ft (14.6 m) wide with a 14-ft (4.3-m) high ceiling. The work done in the shop is classified as rough bench and machine operations. The reflection factor of the ceiling is 80 percent, the walls 50 percent. Maintenance of the fixtures and surfaces will be good.

TABLE 26 Locked-Rotor kVA/hp

NEMA code-letter designation of motor	Locked rotor	
	kVA/hp	kVA/kW
A	0–3.15	0–2.35
B	3.15–3.55	2.35–2.65
C	3.55–4.0	2.65–2.98
D	4.0–4.5	2.98–3.36
E	4.5–5.0	3.36–3.73
F	5.0–5.6	3.73–4.18
G	5.6–6.3	4.18–4.70
H	6.3–7.1	4.70–5.29
J	7.1–8.0	5.29–5.97
K	8.0–9.0	5.97–6.71
L	9.0–10.0	6.71–7.46
M	10.0–11.2	7.46–8.35
N	11.2–12.5	8.35–9.32
P	12.5–14.0	9.32–10.44
R	14.0–16.0	10.44–11.93
S	16.0–18.0	11.93–13.42
T	18.0–20.0	13.42–14.91
U	20.0–22.4	14.91–16.70
	22.4 and up	16.70 and up
V		

TABLE 27 Constants for Motor Starting-Current Determination

Power supply	Constant
Three-phase:	
208 V	0.360
220 V	0.381
440 V	0.762
550 V	0.952
2300 V	3.99
Two-phase:	
Three-wire, 220 V	0.311
Three-wire, 440 V	0.622

Calculation Procedure:

1. Determine the required illumination

Table 28 shows the typical lamp manufacturer's and IES recommended illumination level for various work areas. Study of this tabulation shows that 100 fc (1076.4 lx) is the recommended illumination level for machine shops in which rough bench and machine work is performed.

2. Select the light sources and luminaires

Table 29 lists the various types of light sources used today. Study of this table indicates that 90-W fluorescent lights will probably be suitable. Since the room ceiling height is 14 ft (4.3 m), there will be enough vertical clearance to install the lights because only a 12-ft (3.7-m) clearance is required.

To choose the luminaire type, consider the general practice prevalent in industry today. Three types of luminaires are currently popular for industrial lighting: diffusing, semidirect, and direct. The choice of a particular type of luminaire depends on the degree of diffusion wanted and on the ability of the ceiling and walls to reflect light. Choose direct luminaires for this lighting task because they will provide the high illumination level desired.

3. Compute the room ratio and room index

Use the relation room ratio = $WL/[H(W + L)]$, where W = room width, ft; L = room length, ft; H = room height, ft. For this room, the room ratio = $(48)(80)/[14(48 + 80)]$ = 2.145.

Table 30 shows that a room ratio of 1.75:2.25 indicates a room index of E, with a center point of 2.00.

4. Determine the coefficient of utilization

The coefficient of utilization (CU) is the ratio of the lumens reaching a working plane [normally assumed to be a horizontal plane 30 in (76.2 cm) above the floor] to the total lumens given out by the lamps. It takes into account the lighting efficiency of the luminaire, the mounting height, the room proportions, and the reflection factors of the ceiling and walls.

Coefficient-of-utilization tables are prepared for specific luminaires by the IES and manufacturers. Table 31 lists CU factors for three common general luminaire types. Enter this table with the luminaire type (step 2), room index (step 3), and the ceiling- and wall-reflection factors. Using these data gives CU = 0.65.

TABLE 28 Typical In-Service Illumination Levels, fc (lx)°

Automobile manufacturing:		Iron and steel		Polishing and burnishing	20
Assembly line	100	manufacturing:			
Frame assembly	30	Hot mill	10	Power plants, engine room	
Body manufacturing:		Hot-strip finishing	15	boilers:	
Parts	30	Box annealing	5	Boiler room (operating	
Assembly	30	Hot-strip process	5	floor)	10
Finishing and		Plate finishing	10	Chemical laboratory	20
inspecting	200	Slab-furnace building	10	Coal bunker (conveyor	
		Continuous pickler	15	floor)	10
Chemical works:		Cold mill	15	Coal-crusher house	10
Hand furnaces, boiling		Roll shop	15	Condenser pit (turbine	
tanks, stationary		Cold-mill finishing	15	room)	10
driers, stationary and		Shear building	10	Control rooms (on verti-	
gravity crystallizers	5	Mill runout building	15	cal plane)	30
Mechanical furnaces,		Mesta pickle	10	Heater gallery	10
generators and stills,		Temper mill	15	Machine shop	50
mechanical driers,		Galvanize pickle	10	Oil-pump house (turbine	
evaporators, filtration,		Annealing	10	room)	10
mechanical crystalliz-		Pipe mill	10	Pump bay (turbine	
ers, bleaching	10	Chipping	50	room)	10
Tanks for cooking,				Switchgear area	15
extractors, percolators,		Loading platforms	5	Turbine room (operating	
nitrators, electrolytic				floor)	30
cells	20	Machine shops:		Water pumps	10
		Rough bench and		Water-treating area	10
		machine work	20		
Elevators, freight and		Medium bench and		Sheet-metal works	
passenger	10	machine work, ordi-		Miscellaneous machines,	
		nary automatic		ordinary bench	
		machines, rough		work	20
Forge shops (and		grinding, medium		Punches, presses, shears,	
welding)	10	buffing, polishing	50	stamps, spinning,	
		Fine bench and machine		medium bench	
		work, fine automatic		work	20–25
Foundries:		machines, medium		Tin-plate inspection	10
Annealing (furnaces)	10	grinding, fine buffing			
Cleaning	20	and polishing	100	Stairways and	
Core making:		Extrafine bench and		passageways	10
Fine	50	machine work, grind-			
Medium	25	ing, fine work	200	Storage and stock rooms:	
Grinding and				Rough bulky material	5
chipping	30	Outdoor storage	1	Medium	10
Inspection:				Fine material requiring	
Fine	100	Paper manufacturing:		care	20
Medium fine	50	Beaters, grinding, and			
Medium	30	calendering	20		
Molding:		Finishing, cutting, and		Warehouse	5
Medium	50	trimming	50		
Large	30			Welding and general	
Pouring	10	Paper machine	25	illumination	30
Cupola	10				
Shakeout	10	Parking areas	2	Yard lighting	0.2–0.8
		Plating	10		

°Note:	fc		lx	fc		lx
	0.2–0.8	=	2.2–8.6	25	=	269.1
	2	=	21.5	30	=	322.9
	5	=	53.8	50	=	538.2
	10	=	107.6	100	=	1076.4
	15	=	161.5	200	=	2152.8
	20	=	215.3			

To convert to SI (lx), find the fc value from the table and look
for the lx value as shown above.

TABLE 29 Light-Source Selection Guide*

Light source	Minimum mounting height, ft (m)	Advantages	Disadvantages
Fluorescent: Preheat 40- and 90-W Rapid-start 40-W Slimline: 4 ft, 6 ft, 8 ft (1.2, 1.8, 2.4 m) T-12	12 (3.7) for 20-W	General: High lamp efficiency, long life, low brightness, good color quality; slimlines and rapid-start lamps start instantly (rapid-start within 1 s) and require no starters	General: High initial installation cost; large number of lamps required, with accompanying maintenance problems; low system efficiency in high, narrow areas
Incandescent: Standard-bulb lamps	Varies with wattage	General: Low initial installation cost, good color quality, start instantly	General: Low lamp efficiency, short lamp life in comparison with other types
Reflector lamps: 550-W narrow-beam R57 800-W narrow-beam R57	20 (6.1) 24 (7.3)	Good system efficiency in high narrow areas, relatively long life (2000 h)	High lamp brightness may cause direct and reflected glare; relatively low vertical-surface illumination
550-W wide-beam R52 800-W wide-beam R52	15 (4.6) 18 (5.5)	Good vertical-surface illumination, relatively long life (2000 h)	High lamp brightness may cause direct and reflected glare; low-system efficiency in high, narrow areas
550-W medium-beam R57 800-W medium-beam R57	16 (4.9) 20 (6.1)	Relatively long life (2000 h), designed for use on 230-V distribution systems	High-voltage lamps are slightly less efficient than standard-voltage lamps

Lamp		General characteristics	Remarks
Mercury:		General: High lamp efficiency, long life, easy to maintain, good system efficiency (except for A-H9 in high, narrow areas)	General: Color deficiency characteristic of mercury arc; lamps do not start at full brightness or restart instantly
400-W A-H1	18 (5.5)	Exceptionally long life, relatively low arc brightness	Not economically competitive with other mercury lamps listed
400-W E-H1	18 (5.5)	Exceptionally long life	
700-W A-H18	26 (7.9)	Low transformer cost and losses	High socket voltage (460 V)
1000-W A-H12	35 (10.7)	High lumen output	
1000-W A-H15	35 (10.7)	Low transformer cost and losses	High socket voltage (460 V)
3000-W A-H9	40 (12.2)	Extremely high lumen output, minimum maintenance	Less efficient than other high-wattage mercury lamps, low system efficiency in high, narrow areas
Reflector lamps:		Relatively low initial installation cost	High lamp brightness may cause glare
400-W K-H1	20 (6.1)	High vertical-surface illumination	Low system efficiency in high, narrow areas
400-W L-H1	24 (7.3)	High system efficiency in high, narrow areas	Low vertical-surface illumination
Color-improved mercury:		General: White light, high lamp efficiency, long life, easy to maintain, good system efficiency when used with proper reflector	General: Lamps do not start at full brightness or restart instantly; slightly less efficient than conventional mercury lamps
400-W J-H1	16 (4.9)	Exceptionally long life, relatively low brightness	
700-W B-H18	22 (6.7)	Relatively low brightness, low transformer cost and losses	High socket voltage (460 V)
1000-W C-H12	28 (8.5)	Minimum number of luminaires required	
1000-W B-H15	28 (8.5)	Low transformer cost and losses; minimum number of luminaires required	High socket voltage (460 V)

° *Power* magazine.

5. *Estimate the lamp-maintenance factor*

Illumination decreases in service because dirt accumulates on the lamps and luminaires, reflection factors are reduced by the aging of paint, etc. With a clean atmosphere and good cleaning of the lamps, the maintenance factor (MF) = 0.70. Use this factor for this installation. Typical values of MF for direct luminaires are: good, 0.70; medium, 0.60; poor, 0.55.

TABLE 30   Room-Ratio Ranges

	Room ratio	
Room index	Range	Center point
J	Less than 0.7	0.6
I	0.7–0.9	0.8
H	0.9–1.12	1.0
G	1.12–1.38	1.25
F	1.38–1.75	1.50
E	1.75–2.25	2.00
D	2.25–2.75	2.50
C	2.75–3.50	3.00
B	3.50–4.50	4.00
A	More than 4.50	5.00

6. *Compute the number of lamps and luminaires required*

Use the relation number of lamps = (fc required)(floor area, ft^2)/(lm per lamp) (CU)(MF). Determine the lumens per lamp from Fig. 22. Thus, a 90-W T-17 preheat lamp is rated at 5150 lm per lamp. Entering the appropriate data yields the number of lamps = $(100)(80 \times 48)/(5150)(0.65)(0.70) = 163.8$, say 164 lamps.

Compute the number of luminaires from number of luminaires = number of lamps/lamps per luminaire. With four lamps per luminaire, number of luminaires = 164/4 = 41.

7. *Choose the location of the luminaires*

As a general rule, equal spacing of the luminaires is used to provide uniform illumination. Such a spacing arrangement would be suitable for this installation.

Related Calculations: The procedure presented above is termed the *lumen method*. It is widely used for the design of lighting systems for general-area indoor illumination in all types of installations—industrial, commercial, residential, marine, aircraft, mobile, etc. Used with the data presented, or with additional manufacturer's engineering data, this method permits a quick, economical choice of a suitable indoor-lighting system.

A refinement of the lumen method, termed the *IES zonal-cavity method*, was recently introduced. With this refinement of the lumen method, the room must be divided into three cavities—ceiling, room, and floor—as indicated in Fig. 23. Then, by using the specific reflectances for the surfaces in each of these three cavities, more accurate calculations can be obtained than is possible by the original lumen method. In addition, with the zonal-cavity method, the effects of room proportions, luminaire suspension length, and work-plane height upon the coefficient of utilization are more accurately accounted for.

The zonal cavities are defined as follows:

Ceiling cavity is the space bounded by the ceiling, upper walls, and an imaginary plane through the luminaires.

Room cavity is the space bounded by the plane through the luminaires, the work plane, and the portion of the walls between these planes.

Floor cavity is the space bounded by the work plane, the lower walls, and the floor.

In applying the zonal-cavity method of calculation, the proportions of each cavity may be represented by a *cavity ratio*. These cavity ratios may be obtained from a table (see fig. 9-2, cavity ratios, *IES Lighting Handbook*), or from a formula, as follows:

$$\text{Ceiling cavity ratio (CCR)} = \frac{5h_{CC}(L + W)}{LW}$$

$$\text{Room cavity ratio (RCR)} = \frac{5h_{RC}(L + W)}{LW}$$

$$\text{Floor cavity ratio (FCR)} = \frac{5h_{FC}(L + W)}{LW}$$

where L = room length, ft; W = room width; ft; h = cavity height, ft.

TABLE 31 Coefficients of Utilization for General Types of Luminaires°

Typical luminaire distribution	Room index	Reflection factors Ceiling 80% Walls 50%	30%	10%	70% Walls 50%	30%	10%	50% Walls 50%	30%	10%	30% Walls 50%	30%	10%
Diffusing	J	0.26	0.21	0.18	0.25	0.21	0.17	0.23	0.19	0.16	0.20	0.17	0.15
	I	0.32	0.27	0.23	0.31	0.26	0.22	0.28	0.24	0.21	0.25	0.22	0.19
	H	0.38	0.33	0.29	0.36	0.32	0.28	0.33	0.29	0.26	0.29	0.26	0.23
	G	0.43	0.38	0.34	0.41	0.36	0.33	0.37	0.33	0.30	0.33	0.30	0.27
	F	0.47	0.42	0.38	0.45	0.40	0.36	0.40	0.36	0.33	0.35	0.32	0.30
	E	0.53	0.48	0.44	0.50	0.46	0.42	0.44	0.41	0.38	0.39	0.36	0.34
	D	0.56	0.52	0.48	0.53	0.49	0.46	0.47	0.44	0.41	0.41	0.39	0.37
	C	0.59	0.55	0.51	0.55	0.52	0.49	0.49	0.46	0.44	0.43	0.41	0.39
	B	0.62	0.59	0.56	0.58	0.55	0.53	0.52	0.49	0.47	0.45	0.43	0.42
	A	0.64	0.61	0.59	0.61	0.58	0.55	0.54	0.51	0.49	0.46	0.45	0.44
Semidirect	J	0.34	0.28	0.24	0.33	0.28	0.24	0.31	0.26	0.24	0.30	0.25	0.22
	I	0.42	0.36	0.32	0.40	0.35	0.31	0.38	0.33	0.30	0.36	0.32	0.29
	H	0.48	0.42	0.38	0.47	0.41	0.37	0.44	0.39	0.36	0.41	0.37	0.34
	G	0.54	0.48	0.44	0.52	0.47	0.43	0.49	0.45	0.41	0.46	0.42	0.39
	F	0.58	0.53	0.48	0.56	0.51	0.47	0.53	0.49	0.45	0.49	0.46	0.43
	E	0.64	0.59	0.55	0.62	0.57	0.54	0.58	0.54	0.51	0.54	0.51	0.48
	D	0.67	0.63	0.59	0.65	0.61	0.58	0.60	0.57	0.54	0.56	0.54	0.52
	C	0.70	0.66	0.62	0.68	0.64	0.61	0.63	0.60	0.57	0.58	0.56	0.54
	B	0.73	0.70	0.67	0.70	0.67	0.65	0.66	0.63	0.61	0.61	0.59	0.57
	A	0.75	0.72	0.70	0.72	0.70	0.68	0.68	0.65	0.63	0.62	0.61	0.60
Direct	J	0.34	0.28	0.24	0.34	0.28	0.23	0.33	0.27	0.24	0.32	0.27	0.23
	I	0.43	0.36	0.31	0.42	0.36	0.31	0.41	0.35	0.31	0.40	0.35	0.31
	H	0.49	0.42	0.38	0.48	0.42	0.38	0.47	0.42	0.37	0.46	0.41	0.37
	G	0.55	0.49	0.44	0.55	0.48	0.44	0.53	0.48	0.44	0.52	0.47	0.44
	F	0.60	0.54	0.49	0.59	0.53	0.49	0.57	0.52	0.48	0.56	0.52	0.48
	E	0.65	0.60	0.56	0.64	0.60	0.55	0.63	0.59	0.55	0.61	0.58	0.55
	D	0.69	0.64	0.60	0.68	0.64	0.60	0.66	0.63	0.59	0.65	0.62	0.59
	C	0.72	0.67	0.64	0.71	0.67	0.63	0.69	0.66	0.63	0.67	0.65	0.62
	B	0.76	0.72	0.69	0.75	0.71	0.69	0.73	0.70	0.68	0.71	0.69	0.67
	A	0.78	0.75	0.72	0.77	0.74	0.72	0.75	0.73	0.71	0.74	0.72	0.70

Diffusing: 40 / 40

Semidirect: 25 / 60

Direct: 0 / 80

°A floor reflection factor of 10% is assumed for values in table. For floors having higher reflection factors, refer to IES tables. Numbers accompanying sketches at left indicate percentage of lamp lumens directed upward and downward. Sum of percentages equals luminaire light efficiency.

Room cavity ratios are always required to obtain coefficients of utilization, just as room ratios or room indices were required by the original lumen method of calculation. The *ceiling* and *floor cavity* ratios are required only to obtain *effective cavity reflectances*, which are not used in coefficient-of-utilization tables for specific luminaires to establish specific effective cavity-reflectance conditions. These effective ceiling- or floor-cavity reflectance percentages for various reflectance combinations are also provided in table form, by manufacturers of lighting equipment or in fig. 9-3 in the *IES Lighting Handbook*.

Manufacturers of lighting equipment provide coefficient-of-utilization tables for each luminaire. These CU tables are for room cavity ratios varying from 1 to 10, based on 20 percent effective floor-cavity reflectance and for various ceiling-cavity and wall-cavity reflectances.

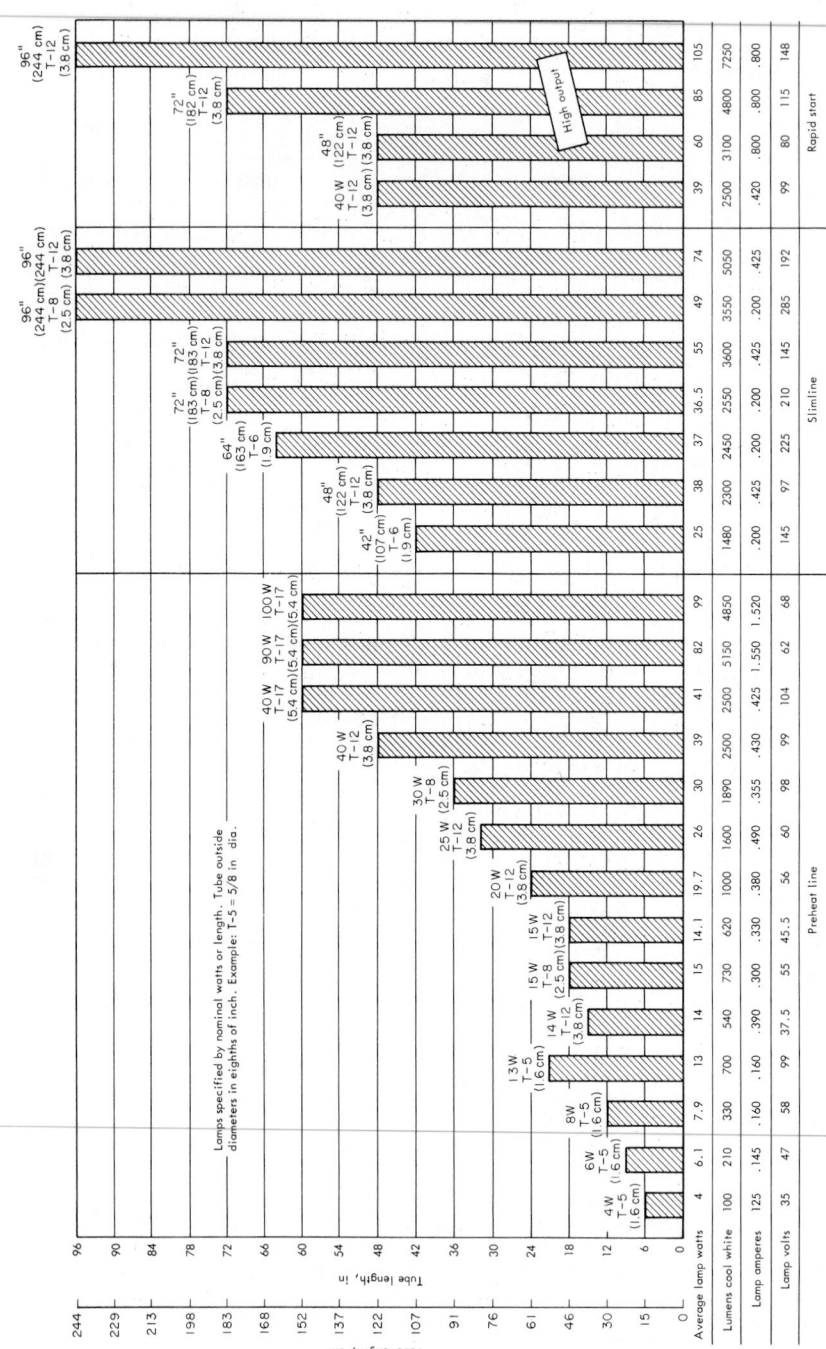

FIG. 22 Performance factors for typical fluorescent lamps.

4.58

When the effective floor-cavity reflectance varies from the 20 percent value normally used in manufacturers' CU tables, correction factors for 10 and 30 percent effective floor-cavity reflectances are provided in table form, either by lighting equipment manufacturers or in fig. 9-5 of the *IES Lighting Handbook*.

A simple procedure may now be followed to obtain the CU value for the specific luminaire. The procedure for a suspended luminaire is typical:

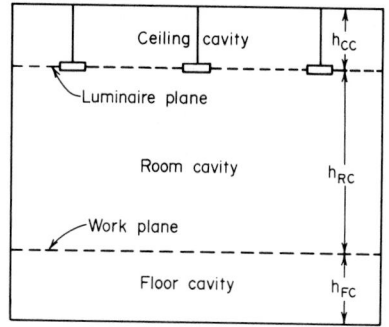

1. Obtain the *room cavity ratio* and *ceiling cavity ratio* by formulas (above), or from the previously mentioned table.

2. Obtain the *effective ceiling-cavity reflectance*, from manufacturers' literature or from fig. 9-3 in the *IES Lighting Handbook* (note that expected *maintained* ceiling and wall reflectances should be used in selecting the proper column).

3. Obtain CU for expected *maintained* wall reflectance and 20 percent effective floor-cavity reflectance from the CU table for the luminaire. (Interpolate as required for the exact RCR and ceiling-cavity reflectance.)

FIG. 23 Three room cavities are used in the IES zonal-cavity method of calculation.

For supplementary lighting (i.e., the lighting of special machines or small areas), the *point-by-point* calculation method is sometimes used. To apply this method, do the following: (1) Select the type of light source that will be used. (2) Find the angle in degrees between the vertical and a line to the point illuminated (Fig. 24). (3) Determine the candlepower in the direction of the point, using the angle θ (Fig. 24) and the candlepower distribution curve of the luminaire, available from the luminaire manufacturer. (4) Compute the horizontal fc = (candlepower of luminaire)(cos θ)/(distance factor, and vertical fc) = (candlepower of luminaire)(cos θ)/distance factor. (The *distance factor* depends on the type of distribution, step 1; for a point source it is the distance squared. Other sources are discussed in the next paragraph.) (5) Multiply each of the results in step 4 by the MF. Use MF similar to those for the lumen method. However, the point-by-point method does not take the reflection of the walls and ceilings into consideration.

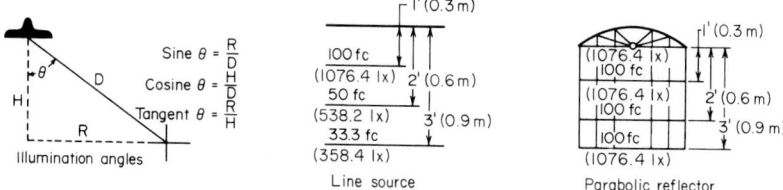

FIG. 24 Variation of illumination with the distance of the lighted area from the light source.

For line light sources, such as a continuous row of fluorescent luminaires, footcandle values near the lamp vary nearly inversely with the distance from it. As the distance increases, the footcandle variation approaches the inverse square (Fig. 24). With parallel beams, such as in spotlights and concentrators, the inverse-square law applies as the distance from the light becomes great.

OUTDOOR-LIGHTING SELECTION AND SIZING

A parking lot in an industrial plant is 400 ft (121.9 m) long and 200 ft (61.0 m) wide. Choose the type and location of the lighting for this lot. Lighting must be provided from sundown to sunrise.

Calculation Procedure:

1. *Choose the type of lamp to use*

Three types of lamps are popular for outdoor lighting: mercury vapor, fluorescent, and filament. The filament-type lamp is usually more economical for installations used less than 1000 lighting hours per year, i.e., under 3 to 4 h per night. Since this installation requires lighting from sundown to sunrise, filament-type lamps would not be economical because they would be used more than 4 h per night.

Thus, the lighting should be provided by either a mercury-vapor or fluorescent lamp. Select mercury-vapor floodlight-type lamps because they provide economical lighting with little maintenance.

2. *Determine the recommended lighting level*

Table 32 shows that the recommended lighting level for parking areas is 0.5 to 2.0 fc (5.4 to 21.5 lx). Use a level of 2.0 fc (21.5 lx) for this area.

To determine the recommended W/ft^2, enter Fig. 25 at the illumination level in footcandles on the left and project horizontally to the appropriate lamp curve. Read at the bottom of the chart the W/ft^2. Thus, with an illumination level of 2.0 fc (21.5 lx), the recommended W/ft^2 is 0.10 (1.1 W/m^2).

3. *Select the lighting-fixture location*

Figure 26 shows five different arrangements of outdoor-lighting fixtures. For large areas with lighting levels of up to 5 fc (53.8 lx), center poles with multiluminaire mountings are economical. Hence, these will be chosen for this parking lot.

4. *Compute the number of lights required*

Use this relation: number of floodlights required = (area lighted, ft^2)(recommended fc)/(floodlight lumen rating)(MF), where MF is 0.165 for open floodlights and 0.75 for enclosed floodlights.

TABLE 32   Recommended Lighting Levels°

Application	Usual recommended level	
	fc	lx
Building exteriors:		
Terra cotta, light marble, plaster	15; 10; 5	161.5; 107.6; 53.8
Buff limestone, buff brick, concrete	20; 15; 10	215.3; 161.5; 107.6
Brownstone, wood shingles, dark finish	50; 35; 20	538.2; 376.7; 215.3
Poster and bulletin boards:		
Bright surroundings, light surfaces	50	538.2
Bright surroundings, dark surfaces	100	1076.4
Dark surroundings, light surfaces	20	215.3
Dark surroundings, dark surfaces	50	538.2
Industrial roadways:		
Between or adjacent to buildings	1.0	10.8
Not bordered by buildings	0.5	5.4
Loading platforms, freight docks	20	215.3
Industrial parking lots	0.5–2.0	5.4–21.5
Smoke stacks and water tanks with advertising signs	Same as poster and bulletin boards	
Storage yards:		
Active	20	215.3
Inactive	1	10.8
Television surveillance:		
To indicate movement	5	53.8
For clear picture	20	215.3

°Adapted from Illuminating Engineering Society and manufacturer's recommendations.

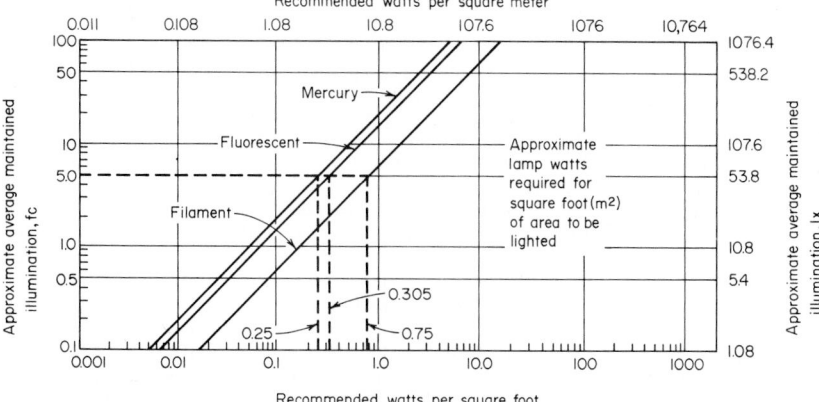

FIG. 25 Typical watts per square foot for different illumination levels.

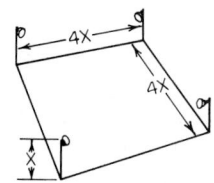

For lighting small areas don't put poles more than four times the mounting height apart. This applies regardless of number floods per pole or fc (lx)

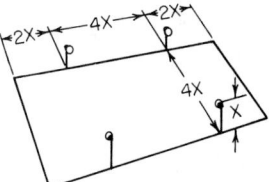

For perimeter poles set in from corners, none should be further from corner than twice mounting nor more than four times mounting height from next pole.

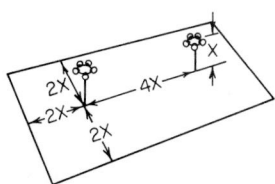

For large areas at up to 5 fc (53.8 lx) center poles with multiluminaire mountings can save. Put outer poles within twice their height from area perimeter.

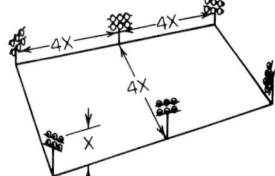

For levels above 5 fc (53.8 lx), poles shouldn't be more than four times mounting height apart. This applies both across area lighted and between poles.

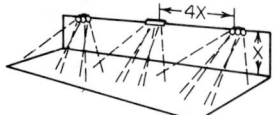

For corridor-type lighting, whether filament or fluorescent sources are used, lamps shouldn't be further apart than four times distance above surface lighted.

FIG. 26 Pole placements recommended for effective outdoor lighting.

Typical mercury-vapor floodlight lumen ratings range from 4200 lm for a 500-W lamp to 18,000 lm for a 1500-W lamp. Assume that 1000-W, 9500-lm, enclosed, clear, general-service floodlights are used. Then, number of floodlights required = $(400 \times 200)(2.0)/(9500)(0.75) = 22.5$, say 24 lamps to obtain an even number of lamps to be divided between two supporting poles.

5.   *Determine the lamp arrangement*

With 24 lamps a two-pole arrangement (Fig. 26) might be suitable. This would lend itself to the mounting of 12 lamps on each pole.

Refer to a lamp manufacturer's engineering data. These data show that a 40° beam floodlight mounted 55 ft (16.8 m) above the ground will illuminate an area 104 ft (31.7 m) long and 65 ft (19.8 m) wide, or an elliptical area of 2650 ft² (246.2 m²). Hence, 24 lamps will illuminate an area of $24(2650) = 63,600$ ft² (5908.6 m²). The area of the parking lot is 400 (200) = 80,000 ft² (7432.2 m²). Thus, the lamps will not illuminate the entire parking lot.

Raising the lamps to 70 ft (21.3 m) above the ground provides an illuminated area of 4700 ft² (436.6 m²) per lamp, or a total area of $24(4700) = 112,800$ ft² (10,479.5 m²). Also, the lamps will be within twice their height of $2(70) = 140$ ft (42.7 m) from the area perimeter because they are mounted 100 ft (30.5 m) from the perimeter (Fig. 27). Since the area covered by the floodlights exceeds the ground area, the beams will overlap. This is a desirable condition because it provides more uniform illumination.

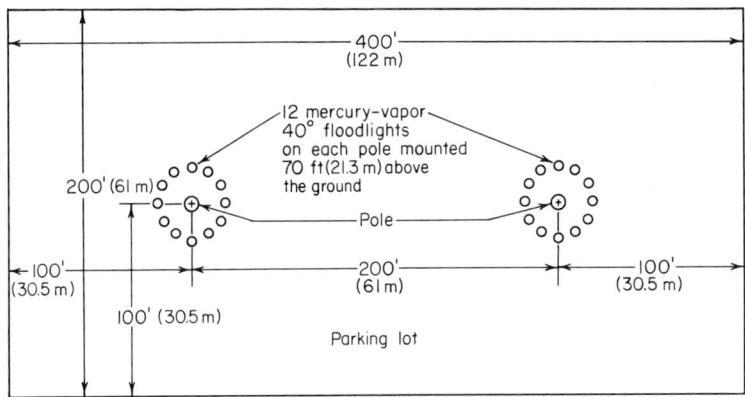

FIG. 27   Parking-lot lighting arrangement.

Related Calculations: Use the method given here for any outdoor-lighting application—building exteriors, catwalks, drill fields, gasoline service stations, piers, prison yards, quarries, railroad yards, shipyards, storage yards, baseball fields, boxing rings, skating rinks, etc. Consult the manufacturer's engineering data for the specific characteristics of the lights that will be used.

Typical output of outdoor lamp in lumens per watt are: filament, 20; mercury-vapor, 50; fluorescent, 70. Useful lamp life is usually: filament, 6 months; fluorescent, 36 months; mercury-vapor, 48 months. These lives are based on all-night every-night operation of the lamps. Give careful consideration to the lamp life. Longer lamp life means fewer maintenance tasks at elevated positions.

In choosing lamps, use the largest-wattage fixtures and the fewest locations that will deliver acceptable uniform lighting. The higher the wattage per fixture, the lower the total cost on a unit-cost-of-light basis. Generally, the spacing between poles should not exceed four times the mounting height of the fixtures. Also, use the widest beam spread fixture available that is consistent with good utilization of light.[1]

[1]John C. Boyter and Robert E. Faucett, General Electric Company, and *Factory* magazine.

To determine the spacing of highway lighting poles, use this relation: spacing between luminaires, ft = (lumens required)(coefficient of utilization)(maintenance factor)/[(fc maintained by the lamp)(width of roadway, curb to curb, ft)].

FEEDER SIZING FOR A COMBINATION ELECTRIC LOAD

Using the *National Electrical Code*, size the feeders for a load consisting of four three-phase 220-V squirrel-cage induction motors designed for a 40°C temperature rise, marked *Code* letter H, started across the line, and made up of one 10-hp (7.5-kW) motor, one 7.5-hp (5.6-kW) motor, and two 1.5-hp (1.1-kW) motors. The feeders will also serve a 20-kW single-phase 115-V lighting load.

Calculation Procedure:

1. Determine the motor average full-load current

Using the motor-current table in the *Code*, we see the average full-load current of the motors is 10 hp (7.5kW), 27 A; 7.5 hp (5.6 kW), 22 A; 1.5 hp (1.1 kW), 5 A.

2. Select the main feeder for the motors

Size the feeder for a current flow of 125 percent of the average full-load current of each motor. Thus, total current flow = 1.25(27) + 1.25(22) + 1.25(5)(2) = 73.7 A. The *Code* motor-feeder table shows that three no. 6 RHW feeders in a 1-in (25.4-mm) conduit are needed.

3. Size the individual branch circuits for each motor

Using the 125 percent current flow computed for each motor in step 2 and *Code* wire data: 10-hp (7.5-kW) motor, 1.25 (27) = 33.75 A; choose three no. 8 TW or RHW feeders in ¾-in (19.1-mm) conduit, as recommended in the *Code*. For the 7.5-hp (5.6-kW) motor, 1.25(22) = 27.5 A; use three no. 10 TW or RHW feeders in ¾-in (19.1-mm) conduit. For each 1.5-hp (1.1-kW) motor, 1.25(5) = 6.25 A; use three no. 14 TW or RHW feeders in ½-in (12.7-mm) conduit.

4. Select the overcurrent protection rating

Since the same 125 percent factor applies for the overcurrent or running protection, the maximum rated thermal elements for each motor will be 10 hp (7.5 kW), 33.75 A; 7.5 hp (5.6 kW), 27.5 A; 1.5 hp (1.1 kW), 6.25 A.

5. Select the motor fuses

The *Code* specifies a maximum of 300 percent for the fuse rating for these *Code* letter H motors, based on the average full-load current of each motor. Or, for the 10-hp (7.5-kW) motor, 3(27 A) = 81 A—use a 100-A block with three 90-A fuses. For the 7.5-hp (5.6-kW) motor, 3(22 A) = 66 A—use a 100-A block with three 70-A fuses. For each 1.5-hp (1.1-kW) motor, 3(5 A) = 15 A—use a 30-A block with three 15-A fuses.

6. Select the main motor-feeder protection

The maximum rating or setting for the motor-feeder protection device must not be greater than the largest rating or setting of a branch-circuit protective device for one of the motors of the group plus the sum of the full-load current of the other motors. Using the largest motor rating plus the others gives 3(27) + 22 + 2(5) = 113 A. Use a 200-A switch with three 125-A fuses.

7. Compute the lighting-load current

Assume the lighting circuit is 115/230 V single-phase. Then, full-load current = load, W/maximum voltage = 20(1000)/230 = 87 A. Use a 100-A switch with two 90-A fuses. If the 87-A load is continuous, a fused switch must be selected such that the load is not over 80 percent of the fuse rating. This means that the minimum fuse size under these conditions would be (87 A)(1.25) = 108.75, or a 110-A fuse (the next *largest* standard rating) for each phase, in a 200-A switch. The factor 1.25 is obtained from an assumed maximum current flow 125 percent of the full-load current. To handle the 87-A flow, use three no. 3 RHW in 1¼-in (31.8-mm) conduit, or three no. 2 RHW for the 108.75-A load.

8. *Select the type of service to use*

If these two loads will be fed from separate services, then the above calculations are complete. But if a combination service is to be used, a four-wire, 240-V delta service would be suitable. The lighting load would then be fed from two of the three phases.

9. *Choose the service feeders and switches*

The motor feeder must handle a starting demand of 113 A, as computed in step 6. The lighting two-phase full-load current is 87 A, step 7. Summing these two current flows gives 113 + 87 = 200 A.

The main switch would be a 200-A size with two 200-A fuses in two phases and one 125-A fuse in the third phase. Two of the main service lines would be 2/0 RHW; one would be no. 6 RHW, and the neutral would be no. 3 RHW. These are the minimum sizes based on the *Code* rules for safe application. Local codes might require larger feeders. Figure 28 shows typical *Code* sizing of feeders and overcurrent devices for motors.

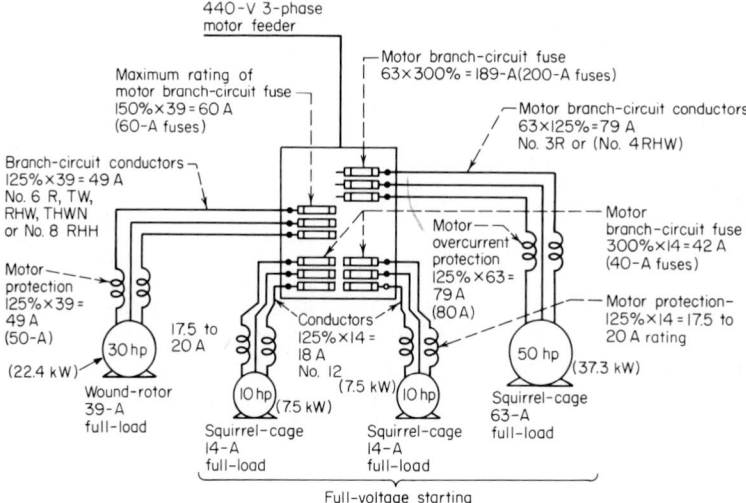

Notes: 1. Full-load current for each motor is taken from NE Code Table 430–150.
2. Running overload protection is sized on basis that nameplate values of motor full-load currents are same as values from NE Code Table 430–150. If nameplate and table values are not the same, overload protection is sized according to nameplate.

FIG. 28   *National Electrical Code* sizing of feeders and overcurrent devices for motors.

Related Calculations: Typical recommended limits for the voltage in various types of circuits are shown in Fig. 29. Application of load demand and diversity factors is shown in Fig. 30, and typical demand factors are listed in Table 33. Fuses of various classes are compared in Table 34. Grounding methods for interior ac wiring systems and equipment enclosures are shown in Fig. 31. The procedure given above is valid for single or combination electric loads in a variety of installations. For specific numerical values of conductor and conduit sizes, refer to the *National Electrical Code* and the local governing code, if any.

SIZING RESIDENTIAL-SERVICE DEMAND LOAD

What size service is required for a 1500-ft² (139.4-m²) house with all-electric utilization having the loads listed in step 1 and sized according to the *National Electrical Code?* Use the *optional method* of calculating the service demand load. The service voltage is 120/230 V, three-wire.

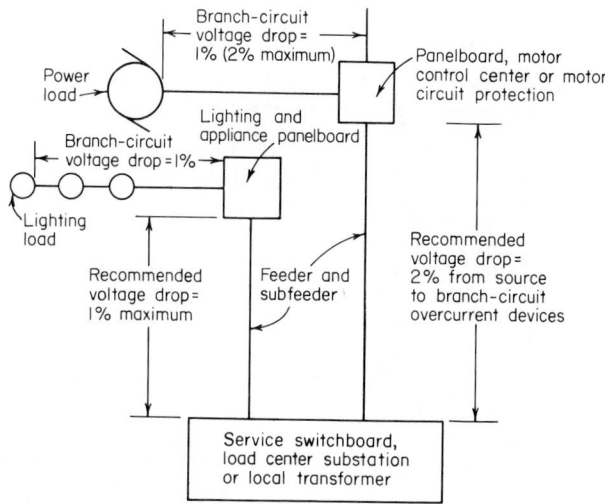

FIG. 29 Recommended limits of voltage drop in various circuits.

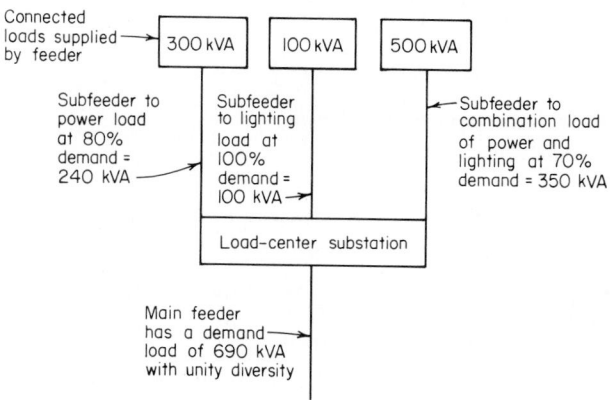

FIG. 30 Typical application of demand and diversity factors.

TABLE 33 Common Demand Factors for Sizing Service and Main Feeders°

Power load devices	Range of common demand factors, %
Motors for pumps, compressors, elevators, machine tools, blowers, etc.	20–60
Motors for semicontinuous operations in various mills and process plants	50–80
Motors for continuous operations, as in textile mills	70–100
Arc furnaces	80–100
Induction furnaces	80–100
Arc welders	30–60
Resistance welders	10–40
Resistance heaters, ovens, and furnaces	80–100

°*National Electrical Code.*

TABLE 34 Comparison of Fuse Classes† (General-Purpose Cartridge-Type)°

UL class	Range, A	Interrupt capacity, A	Maximum let-through	Time delay†	Dimensions
H	0–600	10K	None	No	Old *NEC*
J	0–600	100K or 200K	Yes‡	No	Special§
K-1	0–600	10K, 25K, 50K, 100K or 200K	Yes‡	No	Old *NEC*
K-5	0–600	10K, 25K, 50K, 100K	Yes‡ Yes	Yes	Old *NEC*
K-9	0–600	10K, 25K, 50K, or 100K	Yes‡	Yes	Old *NEC*
L	601–6000	100K or 200K	Yes‡	No	Present NEMA sizes

°*National Electrical Code.*
†NEMA standards call for a minimum of 10-s delay at 500 percent of fuse rating. No UL standards adopted.
‡UL standards state maximum peak let-through in amperes and energy let-through (I^2T) for each size and type of fuse. In 600-A sizes, lowest let-through is class J, increasing slightly through classes K-1, K-5, and K-9.
§Smaller than the noninterchangeable with *NEC*-size fuses.

Calculation Procedure:

1. *Compute the maximum possible demand*

This house has the following loads:

1,500 W for each of two (minimum of two required) kitchen appliance circuits	2(1,500)	= 3,000 W
1,500 ft² (457 m²) of floor area at 3 W/ft² (32.3 W/m²) for general lighting and receptacles	3(1,500)	= 4,500
14 kW of electric space heating from more than four separately controlled units	14,000	
12-kW electric range	12,000	
3-kW water heater	3,000	
5-kW clothes dryer	5,000	
3-kW load of unit air conditioners (Because this load is less than the space-heating load and will *not* be operated simultaneously with it, no load need be added.)		
Hence, maximum possible demand	41,500 W	

2. *Compute the probable demand load*

Table 35, from the *National Electrical Code*, shows that under the optional method, the first 10 kW of all other loads, as defined in Table 35, should be taken at 100 percent, or 10,000 W. The remainder of the load is taken at 40 percent, or $0.40(41,500 - 10,000) = 12,600$ W. Hence, the total probable demand $= 10,000 + 12,600 = 22,600$ W.

3. *Compute the size of the service*

Use the relation, size of service, A = demand load, W/maximum service voltage, or 22,600/230 = 98 A. Use the next larger standard service, or 100 A.

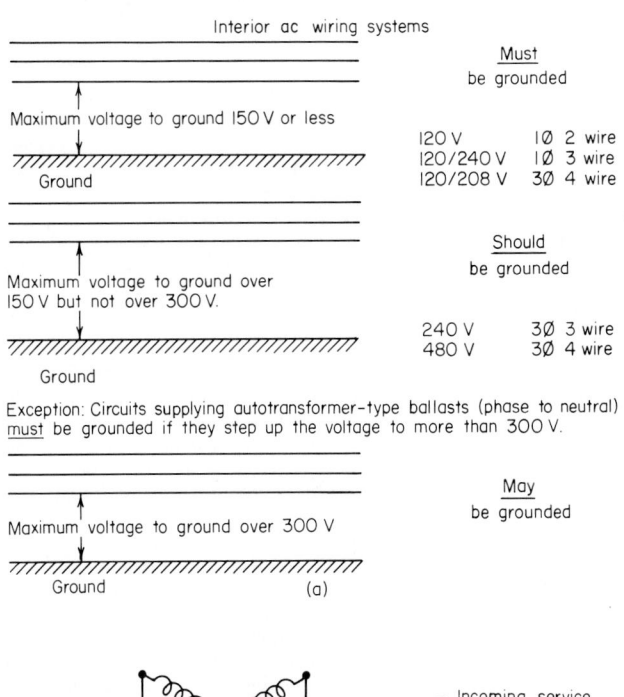

Interior ac wiring systems

<u>Must</u>
be grounded

Maximum voltage to ground 150 V or less

Ground

120 V	1∅	2 wire
120/240 V	1∅	3 wire
120/208 V	3∅	4 wire

<u>Should</u>
be grounded

Maximum voltage to ground over 150 V but not over 300 V.

Ground

| 240 V | 3∅ | 3 wire |
| 480 V | 3∅ | 4 wire |

Exception: Circuits supplying autotransformer-type ballasts (phase to neutral) <u>must</u> be grounded if they step up the voltage to more than 300 V.

<u>May</u>
be grounded

Maximum voltage to ground over 300 V

Ground

(a)

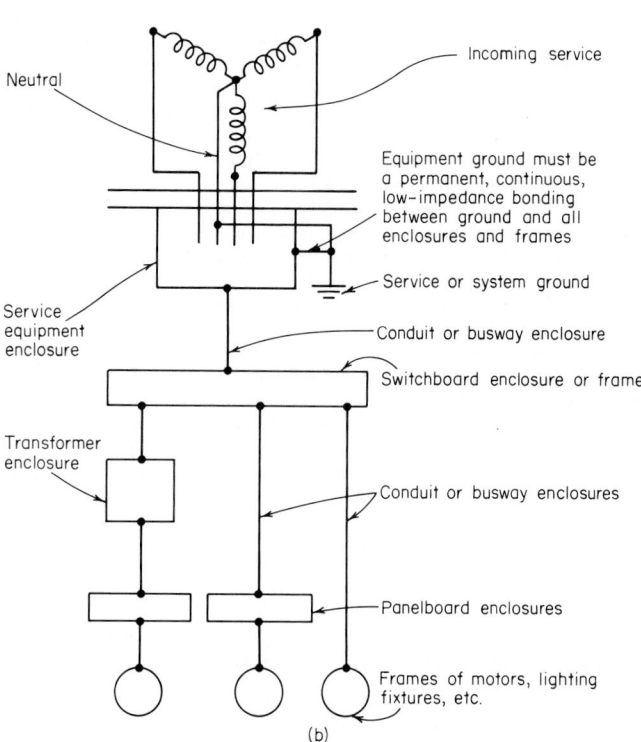

Neutral

Incoming service

Equipment ground must be a permanent, continuous, low-impedance bonding between ground and all enclosures and frames

Service or system ground

Service equipment enclosure

Conduit or busway enclosure

Switchboard enclosure or frame

Transformer enclosure

Conduit or busway enclosures

Panelboard enclosures

Frames of motors, lighting fixtures, etc.

(b)

FIG. 31 (a) When to ground a system conductor; (b) grounding methods for equipment enclosures.

TABLE 35   Optional Calculation for One-Family Residence

Load, kW or kVA	Percentage of load
Air conditioning and cooling including heat pump compressor	100
Central electric space heating	100
Less than four separately controlled electric space-heating units	100
First 10 kW of all other load	100
Remainder of other load	40

All other load shall include 1500 W for each 20-A appliance outlet circuit; lighting and portable appliances at 3 (32.3 W/m^2); all fixed appliances (including electric space heating when there are four or more separately controlled units, ranges, wall-mounted ovens, and counter-mounted cooking units) at nameplate rated load (kVA for motors and other low power-factor loads).

Under certain load conditions, the calculation may indicate a required service capacity substantially less than 100 A. In such cases, however, 100 A is the minimum service that can be used, according to the *Code*. When using the alternative calculation method of *Code* Sec. 220-4, note that a calculated demand load of 10 kW or more requires that the minimum size service be 100-A, three-wire. Further, the optional method may be used instead of the *standard method* under the following conditions: (1) it is for a one-family residence only, (2) it is served by a 115/230-V three-wire 100-A or larger service, and (3) the total load is supplied by one set of service conductors.

Related Calculations: To use the standard method of computing the service entrance for a residence, apply the following steps. (1) Multiply the floor area of the house by 4 W/ft^2 (43.1 W/m^2) to obtain the general-lighting and general-purpose outlet electric load. (2) Add the total circuit capacity, in watts, allowed for the appliance load in the kitchen, dining room, pantry, laundry, and utility area that will be served by 120-V appliance circuits. Find the total circuit capacity by multiplying the number of such circuits laid out in branch-circuit design by 2000 W per circuit. Or assume a load of 4000 W (i.e., two appliance circuits) when the exact number of such circuits is not known. Table 36 lists the typical loads and characteristics of modern circuits for appliances. (3) Take 3000 W of the sum of steps 1 and 2 at 100 percent demand. (4) To the load in step 3 add 35 percent demand of the remaining load above 3000 W computed in the first three steps. (5) Take the sum of the values computed in steps 4 and 5. This is the capacity that must be provided in the service-entrance conductors to supply the general-lighting and general-purpose receptacle loads. (6) Add 8000 W for an electric range if this is rated 12 kW or less. Consult the *Code* if the electric cooking appliances consist of a built-in oven and rangetop. (7) Add the rated wattage of all fixed appliances to be served by individual circuits not previously included in the calculation. If both electric heating and air conditioning will be used in the house, include only the wattage of the larger unit because the two units will not be used simultaneously. (8) Take the sum of steps 5, 6, and 7. (9) Divide the sum found in step 8 by 240 V (for a 120-240-V three-wire single-phase service) to obtain the required ampere rating of the service conductors.

This procedure can also be used to compute the general-lighting, general-purpose receptacle and appliance load for apartments in multiple-dwelling buildings.

ELECTRIC COMFORT HEATING LOAD DETERMINATION

The residence in Fig. 32 is to be heated electrically. What is the heating load for an inside design temperature of 70°F (21.1°C), outside design temperature of 0°F (−17.8°C), a groundwater temperature of 50°F (10.0°C), and a ceiling height of 8 ft (2.4 m) in the structure if (1) the single-floor residence is built over a ventilated crawl space, (2) the same structure is built over a heated basement, and (3) the same structure is mounted on a concrete slab? Use the data given in Tables 37 to 48.

TABLE 36 Modern Circuits for Appliance Loads

Load devices	Typical load, W	Volts	Wires	Circuit breaker or fuse	Number of outlets	Notes
For laundry areas						
Ironer	1,650	120	Two no. 12	20 A	One	Grounding type receptacle required
Washing machine	1,200	120	Two no. 12	20 A	One	Grounding type receptacle required
Dryer	5,000	120/240	Three no. 10	30 A	One	Appliance may be direct-connected—must be grounded
For other loads						
Hand iron	1,000	120	Two no. 12	20 A	Two or more	
Water heater	3,000	. . .	. . .	. . .	. . .	Consult utility code for load requirements
Workshop	1,500	120	Two no. 12	20 A	Two or more	Separate circuit recommended
Portable heater	1,300	120	Two no. 12	20 A	One	Should not be connected to circuit serving other heavy-duty loads
Television	300	120	Two no. 12	20 A	Two or more	Should not be connected to circuit serving appliances
Range	12,000	120/240	Three no. 6	50–60 A	One	Use of more than one outlet is permitted, but not recommended
Oven (built-in)	4,500	120/240	Three no. 10	30 A	One	Appliance may be direct-connected
Rangetop	6,000	120/240	Three no. 10	30 A	One	Appliance may be direct-connected
Rangetop	3,300	120/240	Three no. 12	20 A	One or more	
Dishwasher	1,200	120	Two no. 12	20 A	One	These appliances may be direct-connected on a single circuit; grounded receptacles required otherwise
Waste disposer	300	120	Two no. 12	20 A	One	

4.69

TABLE 36 Modern Circuits for Appliance Loads (*Continued*)

Load devices	Typical load, W	Volts	Wires	Circuit breaker or fuse	Number of outlets	Notes
			For other loads			
Broiler	1,500	120	Two no. 12 At least two kitchen appliance circuits	20 A	Two or more	Heavy-duty appliances regularly used at one location should have a separate circuit, only one such unit should be attached to a single circuit at a time
Fryer	1,300	120				
Coffeemaker	1,000	120				
Refrigerator	300	120	Two no. 12	20 A	Two	Separate circuit serving only refrigerator and freezer is recommended
Freezer	350	120	Two no. 12	20 A	Two	

Individual circuits for unit air conditioners

Size of air conditioner	Average wattage	Circuits required	Size of circuit	Number of outlets	Remarks
¾ hp (0.6 kW)	1,200	Separate circuit	Two no. 12, 120-V	One	Use of three-wire, 120/240-V circuits to unit conditioners offers circuit flexibility for 120 or 240 V
1½ hp (1.1 kW)	2,400	Separate circuit	Three no. 12, 120/240-V	One	

TABLE 37 Heat-Loss Factors

Building section	Applicable table	Factor		Heat loss	
		$W/(ft^2 \cdot deg\ TD)$	$W/(m^2 \cdot deg\ TD)$	At 70° TD, W/ft^2	At 21.1° TD, W/m^2
Walls: Wood siding and sheathing: gypsum board inside; R-11 insulation; 8-ft (2.4-m) ceiling height	2	0.025	0.48	1.75	18.8
Ceiling: Gypsum board; ventilated attic above; R-19 insulation	1	0.018	0.35	1.26	13.6
Floor: Hardwood floor on subfloor; ventilated crawl space below; R-13 insulation	3	0.019	0.32	1.33	14.3
Windows: Tightly fitted storm sash	5	0.132	2.56	9.24	99.5
Doors: 1½-in (3.8-cm) solid wood with storm door	6	0.094	1.82	6.58	70.8
Infiltration: (¾ air change per h)	9	0.00396°	0.25°	0.28†	9.9†

°$W/(ft^3 \cdot deg\ TD)$.
†W/ft^3.

4.70

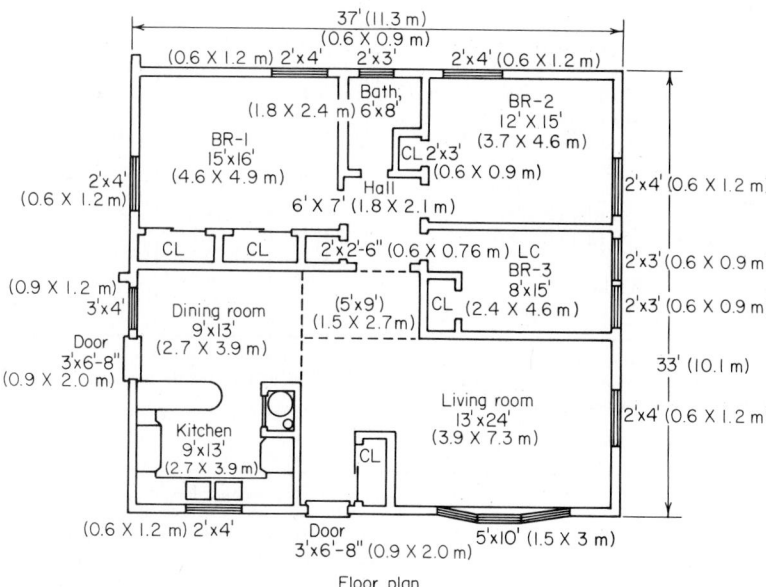

FIG. 32   Floor plan of structure whose heat loss is to be determined.

Calculation Procedure:

1. *Compute the structure heat loss in watts*

For walls, ceilings, doors, windows, and floors over an unheated crawl space or basement, use the relation H = heat loss, $W = WA\Delta T$, where W = heat-loss factor, watts per square foot per degree temperature difference (TD) from Tables 41 through 46; A = surface area—wall, floor, ceiling, etc., ft^2, ΔT = temperature difference between the outside and inside air, °F.

Where the structure has a concrete floor laid at or near the grade level, use the relation $H = WL$, where W = heat-loss factor, W/ft of exposed edge, from Table 45; L = total length of slab edge exposed to the outdoors at the foundation, ft.

TABLE 38   Wall Areas°

Room	Calculations	Glass wall area		Window area		Door area		Net wall area	
		ft^2	m^2	ft^2	m^2	ft^2	m^2	ft^2	m^2
BR-1	$(15 + 16) \times 8$	248	23.0	16	1.5	0	0	232	21.6
BR-2	$(12 + 15) \times 8$	216	20.1	16	1.5	0	0	200	18.6
BR-3	8×8	64	5.9	12	1.1	0	0	52	4.8
LR	$(13 + 24) \times 8$	296	27.5	78	7.2	20	1.9	198	18.4
DR	9×8	72	6.7	12	1.1	20	1.9	40	3.7
K	$(9 + 13) \times 8$	176	16.4	8	0.7	0	0	168	15.6
Bath	6×8	48	4.5	6	0.6	0	0	42	3.9

°Length of outside wall is multiplied by ceiling height to get gross wall area; net wall area is obtained by subtracting the window and door areas from the gross wall area.

TABLE 39 Floor and Ceiling Areas: Volume°

Room	Calculations	Area ft²	Area m²	Volume ft³	Volume m³
BR-1	240 − 5 (linen closet) + 17 (hall and linen closet)	252	23.4	2016	57.1
BR-2	180 + 6 (closet) + 15 (hall and linen closet)	201	18.7	1608	45.5
BR-3	120 + 15 (hall and linen closet)	135	12.5	1080	30.6
LR	312 + 25 (dotted area)	337	31.3	2696	76.3
DR	117 + 20 (dotted area)	137	12.7	1096	31.0
K		117	10.9	936	36.5
Bath	48 − 6 (closet	42	3.9	336	9.6
	Total	1221	113.4	9768	276.6

°Hall and linen closet areas were divided among bedrooms. Dotted area [5 × 9 ft (1.5 × 2.7 m)] was divided between living room and dining room. Room volume is obtained by multiplying the floor area by the ceiling height. Heat loss is the same, regardless of units used for area.

TABLE 40 Room-by-Room Calculations°

Room	ft²	×	W/ft²	=	W	m²	×	W/m²	=	W
Living room:										
Walls	198	×	1.75	=	347	18.4	×	18.83	=	347
Ceiling	337	×	1.26	=	425	31.3	×	13.56	=	425
Floor	337	×	1.33	=	448	31.3	×	14.32	=	448
Windows	78	×	9.24	=	721	7.2	×	99.46	=	721
Door	20	×	6.58	=	132	1.6	×	70.83	=	132
Infiltration	2696†	×	0.28‡	=	755	250.5	×	3.01	=	755
Total heat loss					2828					2828
Suggested heater rating					3.0 kW					3.0 kW
Dining room:										
Walls	40	×	1.75	=	70	3.7	×	18.83	=	70
Ceiling	137	×	1.26	=	173	12.7	×	13.56	=	173
Floor	137	×	1.33	=	182	12.7	×	14.32	=	182
Windows	12	×	9.24	=	111	1.1	×	99.46	=	111
Door	20	×	6.58	=	132	1.9	×	70.83	=	132
Infiltration	1096	×	0.28‡	=	307	101.8	×	3.01	=	307
Total heat loss					975					975
Suggested heater rating					1.0 kW					1.0 kW
Kitchen:										
Walls	168	×	1.75	=	294	15.6	×	18.83	=	294
Ceiling	117	×	1.26	=	147	10.9	×	13.56	=	147
Floor	117	×	1.33	=	156	10.9	×	14.32	=	156
Windows	8	×	9.24	=	74	0.74	×	99.46	=	74
Infiltration	936 †	×	0.28‡	=	262	86.9	×	3.01	=	262
Total heat loss					933					933
Suggested heater rating					1.0 kW					1.0 kW

TABLE 40 Room-by-Room Calculations° (*Continued*)

Room	ft²	×	W/ft²	=	W	m²	×	W/m²	=	W
								Area × heat-loss factor = heat loss		
Bath:										
Walls	42	×	1.75	=	74	3.9	×	18.83	=	74
Ceiling	42	×	1.26	=	53	3.9	×	13.56	=	53
Floor	42	×	1.33	=	56	3.9	×	14.32	=	56
Windows	6	×	9.24	=	55	0.56	×	99.46	=	55
Infiltration	336 †	×	0.28‡	=	94	31.2	×	3.01	=	94
Total heat loss					332					332
Suggested heater rating										

A 500-W heater will satisfy heating requirements; but since quick pickup is often desirable, a 1-kW heater is recommended.

Room	ft²	×	W/ft²	=	W	m²	×	W/m²	=	W
Bedroom 1:										
Walls	232	×	1.75	=	406	21.6	×	18.83	=	406
Ceiling	252	×	1.26	=	318	23.4	×	13.56	=	318
Floor	252	×	1.33	=	335	21.6	×	14.32	=	335
Windows	16	×	9.24	=	148	1.5	×	99.46	=	148
Infiltration	2016†	×	0.28‡	=	564	187.3	×	3.01	=	564
Total heat loss					1771					1771
Suggested heater rating					1.75 kW					1.75 kW
Bedroom 2:										
Walls	200	×	1.75	=	350	18.6	×	18.83	=	350
Ceiling	201	×	1.26	=	253	18.7	×	13.56	=	253
Floor	201	×	1.33	=	267	18.7	×	14.32	=	267
Windows	16	×	9.24	=	148	1.5	×	99.46	=	148
Infiltration	1608†	×	0.28‡	=	450	149.4	×	3.01	=	450
Total heat loss					1468					1468
Suggested heater rating					1.5 kW					1.5 kW
Bedroom 3:										
Walls	52	×	1.75	=	91	4.8	×	18.83	=	91
Ceiling	135	×	1.26	=	170	12.5	×	13.56	=	170
Floor	135	×	1.33	=	180	12.5	×	14.32	=	180
Windows	12	×	9.24	=	111	1.1	×	99.46	=	111
Infiltration	1080†	×	0.28‡	=	302	100.3	×	3.01	=	302
Total heat loss					854					854
Suggested heater rating					1.0 kW					1.0 kW

Room	Walls	Ceiling	Floor	Wind	Doors	Infiltration	Total
			Heat-loss summary, kW				
Living room	347	425	448	721	132	755	2828
Dining room	70	173	182	111	132	307	975
Kitchen	294	147	156	74	. . .	262	933
Bath	74	53	56	55	. . .	94	332
Bedroom 1	406	318	335	148	. . .	564	1771
Bedroom 2	350	253	267	148	. . .	450	1468
Bedroom 3	91	170	180	111	. . .	302	854
Total	1632	1539	1624	1368	264	2734	9161

TABLE 40   Room-by-Room Calculations° (*Continued*)

Basement heat loss with fully heated basement instead of crawl space

Assume groundwater temperature = 50°F (10°C).
Basement walls are 1 ft (0.3 m) above grade, 6 ft (1.8 m) below.
Walls are furred in and insulated to R-5.
There are three 2-ft^2 (0.19-m^2), wood-frame, single-glass windows in above-grade walls.
Infiltration:
    Volume = 1221 × 7 = 8547 ft^3 (242.0 m^3)
    At ¼ air change, heat loss = 8547 × 0.00132 × 70 = 790 W
Windows:
    Area = 2 × 3 = 6 ft^2 (0.56 m^2)
    Temperature difference = 70 − 0 = 70°F (38.9°C)
    Heat loss = 0.331 × 6 × 70 = 139 W

Walls (above grade):
    Length of outside wall = 2(33 + 37) = 140 ft (42.7 m)
    Area = 1 × 40 − windows = 140 − 6 = 134 ft^2 (12.5 m^2)
    Temperature difference = 70 − 0 = 70°F (38.9°C)
    Heat loss = 0.037 × 134 × 70 = 347 W
Walls (below grade):
    Area = 6 × 140 = 840 ft^2 (78.0 m^2)
    Temperature difference = 70 − ½(0 + 50) = 45°F (25.0°C)
    Heat loss = 0.037 × 840 × 45 = 1399 W
Floor:
    Area = 37 × 33 = 1221 ft^2 (113.4 m^2)
    Temperature difference = 70 − 50 = 20°F (11.1°C)
    Heat loss = 0.0293 × 20 × 1221 = 716 W
Total basement heat loss = 3391 W

Floor loss with house on slab at grade level instead of over crawl space

Room	ft	×	W/ft	=	W	m	×	W/m	=	W
						Assumed edge of slab insulated to R-6				
Living room	13 + 24 = 37	×	5.3	=	196	11.3	×	17.4	=	196
Dining room	9 + 0 = 9	×	5.3	=	48	2.7	×	17.4	=	48
Kitchen	9 + 13 = 22	×	5.3	=	117	6.7	×	17.4	=	117
Bath	6 + 0 = 6	×	5.3	=	32	1.8	×	17.4	=	32

Length of exposed slab × heat-loss factor = heat loss

Room	ft	×	W/ft	=	W	m	×	W/m	=	W
Bedroom 1	16 + 15 = 31	×	5.3	=	164	9.4	×	17.4	=	164
Bedroom 2	15 + 12 = 27	×	5.3	=	143	8.2	×	17.4	=	143
Bedroom 3	8 + 0 = 8	×	5.3	=	42	2.4	×	17.4	=	42
Total slab loss					742					742

°Heat loss is the same, regardless of area units used.
†ft^3.
‡W/(ft^3·deg TD) [W/(m^3·deg TD)].

TABLE 41   Heat Loss through Residential Ceilings°

Installed resistance of insulation R	Heat-loss factor, $W/(ft^2 \cdot deg\ TD)$ $[W/(m^2 \cdot deg\ TD)]$	
	Ceilings using plaster or gypsum board products	Ceilings using acoustical tile or insulating board products
15	0.022 (0.426)	0.020 (0.388)
16	0.021 (0.407)	0.019 (0.368)
17	0.020 (0.388)	0.018 (0.349)
18	0.019 (0.368)	0.017 (0.329)
19–20	0.018 (0.349)	0.016 (0.310)
21–22	0.016 (0.310)	0.015 (0.291)
23–24	0.015 (0.291)	0.014 (0.271)
25	0.014 (0.271)	0.013 (0.252)
30	0.011 (0.213)	0.010 (0.194)
35	0.009 (0.174)	0.009 (0.174)
40	0.008 (0.155)	0.008 (0.155)
45	0.007 (0.136)	0.007 (0.136)
50	0.006 (0.116)	0.006 (0.116)

°Assuming space above insulation is ventilated and wood framing covers 15 percent of ceiling area.

TABLE 42   Heat Loss through Frame Walls

Installed resistance of insulation R	Heat-loss factor, $W/(ft^2 \cdot deg\ TD)$ $[W/(m^2 \cdot deg\ TD)]$		
	Masonry walls of low-density concrete 80 lb/ft^3 (1280.8 kg/m^3) or less	Frame walls using insulating board and insulating lath products	Frame walls using wood or metal lath and wood sheathing or masonry walls of stone, concrete block, or high-density concrete
5	0.021 (0.407)	0.030 (0.581)	0.037 (0.717)
6	0.020 (0.388)	0.028 (0.543)	0.034 (0.359)
7	0.019 (0.368)	0.027 (0.523)	0.031 (0.601)
8	0.018 (0.349)	0.025 (0.484)	0.029 (0.562)
9	0.017 (0.329)	0.024 (0.465)	0.027 (0.523)
10	0.016 (0.310)	0.022 (0.426)	0.026 (0.504)
11–12	0.015 (0.291)	0.021 (0.407)	0.025 (0.484)
13–14	0.014 (0.271)	0.019 (0.368)	0.022 (0.426)
15°	0.013 (0.252)	0.017 (0.329)	0.019 (0.368)
20–21°	0.012 (0.233)	0.014 (0.271)	0.016 (0.291)
22–23°	0.011 (0.213)	0.013 (0.252)	0.015 (0.290)

°Using 2 × 6 in (51 × 152 mm) studs.
Note: Values in table assume wood framing covers 20 percent of wall area.

When the structure has a concrete basement floor below the grade level, use the relation $H = 0.0293 A_f\ \Delta T$, where A_f = floor area, ft^2.

To determine the infiltration loss I W, use the relation $I = WV$, where W = heat-loss factor, $W/(ft^3 \cdot °F)$, from Table 44—usual practice is to assume three-fourths air change per hour; V = volume of space to be heated, ft^3.

Assemble and tabulate the heat-loss factors as shown in Table 37. Use the data presented in Tables 40 to 45.

TABLE 43   Heat Loss through Wood Floors

(Assuming wood framing covers 15 percent of floor area. Floor consists of wood subfloor on joists and hardwood floor or tile or linoleum on suitable base.)

Installed resistance of insulation R	Heat-loss factor	
	W/(ft^2·deg TD)	W/(m^2·deg TD)
5	0.035	0.678
6	0.031	0.601
7	0.028	0.543
8	0.026	0.504
9	0.024	0.465
10	0.022	0.426
11	0.021	0.407
12	0.020	0.388
13	0.019	0.368
14	0.018	0.349
15	0.017	0.329
20	0.014	0.271
25	0.012	0.233
30	0.010	0.194
35	0.009	0.174
40	0.008	0.155

TABLE 44   Heat Loss due to Infiltration

No. of air changes per hour	Heat-loss factor	
	W/(ft^3·deg TD)	W/(m^3·deg TD)
⅒	0.00053	0.03769
⅛	0.00066	0.04195
¼	0.00132	0.08391
½	0.00264	0.16782
¾	0.00396	0.25172
1	0.00527	0.33499
1½	0.00791	0.50281

Next, compute the wall areas of the structure, as shown in Table 38. Then compute the floor and ceiling areas as shown in Table 39.

With these data available, the heat loss for each room can be computed by using the equations given above. Table 40 shows the heat-loss computation for each room in this structure when the building is mounted above a ventilated crawl space. The lower portion of Table 40 shows the heat-loss computations for the same structure over a heated basement and mounted on a concrete floor slab.

2. *Summarize the heat losses in the structure*

Table 40 shows the summary heat loss totaled horizontally and vertically for the listed areas and rooms. This summary shows that the heating capacity required for this structure is 9161 kW when mounted over a ventilated crawl space.

When the structure is mounted over a heated basement, the floor heat loss is deleted and the basement heat loss is substituted in its place. Thus, the heating capacity needed for a basement-mounted structure is 10,928 kW. With this structure mounted on a concrete slab, the total heat

TABLE 45 Heat Loss through Concrete Slab on Grade° [W/ft (W/m) of exposed edge]

Outdoor design temperature, °F (°C)	Unheated slab				Heated slab			
	R = 6 to 7	R = 5.0	R = 3.33	R = 2.50	R = 6 to 7	R = 5.0	R = 3.33	R = 2.50
−30 and colder (−34.4 and colder)	7.5 (24.6)	10.0 (32.8)	14.9 (48.9)	19.6 (64.3)	10.1 (33.1)	13.5 (44.3)	20.2 (85.9)	27.0 (88.6)
−25 to −29 (−31.7 – −33.9)	7.0 (23.0)	9.4 (30.8)	14.0 (45.9)	18.8 (61.7)	9.7 (31.8)	12.9 (42.3)	19.3 (63.3)	25.8 (84.6)
−20 to −24 (−28.9 – −31.1)	6.6 (21.7)	8.8 (28.9)	13.1 (43.0)	17.6 (57.7)	8.9 (29.2)	12.0 (39.4)	17.8 (58.4)	24.0 (78.7)
−15 to −19 (−26.1 – −28.3)	6.3 (20.7)	8.2 (26.9)	12.6 (41.3)	16.7 (54.8)	8.1 (26.6)	11.4 (37.4)	17.3 (56.8)	22.8 (74.8)
−10 to −14 (−23.3 – −25.6)	5.9 (19.4)	7.9 (25.9)	11.7 (38.4)	15.8 (51.8)	7.6 (24.9)	10.8 (35.4)	16.1 (52.8)	21.7 (71.2)
−5 to −9 (−20.6 – −22.8)	5.6 (18.4)	7.3 (24.0)	11.1 (36.4)	14.9 (48.9)	7.0 (23.0)	10.2 (33.5)	15.2 (49.9)	20.5 (67.3)
0 to −4 (−17.8 – −20.0)	5.3 (17.4)	7.0 (23.0)	10.5 (34.4)	14.1 (46.3)	7.0 (23.0)	9.4 (30.8)	14.0 (45.9)	18.8 (61.7)
+5 to +1 (−15.0 – −17.2)	4.9 (16.1)	6.5 (21.3)	9.7 (31.8)	12.9 (42.3)	6.6 (21.7)	8.8 (28.9)	13.1 (43.0)	17.6 (57.7)
+10 to +6† (−12.2 – −14.4)	4.6 (15.1)	6.2 (20.3)	9.1 (29.9)	12.3 (40.4)	5.6 (18.4)	7.3 (24.0)	11.1 (36.4)	14.6 (47.9)
+15 to +11† (−9.4 – −11.7)	4.6 (15.1)	6.2 (20.3)	9.1 (29.9)	12.3 (40.4)	5.6 (18.4)	7.3 (24.0)	11.1 (36.4)	14.6 (47.9)
+20 to +16‡ (−6.2 – −8.9)	4.6 (15.1)	6.2 (20.3)	9.1 (29.9)	12.3 (40.4)	5.6 (18.4)	7.3 (24.0)	11.1 (36.4)	14.6 (47.9)

°If no edge insulation is used, calculate heat loss at 0.237 W/ft of exposed edge per degree [1.4 W/(m·deg TD)].
†Factors assume only 12 in (30.5 cm) of edge insulation.
‡Factors assume edge insulation extends down only to bottom of slab.

TABLE 46 Heat Loss through Windows

Number of glass panes	Description	No. of air spaces	Width of each space, in (mm)	Heat loss factor, W/(ft²·deg TD) [W/(m²·deg TD)]
1	Single glass	None	. . . .	0.331 (6.41)
2	Window with usual storm sash		1½ (38.1)	0.220 (4.26)
	Sealed unit	1	¼ (6.4)	0.185 (3.58)
	Sealed unit		½ (12.7)	0.167 (3.24)
	Very tightly fitted storm sash, no vents		1½ (38.1)	0.132 (2.56)
3	Sealed unit	5	¼ (6.4)	0.126 (2.44)
	Sealed unit		½ (12.7)	0.111 (2.15)

TABLE 47 Heat Loss through Solid Wood Doors

Nominal thickness, in (cm)	Actual thickness, in (cm)	Heat-loss factor W/(ft²·deg TD) [W/(m²·deg TD)] Exposed door	With storm door°
1 (2.5)	²⁵⁄₃₂ (2.0)	0.188 (3.64)	0.108 (2.09)
1¼ (3.2)	1¹⁄₁₆ (2.7)	0.161 (3.12)	0.100 (1.94)
1½ (3.8)	1⁵⁄₁₆ (3.3)	0.144 (2.79)	0.094 (1.82)
1¾ (4.4)	1⅜ (3.5)	0.141 (2.73)	0.091 (1.76)
2 (5.1)	1⅝ (4.1)	0.126 (2.44)	0.082 (1.59)
2½ (6.4)	2⅛ (5.4)	0.106 (2.05)	0.076 (1.47)
3 (7.6)	2⅝ (6.7)	0.091 (1.76)	0.067 (1.30)

°50 percent glass and thin wood panels.

TABLE 48 Heat Loss through Exposed-Beam Ceilings with Built-up Roofing

Preformed roof insulation above deck, in (cm)	Flat metal	Type of roof deck, heat loss W/(ft²·deg TD) [W/(m²·deg TD)] — Wood			Preformed slab; wood fiber and cement binder	
		1 in (2.5 cm)	2 in (5.1 cm)	3 in (7.6 cm)	2 in (5.1 cm)	3 in (7.6 cm)
0 (0)	0.264 (5.115)	0.141 (2.732)	0.094 (1.821)	0.067 (1.298)	0.061 (1.182)	0.044 (0.853)
½ (1.3)	0.117 (2.267)	0.085 (1.647)	0.064 (1.24)	0.050 (0.969)		
1 (2.5)	0.076 (1.473)	0.061 (1.182)	0.050 (0.969)	0.041 (0.794)		
1½ (3.8)	0.056 (1.085)	0.047 (0.811)	0.041 (0.794)	0.035 (0.698)	Insulation not used	
2 (5.1)	0.047 (0.911)	0.041 (0.794)	0.035 (0.678)	0.029 (0.562)		
2½ (6.4)	0.038 (0.736)	0.032 (0.62)	0.029 (0.562)	0.026 (0.504)		
3 (7.6)	0.032 (0.62)	0.029 (0.562)	0.026 (0.504)	0.023 (0.446)		

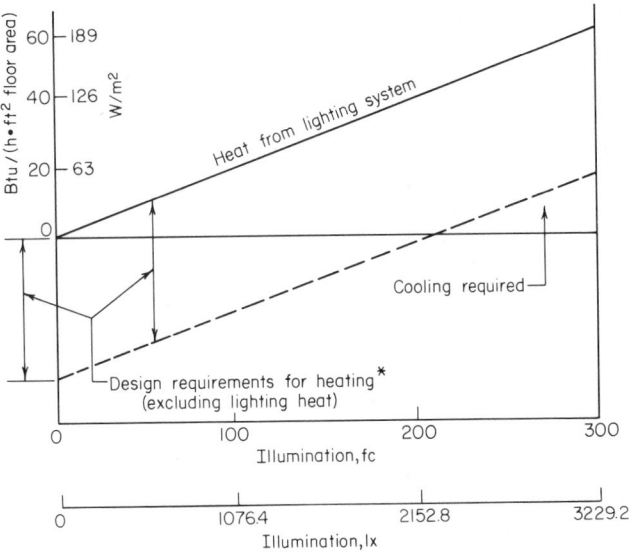

*Prototype example; varies from building to building based on existing climatic and other variable conditions.

FIG. 33   Heat recoverable from lighting systems.

loss is 8279 kW, by the data in Table 40, after the floor loss computed for the crawl-space structure is deducted. The heating capacity must at least equal this heat loss.

Related Calculations: Use this general procedure for any type of structure—industrial, commercial, civic, etc. Where the data given are insufficient, refer to the ASHRAE *Guide and Data Book.*

In newer buildings the heat given off by lighting fixtures is often used to reduce the load on the structure's heating system. Heated air is drawn from lighting troffers and directed to the return-air system. Figure 33 shows the heat given off by a typical modern lighting system. When the lighting level exceeds 100 fc (1076.4 lx), the heat from the lights is often sufficient to heat the entire building. Currently, integrated systems of lighting, heating, and air conditioning are being planned so that the heating and cooling requirements of a structure are reduced.

Definitions: In electric heating, the following definitions are in common use. *R value:* The amount of thermal resistance contributed by insulation when the insulation is installed in a ceiling, wall, or floor. *Outdoor design temperature:* The lowest outdoor temperature at which the heating system is expected to maintain the indoor design temperature. This outdoor design temperature is usually considerably higher than the lowest temperature on record for the area; the outdoor design temperature can be obtained from the local electric utility. *Indoor design temperature:* The temperature which is to be maintained in the heated space, usually 70°F (21.1°C). *Design temperature difference* (TD): The difference between the outdoor and indoor design temperatures.

The data and tables presented in this calculation procedure are based on information presented in *Electrical Construction and Maintenance* magazine.

ELECTRIC-MOTOR CHOICE BASED ON TOTAL COST

Compare two 4000-hp (2984-kW) electric motors when motor A efficiency = 0.955, power factor = 0.90; motor B efficiency = 0.970, power factor = 0.92; energy cost (EC) = \$0.03/kWh; power-cost escalation factor e = 6 percent per year = 1.06; life of each motor for this cost evaluation = 20 years; annual discount factor is based on a 10 percent interest charge for money invested in the motor; usage factor U = 60 percent or 0.6(8760 h/yr) = 5256-h usage per year; demand charge DC = $5 per kVA per month = $60 per kVA per year.

Calculation Procedure:

1. Compute the present value of the energy charge

Set up a listing like that in Table 49 which shows the years in the life of each motor, the cost escalation factor, the discount factor, and the present-value *(pv)* multiplier. Compute each of the factors separately in the following way:

Cost escalation factor e = 1.06 for the first year, or $1 + e$, where e = escalation factor percentage, expressed as a decimal. For the second year, $e = (1.06)(1.06) = 1.124$; for the third year, $e = (1.06)(1.06)(1.06) = 1.191$; for the fourth year, $e = 1.262$, etc. Compute this factor for each of the remaining years, using the same sequence, and list the results in Table 49.

Discount factor d $= 1/(1 + i)^n$, where n = the year number. Thus, for the second year, $d = 1/(1 + 0.10)^2 = 0.826$. Continue calculating the discount factor for each year from 1 through 20, and list the results in Table 49.

The *present-value multiplier* is found from PV $= [(1 + e)/(1 + i)]^n$. Compute the PV for each year, and enter the result in Table 49.

TABLE 49   Computation of PV°

Year	Cost escalation factor	Discount factor	PV multiplier
1	1.06	0.909	0.963
2	1.124	0.826	0.928
3	1.191	0.751	0.894
4	1.262	0.683	0.862
5	1.338	0.621	0.831
6	1.418	0.564	0.780
7	1.504	0.513	0.771
8	1.594	0.466	0.743
9	1.689	0.424	0.716
10	1.791	0.385	0.689
11	1.898	0.350	0.664
12	2.012	0.318	0.640
13	2.133	0.289	0.616
14	2.261	0.263	0.595
15	2.396	0.239	0.573
16	2.540	0.217	0.552
17	2.693	0.197	0.530
18	2.854	0.179	0.511
19	3.025	0.163	0.493
20	3.207	0.148	0.475

Cumulative total of 20 annual PV multipliers =   13.826

° *Power.*

2. Compute the cumulative total of the annual PV multipliers and the PVEC

Sum the PV multipliers in Table 49, to give a total of 13.826. Then PVEC $= \sum_{1}^{n}$ PV EC = 13.826(0.03) = $0.4148 per kWh.

3. Calculate dollars per kilowatt for the units

For a study of this type, $/kW = PVEC$(U)$ = $0.4148(5256) = $2180.

4. *Determine the kilowatt input difference*

For each motor, kilowatt input = 0.746(motor hp)/motor efficiency. Hence, for motor A, input = 0.746(4000)/0.955 = 3124.6 kW; for motor B, input = 0.746(4000)/0.97 = 3076.3 kW. The difference is thus 3124.6 − 3076.3 = 48.3 kW.

5. *Evaluate the efficiency benefits*

The efficiency benefit = (kilowatt difference)($/kW) = 48.3(2180) = $105,294 in favor of motor B.

6. *Evaluate the power-factor benefits*

For each motor, $kVA = kW/pf$, where pf = power factor of the motor, expressed as a decimal. Thus, for motor A, kVA = 3124.6/0.90 = 3471.8; for motor B, kVA = 3076.3/0.92 = 3343.8. The difference is 127.9 kVA.

Then, by using the same procedure as in the efficiency evaluation, demand-charge savings = (127.9 kVA)($60/year)(13.826) = $106,100 in savings in favor of motor B.

Thus, based on the efficiency and demand-charge savings, motor B clearly has the lower total cost and is the more economical choice for this application.

Related Calculations: For many years ac motors were usually chosen only on the basis of first cost. But with the rise in interest rates and the steady escalation of power (energy) costs, careful selection among the available motor choices is now a necessity.

To find out whether the motor pays its way, we must evaluate the efficiency and power factor for each of the various motor choices available. The above calculation procedure shows that in times of escalating power (energy) costs and relatively high interest rates, an extra 1.5 percent in motor efficiency would allow the spending of $105,000 more for a motor of larger horsepower rating.

Power factor also is very important from an economic standpoint. In the above situation, a 2 percent difference in power factor results in an economic difference greater than that from the efficiency difference. For this reason, a power-factor-worth study should always be made, particularly for larger motors or in areas with significant demand penalties.

Although only two evaluations are made above, the concept can be expanded to include the results of other differences or of alternative investments to offset the differences. Thus, the effect of changes in taxes or insurance can be evaluated. Or the power factor of a motor can be improved through an investment in capacitors. Alternatively, investment in a synchronous motor can provide unity or even leading power factor, reducing overall plant demand charge. This could make a lower investment possible or postpone the need for additional switchgear, breakers, conductors, or primary substation equipment. This approach, called *worth analysis*, can aid in decision making in many other areas such as transformer, converter, and generator selection. Data presented in this procedure are the work of K. Lyle Hanson, Electric Machinery Manufacturing Division, Turbodyne Corporation, as reported in *Power* magazine.

DIRECT-CURRENT PERMANENT-MAGNET MOTOR ANALYSIS

A high-efficiency (85 percent minimum) permanent-magnet (PM) motor operating from a 290-V dc line is required to deliver 1 hp (0.75 kW) at 11,500 r/min. The ambient temperature varies from −65 to + 85°C (−85 to 185°F). Determine the motor's performance characteristics.

Calculation Procedures:

1. *Determine the rated operating conditions*

Find the rated operating torque from $T_R = P_{OR}(1352)/(\text{r/min})$, where P_{OR} = power delivered, W; rpm = operating revolutions per minute at rated output, r/min. Solving gives T_R = 746(1352)/11,500 = 87.7 oz·in (0.62 N·m).

2. *Find the rated power input to the motor*

The rated power input $P_{iR} = P_{OR}/\eta_{max}$, where η_{max} = maximum motor efficiency, expressed as a decimal. Or, P_{iR} = 746/0.85 = 877.6 W.

3. Compute the rated input current

The rated input current $I_R = P_{iR}/V$, where V = supply-current voltage. Or, $I_R = 877.6/290 = 3.026$ A.

4. Find the motor parameters, using equations from Table 50

The motor parameters important in dc PM motor analysis are (a) the dimensionless ratio of motor drag torque due to friction, iron, and commutation, to developed stall torque, α; (b) dimensionless speed for maximum efficiency ν_{max}; (c) theoretical no-load speed for $\alpha = 0$ r/min, N_0; (d) no-load speed N_{NL}, r/min; (e) stall torque, W P_{ST}, (f) stall current I_{ST}, A; (g) armature resistance A_R, Ω. Solving for these parameters by using the equations from Table 50 yields the following:

$\quad$ (a) $\alpha = [2 - (\eta_{max})^{0.5}]^2 = [1 - (0.85)^{0.5}]^2 = 0.006091$. (b) $\nu_{max} = 1/(2 - \sqrt{\eta_{max}}) = 0.928$.
(c) $N_0 = N_R/(\eta_{max})^{0.5} = 11{,}500/(0.85)^{0.5} = 12{,}474$ r/min. (d) $N_{NL} = N_0 (\eta_{max})^{0.5} [2 - \eta_{max})^{0.5}] =$

TABLE 50 Summary of Formulas for Universal Performance

(1) $\alpha = \dfrac{T_L}{T_{STD}} = \dfrac{I_{NL}}{I_{ST}}$	(11) $\eta = \dfrac{P_0}{P_i} = \dfrac{(1-\alpha)^2 \nu(1-\nu)}{[1-\nu(1-\alpha)]}$
(2) $\eta_{max} = (1 - \sqrt{\alpha})^2; \alpha = (1 - \sqrt{\alpha_{max}})^2$	(12) $N_{NL} = N_0 \sqrt{\eta_{max}}\,(2 - \sqrt{\eta_{max}})$
(3) $N_0 = \dfrac{N_{NL}}{(1-\alpha)}$	(13) $N_{NL} = N_E(2 - \sqrt{\eta_{max}})$
(4) $\eta_{max} = \left(2 - \dfrac{1}{\nu_{max}}\right)^2$	(14) $N_0 = \dfrac{N_E}{\sqrt{\eta_{max}}}$
(5) $\nu_{max} = \dfrac{1}{1 + \sqrt{\alpha}}$	(15) $T_{STD} = \dfrac{1352}{N_0} P_{ST}$
(6) $\nu_{max} = \dfrac{1}{2 - \sqrt{\eta_{max}}}$	(16) $P_{ST} = \dfrac{1}{1352} N_E T_E \dfrac{1}{\sqrt{\eta_{max}}} \left(\dfrac{2 - \sqrt{\eta_{max}}}{1 - \sqrt{\eta_{max}}}\right)$
(7) $T_{STD} = \dfrac{T_{ST}}{(1-\alpha)}$	(17) $P_{ST} = P_{OE} \dfrac{1}{\sqrt{\eta_{max}}} \left(\dfrac{2 - \sqrt{\eta_{max}}}{1 - \sqrt{\eta_{max}}}\right)$
(8) $P_O = \dfrac{T_{STD}N_0}{1352} (1-\alpha)^2\nu(1-\nu)$	(18) $\dfrac{T_{ST}}{T_E} = \dfrac{2 - \sqrt{\eta_{max}}}{1 - \sqrt{\eta_{max}}}$
$\quad\quad = P_{ST}(1-\alpha)^2\nu(1-\nu)$	(19) $D = \dfrac{1352}{N_{NL^2}} (1-\alpha)^2 P_{ST}$
(9) $P_i = VI_{ST}[1 - \nu(1-\alpha)]$	
(10) $P_i = P_{ST}[1 - \nu(1-\alpha)]$	(20) $\nu = \dfrac{(1+\eta-\alpha) \pm \sqrt{(1-\eta)^2(2+2\eta-\alpha)\,\alpha}}{2(1-\alpha)}$

Nomenclature

α = dimensionless ratio of motor drag torque due to friction, iron, and commutation, to developed stall torque
η = per unit efficiency
N_{NL} = no-load speed, r/min
N_0 = theoretical no-load speed for $\alpha = 0$ r/min
ν = dimensionless ratio of speed N to no-load speed N_{NL}
T_{ST} = stall torque, available at the output 1 oz·in (0.007 N·m)
T_{STD} = developed stall torque for $\alpha = 0$ oz·in (N·m)
P_O = output power, W

P_i = input power, W
η_{max} = maximum efficiency
ν_{max} = dimensionless speed for maximum efficiency
T_E = torque at maximum efficiency point, oz·in (N·m)
N_E = speed at maximum efficiency, r/min
P_{OE} = power output at maximum efficiency
D = slope of torque-speed curve, oz·in·min/r
I_{NL} = no-load current, A
I_{ST} = stall current, A
V = rated voltage, V

$12,474 \ (0.85)^{0.5} \ [2 - (0.85)^{0.5}] = 12,398$ r/min. (e) $P_{ST} = P_{OR} \ (1/\eta_{max})^{0.5} \ (2 - \eta_{max}^{0.5})/(1 - \eta_{max}^{0.5})$ $= 746 \ (1/[0.85]^{0.5}) \ [2 - (0.85)^{0.5}]/[1 - (0.85)^{0.5}] = 11,177$ W. (f) $I_{ST} = P_{ST}/(\text{V dc}) = 11,177/$ $290 = 38.54$ A. (g) $A_R = (\text{V dc})/I_{ST} = 290/38.54 = 7.525 \ \Omega$.

These parameters show the high stall power P_{ST} inherent in high efficiency motors, as shown in Fig. 34. The maximum stall power is even higher if temperature effects are included. To deliver the required output at the specified minimum efficiency, the winding resistance must not exceed the value determined above, 7.525 Ω.

However, with a 7.525-Ω resistance at the maximum operating temperature of the motor, a resistance of about half this must prevail when the motor is started at the coldest temperature. Hence, the peak instantaneous input current may reach twice the value of $I_{ST} = 38.54$ A, or 75 to 80 A. This is an extremely large current for a 290-V dc line [approximately 23-kW peak for a 1-hp (0.75-kW) continuous rating].

5. *Determine the no-load current and power*

For $\alpha = 0.00691$ and $I_{ST} = 38.54$, the no-load current $I_{NL} = \alpha \ I_{ST} = 0.006091(38.54) = 0.235$ A. The no-load power $P_{NL} = (\text{V dc})(I)_{NL} = 290(0.235) = 68.15$ W.

6. *Find the developed stall, actual stall, and drag torques*

The developed stall torque is $T_{STDEV} = (1352/N_0)P_{ST} = (1352/12,474)(11,177) = 1211.4$ oz·in (8.55 N·m). The actual stall torque $T_{ASTDEV} = T_{STDEV} \ (1 - \alpha) = 1211.4(1 - 0.006091) = 1204$ oz-in (8.5 N·m). Last, the drag torque $T_D = \alpha T_{STDEV} = 0.006091(1211.4) = 7.379$ oz·in (0.052 N·m).

The above calculated parameters, determined by the relatively few specified characteristics, provide a complete picture of motor performance.

In an actual application, in-rush current at stall can be reduced in several ways. Current limiting may be added to the input line, corresponding to an acceptable level of stall torque. For example, if the stall torque must exceed 400 percent of the rated torque, the input current may be limited at $4I_R = 12.11$ A. Alternatively, a lower efficiency might be acceptable in a design

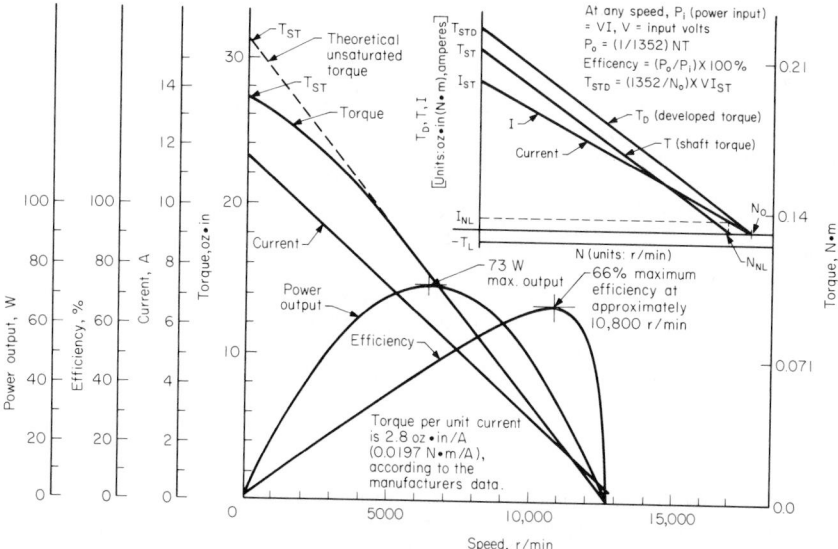

FIG. 34 Performance characteristics of a permanent-magnet dc motor are shown on the large graph; the inset gives basic relations for a linear permanent-magnet dc motor. *Note:* Power output and efficiency are calculated from the theoretical unsaturated torque-speed characteristic, to permit a check of theoretical formulas. Armature resistance = 2.15 Ω, line voltage = 27 V dc. Torque constant = 2.8 oz·in/A (0.0197 N·m/A) (manufacturer's data). *(Product Engineering.)*

tradeoff. By calculation, the ratio of stall input to rated power output decreases by 21.8 percent, as $\eta_{\max}$ goes from 85 to 80 percent.

A third approach would be to accept an efficiency reduction by operating lower on the efficiency curve, at a slower than maximum efficiency speed. Substitute in the efficiency formula of Table 50: $\alpha = 0.006091$; $\eta = 0.80$; $\nu = 0.8348$; $P_E/P_{ST} = 0.1362$; $P_{ST}/P_E = 7.34$.

This large reduction in stall power is obtained by trading off five percentage points of rated efficiency.

The simple analytic procedures and universal curves given here permit the equipment designer to predict complete permanent-magnet dc motor performance from some of its parameters and to optimize operation in specific applications.

Related Calculations: Because of the wide variety of motor applications, it is impossible for the motor manufacturer to give sufficient information on products to cover all operating conditions. So it is important that the designer of motor-driven equipment be able to interpret and extend limited manufacturer's data to cover the specific requirements of the application. Fortunately, for the widely used permanent-magnet dc motor, it is feasible to extend the limited motor data to a complete performance description. The method of doing this is theoretically sound and easy to apply.

The universal motor performance curves in this procedure are readily derived from simple motor performance equations. These equations relate input voltage, current, speed, power, and developed torque. Figure 34 shows standard performance curves of a permanent-magnet dc motor. Motor manufacturers do not usually provide as complete a curve set. Note that drag torque may be estimated from no-load input power VI_{NL} to the power associated with developed torque at no load $T_L N_{NL}/1352$. (Shaft power is zero at no load.) Solving for developed torque (equal to internal drag torque) yields $T_L = 1352 VI_{NL}/N_{NL}$.

For a constant drag torque over the speed range, the curve of developed torque is translated from the shaft torque curve. Developed stall torque therefore exceeds available-at-the-shaft stall torque, by the amount of drag torque. The ratio of drag torque to developed stall torque is an important figure.

To minimize error while retaining analytical simplicity, the drag torque is assumed to be constant at the value corresponding to the speed where motor efficiency is maximum. This assumption provides highest accuracy in the important region of high motor efficiency. No other assumptions are required for the derivations of the universal curve.

Figure 35 shows the starting point for the derivations, introducing the nomenclature and procedures. It also lists the relationships which are to be presented as universal curves. These relationships (together with many others collated for convenience) are summarized in Table 50.

These brief descriptions only hint at the value of the universal curves to the equipment designer. Their real utility can best be demonstrated by worked-out examples of their application.

To show use of the universal curve, the above formulas may be applied to the curves of Fig. 34, which are devised from actual motor data.

From the figure, $N_{NL} = 12,800$ r/min, $\eta_{\max} = 0.66$, armature resistance is 2.15 Ω, and line voltage is 27 V dc. From these data all other parameters of the motor may be derived by standard formulas.

To calculate stall current, assume as a standard approximation a drop of 1 V across each brush, so that $27 - 2 = 25$ V appears across the motor winding. And $I_{ST} = 25/2.15 = 11.6$ A, which agrees with the manufacturer's curve. Stall power is $27(11.6) = 313.2$ W. Calculate $\alpha = (1 - \sqrt{\eta_{\max}})^2 = 0.0352$. Calculate the slope of the torque-versus-speed characteristic, using P_{ST} and α as determined above:

$$D = \frac{1352}{(N_{NL})^2}(1 - \alpha)^2 P_{ST} = 0.00241 \text{ oz}\cdot\text{in}\cdot\text{min/r} \ (1.7 \times 10^{-5} \text{ N}\cdot\text{m}\cdot\text{min/r})$$

This slope applies in the lower-current, higher-speed, unsaturated region. Using $N_{NL} = 12,800$ gives the unsaturated stall torque as $T_{ST} = 0.00241(12,800) = 30.8$ oz$\cdot$in (0.22 N$\cdot$m). The ratio of T_{ST} to I_{ST}, the unsaturated torque constant, is 2.65 oz$\cdot$in/A (0.019 N$\cdot$m/A) compared with the manufacturer's figure of 2.8 (0.0197). The agreement accuracy is to within 5 percent. Calculate $\nu_{\max} = 1/(2 - \sqrt{\eta_{\max}}) = 0.842$. The corresponding speed is $0.842(12,800) = 10,780$ r/min. This checks very accurately with the efficiency curve of Fig. 34.

This procedure is the work of Sid Davis, Consulting Electrical Engineer, East Norwich, New York, as reported in *Product Engineering*.

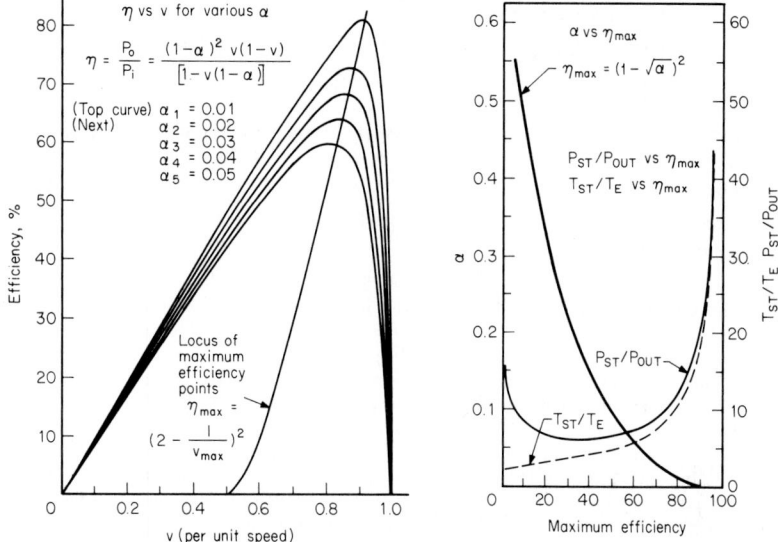

FIG. 35   Dimensionless efficiency curves for permanent-magnet dc motors are shown at left . At right are shown the relationships between per unit drag torque, power, and maximum efficiency (η_{max}). The high stall power of high-efficiency motors is indicated. *(Product Engineering.)*

ELECTRICAL-MEASUREMENT ANALYSIS OF PERMANENT-MAGNET MOTORS

At its rated 20 V, a permanent-magnet motor has a no-load current of $I_O = 1$ A and a no-load speed of $S_0 = 9600$ r/min (9.6 kr/min). When the motor is loaded to a point where its speed $S = 7600$ r/min (7.6 kr/min), the input current $I = 6$ A. Determine the factor of merit M for this motor, maximum motor efficiency, input current at maximum efficiency, and power output and speed at maximum efficiency. Also find the motor drag torque, total stall torque, and stall input current.

Calculation Procedure:

1.   Find the factor of merit M

Use the relation $M_0 = [(S_0/I_0)(\Delta I/\Delta S) + 1]^{0.5}$ where $\Delta I = I - I_0 = 6 - 1 = 5$; $\Delta S = S_0 - S = 9.6 - 7.6 = 2$. Substituting gives $M = [(9.6/1)(5/2) + 1]^{0.5} = 5$.

2.   Compute the maximum motor efficiency

The maximum motor efficiency is $EF_{max} = (1 - 1/M)^2 = (1 - 1/5)^2 = 0.64$, or 64 percent.

3.   Determine the input current at maximum efficiency

The input current at maximum efficiency is given by $I_m = MI_0 = 5(1) = 5$ A.

4.   Compute the power output at maximum efficiency

The power output at maximum efficiency is given by $P_{out} = EI_mEF_{max}$, where $E =$ motor rated voltage. Substituting gives $P_{out} = 20(5)(0.64) = 64$ W.

5.   Find the motor speed at maximum efficiency

The speed at maximum efficiency is given by $S_m = S_0[M/(M + 1)]$ kr/min. Or $S_m = 9.6(5/6) = 8$ kr/min (8000 r/min).

6.   Compute the output torque at maximum efficiency

Use the relation $T_m = 1.35 P_{out}/S_m$, or $T_m = 1.35(64)/8 = 10.8$ oz·in (0.076 N·m).

7. Determine the motor drag torque

The motor drag torque at the speed range from S_0 to S_m is $T_0 = T_m/(M - 1) = 10.8/(5 - 1) = 2.7$ oz·in (0.019 N·m).

8. Find the total stall torque

The total stall torque T_t is the sum of the output stall torque and the drag torque at near-zero speed. Or, $T_t = M(T_m + T_0) = 5(10.8 + 2.7) = 67.5$ oz·in (0.477 N·m).

9. Compute the stall torque

The stall torque $T_s = T_t - T_0 - 67.5 = 2.7 = 64.8$ oz·in (0.458 N·m).

10. Find the stall input current

The stall input current is $I_s = M^2 I_0 = (5)^2(1) = 25$ A

Related Calculations: Because of improved magnetic materials, permanent-magnet (PM) motors no longer are restricted to use in toys and novelty products. New motors with superior performance now regularly show up in more sophisticated applications such as tape and disk drives, machine tools, and instruments. Nevertheless, PM motors have not received much analytical attention in the technical literature. Performance is still determined largely by plotting torque against speed, input current, output power, and motor efficiency. Although this is the traditional procedure for determining maximum efficiency and the most efficient operating point, it is also time-consuming and requires elaborate test equipment, such as dynamometers and torquemeters.

A simpler and quicker analysis can be made by developing an equivalent motor circuit based on plotted performance information (Fig. 36). Equations developed from this circuit can be used with any PM motor to find a factor of merit, which in turn predicts performance capabilities, maximum efficiency, and best operating point of the motor.

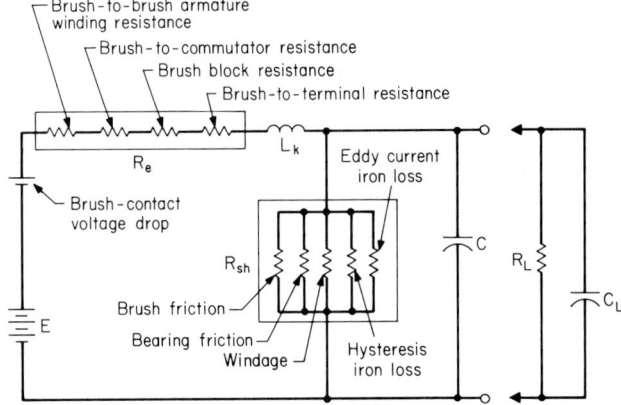

FIG. 36   Equivalent motor circuit for a permanent-magnet motor. *(Machine Design.)*

All that is required to find the factor of merit is a regulated dc power source, a voltmeter, an ammeter, and a tachometer or strobotac. If a good dc source is not available, at least the input voltage should be kept constant during current readings.

After the motor has warmed up, read applied voltage E, no-load current I_0, and no-load speed S_0 (in kr/min). Next decrease the motor speed by loading the shaft. If the actual load is not available, a hand-held rag or wood blocks can be used as a friction load. Speed should be decreased to about 80 percent of no-load speed, although this value is not critical. Now read the loaded speed S and the corresponding input current I. The factor of merit M is found as given in step 1.

This motor analysis method is probably more accurate than most torque or dynamometer readings, since most electrical measurements are more accurate than their mechanical counterparts. Step 2 understates the efficiency by about 1 percent by assuming that the shunt current stays constant over the speed change from S_0 to S. However, by not including the nonohmic brush drop, the equation overstates efficiency by about the same amount. Accuracy, then, is more than adequate for most applications.

This procedure is the work of Joseph A. Mas, Consulting Engineer, Woodbury, New York, as reported in *Machine Design* magazine.

AIR COOLING OF ELECTRIC-MOTOR DRIVES

How much air is required to keep the temperature of an unattended motor room from exceeding 104°F (40°C) if the motor is connected to a 200-hp (149-kW) load, the losses in the motor and drive reduce the overall efficiency to 87 percent, and the entering air temperature is 80°F (27°C)?

Calculation Procedure:

1. Determine how much heat must be removed

The heat generated by the motor and drive as a result of inefficiency must be removed from the room. This heat can be found from $H = 2545 \, \text{bhp}_c (1.0 - E)/E$, where H = heat to be removed, Btu/h (W); bhp_c = brake horsepower of connected load, W; E = combined motor and drive efficiency, expressed as a decimal. Substituting, we find $H = 2545(200)(1 - 0.87)/0.87 = 76{,}057$ Btu/h (22.3 kW).

2. Compute the airflow required to remove the heat

Use the relation $Q = 2358 \, \text{bhp}_c \, (1.0 - E)/(\Delta t \, E)$, where Q = airflow required to dissipate the heat, ft³/min; Δt = temperature rise of the air, °F (°C). Substituting yields $Q = 2358(200)(1 - 0.87)/(104 - 80)0.87 = 2936 \, \text{ft}^3/\text{min}$ (83.1 m³/min).

Related Calculations: Whenever electric motors are operated unattended inside buildings or other enclosures, air circulation must be provided to ensure that the room or enclosure temperature not exceed that required by the motor. Many electric motors are rated for a temperature rise of 40°C (72°F) in an ambient temperature not exceeding 40°C (104°F). Any heat generated by inefficiency of the motor and any drives to which it is connected must be removed to prevent the motor temperature from rising above the rated value.

The airflow required to remove the heat generated by motor and drive inefficiency can be quickly determined as shown above. The required airflow can be provided by either natural or forced ventilation. To determine the maximum airflow required, be certain to use the highest ambient temperature expected in the vicinity of the motor room. The method given here is valid for motors and drives in industrial plants, commercial buildings, ships, aircraft, and similar installations. This procedure is the work of Bill Sisson, Nipak Inc., as reported in *Chemical Engineering* magazine.

FLYWHEEL SELECTION FOR ELECTRIC-MOTOR DRIVES

A traveling crane having a total maximum loaded weight of $W = 90{,}000$ lb (40,500 kg) traverses an overhead track at a speed of $L = 160$ ft/min (48.8 m/min). The crane has a load torque $T_L = 14$ lb·ft (19 N·m) and is driven by a 10-hp (7.5-kW) motor with inertia of $I_r = 3.5$ lb·ft² (0.147 kg·m²), a starting torque of $T_s = 74$ lb·ft (100.3 N·m), a speed of $n = 1800$ r/min, and a braking torque of $T_b = 50$ lb·ft (67.7 N·m). The gear ratio is sufficiently high for the load rotational inertia to be ignored. Required starting and stopping time is 4 s for the load to accelerate and decelerate without swinging. Determine whether a flywheel is required and, if so, what size it should be.

Calculation Procedure:

1. Compute the linear inertia

Use the relation $I_a = WL^2/40n^2$, where I_a = linear inertia, lb·ft² (kg·m²), other symbols as given earlier. Substituting gives $I_a = 90{,}000(160)^2/40(1800)^2 = 17.8$ lb·ft² (0.75 kg·m²).

2. Find the total inertia of the system

Since the rotational inertia of the load is negligible, the motor inertia is the only component of rotational inertia that will enter into this computation. Thus, $I_r = 3.5$ lb·ft^2 (0.147 kg·m^2), and the total inertia in the system is 3.5 lb·ft^2 + 17.8 lb·ft^2 = 21.3 lb·ft^2 (0.898 kg·m^2). (Note that if the rotational inertia of the load were significant, it would be computed and included in the above summation.)

3. Determine the motor system starting and stopping times

The starting time, in seconds, without a flywheel is found from $t_s = nI_i/307(T_s - T_L)$, where I_i = total inertia of the system, lb·ft^2 (kg·m^2), other symbols as given earlier. Substituting, we have $t_s = 1800(21.3)/307(74 - 14) = 2.08$ s.

The stopping time, in seconds, of the motor system without a flywheel is given by $t_b = nI_i/307(0.9T_b + T_L)$, where all the symbols are as given earlier. Thus $t_b = 1800(21.3)/307[(0.9)50 + 14] = 2.1$ s.

The required starting and stopping time for the load to accelerate and decelerate without swinging is 4 s. Since the computed starting and stopping times are 2.08 and 2.1 s, respectively, additional inertia must be added to the system to increase the starting and stopping time.

4. Find the additional inertia required

Use the starting-time equation to find the additional inertia I_f, starting with $t_s = 4$ s, the required starting time, and $I_i = 21.3$ lb·ft^2 + I_f; $I_i = 0.898$ kg·m^2 + I_f in SI. Or, 4 = 1800(21.3 + I_f)/307(74 - 14); $I_f = 20$ lb·ft^2 (0.843 kg·m^2).

The required flywheel inertia of 20 lb·ft^2 (0.843 kg·m^2) can be obtained from any rotational component in the system. To be effective, however, the additional inertia must be connected directly to the motor shaft and have no gearing between the component and the motor.

5. Choose the flywheel to use

A total of 20 lb·ft^2 (0.843 kg·m^2) of inertia is needed for this system. Choose a heavy, iron brake disk which would add 10 lb·ft^2 (0.421 kg·m^2) of inertia. The remaining 10 lb·ft^2 (0.421 kg·m^2) required inertia can be obtained from a flywheel. Assuming that the flywheel is made from steel bar stock of 1-ft (0.3-m) diameter, we know that the required flywheel width w, ft, is given by $w = I_r R^2/0.1pD^4$, where R = gear ratio; p = flywheel material specific weight, lb/ft^3 (kg/m^3); D = outside diameter, ft (m), of the flywheel. Substituting, we have $w = 10(1)^2/0.1(490)(1)^4$; $w = 0.2$ ft (0.06 m).

Caution should be exercised in arbitrarily increasing rotational inertia in this manner to increase starting time. Most motors have maximum startup times that, if exceeded, can severely overheat the motor. These maximum startup times are typically under 5 s, so the motor data sheet should be consulted before a flywheel is used. Extending stopping times is usually safe, however, if the motor brake is designed to dissipate the heat generated in long stopping times.

Related Calculations: In many motor-driven systems, starting and stopping must be done slowly; fast acceleration and deceleration can damage drives or cause loads to swing uncontrollably. Often in such applications, the best choice is an electric or electronic speed control to limit motor acceleration and deceleration, particularly where knob-adjustable motor control is required. But this solution is not cost-effective in many other applications with fixed start-and-stop requirements; electric and electronic controls are expensive to install and maintain, are difficult to protect in harsh factory environments, and often require frequent adjustment to compensate for changing loads.

A much simpler and less expensive way to control acceleration and deceleration in fixed-time applications is with a flywheel connected to the motor to increase the rotational inertia of the system. Although start and stop times are not readily changed once the flywheel is sized and installed, the flywheel is rugged and easy to maintain. In addition, it is inexpensive and helps maintain consistent start and stop times regardless of load changes.

The flywheel is usually a heavy, metal disk placed on the motor shaft expressly for increasing rotational inertia. Sometimes rotational inertia is increased more conveniently through a combination of system components rather than a single flywheel. Gear couplings and pulleys, for example, may be specified in larger sizes. And iron disk brakes may be used instead of the usual aluminum. To be effective, these components must be placed directly on the motor shaft, since gearing reduces rotational inertia by the gear ratio squared. However, heavier rotating components alone are not always enough to increase rotational inertia sufficiently. Then, additional inertia must be provided by a flywheel.

Typically, flywheels are made of steel or cast iron because these materials are heavy enough to increase the rotational inertia with a relatively small flywheel diameter. They are also strong enough to withstand high rotational speeds.

Flywheel disks are made from round bar stock cut to the desired width. Balancing counterweights are rarely used because of the danger of their flying off at high speeds. Rather, small holes are drilled near the perimeter of the disk at appropriate positions. Balancing may be done to a fairly high accuracy on a lathe by an experienced machinist, but it is more accurately performed on a spin-balancing machine.

Flywheels are sized by calculating total system inertia and then determining the additional flywheel inertia needed to produce the required start and stop times, as shown in the procedure.

Figure 37 shows the effect of a 4-in (10.2-cm) thick flywheel on a typical 10-hp (7.5-kW) motor driving a 17 lb·ft^2 (0.716 kg·m^2) load. With no flywheel, the motor accelerates from rest to full speed in less than 1 s. By increasing the flywheel diameter to 1 ft (0.3 m), the startup time can be stretched to more than 5 s.

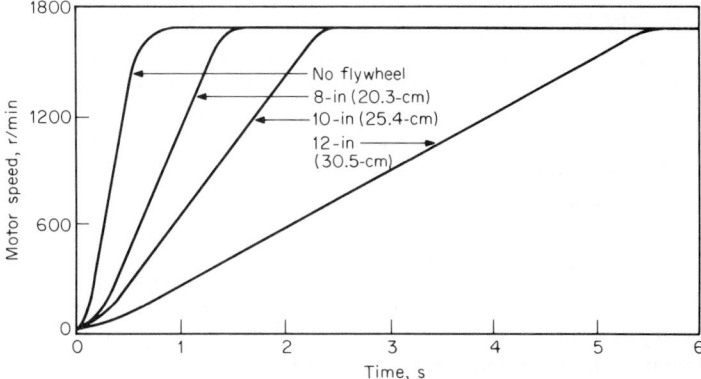

FIG. 37 Effect of flywheel width on motor starting time is shown by these acceleration curves. (*Machine Design.*)

For some applications, high-inertia part-time flywheel control of acceleration and deceleration from high speeds is required along with the positioning accuracy and quick response of a low-inertia motor. This dual part-time flywheel capability is often needed in positioning applications such as automatic warehousing, where a material-transfer conveyor must move quickly between work stations and position very accurately when it gets to the stations. In such dual-mode applications, "part-time" flywheels are used to accelerate and decelerate for high-speed operation, with a simple clutching arrangement to disconnect the flywheel for low-speed final positioning.

In the clutching arrangement (Fig. 38), two motors are used: a main motor with a heavy, iron brake disk flywheel and a smaller secondary motor with a low-inertia aluminum brake. The main motor accelerates the load to high speed. Then, for final positioning, the main-motor brake engages a clutch and the output shaft is driven by the secondary motor until the final position is attained. The flywheel effect of the iron disk brake on the main motor limits high-speed acceleration and deceleration. Final-positioning accuracy is unaffected by the flywheel effect because the rotational inertia of the main-motor brake is reduced by the square of the secondary-motor gear ratio.

This procedure is the work of Peter A. Begley, Product Supervisor, DEMAG Corp., Drives Division, Cleveland, Ohio, as reported in *Machine Design* magazine. The method given is valid for motors used in a variety of industrial, commercial, and process applications.

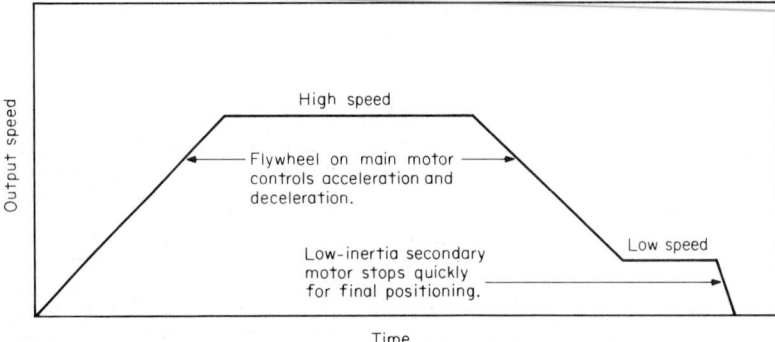

FIG. 38   Speed profile for a part-time flywheel application. *(Machine Design.)*

LIGHTNING PROTECTION OF INDUSTRIAL PLANTS

Design a lightning protection system for an industrial plant located in Omaha, Nebraska, after deciding whether such protection is needed. Choose the protection method to use for the plant if the ground area occupied by the one-story building is 10,000 ft² (929 m²).

Calculation Procedure:

1. Determine whether lightning protection is needed

Figure 39 shows a map of the United States with the average number of thunderstorm days per year in various parts of the country. This map shows that Omaha, Nebraska, has an average of 50 thunderstorm days per year. While this is not the highest for the country, it is in the midrange between 5 and 100 days per year shown on the map. Hence, lightning protection is advisable for this industrial plant.

2. Choose the type of protection system to use

In an industrial electric power system, the two principal sources of overvoltages are lightning and switching surges. Lightning voltage has been estimated by various authorities as between 100×10^6 and 10^9 V just before the stroke. The voltage wavefront (Fig. 40) travels along a conductor and can cause insulation failures and flashovers when it is greater than the basic impulse level (BIL) of a piece of equipment.

Lightning striking a porcelain-insulated power line will travel along the line with an extremely sharp wavefront, such as that in Fig. 40, until it is bled from the system by lightning arrestors or equipment failure.

Switching surges are caused by quick interruption of current, and they cause traveling waves similar to lightning but normally do not have the voltage magnitude of a lightning discharge.

Protection of electric equipment from these surges can be provided with lightning arrestors and with high BILs for equipment. Note that it is difficult to protect fully against all surges—even when the power system is completely insulated with shield conductors and underground cables. Lightning causes extremely high voltages in any equipment near a stroke. Further, all electrical systems have switching operations.

Protecting equipment, buildings, and personnel from direct lightning strokes in the typical industrial plant is usually done by using lightning rods. A single rod will provide a cone of protection (Fig. 41a). The area protected is a function of a radius scribed from a point 100 ft (30 m) above the ground (Fig. 41a).

To protect this 10,000 ft² (929 m²) one-story building, four masts with interconnected wires (Fig. 41b) could be chosen. These masts, when higher than 50 ft (15 m), will provide the protection shown. The area to be protected should be wholly within a cone of protection having a base radius of about 2 times the height of the axis (mast height) of the cone, according to Fink in the *Standard Handbook for Electrical Engineers*. Thus, four 75-ft (23-m) high masts arranged as in Fig. 41b

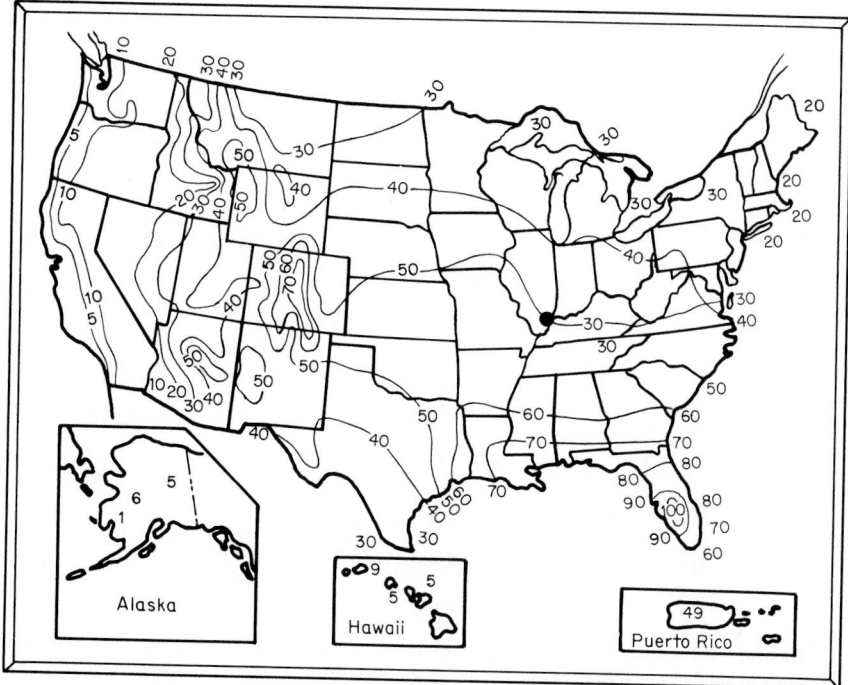

FIG. 39 Average incidence of thunderstorm days per year in the United States. *(Chemical Engineering.)*

will provide adequate protection for this 100×100 ft (31×31 m) building. Individual lightning rods might also be installed on this building to provide more protection.

As Fink notes, the wires and masts in Fig. 41b must all be suitably grounded to an interconnected ground system. This arrangement gives a subdivision of the lightning currents down any one mast and provides further subdivision of currents in the ground.

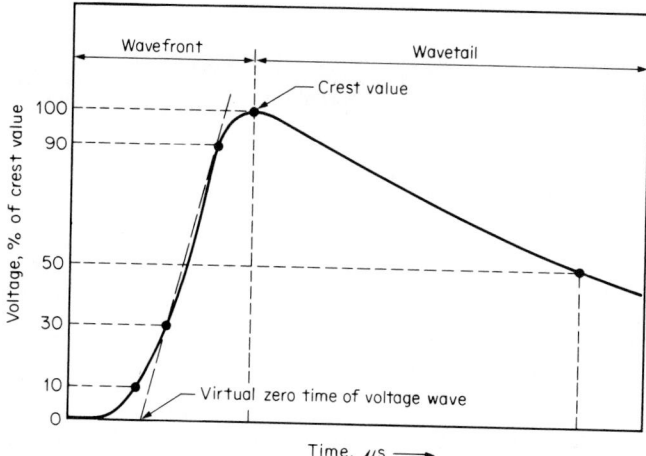

FIG. 40 Voltage wave from surge in an electric conductor. (Beeman and *Chemical Engineering.*)

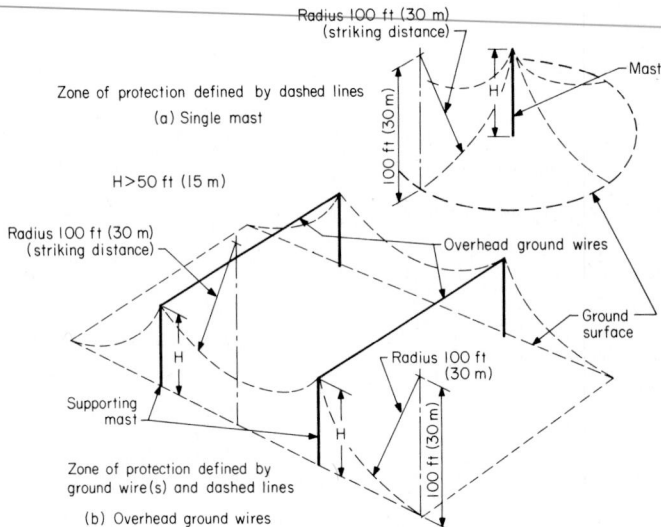

Zone of protection defined by dashed lines
(a) Single mast

Radius 100 ft (30 m)
(striking distance)

100 ft (30 m)

H

Mast

H > 50 ft (15 m)

Radius 100 ft (30 m)
(striking distance)

Overhead ground wires

Ground surface

H

Radius 100 ft (30 m)

Supporting mast

H

100 ft (30 m)

Zone of protection defined by
ground wire(s) and dashed lines

(b) Overhead ground wires

FIG. 41   Zone of protection from towers and masts for lightning strokes. *(Chemical Engineering.)*

Related Calculations: An earlier procedure in this section shows how to select lightning arrestors to protect electric equipment indoors and outdoors. That procedure can be combined with this one to provide overall protection for an industrial plant of any kind.

Other valuable design data can also be obtained from the NFPA publication no. 78 and the *National Electrical Code*. Be certain to follow *NEC* or local code requirements when you design lightning protection systems. Much of the data given here were reported by Thad Brown, Electro Technology Laboratories, Inc., and John L. Cadick, Multi-Amp Institute, in *Chemical Engineering* magazine.

SECTION 5

ELECTRONICS ENGINEERING

CHARLES F. HAFER, P.E.

FREDERICK S. BARTON
DIRECTOR
HEWLETT-PACKARD LIMITED

JOSEPH MITTLEMAN
CONSULTING ENGINEER

ZVI PRIHAR
SCIENTIFIC ADVISOR
OPERATIONS RESEARCH, INC.

ROBERT T. CHIEN
PROFESSOR OF ELECTRICAL ENGINEERING
UNIVERSITY OF ILLINOIS — URBANA-CHAMPAIGN

WILLIAM J. HANNAN
VICE PRESIDENT
RESEARCH AND TECHNOLOGY DEVELOPMENT
ITEK OPTICAL SYSTEMS CORP.

MICHAEL K. STAFFORD
MANAGER, MAGNETICS DEPARTMENT
DOSK-DEVELOPMENT DIVISION, DYSAN CORPORATION

ELECTRONICS—DESIGN AND APPLICATIONS 5.2
   Solid-State-Device Evaluation 5.3
   Transistor Selection and Circuit Arrangement. 5.5
   Vacuum-Tube Plate Resistance, Transconductance, Amplification Factor, and
      Load Line 5.6
   Vacuum-Tube-Amplifier Selection 5.11
   Transistor-Amplifier Selection and Analysis 5.13
   Oscillator Selection and Application 5.14
   Integrated-Circuit Selection and Application 5.14
   Transistor Worst-Case Leakage Current 5.17
   Power-Supply Analysis and Selection 5.19
   Reliability Analysis of Electronic Circuits and Equipment 5.20
   Maintainability Analysis of Electronic Equipment 5.22
   Selection of Electronic-Equipment Cooling Systems 5.25
   Determination of Tuned-Circuit Q Values 5.27
   Antenna Selection and Analysis 5.28
   Mismatch Efficiency and Power Loss 5.29
   Public-Address-System Design and Layout 5.30
   Transistor Hybrid-Parameter Conversions 5.31
   Large-Scale Integration Analysis 5.35
   Frequency-Changer Selection and Application 5.37
   Uses of the Smith Chart in Electronics 5.39
   Ultrasonic-Generator Selection, Crystal Thickness, and Output 5.40
   Applications of Sonar Equations 5.42
   Voltage and Current Gain for Transistorized Amplifier 5.44
   Determination of Load Line and Operating Point of a Transistorized Circuit . . 5.47
   Buffer Amplifier Analysis 5.51
   Transistor Amplifier Gain Control 5.52
   Synthesis of an ac Equivalent Circuit for an Operational Amplifier 5.53
   Dc Output of Operational-Amplifier Chain 5.55
   Gain-Control Circuit Analysis 5.57
   Common-Base Bipolar Amplifier Analysis 5.59
   Low- and High-Pass Filter Design 5.60
   Filter Distortion Determination 5.62
   Network Synthesis by Using an Operational Amplifier 5.64
   Satellite Communication Link Analysis 5.66
   Microwave Transmitter Analysis 5.68
   Selection of Thermoelectric Heat Pump to Cool Electronic Devices 5.71
OPTICS—MIRROR AND LENS SYSTEMS 5.73
   Analysis of Image Produced by Concave, Convex, or Plane Mirrors 5.73
   Determination of Images Produced by Converging and Diverging Lenses . . . 5.76
   Compound Thin-Lens Analysis 5.78
   Thick Compound Lens Systems Analysis 5.78

Electronics—Design and Applications

REFERENCES: Horowitz—*Practical Design with Solid State Devices*, Reston; Ross—*Optoelectronic Devices and Optical Imaging Techniques*, Macmillan; Seymour—*Electronic Devices and Components*, Halsted Press; Grossner—*Transformers for Electronic Circuits*, McGraw-Hill; Alvarez and Tontsch—*Fundamental Circuit Analysis*, SRA; Bonebreak—*Practical Techniques of Electronic Circuit Design*, Wiley; Camenzind—*Electronic Integrated Systems Design*, Krieger; Breuer and Friedman—*Diagnosis and Reliable Design of Digital Systems*, Computer Science Press; Comer—*Electronic Design with Integrated Circuits*, Addison-Wesley; Daryanani—*Principles of Active Network Synthesis and Design*, Wiley; Weller—*Handbook of Electronic Systems Design*, Reston; Hilburn and Johnson—*Manual of Active Filter Design*, McGraw-Hill; Johnson and Jayakumar—*Operational Amplifier Circuits Design and Applications*, Prentice-Hall; Mazda—*Discrete Electronic Components*, Cambridge University Press; Olesky—*Practical Solid-State Circuit Design*, Sams;

Sheingold—*Nonlinear Circuits Handbook*, Analog Devices; Siliconix, Inc.—*Designing with Field Effect Transistors*, McGraw-Hill; Svoboda and White—*Advanced Logical Circuit Design Techniques*, Garland Publications; Wobschall—*Circuit Design for Electronic Instrumentation*, McGraw-Hill; Wolfendale—*Computer-Aided Design Techniques*, Butterworth; Bylander—*Electronic Displays*, McGraw-Hill; Chirlian—*Analysis and Design of Integrated Electronic Circuits*, Harper-Row; Chua and Lin—*Computer-Aided Analysis of Electronic Circuits*, Prentice-Hall; Ghaznavi and Seidman—*Electronic Circuit Analysis*, Macmillan; Graeme—*Designing wiht Operational Amplifiers*, McGraw-Hill; Grob—*Electronic Circuits and Applications*, McGraw-Hill; Pugh—*Robot Vision*, IFS Pubs; Lenk—*Handbook of Microprocessors, Microcomputers and Minicomputers*, Prentice-Hall;.Chang—*Electrical and Computer Engineering*, Wiley; Golde—*Lightning*, Academic Press; Sawin—*Microprocessors and Microcomputer Systems*, Lexington Books; Melen and Buss—*Charge-Coupled Devices: Technology and Applications*, IEEE Press; House—*Laser Beam Information Systems*, PBI-Petrocelli Books; Schiller—*Electron Beam Technology*, Wiley-Interscience; Buchsbaum—*Encyclopedia of Integrated Circuits*, Prentice-Hall; Owen—*PCM and Digital Transmission Systems*, McGraw-Hill; Matisoff—*Handbook of Electronics Packaging Design and Engineering*, Van Nostrand Reinhold; Ewell—*Radar Transmitters*, McGraw-Hill; Einspruch—*VLSI Electronics*, Academic Press; Balanis—*Antenna Theory: Analysis and Design*; Harper & Row; Losev—*Gasdynamic Laser*, Springer-Verlag; Owyang—*Foundations of Optical Waveguides*, G. Elsevier North-Holland; Chryssis—*High-Frequency Switching Power Supplies*, McGraw-Hill.

SOLID-STATE-DEVICE EVALUATION

Select solid-state devices suitable for converting ac to dc (i.e., rectification) and for power switching service. The voltage drop during rectification must not exceed 10 percent with a 12-V power supply and a 0.5-A current rating. For power switching, a control voltage of 40 V is available.

Calculation Procedure:

1. *Compute the actual allowable voltage drop*

The allowable voltage drop = (supply voltage)(allowable drop, percent) = (12)(0.10) = 1.2 V.

2. *Choose the type of rectifier to use*

Junction (i.e., solid-state) rectifier diodes are superior to vacuum rectifier diodes in many ways and are rapidly supplanting them. Hence, only the solid-state diodes are considered here.

Three types of solid-state rectifier diodes used today are selenium, silicon, and germanium. The voltage drop during conducting is approximately 1.5 V for selenium, 0.8 V for silicon, and 0.5 V for germanium. Of the three, the silicon diode is the most commonly used. Hence, a silicon diode will be the first choice.

3. *Analyze the diode voltage drop*

Step 2 shows that the usual voltage drop for a silicon rectifier diode is 0.8 V. The allowable voltage drop, step 1, is 1.2 V for this application. Hence, a silicon rectifier diode will be acceptable from a voltage-drop standpoint.

4. *Check the rectifier current capacity*

Examine several specification sheets for silicon rectifier diodes. These specifications will show that a 0.5-A rating is available from many manufacturers. Also, the ratio of reverse to forward resistance is 100:1 or more. With this high a ratio, the reverse leakage current is negligible.

To provide higher peak-inverse voltage ratings, several solid-state diodes can be stacked in series. When they are installed in matched pairs, full-wave rectification can be obtained. In the usual solid-state rectifier, the current rating depends on the temperature rise caused by the voltage drop and the ability of the diode to drain away the resulting heat. A typical upper operating temperature for a silicon diode is 135°C.

5. *Select the diode case*

Usual cases for solid-state diodes are glass, top-hat, and stud (Fig. 1). Select the case giving the best shock resistance for the service intended. As a general guide, the glass case is the least shock resistant and the stud case has the greatest shock resistance.

6. *Choose a suitable power switching device*

Table 1 summarizes the switching characteristics of a variety of solid-state devices. Use this tabulation of characteristics as a general guide to preliminary switch selection. Figure 2 shows the general arrangement of the solid-state devices listed in Table 1.

With a control voltage of 40 V available, a DIAC is suitable for triggering an SCR. With this arrangement, the DIAC handles the small control current, and the SCR handles the large power current.

Related Calculations: Zener diodes, another type of solid-state device, conduct on reverse voltage when the voltage reaches a set level. The reverse voltage is constant, regardless of the current. This characteristic makes the zener diode a good voltage regulator for levels from about 3 to 200 V or more.

The thyrector consists of two selenium rectifiers connected in opposition. It is useful for dissipating relatively high ac power for a very short time, such as when inductive circuits are energized or opened. Breakdown voltages are not as sharply defined as for zener diodes, but thyrectors are cheaper.

A recently developed process for manufacturing selenium rectifiers offers substantial improvements over conventional methods. Selenium recti-

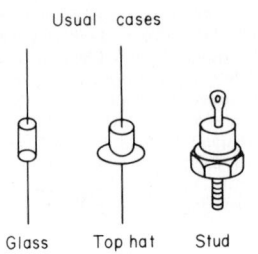

Usual cases

Glass   Top hat   Stud

FIG. 1 Typical cases used for solid-state diodes.

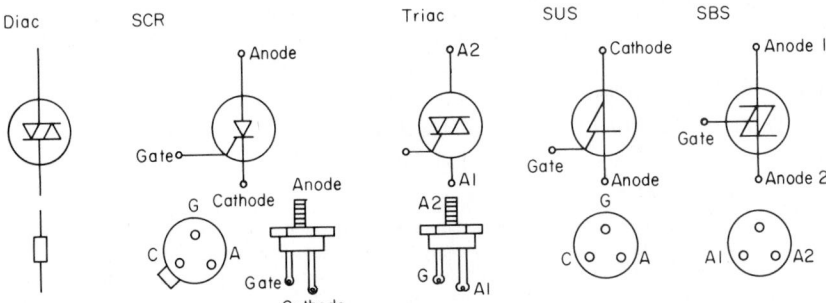

FIG. 2 General arrangement of solid-state devices.

TABLE 1 Solid-State Switch Characteristics

Switch type	Switching characteristics
SCR (silicon-controlled rectifier)	Remote-controlled dc switch; once switched on, it stays on until power to anode is interrupted or polarity is reversed. Switch is activated by a pulse current applied to the gate. Power applied too rapidly may trigger conduction.
DIAC (two-terminal ac diode)	Will not conduct until potential reaches the breakdown voltage of about 35 V. At this voltage, the device offers a low-resistance path useful in generating high-speed pulses for triggering SCRs or TRIACs.
TRIAC (three-terminal device for switching ac power)	May be triggered into conduction in either direction by a gate current of either polarity. Conducts a large current; when the gate current is turned off, conduction stops.
SUS (silicon unilateral switch)	Voltage-controlled dc switch for triggering SCRs. It is a small integrated circuit. When a positive potential of 8 V is applied between the anode and cathode, the SCR is triggered and the device conducts current.
SBS (silicon bilateral switch)	Voltage-sensitive switch for alternating current; works with applied voltages of either polarity. It is the ac equivalent of the SUS and is useful in triggering TRIACs.

TABLE 2   Selenium Rectifier Characteristics°

	Conventional selenium	New selenium	Silicon
Potential drop, V	2.2	2.2	2.4
Aging life, h	20,000	100,000	100,000
Needs transient protection	No	No	Yes
Overload capacity—rating × s	2	3	1
Maximum ambient temperature, °C	35	45	55
Cost, $:			
For 1-A rating	2.00	0.80	1.00
For 10-A rating	6.00	4.00	7.50
Volume, in^3:			
1 A	8.0	0.4	0.35
10 A	100	16	32

°*Product Engineering.*
Note: Size, performance, and cost for 28-V dc bridge rectifiers for 1 and 10 A.

fiers produced by this new method also have advantages over silicon rectifiers for popular control-circuit voltages in cost, size, and overload capacity. Electrical performance is comparable, and the new rectifiers are not subject to damage from transient voltages. Table 2 summarizes the characteristics of these rectifiers.

TRANSISTOR SELECTION AND CIRCUIT ARRANGEMENT

Select a transistor suitable for producing a power gain from 10 to 300 dB without a phase shift of the signal voltage and with a high output resistance.

Calculation Procedure:

1. *Compute the power gain of the transistor*

Use the basic power-gain relation G_o = power gain of the transistor = output power/input power = 300 dB/10 dB = 30.

2. *Select the type of transistor to use*

Table 3 shows the principal characteristics of silicon and germanium transistors—the two types of transistors in common use today. Study of this table shows that grown silicon transistors have power gains in the range of 35. Since the desired power gain is in the range of 30, and germanium transistors normally provide a higher power gain than required here, as shown by Table 3, a grown-silicon transistor will be the initial choice for this device.

3. *Investigate the transistor characteristics*

Table 4 shows the general characteristics of three amplifier configurations. Study of this tabulation shows that a common-base amplifier has no signal phase shift between output and input. Since a phase shift is not acceptable in this transistor and common-emitter and common-collector amplifier configurations do have a phase shift, a common-base transistor will be used.

4. *Check the transistor current characteristics*

Refer to a manufacturer's transistor specification or data sheet (Fig. 3). Typical specification or data sheets list transistor-type number, absolute maximum voltage and current ratings of the transistor, power rating of the transistor, and electrical characteristics of the transistor including the small-signal, high-frequency, dc, cutoff, and switching characteristics.

    The usual transistor specification or data sheet also may include characteristic curves (Fig. 4). Use these curves to determine the current in the transistor. Thus, in Fig. 4*a*, with a collector-base voltage V_{CB} of 20 V and a collector current of I_C = 3 mA, the emitter current I_E = −3.1 mA.

TABLE 3 ~~Characteristics of Silicon and Germanium~~
Transistors°

	Transistor type		
	Silicon grown	Germanium	
		Grown	Alloy
Collector:			
Voltage (maximum), V	40	40	25
Dissipation (maximum), mW	150	50	50
Cutoff current, μA	0.02	2	10
Capacitance, pF	7	14	40
Conductance, parallel, μS†	0.3	0.2	1.0
Emitter:			
Current (minimum usable), mA	1	0.01	0.1
Reverse voltage (maximum), V	2	10	5
Bias voltage, mV	500	160	160
Resistance, Ω	100	25	25
Base, resistance, Ω	500	150	300
Gain:			
Power, dB	35	47	40
Current	26	35	40

° *Electronic Design*, and Kiver—*Transistors*, McGraw-Hill.
† S is the abbreviation for siemens, the SI unit of conductance. It is equivalent to mho.

TABLE 4 General Characteristics of Transistor° Amplifiers

	Amplifier configuration		
Characteristic	Common-emitter	Common-base	Common-collector
Current gain	Large	1, approx.	Large
Voltage gain	Large	Large	1, approx.
Power gain	Largest	Large	Lowest
Input resistance	Low	Lowest	Highest
Output resistance	High	Highest	Lowest
Signal phase shift between output and input	180°	None	180°

° Kiver—*Transistors*, McGraw-Hill.

Related Calculations: Use this general method to make a preliminary selection of a transistor for power, switching, detection, mixing, rectification, oscillation, amplification, and similar applications.

VACUUM-TUBE PLATE RESISTANCE, TRANSCONDUCTANCE, AMPLIFICATION FACTOR, AND LOAD LINE

A 6J5 triode vacuum tube is operated with a grid bias of −6 V. What is the plate resistance of this tube if the plate voltage is raised from 160 to 220 V? What is the tube transconductance when the grid bias is increased from −6 to −8 V while the plate voltage is held at 220 V? What is the tube amplification factor if the grid bias must be changed from −6 to −10 V to maintain a

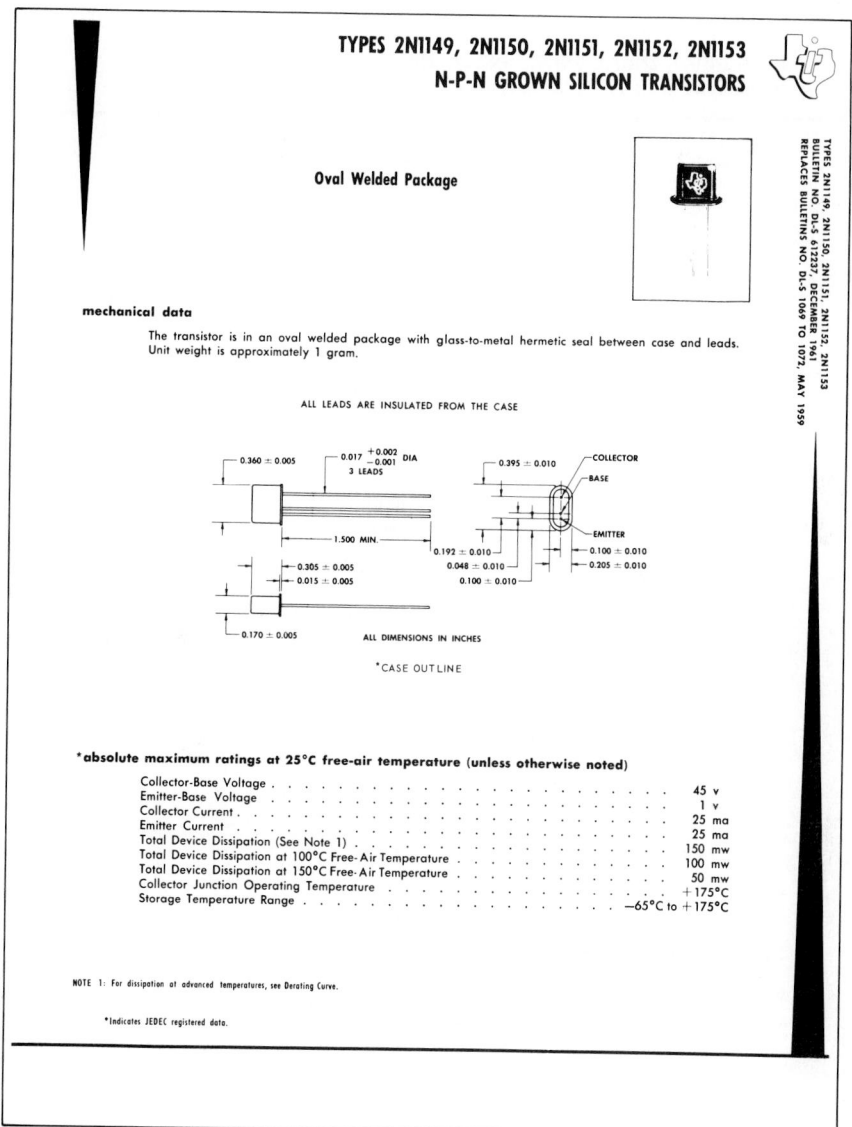

TYPES 2N1149, 2N1150, 2N1151, 2N1152, 2N1153
N-P-N GROWN SILICON TRANSISTORS

Oval Welded Package

mechanical data

The transistor is in an oval welded package with glass-to-metal hermetic seal between case and leads. Unit weight is approximately 1 gram.

ALL LEADS ARE INSULATED FROM THE CASE

*CASE OUTLINE

ALL DIMENSIONS IN INCHES

*absolute maximum ratings at 25°C free-air temperature (unless otherwise noted)

Collector-Base Voltage	45 v
Emitter-Base Voltage	1 v
Collector Current	25 ma
Emitter Current	25 ma
Total Device Dissipation (See Note 1)	150 mw
Total Device Dissipation at 100°C Free-Air Temperature	100 mw
Total Device Dissipation at 150°C Free-Air Temperature	50 mw
Collector Junction Operating Temperature	+175°C
Storage Temperature Range	−65°C to +175°C

NOTE 1: For dissipation at advanced temperatures, see Derating Curve.

*Indicates JEDEC registered data.

FIG. 3   Typical transistor data sheet.

constant plate current when the plate voltage is changed from 160 to 240 V? Plot the tube load line and operating point for a load of 28,000 Ω and a plate-to-cathode voltage of 280 V.

Calculation Procedure:

1. *Determine the tube voltage and current changes*

Obtain a plate characteristic curve for the tube in question. Figure 5 shows a typical set of plate characteristic curves. To determine the tube voltage and current changes, the tube characteristic

---

TYPES 2N1149, 2N1150, 2N1151, 2N1152, 2N1153
N-P-N GROWN SILICON TRANSISTORS

---

electrical characteristics at 25°C free-air temperature (unless otherwise noted)

parameter		test conditions	types	min*	typ	max*	unit
I_{CBO}	Collector Cutoff Current	$V_{BC} = 30$ v $I_E = 0$	ALL			2	μa
		$V_{CB} = 30$ v $I_E = 0$ $T_A = 150°C$	ALL		3		μa
		$V_{CB} = 5$ v $I_E = 0$ $T_A = 100°C$	ALL			10	μa
		$V_{CB} = 5$ v $I_E = 0$ $T_A = 150°C$	ALL		0.5	50	μa
BV_{CBO}	Collector-Base Breakdown Voltage	$I_C = 50 \mu a$ $I_E = 0$	ALL	45			v
$r_{CE(sat)}$	DC Collector-Emitter Saturation Resistance	$I_B = 2.2$ ma $I_C = 5$ ma	ALL		100	200	ohm
C_{ob}	Common-Base Output Capacitance	$V_{CB} = 5$ v $I_E = 0$ $f = 1$ mc	ALL		7		pf
f_{hfb}	Common-Base Alpha Cutoff Frequency	$V_{CB} = 5$ v $I_E = -1$ ma	2N1149	—	12		mc
			2N1150	—	13		
			2N1151	8	14		
			2N1152	—	15		
			2N1153	—	16		
h_{fb}	AC Common-Base Forward Current Transfer Ratio	$V_{CB} = 5$ v $I_E = -1$ ma $f = 1$ kc	2N1149	−0.9	−0.925	−0.953	—
			2N1150	−0.948	−0.96	−0.976	
			2N1151	−0.948	−0.975	−0.989	
			2N1152	−0.9735	−0.98	−0.989	
			2N1153	−0.987	−0.99	−0.997	
h_{ib}	AC Common-Base Input Impedance	$V_{CB} = 5$ v $I_E = -1$ ma $f = 1$ kc	ALL	30	42	80	ohm
h_{ob}	AC Common-Base Output Admittance	$V_{CB} = 5$ v $I_E = -1$ ma $f = 1$ kc	ALL	0	0.4	1.2	μmho
h_{rb}	AC Common-Base Reverse Voltage Transfer Ratio	$V_{CB} = 5$ v $I_E = -1$ ma $f = 1$ kc	2N1149	0	120×10^{-6}	500×10^{-6}	—
			2N1150	0	250×10^{-6}	1000×10^{-6}	
			2N1151	0	400×10^{-6}	1000×10^{-6}	
			2N1152	0	400×10^{-6}	1000×10^{-6}	
			2N1153	0	400×10^{-6}	1000×10^{-6}	

functional tests at 25°C free-air temperature

parameter		test conditions	types	min	typ	max	unit
G_{pe}	Common-Emitter Power Gain	$V_{CE} = 20$ v $I_E = -2$ ma $R_G = 1$ KΩ $R_L = 20$ KΩ $f = 1$ kc $V_g = 0.02$ v	2N1149		35		db
			2N1150		39		
			2N1151		39		
			2N1152		42		
			2N1153		42.5		
NF	Spot Noise Figure	$V_{CE} = 5$ v $I_E = -1$ ma $R_G = 1$ KΩ $f = 1$ kc $BW = 1$ cycle/sec	ALL		20		db

POWER GAIN TEST CIRCUIT

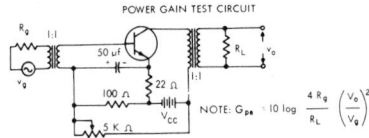

NOTE: $G_{pe} = 10 \log \dfrac{4 R_g}{R_L} \left(\dfrac{V_o}{V_g}\right)^2$

FIG. 3 Typical transistor data sheet *(Continued)*. *(Texas Instruments Incorporated.)*

curve (Fig. 5) must be used. It is known that the plate voltage E_b changes from 160 to 220 V, and the grid bias is −6 V. At a plate voltage of 160 V and a grid bias of $E_c = -6$ V, Fig. 5 shows that the plate current $I_b = 3.8$ mA. Find this value of I_b by projecting vertically upward from $E_b = 160$ V to the curve $E_c = -6$ V. From the intersection project horizontally to the left to read $I_b = 3.8$ mA. By the same procedure, $I_b = 11.4$ mA at $E_b = 220$ V.

2. *Compute the tube plate resistance*

Use the relation $R_p = \Delta E_b / \Delta I_b$, where R_p = plate resistance, Ω; ΔE_b = plate voltage change, V; ΔI_b = plate current with E_b and I_b at a constant grid bias. For this tube at the stated conditions,

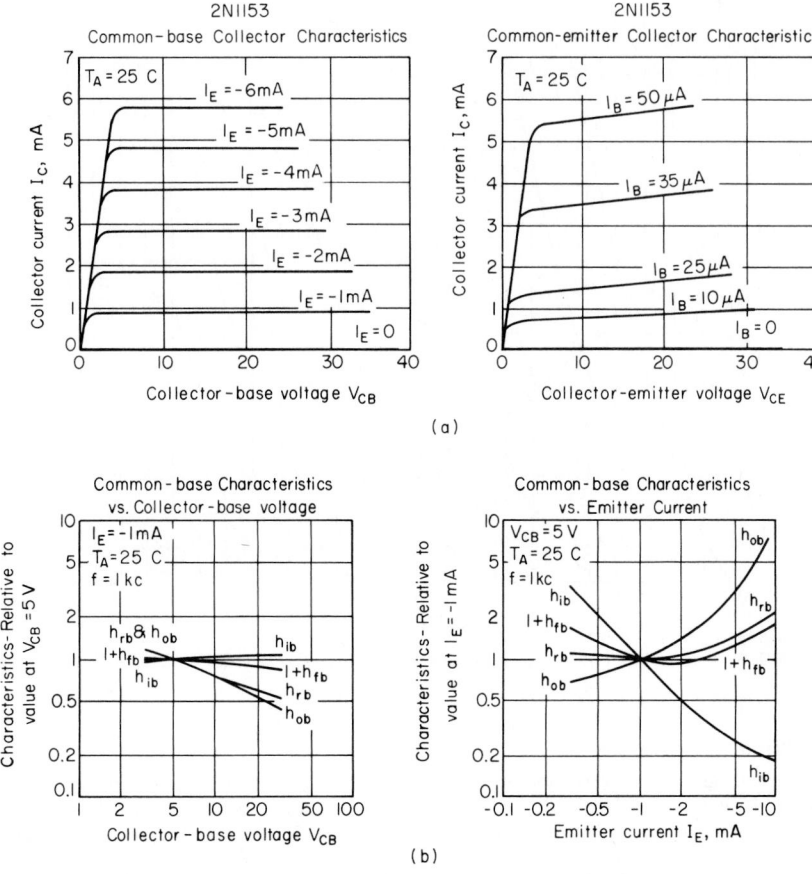

FIG. 4 Typical transistor characteristic curves. *(Texas Instruments Incorporated.)*

$R_p = (220 - 160)/(0.0114 - 0.0038) = 7890\ \Omega$. In the denominator of this expression, the plate currents determined in step 1 were changed from milliamperes to amperes; that is, 1 mA = 0.001 A.

3. *Determine the plate current change*

Using the plate characteristic curve (Fig. 5), find the plate current at $E_b = 220$ V and $E_c = -6$ V as $I_b = 11.4$ mA, using the procedure described in step 2. Also, at $E_b = 220$ V and $E_c = -8$ V, $I_b = 6.0$ mA.

4. *Compute the tube transconductance*

Use the relation $g_m = \Delta I_b/\Delta E_c$, where g_m = tube transconductance, S; ΔI_b = change in plate current, A; ΔE_c = change in grid voltage, V. Thus, $g_m = (0.0114 - 0.006)/[-6 - (-8)] = 0.0027$ S.

5. *Compute the tube amplification factor*

Use the relation $\mu = \Delta E_b/\Delta E_c$, where μ = tube amplification factor; ΔE_b = change in plate voltage, V; ΔE_c = change in grid voltage, V, while the plate current I_b remains constant. Thus, $\mu = (240 - 160)/[-6 - (-10)] = 20$.

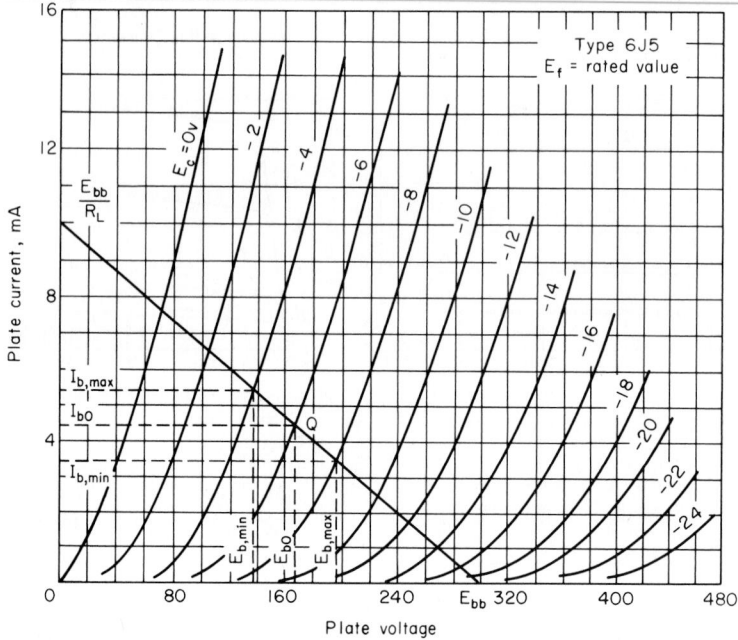

FIG. 5   Vacuum-tube plate characteristics and load line. *(General Electric Company.)*

6. *Plot the tube load line*

The tube load line is plotted on the plate characteristic curve as a straight line between a point corresponding to $I_b = 0$, $E_b = E_{bb}$, where E_{bb} = plate-to-cathode voltage. Since E_{bb} = 280 V, the load line intersects the horizontal axis of the plate characteristic curve at $I_b = 0$, $E_{bb} = 280$ V. Plot this point (Fig. 5).

The load line intersects the vertical axis at $E_b = 0$ V and $I_b = E_{bb}/R_L$. Or $I_b = 280/28,000$ = 0.01 A, or 10 mA. Plot this intersection of the load line on the vertical axis of the plate characteristic curve at $E_b = 0$, $I_b = 10$ mA. Draw a straight line between this and the previously plotted point on the horizontal axis. The resulting line, Fig. 5, is termed the *tube load line*.

The intersection of the load line with any grid-voltage curve is the tube *operating point*. Thus, with a grid voltage of $E_c = -4$ V, $I_b = 5.2$ mA, $E_b = 118$ V.

Related Calculations: Use the general procedure given here to analyze the operating characteristics of diode, triode, tetrode, and pentode vacuum tubes. In determining the operating point of a pentode, the usual procedure is to assume that the screen current is a fixed percentage of the plate current and proceed in the same manner as for a triode. A safe assumption for usual pentodes is that the total cathode current = $1.3I_b$ at the tube operating point.

Brophy[1] states that in many cases it is simpler to resort to the following cut-and-try procedure for determining the operating point for a pentode. Choose a point on the load line corresponding to an arbitrary grid-bias voltage and plate current. Determine the screen current from the tube-screen characteristic curve for these values of E_c and E_b. Then compare the product $(I_b + I_s)R_k$ with the chosen value of E_c, where I_s = screen current, mA; R_k = cathode-bias resistor resistance, Ω. If the two values being compared are equal, the original operating-point choice is satisfactory. If the values are not equal, repeat the process until the desired accuracy is obtained.

[1]Brophy—*Basic Electronics for Scientists*, McGraw-Hill.

VACUUM-TUBE-AMPLIFIER SELECTION

Choose a vacuum-tube amplifier to amplify a 0.5-V input signal to a 450-V output signal. Select the class of amplifier to use.

Calculation Procedure:

1. *Compute the amplification required*

The amplification, or gain, required is amplification = output or plate voltage/input or grid voltage. For this amplifier, amplification = 450/0.5 = 900 times.

2. *Determine the number of amplification stages required*

The usual upper limit on the amplification available from one vacuum tube is 200 times. With this rule as a guide, a 0.5-V input signal could be amplified to (0.5)(200) = 100 V by one tube. Since a 450-V output signal is required, one tube or *stage* of amplification would be unsatisfactory.

   If the 100-V output signal were fed to the grid of a second tube or stage that amplified 200 times, the output voltage would be (100)(200) = 20,000 V. Since this is much greater than the 900-V output desired, a two-stage amplifier will be satisfactory.

3. *Choose the amplification per stage*

Where possible, in multistage vacuum-tube amplifiers, an equal amplification per tube or per stage is provided. Thus, an amplification of 30 times per tube would provide 30 × 30 = 900 times in this two-stage amplifier. As a check, the 0.5-V input signal would be amplified to 0.5 × 30 = 15 V in the first stage. Feeding this output into the second stage gives a voltage of 15 × 30 = 450 V. This is the desired output voltage.

4. *Select the class of amplifier to use*

Table 5 summarizes vacuum-tube-amplifier characteristics. Study of this table shows that class A operation is common in voltage-amplification applications. Hence, a class A amplifier will be chosen for this application.

5. *Select the type of coupling to use*

Resistance-capacitance *(RC)* coupling of two or more amplifier stages provides a flat frequency response in the midband region. Transformer coupling provides a flat frequency response for only about one-half the midband region. Hence, if a flat-frequency-response characteristic throughout the midband region of the output is required, use an *RC*-coupled amplifier. If such a requirement does not exist, use a transformer-coupled amplifier.

   Related Calculations: When a transformer-coupled amplifier is used, the amplification obtainable = tube amplification factor × transformer step-up turns ratio. Thus, the transformer provides added amplification of the vacuum-tube input signal.

TABLE 5   Vacuum-Tube Amplifier Characteristics

Amplifier class	Typical applications	Important characteristics
A	Voltage amplification with small current capacity; power amplification for low-power applications	Small input signal; output current variation exactly duplicates input current variation—i.e., linear operation and little or no distortion; small amplification of input
B	Power amplification with large current capacity	Larger input signal than class A; positive half of input signal is amplified and appears in output; amplification much greater than in class A
C	Radio-frequency power amplification with large current capacity	Largest input signal; top of output curve is flat; largest amplification of the three types

Where desired, class A and B amplifiers can be combined to create class AB_1 and AB_2 amplifiers. The resulting output combines the features of both class A and class B amplifiers. Figure 6 shows the output current waveforms of class A, B, and C amplifiers.

Two amplifiers can be connected so that their input voltages are 180° out of phase. This is termed *push-pull*, and the plate circuit of each amplifier is connected to opposite ends of an output transformer. Push-pull vacuum-tube amplifiers are often used for power amplification. To obtain maximum power amplification, the output of a push-pull amplifier must be matched to the amplifier load. This is termed *impedance matching*. To match impedances, use the relation $Z_L = (N_s/N_p)^2 Z_p$, where Z_L = load impedance, Ω; N_s = number of turns in the output-transformer secondary winding; N_p = number of turns in the output-transformer primary winding; Z_p = impedance of output-transformer primary winding, Ω. The primary winding of the output transformer is connected in series in the amplifier-tube plate circuit, and the secondary winding of the output transformer is connected in series in the load circuit. Usually, the required secondary or load impedance Z_L is fixed by the device served by the amplifier. Given the required load

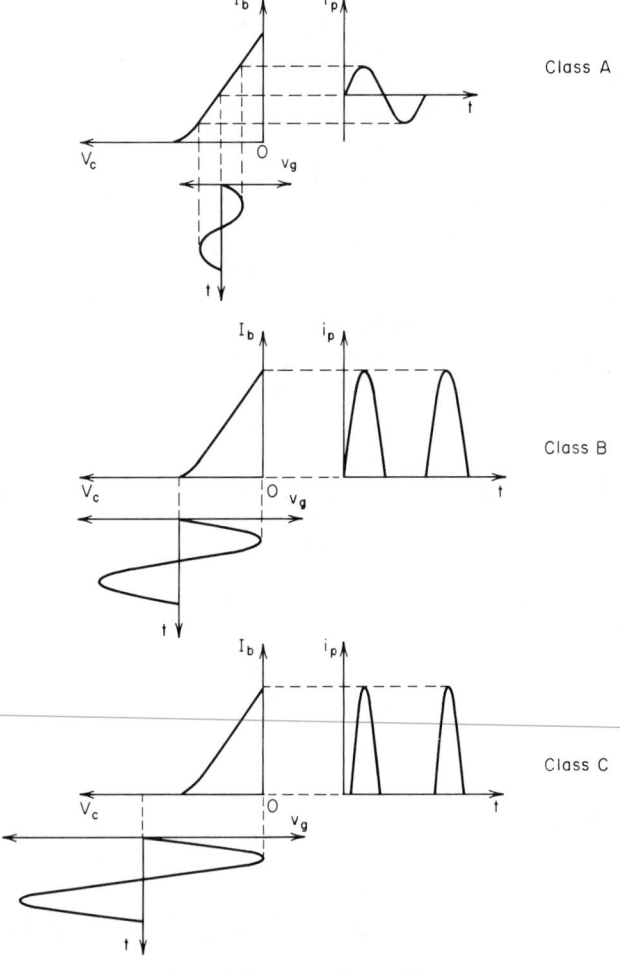

FIG. 6   Output waveforms of class A, B, and C amplifiers.

impedance, a transformer can be selected and the primary impedance computed by using the above relation.

Vacuum tubes used for voltage and power amplifiers include the triode and pentode types. The tetrode is seldom used because it has a secondary-emission disadvantage. A pentode provides more amplification than a triode vacuum tube.

TRANSISTOR-AMPLIFIER SELECTION AND ANALYSIS

Select a transistorized amplifier to amplify a 0.10-V input signal to a 29-V output signal with a 9000-Ω load. The output signal must not undergo a phase reversal. What is the voltage gain with this load and a 30-Ω input resistance? What is the amplifier power gain? Compute the current, voltage, and power gains for a grounded-emitter amplifier configuration with a 9000-Ω load and 3000-Ω input resistance.

Calculation Procedure:

1. *Compute the amplifier voltage gain*

Use the relation gain = output voltage/input voltage = 29/0.10 = 290.

2. *Select the amplifier circuit arrangement*

Table 6 shows typical current, voltage, and power gains for various arrangements of transistor amplifiers. Study of this table shows that either a common-base or a common-emitter circuit configuration will give the desired voltage amplification. Further study of the table shows that a common-emitter amplifier has a 180° phase reversal, or shift. Since a phase reversal cannot be tolerated, a common-base amplifier should be used.

3. *Compute the actual voltage gain*

For a common-base transistor amplifier, voltage gain = output current (A) × load resistance (Ω)/input current (A) × input resistance (Ω).

However, the ratio output current/input current = α = current gains. Table 6 shows that the current gain for a typical common-base transistor amplifier = 0.98. Hence, voltage gain = 0.98 × load resistance/input resistance = 0.98 × 9000/30 = 294. Since a voltage gain of 290 is desired, the amplifier is suitable.

4. *Compute the amplifier power gain*

For an amplifier, power gain = power output in watts/power input in watts, or (output current, A)^2(load resistance, Ω)/[input current, A)^2(input resistance, Ω)]. As in step 3 above, the current gain for a typical common-base transistor amplifier = 0.98. Thus, since current gain = output current/input current, the amplifier power gain using the second relation above = $(0.98)^2(9000/30)$ = 288.

5. *Compute the grounded- or common-emitter current gain*

In a common-emitter amplifier configuration, the input signal is connected to the base. Hence, the base current = input current. Then the current gain = β = output current/input current.

TABLE 6 Typical Transistor Amplifier Characteristics

Circuit configuration	Current gain	Voltage gain	Power gain	Phase reversal	Input resistance	Output resistance
Common base	Low 1.0 (0.98)	Large (200)°	Large (200)	No	Lowest (15 Ω)	Highest (1 MΩ)
Common collector	Large (75)	Low 1.0 (0.98)	Lowest	Yes, 180°	Highest (50,000 Ω)	Lowest (100 Ω)
Common emitter	Large (75)	Large (250)	Largest	Yes, 180°	Medium (600 Ω)	High (50,000 Ω)

°Typical gain value is shown in parentheses.

With a typical emitter current of 6 mA, the output or collector current will be emitter current − base current. Given a base current of 0.15 mA, output or collector current = 6.0 − 0.15 = 5.85 mA. Hence, power gain = β = 5.85/0.15 = 39.

6. *Compute the common-emitter voltage gain*

In a common-emitter amplifier, voltage gain = current gain × resistance gain, or β × load resistance, ohms/input resistance, ohms. With a 9000-Ω load and a 3000-Ω input resistance, voltage gain = (39)(9000/3000) = 117.

7. *Compute the common-emitter power gain*

As in step 4, power gain = (current gain)^2 load resistance, ohms/input resistance, ohms = (39)^2(9000/3000) = 4560.

   Related Calculations: The current gain of a grounded- or common-collector amplifier = β + 1, whereas the voltage gain is less than unity, usually above 0.98. The power gain is computed in the same way as described in steps 4 and 7 above.

   Since the common-emitter amplifier has the largest power gain and large current and voltage gains, this configuration is widely used in transistorized electronic equipment. Even though the current and gain notations are presented in word form in this calculation procedure for each transistor considered, the hybrid symbols are used for hybrid characteristics or parameters in transistor manufacturers' data sheets. Thus, $\alpha = h_{FB}$, $\beta = h_{FE}$ in hybrid notation. The current values given in step 5 are typical for common-emitter transistors.

OSCILLATOR SELECTION AND APPLICATION

Select an oscillator to operate in the 200- to 400-Hz range. At full load the oscillator output must not vary by more than 2 Hz. Ease of frequency change is required for the application. [*Note:* 1 cycle/s = 1 hertz (Hz).]

Calculation Procedure:

1. *Compute the required frequency stability*

The oscillator operating range is 200 to 400 Hz, or 400 − 200 = 200 Hz. With an allowable frequency variation of 2 Hz, the frequency stability must be allowable frequency variation/frequency operating range, or 2/200 = 0.01, or 1 percent. This is usually considered to be a high-frequency stability. Hence, an amplifier having a high-frequency stability is required.

2. *Select the oscillator type*

Table 7 summarizes typical oscillator characteristics. Study of this table shows that a Wein-bridge oscillator has the desired characteristics—excellent frequency stability, easy frequency changing, and a frequency range of 200 to 400 Hz. Hence, this type of oscillator will be chosen for the application.

   Related Calculations: As a general guide, the typical characteristics of oscillators listed in Table 7 apply whether the oscillator uses one or more vacuum tubes or transistors. Hence, the data listed can be used in preliminary selection of oscillators using either type of component. However, transistor oscillators are smaller in physical size than are vacuum-tube oscillators.

INTEGRATED-CIRCUIT SELECTION AND APPLICATION

Choose the type of integrated circuit to use for a wideband amplifier to supply a 10-V 15-mA 100-kHz output. Specify the type of integrated circuit to use.

Calculation Procedure:

1. *Compute the circuit power output*

Use the relation $P = IE$, where P = integrated-circuit power output, W; I = output current, A; E = output voltage. For this circuit, P = (0.015)(10) = 0.15 W.

TABLE 7   Typical Oscillator Characteristics

Oscillator type	Typical characteristics
RC (general)	Good frequency stability below 100 kHz; wide tuning range with constant power output.
LC (Hartley and Colpitts)	Widely used from 100 kHz to 500 MHz. At low power, frequencies to 4000 MHz can be obtained. In Hartley, amplifier portion operates class C; in Colpitts, class A, although class C is possible in the latter, if desired. Colpitts is adaptable to low-impedance loads; frequency stability is moderate.
Phase-shift (uses RC feedback)	Useful at medium and low frequencies to 1 Hz where frequency stability is not critical; frequency changing is cumbersome.
Wein-bridge	Excellent frequency stability; easy frequency changing; typical frequency range is 10:1 or 5 Hz to 1 MHz.
Tickler (Armstrong)	Class C operation used to develop large power output at high frequency. Frequency stability is less than in other oscillator types.
Crystal	Popular for use in fixed-frequency applications; frequency is constant, regardless of load.
Relaxation	Use nonlinear active elements; poor frequency stability.
Blocking	Used to generate pulse waveforms in digital-computer circuits.
Electron-coupled	Excellent frequency stability; can use Armstrong, Colpitts, or Hartley circuit.
Magnetron	Used for frequencies above 1000 MHz.
Klystron	Used for frequencies above 1000 MHz.

2.   *Select the type of integrated circuit to use*

Use Fig. 7 to select the type of integrated circuit. Enter at the power output, watts, and project horizontally until the vertical line at 100 kHz, the output frequency, is intersected. The point of intersection is in the monolithic silicon integrated-circuit area. Hence, this type of circuit should be used.

An integrated circuit consists of two or more electronic components associated on, or within, a substrate to form an electric network. A monolithic integrated circuit has a single semiconductor body, such as a silicon crystal, as its substrate. Monolithic integrated circuits are popular for many applications.

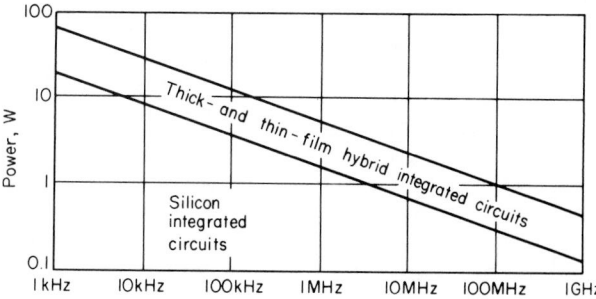

FIG. 7   Approximate performance envelopes for monolithic integrated circuits and thick-film and thin-film hybrid circuits. *(Electronics.)*

TABLE 8 Guide for Integrated-Circuit Selection[a]

Integrated circuits	Monolithic		Thin film		Thick film
	Bipolar silicon	Metal-oxide silicon	Evaporated thin film	Sputtered tantalum thin film	Screen printed thick film
Advantages	Reliability and long life; low cost; high density; low weight; standardization; amenable to high production	Low cost; simple design; high density; resistivities to 20 kΩ/□; amenable to redundant design; direct coupling avoids capacitors; symmetric switching capabilities	Good line definition; high reliability; high precision; wide choice of materials; permit 1:1 translation from discrete component design; amenable to production of interconnections	Good line definition; high reliability; high precision by anodizing	Low cost; simple production techniques; amenable to production of interconnections; resistivities of 1 to 20 kΩ/□; capacitance of 500 to 500,000 pF/in² (77.5 to 77,500.5 pF/cm²); high power capability
Limitations	Power-frequency product; high development cost; loose tolerances; diffused resistivities are 100–200 Ω/□	Limited frequency capability	Active devices must be added; exhibit unique failure modes; low sheet resistivity; production procedures can be complex	Active devices must be added; specialized production processes needed	Active devices must be added; trimming required for close tolerances; high voltage coefficient; nonlinear temperature characteristic

[a] *Electronics.*
Note: Selection of integrated-circuit types is guided by process capability.

TABLE 9   Typical Commercial Integrated Circuits°

Circuit function	Circuit input	Circuit output	Manufacturer's circuit number
Intermediate-frequency amplifier	Differential	Single-ended	μA7703C
Operational amplifier	Differential	Single-ended	μA7709C
Wideband amplifier	Differential	Single-ended	μA7712C
Radio-frequency amplifier	Differential	Single-ended	CA3004
Audio-frequency amplifier	Single-ended or differential	PP emitter-follower	CA3007
Wideband intermediate-frequency amplifier	Differential	Single-ended	CA3011
Half-adder	Digital binary	. . . .	HEP553
Bias driver	Storage element	. . . .	HEP558
Flip-flop	Voltage regulator	. . . .	HEP554

° *Electronics.*

3.   *Check the circuit selection*

Use Table 8 as a guide to the suitability of the type of circuit selected. Study of Table 8 shows that the metal-oxide silicon (MOS) monolithic integrated circuit has a limited frequency capability. Thus, if a wide frequency capability is desired, either a bipolar silicon or other type of circuit has to be used. Since a wide frequency capability is not stated as a requirement for this circuit, the monolithic MOS circuit probably will be acceptable.

4.   *Choose the commercial integrated circuit*

Table 9 shows a condensed listing of typical commercially available integrated circuits. Enter Table 9 at the circuit function and project horizontally to read the circuit input and output. Thus, a μ7712C wideband amplifier has a differential input (i.e., the *difference* between two input signals) and a single-ended output.

Match the desired circuit function with the desired output. When the needed function and required output are obtained from a commercially available integrated circuit, list that circuit as a tentative choice. Once several such tentative choices have been made, the final selection can be made after analysis of the available circuits.

Related Calculations: Use this general method to select integrated circuits for computers, communications equipment, amplifiers (audio, video, radio-frequency, operational, wideband, etc.), oscillators, limiters, mixers, servos, etc. For best results, keep several manufacturers' engineering data tabulations available so that comparisons of the suitable circuits can be quickly made.

TRANSISTOR WORST-CASE LEAKAGE CURRENT

What is the worst-case transistor leakage current in a transistor circuit that is part of a complex equipment if the transistor specification sheet gives the following data for an ambient temperature of 25°C? $I_{CBO,max} = 4\ \mu A(V_{CB} = 2$ V dc); $I_{CBO,max} = 5\ \mu A(V_{CB} = 6$ V dc). The maximum allowable junction temperature = 85°C; maximum allowable power dissipation at 25°C = 300 mW; $h_{FE,min} = 5$; $V_{CB} = 10$ V dc; $V_{CE} = 12$ V dc; $V_{BE} = 2$ V dc; $I_C = 6$ mA. The power sources for the circuit have 1 percent total tolerance ratings, whereas the resistive components have 5 percent total tolerance ratings. The equipment performance specification indicates a maximum ambient temperature requirement of 50°C. The rise in temperature caused by power dissipation within the equipment is 10°C.

Calculation Procedure:

1.   *Compute the transistor power dissipation*

Use the relation $P_d = I_C[/V_{CE} + (V_{BE}/h_{FE})]$, where P_d = power dissipation, mW; I_C = dc collector current, mA; V_{CE} = dc collector-to-emitter voltage; $h_{FE} = I_C/I_B$, where I_B = dc base current, mA. Substituting gives $P_d = 6(12 + 2/5) = 74.4$ mV.

2.   *Compute the thermal resistance factor*

The thermal resistance factor K is computed from $K = (T_{\text{J,max}} - T_{\text{spec}})/P_{\text{dst}}$, where $T_{\text{J,max}} =$ maximum allowable junction temperature, °C; $T_{\text{spec}} =$ specified temperature, °C, at which the maximum allowable power dissipation P_{dst} takes place. Substituting yields $K = (85 - 25)/300 = 0.2$ mW.

3.   *Compute the junction-temperature increase*

Use the relation $\Delta T_{pd} = K P_d$, where $\Delta T_{pd} =$ junction temperature increase, °C. Or $\Delta T_{pd} = 0.2(74.4) \cong 15$°C.

4.   *Compute the junction operating temperature*

The junction operating temperature is $T_{\text{J}} = T_{\text{A}} + T_{\text{E}} + T_{pd}$, where $T_{\text{A}} =$ maximum ambient temperature given in the performance specification, °C; $T_{\text{E}} =$ maximum expected rise in temperature due to the power dissipation within the equipment, °C. Substituting, we have $T_{\text{J}} = 50 + 10 + 15 = 75$°C.

5.   *Compute the transistor thermal constant*

Use the relation $M_{\text{T}} = (T_{\text{J}} - 25°C)/10$, where $M_{\text{T}} =$ transistor thermal constant and the other symbols are as before. Thus, $M_{\text{T}} = (75 - 25)/10 = 5$. Note in this relation that 25°C = the transistor-specification-sheet ambient temperature.

6.   *Compute the segregated-leakage components*

At the specified transistor ambient temperature of 25°C, $I_{\text{TO}(25°C)} + I'_{\text{SO}(25°C)} = 4$ μA, and $I_{\text{TO}(25°C)} + 3I'_{\text{SO}(25°C)} = 5$ μA, where $I_{\text{TO}} =$ the segregated thermal component derived from the transistor-specification sheet for 25°C; $I'_{\text{SO}(25°C)} =$ the zero-life, voltage-dependent, surface-leakage component.

Solve these simultaneous equations by subtracting the first equation from the second to obtain $I'_{\text{SO}(25°C)} = 1$; $I'_{\text{SO}(25°C)} = 0.5$ μA. At $V_{\text{CB}} = 2$ V dc, $I_{\text{TO}} = 3.5$ μA, which is obtained by substituting the value of $I'_{\text{SO}(25°C)} = 0.5$ in the first equation and solving for I_{TO}.

7.   *Compute the zero-life thermal-leakage component*

A general rule of thumb used in transistor analysis is that the thermal-leakage component doubles for every 10°C increase above the ambient temperature given in the transistor-specification sheet. Compute the thermal-leakage component from $I_{\text{TO}} = I_{\text{TO}(25°C)} \times 2^{M(T)}$, where $I_{\text{TO}} =$ the zero-life thermal-leakage component; $I_{\text{TO}(25°C)} =$ the segregated thermal component from the transistor-specification sheet; $2^{M(T)} =$ multiplying factor for the surface-leakage current $I'_{\text{SO}(25°C)}$. This multiplying factor is based on the rule of thumb given above.

Solve for I_{TO} by using the value of $I_{\text{TO}(25°C)}$ from step 6 and applying the rule of thumb to $I'_{\text{SO}(25°C)}$. Thus, with a junction temperature of 75°C, the multiplying factor $2^{M(T)}$ becomes $(75 - 25)/10 = 5$; for example, $I'_{\text{SO}(25°C)} = 0.5$ is doubled five times, or $0.5 \times 2 = 1$; $1 \times 2 = 2$; $2 \times 2 = 4$; $4 \times 2 = 8$; $8 \times 2 = 16$. Then multiplying by 2 and substituting, we have $I_{\text{TO}} = 3.5 \times 32 = 112$ μA.

8.   *Compute the zero-life, voltage-dependent, surface-leakage component*

Use the relation $I_{\text{SO}} = I'_{\text{SO}} \times 2^{M(T)}/2$, where $I_{\text{SO}} =$ zero-life surface-leakage current, μA; other symbols as before. Thus, $I_{\text{SO}} = 2.5 \times 16 = 40$ μA.

9.   *Compute the zero-life leakage current*

Use the relation $I_{\text{LO}} = I_{\text{TO}} + I_{\text{SO}}$, where $I_{\text{LO}} =$ zero-life leakage current, μA; other symbols as before. Thus, $I_{\text{LO}} = 112 + 40 = 152$ μA.

10.   *Compute the effect of transistor aging*

Where the increase in the leakage current caused by aging is not available from physical measurements of the transistor, assume an increase of at least 40 percent. Thus, the total leakage current I_{L} μA $= 1.4I_{\text{LO}}$. Or, $I_{\text{L}} = 1.4 \times 152 = 213$ μA.

11.   *Compute the total worst-case transistor leakage*

With resistive components having 5 percent tolerance ratings and power sources having a 1 percent tolerance rating, the worst-case transistor leakage is $I_{\text{L}}(1.0 + 0.05 + 0.01) = 213 \times 1.06 = 226$ μA.

Related Calculations: Use this method for any equipment having a large number of transistors as well as single isolated transistor circuits. This method and calculation procedure was developed by E. D. Peterson, Military Electronics Division, Motorola, Inc., and published in *Electronics*.

POWER-SUPPLY ANALYSIS AND SELECTION

Choose the power supply for a vacuum-tube electronic equipment to be used on a 120-V power system. High reliability is desired with a full-wave output of 400 mA. The filter ripple must not exceed 0.5 percent. What is the power-supply ripple voltage?

Calculation Procedure:

1. Choose the type of power supply to use

Four types of power supplies are available for vacuum-tube electronic equipment: transformer rectifier-filter voltage-divider assembly; transformerless supply using only a rectifier and filter; vibrator—synchronous or nonsynchronous; and dynamotor.

Of these four, the first is the most popular for high-reliability service. Transformerless power supplies are used for small equipments designed to operate on ac and dc. Vibrators find some use for B + voltage supply for radio receivers. Dynamotors deliver 1000 V or more dc output for aircraft electronic equipment. Hence, assume that a transformer-type power supply is tentatively chosen for this application.

2. Select the transformer for the power supply

When a full-wave output is desired, a center-tap transformer is generally used. Transformers with a center-tapped high-voltage secondary winding are rated by the voltage at each end with respect to the center tap. This rating is usually 350-0-350 V at 100 mA. Low-voltage secondary windings that deliver power to tube heaters are rated by the voltage and current they can deliver, for example, 6.3 V at 2 A.

Since a full-wave power output is desired, assume that a center-tap transformer is used.

3. Select the type of rectifier to use

Table 10 summarizes the characteristics of several types of rectifiers. This table shows that a full-wave output is obtainable from several different types of rectifiers: diodes, gas-filled tubes, selenium, silicon, and bridge rectifiers. However, the vacuum-tube diode is restricted to an output of 300 mA or less. Therefore, one of the other types of rectifiers must be used.

For high-reliability service either a selenium or silicon diode is a good choice. Assume that a silicon-diode rectifier is used because it can be built for lower current ratings—in the range of 0.5 A. Also, the voltage drop of the rectifier is small.

4. Compute the power-supply ripple voltage

With a transformer output of 350 V ac and a silicon-diode rectifier, the dc output voltage will be about 325 V. Compute the dc ripple voltage from $E_r = E_{dc}E_r/100$, where E_r = dc ripple voltage; E_{dc} = rectifier output, dc V; $E_{r\%}$ = percent ripple. Thus $E_r = 325(0.05)/100 = 0.1625$ V.

TABLE 10   Rectifier Characteristics

Rectifier type	Output waveform	Current output	Voltage drop, V
Vacuum-tube diode (one tube)	Half-wave	300 mA or less	15–60
Vacuum-tube diode (two diodes or tubes)	Full wave	300 mA or less	15–50
Gas-filled tube	Half-wave or full wave	300 mA or more	15
Selenium (dry metal)	Half-wave or full wave	1 A or more	Low
Silicon diode	Half-wave or full wave	0.5 A or more	Low
Bridge (vacuum-tube or silicon diodes)	Full wave	0.1 A or more	Low

In general, the maximum allowable ripple in a power supply is less than 1 percent. This power supply meets that requirement.

Related Calculations: Standard, preassembled, off-the-shelf power supplies are available from a number of manufacturers. Use the method given here to check the suitability of such a power supply for a given load. Either semiconductors or vacuum tubes may be used as the diode in these power supplies. For large power outputs, the vacuum tube continues to be popular, but semiconductors are finding ever wider use for this service.

Note that Table 10 is also a useful guide for selecting rectifiers for any electronic-circuit application within the tabulated ranges.

RELIABILITY ANALYSIS OF ELECTRONIC CIRCUITS AND EQUIPMENT

Determine the reliability of the transistor control circuit shown in Fig. 8 when it is used in an aircraft electronic device having a mission time of 200 h. The circuit contains 3 paper Mylar capacitors, 1 silicon diode, 2 silicon amplifier transistors, 1 solenoid, 1 toggle switch, 6 carbon-deposited resistors, 1 potentiometer, and 22 printed-circuit solder joints.

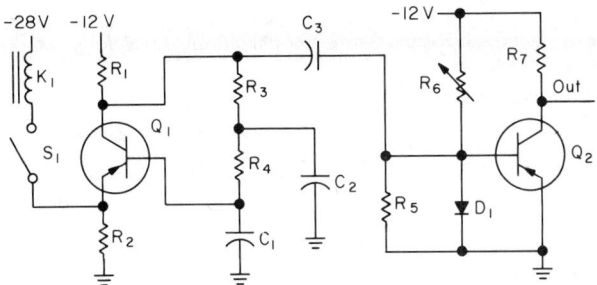

FIG. 8   Transistor control circuit. *(Electronics.)*

Calculation Procedure:

1.   List the circuit components that could cause failure

With respect to reliability, a failure occurs when a circuit no longer performs within the design limits. Any of the components in this circuit could cause a failure. Hence, all the components should be listed as shown in Table 11, along with the number of each type of component in the circuit.

TABLE 11   Reliability Calculation

Component	No. $\times$ F_r
Capacitors	$3 \times 0.01 = 0.03$
Diode	$1 \times 0.20 = 0.20$
Potentiometers	$1 \times 0.25 = 0.25$
Resistors	$6 \times 0.25 = 1.50$
Solenoid	$1 \times 0.05 = 0.05$
Switch	$1 \times$ (negligible)
Transistors	$2 \times 0.50 = 1.00$
Printed-circuit solder joints	$22 \times 0.008 = 0.18$
Total	$= 3.21$

$$F_t = \Sigma F_r K = (3.21)(150) = 481.5$$

2.   Determine the component failure rates

Table 12 lists typical failure rates per million hours for a variety of components. Use the mean failure rate listed in the center column for general-reliability analysis. Enter the respective failure rates in Table 11 opposite the name of the component.

3.   Compute the total, or overall, failure rate

Multiply the number of each component in the circuit by the failure rate for that component. Find the sum of these products, 3.21 (Table 11). This is the total, or overall, failure rate of the circuit per million hours of operation.

TABLE 12   Typical Electronic-Component Failure Rates

Component	Failures per 10^6 h		
	Upper	Mean	Lower
Capacitor, paper Mylar	0.014	0.010	0.006
Diode, silicon	0.250	0.200	0.150
Joint, solder, printed circuit	0.080	0.008	0.004
Potentiometer, carbon-deposited	0.750	0.250	0.100
Resistor, carbon-deposited	0.570	0.250	0.110
Solenoid	0.910	0.050	0.036
Switch toggle°	0.123	0.060	0.015
Transistor, silicon, amplifier	0.840	0.500	0.310

°Failures per 10^6 cycles.

4. Apply the environmental weighting factor

Table 13 lists typical environmental weighting factors for various applications. Inspection of this table shows that for aircraft application $K = 150$. To apply this factor, take the product of it and the total failure rate, or $3.21 \times 150 = 481.5$ per million hours (Table 11). This is called the *total*, or *final*, failure rate.

5. Compute the mean time between failures

Use the relation mean time between failures (MTBF) $= 10^6/F_t K$, where F_t = total or overall failure rate of the circuit per 10^6 h; K = environmental weighting factor. Thus, MTBF $= 10^6/481.5 = 2077$ h.

6. Compute the circuit reliability

Use the relation $R = e^{-t/m}$, where R = circuit reliability, i.e., the probability that a unit or a part will perform its intended function under design conditions for a specified period of time; t = mission time, h; m = MTBF. Substituting gives $R = e^{-200/2077} = 0.908$, or 90.8 percent.

Related Calculations: Use this general method for computing the reliability of single components or complete circuits. Note that Table 12 lists upper, lower, and mean failure rates. For usual circuits, the mean failure rate is a safe value to use. Where a larger number of failures are anticipated or must be provided for, use the upper failure rate. For fewer failures, use the lower failure rate.

For quick reliability analyses, an initial reliability estimate for electronic equipment can be made by (1) counting the number and type of active circuit elements in the electronic package and specifying the environment in which it will operate; and (2) entering Fig. 9 on the horizontal scale at the number of active elements and projecting vertically upward to the appropriate environment curve and reading the MTBF on the left-hand scale. Compute the reliability, using the equation in step 6.

TABLE 13   Typical Environmental Weighting Factors

Environment	Factor
Laboratory	1
Ground	10
Shipboard	20
Trailer	30
Rail	40
Bench	60
Aircraft	150
Missile	1000

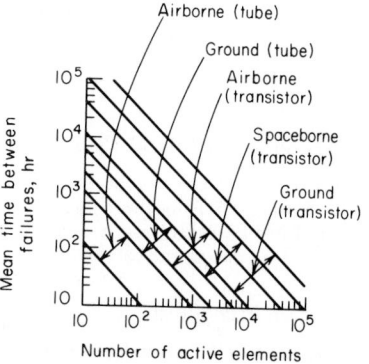

FIG. 9   Mean time between failure of circuit active elements. *(Electronics.)*

The curves in Fig. 9 apply to any heterogeneous electronic equipment using 10 or more active elements. Active elements include transistors, electron tubes, relays, capacitors, and diodes.

As a further guide, Fig. 10 shows how the total reliability R_T can be determined for series, parallel, and time-sequenced redundancy circuit arrangements. Also Table 14 lists values of R for various values of the exponent $-t/m$.

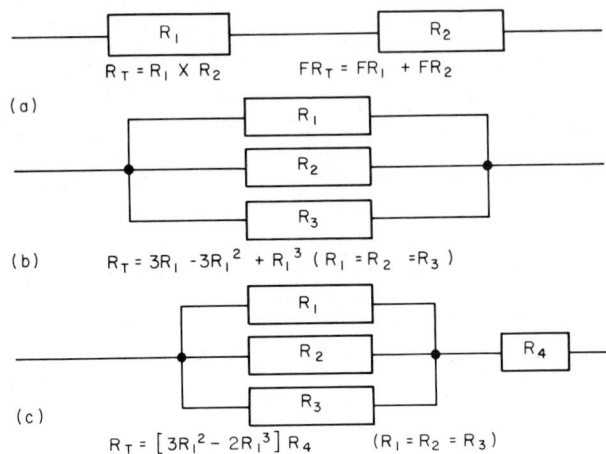

FIG. 10   Failure rates of components in series, parallel, and parallel-series. *(Electronics.)*

TABLE 14   Exponential Values for Reliability Calculations

x	e^{-x}	x	$\ln x$
0.01	0.99005	0.70	$-10 + 9.643$
0.02	0.98020	0.80	$-10 + 9.777$
0.05	0.95123	0.90	$-10 + 9.895$
0.07	0.93239	0.95	$-10 + 9.949$
0.09	0.91393	0.99	$-10 + 9.990$
0.10	0.90484		

MAINTAINABILITY ANALYSIS OF ELECTRONIC EQUIPMENT

An electronic equipment consists of five subsystems as shown in Table 15. Determine the mean time to repair (MTTR) of this equipment if the number of subsystems and failure rates are as shown in Table 15. What is the system MTTR goal if the equipment inherent availability is 99.54 percent and the mean time between failures (MTBF) is 100 h?

Calculation Procedure:

1. *Prepare a maintainability analysis tabulation*

Set up a table listing each subsystem in the equipment. Designate each subsystem by an identifying name, number, or letter (Table 15, column 1). Next, list the number N of subsystems of

TABLE 15 Computation of Mean Time to Repair[*]

(1) Subsystem	(2) Quantity Q	(3) Failure rate F_r	(4) Contribution to total failures C_f $C_f = QF_r$	(5) Percentage of contribution to total failures, R $R = C_f/C_{ft}$	(6) Average corrective- action time M_{ct}	(7) Contribution to total corrective-action time $C_m = C_f M_{ct}$
A	1	0.0030	0.0030	0.30	0.35	0.00105
B	1	0.0010	0.0010	0.10	0.75	0.00075
C	1	0.0035	0.0035	0.35	0.40	0.00140
D	2	0.0005	0.0010	0.10	0.50	0.00050
E	1	0.0015	0.0015	0.15	0.60	0.00090
			$C_{ft} = 0.0100$	1.00		0.00460

[*]Martin Marietta Corporation.

5.23

each type, column 2. Then list the failure rate F_r of each subsystem. Compute this failure rate as described in the previous calculation procedure, or use a value based on past experience with the same or similar subsystems. Note that the F_r value used here is the percentage of failures per 1000 h of operation of each subsystem.

2.   Compute the contribution to total failures

For any subsystem, the contribution that it makes to the overall failure rate of the electronic equipment is $C_f = NF_r$. To obtain the C_f for each subsystem, find the product $C_f = NF_r$ and enter the value in column 4, Table 15. Take the sum of the individual C_f values, and enter it in the last line of Table 15, column 4, as C_{ft}.

3.   Compute the subsystem failure-rate ratio

Set up, for each subsystem, the ratio $R = C_f/C_{fT}$, that is, the ratio of each subsystem's failure-rate contribution to the overall failure rate of the system. Enter the ratio for each system in column 5, Table 15. Find the sum of these ratios and enter it in the last line of Table 15.

4.   Apply the mean corrective-action time for each subsystem

The mean corrective-action time M_{ct} is the number of hours required to repair a defect in a subsystem. To estimate the mean corrective-action time for a subsystem when there is no previous experience to serve as a guide, use the subsystem failure rate as a source of information. Thus, subsystems contributing the highest percentage to the total failures will, ideally, have a low M_{ct}; subsystems with low contributions will have a higher M_{ct}. Using the failure rates as a guide, enter the M_{ct} value for each subsystem in column 6 of Table 15. (Typical values are shown to permit illustration of the method.)

   Compute the contribution of each subsystem to the total corrective-action time C_m by taking the product $C_m = C_f M_{ct}$. Enter the result in column 7, Table 15, for each subsystem. Last, find the sum of the individual C_m values and enter this sum in the last line of column 7.

5.   Compute the mean time to repair

Use the relation mean time to repair (MTTR) $= \Sigma C_f M_{ct}/\Sigma C_f = 0.0046/0.0100 = 0.46$ h.

6.   Compare the actual and desired MTTR values

Compute the desired MTTR from the relation MTTR $=$ MTBF$(1 - A_i)/A_i$, where $A_i =$ equipment inherent reliability expressed as a decimal; other symbols as before. Thus, MTTR $= 100(1 - 0.9954)/0.9954 = 0.462$ h.

   The actual MTTR computed in step 5 is 0.46 h. This compares favorably with the MTTR goal of 0.462 h. Hence, the M_{ct} times assumed in Table 15 are suitable.

   Related Calculations: Should the computed MTTR for the equipment exceed the equipment goal MTTR, three steps can be taken to alter the computed MTTR: (1) Decrease the subsystem failure rates; (2) decrease the mean corrective-maintenance times for the subsystems; (3) decrease either or both items 1 and 2 on a trade-off basis. Should it be decided to change the MTTR by decreasing the failure rate, be certain to select those subsystems where the failure rates can be reduced.

   As a first choice, the subsystem having the highest failure rate appears to be the logical choice for failure-rate reduction. However, such a selection could increase the MTTR rather than reduce it. Thus, reducing the failure rate of subsystem C to 0.0028, that is, by 20 percent in the above procedure, would increase the computed MTTR.

   Study of the typical reliability problem shows that the computed MTTR can be reduced by decreasing the failure rate in the subsystem having the largest M_{ct}. Another course of action is to reduce the M_{ct} of one or more of the subsystems by addition of other desirable maintainability features in the system design. Thus, reducing the M_{ct} of subsystem B by 20 percent in the above equipment will decrease the system computed MTTR.

   Where a large reduction must be made in the computed MTTR, a combination of reducing F_r and M_{ct}, working by trade-off, offers good possibilities of MTTR reduction.

   In a complete reliability analysis, apply the MTTR procedure described here to each subsystem. List the subsystem parts in the first column of the calculation table. Use the subsystem's estimated M_{ct} as the subsystem's MTTR goal adapted from the Martin Marietta Corporation publication *Maintainability Engineering*.

SELECTION OF ELECTRONIC-EQUIPMENT COOLING SYSTEMS

Choose the type of cooling system for an electronic package $4 \times 5 \times 8$ in ($10.5 \times 12.7 \times 20.3$ cm) that requires dissipation of 50 W during operation.

Calculation Procedure:

1. Compute the package external surface area

The package is a six-sided box. Thus, external surface area $= 2(L \times W) + 2(L \times H) + 2(H \times W) = 2(4 \times 5) + 2(4 \times 8) + 2(5 \times 8) = 184$ in^2 (1118.7 cm^2). In this relation, L, W, and H represent the length, width, and height, respectively, of the package.

2. Compute the package volume

Use the relation volume $= L \times H \times W = 4 \times 5 \times 8 = 160$ in^3 (2621.9 cm^3).

3. Compute the surface heat dissipation

The unit dissipation Q_U is an approximate measure of the quantity of heat w per square inch that can be effectively dissipated from the surface of the enclosure. Compute Q_U from $Q_U = w/A$, where w = heat that must be dissipated, W; A = external surface area of the package, in^2. For this package, $Q_U = 50/184 = 0.272$ W/in^2 (0.0422 W/cm^2).

4. Choose the type of cooling system to use

Enter Fig. 11*a* at $Q_U = 0.272$ W/in^2 (0.0422 W/cm^3), and project upward to the highest bar covering the required heat dissipation. Thus, Fig. 11*a* shows that natural cooling of this package

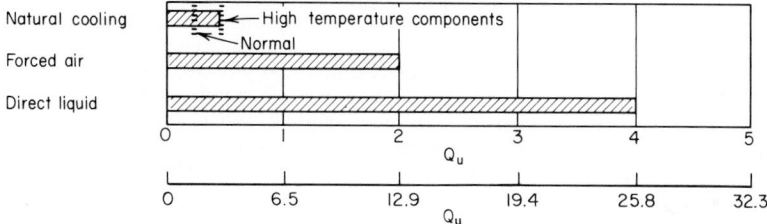

(a) Surface cooling of overall package [unit dissipation Q_U, W/in^2(W/cm^2)]

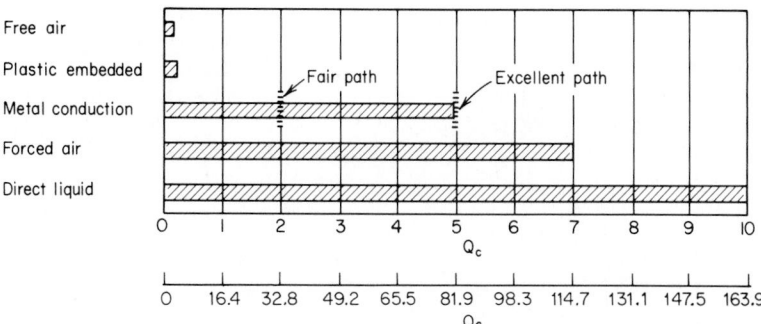

(b) Heat dissipation from crowded components [heat concentration Q_E, W/in^3(W/cm^3)]

FIG. 11 (*a*) Surface cooling of overall electronic package; (*b*) heat dissipation from crowded components. (*Product Engineering.*)

will be satisfactory if the components (vacuum tubes, transistors, resistors, etc.) can withstand fairly high temperatures. If this is not the case, then either forced-air or direct-liquid cooling must be used.

5. Compute the package heat concentration

The package heat concentration Q_c W/in³ is an approximate measure of how many electronic components can be safely packaged inside the enclosure. This measure is based on how effectively each of the components is mounted and what modes of heat transfer there are to conduct the heat out of the component and into the walls of the enclosure. In the method presented here, the temperature of the surroundings is assumed to be 40°C (104°F), and the average temperature rise of the components inside the enclosure is not more than 40°C (72°F). These are typical temperature ranges for a variety of electronic applications.

Compute the heat concentration in the electronic equipment from $Q_c = w/V$, where V = package volume, in³. Thus, $Q_c = 50/160 = 0.313$ W/in³ (0.0191 W/cm³).

6. Select the type of mounting for the electronic components

Refer to Fig. 11b to determine the heat dissipation from crowded electronic components. Study of Fig. 11b shows that plastic embedment of the components is unsuitable because the heat concentration Q_c exceeds the probable heat dissipation. Were plastic embedment used, the package would probably overheat.

Metal conduction (i.e., mounting of the components so that a direct metal conduction path is provided from the components through the mounting surfaces and into the wall of the enclosure) appears better suited for this package. Hence, this type of mounting should be specified.

Related Calculations: This procedure and the charts are based on a hypothetical electronic black-box enclosure (Fig. 12) about the size of a shoe box containing heat-producing miniature components such as resistors, electron (vacuum) tubes, and power transistors. If the overall package size differs markedly from this, make more rigorous estimates of the package heat flow.

In the charts, *natural cooling* refers to an enclosure in an ordinary room, with the usual convection, conduction, and radiation to the surrounding cooler surfaces. *Free air* is assumed for natural cooling and refers to air inside the enclosure, moved by natural draft.

Forced air has two meanings in this procedure. In *unit dissipation*, it means air blown against the outside of the enclosure. In *heat concentration*, it means air blown inside the enclosure against the components. One mode of cooling affects the other, and the distinction is sometimes difficult to make.

Direct-liquid cooling means that the enclosure is partly immersed in liquid, or that the components inside the enclosure are immersed in liquid.

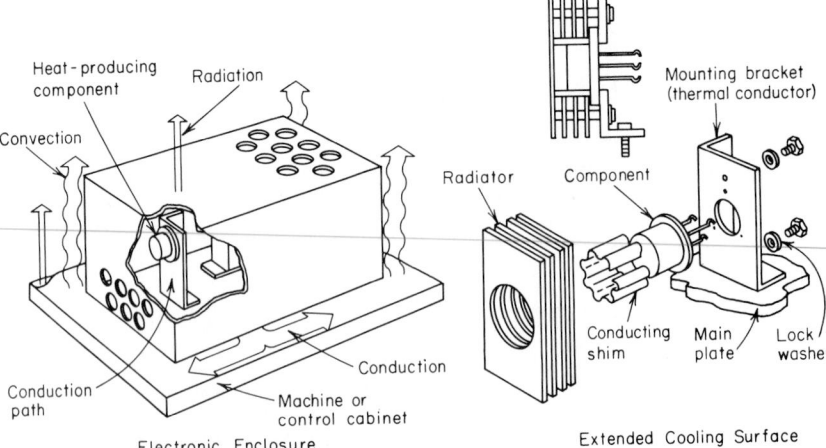

FIG. 12 Typical mounting and heat-dissipation paths in electronic equipment. *(Product Engineering.)*

Plastic-embedded means that each component is encased in thermally conductive plastic. Although this plastic is twice as conductive as free air, it is not particularly efficient as a heat dissipator.

Any improvements on these assumptions will increase the heat-dissipating ability of the overall package components plus the enclosure. For instance, if the package surface area is inadequate, it can be increased by corrugating the surface or by adding fins.

This calculation procedure was developed by B. Mastisoff, electromechanical engineer, and reported in *Product Engineering.*

DETERMINATION OF TUNED-CIRCUIT Q VALUES

What is the Q of a two-stage triode amplifier (Fig. 13) with a frequency response having 3-dB points at frequencies of 21.0 and 21.5 MHz? What Q value is necessary if one stage has a minimum Q of 35?

Calculation Procedure:

1. Determine the amplifier Q $\Delta f/f_0$ ratio

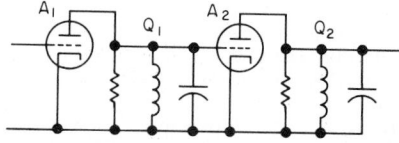

Use Fig. 14, entering at the 3-dB point for a single-tuned circuit on the vertical scale and projecting to the dashed expanded-scale curve. Project vertically downward from the intersection to read the Q $\Delta f/f_0$ ratio as 0.5 closely. In

FIG. 13 Tuned two-stage vacuum-tube amplifier. *(Electronics.)*

this ratio, Q = ratio of the energy stored to the energy dissipated in the circuit each half-cycle, or tuned-coil reactance, ohms/tuned-circuit coil resistance, Ω; Δf = amplifier half-bandwidth, MHz; f_0 = resonant frequency of the circuit.

2. Determine the amplifier resonant frequency

If the 3-dB points are chosen at 21.0 and 21.5 MHz, then the resonant frequency is $f_0 = (21.5 + 21.0)/2 = 21.25$ MHz. Also, $\Delta f = 21.25 - 21.0 = 0.25$ MHz.

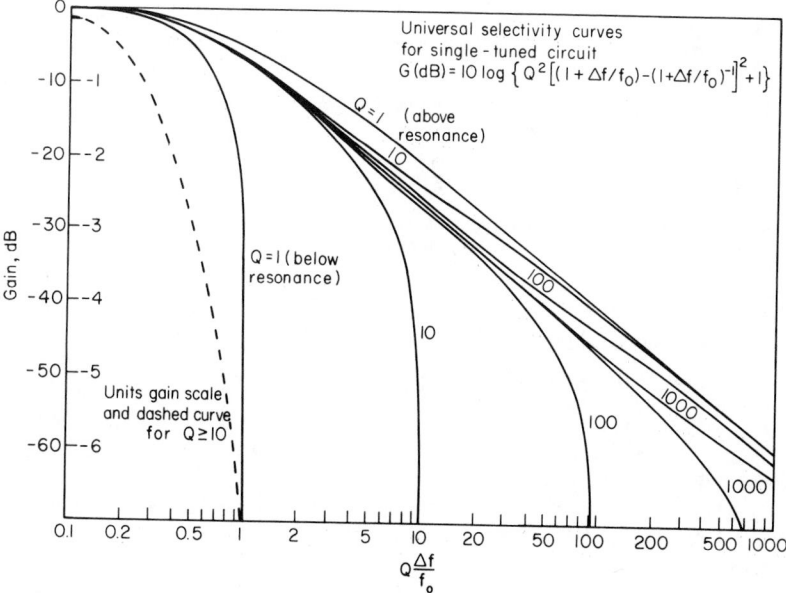

FIG. 14 Universal selectivity curves for single-tuned circuit. *(Electronics.)*

3. Compute the first-stage Q

If all the selectivity were provided by the first stage $Q \, \Delta f/f_0 = 0.5$, then $Q = 0.5 f_0/\Delta f = 0.5(21.25)/0.25 = 42.5$.

With all the selectivity provided by the first stage, the Q of the second stage would be zero because the second stage would not provide any selectivity. This is an impractical case, because the total amplifier gain is zero.

4. Compute the Q of the other stage to meet the bandpass requirements when the Q of one stage is restricted for some reason (for example, from maintaining a specific operating point under conditions of large stray capacitance)

With one stage restricted to a minimum Q or 35 with the same f_0 and Δf used in step 2, $Q \, \Delta f/f_0 = 35(0.25)/21.25 = 0.412$. Entering Fig. 14 with this value of the ratio and projecting vertically upward to the dashed curve show that this stage will contribute about 2.2 dB of attenuation to the total amplifier.

With 3-dB points, the other stage must then roll off $3.0 - 2.2 = 0.8$ dB at the band edges. At this attenuation, the second stage requires a $Q \, \Delta f/f_0 = 0.225$, found from Fig. 14 by entering at 0.8 dB. Then $Q = 0.225 f_0/\Delta f$ for the second state, or $0.225 \, (21.25)/0.25 = 19.1$.

Related Calculations: Figure 14 is equally useful for transistor amplifiers. Use a procedure similar to that given above for transistor amplifiers such as that shown in Fig. 15. Note that the ratio $\Delta f/f_0$ is unity for the second harmonic, 2 for the third harmonic, 3 for the fourth harmonic, etc. In a multistage transistor amplifier, the individual stage Q's must be equal for maximum out-of-band rejection. This procedure was developed by John D. Dunean, Montana State University, and published in *Electronics*.

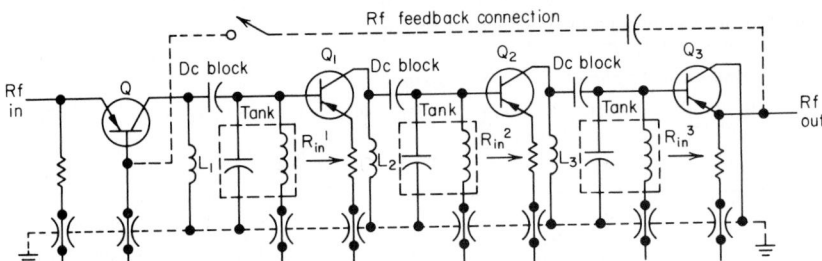

FIG. 15   Three-stage tuned transistor amplifier. Dashed line indicates the feedback connection necessary for a linear radio-frequency amplifier. *(Electronics.)*

ANTENNA SELECTION AND ANALYSIS

Choose a suitable antenna for a communications system operating at 100 MHz. What dimensions should the antenna have? Determine the system wavelength.

Calculation Procedure:

1. Select the type of antenna to use

Table 16 summarizes the names, types, uses, bandwidths, and wavelengths of a variety of antennas in common use. Study of this table shows that the half-wave dipole is a suitable antenna for communications in the 10-MHz to 5-GHz wavelength range. Hence, this type of antenna will be chosen.

2. Compute the antenna dimensions

Use the relation dipole length, ft $= 467.4/$frequency, MHz. Or, dipole length $= 467.4/100 = 4.67$ ft (1.4 m) for half-wave operation.

3. Compute the system wavelength

Use the relation $\lambda = 300 \times 10^6/f$, where $\lambda =$ system wavelength, m; $f =$ system frequency, Hz. Thus, $\lambda = 300 \times 10^6/100 \times 10^6 = 3$ m.

TABLE 16   Antenna Design Characteristics°

Antenna name	Antenna type	Typical uses	Bandwidth	Typical wavelengths
Dielectric rod	Surface-wave	Radar feeds and arrays	10%	1–6 GHz
Yagi	Surface-wave	TV/FM reception	10%	1–5 GHz
Half-wave dipole	Resonant	Communications, navigation, radar, etc.	5%–40%	10 MHz–5 GHz
Half-wave slot	Resonant	Aircraft and missiles	. . .	100 MHz–35 GHz
Rhombic	Traveling-wave	Short-wave transmitting and receiving for long ranges	2–1	2–30 MHz
Axial mode helix	Traveling-wave	Tracking, telemetry, aerospace, ground stations	1.7–1	100 MHz–3 GHz
Log-periodic	Frequency-independent	Ecm and direction finding	10–1	10 MHz–12 GHz
Equiangular spiral	Frequency-independent	Ecm, telemetry, aircraft, missiles, arrays,	10–1	100 MHz–35 GHz
Paraboloidal reflector	Aperture	Radar, communications, radio astronomy; other high-gain uses	Determined by feed	300 MHz–70 GHz
Conical horn	Aperture	Radar, communications	1.6–1	300 MHz–70 GHz
Pyramidal horn	Aperture	Radar, communications	1.6–1	300 MHz–70 GHz
Dielectric lens	Aperture	Radar, communications, radio, astronomy; other high-gain uses	Determined by feed	300 MHz–70 GHz

°For additional data see R. S. Gordon and K. W. Duncan, "Ready-Reference Data Simplifies Antenna Design," *Electronics*, Dec. 21, 1962.

Related Calculations: The length of a quarter-wave antenna is $\lambda/4.2$ ft (1.3 m). A grounded quarter-wave antenna is termed a basic *Marconi antenna*.

To compute the power input to an antenna, use the relation $P = I^2R$, where P = power input to the antenna, W; I = antenna current, A; R = antenna resistance, Ω. Note that a properly tuned antenna is considered to be a pure resistance. Hence, the power, voltage, current, and resistance of a tuned antenna can be computed by using Ohm's law and power equations.

MISMATCH EFFICIENCY AND POWER LOSS

An electronic signal source has an output impedance of 100 Ω. In attempting an impedance match with the load, the closest match that can be achieved is a load impedance of 200 Ω. What effect will this mismatch have on the power output and efficiency of the system?

Calculation Procedure:

1. *Compute the mismatch ratio*

The mismatch ratio for any electronic source-load combination is $R_m = Z_L/Z_s$, where R_m = mismatch ratio; R_L = load impedance, Ω; R_S = source impedance, Ω. For this system, R_m = $200/100 = 2$.

2. *Compute system power loss*

Use the relation $P_L = 10[0.602 + \log R_m - 2\log(1 + R_m)]$, where P_L = power loss, dB; log = logarithm to the base 10; other symbols as before. For this system, $P_L = 10(0.602 + \log 2 - 2\log 3) = -0.51$ dB. Hence the mismatch ratio of $R_m = 2$ causes a 0.51-dB power loss.

3. *Compute the system efficiency*

Use the relation $E = 100\,[R_m/(1 + R_m)]$, where E = system efficiency, percent; other symbols as before. Thus, $E = 100\,[2/(1 + 2)] = 66.6$ percent.

   Related Calculations: Use this procedure for any electronic equipment where an impedance match, or mismatch, is desired between a signal source and a load. If the match is between a source resistance and a load resistance, use the same method but substitute resistance for impedance in the above steps.

   Where the mismatch ratio is unity (i.e., the source and load impedances match exactly), the power output and efficiency are the maximum. As the mismatch increases on either side of unity (i.e., less than or greater than unity), the power loss increases and the efficiency decreases.

PUBLIC-ADDRESS-SYSTEM DESIGN AND LAYOUT

Choose the number, type, and location of the speakers as well as the amplifier rating for a factory sound system. The factory floor served by the system is 80 ft (24.4 m) long and 100 ft (30.5 m) wide. It is desired to use the public-address system for both voice and music amplification.

Calculation Procedure:

1. *Compute the number of speakers required*

The factory floor area is $L \times W = 80 \times 100 = 8000$ ft^2 (743.2 m^2). Using this area, enter Table 17 at the line marked Factories. This line shows that an 8000-ft^2 (743.2-m^2) factory requires four speakers.

2. *Select the type of speaker to use*

Table 17 shows that the reentrant-type horn speaker is recommended for factories having floor areas of 8000 or more ft^2 (743.2 m^2). With only four speakers, each can operate at high output in a high-level speaker system. With a large number of speakers, eight or more, they can be the low-output type. This is termed a *low-level* speaker system. For this factory having four speakers, a high-level speaker system probably would be best.

3. *Select the public-address-system amplifier rating*

Table 17 shows that a 50-W amplifier would be suitable for this factory. The output required for a public-address-system amplifier depends on the size and type of area served by the sound system. Where music is to be played over the sound system, a record player can be built into the amplifier housing. Where clarity of the amplified sound is critical, antifeedback controls are available on most standard amplifiers.

4. *Select the amplifier power source*

Standard sound amplifiers can be operated from 110- to 125-V 60-Hz or 115-V 25-Hz alternating current, or 115-, 4.6-, or 12-V direct current.

5. *Choose the amplifier input devices*

There are five major types of input devices: (*a*) Microphones—crystal, dynamic, or velocity, omnidirectional, bidirectional, or unidirectional (cardioid); (*b*) record player: automatic or manual; (*c*) tape player: single- or multiple-track; (*d*) radio tuner: AM or FM; (*e*) tone generator: produces a tone signal for factory work shifts, lunch periods, etc.; also used as an electronic siren for alarm applications; bell sounds for church belfry.

6. *Choose the output-tap impedance values*

The output-tap impedance value depends on the type of output devices selected in step 5 because there must be an impedance match. In general, three types of output taps are used: (*a*) Direction connection—4-, 8-, and 16-Ω taps; (*b*) constant-voltage line-transformer connection with 70-, 100-, or 140-V taps; (*c*) constant-impedance line-transformer connection with 250- and/or 500-Ω taps. Select one or more taps to give the desired impedance match with the output device or devices.

7. *Choose the type of speaker connection*

There are three main types of speaker connections: (*a*) Direct connection to the amplifier output taps corresponding in impedance value (ohms) to the impedance value (ohms) of a single speaker

TABLE 17 Selection Guide for Public-Address Systems°

Application	Area, ft² (m²)	Amplifier rating, W	Number of speakers	Types of speakers†
Auditoriums	2,000 (186)	15	2	12-in (30.5-cm) cone in wall baffles‡
	5,000 (465)	30	2	12-in (30.5-cm) cone in wall baffles or
	15,000 (1,394)	50	4	12-in (30.5-cm) projector horns
Ballrooms	2,000 (186)	15	4	
	4,000 (372)	30	4	12-in (30.5-cm) cone in wall baffles
	10,000 (929)	50	6	
Churches	1,000 (93)	10	2	10-in (25.4-cm) cone in wall baffles
	4,000 (372)	15	2	
	15,000 (1,394)	30	4	12-in (30.5-cm) cone in wall baffles
Classrooms,	500 (46)	10	1	8-in (20.3-cm) cone in wall baffles
offices, and	2,000 (186)	15	2	
stores	8,000 (743)	30	4	10-in (25.4-cm) cone in wall baffles
Factories	1,000 (93)	15	2	
	4,000 (372)	30	4	12-in (30.5-cm) projector horns
	8,000 (743)	50	4	
	40,000 (3,716)	100	10	Reentrant horns
Funeral	1,000 (93)	10	1	
parlors	4,000 (372)	15	4	12-in (30.5-cm) cone in wall baffles
	10,000 (929)	30	8	
Restaurants	1,000 (93)	15	2	
and	5,000 (465)	30	6	12-in (30.5-cm) projector horns
nightclubs	10,000 (929)	50	12	
Stadiums and	3,000 (279)	15	2	12-in (30.5-cm) cone in wall baffles
gymnasiums	10,000 (929)	30	4	
	50,000 (4,645)	100	8	Reentrant horns

° Values given in table are averages—not minimums or maximums.
† Number of speakers and amplifier power rating should be increased where background noise is higher than normal for the type of area. Acoustically "live" areas generally require lower speaker sound levels. Number of speakers will vary with shape of the plan view of the area.
‡ Although wall baffles are indicated for cone speakers, ceiling-recessed or suspended baffles are frequently advantageous.

or a number of speakers in series, parallel, or series-parallel; (*b*) connection to the amplifier constant-voltage output taps at 70, 100, 140 V, etc., through constant-voltage line-matching transformers; (*c*) connection to amplifier high-impedance output taps of 250- or 500-Ω impedance through constant-impedance line-matching transformers.

8.   *Select the speaker locations*

Locate the speakers on a plan of the area served. In churches, theaters, and auditoriums, place the speakers well forward of the microphones to prevent feedback (squealing).

   Related Calculations: Use this general method to choose sound systems for auditoriums, ballrooms, churches, classrooms, offices, stores, factories, funeral parlors, restaurants, nightclubs, stadiums, gymnasiums, etc. For impedance values of the equipment selected, consult the manufacturer's engineering data. The method presented here is one recommended by *Electrical Construction and Maintenance* magazine.

TRANSISTOR HYBRID-PARAMETER CONVERSIONS

Show how to convert the known parameter values for a common-base connected transistor to the unknown parameter values of a common-emitter transistor. How are the T-equivalent parameters related to the hybrid parameters? List the T-equivalent circuit equations for the three types of transistor amplifiers most commonly used. Compare the transistor amplifier parameters found by means of the T-equivalent circuit equations with the test-determined parameters shown in Fig. 16 for a transistor voltage amplifier.

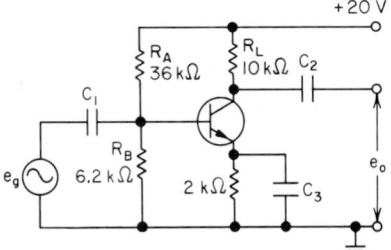

$\alpha = 0.937$
$\beta = 15$
$r_e = 33.5\ \Omega$
$r_b = 750\ \Omega$
$r_c = 1.67\ \mathrm{M}\Omega$

FIG. 16   Transistor voltage amplifier.

Calculation Procedure:

1. List the transistor hybrid-parameter conversions

Table 18 lists the usual symbols employed in transistor calculations and analysis. In actual transistor circuit design, a T-equivalent circuit is often used in circuit analysis in about the same way that the vacuum-tube equivalent circuit is used. The T-equivalent circuit is particularly appropriate for circuit analysis because its parameters are directly related to the basic physical structure of the transistor.

Although the T-equivalent circuit is a satisfactory representation of transistor operation, difficulty is met in determining the various resistance parameters by direct measurements on actual transistors. For this reason an equivalent circuit using *hybrid parameters* is often employed in circuit analysis.[1] Table 19 lists the usual subscript notation for hybrid, or *h*, parameters.

Transistor specification or data sheets may specify hybrid parameters for only one transistor configuration, for example, common-base or common-emitter. When this occurs, the designer may have to convert from one set of parameters to another. Table 20 lists a number of hybrid-parameter conversions, including the T-equivalent parameters for three configurations—common-base, common-emitter, and common-collector.

2. List the transistor amplifier T-equivalent equations

Table 20 also lists the T-equivalent circuit equations for common-base, common-emitter, and common-collector transistor amplifiers. The type of transistor chosen for an amplifier may depend on the types of transistors used in other parts of the equipment. The reason is that most designers try to restrict the number and types of transistors used in electronic equipment because this reduces the initial and maintenance costs of the equipment.

3. Compute the T-equivalent amplifier parameters

Use the equations given in Table 21 to determine the input resistance, voltage gain, and output resistance for the common-emitter amplifier shown in Fig. 16. Thus, $R_{in} \cong r_b + \beta r_e$, where R_{in} = input resistance, Ω; other symbols as listed in Table 18. Substituting gives $R_{in} \cong 750 + (15)(33.5) = 1250\ \Omega$.

The actual amplifier input resistance R'_{in} ohms equals the parallel combination of R_{in} ohms and the two base-biasing resistors of 6.2 and 36 kΩ (Fig. 16). Or, $R' = R_{in} = R_L/(1/R_{in} + 1/R_A + 1/R_B) = 10{,}000/(1/1.25 + 1/6.2 + 1/36) = 1.0\ \mathrm{k}\Omega$.

The voltage gain $A_V \cong -\beta R_L/(r_b + \beta r_e) = -15(10{,}000)/1250 = -120$, compared with a measured voltage gain of -112. Note in Table 21 that A_i = the amplifier current gain.

The output resistance R_o ohms is approximately equal to r_d in parallel with the 10-kΩ load resistance, where $r_d = (1 - \alpha)r_c$. Using the given value of $r_c = 1.67$ MΩ, we have $r_d = (1 - 0.937)1{,}670{,}000 = 105{,}000\ \Omega$. Then $R_o = r_d R_L/(r_d + R_L) = (105)(10\ \mathrm{K})/115 = 9.15\ \mathrm{k}\Omega$. The measured value of $R_o = 9.1$ kΩ.

Under ordinary design conditions one would not expect better than a 5 percent correlation between calculated and measured or experimental results. Resistor tolerances are usually ± 5 percent, depending on the resistor specifications. Further, transistor parameters are not known

---

[1]Brophy—*Basic Electronics for Scientists*, McGraw-Hill.

TABLE 18   Usual Transistor Parameter Symbols[*]

h_{FE}, B	Common-emitter dc current gain
h_{fe}, β	Common-emitter small-signal current gain with output ac short-circuited
h_{fb}, α	Common-base small-signal current gain with output ac short-circuited
h_{fc}	Common-collector small-signal current gain with output ac short-circuited
h_{ib}	Common-base small-signal input impedance with output ac short-circuited
h_{ie}	Common-emitter small-signal input impedance with output ac short-circuited
h_{ic}	Common-collector small-signal input impedance with output ac short-circuited
h_{ob}	Common-base small-signal output admittance with input ac open circuited
h_{oe}	Common-emitter small-signal output admittance with input ac open-circuited
h_{oc}	Common-collector small-signal output admittance with input ac open-circuited
h_{rb}	Common-base small-signal reverse voltage transfer ratio with input ac open-circuited
h_{re}	Common-emitter small-signal reverse voltage transfer ratio with input ac open-circuited
h_{rc}	Common-collector small-signal reverse voltage transfer ratio with input ac open-circuited
f_α, f_{hfb}	Common-base small-signal current gain cutoff frequency
f_β, f_{hfe}	Common-emitter small-signal current gain cutoff frequency
f_{max}	Maximum frequency of oscillation
f_t	Gain bandwidth product where h_{fe} equals 1
I_{CBO}, I_{co}	Dc collector current where emitter is open-circuited (leakage current)
I_{CEO}	Dc collector current where base is open-circuited
I_{CER}	Dc collector current where a resistor is connected from base to emitter
I_{CES}	Dc collector current where the base is shorted to the emitter
BV_{CBO}	Dc breakdown voltage collector to base with emitter open-circuited
BV_{CEO}	Dc breakdown voltage collector to emitter with base open-circuited
BV_{CER}	Dc breakdown voltage collector to emitter with a resistor connected from base to emitter
BV_{CES}	Dc breakdown voltage collector to emitter with base short-circuited to emitter
V_{BE}	Base-to-emitter voltage
$V_{CE(\mathrm{sat})}$	Collector-to-emitter saturation voltage
r_e	Small-signal emitter resistance
r_b	Small-signal base resistance
r_c	Small-signal collector resistance
r_m	αr_c
r_d	$(1 - \alpha)r_c$
r_{sat}	Collector-to-emitter saturation resistance
C_{ob}	Output capacitance
C_c	Collector-to-base capacitance
θ_{JC}	Thermal resistance junction to case (C/W)
t_s	Storage time
t_{on}	Turn-on time
t_{off}	Turn-off time
t_r	Rise time
t_f	Fall time

[*]Long—*Modern Electronic Circuit Design*, McGraw-Hill.

exactly. The results of this calculation show that the computed values for a particular amplifier may vary appreciably from measured or experimental results unless a method of gain stabilization is employed. Where more complete voltage-gain equations are used instead of the approximations in Table 21, a greater degree of accuracy is obtained. However, the approximate equations yield acceptable results where gain stabilization is employed.

TABLE 19 Subscript Notation for h Parameters

Subscript	Usual meaning
i	Input parameter
r	Reverse parameter
f	Forward parameter
o	Output parameter
e	Common emitter
b	Common base
c	Common collector

TABLE 20 Transistor Parameter Conversions°

	Common base	Common emitter	Common collector	Approximate equivalent
α	$\lvert h_{fb} \rvert$	$\dfrac{h_{fe}}{1 + h_{fe}}$	$\dfrac{\lvert h_{fc} \rvert - 1}{\lvert h_{fc} \rvert}$	$\dfrac{\beta}{\beta + 1}$
r_e	$h_{ib} - \dfrac{h_{rb}(1 - \lvert h_{fb} \rvert)}{h_{ob}}$	$\dfrac{h_{re}}{h_{oe}}$	$\dfrac{1 - h_{rc}}{h_{oc}}$	
r_b	$\dfrac{h_{rb}}{h_{ob}}$	$h_{ie} - \dfrac{h_{re}(1 + h_{fe})}{h_{oe}}$	$\dfrac{h_{ic} - \lvert h_{fc} \rvert (1 - h_{rc})}{h_{oc}}$	
r_c	$\dfrac{1 - h_{rb}}{h_{ob}}$	$\dfrac{1 + h_{fe}}{h_{oe}}$	$\dfrac{\lvert h_{fc} \rvert}{h_{oc}}$	
h_{fc}	$\dfrac{\lvert h_{fb} \rvert}{1 - \lvert h_{fb} \rvert}$	....	$\lvert h_{fc} \rvert - 1$	β
h_{ic}	$\dfrac{h_{ib}}{1 - \lvert h_{fb} \rvert}$	....	h_{ic}	$r_b + (\beta + 1)r_e$
h_{rc}	$\dfrac{h_{ib}h_{ob}}{1 - \lvert h_{fb} \rvert} - h_{rb}$	....	$1 - h_{rc}$	$\dfrac{r_e(\beta + 1)}{r_c}$
h_{oc}	$\dfrac{h_{ob}}{1 - \lvert h_{fb} \rvert}$	....	h_{oc}	$\dfrac{1}{r_d}$
$\lvert h_{fb} \rvert$		$\dfrac{h_{fe}}{1 + h_{fe}}$	$\dfrac{\lvert h_{fc} \rvert - 1}{\lvert h_{fc} \rvert}$	α
h_{ib}	....	$\dfrac{h_{ie}}{1 + h_{ie}}$	$\dfrac{h_{ic}}{\lvert h_{fc} \rvert}$	$r_e + \dfrac{r_b}{\beta + 1}$
h_{rb}	....	$\dfrac{h_{ie}h_{oe}}{1 + h_{fe}} - h_{re}$	$(h_{rc} - 1) + \dfrac{h_{ic}h_{oc}}{\lvert h_{fc} \rvert}$	$\dfrac{r_b}{r_c}$
h_{ob}	....	$\dfrac{h_{oe}}{1 + h_{fe}}$	$\dfrac{h_{oc}}{\lvert h_{fc} \rvert}$	$\dfrac{1}{r_c}$

°Long—*Modern Electronic Circuit Design*, McGraw-Hill.

TABLE 21   Transistor Amplifier Parameters°

	Common base	Common emitter	Common collector
A_v	$\dfrac{\beta R_L}{r_b + \beta r_e}$	$\dfrac{-\beta R_L}{r_b + \beta r_e}$	1
A_i	α	$-\beta$	$\beta + 1$
R_o	r_c	r_d	$\dfrac{r_b + R_g}{\beta}$
R_{in}	$r_e + \dfrac{r_b}{\beta}$	$r_b + \beta r_e$	$r_b + \beta R_L$

°Long—*Modern Electronic Circuit Design*, McGraw-Hill.

Related Calculations: Other transistor parameters used in circuit analysis are the hybrid-pi parameters. For a complete discussion of hybrid-pi parameters, see Hunter—*Handbook of Semiconductor Electronics*, McGraw-Hill.

The black-box amplifier (Fig. 17) is useful in deriving the h parameters. Thus, in Fig. 17, $v_1 = i_1 h_{11} + v_2 h_{12}$, and $i_2 = i_1 h_{21} + v_2 h_{22}$. These equations can be solved for the short-circuited case for input impedance, $h_{11} = v_1/i_1$, forward current gain, $h_{21} = i_2/i_1$. With the input ac open-circuited, that is, $i_1 = 0$, $h_{12} = v_1 v_2$, and $h_{22} = i_2/v_2$.

The general subscripts associated with h_{11}, h_{12}, h_{21}, and h_{22} are usually changed to indicate the transistor configuration. Then h_{ie} indicates the common-emitter input impedance, h_{fe} the common-emitter current gain, h_{fe} the common-emitter reverse-voltage transfer ratio, h_{oe} the common-emitter output admittance. In the case of the common-base configuration, the parameters are h_{ib}, h_{fb}, h_{rb}, and h_{ob}. Similarly, for the common-collector configuration, the parameters are h_{ic}, h_{fc}, h_{rc}, and h_{oc}. All these parameters are summarized in Table 20.

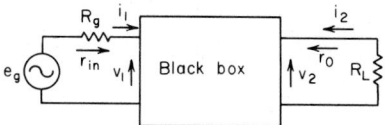

FIG. 17   Transistor "black-box" amplifier.

Transistor parameters are particularly useful in transistor circuit analysis. For a lucid discussion of transistor parameters, see Long—*Modern Electronic Circuit Design*, McGraw-Hill.

LARGE-SCALE INTEGRATION ANALYSIS

How many input-output pins are required for a large-scale integration package having 27 circuits if the average fan-in of the circuits in the package is 2.5? Compare the computed number of pins with actual design practice. Show how the number of pins can be reduced during circuit design.

Calculation Procedure:

1. *Compute the number of input-output pins required*

Use the relation $P = kN^{2/3}$, where P = number of input-output pins required; k = average fan-in of circuits plus 1 = 2.5 + 1 = 3.5 for this package; N = number of circuits in package. So $P = 3.5(27)^{2/3} = 31.5$ input-output pins.

2. *Compare the computed pins with design practice*

Table 22 compares the computed number of input-output pins as found in step 1 with the number of pins used in actual designs. Thus, with 27 circuits in partitions, actual designs use 12 to 24 pins. Figure 18 shows how actual designs for large and small systems deviate from the computed number of pins.

Related Calculations: If the number of chips (i.e., the first-level package count) can be reduced without increasing the total number of input-output pins, then the wiring on the second-

TABLE 22   Actual versus Computed Partitions[°]

Number of circuits in partitions	Computed number of pins	Number of pins in actual designs
1	3.5	
8	14.0	
27	31.0	12–24
64	55.0	24–41
125	87.0	39–52
216	125.0	
343	170.0	
512	220.0	45–96
729	285.0	45–92
1000	350.0	62–120

[°]*Electronics.*

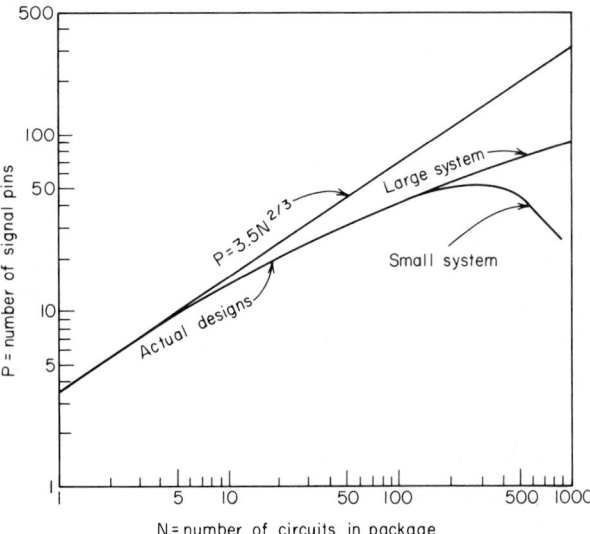

FIG. 18   Comparison of computed and actual design signal pins in large-scale integration. *(Electronics.)*

level package will be reduced. The trade-off here presents a choice between either increasing the unit package cost at the first level or cutting the cost of the second-level package.

To achieve a minimum first-level package count while at the same time reducing the second-level package wiring, a new set of rules has been devised to augment those in general practice, which are applied at low levels of integration to partition logic into small units with few interconnections. These four rules are as follows:

1. In general, to reduce the number of first-level packages and at the same time achieve high circuit-to-pin ratios, begin looking at larger collections of logic circuits as candidates for the partitions, or organize the logic into units with high circuit-to-pin ratios in mind.

2. Where possible, add supernumerary circuits to encode all nonindependent signals leaving the chip and to decode those signals at the destination chips.

3. Partition at the outputs of bit-storage cells in both data-flow sections and sequential-circuit control sections, because here the information is usually completely coded. Try to avoid cutting within the sections of signal-combining circuitry that lie between sets of storage cells; the inter-circuit connection density here is often quite high.

4. Choose as outputs from a chip those signals having large fan-out, and keep all the destination points together—on one chip if possible. The signals with large fan-out can be replicated on the destination chip as required.

FREQUENCY-CHANGER SELECTION AND APPLICATION

Choose a frequency changer suitable for a 60-Hz input and a 400-Hz 1-kVA output with a voltage regulation of 0.3 percent and an output voltage of 220 V. What harmonic distortion, frequency regulation, and response time can be expected in the changer selected? Input and output are both single-phase. What are the weight and volume of the changer?

Calculation Procedure:

1.   *Make a preliminary choice of the type of changer*

Figure 19 summarizes the frequency and power output ranges of three popular types of frequency changers.[1] Study of Fig. 19 shows that either a motor-generator set or a solid-state frequency changer would be suitable from the frequency and power standpoints. Hence, the final choice must be based on additional performance requirements.

2.   *Analyze the changer electrical performance*

Table 23 summarizes the electrical and mechanical aspects of the three popular types of frequency changers—motor-generator sets, vibrators, and solid-state changers. The voltage regulation of a motor-generator set varies between 0.5 and 3 percent, and the voltage regulation of a solid-state frequency changer is 0.5 percent. Since the desired regulation is 0.3 percent, the solid-state changer is more suitable because it has a smaller voltage-regulation range.

Further, Table 23 shows that an output voltage of 120 to 220 V is obtainable from solid-state frequency changers at power outputs up to 2 kVA. A harmonic distortion of 5 percent, frequency regulation of 0.01 to 1 percent, and a response time of 0.1 s can be expected from the solid-state frequency changer.

[1]Ante Lujic and Michael S. Inoue, "Frequency Changers," *Product Engineering.*

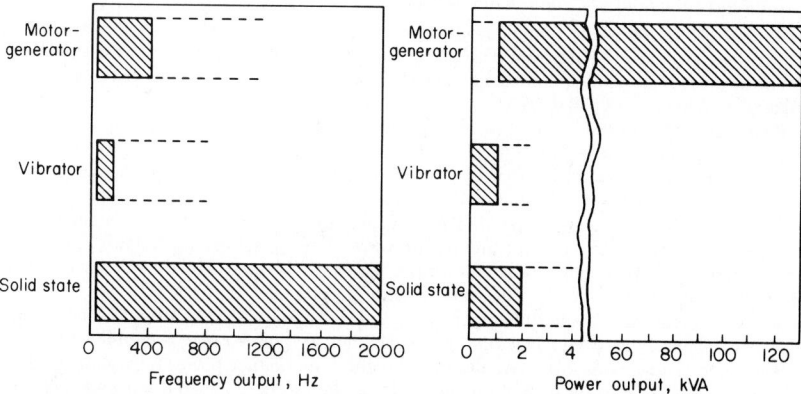

FIG. 19   Preliminary selection chart for frequency changers. (*Product Engineering.*)

TABLE 23   Frequency-Changer Characteristics

Type	Motor-generator set	Vibrator	Solid-state
Electrical:			
Input:			
Power, kVA	Up to 120	Up to 2	Up to 2
Voltage	120–550	10–200	12–220
Power factor	0.8		
Phase	One, two, three	One, two, three	One, two, three
Waveform	Sine	Square	Sine or square
Frequency, Hz	50/60	Dc or ac	Dc or ac
Output:			
Power, kVA	Up to 120	Up to 2	Up to 2
Voltage	120–550	10–200	120–220
Variable voltage	Possible	Possible	Possible
Voltage regulation, %	0.5–3	5	0.5
Efficiency, %	85	50 to 80	80
Current, A	Up to 450	Up to 10	Up to 10
Overload, %	125	. . .	Very low
Phase	One, two, three	One	One, two, three
Waveform	Sine	Square	Sine or square
Frequency, Hz	Up to 420	Up to 150	Up to 2000
Harmonic distortion, %	2–3	5	5
Variable frequency	Possible	yes	yes
Variable regulation, %	1	1–5	0.01-1
Response time, s	0.25	. . .	0.1
Mechanical:			
Weight, lb (kg)	Up to 3000 (1350)	Up to 100 (45)	0.2 lb/VA (0.09 kg/VA) output
Temperature, °C	−40–120	−40–60	−40–60
MIL specification construction	Available	. . .	Available

3.   *Compute the changer volume and weight*

The weight of this solid-state frequency changer will be about 0.2(1000 VA) = 200 lb (90.7 kg), by using the mechanical data from Table 23. Further, the volume of the usual solid-state frequency changer is about 30 in^3/lb (1083.8 cm^3/kg). Hence, the volume of this changer would be (30)(200) = 6000 in^3 (98,322.4 cm^3).

Related Calculations: Note that for higher-frequency outputs or higher-power outputs (above 500 Hz and 2 kVA, respectively), Fig. 19 shows that the choice of changer type is restricted. Thus, Fig. 19 indicates that a solid-state changer is needed for high frequencies and a motor-generator set is needed for high power outputs.

Vibrator frequency changers operate by rectifying alternating current to direct current, then chopping (switching) direct current into square-wave alternating current, followed possibly by wave-shaping for a sinusoidal output. The output power can be used directly at low frequencies; but at frequencies above 150 Hz, the output must be amplified.

Vibrator operation at frequencies much above 150 Hz is generally unreliable. Special choppers have been developed for frequencies up to 2000 Hz, but the power range is in milliwatts.

Solid-state changers use silicon diodes in the rectifier and either power transistors or silicon-controlled rectifiers (SCRs) as switching components in the inverter. The selection and application procedure given here was developed by Lujic and Inoue.

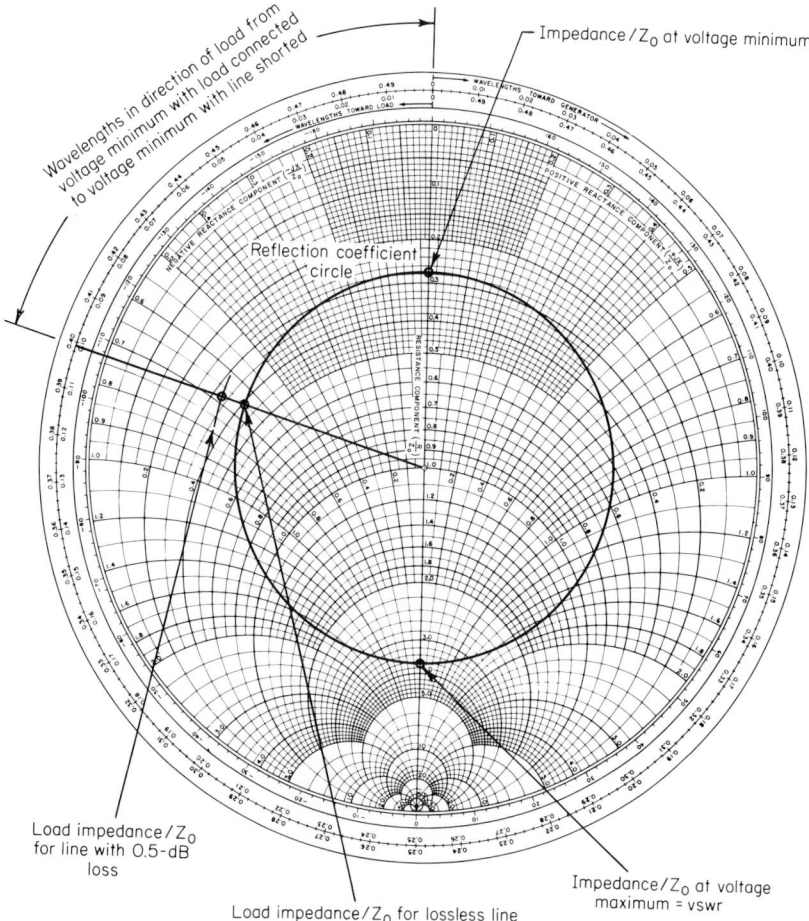

FIG. 20 Smith-chart solution of a transmission-line problem.

USES OF THE SMITH CHART IN ELECTRONICS

A lossless transmission line was tested and the voltage-standing-wave ratio measured as 3.7. With the load connected, a voltage minimum appeared at 62.7 mm; with the load shorted, the minimum moved to 79.6 mm. With the load still shorted, the next minimum toward the load was found at 159.6 mm. Use the Smith chart to analyze this transmission line. Next, consider the case where the cable between the slotted line and the load has an attentuation of 1 dB per 100 ft (30.5 m) and the cable length is 50 ft (15.2 m). Determine from the Smith chart the new conditions for this transmission line.

Calculation Procedure:

1. *Draw the reflection coefficient circle*

The Smith chart (Fig. 20) has a special impedance coordinate system that portrays the impedance at any point along a transmission line in relation to the impedance at any other point.[1] The scales

[1] Gray and Graham—*Radio Transmitters*, McGraw-Hill.

around the outside of the chart are calibrated in fractions of a wavelength, and the full circle (360°) corresponds to one-half wavelength. Resistances and reactances are plotted as fractions of the impedance of the transmission line. The resistance lines are circles that are tangent to the bottom of the chart, and the reactance lines are portions of circles that are tangent to the vertical line through the chart, which is called the *axis of reals*.

Draw the reflection-coefficient circle with its center at $R = 1$, $X = 0$ (Fig. 20), with a radius of 3.7, that is, the circumference of the circle cuts the axis of reals at 3.7.

2. Compute the full wavelength of the signal generator

The half-wavelength of this line is, from the measured values, $159.6 - 79.6 = 80.0$ mm. Hence, the full wavelength $= 2(80.0) = 160$ mm.

3. Compute the wavelength fraction

Use the relation wavelength fraction from the "load" minimum to the " short" minimum = (load-shorted minimum, mm − voltage minimum, mm)/wavelength, mm = $(79.6 - 62.7)/160.0 = 0.1057$.

4. Determine the load impedance

Draw a line from the center of the reflection-coefficient circle through a value of wavelengths *toward* the load of 0.1057 on the chart circumference (Fig. 20). At the intersection of this line with the circumference of the reflection-coefficient circle, read the load impedance as $0.41 - j0.65\ \Omega$.

If this slotted transmission line had an impedance of 50 Ω, then the actual load impedance would be $50(0.41 - j0.65) = 20.5 - j32.5\ \Omega$.

5. Compute the transmission-line attenuation

Use the relation total attenuation, dB = transmission-line length, ft/length, ft, for 1-dB attenuation = 50/100 or 0.5 dB.

6. Determine the new load impedance

The voltage-standing-wave ratio for the lossless line was 3.7. With a transmission-line attenuation of 0.5 dB toward the load, the voltage-standing-wave ratio will increase to about 4.5, as determined from a standard engineering source, such as *Reference Data for Radio Engineers*, ITT.

Plot the new reflection-coefficient circle, Fig. 20. The intersection of the previously drawn line with the new circle shows that the impedance is $0.33 - j0.68\ \Omega$.

Related Calculations: The Smith chart has numerous applications, including general impedance transformations, stub-tuner design, finding of the input impedance, etc. For a full discussion of the chart and its uses, see Smith—*Electronic Applications of the Smith Chart*, McGraw-Hill.

ULTRASONIC-GENERATOR SELECTION, CRYSTAL THICKNESS, AND OUTPUT

What type of ultrasonic generator should be used for an industrial-cleaning operation requiring a frequency of 40,000 Hz? Determine the power output required if the cleaning is done in a 30-gal (113.6-L) bath. What crystal thickness is required for the generator if the crystal is an *x*-cut or a *y*-cut? Sketch a typical vacuum-tube ultrasonic-generator circuit. What wave amplitude should be used for this cleaning operation?

Calculation Procedure:

1. Select the type of ultrasonic generator to use

Magnetostriction is used for frequencies of the order of 30,000 Hz; crystals are used for most frequencies above 30,000 Hz up to about 15 MHz. Higher frequencies can be obtained by vibrating the crystal at one of its harmonics. Motor-generator sets are used for relatively low frequency applications. Since this cleaning operation requires a frequency of 40,000 Hz, a crystal generator will be used.

Figure 21 shows the circuit of a typical vacuum-tube crystal generator. Ultrasonic generators are essentially high-power oscillators such as are commonly used in radio[1] communications.

---

[1]Carlin—*Ultrasonics*, McGraw-Hill.

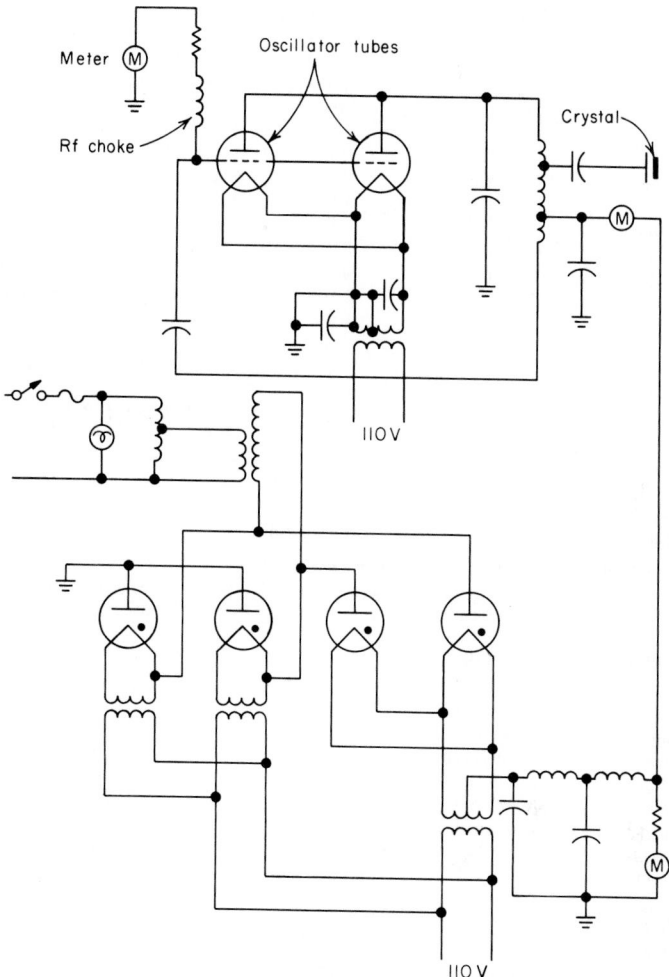

FIG. 21   Vacuum-tube ultrasonic generator.

2. Determine the wave amplitude required

Table 24 shows that a high-amplitude wave is required for cleaning. The energy may be continuous, pulsed, or modulated, in various ways.

3. Compute the cleaning-bath power level required

The usual power level used for ultrasonic cleaning is 50 W/gal (13.2 W/L) of cleaning solution. With a 30-gal (113.6-L) cleaning solution, the power level required is (50 W/gal)(30 gal) = 1500 W.

4. Compute the required crystal thickness

For an x-cut crystal use the relation $t = 0.1126/f$, where t = crystal thickness, in (cm); f = ultrasonic frequency, MHz. Substituting gives $t = 0.1126/0.04 = 2.815$ in (7.2 cm). For a y-cut crystal use the relation $t = 0.0771/f = 0.0771/0.04 = 1.927$ in (4.9 cm). Thus, a y-cut crystal would be thinner than an x-cut crystal. Table 24 lists the required crystal thickness for various common frequencies; Table 25 lists the usual frequencies for various ultrasonic applications.

TABLE 24 Ultrasonic Wave Amplitude and Crystal Thickness

Operation	Wave amplitude	Frequency, MHz	Thickness, in (mm)	
			x-cut	y-cut
Cleaning	High	0.1	1.126 (28.6)	0.771 (19.6)
Welding	High	0.5	0.224 (5.7)	0.154 (3.9)
Drilling	High	1.0	0.113 (2.9)	0.0788 (2.0)
Emulsification	High	2.0	0.0576 (1.5)	0.0394 (1.0)
Soldering	High	3.0	0.0377 (0.96)	0.026 (0.66)
Medical therapy	High	4.0	0.0282 (0.72)	0.0197 (0.50)
Sonar	High	5.0	0.0226 (0.57)	0.016 (0.41)
Chemical	High	6.0	0.0188 (0.48)	0.013 (0.33)
Biological	High	7.0	0.0160 (0.41)	0.0112 (0.28)
Materials testing	Low	8.0	0.0140 (0.36)	0.00964 (0.24)
Burglar alarms	Low	9.0	0.0124 (0.31)	0.00856 (0.22)
Delay lines	Low	10.0	0.0113 (0.29)	0.00771 (0.20)
Medical diagnoses	Low	Varies	Varies	Varies

TABLE 25 Frequencies for Ultrasonic Applications[*]

Applications	Usual frequency, kHz
Sonic altimeter	1.0
Drilling, soldering, cleaning	16–20
Agitation of liquids, aerosol reactions	16–20
Burglar alarms	19.2
Control apparatus and door opening	25
Cleaning (most common types)	40
Blind-guidance devices (upper limit for magnetostriction)	60
Galton whistle (upper limit)	100
Gas whistles in air (upper limit)	120
Resonance testing; emulsion formation	300
Emulsion; agitation	400
Pulsed material testing (lower limit)	500
High-polymer reactions	600
Experimental biological work	750
Material tests; medical therapy; mixing; cleaning	1000
Testing of fine, homogeneous material	5000–25,000

[*]After Carlin—*Ultrasonics*, McGraw-Hill.

Related Calculations: Use this general method to choose the frequency, crystal thickness, and power output for any of a large number of other ultrasonic applications—drilling, soldering, blind-guidance devices, resonance testing, emulsion production, liquid agitation, high-polymer reactions; biological work, medical therapy, etc. Tables 24 and 25 are useful guides to the amplitude, crystal thickness, and frequencies used in various ultrasonic applications.

APPLICATIONS OF SONAR EQUATIONS

What is the detection range of an active sonar gear having a source level of 100 dB and a receiving directivity index of 20 dB if it uses processing that produces a detection threshold of +2 dB and is used against a target having a strength of 12 dB when the noise level is −8 dB? The gear produces a spherical spreading signal, and there is no absorption of the signal. Give the sonar equations for active (monostatic) and passive sonar gears. What is the typical target strength for a bow-stern aspect of a fleet submarine?

Calculation Procedure:

1. Compute the sonar-gear transmission loss

Use the relation TL = 0.5(SL + TS − NL + DI − DT), where TL = transmission loss, dB; SL = source level, dB; TS = target strength, dB; NL = noise level, dB; DI = receiving directivity index, dB; DT = detection threshold, dB. Substituting gives TL = 0.5(100 + 12 − (−8) + 20 − 2) = 69 dB.

2. Compute the detection range

When the gear emits a signal that spreads spherically, the relation between the range r and the transmission loss is TL = 20 log r. Substituting gives 69 = 20 log r; r = 2820 yd (2578.6 m).

Figure 22a shows the range of a sonar gear obtained under ideal conditions (i.e., a constant water temperature at all depths in which the gear is used). However, the temperature of seawater is seldom constant, and refraction of the signal usually occurs. The range might then be reduced to the area shown in Fig. 22b, regardless of the power source. Refraction of the signal may reduce the range at which a submarine can be detected to almost zero.

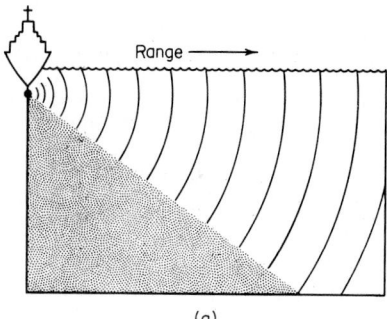

(a)

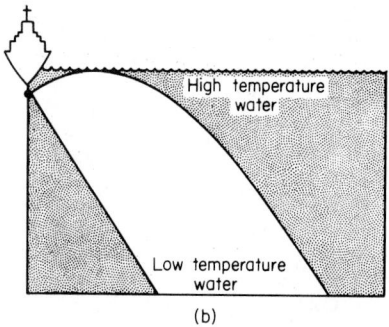

(b)

FIG. 22   (a) Sonar range under ideal conditions; (b) sonar range under average conditions.

3. List the sonar equations

The basic equations used in system design and analysis are as follows for *active* sonars: noise background, SL − 2TL + TS = NL + DI + DT; reverberation background, SL − 2TL + TS = RL + DT; passive sonars, SL − TL = NL − DI + DT.

4. Determine the vessel target strength

Table 26 lists nominal values of target strength for various vessels, torpedoes, and marine life. Study of Table 26 shows that a bow-stern aspect of a submarine will produce a target strength of +10 dB.

TABLE 26   Nominal Values of Target Strength°

Target	Aspect	Target strength, dB
Submarines	Beam	+25
	Bow-stern	+10
	Intermediate	+15
Surface ships	Beam	+25 (highly uncertain)
	Off-beam	+15 (highly uncertain)
Mines	Beam	+10
	Off-beam	+10 to −25
Torpedoes	Bow	−20
Fish of length L	Dorsal view	$-31 + \log L$

°Urick—*Principles of Underwater Sound*, McGraw-Hill.

Related Calculations: Sonar calculations are less certain than many other types of engineering computations because the influencing external factors are so unpredictable. For a complete analysis of sonar calculations, see Urick—*Principles of Underwater Sound for Engineers,* McGraw-Hill.

VOLTAGE AND CURRENT GAIN FOR TRANSISTORIZED AMPLIFIER

Determine the voltage gain $A_v = e_o/e_s$ and the current gain $A_i = i_c/i_s$ for the transistorized amplifier shown in Fig. 23a if the amplifier has the component values shown.

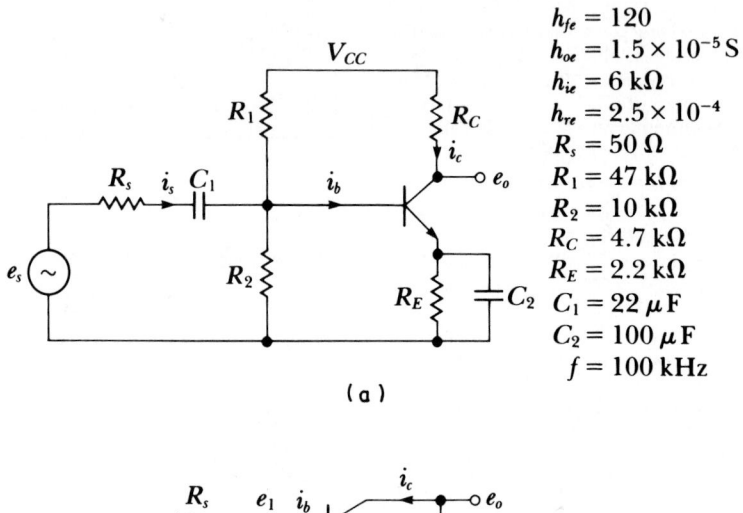

$$h_{fe} = 120$$
$$h_{oe} = 1.5 \times 10^{-5}\,\text{S}$$
$$h_{ie} = 6\,\text{k}\Omega$$
$$h_{re} = 2.5 \times 10^{-4}$$
$$R_s = 50\,\Omega$$
$$R_1 = 47\,\text{k}\Omega$$
$$R_2 = 10\,\text{k}\Omega$$
$$R_C = 4.7\,\text{k}\Omega$$
$$R_E = 2.2\,\text{k}\Omega$$
$$C_1 = 22\,\mu\text{F}$$
$$C_2 = 100\,\mu\text{F}$$
$$f = 100\,\text{kHz}$$

(a)

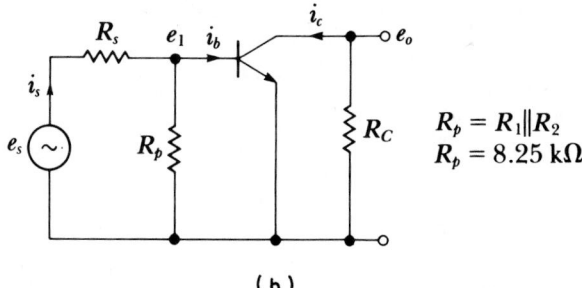

$$R_p = R_1 \| R_2$$
$$R_p = 8.25\,\text{k}\Omega$$

(b)

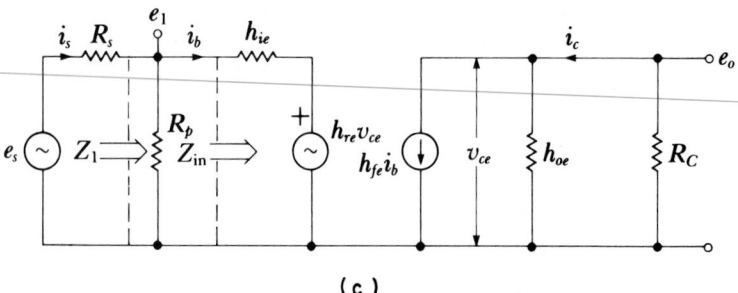

(c)

FIG. 23   (*a*) Transistorized amplifier circuit. (*b*) Ac equivalent circuit for (*a*). (*c*) Linear ac equivalent circuit for (*a*).

Calculation Procedure:

1. Draw the ac equivalent circuit

Check to see that C_1 and C_2 are short circuits at 100 kHz. Next, draw the ac equivalent circuit by short-circuiting C_1 and C_2 and making the power supply ac ground. Then

$$X_{C_1} = \frac{1}{2\pi f C_1} = \frac{1}{2\pi(10^5)(22)(10^{-6})} = 0.0723 \ \Omega$$

$$X_{C_2} = 0.0159 \ \Omega$$

The ac equivalent circuit is shown in Fig. 23b, and the above calculations show that C_1 and C_2 are negligible at 100 kHz.

2. Draw the linear equivalent ac circuit

The linear ac equivalent circuit is shown in Fig. 23c. This is a common-emitter (CE) amplifier.

3. Designate the base and determine the voltage and current gains

Designate e_1 as the base, and solve for the gain expressions as follows, using the symbols given in Table 27:

$$A_v = (A_{v1})(A_{v2})$$

where
$$A_{v2} = \frac{e_o}{e_1} \quad \text{and} \quad A_{v1} = \frac{e_1}{e_s}$$

and
$$A_i = (A_{i1})(A_{i2})$$

where
$$A_{i1} = \frac{i_b}{i_s} \quad \text{and} \quad A_{i2} = \frac{i_c}{i_b}$$

$$\Delta h_e = h_{ie}h_{oe} - h_{fe}h_{re}$$

$$\Delta h_e = (6 \times 10^3)(1.5 \times 10^{-5}) - (120)(2.5 \times 10^{-4}) = 0.06$$

$$A_{v2} = \frac{-h_{fe}R_C}{h_{ie} + R_C \, \Delta h_e} = \frac{(-120)(4.7 \times 10^3)}{6 \times 10^3 + (4.7 \times 10^3)(0.06)} = -89.78$$

$$A_{i2} = \frac{h_{fe}}{1 + R_C h_{oe}} = \frac{120}{1 + (4.7 \times 10^3)(1.5 \times 10^{-5})} = 112.097$$

$$Z_{in} = \frac{h_{ie} + R_C \, \Delta h_e}{1 + R_C h_{oe}} = \frac{6 \times 10^3 + (4.7 \times 10^3)(0.06)}{1 + (4.7 \times 10^3)(1.5 \times 10^{-5})} = 5.868 \ \text{k}\Omega$$

$$Z_1 = R_1 \| R_2 \| Z_{in} = 3.428 \ \text{k}\Omega$$

$$A_{v1} = \frac{Z_1}{R_S + Z_1} = \frac{3.428 \ \text{k}\Omega}{50 + 3.428 \ \text{k}\Omega} = 0.9856$$

Since $i_b/i_s = A_{i1}$ and $i_b Z_{in}/[i_s(R_s + Z_1)] = A_{v1}$,

$$A_{i1} = A_{v1}\left(\frac{R_S + Z_1}{Z_{in}}\right) = \frac{0.9856(3.478 \times 10^3)}{5.868 \times 10^3} = 0.5842$$

$$A_v = (A_{v1})(A_{v2}) = (0.9856)(-89.78) = -88.487$$

$$A_i = (A_{i1})(A_{i2}) = (0.5842)(112.097) = 65.487$$

This procedure is the work of Charles R. Hafer, P.E.

TABLE 27 Symbols and Abbreviations for Calculation Procedures

$\downarrow$	Symbol for ground
$\parallel$	In parallel with
α	Dc current gain, I_C/I_E
β	h_{FE}; dc current gain, I_C/I_B
Δ	Incremental change
Δh	$h_i h_o - h_r h_f$
η	Efficiency
ϕ_m	Phase margin
ω	Angular frequency, $2\pi f$
Ω	Ohms
A_V	Voltage gain
A_I	Current gain
CD, CG, CS	JFET amplifier configurations: common drain, common gate, and common source
CR	Diode designator
C_π, C_μ	Capacitor designations used in hybrid-π model
CB, CC, CE	Transistor configurations: common base, common collector, common emitter
f_H	Upper 3-dB frequency
f_L	Lower 3-dB frequency
f_M	Midband frequency
$G(s)$	Gain expressed in Laplace notation
$G(j\omega)$	Gain expressed as a function of ω
$G(s)H(s)$	Loop gain expressed in Laplace notation
$GH(j\omega)$	Loop gain expressed as a function of ω
g_{fs}, g_{fso}, g_o	Small-signal JFET parameters
g_m	Hybrid-π small-signal bipolar transistor parameter ($I_E/0.026$)
$h_{ib}, h_{ob}, h_{fb}, h_{rb}$	Small-signal common-base transistor parameters
$h_{ic}, h_{oc}, h_{fc}, h_{rc}$	Small-signal common-collector transistor parameters
$h_{ie}, h_{oe}, h_{fe}, h_{re}$	Small-signal common-emitter transistor parameters
h_{fe}	Dc transistor current gain; h_{FE}; I_C/I_B
I_B	Bipolar dc base current
I_C	Bipolar dc collector current
I_D	JFET dc drain current
I_E	Bipolar dc emitter current
I_{DSS}	JFET dc saturation drain current
I_{GSS}	JFET gate leakage current
I_E	Bipolar dc emitter current
I_G	JFET dc gate current
I_S	JFET dc source current
JFET	Junction field-effect transistor
K	Boltzmann's constant (1.3×10^{-23} J/K)
KT/q	Intrinsic transistor parameter; equals 0.026 V at 25°C
r_d	Small-signal diode ac resistance
r_{DS}	JFET on resistance in triode region
r_{DSO}	Minimum value of r_{DS} occurring at zero values for V_{GS} and V_{DS}
r_o, r_x, r_π	Bipolar hybrid-π representation for resistance parameters
S	Siemens (formerly mhos)
S_I	Current stability factor
S_V	Voltage stability factor
V_B	Bipolar dc base voltage
V_{BE}	Bipolar dc base-emitter voltage
V_C	Bipolar dc collector voltage
$V_{CC}, V_{DD}, V_{EE}, \ldots$	Dc power-supply voltages
V_D	JFET dc drain voltage
V_{DS}	JFET dc drain-to-source voltage
V_E	Bipolar dc emitter voltage
V_G	JFET dc gate voltage

TABLE 27   Symbols and Abbreviations for Calculation Procedures (*Continued*)

V_{GS}	JFET dc gate-to-source voltage
V_{GS} (off)	JFET gate-to-source voltage required to reduce I_D to zero; sometimes referred to as pinchoff voltage
V_R	Reference diode designation
V_P	JFET pinchoff voltage
V_S	JFET dc source voltage
V_{TH}	Thevenin's equivalent voltage
X_C	Capacitance reactance, $1/(2\pi fC)$
X_L	Inductance reactance, $2\pi fL$
Z_{IN}	Input impedance
Z_o	Output impedance
Z_T	Transfer impedance, e_{in}/i_o
Z_{TH}	Thevenin's equivalent impedance

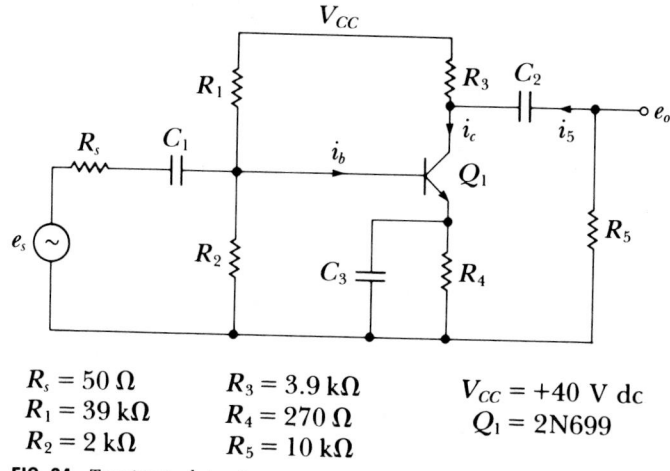

$$R_s = 50\ \Omega \qquad R_3 = 3.9\ \text{k}\Omega \qquad V_{CC} = +40\ \text{V dc}$$
$$R_1 = 39\ \text{k}\Omega \qquad R_4 = 270\ \Omega \qquad Q_1 = 2\text{N}699$$
$$R_2 = 2\ \text{k}\Omega \qquad R_5 = 10\ \text{k}\Omega$$

FIG. 24   Transistorized circuit.

DETERMINATION OF LOAD LINE AND OPERATING POINT OF A TRANSISTORIZED CIRCUIT

The circuit in Fig. 24 has the characteristic curves shown in Fig. 25. Assuming that all capacitors are short circuits at the frequency of interest and that $V_{BE} = 0.7$ V and $H_{FE} = \infty$, find (*a*) V_B, V_E, and V_C, which are the dc values of the base, emitter, and collector voltages, respectively; (*b*) the dc load line and the operating point; (*c*) the ac load line and R_{ac}, the ac equivalent load; (*d*) the approximate value of h_{fe} at this operating point; (*e*) the approximate value of h_{oe} at this operating point; and (*f*) the peak value of i_5 if i_b has a peak value of 0.01 mA.

Calculation Procedure:

1. Find V_B first for this circuit

If H_{FE} is assumed to be ∞, then

$$V_B = \frac{R_2(V_{CC})}{R_1 + R_2} = \frac{2\ \text{k}\Omega(40)}{39\ \text{k}\Omega + 2\ \text{k}\Omega} = 1.95\ \text{V dc}$$

$$V_E = V_B - V_{BE} = 1.95 - 0.7 = 1.25\ \text{V dc}$$

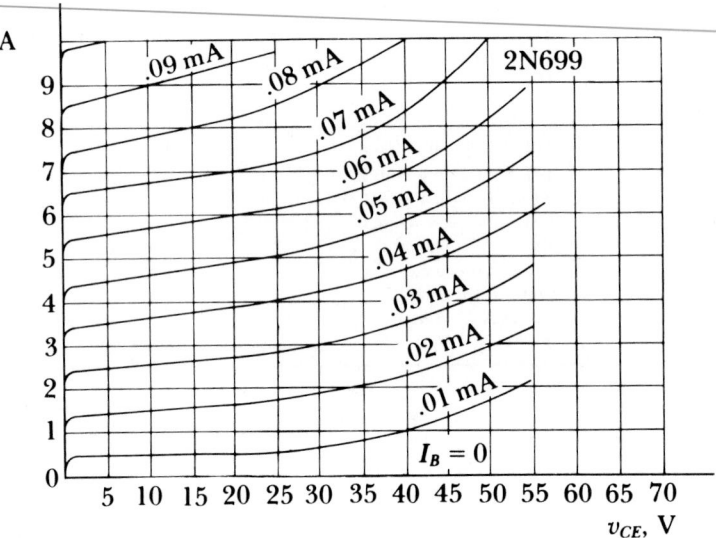

FIG. 25   Characteristic curves for transistor.

and

$$I_E = \frac{V_E}{R_4} = \frac{1.25}{270} = 4.63 \text{ mA}$$

Also, since

$$H_{FE} = \infty \qquad I_C = I_E \qquad I_C = \frac{H_{FE}I_E}{H_{FE} + 1}$$

and

$$V_C = V_{CC} - I_C R_3 = V_{CC} - I_E R_3$$

$$V_C = 40 - (4.63 \times 10^{-3})(3.9 \text{ k}\Omega) = 21.94 \text{ V dc}$$

Note: This would be an acceptable design because the collector voltage is approximately at the midpoint between V_{CC} and V_E, which allows for maximum collector signal swing.

2. *Determine the dc load line*

Determine the dc load line as follows: The minimum collector current flows when $V_C = V_{CC}$; there $I_C = 0$. This locates one point on the abscissa. The maximum current flows when Q_1 is saturated. Assume $V_{CE(sat)} = 0.2$ V dc. This locates I_C on the ordinate, and its maximum value can be determined from

$$I_C \approx \frac{V_{CC} - V_{CE}}{R_4 + R_3} = \frac{39.8}{4.17 \times 10^3} = 9.54 \text{ mA}$$

Now, locate on this load line the operating point of

$$V_{CE} = V_C - V_E = 21.94 - 1.25 \approx 20.7 \text{ V dc}$$

and $I_C = 4.63$ mA from before. The dc load line is drawn on Fig. 26.

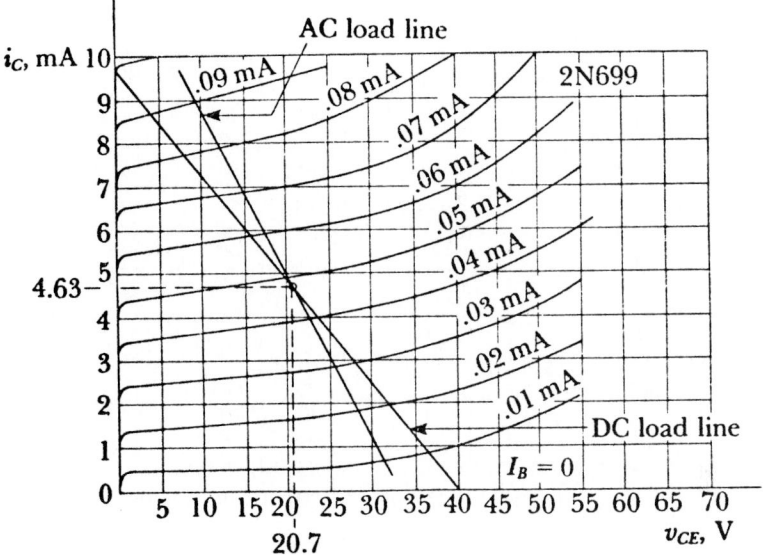

FIG. 26 Determining the ac and dc load lines.

3. *Determine the ac load line*

The ac load line is determined thus: Since the emitter resistor is shunted with zero ac impedance, the ac resistance is

$$R_{ac} = R_3 \| R_5 = 3.9 \text{ k}\Omega \| 10 \text{ k}\Omega = 2.806 \text{ k}\Omega$$

and the slope is

$$m = \frac{1}{R_{ac}} = \frac{1}{2.806 \text{ k}\Omega} = 0.3564 \text{ mA/V}$$

or

$$m = \frac{3.564 \text{ mA}}{10 \text{ V}}$$

and is drawn through the dc operating point as shown in Fig. 26.

4. *Find the small-signal current gain h_{fe}*

Refer to Fig. 27. The small-signal current gain can be determined by taking small increments at the operating point:

$$h_{fe} = \frac{\Delta i_c}{\Delta i_b}\bigg|_{v_{CE}=\text{const}} = \frac{6 \text{ mA} - 3.8 \text{ mA}}{0.06 \text{ mA} - 0.04 \text{ mA}} = 110$$

5. *Find the small-signal output conductance h_{oe}*

The small-signal output conductance h_{oe} is also found by taking small increments about the operating point. Refer to Fig. 28, and look at the slope where $i_b = 0.05$, which is the closest slope to the operating point. Then

$$h_{oe} = \frac{\Delta i_c}{\Delta v_{CE}}\bigg|_{i_b=\text{const}} = \frac{5.15 \text{ mA} - 4.8 \text{ mA}}{27 \text{ V} - 14.5 \text{ V}} = 2.8 \times 10^{-5}$$

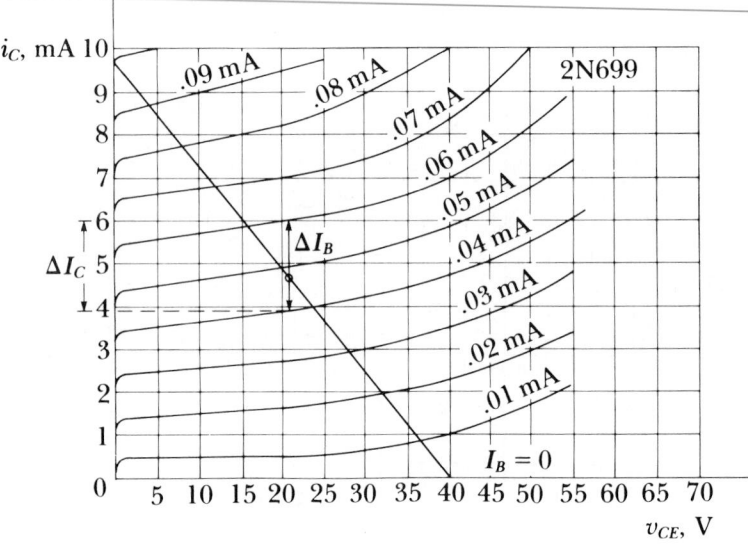

FIG. 27   Determining h_{fe}.

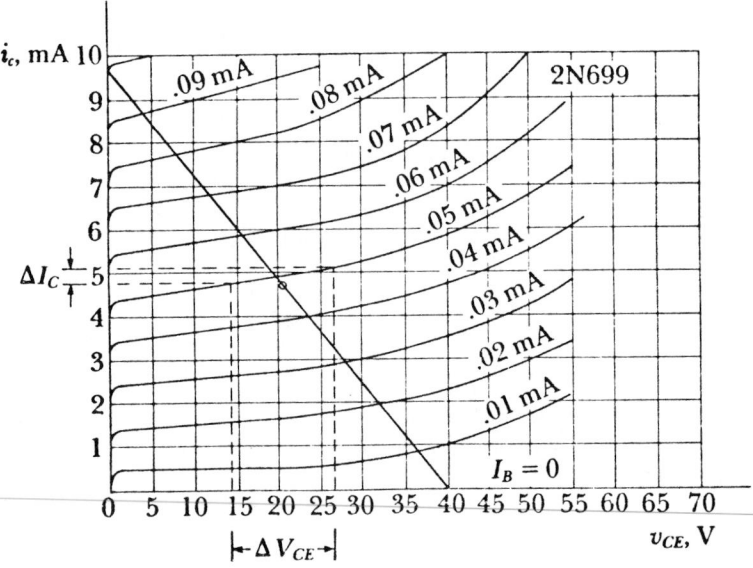

FIG. 28   Determining h_{oe}.

6. Find the peak value of i_5

If the peak value of $i_b = 0.01$ mA, the peak value of i_c can be determined from h_{fe} as follows:

$$i_c = h_{fe}i_b$$
$$i_c = (110)(0.01) = 1.1 \text{ mA}$$

but we want i_5:

$$i_5 = \frac{i_c\,(R_3)}{R_3 + R_5}$$

$$i_5 = \frac{1.1 \text{ mA } (3.9 \text{ k}\Omega)}{3.9 \text{ k}\Omega + 10 \text{ k}\Omega} = 0.309 \text{ mA peak}$$

This procedure is the work of Charles R. Hafer, P.E.

BUFFER AMPLIFIER ANALYSIS

The buffer amplifier shown in Fig. 29 has a minimum input signal of -5 V dc, and the maximum is 0 V dc. If the JFETs are perfectly matched with the parameters shown, what is the maximum output error contributed by the JFETs?

Calculation Procedure:

1. *Evaluate the output voltage and drain current*

Since the JFETs are perfectly matched, $V_{GS1} = V_{GS2}$ as long as the drain currents are the same, and the output voltage will equal the input voltage. When the drain currents become mismatched, an offset error is introduced because $V_{GS1} \neq V_{GS2}$. When $e_o = 0$ V dc, $I_{D1} = I_{D2}$, and $V_{GS1} = V_{GS2} = 0$. The worst-case error is introduced when e_o is maximum negative. Then the current through R_L is

$$I_{RL} = \frac{5}{100 \times 10^3} = 0.05 \text{ mA}$$

$$I_{D2} = I_{DSS2} = 1 \text{ mA}$$

$$I_{D1} = I_{D2} - I_{RL} = 1 - 0.05 = 0.95 \text{ mA}$$

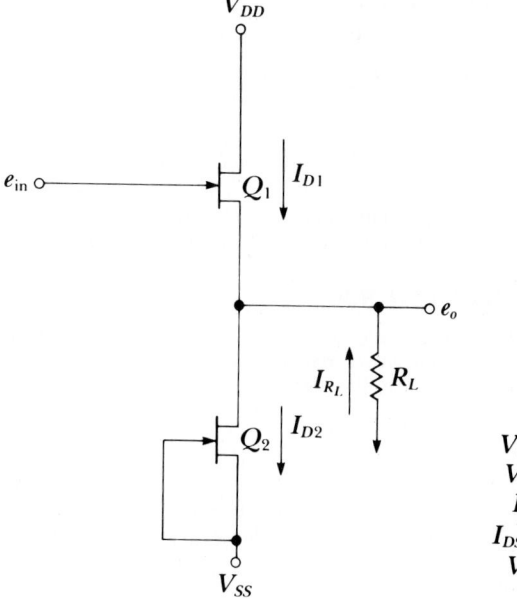

$$V_{CC} = +15 \text{ V dc}$$
$$V_{SS} = -15 \text{ V dc}$$
$$R_L = 100 \text{ k}\Omega$$
$$I_{DSS1} = I_{DSS2} = 1 \text{ mA}$$
$$V_{p1} = V_{p2} = -3 \text{ V dc}$$

FIG. 29 Buffer amplifier.

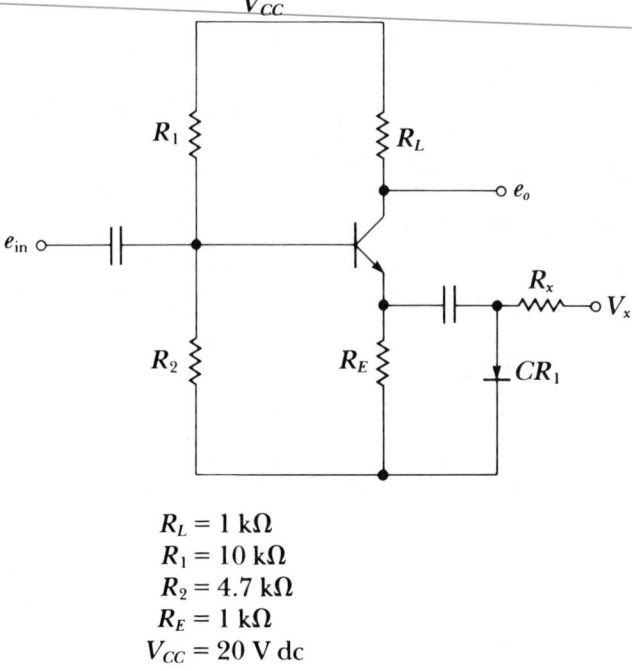

$$R_L = 1 \text{ k}\Omega$$
$$R_1 = 10 \text{ k}\Omega$$
$$R_2 = 4.7 \text{ k}\Omega$$
$$R_E = 1 \text{ k}\Omega$$
$$V_{CC} = 20 \text{ V dc}$$

FIG. 30   Gain-controlled amplifier.

2.   Find the change in V_{GS} for a change in I_D

This change can be found by finding the g_{fs} at the operating point of $I_D = I_{DSS}$. Or

$$g_{fso1} = \frac{\Delta I_{D1}}{\Delta V_{GS1}} = \frac{2 I_{DSS1}}{V_{p1}} = \frac{2 \text{ mA}}{3 \text{ V}} = 667 \ \mu S °$$

$$\Delta V_{GS1} = \frac{\Delta I_{D1}}{g_{fso1}} = \frac{-50 \times 10^{-6}}{667 \times 10^{-6}} = -0.075 \text{ V dc}$$

This is the maximum output error. The absolute output voltage is then $e_o = e_{in} - V_{GS1} = -5 - (0.075) = -4.925$ V dc. This procedure is the work of Charles R. Hafer, P.E.

TRANSISTOR AMPLIFIER GAIN CONTROL

The gain of the transistor amplifier in Fig. 30 is to be varied by controlling the dynamic resistance of the forward-biased diode CR_1. If a gain change from $A_v = 1$ to 10 is wanted, what value must R'_x be for a control voltage range of $V_x = 0$ to 10 V dc?

Calculation Procedure:

1.   Find the emitter current I_E

This is a common-emitter gain-controlled amplifier. Assume the reactance of the capacitor to be $0 \ \Omega$. Then the approximate gain of a common-emitter amplifier is $A_v \approx -R_C/(h_{ib} + R_E)$. In this

---

° S is the abbreviation for siemens, the unit of conductance. It is equivalent to mho.

approximate relation, h_{ib} is a function of the emitter dc current I_E. Using the approximate analysis to determine I_E, we get

$$V_B \approx \frac{(R_2)(V_{CC})}{R_1 + R_2} = 6.39 \text{ V dc}$$

$$V_E \approx V_B - V_{BE} = 6.39 - 0.7 = 5.69 \text{ V dc}$$

$$I_E = \frac{5.69 \text{ V}}{1 \text{ k}\Omega} = 5.69 \text{ mA}$$

$$h_{ib} = \frac{0.026}{5.69 \text{ mA}} = 4.57 \text{ }\Omega$$

2. Find the current flowing through the diode

Let $R'_E = R_E \| r_d$, where r_d is the dynamic impedance of the diode and $r_d = 0.026/I_D$ and I_D is the direct current flowing through the diode. For a gain of 1,

$$1 = \frac{R_C}{h_{ib} + R'_{E(\text{max})}}$$

$$R'_{E(\text{max})} = 1 \text{ k}\Omega - 4.57 = 995.4 \text{ }\Omega$$

$$r_{d(\text{max})} \| R_E = 995.4 \text{ }\Omega$$

Solving, we find

$$r_{d(\text{max})} = 216.4 \text{ k}\Omega$$

$$I_{D(\text{min})} = \frac{0.026}{r_{d(\text{max})}} = \frac{0.026}{216.4} \text{ k}\Omega = 0.12 \text{ }\mu\text{A}$$

3. Investigate the leakage currents because of the small value of I_D

For a gain of 10,

$$10 = \frac{R_C}{h_{ib} + R'_{E(\text{min})}}$$

$$R'_{E(\text{min})} = 95.43 \text{ }\Omega$$

$$r_d \| R_E = 95.43 \text{ }\Omega$$

$$r_{d(\text{min})} = 105.5 \text{ }\Omega$$

$$I_{D(\text{max})} = \frac{0.026}{r_{d(\text{min})}} = \frac{0.026}{105.5} = 0.246 \text{ mA}$$

4. Determine R_x; assume $V_{CR1} = 0.7$ V dc

Solving, we find

$$R_x = \frac{V_x - V_{CR1}}{0.246 \text{ mA}} = 37.8 \text{ k}\Omega$$

Use the nearest standard value of 37.4 kΩ. This procedure is the work of Charles R. Hafer, P.E.

SYNTHESIS OF AN AC EQUIVALENT CIRCUIT FOR AN OPERATIONAL AMPLIFIER

It is desired to synthesize an ac equivalent circuit for an operational amplifier by using a program similar to ECAP. This program has only current generators for its dependent sources. The circuit

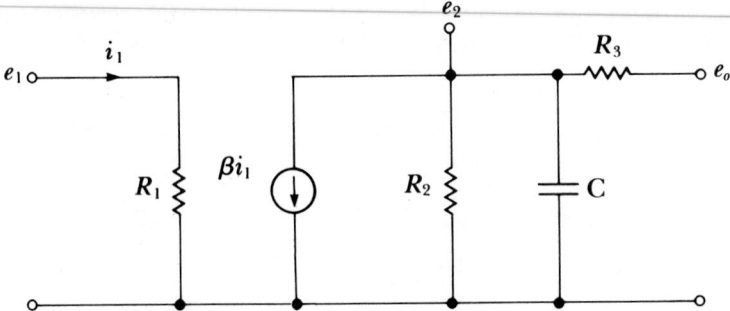

FIG. 31   Operational amplifier ac equivalent circuit.

of Fig. 31 can be used to synthesize the operational amplifier. Determine all circuit values if the amplifier has the characteristics shown:

$$R_{in} = 10^6 \ \Omega \qquad \text{(amplifier input impedance)}$$

$$A_v = 92 \ dB \qquad \text{(amplifier voltage gain)}$$

$$f_c = 100 \ Hz \qquad \text{(3-dB pole)}$$

$$r_o = 100 \ \Omega \qquad \text{(output impedance)}$$

Calculation Procedure:

1. Find R_3 for the amplifier

Since R_1 is equal to the input impedance, $R_1 = 10^6 \ \Omega$. Next, select R_3 as the output impedance and select $R_2 \ll R_3$ so that it will cause $(\beta i_1)(R_2)$ to look like a voltage source. Then $R_3 = r_o = 100 \ \Omega$. Pick $R_2 = 0.001 \ \Omega$.

2. Find β for this amplifier

Since no current is flowing in the output, $e_o = e_2$. Therefore,

$$A_v = \frac{e_o}{e_1} = \frac{e_2}{e_1}$$

$$20 \log A_v = 92 \ dB$$

$$A_v = 10^{92/20}$$

$$= 39{,}811$$

$$e_2 = (\beta i_1)(R_2)$$

$$e_1 = i_1 R_1$$

$$\frac{e_2}{e_1} = \frac{\beta i_1 R_2}{i_1 R_1} = \frac{\beta R_2}{R_1} = A_v$$

$$\beta = \frac{A_v R_1}{R_2}$$

$$= \frac{(39{,}811)(10^6)}{10^{-3}} = 3.9811 \times 10^{13}$$

3. *Determine the breakpoint at 100 Hz*

The breakpoint at 100 Hz is determined by C and R_2:

$$e_2 = (\beta i_1)(Z_2)$$

where

$$Z_2 = \frac{R_2[1/(Cs)]}{R_2 + 1/(Cs)} = \frac{1/C}{s + 1/(R_2C)}$$

Therefore, at $s = j\omega$

$$\omega = \frac{1}{R_2C}$$

and

$$C = \frac{1}{\omega R_2} = \frac{1}{2\pi(100)(10^{-3})} = 1.59 \text{ F}$$

This procedure is the work of Charles R. Hafer, P.E.

DC OUTPUT OF OPERATIONAL-AMPLIFIER CHAIN

Determine the output V_o dc of the amplifier chain shown in Fig. 32. All amplifiers are powered by $+15$ and -15 V, and the useful output swing is ± 10 V dc. What happens if R_6 is changed to 40 kΩ?

Calculation Procedure:

1. *Establish the voltage designations for the chain*

The important point to remember in dealing with dc operation of operational amplifiers is that the forcing voltage is that voltage present at the positive terminal of the amplifier. Whatever voltage is present on the positive terminal is present on the negative terminal. Establish the voltage designations as shown in Fig. 33.

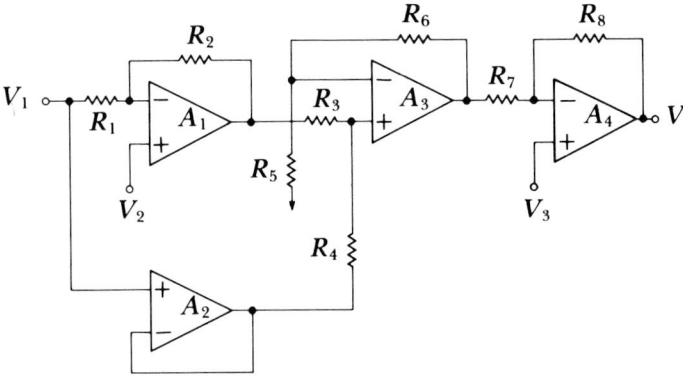

$V_1 = +5$ V dc	$R_1 = 10$ kΩ	$R_4 = 30$ kΩ	$R_7 = 5$ kΩ
$V_2 = +4$ V dc	$R_2 = 20$ kΩ	$R_5 = 10$ kΩ	$R_8 = 10$ kΩ
$V_3 = +5$ V dc	$R_3 = 10$ kΩ	$R_6 = 22$ kΩ	

FIG. 32 Operational amplifier chain.

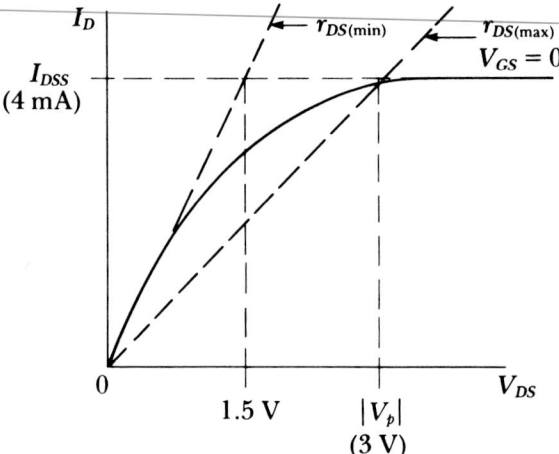

FIG. 35 Determining $r_{DS(max)}$ for JFET in Fig. 34.

Thus,

$$r_{DS} = 667 \ \Omega$$

2. Find out whether the needed resistance can be obtained

Determine the limits of the JFET resistance for the values of V_p and I_{DSS} given. The maximum value of r_{DS} can be found as shown in Fig. 35:

$$r_{DS(max)} = \frac{|V_p|}{I_{DSS}} = \frac{3 \ V}{4 \ mA} = 750 \ \Omega$$

The minimum value of $r_{DS(min)}$ is found from

$$r_{DS(min)} = \frac{0.5|V_p|}{I_{DSS}} = \frac{0.5(3)}{4 \ mA} = 375 \ \Omega$$

3. Find the exact value of the JFET resistance r_{DS}

The value of r_{DS} will lie between 375 and 750 Ω. When $A_v = 0.4$, there is 0.8 V ($= 2 \times 0.4$) across the JFET, and $V_{DS} = 0.8$ V dc. Also, $V_{GS} = 0$ V dc. Solve for I_D thus:

$$I_D = \frac{2V_{DS}I_{DSS}}{V_p} \left(\frac{V_{GS}}{V_p} - \frac{V_{DS}}{2V_p} - 1 \right)$$

$$= \frac{2(0.8)(4 \ mA)}{-3} \left(0 - \frac{0.8}{-6} - 1 \right)$$

$$= 1.85 \ mA$$

and r_{DS} at this value of I_D and V_{DS} is

$$r_{DS} \leq \frac{0.8 \ V}{1.85 \ mA} = 432 \ \Omega$$

This meets the requirements because the resistance can be as low as 432 Ω, as shown in Fig. 35. But the JFET would not be turned on so hard, since only 667 Ω is needed. If a range of V_p and I_{DSS} had been given, the maximum value of V_p and the minimum value of I_{DSS} would be used. This procedure is the work of Charles R. Hafer, P.E.

3. *Determine the breakpoint at 100 Hz*

The breakpoint at 100 Hz is determined by C and R_2:

$$e_2 = (\beta i_1)(Z_2)$$

where

$$Z_2 = \frac{R_2[1/(Cs)]}{R_2 + 1/(Cs)} = \frac{1/C}{s + 1/(R_2 C)}$$

Therefore, at $s = j\omega$

$$\omega = \frac{1}{R_2 C}$$

and

$$C = \frac{1}{\omega R_2} = \frac{1}{2\pi(100)(10^{-3})} = 1.59 \text{ F}$$

This procedure is the work of Charles R. Hafer, P.E.

DC OUTPUT OF OPERATIONAL-AMPLIFIER CHAIN

Determine the output V_o dc of the amplifier chain shown in Fig. 32. All amplifiers are powered by $+15$ and -15 V, and the useful output swing is ± 10 V dc. What happens if R_6 is changed to 40 kΩ?

Calculation Procedure:

1. *Establish the voltage designations for the chain*

The important point to remember in dealing with dc operation of operational amplifiers is that the forcing voltage is that voltage present at the positive terminal of the amplifier. Whatever voltage is present on the positive terminal is present on the negative terminal. Establish the voltage designations as shown in Fig. 33.

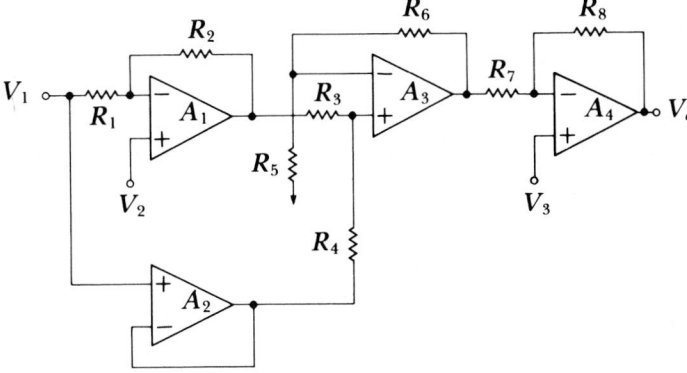

$V_1 = +5$ V dc	$R_1 = 10$ kΩ	$R_4 = 30$ kΩ	$R_7 = 5$ kΩ
$V_2 = +4$ V dc	$R_2 = 20$ kΩ	$R_5 = 10$ kΩ	$R_8 = 10$ kΩ
$V_3 = +5$ V dc	$R_3 = 10$ kΩ	$R_6 = 22$ kΩ	

FIG. 32 Operational amplifier chain.

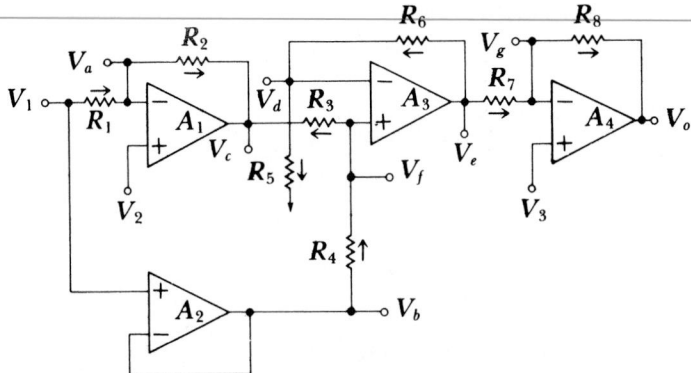

FIG. 33   Node-defined amplifier of Fig. 32.

2.   Compute V_c for this chain

Assume an ideal operational amplifier. Then $V_a = V_2$; $V_d = V_f$; $V_b = V_1$; $V_g = V_3$; $I_{R1} =$ current through R_1; $I_{R2} =$ current through R_2; etc. Do A_1 first, thus:

$$\frac{V_1 - V_a}{R_1} = \frac{V_1 - V_2}{R_1} = I_{R1}$$

$$V_c = V_a - I_{R1}R_2 = V_2 - I_{R1}R_2$$

Numerically,

$$I_{R1} = \frac{5\text{ V} - 4\text{ V}}{10\text{ k}\Omega} = 10^{-4}\text{ A}$$

$$V_c = 4\text{ V} - (10^{-4}\text{ A})(20\text{ k}\Omega) = 2\text{ V dc}$$

3.   Find V_f for this chain

Do A_2. Note that A_2 is simply a buffer amplifier, and $V_b = V_1$. Do A_3. (*Note:* V_f is the independent voltage.) Then

$$V_f = \frac{(V_c - V_b)R_4}{R_3 + R_4} + V_b$$

$$= \frac{(2-5)30\text{ k}\Omega}{40\text{ k}\Omega} + 5$$

$$= -2.25 + 5 = +2.75\text{ V dc}$$

4.   Find the value of V_e for this chain

As before, $V_d = V_f = 2.75$. Then

$$V_d = V_f = 2.75$$

$$I_{R6} = \frac{V_d}{R_5} = I_{R5} \qquad \text{(since the current is zero into negative input of } A_3\text{)}$$

$$\therefore V_e = V_d + I_{R5}R_6$$

$$= V_d + \frac{V_d R_6}{R_5}$$

$$= 2.75 + \frac{(2.75)(22\text{ k}\Omega)}{10\text{ k}\Omega}$$

$$= +8.8\text{ V dc}$$

5. *Find the value of V_o for this chain*

Do A_4, as in the earlier steps:

$$V_g = V_3 = +5\text{ V dc}$$

$$I_{R7} = \frac{V_e - V_g}{R_7} = \frac{8.8 - 5}{5 \times 10^3} = 0.76\text{ mA}$$

and $I_{R8} = I_{R7}$ (since the current is zero into negative input of A_4)

$$V_o = V_g - I_{R8}R_8$$

$$= 5 - 0.76\text{ mA}(10\text{ k}\Omega) = -2.6\text{ V dc}$$

6. *Compute what happens when R_6 is changed to 40 kΩ*

Compute V_e thus:

$$V_c = 2.75 + \frac{40}{10}(2.75) = +13.75\text{ V dc}$$

But this cannot happen because the swing on the output of the operational amplifier limits the positive voltage extreme to $+10$ V dc. This procedure is the work of Charles R. Hafer, P.E.

GAIN-CONTROL CIRCUIT ANALYSIS

The circuit in Fig. 34 is used in a gain-control application; its input signal level e_i varies from 0 to $+2$ V dc. It is desired to control the gain e_o/e_i from 0.4 to 1. Assuming that V_{AGC} controlling the JFET can bias the gate from full on to full off, can the gain-control range desired be obtained from the JFET with the parameters shown?

Calculation Procedure:

1. *Determine the value of the JFET resistance needed*

If the JFET is turned fully off, we can assume that we get a gain of 1. The JFET must now be turned on, which now becomes the condition to be concerned with. What value of JFET resistance is needed? Compute from

$$\frac{r_{DS}}{r_{DS} + R} = \frac{e_o}{e_i} = 0.4$$

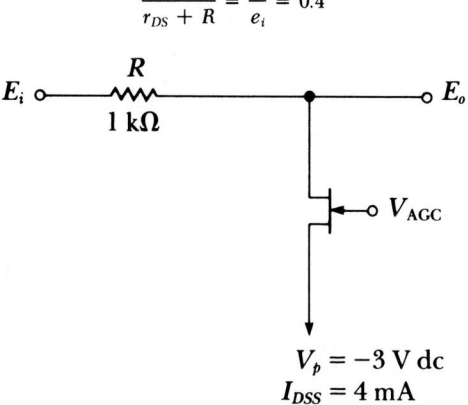

FIG. 34 Automatic gain-control circuit.

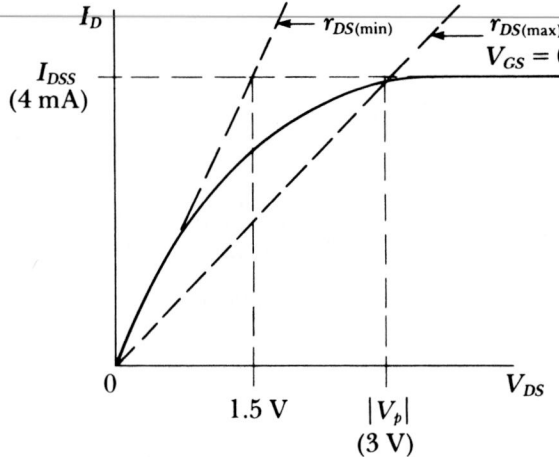

FIG. 35   Determining $r_{DS(\text{max})}$ for JFET in Fig. 34.

Thus,

$$r_{DS} = 667 \ \Omega$$

2.   *Find out whether the needed resistance can be obtained*

Determine the limits of the JFET resistance for the values of V_p and I_{DSS} given. The maximum value of r_{DS} can be found as shown in Fig. 35:

$$r_{DS(\text{max})} = \frac{|V_p|}{I_{DSS}} = \frac{3 \ V}{4 \ \text{mA}} = 750 \ \Omega$$

The minimum value of $r_{DS(\text{min})}$ is found from

$$r_{DS(\text{min})} = \frac{0.5|V_p|}{I_{DSS}} = \frac{0.5(3)}{4 \ \text{mA}} = 375 \ \Omega$$

3.   *Find the exact value of the JFET resistance r_{DS}*

The value of r_{DS} will lie between 375 and 750 Ω. When $A_v = 0.4$, there is 0.8 V ($= 2 \times 0.4$) across the JFET, and $V_{DS} = 0.8$ V dc. Also, $V_{GS} = 0$ V dc. Solve for I_D thus:

$$I_D = \frac{2V_{DS}I_{DSS}}{V_p}\left(\frac{V_{GS}}{V_p} - \frac{V_{DS}}{2V_p} - 1\right)$$

$$= \frac{2(0.8)(4 \ \text{mA})}{-3}\left(0 - \frac{0.8}{-6} - 1\right)$$

$$= 1.85 \ \text{mA}$$

and r_{DS} at this value of I_D and V_{DS} is

$$r_{DS} \le \frac{0.8 \ V}{1.85 \ \text{mA}} = 432 \ \Omega$$

This meets the requirements because the resistance can be as low as 432 Ω, as shown in Fig. 35. But the JFET would not be turned on so hard, since only 667 Ω is needed. If a range of V_p and I_{DSS} had been given, the maximum value of V_p and the minimum value of I_{DSS} would be used. This procedure is the work of Charles R. Hafer, P.E.

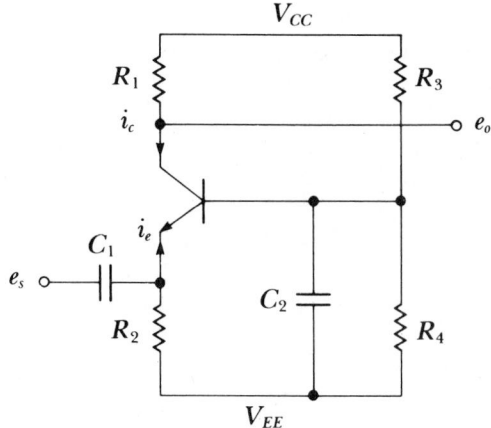

$$R_1 = 4.7 \text{ k}\Omega \qquad h_{ib} = 10 \ \Omega \qquad C_1 = 22 \ \mu\text{F}$$
$$R_2 = 100 \ \Omega \qquad h_{ob} = 10^{-6} \text{ S} \qquad C_2 = 47 \ \mu\text{F}$$
$$R_3 = 10 \text{ k}\Omega \qquad h_{rb} = 5 \times 10^{-4}$$
$$R_4 = 1 \text{ k}\Omega \qquad h_{fb} = -.99$$

FIG. 36   Amplifier circuit.

COMMON-BASE BIPOLAR AMPLIFIER ANALYSIS

The common-base (CB) amplifier in Fig. 36 has the circuit values as shown. Find the input imped-ance as seen by the emitter Q_1; the current gain i_c/i_e; the voltage gain e_o/e_s; and the output impedance as seen at the collector. Assume that the capacitors are short circuits at the frequency of interest.

Calculation Procedure:

1.   Draw the linear ac equivalent circuit

Figure 37 shows the linear ac equivalent circuit. For this circuit, $R_C = R_1 = 4.7 \text{ k}\Omega$.

2.   Find the input impedance as seen at the emitter

First find Δh_b; then Z'_{in}. Or,

$$\Delta h_b = h_{ib} h_{ob} - h_{fb} h_{rb}$$

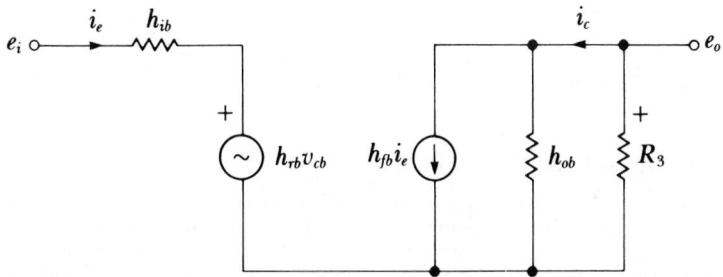

FIG. 37   Linear equivalent circuit of Fig. 36.

$$= (10)(10^{-6}) - (-0.99)(5 \times 10^{-4})$$

$$= 10^{-5} + 4.95 \times 10^{-4} = 5.05 \times 10^{-4}$$

$$Z_{in} = \frac{h_{ib} + R_C \, \Delta h_b}{1 + R_C h_{ob}}$$

$$= \frac{10 + 4.7 \times 10^3 \times 5.05 \times 10^{-4}}{1 + 4.7 \times 10^3 \times 10^{-6}} = 12.316 \ \Omega$$

The Z_{in} just calculated does not include R_2. Since R_2 is in parallel with Z_{in}, this gives Z'_{in}, which is what is sought:

$$Z'_{in} = Z_{in} \| R_2 = 12.316 \| 100$$

$$= 10.97 \ \Omega$$

3.   Determine the current gain A_i

Use the relation

$$A_i = \frac{i_c}{i_e} = \frac{h_{fb}}{1 + R_C h_{ob}} = \frac{-0.99}{1 + 4.7 \times 10^3 \times 10^{-6}}$$

$$= -0.9854$$

4.   Find the voltage gain

The voltage gain is

$$A_v = \frac{-h_{fb} R_C}{h_{ib} + R_C \, \Delta h_b}$$

$$= \frac{(0.99)(4.7 \times 10^3)}{10 + (4.7 \times 10^3)(5.05 \times 10^{-4})} = 376$$

5.   Determine the output impedance looking into the collector

This impedance is given by

$$Z_{ot} = R_C \| Z_o = R_C \left\| \left(\frac{h_{ib} + R_g}{\Delta h_b + R_g h_{ob}} \right) \right.$$

$$= R_C \left\| \frac{h_{ib}}{\Delta h_b} \right. \qquad \text{since } R_g = 0$$

$$= 4.7 \times 10^3 \left\| \left(\frac{10}{5.05 \times 10^{-4}} \right) \right. = 3.8 \text{ k}\Omega$$

This procedure is the work of Charles R. Hafer, P.E.

LOW- AND HIGH-PASS FILTER DESIGN

It is desired to construct a low-pass filter whose half-power point is 25 kHz by using a 1000-pF capacitor and standard 1 percent resistor values. The designer also would like to construct a high-pass filter, using the same capacitor and 1 percent resistors with a half-power frequency of 5 kHz. Design these circuits and show the values. Can these circuits be cascaded to obtain a bandpass with half-power points at 5 and 25 kHz?

Calculation Procedure:

1. *Find the function needed for the low-pass filter*

For the low-pass filter at 25 kHz, we need a function of the form $G(s) = 1/(\tau s + 1)$. This function can be found in a table of transforms. Draw the circuit of Fig. 38a. Then

$$G_1(s) = \frac{1}{T_1 s + 1}$$

$$T_1 = R_1 C = \frac{1}{2\pi f} = 6.37 \times 10^{-6}$$

$$R_1 = 6.37 \text{ k}\Omega$$

(Use the nearest standard value of 6.34 kΩ from a table of standard component values.)

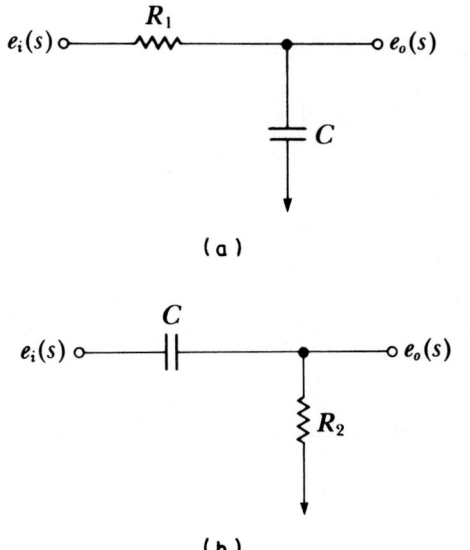

(a)

(b)

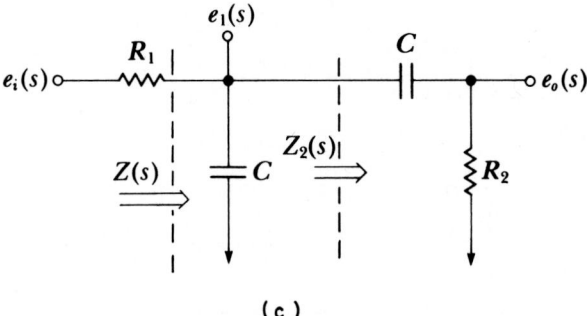

(c)

FIG. 38   (a) Low-pass filter. (b) High-pass filter. (c) Circuit for analysis of Fig. 44a and b.

2. Design the high-pass filter

For the high-pass filter of 5 kHz we use the circuit shown in Fig. 38b. Then

$$G_2(s) = \frac{T_2 s}{T_2 s + 1}$$

$$T_2 = R_2 C = \frac{1}{2\pi f} = 31.83 \times 10^{-6}$$

$$R_2 = 31.83 \text{ k}\Omega$$

(Use the nearest standard value of 31.6 kΩ from a table of standard component values.)

3. Determine whether the two filters can be cascaded for a bandpass

The two filters cannot be cascaded because one circuit loads the other and gives a transfer function altered from that desired. Deriving the expression for $e_o(s)/e_j(s) = G_3(s)$ for the circuit of Fig. 38c shows that $G_3(s) \neq G_1(s)G_2(s)$. This procedure is the work of Charles R. Hafer, P.E.

FILTER DISTORTION DETERMINATION

The filter in Fig. 39a has the voltage square wave in Fig. 39b applied to its input. It is desired to extract the fundamental and attenuate the harmonics. What percentage of distortion is obtained from the first three contributing harmonics?

Calculation Procedure:

1. Write the transfer function for this circuit

Standard filter references show the transfer function for this circuit as

$$G(s) = \frac{1}{(T_3 s + 1)(T_4 s + 1)} = \frac{1}{T_3 T_4 s^2 + (T_3 + T_4)s + 1}$$

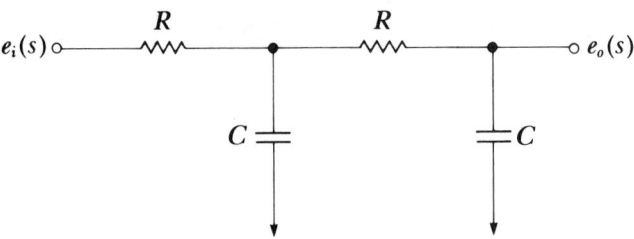

$$R = 10 \text{ k}\Omega$$
$$C = 0.01 \ \mu\text{F}$$

(a)

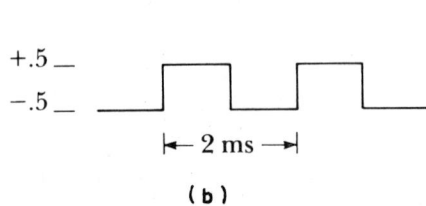

(b)

FIG. 39 (a) Filter circuit. (b) Input voltage waveform.

$$T_3 T_4 = R_1 R_2 C_1 C_2 = R^2 C^2 = 10^{-8}$$

$$T_3 + T_4 = R_1(C_1 + C_2) + R_2 C_2 = 3RC = 3 \times 10^{-4}$$

$$G(s) = \frac{1}{10^{-8} s^2 + 3 \times 10^{-4} s + 1}$$

2.   Write the Fourier expression for the square wave

Again, from standard references, the Fourier transform is as follows (there is no dc term because the waveform is symmetrical about the abscissa):

$$e_o(t) = \frac{4 \sin \omega t}{\pi} + \frac{4 \sin 3\omega t}{3\pi} + \frac{4 \sin 5\omega t}{5\pi} + \frac{4 \sin 7\omega t}{7\pi} + \cdots$$

3.   Find the magnitude of each harmonic as it exits the filter

$$G(j\omega) = \frac{1}{(1 - 10^{-8} \omega^2) + j3 \times 10^{-4} \omega}$$

For $f = 500$ Hz, $\omega = 3142$:

$$G(j3142) = \frac{1}{(1 - 0.099) + j0.94} = 0.768\underline{/-46.2°}$$

For $f = 1500$ Hz, $\omega = 9425$:

$$G(j9425) = \frac{1}{(1 - 0.89) + j2.83} = 0.353\underline{/-87.8°}$$

For $f = 2500$, $\omega = 1.57 \times 10^4$:

$$G(j1.57 \times 10^4) = \frac{1}{(1 - 2.465) + j4.71} = 0.203\underline{/-107.3°}$$

For $f = 3500$, $\omega = 2.2 \times 10^4$:

$$G(j2.2 \times 10^4) = \frac{1}{(1 - 4.84) + j6.6} = 0.131\underline{/-120.2°}$$

4.   Compute the magnitudes of the fundamental and harmonics

$$A_1 = G(j3142) \left(\frac{4}{\pi} \right) = (0.768)(1.273) = 0.978$$

$$A_3 = G(j9425) \left(\frac{4}{3\pi} \right) = (0.353)(0.424) = 0.15$$

$$A_5 = G(j1.57 \times 10^4) \left(\frac{4}{5\pi} \right) = (0.203)(0.255) = 0.052$$

$$A_7 = G(j2.2 \times 10^4) \left(\frac{4}{7\pi} \right) = (0.152)(0.131) = 0.02$$

5.   Determine the total distortion from the first three unwanted harmonics

$$D = (A_3^2 + A_5^2 + A_7^2)^{1/2}$$

$$= 0.16$$

$$\text{Percentage of distortion} = \frac{0.16}{0.978} = 16.4\%$$

This procedure is the work of Charles R. Hafer, P.E.

NETWORK SYNTHESIS BY USING AN OPERATIONAL AMPLIFIER

Synthesize a network, using an operational amplifier, resistors, and capacitors arranged as shown in Fig. 40a, which contains only part of the network. The response desired is shown in Fig. 40b. Find what frequency is f_b, where capacitor C_1 would be located to create f_b, and the values R_1, C_1, R_2, and C_2 if the low-frequency input impedance is 10 kΩ.

Calculation Procedure:

1. Find f_b for this circuit

Bode analysis may be used in the synthesis of this network. Since the gain decreases at 20 db/decade (factor of 10) or 6 dB/octave (a factor of 2), f_b can be found thus: The first decrease of 20

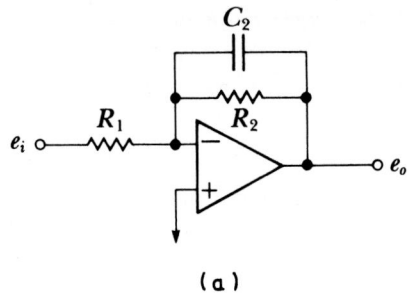

(a)

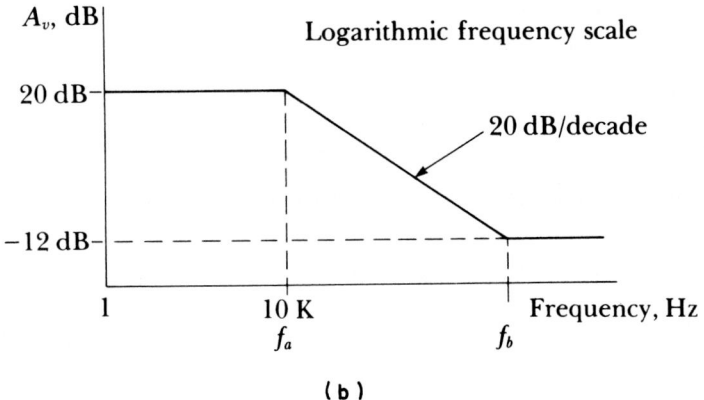

(b)

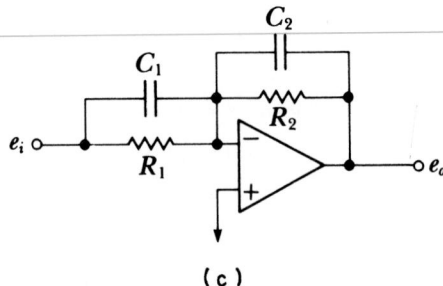

(c)

FIG. 40   (a) Given circut components. (b) Desired response. (c) Final circuit.

dB of gain puts the frequency at 100 kHz, while the next 6-dB decrease puts the frequency at 200 kHz, followed by another 6-dB decrease which puts the frequency at 400 kHz—a total of 32 dB [20 dB $-$ ($-$12 dB)].

2. Determine where C_1 should be located to create f_b

A pole exists at f_a and surely must be caused by the $R_2 = C_2$ combination, because

$$\frac{e_o}{e_{\text{in}}} = \frac{Z_2}{R_1}$$

where

$$Z_2 = \frac{R_2/(1/C_2 s)}{R_2 + 1/C_2 s}$$

$$Z_2 = \frac{1/C_2}{s + 1/R_2 C_2}$$

Therefore,

$$\frac{e_0}{e_{\text{in}}} = \frac{1}{R_1 C_2} \frac{1}{s + 1/R_2 C_2} \quad \text{and} \quad \omega_a = \frac{1}{R_2 C_2}$$

The zero which occurs at f_b can be synthesized similarly by locating a capacitor in shunt with R_1, as shown in Fig. 40c.

3. Find the value of C_1

The zero occurs at f_b; therefore

$$\omega_b = \frac{1}{R_1 C_1}$$

Also, $R_1 = 10 \text{ k}\Omega$ since R_1 is the low-frequency input impedance. Knowing R_1, we solve for C_1:

$$C_1 = \frac{1}{2\pi f R_1} = \frac{1}{2\pi (400 \times 10^3)(10 \times 10^3)}$$

$$= 3.98 \times 10^{-11}$$

Check: At $f = 400$ kHz, X_{C1} should equal 10 kΩ, since $X_{C1} = R$.

$$X_{C1} = \frac{1}{2\pi f C_1} = \frac{1}{2\pi (400 \times 10^3)(3.98 \times 10^{-11})}$$

$$= 10 \text{ k}\Omega$$

4. Synthesize R_2 and C_2

Since the dc gain is 10(20 dB), find R_2 thus:

$$A_v = \frac{R_2}{R_1} \quad \text{at dc}$$

$$R_2 = A_v R_1 = (10)(10 \times 10^3) = 100 \text{ k}\Omega$$

From before, it is known that the pole of the circuit is determined by the $R_2 = C_2$ combination. Then

$$\omega_a = \frac{1}{R_2 C_2}$$

$$C_2 = \frac{1}{\omega_a R_2} = \frac{1}{2\pi (10 \times 10^3)(100 \times 10^3)}$$

$$= 1.59 \times 10^{-10}$$

Check: At $f = 10$ kHz, $X_{C2} = 100$ kΩ, since $X_{C2} = R$.

$$X_{C2} = \frac{1}{2\pi f C_2} = \frac{1}{2\pi (10 \times 10^3)(1.59 \times 15^{-10})}$$

$$= 100 \text{ kΩ}$$

This procedure is the work of Charles R. Hafer, P.E.

SATELLITE COMMUNICATION LINK ANALYSIS

A communication link between a satellite and a ground station is separated by a distance of $d = 30,000$ km (18,000 mi). Find the following for this link:

1. The maximum power density reaching the ground station if the directive transmitter power is $P_T = 15$ W and has an antenna of cross-sectional area $A_T = 1.15$ m². Assume isotropic transmission. The frequency of transmission is 500 MHz, and the atmospheric attenuation is determined to be 50 dB.

2. The power received by the ground receiver if its antenna has a directive gain $G_R = 30$ dB, and its area is 300 m².

3. The background noise power received by the ground antenna given the bandwidth of the receiver is 10 kHz, and the space noise effective temperature is $T_e = 1000$ K.

4. The signal-to-noise ratio in decibels.

Calculation Procedure

1. *Draw a block diagram of the system*

The model for this sysem can be described by the block diagram in Fig. 41.

2. *Develop the equations for each power*

Develop the equations for each power, or power density, with magnitudes referenced with respect to the receiver. Here P_1 represents the power density at the receiver as a result of isotropic radiation, with the ground site a part of a sphere of radius d. Then

$$P_1 = \frac{P_T}{4\pi d^2} = G_1 P_T \text{ W/m}^2 \qquad G_1 = \frac{1}{4\pi d^2}$$

There is a directional gain of the transmitter represented by

$$P_2 = G_2 P_1 = \frac{4\pi A_T P_1}{\lambda^2} \text{ W/m}^2 \qquad G_2 = \frac{4\pi A_T}{\lambda^2}$$

Atmospheric attenuation can be included as follows:

$$G_3 = -50 \text{ dB} = 10 \log \frac{P_3}{P_2}$$

$$-5 = \log \frac{P_3}{P_2}$$

$$\frac{P_3}{P_2} = 10^{-5}$$

$$G_3 = 10^{-5}$$

The power density reaching the ground excludes the receiver directional gain and receiver antenna gain and can be calculated as follows [*note:* $\lambda = (3 \times 10^8)/f = (3 \times 10^8)/(5 \times 10^8) = 0.6$ m]:

$$P_3 = G_3 P_2 = G_3 G_2 P_1 = G_1 G_2 G_3 P_T$$

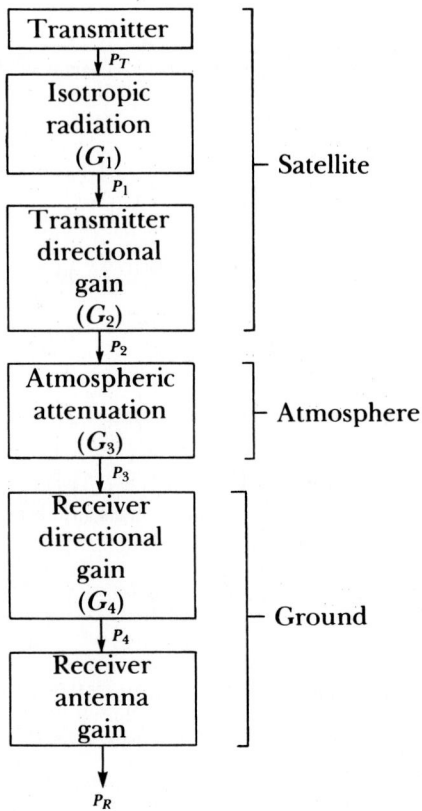

FIG. 41 Block diagram of a satellite-ground communication link.

$$= \left(\frac{1}{4\pi d^2}\right)\left(\frac{4\pi A_T}{\lambda^2}\right)(10^{-5})P_T$$

$$= \frac{10^{-5}A_T P_T}{\lambda^2 d^2} \quad \text{W/m}^2$$

$$= \frac{(10^{-5})(1.15 \text{ m}^2)(15 \text{ W})}{(0.6 \text{ m})^2(3 \times 10^7 \text{ m})^2} = 5.32 \times 10^{-19} \text{ W/m}^2$$

3. Compute the power input to the receiver

The power input to the receiver is:

$$P_R = A_R P_4 = A_R G_4 P_3$$

$$= (300 \text{ m}^2)(10^3)(5.32 \times 10^{-19})$$

$$= 1.6 \times 10^{-13} \text{ W}$$

4. Find the background noise power received by the ground antenna

The expression for the background noise power is

$$P_N = KT_e B$$

where

K = Boltzmann's constant; 1.3×10^{-23} J/K
B = noise bandwidth; assume the bandwidth given is the noise bandwidth, since no other information is given
T_e = effective noise temperature

So

$$P_N = (1.38 \times 10^{-23})(10^3)(10^4) = 1.38 \times 10^{-16} \text{ W}$$

5. **Find the signal-to-noise ratio**

$$\frac{S}{N} = \frac{P_R}{P_N} = \frac{1.6 \times 10^{-13}}{1.38 \times 10^{-16}} = 1159.4 \text{ W/W}$$

$$\frac{S}{N} = 10 \log 1159.4 = 30.64 \text{ dB}$$

This procedure is the work of Charles R. Hafer, P.E.

MICROWAVE TRANSMITTER ANALYSIS

A microwave transmitter, shown in Fig. 42, operates in an assigned band from 5.5 to 6.0 GHz. Each component block possesses the following characteristics:

Designation	Component	Characteristics
A_1	Modulating amplifier	f_m = 0 to 5 MHz
A_2	Gunn oscillator 1	f_1 = 5.15 GHz; modulation sensitivity is 0.25 MHz/V
A_3	Gunn oscillator 2	f_2 = 5.1 GHz
A_4	Mixer 1	$e_4 = a_1 e_1 + a_1 e_1^2$
A_5	IF amplifier	
A_6	Mixer 2	$e_6 = a_1 e_1 + a_1 e_1^2$
A_7	Gunn oscillator 3	f_3 = 5.5 GHz
A_8	Traveling-wave-tube amplifier	A_v = K V/V
A_9	Bandpass filter	

Determine (a) e_1 if the transmitter deviation is to be 2.5 MHz; (b) the frequencies present in the output e_4; (c) the frequencies present at e_5 and e_6 with no modulation; (d) the minimum bandwidth of the IF amplifier with modulation. What is the frequency range of interest?

Calculation Procedure:

1. **Determine the voltage magnitude of e_1**

Since the modulation sensitivity of A_2 is 0.25 MHz/V, the magnitude of the voltage e_1 is

$$e_1 = (2.5 \text{ MHz}) \left(\frac{1 \text{ V}}{0.25 \text{ MHz}} \right) = 10 \text{ V}$$

and the complete modulating waveform can be represented by

$$e_1 = 10 \sin \omega_m t = 10 \sin 2\pi f_m t$$

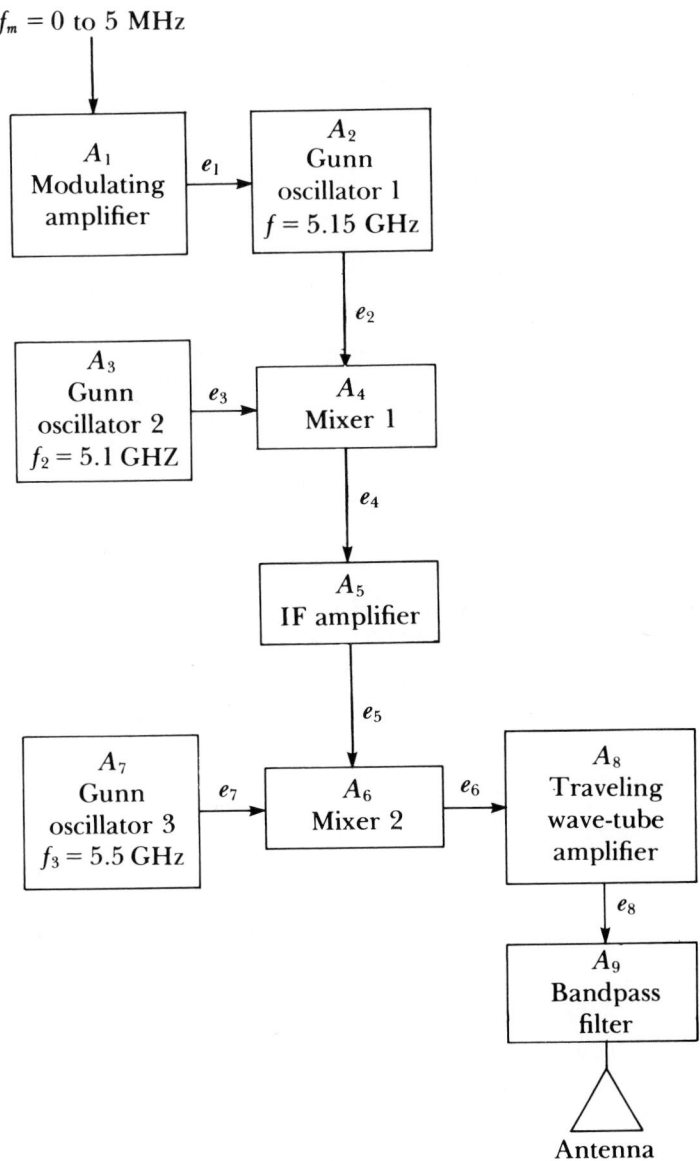

$f_m = 0$ to 5 MHz

FIG. 42 Microwave transmitter block diagram.

2. *Determine the frequencies present in output e_4*

Here e_4 represents the output of mixer 1. With no modulation, we can represent e_2 and e_3 as follows:

$$e_2 = A_1 \sin \omega_1 t \qquad e_3 = A_2 \sin \omega_2 t$$

and the output e_4 is represented by

$$e_4 = a_1(A_1 \sin \omega_1 t + A_2 \sin \omega_2 t) + a_1(A_1 \sin \omega_1 t + A_2 \sin \omega_2 t)^2$$

$$e_4 = a_1(A_1 \sin \omega_1 t + A_2 \sin \omega_2 t) + a_1[A_1^2 \sin^2 \omega_1 t + (2A_1 A_2 \sin \omega_1 t)(\sin \omega_2 t) + A_2^2 \sin^2 \omega_2 t]$$

Using trigonometric identities, we find

$$e_4 = a_1(A_1 \sin \omega_1 t + A_2 \sin \omega_2 t) + a_1[0.5A_1^2(1 - \cos 2\omega_1 t)$$

$$+ 0.5A_2^2(1 - \cos 2\omega_2 t) + A_1 A_2 \sin (\omega_1 + \omega_2)t$$

$$+ A_1 A_2 \sin (\omega_1 - \omega_2)t]$$

Therefore the frequencies present are

Term	Frequency, GHz
ω_1	5.15
ω_2	5.10
$2\omega_1$	10.30
$2\omega_2$	10.20
$\omega_1 + \omega_2$	10.25
$\omega_1 - \omega_2$	0.05

3. Find the frequencies at e_5 and e_6

The selected signal out of the IF amplifier (e_5) is the difference frequency which contains the needed information. Of course, the lower frequency is selected because of the less stringent design procedures. Therefore f_0 is selected to be 0.05 GHz.

The signal out of the second mixer (e_6) can be determined as in step a, except the two frequencies of interest are $f_0 = 0.05$ GHz and $f_3 = 5.5$ GHz.

Term	Frequency, GHz
f_0	0.05
f_5	5.50
$2f_0$	0.10
$2f_5$	11.00
$f_5 + f_0$	5.55
$f_5 - f_0$	5.45

4. Find the minimum bandwidth

Determine the minimum bandwidth by first finding the deviation ratio:

$$\delta = \frac{f_d}{f_m} = \frac{2.5 \text{ MHz}}{5 \text{ MHz}} = 0.5$$

With this value of deviation ratio, look at the Bessel tables given in many engineering books to determine the significant contributors to the output for $\delta = 0.5$. Data in the table below are reconstructed for $\delta = 0.5$. Assume that any contributions less than 0.001 are negligible. Clearly all spectral contributors up to and including $n = 3$ are significant. Therefore, the minimum bandwidth (BW) can be determined to be

$$\text{BW} = 4f_m = 4(5 \text{ MHz}) = 20 \text{ MHz}$$

This is centered at the IF of 50 MHz, so the range of interest is 40 to 60 MHz.

The assigned bandwidth is in the range of 5.5 to 6.0 GHz. Therefore, select a frequency in this range from step 4. Try $f_5 + f_0 = 5.55$ GHz. The minimum required bandwidth for the bandpass filter would be the same as for the IF stage, or 20 MHz. This procedure is the work of Charles R. Hafer, P.E.

n	$J_n(0.5)$
0	0.9385
1	0.2423
2	0.0306
3	0.0026
4	0.0002

SELECTION OF THERMOELECTRIC HEAT PUMP TO COOL ELECTRONIC DEVICES

Select a thermoelectric heat pump (Peltier-effect device) to cool an electronic package which generates 10 W of heat with the package temperature being maintained at 30°C (111.6°F), or lower, with an ambient temperature of 40°C (129.6°F). The heat leak for the package is 2 W. Space constraints allow the use of only two cooling modules. The heat-sink base temperature = 50°C (147.6°F).

Calculation Procedure:

1. Determine the total heat load

In any electronic cooling application of the total heat load, Q_{cT} is the sum of the active heat sources and the heat leak, or $Q_{cT} = 10 + 2 = 12$ W for this application. *Active heat sources* are components on the cold mounting surface that generate heat—those that dissipate power, for example. Heat leak is the passive absorption of heat into the cold surface from the warmer surrounding ambient. This source of heat often represents a significant portion of the total heat load of the cooling system.

2. Find the heat load per cooling module

With two modules for cooling, the heat load per module $Q_c = Q_{cT}/N$, where N = number of cooling modules. Or, $Q_c = 12/2 = 6$ W per cooling module.

3. Determine the required operating current

Allow a design margin for Q_c for each module. A suitable design margin for this package is 0.5 W. Hence, $Q_c = 6 + 0.5 = 6.5$ W.

To find the operating current per module, enter Fig. 43 at the difference between the desired package temperature, 30°C, and the heat-sink base temperature, 50°C, or $T = 50 - 30 = 20$°C, and the module heat load, $Q_c = 6.5$ W. At the bottom of Fig. 43 read the operating current as 3.4 A per module, point 5 on the curve.

4. Find the coefficient of performance (COP)

The COP is defined as the heat in watts absorbed at the cold side of the thermoelectric heat pump, Q_c, divided by the electric power input. Thus, at 3.4 A and a ΔT of 20°C, the COP is found from Fig. 43 to be 105 percent.

5. Find the total power consumed by the modules and the voltage across each

The power P, in watts, consumed by the module is $P = (Q_c/\text{COP})N$, where N = number of cooling units in the module. Substituting gives $P = (6.5/1.05)2 = 12.4$ W. Then with two modules connected in series, the voltage across each will be $V = P/I = 12.4/3.4 = 3.65$ V.

6. Determine the total heat rejection of the unit

The total heat rejection is $Q_h = Q_cN + P = 6.5(2) + 12.4 = 25.4$ W.

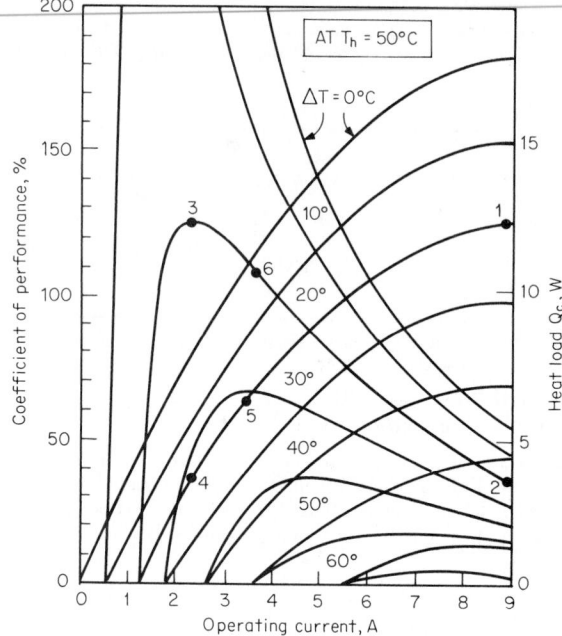

FIG. 43   Thermal performance curves for a typical thermoelectric module. *(Electronics.)*

7.   Find the required thermal resistance of the heat sink

The required thermal resistance of the heat sink is given by $\theta = (T_h - T_a)/Q_h$, where T_h = hot-side temperature, °C; T_a = lower or ambient temperature, °C; other symbols as before. Substituting yields $\theta = (40 - 30)/25.4 = 0.39$ C/W.

Related Calculations: Thermoelectric heat pumps are solid-state devices with no moving parts. They use the Peltier effect to reduce the temperature of the parts they are designed to cool. Typical cooling applications for which the above procedure can be used include electronic instrument cabinets; controlling temperature-sensitive electronic parameters such as noise or bias current in instrumentation operational amplifiers; cooling infrared and charge-coupled device detectors; maintaining sample temperatures in medical and laboratory instruments; laser tuning through temperature control; cooling photomultiplier tubes; semiconductor testing; and cooling microprocessors and other devices operating in industrial environments having high ambient temperatures.

Thermoelectric heat pumps can operate at 150°C (327.6°F) or higher, even in a vacuum. Compared with electromechanical cooling systems, the solid-state modules combine reduced size and weight with long-term reliability approaching that of other solid-state devices.

One of the better ways of designing with a thermoelectric module is to work with parametric performance curves such as those in Fig. 43. These typical transfer functions relate the important input and output parameters of the module, including operating current, heat absorbed at the cold side, the temperature difference between the hot and cold sides (ΔT), and the module's thermoelectric efficiency. Table 28 shows three design alternatives for this electronic package. The "practical design" is the alternative for which the steps are given here.

Temperature differences between hot and cold sides of up to 60°C (166°F) can be obtained routinely. However, as the temperature difference increases, both the heat absorbed at the cold side Q_c and the COP decrease. The designer should be certain that the performance curves for competing units are based on the same test conditions, such as in a vacuum or in the air, mounting by soldering to the test fixture heat sink, etc.

TABLE 28 Summary of Results for Three Thermoelectric Design Alternatives

Parameter	Maximum power dissipation	Maximum coefficient of performance	Practical design
Input current I, A	9	2.3	3.4
Input voltage V, V	3.5	4.6	3.6
Input power P, W	31.3	10.6	12.4
Heat rejection O_h, W	43.8	23.8	25.4
Number of TE modules	1	4	2
Required heat-sink thermal conductivity, C/W	0.23	0.42	0.39

The procedure described here is the work of Dale A. Zeskind, consultant to Cambridge Thermionic Corporation, as reported in *Electronics* magazine.

Optics—Mirror and Lens Systems

REFERENCES: Beiser—*Applied Physics*, McGraw-Hill; Hecht—*Optics*, McGraw-Hill; Born—*Principles of Optics*, Pergamon; Brown—*Modern Optics*, Krieger; Carlson—*Introduction to Applied Optics for Engineers*, Academic Press; Fincham—*Optics*, Butterworths; Francon—*Optical Image Formation and Processing*, Academic Press; Garmire—*Integrated Optics*, Springer-Verlag, Ghatak—*Contemporary Optics*, Plenum; Jenkins—*Fundamentals of Optics*, McGraw-Hill; Kingslake—*Applied Optics and Optical Engineering*, Academic Press; Levi—*Applied Optics: A Guide to Optical Systems Design*, Wiley-Interscience; Midwinder—*Optical Fibers for Transmission*, Wiley; Nelkon—*Optics, Sound and Waves*, Heineman; Nussbaum—*Contemporary Optics for Scientists and Engineers*, Prentice-Hall; Pressley—*Handbook of Lasers*, CRC; Robertson—*Engineering Uses of Coherent Optics*, Cambridge University Press; Van Heel—*Advanced Optical Techniques*, Elsevier; Young—*Applied Optics*, Springer-Verlag.

ANALYSIS OF IMAGE PRODUCED BY CONCAVE, CONVEX, OR PLANE MIRRORS

An object 5 in (12.7 cm) high is placed 20 in (50.8 cm) in front of a concave mirror whose focal length is 15 in (38.1 cm). Find the location, size, and nature of the image.

Calculation Procedure:

1. Compute the image distance

Use the mirror equation $1/p + 1/q = 1/f$, or 1/object distance + 1/image distance = 1/focal length, all measurements being in the same units (inches, centimeters, feet, etc.). This equation is valid for both concave and convex mirrors. In words, the mirror equation is stated thus: When an object is a distance p from a mirror of focal length f, the image will be located a distance q from the mirror. Substituting for this mirror, we find $q = pf/(p - f) = (20 \times 15)/(20 - 15) = 60$ in (152.4 cm).

2. Determine the type of image produced

A positive value of p or q in the mirror equation denotes a *real* image or object; a negative value denotes a *virtual* image or object. A *real object* is one that is in *front* of a mirror; a *virtual* object appears to be located *behind* the mirror.

A *real image* is formed by light rays that actually pass through the image; so a real image will appear on a screen placed at the position of the image. (This characteristic is important in the design of copying machines, cameras, telescopes, and other optical equipment.)

A *virtual image* can be seen only by the eye since the light rays that appear to come from the image actually do not pass through it. Real images are located in *front* of a mirror; virtual images, *behind* it.

Draw the object and mirror (Fig. 1) to scale, showing two different light rays, one parallel to the axis of the mirror and the other leaving the object along a radius of the convex mirror. Extend

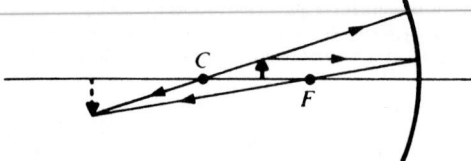

FIG. 1 Concave mirror image location. (*Beiser*—Applied *Physics, McGraw-Hill.*)

these rays until they intersect beyond the mirror. Thus, in Fig. 1, the radial ray passes through the top end of the object and the center of curvature of the concave mirror C, while the horizontal ray passes through the real focal point F, when reflected by the mirror. The dotted line shows that the image is real and on the same side of the mirror as the object.

3. Compute the height of the image and its nature

The *linear magnification m* of any optical system is the ratio between the size (height, width, or other transverse linear dimension) of the image and the size of the object. In the case of a mirror, $m = h'/h = -q/p$. In words, linear magnification m = image height h'/object height h = $-$(image distance q/object distance p). As before, the same units are used throughout.

For this concave mirror, solving for the image height gives $h' = -h(q/p) = -5(60/20) = -15$ in (-38.1 cm), which is 3 times greater than the height of the 5-in (12.7-cm) object. The minus sign indicates an inverted image.

Related Calculations: For a convex mirror, an object placed between the focal point F and the center of curvature C will, in general, have a real, inverted image that is larger than the object, that is, with p greater than f but less than $2f$.

A positive magnification for a mirror signifies an erect image; a negative magnification signifies an inverted image. Table 1 summarizes the sign conventions used for spherical mirrors.

TABLE 1 Sign Conventions for Spherical Mirrors°

Quantity	Positive	Negative
Focal length f	Concave mirror	Convex mirror
Object distance p	Real object	Virtual object
Image distance q	Real image	Virtual image
Magnification m	Erect image	Inverted image

°Beiser—*Applied Physics*, McGraw-Hill.

With plane (flat) mirrors, the angle of light reflection equals the angle of incidence (Fig. 2). The image of an object in a plane mirror has the same size and shape as the object but with left and right reversed. And the image is the same distance behind the mirror as the object is in front of the mirror.

Two different light rays traced from each point of interest in an object to where they (or their extensions, in the case of a virtual image) intersect after being reflected by the mirror give the

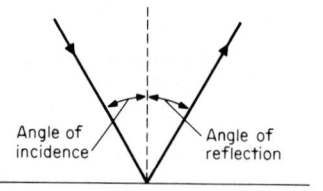

Angle of incidence   Angle of reflection

FIG. 2 Reflection of light from a plane surface. (*Beiser*—Applied Physics, *McGraw-Hill.*)

position and size of the image formed by a spherical mirror. Three rays are especially useful (Fig. 3) for this purpose; any two of the rays are sufficient. The three rays (any two of which can be used) are as follows:

1. A ray that leaves the object parallel to the axis of the mirror. After reflection, this ray passes through the focal point of a concave mirror or seems to come from the focal point of a convex mirror.

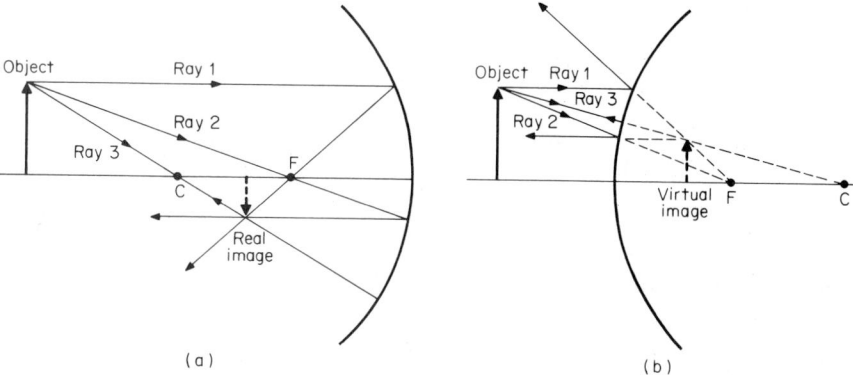

FIG. 3 Spherical-mirror images. (*Beiser*—Applied Physics, *McGraw-Hill.*)

2. A ray that passes through the focal point of a concave mirror. After reflection, this ray travels parallel to the axis of the mirror.

3. A ray that leaves the object along a radius of the mirror. After reflection this ray returns along the same radius.

Spherical aberration in a spherical mirror refers to the condition in which light rays from a point on an object are reflected at different distances from the mirror axis and do not converge (or appear to diverge from) a single point. This effect is shown in Fig. 4 for parallel rays reaching a concave mirror. Rays reflected from the outer parts of the mirror converge at focal points closer to the mirror than those reflected near the mirror's axis. This effect causes a spherical mirror to produce sharp images only when its diameter is small compared with its focal length.

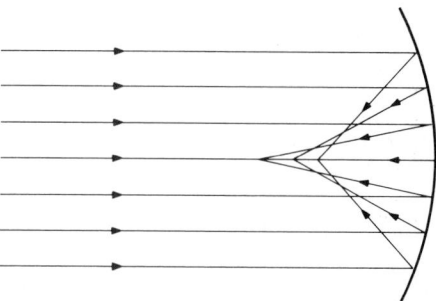

FIG. 4 Spherical aberration. (*Beiser*—Applied Physics, *McGraw-Hill.*)

Note that for a concave mirror the focal length $f = +R/2$, where R = mirror radius of curvature. But for a convex mirror $f = -R/2$.

When the mirror equation is used for a convex mirror, normally the value of f will be negative. This usually produces a negative image distance, meaning that the image is a virtual one, behind the mirror. Thus, an object 5 cm (2 in) long in front of a convex mirror whose curvature is 80 cm (31.5 in) is analyzed as follows: (1) Focal length of the mirror $f = -R/2 = -80/2 = -40$ cm (-15.75 in); (2) the image distance is $q = pf/(p - f) = 20(-40)/[25 - (-40)] = -15.4$ cm (-6.06 in) (the minus sign indicates a virtual image located behind the mirror); and (3) the length of the image is $h' = -hq/p = -5(-15.4)/25 = 3.1$ cm (1.22 in). A positive value of h' signifies an erect image.

The data contained in this calculation procedure are sufficient to enable an engineer or designer to analyze—in a preliminary way—the plane and spherical mirrors used in various optical devices such as cameras, telescopes, binoculars, microscopes, infrared military devices, copying machines, etc. It is the work of Arthur Beiser—*Applied Physics*, McGraw-Hill.

DETERMINATION OF IMAGES PRODUCED BY CONVERGING AND DIVERGING LENSES

An object 8 cm (3.15 in) long is 30 cm (11.8 in) away from a converging lens whose focal length is 15 cm (5.91 in). Find the location, size, and nature of the image.

Calculation Procedure:

1. *Find the image distance for this lens*

The object distance p, image distance q, and focal length f of a lens (Fig. 5) are related by the lens equation, $1/p + 1/q = 1/f$. Note that this equation is similar to the mirror equation in the previous calculation procedure. This equation is valid for both converging and diverging lenses. The same units of measurement should be used throughout the equation.

As in the case of mirrors, a positive value of p or q denotes a real object or image; a negative value denotes a virtual object or image. A real image or a real object is always on the opposite side of the lens from the object. A virtual image is on the same side of the lens. Thus, if a real object is on the left of a lens, a positive image distance q signifies a real image to the right of the lens whereas a negative image distance q denotes a virtual image to the left of the lens.

Substituting in the lens equation, we have $q = pf/(p - f) = 30(15)/(30 - 15) = 30$ cm (11.81 in). The image is real since q is positive (Fig. 6).

2. *Compute the length of the image*

Use the lens magnification equation $m = h'/h = q/p$, where m = linear magnification, h' = image height, h = object height, q = image distance, and p = object distance, with distances being expressed in the same units. A positive magnification signifies an erect image; a negative magnification indicates an inverted image. Table 2 summarizes the sign conventions used for lenses.

Substituting in the magnification equation gives $h' = -hq/p = -8(30/30) = -8$ cm (−3.15 in). The image is inverted since h' is negative and is the same size as the object. In general,

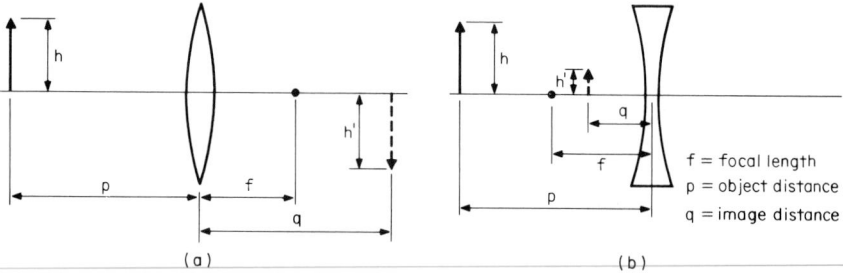

FIG. 5 Distances in lens equation. (*Beiser—Applied Physics, McGraw-Hill.*)

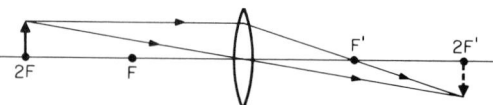

FIG. 6 Converging-lens diagram. (*Beiser—Applied Physics, McGraw-Hill.*)

TABLE 2 Sign Conventions for Lenses°

Quantity	Positive	Negative
Focal length f	Converging lens	Diverging lens
Object distance p	Real object	Virtual object
Image distance q	Real image	Virtual image
Magnification m	Erect image	Inverted image

°Beiser—*Applied Physics*, McGraw-Hill.

an object that is a distance $2f$ from a converging lens will have a real, inverted image the same size as the object with an image distance equal to $2f$.

   Related Calculations: The position and size of the image of an object formed by a lens can be found by constructing a scale drawing. Two different light rays are traced from each point of interest in the object to where they (or their extensions, in the case of a virtual image) intersect after being refracted by the lens. Three rays are especially useful for this purpose (Fig. 7):

1. A ray that leaves the object parallel to the axis of the lens. After refraction (defined as the bending, or deflection, of light away from or toward the normal, i.e., a perpendicular to the center of the lens), the ray that leaves an object parallel to the axis of the lens passes through the far focal point of a converging lens, or seems to come from the near focal point of a diverging lens. (A positive focal length corresponds to a converging lens and a negative focal length to a diverging lens.)

2. A ray that passes through the near focal point of a converging lens or is directed toward the far focal point of a diverging lens. After refraction, this ray travels parallel to the axis of the lens.

3. A ray that leaves the object and proceeds toward the center of the lens. This ray is not deviated by refraction.

   In general, a converging lens brings a parallel beam of light to a real focal point while a diverging lens spreads out a parallel beam of light so that the refracted rays appear to come from a virtual focal point. *Thin lenses* (those whose thickness can be neglected, as far as optical effects are concerned) can be analyzed by the *lensmaker's equation*

$$\frac{1}{f} = (n - 1)\left(\frac{1}{R_1} + \frac{1}{R_2}\right)$$

where n = index of refraction of the lens material relative to the medium (air, water, etc.) and R_1 and R_2 are the radii of curvature of the two surfaces of the lens. Both R_1 and R_2 are considered to be positive for a convex (curved outward) surface and negative for a concave (curved inward) surface; it does *not* matter which surface is taken as 1 and which as 2.

   When a system of lenses is used to produce an image of an object, as in a telescope or microscope, the procedure for finding the position and nature of the final image is to let the image formed by each lens in turn be the object for the next lens in the system. To find the image

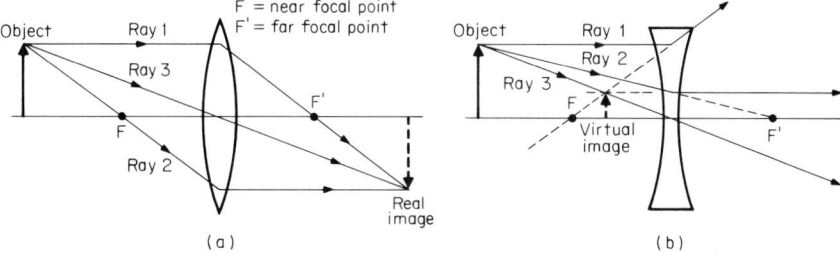

FIG. 7 Ray tracing for (*a*) converging lens, (*b*) diverging lens. (*Beiser*—Applied Physics, *McGraw-Hill.*)

produced by a system of two lenses, first determine the image formed by the lens nearest the object. This image then serves as the object for the second lens, with the usual sign convention. Thus, if the image is on the front side of the second lens, the object distance is considered positive; if the image is on the back side, the object distance is considered negative.

The total magnification produced by a system of lenses is equal to the product of the magnifications of the individual lenses. Thus, if the magnification of the objective lens of a microscope or telescope is m_1 and that of the eyepiece is m_2, the total magnification is $m_T = m_1 m_2$.

When a lens has one flat or plane surface and one curved surface, it is termed a *plano-convex lens* or a *plano-diverging lens*, depending on the curvature of the curved portion. A *meniscus* lens has one concave and one convex surface.

When a lens made of glass is immersed in water, such as in a still camera or an underwater TV monitoring camera, the index of refraction n' of glass relative to water is $n' =$ index of refraction of glass/index of refraction of water. For the usual lens, $n' = 1.60/1.33 = 1.20$.

When $R_1 = R_2$ in the lensmaker's equation, as might happen in water, the focal length f' of a lens in water is found from $f'/f = (n - 1)/(n' - 1)$. Substituting for a lens will show that the focal length of a lens having an index refraction of 1.6 and a focal length of $+20$ in air will have a focal length 3 times longer in water than in air.

This procedure is useful for analyzing and selecting lenses for cameras, telescopes, microscopes, projectors, telephoto devices, scanning instruments, etc. It is based on the work of Arthur Beiser—*Applied Physics*, McGraw-Hill.

COMPOUND THIN-LENS ANALYSIS

A compound lens consists of two thin biconvex lenses L_1 and L_2 of focal lengths of 10 cm (3.94 in) and 20 cm (7.87 in), separated by a distance d of 80 cm (31.5 in). What will be the image produced by an object 5 cm (2 in) tall that is 15 cm (5.91 in) from the first lens?

Calculation Procedure:

1. *Find the image distance for the compound lens*

As noted earlier, in a compound lens system assume that the image formed by the first lens serves as the object for the second lens, and so on. For any two thin lenses, $q = [f_2 d - f_1 f_2 p/(p - f_1)]/[d - f_2 - f_1 p/(p - f_1)]$, where $d =$ distance between lenses; other symbols as before with the subscripts 1 and 2 referring to lens L_1 and lens L_2, respectively. Substituting in the given values, we find $q = [20(80) - 10(20)(15)/(15 - 10)]/[80 - 20 - 10(15)/(15 - 10)] = 33.3$ cm (13.1 in). Since q is positive, the image is real and is located 33.3 cm (13.1 in) beyond the last lens, L_2.

2. *Find the magnification produced by the lenses*

Use the relation $m_T = f_1 q/[d(p - f_1) - pf_1]$. Or, the total magnification $m_T = 10(33.3)/[80(15 - 10) - 15(10)] = 1.3$. Thus, the image is magnified 30 percent and is erect.

Related Calculations: This procedure is valid for any compound lens system using two thin lenses. It is the work of Eugene Hecht—*Optics*, McGraw-Hill.

THICK COMPOUND LENS SYSTEMS ANALYSIS

Two identical biconvex thick lens are placed in line with a separation distance of 25.7 mm (10.12 in), as in Fig. 8. Each lens has radii of 60 mm (23.62 in) and 40 mm (15.75 in), a thickness of 20 mm (7.87 in), and an index of refraction of $n = 1.5$. Determine the focal length of each lens, and locate the principal points of the lenses in Fig. 8. Find the effective focal length of the system in air.

Calculation Procedure:

1. *Find the reciprocal of the focal length of each lens in the system*

Compute the reciprocal of the focal length of each lens from $1/f = (n - 1)[1/R_1 - 1/R_2 + (n - 1)d/nR_1R_2]$, where all the symbols are as given in earlier procedures in this section. Substituting the given data, we get $1/f = (1.5 - 1)[1/60 - 1/(-40) + (1.5 - 1)20/1.5(60)(-40)] = 7/360$.

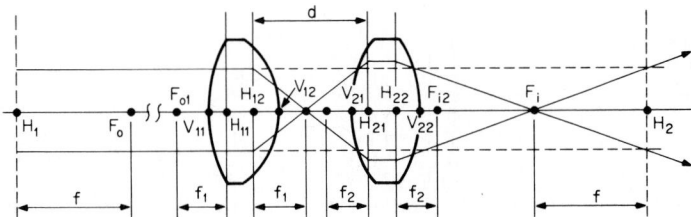

FIG. 8 Compound lens system. (*Hecht—Optics, McGraw-Hill.*)

2. *Find the focal length of each thick lens*

The focal length of a composite system of lenses such as that in Fig. 8 is a function of the focal length of each of the thick lenses. Since the lenses are identical and for each lens $1/f = 7/360$, $f_1 = f_2 = 51.43$ mm (2.02 in).

3. *Find the principal point for each lens*

If the incoming and outgoing rays of a thick lens are extended, as shown by the dotted lines in Fig. 9, each pair of lines will intersect on a surface. In the paraxial approximation, the surfaces reduce to planes known as the *first* and *second principal planes;* their points of intersection with the central axis, H_1 and H_2, are the *first* and *second principal points*, respectively. As a rule of thumb, for glass lenses in air the distance $\overline{H_1H_2}$ is roughly equal to one-third the thickness ($d = \overline{V_1V_2}$) of the lens. Note that the principal planes need not lie within the lens itself.

The gaussian lens formula, also called the *thin-lens equation,* $1/p - 1/q = 1/f$, can be used to solve for the principal planes and principal points to give $\overline{V_1H_1} \equiv h_1 = -f(n_l - 1)d/R_2n_l$ and $\overline{V_2H_2} \equiv h_2 = -f(n_l - 1)d/R_1n_l$, where $n_l =$ index of refraction of lens material. Both h_1 and h_2 will be *positive* when the principal planes are to the *right* of their respective vertices V_1 and V_2. The relationships between the various distances are shown in Fig. 10.

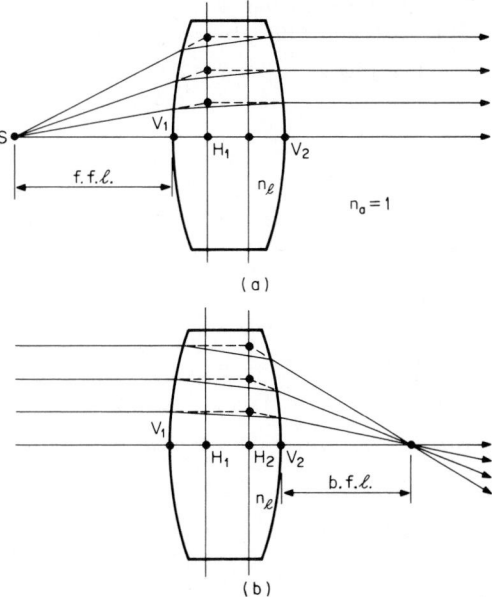

FIG. 9 Thick-lens rays. (*Hecht—Optics, McGraw-Hill.*)

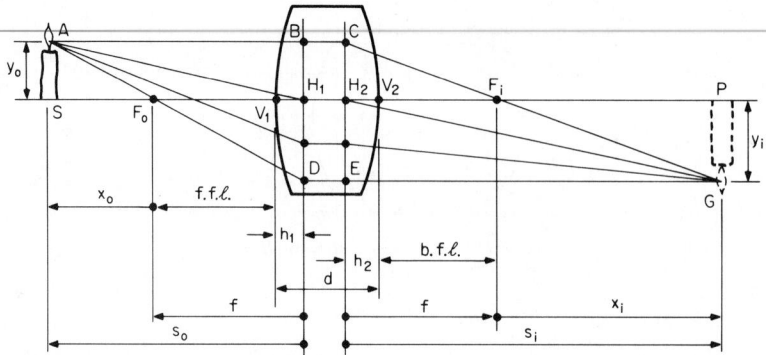

FIG. 10   Principal planes for a thick lens. (*Hecht—Optics, McGraw-Hill.*)

Substituting for this lens system gives $h_1 = -51.4(1/2)(20)/(-40)(3/2) = 8.6$ mm (0.34 in) and $h_2 = -51.4(1/2)(20)/60(3/2) = -5.7$ mm (-0.22 in). Since the lenses are identical, these values (h_1 and h_2) fix the positions of the principal planes with respect to the vertices for both lenses.

4.   *Find the focal length of the compound lens*

The compound lens has a focal length of $1/f = 1/f_1 + 1/f_2 - d/f_1f_2 = 1/51.4 + 1/51.4 - 25.7/51.4(51.4) = 3/102.8$; so $f = 34.3$ mm (1.35 in).

Related Calculations: Use this procedure to analyze any thick-lens system. The procedure given here combines two lenses (thick or thin) into one. Thus, if you have, say, six lenses in a centered system, you apply the procedure, replacing two lenses at a time until you have one equivalent lens representing the entire system.

The reciprocal of the focal length is defined as the *dioptric power D* of a lens. When the focal length f is measured in meters, D has the unit of m^{-1}, or diopters. For the case of two thin lenses in contact, $D = D_1 + D_2$ yields the combined power of the individual elements. This procedure is the work of Eugene Hecht—*Optics*, McGraw-Hill.

SECTION 6

CHEMICAL AND PROCESS PLANT ENGINEERING

ROBERT L. DAVIDSON
CONSULTING ENGINEER

JOHN S. REARICK, P.E.
CONSULTING ENGINEER

TYLER G. HICKS, P.E.
INTERNATIONAL ENGINEERING ASSOCIATES

CHEMICAL ENGINEERING 6.2
   Analysis of a Saturated Solution 6.2
   Ternary Liquid System Analysis 6.3
   Determining the Heat of Mixing of Chemicals 6.5
   Chemical Equation Material Balance 6.6
   Batch Physical Process Balance 6.7
   Steady-State Continuous Physical Balance with Recycle and Bypass 6.7
   Steady-State Continuous Physical Process Balance 6.8
   Determining the Characteristics of an Immiscible Solution 6.10
   Pump Selection for Chemical Plants 6.10
   Crusher Power Input Determination 6.12
   Cooling-Water Flow Rate for Chemical-Plant Mixers 6.12
   Liquid-Liquid Separation Analysis 6.13
PROCESS PLANT ENGINEERING 6.14
   Designing Steam Tracing for Piping 6.14
   Steam Tracing a Vessel Bottom to Keep the Contents Fluid 6.15

For additional calculation procedures useful in chemical engineering, please refer to the following sections of this handbook: Sec. 3, Mechanical Engineering; Sec. 4, Electrical Engineering; Sec. 7, Control Engineering; Sec. 11, Sanitary Engineering; Sec. 12, Engineering Economics. Each section contains a number of calculation procedures pertinent to the content of Sec. 6, Chemical Engineering, but size limitations prevent their repetition in Sec. 6.

Designing Steam-Transmission Lines without Steam Traps 6.17
Line Sizing for Flashing Steam Condensate 6.20
Saving Energy Loss from Storage Tanks and Vessels 6.22
Saving Energy Costs by Relocating Heat-Generating Units 6.25
Energy Savings from Vapor Recompression 6.27
Effective Stack Height for Disposing Plant Gases and Vapors 6.29
Savings Possible from Using Low-Grade Waste Heat for Refrigeration 6.31
Excess-Air Analysis to Reduce Waste-Heat Losses 6.33
Estimating Size and Cost of Venturi Scrubbers 6.34
Sizing Desuperheater-Condensers Economically 6.39
Sizing Vertical Liquid-Vapor Separators. 6.41
Sizing a Horizontal Liquid-Vapor Separator 6.43
Sizing Rupture Disks for Gases and Liquids 6.45
Time Needed to Empty a Storage Vessel without Dished Ends. 6.48
Cost Estimation of Heat Exchangers and Storage Tanks via Correlations . . .   6.49
Estimating Centrifugal-Pump and Electric-Motor Cost by Using Correlations . .   6.51
Determining the Friction Factor for Flow of Bingham Plastics 6.56
Time Needed to Empty a Storage Vessel with Dished Ends 6.59
Checking the Vacuum Rating of a Storage Vessel 6.61
Designing Prismatic Pressure Vessels 6.63
Minimum-Cost Pressure Vessels 6.67

REFERENCES: Teja—*Chemical Engineering and the Environment*, Halsted Press; Allen—*A Guide to the Economic Evaluation of Projects*, Institute of Chemical Engineers (England); Ginn—*Fundamentals of Computerized Chemical Process Optimization and Control*, Gulf Publishing; Nonhebel and Barry—*Chemical Engineering in Practice*, Crane-Russak; Schweitzer—*Handbook of Separation Techniques for Chemical Engineers*, McGraw-Hill; Sturbacek—*Calculation of Chemical Engineering Properties Using Corresponding State Methods*, Elsevier; Davidson—*Handbook of Water-Soluable Gums and Resins*, McGraw-Hill; Perry—*Chemical Engineers' Handbook*, McGraw-Hill; Kraus—*Pneumatic Conveying of Bulk Material*, Ronald; Box and Draper—*Evolutionary Operation*, Wiley; McCabe and Smith—*Unit Operations of Chemical Engineering*, McGraw-Hill; Warn—*Concise Chemical Thermodynamics in SI Units*, Van Nostrand Reinhold; Weast and Selby—*Handbook of Chemistry and Physics*, Chemical Rubber; Erskine—*Chemical Conversion Factors and Yields*, Chemical Information Services; Anderson and Wenzel—*Introduction to Chemical Engineering*, McGraw-Hill; Levenspiel—*Chemical Reaction Engineering*, Wiley; Henley and Bieber—*Chemical Engineering Calculations*, McGraw-Hill; Brown—*Unit Operations*, Wiley; Peters—*Plant Design and Economics for Chemical Engineers*, McGraw-Hill; Vilbrandt and Dryden—*Chemical Engineering Plant Design*, McGraw-Hill; Foust et al.—*Principles of Unit Operations*, Wiley; Williams—*Systems Engineering for the Process Industries*, McGraw-Hill.

Chemical Engineering

ANALYSIS OF A SATURATED SOLUTION

If 1000 gal (3785.4 L) of water is saturated with potassium chlorate ($KClO_3$) at 80°C (176°F), determine (*a*) the weight, lb, of $KClO_3$ that will precipitate if the solution is cooled to 30°C (86°F) and (*b*) the weight of $KClO_3$ that will precipitate if one-half the 1000 gal (3785.4 L) of water is evaporated at 100°C (212°F).

Calculation Procedure:

1. *Compute the precipitate when the solution is cooled*

When a solid is dissolved in water (or any other solvent liquid), the resulting solution is termed *saturated* when at a given temperature the solvent cannot dissolve any more of the solid. Most solvents dissolve (hold) more solids at higher temperatures than at lower temperatures. Thus, when the solution temperature is lowered or a portion of the solvent is evaporated, the solution becomes *supersaturated* and solid material may precipitate. This is the basis of *crystallization*, a chemical engineering operation frequently used to produce a purer or more crystalline product.

Referring to Fig. 1, obtain these solubilities: at 80°C (176°F), $KClO_3$ solubility = 38.5 g per 100 g H_2O; at 30°C (86°F), $KClO_3$ solubility = 10.5 g per 100 g of H_2O.

The weight of the water at 80°C (176°F) = (1000 gal H_2O)(0.97183 g H_2O per cm^3 H_2O) ×

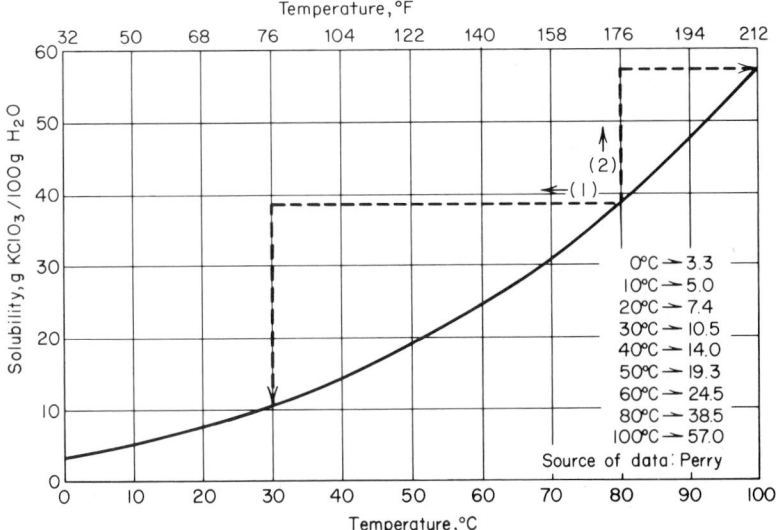

FIG. 1 Solubility of KClO₃.

(1 lb/454 g) = 8103 lb (3683.2 kg). Now, the weight of KClO₃ that any solvent can dissolve at a given temperature = weight of solvent at the given temperature, lb (solubility of KClO₃ at the given temperature, g per 100 g of the solvent). Or, at 80°C (176°F), weight of KClO₃ dissolved by the water = (8103 lb of water)(38.5 g KClO₃ per 100 g of H₂O) = 3119 lb (1417.7 kg) of KClO₃. And at 30°C (86°F) with the same quantity of water but the reduced solubility, the weight of KClO₃ that can be dissolved = (8103)(10.5 g per 100 g) = 851 lb (386.8 kg) of KClO₃.

When the temperature of the water (solvent) is reduced from 80 to 30°C (176 to 86°F), the weight of KClO₃ precipitated = weight of KClO₃ dissolved at 80°C (176°F) − weight of KClO₃ dissolved at 30°C (86°F), or 3119 − 851 = 2271 lb (1032.3 kg) of KClO₃ precipitated.

Note that the same procedure can be followed for any similar solution, i.e., any similar solvent and solid. Neither the solvent nor the solid need be the ones considered here.

2. *Compute the precipitate when a portion of the solvent is evaporated*

Since half the solvent (water in this case) is evaporated, the weight of water remaining = 8103/2 = 4051.5 lb (1841.6 kg). Using the solubility of KClO₃ as before, except that the solvent temperature is 100°C (212°F), we see the weight of KClO₃ dissolved = 4051.5(57 g KClO₃ per 100 g H₂O) = 2309 lb (1047.3 kg) of KClO₃. Then the weight of KClO₃ precipitated by the evaporation = weight of KClO₃ dissolved in 1000 gal (3785.0 L) of water at 80°C (176°F) − weight of KClO₃ dissolved in 500 gal (1892.5 L) of water at 100°C (212°F) = 3119 − 2309 = 810 lb (367.4 kg) of KClO₃ precipitated.

TERNARY LIQUID SYSTEM ANALYSIS

For a liquid mixture of 20 weight percent water, 30 weight percent acetic acid, and 50 weight percent isopropyl ether, determine the composition of the two phases (e.g., the ether layer and the water layer) and the amount of acetic acid that must be added to the system to form a one-phase (single-layer) solution.

Calculation Procedure:

1. *Compute the composition of the two layers*

When two pure liquids are mixed, they will dissolve in each other to some degree. If they are completely soluble in each other, such as water and acetic acid, they are *miscible*.

If their mutual solubilities are zero, they are called immiscible. Between these extremes, liquids are partially miscible.

Addition of a third liquid component often affects the mutual solubilities of the two original liquids. The third liquid may be more soluble in one liquid than in another. This difference in solubilities is the basis of the chemical engineering operation termed *liquid-liquid extraction*.

The third liquid may cause immiscible liquids to become completely miscible, or the third liquid may produce miscibility only in certain concentration ranges. Such interrelationships can be shown graphically, as with the two parts in Fig. 2.

The *phase envelope*, Fig. 2, separates the two-phase region from the one-phase region. Note, Fig. 2, that the acetic acid and water are completely miscible, as indicated by the phase envelope not touching the horizontal axis at any point. Likewise, the isopropyl ether and acetic acid are completely miscible. But water and isopropyl ether are virtually immiscible, as indicated by little of the vertical axis being free of the two-phase region of the phase envelope, Fig. 2.

The composition of the two phases, for the mixture in the two-phase region, is found on the phase envelope line itself, Fig. 2. Toward the lower part of the phase envelope, Fig. 2, is the water-rich layer, and toward the top of the phase envelope line is the ether-rich layer.

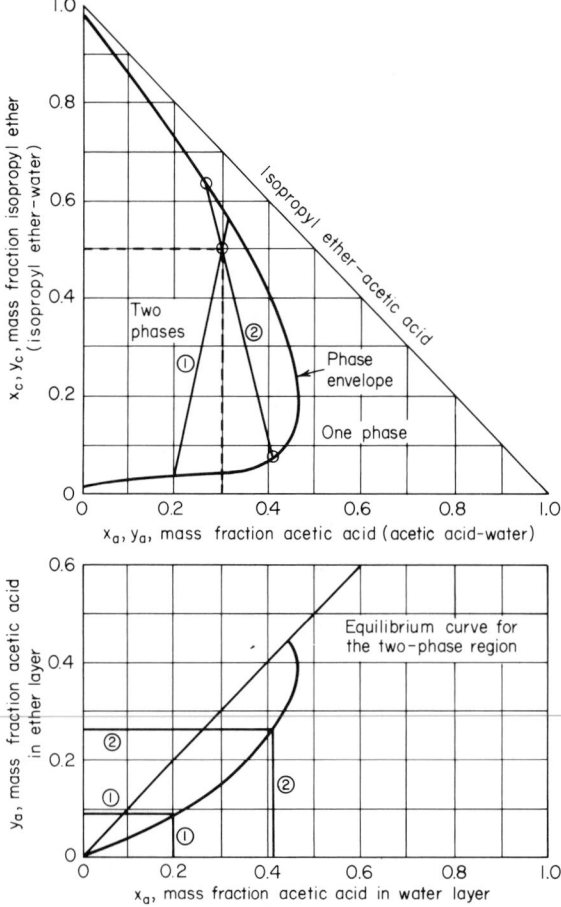

FIG. 2   Liquid-system phase-envelope plot. (*After Anderson and Wenzel*—Introduction to Chemical Engineering, *McGraw-Hill*.)

Plot on the upper portion of Fig. 2 the given values of acetic acid and isopropyl ether, that is, 30 weight percent, 50 weight percent. Through this point, draw a tie line to intersect the phase envelope at two points, line 1, Fig. 2. Read the values: *lower point*—acetic acid in water layer = 20 weight percent; *upper point*—acetic acid in isopropyl ether layer = 31.5 weight percent.

Transferring the lower intersection point to the bottom diagram for tie line 1 shows that equilibrium exists between a layer that is 20 weight percent acetic acid in water and a layer that is 9 weight percent (not 31.5 weight percent) acetic acid in isopropyl ether.

Draw a second tie line, 2, Fig. 2, as shown. Line 2 gives a check between the upper and lower diagrams, giving *water layer* — $x_a = 0.415$; $x_c = 0.065$; $x_w = 0.520$; *ether layer* — $y_a = 0.270$; $y_c = 0.650$; $x_w = 0.080$.

2. Compute the amount of acetic acid that must be added to form a one-phase system

The water/ether ratio remains unchanged at water/ether = 0.20/0.50 = 0.40. Then the total system is: water + ether + acid = 1.000; ∴ ether (weight percent) = [1.000 − acid (weight percent)]/1.40.

Assume that the acid = 0.350. Then ether = (1.000 − 0.350)/1.40 = 0.464. Checking against the upper diagram in Fig. 2, this point ($x_a = 0.350$, $x_c = 0.464$) falls inside the two-phase region. Hence, the assumption was incorrect.

As a second trial, assume acid = 0.380. Then ether = (1.000 − 0.380)/1.40 = 0.443. Checking $x_a = 0.380$, $x_c = 0.443$ in the upper diagram of Fig. 2 shows that the point falls exactly on the phase envelope line. Hence, it is at the minimum one-phase region.

DETERMINING THE HEAT OF MIXING OF CHEMICALS

How many Btu's of heat are released (generated) when 1000 lb (453.6 kg) of water at 80°C (176°F) is mixed with (1) 500 lb (226.8 kg) of aluminum bromide, $AlBr_3$; (2) 750 lb (340.3 kg) of barium nitrate, $Ba(NO_3)_2$; and (3) 1000 lb (453.6 kg) of dextrin, $C_{12}H_{20}O_{10}$?

Calculation Procedure:

1. Compute the heat released when $AlBr_3$ is dissolved in water

When two or more substances are mixed, heat is usually generated or absorbed. The heat released or absorbed may be small when two similar organic liquids are mixed or very large when strong acids are mixed in water. The heat evolved (or absorbed) during the mixing of liquids is often called the *heat of dilution*, whereas the heat from the mixing of solids is often termed the *heat of solution*. Data for heats of solution for both organic and inorganic liquids and solids are given in Perry, Lange, and similar reference works.

Thus, at 80°C (176°F) the solubility of $AlBr_3$ in water is 126 g per 100 g of water. The weight of $AlBr_3$ that will dissolve in 1000 lb (454.5 kg) of water = (126/100)1000 = 1260 lb (572.7 kg). Standard references show that the heat of solution for $AlBr_3$ is 85.3 kg·cal per g·mol $AlBr_3$. Since the total Btu/(lb·mol) = 1.8 g·cal/g, the $AlBr_3$ in this solution can release [85.3 kg·cal/(g·mol)][1000 cal/(kg·cal)]1.8 = 153,540 Btu/(lb·mol) (357.1 kJ/mol).

The weight of 1 lb·mol of $AlBr_3$ = (27 + 79.9 × 3) = 266.7 lb (121.2 kg). Hence, the heat evolved when 500 lb (227.3 kg) of $AlBr_3$ is dissolved in water is {500 lb $AlBr_3$/[266.7 lb/(lb·mol)]} [153,540 Btu/(lb·mol)] = 287,800 Btu (303.6 kJ).

2. Compute the heat released in dissolving $Ba(NO_3)_2$ in water

At 80°C (176°F), the solubility of $Ba(NO_3)_2$ in water is 27.0 lb/lb (12.3 kg/kg) of water. The weight of $Ba(NO_3)_2$ that will dissolve in 1000 lb (454.5 kg) of water = (27/100)1000 = 270 lb (122.7 kg). Since 750 lb (340.9 kg) of $Ba(NO_3)_2$ is available for dissolving, the weight that will not dissolve is 750 − 270 = 480 lb (218.2 kg).

The heat of solution of $Ba(NO_3)_2$ is − 10.2 kg·cal/(g·mol) of $Ba(NO_3)_2$. As in step 1, the weight of 1 lb·mol of $Ba(NO_3)_2$ = 137.34 + 2(14.0 + 3 × 16) = 261.4. Then, as in step 1, the heat released = [480 lb $Ba(NO_3)_2$/261.4 lb/(lb·mol)][− 10.2 kg·cal/(g·mol) × 1000 g·cal/(kg·cal) × 1.8 [Btu/(lb·mol)/[g·cal/(g·mol)] = −33,300 Btu (−35,131.5 J).

The negative heat release means that 33,300 Btu (35,131.5 J) of heat must be added to the system to maintain the solution temperature at 80°C (176°F) because a fall in temperature would reduce the solubility of the $Ba(NO_3)_2$ in water and thus change the resulting solution.

3. *Compute the heat released in dissolving $C_{12}H_{20}O_{10}$ in water*

Perry indicates that there is no solubility limit for $C_{12}H_{20}O_{10}$ in water. By following the same procedure as in step 1, the heat released = [1000 lb $C_{12}H_{20}O_{10}$/324.2 lb/(lb·mol)] × {268 g·cal/(g·mol) × 1.8[Btu/(lb·mol)]/[g·cal/(g·mol)]} = 1488 Btu (1569.8 J) released.

Related Calculations: Use the general procedure to determine the heat of mixing of any material dissolved in another.

CHEMICAL EQUATION MATERIAL BALANCE

Ethylene oxide is produced by the catalytic reaction of ethylene and oxygen: $C_2H_4 + \frac{1}{2}O_2 \rightarrow (CH_2)_2O$. For each 100 lb (45.5 kg) of ethylene, (1) how much ethylene oxide is produced, (2) how much oxygen is required, and (3) what are the quantities of ethylene oxide and ethylene in the product if there is a 20 percent deficiency of oxygen?

Calculation Procedure:

1. *Compute the quantity of ethylene oxide produced*

The two most frequently met calculations in day-to-day chemical engineering are the *material balance* (discussed here) and the *energy balance*, discussed later. In a chemical process, a balance is the same as any other type of balance, i.e., an equating of input, output, and accumulation or loss: Input − output = ± accumulation.

Such a balance may be written around a single item of chemical process equipment, a portion of a process, or an entire chemical plant. A balance may be used to check experimental data or to determine an unknown quantity of some process stream.

For purposes of balance calculations, chemical processes are classified as *steady-state*, i.e., input = output, no accumulation; *unsteady-state*, i.e., input ≠ output, a ± accumulation; *batch process*, i.e., system is loaded, no further ± accumulation; continuous process, i.e., continuous input and output. Chemical processes may be further classed as physical, in which there is no chemical reaction, or chemical, in which a chemical change occurs. To analyze chemical reactions, the principles of chemical equation balances must be understood.

A *stoichiometrically balanced reaction* is one in which the reactants are exactly proportioned to give a product free of excess reactants, as in $C_2H_4 + \frac{1}{2}O_2 \rightarrow (CH_2)_2O$. An *excess reactant* is one present in excess of the stoichiometric quantity, such as if there were more than 0.5 mol of oxygen in the above equation.

The *degree of completion* is the percentage of the limiting reactant that reacts. The *limiting reactant* is the one present in less than stoichiometric proportion, so that the other reactant is in excess.

To determine how much ethylene oxide is produced, find the molecular weight of $C_2H_4 = 2(12) + 4(1) = 28$. The moles of $C_2H_4 = 100/28 = 3.571$.

Referring to the reaction equation shows that for each mole of C_2H_4, 1 mol of $(CH_2)_2O$ is produced, having a molecular weight of $2(12) + 4(1) + 16 = 44$. Then the weight of $(CH_2)_2O$ = 44(3.571) = 157.14 lb (71.4 kg).

2. *Compute the amount of oxygen required*

The molecular weight of $O_2 = 16(2) = 32$. Referring to the reaction equation shows that $\frac{1}{2}$ mol of oxygen is needed for each mole of ethylene, C_2H_4. Hence, the weight of oxygen needed = $\frac{1}{2}(32)(3.571) = 57.14$ lb (25.9 kg).

3. *Compute the product mix for a reactant deficiency*

Referring to step 2, we see that a 20 percent oxygen deficiency means that there was 0.80 ($\frac{1}{2}$) = 0.40 mol of oxygen available. Rewriting the equation gives $0.2C_2H_4 + 0.8C_2H_4 + 0.4O_2 \rightarrow 0.8(CH_2)O + 0.2C_2H_4$. Hence, the ethylene oxide $(CH_2)_2O$ in the product = 0.8(157.14) = 125.71 lb (57.14 kg). And the ethylene, C_2H_4, in the product = 0.2(100) = 20 lb (9.1 kg).

Related Calculations: Use this general procedure for any chemical equation balance similar to that analyzed here.

BATCH PHYSICAL PROCESS BALANCE

A load of clay containing 35 percent moisture on a wet basis weighs 2000 lb (909.1 kg). If the clay is dried to a 15 percent moisture content (on a wet basis), how much water is evaporated in the drying process?

Calculation Procedure:

1. *Compute the initial moisture content*

The 2000 lb (909.1 kg) of wet clay contains 35 percent moisture, or 2000 (0.35) = 700 lb (318.2 kg) of water. Thus, the dry clay weighs 2000 − 700 = 1300 lb (590.9 kg).

2. *Compute the weight after drying*

Set up the relation y lb of wet clay + x lb of water = 1300 lb (590.9 kg) of dry clay. But the final batch contains 15 percent moisture. Hence, the water = $0.15y$. Therefore, the dry clay = $(1.00 - 0.15)y = 0.85y = 1300$. Solving, we find y = 1529 lb (694.9 kg) of wet (15 percent moisture) clay. And since $y + x$ = 2000 lb (909.1 kg), x = 2000 − y = 2000 − 1529 = 471 lb (214.1 kg) of water evaporated.

Related Calculations: Use this general procedure for any batch physical process balance involving evaporation or drying of a solid.

Where the rate of feed is given, a steady-state physical process balance can be analyzed. Thus, if the 2000 lb (909.1 kg) of clay in the above process were fed to the dryer in 1 h, the rate of evaporation would be 471 lb/h (214.1 kg/h) of water.

STEADY-STATE CONTINUOUS PHYSICAL BALANCE WITH RECYCLE AND BYPASS

Feed to a distillation tower is 1000 lb·mol/h (0.126 kg·mol/s) of a solution of 35 mole percent ethylene dichloride (EDC) in xylene. There is not any accumulation in the tower. The overhead distillate stream contains 90 mole percent ethylene dichloride, and the bottoms stream contains 15 mole percent ethylene dichloride. Cooling water to the overhead condenser is adjusted to give a reflux ratio of 10:1 (10 mol reenters the column for each mole of overhead product). Heat to the reboiler, Fig. 3, is adjusted so that the recycle ratio is 5:1 (5 mol reenters the column for each mole of bottom product), with a 2:15 bypass (2 mol bypasses the reboiler for each 15 mol that passes through the reboiler). Determine the flow rate of the overhead product, bottoms product, overhead reflux reentering the column, bottoms recycle reentering the column, bottoms bypassing the reboiler, and the total bottoms.

Calculation Procedure:

1. *Compute the bottoms product*

Since this is a physical system with no change in chemical composition and no accumulation, any component may be followed through the system. Having been given the important values of the ethylene dichloride (X_F, X_D, X_W), use them as the basis of the calculation, with X representing the moles of ethylene dichloride in each stream.

Set up a total material balance thus: $F = D + W$ = 1000 lb·mol/h (0.126 kg·mol/s), Eq. 1, where F = feed; D = product (i.e., distillate); W = product (i.e., bottoms), all expressed in lb·mol/h as shown in Fig. 3.

An ethylene dichloride balance is $FX_F = DX_D + WX_W$. Substituting given values, we have $1000(0.35) = 0.90D + 0.15W = 350$, Eq. 2. Solving Eqs. 1 and 2 simultaneously gives $D = 1000 - W$; $W = (350 - 0.90D)/0.15$; $D = 1000 - (350 - 0.90D)/0.15$; D = 266.67 lb·mol/h (0.034 kg·mol/s) distillate product; W = 733.33 lb·mol/h (0.092 kg·mol/s) bottoms product.

2. *Compute the reflux flow rate*

Taking the tower overhead as a separate system, Fig. 3, we find $X_L/X_D = 10 = L/D$. Hence, $L = 10D = 10(266.67)$ = 2666.7 lb·mol/h (0.34 kg·mol/s) reflux.

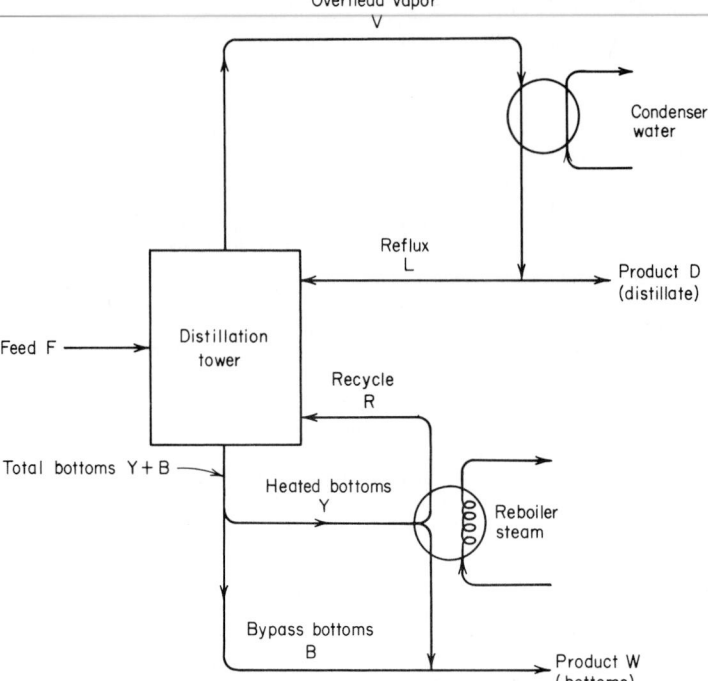

FIG. 3   Distillation tower flow.

3. *Analyze the condenser*

A total material balance around the condenser is input = output; or $V = D + L$, Fig. 3. Hence, $V = 266.67 + 2666.7 = 2933.37$ lb·mol/h (0.37 kg·mol/s) overhead vapor.

4. *Analyze the tower reboiler*

Taking the tower reboiler as a separate system, Fig. 3, gives $X_R/X_W = 5 = R/W$; hence, $R = 5W = 5(733.33) = 3666.7$ lb·mol/h (0.46 kg·mol/s) bottoms recycle.

Also, $X_B/X_Y = 2/15 = B/Y$; $2Y = 15B$, Eq. 3. And a total material balance around the reboiler is input = output, or $Y = R + (W - B)$, Eq. 4. Solving Eqs. 3 and 4 simultaneously gives $B = 2/15(Y)$; $Y = R + [W - 2/15(Y)]$; $Y = [15(3666.7) + 15(733.3)]/17 = 3882.35$ lb·mol/h (0.49 kg·mol/s) reboiled bottoms. Then $Y + B = 4399.97$ lb·mol/h (0.55 kg·mol/s) total bottoms.

Related Calculations: Use this general procedure to analyze distillation towers handling liquids similar to those considered here.

STEADY-STATE CONTINUOUS PHYSICAL PROCESS BALANCE

The distillation tower of the previous calculation procedure has the temperature and thermal conditions shown in Fig. 4. The reboiler is heated by steam that condenses at 280°F (137.8°C). Cooling water enters the overhead condenser at 70°F (21.1°C) and leaves at 120°F (48.9°C). In the condenser, the overhead vapor condenses at 184°F (84.4°C) before being cooled at 175°F (79.4°C), the temperature of the liquid reflux and distillate product. The heat of condensation ΔH of the overhead vapor is 14,210 Btu/(lb·mol) [33.1 kJ/(kg·mol)], as given in a standard reference work, and 14,820 Btu/(lb·mol) [34.5 kJ/(kg·mol)] for the tower bottoms. The heat capac-

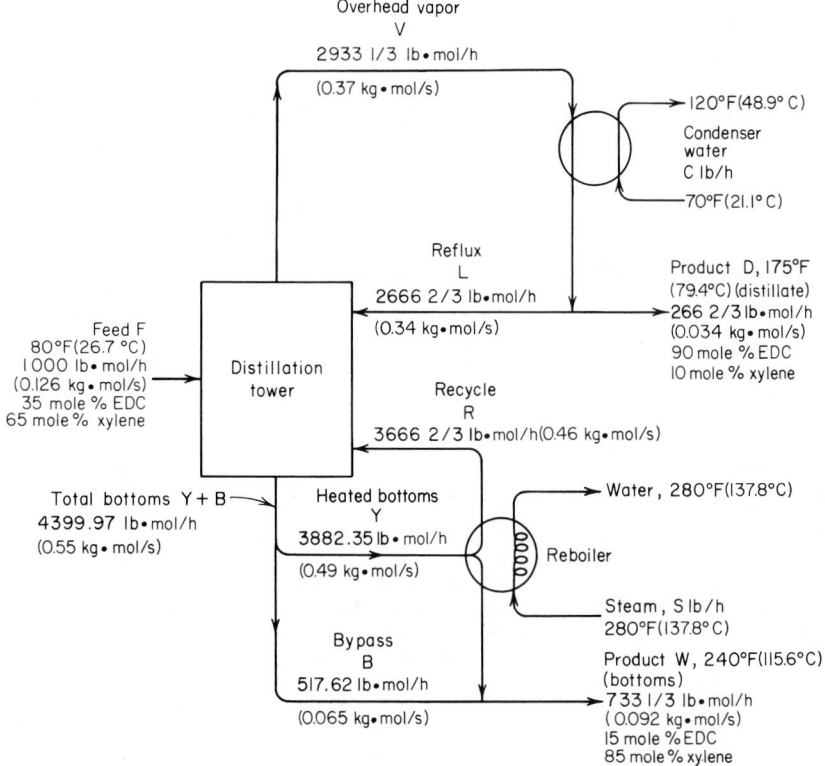

FIG. 4 Distillation-tower flow quantities and flow rates.

ity of all liquid streams in this installation is 40 Btu/(lb·mol·°F) [167.4 kJ/(kg·mol·K)]. Determine the steam and cooling-water flow rates required.

Calculation Procedure:

1. Set up a heat balance for the column

Thus, heat in = heat in feed + heat in steam. Let the temperature basis for the calculation = 80°F (26.7°C) = t_b. Then, the heat in the feed = ΔH_F = (feed rate, lb/h)[heat capacity of the feed, Btu/(lb·mol·°F)](feed temperature, °F − temperature basis for the calculation, °F) = 1000(40)(80 − 80) = 0.

The enthalpy of vaporization of the steam is, from the steam tables, 924.74 Btu/lb (2151 kJ/kg) when there is complete condensation of the steam and the condensate leaves the reboiler at 280°F (137.8°C). Then the heat given up by the condensation of S lb (kg) of steam is ΔH_S = 924.74S Btu/h (271.0 W).

The heat out = heat in distillate + heat in bottoms + heat in water, all expressed in Btu/h (W). Using the same procedure as for the heat in the feed, we see the heat in the distillate = $\Delta H_D/DC\Delta t_d$, where C = distillate heat capacity, Btu/(lb·mol·°F) [kJ/(kg·°C)]; Δt_d = temperature change of the distillate, °F (°C). Or, ΔH_D = 266.7(40)(175 − 80) = 1,010,000 Btu/h (295.9 kW). Likewise, for the bottoms, ΔH_W = 733.3(40)(240 − 80) = 4,690,000 Btu/h (1374.2 kW).

For the water, ΔH_w = heat to condense overhead vapor + heat absorbed when cooling the condensed vapor from 184 to 175°F (84.4 to 79.4°C), all expressed in Btu/h (W). With a flow of

2933.3 lb/h (0.37 kg/s) of vapor, $\Delta H_w = 2933.3\,[14,210 + 40(184 - 175)] = 42,700,000$ Btu/h (12.5 MW).

2. Analyze the heat balance

The heat balance is heat in = heat out, or $\Delta H_S = \Delta H_D + \Delta H_W + \Delta H_w$. Thus, $924.74S = 1,010,000 + 4,690,000 + 42,700,000$; $S = 52,300$ lb/h (6.59 kg/s) of steam.

For the water, $\Delta H_W = Cc\Delta t_c$, where C = water flow rate, lb/h; c = specific heat of water = 1.0 Btu/(lb·°F). Substituting gives $\Delta H_W = C(1)(120 - 70) = 42,700,000$; $C = 854,000$ lb/h (107.6 kg/s) of water.

DETERMINING THE CHARACTERISTICS OF AN IMMISCIBLE SOLUTION

For steam distillation of 2-bromoethylbenzene, the vapor temperature is 222.4°F (105.8°C). Analysis shows 0.16 lb (0.073 kg) of 2-bromoethylbenzene (BB) per lb (kg) of vapor. Saturated steam is used in the distillation process. Determine the pressure in the still and how far from ideal the actual conditions are.

Calculation Procedure:

1. Compute the pressure in the still

Each component of an immiscible mixture of liquids exerts a vapor pressure that is independent of its concentration and equal to the vapor pressure of the pure substance—but only if stratification is avoided by vigorous mixing or boiling. The major industrial application of immiscible systems is in steam distillation of high-molecular-weight heat-sensitive organic materials. The mixture of water (steam) and an organic substance will boil when the total solution pressure equals atmospheric pressure. Since the organic material must exert some vapor pressure, it vaporizes with the steam, at a greatly reduced temperature.

The relationship for immiscible components A and B is $w_A/w_B = y_A M_A/y_B M_B = P_{VA}M_A/P_{VB}M_B$, where $w_{A,B}$ = weight of component A, B in vapor; $M_{A,B}$ = molecular weight of component A, B; $y_{A,B}$ = vapor-phase mole fraction of component A, B; $P_{VA,VB}$ = vapor pressure of component A, B.

The vapor pressure of BB at 222.4°F (105.8°C) is, from Perry—*Chemical Engineers' Handbook*, 20 mmHg, and the vapor pressure of water at 222.4°F (105.8°C) is 938 mmHg. Hence, the total pressure (ideal) in the still is $938 + 20 = 958$ mmHg.

2. Compare the ideal to the actual conditions

If conditions in the still were ideal (i.e., exactly according to theory), the weight of the BB in the vapor would be, according to step 1, $w_{BB} = (P_{V,BB}/P_{V,H_2O})(M_{BB}/M_{H_2O}) = (20/938)(185/18) = 0.219$ lb (0.0995 kg), versus 0.16 lb (0.073 kg) actual, as given.

Or, by computing the ideal BB vapor pressure for 0.16 lb of BB per lb (0.07 kg/kg) of vapor, from the relation in step 1, $(P_{V,BB}/P_{H_2O})(185/18) = (P_{V,BB}/938)(185/18) = 0.16$ lb/lb (0.07 kg/kg). Solving, we find $P_{V,BB} = (0.16)(938)(18/185) = 14.6$ mmHg versus 20 mmHg actual.

The divergence between the actual and ideal most likely means that the time of contact between the steam and the BB is insufficient to reach equilibrium. Also, the total pressure should be $938 + 14.6 = 952.6$ mmHg, not the 958 mmHg of the ideal case.

Related Calculations: This procedure is valid for immiscible solutions of all types resembling the one considered here.

Plant engineers and designers in the chemical processing industry must be extremely careful about making changes in chemical processes or waste disposal. Seemingly routine decisions changing a process or disposal method can run into trouble under the Toxic Substance Control Act (TSCA). This act gives the U.S. Environmental Protection Agency (EPA) information and control over commercial chemicals.

Fines as high as $23,000 a day can be levied when TSCA rules are not obeyed. EPA applies rigid formulas when enforcing the act. Violation can result in million-dollar assessments.

It is important that engineers submit a premanufacture notification (PMN) in accordance with Section 5 of TSCA *before* manufacturing or importing a new chemical substance. A new chemical substance, as defined by Matthew Kuryla in *Chemical Engineering* magazine, is one that does not appear on an EPA list known as the TSCA Inventory. This list is constantly changing. Further, a portion of the list is confidential. To make a comprehensive search of the list requires a written request to EPA certifying a bona fide intent to manufacture or import a chemical.

Unless an exemption applies, a manufacturer must file a PMN with EPA *before* commencing production or importation of a new chemical. (In certain instances, a PMN must also be filed for existing chemical production that falls under a regulation known as the "significant new use rule.") The PMN must include the identity of the chemical, information about its proposed use and quantity, its by-products, and all available data concerning potential worker exposure and environmental or public-health effects.

Ninety days after filing a PMN, a company may commence manufacture or import of the chemical if Notice of Commencement (NOC) is filed. When EPA receives an NOC, it places the chemical on the TSCA Inventory. The chemical is then no longer considered new, but an "existing" chemical subject to other TSCA rules.

A number of chemicals and processes are exempt from the PMN requirements of TSCA, including: (1) foods, drugs, and cosmetics (including their intermediates); (2) pesticides (but not their intermediates); (3) chemicals used solely for research and development purposes, in small quantities; (4) chemicals manufactured solely for export; (5) impurities unintentionally present in another chemical; (6) by-products whose only commercial purpose is for burning as fuel, disposal as waste, or reclamation; (7) nonisolated intermediates (i.e., those mixed with other products and reactants) or incidental reaction by-products.

Besides these exemptions, TSCA provides a specialized PMN process for certain limited uses of a new chemical. These rules are known as the "test market," "low volume," and "polymer" exemptions. Such exemptions are subject to detailed rules of their own. They do not apply across the board, and frequently do not have specific, quantitative limits, notes Matthew Kuryla in *Chemical Engineering*.

PUMP SELECTION FOR CHEMICAL PLANTS

Choose a pump to handle 26,000 gal/min (1640 L/s) of water at 60°F (15.6°C) in a chemical plant when the total dynamic head is 37 ft (11.3 m) of water. What is the required hp input to the pump if the pump efficiency is 85 percent? What type of pump should be used if the rotational speed is limited to 880 r/min?

Calculation Procedure:

1. Determine the required power input to the pump

A quick way to determine the power input to a pump handling water at normal atmospheric temperatures is to use Fig. 5. Enter on the left at the total dynamic head, 37 ft (11.3 m), and project to the right to the required pump capacity, 26,000 gal/min (1640 L/s). At the intersection with the hp stem, read the required power input as 285 hp (212.6 kW).

2. Select the type of pump to use

From the rotational speed, 880 r/min on the bottom stem, draw a straight line at right angles to the first construction line, as shown. At the intersection with the top stem, read the type of pump as a propeller pump having a specific speed of 9500 r/min.

Related Calculations: Note that this pump application chart applies to rotating-type centrifugal pumps. Where a reciprocating pump is desired, use the methods given in Sec. 3 of this handbook. The chart in Fig. 5 was developed by H. W. Hamm and was first presented in *Power* magazine.

Pending environmental regulations will strictly limit pump leakage in chemical and process plants of all kinds. Today's laws require plant operators to report leakage of toxic substances of 0.0001% of the pump's capacity.

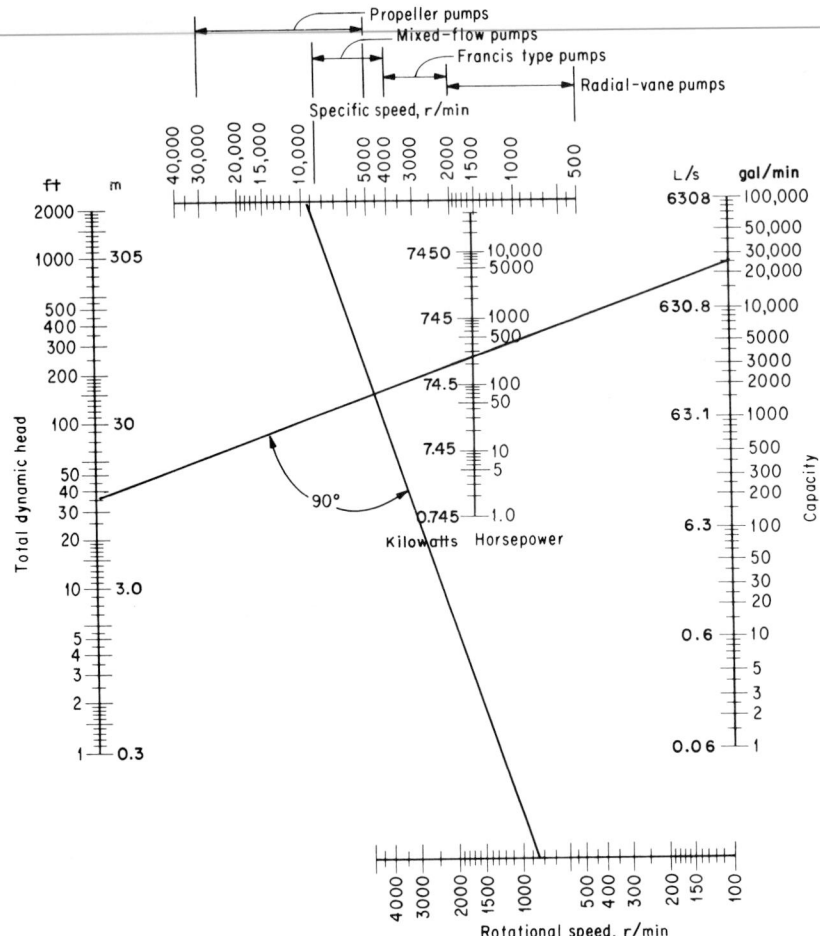

FIG. 5 Pump hp and type selection chart. *(Power.)*

There are both national (EPA) and state laws controlling pump leakage. For example, the state of New Jersey has a Toxic Catastrophe Prevention Act (TCPA) which strictly controls pump leakage. This and similar state environmental laws controlling pump seal leakage of toxic materials will probably become stricter in the future. For this reason, careful selection of pump shaft seals is important to every engineer working with toxic materials.

Typical toxic materials whose leakage must be prevented from pumps are sulfuric and nitric acid. Where water-flushed seals are used to contain leakage of such materials, the acidic flush water must be treated before disposal. Leakage of toxic materials must be prevented both while a pump is operating and while it is idle. Often, a seal that prevents leakage while the pump is operating will allow leakage when the pump is shut down. Such leakage is just as unlawful as leakage while the pump is operating.

For these reasons, engineers must carefully specify leak-free seals when choosing pumps handling toxic materials. The best seals can only be chosen after thorough study and consultation with both the pump and seal manufacturers.

CRUSHER POWER INPUT DETERMINATION

A chemical process requires the crushing of 240 tons/h (217.7 t/h) of quartz. The quartz feed used is such that 80 percent passes a 3-in (7.6-cm) screen and 80 percent of the product must pass a ¼-in (0.64-cm) screen. Determine the power input to the crusher.

Calculation Procedure:

1. Compute the crusher capacity in tons/min

Use the relation $t_m = t_h/60$, where t_m = crusher capacity, tons/min; t_h = crusher capacity, tons/h. Substituting yields $t_m = 240/60 = 4$ tons/min (3.63 t/min).

2. Determine the material work index

The work index for any material that will be crushed is the total energy, kWh/ton, needed to reduce the feed to a size so that 80 percent of the product will pass through a 100-μm screen. Standard references such as Perry—*Chemical Engineers' Handbook* list work indexes for various materials. For quartz having a specific gravity of 2.65, Perry gives the work index $W_i = 13.57$ kWh/ton (14.96 kWh/t).

3. Compute the raw-material and product mesh sizes

Use the relation $d_r = s/12$, where d_r = mesh size, ft, for feed; s = mesh opening measure used, in. For the product, $d_p = s/12$, where the symbols are the same as before except that the mesh opening is that used for the product. Substituting gives $d_r = 3/12 = 0.25$; $d_p = 0.25/12 = 0.0208$ (0.0064 m).

4. Compute the required power input to the crusher

Use the relation $hp = 1.46t_m W_i(1/d_p^{0.5} - 1/d_r^{0.5})$, where the symbols are as given earlier. Substituting gives $hp = 1.46(4)(13.57)(1/0.0208^{0.5} - 1/0.25^{0.5}) = 391$ hp (291.6 kW). A 400-hp (298.3 kW) motor would be used to drive this crusher.

Related Calculations: Use this general procedure, known as the bond crushing law and work index, to determine the power input required for commercially available grinders and crushers of all types. The result obtained is valid for all usual preliminary calculations.

COOLING-WATER FLOW RATE FOR CHEMICAL-PLANT MIXERS

A kneader used in a chemical plant requires 300-hp (223.7-kW) input per 1000 gal (3785.0 L) of material kneaded. If this kneader handles 3000 lb (1360.8 kg) of a chemical having a density of 65 lb/ft³ (1041.2 kg/m³), determine the quantity of cooling water required in gal/min and gal/h if the maximum allowable temperature rise of the water during passage through the kneader is 25°F (13.9°C).

Calculation Procedure:

1. Convert the kneader load to gallons

Since the power-input requirements of chemical mixers are normally stated in hp/gal, the kneader load must be converted to gal. Use the relation, load, gal = load weight, lb (7.48 gal/ft³ water)/load density, lb/ft³. For this kneader, load, gal = 3000(7.48)/65 = 345 gal (1306.0 L).

2. Compute the required power input

Use this relation: power input hp = hp input per 1000 gal (load,gal)/1000. For this kneader, power input $hp = 300(345)/1000 = 103.5$ hp (77.2 kW).

3. Compute the heat that must be removed

Since 1 hp = 2545 Btu/h (745.9 W), the heat that must be removed = (103.5 hp)(2545) = 263,407.5 Btu/h (77,145.5 W).

4. Compute the cooling-water flow rate

With an allowable temperature rise of 25°F (−3.9°C), and a specific heat of 1 Btu/(h·°F) (0.293 W), the cooling-water flow rate required = (263,407.5 Btu/h)/[(25°F)(1.0)(8.33 lb/gal of water)(60 min/h)] = 21.1 gal/min, or 21.2(60 min/h) = 1265 gal/h (4788 L/h).

Related Calculations: Use the general procedure given here for any of the usual chemical mixers, such as paddles, turbines, propellers, disks, cones, change cans, dispersers, tumbling mixers, mixing rolls, masticators, pug mills, and mixer-extruders. Consult Perry—*Chemical Engineers' Handbook* for suitable power-input data for mixers of various types.

LIQUID-LIQUID SEPARATION ANALYSIS

Size a liquid-liquid separator or decanter by using gravitational force for continuous separation of two liquids, the first of which has a density of 47 lb/ft³ (752.5 kg/m³) and the second liquid a density of 81 lb/ft³ (1296.8 kg/m³). Both liquids flow into the separator at a rate of 50 gal/min (189.3 L/min). The time required for settling is 35 min. What size separator is required to handle this flow? How far above the separator bottom should overflow of the heavier liquid be located?

Calculation Procedure:

1. Compute the liquid holdup volume

Since there are two liquids, a light one and a heavy one, entering the separator, the holdup volume = (number of liquids entering)(liquid flow rate into the separator, gal/min)(holdup time, min). Or for this separator, total holdup volume = 2(50)(35) = 3500 gal (13,247.5 L).

2. Determine the separator tank volume

Usual design practice is to make the separator tank volume 10 to 25 percent greater than the required holdup volume. Using a volume 20 percent greater than the required holdup volume gives a required tank volume of 1.20(3500 gal) = 4200 gal (15,897.0 L).

3. Size the separator tank

Most decanter-type separator tanks are sized so that the tank diameter and height are approximately equal. Selecting a 10-ft (3.05-m) diameter and 10-ft (3.05-m) high tank gives a total tank volume of (head area, ft²)(height, ft) = $(d^2\pi/4)h$, where d and h are the diameter and height of the tank, ft, respectively. Or, volume = $(10^2\pi/4)(10)$ = 785.4 ft³ (2.22 m³). Since 1 gal (3.8 L) of liquid occupies 0.13 ft³, the capacity of this tank = 785.4/0.13 = 5850 gal (22,142.2 L). This is sufficient to store the holdup liquid but somewhat oversize.

Try a 9-ft (2.74-m) diameter and high tank. By the same method, the tank capacity is 4250 gal (16,086.3 L). This is closer to the required holdup capacity. Hence a 9-ft (2.74-m) tank will be used.

4. Compute the liquid depth in the tank

Use the relation D_1 ft = 4(holdup volume, gal)/$7.48\pi d^2$, where D_1 = liquid depth, ft. So D_1 = $4(3500)/7.48\pi 9^2$ = 7.34 ft (2.24 m).

5. Determine the height of the heavy-liquid overflow

Assume that the two liquids interface midway between the vessel bottom and the liquid surface. Then the height of the heavy liquid = 7.34/2 = 3.67 ft (1.12 m).

To find the height of the heavy-liquid overflow, solve $H_h = H_1 + (D_1 - H_1)$(density of lighter liquid, lb/ft³)/(density of heavier liquid, lb/ft³), where H_h = height of heavy-liquid overflow above tank bottom, ft; H_1 = height of heavy liquid in tank, ft; other symbols as before. Solving gives H_h = 3.67 + (7.34 − 3.67)(47/81) = 5.80 ft (1.77 m). This is the distance measured to the inside lower surface of the overflow pipe from the tank bottom.

The *continuous decanter* is a popular type of static separator for immiscible liquids of many types. This type of separator is fed from the top and vented to the open air through both the light- and heavy-liquid overflow lines.

Process Plant Engineering

DESIGNING STEAM TRACING FOR PIPING

A 10-in (25-cm) stainless-steel pipe is conveying phthalic anhydride at 300°F (148.9°C). The line is to be steam-traced by using heat-transfer cement to attach the tracing line to the main pipe which is insulated with 1.5 in (3.81 cm) of calcium silicate insulation to maintain the bulk temperature at 300°F (148.9°C). The process fluid is stagnant in the pipe with an average, inside, natural convection coefficient of 20 Btu/(h·ft²·°F) [113.6 W/(m²·K)]. Determine the number of parallel tracers required, the heat transferred to the process fluid, and the steam consumption, using 150 lb/in² (gage) (1034 kPa) saturated steam. Other key data are as follows: ambient temperature $T_{amb} = -10°F$ ($-23.3°C$); supply-steam temperature $T_s = 366°F$ (185.6°C); process fluid temperature $t_p = 300°F$ (148.9°C); thermal conductivity of stainless-steel pipe wall $K = 9.8$ Btu/(h·ft²·°F·in) [1.4 W/(m·K)]; pipe-wall thickness $= 0.165$ in (0.413 cm); ID $= 0.43$ in (1.08 cm); heat-transfer coefficient between the insulation and ambient atmosphere $h_{air} = 2$ Btu/(h·ft²·°F) [11.4 W/(m²·K)].

Calculation Procedure:

1. Compute the overall heat-transfer coefficient between the wall and the ambient atmosphere

Use the relation

$$\frac{1}{h_o} = \frac{1}{h_{air}} + \frac{x_{ins}}{k_{ins}}$$

where the symbols are h_o = overall heat-transfer coefficient between the pipe wall and the air, Btu/(h·ft²·°F) [W/(m²·K)]; h_{air} = heat-transfer coefficient between the ambient air and insulation, Btu/(h·ft²·°F) [W/(m²·K)]; x_{ins} = insulation thickness, in (cm); k_{ins} = thermal conductivity of insulation, Btu/(h·ft²·°F·in) [W/(m²·K·cm)]. Entering the given values, we find $1/h_o = 1/2.0 + 1.5/0.3$; $h_o = 0.182$ Btu/(h·ft²·°F) [1.03 W/(m²·K)].

2. Determine the constants A and B and the ratio B/A

The dimensionless constant $A = (h_o + h_i)/Kt$, where h_i = process-fluid heat-transfer coefficient, Btu/(h·ft²·°F) [W/(m·K)]; K = thermal conductivity of pipe or vessel wall, Btu/(h·ft²·°F·ft) [W/(m²·K)], other symbols as before. Substituting gives $A = (0.182 + 20)/(9.8)(0.165/12) = 150$; $\sqrt{A} = (150)^{1/2} = 12.25$. And $B = (h_i T_p + h_o T_{amb})/Kt$, where T_p = pipe or vessel-wall temperature, °F (°C); other symbols as before. Substituting, we find $B = [(20)(300) + (0.182)(-10)]/9.8(0.165/12) = 44,510$. Then $B/A = 44,510/150 = 296.7$. This corresponds to the equilibrium temperature, 296.7°F (147.1°C), of the pipe wall in the absence of steam tracing.

3. Determine the temperature of the pipe wall at the tracer

The temperature of the pipe wall at the tracer T_o is usually set equal to the lowest temperature in the tracer, namely the saturated steam temperature at the tracer outlet. Normally, trapping distances are based on a 10 percent or 10-lb/in² (gage) (68.9-kPa) pressure drop, whichever is greater.

Assume a 10 percent pressure drop for this tracer circuit. The outlet pressure = 135 lb/in² (gage) (930.8 kPa), corresponding to a saturated steam temperature of $T_o = 358°F$ (181.1°C) from the steam tables.

4. Calculate the steam tracing half-pitch

Use a value of heat transferred to the process fluid, Btu/h (W)/length of the tracer, equivalent ft (m) of 100 Btu/(h·ft) (8.92 W/m) of tracer to compute L = one-half the steam tracing pitch, ft (m). Then, $L = [Q/z - 2(T_o - B/A)h_i/\sqrt{A}]/2h_i(B/A - T_p)$, where the symbols are as given earlier. Substituting, we find $L = [100 - 2(358 - 296.7)(20)/12.25]/2(20)(2.967 - 300) = 0.76$ ft (0.23 m).

Checking the assumption of the 10 percent pressure drop, we see tanh $(\sqrt{A}L) = $ tanh(12.25 × 0.76) = tanh 9.3 ≈ 1.0. Therefore, the assumption was valid and $L = 0.76$ is correct. If it were not correct, another pressure drop would have to be assumed and L computed again until a suitable value were obtained.

For certain applications, another value for Q/z different from 100 may be preferable, depending on the reheat time required if the tracing steam supply were lost. Hence, the designer must verify that the heatup time possibly meets this requirement while not producing an uneconomical tracing design.

5. Compute the minimum required pitch and distance between tracers

The minimum required pitch $= 2L = 2(0.76) = 1.52$ ft (0.46 m). Determine the number of parallel tracers from $\pi D/2L$, where the symbols are as defined earlier. For this installation, $N = \pi(10.75/2)/1.52 = 1.85$. With two parallel tracers the resultant distance between tracers is $\pi D/2 = \pi(10.75/12)/2 = 1.4$ ft (0.42 m).

6. Determine the total heat transferred by the tracer

The total heat transferred by the tracer $H_T = Q_T/2 = [2(20.182)/12.25](358 - 296.7)$ tanh $(12.25 \times 0.7) = 202$ Btu/(h·ft) (18 W/m).

7. Determine the tracer steam consumption

The steam consumption is $w' = (Q_T/2)\Delta H$, where $\Delta H =$ change in enthalpy of the steam from the tracer inlet to the tracer outlet. Or, $w' = 202/865 = 0.234$ lb/(h·ft) [0.0000967 kg/(s·m)].

8. Find the maximum equivalent feet of tracing run per steam trap

Using the tracer circuit pressure-drop assumed in step 3, compute the equivalent length, ft, from $z = [1.48(10^{11})(D^5)(\Delta P)/fw'^2(V_V - V_L)]^{13} = (1.48)(10^{11})(0.43/12)^5(15)/(0.012)(0.234)^2(3.02 - 0.018) = 405$ maximum equivalent ft (123.4 m).

The equation used in this step is derived by integrating the Darcy equation for a fluid with a changing specific volume. No account is taken of the fact that that fluid has two phases, since the largest portion of the total pressure drop is taken where the fluid is nearly all vapor. Further, the steam pressure is taken to be that at the outlet of the tracer circuit, thereby a somewhat conservative steam specific volume for the circuit.

Related Calculations: This procedure can be used for piping in a variety of applications, including chemical, petroleum, food, textile, marine, steel, etc. The ultimate use of the medium in the pipe has little or no effect on the calculation. This procedure is the work of Carl G. Bertram, Vikram J. Desai, and Edward Interess, the Badger Co., as reported in *Chemical Engineering* magazine.

STEAM TRACING A VESSEL BOTTOM TO KEEP THE CONTENTS FLUID

The bottom of a 4-ft (1.22-m) diameter, stainless-steel, solvent-recovery column holds a liquid that freezes at 320°F (160°C) and polymerizes at 400°F (204.4°C). The bottom head must be traced to keep the material fluid after a shutdown. Determine the required pitch of the tracing, using 150 lb/in² (gage) (1034 kPa) saturated steam. The ambient temperature is −20°F (−28.9°C), the supply steam temperature $T_s = 366°F$ (185.6°C), thermal conductivity of stainless steel = 9.8 Btu/(h·ft²·°F·ft) [16.95 W/(m·K)], insulation thickness = 2 in (5.1 cm), thermal conductivity of insulation = 0.3 Btu/(h·ft²·°F·ft) [0.52 W/(m·K)], and wall thickness = 0.375 in (0.95 cm). The heat-transfer coefficient between the insulation and the air is 2.0 Btu/(h·ft²·°F) [11.4 W/(m²·K)], and the inside convection coefficient is 20.

Calculation Procedure:

1. Compute the process-fluid heat-transfer coefficient and the overall heat-transfer coefficient

Use the relation $1/h_o = 1/h_{air} + x_{ins}/k_{ins}$, where the symbols are as defined in the previous calculation procedure. Substituting, we get $1/h_o = \frac{1}{2} + 2/0.3$; $h_o = 0.14$ Btu/(h·ft²·°F) [0.79 W/(m²·K)].

The process-fluid heat-transfer coefficient $h_i = 20$ Btu/(h·ft²·°F) [113.6 W/(m²·K)], assumed.

2. Determine the constants A and B and the ratio B/A

To determine the value of A, solve $A = (h_o + h_i)/Kt = (0.14 + 20)/(9.8)(0.375/12) = 65.6$, dimensionless. Also, $B = (h_iT_p + h_oT_{amb})/Kt = [(20)(320) + (0.14)(-20)]/(9.8)(0.375/12) = 20,900$. Then $B/A = 20,900/65.6 = 318$.

3. Compute the tracer-steam outlet temperature

The tracer steam is supplied at 50 lb/in² (gage) (344.7 kPa). Assuming a 15-lb/in² (gage) (103.4-kPa) pressure drop in the tracer system, we see the outlet pressure = 150 − 15 = 135 lb/in² (gage) (930.8 kPa). The corresponding saturated-steam temperature is, from the steam tables, $T_o = 358°F$ (181°C). This is the tracer outlet steam temperature.

4. Calculate the adjusted temperature ratio

Use the relation $(T_{mid} - B/A)/(T_o - B/A) = (320 - 318)/(358 - 318) = 0.05$.

5. Determine the required tracing pitch

From Fig. 6, with the adjusted temperature ratio of 0.05, $\sqrt{A}(L) = 3.7$ when $\alpha = 1$. (Here $\alpha =$ a parameter $= x/L$, where $x =$ distance along the pipe or vessel wall, ft.) Then, by solving for $L = 3.7/65.6^{1/2}$, $L = 0.46$ ft (0.14 m) = 5.5 in (13.97 cm).

The maximum allowable pitch for tracing the bottom of the column, Fig. 7, is 21, or 2(5.5) = 11 in (27.9 cm). A typical tracing layout is shown in Fig. 7.

Related Calculations: This procedure can be used to design steam tracing for a variety of tanks and vessels used in chemical, petroleum, food, textile, utility, and similar industries. The medium heated can be liquid, solid, vapor, etc. As with the previous calculation procedure, this procedure is the work of Carl G. Bertram, Vikram J. Desai, and Edward Interess, the Badger Company, as reported in *Chemical Engineering* magazine.

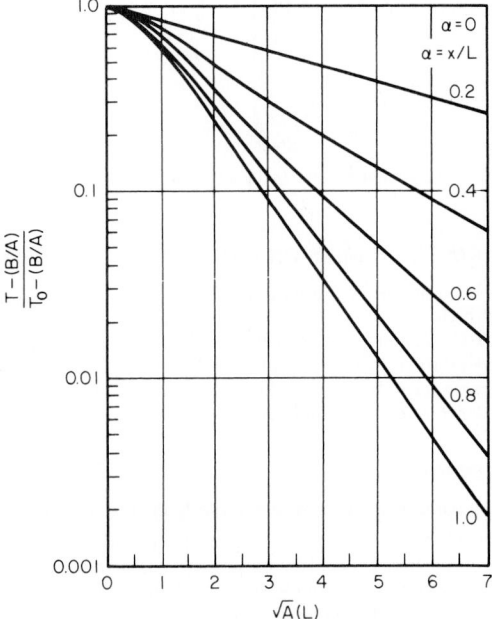

FIG. 6   Graphical solution for steam-tracing design. (*Chemical Engineering.*)

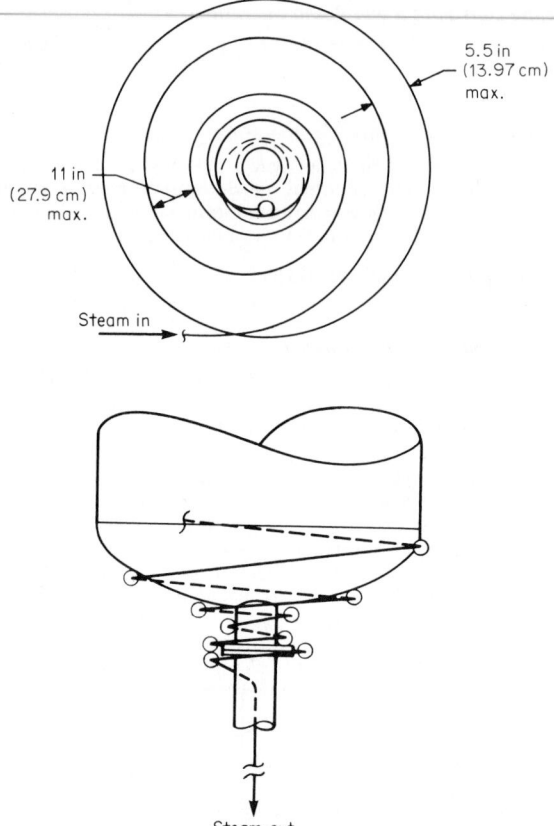

5.5 in
(13.97 cm)
max.

11 in
(27.9 cm)
max.

Steam in

Steam out

FIG. 7   Steam-traced vessel bottom. *(Chemical Engineering.)*

DESIGNING STEAM-TRANSMISSION LINES WITHOUT STEAM TRAPS

Design a steam line for transporting a minimum of 6.0×10^5 lb/h (2.7×10^5 kg/h) and a maximum of 8.0×10^5 lb/h (3.6×10^5 kg/h) of saturated steam at 205 lb/in² (gage) and 390°F (1413 kPa and 198.9°C). The line is 3000 ft (914.4 m) long, with eight 90° elbows and one gate valve. Ambient temperatures range from −40 to 90°F (−40 to 32.2°C). The line is to be designed to operate without steam traps. Insulation 3-in (7.6-cm) thick with a thermal conductivity of 0.48 Btu·in/(h·ft²·°F) [0.069 W/(m·K)] will be used on the exterior of the line.

Calculation Procedure:

1.   *Size the pipe by using a suitable steam velocity for the maximum flow rate*

The minimum acceptable steam velocity in a transmission line which is not fitted with steam traps is 110 ft/s (33.5 m/s). Assuming, for safety purposes, a steam velocity of 160 ft/s (48.8 m/s) to use in sizing this transmission line, compute the pipe diameter in inches from $d = 0.001295 f\rho LV^2/\Delta P$, where f = friction factor for the pipe (= 0.0105, assumed); ρ = density of the steam, lb/ft³ (kg/m³) [= 0.48 (7.7) for this line]; L = length of pipe, ft, including the equivalent length of fittings [= 3500 ft (1067 m) for this pipe]; V = steam velocity, ft/s [= 160 ft/s (48.8 m/s) for this line]; ΔP = pressure drop in the line between inlet and outlet, lb/in^2 [= 25 lb/in^2 (172.4 kPa) assumed for this line]. Substituting yields d = 0.001295(0.0105)(0.48)(3500)(160)^2/25 = 22.94 in (58.3 cm); use 24-in (61-cm) schedule 40 pipe, the nearest standard size.

2. Check the actual steam velocity in the pipe chosen

The actual velocity of the steam in the pipe can be found from $V = Q/A$, where V = steam velocity, ft/s (m/s); Q = flow rate of steam, lb/s (kg/s); A = cross-sectional area of pipe, ft^2 (m^2). Substituting gives V = (800,000 lb/h ÷ 3600 s/h)(2.08 ft^3/lb for steam at the entering pressure)/ 2.94 ft^2 = 157.6 ft/s (48.0 m/s) for maximum-flow conditions; V = (600,000/3600)(2.08)/2.94 = 117.9 ft/s (35.9 m/s) for minimum-flow conditions.

3. Compute the pressure drop in the pipe for each flow condition

Use the relation $\Delta P = 0.001295 f\rho L V^2/D$, where the symbols are the same as in step 1. Substituting, we find ΔP = 0.001295(0.0105)(0.48)(3500)(157.6)^2/24 = 23.2 lb/in^2 (159.9 kPa) for maximum-flow conditions. For minimum-flow conditions by the same relation, ΔP = 0.001295 (0.0105)(0.48)(3500)(117.9)^2/24 = 13.23 lb/in^2 (91.2 kPa). The pressure at the line outlet will be 220.0 − 23.2 = 196.5 lb/in^2 (1356.9 kPa) for the maximum-flow condition and 220.0 − 13.2 = 206.8 lb/in^2 (1425.9 kPa) for minimum-flow conditions.

4. Compute the steam velocity at the pipe outlet

Use the velocity relation in step 1. Hence, for maximum-flow conditions, V = (800,000/ 3600)(2.30)/2.94 = 173.8 ft/s (52.9 m/s). Likewise, for minimum-flow conditions, V = (600,000/ 3600)(2.19)/2.94 = 124.1 ft/s (37.8 m/s).

5. Determine the enthalpy change in the steam at maximum temperature-difference conditions

First, the heat loss from the insulated pipe must be determined for the maximum temperature-difference condition from $Q_m = h\Delta t A$, where Q_m = heat loss at maximum flow rate, Btu/h (W); h = overall coefficient of heat transfer for the insulated pipe, Btu·in/(h·ft^2·°F) [W·cm/(m^2· °C)]; Δt = temperature difference when the minimum ambient temperature prevails, °F (°C); A = insulated area of pipe exposed to the outdoor air, ft^2 (m^2). Substituting yields Q_m = 0.16 (430)(3362)(6.28) = 1,452,599 Btu/lb (3378.7 MJ/kg). In this relation, 430°F = 390°F steam temperature ± (−40°F) ambient temperature; 3362 = pipe length including elbows and valves, ft; 6.28 = area of pipe per ft of pipe length, ft^2.

The enthalpy change for the maximum temperature difference will be the largest with the minimum steam flow. This change, in Btu/lb (J/kg) of steam, is $\Delta h_{max} = Q_m/F$, where F = flow rate in the line, lb/h, or Δh_{max} = 1,452,599/600,000 = 2.42 Btu/lb (5631 J/kg).

The minimum enthalpy at the pipe line outlet = inlet enthalpy − enthalpy change. For this pipe line, h_{0min} = 1199.60 − 2.42 = 1197.18 Btu/lb (2784.6 kJ/kg).

6. Determine the enthalpy change in the steam at the minimum temperature-difference conditions

As in step 5, $Q_{min} = h\Delta t A$, or Q_{min} = 0.16(300)(3362)(6.28) = 1,013,441 Btu/lb (2357.3 mJ/kg). Then Δh_{min} = 1,013,441/800,000 = 1.26 Btu/lb (2946.6 J/kg). Also, h_{2max} = 1199.60 − 1.26 = 1198.34 Btu/lb (2787.3 kJ/kg).

7. Determine the steam conditions at the pipe outlet

From step 3, the pressure at the transmission line outlet at minimum flow and lowest ambient temperature is 206.8 lb/in^2 (1425.9 kPa), and the enthalpy is 1197.18 Btu/lb (2784.6 kJ/kg). Checking this condition on a Mollier chart for steam, we find that the steam is wet because the condition point is below the saturated-vapor line.

From steam tables, the specific volume of the steam is 2.22 ft^3/lb of total mass, while the specific volume of the condensate is 0.0000342 ft^3/lb of total mass. Thus, the percentage of condensate per volume = 100(0.0000342)/2.22 = 0.00154 percent condensate per volume. The percentage volume of dry steam therefore = 100(1.00000 − 0.00154) = 99.99846 percent dry steam per volume.

Since the velocity under these steam conditions is 124.1 ft/s (37.8 m/s), the steam will exist as a fine mist because such a status prevails when the steam velocity exceeds 110 ft/s (33.5 m/s). In the fine-mist condition, the condensate cannot be collected by a steam trap. Hence, no steam traps are required for this transmission line as long as the pressure and velocity conditions mentioned above prevail.

Related Calculations: Some energy is lost whenever a steam trap is used to drain condensate from a steam transmission line. This energy loss continues for as long as the steam trap is draining the line. Further, a steam-trap system requires an initial investment and an ongoing cost for routine maintenance. If the energy loss and trap-system costs can be reduced or eliminated, many designers will take the opportunity to do so.

A steam transmission line carries energy from point 1 to point 2. This energy is a function of temperature, pressure, and flow rate. Along the line, energy is lost through the pipe insulation and through steam traps. A design that would reduce the energy loss and the amount of required equipment would be highly desirable.

The designer's primary concern is to ensure that steam conditions stay as close to the saturated line as possible. The steam state in the line changes according to the change in pressure due to a pressure drop and the change in enthalpy due to a heat loss through insulation. These changes of condition are plotted in Fig. 8, a simplified Mollier chart for steam. Point 1 is defined by P_1 and T_1 steam conditions. Because of the variability of such parameters as flow rate and ambient temperature, the designer should consider extreme conditions. Thus, P_2 would be defined by the minimum pressure drop produced by the minimum flow rate. Similarly, h_2 would be defined by the maximum heat loss produced by the lowest ambient temperature. Point 2 on the h-s diagram is defined by the above P_2 and h_2. If point 2 is above or on the saturated-steam line, no condensate is generated and steam traps are not required.

In some cases (small pressure drop, large heat loss), point 2' is below the saturated-steam line, and some condensate is generated. The usual practice has been to provide trap stations to collect this condensate and steam traps to remove it. However, current research in two-phase flow demonstrates that the turbulent flow, produced by normal steam velocities and reasonable steam qualities, disperses any condensate into a fine mist equally distributed along the flow profile. The trap stations do not collect the condensate, and once again, the steam traps are not required. For velocities greater than 110 ft/s (33.5 m/s) and a steam fraction more than 98 percent by volume, the condensate normally generated in a transmission line exists as a fine mist that cannot be collected by steam-trap stations.

A few basic points should be followed when a steam line is operated without steam traps. All lines must be sloped. If a line is long, several low points may be required. Globe valves are used on drains for each low point and for a drain at the end of the line. Since trap stations are not required, drain valves should be located as close to the line as possible to avoid freezing. Vents are placed at all high points. All vents and drains are opened prior to warming the line. Once steam is flowing from all vent valves, they are closed. As each drain valve begins to drain steam only, it is partially closed so that it may still bleed condensate if necessary. When full flow is established, all drain valves are shut. If the flow is shut down, all valves are opened until the pipe cools and are then closed to isolate the line from the environment. The above procedure would be the same if the steam traps were on the line.

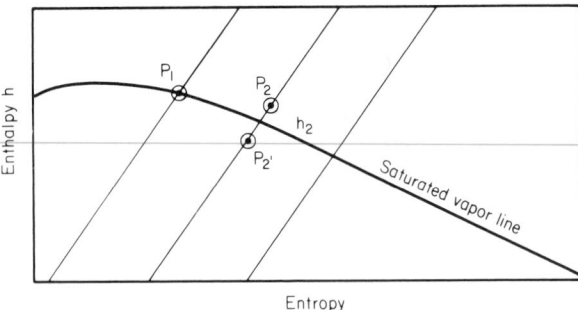

FIG. 8 Simplified Mollier chart showing changes in steam state in a steam-transmission line.

In summary, it can be demonstrated that steam traps are not required for steam transmission lines, provided that one of the following parameters is met:

1. Steam is saturated or superheated.

2. Steam velocity is greater than 110 ft/s (33.5 m/s), and the steam fraction more than 98 percent by volume.

By using the above design, the steam energy normally lost through traps is saved, along with the construction, maintenance, and equipment costs for the traps, drip leg, strainers, etc., associated with each trap station.

This calculation procedure can be used for steam transmission lines in chemical plants, petroleum refineries, power plants, marine installations, factories, etc. The procedure is the work of Mileta Mikasinovic and David R. Dautovich, Ontario Hydro, and reported in *Chemical Engineering* magazine.

Leaks of hazardous materials from underground piping and tanks can endanger lives and facilities. To reduce leakage dangers, the EPA now requires all underground piping through which hazardous chemicals or petrochemicals flow to be designed for double containment. This means that the inner pipe conveying the hazardous material is contained within an outer pipe, giving the "double-containment" protection.

Likewise, underground tanks are governed by the new Underground Storage Tank (UST) laws. The UST laws also cover underground piping. By December 1998, all existing underground piping conveying hazardous materials will have to be retrofitted to double-containment systems to comply with EPA requirements.

Double containment of piping brings a host of new problems for the engineering designer. Expansion of the inner and outer pipes must be accommodated so that there is no interference between the two. While prefabricated double-containment piping can solve some of these problems, engineers are still faced with considerations of soil loading, pipe expansion and contraction, and fluid flow. Careful study of the EPA requirements is needed before any double-containment design is finalized. Likewise, local codes and laws must be reviewed prior to starting and before finalizing any design.

LINE SIZING FOR FLASHING STEAM CONDENSATE

REFERENCES: [1] O. Baker, *Oil & Gas J.*, July 26, 1954; [2] S. G. Bankoff, *Trans. ASME*, vol. C82, 265 (1960); [3] M. W. Benjamin and J. G. Miller, *Trans. ASME*, vol. 64, 657 (1942); [4] J. M. Chenoweth and M. W. Martin, *Pet. Ref.*, vol. 34, 151 (1955); [5] A. E. Dukler, M. Wickes, and R. G. Cleveland, *AIChE J.*, Vol. 10, 44 (1964); [6] E. C. Kordyban, *Trans. ASME*, Vol. D83, 613 (1961); [7] R. W. Lockhart and R. C. Martinelli, *Chem. Eng. Prog.*, Vol. 45, 39 (1949); [8] P. M. Paige, *Chem. Eng.*, p. 159, Aug. 14, 1967.

A reboiler in an industrial plant is condensing 1000 lb/h (0.13 kg/s) of steam of 600 lb/in^2 (gage) (4137 kPa) and returning the condensate to a nearby condensate return header nominally at 200 lb/in^2 (gage) (1379 kPa). What size condensate line will give a pressure drop of (1 lb/in^2)/100 ft (6.9 kPa/30.5 m) or less?

Calculation Procedure:

1. *Use a graphical method to determine a suitable pipe size*

Flow in condensate-return lines is usually two-phase, i.e., comprised of liquid and vapors. As such, the calculation of line size and pressure drop can be done by using a variety of methods, a number of which are listed below. Most of these methods, however, are rather difficult to apply because they require extensive physical data and lengthy computations. For these reasons, most design engineers prefer a quick graphical solution to two-phase flow computations. Figure 9 provides a rapid estimate of the pressure drop of flashing condensate, along with a determination of fluid velocity. To use Fig. 9, take these steps.

Enter Fig. 9 near the right-hand edge at the steam pressure of 600 lb/in^2 (gage) (4137 kPa) and project downward to the 200-lb/in^2 (gage) (1379-kPa) end-pressure curve.

From the intersection with the end-pressure curve, project horizontally to the left to intersect

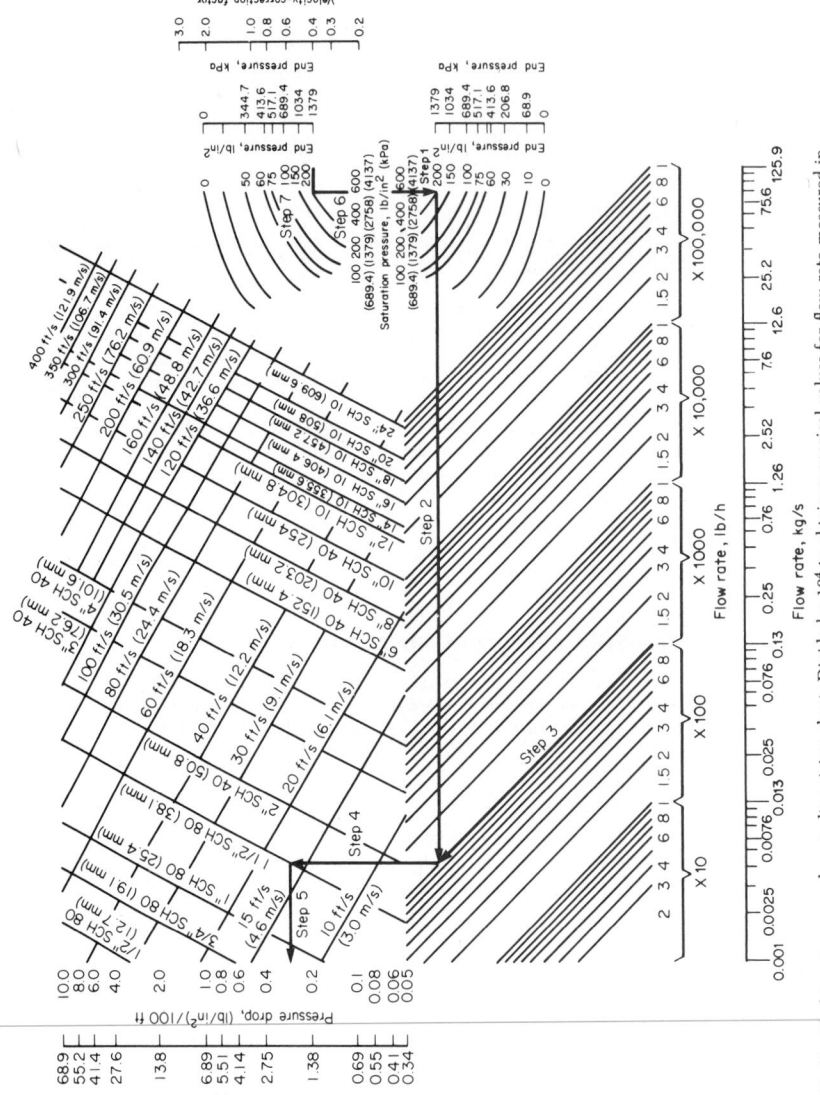

FIG. 9 Flashing steam condensate line-sizing chart. Divide by 10⁴ to obtain numerical values for flow rate measured in kg/s. (*Chemical Engineering.*)

the 1000-lb/h (0.13-kg/s) curve. Project vertically from this intersection to one or more trial pipe sizes to find the pressure loss for each size.

Trying the 1-in (2.5-cm) pipe diameter first shows that the pressure loss—[3.0 lb/in^2 (gage)]/100 ft (20.7 kPa/30.5 m) exceeds the desired [1 lb/in^2 (gage)]/100 ft (6.9 kPa/30.5 m). Projecting to the next larger standard pipe size, 1.5 in (3.8 cm), gives a pressure drop of [2 lb/in^2 (gage)]/100 ft (1.9 kPa/30.5 m). This is within the desired range. The velocity in this size pipe will be 16.5 ft/s (5.0 m/s).

2. *Determine the corrected velocity in the pipe*

At the right-hand edge of Fig. 9, project upward from 600-lb/in^2 (gage) (4137-kPa) to 200-lb/in^2 (gage) (1379-kPa) end pressure to read the velocity correction factor as 0.41. Thus, the actual velocity of the flashing mixture in the pipe = 0.41 (16.5) = 6.8 ft/s (2.1 m/s).

Related Calculations: This rapid graphical method provides pressure-drop values comparable to those computed by more sophisticated techniques for two-phase flow [1–8]. Thus, for the above conditions, the Dukler [5] no-slip method gives [0.22 lb/in^2 (gage)]/100 ft (1.52 kPa/30.5 m), and the Dukler constant-slip method gives [0.25 lb/in^2 (gage)]/100 ft (1.72 kPa/30.5 m).

The chart in Fig. 9 is based on the simplifying assumption of a single homogeneous phase of fine liquid droplets dispersed in the flashed vapor. Pressure drop is computed by Darcy's equation for single-phase flow. Steam-table data were used to calculate the isenthalpic flash of liquid condensate from a saturation pressure to a lower end pressure; the average density of the resulting liquid-vapor mixture is used as the assumed homogeneous fluid density. Flows within the regime of Fig. 9 are characterized as either in complete turbulence or in the transition zone near complete turbulence.

Pressure drops for steam-condensate lines can be determined by assuming that the vapor-liquid mix throughout the lines is represented by the mix for conditions at the end pressure. This assumption conforms to conditions typical of most actual condensate systems, since condensate lines are sized for low-pressure drop, with most flashing occurring across the steam trap or control valve at the entrance.

If the condensate line is to be sized for a considerable pressure drop, so that continuous flashing occurs throughout its length, end conditions will be quite different from those immediately downstream of the trap. In such cases, an iterative calculation should be performed, involving a series of pressure-drop determinations across given incremental lengths.

This iteration is begun at the downstream end pressure and worked back to the trap, taking into account the slightly higher pressure, and thus the changing liquid-vapor mix, in each successive upstream incremental pipe length. The calculation is complete when the total equivalent length for the incremental lengths equals the equivalent length between the trap and the end-pressure point. This operation can be performed by using Fig. 9.

Results from Fig. 9 have also been compared to those calculated by a method suggested by a Paige [8] and based on the work of Benjamin and Miller [3]. Paige's method assumed a homogeneous liquid-vapor mixture with no liquid holdup, and thus it is similar in approach to the present method. However, Paige suggests calculation of the liquid-vapor mix based on an isentropic flash, whereas Fig. 9 is based on an isenthalpic flash; and this is believed to be more representative of steam-condensate collecting systems.

For the example, the Paige method gives (0.26 lb/in^2)/100 ft (1.79 kPa/30.5 m) at the terminal pressure and (0.25 lb/in^2)/100 ft (1.72 kPa/30.5 m) at a point 1000 ft (305 m) upstream of the terminal pressure, owing to the slightly higher pressure, which suppresses flashing.

The method given here is valid for sizing lines conveying flashing steam used in power plants, factories, air-conditioning systems, petroleum refineries, ships, heating systems, etc. Further, Fig. 9 is designed so that it covers the majority of steam-condensate conditions met in these applications. This calculation procedure is the work of Richard P. Ruskin, Process Engineer, Arthur G. McKee & Co., as reported in *Chemical Engineering* magazine.

SAVING ENERGY LOSS FROM STORAGE TANKS AND VESSELS

Fuel oil at 12° API with a viscosity of 50 SSF (0.01068 mm^2/s) at 122°F (50°C) is stored at 300°F (148.9°C) in a 20-ft (6.1-m) diameter by 30-ft (9.1-m) high carbon-steel tank at atmospheric pressure. The oil level in the tank is 18 ft (5.4 m); the air temperature is 70°F (21.1°C). Determine the heat loss to the environment for two situations: (a) Total surface of the tank is uninsulated and black in color. Wind velocity is 0 mi/h (0 km/h). Surface emissivity of the tank is 0.9. Thermal conductivity of the ground under the tank is 0.8 Btu/(h·ft²·°F·ft) [1.38 W/(m·K)]. (b) Roof of tank is uninsulated and is coated with aluminum paint. Sidewall is insulated with calcium silicate, or equivalent, and has a surface emissivity of 0.8. Wind velocity is 30 mi/h (48.0 km/h). The tank contents are not agitated. Thermal conductivity of the ground is 0.8 Btu/(h·ft²·°F·ft) [1.38 W/(m·K)].

Calculation Procedure:

1.   Determine the heat loss from the wetted surface inside the tank

For situation (a), the wetted area inside the tank is $A_L = \pi D H_L$, where A_L = wetted area, ft² (m²); D = tank diameter, ft (m); H_L = liquid height, ft (m). For this tank, $A_L = \pi 20(18) = 1130.9$ ft² (105.1 m²).

2.   Find the temperature difference between the stored liquid and the atmospheric air

The oil temperature T_i is 300°F (148.9°C), and the air temperature T_A is 70°F (21.1°C). Hence, $\Delta T_W = \Delta(T_i - T_A) = 300 - 70 = 230$°F (127.8°C).

3.   Determine the heat loss from the tank

Enter Fig. 10 at the bottom with the temperature difference of 230°F (127.8°C), project vertically upward to the unit heat loss curve q_T, and read $q_T = 664$ Btu/(h·ft²) (2086.6 W/m²). Then the total heat loss from the tank $Q_L = A_L q_T = 1130(664) = 750,320$ Btu/h (219,896.3 W).

4.   Compute the heat loss from the dry inside surface and tank roof surface

The area of the vessel $A_V = \pi D H_V + \pi D^2/4$, where the symbols are as above except that they apply to the dry surfaces of the tank. Then $A_V = \pi(20)(30 - 18) + \pi(20)^2/4 = 1068$ ft² (99.2 m²), where A_V = area.

The temperature difference, by using the correction factor W from Fig. 10 for a noncondensing vapor and the temperature differences in step 2, is $\Delta T_W = (T_i - T_A)W$, or $\Delta T_W = (300 - 70)(0.2) = 46$°F (7.7°C).

Next, the unit heat loss is found from Fig. 10 to be 84.3 Btu/(h·ft²) (265.9 W/m²) for a temperature difference of 46°F (7.7°C). Then the total heat loss from the dry inside surface in contact with the vapor is $Q_V = 1068(84.3) = 90,032.4$ Btu/h (26,385.8 W).

5.   Compute the heat loss through the tank bottom to the ground

The heat loss to the ground through the tank bottom, in Btu/h is $O_G = 2dk_G(T_L - T_G)$, where k_G = thermal conductivity of the ground, Btu/(h·ft²·°F·ft); T_L = liquid temperature, °F; T_G = ground temperature, °F; other symbols as before. Assuming that the ground temperature equals the air temperature, and with $k_G = 0.8$, we see $O_G = 2(20)(0.8)(300 - 70) = 7360$ Btu/h (2156.4 W).

6.   Compute the total heat loss from the tank

The total heat loss from the tank will be the sum of the losses from the liquid, vapor, and ground areas of the tank, or $Q_T = Q_L + Q_V + Q_G = 750,320 + 90,032 + 7360 = 847,712$ Btu/h (248,438.9 W).

7.   Compare the results by using exact equations

Figure 10 gives the exact algebraic equations for the total heat loss from uninsulated tanks. Substituting in these equations gives an exact total heat loss of 894,122 Btu/h (261,977 W). This is a difference of 4.7 percent from the approximate solution obtained by using Fig. 10. Most working engineers would be willing to accept such a difference in view of the savings in time and labor obtained by using the graphic solution.

8.   Determine the insulation thickness needed for the tanks

For situation (b), the wetted surface of the tank $A_L = 1130$ ft² (105.1 m²). This represents the interior circumferential area of the tank wetted by the fuel oil to a height of 18 ft (5.4 m).

Heat loss from the tank to the ambient air is a function of the temperature difference between the tank wall and the air, or $\Delta T_I = T_W - T_A = 300 - 70 = 230$°F (110°C).

From Fig. 11, find the recommended insulation thickness as 1.5 in (3.8 cm) and the wind correction factor for a 30-mi/h (48-km/h) wind and 1.5-in (3.8-cm) insulation as 1.10.

9. Correct the unit heat loss for wind velocity

From step 4, $\Delta T_W = 46$; then $q_L = 46(1.10) = 50.6$ Btu/(h·ft²) (159.6 W/m²).

10. Compute the heat loss from liquid and vapor in the tank

Heat loss from the liquid is $Q_L = A_L q_L = 1130(50.6) = 57,178$ Btu/h (16,757.2 W).
Now the heat loss from the insulated dry-side surface which contacts the vapor is computed

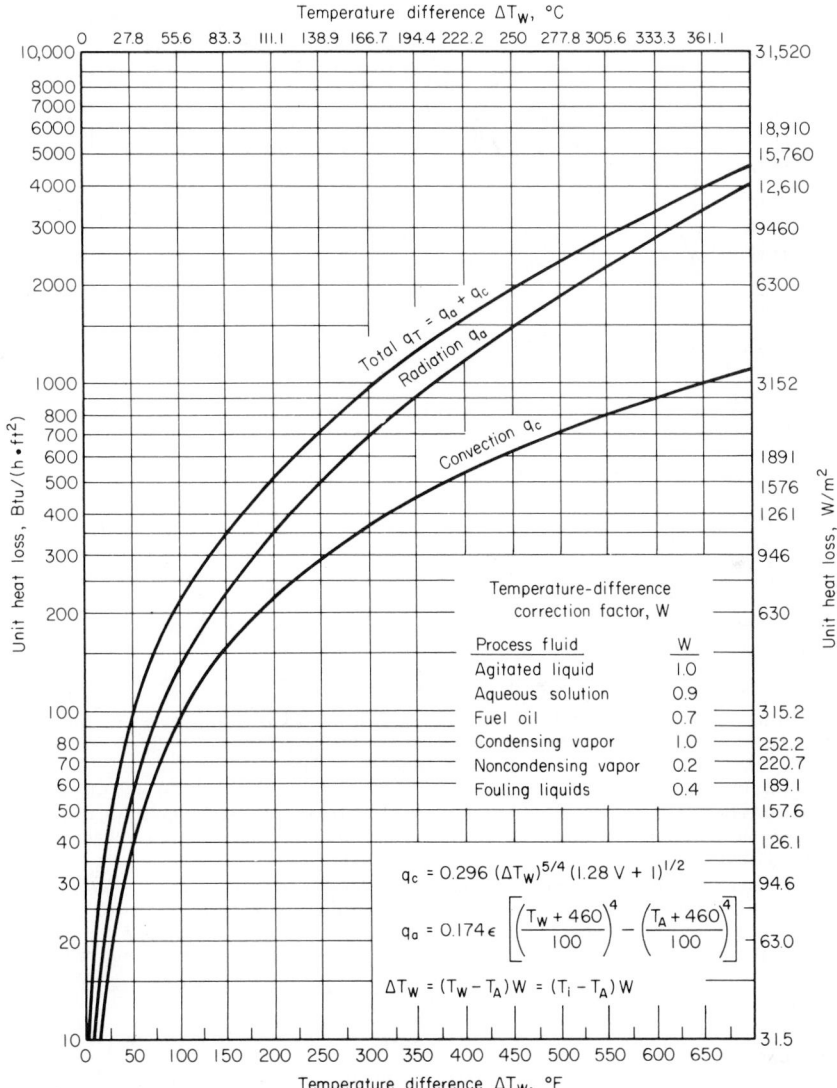

FIG. 10 Heat losses from uninsulated tanks depend on nature of tank contents. Values in chart are for wind velocity of zero, surface emissivity of 0.9, and ambient air temperature of 70°F (21.1°C). *(Chemical Engineering.)*

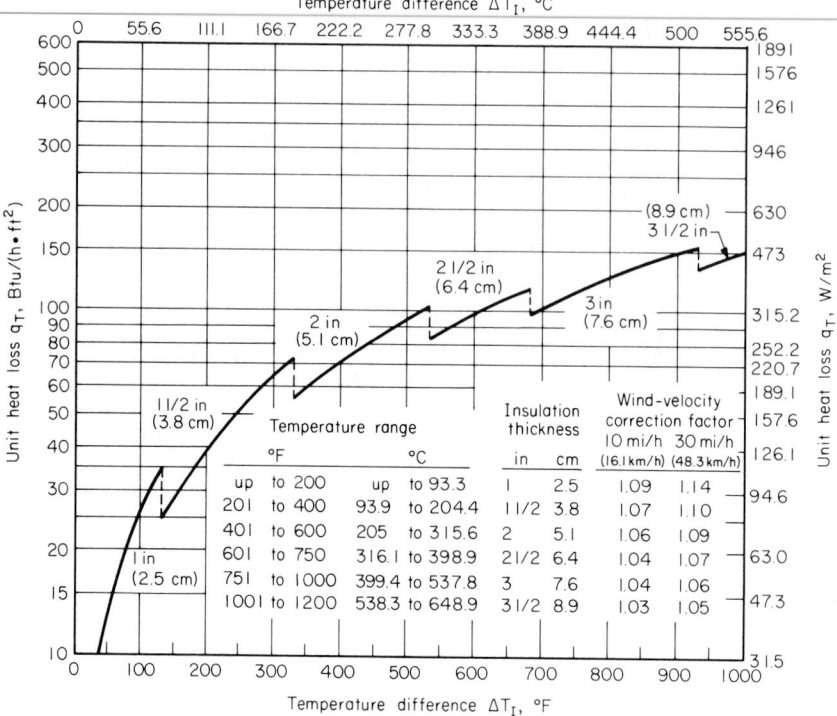

FIG. 11 Insulated tanks, covered with calcium silicate, have heat losses based on negligible resistance to heat flow on process side. Values in chart are for wind velocity of zero, emissivity of 0.8, ambient air temperature of 70°F (21.1°C). *(Chemical Engineering.)*

from $A_V = \pi(20)(30 - 18) = 754$ ft^2 (70.0 m^2). Then $Q_V = A_V q_V = 754(50.6) = 38,152$ Btu/h (11,181.2 W).

11. Determine the heat loss from the uninsulated roof

The area of the roof is $A_R = \pi D^2/4 = \pi(20)^2/4 = 314$ ft^2 (28.3 m^2). Correcting the temperature difference for a noncondensing vapor by using Fig. 10, we find $\Delta T_W = (T_i - T_A)W = 230(0.20) = 46°F$ (7.77°C), where T_i = tank contents temperature, °F.

Using this corrected temperature difference and Fig. 10, we see the unit heat loss for radiation and convection can be found after the emissivity is determined. From the ASHRAE *Guide*, the surface emissivity is 0.40 for bright aluminum-painted surfaces at temperatures in the 50 to 100°F (10 to 37.7°C) range.

The unit heat loss from the roof for radiation and convection is found by using the corrected temperature difference and Fig. 10 and correcting the radiation loss for an emissivity of 0.4 and the convection loss for the wind velocity: $q_a = 54.2(0.4) = 21.7$ Btu/(h·ft^2) (68.4 W/m^2); $q_c = 35.5[(1.28)(30) + 1]^{0.5} = 222.8$ Btu/(h·ft^2) (702.3 W/m^2). Then $q_R = 21.7 + 222.8 = 244.5$ Btu/(h·ft^2) (770.7 W/m^2). Hence, for the roof, $Q_R = A_R q_R$, where A_R = roof area, ft^2. Or, $Q_R = 314(244.5) = 76,773$ Btu/h (22,499.9 W).

12. Determine the total heat loss from the tank

The total heat loss from the tank is the sum of the component heat losses. Or, since the heat loss to the ground is the same as in situation (*a*), $Q_T = 57,543 + 38,529 + 76,773 + 7360 = 180,405$ Btu/h (52,871.3 W).

TABLE 1   Air-Pollution Control Criteria

Substance	Maximum ground concentration, ppm°	Lower explosive limit, ppm	Odor threshold, ppm
Acetylene	—	2.5	—
Ammonia	100	15.5	53
Amylene	—	1.7	2.3
Benzene	50	1.4	1.5
Butane	—	1.9	5,000
Carbon monoxide	100	12.5	Odorless
Ethylene	—	2.8	—
Hydrogen sulfide	30	4.3	0.1
Methanol	200	6.7	410
Propane	—	2.1	20,000
Sulfur dioxide	10	—	3.0

° 8-h exposure.

Using the algebraic equations gives Q_T = 147,945 Btu/h (43,358.2 W). This is a difference greater than in the first situation, but the time savings accrued from using the approximations are significant.

Related Calculations: This procedure can be used for a variety of insulated and uninsulated tanks and vessels used to store oil, chemicals, food, water, and similar liquids in almost any industry. Table 1 lists the typical conditions encountered with such tanks and vessels, the factors which can be neglected in the insulation calculations, and the exact procedure to follow for both graphical and algebraic methods. The methods given here are the work of Richard Hughes and Victor Deumango of the Badger Company and reported in *Chemical Engineering* magazine.

SAVING ENERGY COSTS BY RELOCATING HEAT-GENERATING UNITS

A vacuum pump is driven by a 10-hp (7.5-kW) electric motor and is located in a refrigerated packing room. Determine the energy saving if the vacuum pump and motor are moved out of the room into a noncooled area. Find the energy cost if only the motor is removed from the room. The refrigeration unit has a coefficient of performance of 2.5, the cost of electricity is $0.05 per kilowatthour, the vacuum pump operates 75 percent of the time, and the packing room is cooled 2000 h/year.

Calculation Procedure:

1. Determine the annual cooling load for the motor and pump

Use Fig. 12 to find the heat gain for continuously operated electric motors and equipment. Entering Fig. 12 at 10 hp (7.5 kW) at the bottom and projecting up to the top curve, we see that the amount of heat generated is 30,000 Btu/h (8790 W).

2. Adjust the cooling load for the actual operating time

The pump operates 75 percent of the packing-room annual schedule of 2000 h. Hence, annual cooling load = 0.75(30,000 Btu/h)(2000 h/year) = 45 × 10⁶ Btu/year (47.5 kJ/year).

3. Compute the energy saved

With a coefficient of performance (COP) of 2.5, the energy saved = (annual cooling load, Btu/year) (2.93 × 10⁻⁴ kWh/Btu)/COP, or (45 × 10⁶ Btu/year) (2.93 × 10⁻⁴ kWh/Btu)/2.5 =

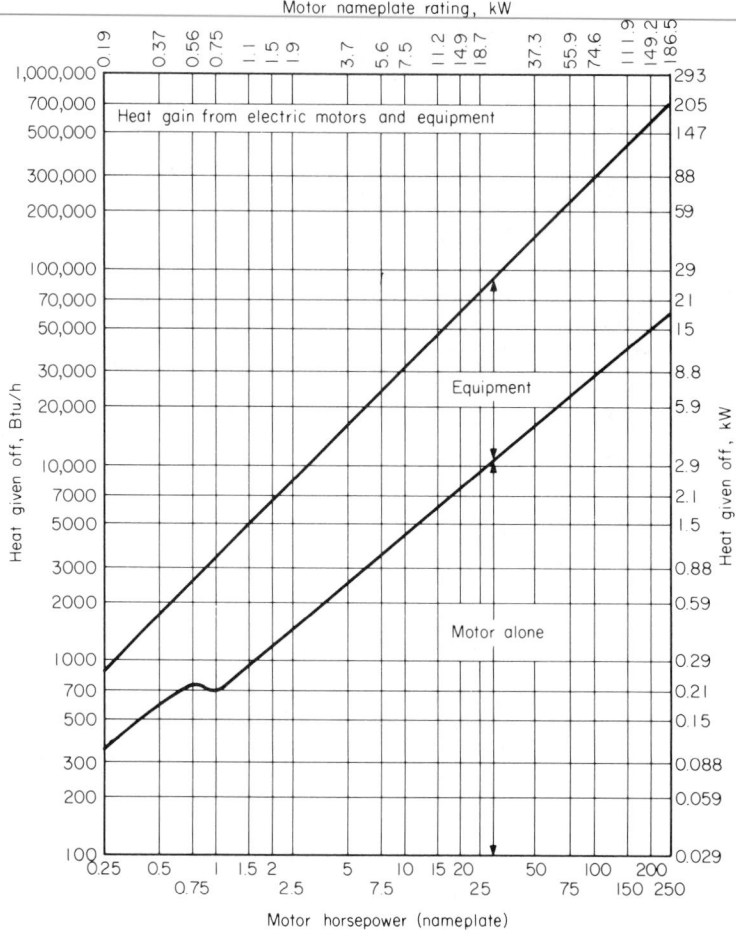

Motor nameplate rating, kW

Heat gain from electric motors and equipment

Equipment

Motor alone

Heat given off, Btu/h

Heat given off, kW

Motor horsepower (nameplate)

FIG. 12   Heat gain from electric motors and equipment. *(Chemical Engineering.)*

5274 kWh/year. At $0.05 per kilowatthour for electric power, the annual saving = ($0.05) (5274) = $263.70.

4.   Determine the saving if only the motor is removed

With only the motor removed from the refrigerated room, the saving is smaller. Figure 12 shows a 4500-Btu/h (1318.5-W) heat gain for a 10-hp (7.5-kW) motor alone. From the assumptions from steps 3 and 4, annual cooling load = 0.75 (4500 Btu/h)(2000 h/year) = 6.75 × 10⁶ Btu/year (7.12 kJ/year). Then, energy saved = (6.75 × 10⁶ Btu/year)(7.12 kJ/year) (2.93 × 10⁻⁴ kWh/Btu)/2.5 = 786 kWh/year. The monetary saving is ($0.05)(786) = $39.30 per year.

   Related Calculations: This method of calculation can be used for any type of motor-driven equipment located in air-conditioned or refrigerated areas. The total heat gain depends on the nameplate (brake) horsepower of the motor and on the motor efficiency. Or, total heat gain, Btu/h (W) = [2545Btu/(hp·h)] (P/E), where P = brake horsepower; E = motor efficiency.

Heat given off by a motor or powered equipment can also be expressed as motor heat gain, Btu/h = [2545 Btu/(hp·h)](P)(1 − E)/E. Also, equipment heat gain, Btu/h = [2545 Btu/(hp·h)]P. Because smaller motors are less efficient, they contribute more of the total heat gain than large motors. For example, 36 percent of the heat gain is contributed by a 0.25-hp (0.19-kW) motor, but only 9 percent by a 250-hp (186.5-kW) motor.

Use this procedure for motor-driven equipment in commercial buildings, factories, ships, aircraft, cold-storage warehouses, and other installations where the heat given off by the motor or equipment will place an extra or unwanted load on the air-conditioning system. The procedure given here is the work of Walter A. Hendrix and William G. Moran, Engineering Experiment Station, Georgia Institute of Technology, as reported in *Chemical Engineering* magazine.

ENERGY SAVINGS FROM VAPOR RECOMPRESSION

Determine the energy savings possible in a plant where 15-lb/in² (gage) (103.4-kPa) steam is vented to the atmosphere while 5000 lb/h (2250 kg/h) of 40-lb/in² (gage) (275.8-kPa) steam is used from the boiler in another process, if the vented steam is recompressed in an electrically driven compressor to the 40-lb/in² (gage) (275.8-kPa) level. The boiler feedwater temperature is 80°F (26.7°C), the boiler efficiency is 80 percent, the boiler operates 8000 h/year, the cost of 150,000-Btu/gal no. 6 fuel oil is $1.00 per gallon, and electricity costs $0.03 per kilowatthour.

Calculation Procedure:

1. *Compute the annual heat input to the boiler*

The annual heat input to the boiler $H = W(\Delta h)T/e$, where H = heat input, Btu/year (W); W = weight of steam used, lb/h (kg/h); Δh = enthalpy change in the boiler = enthalpy of steam − enthalpy of the feedwater, both expressed in Btu/lb (J/kg); T = annual operating time of the boiler, h; e = boiler efficiency, expressed as a percentage. Substituting, we find H = 5000 (1176 − 48)(8000)/0.80 = 5.6 × 10¹⁰ Btu/year (5.91 × 10¹⁰ J/year).

2. *Find the annual fuel cost for generating the steam in the boiler*

The annual fuel cost $C = HP/h_v$, where C = annual fuel cost; P = price per gallon of fuel oil; h_v = fuel heating value, Btu/gal; other symbols as before. Substituting gives C = (5.6 × 10¹⁰)($1.00)/150,000 = $373,333 per year.

3. *Determine the recompression energy input*

When 15-lb/in² (gage) (103.4-kPa) waste steam is compressed to 40 lb/in² (gage) (275.8 kPa) by an electrically driven compressor, the compressor ratio is $c_r = P_d$ = discharge pressure, lb/in² (abs) (kPa)/p_i = inlet pressure, lb/in² (abs) (kPa). Substituting gives c_r = (40 + 14.7)/(15 + 14.7) = 1.84. To find the recompression energy input, enter Fig. 13 at the computed compression ratio and project vertically to the inlet pressure curve. At the left read the energy input as 66 Btu/lb (153.5 kJ/kg) of steam recompressed.

4. *Compute the energy cost of recompression*

The energy cost of recompression = (Btu/lb for recompression) (WT)(0.000293 kWh/Btu) ($0.03 per kWh) = $23,000 per year.

5. *Find the annual energy cost saving from recompression*

The annual energy cost saving for this installation would be $373,333 − $23,200 = $350,133.

Related Calculations: It is common practice in many industrial plants to vent any steam at pressures below 20 lb/in² (gage) (137.9 kPa) to the atmosphere. At the same time, there may be several users that require somewhat higher-pressure steam, 30 to 50 lb/in² (gage) (206.9 to 344.6 kPa). Rather than reducing high-pressure boiler steam to supply these needs, it is possible to compress the waster low-pressure vapor to a higher pressure so that it can be reused. Although energy must be supplied to the compressor to raise the steam pressure, this operation typically requires only 5 to 10 percent of the energy necessary to generate the same steam in a boiler. In practice,

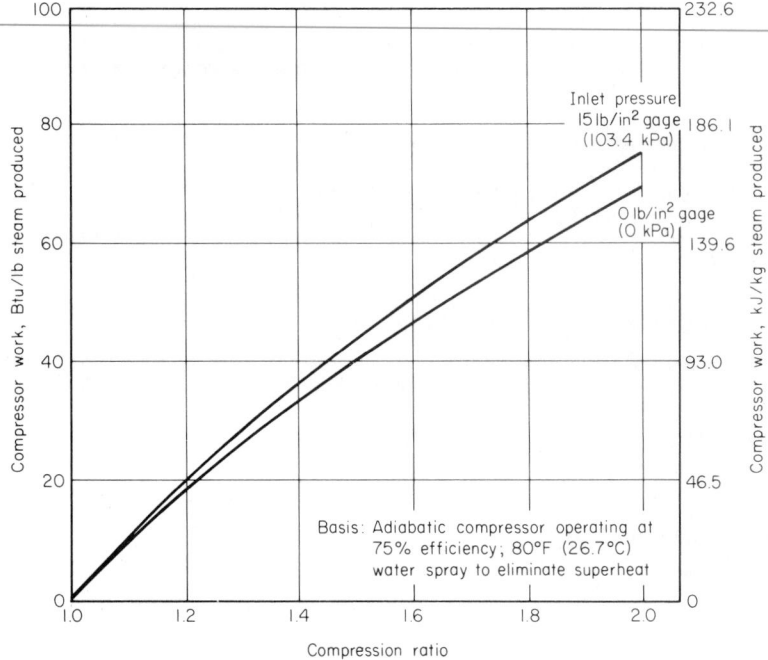

FIG. 13   Recompression work input. *(Chemical Engineering.)*

this principle is limited to situations where the compressor inlet pressure is above 14.7 lb/in² (abs) (101.4 kPa) and the compression ratio is less than about 2.0, owing to physical limitations of the compressors used.

Vapor recompression has been used for years as a means of lowering the steam requirements of evaporators. In this application, the overhead vapors are compressed and recycled to an evaporator steam chest where they evaporate more liquid. In this way, a single-effect evaporator can achieve a steam economy equivalent to an evaporator with up to 15 effects.

Figure 13 indicates the energy required to compress steam as a function of the compression ratio and inlet pressure. It is based on adiabatic compression with a compressor efficiency of 75 percent. Since the steam leaving the compressor is superheated, it is also assumed that water at 80°F (26.7°C) is sprayed into the steam to eliminate the superhead. The graph can be used in conjunction with standard steam tables to estimate the energy saving possible from employing vapor recompression.

The procedure given here can be used for any application—industrial, commercial, residential, marine, etc.—in which recompression of steam might prove economical. This procedure is the work of George Whittlesey and John D. Muzzy, School of Chemical Engineering, Georgia Institute of Technology, as reported in *Chemical Engineering* magazine.

EFFECTIVE STACK HEIGHT FOR DISPOSING PLANT GASES AND VAPORS

Acetylene (molecular weight 26) is being emitted from a process at 100 lb/min (0.76 kg/s). What stack height is needed to achieve an allowable downwind concentration of 40 percent of the lower explosive limit of 2.5?

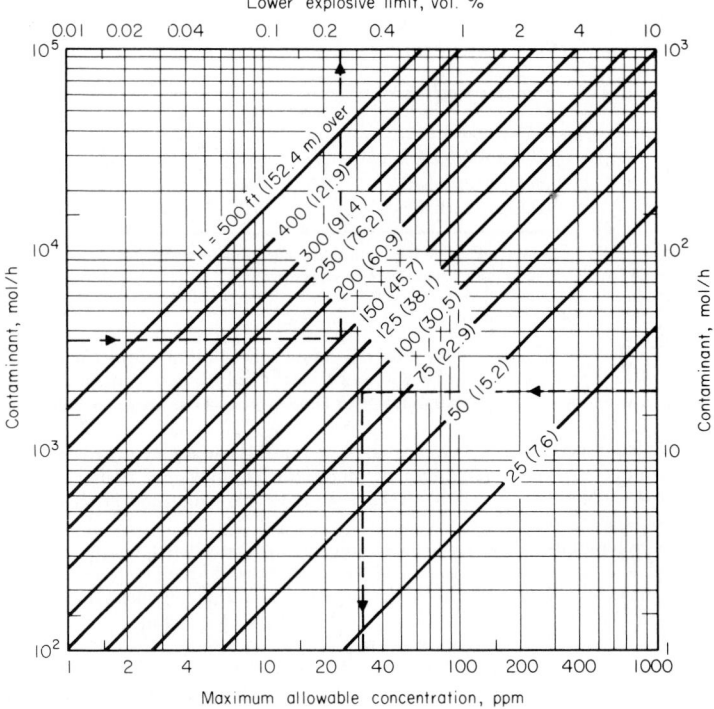

FIG. 14   Stack height needed to dispose of plant gases and vapors. *(Chemical Engineering.)*

Calculation Procedure:

1.   *Convert the flow rate to moles*

To convert from lb/min to mol/h, use the expression $M = 60L/m$, where M = flow rate, mol/h; L = flow rate, lb/min; m = molecular weight of the flowing gas or vapor. Substituting gives $M = 60 (100)/26 = 230$ mol/h.

2.   *Determine the allowable concentration of the vapor*

Use the relation $A = E_L L_e$, where A = allowable concentration downwind of the stack, in percent. Or, $A = 2.5(0.40) = 1.00$ volume percent.

3.   *Find the required stack height*

Use Fig. 14 to determine the stack height, entering at the top of the chart at the lower explosive limit of 1 volume percent and projecting vertically downward to the contaminant flow rate of 230 mol/h. The intersection is just below the stack height of 25 ft (7.6 m). Use a height of 25 ft (7.6 m) because this height is accepted in industry as the minimum allowable. This is the effective height for a 1-mi/h (1.6-km/h) wind.

4.   *Find the distance downwind from the stack*

The distance downwind from the stack where the maximum concentration will be found varies with the turbulence conditions in the area. Turbulence parameters as given by Bosanquet-Pearson are as follows:

	p	q	p/q
Low turbulence	0.02	0.04	0.50
Average turbulence	0.05	0.08	0.63
Moderate turbulence	0.10	0.16	0.63

Note: p and q are the vertical and horizontal dimensionless diffusion coefficients, respectively.

The distance downwind from the stack for maximum concentration of the effluent is given by $d = H/2p$, where d = distance, ft (m); H = stack height, ft (m); p = vertical diffusion coefficient. For low turbulence, $d = 25/2(0.02) = 625$ ft (190.6 m). For moderate turbulence, $d = 25/2(0.1) = 125$ ft (38.1 m).

Related Calculations: When designing emission control systems, be sure to consult the local air-pollution control ordinance (if any) for the criteria which must be met. If an odorous pollutant is being emitted, the design basis will be the concentration below the odor threshold (Table 1) at ground level outside the plant.

Do not design for emission directly to the atmosphere in areas where atmospheric temperature inversions are known to occur. Use, instead, a closed system to rid the plant of the gas or vapor. Extreme care in design is required to avoid the possibility of a legal nuisance suit. Be sure that all applicable ordinances are reviewed before any final design work is begun.

Figure 14 is based on the well-known Bosanquet-Pearson formula. Two solutions plotted on Fig. 14 are for a contaminant flow rate of 3800 mol/h and a lower explosive limit of 0.25 volume percent, requiring a stack height of 150 ft (45.7 m); and a contaminant flow rate of 20 mol/h and a maximum allowable concentration of 31 ppm, requiring a stack height of 75 ft (22.9 m).

As a general guide for stack design, the following relations are given by the above formula:

1. Concentration of an effluent downwind from a source is directly proportional to the discharge quantity.

2. It is impossible to alter materially the downwind ground-level concentration of a contaminant by diluting the effluent.

3. Concentrations downwind of a stack are inversely proportional to wind speed; doubling the wind speed cuts pollutant concentration by half.

4. Pollutant concentration is inversely proportional to the square of the stack height. Doubling the stack height reduces the maximum ground-level concentration to one-fourth the previous level.

5. Location of the maximum ground-level concentration depends on atmospheric stability. When atmospheric conditions are unstable—i.e., wind speeds are low or there is an inversion—the maximum concentration occurs close to the stack. As the wind speed increases and the inversion disappears, the maximum ground-level concentrations move farther away from the stack.

6. Figure 14 is based on a wind speed of 1 mi/h (1.6 km/h) and $p/q = 1$. Further, a smooth, level terrain was ensured for the equation from which the chart is plotted.

7. The effective height to which the plume from a stack rises before it begins to turn downward is the actual height of the stack plus the plume rise created by the sum of the exit velocity and the difference in density above the plume. Use the relation $H_V = 4.77(Q_1 V/1.5)^{0.5}$, where H_V = plume rise due to exit velocity, ft; V = stack exit velocity, ft/s; $Q_1 = Q_V T_1/530$, in which Q_V = stack exit volume, ft^3/s; $T_1 = 18.3$ (molecular weight of contaminant).

This procedure can be used for a variety of gases and vapors, including acetylene, ammonia, amylene, benzene, butane, carbon monoxide, ethylene, hydrogen sulfide, methanol, propane, and sulfur dioxide. Pertinent data for these effluents are given in Table 1.

This calculation procedure is the work of John D. Constance, Consultant, as reported in *Chemical Engineering* magazine.

Where plant gases or vapors pollute the local environment, expensive pollution-abatement equipment may be required. On the west coast of the United States certain chemical and refining plants are gaining emissions credits for their stacks by eliminating pollution elsewhere.

Thus, at the time of writing, one chemical company is paying $700 each for pre-1972 cars. Each car and light truck purchased is cut up for scrap, thereby eliminating the smoke and pollution such vehicles emit. Emissions credits are issued to the chemical plant for each vehicle scrapped. This approach to emissions control is in accord with federal regulations requiring companies to either reduce their own pollution or obtain emissions credits by reducing other pollutants generated in the area.

SAVINGS POSSIBLE FROM USING LOW-GRADE WASTE HEAT FOR REFRIGERATION

An industrial plant presently exhausts low-pressure steam to the atmosphere. What would the annual savings be if a mechanical chiller having a coefficient of performance (COP) of 4.0 producing an average of 150 tons/year (527.4 kW) of refrigeration for 4000 h were replaced by an absorption refrigeration unit using the exhaust steam? The cost of electricity is 3.0 cents per kilowatthour.

Calculation Procedure:

1. *Sketch the refrigeration system being considered*

Figure 15 shows the absorption refrigeration unit being considered. The heat input from the exhaust steam is indicated as Q_G.

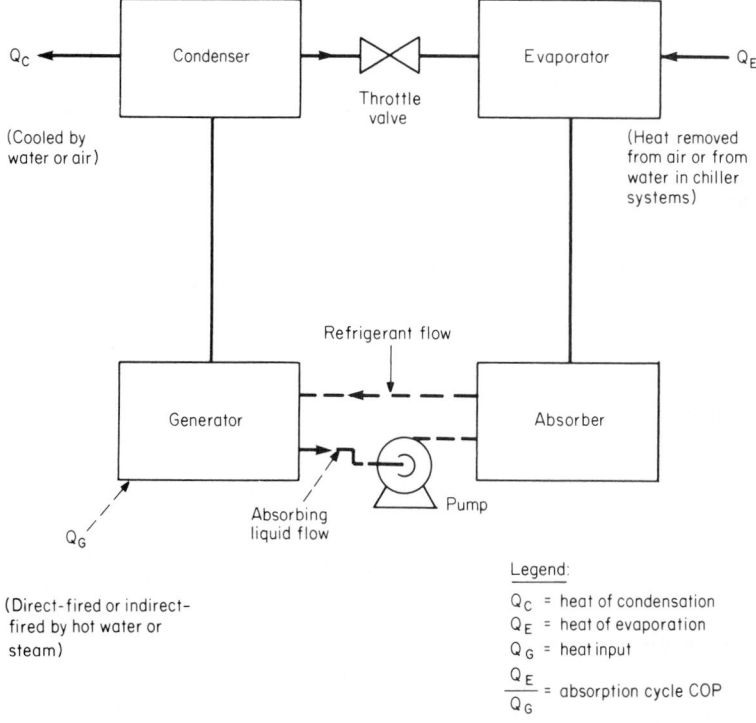

FIG. 15   Absorption refrigeration system. *(Chemical Engineering.)*

In an absorption refrigeration unit, the mechanical vapor compressor is replaced by the generator which uses steam or hot water to revaporize the refrigerant. But absorption refrigeration units are characterized by a low COP, compared to the mechanical type. Hence, to be competitive, an absorption unit must use low-grade waste heat to power the generator. The low-pressure steam available in this plant would be ideal for this purpose.

2. Determine the hourly savings possible

Use Fig. 16 to find the hourly savings. Enter at the refrigeration load of 150 tons (527.4 kW) on the left, and project vertically to the COP value of 4. From the intersection with this curve, project horizontally to the right to intersect the electricity cost curve of 3 cents per kilowatthour. At the bottom read the saving as $4 per hour.

3. Compute the annual savings

Use this relation: annual savings, $ = (hourly savings)(annual number of operating hours) = ($4)(4000) = $16,000 per year.

Related Calculations: This procedure can be used for an absorption refrigeration system by using waste heat in the generator. The heat can be in the form of exhaust steam, hot waste liquids, warm air, etc. The source of the heat is not important provided (1) the temperature of the heating medium is high enough for use in the generator, (2) the supply of heat is steady, and (3) the heat is not chargeable to the refrigeration process.

Given the above criteria as guidelines, the procedure given here can be used for absorption refrigeration machines in industrial plants, commercial building, ships, and domestic applications. Where the COP of an equivalent mechanical refrigeration system is not known, it can be approx-

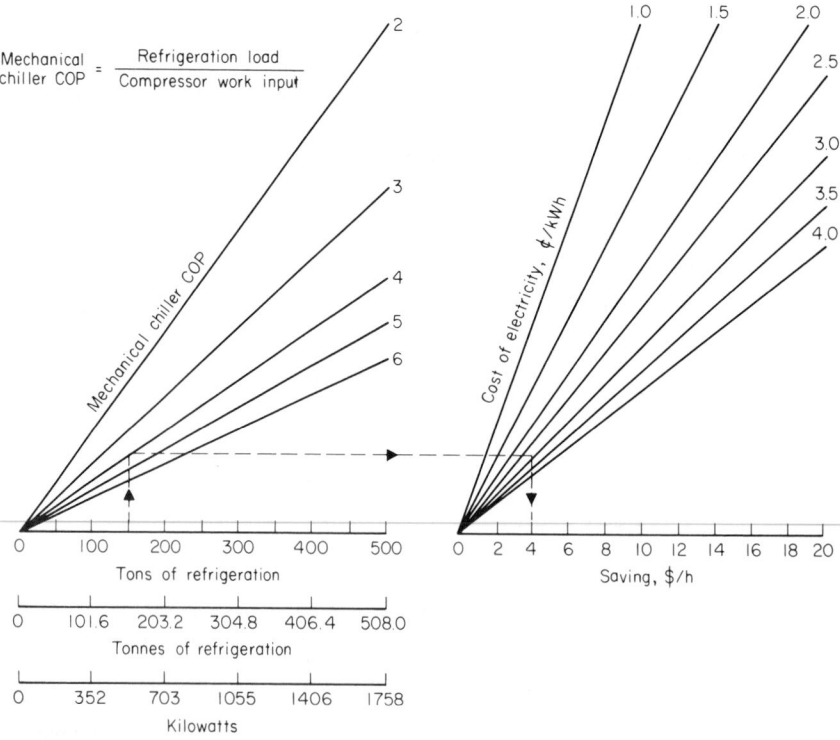

FIG. 16 Savings using low-grade heat. *(Chemical Engineering.)*

imated by applying data from known similar installations. Where the electricity cost exceeds the plotted values, use half the actual cost and multiply the result by 2.

This procedure is the work of Guillermo H. Hoyos, Universidad de los Andes, and John D. Muzzy, School of Chemical Engineering, Georgia Institute of Technology, as reported in *Chemical Engineering* magazine.

EXCESS-AIR ANALYSIS TO REDUCE WASTE-HEAT LOSSES

The fuel input to the primary reformer of an ammonia plant is a mixture of natural gas, tail gas, and naphtha with these flow rates and compositions:

	Natural gas	Tail gas	Naphtha
Fuel no.	1	2	3
Flow rate, kg/h	6800	3420	7800
C:H weight ratio	3.20	2.40	5.35
Weight % carbon	72.3	17.4	84.2
Volume % incombustibles	2.8	53.0	nil

How much excess air is being fed to the reformer if the flue gas analysis shows 3.5 percent O_2 and 11 percent CO_2 on a dry basis?

Calculation Procedure:

1. *Calculate the hydrogen:carbon weight ratio*

The given ratios are for carbon:hydrogen, or the inverse of the required ratios. Hence, the carbon:hydrogen ratio r for each fuel can be found by inverting the given values. Or, for fuel no. 1, natural gas, $r_1 = 1/3.20 = 0.3125$. For fuel no. 2, tail gas, $r_2 = 1/2.40 = 0.4167$; for fuel no. 3, naphtha, $r_3 = 1/5.35 = 0.1869$.

2. *Determine the weight of carbon in the fuel*

The weight of carbon in the fuel c = fraction of carbon × mass flow rate. Or, for natural gas, fuel no. 1, $c_1 = 0.723 \times 6800 = 4915$ kg/h (10,837.7 lb/h), where 0.723 is the weight percent of carbon in the natural gas, expressed as a decimal and given in the table above. Then, for fuel no. 2, tail gas, $c_2 = 0.174 \times 3450 = 600.3$ kg/h (1323.4 lb/h). For fuel no. 3, naphtha, $c_3 = 0.842 \times 7800 = 6568$ kg/h (14,478.8 lb/h).

3. *Determine the mole fraction (volume fraction for gases) of incombustibles in the fuel*

Divide the given percentage of incombustibles by 100 to get the mole fraction for each fuel. Or, for natural gas, $a_i = 2.8/100 = 0.028$; for tail gas, $a_i = 53.0/100 = 0.53$; for naphtha, $a_i = $ nil/100 = nil.

4. *Compute the value R for the fuel*

Here R is defined as $R = \Sigma(1 - a_i)r_i c_i / \Sigma(1 - a_i)c_i$, where the summations are over all fuels in the mixture. Substituting yields $R = [(1 - 0.028)(0.3125)(4916) + (1 - 0.53)(0.4167)(600) + (1 - 0)(0.1869)(6568)]/[(1 - 0.028)(4916) + (1 - 0.53)(600) + (1 - 0)(6568)] = 0.2441$.

5. *Calculate the percentage of excess air*

Use the relation, percentage of excess air = $[100(\%O_2$ in dry flue gas)$]/(\%CO_2$ in dry flue gas)$(1 + 3R)$. Substituting the previously calculated values gives percentage of excess air = $100(3.5)/11[1 + 3(0.2441)] = 18.36$, say 18.4 percent.

Related Calculations: Combustion processes require at least stoichiometric air to get complete fuel utilization, but excess air should be limited to keep the waste heat carried away by flue gases to a minimum. Because it is linked to energy usage, excess air is a key factor in evaluating combustion processes.

The method given here is useful for determining the percentage of excess air being fed to a combustion process. It can be used for both liquid and gaseous fuels, singly or in mixtures, as well as for fuels that contain incombustibles. The method only requires readily available information such as fuel characteristics, flow rates, and flue-gas analysis and avoids the laborious solution of simultaneous material-balance equations that is usually needed.

This calculation procedure is the work of S. Michael Antony, IFFCO, Ltd., as reported in *Chemical Engineering* magazine.

ESTIMATING SIZE AND COST OF VENTURI SCRUBBERS

Determine the size and cost of a venturi scrubber to handle 100,000 actual ft^3/min (47.2 m^3/s) of gas entering the venturi in an air-pollution control system. The scrubber, Fig. 17, is to remove all particles larger than 0.6 μm.

Calculation Procedure:

1. *Determine the scrubber base cost*

The cost of a venturi scrubber depends on the volumetric flow, operating pressure, and materials of construction. Figure 18 gives flange-to-flange costs—covering the venturi, elbow, separator, pumps, and controls—for base systems [i.e., constructed of 0.125-in (0.32-cm) carbon steel] for different volumetric flow rates. Do not extend the curve or equation beyond 200,000 actual ft^3/min (94.4 m^3/s).

Whether a thickness other than 0.125 in (0.32 cm) is required can be determined from Fig. 19, the thickness being a function of design operating pressure and shell diameter. The curves in Fig. 19 include a safety factor of 2, but no allowance is made for corrosion and erosion. Figure 19 is used after the base cost is determined from Fig. 18.

Entering Fig. 18 at 100,000 ft^3/min (47.2 m^3/s) at the bottom and projecting vertically upward to the curve, we find a base cost of $39,400. From the equation, the exact cost would be $39,417.

2. *Find the pressure drop in the scrubber*

For an efficiency that removes all particles larger than 0.6 μm, Fig. 20 shows that a 35-in water gage (w.g.) (88.9-cm w.g.) pressure drop is needed. Note that as the particle size removed decreases, the pressure drop required for particle removal increases.

3. *Compute the metal thickness and scrubber cost*

At the 35-in w.g. (88.9-cm w.g.) pressure drop, Fig. 19 shows that a scrubber metal thickness of 0.25 in (0.64 cm) is required. (Always round *up* to the next standard metal thickness when you

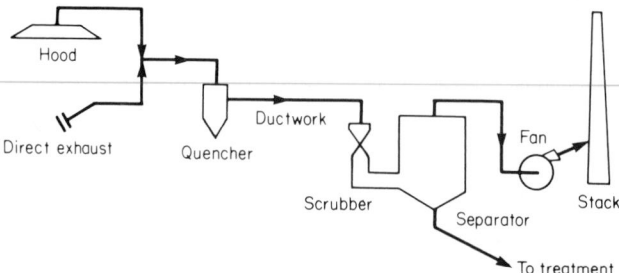

FIG. 17 Scrubber-separator in an air-pollution control system. *(Chemical Engineering.)*

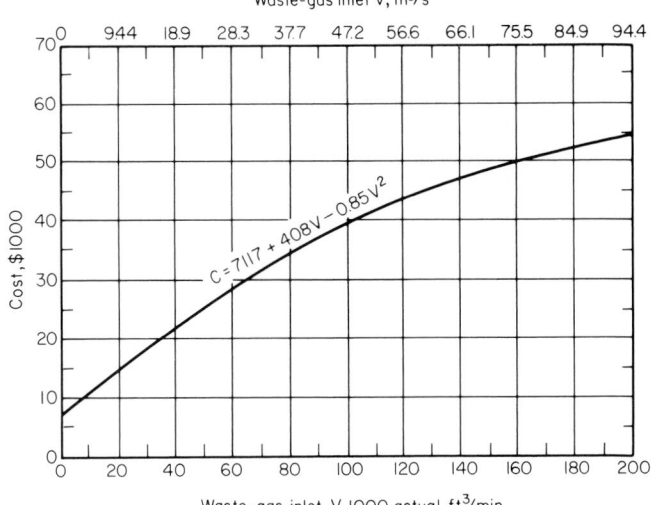

FIG. 18 Base scrubber cost, flange-to-flange construction of 0.125-in (0.32-cm) carbon steel. (*Chemical Engineering magazine and Fuller Company.*)

use this procedure.) For 0.25-in (0.64-cm) carbon steel, Fig. 21 gives a cost adjustment factor of 1.6. Hence, the scrubber cost will be 1.6 × $39,400 = $63,040.

If the scrubber metal is to be 304 or 316 stainless steel, multiply the above cost estimate, $63,040, by 2.3 or 3.2, respectively. And if the scrubber is to be equipped with a fiberglass lining to reduce wear of the metal, multiply the Fig. 18 estimate of $39,400 by 0.15 to obtain $5910, and add this to $63,040 to arrive at an estimate of $68,950.

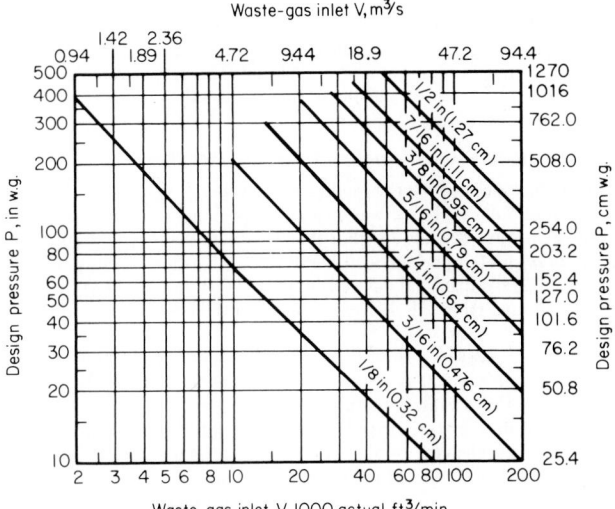

FIG. 19 Flow rate and design pressure dictate scrubber metal thickness. (*Chemical Engineering magazine and Fuller Company.*)

Pressure drop required, cm w.g.

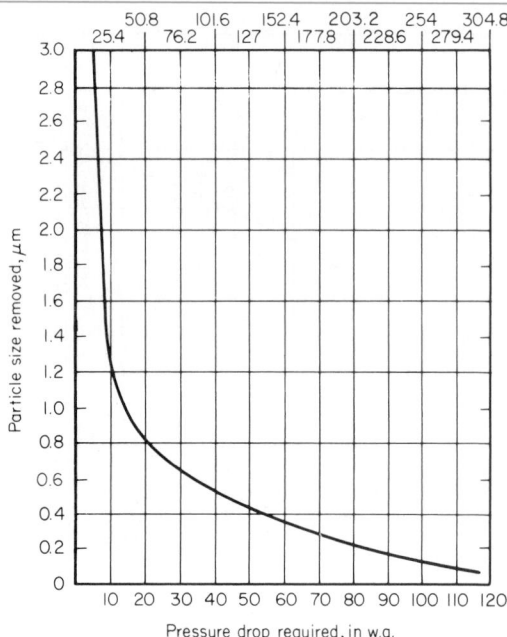

FIG. 20 Correlation gives efficiency performance of venturi scrubbers. *(Chemical Engineering magazine and AIChE.)*

4. *Determine the cost of a rubber-lined scrubber*

If the scrubber is to be lined with 0.1875-in (0.476-cm) rubber, or any other thickness rubber, determine the internal surface area of the scrubber from Fig. 22. For 100,000 actual ft³/min (47.2 m³/s), the scrubber internal area is 1500 ft² (139.4 m²). The unit cost for lining a scrubber with this thickness rubber is \$4.69 per ft² (\$50.48 per m²). Hence, the total cost = (1500 ft²)(\$4.69/ft²) = \$7035. Adding this to the \$63,040 gives an estimate of \$70,075.

5. *Determine the number and diameter of the scrubber trays*

If the separator is to be equipped for gas cooling, the number of trays needed must be determined, based on an average of 5 lb (2.3 kg) of water removed per square foot of tray area, with an outlet gas temperature about 40°F (22.2°C) higher than the inlet water temperature. [This is valid for typical scrubber outlet-gas temperatures of 200°F (93.3°C) or less, cooling-water temperature of about 70°F (21.1°C), and superficial gas velocities of 600 ft/min (3.05 m/s).] The total water to be removed is determined from the difference between the absolute humidities of the inlet and outlet gas streams.

Find the diameter of each tray from Fig. 22 for a flow of 100,000 ft³/min as 13.5 ft (4.1 m). Figure 23 gives a cost of \$14,000 per tray. This includes the cost of the tray plus the cost of additional separator height to contain the tray.

If the separator requires six trays to achieve the dehumidification required, the \$70,075 estimate will be increased by \$84,000 (= 6 × \$14,000) to \$154,075, say \$154,000. (If the chart in Fig. 18 were used instead of the equation, the total estimate would be \$153,000 after rounding.) To update the estimate to current costs, use a suitable cost index, as detailed below.

Related Calculations: Venturi scrubbers are highly efficient in removing submicron dust particles from gas streams.

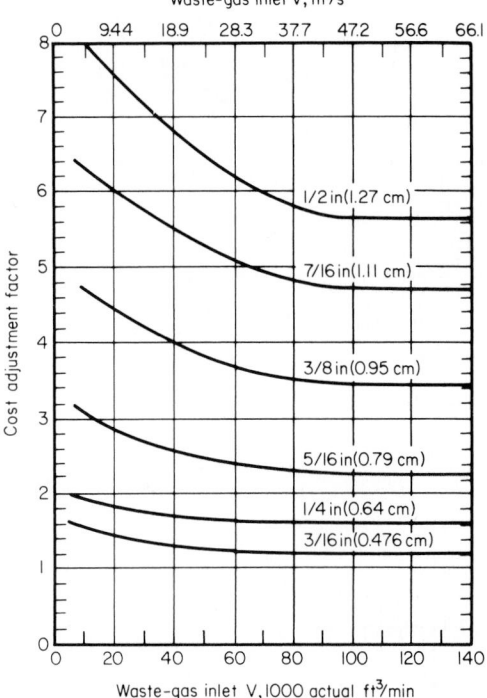

FIG. 21 Factors adjust scrubber cost for flow rate and metal thickness. *(Chemical Engineering magazine and Fuller Company.)*

Basically, the gas stream accelerates in the converging section of the venturi to maximum velocity in the throat, where it is sprayed by a scrubbing liquor. The faster velocity of the gas stream atomizes the liquor and promotes collisions between the particles and the droplets. Agglomeration in the diverging section produces droplets, with entrapped particles, of a size easily removed by mechanical means.

Collection efficiency depends on the venturi pressure drop, which is a function of gas-stream throat velocity and scrubbing-liquor flow rate. The smaller the particles, the higher the pressure drop required. Venturis are normally operated at pressure drops between 6 in (15.2 cm) and 80 in (203.2 cm) water gage, depending on the characteristics of the dust, and at liquor flow rates of 3 to 20 gal/min (0.19 to 1.26 L/s) per 1000 actual ft^3/min (0.47 m^3/s). Collection efficiencies range from 99+ percent for 1-μm and larger particles to between 90 and 99 percent for those less than 1 μm.

Precise pressure drops can be obtained from Fig. 20, which represents 100 percent removal of particle sizes indicated for a particular pressure drop. For instance, to remove all particles 0.4 μm (10^{-6} m) and larger requires a pressure drop of 55 in (139.7 cm) w.g. (Also $P_d = 15.4d^{-1.39}$, with d the diameter in micrometers, also gives the pressure drop.)

A separator—normally a cylindrical tank having a low tangential inlet and a centered top gas outlet—located immediately downstream of the scrubber removes the agglomerated liquor drops from the gas stream by a cyclonic motion that forces them to impinge on the tank wall. Slurry settles into a bottom cone, from which most of it is sent to the water treatment facility, with the cleaner liquid above the sediment being recycled to the venturi (Fig. 17).

In hot processes, a considerable amount of water is vaporized in the scrubber and upstream equipment (particularly the quencher). Unless this vapor is removed, it must be handled by the

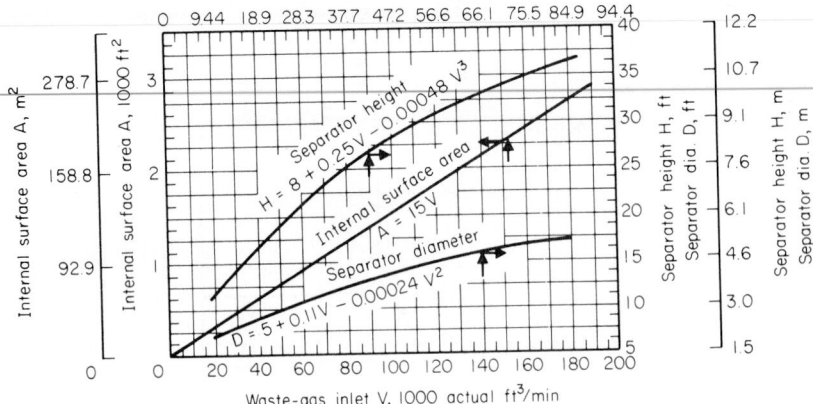

FIG. 22 Scrubber internal surface area and separator dimensions. *(Chemical Engineering magazine and Fuller Company.)*

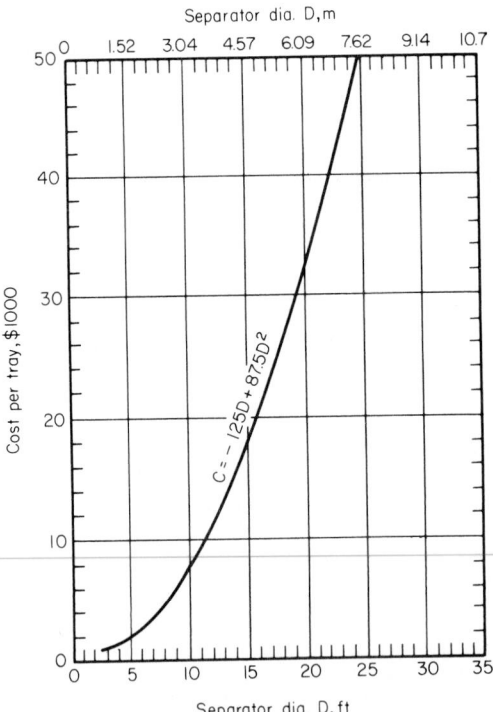

FIG. 23 Cost per tray of separator internal cooler. *(Chemical Engineering magazine and Fuller Company.)*

fan (commonly a radial-tip fan), which therefore must be of higher horsepower and so is more costly to operate.

A gas cooler can be incorporated into the separator to cool and dehumidify the gas stream. Such a cooler can be one of several types, including one in which the gas stream passes through spray banks of cooling water and then impinges on baffles, and another in which the stream rises through perforated holes or bubble caps in trays flooded with cooling water.

The cost data in this procedure are based on 1977 values. They can be easily updated to the current year by using a suitable cost index, for the year in which the estimate is being made. Thus, if the cost index (*Chemical Engineering, EN-R, Marshall & Stevens*, etc.) is 229 for the year 1977 and 245 for the design year, the cost of the scrubber in the design year will be $154,000(245/229) = $164,759, say $165,000.

This procedure is the work of William M. Vatavuk of the U.S. Environmental Protection Agency and Robert B. Neveril of Gard, Inc., as reported in *Chemical Engineering* magazine, using data from an AIChE paper and Fuller Company.

Toxic air pollutants are now strictly controlled by the eleven titles of the U.S. Clean Air Act Amendments (CAAA) passed in 1990. Title III deals with 189 chemicals designated as hazardous air pollutants (HAP), which may be emitted by chemical process plants. EPA is required to issue control standards based on maximum achievable control technologies (MACT) for sources designated to be "major" or "area" generators of hazardous pollutants.

Any facility that emits 10 or more tons per year of any single HAP, or 25 or more tons per year of any combination of HAPs is considered a major source. Alternatively, an area source is any facility that routinely emits HAPs but is not classified a major source. EPA is required to ensure that 90 percent of the emissions from the 30 most serious area-source pollutants are regulated by the year 2000. For these reasons, scrubbers like those discussed above are becoming more important to chemical engineers worldwide.

Title IV of the act will impose strict SO_2 and NO_x control requirements on electric utilities. Title V requires operators of plants to apply for, and obtain, permits. These operating permits, issued by the states, will establish emission limits, permit fees, and monitoring and reporting requirements.

Volatile organic compounds (VOC) comprise half of all regulated toxic air pollutants. VOC are regulated by Title I, which imposes strict compliance requirements on particulate matter and ozone precursors, including some VOC. These regulations may exceed MACT requirements under Title III. Data on toxic air pollutants as given here are from K. Sampeth Kumar, Rodney L. Pennington, and Jan T. Zmuda, Research-Cottrell Co., a subsidiary of Air & Water Technologies Corp., as reported in *Chemical Engineering*.

SIZING DESUPERHEATER-CONDENSERS ECONOMICALLY

A reactor exhausts 27,958 lb/h (3.52 kg/s) of isobutane with a small amount of *n*-butane at 200°F (93.3°C) and 85 lb/in² (gage) (586.0 kPa). This gas becomes saturated at 130°F (54.4°C) and condenses completely at 125°F (51.7°C). The gas is to be cooled and condensed by a horizontal counterflow heat exchanger like that in Fig. 24 using well water at 65°F (18.3°C) inlet temperature and an outlet temperature of 100°F (37.8°C). How much heat-transfer area is required in this exchanger?

Calculation Procedure:

1. *Check for condensation at the hot end*

Desuperheater-condensers are widely used in the process, petrochemical, chemical, and power industries. Figure 24 shows a horizontal in-shell design which might be used as a high-pressure feed heater, an inter- or aftercondenser in a steam-jet ejector system, or a gas cooler in a compressor system.

Conventional design practice splits the heat load and sizes the desuperheating and condensing zones separately. This assumes that the superheated vapor cools as if it were a dry gas, which is

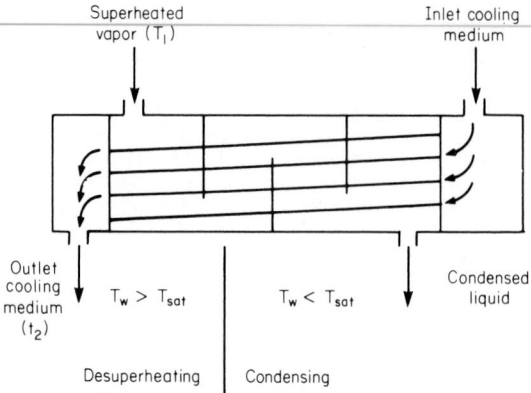

FIG. 24 Horizontal in-shell desuperheater-condenser. *(Chemical Engineering.)*

true only when the tube-wall temperature T_w in the desuperheating zone is greater than the vapor's saturation temperature T_{sat}.

When T_w is less than T_{sat}, the superheated vapor condenses directly, in the same way as saturated vapor. Because the heat flux in condensation is much greater than in desuperheating, this situation requires less heat-transfer area than the conventional design would prescribe.

For a counterflow desuperheater-condenser with vapor on the shell side like that in Fig. 24, the energy balance for the desuperheating zone is $h_d(T_1 - T_w) = U_d(T_1 - t_2)$, where h_d is the desuperheating heat-transfer coefficient; U_d is the overall heat-transfer coefficient; T_1 is the vapor inlet temperature; t_2 is the cooling-medium outlet temperature. Using this energy-balance equation and the data from Fig. 25, we find the tube-wall temperature at the hot end of the unit is $45.2(200 - T_w) = 42.3(200 - 100)$; $T_w = 106.4°F$ (41.3°C). Since T_w is less than T_{sat}, condensation does take place as soon as the superheated vapor enters the shell.

2. Determine the area required for the condensing load

Since condensation does take place, the entire desuperheating load, 860,000 Btu/h (251.9 W), should be treated as a condensing load, by using the condensing heat-transfer coefficient shown in Fig. 25. The heat-transfer area required for this load is (Btu/h)/U_c(LMTD); or $A = 860,000/158(33) = 165$ ft² (15.3 m²).

3. Compare the conventional approach to this approach

Using the conventional approach, such as that in Kern—*Process Heat Transfer*, McGraw-Hill, and in the heat-exchanger calculation procedures given elsewhere in this handbook (see index), shows that the heat duty for the exchanger is split into two zones—condensing and desuperheating. The area required for the condensing zone is 523 ft² (48.6 m²); for the desuperheating zone it is 323 ft² (30 m²). Important data for the desuperheating zone are shown in Fig. 25.

The total heat-transfer area, when the approach given here is used, will be the condensing-zone area from the conventional approach + desuperheating area computed in step 2. Or, $A_{total} = 523 + 165 = 688$ ft² (63.9 m²). This compares with $523 + 323 = 846$ ft² (78.6 m²) for the conventional approach, or 23 percent greater area.

Note that the desuperheat was a sizable fraction (over 20 percent) of the total heat load in this exchanger, which is cooling an organic vapor. With steam, the importance of desuperheating is generally less.

Related Calculations: Energy conservation studies often show a longer breakeven period and a smaller payout because equipment costs are excessive. Higher equipment costs lead to greater cost of money for the heat-recovery unit or units. Hence, it is important that any equipment chosen to conserve heat be sized properly. The procedure given here shows how a saving of

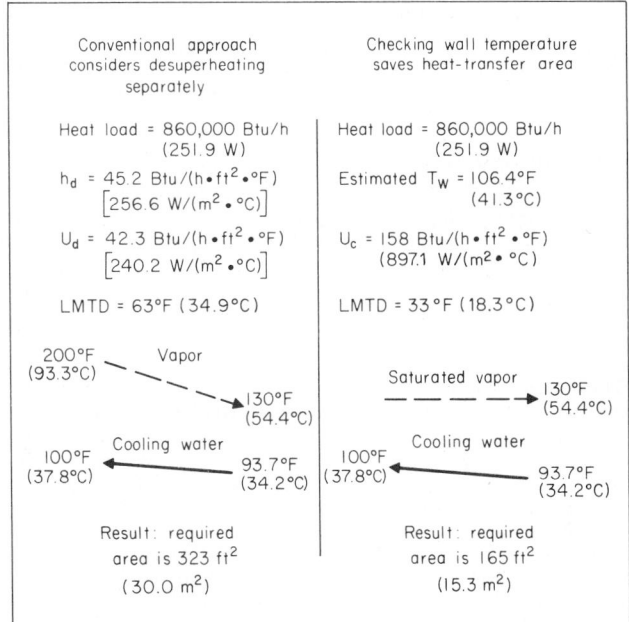

FIG. 25 Example shows importance of checking wall temperature. *(Chemical Engineering.)*

nearly 25 percent can be made in the area of certain types of heat exchangers. Such savings can significantly reduce the required investment, leading to an earlier breakeven and higher payout. Thus, energy conservation will be easier to justify when this procedure is used.

Since this procedure is relatively simple, it should be applied in selecting a heat exchanger which involves desuperheating. The method is applicable in land, marine, chemical, petrochemical, and process heat-exchanger selection. This procedure is the work of P. S. V. Kurmarao, Ph.D., EDC (Heat Exchangers) Bharat Heavy Electricals Ltd., as reported in *Chemical Engineering* magazine.

SIZING VERTICAL LIQUID-VAPOR SEPARATORS

Find the diameter needed for a vertical vessel to separate a liquid having a density $d = 58.0$ lb/ft^3 (928.6 kg/m^3) from 2000 mol/h of vapor having a molecular weight of 25.0 at an operating temperature of 300°F (148.7°C) and 250 lb/in^2 (gage) (1723.5 kPa). The compressibility factor $Z = 1.0$ for the vapor.

Calculation Procedure:

1. Find the vapor volumetric flow rate V

Using the gas law $PV = nRTZ$, let $V = $ ft^3/s and $n = 2000/3600$ mol/s. Solving yields $V = nRTZ/P = (2000/3600)(10.73)(760)(1.0)/264.7 = 17.1$ ft^3/s (48,393 m^3/s). In this equation, R = the gas constant; T = absolute temperature, °R = 460 + operating temperature, °F; P = absolute pressure of the vapor, lb/in^2 (abs).

2. Determine the density of the vapor d_v

Use the relation $d_v = $ (mol/h) (molecular weight)/volumetric flow rate, lb/h. Or, $d_v = (2000)(25)/(17.1)(3600$ s/h$) = 0.812$ lb/ft^3 (13.0 kg/m^3).

3. Compute the terminal vapor velocity v_t

Use the relation $v_t = K'[(d - d_v)/d_v]^{0.5}$, where K' is a constant which ranges between 0.1 and 0.35, with 0.227 being the value for many satisfactory designs and recommended except when special considerations are warranted. Substituting, we find $v_t = 0.227[(58.0 - 0.812)/0.812]^{0.5}$ = 1.91 ft/s (0.58 m/s).

4. Find the allowable vapor velocity v_a

Use the relation $v_a = 0.15v_t$, where the constant 0.15 is based on an allowable vapor velocity of 15 percent of v_t to ensure good liquid disentrainment during the normal flow surges. For usual designs, researchers have determined that v_a should be 15 percent of v_t. By substituting, $v_a = 0.15(1.91) = 0.286$ ft/s (0.087 m/s).

5. Determine the separator cross-sectional area and diameter

The separator cross-sectional area $A = V/v_a = 17.1/0.286 = 59.8$ ft^2 (5.6 m^2). Then the separator diameter $D = [(4)(59.8)/\pi]^{0.5} = 8.7$ ft (2.65 m). A diameter of 9 ft (2.74 m) would be chosen.

Related Calculations: Vertical liquid-vapor separators are used primarily to disengage a liquid from a vapor when the volume of the first is small compared with that of the second. The separation is accomplished by providing an environment (i.e., a vessel) in which the liquid particles are directed by the force of gravity rather than the force of the flowing vapor.

Devices have been developed to agglomerate liquid particles in a vapor stream and enhance disentrainment. Some act as baffles, causing multiple changes in the direction of vapor flow. Inertia keeps the liquid particles from changing direction, and they impinge on the baffles. As the particles coalesce on the baffles, they agglomerate into droplets, which fall because of gravity.

Other devices, such as packing and grids, provide a large surface area for liquid coalescence and agglomeration. One such device that has gained wide acceptance—because it is highly efficient and relatively inexpensive and causes negligible pressure drop—is the mist elimination pad. Usually a mesh formed by knitting metal wire, it comes in a variety of standard thicknesses and densities. For general process-separator and compressor-suction knockout-pot services, a stainless-steel pad of 4-in (10.2-cm) thickness and nominal 9-lb/ft^3 (144.1-kg/m^3) density is the most economical.

A separator equipped with a mist eliminator can be considerably smaller in diameter than one not having it. Indeed, design practice permits ignoring the 15 percent safety factor and letting the allowable vapor velocity be equal to the terminal vapor velocity (that is, $v_a = v_t$).

In the separator above, therefore, the required cross-sectional area of the vertical separator A now becomes $A = 18.1/1.91 = 9.0$ ft^2 (2.74 m^2). And the separator diameter D becomes $D = [(4 \times 9.0)/\pi]^{0.5} = 3.4$ ft (1.04 m). A diameter of 3 ft 6 in (1.07 m) would now be chosen.

The height of the liquid level in a vertical separator, Fig. 26, depends primarily on the residence time dictated by process considerations. Suppose that the residence time for the above separator is chosen as 5 min. For the 3.5-ft (1.07-m) diameter separator, the cross-sectional area is $A = (\pi/4)(3.5)^2 = 9.62$ ft^2 (0.89 m^2).

Assuming that liquid is entering the separator at a rate of 2000 gal/h (2.1 L/s), or 4.64 ft^3/min (0.00219 m^3/s), we find the liquid level for a 5-min residence time is L = (ft^3/min)(residence time, min)/A = 4.46(5)/9.62 = 2.32 ft (0.71 m). Choose a sump height of 2.5 ft (0.76 m). A vertical separator is usually specified when a short liquid holdup time is permitted.

The following procedure is standard in the process design of vertical liquid-vapor separators. A standardized design procedure and vessel configuration saves much engineering time. A separator is usually relatively inexpensive, and the application of a rigorous, sophisticated procedure to achieve an optimum design is seldom warranted. Only in special cases, such as when a separator is built of extra thick laminated shells, does it become economical to attain an optimum design, because the saving in fabrication cost can be significant.

The standard procedure stipulates the following:

1. The allowable vapor velocity v_a in a separator shall be equal to the terminal velocity v_t, calculated by rounding up the vessel diameter to the nearest 6 in (15.2 cm), when a mist eliminator is used. However, v_a shall be no greater than 15 percent of v_t when the separator is not equipped with a mist eliminator.

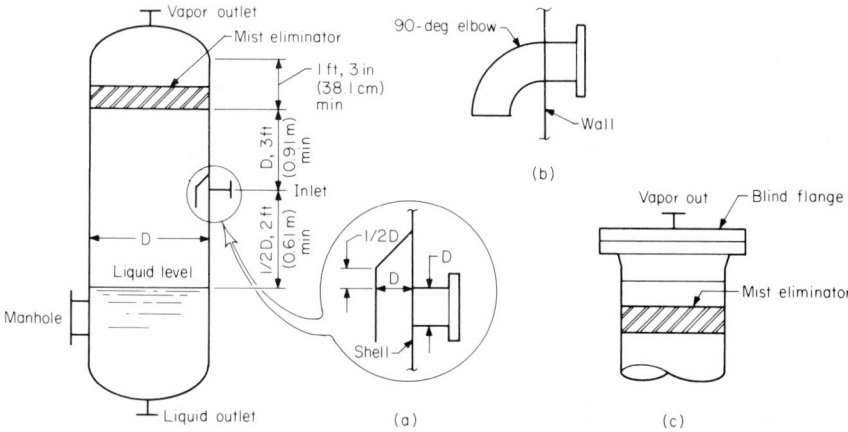

FIG. 26 (*a*) Inlet configuration, diameter $\geq$ 30 in (9.1 cm). (*b*) Inlet configuration, diameter < 30 in (9.1 cm). (*c*) Top head configuration of pipe separators, diameter < 30 in (9.1 cm). *(Chemical Engineering.)*

2. The disengaging space, the distance between any inlet and the bottom of the mist elimination pad (see Fig. 26), shall be equal to the diameter of the separator. However, when the diameter of the separator is less than 3 ft 0 in (0.91 m), the height of the disengaging space shall be a minimum of 3 ft (0.91 m).

3. The distance between the inlet nozzle and the maximum liquid level shall be equal to one-half the vessel diameter, or a minimum of 2 ft (0.61 m).

4. The dimension between the top tangent line of the separator and the bottom of the mist elimination pad shall be a minimum of 1 ft 3 in (38.1 cm) (Fig. 26).

5. Vessel diameters 3 ft 0 in (0.91 m) and larger shall be specified in increments of 6 in (15.2 cm). Diameters of shell plate vessels shall be specified as inside diameters. Vessel lengths shall be specified in 3-in (7.6-cm) increments.

6. Separators of 30-in (76.2-cm) diameter and smaller shall be specified as fabricated from pipe. Diameter dimensions shall represent pipe outside diameters. Top heads shall be specified as full-diameter flanges, with blind flange covers (Fig. 26). Bottom heads shall be standard heads or pipe caps.

7. Inlets shall have an internal arrangement to divert flow downward. Vessels 3 ft 0 in (0.91 m) and larger shall have a hood, attached to the shell, covering the inlet nozzle (Fig. 26).

8. Outlets shall have antivortex baffles.

9. Mist elimination pads shall be specified as 4-in (10.2-cm) thick, nominal 9-lb/ft^3 (144.1-kg/m^3) density and stainless steel. Spiral-wound pads are not acceptable.

The method given here is valid for vertical separators used in process, chemical, petrochemical, power, marine, and a variety of other plants. This procedure is the work of Arthur Gerunda, Vice President of Commercial Development, The Heyward-Robinson Co., as reported in *Chemical Engineering* magazine.

SIZING A HORIZONTAL LIQUID-VAPOR SEPARATOR

Design a horizontal vessel to separate 7000 gal/h (7.36 L/s) of liquid having a density of 60 lb/ft^3 (960.6 kg/m^3) from 1000 mol/h of vapor having a molecular weight of 28 if the holding time for the liquid is to be 8 min when the operating temperature is 100°F (37.8°C), the operating pressure is 300 lb/in^2 (gage) (2068.2 kPa), and Z = 1.0.

Calculation Procedure:

1. Find the volumetric flow rate V

Using the gas law as in the previous calculation procedure with $n = 1000/3600$ mol/s and $A = 1.0$, we get $V = (1000/3600)(10.73)(560)(1.0) = 5.3$ ft³/s (15,004 m³/s).

2. Compute the density of the vapor d_v

As in the previous calculation procedure, $d_v = $ (mol/h) (molecular weight)/(V) (3600) $= 1.47$ lb/ft³ (23.5 kg/m³).

3. Determine the terminal vapor velocity v_t

Use the relation $v_t = 0.227[(d_l - d_v)/d_v]^{0.5}$, where $d_l = $ density of liquid. Or $v_t = 0.227 [(60 - 1.47)/1.47]^{0.5} = 1.43$ ft/s (0.44 m/s). The constant 0.227 is obtained in the same way as described in the previous calculation procedure.

4. Decide what sets the size of the separator

Either the rate of liquid separation from the vapor or the liquid holding time will set the size of the separator. *By liquid separation,* $D = [V/3(\pi/4)(0.15)(V_t)]^{0.5}$ where $D = $ separator diameter, ft. Or, $D = [5.3/(4\pi/4)(0.15 \times 1.43)]^{0.5} = 2.8$ ft (0.85 m).

By holding time, $D = [t_h V_i/3(\pi/4)f]^{0.333}$, where $t_h = $ holding time, min; $V_i = $ volumetric flow rate of the liquid, ft³/min; $f = $ fraction of the separator area occupied by the liquid. Assuming an L/D ratio, i.e., separator length/diameter, of 4, and with 7000 gal/h = 15.6 ft³/min (0.01 m³/s), make a first approximation with an assumed liquid-space area f of 0.70: $D = [8 (15.6)/4(\pi/4) (0.70)]^{0.333} = 3.84$ ft (1.17 m).

Since the larger diameter is set by the holding time, this is the determining factor in the sizing of the separator.

Next, examine a 4-ft (1.22-m) diameter vessel with $f = 0.70$. With $L/D = 4$, $L = 4D = 4(4) = 16$ ft (4.9 m).

With the area of the vapor space = 30 percent (that is, $1.00 - 0.70$), the fractional height of the vapor space $f_{hv} = 0.342$, from the geometry of the tank. The height of the vapor space then is 0.342 (46 in) $= 15.7$ in (39.9 cm), where the 46 in (116.8 cm) is the approximate actual internal diameter of the vessel. Make the vapor-space height 18 in (45.7 cm). This gives a vapor-space area fraction f_{av} of 0.36 and a liquid-space area fraction f_{al} of 0.64.

The holding time t_h now is $t_h = [(\pi/4)(4^2)(0.64)(16)]/15.6 = 8.25$ min. Hence, the final separator size is as follows: diameter = 4 ft (1.22 m), length = 16 ft (4.88 m), and liquid height = 2.5 ft (0.76 m). As with a vertical separator, when a horizontal separator is equipped with a mist elimination pad, the allowable vapor velocity v_a can be taken to be the same as the terminal velocity v_t in the vessel diameter calculations. Figure 27 shows some typical arrangements of mist eliminators in horizontal liquid-vapor separators.

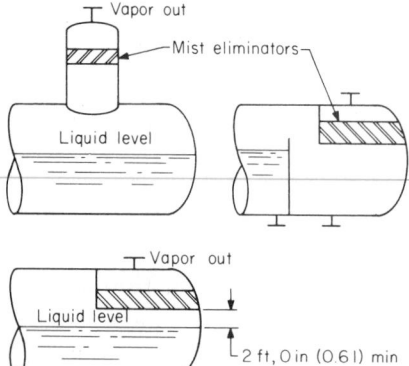

FIG. 27 Mist eliminators in horizontal separators. *(Chemical Engineering.)*

Related Calculations: The chief concern in designing a horizontal liquid separator is to have the vapor velocity sufficiently low to give the liquid particles just enough time to settle out before the vapor leaves the vessel. Figure 28 shows the approximate traverse of a liquid particle for which the minimum time has been allowed for its disentrainment from the vapor. Indicated in the cross section are the fraction of area f_{av} and height f_{hv} taken up by the vapor space.

As with a vertical separator, empirical findings have shown that for safe design the allowable vapor velocity v_a in a horizontal separator should be no greater than 15 percent of the calculated terminal velocity v_t. Another restriction found necessary is that f_{av} be no less than 15 percent of the cross-sectional area.

For horizontal separators L/D ratios are dictated by economics and plot restrictions. As a general guide, the following provides economic designs:

Operating pressure		L/D ratio
lb/in² (gage)	kPa	
0–250	0–1723.5	3.0
251–500	1730.4–3447.0	4.0
501 and higher	3423.9 and higher	5.0

For a first trial size, set the liquid level at the centerline of the separator, so that $f_h = f_a = 0.5$.

Now, $D = [V/(3)(\pi/4)(0.15 \times v_t)]^{0.5}$, or $D = [V/(0.35 \times v_t)]^{0.5}$.

These equations provide a good starting point for trial-and-error calculations to determine the size of a liquid-vapor horizontal separator operating at less than 251 lb/in² (gage) (1730.4 kPa). Note that the terminal vapor velocity v_t has, in effect, been replaced by the allowable vapor velocity v_a, because v_t is multiplied by the safety design factor, 0.15.

The following specifications are generally standard in the design of horizontal separators: (1) The maximum liquid level shall provide a minimum vapor space height of 15 in (38.1 cm) but not be below the centerline of the separator. (2) The volume of dished heads is not taken into account in vessel sizing calculations. (3) Inlet and outlet nozzles shall be located as closely as practical to the vessel tangent lines. (4) Liquid outlets shall have antivortex baffles.

When the size of a horizontal separator is set by the holdup time for the liquid, the diameter of the vessel must be determined by trial-and-error calculations. If f_{al} = the fraction of area occupied by the liquid, the holdup time t_h is given by $t_h = [(\pi/4)D^2 f_{al}L]V_t$. Here, L = vessel

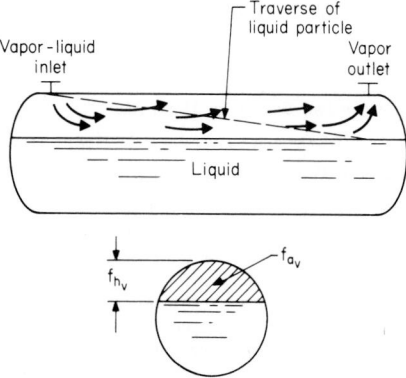

FIG. 28 Traverse of liquid particle. *(Chemical Engineering.)*

length and V_t = volumetric flow rate of the liquid. When the operating pressure is below 251 lb/in^2 (gage) (1730.4 kPa), L/D = 3.0. Solving for D yields $D = [t_h V_t/3(\pi/4)f_{al}]^{0.333}$.

This procedure is the work of Arthur Gerunda, Vice-President, Commercial Development, Heyward-Robinson Co., as reported in *Chemical Engineering* magazine.

SIZING RUPTURE DISKS FOR GASES AND LIQUIDS

What diameter rupture disk is required to relieve 50,000 lb/h (6.3 kg/s) of hydrogen to the atmosphere from a pressure of 80 lb/in^2 (gage) (551.5 kPa)? Determine the diameter of a rupture disk required to relieve 100 gal/min (6.3 L/s) of a liquid having a specific gravity of 0.9 from 200 lb/in^2 (gage) (1378.8 kPa) to atmosphere.

Calculation Procedure:

1. Determine the rupture disk diameter for the gas

For a gas, use the relation $d = (W/146P)^{0.5} (1/Mw)^{0.25}$, where d = minimum rupture-disk diameter, in; W = relieving capacity, lb/h; P = relieving pressure, lb/in^2 (abs); Mw = molecular weight of gas being relieved. By substituting, $d = [50,000/146(94.7)]^{0.5} (1/2)^{0.25} = 1.60$ in (4.1 cm).

2. Find the rupture-disk diameter for the liquid

Use the relation $d = 0.236(Q)^{0.5} (Sp)^{0.25}/P^{0.25}$, where the symbols are the same as in step 1 except that Q = relieving capacity, gal/min; Sp = liquid specific gravity. So $d = 0.236(100)^{0.5}(0.0)^{0.25}/(214.7)^{0.25} = 0.60$ in (1.52 cm).

Related Calculations: Rupture disks are used in a variety of applications—process, chemical, power, petrochemical, and marine plants. These disks protect pressure vessels from pressure surges and are used to separate safety and relief valves from process fluids of various types.

Pressure-vessel codes give precise rules for installing rupture disks. Most manufacturers will guarantee rupture disks they size according to the capacities and operating conditions set forth in a purchase requisition or specification.

Designers, however, often must know the needed size of a rupture disk long before bids are received from a manufacturer so the designer can specify vessel nozzles, plan piping, etc.

The equations given in this procedure are based on standard disk sizing computations. They provide a quick way of making a preliminary estimate of rupture-disk diameter for any gas or liquid whose properties are known. The procedure is the work of V. Ganapathy, Bharat Heavy Electricals Ltd., as reported in *Chemical Engineering* magazine.

TIME NEEDED TO EMPTY A STORAGE VESSEL WITHOUT DISHED ENDS

How long will it take to empty a 10-ft (3-m) diameter spherical tank filled to a height of 8 ft (2.4 m) with ethanol, a newtonian fluid, if the drain is a short 2-in (5.1-cm) diameter tube of double extra-strong pipe?

Calculation Procedure:

1. Determine the discharge coefficient for the drain

Figure 29 shows that the discharge coefficient is $C_d = 0.80$ for a short, flush-mounted tube.

2. Compute the discharge time

Substitute the appropriate values in the equation in Fig. 29 for spherical storage tanks. Or, $t = (2)^{0.5}(\pi)(8)^{1.5}[10 - (0.6 \times 8)]/[3(0.8)(1.774/144)(32.2)^{0.5}] = 3116$ s, or 51.9 min.

Related Calculations: Figure 29 gives the equations for computing the emptying time for four common tank geometrics. The discharge coefficient C_d is constant for newtonian fluids in turbulent flow, but the coefficient depends on the shape of the orifice. Water flowing through sharp-edged orifices of 0.25-in (0.64-cm) diameter, or larger, is always turbulent. Thus, the assumption of a constant C_d is valid for most practical applications. Figure 29 lists accepted C_d values.

The relations given here are valid for storage tanks used in a variety of applications—chemical and petrochemical plants, power plants, waterworks, ships and boats, aircraft, etc. This procedure is the work of Thomas C. Foster, as reported in *Chemical Engineering* magazine.

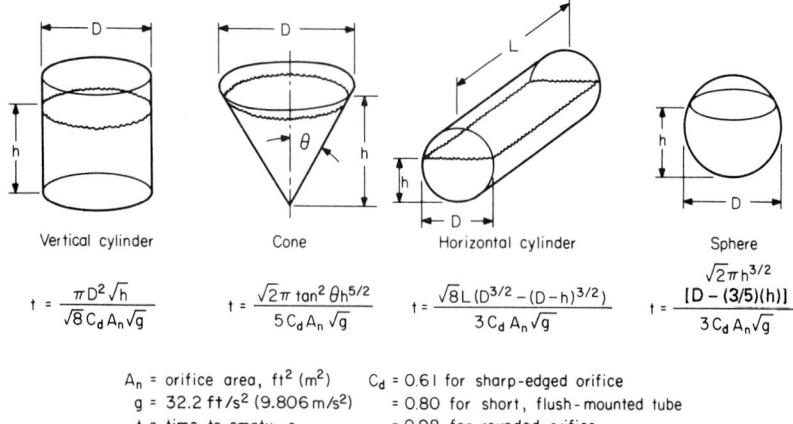

$$t = \dfrac{\pi D^2 \sqrt{h}}{\sqrt{8}\,C_d A_n \sqrt{g}}$$

$$t = \dfrac{\sqrt{2\pi}\,\tan^2\theta\, h^{5/2}}{5\,C_d A_n \sqrt{g}}$$

$$t = \dfrac{\sqrt{8}\,L\,(D^{3/2}-(D-h)^{3/2})}{3\,C_d A_n \sqrt{g}}$$

$$t = \dfrac{\sqrt{2}\,\pi\, h^{3/2}\,[D-(3/5)(h)]}{3\,C_d A_n \sqrt{g}}$$

A_n = orifice area, ft² (m²) C_d = 0.61 for sharp-edged orifice
g = 32.2 ft/s² (9.806 m/s²)   = 0.80 for short, flush-mounted tube
t = time to empty, s   = 0.98 for rounded orifice

FIG. 29   Time to empty tanks. *(Chemical Engineering.)*

COST ESTIMATION OF HEAT EXCHANGERS AND STORAGE TANKS VIA CORRELATIONS

Using correlations, estimate the cost of a fixed-head, carbon-steel heat exchanger rated for 150 lb/in² (gage) (1034 kPa) having a total heat-transfer area of 1500 ft² (139.4 m²). Using the same approach, estimate the cost of a cone-roof storage tank made of carbon steel having a total capacity of 677,000 gal (2,562,445 L). Show how to update the costs from the base year (cost index = 200.8) to a year in which the cost index is 265.

Calculation Procedure:

1.   Compute the base cost of the heat exchanger

Using Table 2, substitute the area A in the relation $C_B -$ exp [8.551 $-$ 0.30863 ln A + 0.06811 (ln $A)^2$]. Or, $C_B =$ exp [8.551 $-$ 0.30863 ln 1500 + 0.06811 (ln 1500)²] = $20,670.

2.   Determine the exchanger-type cost factor for the heat exchanger

Again, by using Table 2 for a fixed-head exchanger, $F_D =$ exp (-1.1156 + 0.090606 ln A), where F_D = exchanger-type cost factor. Substituting yields $F_D =$ exp (-1.1156 + 0.090606 ln 1500) = 0.6357.

3.   Find the design-pressure cost factor for the exchanger

From Table 2, the design-pressure cost factor for a pressure in the 100 to 300-lb/in² (gage) range (700 to 2100-kPa range) is $F_P =$ 0.7771 + 0.04981 ln A. Substituting, we find $F_P =$ 0.7771 + 0.04981 ln 1500 = 1.1414.

4.   Find the materials-of-construction cost factor

The materials-of-construction cost factor, F_M, for carbon steel is unity, or 1.0. Factors for other materials of construction are shown in Table 3.

5.   Compute the heat-exchanger cost

Use the relation $C_E = C_B F_D F_P F_M$, where C_E = exchanger cost. Or, $C_E =$ ($20,670)(0.6357) (1.1414)(1.0) = $15,000.

6.   Update the heat-exchanger cost

The base-year cost index—for 1976, the year on which the above costs are based—is 200.8. For the year in which the cost estimate is being made, the cost index is 265 (obtained from any of the standard, widely accepted cost indices). Updating the heat-exchanger cost reveals $C_{EU} =$ $15,000(265/200.8) = $19,796. In this relation, the updated cost is $C_{EU} = C_E$(current-year equipment cost index/base-year cost index).

TABLE 2 Correlations for Costs of Heat Exchangers[a]

USCS units	SI units
Base cost for carbon-steel, floating-head, 100 lb/in² (gage) exchanger:	Base cost for carbon-steel, floating-head, 700-kN/m² exchanger:
$C_B = \exp[8.551 - 0.30863 \ln A + 0.06811 (\ln A)^2]$	$C_B = \exp[8.202 + 0.01506 \ln A + 0.06811 (\ln A)^2]$
Exchanger-type cost factor:	Exchanger-type cost factor:
Fixed-head: $F_D = \exp(-1.1156 + 0.0906 \ln A)$	Fixed-head: $F_D = \exp(-0.9003 + 0.0906 \ln A)$
Kettle reboiler: $F_D = 1.35$	Kettle reboiler: $F_D = 1.35$
U-tube: $F_D = \exp(-0.9816 + 0.0830 \ln A)$	U-tube: $F_D = \exp(-0.7844 + 0.0830 \ln A)$
Design-pressure cost factor:	Design-pressure cost factor:
100 to 300 lb/in² (gage): $F_P = 0.7771 + 0.04981 \ln A$	700–2100 kN/m²: $F_P = 0.8955 + 0.04981 \ln A$
300 to 600 lb/in² (gage): $F_P = 1.0305 + 0.07140 \ln A$	2100–4200 kN/m²: $F_P = 1.2002 + 0.07140 \ln A$
600 to 900 lb/in² (gage): $F_P = 1.1400 + 0.12088 \ln A$	4200–6200 kN/m²: $F_P = 1.4272 + 0.12088 \ln A$
A in ft², lower limit: 150 ft², upper limit: 12,000 ft²	A in m², lower limit: 14 m², upper limit: 1100 m².

[a] *Chemical Engineering.*

TABLE 3   Material-of-Construction Cost Factors for Heat Exchangers[*]

Material	USCS units, A in ft² $F_M = g_1 + g_2 \ln A$		SI units, A in m² $F_M = g_1 + g_2 \ln A$	
	g_1	g_2	g_1	g_2
Stainless steel 316	0.8608	0.23296	1.4144	0.23296
Stainless steel 304	0.8193	0.15984	1.1991	0.15984
Stainless steel 347	0.6116	0.22186	1.1388	0.22186
Nickel 200	1.5092	0.60859	2.9553	0.60859
Monel 400	1.2989	0.43377	2.3296	0.43377
Inconel 600	1.2040	0.50764	2.4103	0.50764
Incoloy 825	1.1854	0.49706	2.3665	0.49706
Titanium	1.5420	0.42913	2.5617	0.42913
Hastelloy	0.1549	1.51774	3.7614	1.51774

[*]*Chemical Engineering.*

7. *Compute the storage-tank cost*

Using Table 4, apply the relation $C_B = \exp\,[11.362 - 0.6104 \ln V + 0.045355 \,(\ln V)^2]$, where C_B = base cost of field-erected tank in carbon steel; V = tank volume, gal. Substituting gives $C_B = \exp\,[11.362 - 0.6104 \ln 677.000 + 0.045355 \,(\ln 677{,}000)^2] = \$84{,}300$. Updating the cost, as before, we find $C_{BU} = \$84{,}300(265/200.8) = \$111{,}252$. Table 5 shows materials of construction cost factors for storage tanks.

Related Calculations: The approach given here correlates the cost of shell-and-tube heat exchangers and heat-transfer area. This contrasts with cost estimation procedures that take into account shell diameter, number and length of the tubes, types of heads, and other construction details. The accuracy of the simple correlation of cost versus area is sufficient for preliminary cost estimates.

Correlations for base cost are given in both USCS and SI units in the accompanying tables. The base-cost basis for the equipment is given in each table. While heat-exchanger costs are based on area, storage-tank costs are based on the total tank volume. The tank volume is calculated (for the base cost) from residence time, a fixed overcapacity factor of 20 percent, and volumetric flow rate.

Omitted from the cost estimation procedure given here are the number and sizes of nozzles and manholes and other design details. These details cause variations in cost that are usually within the accuracy of preliminary estimates.

Data on the cost of shell-and-tube heat exchangers in a wide range of heat-transfer areas and design pressures were used in developing the correlations for 10 different materials of construction and three design types. PDQ\$, Inc. supplied these and the cost data for cylindrical carbon-steel tanks having cone roofs and flat bottoms in a wide range of volumes. The cost of field-erected tanks includes the cost of platforms and ladders, but not of foundations and other installation materials (piping, electric instrumentation, etc.). The cost of the shop-fabricated tanks does not include any of the installation materials.

This procedure is the work of Armando B. Corripio, Louisiana State University, and Katherine S. Chrien and Lawrence B. Evans, both of the Massachusetts Institute of Technology, as reported in *Chemical Engineering* magazine.

ESTIMATING CENTRIFUGAL-PUMP AND ELECTRIC-MOTOR COST BY USING CORRELATIONS

Determine the cost of a ductile-steel pump to deliver 1430 gal/min (90.2 L/s) at a differential head of 77 ft·lbf/lb (230.2 J/kg). A horizontally split case one-stage pump running at 3550 r/min is specified. The specific gravity of the fluid being pumped is 0.952.

TABLE 4 Correlations for Costs of Storage Tanks[a]

USCS units	SI units
Base cost for carbon-steel, shop-fabricated tanks:	Base cost for carbon-steel, shop-fabricated tanks:
$C_B = \exp[2.331 + 1.3673 \ln V - 0.063088 (\ln V)^2]$	$C_B = \exp[7.994 + 0.6637 \ln V - 0.063088 (\ln V)^2]$
V in gallons; lower limit: 1300 gal, upper limit: 21,000 gal	V in m³, lower limit: 5 m³, upper limit: 80 m³
Base cost for carbon-steel, field-erected tanks:	Base cost for carbon-steel, field-erected tanks:
$C_B = \exp[11.362 - 0.6104 \ln V + 0.045355 (\ln V)^2]$	$C_B = \exp[9.369 - 0.1045 \ln V + 0.045355 (\ln V)^2]$
V in gallons; lower limit: 21,000 gal, upper limit: 11,000,000 gal	V in m³, lower limit: 80 m³, upper limit: 45,000 m³

[a]*Chemical Engineering.*

TABLE 5 Material-of-Construction Cost Factors for Storage Tanks°

Material of construction	Cost factor F_M
Stainless steel 316	2.7
Stainless steel 304	2.4
Stainless steel 347	3.0
Nickel	3.5
Monel	3.3
Inconel	3.8
Zirconium	11.0
Titanium	11.0
Brick-and-rubber- or brick-and-polyester-lined steel	2.75
Rubber- or lead-lined steel	1.9
Polyester, fiberglass-reinforced	0.32
Aluminum	2.7
Copper	2.3
Concrete	0.55

° *Chemical Engineering.*

Calculation Procedure:

1. Determine the size parameter S

The size parameter is defined as $S = QH^{0.5}$, where Q = design capacity of the pump, gal/min (m³/s), and H is the required head for the pump, ft·lb/lb or J/kg. Substituting, we get $S = 1430(77)^{0.5} = 12,550$, closely.

2. Find the pump base cost C_B

Use the base cost relation from Table 6, or $C_B = \exp[8.3949 - 0.6019 \ln S + 0.0519 (\ln S)^2]$. So $C_B = \exp[8.3949 - 0.6019 \ln 12,550 + 0.0519 (\ln 12,550)^2] = \$1536$.

3. Compute the pump design-type factor F_T

Table 6 shows that the design-type factor for a one-stage, 3550-r/min HSC pump is found from $F_T = \exp[b_1 + b_2 \ln S + f_3 (\ln S)^2]$. Substituting the values given in the table, we see $F_T = \exp[0.0632 + 0.2744 \ln 12,550 - 0.0253 (\ln 12,550)^2] = 1.491$.

4. Find the materials-of-construction factor F_M

From Table 7 for ductile iron, $F_M = 1.15$.

5. Compute the pump cost C_P with base plate and coupling

Use the relation $C_P = C_B F_T F_M$, where the symbols are as given above. Thus, $C_P = (\$1536)(1.491)(1.15) = \$2630$.

6. Determine the required horsepower for the motor

Use the relation $P_B = pQH/33,000N_P$, where P_B = bhp input to pump; p = fluid density, lb/gal; N_P = pump efficiency, percent; other symbols as given earlier. In this method of cost estimating, $N_Q = -0.316 + 0.24015 (\ln Q) - 0.01199 (\ln Q)^2$.

Find the fluid density from p = specific gravity (8.33 lb/gal) = 0.952(8.33) = 7.93 lb/gal (0.94 kg/L). The pump efficiency, from the above relation, is $N_P = -0.316 + 0.24015 (\ln 1430) - 0.01199 (\ln 1430)^2 = 0.796$.

Substituting in the power relation yields $P_B = 7.93 (1430)(77)/33,000(0.796) = 33.2$ hp (24.8 kW). A 40-hp (29.8 · kW) motor is required for this pump.

7. Compute the cost of the electric motor

Use the appropriate correlation from Table 8. Assume a 3600-r/min totally enclosed fan-cooled motor is needed. Then the motor cost $C_M = \exp[3.8544 + 0.8331 (\ln P_B) + 0.02399 (\ln P_B)^2] = \exp[3.8544 + 0.8331 (\ln 40) + 0.02399 (\ln 40)^2] = \$1410$.

8. Determine the total cost of the pump and motor

Find the sum of $C_P + C_M$. Or, $C_P + C_M = \$2630 + \$1410 = \$4040$.

9. Compute the pump power consumption

Use the relation $U_M + 0.80 = 0.0319 (\ln P_B) - 0.00182 (\ln P_B)^2$ to find the efficiency of the motor. By substituting, $N_M = 0.80 + 0.0319 (\ln 33.2) - 0.00182 \ln (33.2)^2 = 0.889$. Then the pump power consumption $P_C = P_B/N_M = 33.2/0.889 = 37.3$ hp (27.8 kW).

TABLE 6 Correlations for Costs of Centrifugal Pumps[°]

USCS units

Base cost for one-stage, 3550 r/min, VSC cast-iron pump:

$$C_B = \exp[8.3949 - 0.6019 \ln S + 0.0519(\ln S)^2]$$

Here, $S = Q\sqrt{H}$, with Q in gal/min and H in ft·lbf/lb (ft of head).

Cost factor for pump type:

$$F_T = \exp[b_1 + b_2 \ln S + b_3 (\ln S)^2]$$

Type	b_1	b_2	b_3
One-stage, 1750-r/min, VSC	5.1029	-1.2217	0.0771
One-stage, 3550-r/min, HSC	0.0632	0.2744	-0.0253
One-stage, 1750-r/min, HSC	2.0290	-0.2371	0.0102
Two-stage, 3550-r/min, HSC	13.7321	-2.8304	0.1542
Multistage, 3550-r/min, HSC	9.8849	-1.6164	0.0834

SI units

Base cost for one-stage, 3550 r/min, VSC cast-iron pump:

$$C_B = \exp[7.2234 + 0.3451 \ln S + 0.0519(\ln S)^2]$$

Here, $S = Q\sqrt{H}$, with Q in m³/s, and H in J/kg or m²/s².

Cost factor for pump type:

$$F_T = \exp[b_1 + b_2 \ln S + b_3 (\ln S)^2]$$

Type	b_1	b_2	b_3
One-stage, 1750-r/min, VSC	0.3740	0.1851	0.0771
One-stage, 3550-r/min, HSC	0.4612	-0.1872	-0.0253
One-stage, 1750-r/min, HSC	0.7147	-0.0510	0.0102
Two-stage, 3550-r/min, HSC	0.7445	-0.0167	0.1542
Multistage, 3550-r/min, HSC	2.0798	-0.0946	0.0834

[°] *Chemical Engineering and Richardson Engineering Services, Inc.*

TABLE 7 Cost Factors for Material of Construction[*]

(Source: Monsanto Co.'s FLOWTRAN pump-costing subprogram.)

Material	Cost factor F_M
Cast steel	1.35
304 or 316 fittings	1.15
Stainless steel, 304 or 316	2.00
Cast Gould's alloy no. 20	2.00
Nickel	3.50
Monel	3.30
ISO B	4.95
ISO C	4.60
Titanium	9.70
Hastelloy C	2.95
Ductile iron	1.15
Bronze	1.90

[*] *Chemical Engineering.*

TABLE 8 Correlation for Cost of Electric Motors[*]

Cost of 60-Hz standard-voltage motor and insulation, discounted:
$$C_M = exp\,[a_1 + a_2\,ln\,P + a_3\,(ln\,P)^2]$$
P is the nominal size in horsepower

	Coefficients				
	a_1	a_2	a_3	hp limits	kW limits
Open, drip-proof:					
3600 r/min	4.8314	0.09666	0.10960	1–7.5	0.75–5.6
	4.1514	0.53470	0.05252	7.5–250	5.6–186.5
	4.2432	1.03251	−0.03595	250–700	186.5–522.2
1800 r/min	4.7075	−0.01511	0.22888	1–7.5	0.75–5.6
	4.5212	0.47242	0.04820	7.5–250	5.6–186.5
	7.4044	−0.06464	0.05448	250–600	186.5–447.6
1200 r/min	4.9298	0.30118	0.12630	1–7.5	0.75–5.6
	5.0999	0.35861	0.06052	7.5–250	5.6–186.5
	4.6163	0.88531	−0.02188	250–500	186.5–373.0
Totally enclosed, fan-cooled:					
3600 r/min	5.1058	0.03316	0.15374	1–7.5	0.75–5.6
	3.8544	0.83311	0.02399	7.5–250	5.6–186.5
	5.3182	1.08470	−0.05695	250–400	186.5–298.4
1800 r/min	4.9687	−0.00930	0.22616	7.5–250	5.6–186.5
	4.5347	0.57065	0.04609		
1200 r/min	5.1532	0.28931	0.14357	1–7.5	0.75–5.6
	5.3858	0.31004	0.07406	7.5–350	5.6–261.1
Explosion-proof:					
3600 r/min	5.3934	−0.00333	0.15475	1–7.5	0.75–5.6
	4.4442	0.60820	0.05202	7.5–200	5.6–149.2
1800 r/min	5.2851	0.00048	0.19949	1–7.5	0.75–5.6
	4.8178	0.51086	0.05293	7.5–250	5.6–186.5
1200 r/min	5.4166	0.31216	0.10573	1–7.5	0.75–5.6
	5.5655	0.31284	0.07212	7.5–200	5.6–149.2

[*] *Chemical Engineering.*

TABLE 9   Flow, Head, and Power Limits for Centrifugal Pumps°

	Flow, gal/min (m³/s)		Head, ft·lbf/lb (J/kg)		Motor hp, upper limit	Motor kW
	Lower limit	Upper limit	Lower limit	Upper limit		
One-stage, 3550 r/min, VSC	50 (0.00315)	900 (0.568)	50 (150)	400 (1200)	75	55.95
One-stage, 1750 r/min, VSC	50 (0.00315)	3500 (0.2208)	50 (150)	200 (600)	200	149.2
One-stage, 3550 r/min, HSC	100 (0.00631)	1500 (0.0946)	100 (300)	450 (1350)	150	111.9
One-stage, 1750 r/min, HSC	250 (0.01577)	5000 (0.3155)	50 (150)	500 (1500)	250	186.5
Two-stage, 3550 r/min, HSC	50 (0.00315)	1100 (0.0694)	300 (900)	1100 (3300)	250	186.5
Multistage, 3550 r/min, HSC	100 (0.00631)	1500 (0.0946)	650 (2000)	3200 (9600)	1450	1081.7

°*Chemical Engineering.*

Related Calculations: This procedure can be used for centrifugal pumps and electric motors in a variety of industries and applications provided the pump and motor are of the type listed in the tables. Typical industries and applications include chemical, petroleum, petrochemical, power, marine, air-conditioning, heating, and food processing.

Data on the cost of centrifugal pumps and electric motors were taken from Vol. 4 of the data book by Richardson Engineering Services (Solana Beach, CA), *Process Plant Construction Estimating Standards.* The material-of-construction cost factors for pumps were taken from Monsanto Co.'s FLOWTRAN pump-costing subprograms.

Although the cost of a pump includes the cost of the driver coupling, cost correlations for belt-, chain- and variable-speed drive couplings were obtained from the U.S. Bureau of Mines equipment-costing program. These correlations were escalated from their original data of 1967 to the first quarter of 1979 by using the chemical engineering pumps and compressors index ratio of $270/11.2 = 2.43$. All other cost data were for the first quarter of 1979, when the pumps and compressors index was 270 and the electrical equipment index was 175.5. To update the costs to the year in which an estimate is being made, simply apply the current index, as detailed in the preceding procedure.

Table 9 gives the flow, head, and power limits for the centrifugal pumps considered in this procedure. Table 10 shows the correlations for the cost of drive couplings for the pumps.

This procedure is the work of Armando B. Corripio, Louisiana State University; Katherine S. Chrien of J. S. Dweck, Consultant, Inc.; and Lawrence B. Evans, Massachusetts Institute of Technology, as reported in *Chemical Engineering* magazine.

DETERMINING THE FRICTION FACTOR FOR FLOW OF BINGHAM PLASTICS

REFERENCES: [1] E. Buckingham, On Plastic Flow Through Capillary Tubes, *ASTM Proc.*, Vol. 21, 1154 (1921); [2] R. W. Hanks and D. R. Pratt, On the Flow of Bingham Plastic Slurries in Pipes and between Parallel Plates, *Soc. Petrol. Eng. J.*, Vol. 1, 342 (1967); [3] R. W. Hanks and B. H. Dadia, Theoretical Analysis of the Turbulent Flow of Non-Newtonian Slurries in Pipes, *AIChE J.*, Vol. 17, 554 (1971); [4] S. W. Churchill, Friction-factor Equation Spans All Fluid-flow Regimes, *Chem. Eng.*, Nov. 7, 1977, pp. 91–92; [5] S. W. Churchill and R. A. Usagi, A General Expression for the Correlation of Rates of Transfer and Other Phenomena, *AIChE J.*, Vol. 18, No. 6, 1121–1128 (1972); [6] R. L. Whitmore, *Rheology of the Circulation*, Pergamon Press, Oxford, 1968; [7] N. Casson, A Flow Equation for Pigment-Oil Dispersions of the Printing Ink Type, Ch. 5 in *Rheology of Disperse Systems*, C. C. Mill (ed.), Pergamon Press, Oxford, 1959; [8] R. Darby and B. A. Rogers, Non-Newtonian Viscous Properties of Methacoal Suspensions, *AIChE J.*, Vol. 26, 310 (1980); [9] G. W. Govier and A. K. Aziz, *The Flow of Complex Mixtures in Pipes*, Van Nostrand Reinhold, New York, 1972; [10] E. H. Steiner, The Rheology of Molten Chocolate, Ch. 9 in C. C. Mill (ed.), *op. cit.*; [11] R. B. Bird, W. 1 . Stewart, and E. N. Lightfoot, *Transport Phenomena*, John Wiley & Sons, New York, 1960.

TABLE 10 Correlations for Cost of Drive Coupling°

Cost of belt-drive coupling:

$$C_C = \exp (3.689 + 0.8917 \ln P)$$

Cost of chain-drive coupling:

$$C_C = \exp (5.329 + 0.5048 \ln P)$$

Cost of variable-speed-drive coupling:

$$C_C = 1/[1.562 \times 10^{-4} + (7.877 \times 10^{-4}/P)]$$

Upper limit = 75 hp; S = nominal motor size in hp

°*Chemical Engineering* and U.S. Bureau of Mines.

A coal slurry is being pumped through a 0.4413-m (18-in) diameter schedule 20 pipeline at a flow rate of 400 m³/h. The slurry behaves as a Bingham plastic, with the following properties (at the relevant temperature): τ_0 = 2 N/m² (0.0418 lbf/ft²); μ_∞ = 0.03 Pa·s (30 cP); ρ = 1500 kg/m³ (93.6 lbm/ft³). What is the Fanning friction factor for this system?

Calculation Procedure:

1. *Determine the Bingham Reynolds number and the Hedstrom number*

Engineers today often must size pipe or estimate pressure drops for fluids that are non-Newtonian in nature—coal suspensions, latex paint, or printer's ink, for example. This procedure shows how to find the friction factors needed in such calculations for the many fluids that can be described by the Bingham-plastic flow mode. The method is convenient to use and applies to all regimes of pipe flow.

A Bingham plastic is a fluid that exhibits a yield stress; that is, the fluid at rest will not flow unless some minimum stress τ_0 is applied. Newtonian fluids, in contrast, exhibit no yield stress, as Fig. 30 shows.

The Bingham-plastic flow model can be expressed in terms of either shear stress τ versus shear rate $\dot\gamma$, as in Fig. 30, or apparent viscosity η versus shear rate:

$$\tau = \tau_0 + \mu_\infty \dot\gamma \tag{1}$$

$$\eta = \frac{\tau}{\dot\gamma} = \frac{\tau_0}{\dot\gamma} + \mu_\infty \tag{2}$$

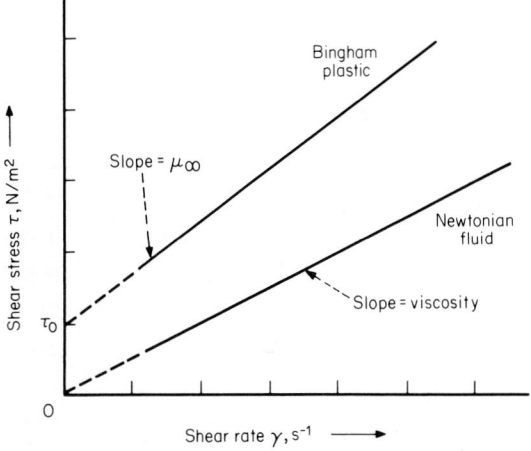

FIG. 30 Bingham plastics exhibit a yield stress. *(Chemical Engineering.)*

Equation 2 means that the apparent viscosity of a Bingham plastic depends on the shear rate. The parameter μ_∞ is sometimes called the coefficient of rigidity, but it is really a limiting viscosity. As Eq. 2 shows, apparent viscosity approaches μ_∞ as shear rate increases indefinitely. Thus the Bingham plastic behaves almost like a newtonian at sufficiently high shear rates, exhibiting a viscosity of μ_∞ at such conditions. Table 11 shows values of τ_0 and μ_∞ for several actual fluids.

For any incompressible fluid flowing through a pipe, the friction loss per unit mass F can be expressed in terms of a Fanning friction factor f:

$$F = \frac{2fLv^2}{D} \tag{3}$$

where L is the length of the pipe section, D is its diameter, and v is the fluid velocity.

An exact description of friction loss for Bingham plastics in fully developed laminar pipe flow was first published by Buckingham [1]. His expression can be rewritten in dimensionless form as follows:

$$f_L = \frac{16}{N_{Re}}\left(1 + \frac{N_{He}}{6N_{Re}} - \frac{N_{He}^4}{3f_L^3 N_{Re}^7}\right) \tag{4}$$

where N_{Re} is the Bingham Reynolds number (Dvp/μ_∞) and N_{He} is the Hedstrom number $(D^2 p\tau_0/\mu^2)$. Equation 4 is implicit in f_L, the laminar friction factor, but can be readily solved either by Newton's method or by iteration. Since the last term in Eq. 4 is normally small, the value of f obtained by omitting this term is usually a good starting point for iterative solution.

For this pipeline

$$N_{Re} = \frac{4Q\rho}{\pi D\mu_\infty} = \frac{4(400)(1/3)(600)(1500)}{\pi(0.4413)(0.03)} = 16,030$$

$$N_{He} = \frac{D^2 \rho\tau_0}{\mu_\infty^2} = \frac{(0.4413)^2(1500)(2)}{(0.03)^2} = 649,200$$

2.   *Find the friction factor f_L for the laminar-flow regime*

Substituting the values for N_{Re} and N_{He} into Eq. 4, we find $f_L = 0.007138$.

3.   *Determine the friction factor f_T for the turbulent-flow regime*

Equation 4 describes the laminar-flow sections. An empirical expression that fits the turbulent-flow regime is

$$f_T = 10^a N_{Re}^{-0.193} \tag{5}$$

where

$$a = -1.378\,[1 + 0.146\,\exp\,(-2.9 \times 10^{-5}\,N_{He})] \tag{6}$$

TABLE 11   Values of τ_0 and μ_∞

Fluid	τ_0, N/m^2	μ_∞, Pa·s	Ref.
Blood (45% hematocrit)	0.005	0.0028	[6]
Printing-ink pigment in varnish (10% by wt.)	0.4	0.25	[7]
Coal suspension in methanol (35% by vol.)	1.6	0.04	[8]
Finely divided galena in water (37% by vol.)	4.0	0.057	[9]
Molten chocolate (100°F)	20	2.0	[10]
Thorium oxide in water (50% by vol.)	300	0.403	[11]

We now have friction-factor expressions for both laminar and turbulent flow. Equation 6 does not apply when N_{He} is less than 1000, but this is not a practical constraint for most Bingham plastics with a measurable yield stress.

When N_{He} is above 300,000, the exponential term in Eq. 6 is essentially zero. Thus $a = -1.378$ here, and Eq. 5 becomes

$$f_T = 10^{-1.378}(16,030)^{-0.193}$$

$$= 0.006463$$

4. Find the friction factor f

Combine the f_L and f_T expressions to get a single friction factor valid for all flow regimes:

$$f = (f_L^m + f_T^m)\,\frac{1}{m} \tag{7}$$

where f_L and f_T are obtained from Eqs. 4 and 5, and the power m depends on the Bingham Reynolds number:

$$m = 1.7 + \frac{40,000}{N_{Re}} \tag{8}$$

The values of f predicted by Eq. 7 coincide with Hanks's values in most places, and the general agreement is excellent. Relative roughness is not a parameter in any of the equations because the friction factor for non-Newtonian fluids, and particularly plastics, is not sensitive to pipe roughness.

Substituting yields $m = 1.7 + 40,000/16,030 = 4.20$, and $f = [(0.007138)^{4.20} + (0.006463)^{4.20}]^{1/4.20} = 0.00805$.

If m had been very large, the bracketed term above would have approached zero. Generally, when N_{Re} is below 4000, Eq. 8 should be solved by taking f equal to the greater of f_L and f_T.

Related Calculations: This procedure is valid for a variety of fluids met in many different industrial and commercial applications. The procedure is the work of Ron Darby, Professor of Chemical Engineering, Texas A & M University, College of Engineering, and Jeff Melson, Undergraduate Fellow, Texas A & M, as reported in *Chemical Engineering* magazine. In their report they cite works by Hanks and Pratt [2], Hanks and Dadia [3], Churchill [4], and Churchill and Usagi [5] as important in the procedure described and presented here.

TIME NEEDED TO EMPTY A STORAGE VESSEL WITH DISHED ENDS

A tank with a 6-ft (1.8-m) diameter cylindrical section that is 16 ft (4.9 m) long has elliptical ends, each with a depth of 2 ft (0.7 m), and is half-full with ethanol, a newtonian fluid. How long will it take to empty the tank if it is set horizontally and fitted at the bottom with a drain consisting of a short tube of 2-in (5.1-cm) double extrastrong pipe? How long will it take to empty the tank if it is set vertically and fitted at the bottom with a drainpipe of 2-in (5.1-cm) double extrastrong pipe? The drain system extends 4 ft (1.2 m) below the dished bottom and has an equivalent length of 250 ft (76.2 m).

Calculation Procedure:

1. Determine the discharge coefficient for, and orifice area of, the drain tube

Figure 29 shows the discharge coefficient is $C_d = 0.80$ for a short, flush-mounted tube. Baumeister, in *Mark's Standard Handbook for Mechanical Engineers*, indicates the internal section area of the tube is $A_n = 1.774$ in² (11.4 cm²).

2. Compute the discharge time for the tank in a horizontal position

Substitute the appropriate values in the equation for t_p shown under the storage tanks in Fig. 31. Thus, $t_p = [(8)^{0.5}/[3(0.80)(1.774/144)(32.2)^{0.5}]\{16[(6)^{1.5} - (6 - 3)^{1.5}] + [2\pi(3)^{1.5}/6][6 - (3/5)(3)]\} = 2948$ s, or 49.1 min.

3. Determine the internal diameter and friction factor for the drainpipe

From Baumeister, *Mark's Standard Handbook for Mechanical Engineers*, the internal diameter of the pipe is $d = 1.503$ in (3.8 cm), or 0.125 ft (0.038 m) and the Moody friction factor is $f = 0.020$ for the equivalent length, $l = 250$ ft (76.2 m), of pipe.

4. Compute the initial and final height above the drainpipe outlet for the cylindrical section

Initial height of the liquid is $H_I = a + b + h_o = [(16/2) + 2 + 4] = 14.0$ ft (4.3 m). Final height is $H_F = b + h_o = 2 + 4 = 6$ ft (1.8 m).

5. Compute the time required to drain the cylindrical section of the tank

Substitute the appropriate values in the equation for t_c shown under the storage tanks in Fig. 31. Hence, $t_c = [(6)^2/(0.125)^2]\{(2/32.2)[1 + (0.020 \times 250/0.125)]\}^{0.5}[(14)^{0.5} - (6)^{0.5}] = 4751$ s, or 79.2 min.

6. Compute the initial and final liquid height above the drainpipe outlet for the elliptically dished head

Initial height of the liquid is $H_1 = b + h_o = 2 + 4 = 6$ ft (1.8 m). Final height is $H_2 = h_o = 4$ ft (1.2 m).

7. Compute how long it will take to empty the dished bottom of the tank

In order to solve for t_e it is necessary to determine the following values: $B = h_o + b = 4 + 2 = 6$ ft (1.8 m); $E^2 = h_o^2 + 2bh_o = (4)^2 + 2(2)(4) = 32$ ft^2 (3.0 m^2); $C = [D/(db)]^2\{[1/(2g)][1 + (f_1/d)]\}^{0.5} = [6/(0.125 \times 2)]^2\{[1/(2 \times 32.2)][1 + ([0.02 \times 250]/0.125)]\}^{0.5} = 459.6$, s/ft$^{5/2}$ (s/m$^{5/2}$).

Then, use the values for B, E^2, C, and other relevant dimensions to find t_e from the equation shown under the storage tanks in Fig. 31. Thus, $t_e = 459.6 [(2 \times 4^2/5) - (4 \times 6 \times 4/3) + 2(32)](4)^{0.5} - [(2 \times 6^2/5) - (4 \times 6 \times 6/3) + 2(32)](6^{0.5}) = 1073$ s, or 17.9 min.

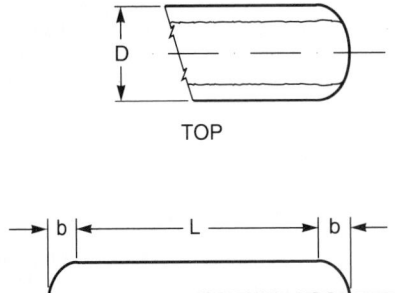
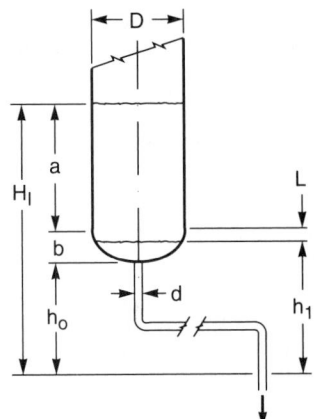

TOP

FRONT
Horizontal cylinder with dished ends

Vertical Cylinder with dished end and drainpipe system

$$t_c = \frac{D^2}{d^2} \{[(2/g)(1 + [fl/d])]^{1/2}(H_I^{1/2} - H_F^{1/2})\}$$

$$t_e = C\{[(2 \times H_2^2/5) - (4 \times B \times H_2/3) + 2E^2\}(H_2^{1/2})$$
$$- [(2 \times H_1^2/5) - (4 \times B \times H_1/3) + 2E^2](H_1^{1/2})\}$$

$$t_p = \frac{\sqrt{8}}{3C_dA_n\sqrt{g}} \{L[D^{3/2} - (D - h)^{3/2}] + \frac{bph^{3/2}}{D}[D - (3h/5)]\}$$

FIG. 31 Time to drain tanks. (*a*) Top and (*b*) front view of horizontal cylinder with dished ends. (*c*) Vertical cylinder with dished-end and drainpipe system. (*Chemical Engineering*).

8. Compute the time it will take to drain the half-full vertical tank

Total time is $t_v + t_c + t_e = 4751 + 1073 = 5824$ s, or 96.1 min.

Related Calculations: Figure 31 shows the equation for computing the emptying time for a horizontal cylindrical tank with elliptically dished ends and equations for calculating the emptying time for a vertical cylindrical tank with an elliptically dished bottom end fitted with a drain system. The symbols A_n, g, t, and C_d are defined as in the previous problem for a storage vessel without dished ends, except that A_n is now the drainpipe internal area.

The term associated with the second pair of brackets in the equation for t_p accounts for the dished ends of the horizontal tank. For hemispherical ends $b = D/2$ and for flat ends, $b = 0$.

When seeking the time required to drain a portion of the cylindrical part of the vertical tank use the formula for t_c with the appropriate values for H_1 and H_F and other pertinent variables. To find the time it takes to drain a portion of the dished bottom of the vertical tank use the formula for t_e with given values of H_1 and H_2 and other applicable variables.

The relations given here are valid for storage tanks used in a variety of applications—chemical and petrochemical plants, power plants, waterworks, ships and boats, aircraft, etc. The procedure for a horizontal cylindrical tank with dished ends is the work of Jude T. Sommerfeld, and the procedure for a vertical cylindrical tank with a dished bottom end is the work of Mahnoosh Shoael and Jude T. Sommerfeld, as reported in *Chemical Engineering* magazine.

CHECKING THE VACUUM RATING OF A STORAGE VESSEL

Check the vacuum rating of a cylindrical flat-ended process tank which is 12.75 ft (3.9 m) tall and 4 ft (1.22 m) in diameter. It contains fresh water at 190°F (87.8°C) and is located where $g = 32.0$ ft/s^2 (9.8 m/s^2) and the atmospheric pressure is 14.7 lb/in^2 (101.3 kPa). What is its maximum vacuum when the tank is gravity-drained? Find both the final tank vacuum and final height of liquid above the tank bottom when the tank is initially 75 percent full, first with gravity drain and then using a pumped drain, each discharging to the atmosphere. The pumped-drain piping system consists of double extra-strong 2-in (5.1-cm) pipe with an equivalent length of 75 ft (22.9 m) and a pump which can discharge 40 gal/min (151 L/min) with its suction centerline 2 ft (0.61 m) below the tank bottom and has a net positive suction head of 4.5 ft (1.4 m).

Calculation Procedure:

1. Find the density of the fresh water

From a suitable source such as Baumeister, *Mark's Standard Handbook for Mechanical Engineers*, freshwater density $\rho = 60.33$ lb/ft^3 (966.7 kg/m^3) at the given conditions.

2. Compute the maximum vacuum rating with gravity drain

A shortcut method gives the maximum vacuum rating by use of the equation for P_s in Fig. 32 where $H =$ overall vertical dimension of the tank; $\rho =$ density of the fresh water; $g =$ acceleration due to gravity at the tank's location; and $g_c = 32.174$ lb · ft/lb · s^2, a conversion factor. Substituting appropriate values gives $P_s = 2.036(12.75)(60.33)(32.0/[144 \times 32.174]) = 10.82$ in Hg (36.64 kPa).

3. Compute the head space volume when the tank is 75 percent full

Airspace volume $V_o = (1.00 - 0.75)(\text{H} \times 0.7854 \text{ D}^2) = 0.25(12.75 \times 0.7854 \times 4^2) = 40.1$ ft^3 (1.14 m^3).

4. Compute the final height of liquid above the tank bottom created by gravity drain

By trial and error, the final height of liquid can be computed by solving the equation for h_{fg} shown in Fig. 32 where the ambient atmospheric pressure $P_o = 14.7$ lb/in^2 (101.3 kPa); tank radius $R = 2$ ft (0.61 m); initial fluid height above tank bottom $h_o = 0.75H = 0.75(12.75) = 9.56$ ft (2.91 m); ratio of molar specific heats, C_p/C_v, is $\gamma = 1.4$ for diatomic gases. Assuming a reasonable initial value of h_{fg} on the right-hand side of the equation and substituting appropriate other values, too, gives $h_{fg} = 144(14.7)(1 - \{40.1/[40.1 + \pi(2^2)(9.56 - h_{fg})]\}1.4)/(60.33 \times 32.0/32.174) = 7.18$ ft (2.19 m), where the left-hand side value is used to repeat the interation process until the h_{fg} values on either side of the equal sign are in close agreement.

5. Compute the final tank vacuum under gravity drain

Use the equation for P_g shown in Fig. 32 to find the final tank vacuum. Thus, $P_g = 29.92 \times 7.18 \times 60.33 \times 32.0/(144 \times 14.7 \times 32.174) = 6.09$ in Hg (20.62 kPa).

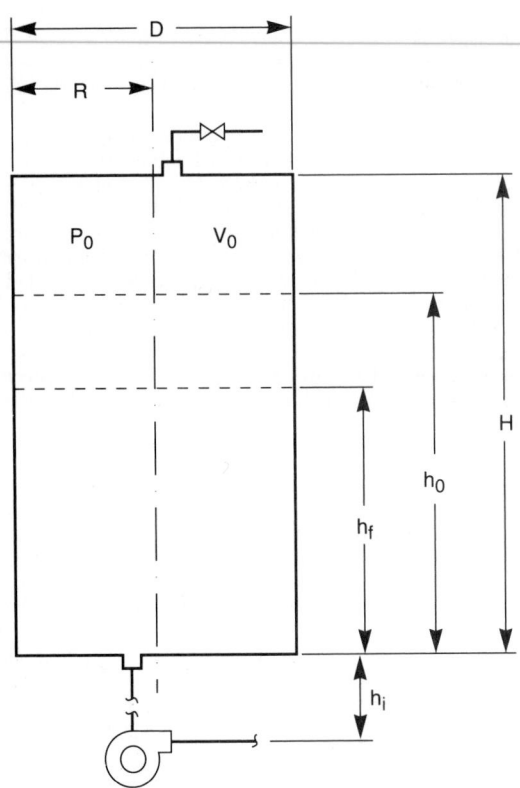

rho $P_s = 2.036Hd(g/[144g_c)$, in Hg|(kPa)

rho $h_{fg} = 144P_0(1 - \{V_0/[V_0 + pR^2(h_o - h_{fg})]\}n)/(eg/g_c)$, ft|(m)

pi gamma $P_g = 29.92h_{fg}eg/(144P_0g_c)$, in Hg|(kPa)

$h_{fg} = NPSH + F_1 + V_p - h_i - 144P_0 \{V_0/[V_0 + pR^2(h_o - h_{fp})]\}n/(eg/g_c)$, ft|(m)

$P_p = 2.036P_0\{1 - [V_0/(V_0 + pR^2h_o)]n\}$, in Hg|(kPa)

FIG. 32   Vacuum rating of tanks.

6.   *Compute the water velocity in the pumped-drain system piping*

Pump's flow rate $q = (40 \text{ gal/min})/([7.48 \text{ gal/ft}^3][60 \text{ s/min}]) = 0.0891 \text{ ft}^3/\text{s}$ (0.0025 m³/s). The sectional area of the 2 in (5.1 cm) double extra-strong pipe is $a = 0.7854d^2 = 0.7854(1.503/12)^2 = 0.0123 \text{ ft}^2$ (0.00114 m²). Thus, water mean velocity $v = q/a = 0.0891/0.0123 = 7.24$ ft/s (2.2 m/s).

7.   *Determine the fluid's viscosity*

From Baumeister, *Mark's Standard Handbook for Mechanical Engineers*, fresh water at 190°F (87.8°C) has a dynamic viscosity of $\mu = 6.75 \times 10^{-6}$ lb · s/ft² (323.2 × 10⁻⁶ Pa · s).

8.   *Compute the Reynold's number for the drain system piping*

Using pertinent values found previously, Reynold's number = Re = $\rho vd/\mu = (60.33 \times 7.24)(1.1503/12)/(6.75 \times 10^{-6}) = 8,088,690$.

9. Find the friction factor for the drain system piping

Baumeister, *Marks' Standard Handbook for Mechanical Engineers*, indicates that the relative roughness factor for the drain piping is $\epsilon/d = 150 \times 10^{-6}/0.12 = 0.0012$, hence the Moody friction factor is $f = 0.02$ for the above value of N_R.

10. Compute the friction loss of the pumped-drain piping system

Friction loss $F_1 = (fL/d)(v^2/2g)$ where F_1 is in feet (m) of fresh water and L is the equivalent length of the piping system in ft (m). Substituting, $F_1 = (0.02 \times 75/0.125)(7.24^2/[2 \times 32.174])$ = 9.77 ft (1.98 m).

11. Compute the final height of liquid above the tank bottom created by the pumped drain

In the equation for h_{fp} shown in Fig. 32, the net positive suction head (NPSH) = 4.5 ft (1.37 m); freshwater vapor pressure $V_p = 22.29$ ft (6.79 m) of water (Baumeister, *Marks' Standard Handbook for Mechanical Engineers*); height between tank bottom and centerline of pump suction $h_i = 2$ ft (0.61 m). Substituting appropriate values gives $h_{fp} = 4 + 9.77 + 22.29 - 2 - 144(14.7)\{40.1/[40.1 + \pi(2^2)(9.56 - h_{fp})]\}1.4/(60.33 \times 32.0/32.174) = 6.94$ ft (2.12 m), by trial and error, as was done for the gravity drain. From 4 to 10 trials should do it.

12. Compute the final tank vacuum for the pumped drain

Final tank vacuum is found by solving the equation for P_p shown in Fig. 32, thus $P_p = 2.036(14.7)\{1 - (40.1/[40.1 + (\pi \times 2^2)(9.56)])1.4\} = 19.44$ in Hg (65.83 kPa).

Related Calculations: Specifying an appropriate vacuum rating could prevent the collapse of a storage vessel as the contents are being drained while the vent is inadvertently blocked. Vacuum ratings range from full vacuum to no vacuum. A full vacuum rating is advisable for tanks such as those for steam-sterilized sanitary service and those with pumped discharge. Tanks with vents that cannot be blocked require no vacuum rating.

That the maximum gravity-drain vacuum rating occurs at 100 percent full capacity was borne out by the above calculations for P_s and P_g. However, the pumped-drain vacuum rating P_p, under more favorable conditions, still turned out to be 3.19 times greater than P_s, the maximum for gravity drain. This varies with pump capacity and the drain system piping size. These calculations assume ideal gas behavior in the head space above the fluid surface and the process is considered isothermal for drain times longer than 5 min. If the initial fluid height is set too low, it is possible for the tank to be emptied by pump drain before maximum vacuum occurs. The calculations presume a centrifugal pump will not deliver if the NPSH requirements are not met and then backflow into the tank starts. Use the equation for P_p to find the final tank vacuum if it is expected that the tank will be emptied before backflow occurs.

The procedure presented here allows the designer to choose a vacuum rating appropriate to the tank. However, it is suggested that the designer perform applicable code calculations before making a final decision on the vacuum rating for a tank. This presentation is based upon an article by Barry Wintner of Life Sciences International, and which appeared in *Chemical Engineering* magazine.

DESIGNING PRISMATIC PRESSURE VESSELS

A closed-top tank filled with fresh water, Fig. 33a, is constructed of 0.3125 in (0.79 cm) medium steel-plate sides. Plate's allowable bending stress = 19,330 lb/in² (133.3 MPa) and modulus of elasticity = 27.9×10^6 lb/in² (192.4 GPa). Where should horizontal stiffeners be located and what size should they be if their allowable bending stress is 16,000 lb/in² (110.32 GPa)?

Calculation Procedure:

1. Check out an initial tentative height for the lowest plate, panel No. 1

Assume the vertical dimension of panel No. 1 is one-third the tank height, or $b_1 = H/3 = 135/3 = 45$ in (1.14 m). Variation of the liquid's horizontal pressure is symbolized by the inclined line shown in Fig. 33b. Average pressure distribution over the entire panel is considered to be represented by the pressure head at a level two-thirds down the panel, as indicated by h_1 in Fig. 33c. Thus, the applied uniform pressure is $p_1 = h_1\gamma$, where p_1 is in lb/in² (Pa); h_1 is in in (m); specific weight γ is in lb/in³ (N/m³). Hence, $p_1 = [H - (b_1/3)][\gamma] = [135 - (45/3)][62.4/1728] = 120(0.0361) = 4.33$ lb/in² (29.85 kPa).

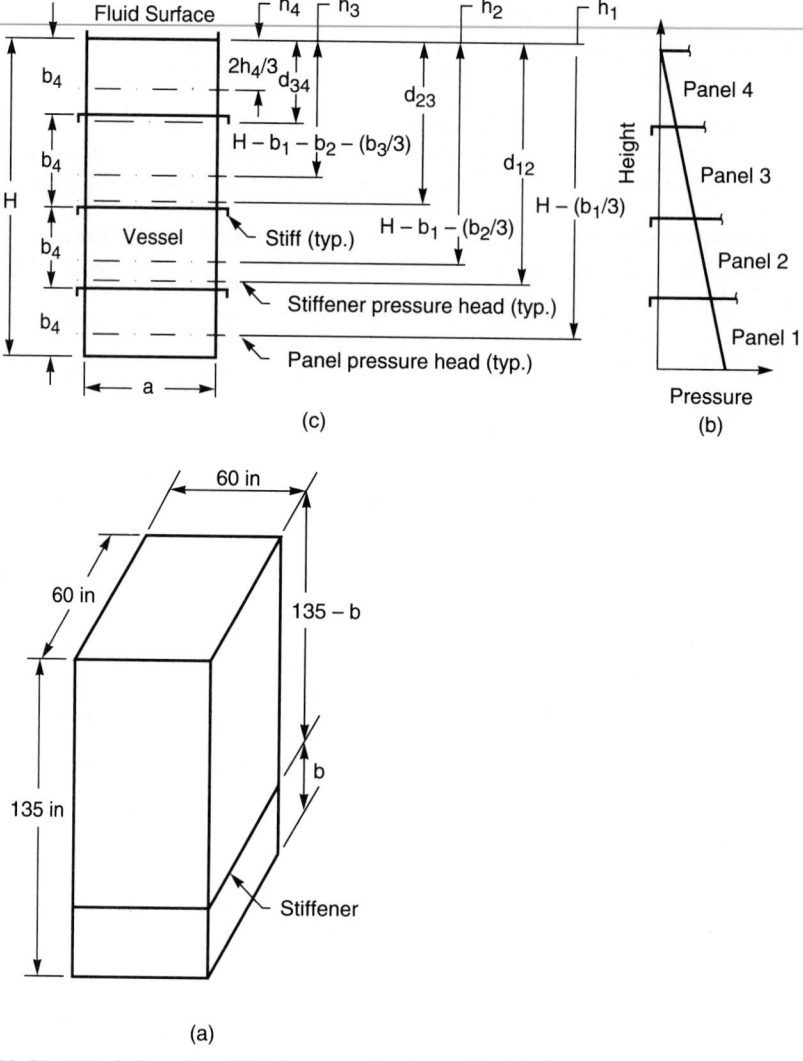

FIG. 33   (a) Tank dimensions. (b) Tank pressure distribution. (c) Tank elevation.

Then, since the panel dimension ratio $a/b = 60/45 = 1.33$ select, by interpolation from Table 12, deflection parameter $\alpha = 0.0213$ and bending parameter $\beta = 0.4173$. The equation for calculating the panel height is $b = (s_b t^2/\beta p)^{1/2}$, where b is in in (m); allowable bending stress $s_b = 19{,}330$ lb/in² (133.3 MPa); plate thickness $t = 0.3125$ in (0.79 × 10⁻² m); other symbols are as determined previously. Substituting, $b_1 = [(19{,}330 \times 0.3125^2)/(0.4173 \times 4.33)]^{1/2} = 32.32$ in (0.82 m). Since the result is much smaller than the assumed value, it is necessary to iterate again. Solutions emerge quickly for low-pressure gradients such as those for water or oil.

2.   *Review a revised tentative height for panel No. 1*

Repeating the previous procedure, but with $b_1 = 29$ in (0.74 m), $p_1 = [H - (b_1/3)](0.0361)$ $= [135 - (29/3)](0.0361) = 4.48$ lb/in² (30.89 kPa). Also, $a/b_1 = 60/29 = 2.07$, $\alpha = 0.0279$,

TABLE 12 Deflection and Bending Parameters (Machine Design).

Dimension ratio a/b	Deflection parameter α	Bending parameter β
1.0	0.0138	0.3078
1.2	0.0188	0.3834
1.4	0.0226	0.4356
1.6	0.0251	0.4680
1.8	0.0267	0.4872
2.0	0.0277	0.4974
2.2	0.0282	0.4999
∞	0.0284	0.5000

and $\beta = 0.4979$. Then, $b_1 = [(19{,}330 \times 0.3125^2)/(0.4979 \times 4.48)]^{1/2} = 29.1$ in (0.74 m), and is tentatively acceptable as 29 in (0.74 m).

3. Check panel No. 1 for bending stress

Although the calculated value is very close to the assumed value of b_1, it is best to check the actual plate stress by $s_b = \beta pb^2/t^2$, or $s_{b1} = 0.4979 \times 4.48 \times 29^2/0.3125^2 = 19{,}210$ lb/in^2 (132.4 MPa). This is less than the allowable bending stress; hence the panel height is still tentatively acceptable as 29 in (0.74 m).

4. Check panel No. 1 for deflection

Although the bending stress is acceptable, the panel should now be checked for deflection, which may not exceed one-half the plate thickness $t/2 = 0.3125/2 = 0.156$ in (0.40 cm). The formula for deflection is $\delta = \alpha pb^4/Et^3$, where δ = plate deflection, in (m); modulus of elasticity $E = 27.9 \times 10^6$ lb/in^2 (192.4 GPa); other variables are as previously determined. Thus, $\delta_1 = 0.0279 \times 4.48 \times 29^4/(27.9 \times 10^6 \times 0.3125^3) = 0.104$ in (0.26 cm). Since the deflection is within the allowable limit, the value of $b_1 = 29$ in (0.74 m) is now acceptable.

5. Size the stiffener for panel No. 1

The stiffener is continuous around the tank; therefore, the ends are practically fixed. Uniform loading along the length a of the stiffener is considered to be the sum of the pressure loading acting on half of the panels on either side of the stiffener. Average pressure exists at the horizontal centerline of the loaded area. This centerline is below the fluid surface at a distance $d_{12} = b_4 + b_3 + (b_2/2) + [(b_2 + b_1)/2] = 33 + 39 + (34/2) + [(34 + 29)/4] = 104.75$ in (2.66 m). Average pressure $p_{a1} = 0.0361d_{12} = 0.0361 \times 104.75 = 3.78$ lb/in^2 (26.06 kPa). Thus, the stiffener is subjected to a uniform load $W_1 = p_{a1}(b_1 + b_2)/2 = 3.78(29 + 34)/2 = 119.07$ lb/in (20.85 kN/m). The required composite section modulus of the stiffener and plate can be calculated from $S = WL^2/(12s_a)$, where S = section modulus, in^3 (m^3); $L = a$ = length of the stiffener, in (m); s_a = stiffener allowable bending stress, lb/in^2 (Pa). Thus, $S_1 = 119.07 \times 60^2/(12 \times 16{,}000) = 2.24$ in^3 (3.79 $\times$ 10^{-5} m^3).

6. Check out an initial tentative height for panel No. 2

Assume $b_2 = (H - b_1)/3 = (135 - 29)/3 = 35.3$ in (0.90 m). Applied pressure $p_2 = h_2(0.0361) = [H - b_1 - (b_2/3)](0.0361) = [135 - 29 - (35.3/3)](0.0361) = 3.40$ lb/in^2 (23.44 kPa). And, $a/b = 60/35.3 = 1.70$, where $\alpha = 0.0259$ and $\beta = 0.4776$. Substituting in the equation for panel height $b_2 = [(19{,}330 \times 0.3125^2)/(0.4776 \times 3.40)]^{1/2} = 34.1$ in (0.87 m). This is so close to the assumed value that a tentative value of $b_2 = 34$ in (0.86 m) can be set.

7. Check panel No. 2 for bending stress

Substituting in the equation for bending stress $s_{b2} = 0.4776 \times 3.40 \times 34^2/0.3125^2 = 19{,}220$ lb/in^2 (132.5 MPa). This is within the allowable limit of the bending stress.

8. Check Panel No. 2 for deflection

Substituting in the equation for deflection $\delta_2 = 0.0259 \times 3.40 \times 34^4/(27.9 \times 10^6 \times 0.3125^3)$ = 0.138 in (0.35 cm). Since this value is less than the allowable deflection of 0.156 in (0.40 cm), the height of panel 2 is now acceptable as 34 in (0.86 m).

9. Size the stiffener for panel No. 2

Pressure head $d_{23} = b_4 + (b_3/2) + [(b_3 + b_2)/4] = 33 + (39/2) + [(39 + 34)/4] = 70.75$ in (1.80 m). Average pressure $p_{a2} = 0.0361d_2 = 0.0361(70.75) = 2.55$ lb/in^2 (17.56 kPa). Uniform load $W_2 = p_{a2}(b_2 + b_3)/2 = 2.55(34 + 39)/2 = 93.08$ lb/in (16.30 kN/m). Composite section modulus $S_2 = W_2a^2/(12s_a) = 93.08 \times 60^2/(12 \times 16,000) = 1.75$ in^3 (2.94 $\times$ 10^{-5} m^3).

10. Check out an initial tentative height for panel No. 3

Following the previous procedures, assume panel height $b_3 = (135 - 29 - 34)/3 = 24$ in (0.61 m). Applied pressure $p_3 = [135 - 29 - 34 - (24/3)](0.0361) = 2.31$ lb/in^2 (15.93 kPa). Then, since $a/b = 60/24 = 2.50$, $\alpha = 0.0284$ and $\beta = 0.5000$ for infinity because corresponding values for $a/b = 2.50$ are not much less. Substituting in the equation for panel height $b_3 = [19,330 \times 0.3125^2/(0.5000 \times 2.31)]^{1/2} = 40.4$ in (1.23 m). Because the result is much greater than the assumed value, it is necessary to iterate again.

11. Check out a revised tentative height for panel No. 3

Repeating the previous procedure, but with $b_3 = 48$ in (1.22 m), $p_3 = h_3(0.0361) = [135 - 29 - 34 - (48/3)](0.0361) = 2.02$ lb/in^2 (13.94 kPa). Also, $a/b = 60/48 = 1.25$ and, by interpolation, $\alpha = 0.0201$ and $\beta = 0.4023$. Again, solving for $b_3 = [19,330 \times 0.3125^2/(0.4023 \times 2.02)]^{1/2} = 48.2$ in (1.22 m). This result is very close to the assumed value; hence $b_3 = 48$ in (1.22 m) is tentatively acceptable.

12. Check panel No. 3 for bending stress

The actual plate stress $s_{b3} = 0.4023 \times 2.02 \times 48^2/(0.3125)^2 = 19,170$ lb/in^2 (132.17 MPa). This stress is less than the allowable 19,330 lb/in^2 (133.3 MPa), hence the height of panel No. 3 may be less, but not more, than 48 in (1.22 m).

13. Check panel No. 3 for deflection

Substituting in the equation for deflection $\delta_3 = 0.0201 \times 2.02 \times 48^4/(27.9 \times 10^6 \times 0.3125^3)$ = 0.253 in (0.64 cm). Since this deflection exceeds 0.156 in (0.40 cm), it is necessary to select a smaller value for b_3.

14. Check panel No. 3 for deflection with its height reduced

With $b_3 = 39$ in (0.99 m), $h_3 = 135 - 29 - 34 - (39/3) = 59$ in (1.5 m) and $p_3 = 59 \times 0.0361 = 2.13$ lb/in^2 (14.69 kPa). Also, for $a/b = 60/39 = 1.54$ the value of $\alpha = 0.0243$. Then, $\delta_3 = 0.0243 \times 2.13 \times 39^4/(27.9 \times 10^6 \times 0.3125^3) = 0.141$ in (0.36 cm). This deflection is close enough to 0.156 in (0.40 cm) to stop iterating for a value of b_3 greater than 39 in (0.99 m) and less than 48 in (1.22 m) wherein the bending stress does not exceed the allowable value.

15. Size the stiffener for panel No. 3

Pressure head $d_{34} = (b_4/2) + [(b_4 + b_3)/4] = (33/2) + [(33 + 39)/4] = 34.5$ in (0.88 m). Average pressure $p_{a3} = 0.0361d_3 = 0.0361 \times 34.5 = 1.25$ lb/in^2 (8.62 kPa). Uniform load $W_3 = p_{a3}(b_3 + b_4)/2 = 1.25(39 + 33)/2 = 45$ lb/in (7.88 kN/m). Composite section modulus $S_3 = W_3a^2/(12s_a) = 45.0 \times 60^2/(12 \times 16,000) = 0.84$ in^3 (1.42 $\times$ 10^{-5} m^3).

16. Determine if there is a need for reinforcing panel No. 4

The height of the uppermost panel is $b_4 = 135 - 29 - 34 - 39 = 33$ in (0.84 m). Applied water pressure $p_4 = h_4(0.0361) = (2 \times 33/3)(0.0361) = 0.79$ lb/in^2 (5.45 kPa). Panel dimension ratio $a/b = 60/33 = 1.82$, hence $\alpha = 0.0268$ and $\beta = 0.4882$. Then, $b_4 = [19,330 \times 0.3125^2/(0.4882 \times 0.79)]^{1/2} = 70.0$ in (1.78 m). Since this calculated height is much greater than the actual height of the panel, it shows that panel No. 4 is adequate without an additional stiffener.

Related Calculations: Stiffeners divide the vessel faces into rectangular plates whose edges are considered to be fixed. Plate thickness can be minimized to an optimum size which provides panel heights within preferred magnitudes while maintaining maximum stress or deflection just within allowable limits. For more accuracy in obtaining values for α and β and to avoid interpolation, values may be obtained from graphs plotted from the listed values.

The solution to this example is based upon the method shown in an article written by R. Jay Smith and G.L.B. Knight in the November 26, 1981, issue of *Machine Design* magazine. Both authors were associated with the Davy McKee Corp., Lakeland, Florida.

MINIMUM-COST PRESSURE VESSELS

Find the diameter that minimizes the cost to construct a pressure tank with the following design characteristics: volume $V = 1400$ ft^3 (40 m^3); allowable stress, $s = 16,000$ lb/in^2 (110 MPa); welding efficiency $e = 85$ percent; internal pressure $p = 400$ lb/in^2 (2760 kPa); corrosion allowance $t_c = 0.05$ in (1.27×10^{-3} m). Also, find the length-to-diameter ratio of the tank.

Calculation Procedure:

1. *Compute the approximate diameter*

Use the equation for the approximate diameter, $D_a = [V(0.2898V^{1/3} + Z_1)/(1.426V^{1/3} + Z_2)]^{1/3}$, where $Z_1 = 0.2829t_c[(se/p) - 0.6]$ and $Z_2 = 0.2175t_c[(se/p) - 0.6]$. Substituting, $Z_1 = 0.2829 \times 0.05[(16,000 \times 0.85/400) - 0.6] = 0.4724$ and $Z_2 = 0.2175 \times 0.05[(16,000 \times 0.85/400) - 0.6] = 0.3632$. Then, $D_a = \{1400[0.2898(1400)^{1/3} + 0.4724]/[1.426(1400)^{1/3} + 0.3632]\}^{1/3} = 6.83$ ft (2.08 m).

2. *Compute the exact diameter*

The equation for the exact diameter is $D = [V(0.4244D + Z_1)/(2.088D + Z_2)]^{1/3}$. For a first trial use the approximate diameter; hence $D = \{1400[(0.4244 \times 6.83) + 0.4724]/[(2.088 \times 6.83) + 0.3632)]\}^{1/3} = 6.87$ ft (2.09 m). Next, try $D = 6.87$ ft (2.09 m). Thus, $D = \{1400[(0.424 \times 6.87) + 0.4724]/[(2.088 \times 6.87) + 0.3632)]\}^{1/3} = 6.86$ ft (2.09 m). Iterating once more, $D = \{1400[(0.4244 \times 6.86) + 0.4724]/[(2.088 \times 6.86) + 0.3632]\}^{1/3} = 6.86$ ft (2.09 m); the exact diameter.

3. *Determine the length-to-diameter ratio of the tank*

The length-to-diameter ratio $L/D = 4V/\pi D^3 = (4 \times 1400)/(\pi \times 6.86^3) = 5.52$.

Related Calculation: The approximate equation directly provides the size of a diameter within 3 percent, or less, of the exact size. Greater accuracy can be obtained by iterating the exact equation which converges rapidly. The length-to-diameter ratio varies accordingly. For metric calculations the coefficient for Z_1 should be 3.420 instead of 0.2829 and for Z_2 it should be 2.629 instead of 0.2175.

This example is based on an article written by J. Zigrang and N.D. Sylvester, professor and dean, respectively, at the University of Tulsa, Tulsa, Oklahoma. The piece appeared in the December 12, 1985, issue of *Machine Design* magazine.

SECTION 7

CONTROL ENGINEERING

CHARLES R. HAFER, P.E.

GEORGE M. MUSCHAMP
CONSULTING ENGINEER
HONEYWELL, INC.

TYLER G. HICKS, P.E.
INTERNATIONAL ENGINEERING ASSOCIATES

Selection of a Process-Control System 7.2
Process-Temperature-Control Analysis 7.4
Computer Selection for Industrial Process-Control Systems 7.6
Control-Valve Selection for Process Control 7.8
Controlled-Volume-Pump Selection for a Control System 7.10
Steam-Boiler-Control Selection and Application 7.11
Control-Valve Characteristics and Rangeability 7.12
Fluid-Amplifier Selection and Application 7.14
Cavitation, Subcritical- and Critical-Flow Considerations in Controller Selection 7.15
Servo System Stability Determination 7.20
Angular-Position Servo System Analysis 7.21
Loop-Gain Function Analysis 7.22
Servo System Overshoot and Settling Time 7.24
Analysis of a Servo System with a Loop Delay 7.25
Servo System Closed-Loop Transfer Function Characteristics 7.26
Developing a Transfer Function for a Given Phase Margin 7.30
Feedback Control System Phase-Lag Compensator Design 7.33
Analysis of Temperature-Measuring Amplifier 7.37
Analysis of Temperature-Measuring Instrument 7.38
Measuring RMS Value of a Waveform 7.40
NAND Gate Circuit Implementation 7.41

Using NAND Gates to Implement Two Levels of Logic. 7.42
Sizing Steam-Control and Pressure-Reducing Valves 7.45
Boolean Algebra 7.48

REFERENCES: Morris—*Control Engineering*, McGraw-Hill; Kuo—*Automatic Control Systems*, Prentice-Hall; Schwartz—*Multivariable Technical Control Systems*, Elsevier North-Holland; Berkovitz—*Optimal Control Theory*, Springer-Verlag; Bryson and Ho—*Applied Optimal Control*, Hemisphere; Craven—*Mathematical Programming & Control Theory*, Chapman & Hall; Fallside—*Control System Design by Pole-Zero Assignment*, Cambridge University Press; Hafer—*Electronics Engineering for Professional Engineers' Examinations*, McGraw-Hill; Considine—*Process Instruments and Controls Handbook*, McGraw-Hill; Merritt—*Hydraulic Control Systems*, Wiley; Considine and Ross—*Handbook of Applied Instrumentation*, McGraw-Hill; Eckman—*Automatic Process Control*, Wiley; Kallen—*Handbook of Instrumentation and Controls*, McGraw-Hill; Farrington—*Fundamentals of Automatic Control*, Wiley; ASME—*Fluid Meters*; Shinskey—*Process-Control Systems*, McGraw-Hill; Graham-McRuer—*Analysis of Nonlinear Control Systems*, Wiley; Harriott—*Process Control*, McGraw-Hill; Mesarovic—*The Control of Multivariable Systems*, Wiley; Savas—*Computer Control of Industrial Processes*, McGraw-Hill; Newton-Gould-Kaiser—*Analytical Design of Linear Feedback Controls*, Wiley; Burr-Brown Research Corp.—*Handbook of Operational Amplifier Active RC Networks*; Bower-Schultheiss—*Introduction to Design of Servomechanisms*, Wiley; Gibson-Tuteur—*Control System Components*, McGraw-Hill; Bode—*Network Analysis and Feedback Amplifier Design*, D. Van Nostrand; Doss—*Information Processing Equipment*, Reinhold; ASME—*Flowmeter Computation Handbook*; ASME—*Flow Measurement*; Dommasch and Laudeman—*Principles Underlying Systems Engineering*, Pitman.

SELECTION OF A PROCESS-CONTROL SYSTEM

A continuous industrial process contains four process centers, each of which has two variables that must be controlled. If a fast process-reaction rate is required with only small to moderate dead time, select a suitable mode of control. The system contains more than two resistance-capacity pairs. What type of transmission system would be suitable for this process?

Calculation Procedure:

1. Compute the number of process capacities

The number of process capacities = (number of process centers)(number of variables per center), or, for this system, $4 \times 2 = 8$ process capacities. This is defined as a *multiple* number of process capacities because the number controlled is greater than unity.

2. Analyze the process-time lags

A small to moderate dead time is allowed in this process-control system. With such a dead-time allowance and with two or more resistance-capacity pairs in the system, a mode of control that provides for any number of process-time lags is desirable.

3. Select a suitable mode of control

Table 1 summarizes the forms of control suited to processes having various characteristics. This table is a *general guide*—it provides, at best, an *approximate* aid in selecting control modes. Hence it is suitable for tentative selection of the mode of control. Final selection must be based on actual experience with similar systems.

Inspection of Table 1 shows that for a multiple number of processes with small to moderate dead time and any number of resistance-capacity pairs, a proportional plus reset mode of control is probably suitable. Further, this mode of control provides for any (i.e., fast or slow) reaction rate. Since a fast reaction rate is desired, the proportional plus reset method of control is suitable because it can handle any process-reaction rate.

4. Select the type of transmission system to use

Four types of transmission systems are used for process control today: pneumatic, electric, electronic, and hydraulic. The first three types are by far the most common.

TABLE 1 Process Characteristics versus Mode of Control°

Number of process capacities	Process reaction rate	Resistance capacity (RC)	Dead time (transportation)	Load changes		Suitable mode of control
				Size	Speed	
Single	Slow	Moderate to large	Small	Any	Any	Two-position; two-position with differential gap
				Moderate	Slow	Multiposition; proportional input
Single (self-regulating)	Fast	Small	Small	Any	Slow	Floating modes: Single speed, multispeed
					Moderate	Proportional-speed floating
Multiple	Slow to moderate	Moderate	Small	Small	Moderate	Proportional position
Multiple	Moderate	Any	Small	Small	Any	Proportional plus rate
Multiple	Any	Any	Small to moderate	Large	Slow to moderate	Proportional plus reset
Multiple	Any	Any	Small	Large	Fast	Proportional plus reset plus rate
Any	Faster than that of the control system	Small or nearly zero	Small to moderate	Any	Any	Wideband proportional plus fast reset

°Considine—*Process Instruments and Controls Handbook*, McGraw-Hill.

Pneumatic transmission systems use air at 3 to 20 lb/in^2 (gage) (20.7 to 137.9 kPa) to convey the control signal through small-bore metal tubing at distances ranging to several thousand feet. The air used in pneumatic systems must be clean and dry. To prevent a process from getting out of control, a constant supply of air is required. Pneumatic controllers, receivers, and valve positioners usually have small air-space volumes of 5 to 10 in^3 (81.0 to 163.9 cm^3). Air motors of the diaphragm or piston type have relatively large volumes: 100 to 5000 in^3 (1639 to 81,935 cm^3).

Pneumatic control systems are generally considered to be spark-free. Hence, they find wide use in hazardous process areas. Also, control air is readily available, and it can be "dumped" to the atmosphere safely. The response time of pneumatic control systems may be slower than that of electric or hydraulic systems.

Electric and electronic control systems are fast-response with the signal conveyed by a wire from the sensing point to the controller. In hazardous atmospheres the wire must be protected against abrasion and breakage.

Hydraulic control systems are also rapid-response. These systems are capable of high power actuation. Slower-acting hydraulic systems use fluid pressures in the 50 to 100 lb/in^2 (344.8 to 689.5 kPa) range; fast-acting systems use fluid pressures to 5000 lb/in^2 (34,475 kPa).

Dirt and fluid flammability are two factors that may be disadvantages in certain hydraulic-control-system applications. However, new manufacturing techniques and nonflammable fluids are overcoming these disadvantages.

Since a fast response is desired in this process-control system, electric, electronic, or hydraulic transmission of the signals would be considered first. With long distances between the sensing points [say 1000 ft (305 m) or more], an electric or electronic system would probably be best.

Next, determine whether the systems being considered can provide the mode of control (step 2) required. If a system cannot provide the necessary mode of control, eliminate the system from consideration.

Before a final choice of a system is made, other factors must be considered. Thus, the relative cost of each type of system must be determined. Should an electric system prove too costly, the slightly slower response time of the pneumatic system might be accepted to reduce the initial investment.

Other factors influencing the choice of the type of a control system include type of controls, if any, currently used in the installation, skill and experience of the operating and maintenance personnel, type of atmosphere in which, and type of process for which, the controls will be used. Any of these factors may alter the initial choice.

Related Calculations: Use this general method to make a preliminary choice of controls for continuous processes, intermittent processes, air-conditioning systems, combustion-control systems, etc. Before making a final choice of any control system, be certain to weigh the cost, safety, operating, and maintenance factors listed above. Last, the system chosen *must* be able to provide the mode of control required.

PROCESS-TEMPERATURE-CONTROL ANALYSIS

A water storage tank (Fig. 1) contains 500 lb (226.8 kg) of water at 150°F (65.6°C) when full. Water is supplied to the tank at 50°F (10.0°C) and is withdrawn at the rate of 25 lb/min (0.19 kg/s). Determine the process-time constant and the zero-frequency process gain if the thermal sensing pipe contains 15 lb (6.8 kg) of water between the tank and thermal bulb and the maximum steam flow to the tank is 8 lb/min (0.060 kg/s). The steam flow to the tank is controlled by a standard linear regulating valve whose flow range is 0 to 10 lb/min (0 to 0.076 kg/s) when the valve operator pressure changes from 5 to 30 lb/in^2 (34.5 to 206.9 kPa).

Calculation Procedure:

1. *Compute the distance-velocity lag*

The time in minutes needed for the thermal element to detect a change in temperature in the storage tank is the *distance-velocity lag*, which is also called the *transportation lag*, or *dead time*. For this process, the distance-velocity lag d is the ratio of the quantity of water in the pipe between the tank and the thermal bulb—that is, 15 gal (57.01 L)—and the rate of flow of water out of the tank—that is, 25 lb/min (0.114 kg/s)—or $d = 15/25 = 0.667$ min.

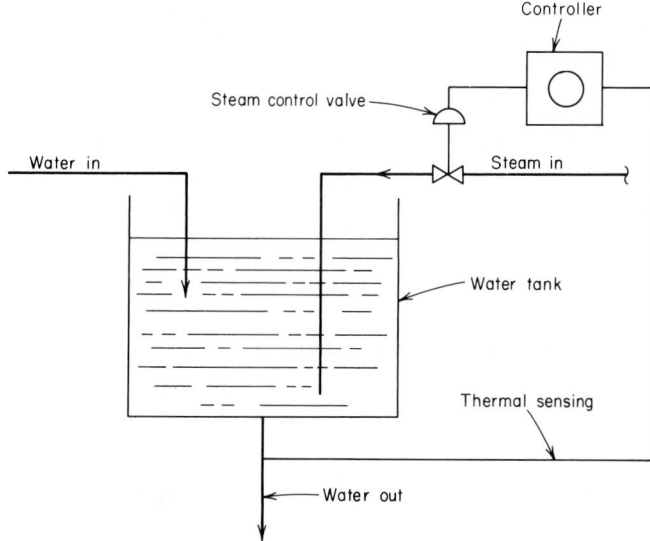

FIG. 1   Temperature control of a simple process.

2.   *Compute the energy input to the tank*

This is a *transient-control process;* i.e., the conditions in the process are undergoing constant change instead of remaining fixed, as in *steady-state conditions.* For transient-process conditions the heat balance is $H_{in} = H_{out} + H_{stor}$, where H_{in} = heat input, Btu/min; H_{out} = heat output, Btu/min; H_{stor} = heat stored, Btu/min.

The heat input to this process is the enthalpy of vaporization h_{fg} Btu/(lb·min) of the steam supplied to the process. Since the regulating valve is linear, its sensitivity s is (flow-rate change, lb/min)/(pressure change, lb/in²). Or, by using the known valve characteristics, $s = (10 - 0)/(30 - 5) = 0.4$ (lb/min)/(lb/in²) [0.00044 kg/(kPa·s)].

With a change in steam pressure of p lb/in² (p' kPa) in the valve operator, the change in the rate of energy supply to the process is $H_{in} = 0.4$ (lb/min)/(lb/in²) $\times p \times h_{fg}$. Taking h_{fg} as 938 Btu/lb (2181 kJ/kg) gives $H_{in} = 375p$ Btu/min (6.6p' kW).

3.   *Compute the energy output from the system*

The energy output H_{out} = lb/min of liquid outflow $\times$ liquid specific heat, Btu/(lb·°F) $\times$ ($T_a - 150°F$), where T_a = tank temperature, °F, at any time. When the system is in a state of equilibrium, the temperature of the liquid in the tank is the same as that leaving the tank or, in this instance, 150° F (65.6°C). But when steam is supplied to the tank under equilibrium conditions, the liquid temperature will rise to $150 + T_r$, where T_r = temperature rise, °F ($T_{r'}$, °C), produced by introducing steam into the water. Thus, the above equation becomes $H_{out} = 25$ lb/min $\times 1.0$ Btu/(lb·°F) $\times T_r = 25T_r$ Btu/min (0.44$T_{r'}$ kW).

4.   *Compute the energy stored in the system*

With rapid mixing of the steam and water, H_{stor} = liquid storage, lb $\times$ liquid specific heat, Btu/(lb·°F) $\times T_r q = 500 \times 1.0 \times T_r q$, where q = derivative of the tank outlet temperature with respect to time.

5.   *Determine the time constant and process gain*

Write the process heat balance, substituting the computed values in $H_{in} = H_{out} + H_{stor}$, or $375p = 25T_r + 500T_r q$. Solving gives $T_r/p = 375/(25 + 500q) = 15/(1 + 20q)$.

The denominator of this linear first-order differential equation gives the process-system time constant of 20 min in the expression $1 + 20q$. Likewise, the numerator gives the zero-frequency process gain of 15°F/(lb/in²) (1.2°C/kPa).

Related Calculations: This general procedure is valid for any liquid using any gaseous heating medium for temperature control with a single linear lag. Likewise, this general procedure is also valid for temperature control with a double linear lag and pressure control with a single linear lag.

COMPUTER SELECTION FOR INDUSTRIAL PROCESS-CONTROL SYSTEMS

Select the type of computer and its speed of operation for use in an industrial control application. The computer will be used to monitor and control two continuous-flow process operations. Budget limitations restrict the investment in the computer to about $100,000 with a typical execution time of 10 ms or better.

Calculation Procedure:

1.   *Analyze the computers available; select a suitable computer*

Four types of computers are available for consideration in any control problem: analog; digital; hybrid—consisting of analog and digital; and special-purpose—analog or digital computers for industrial control.

Digital computers (Fig. 2) find wide use for controlling continuous-flow processes. When used to control continuous-flow processes, the digital computer is connected *online;* i.e., information reflecting the activity in the process being controlled is introduced to the data-processing system as soon as it occurs, and action is immediately initiated by the system to make any needed adjustments. Since digital computers are proven machines for process control, this type will be tentatively chosen for this process and its suitability investigated further.

The usual digital computer used in process control is a general-purpose one that can receive and transmit analog signals. A magnetic-drum-type stored memory is often used in control applications.

2.   *Determine the computer operating time*

The speed of a computer depends on the actions needed to perform a calculation. Thus, a drum or disk memory may have a 150-μs add time with a memory access time of 16,000 μs. In such a computer the memory-access time is the controlling speed factor.

Figure 3 shows the execution times for a variety of digital computers of different makes in solving the same *bench-mark* or *test* problems. A typical bench-mark problem consists of a pair of simultaneous equations having two unknowns.

Study of Fig. 3 shows that three machines—*A*, *B*, and *D*—meet the general requirements for

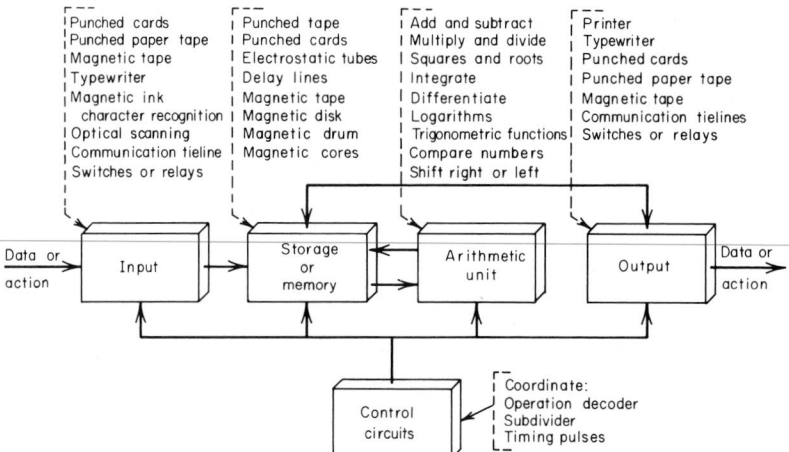

FIG. 2   Digital-computer elements used in process control.

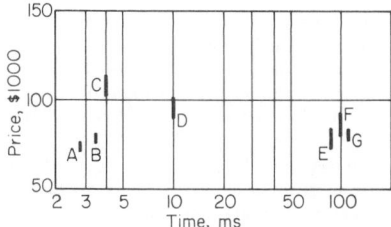

FIG. 3 Execution time for an arithmetic bench-mark problem shows the solution time required by different makes of computers.

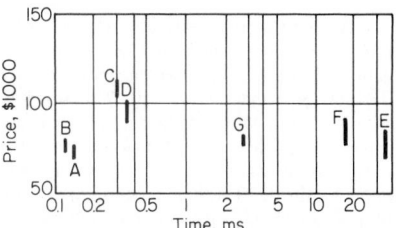

FIG. 4 Execution time for a bench-mark problem shows no clear relationships between solution time and computer cost.

this process with respect to speed and cost. Since each of these computers is produced by a different manufacturer, a fairly wide choice of units is available. When a chart such as Fig. 3 is not available, prepare one after computing the costs of machines available from various manufacturers.

3. *Check the computer performance*

Computer performance is rated according to three factors: speed margin, i.e., how the computer copes with the worst-case time combination of events in the process; memory-storage margin, that is, provisions for worst-case data storage, retrieval, and working space; and reliability or consistency of performance.

Word length affects computer speed. Although a computer handling 24-bit words may seem to have an operation rate identical to a 12-bit-word computer, there may be as much as a 2:1 variation in the speed to the same degree of precision. To compare two computers, use their relative *problem-solving speeds* as a guide.

Manufacturers publish the relative problem-solving speeds of each model of computer. List these speeds for each of the suitable makes—A, B, and D, step 2. Select the computer giving the fastest speed for the smallest investment.

4. *Investigate the computer logic function*

In some control applications, the logical-problem-solving ability of the computer may be more important than the arithmetic ability. In the logic function, the computer uses information transfers, manipulations, and comparisons, all of which take time. Using manufacturer's data, plot the logic speed against cost for each computer for a specific bench-mark or test problem. Usually the execution time for a logical bench-mark problem shows no clear relation to price (Fig. 4). However, the plot does indicate the price range for various operating speeds.

5. *Evaluate the computer-memory size*

An online-control computer cannot deal with a real-time problem unless all instructions defining the action are stored and available when needed. Distribution of the storage between high-speed random-access memory sections and lower-cost cyclic-access sections, such as drums and disks, influences computer speed. Good practice stores an image of the high-speed working memory in one of the slower-speed backup memories.

In machines with 12- to 15-bit words, more than one memory word may be needed to store an instruction or data item. Compare bench-mark problems to determine the real-time need of memory addresses between different computers. In general, more process-control applications are better served by machines having expandable memories. The tendency of many control-system designers is to underestimate the memory capacity needed. So choose a machine having the largest memory possible within the prevailing financial constraints.

6. *Evaluate the computer reliability*

The mean time between failure (MTBF) is a good measure of computer reliability. For a typical control computer, the MTBF should be in the 1000+-h category.

Compare the MTBF for each of the computers in step 2. Choose the machine having the highest MTBF at the desired speed. Further, the computer should be capable of operating in the 50 to 120°F (10 to 48.9° C) ambient temperature range.

Related Calculations: Use this general method to choose a control computer for any process-type application. Where high-speed operations are involved, such as in missile-guidance applications, computers operating at nanosecond speeds are generally required. The selection procedure for such machines is different from that for process-control computers where millisecond speeds are usually satisfactory. Note that the computer prices cited here are relative; for actual prices consult the manufacturers concerned.

The program or software prepared for automatic process control can cost as much as the computer hardware. The first process applications of computers in an industry often show that the programs needed cost more than anticipated. Thus, computer programming to perform startup and shutdown sequence monitoring in a process system may require 4000 instructions. For automatic sequence control, as many as 20,000 instructions may be needed. Further, each plant is unique, and little of the programming work done for one process can be applied to another.

Programming still uses the most time of any phase of total computer operation for problem solving. So even if the computer operating time were cut to nearly zero, the saving in machine time would not be significant when compared with the programming time. On the other hand, saving in computer machine time is important in process-control applications because processes can be held more closely to optimum levels with reduced machine time.

Direct digital control (DDC) is replacing analog control in some process applications. The major advantage of DDC is that it removes the need for digital-to-analog converters and gives the computer more direct control over plant equipment while removing a source of spurious signals.

In process control, certain aspects of programming warrant special consideration. These aspects include real-time operation, memory capability, and operator misuse. Because a process computer functions in a real-time environment, a certain amount of "free" time must be available to allow for emergency reactions and special operator requests. A good rule of thumb is to allow 40 percent of free time within the computer; any less may cause the computer program to fall out of step with the process situation.

Most process-control computers use one of the variants of IBM's original Fortran. This language finds its principal applications in relatively infrequent procedures, such as plant startup and shutdown. Minute-by-minute scanning is usually handled by a standard "scan, monitor, and alarm" program prepared manually for the particular scanning sequence dictated by operating requirements.

A self-checking program is an important feature of a process-control computer system. Unlike a scientific program in which the results obtained are printed out for perusal by a scientist or engineer, a closed-loop control system utilizes the computer's calculations to act directly on the process plant itself. Should either the input to the computer or the data handling within it be in error, the resulting calculated control points and output signals will be incorrect and could result in hazardous operation. Double and triple checks may be necessary to ensure that the operating data are valid. As more and more input signals are utilized, the programming necessary to provide such validity checks becomes increasingly more complicated. Continuous calculation of a process heat balance can, for example, be useful in determining the validity of the input information, since an extensive range of input data figures in the calculation.

CONTROL-VALVE SELECTION FOR PROCESS CONTROL

Select a steam control valve for a heat exchanger requiring a flow of 1500 lb/h (0.19 kg/s) of saturated steam at 80 lb/in^2 (gage) (551.6 kPa) at full load and 300 lb/h (0.038 kg/s) at 40 lb/in^2 (gage) (275.8 kPa) at minimum load. Steam at 100 lb/in^2 (gage) (689.5 kPa) is available for heating.

Calculation Procedure:

1. Compute the valve flow coefficient

The valve flow coefficient C_v is a function of the maximum steam flow rate through the valve and the pressure drop that occurs at this flow rate. In choosing a control valve for a process-control system, the usual procedure is to assume a maximum flow rate for the valve based on a considered judgment of the overload the system may carry. Usual overloads do not exceed 25 percent of the maximum rated capacity of the system. Using this overload range as a guide, assume that the valve must handle a 20 percent overload, or 0.20 (1500) = 300 lb/h (0.038 kg/s). Hence, the rated capacity of this valve should be 1500 + 300 = 1800 lb/h (0.23 kg/s).

The pressure drop across a steam control valve is a function of the valve design, size, and flow rate. The most accurate pressure-drop estimate usually available is that given in the valve manufacturer's engineering data for a specific valve size, type, and steam-flow rate. Without such data, assume a pressure drop of 5 to 15 percent across the valve as a first approximation. This means that the pressure loss across this valve, assuming a 10 percent drop at the maximum steam-flow rate, would be 0.10 × 80 = 8 lb/in² (gage) (55.2 kPa).

With these data available, compute the valve flow coefficient from $C_v = WK/3(\Delta p\ P_2)^{0.5}$, where W = steam flow rate, lb/h; $K = 1 + (0.0007 \times °F$ superheat of the steam); p = pressure drop across the valve at the maximum steam flow rate, lb/in²; P_2 = control-valve outlet pressure at maximum steam flow rate, lb/in² (abs). Since the steam is saturated, it is not superheated and $K = 1$. Then $C_v = 1500/3(8 \times 94.7)^{0.5} = 18.1$.

2. *Compute the low-load steam flow rate*

Use the relation $W = 3(C_v\ \Delta p\ P_2)^{0.5}/K$, where all the symbols are as before. Thus, with a 40-lb/in² (gage) (275.8-kPa) low-load heater inlet pressure, the valve pressure drop is 80 − 40 = 40 lb/in² (gage) (275.8 kPa). The flow rate through the valve is then $W = 3(18.1 \times 40 \times 54.7)^{0.5}/1 = 598$ lb/h (0.75 kg/s).

Since the heater requires 300 lb/h (0.038 kg/s) of steam at the minimum load, the valve is suitable. Had the flow rate of the valve been insufficient for the minimum flow rate, a different pressure drop, i.e., a larger valve, would have to be assumed and the calculation repeated until a flow rate of at least 300 lb/h (0.038 kg/s) was obtained.

Related Calculations: The flow coefficient C_v of the usual 1-in (2.5-cm) diameter double-seated control valve is 10. For any other size valve, the approximate C_v valve can be found from the product $10 \times d^2$, where d = nominal body diameter of the control valve. Thus, for a 2-in (5.1-cm) diameter valve, $C_v = 10 \times 2^2 = 40$. By using this relation and solving for d, the nominal diameter of the valve analyzed in steps 1 and 2 is $d = (C_v/10)^{0.5} = (18.1/10)^{0.5} = 1.35$ in (3.4 cm); use a 1.5-in (3.8-cm) valve because the next smaller standard control valve size, 1.25 in (3.2 cm), is too small. Standard double-seated control-valve sizes are ¾, 1, 1¼, 1½, 2, 2½, 3, 4, 6, 8, 10, and 12 in (1.9, 2.5, 3.2, 3.8, 5.1, 6.4, 7.6, 10.2, 15.2, 20.3, 25.4, 30.5 cm). Figure 5 shows typical flow-lift characteristics of popular types of control valves.

To size control valves for liquids, use a similar procedure and the relation $C_v = V(G/\Delta p)$, where V = flow rate through the valve, gal/min; Δp = pressure drop across the valve at maxi-

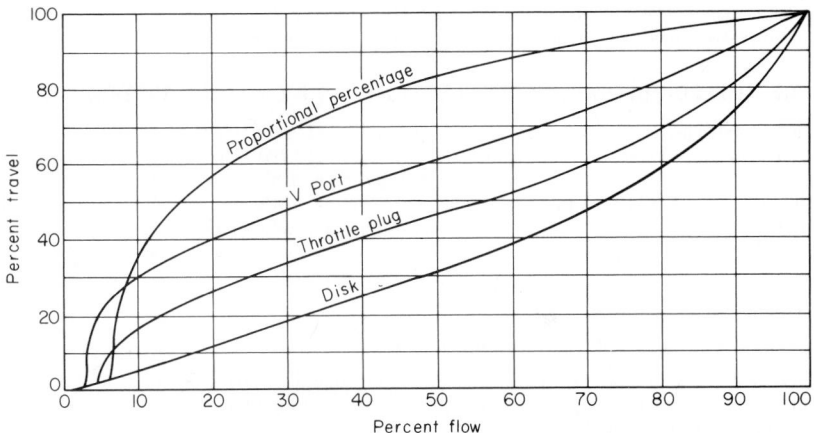

FIG. 5 Flow-lift characteristics of control valves. *(Taylor Instrument Process Control Division of Sybron Corporation.)*

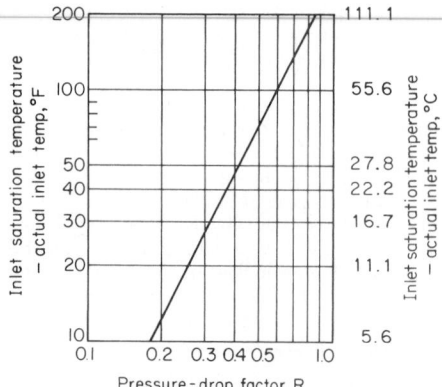

FIG. 6   Pressure-drop correction factor for water in the liquid state. *(International Engineering Associates.)*

mum flow rate, lb/in²; G = specific gravity of the liquid. When a liquid has a specific gravity of 100 SSU or less, the effect of viscosity on the control action is negligible.

To size control valves for gases, use the relation $C_v = Q(GT_a)^{0.5}/1360(\Delta p\ P_2)^{0.5}$, where Q = gas flow rate, ft³/h at 14.7 lb/in² (abs) (101.4 kPa) and 60°F (15.6°C); T_a = temperature of the flowing gas, °F abs = 460 + °F; other symbols as before. When the valve outlet pressure P_2 is less than $0.5P_1$, where P_1 = valve inlet pressure, use the value of $P_1/2$ in place of $(\Delta p\ P_2)^{0.5}$ in the denominator of the above relation.

To size control valves for vapors other than steam, use the relation $C_v = W(v_2/\Delta p)^{0.5}/63.4$, where W = vapor flow rate, lb/h; v_2 = specific volume of the vapor at the outlet pressure P_2, ft³/lb; other symbols as before. When P_2 is less than $0.5P_1$, use the value of $P_1/2$ in place of Δp and use the corresponding value of v_2 at $P_1/2$.

When the control valve handles a flashing mixture of water and steam, compute C_v by using the relation for liquids given above after determining which pressure drop to use in the equation. Use the *actual* pressure drop or the *allowable* pressure drop, whichever is smaller. Find the allowable pressure drop by taking the product of the supply pressure and the correction factor R, where R is obtained from Fig. 6. For a further discussion of control-valve sizing, see Considine—*Process Instruments and Controls Handbook*, McGraw-Hill, and G. F. Brockett and C. F. King—"Sizing Control Valves Handling Flashing Liquids," Texas A & M Symposium.

CONTROLLED-VOLUME-PUMP SELECTION FOR A CONTROL SYSTEM

Select a controlled-volume pump to deliver 80 gal/h (0.084 L/s) of 100°F (37.8°C) distilled water to a chemical-feed system operating at 2000 lb/in² (abs) (13,790 kPa). What is the net positive suction head (NPSH) at the beginning of the pump suction stroke if the supply tank produces a 2-lb/in² (gage) (13.8-kPa) suction head at the pump centerline? Compute the minimum allowable NPSH for this pump if the length of the 1.5-in (3.8-cm) pipe between the pump and suction tank is 30 ft (9.1 m).

Calculation Procedure:

1.   *Choose the general type of pump to use*

Controlled-volume pumps serve two functions when used as the final control elements in a control loop: to deliver liquid at the required pressure and to deliver liquid in the required quantities. In its second role, the pump also serves as a meter.

Two types of controlled-volume pumps are popular: plunger and diaphragm. The plunger pump is of somewhat simpler construction and is often used where contact of the plunger and liquid handled is not objectionable. Since distilled water is a relatively bland liquid, a plunger pump will be the tentative first choice for this control application.

2.   Determine the pump dimensions and speed

The capacity, dimensions, speed, and efficiency of plunger-type controlled-volume pumps are given by $Q = D^2LNE/K$, where Q = pump capacity, gal/h (L/s); D = plunger diameter, in (cm); L = plunger stroke length, in (cm); N = number of strokes per minute, i.e., the pump speed; E = volumetric efficiency of the pump; K = dimensional constant = 4.92 for Q gal/h (0.0052 L/s), 295 for pump capacity in gal/min (18.6 L/s), and 0.0013 for pump capacity in mL/h.

Assume a pump speed of 50 strokes/min. This is a typical speed for plunger-type controlled-volume pumps. The usual efficiency of such a pump is 90 percent. With a 3-in (7.6-cm) stroke, the plunger diameter is $D = (QK/LNE)^{0.5} = [(80 \times 4.92)/(3 \times 50 \times 0.9)]^{0.5} = 1.71$ in, say 1.75 in (4.4 cm). This is a standard pump-plunger diameter.

3.   Compute the pump NPSH

Use the relation NPSH = $P_a \pm P_h - P_v$, where NPSH = pump net positive suction head, lb/in^2 (abs) (kPa); P_a = atmospheric pressure at pump location = 14.7 lb/in^2 (abs) (101.4 kPa) at sea level; P_h = pressure head of liquid column above (+) or below (−) the centerline of the pump suction, lb/in^2 (gage) (kPa); P_v = vapor pressure of the liquid at the pumping temperature, lb/in^2 (abs) (kPa).

From the steam tables, P_v = 0.949 lb/in^2 (abs) (6.54 kPa) for water at 100°F (37.8°C). With an atmospheric pressure of 14.7 lb/in^2 (abs) (101.4 kPa), NPSH = 14.7 + 2.0 − 0.949 = 15.751 lb/in^2 (abs) (108.6 kPa).

4.   Compute the pump minimum NPSH

Use the relation NPSH$_{min}$ = $sL_pLN^2D^2/120,000D_p^2$, where NPSH$_{min}$ = minimum net positive suction head with which the pump can operate, lb/in^2 (abs) (kPa); s = specific gravity of liquid handled; L_p = length of suction pipe, ft (m); D_p = suction pipe inside diameter, in (cm); other symbols as given before. Assuming a specific gravity of 1.0 (NPSH$_{min}$) = $(1.0 \times 30 \times 3 \times 50 \times 50 \times 1.5 \times 1.5)/(120,000 \times 1.5 \times 1.5)$ = 1.87 lb/in^2 (abs) (12.9 kPa).

Since the available NPSH [15.751 lb/in^2 (abs) (108.6 kPa), step 3] is greater than the minimum NPSH required [1.87 lb/in^2 (abs) (12.9 kPa), step 4], the pump will operate satisfactorily and without cavitation.

Related Calculations: Use this general procedure to choose controlled-volume metering pumps for control systems requiring flows ranging from 1 mL/h to 20 gal/min (1 mL/h to 1.3 L/s) or more, at pressures ranging to 50,000 lb/in^2 (344,750 kPa). Typical applications for which this procedure is valid include chemical feed, ratioing, proportioning, and control of process variables.

STEAM-BOILER-CONTROL SELECTION AND APPLICATION

Choose a suitable feedwater regulator and combustion control for an industrial boiler serving the following loads: heating, 18,000 lb/h (2.3 kg/s); process 100,000 lb/h (12.6 kg/s); miscellaneous uses, 12,000 lb/h (1.5 kg/s). The boiler will have a maximum overload of 20 percent, and wide load fluctuations are expected at frequent intervals during operation. Pulverized-coal fuel is used to fire the boiler.

Calculation Procedure:

1.   Determine the required boiler rating

Find the sum of the individual loads on the boiler, or 18,000 + 100,000 + 12,000 = 130,000 lb/h (16.4 kg/s). With a 20 percent overload, the boiler rating must be 1.2(130,000) = 156,000 lb/h (19.7 kg/s). With a 10 percent additional reserve capacity to provide for unusual loads, the rated boiler capacity should be 1.1(156,000) = 171,500 lb/h, say 175,000 lb/h (22.0 kg/s) for selection purposes.

2.   Choose the type of feedwater regulator to use

Table 2 summarizes typical feedwater regulators used for boilers of various capacities. Study of Table 2 shows that a boiler in the 75,000 to 200,000 lb/h (9.4 to 25.2 kg/s) capacity range can use a relay-operated regulator with one or two elements when the load fluctuations are reasonable. With wide load swings, the relay-operated three-element regulator is a better choice. Since this boiler will encounter wide load swings, a three-element regulator is a wise and safe choice.

TABLE 2   Boiler-Feedwater-Regulator Selector Chart[°][†]

Boiler capacity	Type of feedwater regulator		
	Self-operated single-element	Relay-operated single- or two-element	Relay-operated three-element
Below 75,000 lb/h (9.4 kg/s)	For steady loads (building heating or continuous processes)	For irregular loads (batch processes, hoists, rolling mills, etc.)	
75,000–200,000 lb/h (9.4–25.2 kg/s)	Use only in special cases	For all steady and fluctuating loads	For extreme load and water conditions and boilers with steaming economizers
Above 200,000 lb/h (25.2 kg/s)	—	Use only on steady loads	For all types of loads

[°] From Kallen—*Handbook of Instrumentation and Controls*, McGraw-Hill.
[†] Excess pressure ahead of feedwater regulator should be at least 50 lb/in^2 (344.8 kPa) and should be controlled by regulation of the feed pump. Use excess-pressure valves only when excess pressure varies more than plus or minus 30 percent. Where drum level is unsteady owing to high solids concentration or boiler feed or other causes, use next-higher-class feed regulator.

3.   Choose the type of combustion-control system

Table 3 summarizes the important selection features of four types of combustion-control systems. Study of Table 3 shows that a stream flow–air flow type of combustion-control system would probably be best for the fuel and load conditions in this plant. Hence, this type of control system will be chosen.

Related Calculations: Any control system selected for a boiler by using this procedure should be checked out by studying the engineering data available from the control-system manufacturer. The procedure given here is valid for heating, industrial, power, marine, and similar boilers.

CONTROL-VALVE CHARACTERISTICS AND RANGEABILITY

A flow control valve will be installed in a process system in which the flow may vary from 100 to 20 percent while the pressure drop in the system rises from 5 to 80 percent. What is the required rangeability of the control valve? What type of control-valve characteristic should be used? Show how the effective characteristic is related to the pressure drop that the valve should handle.

Calculation Procedure:

1.   Compute the required valve rangeability

Use the relation $R = (Q_1/Q_2)(\Delta P_2/\Delta P_1)^{0.5}$, where R = valve rangeability; Q_1 = valve initial flow, percentage of total flow; Q_2 = valve final flow, percentage of total flow; P_1 = initial pressure drop across the valve, percentage of total pressure drop; P_2 = percentage of final pressure drop across the valve.

Substituting gives $R = (100/20)(80/5)^{0.5} = 20$.

2.   Select the type of valve characteristic to use

Table 4 lists the typical characteristics of various control valves. Study of Table 4 shows that an equal-percentage valve must be used if a rangeability of 20 is required. Such a valve has equal stem movements for equal-percentage changes in flow at a constant pressure drop based on the

TABLE 3 Classification of Combustion-Control Systems[a]

	A, series-fuel	B, series-air	C, parallel	D, calorimeter or steam flow–air flow
Action	Temperature- or pressure-actuated master adjusts fuel rate; fuel meter adjusts air flow	Temperature- or pressure-actuated master adjusts air flow; air-flow meter adjusts fuel flow	Temperature- or pressure-actuated master adjusts fuel flow and air flow simultaneously	Pressure-actuated master adjusts fuel flow; steam flow adjusts air flow
Relative speed of control	Master adjusted for fast response because fuel-rate fluctuations caused by fluctuating pressure or temperature on master do not have correspondingly fast effect on that controlled variable	Master adjusted for slow response, because air-flow fluctuations following fast fluctuating master signal have a rapid effect on controlled variable and may cause hunting action if air-flow response is too fast	Master adjusted for slow response for same reason as in series-air	Master adjusted for fast response for same reason as in series-fuel. Steam flow–air flow control can be relatively rapid since steam-flow fluctuations are not so rapid as pressure variations
Used on fuels	Easily metered fuels such as oil and gas	All fuels. Oil, gas, and coal, either solid or burned in suspension	Primarily on solid fuels (grate firing)	Fuels hard to meter or fuels burned simultaneously. Commonly used on pulverized-coal-fired boilers
Advantages	When fuel may be in short supply, eliminates possibility of carrying high excess air for long period	Eliminates possibility of explosive mixture in combustion space when air fails. Eliminates need of fuel cutback for this purpose	Relatively inexpensive control system. No metering necessary	Ensures proper air-fuel ratio, even though fuel cannot be accurately metered or is of varying heat content. Ensures this condition even when burning a mixture of different fuels at the same time

[a] From Kallen—*Handbook of Instrumentation and Controls*, McGraw-Hill.

TABLE 4   Control-Valve Characteristics

Valve type	Typical flow rangeability	Stem movement
Linear	12-1	Equal stem movement for equal flow change
Equal-percentage	30-1 to 50-1	Equal stem movement for equal-percentage flow change°
On-off	Linear for first 25% of travel; on-off thereafter	Same as linear up to on-off range

°At constant pressure drop.

flow occurring just before the change is made.[1] The equal-percentage valve finds use where large rangeability is desired and where equal-percentage characteristics are necessary to match the process characteristics.

3.   *Show how the valve effective characteristic is related to pressure drop*

Figure 7 shows the inherent and effective characteristics of typical linear, equal-percentage, and on-off control valves. The inherent characteristic is the theoretical performance of the valve.[1] If a valve is to operate at a constant load without changes in the flow rate, the characteristic of the valve is not important, since only one operating point of the valve is used.

Figure 7b and c gives definite criteria for the amount of pressure drop the control valve should handle in the system. This pressure drop is not an arbitrary value such as 5 lb/in^2 (34.4 kPa) but rather a percentage of the total dynamic drop. The control valve should take at least 33 percent of the total dynamic system pressure drop[1] if an equal-percentage valve is used and is to retain its inherent characteristics. A linear valve should not take less than a 50 percent pressure drop if its linear properties are desired.

There is an economic compromise in the selection of every control valve. Where possible, the valve pressure drop should be as high as needed to give good control. If experience or an economic study dictates that the requirement of additional horsepower to provide the needed pressure is not worth the investment in additional pumping or compressor capacity, the valve should take less pressure drop with the resulting poorer control.

FLUID-AMPLIFIER SELECTION AND APPLICATION

Select a fluid amplifier to amplify the output of a fluidic sensor in a control system having a sensor output of 1 lb/in^2 (6.9 kPa) and requiring a control-valve operating pressure for modulating purposes of at least 10 lb/in^2 (68.9 kPa).

[1]E. Ross Forman, "Fundamentals of Process Control," *Chemical Engineering*, June 21, 1965.

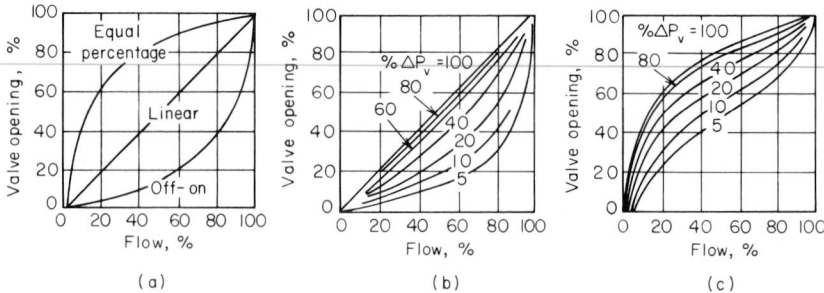

FIG. 7   (a) Inherent flow characteristics of valves at constant pressure drop; (b) effective characteristics of a linear valve; (c) effective characteristics of a 50:1 equal-percentage valve.

Calculation Procedure:

1. *Compute the required amplification ratio*

The amplification ratio μ = the gain of the amplifier = amplifier output, lb/in^2 (kPa)/amplifier input, lb/in^2 (kPa). For this amplifier, $\mu = 10/1 = 10$. Hence, this fluid amplifier must increase a 1-lb/in^2 (6.9-kPa) input signal from the sensor to at least a 10-lb/in^2 (68.9-kPa) signal at the amplifier output.

2. *Select a suitable fluid amplifier*

Refer to a manufacturer's engineering data for the characteristic of a suitable fluid amplifier. Figure 8 shows the characteristics of a typical commercially available fluid amplifier. This amplifier is available for two gains: 9:1 and 18:1.

Since the first gain is lower than required, the second, or 18:1, gain would be used if this amplifier were chosen. With this gain, a 1-lb/in^2 (6.9-kPa) input signal would be amplified in the device to an output of an $18(1) = 18$-lb/in^2 (124.4-kPa) signal. Since a 10-lb/in^2 (68.9-kPa) or larger signal is required to operate the control valve, this amplifier is acceptable.

Related Calculations: Many fluid amplifiers of the proportional type are packaged into fully operational modules, just as transistors and linear integrated circuits are. When so packaged, the device is termed an *operational amplifier*. An operational amplifier[1] is a module that contains within itself all the elements of a high-gain, accurate, and repeatable analog amplifier. Called *op-amps* for short, these devices are useful because they can amplify low-level signals, such as the outputs of fluidic sensors, proportionally to levels high enough to modulate control valves.

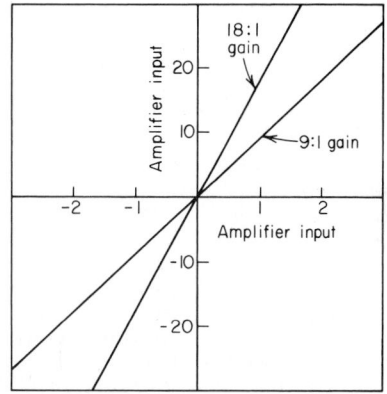

FIG. 8   Operational-amplifier performance for two gains.

A fluid amplifier can produce a gain of pressure as, in this instance, power, or flow. Compute the gain for any of these outputs by setting up the ratio of output to input, as in step 1. Be certain to use consistent units for the output and input variables.

Where an amplifier is *not* linear (Fig. 9), the gain is different for each output. Hence, the operating points (i.e., input and output variables) must be stated for the gain being considered. Table 5 lists the selection characteristics for fluid amplifiers, control valves, and integrated circuits. Data presented in this tabulation are useful in choosing the most suitable control for a given application.

CAVITATION, SUBCRITICAL-, AND CRITICAL-FLOW CONSIDERATIONS IN CONTROLLER SELECTION

Given the sizing formulas of the Fluid Controls Institute (FCI), size control valves for the cavitation, subcritical-, and critical-flow situations described below. Show how accurate the FCI formulas are.

Cavitation: Select a control valve for a situation where cavitation may occur. The fluid is steam condensate; inlet pressure P_1 is 167 lb/in^2 (abs) (1151.5 kPa); $\Delta P = 105$ lb/in^2 (724.0 kPa); inlet temperature T_1 is 180°F (82.2°C); vapor pressure P_v is 7.5 lb/in^2 (abs) (51.7 kPa).

Subcritical gas flow: Determine the valve capacity required at these conditions: fluid is air; flow Q_g is 160,000 standard ft^3/h (1.3 standard m^3/s); inlet pressure P_1 is 275 lb/in^2 (abs) (1896.1 kPa); $\Delta P = 90$ lb/in^2 (620.4 kPa); gas temperature T_1 is 60°F (15.6°C).

Critical vapor flow: A heavy-duty angle valve is suggested for a steam pressure-reducing application. Determine the capacity required, and compare an alternate valve type. The fluid is

---

[1]Frank Yeaple, "Analog Fluidic Amplifiers Are Waiting in the Wings," *Product Engineering*, Oct. 23, 1967.

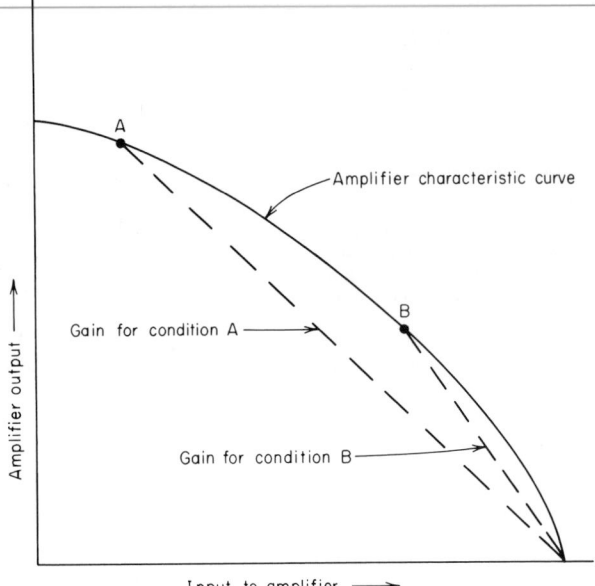

FIG. 9 Fluid-amplifier operating characteristic curve and gain for two operating conditions.

saturated steam; flow W is 78,000 lb/h (9.8 kg/s); inlet pressure P_1 is 1260 lb/in² (abs) (8686.4 kPa); outlet pressure P_2 is 300 lb/in² (abs) (2068.5 kPa).

Calculation Procedure:

1. *Choose the valve type and determine its critical flow factor for the cavitation situation*

If otherwise suitable (i.e., with respect to size, materials, and space considerations), a butterfly control valve is acceptable on a steam-condensate application. Find, from Table 6, the value of the critical flow factor $C_f = 0.68$ for a butterfly valve with 60° operation.

2. *Compute the maximum allowable pressure differential for the valve*

Use the relation $\Delta P_m = C_f^2(P_1 - P_v)$, where ΔP_m = maximum allowable pressure differential, lb/in² (kPa); P_1 = inlet pressure, lb/in² (abs) (kPa); P_v = vapor pressure, lb/in² (abs) (kPa). Substituting gives $\Delta P_m = (0.68)^2(167 - 7.5) = 74$ lb/in² (510.2 kPa). Since the actual pressure drop, 105 lb/in² (724.0 kPa), exceeds the allowable drop, 74 lb/in² (510.2 kPa), cavitation *will* occur.

3. *Select another valve and repeat the cavitation calculation*

For a single-port top-guided valve with flow to open plug, find $C_f = 0.90$ from Table 6. Then $\Delta P_m = (0.90)^2(167 - 7.5) = 129$ lb/in² (889.5 kPa).

In the case of the single-port top-guided valve, the allowable pressure drop, 129 lb/in² (889.5 kPa) exceeds the actual pressure drop, 105 lb/in² (724.0 kPa) by a comfortable margin. This valve is a better selection because cavitation will be avoided. A double-port valve also might be used, but the single-port valve offers lower seat leakage. However, the double-port valve offers the possibility of a more economical actuator, especially in larger valve sizes. This concludes the steps for choosing the valve where cavitation conditions apply.

4. *Apply the FCI formula for subcritical flow*

The FCI formula for subcritical gas flow is $C_v = Q_g/1360(\Delta P/GT)^{0.5}[(P_1 + P_2)/2]^{0.5}$, where C_v = valve flow coefficient; Q_g = gas flow, standard ft³/h (standard m³/s); ΔP = pressure differ-

TABLE 5   Control Selection Guide° †

	Fluid amplifiers	Spools, poppets	Solid-state integrated circuits
Characteristics of device needed			
Is environment-proof:			
Can be designed to operate at extremely high			
temperature	✓		
Can be made tolerant of any atmosphere	✓	✓	✓
Is unhurt by nuclear radiation	✓	✓	
Is unhurt by heavy vibration or shock	✓	. . .	✓
Has no moving parts:			
No stiction, hysteresis, dead zone, or jamming; also,			
no mechanical blocking, so no fluid hammer	✓	. . .	✓
Is tiny, stackable, monolithic:			
Whole circuits can be made in one integrated block,			
all permanently sealed, and with no moving parts	✓	. . .	✓
Can be supplied from any fluid source:			
Air, gas, water, oil, or process fluids will work; even			
the water rushing by the hull of a boat, or the air			
slipping past an airfoil, can be exploited	✓		
Needs no electricity:			
Is not affected in performance by radio or electrical			
interference	✓	✓	
Is responsive to extremely small inputs:			
Breaths of air; proximity of anything, such as tiny			
threads, specks, liquid surfaces, and air bubbles;			
motion of housing (fluid inertia effect); shock			
waves; fluid disturbances; spark discharges; sound			
waves; controlled vibration; localized heat	✓	. . .	✓
Is fast (millisecond or better)	✓	. . .	✓
Characteristics of device sought			
Has high energy output:			
Easily transduced to mechanical movement	Some	✓	
Is widely available from many sources:			
For proportional control	Some	Some	✓
For on-off control	✓	✓	✓
Is ultrafast (μsecond)	. . .	. . .	✓
Can be shut off individually when not in use:			
Power requirement drops drastically when device is			
switched off	. . .	✓	✓

°Checks show which control to use.
†*Product Engineering.*

ential, lb/in² (kPa); G = specific gravity of gas at 14.7 lb/in² (abs) (101.4 kPa) and 60°F (15.6°C); T = absolute temperature of the gas, R; other symbols as given earlier. Substituting yields C_v = 160,000/1360 $(90/520)^{0.5}$ $[(275+185)/2]^{0.5}$ = 18.6. Note that G = 1.00 for air.

5.   *Compute C_v, using the unified gas-sizing formula*

For greater accuracy, many engineers use the unified gas-sizing formula. Assuming a single-port top-guided valve installed open to flow, Table 6 shows C_f = 0.90. Then $Y = (1.63/C_f)(\Delta P/P_1)^{0.5}$, where Y is defined by the equation and the other symbols are as given earlier. Substituting gives $Y = (1.63/0.90)(90/275)^{0.5}$ = 1.04. Figure 10 shows the flow correlation established from actual

TABLE 6 Critical Flow Factors for Control Valves at 100 Percent Lift[*]

Split body

A	Flow to close plug	0.80
	Flow to open plug	0.75
	Parabolic plug only	
B	Flow to close plug	0.50
	Flow to open plug	0.90
	Parabolic plug only	

Double-port, globe body

A	Parabolic plug	0.90
	V-port plug	1.00
B	Parabolic plug	0.62
	V-port plug	0.95

Single-port, globe body

A	Flow to close plug	0.85
	Flow to open plug	0.90
	Parabolic plug only	
B	Flow to close plug	0.50
	Flow to open plug	0.90
	Parabolic plug only	

Butterfly

		$\alpha = 60°$	$\alpha = 90°$
D/d = 1		0.68	0.58
D/d = 2		0.62	0.50

Angle body

A	Flow to close plug	0.40
	Flow to open plug	0.90
	Parabolic plug only	
B	Flow to close plug	0.55
	Flow to open plug	0.95
	Parabolic plug only	

(A) Full-capacity trim, orifice diameter ~ 0.8 valve diameter

(B) Reduced capacity trim, 50% of (A) and less.

NOTE: The listed values apply for equal port-area valves only and do not include corrections for pipe friction.

[*] Henry W. Boger and *Chemical Engineering.*

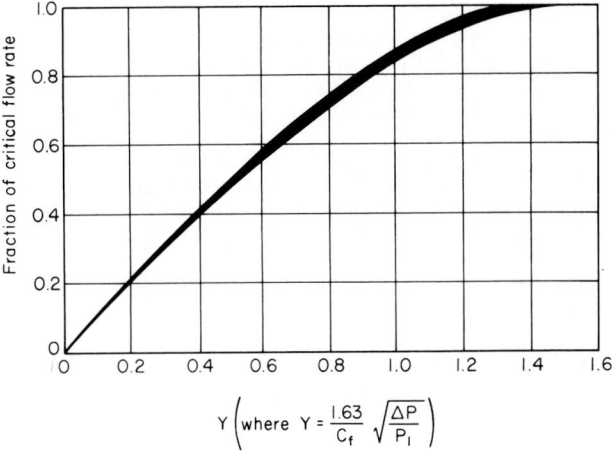

$$Y \left(\text{where} \quad Y = \frac{1.63}{C_f} \sqrt{\frac{\Delta P}{P_1}} \right)$$

FIG. 10 Flow correction established from actual data for many valve configurations at maximum valve opening.

test data for many valve configurations at maximum valve opening, and relates Y and the fraction of the critical flow rate.

Find from Fig. 11 the value of $Y - 0.148Y^3 = 0.87$. Compute $C_v = Q_g(GT)^{0.5}/834C_f(Y - 0.148Y^3)$, where all the symbols are as given earlier. Or, $C_v = 160,000(520)^{0.5}/[834(0.90)(275)(0.87)] = 20.4$. This value represents an error of approximately 10 percent in the use of the FCI formula.

6. Determine C_f for critical vapor flow

Assuming reduced valve trim for a heavy-duty angle valve, we find $C_f = 0.55$ from Table 6.

7. Compute the critical pressure drop in the valve

Use $\Delta P_c = 0.5(C_f)^2 P_1$, where P_c = critical pressure drop, lb/in² (kPa); other symbols as given earlier. So $\Delta P_c = 0.5(0.55)^2(1260) = 191$ lb/in² (1316.9 kPa).

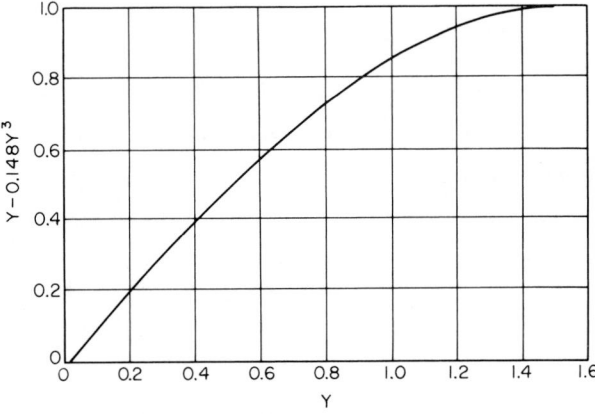

FIG. 11 Correction-factor values.

8.   *Determine the value of C_v*

Use the relation $C_v = W/1.83 C_f P_1$, where the symbols are as given earlier. Substituting yields $C_v = 78{,}000/[1.83(0.55)(1260)] = 61.5$. A lower C_v could be attained by using the valve flow to open, but a more economical choice is a single-port top-guided valve installed open to flow.

For a single-port top-guided valve flow to open, $C_f = 0.90$ from Table 6. Hence, $C_v = 78{,}000/[1.83(0.90)(1260)] = 37.6$.

A lower capacity is required at critical flow for a valve with less pressure recovery. Although this may not lead to a smaller body size because of velocity and stability considerations, the choice of a more economical body type and a smaller actuator requirement is attractive. The heavy-duty angle valve finds its application generally on flashing-hydrocarbon liquid service with a coking tendency.

This calculation procedure is the work of Henry W. Boger, Engineering Technical Group Manager, Worthington Controls Co.

SERVO SYSTEM STABILITY DETERMINATION

The servo system shown in Fig. 12 is to be used in an industrial application. Determine the range of K for the system to be stable.

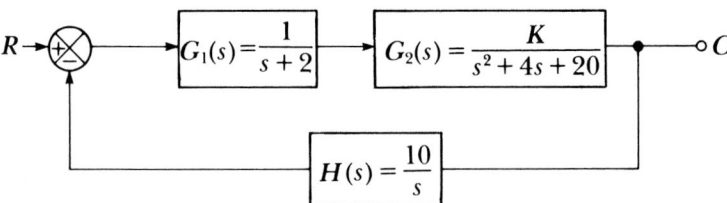

FIG. 13   Angular-position servo system.

Calculation Procedure:

1.   *Solve, using the Routh criterion*

Using the Routh criterion gives

$$G(s) = G_1(s)G_2(s) = \frac{K}{(s + 2)(s^2 + 4s + 20)} = \frac{K}{D}$$

We set $1 + G(s)H(s) = 0$:

$$1 + \frac{K}{(s + 2)(s^2 + 4s + 20)}\frac{10}{s} = 0$$

$$s(s + 2)(s^2 + 4s + 20) + 10K = 0$$

$$s^4 + 6s^3 + 28s^2 + 40s + 10K = 0$$

2.   *Set up the Routh array*

s^4	1	28	$10K$
s^3	6	40	0
s^2	$\dfrac{(6)(28) - (1)(40)}{6} = 21.33$	$\dfrac{60K - 1(0)}{6} = 10K$	0
s^1	$\dfrac{(21.33)40 - 60K}{21.33} = 40 - 2.813K$	0	
s^0	$10K$		

3. *Solve for the range of K*

$$K = 0 \qquad \text{(for the } s^0 \text{ term)}$$

$$40 - 2.813K = 0 \qquad \text{(for the } s^1 \text{ term)}$$

$$2.813K = 40$$

$$K \geq 14.22$$

Therefore K must lie between 0 and 14.22. This procedure is the work of Charles R. Hafer, P.E.

ANGULAR-POSITION SERVO SYSTEM ANALYSIS

An angular-position servo system uses potentiometer feedback and has these gain-transfer functions:

Amplifier: $\qquad G_1(s) = \dfrac{100}{(s + 5)} \quad \dfrac{\text{V}}{\text{V}}$

Motor mechanical transfer function: $\qquad G_2(s) = \dfrac{1}{40s} \quad \dfrac{\text{rad}}{\text{A}}$

Motor electrical transfer function: $\qquad G_3(s) = \dfrac{1}{s + 4} \quad \dfrac{\text{A}}{\text{V}}$

Potentiometer gain constant: $\qquad G_4(s) = K \quad \dfrac{\text{V}}{\text{rad}}$

Draw a block diagram for this system, and determine the open-loop transfer function and the closed-loop transfer function. If $K = 100$, is the system stable?

Calculation Procedure:

1. *Write the open-loop transfer function*

The block diagram is drawn as shown in Fig. 13. The open-loop transfer function is

$$G(s) = G_1(s)G_2(s)G_3(s) = \frac{100}{40s(s + 5)(s + 4)} = \frac{2.5}{s(s + 5)(s + 4)}$$

2. *Write the closed-loop transfer function*

The closed-loop transfer function requirements suggest that the system be put in the canonical form as

$$\frac{C}{R} = \frac{G(s)}{1 + G(s)H(s)} = \frac{2.5}{(s)(s + 5)(s + 4) + 2.5K}$$

3. *Determine whether the system is stable*

The characteristic equation for this system is

$$(s)(s + 5)(s + 4) + 2.5K = 0$$

$$s^3 + 9s^2 + 20s + 2.5K = 0$$

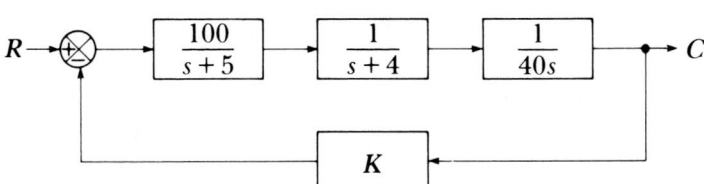

FIG. 12 Servo system block diagram.

Set up a Routh array to find the gain range of K for a stable system:

$$
\begin{array}{c|cc}
s^3 & 1 & 20 \quad 0 \\
s^2 & 9 & 2.5K \ 0 \\
s^1 & \dfrac{180 - 2.5K}{9} & 0 \\
s^0 & 2.5K & 0
\end{array}
$$

Setting the s^1 and s^0 terms in the first column ≥ 0, solve for the range of K that makes the system stable. Thus,

For $180 - 2.5K \geq 0$: $K \leq 72$

For $2.5K \geq 0$: $K \geq 0$

So the range of K for a stable system is $0 \leq K \leq 72$ V/rad. So the system is not stable for $K = 100$. This procedure is the work of Charles R. Hafer, P.E.

LOOP-GAIN FUNCTION ANALYSIS

The loop-gain function $GH(s) = K/s(s + \omega_1)(s + \omega_2)$ is represented by the straight-line approximation Bode plot in Fig. 14. The solid line represents the gain, and the dashed line represents phase. Find ω_1 and ω_2, K, phase margin ϕ_m, and the value of the K adjustment for a phase margin of $45°$.

Calculation Procedure:

1.   *Determine the first pole and the other two breakpoints*

The first pole determined by $1/s$ is at zero frequency. The other two breakpoints can be determined from Fig. 14 to be

$$\omega_1 = 10 \text{ rad/s} \qquad T_1 = \frac{1}{\omega_1} = 0.1 \text{ s}$$

$$\omega_2 = 100 \text{ rad/s} \qquad T_2 = \frac{1}{\omega_2} = 0.01 \text{ s}$$

2.   *Write the transfer function*

The transfer function is now

$$GH(s) = \frac{K'}{s(0.1s + 1)(0.01s + 1)} = \frac{K}{s(s + 10)(s + 100)}$$

$$GH(j\omega) = \frac{K}{j\omega(j\omega + 10)(j\omega + 100)}$$

Since the approximation is a straight line, the maximum gain error occurs at the breakpoints. To solve for K with minimum error, move as far from a break as possible. Choose $\omega = 0.1$ rad/s. Then

$$|GH(j\omega)| = 60 \text{ dB} = 1000 = \left| \frac{K}{(0.1j)(0.1j + 10)(0.1j + 100)} \right|$$

$$\left| \frac{K}{(0.1\underline{/90°})(10\underline{/0.6°})(100\underline{/0.06°})} \right| = 1000$$

$$\frac{K}{100} = 1000$$

$$K = 10^5$$

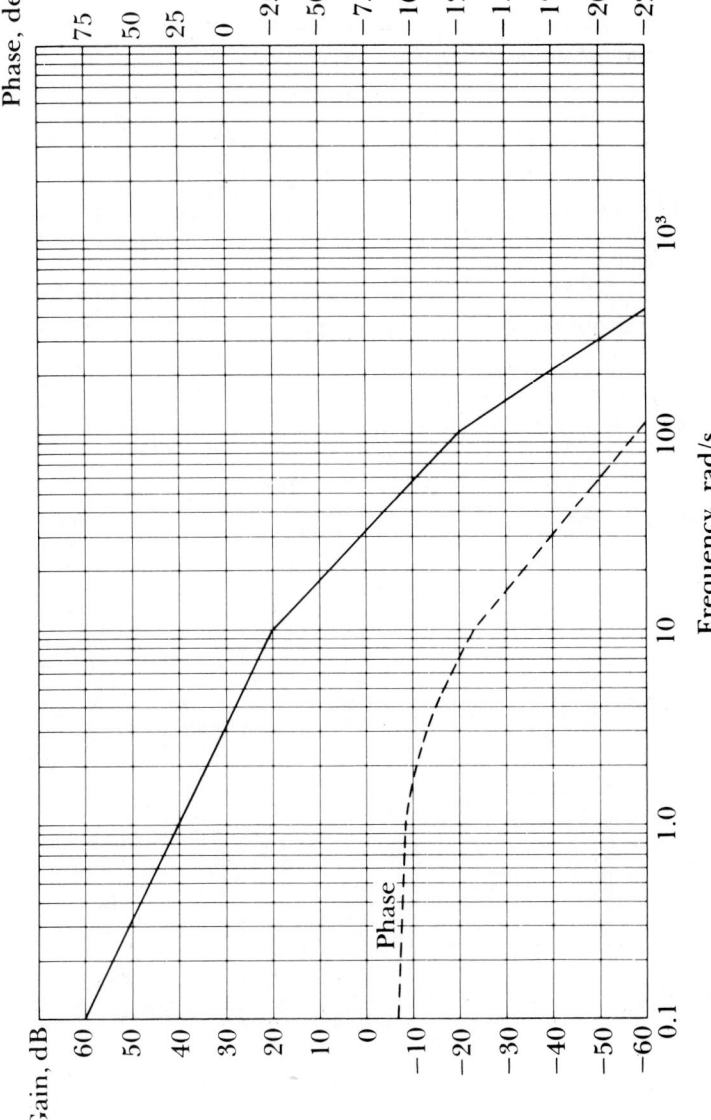

FIG. 14 Straight-line Bode plot approximation.

7.23

3. *Find the phase margin*

From Fig. 14, the phase is 175° at the point where the gain is 0 dB. Hence, the phase margin is

$$\phi_m = 180° - 175° = 5°$$

This would be very marginal for stability, since 45° would be preferred.

4. *Determine the phase plot for a phase margin of 45°*

For a phase margin of 45°, the phase plot has a value of

$$\phi = -180° + \phi_m = -180° + 45° = -135°$$

At this value of ϕ on Fig. 14, the gain is 20 dB at a frequency of 10 rad/s. It is desired that the gain be 0 dB at this magnitude of phase and frequency. Therefore, it would be necessary to reduce the gain by 20 dB (a factor of 10). Now K becomes 10^4. This procedure is the work of Charles R. Hafer, P.E.

SERVO SYSTEM OVERSHOOT AND SETTLING TIME

A servo system has the block diagram shown in Fig. 15a. Determine the values of A and a from the expression $A/s(s + a)$ for a 15 percent overshoot and a 10-s settling time.

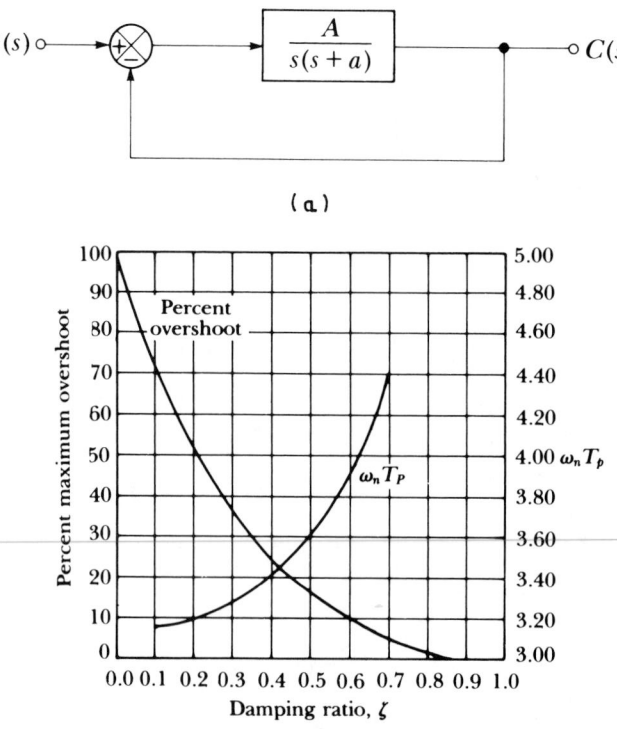

FIG. 15   (a) Block diagram of servo system; (b) percentage of overshoot and peak time versus damping ratio for a second-order system.

Calculation Procedure:

1. Write the transfer function for the closed-loop system

The transfer function is

$$F(s) = \frac{C(s)}{R(s)} = \frac{G(s)}{1 + G(s)H(s)} = \frac{A}{s^2 + as + A}$$

This is the classic form of

$$F(s) = \frac{\omega_n^2}{s^2 + 2\xi\omega_n s + \omega_n^2}$$

2. Solve for the settling-time criteria

Assume that settling time means less than 5 percent, which it normally implies. The equation for a 5 percent settling time is

$$t_s = \frac{3}{\xi\omega_n}$$

$$PO = \frac{100 \exp(-\xi\pi)}{(1 - \xi^2)^{1/2}}$$

Solving for the settling-time criteria, we find

$$\xi\omega_n = \frac{3}{10} = 0.3$$

3. Find the damping ratio which satisfies the requirements

From the curve of Fig. 15b, the value of the damping ratio which satisfies the requirements is approximately 0.53 for 15 percent of overshoot. Therefore,

$$\omega_n = \frac{0.3}{\xi} = 0.57$$

$$\omega_n^2 = 0.32$$

4. Write the closed-loop expression and solve it

$$F(s) = \frac{0.32}{s^2 + 0.6s + 0.32}$$

$$A = 0.32$$

$$a = 0.6$$

This procedure is the work of Charles R. Hafer, P.E.

ANALYSIS OF A SERVO SYSTEM WITH A LOOP DELAY

A servo system with a loop delay is represented by the block diagram in Fig. 16 and the function e^{-sT}. Find the maximum value of T for a stable system.

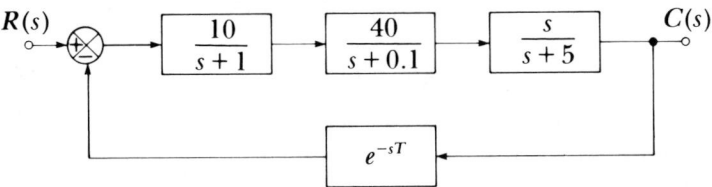

FIG. 16   Block diagram of servo system.

Calculation Procedure:

1. Find the loop gain first

The loop gain is

$$GH(s) = G(s)H(s) = G_1(s)G_2(s)G_3(s)e^{-sT}$$

Neglecting the phase shift, we obtain $GH'(s)$:

$$GH'(s) = \frac{10}{s+1}\frac{40}{s+0.1}\frac{s}{s+5} = \frac{400s}{(s+0.1)(s+1)(s+5)}$$

$$GH'(j\omega) = \frac{400j\omega}{(0.1+j\omega)(1+j\omega)(5+j\omega)}$$

2. Plot the function

For a starting point, find $G(j\omega)$ far from a breakpoint (for accuracy purposes), and plot the function. Pick $\omega = 0.01$ rad/s, and solve for $GH(j\omega)$:

$$GH(j0.01) = \frac{4j}{(0.1)(1)(5)} = 8j$$

$$|GH(j0.01)| = 20\log_{10} 8 = 18.6 \text{ dB}$$

Now plot the straight-line approximation as shown in Fig. 17.

3. Determine the phase at the crossover point

It appears as though the plot crosses 0 dB at 20 rad/s. The phase at this crossover point is

$$GH(j20) = \frac{400(j20)}{(0.1+j20)(1+j20)(5+j20)} \approx \frac{400}{(20\underline{/87.1°})(20.6\underline{/76°})}$$

$$GH(j20) = 0.97\underline{/-163.1°}$$

The gain checks since it is close to 1. The phase margin is only $16.9°$ ($= 180° - 163.1°$) and is already potentially unstable. It will surely be unstable if another $16.9°$ is added. Therefore, at $\omega = 20$ rad/s

$$\omega T = 16.9° \times \frac{2\pi \text{ rad}}{360°}$$

$$\omega T = 0.294 \text{ rad}$$

$$T = \frac{0.294}{20} = 14.75 \text{ ms}$$

This procedure is the work of Charles R. Hafer, P.E.

SERVO SYSTEM CLOSED-LOOP TRANSFER FUNCTION CHARACTERISTICS

The servo system in Fig. 18 is to have the closed-loop transfer function $G(s) = C(s)/R(s)$ to possess the following characteristics: $\xi = 0.707$, $\omega_n = 15$. Find the transfer function of a compensating network to accomplish this, using cancellation compensation. Determine the network that will provide this function.

Calculation Procedure:

1. Describe how to obtain the desired function

Cancellation compensation is accomplished by removing a pole or zero and replacing it with another pole or zero. For example, if we have a function of the form $G(s) = 1/(s + a)$ and we

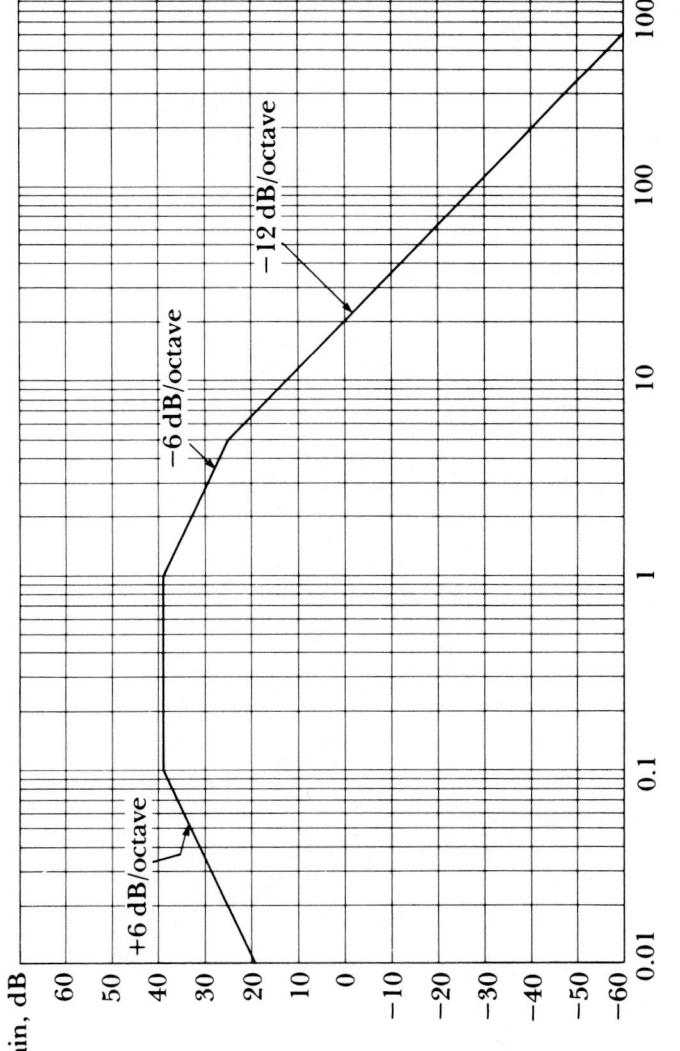

FIG. 17 Bode plot straight-line approximation for $GH(s)$.

7.27

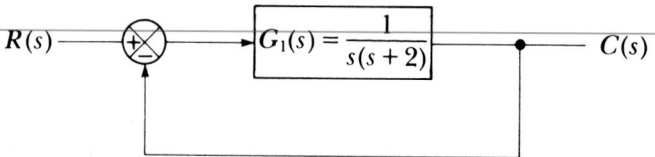

FIG. 18   Block diagram of servo system.

desire $G'(s) = 1/(s + b)$, we can multiply our original function by $G_c(s) = (s + a)/(s + b)$, the compensating network. Thus, we can get our desired function $G'(s)$ as follows:

$$G'(s) = G(s)G_C(s) = \frac{[1/(s + a)](s + a)}{s + b} = \frac{1}{s + b}$$

2.   Find the closed-loop transfer function

The closed-loop transfer function can be found from

$$G(s) = \frac{C(s)}{R(s)} = \frac{G_1(s)}{1 + G_1(s)H(s)}$$

Since $H(s) = 1$ and $G_1(s) = 1/s(s + 2)$,

$$G(s) = \frac{1}{s^2 + 2s + 1}$$

This is the general form of a second-order system described by

$$G(s) = \frac{\omega_n^2}{s^2 + 2\xi\omega_n^s + \omega_n^2}$$

For the circuit of Fig. 18 we have

$$\omega_n = 1 \qquad \xi = 1$$

For the given requirements of $\xi = 0.707$ and $\omega_n = 15$ we have $\omega_n^2 = 225$ and $2\xi\omega_n = 21.21$, and our desired closed-loop function becomes

$$G'(s) = \frac{225}{s^2 + 21.21s + 225}$$

3.   Find the compensation network

Now take this expression back to open-loop form and obtain $G_1'(s)$ from

$$G'(s) = \frac{G'_1(s)}{1 + G_1'(s)H(s)} = \frac{N_1'(s)}{H(s)N_1'(s) + D_1'(s)}$$

where

$$G_1'(s) = \frac{N_1'(s)}{D_1'(s)}$$

$$= \frac{225}{s^2 + 21.21s} = \frac{225}{s(s + 21.21)}$$

If we use cancellation techniques, we should cancel out our original pole with a zero and insert a new pole as described by $G_1'(s)$. Since our gain is 225, our compensation network now becomes

$$G_C(s) = \frac{225(s + 2)}{s + 21.21}$$

The new forward-loop transfer function is now

$$G_1'(s) = G_1(s)G_C(s)$$

$$= \frac{1}{s(s+2)} \frac{225(s+2)}{s+21.21} = \frac{225}{s(s+21.21)}$$

4.   *Use networks for an active network solution*

Use an active solution because good isolation (low output impedance) is obtained. The gain of the compensating network will be negative because of the inverting amplifier. Or,

$$G_C(s) = \frac{-225(s+2)}{s+21.21} = \frac{-21.21(0.5s+1)}{0.047s+1}$$

These requirements can be satisfied by circuits from a table of networks. The transfer-impedance function is

$$G_C(s) = \frac{A(1+sT)}{(1+s\theta T)}$$

where

$$A = 21.22$$

$$T = 0.5$$

$$\theta T = 0.047$$

$$\theta = 0.094$$

From a table of transfer functions,

$$R_1' = \frac{A}{2} = \frac{21.22}{2} = 10.61$$

$$R_2' = \frac{A\theta}{4(1-\theta)} = 0.55$$

$$C = \frac{4T(1-\theta)}{A} = 0.0854$$

Choose $C = 100\ \mu\text{F}$ and scale the other values accordingly. Then

$$\alpha = \frac{0.0854}{10^{-4}} = 854$$

$$R_1 = \alpha R_1' = 9.06\ \text{k}\Omega$$

$$R_2 = \alpha R_2' = 470\ \Omega$$

Pick the nearest standard value of R_1 of 9.09 kΩ.

5.   *Find the dc gain function*

The dc gain for the compensator is

$$\underset{s\to 0}{G_C(s)} = -21.22$$

Therefore, since C is an open circuit at dc, our dc gain function is

$$|A| = \frac{Z_f}{Z_i}$$

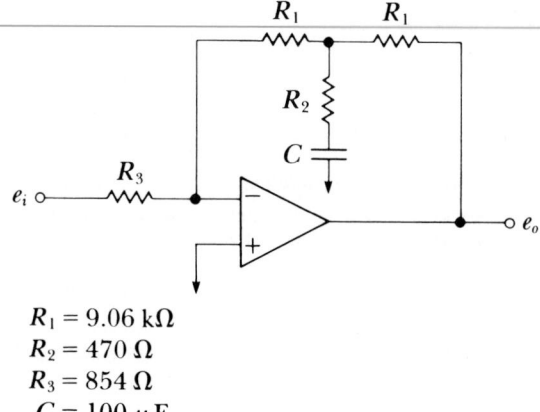

$R_1 = 9.06 \text{ k}\Omega$
$R_2 = 470 \, \Omega$
$R_3 = 854 \, \Omega$
$C = 100 \, \mu\text{F}$

FIG. 19   Cancellation compensation circuit.

$$\frac{2R_1}{R_3} = \frac{18.12 \times 10^3}{R_3}$$

$$R_3 = 854 \, \Omega$$

Choose the nearest standard value of 845 Ω from a table of standard component values. Note that 180° of phase reversal has been introduced when an inverting amplifier is used. The sign of the summer must therefore be changed at the input error circuit. The circuit implementation is shown in Fig. 19 and the system block diagram in Fig. 20. This procedure is the work of Charles R. Hafer, P.E.

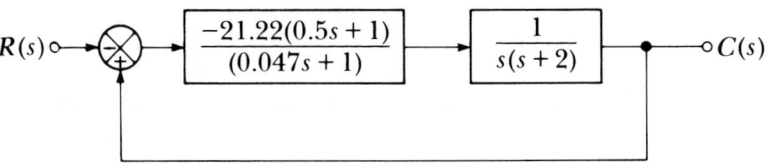

FIG. 20   Block diagram of servo-system solution.

DEVELOPING A TRANSFER FUNCTION FOR A GIVEN PHASE MARGIN

Using a lead compensator, find the transfer function for the system shown in Fig. 21 when the gain at dc remains the same as for the uncompensated case. Provide a circuit arrangement that will satisfy the transfer function.

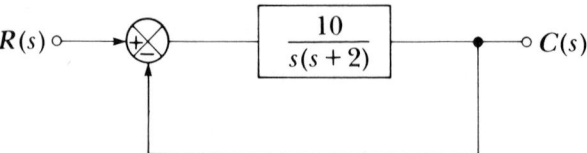

FIG. 21   Block diagram of system.

Calculation Procedure:

1. Plot the magnitude and phase in Bode form

Start with $\omega = 0.1$ rad/s because it is far from a breakpoint and will not introduce much error. From this plot (Fig. 22), the phase ϕ_x at 0-dB gain is approximately $-146°$. This shows that the phase margin is $+34(180° - 146°)$.

2. Determine the phase shift needed

For a phase margin $+60°$, a phase shift of $\phi_m =$ phase margin $- 180° - \phi_x = 60 - 180 + 146 = 26°$. The normal procedure is to add a few degrees because the crossover shifts on the axis in a way to introduce more lag. Use $30°$. Then

$$\sin \phi_m = \frac{\alpha - 1}{\alpha + 1}$$

$$\frac{\alpha - 1}{\alpha + 1} = 0.5$$

$$\alpha = 3$$

$$GH(\omega_m) = -10 \log \alpha = -10 \log 3 = -4.8 \text{ dB}$$

3. Find T_1 for the system

Refer to Fig. 22 and find $GH(\omega_n) = -4.8$ dB. Then ω_m is approximately 4 rad/s, which will be the new crossover. Now T_1 and $G_C(s)$ can be calculated.

$$T_1 = \frac{\alpha^{-1/2}}{\omega_m} = \frac{1}{(1.73)(4)} = 0.145s$$

and

$$G_C(s) = \frac{1 + \alpha T_1 s}{\alpha(1 + T_1 s)} = \frac{1}{3} \frac{1 + 0.43s}{1 + 0.145s}$$

$$= \frac{s + 2.3}{s + 6.9}$$

4. Choose the network that will provide the above function

Review of a table of Bode plots shows that the circuit of Fig. 23 followed by a buffer amplifier to provide isolation and gain will be suitable. The transfer function desired is

$$G_C(s) = \frac{1 + 0.43s}{3(1 + 0.145s)} = \frac{A(T_1 s + 1)}{T_3 s + 1}$$

Pick $C_1 = 100 \ \mu F$ and $R_3 = 10$ kΩ and solve for our other values:

$$R_1 = \frac{T_1}{C_1} = \frac{0.435}{10^{-4}} = 4.35 \text{ k}\Omega$$

$$A = \frac{R_2}{R_1 + R_2}$$

$$R_2 = 2.15 \text{ k}\Omega$$

In order that the gain at dc remain the same, the gain of the amplifier must equal α. So

$$\frac{R_4 + R_3}{R_3} = \alpha = 3$$

$$R_4 = 20 \text{ k}\Omega$$

This procedure is the work of Charles R. Hafer, P.E.

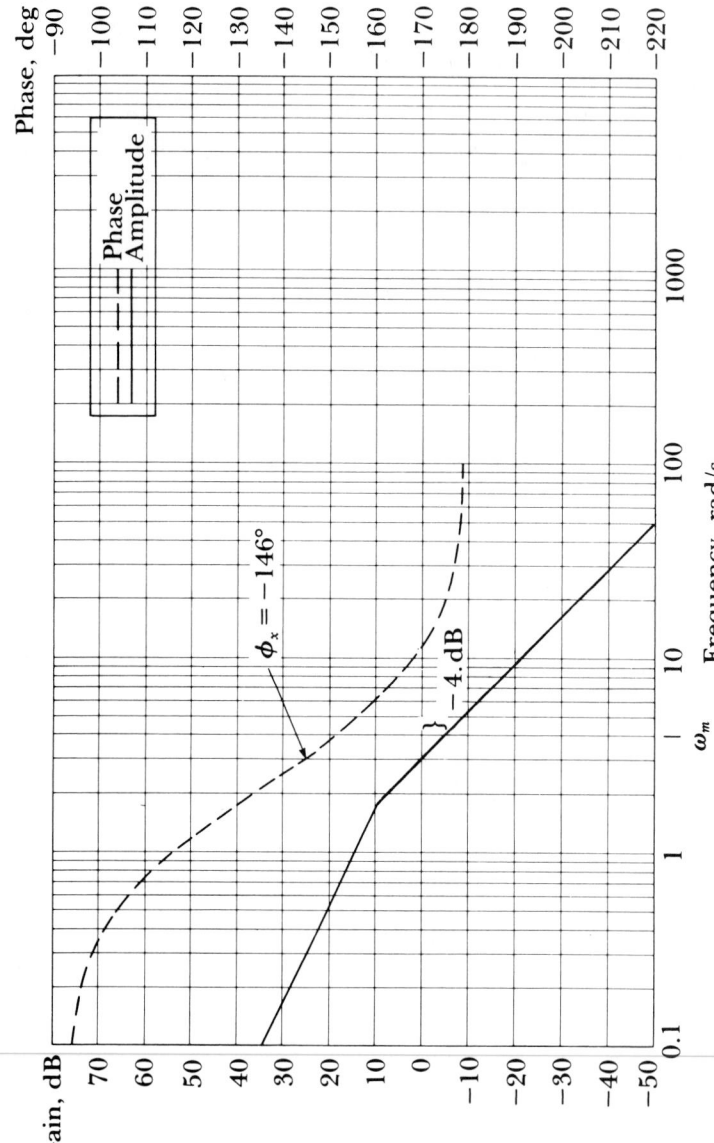

FIG. 22 Bode plot for system having 60° phase margin.

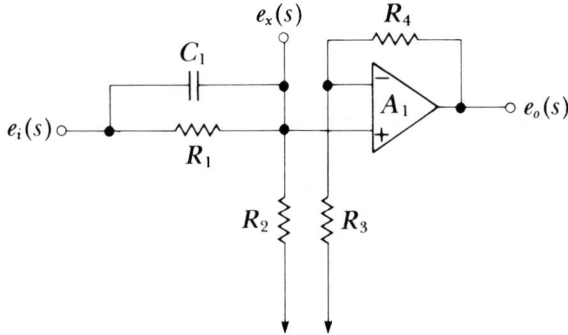

FIG. 23 Lead-compensation circuit.

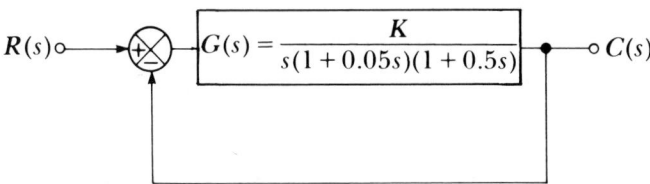

FIG. 24 Feedback control system.

FEEDBACK CONTROL SYSTEM PHASE-LAG COMPENSATOR DESIGN

For the feedback control system shown in Fig. 24, find a phase-lag compensator that will fulfill the requirements of $K_v = 20\ \text{s}^{-1}$, the velocity = error constant, and $\phi_m = 45°$, phase margin. Design a circuit that will provide the compensating network.

Calculation Procedure:

1. Plot the Bode diagram for GH(s) for this system

Place $GH(s)$ in the proper format and plot, as in Fig. 25. Note that $H(s) = 1$. Then

$$GH(s) = \frac{K}{s(1 + 0.05s)(1 + 0.5s)}$$

$$= \frac{40K}{s(s + 20)(s + 2)}$$

2. Develop the transfer function for the velocity-error constant

To meet the first requirement (that the velocity-error constant be 20 s^{-1}), solve for K_v thus:

$$K_v = \lim_{s \to 0} sG(s) = \frac{40K}{(20)(2)} = 20$$

Therefore, $K = 20$ and the transfer function becomes

$$GH(s) = \frac{800}{s(s + 2)(s + 20)}$$

$$GH(j\omega) = \frac{800}{j\omega(2 + j\omega)(20 + j\omega)}$$

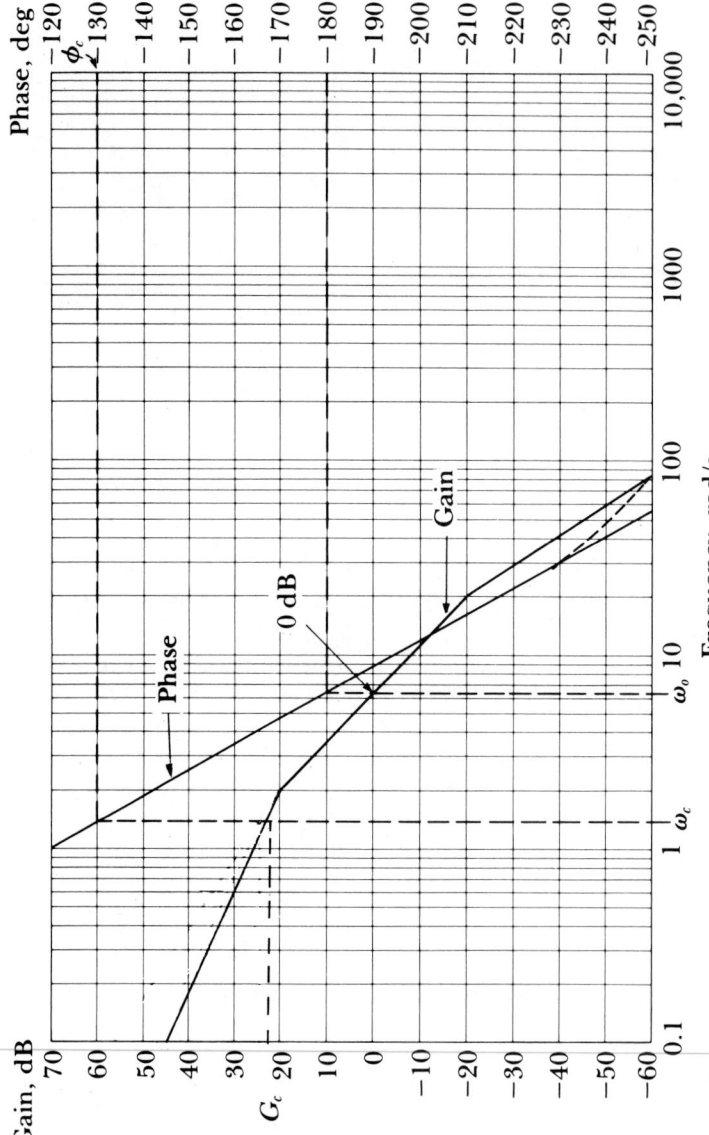

FIG. 25 Bode plot of $GH(s)$ for feedback control system.

7.34

3. *Compute the phase from the ω_0 value*

The Bode plot is shown in Fig. 25 for the function in step 2. From this plot, determine that $\omega_0 = 6.3$ rad/s. Calculate the phase from this frequency. This is more accurate than obtaining the phase from the plot. Thus

$$GH(j6.3) = \frac{800}{j6.3(2 + j6.3)(20 + j6.3)}$$

$$= \frac{800}{(6.3\underline{/90°})(6.61\underline{/72.4°})(20.97\underline{/17.5°})} = 0.916\underline{/-179.9°}$$

The gain is close to 0 dB (-0.8 dB), and the phase checks. We have 0° phase margin and need to obtain a phase of at least $-135°$ to obtain a phase margin of 45°. Add an additional 5° for a safety margin, and ϕ_c is now $-130°$. This occurs at a frequency ω_c of approximately 1.5 rad/s. The gain G_c at this frequency is 22 dB. To cross 0 dB at 1.5 rad/s, the lag compensator must reduce the gain by this amount. The transfer function of a lag compensator is represented by

$$G_c(s) = \frac{1 + \alpha T_1 s}{1 + T_1 s}$$

The constants can be determined as follows:

$$\alpha = 10^{-G_c/20} = 10^{-22/20} = 0.0794$$

$$\alpha T_1 = \frac{10}{\omega_c} = \frac{10}{1.5} = 6.67 \text{ s}$$

$$T_1 = 84 \text{ s}$$

The transfer function becomes

$$G_c(s) = \frac{1 + 6.67s}{1 + 84s}$$

The total compensated loop transfer function is now:

$$GH_c(s) = GH(s)G_c(s) = \frac{63.5(s + 0.015)}{s(s + 0.012)(s + 2)(s + 20)}$$

$$GH_c(j\omega) = \frac{63.5(0.15 + j\omega)}{j\omega(0.012 + j\omega)(2 + j\omega)(20 + j\omega)}$$

This function is plotted in Fig. 26.

4. *Select a circuit to meet the G_c (s) transfer function*

Study circuits and their related transfer functions in a standard reference; this study will show that the circuit of Fig. 27 meets the $G_c(s)$ transfer function requirements. Follow this circuit with a buffer amplifier to prevent loading. Then

$$G_c(s) = \frac{1 + 6.67s}{1 + 84s} = \frac{T_1 s + 1}{T_3 s + 1}$$

$$T_1 = R_1 C_1 = 6.67$$

$$T_3 = (R_1 + R_2)C_1 = 84$$

Pick $C_1 = 150 \ \mu$F and solve for the remaining component values:

$$R_1 = 44.5 \text{ k}\Omega \qquad R_2 = 516 \text{ k}\Omega$$

Pick the nearest standard values from a table of component characteristics. The circuit is shown in Fig. 27. This procedure is the work of Charles R. Hafer, P.E.

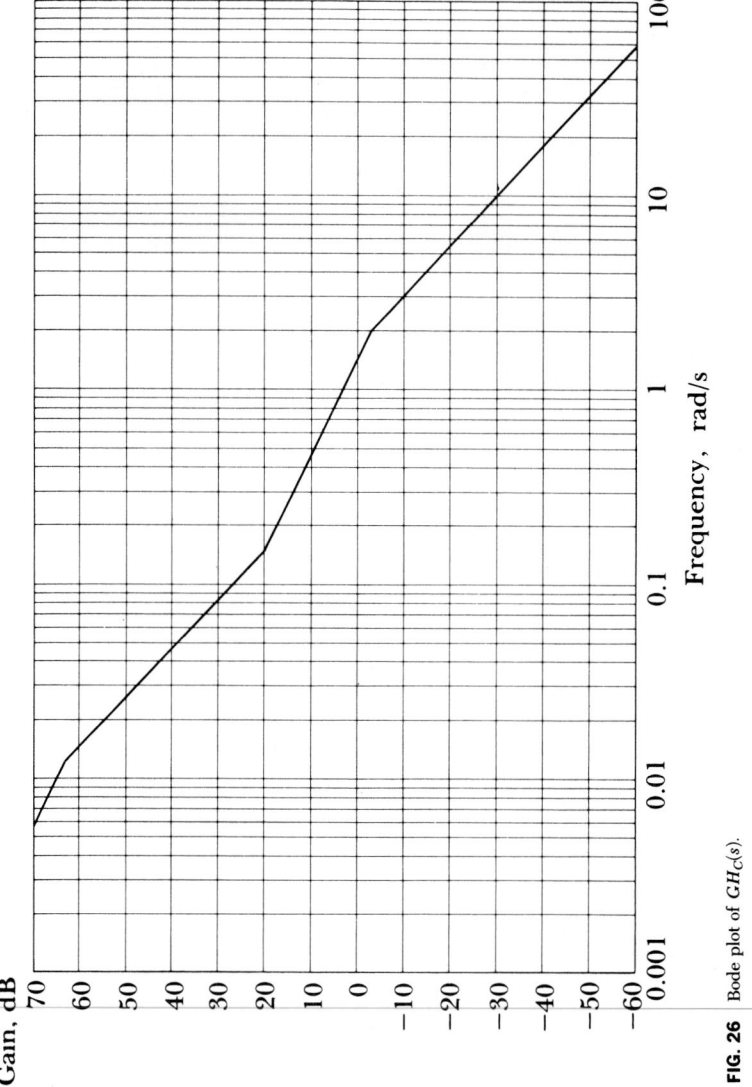

Gain, dB

Frequency, rad/s

FIG. 26 Bode plot of $GH_C(s)$.

7.36

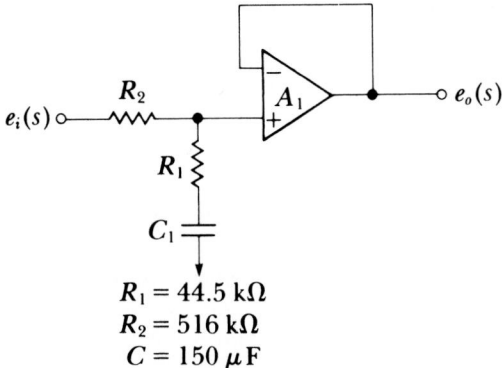

FIG. 27 Circuit lag compensator.

ANALYSIS OF TEMPERATURE-MEASURING AMPLIFIER

The operational amplifier shown in Fig. 28 is used to measure temperature. Here R_T is a temperature-sensitive resistor that varies with temperature as follows: $R_T = 1000e^{-T/25°C}$. Determine R_2 if E_o is to be 0 V at $-55°C$. What value must R_5 be if full-scale deflection at 125°C is required? The meter resistance R_M is 1 kΩ and the meter has a full-scale deflection of 1 mA.

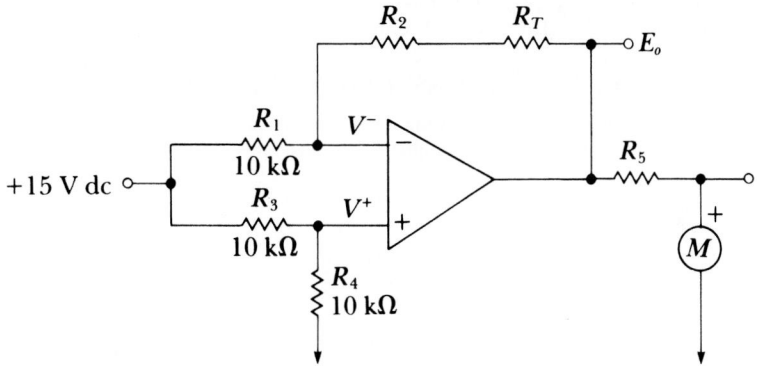

FIG. 28 Amplifier circuit.

Calculation Procedure:

1. Find the values of R_T at minimum and maximum temperatures

At $-55°C$, the minimum temperature is $R_T = 1000e^{-55/(-25)} = 9.025$ kΩ. At the maximum temperature of 125°C, $R_T = 1000e^{-125/25} = 6.74$ Ω.

2. Find R_2 for this amplifier

Using the generalized equation for the analysis of dc amplifiers, we get

$$E_o = \frac{(V_1 - V_2)(R_3 + R_4)R_2}{(R_1 + R_2)R_3} + \frac{(R_3 + R_4)V_2}{R_3} - \frac{V_3R_4}{R_3}$$

We substitute the terms for this amplifier, allowing $V = 15$ V dc:

$$E_o = \frac{V(R_1 + R_2 + R_T)R_4}{(R_3 + R_4)R_1} - \frac{V(R_2 + R_T)}{R_1}$$

Solve for R_2 with $E_o = 0$ V and $R_T = 9.025$ kΩ:

$$0 = \frac{15(10 \times 10^3 + R_2 + 9.025 \times 10^3)10 \times 10^3}{(20 \times 10^3)10 \times 10^3} - \frac{15(R_2 + 9.025 \times 10^3)}{10 \times 10^3}$$

So

$$0.5R_2 + 9.5125 \times 10^3 - R_2 - 9.025 \times 10^3 = 0$$

$$R_2 = 975 \ \Omega$$

From a table of standard metal-film resistor values, select $R_2 = 976$ Ω.

3. *Determine the values of E_o and R_5*

At maximum temperature of 125°C, full-scale deflection is required. Calculate the value of E_o at 125°C ($R_T = 6.74$ Ω from step 1):

$$E_o = \frac{15(10 \times 10^3 + 975 + 6.74)10 \times 10^3}{(20 \times 10^3)(10 \times 10^3)} - \frac{15(975 + 6.74)}{10 \times 10^3}$$

$$= 6.764 \text{ V dc}$$

With this value of E_o, 1 mA is to flow into the meter. Therefore,

$$R_5 + R_M = \frac{E_o}{1 \text{ mA}} = \frac{6.764}{1 \text{ mA}} = 6.764 \text{ k}\Omega$$

$$R_5 = 5.764 \text{ k}\Omega$$

Choose the nearest 1% value; let $R_5 = 5.76$ kΩ.
    This procedure is the work of Charles R. Hafer, P.E.

ANALYSIS OF TEMPERATURE-MEASURING INSTRUMENT

Figure 29 shows the schematic of a measuring system. In this system the motor drives a recording stylus which indicates temperature and simultaneously moves the wiper of the potentiometer that

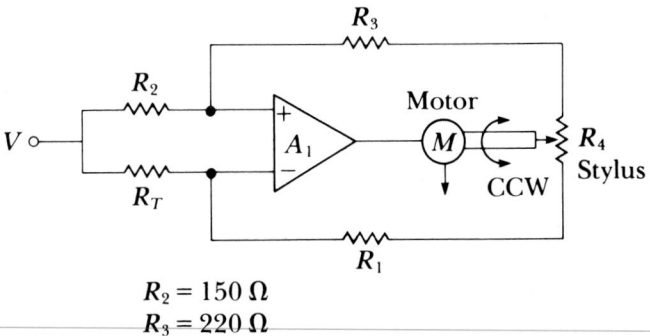

$$R_2 = 150 \ \Omega$$
$$R_3 = 220 \ \Omega$$

FIG. 29   Measuring system schematic.

indicates the position of the stylus. Here R_1 and R_2 are internal to the recorder; R_T is the transducer and has a resistance value of 330 Ω at 25°C and a temperature coefficient of +3 Ω/°C. Find the values of R_1 and R_4 if we want to measure a temperature range of −25 to 125°C and the wiper of the potentiometer is returned to ground potential. At −25°C, the stylus is fully counterclockwise (CCW).

Calculation Procedure:

1. Determine the value of R_T at -25 and $+125°C$

At $-25°C$: $R_T' = 330 + (3\Omega/°C)(-50°C) = 180\ \Omega$

At $125°C$: $R_T'' = 330 + (3\ \Omega/°C)(100°C) = 630\ \Omega$

2. Draw the equivalent circuit, and write the equations for it at $-25°C$

Draw the equivalent circuit for $-25°C$ as shown in Fig. 30. The potentiometer is full CCW. Since the voltage at the negative terminal must equal the voltage at the positive terminal, the following equations apply:

$$V^+ = V^-$$

$$\frac{(R_3 + R_4)V}{R_2 + R_3 + R_4} = \frac{R_1 V}{R_T' + R_1}$$

$$\frac{220 + R_4}{150 + 220 + R_4} = \frac{R_1}{180 + R_1}$$

Solving gives

$$R_4 = 0.833R_1 - 220$$

3. Draw the equivalent circuit, and write the equations for it at $125°C$

For the $125°C$ condition, the potentiometer is fully clockwise, and the circuit of Fig. 31 is applicable. Write these equations for $125°C$:

$$V^+ = V^-$$

$$\frac{(R_1 + R_4)V}{R_T'' + R_1 + R_4} = \frac{R_3 V}{R_2 + R_3}$$

$$\frac{R_1 + R_4}{630 + R_1 + R_4} = \frac{220}{370}$$

Thus

$$R_1 + R_4 = 925\ \Omega$$

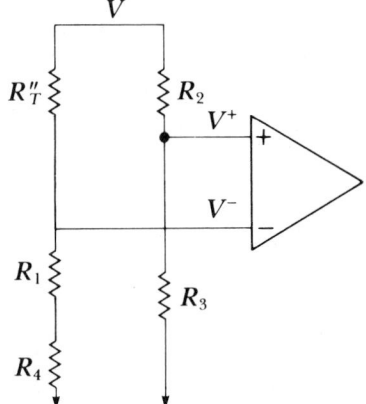

FIG. 30 Fully counterclockwise equivalent circuit.

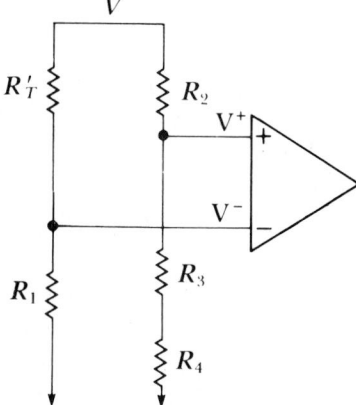

FIG. 31 Fully clockwise equivalent circuit.

And substituting from our previous solution for R_4 gives

$$R_1 + 0.833R_1 - 220 = 925 \ \Omega$$

$$R_1 = 625 \ \Omega$$

$$R_4 = 300 \ \Omega$$

Choose the nearest standard value for R_1 from a table of component values. So $R_1 = 619$; R_4 would be a potentiometer of $300 \ \Omega$.

This procedure is the work of Charles R. Hafer, P.E.

MEASURING RMS VALUE OF A WAVEFORM

A dc ammeter is used to measure the rms values of the waveform shown in Fig. 32. Determine the value of R_1 for the circuit in Fig. 33 if the internal resistance of the meter is 1 kΩ and the meter has a full-scale deflection of 1 mA. The peak input signal level is 10 V.

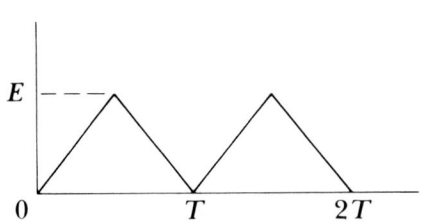

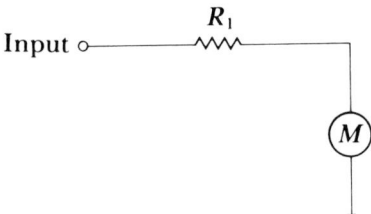

FIG. 32 Input waveform.   **FIG. 33** Ammeter circuit.

Calculation Procedure:

1.   *Calculate the average value of the waveform voltage*

A dc ammeter reads the average value of a waveform. Since we need rms values, we must find the rms and average values for the waveform and design the resistor accordingly. We must make the assumption that the natural time constant of the meter movement is much larger than the period of the waveform, or the meter movement will follow the profile of the waveform.

Calculate the average value for the waveform first. The equation for the waveform from the origin to $T/2$ is $E_1(t) = Et/(T/2) = 2Et/T$. The equation for the waveform from $T/2$ to T is $E_2(t) = 2E - 2Et/T$. The average value can be calculated as follows:

$$E_{av} = \frac{1}{T} \int_0^T e \, dt = \frac{1}{T} \left[\int_0^{T/2} \frac{2Et}{T} \, dt + \int_{T/2}^T \left(2E - \frac{2Et}{T} \right) dt \right]$$

$$= \frac{1}{T} \left(\frac{Et^2}{T} \Big|_0^{T/2} + 2Et - \frac{Et^2}{T} \Big|_{T/2}^T \right)$$

$$= \frac{1}{T} \left(\frac{ET}{4} + 2ET - ET - ET + \frac{ET}{4} \right) = \frac{E}{2} = 5 \text{ V}$$

2.   *Compute the rms value of the waveform*

The rms value of the triangular waveform can be calculated as follows:

$$E_{\text{rms}} = \left(\frac{1}{T} \int_0^T e^2 \, dt \right)^{1/2}$$

$$= \left[\frac{1}{T} \int_0^{T/2} \left(\frac{2Et}{T} \right)^2 dt + \frac{1}{T} \int_{T/2}^T \left(2E - \frac{2Et}{T} \right)^2 dt \right]^{1/2}$$

$$= \left[\frac{1}{T} \left(\frac{4E^2 t^3}{3T^2} \right)_0^{T/2} + \frac{1}{T} \left(4E^2 t - \frac{4E^2 t^2}{T} + \frac{4E^2 t^3}{3T^2} \right)_{T/2}^{T} \right]^{1/2}$$

$$= \left[\frac{1}{T} \left(\frac{E^2 T}{6} \right) + \frac{1}{T} \left(4E^2 T - 2E^2 T - 4E^2 T + E^2 T + \frac{4E^2 T}{3} - \frac{E^2 T}{6} \right) \right]^{1/2}$$

$$= \left(\frac{E^2}{6} + 4E^2 - 2E^2 - 4E^2 + E^2 + \frac{4E^2}{3} - \frac{E^2}{6} \right)^{1/2} = \left(\frac{E^2}{3} \right)^{1/2}$$

$$= 5.774 \text{ V}$$

3. Determine the value of R_1

For a full-scale deflection of 10 V, we would want the meter resistance plus R_1 to equal 10 kΩ for a 1-mA full-scale deflection. Thus the meter would read 5 V for the average value. But we want an rms reading which is 5.774 V, so we need to increase our current by 5.774/5, which yields a current increase to 1.155 mA. Therefore,

$$R_M + R_1 = \frac{10 \text{ V}}{1.155 \text{ mA}} = 8.66 \text{ k}\Omega$$

$$R_1 = 7.66 \text{ k}\Omega \quad \text{or the nearest standard value}$$

This procedure is the work of Charles R. Hafer, P.E.

NAND GATE CIRCUIT IMPLEMENTATION*

A 4-bit binary-coded decimal (BCD) code is shown in Table 7, appearing as inputs A, B, C, and D, with A being the least significant bit (LSB). Find a NAND gate implementation to convert from this BCD to the Gray code shown in Table 7.

Calculation Procedure:

1. Set up the standard basis for the designation numbers of inputs and outputs

Set up the standard basis which gives the designation numbers of the inputs and then those of the outputs (W, X, Y, Z) desired. This is done as shown in Table 8. The table is set up as follows: For the BCD portion of the table, column 1, we have a zero code; in the bottom part of the table we also want a zero code. This procedure is continued for all columns and corresponds to the desired results shown in Table 7. Now, since W, X, Y, and Z are functions of A, B, C, and D, we can use a Karnaugh map to

TABLE 7 BCD Input, Gray Code Output

Input				Output			
A	B	C	D	W	X	Y	Z
0	0	0	0	0	0	0	0
1	0	0	0	1	0	0	0
0	1	0	0	1	1	0	0
1	1	0	0	1	1	1	0
0	0	1	0	1	1	1	1
1	0	1	0	0	1	1	1
0	1	1	0	0	0	1	1
1	1	1	0	0	0	0	1
0	0	0	1	1	0	0	1
1	0	0	1	1	1	0	1

simplify each of our expressions W, X, Y, and Z. The numbers in each square correspond to the minterm representation in the designation number. For example,

$$0 \to \overline{ABCD} \qquad 2 \to \overline{AB}C\overline{D}$$

2. Use the Karnaugh map to simplify each expression

Go through the complete procedure for determining W, and the solutions for X, Y, and Z can be done similarly. The designation number for W is determined from Table 8.

$$W = 0111 \qquad 1000 \qquad 11XX \qquad XXXX$$

The X's denote constrained states or "don't care" conditions. The Karnaugh map of W is shown in Fig. 34. In all boxes where we want a 1 we put a diagonal line. In all squares where we don't

---

*See pg 7.48ff for a discussion of boolean algebra with a number of pertinent calculation procedures.

TABLE 8 Standard Basis

A	0101	0101	01	(LSB)		
B	0011	0011	00		BCD	
C	0000	1111	00		code	(Input)
D	0000	0000	11			
W	0111	1000	11	(LSB)		
X	0011	1100	01		Gray	
Y	0001	1110	00		code	(Output)
Z	0000	1111	11			

care, we put an X. The X means that the square can be occupied by either a 1 or a 0 at the convenience of the designer. Now minimization can be achieved. All X's are taken to be 1 for this variable. Grouping as shown yields

Squares	Representation
3, 11, 9, 1	$A\overline{C}$
2, 3, 10, 11	$B\overline{C}$
8, 9, 10, 11, 12, 13, 14, 15	D
4, 12	$\overline{A}\overline{B}C$

Similar solutions for X, Y, and Z are shown in Figs. 35 through 37, respectively. The circuit implementation is shown in Fig. 38. If $\overline{A}$, $\overline{B}$, $\overline{C}$, and $\overline{D}$ are not available, inverters can be placed on A, B, C, and D.

This procedure is the work of Charles R. Hafer, P.E.

USING NAND GATES TO IMPLEMENT FUNCTIONS WITH TWO LEVELS OF LOGIC

The boolean expression

$$F_1 = AB + ABC + BC$$

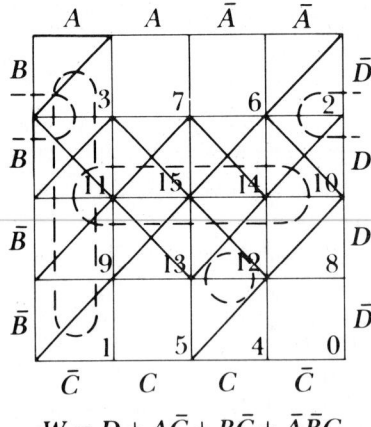

$$W = D + A\overline{C} + B\overline{C} + \overline{A}\overline{B}C$$

FIG. 34 Karnaugh map of W.

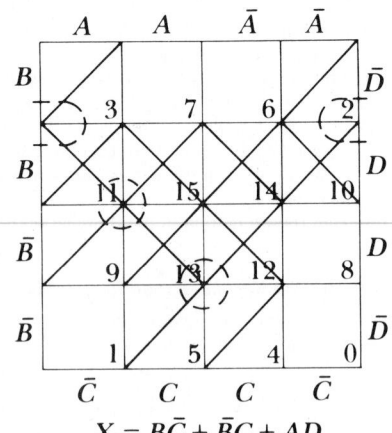

$$X = B\overline{C} + \overline{B}C + AD$$

Fig. 35 Karnaugh map of X.

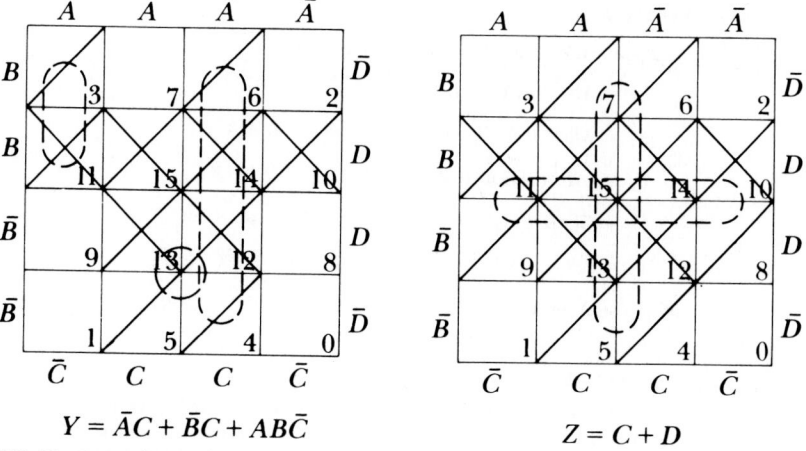

$$Y = \bar{A}C + \bar{B}C + AB\bar{C}$$

FIG. 36 Karnaugh map of Y.

$$Z = C + D$$

FIG. 37 Karnaugh map of Z.

$$F_2 = (A\bar{B} + C)(A + \bar{B})C$$

$$F_3 = AB + (\bar{B} + \bar{C}) + \bar{A}C$$

is to be simplified by using boolean algebra and to be implemented by using a maximum of two levels of logic. Use NAND gates to implement these functions, and show the diagrams.

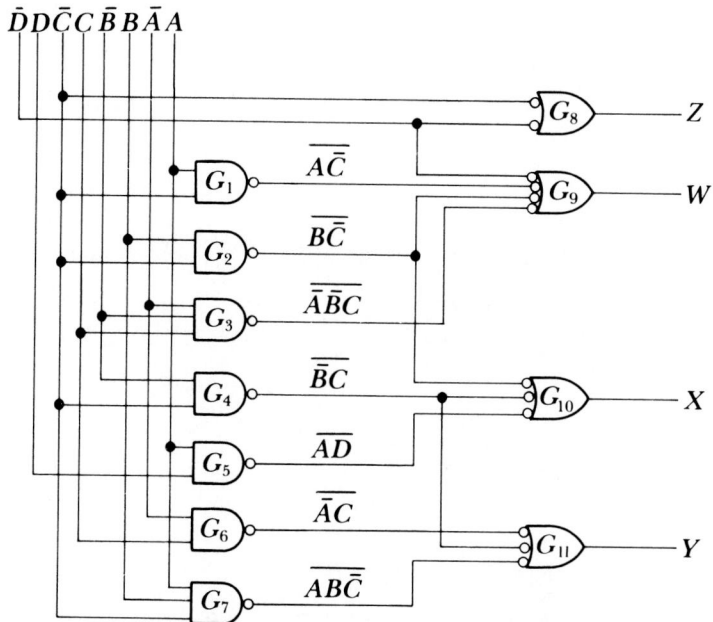

FIG. 38 Circuit implementation of W, X, Y, and Z.

Calculation Procedure:

1. Simplify the given expression, using boolean algebra

$$F_1 = AB + ABC + BC$$

$$F_1 = AB(1 + C) + BC = AB + BC = B(A + C)$$

$$F_2 = (A\overline{B} + C)(A + \overline{B})C = (A\overline{B}C + C)(A + \overline{B})$$

$$F_2 = C(A\overline{B} + 1)(A + \overline{B}) = C(A + \overline{B}) = AC + \overline{B}C$$

$$F_3 = AB + (\overline{B} + \overline{C}) + \overline{A}C = AB + A\overline{B} + \overline{A}\,\overline{C} + \overline{A}C$$

$$F_3 = A + \overline{A} = 1$$

(Note: $\overline{B}$ implies $A\overline{B}$ and $\overline{C}$ implies $\overline{A}\,\overline{C}$.)

2. Implement, using NAND gates; show the diagram

The method for NAND gate implementation is to use AND gates and OR gates (no inversions) to implement the function and to place an inversion "ball" at the output of all AND gates and an inversion "ball" at the input of all AND gates used as OR functions. NAND gates may be added where required to provide an inverter function. Figure 39 shows how F_1 is obtained by using NAND logic, while Fig. 40 demonstrates the technique for F_2. Here F_3 is a trivial solution since it is a logic 1 level only.

This procedure is the work of Charles R. Hafer, P.E.

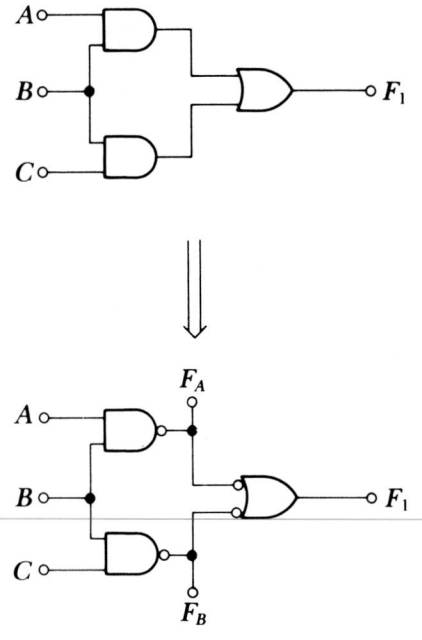

$$F_A = \overline{AB} \qquad F_B = \overline{BC}$$

$$F_1 = \overline{F_A} + \overline{F_B} = \overline{\overline{AB}} + \overline{\overline{BC}} = AB + BC$$

FIG. 39   Circuit for F_1 implementation.

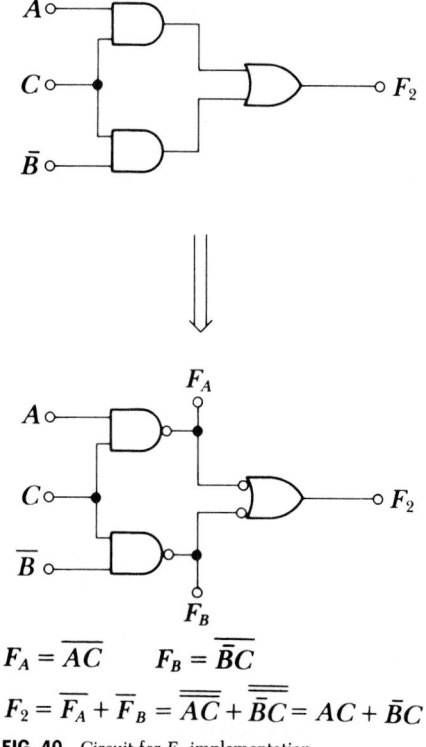

$$F_A = \overline{AC} \qquad F_B = \overline{\bar{B}C}$$

$$F_2 = \overline{F_A} + \overline{F_B} = \overline{\overline{AC}} + \overline{\overline{\bar{B}C}} = AC + \bar{B}C$$

FIG. 40 Circuit for F_2 implementation.

SIZING STEAM-CONTROL AND PRESSURE-REDUCING VALVES

Dry saturated steam at 30 lb/in^2 (abs) (206.9 kPa) will flow at the rate of 1000 lb/h (0.13 kg/s) through a single-seat pressure-reducing throttling valve. The desired exit pressure is 20 lb/in^2 (abs) (137.9 kPa) at the valve outlet. Select a valve of suitable size.

Calculation Procedure:

1. Determine the critical pressure for the valve

Critical pressure exists in a valve and piping system when the pressure at the valve outlet is 58 percent, or less, of the absolute inlet pressure for saturated steam (55 percent for hydrocarbon vapors and superheated steam). Thus, for this system the critical outlet pressure is $P_c = 0.58P_i$, where P_c = critical pressure for the system, lb/in^2 (abs) (kPa); P_i = inlet pressure, lb/in^2 (abs) (kPa). Or, $P_c = 0.58(30) = 17.4$ lb/in^2 (abs) (119.9 kPa).

Since the outlet pressure, 20 lb/in^2 (abs) (137.9 kPa), is greater than the critical pressure, the flow through the valve is noncritical.

2. Find the density of the outlet steam

Assume adiabatic expansion of the steam from 30 lb/in^2 (abs) (206.9 kPa) to 20 lb/in^2 (abs) (137.9 kPa). (This is a valid assumption for a throttling process such as that which takes place in a pressure-reducing valve.) Using the steam tables, we find the density of the steam at the outlet pressure of 20 lb/in^2 (abs) (137.9 kPa) is 0.05 lb/ft^3 (0.8 kg/m^3).

3. Compute the valve flow coefficient c_v

Use the relation $c_v = W/63.5\sqrt{(P_i - P_2)\rho}$, where c_v = valve flow coefficient, dimensionless; W = steam (or vapor or gas) flow rate, lb/h (kg/s); P_i = valve inlet pressure, lb/in^2 (abs); ρ =

density of the vapor or gas flowing through the valve, lb/ft^3 (kg/m^3). For this valve, $c_v = 1000/$ $63.5\sqrt{(30-20)0.05} = 22.3$, say 22.0 because c_v valves are usually stated in even numbers for larger-size valves.

4. Select the control valve to use

At normal operating conditions, most engineers recommend that the flow through the valve not exceed 80 percent of the maximum flow possible. Thus the valve selected should have a c_v equal to or greater than the computed $c_v/0.80$. Thus, for this valve, choose a unit having a c_v equal to or greater than $22/0.80 = 27.5$. From Table 9 choose a 2-in (5.08-cm) single-seat valve having a c_v of 36.

The operating c_v of any valve is $c_{vo} = c_{vf}/c_{vs}$, where $c_{vf} = c_v$ value computed by the formula in step 3 and $c_{vs} = c_v$ of actual valve selected. Or, for this valve, $c_{vo} = 22/36 = 0.61$.

To avoid wire drawing which occurs when the valve plug operates too close to the valve seat, c_{vo} values of less than 0.10 should not be used. Since $c_{vo} = 0.61$ for this valve, wire drawing will not occur.

Related Calculations: To speed up the determination of c_v, Fig. 41 can be used instead of the formula in step 3. This is a performance-tested chart valid for steam control valves for blast-heating coils, tank heaters, pressure-reducing stations, and any other installations—stationary, mobile, or marine—where steam flow and pressure are to be regulated. The approach can also be used for valves handling gases other than steam.

The valve coefficient c_v is conventional; it equals the gallons per minute (liters per second) of clear cold water at 60°F (15.6°C) that will pass through the flow restriction (valve or orifice) while undergoing a pressure drop of 1 lb/in^2 (7.0 kPa). The c_v value is the same for liquids, gases, and steam. Tables listing c_v values versus valve size and type are published by the various valve manufacturers. General c_v values not limited to any manufacturer are given in Table 9 for a variety of valve types and sizes.

Note in Fig. 41 the relations for the density of the steam of various valve outlet pressures. In these relations P_c = critical pressure, lb/in^2 (abs) (kPa), as defined earlier. The solution given in

TABLE 9 Flow Coefficients for Steam-Control Valves°

Size		Straight-through throttling		Straight-through on-off	Straight-through regulators	
in	cm	Single seat	Double seat	Single seat	Single seat	Double seat
⅛	0.32	0.23				
¼	0.64	0.78				
⅜	0.95	1.7				
½	1.27	3.2				
¾	1.91	5.4	7.2	7	3.6	4.3
1	2.54	9	12	12	6	7.2
1¼	3.18	14	18	18	9	10.8
1½	3.81	21	28	27	14	16.8
2	5.08	36	48	42	24	28.8
2½	6.35	54	72	65	36	43.2
3	7.62	75	100	93	50	60
4	10.2	124	165	170	83	99
6	15.2	270	360	380	180	216
8	20.3	480	640	660	320	384
10	25.4	750	1000	1100	500	600
12	30.5	1080	1440	1550	720	864

°*Chemical Engineering.*

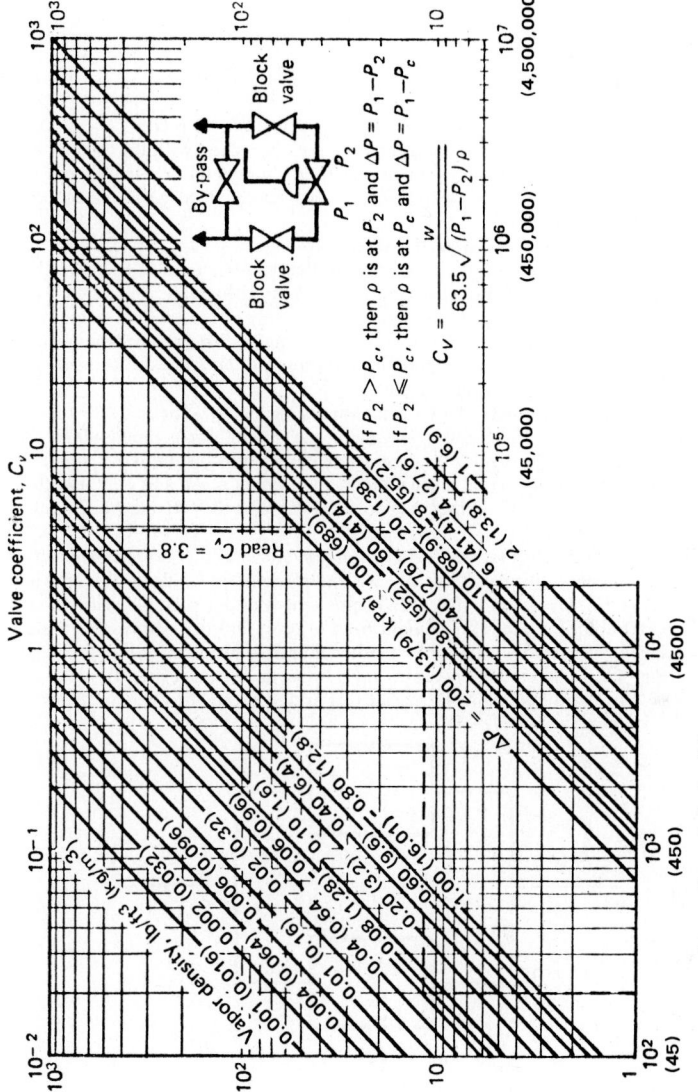

FIG. 41 c_v values for steam-control and pressure-reducing valves. *(Chemical Engineering.)*

7.47

Fig. 41 is for a flow rate of 200 lb/h (0.025 kg/s) of steam having a density of 0.08 lb/ft^3 (1.25 kg/m^3) at the valve outlet with a 10-lb/in^2 (abs) (68.9-kPa) pressure drop through the valve, giving a c_v of 3.8.

This procedure is the work of John D. Constance, P.E., as reported in *Chemical Engineering* magazine.

BOOLEAN ALGEBRA

We shall develop the basic concepts and laws of boolean algebra by relating them directly to a set of tangible objects. In this system of algebra, we are concerned with the *state* of an object, and each object has two possible states: open or closed, horizontal or vertical, red or green, etc. These objects, which are termed *components*, are arranged to form a *system*, and the system itself has two possible states.

The states of the components and of the system are expressed by use of a code consisting of the numerals 0 and 1. For example, 0 can denote that a component is open and 1 that it is closed. The expression $A = 0$ states that component A is at the state represented by 0. For brevity, we say that A has the *value* 0, although we are expressing a state of being rather than a numerical value. When referring to the system, the numeral 1 represents the required state of the system; when referring to a component, 1 represents the state of the component that enables the system to attain its required state.

Assume that a system consists of two components: A and B. The state of the system is determined by the states of A and B and by the manner in which these components are arranged. Two arrangments are possible. Under the first arrangement, the system is 1 if *either A or B is* 1 (or if both are 1); under the second arrangement, the system is 1 if and only if *both A and B* are 1. These arrangements of the components are referred to as the OR and AND relationships, respectively.

As an illustration, refer to Fig. 42a, where components A and B are arranged in parallel. Assume that we are currently at point m and wish to reach point n. To do this, we must pass through either A or B. Now assume that each component is either passable or impassable. The system itself is passable if it allows movement from m to n, and it is impassable if such is not the case. Manifestly, the system is passable if either A or B is passable (or if both are passable). Now refer to Fig. 42b, where components A and B are arranged in series. To move from m to n, we must pass through both A and B. Under this arrangement, the system is passable only

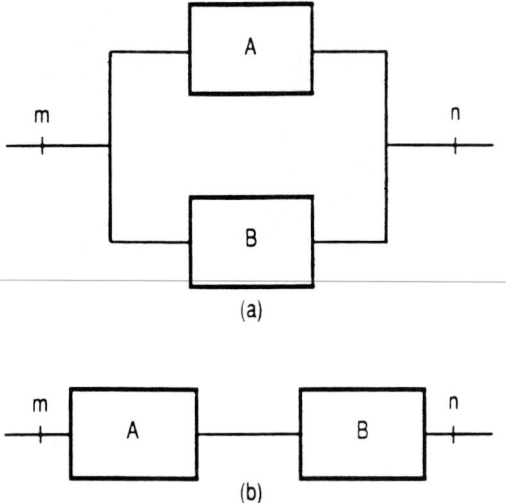

(a)

(b)

FIG. 42 Passage through system. (*a*) Components in parallel; (*b*) components in series.

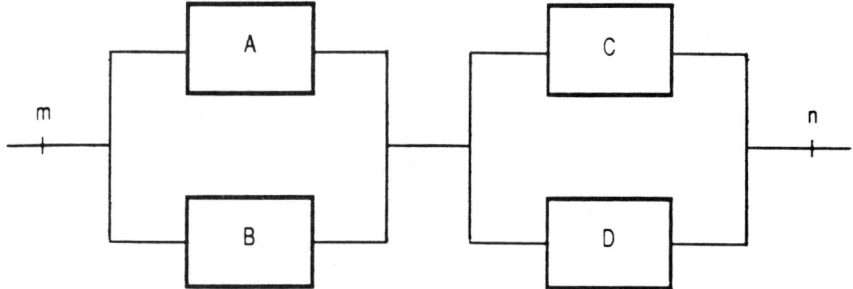

FIG. 43   System consisting of subsystems arranged in series.

if both A and B are passable. In summary, the OR relationship corresponds to the parallel arrangement of the components, and the AND relationship corresponds to the series arrangement. This concept of movement across the system provides a simple means of arriving at the laws of boolean algebra.

The OR and AND relationships are denoted in this manner: The expression $A + B$ (read "A or B") signifies that the system is 1 if either A or B is 1 (or if both are 1); the expression AB (read "A and B") signifies that the system is 1 if both A and B are 1. However, it is to be emphasized that these expressions have nothing to do with addition and multiplication in ordinary algebra, despite the similarity of notation.

A system consisting of three or more components may be considered to contain *subsystems*, and the composition of a subsystem can be described by the use of parentheses. As an illustration, refer to Fig. 43. This system contains two subsystems. One consists of A and B in parallel, and the other consists of C and D in parallel. The two subsystems are arranged in series. Thus, to move from m to n, we must first pass through either A or B, and then through either C or D. The expression for this system is $(A + B)(C + D)$.

A component is described as a *variable* or a *constant* according to whether its state is alterable or unalterable, respectively. The expression $A0$ describes a system in which A is in series with a component that is always at state 0. Similarly, the expression $A + 1$ describes a system in which A is in parallel with a component that is always at state 1. Two or more components are said to be *identical* if they are always at the same state, and they are represented by an identical symbol. For example, the expression AA describes a system in which two identical components are arranged in series.

Equivalence:

Two systems are *equivalent* to each other if they are always at the same state, and equivalence is expressed by use of an equals sign. Thus, the statement $AA = A$ means that a system consisting of two components A in series is equivalent to a system having a single component A. A statement of equivalence is called an *equation*. The following equations are self-evident:

$$A + 0 = A \tag{1}$$

$$A + 1 = 1 \tag{2}$$

$$A + A = A \tag{3}$$

$$A0 = 0 \tag{4}$$

$$A1 = A \tag{5}$$

$$AA = A \tag{6}$$

$$A + B = B + A \tag{7}$$

$$AB = BA \tag{8}$$

Equations 7 and 8 state that components A and B are *commutative*.

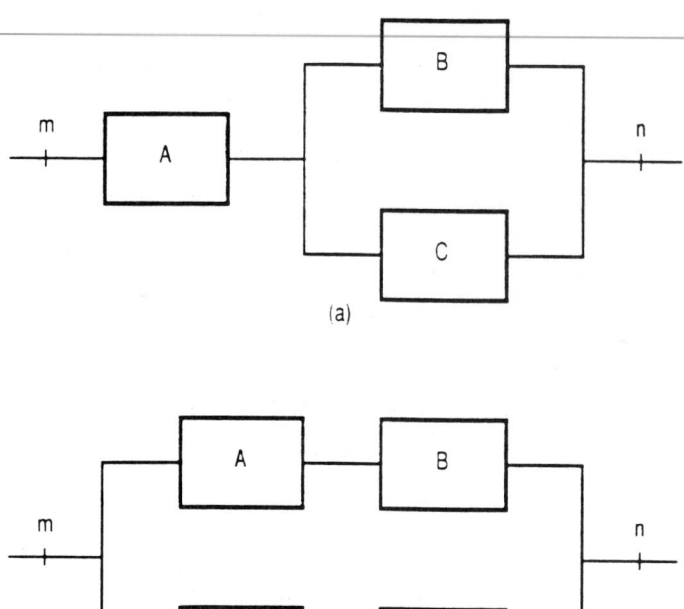

FIG. 44   Equivalent systems. (*a*) System for $A(B + C)$; (*b*) system for $AB + AC$.

If three components are arranged in parallel or in series, any two may be considered to form a subsystem. Therefore, the components are *associative*. Expressed symbolically,

$$(A + B) + C = A + (B + C) \tag{9}$$

$$(AB)C = A(BC) \tag{10}$$

Consider the expression $A(B + C)$, which describes the system in Fig. 44*a*. In moving from *m* to *n*, we have two alternative paths: *A* and *B*, and *A* and *C*. Therefore, the system is passable if both *A* and *B* are passable or if both *A* and *C* are passable. It follows that the system in Fig. 44*a* is equivalent to that in Fig. 44*b*, and we have

$$A(B + C) = AB + AC \tag{11}$$

Therefore, in the expression at the left, *A* is *distributive*. Reversing the procedure, we see that it is possible to *factor* the repeating variable *A* in the expression at the right, thereby obtaining the expression at the left.

Equation 11 is extendible. Consider the expression $(A + B)(C + D)$, which describes the system in Fig. 43. In moving from *m* to *n*, we have found alternative paths: *A* and *C*, *A* and *D*, *B* and *C*, and *B* and *D*. Therefore,

$$(A + B)(C + D) = AC + AD + BC + BD$$

Generalized Terminology:

Having developed the basic laws of boolean algebra by use of a specific application, we can now modify our terminology to make it more general. In our new terminology, a component becomes a constant or a variable, a system becomes a function, and a subsystem becomes a term in an expression. Thus, Eq. 11 now acquires the following meaning: The expression $A(B + C)$ is equivalent to the expression $AB + AC$.

A function is denoted by the letter f. The statement $f = AB + CD$ means that f assumes the value 1 if either of these conditions exists: A and B are both 1; C and D are both 1. Thus, the expression at the right is a set of specifications; it lists the requirements that must be satisfied for f to become 1. In the following material the word *expression* refers to the expression for a function, and the value of the expression is the value of the function itself.

Complementary Variables and Expressions:

The symbol $\overline{A}$ (read "not A") denotes a variable that is always at the state different from that of A. Thus, if $A = 0$, $\overline{A} = 1$, and vice versa. (The symbol A' is sometimes used to denote "not A".) It follows at once that

$$A + \overline{A} = 1 \tag{12}$$

For this reason, A and $\overline{A}$ are said to be *complementary* to each other. Similarly, we have

$$A\overline{A} = 0 \tag{13}$$

The symbol $\overline{\overline{A}}$ denotes the complement of $\overline{A}$. Then

$$\overline{\overline{A}} = A \tag{14}$$

We shall consider A to be the independent variable and $\overline{A}$ the dependent variable.

The NOT notation also applies to expressions as well as single variables. For example, the expression $\overline{AB}$ is always at the state different from that of AB. Thus, if $AB = 0$, $\overline{AB} = 1$, and vice versa. The expressions AB and $\overline{AB}$ are also said to be complementary to each other.

The conditions that make a given expression 1 are those that make its complementary expression 0. By establishing the conditions that make $AB = 0$, we arrive at this result:

$$\overline{AB} = \overline{A} + \overline{B} \tag{15}$$

By extension,

$$\overline{ABC} = \overline{A} + \overline{B} + \overline{C}$$

Similarly,

$$\overline{A + B} = \overline{A}\,\overline{B} \tag{16}$$

By extension,

$$\overline{A + B + C} = \overline{A}\,\overline{B}\,\overline{C}$$

Equations 15 and 16 are referred to as *De Morgan's laws*. We also have the following:

$$\overline{AB + AC + BC} = \overline{A}\overline{B} + \overline{A}\overline{C} + \overline{B}\overline{C} \tag{17}$$

and

$$\overline{AB + AC + BC} = (\overline{A} + \overline{B})(\overline{A} + \overline{C})(\overline{B} + \overline{C}) \tag{18}$$

Equations 17 and 18 stem from the fact that $AB + AC + BC = 0$ if any two of the three variables are 0.

An expression may contain both a given variable and its complementary variable. The expression $A + \overline{A}B$ is an illustration. This expression is 1 if either of the following conditions exists: $A = 1$; $A = 0$ and $B = 1$. Thus, it is simply necessary that either A or B be 1, and

$$A + \overline{A}B = A + B \tag{19}$$

The following equations can be derived by simple logic or by applying the preceding equations:

$$A + AB = A \tag{20}$$

$$A(A + B) = A \tag{21}$$

$$(A + B)(A + C) = A + BC \tag{22}$$

$$A(A + B) = AB \tag{23}$$

$$AB + A\overline{B} = A \tag{24}$$

The terms NOR and NAND are contractions for "Not OR" and "Not AND," respectively. Thus, $\overline{A + B + C}$ is a NOR term, and $\overline{ABC}$ is a NAND term.

The EXCLUSIVE OR Relationship:

The expression $A + B$ is 1 if either A or B is 1, or if both are 1. However, in some applications of boolean algebra it is necessary to impose a more stringent requirement: The expression is to be 1 if *one and only one* variable is 1. This requirement is referred to as the EXCLUSIVE OR relationship, and it is denoted by the symbol $A \oplus B$.

Since the expression $A \oplus B$ is 1 if one variable is 1 and the other is 0, it follows that

$$A \oplus B = (A + B)(\overline{A} + \overline{B}) \tag{25}$$

and

$$A \oplus B = A\overline{B} + \overline{A}B \tag{26}$$

We also have the following:

$$A \oplus 0 = A \tag{27}$$

$$A \oplus 1 = \overline{A} \tag{28}$$

$$A \oplus A = 0 \tag{29}$$

Consider the expression $\overline{A \oplus B}$. The complementary expression $A \oplus B$ is 0 if both variables are 0, or if both variables are 1. Therefore,

$$\overline{A \oplus B} = AB + \overline{A}\overline{B} \tag{30}$$

and

$$\overline{A \oplus B} = (A + \overline{B})(\overline{A} + B) \tag{31}$$

Inclusion:

A given AND term is said to *include* a longer AND term if the variables that appear in the former also appear in the latter. For example, the term $A\overline{C}D$ includes the term $A\overline{B}\overline{C}D$ because the latter contains A, $\overline{C}$, and D. The significance of inclusion is this: The term $A\overline{C}D$ requires that $A = 1$, $C = 0$, $D = 1$, and B can be 0 or 1. The term $A\overline{B}\overline{C}D$ requires that $A = 1$, $C = 0$, $D = 1$, and $B = 0$. Thus, the second requirement is encompassed in the first. It is convenient to extend the definition of inclusion by saying that a given AND term includes itself.

Compatible Terms:

Two AND terms are said to be *compatible* with each other if one term can be obtained from the other by replacing one and only one variable with its complement. For example, the terms $A\overline{B}DG$ and $ABDG$ are compatible because the second term can be obtained from the first by replacing $\overline{B}$ with B.

Compatible terms that are connected by the OR relationship can be combined in accordance with Eq. 24 or an extension of it. For example, let

$$f = \overline{A}BC + \overline{A}\overline{B}C$$

The first term states that $f = 1$ if $A = 0$, $B = 1$, $C = 1$; the second term states that $f = 1$ if $A = 0$, $B = 0$, $C = 1$. These terms can be combined to form $f = \overline{A}C$. This condensed equation states that $f = 1$ if $A = 0$ and $C = 1$, and the value of B is irrelevant.

If a given term is incompatible with any other term in the expression, it is called a *prime implicant*.

Standard Expressions:

Assume that f is a function of n independent variables. In the expression for f, a term is said to be in *standard form* if it is either an OR or AND term and each variable (or its complement) appears in the term once and only once. The following are standard terms for a four-variable function: $ABCD$, $\overline{A}BCD$, $A + B + C + \overline{D}$, $\overline{A} + \overline{B} + C + D$. A standard AND term is also called a *minterm*, and a standard OR term is also called a *maxterm*.

The expression for f is said to be in *standard, elemental,* or *canonical form* if it's composed of standard terms. The following are standard four-variable expressions:

$$A\overline{B}C\overline{D} + \overline{A}B\overline{C}D + A\overline{B}\overline{C}D \qquad ABCD + A\overline{B}\overline{C}D$$

$$(A + \overline{B} + C + D)(\overline{A} + B + C + \overline{D})$$

An AND-to-OR expression for f is one where each term is an AND expression, and these terms are linked by the OR relationship. Thus, $ABC + \overline{A}B\overline{C}$ is an AND-to-OR expression. Similarly, an OR-to-AND expression for f is one where each term is an OR expression, and these terms are linked by the AND relationship. Thus, $(A + \overline{B} + C)(\overline{A} + B + \overline{C})$ is an OR-to-AND expression. Standard AND-to-OR expressions are also called *minterm forms* or *disjunctive expressions*, and standard OR-to-AND expressions are also called *maxterm forms* or *conjunctive expressions*.

Let n denote the number of independent variables that are present. Since each standard term contains either a given independent variable or its complement, the number of standard terms that can be formed is 2^n. If we expand our definition of a standard expression to include the degenerate cases where f is a constant, we may say that a standard expression is a combination of m standard terms, where $0 < m \leq 2^n$. It follows that the number of possible standard expressions is 2 raised to the 2^n power.

As an illustration, let $n = 2$. The number of possible standard expressions is $2^4 = 16$, and the AND-to-OR forms are recorded in Table 10. This table is constructed by starting with the basic forms $\overline{A}\overline{B}$, $\overline{A}B$, $A\overline{B}$, and AB and then combining them two at a time, three at a time, and four at a time. Since the last expression encompasses all possible terms, it corresponds to the case $f = 1$.

TABLE 10 Standard AND-to-OR Expressions (Minterm Forms) for Two-Variable Functions

0	$\overline{A}\overline{B} + A\overline{B}$
$\overline{A}\overline{B}$	$\overline{A}B + AB$
$\overline{A}B$	$A\overline{B} + AB$
$A\overline{B}$	$\overline{A}\overline{B} + \overline{A}B + A\overline{B}$
AB	$\overline{A}\overline{B} + \overline{A}B + AB$
$\overline{A}\overline{B} + \overline{A}B$	$\overline{A}\overline{B} + A\overline{B} + AB$
$\overline{A}\overline{B} + A\overline{B}$	$\overline{A}B + A\overline{B} + AB$
$\overline{A}B + AB$	$\overline{A}\overline{B} + \overline{A}B + A\overline{B} + AB$

Reduction of Expressions:

A letter that represents a variable is termed a *literal*. In counting literals, we include duplicates. For example, the expression $\overline{A}B + \overline{A}BCD + CD\overline{E}$ contains nine literals.

A given expression is said to be *reduced to simpler form* when it is transformed to an equivalent expression with fewer literals. There are three basic methods of reducing an expression to simpler form. They are as follows:

1. *Consolidation of terms:* As previously stated, two compatible terms that are connected by the OR relationship can be combined. For example,

$$A\overline{B}D\overline{G} + ABD\overline{G} = AD\overline{G}(\overline{B} + B) = AD\overline{G}1 = AD\overline{G}$$

Thus, the combined term contains solely the variables that are common to the compatible terms.

2. *Elimination of redundancies:* For example, Eq. 19 reduces the expression $A + \overline{A}B$, which contains three literals, to $A + B$, which contains only two. The variable $\overline{A}$ in the original expression was redundant.

3. *Factoring:* For example, the equation $AB + AC = A(B + C)$ reduces an expression with four literals to an expression with three.

A given term in an AND-to-OR expression may be compatible with several terms in the expression. It is permissible to combine each compatible pair of terms. The justification is that $A + A = A$, and therefore this duplication of terms does not inject any error.

When a given expression has been reduced to the fullest extent possible by consolidating terms and by eliminating redundancies, it is said to be in *optimal form*. If the resulting expression is then reduced by factoring, the original expression is said to be in its *minimal form*. In many applications of boolean algebra, it is necessary to reduce a given expression to its minimal form, and we shall discuss systematic methods of accomplishing this objective.

Laws of Duality:

Boolean algebra is characterized by a duality that imparts an image to each expression and each equation. This duality can be exploited to considerable advantage. There are two laws of duality, as follows:

Theorem 1. A given expression is transformed to its complementary expression if the OR and AND relationships are interchanged and each variable is replaced with its complementary variable.

This principle stems from Eqs. 15 and 16. As an illustration, let

$$f = A\overline{B}(C + \overline{D}E)$$

Then

$$\overline{f} = \overline{A} + B + \overline{C}(D + \overline{E})$$

The expression for $\overline{f}$ is clearly valid because $f = 0$ if any of these conditions exist: $A = 0$; $B = 1$; $C = 0$, and either $D = 1$ or $E = 0$.

Theorem 2. A given equation is transformed to a corresponding equation if the OR and AND relationships are interchanged and the constants 0 and 1 are interchanged.

The corresponding equation is called the *dual* of the first. Thus, Eq. 5 is the dual of Eq. 1.

ALGEBRAIC PROOF OF AN EQUATION

Applying solely the equations of boolean algebra, prove the following:

$$\overline{A} + A\overline{B}C + \overline{D(\overline{A} + E)} = \overline{A} + \overline{B}C + \overline{D} + \overline{E} \tag{32}$$

Calculation Procedure:

1.   *Apply Eq. 19 to the first two terms and Eq. 15 to the third term*

Let f denote the expression at the left. The specified equations transform f to the following:

$$f = \overline{A} + \overline{B}C + \overline{D} + \overline{\overline{A} + E}$$

2.   *Apply Eqs. 16 and 14 in turn*

The result is

$$f = \overline{A} + \overline{B}C + \overline{D} + A\overline{E}$$

3.   *Apply Eq. 19 to the first and fourth terms*

The result is

$$f = \overline{A} + \overline{B}C + \overline{D} + \overline{E}$$

Equation (32) is thus proved.

PROVING AN EQUATION BY IDENTIFYING THE SATISFACTORY CONDITIONS

With reference to the preceding Calculation Procedure, prove Eq. 32 by identifying the conditions that make the expression on each side of the equation equal to 1.

Calculation Procedure:

1.   *Identify and record these conditions*

Let $f1$ and $f2$ denote, respectively, the expression on the left side and on the right side of Eq. 32. In Table 11, record the alternative conditions that make $f1 = 1$ and $f2 = 1$. Number the conditions in the manner shown.

2. Eliminate any redundancies that may exist

Conditions 2 and 3 are relevant solely if condition 1 does not exist; i.e., if $A = 1$. Therefore, in conditions 2 and 3, the requirement $A = 1$ is redundant and should be eliminated.

3. Compare the two sets of conditions as they now exist

With the redundancy eliminated, the conditions that make $f1 = 1$ coincide with those that make $f2 = 1$. Equation 32 is thus proved.

USE OF TRUTH TABLES

Construct a truth table to confirm the following equation, which stems from the first law of duality:

$$\overline{A + \overline{B}C} = \overline{A}(B + \overline{C})$$

TABLE 11

	Conditions for $f1 = 1$		Conditions for $f2 = 1$
1	$A = 0$	4	$A = 0$
2	$A = 1$ and $B = 0$ and $C = 1$	5	$B = 0$ and $C = 1$
		6	$D = 0$
3 or	$D = 0$ $A = 1$ and $E = 0$	7	$E = 0$

Calculation Procedure:

1. Number the rows and assign values to the variables in the conventional manner

A *truth table* is used to confirm an equation. In this table, we assign a column to each independent variable and to each term that appears in the equation. We then record every possible combination of values of the independent variables and the corresponding value of a given term and expression. If the expression contains n independent variables, the number of combinations of values is 2^n. In the present case, $n = 3$ and the number of combinations is $2^3 = 8$.

Refer to Table 12. By convention, the rows are numbered consecutively, starting with 0. Values are assigned to the variables in such manner that the digits form the row number in the binary system. For example, since 5 is 101 in binary form, row 5 contains these values: $A = 1$, $B = 0$, $C = 1$.

2. Establish the values of the terms $\overline{B}C$, $B + \overline{C}$, $A + \overline{B}C$, and $\overline{A}(B + \overline{C})$

Consider the term $\overline{B}C$. This is 1 if $B = 0$ and $C = 1$. Therefore, record the value 1 in rows 1 and 5 and the value 0 in all other rows. Now consider the term $B + \overline{C}$. This is 1 if $B = 1$ or $C = 0$; conversely, it is 0 if $B = 0$ and $C = 1$. Therefore, record the value 0 in rows 1 and 5 and the value 1 in all other rows. Continue in this manner to obtain the results recorded in Table 12.

3. Compare the values of $A + \overline{B}C$ and of $\overline{A}(B + \overline{C})$

In all instances, these two terms differ in value. Therefore, the given equation is confirmed.

TABLE 12 Truth Table

Row	A	B	C	$\overline{B}C$	$B + \overline{C}$	$A + \overline{B}C$	$\overline{A}(B + \overline{C})$
0	0	0	0	0	1	0	1
1	0	0	1	1	0	1	0
2	0	1	0	0	1	0	1
3	0	1	1	0	1	0	1
4	1	0	0	0	1	1	0
5	1	0	1	1	0	1	0
6	1	1	0	0	1	1	0
7	1	1	1	0	1	1	0

EXPANSION OF NONSTANDARD EXPRESSION TO STANDARD FORM

The expression for f is

$$f = \overline{A}BC + A\overline{B} + B$$

Expand this expression to standard form.

Calculation Procedure:

1. *Examine each term of the expression to determine which variables must be added to place the term in standard form*

The first term is standard, the second term requires the addition of C (or $\overline{C}$), and the third term requires the addition of A and C (or $\overline{A}$ and $\overline{C}$).

2. *Fill in the gaps by applying Eqs. 5 and 12*

The result is

$$f = \overline{A}BC + A\overline{B}(C + \overline{C}) + (A + \overline{A})B(C + \overline{C})$$

3. *Remove the parentheses by applying Eq. 11*

The result is

$$f = \overline{A}BC + A\overline{B}C + A\overline{B}\overline{C} + ABC + AB\overline{C} + \overline{A}BC + \overline{A}B\overline{C}$$

4. *Eliminate duplications*

Since the first and sixth terms in the last expression are identical and $A + A = A$, delete the sixth term. The final expression is

$$f = \overline{A}BC + A\overline{B}C + A\overline{B}\overline{C} + ABC + AB\overline{C} + \overline{A}B\overline{C}$$

As the previous Calculation Procedure illustrates, every nonstandard AND-to-OR expression can be expanded to standard form. From the principle of duality, it follows that every non-standard OR-to-AND expression can also be expanded to standard form.

ALGEBRAIC METHOD OF REDUCTION

Applying purely algebraic operations, reduce the following expression to its minimal form:

$$f = A\overline{B}\overline{C}D + \overline{A}B\overline{C}D + \overline{A}\overline{B}CD + ABC\overline{D} + \overline{A}\overline{B}\overline{C}D + A\overline{B}\overline{C}\overline{D} + \overline{A}BCD \qquad (a)$$

Calculation Procedure:

1. *Perform cycle 1 in the consolidation of terms*

We shall follow the formalized algebraic procedure for reducing a standard AND-to-OR expression to optimal form that was developed by W. V. Quine. The given expression is reduced by combining compatible terms, proceeding in cycles until all possibilities have been exhausted.

Number the terms in (a) from left to right, and record them in columnar form in Table 13. Starting with term 1, compare it with each subsequent term for compatibility. Term 1 is compatible with both term 3 and term 6. Identify each compatible pair of terms in Table 13 by placing the mark X in a unique column and on the same row as the term. Thus, the first column has an X on rows 1 and 3, and the second column has an X on rows 1 and 6. Now take term 2; it is compatible with both term 3 and term 7. Thus, the third column has an X on rows 2 and 3, and the fourth column has an X on rows 2 and 7. Similarly, term 3 is compatible with term 5, and term 5 is compatible with term 6. Term 4 is a prime implicant, and it is carried into the subsequent expression for f.

In Table 14, record the numbers of the terms that are compatible and the combined term they form. For example, terms 1 and 3 form $\overline{B}\overline{C}D$, and terms 1 and 6 form $A\overline{B}\overline{C}$. At the end

TABLE 13 Identity of Compatible Pairs of Terms

	Cycle 1								Cycle 2		
1 $A\overline{B}\overline{C}D$	X	X					1 $\overline{B}CD$	X	
2 $A\overline{B}CD$			X	X			2 $A\overline{B}\overline{C}$		X
3 $AB\overline{C}D$	X		X		X		3 $\overline{A}CD$		
4 $ABC\overline{D}$							4 $\overline{A}BD$		
5 $\overline{A}BCD$					X	X	5 $\overline{A}B\overline{C}$		X
6 $\overline{A}B\overline{C}D$		X				X	6 BCD	X	
7 $ABCD$				X					

of cycle 1, the expression for f has been reduced to

$$f = ABC\overline{D} + \overline{B}CD + A\overline{B}\overline{C} + \overline{A}CD + \overline{A}BD + \overline{A}B\overline{C} + BCD \tag{b}$$

However, the expression in (b) also contains compatible terms; therefore, a second cycle of consolidation is needed.

2. *Perform cycle 2 in the consolidation of terms*

Number all terms in (b) beyond the first and record them in Table 13. Again investigate for compatibility. Thus, term 1 is compatible with term 6, term 2 is compatible with term 5, and terms 3 and 4 are prime implicants. Complete Table 14. This table shows that both compatible

TABLE 14

Cycle	Compatible terms	Combined term
1	1 and 3	$\overline{B}CD$
	1 and 6	$A\overline{B}\overline{C}$
	2 and 3	$\overline{A}CD$
	2 and 7	$\overline{A}BD$
	3 and 5	$\overline{A}B\overline{C}$
	5 and 6	BCD
2	1 and 6	$\overline{B}\overline{C}$
	2 and 5	$\overline{B}\overline{C}$

pairs form $\overline{B}\overline{C}$. Since $A + A = A$, take $\overline{B}\overline{C}$ only once. At the end of cycle 2, the expression for f has been reduced to the following:

$$f = ABC\overline{D} + \overline{A}CD + \overline{A}BD + \overline{B}\overline{C} \tag{c}$$

The terms in (c) are incompatible with one another, and the consolidation process is now complete.

3. *Examine the expression in (c) for redundancies and eliminate any that may exist*

Ascertain how the terms in Eq. *a* are included in those in Eq. *c*. In Table 15, again record the terms in Eq. *a* in columnar form at the left, and record the terms in Eq. *c* in a horizontal row across the top. Take each term in Eq. *c* and compare it with each term in Eq. *a* with respect to inclusion. Thus, $ABC\overline{D}$ includes only term 4. Indicate this condition by placing the mark X at the indicated location. Also, $\overline{A}CD$ includes terms 2 and 3, $\overline{A}BD$ includes terms 2 and 7, etc.

Now proceed to construct the optimal expression for f. Each term in the original expression must be included at least once in the optimal expression. Therefore, the optimal expression

TABLE 15

Terms in Eq. a	$ABC\overline{D}$	$\overline{A}CD$	$\overline{A}BD$	$\overline{BC}$
		Terms in Eq. c		
1 $A\overline{B}C\overline{D}$				X
2 $A\overline{B}CD$		X	X	
3 $ABCD$		X		X
4 $ABC\overline{D}$	X			
5 $\overline{A}\overline{B}CD$				X
6 $\overline{A}BCD$				X
7 $\overline{A}BCD$			X	

must contain $ABCD$ because that is the only term in Eq. c that includes term 4 in Eq. a. Similarly, the optimal expression must contain $\overline{BC}$ because that is the only term in Eq. c that includes terms 1, 5, and 6 in Eq. a. Finally, the optimal expression must contain $\overline{A}BD$ because that is the only term in Eq. c that contains term 7 in Eq. a. Term $\overline{A}CD$ is redundant because terms 2 and 3 are already embodied in the optimal expression. Therefore, the optimal expression is

$$f = ABC\overline{D} + \overline{BC} + \overline{A}BD$$

(This expression will be obtained in a subsequent Calculation Procedure by a graphical method.)

4.   *Factor the optimal expression to obtain the minimal expression*

The result is

$$f = B(AC\overline{D} + \overline{A}D) + \overline{BC}$$

CONSTRUCTION OF KARNAUGH MAP FOR THREE-VARIABLE EXPRESSION

Construct the Karnaugh map for the expression

$$f = ABC + A\overline{BC} + \overline{A}BC$$

Calculation Procedure:

1.   *Establish the size of the array and assign a cell to each possible term*

A *Karnaugh map* (or *truth map*) affords a graphical method of reducing a given boolean AND-to-OR expression to optimal form through the combination of compatible terms. The map is a rectangular array of squares or *cells*. Each cell corresponds to a specific standard AND term, and a cell is provided for each possible term. Therefore, if n denotes the number of independent variables, the number of cells required is 2^n. In the present case, $n = 3$, and the number of cells is 8.

The basic requirement of a Karnaugh map is this: As we move from one row or column to an adjacent row or column, only one variable can change. In the Karnaugh map, the horizontal rows are numbered from the top down, and the vertical columns are numbered from left to right, starting with the number 1 in both cases. A cell is identified by specifying its row and column, in that order. For example, cell 25 (read "two five") lies in row 2 and column 5.

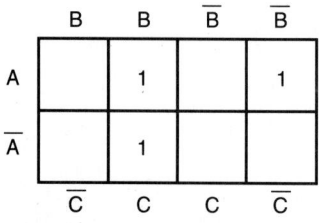

FIG. 45   Karnaugh map for three-variable expression.

Refer to Fig. 45, which is an array consisting of 2 rows and 4 columns. Each variable, both independent and dependent, must be represented by at least 1 row or column. Make the following assignments: row 1 to A; row 2 to $\overline{A}$; columns 1 and 2 to B; columns 3 and 4 to $\overline{B}$; columns 2 and 3 to C; columns 1 and 4 to $\overline{C}$. Record these assignments by placing the appropriate labels along the periphery of the array, as shown. Visualize that the map is inscribed on a torus rather than a plane, the result being that it is circular in both directions. Thus, columns 1 and 4 are adjacent to each other.

The term that corresponds to a given cell is found by taking the row and column designations of that cell. For example, in Fig. 45, cell 11 corresponds to $AB\overline{C}$, cell 21 corresponds to $\overline{AB}\,\overline{C}$, and cell 23 corresponds to $\overline{A}BC$. Moreover, the manner in which rows and columns have been assigned satisfies the basic requirement of a Karnaugh map.

In a Karnaugh map, adjacent cells correspond to compatible terms. For example, cells 12 and 22 correspond, respectively, to ABC and $\overline{A}BC$, and these terms are compatible. Similarly, cells 21 and 24 (which are adjacent) correspond, respectively, to $\overline{AB}\,\overline{C}$ and $\overline{A}B\overline{C}$, and these terms are compatible.

2. Complete the Karnaugh map by inserting 1's

The terms in the expression $ABC + A\overline{BC} + \overline{A}BC$ are represented by cells 12, 14, and 22 in Fig. 45. Therefore, to record the expression in the map, place the numeral 1 in each of these cells. The cells marked 1 are referred to as the p cells.

CONSTRUCTION OF KARNAUGH MAP FOR FOUR-VARIABLE EXPRESSION

Construct the Karnaugh map for the expression

$$f = \overline{ABC}\overline{D} + \overline{AB}C\overline{D} + ABCD + A\overline{BC}D + \overline{A}BCD$$

Calculation Procedure:

1. Establish the size of the array and assign a cell to each possible term

Since there are now 4 independent variables, the number of cells is $2^4 = 16$. Refer to Fig. 46. Again, visualize the map to be circular in both directions. As a result, rows 1 and 4 are adjacent, and columns 1 and 4 are adjacent. Assign the rows and columns in the manner shown.

2. Complete the Karnaugh map by inserting 1's

The terms in the given expression are represented by cells 41, 34, 22, 14, and 43. Therefore, to record the given expression in the map, place the numeral 1 in each of these cells.

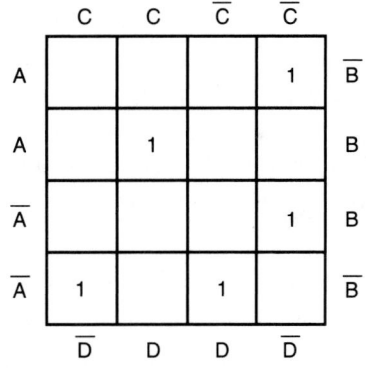

FIG. 46 Karnaugh map for four-variable expression.

CONSTRUCTION OF KARNAUGH MAP FOR FIVE-VARIABLE EXPRESSION

Construct the Karnaugh map for the expression

$$f = \overline{ABCD}\overline{E} + A\overline{BC}DE + A\overline{BCDE}$$

Calculation Procedure:

1. Establish the size of the array and assign a cell to each possible term

Since there are 5 independent variables, the number of cells is $2^5 = 32$. Refer to Fig. 47, where the array consists of 8 rows and 4 columns. Since the map is considered to be circular in both directions, rows 1 and 8 are adjacent, and columns 1 and 4 are adjacent. Assign the rows and

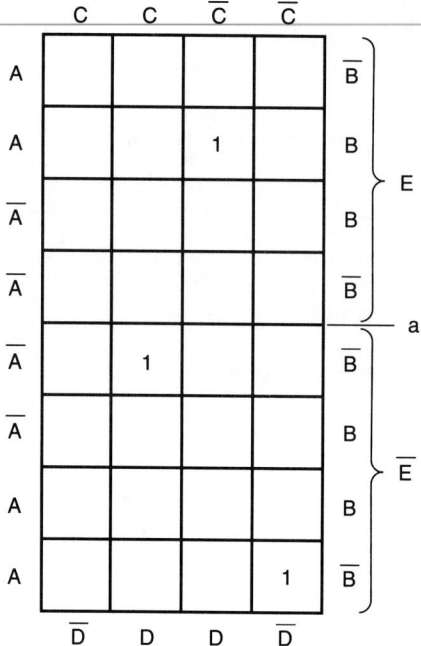

FIG. 47   Karnaugh map for five-variable expression.

columns in the manner shown, and visualize that the subarray for $\overline{E}$ is obtained by revolving the subarray for E about line a. The basic requirement is satisfied. For example, when we move from row 4 to row 5, the sole change is from E to $\overline{E}$. Similarly, when we move from row 8 to row 1, the sole change is from $\overline{E}$ to E.

2. Complete the Karnaugh map by inserting 1's

The terms in the given expression are represented by cells 52, 23, and 84. Therefore, place the numeral 1 in each of these cells.

REDUCTION OF EXPRESSION TO OPTIMAL FORM BY USE OF KARNAUGH MAP

Reduce the following expression to optimal form by constructing a Karnaugh map:

$$f = ABC\overline{D} + ABCD + \overline{A}BC\overline{D} + \overline{A}BCD$$

Then verify the result by reducing the given expression algebraically.

Calculation Procedure:

1. Construct the Karnaugh map

Assume that a Karnaugh map contains 2^m adjacent p cells, where m is a positive integer. These cells are said to form an *mth-order block*, and the extent of the block is shown by enclosing it with light lines. As we shall demonstrate, the terms that correspond to the cells in the block can be combined into a single term that contains solely the variables within which the block is confined.

In general, if the given expression contains n independent variables and the block is of the mth order, the combined term corresponding to this block contains $n - m$ variables. Therefore, in forming blocks, the guiding principle is to make the blocks as large as possible. It is convenient to view an isolated p cell as a zero-order block, the justification being that $2^0 = 1$.

The given expression is mapped in Fig. 48.

2. Establish the optimal expression

In Fig. 48, the p cells form a second-order block, and this block is confined within the rows for B and the columns for C. Therefore, the given expression reduces to $f = BC$. This condensed equation signifies that the values assumed by A and D are irrelevant.

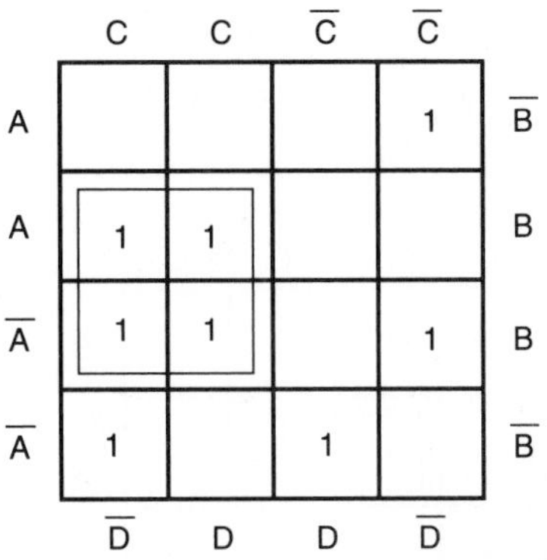

3. Verify the result

To reduce the given expression algebraically, factor recurrently and then apply Eq. 12, in this manner: Since each term in the given expression contains B and C, factor these variables in the first step. Then

$$f = BC(A\overline{D} + AD + \overline{A}\,\overline{D} + \overline{A}D)$$
$$= BC[A(\overline{D} + D) + \overline{A}(\overline{D} + D)]$$
$$= BC(A + \overline{A}) = BC$$

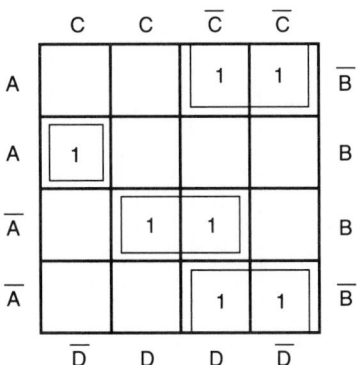

FIG. 48 Karnaugh map with second-order block.

USE OF KARNAUGH MAP WITH DISTINCTIVE BLOCKS

Reduce the following expression to optimal form by constructing a Karnaugh map:

$$f = A\overline{B}\overline{C}D + \overline{A}B\overline{C}D + \overline{A}\,\overline{B}CD + ABC\overline{D} + \overline{A}\overline{B}CD + A\overline{B}C\overline{D} + \overline{A}BCD$$

(This expression was reduced algebraically in an earlier Calculation Procedure.)

Calculation Procedure:

1. Construct the Karnaugh map

The given expression is mapped in Fig. 49.

2. Establish the optimal expression

Since rows 1 and 4 are adjacent, cells 13, 14, 43, and 44 form a second-order block, and their terms combine to form $\overline{BC}$. Cells 32 and 33 form a first-order block, and their terms combine to form $\overline{A}BD$. Cell 21 is a zero-order block, and its term is $AB\overline{C}D$. At this point, all p cells have been taken into account, and the optimal expression is complete. That cells 33 and 43 constitute a block is not relevant because both these cells have already been taken into account. Therefore, carrying the term $A\overline{C}D$ into the reduced expression would be redundant. Thus, the optimal expression is

FIG. 49 Map with distinctive blocks.

$$f = \overline{BC} + \overline{A}BD + AB\overline{C}D$$

This result agrees with that obtained algebraically in the earlier Calculation Procedure.

USE OF KARNAUGH MAP WITH OVERLAPPING BLOCKS

A function f has the expression that is mapped in Fig. 50. Establish the optimal form of the expression.

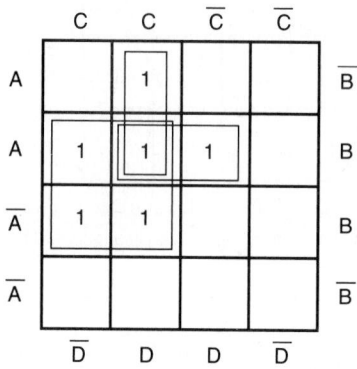

FIG. 50   Map with overlapping blocks.

Calculation Procedure:

1. Form blocks of the largest possible size

In forming blocks, it is permissible to include a given cell in more than one block. As an illustration, consider the expression

$$f = \overline{AB}\,\overline{C} + \overline{A}\overline{B}C + A\overline{B}C$$

The first and second terms combine to form $\overline{A}\overline{B}$, and the second and third terms combine to form $\overline{B}C$. Then

$$f = \overline{A}\overline{B} + \overline{B}C$$

Thus, $f = 1$ if either of these conditions exist: $A = 0$, $B = 0$, and C can have either value; $B = 0$, $C = 1$, and A can have either value.

In the present case, form the second-order block and the two first-order blocks shown.

2. Formulate the optimal expression

The second-order block corresponds to BC, and the first-order blocks correspond to ACD and ABD. Therefore, the optimal expression is

$$f = BC + ACD + ABD$$

USE OF LARGE BLOCKS IN A KARNAUGH MAP

A function f has the expression that is mapped in Fig. 51. Establish the minimal form of the expression.

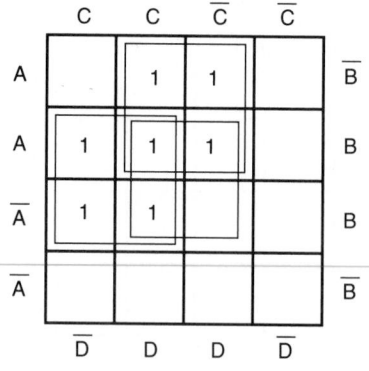

FIG. 51   Map with overlapping second-order blocks.

Calculation Procedure:

1. Form blocks in the most effective manner

It is possible to combine the p cells to form one second-order block and two first-order blocks. The resulting expression would have $2 + 3 + 3 = 8$ literals. On the other hand, by using overlapping blocks, it is possible to form second-order blocks exclusively, in the manner shown. The resulting expression would have $2 \times 3 = 6$ literals. Therefore, choose the second method.

2. Formulate the optimal expression

The result is

$$f = AD + BC + BD$$

3. Formulate the minimal expression by factoring

We can factor either B or D, and the following alternative forms result:

$$f = AD + B(C + D) \qquad f = D(A + B) + BC$$

USE OF KARNAUGH MAP WITH AN INCOMPLETE BLOCK

A function f has the expression that is mapped in Fig. 52. Establish the minimal form of the expression.

Calculation Procedure:

1. Form the second-order and first-order blocks shown, and obtain the optimal expression

The expression is

$$f = CD + \overline{BCD}$$

2. Obtain the minimal expression

Factor the foregoing expression and then apply Eq. 19. The result is

$$f = D(C + \overline{BC}) = D(\overline{B} + C)$$

The last expression is the minimal one.

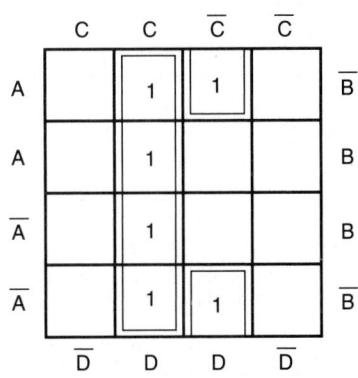

FIG. 52   Map with an incomplete third-order block.

3. Alternatively, obtain the minimal expression by using an incomplete block

Tentatively transform cells 23 and 33 to p cells, thereby forming a third-order block having D as its expression. Cells 23 and 33 form a first-order block having $B\overline{C}D$ as its term. Therefore, impose the additional requirement that $B\overline{C}D$ be 0. Then

$$f = D(\overline{B\overline{C}D})$$

Now apply Eqs. 15 and 13 to obtain

$$f = D(\overline{B} + C + \overline{D}) = D(\overline{B} + C)$$

Thus, use of an incomplete block yields the minimal expression in a more direct manner.

USE OF IRRELEVANT (DON'T CARE) TERMS

The variables A, B, and C determine a function f. Table 16 exhibits the alternative sets of values that make $f = 1$ and those that make $f = 0$. The two sets of values not shown in the table have no effect on the value of f. Formulate the minimal expression for f (a) without using irrelevant terms and (b) by using these terms.

Calculation Procedure:

1. Write the expression for f as given by Table 16

The table states that $f = 1$ if any of these conditions exist: $A B \overline{C} = 1$, $A \overline{B} C = 1$, $\overline{A} B \overline{C} = 1$. Therefore,

$$f = A B \overline{C} + A \overline{B} C + \overline{A} B \overline{C}$$

TABLE 16

Condition	A	B	C
$f = 1$	1	1	0
	1	0	1
	0	1	0
$f = 0$	1	1	1
	1	0	0
	0	1	1

2. Construct the Karnaugh map and place the appropriate mark in each cell

Figure 53 is the Karnaugh map. As before, place the numeral 1 in the cells corresponding to the terms that make $f = 1$. In addition, place the numeral 0 in the cells corresponding to the

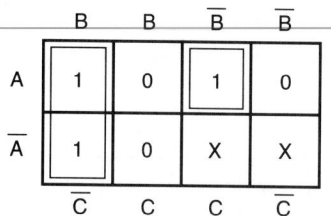

FIG. 53 Method of identifying irrelevant terms on a map.

terms that make $f = 0$. Now place an X in the two remaining cells. Where certain combinations of values cannot occur in practice or have no effect on the value of the function, the corresponding terms are called *irrelevant* or *don't-care terms*. Thus, the X's in the Karnaugh map identify the irrelevant terms.

3. Formulate the optimal expression by combining solely the p cells

Combine the p cells in the manner shown to obtain

$$f = A\overline{B}C + B\overline{C}$$

4. Now formulate the optimal expression by incorporating the irrelevant terms

Where irrelevant terms are present, it is permissible to enlarge the original expression for the function to include these terms if doing so yields a briefer final expression for the function.

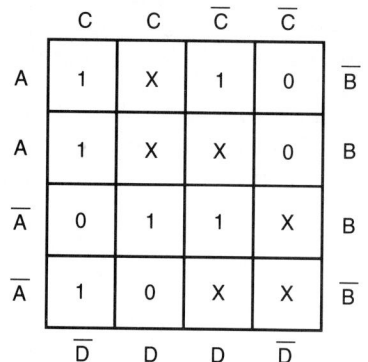

FIG. 54 Formation of blocks containing irrelevant terms.

Since irrelevant terms cannot become 1 or their values are of no consequence, their inclusion in the expression does not inject any error. In Fig. 53, combine cell 23 with cell 13. We now have two first-order cells, and the resulting expression is

$$f = \overline{B}C + B\overline{C}$$

This expression is preferable to that in step 3 because it contains four literals instead of five.

ALTERNATIVE OPTIMAL EXPRESSIONS RESULTING FROM USE OF IRRELEVANT TERMS

A function f has the Karnaugh map in Fig. 54. Formulate the optimal expression for f.

Calculation Procedure:

1. Combine the terms in a suitable manner and write the corresponding expression

Use the irrelevant terms to obtain the blocks shown in Table 17. The corresponding expression is

TABLE 17

Cells in block	Term
11, 12, 21, 22	AC
12, 13, 22, 23	AD
22, 23, 32, 33	BD
11, 41	$\overline{BCD}$

$$f = AC + AD + BD + \overline{B}C\overline{D}$$

2. Combine the terms in an alternative manner and write the corresponding expression

Now combine cell 41 with cell 44 rather than cell 11. The expression now becomes

$$f = AC + AD + BD + \overline{A}B\overline{D}$$

As the foregoing Calculation Procedure demonstrates, there may be alternative ways of forming blocks in the Karnaugh map where irrelevant terms are present. If such is the case, the problem of formulating the optimal expression for the function is ambiguous.

SECTION 8

AERONAUTICAL AND ASTRONAUTICAL ENGINEERING

HAROLD BECHER
PRESIDENT
STRATO MISSILES, INC.

JOHN P. ROEDEL
RESEARCH ENGINEER
THE BOEING COMPANY

TYLER G. HICKS, P.E.
INTERNATIONAL ENGINEERING ASSOCIATES

Aircraft Stall or Landing Speed 8.2
Aircraft True Airspeed, Drift, and Groundspeed 8.3
Aircraft Rate of Climb, Climbing Time, Speed, and Range 8.4
Aircraft Engine Thrust Determination 8.5
Air-Cushion Vehicle Proportions and Power Requirements 8.6
Power Required for Vertical-Takeoff Aircraft 8.9
Commercial Aircraft Operating-Cost Analysis 8.10
Lifting Power and Maximum Altitude of Balloons 8.12
Rocket Flight Velocity 8.12
Missile Maximum Range and Launch Angle 8.13
Satellite Flight Velocity, Escape Velocity, and Period 8.14
Interplanetary Flight Launch Velocity and Flight Time 8.14
Space Vehicle Burnout Velocity and Fuel Selection 8.16
Observation-Satellite Detail Detection 8.18
Aircraft Cargo-Carrying Capacity Analysis 8.20
Ramjet Engine Diffuser Analysis 8.21
Jet Plane Speed Estimate 8.23
Supersonic Wind Tunnel Back Pressure 8.24
Rocket Combustion Chamber Nozzle Throat Area 8.25
Mach 2.5 Air-Stream Conditions 8.26

REFERENCES: Bertin and Smith—*Aerodynamics for Engineers*, Prentice-Hall; Schlichting and Trucken-brodt—*Aerodynamics of the Airplane*, McGraw-Hill; Clancy—*Aerodynamics*, Halsted Press; Kuchemann—*The Aerodynamic Design of Aircraft in SI/Metric Units*, Pergamon; Carafoli—*Wing Theory in Supersonic Flow*, Pergamon; Collins—*Takeoffs and Landings*, Delacorte; Cameron—*Ballooning Handbook*, Merrimack Book; Carroll—*The Aerodynamics of Powered Flight*, Wiley; Shields—*Air Pilot Training*, McGraw-Hill; Elsley and Devereux—*Hovercraft Design and Construction*, David & Charles (London); Koelle—*Handbook of Astronautical Engineering*, McGraw-Hill; Corliss—*Propulsion Systems for Space Flight*, McGraw-Hill; Millikan—*Aerodynamics of the Airplane*, Wiley; *Space Handbook: Astronautics and Its Applications*, GPO; Allen—*Astrophysical Quantities*, Athlone Press (London); Carter—*Realities of Space Travel*, McGraw-Hill; Clarke—*Interplanetary Flight*, Temple Press (London); Puckett and Ramo—*Guided Missile Engineering*, McGraw-Hill; Sutton—*Rocket Propulsion Elements*, Wiley; von Braun—*The Mars Project*, University of Illinois Press; Herrick—*Astrodynamics*, Van Nostrand; Moulton—*Celestial Mechanics*, Macmillan; Van Allen—*Scientific Uses of Earth Satellites*, University of Michigan Press; Stephenson—*Introduction to Nuclear Engineering*, McGraw-Hill; Cowling—*Magnetohydrodynamics*, Interscience; Blasingame—*Astronautics*, McGraw-Hill; Berkner and Odishaw—*Science in Space*, McGraw-Hill; Nokilayev—*Thermodynamic Assessment of Rocket Engines*, Pergamon; Boehm—*ROCKET: Rand's Omnibus Calculator of the Kinematics of Earth Trajectories*, Prentice-Hall; Bell—*Cryogenic Engineering*, Prentice-Hall; Pogorelov—*Fundamentals of Orbital Mechanics*, Holden-Day; Casamassa and Bent—*Jet Aircraft Power Systems*, McGraw-Hill; McClintock—*Cryogenics*, Reinhold; Henshaw—*Supersonic Engineering*, Wiley; Wolverton—*Flight Performance Handbook for Orbital Operations*, Wiley; Loomis—*High-Speed Commercial Flight: The Coming Era*, Battelle Press; Ashley and Landahl—*Aerodynamics of Wings and Bodies*, Dover; Rae and Pope—*Low-Speed Wind-Tunnel Testing*, Wiley.

AIRCRAFT STALL OR LANDING SPEED

What is the sea-level stall or landing speed of a 9300-lb (4218-kg) airplane having a wing area of 361 ft^2 (33.5 m^2) if a Clark wing with a fixed slot is used? Determine these speeds for a Clark wing with a fixed slot and slotted flap. What is the stall angle of attack for each wing?

TABLE 1   Typical Aircraft Wing Characteristics

Wing type	C_{Lm}	$\alpha°$ at C_{Lm}
Clark, fixed-slot	1.77	24
Clark, fixed-slot, slotted-flap	2.26	19
Clark, most complex	3.37	16
NACA 4415	1.6	20

Calculation Procedure:

1.   *Determine the wing lift coefficient*

The maximum value of the wing lift coefficient C_{Lm} is used to compute the stall or landing speed of an aircraft. Values of C_{Lm} are published in various handbooks and other references. Table 1 lists a few typical wings and the value of C_{Lm} for each wing. Thus, for the basic Clark wing with a fixed slot, $C_{Lm} = 1.77$. With a fixed slot and slotted flap, $C_{Lm} = 2.26$.

2.   *Compute the wing stall or landing speed*

Use the relation $v_s = (2W/\rho A C_{Lm})^{0.5}$, where v_s = stall or landing speed, ft/s; W = aircraft weight, lb; ρ = density of air at sea level = 0.00233 slugs/ft^3 (1.2 kg/m^3); A = aircraft wing area, ft^2 (m^2). Substituting for the fixed-slot Clark wing gives $v_s = [2 \times 9300/(0.00233 \times 361 \times 1.77)]^{0.5} = 118$ ft/s (34.1 m/s), or 76.2 mi/h (122.6 km/h). For the fixed-slot, slotted-flap wing, $v_s = [2 \times 9300/(0.00233 \times 361 \times 2.26)]^{0.5} = 98.9$ ft/s (30.1 m/s), or 67.4 mi/h (108.4 km/h). Thus, the greater the number of high-lift devices (slots, flaps, etc.) used on a wing, the lower the landing speed. A low landing speed is desirable because it reduces the required runway length, makes for easier control of the aircraft, and lowers the impact forces between the aircraft and the runway. The most complex Clark wing has a C_{Lm} of 3.37, giving a stall or landing speed of 55.2 mi/h (88.8 km/h).

3.   *Determine the stall or landing speed angle of attack*

Table 1 shows that the angle of attack at stall for the fixed-slot wing is 24° and for the fixed-slot, slotted-flap wing 19°. For the most complex Clark wing, the stall angle of attack is 16°. Thus, high-lift devices also reduce the stall or landing angle of attack. This is a desirable feature in a wing.

   Related Calculations: Use this general method to compute the minimum horizontal speed of any winged aircraft. Wing characteristics are tabulated in aeronautical engineering handbooks and wind-tunnel reports.

AIRCRAFT TRUE AIRSPEED, DRIFT, AND GROUNDSPEED

An aircraft is flying at 8000-ft (2438.3-m) altitude at an indicated airspeed of 300 mi/h (482.7 km/h). What is its true airspeed? If the aircraft is on a true heading of 300° and there is a 200° wind blowing at 100 mi/h (160.9 km/h), find the track of the aircraft and its groundspeed.

Calculation Procedure:

1. *Compute the aircraft true airspeed*

Use the relation $t_a = i_a + (0.02i_a h/1000)$, where t_a = aircraft true airspeed, mi/h; i_a = aircraft indicated airspeed, mi/h; h = flight altitude, ft. Substituting gives $t_a = 300 + (0.02 \times 300 \times 8000/1000) = 348$ mi/h (559.9 km/h). The constant of 0.02 represents the average altitude error of the usual airspeed indicator of 2 percent per 1000 ft (304.8 m) of altitude. If the indicator being used had a different error percentage, the actual error would be substituted in the relation above.

2. *Compute the aircraft groundspeed*

The aircraft groundspeed, i.e., its speed on the ground, will be greater or less than the true airspeed, depending on whether the wind, if any, at the flight altitude aids or retards the aircraft movement.

To determine the groundspeed t_g mi/h, draw a vector to scale representing the true heading and true airspeed of the aircraft, Fig. 1. Label this vector TH for true heading, and indicate on it the angular measure of the true heading, 300°, and the true airspeed, 348 mi/h (559.9 km/h).

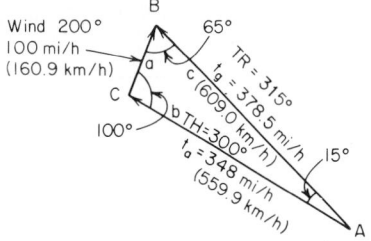

From the end of the true-heading vector lay off the wind vector at 200° with length equal to the wind velocity, or 100 mi/h (160.9 km/h). Connect the origin of the true-heading vector with the terminal end of the wind vector. The resulting vector is the track TR and ground-speed t_g vector of the aircraft.

Determine the groundspeed graphically by scaling the t_g vector, or solve for this value by computation. To solve for t_g by computation, determine the included angle between the TH

FIG. 1 Velocity diagram for determining the track and groundspeed of an aircraft.

and the wind vector by geometry. Then, by the law of cosines, $t_g^2 = c^2 = a^2 + b^2 - 2ab \cos C$, where the letters a, b, and c represent the sides of the triangle (Fig. 1). Or $c^2 = 100^2 + 348^2 - 2(100)(348)(\cos 100°)$; $c = t_g = 378.5$ mi/h (609.0 km/h). Thus, the wind increases the speed of the aircraft over the ground.

3. *Compute the aircraft track*

Solve for angle A in Fig. 1, using the relation $\tan A/2 = r/s - a$, where $r = [(s - a) \times (s - b)(s - c)/s]^{0.5}$, and $s = (a + b + c)/2$. Solving gives $A = 15°$. Hence, the aircraft track = 300 + 15° = 315°. This is the course made good over the ground.

Related Calculations: The flight condition met in this calculation procedure is termed *right drift*; i.e., the aircraft drifts to the right of its true heading because the wind is from the left. A general rule applicable to right drift is: *When the wind is from the left or the drift is to the right, the track vector will always be to the right of the heading vector and will have a larger directional (i.e., angular) value. Left drift* results from a right wind, and the track vector is always to the left of the heading vector and has a lower directional value.

Use the same general procedure to find the wind velocity, wind direction, true airspeed, or true heading. This same general procedure also can be used for determining the drift of a ship or boat resulting from the water current or surface wind, or both. The only difference is that the speed quantities involved are smaller than for aircraft. However, the difference in numerical quantities does not influence the graphical or computation procedure in any way.

AIRCRAFT RATE OF CLIMB, CLIMBING TIME, SPEED, AND RANGE

Figure 2 shows a simplified performance diagram for a 5000-lb (2268.0-kg) airplane. Using this chart, determine the rate of climb at sea level and at an 8000-ft (2438.4-m) altitude. What is the aircraft speed at sea level and at 8000 ft (2438.4 m)? How long will it take the aircraft to reach an altitude of 8000 ft (2438.4 m) after takeoff? What is the absolute ceiling of the aircraft? Determine the aircraft range if it consumes 50 gal/h of fuel at full throttle at 8000 ft (2438.4 m) and its total fuel capacity is 500 gal (1842.5 L).

Calculation Procedure:

1. Compute the sea-level rate of climb of the aircraft

Determine the largest difference between the power available at sea level and the power required at sea level, Fig. 2. Do this by studying the performance diagram and by using a scale to find the largest power difference at sea level on the performance diagram plot of the power available.

For any airplane, the power available for climbing $hp_{ac} = hp_a - hp_r$, where hp_a = maximum available engine-horsepower output at the altitude in question; hp_r = horsepower required to overcome the aircraft drag at the altitude in question. For this airplane the maximum value of hp_{ac} at sea level is 120 hp (89.5 kW). This value is found by inspection and scaling of Fig. 2.

Knowing hp_{ac} and the weight of the airplane, we compute the maximum rate of climb at sea level from $r_c = 33,000\ hp_{ac}W$, where r_c = rate of climb, ft/min, and W = aircraft weight, lb. So $r_c = 33,000(120)/5000 = 794$ ft/min (4.0 m/s).

2. Determine the maximum speed of the aircraft at sea level

Find the intersection of the power-available and power-required curves at sea level, Fig. 1. Read the velocity at this point as 320 ft/s (97.5 m/s), or $v = (320$ ft/s$)(3600$ s/h$)/(5280$ ft/mi$) = 218.2$ mi/h (351.1 km/h).

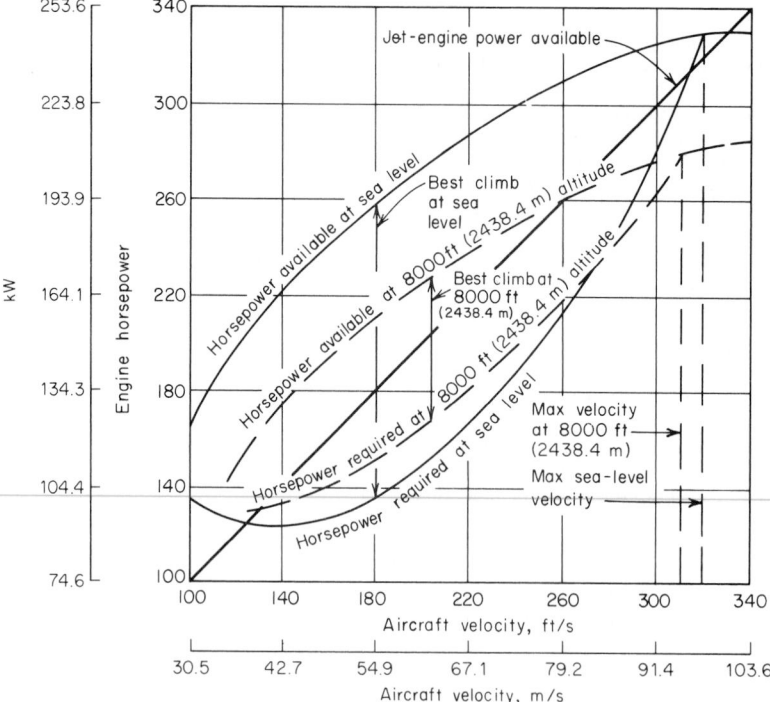

FIG. 2   Performance chart for a typical propeller aircraft. The power curve for a typical jet engine is also shown. The jet power curve is used in the same way as the propeller power curve.

3. *Determine the altitude rate of climb of the aircraft*

Use the same procedure as in step 1 to find the maximum difference between hp_a and hp_r, Fig. 2, at the altitude in question, 8000 ft (2438.4 m). Note that there are separate hp_a and hp_r curves for the 8000-ft (2438.4-m) altitude. These performance curves are different and distinct from the corresponding sea-level curves. Figure 2 shows that the maximum difference between hp_a and hp_r at 8000 ft (2438.4 m) is 60 hp (44.7 kW). Hence, the maximum rate of climb at 8000 ft (2438.4 m) is $r_c = 33,000(60)/5000 = 396$ ft/min (2.0 m/s), or about half the maximum rate of climb at sea level. The *service ceiling* of an airplane is the altitude at which the aircraft rate of climb is 100 ft/min (0.5 m/s).

4. *Determine the maximum aircraft speed at altitude*

Use the same procedure as in step 2, and read the maximum speed at 8000-ft (2438.4-m) altitude as 310 ft/s (94.5 m/s), or $(310)(3600)/5280 = 211.5$ mi/h (340.3 km/h), closely.

5. *Compute the absolute ceiling of the aircraft*

At the *absolute ceiling* of an aircraft the rate of climb is zero. To compute the absolute ceiling, use the relation $H = r_s h/(r_s - r)$, where H = absolute ceiling of aircraft, ft; r_s = rate of climb at sea level, ft/min; h = altitude for which another rate of climb is known, ft; r = rate of climb, ft/min, at altitude h. Using the sea-level and 8000-ft (2438.4-m) values computed above, we get $H = 794(8000)/(794 - 396) = 15,950$ ft (4861.1 m).

6. *Compute the time for the aircraft to climb to altitude*

Use the relation $t = 2.303(H/r_s) \log H/(H - h)$, where t = time to climb to altitude h, min; other symbols as before. So $t = 2.303(15,950/794)\log 15,950/(15,950 - 8000) = 14.0$ min.

7. *Compute the aircraft range*

Allow 20 percent of the fuel capacity for the takeoff, climbing, and landing maneuvers, including holding patterns before landing. Hence, fuel available for cruise $= 500 - 0.20(500) = 400$ gal (1514.0 L).

At 8000 ft (2438.4 m) the speed of the aircraft, from step 4, is 211.5 mi/h (340.3 km/h) with a fuel consumption of 50 gal/h (0.053 L/s). Use the relation $R = vC/f$, where R = aircraft range, mi; v = aircraft speed at the altitude in question, mi/h; C = fuel available for cruising, gal; f = fuel consumption at cruise altitude, gal/h. Substituting gives $R = 211.5(400/50) = 1692$ mi (2723.0 km).

The range computed here is the distance the aircraft could travel in still air, i.e., without head, tail, or off-course winds. In actual operation of the aircraft, the pilot would make allowance for winds in computing true airspeed and range.

Related Calculations: The performance curve shown in Fig. 2 is typical for a variety of aircraft. For each altitude at which the aircraft will fly, a power-available and power-required curve can be plotted. With these curves available, the maximum rate of climb and airspeed can be determined as described in steps 1 and 2 for each altitude plotted.

AIRCRAFT ENGINE THRUST DETERMINATION

An aircraft jet engine has an air inlet velocity v_i of 1000 ft/s (304.8 m/s) and a jet exit velocity v_e of 3000 ft/s (914.4 m/s). What thrust will this engine develop when the airflow rate through it is 300 lb/s (136.1 kg/s)? What is the equivalent hp of a reciprocating engine driving a propeller having an efficiency of 70 percent at a true airspeed of 400 mi/h (643.6 km/h)? What is the rate of climb with the reciprocating power plant if the airplane weighs 100,000 lb (45,359 kg) and the power required at sea level is 27,000 hp (20,133 kW)?

Calculation Procedure:

1. *Compute the jet-engine thrust*

Use the relation $T = W\Delta v/g$, where T = jet-engine thrust, lb; W = rate of airflow through the engine, lb/s; Δv = change in velocity of the air during passage through the engine, ft/s; g = 32.2 ft/s^2 (9.8 m/s^2). Substituting, $T = 300(3000 - 1000)/32.2 = 18,620$ lb (8445.4 kg).

2. *Compute the equivalent reciprocating engine hp*

Use the relation $hp_e = Tv/375e$, where hp_e = equivalent hp of the reciprocating engine; v = true airspeed at which the comparison is made; e = propeller efficiency. So $hp_e = 18,620(400)/$

[375(0.70)] = 28,600 hp (21,327 kW). This is much greater than the takeoff hp rating of any standard reciprocating aircraft engine. However, several engines whose hp sum equals or exceeds the required rating could be used if the weight conditions were acceptable.

3. Calculate the rate of climb of the aircraft

Use the relation $r_c = 33,000\ hp_{ac}/W$, where r_c = rate of climb of the airplane, ft/min; hp_{ac} = available power for climbing = maximum rated hp − hp required, both at sea level; W = weight of loaded aircraft, lb. Thus $r_c = 33,000(28,600 − 27,000)/100,000 = 528$ ft/min (2.7 m/s).

Related Calculations: Use this general procedure for preliminary analysis of any jet engine, whose overall characteristics are known. The equivalent hp relation, step 2, is useful in comparing piston-engine and jet-engine aircraft.

AIR-CUSHION VEHICLE PROPORTIONS AND POWER REQUIREMENTS

Select the required power, fan airflow, and natural frequency of a plenum-type air-cushion vehicle having a total weight, including the payload, of 5000 lb (2268.0 kg). The vehicle must have an emergency load capacity for extra passengers or dynamic loads of 1000 lb (453.6 kg). The available area for the plenum is 8 × 16 ft (2.4 × 4.9 m), and the floating height is to be 2 in (5.1 cm) above the surface.

Calculation Procedure:

1. Compute the vehicle force increment

The vehicle force increment F is defined as the ratio SF/F_e, where SF = emergency load capacity for extra passengers or dynamic loads, lb; F_e = vehicle total weight including payload, lb. Thus, $F = SF/F_e = 1000/5000 = 0.2$.

2. Compute the cushion gage pressure

Use the relation $p_c = F_e/A_p$, where p_c = cushion gage pressure, lb/ft²; A_p = plenum area, ft²; other symbols as before. Thus, $p_c = 5000/128 = 39$ lb/ft² (gage) (268.9 kPa).

3. Compute the sealing perimeter

The sealing perimeter L is the distance around the plenum exterior, or $L = 8 + 8 + 16 + 16 = 48$ ft (14.6 m).

4. Compute the plenum sealing area

Use the relation $A_s = Lh_e$, where A_s = perimeter sealing area, ft²; h_e = floating height, ft; other symbols as before. Thus, $A_s = 48(2/12) = 8$ ft² (0.7 m²).

5. Compute the power required to drive the vehicle

Enter Fig. 3 at the force increment value $F = 0.2$ from step 1, and project vertically upward to the power curve. From the intersection with the power curve, project horizontally to the right to read the value $P/[A_s(p_c)^{1.5}] = 0.042$, where P = required power input, hp. So $P = 0.042\,[8(39)^{1.5}] = 81.2$ hp (60.6 kW).

6. Compute the required airflow rate

Use the relation $W = 1.4A_s p_c^{0.5}$, where W = required airflow rate to support the vehicle, lb/s. Substituting yields $W = 1.4(8)(39)^{0.5} = 70$ lb/s (31.8 kg/s).

7. Determine the vehicle natural frequency

Enter Fig. 3 at $f = 0.2$ and project vertically upward to the top scale. From here project through $h_e = 2$ in (5.1 cm) to the natural frequency $f = 1.3$ Hz.

Related Calculations: Use Fig. 4 to analyze peripheral-jet-type air-cushion vehicles. The power equation for the peripheral-jet-type vehicle is P = chart factor $[LH(p_c)^{1.5}]$, and the airflow-rate equation is $W = PW/Lh$, where the value of W/Lh is obtained from the extreme left-hand scale of Fig. 4. The peripheral-jet air-cushion vehicle is more efficient than the simpler and more reliable plenum type.

For a peripheral-jet air-cushion vehicle with $F_e = 40,000$ lb (177,929 kN), $SF = 18,000$ lb (80,068 kN), $h = 2$ in (5.1 cm), length = 50 ft (15.2 m), width = 10 ft (3.0 m), and 80 percent of the base area available for suspension $SF/F_e = 18,000/40,000 = 0.45$. The available suspen-

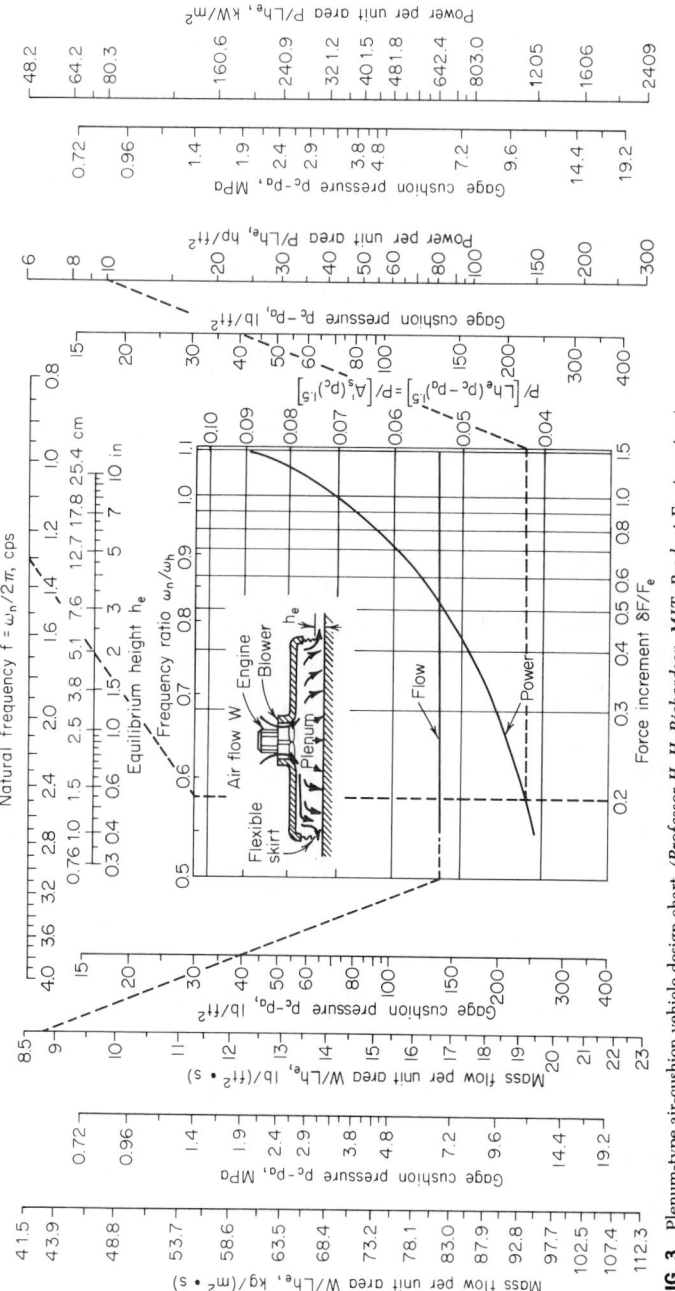

FIG. 3 Plenum-type air-cushion vehicle design chart. (*Professor H. H. Richardson, MIT; Product Engineering.*)

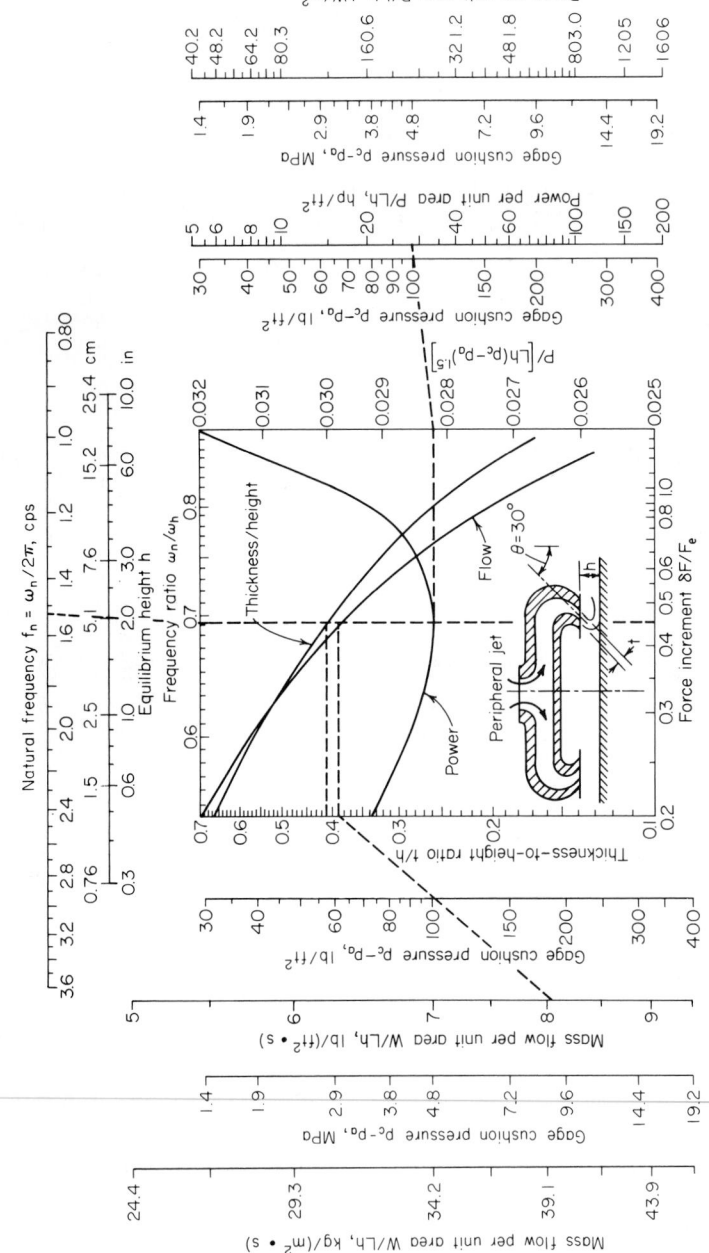

FIG. 4 Peripheral-jet type of air-cushion vehicle design chart. *(Professor H. H. Richardson, MIT; Product Engineering.)*

8.8

sion area $A_a = 50(10)(0.80) = 400$ ft^2 (37.2 m^2), and the gage cushion pressure $= F_e/A_a = 40,000/400 = 100$ lb/ft^2 (41.3 N/m^2).

Assume that two suspension pads, each 10×20 ft (3.0×6.1 m), are used, giving a total suspension area of $2(10 \times 20) = 400$ ft^2 (37.2 m^2) $= A_a$. The perimeter length of each pad is $10 + 10 + 20 + 20 = 60$ ft (18.3 m); for two pads, the perimeter length $= 2(60) = 120$ ft (36.6 m). The sealing area $= A_s = Lh = 120(2/12) = 20$ ft^2 (1.9 m^2).

To determine the power required per square foot of sealing area and the total power required for the vehicle, enter the bottom of Fig. 4 at $SF/F_e = 0.45$ and project vertically to the power curve. From the intersection, project horizontally to the right-hand chart border (Fig. 4); then draw a straight line through $p_c - p_a = 100$ lb/ft^2 (488.2 kg/m^2) and extend this line to $P/Lh = 28$ hp/ft^2 (224.7 kW/m^2) of sealing area. Since the total sealing area $A_s = 20$ ft^2 (1.9 m^2), $P = 28(20) = 560$ hp (417.6 kW), the total power required for the vehicle. Note that the chart factor in the power equation given earlier is plotted on the right-hand border of the chart (Fig. 4).

To determine the air-mass flow required for the blower, enter Fig. 4 as before and project vertically to the flow curve. From the intersection project horizontally to the left-hand chart border, then through $p_c - p_a = 100$ lb/ft^2 (488.2 kg/m^2) to $W/Lh = 8.05$ lb/(ft^2·s) [39.3 kg/(m^2·s)]. Since $Lh = 20$ ft^2 (1.9 m^2), $W = 8.05(20) = 161$ lb/s (73.0 kg/s).

To determine the jet thickness/height ratio $= t/h$, enter Fig. 4 as before and project to the thickness/height curve. From the intersection, project to the left-hand chart border to read $t/h = 0.41$. Since $h = 2$ in (5.1 cm), $t = 2(0.41) = 0.82$ in (2.1 cm) $=$ jet thickness. To determine the vehicle natural frequency, project from the $SF/F_e = 0.45$ starting point to the frequency ratio (Fig. 4), then through $h = 2.0$ in (5.1 cm) to the natural frequency $f_n = 1.52$ Hz. Where desired, Fig. 3 can be used in a similar manner instead of the equations given earlier.

The methods given here are valid for a variety of loadings on the types of vehicles considered.

POWER REQUIRED FOR VERTICAL-TAKEOFF AIRCRAFT

Which requires a larger power output to produce a vertical takeoff—a 30,000-lb (133.4-kN) helicopter whose blades develop an air downwash velocity of 75 ft/s (22.9 m/s) or a jet-propelled aircraft of the same weight whose jet engine produces an 800-ft/s (243.8-m/s) downwash velocity?

Calculation Procedure:

1. Compute the helicopter takeoff power

Use the relation $hp = Wv_d/550$, where $hp =$ takeoff hp (kW) of the aircraft; $W =$ weight of aircraft fully loaded and fully fueled, lb (N); $v_d =$ air downwash velocity, ft/s (m/s).

Substituting gives $hp = (30,000)(75)/550 = 4090$ hp (3049.9 kW). The installed hp would probably be 4200 or 4500 hp (3131.9 to 3355.6 kW) to provide for reduced efficiency of the rotor blades, gear mechanism, or engine.

2. Compute the jet-engine power requirement

Use the same relation as in step 1, or $hp = 30,000(800)/550 = 43,600$ hp (32,512.5 kW). Thus, the jet engine requires about 10 times the power that the helicopter requires.

Related Calculations: This method is also suitable for analyzing the landing hp required by either type of aircraft. Thus, the power that must be developed during landing initially is the same as for takeoff but is gradually reduced as the aircraft approaches the ground.

COMMERCIAL AIRCRAFT OPERATING-COST ANALYSIS

Compare the direct operating costs in dollars per airplane mile for ranges up to 6000 statute mi (9656.1 km) of three proposed commercial passenger aircraft: a jumbo jet, a tri-jet, and a stretched medium jet. Determine the tri-jet direct operating costs in cents per seat-mile for three cabin seating configurations: 256, 295, and 330 passengers. Plot the range-payload tradeoffs for the tri-jet. A normal flight range of 3000 statute mi (4828.0 km) is required for the contemplated service of the aircraft.

Calculation Procedure:

1. Summarize the direct operating costs for each aircraft

The *direct operating costs* of a commercial aircraft consist of the out-of-pocket costs required to operate the aircraft between two points a known distance apart. These costs are fuel, lubricants, crew wages, flight taxes and airport fees, direct maintenance, and aircraft depreciation. Figure 5a summarizes the typical direct operating costs for commercial aircraft of the jumbo jet, tri-jet, and stretched-jet types, as supplied by the manufacturers of these aircraft. Alternatively, the direct operating cost per mile can be computed by summing the direct costs listed above for each flight range—300, 1000, 2000, 3000, etc., statute mi (482.8, 1609.4, 3218.7, 4828.0 km).

2. Compute the seat-mile operating costs

With a 3000-mi (4828.0-km) operating range, the aircraft choice is between the jumbo jet and the tri-jet, because the maximum range of the stretched jet is only 2000 mi (3218.7 km), Fig. 5a. To compute the seat-mile direct operating cost s_m, use the relation $s_m = c_m/S$, where $c_m =$ direct operating cost, \$/mi; $S =$ number of seats in the aircraft. For the 256-seat configuration in the tri-jet at a 3000-mi (4828.0 km) range, $s_m = 2.1/256 = 0.82$ cent per seat-mile (0.51 cent per km), using the direct operating cost per mile plotted in Fig. 5a.

Compute the seat-mile cost for the tri-jet for flight ranges varying from the maximum of 3000 mi (4828.0 km) to the minimum of 250 mi (402.3 km). Plot the results as shown in Fig. 5b. The jumbo jet can be ignored because its per-mile operating cost is considerably higher than that of the tri-jet, Fig. 5a. However, if the jumbo jet had a seating capacity approximately twice that of the tri-jet, a seat-mile cost analysis would be worth making to determine the comparable cost. But the jumbo jet being considered here has the same seating capacity as the tri-jet. Hence, the cost *differences* shown in Fig. 5a on a per-mile basis would not change significantly when converted to a seat-mile basis.

3. Plot the range-payload tradeoffs for the tri-jet

A range-payload tradeoff correlates the distance a commercial aircraft can fly carrying a given payload and the distance flown. As the range of an aircraft is increased, its payload decreases because more fuel must be carried to provide the greater flight range.

Plot, or obtain from the aircraft manufacturer, the range-payload curve. Figure 5c shows the range-payload curve for the tri-jet aircraft being considered here. Note in Fig. 5c that this aircraft has a maximum design payload of 87,381 lb (388,690 N) when its range is 2000 nmi (3704.0 km). The upper knee of this range-payload curve, A in Fig. 5c, provides optimum operating costs for a carrier having a maximum route flight range of approximately 2000 nmi (3704.0 km). The lower knee, point B in Fig. 5c, provides optimum operating costs for a carrier having a maximum route flight range of approximately 3500 nmi (6428.0 km). Between these two ranges, a variety of payloads and flight ranges are obtainable, as Fig. 5c shows.

The tri-jet analyzed in Fig. 5c would be suitable for a U.S. domestic carrier with routes east of the Mississippi River if operated at point A with respect to payload and range. When operated at point B, Fig. 5c, this tri-jet would be suitable for intercontinental service, such as between the United States and Europe. The aircraft was actually chosen for these two services.

Related Calculations: The three curves shown above are typical for a variety of aircraft—large and small, pure jet, prop jet, and propeller. Hence, the procedure used here can be applied to corporate, private, air taxi, commercial, and similar aircraft carrying passengers for hire. Of course, many other factors enter the final choice of an aircraft. These factors include loading facilities, cabin arrangement, engine arrangement, type of tail, gross takeoff weight, takeoff distance, landing distance, noise level, etc. However, the economic factors analyzed here almost always lead the list of factors considered. Figure 5d and e shows two other versions of range-payload charts for modern jet aircraft.

Design studies are currently underway for a Very Large Commercial Transport (VLCT) capable of carrying some 600 passengers, compared to the 421 passengers carried by the Boeing 747-400 jumbo jet. There is an estimated need of 400 to 500 VLCTs by the world's airlines. An international partnership is believed to be the only way that such a large aircraft could be successfully produced. The first version of the VLCT would have a range of 7000 nm (11,263 km). Later versions might have a greater range and a larger passenger-carrying capacity. Operation of the VLCT from current international airports is thought to be feasible.

Recent internal environmental studies of passenger aircraft used for commercial purposes have focused on the quality of air in passenger cabins. Aircraft built before the mid-80s furnished 100 percent outside fresh air to the passenger cabins. This air was circulated through the cabin in sufficient volume to provide a complete air change every 3 min, or 20 air changes per hour.

Newer aircraft provide 50 percent outside air and 50 percent recirculated air with an air change every 6 min, or 10 air changes per hour. Using recirculated air produces a fuel saving

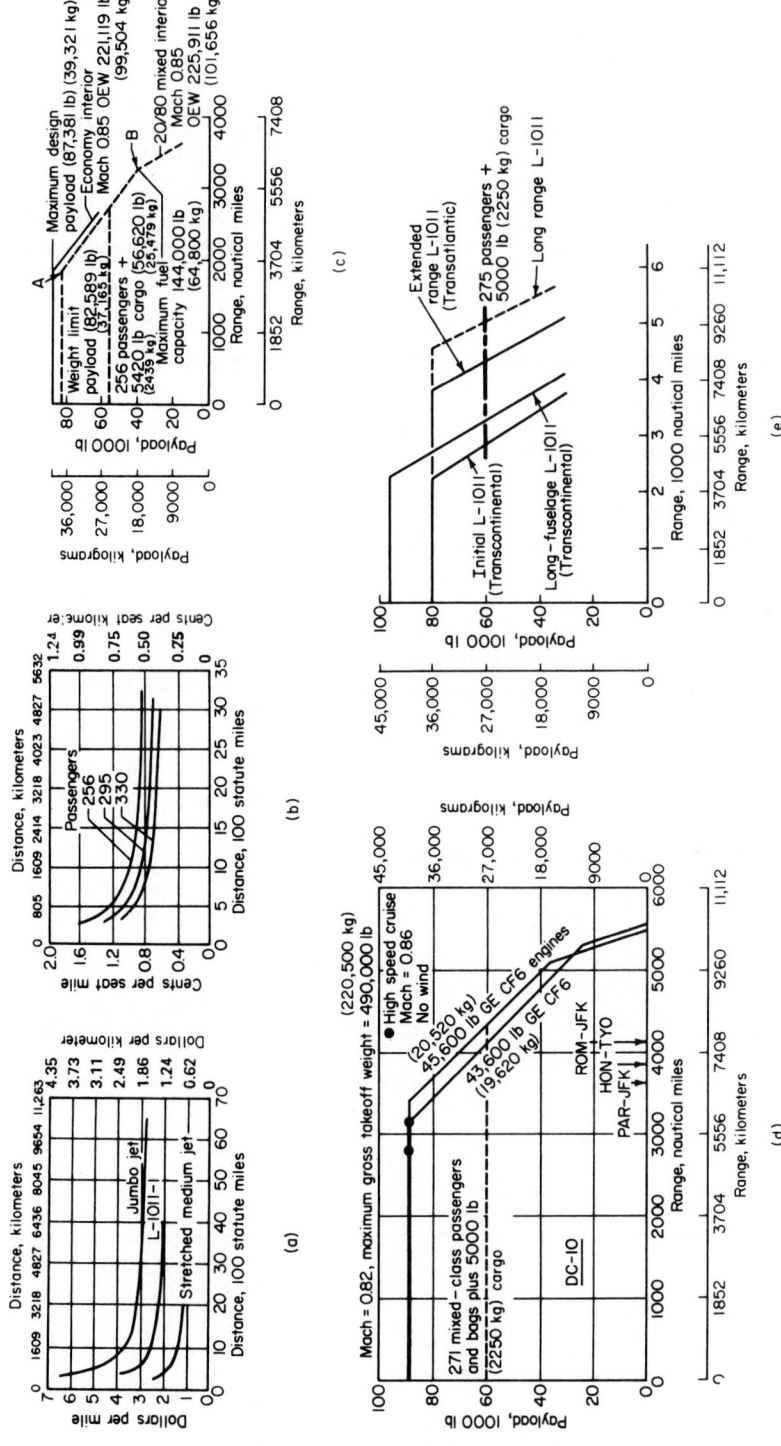

FIG. 5 (a) Aircraft direct operating cost, dollars per mile. (b) Aircraft direct operating cost, cents per seat-mile. (c) Range-payload tradeoffs for tri-jet aircraft. (d), (e) Range-payload charts for modern jet aircraft. (*Aviation Week.*)

8.11

of about $60,000 per aircraft per year. This is a significant saving to an airline that is operating hundreds of aircraft in its fleet.

Complaints of poor air quality and various symptoms—headaches, dried out nasal passages, dizziness, and nausea—by passengers and crew members has led to a questioning of ventilation system design. Inquiries are being conducted by the Transportation Department and the Centers for Disease Control and Prevention.

While minor discomfort is undesirable, a greater concern is the possibility of tuberculosis transmission between passengers during long flights. It is possible that air recirculation in the passenger cabin may spread bacteria throughout the cabin. The federal studies mentioned above seek to isolate the cause of disease transmission, if any, in aircraft cabins. If the air circulation system is responsible, design changes may be necessary.

LIFTING POWER AND MAXIMUM ALTITUDE OF BALLOONS

What is the gross and net lifting power of a 100-ft (30.5-m) diameter balloon containing pure helium if the balloon and its equipment weigh 1500 lb (680.4 kg)? How high will this balloon rise if the temperature of the air and lifting gas averages 32°F (0.0°C)?

Calculation Procedure:

1. *Compute the unit lifting power of the balloon*

Use the relation $u = w_a - w_g$, where u = unit lifting power of balloon, lb/ft^3; w_a = weight of air, lb/ft^3; w_g = weight of gas used, lb/ft^3. For this balloon, $u = 0.07658 - 0.01058 = 0.066$ lb/ft^3 (0.0041 kg/m^3). The weights of various gases used in balloons are listed in Table 2.

2. *Compute the gross lifting power of the balloon*

Use the relation $G = uV$, where G = gross lifting power of balloon, lb; V = volume of balloon, ft^3 = $\pi d^3/6$, where d = balloon diameter, ft. Substituting yields $G = 0.066(\pi \times 100^3/6)$ = 34,558 lb (15,551 kg).

3. *Compute the net lifting power of the balloon*

Use the relation $N = G - W$, where N = net lifting power of balloon, lb; W = weight of the balloon and its equipment. So $N = 34,558 - 1500 = 33,058$ lb (14,876 kg).

4. *Compute the height to which the balloon will rise*

Use the relation $h = 60,350 \log (100 \ W/aV)$, where h = height, ft above sea level to which the balloon will rise; a = ratio from Table 2. Substituting gives $h = 60,350 \log [100 \times 1500/(0.0695 \times 523,600)] = 37,109$ ft (11,311 m).

Related Calculations: Use this general procedure to compute the weight a balloon can lift when used to transport logs, boats, and other objects. If the balloon gondola contains ballast, the balloon will rise approximately 260 ft (79.2 m) for each percent of the balloon weight W that is discarded. A rise of 1°F (0.6°C) in the air temperature increases the balloon altitude by about 55 ft (16.8 m) and the lifting power by 0.2 percent. A decrease in the air temperature of 1°F (0.6°C) causes the reverse effect, i.e., a decrease in altitude of about 55 ft (16.8 m) and in lifting power of 0.2 percent.

TABLE 2 Balloon Gas Weights

Gas	Weight lb/ft^3	Weight kg/m^3	Factor a
Helium, pure	0.01058	0.00066	0.0695
Hydrogen	0.00530	0.00033	0.0713
Coal gas	0.1166	0.0073	0.0400
Atmospheric air	0.07658	0.0048	

ROCKET FLIGHT VELOCITY

What is the final velocity of a rocket in which 75 percent of the rocket mass is discharged as propellant at a velocity of 10,000 ft/s (3048.0 m/s)? How does this final velocity compare with that of a rocket in which the propellant velocity is 170,000 ft/s (51,816 m/s)? How much of a velocity increase could be produced by increasing the discharged mass to 80 percent?

Calculation Procedure:

1. Compute the rocket mass ratio

Use the relation $m_r = m_i/m_f$, where m_r = rocket mass ratio; m_i = initial mass of the rocket, lb (kg); m_f = final mass of the rocket, lb (kg). Where the mass ratios are given instead of the actual mass, use the appropriate ratio in the above relation. Thus, the initial mass of the rocket is 1.00, and the final mass = $m_f = m_i$ − propellant ejected, or $1.00 - 0.75 = 0.25$. Hence, $m_r = 1.00/0.25 = 4.0$.

2. Compute the final velocity of the rocket

Use the relation $v_f = v_e \ln m_r$, where v_f = final velocity of the rocket ft/s (m/s); v_e = ejection velocity of the propellant, ft/s (m/s); ln = logarithm to the base e, or 2.71828; other symbols as before. Substituting for the 10,000-ft/s (3048-m/s) propellant velocity we get $v_f = 10,000 \ln 4.0 = 13,863$ ft/s (4225 m/s). For the 170,000-ft/s (51,816-m/s) propellant velocity, $v_f = 170,000 \ln 4.0 = 235,700$ ft/s (71,841 m/s).

3. Compute the velocity increase produced by the larger mass

Using the same procedure as in step 1, we find $m_r = 1.00/0.20 = 5.0$. Hence, $v_f = 10,000 \ln 5.0 = 16,094$ ft/s (4905.5 m/s), or a velocity increase of $16,094 - 13,863 = 2231$ ft/s (680.0 m/s) for a 10 percent increase in the discharged mass. For the second propellant, $v_f = 170,000 \ln 5.0 = 273,600$ ft/s (83,393 m/s), or a velocity increase of $273,600 - 235,700 = 37,900$ ft/s (11,551 m/s) for a 10 percent increase in the discharged mass.

Related Calculations: In designing a rocket, a greater final velocity of the vehicle can be obtained by using a propellant having a higher discharge velocity or by ejecting a larger percentage of the initial mass of the rocket. In general, rocket designers prefer to use propellants having a higher exit velocity instead of discharging a larger portion of the original rocket mass.

MISSILE MAXIMUM RANGE AND LAUNCH ANGLE

What is the minimum-energy maximum range of a ballistic missile having a burnout velocity of 12,000 ft/s (3657.6 m/s)? What range angle must be used to obtain this range? What elevation angle should be used for the minimum-energy trajectory?

Calculation Procedure:

1. Compute the missile half-range angle

Use the relation $\cos \theta_{max} = (u_e^2 - r_e u_e v_b^2)^{0.5}/(u_e - r_{bo}v^2/2)$, where θ_{max} = maximum half-range angle, deg; $u_e = 14.05 \times 10^{15}$ ft^3/s^2 (397.9 m^3/s^2) = $g_c r_e$, where g_c = acceleration of gravity at sea level = 32.17 ft/s^2 (9.8 m/s^2), and r_e radius of the earth = 3963 statute mi (6377.8 km) = 20.90×10^6 ft (6.38 $\times 10^6$ m); r_{bo} = missile burnout radius, ft (m); v_b = missile burnout velocity, ft/s (m/s). Substituting gives $\cos \theta_{max} = (14.05^2 \times 10^{30} - 20.9 \times 10^6 \times 14.05 \times 10^{15} \times 12^2 \times 10^6)^{0.5}/(14.05 \times 10^{15}) - (20.9 \times 10^6 \times 12^2 \times 10^6)/2 = 0.895$; $\theta = 26.5° = 0.4625$ rad.

2. Compute the missile maximum range

Use the relation $R_m = r_e \theta_r$, where r_m = maximum range of the missile, nmi; r_e radius of earth, nmi; θ_r = half-range angle, expressed in radians. Since the radius of the earth $r_e = 3440$ nmi (6370.9 km), $R_m = 3440(0.4625) = 1590$ nmi (2944.7 km).

3. Compute the required missile elevation angle

Use the relation $\cos \phi = (u_e/2u_e - r_{bo}v_b^2)^{0.5}$, where ϕ = missile elevation angle; other symbols as before. By substituting, $\cos \phi = [14.05 \times 10^{15}/(2 \times 14.05 \times 10^{15}) - (20.9 \times 10^6 \times 12^2 \times 10^6)]^{0.5} = 0.749$; $\phi = 41.5°$.

Related Calculations: Reducing the missile elevation angle below that computed in step 3 shortens the range of the missile below that computed in step 2. Increasing the elevation angle above that computed in step 3 also reduces the missile range below that computed in step 2.

SATELLITE FLIGHT VELOCITY, ESCAPE VELOCITY, AND PERIOD

An unstaffed satellite is placed in a circular orbit 800 mi (1287.2 km) above the earth. What is the velocity of the satellite? What is the escape velocity of this satellite at this altitude? How long will it take this satellite to make one revolution of the earth?

Calculation Procedure:

1. Compute the satellite velocity

Use the relation $v_s = (u_e/r_e + h)^{0.5}$, where v_s = satellite velocity, ft/s (m/s); $u_e = 14.05 \times 10^{15}$ ft^3/s^2 (397.9 $\times$ 10^{12} m^3/s^2), as defined in the previous calculation procedure; r_e = radius of the earth = 20.90 $\times$ 10^6 ft (6.4 $\times$ 10^6 m); h = satellite altitude, ft (m). Substituting gives v_s = [14.05 $\times$ 10^{15}/(20.90 $\times$ 10^6 + 800 $\times$ 5280)]$^{0.5}$ = 23,650 ft/s (7208.5 m/s).

2. Compute the satellite escape velocity

Use the relation $v_e = v_s\sqrt{2}$, where v_e = satellite escape velocity, ft/s (m/s); other symbols as before. So $v_e = 23,650\sqrt{2} = 33,450$ ft/s (10,196 m/s).

3. Compute the time required for one revolution

The satellite revolves around the center of the earth. Since the earth has a radius of 3963 statute mi (6377.8 km), the radius of rotation of this satellite, which is in orbit 800 mi (1287.5 km) above the earth, is r_s = 3963 + 800 = 4763 mi (7665.3 km). The circumference of this orbit is the distance d_s traveled by the satellite during one revolution about the earth, or $d_s = 2\pi r = 2\pi (4763) = 29,990$ mi (48,264 km). Since $t = 5280d_s/v_s$, where t = the time required for one revolution, or the *period* of the satellite, $t = 5280(29,990)/23,650 = 6690$ s, or 111.5 min.

Related Calculations: When the satellite escape velocity is known, compute the satellite velocity from $v_s = v_e/\sqrt{2}$. To compute the period of a satellite having an elliptical orbit with a major axis of length a, use the relation $t = 2\pi a^{3/2}/u_e^{1/2}$. Be certain to use consistent units—feet or miles—in both the numerator and denominator of this expression.

INTERPLANETARY FLIGHT LAUNCH VELOCITY AND FLIGHT TIME

What minimum launch velocity is required for an interplanetary satellite launched from Earth and traveling to Saturn? How long will the flight take? Compare the computed velocity and flight time with published values. Figure 6 shows the relationship of the various planets in the solar system.

Calculation Procedure:

1. Compute the major axis of the flight path

The flight path from Earth to another planet can be an ellipse. Assuming that the flight path of this satellite is elliptical (Fig. 7), use the relation $2m_a = d_1 + d_t$, where m_a = length of major axis of flight-path ellipse, mi; d_1 = distance of launch body from Sun, mi; d_t = distance of target body from Sun, mi. Since Earth is the launch body and is 93 $\times$ 10^6 mi (150 $\times$ 10^6 km) from the Sun and Saturn, the target body is 886 $\times$ 10^6 mi (1.4 $\times$ 10^9 km) from the Sun, $2m_a$ = 93 $\times$ 10^6 + 886 $\times$ 10^6 = 979 $\times$ 10^6 mi (1.6 $\times$ 10^9 km); m_a = 489.5 $\times$ 10^6 mi (788 $\times$ 10^6 km).

2. Compute the specific mechanical energy of the system

Use the relation $m_a = -u_s/2E$, where $u_s = g_s r_s^2$, in which g_s = acceleration of gravity at the surface of the Sun = 900.0 ft/s^2 (274.3 m/s^2); r_s = radius of sun = 2.285 $\times$ 10^9 ft (697 $\times$ 10^6 m); thus u_s = (900)(2.285 $\times$ 10^9) = 4.69 $\times$ 10^{21} ft^3/s^2 (1.33 $\times$ 10^{18} m^3/s^2); E = specific mechanical energy of the system, per unit mass. Substituting and solving, we find $E = -u_s/2m_a = -4.69 \times 10^{21}/2(489.5 \times 10^6 \times 5280) = -0.906 \times 10^9$.

3. Compute the satellite orbit perigee velocity

Use the relation $v_p = (2E + 2u_s/r_p)^{0.5}$, where v_p = velocity at orbit perigee, ft/s; r_p = radius at perigee = earth radius from Sun, ft; other symbols as before. Substituting, we find $v_p = [2(-0.906 \times 10^9) + 4.69 \times 10^{21}/(93 \times 10^6 \times 5280)]^{0.5} = 131,800$ ft/s (40,172.6 m/s).

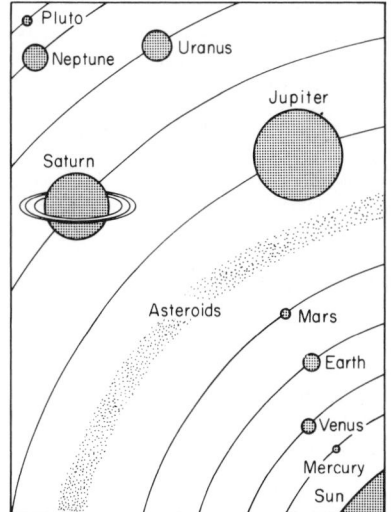

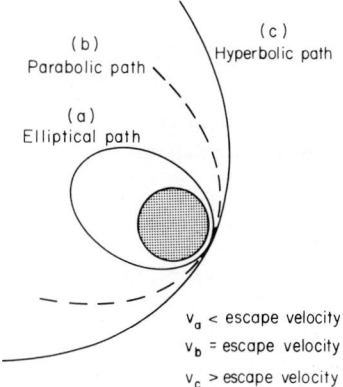

FIG. 7 Types of space-vehicle paths. *(RAND Corporation.)*

FIG. 6 The solar system. *(RAND Corporation.)*

4. *Compute the required velocity increment*

Far from the Earth, the velocity increment required from a satellite is reduced by the orbital velocity of the Earth if the satellite is orbited to take advantage of the Earth's velocity. Since the orbital velocity of the Earth is 97,800 ft/s (29,809 m/s), the required velocity increment is $v_i =$ 131,800 − 97,800 = 34,000 ft/s (10,363 m/s).

5. *Compute the velocity required at the Earth's surface*

Use the relation $E = 1/2v_i^2 - u_e/\infty = 1/2v_e^2 - u_e/r_e$, where $u_e = g_e r_e^2$, as defined above for the Sun; ∞ = infinity; v_e = velocity required at the Earth's surface. Thus $E = 1/2(34,000)^2 - u_e/\infty$; $E = 11.56 \times 10^8 = v_e^2 - 1.344 \times 10^9$; $v_e^2 = 24.96 \times 10^8$; $v_e = 49,900$ ft/s (15,209 m/s). This compares favorably with the actual escape velocity shown in Table 3.

TABLE 3 Minimum Launch Velocities and Transit Times to Reach the Planets from the Earth°

Planet	Planet surface escape velocity		Minimum Earth launching velocity		Transmit time one way
	ft/s	m/s	ft/s	m/s	
Mercury	13,600	4,145.3	44,000	13,411.2	110 days
Venus	33,600	10,241.3	38,000	11,582.4	150 days
Mars	16,700	5,090.2	38,000	11,582.5	260 days
Jupiter	197,000	60,045.6	46,000	14,020.8	2.7 years
Saturn	119,500	36,423.6	49,000	14,935.2	6 years
Uranus	72,400	22,067.5	51,000	15,544.8	16 years
Neptune	82,100	25,024.1	52,000	15,849.6	31 years
Pluto	31,200†	9,509.8	53,000	16,154.4	46 years
Earth	36,700	11,186.2			

° NASA data.
† Approximate.

6. *Compute the period of the flight*

Use the relation $t = 2\pi m_a^{1.5}/u_s^{0.5}$, where t = flight time for a round trip, s. Thus $t = 2\pi(489.5 \times 10^6 \times 5280)^{1.5}/(4.69 \times 10^{21})^{0.5} = 3.81 \times 10^8$ s = 4405 days = 12.05 years. This is the time for a round-trip flight. A one-way flight would take $12.05/2 = 6.025$ years.

Table 3 lists minimum launching velocities and transit times for all planets. From Table 3, the computed one-way flight time agrees well with the published flight time for the planet Saturn.

Related Calculations: Use this general method to compute the velocities and flight time for any interplanetary probe or satellite flight. Table 3 lists the surface escape velocities for the planets. This is the velocity an interplanetary spaceship would have to develop to return to the earth from a given planet.

SPACE VEHICLE BURNOUT VELOCITY AND FUEL SELECTION

What is the burnout velocity v_b of a single-stage rocket having a mass ratio m_r of 9.25 when it uses a fuel having a specific impulse s_i of 250 s? Would this rocket be suitable for a Saturn-probe launching? If the rocket payload were reduced by 200 lb (90.7 kg), would the rocket be suitable for this probe? The rocket weighs 50,000 lb (22,679 kg) before payload reduction. What specific impulse is required for a Saturn probe when the rocket mass ratio is 9.25? Choose a suitable fuel for the rocket.

Calculation Procedure:

1. *Compute the rocket burnout velocity*

Use the relation $v_b = s_i g \ln m_r$, where v_b = burnout velocity, ft/s; s_i = fuel specific impulse, s; g = acceleration due to gravity = 32.2 ft/s² (9.8 m/s²); ln = log to the base e; m_r = rocket mass ratio. Substituting gives $v_b = 250(32.2) \ln 9.25 = 17.880$ ft/s (5449.8 m/s).

The minimum launching velocity required for a Saturn probe is 49,000 ft/s (14,935 m/s).

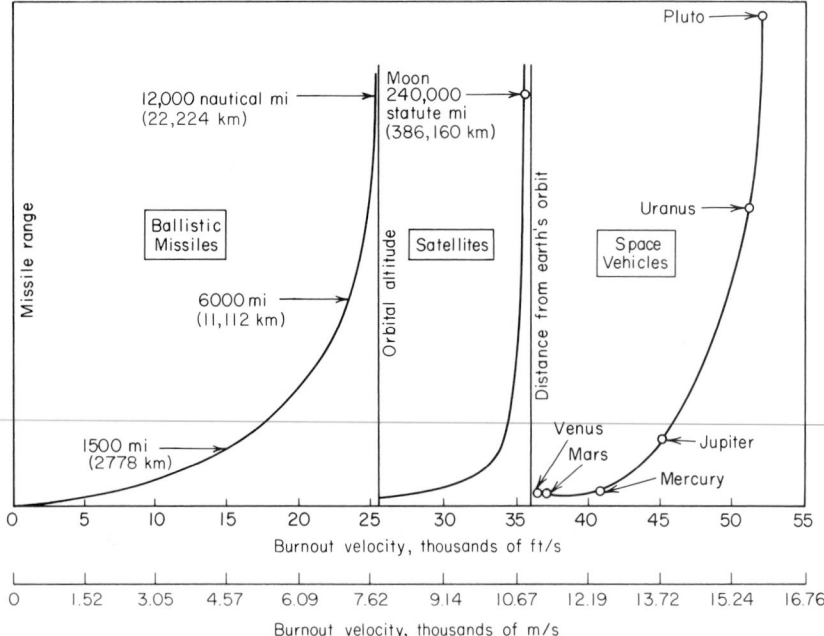

FIG. 8 Velocity requirements for ballistic missile and space flights. *(RAND Corporation.)*

Hence, a velocity of 17,880 ft/s (5449.8 m/s) is completely unsatisfactory for such a probe because it is less than half the required velocity. To overcome this velocity deficiency, a multistage rocket is needed. Figure 8 shows the burnout velocity required for ballistic missiles, satellites, and space vehicles.

2. *Compute the effect of a reduced payload*

The new mass ratio $\Delta m_r = m_{rf} - m_{ri}$, where Δm_r = change in rocket mass ratio; m_{rf} = final mass ratio after payload change; m_{ri} = initial mass ratio of the rocket, However, $m_r = m_i/m_f$, where m_i = initial mass of rocket, lb; m_f = final mass of rocket, lb. Substituting for the initial condition, we find $m_f = m_i/m_r = 50.000/9.25 = 5400$ lb (2449.4 kg). With a 200-lb (90.7 kg) reduction in payload, $m_i = 50,000 - 200 = 49,800$ lb (22,588.9 kg); $m_f = 5400 - 200 = 5200$ lb (2358.7 kg), $m_r = m_i/m_f = 49,800/5200 = 9.57$. Then $\Delta m_r = m_{rf} - m_{ri} = 9.57 - 9.25 = 0.32$.

Compute the change in burnout velocity from $\Delta v_b = \Delta m_r s_i g/m_{ri} = 0.32(250)(32.2)/9.25 = 279$ ft/s (85.0 m/s). Hence the new burnout velocity is $17,880 + 279 = 18,159$ ft/s (5534.9 m/s). This is still far below the required burnout velocity. Figure 9 shows the relationship among the propellant fraction, mass ratio, and rocket velocity for single-stage vehicles. Study of this chart shows that the high propellant fractions associated with high rocket velocities can be achieved only by severely reducing to a minimum all components of the rocket that contribute to the weight at propellant exhaustion, including the payload. Once again, a multistage rocket is needed to overcome the velocity deficiency.

Use the equation in step 1 and solve for $s_i = v_b/g \ln m_r$. Since $v_b = 49,000$ ft/s (14,935.2 m/s) for a Saturn prober, as shown in the previous calculation procedure (Table 3), $s_i = 49,000/(32.2 \ln 9.5) = 676$ s.

3. *Select a suitable fuel for the rocket*

Table 4 lists the characteristics of typical common propulsion systems. Study of this tabulation shows that liquid fuels are unsuitable for this single-stage rocket because their specific impulse is too low. Only a nuclear or free-radical propellant gives the desired specific impulse for a single-stage vehicle. However, two or more stages of liquid fuel would give the desired velocity to this

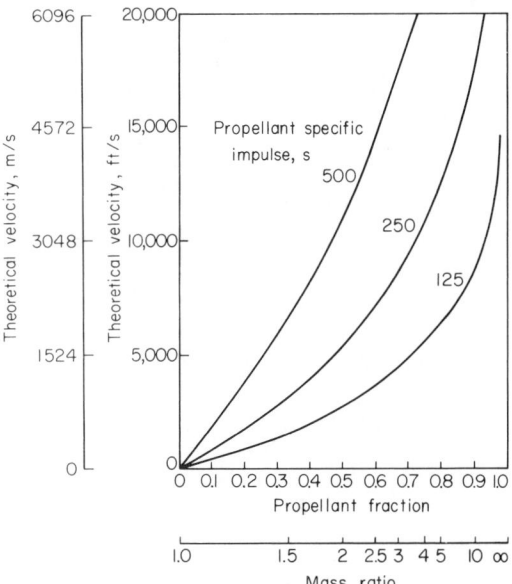

FIG. 9 Velocity characteristics of single-stage rocket vehicles. *(RAND Corporation.)*

TABLE 4 Typical Characteristics of Propulsion Systems

System	Specific impulse, s	Ratio of thrust to engine weight
Liquid propellants	200–300	50 to 80
High-energy liquid propellants	340–440	
		Less than 50 to 80 but of the same order of magnitude
Nuclear energy	400–900	
Free radicals	400–1800	
Solar heat transfer	400–500	0.05
Ion	5000–20,000	0.0005 to 0.00005
Thiokol perchlorate (solid)	200–215	
Rubber nitrate (solid)	180–195	

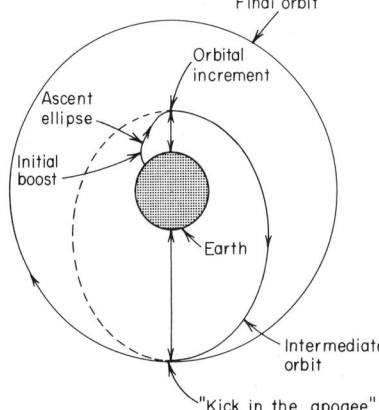

FIG. 10 "Kick in the apogee" technique of satellite launching. *(RAND Corporation.)*

rocket because the specific impulse of the second stage adds to the specific impulse of the first stage. For these reasons, the type of fuel chosen depends on the rocket mission, the propulsion system availability, economics, and similar factors.

Related Calculations: Use the procedure given here to determine the burnout velocity and fuel for ballistic missiles, interplanetary launch vehicles, and similar applications.

To achieve the needed escape velocity required for interplanetary flights with liquid propellants, two or more rocket stages are required. Thus, if the burnout velocity of the first stage of a rocket is 17,880 ft/s (5449.8 m/s), the second stage begins accelerating from this initial velocity. In an efficient two-stage rocket, the mass ratios of each stage are equal, or nearly so.

Figure 10 shows how the second stage of a rocket can produce a "kick in the apogee." If the second stage is fired with the rocket correctly oriented, the final orbit that can be achieved is not limited by the projection altitude of the basic booster or first-stage rocket.

OBSERVATION-SATELLITE DETAIL DETECTION

An observation satellite carries a precision camera having a focal length of 6 in (15.2 cm). What is the scale on pictures taken by this camera from an altitude of 150 mi (241.4 km)? What focal length is needed for a camera at an altitude of 1000 mi (1609.3 km) to produce a ground resolution of 20 ft (6.1 m)? Determine the ground resolution obtainable with the 6-in (15.2-cm) focal-length camera if the film resolution is 100 lines per minute.

Calculation Procedure:

1. *Compute the scale number of the photograph*

Factors that enter an estimate of the degree of detail that can be detected or identified by a camera include the distance between the camera and the object photographed and the focal length of the viewing lens. Using these two factors, we see a *scale number S* is computed from S

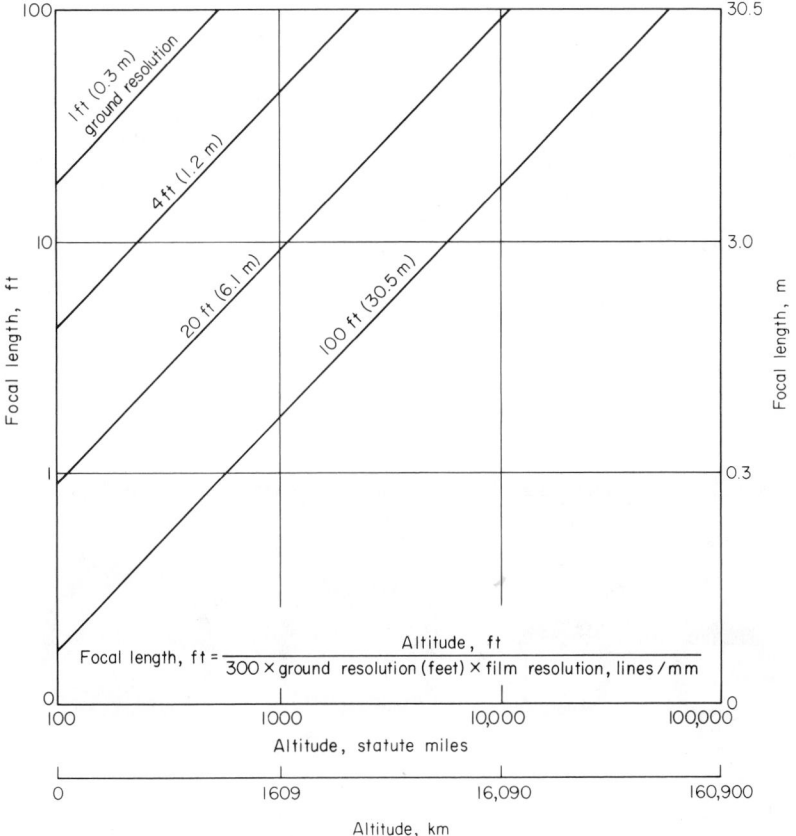

FIG. 11 Required focal-length variation with altitude for various ground resolutions. *(RAND Corporation.)*

= camera altitude, ft (m)/lens focal length, ft (m). For this camera at an altitude of 150 mi (241.4 km), $S = (150 \text{ mi})(5280 \text{ ft/mi})/0.5 = 1,584,000$.

A scale number of 1,584,000 means that 1 in (2.54 cm) on the photograph taken by this camera corresponds to 1,584,000 in (4,023,360 cm) on the ground area photographed. Since scale numbers are usually in miles, 1 in (2.54 cm) on the photograph = 1,584,000 in/[(5280 ft/mi)(12 in/ft)] = 25.05 mi (40.31 km).

In general, the larger the scale number, the more difficult it is to detect fine details in the photograph.

2.   *Determine the required focal length of the camera*

Assume a film resolution of 100 lines per minute. This is a typical resolution used for high-altitude photography.

Enter Fig. 11 at the satellite altitude, 1000 mi (1609.3 km), on the horizontal axis, and project vertically upward to the desired ground resolution, 20 ft (6.1 m). From the intersection, project horizontally to the left to read the required focal length of 9.5 ft (2.9 m). With the high resolution required in cameras carried aboard observation satellites, a speed of at least $f/8$, and preferably faster, is desirable.

3.   *Compute the ground resolution for the given scale number*

Use the relation $G_r = S/(300R)$, where G_r = ground resolution for a given scale number, ft; S = photograph scale number, inches on the ground per inch on the photograph; R = lines per

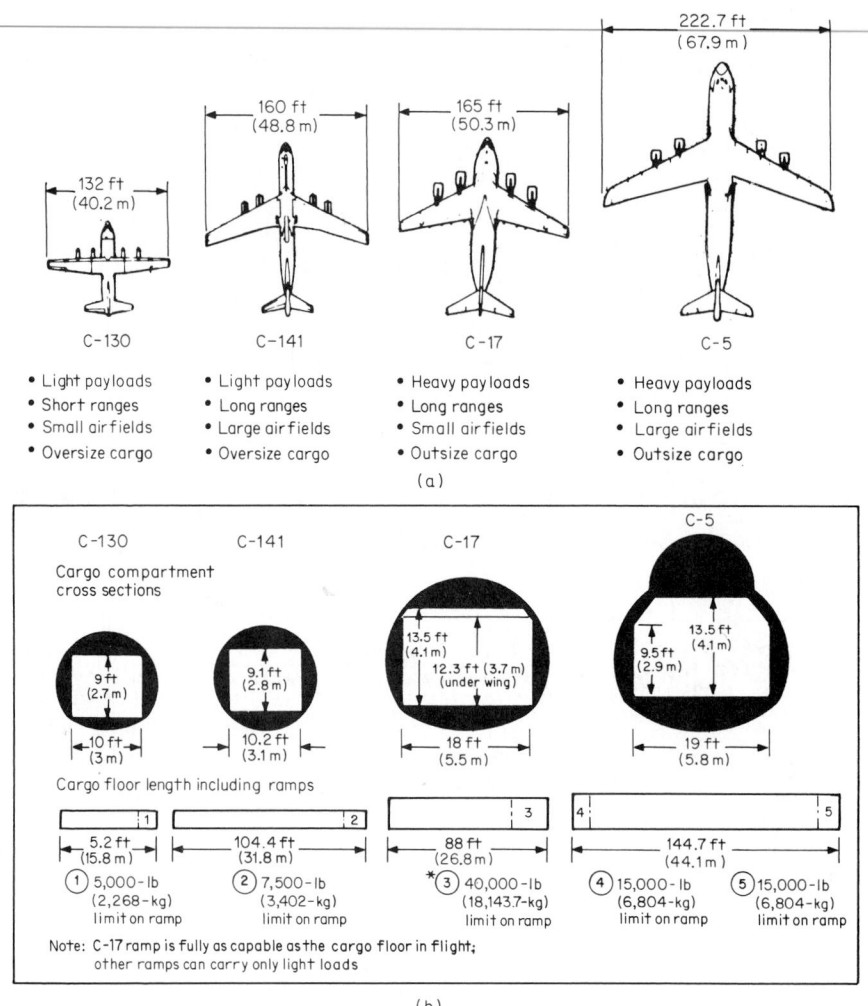

FIG. 12 Comparison of four cargo aircraft. (*Aviation Week & Space Technology*)

mm produced by the film-lens combination. By substituting in the relation the scale number from step 1, $G_r = 1,584,000/[300(100)] = 56.1$ ft (17.1 m).

Related Calculations: Four levels of photographic detail are generally used to define the ground resolution obtainable with observation satellites.[1] In terms of ground resolution these levels are A, 50 to 200 ft (15.2 to 61.0 m); B, 10 to 40 ft (3.0 to 12.2 m); C, 2 to 8 ft (0.6 to 2.4 m); D, 0.5 to 2 ft (0.2 to 0.6 m). The range over a factor of 4 within each level arises from a practical inability to measure and interpret ground resolution as a fixed number and from additional detailed factors, such as the graininess of photographic emulsions.

[1]Staff Report of the Select Committee on Astronautics and Exploration—*Space Handbook: Astronautics and Its Applications*, House Document 86, GPO.

AIRCRAFT CARGO-CARRYING CAPACITY ANALYSIS

Analyze the cargo-carrying capacity of several aircraft with respect to payload, range, airfield requirements, and cargo size.

Calculation Procedure:

1. *Obtain pertinent aircraft data*

Contact the appropriate aircraft manufacturer and obtain from them key data on the aircraft being considered. Assemble this information in a form such as that in Fig. 12*a*.

Below an outline diagram of each aircraft, list the relative features of the particular configuration being considered. Thus, for the C-130, the first aircraft shown in Fig. 12*a*, it can: handle light payloads, short flight ranges, land on small air fields, and accommodate oversize cargo. The listings below the three other aircraft being studied show the characteristics of each.

2. *Plot the pertinent aircraft cross-sectional data*

Using the data supplied by the aircraft manufacturers, plot the maximum interior fuselage width, the maximum cargo width, and the maximum cargo height, to a suitable scale (Fig. 12*b*). Next, plot the maximum cargo floor length to scale, including the ramp. Locate these scale lengths beneath the cross section of each aircraft. Below the cargo floor length list the maximum load limit for the ramp for each aircraft because this can be a limiting factor in aircraft use, as Fig 12*b* shows.

3. *Analyze the pertinent features of each choice*

With the pertinent features of each aircraft graphically plotted as shown, a preliminary comparison can be made. Knowing the characteristics of the planned cargo (typical length, width, height, and weight), you an quickly determine which aircraft would be suitable from a dimension and ramp standpoint. Then, other factors must be considered—range, airfield size requirement, economy of operation, etc. These factors will normally require additional economic analyses before a final choice is made. Hence, the procedure given here is simply a "first-cut" approach that permits the engineer to narrow a choice to fewer possibilities, thereby saving time, energy, and money.

   Related Calculations: Use this general approach for analyzing a variety of possibilities in aircraft choice where the variables can be plotted as shown. Remember, however, that far more extensive analyses will be required before a final selection is made. This procedure is applicable to commercial, military, private, agricultural, fire fighting, and a variety of other aircraft where several possibilities exist for accomplishing a given task.

RAMJET ENGINE DIFFUSER ANALYSIS

A ramjet aircraft flies at 800 mi/h (1287.5 km/h) at an altitude at which still air has a pressure of 12 lb/in² (82.7 kPa) absolute and a temperature of 0°F (-17.8°C). Plane normal shock occurs at entrance to the ram. If isentropic diffusion is assumed to take place behind the shock plane to a final velocity of 500 ft/s (152.4 m/s), calculate the final static pressure and temperature.

Calculation Procedure:

1. *Compute the acoustic velocity of the still air*

Acoustic velocity $a = (gkRT)^{1/2}$, for dry air considered as an ideal diatomic gas. The gravitational factor $g = 32.174$ ft/s² (9.8066 m/s²); dimensionless gas constant $k = c_p/c_v = 1.4$; gas constant $R = 53.34$ ft·lb/lb·°F (287.0 N·m/kg·°K); temperature T in °R (K), as identified by a subscript such as x. Then, $a_o = (32.174 \times 1.4 \times 53.34)^{1/2}(T_o)^{1/2} = 49.016(T_o)^{1/2}$, where $T_o = 0$°F $+ 459.67 = 459.67$°R (255.4 K). Thus, $a_o = 49.016(459.67)^{1/2} = 1050.9$ ft/s (313.8 m/s).

2. *Compute the Mach number of the aircraft's velocity*

The aircraft Mach number is $M = V_1/a_o$, where the aircraft velocity $V_1 = $ (mi/h)(ft/mi)/(s/h) $= 800 \times 5280/3600 = 1173.3$ ft/s (357.6 m/s) and $a_o = $ acoustic velocity, from step 1 above. Then, $M_1 = 1173.3/1050.9 = 1.1165$.

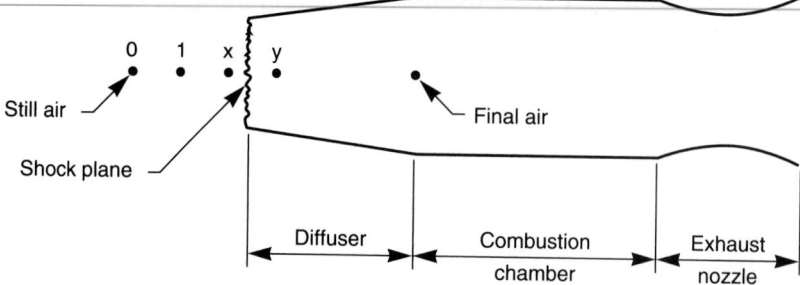

FIG. 13 Ram diffuser configuration.

3.   *Determine the static pressure, static temperature, and Mach number of the air at the upstream face of the shock plane*

During supersonic aircraft velocity, air at point 1 in Fig. 13 is at the brink of entering the upstream face of the shock plane. It has the same physical properties as the atmosphere's still air at point 0 in Fig. 13 and is moving relative to the ram diffuser at the aircraft's speed. Therefore, at the upstream face of the shock plane the static pressure p_x = 12.0 lb/in² (82.7 kPa); static temperature T_x = 459.67°R (255.4 K); Mach number M_x = 1.1165.

4.   *Determine the static pressure, static temperature, and Mach number of the air at the downstream face of the shock plane*

From the Gas Tables mentioned under Related Calculations of this procedure, use Table 48, One-Dimensional Normal-Shock Functions for a Perfect Gas with Constant Specific Heat and Molecular Weight, k = 1.4 to find required values corresponding to M_x = 1.1165. By interpolation, at the downstream face of the shock plane the Mach number M_y = 0.89919; p_y/p_x = 1.2877; T_y/T_x = 1.07542. From these values and previous results, P_y = 1.2877 × 12.0 = 15.452 lb/in² (106.54 kPa); T_y = 1.07542 × 459.67 = 494.34°R (274.6 K).

5.   *Compute the acoustic velocity and air velocity at the downstream face of the shock plane*

Acoustic velocity a_y = 49.016$(T_y)^{1/2}$ = 49.016(494.34)$^{1/2}$ = 1089.8 ft/s (332.2 m/s). Air velocity V_y = $a_y M_y$ = 1.089.8 × 0.89919 = 979.94 ft/s (298.7 m/s).

6.   *Determine the isentropic critical Mach number and related local-to-stagnation ratios of the pressure and temperature corresponding to the Mach number at the downstream face of the shock plane*

From the Gas Tables, use Table 30, One-Dimensional Isentropic Compressible-Flow Functions for a Perfect Gas with Constant Specific Heat and Molecular Weight, k = 1.4 to obtain required values corresponding to M = M_y = 0.89919. By interpolation, the isentropic critical Mach number M_y^* = 0.91389; local-to-stagnation, or static-to-total, pressure ratio P_y/P_o = 0.59178; local-to-stagnation temperature ratio T_y/T_o = 0.86079.

7.   *Compute the critical air velocity for the critical Mach number at the shock plane*

Critical Mach number M_y^* = V_y/V_y^*. Hence, 0.91389 = 979.94/V_y^* and the critical air velocity V_y^* = 979.94/0.91389 = 1072.3 ft/s (326.8 m/s).

8.   *Determine the critical Mach number downstream where the final velocity is 500 ft/s (152.4 m/s)*

Since, at point 2 on Fig. 13, the final critical Mach number M_f^* = V_f/V_y^*, where V_f is the given final velocity and V_y^* is now known, hence M_f^* = 500/1072.3 = 0.46629.

9.   *Determine the local-to-stagnation ratio of the pressure and temperature corresponding to the final critical Mach number*

Table 30 of the Gas Tables provides, by interpolation, the required ratios corresponding to M_f^* = 0.46629 as p_f/p_o = 0.87880 and T_f/T_o = 0.96376.

SECTION 9

MARINE ENGINEERING

BERNARD TICHAZ
DIRECTOR
GEORGE G. SHARP INC.

TYLER G. HICKS, P.E.
INTERNATIONAL ENGINEERING ASSOCIATES

JOHN C. STERLING
CONSULTANT
NEWPORT NEWS SHIPBUILDING & DRYDOCK CO.

Form Coefficients for Various Vessel Types 9.2
Vessel Wetted Area and Shallow-Water Speed 9.3
Power Required to Propel a Vessel 9.4
Marine Propeller-Shaft Diameter and Propeller Slip 9.6
Propeller Selection for Vessels 9.7
Propeller Revolutions and Vessel Speed 9.9
Vessel Immersion and Flooding Effects 9.10
Marine Pump Selection for Commercial Vessels 9.10
Marine Power-Plant Selection 9.13
Nuclear Propulsion for Oceangoing Vessels 9.16
Tanker Capacity and Cargo-Handling Characteristics 9.17
Marine Refrigeration and Air-Conditioning Systems 9.19
Choice of Vessel Type for Seagoing Service 9.22

REFERENCES: Bhattacharya—*Dynamics of Marine Vehicles*, Wiley-Interscience; Gritzen—*Introduction to Naval Engineering*, Naval Institute Press; Osbourne and Neild—*Modern Marine Engineer's Manual*, Cornell Maritime Press; Weddle—*Marine Engineering Systems*, Sheridan; Baxter—*Naval Architecture*, Charles Griffin; Gillmer—*Modern Ship Design*, Naval Institute Press; Hutchinson—*A Treatise on Naval Architecture*, Conway Maritime; Jacobsson—*Computer Applications in the Automation of Shipyard Operation and Ship Design*, Elsevier; Taggart—*Marine Propulsion*, Gulf Publishing; Bishop and Price—*Hydroelasticity of Ships*, Cambridge University Press; Baker—*Introduction to Steel Shipbuilding*, McGraw-Hill; Evans—*Ship Structural Design Concepts*, Cornell Maritime Press; MacBride—*Handbook of Practical Shipbuilding*, Gordon Press; Vossers—*Behavior of Ships in Waves*, Stam Press; Munro-Smith—*Applied Naval Architecture*, Longmans; Friedman—*Modern Warship: Design and Development*, Mayflower; Labberton—*Marine Engineering*, McGraw-Hill; Bell—*Petroleum Transportation Handbook*, McGraw-Hill; Seward—*Marine Engineering*, Society of Naval Architects and Marine Engineers; U.S. Coast Guard—*Regulations for Commercial Vessels*, Government Printing Office; Lammeren—*Resistance, Propulsion, and Steering of Ships*, Technical Publishing Co, Haarlem, Holland; Crouch—*Nuclear Ship Propulsion*, Cornell Maritime Press; Quinn—*Design and Construction of Ports and Maritime Structures*, McGraw-Hill.

FORM COEFFICIENTS FOR VARIOUS VESSEL TYPES

Determine the form coefficients for a proposed 15.4-kn (28.5-km/h) ocean-going freighter having a waterline length of 441 ft (134.4 m), a molded beam of 56.9 ft (17.3 m), a molded depth of 37.3 ft (14.4 m), a load draft of 26.0 ft (7.92 m), and a displacement of 13,570 tons (12,309 t). The area of the midship section below the water plane is 1450 ft^2 (134.7 m^2). Do the form coefficients of this vessel agree with generally accepted design practice?

Calculation Procedure:

1. Compute the vessel speed-length ratio

Use the relation $S = V/L^{0.5}$, where S = vessel speed-length ratio, V = vessel normal speed, kn; L = vessel waterline length, ft. For this vessel, $S = 15.4/(441)^{0.5} = 0.733$.

2. Compute the vessel block coefficient of fineness

Use the relation $b = 35D/LBH$, where b = block coefficient of fineness; D = displacement, tons of seawater; B = beam, ft; H = draft, ft. For this vessel, $b = 35(13,570)/(441 \times 56.9 \times 26.0) = 0.727$.

3. Compute the vessel midship section coefficient

Use the relation $m = M/BH$, where m = midship section coefficient; M = area of midship section below water plane, ft^2; other symbols as before. For this vessel, $m = 1450/(56.9 \times 26) = 0.98$.

4. Compute the longitudinal prismatic coefficient

Use the relation $l = 35D/ML$, where l = longitudinal prismatic or mean-length coefficient; other symbols as before. Thus, $l = 35 \times 13,570/(1450 \times 441) = 0.741$.

5. Compute the water-plane coefficient

Use the relation $\alpha = 0.667b + 0.333$, where α = water-plane coefficient; other symbols as before. Thus, $\alpha = 0.667(0.727) + 0.333 = 0.818$. Alternatively, $\alpha = A/BL$, where A = area of water plane at the surface of the water, ft^2; other symbols as before.

6. Compute the displacement length coefficient

Use the relation $c_d = D/(L/100)^3$, where c_d = displacement length coefficient; other symbols as before. Thus, $c_d = 13,570/[(441/100)^3] = 158$.

7. Compare the form coefficients with typical values

Table 1 lists form coefficients for a variety of typical vessels. This table shows that the form coefficients computed for the freighter under consideration agree with those listed for moderate-speed freighters in all instances except one—the displacement length coefficient. However, the difference here, 158 versus 165, is insignificant. The displacement length coefficient is the displacement in tons of a mechanically similar ship that is 100 ft (30 m) in length.

TABLE 1 Typical Coefficients of Form for Various Types of Vessels

Vessel type	$V/L^{0.5}$	b	m	l	α	$D/(L/100)^3$
Great Lakes ore ships	0.39–0.43	0.85–0.87	0.99–0.995	0.86–0.88	0.89–0.92	70–95
Slow ocean freighters	0.45–0.50	0.77–0.82	0.99–0.995	0.78–0.83	0.85–0.88	180–200
Moderate-speed freighters	0.55–0.75	0.67–0.76	0.98–0.99	0.68–0.78	0.78–0.84	165–195
Fast passenger liners	0.70–1.05	0.56–0.65	0.94–0.985	0.59–0.67	0.71–0.76	75–105
Fast cruisers	1.30–1.70	0.45–0.53	0.80–0.90	0.55–0.60	0.60–0.65	
Destroyers	1.8–2.5	0.44–0.53	0.72–0.83	0.62–0.71	0.67–0.73	40–65
Tugs	0.9–1.2	0.45–0.53	0.71–0.83	0.61–0.66	0.71–0.77	200–420

Related Calculations: Form coefficients are valuable to ship designers because the coefficients are dimensionless. Hence, the coefficients apply to ship forms of all sizes. A designer can thus easily compare two ships of widely different sizes. To compare two ships, simply examine their speed-length ratios and their prismatic coefficients. Using this procedure, a 300-ft (91.4-m) vessel with a 1.04 speed-length ratio and a prismatic coefficient of 0.68 is similar to a 900-ft (274-m) vessel having the same speed-length ratio and prismatic coefficient. This means that the sectional area curves of the two vessels are also similar.

The form coefficients are also useful in determining the dimensions of a certain class of ship. To use the form coefficients in this manner, select a coefficient from within the range given in Table 1. Assume one or more dimensions of the vessel, and solve for the unknown dimension. By using this procedure, a vessel of desired dimensions can be developed for a given service.

VESSEL WETTED AREA AND SHALLOW-WATER SPEED

What are the wetted area of and the frictional hp required for a 600-ft (182.9 m) long 16-kn (29.6-km/h) vessel displacing 17,350 tons (15,740 t)? At what water depth will there be no increase in the resistance of the vessel if its loaded draft is 32 ft (9.8 m)? What percentage increase in resistance to vessel movement is there in 40 ft (12.2 m) of water?

Calculation Procedure:

1. Compute the wetted area of the vessel

Use the Taylor relation $A = k(DL)^{0.5}$, where A = vessel wetted hull area, ft²; k = the Taylor constant = 15.0 to 16.3, with the lower values of l corresponding to finely shaped vessels (destroyers, cruisers, and some yachts) and higher values of k corresponding to less finely shaped vessels (barges, bulk-cargo freighters, and small tankers); D = vessel displacement, tons; L = length of vessel on the waterline, ft. Thus, with k = 15.6, the approximate midpoint value is A = $15.6(17,350 \times 600)^{0.5}$ = 50,400 ft² (4682.3 m²).

2. Compute the frictional resistance of the vessel

The frictional hp required to propel a vessel is the shaft horsepower (shp) needed to overcome frictional resistance to movement of the hull through the water. Use the relation $F_r = fAV^{1.825}$, where F_r = vessel frictional resistance, lb; f = coefficient of friction for the vessel in saltwater; A = wetted hull area, ft²; V = vessel speed, kn. Values of f for vessels of various lengths operating in saltwater are given in Table 2. With f = 0.008726 for a 600-ft (182.9-m) vessel, F_r = $0.008726(50,400)(16)^{1.825}$ = 69,000 lb (306,927 kN).

3. Compute the frictional hp required

Use the relation $H_f = 101.3F_rV/33,000$, where H_f = frictional hp required; other symbols as before. The constant 101.3 converts kn to ft/min (0.514 converts kn to m/s); the constant 33,000

TABLE 2 Coefficient of Friction for Vessels in Salt Water

Vessel length, ft (m)	Coefficient of friction, f
100 (30.5)	0.009207
200 (60.9)	0.008992
300 (91.4)	0.008902
400 (121.9)	0.008832
500 (152.4)	0.008776
600 (182.9)	0.008726
700 (213.4)	0.008680
800 (243.8)	0.008639
900 (274.3)	0.008608
1000 (304.8)	0.008574

converts ft·lb/min to hp (1,088,965 converts ft·lb/min to kW). So $H_f = 101.3(69,000)(16)/33,000 = 3390$ hp (2527.9 kW).

As an alternate solution for frictional hp, use the appropriate Schoenherr curve from Fig. 1. This curve shows that a 600-ft (182.9-m) 16-kn (29.6-km/h) vessel requires 69 hp/1000 ft² (553.8 kW/1000 m²) of wetted hull area to overcome the frictional resistance. Thus, $H_f = 69(50,400/1000) = 3475$ hp (2591.3 kW). This value agrees closely with the earlier computed value of 3390 hp (2527.9 kW). If desired, the Schoenherr curves in Fig. 1 can be interpolated or extrapolated for vessel lengths that are not plotted.

4. Compute the minimum depth of water for zero resistance increase

The resistance to the movement of a vessel through the water increases markedly when $V = 2H^{0.5}$, where V = vessel speed, kn; H = water depth, ft. Solving yields $H = V^2/4$. Or, the minimum depth of water for zero resistance increase for a 16-kn (29.6-km/h) vessel is $H = 16^2/4 = 64$ ft (19.5 m).

The maximum vessel speed v_{md} kn for a nonplaning hull is $V_{md} = 3.36H^{0.5}$ in deep water. In shallow water the maximum speed in knots is $V_{ms} = 2.5H^{0.5}$. For trial runs, Admiral Taylor recommends a least-water depth, ft = 10(vessel draft, ft) $V/L^{0.5}$, where L = load waterline length of vessel, ft.

5. Compute the percentage increase in vessel resistance

In shallow water the percentage increase in resistance is $p = 50H/d$, where H = vessel draft, ft; d = water depth, ft. Thus, at a depth of 40 ft (12.2 m), $p = 50(32)/40 = 40$ percent. An increase in resistance of this magnitude will appreciably reduce the speed of the vessel.

Related Calculations: The total resistance of a vessel is the sum of the frictional resistance and the residual resistance. This latter resistance is comprised of the sum of eddy resistance and wave resistance. Residual resistance can be determined by test of a model of the vessel in a towing tank or from published systematic model test data such as Taylor's Standard Series or Ayre's method.[1] In model tests, residual resistance of model/residual resistance of ship = (length of model, ft)³/(length of ship, ft)³.

One modern way to reduce the overall water resistance of large vessels is the use of a bulbous bow. Such a bow creates a secondary wave that partially cancels the primary wave system caused by movement of the vessel. A properly designed bulbous bow can (1) increase vessel speed 4 to 6 percent, (2) increase cargo-carrying capacity 4.5 percent without increasing shaft power or decreasing speed, and (3) reduce power and fuel consumption 10 to 15 percent without sacrificing speed or cargo capacity.

POWER REQUIRED TO PROPEL A VESSEL

What is the hp required to drive a 600-ft (182.9-m), tanker at a cruising speed of 17.4 kn (32.2 km/h) if the displacement of the vessel is 34,650 tons (31,434 t) and the propulsion machinery is a geared turbine?

[1]Van Lammeren—*Resistance, Propulsion, and Steering of Ships*, Technical Publishing Co., Haarlem, Holland.

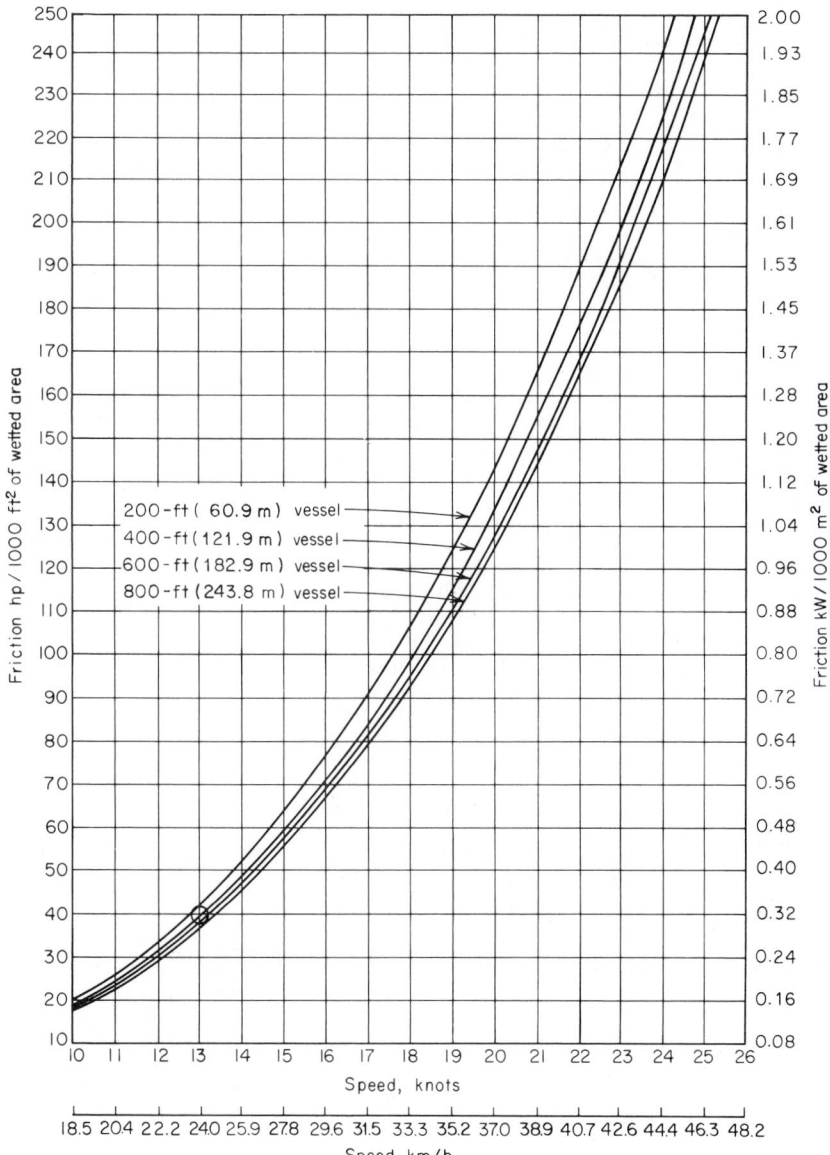

FIG. 1 Frictional hp required per 1000 ft² (92.9 m²) of wetted hull surface of vessels of various lengths.

Calculation Procedure:

1. *Determine the admiralty coefficient for the vessel*

When a vessel is being designed, its admiralty coefficient must be approximated from known values for vessels of about the same overall length. Table 3 lists typical admiralty coefficients for various modern vessels propelled by geared turbines or diesel engines. Table 3 shows that the

TABLE 3 Admiralty Coefficients for Modern Vessels

Vessel LWL, ft (m)	30 (9.1)	50 (15.2)	75 (22.9)	100 (30.5)	150 (45.7)	200 (60.9)	300 (91.4)	400 (121.9)	500 (152.4)	600 (182.9)	800 (243.8)
Admiralty coefficient	70	95	120	150	180	200	250	345	360	400	500

admiralty coefficient $K = 400$ for a vessel having a length on the load waterline (LWL) of 600 ft (182.9 m).

2. Compute the required power to drive the vessel

Use the relation $P = D^{2/3}V^3/K$, where P = required shp to drive the vessel; D = vessel displacement, long tons; V = vessel cruising speed, kn; K = admiralty coefficient. Using the given data, we get $P = 34,650^{2/3} \times 17.4^3/400 = 14,000$ hp (10,440 kW), closely.

3. Check the power, using Froude's law of comparison

Froude's law of comparison states: For ships of similar geometric form and proportion, the shaft horsepowers required at corresponding speeds will be in the ratio of the products of the displacements by the speeds, or the power ratio will equal the product of the displacements ratio by the speed ratio, or power is proportional to $L^{7/2}$ or to $D^{7/6}$. Also, corresponding speeds are in the ratio of the square roots of the vessel lengths.

To apply Froude's law of comparison, obtain the speed, LWL, and P of a similar vessel. Thus, a 445-ft (135.6-m) tanker of modern design has a speed of 13.7 kn (25.3 km/h) when developing 4100 shp (3057.4 kW). By Froude's speed law, the corresponding speed of the vessel being analyzed here is $13.7(600/445)^{1/2} = 15.9$ kn (29.4 km/h).

Next apply Froude's power law: P at 15.9 kn (29.4 km/h) $= 4100(600/445)^{7/2} = 11,000$ hp (8202.7 kW). Then the shp at the design speed of 17.4 kn (32.2 km/h) is, if the power varies as the cube of the speed, $11,000(17.4/15.9)^3 = 14,350$ hp (10,700 kW). This agrees closely with the admiralty coefficient computed shp of 14,000 hp, computed in step 2.

Related Calculations: Use this method for any size vessel, from 30 ft (9.1 m) to 1000 ft (304.8 m) or more LWL, of any type—cargo, passenger, tanker, yacht, tug, naval vessel, etc. Be certain to use admiralty coefficients developed for modern vessels. Typical up-to-date values are published in *Marine Engineering/Log* (New York), *Transactions of the Institute of Marine Engineers* (London), and *Journal of the American Society of Naval Engineers* (Washington, D.C.).

For yachts, powerboats, and motorboats, the power required can be computed from $P = (VB/C)^3/L$, where P = bhp (brake horsepower) to propel the boat or yacht at V statute mi/h; B = maximum waterline beam, ft; L = load waterline length, ft; C = a constant from Table 4.

MARINE PROPELLER-SHAFT DIAMETER AND PROPELLER SLIP

What diameter line and tail or propeller shafts should be used for a 15,000-shp (11,185 kW) single-screw 17-kn (31.5-km/h) ocean vessel if the propeller turns at 110 r/min at the service speed of the ship? The tail shaft will be fitted with a continuous bronze liner, and the propeller is 22.5 ft (6.9 m) in diameter. What is the twisting moment on the shaft if the ship is driven by a geared steam turbine? What are the apparent and true slip of the propeller at normal speed if the propeller pitch is 15.75 ft (4.8 m)?

TABLE 4 Coefficient for Powerboat Power Relation°

Powerboat type	C	Powerboat type	C
Heavy cruiser	8–9	Heavy runabout	11.7–14.3
Medium cruiser	8.4–10	Average runabout	14–16.3
Light cruiser	9.2–11.3	Racing runabout	15.8–18.2
Express cruiser	9.8–12.6	Hydroplane	18–20

°White—*Yachting.*

Calculation Procedure:

1. *Compute the line-shaft diameter*

Use the American Bureau of Shipping relation $d_1 = c(KP/N)^{1/3}$, where d_1 = line-shaft diameter, in; c = 1.0 for line shafts, 1.05 for thrust shafts transmitting torque; K = 64 for ocean and coastwise service, 58 for river and harbor service; P = ship shp; N = propeller r/min. For this ship, d_1 = $1.0(64 \times 15,000/110)^{1/3}$ = 20.6 in, say 20.75 in (52.7 cm).

2. *Compute the tail- or propeller-shaft diameter*

Use the relation $d_p = d_1 + d/b$, where d_p = propeller- or tail-shaft diameter, in; d_1 = line-shaft diameter, in; d = propeller diameter, ft; b = 12 for a continuous bronze liner, 8.3 if liners are fitted only at the bearings. For this vessel, d_p = 20.75 + 22.5/12 = 22.62 in, say 22.75 in (57.8 cm). Thus, the tail shaft is 2 in (5.1 cm) larger in diameter than the line shaft.

3. *Compute the shaft twisting moment*

Use the relation $M_t = 63.024(P/N)$, where M_t = shaft twisting moment, in·lb; other symbols as before. For this vessel, M_t = 63.024(15.000/110) = 8600 in·lb (971.7 N·m).

4. *Compute apparent slip of the propeller*

Use the relation $s_a = (pN - 101.3V)/pN$, where s_a = apparent slip = slip ratio; p = propeller pitch, ft; N = propeller r/min; V = ship speed, kn. So s_a = $(15.75 \times 110 - 101.3 \times 17)/(15.75 \times 110)$ = 0.006, or 0.6 percent.

5. *Compute the true slip of the propeller*

Use the relation $s_t = (pN - 101.3V_a)/pN$, where s_t = propeller true slip; V_a = speed of advance of the propeller, kn = $V(1 - w)$, where w = wake fraction for the vessel; other symbols as before. For w = 0.35, V_a = 17(1 − 0.35) = 11.05 kn (20.44 km/h). Then s_t = $(15.75 \times 110 - 11.05 \times 101.3)/(15.75 \times 110)$ = 0.352, or 35.2 percent.

Related Calculations: The apparent slip of a propeller can be negative, particularly if the vessel is a single-screw ship at light draft. Apparent slip varies from 5 to 30 percent, with 10 percent being a typical value.

Real or true slip is based on the pitch speed of the propeller through the moving water or wake. The real or true slip is greater than the apparent slip.

Table 5 lists typical values of the wake fraction w for various types of vessels.

PROPELLER SELECTION FOR VESSELS

Choose a three-bladed propeller for a 15,000-shp (11,185 kW) single-screw 17-kn (31.5-km/h) ocean vessel if the maximum allowable diameter of the screw is limited to 22.5 ft (6.9 m). What is the thrust on the propeller blades? Will this propeller cavitate?

Calculation Procedure:

1. *Compute the propeller speed of advance through the wake*

Use the relation $V_a = V(1 - w)$, where V_a = speed of advance of the propeller through the wake, kn; V = vessel speed, kn; w = wake fraction = wake velocity, kn/vessel velocity, kn.

TABLE 5   Typical Wake-Fraction Values*

Vessel type	Single-screw	Twin- or quadruple-screw
Tankers and slow cargo vessels	0.35	0.20
Passenger and fast cargo vessels	0.30	0.15
Yachts and bay steamers	0.25	0.10
Scout cruisers and tugs	0.20	0.05
Destroyers and motorboats	0.15	0

*Baker—*Introduction to Steel Shipbuilding*, McGraw-Hill.

TABLE 6   Best Combinations of Propeller Diameter and Revolutions per Minute° †

Combination‡	Pitch-diameter ratio, $a = p/d$	Diameter factor	R/min factor	Efficiency e
1	1.4	19.3	4.5	0.78
2	1.3	17.4	5.5	0.765
3	1.2	16.0	6.5	0.75
4	1.1	14.5	8.0	0.735
5	1.0	13.0	10.0	0.72
6	0.9	11.4	13.0	0.695
7	0.8	9.0	20.0	0.65
8	0.7	6.8	33.0	0.59
9	0.6	4.9	60.0	0.51

°Baker—*Introduction to Steel Shipbuilding*, McGraw-Hill.
†For even pitch ratio p/d.
‡Since these are the best combinations, use them where circumstances permit. For other combinations, use the Taylor or Baker curves.

Table 5 lists typical values of the wake fraction for single- and multiple-screw ships of various types. Thus, for a single-screw tanker, $w = 0.35$, and $V_a = 17(1 - 0.35) = 11.05$ kn (20.44 km/h).

2. Compute the propeller diameter factor

The diameter factor is used in determining the best combinations of propeller diameter and pitch for a given pitch ratio. Use the relation $D = dV_a^{3/2}/P^{1/2}$, where D = propeller diameter factor; d = propeller diameter, ft; P = propeller shp. For this ship, $D = 22.5(11.05)^{3/2}/15,000^{1/2} = 6.79$, say 6.8

3. Determine the propeller r/min factor

Refer to Table 6 and read the r/min factor B as 33.0 opposite the diameter factor $D = 6.8$.

4. Compute the propeller r/min

Use the relation r/min $= BV_a^{5/2}/P^{1/2} = 33(11.05)^{5/2}/15,000^{1/2} = 109.9$, say 110 r/min.

5. Compute the propeller pitch

Read from Table 6 the pitch-diameter ratio $a = p/d = 0.7$, where p = propeller pitch, ft; d = propeller diameter, ft. So $p = da = 22.5(0.7) = 15.75$ ft (4.8 m), closely.

6. Determine the propeller efficiency

Read the propeller efficiency in Table 6 as 59 percent. This is an acceptable efficiency for a propeller of this type and size.

7. Compute the thrust on the propeller blades

Use the relation $T = 33,000eP/(101.3V_a)$, where T = thrust on propeller blades, lb; e = propeller efficiency; other symbols as before. For this vessel, $T = 33,000(0.59)(15,000)/[101.3(11.05)] = 260,500$ lb (1159 kN).

8. Compute the thrust on the propeller blades

Propeller thrust is also expressed in pressure terms, i.e., pounds per square inch of actual blade area. When the developed area ratio a_r is known (0.40 is a typical value for slow- and moderate-speed vessels), the thrust is $T_p = T/(36\pi a_r d^2)$ lb/in². With $a_r = 0.40$, $T_p = 260,500/[36\pi(0.40)(22.5)^2] = 11.36$ lb/in² (78.3 kPa).

9. Determine whether the propeller will cavitate

Compute the propeller tip speed from $t = \pi dN$, where t = propeller tip speed, ft/min; N = propeller r/min; other symbols as before. So $t = \pi(22.5)(110) = 7800$ ft/min (39.6 m/s).

Table 7 lists tip speeds and pressure thrusts of cavitating propellers. To prevent cavitation, the actual pressure thrust should be about 10 percent less than the tabulated pressure thrust. Since the

TABLE 7 Critical Thrust of Propellers at Various Tip Speeds°

Propeller tip speed							
ft/min	2,000	4,000	6,000	8,000	10,000	12,000	14,000
m/s	10.2	20.3	30.5	40.6	50.8	61.0	71.1
Thrust							
lb/in²	1.2	5.6	12.0	18.2	23.6	28.5	33.0
kPa	8.3	38.6	32.7	125.5	162.7	196.5	227.5

°Developed by Commander Irish.

actual pressure thrust, 11.36 lb/in² (78.3 kPa), is 10 percent less than the cavitating pressure thrust at a tip speed of 8000 ft/min (40.6 m/s), the nearest tabulated tip speed, this propeller will *not* cavitate. Hence, it is an acceptable propulsion unit for this vessel. Also, the assumed developed area ratio, 0.40, is safe, as far as cavitation is concerned.

Related Calculations: Use this general method for choosing a propeller to run at a limited r/min or limited r/min and diameter. If desired, Taylor's diagram (Taylor—*The Speed and Power of Ships*, Ransdell, Inc., Washington, D.C.) for three-bladed propellers can be used in place of Table 6. Where desired, the wake factor or fraction can be approximated from the Taylor relation: $w = 0.5b - 0.05$ for single-screw ships, and $w = 0.55b - 0.2$ for twin-screw ships, where b = vessel block coefficient of fineness.

To determine the *approximate* optimum diameter of a three-bladed propeller, use the relation $d = 50P^{0.2}/(r/min)^{0.6}$. By this relation, the optimum diameter for the above propeller is 20.25 ft (6.17 m). A four-bladed propeller of 0.97 × the diameter of a three-bladed propeller, the same pitch ratio, and 1.33 × the area will absorb the same shp at the same r/min as the three-bladed propeller. Likewise, a two-bladed propeller of 5 percent greater diameter is about equivalent to a three-bladed propeller.

The *developed area ratio* of a propeller = developed area of blades, in² (cm²)/area of circle of same diameter as propeller, in² (cm²). Even though a value of 0.40 was assumed for a_r, this value can range up to 1.0 for high-speed, high-power vessels.

The propulsive coefficient (PC) of a marine propeller = ehp/shp (ekW/skW), where *ehp* = the effective or tow-rope hp (kW) = $0.00307VR$ ($0.00229VR$), where R = total resistance of the vessel, lb (kg). Typical values of PC are 0.60 for a single-screw vessel and 0.57 for a twin-screw vessel; these values can increase to 0.80 and 0.70, respectively, for moderate-power (i.e., merchant) ships with well-designed propellers and sterns.

Use the same procedure as that given above to select propellers for small craft (tugs, powerboats, auxiliaries, etc.).

PROPELLER REVOLUTIONS AND VESSEL SPEED

At what r/min should a propeller having a 25-ft (7.6-m) pitch be turned to drive a vessel at 22 kn (40.7 km/h) if the apparent slip of the propeller is 12 percent? How many revolutions must the propeller turn to drive the vessel 50 nmi (92.6 km) in calm weather? What is the speed of this vessel when the propeller is turning at 70 r/min?

Calculation Procedure:

1. *Compute the required r/min of the propeller*

Use the relation $N = 101.3V/[p(1 - s_a)]$, where N = required propeller r/min; V = vessel speed, kn; p = propeller pitch, ft; s_a = propeller apparent slip, expressed as a decimal. So $N = 101.3(22)/[25(1 - 0.12)] = 101.3$, say 102 r/min.

2. *Compute the required number of propeller revolutions*

Use the relation $n = 6080d/p(1 - s_a)$, where n = total number of turns of the propeller to drive the vessel d nmi; other symbols as before. Thus, $n = 6080(50)/[25(1 - 0.12)] = 13,820$ r.

To check this result, divide the distance traveled by the vessel speed to determine the running time in hours. Or, T = distance, nmi/speed, kn = $50/22$ = 2.27 h. Since there are 60 min/h and the required propeller r/min is 101.3, n = 2.27(60)(101.3) = 13,820 r.

3. Compute the vessel speed at the different r/min

Use the relation $N_1/N_2 = V_1/V_2$, where subscripts 1 and 2 refer to different propeller and vessel speeds, respectively. Substituting gives $101.3/70 = 22/V_2$, V_2 = 15.2 kn (7.8 m/s).

Related Calculations: Operating engineers aboard seagoing vessels generally use the apparent slip when computing the performance of their vessel at sea. Hence, the procedures given here are useful in propeller design and selection and vessel operation. By using the method of step 3, the vessel speed at various r/min, or the r/min for various speeds, is easily determined. In using this method, the apparent slip is assumed to be constant. This is a valid assumption for practical calculation purposes.

VESSEL IMMERSION AND FLOODING EFFECTS

A 600-ft (182.9-m) tanker has an 82.5-ft (25.1-m) beam and a water-plane coefficient of fineness of 0.76 at a loaded draft of 31.9 ft (9.7 m). How many tons of cargo must be placed aboard this vessel to make it sink 1 in (2.5 cm)? How many tons of water will enter this vessel if a 2 × 2 ft (0.6 × 0.6 m) hole is stove in the hull 16 ft (4.9 m) below the waterline? What effect will this hole have on the vessel?

Calculation Procedure:

1. Compute the vessel water-plane area

Use the relation $A_w = LB\alpha$, where A_w = waterplane area, ft²; L = vessel length on waterline, ft; B = vessel beam at waterline, ft; α = water-plane coefficient of fineness. For this vessel, A_w = 600(82.5)(0.76) = 37,600 ft² (3493.2 m²).

2. Compute the weight required to increase immersion

The weight required to increase the immersion of the vessel 1 in (2.5 cm) is $W = A_w/420$ in saltwater, where W = weight, tons. Thus, for this vessel, W = 37,600/420 = 89.6 tons (81.3 t).

3. Compute the quantity of water entering the vessel

Use the relation $Q = 13.7(H)^{0.5}A$, where Q = weight of water entering the vessel, tons/min; H = distance of centerline of hole below the waterline, ft; A = area of hole, ft². For this vessel, Q = $13.7(16)^{0.5}(4)$ = 219.5 tons/min (199.1 t/min).

4. Compute the rate of vessel settling

Step 2 shows that the vessel will settle 1 in (2.5 cm) for every 89.6 tons (81.3 t) taken aboard. With water entering at the rate of 219.5 tons/min (199.1 t/min), the vessel will sink 219.5/89.6 = 2.44 in/min (0.10 cm/s). Assuming a constant water-plane shape (which is not quite true), the vessel will sink, in 1 h, 60(2.44)/12 = 12.20 ft (3.7 m). With a freeboard of 12 ft (3.7 m), to the main deck, this deck would be awash about 1 h after the hole were stove in the hull.

Related Calculations: Use this general procedure to determine the effect of loading cargo aboard any vessel whose dimensions are known. When the vessel operates in fresh water, as in the Great Lakes, compute the weight required to increase the draft 1 in (2.5 cm) from $W = A_w/409$.

Environmental considerations impact channels, waterways, and berths of deep-draft vessels in ports throughout the world. Thus, when dioxin-contaminated sediment is removed from a channel or berth sea bed, its dumping at sea or other disposal may violate environmental laws.

Where burial of contaminants in offshore sites is chosen as a disposal method, capping with clean sand is often required. The clean sand is poured on top of the sediment after the contaminated materials reach the bottom of the ocean. Thus, in a recent capping operation, some 2.2 million yd³ (1.68 million m³) of clean sand was used to cap 450,000 yd³ (344,045 m³) of dioxin-contaminated sediment buried offshore.

Removal of the dioxin-contaminated sediment allowed the deepening of berths and the channel leading to them in a large port. Dredge vessels are used to remove sediment and cap it. Availability of such vessels may be limited, delaying much-needed deepening of harbors and channels. Environmental agencies will often ban the disposal of contaminated sediment near

TABLE 8 Typical Pumps Used for Marine Service

Service	Usual pump type	Typical total head lb/in²	kPa	Capacity gal/min	L/st
Lube-oil	Rotary	50	345	$36 + (7.5 \text{ shp} + 1300)^{0.5}$	$0.0631 [36 + (7.5 \text{ shp} + 1300)^{0.5}]$
Sanitary or flushing	Centrifugal	100	689	(1.6 gal/min)/1000 td°	(0.11 L/s)/1000-t displacement
Fire protection	Centrifugal	125	862	At least 800 gal/min in two pumps	At least 50 L/s in two pumps
Ballast	Centrifugal	50	345	(35 gal/min)/1000 td	2.4 L/s per 1000-t displacement
Bilge	Centrifugal	50	345	(35 gal/min)/1000 td	2.4 L/s per 1000-t displacement
Fuel-oil transfer	Rotary	50	345	At least one 225-gal/min pump	At least one 14.2-L/s pump
Fuel oil	Rotary	400	2758	Depends on steam capacity of boilers	Depends on steam capacity of boilers
Refrigeration condenser	Centrifugal	25	172	5 gal/(min·ton) of refrigeration	0.39 L/(s·t) of refrigeration
Fresh water	Centrifugal	75	517	(3 gal/min)/1000 td	0.2 L/s per 1000-t displacement
Ice water	Centrifugal	25	172	(0.5 gal/min)/1000 td	0.03 L/s per 1000-t displacement
Condenser circulating	Centrifugal	50	345	1 gal/min per 5 lb/h of steam condensed	0.06 L/s per 2.3 kg of steam condensed

°td = tons displacement of the vessel.
†All tons in this column are metric tons.

fishing grounds, further complicating the problem of deepening channels and berths.

With container ships growing in size and tonnage, deeper channels and berths are a necessity in harbors that want to continue attracting marine traffic. A 40-ft (12.2-m) depth is often cited as the minimum for busy modern ports. Where less depth is available, vessel sailings and arrivals must be coordinated with the times of high tides.

Disposal costs vary with the type of material being handled. Thus, clean clay and sand dumped at sea can be handled for $6 per yd³ (0.76 cu m³). Badly contaminated sediment can cost $100 per yd³ (0.76 m³) when dumped in landfills. An environmental impact statement must often be prepared before any disposal can begin. This requirement will often delay the deepening of harbors and berths for the larger tankers and container vessels plying trade routes today.

MARINE PUMP SELECTION FOR COMMERCIAL VESSELS

Choose the capacity, head, and types of pumps for a 10,000-ton (9071.8-t) cargo vessel for the following shipboard services: lube oil, sanitary or flushing, fire protection, ballast, bilge, fuel-oil transfer, and refrigeration condenser, fresh-water, ice-water-, main condenser and auxiliary condenser circulation. The shp of this steam-turbine-driven vessel is 6000 hp (4474.2 kW) with a steam flow of 37,500 lb/h (4.7 kg/s). What is the power input to the main-condenser circulating pump?

Calculation Procedure:

1. *Compute the lube-oil pump size*

Table 8 lists the usual type, total pressure, and capacity for various pumps used in marine service.

Thus the lube-oil pump capacity $C = 36 + (7.5\ shp + 1300)^{0.5}$, where C = pump capacity, gal/min; shp = installed shp in the vessel. So $C = 36 + (7.5 \times 6000 + 1300)^{0.5} = 251.5$ gal/min (15.9 L/s). A 260-gal/min (16.4-L/s) rotary pump (gear, screw, lobe, etc.) rated at a total head of 50 lb/in² (344.8 kPa) would be chosen for this ship.

2.   Compute the sanitary pump size

Table 8 shows that sanitary or flushing pumps should have a capacity of 1.6 gal/min (0.10 L/s) per 1000-ton (907.1-t) displacement of the vessel. Since this is a 10,000-ton (9071.8-t) vessel, the required capacity of the sanitary pump is $C = 1.6(10,000/1000) = 16$ gal/min (1.0 L/s). A 20-gal/min (1.6-L/s) centrifugal pump developing a total pressure of 100 lb/in² (689.5 kPa) would probably be chosen for this vessel.

3.   Select the fire-pump size

The American Bureau of Shipping and other maritime agencies publish recommendations or requirements for marine fire protection. Whereas the recommendations or requirements vary with the type of vessel and its intended service, a fire-protection pump capacity of at least 800 gal/min (50.5 L/s) in two pumps—usually 400 gal/min (25.2 L/s) each—is generally needed. Centrifugal pumps developing a total head of 125 lb/in² (861.9 kPa) or more are almost universally used for marine fire protection.

4.   Compute the ballast and bilge pump sizes

Table 8 shows that ballast and bilge pump capacity are based on the same relation. Or $C = 35(10,000/1000) = 350$ gal/min (22.1 L/s). Likewise, the total head developed is usually 50 lb/in² (344.8 kPa) or more. Modern vessels use centrifugal pumps for ballast and bilge service.

5.   Select the fuel-transfer and fuel-service pumps

Many steam vessels use one or more 225-gal/min (14.2-L/s) rotary pumps for fuel-oil transfer. The total pressure developed by the pump is 50 lb/in² (344.8 kPa) or more.

   To determine the capacity of the fuel-service pump, use the relation $C = W/486$, where W = weight of fuel oil pumped per hour, lb. If the pump handles 6000 lb/h (0.8 kg/s) of fuel, $C = 6000/486 = 12.34$ gal/min (0.8 L/s). With a 30 percent reserve capacity for overloads, $C = 1.3(12.34) = 16$ gal/min (1.0 L/s), closely.

   Install two fuel-service pumps, each rated at 16-gal/min (0.1-L/s) capacity and 400-lb/in² (2758.0-kPa) total pressure. Two full-capacity fuel-service pumps are needed so that one pump is available while the other is being overhauled. On some vessels a third pump rated at one-half capacity may also be installed. In modern ships, fuel-service pumps are almost always the rotary type.

6.   Compute the size of the refrigeration-condenser pump

Table 8 shows that the refrigeration condenser pump must supply 5 gal/(min·ton) [0.3 L/t·s)] of installed refrigeration capacity. With a 500-ton (453.6-t) refrigerating plant, $C = 5(500) = 2500$ gal/min (157.7 L/s). Use a centrifugal pump developing a total head of 25 lb/in² (172.4 kPa) or more.

7.   Compute the size of the fresh- and ice-water pumps

Table 8 shows that the fresh-water pump should deliver 3 gal/min per 1000-ton (0.2 L/s per 1000-t) displacement of the vessel. Or, $C = 3(10,000/1000) = 30$ gal/min (1.9 L/s). Use a centrifugal pump rated for a total head of 75 lb/in² (517.1 kPa) or more.

   Likewise, by using data from Table 8 for the ice-water pump, $C = 0.5(10,000/1000) = 5$ gal/min (0.32 L/s). Use a centrifugal pump rated for a total head of 25 lb/in² (172.4 kPa) or more.

8.   Compute the size of the steam-condenser circulating pumps

Table 8 shows that the main- and auxiliary-condenser circulating pumps must supply 1 gal/min of seawater per 5 lb (1 L/s per 36 kg) of steam condensed. This flow rate is based on a 10°F (5.6°C) rise in the seawater temperature during passage through the condenser, a specific heat of the seawater of 0.94 Btu/(lb·°F) [3.94 kJ/(kg·°C)], a seawater weight of 8.58 lb/gal (1.03 kg/L) under average conditions, and a reserve capacity of 20 percent of the required capacity.

   With a steam flow of 37,500 lb/h (17,045.5 kg/h) to the main condenser, the circulating pump

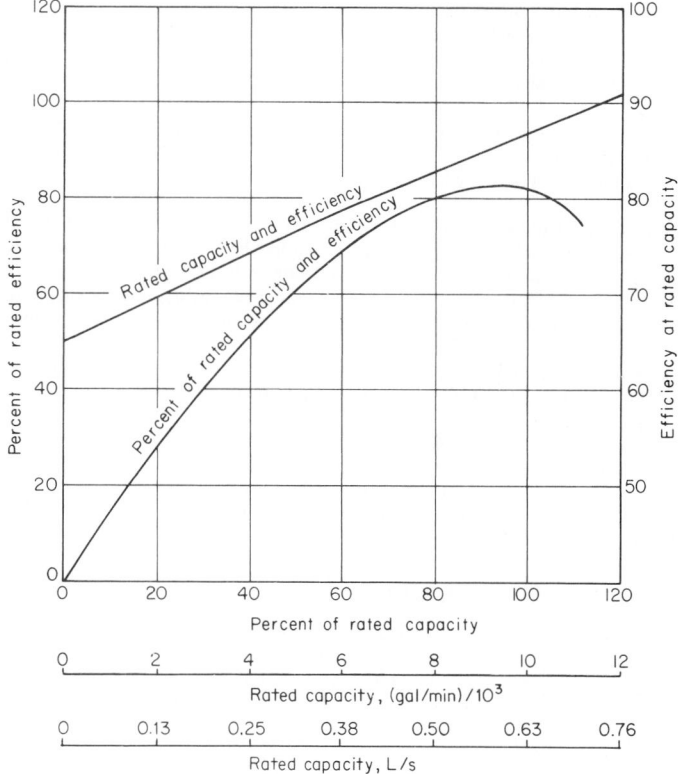

FIG. 2 Hydraulic characteristics of a typical centrifugal pump for marine service.

should have a capacity of $C = 37,500/5 = 7500$ gal/min (473.2 L/s). To provide for overloads and reduced heat transfer, use a 20 percent larger capacity, or 1.2(7500) = 9000 gal/min (567.8 L/s). Choose a centrifugal pump developing 50-lb/in² (344.7-kPa) total head. Install two full-capacity pumps per main condenser.

Compute the required capacity of the auxiliary-condenser circulating pump in the same way, using the auxiliary steam flow as the basis of the required seawater flow rate and pump capacity. Choose a pump having a 20 percent larger capacity, as was done above.

9. *Compute the power input to the circulating pump*

As a guide to computation of the power input to any marine pump, the main-condenser circulating pump will be analyzed.

The rated capacity of this pump is 9000 gal/min (567.8 L/s), including a 20 percent reserve. Refer to the characteristic curve of the pump, Fig. 2, to determine the efficiency of the pump at this rated flow rate. Figure 2 shows that the efficiency is 85 percent at a rated capacity of 9000 gal/min (567.8 L/s).

Compute the required power input to the pump from hp = gal/min(head, lb/in²)/1715(efficiency), or hp = 9000 (50)/1715(0.85) = 309 hp (230.5 kW). Use a 350-hp (261.1-kW) motor to drive the pump, because this is the next largest standard-size motor.

Next, compute the percent operating capacity of the pump from actual gal/min required/pump rating, gal/min. Since the pump has a 20 percent reserve capacity, actual gal/min required = 9000/1.20 = 7500 gal/min (473.2 L/s). Hence, operating capacity = 7500/9000 = 0.833, or

~~83.3 percent of the pump rated capacity.~~

Enter Fig. 2 at 83.3 percent of the pump rated capacity, and project to the percent of rated capacity and efficiency curve. At the left read percentage of rated efficiency of the pump = (percentage of rated efficiency at actual capacity)(percentage efficiency at pump rated capacity) = (0.81)(0.85) = 0.688, or 68.8 percent.

Compute the pump operating hp from hp = (actual gal/min)(head, lb/in^2)/1715(pump actual operating efficiency), or hp = (7500)(50)/[1715(68.8)] = 318 hp (237.2 kW). With a 350-hp (261.1-kW) motor the operating load is 318/350 = 0.907, or 90.7 percent of the rated motor capacity.

To convert the mechanical power input to the pump to the electric power input to the motor, i.e., the electrical load on the ship's generator caused by the main-condenser circulating-pump drive motor: (*a*) Determine the motor efficiency at full load from the motor characteristic curve; (*b*) determine the motor efficiency at the actual operating rating, again using the motor characteristic curve; (*c*) take the product of these two efficiencies; and (*d*) convert the motor operating hp to kilowatts and divide by the product *c*.

Thus, with *a* = 0.96 and *b* = 0.85, the input to this motor = 318 hp (0.746 kW/hp)/ [(0.96)(0.85)] = 291 kW, say 300 kW. The power input to any of the other pumps is determined in the same way.

Related Calculations: Use this general method for preliminary sizing of pumps for steam-propelled vessels of any type—tankers, dry-cargo, passenger, etc. For best results, assemble a set of typical pump and motor characteristic curves, such as Fig. 2. These will shorten the time spent on pump sizing. Note that the discharge pressures listed in Table 8 may vary if a vessel is used for special services.

MARINE POWER-PLANT SELECTION

Choose a power plant suitable for a 700-ft (213.4-m) 30,000-ton (27,213 t) oceangoing tanker that must have a service speed of 20 kn (37 km/h). Indicate the factors entering the decision. The owners require a low fuel rate, small investment, and minimum maintenance cost. List the typical installation and capacity factors for the type of power plant selected.

TABLE 9   Typical Marine Power-Plant Choices

Power plant type	Typical application
Geared steam turbine	Ocean and Great Lakes vessels of 6000 shp (4474.2 kW) or more
Turboelectric drive	Ocean and Great Lakes vessels of 2000 shp (1491.4 kW) or more in service requiring much maneuvering
Reciprocating steam engine	Rarely used in new vessels today except where simple machinery is required in power ranges up to 5000 bhp (3728.5 kW) per shaft
Direct diesel drive	Ocean and Great Lakes vessels with power requirements up to 30,000 bhp (22,371 kW) per shaft, although usual installations range up to about 15,000 bhp (11,185.5 kW) per shaft; also popular for harbor craft—tugs, sewage vessels, etc.
Diesel-electric drive	Popular for harbor craft (tugs, ferries, etc.) where maneuvering is a primary consideration; also fitted to ocean and Great Lakes vessels
Geared diesel drive	Popular for harbor craft but is also finding use in ocean and Great Lakes vessels
Nuclear power	Ocean surface and undersea vessels
Gasoline engine	Limited to relatively small vessels up to 60 ft (18.3 m) plying the coastal and inland waters
Gas turbine	Limited use on large and small vessels in ocean and coastal service

TABLE 10 Comparative Costs of Marine Power Plants°

Type of plant	Typical fuel rate, lb/(shp·h) (kg/kWh)	Lube-oil cost	Maintenance cost	Relative cost
Geared steam turbine	0.50–0.56 (0.30–0.34)	Nominal	Nominal	190
Turboelectric drive	0.60 (0.36)	Nominal	Moderate	210
Reciprocating steam engine	0.75–0.90 (0.46–0.55)	Nominal	Nominal	150
Direct diesel drive	0.38 (0.23)	High	High	220
Diesel-electric drive	0.40 (0.24)	High	High	230
Geared diesel drive	0.40 (0.24)	High	High	225
Nuclear power	. . .	Nominal	High	300
Gasoline engine	0.5 (0.30)	High	High	
Gas turbine	0.39–0.60 (0.24–0.36)	High	High	210

°Output of 2000 shp (1491.4 kW) or more.

Calculation Procedure:

1. *Compute the power required to propel the vessel*

Use the admiralty method given earlier. Or, $P = D^{2/3}V^3/K$, where P = shp required to drive the vessel; D = vessel displacement, long tons; V = vessel cruising or service speed, kn; K = admiralty coefficient from Table 3. With $K = 450$, by interpolation in Table 3, $P = (30,000)^{2/3}(20)^3/450 = 17,150$ hp (12,794 kW). This power requirement is a typical value for vessels of this size.

2. *Select the type of power plant to use*

In the power range needed for this vessel—17,000 hp (12,682 kW)—the geared steam turbine (Tables 9 and 10) will provide the lowest fuel rate and lowest investment if the reciprocating steam engine is ignored. Few large vessels being built today use reciprocating steam engines.

Diesel propulsion, although suitable for this vessel, would be more expensive because (*a*) fuel cost is higher if diesel oil is used; (*b*) lube-oil cost is higher; (*c*) the initial investment for the engines is higher; (*d*) the weight of the engines is about 40 percent greater than that of a geared steam turbine of equal power output; (*e*) diesel-engine maintenance costs are substantially higher than are steam-turbine maintenance costs.

3. *List typical installation and capacity factors*

Watertube three-pass boilers weighing 24 to 30 lb/ft² (117.4 to 146.7 kg/m²) of heating surface are used in modern oceangoing merchant vessels. When an air heater is installed, as is often done today, it weighs about 6.5 lb/ft² (31.8 kg/m²). D-type watertube boilers weigh 20 to 30 lb/ft² (97.8 to 146.7 kg/m²) of heating surface. Where an economizer is used, add its weight to that of the boiler. The economizer feedwater temperature is usually in the 400°F (204.4°C) range. Currently, 600-lb/in² (4136-kPa) 875°F (468°C) boilers are popular for standard merchant vessels.

Usual geared steam turbine installations use two turbines: a high-pressure and a low-pressure turbine driving one propeller shaft. Large ships, and high-speed vessels, may use three turbines per propeller shaft. With this arrangement the turbine receiving steam from the boiler is called the *cruising turbine*. Monel and low-carbon steel blades are often used in modern marine turbines.

TABLE 11 Ocean-Going Vessel Performance

	Vessel Type					
	Liquid carrier		Bulk cargo		General cargo	
Propulsion	Oil	Nuclear	Oil	Nuclear	Oil	Nuclear
Machinery weight, tons (t)	250 (227)	2,000 (1,814)	250 (227)	2,000 (1,814)	250 (227)	2,000 (1,814)
Fuel used, tons/nmi (t/km)	0.5 (0.25)	. . .	0.5 (0.25)	. . .	0.5 (0.25)	
Cargo weight, tons (t)	25,000 (22,750)	25,000 (22,750)	25,000 (22,750)	25,000 (22,750)	25,000 (22,750)	25,000 (22,750)
Voyage distance, mi (km)	5,000 (8,045)	5,000 (8,045)	10,000 (16,090)	10,000 (16,090)	15,000 (24,135)	15,000 (24,135)
CF ratio	9.1	12.5	4.77	12.5	3.23	12.5

Reduction gears generally provide an 8:1 speed reduction between the turbine and the propeller shaft. Double-helical gears are used to neutralize unbalanced end thrust in the gears. The tooth opening resulting from torsional and bending deflection is limited to 0.001 in (0.03 mm). Typical weights of complete reduction gears, lb = 1300(shp/propeller r/min), according to J. F. Nace. The usual efficiency of large reduction gears ranges from 97 to 98.5 percent.

Typical marine surface condensers have 0.9 to 1.6 ft^2/shp (0.11 to 0.20 m^2/kW) of heat-transfer surface. The temperature rise of the seawater during passage through a condenser serving a geared steam turbine is 6 to 10°F (3.3 to 5.6°C) when the inlet temperature is 75°F (23.9°C).

Size the condensate-pump inlet pipe for a flow velocity of less than 2 ft/s (0.61 m/s). Choose the main boiler feed pump such that it has a capacity that will permit the rated cruising speed when the pump is delivering 70 percent of its rated capacity.

When a vessel is equipped with all electrically driven auxiliaries and an electric galley, the generator load, kW = 0.75 (number of persons aboard the vessel) + 0.025 (shp of the vessel). With a turbine-driven main feed pump, kW = 0.75 (number of persons aboard the vessel) + 0.017 (shp of the vessel).

In air ejectors, the quantity of dry air removed, lb/h = 7.5 + 0.00025 (condensate flow, lb/h). With a 28.5-in (723.9-mm) vacuum in the condenser, the weight of air and vapor removed by the air ejector is 2 to 2.5 times that given above. Usual air ejectors use 5 lb (2.3 kg) of steam per lb (kg) of air and vapor mixture removed.

Steam soot blowers use, in lb of steam, 0.008 (total heating surface of boiler, ft^2) (number of blows per day), if each soot blower blows for 45 s.

Related Calculations: Use this general method to choose the type of power plant for any vessel, large or small. When two or more types of power plants will provide equal service, the final choice must be based on an economic comparison of the alternative plants.

Some of the newest steam-propelled vessels, such as the 25-kn (46.5-km/h) roll-on roll-off trailership *Ponce de Leon*, use only one steam boiler. This boiler is usually of the reheat type. Other vessels using only one boiler are the tankers *Esso Houston, Esso New Orleans*, and the 206,000-deadweight-ton (186,842-t) *Idemitsu Maru*.

Fuel consumption when using reheat is 0.43 lb/(shp·h) (0.26 kg/kWh) of oil. The initial cost of a reheat boiler plant is 15 to 20 percent higher than a nonreheat plant. The reheat section of the boiler is used not when the vessel is maneuvering, only when it is at sea.

NUCLEAR PROPULSION FOR OCEANGOING VESSELS

A transportation firm is considering replacing its oil-fueled vessels with nuclear-powered vessels. Compare the performance of the vessels listed in Table 11 on the basis of the cargo-fuel ratios for the routes listed. What magnitude of tank-top loading can be expected in nuclear-powered vessels? List the types of reactors suitable for marine applications of nuclear power.

TABLE 12 Types of Marine Reactors

Reactor type	Steam conditions	Advantages	Disadvantages
Pressurized-water (used on the N.S. *Savannah*)	460 lb/in² saturated (3171 kPa saturated)	Simple; easy to control	Requires large heat exchangers; boiling takes place outside the reactor; large pumps needed
Boiling-water	460 lb/in² saturated (3171 kPa saturated)	Does not need heat exchangers; requires little pumping power; boiling occurs in reactor	Precise pressure and water-level control needed
Moderated	460 lb/in² superheated (3171 kPa superheated)	Simplest reactor design; produces superheated steam at low reactor pressures	Coolant may decompose, requiring makeup
Liquid-sodium	650 lb/in², 850°F (4481 kPa, 454°C)	Reactor operates at 14.7 lb/in² (abs) (101.3 kPa)	Uses more complex coolant piping; needs more pumps; requires moderator
Gas-cooled	High pressures and temperatures	Simple; safer than other types; higher steam pressures and temperature possible	Needs gas blowers; uses large pumping power

Calculation Procedure:

1. Compute the vessel cargo-fuel ratios

The cargo-fuel (CF) ratio of any vessel is the ratio of cargo capacity, tons/(fuel capacity, tons + steam-generating apparatus weight, tons). Table 11 shows the fuel and machinery weights for each of three ships propelled by oil and nuclear power.

For the data in Table 11 for the tanker or liquid carrier, CF = 25,000 tons/(0.5 tons/nmi × 5000 nmi + 250 tons machinery weight) = 9.1. For the nuclear-powered tanker, CF = 12.5. Compute the CF values for the other vessels, and list the results in Table 11.

2. Evaluate the cargo-fuel ratios computed

For any ocean trade route, the higher the CF ratio, the more economical the vessel. Table 11 indicates that the nuclear-propelled vessel has a higher CF ratio for each type of cargo carried and for each voyage length. This tabulation also shows that the nuclear-propelled vessel has a greater advantage from the CF standpoint for the longer voyages. The reason is that the oil-fueled vessel CF decreases as the voyage length increases, whereas the nuclear-propelled vessel CF remains constant. Hence, the nuclear-propelled vessel is more economical for all the routes considered.

3. Compute the nuclear-propelled vessel tank loading

A modern pressurized-water marine reactor plant developing 20,000 shp (19,914 kW) weighs about 2000 tons (1820 t) including its fuel, shield, and related apparatus. Such a reactor will require some 2000 ft² (185.8 m²) of tank-top area. Thus, the load on the tank top is (2000 tons × 2000 lb/ton)/(2000 ft²) = 2000 lb/ft² (9765 kg/m²).

Usual modern merchant vessels have a tank-top loading of 1500 to 1800 lb/ft² (7324 to 8788 kg/m²). Hence, the hull of a nuclear-propelled vessel must be strengthened to carry the extra load of the reactor. This strengthening will, of course, increase the cost of the hull.

4. List the types of reactors suitable for the vessels

Table 12 shows the types of nuclear reactors suitable for marine propulsion systems. The major

TABLE 13 Typical Tanker Cargo-Discharge Times

Tanker type or name	DWT, tons (t)	Cargo capacity of 42-gal bbl (L)	Pump capacity, bbl/h (L/s)	Discharge time, h
T-1	16,800 (15,241)	141,000 (22,414,770)	8,600 (379.8)	16.5
T-5	26,500 (24,040)	204,000 (32,429,880)	17,600 (777.2)	11.6
T-5-5	25,000 (22,680)	190,000 (30,204,300)	23,400 (1,033.3)	8.1
Pennsylvania	28,170 (25,555)	241,500 (38,391,255)	23,400 (1,033.3)	10.5
Alton Jones	38,000 (34,473)	336,000 (53,413,920)	30,600 (1,351.2)	11.0
World Glory	45,500 (41,277)	396,000 (62,952,120)	20,000 (883.2)	19.8
Barracuda	60,000 (54,431)	430,000 (68,357,100)	34,300 (1,514.6)	12.5
Victory tankers	100,000 (90,719)	825,000 (131,151,000)	40,000 (1,766.3)	20.6
Niarchos tankers	106,000 (96,162)	821,000 (130,514,000)	40,000 (1,766.3)	20.5
Super tanker	110,000 (99,790)	835,000 (132,740,000)	40,000 (1,766.3)	21
Universe Ireland	312,000 (283,042)	2,400,000 (381,528,000)	125,000 (5,519.8)	20
Ultratanker	470,000 (426,377)	3,620,000 (575,471,000)	200,000 (8,831.7)	19

advantages and disadvantages of each type of reactor are listed.

Related Calculations: Nuclear propulsion is still in the development stage for commercial vessels. However, the data presented here are appropriate for preliminary selection of the type of reactor suitable for a given vessel.

Nuclear vessels constructed thus far are able to operate at higher speeds over longer trade routes than are oil-fueled vessels. The increased speed produces a substantial increase in a vessel's cargo-carrying capacity over a given time. In a conventional merchant vessel, the hp required for higher speeds is increased by the third power of the speed ratio. To produce the higher speed, the vessel must have larger engines and a greater fuel-storage capacity. The nuclear vessel does not require a larger fuel storage capacity and can thus operate continually at the maximum level of its hull form without penalty in its cargo-carrying capacity.

TANKER CAPACITY AND CARGO-HANDLING CHARACTERISTICS

What is the T-2 equivalent of a 110,000-deadweight-ton (DWT) 18-kn (33.3-km/h) tanker having a capacity of 835,000 bbl (1.32×10^8 L) of oil? How long will it take to unload such a tanker if it is fitted with the usual cargo pumps? What is the energy that must be absorbed by a dock if this ship moves at a velocity of 30 ft/min (0.15 m/s) during normal docking and strikes the dock at this velocity?

Calculation Procedure:

1. Compute the T-2 equivalent of the vessel

The T-2 equivalent of a tanker is a quick means of comparing a new vessel with the well-known capacity of the T-2 type of tanker. To compute the T-2 equivalent of any tanker, use the relation T-2 equivalent = (new tanker DWT)(new tanker speed, kn)/(16,000 × 14.6), where DWT = tanker deadweight tons = displacement of the tanker fully loaded, ready for sea, less the weight of the ship itself, or displacement, light, expressed in tons of 2240 lb. The constants in the denominator of this expression are the DWT and speed, respectively, of the standard T-2 tanker. So T-2 equivalent = (110,000)(18)/[16,000(14.6)] = 8.46.

This equivalent means that the new tanker has a carrying capacity of nearly 8½ times that of a standard T-2 tanker when the larger capacity and higher speed of the new vessel are taken into consideration. The equivalent provides an easily understood comparison for all tanker personnel afloat and ashore.

2.   *Determine the vessel unloading time*

Modern tankers are fitted with cargo pumps having sufficient capacity to unload the vessel in 12 to 24 h. An average unloading time for a new tanker discharging at a modern terminal is 16 h.

Table 13 shows the pumping rates for a variety of tankers in current use. A study of this list shows that a 110,000-ton (99,781-t) tanker can be unloaded in about 21 h.

In some terminals the tankage or piping capacity may be small. This condition can increase the unloading time. Or if the tanker is transporting a "dirty" cargo (i.e., asphalt, heavy residual oil, fuel oil, or certain crude oils), unloading may take longer because the cargo is too viscous for the ship's pumps to handle at normal temperature. When this situation occurs, the cargo is heated to a temperature high enough to produce a viscosity suitable for pumping. The required temperature usually ranges between 125 and 150°F (51.7 and 65.6°C). Steam coils located in the cargo tanks are used to heat the cargo.

3.   *Compute the energy absorbed by the tanker dock*

Use the relation $E = Wv^2/4g$, where E = energy absorbed by the dock, ft·lb, when the vessel strikes it at an assumed angle of 10° between the face of the dock and the vessel hull; W = displacement of the loaded vessel, lb; v = velocity of the vessel at impact, ft/s; g = 32.2 ft/s² (9.8 m/s²). Substituting yields $E = 110,000(2240 \text{ lb/long ton})(0.5)^2/4(32.2) = 478,000$ ft·lb (648.1 kN·m).

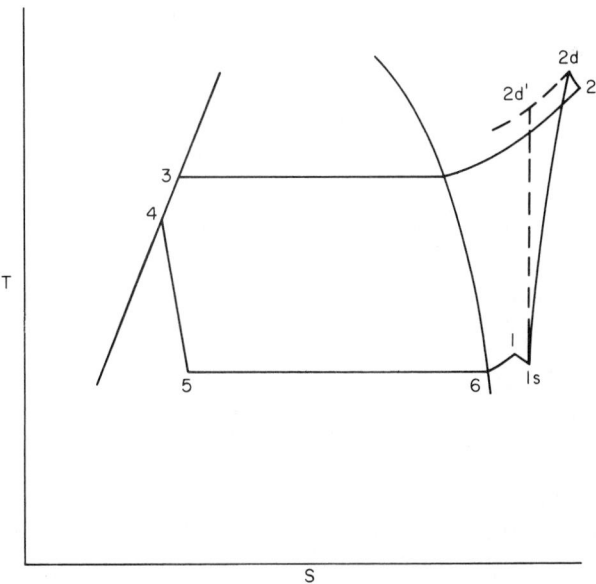

FIG. 3   Temperature-entropy diagram for a refrigeration cycle.

Related Calculations: A loaded tanker approaching a dock moves at a velocity of 0.15 to 1.0 ft/s (0.046 to 0.31 m/s). When tugs are used to dock the tanker, the velocity of approach is usually less than 0.5 ft/s. (0.15 m/s). In exposed locations where the tanker docks without the aid of tugs, the velocity of approach may range between 0.5 and 1.0 ft/s (0.15 and 0.31 m/s). Note that the energy absorbed by the dock is a function of the vessel displacement and the velocity of approach. Hence, the relation presented in step 3 can be used for any type of vessel.

MARINE REFRIGERATION AND AIR-CONDITIONING SYSTEMS

A marine refrigeration system has a load of 30 tons (27.2 t) and uses a Genetron refrigerant that evaporates at 5°F (-15.0°C) and condenses at 105°F (40.6°C). The temperature of the refrigerant leaving the evaporator is 10°F (-12.2°C), resulting from a slight superheating in the evaporator. Wiredrawing in the compressor suction valves causes a 5-lb/in^2 (34.5-kPa) pressure loss; in the discharge valves, wiredrawing causes a 10-lb/in^2 (69.0-kPa) pressure loss. There is a 10 percent increase in isentropic work during compression resulting from turbulence in the compressor cylinders. The volumetric efficiency of the compressor is 70 percent; its mechanical efficiency is 80 percent. Determine the quantity of refrigerant that must be circulated, the work done on the refrigerant in the compressor, the rating of the compressor driving motor, and the required capacity of the condenser circulating pump.

Calculation Procedure:

1. Compute the quantity of refrigerant circulated

Using a tabulation or a plot of Genetron refrigerant properties, list the following values for this cycle, as shown in Fig. 3: Condensing pressure = p_3 = 141 lb/in^2 (abs) (972.2 kPa); cylinder discharge pressure = p_{2d} = 141 + 10 = 151 lb/in^2 (abs) (1041.1 kPa); compressor suction pressure = p_5 = 26.5 lb/in^2 (abs) (182.7 kPa); cylinder suction pressure = p_{1s} = 26.5 $-$ 5 = 21.5 lb/in^2 (abs) (148.2 kPa); suction vapor enthalpy = h_1 = h_{1s} = 81.1 Btu/lb (188.6 kJ/kg); suction vapor specific volume = V_1 = 1.576 ft^3/lb (0.1 m^3/kg); suction entropy in cylinder = S_2 = S_{2d} = 0.17568; discharge enthalpy, isentropic = h_{2d} = 95.18 Btu/lb (221.4 kJ/kg); evaporator liquid temperature = t_4 = 100°F (37.8°C); liquid enthalpy = h_4 = h_5 = 31.16 Btu/lb (72.5 kJ/kg); evaporator exit enthalpy = h_6 = 80.99 Btu/lb (188.4 kJ/kg).

The refrigeration effect of Genetron circulated = $h_6 - h_5$ = 80.99 $-$ 31.16 = 49.83 Btu/lb (115.9 kJ/kg). With a load of 30 tons (27.2 t), the refrigerant flow rate required = F = (30-ton load)[200 Btu/(min·ton)]/49.83 = 120.5 lb/min (54.2 kg/min).

2. Compute the work done on the refrigerant

With a 10 percent work-input increase resulting from turbulence, the work input = W = 1.1($h_{2d} - h_1$) = 1.1(95.18 $-$ 81.1) = 15.5 Btu/lb (36.1 kJ/kg) of refrigerant circulated.

Compute the compressor indicated hp from ihp = WF/42.4, where the constant in the denominator = Btu/(hp·min). Substituting gives ihp = 15.5(120.5)/42.4 = 44.1 ihp (32.9 kW).

3. Select the size of the compressor drive motor

The mechanical efficiency of the compressor is 80 percent. Hence the power input to the compressor = ihp/mechanical efficiency = 44.1/0.80 = 55.1 hp (41.1 kW). Use the next larger size standard motor, or 60 hp (44.8 kW).

4. Compute the compressor displacement required

Use the relation D = 1728 FV_1/e_v, where D = compressor displacement, in^3; e_v = compressor volumetric efficiency; other symbols as before. So D = 1728 $\times$ (120.5)(1.576)/0.70 = 469,000 in^3/min (128,084 cm^3/s).

5. Compute the condenser heat load

The heat that must be removed in the condenser is h = $h_2 - h_4$, where h = Btu/lb of refrigerant circulated; other symbols as before. Thus h = 96.6 $-$ 31.16 = 65.44 Btu/lb (155.21 kJ/kg). With a refrigerant flow rate of 120.5 lb/min (54.6 kg/min), the total heat removed in the condenser H = 65.44(120.5) = 7880 Btu/min (38.5 W).

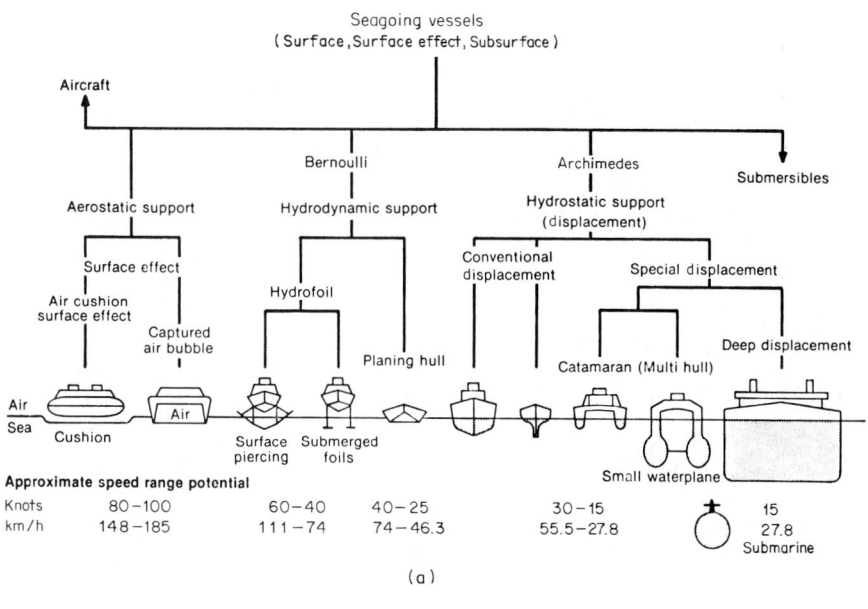

Approximate speed range potential

Knots	80–100	60–40	40–25	30–15	15
km/h	148–185	111–74	74–46.3	55.5–27.8	27.8
					Submarine

(a)

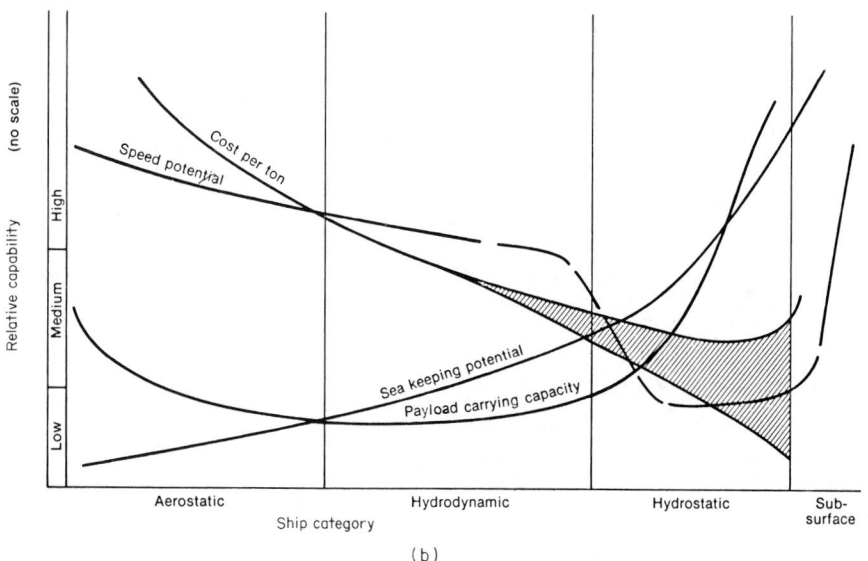

(b)

FIG. 4 (a) Seagoing vessels arranged according to their mode of support on or in the sea. (b) Relative performance and associated capabilities of ship categories in (a). *(United States Naval Institute.)*

6.   Compute the condensing water flow rate

With a rise in the seawater temperature of Δt during passage through the condenser, the quantity of heat absorbed per gallon of water is $H_a = \Delta t \, sw$, where H_a = heat absorbed, Btu/gal; s = specific heat of seawater, Btu/(lb·°F) = 0.94 [3.9 kJ/(kg·°C)]; w = weight of seawater, lb/gal = 8.58 (1.03 kg/L). Assuming a 10°F (5.6°C) temperature rise in the water during passage through the condenser, $H_a = 10(0.94)(8.58) = 80.6$ Btu/gal (22.5 MJ/L). With a total heat load of H, the quantity of water that must be circulated is $H/H_a = 7880/80.6 = 97.8$ gal/min (6.2 L/s).

7.   Select the condenser circulating water pump

Use a pump with a 50 percent larger capacity to guard against overloads and fouling. Thus, pump capacity = 1.5(97.8) = 146.8 gal/min, say 150 gal/min (9.5 L/s).

Compute the total head loss through the piping and condenser. This seldom exceeds 25 lb/in² (172.4 kPa). For usual marine air-conditioning and refrigeration service, a centrifugal circulating pump is best because it is less subject to fouling by dirty harbor water or marine growths.

8.   Compute the power input to the pump

Use the relation hp = gal/min (discharge pressure, lb/in²)/1715 (pump efficiency). With a pump efficiency of 60 percent, hp = 150(25)/1715(0.60) = 3.64 hp (2.72 kW); use a 5-hp (3.73-kW) motor.

Related Calculations: Use this method to analyze refrigeration systems for cargo holds, ice-water service, air conditioning of passenger and crew quarters, and other marine applications.

Note that the general design procedures for marine refrigeration and air-conditioning systems resemble those for land service. The outdoor design conditions are, of course, determined by the areas of the world in which the vessel will sail. Carrier Air Conditioning Company recommends summer outdoor design conditions of 95°F (35°C) db and 82°F (27.8°C) wb, unless the vessel will be sailing in predominantly tropical areas. If the latter is the case, then the average warm port of call defines the summer outdoor design condition. The coldest port of call determines the winter outdoor design condition. Besides the usual sun load on the glass area of a ship, there is the added load of the radiant energy of the sun diffusely reflected from the water surface.

Ventilation needs on shipboard require a minimum of 12.5 ft³/min (0.35 m³/min) per person or 2.5 air changes per hour, whichever is greater. Deluxe passenger vessels require more ventilation for maximum comfort. Air-conditioning systems for passenger staterooms and crew quarters are limited to air-water induction, all-air reheat, and dual-duct systems. The public spaces are air-conditioned by field- or factory-assembled conventional all-air central-station fan-coil systems.

CHOICE OF VESSEL TYPE FOR SEAGOING SERVICE

Select the hull type for a seagoing cargo vessel for fast delivery of small quantities of boxed cargo on a route having a 200-mi (120-km) radius from the vessel's home port. Seakeeping potential of the vessel must be of medium capability; payload capability must be low; cost per ton of the vessel is critical in hull choice.

Calculation Procedure:

1.   Determine the hull type suitable for this application

Figure 4 shows the hull body profiles for vessels typed according to the method of hull support—aerostatic, hydrodynamic, or hydrostatic support. As a general guide, the surface speed of a vessel decreases as the hull support changes from aerostatic to hydrostatic. Thus, in Fig. 4, the hull body profiles are arranged, from left to right, from high- to low-speed types. The one exception is the multihull type which may be either high- or low-speed, depending on the vessel use.

The vessel being considered here must provide fast delivery of small quantities of freight with medium seakeeping potential. Studying Fig. 4 shows that a planing hull can provide speeds of 25 to 40 kn (46.3 to 74.0 km/h); both values are considered "fast" in terms of cargo-vessel speeds. Also, from the same chart, a conventional displacement hull can provide a speed in the range of 15 to 30 kn (27.8 to 55.5 km/h). Again, this range is considered "reasonably fast" for cargo service.

2.   Determine the seakeeping potential of the vessels being studied

Figure 4 shows that the seakeeping potential of both the planing and the conventional displacement hulls is in the medium range, as required by this application.

3.   Determine the relative payload carrying capacity

Again, from Fig. 4, the payload carrying capacity of both hulls is about the same. With a low payload requirement (carrying capacity), the planing hull has an advantage because it is faster than the displacement hull for equal fuel consumption.

4.   Analyze the relative cost per hull ton for each option

Figure 4 shows that the planing hull has a distinct cost advantage over the displacement hull when compared on a cost-per-ton basis. Since the other factors being considered—speed, seakeeping potential, and payload—appear to be approximate tradeoffs for both hull types, the cost per ton of the hull would be the deciding factor. Hence, a planing hull appears to be the best choice for this application.

Were the cargo to be carried extremely fragile and the route to be traveled characterized by extremely rough seas, the hull choice might be the displacement type because the planing hull is known for its slamming characteristics in rough waters.

Related Calculations: The vessel considered above is an inter-island trading ship having a relatively short voyage length. Hence, the hull configuration is one which provides maximum speed at minimum first cost. For longer voyage lengths, other factors will influence the hull choice. The general procedure given here is valid for vessels for a variety of services—naval, merchant, rescue, pleasure, fire, commuter, etc. Figure 4 is the work of Professor Thomas C. Gillmer, Naval Architect and formerly a member of the Engineering Faculty, U.S. Naval Academy.

New federal requirements for deep-displacement tanker vessels mandate that double hulls and double skins be used to prevent environmentally damaging oil spills. Estimates predict that some 30 percent of the world's single-hull tanker fleet will be scrapped during the next 5 years, to be replaced by double-hull vessels. While there is ongoing discussion amongst designers as to the efficacy of double-hull/double-skin vessels, environmental regulations will eventually prohibit single-hull tankers from entering the waters of the United States when loaded with an oil cargo.

Other areas where environmental requirements are impacting vessel design are in waste and trash disposal. Recent estimates show that annually some 7 million tons of trash are discharged illegally by ships into ocean waters. While federal regulations allow dumping of certain trash outside the 12-mi (19.2-km) limit, the items being dumped must meet certain physical requirements. Thus, the trash must be heavy enough to sink of its own weight, must not be larger than 1 in^2 (6.45 cm^2), and may not be made of, nor contained in, plastic. These requirements mean that such trash must be processed in a suitable shredder/macerator before being discharged overboard. Designers must equip each vessel with such suitable equipment prior to its first voyage.

With greater emphasis on environmental control of the seas, many countries are punishing violators who dump trash within their exclusive economic zone, which is defined as 200 nm (320 km) from their shore. Territorial waters are defined as 3 nm (4.8 km) from the shore.

Cruise ships, which are popular throughout the world, are thought to be responsible for some 15 percent of the illegal ocean dumping. This estimate was made by the Center for Marine Conservation. With the number of cruise ships increasing, along with their passenger-carrying capacity (as high as 3,500 persons for larger ships), greater attention is being given to designing nonpolluting trash- and waste-disposal systems for these ships.

The center is especially concerned with plastic refuse from ships. Plastics have a long life, taking more than 300 years to return to a nonpolluting state. Thus, the dumping of plastics can have a long-term impact on the environment because birds, fish, mammals, and turtles can be adversely affected. Estimates by the center show that 1 million birds are killed each year by plastic dumped at sea.

Environmental concerns will continue to influence marine designs in all types of ships. The examples cited above are just a few of the many that will face engineers as they design for a pollution-free marine environment.

A U.S. Maritime Administration projection of commercial vessel needs for the 10-year period

TABLE 14   Projected Future Vessel Needs*

Vessel type	Replacement vessels	New vessels	Total vessels
Chemical transport	115	180	295
Containerships	348	665	1013
Dry bulk cargo	2120	524	2644
Gas transport (LNG/LPG)	234	273	507
Tankers	3550	235	3785
Other types	256	500	756
Total	6623	2377	9000

*U.S. Maritime Administration

of 1992 to 2001 shows that a total of 9000 new vessels will be required. Table 14 shows the numbers of specific vessel types that will be required. As can be seen from this tabulation, the most severely regulated vessels—tankers—represent the largest percentage of both new and replacement ships, i.e., 42.1 percent. Included in this future tanker fleet will be a number of Panmax and Suezmax vessels—i.e., tankers of a size (beam, loaded draft) that is the maximum that can be handled by either the Panama (Panmax) or Suez (Suezmax) Canal. Designers can use the projections in Table 14 to plan their future engineering efforts.

SECTION 10

NUCLEAR ENGINEERING

B. G. A. SKROTZKI, P.E.
POWER MAGAZINE

SAMUEL C. LIND
CONSULTANT
UNITED STATES ATOMIC ENERGY COMMISSION

Nuclear Power Reactor Selection 10.2
Nuclear Power-Plant Cycle Analysis 10.2
Reactor Fuel Consumption, Atom Burnup, and Neutron Flux 10.6
Value of Fissionable Material for Power Generation 10.7
Effect of Nuclear Radiation on Human Beings 10.8
Analysis of Nuclear Power and Desalting Plants 10.9

REFERENCES: Klema and West—*Public Regulation of Site Selection for Nuclear Power Plants,* Johns Hopkins University Press; Hagel—*Alternative Energy Strategies: Constraints and Opportunities,* Holt, Rinehart and Winston; Komanoff—*Power Plant Cost Escalation,* Van Nostrand Reinhold; Seeley—*Elements of Thermal Technology,* Marcel Dekker; Hunt—*Handbook of Energy Technology,* Van Nostrand Reinhold; Munn—*Environmental Impact Assessment,* Wiley; Canapathy—*Applied Heat Transfer: A Complete Handbook for Power and Process Engineers,* PennWell Books; El-Wakil—*Nuclear Power Engineering,* McGraw-Hill; Sachs—*Nuclear Theory,* Addison-Wesley; Hoegerton and Grass—*Reactor Handbook,* U.S. Atomic Energy Commission; Schwenk and Shannon—*Nuclear Power Engineering,* McGraw-Hill; Murphy—*Elements of Nuclear Engineering,* Wiley; Rockwell—*Reactor Shielding Design Manual,* Van Nostrand; Price—*Radiation Shielding,* Pergamon; Hollaender—*Radiation Biology,* McGraw-Hill; Glasstone and Edlund—*The Elements of Nuclear Reactor Theory,* Van Nostrand; Murray—*Nuclear Reactor Physics,* Prentice-Hall; Etherington—*Nuclear Engineering Handbook,* McGraw-Hill; Glasstone—*Principles of Nuclear Reactor Engineering,* Van

Nostrand; Bonilla—*Nuclear Engineering*, McGraw-Hill; Glasstone and Lovberg—*Controlled Thermonuclear Reactions*, Van Nostrand; Schultz—*Control of Nuclear Reactors and Power Plants*, McGraw-Hill; International Atomic Energy Agency—*Directory of Nuclear Reactors*; Henley—*Advances in Nuclear Science and Technology*, Academicol'denblat—*Calculation of Thermal Stresses in Nuclear Reactors*, Consultants Bureau; Greenspan—*Computing Methods in Reactor Physics*, Gordon and Breach; International Atomic Energy Agency—*Programming and Utilization of Research Reactors*; Marchuk—*Theory and Methods of Nuclear Reactor Calculations*, Consultants Bureau.

NUCLEAR POWER REACTOR SELECTION

Select a nuclear power reactor to generate 60,000 kW at a thermal efficiency of 35 percent or more. If the selected unit is a 10-ft (3.0-m) diameter reactor that uses a fluidized bed containing 20×10^6 fuel pellets each 0.375 in (9.5 mm) in diameter with a density of 700 lbm/ft^3 (11,213 kg/m^3) and the reactor fluid is pressurized water at 600°F (315.6°C), determine the bed pressure drop when fluidized. Also, compute the reactor fuel volume, the collapsed fuel bed height, and the density of the pressurized water.

Calculation Procedure:

1.   *Select the type of reactor to use*

Table 1 summarizes the operating characteristics of six types of power reactors. Study shows that a pressurized-water reactor will provide the desired thermal efficiency. Further, this type of reactor is successfully used for large-scale power generation. Hence, a pressurized-water reactor will be the first tentative choice for this plant.

2.   *Compute the reactor fuel volume*

Use the relation $v_f = nv_p$, where v_f = fuel volume, ft^3; n = number of fuel pellets in the reactor; v_p = volume of each pellet, ft^3. Substituting yields $v_f = 20 \times 10^6\pi(0.375)^3/[6(1728)] = 320$ ft^3 (9.1 m^3).

3.   *Compute the fuel volume in the collapsed form*

With the fuel bed not fluidized, the porosity P with packed spheres is about 0.40. Then collapsed volume $v_c = v_f/(1 - P) = 320/0.60 = 534$ ft^3 (15.1 m^3).

4.   *Compute the collapsed fuel-bed height*

Use the relation $h = v_c/A_r$, where h = collapsed height of fuel bed, ft; A_r = reactor fuel bed area, ft^2. So $h = 534/(\pi 10^2/4) = 6.78$ ft (2.1 m).

5.   *Determine the density of the pressurized water*

Using the steam tables shows $d_w = 42.45$ lb/ft^3 (680.0 kg/m^3) at 600°F (315.6°C) for saturated liquid.

6.   *Compute the pressure loss through the fluidized bed*

Use the relation $p = 2.9h[(1 - P)d_f + Pd_w]$, where p = pressure loss through fluidized fuel bed, lb/ft^2; d_f = fuel density, lbm/ft^3; other symbols as before. Substituting, we find $p = 2.9\,[(1 - 0.4)700 + 0.4 \times 42.45] = 1268$ lb/ft^2 or 8.79 lb/in^2 (60.6 kPa).

   Related Calculations: This general procedure is valid for preliminary selection of the type of nuclear reactor to use for a given power application. Since reactors are expensive, a complete economic analysis must be made of the alternatives available before the final choice is made.

NUCLEAR POWER-PLANT CYCLE ANALYSIS

A nuclear power plant using two coolants, Na and NaK, is arranged as shown in Fig. 1. Sodium, the first coolant, enters the reactor at 600°F (315.6°C) and leaves at 1000°F (537.8°C); NaK, the second coolant, enters the intermediate heat exchanger at 550°F (287.8°C) and leaves at 950°F (510.0°C). Neglecting heat and pressure losses in the piping, plot the enthalpy-temperature diagram for the plant if steam leaves the boiler at 1200 lb/in^2 (8273 kPa). What are the Na and NaK flow rates with the cycle arrangement shown in Fig. 1, a reactor capacity of 400,000 kW of heat energy, and a 155,000-kW turbine output? Determine the plant thermal efficiency if the auxiliary-power needs = 12,000 kW.

TABLE 1 Nuclear-Power Reactor Characteristics

Reactor type	Typical thermal efficiency, %	Typical power density, thermal, kW/ft^3 (MW/m^3)	Typical reactor pressure, lb/in^2 (gage) (kPa)	Average heat flux, Btu/(h·ft^2) (MW/m^2)	Typical fuel enrichment, %	Reactor coolant
Pressurized-water	36	1,600 (56.5)	1,500 (10,341)	300,000 (945.6)	1.5–3.0	Light water
Boiling-water	22–30	800 (28.3)	1,000 (6,894)	100,000 (315.2)	1.5	Light water
Gas-cooled	30	200 (7.1)	600–1,000 (4,136–6,894)	... ...	0.70–2.5	Carbon dioxide
Liquid-metal	33	300 (10.6)	100 (689.4)	... ...	...	Sodium, bismuth, lead, etc.
Fast-breeder	32	20,000 (706.5)	100 (689.4)	650,000 (2,049)	...	Sodium
Fluid-fueled	30	400 (14.1)	1,000–2,000 (6,894–13,788)	Varies (varies)	Varies	Reactor fuel solution

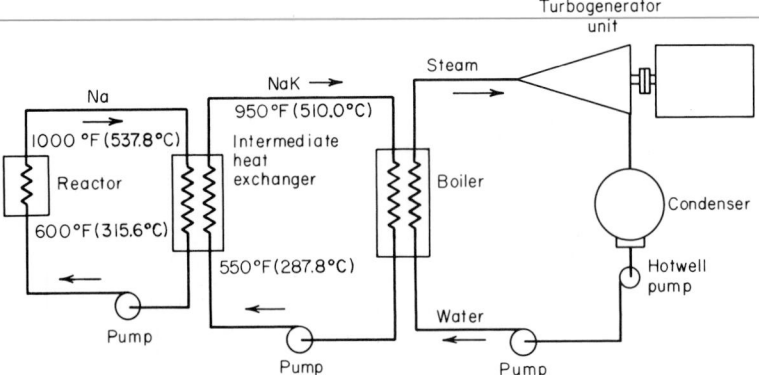

FIG. 1   Reactor plant with two-coolant system uses Na in the reactor circuit and transfers heat to the intermediate NaK circuit, which acts as a buffer against making the steam circuit radioactive.

Calculation Procedure:

1. *Determine the steam outlet and saturation temperature*

Figure 1 shows that NaK enters the boiler at 950°F (510.0°C). Draw a horizontal line on the enthalpy-temperature *(h-t)* diagram (Fig. 2), indicating the 950°F (510.0°C) NaK temperature entering the boiler. Also draw a horizontal line on the *h-t* diagram, Fig. 2, at 1000°F (537.8°C), indicating the Na temperature leaving the reactor.

The steam outlet temperature from the boiler will be less than 950°F (510.0°C) because transfer of heat between the NaK and the water and steam in the boiler provides the energy required to convert the water to steam. A temperature difference between the NaK and the steam is needed to produce the desired heat transfer.

Assume a 50°F (27.8°C) temperature difference between the boiler outlet steam and the NaK, which is a typical temperature difference for this type of cycle. With such a temperature difference the outlet steam temperature = 950 − 50 = 900°F (482.2°C). From the steam tables find the saturation temperature of steam at 1200 lb/in^2 (abs) (8273 kPa) as 567.2°F (297.3°C). Hence the steam will be superheated when it leaves the boiler.

2. *Compute the boiler evaporator coolant outlet temperature*

Incoming feedwater enters the boiler evaporator section where it is heated by the NaK before entering the boiler steam section. To provide heat transfer between the NaK leaving the evaporator section of the boiler and the incoming boiler feedwater, a temperature difference between the two fluids is necessary. Assume that the NaK coolant leaves the boiler evaporator section at a temperature 40°F (22.2°C) higher than the incoming feedwater. With the incoming feedwater at the saturation temperature, or 567.2°F (297.3°C), the NaK coolant outlet temperature from the boiler evaporator = 567.2 + 40 = 607.2, say 607°F (319.4°C).

3. *Plot the boiler coolant temperature path*

Locate the boiler outlet steam state on the *h-t* diagram, Fig. 2, on the 1200-lb/in^2 (abs) (8273-kPa) pressure curve and the 900°F (482.2°C) temperature horizontal. From this point, project vertically upward to the 950°F (510°C) NaK temperature horizontal to locate point 1, the temperature of the NaK entering the boiler, Fig. 2.

Next, locate the point 1*a* where the liquid enthalpy line of the *h-t* diagram, Fig. 2, intersects the 1200-lb/in^2 (abs) (8273-kPa) evaporation enthalpy line. From point 1*a*, project vertically upward to 607°F (319.4°C), point 2, the temperature of the NaK coolant leaving the boiler evaporator section.

Points 1 and 2 are the NaK *temperature path* in the boiler evaporator and steam-generating sections. Assuming that the NaK has a constant specific heat while flowing through the boiler evaporator and steam-generating sections (a completely valid assumption), draw a straight line

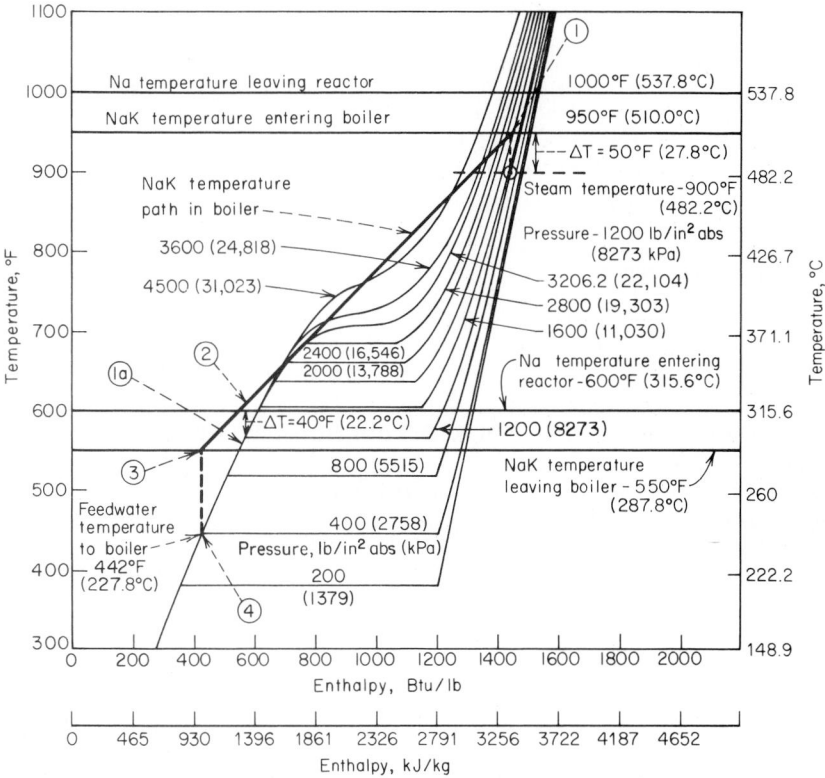

FIG. 2 Steam-water enthalpy-temperature diagram shows the relation between NaK circuit and steam circuit. Keeping the steam temperature high raises the thermal efficiency of the plant.

between points 1 and 2 and extend it to intersect the 550°F (287.8°C) temperature line at point 3. Note that point 3 represents the temperature of the NaK entering the intermediate heat exchanger.

4. Determine the boiler feedwater inlet temperature

Feedwater enters the boiler at a yet unknown temperature. During passage between the boiler inlet and the evaporator section inlet, the feedwater absorbs heat from the NaK coolant, leaving the evaporator at 607°F (319.4°C).

Draw a line vertically downward from point 3 until the liquid enthalpy curve is intersected, point 4. Point 4 represents the boiler feedwater inlet temperature, or 442°F (227.8°C), based on the valid assumption that the feedwater leaving the condenser hot well is in the saturated state.

5. Compute the reactor coolant flow rate

Sodium enters the reactor at 600°F (315.6°C) and leaves at 1000°F (537.8°C), Fig. 1. Thus, the temperature rise of the Na during passage through the reactor is $1000 - 600 = 400$°F (222.2°C). Also, the average specific heat of Na is 0.306 Btu/(lb·°F) [1.28 kJ/(kg·°C)], found from a tabulation of Na properties in an engineering handbook.

Compute the Na flow from $f = 3413 \, kw/\Delta tc$, where f = Na flow rate, lb/h; kw = reactor heat rating, kW; Δt = Na temperature rise during passage through the reactor, °F; c = specific heat of the Na coolant, Btu/(lb·°F). Substituting gives us $f = 3413(400,000)/[400(0.306)] = 11,130,000$ lb/h (1402.4 kg/s).

6. Compute the boiler heating liquid flow rate

Use the same relation as in step 5, substituting the temperature change and specific heat of NaK. Since the NaK enters the boiler at 950°F (510.0°C) and leaves at 550°F (287.8°C), its temperature change is $950 - 550 = 400°F$ (222.2°C). Also, the specific heat of NaK is 0.251 Btu/(lb·°F) [1.05 kJ/(kg·°C)], as found from NaK properties tabulated in an engineering handbook. So $f = 3413(400,000)/[400(0.251)] = 13,600,000$ lb/h (1713.6 kg/s).

7. Compute the plant thermal efficiency

The net station output kw = gross output of turbine, kW, minus the total plant auxiliary demand, kW = $155,000 - 12,000 = 143,000$ kW. Then overall plant thermal efficiency = net station output, kW/reactor heat output, kW = $143,000/400,000 = 0.357$, or 35.7 percent.

Related Calculations: This analysis is valid for a cycle in which the reactor coolant does not do work in the turbine. In general, designers prefer to avoid using the reactor coolant in the turbine. Although the thermodynamic aspects of a nuclear cycle are important, the cost of the plant must also be considered before a final choice of a cycle is made. The method presented is the work of Henry C. Schwenk and Robert H. Shannon, as reported in *Power* magazine.

REACTOR FUEL CONSUMPTION, ATOM BURNUP, AND NEUTRON FLUX

Determine the amount of fissionable material used in a 500-mW reactor having 3×10^{10} fissions per watt-second. The reactor core has a volume of 1360 ft³ (38.5 m³) and the fuel (99.3 percent U 238 plus 0.7 percent U 235) occupies 6 percent of the reactor volume. How much fissionable material is consumed if the plant operates 8760 h/year at an 80 percent load factor and the capture cross section/fission cross section ratio = 1.2? What are the maximum allowable atom burnup, the average fuel-cycle time, and the reactor neutron flux?

Calculation Procedure:

1. Compute the reactor fission rate

Use the relation $F_r = P_T C$, where F_r = reactor fission rate, fissions/(W·s); P_T = total reactor power, W; C = fissions (W·s). So $F_r = 500 \times 10^6 (3 \times 10^{10}) = 1.5 \times 10^{19}$ fissions/s.

2. Compute the total volume of the fuel

Since the fuel occupies 6 percent of the reactor volume, the fuel volume $V_f = 0.06 \times 1360 = 81.6$ ft³ (2.3 m³). Since reactor fuel quantities are often expressed in cubic centimeters, convert the fuel volume in cubic feet by multiplying by the conversion factor 2.832×10^4, or $V_{fc} = 2.832 \times 10^4 (81.6) = 2.31 \times 10^6$ cm³.

3. Compute the U 235 nuclei in the reactor

First determine the uranium nuclei per cm³ N_U, using the relation N_U = [(uranium density, g/cm³)/uranium atomic weight] (Avogadro's constant) = $(18.68/238.07)(6.023 \times 10^{23}) = 0.0472 \times 10^{24}$ nuclei/cm³. In this relation the following constants are used: uranium density = 18.68 g/cm³; uranium atomic weight = 238.07; Avogadro's constant = $N_m = 6.023 \times 10^{23}$ atoms/(g·atom).

With the uranium nuclei per cm³ known, compute the U 235 nuclei in the reactor from $N_{U\ 235} = 0.007 N_U V_{fc} = 0.007(0.0472 \times 10^{24})(2.31 \times 10^6) = 7.64 \times 10^{26}$ U 235 nuclei in the reactor.

4. Compute the U 235 fissionable material consumed

Use the relation $F_{U\ 235} = F_r G_m / N_m$, where $F_{U\ 235}$ = fissionable U 235 material consumed or burned up for power only, g/s; G_m = g/mol of the fissionable material; other symbols as before. Substituting gives $F_{U\ 235} = (1.5 \times 10^{19})(235)/6.023 = 5.85 \times 10^{-3}$ g/s.

5. Compute the annual consumption of fissionable material

Use the relation $A_c = F_{U\ 235}\ YL/1000$, where A_c = annual consumption of fissionable material, kg; Y = s/year; L = load factor; other symbols as before. Substituting reveals $A_c = 5.85 \times 10^{-3}(3600 \times 8760)(0.8)/1000 = 147.4$ kg/year.

6. *Compute the U 235 annual consumption*

The U 235 is consumed by fissioning for power and is also lost by absorption. The proportion of these two forms of consumption is expressed by $\alpha = $ U 235 total capture cross section/U 235 fission cross section. With $\alpha = 1.2$ for a typical reactor, the total annual U 235 consumption $= 1.2(147.4) = 177$ kg/year.

7. *Compute the maximum allowable atom burnup*

Both U 235 and U 238 are regarded as reactor fuel. The allowable percentage of burnup depends on the total integrated radiation dosage and radiation energy level, and the effect on fuel material dimensional stability, thermal conductivity, and reduction in effective multiplication factor. Assuming a maximum allowable burnup of 20 percent, which is a typical value, compute $B_{ma} = $ (percentage of burnup)(fuel atoms per cm^3)(total cm^3 of fuel), where $B_{ma} = $ maximum allowable atom burnup, atoms. Substituting gives us $B_{ma} = (0.002)(0.0472 \times 10^{24})(2.31 \times 10^6) = 2.18 \times 10^{26}$ atoms.

8. *Compute the average fuel-cycle time*

Use the relation $A_f = B_{ma}/F_r$, where $A_f = $ average fuel-cycle time, s. Thus $A_f = 2.18 \times 10^{26}/(1.5 \times 10^{19}) = 1.45 \times 10^7$ s $= 4040$ h $= 30$ weeks, approximately.

9. *Compute the reactor neutron flux*

Use the reaction $N_f = P_T C / \sum f V_f$, where $N_f = $ reactor neutron flux; $\sum f = N_{U\ 235} \times \sigma_{f\ 235}$, where $\sigma_{f\ 235} = $ total microscopic absorption cross section for U 235; other symbols as before. So $N_f = 500 \times 10^6 (3 \times 10^{10})/(0.00033 \times 10^{24})(549 \times 10^{-24})(2.31 \times 10^6) = 3.57 \times 10^{13}$. Note that values of $\sigma_{f\ 235}$ are obtained from nuclear data sources.

Related Calculations: Use this general method for any reactor designed to generate power. The method presented is the work of Henry C. Schwenk and Robert H. Shannon, as reported in *Power* magazine.

At the time of this writing (1994) the United States is generating more than 22 percent of its power requirements in nuclear plants. Thus, nuclear stations are number two in generating electricity for the United States.

Nuclear power does not pollute the air. Recent studies show that the nuclear plants currently operating annually reduce the amount of carbon dioxide that would be emitted to the atmosphere by some 500 million tons. Likewise, these plants reduce atmospheric pollution by 3.6 million tons of methane and some 2 million tons of nitrous oxides annually. The NO_x reduction closely approximates the requirements of amendments to the 1990 Clean Air Act.*

VALUE OF FISSIONABLE MATERIAL FOR POWER GENERATION

How many tons of coal are required to produce the heat equivalent of 1 lb (0.45 kg) of fissionable U 235? If heat is worth 40 cents per million Btu (37.9 cents per 10^6 kJ), what is 1 g of fissionable U 235 worth? One ton of coal contains 24×10^6 Btu (25.3×10^6 kJ).

Calculation Procedure:

1. *Compute the heat produced by 1 lb (0.45 kg) of fissionable material*

When all the nuclei in the atoms of 1 lb (0.45 kg) of fissionable U 235 fission, about 0.001 lb (0.45 g) of material converts to heat energy. Since by Einstein's mass-energy equation, 1 lbm $= 11.3 \times 10^9$ kWh of energy, 1 lb (0.45 kg) of fissioning U 235 produces $0.001(3413)(11.3 \times 10^9) = 39.5 \times 10^9$ Btu/lb (91.9×10^9 kJ/kg). In this relation, the constant 3413 (3600.9) converts kW to Btu (kJ).

2. *Compute the heat equivalent of the fissionable material*

Use the relation equivalent tons of coal per pound of U 235 $= $ heat released per pound of U 235, Btu/heat released by 1 ton of coal, Btu $= 39.5 \times 10^9/(24 \times 10^6) = 1645$ tons of coal per pound of U 235 (3290.0 t of coal per 1 kg U 235). Thus, it takes 1645 tons of coal to equal the potential heat produced by 1 lb (3290 t/kg) of U 235 in a nuclear reactor.

*Orval Hansen, President, Columbia Institute.

3.   *Compute the monetary worth of the nuclear material*

Since heat is worth 40 cents per million Btu in this plant, the value of 1 lb (0.45 kg) of U 235 is $(39.5 \times 10^9)(0.4/10^6)$ = \$15,800, or about \$34.80 per gram of U 235.

Related Calculations: Use this general procedure for other fissionable materials used for fuels in nuclear plants. The method presented is the work of Henry C. Schwenk and Robert H. Shannon, as reported in *Power* magazine.

With nuclear-power generation there is always the consideration of what to do with spent fuel. Spent nuclear fuel is still radioactive and hazardous to humans.

Spent nuclear fuel is a waste material that requires much more care than ash from coal or SO_2 emitted by a power-plant stack for a coal-fired generating plant. Environmental regulations are equally strong in their control of nuclear waste, stack and boiler-grate effluent, and internal-combustion engine exhausts.

But fossil-fuel-fired generating plants have an option nuclear plants do not have. A fossil-fuel plant can purchase allowances to emit SO_2, as sanctioned by Title IV, the acid-rain provisions, of the 1990 Clean Air Act Amendments (CAAA). No such allowances are permitted for nuclear-fuel waste because spent fuel is much more lethal than SO_2.

At a recent auction one low-sulfur-coal-burning utility bought 85,103 allowances for over \$11 million. Each allowance permits a plant to emit 1 ton of SO_2 per calendar year. The utility justified its purchases of the allowances at \$135 per ton by comparing it to the cost of installing scrubbers to provide similar reductions, namely \$500 per ton. By buying the allowances now the utility believes it can postpone large capital outlays until less costly controls become available in the marketplace.

Under Title IV, the Environmental Protection Agency (EPA) seeks to establish a nationwide limit of 8.9 million tons of SO_2 emitted per year by the year 2000. If this goal is achieved it will be a 50 percent reduction of the 1980 SO_2 emission level. By Jan. 1, 1995, 110 fossil-fuel-fired power plants with the highest SO_2 emission must meet Phase I of CAAA emission limits of 2.5 lb SO_2/million Btu (1.1 kg SO_2/1055 kJ). By Jan. 1, 2000, all other 25-MW or larger utilities, and certain industrial cogenerators, must meet Phase II emission limits of 1.2 lb SO_2/million Btu (0.54 kg SO_2/1055 kJ). The above data are from a summary presented in *Environmental Engineering*.

With no allowances available to nuclear plants, the designer must give thought to the eventual disposal of spent fuel. Two approaches can be used in the handling of spent nuclear fuel: (1) storage, (2) reprocessing.

In the first approach, storage, both the heat and radiation of the spent fuel must be contended with during the long-term storage period required. Underground storage of spent fuel is the most common way of handling the waste. Today most spent fuel is buried intact, with no processing before storage. Handling spent nuclear fuel is an ongoing problem for which no final solutions appear available at this time.

In the second approach, reprocessing, a number of usable by-products—plutonium, uranium, and radioisotopes—are obtained. These can be used in agriculture, industry, and medicine to perform beneficial tasks. But even after reprocessing there is a residue of high-level nuclear waste. This residue must be stored in stainless-steel tanks or in solid form. Many different storage options are being studied.

With increasing attention on environmental aspects of nuclear-power generation, the designer has much to contend with. Between federal and state regulators, the environmental demands are enormous. The environment must be "factored into" every engineering cost estimate today. If the environmental aspects are overlooked, the cost estimate will be completely unrealistic, and the time schedule may be off by years.

EFFECT OF NUCLEAR RADIATION ON HUMAN BEINGS

What is the total radiation dose in rems for a worker exposed to 0.3 rad of 1.0 MeV beta particles and 0.05 rad of 1.0 MeV neutrons each day? Is the total dose dangerous to this worker? Use National Bureau of Standards data (Tables 2 to 4) in the analysis.

Calculation Procedure:

1. *Compute the total radiation dose*

Use the relation, total dose, rem $= \sum(\text{dose, rad})(\text{RBE})$, where rem = roentgen equivalent per man; rad = radiation absorbed dose; RBE = relative biological effectiveness. Table 2 lists the RBE values for various types of radiation. By substituting the appropriate values from Table 2, total dose $= (0.3)(1.0) + 0.05(10.5) = 0.825$ rem.

2. *Determine whether the dose is dangerous*

Table 3 lists the exposure tolerance of the human body. This listing shows that a dose of 1 rem/day is believed to cause debilitation within 3 to 6 months and death within 3 to 6 years. Since the daily dose to which this worker is exposed—0.825 rem—is close to the 1.0-rem danger level, the dose is excessive and dangerous.

Table 4 lists the recommended weekly maximum dosage for various types of radiation on different parts of the body. Study of this list also indicates that the radiation to which this worker is exposed is dangerous.

Related Calculations: The effects of radiation can be fatal to all living organisms. Hence, extreme care must be used in computing the dose received by anyone exposed to radiation. Since the allowable dose and the effects of various doses are under constant study, be certain to refer to the latest available data from the Nuclear Regulatory Commission before permitting exposure of any worker to radiation of any kind.

Environmental cleanup after a nuclear power plant accident can be expensive and time consuming. Thus, it took some 14 years to clean up the contamination at Three Mile Island 2 after the accident in March, 1979. An electric evaporator operated for 2 years to boil off some 2.23 10⁶ gal (8440 m³) of contaminated water from Reactor No. 2. Although some contamination still remains in the reactor building, it is confined to the walls and is not thought to pose any danger to the environment.

TABLE 2 Conversion: Rad to Rem° †

Radiation effects on humans: Definitions
One r (*roentgen*) is the quantity of gamma or x-radiation that produces an energy absorption of 83 ergs/g of dry air.
One rep (*roentgen equivalent physical*) is the quantity of radiation that produces an energy absorption of 93 ergs/g of aqueous tissue.
One rad (*radiation absorbed dose*) is required to deposit 100 ergs/g in any material by any kind of radiation.
One rem (*roentgen equivalent man*) is the unit of particulate radiation that produces tissue damage in humans.
The conversion factor from rad to rem is the RBE (relative biological effectiveness), i.e., dose in rem = dose in rad × RBE.

Type of radiation	RBE°	Type of radiation	RBE°
X-rays	1	Neutrons, 0.5 MeV	10.2
Gamma rays	1	Neutrons, 1.0 MeV	10.5
Beta particles, 1.0 MeV	1	Neutrons, 10 MeV	6.4
Beta particles, 0.1 MeV	1.08	Protons, 100 MeV	1–2
Neutrons, thermal	2.8	Protons, 1 MeV	8.5
Neutrons, 0.0001 MeV	2.2	Protons, 0.1 MeV	10
Neutrons, 0.005 MeV	2.4	Alpha particles, 5 MeV	15
Neutrons, 0.02 MeV	5	Alpha particles, 1 MeV	20

°Example for total dose: For a given exposure time, a dose of 0.2 rad of γ radiation plus 0.04 rad of thermal neutrons gives a total dose of $(0.2 \times I\text{RBE}) + (0.04 \times 2.8 \text{ RBE}) = 0.312$ rem.
†Based on most detrimental chronic biological effects for continuous low-dose exposures.

TABLE 3 Exposure Tolerance Values for Humans°

0.001 rem/day	Natural background radiation
0.01 rem/day	Permissible dose range, 1957
0.1 rem/day	Permissible dose range, 1930 to 1950
1 rem/day	Debilitation 3 to 6 months; death 3 to 6 years (projected from animal data)
10 rem/day	Debilitation 3 to 6 weeks; death 3 to 6 years (projected from animal data)

100 rem—1 day
  150 rem—1 week
  300 rem—1 month
	Survivable emergency exposure dose but permitting no further exposure for life

25 rem	Single emergency exposure
100 rem	Twenty-year career allowance
500 rem	Maximum permissible 20-year-career allowance

° Whole-body radiation doses.

Since the reactor wall of TMI No. 2 was not punctured, it was possible to leave unreachable radioactive materials inside the reactor. Over time, the radiation level of these materials will decline. Some 150 tons of radioactive wreckage was removed from TMI No. 2 reactor and deposited at the National Engineering Laboratory of the Department of Energy in Idaho. A cooling system malfunction damaged TMI No. 2 reactor core, leading to leakage of radioactive gases. The cleanup has been completed.

ANALYSIS OF NUCLEAR POWER AND DESALTING PLANTS

Analyze the feasibility of building and operating nuclear-powered combined electric generating and water-desalting plants. Sketch the different types of cycles that might be used. Determine the cycle to use for a water production of 100×10^6 gal/day (4.4×10^3 L/s), electric power net output of 500 mW, and a desalting heat performance of 100.

Calculation Procedure:

1. *Draw the cycle diagrams*

Three cycles will be considered: the back-pressure, extraction, and multishaft cycles.

Figure 3 shows the back-pressure cycle in which the entire exhaust steam flow from the turbine is used to heat brine in the water-desalting system. For a given amount of water produced, this cycle generates large quantities of electric power.

In the extraction cycle (Fig. 4), the steam for brine heating is removed from the turbines at some midpoint during expansion. The exhaust steam goes to a standard condenser. This cycle can

TABLE 4 Maximum Weekly Dosage°

	Skin		Lens of		Blood-forming	Intermediate tissue
Radiation	Total body	Appendages	eye	Gonads	organs	(0.07–5.0 cm depth)
X-rays or γ-rays < 3 MeV	0.45	1.5	0.45	0.3	0.4	0.4–0.45
Electrons or β	0.6	1.5	0.3	0.3	0.3	0.3–0.6
Protons	0.6	1.5	0.3	0.3	0.3	0.3–0.6
Fast neutrons	0.3–0.6	0.75–1.5	0.3	0.3	0.3	0.3–0.6
Thermal neutrons	0.5	1.2	0.3	0.1	0.17	0.17–0.5
Alpha particles	1.5	1.5	0.3	0.3	0.3	0.3–1.5
Heavy nuclei (O, N, C, locally generated)	1.5	1.5	0.3	0.3	. . .	0.3–1.6

° Rems per week.

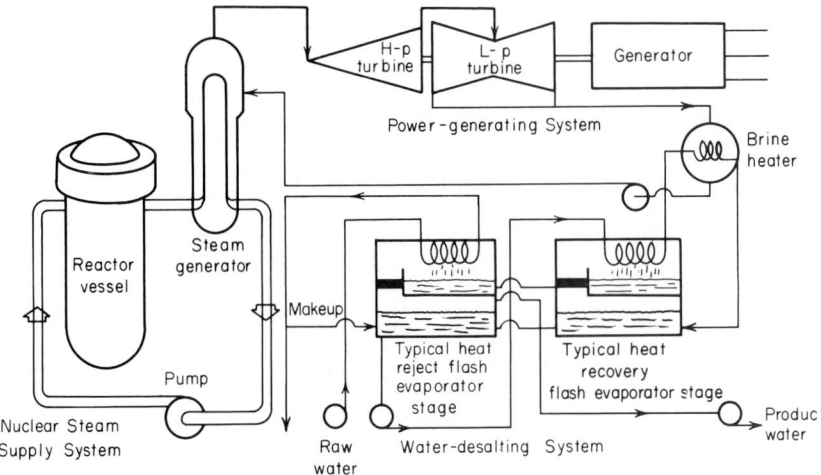

FIG. 3   Back-pressure cycle in which the entire exhaust steam from the turbines is used to heat brine in the water-desalting system.

have a high product ratio (PR), that is, the ratio of the electric power to desalted water. If desired, large amounts of water can be produced when needed.

The multishaft cycle (Fig. 5) is fundamentally the same as Fig. 3, but it uses parallel condensing and noncondensing turbines. The electric output can vary over a wide range without changing the water-desalting production. Although many other cycles are possible, all are variations of the three basic arrangements described above.

2.   *Choose the type of cycle and reactor size to use*

Figure 6 allows quick *estimates* of the type of cycle and reactor size. Any of the four plotted quantities can be determined from Fig. 6 when the other three are known.

Enter Fig. 6 at the bottom at the water production rate of 100×10^6 gal/day (4.4×10^3 L/s), and project vertically upward (1) to the desalting heat performance of 100. From the intersection with the appropriate curve, project horizontally to the left-hand scale of Fig. 6. Next project upward (2) parallel to the index scale. Then project vertically downward (3) from the net electric

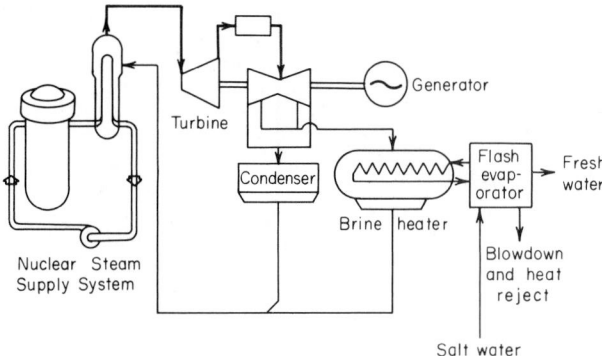

FIG. 4   Extraction cycle in which the steam for brine heating is removed from the turbines at some midpoint during expansion.

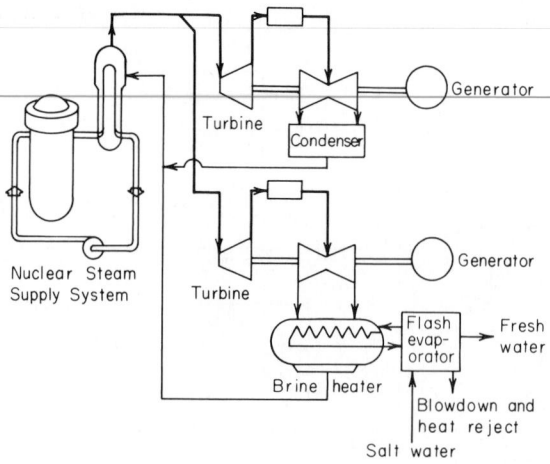

FIG. 5 Multishaft cycle is the same as the back-pressure cycle, but it uses parallel condensing and noncondensing turbines.

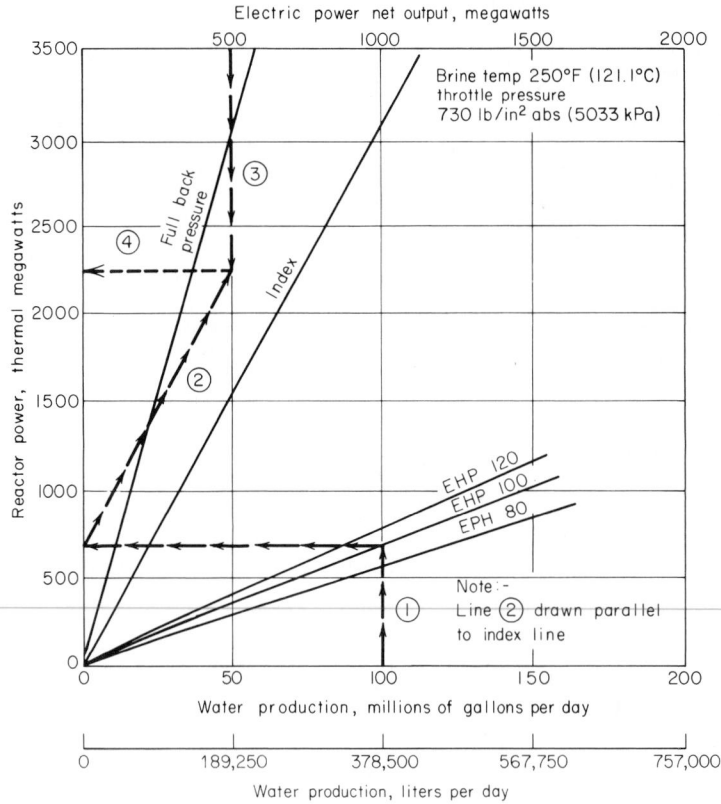

FIG. 6 Nomogram for plant ratings relates the four variables important in desalting when combined with power generation.

power output, 500 mW, on the top scale. From the intersection between lines 2 and 3, draw line 4 horizontally to the left-hand scale. At the intersection, read the reactor power as 2250 thermal mW.

The *type of cycle* is determined by the location of the point of intersection between lines 2 and 3. If lines 2 and 3 intersect to the *right* of the full back-pressure (FBP) line, the cycle used is the extraction or multishaft type. When the intersection falls directly on the FBP line, a back-pressure cycle is indicated. An intersection to the left of the FBP line indicates that some of the steam to the brine heater is bypassed around the turbine regardless of the cycle used. Since the intersection in Fig. 6 occurs to the right of the FBP line, either an extraction or multishaft type of cycle could be used. The final choice of a cycle would depend on the water output required.

Related Calculations: The data presented here were developed by W. H. Comtois, Westinghouse Electric Corp., and were reported in *Mechanical Engineering*. Studies made at Westinghouse show that:

1. The fixed-annual-charge rate exerts the greatest single influence on water cost, increasing the cost by about two-thirds for a factor of 2 increase in the rate. This effect is moderated somewhat for large plant sizes.

2. The plant load factor gives the expected result of decreasing product costs with increasing load factor. The effect is a 1 to 2 percent decrease (increase) for every percentage increase (decrease) in load factor in the range from 75 to 95 percent.

3. Plant design life is of little consequence in the range normally considered (30 to 40 years).

4. The range of maximum brine temperatures studied was 200 to 250°F (93.3 to 121.1°C). Without exception, the computed optimum brine temperature was 250°F (121.1°C).

5. The single-shaft cycles (backpressure or extraction) enjoy a small (5 to 10 percent) water cost advantage over the multishaft cycle.

SECTION 11

SANITARY ENGINEERING

EDMUND B. BESSELIEVRE, P.E.
CONSULTANT
FORREST & COTTON, INC.

TYLER G. HICKS, P.E.
INTERNATIONAL ENGINEERING ASSOCIATES

MAX KURTZ, P.E.
CONSULTING ENGINEER

Water-Supply System Flow-Rate and Pressure-Loss Analysis 11.2
Water-Supply System Selection 11.7
Selection of Treatment Method for Water-Supply System 11.10
Storm-Water Runoff Rate and Rainfall Intensity 11.12
Sizing Sewer Pipes for Various Flow Rates 11.14
Sewer-Pipe Earth Load and Bedding Requirements 11.17
Sanitary Sewer System Design. 11.20
Storm-Sewer Inlet Size and Flow Rate 11.24
Storm-Sewer Design 11.25
Selection of Sewage-Treatment Method 11.26

REFERENCES: McClelland and Evans—*Individual Onsite Wastewater Systems*, National Sanitation Foundation; Harbold—*Sanitary Engineering Problems and Calculations for Professional Engineers*, Ann Arbor Science; Feachem—*Water, Waste and Health in Hot Climates*, Wiley; Noble—*Sanitary Land Fill Design*,

Technomic; Rich—*Environmental Systems Engineering*, McGraw-Hill; Kalbermatten—*Appropriate Sanitation Alternatives: A Technical and Economic Appraisal*, John Hopkins; *Sanitation Details in SI Metric*, International Ideas; Sawyer and McCarty—*Chemistry for Environmental Engineering*, McGraw-Hill; Fair—*Sewage Treatment*, Wiley; Steel—*Water Supply and Sewerage*, McGraw-Hill; Gurnham—*Principles of Industrial Waste Treatment*, Wiley; Babbitt, Dolan, and Cleasby—*Water Supply Engineering*, McGraw-Hill; Wright—*Rural Water Supply and Sanitation*, Wiley; Ehlers and Steel—*Municipal and Rural Sanitation*, McGraw-Hill; Babbitt and Baumann—*Sewerage and Sewage Treatment*, Wiley; Chow—*Handbook of Applied Hydrology*, McGraw-Hill; American Society of Civil Engineers—*Design and Construction of Sanitary and Storm Sewers*; King and Brater—*Handbook of Hydraulics*, McGraw-Hill; Woods—*Highway Engineering Handbook*, McGraw-Hill; Hicks—*Pump Application Engineering*, McGraw-Hill; Fair, Geyer, and Okun—*Water Supply and Wastewater Engineering*, Wiley; Besselievre—*Industrial Waste Treatment*, McGraw-Hill; Federation of Sewage and Industrial Wastes Association—*Chlorination of Sewage and Industrial Wastes*; American Society of Civil Engineers—*Sewage Treatment Plant Design*; Imhoff and Fair—*Sewage Treatment*, Wiley; American Society of Civil Engineers—*Filtering Materials for Sewage Treatment Plants*; Mahlie—*Manual for Sewage Plant Operators*, Texas Water & Sewage Works Association; Escritt—*Sewerage and Sewage Disposal*, Contractors Record, Ltd. (London); Escritt—*Pumping Station Equipment and Design*, C. R. Books, Ltd.

WATER-SUPPLY SYSTEM FLOW-RATE AND PRESSURE-LOSS ANALYSIS

A water-supply system will serve a city of 100,000 population. Two water mains arranged in a parallel configuration (Fig. 1a) will supply this city. Determine the flow rate, size, and head loss of each pipe in this system. If the configuration in Fig. 1a were replaced by the single pipe shown in Fig. 1b, what would the total head loss be if $C = 100$ and the flow rate were reduced to 2000 gal/min (126.2 L/s)? Explain how the Hardy Cross method is applied to the water-supply piping system in Fig. 3.

Calculation Procedure:

1. *Compute the domestic water flow rate in the system*

Use an average annual domestic water consumption of 150 gal/day (0.0066 L/s) per capita. Hence, domestic water consumption = (150 gal per capita per day)(100,000 persons) = 15,000,000 gal/day (657.1 L/s). To this domestic flow, the flow required for fire protection must be added to determine the total flow required.

2. *Compute the required flow rate for fire protection*

Use the relation $Q_f = 1020(P)^{0.5}[1 - 0.01(P)^{0.5}]$, where Q_f = fire flow, gal/min; P = population in thousands. Substituting gives $Q_f = 1020(100)^{0.5}[1 - 0.01(100)^{0.5}] = 9180$, say 9200 gal/min (580.3 L/s).

3. *Apply a load factor to the domestic consumption*

To provide for unusual water demands, many design engineers apply a 200 to 250 percent load factor to the average hourly consumption that is determined from the average annual consump-

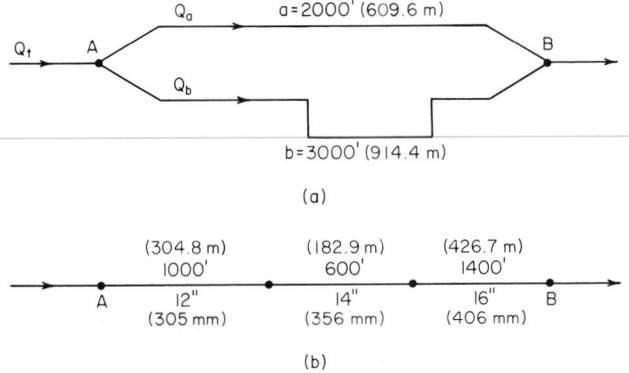

(a)

(b)

FIG. 1 (*a*) Parallel water distribution system; (*b*) single-pipe distribution system.

tion. Thus, the average daily total consumption determined in step 1 is based on an average annual daily demand. Convert the average daily total consumption in step 1 to an average hourly consumption by dividing by 24 h or 15,000,000/24 = 625,000 gal/h (657.1 L/s). Next, apply a 200 percent load factor. Or, design hourly demand = 2.00(625,000) = 1,250,000 gal/h (1314.1 L/s), or 1,250,000/60 min/h = 20,850, say 20,900 gal/min (1318.6 L/s).

4. Compute the total water flow required

The total water flow required = domestic flow, gal/min + fire flow, gal/min = 20,900 + 9200 = 30,100 gal/min (1899.0 L/s). If this system were required to supply water to one or more industrial plants in addition to the domestic and fire flows, the quantity needed by the industrial plants would be added to the total flow computed above.

5. Select the flow rate for each pipe

The flow rate is not known for either pipe in Fig. 1a. Assume that the shorter pipe a has a flow rate Q_a of 12,100 gal/min (763.3 L/s), and the longer pipe b a flow rate Q_b of 18,000 gal/min (1135.6 L/s). Thus, $Q_a + Q_b = Q_t$ = 12,100 + 18,000 = 30,100 gal/min (1899.0 L/s), where Q = flow, gal/min, in the pipe identified by the subscript a or b; Q_t = total flow in the system, gal/min.

6. Select the sizes of the pipes in the system

Since neither pipe size is known, some assumptions must be made about the system. First, assume that a friction-head loss of 10 ft of water per 1000 ft (3.0 m per 304.8 m) of pipe is suitable for this system. This is a typical allowable friction-head loss for water-supply systems.

Second, assume that the pipe is sized by using the Hazen-Williams equation with the coefficient C = 100. Most water-supply systems are designed with this equation and this value of C.

Enter Fig. 2 with the assumed friction-head loss of 10 ft/1000 ft (3.0 m/304.8 m) of pipe on the right-hand scale, and project through the assumed Hazen-Williams coefficient C = 100. Extend this straight line until it intersects the pivot axis. Next, enter Fig. 2 on the left-hand scale at the flow rate in pipe a, 12,100 gal/min (763.3 L/s), and project to the previously found intersection on the pivot axis. At the intersection with the pipe-diameter scale, read the required pipe size as 27-in (686-mm) diameter. Note that if the required pipe size falls between two plotted sizes, the next *larger* size is used.

Now in any parallel piping system, the friction-head loss through any branch connecting two common points equals the friction-head loss in any other branch connecting the same two points. Using Fig. 2 for a 27-in (686-mm) pipe, find the actual friction-head loss at 8 ft/1000 ft (2.4 m/304.8 m) of pipe. Hence, the total friction-head loss in pipe a is (2000 ft long)(8 ft/1000 ft) = 16 ft (4.9 m) of water. This is also the friction-head loss in pipe b.

Since pipe b is 3000 ft (914.4 m) long, the total friction-head loss per 1000 ft (304.8 m) is total head loss, ft/length of pipe, thousands of ft = 16/3 = 5.33 ft/1000 ft (1.6 m/304.8 m). Enter Fig. 2 at this friction-head loss and C = 100. Project in the same manner as described for pipe a, and find the required size of pipe b as 33 in (838.2 mm).

If the district being supplied by either pipe required a specific flow rate, this flow would be used instead of assuming a flow rate. Then the pipe would be sized in the same manner as described above.

7. Compute the single-pipe equivalent length

When we deal with several different sizes of pipe having the same flow rate, it is often convenient to convert each pipe to an *equivalent length* of a common-size pipe. Many design engineers use 8-in (203-mm) pipe as the common size. Table 1 shows the equivalent length of 8-in (203-mm) pipe for various other sizes of pipe with C = 90, 100, and 110 in the Hazen-Williams equation.

From Table 1, for 12-in (305-mm) pipe, the equivalent length of 8-in (203-mm) pipe is 0.14 ft/ft when C = 100. Thus, total equivalent length of 8-in (203-mm) pipe = (1000 ft of 12-in pipe)(0.14 ft/ft) = 140 ft (42.7 m) of 8-in (203-mm) pipe. For the 14-in (356-mm) pipe, total equivalent length = (600)(0.066) = 39.6 ft (12.1 m), using similar data from Table 1. For the 16-in (406-mm) pipe, total equivalent length = (1400)(0.034) = 47.6 ft (14.5 m). Hence, total equivalent length of 8-in (203-mm) pipe = 140 + 39.6 + 47.6 = 227.2 ft (69.3 m).

8. Determine the friction-head loss in the pipe

Enter Fig. 2 at the flow rate of 2000 gal/min (126.2 L/s), and project through 8-in (203-mm) diameter to the pivot axis. From this intersection, project through C = 100 to read the friction-

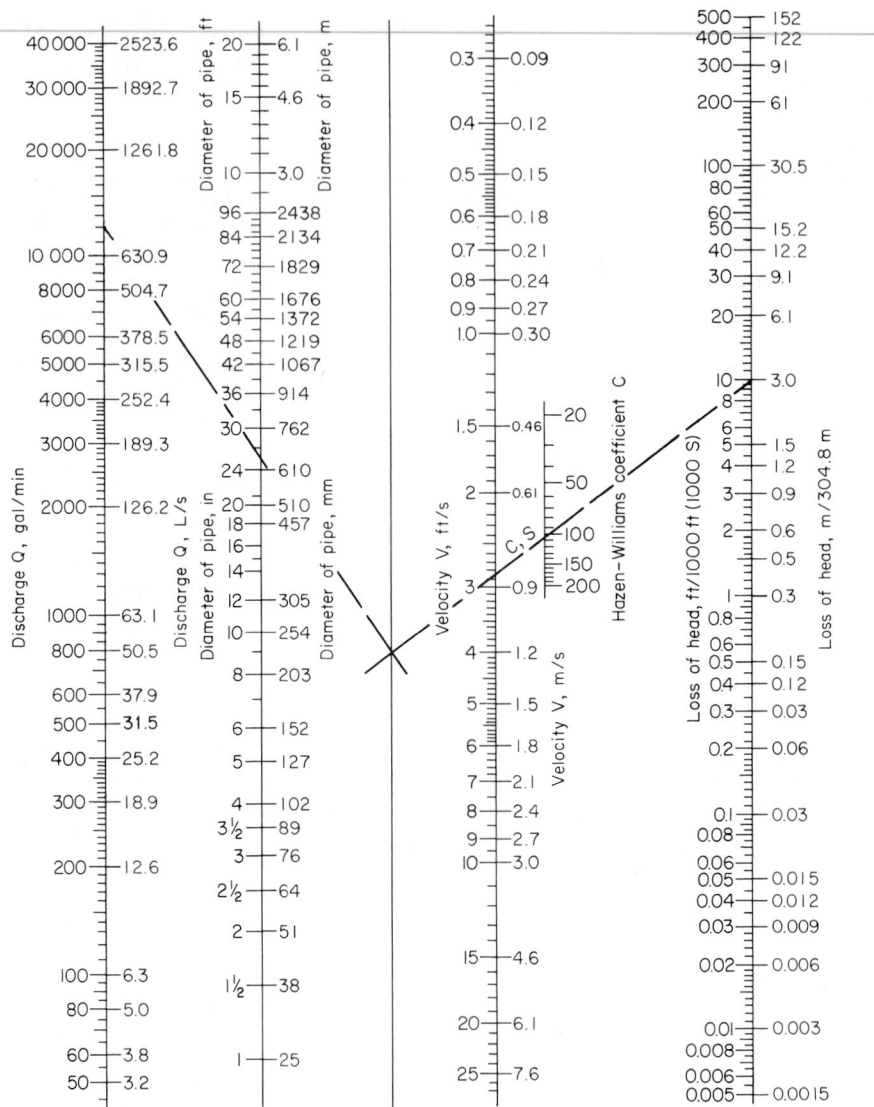

FIG. 2 Nomogram for solution of the Hazen-Williams equation for pipes flowing full.

head loss as 100 ft/1000 ft (30.5 m/304.8 m), due to the friction of the water in the pipe. Since the equivalent length of the pipe is 227.2 ft (69.3 m), the friction-head loss in the compound pipe is $(227.2/1000)(110) = 25$ ft (7.6 m) of water.

Related Calculations: Two pipes, two piping systems, or a single pipe and a system of pipes are said to be *equivalent* when the losses of head due to friction for equal rates of flow in the pipes are equal.

To determine the flow rates and friction-head losses in complex waterworks distribution systems, the Hardy Cross method of network analysis is often used. This method[1] uses trial and error

---

[1]O'Rourke—*General Engineering Handbook*, McGraw-Hill.

TABLE 1   Equivalent Length of 8-in (203-mm) Pipe for $C = 100$

Pipe diameter				
in	mm	$C = 90$	$C = 100$	$C = 110$
2	51	1012	851	712
4	102	34	29	24.3
6	152	4.8	4.06	3.4
8	203	1.19	1.00	0.84
10	254	0.40	0.34	0.285
12	305	0.17	0.14	0.117
14	356	0.078	0.066	0.055
16	406	0.040	0.034	0.029
18	457	0.023	0.019	0.016
20	508	0.0137	0.0115	0.0096
24	610	0.0056	0.0047	0.0039
30	762	0.0019	0.0016	0.0013
36	914	0.00078	0.00066	0.00055

to obtain successively more accurate approximations of the flow rate through a piping system. To apply the Hardy Cross method: (1) Sketch the piping system layout as in Fig. 3. (2) Assume a flow quantity, in terms of percentage of total flow, for each part of the piping system. In assuming a flow quantity note that (*a*) the loss of head due to friction between any two points of a closed circuit must be the same by any path by which the water may flow, and (*b*) the rate of inflow

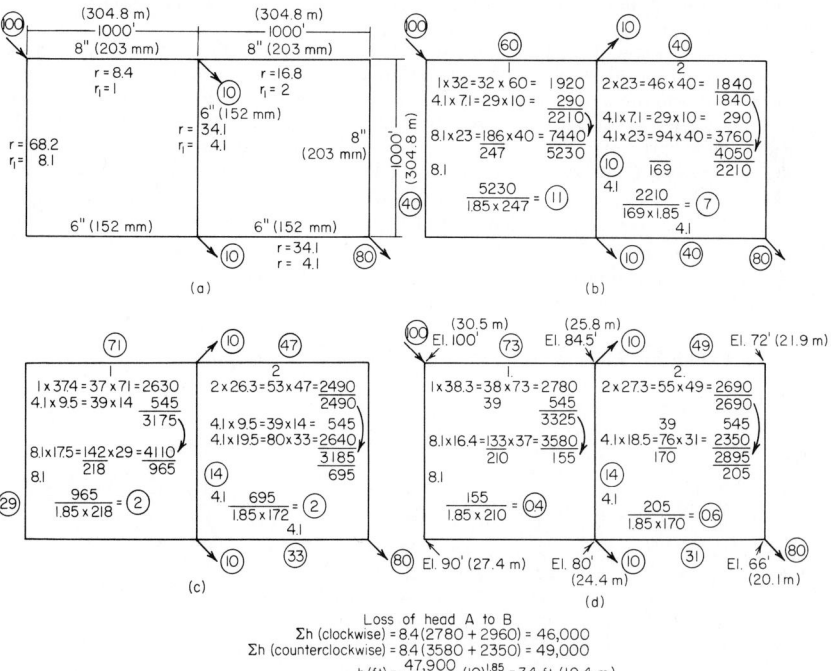

Loss of head A to B
Σh (clockwise) = 8.4(2780 + 2960) = 46,000
Σh (counterclockwise) = 8.4(3580 + 2350) = 49,000
$h(\text{ft}) = \dfrac{47,900}{100,000}(10)^{1.85} = 34$ ft (10.4 m)

FIG. 3   Application of the Hardy Cross method to a water distribution system.

TABLE 2 Values of r for 1000 ft (304.8 m) of Pipe Based on the Hazen-Williams Formula[*]

d, in (mm)	$C = 90$	$C = 100$	$C = 110$	$C = 120$	$C = 130$	$C = 140$
4 (102)	340	246	206	176	151	135
6 (152)	47.1	34.1	28.6	24.3	21.0	18.7
8 (203)	11.1	8.4	7.0	6.0	5.2	4.6
10 (254)	3.7	2.8	2.3	2.0	1.7	1.5
12 (305)	1.6	1.2	1.0	0.85	0.74	0.65
14 (356)	0.72	0.55	0.46	0.39	0.34	0.30
16 (406)	0.38	0.29	0.24	0.21	0.18	0.15
18 (457)	0.21	0.16	0.13	0.11	0.10	0.09
20 (508)	0.13	0.10	0.08	0.07	0.06	0.05
24 (610)	0.052	0.04	0.03	0.03	0.02	0.02
30 (762)	0.017	0.013	0.011	0.009	0.008	0.007

Example: r for 12-in (305-mm) pipe 4000 ft (1219 m) long, with $C = 100$, is $1.2 \times 4.0 = 4.8$.
[*] Head loss in ft (m) $= r \times 10^{-5} \times Q^{1.85}$ per 1000 ft (304.8 m), Q representing gal/min (L/s).

into any section of the piping system must equal the outflow. (3) Compute the loss of head due to friction between two points in each part of the system, based on the assumed flow in (*a*) the clockwise direction and (*b*) the counterclockwise direction. A difference in the calculated friction-head losses in the two directions indicates an error in the assumed direction of flow. (4) Compute a counterflow correction by dividing the difference in head, Δh ft, by $n(Q)^{n-1}$, where $n = 1.85$ and $Q =$ flow, gal/min. Indicate the direction of this counterflow in the pipe by an arrow starting at the right side of the smaller value of h and curving toward the larger value, Fig. 3. (5) Add or subtract the counterflow to or from the assumed flow, depending on whether its direction is the same or opposite. (6) Repeat this process on each circuit in the system until a satisfactory balance of flow is obtained.

To compute the loss of head due to friction, step 3 of the Hardy Cross method, use any standard formula, such as the Hazen-Williams, that can be reduced to the form $h = rQ^nL$, where h = head loss due to friction, ft of water; r = a coefficient depending on the diameter and roughness of the pipe; Q = flow rate, gal/min; $n = 1.85$; L = length of pipe, ft. Table 2 gives values of r for 1000-ft (304.8-m) lengths of various sizes of pipe and for different values of the Hazen-Williams coefficient C. When the percentage of total flow is used for computing $\sum h$ in Fig. 3, the loss of head due to friction in ft between any two points for any flow in gal/min is computed from $h = [\sum h \text{ (by percentage of flow)}/100,000]\,(\text{gal/min}/100)^{0.85}$. Figure 3 shows the details of the solution using the Hardy Cross method. The circled numbers represent the flow quantities. Table 3 lists values of numbers between 0 and 100 to the 0.85 power.

TABLE 3 Value of the 0.85 Power of Numbers

N	0	1	2	3	4	5	6	7	8	9
0	0	1.0	1.8	2.5	3.2	3.9	4.6	5.2	5.9	6.5
10	7.1	7.7	8.3	8.9	9.5	10.0	10.6	11.1	11.6	12.2
20	12.8	13.3	13.8	14.4	14.9	15.4	15.9	16.4	16.9	17.5
30	18.0	18.5	19.0	19.5	20.0	20.5	21.0	21.5	22.0	22.5
40	23.0	23.4	23.9	24.3	24.8	25.3	25.8	26.3	26.8	27.3
50	27.8	28.2	28.7	29.1	29.6	30.0	30.5	31.0	31.4	31.9
60	32.4	32.9	33.3	33.8	34.2	34.7	35.1	35.6	36.0	36.5
70	37.0	37.4	37.9	38.3	38.7	39.1	39.6	40.0	40.5	41.0
80	41.5	42.0	42.4	42.8	43.3	43.7	44.1	44.5	45.0	45.4
90	45.8	46.3	46.7	47.1	47.6	48.0	48.4	48.8	49.2	49.6

WATER-SUPPLY SYSTEM SELECTION

Choose the type of water-supply system for a city having a population of 100,000 persons. Indicate which type of system would be suitable for such a city today and 20 years hence. The city is located in an area of numerous lakes.

Calculation Procedure:

1. Compute the domestic water flow rate in the system

Use an average annual domestic water consumption of 150 gal per capita day (gcd) (6.6 mL/s). Hence, domestic water consumption = (150 gal per capita day)(100,000 persons) = 15,000,000 gal/day (657.1 L/s). To this domestic flow, the flow required for fire protection must be added to determine the total flow required.

2. Compute the required flow rate for fire protection

Use the relation $Q_f = 1020(P)^{0.5}[1 - 0.01(P)^{0.5}]$, where Q_f = fire flow, gal/min; P = population in thousands. So $Q_f = 1020(100)^{0.5}[1 - 0.01 \times (100)^{0.5}] = 9180$, say 9200 gal/min (580.3 L/s).

3. Apply a load factor to the domestic consumption

To provide for unusual water demands, many design engineers apply a 200 to 250 percent load factor to the average hourly consumption that is determined from the average annual consumption. Thus, the average daily total consumption determined in step 1 is based on an average annual daily demand. Convert the average daily total consumption in step 1 to an average hourly consumption by dividing by 24 h, or 15,000,000/24 = 625,000 gal/h (657.1 L/s). Next, apply a 200 percent load factor. Or, design hourly demand = 2.00(625,000) = 1,250,000 gal/h (1314.1 L/s), or 1,250,000/(60 min/h) = 20,850, say 20,900 gal/min (1318.4 L/s).

4. Compute the total water flow required

The total water flow required = domestic flow, gal/min + fire flow, gal/min = 20,900 + 9200 = 30,100 gal/min (1899.0 L/s). If this system were required to supply water to one or more industrial plants in addition to the domestic and fire flows, the quantity needed by the industrial plants would be added to the total flow computed above.

5. Study the water supplies available

Table 4 lists the principal sources of domestic water supplies. Wells that are fed by groundwater are popular in areas having sandy or porous soils. To determine whether a well is suitable for supplying water in sufficient quantity, its specific capacity (i.e., the yield in gal/min per foot of drawdown) must be determined.

Wells for municipal water sources may be dug, driven, or drilled. Dug wells seldom exceed 60 ft (18.3 m) deep. Each such well should be protected from surface-water leakage by being lined with impervious concrete to a depth of 15 ft (4.6 m).

Driven wells seldom are more than 40 ft (12.2 m) deep or more than 2 in (51 mm) in diameter when used for small water supplies. Bigger driven wells are constructed by driving large-diameter casings into the ground.

TABLE 4 Typical Municipal Water Sources

Source	Collection method	Remarks
Groundwater	Wells (artesian, ordinary, galleries)	30 to 40 percent of an area's rainfall becomes groundwater
Surface freshwater (lakes, rivers, streams, impounding reservoirs)	Pumping or gravity flow from submerged intakes, tower intakes, or surface intakes	Surface supplies are important in many areas
Surface saltwater	Desalting	Wide-scale application under study at present

Drilled wells can be several thousand feet deep, if required. The yield of a driven well is usually greater than any other type of well because the well can be sunk to a depth where sufficient groundwater is available. Almost all wells require a pump of some kind to lift the water from its subsurface location and discharge it to the water-supply system.

Surface freshwater can be collected from lakes, rivers, streams, or reservoirs by submerged-, tower-, or crib-type intakes. The intake leads to one or more pumps that discharge the water to the distribution system or intermediate pumping stations. Locate intakes as far below the water surface as possible. Where an intake is placed less than 20 ft (6.1 m) below the surface of the water, it may become clogged by sand, mud, or ice.

Choose the source of water for this system after studying the local area to determine the most economical source today and 20 years hence. With a rapidly expanding population, the future water demand may dictate the type of water source chosen. Since this city is in an area of many lakes, a surface supply would probably be most economical, if the water table is not falling rapidly.

6.   *Select the type of pipe to use*

Four types of pipes are popular for municipal water-supply systems: cast iron, asbestos cement, steel, and concrete. Wood-stave pipe was once popular, but it is now obsolete. Some communities also use copper or lead pipes. However, the use of both types is extremely small when compared with the other types. The same is true of plastic pipe, although this type is slowly gaining some acceptance.

In general, cast-iron pipe proves dependable and long-lasting in water-supply systems that are not subject to galvanic or acidic soil conditions.

Steel pipe is generally used for long, large-diameter lines. Thus, the typical steel pipe used in water-supply systems is 36 or 48 in (914 or 1219 mm) in diameter. Use steel pipe for river crossings, on bridges, and for similar installations where light weight and high strength are required. Steel pipe may last 50 years or more under favorable soil conditions. Where unfavorable soil conditions exist, the life of steel pipe may be about 20 years.

Concrete-pipe use is generally confined to large, long lines, such as aqueducts. Concrete pipe is suitable for conveying relatively pure water through neutral soil. However, corrosion may occur when the soil contains an alkali or an acid.

Asbestos-cement pipe has a number of important advantages over other types. However, it does not flex readily, it can be easily punctured, and it may corrode in acidic soils.

Select the pipe to use after a study of the local soil conditions, length of runs required, and the quantity of water that must be conveyed. Usual water velocities in municipal water systems are in the 5-ft/s (1.5-m/s) range. However, the velocities in aqueducts range from 10 to 20 ft/s (3.0 to 6.1 m/s). Earthen canals have much lower velocities—1 to 3 ft/s (0.3 to 0.9 m/s). Rock- and concrete-lined canals have velocities of 8 to 15 ft/s (2.4 to 4.6 m/s).

In cold northern areas, keep in mind the occasional need to thaw frozen pipes during the winter. Nonmetallic pipes—concrete, plastic, etc., as well as nonconducting metals—cannot be thawed by electrical means. Since electrical thawing is probably the most practical method available today, pipes that prevent its use may put the water system at a disadvantage if subfreezing temperatures are common in the area served.

7.   *Select the method for pressurizing the water system*

Water-supply systems can be pressurized in three different ways: by gravity or natural elevation head, by pumps that produce a pressure head, and by a combination of the first two ways.

Gravity systems are suitable where the water storage reservoir or receiver is high enough above the distribution system to produce the needed pressure at the farthest outlet. The operating cost of a gravity system is lower than that of a pumped system, but the first cost of the former is usually higher. However, the reliability of the gravity system is usually higher because there are fewer parts that may fail.

Pumping systems generally use centrifugal pumps that discharge either directly to the water main or to an elevated tank, a reservoir, or a standpipe. The water then flows from the storage chamber to the distribution system. In general, most sanitary engineers prefer to use a reservoir or storage tank between the pumps and distribution mains because this arrangement provides greater reliability and fewer pressure surges.

Surface reservoirs should store at least a 1-day water supply. Most surface reservoirs are designed to store a supply for 30 days or longer. Elevated tanks should have a capacity of at least

25 gal (94.6 L) of water per person served, *plus* a reserve for fire protection. The capacity of typical elevated tanks ranges from a low of 40,000 gal (151 kL) for a 20-ft (6.1-m) diameter tank to a high of 2,000,000 gal (7.5 ML) for an 80-ft (24.4-m) diameter tank.

Choose the type of distribution system after studying the topography, water demand, and area served. In general, a pumped system is preferred today. To ensure continuity of service, duplicate pumps are generally used.

8.   *Choose the system operating pressure*

In domestic water supply, the minimum pressure required at the highest fixture in a building is usually assumed to be 15 lb/in^2 (103.4 kPa). The maximum pressure allowed at a fixture in a domestic water system is usually 65 lb/in^2 (448.2 kPa). High-rise buildings (i.e., those above six stories) are generally required to furnish the pressure increase needed to supply water to the upper stories. A pump and overhead storage tank are usually installed in such buildings to provide the needed pressure.

Commercial and industrial buildings require a minimum water pressure of 75 lb/in^2 (517.1 kPa) at the street level for fire hydrant service. This hydrant should deliver at least 250 gal/min (15.8 L/s) of water for fire-fighting purposes.

Most water-supply systems served by centrifugal pumps in a central pumping station operate in the 100-lb/in^2 (689.5-kPa) pressure range. In areas of one- and two-story structures, a lower pressure, say 65 lb/in^2 (448.2 kPa), is permissible. Where the pressure in a system falls too low, auxiliary or booster pumps may be used. These pumps increase the pressure in the main to the desired level.

Choose the system pressure based on the terrain served, quantity of water required, allowable pressure loss, and size of pipe used in the system. Usual pressures required will be in the ranges cited above, although small systems serving one-story residences may operate at pressures as low as 30 lb/in^2 (206.8 kPa). Pressures over 100 lb/in^2 (689.5 kPa) are seldom used because heavier piping is required. As a rule, distribution pressures of 50 to 75 lb/in^2 (344.7 to 517.1 kPa) are acceptable.

9.   *Determine the number of hydrants for fire protection*

Table 5 shows the required fire flow, number of standard hose streams of 250 gal/min (15.8 L/s) discharged through a 1⅛-in (28.6-mm) diameter smooth nozzle, and the average area served by a hydrant in a high-value district. A standard hydrant may have two or three outlets.

Table 5 indicates that a city of 100,000 persons requires 36 standard hose streams. This means that 36 single-outlet or 18 dual-outlet hydrants are required. More, of course, could be used if better protection were desired in the area. Note that the required fire flow listed in Table 5 agrees closely with that computed in step 2 above.

Related Calculations: Use this general method for any water-supply system, municipal or industrial. Note, however, that the required fire-protection quantities vary from one type of

TABLE 5   Required Fire Flow and Hydrant Spacing°

Population	Required fire flow, gal/min (L/s)	Number of standard hose streams	Average area served per hydrant, ft^2 (m^2)†	
			Direct streams	Engine streams
22,000	4,500 (284)	18	55,000 (5,110)	90,000 (8,361)
28,000	5,000 (315)	20	40,000 (3,716)	85,000 (7,897)
40,000	6,000 (379)	24	40,000 (3,716)	80,000 (7,432)
60,000	7,000 (442)	28	40,000 (3,716)	70,000 (6,503)
80,000	8,000 (505)	32	40,000 (3,716)	60,000 (5,574)
100,000	9,000 (568)	36	40,000 (3,716)	55,000 (5,110)
125,000	10,000 (631)	40	40,000 (3,716)	48,000 (4,459)
150,000	11,000 (694)	44	40,000 (3,716)	43,000 (3,995)
200,000	12,000 (757)	48	40,000 (3,716)	40,000 (3,716)

°National Board of Fire Underwriters.
†High-value districts.

TABLE 6   Selected Industrial Water and Steam Requirements[°]

	Water	Steam
Air conditioning	6000 to 15,000 gal (22,700 to 57,000 L) per person per season	. . .
Aluminum	1,920,000 gal/ton (8.0 ML/t)	. . .
Cement, portland	750 gal/ton cement (3129 L/t)	. . .
Coal, by-product coke	1430 to 2800 gal/ton coke (5967 to 11,683 L/t)	570 to 860 lb/ton (382 to 427 kg/t)
Rubber (automotive tire)	. . .	120 lb (54 kg) per tire
Electricity	80 gal/kW (302 L/kW) of electricity	. . .

[°]Courtesy of American Society for Testing and Materials.

municipal area to another and among different industrial exposures. Refer to *NFPA Handbook of Fire Protection*, available from NFPA, 60 Batterymarch Street, Boston, Massachusetts 02110, for specific fire-protection requirements for a variety of industries. In choosing a water-supply system, the wise designer looks ahead for at least 10 years when the water demand will usually exceed the present demand. Hence, the system may be designed so it is oversized for the present population but just adequate for the future population. The American Society for Testing and Materials (ASTM) publishes comprehensive data giving the usual water requirements for a variety of industries. Table 6 shows a few typical water needs for selected industries.

To determine the storage capacity required at present, proceed as follows: (1) Compute the flow needed to meet 50 percent of the present domestic daily (that is, 24-h) demand. (2) Compute the 4-h fire demand. (3) Find the sum of (1) and (2).

For this city, procedure (1) = (20,900 gal/min)(60 min/h)(24 h/day)(0.5) = 15,048,000 gal (57.2 ML) with the data computed in step 3. Also procedure (2) = (4 h)(60 min/h)(9200 gal/min) = 2,208,000 gal (8.4 ML), using the data computed in step 2, above. Then, total storage capacity required = 15,048,000 + 2,208,000 = 17,256,000 gal (65.3 ML). Where one or more reliable wells will produce a significant flow for 4 h or longer, the storage capacity can be reduced by the 4-h productive capacity of the wells.

SELECTION OF TREATMENT METHOD FOR WATER-SUPPLY SYSTEM

Choose a treatment method for a water-supply system for a city having a population of 100,000 persons. The water must be filtered, disinfected, and softened to make it suitable for domestic use.

Calculation Procedure:

1. *Compute the domestic water flow rate in the system*

When water is treated for domestic consumption, only the drinking water passes through the filtration plant. Fire-protection water is seldom treated unless it is so turbid that it will clog fire pumps or hoses. Assuming that the fire-protection water is acceptable for use without treatment, we consider only the drinking water here.

Use the same method as in steps 1 and 3 of the previous calculation procedure to determine the required domestic water flow of 20,900 gal/min (1318.6 L/s) for this city.

2. *Select the type of water-treatment system to use*

Water supplies are treated by a number of methods including sedimentation, coagulation, filtration, softening, and disinfection. Other treatments include disinfection, taste and odor control, and miscellaneous methods.

Since the water must be filtered, disinfected, and softened, each of these steps must be considered separately.

TABLE 7   Typical Limits for Impurities in Water Supplies

Impurity	Limit, ppm	Impurity	Limit, ppm
Turbidity	10	Iron plus manganese	0.3
Color	20	Magnesium	125
Lead	0.1	Total solids	500
Fluoride	1.0	Total hardness	100
Copper	3.0	Ca + Mg salts	

3. *Choose the type of filtration to use*

Slow sand filters operate at an average rate of 3 million gal/(acre·day) [2806.2 L/(m^2·day)]. This type of filter removes about 99 percent of the bacterial content of the water and most tastes and odors.

Rapid sand filters operate at an average rate of 150 million gal/(acre·day) [1.6 L/(m^2·s)]. But the raw water must be treated before it enters the rapid sand filter. This preliminary treatment often includes chemical coagulation and sedimentation. A high percentage of bacterial content—up to 99.98 percent—is removed by the preliminary treatment and the filtration. But color and turbidity removal is not as dependable as with slow sand filters. Table 7 lists the typical limits for certain impurities in water supplies.

The daily water flow rate for this city is, from step 1, (20,900 gal/min)(24 h/day)(60 min/h) = 30,096,000 gal/day (1318.6 L/s). If a slow sand filter were used, the required area would be (30.096 million gal/day)/[3 million gal/(acre·day)] = 10+ acres (40,460 m^2).

A rapid sand filter would require 30.096/150 = 0.2 acre (809.4 m^2). Hence, if space were scarce in this city—and it usually is—a rapid sand filter would be used. With this choice of filtration, chemical coagulation and sedimentation are almost a necessity. Hence, these two additional steps would be included in the treatment process.

Table 8 gives pertinent data on both slow and rapid sand filters. These data are useful in filter selection.

TABLE 8   Typical Sand-Filter Characteristics

Slow sand filters	
Usual filtration rate	2.5 to 6.0 × 10^6 gal/(acre·day) [2339 to 5613 L/(m^2·day)]
Sand depth	30 to 36 in (76 to 91 cm)
Sand size	35 mm
Sand uniformity coefficient	1.75
Water depth	3 to 5 ft (0.9 to 1.5 m)
Water velocity in underdrains	2 ft/s (0.6 m/s)
Cleaning frequency required	2 to 11 times per year
Units required	At least two to permit alternate cleaning

Fast sand filters	
Usual filtration rate	100 to 200 × 10^6 gal/(acre·day) [24.7 to 49.4 kL/(m^2·day)]
Sand depth	30 in (76 cm)
Gravel depth	18 in (46 cm)
Sand size	0.4 to 0.5 mm
Sand uniformity coefficient	1.7 or less
Units required	At least three to permit cleaning one unit while the other two are operating

4. Select the softening process to use

The principal water-softening processes use: (*a*) lime and sodium carbonate followed by sedimentation or filtration, or both, to remove the precipitates and (*b*) zeolites of the sodium type in a pressure filter. Zeolite softening is popular and is widely used in municipal water-supply systems today. Based on its proven usefulness and economy, zeolite softening will be chosen for this installation.

5. Select the disinfection method to use

Chlorination by the addition of chlorine to the water is the principal method of disinfection used today. To reduce the unpleasant effects that may result from using chlorine alone, a mixture of chlorine and ammonia, known as chloramine, may be used. The ammonia dosage is generally 0.25 ppm or less. Assume that the chloramine method is chosen for this installation.

6. Select the method of taste and odor control

The methods used for taste and odor control are: (*a*) aeration, (*b*) activated carbon, (*c*) prechlorination, and (*d*) chloramine. Aeration is popular for groundwaters containing hydrogen sulfide and odors caused by microscopic organisms.

Activated carbon absorbs impurities that cause tastes, odors, or color. Generally, 10 to 20 lb (4.5 to 9.1 kg) of activated carbon per million gallons of water is used, but larger quantities— from 50 to 60 lb (22.7 to 27.2 kg)—may be specified. In recent years, some 2000 municipal water systems have installed activated carbon devices for taste and odor control.

Prechlorination and chloramine are also used in some installations for taste and odor control. Of the two methods, chloramine appears more popular at present.

Based on the data given for this water-supply system, method *b*, *c*, or *d* would probably be suitable. Because method *b* has proven highly effective, it will be chosen tentatively, pending later investigation of the economic factors.

Related Calculations: Use this general procedure to choose the treatment method for all types of water-supply systems where the water will be used for human consumption. Thus, the procedure is suitable for municipal, commercial, and industrial systems.

Hazardous wastes of many types endanger groundwater supplies. One of the most common hazardous wastes is gasoline which comes from the estimated 120,000 leaking underground gasoline-storage tanks. Major oil companies are replacing leaking tanks with new noncorrosive tanks. But the soil and groundwater must still be cleaned to prevent pollution of drinking-water supplies.

Other contaminants include oily sludges, organic (such as pesticides and dioxins), and nonvolatile organic materials. These present especially challenging removal and disposal problems for engineers, particularly in view of the stringent environmental requirements of almost every community.

A variety of treatment and disposal methods are in the process of development and application. For oily waste handling, one process combines water evaporation and solvent extraction to break down a wide variety of hazardous waste and sludge from industrial, petroleum-refinery, and municipal-sewage-treatment operations. This process typically produces dry solids with less than 0.5 percent residual hydrocarbon content. This meets EPA regulations for nonhazardous wastes with low heavy-metal contents.

Certain organics, such as pesticides and dioxins, are hydrophobic. Liquified propane and butane are effective at separating hydrophobic organics from solid particles in tainted sludges and soils. The second treatment method uses liquified propane to remove organics from contaminated soil. Removal efficiencies reported are: polychlorinated biphenyls (PCBs) 99.9 percent; polyaromatic hydrocarbons (PAHs) 99.5 percent; dioxins 97.4 percent; total petroleum hydrocarbons 99.9 percent. Such treated solids meet EPA land-ban regulations for solids disposal.

Nonvolatile organic materials at small sites can be removed by a mobile treatment system using up to 14 solvents. Both hydrophobic and hydrophilic solvents are used; all are nontoxic; several have Food and Drug Administration (FDA) approval as food additives. Used at three different sites (at this writing) the process reduced PCB concentration from 500 to 1500 ppm to less than 100 ppm; at another site PCB concentration was reduced from an average of 30 to 300 ppm to less than 5 ppm; at the third site PCBs were reduced from 40 ppm to less than 3 ppm.

TABLE 9   Coefficient of Runoff for Various Surfaces

Surface	Coefficient
Parks, gardens, lawns, meadows	0.05–0.25
Gravel roads and walks	0.15–0.30
Macadamized roadways	0.25–0.60
Inferior block pavements with uncemented joints	0.40–0.50
Stone, brick, and wood-block pavements with tightly cemented joints	0.75–0.85
Same with uncemented joints	0.50–0.70
Asphaltic pavements in good condition	0.85–0.90
Watertight roof surfaces	0.70–0.95

TABLE 10   Coefficient of Runoff for Various Areas

Area	Coefficient
Business:	
Downtown	0.70–0.95
Neighborhood	0.50–0.70
Residential:	
Single-family	0.30–0.50
Multiunits, detached	0.40–0.60
Multiunits, attached	0.60–0.75
Residential (suburban)	0.25–0.40
Apartment dwelling	0.50–0.70
Industrial:	
Light industry	0.50–0.80
Heavy industry	0.60–0.90
Playgrounds	0.20–0.35
Railroad yards	0.20–0.40
Unimproved	0.10–0.30

STORM-WATER RUNOFF RATE AND RAINFALL INTENSITY

What is the storm-water runoff rate from a 40-acre (1.6-km^2) industrial site having an imperviousness of 50 percent if the time of concentration is 15 min? What would be the effect of planting a lawn over 75 percent of the site?

Calculation Procedure:

1. *Compute the hourly rate of rainfall*

Two common relations, called the *Talbot formulas,* used to compute the hourly rate of rainfall R in/h are $R = 360/(t + 30)$ for the heaviest storms and $R = 105/(t + 15)$ for ordinary storms, where t = time of concentration, min. Using the equation for the heaviest storms because this relation gives a larger flow rate and produces a more conservative design, we see $R = 360/(15 + 30) = 8$ in/h (0.05 mm/s).

2. *Compute the storm-water runoff rate*

Apply the *rational method* to compute the runoff rate. This method uses the relation $Q = AIR$, where Q = storm-water runoff rate, ft^3/s; A = area served by sewer, acres; I = coefficient of runoff or percentage of imperviousness of the area; other symbols as before. So $Q = (40)(0.50)(8) = 160$ ft^3/s (4.5 m^3/s).

3. Compute the effect of changed imperviousness

Planting a lawn on a large part of the site will increase the imperviousness of the soil. This means that less rainwater will reach the sewer because the coefficient of imperviousness of a lawn is lower. Table 9 lists typical coefficients of imperviousness for various surfaces. This tabulation shows that the coefficient for lawns varies from 0.05 to 0.25. Using a value of $I = 0.10$ for the $40(0.75) = 30$ acres of lawn, we have $Q = (30)(0.10)(8) = 24$ ft^3/s (0.68 m^3/s).

The runoff for the remaining 10 acres (40,460 m^2) is, as in step 2, $Q = (10)(0.5)(8) = 40$ ft^3/s (1.1 m^3/s). Hence, the total runoff is $24 + 40 = 64$ ft^3/s (1.8 m^3/s). This is $160 - 64 = 96$ ft^3/s (2.7 m^3/s) less than when the lawn was not used.

Related Calculations: The time of concentration for any area being drained by a sewer is the time required for the maximum runoff rate to develop. It is also defined as the time for a drop of water to drain from the farthest point of the watershed to the sewer.

When rainfall continues for an extended period T min, the coefficient of imperviousness changes. For impervious surfaces such as watertight roofs, $I = T/(8 + T)$. For improved pervious surfaces, $I = 0.3T/(20 + T)$. These relations can be used to compute the coefficient in areas of heavy rainfall.

Equations for R for various areas of the United States are available in Steel—*Water Supply and Sewerage*, McGraw-Hill. The Talbot formulas, however, are widely used and have proved reliable.

The time of concentration for a given area can be approximated from $t = I(L/Si^2)^{1/3}$, where L = distance of overland flow of the rainfall from the most remote part of the site, ft; S = slope of the land, ft/ft; i = rainfall intensity, in/h; other symbols as before. For portions of the flow carried in ditches, the time of flow to the inlet can be computed by using the Manning formula.

Table 10 lists the coefficient of runoff for specific types of built-up and industrial areas. Use these coefficients in the same way as shown above. Tables 9 and 10 present data developed by Kuichling and ASCE.

SIZING SEWER PIPES FOR VARIOUS FLOW RATES

Determine the size, flow rate, and depth of flow from a 1000-ft (304.8-m) long sewer which slopes 5 ft (1.5 m) between inlet and outlet and which must carry a flow of 5 million gal/day (219.1 L/s). The sewer will flow about half full. Will this sewer provide the desired flow rate?

Calculation Procedure:

1. Compute the flow rate in the half-full sewer

A flow of 1 million gal/day = 1.55 ft^3/s (0.04 m^3/s). Hence, a flow of 5 million gal/day = 5(1.55) = 7.75 ft^3/s (219.1 L/s) in a *half-full* sewer.

2. Compute the full-sewer flow rate

In a *full sewer*, the flow rate is twice that in a half-full sewer, or $2(7.75) = 15.50$ ft^3/s (0.44 m^3/s) for this sewer. This is equivalent to $15.50/1.55 = 10$ million gal/day (438.1 L/s). Full-sewer flow rates are used because pipes are sized on the basis of being full of liquid.

3. Compute the sewer-pipe slope

The pipe slope S ft/ft $= (E_i - E_o)/L$, where E_i = inlet elevation, ft above the site datum; E_o = outlet elevation, ft above site datum; L = pipe length between inlet and outlet, ft. Substituting gives $S = 5/1000 = 0.005$ ft/ft (0.005 m/m).

4. Determine the pipe size to use

The Manning formula $v = (1.486/n)R^{2/3}S^{1/2}$ is often used for sizing sewer pipes. In this formula, v = flow velocity, ft/s; n = a factor that is a function of the pipe roughness; R = pipe hydraulic radius = 0.25 pipe diameter, ft; S = pipe slope, ft/ft. Table 11 lists values of n for various types of sewer pipe. In sewer design, the value $n = 0.013$ for pipes flowing full.

Since the Manning formula is complex, numerous charts have been designed to simplify its solution. Figure 4 is one such typical chart designed specifically for sewers.

Enter Fig. 4 at 15.5 ft^3/s (0.44 m^3/s) on the left, and project through the slope ratio of 0.005. On the central scale between the flow rate and slope scales, read the *next larger* standard sewer-pipe diameter as 24 in (610 mm). When using this chart, always read the next larger pipe size.

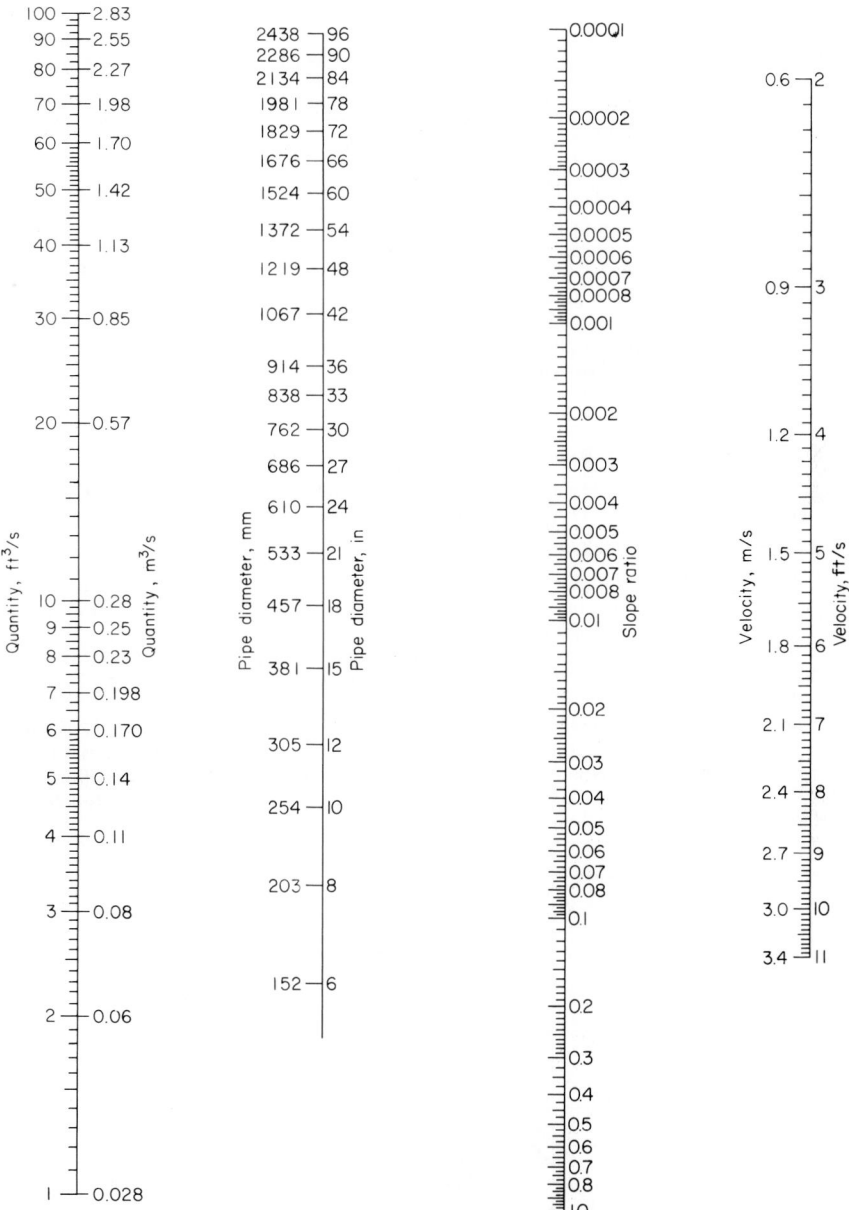

FIG. 4 Nomogram for solving the Manning formula for circular pipes flowing full and $n = 0.013$.

TABLE 11   Values of n for the Manning Formula

Type of surface of pipe	n
Ditches and rivers, rough bottoms with much vegetation	0.040
Ditches and rivers in good condition with some stones and weeds	0.030
Smooth earth or firm gravel	0.020
Rough brick; tuberculated iron pipe	0.017
Vitrified tile and concrete pipe poorly jointed and unevenly settled; average brickwork	0.015
Good concrete; riveted steel pipe; well-laid vitrified tile or brickwork	0.013°
Cast-iron pipe of ordinary roughness; unplaned timber	0.012
Smoothest pipes; neat cement	0.010
Well-planed timber evenly laid	0.009

°Probably the most frequently used value.

5.   *Determine the fluid flow velocity*

Continue the solution line of step 4 to read the fluid flow velocity as 5 ft/s (1.5 m/s) on the extreme right-hand scale of Fig. 4. This is for a sewer flowing *full*.

6.   *Compute the half-full flow depth*

Determine the full-flow capacity of this 24-in (610-mm) sewer by entering Fig. 4 at the slope ratio, 0.005, and projecting through the pipe diameter, 24 in (610 mm). At the left read the full-flow capacity as 16 ft³/s (0.45 m³/s).

The required half-flow capacity is 7.75 ft³/s (0.22 m³/s), from step 1. Determine the ratio of the required half-flow capacity to the full-flow capacity, both expressed in ft³/s. Or 7.75/16.0 = 0.484.

Enter Fig. 5 on the bottom at 0.484, and project vertically upward to the discharge curve. From the intersection, project horizontally to the left to read the depth-of-flow ratio as 0.49. This means that the depth of liquid in the sewer at a flow of 7.75 ft³/s (0.22 m³/s) is 0.49(24 in) = 11.75 in (29.8 cm). Hence, the sewer will be just slightly less than half full when handling the designed flow quantity.

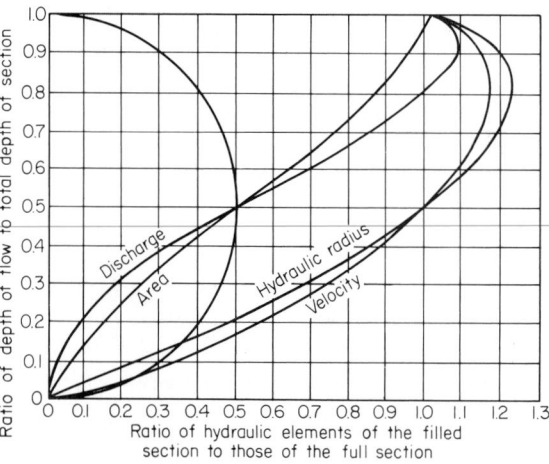

FIG. 5   Hydraulic elements of a circular pipe.

7. Compute the half-full flow velocity

Project horizontally to the right along the previously found 0.49 depth-of-flow ratio until the velocity curve is intersected. From this intersection, project vertically downward to the bottom scale to read the ratio of hydraulic elements as 0.99. Hence, the fluid velocity when flowing half-full is 0.99(5.0 ft/s) = 4.95 ft/s (1.5 m/s).

Related Calculations: The minimum flow velocity required in sanitary sewers is 2 ft/s (0.6 m/s). At 2 ft/s (0.6 m/s), solids will not settle out of the fluid. Since the velocity in this sewer is 4.95 ft/s (1.5 m/s), as computed in step 7, the sewer meets, and exceeds, the minimum required flow velocity.

Certain localities have minimum slope requirements for sanitary sewers. The required slope produces a minimum flow velocity of 2 ft/s (0.6 m/s) with an n value of 0.013.

Storm sewers handling rainwater and other surface drainage require a higher flow velocity than sanitary sewers because sand and grit often enter a storm sewer. The usual minimum allowable velocity for a storm sewer is 2.5 ft/s (0.76 m/s); where possible, the sewer should be designed for 3.0 ft/s (0.9 m/s). If the sewer designed above were used for storm service, it would be acceptable because the fluid velocity is 4.95 ft/s (1.5 m/s). To prevent excessive wear of the sewer, the fluid velocity should not exceed 8 ft/s (2.4 m/s).

Note that Figs. 4 and 5 can be used whenever two variables are known. When a sewer flows at 0.8, or more, full, the partial-flow diagram, Fig. 5, may not give accurate results, especially at high flow velocities.

SEWER-PIPE EARTH LOAD AND BEDDING REQUIREMENTS

A 36-in (914-mm) diameter clay sewer pipe is placed in a 15-ft (4.5-m) deep trench in damp sand. What is the earth load on this sewer pipe? What bedding should be used for the pipe? If a 5-ft (1.5-m) wide drainage trench weighing 2000 lb/ft (2976.3 kg/m) of length crosses the sewer pipe at right angles to the pipe, what load is transmitted to the pipe? The bottom of the flume is 11 ft (3.4 m) above the top of the sewer pipe.

Calculation Procedure:

1. Compute the width of the pipe trench

Compute the trench width from $w = 1.5d + 12$, where w = trench width, in; d = sewer-pipe diameter, in. So $w = 1.5(36) + 12 = 66$ in (167.6 cm), or 5 ft 6 in (1.7 m).

2. Compute the trench depth-to-width ratio

To determine this ratio, subtract the pipe diameter from the depth and divide the result by the trench width. Or, $(15 - 3)/5.5 = 2.18$.

3. Compute the load on the pipe

Use the relation $L = kWw^2$, where L = pipe load, lb/lin ft of trench; k = a constant from Table 12; W = weight of the fill material used in the trench, lb/ft^3; other symbol as before.

Enter Table 12 at the depth-to-width ratio of 2.18. Since this particular value is not tabulated, use the next higher value, 2.5. Opposite this, read $k = 1.70$ for a sand filling.

Enter Table 13 at damp sand, and read the weight as 115 lb/ft^3 (1842.1 kg/m^3). With these data the pipe load relation can be solved.

Substituting in $L = kWw^2$, we get $L = 1.70(115)(5.5)^2 = 5920$ lb/ft (86.4 N/mm). Study of the properties of clay pipe (Table 14) shows that 36-in (914-mm) extra-strength clay pipe has a minimum average crushing strength of 6000 lb (26.7 kN) by the three-edge-bearing method.

4. Apply the loading safety factor

ASTM recommends a factor of safety of 1.5 for clay sewers. To apply this factor of safety, divide it into the tabulated three-edge-bearing strength found in step 3. Or, $6000/1.5 = 4000$ lb (17.8 kN).

5. Compute the pipe load-to-strength ratio

Use the strength found in step 4. Or pipe load-to-strength ratio (also called the *load factor*) = $5920/4000 = 1.48$.

TABLE 12   Values of *k* for Use in the Pipe Load Equation°

Ratio of trench depth to width	Sand and damp topsoil	Saturated topsoil	Damp clay	Saturated clay
0.5	0.46	0.46	0.47	0.47
1.0	0.85	0.86	0.88	0.90
1.5	1.18	1.21	1.24	1.28
2.0	1.46	1.50	1.56	1.62
2.5	1.70	1.76	1.84	1.92
3.0	1.90	1.98	2.08	2.20
3.5	2.08	2.17	2.30	2.44
4.0	2.22	2.33	2.49	2.66
4.5	2.34	2.47	2.65	2.87
5.0	2.45	2.59	2.80	3.03
5.5	2.54	2.69	2.93	3.19
6.0	2.61	2.78	3.04	3.33
6.5	2.68	2.86	3.14	3.46
7.0	2.73	2.93	3.22	3.57
7.5	2.78	2.98	3.30	3.67

°Iowa State Univ. Eng. Exp. Sta. Bull. 47.

TABLE 13   Weight of Pipe-Trench Fill

Fill	lb/ft^3	kg/m^3
Dry sand	100	1601
Damp sand	115	1841
Wet sand	120	1921
Damp clay	120	1921
Saturated clay	130	2081
Saturated topsoil	115	1841
Sand and damp topsoil	100	1601

TABLE 14   Clay Pipe Strength

Pipe size, in (mm)	Minimum average strength, lb/lin ft (N/mm)	
	Three-edge-bearing	Sand-bearing
4 (102)	1000 (14.6)	1500 (21.9)
6 (152)	1100 (16.1)	1650 (24.1)
8 (203)	1300 (18.9)	1950 (28.5)
10 (254)	1400 (20.4)	2100 (30.7)
12 (305)	1500 (21.9)	2250 (32.9)
15 (381)	1750 (25.6)	2625 (38.3)
18 (457)	2000 (29.2)	3000 (43.8)
21 (533)	2200 (32.1)	3300 (48.2)
24 (610)	2400 (35.0)	3600 (52.6)
27 (686)	2750 (40.2)	4125 (60.2)
30 (762)	3200 (46.7)	4800 (70.1)
33 (838)	3500 (51.1)	5250 (76.7)
36 (914)	3900 (56.9)	5850 (85.4)

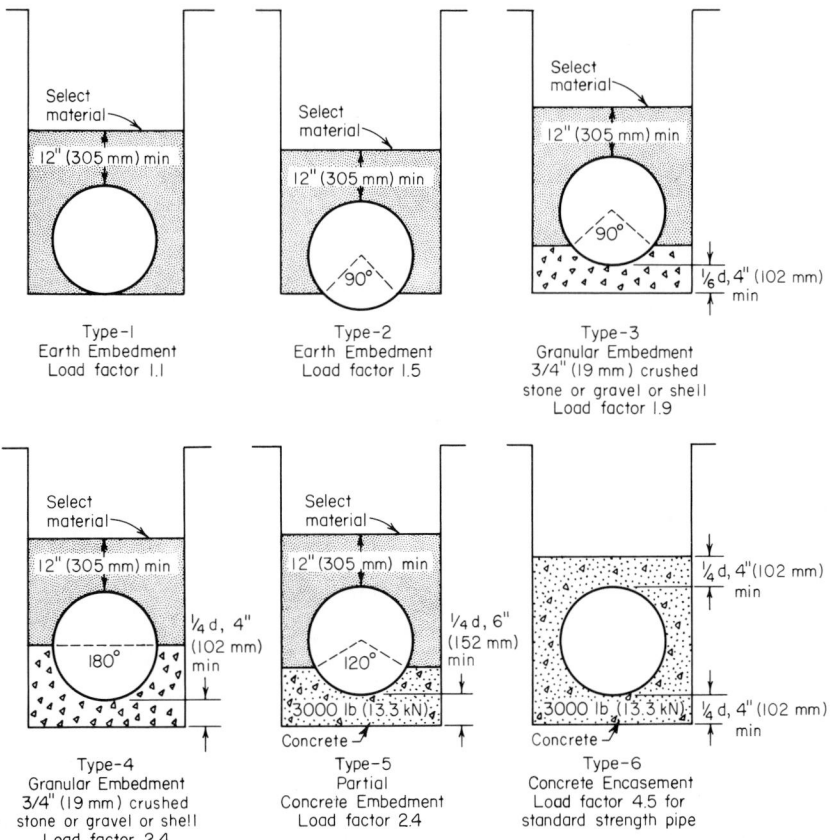

FIG. 6 Strengths developed for various methods of bedding sewer pipes. (*W. S. Dickey Clay Manufacturing Co.*)

6. Select the bedding method for the pipe

Figure 6 shows methods for bedding sewer pipe and the strength developed. Thus, earth embedment, type 2 bedding, develops a load factor of 1.5. Since the computed load factor, step 5, is 1.48, this type of bedding is acceptable. (In choosing a type of bedding be certain that the load factor of the actual pipe is less than, or equals, the developed load factor for the three-edge-bearing strength.)

The type 2 earth embedment, Fig. 6, is a highly satisfactory method, except that the shaping of the lower part of the trench to fit the pipe may be expensive. Type 3 granular embedment may be less expensive, particularly if the crushed stone, gravel, or shell is placed by machine.

7. Compute the direct load transmitted to the sewer pipe

The weight of the drainage flume is carried by the soil over the sewer pipes. Hence, a portion of this weight may reach the sewer pipe. To determine how much of the flume weight reaches the pipe, find the weight of the flume per foot of width, or 2000 lb/5 ft = 400 lb/ft (5.84 kN/mm) of width.

Since the pipe trench is 5.5 ft (1.7 m) wide, step 1, the 1-ft (0.3-m) wide section of the flume imposes a total load of 5.5(400) = 2200 lb (9.8 kN) on the soil beneath it.

To determine what portion of the flume load reaches the sewer pipe, compute the ratio of the depth of the flume bottom to the width of the sewer-pipe trench, or 11/5.5 = 2.0.

TABLE 15   Proportion of Short Loads Reaching Pipe in Trenches

Depth-to-width ratio	Sand and damp topsoil	Saturated topsoil	Damp clay	Saturated clay
0.0	1.00	1.00	1.00	1.00
0.5	0.77	0.78	0.79	0.81
1.0	0.59	0.61	0.63	0.66
1.5	0.46	0.48	0.51	0.54
2.0	0.35	0.38	0.40	0.44
2.5	0.27	0.29	0.32	0.35
3.0	0.21	0.23	0.25	0.29
4.0	0.12	0.14	0.16	0.19
5.0	0.07	0.09	0.10	0.13
6.0	0.04	0.05	0.06	0.08
8.0	0.02	0.02	0.03	0.04
10.0	0.01	0.01	0.01	0.02

TABLE 16   Proportion of Long Loads Reaching Pipe in Trenches

Depth-to-width ratio	Sand and damp topsoil	Saturated topsoil	Damp yellow clay	Saturated yellow clay
0.0	1.00	1.00	1.00	1.00
0.5	0.85	0.86	0.88	0.89
1.0	0.72	0.75	0.77	0.80
1.5	0.61	0.64	0.67	0.72
2.0	0.52	0.55	0.59	0.64
2.5	0.44	0.48	0.52	0.57
3.0	0.37	0.41	0.45	0.51
4.0	0.27	0.31	0.35	0.41
5.0	0.19	0.23	0.27	0.33
6.0	0.14	0.17	0.20	0.26
8.0	0.07	0.09	0.12	0.17
10.00	0.04	0.05	0.07	0.11

Enter Table 15 at a value of 2.0, and read the load proportion for sand and damp top soil as 0.35. Hence, the load of the flume reaching each foot of sewer pipe is 0.35(2200) = 770 lb (3.4 kN).

Related Calculations: A load such as that in step 7 is termed a *short load;* i.e., it is shorter than the pipe-trench width. Typical short loads result from automobile and truck traffic, road rollers, building foundations, etc. *Long loads* are imposed by weights that are longer than the trench is wide. Typical long loads are stacks of lumber, steel, and poles, and piles of sand, coal, gravel, etc. Table 16 shows the proportion of long loads transmitted to buried pipes. Use the same procedure as in step 7 to compute the load reaching the buried pipe.

When a sewer pipe is placed on undisturbed ground and covered with fill, compute the load on the pipe from $L = kWd^2$, where d = pipe diameter, ft; other symbols as in step 3. Tables 15 and 16 are the work of Prof. Anson Marston, Iowa State University.

To find the total load on trenched or surface-level buried pipes subjected to both fill and long or short loads, add the proportion of the long or short load reaching the pipe to the load produced by the fill.

Note that sewers may have several cross-sectional shapes—circular, egg, rectangular, square, etc. The circular sewer is the most common because it has a number of advantages, including economy. Egg-shaped sewers are not as popular as circular and are less often used today because of their higher costs.

Rectangular and square sewers are often used for storm service. However, their hydraulic characteristics are not as desirable as circular sewers.

SANITARY SEWER SYSTEM DESIGN

What size main sanitary sewer is required for a midwestern city 30-acre (1.21×10^5 m^2) residential area containing six-story apartment houses if the hydraulic gradient is 0.0035 and the pipe roughness factor $n = 0.013$? One-third of the area is served by a branch sewer. What should the size of this sewer be? If the branch and main sewers must also handle groundwater infiltration, determine the required sewer size. The sewer is below the normal groundwater level.

Calculation Procedure:

1. *Compute the sanitary sewage flow rate*

Table 17 shows the typical population per acre for various residential areas and the flow rate used in sewer design. Using the typical population of 500 persons per acre (4046 m^2) given in Table 17, we see the total population of the area served is (30 acres)(500 persons per acre) = 15,000 persons.

Since this is a midwestern city, the sewer design basis, per capita, used for Des Moines, Iowa, 200 gal/day (8.76 mL/s), Table 17, appears to be an appropriate value. Checking with the minimum flow recommended in Table 17, 100 gal/day (4.38 mL/s), we see the value of 200 gal/day

TABLE 17   Sanitary Sewer Design Factors

Population data		
	Typical population	
Type of area	Per acre	Per km^2
Light residential	15	3,707
Closely built residential	55	13,591
Single-family residential	100	24,711
Six-story apartment district	500	123,555

Sewage-flow data		
	Sewer design basis, per capita	
City	gal/day	mL/s
Berkeley, California	92	4.03
Cranston, Rhode Island	167	7.32
Des Moines, Iowa	200	8.76
Las Vegas, Nevada	250	10.95
Little Rock, Arkansas	100	4.38
Shreveport, Louisiana	150	6.57

Typical sewer design practice		
	Design flow, per capita	
Sewer type	gal/day	mL/s
Laterals and submains	400	17.5
Main, trunks, and outfall	250	10.95
New sewers	Never 100	Never 4.38

TABLE 18   Manning Formula Conveyance Factor

Pipe diameter, in (mm)	Pipe cross-sectional area, ft² (m²)	n 0.011	0.013	0.015	0.017
6 (152)	0.196 (0.02)	6.62	5.60	4.85	4.28
8 (203)	0.349 (0.03)	14.32	12.12	10.50	9.27
10 (254)	0.545 (0.05)	25.80	21.83	18.92	16.70
12 (305)	0.785 (0.07)	42.15	35.66	30.91	27.27
15 (381)	1.227 (0.11)	76.46	64.70	56.07	49.48
18 (457)	1.767 (0.16)	124.2	105.1	91.04	80.33
21 (533)	2.405 (0.22)	187.1	158.3	137.2	121.1

(8.76 mL/s) seems to be well justified. Hence, the sanitary sewage flow rate that the main sewer must handle is (15,000 persons)(200 gal/day) = 3,000,000 gal/day.

2.   *Convert the flow rate to cfs*

Use the relation $cfs = 1.55 \, (gpd/10^6)$, where cfs = flow rate, ft³/s; gpd = flow rate, gal/24 h. So $cfs = 1.55(3,000,000/1,000,000) = 465$ ft³/s (0.13 m³/s).

3.   *Compute the required size of the main sewer*

Size the main sewer on the basis of its flowing full. This is the usual design procedure followed by experienced sanitary engineers.

Two methods can be used to size the sewer pipe. (*a*) Use the chart in Fig. 4 for the Manning formula, entering with the flow rate of 4.65 ft³/s (0.13 m³/s) and projecting to the slope ratio or hydraulic gradient of 0.0035. Read the required pipe diameter as 18 in (457 mm).

(*b*) Use the Manning formula and the appropriate *conveyance factor* from Table 18. When the conveyance factor C_f is used, the Manning formula becomes $Q = C_f S^{1/2}$, where Q = flow rate through the pipe, ft³/s; C_f = conveyance factor corresponding to a specific n value listed in Table 18; S = pipe slope or hydraulic gradient, ft/ft. Since Q and S are known, substitute and solve for C_f, or $C_f = Q/S^{1/2} = 4.65/(0.0035)^{1/2} = 78.5$. Enter Table 18 at $n = 0.013$ and $C_f = 78.5$, and project to the exact or next higher value of C_f. Table 18 shows that C_f is 64.70 for 15-in (381-mm) pipe and 105.1 for 18-in (457-mm) pipe. Since the actual value of C_f is 78.5, a 15-in (381-mm) pipe would be too small. Hence, an 18-in (457-mm) pipe would be used. This size agrees with that found in procedure *a*.

4.   *Compute the size of the lateral sewer*

The lateral sewer serves one-third of the total area. Since the total sanitary flow from the entire area is 4.65 ft³/s (0.13 m³/s), the flow from one-third of the area, given an even distribution of population and the same pipe slope, is 4.65/3 = 1.55 ft³/s (0.044 m³/s). Using either procedure in step 3, we find the required pipe size = 12 in (305 mm). Hence, three 12-in (305-mm) laterals will discharge into the main sewer, assuming that each lateral serves an equal area and has the same slope.

5.   *Check the suitability of the main sewer size*

Compute the value of $d^{2.5}$ for each of the lateral sewer pipes discharging into the main sewer pipe. Thus, for one 12-in (305-mm) lateral line, where d = smaller pipe diameter, in, $d^{2.5} = 12^{2.5}$ = 496. For three pipes of equal diameter, $3d^{2.5} = 1488 = D^{2.5}$, where D = larger pipe diameter, in. Solving gives $D^{2.5} = 1488$ and $D = 17.5$ in (445 mm). Hence, the 18-in (457-mm) sewer main has sufficient capacity to handle the discharge of three 12-in (305-mm) sewers. Note that Fig. 4 shows that the flow velocity in both the lateral and main sewers exceeds the minimum required velocity of 2 ft/s (0.6 m/s).

6.   *Compute the sewer size with infiltration*

Infiltration is the groundwater that enters a sewer. The quantity and rate of infiltration depend on the character of the soil in which the sewer is laid, the relative position of the groundwater level and the sewer, the diameter and length of the sewer, and the material and care with which the sewer is constructed. With tile and other jointed sewers, infiltration depends largely on the type of joint used in the pipes. In large concrete or brick sewers, the infiltration depends on the type of waterproofing applied.

Infiltration is usually expressed in gallons per day per mile of sewer. With very careful construction, infiltration can be kept down to 5000 gal/(day·mi) [0.14 L/(km·s)] of pipe even when the groundwater level is above the pipe. With poor construction, porous soil, and high groundwater level, infiltration may amount to 100,000 gal/(day·mi) [2.7 L/(km·s)] or more. Sewers laid in dense soil where the groundwater level is below the sewer do not experience infiltration except during and immediately after a rainfall. Even then, the infiltration will be in small amounts.

Assuming an infiltration rate of 20,000 gal/(day·mi) [0.54 L/(km·s)] of sewer and a sewer length of 1.2 mi (1.9 km) for this city, we see the daily infiltration is 1.2(20,000) = 24,000 gal (90,850 L).

Checking the pipe size by either method in step 3 shows that both the 12-in (305-mm) laterals and the 18-in (457-mm) main are of sufficient size to handle both the sanitary and infiltration flow.

Related Calculations: Where a sewer must also handle the runoff from fire-fighting apparatus, compute the quantity of fire-fighting water for cities of less than 200,000 population from $Q = 1020(P)^{0.5}[1 - 0.01(P)^{0.5}]$, where Q = fire demand, gal/min; P = city population in thousands. Add the fire demand to the sanitary sewage and infiltration flows to determine the maximum quantity of liquid the sewer must handle. For cities having a population of more than 200,000 persons, consult the fire department headquarters to determine the water flow quantities anticipated.

Some sanitary engineers apply a demand factor to the average daily water requirements per capita before computing the flow rate into the sewer. Thus, the maximum monthly water consumption is generally about 125 percent of the average annual demand but may range up to 200 percent of the average annual demand. Maximum daily demands of 150 percent of the average annual demand and maximum hourly demands of 200 to 250 percent of the annual average demand are commonly used for design by some sanitary engineers. To apply a demand factor, simply multiply the flow rate computed in step 2 by the appropriate factor. Current practice in the use of demand factors varies; sewers designed without demand factors are generally adequate. Applying a demand factor simply provides a margin of safety in the design, and the sewer is likely to give service for a longer period before becoming overloaded.

Most local laws and many sewer authorities recommend that no sewer be less than 8 in (203 mm) in diameter. The sewer should be sloped sufficiently to give a flow velocity of 2 ft/s (0.6 m/s) or more when flowing full. This velocity prevents the deposit of solids in the pipe. Manholes serving sewers should not be more than 400 ft (121.9 m) apart.

Where industrial sewage is discharged into a sanitary sewer, the industrial flow quantity must be added to the domestic sewage flow quantity before the pipe size is chosen. Swimming pools may also be drained into sanitary sewers and may cause temporary overflowing because the sewer capacity is inadequate. The sanitary sewage flow rate from an industrial area may be less than from a residential area of the same size because the industrial population is smaller.

Many localities and cities restrict the quantity of commercial and industrial sewage that may be discharged into public sewers. Thus, one city restricts commercial sewage from stores, garages, beauty salons, etc., to 135 gal/day per capita. Another city restricts industrial sewage from factories and plants to 50,000 gal/(day·acre) [0.55 mL/(m·s)]. In other cities each proposed installation must be studied separately. Still other cities prohibit any discharge of commercial or industrial sewage into sanitary sewers. For these reasons, the local authorities and sanitary codes, if any, must be consulted before the design of any sewer is begun.

Before starting a sewer design, do the following: (a) Prepare a profile diagram of the area that will be served by the sewer. Indicate on the diagram the elevation above grade of each profile. (b) Compile data on the soil, groundwater level, type of paving, number and type of foundations, underground services (gas, electric, sewage, water supply, etc.), and other characteristics of the area that will be served by the sewer. (c) Sketch the main sewer and lateral sewers on the profile diagram. Indicate the proposed direction of sewage flow by arrows. With these steps finished, start the sewer design.

To design the sewers, proceed as follows: (a) Size the sewers using the procedure given in steps 1 through 6 above. (b) Check the sewage flow rate to see that it is 2 ft/s (0.6 m/s) or more. (c) Check the plot to see that the required slope for the pipes can be obtained without expensive blasting or rock removal.

Where the outlet of a building plumbing system is below the level of the sewer serving the building, a pump must be used to deliver the sewage to the sewer. Compute the pump capacity, using the discharge from the various plumbing fixtures in the building as the source of the liquid flow to the pump. The head on the pump is the difference between the level of the sewage in the pump intake and the centerline of the sewer into which the pump discharges, plus any friction losses in the piping.

STORM-SEWER INLET SIZE AND FLOW RATE

What size storm-sewer inlet is required to handle a flow of 2 ft³/s (0.057 m³/s) if the gutter is sloped ¼ in/ft (2.1 cm/m) across the inlet and 0.05 in/ft (0.4 cm/m) along the length of the inlet? The maximum depth of flow in the gutter is estimated to be 0.2 ft (0.06 m), and the gutter is depressed 4 in (102 mm) below the normal street level.

Calculation Procedure:

1. *Compute the reciprocal of the gutter transverse slope*

The *transverse slope* of the gutter across the inlet is ¼ in/ft (2.1 cm/m). Expressing the reciprocal of this slope as r, compute the value for this gutter as $r = 4 \times 12/1 = 48$.

2. *Determine the inlet capacity per foot of length*

Enter Table 19 at the flow depth of 0.2 ft (0.06 m), and project to the depth of depression of the gutter of 4 in (102 mm). Opposite this depth, read the inlet capacity per foot of length as 0.50 ft³/s (0.014 m³/s).

3. *Compute the required gutter inlet length*

The gutter must handle a maximum flow of 2 ft³/s (0.057 m³/s). Since the inlet has a capacity of 0.50 ft³/(s·ft) [0.047 m³/(m·s)] of length, the required length, ft = maximum required capacity, ft³/s/capacity per foot, ft³/s = 2.0/0.50 = 4.0 ft (1.2 m). A length of 4.0 ft (1.2 m) will be satisfactory. Were a length of 4.2 or 4.4 ft (1.28 or 1.34 m) required, a 4.5-ft (1.37-m) long inlet would be chosen. The reasoning behind the choice of a longer length is that the extra initial investment for the longer length is small compared with the extra capacity obtained.

4. *Determine how far the water will extend from the curb*

Use the relation $l = rd$, where l = distance water will extend from the curb, ft; d = depth of water in the gutter at the curb line, ft; other symbols as before. Substituting, we find $l = 48(0.2) = 9.6$ ft (2.9 m). This distance is acceptable because the water would extend out this far only during the heaviest storms.

   Related Calculations: To compute the flow rate in a gutter, use the relation $F = 0.56(r/n)s^{0.5}d^{8/3}$, where F = flow rate in gutter, ft³/s; n = roughness coefficient, usually taken as 0.015; s = gutter slope, in/ft; other symbols as before. Where the computed inlet length is 5 ft (1.5 m) or more, some engineers assume that a portion of the water will pass the first inlet and enter the next one along the street.

TABLE 19   Storm-Sewer Inlet Capacity per Foot (Meter) of Length

Flow depth in gutter, ft (mm)	Depression depth, in (mm)	Capacity per foot length, ft³/s (m³/s)
0.2 (0.06)	0 (0)	0.062 (5.76)
	1 (25.4)	0.141 (13.10)
	2 (50.8)	0.245 (22.76)
	3 (76.2)	0.358 (33.26)
	4 (101.6)	0.500 (46.46)
0.3 (0.09)	0 (0)	0.115 (10.69)
	1 (25.4)	0.205 (19.05)
	2 (50.8)	0.320 (29.73)
	3 (76.2)	0.450 (41.81)
	4 (101.6)	0.590 (54.82)

STORM-SEWER DESIGN

Design a storm-sewer system for a 30-acre $(1.21 \times 10^5\text{-m}^2)$ residential area in which the storm-water runoff rate is computed to be 24 ft³/s (0.7 m³/s). The total area is divided into 10 plots of equal area having similar soil and runoff conditions.

Calculation Procedure:

1. *Sketch a plan of the sewer system*

Sketch the area and the 10 plots as in Fig. 7. A scale of 1 in = 100 ft (1 cm = 12 m) is generally suitable. Indicate the terrain elevations by drawing the profile curves on the plot plan. Since the profiles (Fig. 7) show that the terrain slopes from north to south, the main sewer can probably be best run from north to south. The sewer would also slope downward from north to south, following the general slope of the terrain.

Indicate a storm-water inlet for each of the areas served by the sewer. With the terrain sloping from north to south, each inlet will probably give best service if it is located on the southern border of the plot.

Since the plots are equal in area, the main sewer can be run down the center of the plot with each inlet feeding into it. Use arrows to indicate the flow direction in the laterals and main sewer.

2. *Compute the lateral sewer size*

Each lateral sewer handles 24 ft³/s/10 plots = 2.4 ft³/s (0.07 m³/s) of storm water. Size each lateral, using the Manning formula with $n = 0.013$ and full flow in the pipe. Assume a slope ratio of 0.05 for each inlet pipe between the inlet and the main sewer. This means that the inlet pipe will slope 1 ft in 20 ft (0.3 m in 6.1 m) of length. In an installation such as this, a slope ratio of 0.05 is adequate.

By using Fig. 4 for a flow of 2.4 ft³/s (0.0679 m³/s) and a slope of 0.05, an 8-in (203-mm) pipe is required for each lateral. The fluid velocity is, from Fig. 4, 7.45 ft/s (2.27 m/s). This is a high

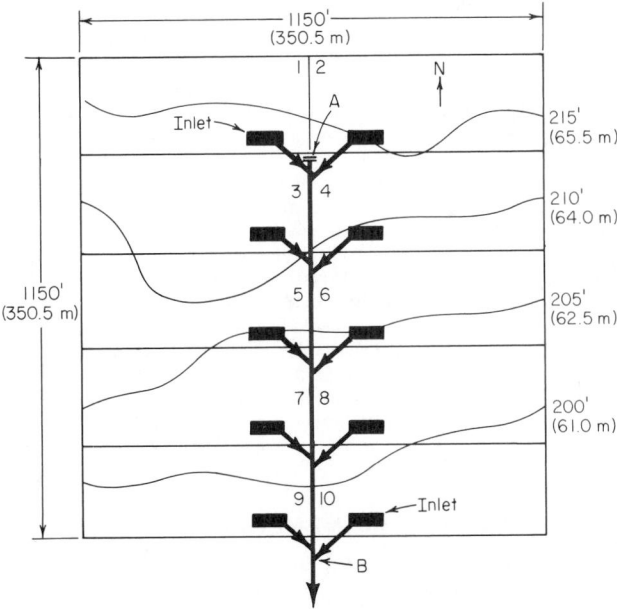

FIG. 7 Typical storm-sewer plot plan and layout diagram.

enough velocity to prevent solids from settling out of the water. [The flow velocity should not be less than 2 ft/s (0.61 m/s).]

3. Compute the size of the main sewer

There are four sections of the main sewer (Fig. 7). The first section, section 3-4, serves the two northernmost plots. Since the flow from each plot is 2.4 ft³/s (0.0679 m³/s), the storm water that this portion of the main sewer must handle is 2(2.4) = 4.8 ft³/s (0.14 m³/s).

The main sewer begins at point A, which has an elevation of about 213 ft (64.9 m), as shown by the profile. At point B the terrain elevation is about 190 ft (57.8 m). Hence, the slope between points A and B is about 213 − 190 = 23 ft (7.0 m), and the distance between the two points is about 920 ft (280.4 m).

Assume a slope of 1 ft/100 ft (0.3 m/30.5 m) of length, or 1/100 = 0.01 for the main sewer. This is a typical slope used for main sewers, and it is within the range permitted by a pipe run along the surface of this terrain. Table 20 shows the minimum slope required to produce a flow velocity of 2 ft/s (0.61 m/s).

TABLE 20 Minimum Slope of Sewers°

Sewer diameter, in (mm)	Minimum slope, ft/ 100 ft (m/30.5 m) of length
4 (102)	1.20 (0.366)
6 (152)	0.60 (0.183)
8 (203)	0.40 (0.122)
10 (254)	0.29 (0.088)
12 (305)	0.22 (0.067)
15 (381)	0.15 (0.046)
18 (457)	0.12 (0.037)
20 (505)	0.10 (0.030)
24 (610)	0.08 (0.024)

°Based on the Manning formula with n = 0.13 and the sewer flowing either full or half full.

Using Fig. 4 for a flow of 4.8 ft³/s (0.14 m³/s) and a slope of 0.01, we see the required size for section 3-4 of the main sewer is 15 in (381 mm). The flow velocity in the pipe is 4.88 ft/s (1.49 m/s). The size of this sewer is in keeping with general design practice, which seldom uses a storm sewer less than 12 in (304.8 mm) in diameter.

Section 5-6 conveys 9.6 ft³/s (0.27 m³/s). Using Fig. 4 again, we find the required pipe size is 18 in (457.2 mm) and the flow velocity is 5.75 ft/s (1.75 m/s). Likewise, section 7-8 must handle 14.4 ft³/s (0.41 m³/s). The required pipe size is 21 in (533 mm), and the flow velocity in the pipe is 6.35 ft/s (1.94 m/s). Section 9-10 of the main sewer handles 19.2 ft³/s (0.54 m³/s), and must be 24 in (609.6 mm) in diameter. The velocity in this section of the sewer pipe will be 6.9 ft/s (2.1 m/s). The last section of the main sewer handles the total flow, or 24 ft³/s (0.7 m³/s). Its size must be 27 in (686 mm), Fig. 4, although a 24-in (610.0-mm) pipe would suffice if the slope at point B could be increased to 0.012.

Related Calculations: Most new sewers built today are the *separate* type, i.e., one sewer for sanitary service and another sewer for storm service. Sanitary sewers are usually installed first because they are generally smaller than storm sewers and cost less. *Combined sewers* handle both sanitary and storm flows and are used where expensive excavation for underground sewers is necessary. Many older cities have combined sewers.

To size a combined sewer, compute the sum of the maximum sanitary and storm water flow for each section of the sewer. Then use the method given in this procedure after having assumed a value for n in the Manning formula and for the slope of the sewer main.

Where a continuous slope cannot be provided for a sewer main, a pumping station to lift the sewage must be installed. Most cities require one or more pumping stations because the terrain does not permit an unrestricted slope for the sewer mains. Motor-driven centrifugal pumps are generally used to handle sewage. For unscreened sewage, the suction inlet of the pump should not be less than 3 in (76 mm) in diameter.

SELECTION OF SEWAGE-TREATMENT METHOD

A city of 100,000 population is considering installing a new sewage-treatment plant. Select a suitable treatment method. Local ordinances require that suspended matter in the sewage be reduced 80 percent, that bacteria be reduced 60 percent, and that the biochemical oxygen demand be reduced 90 percent. The plant will handle only domestic sanitary sewage. What are the daily oxygen demand and the daily suspended-solids content of the sewage? If an industrial plant discharges into this system sewage requiring 4500 lb (2041.2 kg) of oxygen per day, determine the population equivalent of the industrial sewage.

Calculation Procedure:

1. *Compute the daily sewage flow*

With an average flow per capita of 200 gal/day (8.8 mL/s), this sewage treatment plant must handle per capita (200 gal/day)(100,000 population) = 20,000,000 gal/day (896.2 L/s).

2. *Compute the sewage oxygen demand*

Usual domestic sewage shows a 5-day oxygen demand of 0.12 to 0.17 lb/day (0.054 to 0.077 kg/day) per person. With an average of 0.15 lb (0.068 kg) per person per day, the daily oxygen demand of the sewage is (0.15)(100,000) = 15,000 lb/day (78.7 g/s).

3. *Compute the suspended-solids content of the sewage*

Usual domestic sewage contains about 0.25 lb (0.11 kg) of suspended solids per person per day. Using this average, we see the total quantity of suspended solids that must be handled is (0.25)(100,000) = 25,000 lb/day (0.13 kg/s).

4. *Select the sewage-treatment method*

Table 21 shows the efficiency of various sewage-treatment methods. Since the desired reduction in suspended matter, biochemical oxygen demand (BOD), and bacteria is known, this will serve as a guide to the initial choice of the equipment.

Study of Table 21 shows that a number of treatments are available which will reduce the suspended matter by 80 percent. Hence, any one of these methods might be used. The same is true for the desired reduction in bacteria and BOD. Thus, the system choice resolves to selection of the most economical group of treatment units.

For a city of this size, four steps of sewage treatment would be advisable. The first step, *pre-*

TABLE 21 Typical Efficiencies of Sewage-Treatment Methods°

Treatment	Percentage reduction		
	Suspended matter	BOD	Bacteria
Fine screens	5–20	. . .	10–20
Plain sedimentation	35–65	25–40	50–60
Chemical precipitation	75–90	60–85	70–90
Low-rate trickling filter, with pre- and final sedimentation	70–90+	75–90	90+
High-rate trickling filter with pre- and final sedimentation	70–90	65–95	70–95
Conventional activated sludge with pre- and final sedimentation	80–95	80–95	90–95+
High-rate activated sludge with pre- and final sedimentation	70–90	70–95	80–95
Contact aeration with pre- and final sedimentation	80–95	80–95	90–95+
Intermittent sand filtration with presedimentation	90–95	85–95	95+
Chlorination:			
Settled sewage	. . .	†	90–95
Biologically treated sewage	. . .	†	98–99

°Steel—*Water Supply and Sewerage*, McGraw-Hill.
†Reduction is dependent on dosage.

TABLE 22 Sludge and Other Products of Sewage-Treatment Processes per Million Gallons of Sewage Treated[*]

Data	Treatment process							
	Racks	Fine screens	Grit chambers	Plain sedimentation	Septic tanks	Imhoff or separate tanks	Activated sludge	Trickling filter humus tanks
Character of product	Screenings	Screenings	Grit	Raw sludge	Digested sludge	Digested sludge	Raw sludge	Raw sludge
Average amount per million gallons	4–8 ft³ (0.11–0.23 m³)	10–30 ft³ (0.28–0.83 m³)	2.5 ft³ (0.07 m³)	2500 gal (9462.5 L)	900 gal (3406.5 L)	500 gal (1892.5 L)	13,500 gal (51,098 L)	500 gal (1892.5 L)
Average moisture content, percent	80	80	15	95	90	85	99	92.5
Specific gravity				1020	1040	1040	1005	1025
Usual disposal methods	Burying, burning, or shredding and digestion with sludge	Burying, burning, or digesting with sludge	Filling land	Processing, digestion, or drying	Drying	Drying	Processing, digestion, or lagooning	Digestion and drying

[*]O'Rourke—*General Engineering Handbook*, McGraw-Hill.

liminary treatment, could include screening to remove large suspended solids, grit removal, and grease removal. The next step, *primary treatment*, could include sedimentation or chemical precipitation. *Secondary treatment*, the next step, might be of a biological type such as the activated-sludge process or the trickling filter. In the final step, the sewage might be treated by chlorination. Treated sewage can then be disposed of in fields, streams, or other suitable areas.

Choose the following units for this sewage-treatment plant, using the data in Table 21 as a guide: rocks or screens to remove large suspended solids, grit chambers to remove grit, skimming tanks for grease removal, plain sedimentation, activated-sludge process, and chlorination.

Reference to Table 21 shows that screens and plain sedimentation will reduce the suspended solids by the desired amount. Likewise, the activated-sludge process reduces the BOD by up to 95 percent and the bacteria up to 95+ percent. Hence, the chosen system satisfies the design requirements.

5. Compute the population equivalent of the industrial sewage

Use the relation $P_e = R/D$, where P_e = population equivalent of the industrial sewage, persons; R = required oxygen of the sewage, lb/day; D = daily oxygen demand, lb per person per day. So $P_e = 4500/0.15 = 30,000$ persons.

Related Calculations: Where sewage is combined (i.e., sanitary and storm sewage mixed), the 5-day per-capita oxygen demand is about 0.25 lb/day (0.11 kg/day). Where large quantities of industrial waste are part of combined sewage, the per-capita oxygen demand is usually about 0.5 lb/day (0.23 kg/day). To convert the strength of an industrial waste to the same base used for sanitary waste, apply the population equivalent relation in step 5. Some cities use the population equivalent as a means of evaluating the load placed on the sewage-treatment works by industrial plants.

Table 22 shows the products resulting from various sewage-treatment processes per million gallons of sewage treated. The tabulated data are useful for computing the volume of product each process produces.

Environmental considerations are leading to the adoption of biogas methods to handle the organic fraction of municipal solid waste (MSW). Burning methane-rich biogas can meet up to 60 percent of the operating cost of waste-to-energy plants. Further, generating biogas avoids the high cost of disposing of this odorous by-product. A further advantage is that biogas plants are exempt from energy or carbon taxes. Newer plants also handle industrial wastes, converting them to biogas.

Biogas plants are popular in Europe. The first anaerobic digestion plant capable of treating unsorted MSW handles 55,000 mt/yr. It treats wastestreams with solids contents of 30 to 35 percent. Automated sorting first removes metals, plastics, paperboard, glass, and inerts from the MSW stream. The remaining organic fraction is mixed with recycled water from a preceding compost-drying press to form a 30 to 35 percent solids sludge which is pumped into one of the plant's three 2400-m³ (84,720 ft³) digesters.

Residence time in the digester is about 3 weeks with a biogas yield of 99 m³/mt of MSW (3495 ft³/t), or 146 m³/mt (5154 ft³/t) of sorted organic fraction. Overflow liquid from the digester is pressed, graded, and sold as compost. Mixtures of MSW, sewage sludges, and animal slurries can also be digested in this process (Valorga) developed by Valorga SA (Vendargues, France). This is termed a *dry* process.

Wet processes handle wastestreams with only 10 to 15 percent solids content. Featuring more than one digestion stage, it is easier to control parameters such as pH and solids concentration than dry fermentation. The first plant to use wet digestion to process MSW is a 20,000-mt/yr installation in Denmark. About two-thirds of the annual operating cost of $2 million is recovered through the sale of biogas. In a 14,000-mt/yr plant in Finland the biogas produced is used to fire a gas turbine. Multiple stages are said to make wet fermentation 65 percent faster than single-stage processes, with a 50 percent higher gas yield.

These developments show that sanitary engineers will be more concerned than ever with the environmental aspects of their designs. With the world population growing steadily every year and the longer lifespan of older individuals, biogas and similar recovery-conversion processes will become standard practice in every major country.

The data on biogas given above was reported in *Chemical Engineering*.

SECTION 12

ENGINEERING
ECONOMICS

MAX KURTZ, P.E.
CONSULTING ENGINEER

CALCULATION OF INTEREST, PRINCIPAL, AND PAYMENTS 12.4
Determination of Simple Interest 12.5
Compound Interest; Future Value of Single Payment 12.5
Present Worth of Single Payment 12.6
Principal in Sinking Fund 12.6
Determination of Sinking-Fund Deposit 12.6
Present Worth of a Uniform Series 12.6
Capital-Recovery Determination 12.7
Effective Interest Rate 12.7
Perpetuity Determination 12.7
Determination of Equivalent Sums 12.7
Analysis of a Nonuniform Series 12.8
Uniform Series with Payment Period Different from Interest Period 12.9
Uniform-Gradient Series: Conversion to Uniform Series 12.9
Present Worth of Uniform-Gradient Series 12.9
Future Value of Uniform-Rate Series 12.10
Determination of Payments under Uniform-Rate Series 12.10
Continuous Compounding 12.11
Future Value of Uniform Series with Continuous Compounding 12.11
Present Worth of Continuous Cash Flow of Uniform Rate 12.11
Future Value of Continuous Cash Flow of Uniform Rate 12.11
DEPRECIATION AND DEPLETION 12.12
Straight-Line Depreciation 12.12
Straight-Line Depreciation with Two Rates 12.12
Depreciation by Accelerated Cost Recovery System 12.12
Sinking-Fund Method: Asset Book Value 12.13
Sinking-Fund Method: Depreciation Charges 12.13
Fixed-Percentage (Declining-Balance) Method 12.14
Combination of Fixed-Percentage and Straight-Line Methods 12.14
Constant-Unit-Use Method of Depreciation 12.14
Declining-Unit-Use Method of Depreciation 12.15
Sum-of-the-Digits Method of Depreciation 12.15

Combination of Time- and Use-Depreciation Methods 12.15
Effects of Depreciation Accounting on Taxes and Earnings 12.16
Depletion Accounting by the Sinking-Fund Method 12.17
Income from a Depleting Asset 12.17
Depletion Accounting by the Unit Method 12.17
COST COMPARISONS OF ALTERNATIVE PROPOSALS 12.18
Determination of Annual Cost of an Asset 12.18
Minimum Asset Life to Justify a Higher Investment 12.19
Comparison of Equipment Cost and Income Generated 12.19
Selection of Relevant Data in Annual-Cost Studies 12.19
Determination of Manufacturing Break-Even Point 12.20
Cost Comparison with Nonuniform Operating Costs 12.21
Economics of Equipment Replacement 12.22
Annual Cost by the Amortization (Sinking-Fund-Depreciation) Method 12.23
Annual Cost by the Straight-Line-Depreciation Method 12.23
Present Worth of Future Costs of an Installation 12.24
Determination of Capitalized Cost 12.25
Capitalized Cost of Asset with Uniform Intermittent Payments 12.25
Capitalized Cost of an Asset with Nonuniform Intermittent Payments 12.26
Stepped-Program Capitalized Cost 12.26
Calculation of Annual Cost on After-Tax Basis 12.27
Cost Comparison with Anticipated Decreasing Costs 12.28
Economy of Replacing an Asset with an Improved Model 12.29
Economy of Replacement under Continuing Improvements 12.30
Economy of Replacement on After-Tax Basis 12.31
EFFECTS OF INFLATION 12.32
Determination of Replacement Cost with Constant Inflation Rate 12.32
Determination of Replacement Cost with Variable Inflation Rate 12.33
Present Worth of Costs in Inflationary Period 12.33
Cost Comparison with Anticipated Inflation 12.34
Endowment with Allowance for Inflation 12.34
EVALUATION OF INVESTMENTS 12.35
Premium-Worth Method of Investment Evaluation 12.35
Valuation of Corporate Bonds 12.36
Rate of Return on Bond Investment 12.36
Investment-Rate Calculation as Alternative to Annual-Cost Calculation 12.37
Allocation of Investment Capital 12.37
Allocation of Capital to Two Investments with Variable Rates of Return . . . 12.39
Allocation of Capital to Three Investments by Dynamic Programming 12.39
Economic Level of Investment 12.41
Relationship between Before-Tax and After-Tax Investment Rates 12.42
Apparent Rates of Return on a Continuing Investment 12.42
True Rate of Return on a Completed Investment 12.43
Average Rate of Return on Composite Investment 12.44
Rate of Return on a Speculative Investment 12.44
Investment at an Intermediate Date (Ambiguous Case) 12.45
Payback Period of an Investment 12.46
Payback Period to Yield a Given Investment Rate 12.46
Benefit-Cost Analysis 12.47
ANALYSIS OF BUSINESS OPERATIONS 12.48
Linear Programming to Maximize Income from Joint Products 12.48
Allocation of Production among Multiple Facilities with Nonlinear Costs . . . 12.49
Optimal Product Mix with Nonlinear Profits 12.51
Dynamic Programming to Minimize Cost of Transportation 12.51
Optimal Inventory Level 12.54
Effect of Quantity Discount on Optimal Inventory Level 12.55
Project Planning by the Critical-Path Method 12.55
Project Planning Based on Available Workforce 12.59

STATISTICS, PROBABILITY, AND THEIR APPLICATIONS 12.62
   Determination of Arithmetic Mean, Median, and Standard Deviation 12.62
   Determination of Arithmetic Mean and Standard Deviation of Grouped Data . . 12.63
   Number of Ways of Assigning Work 12.65
   Formation of Permutations Subject to a Restriction 12.65
   Formation of Combinations Subject to a Restriction 12.65
   Probability of a Sequence of Events 12.66
   Probability Associated with a Series of Trials. 12.67
   Binomial Probability Distribution 12.67
   Pascal Probability Distribution 12.68
   Poisson Probability Distribution 12.69
   Composite Event with Poisson Distribution 12.70
   Normal Distribution 12.70
   Application of Normal Distribution. 12.71
   Negative-Exponential Distribution 12.72
   Sampling Distribution of the Mean 12.74
   Estimation of Population Mean on Basis of Sample Mean 12.75
   Decision Making on Statistical Basis. 12.76
   Probability of Accepting a False Null Hypothesis 12.78
   Decision Based on Proportion of Sample 12.78
   Probability of Accepting an Unsatisfactory Shipment 12.80
   Device with Negative-Exponential Life Span 12.81
   Correspondence between Poisson Failure and Negative-Exponential Life Span 12.82
   Probability of Failure during a Specific Period 12.82
   System with Components in Series 12.83
   System with Components in Parallel 12.83
   System with Identical Components in Parallel 12.84
   Analysis of Composite System by Conventional Method 12.84
   Analysis of Composite System by Alternative Method. 12.85
   Analysis of System with Safeguard by Conventional Method 12.87
   Analysis of System with Safeguard by Alternative Method 12.88
   Optimal Inventory to Meet Fluctuating Demand 12.89
   Finding Optimal Inventory by Incremental-Profit Method 12.90
   Simulation of Commercial Activity by the Monte Carlo Technique. 12.91
   Linear Regression Applied to Sales Forecasting. 12.94
   Standard Deviation from Regression Line 12.96
   Short-Term Forecasting with a Markov Process 12.96
   Long-Term Forecasting with a Markov Process 12.97
   Verification of Steady-State Conditions for a Markov Process 12.99

REFERENCES: Kurtz—*Handbook of Engineering Economics*, McGraw-Hill; Barish and Kaplan—*Economic Analysis for Engineering and Managerial Decision Making*, McGraw-Hill; DeGarmo et al.—*Engineering Economy*, Macmillan; Grant and Leavenworth—*Principles of Engineering Economy*, Ronald Press; Kasmer—*Essentials of Engineering Economics*, McGraw-Hill; Smith—*Engineering Economy*, Iowa State University Press; Cissell—*Mathematics of Finance*, Houghton Mifflin; Clifton and Fyffe—*Project Feasibilty Analysis*, Wiley; Sullivan and Claycombe—*Fundamentals of Forecasting*, Reston; Weston and Brigham—*Essentials of Managerial Finance*, Dryden Press; Lock—*Engineer's Handbook of Management Techniques*, Grove Press (London, England); Jelen—*Project and Cost Engineers' Handbook*, American Association of Cost Engineers; Kharbanda—*Process Plant and Equipment Cost Estimation*, Vivek Enterprises (Bombay, India); Johnson and Peters—*A Computer Program for Calculating Capital and Operating Costs*, Bureau of Mines Information Circular 8426, U.S. Department of Interior; Ostwald—*Cost Estimation for Engineering and Management*, Prentice-Hall; American Association of Cost Engineers—*Cost Engineers' Notebook*; Gass—*Linear Programming: Methods and Applications*, McGraw-Hill; Hadley—*Linear Programming*, Addison-Wesley; Bellman and Dreyfus—*Applied Dynamic Programming*, Princeton University Press; Hadley—*Nonlinear and Dynamic Programming*; Addison-Wesley; Allen—*Probability and Statistics, and Queuing Theory*, Academic Press; Gross and Harris—*Fundamentals of Queuing Theory*, Wiley; Beightler—*Foundations of Optimization*, Prentice-Hall; Blum and Rosenblatt—*Probability and Statistics*, W. B. Saunders; Brownlee—*Statistical Theory and Methodology in Science and Engineering*, Wiley; Quinn—*Probability and Statistics*, Harper & Row; Newnan—*Engineering Economic Analysis*, Engineering Press; Park —*Cost Engineering*, Wiley; Taylor—*Managerial and Engineering Economy*, VNR; Mishan—*Cost-Benefit Analysis*, Praeger; Jelen and Black—*Cost and Optimization Engineering*, McGraw-Hill; White et al.—*Principles of Engineering Economic Analysis*, Wiley; Riggs—*Engineering Economics*, McGraw-Hill; Guenther—*Concepts of Statistical Inference*, McGraw-Hill; Lindgren—*Statistical Theory*, Macmillan; Meyer—*Introductory Probability and Statistical Applications*, Addison-Wesley; Renwick—*Introduction to Investments and Finance*, Macmillan; O'Brien—*CPM in Construction Management*, McGraw-Hill; Gupta and Cozzolino—*Fundamentals of Operations Research for Management*, Holden-Day.

Calculation of Interest, Principal, and Payments

Symbols and Abbreviations

General: With discrete compounding, i = interest rate per period, percent; n = number of interest periods. With continuous compounding, j = nominal annual interest rate, percent; interest period = 1 year.

Simple and compound interest—single payment: P = value of payment at beginning of first interest period, also termed *present worth* of payment; S = value of payment at end of nth interest period, also termed *future value* of payment.

Compound interest—uniform-payment series: R = sum paid at end of each interest period for n periods; P = value of payments at beginning of first interest period, also termed *present worth* of payments; S = value of payments at end of nth interest period, also termed *future value* of payments.

Compound interest—uniform-gradient series: R_m = payment at end of mth interest period; g = constant difference between given payment and preceding payment, also termed *gradient* of series. Then $R_m = R_1 + (m - 1)g$. Also, P and S have the same meaning as for uniform-payment series.

Compound interest—uniform-rate series: R_m = payment at end of mth interest period; r = constant ratio of given payment to preceding payment. Then $R_m = R_1 r^{m-1}$, and P and S have the same meaning as for uniform-payment series.

Compound-interest factors: *Single payment*—S/P = single-payment compound-amount (SPCA) factor; P/S = single-payment present-worth (SPPW) factor. *Uniform-payment series*—S/R = uniform-series compound-amount (USCA) factor; R/S = sinking-fund-payment (SFP) factor; P/R = uniform-series present-worth (USPW) factor; R/P = capital-recovery (CR) factor. *Uniform-rate series*—S/R_1 = uniform-rate-series compound-amount (URSCA) factor; P/R_1 = uniform-rate-series present-worth (URSPW) factor.

Basic Equations

Simple interest, single payment

$$S = P(1 + ni) \qquad (1)$$

Compound interest with discrete compounding

$$\text{SPCA} = (1 + i)^n \qquad (2)$$

$$\text{SPPW} = (1 + i)^{-n} \qquad (3)$$

$$\text{USCA} = \frac{(1 + i)^n - 1}{i} \qquad (4)$$

$$\text{SFP} = \frac{i}{(1 + i)^n - 1} \qquad (5)$$

$$\text{USPW} = \frac{(1 + i)^n - 1}{i(1 + i)^n} \tag{6}$$

$$\text{CR} = \frac{i(1 + i)^n}{(1 + i)^n - 1} \tag{7}$$

$$\text{URSCA} = \frac{r^n - (1 + i)^n}{r - i - 1} \tag{8}$$

$$\text{URSPW} = \frac{[r/(1 + i)]^n - 1}{r - i - 1} \tag{9}$$

A uniform-payment series that continues indefinitely is termed a *perpetuity*. For this case,

$$\text{USPW} = \frac{1}{i} \tag{6a}$$

$$\text{CR} = i \tag{7a}$$

Compound interest with continuous compounding

$$\text{SPCA} = e^{jn} \tag{10}$$

where e = base of natural logarithms = 2.71828 . . .

$$\text{SPPW} = e^{-jn} \tag{11}$$

$$\text{USCA} = \frac{e^{jn} - 1}{e^j - 1} \tag{12}$$

$$\text{USPW} = \frac{1 - e^{-jn}}{e^j - 1} \tag{13}$$

The compound-interest factors for a single payment and for a uniform-payment series can be found by referring to compound-interest tables or by solving the relevant equations by calculator.

DETERMINATION OF SIMPLE INTEREST

A company borrows $4000 at 6 percent per annum simple interest. What payment must be made to retire the debt at the end of 5 years?

Calculation Procedure:

Apply the equation for simple interest

This equation is $S = P(1 + ni) = \$4000(1 + 5 \times 0.06) = \$5200$.
  Note: See the introduction to this section for the symbols used.

COMPOUND INTEREST; FUTURE VALUE OF SINGLE PAYMENT

The sum of $2600 was deposited in a fund that earned interest at 8 percent per annum compounded quarterly. What was the principal in the fund at the end of 3 years?

Calculation Procedure:

1. Compute the true interest rate and number of interest periods

Since there are four interest periods per year, the interest rate i per period is i = 8 percent/4 = 2 percent per period. With a 3-year deposit period, the number n of interest periods is $n = 3 \times 4 = 12$.

2. Apply the SPCA value given in a compound-interest table

Look up the SPCA value for the interest rate, 2 percent, and the number of interest periods, 12. Then substitute in $S = P(\text{SPCA}) = \$2600(1.268) = \$3296.80$.

PRESENT WORTH OF SINGLE PAYMENT

On January 1 of a certain year, a deposit was made in a fund that earns interest at 6 percent per annum. On December 31, 7 years later, the principal resulting from this deposit was $1082. What sum was deposited?

Calculation Procedure:

Apply the SPPW relation

Obtain the SPPW factor for $i = 6$ percent, $n = 7$ years from the interest table. Thus $P = S(\text{SPPW}) = \$1082(0.6651) = \$719.64$.

PRINCIPAL IN SINKING FUND

To accumulate capital for an expansion program, a corporation made a deposit of $200,000 at the end of each year for 5 years in a fund earning interest at 4 percent per annum. What was the principal in the fund immediately after the fifth deposit was made?

Calculation Procedure:

Apply the USCA factor

Obtain the USCA factor for $i = 4$ percent, $n = 5$ from the interest table. Substitute in the relation $S = R(\text{USCA}) = \$200,000(5.416) = \$1,083,200$.

DETERMINATION OF SINKING-FUND DEPOSIT

The XYZ Corporation borrows $65,000, which it is required to repay at the end of 5 years at 8 percent interest. To accumulate this sum, XYZ will make five equal annual deposits in a fund that earns interest at 3 percent, the first deposit being made 1 year after negotiation of the loan. What is the amount of the annual deposit required?

Calculation Procedure:

1. Compute the sum to be paid at the expiration of the loan

Obtain the SPCA factor from the interest table for $i = 8$ percent, $n = 5$. Then substitute in the relation $S = P(\text{SPCA}) = \$65,000(1.469) = \$95,485$.

2. Compute the annual deposit corresponding to this future value

Obtain the SFP factor from the interest table for $i = 3$ percent, $n = 5$ and substitute in the relation $R = S(\text{SFP}) = \$95,485(0.18835) = \$17,985$.

PRESENT WORTH OF A UNIFORM SERIES

An inventor is negotiating with two firms for assignment of rights to a patent. The ABC Corp. offers an annuity of 12 annual payments of $15,000 each, the first payment to be made 1 year after sale of the patent. The DEF Corp. proposes to buy the patent by making an immediate lump-sum payment of $120,000. If the inventor can invest the capital at 10 percent, which offer should be accepted?

Calculation Procedure:

Compute the present worth of the annuity, using an interest rate of 10 percent

Obtain the USPW factor from an interest table for $i = 10$ percent, $n = 12$ and substitute in the relation $P = R(\text{USPW}) = \$15,000(6.814) = \$102,210$. Since the DEF Corp. offered an immediate payment of $120,000, its offer is more attractive than the offer made by ABC Corp.

CAPITAL-RECOVERY DETERMINATION

On January 1 of a certain year a company had a bank balance of $58,000. The company decided to allot this money to an improvement program by making a series of equal payments 4 times a year for 5 years, beginning on April 1 of the same year. If the account earned interest at 4 percent compounded quarterly, what was the amount of the periodic payment?

Calculation Procedure:

1. *Compute the true interest rate and number of interest periods*

Since the annual rate = 4 percent and there are four interest periods per year, the rate per period is i = 4 percent/4 = 1 percent. And with a 5-year pay period, the number of interest periods = 5 years (4 periods per year) = 20 periods.

2. *Compute the uniform payment, i.e., capital recovery*

The present worth of the sum is $58,000. Obtain the CR factor from an interest table for i = 1 percent, n = 20 and substitute in the relation $R = P(\text{CR}) = \$58,000(0.05542) = \$3214.36$.

EFFECTIVE INTEREST RATE

An account earns interest at the rate of 6 percent per annum, compounded quarterly. Compute the effective interest rate to four significant figures.

Calculation Procedure:

Compute the interest earned by $1 per year

With four interest periods per year, the interest rate per period = i = 6 percent/4 = 1.5 percent. In 1 year there are four interest periods for this account.

Find the compounded value of $1 at the end of 1 year from $S = (1 + i)^n = (1 + 0.015)^4 = \$1.06136$. Thus, the interest earned by $1 in 1 year = $1.06136 − 1.00000 = $0.06136. Hence, the effective interest rate = 6.136 percent.

PERPETUITY DETERMINATION

What sum must be deposited to provide annual payments of $10,000 that are to continue indefinitely if the endowment fund earns interest of 4 percent compounded semiannually?

Calculation Procedure:

1. *Compute the effective interest rate*

Using the same procedure as in the previous calculation procedure for $1, we find the effective interest rate $i_e = (1.02)^2 − 1 = 0.04040$, or 4.04 percent.

2. *Apply the USPW relation*

The endowment or principal required = $P = \text{payment}/i_e$, or $P = \$10,000/0.0404 = \$247,525$.

DETERMINATION OF EQUIVALENT SUMS

Jones Corp. borrowed $900 from Brown Corp. on January 1 of year 1 and $1200 on January 1 of year 3. Jones Corp. made a partial payment of $700 on January 1 of year 4. It was agreed that the balance of the loan would be discharged by two payments, one on January 1 of year 5 and the other on January 1 of year 6, with the second payment being 50 percent larger than the first. If the interest rate is 6 percent, what is the amount of each payment?

Calculation Procedure:

1. *Construct a line diagram indicating the loan data*

Figure 1 shows the line diagram for these loans and is typical of the diagrams that can be prepared for any similar set of loans.

2. *Select a convenient date for evaluating all the sums*

For this situation, select January 1 of year 6. Mark the valuation date on Fig. 1, as shown.

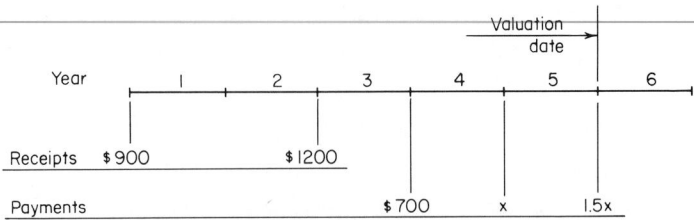

FIG. 1   Time, receipt, and payment diagram.

3.   *Evaluate each sum at the date selected*

Use the applicable interest rate, 6 percent, and the equivalence equation, value of money borrowed = value of money paid. Substituting the applicable SPCA factor from the interest table for each of the interest periods involved, or $n = 5$, $n = 3$, $n = 2$, and $n = 1$, respectively, gives $900(\text{SPCA}) + \$1200(\text{SPCA}) = \$700(\text{SPCA}) + x(\text{SPCA}) + 1.5x$, where $x =$ payment made on January 1 of year 5 and $1.5x =$ payment made on January 1 of year 6. Substituting, we get $900(1.338) + \$1200(1.191) = \$700(1.124) + 1.06x + 1.5x$; $x = \$721.30$. Hence, $1.5x = \$1081.95$.

Related Calculations: Note that this procedure can be used for more than two loans and for payments of any type that retire a debt.

ANALYSIS OF A NONUNIFORM SERIES

On January 1 of a certain year, ABC Corp. borrowed $1,450,000 for 12 years at 6 percent interest. The terms of the loan obliged the firm to establish a sinking fund in which the following deposits were to be made: $200,000 at the end of the second to the sixth years; $250,000 at the end of the seventh to the eleventh years; and one for the balance of the loan at the end of the twelfth year. The interest rate earned by the sinking fund was 3 percent. Adverse financial conditions prevented the firm from making the deposit of $200,000 at the end of the fifth year. What was the amount of the final deposit?

Calculation Procedure:

1.   *Prepare a money-time diagram*

Figure 2 shows a money-time diagram for this situation, where $x =$ deposit made at end of twelfth year.

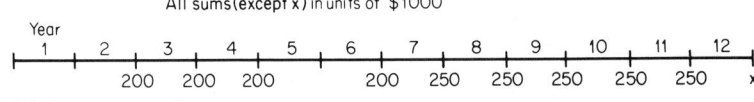

FIG. 2   Money-time diagram.

2.   *Compute the principal of the loan at the end of the twelfth year*

Use the relation $S = P(\text{SPCA})$ for $i = 6$ percent, $n = 12$. Obtain the SPCA value from an interest table, and substitute in the above relation, or $S = \$1,450,000(2.012) = \$2,917,400$.

3.   *Set up an expression for the principal in the sinking fund at the end of the twelfth year*

From Fig. 2, principal $= \$200,000(\text{USCA}, n = 3)(\text{SPCA}, n = 8) + \$200,000(\text{SPCA}, n = 6) + \$250,000(\text{USCA}, n = 5)(\text{SPCA}, n = 1) + x$. With an interest rate of 3 percent, principal $= \$200,000(3.091)(1.267) + \$200,000(1.194) + \$250,000(5.309)(1.030) + x$, or principal $= \$2,389,100 + x$.

4.   *Compute the final deposit*

Equate the principal in the sinking fund to the principal of the loan: $\$2,389,100 + x = \$2,917,400$. Thus, $x = \$528,300$.

UNIFORM SERIES WITH PAYMENT PERIOD DIFFERENT FROM INTEREST PERIOD

Deposits of $2000 each were made in a fund earning interest at 4 percent per annum compounded quarterly. The interval between deposits was 18 months. What was the balance in the account immediately after the fifth deposit was made?

Calculation Procedure:

1. Compute the actual interest rate

Replace the interest rate i_3 for the quarterly period with an equivalent rate i_{18} for the 18-month period. Or, $i_{18} = (1 + i_3)^n - 1 = (1.01)^6 - 1 = 6.15$ percent.

2. Compute the USCA value

Apply the equation USCA $= [(1 + i)^n - 1]/i$, or USCA $= [(1.0615)^5 - 1]/0.0615 = 5.654$.

3. Compute the principal in the fund

Use the relation $S = R(\text{USCA}) = \$2000(5.654) = \$11,308$.

UNIFORM-GRADIENT SERIES: CONVERSION TO UNIFORM SERIES

A loan was to be amortized by a group of six end-of-year payments forming an ascending arithmetic progression. The initial payment was to be $5000, and the difference between successive payments was to be $400, as shown in Fig. 3. But the loan was renegotiated to provide for the payment of equal rather than uniformly varying sums. If the interest rate of the loan was 8 percent, what was the annual payment?

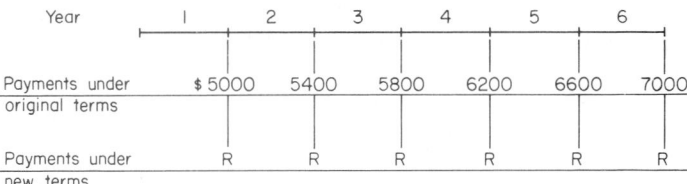

FIG. 3   Diagram showing changed payment plan.

Calculation Procedure:

1. Apply the equivalent-uniform-series equation

Let R_1 = initial payment in a uniform-gradient series; g = difference between successive payments; n = number of payments; R_e = periodic payment in an equivalent uniform series. Then $R_e = R_1 + (g/i)(1 - n\text{SFP})$. Substituting with $R_1 = \$5000$, $g = \$400$, $n = 6$, and $i = 8$ percent, we find $R_e = \$5000 + (\$400/0.08)(1 - 6 \times 0.13632) = \$5911$.

2. As an alternative, use the uniform-gradient conversion (UGC) factor

With $n = 6$ and $i = 8$ percent, UGC $= 2.28$. Then, $R_e = \$5000 + \$400(2.28) = \$5912$.

PRESENT WORTH OF UNIFORM-GRADIENT SERIES

Under the terms of a contract, Brown Corp. was to receive a payment at the end of each year from year 1 to year 7, with the payments varying uniformly from $8000 in year 1 to $5000 in year 7. At the beginning of year 1, Brown Corp. assigned its annuity to Edwards Corp. at a price that yielded Edwards Corp. a 6 percent investment rate. What did Edwards Corp. pay for the annuity?

Calculation Procedure:

1. Apply the relation of step 1 of the previous calculation procedure

The relation referred to converts a uniform-gradient series to an equivalent uniform series. Thus, with $g = -\$500$, $n = 7$, and $i = 6$ percent, we get $R_e = \$8000 + (-\$500/0.06) \times (1 - 7 \times 0.11914) = \$6617$.

2.   Compute the present worth of the equivalent annuity

Use the relation $P = R_e(\text{USPW}) = \$6617(5.582) = \$36,936$.

FUTURE VALUE OF UNIFORM-RATE SERIES

A deposit was made in a fund at the end of each year for 8 consecutive years. The first deposit was $1000, and each deposit thereafter was 25 percent more than the preceding deposit. If the interest rate of the fund was 7 percent per annum, what was the principal in the fund immediately after the eighth deposit was made?

Calculation Procedure:

1.   Compute the URSCA value

A uniform-rate series is a set of payments made at equal intervals in which the payments form a geometric progression (i.e., the ratio of a given payment to the preceding payment is constant). In this instance, the deposits form a uniform-rate series because each deposit is 1.25 times the preceding deposit. Apply Eq. 8: $\text{URSCA} = [r^n - (1 + i)^n]/(r - i - 1)$, where r = ratio of given payment to preceding payment, n = number of payments, and i = interest rate for the payment period. With $r = 1.25$, $n = 8$, and $i = 7$ percent, $\text{URSCA} = [(1.25)^8 - (1.07)^8]/(1.25 - 0.07 - 1) = 23.568$.

2.   Compute the future value of the set of deposits

Use the relation $S = R_1(\text{URSCA})$, where R_1 = first payment. Then $S = \$1000(23.568) = \$23,568$.

DETERMINATION OF PAYMENTS UNDER UNIFORM-RATE SERIES

At the beginning of year 1, the sum of $30,000 was borrowed with interest at 9 percent per annum. The loan will be discharged by payments made at the end of years 1 to 6, inclusive, and each payment will be 95 percent of the preceding payment. Find the amount of the first and sixth payments.

Calculation Procedure:

1.   Compute the URSPW value of the uniform-rate series

Apply Eq. 9: $\text{URSPW} = \{[r/(1 + i)]^n - 1\}/(r - i - 1)$, where the symbols are as defined in the previous calculation procedure. With $r = 0.95$, $n = 6$, and $i = 9$ percent, $\text{URSPW} = [(0.95/1.09)^6 - 1]/(0.95 - 0.09 - 1) = 4.012$.

2.   Find the amount of the first payment

Use the relation $P = R_1(\text{URSPW})$, where R_1 = first payment. Then $\$30,000 = R_1(4.012)$, or $R_1 = \$7477.60$.

3.   Find the amount of the sixth payment

Use the relation $R_m = R_1 r^{m-1}$, where R_m = mth payment. Then $R_6 = \$7477.60(0.95)^5 = \$5786.00$.

CONTINUOUS COMPOUNDING

If $1000 is invested at 6 percent per annum compounded continuously, what will it amount to in 5 years?

Calculation Procedure:

Apply the continuous compounding equation

Use the relation $\text{SPCA} = e^{jn}$, where e = base of the natural logarithm system = $2.71828 \ldots$, j = nominal interest rate, n = number of years. Substituting gives $\text{SPCA} = (2.718)^{0.30} = 1.350$. Then $S = P(\text{SPCA}) = \$1000(1.350) = \$1350$.

FUTURE VALUE OF UNIFORM SERIES WITH CONTINUOUS COMPOUNDING

An inventor received a royalty payment of $25,000 at the end of each year for 7 years. The royalties were invested at 12 percent per annum compounded continuously. What was the inventor's capital at the expiration of the 7-year period?

Calculation Procedure:

1. *Compute the USCA value*

Apply Eq. 12: USCA $= (e^{jn} - 1)/(e^{j} - 1)$, where $n =$ number of annual payments in uniform-payment series and e and j are as defined in the previous calculation procedure. Thus, with $n = 7$ and $j = 12$ percent, USCA $= (e^{0.84} - 1)/(e^{0.12} - 1) = 10.325$.

2. *Compute the future value of the series*

Set $S = R(\text{USCA}) = \$25,000(10.325) = \$258,100$.

Related Calculations: Note that if interest were compounded annually at 12 percent, the USCA value would be 10.089.

PRESENT WORTH OF CONTINUOUS CASH FLOW OF UNIFORM RATE

An investment syndicate is contemplating purchase of a business that is expected to yield an income of $200,000 per year continuously and at a constant rate for the next 5 years. If the syndicate wishes to earn 18 percent on its investment, what is the maximum price it should offer for the business?

Calculation Procedure:

1. *Compute the present-worth factor*

Apply the equation CFPW $= (1 - e^{-jn})/j$, where CFPW $=$ present-worth factor for a continuous cash flow of uniform rate and $n =$ number of years of the flow. Where the cash flow is continuous, it is understood that the given interest or investment rate is based on continuous compounding. Thus, with $n = 5$ and $j = 18$ percent, CFPW $= (1 - e^{-0.90})/0.18 = 3.297$.

2. *Compute the present worth of the income*

Set $P = R(\text{CFPW})$, where $R =$ annual cash-flow rate. Then $P = \$200,000(3.297) = \$659,400$.

FUTURE VALUE OF CONTINUOUS CASH FLOW OF UNIFORM RATE

The sum of $30 will be invested daily in a venture that yields 14 percent per annum. What will be the accumulated capital at the expiration of 18 months?

Calculation Procedure:

1. *Compute the cash-flow rate*

Where money is invested daily, the cash flow may be considered to be continuous for all practical purposes. Assume that deposits are made every day of the year. The cash-flow rate $R = \$30(365) = \$10,950$ per year.

2. *Compute the future-value factor*

Apply the equation CFFV $= (e^{jn} - 1)/j$, where CFFV $=$ future-value factor for a continuous cash flow of uniform rate. Thus, with $n = 1.5$ and $j = 14$ percent, CFFV $= (e^{0.21} - 1)/0.14 = 1.669$.

3. *Compute the future value of the money invested*

Set $S = R(\text{CFFV}) = \$10,950(1.669) = \$18,280$.

Depreciation and Depletion

Notational System

Here $D_U =$ depreciation charge for Uth year; $D =$ annual depreciation; $\Sigma D_U =$ cumulative depreciation at end of Uth year $= D_1 + D_2 + D_3 + \cdots + D_U$, where the subscript numbers refer to the year numbers; P_0 = original cost of asset; P_U = book value of asset at end of Uth year = $P_0 - \Sigma D_U$; IRS = Internal Revenue Service; L = salvage value; W = wearing value, or total depreciation = $P_0 - L$; N = longevity or life of asset, years.

STRAIGHT-LINE DEPRECIATION

The initial cost of a machine, including its installation, is $15,000. The IRS life of this machine is 10 years. The estimated salvage value of the machine is $1000, and the cost of dismantling the machine is estimated to be $200. Using straight-line depreciation, what is the annual depreciation charge? What is the book value of the machine at the end of the seventh year?

Calculation Procedure:

1. Compute the annual depreciation charge

When straight-line depreciation is used, the annual depreciation charge is constant, $D = W/N$. Since $P_0 = \$15,000$, $L = \$1000 - \$200 = \$800$, $W = \$15,000 - \$800 = \$14,200$, $N = 10$. Then $D = \$14,200/10 = \$1420$.

2. Compute the book value of the machine at the end of the seventh year

$\Sigma D_7 = 7D = 7(\$1420) = \$9940$. Then $P_7 = \$15,000 - \$9940 = \$5060$.

STRAIGHT-LINE DEPRECIATION WITH TWO RATES

An asset having an initial cost of $30,000 has a life expectancy of 15 years and an estimated salvage value of $5000. What are the depreciation charges under a modified straight-line method in which 60 percent of the total depreciation is considered to occur during the first 5 years of the life of the asset?

Calculation Procedure:

1. Proportion the total wearing value of the asset

Divide the asset's life span into the two specified intervals, and proportion the total wearing value between them. Thus, $W = \$30,000 - \$5000 = \$25,000$; $N = 15$; W_1 = first-period wearing value = $0.60(\$25,000) = \$15,000$; W_2 = second-period wearing value = $0.40(\$25,000) = \$10,000$.

2. Compute the annual depreciation charge

For the first 5 years, $D = \$15,000/5 = \$3000$. For the next 10 years, $D = \$10,000/10 = \$1000$.

DEPRECIATION BY ACCELERATED COST RECOVERY SYSTEM

An asset having a first cost of $120,000 is to be depreciated by the Accelerated Cost Recovery System (ACRS). This asset is assigned a 5-year cost-recovery period, and the following depreciation factors are to be applied: year 1, 20.0 percent; year 2, 32.0 percent; year 3, 19.2 percent; years 4 and 5, 11.5 percent; year 6, 5.8 percent. Compute the depreciation charges and the book value of the asset during the cost-recovery period.

Calculation Procedure:

1. Compute the depreciation charges

The ACRS for allocating depreciation was adopted by the federal government in 1981, but it subsequently underwent several modifications. ACRS was designed to allow a firm to write off an asset rapidly, the expectation being that industry would thus be encouraged to modernize its plants and facilities.

The salient features of ACRS are as follows: Each asset is assigned a *cost-recovery period* during which depreciation is to be charged, and this period is independent of its estimated longevity; the estimated salvage value is ignored; the depreciation for a given year is computed by multiplying the first cost of the asset by a specified *depreciation factor;* the initial depreciation charge occurs in the year the asset is placed in service, and the depreciation charge for that year is independent of the specific date at which this placement occurs. Since many assets are placed in service relatively late in the year, the allowable depreciation charge for the first year is low compared with that for the second year, and depreciation is charged for one year beyond the cost-recovery period. If the salvage value that accrues from disposal of the asset exceeds the book value of the asset at that date, the excess is subject to taxation.

The depreciation charges are recorded in the accompanying table.

2. Compute the end-of-year book values

The results are recorded in the accompanying table. Currently, the federal government recognizes only the straight-line method and ACRS for allocating depreciation. However, many state governments still recognize other methods. Moreover, a firm may wish to compute depreciation by some other method for its private records as a means of obtaining a more accurate appraisal of its annual profit.

Year	Depreciation charge, $	Book value at year end, $
1	$120,000(0.200) = 24,000$	$120,000 - 24,000 = 96,000$
2	$120,000(0.320) = 38,400$	$96,000 - 38,400 = 57,600$
3	$120,000(0.192) = 23,040$	$57,600 - 23,040 = 34,560$
4	$120,000(0.115) = 13,800$	$34,560 - 13,800 = 20,760$
5	$120,000(0.115) = 13,800$	$20,760 - 13,800 = 6,960$
6	$120,000(0.058) = 6,960$	$6,960 - 6,960 = 0$

SINKING-FUND METHOD: ASSET BOOK VALUE

A factory constructed at a cost of $9,000,000 has an anticipated salvage value of $400,000 at the end of 30 years. What is the book value of this factory at the end of the tenth year if depreciation is charged by the sinking-fund method with an interest rate of 5 percent?

Calculation Procedure:

1. Compute the cumulative depreciation

This method of depreciation accounting assumes that when the asset is retired, it is replaced by an exact duplicate and that replacement capital is accumulated by making uniform end-of-year deposits in a reserve fund. The cumulative depreciation ΣD_U is therefore equated to the principal in the fund at the end of the Uth year. Or, $\Sigma D_U = W(SFP)(USCA)$. So $W = \$9,000,000 - \$400,000 = \$8,600,000$, SFP $= 0.01505$ for 30 years, $U = 10$ years, and $i = 5$ percent, $\Sigma D_{10} = \$8,600,000(0.01505)(12.578) = \$1,628,000$.

2. Compute the book value

At the end of 10 years, the book value $P_{10} = P_0 - \Sigma D_{10} = \$9,000,000 - \$1,628,000 = \$7,372,000$.

SINKING-FUND METHOD: DEPRECIATION CHARGES

An asset costing $20,000 is expected to remain serviceable for 5 years and to have a salvage value of $3000. Compute the depreciation charges, using the sinking-fund method and an interest rate of 4 percent.

Calculation Procedure:

1. Compute the annual sinking-fund payment

Use the relation $R = W(SFP)$. With $W = \$20,000 - \$3000 = \$17,000$, $N = 5$ years, and $i = 4$ percent, SFP $= 0.18463$. Then $R = \$17,000(0.18463) = \$3139$.

2. Compute the annual depreciation charges

Use the relation $D_U = R(SPCA)$, or $D_1 = \$3139(1.000) = \$3139$; $D_2 = \$3139(1.040) = \$3265$; $D_3 = \$3139(1.082) = \$3396$; $D_4 = \$3139(1.125) = \$3531$; $D_5 = \$3139(1.170) = \$3673$. Then $\Sigma D_5 = \$17,004$.

FIXED-PERCENTAGE (DECLINING-BALANCE) METHOD

An asset cost $5000 and has a life expectancy of 6 years and an estimated salvage value of $800. Construct a depreciation schedule for this asset, using the fixed-percentage method.

Calculation Procedure:

1. Compute the rate of depreciation

Use the relation $h = 1 - (L/P_0)^{1/N}$, where h = rate of depreciation. Substituting gives $h = 1 - (800/5000)^{1/6} = 0.2632$, or 26.32 percent.

2. Compute the end-of-year book value

Use the relation $D_1 = hP_0 = 0.2632(\$5000) = \$1316$. Then $P_1 = P_0 - D_1 = \$5000 - \$1316 = \$3684$. Likewise, $D_2 = 0.2632(\$3684) = \$969.63$; $P_2 = \$3684 - \$969.63 = \$2714.37$. In a similar manner, $D_3 = \$714.42$, $P_3 = \$1999.95$, $D_4 = \$526.39$, $P_4 = \$1473.56$, $D_5 = \$387.84$, $P_5 = \$1085.72$, $D_6 = \$285.76$, $P_6 = \$799.96$.

COMBINATION OF FIXED-PERCENTAGE AND STRAIGHT-LINE METHODS

An asset cost $20,000 and has a life of 8 years and a salvage value of $1000. The IRS permits use of the double-declining-balance method to charge depreciation. Compute the depreciation charges.

Calculation Procedure:

1. Compute the rate of depreciation

Under the double-declining-balance method, depreciation is initially charged on a fixed-percentage basis, with $2/N$ as the rate of depreciation. Thus, rate of depreciation = $2/8 = 0.25$, or 25 percent.

2. Compute the depreciation charge for each year by the fixed-percentage method

For the first year, depreciation charge = $0.25(\$20,000) = \$5000$. Then the book value at the end of the first year = $\$20,000 - \$5000 = \$15,000$. Following this procedure, construct Table 1.

3. Compute the depreciation for the transfer study

Assume that the transfer in depreciation accounting from the fixed-percentage to the straight-line method is made at the end of a particular year. Calculate the annual depreciation charge D' that applies for the remaining life of the asset.

For example, at the end of the third year the book value is $8437, and the depreciation that remains to be charged during the last 5 years is $7437. Then $D' = \$7437/5 = \$1487$. Record the values found in this manner in Table 1.

TABLE 1 Depreciation by the Double-Declining-Balance Method

Year	Depreciation charge, $	Book value at year end, $	D', $
0	. . .	20,000	
1	5,000	15,000	2,000
2	3,750	11,250	1,708
3	2,813	8,437	1,487
4	2,109	6,328	1,332
5	1,582	4,746	1,249
6	1,187	3,559	1,280
7	890	2,669	1,669
8	667	2,002	

4. Determine the transfer date

To establish the transfer date, compare each value of D' with the depreciation charge that will occur in the following year if the fixed-percentage method is used. This comparison shows that the method should be revised at the end of the fifth year because after that time the fixed-percentage method results in a smaller depreciation charge. The depreciation charges (Table 1) are thus $D_1 = \$5000$; $D_2 = \$3750$; $D_3 = \$2813$; $D_4 = \$2109$; $D_5 = \$1582$; $D_6 = D_7 = D_8 = \$1249$.

CONSTANT-UNIT-USE METHOD OF DEPRECIATION

A machine cost $38,000 and has a life of 5 years and a salvage value of $800. The production output of this machine in units per year is: first year, 2000; second year, 2500; third year, 2250;

fourth year, 1750; fifth year, 1500 units. If the depreciation is ascribable to use rather than the effects of time, and the units produced are of uniform quality, what are the annual depreciation charges?

Calculation Procedure:

1. Determine the depreciation charge per production unit

Proportion the wearing value on the basis of annual production. Since W = $38,000 − $800 = $37,200 and 10,000 units are produced in 5 years, the depreciation charge per production unit = $37,200/10,000 = $3.72.

2. Compute the annual depreciation charge

Since the annual depreciation charge is a function of the production rate, take the product of the depreciation charge per production unit and the annual production. Or, D_1 = $3.72(2000) = $7440; D_2 = $9300; D_3 = $8370; D_4 = $6510; D_5 = $5580.

DECLINING-UNIT-USE METHOD OF DEPRECIATION

Using the same data as in the previous calculation procedure, assume that depreciation will be charged by weighting the units produced according to their relative quality. This method reflects the quality loss resulting from increased use of the machine. The quality weights assigned this machine are: first 4000 units produced, 2.0; next 3000 units, 1.5; remainder, 1.0. Compute the depreciation charges for this machine.

Calculation Procedure:

1. Compute the number of depreciation units

The depreciation units are related to the annual production by applying the assigned quality rates. Thus

Year	Depreciation units
1	$2,000 \times 2 = 4,000$
2	$\begin{cases} 2,000 \times 2 = 4,000 \\ 500 \times 1.5 = 750 \end{cases}$
3	$2,250 \times 1.5 = 3,375$
4	$\begin{cases} 250 \times 1.5 = 375 \\ 1,500 \times 1 = 1,500 \end{cases}$
5	$1,500 \times 1 = 1,500$
	Total 15,500

2. Proportion the wearing value

Consider the number of depreciation units as the criterion. Or, depreciation charge per depreciation unit = $37,200/15,500 = $2.40.

3. Compute the annual depreciation

Take the product of the depreciation charge per depreciation unit and the annual depreciation units. Or, D_1 = $2.40(4000) = $9600; likewise, D_2 = $11,400; D_3 = $8100; D_4 = $4500; D_5 = $3600. Taking the sum of these charges, we see that the total depreciation = $37,200.

SUM-OF-THE-DIGITS METHOD OF DEPRECIATION

A machine costing $15,000 is expected to remain serviceable for 7 years. The machine will have a salvage value of $1000. What are the annual depreciation charges based on the sum-of-the-digits method?

Calculation Procedure:

1.   Compute the machine wearing value

The wearing value W, or total depreciation $= \$15,000 - \$1000 = \$14,000$.

2.   Compute the annual depreciation

Use the relation $D_U = W(N - U + 1)/0.5[N(N + 1)]$, where $U =$ year number. Thus, for $U = 1$, $D_1 = \$3500$. Likewise, for $U = 2$, $D_2 = \$3000$; for $U = 3$, $D_3 = \$2500$; for $U = 4$, $D_4 = \$2000$; for $U = 5$, $D_5 = \$1500$; for $U = 6$, $D_6 = \$1000$; for $U = 7$, $D_7 = \$500$.

COMBINATION OF TIME- AND USE-DEPRECIATION METHODS

A machine cost \$38,000 and has a life of 5 years and a salvage value of \$800. Studies show that one-third of the total depreciation stems from the effects of time and two-thirds stems from use. Compute the annual depreciation charges if time depreciation is based on sum of the digits and use depreciation on a production basis with all units of equal quality. Use the same production as in the third previous procedure.

Calculation Procedure:

1.   Divide the wearing value into its two elements

Knowing the respective depreciation proportions, let the subscripts t and u refer to time and use, respectively. Also, $W = \$38,000 - \$800 = \$37,200$, and $W_t = \frac{1}{3}(\$37,200) = \$12,400$; $W_u = \frac{2}{3}(\$37,200) = \$24,800$.

2.   Compute the annual depreciation charge

For the first year, $D_{t1} = W_t N/[N(N + 1/2] = \$12,400(5)/[5(6/2)] = \$4133$. Also, $D_{u1} = (\$24,800/10,000 \text{ units})(2000 \text{ units the first year}) = \$4960$. Thus, the total depreciation for the first year is $D_1 = \$4133 + \$4960 = \$9093$.

EFFECTS OF DEPRECIATION ACCOUNTING ON TAXES AND EARNINGS

The QRS Corp. purchased capital equipment for use in a 5-year venture. The equipment cost \$240,000 and had zero salvage value. If the income tax rate was 52 percent and the annual income from the investment was \$83,000 before taxes and depreciation, what was the average rate of earnings if the profits after taxes were invested in tax-free bonds yielding 3 percent? Compare the results obtained when depreciation is computed by the straight-line and sum-of-the-digits methods.

Calculation Procedure:

1.   Compute the taxable income

With straight-line depreciation, the depreciation charge is $\$240,000/5 = \$48,000$ per year. Then the taxable income $= \$83,000 - \$48,000 = \$35,000$, because depreciation is fully deductible from gross income.

2.   Compute the annual tax payment

With a tax rate of 52 percent, the annual tax payment, excluding other deductions, is $0.52(\$35,000) = \$18,200$.

3.   Compute the net income

The net cash income $=$ gross income $-$ tax payment, if there are no other expenses. Or, net income $= \$83,000 - \$18,200 = \$64,800$.

4.   Determine the capital accumulated by investing the net income in bonds

Use the USCA factor for $i = 3$ percent, $n = 5$ years. Or, $S = R(USCA) = \$64,800(5.309) = \$344,000$.

5.   Compute the average earnings rate on the venture

Use the relation $SPCA = (1 + i)^n$, where $SPCA = \$344,000/\$240,000 = (1 + i)^5$; $i = 7.47$ percent.

6. *Compute the sum-of-the-digits annual depreciation*

Using the previously developed procedure for sum-of-the-digits depreciation charges gives $D_1 = \$80,000$; $D_2 = \$64,000$; $D_3 = \$48,000$; $D_4 = \$32,000$; $D_5 = \$16,000$.

7. *Compute the annual tax and net income*

Using the same method as in steps 2 and 3, we find the annual net income R is $R_1 = \$81,440$; $R_2 = \$73,120$; $R_3 = \$64,800$; $R_4 = \$56,480$; $R_5 = \$48,160$.

8. *Determine the capital accumulated*

Use the respective SPCA values for $i = 3$ percent and years 1 through 5 for the income earned in each year. Or, $S = \$81,440(1.126) + \$73,120(1.093) + \$64,800(1.061) + \$56,480(1.030) + \$48,160 = \$346,700$.

9. *Compare the average earnings rate on the venture*

By the method of step 5, $\$346,700/\$240,000 = (1 + i)^5$; $i = 7.63$ percent.

The computed interest rates apply to a composite investment—the purchase and operation of the capital equipment and the purchase of bonds. The total income accruing from the first element is $324,000, regardless of the depreciation method used. However, the *timing* as well as the amount of this income is important.

The straight-line method produces a uniform annual depreciation charge, tax payment, and net income. Under the sum-of-digits method, these amounts are nonuniform; the net income is highest in the first year and then gradually declines. Therefore, the interest earned through the purchase of bonds is higher if the firm adopts the sum-of-the-digits method.

If the interest rate associated with the second element of this composite investment had been higher, say 4 or 5 percent, the disparity between the two average returns would have been correspondingly higher.

DEPLETION ACCOUNTING BY THE SINKING-FUND METHOD

An oil field is anticipated to yield an annual income, before depletion allowances, of $120,000. The field will be dry after 5 years, at which time the land will have a residual value of $60,000. If a firm desires a return of 10 percent on its investment, what is the maximum amount it should invest in this oil field? Use a 4 percent interest rate for the sinking fund.

Calculation Procedure:

1. *Determine the replacement cost of the asset*

In this method of depletion accounting, it is assumed that the firm deposits a portion of the annual income in a reserve fund to accumulate the capital needed to replace the asset. Let C denote the investment required. Then the replacement cost $r = C - \$60,000$ for this venture.

2. *Compute the annual deposit required*

Let $d = $ annual deposit required. Then $d = r(\text{SFP})$ for this venture, or any similar situation. With $i = 4$ percent, $n = 5$, $d = (C - \$60,000)(0.18463) = 0.18463C - 11,077.80$.

3. *Compute the investment required*

Set the residual income equal to 10 percent of the investment and solve for C. Or, $\$120,000 - (0.18463C - 11,077.80) = 0.10C$; $C = \$460,520$.

Related Calculations: Note that this method can be applied to any situation where there is a gradual depletion of a valuable, profit-generating asset. Further, the method given here is homologous to the sinking-fund method of depreciation accounting.

INCOME FROM A DEPLETING ASSET

An oil field purchased for $800,000 is expected to be dry at the end of 4 years. If the resale value of the land is $20,000, what annual income is required to yield an investment rate of 8 percent? Use a sinking-fund rate of 3 percent.

Calculation Procedure:

1. *Compute the annual deposit required to accumulate the replacement capital*

The replacement cost $= \$800,000 - \$20,000 = \$780,000 = r$. Use the relation annual deposit $d = r(\text{SFP})$. With $i = 3$ percent, $n = 4$, $d = \$780,000(0.23903) = \$186,440$.

2. *Compute the annual income required*

Combine the annual return on the invested capital with the reserve-fund deposit to obtain the required annual income from the asset. Or, annual return on investment = 0.08($800,000) = $64,000. Then the required annual income = $64,000 + $186,440 = $250,440.

DEPLETION ACCOUNTING BY THE UNIT METHOD

The sum of $500,000 was expended in purchasing and developing a mine. During the first 2 years, ore was extracted at these rates: first year, 20,000 tons; second year, 18,000 tons. Originally, the mine was estimated to have a capacity of 230,000 tons, but at the beginning of the second year the remaining capacity was estimated to be only 170,000 tons. Compute the depletion allowance for the first 2 years by the unit method.

Calculation Procedure:

1. *Compute the depletion allowance for the first year*

Under the unit method, it is assumed that the entire capital invested in a depleting asset is consumed in the venture, and the loss of capital is prorated over the life of the venture on the basis of the amount of mineral extracted each year. Thus, for the first year, depletion = $500,000(20,000/230,000) = $43,480.

2. *Compute the depletion allowance for the second year*

At the beginning of the second year, the unrecovered capital is $500,000 − $43,480 = $456,520, and the estimated amount of ore remaining is 170,000 tons. Thus, for the second year, depletion = $456,520(18,000/170,000) = $48,340.

Cost Comparisons of Alternative Proposals

Annual Cost

For analytical purposes, it is desirable to convert the estimated costs associated with a proposed scheme to an equivalent series of uniform annual payments. The annual payment thus obtained is termed the *annual cost* of the scheme. The interest rate applied in making this conversion is the minimum investment rate that is considered acceptable by the organization making the investment or incurring the costs.

Where alternative schemes are being evaluated on the basis of their annual cost, the usual procedure is to exclude those expenses which are identical for all schemes, since they do not affect the comparison.

Notational System

Here P = initial cost of asset acquired in the proposed scheme; L = salvage value of the asset; N = life of asset, years; i_1 = interest rate; c = sum of annual costs of operation, maintenance, etc., that are assumed to remain constant for the asset life; A = annual cost = $(P - L)(\text{CR}) + Li_1 + c$, where CR is the capital-recovery factor from the compound-interest tables for $i = i_1$, $n = N$; other symbols are as defined earlier. Also, $A = (P - L)(\text{SFP}) + Pi_1 + c$, for the same interest and life as the above annual-cost relation.

DETERMINATION OF ANNUAL COST OF AN ASSET

A firm contemplates building a new warehouse. A choice is to be made between a brick and a galvanized-iron structure. The cost data associated with each structure are as follows:

	Brick	Galvanized iron
First cost, $	80,000	36,000
Salvage value, $	15,000	4,000
Life, years	40	15
Annual maintenance cost, $	1,000	2,300
Annual taxes, $/$100	1.30	1.30
Annual insurance, $/$1000	2	5

If this firm earns 6 percent on its invested capital, which type of structure is the more economical one?

Calculation Procedure:

1. *Compute the operating and maintenance costs*

For the brick building, the annual operating and maintenance cost $= c =$ maintenance cost per year, $ + annual taxes, $ + annual insurance cost, $, or $1000 + 0.013($80,000) + 0.002($80,000) = $2200.

For the galvanized-iron building, $c = $2300 + 0.013($36,000) + 0.005($36,000) = $2948.

2. *Compute the annual cost of each building*

Use the capital-recovery equation. Thus, for the brick building, $A = ($80,000 - $15,000)(0.06646) + $15,000)(0.06) + $2200 = $7420.

For the galvanized-iron building, $A = ($36,000 - $4000)(0.10296) + $4000(0.06) + $2948 = $6483.

Since the galvanized-iron building has a lower annual cost, it is the more economical structure.

Related Calculations: This general method of computing annual costs can be used for any number of industrial or commercial assets regardless of whether they are stationary, moving, or water- or air-borne. The key fact is that accurate costs are required if the annual cost comparison is to have validity.

MINIMUM ASSET LIFE TO JUSTIFY A HIGHER INVESTMENT

The timber floor of a bridge is to be replaced, and consideration is being given to treating the timber to prolong its life and reduce maintenance costs. An untreated timber floor costs $5000 and has an annual maintenance cost of $500 and a life of 10 years. A treated timber floor costs $8500 and has an annual maintenance cost of $300. How long should the treated timber last to make it more economical than the untreated timber? Use an interest rate of 5 percent.

Calculation Procedure:

1. *Compute the annual cost of the untreated timber floor*

Using the capital-recovery factor, we see the annual cost is $5000(0.12950) + $500 = $1147.50.

2. *Set up an expression for the annual cost of the treated timber floor*

The annual cost is $8500(CR) + $300.

3. *Compute the minimum life required to justify treating the timber*

Equate the annual costs, giving $8500(CR) + $300 = $1147.50, or CR = 0.09971. Interpolating in the compound-interest table for 5 percent, we find $N = 14.3$ years. The life of the treated timber floor must exceed 14.3 years to make it more economical.

COMPARISON OF EQUIPMENT COST AND INCOME GENERATED

A firm is considering purchasing equipment that will reduce annual labor costs by $4000. The equipment costs $30,000 and has a salvage value of $5000 and a life of 7 years. The annual maintenance cost is $600. While not in use by the firm, the equipment can be rented to others to generate an income of $1000 per year. If money can be invested for an 8 percent return, is the firm justified in buying the equipment?

Calculation Procedure:

1. *Compute the annual cost of using the equipment*

Using the capital-recovery-factor annual cost, we get $A = ($30,000 - $5000)(0.19207) + $5000(0.08) + $600 = $5802.

2. *Compute the annual cost of not purchasing the equipment*

If the equipment is not purchased, the firm will incur an extra labor cost of $4000 over that with the equipment. Also, the rental income that would be obtained from the equipment will be lost. Hence, the total annual cost without the equipment would be $A = $4000 + $1000 = $5000.

Since the annual cost with the equipment would be $5802, the firm should not purchase the equipment because without it the annual cost is only $5000.

SELECTION OF RELEVANT DATA IN ANNUAL-COST STUDIES

An existing factory must be enlarged or replaced to accommodate new production machinery. The structure was built at a cost of $130,000. Its present book value, based on straight-line depreciation, is $35,000, but it has been appraised at $40,000. If the structure is altered, the cost will be $80,000 and its service life will be extended 8 years, with a salvage value of $30,000. A new factory could be purchased for $250,000. It would have a life of 20 years and a salvage value of $35,000. Annual maintenance costs of the new building would be $8000, compared with $5000 in the enlarged structure. However, the improved layout in the new building would reduce annual production costs by $12,000. All other expenses for the two structures are estimated as being equal. Using an investment rate of 8 percent, determine which is the more attractive investment for this firm.

Calculation Procedure:

1.   *Segregate the relevant data for the existing structure*

Relevant data—present resale value. Irrelevant data—cost of construction and present book value.

2.   *Record the pertinent cost data for each scheme*

Classify the income that would accrue from one scheme as a "cost" of its alternative. Thus

	Enlarged building	New building
Initial cost or payment, $	80,000	250,000
Resale value existing building, $	40,000	
Total first cost, $	120,000	250,000
Salvage value, $	30,000	35,000
Life, years	8	20
Operating cost, $	5,000	8,000
Production "cost," $	12,000	

3.   *Compute the annual cost of the enlarged building*

Using the capital-recovery factor for $i_1 = 8$ percent, $n = 8$ years, we have $A = ($120,000 - $30,0000)(0.17401) + $30,000(0.08) + $5000 + $12,000 = $35,061.$

4.   *Compute the annual cost of the new building*

Using the capital-recovery factor for $i = 8$ percent, $n = 20$ years gives $A = ($250,000 - $35,000)(0.10185) + $35,000(0.08) + $8000 = $32,698.$

Since the new building has an annual cost almost $2400 less than the enlarged existing structure, the new building is the more economical choice.

Related Calculations: This general procedure can be used to compare any two or more alternatives having characteristics similar to those described above.

DETERMINATION OF MANUFACTURING BREAK-EVEN POINT

A manufacturing firm has a choice between two machines to produce a product. The relevant data are as follows:

	Machine A	Machine B
First cost, $	20,000	28,000
Salvage value, $	2,000	
Life, years	10	6
Annual operating cost, $	3,000 + 5.00 per unit	2,500 + 1.50 per unit

If money is worth 7 percent, what annual production is required to justify purchase of machine B?

Calculation Procedure:

1. Compute the annual cost of the first machine

Let x denote the number of units produced annually. Then, by using the capital-recovery factor, $A = (\$20,000 - \$2000)(0.14238) + \$2000(0.07) + \$3000 + 5x = \$5703 + 5x$ for machine A.

2. Compute the annual cost of the second machine

Using the same procedure for machine B gives $A = (\$28,000)(0.20980) + \$2500 + 1.5x = \$8374 + 1.5x$.

3. Equate the annual costs, and solve for the unknown

Substituting the annual costs from steps 1 and 2 yields $\$5703 + 5x = \$8374 + 1.5x$; $x = 763$ units.

This is the break-even point at which the costs of each machine are equal. If production is expected to exceed this volume, machine B is the economical choice.

COST COMPARISON WITH NONUNIFORM OPERATING COSTS

Two alternative machines have the following data:

	Machine A	Machine B
First cost, $	6,800	12,000
Salvage value, $	. . .	1,000
Life, years	6	10

For machine A, the estimated annual operating cost is $1240. For machine B, it is $800 for the first 4 years, $1200 for the next 3 years, and $1500 for the remaining 3 years. Determine which machine is more economical, using an 8 percent interest rate.

Calculation Procedure:

1. Compute the annual cost of machine A

By using the capital-recovery factor, $A = \$6800(0.21632) + \$1240 = \$2711$.

2. Construct a money-time diagram for machine B

Figure 4 shows the annual operating costs for machine B.

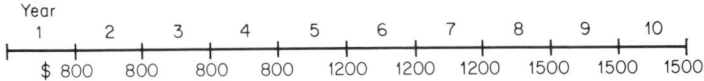

Year

| 1 | 2 | 3 | 4 | 5 | 6 | 7 | 8 | 9 | 10 |

$ 800   800   800   800   1200   1200   1200   1500   1500   1500

FIG. 4   Annual operating costs for machine B.

3. Convert the operating costs for machine B to an equivalent uniform series

The value S of these costs as of the end of the tenth year is $S = \$800(\text{USCA}, n = 4)(\text{SPCA}, n = 6) + \$1200(\text{USCA}, n = 3)(\text{SPCA}, n = 3) + \$1500(\text{USCA}, n = 3)$, or $S = \$800(4.506)(1.587) + \$1200(3.246)(1.260) + \$1500(3.246) = \$15,498$. Now apply the relationship $R = S(\text{SFP}, n = 10)$, where R = annual payment of a uniform series that is equivalent to the actual operating costs. Then $R = \$15,498(0.06903) = \$1070$. This is the equivalent uniform annual operating cost for machine B.

4. Compute the annual cost of machine B, and compare the two machines

Using the capital-recovery factor, we find $A = (\$12,000 - \$1000)(0.14903) + \$1000(0.08) + \$1070 = \$2789$. Machine A has a lower annual cost, and so it is more economical.

Related Calculations: As an alternative method in step 3, compute the value P' of the operating costs as of the purchase date. Then $P' = \$800(\text{USPW}, n = 4) + \$1200(\text{USPW}, n = 3)(\text{SPPW}, n = 4) + \$1500(\text{USPW}, n = 3)(\text{SPPW}, n = 7)$, or $P' = \$800(3.312) + \$1200(2.577)(0.7350) + \$1500(2.577)(0.583) = \$7176$. Now apply the relationship $R = P'(\text{CR}, n = 10)$, or $R = \$7176(0.14903) = \$1069$. Note that the *arithmetic mean* of the annual operating costs for machine B is $1130. However, since the costs increase with time and the earlier payments in a series have a more pronounced effect than the later payments, the *equivalent* annual operating cost is less than $1130.

ECONOMICS OF EQUIPMENT REPLACEMENT

A machine having an installed cost of $10,000 was used for 5 years. During that time its trade-in value and operating costs changed as follows:

End of year	Salvage value, $	Operating cost, $/year
1	6000	2300
2	4000	2500
3	3200	3300
4	2500	4800
5	2000	6800

If the cost of a new machine remained constant during this time, at what date would it be most economical to replace the machine with a duplicate? Use a 7 percent interest rate.

Calculation Procedure:

1. Compute the present worth of all payments on the asset

Let P' denote the present worth (i.e., the value at the date of purchase) of all expenditures ascribable to an asset. In the capital-recovery annual-cost equation, substitute P' for P and set $c = 0$ to obtain the following alternative equation: $A = (P' - L)(\text{CR}) + Li_1$. Using $i_1 = 7$ percent, compute the present worth of the operating costs. Or,

Year	
1	PW = ($2300)(0.9346) = $2150
2	PW = ($2500)(0.8734) = $2184
3	PW = ($3300)(0.8163) = $2694
4	PW = ($4800)(0.7629) = $3662
5	PW = ($6800)(0.7130) = $4848

2. Determine the present worth for each life span

Take the sum of the installed cost, $10,000, and the present worth of the operating cost found in step 1. Or,

Life, years	
0	$P' = \$10,000 + \quad \$0 = \$10,000$
1	$P' = \$10,000 + \$2,150 = \$12,150$
2	$P' = \$12,150 + \$2,184 = \$14,334$
3	$P' = \$14,334 + \$2,694 = \$17,028$
4	$P' = \$17,028 + \$3,662 = \$20,690$
5	$P' = \$20,690 + \$4,848 = \$25,538$

3. Apply the annual-cost equation developed in step 1

Life, years	
1	$A = \$6{,}150(1.07000) + \$6{,}000(0.07) = \$7{,}001$
2	$A = \$10{,}334(0.55309) + \$4{,}000(0.07) = \$5{,}996$
3	$A = \$13{,}828(0.38105) + \$3{,}200(0.07) = \$5{,}493$
4	$A = \$18{,}190(0.29523) + \$2{,}500(0.07) = \$5{,}545$
5	$A = \$23{,}538(0.24389) + \$2{,}000(0.07) = \$5{,}881$

Inspect these annual costs to determine when the minimum annual cost occurs. Since the annual cost is a minimum when $N = 3$, the asset should be retired at the end of the third year.

ANNUAL COST BY THE AMORTIZATION (SINKING-FUND-DEPRECIATION) METHOD

A machine costs $30,000 and will be retired at the end of 8 years with a salvage value of $5000. The annual operating cost is $3200. Determine the annual cost by the amortization method if the interest rate on the loan is 6 percent and that of the sinking fund is 3 percent.

Calculation Procedure:

Compute the annual cost of the asset

The amortization method is based on the following assumptions: The asset is purchased with borrowed funds; interest on the loan is paid annually; the loan principal is paid as a lump sum at the retirement of the asset; the funds required to retire the debt are accumulated by uniform annual deposits in a reserve fund. This assumed method of financing is unrealistic; the amortization method is therefore approximate.

Let i_1 = interest rate on loan; i_2 = interest rate on sinking fund. Then $A = (P - L)(\text{SFP}) + Pi_1 + c$. In this equation the SFP factor is based on i_2. Apply this equation, using $P = \$30{,}000$, $L = \$5000$, $N = 8$, $c = \$3200$, $i_1 = 6$ percent, $i_2 = 3$ percent. Thus, $A = \$7812$.

ANNUAL COST BY THE STRAIGHT-LINE-DEPRECIATION METHOD

The director of a corporation recommends that a firm buy a computer instead of renting at the rate of $50 per hour. A new computer costs $120,000; annual operating, maintenance, and insurance costs total $8500. The computer will be traded at the end of 10 years for $30,000. The director forecasts computer usage for 480 h/year. He bases his calculation of annual cost on the straight-line-depreciation method with an interest rate of 6 percent. Is his recommendation sound?

Calculation Procedure:

1. Compute the annual cost of owning the asset

The straight-line method is an approximate one which assumes that the asset is purchased with borrowed funds. However, the method disregards the timing of payments and considers only their arithmetic average. Thus, the annual cost $A = (P - L)/N + (P - L)i_1(N + 1)/2N + Li_1 + c$. So $A = \$90{,}000/10 + \$90{,}000(0.06)(11/20) + \$30{,}000(0.06) + \$8500 = \$22{,}270$.

2. Compute the annual cost of renting the asset

The annual cost of renting the asset = (hourly rate, $)(annual use, hr) = ($50)(480) = $24,000.

Since the annual cost of owning the asset is less than the annual cost of renting it, the firm would save money by owning the asset. Note that this is an approximate method.

Present Worth of Future Costs

A cost analysis of alternative schemes may be performed by computing the present worth of all expenses incurred in each scheme during a stipulated period called the *analysis period*. This period should encompass an integral number of lives of each asset required under the alternative schemes.

PRESENT WORTH OF FUTURE COSTS OF AN INSTALLATION

A city contemplates increasing the capacity of existing water-transmission lines. Two plans are under consideration: Plan A requires construction of a parallel pipeline, flow being maintained by gravity. The initial cost is $800,000, and the life is 60 years with an annual operating cost of $1000. Plan B requires construction of a booster pumping station costing $210,000 with a life of 30 years. The pumping equipment costs an additional $50,000; it has a life of 15 years and a salvage value of $10,000. The annual operating cost is $35,000. Which is the more economical plan if the interest rate is 6 percent?

Calculation Procedure:

1. Construct a money-time diagram of the situation

Figure 5 shows the money-time diagram. Note that this diagram uses 60 years as the analysis period. Record on the money-time diagram the capital expenditures during this 60-year period.

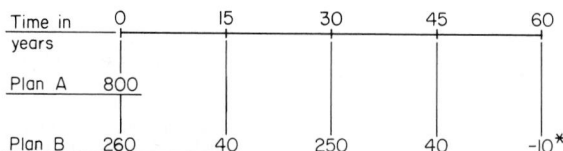

*Income from disposal of equipment.
All sums in units of $1000

FIG. 5   Money-time diagram.

2. Compute the total present worth of the payments

For plan A, using the USPW factor for $n = 60$ years, we get PW = $800,000 + $1000(16.161) = $816,160. For plan B, by using the SPPW factor for the payments shown in Fig. 5, and the uniform series present-worth factor for the operating cost, PW = $260,000 + $40,000(SPPW) + $250,000(SPPW) + $40,000(SPPW) + $35,000(USPW) − $10,000(SPPW) = $260,000 + $40,000(0.4173) + $250,000(0.1741) + $40,000(0.0727) + $35,000(16.161) − $10,000(0.0303) = $888,460.

Since the present worth of plan A is less than that of plan B, the scheme for plan A should be adopted because it is more economical.

Capitalized Cost

In computing the present worth of the costs associated with a proposed scheme, it is often advantageous to select an analysis period of infinite duration. The present worth of the future costs is then referred to as the *capitalized cost* of the scheme.

Since each expenditure recurs indefinitely during the analysis period, the various costs constitute a group of perpetuities. Thus, the capitalized cost C_c is $C_c = [(P − L)/i_1](CR) + L + c/i_1$, or $C_c = [(P − L)/i_1](SFP) + P + c/i_1$.

If an asset is considered to have an infinite life span, these equations reduce to $C_c = P + c/i_1$. In these equations, $i = i_1$, $n = N$.

DETERMINATION OF CAPITALIZED COST

Two methods of conveying water for an industrial plant are being analyzed. Method A uses a tunnel, and method B a ditch and flume. The costs are as follows:

	Method A	Method B	
	Tunnel	Ditch	Flume
First cost, $	180,000	50,000	40,000
Salvage value, $	. . .	. . .	5,000
Life, years	Infinite	50	15
Operating cost, $/year	2,300	2,000	3,600

Evaluate these two alternatives on the basis of capitalized cost, using a 5 percent interest rate.

Calculation Procedure:

1. Compute the capitalized cost of the first alternative

Since the tunnel has an infinite life, $C_c = P + c/i_1 = \$180,000 + \$2300/0.05 = \$226,000$.

2. Compute the capitalized cost of the second alternative

Using the capital-recovery factor for $n = 50$ years, $i = 5$ percent, we find for the ditch $C_c = (\$50,000/0.05)(0.05478) + \$2000/0.05 = \$94,780$.

Using a similar procedure for the flume, which has a 15-year life, gives $C_c = (\$35,000/0.05)(0.09634) + \$5000 + \$3600/0.05 = \$144,440$. The total capitalized cost for method B = the sum of the flume and ditch costs, or $239,220. Since method A costs less, it is more economical.

CAPITALIZED COST OF ASSET WITH UNIFORM INTERMITTENT PAYMENTS

What is the capitalized cost of a bridge costing $85,000 and having a 25-year life, a $10,000 salvage value, $400 annual maintenance cost, and repairs at 5-year intervals of $2000, if the interest rate is 5 percent?

Calculation Procedure:

1. Convert the assumed repair costs to an equivalent series of uniform annual payments

Assume that the repairs are made at the end of every 5-year interval, including the replacement date. Using the SFP factor, we see the equivalent series of uniform annual payments $R_1 = \$2000(SFP)$ for $i = 5$ percent, $n = 5$ years, or $R_1 = \$2000(0.18097) = \$362$.

2. Convert the true repair costs to an equivalent series of uniform annual payments

Repairs are omitted when the bridge is scrapped at the end of 25 years, thereby saving $2000 in the final 5-year period. Convert this amount to an equivalent series of uniform annual payments (i.e., savings) and subtract from the result in step 1. Or, $R_2 = \$2000(SFP)$ for $i = 5$ percent, $n = 25$ years. Or, $R_2 = \$2000(0.02095) = \$42$. Thus the annual cost of the repairs = $362 − $42 = $320.

3. Compute the capitalized cost

Using the capital-recovery factor for $i = 5$ percent, $n = 25$ years, we get $C_c = (\$75,000/0.05)(0.07095) + \$10,000 + \$400/0.05 + \$320/0.05 = \$130,830$.

Related Calculations: An alternative solution could be worked as follows: Since the $2000 saving at the end of every 25-year interval coincides in timing with the income of $10,000 from the sale of the old bridge as scrap, this saving can be combined with the salvage value to obtain an effective value of $12,000 for salvage. The annual cost of repairs is therefore taken as $362, the value of R_1, step 1. Applying the capital-recovery factor gives $C_c = [(\$85,000 − \$12,000)/$

~~0.05](0.07095) + $12,000 + $400/0.05 + $362/0.05 = $130,830. This agrees with the previously~~
determined value.

CAPITALIZED COST OF AN ASSET WITH NONUNIFORM INTERMITTENT PAYMENTS

A bridge has the same cost data as in the previous calculation procedure except for the repairs, which are as follows:

End of year	Repair cost, $
10	2000
15	3500
20	1500

What is the capitalized cost of the bridge if the interest rate is 5 percent?

Calculation Procedure:

1. Compute the present worth of the repairs for one life span

Use the single-payment present-worth factor for each of the repair periods. Or, PW = $2000(0.6139) + $3500(0.4810) + $1500(0.3769) = $3477.

2. Convert the result of step 1 to an equivalent series of uniform annual payments

Using the capital-recovery factor, we find the annual cost of repairs = $3477(CR), where i = 5 percent, n = 25 years. Or, c_a = $3477(0.07095) = $247.

3. Compute the capitalized cost

Using the same method as in step 3 of the previous calculation procedure gives C_c = $106,430 + $10,000 + $8000 + $247/0.05 = $129,370.

Related Calculations: An alternative way of solving this problem is to combine the present worth of the payments for repairs ($3477) with the initial cost ($85,000) to obtain an equivalent initial cost P'. Then, P' = $88,477, and $P' - L$ = $88,477 - $10,000 = $78,477. By applying the capital-recovery factor, C_c = $129,370 as before.

STEPPED-PROGRAM CAPITALIZED COST

A firm plans to build a new warehouse with provision for anticipated growth. Two alternative plans are available.

	Plan A	Plan B
First cost, $	100,000	80,000
Salvage value, $	10,000	15,000
Life, years	25	30
Annual maintenance, $	1,400	1,200 first 10 years, 1,800 thereafter
Cost of enlarging structure 10 years hence, $	. . .	40,000

If money is worth 10 percent, which is the more economical plan?

Calculation Procedure:

1. Compute the total present worth of the second plan costs

Let P' represent the total present worth of the costs associated with plan B for one life span. Using the SPPW for $i = 10$ percent, $n = 10$ years, for the cost of enlarging the structure, and the USPW for the annual maintenance *after* expansion, and the *difference* between the annual maintenance costs of this structure and the original structure, we get $P' = \$80,000 + \$40,000(0.3855) + \$1800(9.427) - \$600(6.144) = \$108,700$.

2. Compute the capitalized cost of each alternative

Using the capital-recovery factor for plan A with $i = 10$ percent, $n = 25$ years yields $C_c = [(\$100,000 - \$10,000)/0.10](0.11017) + \$10,000 + \$1400/0.10 = \$123,150$.

   For plan B, by using the present worth from step 1, $C_c = [(\$108,700 - \$15,000)/0.10](0.10608) + \$15,000 = \$114,400$. Note that the capital-recovery factor for plan B is for 30 years.

   Since plan B has the lower capitalized cost, it is more economical.

Cost Comparisons with Taxation and Technological Advances

In the preceding material, the costs of alternative proposals were compared by disregarding taxation and assuming that financial and technological conditions remain static. The cost analysis is now made more realistic by including the effects of taxation and technological advances. Later the effects of inflation also are included.

CALCULATION OF ANNUAL COST ON AFTER-TAX BASIS

An asset has the following cost data: First cost, $80,000; life, 10 years; salvage value, $5000; annual operating cost, $3600. The firm that owns the asset is subject to a tax rate of 47 percent, and its investment rate is 8 percent after payment of taxes. Compute the after-tax annual cost of this asset if depreciation is allocated by (a) the straight-line method and (b) the sum-of-digits method.

Calculation Procedure:

1. Compute the annual depreciation charge under the straight-line method

The charge is $D = (\$80,000 - \$5000)/10 = \$7500$.

2. Compute the annual cost under straight-line depreciation

Most income earned by a corporation is subject to the payment of corporate income tax. The *effective* (or *after-tax*) income is the difference between the original income and the tax payment pertaining to that income. The before-tax investment rate i_b = rate of return on an investment as calculated on the basis of original income; the after-tax investment rate i_a = rate of return as calculated on the basis of effective income. Every cost incurred in operating an asset serves to reduce taxable income and thus the tax payment. The *effective* cost is the difference between the actual expenditure and the tax savings that results from the expenditure. The cost of an asset is said to be computed on an after-tax basis if all calculations are based on effective costs and the after-tax investment rate.

   Let t = tax rate and D = annual depreciation charge. Where annual operating costs and depreciation charges are uniform, the annual cost $A = (P - L)(\text{CR}, n = N, i = i_a) + Li_a + c(1 - t) - Dt$. The last term represents the tax savings that accrues from the depreciation charge. With $n = 10$, $i_a = 8$ percent, and $t = 47$ percent, $A = (\$80,000 - \$5000)(0.14903) + \$5000(0.08) + \$3600(0.53) - \$7500(0.47) = \$9960$.

3. Compute the annual depreciation charges under the sum-of-digits method

As given in an earlier calculation procedure, $D_U = W(N - U + 1)/0.5[N(N + 1)]$, where D_U = depreciation charge for Uth year and W = total depreciation. With $W = \$75,000$ and $N = 10$, $D_1 = \$13,636$, and every depreciation charge thereafter is $1363.64 less than the preceding charge.

4.   *Convert the depreciation charges under the sum-of-digits method to an equivalent uniform depreciation charge, using an 8 percent interest rate*

Refer to an earlier calculation procedure for converting a uniform-gradient series to an equivalent uniform series. The equivalent uniform depreciation charge $D = D_1 + (g/i)(1 - n\text{SFP})$. With $D_1 = \$13,636$, $g = -\$1363.64$, $i = 8$ percent, and $n = 10$, $D = \$13,636 + (-\$1363.64/0.08)[1 - 10(0.06903)] = \$8357$.

5.   *Compute the annual cost under sum-of-digits depreciation*

Referring to step 2 and taking the difference between the equivalent depreciation charge in the present case and the depreciation charge under the straight-line method, we determine the annual cost $A = \$9960 - (\$8357 - \$7500)(0.47) = \$9557$.

   Related Calculations: A comparison of the two values of annual cost—$9960 when straight-line depreciation is used and $9557 when sum-of-digits depreciation is used—confirms the statement made in an earlier calculation procedure. Since tax savings accrue more quickly under sum-of-digits depreciation than under straight-line depreciation, the former method is more advantageous to the firm. In general, a firm seeks to write off an asset rapidly in order to secure tax savings as quickly as possible, thus allowing it to retain more capital for investment. For this reason depreciation accounting is subject to stringent regulation by the IRS.

COST COMPARISON WITH ANTICIPATED DECREASING COSTS

Two alternative machines, A and B, are available for a manufacturing operation. The life span is 4 years for machine A and 6 years for machine B. The equivalent uniform annual cost is estimated to be $16,000 for machine A and $15,000 for machine B. However, as a result of advances in technology, the annual cost is expected to decline at a constant rate from one life to the next, the rate of decline being 10 percent for machine A and 6 percent for machine B. Applying an investment rate of 12 percent, determine which machine is preferable.

Calculation Procedure:

1.   *Select the analysis period, and compute annual costs for this period*

The cost comparison will be made by the present-worth method. The analysis period is 12 years, since this is the lowest common multiple of 4 and 6. The annual costs are as follows: Machine A: first life, $16,000; second life, $16,000(0.90) = $14,400; third life, $14,400(0.90) = $12,960. Machine B: first life, $15,000; second life, $15,000(0.94) = $14,100.

2.   *Construct a money-time diagram*

The equivalent uniform annual payments are shown in Fig. 6.

3.   *Compute the present worth of costs for the first analysis period, and identify the more economical machine*

For machine A, PW = $16,000(USPW, $n = 4$) + $14,400(USPW, $n = 4$)(SPPW, $n = 4$) + $12,960(USPW, $n = 4$)(SPPW, $n = 8$). With $i = 12$ percent, PW = $16,000(3.037) + $14,400(3.037)(0.6355) + $12,960(3.037)(0.4039) = $92,280. For machine B, PW = $15,000(USPW, $n = 6$) + $14,100(USPW, $n = 6$)(SPPW, $n = 6$), or PW = $15,000(4.111) + $14,100(4.111)(0.5066) = $91,030. Machine B should be used for the first 12 years because it costs less.

4.   *Compute the present worth of costs for the second analysis period*

The second 12-year period encompasses the fourth, fifth, and sixth lives of machine A and the third and fourth lives of machine B. The "present" is the beginning of the second 12-year period. The annual cost of machine A during its fourth life is $(0.90)^3$ times the annual cost during its first life. Therefore, the results of step 3 can be applied. For machine A, PW = $92,280(0.90)^3 = $67,270. For machine B, PW = $91,030(0.94)^2 = $80,430. Thus, machine A should be used after the first 12 years. Realistically, since the transfer from machine B to machine A can be made at the end of the first 6-year period, the decision should be reviewed at that time in the light of currently available forecasts.

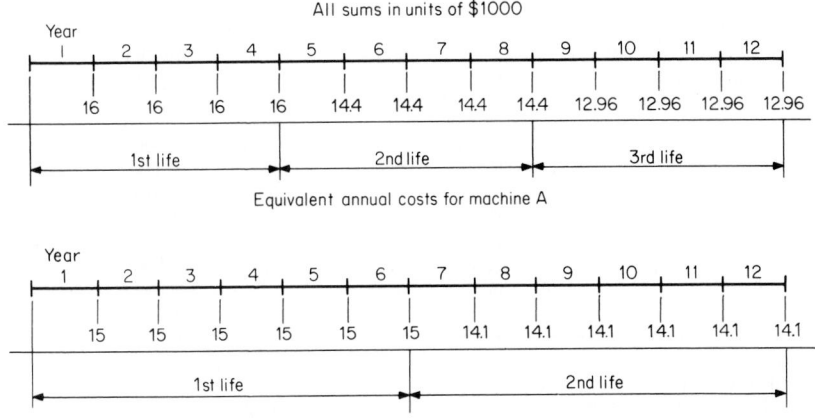

FIG. 6   Equivalent payments for 12-year analysis period.

ECONOMY OF REPLACING AN ASSET WITH AN IMPROVED MODEL

A machine has been in use for 3 years, and its cost data for the next 8 years are shown in Table 2, where the years are counted from the present. An improved model of this machine has just appeared on the market, and according to estimates it has an optimal life of 6 years with an equivalent uniform annual cost of $9400. No additional improvements are anticipated in the near future. If money is worth 10 percent, when will it be most economical to retire the existing machine?

Calculation Procedure:

1. *Compute an equivalent single end-of-life payment for each prospective remaining life*

Let R = remaining life of machine, years. The annual cost corresponding to every possible value of R will be found by a method that is a variation of that used in an earlier calculation

TABLE 2

Year	Salvage value at end, $	Annual operating cost, $
0	12,000	
1	8,000	4,700
2	5,000	5,200
3	4,000	5,800
4	3,500	6,600
5	3,000	7,500
6	2,700	8,500
7	2,600	9,700
8	2,500	10,900

procedure. Let c_R = operating cost at end of Rth year and F_R = equivalent single payment at end of Rth year. Then $F_R = F_{R-1}(1 + i) + c_R$. By retaining the existing machine, the firm forfeits an income of $12,000, and this is equivalent to making a payment of that amount now. Then F_0 = $12,000; F_1 = $12,000(1.10) + $4700 = $17,900; F_2 = $17,900(1.10) + $5200 = $24,890; F_3 = $24,890(1.10) + $5800 = $33,179; etc. The results are shown in Table 3.

2. *Compute the annual cost for every prospective remaining life*

Let A_R = annual cost for a remaining life of R years, and L_R = salvage value at end of Rth year. Then $A_R = (F_R - L_R)(\text{SFP}, n = R)$. Thus, with i = 10 percent, A_1 = ($17,900 - $8000)1 = $9900; A_2 = ($24,890 - $5000)(0.47619) = $9471; A_3 = ($33,179 - $4000)(0.30211) = $8815; etc. The results are shown in Table 3.

3. *Determine whether the existing machine should be retired now*

When an asset is purchased and installed, its resale value drops sharply during the early years of its life, and thus the firm incurs a rapid loss of capital during those years. The result is that the

TABLE 3 Calculation of Annual Cost

Remaining life, years	F, $	L, $	SFP	Annual cost, $
0	12,000			
1	17,900	8,000	1.00000	9,900
2	24,890	5,000	0.47619	9,471
3	33,179	4,000	0.30211	8,815
4	43,097	3,500	0.21547	8,532
5	54,907	3,000	0.16380	8,502
6	68,897	2,700	0.12961	8,580
7	85,487	2,600	0.10541	8,737
8	104,936	2,500	0.08744	8,957

annual cost for the remaining life of an existing asset is considerably less than the annual cost when the asset was first purchased.

Table 3 shows that the optimal remaining life of the existing machine is 5 years and the corresponding annual cost is $8502. Since this is less than the annual cost of the new machine ($9400), the existing machine should be retained for at least 5 years.

4. Determine precisely when the existing machine should be replaced

Since costs increase beyond the fifth year, the existing machine and the improved model will be compared on a year-by-year basis. Let B_R = cost of retaining existing machine 1 year beyond the Rth year. Then $B_R = L_R(1 + i) - L_{R+1} + c_{R+1}$. Thus, $B_5 = \$3000(1.10) - \$2700 + \$8500 = \$9100$. Since this is less than the annual cost of the new model, the existing machine should not be retired 5 years hence. Continuing, we find $B_6 = \$2700(1.10) - \$2600 + \$9700 = \$10,070$, which exceeds $9400. Therefore, the existing machine should be retired 6 years hence.

ECONOMY OF REPLACEMENT UNDER CONTINUING IMPROVEMENTS

A newly acquired machine costs $40,000, and it has the salvage values and annual operating costs shown in Table 4. It is anticipated that a new model will become available at the end of each year. All future models will have first costs and salvage values identical with those of the present model, but the annual operating cost for a given model will be $600 lower than the corresponding annual operating cost of the preceding model. For example, the model that becomes available 1 year hence will have an operating cost of $11,400 for the first year, $12,400 for the second year, etc. Applying an interest rate of 10 percent, determine how long this machine should be held.

TABLE 4

Year	Salvage value at end, $	Annual operating cost, $
1	25,000	12,000
2	20,000	13,000
3	17,000	14,600
4	15,000	16,500
5	13,500	18,800
6	12,000	21,500
7	11,000	24,500
8	10,000	28,000

Calculation Procedure:

1. Establish the excess operating costs in relation to a 1-year life

When the machine is retired, it will be replaced with the model that becomes available at that date. First assume that the machine is retired at the end of each year. The operating costs for the next 8 years are shown in Table 5. Now assume that the machine is held for 8 years. Subtracting the values just found from the values in Table 4 gives the *excess* operating costs for an 8-year life; these are shown in Table 5. This table also gives the excess operating costs for every

TABLE 5

Year	Operating cost for 1-year life, $	Excess operating cost for 8-year life, $
1	12,000	0
2	11,400	1,600
3	10,800	3,800
4	10,200	6,300
5	9,600	9,200
6	9,000	12,500
7	8,400	16,100
8	7,800	20,200

prospective life of the machine. For example, if the machine is held 3 years, the excess operating cost is $1600 for the second year and $3800 for the third year, and these values apply to each subsequent life. Since only *differences* in cost are significant, the prospective lives of the machine will be compared by applying the excess rather than the actual operating costs.

2. Compute an equivalent single end-of-life payment for every prospective life

Annual costs will be computed by using the same method as in the previous calculation procedure, but applying excess operating costs. Thus, with i = 10 percent, F_1 = $40,000(1.10) = $44,000; F_2 = $44,000(1.10) + $1600 = $50,000; F_3 = $50,000(1.10) + $3800 = $58,800; etc. The results are shown in Table 6.

3. Compute the annual cost for every prospective life

Proceeding as in the previous calculation procedure gives A_1 = ($44,000 − $25,000)1 = $19,000; A_2 = ($50,000 − $20,000)(0.47619) = $14,286; A_3 = ($58,800 − $17,000)(0.30211) = $12,628; etc. The results are shown in Table 6.

TABLE 6 Calculation of Annual Cost

Life, years	F, $	L, $	SFP	Annual cost, $
1	44,000	25,000	1.00000	19,000
2	50,000	20,000	0.47619	14,286
3	58,800	17,000	0.30211	12,628
4	70,980	15,000	0.21547	12,062
5	87,278	13,500	0.16380	12,085
6	108,506	12,000	0.12961	12,508
7	135,456	11,000	0.10541	13,119
8	169,203	10,000	0.08744	13,921

4. Identify the most economical life of the machine

Table 6 reveals that a 4-year life has the minimum annual cost.

Related Calculations: Each excess annual operating cost shown in Table 5 consists of two parts: a *deterioration cost*, which is the increase in operating cost due to aging of the machine, and an *obsolescence cost*, which results from the development of an improved model. For example, at the end of the fourth year the deterioration cost is $16,500 − $12,000 = $4500, and the obsolescence cost is $600 × 3 = $1800. If the quality of the product declines as the machine ages, the resulting loss of income can be added to the deterioration cost.

ECONOMY OF REPLACEMENT ON AFTER-TAX BASIS

A machine was purchased 3 years ago at a cost of $45,000. It had a life expectancy of 7 years and anticipated salvage value of $3000. It has been depreciated by the sum-of-digits method. The net resale value of the machine is $13,000 at present and is expected to be $9000 a year hence. The operating cost during the coming year will be $2600. A newly developed machine can be substituted for the existing one. According to estimates, this machine will have an optimal life of 5 years with an annual cost of $4800 on an after-tax basis. The tax rate is 45 percent for ordinary income and 30 percent for long-term capital gains. The desired investment rate on an after-tax basis is 8 percent. Determine whether the existing machine should be replaced at present.

Calculation Procedure:

1.   *Compute the depreciation charges for the first 4 years*

Refer to an earlier calculation procedure for sum-of-digits depreciation. The charges are $D_1 = \$10,500$; $D_2 = \$9000$; $D_3 = \$7500$; $D_4 = \$6000$.

2.   *Compute the book value at the end of the third and fourth years*

Let B_R = book value at end of Rth year. Then $B_3 = \$45,000 - (\$10,500 + \$9,000 + \$7,500) = \$18,000$ and $B_4 = \$18,000 - \$6000 = \$12,000$.

3.   *Compute the cost of retaining the machine through the fourth year*

Income that accrues from normal business operations is called *ordinary income;* other forms of income are called *capital gains.* The difference between the net income that accrues from selling an asset and its book value at the date of sale is a capital gain (or loss). If the asset was held for a certain minimum amount of time, this capital gain (or loss) is subject to a tax rate different from that for ordinary income.

Let R = age of asset, years. The after-tax cost of retaining the asset through the $(R + 1)$st year, as evaluated at the end of that year, is $[L_R(1 - t_c) + B_R t_c](1 + i_a) + c_{R+1}(1 - t_o) - L_{R+1}(1 - t_c) - B_{R+1}t_c - D_{R+1}t_o$, where i_a = after-tax investment rate; t_o and t_c = tax rate on ordinary income and long-term capital gains, respectively; L = true salvage value; c = annual operating cost; and the subscript refers to the age of the asset. The expression in brackets is the income that would be earned if the asset were sold at the end of the Rth year; if the asset is retained, this income is forfeited and becomes part of the cost of retention. With $i_a = 8$ percent, $t_o = 45$ percent, $t_c = 30$ percent, $L_3 = \$13,000$, $L_4 = \$9000$, and $c_4 = \$2600$, the cost of retaining the machine through the fourth year is $[\$13,000(0.70) + \$18,000(0.30)](1.08) + \$2600(0.55) - \$9000(0.70) - \$12,000(0.30) - \$6000(0.45) = \$4490$.

4.   *Determine whether the existing machine should be retired now*

Since the cost of retaining the machine for 1 additional year ($4490) is less than the annual cost of the new machine ($4800), the existing machine should not be retired at present.

Effects of Inflation

Notational System

Here C_0 and C_1 are the costs of a commodity now and 1 year hence, respectively, and f = annual rate of inflation during the coming year with respect to this commodity. Then $f = (C_1 - C_0)/C_0$, or $C_1 = C_0(1 + f)$. Also, C_n = cost of the commodity n years hence. If the annual rate of inflation remains constant at f, then $C_n = C_0(1 + f)^n = C_0(\text{SPCA}, i = f)$. In the subsequent material, it is understood that the given inflation rate applies to the asset under consideration.

DETERMINATION OF REPLACEMENT COST WITH CONSTANT INFLATION RATE

A machine has just been purchased for $60,000. It is anticipated that the machine will be held 5 years, that it will have a salvage value of $4000 as based on current prices, and that the annual rate of inflation during the next 5 years will be 7 percent. The machine will be replaced with a duplicate, and the firm will accumulate the necessary capital by making equal end-of-year deposits in a reserve fund that earns 6 percent per annum. Determine the amount of the annual deposit.

Calculation Procedure:

1.   *Compute the required replacement capital*

Both the cost of a new machine and the salvage value of the existing machine increase at the given rate. Thus, the amount of money the firm must accumulate to buy a new machine is $(\$60,000 - \$4000)(1.07)^5 = \$56,000(1.403) = \$78,568$.

2. Compute the annual deposit

Use this relation: Annual deposit $R = S(SFP)$. With $i = 6$ percent and $n = 5$, $R = \$78,568(0.17740) = \$13,938$.

DETERMINATION OF REPLACEMENT COST WITH VARIABLE INFLATION RATE

In the preceding calculation procedure, determine the amount of the annual deposit if the annual rate of inflation is expected to be 7 percent for the next 3 years and 9 percent thereafter.

Calculation Procedure:

1. Compute the required replacement capital

Replacement capital $= \$56,000(1.07)^3(1.09)^2 = \$56,000(1.225)(1.188) = \$81,497$.

2. Compute the annual deposit

From the preceding calculation procedure, annual deposit $= \$81,497(0.17740) = \$14,458$.

PRESENT WORTH OF COSTS IN INFLATIONARY PERIOD

An asset with a first cost of $70,000 is expected to last 6 years and to have the following additional cost data as based on present costs: salvage value, $5000; annual maintenance, $8400; major repairs at the end of the fourth year, $9000. The asset will be replaced with a duplicate when it is retired. Using an interest rate of 12 percent and an inflation rate of 8 percent per year, find the present worth of costs of this asset for the first two lives (i.e., for 12 years).

Calculation Procedure:

1. Compute the present worth of the capital expenditures for the first life

The "present" refers to the beginning of the first life. The payment for repairs will be $9000(1.08)^4$, and the present worth of this payment is $9000(1.08)^4(SPPW, n = 4, i = 12$ percent$) = \$9000(1.08)^4/(1.12)^4 = \$7780$. Similarly, the present worth of the salvage value is $\$5000(1.08)^6/(1.12)^6 = \$4020$. Thus, the present worth of capital expenditures for the first life is $\$70,000 + \$7780 - \$4020 = \$73,760$.

2. Compute the present worth of maintenance for the first life

The annual payments for maintenance constitute a uniform-rate series in which the first payment $R_1 = \$8400(1.08) = \$9072$ and the ratio of one payment to the preceding payment is $r = 1.08$. By Eq. 9, the present-worth factor of the series is URSPW $= [(1.08/1.12)^6 - 1]/(1.08 - 1.12) = 4.901$. Then present worth of series $= R_1(URSPW) = \$9072(4.901) = \$44,460$.

3. Compute the present worth of costs for the first life

Summing the results, we see that present worth $= \$73,760 + \$44,460 = \$118,220$.

4. Compute the present worth of costs for the second life

Since each payment in the second life is $(1.08)^6$ times the corresponding payment in the first life, the value of all payments in the second life, evaluated at the beginning of that life, is $(1.08)^6$ times that for the first life, or $\$118,220(1.08)^6$. The present worth of this amount is $\$118,220(1.08)^6/(1.12)^6 = \$95,040$.

5. Compute the present worth of costs for the first two lives

Summing the results yields PW $= \$118,220 + \$95,040 = \$213,260$.

   Related Calculations: Let $h = [(1 + f)/(1 + i)]^N$, where $N =$ life of asset, years. In the standard case, where all annual payments as based on present costs are equal and no extraordinary intermediate payments occur, the present worth of costs for the first life is $P - Lh + c(1 + f)(h - 1)/(f - i)$, where $P =$ initial cost; $L =$ salvage value as based on present costs; $c =$ annual payment for operation, maintenance, etc., as based on present costs. Where extraordinary payments occur, simply add the present worth of these payments, as was done in the present case with respect to repairs at the end of the fourth year. In the special case where $f = i$, the present worth of costs for the first life is $P - L + Nc$.

COST COMPARISON WITH ANTICIPATED INFLATION

Two alternative machines have the following cost data as based on present costs:

	Machine A	Machine B
First cost, $	45,000	80,000
Salvage value, $	3,000	2,000
Life, years	4	6
Annual maintenance, $	8,000	6,000

Determine which machine is more economical, using an interest rate of 10 percent and annual inflation rate of 7 percent.

Calculation Procedure:

1. Establish the method of cost comparison

The present-worth method is suitable here. Select an analysis period of 12 years, which encompasses three lives of machine A and two lives of machine B.

2. Compute the present worth of costs of machine A for the first life

Refer to the equation given at the conclusion of the preceding calculation procedure. Set $h = (1.07/1.10)^4 = 0.89529$. By reversing the sequence in the last two terms of the equation, present worth $= \$45,000 - \$3000(0.89529) + \$8000(1.07)(1 - 0.89529)/(0.10 - 0.07) = \$72,190$.

3. Compute the present worth of costs of machine A for the first three lives

Refer to step 4 of the preceding calculation procedure. Thus, PW $= \$72,190[1 + (1.07/1.10)^4 + (1.07/1.10)^8] = \$72,190(2.69684) = \$194,680$.

4. Compute the present worth of costs of machine B for the first life

Set $h = (1.07/1.10)^6 = 0.84712$. The present worth of costs for the first life $= \$80,000 - \$2000(0.84712) + \$6000(1.07)(1 - 0.84712)/(0.10 - 0.07) = \$111,020$.

5. Compute the present worth of costs of machine B for the first two lives

PW $= \$111,020(1 + 0.84712) = \$205,070$.

6. Determine which machine is preferable

Machine A has the lower cost and so is preferable.

ENDOWMENT WITH ALLOWANCE FOR INFLATION

An endowment fund is to provide perpetual annual payments to a research institute. The first payment, to be made 1 year hence, will be $10,000. Each subsequent payment will be 2 percent more than the preceding payment, to allow for inflation. If the interest rate of the fund is 7 percent per annum, what amount must be deposited in the fund now? Verify the result.

Calculation Procedure:

1. Compute the amount to be deposited

The payments form a uniform-rate series, and the amount to be deposited $= P =$ present worth of series. Refer to Eq. 9 for the present-worth factor. When $r < 1 + i$ and n is infinite, URSPW $= 1/(1 + i - r)$. With $i = 7$ percent and $r = 1.02$, URSPW $= 1/(1.07 - 1.02) = 20$. Then $P = R_1(\text{URSPW}) = \$10,000(20) = \$200,000$.

2. *Prove that this deposit will provide an endless stream of payments*

The proof consists in finding the rate at which the principal in the fund is growing. At the end of the first year, principal = $200,000(1.07) − $10,000 = $204,000. The rate of increase in principal = ($204,000 − $200,000)/$200,000 = 2 percent per year. Similarly, at the end of the second year, principal = $204,000(1.07) − $10,000(1.02) = $208,080. The rate of increase in principal = ($208,080 − $204,000)/$204,000 = 2 percent per year. Thus, the end-of-year principal expands at the same rate as the payments, and so the payments can continue indefinitely.

Related Calculations: If the interest period of the fund differs from the payment period, it is necessary to use the interest rate corresponding to the payment period. For example, assume that the interest rate is 7 percent per annum compounded quarterly. The corresponding annual (or effective) rate is $i = (1.0175)^4 − 1 = 7.186$ percent, and URSPW = $1/(1.07186 − 1.02) = 19.283$. The amount to be deposited = $192,830. Note that if $r \geq 1 + i$, URSPW becomes infinite as n becomes infinite. Thus, if the interest rate of the fund is 7 percent per annum, it is impossible to allow the payments to increase by 7 percent or more.

Evaluation of Investments

PREMIUM-WORTH METHOD OF INVESTMENT EVALUATION

A firm contemplates investing in a depleting asset and has a choice between two enterprises. Project A requires the investment of $57,500; project B requires the investment of $63,000. The forecast end-of-year dividends are as follows:

Year	Project A, $	Project B, $
1	10,000	15,000
2	15,000	25,000
3	25,000	30,000
4	20,000	20,000
5	10,000	

After weighing the risks involved, the firm decides that the minimum acceptable rate of return on project A is 10 percent; on project B, 12 percent. Evaluate these investments by the premium-worth method. If both investments are satisfactory, determine which is more satisfactory.

Calculation Procedure:

1. *Compute the present worth of the dividends from both investments*

The *present* generally refers to the date on which the investment is made. Where the present worth is greater than the sum invested, the excess is termed the *premium worth*. Such a result signifies that the true investment rate exceeds the minimum acceptable rate.

For any year, PW = (dividend, $)(PW factor for 10 percent and the number of years involved). Thus, for project A:

Year	PW
1	$10,000(0.9091) = $9,091
2	15,000(0.8264) = 12,396
3	25,000(0.7513) = 18,783
4	20,000(0.6830) = 13,660
5	10,000(0.6209) = 6,209
Total	$60,139

Then the premium worth = $60,139 − 57,500 = $2639.

By using a similar procedure, the present worth of project B at 12 percent is as follows:

Year	PW
1	$15,000(0.8929)= $13,394
2	25,000(0.7972)= 19,930
3	30,000(0.7118)= 21,354
4	20,000(0.6355)= 12,710
Total	$67,388

Then the premium worth = $67,388 − $63,000 = $4388.

2.   Determine the relative values of the investments

Since both investments satisfy the minimum requirements, determine their relative values by computing the premium-worth percentage (i.e., the ratio of the premium worth to the capital invested). Thus, for project A the premium-worth percentage is $2639(100)/$57,500 = 4.6 percent. For project B the premium-worth percentage is $4388(100)/$63,000 = 7.0 percent. Thus, project B is the more attractive because it has a higher premium-worth percentage.

VALUATION OF CORPORATE BONDS

A $10,000, 4 percent corporation bond paying semiannual dividends is redeemable at 102 at the end of 15 years. What is the maximum price an investor should pay for this bond if he desires a return of 6 percent compounded semiannually?

Calculation Procedure:

1.   Determine the semiannual dividend and redemption payment

The dividend = (principal, $)$i/2$ = $10,000(0.04/2) = $200. Also, the redemption payment = (redemption price/100)(principal) = (102/100)($10,000) = $10,200.

2.   Compute the purchase price

Using an interest rate of 6/2 = 3 percent per semiannual period, compute the present worth of the dividends and the redemption payment. Equate the present worth to the purchase price of the bond. Or, purchase price = (dividend, $)(USPW) + (redemption payment, $)(SPPW), for i = 3 percent, n = 30. Hence, purchase price = ($200)(19.60) + ($10,200)(0.4120) = $8122.

RATE OF RETURN ON BOND INVESTMENT

A $10,000, 6 percent, 20-year bond paid dividends semiannually and was redeemed at par. An investor bought the bond for $11,500 at its date of issue and held it to maturity. What interest rate did the holder earn?

Calculation Procedure:

1.   Record the payment and receipts associated with the investment

The *payment* was $11,500 at the date of issue. The *receipt* for each semiannual interest period was (6 percent/2)($10,000) = $300 for 40 periods. Also, $10,000 was received at the end of the 40 periods. The correct interest rate is that which will make the payment equal the receipts.

2.   Select a trial interest rate and compute the results

Selecting an interest rate of 2.5 percent as a trial, compute the value of the receipts at the date of issue, using the USPW factor for the dividends and the SPPW factor for the principal repayment. Or, ($300)(25.103) + $10,000(0.3724) = $11,255. Since the purchase price exceeded this value, the true interest rate was less than 2.5 percent.

3. Select another trial interest rate

Repeat the previous calculation, using a 2 percent rate. Or, $300(27.355) + $10,000(0.4529) = $12,736.

4. Interpolate linearly between the trial values

Interest rate, %	Purchase price, $
2.5	11,255
i	11,500
2	12,736

This interpolation gives $i = 2.42$ percent per semiannual period, or 4.84 percent per annum compounded semiannually.

INVESTMENT-RATE CALCULATION AS ALTERNATIVE TO ANNUAL-COST CALCULATION

In the Comparison of Equipment Cost and Income Generated procedure in this section, it was concluded that the proposed investment in labor-saving equipment could not be justified because it failed to yield the minimum acceptable rate of 8 percent. Determine the actual rate of return for this investment.

Calculation Procedure:

1. Compute the net annual dividend

Labor saving	$4000
Rental income	1000
Total	$5000
Less maintenance	600
Net dividend	$4400

2. Select a trial interest rate

Using an interest rate of 5 percent, determine the present worth of the dividends and the equipment salvage value. Thus, (net dividend, $)(USPW) + (salvage value, $)(SPPW), for $i = 5$ percent, $n = 7$ years. Or, $4400(5.786) + $5000(0.7107) = $29,012. Since the investment was $30,000, the actual interest rate is smaller.

3. Test another trial interest rate

Using a 4 percent interest rate and repeating the calculation in step 2, we get $4400(6.002) + $5000(0.7599) = $30,208.

4. Interpolate linearly to obtain the actual interest rate

Linear interpolation yields a rate of $i = 4.2$ percent. This verifies that the earlier results were valid.

ALLOCATION OF INVESTMENT CAPITAL

In devising a program for investment of $8000 in surplus funds, a firm has a choice between two plans. Each plan pays an annual dividend and repayment of the invested capital when the venture terminates. Under plan A the dividend varies with the sum invested in the manner shown below. Under plan B the dividend rate is 10 percent, irrespective of the sum invested. In what manner should this firm divide its investment capital to secure the maximum return?

Calculation Procedure:

1.   *List the annual dividends obtainable*

The table below shows the dividend that can be expected under plan A.

Investment, $	Annual dividend, $	Dividend rate, %
1000	300	30.0
2000	540	27.0
3000	720	24.0
4000	900	22.5
5000	950	19.0
6000	1020	17.0
7000	1220	17.4
8000	1300	16.3

2.   *Construct a dividend-investment diagram*

Figure 7 shows the dividend-investment diagram for this situation. Points A to H represent the sets of values under plan A. The slope of a line connecting any two points represents the rate of return on the incremental investment. For example, the slope of line $EG = \$270/\$2000 = 0.135$, or 13.5 percent represents the rate obtained on the $2000 investment added in going from E to G, Fig. 7.

3.   *Determine the investments to make*

Draw line OJ, Fig. 7, having a slope of 10 percent, the dividend rate under plan B. Next, determine which of the points, A to H, is most distant from line OJ. Do this by scaling the vertical offsets or by drawing lines through these points parallel to OJ.

   Point G, which has a vertical offset of $520, is the most distant one. Therefore, $7000 is the appropriate sum to invest in plan A because of the following: (*a*) When the investment is extended from some lower level, such as $5000, to the stipulated level, the rate of return on this incremental investment, which is represented by the slope of line EG, exceeds 10 percent. (*b*) If the investment is carried beyond G, the rate of return on this incremental investment, represented by the slope

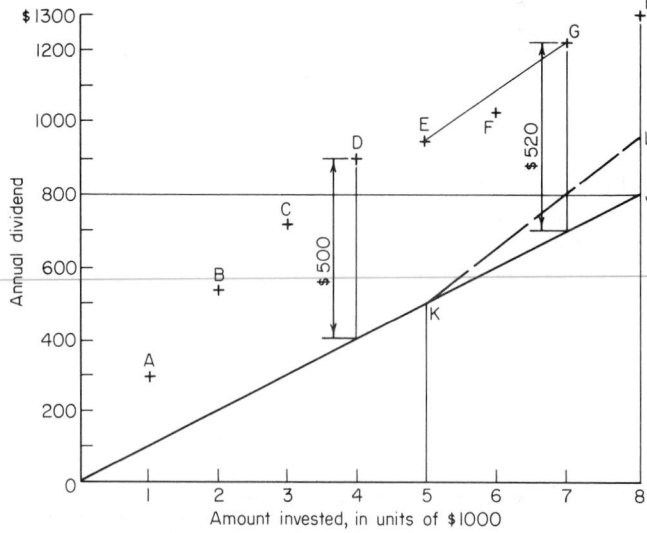

FIG. 7   Dividend-investment diagram.

of line *GH*, is less than 10 percent. Hence, this firm should allocate $7000 to plan A and $1000 to plan B.

Related Calculations: As an alternative, construct the following tabulation to determine the total annual dividend corresponding to every possible division of the capital. Study of this table shows that the maximum dividend of $1320 accrues when $7000 is allocated to plan A and $1000 to plan B.

Investment, $		Dividend, $		Total dividend, $
Plan A	Plan B	Plan A	Plan B	
...	8000	...	800	800
1000	7000	300	700	1000
2000	6000	540	600	1140
3000	5000	720	500	1220
4000	4000	900	400	1300
5000	3000	950	300	1250
6000	2000	1020	200	1220
7000	1000	1220	100	1320
8000	...	1300	...	1300

ALLOCATION OF CAPITAL TO TWO INVESTMENTS WITH VARIABLE RATES OF RETURN

Suppose that the dividend under plan B in the previous procedure is 15 percent of the first $3000 invested and 10 percent of the excess. Determine the optimal division of the $8000 investment between the two plans.

Calculation Procedure:

1. *Construct a dividend-investment diagram*

Use Fig. 7 and draw line *OK* having a slope of 10 percent and line *KL* having a slope of 15 percent, where *K* has an abscissa of $5000. The ordinate of each point on the line *OKL* represents the prospective plan B dividend that is forfeited by allocating part of the investment capital to plan A. The optimal division of the investment capital is that for which the excess of plan A dividends over forfeited plan B dividends is the maximum.

2. *Determine the investment allocation*

Find which of the points from *A* to *H* is most distant from the line *OKL*. This is point *D*, which has a vertical offset of $500. Therefore, the firm should allocate $4000 to plan A and $4000 to plan B.

Related Calculations: Alternatively, calculate the total dividend corresponding to every possible division of the available capital. For example, if $1000 is allocated to plan A and $7000 to plan B: Dividend under plan A = $300; dividend under plan B = $3000(0.15) + $4000(0.10) = $850; total dividend = $1150. Using this technique, we find the maximum total dividend to be $1450; this occurs when the capital is divided equally between two plans. Thus, the previous findings are verified.

This procedure and the previous one show two methods of establishing the optimal division of available capital, i.e., computing the rate of return on an incremental investment and computing the total dividend. The latter represents a more straightforward approach, particularly where both alternative investments yield a variable rate of return.

ALLOCATION OF CAPITAL TO THREE INVESTMENTS BY DYNAMIC PROGRAMMING

A syndicate has $600,000 available for investment; and three investment plans, A, B, and C, are under consideration. Under each plan, the amount that can be invested is a multiple of $100,000,

TABLE 7

Amount invested, $	Annual dividend, $		
	Plan A	Plan B	Plan C
100,000	25,000	10,000	15,000
200,000	44,000	32,000	31,000
300,000	63,000	60,000	48,000
400,000	80,000	91,000	56,000
500,000	89,000	93,000	79,000
600,000	95,000	94,000	102,000

TABLE 8 Combinations of Plans A and B with Total Investment of $600,000 (multiply all values in table by 1000)

Individual investment, $		Annual dividend, $
Plan A	Plan B	
600	0	95 + 0 = 95
500	100	89 + 10 = 99
400	200	80 + 32 = 112
300	300	63 + 60 = 123
200	400	44 + 91 = 135
100	500	25 + 93 = 118
0	600	0 + 94 = 94

and the investors receive annual dividends and recover their capital when the venture terminates at the expiration of 5 years. The annual dividends corresponding to the various levels of investment are shown in Table 7. The investments can be combined in any manner whatever. Devise the most profitable composite investment.

Calculation Procedure:

1. Identify the most profitable combination of plans A and B and determine the corresponding annual dividend if $600,000 is placed in this combination

The problem of identifying the most profitable combination of all three plans can be solved by dynamic programming. By this technique, the most profitable combination will be formed in stages, starting with combinations of A and B, identifying the most profitable ones, and then expanding these to include C.

The possible combinations of plans A and B corresponding to a total investment of $600,000 are shown in Table 8, and their corresponding annual dividends are computed by applying the values given in Table 7. The most profitable combination for the stipulated total investment is one in which $200,000 is placed in plan A and $400,000 is placed in plan B. The corresponding annual dividend is $135,000.

2. Identify the most profitable combination of plans A and B and determine the corresponding annual dividend if any amount is placed in this combination

The amount that can be placed in this combination is a multiple of $100,000 with an upper limit of $600,000. Repeat the procedure in step 1 to obtain the results shown in Table 9. The combi-

TABLE 9 Optimal Combinations of Plans A and B (multiply all values in table by 1000)

Total investment, $	Individual investment, $		Annual dividend, $
	Plan A	Plan B	
600	200	400	44 + 91 = 135
500	100	400	25 + 91 = 116
400	0	400	0 + 91 = 91
300	300	0	63 + 0 = 63
200	200	0	44 + 0 = 44
100	100	0	25 + 0 = 25

nations listed in this table are candidates for the most profitable combination of all three plans, and the other combinations of A and B are now discarded.

3. Compute the maximum dividend that can be obtained from a combination of all three plans

The calculations are shown in Table 10. As an illustration, assume that $400,000 will be placed in the A-B combination, leaving $600,000 − $400,000 = $200,000 for plan C. From Table 9, the

TABLE 10 Combinations of Plans A, B, and C (multiply all values in table by 1000)

Individual investment, $		
Combination of plans A and B	Plan C	Annual dividend, $
600	0	135+ 0 = 135
500	100	116+ 15 = 131
400	200	91+ 31 = 122
300	300	63+ 48 = 111
200	400	44+ 56 = 100
100	500	25+ 79 = 104
0	600	0+ 102 = 102

annual dividend from the A-B combination is $91,000; from Table 7, the dividend from plan C is $31,000. Thus, the total dividend is $91,000 + $31,000 = $122,000. Table 10 shows that the maximum possible dividend is $135,000.

4. Identify the most profitable combination of all three plans

From Table 10, the most profitable combination results from placing $600,000 in the A-B combination and nothing in plan C. From Table 9, the A-B combination consists in placing $200,000 in A and $400,000 in B. Thus, the most profitable way of dividing the capital is $200,000 in plan A, $400,000 in plan B, and nothing in plan C.

Related Calculations: This problem can be solved directly by forming all possible combinations of plans A, B, and C and computing their annual dividends; the number of combinations is 28. However, where the number of combinations is very large, the direct method becomes unwieldy. Dynamic programming provides a systematic way of solving the problem, and it reduces the number of calculations. It will be applied again in a later calculation procedure.

ECONOMIC LEVEL OF INVESTMENT

A firm planned to purchase and improve property in the expectation that land values in the area would appreciate in the near future. The question arose as to how large an investment should be made. The following data were compiled for five alternative plans, each representing a different level of investment.

	Plan				
	A	B	C	D	E
Investment, $	200,000	270,000	340,000	410,000	460,000
Rate of return, %	14.1	13.8	12.5	11.6	12.3

If 12 percent is considered the minimum acceptable rate of return, determine the most attractive plan.

Calculation Procedure:

1. Establish a basis of comparison for the investments

To establish a basis of comparison, assume that each investment pays an annual dividend and that the invested capital remains intact until the venture terminates. (Although these assumptions are not realistic, they are entirely valid for comparative purposes.)

2. Calculate the annual dividend under each plan

Thus, for plan A, the annual dividend = $200,000(0.141) = $28,200. Compute the dividends for the other plans in the same manner.

3. Construct a dividend-investment diagram

Use the same procedure as for Fig. 7. Plot the points representing the sets of values associated with the five plans.

Draw a line through the origin of the dividend-investment diagram with a slope of 12 percent. Determine which point is most distant from this line. This point corresponds to the most profitable rate of return. Study of the plot shows that plan B should be adopted.

Related Calculations: To compare these five plans algebraically, assume that the firm has a total available capital of $460,000 (the investment required under plan E) and that the amount remaining after investment in one of the five plans will be allocated to another investment yielding an annual dividend of 12 percent.

Next, calculate the total annual dividend corresponding to each plan.

Plan	Dividend, $
A	200,000(0.141) + 260,000(0.12)= 59,400
B	270,000(0.138) + 190,000(0.12)= 60,060
C	340,000(0.125) + 120,000(0.12)= 56,900
D	410,000(0.116) + 50,000(0.12)= 53,560
E	460,000(0.123) = 56,580

Since plan B yields the highest total dividend, it is the best choice.

RELATIONSHIP BETWEEN BEFORE-TAX AND AFTER-TAX INVESTMENT RATES

A corporation is investigating a proposed investment under which it will receive annual dividends and recover its capital when the venture terminates in 12 years. Since the venture is highly speculative, the firm wishes to earn a minimum of 15 percent on its capital as calculated after the payment of taxes. If income from the investment will be taxed at 56 percent, what must be the minimum rate of return as calculated before the payment of taxes?

Calculation Procedure:

Compute the minimum acceptable before-tax investment rate

Use the relation $i_b = i_a/(1 - t)$, where i_b and i_a = before-tax and after-tax investment rates, respectively; t = tax rate. Thus, $i_b = 0.15/0.44 = 34.1$ percent.

APPARENT RATES OF RETURN ON A CONTINUING INVESTMENT

A firm leasing construction equipment purchased an asset for $24,000, charging depreciation on a straight-line basis. The life used was 4 years; salvage value, zero. The asset was used for 6 years and scrapped for $800. Net revenues obtained from this asset are listed in Table 11. The firm's normal income was taxed at 50 percent, but the proceeds from the salvage sale were taxed at 25 percent. What were the apparent rates of return on this asset investment, after taxes, computed during the life of the asset?

TABLE 11   Determination of Apparent Rates of Return

Year	Net revenue, $	Depreciation charge, $	Net profit before tax, $	Net profit after tax, $	Book value beginning of year, $	Apparent rate of return, %
1	10,000	6,000	4,000	2,000	24,000	8.3
2	9,600	6,000	3,600	1,800	18,000	10.0
3	8,000	6,000	2,000	1,000	12,000	8.3
4	6,400	6,000	400	200	6,000	3.3
5	4,400	. . .	4,400	2,200	. . .	Infinite
6	2,400	. . .	2,400	1,200	. . .	Infinite
	800°	. . .	800	600		

°Income from sale of asset.

Calculation Procedure:

1.   *Compute the annual depreciation charge*

Using the straight-line method and a 4-year life, we get the annual depreciation = $24,000/4 = $6000, assuming zero salvage. Record the depreciation charge in the third column of Table 11.

2.   *Compute the net profit before taxes*

Deduct from the annual net revenue the annual depreciation charge, Table 11, and enter the result in column 4. Thus, for the first year with a revenue of $10,000, the net income before taxes = $10,000 − $6000 = $4000.

3.   *Compute the after-tax profit*

With a tax of 50 percent of the profit before taxes, multiply the value in Table 11, column 4, by 0.50 to determine the after-tax profit. Thus, for year 1, the after-tax profit = $4000(0.50) = $2000.

4.   *Record the asset book value at the beginning of the year*

In this type of calculation, the book value = the unrecovered capital investment for that year. Or, for year 1, the book value = $24,000. For year 2, the book value = $24,000 − $6000 = $18,000. In this relation, $6000 is the depreciation during year 1.

5.   *Compute the apparent rate of return*

Divide the after-tax profit for any year by the book value of the asset at the beginning of the year to determine the apparent rate of return. Or, for year 2, apparent rate of return = $1800/$18,000 = 0.10, or 10.0 percent.

TRUE RATE OF RETURN ON A COMPLETED INVESTMENT

Refer to the previous calculation procedure. What was the actual after-tax rate of return yielded by this investment, computed at the conclusion of the venture?

Calculation Procedure:

1.   *Determine the after-tax income*

In this situation, only the actual disbursements and receipts, as well as their timing, are pertinent. The depreciation charges, which arise from bookkeeping entries, are irrelevant.

   To determine the after-tax income, deduct the tax payment from the net revenue listed in Table 11 to obtain the after-tax income. List this income in Table 12.

2.   *Determine the present worth of the annual receipts*

Compute the present worth of each year's after-tax income for years 1 through 6, and take the sum. To perform this computation, assume an interest rate that is believed to approximate the actual rate of return on the $24,000 investment.

TABLE 12 Determination of True Rate of Return

Year	Net revenue, $	Tax payment, $	After-tax income, $
1	10,000	2,000	8,000
2	9,600	1,800	7,800
3	8,000	1,000	7,000
4	6,400	200	6,200
5	4,400	2,200	2,200
6	3,200	1,400	1,800

At 12 percent, present worth of the after-tax income = $24,444. At 15 percent, present worth of the after-tax income = $22,874. By linear interpolation for a present worth of $24,000, i = rate of return = 12.8 percent.

AVERAGE RATE OF RETURN ON COMPOSITE INVESTMENT

Suppose that the income in the previous calculation procedure were reinvested at 8 percent after taxes, until the end of the fourth year. Thereafter, the income received was reinvested at 10 percent. What was the average rate of return on the $24,000 capital during the 6-year period?

Calculation Procedure:

1. Compute the value of the original capital at the end of the sixth year

Thus, by using the SPCA factor for the after-tax income listed for each year in Table 12 for i = 8 or 10 percent, the value of the original capital at the end of the sixth year = $8000(1.469) + $7800(1.360) + $7000(1.260) + $6200(1.166) + $2200(1.100) + $1800 = $42,629.

2. Compute the average investment rate

Let i' = average investment rate. Equate the original investment to $42,629 at a date 6 years in the future, and solve for i'. Thus, $24,000(SPCA for i') = $42,629; i' = 10.1 percent.

RATE OF RETURN ON A SPECULATIVE INVESTMENT

A firm purchased a parcel of land for $25,000 and spent $600 during the first year to improve the property. (This investment for improvements should be considered a lump-sum end-of-year payment.) The expenses for real estate tax, insurance, and maintenance totaled $1200 per year. At the end of 5 years, the firm sold the property at a price that yielded $48,700 after payment of legal fees and commissions. In computing the federal income tax, the firm deducted the ordinary expenses of holding this property from the income derived from other sources. This income was subject to a 53 percent tax rate. The profit on the sale of the land was taxed at the 25 percent capital-gains rate. What was the rate of return on the investment?

Calculation Procedure:

1. Determine the effective annual payment

The expenses related to possession of the land served to reduce the income tax payments. Therefore, the *effective* cost of holding the property (or any similar asset) was less than the actual expenses. To obtain the effective annual payment, deduct the annual income tax saving from the annual payment related to the asset. Thus, effective annual payment = $1200(1.00 − 0.53) = $564.

2. Compute the net proceeds from the sale of the asset

Deduct the capital-gains tax from the selling price of the asset to obtain the net proceeds. This is often called the *effective selling price*. Thus, capital gains = $48,700 − ($25,000 + $600) = $23,100. The capital-gains tax = $23,100(0.25) = $5775. Hence, net proceeds = $48,700 − $5775 = $42,925.

3. Set up an equation for the rate of return

Selecting the date at which the asset was sold as the reference date, express the value of every sum of money, and equate the total effective payments to the income. Thus, \$25,000(SPCA for n = 5 years, i = ?) + \$600(SPCA for n = 4 years, i = ?) + \$564(USCA for n = 5 years, i = ?) = \$42,925.

4. Solve the rate-of-return equation, using trial values

As a trial, set i = 10 percent, and evaluate the left-hand side of the relation in step 3. Thus, \$25,000(1.611) + \$600(1.464) + \$564(6.105) = \$44,597.

Since the actual income, \$42,925, was less than \$44,597, the assumed rate of return is too high. Try 8 percent. Then \$25,000(1.469) + \$600(1.360) + \$564(5.867) = \$40,850. This is less than the actual income. Interpolating linearly between the two trial values yields i = 9.1 percent.

Related Calculations: As a general guide for selecting trial rate-of-return values, choose a higher value and a lower value around the estimated true rate of return. Check the result by computing the dollar return. Interpolate linearly when higher and lower dollar returns are obtained.

INVESTMENT AT AN INTERMEDIATE DATE (AMBIGUOUS CASE)

A firm purchased an oil-producing property under terms which did not require an immediate payment to the seller but which did require payment of royalties on income from sale of the oil.

TABLE 13 Income from an Asset

Year	Net income, $
1	600,000
2	300,000
3	100,000
4	1,220,000
5	500,000
6	200,000

TABLE 14 Trial Calculations for Rate-of-Return Equation

Interest rate, %	Value of polynomial, $
8	10,600
10	-1,400
15	-21,300
20	-26,400
25	-19,400
30	-700
40	65,400

By the end of the third year, the primary reserves were nearly exhausted, and the firm spent \$2,830,000 on a water-injection program to extend the oil yield. Operations were continued until the end of the sixth year. The income from the venture is listed in Table 13. Compute the rate of return on this investment. Is more than one solution obtained? How may the ambiguity inherent in this type of investment be resolved?

Calculation Procedure:

1. Set up an equation for the rate of return

Selecting the end of the third year as the reference date, express the value of every sum of money. Consider receipts to be positive and expenditures to be negative. Then \$600,000(SPCA for n = 2, i = ?) + \$300,000(SPCA for n = 1, i = ?) + \$100,000 + \$1,220,000(SPPW for n = 1, i = ?) + \$500,000(SPPW for n = 2, i = ?) + \$200,000(SPPW for n = 3, i = ?) − \$2,830,000 = 0.

2. Solve the rate-of-return equation, using trial values

Assign a series of trial values for i in the equation in step 1. Record the results in Table 14. Then, by linear interpolation, i = 9.8 percent, or i = 30.1 percent.

3. Evaluate the rates of return obtained

The polynomial in step 1 resembles a quadratic polynomial since it contains either two real roots or none. That there are two values of i which satisfy this equation is explained as follows.

First, consider that $i = 9.8$ percent, causing the polynomial to assume the value of zero. Then replace 9.8 percent with the higher rate of 30.1 percent. Second, when this substitution is made, the value of the income received prior to the end of the third year is increased by a certain amount. The value of the income received after that date is decreased by the same amount. Hence, the value of the polynomial remains zero.

4. *Make a realistic appraisal of the investment*

A realistic appraisal of an investment of this type requires consideration of the reinvestment rate earned by either the entire income or that part of the income received prior to the expenditure. In the present instance, assume that the income received up to the end of the third year was reinvested at 8 percent. Its value at the date of the expenditure for water injection is, by the equation from step 1, $600,000(1.166) + $300,000(1.080) + $100,000 = $1,123,600.

Then the effective investment = $2,830,000 − $1,123,600 = $1,706,400. To determine the rate of return, set the effective investment $1,706,400 = $1,220,000(SPPW for $n = 1$) + $500,000(SPPW for $n = 2$) + $200,000(SPPW for $n = 3$) and solve for i. The result of this solution is $i = 8.5$ percent.

Related Calculations: This procedure illustrates the fact that in financial analyses it is not possible to place exclusive reliance on mathematical results. However rigorous the mathematical solution may appear to be, the results must be interpreted in a practical manner. Note that this procedure may be used for any type of asset.

PAYBACK PERIOD OF AN INVESTMENT

A firm has a choice of two alternative investment plans, A and B. Each plan requires an immediate expenditure of $2,000,000, lasts 10 years, and yields an income at the end of each year. Under plan A, the annual income is expected to be $450,000 for the first 5 years and $33,000 for the remaining 5 years. Under plan B, the annual income is expected to be $150,000 for the first 3 years, $250,000 for the next 3 years, and $650,000 for the last 4 years. If the decision is to be based on a short payback period, which investment plan should the firm adopt?

Calculation Procedure:

1. *Compute the payback period under plan A*

For various reasons, a firm often prefers an investment that allows it to recover its capital quickly. The speed with which capital is recovered is measured by the *payback period*, defined thus: Assume that all income accruing from the investment initially represents recovered capital, and all income accruing after capital has been fully recovered represents interest. The time required for completion of capital recovery is called the payback period.

Under plan A, the first four payments total $1,800,000, and the first five payments total $2,250,000. Thus, the fifth payment completes capital recovery, and the payback period is 5 years.

2. *Compute the payback period under plan B*

The first seven payments total $1,850,000, and the first eight payments total $2,500,000. Thus, the payback period is 8 years.

3. *Select the investment plan*

The firm should adopt plan A because it has the lower payback period.

Related Calculations: The investment rate is 6.0 percent for plan A and 10.0 percent for plan B. However, since the income under plan B is largely deferred, use of the payback period as a criterion in investment appraisal places plan B at a disadvantage.

PAYBACK PERIOD TO YIELD A GIVEN INVESTMENT RATE

An asset has a first cost of $40,000 and maximum life of 10 years. Its resale value will be $3500 at the end of the first year, and then it will diminish by $500 per year, becoming zero at the end of the eighth year. The end-of-year income that accrues from use of this asset will be $12,000 for the first 2 years, $8000 for the next 3 years, $6000 for the next 3 years, and $2000 for the last 2 years. Determine how long this asset must be held to secure a 10 percent return on the investment.

Calculation Procedure:

1. *Establish the criterion for finding the life of the asset*

As the preceding calculation procedure showed, a firm that considers solely how long it takes an investment to restore the sum invested is apt to undertake investments of relatively low yield. A more logical approach is to consider how long it takes an investment to restore the sum invested and yield a certain minimum rate of return.

Let N = life of asset, years; i and i' = required investment rate and true investment rate, respectively; V_{exp} and V_{inc} = value of expenditures and value of income, respectively, where all sums of money are evaluated at a specific date and by using i as the interest rate. If $V_{exp} = V_{inc}$, then $i' = i$. Thus, the problem is to find the value of N at which this equality of expenditures and income becomes a fact.

2. *Perform the calculations*

Evaluate all sums of money at the date of purchase, using an interest rate of 10 percent. Set N = 1. Then V_{exp} = \$40,000 − \$3500(SPPW, n = 1) = \$40,000 − \$3500(0.90909) = \$36,818. Also, V_{inc} = \$12,000(0.90909) = \$10,909. Since $V_{exp} > V_{inc}$, $i' <$ 10 percent.

Now set N = 2. Then V_{exp} = \$40,000 − \$3000(SPPW, n = 2) = \$40,000 − \$3000(0.82645) = \$37,521. Applying the previous result and adding the income at the end of the second year, we find V_{inc} = \$10,909 + \$12,000(0.82645) = \$20,826. Thus, $i' <$ 10 percent.

Set N = 3. Then V_{exp} = \$40,000 − \$2500(0.75131) = \$38,122. Also, V_{inc} = \$20,826 + \$8000(0.75131) = \$26,836.

Continue these calculations to obtain the results in Table 15. This table shows that $i' <$ 10 percent when N = 5 and $i' >$ 10 percent when N = 6. Thus, the asset must be held 6 years to secure a 10 percent rate of return.

TABLE 15   Value of Expenditures and Income

Life of asset, years	Value of expenditures, \$	Value of income, \$
1	36,818	10,909
2	37,521	20,826
3	38,122	26,836
4	38,634	32,300
5	39,069	37,267
6	39,435	40,654

BENEFIT-COST ANALYSIS

A proposed flood-control dam is expected to have an initial cost of \$5,000,000 and to require annual maintenance of \$24,000. It will also require major repairs and reconstruction costing \$120,000 at the end of every 10-year period. The life of the dam may be assumed to be infinite. The reduction in losses due to flood damage is estimated to be \$300,000 per year. However, there will be an immediate loss of \$100,000 in the value of the property surrounding the dam, and this loss will be borne by the public. Applying an interest rate of 6 percent, determine whether the proposed dam is feasible.

Calculation Procedure:

1. *Compute the present worth of costs*

With reference to a federal project, any income or reduction in loss that accrues to the public is called a *benefit*, any loss that accrues to the public is called a *disbenefit*, and the difference between the benefits and disbenefits is called the *net benefit*. The ratio of net benefit to costs is called the *benefit-cost (B/C) ratio*. If this ratio exceeds 1, the project is considered desirable.

A uniform-payment series that continues indefinitely is called a *perpetuity*. The value of a perpetuity at its origin date (i.e., one payment period before the first payment) is $R/[(1 + i)^m - 1]$, where R = periodic payment, i = interest rate, and m = number of interest periods in one payment period.

The present worths are: First cost, PW = \$5,000,000; annual maintenance, PW = \$24,000/0.06 = \$400,000; repairs and reconstruction, PW = \$120,000/[(1.06)^{10} − 1] = \$151,700. Then PW of costs = \$5,000,000 + \$400,000 + \$151,700 = \$5,551,700.

2. *Compute the present worth of net benefit*

The present worths are: Savings, PW = \$300,000/0.06 = \$5,000,000; devaluation of property, PW = \$100,000. Then PW of net benefit = \$5,000,000 − \$100,000 = \$4,900,000.

3. *Determine whether the dam is feasible*

Since B/C ratio = \$4,900,000/\$5,551,700 < 1, the dam is not feasible.

Analysis of Business Operations

LINEAR PROGRAMMING TO MAXIMIZE INCOME FROM JOINT PRODUCTS

A firm manufactures two articles, A and B. The unit cost of production, exclusive of fixed costs, is \$10 for A and \$7 for B. The unit selling price is \$16 for A and \$13.50 for B. The estimated maximum monthly sales potential of A is 9000 units; of B, 7000 units. It is the policy of the firm to produce only as many units as can readily be sold. If production is restricted to one article, the factory can turn out 13,000 units of A or 8500 units of B per month. The capital allotted to monthly production after payment of fixed costs is \$100,000. What monthly production of each article will yield the maximum profit?

Calculation Procedure:

1. *Express the production constraints imposed by sales and capital*

Let N_A and N_B denote the number of articles A and B, respectively, produced monthly. Then potential sales: $N_A \le 9000$, Eq. *a*; $N_B \le 7000$, Eq. *b*. Available capital: $10N_A + 7N_B \le \$100,000$, Eq. *c*.

2. *Determine the production constraint imposed by the plant capacity*

The number of months required to produce N_A units of A is $N_A/13,000$. Likewise, to produce N_B units of B would be $N_B/8500$. Then, $N_A/13,000 + N_B/8500 \le 1$, or $8.5N_A + 13N_B \le 110,500$, Eq. *d*.

3. *Express the monthly profit in equation form*

Before fixed costs are deducted, the profit $P = (16 - 10)N_A + (13.5 - 7)N_B$, or $P = 6N_A + 6.5N_B$, Eq. *e*.

4. *Construct a monthly production chart*

Considering the expressions *a* to *d* above to be equalities, plot the straight lines representing them (Fig. 8).

Since these expressions actually establish upper limits to the values of N_A and N_B, the point representing the joint production of articles A and B must lie either within the shaded area, which is termed the *feasible region*, or on one of its boundary lines.

5. *Plot an equal-profit line*

Assign the arbitrary value of \$30,000 to P, and plot the straight line corresponding to Eq. *e* above. Every point on this line (Fig. 8) represents a set of values for N_A and N_B for which the profit is \$30,000. This line is therefore termed an *equal-profit line*.

Next, consider that P assumes successively greater values. As P does so, the equal-profit line moves away from the origin while remaining parallel to its initial position.

6. *Maximize the profit potential*

To maximize the profit, locate the point in Fig. 8 at which the equal-profit line, in its outward displacement, is on the verge of leaving the feasible region. This is point Q, which lies at the intersection of the lines representing the equalities *c* and *d*.

7. *Determine the number of units for maximum profit*

Establish the coordinates of the maximum-profit point Q either by reading them from the chart, Fig. 8, or by solving the equalities *c* and *d* simultaneously. The results are $N_A = 7468$ units; $N_B = 3617$ units.

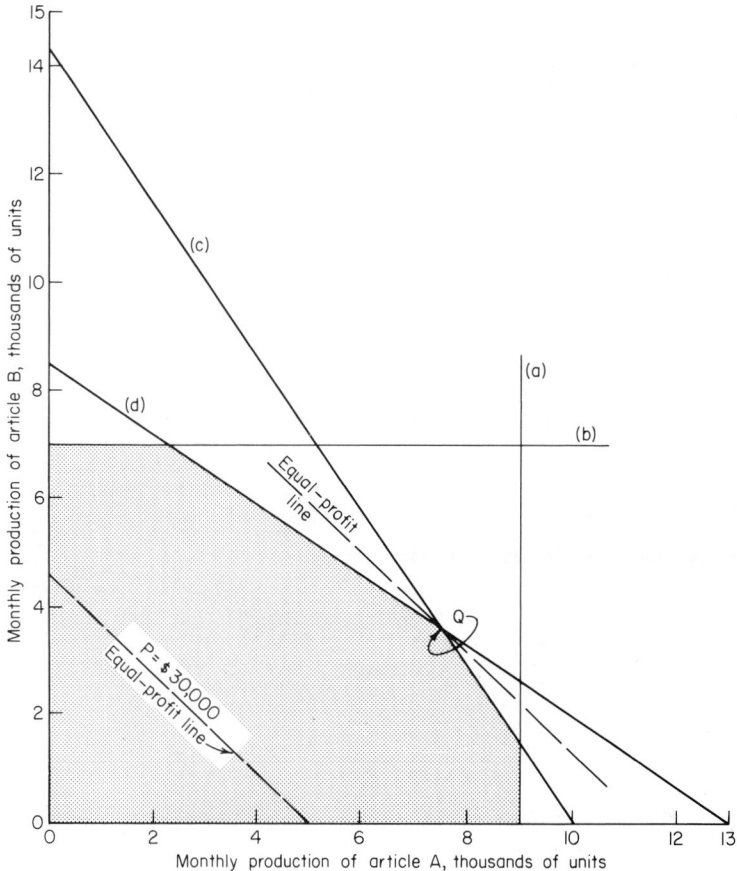

FIG. 8 Linear programming solution.

Related Calculations: Note that this method can be used for any type of product manufactured by any means.

ALLOCATION OF PRODUCTION AMONG MULTIPLE FACILITIES WITH NONLINEAR COSTS

A firm must produce 700 units per month of a commodity, and three machines, A, B, and C, are available for this purpose. Production costs are $C_A = 0.28N_A^{1.42} + 400$; $C_B = 0.36N_B^{1.47} + 500$; $C_C = 0.30N_C^{1.53} + 420$, where N = number of units produced monthly; C = monthly cost of production, \$; and the subscript refers to the machine. Find the most economical manner of allocating production among the three machines.

Calculation Procedure:

1. *Write the equations of incremental costs*

The objective is to minimize the total cost of production. Since costs vary nonlinearly, this situation does not lend itself to linear programming.

Assume that N units have been produced on a given machine. The *incremental* (or *marginal*) *cost* at that point is the cost of producing the $(N + 1)$th unit. Let I = incremental cost, $. If N is large, $I \simeq dC/dN$. By differentiating the foregoing expressions, this approximation gives $I_A = 0.3976N_A^{0.42}$, $I_B = 0.5292N_B^{0.47}$, $I_C = 0.4590N_C^{0.53}$, where the subscript refers to the machine. Also, if N is large, cost of producing Nth unit $\simeq$ cost of producing $(N + 1)$th unit.

2.   *Establish the condition at which the total cost of production is minimum*

Arbitrarily set $N_A = 150$, $N_B = 250$, $N_C = 300$, which gives a total of 700 units. The incremental costs are $I_A = \$3.2614$, $I_B = \$7.0901$, and $I_C = \$9.4338$. Also, when $N_A = 151$, then $I_A = \$3.2705$. The total cost of production can be reduced by shifting 1 unit from machine B to machine A and 1 unit from machine C to machine A, with the reduction being approximately $\$7.0901 + \$9.4338 - (\$3.2614 + 3.2705) = \$9.9920$. Thus, the arbitrary set of N values given above does not yield the minimum total cost.

Clearly the total cost of production is minimum when all three incremental costs are equal (or as equal as possible, since N is restricted to integral values).

3.   *Find the most economical allocation of production*

At minimum total cost, $I_A = I_B = I_C$, or $0.3976N_A^{0.42} = 0.5292N_B^{0.47} = 0.4590N_C^{0.53}$, Eq. *a*; and $N_A + N_B + N_C = 700$, Eq. *b*. By a trial-and-error solution, $N_A = 468$, $N_B = 132$, and $N_C = 100$.

Alternatively, proceed as follows: From Eq. *a*, $N_B = 0.5442N_A^{0.8936}$ and $N_C = 0.7627N_A^{0.7925}$. Substituting in Eq. *b* gives $N_A + 0.5442N_A^{0.8936} + 0.7627N_A^{0.7925} = 700$. Assign trial values to N_A until this equation is satisfied. The solution is $N_A = 468$, and the remaining values follow.

4.   *Devise a semigraphical method of solution*

In Fig. 9, plot the incremental-cost curves. Pass an arbitrary horizontal line L through these curves to obtain a set of N values at which $I_A = I_B = I_C$. Scale the N values, and find their sum. Now displace the horizontal line until the sum of the N values is 700.

Related Calculations: Allocation problems of this type usually are solved by applying *Lagrange multipliers*. However, as the previous solution demonstrates, the use of simple economic logic can circumvent the need for abstract mathematical concepts.

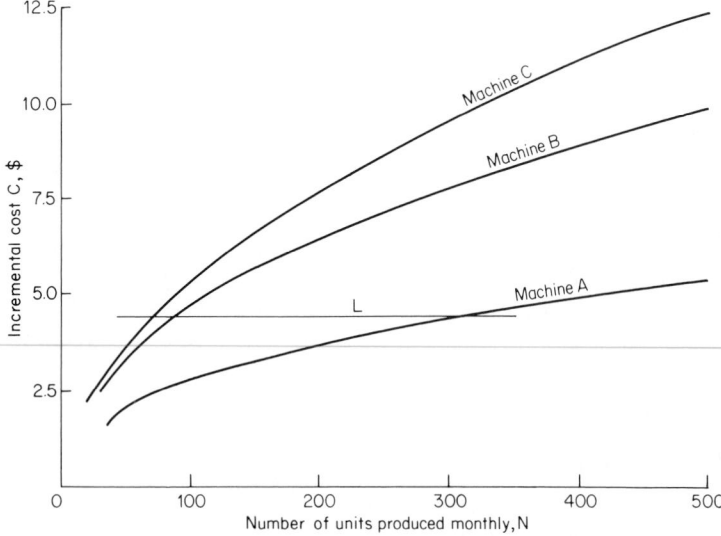

FIG. 9   Incremental-cost curves.

OPTIMAL PRODUCT MIX WITH NONLINEAR PROFITS

A firm manufactures three articles, A, B, and C, and it can sell as many units as it can produce. The monthly profits, exclusive of fixed costs, are $P_A - 4.75N_A - 0.0050N_A^2$, $P_B = 2.60N_A - 0.0014N_B^2$, and $P_C = 2.25N_C - 0.0010N_C^2$, where N = number of units produced monthly, P = monthly profit, and the subscript refers to the article. If production is restricted to one article, the firm can produce 1000 units of A, 1500 units of B, and 1800 units of C per month. What monthly production of each article will yield the maximum profit?

Calculation Procedure:

1.   *Express the constraint imposed on production*

Let T = number of months required to produce N units of an article, with a subscript to identify the article. Then $T_A = N_A/1000$; $T_B = N_B/1500$; and $T_C = N_C/1800$. Since 1 month is available, $N_A/1000 + N_B/1500 + N_C/1800 = 1$, or $1.8N_A + 1.2N_B + N_C = 1800$, Eq. *a*.

2.   *Determine how the values of N can vary*

Assume for simplicity that N_A is restricted to integral values but N_B and N_C can assume nonintegral values. Equation *a* reveals that if N_A increases by 1 unit, N_B must decrease by $1.8/1.2 = 1.5$ units, or N_C must decrease by 1.8 units. Expressed formally, the partial derivatives are $\partial N_B/\partial N_A = -1.5$ and $\partial N_C/\partial N_A = -1.8$.

3.   *Write the equations of incremental profits*

If N units of an article have been produced, the incremental profit at that point is the profit that accrues from producing the $(N + 1)$th unit. Let I = incremental profit. If N is large, $I \simeq dP/dN$. By differentiating the foregoing expressions, the incremental profits are $I_A = 4.75 - 0.0100N_A$, $I_B = 2.60 - 0.0028N_B$, and $I_C = 2.25 - 0.0020N_C$, where the subscript refers to the article. Also, if N is large, the profit from the Nth unit $\simeq$ profit from the $(N + 1)$th unit.

4.   *Establish the condition at which the total profit is maximum*

Arbitrarily set $N_A = 300$, $N_B = 400$, $N_C = 780$, satisfying Eq. *a*. The incremental profits are $I_A = \$1.750$; $I_B = \$1.480$ and $1.5I_B = \$2.220$; $I_C = \$0.690$ and $1.8I_C = \$1.242$. Also, when $N_B = 401.5$, then $1.5I_B = \$2.214$. The total profit can be increased by reducing N_A by 1 unit, reducing N_C by 1.8 units, and increasing N_B by $2(1.5) = 3$ units, with the increase in profit being approximately $\$2.220 + \$2.214 - (\$1.750 + \$1.242) = \$1.442$. Thus, the arbitrary set of N values given above does not yield the maximum profit.

So the total profit is maximum when $I_A = 1.5I_B = 1.8I_C$, or $4.75 - 0.0100N_A = 3.90 - 0.0042N_B = 4.05 - 0.0036N_C$, Eq. *b*.

5.   *Find the production that will maximize profit*

Applying Eq. *b*, express N_B and N_C in terms of N_A. Substitute these expressions in Eq. *a*, and solve the resulting equation for N_A. Then calculate N_B and N_C. The results are $N_A = 301$, $N_B = 514$, and $N_C = 641$.

6.   *Devise a semigraphical method of solution*

In Fig. 10, plot the straight lines that represent I_A, $1.5I_B$, and $1.8I_C$. Pass an arbitrary horizontal line L through these lines to obtain a set of N values at which $I_A = 1.5I_B = 1.8I_C$. Scale the N values, and determine whether they satisfy Eq. *a*. Now displace the horizontal line until the N values do satisfy this equation.

DYNAMIC PROGRAMMING TO MINIMIZE COST OF TRANSPORTATION

A firm must ship merchandise by truck from town A to town E, and the trip will last 4 days. The driver will stop in district B the first night, district C the second night, and district D the third night. The number of towns in each district is: district B, three; district C, two; district D, three. The driver can stay overnight in any of these towns, and the cost of lodging is the same in all. The relative cost of traveling from one town to another is recorded in Fig. 11. Design the most economical route.

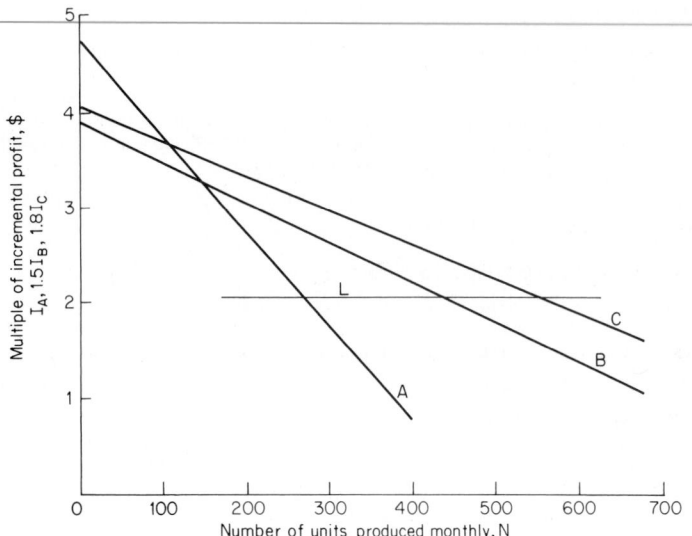

FIG. 10 Plotting of indicated multiples of incremental profit.

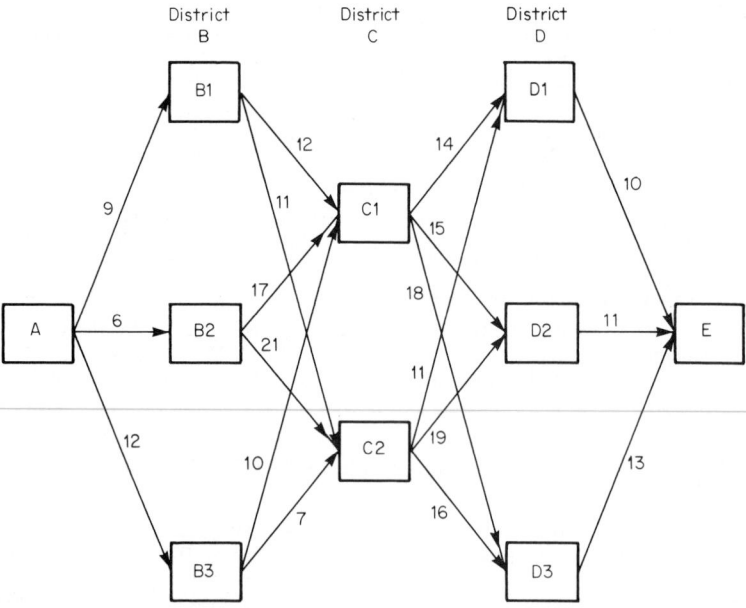

FIG. 11 Relative cost of transportation between successive stops.

12.52

Calculation Procedure:

1. Determine the minimum cost of transportation from towns C1 and C2 to town E

The design will be executed by dynamic programming, which was applied in an earlier calculation procedure to identify the most lucrative combination of investments. The most economical route from A to E will be constructed in stages, in reverse order.

The cost of transportation from C1 to E is as follows: for C1-D1-E, 14 + 10 = 24; for C1-D2-E, 15 + 11 = 26; for C1-D3-E, 18 + 13 = 31. Thus, the minimum cost is 24. The cost of transportation from C2 to E is: for C2-D1-E, 11 + 10 = 21; for C2-D2-E, 19 + 11 = 30; for C2-D3-E, 16 + 13 = 29. Thus, the minimum cost is 21. Record these minimum costs and their corresponding towns in district D in Fig. 12; they are the only costs that are relevant from now on.

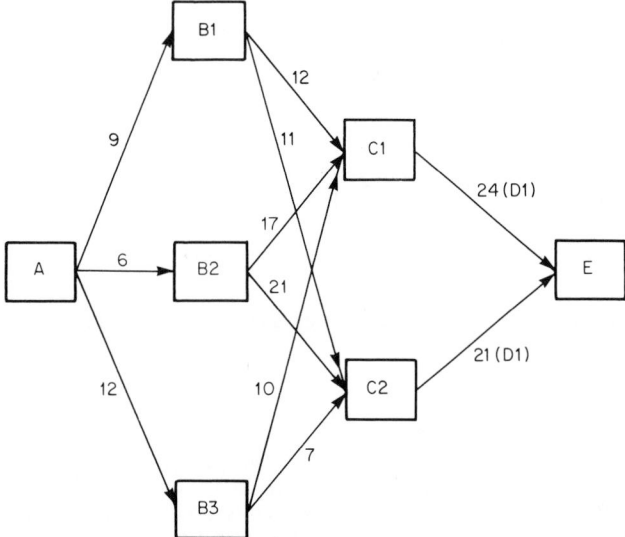

FIG. 12 First stage in finding most economical route.

2. Determine the minimum cost of transportation from towns B1, B2, and B3 to town E

Refer to Fig. 12. The cost of transportation from B1 to E is: for B1-C1-E, 12 + 24 = 36; for B1-C2-E, 11 + 21 = 32. Thus, the minimum cost is 32. The cost of transportation from B2 to E is: for B2-C1-E, 17 + 24 = 41; for B2-C2-E, 21 + 21 = 42. Thus, the minimum cost is 41. The cost of transportation from B3 to E is: for B3-C1-E, 10 + 24 = 34; for B3-C2-E, 7 + 21 = 28. Thus, the minimum cost is 28. Record these minimum costs and their corresponding towns in district C in Fig. 13.

3. Determine the minimum cost of transportation from town A to town E

Refer to Fig. 13. The cost of transportation from A to E is: for A-B1-E, 9 + 32 = 41; for A-B2-E, 6 + 41 = 47; for A-B3-E, 12 + 28 = 40. Thus, the minimum cost is 40, and the corresponding town in district B is B3.

4. Identify the most economical route

Refer to step 3 and Figs. 13 and 12, in that order. The most economical route is A-B3-C2-D1-E. From Fig. 11, the cost of transportation corresponding to this route is 12 + 7 + 11 + 10 = 40, which agrees with the result in step 3.

Related Calculations: The number of alternative routes from town A to town E is $3 \times 2 \times 3 = 18$. Therefore, the most economical route can be found by listing all 18 routes and computing

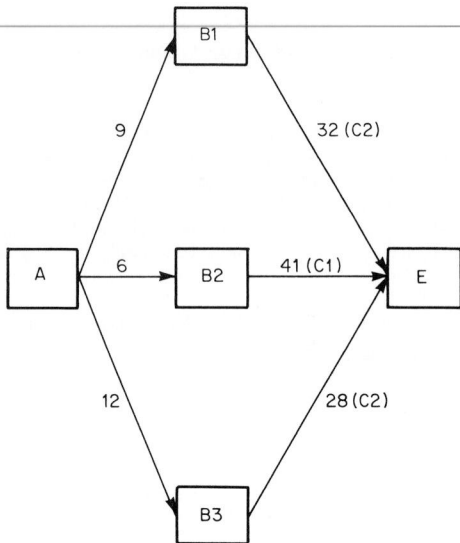

FIG. 13   Second stage in finding most economical route.

their respective costs. However, the solution by dynamic programming given above simplifies the work.

OPTIMAL INVENTORY LEVEL

A firm is under contract to supply 41,600 parts per year and plans to produce them in equal lots spaced at equal intervals. The production capacity is 800 parts per day. Setup and teardown cost for the production machines is $550 for each run. The cost of storage, insurance, and interest on the investment is $1.40 per part for each year the part is carried in inventory. The regular production cost, exclusive of setup and teardown, is $5 per part. A reserve stock of parts is not needed. Determine the most economical lot size and the corresponding cost of production.

Calculation Procedure:

1.   *Compute the parts delivery rate*

Assume that the parts are delivered to the buyer at a uniform rate, and compute the daily delivery rate. Since there are approximately 260 working days per year, the rate of delivery = 41,600 parts/260 days = 160 parts per day.

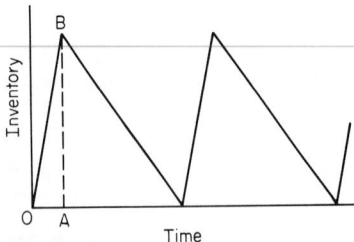

FIG. 14   Variation in inventory level.

2.   *Construct an inventory-time diagram*

Figure 14 shows such a diagram, starting with zero inventory.

3.   *Compute the peak inventory for a lot size of N*

The time OA required to produce 1 lot, Fig. 14, = lot size/maximum production rate = $N/800$, days. The slope of OB = rate of production − rate of delivery = 800 − 160 = 640 parts per day. Then $AB = (N/800)(640) = 0.8N$ parts.

4.   Compute the total annual cost in terms of N

The number of runs per year = 41,600/N. Also, the annual cost of setup and teardown = $550(41,600/$N$) = 22,880,000/$N$. Further, the average inventory = 0.5AB = 0.4N parts. By taking the product of the carrying cost and the average inventory, the annual cost of carrying the inventory = $1.40(0.4$N$) = 0.56$N$. With a $5 per unit regular cost, the annual regular cost = $5(41,600 units) = $208,000. Then the annual total cost C = (22,880,000/N) + 0.56N + 208,000.

5.   Find the economical lot size

To minimize C, set the derivative of C with respect to N equal to zero; solve for N to find the economical lot size. Thus, $dC/dN = -(22,880,000/N^2) + 0.56 = 0$; N = 6392 parts per lot.

6.   Compute the total annual cost

Substitute the number of parts from step 5 in the total annual-cost equation in step 4. Or, C = 3580 + 3580 + 208,000 = $215,160.

EFFECT OF QUANTITY DISCOUNT ON OPTIMAL INVENTORY LEVEL

Given the data from the previous calculation procedure, the firm finds that it can obtain quantity discounts if the parts are produced in lots of 7500 or more. These discounts reduce the regular production cost from $5 to $4.80 per part; this saving reduces the interest cost on inventory by $0.02 per part. Determine the most economical lot size under these conditions.

Calculation Procedure:

1.   Determine the number of parts for minimum cost

Assume that $N \geq 7500$. Proceeding as before, express C in terms of N, and set $dC/dN = 0$.

The annual cost of carrying the inventory = $1.38(0.4$N$) = 0.552$N$. Also, the annual regular cost = 41,600($4.80) = $199,680. Also, C = (22,880,000/N) + 0.552N + 199,680. And $dC/dN = -(22,880,000/N^2) + 0.552 = 0$. For minimum cost, N = 6438 parts. However, since discounts are not obtained until N reaches 7500, the last calculation lacks significance for this situation.

2.   Compute the economical lot size

Set N = 7500 and substitute in the cost equation above. Then C = 3051 + 4140 + 199,680 = $206,871. Since this result is less than the value of $215,160 computed in the previous calculation procedure corresponding to 6392 parts, the economical lot size is 7500 parts.

PROJECT PLANNING BY THE CRITICAL-PATH METHOD

Table 16 lists the activities performed in preparing a building site and installing the utilities. Assuming that the estimated durations are precise, determine the minimum time needed to complete the project. Identify the critical path. Upon completion of the project, it was found that each activity was undertaken at the earliest possible date and that its duration coincided with the estimate, except for the following: activity D was started 3 days late; activity H required 9 days instead of 6; activity I required 5 days instead of 4. Determine the duration of the project.

Calculation Procedure:

1.   Identify the predecessor(s) of each activity; tabulate results

The critical-path method (CPM) offers a systematic means of scheduling activities in a project and analyzing the consequences of departures from the schedule. The procedure consists of devising a logical concatenation of activities after ascertaining the relationships that exist among them. For example, activities D and E, Table 16, are independent of each other and therefore may be performed concurrently. But activities D and F are sequentially related—F cannot commence until D is finished. Thus, D is the immediate predecessor of F. Using these principles, list the related activities as shown in Table 17.

TABLE 16   Project Activities

Mark	Activity	Estimated duration, days
A	Clear site	4
B	Survey and lay out site	3
C	Rough grade	3
D	Excavate for sewer	8
E	Excavate for electrical manholes	1
F	Install sewer and backfill	4
G	Install electrical manholes	6
H	Install overhead pole line	6
I	Install electrical duct bank	4
J	Pull in power feeder	5
K	Construct foundations for water tank	3
L	Erect water tank	8
M	Install piping and valves for water tank	12
N	Drill well	14
O	Install well pump	2
P	Install underground water piping	9
Q	Connect all piping	2

TABLE 17   Related CPM Activities

Activity	Predecessor	Activity	Predecessor
B	A	J	H and I
C	B	K	C
D	C	L	K
E	C	M	L
F	D	N	C
G	E	O	N
H	C	P	O
I	F and G	Q	M and P

2. *Construct the network for the project, Fig. 15*

The network is a delineation of the sequence in which the activities are to be performed. Each activity is represented by a horizontal arrow, which may or may not be to scale. The arrow representing a given activity is placed to the right of its immediate predecessor activity. Where there are multiple predecessors or successors, broken arrows are used to transfer from one activity to another. The duration of each activity is recorded under its corresponding arrow.

Completion of an activity and the start of its successor constitute an *event.* Commencement of a given activity is termed its *i event;* completion of an activity is termed its *j event.* A number is assigned to each event and is recorded in the network in a circle between consecutive arrows. Each activity is identified by the events it separates. For instance, 6-7 designates *erect water tank.* A chain of activities extending from inception to completion of the project is termed a *path.* In this project there are the following five paths:

0-1-2-3-4-5-8-13

0-1-2-3-6-7-8-13

0-1-2-3-9-11-12-13

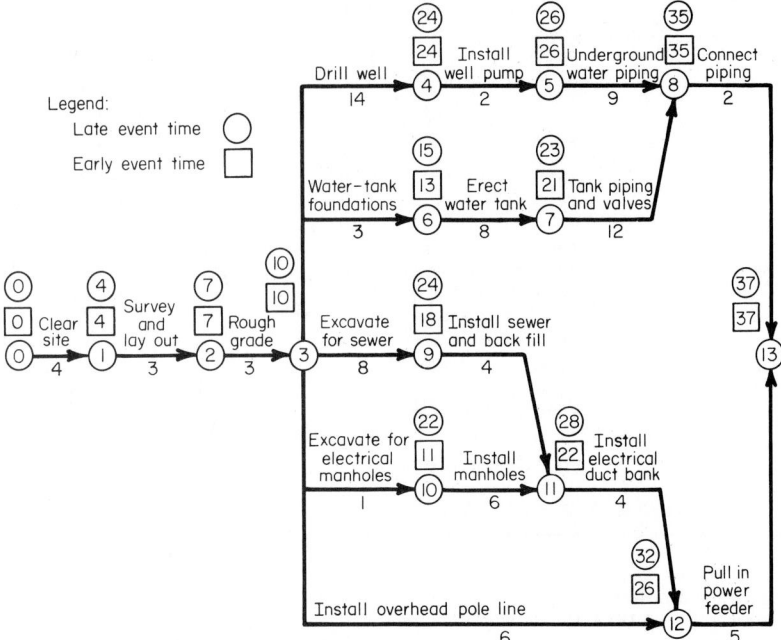

FIG. 15   CPM network for site-preparation project.

0-1-2-3-10-11-12-13

0-1-2-3-12-13

3.   Compute the early event time T_E of each event

The *early event time* is the earliest possible date at which the event may occur. Compute the early event time from $T_{E(n)} = T_{E(n-1)} + D$, where $T_{E(n)}$ = early event time of a given event; $T_{E(n-1)}$ = early event time of preceding event; D = duration of intervening activity. Where an event has multiple immediate predecessors, this equation yields multiple values of T_E; the correct value is the maximum value.

Table 18 shows the calculations for the early event times. In the calculations, the starting date of the project was used as the datum. Record the early event times on the network by entering the time (usually in days) in a square above each event. From Table 18, the minimum duration of this project is 37 days.

4.   Compute the late event time T_L of each event

The late event time of each event is the latest date at which the event may occur without extending the duration of the project beyond the minimum time. Use the relation $T_{L(n)} = T_{L(n+1)} - D$, where $T_{L(n)}$ = late event time of a given event; $T_{L(n+1)}$ = late event time of succeeding event; D = duration of intervening activity. Both early and late event times are usually measured in days, but on unusually long projects they may be measured in months. Where an event has multiple immediate successors, the equation above yields multiple values of T_L; the correct result is the minimum value.

Table 19 shows the calculations of the late event times. Enter the late event time on the network in a circle above each event.

5.   Compute the float of each activity

Float is the time, usually in days, that the completion of each activity may be delayed without extending the duration of the project, with the understanding that no other delays will occur.

TABLE 18 Calculation of Early Event Times

Event	T_E, days
0	0
1	4
2	4 + 3 = 7
3	7 + 3 = 10
4	10 + 14 = 24
5	24 + 2 = 26
6	10 + 3 = 13
7	13 + 8 = 21
8	26 + 9 = 35
	or 21 + 12 = 33 (disregard)
9	10 + 8 = 18
10	10 + 1 = 11
11	18 + 4 = 22
	or 11 + 6 = 17 (disregard)
12	22 + 4 = 26
	or 10 + 6 = 16 (disregard)
13	35 + 2 = 37
	or 26 + 5 = 31 (disregard)

TABLE 19 Calculation of Late Event Times

Event	T_L, days
13	37
12	37 − 5 = 32
11	32 − 4 = 28
10	28 − 6 = 22
9	28 − 4 = 24
8	37 − 2 = 35
7	35 − 12 = 23
6	23 − 8 = 15
5	35 − 9 = 26
4	26 − 2 = 24
3	24 − 14 = 10
	or 15 − 3 = 12 (disregard)
	or 24 − 8 = 16 (disregard)
	or 22 − 1 = 21 (disregard)
	or 32 − 6 = 26 (disregard)
2	10 − 3 = 7
1	7 − 3 = 4
0	4 − 4 = 0

Compute the float from $F = T_{L(j)} - (T_{E(i)} + D)$, where F = float, usually in days; $T_{L(j)}$ = late event time of completion in the same time units as F; $T_{E(i)}$ = early event starting time, in the same time units as F; D = activity duration, in the same time units as F. The expression in parentheses represents the earliest possible date at which the activity may be completed. Table 20 shows the float calculations.

TABLE 20 Calculation of Project Float°

Activity	$T_{L(j)}$	$T_{E(i)}$	D	F
0-1	4	0	4	0
1-2	7	4	3	0
2-3	10	7	3	0
3-4	24	10	14	0
4-5	26	24	2	0
5-8	35	26	9	0
3-6	15	10	3	2
6-7	23	13	8	2
7-8	35	21	12	2
8-13	37	35	2	0
3-9	24	10	8	6
9-11	28	18	4	6
3-10	22	10	1	11
10-11	28	11	6	11
11-12	32	22	4	6
3-12	32	10	6	16
12-13	37	26	5	6

°Measured in days.

6.   Identify the critical path

An activity is *critical* if any delay in its completion will extend the duration of the project. The path on which the critical activities are located is termed the *critical path*. (There may be several critical paths associated with a project.) In the terminology of CPM, a critical activity is one having zero float. The critical path for this project is therefore 0-1-2-3-4-5-8-13.

7.   Verify the results of step 6

Plot the project activities on a time scale, Fig. 16. This diagram was constructed under the assumption that each activity commences at the earliest possible date.

Note in Fig. 16 that the float of a given activity equals the total gap in the chain extending from the completion of that activity to the completion of the project. For instance, 6-7 has a float of 2 days, and 10-11 has a float of 5 + 6 = 11 days.

8.   Indicate where the actual schedule departed from the forecast

List the data as follows:

Old mark	New mark	Delay in completion, days
D	3-9	3
H	3-12	3
I	11-12	1

9. *Determine the true duration of the project*

Treating each departure individually, deduce the effects of that departure by referring to Fig. 16. Combine these effects where they are cumulative. On the basis of these results, establish the true duration of the project. Thus activity 3-9: events 11 and 12 would be delayed 3 days; activity 3-12: event 12 would not be delayed, since there is a latitude of 10 days along this path; activity 11-12: event 12 would be delayed 1 day. *Summary:* Event 12 is delayed 4 days; event 13 is not delayed, since there is a latitude of 6 days along this path. Therefore, the true duration of the project = 37 days, as forecast.

PROJECT PLANNING BASED ON AVAILABLE WORKFORCE

A manufacturing firm has a contract to build a pilot model of a newly invented machine. To plan the work, the firm constructed the CPM network in Fig. 17 and compiled the data in Table 21. The following activities must be performed as a unit, without loss of continuity: 0-1-2-4; 2-3-4; 5-7-8. Activities 4-8 and 6-8 may be performed piecemeal, if this proves convenient. The workforce available for assignment to this project is 15 workers for the first 8 days, 25 for the remaining time. Each employee is capable of performing all 11 activities. But the constraints imposed by the available facilities limit the number of workers for each activity to that shown in Table 21. Overtime is not permissible. Devise a schedule that will allow completion of this project at the earliest possible date.

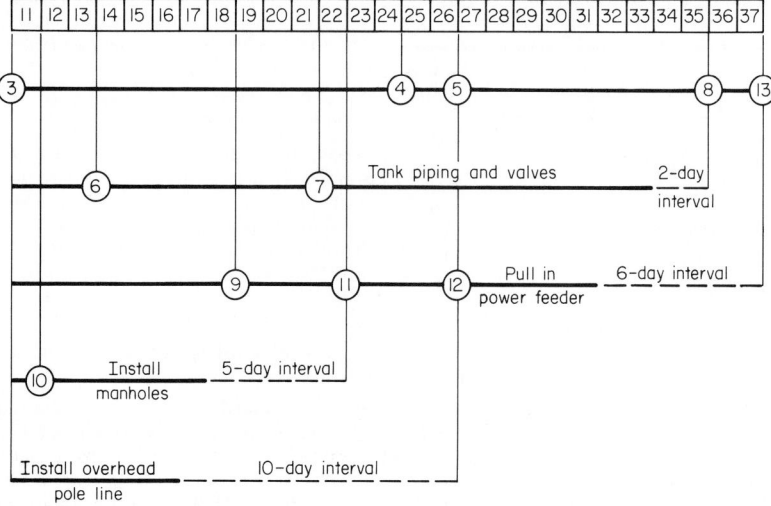

FIG. 16 Activity-time diagram.

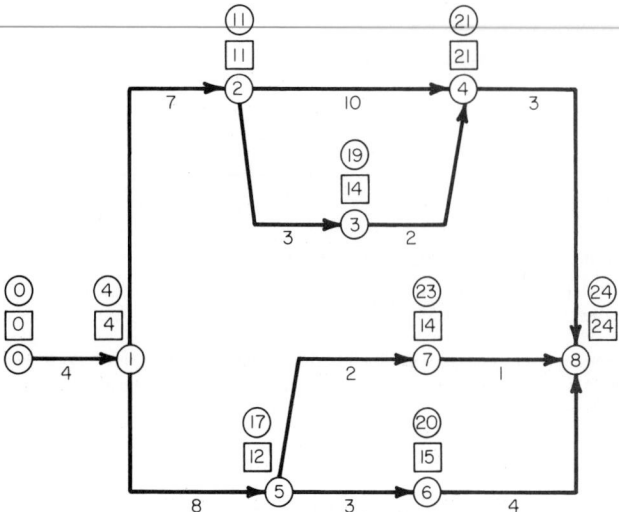

FIG. 17 CPM network for construction of pilot model.

Calculation Procedure:

1. *Compute the early and late event times*

The results of these calculations, made in accordance with the previous calculation procedure, are shown in Fig. 17.

2. *Identify the critical path*

Since the critical path is the longest path through the network, Fig. 17 shows that this is 0-1-2-4-8. Note that the critical path sets a lower limit of 24 days on the duration of the project. But the workforce limitations may lengthen the project beyond 24 days. Thus, the objective will be to devise a schedule that fits noncritical activities into the 24-day period while satisfying the workforce availability.

TABLE 21 Time and Workforce Requirements for a Technical Project

Activity, Fig. 17	Duration, days	Workers required
0-1	4	3
1-2	7	5
2-4	10	9
2-3	3	9
3-4	2	12
4-8	3	4
1-5	8	6
5-6	3	10
6-8	4	8
5-7	2	6
7-8	1	7

3. *Schedule the project, assuming unlimited workforce*

As a first trial, assume that unlimited workforce is available, and schedule each activity to start at the *earliest* possible date. Construct the workforce-time diagram for this condition, as shown in Fig. 18a. Study of the diagram shows that this schedule is unsatisfactory because the workforce requirements exceed the available workers on days 12 to 15, inclusive.

4. *Schedule the project with the latest possible start*

Assume unlimited workforce again, and schedule each activity to start at the *latest* possible date. Construct the corresponding workforce-time diagram, Fig. 18b. Study of the diagram shows that this schedule is also unsatisfactory, but it is an improvement over the schedule in step 3.

Although both these schedules are unsatisfactory, they are useful because their workforce-time diagrams reveal the boundaries of each activity or chain of activities based on a project duration

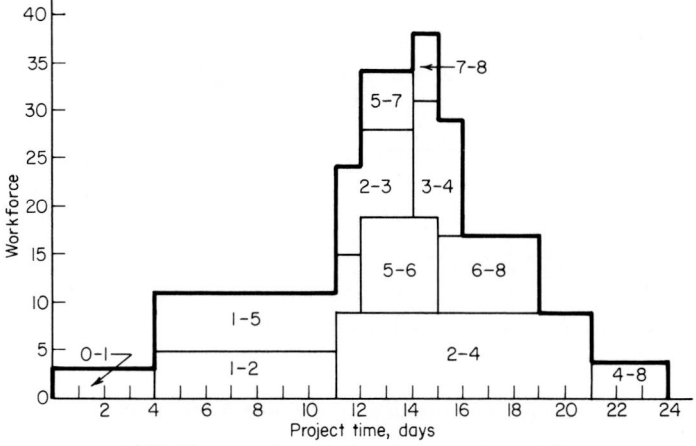

(a) Workforce requirements based upon early event times

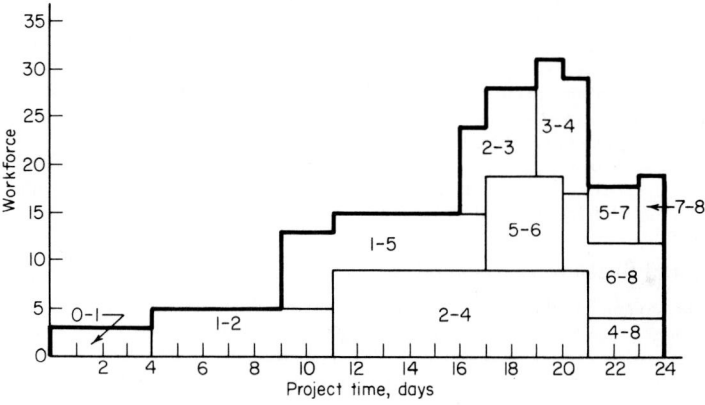

(b) Workforce requirements based upon late event times

FIG. 18

of 24 days. For example, the chain 5-7-8 may start at any time between days 12 and 21, Fig. 18. This information is not explicitly supplied by the network, since the late event time for event 5 is determined by the chain 5-6-8.

5. *Shift noncritical activities to obtain a suitable schedule*

Using the allowable workforce limits, shift noncritical activities such that the project can be completed in 24 days using the available workers.

Construct the schedule shown in Fig. 19. Note that the workforce requirements are less than the available personnel. Further, the schedule preserves the integrity of the three chains of activities mentioned earlier. Although fragmentation of activity 6-8 is permissible, it proved unnecessary. With the schedule shown, the project will be finished in 24 days.

Related Calculations: Note that this procedure can be used for any type of project requiring allocation of workforce or other resources.

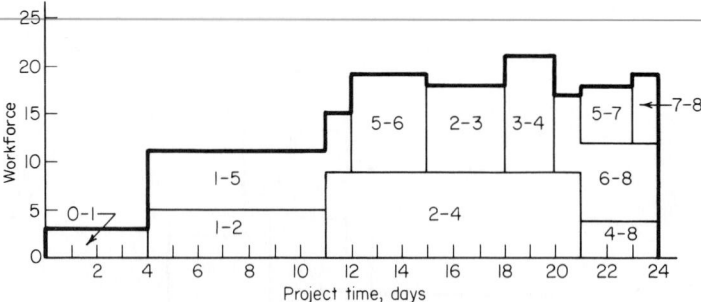

FIG. 19   Workforce requirements based on project schedule.

Statistics, Probability, and Their Applications

Basic Statistics

If the value assumed by a variable on a given occasion cannot be predicted because it is influenced by chance, the variable is known as a *random*, or *stochastic*, variable. The number of times the variable assumes a given value is called the *frequency* of that value. A number that may be considered representative of all values of the variable is called an *average*. There are several types of averages, such as arithmetic mean, geometric mean, harmonic mean, median, mode, etc. Each is used for a specific purpose. The *standard deviation* is a measure of the dispersion, or scatter, of the values of the random variable; it is approximately equal to the amount by which a given value of the variable may be expected to differ from the arithmetic mean, in either direction. The *variance* is the square of the standard deviation. Where no mention of frequency appears, it is understood that all frequencies are 1.

Notational System

Here X = random variable; X_i = ith value assumed by X; f_i = frequency of X_i; n = sum of frequencies; $\overline{X}$ = arithmetic mean of values of X; X_{med} = median of these values; d_i = deviation of X_i from $\overline{X} = X_i - \overline{X}$; A = an assumed arithmetic mean; $d_{A,i}$ = deviation of X_i from $A = X_i - A$; s = standard deviation; s^2 = variance. In the following material, the subscript i will be omitted.

DETERMINATION OF ARITHMETIC MEAN, MEDIAN, AND STANDARD DEVIATION

Column 2 of Table 22 presents the number of units of a commodity that were sold monthly by a firm for 7 consecutive months. Find the arithmetic mean, median, and standard deviation of the number of units sold monthly.

Calculation Procedure:

1. *Compute the arithmetic mean*

Let X = number of units sold monthly. Find the sum of the values of X, which is 301. Then set $\overline{X} = (\Sigma X)/n$, or $\overline{X} = 301/7 = 43$.

2. *Find the median*

Consider that all values of X are arranged in ascending order of magnitude. If n is odd, the value that occupies the central position in this array is called the *median*. If n is even, the median is taken as the arithmetic mean of the two values that occupy the central positions. In either case,

TABLE 22

(1) Month	(2) Number of units sold X	(3) $d = X - 43$	(4) d^2	(5) $d_A = X - 40$	(6) d_A^2
1	32	-11	121	-8	64
2	49	6	36	9	81
3	51	8	64	11	121
4	44	1	1	4	16
5	37	-6	36	-3	9
6	41	-2	4	1	1
7	47	4	16	7	49
Total	301	0	278	21	341

the total frequency of values below the median equals the total frequency of values above the median. The median is useful as an average because the arithmetic mean can be strongly influenced by an extreme value at one end of the array and thereby offer a misleading view of the data.

In the present instance, the array is 32, 37, 41, 44, 47, 49, 51. The fourth value in the array is 44; then $X_{med} = 44$.

3. Compute the standard deviation

Compute the deviations of the X values from $\overline{X}$, and record the results in column 3 of Table 22. The sum of the deviations must be 0. Now square the deviations, and record the results in column 4. Find the sum of the squared deviations, which is 278. Set the variance $s^2 = (\Sigma d^2)/n = 278/7$. Then set the standard deviation $s = \sqrt{278/7} = 6.30$.

4. Compute the arithmetic mean by using an assumed arithmetic mean

Set $A = 40$. Compute the deviations of the X values from A, record the results in column 5 of Table 22, and find the sum of the deviations, which is 21. Set $\overline{X} = A + (\Sigma d_A)/n$, or $\overline{X} = 40 + 21/7 = 43$.

5. Compute the standard deviation by using the arithmetic mean assumed in step 3

Square the deviations from A, record the results in column 6 of Table 22, and find their sum, which is 341. Set $s^2 = (\Sigma d_A^2)/n - [(\Sigma d_A)/n]^2 = 341/7 - (21/7)^2 = 341/7 - 9 = 278/7$. Then $s = \sqrt{278/7} = 6.30$.

Related Calculations: Note that the equation applied in step 5 does not contain the true mean $\overline{X}$. This equation serves to emphasize that the standard deviation is purely a measure of dispersion and thus is independent of the arithmetic mean. For example, if all values of X increase by a constant h, then $\overline{X}$ increases by h, but s remains constant. Where $\overline{X}$ has a nonintegral value, the use of an assumed arithmetic mean A of integral value can result in a faster and more accurate calculation of s.

DETERMINATION OF ARITHMETIC MEAN AND STANDARD DEVIATION OF GROUPED DATA

In testing a new industrial process, a firm assigned a standard operation to 24 employees in its factory and recorded the time required by each employee to complete the operation. The results are presented in columns 1 and 2 of Table 23. Find the arithmetic mean and standard deviation of the time of completion.

Calculation Procedure:

1. Record the class midpoints

Where the number of values assumed by a variable is very large, a comprehensive listing of these values becomes too cumbersome. Therefore, the data are presented by grouping the values in

TABLE 23

(1) Time of completion, min (class interval)	(2) Number of employees (frequency f)	(3) Midpoint X	(4) Code c
20 to less than 24	3	22	-2
24 to less than 28	9	26	-1
28 to less than 32	7	30	0
32 to less than 36	5	34	1
Total	24		

classes and showing the frequency of each class. The range of values of a given class is its *class interval*, and the end values of the interval are the *class limits*. The difference between the upper and lower limits is the *class width*, or *class size*. Thus, in Table 23, all classes have a width of 4 min. The arithmetic mean of the class limits is the *midpoint*, or *mark*. In analyzing grouped data, all values that fall within a given class are replaced with the class midpont. The midpoints are recorded in column 3 of Table 23, and they are denoted by X.

2.   Compute the arithmetic mean

Set $\overline{X} = (\Sigma fX)/n$, or $\overline{X} = (3 \times 22 + 9 \times 26 + 7 \times 30 + 5 \times 34)/24 = 680/24 = 28.33$ min.

3.   Compute the standard deviation

Set $s^2 = (\Sigma fd^2)/n$, or $s^2 = [3(-6.33)^2 + 9(-2.33)^2 + 7(1.67)^2 + 5(5.67)^2]/24 = 14.5556$. Then $s = \sqrt{14.5556} = 3.82$ min.

4.   Compute the arithmetic mean by the coding method

This method simplifies the analysis of grouped data where all classes are of uniform width, as in the present case. Arbitrarily selecting the third class, assign the interger 0 to this class, and then assign intergers to the remaining classes in consecutive and ascending order, as shown in column 4 of Table 23. These integers are the *class codes*. Let c = class code, w = class width, and A = midpoint of class having the code 0. Compute Σfc, or $\Sigma fc = 3(-2) + 9(-1) + 7(0) + 5(1) = -10$. Now set $\overline{X} = A + w(\Sigma fc)/n$, or $\overline{X} = 30 + 4(-10)/24 = 28.33$ min.

5.   Compute the standard deviation by the coding method

Using the codes previously assigned, compute Σfc^2, or $\Sigma fc^2 = 3(-2)^2 + 9(-1)^2 + 7(0)^2 + 5(1)^2 = 26$. Now set $s^2 = w^2 \{(\Sigma fc^2)/n - [(\Sigma fc)/n]^2\}$. Then $s^2 = 16[26/24 - (-10/24)^2] = 14.5556$, and $s = \sqrt{14.5556} = 3.82$ min.

Permutations and Combinations

An arrangement of objects or individuals in which the order or rank is significant is called a *permutation*. A grouping of objects or individuals in which the order or rank is not significant, or in which it is predetermined, is called a *combination*. Assume that n objects are available and that r of these objects are selected to form a permutation or combination. If interest centers on only the *identity* of the r objects selected, a combination is formed; if interest centers on both the identity and the order or rank of the r objects, a permutation is formed. In the following material, the n objects all differ from one another.

Where necessary, the number of permutations or combinations that can be formed is computed by applying the following law, known as the *multiplication law:* If one task can be performed in m_1 different ways and another task can be performed in m_2 different ways, the set of tasks can be performed in $m_1 m_2$ different ways.

Notational System

Here $n!$ (read "n factorial" or "factorial n") = product of first n integers, and the integers are usually written in reverse order. Thus, $5! = 5 \times 4 \times 3 \times 2 \times 1 = 120$. For mathematical consistency, $0!$ is taken as 1.

Also, $P_{n,r}$ = number of permutations that can be formed of n objects taken r at a time; and $C_{n,r}$ = number of combinations that can be formed of n objects taken r at a time.

NUMBER OF WAYS OF ASSIGNING WORK

A firm has three machines, A, B, and C, and each machine can be operated by only one individual at a time. The number of employees who are qualified to operate a machine is: machine A, five; machine B, three; machine C, seven. In addition to these 15 employees, Smith is qualified to operate all three machines. In how many ways can operators be assigned to the machines?

Calculation Procedure:

1. *Compute the number of possible assignments if Smith is excluded*

Apply the multiplication law. The number of possible assignments = $5 \times 3 \times 7 = 105$.

2. *Compute the number of possible assignments if Smith is selected*

If Smith is assigned to machine A, the number of possible assignments to B and C = $3 \times 7 = 21$. If Smith is assigned to machine B, the number of possible assignments to A and C = $5 \times 7 = 35$. If Smith is assigned to machine C, the number of possible assignments to A and B = $5 \times 3 = 15$. Thus, the number of possible assignments with Smith selected = $21 + 35 + 15 = 71$.

3. *Compute the total number of possible assignments*

By summation, the number of ways in which operators can be assigned to the three machines = $105 + 71 = 176$.

FORMATION OF PERMUTATIONS SUBJECT TO A RESTRICTION

Permutations are to be formed of the first seven letters of the alphabet, taken four at a time, with the restriction that d cannot be placed anywhere to the left of c. For example, the permutation edgc is unacceptable. How many permutations can be formed?

Calculation Procedure:

1. *Compute the number of permutations in the absence of any restriction*

Use the relation $P_{n,r} = n!/(n - r)!$, or $P_{7,4} = 7!/3! = 7 \times 6 \times 5 \times 4 = 840$.

2. *Compute the number of permutations that violate the imposed restriction*

Form permutations that violate the restriction. Start by placing d in the first position. Letter c can be placed in any of the three subsequent positions. Two positions now remain unoccupied, and five letters are available; these positions can be filled in $5 \times 4 = 20$ ways. Thus, the number of permutations in which d occupies the first position and c some subsequent position is $3 \times 20 = 60$. Similarly, the number of permutations in which d occupies the second position and c occupies the third or fourth position is $2 \times 20 = 40$, and the number of permutations in which d occupies the third position and c occupies the fourth position is $1 \times 20 = 20$.

By summation, the number of unacceptable permutations = $60 + 40 + 20 = 120$.

3. *Compute the number of permutations that satisfy the requirement*

By subtraction, the number of acceptable permutations = $840 - 120 = 720$.

FORMATION OF COMBINATIONS SUBJECT TO A RESTRICTION

A committee is to consist of 6 individuals of equal rank, and 15 individuals are available for assignment. However, McCarthy will serve only if Polanski is also on the committee. In how many ways can the committee be formed?

Calculation Procedure:

1. *Compute the number of possible committees in the absence of any restriction*

Since the members will be of equal rank, each committee represents a combination. Use the relation $C_{n,r} = n!/[r!(n-r)!]$, or $C_{15,6} = 15!/(6!9!) = (15 \times 14 \times 13 \times 12 \times 11 \times 10)/(6 \times 5 \times 4 \times 3 \times 2) = 5005$.

2. *Compute the number of possible committees that violate the imposed restriction*

Assign McCarthy to the committee, but exclude Polanski. Five members remain to be selected, and 13 individuals are available. The number of such committees $= C_{13,5} = 13!/(5!8!) = (13 \times 12 \times 11 \times 10 \times 9)/(5 \times 4 \times 3 \times 2) = 1287$.

3. *Compute the number of possible committees that satisfy the requirement*

By subtraction, the number of ways in which the committee can be formed $= 5005 - 1287 = 3718$.

Probability

If the outcome of a process cannot be predicted because it is influenced by chance, the process is called a *trial*, or *experiment*. The outcome of a trial or set of trials is an *event*. Two events are *mutually exclusive* if the occurrence of one excludes the occurrence of the other. Two events are *independent* of each other if the occurrence of one has no effect on the likelihood that the other will occur.

Assume that a box contains 17 objects, 12 of which are spheres. If an object is to be drawn at random and all objects have equal likelihood of being drawn, then the probability that a sphere will be drawn is 12/17. Thus, the probability of a given event can range from 0 to 1. The lower limit corresponds to an impossible event, and the upper limit corresponds to an event that is certain to occur. If two events are mutually exclusive, the probability that *either* will occur is the sum of their respective probabilities. If two events are independent of each other, the probability that *both* will occur is the product of their respective probabilities.

Assume that a random variable is discrete and the number of values it can assume is finite. A listing of these values and their respective probabilities is called the *probability distribution* of the variable. Where the number of possible values is infinite, the probability distribution is expressed by stating the functional relationship between a value of the variable and the corresponding probability. Where the random variable is continuous, the method of expressing its probability distribution is illustrated in the calculation procedure below pertaining to the normal distribution.

Notational System

Here E = given event; X = random variable; $P(E)$ = probability that event E will occur; $P(X_i)$ = probability that X will assume the value X_i; μ and σ = arithmetic mean and standard deviation, respectively, of a probability distribution.

PROBABILITY OF A SEQUENCE OF EVENTS

A box contains 12 bolts. Of these, 8 have square heads and 4 have hexagonal heads. Seven bolts will be removed from the box, individually and at random. What is the probability that the second and third bolts drawn will have square heads and the sixth bolt will have a hexagonal head?

Calculation Procedure:

1. *Compute the total number of ways in which the bolts can be drawn*

The sequence in which the bolts are drawn represents a permutation of 12 bolts taken 7 at a time, and each bolt is unique. The total number of permutations $= P_{12,7} = 12!/5!$.

2. *Compute the number of ways in which the bolts can be drawn in the manner specified*

If the bolts are drawn in the manner specified, the second and third positions in the permutation are occupied by square-head bolts and the sixth position is occupied by a hexagonal-head bolt.

Construct such a permutation, in these steps: Place a square-head bolt in the second position; the number of bolts available is 8. Now place a square-head bolt in the third position; the number of bolts available is 7. Now place a hexagonal-head bolt in the sixth position; the number of bolts available is 4. Finally, fill the four remaining positions in any manner whatever; the number of bolts available is 9.

The second position can be filled in 8 ways, the third position in 7 ways, the sixth position in 4 ways, and the remaining positions in $P_{9,4}$ ways. By the multiplication law, the number of acceptable permutations is $8 \times 7 \times 4 \times P_{9,4} = 224(9!/5!)$.

3.   *Compute the probability of drawing the bolts in the manner specified*

Since all permutations have an equal likelihood of becoming the true permutation, the probability equals the ratio of the number of acceptable permutations to the total number of permutations. Thus, probability $= 224(9!/5!)/(12!/5!) = 224(9!)/12! = 224(12 \times 11 \times 10) = 224/1320 = 0.1697$.

4.   *Compute the probability by an alternative approach*

As the preceding calculations show, the exact positions specified (second, third, and sixth) do not affect the result. For simplicity, assume that the first and second bolts are to be square-headed and the third bolt hexagonal-headed. The probabilities are: first bolt square-headed, 8/12; second bolt square-headed, 7/11; third bolt hexagonal-headed, 4/10. The probability that all three events will occur is the product of their respective probabilities. Thus, the probability that bolts will be drawn in the manner specified $= (8/12)(7/11)(4/10) = 224/1320 = 0.1697$. Note also that the precise number of bolts drawn from the box (7) does not affect the result.

PROBABILITY ASSOCIATED WITH A SERIES OF TRIALS

During its manufacture, a product passes through five departments, A, B, C, D, and E. The probability that the product will be delayed in a department is: A, 0.06; B, 0.15; C, 0.03; D. 0.07; E, 0.13. These values are independent of one another in the sense that the time for which the product is held in one department has no effect on the time it spends in any subsequent department. What is the probability that there will be a delay in the manufacture of this product?

Calculation Procedure:

1.   *Compute the probability that the product will be manufactured without any delay*

Since it is certain that the product either will or will not be delayed in a department and the probability of certainty is 1, probability of no delay $= 1 -$ probability of delay. Thus, the probability that the product will pass through department B without delay $= 1 - 0.15 = 0.85$. The probability that the product will pass through every department without delay is the product of the probabilities of these individual events. Thus, probability of no delay in manufacture $= (0.94)(0.85)(0.97)(0.93)(0.87) = 0.6271$.

2.   *Compute the probability of a delay in manufacture*

Probability of delay in manufacture $= 1 -$ probability of no delay in manufacture $= 1 - 0.6271 = 0.3729$.

Related Calculations: This method of calculation can be applied to any situation where a series of trials occurs, either simultaneously or in sequence, and any trial can cause the given event. Thus, assume that several projectiles are fired simultaneously and the probability of landing in a target area is known for each projectile. The above method can be used to find the probability that at least one projectile will land in the target area.

BINOMIAL PROBABILITY DISTRIBUTION

A case contains 14 units, 9 of which are of type A. Five units will be drawn at random from the case; and as a unit is drawn, it will be replaced with one of identical type. If X denotes the number of type A units drawn, find the probability distribution of X and the average value of X in the long run.

Calculation Procedure:

1. Compute the probability corresponding to a particular value of X

Consider that n independent trials are performed, and let X denote the number of times an event E occurs in these n trials. The probability distribution of X is called *binomial*. In this case, since each unit drawn is replaced with one of identical type, each drawing is independent of all preceding drawings; therefore, X has a binomial probability distribution. The event E consists of drawing a type A unit.

With respect to every drawing, probability of drawing a type A unit = 9/14, and probability of drawing a unit of some other type = 5/14. Arbitrarily set $X = 3$, and assume that the units are drawn thus: A-A-A-N-N, where N denotes a type other than A. The probability of drawing the units in this sequence = $(9/14)(9/14)(9/14)(5/14)(5/14) = (9/14)^3(5/14)^2$. Clearly this is also the probability of drawing 3 type A units in any other sequence. Since the type A units can occupy any 3 of the 5 positions in the set of drawings and the exact positions do not matter, the number of sets of drawings that contain 3 type A units = $C_{5,3}$. Summing the probabilities, we find $P(3)$ = $C_{5,3}(9/14)^3(5/14)^2 = [5!/(3!2!)](9/14)^3(5/14)^2 = 0.3389$.

2. Write the equation of binomial probability distribution

Generalize from step 1 to obtain $P(X) = C_{n,X}P^X(1 - P)^{n-X}$, where P = probability event E will occur on a single trial. Here $P = 9/14$.

3. Apply the foregoing equation to find the probability distribution of X

The results are $P(0) = 1(9/14)^0(5/14)^5 = 0.0058$. Similarly, $P(1) = 0.0523$; $P(2) = 0.1883$; $P(3)$ = 0.3389 from step 1; $P(4) = 0.3050$; $P(5) = 0.1098$.

4. Verify the values of probability

Since it is certain that X will assume some value from 0 to 5, inclusive, the foregoing probabilities must total 1. There sum is found to be 1.0001, and the results are thus confirmed.

5. Compute the average value of X in the long run

Consider that there are an infinite number of cases of the type described and that 5 units will be drawn from each case in the manner described, thereby generating an infinite set of values of X. Since the chance of obtaining a type A unit on a single drawing is 9/14, the average number of type A units that will be obtained in 5 drawings is $5(9/14) = 45/14 = 3.21$. Thus, the arithmetic mean of this infinite set of values of X is 3.21.

Alternatively, find the average value of X by multiplying all X values by their respective probabilities, to get $0.0523 + 2(0.1883) + 3(0.3389) + 4(0.3050) + 5(0.1098) = 3.21$. The arithmetic mean of an infinite set of X values is also called the *expected value* of X.

PASCAL PROBABILITY DISTRIBUTION

Objects are ejected randomly from a rotating mechanism, and the probability that an object will enter a stationary receptacle after leaving the mechanism is 0.35. The process of ejecting objects will continue until four objects have entered the receptacle. Let X denote the number of objects that must be ejected. Find (*a*) the probability corresponding to every X value from 4 to 10, inclusive; (*b*) the probability that more than 10 objects must be ejected; (*c*) the average value of X in the long run.

Calculation Procedure:

1. Compute the probability corresponding to a particular value of X

Consider that a trial is performed repeatedly, each trial being independent of all preceding trials, until a given event E has occurred for the kth time. Let X denote the number of trials required. The variable X is said to have a *Pascal* probability distribution. (In the special case where $k = 1$, the probability distribution is called *geometric*.) In the present situation, the given event is entrance of the object into the receptacle, and $k = 4$.

Use this code: A signifies the object has entered; B signifies it has not. Arbitrarily set $X = 9$, and consider this sequence of events: A-B-B-A-B-B-B-A-A, which contains four A's and five B's. The probability of this sequence, and of every sequence containing four A's and five B's, is

$(0.35)^4(0.65)^5$. Other arrangements corresponding to $X = 9$ can be obtained by holding the fourth A in the ninth position and rearranging the preceding letters, which consist of three A's and five B's. Since the A's can be assigned to any 3 of the 8 positions, the number of arrangements that can be formed is $C_{8,3}$. Thus, $P(9) = C_{8,3}(0.35)^4(0.65)^5 = 56(0.35)^4(0.65)^5 = 0.0975$.

2.   *Write the equation of Pascal probability distribution*

Generalize from step 1 to obtain $P(X) = C_{X-1,k-1}P^k(1 - P)^{X-k}$, where $P =$ probability that event E will occur on a single trial. Here $P = 0.35$.

3.   *Apply the foregoing equation to find the probabilities corresponding to the given X values*

The results are $P(4) = 1(0.35)^4(0.65)^0 = 0.0150$; $P(5) = 4(0.35)^4(0.65)^1 = 0.0390$; $P(6) = 10(0.35)^4(0.65)^2 = 0.0634$. Similarly, $P(7) = 0.0824$; $P(8) = 0.0938$; $P(9) = 0.0975$ from step 1; $P(10) = 0.0951$.

Thus, as X increases, $P(X)$ increases until $X = 9$, and then it decreases. The variable X can assume an infinite number of values in theory, and the corresponding probabilities form a converging series having a sum of 1.

4.   *Compute the probability that 10 or fewer ejections will be required*

Sum the values in step 3; $P(X \leq 10) = 0.4862$.

5.   *Compute the probability that more than 10 ejections will be required*

Since it is certain that X will assume a value of 10 or less or a value of more than 10, $P(X > 10) = 1 - 0.4862 = 0.5138$.

6.   *Compute the average number of ejections required in the long run*

Consider that the process of placing a set of four objects in the receptacle is continued indefinitely, thereby generating an infinite set of values of X. Since there is a 35 percent chance that a specific object will enter the receptacle after being ejected from the mechanism, it will require an average of $1/0.35 = 2.86$ ejections to place one object in the receptacle and an average of $4(1/0.35) = 11.43$ ejections to place four objects in the receptacle. Thus, the infinite set of values of X has an arithmetic mean of 11.43.

POISSON PROBABILITY DISTRIBUTION

A radioactive substance emits particles at an average rate of 0.08 particles per second. Assuming that the number of particles emitted during a given time interval has a Poisson distribution, find the probability that the substance will emit more than three particles in a 20-s interval.

Calculation Procedure:

1.   *Compute the average number of particles emitted in 20 s*

Let T denote an interval of time, in suitable units. Consider that an event E occurs randomly in time but the average number of occurrences of E in time T, as measured over a relatively long period, remains constant. Let $m =$ average (or expected) number of occurrences of E in T, and $X =$ true number of occurrences of E in T. The variable X is said to have a *Poisson* probability distribution.

In the present case, $X =$ number of particles emitted in 20 s, and $m = 20(0.08) = 1.6$.

2.   *Compute the probability that X ≤ 3*

Use the relation $P(X) = m^X/[e^m(X!)]$, where $e =$ base of natural logarithms $= 2.71828. \ldots$ Thus, $e^m = e^{1.6} = 4.95303$. Then $P(0) = (1.6)^0/(4.95303 \times 1) = 0.2019$; $P(1) = (1.6)^1/(4.95303 \times 1) = 0.3230$; $P(2) = (1.6)^2/(4.95303 \times 2) = 0.2584$; $P(3) = (1.6)^3/(4.95303 \times 6) = 0.1378$. Sum these results to obtain $P(X \leq 3) = 0.9211$.

3.   *Compute the probability that X > 3*

$P(X > 3) = 1 - 0.9211 = 0.0789$.

Related Calculations: The probabilities in step 2 also can be found by referring to a table of Poisson probability. The foregoing discussion pertains to an event that occurs in *time*, but analogous comments apply to an event that occurs in *space*. For example, assume that a firm manufactures long rolls of tape. Defects in the tape occur randomly, but the average number of defects in a 300-m length, as measured across long distances, is constant. The number of defects in a given length of tape has a Poisson distribution. The Poisson distribution is an extreme case of the binomial distribution. As the probability that event E will occur on a single trial becomes infinitesimally small and the number of trials becomes infinitely large, the binomal distribution approaches the Poisson distribution as a limit.

COMPOSITE EVENT WITH POISSON DISTRIBUTION

With reference to the preceding calculation procedure, a counting device is installed to determine the number of particles emitted. The probability that the device will actually count an emission is 0.90. Find the probability that the number of emissions counted in a 20-s interval will be 3.

Calculation Procedure:

1.   *Compute the average number of emissions counted in 20 s*

In the present case, event E is that an emission is counted. This event is a composite of two basic events: a particle is emitted, and the device functions properly. Thus, $m = 20(0.08)(0.90) = 1.44$.

2.   *Compute the probability that X = 3*

Use the equation given in the preceding calculation procedure, or $P(3) = (1.44)^3/[e^{1.44}(3!)] = 0.1179$.

NORMAL DISTRIBUTION

A continuous random variable X has a normal probability distribution with an arithmetic mean of 14 and a standard deviation of 2.5. Find the probability that on a given occasion X will assume a value that (*a*) lies between 14 and 17; (*b*) lies between 12 and 16.2; (*c*) is less than 10.

Calculation Procedure:

1.   *Compute the values of z corresponding to the specified boundary values of X*

If a random variable X is continuous, the probability that X will assume a value between X_j and X_k is represented graphically by constructing a *probability diagram* in this manner: Plot values of X on the horizontal axis; then construct a curve such that $P(X_j < X < X_k) =$ area bounded by the curve, the horizontal axis, and vertical lines at X_j and X_k. The ordinate of this curve is denoted by $f(X)$ and is called the *probability density function*. The total area under the curve is 1, the probability of certainty.

    A continuous random variable has a *normal* or *gaussian* probability distribution if the range of its possible values is infinite and its probability curve has an equation of this form: $f(X) = (1/b\sqrt{2\pi})e^{-(X-a)^2/2b^2}$, where a and b are constants and $e =$ base of natural logarithms. Figure 20 is the probability diagram; the curve is bell-shaped and symmetric about a vertical line through the summit.

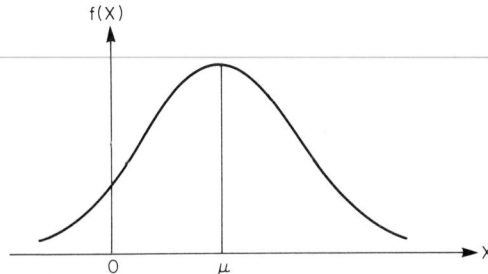

FIG. 20   Curve of normal probability distribution.

Consider that the trial that yields a value of X is repeated indefinitely, generating an infinite set of values of X. Let μ and σ = arithmetic mean and standard deviation, respectively, of this set of values. The summit of the probability curve lies at $X = \mu$. By symmetry, the area under the curve to the left and to the right of $X = \mu$ is 0.5.

The deviation of X_i from μ is expressed in *standard units* in this form: $z_i = (X_i - \mu)/\sigma$. Thus, for $X = 14$, $z = 0$; for $X = 17$, $z = (17 - 14)/2.5 = 1.20$; for $X = 12$, $z = (12 - 14)/2.5 = -0.80$; etc. Record the z values in Table 24.

TABLE 24

X	z	$A(z)$
14	0	0
17	1.20	0.38493
12	-0.80	0.28814
16.2	0.88	0.31057
10	-1.60	0.44520

2. Find the values of A(z)

Let $A(z_i)$ = area under probability curve from centerline (where $X = \mu$) to $X = X_i$; this area = $P(\mu < X < X_i)$. Obtain the values of $A(z)$ from the table of areas under the normal probability curve. Refer to Table 25, which is an excerpt from this table. Thus, if $z = 1.60$, $A(z) = 0.44520$; if $z = 1.78$, $A(z) = 0.46246$. Note that $A(-z_i) = A(z_i)$ by symmetry. Record the values of $A(z)$ in Table 24.

TABLE 25   Area under the Normal Curve

z	.00	.01	.02	.03	.04
1.5	.43319	.43448	.43574	.43699	.43822
1.6	.44520	.44630	.44738	.44845	.44950
1.7	.45543	.45637	.45728	.45818	.45907

z	.05	.06	.07	.08	.09
1.5	.43943	.44062	.44179	.44295	.44408
1.6	.45053	.45154	.45254	.45352	.45449
1.7	.45994	.46080	.46164	.46246	.46327

3. Compute the required probabilities

Refer to Fig. 21. Apply the areas in Table 24 to obtain these results: $P(14 < X < 17) = 0.38493$; $P(12 < X < 16.2) = 0.28814 + 0.31057 = 0.59871$; $P(X < 10) = 0.5 - 0.44520 = 0.05480$.

Related Calculations: Many random variables that occur in nature have a normal probability distribution. For example, the height, weight, and intelligence of members of a species have normal distributions. Although in theory this distribution applies solely where the range of X values is infinite, in practice the distribution is applied as a valid approximation where the range of X values is finite.

APPLICATION OF NORMAL DISTRIBUTION

The time required to perform a manual operation is assumed to have a normal distribution. Studies of past performance disclose that the average time required is 5.80 h and the standard deviation is 0.50 h. Find the probability (to three decimal places) that the operation will be performed within 5.25 h.

Calculation Procedure:

1. Compute the value of z corresponding to the boundary value of X

Let X = time required to perform the operation, and refer to the preceding calculation procedure for the definition of z. For $X = 5.25$, $z = (5.25 - 5.80)/0.50 = -1.10$.

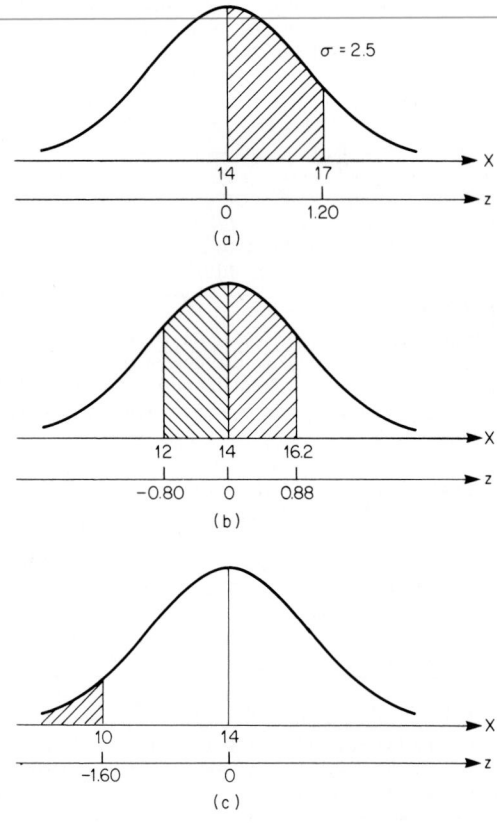

FIG. 21

2. *Find the value of A(z)*

Refer to the table of areas under the normal probability curve, and take the absolute value of z. If $z = 1.10$, $A(z) = 0.364$.

3. *Compute the required probability*

The area under consideration lies to the *left* of a vertical line at $X = 5.25$, and the area found in step 2 lies between this line and the centerline. Then $P(X < 5.25) = 0.5 - 0.364 = 0.136$.

NEGATIVE-EXPONENTIAL DISTRIBUTION

The mean life span of an electronic device that operates continuously is 2 months. If the life span of the device has a negative-exponential distribution, what is the probability that the life span will exceed 3 months?

Calculation Procedure:

1. *Write the equation of cumulative probability*

Refer to the calculation procedure on the normal distribution for definitions pertaining to a continuous random variable. A variable X is said to have a *negative-exponential* (or simply *expo-*

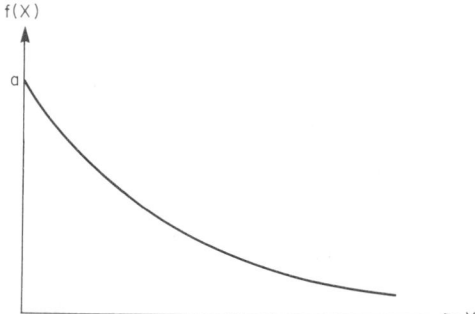

FIG. 22   Negative-exponential probability distribution.

nential) probability distribution if its probability density function is of this form: $f(X) = 0$ if $X < 0$ and $f(X) = ae^{-aX}$ if $X \geq 0$, Eq. *a*, where a = positive constant and e = base of natural logarithms. Figure 22 shows the probability diagram. The arithmetic mean of X is $\mu = 1/a$, Eq. *b*.

Let X = life span of device, months, and let K denote any positive number. Integrate Eq. *a* between the limits of 0 and K, giving $P(X \leq K) = 1 - e^{-aK}$, Eq. *c*. Then $P(X > K) = e^{-aK}$, Eq. *d*.

2.   *Compute the required probability*

Compute a by Eq. *b*, giving $a = 1/\mu = 1/2 = 0.5$. Set $K = 3$ months. By Eq. *d*, $P(X > 3) = e^{-1.5} = 1/e^{1.5} = 0.2231$.

Statistical Inference

Consider that there exists a set of objects, which is called the *population*, or *universe*. Also consider that interest centers on some property of these objects, such as length, molecular weight, etc., and that this property assumes many values. Thus, associated with the population is a set of *numbers*. This set of numbers has various characteristics, such as arithmetic mean and standard deviation. A characteristic of this set of numbers is called a *parameter*. For example, assume that the population consists of five spheres and that they have the following diameters: 10, 13, 14, 19, and 21 cm. The diameters have an arithmetic mean of 15.4 cm and standard deviation of 4.03 cm, and these values are parameters of the given population.

Now consider that a subset of these objects is drawn. This subset is called a *sample*, and a characteristic of the sample is called a *statistic*. Thus, using the previous illustration, assume that the sample consists of the spheres having diameters of 14, 19, and 21 cm. These diameters have an arithmetic mean of 18 cm and standard deviation of 2.94 cm, and these values are statistics of the sample drawn. The number of objects in the sample is the sample *size*.

In many instances, it is impossible to evaluate a parameter precisely, for two reasons: The population may be so large as to preclude measurement of every object, and measurement may entail destruction of the object, as in finding the breaking strength of a cable. In these cases, it is necessary to *estimate* the parameter by drawing a representative sample and evaluating the corresponding statistic. The process of estimating a parameter by means of a statistic is known as *statistical inference*.

Since a statistic is a function of the manner in which the sample is drawn and thus is influenced by chance, the statistic is a random variable. The probability distribution of a statistic is called the *sampling distribution* of that statistic. Consider that all possible samples of a given size have been drawn and the statistic S corresponding to each sample has been calculated. A characteristic of this set of values of S, such as the arithmetic mean, is referred to as a characteristic of the sampling distribution of S. In the subsequent material, the term *mean* refers exclusively to the arithmetic mean.

TABLE 26   Notation

Characteristic	Sample	Population	Sampling distribution of a statistic S
Mean	$\overline{X}$	μ	μ_S
Standard deviation	s	σ	σ_S
Number of items	n	N	

Notational System

Table 26 is presented for ease of reference. Here N = number of objects in the population; μ and σ = arithmetic mean and standard deviation, respectively, of the population; n = number of objects in the sample; $\overline{X}$ and s = arithmetic mean and standard deviation, respectively, of the sample; μ_S and σ_S = arithmetic mean and standard deviation, respectively, of the sampling distribution of the statistic S.

Basic Equations

The mean and standard deviation of the sampling distribution of the mean are

$$\mu_{\overline{X}} = \mu \tag{14}$$

$$\sigma_{\overline{X}} = \sigma \sqrt{\frac{N-n}{n(N-1)}} \tag{15}$$

If the population is infinite, Eq. 15 reduces to

$$\sigma_{\overline{X}} = \frac{\sigma}{\sqrt{n}} \tag{15a}$$

The quantity $\sigma_{\overline{X}}$ is an index of the diversity of the sample means. As the sample size increases, the samples become less diverse.

Since a sample represents a combination of N objects taken n at a time, the number of samples that may be drawn is $C_{N,n} = N!/[n!(N-n)!]$.

SAMPLING DISTRIBUTION OF THE MEAN

The population consists of 5 objects having the numerical values 15, 18, 27, 36, and 54; the sample size is 3. Find the mean and standard deviation of the sampling distribution of the mean.

Calculation Procedure:

1.   Compute the mean and variance of the population

Mean $\mu = (15 + 18 + 27 + 36 + 54)/5 = 30$; variance $\sigma^2 = [(15 - 30)^2 + (18 - 30)^2 + (27 - 30)^2 + (36 - 30)^2 + (54 - 30)^2]/5 = 198$.

2.   Compute the properties of the sampling distribution

Apply Eq. 14 to find the mean of the sampling distribution of the mean, or $\mu_{\overline{X}} = 30$. Apply Eq. 15 to find the variance of the sampling distribution, or $\sigma_{\overline{X}}^2 = 198(5 - 3)/(3 \times 4) = 33$. Then $\sigma_{\overline{X}} = \sqrt{33} = 5.74$.

3.   Compute the required properties without recourse to any set equations

If the population is finite, the number of possible samples is finite. Since all samples have an equal likelihood of becoming the true sample, the sampling distribution of a statistic can be found by forming all possible samples and computing the statistic under consideration for each.

Record all possible samples in the first column of Table 27; the number of these samples is

$C_{5,3} = 10$. Now compute the mean $\overline{X}$ of each sample, record the results in the second column, and total them. This column contains full information concerning the sampling distribution of the mean. Thus, since no duplications occur, $P(\overline{X} = 20) = 1/10$; $P(\overline{X} = 23) = 1/10$; etc. Compute the mean of the possible values of $\overline{X}$, or $\mu_{\overline{X}} = 300/10 = 30$. Record the deviations from 30 in Table 27, square the deviations, and total the results. Compute the variance of the possible values of $\overline{X}$, or $\sigma_{\overline{X}}^2 = 330/10 = 33$. Then $\sigma_{\overline{X}} = \sqrt{33} = 5.74$. These results are consistent with those in step 2.

TABLE 27 Properties of Sampling Distribution of the Mean

Sample	Sample mean $\overline{X}$	Deviation d $= \overline{X} - 30$	d^2
15, 18, 27	20	−10	100
15, 18, 36	23	−7	49
15, 18, 54	29	−1	1
15, 27, 36	26	−4	16
15, 27, 54	32	2	4
15, 36, 54	35	5	25
18, 27, 36	27	−3	9
18, 27, 54	33	3	9
18, 36, 54	36	6	36
27, 36, 54	39	9	81
Total	300	0	330

ESTIMATION OF POPULATION MEAN ON BASIS OF SAMPLE MEAN

A firm produces rods, and their lengths vary slightly because of unavoidable differences in manufacture. Assume that the lengths are normally distributed. One hundred rods were selected at random, and they were found to have a mean length of 1.856 m and a standard deviation of 0.074 m. Estimate the mean length of all rods manufactured by this firm, using a 95 percent confidence level.

Calculation Procedure:

1. Compute the z value corresponding to the given confidence level

Let X = length of a rod. The population consists of all rods manufactured by the firm, and it may be considered infinite. There are two types of estimates: a *point estimate*, which assigns a specific value to X, and an *interval estimate*, which states that X lies within a given interval. This interval is called a *confidence interval*, its boundaries are called the *confidence limits*, and the probability that the estimate is correct is called the *confidence level*, or *confidence coefficient*. Statistical inference can supply only interval estimates.

The *central-limit theorem* states: (*a*) If the population is extremely large and the probability distribution of X is normal, then the sampling distribution of the sample mean $\overline{X}$ is also normal; (*b*) if the population is extremely large but the probability distribution of X is not normal, the sampling distribution of $\overline{X}$ is approximately normal if the sample size is 30 or more. Thus, in the present case, the sampling distribution of $\overline{X}$ is considered to be normal.

Figure 23 is the sampling distribution diagram of the sample mean $\overline{X}$. Let M = area under

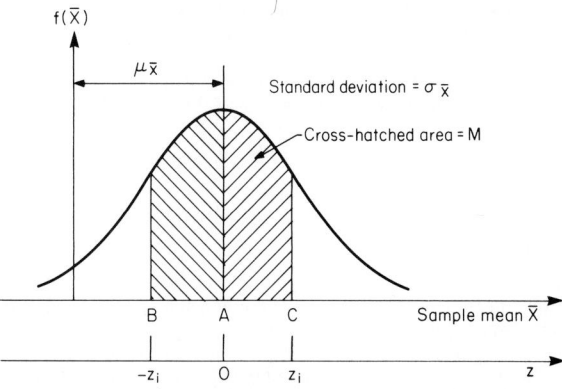

FIG. 23 Sampling distribution of the mean.

curve from B to C. If a sample is drawn at random, there is a probability M that the true sample mean $\overline{X}_j$ lies within the interval BC, or $P(\mu_{\overline{X}} - z_i\sigma_{\overline{X}} < \overline{X}_j < \mu_{\overline{X}} + z_i\sigma_{\overline{X}}) = M$. This equation can be transformed to $P(\overline{X}_j - z_i\sigma_{\overline{X}} < \mu_{\overline{X}} < \overline{X}_j + z_i\sigma_{\overline{X}}) = M$, Eq. a.

In the present case, $\overline{X}_j = 1.856$ m, $s = 0.074$ m, $n = 100$, and $M = 0.95$. Then area under curve from A to C $= (0.50)(0.95) = 0.475$. From the table of areas under the normal curve, if $A(z_i) = 0.475$, $z_i = 1.96$.

2. Set up expressions for the mean and standard deviation of the sampling distribution of the mean

By Eq. 14, $\mu_{\overline{X}} = \mu$, where μ is the population mean to be estimated. Use the standard deviation s of the sample as an estimate of the standard deviation σ of the population. Then $\sigma = 0.074$ m, and by Eq. 15a, $\sigma_{\overline{X}} = 0.074/\sqrt{100} = 0.0074$ m.

3. Estimate the mean length of the rods

Refer to Eq. a, and compute $z_i \sigma_{\overline{X}} = (1.96)(0.0074) = 0.015$ m. Compute the confidence limits in Eq. a, or $1.856 - 0.015 = 1.841$ m and $1.856 + 0.015 = 1.871$ m. Equation a becomes $P(1.841 < \mu < 1.871) = 0.95$. Thus, there is a 95 percent probability that the mean length of all rods lies between 1.841 and 1.871 m.

Related Calculations: Note that the confidence interval is a function of the degree of probability that is demanded, and the two quantities vary in the same direction. For example, if the confidence level were 90 percent, the confidence interval would be 1.844 to 1.868 m.

DECISION MAKING ON STATISTICAL BASIS

Units of a commodity are produced individually, and studies have shown that the time required to produce a unit has a mean value of 3.50 h and a standard deviation of 0.64 h. An industrial engineer claims that a modification of the production process will substantially reduce production time. The proposed method was tested on 40 units, and it was found that the mean production time was 3.37 h per unit. Management has decided that it will make the proposed modification only if there is a probability of 95 percent or more that the engineer's claim is valid. What is your recommendation?

Calculation Procedure:

1. Formulate the null and alternative hypotheses

The population consists of all units that will be produced under the modified method if it is adopted, and the sample consists of the 40 units actually produced under this method. An assumption based on conjecture is termed a *hypothesis*. A hypothesis that is formulated merely to provide a basis for investigation is a *null* hypothesis, and any hypothesis that contradicts the null hypothesis is an *alternative* hypothesis. However, interest centers on the particular alternative hypothesis that is significant in the given case. The null and alternative hypotheses are denoted by H_0 and H_1, respectively.

Let X = time required to produce 1 unit, h. Place the burden of proof on the industrial engineer by assuming that production time under the modified method is identical with that under the present method. Thus, the hypotheses are H_0: $\mu = 3.50$ h and $\sigma = 0.64$ h; H_1: $\mu < 3.50$ h.

2. Compute the properties of the sampling distribution of the mean as based on the null hypothesis

Apply Eqs. 14 and 15a, giving $\mu_{\overline{X}} = 3.50$ h and $\sigma_{\overline{X}} = 0.64/\sqrt{40} = 0.101$ h.

3. Compute the critical value of $\overline{X}$

By the central-limit theorem given in the preceding calculation procedure, the sampling distribution of the sample mean $\overline{X}$ may be considered normal, and the sampling distribution diagram is shown in Fig. 24a. Management has imposed a requirement of 95 percent probability. Therefore, the null hypothesis is disproved and the alternative hypothesis validated if the true sample mean has a value less than that corresponding to 95 percent of all possible samples. In Fig. 24a, locate B such that the area to the right of B $= 0.95$; then area from A to B $= 0.95 - 0.50 = 0.45$. The null hypothesis is to be accepted or rejected according to whether the true value of $\overline{X}$

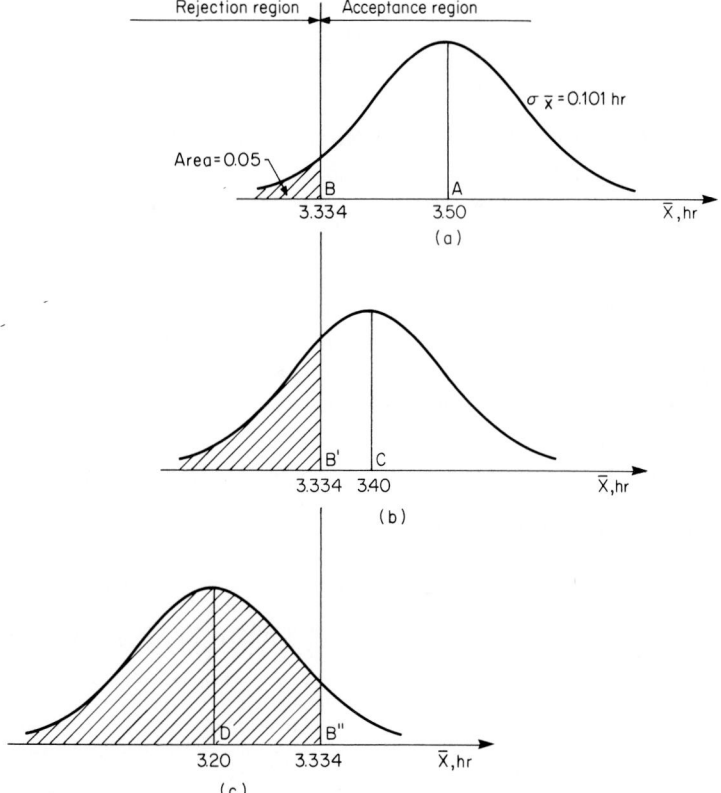

FIG. 24 Sampling distribution of mean production time corresponding to three distinct values of the population mean.

lies to the right or left of B, respectively, and the regions are labeled as shown. The values of $\overline{X}$ and z at the boundary of the acceptance and rejection regions are called the *critical* values.

At B, $A(z) = 0.45$. From the table of areas under the normal curve, $z = -1.645$. Thus, at B, $\overline{X} = 3.50 + (-1.645)(0.101) = 3.334$ h.

4. Make a recommendation

Since the true sample mean of 3.37 h falls to the right of B, the null hypothesis stands. Therefore, we must recommend that the modified method of production be disapproved and the present method retained.

This recommendation does not necessarily imply that the industrial engineer's claim is invalid. The decision must be based on probability rather than certainty, and the test results have failed to demonstrate a 95 percent probability that the modified method is superior to the present method. The difference between the assumed population mean of 3.50 h and the sample mean of 3.37 h can be ascribed to chance.

Related Calculations: If the null hypothesis is rejected when, in fact, it is true, a *Type I error* has been committed. The probability of committing this error is denoted by α, and the acceptable value of α is termed the *level of significance*. In this case, if the null hypothesis is correct, the sampling distribution of $\overline{X}$ is as shown in Fig. 24a. The hypothesis will be rejected if $\overline{X}$ assumes a value to the left of B, and the probability of this event is $1 - 0.95 = 0.05$. Thus, the level of significance is 0.05.

Whether a null hypothesis is accepted or rejected depends largely on the level of significance imposed. Therefore, selecting an appropriate level of significance is one of the crucial problems that arise in statistical decision making. The selection must be based on the amount of the loss that would result from a false decision.

PROBABIILTY OF ACCEPTING A FALSE NULL HYPOTHESIS

With reference to the preceding calculation procedure, the time required to produce 1 unit under the modified method has these characteristics: The standard deviation remains 0.64 h, but the arithmetic mean is (*a*) 3.40 h; (*b*) 3.20 h. Determine the probability that the industrial engineer's proposal will be vetoed despite its merit.

Calculation Procedure:

1. Compute the critical value of z in each case

Since σ remains 0.64 h and the sample size is still 40, $\sigma_{\bar{X}}$ remains 0.101 h. Refer to Fig. 24*b* and *c*, which gives the sampling distributions of $\bar{X}$ when $\mu = 3.40$ h and $\mu = 3.20$ h, respectively. The null hypothesis will be accepted if $\bar{X} > 3.334$, and it is necessary to calculate the probability of this event.

In Fig. 24*b*, at *B'*, $z = (3.334 - 3.40)/0.101 = -0.653$. In Fig. 24*c*, at *B''*, $z = (3.334 - 3.20)/0.101 = 1.327$.

2. Compute the required probabilities

Refer to the table of areas under the normal curve. When $z = -0.653$, $A(z) = 0.243$. In Fig. 24*b*, area to right of *B'* = $0.243 + 0.5 = 0.743$. Thus, when $\mu = 3.40$ h, there is a probability of 74.3 percent that the proposal will be vetoed. Similarly, when $z = 1.327$, $A(z) = 0.408$. In Fig. 24*c*, area to right of *B''* = $0.5 - 0.408 = 0.092$. Thus, when $\mu = 3.20$ h, there is a probability of 9.2 percent that the proposal will be vetoed.

Related Calculations: If the null hypothesis is accepted when, in fact, it is false, a *Type II error* has been committed, and the probability of committing this error is denoted by β. Thus, this calculation procedure involves the determination of β. It follows that $1 - \beta$ is the probability that a false null hypothesis will be rejected. The process of drawing and analyzing a sample represents a *test* of the null hypothesis, and the quantities β and $1 - \beta$ are called the *operating characteristic* and *power*, respectively, of the test. Thus, the power of a test is its ability to detect that the null hypothesis is false if such is truly the case.

Consider that a diagram is constructed in which assumed values of the parameter are plotted on the horizontal axis and the corresponding values of β resulting from the null hypothesis are plotted on the vertical axis. The curve thus obtained is called an *operating-characteristic curve*. Similarly, the curve obtained by plotting values of $1 - \beta$ against assumed values of the parameter is called a *power curve*.

DECISION BASED ON PROPORTION OF SAMPLE

A firm receives a large shipment of small machine parts, and it must determine whether the number of defectives in a shipment is tolerable. Its policy is as follows: A shipment is accepted only if the estimated incidence of defectives is 3 percent or less, the decision is based on an inspection of 250 parts selected at random, and a shipment is rejected only if there is a probability of 90 percent or more that the incidence of defectives exceeds 3 percent. What is the highest incidence of defectives in the sample if the shipment is to be considered acceptable?

Calculation Procedure:

1. Formulate the null hypothesis

Consider that a set of objects consists of type A and type B objects. The ratio of the number of type A objects to the total number of objects is called the *proportion* of type A objects. Let *P* and *p* = proportion of type A objects in the population and sample, respectively.

In this case, the population consists of all machine parts in the shipment, the sample consists of the 250 parts that are inspected, and interest centers on the proportion of defective parts. To provide a basis for investigation, assume that the proportion of defectives in the shipment is precisely 3 percent. Thus, the null hypothesis is H_0: $P = 0.03$.

2. *Compute the properties of the sampling distribution of the proportion as based on the null hypothesis*

Consider that all possible samples of a given size are drawn and their respective values of p determined, thus obtaining the sampling distribution of p. As before, let N and n = number of objects in the population and sample, respectively. The sampling distribution of p has these values: the mean $\mu_p = P$, Eq. *a*; the variance $\sigma_p^2 = P(1 - P)(N - n)/[n(N - 1)]$, Eq. *b*. Where the population is infinite, $\sigma_p^2 = P(1 - P)/n$, Eq. *c*. In the present case, $P = 0.03$, N may be considered infinite, and $n = 250$. By Eq. *a*, $\mu_p = 0.03$; by Eq. *c*, $\sigma_p^2 = (0.03)(0.97)/250 = 0.0001164$. Then standard deviation $\sigma_p = 0.0108$.

3. *Compute the critical value of p*

For simplicity, treat the number of defective parts as a continuous rather than a discrete variable; then P and p are also continuous. Since the sample is very large, the sampling distribution of p is approximately normal, and it is shown in Fig. 25*a*. The null hypothesis is to be rejected if p assumes a value greater than that corresponding to 90 percent of all possible samples. In Fig. 25*a*, locate B such that area to left of $B = 0.90$; then area from A to $B = 0.90 - 0.50 = 0.40$. At B, $A(z) = 0.40$. From the table of areas under the normal curve, $z = 1.282$. Thus, at B, $p = 0.03 + (1.282)(0.0108) = 0.044$.

4. *State the decision rule*

If the proportion of defectives in the sample is 4.4 percent or less, accept the shipment; if the proportion is greater, reject the shipment.

By setting the limiting proportion of defectives in the sample at 4.4 percent as compared with

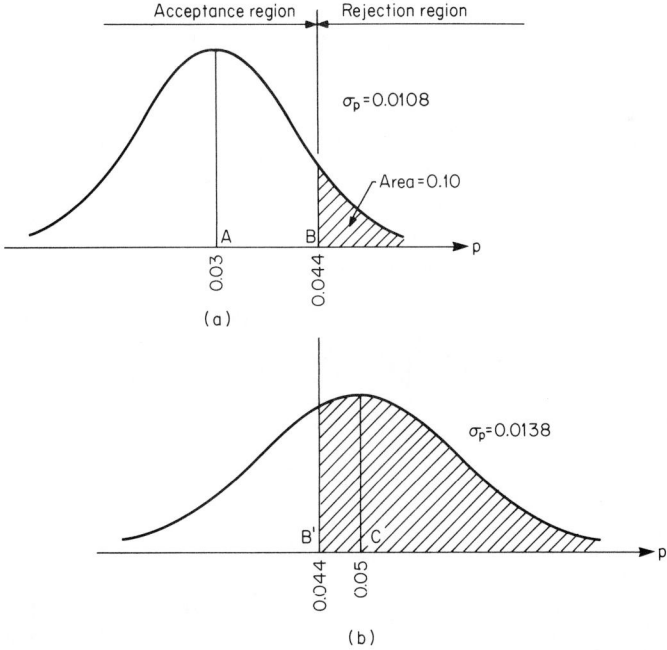

FIG. 25 Sampling distribution of proportion of defective units.

the limiting proportion of 3 percent in the population, allowance is being made for the random variability of sample results.

PROBABILITY OF ACCEPTING AN UNSATISFACTORY SHIPMENT

With reference to the preceding calculation procedure, what is the probability that a shipment in which the incidence of defectives is 5 percent will nevertheless be accepted?

Calculation Procedure:

1. *Compute the true properties of the sampling distribution of the proportion*

Apply Eqs. *a* and *c* of the previous calculation procedure, giving $\mu_p = 0.05$ and $\sigma_p^2 = (0.05)(0.95)/250 = 0.0001900$. Then $\sigma_p = 0.0138$. The sampling distribution diagram appears in Fig. 25*b*.

2. *Compute the probability that the shipment will be accepted*

The shipment will be accepted if $p < 0.044$, and it is necessary to calculate the probability of this event. In Fig. 25*b*, at B', $z = (0.044 - 0.05)/0.0138 = -0.435$. From the table of areas under the normal curve, $A(z) = 0.168$. Then the area to left of $B' = 0.5 - 0.168 = 0.332$. Thus, there is a probability of 33.2 percent that the shipment will be accepted.

Reliability

Consider that a device operates continuously and fails abruptly. The life span of the device is a continuous random variable. The reliability of the device corresponding to a given length of time t, denoted by $R(t)$, is the probability that its life span will exceed t. Let T = life span. Then $R(t) = P(T > t)$.

Refer to Fig. 26, which is the assumed *life-span curve* of a device. The diagram is constructed

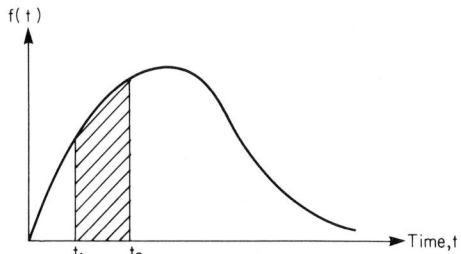

FIG. 26   Life-span curve.

so that $P(t_1 < T \leq t_2)$ = area under curve from t_1 to t_2. Thus, a life-span curve is the probability curve of the continuous variable T. Left $f(t)$ = ordinate of life-span curve = probability-density function. From the definition of reliability, it follows that $R(t)$ = area under curve to right of t, or

$$R(t) = \int_t^\infty f(t)\, dt \tag{16a}$$

and

$$f(t) = -\frac{dR(t)}{dt} \tag{16b}$$

A mechanism formed by the assemblage of devices is called a *system*, and the individual device is called a *component* of the system. Assume that a system consists of two components, C_1 and C_2, and let the subscripts 1, 2, and S refer to C_1, C_2, and the system, respectively. If the components are arranged in series, as shown in Fig. 27, the system is operating only if *both* com-

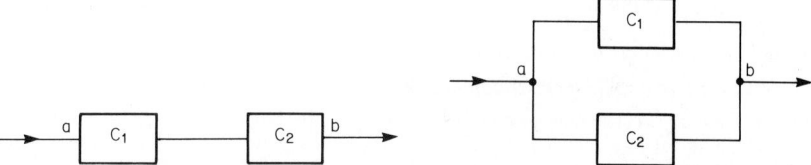

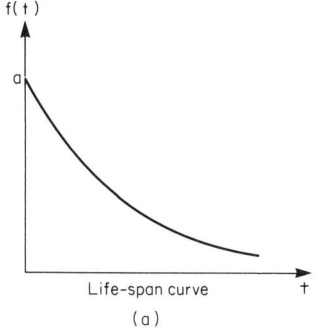

FIG. 27   System with components in series.

FIG. 28   System with components in parallel.

ponents are operating. Thus, the reliability of the system is

$$R_S(t) = R_1(t)R_2(t) \qquad (17)$$

If the components are arranged in parallel, as shown in Fig. 28, the system is operating if *either* component is operating. To express it another way, the system fails if both components fail, and

$$1 - R_S(t) = [1 - R_1(t)][1 - R_2(t)]$$

or

$$R_S(t) = 1 - [1 - R_1(t)][- R_2(t)] \qquad (18)$$

Equations 17 and 18 can be generalized to include any number of components. A system having components in series has a reliability less than that of any component; a system having components in parallel has a reliability greater than that of any component.

DEVICE WITH NEGATIVE-EXPONENTIAL LIFE SPAN

A certain type of earth satellite has a negative-exponential life span with a mean value of 15 months. Four satellites of this type are launched simultaneously. If X denotes the number of satellites that remain in orbit at the expiration of 1 year, establish the probability distribution of X.

Calculation Procedure:

1. *Compute the probability that a particular satellite will survive the first year*

Refer to the earlier calculation procedure pertaining to the negative-exponential probability distribution. A device has a negative-exponential life span if its life-span curve has an equation of this form: $f(t) = ae^{-at}$, Eq. *a*, where a = positive constant and e = base of natural logarithms. From Eq. 16a, $R(t) = e^{-at}$, Eq. *b*. Figure 29 presents the life-span and reliability diagrams. The mean life span $\mu = 1/a$, Eq. *c*.

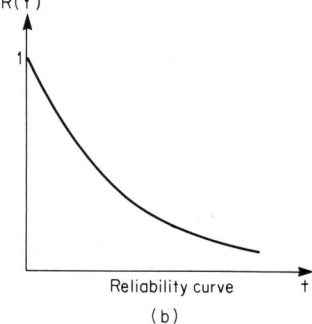

FIG. 29

Take 1 month as the unit of time. In this case, $\mu = 15$; then $a = 1/15$. By Eq. b, $R(t) = e^{-t/15}$, or $R(12) = e^{-12/15} = e^{-0.8} = 1/e^{0.8} = 0.4493$. Thus, there is a probability of 0.4493 that a particular satellite will survive the first year.

2. Establish the probability distribution of X

Refer to the earlier calculation procedure pertaining to the binomial probability distribution. Launching a satellite may be viewed as a trial to determine whether the satellite will survive the first year. Since all satellites operate independently, the trials are independent of one another, and therefore X has a binomial probability distribution.

Apply the equation in step 2 of the procedure for the binomial distribution, with $n = 4$, $P = 0.4493$, and $1 - P = 0.5507$. Now, $C_{4,0} = C_{4,4} = 1$; $C_{4,1} = C_{4,3} = 4$; $C_{4,2} = 6$. Then $P(0) = 1(0.5507)^4 = 0.092$; $P(1) = 4(0.4493)(0.5507)^3 = 0.300$; $P(2) = 6(0.4493)^2(0.5507)^2 = 0.367$; $P(3) = 4(0.4493)^3(0.5507) = 0.200$; $P(4) = 1(0.4493)^4 = 0.041$. These probabilities total 1, as they must.

Related Calculations: It can readily be shown that a device that has a negative-exponential life span and has been operating for some time has the same probability of surviving the next unit of time as one that was just activated. Thus, the age of the device is completely irrelevant in predicting its remaining life. Thus failure is caused not by cumulative damage resulting from use but by a sudden accidental occurrence, and the probability of this occurrence during the next unit of time is independent of the age of the device. To apply an analogy, the probability that an individual who travels extensively by plane will become the victim of a plane crash during the next year has no relation to the amount of air travel that person has done in the past.

CORRESPONDENCE BETWEEN POISSON FAILURE AND NEGATIVE-EXPONENTIAL LIFE SPAN

The failure of a certain type of device has a Poisson probability, and the mean number of failures in 1500 h of operation is 6. What is the reliability of the device for 270 h of operation?

Calculation Procedure:

1. Compute the mean life span

Refer to the previous calculation procedure pertaining to the Poisson probability distribution. Consider the following: A device is set in operation. When this device fails, a device of identical type is set in operation, and this replacement process continues indefinitely. Let X denote the number of failures in time t, and assume that X has a Poisson probability distribution. Thus the probability that exactly 1 failure will occur in 1 h remains constant as time elapses. It can readily be shown that the life span of the device, which is the time interval between successive failures, has a negative-exponential probability distribution.

Select 1 h as the unit of time. Then mean life span $\mu = 1500/6 = 250$ h.

2. Compute the reliability

Apply Eq. c of the preceding calculation procedure, giving $a = 1/\mu = 1/250$. Now apply Eq. b, giving $R(270) = e^{-270/250} = e^{-1.08} = 0.340$.

PROBABILITY OF FAILURE DURING A SPECIFIC PERIOD

The life span of a device is negative-exponential, and its mean value is 8 days. What is the probability that the device will fail on the fifth day?

Calculation Procedure:

1. Set up the probability equation

Set $T = $ life span, and refer to Fig. 26. As previously stated, $P(t_1 < T \le t_2) = $ area under curve from t_1 to t_2. It follows that $P(t_1 < T \le t_2) = R(t_1) - R(t_2)$. Take 1 day as the unit of time. So $P(4 < T \le 5) = R(4) - R(5)$.

2. *Compute the probability*

Set $a = 1/\mu = 1/8$. Then $R(t) = e^{-at} = e^{-t/8}$, or $R(4) = e^{-4/8} = 0.607$ and $R(5) = e^{-5/8} = 0.535$. Thus, $P(4 < T \leq 5) = 0.607 - 0.535 = 0.072$.

Related Calculations: Assume that the device has been operating for some time. In accordance with the statement previously made, the probability that this device will fail on the fifth day from the present is 0.072, regardless of the present age of the device.

SYSTEM WITH COMPONENTS IN SERIES

A system consists of three type A components and two type B components, all arranged in series. These components have negative-exponential life spans, and the mean life span is 30 h for type A and 36 h for type B. Find the reliability of the system for 9 h of operation.

Calculation Procedure:

1. *Compute the reliability of each component for 9 h*

Let the subscripts A and B refer to the type of component, and take 1 h as the unit of time. Then $a_A = 1/\mu_A = 1/30$, and $R_A(t) = e^{-t/30}$, or $R_A(9) = e^{-9/30} = 0.7408$. Similarly, $R_B(9) = e^{-9/36} = 0.7788$.

2. *Compute the reliability of the system for 9 h*

Apply Eq. 17: $R_S(9) = (0.7408)^3(0.7788)^2 = 0.247$.

3. *Compute the reliability of the system by an alternative method*

Apply Eq. 17 to prove that the system also has a negative-exponential life span. Now assume that when a component fails, it is instantly replaced with one of identical type, thus maintaining continuous operation. In 180 h of operation, the mean number of failures of an individual component is $180/30 = 6$ for type A and $180/36 = 5$ for type B. Since failure of any component causes failure of the system, the mean number of failures of the system in 180 h is $3 \times 6 + 2 \times 5 = 28$. Thus, the mean life span of the system is $\mu_S = 180/28$, and $a_S = 28/180$. Then $a_S t = (28/180)9 = 1.4$, and $R_S(9) = e^{-1.4} = 0.247$.

In the foregoing calculations, 180 h was selected for convenience, since 180 is the lowest common multiple of 30 and 36. However, a period of any length whatever can be selected.

SYSTEM WITH COMPONENTS IN PARALLEL

With reference to the system in Fig. 30, the reliability of each component for a 60-day period is as indicated. For example, the reliability of C_1 is 0.18. Determine the probability that the system will be operating at the expiration of 60 days as a result of each of the following causes: (*a*) only one component survives; (*b*) only two components survive; (*c*) all three components survive.

Calculation Procedure:

1. *Compute the probability that a component fails during the 60-day period*

The probability of failure is: for C_1, $1 - 0.18 = 0.82$; for C_2, $1 - 0.23 = 0.77$; for C_3, $1 - 0.15 = 0.85$.

2. *Compute the probability that the system survives because one and only one component survives*

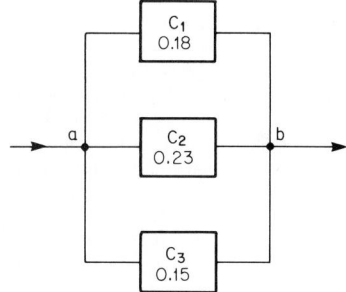

FIG. 30

Multiply the probabilities of the individual events. Thus, P(only C_1 survives) $= (0.18)(0.77)(0.85) = 0.11781$; P(only C_2 survives) $= (0.82)(0.23)(0.85) = 0.16031$; P(only C_3 survives) $= (0.82)(0.77)(0.15) = 0.09471$. Sum the results: $0.11781 + 0.16031 + 0.09471 = 0.37283$. This is the probability that only one component survives.

3. *Compute the probability that the system survives because two and only two components survive*

Proceed as in step 2. Thus, P(only C_1 and C_2 survive) = $(0.18)(0.23)(0.85)$ = 0.03519; P(only C_1 and C_3 survive) = $(0.18)(0.77)(0.15)$ = 0.02079; P(only C_2 and C_3 survive) = $(0.82)(0.23)(0.15)$ = 0.02829. Sum the results: 0.03519 + 0.02079 + 0.02829 = 0.08427 = the probability that only two components survive.

4. *Compute the probability that the system survives because all three components survive*

P(all survive) = $(0.18)(0.23)(0.15)$ = 0.00621.

5. *Verify the foregoing results*

Sum the results in steps 2, 3, and 4: $R_S(60)$ = 0.37283 + 0.08427 + 0.00621 = 0.46331. Now apply Eq. 18, giving $R_S(60)$ = 1 − $(0.82)(0.77)(0.85)$ = 0.46331. The equality of the two values confirms the results obtained in the previous steps.

SYSTEM WITH IDENTICAL COMPONENTS IN PARALLEL

A certain type of component has a reliability of 0.12 for 20 days. How many such components must be connected in parallel if the reliability of the system is to be at least 0.49 for 20 days?

Calculation Procedure:

1. *Write the equation for the reliability of the system*

Let n = number of components required. Apply Eq. 18, giving $R_S(20)$ = 1 − $(1 − 0.12)^n$ = 1 − $(0.88)^n$, Eq. a.

2. *Determine the number of components*

Set $R_S(20)$ = 0.49 and solve Eq. a for n, giving n = $(\log 0.51)/(\log 0.88)$ = 5.3. Use six components.

ANALYSIS OF COMPOSITE SYSTEM BY CONVENTIONAL METHOD

A system is constructed by arranging the components in the manner shown in Fig. 31a, and the reliability of each component for a given time t is recorded in the drawing. For example, the reliability of C_1 is 0.58. Find the reliability of the system for time t.

Calculation Procedure:

1. *Perform the first cycle in transforming the system to a simpler one*

A system in which the components are arranged solely in series or in parallel is a *simple* system, and one that combines series and parallel arrangements is a *composite* system. A composite system may be regarded as composed of *subsystems*, with a subsystem being a set of components arranged solely in series or parallel formation. A subsystem can be replaced with its *resultant*, which is a single component that has a reliability equal to that of the subsystem. Therefore, the conventional method of analyzing a composite system consists of resolving the system into subsystems and replacing the subsystems with their resultants, continuing the process until the given system has been transformed to an equivalent simple system.

Let D_1 and D_2 denote the resultants of C_3 and C_4 and of C_5 and C_6, respectively. Apply Eq. 18 to obtain these reliabilities: for D_1, 1 − $(0.85)(0.83)$ = 0.29450; for D_2, 1 − $(0.79)(0.70)$ = 0.44700. Replace these components with their resultants, producing the equivalent composite system shown in Fig. 31b.

2. *Perform the second cycle*

Let D_3 denote the resultant of C_1, C_2 and D_1. Apply Eq. 17: reliability of D_3 = $(0.58)(0.49)(0.29450)$ = 0.08370. Replace these components with D_3, producing the equivalent simple system shown in Fig. 31c.

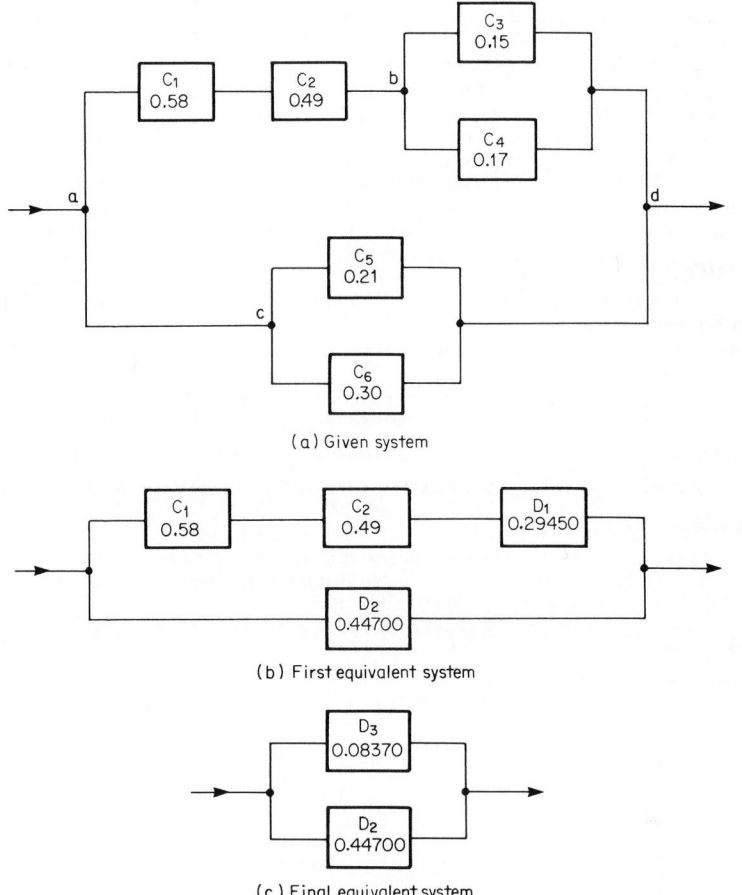

(a) Given system

(b) First equivalent system

(c) Final equivalent system

FIG. 31

3. Compute the reliability of the system

Apply Eq. 18. The reliability of the system in Fig. 31c is $R_S(t) = 1 - (0.91630)(0.55300) = 0.49329$. This is also the reliability of the original system.

ANALYSIS OF COMPOSITE SYSTEM BY ALTERNATIVE METHOD

With reference to the preceding calculation procedure, find the reliability of the system by the moving-particle method.

Calculation Procedure:

1. Compute the number of particles that traverse the system by way of C_1, C_2, and either C_3 or C_4

The alternative method of analyzing a composite system is based on this conception: During a given interval, a certain number of particles enter the system at one terminal and seek to move through the system to the other terminal. Each component offers resistance to the movement of these particles, and the proportion of particles that penetrate a component is equal to the reliability of that component. If a particle is obstructed at any point along its path, it returns to an earlier point and then proceeds along an alternative path if one is available. However, a particle can enter a component in a given direction only once. The reliability of the system is equal to the proportion of particles that traverse the system.

Consider that during a given interval 1,000,000 particles arrive at point a in Fig. 31a, seeking a path to d. They can reach d by passing through C_1, C_2, and either C_3 or C_4, or by passing through C_5 or C_6. Assume that the particles attempt passage by the first route, and refer to the schematic drawing in Fig. 32. The number of particles that penetrate C_1 = 1,000,000(0.58) = 580,000, and the number that then penetrate C_2 = 580,000(0.49) = 284,200. Assume that these particles now enter C_3. The number of particles that penetrate C_3 = 284,200(0.15) = 42,630. The number that fail to penetrate C_3 = 284,200 − 42,630 = 241,570; these particles return to b in Fig. 31a and enter C_4. The number that penetrate C_4 = 241,570(0.17) = 41,067. Thus, the number of particles that reach d by way of C_1, C_2, and either C_3 or C_4 = 42,630 + 41,067 = 83,697.

As an alternative calculation, the number of particles that fail to penetrate C_3 = 284,200(0.85) = 241,570.

Note that the *proportion* of particles that traverse the system by the indicated route is 83,697/1,000,000 = 0.08370 (to five decimal places), and this is the reliability of D_3 in Fig. 31c.

2.   *Compute the number of particles that traverse the system by way of either C_5 or C_6*

Refer to Fig. 32. The number of particles that fail to penetrate a component is: for C_1, 1,000,000 − 580,000 = 420,000; for C_2, 580,000 − 284,200 = 295,800; for C_4, 241,570 − 41,067 = 200,503. The total is 420,000 + 295,800 + 200,503 = 916,303. These particles return to a in Fig. 31a and then proceed to c. Assume that they now enter C_5. The number of particles that penetrate C_5 = 916,303(0.21) = 192,424, and the number that fail to penetrate C_5 = 916,303 − 192,424 = 723,879. The latter enter C_6, and the number that penetrate C_6 = 723,879(0.30) = 217,164. Thus, the number of particles that reach d by way of either C_5 or C_6 = 192,424 + 217,164 = 409,588.

Note that the *proportion* of particles that traverse this route is 409,588/916,303 = 0.44700, and this is the reliability of D_2 in Fig. 31c.

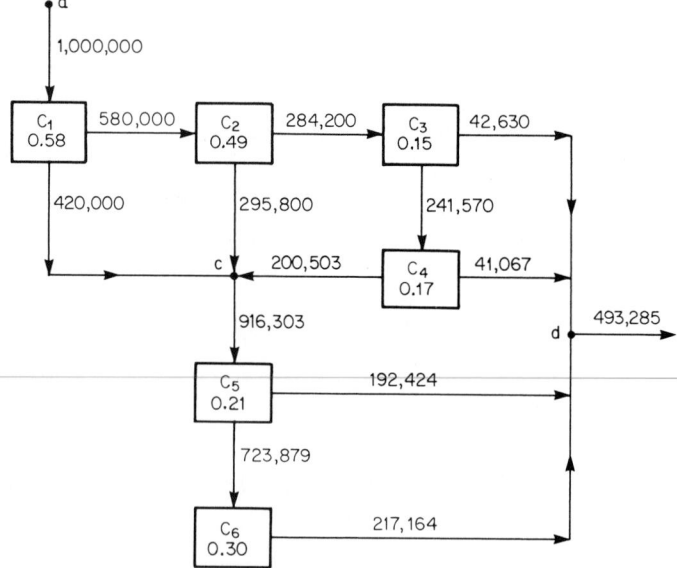

FIG. 32

3. *Compute the reliability of the system*

From steps 1 and 2 (or from Fig. 32), the number of particles that traverse the system from a to $d = 83,697 + 409,588 = 493,285$. The *proportion* of particles that traverse the system $= 493,285/1,000,000 = 0.49329$, and this is the reliability of the system. This result is consistent with that in the preceding calculation procedure.

Alternatively, find the number of particles that traverse the system thus: A particle fails to traverse the system if it fails to penetrate C_6. From Fig. 32, the number of such particles is 723,879 $- 217,164 = 506,715$. Thus, the number of particles that traverse the system $= 1,000,000 - 506,715 = 493,285$.

Related Calculations: The moving-particle method discloses certain principles very clearly. For example, assume that a simple composite system has components in series. To traverse the system, a particle must penetrate *all* components, and therefore the resistance of each component contributes to the resistance of the system. Thus, the resistance of the system exceeds that of any component, and the *reliability* of the system is less than that of any component. Now assume that a simple composite system has components in parallel. These components offer alternative paths for the moving particles, and therefore the reliability of the system exceeds that of any component.

ANALYSIS OF SYSTEM WITH SAFEGUARD BY CONVENTIONAL METHOD

A system is constructed by arranging the components in the manner shown in Fig. 33, and the reliability of each component for a given time t is recorded in the drawing. Find the reliability of the system for time t.

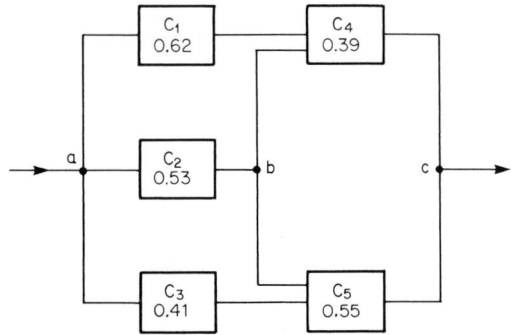

FIG. 33

Calculation Procedure:

1. *Identify the types of failure*

The system operates if any of the following pairs of components operate: C_1 and C_4; C_2 and C_4; C_2 and C_5; C_3 and C_5. Thus, if C_1 and C_3 both fail, the system continues to operate through C_2, and therefore C_2 is a *safeguard*. The reliability of the system can be found most simply by determining the probability that the system will fail.

There are several modes of potential failure, but they can all be encompassed within two broad types. A *type 1 failure* occurs if each of the following events occurs: (*a*) C_2 fails; (*b*) either C_1 or C_4 fails, or both fail; (*c*) either C_3 or C_5 fails, or both fail. A *type 2 failure* occurs if each of the following events occurs: (*a*) C_2 operates; (*b*) C_4 fails; (*c*) C_5 fails.

2. *Compute the probability of a type 1 failure*

To simplify the notation, let R = reliability of a component, with a subscript identical with that of the component, and R_S = reliability of the system.

As previously stated, if two events are independent of each other, the probability that *both* will occur is the product of their respective probabilities. Consider a type 1 failure. The probability that C_2 fails is $1 - R_2$. The probability that both C_1 and C_4 operate is R_1R_4; thus, the probability that either or both components fail is $1 - R_1R_4$. Similarly, the probability that C_3 or C_5 fails or both fail is $1 - R_3R_5$. Thus, the probability of a type 1 failure is $P(\text{type 1}) = (1 - R_2)(1 - R_1R_4)(1 - R_3R_5)$, or $P(\text{type 1}) = (0.47)[1 - (0.62)(0.39)][1 - (0.41)(0.55)] = 0.27600$.

3.   Compute the probability of a type 2 failure

Multiply the probabilities of the three specified events, giving $P(\text{type 2}) = R_2(1 - R_4)(1 - R_5)$, or $P(\text{type 2}) = (0.53)(0.61)(0.45) = 0.14549$.

4.   Compute the reliability of the system

The probability that either of two mutually exclusive events will occur is the sum of their respective probabilities. From steps 2 and 3, the probability that the system will fail is $0.27600 + 0.14549 = 0.42149$. Then $R_S = 1 - 0.42149 = 0.57851$.

ANALYSIS OF SYSTEM WITH SAFEGUARD BY ALTERNATIVE METHOD

With reference to the preceding calculation procedure, find the reliability of the system by the moving-particle method.

Calculation Procedure:

1.   Compute the number of particles that traverse the system by way of C_2

Consider that during a given interval 1,000,000 particles arrive at point a in Fig. 33, seeking a path to c. They can reach c by any of three routes: C_2 and either C_4 or C_5; C_1 and C_4; C_3 and C_5. Assume that the particles attempt passage by the first route, and refer to Fig. 34. The number of particles that penetrate $C_2 = 1,000,000(0.53) = 530,000$. Assume that these particles now enter C_4. The number of particles that penetrate $C_4 = 530,000(0.39) = 206,700$, and the number that

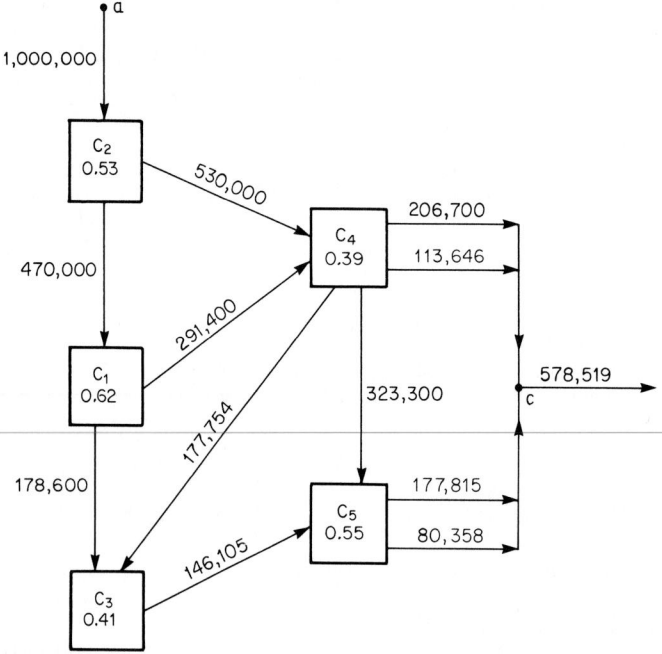

FIG. 34

fail to penetrate $C_4 = 530,000 - 206,700 = 323,300$. The latter return to b in Fig. 33 and then enter C_5. The number that penetrate $C_5 = 323,300(0.55) = 177,815$. Thus, the number of particles that reach c by way of C_2 and either C_4 or $C_5 = 206,700 + 177,815 = 384,515$.

2.   Compute the number of particles that traverse the system by way of C_1 and C_4

Refer to Fig. 34. The number of particles that fail to penetrate $C_2 = 1,000,000 - 530,000 = 470,000$. These particles return to a in Fig. 33; assume that they then enter C_1. The number of particles that penetrate $C_1 = 470,000(0.62) = 291,400$, and the number that then penetrate $C_4 = 291,400(0.39) = 113,646$. Thus, 113,646 particles reach c by way of C_1 and C_4.

3.   Compute the number of particles that traverse the system by way of C_3 and C_5

From step 2, the number of particles that fail to penetrate $C_1 = 470,000 - 291,400 = 178,600$, and the number that fail to penetrate $C_4 = 291,400 - 113,646 = 177,754$. These particles return to a in Fig. 33 and then enter C_3. Refer to Fig. 34. The number of particles that penetrate $C_3 = (178,600 + 177,754)(0.41) = 146,105$, and the number that then penetrate $C_5 = 146,105(0.55) = 80,358$. Thus, 80,358 particles reach c by way of C_3 and C_5.

4.   Compute the reliability of the system

From steps 1, 2, and 3 (or from Fig. 34), the number of particles that traverse the system from a to $c = 384,515 + 113,646 + 80,358 = 578,519$. The *proportion* of particles that traverse the system $= 578,519/1,000,000 = 0.57852$, and this is the reliability of the system. This result is consistent with that in the preceding calculation procedure.

Making Business Decisions under Uncertainty

The industrial world is characterized by uncertainty, and so many business decisions must be based on considerations of probability. The calculation procedures that follow illustrate several techniques that have been developed for making decisions of this type.

OPTIMAL INVENTORY TO MEET FLUCTUATING DEMAND

A firm sells a commodity that is used only during the winter. Because the commodity deteriorates with age, units of the commodity that remain unsold by the end of the season cannot be carried over to the following winter. To allow time for manufacture, the firm must place its order for the commodity before July 1. Thus, the firm must decide how many units to stock. For simplicity, the firm orders only in multiples of 10, and it assumes that the number of units demanded by its customers is also a multiple of 10.

A study of past records reveals that the number of units demanded per season ranges from 150 to 200, and the probabilities are as shown in Table 28. The cost of the commodity, including

TABLE 28   Demand Probabilities

Number of units demanded	150	160	170	180	190	200
Probability, percent	8	13	20	32	18	9

purchase price and allowance for handling, storage, and insurance, is $50 per unit; the selling price is $75 per unit. Units that are not sold can be disposed of as scrap for $6 each. If the firm is unable to satisfy the demand, it suffers a loss of goodwill because there is some possibility of permanently losing customers to a competitor; this loss of goodwill is assigned the value of $4 per unsold unit. How many units of this commodity should the firm order?

Calculation Procedure:

1.   Set up the equations for profit

A firm that sells a perishable commodity with a widely fluctuating demand runs a risk at each end of the spectrum. If its stock is excessive, it suffers a loss on the unsold units; if its stock is inadequate, it forfeits potential profits and suffers a loss of goodwill. So it must determine how large a stock to maintain to maximize profits in the long run, applying past demand as a guide.

Let X = number of units ordered; Y = number of units demanded; P = profit (exclusive of fixed costs), \$. If $X = Y$, then $P = (75 - 50)X$, or $P = 25X$, Eq. a. If $X > Y$, then $P = 75Y - 50X + 6(X - Y)$, or $P = -44X + 69Y$, Eq. b. If $X < Y$, then $P = (75 - 50)X - 4(Y - X)$, or $P = 29X - 4Y$, Eq. c.

2. *Construct the profit matrix*

In Table 29, list all possible values of X in the column at the left and all possible values of Y in the row across the top. Compute the value of P for every possible combination of X and Y, and

TABLE 29   Values of P

Number of units ordered X	Number of units demanded Y					
	150	160	170	180	190	200
150	3750	3710	3670	3630	3590	3550
160	3310	4000	3960	3920	3880	3840
170	2870	3560	4250	4210	4170	4130
180	2430	3120	3810	4500	4460	4420
190	1990	2680	3370	4060	4750	4710
200	1550	2240	2930	3620	4310	5000
Probability of Y	0.08	0.13	0.20	0.32	0.18	0.09

record the value in the table. Thus, assume $X = Y = 160$; by Eq. a, $P = 25 \times 160 = \$4000$. Now assume $X = 180$ and $Y = 160$; by Eq. b, $P = -44 \times 180 + 69 \times 160 = \$3120$. Finally, assume $X = 160$ and $Y = 200$; by Eq. c, $P = 29 \times 160 - 4 \times 200 = \$3840$. Table 29 shows that P can range from \$1550 (when the stock is highest and the demand is lowest) to \$5000 (when the demand is highest and the stock is adequate for the demand).

Alternatively, find the values of P thus: In Table 29, insert all values lying on the diagonal from the upper left-hand corner to the lower right-hand corner by applying Eq. a. In each column, proceed upward from this diagonal by successively deducting \$290, in accordance with Eq. c. Then proceed downward from the diagonal by successively deducting \$440, in accordance with Eq. b.

3. *Compute the expected profit corresponding to each value of X*

As stated earlier, if all possible values of a random variable are multiplied by their respective probabilities and the products are added, the result equals the arithmetic mean of the variable in the long run, and it is also called the *expected value* of the variable. For convenience, repeat the probability corresponding to every possible value of Y at the bottom of Table 29. Let $E(P)$ = expected value of P. When $X = 150$, $E(P) = \$3750(0.08) + \$3710(0.13) + \$3670(0.20) + \$3630(0.32) + \$3590(0.18) + \$3550(0.09) = \$3643.60$. When $X = 160$, $E(P) = \$3310(0.08) + \$4000(0.13) + \$3960(0.20) + \$3920(0.32) + \$3880(0.18) + \$3840(0.09) = \$3875.20$. Continue these calculations to obtain: when $X = 170$, $E(P) = \$4011.90$; when $X = 180$, $E(P) = \$4002.60$; when $X = 190$, $E(P) = \$3759.70$; when $X = 200$, $E(P) = \$3385.40$.

4. *Determine how many units the firm should order*

The results in step 3 show that the expected profit is maximum when $X = 170$. Therefore, the firm should order 170 units. If the fluctuation in demand follows the same pattern as in the past, the firm will maximize its profits in the long run by maintaining a stock of this size.

FINDING OPTIMAL INVENTORY BY INCREMENTAL-PROFIT METHOD

With reference to the preceding calculation procedure, determine how many units the firm should order by applying incremental analysis.

Calculation Procedure:

1. *Set up the equation for expected incremental profit*

Consider that the firm increases the number of units ordered by 10. In doing this, the firm has undertaken an *incremental investment,* and the profit that accrues from this incremental investment is called the *incremental profit.* Since the objective is to maximize profits from the sale of this commodity without reference to the rate of return that the firm earns on invested capital, the incremental investment is justified if the incremental profit has a positive value.

If a demand for these 10 additional units exists, the firm earns a direct profit of $10(\$75 - \$50) = \$250$, and it reduces its loss of goodwill by $10(\$4) = \$40$. Thus, the *effective* profit $= \$250 + \$40 = \$290$. If a demand for the 10 additional units does not exist, the firm incurs a loss of $10(\$50 - \$6) = \$440$. Let P(sold) and P(not sold) = probability the 10 additional units will be sold and will not be sold, respectively, and $E(\Delta P)$ = expected incremental profit, \$. Then $E(\Delta P) = 290$ $[P(\text{sold})] - 440 [P(\text{not sold})]$. Set $P(\text{not sold}) = 1 - P(\text{sold})$, giving $E(\Delta P) = 730 [P(\text{sold})] - 440$, Eq. *a.*

2. *Apply this equation to find the optimal inventory*

From the preceding calculation procedure, $E(P)$ is positive if $X = 150$; thus, the firm should order at least 150 units. Assume X increases from 150 to 160. From Table 28, $P(\text{sold}) = 1 - 0.08 = 0.92$. By Eq. *a,* $E(\Delta P) = 730(0.92) - 440 = \$231.60 > 0$, and the incremental investment is justified. Assume X increases from 160 to 170. Then $P(\text{sold}) = 1 - (0.08 + 0.13) = 0.79$. By Eq. *a,* $E(\Delta P) = 730(0.79) - 440 = \$136.70 > 0$, and the incremental investment is justified. Assume X increases from 170 to 180. Then $P(\text{sold}) = 1 - (0.08 + 0.13 + 0.20) = 0.59$. By Eq. *a,* $E(\Delta P) = 730(0.59) - 440 = -\$9.30 < 0$, and the incremental investment is not justified. Thus, the firm should order 170 units.

3. *Devise a direct method of solution*

Determine when $E(\Delta P)$ changes sign by setting $E(\Delta P) = 730 [P(\text{sold})] - 440 = 0$, giving $P(\text{sold}) = 440/730 = 0.603$. This is the lower limit of $P(\text{sold})$ if the incremental investment is to be justified. Now, $P(\text{sold})$ first goes below this value when $X = 180$; thus, the expected profit is maximum when $X = 170$.

Related Calculations: From the preceding calculation procedure, when $X = 150$, $E(P) = \$3643.60$; when $X = 160$, $E(P) = \$3875.20$. Thus, when X increases from 150 to 160, $E(\Delta P) = \$3875.20 - \$3643.60 = \$231.60$, and this is the result obtained in step 2. Similarly, from the preceding calculation procedure, when X increases from 160 to 170, $E(\Delta P) = \$4011.90 - \$3875.20 = \$136.70$, and this is the result obtained above. The two methods of solution yield consistent results.

The incremental-profit method is less time-consuming than the method followed in the preceding calculation procedure, and it is particularly appropriate when the firm sets a minimum acceptable rate of return. Thus, assume that the firm will undertake an investment only if the expected rate of return is 15 percent or more. When the firm orders 10 additional units, it undertakes an incremental investment of \$440. This incremental investment is justified only if the expected incremental profit is at least $\$440(0.15)$, or \$66.

SIMULATION OF COMMERCIAL ACTIVITY BY THE MONTE CARLO TECHNIQUE

A firm sells and delivers a standard commodity. The terms of sale require that the firm deliver the product within 1 day after an order is placed. In the past, the volume of orders received averaged 3315 units per week, with the variation in volume shown in Table 30.

The firm currently employs a trucking company. But the firm contemplates purchasing its own fleet of trucks to make deliveries. It is therefore necessary to decide how many trucks are to be purchased. Several plans are under consideration. The shipping facilities under plan A have an estimated average capacity of 3405 units per week. Experience indicates that this capacity may be expected to vary in the manner shown in Table 30.

When the volume of daily orders exceeds the shipping capacity, sales will be lost; when the reverse condition occurs, trucks will be idle. Lost sales are valued at \$2.40 per unit, which includes

TABLE 30   Frequency Distribution

Orders received per week			Weekly shipping capacity		
Number of units	Relative frequency	Median value	Number of units	Relative frequency	Median value
3000–3099	0.05	3050	3300–3349	0.15	3325
3100–3199	0.10	3150	3350–3399	0.30	3375
3200–3299	0.35	3250	3400–3449	0.35	3425
3300–3399	0.25	3350	3450–3499	0.20	3475
3400–3499	0.15	3450			
3500–3599	0.10	3550	Total	1.00	
Total	1.00				

an allowance for partial loss of goodwill. Unused shipping capacity is valued at $1.10 per unit. Applying the Monte Carlo technique, estimate the amount of these losses if plan A is adopted.

Calculation Procedure:

1.   *Determine the average weekly losses*

In Table 30, record the *median* value for each range, as shown in the third and sixth columns. For convenience, apply only these median values in the calculations. This procedure is equivalent to assuming, for example, that the volume of prospective sales varies discretely from 3050 to 3550 units with an interval of 100 units between consecutive values.

Analysis of Table 30 reveals that the excess of weekly shipping capacity over delivery requirements may range between 425 units ($3475 - 3050$) and -225 units ($3325 - 3550$), and that it may assume any of the following values:

425	375	325	275	225
175	125	75	25	-25
-75	-125	-175	-225	

To evaluate the average weekly losses, it is necessary to evaluate the frequency with which these values are likely to exist. The Monte Carlo technique is a probabilistic device that circumvents the mathematical complexity inherent in a rigorous solution by resorting to a set of numbers generated in a purely random manner. Tables of random numbers are published in books listed in the references for this section.

2.   *Compute the cumulative frequency of the prospective sales*

The cumulative frequency of each value of prospective sales is the relative frequency with which orders of the designated magnitude, or less, are received. The results of this calculation appear in Table 31.

TABLE 31   Cumulative Frequency of Prospective Sales

Number of units demanded	Cumulative frequency
3050	0.05
3150	0.15
3250	0.50
3350	0.75
3450	0.90
3550	1.00

3.   *Prepare a histogram of the frequency distributions*

Plot the cumulative-frequency values in Fig. 35. Draw horizontal and vertical lines as shown. The relative frequency of a given value of the prospective sales is represented by the length of the vertical line directly above the value.

4.   *Select random numbers for the solution*

Refer to a table of random numbers. Select the first 10 numbers found in the table. Enter these

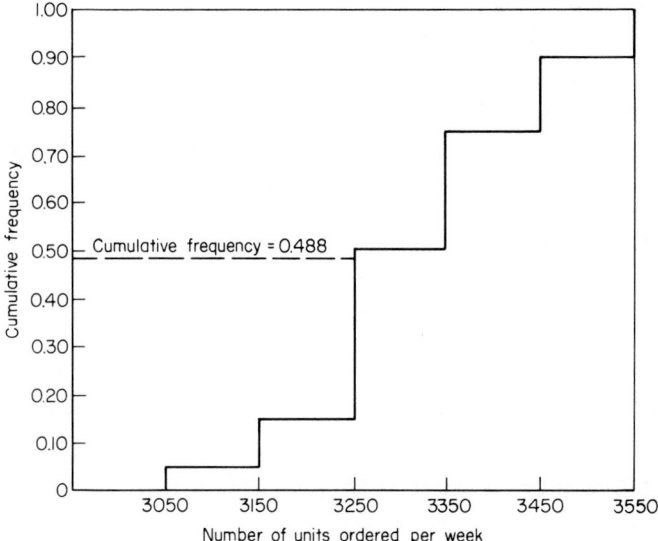

FIG. 35 Cumulative-frequency histogram of prospective sales.

numbers in the second column of Table 32. (In actual practice, a larger quantity of random numbers would be selected.)

5. Use the random numbers in the solution

Consider each random number as a cumulative frequency. Refer to the histogram, Fig. 35, to find the volume of orders corresponding to this value of the random number. Then, draw a horizontal through the random-number value of 0.488 on the vertical axis of Fig. 35. This line intersects the vertical that lies above the value of 3250 on the horizontal axis. Therefore, enter in Table 32 the value of 3250 opposite the random number 0.488.

6. Repeat steps 3 to 5 for the shipping capacity

Enter the results in Table 32 in the same manner as for the units demanded, step 5.

TABLE 32 Simulated Values of Prospective Sales and Shipping Capacity

Week	Random number	Number of units demanded	Random number	Shipping capacity
1	0.488	3250	0.339	3375
2	0.322	3250	0.697	3425
3	0.274	3250	0.031	3325
4	0.557	3350	0.052	3325
5	0.931	3550	0.506	3425
6	0.986	3550	0.865	3475
7	0.682	3350	0.948	3475
8	0.179	3250	0.308	3375
9	0.881	3450	0.218	3375
10	0.834	3450	0.367	3375

7.   *Evaluate the loss on sales and unused capacity*

Compare the simulated prospective sales with the simulated capacity. For example, during week 1, the loss on unused capacity = $1.10(3375 - 3250) = \$137.50$, given the data from Table 32. Likewise, during week 4, loss on lost sales = $2.40(3350 - 3325) = \$60.00$.

8.   *Determine the average weekly losses*

Total the computed losses obtained in step 7, and divide by 10 to obtain the following average weekly values:

Loss on unused capacity	$68.75
Loss on forfeited sales	90.00
Total	$158.75

If more trucking facilities are procured, the forfeited sales will be reduced, but the unused capacity will be increased. The optimal number of trucks to be purchased is that for which the total loss is a minimum.

LINEAR REGRESSION APPLIED TO SALES FORECASTING

A firm had the following sales for 5 consecutive years:

Year	Sales, $000
19AA	348
19BB	377
19CC	418
19DD	475
19EE	500

In 19FF, the firm decided to expand its production facilities in anticipation of future growth, and therefore it required a forecast of future sales. Apply linear regression to discern the sales trend. What is the projected sales volume for 19JJ?

Calculation Procedure:

1.   *Plot a scatter diagram for the given data*

Regression analysis is applied where a causal relationship exists between two variables, although the relationship is obscured by the influence of random factors. The problem is to establish the relationship on the basis of observed data. Here we assume that the sales volume is a linear function of time.

Consider the annual sales income to be a lump sum received at the end of the given year, and plot the sales data as shown in Fig. 36. The aggregate of points is termed a *scatter diagram*. This diagram will be replaced by a straight line that most closely approaches the plotted points; this straight line is called the *regression line*, or *line of best fit*.

2.   *Set up the criterion for the regression line*

Draw the arbitrary straight line in Fig. 36, and consider the vertical deviation e of a point in the scatter diagram from this arbitrary line. The regression line is taken as the line for which the sum of the squares of the deviations is minimum.

Let Y denote the ordinate of a point in the scatter diagram and Y_R the corresponding ordinate on the regression line. By definition, $\Sigma e^2 = \Sigma(Y - Y_R)^2 = $ minimum.

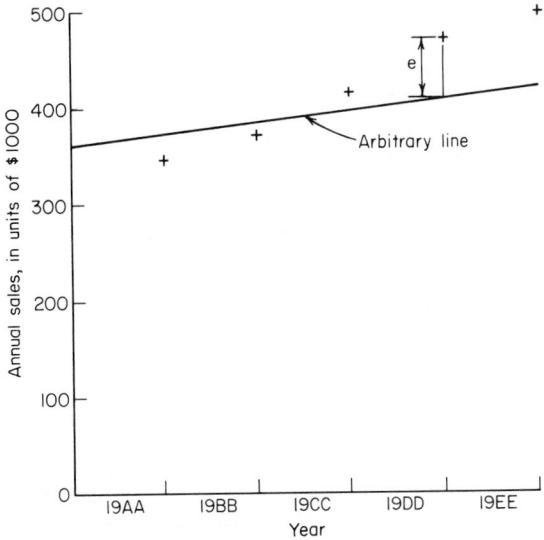

FIG. 36   Regression line, or line of best fit.

3. Write the equation of the regression line

Let n denote the number of points in the scatter diagram, and let $Y_R = a + bX$ be the equation of the regression line, where X denotes the year number as measured from some convenient datum. To find the regression line, parameters a and b must be evaluated.

Since Σe^2 is to have a minimum value, express the partial derivatives of Σe^2 with respect to a and b, and set these both equal to zero. Then derive the following simultaneous equations containing the unknown quantities a and b:

$$\Sigma Y = an + b\Sigma X$$

$$\Sigma XY = a\Sigma X + b\Sigma X^2$$

4. Simplify the calculations

Select the median date (the end of 19CC) as a datum. This selection causes the term ΣX to vanish.

5. Determine the values of ΣX^2 and ΣXY

Prepare a tabulation such as Table 33. Use the data for each year in question.

TABLE 33   Locating a Regression Line

Year	X	Y	X²	XY	Y_R
19AA	−2	348	4	−696	343.2
19BB	−1	377	1	−377	383.4
19CC	0	418	0	0	423.6
19DD	1	475	1	475	463.8
19EE	2	500	4	1000	504.0
Total	0	2118	10	402	

6. Solve for parameters a and b

Substitute in the equations in step 3, and solve for a and b. Thus $2118 = 5a$; $402 = 10b$; $a = 423.6$; $b = 40.2$.

7.   Write the regression equation; extrapolate for the year in question

Here $Y_R = 423.6 + 40.2X$. For 19JJ: $X = 7$; $Y_R = 423.6 + 40.2$ (7) $= 705$. Hence, the forecast sales for 19JJ $= \$705,000$. For comparative purposes, the past sales volumes as determined by the regression line are listed in Table 33.

STANDARD DEVIATION FROM REGRESSION LINE

Using the data in the previous calculation procedure, appraise the reliability of the regression line in forecasting future sales by computing the standard deviation of the points in the scatter diagram, using the regression line as the datum from which the deviation is measured.

Calculation Procedure:

1.   Calculate the deviation of each point; square the result

The standard deviation serves as an index of the dispersion of the points in the scatter diagram. The standard deviation $\sigma = \sqrt{\Sigma e^2 / n}$.

Calculate the value of e for each point, Fig. 36. Enter the results for each year in a tabulation such as Table 34. Then, $\sigma = \sqrt{236.8/5} = 6.9$.

TABLE 34   Determining the Standard Deviation

Year	$Y - Y_R = e$	e^2
19AA	$348 - 343.2 = 4.8$	23.0
19BB	$377 - 383.4 = -6.4$	41.0
19CC	$418 - 423.6 = -5.6$	31.4
19DD	$475 - 463.8 = 11.2$	125.4
19EE	$500 - 504.0 = -4.0$	16.0
Total		236.8

TABLE 35   Probabilities for Two Successive Purchases

Present model	Next model		
	A	B	C
A	.4167	.3333	.2500
B	.5000	.3000	.2000
C	.1538	.2308	.6154

2.   Determine the monetary value represented by the standard deviation

Since the given monetary values are expressed in thousands of dollars, the value of the standard deviation $= 6.9$ ($\$1000$) $= \$6900$.

SHORT-TERM FORECASTING WITH A MARKOV PROCESS

The XYZ Company manufactures a machine that is available in three models, A, B, and C. There are currently 1200 such machines in use, divided as follows: model A, 460; model B, 400; model C, 340. On the basis of a survey, the XYZ Company has established probabilities corresponding to two successive purchases, and they are recorded in Table 35. Thus, if a firm currently owns model B, there is a probability of 0.5000 that its next model will be A; if a firm currently owns model C, there is a probability of 0.6154 that its next model will also be C. Assume that each machine will remain in service for precisely 1 year, after which it will be replaced with another machine manufactured by the XYZ Company. Also assume that the XYZ Company will not acquire any new customers in the foreseeable future. Estimate the number of units of each model that will be in use 1 year, 2 years, and 3 years hence.

Calculation Procedure:

1.   Set up the basic equations that link two successive years

Assume that a trial will be performed repeatedly and that the outcome of one trial directly influences the outcome of the succeeding trial. A trial of this type is called a *Markov process*. In this situation, the purchase of a machine is a Markov process because the model that a firm selects on one occasion has a direct bearing on the model it selects on the following occasion. The probabilities in Table 35 are termed *transition probabilities*, and the table itself is called a *transition matrix*.

Let $X_{A,n}$ = expected number of units of model A that will be in use n years hence. Multiply the expected values for n years hence ' y their respective probabilities to obtain the expected values for $n + 1$ years hence, giving

$$X_{A,n+1} = 0.4167X_{A,n} + 0.5000X_{B,n} + 0.1538X_{C,n} \qquad (a)$$

$$X_{B,n+1} = 0.3333X_{A,n} + 0.3000X_{B,n} + 0.2308X_{C,n} \qquad (b)$$

$$X_{C,n+1} = 0.2500X_{A,n} + 0.2000X_{B,n} + 0.6154X_{C,n} \qquad (c)$$

2. Calculate the expected values for 1 year hence

Apply Eqs. *a*, *b*, and *c* with $n = 0$ and $X_{A,0} = 460$, $X_{B,0} = 400$, $X_{C,0} = 340$. Then $X_{A,1} = (0.4167)460 + (0.5000)400 + (0.1538)340 = 444$; $X_{B,1} = (0.3333)460 + (0.3000)400 + (0.2308)340 = 352$; $X_{C,1} = (0.2500)460 + (0.2000)400 + (0.6154)340 = 404$. Record the results in Table 36.

3. Calculate the expected values for 2 years hence

Apply Eqs. *a*, *b*, and *c* with $n = 1$ and the values of $X_{A,1}$, $X_{B,1}$, and $X_{C,1}$ shown in Table 36. Then $X_{A,2} = (0.4167)444 + (0.5000)352 + (0.1538)404 = 423$; $X_{B,2} = (0.3333)444 + (0.3000)352 + (0.2308)404 = 347$; $X_{C,2} = (0.2500)444 + (0.2000)352 + (0.6154)404 = 430$. Record the results in Table 36.

4. Calculate the expected values for 3 years hence

Apply Eqs. *a*, *b*, and *c* for the third cycle. Then $X_{A,3} = (0.4167)423 + (0.5000)347 + (0.1538)430 = 416$; $X_{B,3} = (0.3333)423 + (0.3000)347 + (0.2308)430 = 344$; $X_{C,3} = (0.2500)423 + (0.2000)347 + (0.6154)430 = 440$. Record the results in Table 36.

TABLE 36 Expected Number of Units in Use

Elapsed time, years	X_A	X_B	X_C
0	460	400	340
1	444	352	404
2	423	347	430
3	416	344	440

5. Determine the expected values with the aid of a diagram

Refer to Fig. 37, which shows the expected manner in which units of a given model will be replaced. Each value of X is recorded in the appropriate box. Multiply the values of $X_{A,0}$, $X_{B,0}$, and $X_{C,0}$ by the corresponding probabilities to find the expected replacements during the first year. Thus, with reference to the 460 units of model A, the expected replacements are as follows: model A, $460(0.4167) = 192$; model B, $460(0.3333) = 153$; model C, $460(0.2500) = 115$. Record all values in Fig. 37. Then $X_{A,1} = 192 + 200 + 52 = 444$; $X_{B,1} = 153 + 120 + 79 = 352$; $X_{C,1} = 115 + 80 + 209 = 404$. Repeat the cycle of calculations for the second and third years. The values in Fig. 37 agree with those in Table 36.

Related Calculations: Matrix multiplication provides a compact procedure for solving problems pertaining to a Markov process. Let P denote the matrix in Table 35 and R_n denote a row vector consisting of $X_{A,n}$, $X_{B,n}$, and $X_{C,n}$. The values of these variables appear in Table 36. Then

$$R_1 = R_0P = [460 \quad 400 \quad 340] \begin{bmatrix} 0.4167 & 0.3333 & 0.2500 \\ 0.5000 & 0.3000 & 0.2000 \\ 0.1538 & 0.2308 & 0.6154 \end{bmatrix} = [444 \quad 352 \quad 404]$$

Similarly, $R_2 = R_1P = R_0P^2$, and $R_3 = R_2P = R_0P^3$. In general, $R_n = R_0P^n$.

LONG-TERM FORECASTING WITH A MARKOV PROCESS

With reference to the preceding calculation procedure, estimate the number of units of each model that will ultimately be in use simultaneously.

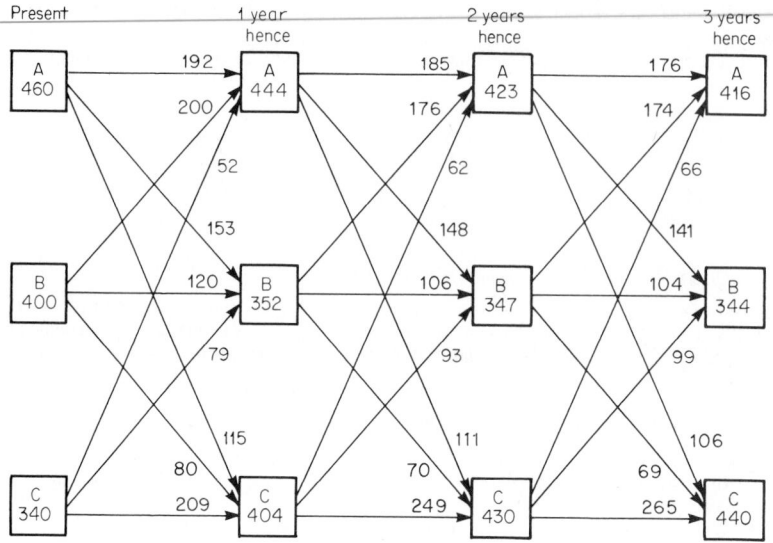

FIG. 37   Replacement diagram.

Calculation Procedure:

1. Form a system of simultaneous equations containing the limits of the expected values

In the long run, the probability that a firm will buy a given model is solely a function of the specific needs of that firm and the characteristics of each model; it is independent of the particular model that the firm happens to own at present. Therefore, $X_{A,n}$, $X_{B,n}$, and $X_{C,n}$ approach definite limits as n increases beyond bound. The values of these variables when n has a finite value constitute *transient conditions*, and the values when n is infinite constitute the *steady-state conditions*. In practice, however, the steady-state conditions may be considered to exist when all differences between transient and steady-state values become less than some specified small number.

Let $X_{A,u} = \lim\limits_{n \to \infty} X_{A,n}$. In Eqs. a and b of the preceding calculation procedure, replace the transient values with their respective limits and rearrange:

$$-0.5833X_{A,u} + 0.5000X_{B,u} + 0.1538X_{C,u} = 0 \qquad (a')$$

$$0.3333X_{A,u} - 0.7000X_{B,u} + 0.2308X_{C,u} = 0 \qquad (b')$$

Also,

$$X_{A,u} + X_{B,u} + X_{C,u} = 1200 \qquad (d)$$

2. Solve the system of equations

The results are $X_{A,u} = 411$; $X_{B,u} = 343$; $X_{C,u} = 446$. Thus, it is expected that there will ultimately be 411 units of model A, 343 units of model B, and 446 units of model C in use simultaneously. Note that the values of $X_{A,3}$, $X_{B,3}$, and $X_{C,3}$ in Table 36 are very close to the limiting values. Thus, the expected values approach their respective limits rapidly.

Related Calculations: Many problems in engineering, economics, and other areas lend themselves to solution as Markov processes. The computational techniques applied in this calculation procedure and the preceding one are entirely general, and they may be applied to any problem where a Markov process exists.

VERIFICATION OF STEADY-STATE CONDITIONS FOR A MARKOV PROCESS

Verify the accuracy of the results obtained in the preceding calculation procedure by devising an alternative method of solution.

Calculation Procedure:

1. Construct a recurring series of outcomes that conforms with the given process

Assume that a Markov process has three possible outcomes, A, B, and C, and that the first 35 outcomes were these:

B-A-A-B-B-A-C-C-C-C-B-A-A-A-C-C-A-A-A-C-C-B-A-B-A-B-C-C-C-C-A-B-C-B-B

This series consists of 12 A's, 10 B's, and 13 C's. Also assume that this series of outcomes will recur indefinitely. Thus, the last outcome in the series will be followed by B. It will be demonstrated that this series is relevant to the preceding calculation procedure.

2. Compute the transition probabilities as established by the recurring series

Count the successors of the outcomes in this series, and then compute the relative frequencies of the various successions. Refer to Table 37 for the calculations. Since the given series of outcomes will recur indefinitely, the relative frequencies in Table 37 equal the *transition probabilities* cor-

TABLE 37

Given outcome	Successor	Frequency of successor	Relative frequency of successor
A	A	5	5/12 = 0.4167
	B	4	4/12 = 0.3333
	C	3	3/12 = 0.2500
		Total 12	
B	A	5	5/10 = 0.5000
	B	3	3/10 = 0.3000
	C	2	2/10 = 0.2000
		Total 10	
C	A	2	2/13 = 0.1538
	B	3	3/13 = 0.2308
	C	8	8/13 = 0.6154
		Total 13	

responding to the present Markov process. Thus, the probability that A will be followed by B is 0.3333, and the probability that C will be followed by A is 0.1538. Since these transition probabilities coincide with those in Table 35, it follows that the present recurring series provides a basis for investigating the Markov process in the preceding calculation procedure.

3. Compute the steady-state probabilities

In the long run, the probability that a given outcome will occur is independent of some outcome in the distant past. In the recurring series, the relative frequencies of the outcomes are: outcome A, 12/35; outcome B, 10/35; outcome C, 13/35. These relative frequencies are the *steady-state probabilities* corresponding to the Markov process.

4. Compute the expected number of units in use at steady-state conditions

In the preceding calculation procedure, the total number of machines that will be in use simultaneously is 1200. Multiply the steady-state probabilities found in step 3 by 1200 to obtain the expected number of units of each model that ultimately will be in use simultaneously. The results are: model A, 1200(12/35) = 411; model B, 1200(10/35) = 343; model C, 1200(13/35) = 446. These results coincide with those obtained in the preceding calculation procedure, and so the latter are confirmed.

Related Calculations: In constructing the recurring series of outcomes, it is necessary to apply the *principle of succession*. Assume that a Markov process has three possible outcomes, A, B, and C; and let $N(A-B)$ = number of times that A is followed by B. The principle is $N(A-B) + N(A-C) = N(B-A) + N(C-A)$. As an illustration, consider the following skeletal series, where each outcome is followed by a different outcome:

$$A-C-A-C-B-A-B-A-C-B-A-C-B-C-A-C$$

The last outcome will be followed by A. Then $N(A-B) + N(A-C) = N(B-A) + N(C-A) = 6$; $N(B-A) + N(B-C) = N(A-B) + N(C-B) = 4$; $N(C-A) + N(C-B) = N(A-C) + N(B-C) = 6$. Now this skeletal series can be expanded to the true series by allowing one outcome to be followed by the same outcome. For example, assume the requirements are $N(A-A) = 3$, $N(B-B) = 5$, and $N(C-C) = 6$. A true recurring series is

$$A-A-C-C-C-C-A-C-C-B-B-B-A-B-B-A-C-C-B-B-B-A-A-C-B-C-A-A-C$$

SECTION 13

ENVIRONMENTAL ENGINEERING

TYLER G. HICKS, P.E.
INTERNATIONAL ENGINEERING ASSOCIATES

JOSEPH LETO, P.E.

General Cost-Benefit Analysis 13.2
Selection of the Most Desirable Project Using Cost-Benefit Analysis 13.3
Economics of Energy-from-Wastes Alternatives 13.4
Flue-Gas Heat Recovery and Emissions Reduction 13.8
Estimating Total Costs of Cogeneration-System Alternatives 13.14
Choosing Steam Compressor for Cogeneration System 13.19
Using Plant Heat-Need Plots for Cogeneration Decisions 13.22
Geothermal and Biomass Power-Generation Analyses 13.27
Estimating Capital Cost of Cogeneration Heat-Recovery Boilers 13.31
"Clean" Energy from Small-Scale Hydro Sites 13.34
Central Chilled-Water System Design to Meet Chlorofluorocarbon (CFC)
  Issues 13.37
Work Required to Clean Oil-Polluted Beaches 13.39
Sizing Explosion Vents for Industrial Structures 13.42
Industrial Building Ventilation for Environmental Safety 13.44
Estimating Power-Plant Thermal Pollution 13.47

REFERENCES: McGraw-Hill *Encyclopedia of Environmental Science and Engineering*, McGraw-Hill; Lesage and Jackson—*Groundwater Contamination and Analysis at Hazardous Waste Sites*, Marcel Dekker; Corbitt—*Standard Handbook of Environmental Engineering*, McGraw-Hill; Woodslide—*Harzardous Materials and Hazardous Waste Management*, Wiley; Fthenakis—*Prevention and Control of Accidental Releases of Hazardous Gases*, Van Nostrand Reinhold; LaGrega, Buckingham and Evans—*Hazardous Waste Management*, McGraw-Hill; Cheremisinoff—*Air Pollution Control and Design for Industry*, Marcel Dekker; Freeman—*Standard Handbook of Hazardous Waste Treatment and Disposal*, McGraw-Hill; Office of Technology Assessment—*Green Products by Design—Choices for a Cleaner Environment*, U.S. Government

Printing Office; Arbuckle—*Environmental Law Handbook*, Government Institute, Inc.; Lund—*The Mc-Graw-Hill Recycling Handbook*, McGraw-Hill; Holmes—*Refuse Recycling and Recovery*, Wiley; Gabor—*Beyond the Age of Waste*—*A Report to the Club of Rome*, Pergamon; Porteous—*Refuse Derived Fuels*, Halsted; Council on Environmental Quality—*Environmental Quality*, U.S. Government Printing Office; Jain et al—*Environmental Assessment*, McGraw-Hill; Morgan—*Renewable Resource Utilization for Development*, Pergamon; White and Plaskett—*Biomass as Fuel*, Academic; Herwan and Boyce—*Gas Turbine Engineering Handbook*, Gulf; Wetzel and Murphy—*Treating Industrial-Waste Interferences at Publicly-Owned Treatment Works*, Noyes Data; Beranek and Ver—*Noise and Vibration Control Engineering*, Wiley; Neporozhny—*Thermal Power Plants and Environmental Control*, MIR Publishing; Payne—*Cogeneration Sourcebook*, Fairmont Press; Fortuna and Lennett—*Hazardous Waste Regulation*—*The New Era*, McGraw-Hill; Hu—*Cogeneration*, Reston; McDermott—*Handbook of Ventilation for Contaminant Control*, Butterworths; Goldstick and Thumann—*Principles of Waste Heat Recovery*, Fairmont Press; Vutukuri and Lama—*Environmental Engineering in Mines*, Cambridge University Press; Hallenbeck and Cunningham—*Quantitative Risk Assessment for Environmental and Occupational Health*, Lewis Publishers; Nejat and Veziroglu—*Alternative Energy Sources*, Hemisphere Publishing; Spiewak—*Cogeneration and Small Power Production*, Fairmont Press; Plunkett—*Handbook of Industrial Toxicology*, Chemical Publishing; Marine Board et al—*Dredging Coastal Ports*, National Academy Press; Thumann and Miller—*Fundamentals of Noise Control Engineering*, Prentice Hall.

Environmental engineering is probably the fastest growing branch of engineering today. Impacting every facet of industry and society, environmental engineering is the answer to a cleaner, safer world. Regardless of where pollution control is exercised—before the pollution occurs, or afterwards—environmental engineering *is* the answer to creating a better environment for everyone, everywhere.

Environmental engineering uses the skills and technologies of almost every other branch of the profession. Thus, the environmental engineer will use methods and solutions from engineering disciplines including mechanical, civil, electrical, chemical, industrial, architectural, sanitary, nuclear, and control engineering. Today a number of engineering schools are offering a major in environmental engineering. Graduates have studied portions of the disciplines just mentioned.

This section of the *Handbook* concentrates on procedures for solving environmental problems of many types. Where procedures in related disciplines are needed, for example pipe sizing, the reader should refer to that discipline in this handbook. By combining the methods given in related sections with those in this section, an engineer should be able to develop solutions to a variety of practical, everyday environmental problems.

GENERALIZED COST-BENEFIT ANALYSIS

An engineering atmospheric control to protect the public against environmental pollution will have an incremental operating cost of $100,000. If the pollution were uncontrolled, the damage to the public would have an estimated incremental cost of $125,000. Would this atmospheric control be a beneficial investment?

Calculation Procedure:

1. *Write the cost-benefit ratio for this investment*

The generalized dimensionless cost-benefit equation is $0 \le C/B \le 1$, where C = incremental operating cost of the proposed atmospheric control, $, or other consistent monetary units; B = benefit to the public of having the pollution controlled, $, or other consistent monetary units.

2. *Compute the cost-benefit ratio for this situation*

Using the values given, $0 \le \$100,000/\$125,000 \le 1$. Or, $0 \le 0.80 \le 1$. This result means that 80¢ spent on environmental control will yield $1.00 in public benefits. Investing in the control would be a wise decision because a return greater than the cost of the control is obtained.

Related Calculations: In the general cost-benefit equation, $0 \le C/B \le 1$, the upper limit of unity means that $1.00 spent on the incremental operating cost of the atmospheric control will deliver $1.00 in public benefits. A cost-benefit ratio of more than unity is uneconomic.

Thus, $1.25 spent to obtain $1.00 in benefits would not, in general, be acceptable in a rational analysis. The decision would be to accept the environmental pollution until a satisfactory cost-benefit solution could be found.

A negative result in the generalized equation means that money invested to improve the environment actually degrades the condition. Hence, the environmental condition becomes worse. Therefore, the technology being applied cannot be justified on an economic basis.

In applying cost-benefit analyses, a number of assumptions of the benefits to the public may have to be made. Such assumptions, particularly when expressed in numeric form, can be open to change by others. Fortunately, by assigning a number of assumed values to one or more benefits, the cost-benefit ratios can easily be evaluated, especially when the analysis is done on a computer.

SELECTION OF MOST DESIRABLE PROJECT USING COST-BENEFIT ANALYSIS

Five alternative projects for control of environmental pollution are under consideration. Each project is of equal time duration. The projects have the cost-benefit data shown in Table 1. Determine which project, if any, should be constructed.

TABLE 1   Project Costs and Benefits

Project	Equivalent uniform net annual benefits, $	Equivalent uniform net annual costs, $	C/B ratio
A	200,000	135,000	0.68
B	250,000	190,000	0.76
C	180,000	125,000	0.69
D	150,000	90,000	0.60
E	220,000	150,000	0.68

Calculation Procedure:

1. Evaluate the cost-benefit (C/B) ratios of the projects

Setting up the C/B ratios for the five projects by the cost by the estimated benefit shows—in Table 1—that all C/B ratios are less than unity. Thus, each of the five projects passes the basic screening test of $0 \leq C/B \leq 1$. This being the case, the optimal project must be determined.

2. Analyze the projects in terms of incremental cost and benefit

Alternative projects cannot be evaluated in relation to one another merely by comparing their C/B ratios, because these ratios apply to unequal bases. The proper approach to analyzing such a situation is: Each project corresponds to a specific *level* of cost. To be justified, *every* sum of money expended must generate at least an equal amount in benefits; the step from one level of benefits to the next should be undertaken only if the incremental benefits are at least equal to the incremental costs.

Rank the projects in ascending order of costs. Thus, Project D costs $90,000; Project C costs $125,000; and so on. Ranking the projects in ascending order of costs gives the sequence D-C-A-E-B.

Next, compute the incremental costs and benefits associated with each step from one level to the next. Thus, the incremental cost going from Project D to Project C is $125,000 − $90,000 = $35,000. And the benefit from going from Project D to Project C is $180,000 − $150,000 = $30,000, using the data from Table 1. Summarize the incremental costs and benefits in a tabulation like that in Table 2. Then compute the C/B ratio for each situation and list it in Table 2. This computation shows that Project E is the best of these five projects because it has the lowest cost—75¢ per $1.00 of benefit. Hence, this project would be chosen for control of environmental pollution in this instance.

Related Calculations: There are some situations in which the minimum acceptable C/B ratio should be set at some value close to 1.00. For example, with reference to the above projects,

TABLE 2 Cost-Benefit Comparison

Step	Incremental benefit, $	Incremental cost, $	C/B ratio	Conclusion
D to C	30,000	35,000	1.17	Unsatisfactory
D to A	50,000	45,000	0.90	Satisfactory
A to E	20,000	15,000	0.75	Satisfactory
E to B	30,000	40,000	1.33	Unsatisfactory

assume that the government has a fixed sum of money that is to be divided between a project listed in Table 1 and some unrelated project. Assume that the latter has a C/B ratio of 0.91, irrespective of the sum expended. In this situation, the step from one level to a higher one is warranted only if the C/B ratio corresponding to this increment is at least 0.91.

Closely related to cost-benefit analysis and an outgrowth of it is *cost-effectiveness analysis*, which is used mainly in the evaluation of military and space programs. To apply this method of analysis, assume that some required task can be accomplished by alternative projects that differ in both cost and degree of performance. The effectiveness of each project is expressed in some standard unit, and the projects are then compared by a procedure analogous to that for cost-benefit analysis.

Note that cost-benefit analysis can be used in any comparison of environmental alternatives. Thus, cost-benefit analyses can be used for air-pollution controls, industrial thermal discharge studies, transportation alternatives, power-generation choices (windmills vs. fossil-fuel or nuclear plants), cogeneration, recycling waste for power generation, solar power, use of recycled sewer sludge as a fertilizer, and similar studies. The major objective in each comparison is to find the most desirable alternative based on the benefits derived from various options open to the designer.

For example, electric utilities using steam generating stations burning coal or oil may release large amounts of carbon dioxide into the atmosphere. This carbon dioxide, produced when a fuel is burned, is thought to be causing a global greenhouse effect. To counteract this greenhouse effect, some electric utilities have purchased tropical rain forests to preserve the trees in the forest. These trees absorb carbon dioxide from the atmosphere, counteracting that released by the utility.

Other utilities pay lumber companies to fell trees more selectively. For example, in felling the 10 percent of the marketable trees in a typical forest, as much as 40 to 50 percent of a forest may be destroyed. By felling trees more selectively, the destruction can be reduced to less than 20 percent of the forest. The remaining trees absorb atmospheric carbon dioxide, turning it into environmentally desirable wood. This conversion would not occur in these trees if they were felled in the usual foresting operation. The payment to the lumber company to do selective felling is considered a cost-benefit arrangement because the unfelled trees remove carbon dioxide from the air. The same is true of the tropical rain forests purchased by utilities and preserved to remove carbon dioxide which the owner-utility emits to the atmosphere.

Recently, a market has developed in the sale of "pollution rights" in which a utility that emits less carbon dioxide because it has installed pollution-control equipment can sell its "rights" to another utility that has less effective control equipment. The objective is to control, and reduce, the undesirable emissions by utilities.

With a potential "carbon tax" in the future, utilities and industrial plants that produce carbon dioxide as a by-product of their operations are seeking cost-benefit solutions. The analyses given here will help in evaluating potential solutions.

ECONOMICS OF ENERGY-FROM-WASTES ALTERNATIVES

A municipality requires the handling of 1500 tons/day (1524 mt/day) of typical municipal solid waste. Determine if a waste-to-energy alternative is feasible. If not, analyze the other means by which this solid-waste stream might be handled. Two waste-to-energy alternatives are being considered—mass burn and processed fuel. The expected costs are shown in Table 3. If earnings of 6 percent on invested capital are required, which alternative is more economical?

TABLE 3   Estimated Costs of Municipal Solid Waste Disposal Facilities

	Mass burn	Processed fuel
First cost, $	15,000,000	22,500,000
Salvage value, $	1,500,000	2,500,000
Life, years	15	25
Annual maintenance cost, $	750,000	400,000
Annual taxes, $	15,000	15,000
Annual insurance, $	100,000	150,000

Calculation Procedure:

1.   Plot the options available for handling typical municipal waste

Figure 1 shows the options available for handling solid municipal wastes or refuse. The refuse enters the energy-from-waste cycle and undergoes primary shredding. Then the shredded material is separated according to its density. Heavy materials—such as metal and glass—are removed for recovery and recycling. Experience and studies show that recycling will recover no more than 35 percent of the solid wastes entering a waste-to-energy facility. And most such facilities today are able to recycle only about 20 percent of municipal refuse. Assuming this 20 percent applies to the plant facility being considered here, the amount of waste that would be recycled would be 0.20 (1500) = 300 tons/day (305 mt/day).

Numerous studies show that complete recycling of municipal waste is uneconomic. Therefore, the usual solution to municipal waste handling today features four primary components: (1) source reduction, (2) recycling, (3) waste to energy, and (4) landfilling. The Environmental Protection Agency (EPA) recently proposed broad policies encouraging recycling and reduction of pollutants at their source.

Using waste-to-energy facilities reduces the volume of wastes requiring disposal while producing a valuable commodity—steam and/or electric power. Combustion control is needed in every waste-to-energy facility to limit the products of incomplete combustion which escape in the flue gas and cause atmospheric pollution. Likewise, limiting the quantities of metal entering the combustor reduces their emission in the ash or flue gas. This, in turn, reduces pollution.

2.   Determine the energy available in the municipal waste

Usual municipalities generate 1 ton (0.91 Mg) of solid waste per year per capita. About 35 percent of this waste is from residences; 65 percent is from industrial and commercial establishments. The usual heating value of municipal waste is 5500 Btu/lb (12×10^6 J/kg). Table 4 shows typical industrial wastes and their average heating values. Municipalities typically spend $25 or more per ton (0.91 Mg) to dispose of solid wastes.

Because municipal wastes have a variety of ingredients, many plants burn the solid waste as a supplement to coal. The heat in the waste is recovered for useful purposes, such as generating steam or electricity. When burned with high-sulfur coal, the solid waste reduces the sulfur content discharged in the stack gases. The solid waste also increases the retention of sulfur compounds in the ash. The result is reduced corrosion of the boiler tubes by HCl. Further, acid-rain complaints are fewer because of the reduced sulfur content in the stack gases.

Where an existing or future plant burns, or will burn, oil, another approach may be taken to the use of solid municipal waste as fuel. The solid waste is first shredded; then it is partially burned in a rotary kiln in an oxygen-deficient atmosphere at 1652°F (900°C). The gas produced is then burned in a conventional boiler to supplement the normal oil fuel.

Estimates show that about 5 percent of the energy needs of the United States could be produced by the efficient burning of solid municipal wastes in steam plants. Such plants must be located within about 100 mi (160 km) of the waste source to prevent excessive collection and transportation costs.

Combustion of, and heat recovery from, solid municipal wastes reduces waste volume considerably. But there is still ash from the combustion that must be disposed of in some manner. If landfill disposal is used, the high alkali content of the typical municipal ash must be considered.

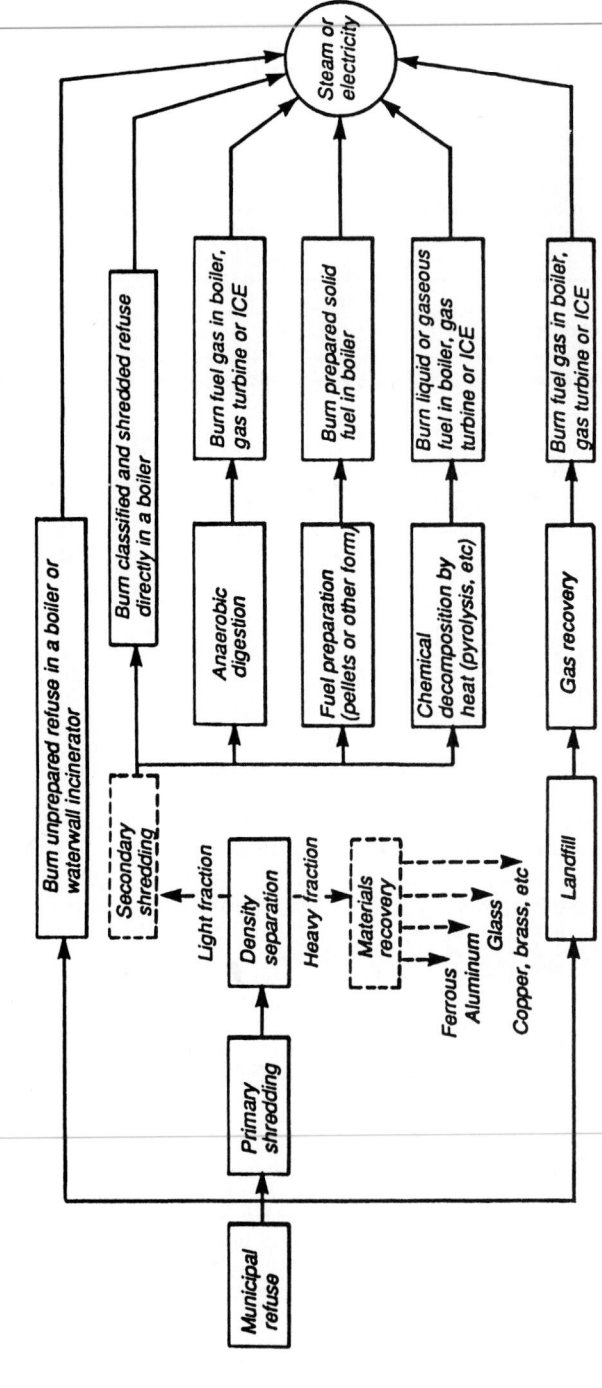

FIG. 1 Several options for energy extraction from municipal waste. Selection should be based on local variables and economics. (*Power.*)

TABLE 4 Typical Industrial Wastes with Significant Fuel Value

	Average heating value (as fired)	
	Btu/lb	kJ/kg
Waste gases:		
Coke-oven	19,700	45,900
Blast-furnace	1,139	2,654
Carbon monoxide	579	1,349
Liquids:		
Refinery	21,800	50,794
Industrial sludge	3,700–4,200	8,621–9,786
Black liquor	4,400	10,252
Sulfite liquor	4,200	9,786
Dirty solvents	10,000–16,000	23,300–37,280
Spent lubricants	10,000–14,000	23,300–32,620
Paints and resins	6,000–10,000	13,980–23,300
Oily waste and residue	18,000	41,940
Solids:		
Bagasse	3,600–6,500	8,388–15,145
Bark	4,500–5,200	10,485–12,116
General wood wastes	4,500–6,500	10,485–15,145
Sawdust and shavings	4,500–7,500	10,485–17,475
Coffee grounds	4,900–6,500	11,417–15,145
Nut hulls	7,700	17,941
Rice hulls	5,200–6,500	12,116–15,145
Corn cobs	8,000–8,300	18,640–19,339

Source: Power; SI units added by editor.

This alkali content often presents leaching and groundwater contamination problems. So, while the solid-waste disposal problem may have been solved, there are still environmental considerations that must be faced. Further, the large noncombustible items often removed from solid municipal waste before combustion—items like refrigerators and auto engine blocks—must still be disposed of in an environmentally acceptable manner. Table 5 shows a number of ash reuse and disposal options available for use today.

3. *Choose between available alternatives*

The two alternatives being considered—*mass burn* and *processed fuel*—have separate and distinct costs. These costs must be compared to determine the most desirable alternative.

In a mass-burn facility the trash is burned as received, after hand removal of large noncombustible items—sinks, bathtubs, engine blocks, etc. The remaining trash is rough-mixed by a clamshell bucket and delivered to the boiler's moving grate. Some 30 to 50 percent by weight and 5 to 15 percent by volume of the waste burned in a mass-burn facility leaves in the form of bottom ash and flyash.

TABLE 5 Ash Reuse and Disposal Options

	Treatment required	Use
Bottom ash	Particle-size screening	Coarse highway aggregate, concrete products
Bottom ash		Asphalt paving
Combined ash	Particle-size screening	Artificial reefs
Combined ash	Particle-size screening	Aggregate for paving
Flyash	Particle-size screening	Aggregate for paving
Flyash	Particle-size screening	Fine cement aggregate

Source: Power.

In a processed-fuel facility [also called a refuse-derived-fuel (RDF) facility], the solid waste is processed in two steps. First, noncombustibles are separated from combustibles. The remaining combustible waste is reduced to uniform-sized pieces in a hammermill-type shredder. The shredded pieces are then delivered to a boiler for combustion.

Using the annual cost of each alternative (see Section 12) as the "first cut" in the choice: Operating and maintenance cost = C = maintenance cost per year, $ + annual taxes, $ + annual insurance cost, $. For mass burn, C = $750,000 + $15,000 + 0.002 ($15,000,000) = $795,000. For processed fuel, C = $400,000 + $15,000 + 0.002 ($22,500,000) = $460,000.

Next, using the capital-recovery equation from Section 12, for mass burn, the annual cost, A = ($15,000,000 − $1,500,000)(0.06646) + $750,000 + $1,500,000(0.06) + $795,000 = $1,737,210. For processed fuel, A = ($22,500,000 − $2,500,000)(0.06646) + $400,000 + $2,500,000(0.06) = $1,879,200. Therefore, mass burn is the more attractive alternative from an annual-cost basis because it is $1,879,000 − $1,737,210 = $141,990 per year less expensive than the processed-fuel alternative.

Several more analyses would be made before this tentative conclusion was accepted. However, this calculation procedure does reveal an acceptable first-cut approach to choosing between different available alternatives for evaluating an environmental proposal.

Related Calculations: Another source of usable energy from solid municipal waste is landfill methane gas. This methane gas is produced by decomposition of organic materials in the solid waste. The gas has a heating value of about 500 Btu/ft^3 (1.1×10^6 J/kg) and can be burned in a conventional boiler, gas turbine, or internal-combustion engine. Using landfill gas to generate steam or electricity can reduce landfill odors. But such burning does *not* reduce the space and groundwater problems produced by landfills. The cost of landfill gas can range from $0.45 to $5/million Btu ($0.45 to $5/1055 kJ). Much depends on the cost of recovering the gas from the landfill.

Methane gas is recovered from landfills by drilling wells into the field. Plastic pipes are then inserted into the wells and the gas is collected by gas compressors. East coast landfills in the United States have a lifespan of 5 to 7 years. West coast landfills have a lifespan of 15 to 18 years.

Data in this procedure were drawn from *Power* magazine and Hicks, *Power Plant Evaluation and Design Reference Guide*, McGraw-Hill.

FLUE-GAS HEAT RECOVERY AND EMISSIONS REDUCTION

A steam boiler rated at 32,000,000 Btu/h (9376 MW) fired with natural gas is to heat incoming feedwater with its flue gas in a heat exchanger from 60°F (15.6°C) to an 80°F (26.7°C) outlet temperature. The flue gas will enter the boiler stack and heat exchanger at 450°F (232°C) and exit at 100°F (37.8°C). Determine the efficiency improvement that might be obtained from the heat recovery. Likewise, determine the efficiency improvement for an oil-fired boiler having a flue-gas inlet temperature of 300°F (148.9°C) and a similar heat exchanger.

Calculation Procedure:

1. *Sketch a typical heat-recovery system hookup*

Figure 2 shows a typical hookup for stack-gas heat recovery. The flue gas from the boiler enters the condensing heat exchanger at an elevated temperature. Water sprayed into the heat exchanger absorbs heat from the flue gas and is passed through a secondary external heat exchanger. Boiler feedwater flowing through the secondary heat exchanger is heated by the hot water from the condensing heat exchanger. Note that the fluid heated can be used for a variety of purposes other than boiler feedwater—process, space heating, unit heaters, domestic hot water, etc.

Flue gas from the boiler can enter the condensing heat exchanger at temperatures of 300°F (148.9°C), or higher, and exit at 100 to 120°F (37.8 to 48.9°C). The sensible and latent heat given up is transferred to the spray water. Since this sprayed cooling water may be contaminated by the flue gas, a secondary heat exchanger (Fig. 2), may be used. Where the boiler fuel is clean-burning natural gas, the spray water may be used directly, without a secondary heat exchanger. Since there may be acid contamination from SO_2 in the flue gas, careful analysis is needed to determine if the contamination level is acceptable in the process for which the heated water or other fluid will be used.

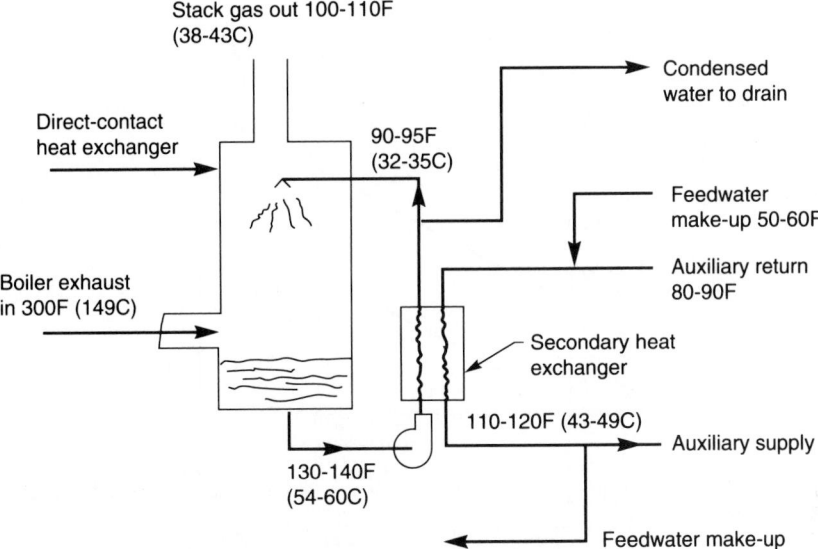

FIG. 2 Effective condensation heat recovery depends on direct contact between flue gas and cooling medium and low gas-side pressure drop. (*Power.*)

2. Determine the efficiency gain from the condensation heat recovery

Efficiency gain is a function of fuel hydrogen content, boiler flue-gas exit temperature, spray (process) water temperature, amount of low-level heat needed, fuel moisture content, and combustion-air humidity. The first four items are of maximum significance for gas-, oil-, and coal-fired boilers. Installations firing lignite or high-moisture-content biomass fuels may show additional savings over those computed here. If combustion-air humidity is high, the efficiency improvement from the condensation heat recovery may be 1 percent higher than predicted here.

The inlet water temperature is normally 20°F (36°C) lower than the flue-gas outlet temperature. And for the usual preliminary evaluation of the efficiency of condensation heat recovery, the flue-gas exit temperature from the heat exchanger is taken as 100°F (37.8°C).

For the natural-gas-burning boiler, flue-gas inlet temperature = 450°F (232°C); cold-water inlet temperature = 60°F (15.6°C); water outlet temperature = 60 + 20 = 80°F (26.7°C); flue-gas outlet temperature = 80°F (26.7°C). Find the *basic* efficiency improvement, ΔEi, from Fig. 3 as ΔEi = 14.5 percent by entering at the bottom at the flue-gas temperature of 450°F (232°C), projecting to the gas-fired curve, and reading ΔEi on the left-hand axis.

Next, find the *actual* efficiency improvement, ΔE, from $\Delta E = F(\Delta Ei)$, where F is a factor depending on the flue-gas outlet temperature. Values of F are shown in Fig. 4 for various outlet gas temperatures. With an outlet gas temperature of 80°F (26.7°C), Fig. 4a shows F = 1.19. Then, ΔE = 14.5 × 1.19 = 17.3 percent. Table 6 details system temperatures.

For the oil-fired boiler, flue-gas inlet temperature = 300°F (148.9°C); cold-water inlet temperature = 70°F (21.1°C); water outlet temperature = 70 + 20 = 90°F (32.2°C). Find the *basic* efficiency improvement from Fig. 3 as Ei = 7.2 percent. Next, find F from Fig. 4b as 1.18. Then, the *actual* efficiency = ΔE = 1.18 (7.2) = 8.5 percent.

The lower efficiency improvement for oil-fired boilers is generally due to the lower hydrogen content of the fuel. Note, however, that where the cost of oil is higher than natural gas, the dollar saving may be greater.

The efficiency-improvement charts given here assume that all of the low-level heat generated can be used. A plant engineer familiar with a plant's energy balance is in the best position to choose the optimum level of heat recovery. Typical applications are: makeup-water preheat, low-temperature process load, space heating, and domestic hot water.

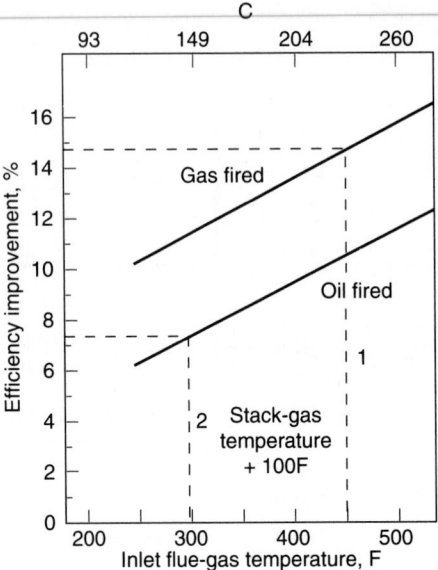

FIG. 3   Efficiency increase depends on fuel and on temperature of flue gas. (*Power.*)

Makeup-water-preheat needs depend largely on the amount of condensate that is returned to the boiler. Generally, there is more heat available in the flue gas than can be used to preheat feedwater. If the boiler is operating at 100 percent makeup, only about 60 percent of the available heat can be transferred to the incoming feedwater. One reason for this is the low temperature of the hot water. This limitation can be handled in two ways: (1) Design the heat-recovery unit to take a slip-stream from the flue gas and only recover as much heat as can be used to heat feedwater; or (2) in multiple-boiler plants, install a heat-recovery system on one boiler only and use it to preheat feedwater for all the boilers.

Process hot water, if it is needed, can provide extremely short payback for a condensation heat-recovery unit. In food and textile processes, the hot-water needs account for 15 percent or more of the total boiler load. Any facility with hot-water requirements between 10 and 15

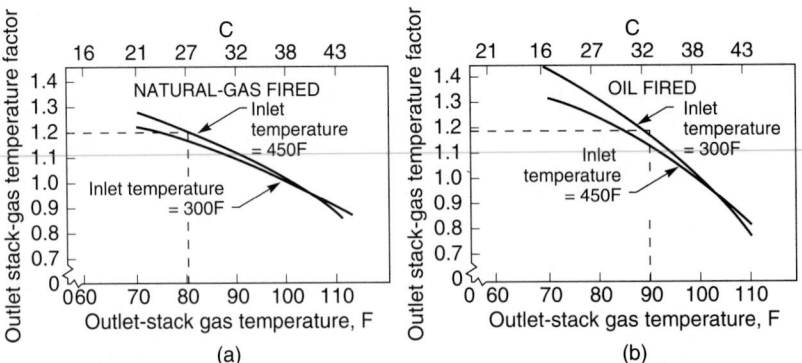

FIG. 4   Use these curves to allow for the effect of variations in the exit temperature from the recovery unit on efficiency increase possible with heat-recovery unit. (*Power.*)

percent of boiler capacity and an operating schedule greater than 4000 h/yr should seriously investigate condensation heat recovery.

Space-heating economics are generally less favorable than makeup or process hot water because of the load variation, limited heating season, and the difficulty of matching demand and supply schedules. The difficulty of retrofitting heat exchangers to an existing heating system also limits the number of useful applications. However, paybacks between 2.5 and 3 years are possible in colder regions and certainly warrant preliminary investigation. Sometimes space heating can be combined with feedwater or process-water heating.

3. *Estimate the cost of the condensation heat-recovery equipment*

Figure 5 shows an approximate range of costs for equipment and installation. Note that the installed cost may be three times the equipment cost because of retrofit difficulties involved with an existing installation.

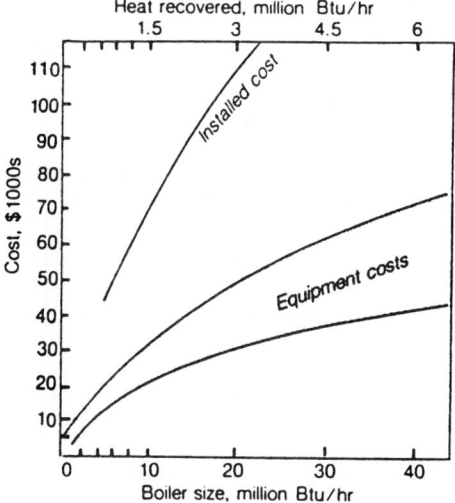

FIG. 5   Installation cost of condensation heat-recovery unit may run as high as three times equipment cost. (*Power.*)

Operating costs are primarily fan and pump power consumption. These generally range from 5 to 10 percent of the value of the recovered heat. The lower figure applies to limited distribution of the hot water, while the higher figure applies to systems where hot water is distributed 100 ft (30.5 m) or more from the boiler or where high-pressure-drop heat-recovery units are used.

Figure 6 shows how a heat-recovery unit can be used to heat the feedwater for one or more boilers. The heat recovered, as noted above, may be more than needed to heat the feedwater for just one boiler.

Corrosion in a condensing heat-recovery unit can usually be prevented by using Type 304 or 316 stainless steel or fiberglass-reinforced plastic for the tower pump and secondary exchanger. If the flue gas is unusually corrosive, it may be advisable to do a chemical analysis before planning the recovery unit.

A unique feature of condensation heat recovery is that it recovers energy while also reducing emissions. In addition, when natural gas is burned, a small percentage of the NO_x emissions are reduced by condensation of oxides of nitrogen. SO_2 emissions can be reduced significantly by using an alkaline water spray in a pH range of 6 to 8. Natural gas depletes the ozone layer less than other fossil fuels.

The potential emission reduction can have a significant effect in nonattainment areas and could increase allowable plant capacity. But it should be pointed out that the SO_2-emission reduction from the scrubbing and condensation have not been substantiated by independent

PARTIAL HEAT RECOVERY

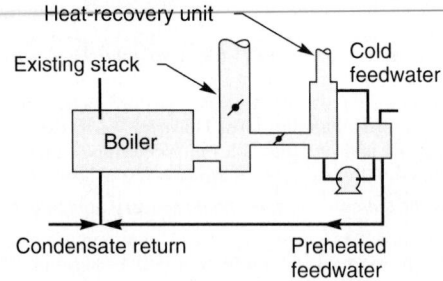

HEAT RECOVERY UNIT SERVING
MULTIPLE BOILERS

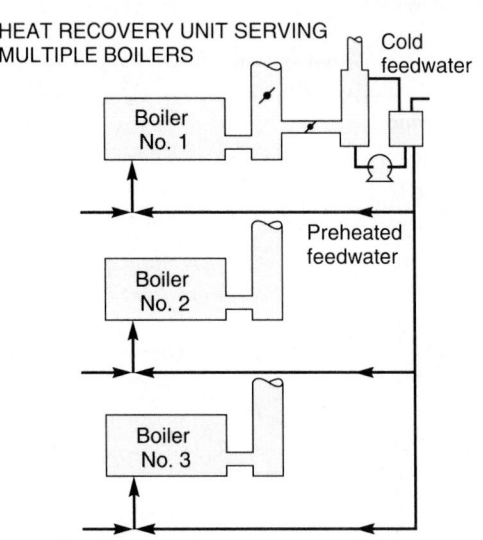

FIG. 6 Heat recovery from the unit is more than enough to
heat feedwater to one boiler. (*Power.*)

TABLE 6 Use of Figs. 3 and 4

1. Natural gas		
Inlet temperature	450°F	232°C
Inlet cold water	60°F	16°C
Outlet temperature = 60 + 20 =	80°F	27°C
ΔEb (from Fig. 3)	14.5%	14.5%
F (from Fig. 4)	1.19	1.19
$\Delta E = (14.5 \times 1.19)$	17.3%	17.3%
2. Fuel oil		
Inlet temperature	300°F	149°C
Inlet cold water	70°F	21°C
Outlet temperature = 70 + 20 =	90°F	32°C
ΔEb (from Fig. 3)	7.2%	7.2%
F (from Fig. 4)	1.18	1.18
ΔE (7.2 × 1.18)	8.5%	8.5%

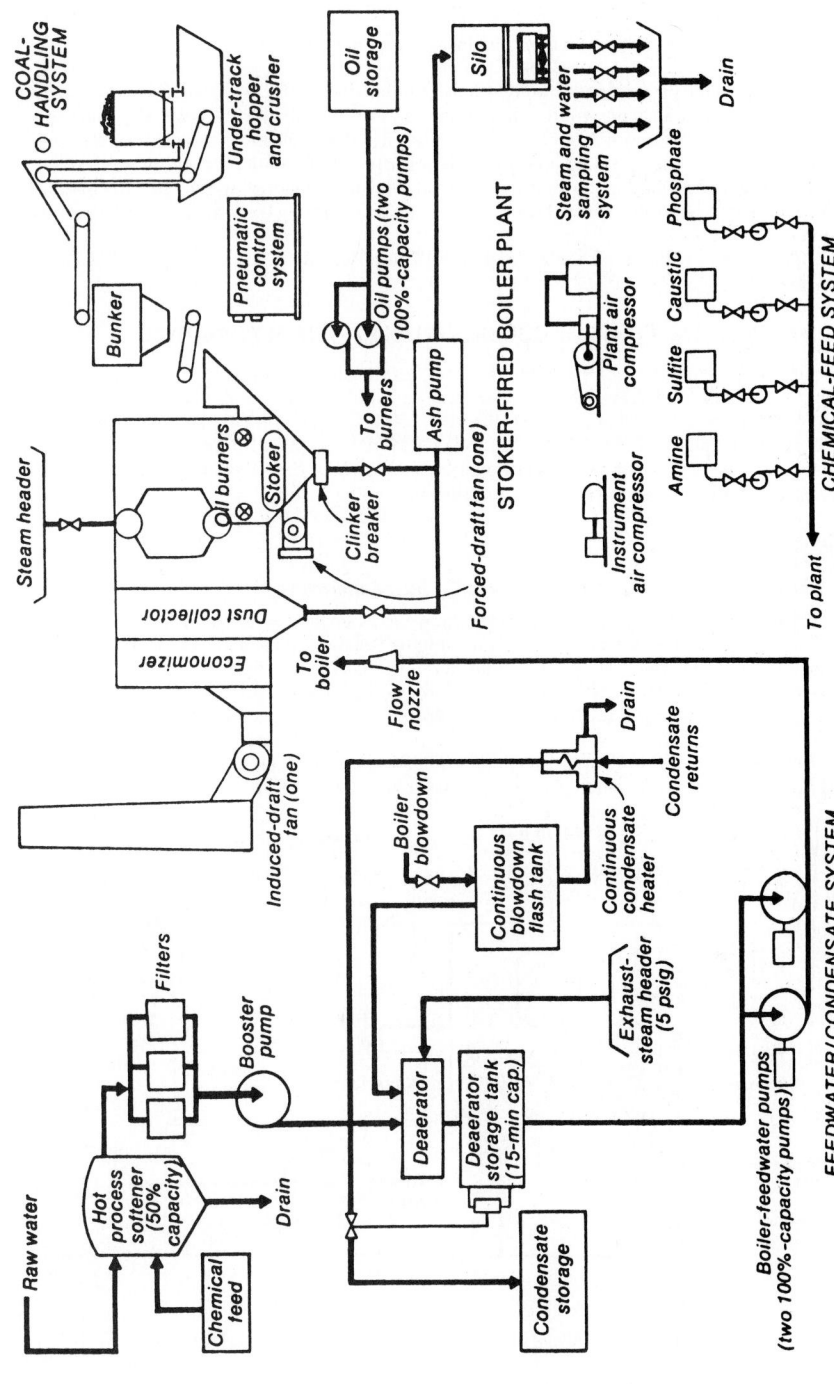

FIG. 7 Stoker-fired steam plant used in the cost analysis includes all the components shown. (*Power.*)

13.13

tests. Such tests should be provided for any installation that depends on emission reduction for its justification.

At the time of the preparation of the revision of this handbook, the Tennessee Valley Authority (TVA) is testing at its Shawnee plant a lime treatment to reduce boiler stack gas SO_2 emissions. Lime, in fine particle form, is suspended in the flue gas before release to the plant smokestack. Sulfur in the flue gas binds to the lime, thereby reducing the potential for acid rain. A cyclone and electrostatic precipitator separate the lime particles from the exiting flue gas. At this time the lime system is believed to have lower equipment and operating costs than other competitive systems.

This procedure is based on the work of R. E. Thompson, KVB, Inc., as reported in *Power* magazine.

ESTIMATING TOTAL COSTS OF COGENERATION-SYSTEM ALTERNATIVES

Compare the capital and operating costs of two cogenerational coal-fired industrial steam plants—Option 1: a stoker-fired (SF) plant, Fig. 7; Option 2: a pulverized-coal-fired (PF) plant, Fig. 8. Both plants operate at 600 lb/in² (gage) (4134 kPa)/750°F (399°C). The SF plant meets a demand of 200,000 lb/h (90,800 kg/h) of 150-lb/in² gage (1034-kPa) steam and 10 MW of electric power with the SF boiler and purchased power from a local utility. For the PF boiler the same demands are met using a nonextraction backpressure turbine.

Calculation Procedure:

1. *Obtain, or develop, the capital costs for the boiler alternatives*

Contact manufacturers of suitable boilers, asking for estimated costs based on the proposed operating capacity, pressure, and temperature. Figure 9 shows a typical plot of the data supplied by manufacturers for the boilers considered here. For coal firing, field-erected boilers are to be used in this plant. (*Note:* The costs given here are for example purposes *only.* Do *not* use the given costs for actual estimating purposes. Instead, obtain current costs from the selected manufacturers.)

From Fig. 9, the SF unit costs $61/lb ($27.70/kg) of steam generated; and the PF unit $72/lb ($32.70/kg) of steam generated per hour. Assuming that SO_2 reduction is not required by

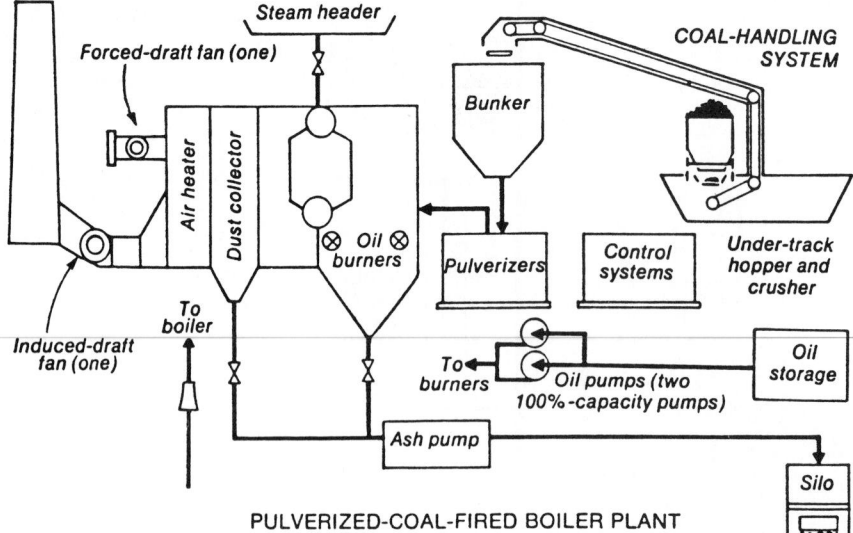

FIG. 8   Pulverized-coal-fired industrial boilers produce from 200,000 lb/h (90,800 kg/h) to 1 million lb/h (454,000 kg/h) of steam. Unit here does not have an economizer. (*Power.*)

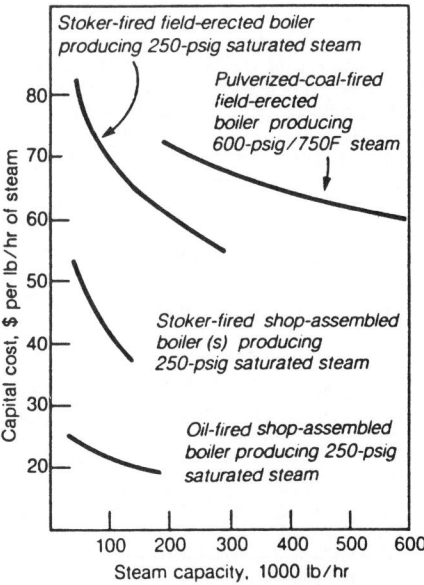

FIG. 9 Capital costs for three boiler alternatives. (*Power.*)

environmental considerations for either unit, a dry-scrubber/fabric-filter combination can be used to remove the total suspended particulates (TSP). To cover the cost of this combination to remove TSP, add $7/lb ($3.20/kg) of steam generated. This brings the cost to $68/lb ($30.90/kg) and $79/lb ($35.90/kg) of steam generated.

2. Compute the capital cost for the turbine installation

Before determining the capital cost of the turbine installation—often called a *turbine island*—estimate the potential electric-power generation from the process-steam flow based on the ASME data in Table 7. At 600 lb/in² (gage) (4134 kPa)/750°F (399°C) throttle conditions, and 150-lb/in² (gage) (1034 kPa) backpressure, 6127 kW is available, determined as follows: Theoretical steam rate from Table 7 = 23.83 lb/kWh (10.83 kg/kWh); turbine efficiency = 73 percent from Table 7. Then, actual steam rate = theoretical steam rate/efficiency. Or, actual steam rate = 23.83/0.73 = 32.64 lb/kWh (14.84 kg/kWh). Then kW available = steam flow rate, lb/h/steam rate, lb/kWh. Or, kW available = 200,000/32.64 = 6127 kW.

Referring to Fig. 10 for a 6-MW nonextraction backpressure turbine shows a capital cost of $380/kW. Summarize the capital costs in tabular form, Table 8. Thus, the SF boiler cost, Option 1 in Table 8, is (200,000 lb/h)($68/lb of steam generated) = $13,600,000. The PC-fired boiler, Option 2, will have a cost of (200,000 lb/h)($79/lb of steam generated) = $15,800,000. Since a turbine is used with the PC-fired unit, its cost must also be included. Or, 6100 kW ($380/kW) = $2,318,000. Computing the total cost for each option shows, in Table 8, that Option 2 costs $18,118,000 − $13,600,000 = $4,518,000 more than Option 1.

3. Compare operating costs of each option

Obtain from the plant owner and equipment suppliers the key data needed to compare operating costs, namely: Operating time, h/yr; boiler efficiency, %; fuel cost, $/ton ($/tonne); electric-power use, kWh/yr; electric power cost, ¢/kWh; maintenance cost as a percent of the capital investment per year; personnel required; personnel cost; ash removal, tons/yr (tonnes/yr); ash removal cost, $/yr. Using these data, compute the operating cost for each option and tabulate the results as shown in Table 9.

With Option 1, all electric power is purchased; with Option 2, 3900 kW must be purchased. The difference in annual operating cost is $6,288,350 − $4,895,000 = $1,393,350. Since Option

TABLE 7 ASME Values for Estimating Turbine Steam Rates*

	Theoretical steam rate, lb/kWh (kg/kWh)					
Exhaust pressure	600 lb/in² (gage) (4134 kPa)		900 lb/in² (gage) (6201 kPa)		1500 lb/in² (gage) (10,335 kPa)	
	750°F	(399°C)	900°F	(482°C)	900°F	(482°C)
4.0 inHg (10.2 cmHg)	7.64	3.47	6.69	3.04	6.48	2.94
5 lb/in² (gage) (34.5 kPa)	11.05	5.02	9.21	4.18	8.53	3.87
15 lb/in² (gage) (103.4 kPa)	12.16	5.52	9.98	4.53	9.06	4.11
150 lb/in² (gage) (1034 kPa)	23.83	10.82	16.91	7.68	14.30	6.49
	Turbine efficiency, %					
4.0 in Hg		77		78		78
5 lb/in² (gage)		73		76		76
15 lb/in² (gage)		74		76		76
150 lb/in² (gage)		73		74		74

*Actual turbine steam rate equals theoretical steam rate divided by efficiency.

2 costs $4,518,000 more than Option 1, but has a $1,393,350-per-year lower operating cost, the simple payback time for the cogeneration option (2), ignoring the cost of money, is: Payback time = larger capital cost, $/annual savings, $, of the higher cost option. Or, $4,518,000/$1,393,350 = 3.24 years. Thus, the cogeneration option (2) is attractive because the payback time is relatively short.

Other economic analyses should also be conducted. For example, higher cycle efficiencies can be obtained with higher throttle conditions, but boiler capital and operating costs will be higher. Plants in the 40- to 60-MW range might benefit from an extraction–condensing-turbine

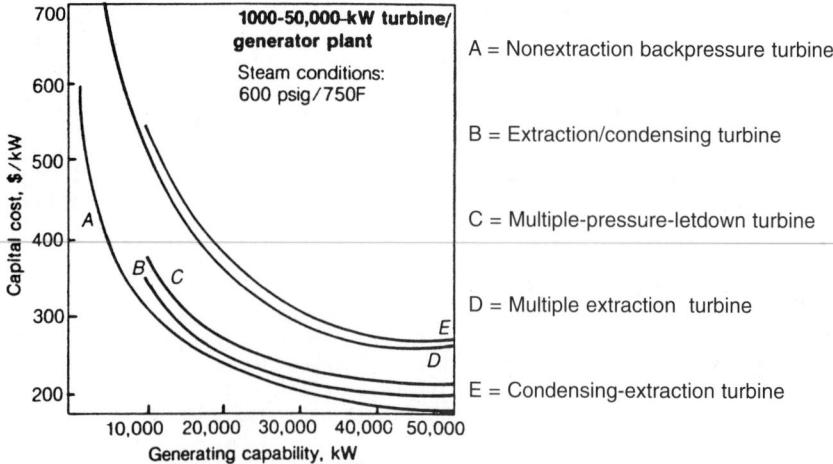

A = Nonextraction backpressure turbine

B = Extraction/condensing turbine

C = Multiple-pressure-letdown turbine

D = Multiple extraction turbine

E = Condensing-extraction turbine

FIG. 10 Turbine capital cost is estimated from this type of graph. (*Power.*)

TABLE 8 Summary of Capital Costs

	Option 1	Option 2
Boiler island		
Stoker-fired unit ($68/lb steam)	$13,600,000	—
($30.90/kg steam)		
PC-fired unit ($79/lb steam)	—	$15,800,000
($35.90/kg steam)		
Turbine island (6100 kW, $380/kW)	—	2,318,000
Total	$13,600,000	$18,118,000

arrangement generating all plant electric-power demand. Although not as efficient as a straight extraction machine, the cost, in $/kW, will be less than purchased power.

   Related Calculations: The EPA and state environmental bodies favor cogeneration of electricity because it reduces atmospheric pollution while conserving fuel. While the options considered here use steam-powered prime movers, cogeneration installations can use diesel-, gasoline-, or natural-gas-fueled prime movers of many different types—reciprocating, gas-turbine, etc. The principal objective of cogeneration is to wrest more heat from available energy streams by the simultaneous generation of electricity and steam (or some other heated medium), thereby saving fuel while reducing atmospheric pollution.

   While the possible choices for boiler fuel are oil, gas, and coal, practical choices are limited to coal for most industrial cogeneration projects. Gas firing in industrial plants is restricted to units 80,000 lb/h (36,364 kg/h) or less by the Industrial Fuel Use Act. Oil firing is allowed in field-erected boilers, but usually gives way to coal on an economic basis. Packaged oil-fired boilers top out at 200,000 lb/h (90,909 kg/h) to permit shipping.

   Coal can be burned in either a stoker-fired unit or a pulverized-coal-fired unit, the essential cost differences being in coal preparation and ash handling. PC-fired units produce about 75 percent flyash and 25 percent bottom ash, and stoker-fired units the reverse. Coal supply

TABLE 9 Comparison of Operating Costs for Coal-Fired Plants

Operating requirements	Option 1 Stoker-fired	Option 2, PC-fired
Equivalent full-power operating time, h/yr	5,400	5,700
Boiler efficiency, %	82	87
Fuel consumption, tons/yr (tonne/yr)	52,700 (47,798)	62,000 (56,234)
Fuel cost, $/ton ($/tonne)	50 (45.35)	42 (38.09)
Total fuel cost, $/yr	2,640,000	2,600,000
Electric-power use, kWh/yr	3,240,000	8,550,000
Electric-power cost, $/yr @ 5¢/kWh	162,000	427,000
Maintenance cost, $/yr (based on 2.5% of capital investment per year)	340,000	395,000
Personnel per shift	2.5	2.75
Personnel cost, $/yr (based on $30,000/yr for each person)	270,000	330,000
Ash removal, tons/yr (tonne/yr)	5,270 (4780)	6,200 (5,623)
Ash-removal cost, $/yr	26,350	31,000
Total operating cost, $/yr	3,438,350	3,783,500
Total operating costs, steam + purchased power		
Steam	3,438,350	3,783,500
Purchased power @ 5¢/kWh	10MW 2,850,000	3.9 MW 1,111,500
Total operating cost	$6,288,350	$4,895,000

specifications also vary between the two. Grindability is important to PC-fired units; top size and fines content is critical to stoker-fired units. Coal for stoker-fired units averages $5/ton ($5.51/tonne) more for coals of similar heating values.

PC-fired boilers always include an air heater for drying coal upstream of the pulverizer, and usually an economizer for flue-gas heat recovery. The reverse is true of stoker-fired units, but, concerning the air heater, more care is needed to ensure that the grate is not overheated during normal operation.

The greatest overlap in choosing between the two exists in the 200,000 to 300,000 lb/h (90,909 to 136,364 kg/h) size range. Above this range, stoker-fired units are limited by grate size. Below this range, PC-fired units usually do not compete economically. Shop-assembled chain-grate stoker-fired units have been shipped up to a capacity of 45,000 lb/h (20,455 kg/h).

No matter what type of firing is used, industrial power plants must meet EPA emission limits for total suspended particulates (TSP), SO_2, and NO_x. If an on-site coal pile is contemplated, water runoff control must meet National Pollution Discharge Elimination System (NPDES) standards.

NO_x formation is typically limited in the combustion process through careful choice of burners. SO_2 and TSP are usually removed from the flue gas. For TSP reduction, an electrostatic precipitator or a fabric filter is used. These will sometimes be preceded by cyclone collectors for stoker-fired boilers to reduce the total load on the final stage of the ash-collection system.

For boilers under 250-million Btu/h (73.3 MW) heat input, SO_2 formation can be limited by burning low-sulfur coal. Where SO_2 emissions reduction is required to meet the National Ambient Air Quality Standards (NAAQS), a dry-scrubber–fabric-filter combination is a satisfactory strategy. Wet scrubbers, though highly effective, create an additional sludge-disposal problem.

Operating costs for stoker- and pulverized-coal-fired plants designed to produce 200,000 lb/h (90,800 kg/h) of steam from one boiler are shown in Table 9.

To understand more about how operating costs were calculated, look at the entries in Table 9 line by line. First, equivalent full-power operating hours are determined by subtracting 336 h (2 weeks) for maintenance from the total number of hours in a year (8760), and by multiplying the result by both unit availability (85 percent for stoker, 90 percent for pulverized coal) and the assumed plant load factor—in this case 75 percent.

Fuel consumption is based on typical operating efficiencies for similar plants and a fuel heating value of 12,500 Btu/lb (29,125 kJ/kg) for coal. Note that a premium is paid for stoker coal because a relatively clean fuel of suitable size is needed to maintain efficient operation.

The general procedure given here is applicable to a variety of options because today's emphasis on industrial cogeneration calls for a method of reasonably estimating costs of the many system alternatives. The approach differs from utility cost-estimating mainly because steam capacity and power-generation capability are separate design objectives. Either the industrial power plant meets the process-steam demand and then generates whatever power that creates, or else it meets the electric-power demand and generates the required steam.

Choice of approach depends on the steam and electric-power requirements of the facility. Ideally, they balance exactly. In practice, most steam requirements will not generate enough electric power to meet the plant load. Conversely, the steam flow can rarely generate more electric power than the plant needs. Variations in steam conditions and turbine-exhaust pressure lead to many ways of matching the loads. More important, regulated buyback of excess electric power by public utilities now eases the problem of load balancing.

A profile of steam and electric-power consumption is necessary to begin evaluating alternatives. Daily, weekly, monthly, and seasonal variations are all important. During initial evaluation of the balance, use an average of 25 lb (11.4 kg) of steam/kWh as the steam rate of a small steam turbine. It is a conservative number, and the actual value will probably be lower—meaning more electric power for the steam flow—but it will give a rough idea of how close the two demands will match.

In the above procedure, capital costs are separated into costs for the boiler island and costs for the turbine island. Absolute accuracy is to within ±25 percent, not of appropriation quality but good enough to compare different plant designs. In fact, relative accuracy is closer to ±10 percent.

This procedure is the work of B. Dwight Coffin, H. K. Ferguson Co., and was reported in *Power* magazine.

CHOOSING STEAM COMPRESSOR FOR COGENERATION SYSTEM

Select a suitable steam compressor to deliver an 80-lb/in² (gage) (551-kPa) discharge pressure for a cogeneration system using two 1500-kW diesel-engine-generator sets operating at 1200 rpm with 1200°F (649°C) exhaust temperature. Each engine exhaust is vented through a waste-heat (heat-recovery) boiler which generates 3800 lb/h (1725 kg/h) of steam at 110-lb/in² (gage) (758-kPa) saturated. The cooling system of each engine generates 5000 lb/h (2270 kg/h) of 15-lb/in² (gage) (103-kPa) steam. Choose the compressor to boost the 15-lb/in² (gage) (103-kPa) steam pressure to 80-lb/in² (gage) (551-kPa) to be used in a distribution system. Steam at 110 lb/in² (758 kPa) is first used in laundry and heat-exchange equipment before being reduced in 80 lb/in² (gage) (551 kPa) and combined with the compressor discharge flow. About 16,500 lb/h (7491 kg/h) of 80-lb/in² (gage) (551-kPa) steam satisfies the distribution system requirements, except in severe weather when the existing boilers are fired to supplement the steaming requirements.

Calculation Procedure:

1. *Determine the amount of steam that can be generated by each waste-heat boiler*

Exhaust gas from each diesel engine enters the waste-heat boiler at 1200°F (649°C). Using the rule of thumb that a diesel-engine exhaust heat boiler can produce 1.9 lb/h (0.86 kg/h) of 100-lb/in² (gage) (689-kPa) saturated steam at full load per rated horsepower, find the amount of steam generated as 1.9 (1500 kW/0.746 hp/kW) = 3820.4 lb/h (1734.5 kg/h). Since the 110-lb/in² (gage) (758 kPa) steam required by the cogeneration system needs slightly more heat input, round off the quantity of steam generated to 3800 lb/h (1725 kg/h).

2. *Select the type of steam compressor to use*

The compression ratios used in cogeneration—higher than for many process applications—dictate use of mechanical compressors. Several different thermodynamic paths may be followed during the compression process (Fig. 11). Of the three processes shown, the highest compressor coefficient of performance (COP) is exhibited by direct two-phase compression (Fig. 12). The task becomes one of selecting a suitable unit to follow this path.

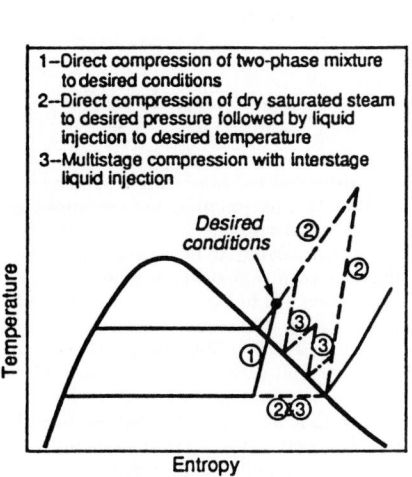

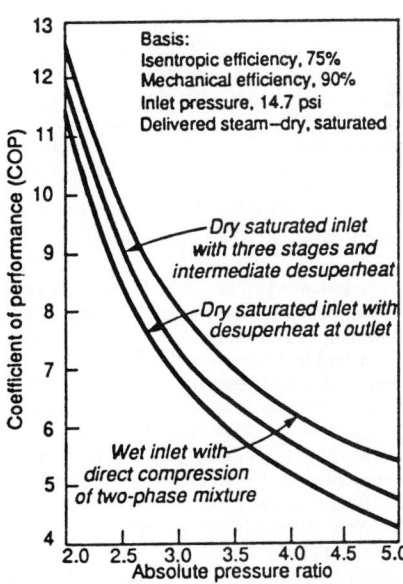

FIG. 11 Compression can follow several different thermodynamic paths. (*Power.*)

FIG. 12 Each compression path exhibits a different coefficient of performance. (*Power.*)

Centrifugal and axial compressors, both of the general category of dynamic compressors, work on aerodynamic principles. Though capable of handling large flow rates, they are sensitive to water droplets that may cause blade erosion. Thus, two-phase flow is undesirable. These compressors are also limited to operation within a narrow range because of surging or low efficiency when conditions deviate from design.

Positive-displacement compressors—such as the screw, lobe, or reciprocating variety—are more suited to cogeneration applications. Both the screw compressor and, to a lesser extent, the reciprocating compressor, achieve high pressure ratios at high COP. Further, the units approach the isothermal condition because work which normally goes into producing sensible heat during compression simply causes additional liquid to evaporate. Intercooling is avoided, and input power requirements are acceptable (Fig. 13).

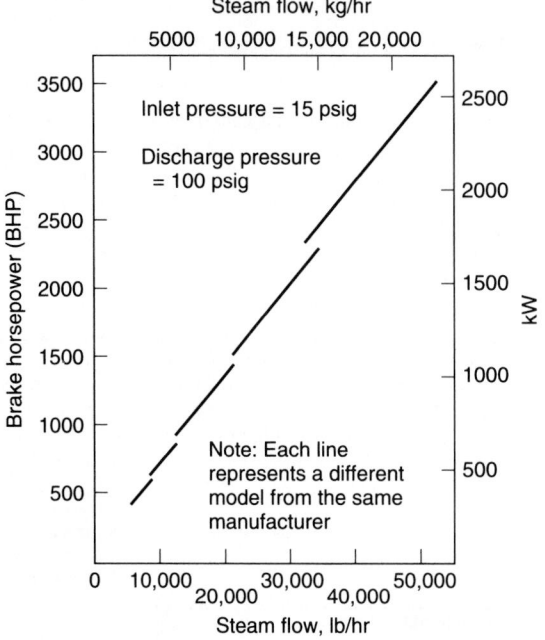

FIG. 13   Energy input requirements for a screw compressor. (*Power.*)

Cost considerations tend to make reciprocating units the second choice after screw and lobe units, but they are comparable on a technical basis. Note that reciprocating units require large foundations and must be driven at slow speeds. Some manufacturers offer carbon-ring units, requiring no lubrication. Even with the more common units requiring lubrication, use of synthetic lubricants keeps the amount of oil small, making steam contamination a negligible concern.

Based on the above information, two screw compressors will be chosen to boost the 15-lb/in² (gage) (103-kPa) steam to 80-lb/in² (gage) (551 kPa). Use Fig. 13 to approximate the required horsepower (kW) input to each screw compressor. With 16,500 lb/h (7491 kg/h) process steam required, Fig. 13 shows that the required horsepower input for each screw compressor will be 1250 hp (933 kW).

Note: Although Fig. 13 applies to an inlet pressure of 15 lb/in² (gage) (103 kPa) and an outlet pressure of 100 lb/in² (gage) (689 kPa), the results are accurate enough for an 80-lb/in² (gage) (551-kPa) outlet pressure. The required horsepower will be slightly less than that shown.

The two screw compressors will be clutch-connected at the generator end of the two engine-generator sets, as shown in Fig. 14. This diagram also shows the piping layout for the three

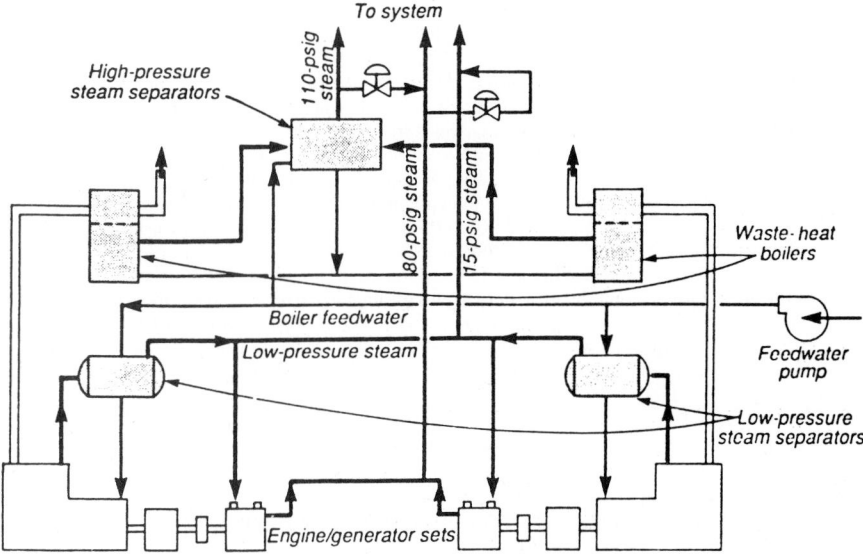

FIG. 14 System uses common high-pressure steam separator. (*Power.*)

steam systems—110 (758), 80 (551), and 15-lb/in² (gage) (103 kPa). Low-pressure steam separators are used to remove water from the low-pressure cogenerated steam.

In an actual system similar to that shown here, 110-lb/in² (gage) (758-kPa) steam is first used in a laundry and in heat-exchange equipment before being reduced to 80 lb/in² (gage) (551 kPa) and combined with the compressor discharge flow. In the summer, 15-lb/in² (gage) (103-kPa) steam is used to drive a large absorption chiller supplying 600 tons (2112 kW) of refrigeration.

Related Calculations: The recent popularity of reciprocating-engine-based cogeneration has caused a new factor to enter the economic evaluation of these systems: the value of the 15-lb/in² (gage) (103-kPa) steam typically recovered from the engine's cooling system. Because uses for 15-lb/in² (gage) (103-kPa) steam are limited, boosting the pressure to about 100 lb/in² (gage) (689 kPa) multiplies the practical uses of the recovered heat. This concept of pressure boosting is compatible with recent trends in cogeneration to maximize the value of the thermal output through closer coupling of the power-process interface.

Steam recompression has long been an accepted practice in the process industries where large quantities of low-pressure steam can be economically upgraded. A pound (0.45 kg) of steam vented to the atmosphere or condensed represents a loss of about 1000 Btu (1055 kJ) of heat energy. Thus, in many applications, it is less expensive (and more environmentally wise) to boost steam pressure than to produce the equivalent amount in a boiler.

Pressure ratios used to satisfy process requirements are relatively low—about 1.5 to 2. Thermocompressors most economically satisfy these ratios. They use high-pressure steam to boost low-pressure steam to a point in between the two.

Practical limitations on thermocompressors for satisfying higher ratios are two: (1) a large quantity of high-pressure steam is needed, and (2) the heat balance must be such that the steam need not be vented or condensed. For example, if 600-lb/in² (gage) (4134-kPa) boiler steam is available to boost 15-lb/in² (gage) (103-kPa) steam to 150 lb/in² (gage) (1034 kPa), about 12 times the quantity of high-pressure steam is required. So if an engine produces 5000 lb/h (2270 kg/h) of 15-lb/in² (gage) (103-kPa) steam, 60,000 lb/h (27,240 kg/h) of high-pressure steam is required to meet a 65,000 lb/h (29,510 kg/h) 150-lb/in² (1034 kPa) steam demand.

The typical reciprocating engine rejects 65 to 70 percent of its heat input to exhaust, engine cooling, lube-oil coolers, intercoolers, and radiation. About 20 percent of this heat is recoverable from the exhaust as steam at pressures up to 150 lb/in² (gage) (1034 kPa) and beyond, representing about 70 percent of the available heat in the exhaust. Another 28 percent represents engine cooling that can be completely recovered as hot water or as steam at a maximum pressure of 15 lb/in² (gage) (103 kPa). Lube-oil heat may also be recoverable, but usually not the intercooler heat.

Although hot water and 15-lb/in² (gage) (103-kPa) steam can be used for space heating, as a heat source for absorption refrigeration, or for domestic water heating, there usually is more demand for higher-pressure steam—such as that produced by the engine's exhaust.

As cogeneration systems maximize heat recovery from an internal-combustion engine, it is important to note that the engine exhaust no longer is a simple pipe protruding through the roof of a building. Exhaust explosions, not uncommon to engine operation, thus can be destructive to the often large and complex exhaust systems of cogeneration installations.

For this reason, it is prudent to tighten engine specifications. Partial failure of the ignition system, for example, should not be able to cause a potentially catastrophic exhaust explosion. Further, cross-limiting should be provided when fuel and air are measured at different locations.

Engine manufacturers may require that the exhaust system resist any explosion. A preferred alternative to this requirement is to insist that the engine manufacturer design to minimize exhaust explosions so that the cogeneration plant designer can confidently specify lightweight preformed exhaust ducting.

The data presented in this procedure were drawn from the work of Paul N. Garay, FMC Associates, a division of Parsons Brinckerhoff Quade & Douglas Inc., as reported in *Power* magazine.

USING PLANT HEAT NEED PLOTS FOR COGENERATION DECISIONS

An industrial process plant's heat needs are dominated by distillation. Its heat needs are represented by the fired-heat composite curve (FHCC) shown in Fig. 15. Five distinct heat sources are used in this process plant: two furnaces and steam supplied at three different pressure levels, as shown in Table 10. A gas turbine with the exhaust profile shown in Fig. 15 can supply all the heat needs of the process. Determine the annual fuel savings and payback time.

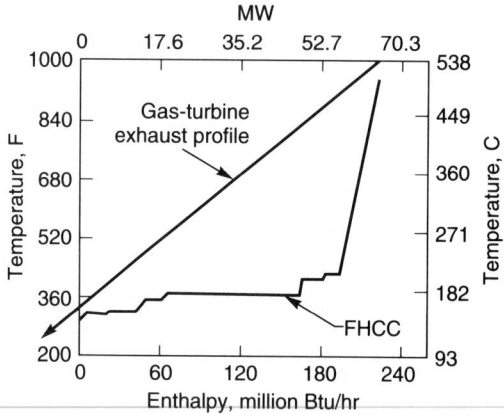

FIG. 15   Fired-heat composite curve for BTX plant matches well with exhaust profile of gas turbine. (*Power.*)

Calculation Procedure:

1. Analyze the fired-heat composite curve

Heat obtained from a cogeneration system generally displaces heat from other sources that can be traced back to direct fuel firing. Even though the fuel may not be fired at the point of use, it is almost always fired somewhere, such as in a boiler.

TABLE 10   Heat Loads

Heat source	Heat load, million Btu/h (MW)
280-lb/in² (gage) (1.93 kPa) steam	25.1 (7.35)
140-lb/in² (gage) (0.96 kPa) steam	88.1 (25.8)
70-lb/in² (gage) (0.48 kPa) steam	$23.9 (7.0)
Furnace duty, 242–954°F (117–512°C) nonlinear	51.8 (15.2)
Furnace duty, 356–360°F (180–182°C)	19.4 (5.7)
Total	208.3 (61.05)

All these heating needs can be represented by a single FHCC. This curve of heat quantity (H) vs. temperature (T) represents the overall heating duties that, as far as possible, must be satisfied by the cogeneration system. The exhaust-heat profile of the cogeneration system can also be represented by a curve of heat quantity vs. temperature.

To see how such curves are developed, consider the three heat-acceptance profiles in Fig. 16. Profile (a) is a simple constant-heat-capacity profile, typical of heating duties in which no phase change occurs. The process stream is heated from its supply temperature to its target temperature and the heat load varies linearly between these points.

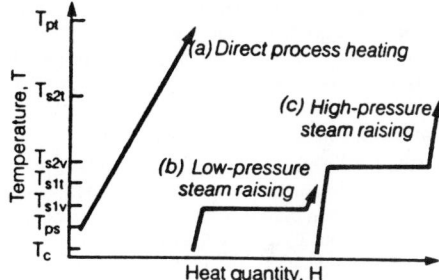

T_c = condensate to boiler
T_{ps} = process supply
T_{s1v} = steam level 1 vaporization
T_{s1t} = steam level 1 target
T_{s2v} = steam level 2 vaporization
T_{s2t} = steam level 2 target
T_{pt} = process target

FIG. 16   Every heating duty has characteristic heat-acceptance profile. (*Power.*)

Profile (b) represents low-pressure steam raising. The first linear part of this curve corresponds to preheat, the horizontal plateau to vaporization, and the final linear section to superheating. Profile (c) represents high-pressure steam raising.

Heat loads, unlike temperature, are additive. Thus it is possible to add the three profiles of Fig. 16 to obtain a combined heat-acceptance profile (Fig. 17). This is the FHCC and it shows total heating needs in terms of the quantity of heat required and the temperature at which it is needed.

The exhaust profile of the proposed cogeneration plant is also shown in Fig. 17. In this case it represents heat in the gas-turbine exhaust and is a straight line, neglecting the effect of

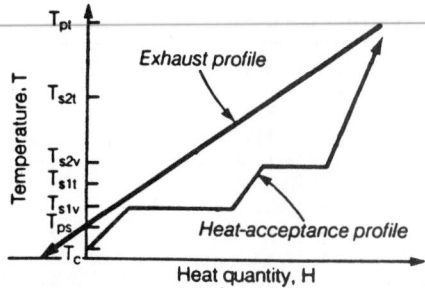

FIG. 17 Total process heat-acceptance profile is matched with prospective exhaust profile. (*Power.*)

condensation. Note that the exhaust profile lies above the heat-acceptance curve, implying that heat can be transferred from the exhaust stream to the process. The vertical separation between the two profiles is a measure of the available thermal driving force for heat transfer. Residual heat in the exhaust system, after the process duties have been satisfied, overhangs the heat-acceptance curve (at the left-hand end) and is lost up the stack.

Composite curves and profile matching provide a convenient way of representing the thermodynamics of heat recovery in cogeneration systems. Implicit within the construction of Fig. 17 are the requirements of the first law of thermodynamics, which demand a heat balance, and those of the second law, which lead to a relationship between the temperatures at which heat is required and the efficiency of the cogeneration system.

Analysis of the FHCC in Fig. 15 shows that all the needed process heat can be supplied by the gas turbine exhaust. Hence, a further evaluation of the proposed cogeneration installation is justified.

2. *Determine the annual fuel saving and payback period*

Assemble the financial data in Table 11 from information available in plant records and estimates. These data show, for *this* proposed cogeneration installation, that the savings that can be obtained are: (a) boiler fuel savings, $4.1 million per year; (b) credit for cogenerated power, $13.1 million per year; total savings = $4.1 million + $13.1 million = $17.2 million per year. The additional cost is that for the cogeneration gas which is burned in the gas turbine, or $12.2 million. Thus, the net savings will be $17.2 million − $12.2 million = $5.0 million per year.

The payback time = installed cost, $/annual savings, $. Or, payback time = $15.8 million/ $5.0 = 3.16, say 3.2 years. This is a relatively short payback time that would be acceptable in most industries.

TABLE 11 Parameters Used to Evaluate Cogeneration Process

Displaced furnace fuel cost, $/million Btu	2
Furnace efficiency, %	85
Boiler fuel savings, $ million/yr	4.1
Displaced or exported power, $/kWh	0.045
Gas for cogeneration system, $/million Btu	3.50
Cogeneration gas cost, $ million/yr	12.2
Operating hours per year	8000
Power output, MW	36.3
Credit for cogenerated power, $ million/yr	13.1
Cogeneration efficiency, %	78.1
Installed cost, $ million	15.8
Total cash benefit, $ million/yr	5
Estimated payback, years	3

Related Calculations: Reciprocating internal-combustion engines are also often considered where gas turbines appear to be a possible choice. The reason for this is that about 20 percent of the heat content of fuel fired in a reciprocating engine is rejected in the exhaust gases and the heat-rejection profile is similar to that of a gas turbine. And even more heat, about 30 percent, is removed in cooling water at a temperature of 160°F (71°C) to 240°F (1116°C). A further 5 percent is available in the lubricating oil, usually below 180°F (82°C). The heat-rejection profile of a reciprocating engine that closely matches the composite curve of the plant's process is also shown in Fig. 18.

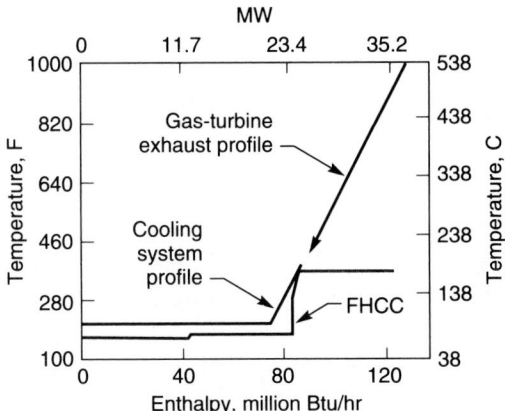

FIG. 18 Exhaust-heat profile of reciprocating engine is good fit with fired-heat composite curve of textile mill. (*Power.*)

A reciprocating engine has a higher overall efficiency than a gas turbine and therefore generates a greater cash benefit for the plant owner. For the scale of operation we are considering here, it would be necessary to use several engines and the capital cost would be substantially greater than that of a single gas turbine. As a result, payback periods for the two systems are about the same.

Gas turbines are often mated with steam turbines in combined-cycle cogeneration plants. In its basic form the combined-cycle power plant has the gas turbine exhausting into a heat-recovery steam generator (HRSG) that supplies a steam-turbine cycle. This cycle is the most efficient system for generating steam and/or electric power commercially available today. The cycle also has significantly lower capital costs than competing nuclear and conventional fossil-fuel-fired steam/electric stations. Other advantages of the combined-cycle plant are low air emissions, low water consumption, reduced space requirements, and modular units which allow phased-in construction. And from an efficiency standpoint, even in a simple-cycle configuration, gas turbines now exhibit efficiencies of between 30 and 35 percent, comparable to state-of-the-art fossil-fuel-fired power stations.

Cogeneration, which is the simultaneous production of useful thermal energy and electric power from a fuel source, or some variant thereof, is a good match for combined cycles. Experience with cogeneration and combined-cycle power plants has been most favorable. Figure 19 shows a variety of combined-cycle cogeneration plants using reheat in an HRSG to provide steam for a steam-turbine generator. Flexibility is extended as gas turbines, steam turbines, and HRSGs are added to a system. Reheat can improve thermal efficiency and performance by several percentage points, depending on how it is integrated into the combined cycle.

Aeroderivative gas turbines, as part of a combined cycle, increasingly are finding application in cogeneration in the under 100-MW capacity range. Cogeneration has the airline and defense industries to thank for the rapid development of high-efficiency, long-running gas turbines at extremely low research cost.

And the new large gas turbines have exhaust temperatures high enough to justify reheat in the steam cycle without supplementary firing in a boiler. Depending on how the reheat cycle

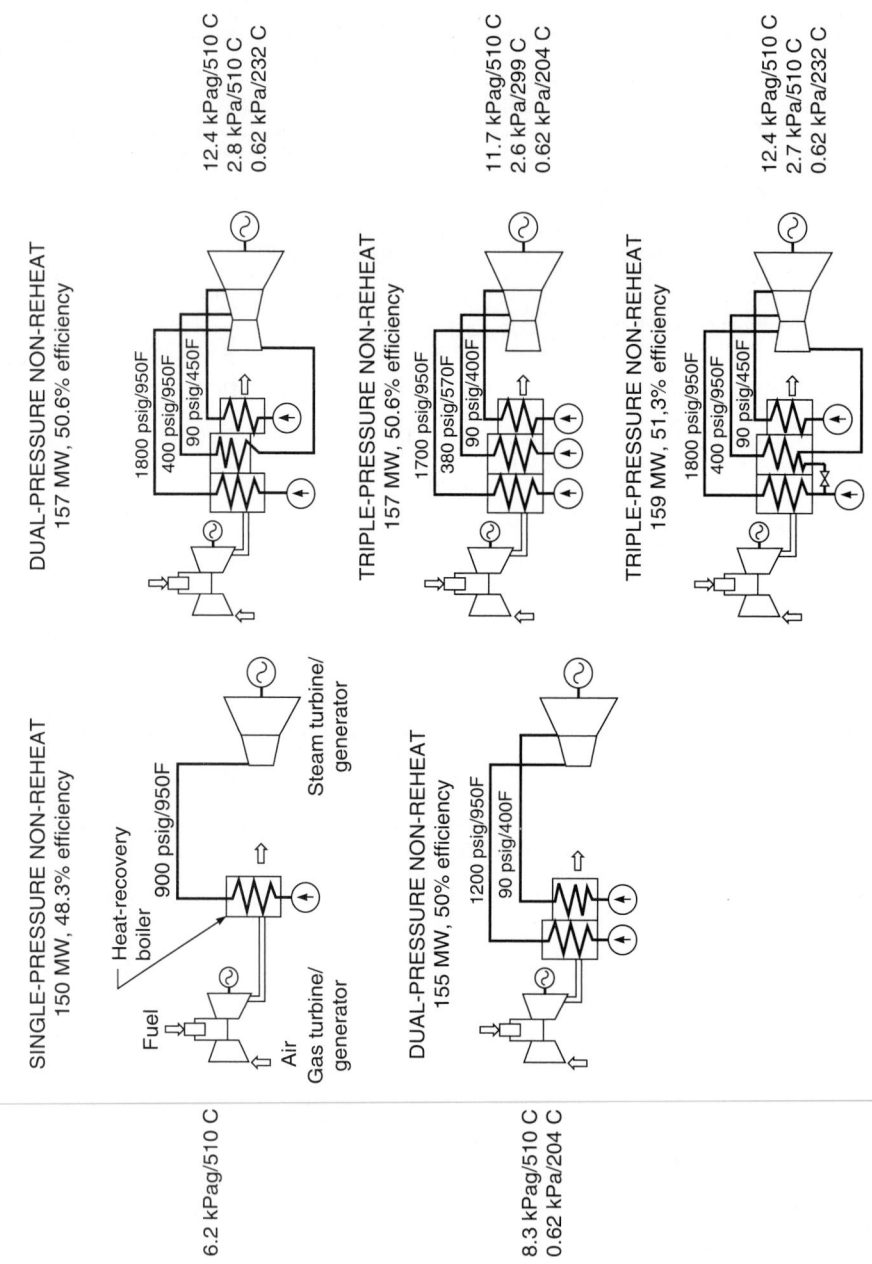

FIG. 19 Combined-cycle gas-turbine cogeneration plants using reheat in an HRSG to provide steam for a steam-turbine generator. (*Power.*)

SINGLE-PRESSURE NON-REHEAT
150 MW, 48.3% efficiency

900 psig/950F

Heat-recovery boiler

Fuel

Air

Gas turbine/ generator

Steam turbine/ generator

6.2 kPag/510 C
0.62 kPa/204 C

DUAL-PRESSURE NON-REHEAT
155 MW, 50% efficiency

1200 psig/950F
90 psig/400F

8.3 kPag/510 C
0.62 kPa/204 C

DUAL-PRESSURE NON-REHEAT
157 MW, 50.6% efficiency

1800 psig/950F
400 psig/950F
90 psig/450F

12.4 kPag/510 C
2.8 kPa/510 C
0.62 kPa/232 C

TRIPLE-PRESSURE NON-REHEAT
157 MW, 50.6% efficiency

1700 psig/950F
380 psig/570F
90 psig/400F

11.7 kPag/510 C
2.6 kPa/299 C
0.62 kPa/204 C

TRIPLE-PRESSURE NON-REHEAT
159 MW, 51.3% efficiency

1800 psig/950F
400 psig/950F
90 psig/450F

12.4 kPag/510 C
2.7 kPa/510 C
0.62 kPa/232 C

is configured, thermal performance at rated conditions can vary by up to three percentage points.

The Public Utilities Regulatory Policies Act (PURPA) passed by Congress to help manage energy includes incentives for efficient cogeneration systems. Cogeneration plants are allowed to sell power to local electric utilities to increase the return on investment earned from cogeneration.

A whole new energy-saving industry—termed nonutility generation (NUG)—has developed. At this writing NUG plants in the 200- to 300-MW range are common. And the pipeline industry which supplies natural-gas fuel for gas turbines is being restructured under the Federal Energy Regulatory Commission (FERC). Lower fuel costs are almost certain to result.

While lower electricity and energy costs are in the offing, these must be balanced against increased environmental requirements. The Clean Air Act Amendments of 1990 require better cleaning of stack emissions to provide a cleaner atmosphere. Yet this same 1990 act allows utilities to meet the required sulfur standard by installing suitable scrubber cleaning equipment, or by switching to a low-sulfur fuel.

A utility may buy—from another utility which exceeds the required sulfur standard—allowances to exhaust sulfur to the atmosphere. Each allowance permits a utility to emit 1 ton (tonne) of sulfur to the atmosphere. Public auctions of these allowances are now being held periodically by the Chicago Board of Trade.

Active discussions are underway at present over the suitability of selling sulfur allowances. Some opponents to sulfur pollution allowances believe that their use will delay the cleanup that ultimately must take place. Further, these opponents say, the pollution allowances delay the installation of sulfur-removal equipment. Meanwhile, sulfuric acid rain (also called acid rain) continues to plague communities in the path of a utility's sulfur effluent.

Challenging the above view is the Environmental Defense Fund. Its view is that there are too few allowances available to prevent the ultimate cleanup required by law.

The calculation data in this procedure are the work of A. P. Rossiter and S. H. Chang, ICI/ Tensa Services as reported in *Power* magazine, along with John Makansi, executive editor, reporting in the same publication. Data on environmental laws are from the cited regulatory agency or act.

GEOTHERMAL AND BIOMASS POWER-GENERATION ANALYSES

Compare the costs—installation and operating—of a 50-MW geothermal plant with that of a conventional fossil-fuel-fired installation of the same rating. Likewise, compare plant availability for each type. Brine available to the geothermal plant free-flows at 4.3 million lb/h (1.95 million kg/h) at 450 lb/in^2 (gage) at 450°F (3100 kPa at 232°C).

Calculation Procedure:

1. *Estimate the cost of each type of plant*

The cost of constructing a geothermal plant (i.e., an electric-generating station that uses steam or brine from the ground produced by nature) is in the $1500 to $2000 per installed kW range. This cost includes all associated equipment and the development of the well field from which the steam or brine is obtained.

Using this cost range, the cost of a 50-MW geothermal station would be in the range of: 50 MW × ($1500/kW) × 1000 = $75 million to 50 MW × ($2000/kW) × 1000 = $100 million. Fossil-fuel-fired installations cost about the same—i.e., $1500 to $2000 per installed kW. Therefore, the two types of plants will have approximately the same installed cost.

Department of Energy (DOE) estimates give the average cost of geothermal power at 5.7¢/ kWh. This compares with the average cost of 2.4¢/kWh for fossil-fuel-based plants. Advances in geothermal technology are expected to reduce the 5.7¢ cost significantly over the next 40 years.

Because of the simplicity of geothermal plant design, maintenance requirements are relatively low. Some modular plants even run unattended; and because maintenance is limited, plant availability is high. In recent years geothermal-plant availability averaged 97 percent. Thus, the maintenance cost of the usual geothermal plant is lower than a conventional fossil-fuel plant. Further, geothermal plants can meet new emission regulations with little or no pollution-abatement equipment.

2. *Choose the type of cycle to use*

Tapping geothermal energy from liquid resources poses a number of technical challenges—from drilling wells in a high-temperature environment to excessive scaling and corrosion in plant equipment. But DOE-sponsored and private-sector R&D programs have effectively overcome most of these problems. Currently, there are more than 35 commercial plants exploiting liquid-dominated resources. Of the 800 MW of power generated by these plants, 620 MW is produced by flash-type plants and 180 MW by binary-cycle units (Fig. 20).

The flashed-steam plant is best suited for liquid-dominated resources above 350°F (177°C). For lower-temperature sources, binary systems are usually more economical.

In flash-type plants, steam is produced by dropping the pressure of hot brine, causing it to "flash." The flashed steam is then expanded through a conventional steam turbine to produce power. In binary-cycle plants, the hot brine is directed through a heat exchanger to vaporize a secondary fluid which has a relatively low boiling point. This working fluid is then used to generate power in a closed-loop Rankine-cycle system. Because they use lower-temperature brines than flash-type plants, binary units (Fig. 20), are inherently more complex, less efficient, and have higher capital equipment costs.

In both types of plants the spent brine is pumped down a well and reinjected into the resource field. This is done for two reasons: (1) to dispose of the brine—which can be mineral-laden and deemed hazardous by environmental regulatory authorities, and (2) to recharge the geothermal resource.

One recent trend in the industry is to collect noncondensable gases (NCGs) purged from the condenser and reinject them along with the brine. Older plants use pollution-abatement devices to treat NCGs, then release them to the atmosphere. Reinjection of NCGs with brine lowers operating costs and reduces gaseous emissions to near zero.

Major improvements in flashed-steam plants over the past decade centered around: (1) improving efficiency through a dual-flash process and (2) developing improved water treatment processes to control scaling caused by brines. The pressure of the liquid brine stream remaining after the first flash is further reduced in a secondary chamber to generate more steam. This two-stage process can generate 20 to 30 percent more power than single-flash systems.

Most of the recent improvements in binary-cycle plants have been made by applying new working fluids. The thermodynamic and transport properties of these fluids can improve cycle efficiency and reduce the size and cost of heat-transfer equipment.

To illustrate: By using ammonia rather than the more common isobutane or isopentane, capital cost can be reduced by 20 to 30 percent. It is also possible to improve the conversion efficiency by using mixtures of working fluids, which in turn reduces the required brine flow rate for a given power output.

A flashed-steam cycle will be tentatively chosen for this installation because the brine free-flows at 450°F (232°C), which is higher than the cutoff temperature of 350°F (177°C) for binary systems. An actual plant (Fig. 21), operating with these parameters uses two flashes. The first flash produces 623,000 lb/h (283,182 kg/h) of steam at 100 lb/in² (gage) (689 kPa). In the second flash an additional 262,000 lb/h (117,900 kg/h) of steam at 10 lb/in² (gage) (68.9 kPa) is produced.

Steam is cleaned in two trains of scrubbers, then expanded through a 54-MW, 3600-rpm, dual-flow, dual-pressure, five-stage turbine-generator to produce 48.9 MW. Of this total, 47.5 MW is sold to Southern California Edison Co. because of transmission losses.

The turbine exhausts into a surface condenser, coupled to a seven-cell cooling tower. About 40,000 lb/h (18,000 kg/h) of the high-pressure steam is required by the plant's air ejectors to remove NCGs from the main condenser at a rate of 6500 lb/h (2925 kg/h).

Because the liquid brine from the flash process is supersaturated, various solid compounds precipitate out of solution and must be removed to avoid scaling and fouling of the pumps, pipelines, and injection wells. This is accomplished as the brine flows to the crystallizer and clarifier tanks where, respectively, solid crystals grow and then are separated. The solids are dewatered and used in construction-grade soil cement. The clarified brine is disposed of by pumping it into three injection wells.

Related Calculations: Geothermal generating plants are environmentally friendly because there are no stack emissions from a boiler. Further, such plants do not consume fossil fuel, so

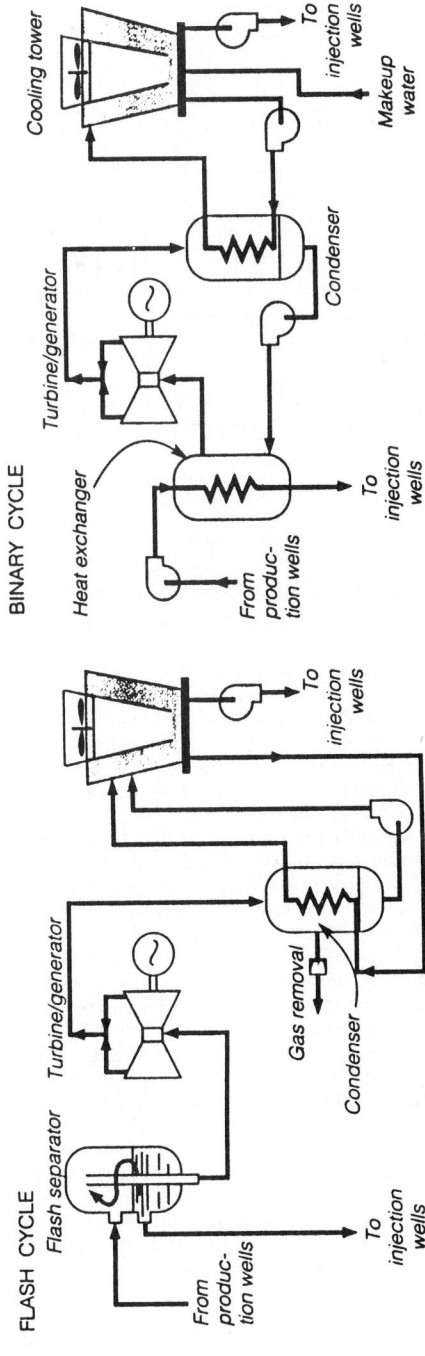

FIG. 20 Energy from hot-water geothermal resources is converted by either a flash-type or binary-cycle plant. (*Power.*)

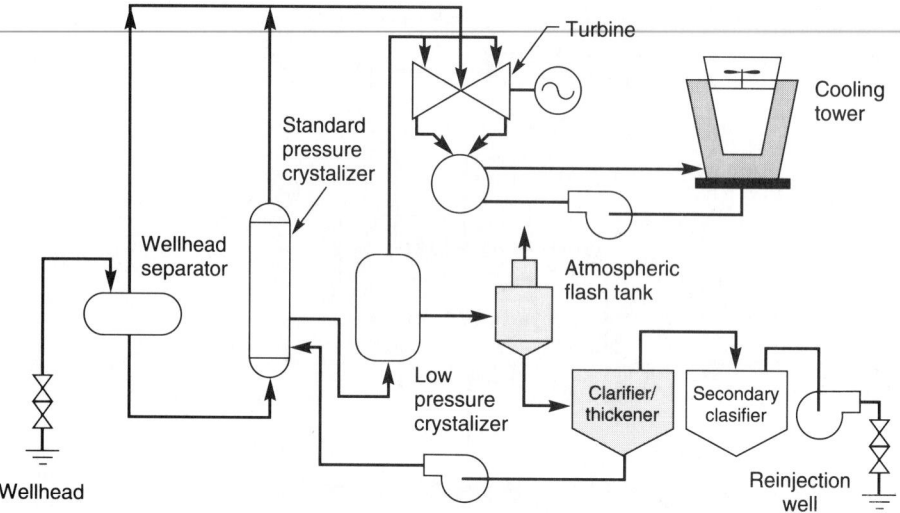

FIG. 21   Dual-flash process extracts up to 30 percent more power than older, single-flash units. (*Power.*)

they are not depleting the world's supply of such fuels. And by using the seemingly unlimited supply of heat from the earth, such plants are contributing to an environmentally cleaner and safer world while using a renewable fuel.

Another renewable fuel available naturally that is receiving—like geothermal power—greater attention today is *biomass*. The most common biomass fuels used today are waste products and residues left over from various industries, including farming, logging, pulp, paper, and lumber production, and wood-products manufacturing. Wooden and fibrous materials separated from the municipal waste stream also represent a major source of biomass.

Although biomass-fueled power plants currently account only for about 1 percent of the installed generating capacity in the United States, or 8000 MW, they play an important role in solving energy and environmental problems. Since the fuels burned in these facilities are considered waste in many cases, combustion yields the double benefits of reducing or eliminating disposal costs for the seller and providing a low-emissions fuel source for the buyer. On a global scale, biomass firing could present even more advantages, such as: (1) there is no net buildup of atmospheric CO_2 and air emissions are lower compared to many coal- or oil-fired plants. (2) Vast areas of deforested or degraded lands in tropical and subtropical regions can be converted to practical use. Because much of the available land is in the developing regions of Latin America and Africa, the fuels produced on these plantations could help improve a country's balance of payments by reducing dependence on imported oil. (3) Industrialized nations could potentially phase out agricultural subsidies by encouraging farmers to grow energy crops on idle land.

The current cost of growing, harvesting, transporting, and processing high-grade biomass fuels is prohibitive in most areas. However, proponents are counting on the successful development of advanced biomass-gasification technologies. They contend that biomass may be a more desirable feedstock for gasification than coal because it is easier to gasify and has a very low sulfur content, eliminating the need for expensive O_2 production and sulfur-removal processes.

One report indicates that integrated biomass-gasification–gas-turbine-based power systems with efficiencies topping 40 percent should be commercially available by year 2000. By 2025, efficiencies may reach 57 percent if advanced biomass-gasification–fuel-cell combinations become viable. Proponents are optimistic because this technology is currently being developed for coal gasification and can be readily transformed to biomass.

Data in this procedure are the work of M. D. Forsha and K. E. Nichols, Barber-Nichols Inc., for the geothermal portion, and Steven Collins, assistant editor, *Power*, for the biomass portion. Data on both these topics was published in *Power* magazine.

ESTIMATING CAPITAL COST OF COGENERATION HEAT-RECOVERY BOILERS

Use the Foster-Pegg* method to estimate the cost of the gas-turbine heat-recovery boiler system shown in Fig. 22 based on these data: The boiler is sized for a Canadian Westinghouse 251 gas turbine; the boiler is supplementary fired and has a single gas path; natural gas is the fuel for both the gas turbine and the boiler; superheated steam generated in the boiler at 1200 lb/in² (gage) (8268 kPa) and 950°F (510°C) is supplied to an adjacent chemical process facility; 230-lb/in² (gage) (1585-kPa) saturated steam is generated for reducing NO_x in the gas turbine; steam is also generated at 25 lb/in² (gage) (172 kPa) saturated for deaeration of boiler feedwater; a low-temperature economizer preheats undeaerated feedwater obtained from the process plant before it enters the deaerator. Estimate boiler costs for two gas-side pressure drops: 14.4 in (36.6 cm) and 10 in (25.4 cm), and without, and with, a gas bypass stack. Table 12 gives other application data. *Note*: Since cogeneration will account for a large portion of future power generation, this procedure is important from an environmental standpoint. Many of the new cogeneration facilities planned today consist of gas turbines with heat-recovery boilers, as does the plant analyzed in this procedure.

Calculation Procedure:

1. Determine the average LMTD of the boiler

The average log mean temperature difference (LMTD) of a boiler is indicative of the relative heat-transfer area, as developed by R. W. Foster-Pegg, and reported in *Chemical Engineering* magazine. Thus, $LMTD_{avg} = Q_t/C_t$, where Q_t = total heat exchange rate of the boiler, Btu/s (W); C_t = conductance, Btu/s · F (W). Substituting, using data from Table 12, $LMTD_{avg} = 81,837/1027 = 79.7°F$ (26.5°C).

2. Compute the gas pressure drop through the boiler

The gas pressure drop, ΔP in H_2O (cmH_2O) = $5C_t/G$, where G = gas flow rate, lb/s (kg/s). Substituting, $\Delta P = 5(1027/355.8)$ with a gas flow of 355.8 lb/s (161.5 kg/s), as given in Fig. 12; then $\Delta P = 14.4$ inH_2O (36.6 cmH_2O). With a stack and inlet pressure drop of 3 inH_2O (7.6 cmH_2O) and a supplementary-firing pressure drop of 3 inH_2O (7.6 cmH_2O) given by the manufacturer, or determined from previous experience with similar designs, the total pressure drop = 14.4 + 3.0 + 3.0 = 20.4 inH_2O (51.8 cmH_2O).

3. Compute the system costs

The conductance cost component, $Cost_{ts}$, is given by $Cost_{ts}$, in thousands of $, = $5.65[(C_{sh}^{0.8} + C_1^{0.8} + \cdots + (C_n^{0.8}) + 2(C_n^{0.8})]$, where C = conductance, Btu/s · F; (W), and the subscripts represent the boiler elements listed in Table 12. Substituting, $Cost_{ts} = 5.65(404.37) = \$2,285,000$ in 1985 dollars. To update to present-day dollars, use the ratio of the 1985 *Chemical Engineering* plant cost index (310) to the current year's cost index thus: Current cost = (today's plant cost index/310)(cost computed above).

The steam-flow cost component, $Cost_w$, in thousands of $ = $4.97(W_1 + W_2 + \cdots + W_n)$, where $Cost_w$ = cost of feedwater, $; W = feedwater flowrate, lb/s (kg/s); the subscripts 1, 2, and n denote different steam outputs. Substituting, $Cost_w = 4.97(59.14) = \$294,000$ in 1985 dollars, with a total feedwater flow of 59.14 lb/s (26.9 kg/s).

The cost for gas flow includes connecting ducts, casing, stack, etc. It is proportional to the sum of the separate gas flows, each raised to the power of 1.2. Or, cost of gas flow, $Cost_g$, in thousands of $ = $0.236(G_1^{1.2} + G_2^{1.2} + \cdots + G_n^{1.2})$. Substituting, $Cost_g = 0.236(355.8)^{1.2} = \$272,000$ with a gas flow of 355.8 lb/s (161.5 kg/s) and no bypass stack.

The cost of a supplementary-firing system for the heat-recovery boiler in 1985 dollars is additional to the boiler cost. Typical fuels for supplementary firing are natural gas or No. 2 fuel oil, or both. The supplementary-firing system cost, $Cost_f$, in thousands of $ = B/1390 + 30N + 20$, where B = boiler firing capacity in Btu (kJ) high heating value; N = number of fuels burned. For this installation with *one* fuel, $Cost_f = 16,980/1390 + 30 + 20 = \$62,000$, rounded off. In this equation the 16,980 Btu/s (17,914 kJ/s) is the high heating value of the fuel and N = 1 since only *one* fuel is used.

The total boiler cost (with base gas ΔP and no gas bypass stack) = total material cost + erection cost, or $2,285,000 + 294,000 + 272,000 + 62,000 = \$2,913,000 for the materials.

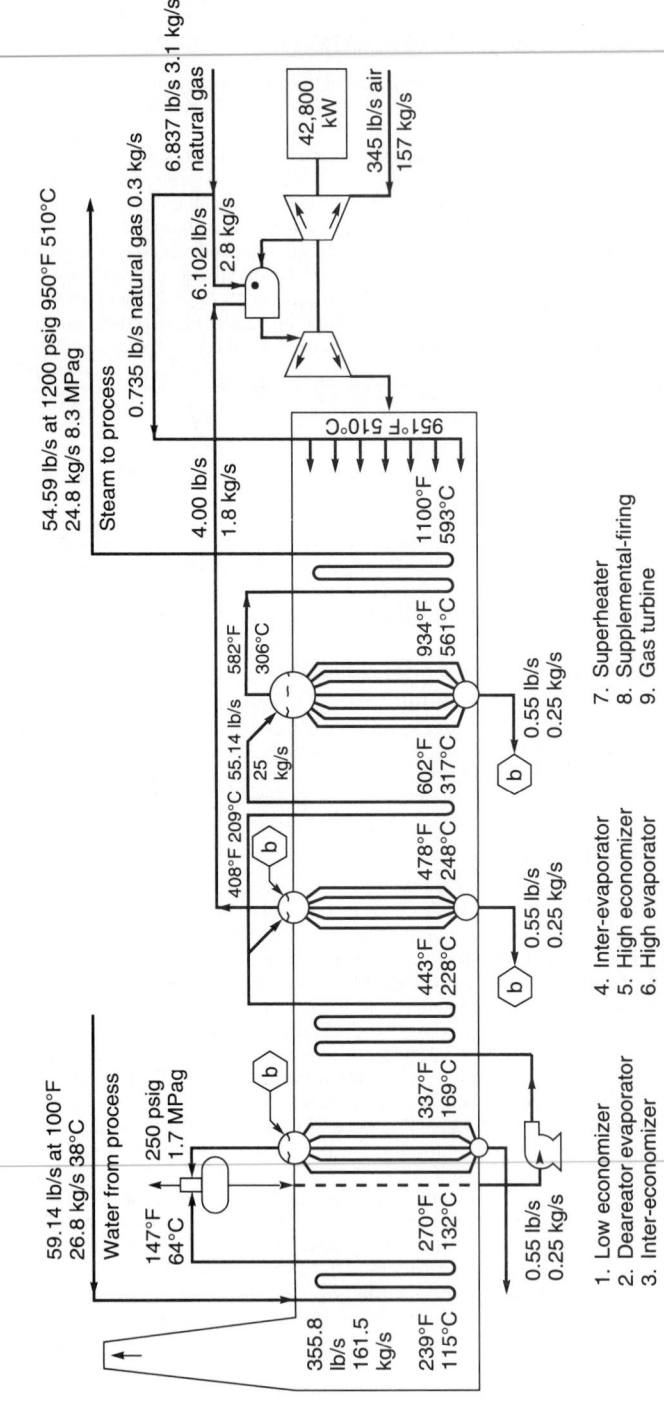

FIG. 22 Gas-turbine and heat-recovery-boiler system.(*Chemical Engineering.*)

1. Low economizer
2. Deareator evaporator
3. Inter-economizer
4. Inter-evaporator
5. High economizer
6. High evaporator
7. Superheater
8. Supplemental-firing
9. Gas turbine

59.14 lb/s at 100°F
26.8 kg/s 38°C

Water from process

147°F   250 psig
64°C   1.7 MPag

355.8 lb/s
161.5 kg/s

239°F   270°F   337°F   443°F   478°F   602°F   934°F   1100°F
115°C   132°C   169°C   228°C   248°C   317°C   561°C   593°C

0.55 lb/s
0.25 kg/s

0.55 lb/s
0.25 kg/s

0.55 lb/s
0.25 kg/s

0.55 lb/s
0.25 kg/s

408°F 209°C   55.14 lb/s   582°F
25 kg/s   306°C

4.00 lb/s
1.8 kg/s

951°F 510°C

Steam to process

54.59 lb/s at 1200 psig 950°F 510°C
24.8 kg/s 8.3 MPag

0.735 lb/s natural gas 0.3 kg/s

6.102 lb/s
2.8 kg/s

6.837 lb/s 3.1 kg/s
natural gas

42,800
kW

345 lb/s air
157 kg/s

13.32

TABLE 12 Data for Heat Recovery Boiler*

	LMTD, °F	Q, Btu/s	C, Btu/s · °F	$C^{0.8}$ Btu/s · °F
Superheater	237	16,098	67.92	29.22
High evaporator	116	32,310	278.53	90.34
High economizer	40	11,583	290.3	93.39
Inter-evaporator	50.5	3,277	64.89	28.17
Inter-economizer	57	9,697	169.82	60.81
Deareator evaporator	46	6,130	134.43	50.44
Low economizer	131	2,742	20.93	11.39
Additional for superheater material				29.22
Additional for low-economizer material				11.39
Total	81,837		1,027	404.37

*See procedure for SI values in this table.
Source: Chemical Engineering.

A *budget estimate* for the cost of erection = 25 percent of the total material cost, or 0.25 × \$2,913,000 = \$728,250. Thus, the budget estimate for the erected cost = \$2,913,000 + \$728,250 = \$3,641,250.

The estimated cost of the entire system—which includes the peripheral equipment, connections, startup, engineering services, and related erection—can be approximated at 100 percent of the cost of the major equipment delivered to the site, but not erected. Thus, the total cost of the boiler ready for operation is approximately twice the cost of the major equipment material, or 2(boiler material cost) = 2 (\$2,913,000) = \$5,826,000.

4. Determine the costs with the reduced pressure drop

The second part of this analysis reduces the gas pressure drop through the boiler to 10 in H_2O (25.4 cm H_2O). This reduction will increase the capital cost of the plant because much of the equipment will be larger.

Proceeding as earlier, the total pressure drop, ΔP = 10 + 3 + 3 = 16 in H_2O (40.6 cm H_2O). The pressure drop for normal solidity (i.e., normal tube and fin spacing in the boiler) is ΔP_1 = 14.4 in H_2O (36.6 cm H_2O). For a different pressure drop, ΔP_2, the surface cost, C_s(\$), is at ΔP_2, C_s = $[1.67(\Delta P_1/\Delta P_2)^{0.28} - 0.67](C_s$ at $P_1)$. Substituting, C_s = $1.67(14.4/10)^{0.28} - 0.67$ = 1.18 × base cost from above. Hence, the surface cost for a pressure drop of 10 in H_2O (25.4 cm H_2O) = 1.18 (\$2,285,000) = \$2,696,300.

The total material cost will then be \$2,696,300 + \$272,000 + \$62,000, using the data from above, or \$3,324,300. Budget estimate for erection, as before = 1.25 (\$3,324,300) = \$4,155,375. And the estimated system cost, ready to operate = 2(\$3,324,300) = \$6,648,600.

Adding for a gas bypass stack, the gas-flow component is the same as before, \$272,000. Then the budget estimate of the installed cost of the gas bypass stack = 1.25(\$272,000) = \$340,000. And the total cost of the boiler ready for operation at a gas-pressure drop of 10 in H_2O (25.4 cm H_2O) with a gas bypass stack = 2(\$3,324,300 + \$272,000) = \$7,192,600.

Related Calculations: To convert the costs in this procedure to current-day costs, assume that the *Chemical Engineering* plant cost index today is 435, compared to the 1985 index of 310. Then, today's cost, \$ = (today's cost index/1985 cost index)(1985 plant or equipment cost, \$). Thus, for the first installation, today's cost = (435/310)(\$5,826,000) = \$8,175,194. And for the second installation, today's cost = (435/310)(\$7,192,600) = \$10,092,842.

Boilers for recovering exhaust heat from gas turbines are very different from conventional boilers, and their cost is determined by different parameters. Because engineers are becoming more involved with cogeneration, the differences are important to them when making design and cost estimates and decisions.

In a conventional boiler, combustion air is controlled at about 110 percent of the stoichiometric requirement, and combustion is completed at about 3000°F (1649°C). The maximum temperature of the water (i.e., steam) is 1000°F (538°C), and the temperature difference between the gas and water is about 2000°F (1093°C). The temperature drop of the gas to the stack is about 2500°F (1371°C), and the gas/water ratio is consistent at about 1.1.

By contrast, the exhaust from a gas turbine is at a temperature of about 1000°F (538°C), and the difference between the gas and water temperatures averages 100°F (56°C). The temperature drop of the gas to the stack is a few hundred degrees, and the gas/water ratio ranges between 5 and 10. Because the airflow to a heat-recovery boiler is fixed by the gas turbine, the air varies from 400 percent of the stoichiometric requirement of the fuel to the turbine (unfired boiler) to 200 percent if the boiler is supplementary fired.

In heat-recovery boilers, the tubes are finned on the outside to increase heat capture. Fins in conventional boilers would cause excessive heat flux and overheating of the tubes. Although the lower gas temperatures in heat-recovery boilers allow gas enclosures to be uncooled internally insulated walls, the enclosures in conventional boilers are water-cooled and refractory-lined.

Because the exhaust from a gas turbine is free of particles and contaminants, gas velocities past tubes can be high, and fin and tube spacings can be close, without erosion or deposition. Because the products of combustion in a conventional boiler may contain sticky residues, carbon, and ash particles, tube spacing must be wider and gas velocities lower. Because of its configuration and absence of refractories, the heat-recovery boiler used with gas turbines can be shop-fabricated to a greater extent than conventional boilers.

These differences between conventional and heat-recovery boilers result in different cost relationships. With both operating on similar clean fuels, a heat-recovery boiler will cost more per pound of steam and less per square foot (m²) of surface area than a conventional boiler. The cost of a heat-recovery boiler can be estimated as the sum of three major parameters, plus other optional parameters. Major parameters are: (1) the capacity to transfer heat ("conductance"), (2) steam flow rate, and (3) gas flow rate. Optional parameters are related to the optional components of supplementary firing and a gas bypass stack.

This procedure is the work of R. W. Foster-Pegg, Consultant, as reported in *Chemical Engineering* magazine. Note that the costs computed by the given equations are in 1985 dollars. Therefore, they *must* be updated to current costs using the *Chemical Engineering* plant cost index.

"CLEAN" ENERGY FROM SMALL-SCALE HYDRO SITES

A newly discovered hydro site provides a potential head of 65 ft (20 m). An output of 10,000 kW (10 MW) is required to justify use of the site. Select suitable equipment for this installation based on the available head and the required power output.

Calculation Procedure:

1. *Determine the type of hydraulic turbine suitable for this site*

Enter Fig. 23 on the left at the available head, 65 ft (20 m), and project to the right to intersect the vertical projection from the required turbine output of 10,000 kW (10 MW). These two lines intersect in the *standardized tubular unit* region. Hence, such a hydroturbine will be tentatively chosen for this site.

2. *Check the suitability of the chosen unit*

Enter Table 13 at the top at the operating head range of 65 ft (20 m) and project across to the left to find that a tubular-type hydraulic turbine with fixed blades and adjustable gates will produce 0.25 to 15 MW of power at 55 to 150 percent of rated head. These ranges are within the requirements of this installation. Hence, the type of unit indicated by Fig. 23 is suitable for this hydro site.

Relation Calculations: Passage of legislation requiring utilities to buy electric power from qualified site developers is leading to strong growth of both site development and equipment suitable for small-scale hydro plants. Environmental concerns over fossil-fuel-fired and nuclear generating plants make hydro power more attractive. Hydro plants, in general, do not pollute the air, do not take part in the acid-rain cycle, are usually remote from populated areas, and run for up to 50 years with low maintenance and repair costs. Environmentalists rate hydro power as "clean" energy available with little, or no, pollution of the environment.

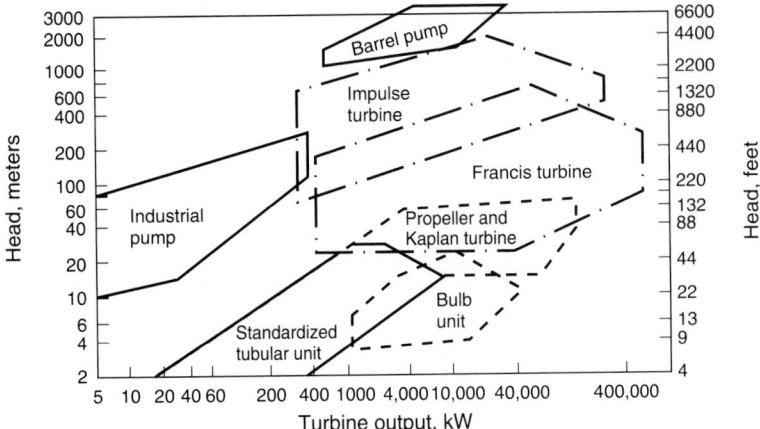

FIG. 23 Traditional operating regimes of hydraulic turbines. New designs allow some turbines to cross traditional boundaries. (*Power.*)

To reduce capital cost, most site developers choose standard-design hydroturbines. With essentially every high-head site developed, low-head sites become more attractive to developers. Table 13 shows the typical performance characteristics of hydroturbines being used today. Where there is a region of overlap in Table 13 or Fig. 23, site-specific parameters dictate choice and whether to install large units or a greater number of small units.

Delivery time and ease of maintenance are other factors important in unit choice. Further, the combination of power-generation and irrigation services in some installations make hydro-turbines more attractive from an environmental view because two objectives are obtained: (1) "clean" power, and (2) crop watering.

Maintenance considerations are paramount with any selection; each day of downtime is lost revenue for the plant owner. For example, bulb-type units for heads between 10 and 60 ft (3 and 18 m) have performance characteristics similar to those of Francis and tubular units, and are often 1 to 2 percent more efficient. Also, their compact and, in some cases, standard design makes for smaller installations and reduced structural costs, but they suffer from poor accessibility. Sometimes the savings arising from the unit's compactness are offset by increased costs for the watertight requirements. Any leakage can cause severe damage to the machine.

To reduce the costs of hydroturbines, suppliers are using off-the-shelf equipment. One way this is done is to use centrifugal pumps operated in reverse and coupled to an induction motor. Although this is not a novel concept, pump manufacturers have documented the capability of many readily available commercial pumps to run as hydroturbines. The peak efficiency as a turbine is at least equivalent to the peak efficiency as a pump. These units can generate up to 1 MW of power. Pumps also benefit from a longer history of cost reductions in manufacturing, a wider range of commercial designs, faster delivery, and easier servicing—all of which add up to more rapid and inexpensive installations.

Though a reversed pump may begin generating power ahead of a turbine installation, it will not generate electricity more efficiently. Pumps operated in reverse are nominally 5 to 10 percent less efficient than a standard turbine for the same head and flow conditions. This is because pumps operate at fixed flow and head conditions; otherwise efficiency falls off rapidly. Thus, pumps do not follow the available water load as well unless multiple units are used.

With multiple units, the objective is to provide more than one operating point at sites with significant flow variations. Then the units can be sequenced to provide the maximum power output for any given flow rate. However, as the number of reverse pump units increases, equipment costs approach those for a standard turbine. Further, the complexity of the site increases with the number of reverse pump units, requiring more instrumentation and auto-mation, especially if the site is isolated.

Energy-conversion-efficiency improvements are constantly being sought. In low-head ap-plications, pumps may require specially designed draft tubes to minimize remaining energy

TABLE 13 Performance Characteristics of Common Hydroturbines

Type	Operating head range		Capacity range	
	Rated head, ft (m)	% of rated head	MW	% of design capacity
Vertical fixed-blade propeller	7–120 (3–54) and over	55–125	0.25–15	30–115
Vertical Kaplan (adjustable blades and guide vanes)	7–66 (3–30) and over	45–150	1–15	10–115
Vertical Francis	25–300 (11–136) and over	50–150 and over	0.25–15	35–115
Horizontal Francis	25–500 (11–227) and over	50–125	0.25–10	35–115
Tubular (adjustable blades, fixed gates)	7–59 (3–27)	65–140	0.25–15	45–115
Tubular (fixed blades, adjustable gates)	7–120 (3–54)	55–150	0.25–15	35–115
Bulb	7–66 (3–30)	45–140	1–15	10–115
Rim generator	7–30 (3–14)	45–140	1–8	10–115
Right-angle-drive propeller	7–59 (3–27)	55–140	0.25–2	45–115
Cross flow	20–300 (9–136) and over	80–120	0.25–2	10–115

Source: Power.

13.36

after the water exists from the runner blades. Other improvements being sought for pumps are: (1) modifying the runner-blade profiles or using a turbine runner in a pump casing, (2) adding flow-control devices such as wicket gates to a standard pump design or stay vanes to adjust turbine output.

Many components of hydroturbines are being improved to reduce space requirements and civil costs, and to simplify design, operation, and maintenance. Cast parts used in older turbines have largely been replaced by fabricated components. Stainless steel is commonly recommended for guide vanes, runners, and draft-tube inlets because of better resistance to cavitation, erosion, and corrosion. In special cases, there are economic tradeoffs between using carbon steel with a suitable coating material and using stainless steel.

Some engineers are experimenting with plastics, but much more long-term experience is needed before most designers will feel comfortable with plastics. Further, stainless steel material costs are relatively low compared to labor costs. And stainless steel has proven most cost-effective for hydroturbine applications.

While hydro power does provide pollution-free energy, it can be subject to the vagaries of the weather and climatic conditions. Thus, at the time of this writing, some 30 hydroelectric stations in the northwestern part of the United States had to cut their electrical output because the combination of a severe drought and prolonged cold weather forced a reduction in water flow to the stations. Purchase of replacement power—usually from fossil-fuel-fired plants—may be necessary when such cutbacks occur. Thus, the choice of hydro power must be carefully considered before a final decision is made.

This procedure is based on the work of Jason Makansi, associate editor, *Power* magazine, and reported in that publication.

CENTRAL CHILLED-WATER SYSTEM DESIGN TO MEET CHLOROFLUOROCARBON (CFC) ISSUES

Choose a suitable storage tank size and capacity for a thermally stratified water-storage system for a large-capacity thermal-energy storage system for off-peak air conditioning for these conditions: Thermal storage capacity required = 100,000 ton-h (35,169 kWh); difference between water inlet and outlet temperatures = T = 20°F (36°C); allowable nominal soil bearing load in one location is 2500 lb/ft^2 (119.7 kPa); in another location 4000 lb/ft^2 (191.5 kPa). Compare tank size for the two locations.

Calculation Procedure:

1. Compute the required tank capacity in gallons (liters) to serve this system

Use the relation C = 1800 $S/\Delta T$, where C = required tank capacity, gal; S = system capacity, ton-h; ΔT = difference between inlet and outlet temperature, °F (°C). For this installation, C = 1800 (100,000)/20 = 9,000,000 gal (34,065 m^3).

2. Determine the tank height and diameter for the allowable soil bearing loads

Depending on the proposed location of the storage tank, either the height or diameter may be a restricted dimension. Thus, tank height may be restricted by local zoning laws or possible interference with aircraft landing or takeoff patterns. Tank diameter may be restricted by the ground area available. And the allowable nominal soil bearing load will determine if the required amount of water can be stored in one tank or if more than one tank will be required.

Starting with 2500-lb/ft^2 (119.7-kPa) bearing-load soil, assume a standard tank height of 40 ft (12.2 m). Then, the required tank volume will be V = 0.134C, where 0.134 = ft^3/gal; or V = 0.134 (9,000,000) = 1,206,000 ft^3 (34,130 m^3). The tank diameter is d = $(4V/\pi h)^{0.5}$, where d = diameter in feet (m). Or d = $[4(1,206,000)/\pi 40]^{0.5}$ = 195.93 ft; say 196 ft (59.7 m). This result is consistent with the typical sizes, heights, and capacities used in actual practice, as shown in Table 14.

Checking the soil load, the area of the base of this tank is A = $\pi d^2/4$ = $\pi(196)^2/4$ = 30,172 ft^2 (2803 m^2). The weight of the water in the tank is W = 8.35C = 75,150,000 lb (34,159 kg). This will produce a soil bearing pressure of p = W/A lb/ft^2 (kPa). Or, p = 75,150,000/30,172 = 2491 lb/ft^2 (119.3 kPa). This bearing load is within the allowable nominal specified load of 2500 lb/ft^2.

TABLE 14 Typical Thermal Storage Tank Sizes, Heights, Capacities*

Rated thermal energy storage capacity, ton-hours			Tank shell height (and nominal soil bearing load)					
			64-ft shell height (4000 lb/ft²soil)		40-ft shell height (2500 lb/ft²)		24-ft shell height (1500 lb/ft² soil)	
ΔT, 10°F	ΔT, 15°F	ΔT, 20°F	Gross volume, gal	Tank diameter, ft	Gross volume, gal	Tank diameter, ft	Gross volume, gal	Tank diameter, ft
40,000	60,000	80,000	6,880,000	135	7,200,000	175	7,880,000	236
50,000	75,000	100,000	8,610,000	151	9,000,000	196	9,850,000	264
60,000	90,000	120,000	10,330,000	166	10,800,000	214	11,820,000	290

Source: Chicago Bridge & Iron Company.
*See calculation procedures for SI values.

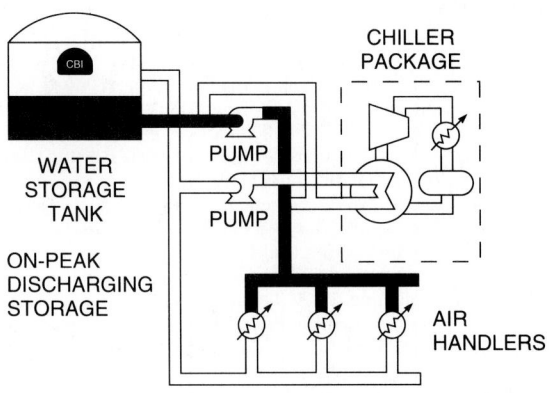

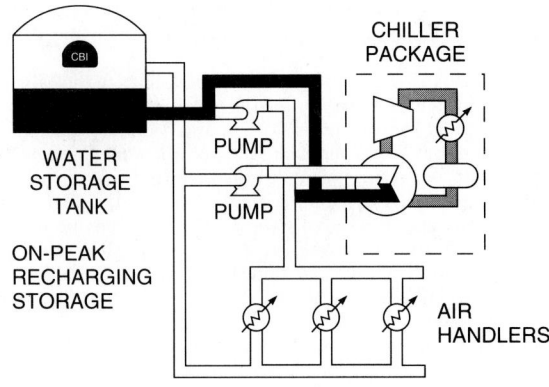

KEY

■■■ REFRIGERANT   ▨▨▨ CHILLED WATER   ▭ WARM WATER

FIG. 24 On-peak and off-peak storage discharging and recharging of thermally stratified water-storage system. (*Chicago Bridge & Iron Company.*)

TYPICAL STRATA-THERM CHILLED WATER
THERMAL ENERGY STORAGE INSTALLATION

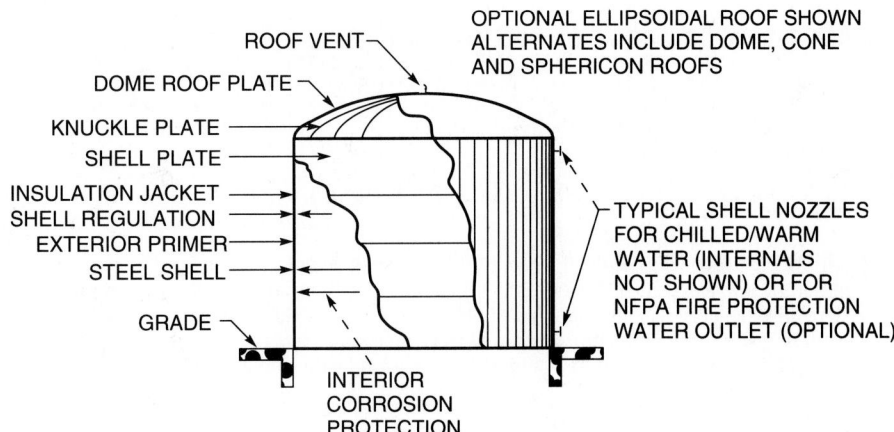

FIG. 25 Typical chilled-water thermal-energy storage tank installation. (*Chicago Bridge & Iron Company.*)

Where a larger soil bearing load is permitted, tank diameter can be reduced as the tank height is increased. Thus, using a standard 64-ft (19.5-m) high tank with the same storage capacity, the required diameter would be $d = [4(1,206,000)/\pi64)]^{0.5} = 154.9$ ft (47.2 m). Soil bearing pressure will then be $W/A = 75,140,000/[\pi(154.9)^2/4] = 3987.8$ lb/ft^2 (190.9 kPa). This is within the allowable soil bearing load of 4000 lb/ft^2 (191.5 kPa).

By reducing the storage capacity of the tank 4 percent to 8,610,000 gal (32,589 m^3), the diameter of the tank can be made 151 ft (46 m). This is a standard dimension for 64-ft (19-m) high tanks with a 4000-lb/ft^2 (191.5-kPa) soil bearing load.

Related Calculations: Thermal energy storage (TES) is environmentally desirable because it uses heating, ventilating, and air-conditioning (HVAC) equipment and a storage tank to store heated or cooled water during off-peak hours, allowing more efficient use of electric generating equipment. The stored water is used to serve HVAC or industrial process loads during on-peak hours.

To keep investment, operating, and maintenance costs low, one storage tank can be used to store both cool and warm water. Thermal stratification permits a smaller investment in the tank, piping, insulation, and controls to produce a higher-efficiency system. Lower-density warm water is thermally stratified from higher-density cool water without any mechanical separation in a full storage tank.

While systems using 200,000 gal (757 m^3) of stored water are feasible, the usual minimum size storage tank is 500,000 gal (1893 m^3). Tanks as large as 4.4 million gal (16,654 m^3) are currently in use in TES for HVAC and process needs. TES is also used for schools, colleges, factories, and a variety of other applications. Where chlorofluorocarbon (CFC)-based refrigeration systems must be replaced with less environmentally offensive refrigerants, TES systems can easily be modified because the chiller (Fig. 24), is a simple piece of equipment.

As an added environmental advantage, the stored water in TES tanks can be used for fire protection. The full tank contents are continuously available as an emergency fire water reservoir. With such a water reserve, the capital costs for fire-protection equipment can be reduced. Likewise, fire-insurance premiums may also be reduced. Where an existing fire-protection-water tank is available, it may be retrofitted for TES use.

Above-ground storage tanks (Fig. 25) are popular in TES systems. Such tanks are usually welded steel, leak-free with a concrete ringwall foundation. Insulated to prevent heat gain or loss, such tanks may have proprietary internal components for proper water distribution and stratification.

Some TES tanks may be installed partially, or fully, below grade. Before choosing partially or fully below-grade storage, the following factors should be considered: (1) system hydraulics may be complicated by a below-ground tank; (2) the tank must be designed for external pressure, particularly when the tank is empty; (3) soil and groundwater conditions may make the tank more costly; (4) local and national regulations for underground tanks may increase costs; (5) the choice of water-treatment methods may be restricted for underground tanks; (6) the total cost of an underground tank may be twice that of an above-ground tank.

The data and illustrations for this procedure were obtained from the Strata-Therm Thermal Systems Group of the Chicago Bridge & Iron Company.

WORK REQUIRED TO CLEAN OIL-POLLUTED BEACHES

How much relative work is required to clean a 300-yd (274-m) long beach coated with heavy oil, if the width of the beach is 40 yd (36.6 m), the depth of oil penetration is 20 in (50.8 cm), the beach terrain is gravel and pebbles, the oil coverage is 60 percent of the beach, and the beach contains heavy debris?

Calculation Procedure:

1. Establish a work-measurement equation from a beach model

After the *Exxon Valdez* ran aground on Bligh Reef in Prince William Sound, a study was made to develop a model and an equation that would give the relative amount of work needed to rid a beach of spilled oil. The relative amount of work remaining, expressed in clydes, is defined as the amount of work required to clean 100 yd (91.4 m) of lightly polluted beach. As the actual cleanup progressed, the actual work required was found to agree closely with the formula-predicted relative work indicated by the model and equation that were developed.

The work-measurement equation, developed by on-the-scene Commander Peter C. Olsen, U.S. Coast Guard Reserve, and Commander Wayne R. Hamilton, U.S. Coast Guard, is $S = (L/100)(EWPTCD)$, where S = standardized equivalent beach work units, expressed in clydes; L = beach-segment length in yards or meters (considered equivalent because of the rough precision of the model); E = degree of contamination of the beach expressed as: light oil = 1; moderate oil = 1.5; heavy oil = 2; random tar balls and very light oil = 0.1; W = width of beach expressed as: less than 30 m = 1; 30 to 45 m = 1.5; more than 45 m = 2; P = depth of penetration of the oil expressed as: less than 10 cm = 1; 10 to 20 cm = 2; more than 30 cm = 3; T = terrain of the beach expressed as: boulders, cobbles, sand, mud, solid rock without vertical faces = 1; gravel/pebbles = 2; solid rock faces = 0.1; C = percent of oil coverage of the beach expressed as: more than 67 percent coverage = 1; 50 to 67 percent = 0.8; less than 50 percent = 0.5; D = debris factor expressed as: heavy debris = 1.2; all others = 1.

2. *Determine the relative work required*

Using the given conditions, $S = (300/100)(2 \times 1.5 \times 1 \times 1 \times 0.8 \times 1.2) = 8.64$ clydes. This shows that the work required to clean this beach would be some 8.6 times that of cleaning 100 yd of lightly oiled beach. Knowing the required time input to clean the "standard" beach (100 yd, lightly oiled), the approximate time to clean the beach being considered can be obtained by simple multiplication. Thus, if the cleaning time for the standard lightly oiled beach is 50 h, the cleaning time for the beach considered here would be 50 (8.64) = 432 h.

Related Calculations: The model presented here outlines—in general—the procedure to follow to set up an equation for estimating the working time to clean any type of beach of oil pollution. The geographic location of the beach will not in general be a factor in the model unless the beach is in cold polar regions. In cold climates more time will be required to clean a beach because the oil will congeal and be difficult to remove.

A beach cleanup in Prince William Sound was defined as eliminating all gross amounts of oil, all migratory oil, and all oil-contaminated debris. This definition is valid for any other polluted beach be it in Europe, the Far East, the United States, etc.

Floating oil in the marine environment can be skimmed, boomed, absorbed, or otherwise removed. But oil on a beach must either be released by (1) scrubbing or (2) steaming and floated to the nearby water where it can be recovered using surface techniques mentioned above.

Where light oil—gasoline, naphtha, kerosene, etc.—is spilled in an accident on the water, it will usually evaporate with little damage to the environment. But heavy oil—No. 6, Bunker C, unrefined products, etc.—will often congeal and stick to rocks, cobbles, structures, and sand. Washing such oil products off a beach requires the use of steam and hot high-pressure water. Once the oil is freed from the surfaces to which it is adhering, it must be quickly washed away with seawater so that it flows to the nearby water where it can be recovered. Several washings may be required to thoroughly cleanse a badly polluted beach.

The most difficult beaches to clean are those comprised of gravel, pebbles, or small boulders. Two reasons for this are: (1) the surface areas to which the oil can adhere are much greater, and (2) extensive washing of these surface areas is required. This washing action can carry away the sand and the underlying earth, destroying the beach. When setting up an equation for such a beach, this characteristic should be kept in mind.

Beaches with larger boulders having a moderate slope toward the water are easiest to clean. Next in ease of cleaning are sand and mud beaches because thick oil does not penetrate deeply in most instances.

Use this equation as is, and check its results against actual cleanup times. Then alter the equation to suit the actual conditions and personnel met in the cleanup.

The model and equation described here are the work of Commander Peter C. Olsen, U.S. Coast Guard Reserve and Commander Wayne R. Hamilton, U.S. Coast Guard, as reported in government publications.

SIZING EXPLOSION VENTS FOR INDUSTRIAL STRUCTURES

Choose the size of explosion vents to relieve safely the maximum allowable overpressure of 0.75 lb/in² (5.2 kPa) in the building shown in Fig. 26 for an ethane/air explosion. Specify how the vents will be distributed in the structure.

Calculation Procedure:

1.   *Determine the total internal surface area of Part A of the building*

Using normal length and width area formulas for Part A, we have: Building floor area = 100 × 25 = 2500 ft² (232.3 m²); front wall area = 12 × 100 = 1200 ft² − 12 × 20 = 960 ft² (89.2 m²); rear wall area = 12 × 100 = 1200 ft² (111.5 m²); end wall area = 2 × 25 × 12 + 2 × 25 × 3/2 = 675 ft² (62.7 m²); roof area = 2 × 3 × 100 = 600 ft² (55.7 m²). Thus, the total internal surface area of Part A of the building is 2500 + 960 + 1220 + 600 = 5935 ft² (551.4 m²).

2.   *Determine the total internal surface area of Part B of the building*

Using area formulas, as before: Floor area = 50 × 20 = 1000 ft² (92.9 m²); side wall area = 2 × 50 × 12 = 1200 ft² (111.5 m²) front wall area = 20 × 12 = 240 ft² (22.3 m²); roof area

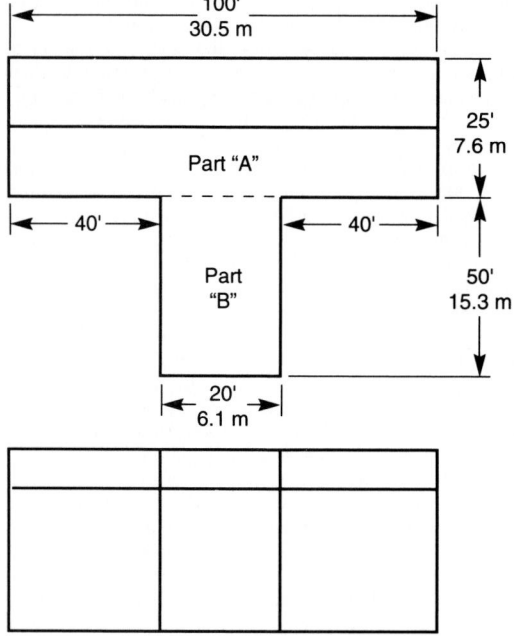

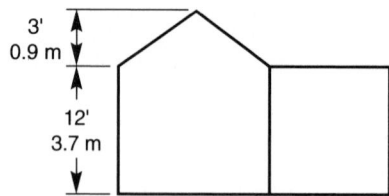

FIG. 26   Typical industrial building for which explosion vents are sized.

$= 50 \times 20 = 1000$ ft^2 (92.9 m^2); total internal surface area of Part B is $1000 + 1200 + 240 + 1000 = 3440$ ft^2 (319.6 m^2).

3. Compute the vent area required

Using the relation $A_v = CA_s/(P_{red})^{0.5}$, where A_v = required vent area, m^2; C = deflagration characteristic of the material in the building, (kPa)$^{0.5}$, from Table 15; A_s = internal surface area of the structure to be protected, m^2. For this industrial structure, $A_v = 0147 \,(551.4 + 319.6)/(5.17)^{0.5} = 180.1$ m^2 (1939 ft^2) total vent area.

The required vent area should be divided proportionately between Part A and Part B of the building, or Part A vent area $= 180.1(551.4/871.0) = 114$ m^2 (1227 ft^2); Part B vent area $= 180.1(319.6/871.0) = 66.1$ m^2 (712 ft^2).

The required vent area should be distributed equally over the external wall and roof areas in each portion of the building. Before final choice of the vent areas to be used, the designer should consult local and national fire codes. Such codes may require different vent areas, depending on a variety of factors such as structure location, allowable overpressure, and gas mixture.

Related Calculations: This procedure is the work of Tom Swift, a consultant, reported in *Chemical Engineering*. In his explanation of his procedure he points out that the word *explosion* is an imprecise term. The method outlined above is intended for those explosions known as deflagrations—exothermic reactions that propagate from burning gases to unreacted materials by conduction, convection, and radiation. The great majority of structural explosions at chemical plants are deflagrations.

The equation used in this procedure is especially applicable to "low-strength" structures widely used to house chemical processes and other manufacturing operations. This equation is useful for both gas and dust deflagrations. It applies to the entire subsonic venting range. Nomenclature for Table 15 is given as follows:

A_s Internal surface area of structure to be protected, m^2

A_v Vent area, m^2

B Dimensionless constant

C Deflagration characteristic, (kPa)$^{1/2}$

C_D Discharge coefficient

G' Maximum subsonic mass flux through vent, kg/m$^2 \cdot$ s

P_f Overpressure, kPa

TABLE 15   Parameters for Vent Area Equation*

Material	$\dfrac{S_u \rho_u}{G'}$	$\dfrac{P_{max}}{P_o}$
Methane	1.1×10^{-3}	8.33
Ethane	1.2×10^{-3}	9.36
Propane	1.2×10^{-3}	9.50
Pentane	1.3×10^{-3}	9.42
Ethylene	1.9×10^{-3}	9.39

Material	C, (kPa)$^{1/2}$
Methane	0.41
Ethane	0.47
Propane	0.48
Pentane	0.51
Ethylene	0.75
ST 1 dusts	0.26
ST 2 dusts	0.30

P_{max} Maximum deflagration pressure in a sealed spherical vessel, kPa

P_o Initial (ambient) pressure, kPa

P_{red} Maximum reduced explosion pressure that a structure can withstand, kPa

S_u Laminar burning velocity, m/s

γ_b Ratio of specific heats of the combustion gases

ρ_u Density of the unburnt gases, kg/m^3

λ Turbulence enhancement factor

With increased interest in the environment by regulatory authorities, greater attention is being paid to proper control and management of industrial overpressures. Explosion vents that are properly sized will protect both the occupants of the building and surrounding structures. Therefore, careful choice of explosion vents is a prime requirement of sensible environmental protection.

INDUSTRIAL BUILDING VENTILATION FOR ENVIRONMENTAL SAFETY

Determine the ventilation requirements to maintain interior environmental safety of a pump and compressor room in an oil refinery in a cool-temperate climate. Floor area of the pump and compressor room is 2000 ft^2 (185.8 m^2) and room height is 15 ft (4.6 m); gross volume = 30,000 ft^3 (849 m^3). The room houses two pumps—one of 150 hp (111.8 kW) with a pumping temperature of 350°F (177°C), and one of 75 hp (55.9 kW) with a pumping temperature of 150°F (66°C). Also housed in the room is a 1000-hp (745.6-kW) compressor and a 50-hp (37.3-kW) compressor.

Calculation Procedure:

1. *Determine the hp-deg for the pumps*

The hp-deg = pump horsepower × pumping temperature. For these pumps, the total hp-deg = (150 × 350) + (75 × 150) = 63,750 hp-deg (19,789 kW-deg). Enter Fig. 27 on the left axis at 63,750 and project to the diagonal line representing the ventilation requirements for pump rooms in cool-temperate climates. Then extend a line vertically downward to the bottom axis to read the air requirement as 7200 ft^3/min (203.8 m^3/min).

The compressors require a total of 1050 hp (782.9 kW). Enter Fig. 27 on the right-hand axis at 1050 and project horizontally to cool-temperate climates for compressor and machinery rooms. From the intersection with this diagonal project vertically to the top axis to read 2200 ft^3/min (62.3 m^3/min) as the ventilation requirement.

Since the ventilation requirements of pumps and compressors are additive, the total ventilation-air requirement for this room is 7200 + 2200 = 9400 ft^3/min (266 m^3/min).

2. *Check to see if the computed ventilation flow meets the air-change requirements*

Use the relation $N = 60 F/V$, where N = number of air changes per hour; F = ventilating-air flow rate, ft^3/min (m^3/min); V = room volume, ft^3 (m^3). Using the data for this room, $N = 60(9400)/30,000 = 18.8$ air changes per hour.

Figure 27 is based on a minimum of 10 air changes per hour for summer and 5 air changes per hour for winter. Since the 18.8 air changes per hour computed exceeds the minimum of 10 changes per hour on which the chart is based, the computed air flow is acceptable.

In preparing the chart in Fig. 27 the climate lines are based on ASHRAE degree-day listings, namely: *Cool, temperate climates*, 5000 degree-days and up; *average climates*, 2000 to 5000 degree-days; *warm climates*, 2000 degree-days maximum.

3. *Select the total exhaust-fan capacity*

An exhaust fan or fans must remove the minimum computed ventilation flow, or 9400 ft^3/min (266 m^3/min) for this room. To allow for possible errors in room size, machinery rating, or temperature, choose an exhaust fan 10 percent larger than the computed ventilation flow. For this room the exhaust fan would therefore have a capacity of 1.1 × 9400 = 10,340 ft^3/min (292.6 m^3/min). A fan rated at 10,500 or 11,000 ft^3/min (297.2 or 311.3 m^3/min), depending on the ratings available from the supplier, would be chosen.

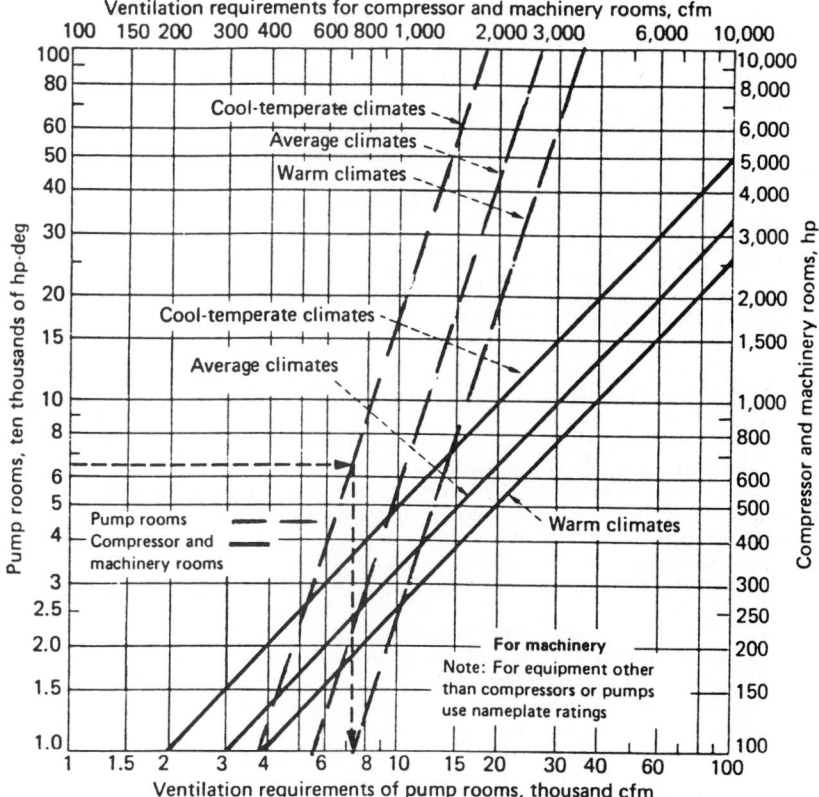

FIG. 27   Chart for determining building ventilation requirements. (*Chemical Engineering.*)

Related Calculations: Ventilation is environmentally important and must accomplish two goals: (1) Removal of excess heat generated by machinery or derived from hot piping and other objects; (2) removal of objectionable, toxic, or flammable gases from process pumps, compressors, and piping.

The usual specifications for achieving these goals commonly call for an arbitrary number of hourly air changes for a building or room. However, these specifications vary widely in the number of air changes required, and use inconsistent design methods for ventilation. The method given in this procedure will achieve proper results, based on actual applications.

Because of health and explosion hazards, workers exposed to toxic or hazardous vapors and gases should be protected against dangerous levels [threshold limit values (TLV)] and explosion hazards [lower explosive limit (LEL)] by diluting workspace air with outside air at adequate ventilation rates. If a workspace is protected by adequate ventilation rates for health (i.e., below TLV) purposes, the explosion hazard (LEL) will not exist. The reason for this is that the health air changes far exceed those required for explosion prevention.

To render a workspace safe in terms of TLV, the number of ft^3/min (m^3/min) of dilution air, A_d, required can be found from: $A_d = [1540 \times S \times T/(M \times \text{TLV})]K$, or in SI, $A_{dm} = $ m^3/min $= 0.0283A_d$, where $S = $ gas or vapor expelled over an 8-h period, lb (kg); $M = $ molecular weight of vapor or gas; TLV $= $ threshold limit value, ppm; $T = $ room temperature, absolute °R(K); $K = $ air-mixing factor for nonideal conditions, which can vary from 3 to 10, depending on actual space conditions and the efficiency of the ventilation-air distribution system.

If the space temperature is assumed to be 100°F (37.8°C) (good average summer conditions), the above equation becomes $A_d = [862,400 \times S/(M \times \text{TLV})]K$.

For every pound (kg) of gas or vapor expelled of an 8-h period, when $S = 1$, the second equation becomes $A_d = [862,400/(M \times \text{TLV})]K$. For values of S less or greater than unity, simple multiplication can be used.

It is only for ideal mixing that $K = 1$. Hence, K must be adjusted upward, depending on ventilation efficiency, operation, and the particular system application.

If mixing is perfect and continuous, then each air change reduces the contaminant concen tration to about 35 percent of that before the air change. Perfect mixing is seldom attainable, however, so a room mixing factor, K, ranging from 3 to 10 is recommended in actual practice.

The practical mixing factor for a particular workspace is at best an estimate. Therefore, some flexibility should be built into the ventilation system in anticipation of actual operations. For small enclosures, such as ovens and fumigation booths, K-values range from 3 to 5. If you are not familiar with efficient mixing within enclosures, use a K-factor equal to 10. Then your results will be on the safe side. Figure 27 is based on a K-value equal to 8 to 10. Table 16 gives K-factors for ventilation-air distribution systems as indicated.

In some installations, heat generated by rotating equipment (pumps, compressors, blowers) process piping, and other equipment can be calculated, and the outside-air requirements for dilution ventilation determined. In most cases, however, the calculation is either too cumbersome and time consuming or impossible.

Figure 27 was developed from actual practice in the chemical-plant and oil-refinery businesses. The chart is based on a closed processing system. Hence, air quantities found from the chart are not recommended if (1) the system is not closed or (2) if abnormal operating conditions prevail that permit the escape of excessive amounts of toxic and explosive materials into the workplace atmosphere.

For these situations, special ventilation measures, such as local exhaust through hoods, are required. Vent the exhaust to pollution-control equipment or, where permitted, directly outdoors.

Figure 27 and the procedure for determining dilution-air ventilation requirements were developed from actual tests of workspace atmospheres within processing buildings. Design and operating show that by supplying outside air into a building near the floor, and exhausting it high (through the roof or upper outside walls), safe and comfortable conditions can be attained. Use of chevron-type stormproof louvers permits outside air to enter low in the room.

The chevron feature causes the air to sweep the floor, picking up heat and diluting gases and vapors on the way up to the exhaust fan, (Fig. 28).

In the system shown in Fig. 28 there are a number of features worth noting. With low-level distribution and adequate high exhaust, only the internal plant heat load (piping, equipment) is of importance in maintaining desirable workspace conditions. Wall and transmission heat loads are swept out of the building and do not reach the work areas. Even the temperature rise caused by the plant load occurs above the work level. Hence, low-level distribution of the supply air maintains the work area close to supply-air temperatures.

For any installation, it is good practice to check the ratio of hp-deg/ft^2 (kW-deg/m^2) of floor area. When this ratio exceeds 100, consider installing a totally enclosed ventilation system for cooling. This should be complete with ventilating fans taking outside air, preferably from a high stack, and discharging through ductwork into a sheetmetal motor housing.

The result is a the greater use of outside air for cooling through a confined system at a lower ventilation rate. Ventilation air flow needs may be obtained from the equipment manufacturer or directly from the chart (Fig. 27). The remainder of the building may be ventilated as usual, based either on the absence of equipment or on any equipment outside the ventilation enclosure.

TABLE 16   *K*-values for Various Ventilation-Air Distribution Systems

K-values	Distribution system
1.2–1.5	Perforated ceiling
1.5–2.0	Air diffusers
2.0–3.0	Duct headers along ceiling with branch headers pointing downward
3.0 and up	Window fans, wall fans, and the like

Chemical Engineering

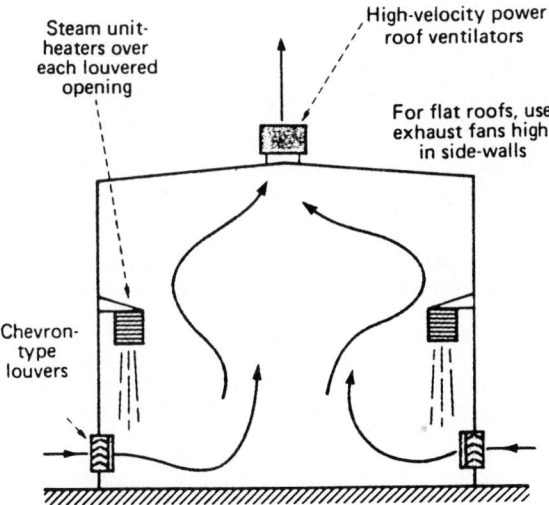

FIG. 28 Ventilation system for effective removal of plant heat loads. (*Chemical Engineering.*)

When designing the duct system, take care to prevent moisture entrainment with the incoming airstream.

This procedure is the work of John A. Constance, P.E., consultant, as reported in *Chemical Engineering.*

ESTIMATING POWER-PLANT THERMAL POLLUTION

A steam power plant has a 1000-MW output rating. Find the cooling-water thermal pollution by this power plant when using a once-through cooling system for the condensers if the plant thermal efficiency is 30 percent.

Calculation Procedure:

1. Determine the amount of heat added during plant operation

The general equation for power-plant efficiency is $E = W/Q_A$, where E = plant net thermal efficiency, %; W = plant net output, MW; Q_A = heat added, MW. For this plant, Q_A = W/E = 1000/0.30 = 3333 MW.

2. Compute the heat rejected by this plant

The general equation for heat rejected is $Q_R = (W/E - W)$, Q_R = heat rejected, MW. For this power plant, Q_R = (1000/0.3 − 1000) = 2333 MW. Thus, this plant will reject 2333 MW to the condenser cooling water.

The heat rejected to the cooling water will be absorbed by the river, lake, or ocean providing the water pumped through the condenser. Depending on the thermal efficiency of the plant, the required cooling-water flow for the condenser will range from 250×10^6 lb/h, or 65,000 ft^3/min (30 m^3/s) to 400×10^6 lb/h or 100,000 ft^3/min (50 m^3/s). The discharged water in a once-through cooling system will be 20 to 25°F (11 to 14°C) higher in temperature than the entering water.

3. Assess the effects of this thermal pollution

Warm water discharged in large volume to a restricted water mass may affect the ecosystem in a deleterious way. Fish and plant life, larvae, plankton, and other organisms can be damaged or have a high mortality rate. If chlorine is used to control condenser scaling, the effect on the ecosystem can be more damaging.

If the warm condenser cooling water is discharged into a large body of water, such as a major river or ocean, the effect on the ecosystem can be more beneficial than deleterious. Thus, well-planned cooling-water outlets can be used to increase fish production in hatcheries. In agriculture, the warm-water discharge can be used to markedly increase the output of greenhouses and open fields in cold climates. Thus, the overall effect of thermal pollution can be positive, if the pollution energy is used in an antipolluting manner.

Related Calculations: Since thermal and atmospheric pollution are associated with the generation of electricity, environmental engineers are seeking ways to reduce electricity use. Personal computers (PCs) are big users of electricity today. At the time of this writing, PCs consume some 5 percent of commercial energy used in the United States.

Typical PCs use 150 to 200 W of power when in use, or just on but not in use. Some 30 to 40 percent of PCs are left on overnight and during weekends. The extra electricity which must be generated to carry this PC load leads to more thermal and air pollution.

New PCs have "sleep" circuitry which reduces the electrical load to 30 W when the computer is not being used. Such microprocessors will reduce the electrical load caused by PCs. This, in turn, will reduce thermal and air pollution produced by power-generating plants.

Internal-combustion engines—diesel, gas, and gas turbines—produce both thermal and air pollution. To curb this pollution and wrest more work from the fuel burned, cogeneration is being widely applied. Heat is extracted from the internal-combustion engine's cooling water for use in process or space heating. In addition, exhaust gases are directed through heat exchangers to extract more heat from the internal-combustion engine exhaust. Thus, environmental considerations are met while conserving fuel. This is one reason why cogeneration is so popular today.

Compute the heat recovery from cogeneration using the many concepts given earlier in this section. Try to combine both heat recovery and pollution reduction; then the required investment will be easier to justify from an economic standpoint.

INDEX

Accumulator, steam, 3.469 to 3.471
Acid Rain Control Program, 3.192
Adjustable-speed drives, 4.16
Admiralty coefficients, 9.5
Aerial photogrammetry, 1.255 to 1.265
  definitions pertaining to, 1.256, 1.257
  flying height in, 1.255 to 1.257
  with oblique photograph, 1.263 to 1.265
  with overlapping photographs, 1.261 to
    1.263
  with tilted photograph, 1.259 to 1.261
  with vertical photograph, 1.255 to 1.258,
    1.261 to 1.263
Aeroderitative gas turbines, 13.25
Aeronautical engineering, 8.1 to 8.20
  air-cushion vehicles, 8.6 to 8.9
  aircraft: climb rate, 8.4
    drift, 8.3
    engine thrust, 8.5
    groundspeed, 8.3
    landing speed, 8.2
    stall speed, 8.2
    true airspeed, 8.3
    vertical-takeoff, 8.9
  commercial aircraft, 8.10
  lifting power of balloons, 8.12
  vertical-takeoff aircraft, 8.9
Affinity laws for pumps, 3.352 to 3.353
Air compressors, 3.325 to 3. 338
  air receivers for, 3.331
  altitude factors for, 3.330
  brake horsepower (kW) input to, 3.328

Air compressors (*Cont.*):
  components for, 3.329 to 3.331
  computation of air requirements, 3.326
  cooling-water requirements for, 3.330
  effect of intake temperature on, 3.330
  selection of, 3.325 to 3.329
Air conditioning, 2.80 to 2.126
  cooling coils for, 2.82 to 2.93
  heat-load determination for, 2.80 to 2.87
  high-velocity ducts for, 2.112 to 2.114
  marine, 9.19 to 9.21
  noise-reduction materials for, 2.122 to
    2.126
  outlet and return grilles for, 2.114 to
    2.118
  selection of system for known load, 2.94
    to 2.96
  sizing low-velocity ducts, 2.95 to 2.105
    branch areas, 2.98
    friction-loss chart for, 2.99
    recommended velocities in, 2.97
  systems and applications, 2.95
  vibration isolators for, 2.120 to 2.122
Air-cooled heat exchanger selection, 3.540
    to 3.543
Air-cushion vehicles, 8.6 to 8.9
Air-snubber springs, 3.92 to 3.96
Air-volume conversion factors, 2.70
Aircraft:
  climb rate of, 8.4
  drift of, 8.3
  engine thrust of, 8.5

Aircraft (*Cont.*):
  groundspeed of, **8.3**
  landing speed of, **8.2**
  stall speed of, **8.2**
  true airspeed of, **8.3**
  vertical-takeoff, **8.8**
Algebra, boolean, **7.48** to **7.64**
Allowable stress, **3.113** to **3.117**
Allowable stresses in pipes, **3.398**
Allowance to emit CO_2, **10.8**
Alternating-current circuit analysis, **4.7**
  alternator-selection in, **4.13**
  motor, selection in, **4.44** to **4.47**
    starting current of, **4.47**
    starting time of, **4.447**
  vector algebra in, **4.8**
Alternative proposals, evaluation of, **12.18** to **12.32**
Alternatives, cogeneration, **13.14**
  energy-from-wastes, **13.4**
Altitude of star, **1.234**
Amplifier, **5.44** to **5.66**
  bipolar, **5.59**
  black-box, **5.35**
  buffer, **5.51**
  operational, **5.53** to **5.57**, **5.64**
    as equivalent circuit synthesis, **5.53**
    chain, **5.55** to **5.57**
    network synthesis with, **5.64**
  transistorized, **5.6**, **5.13**, **5.44**
    equivalent circuit for, **5.45**
    gain control in, **5.52**
Analysis of business operations, **12.48** to **12.62**
  linear programming in, **12.48**
  optimal inventory level, **12.54**
  project planning using CPM/PERT, **12.55** to **12.59**
Anchor force of pipes, **3.441**
Antenna selections, **5.28**
Appliances, electric, **4.69**
Arch, reactions of three-hinged, **1.14**
Architectural engineering, **2.1** to **2.150**
  air conditioning, **2.80** to **2.126**
    branch ducts, **2.98**
    cooling coils for, **2.82** to **2.93**
    friction-loss chart for low-velocity ducts, **2.99**
    heat-load determination for, **2.80** to **2.87**
    high-velocity ducts for, **2.112** to **2.114**
    noise-reduction materials for, **2.122** to **2.126**

Architectural engineering, air conditioning (*Cont.*):
    outlet and return grilles for, **2.114** to **2.118**
    recommended velocities in ducts, **2.97**
    selection of system for known load, **2.94** to **2.96**
    sizing low-velocity ducts, **2.96** to **2.105**
    systems and applications, **2.95**
    vibration isolators for, **2.120** to **2.122**
  beam connection: riveted moment, **2.7** to **2.9**
    semirigid, **2.6**
    welded flexible, **2.9**, **2.10**
    welded moment, **2.12**
    welded seated, **2.10**, **2.11**
  column base: for axial load, **2.15**
    for end moment, **2.15**
    grillage type, **2.16** to **2.19**
  composite steel-and-concrete beam, **2.32** to **2.34**
  connection: beam-to-column (*see* Beam connection)
    to resist horizontal shear, **2.31** to **2.34**
    of truss members, **2.5**
  eyebar, **2.3**
  gusset plate, **2.5**
  hanger, steel, **2.4**
  heating systems: applications of, **2.63**
    coils for, **2.72**
    condensate formation rate of coils, **2.73**
      final temperatures in coils, **2.73**
      fuel-consumption factors for, **2.64**
      passive solar, **2.145** to **2.150**
      selection of, **2.62** to **2.65**
      solar, **2.134** to **2.141**, **2.145** to **2.150**
  humidifiers, selection of, **2.105** to **2.110**
    capacities of, **2.107**
    location of, **3.108**
    piping for, **2.110**
    recommended humidities for, **2.109**
    steam required by, **2.106**
  joists, reinforced-concrete, **2.34ff.**
  knee: curved, **2.14**
    rectangular, **2.13**
  plumbing and drainage, **2.37** to **2.61**
    cold- and hot-water piping, **2.44** to **2.51**
    gas piping, **2.55 to 2.57**
    pipe-size determination for, **2.37** to **2.42**
    roof and yard rainwater systems, **2.42** to **2.44**

Architectural engineering, heating systems:
    applications of (*Cont.*):
    sprinkler systems, **2.**52 to **2.**54
    swimming-pool sizing, **2.**58
    solar energy, **2.**126 to **2.**150
        ASHRAE, **2.**132
        average annual amount, **2.**126
        collector choice for, **2.**127, **2.**146
        collector sizing for, **2.**134, **2.**145 to
            2.147
        domestic hot-water heater, **2.**141 to
            2.145
        F chart method, **2.**135 to **2.**141
        flat-plate collector, **2.**127
        heat exchangers for flat-plate collector,
            2.130
        heating contribution of passive system,
            2.149
        heating systems, **2.**134 to **2.**141, **2.**145
            to **2.**150
        insolation determination, **2.**131 to
            2.133, **2.**147
        key system components with flat-plate
            collector, **2.**129
        orientation of collectors, **2.**128
        passive solar heating, **2.**145 to **2.**150
    stair slab, **2.**35 to **2.**37
    steel beam, **2.**27 to **2.**32
        encased in concrete, **2.**30 to **2.**32
        light-gage, **2.**27 to **2.**30
    unit heaters: blow distances of, **2.**69
        capacity of, **2.**65 to **2.**70
        conversion factors for, **2.**68
        diameter of pipes for, **2.**70
        outlet velocities of, **2.**69
        vertical types, **2.**67
    ventilators, roof, **2.**118 to **2.**120
    vibration of bent, **2.**37
    wind drift, **2.**25
        reduction with diagonal bracing, **2.**26
    wind-stress analysis, **2.**19 to **2.**25
        cantilever method, **2.**21
        portal method, **2.**19 to **2.**21
        slope-deflection method, **2.**23 to **2.**25
Area:
    calculation of, **1.**227 to **1.**230
    geometric properties of, **1.**22 to **1.**24
Ash reuse, **13.**7
ASHRAE, **2.**132
Astronautical engineering, **8.**12 to **8.**20
    interplanetary launch velocity, **8.**14
    missile range, **8.**13
    observation satellites, **8.**18

Astronautical engineering (*Cont.*):
    rocket flight velocity, **8.**12
    satellite velocity, **8.**14
    space vehicle burnout velocity, **8.**16
Astronomy, field, **1.**233 to **1.**235
Atmospheric pollution, **13.**17
Audio pollution, **2.**125
Average-end-area method, **1.**232, **1.**233
Average-grade method, **1.**245 to **1.**247
Azimuth of star, **1.**233 to **1.**235

Balanced design:
    of prestressed-concrete beam, **1.**166
    of reinforced-concrete beam, **1.**126, **1.**139,
        1.140
    of reinforced-concrete column, **1.**149
Ball bearings, **3.**60ff.
Barometric-condenser selection, **3.**247
    heat transfer in, **3.**470
Batch physical process balance, **6.**7
Batteries, industrial, **4.**40
    characteristics of, **4.**42
Beach cleaning, work required for, **13.**40
Beam(s):
    bending moment in, **1.**37 to **1.**39
    bending stress in, **1.**39 to **1.**42
    composite steel-and-concrete, **1.**200 to
        1.204, **2.**32 to **2.**34
    composite steel-and-timber, **1.**41, **1.**42
    compound, **1.**40
    conjugate, **1.**50
    continuous, **1.**53 to **1.**58
        of prestressed concrete, **1.**187 to **1.**197
        of reinforced concrete, **1.**134 to **1.**136,
            1.148
        of steel, **1.**82, **1.**102 to **1.**105
    deflection of, **1.**48 to **1.**53
    on movable supports, **1.**40
    with moving loads, **1.**59, **1.**61, **1.**67
    prestressed-concrete (*see* Prestressed-con-
        crete beam)
    reinforced-concrete (*see* Reinforced-con-
        crete beam)
    shear center of, **1.**43, **1.**44
    shear flow in, **1.**42
    shearing stress in, **1.**42
    statically indeterminate, **1.**53 to **1.**59
    steel (*see* Steel beam)
    timber (*see* Timber beam)
    vertical shear in, **1.**37 to **1.**39
Beam column:
    pile group as, **1.**48

Beam column (*Cont.*):
  soil prism as, **1**.47
  steel, **1**.94, **1**.95, **1**.114, **1**.115
Beam connection:
  riveted moment, **2**.7, to **2**.9
  semirigid, **2**.6
  welded flexible, **2**.9, **2**.10
  welded moment, **2**.12
  welded seated, **2**.10, **2**.11
Bearings, shaft, **3**.56 to **3**.77
  ball, **3**.60ff.
  characteristics of, **3**.57
  gas-type, **3**.75 to **3**.77
  heat generation of, **3**.64
  hydrostatic journal analysis, **3**.69 to **3**.72
    constants for, **3**.72
  hydrostatic thrust, **3**.67 to **3**.69
    constants for equations, **3**.70, **3**.71
    oil-film thickness, **3**.69
  length of, **3**.62
  load capacity of, **3**.79
  materials for sleeve-type, **3**.58
  multidirection-type analysis, **3**.72 to **3**.75
    constants for, **3**.72
  oil-film, **3**.60
    design load for, **3**.60
  porous-metal capacity, friction, **3**.65
    capacity factors, **3**.66
    friction factors, **3**.66
  roller-type, **3**.63 to **3**.65
    capacity and reliability of, **3**.65
    operating-life analysis, **3**.63
    radial load rating of, **3**.64
  rolling-type, capacity, cost, and size of,
    3.61, **3**.65
  selection of type of, **3**.56 to **3**.62
  speed limits for ball and roller, **3**.61, **3**.63
  temperature limits of, **3**.60
Belleville springs, **3**.88
Belts, **3**.16 to **3**.54
  leather, **3**.16ff.
    capacity factors for, **3**.16
    correction factors for, **3**.17
  rubber, **3**.17ff.
    arc-of-contact factor, **3**.18
    correction factors, **3**.17, **3**.21
    horsepower (kW) ratings, **3**.19, **3**.21ff.
    length correction factors, **3**.23
    minimum pulley diameters, **3**.19
    multiple V, **3**.22
    service factors, **3**.17, **3**.18
    sheave dimensions, **3**.20

Belts, rubber (*Cont.*):
    small-diameter factors, **3**.23
   V (*see* V belts)
  timing, **3**.51 to **3**.54
Bending flat plate, **1**.45
Bending metal parts, **3**.180 to **3**.182
Bending moment:
  in beam, **1**.37 to **1**.39
  in column footing, **1**.156 to **1**.161
  modified, in continuous beam, **1**.82
  in three-hinged arch, **1**.66, **1**.67
  in truss, **1**.64
Bending-moment diagram:
  for beam, **1**.37 to **1**.39, **1**.81, **1**.82
  for combined footing, **1**.160
  for gable frame, **1**.111
Bending stress:
  in beam, **1**.39 to **1**.42
  in curved member, **1**.46
Bends, pipe, **3**.443 to **3**.455
Benefit-cost analysis, **12**.47
Bernoulli's theorem, **1**.208, **1**.209
Bingham plastics, friction factor for, **6**.54 to
    6.57
Biomass power generation, **13**.27, **13**.30
Bipolar amplifier, **5**.59
Black-box amplifier, **5**.35
Blank diameters, **3**.182
Blanking metals, **3**.165
Bleed-steam cycles, **3**.227 to **3**.233
Blowdown, boiler heat recovery from, **3**.538
Bode plot, **7**.22, **7**.27, **7**.31, **7**.34, **7**.36
Boiler:
  blowdown heat recovery, **3**.538 to **3**.540
    percentage, **3**.540
  controls, **7**.11 to **7**.12
    combustion, **7**.12
  conversion to coal, **3**.533 to **3**.535
  feedwater regulators, **7**.11
  furnace size comparisons, **8**.531
  heat balance, **3**.204 to **3**.206
    draft fans, **3**.281 to **3**.285
    ducts and uptakes, **3**.285 to **3**.290
    efficiency, **3**.268
    fire-tube, analysis of, **3**.270
    steam, selection of, **3**.278 to **3**.281
  safety valves, **3**.271 to **3**.277
    capacity of, **3**.271
    selection of, **3**.271 to **3**.277
  scale, savings from reduced, **3**.534
Bolt diameter selection, **3**.121 to **3**.125
  endurance limit of, **3**.124

Bolt diameter selection (*Cont.*):
  for pressurized joint, **3**.125
  torque tightening factor, **3**.122
Boolean algebra, **7**.48 to **7**.64
Boring, time and power for, **3**.152
Boussinesq equation, **1**.269, **1**.270
Bracing, diagonal, **2**.26
Brakes:
  characteristics of, **3**.103
  cooling time of, **3**.106
  electric, **3**.104
  performance characteristics of, **3**.104
  selection of, **3**.102
  surface area of, **3**.103 to **3**.106
Branching pipes, **1**.217
Breakdown in manufacturing, **3**.184 to **3**.186
Bridge, highway, **1**.197 to **1**.204
  composite steel-and-concrete, **1**.200 to
      1.204
  concrete, **1**.198 to **1**.200
Bridge truss, **1**.60 to **1**.65
Brinell hardness, **3**.173
Broaching time and production rate, **3**.160
Bucket elevators, **3**.326, **3**.327
Buffer amplifier, **5**.51
  current drain, **5**.51
  output voltage, **5**.51
Bulk material conveying, **3**.326 to **3**.330
Bulkhead, thrust on, **1**.277 to **1**.280
Buoyancy, **1**.204 to **1**.208
  center of, **1**.207
Business operations, analysis of, **12**.48 to
      12.62
  linear programming in, **12**.48
  optimal inventory level, **12**.54
  project planning using CPM/PERT, **12**.55
      to **12**.59
Bypass cooling system for engines, **3**.301 to
      3.305

Cable(s):
  catenary, **1**.17
  with concentrated loads, **1**.15, **1**.16, **1**.28
  parabolic, **1**.17
  voltage regulation of, **4**.28
Calorimeter, analysis of, **3**.267
Cam clutch selection, **3**.50
Cantilever method of wind-stress analysis,
      2.21
Capacity-reduction factor, **1**.126

Capacity tables, fan, **3**.275
Capital, recovery of, **12**.7
Carbon dioxide (CO_2), **13**.4
Cash flow calculations, **12**.11
Celestial sphere, **1**.234, **1**.235
Centrifugal compressor, refrigeration, **3**.505
Centrifugal pumps, **3**.352 to **3**.382
  affinity laws for, **3**.352, **3**.353
  analysis of characteristic curves of, **3**.368
      to **3**.374
  condensate, selection of, **3**.374 to **3**.378
  critical speed of, **3**.381
  as hydraulic turbines, **3**.387 to **3**.392
    cavitation constant, **3**.389
    constant-head curves, **3**.390
    constant-speed curves, **3**.389
    converting design conditions, **3**.387
    number of stages, **3**.387
    specific speed of, **3**.387
    turbine performance, **3**.388
  minimum safe flow for, **3**.378
  net positive suction head of, **3**.374
  selecting for viscous liquid, **3**.379
  selection of, **3**.362 to **3**.368
  shaft deflection of, **3**.381
  similarity laws for, **3**.352, **3**.353
  sizing impellers, for safety service, **3**.342
  specific speed of, **3**.352 to **3**.353
  total head on, **3**.357 to **3**.361
Centroid of area, **1**.22 to **1**.24
Chain drives, **3**.48 to **3**.50
  inverted-tooth (silent), **3**.48
  roller, **3**.48
    horsepower (kW) rating of, **3**.49
    length factors for, **3**.50
    loads and service factors for, **3**.48
Channel:
  nonuniform flow in, **1**.221
  uniform flow in, **1**.217
Characteristic curves, pump, **3**.356 to **3**.362
Chemical engineering, **6**.1 to **6**.57
  batch physical process balance, **6**.7
  crusher power input, **6**.12
  heat of mixing, **6**.5
  immiscible-solution characteristics, **6**.10
  liquid-liquid separation, **6**.13
  material balance, **6**.6
  mixer cooling-water flow rate, **6**.12
  process plant engineering (*see* Process
      plant engineering)
  pump selection for, **6**.10
  saturated-solution analysis, **6**.2

Chemical engineering (*Cont.*):
  steady-state continuous physical balance,
    6.7 to 6.10
  ternary liquid system, 6.3
Chemical reactions in combustion, 3.193
Chicago Board of Trade, 13.27
Chilled-water system design, 13.37
Civil engineering, 1.1 to 1.292
Clean Air Act (CAA), 3.192
Clean Air Act Amendments (CAAA), 6.41,
    10.8, 13.27
"Clean" energy hydro, 13.34
Cleaning polluted beaches, 13.40
Clothoid, 1.241
Clutch selection, 3.100 to 3.102
Clutches:
  cam, 3.50
  characteristics of, 3.100
  ratings of, 3.102
  service factors for, 3.101
Coal-burning plant, ground area for, 3.537
Coal fuel in a furnace, combustion of, 3.189
    to 3.192
Coal storage capacity of piles and bunkers,
    3.295
Coefficient of performance, 13.19
Cogeneration, choosing steam compressor
    for, 13.19
  costs of, 3.526 to 3.533
  defined, 13.25
Cold spring of pipes, 3.442
Cold-water pipe sizes, 2.44
Collectors, solar, 2.127 to 2.130, 2.134,
    2.145 to 2.150
Column:
  reinforced-concrete (*see* Reinforced-con-
    crete column)
  steel (*see* Steel column)
  timber, 1.119, 1.120
Column base:
  for axial load, 2.15
  for end moment, 2.15
  grillage-type, 2.16 to 2.19
Combined bending and axial loading, 1.45
Combustion, 3.189 to 3.208
  chemical reactions in, 3.197
  of coal, 3.189 to 3.192
  of fuel oil, 3.193
  molal conversion factors for, 3.201 to
    3.203
  of natural gas, 3.191 to 3.199

Combustion (*Cont.*):
  products, final, 3.203
  properties of elements in, 3.193
  of wood fuel, 3.199 to 3.201
Combustion controls, boiler, 7.12
Commercial aircraft, 8.10
Common-base bipolar amplifier, 5.59
Communicating vessels, discharge between,
    1.221
Composite mechanisms, theorem of, 1.108
Composite steel-and-concrete beam, 1.200
    to 1.204, 2.32 to 2.34
Composite steel-and-timber beam, 1.41,
    1.42
Compound thin lens, 5.78
Compressed-air requirements, 3.325 to
    3.329
  and pressure loss, 3.433
  receiver volume for, 3.331
  of tools, 3.326
Compression index, 1.285
Compression ratio, engine, 3.294
Compression test:
  triaxial, 1.272
  unconfined, 1.270 to 1.272
Compressors, 13.20
  cost considerations, 13.20
Computers for process control, 7.6 to 7.8
Concave mirrors, 5.73
Concordant trajectory, 1.190
Concrete joist, 2.34
Condensate heat recovery, 13.9
Condensate pump selection, 3.374 to 3.378
Condensers, steam, 3.246 to 3.254
Condensing turbine output, 3.238 to 3.240
Conjugate-beam method, 1.50
Connection:
  beam-to-column (*see* Beam connection)
  for pipe joint, 1.71
  to resist horizontal shear, 1.84, 1.85,
    1.203, 2.31 to 2.34
  riveted, 1.68 to 1.75
  riveted moment, 2.7
  semirigid, 2.6
  timber, 1.121 to 1.125
  of truss members, 2.5, 2.6
  welded, 1.74, 1.75, 2.9 to 2.13
Connectors, shear, 1.203, 2.34
Conservation of energy (*see* Energy conser-
    vation)
Constant-entropy steam process, 3.219
Constant-pressure steam process, 3.214

Constant-temperature steam process, **3.218**
Constant-volume steam process, **3.215 to 3.217**
Control engineering, **7.1 to 7.48**
  Bode plot, **7.22, 7,27, 7.31, 7.34, 7.36**
  computers for, **7.6 to 7.8**
  developing a transfer function, **7.30**
  feedback control system, **7.33**
  fluid amplifiers for, **7.14**
  NAND gate circuit, **7.41 to 7.45**
    Karnaugh map for, **7.41 to 7.43**
    with two logic levels, **7.42 to 7.45**
  phase-lag compensator design, **7.31**
  process system, **7.2 to 7.4**
  process temperature, **7.4**
  pump selection to, **7.10**
  servo systems, **7.20 to 7.30**
    active network solution, **7.29**
    angular-position system analysis, **7.21**
    closed-loop transfer function, **7.21, 7.26**
    compensation network, **7.28**
    dc gain function, **7.29**
    with loop delay, **7.25**
    loop-gain function analysis, **7.22**
    open-loop transfer function, **7.21**
    overshoot and settling time, **7.24**
    Routh array for stability determination, **7.20**
    Routh criterion for stability determination, **7.20**
    stability determination, **7.20**
  sizing steam-control valves, **7.45 to 7.48**
    pressure-reducing valves, **7.45 to 7.48**
  for steam boilers, **7.110**
  temperature-measuring amplifier, **7.37**
    equivalent circuit for, **7.39**
    instrument, analysis of, **7.38**
  transfer function, **7.21, 7.26, 7.30, 7.35**
    phase shift, **7.31**
  valve selection for, **7.8 to 7.10**
  waveform rms value, **7.40**
    average value of, **7.40**
Control valves, **7.8 to 7.10, 7.12 to 7.20**
  characteristics and rangeability of, **7.12 to 7.14**
  flow considerations of, **7.15 to 7.20**
  for process control, **7.8 to 7.10**
  steam-control and pressure-reducing, **7.45 to 7.48**
Controlled-volume pump, **7.10**
Converse-Labarre equation, **1.288**
Convex mirror, **5.73**

Conveyor characteristics, **3.340**
  horsepower (kW) requirements, **3.341**
  maximum belt speeds, **3.342**
  minimum belt width, **3.341**
  screw-type, **3.342 to 3.344**
    capacities and speeds of, **3.342**
    material factors for, **3.343**
    size factors for, **3.341**
Cooling of electric motors, **4.87**
Cooling coils, **2.82 to 2.93**
  bypass factors for, **2.82**
  characteristics of, **2.89ff.**
  selection of, **2.88 to 2.93**
Cooling ponds, **3.248 to 3.250**
Cooling systems for electronics, **5.25, 5.71**
  heat pump selection, **5.71 to 5.73**
Cooling water:
  for air compressors, **3.330**
  engine, **3.306 to 3.311**
Cost estimates for plant and equipment, **6.34 to 6.54**
  centrifugal-pump and electric-motor, **6.50 to 6.54**
  correlations in, **6.46 to 6.54**
  heat exchangers and storage tanks, **6.46**
Cost of alternatives, **13.14**
  capital cost of, **13.31**
  energy plots for, **13.22**
  and gas turbines, **13.25**
Cost of heat loss, **3.521**
Cost separation in cogeneration, **3.526 to 3.529**
Cost vs. benefit, analysis of, **12.47**
  selection of profit with, **13.3**
Coulomb's theory, **1.274**
Countersinking, time and power for, **3.152**
Couplings, shaft, **3.39 to 3.48**
  flexible, **3.42 to 3.44**
    allowable misalignment of, **3.42**
    horsepower (kW) ratings of, **3.43**
    service factors for, **3.43**
  functional characteristics of, **3.46**
  high-speed, **3.44**
  operating characteristics of, **3.45**
  rigid flange-type, **3.39**
  selection for torque and thrust loads, **3.44**
  universal-joint output variations, **3.47**
Cover plates:
  for highway girder, **1.200 to 1.204**
  for plate girder, **1.84, 1.85**
  for rolled section, **1.79 to 1.82**
  for steel-and-concrete beam, **2.32 to 2.34**

Crane, traveling, 4.87
Critical depth of fluid flow, 1.218 to 1.220
Critical-path method (CPM) in project planning, 12.55 to 12.59
Critical speed, pump, 3.381
Cross-border pollution, 3.206
Crusher power input, 6.12
Culmination of star, 1.235
Curve:
   circular, 1.236 to 1.241
   compound, 1.239 to 1.241
   transition, 1.241 to 1.244
   vertical (see Vertical parabolic curve)
Curved springs, 3.82 to 3.84
Cut, length and angle of, 3.150
Cutting speeds:
   economical, 3.176
   for lowest-cost machining, 3.181
   for materials, 3.149
Cutting time:
   keyway, 3.150
   tool feed rate and, 3.152

Darcy-Weisbach formula, 1213 to 1.215
Declination of star, 1.234
Deflection:
   of beam, 1.48 to 1.53
   of cantilever frame, 1.52
   by conjugate-beam method, 1.50
   by double-integration method, 1.49
   by moment-area method, 1.49
   under moving loads, 1.67
   of prestressed-concrete beam, 1.178
   of reinforced-concrete continuous beam,
      1.148, 1.149
   of shafts, 3.13 to 3.15
      pump, 3.381
   of spring, 3.77
   by unit-load method, 1.51
Deformation:
   of built-up member, 1.26
   of member under axial load, 1.26
Degree of saturation, 1.265, 1.266
Departure of line, 1.226
Depletion and depreciation (see Depreciation and depletion)
Depreciation and depletion, 12.12 to 12.18
   accounting for, 12.17
   combination, 12.13, 12.15
   constant-unit use, 12.15
   declining-balance, 12.13

Depreciation and depletion (Cont.):
   declining-unit-use, 12.14
   income from, 12.17
   sinking-fund, 12.12
   straight-line, 12.18
   sum-of-the-digits, 12.15
   taxes and earnings, 12.16
   unit method, 12.17
Depth factor, 1.117
Designing parts, 3.110 to 3.112
Desuperheater, steam, 3.468, 3.473
Desuperheater condensers, 6.39
Diesel engines, 3.301 to 3.325
   cooling water for, 3.306 to 3.311
   displacement of, 3.302
   horsepower (kW) of, 3.302
   mean effective pressure of, 3.302
   output of, 3.304
   selection of, 3.303
Differential leveling, 1.230, 1.231
Digital control, fuel savings from, 3.543
Dimensional analysis, 1.224, 1.225
Dimpling metal parts, 3.183 to 3.185
Direct-current circuit analysis, 4.2 to 4.7
   generator selection in, 4.12
   Kirchhoff's laws for, 4.5
Direct-current permanent-magnet motors,
      4.81 to 4.87
Displacement of truss joint, 1.29
Dissolved gas, effect of, on rotary pump,
      3.384 to 3.386
Distance:
   double meridian, 1.227, 1.228
   sight, 1.249
Double-integration method, 1.49
Dovetails, dimensions of, 3.152
Dowel pins, sizing of, 3.147
Dowels:
   in column footing, 1.157 to 1.159
   in retaining wall, 1.165
Drains:
   building, 2.41
   roof, 2.42
   (See also Plumbing and drainage)
Drawing of metals, 3.168, 3.183 to 3.185
Drill penetration rate, 3.183
Drilling, time and power for, 3.155
Dryers, coal, 3.293
Duct sizing, 2.96 to 2.105
   equal-friction method, 2.96 to 2.102
   friction chart for, 2.99
   static-regain method, 2.102 to 2.105
Dummy pile, 1.291

Earth thrust:
  on bulkhead, 1.277 to 1.280
  on retaining wall, 1.273 to 1.275
  on timbered trench, 1.275 to 1.277
Earthwork, volume of, 1.232, 1.233
Economical cutting speeds and production
    rates, 3.176
Economics, engineering, of energy-from-
    wastes, 13.4 (*See also* Engineering
    economics)
Economizer, heat transfer, 3.489
Efficiency, internal-combustion engines,
    3.289, 3.290
Elasticity, modulus of, 1.25
Electric-arc welding, 3.165
Electric brakes, 3.104
Electric comfort-heating load determination,
    4.68 to 4.79
Electric heaters, 3.488
Electric-motor controls, 4.14ff.
  characteristics and applications of, 4.48 to
    4.50
  constants for starting current, 4.52
  dimensions of, 4.47
  frame numbers for, 4.51
  full-load current of, 4.32
  locked-rotor kVA, 4.52
  power ratings of, 4.45
  selection of, 4.44 to 4.47
  speed- and power-range applications of,
    4.46
  speed-torque characteristics of, 4.45, 4.47
Electric motors, 4.2 to 4.14, 4.44 to 4.51,
    4.79 to 4.89
  benefits, 4.81
  cooling of, 4.87
  cost of, 4.79
  energy charge, present value, 4.80
  flywheel selection for, 4.87 to 4.89
  permanent-magnet type, 4.81 to 4.87
Electric transformers, 4.20 to 4.24
  characteristics of, 4.22ff.
  coolant characteristics of, 4.22
  load-center units, 4.23
  selection of, 4.21
Electrical engineering, 4.1 to 4.92
  circuit lengths in, 4.34 to 4.35
  crane, traveling, 4.87
  direct-current permanent-magnet motors,
    4.81 to 4.87
  electric motors, 4.79 to 4.89
    cooling of, 4.87
    cost of, 4.79

Electrical engineering, electric motors
    (*Cont.*):
  efficiency benefits, 4.81
  energy charge, present value, 4.80
  flywheel selection for, 4.87 to 4.89
  permanent-magnet type, 4.81 to 4.87
  power-factor benefits, 4.81
flywheels for electric motors, 4.87 to 4.89
  effect of width on, 4.89
  linear inertia of, 4.87
  starting and stopping times, 4.88
  total inertia of, 4.88
lightning protection, 4.9 to 4.12, 4.90 to
    4.92
  arrester selection, 4.9 to 4.12
  incidence of thunderstorms, 4.91
  type of system to use, 4.90
  zone of protection, 4.92
load estimating in, 4.33 to 4.37
permanent-magnet motors, 4.81 to 4.87
  analysis, 4.81
  dimensionless efficiency curves, 4.85
  electrical measurement analysis, 4.85
  equivalent motor circuit, 4.86
  factor of merit, 4.85
  performance characteristics, 4.83
  stall and drag torques, 4.83, 4.86
  universal performance formulas, 4.82
thunderstorms, incidence of, 4.91
transformers, 4.20 to 4.24
Electronics engineering, 5.1 to 5.80
  electronics—design and applications, 5.2
    to 5.73
    ac equivalent circuit synthesis for oper-
      ational amplifier, 5.53
    amplifiers, 5.44 to 5.66
    antenna selection, analysis of, 5.28
    bipolar amplifier, 5.59
    black-box amplifier, 5.35
    buffer amplifier, 5.51
    cooling of equipment, 5.25, 5.71 to 5.73
    current drain, 5.51
    filter distortion, 5.62
    Fourier expression, 5.63
    frequency-changer selection and appli-
      cation, 5.37
    gain-control circuit, 5.57
    heat pump, 5.71 to 5.73
    high-pass filter design, 5.60
    hybrid-parameter conversions, 5.31 to
      5.35
    integrated-circuit selection, applications
      of, 5.14 to 5.17

Electronics engineering, electronics—
  design and applications (*Cont.*):
  JFET resistance of gain-control circuit,
    5.58
  large-scale integration, **5.**35
  low-pass filter design, **5.**60
  maintainability analysis in, **5.**22 to **5.**24
  microwave transmitter analysis, **5.**68 to
    5.71
  mismatch efficiency, **5.**29
  network synthesis, **5.**64
  operational amplifier, **5.**53 to **5.**57, **5.**64
  operational amplifier chain, **5.**55 to **5.**57
  oscillator selection, application of, **5.**14
  output voltage of buffer amplifier, **5.**51
  power-supply analysis, selection of, **5.**19
  public-address systems, **5.**30
  reliability analysis in, **5.**20 to **5.**22
  satellite communications, **5.**66
  Smith chart use, **5.**39
  solid-state device evaluation, **5.**3 to **5.**5
  sonar equations, **5.**42
  transistor selection, **5.**5
    amplifiers, **5.**6, **5.**13
  transistorized amplifier, **5.**44
    equivalent circuit for, **5.**45
    gain control in, **5.**52
  transistorized circuit, **5.**47
    load line, **5.**49
    operating point, **5.**48
    small-signal, **5.**49
  tuned-circuit *Q* values, **5.**27
  ultrasonic-generators, **5.**40
  vacuum-tube: amplifier selection for,
    5.11
    characteristics of, **5.**8, **5.**11
  worst-case transistor leakage current,
    5.17
optics—mirror and lens systems, **5.**73 to
    5.80
  compound thin-lens analysis, **5.**78 to
    5.80
  lenses, image produced by, **5.**76 to **5.**78
  mirrors, image produced by, **5.**73 to
    5.76
  thick compound lens systems analysis,
    5.78
Embankment, stability of, **1.**280 to **1.**285
Emissions, credits, **6.**33
  reduction of, **13.**8
Emptying vessel, time needed for, **6.**59
Endowment fund, allowance for inflation,
    12.34

Energy, solar (*see* Solar energy)
Energy conservation, **3.**509 to **3.**562
  air-cooled heat exchanger, preliminary se-
    lection, **3.**542 to **3.**544
  boiler conversion to coal, **3.**533 to **3.**535
  boiler furnace size comparisons, **3.**531
  cogeneration costs, **3.**525 to **3.**530
  flash tanks, sizing, **3.**548 to **3.**551
    from waste alternatives, **13.**4
  fuel savings from direct digital control,
    3.543 to **3.**545
  ground area, for coal burning plants,
    3.537
  heat exchanger, air-cooled, **3.**542
  heat-loss cost, **3.**521
  heat-rate improvement, **3.**522 to **3.**526
  heat recovery from boiler blowdown,
    3.538
  high-temperature hot-water heating, **3.**516
    to **3.**519
  produced by heat recovery, **3.**519
    pump choice for, **3.**394 to **3.**395
  from reduced boiler scale, **3.**536
  return on investment, **3.**535
  small hydro power analysis, **3.**546
  uninsulated pipes, cost of heat loss, **3.**521
  wind-energy systems for, **3.**509 to **3.**516
    capital costs of machines, **3.**514
    electric power output, **3.**510
    kites for power generation, **3.**514
    machine performance curves, **3.**513
    machine selection, **3.**509
    types of machines, **3.**511 to **3.**514
Energy gradient, **1.**221
Energy requirements of screw compressors,
    13.20
Energy savings:
  in industrial hydraulic systems, **3.**146
  in plants, **6.**22
  plots of, **13.**22
  from recompression, **6.**27 to **6.**29
  by relocating units, **6.**25
  from storage tanks and vessels, **6.**22 to
    6.25
  from waste heat for refrigeration, **6.**31
Engine lathe, **3.**155
Engineering economics, **12.**1 to **12.**100
  alternative proposals, **12.**18 to **12.**32
    annual cost, after-tax basis, **12.**27
    annual cost of asset, **12.**18, **12.**23
    annual-cost studies, **12.**20
    asset replacement, **12.**29 to **12.**32
    capitalized cost, **12.**24 to **12.**27

Engineering economics, alternative pro-
    posals (*Cont.*):
  cost and income, **12.**19
  equipment replacement, **12.**22
  manufacturing break-even point, **12.**20
  minimum asset life, **12.**19
  nonuniform operating costs, **12.**21
  present worth of future costs of an in-
    stallation, **12.**24
  analysis of business operations, **12.**48 to
    12.62
    linear programming in, **12.**48
    optimal inventory level, **12.**54
    project planning using CPM/PERT,
      12.55 to **12.**59
  capital recovery, **12.**7
  cost comparison, **12.**18 to **12.**22
  depreciation and depletion, **12.**12 to **12.**18
    accelerated cost recovery, **12.**12
    accounting for, **12.**17
    combination, **12.**13, **12,**16
    constant-unit use, **12.**14
    declining-balance, **12.**14
    declining-unit-use, **12.**14
    income from, **12.**17
    sinking-fund, **12.**13
    straight-line, **12.**12
    sum-of-the-digits, **12.**15
    taxes and earning, **12.**16
    unit method, **12.**17
  effects of inflation (*see* inflation, effects of,
    below)
  equivalent sums, **12.**7
  evaluation of investments, **12.**35 to **12.**48
    allocation of capital, **12.**37 to **12.**41
    apparent rates of return, **12.**42
    average rate of return, **12.**44
    benefit-cost analysis, **12.**47
    corporate bonds, **12.**36
    economic level of investment, **12.**41
    investment at an intermediate date,
      12.45
    investment-rate calculations, **12.**37,
      12.42
    payback period, **12.**46
    premium worth method, **12.**35
    true rate of return, **12.**43
  inflation, effects of, **12.**32 to **12.**35
    anticipated, **12.**34
    at constant rate, **12.**32
    on endowment fund, **12.**34
    on present worth of costs, **12.**33
    on replacement cost, **12.**32
    at variable rate, **12.**33

Engineering economics (*Cont.*):
  interest calculations, **12.**4 to **12.**7
    compound, **12.**5
    effective rate, **12.**7
    simple, **12.**5
  nonuniform series, **12.**8
  perpetuity determination, **12.**7
  present worth, **12.**6ff.
    of continuous cash flow of uniform rate,
      12.11
    of costs in inflationary period, **12.**33
    of future costs of installation, **12.**24
    of single payment, **12.**6
    of uniform-gradient series, **12.**9
    of uniform series, **12.**6
  probability, **12.**66 to **12.**90
  sinking fund, principal in, **12.**6
  sinking-fund deposit, **12.**6
  statistical inference, **12.**73ff.
  statistics and probability, **12.**62 to **12.**100
    arithmetic mean and median, **12.**62 to
      12.64
    binomial distribution, **12.**67
    composite event, **12.**70
    decision making, **12.**76, **12.**89 to **12.**100
    of failure, **12.**82
    of false null hypothesis, **12.**79
    forecasting with a Markov process,
      12.96 to **12.**98
    life span of devices, **12.**81 to **12.**83
    life span of systems, **12.**83 to **12.**89
    Markov process, **12.**96 to **12.**100
    Monte Carlo simulation of commercial
      activity, **12.**91 to **12.**94
    negative-exponential distribution, **12.**72
    normal distribution, **12.**70 to **12.**72
    number of ways of assigning work,
      12.65
    optimal inventory, **12.**89 to **12.**91
    Pascal distribution, **12.**68 to **12.**69
    permutations, and combinations, **12.**64
      to **12.**73
    Poisson distribution, **12.**69 to **12.**71
    population mean, **12.**75
    proportion of sample, **12.**78 to **12.**80
    sales forecasting by linear regression,
      12.94
    of sequence of events, **12.**66
    series of trials, **12.**67
    standard deviation, **12.**62 to **12.**64
    standard deviation from regression line,
      12.96
    uniform series, **12.**9 to **12.**11
    of unsatisfactory shipment, **12.**80

Enlargement of pipe, 1.215 to 1.217
Environmental cleanup, 10.9
Environmental Protection Agency (EPA), 6.10, 10.8, 13.5, 13.17
Epicyclic gears, 3.37
Equal-friction duct sizing, 2.96 to 2.102
Equilibrant of force system, 1.7, 1.8
Equilibrium, equations of, 1.7
Equipotential line, 1.267
Equivalent-beam method, 1.279
Equivalent length of pipes, 3.410, 3.411
Estimating, cost of cogeneration alternatives, 13.14
  cost of waste disposal, 13.5
  electrical loads, 4.33 to 4.37
Euler equation for column, 1.89
Evaluation of investments, 12.35 to 12.48
Excess-air analysis, 6.33
Expansion, pipe, 3.453ff.
  bends, 3.453
  joints, 3.455
Expansion fits, 3.170
Expected life, designing for, 3.110 to 3.112
Explosion vent, sizing of, 13.42
Extraction turbine, 3.298 to 3.300
Eyebar, 2.3

F chart method, 2.135 to 2.141
Facing, time for, 3.155
Factor of safety, 3.113 to 3.116
Fans, draft:
  controls for, 3.290 to 3.292
  selection of, 3.281 to 3.285
Fatigue loading, 1.93, 1.94
Federal Energy Regulatory Commission (FERC), 13.27
Feed rate, tool, 3.152
Feedback control system, 7.33
Feeders, electrical, 4.63
Feedwater heaters, 3.258 to 3.267
  closed, 3.259 to 3.264
  direct-contact, 3.258
  extraction-cycle, 3.264 to 3.268
Feedwater regulators, boiler, 7.11
Filter distortion, 5.62
Finned-tube heat exchangers, 3.484 to 3.486
Fired-heat composite curve (FHCC), 13.22
Fixed-end moment, 1.56
Fixture, plumbing, 2.39ff.
  demand weight of, 2.47ff.
Flash tank, sizing, 3.548 to 3.551

Flashing condensate, line sizing for, 6.20 to 6.22
Flexural stress (see Bending stress)
Flow line, 1.267
Flow set, 1.267, 1.268
Flue gas, 3.560
Flue-gas heat recovery, 13.8
Fluid amplifiers, 7.14
Fluid velocity in pipes, 3.397
Flying height for aerial photographs, 1.255 to 1.257
Flywheels, 3.7, 4.87 to 4.89
  for electric motors, 4.87 to 4.89
    effect of width, 4.89
    linear inertia of, 4.87
    starting and stopping times, 4.88
    total inertia of, 4.88
Footing:
  combined, 1.159 to 1.161
  isolated square, 1.157 to 1.159
  settlement of, 1.286
  sizing of, by Housel's method, 1.287
  stability of, 1.284, 1.285
Force, hydrostatic, 1.204 to 1.206
Force fit, 3.117
Force polygon, 1.10, 1.289
Form coefficients, vessel, 9.2
Form milling, 3.160
Fouling factors, heat-exchanger, 3.482
Fourier expression, 5.63
Francis equation, 1.212
Frequency of vibrating bent, 2.37
Frequency changers, 5.37
Friction, static, 1.8, 1.9
Friction damping, 3.108 to 3.110
Friction factor for Bingham plastics, 6.54 to 6.57
Friction head, pipe, 3.416 to 3.420
Fuel, industrial waste, 3.281
  low-sulfur Diesel, 3.319
Fuel-consumption factors, 2.64
Fuel oil, combustion of, 3.193
Fuel savings:
  from direct digital control, 3.543 to 3.545
  from heat recovery, 3.503
Fuse, classes of, 4.66

Gain-control circuit, 5.57
  JFET resistance of, 5.58
Gang milling, 3.157
Gas, flow rate of, 3.422
Gas and vapor disposal for plants, 6.29

Gas bearings, **3.**75 to **3.**77
Gas engines, **3.**289 to **3.**313
Gas pressure loss in pipes, **2.**55ff., **3.**421, **3.**422
Gas turbines, **3.**297 to **3.**298
   inlet air temperature, **3.**206
Gases, properties of mixtures of, **3.**286
Gear drives, **3.**33 to **3.**37
   bearing loads in, **3.**34
   force ratio of, **3.**35
   moment of inertia of, **3.**33
   transmission gear ratio, **3.**37
Gear trains, **3.**37
   epicyclic, **3.**37
   planetary, **3.**38
Geared speed reducer, **3.**54
Gears, **3.**24 to **3.**39
   bearing loads of, **3.**34
   bore diameter of, **3.**36
   dimensions of, **3.**31
   epicyclic, **3.**37
   force ratio of, **3.**35
   horsepower (kW) rating of, **3.**32
   for light loads, **3.**29 to **3.**31
   moment of inertia of, **3.**33
   pitch selection of, **3.**29ff.
   planetary, **3.**38
   plastics, **3.**125 to **3.**128
   ratio for, **3.**37
   selection of: by application, **3.**26
      by arrangement of equipment, **3.**28
      by convenience, **3.**27
   size of, **3.**25
   speed of, **3.**24
   type of, **3.**25
General wedge theory, **1.**275 to **1.**277
Geothermal power generation, **13.**27
Graphical analysis:
   of cable with concentrated loads, **1.**15 to **1.**16
   of force system, **1.**7, **1.**8
   of pile roup, **1.**288 to **1.**290
   of plane truss, **1.**10 to **1.**12
Grillage under column, **2.**16 to **2.**19
Grille, outlet and return, **2.**114 to **2.**118
Grinding feed and work time, **3.**159
Gusset plate, **2.**5
Gutter sizes, **2.**44

Hanger:
   pipe, **3.**434 to **3.**442
   steel, **2.**4

Hankinson's equation, **1.**121
Hardy Cross method, **11.**5
Hazardous air pollutants (HAP), **6.**41
Hazen-Williams formula, **3.**415, **11.**4
Head:
   friction, loss in water pipe, **3.**415
   total, on a pump, **3.**357 to **3.**361
Heat, waste, **6.**33
Heat of mixing, **6.**5
Heat balance, boiler, **3.**204 to **3.**206
   draft fans, **3.**281 to **3.**285
   ducts and uptakes, **3.**285 to **3.**290
   efficiency of, **3.**268
   fire-tube, analysis of, **3.**270
   steam, selection of, **3.**278 to **3.**281
Heat exchangers:
   costs, **6.**46 to **6.**50
   selection of air-cooled, **3.**542
Heat insulation for pipes, **3.**410
Heat loss, cost of, **3.**521
Heat-loss determination, **2.**61
Heat-loss factors, **4.**7 to **4.**78
   for concrete slabs, **4.**77
   for frame walls, **4.**75
   for infiltration, **4.**76
   for residential ceilings, **4.**75
   for roofing, **4.**78
   for windows, **4.**78
   for wood floors, **4.**76
   for wooden doors, **4.**78
Heat pump:
   analysis of, **3.**504 to **3.**507
   cooling electronic devices, **5.**71 to **5.**73
Heat-rate improvement, **3.**523 to **3.**526
Heat recovery:
   from boiler blowdown, **3.**542
   boilers, **13.**31
   engine, **3.**317
   from lighting systems, **2.**79, **4.**79
   steam generator (HRSG), **13.**25
   unit, **13.**12
Heat transfer, **3.**474 to **3.**491
   actual temperature difference of, **3.**480 to **3.**482
   air-cooled, selection of, **3.**540 to **3.**543
   in barometric condensers, **3.**483
   costs, **6.**46 to **6.**50
   electric, **3.**488
   exchangers for, **3.**475 to **3.**489
   finned-tube type, **3.**484 to **3.**486
   fouling factors in, **3.**482
   quick design and evaluation, **3.**552 to **3.**560

Heat transfer (*Cont.*):
  selecting type of, **3.**475
  shell-and-tube type, **3.**475 to **3.**480
  spiral type, **3.**486
Heating, electric comfort, **4.**68 to **4.**79
Heating systems:
  applications of, **2.**63
  coils for, **2.**72
    condensate formation rate of, **2.**73
    final temperatures in, **2.**73
  fuel-consumption factors for, **2.**64
  passive solar, **2.**145 to **2.**150
  selection of, **2.**62 to **2.**65
  solar, **2.**134 to **2.**141, **2.**145 to **2.**150
  steam consumption of, **2.**70 to **2.**72
Heaters:
  electric, **3.**488
  feedwater, **3.**258 to **3.**267
    closed, **3.**259 to **3.**264
    direct-contact, **3.**258
    extraction-cycle, **3.**264 to **3.**268
Helical springs, **3.**78 to **3.**81
High-pass filter design, **5.**60
High-temperature hot-water heating, **3.**516
    to **3.**519
High-vacuum systems, **3.**332 to **3.**338
  pipe size for, **3.**337
  pumping speed for, **3.**327
  pumps for, **3.**332 to **3.**338
  system factors for, **3.**336
Hobbing time, **3.**163
Horsepower (kW) for metalworking, **3.**179 to
    3.181
Hot water:
  domestic hot-water heater, **2.**141 to **2.**145
  heating with high-temperature, **3.**516 to
    3.519
  pipe sizes for, **2.**44 to **2.**51
  piping systems for, **2.**51
  temperatures for various services, **2.**51
Housel's method, **1.**287
Hull type, marine vessel, **9.**22
Humidifiers, selection of, **2.**105 to **2.**110
  capacities of, **2.**107
  location of, **2.**108
  piping for, **2.**110
  recommended humidities for, **2.**109
  steam required by, **2.**106
Hybrid-parameter conversions, **5.**31 to **5.**35
Hydraulic gradient, **1.**215, **1.**266
Hydraulic jump, **1.**220, **1.**221
Hydraulic piston, **3.**138ff.
  accelerating force, **3.**139

Hydraulic piston (*Cont.*):
  cushioning pressure, **3.**134
  fluid flow required, **3.**139
Hydraulic radius, **1.**215
  pipe, **3.**414
Hydraulic similarity, **1.**225
Hydraulic systems, **3.**117 to **3.**121
  energy savings in industrial, **3.**145
  valving and piping for, **3.**121
Hydraulic turbines from centrifugal pumps,
    3.376 to **3.**381
Hydro sites, **13.**34
  small, **3.**546
Hydropneumatic accumulator, **3.**144
  final gas pressure, **3.**144
  final volume, **3.**145
  kinetic energy absorbed, **3.**144
Hydropneumatic storage tank, sizing, **3.**375
Hydrostatic bearings, **3.**69 to **3.**75
  journal, **3.**69 to **3.**72
  multidirection-type, **3.**72 to **3.**75
  thrust, **3.**67 to **3.**69
Hydrostatic force:
  on curved surface, **1.**206
  on plane surface, **1.**204, **1.**205

Immiscible-solution characteristics, **6.**10
Impact load, **1.**29
Indicators, engine, **3.**294
Indoor lighting, **4.**51 to **4.**59
Industrial building ventilation, **13.**44
Industrial Fuel Use Act (IFUA), **13.**17
Industrial waste fuel, **13.**7
Inflation, effects of, **12.**32 to **12.**35
  anticipated, **12.**34
  at constant rate, **12.**32
  on endowment fund, **12.**34
  on present worth of costs, **12.**33
  on replacement cost, **12.**32
  at variable rate, **12.**33
Influence line:
  for bridge truss, **1.**60 to **1.**62
  for three-hinged arch, **1.**66
Initial yielding, **1.**97
Insolation determination, solar, **2.**132 to
    2.134, **2.**147
Integrated circuits, **5.**14ff.
Interaction diagram, **1.**150 to **1.**156
Interest calculations, **12.**4 to **12.**7
Interior lighting, **4.**51 to **4.**59
Internal-combustion engines, **3.**201 to **3.**325
  bypass cooling system for, **3.**312 to **3.**317

Internal-combustion engines (*Cont.*):
  compression ratio of, **3.306**
  cooling-water requirements of, **3.306** to
        3.311
    hookups for, **3.308**
    slant diagrams for, **3.309**
  displacement of, **3.302**
  efficiency of, **3.301**
  fuel storage capacity and cost, **3.318**
  horsepower (kW) of, **3.302**
  hot-water heat recovery, **3.317**
  indicator, use for, **3.305**
  low-sulfur fuel, **3.319**
  mean effective pressure of, **3.302**
  oil coolers for, **3.320**
  output at high temperatues and high alti-
        tudes, **3.304**
  performance factors for, **3.321** to **3.325**
  piston speed of, **3.306**
  power input to pumps for, **3.319**
  selection of, **3.303**
  solids entering, **3.321**
  torque of, **3.306**
  vent system for, **3.311** to **3.312**
Interplanetary flight, **8.14**
Inventory, optimal level of, **12.54**, **12.89** to
        12.91
Inverted-tooth (silent) chain drive, **3.48**
Investments:
  evaluation of, **12.35** to **12.48**
  return on, for energy savings, **3.518**
Involute splines, **3.106** to **3.108**
  face width of, **3.107**
  number of teeth for, **3.108**
  size of, **3.107**

Jet-plane speed, **8.23**
Joists, reinforced-concrete, **2.34ff.**

Karnaugh map, construction of, **7.58** to **7.60**
  with distinctive blocks, **7.61**
  incomplete block, use of, **7.63**
  large block, use of, **7.62**
  reduction of expression, **7.60**
Kern distance, **1.176**
Keyway, cutting of, **3.148**
Kirchhoff's laws, **4.5**
Kites for wind energy, **3.500**
Knee:
  curved, **2.14**
  rectangular, **2.13**
Krey ϕ-circle method of analysis, **1.282**

Laminar flow, **1.212**, **1.213**
  in pipes, **3.426**
Landfills, mining of, **1.286**
Laplace equation, **1.267**
Large-scale integration, **5.35**
Latitude of line, **1.226**
Learning curves, **3.173** to **3.176**
Lenses, **5.76** to **5.80**
  compound thin analysis, **5.78** to **5.80**
  images produced by, **5.76** to **5.78**
  thick compound systems analysis, **5.78**
Leveling, differential, **1.230**, **1.231**
Life of springs, **3.98**
  designing parts for expected, **3.110** to
        3.112
Light-gage steel beam:
  with stiffened flange, **2.28** to **2.30**
  with unstiffened steel flange, **2.27**
Lighting systems, **4.51** to **4.63**
  heat recoverable from, **2.79**, **4.79**
  indoor, **4.51** to **4.59**
    coefficients of utilization of, **4.57**
    illumination level of, **4.52**
    light sources for, **4.54**
    performance factors in, **4.58**
  outdoor, **4.59** to **4.63**
    pole placements for, **4.61**
    recommended lighting levels for, **4.60**
Lightning protection, **4.9** to **4.12**, **4.90** to
        4.92
  arrester selection, **4.9** to **4.12**
  incidence of thunderstorms, **4.91**
  type of system to use, **4.90**
  zone of protection, **4.92**
Line sizing for flashing condensate, **6.20** to
        6.22
Linear transformation, principle of, **1.188** to
        1.190
Liquid, **3.424ff.**
  siphon, **3.424**
  specific gravity of, **3.425**
  viscosity of, **3.425**
Liquid-liquid separation, **6.13**
Liquid springs, **3.91**
Liquid-vapor separators, **6.41** to **6.45**
  horizontal, **6.43** to **6.45**
  vertical, **6.41** to **6.43**
Load, structural, **1.48ff.**, **1.72ff.**
  eccentric, **1.48ff.**, **1.72ff.**
    on pile group, **1.48**
    on rectangular section, **1.45**
    on riveted connection, **1.72** to **1.74**
    on welded connection, **1.75**
  moving (*see* Moving-load system)

Load factor, 1.95
Load-stress factors, 3.113
Looping pipes, 1.216
Lot size in manufacturing, 3.174
Low-pass filter design, 5.60
Lube-oil coolers, engine, 3.308

Mach 2.5 air stream, 8.26
Machine design and analysis, 3.7 to 3.146
Magnel diagram, 1.176 to 1.178, 1.193 to
    1.195
Maintainability, electronic, 5.22 to 5.24
Manning formula factor, 1.215, 11.22
Maps, Karnaugh, 7.58 to 7.64
Marine engineering, 9.1 to 9.22
    choice of vessel, 9.22
    hull type, 9.22
    immersion and flooding effects, 9.10
    marine propeller shafts, 9.6
    nuclear propulsion, 9.15 to 9.17
    power-plant selection in, 9.13 to 9.16
    propeller selection, 9.7 to 9.9
    pump selection in, 9.10 to 9.13
    refrigeration and air conditioning, 9.19 to
        9.21
    revolutions of propeller, 9.9
    tanker capacity, 9.17 to 9.18
    vessel choice, 9.22
    vessel form coefficients, 9.2
        power for propulsion, 9.4 to 9.6
        shallow-water speed, 9.3 to 9.5
        speed, 9.9
        wetted area, 9.3
    vessel seakeeping potential, 9.22
Marine refrigeration and air conditioning,
    9.19 to 9.21
Markov process in sales forecasting, 12.96 to
    12.100
Material balance, 6.6
Materials handling, 3.338 to 3.352
    bulk elevators and conveyors, 3.338 to
        3.342
        conveyor characteristics, 3.340
        horsepower (kW) required for, 3.341
        maximum belt speeds for, 3.342
        minimum belt width for, 3.341
    pneumatic conveying systems, 3.344 to
        3.352
        air quantities for, 3.348
        design calculations for, 3.346
        duct diameters for, 3.348

Materials handling, pneumatic conveying
        systems (*Cont.*):
        duct gages for, 3.351
        duct resistance in, 3.349
        entrance losses of, 3.350
    screw conveyors, 3.342 to 3.344
        capacity and speed of, 3.342
        material factors for, 3.343
        size factors for, 3.343
Maxwell's theorem, 1.68
Mean effective pressure, 3.000
Mechanical-drive turbines, 3.000 to 3.000
Mechanical engineering, 3.1 to 3.562
Mechanism method of plastic design, 1.101
Membrane vibration, 3.145
Meridian, 1.234
Metacenter, 1.208
Metalworking, 3.148 to 3.189
    angle and length of cut, 3.152
    bending, dimpling, and drawing, 3.183 to
        3.185
    blank diameters for, 3.185
    blanking, drawing, and necking, 3.168
    breakeven in, 3.187 to 3.189
    Brinnel hardness, 3.176
    broaching, 3.163
    centerless grinder, 3.183
    cutting speed for lowest cost, 3.181
    cutting speeds in, 3.149
    cutting time in, 3.152
    drill penetration rate, 3.183
    economical cutting speeds and production
        rates, 3.176
    element time in, 3.149
    grinding, 3.162
    hobbing, 3.163
    horsepower (kW) required for, 3.179 to
        3.181
    keyway cutting, 3.150
    learning curves in, 3.173 to 3.176
    mechanical-press capacity, 3.167
    milling-machine feed and approach, 3.351
    milling operations, 3.159 to 3.161
    minimum lot size for, 3.153
    operator time of, 3.149
    optimum lot size in, 3.177
    oxyacetylene cutting, 3.165
    planer operations, 3.161
    plating of metals, 3.169
    precision dimensions, 3.178
    press fits, 3.170
    presswork force, 3.167
    reorder quantity for, 3.181

Metalworking (*Cont.*):
  savings with more machinable materials 3.182
  sawing time, 3.164
  serrating time, 3.163
  shaper operations, 3.161
  shrink fits, 3.170
  splining, 3.163
  tapers and dovetails, 3.151
  tapping time, 3.156
  thread milling, 3.182
  threading time, 3.156, 3.157
  time and power for, 3.154
    boring, 3.154
    countersinking, 3.154
    drilling, 3.154
    facing, 3.155
    reaming, 3.154
  time to tap, 3.156, 3.158
  time to thread, 3.158
  tool-change time in, 3.153
  tool feed rate in, 3.152
  true unit time of, 3.153
  turning time for, 3.153
  turret-lathe power, 3.157
    constants for, 3.158
Methane gas, 13.8
Method:
  of joints, 1.12 to 1.14
  of sections, 1.14
  of slices, 1.280 to 1.282
Microwave transmitter analysis, 5.68 to 5.71
Milling machine:
  cutting speed of, 3.159
  feed for, 3.148, 3.159
  horsepower (kW) of, 3.159
  teeth number for, 3.159
Mine surveying, 1.250 to 1.255
Minimum lot size, 3.153
Minimum safe flow for pump, 3.378
Mining landfills, 1.286
Mirrors, analysis of image produced by, 5.73 to 5.76
  concave, 5.73
  convex, 5.73
  plane, 5.73
Mismatch, efficiency, 5.29
Missiles, range of, 8.13
Mixer cooling-water flow rate, 6.12
Mixing of two airstreams, 2.93
Modulus:
  of elasticity, 1.25
  of rigidity, 1.36

Mohr's circle of stress, 1.30, 1.31, 1.270 to 1.273
Moisture content:
  of saturated air, 2.107
  of soil, 1.265, 1.266
Molal method of combustion analysis, 3.201 to 3.203
Mollier diagram, steam, 3.209 to 3.211
Moment:
  bending (*see* Bending moment)
  of inertia, 1.22 to 1.25
    polar, 1.22 to 1.25
  statical, 1.23
Moment-area method, 1.49ff.
Moment distribution, 1.57
Monte Carlo simulation, 12.91 to 12.94
Motors, electric (*see* Electric motors)
Moving-load system:
  on beam, 1.59, 1.61, 1.67
  on bridge truss, 1.60 to 1.64
Multiple milling, 3.160
Multirate helical springs, 3.87

Nadir of observer, 1.234
NAND gate circuit, 7.41 to 7.45
  Karnaugh map for, 7.41 to 7.43
  with two logic levels, 7.42 to 7.45
National Pollution Discharge Elimination System (NPDES), 13.18
Natural gas, combustion of, 3.194 to 3.199
Necking of metals, 3.165
Net positive suction head for pumps, 3.362
Network synthesis, 5.64
Neutral axis in composite beam, 1.41
Neutral point, 1.63
NO_x formation, 6.41, 13.18
Noise-reduction material, 3.125 to 3.126
Noncircular shafts, 3.130 to 3.138
  composite, 3.132
  cross-shaped, 3.138
  four-keyway, 3.135
  four-spline, 3.133
  milled, 3.133
  pinned-shaft, 3.137
  rectangular, 3.137
  single-keyway, 3.134
  single-spline, 3.136
  two-keyway, 3.135
  two-spline, 3.136
Noncondensable gases (NCGs), 13.28
Nonuniform series, 12.8
Notice of commencement (NOC), 6.11

Nonutility generation (NUG), 13.27
Nuclear power plants, 9.16, 10.2 to 10.12
    cycle analysis, 10.3 to 10.6
    and desalting plants, 10.10 to 10.13
    fissionable material for, 10.7
    marine, 9.16 to 9.17
    power reactor selection, 10.2
    reactor characteristics, 10.3
    reactor fuel consumption, 10.6
    spent fuel care, 10.8
Nuclear radiation, effects of, on humans,
    10.8

Observation satellites, 8.18
Oil-film bearings, 3.60
Oil pipes, pressure loss in, 3.427 to 3.433
Operating costs, 13.17
Operating speed for pumps, 3.343 to 3.345
Operational amplifier, 5.53 to 5.57, 5.64
    ac equivalent circuit synthesis, 5.53
    chain, 5.55 to 5.57
        dc output of, 5.55
        node-defined, 5.56
    network synthesis with, 5.64
Optics, 5.73 to 5.80
    compound thin-lens analysis, 5.78 to 5.80
    lenses, image produced by, 5.76 to 5.78
    mirrors, image produced by, 5.73 to 5.76
    thick compound lens system analysis, 5.78
Optimal inventory level, 12.54, 12.89 to
    12.91
Optimum lot size, 3.153
Orifice:
    flow through, 1.211
    meter, 3.411
    variation in head on, 1.222
Orifice meter, pipe, 3.411
Oscillator, selection of, 5.14
Outdoor lighting, 4.59 to 4.63
Oxyacetylene cutting, 3.165 to 3.166
Ozone Transport Commission, 3.319

Parabolic curve:
    coordinates of, 1.186
    vertical (see Vertical parabolic curve)
Pascal probability distribution, 12.68 to
    12.69
Passive solar heating, 2.145 to 2.151
Payback period of investments, 12.46
Performance curves, wind-energy machines,
    3.513

Permanent-magnet motors, 4.81 to 4.87
    analysis, 4.81
    dimensionless efficiency curves, 4.85
    electrical measurement analysis, 4.85
    equivalent motor circuit, 4.86
    factor of merit, 4.85
    performance characteristics, 4.83
    stall and drag torques, 4.83, 4.86
    universal performance formulas, 4.82
Permeability of soil, 1.267
PERT in project planning, 12.55 to 12.59
Phase-lag compensator design, 7.31
ϕ-circle method, 1.282
Photogrammetry, aerial (see Aerial photo-
    grammetry)
Pile-driving formula, 1.288
Pile group:
    under concentric load, 1.288
    under eccentric load, 1.48
    load distribution within, 1.289 to 1.292
Piping (and fluid flow), 3.396 to 3.474
    allowable stresses in, 3.398
    anchor force in, 3.453
    chart determination of friction head, 3.416
        to 3.420
    cold-spring effect, 3.442
    compressed-air pressure loss, 3.433
    corrugated expansion joint, 3.457 to 3.459
    equivalent length: of parallel pipeline,
        3.423
        of series pipeline, 3.422
    expansion-bend stress, deflection of, 3.453
    flow of water in pipes, 1.209 to 1.217
    fluid velocities in 3.397
    friction loss, for solids in suspension,
        3.472
    gas pressure loss, 2.55ff., 3.434
        flow rate of, 3.434
    hanger selection for, 3.434 to 3.442
        spacing, 3.442
    Hazen-Williams formula, 3.415
    head loss in water piping, 3.415
    heat insulation for, 3.410
    heat loss in uninsulated, 3.521
    hydraulic radius of, 3.411
    laminar flow, pressure loss with, 3.426
    liquid siphon height, 3.424
    liquid specific gravity in, 3.425
    liquid velocity in, 3.414
    liquid viscosity in, 3.425
    orifice meter for, 3.411
    pipe bends for, 3.443 to 3.454
    pipe slope for, 3.442

Piping (*Cont.*):
  plastic piping, **3.471**
  plumbing (*see* Plumbing and drainage)
  pressure loss: in oil pipes, **3.427** to **3.433**
    in steam piping, **3.399** to **3.403**
  pressure-reducing valves for water, **3.420**
    to **3.421**
  pressure-regulating valve for, **3.412** to
    3.414
  relative carrying capacity of, **3.420**
  schedule number of, **3.396**
  single-plane bends in, **3.443** to **3.446**
  slip-type expansion joint, **3.455** to **3.457**
  steam transmission lines, **3.460** to **3.468**
    accumulator selection and sizing, **3.469**
      to **3.471**
    desuperheater analysis, **3.468**
  steam trap selection for, **3.404** to **3.410**
    bucket-trap capacity, **3.409**
    coefficient of heat transfer for, **3.406**
    condensate formed, **3.407**
    factors in, **3.404**
    impulse-trap capacities, **3.409**
    specific heats for, **3.405**
    unit-heater correction factors for, **3.408**
  tabular determination of friction head,
    3.416 to **3.420**
  three-plane bends in, **3.451** to **3.453**
  two-plane bends in, **3.446** to **3.449**
  wall thickness of, **3.396**ff.
  wall thickness determination by ANSI
    piping code formula, **3.396** to **3.399**
  warm-up condensate load of, **3.403**
  water-hammer effects, **3.425**
  water-meter, sizing of, **3.422**
Plane mirror, **5.73**
Planer characteristics, **3.158** to **3.159**
Planetary gears, **3.38**
Plant gas and vapor disposal, **6.29**
Plastic design, **1.97** to **1.116**
  of beam-column, **1.114**
  of continuous beam, **1.102** to **1.105**
  definitions relating to, **1.97**
  of gable frame, **1.110** to **1.114**
  mechanism method of, **1.101**
  of rectangular frame, **1.105** to **1.110**
  shape factor in, **1.99**
  statical method of, **1.100**
Plastic gears, **3.125** to **3.128**
  horsepower (kW) equations, **3.126**
  materials for, **3.126**
  safe stress for, **3.125**
Plastic hinge, **1.98**

Plastic modulus, **1.98**
Plastic moment, **1.98**
Plastic piping, **3.459**
Plate girder, **1.84** to **1.88**
Plating of metals, **3.169**
Plumbing and drainage, **2.37** to **2.61**
  cold- and hot-water piping, **2.44** to **2.51**
  gas piping, **2.55** to **2.57**
  pipe-size determination for, **2.37** to **2.42**
  roof and yard rainwater systems, **2.42** to
    2.44
  sprinkler systems, **2.52** to **2.54**
  swimming-pool sizing, **2.58**
Pneumatic conveying systems, **3.344** to
    3.352
  design calculations for, **3.346**
  duct diameters and areas for, **3.348**
  duct gages for, **3.351**
  duct resistance chart for, **3.349**
  entrance losses for, **3.350**
  exhaust air quantities in, **3.348**
Poisson probability distribution, **12.69** to
    12.70, **12.82**
Poisson's ratio, **3.119**
Polar moment of inertia, **1.22** to **1.25**
Polluted beach, cleaning of, **13.40**
Pollution:
  air, reduction of, **2.145**, **2.150**
  audio, **2.125**
  cross-border, **3.206**
  by tabacco smoke, **2.120**
Porosity of soil, **1.265**
Portal method of wind-stress analysis, **2.19**
    to **2.21**
Power:
  of flowing liquid, **1.211**
  generation of (*see* Power generation)
  marine nuclear, **9.16**
  to propel vessels, **9.4** to **9.6**
Power distribution for plants, **4.37** to **4.40**
Power factor:
  analysis of, **4.28**
  determination of, **4.26**
  improvement of, **4.26**
Power generation, **3.208** to **3.300**
  air-ejector analysis, **3.252** to **3.254**
  barometric condenser analysis, **3.255**
  with biomass, **13.27**
  bleed-stream regenerative cycle, **3.227** to
    3.233
  boiler efficiency in, **3.268** to **3.270**
  boiler selection, **3.278** to **3.281**
    air ducts for, **3.285** to **3.290**

Power generation, boiler selection (*Cont.*):
    fan control selection, 3.290 to 3.292
    fan selection in, 3.281 to 3.285
    coal-dryer analysis, 3.293
    coal storage capacity, 3.295
    condenser: water pressure loss in, 3.254
        weight analysis of, 3.254
    condensing turbine analysis, 3.238 to
        3.240
    constant-entropy steam process, 3.219
    constant-pressure steam process, 3.214
    constant-temperature steam process,
        3.218
    constant-volume steam process, 3.215 to
        3.217
    cooling-pond sizing, 3.257 to 3.258
    feedwater heaters, 3.258 to 3.267
        closed, 3.259 to 3.264
        direct-contact, 3.258
        extraction-cycle, 3.264 to 3.268
    fire-tube boiler analysis, 3.270
    geothermal, 13.27
    internal-combustion engines for, 3.301 to
        3.325
    irreversible adiabatic compression, 3.221
        to 3.222
    irreversible adiabatic expansion, 3.222
    mechanical-drive turbine analysis, 3.236
        to 3.238
    nuclear (*see* Nuclear power plants)
    properties of gas mixtures, 3.234
    regenerative-cycle performance, steam
        turbine, 3.240 to 3.242
    regenerative-gas turbine, 3.297 to 3.298
    reheat cycle performance, 3.233 to 3.236
    reheat-regenerative heat rate, 3.243
    reversible heating process, 3.225 to 3.227
    safety valve: capacity of, 3.271
        selection of, for boiler, 3.272 to 3.277
    smokestack sizing, 3.292
    steam condenser: air-ejector analysis for,
        3.252 to 3.254
        circulating-water pressure loss in, 3.254
        performance of, 3.246 to 3.249
        selection of, 3.250
        weight analysis of, 3.254
    steam injection, 3.296
    steam Mollier diagram for, 3.209 to 3.214
    steam-quality determination, 3.277
    steam table use for, 3.209 to 3.214
        interpolation of values in, 3.211 to
        3.214
    steam turbine-gas turbine cycles, 3.245

Power generation (*Cont.*):
    superheater pressuredrop, 3.277
    throttling processes, 3.223
Power-plant thermal pollution, 13.47
Power plants, marine, 9.13 to 9.16
Power savings in industrial hydraulic sys-
    tems, 3.146
Power-supply analysis, 5.19
Power transmission, 3.56
Powerboats, coefficients for, 9.6
Precision dimensions in metalworking, 3.178
Premanufacture notificaton (PMN), 6.10
Present worth:
    of continuous cash flow of uniform rate,
        12.11
    of costs in inflationary period, 12.33
    of future costs of an installation, 12.24
    of single payment, 12.6
    of uniform-gradient series, 12.9
    of uniform series, 12.6
Pressing, 3.167, 3.170
Pressure, soil, 1.47, 1.269, 1.270
Pressure center, 1.204
Pressure loss:
    in oil pipes, 3.427 to 3.433
    in reducing valves for water piping, 3.420
    in regulating valves, 3.412 to 3.414
    for solids in suspension, 3.472
    in steam pipes, 3.399
Pressure prism of fluid, 1.205
Pressure-reducing valves, 7.45 to 7.48
    for water piping, 3.420
Pressure vessel:
    prestressed, 1.32
    thick-walled, 1.33
    thin-walled, 1.32
Pressurized joint, bolt diameter, 3.121 to
    3.125
    endurance limit of, 3.124
    loss of clamping force, 3.124
    required bolt area, 3.122
    spring rate of, 3.123
    tightening torque applied, 3.125
    tightening torque required, 3.124
Presswork force, 3.167
Prestressed-concrete beam, 1.165 to 1.197
    in balanced design, 1.166
    concordant trajectory in, 1.190
    continuity moment in, 1.188
    continuous, 1.187 to 1.197
        with nonprestressed reinforcement,
        1.191 to 1.197
        reactions for, 1.197

Prestressed-concrete beam (*Cont.*):
  deflection of, 1.178
  guides in design of, 1.175
  kern of, 1.176
  linear transformation in, 1.188 to 1.190
  loads carried by, 1.166
  Magnel diagram for, 1.176 to 1.178, 1.193
    to 1.195
  notational system for, 1.166, 1.167
  posttensioned, design of, 1.182 to 1.186
  posttensioning of, 1.166
  prestress shear and bending moment in,
    1.167
  pretensioned, design of, 1.179 to 1.182
  pretensioning of, 1.165, 1.166
  radial forces in, 1.187
  stress diagrams for, 1.168 to 1.172
  trajectory of prestressing force in, 1.166
  transmission length of, 1.166
  web reinforcement of, 1.181
Principal axis, 1.89
Principal plane, 1.30 to 1.32
Principal stress, 1.31, 1.32
Prismoidal method, 1.233
Probability, 12.66 to 12.90
  binomial distribution, 12.67
  composite event, 12.70
  of failure, 12.82
  of false null hypothesis, 12.79
  negative-exponential distribution, 12.72
  negative-exponential life span, 12.81 to
    12.83
  normal distribution, 12.70 to 12.72
  Pascal distribution, 12.68
  Poisson distribution, 12.69 to 12.70, 12.82
  of sequence of events, 12.66
  series of trials, 12.67
  of unsatisfactory shipment, 12.80
Process-control systems, 7.2 to 7.11
  computers for, 7.6 to 7.8
  mode of control for, 7.3
  for temperature, 7.4
  valves for, 7.8
Process plant engineering, 6.14 to 6.57
  Bingham plastics, friction factor for, 6.54
    to 6.57
  cost estimates, 6.34 to 6.54
    centrifugal-pump and electric-motor,
      6.50 to 6.54
    correlations in, 6.46 to 6.54
    heat exchangers and storage tanks, 6.46
  desuperheater condensers, 6.39 to 6.41

Process plant engineering (*Cont.*):
  energy loss savings, 6.22
    by relocating units, 6.25
    from storage tanks and vessels, 6.22 to
      6.25
    from waste heat for refrigeration, 6.31
  energy savings, 6.22 to 6.29
    from recompression, 6.27 to 6.29
  excess-air analysis, 6.33
  flashing condensate, line sizing for, 6.20
    to 6.22
  friction factor for Bingham plastics, 6.54
    to 6.57
  gas and vapor disposal, plant, 6.29
  heat exchanger and storage tank costs,
    6.46 to 6.50
  line sizing for flashing condensate, 6.20 to
    6.22
  liquid-vapor separators, 6.41 to 6.45
    horizontal, 6.43 to 6.45
    vertical, 6.41 to 6.43
  rupture disks, 6.45
  size and cost estimates, process equip-
    ment, 6.34 to 6.54
  stack height, effective, 6.29
  steam tracing: for piping, 6.14
    vessel bottom, 6.15
  steam transmission line, 6.17
    without steam traps, 6.17 to 6.20
  storage vessel, time to empty, 6.46
  Venturi scrubbers, 6.34 to 6.39
  waste-heat loss reduction, 6.33
Product of inertia, 1.24
Product cooling, 3.493 to 3.498
Project planning, 12.55 to 12.62
  selection, 13.3
Propellers, marine, 9.6 to 9.10
  best combinations of, 9.8
  critical thrust of, 9.9
  diameter of shaft for, 9.6
  revolutions of, 9.9
  selection of, 9.7 to 9.9
  slip of, 9.6
  and vessel speed, 9.9
Properties of combustion elements, 3.192
Psychrometric charts, 2.90, 2.110 to 2.112
Public-address systems, 5.30
Public Utilities Regulatory Policies Act
    (PURPA), 13.27
Pulley, load on shaft, 3.9
Pump-down time, vacuum systems, 3.352

Pumps (and pumping systems), 3.352 to 3.395
  affinity laws for, 3.352, 3.353
  analysis of characteristic curves, 3.368 to 3.374
  centrifugal, 3.352 to 3.383
    as hydraulic turbines, 3.387 to 3.392
  characteristics of modern, 3.365
  chemical-plant, 6.12
  choice to reduce energy consumption, 3.394
  classes and types of, 3.365
  composite chart for, 3.366
  condensate, 3.374 to 3.378
  critical speed of, 3.381
  effect of liquid viscosity, 3.382 to 3.386
    on reciprocating pumps, 3.383
    on regenerative pumps, 3.382
    on rotary pumps, 3.384 to 3.386
  energy efficiency, 3.395
  engine, 3.316
  essential data for, 3.364
  hydraulic system, 3.117 to 3.121
  hydropneumatic tank sizing, 3.386
  impeller shapes in, 3.356
  marine, 9.10 to 9.13
  minimum safe flow for, 3.378
  net positive suction head for, 3.374
  pipe friction loss for water, 3.361
  piping arrangements for, 3.358
  rotary pump capacity ranges, 3.367
  selecting best operating speed of, 3.355 to 3.357
  selection of: for any system, 3.362 to 3.368
    for chemical plants, 6.10
    materials for, 3.386
    for reduced energy consumption and loss, 3.394 to 3.395
  shaft deflection of, 3.381
  similarity laws for, 3.352 to 3.353
  specific-speed considerations for, 3.353 to 3.355
  suction specific speed ratings, 3.357
  total head on, 3.357 to 3.361
  types of, by specific speeds, 3.357
  typical rating table, 3.356
  for viscous liquids, 3.379 to 3.381

Quick design of heat exchangers, 3.552
Quicksand conditions, 1.266

Radiant-heating panels, 2.75 to 2.78
Radiation, nuclear, 10.8
Rain forests, 13.4
Ramjet engine, 8.21
Rankine's theory, 1.273
Rate of springs, 3.99
Reactions at elastic supports, 1.27, 1.53 to 1.55
Reaming, time and power for, 3.152
Rebhann's theorem, 1.275
Reciprocal deflections, theorem of, 1.67
Reciprocating-pump performance, 3.383
Reciprocating refrigeration compressors, 3.503ff.
Rectangular springs, 3.81
Redtenbacker's formula, 1.287
Refrigeration, 3.491 to 3.509
  centrifugal-machine load analysis, 3.505
  compressor cycle analysis for, 3.499 to 3.503
  energy for steam-jet, 3.498 to 3.499
  heat-pump cycle, 3.506 to 3.509
  marine, 9.19 to 9.21
  for product cooling, 3.493 to 3.498
  reciprocating, compressor selection, 3.503 to 3.505
  system selection for, 3.491 to 3.493
Regenerative cycle, steam, 3.240 to 3.242
Regenerative-pump performance, 3.358
Reheat cycle, steam, 3.233 to 3.236
Reheat-regenerative cycle, steam, 3.243
Reinforced-concrete beam, 1.125 to 1.149
  in balanced design, 1.126, 1.139, 1.140
  bond stress in, 1.134
  with compression reinforcement, 1.131, 1.147
  continuous: deflection of, 1.148
    design of, 1.134 to 1.136
  equations of, 1.127, 1.139
  of joist construction, 2.34
  of rectangular section, 1.127 to 1.132, 1.140 to 1.148
  shearing stress in, 1.132, 1.143
  supporting stair, 2.35
  of T section, 1.129, 1.130, 1.144 to 1.147, 1.198 to 1.200
  transformed section of, 1.141, 1.142
  with two-way reinforcement, 1.136 to 1.138
  ultimate-strength design of, 1.126 to 1.138
  web reinforcement of, 1.132 to 1.134, 1.143
  working-stress design of, 1.138 to 1.149

Reinforced-concrete column, 1.149 to 1.157
  in balanced design, 1.149
  interaction diagram for, 1.150, 1.154
  ultimate-strength design of, 1.149 to 1.153
  working-stress design of, 1.153 to 1.157
Reinforced-plastic springs, 3.96 to 3.98
Relative carrying capacity of pipes, 3.420
Reliability, electronic, 5.20 to 5.22
Reorder quantity for metalworking, 3.181
Repowering of plants, 3.532
Residential-service demand load, 4.64 to
    4.68
Resource Conservation and Recovery Act,
    3.281
Retaining wall:
  cantilever, design of, 1.162 to 1.165
  earth thrust on, 1.273 to 1.275
  gravity, stability of, 1.18
Return on investment for energy saving,
    3.536
Reversible heating processes, 3.225 to 3.227
Reynolds number, 1.213, 3.560
Ribbed floor, 2.34
Right ascension of star, 1.234
Rigidity, modulus of, 1.36
Ring springs, 3.89 to 3.91
Riveted connection, 1.68 to 1.75, 2.7
  eccentric load on, 1.72 to 1.74
  moment on, 1.72, 2.7
Robots, 3.140 to 3.144
  arcs of, 3.140
  three-link, 3.141
  types of bodies, 3.143
  workspace of, 3.141
Rocket combustion chamber, 8.25
Rocket flight velocity, 8.13
Roller bearings, 3.63 to 3.65
  capacity: and cost and size of, 3.63
    and reliability of, 3.65
  operating-life analysis of, 3.63
  radial load rating of, 3.64
  speed limits for, 3.61, 3.63
Roller chain drives, 3.48 to 3.50
Rolling surface wear life, 3.112
Roof ventilators, 2.118 to 2.120
Rotary pump performance, 3.372 to 3.374
Roughness coefficient, 1.215
Rupture disks, 6.45
Rupture factor, 3.116

Safety service, sizing centrifugal pumps for,
    3.392

Safety valves, boiler, 3.271 to 3.277
  capacity of, 3.271
  selection of, 3.271 to 3.277
Sanitary engineering, 11.1 to 11.29
  Manning formula factor, 11.22
  sanitary sewer design, 11.20 to 11.23
  sewage-treatment methods: products of,
      11.28
    selection of, 11.26 to 11.29
  sewer-pipe earth load, 11.17 to 11.20
  sizing sewer pipes, 11.14 to 11.16
  storm-sewer design, 11.25
  storm-sewer inlet size and flow rate, 11.24
  storm-water runoff rate, 11.12
  water supply: flow rate, 11.2 to 11.6
    Hardy Cross method, 11.5
    Hazen-Williams equation for, 11.4
    industrial water requirements, 11.10
    pressure loss of, 11.2 to 11.6
    system selection for, 11.7 to 11.10
    treatment methods for, 11.10 to 11.12
Satellite:
  communications link analysis, 5.66
  flight velocity of, 8.14
  observation, 8.18
Saturated-air moisture content, 2.107
Saturated-solution analysis, 6.2
Scale, boiler, savings from reduced, 3.534
Scale of aerial photograph:
  definition of, 1.257
  determination of, 1.263 to 1.265
  in relation to flying height, 1.255 to 1.257
Schedule number, pipe, 3.396
Screw compressor, 13.20
Selection of project, 13.3
Serrating time, 3.163
Servo system(s), 7.20 to 7.30
  active network solution, 7.29
  angular-position system analysis, 7.21
    closed-loop transfer function, 7.21, 7.26
    open-loop transfer function, 7.21
  compensation network, 7.28
  dc gain function, 7.29
  with loop delay, 7.25
  loop-gain function analysis, 7.22
  overshoot and settling time, 7.24
  stability determination, 7.20
    Routh array, 7.20
    Routh criterion, 7.20
Sewage treatment, 11.26 to 11.29
Sewer pipes:
  earth loads of, 11.17 to 11.20
  sanitary system design, 11.20 to 11.23

Sewer pipes (*Cont.*):
  sizing of, **11**.14 to **11**.16
  storm-sewer design, **11**.25
  storm-sewer inlet size and flow rate, **11**.24
Shaft bearings (*see* Bearings, shaft)
Shaft vibration damping, **3**.108 to **3**.110
Shafts:
  bearings for (*see* Bearings, shaft)
  bending of, **3**.11, **3**.12
  bending moment of, **3**.9, **3**.10
  concentrated loads on, **3**.14
  couplings for, **3**.39 to **3**.48
    flexible, **3**.42 to **3**.44
  deflection of, **3**.13 to **3**.15
    pump, **3**.369
  driver efficiency of, **3**.8
  equivalent bending moment of, **3**.13
  hollow, **3**.10
  horsepower (kW) of, **3**.8
  ideal torque for, **3**.13
  keys for, **3**.15
  marine propeller, **9**.6 to **9**.10
  noncircular (*see* Noncircular shafts)
  pump, deflection of, **3**.369
  reactions of, **3**.9
  solid, **1**.36, **3**.10, **3**.11
  torque of, **3**.8, **3**.13
  torsion of, **1**.36, **3**.10, **3**.11
  uniform loads on, **3**.14
Shape factor, **1**.97 to **1**.99
Shaper characteristics, **3**.158
Shear:
  in beam, **1**.37 to **1**.39
  in column footing, **1**.157 to **1**.161
  punching, **1**.157
  pure, **2**.31, **2**.32
  in truss, **1**.60 to **1**.62
Shear center, **1**.43, **1**.44
Shear connectors, **1**.203, **2**.34
Shear diagram:
  for beam, **1**.37 to **1**.39
  for combined footing, **1**.160
Shell-and-tube heat exchangers, **3**.475 to
    3.480
Short-circuit current, **4**.17
  in power systems, **4**.18
Shrink-fit stress, **1**.35
Shrink fits, **3**.117, **3**.170
Similarity, hydraulic, **1**.225
Similarity laws for pumps, **3**.352 to **3**.353
Sinking fund, **12**.6

Size and cost estimates, process equipment,
    6.34 to **6**.54
  desuperheater condensers, **6**.39 to **6**.41
  heat exchangers and storage tanks, **6**.46 to
    6.50
  horizontal liquid-vapor separators, **6**.43 to
    6.45
  rupture disks, **6**.45
  Venturi scrubbers, **6**.34 to **6**.39
  vertical liquid-vapor separators, **6**.41 to
    6.43
Slenderness ratio, **1**.90
Slices, method of, **1**.280 to **1**.282
Slip of propellers, **9**.6
Slope of pipes, **3**.442
Slope-deflection method of wind-stress anal-
    ysis, **2**.23 to **2**.25
Small hydro power, **3**.546, **13**.34
Smith chart, use of, **5**.39
Smoke, tobacco, **2**.120
Smokestacks, **3**.292
  diameter of, **3**.292
  height of, **3**.292
Snow-melting heating panels, **2**.78
Soil:
  composition of, **1**.265
  compression index of, **1**.285
  consolidation of, **1**.285
  contaminated, **1**.266
  moisture content of, **1**.265, **1**.266
  permeability of, **1**.267
  porosity of, **1**.265
  shearing capacity of, **1**.270
Soil pressure:
  caused by point load, **1**.269
  caused by rectangular loading, **1**.271
  under dam, **1**.47
Solar energy, **2**.126 to **2**.151
  ASHRAE, **2**.132
  average annual amount, **2**.126
  collectors for: choice of, **2**.127
    flat-plate, **2**.127
    heat exchangers for, **2**.130
    key system components, **2**.129
    orientation, **2**.128
    for passive solar heating, **2**.145 to **2**.151
    sizing of, **2**.134
  domestic hot-water heater, **2**.141 to **2**.145
  F chart method, **2**.135 to **2**.141
  heating systems, **2**.134 to **2**.141, **2**.145 to
    2.150

Solar energy (*Cont.*):
  insolation determination, 2.132 to 2.133,
    2.147
  passive solar heating, 2.145 to 2.150
    collector for, 2.146
    heating contribution, 2.149
    insolation, 2.147
Solid-state device evaluation, 5.3 to 5.5
Sonar equations, 5.42
Sound levels for various occupancies, 4.26
Space frame, 1.19 to 1.22
Space vehicles, 8.16 to 8.17
Specific gravity of liquid, 3.425
Specific speed of pumps, 3.353
Speed reducer:
  selection of, 3.54
  torque rating of, 3.55
Spiral, transition, 1.241 to 1.244
Spiral-type heat exchangers, 3.486
Splines, involute, 3.106 to 3.108
Splining time, 3.163
Spring constant of bent, 2.37
Springs, 3.77 to 3.100
  air-snubber, 3.92 to 3.96
  Belleville, 3.88
  curved, 3.82 to 3.84
  helical compression and tension, 3.78 to
    3.80
  life of, 3.98
    designing parts for expected, 3.110 to
      3.112
  liquid, 3.91
  multirate helical, 3.87
  rate of, 3.99
  reinforced-plastic, 3.96 to 3.98
  ring, 3.89 to 3.91
  round- and square-wire helical, 3.84 to
    3.86
  selection for load and deflection, 3.77
  sizing helical, 3.80
  square- and rectangular-wire helical, 3.81
    to 3.82
  stripping, 3.168
  torsion, 3.84 to 3.86
  torsion-bar, 3.86
  wire length and weight, 3.78
Sprinkler systems, 2.52 to 2.54
  pipe sizes for, 2.53
  piping for, 2.55
Square springs, 3.84 to 3.86
Stability:
  of embankment, 1.280 to 1.285

Stability (*Cont.*):
  of footing, 1.284
  of vessel, 1.206 to 1.208
Stack height, effective, 6.29
Stadia surveying, 1.231
Stair slab, 2.35 to 2.37
Star strut, 1.90
Static-regain duct sizing, 2.102 to 2.105
Statically indeterminate structures, 1.53 to
    1.59
Statistical inference, 12.73
Statistics and probability, 12.62 to 12.100
Steady-state continuous physical balance,
    6.7 to 6.10
Steam boiler:
  heat balance determination, 3.204 to
    3.206
  selection of, 3.278 to 3.281
Steam compressor, choice of, 13.19
Steam condenser:
  air-ejector analysis for, 3.252 to 3.254
  circulating-water pressure loss in, 3.254
  performance of, 3.246 to 3.249
  selection of, 3.250
  weight analysis of, 3.254
Steam consumption of heating systems, 2.70
    to 2.72
Steam-control valves, 7.45 to 7.48
Steam-jet refrigeration, 3.498 to 3.499
Steam piping, pressure loss in, 3.399 to
    3.403
Steam pressure drop in superheater, 3.277
Steam processes, 3.214 to 3.219
  constant-entropy, 3.219
  constant-pressure, 3.214
  constant-temperature, 3.218
  constant-volume, 3.215 to 3.217
Steam-quality determination, 3.277
Steam tables, use of, 3.208 to 3.211
  interpolation of values in, 3.211 to 3.214
Steam tracing:
  for piping, 6.14
  vessel bottom, 6.15
Steam transmission line, 6.17
  without steam traps, 6.17 to 6.20
Steam transmission piping, 3.460 to 3.468
  accumulator selection and sizing, 3.469 to
    3.471
  desuperheater analysis, 3.468, 3.473
Steam trap selection, 3.406 to 3.410
  bucket-trap capacity, 3.409
  coefficient of heat transfer for, 3.406

Steam trap selection (*Cont.*):
  condensate formed, 3.407
  factors in, 3.404
  impulse-trap capacities, 3.409
  specific heats for, 3.405
  unit-heater correction factors for, 3.408
Steam turbine–gas turbine cycle analysis,
    3.245
Steel beam, 1.76 to 1.88, 2.27 to 2.32
  continuous, 1.82, 1.102 to 1.105
    elastic design of, 1.82, 1.101 to 1.105
    plastic design of, 1.102 to 1.105
  with continuous lateral support, 1.76
  cover-plated, 1.79 to 1.82
  encased in concrete, 2.30 to 2.32
  with intermittent lateral support, 1.76
  light-gage, 2.27 to 2.30
  with reduced allowable stress, 1.78
  shearing stress in, 1.83
Steel beam-column, 1.94, 1.95, 1.114, 1.115
Steel column, 1.88 to 1.95
  base for, 2.15 to 2.19
  built-up, 1.89
  effective length of, 1.88
  with end moments, 1.94, 1.95
  under fatigue loading, 1.93
  with intermediate loading, 1.93
  lacing of, 1.92
  with partial restraint, 1.91
  of star-strut section, 1.90
  with two effective lengths, 1.91
Storage tank costs, 6.46 to 6.50
Storage vessel:
  energy savings from, 6.22 to 6.25
  time to empty, 6.46
Storm sewer:
  design of, 11.5
  inlet size for, 11.24
Storm-water runoff rate, 11.12
Storm-water systems, 11.24 to 11.26
Strain, axial, 1.25, 1.26
Stress(es):
  allowable, in design, 3.113 to 3.117
  axial, 1.26
  bending, 1.39 to 1.42, 1.46
  bond, 1.134, 1.157 to 1.165
  hoop, 1.32 to 1.35
  for plastic gears, 3.125 to 3.128
  principal, 1.31, 1.32
  shearing, 1.42, 1.83, 1.117, 1.132, 1.143,
    1.289
    in homogeneous beam, 1.31, 1.42

Stress(es), shearing (*Cont.*):
    in reinforced-concrete beam, 1.133,
      1.139
    in steel beam, 1.83
    in timber beam, 1.117
  shrink-fit, 1.35
  thermal, 1.34
String polygon, 1.7, 1.287
Structural engineering, 2.2 to 2.37
  beam connection: riveted moment, 2.7 to
    2.9
    semirigid, 2.6
    welded flexible, 2.9, 2.10
    welded moment, 2.12
    welded seated, 2.10, 2.11
  column base: for axial load, 2.15
    for end moment, 2.15
    grillage-type, 2.16 to 2.19
  composite steel-and-concrete beam, 2.32
    to 2.34
  connection: beam-to-column (*see* Beam
      connection)
    to resist horizontal shear, 2.31 to 2.34
    of truss members, 2.5, 2.6
  eyebar, 2.3
  gusset plate, 2.5
  hanger, steel, 2.4
  joists, reinforced-concrete, 2.34
  knee: curved, 2.14
    rectangular, 2.13
  stair slab, 2.35 to 2.37
  steel beam, 2.27 to 2.32
    encased in concrete, 2.30 to 2.32
    light-gage, 2.27 to 2.30
  vibration of bent, 2.37
  wind drift, 2.25
    reduction with diagonal bracing, 2.26
  wind-stress analysis, 2.19 to 2.25
    cantilever method, 2.21
    portal method, 2.19 to 2.21
    slope-deflection method, 2.23 to 2.25
Sulfur dioxide, 3.192, 6.41, 13.8, 13.11,
    13.14, 13.18
Superheater, boiler, 3.277
Supersonic wind tunnel, 8.24
Surveying, mine, 1.250 to 1.255
Swedish method for analysis of slope stabil-
    ity, 1.280 to 1.282
Swimming pool, sizing of, 2.58

T beam reinforced-concrete, 1.129, 1.130,
    1.144, to 1.147, 1.198 to 1.200

Tables, truth, **7.55**
Tangent-offset method, **1.245** to **1.247**
Tangential deviation, **1.49**
Tankers, marine, **9.17** to **9.18**
  capacity of, **9.17**
  discharge time of, **9.18**
  T-2 equivalent of, **9.17**
Tapers, dimensions of, **3.149**
Tapping time, **3.156**
Temperature control, **7.4**
Temperature difference in heat exchangers, **3.480**
Temperature-measuring amplifier, **7.37**
  instrument, analysis of, **7.38**
    equivalent circuit for, **7.39**
Temperature reinforcement, **1.165**, **1.198**, **2.35**
Tennessee Valley Authority (TVA), **13.14**
Tension member, steel, **1.96**
Ternary liquid system, **6.3**
Terzaghi general wedge theory, **1.275**
Terzaghi theory of consolidation, **1.285**
Thermal effects, **1.34**, **1.35**
Thermal pollution, estimating, **13.47**
Threading time, **3.156**, **3.157**
Three moments, theorem of, **1.55**
Throttling processes, steam and water, **3.223**
Thrust, aircraft engine, **8.5**
Thunderstorms, incidence of, **4.91**
Timber beam, **1.116** to **1.119**
  bending stress in, **1.116**
  built-up, **1.117** to **1.119**
  depth factor of, **1.117**
  moving load on, **1.117**
  shearing stress in, **1.117**
Timber column, **1.119**, **1.120**
Timber connection, **1.121** to **1.125**
Timber engineering, **1.116** to **1.125**
Timber member under oblique force, **1.121**
Time:
  to bore: **3.154**
  for broaching, **3.163**
  to countersink, **3.154**
  to drill, **3.154**
  to empty dished-end vessel, **6.59**
  to empty storage vessel, **6.46**
  to face, **3.155**
  for grinding, **3.162**
  for hobbing, **3.163**
  for oxyacetylene cutting, **3.165**
  to plate metals, **3.169**
  to ream, **3.154**
  for sawing, **3.164**

Time (*Cont.*):
  for serrating, **3.163**
  for splining, **3.163**
  to tap, **3.156**
  to thread, **3.156**
  for thread milling, **3.182**
  for tool change, **3.153**
  to turn, **3.153**
Timing belts:
  center distances of, **3.53**
  horsepower (kW) ratings of, **3.54**
  number of sprocket teeth for, **3.53**
  pitch of, **3.52**
  selection of, **3.51** to **3.54**
  service factors for, **3.52**
Tool-change time, **3.153**
Torsion of shaft, **1.36**, **3.10**, **3.11**
Torsion-bar springs, **3.86**
Torsion springs, **3.84** to **3.86**
Total costs, cogeneration alternatives, **13.47**
Total element time, **3.147**
Total head on a pump, **3.357** to **3.361**
Total operation time, **3.150**
Toxic Substance Control Act (TSCA), **6.10**
Tracing, steam:
  for piping, **6.14**
  tank bottom, **6.15**
Tract:
  area of, **1.227** to **1.230**
  partition of, **1.228** to **1.230**
Trajectory:
  concordant, **1.190**
  linear transformation of, **1.188** to **1.190**
Transfer function, **7.21**, **7.26**, **7.30**, **7.35**
  phase shift, **7.31**
Transformed section, **1.41**, **1.42**
Transformers, electric, **4.20** to **4.24**
  characteristics of, **4.22ff**.
  coolant characteristics of, **4.22**
  load-center units, **4.23**
  selection of, **4.21**
Transistorized amplifier, **5.6**, **5.13**, **5.44**
  equivalent circuit for, **5.45**
  gain control in, **5.52**
Transistorized circuit, **5.47**
  load line, **5.49**
  operating point, **5.48**
  small-signal, **5.49**
Transistors, selection of, **5.5**
Transition spiral, **1.241** to **1.244**
Traverse, closed, **1.225** to **1.227**
Treatment, sewage, **11.26** to **11.29**
Trench, earth thrust on, **1.275** to **1.277**

True unit time, 3.153
Truth tables, 7.55
Truss:
    force analysis of, 1.10 to 1.14
        by graphical method, 1.10 to 1.12
        by method of joints, 1.12 to 1.14
        by method of sections, 1.14
        with moving loads, 1.60 to 1.65
        statically indeterminate, 1.58
    gusset plate of, 2.5
Tuned-circuit Q values, 5.27
Turbines:
    gas, 3.297 to 3.300
    hydraulic, 3.546, 13.34
    mechanical-drive, 3.236
    steam rates, 13.16
    steam turbine–gas turbine cycle analysis,
        3.245
Turbulent flow, 1.213 to 1.215
Turning, time for, 3.153
Turret lathe:
    power constant, 3.158
    power input, 3.157
Two-way slab, 1.136 to 1.138

Ultimate load, 1.98, 1.99
Ultrasonic generators, 5.40
Underground storage tank (UST) laws, 6.21
Uniform series, 12.9 to 12.11
Uninsulated pipes, costs of heat loss, 3.521
Unit heaters:
    blow distances of, 2.69
    capacity of, 2.65 to 2.70
    conversion factors for, 2.68
    diameter of pipes for, 2.70
    outlet velocities of, 2.69
    vertical types, 2.67
Unit-load method, 1.51
Universal joints, 3.47

V belts, 3.19 to 3.23
    arc of contact factor for, 3.18
    correction factors for, 3.17, 3.21
    horsepower (kW) ratings of, 3.19, 3.21ff.
    length correction factors, 3.23
    minimum pulley diameters for, 3.19
    multiple, 3.22
    service factors for, 3.17, 3.18
    sheave dimensions for, 3.20
    small-diameter factors, 3.23

Vacuum systems, 3.332 to 3.338
    pipe size for, 3.337
    pump-down time of, 3.332 to 3.345
    pump selection for, 3.334 to 3.337
    rating, checking, 6.61
    system factors for, 3.336
Vacuum-tube amplifier, 5.11
    characteristics of, 5.8, 5.11
Valves, safety, 3.272 to 3.277
Variable-speed drive, 3.56
Vector algebra, 4.8
Vent, explosion, sizing of, 13.42
Vent system, engine, 3.298 to 3.301
Ventilation, building, 13.44
Ventilators, roof, 2.118 to 2.120
Venturi meter, flow through, 1.210
Venturi scrubbers, 6.34 to 6.39
Vertical parabolic curve:
    containing given point, 1.248
    plotting of, 1.245 to 1.248
    sight distance on, 1.249
Vertical shear (see Shear)
Vertical-takeoff aircraft, 8.9
Very large commercial transport, 8.10
Vessel(s), marine, 9.2ff.
    choice of, 9.22
    form coefficients for, 9.2
    hull type, 9.22
    immersion and flooding of, 9.10
    nuclear propulsion for, 9.16 to 9.17
    power-plant selection for, 9.13 to 9.15
    power required to propel, 9.4 to 9.6
    pressure, minimum-cost, 6.67
    prismatic, designing, 6.63ff.
    propeller revolutions and speed, 9.9
    propellers for, 9.7 to 9.9
    pumps for, 9.10 to 9.13
    refrigeration and air conditioning for, 9.19
        to 9.21
    seakeeping potential, 9.22
    shallow-water speed of, 9.3 to 9.5
    vacuum rating, checking, 6.61
    wetted area of, 9.3
Vibration of bent, 2.37
Vibration isolators, 2.121 to 2.125
Virtual displacements, theorem of, 1.112
Viscosity:
    of fluid, 1.208
    of liquid, 3.425
    liquid, effect of, on pump, 3.382 to 3.386
Viscous liquid, selecting pump for, 3.379 to
        3.381

Void ratio, 1.265, 1.285
Volatile organic compounds (VOC), 6.41
Voltage drop, tables of, 4.34 to 4.35
Voltage ratings for power equipment, 4.37
Voltage regulation of cables, 4.28

Wall thickness of pipe, 3.396ff.
Warm-up condensate load, piping, 3.403
Waste-heat boiler, 3.551
Waste-heat loss reduction, 6.33
Water-hammer effects in pipes, 3.425
Water meters, sizing of 3.422
Water pressure, loss in, 2.48
Water-supply piping, 2.44 to 2.51
Water-supply systems, 11.2 to 11.12
    flow rates in, 11.2 to 11.6
    Hardy Cross method for, 11.5
    Hazen-Williams equation for 11.4
    industrial water requirements, 11.10
    pipe sizes for, 11.3
    pressure losses in, 11.2 to 11.6
    selection of, 11.7 to 11.10
    treatment methods for, 11.10 to 11.12
    typical sources for, 11.7
Waveform rms value, 7.40
Wear life of rolling surfaces, 3.112
Web reinforcement:
    of prestressed-concrete beam, 1.181
    of reinforced-concrete beam, 1.132 to
        1.134, 1.143
Weir:
    discharge over, 1.212
    variation in head on, 1.222 to 1.224

Welded connection, 1.74, 1.75, 2.9, to 2.13
Welding:
    electric-arc, 3.165
    oxyacetylene, 3.165
Westergaard construction, 1.288 to 1.290
Wetted perimeter, 1.215
Williott displacement diagram, 1.29
Wind drift, 2.25, 2.26
Wind-energy systems, 3.509 to 3.516
    capital costs of, 3.514
    electric power output, 3.510
    kites in, 3.514
    machine performance curves, 3.513
    machine selection, 3.509
    types of machines, 3.511
Wind-stress analysis, 2.19 to 2.25
    by cantilever method, 2.21
    by portal method, 2.19 to 2.21
    by slope-deflection method, 2.23 to 2.25
Wind tunnel, supersonic, 8.24
Wire length for springs, 3.78
Wire-rope drives, 3.23
Wire size, choice of, 4.26 to 4.33
Wood, combustion of, 3.199 to 3.201
Wood-plywood girder, 1.118
Worst-case transistor leakage current, 5.17

Yield-line theory, 1.136 to 1.138
Yield moment, 1.98

Zenith of observer, 1.236

About the Editor

Tyler G. Hicks is a consulting engineer with International Engineering Associates. He has worked in plant design and operation in a variety of industries, taught at several engineering schools, and lectured in the United States and abroad. He is the author or coauthor of numerous engineering reference books, including *Standard Handbook of Consulting Engineering Practice, The McGraw-Hill Handbook of Essential Engineering Information and Data,* and *Handbook of Effective Technical Communications,* available from McGraw-Hill.